Teacher's Edition

Geometry

Explorations and Applications

$\angle BCD \cong \angle CDE$

Authors

Douglas B. Aichele

Patrick W Hopfensberger

Miriam A. Leiva

Marguerite M. Mason

Stuart J. Murphy

Vicki J. Schell

Matthias C. Vheru

McDougal Littell

A HOUGHTON MIFFLIN COMPANY

Evanston, Illinois • Boston • Dallas

Authors

Douglas B. Aichele Regents Professor of Mathematics, Oklahoma State University, Stillwater, Oklahoma

Patrick W. Hopfensperger Mathematics Teacher, Homestead High School, Mequon, Wisconsin

Miriam A. Leiva Cone Distinguished Professor for Teaching and Professor of Mathematics, University of North Carolina at Charlotte

Marguerite M. Mason Assistant Professor of Mathematics Education, University of Virginia, Charlottesville, Virginia

Stuart J. Murphy Visual Learning Specialist, Evanston, Illinois

Vicki J. Schell Professor of Mathematics and Mathematics Education, Lenoir-Rhyne College, Hickory, North Carolina

Matthias C. Vheru Mathematics Teacher, Locke High School, Los Angeles, California

The authors wish to thank **Jane Pflughaupt**, Mathematics Teacher, Pioneer High School, San Jose, California, and **Martha E. Wilson**, Preparatory Mathematics Specialist, Mathematical Sciences Teaching and Learning Center, University of Delaware, Newark, Delaware, for their contributions to this Teacher's Edition.

ISBN: 0-395-72286-1

123456789—VH—01 00 99 98 97

Internet Web Site: http://www.hmco.com

Contents of the Teacher's Edition

T4   Philosophy of *Geometry: Explorations and Applications*

T6   Contents of the Student Edition with chapter overviews

T19   Program Overview

- **Teaching Resource Materials**   T20-T24
- **Features of the Student Edition**   T25-T33

T34   Using the Teacher's Edition

- **Planning Pages for Every Chapter**   T34
- **A Teaching Plan for Every Section**   T36
- **Pacing and Making Assignments**   T38

1   Student Edition with answers and teaching suggestions

A1   Additional Answers

Philosophy of Geometry

Explorations and Applications

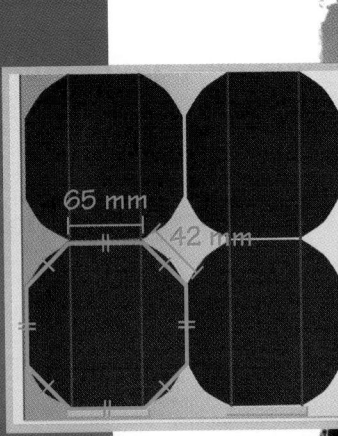

65 mm

42 mm

Debby Tewa, a solar energy expert, displays a solar module used to convert sunlight into electricity.

GOALS OF THE COURSE

This course has been designed to make mathematics accessible and inviting to the wide range of students who are studying geometry today. It helps you prepare today's students for tomorrow's world by:

- Building understanding of the concepts that provide a strong foundation for future courses and careers
- Connecting geometry to the real world and to other subjects and math topics
- Involving students in exploring and discovering math concepts
- Assessing students' progress in ways that support learning

MATHEMATICAL CONTENT

The content and the teaching strategies in the textbook reflect the curriculum, teaching, and assessment standards of the National Council of Teachers of Mathematics. The fresh, new course outline:

- Emphasizes reasoning and logical thinking
- Balances coordinate, synthetic, and transformational geometry
- Integrates technology as a problem-solving tool
- Connects geometry to algebra, data analysis, probability, trigonometry, and discrete mathematics
- Incorporates two-column, paragraph, flow, and coordinate proof

TEACHING STRATEGIES

The flexible course design offers frequent opportunities for you to incorporate:

- Exploratory activities that build conceptual understanding
- Applications that strengthen problem-solving skills
- Writing questions and proofs that develop communication skills
- Strategies for using technology to visualize and solve problems

ASSESSMENT

You and your students can measure their mathematical growth throughout the course in a variety of ways, including:

- Cooperative learning activities
- Open-ended problems
- Journal writing
- Portfolio projects

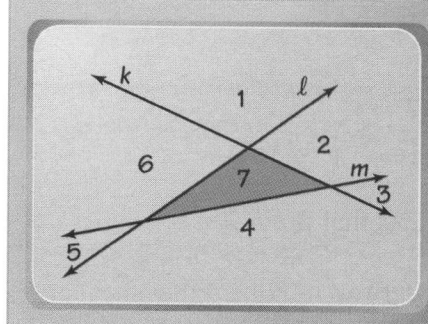

Program Overview

Pages T6-T39 give an overview of *Geometry: Explorations and Applications* and the teaching materials that support it. These pages provide information about:

- **Table of contents of the textbook** pp. T6-T18
- **Teaching resource materials** pp. T19-T24
- **Features of the textbook** pp. T25-T33
- **Content and organization of the Teacher's Edition** pp. T34-T39

Contents

CHAPTER

1

Patterns, Lines, and Planes

Interview *Latecia Leavy*		xxxii
1.1	**Patterns and Reasoning**	3
	Exploration: *Analyzing Patterns*	3
1.2	**Transformations and Symmetry**	8
1.3	**Making Conjectures**	14
	Exploration: *Tracing Networks*	14
1.4	**Modeling Points, Lines, and Planes**	21
	Exploration: *Representing Points, Lines, and Planes*	21
1.5	**Segments and Their Measures**	28
1.6	**Working with Angles**	34
1.7	**Bisecting Segments and Angles**	40
Review	**Chapter Review**	50

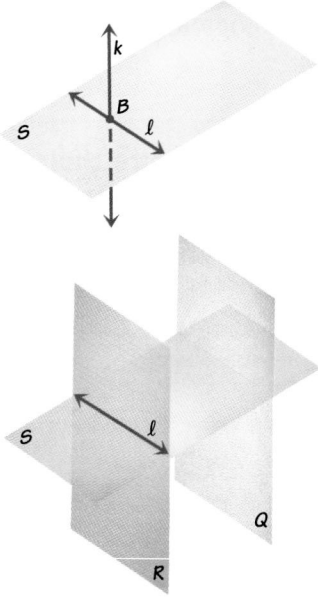

1.4 *Representing planes* 23

Assessment	**Ongoing Assessment**	7, 13, 19, 27, 32, 39, 46
	Assess Your Progress	20, 33, 47
	Journal writing	20, 33, 47
	Portfolio Project: *Investigating Symmetry*	48
	Chapter Assessment	52

Applications

Interview:
Latecia Leavy
Tumbler
12, 17, 37

Connection			
Astronomy	7	**Manufacturing**	31
Communication	13	**Science**	39
History	19	**Crafts**	45
Art	26		

Applying Algebra 16, 32, 41, 43, 44, 46

Additional applications: chemistry, travel, social studies, history, skiing, racing, boating

CHAPTER 1

This chapter introduces **logical reasoning** and lays the foundation for a formal study of geometry. Students gain experience with **inductive reasoning** as they make **conjectures** about real-world objects and mathematical patterns. Students also learn the **definitions** of essential terms such as angle, segment, and bisector.

Triangles and Polygons

Interview *Virgil Lueth*		54
2.1	**Types of Angles**	57
2.2	**Classifying Triangles**	64
	Exploration: *Sorting Triangles*	64
2.3	**Types of Polygons**	72
2.4	**Angles in Polygons**	79
	Exploration: *Finding Angle Measures in Polygons*	79
2.5	**Parallelograms**	87
	Exploration: *Analyzing Parallelograms*	87
2.6	**Building Prisms**	93
Review	**Chapter Review**	102
Algebra	**Review/Preview**	106

Assessment	**Ongoing Assessment**	63, 71, 78, 86, 92, 98
	Assess Your Progress	71, 86, 99
	Journal writing	71, 86, 99
	Portfolio Project: *Building the Platonic Solids*	100
	Chapter Assessment	104

2.2 Sorting Triangles 64

Applications

Interview:
Virgil Lueth
Mineralogist
68, 82, 85, 98

——————— Connection ———————

Biology	61	**Crafts**	77
Communications	62	**Art**	83
History	70	**Optics**	91

Applying Algebra 66, 67, 68, 80, 84, 86, 88, 104

Additional applications: architecture, bicycling, sports, computer-aided design, games

CHAPTER 2

This chapter presents important **properties** of triangles and polygons. Students **classify** polygons and **explore properties** of their sides and angles. Students develop their **visualization skills,** and the chapter concludes with a study of **prisms**.

Reasoning in Geometry

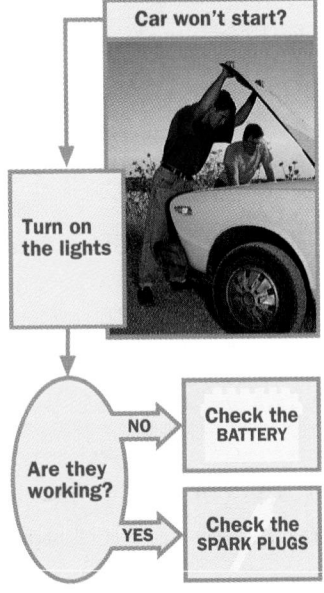

Car won't start?

Turn on the lights

NO → Check the BATTERY

Are they working?

YES → Check the SPARK PLUGS

3.7 *Auto Maintenance* 151

Interview *Mary-Jacque Mann* 108

3.1 **Inductive and Deductive Reasoning** 111

3.2 **Postulates, Definitions, and Properties** 117

3.3 **Paragraph Proof** 124
    Exploration: *Measuring Exterior Angles* 125

3.4 **Two-Column Proof** 130

3.5 **Converses of Statements** 136

3.6 **The Pythagorean Theorem** 141
    Exploration: *Proving the Pythagorean Theorem* 141

3.7 **Negations and Contrapositives** 148
    Exploration: *Analyzing Triangles* 148

Review **Chapter Review** 156

Assessment **Ongoing Assessment** 116, 123, 129, 134, 140, 147, 152

    Assess Your Progress 123, 135, 153

    Journal writing 123, 135, 153

    Portfolio Project: *Classifying Information* 154

    Chapter Assessment 158

    Cumulative Assessment *Ch. 1–3* 160

Applications

Interview: **Mary-Jacque Mann**
    Forensic Scientist 128, 133

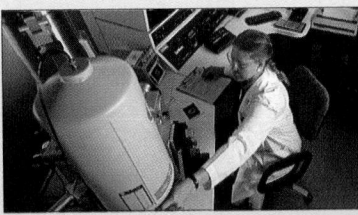

--- Connection ---

Zoology	115	**Language**	139
Navigation	121	**History**	146
Measurement	122	**Auto Maintenance**	151
Literature	127		

Applying Algebra 112, 115, 142, 144

Additional applications: personal finance, marketing, carpentry, archeology, architecture, design, travel

CHAPTER 3

In this chapter, students expand their knowledge of **reasoning strategies** to include deductive methods. Students develop skill in understanding and writing convincing arguments as **paragraph proofs** and **two-column proofs**. Logical statements such as converse, negative, and contrapositive are studied.

Coordinates in Geometry

Interview *Marc Hannah*		162
4.1	**Finding Distances and Midpoints**	165
4.2	**Equations of Lines**	173
4.3	**Exploring Parallels and Perpendiculars**	180
	Exploration: *Comparing Slopes of Lines*	180
4.4	**Equations of Circles**	187
4.5	**Coordinates and Proof**	194
	Exploration:	
	Placing a Parallelogram on a Coordinate Plane	194
4.6	**Coordinates in Three Dimensions**	201
Review	**Chapter Review**	210
Algebra	**Review/Preview**	214

Assessment	**Ongoing Assessment**	172, 179, 186, 193, 200, 206
	Assess Your Progress	186, 207
	Journal writing	186, 207
	Portfolio Project:	
	Exploring Taxicab Geometry	208
	Chapter Assessment	212

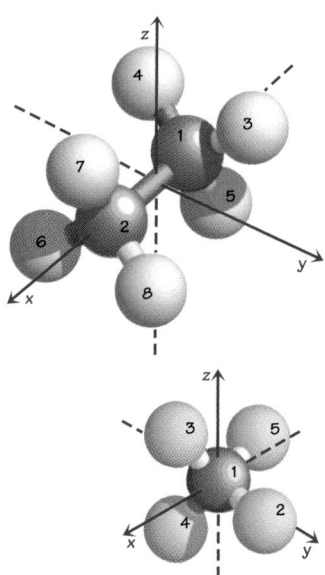

4.6 *Ethane and methane molecules* 201, 205

Applications

Interview: *Marc Hannah*
Computer-system Designer 170, 198, 206

——— Connection ———

Puzzles	171	**Quilt Patterns**	185
Engineering	177	**Architecture**	192
City Planning	178		

Applying Algebra 170, 177, 185, 193, 196, 200

Additional applications: sports, recreation, accessibility, city planning, carpentry, printing, chemistry

CHAPTER 4

Central to the work throughout the book **linking algebra and geometry**, this chapter presents figures and their properties on the **coordinate plane**. Students use coordinates to study distance, midpoints, slopes, lines, and polygons. Students use **coordinate proof** to justify conjectures.

Interview *José Saínz*   216

5.1 **Parallel Lines and Transversals**   219
     Exploration: *Angles and Transversals*   221

5.2 **Properties of Parallel Lines**   226

5.3 **Types of Proofs**   234

5.4 **Conditions for Parallel Lines**   242
     Exploration: *Drafting Parallel Lines*   242

5.5 **Proving Theorems About Parallels**   249

5.6 **Parallels in Space**   256
     Exploration:
     Investigating Planes and Their Intersections   256

5.7 **Constructing Parallels and Perpendiculars**   263

Review **Chapter Review**   272

Assessment **Ongoing Assessment**   225, 233, 241, 248,
                                                        254, 262, 269

                Assess Your Progress   233, 255, 269

                Journal writing   233, 255, 269

                Portfolio Project:
                Creating Technical Drawings   270

                Chapter Assessment   274

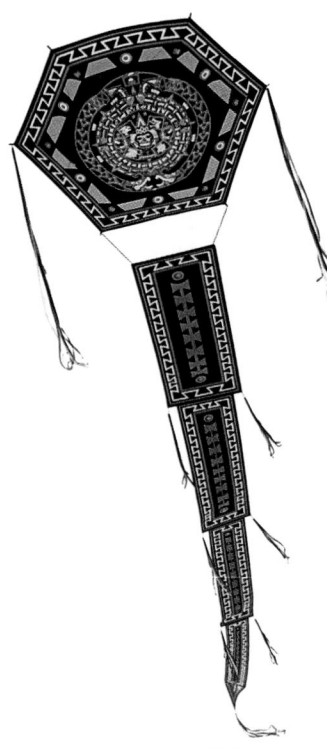

5.2 *Kite design*   231

Applications

Interview: *José Saínz*
         Kite Designer 231, 252, 262

─────── Connection ───────

Botany   224   Design   254
Architecture   232   Electronics   259
Engineering   240   Technical Drawing   261
Horticulture   246   Paper Folding   266
Dance   247

Applying Algebra 222, 224, 227, 230, 232, 243, 245

Additional applications: biology, automobiles, art, design, engineering, city planning, aviation

CHAPTER 5

In this chapter, students **investigate and justify** properties of parallel lines and transversals. Students are introduced to **flow proof** and then compare the **methods of proof** they have learned. Students also continue their study of **constructions** begun in Chapter 2.

Conjectures About Triangles

Interview *Madeleine Fleming*   276

6.1   **Triangle Inequalities**   279
    Exploration: *Comparing Sides of Triangles*   279

6.2   **Exploring Congruence**   285
    Exploration: *Investigating Overlapping Triangles*   287

6.3   **Congruent Triangles: SSS and SAS**   292

6.4   **Congruent Triangles: ASA, AAS, and HL**   299

6.5   **Applying Congruence**   306

6.6   **Properties of Isosceles Triangles**   313

6.7   **Altitudes, Medians, and Bisectors**   319

Review   **Chapter Review**   328

Algebra **Review/Preview**   332

Assessment   **Ongoing Assessment**   284, 291, 298, 305, 312, 318, 324

       Assess Your Progress   291, 312, 325

       Journal writing   291, 312, 325

       Portfolio Project: *Building a Mobile*   326

       Chapter Assessment   330

       Cumulative Assessment *Ch. 4–6*   334

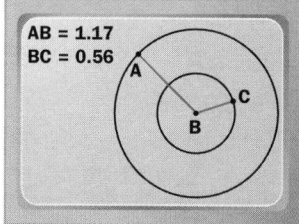

AB = 1.17
BC = 0.56

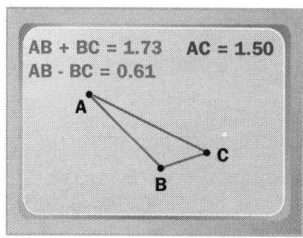

AB + BC = 1.73   AC = 1.50
AB - BC = 0.61

6.1 *Exploring using technology*   279

Applications

Interview:
Madeleine Fleming
Optical Physicist
290, 295, 317

——— Connection ———

Probability	283
Interior Design	284
Rescue Safety	297
Astronomy	305

Aerial Photography	310
Physics	323

Applying Algebra   286, 289, 310, 315, 317, 322, 323, 325, 330

Additional applications: art, pottery, biology, cars, rock climbing, architecture

Contents   **ix**

CHAPTER 6

In this chapter, **triangle congruence** relationships are applied and proved. Explorations with **manipulatives** and **technology** are used to highlight and reinforce proofs involving congruent triangles. Properties of **isosceles** triangles, altitudes, medians, and bisectors are also explored.

Quadrilaterals, Areas, and Volumes

Interview *Walt Stone*   336

7.1 Classifying Quadrilaterals   339
  Exploration: *Investigating Diagonals*   340

7.2 Identifying Parallelograms   346

7.3 Conditions for Special Parallelograms   353

7.4 Areas of Triangles and Quadrilaterals   360
  Exploration: *Discovering Area Formulas*   361

7.5 Areas of Regular Polygons and Circles   367

7.6 Prisms and Cylinders   374
  Exploration: *Comparing Volumes*   374

Review Chapter Review   384

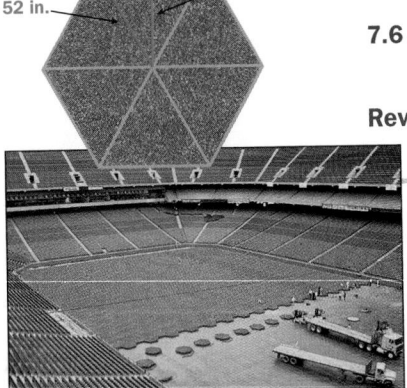

26 in.
26√3 in.
52 in.

Assessment   **Ongoing Assessment**   345, 352, 358, 366, 373, 381

   Assess Your Progress   359, 381

   Journal writing   359, 381

   Portfolio Project: *Designing a Cottage*   382

   Chapter Assessment   386

7.5 *Grass for the Silverdome*   367

Applications

Interview:
Walt Stone
Alderman
366, 380

———— Connection ————

Engineering   344   Biology   371
Machinery   351   Consumer
Social Studies   357   Economics   379
Gardening   365

Applying Algebra   341, 351, 354, 357, 365, 380

Additional applications: auto repair, carpentry, drafting, crafts, quilting, history, clothing, sports, architecture, package design, nutrition

CHAPTER 7

In this chapter, quadrilateral relationships introduced in Chapter 2 are extended and formalized. Students identify types of **parallelograms** and expand their proof skills by investigating their properties. Students discover **area formulas** and continue their work with figures in three dimensions by exploring **volumes** of prisms and cylinders.

Using Transformations

Interview *Terri Johnson*   388

8.1 Using Reflections   391
    Exploration: *Reflecting a Polygon Over a Line*   392

8.2 Reflections with Coordinates   398

8.3 Translations   405
    Exploration: *Comparing Translated Polygons*   405

8.4 Applying Rotations   412

8.5 Glide Reflections   419

8.6 Dilations   426
    Exploration: *Making an Enlargement*   427

Review Chapter Review   436

Algebra Review/Preview   440

Assessment **Ongoing Assessment**   397, 404, 411, 418, 425, 432

        Assess Your Progress   411, 433

        Journal writing   411, 433

        Portfolio Project:
        Tessellating the Plane   434

        Chapter Assessment   438

8.4 *Hawaiian quilts*   417

Applications

Interview: Terri Johnson
    Architect   395, 409, 418

——————— Connection ———————

Audio-Visual Presentation   396   **Hawaiian Quilts**   417

Miniature Golf   403   **Computer Simulation**   424

Architecture   410   **Literature**   431

Applying Algebra 395, 402

Additional applications: periscopes, optics, computers, furniture design, fabric design, animal tracks, manufacturing

CHAPTER 8

This chapter connects symmetry and coordinates with a look at transformations such as **reflections, rotations,** and **translations**. Students use coordinates to examine the effects of transformations on figures and their properties. Students expand their knowledge of transformations by investigating **glide reflections** and **dilations**.

Similar Polygons

Interview *Loy Arcenas*   442

9.1   **Properties of Similar Figures**   445

9.2   **Similar Triangles**   453

9.3   **Proportions and Similarity**   461
      Exploration:
      Finding Proportions in Triangles   461

9.4   **Areas and Volumes of Similar Figures**   468

9.5   **Geometric Probability**   475
      Exploration:
      Investigating Probability Based on Area   475

Review **Chapter Review**   484

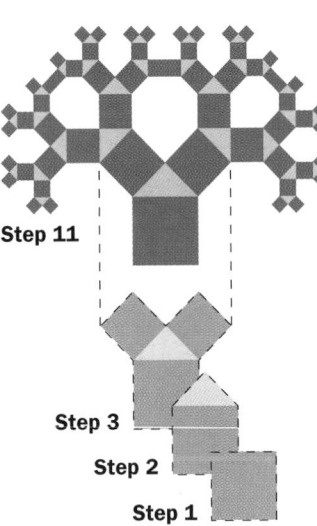

Step 11

Step 3

Step 2

Step 1

Assessment   **Ongoing Assessment**   452, 460, 466, 474, 480

             Assess Your Progress   467, 481

             Journal writing   467, 481

             Portfolio Project:
             Scaling the Planets   482

             Chapter Assessment   486

             Cumulative Assessment *Ch. 7–9*   488

9.1 *Self-similarity*   452

Applications

Interview:
Loy Arcenas
Set Designer
465, 468, 472

——————— Connection ———————

Astronomy	450	**Consumer**	
Surveying	457	**Economics**	473
Electronics	459	**Zoology**	479

Applying Algebra 448, 449, 462, 464, 466, 470, 474

Additional applications: graphic design, technical drawing, movies, geography, fractals, sculpture, ballooning, advertising

CHAPTER 9

In this chapter, the connection of many **strands of mathematics** is demonstrated in the use of ratios to analyze similarity, scale factors, and geometric probability. Using **spreadsheets** and **geometry software** as appropriate, students discover patterns and investigate applications. Students compare areas and volumes of similar figures in three dimensions.

Applying Right Triangles

Interview *Debby Tewa*		490
10.1 Similar Right Triangles		493
Exploration: *Comparing Right Triangles*		493
10.2 Special Right Triangles		500
10.3 The Tangent Ratio		507
10.4 Sine and Cosine Ratios		514
Exploration: *Analyzing Ratios in Triangles*		514
10.5 Using Vectors		521
10.6 Areas and Trigonometry		529
10.7 Pyramids and Cones		535
Review Chapter Review		544
Algebra Review/Preview		548

Assessment	**Ongoing Assessment**	499, 506, 513, 520, 528, 534, 541
	Assess Your Progress	506, 528, 541
	Journal writing	506, 528, 541
	Portfolio Project: *Applying Solar Geometry*	542
	Chapter Assessment	546

Angles of grain piles

	wheat	corn	oat
$m \angle A$	27.0	27.5	28.0
tan A	0.5095	0.5206	0.5317

10.3 *Angles of grain piles* 507, 513

Applications

Interview:
Debby Tewa
Solar Engineer
503, 519

—————— Connection ——————

Literature	498	**Geology**	526
Architecture	504	**History**	533
Transportation	512	**Architecture**	539
Astronomy	518		

Applying Algebra 494, 497, 499, 505, 520, 534

Additional applications: home repair, music, sports, history, forestry, agriculture, navigation, paleontology, theater, aviation, orienteering, archaeology, packaging

Contents **xiii**

CHAPTER 10

In this chapter, many applications of right triangles are presented and extended. Students develop an understanding of right triangle relationships, **trigonometry**, and **vectors**. Students also continue their work with figures in three dimensions by learning about the areas and volumes of **pyramids** and **cones**.

Circles and Spheres

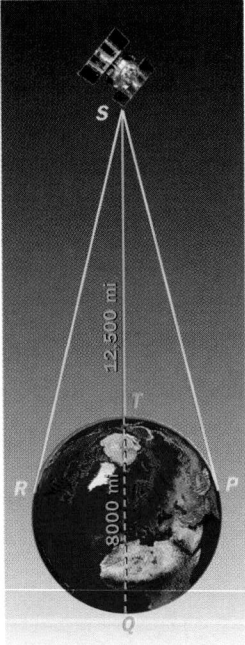

Interview *Ron Courson*   550

11.1 Angles and Circles   553
      Exploration: *Comparing Central and Inscribed Angles*   553

11.2 Tangents, Secants, and Chords   560

11.3 Applying Properties of Chords   567
      Exploration: *Circumscribing a Triangle*   567

11.4 Segment Lengths in Circles   573

11.5 Sectors and Arc Lengths   579

11.6 Surface Areas and Volumes of Spheres   586

11.7 Volumes of Similar Solids   592

11.8 Spherical Geometry   597
      Exploration: *Triangles on a Sphere*   597

Review Chapter Review   606

Assessment   **Ongoing Assessment**   559, 565, 572, 578,
                                                  585, 591, 596, 603

                 Assess Your Progress   566, 585, 603

                 Journal writing   566, 585, 603

                 Portfolio Project: *Modeling Cavalieri's Principle*   604

                 Chapter Assessment   608

11.4 *Global positioning* 575

Applications

Interview:
Ron Courson
Athletic Trainer/
Physical Therapist
557, 589

——— Connection ———

Biology	564	**Biology**	595
Highway Safety	571	**Cooking**	596
Earth Science	577	**Geography**	601
Rock Climbing	584	**Hyperbolic**	
Sports	590	**Geometry**	602

Applying Algebra 558, 562, 568, 574, 576, 587, 595

Additional applications: global positioning, sports,
irrigation, catering, architecture, astronomy, physics

CHAPTER 11

Investigating properties of circles, arcs, chords, tangents,
secants, and sectors is the focus of this chapter. Work with
three-dimensional figures is expanded as students study
surface areas and volumes of **spheres.** Students explore
non-Euclidean geometry by examining relationships
among figures on a sphere.

Coordinates for Transformations

Interview *Adriana Ocampo*		610
12.1	**Dilations with Matrices**	613
12.2	**Translations with Matrices**	619
12.3	**Multiplying Matrices**	626
12.4	**Reflections with Matrices**	634
	Exploration: *Reflections Using Matrix Multiplication*	634
12.5	**Rotations with Matrices**	641
	Exploration: *Matrices with Zeros on a Diagonal*	641
Review	**Chapter Review**	652

Assessment	**Ongoing Assessment**	618, 625, 633, 640, 647
	Assess Your Progress	625, 647
	Journal writing	625, 647
	Portfolio Project: *Creating Fractals*	648
	Chapter Assessment	654
	Cumulative Assessment *Ch. 10–12*	656

12.3 *Morphing*   626

Applications

Interview: *Adriana Ocampo*
Planetary Geologist 623, 645

Connection

Digital Maps	617	**Art Photography**	637
Color Photocopiers	624	**Virtual Reality**	639
Special Effects	631	**Video Games**	646

Applying Algebra 630, 640

Additional applications: computers, morphing

Contents   **xv**

CHAPTER 12

This chapter establishes another important **link** between **algebra and geometry** by using matrices to describe transformations of points, segments, and polygons. Students apply **matrix operations** to transformations of geometric figures, using **technology** where appropriate.

Student Resources

Extra Practice   659–686

Toolbox   687–711

Using Geometric Tools and Transformations   687
Operations with Variable Expressions   690
Linear Equations   694
Graphing on the Coordinate Plane   696
Quadratic Equations   700
Formulas for Geometric Figures   702
Inequalities   705
Ratio and Proportion   707
Data Analysis and Probability   708
Matrices   711

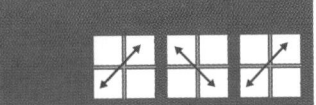

Postulates, Properties, Theorems, and Constructions   712–720

List of Postulates and Properties   712
List of Theorems   714
List of Constructions   720

Technology Handbook   721–729

Using a Graphing Calculator   721
Using a Spreadsheet   725
Using Geometry Software   726

Tables   730–733

Table of Measures   730
Table of Symbols   731
Table of Squares and Square Roots   732
Table of Trigonometric Ratios   733

Appendices   734–741

Appendix 1: A Brief History of Geometric Systems   734
Appendix 2: Truth Tables and Logic   736

Glossary   742–747

Index   748–758

Selected Answers

Student Resources

These resources include practice, review, and reference material to help students with the course. The Toolbox provides a quick brush-up on prerequisite skills. It includes worked examples and practice exercises. The Technology Handbook introduces students to useful features provided by graphing calculators and geometry software.

Geometry

Explorations and Applications

This new program helps you prepare today's students for tomorrow's world.

Support Materials
for all teaching and learning needs

Course Guides
- Standard Courses
- Block Schedule
- Prerequisite Skills Review

Lesson Plans

Assessment Book
- Chapter Tests—2 Forms
- Quizzes
- Performance-Based Assessment

Explorations Lab Manual
- Additional Explorations
- Diagram Masters
- Recording Sheets

Professional Handbook
- Professional Articles
- Support for New Approaches
- Sample Classroom Activities

Portfolio Project Book
- Additional Projects
- Scoring Rubrics
- Recording Sheets

Libro de evaluación
- Exámenes de los capítulos– 2 versiones
- Pruebas
- Evaluación por rendimiento

Multi-Language Glossary
- Spanish
- Vietnamese
- Laotian
- Chinese
- Cambodian
- Arabic

Of Special Interest . . .

Assessment Book
- Chapter Tests—2 Forms
- Quizzes
- Performance-Based Assessment

Study Guide
- Study Guide lesson for each section of the textbook
- Examples and key terms
- Practice and review

Also available in Spanish

Assessment Book
- Chapter tests — 2 versions
- Performance-based assessment
- Short quizzes and cumulative tests

Also available in Spanish

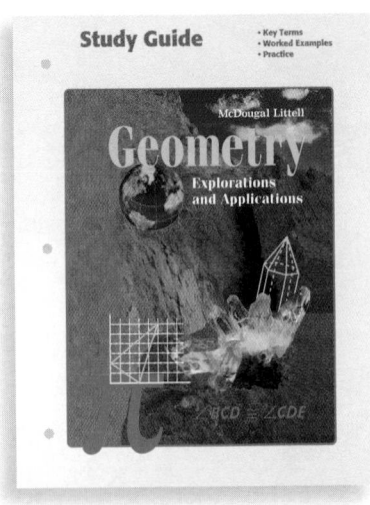

Study Guide
- Key Terms
- Worked Examples
- Practice

. . . for You and your Students

Preparing for College Entrance Tests

- Student handbook with test-preparation strategies
- Preview tests with questions in standardized-test formats

Challenge Problems

- Additional challenging exercises and problems for each section of the textbook
- Chapter-review challenge sets, including Extension problems

Technology Support

McDougal Littell Mathpack

Geometry Inventor

Geometry Inventor is one of four software packages included in *McDougal Littell Mathpack*. This software for exploring and applying geometry enables students to:

- use properties, coordinates, or transformations to create geometric diagrams
- measure figures
- discover the properties of figures

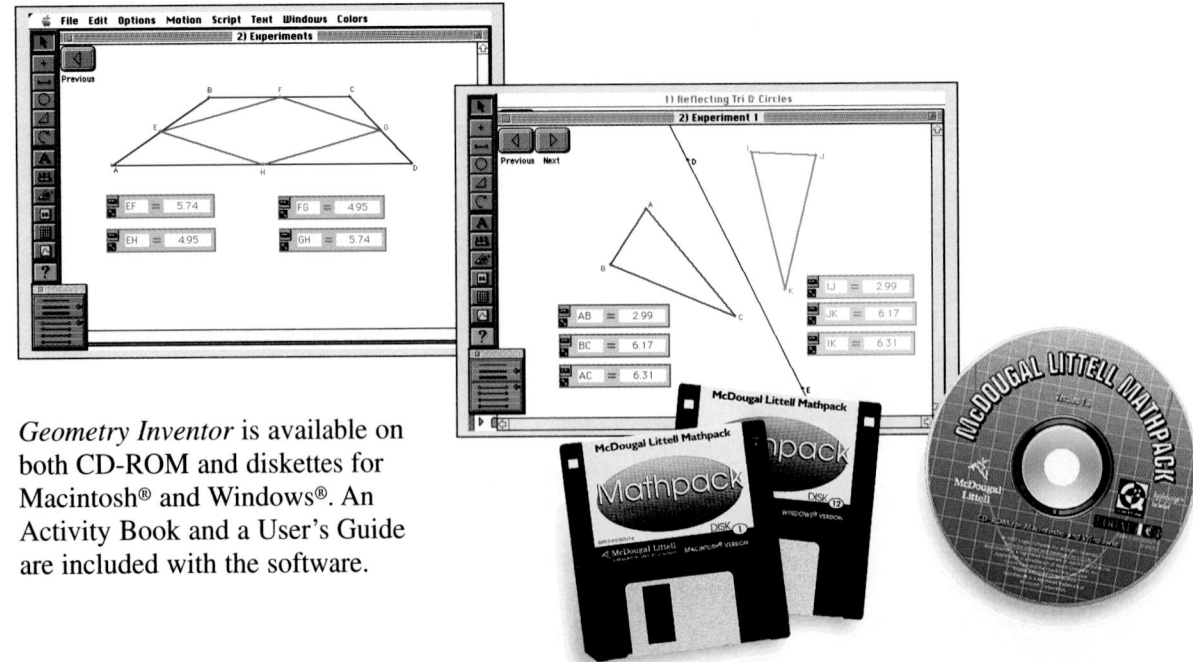

Geometry Inventor is available on both CD-ROM and diskettes for Macintosh® and Windows®. An Activity Book and a User's Guide are included with the software.

Probability Constructor, Function Investigator, and Stats!

These three software packages are also available, separately and as part of *McDougal Littell Mathpack*. Graphics from each of these can be combined with text using *Portfolio Processor*.

Function Investigator
- examine functions
- perform matrix calculations

Probability Constructor
- simulate experiments
- analyze results

Stats!
- collect and record data
- represent data in several ways

Other Technology Support

Test and Practice Generator
- Create worksheets, quizzes, practice tests, and more
- Available for Macintosh and IBM
- Test bank with user's guide

Technology Book
Includes activities for:
- TI-81, TI-82, TI-83, and TI-92 graphing calculators
- Spreadsheets

Complete Teaching Resources

Teacher's Edition includes complete support for planning and teaching your lessons.

- Student pages with complete answers
- Point-of-use teaching and exercise notes
- Planning pages for each chapter, with assignment guides for standard and block scheduling

Teacher's Resources provides a comprehensive selection of support materials, with suggestions for implementing new ideas for teaching and assessment.

- Lesson Plans
- Study Guide
- Practice Bank
- Assessment Book
- Challenge Problems
- Portfolio Project Book
- Explorations Lab Manual
- Preparing for College Entrance Tests
- Warm-Up Transparencies
- Professional Handbook
- Technology Book
- Course Guides
- Solution Key
- Overhead Visuals

Also available:

- Multi-Language Glossary
- Assessment Book, Spanish Edition
- Study Guide, Spanish Edition
- Test and Practice Generator (Macintosh and IBM)
- McDougal Littell Mathpack software

Geometry Explorations and Applications

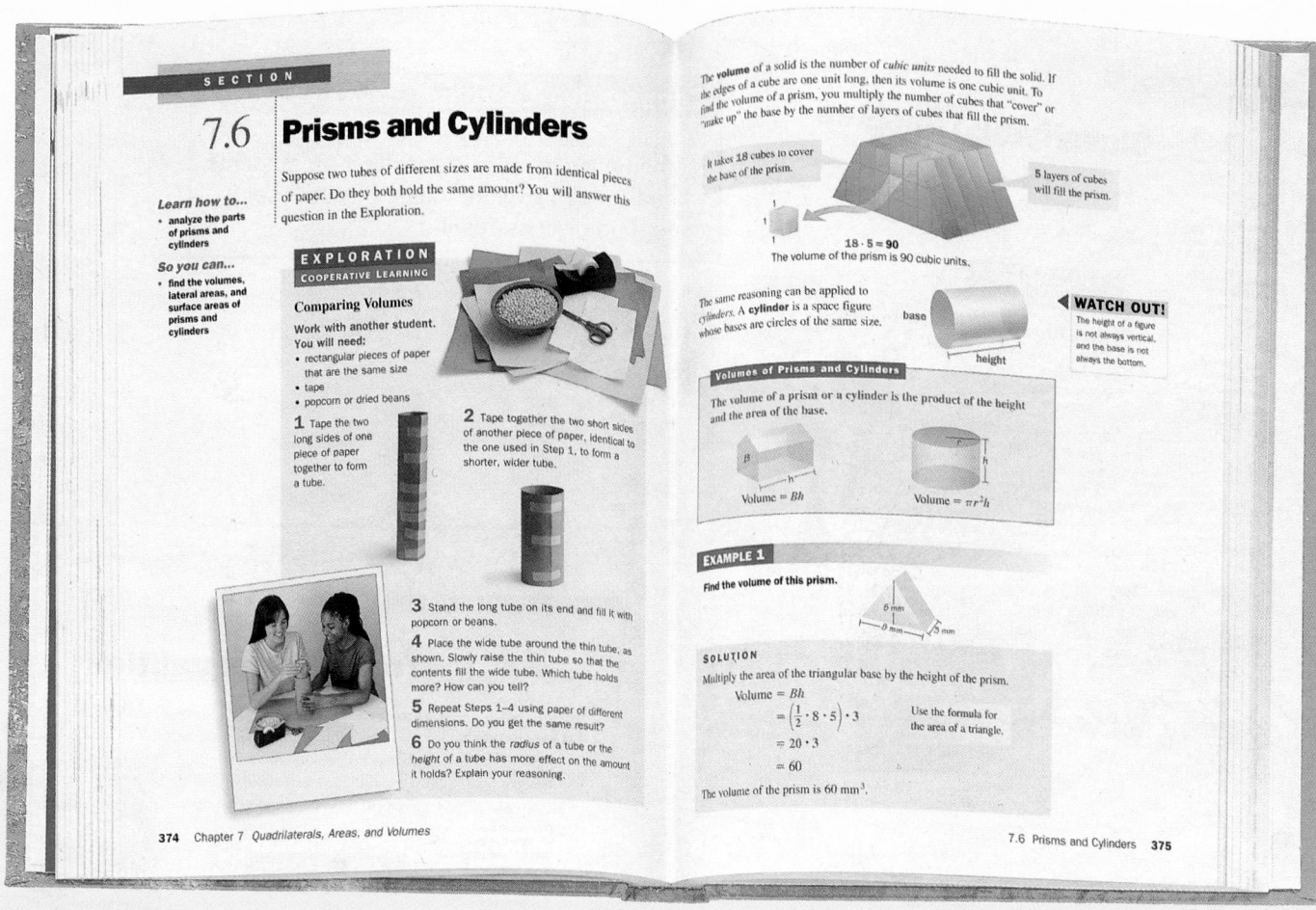

Turn to pages T26–T33 to see how this book integrates these teaching strategies:

- Involve students in exploring and discovering math concepts.
- Connect geometry to the real world and to other subjects and math topics.
- Build understanding of the concepts that provide a strong foundation for future courses and careers.
- Assess students' progress in ways that support learning.

Involve *students in exploring and discovering math concepts.*

2.2 Classifying Triangles

A **triangle** is the figure formed by the segments whose endpoints are three noncollinear points. The shape of a triangle depends on the measures of its angles and the lengths of its sides.

Learn how to...
- name, label, and classify triangles
- sketch triangles and find the measures of their angles

So you can...
- use types of triangles to describe objects such as crystals

EXPLORATION
COOPERATIVE LEARNING

Sorting Triangles

Work in a group of four students.
You will need:
- a set of triangles like the ones at the right

1 One student should write down a "secret rule" that a second student will use to sort the triangles.

2 The second student, the sorter, should choose triangles that fit the rule as the other students watch. At any time, another student can stop the sorter and guess the rule.

One rule might be "triangles with an obtuse angle."

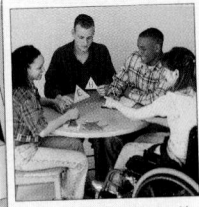

Another possible rule is "triangles with three congruent sides."

3 Repeat Steps 1 and 2 until five rules have been used. Make sure each student has a chance to write a rule, sort the triangles, and guess a rule.

4 Make a list of the rules your group used. What other rules can you think of? What rules did the other groups in your class use?

Cooperative Learning   *A Team Approach*

Mathematics is more meaningful for students when they can explore ideas, discover solutions, and discuss results.

6.1 Triangle Inequalities

Can a triangle have sides of lengths 3 in., 4 in., and 5 in.? How about 3 in., 4 in., and 10 in.? If you know the lengths of two sides of a triangle, are there any limits on how long the third side can be? In the Exploration you will investigate the lengths of sides in triangles.

Learn how to...
- apply the Triangle Inequality Theorems

So you can...
- determine if a triangle can be formed from three given lengths
- use the lengths of two sides of a triangle to describe the length of the third side

EXPLORATION
COOPERATIVE LEARNING

Comparing Sides of Triangles

Work with another student.
You will need:
- geometry software

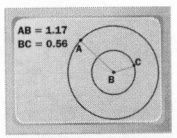

1 Construct two circles with the same center. Label the center *B*. Construct and label point *A* on the larger circle and point *C* on the smaller circle.

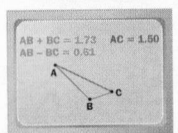

2 Construct and measure $\overline{AB}$ and $\overline{BC}$. Hide the circles so only ∠ *ABC* is showing.

3 Calculate $AB + BC$ and $AB − BC$.

4 Using a different type or color of line, construct $\overline{AC}$. Measure *AC*.

5 Compare *AC* with $AB + BC$ as you move points *A* and *C* to different positions on the circles. Compare *AC* with $AB − BC$. What do you notice?

6 When *AC* is at its largest value, do $\overline{AB}$, $\overline{BC}$, and $\overline{AC}$ form a triangle? Explain why or why not. Do the segments form a triangle when *AC* is at its smallest value?

7 If segments $\overline{AB}$, $\overline{BC}$, and $\overline{AC}$ form a triangle, write two inequalities to describe *AC* in terms of *AB* and *BC*.

Exploring with Technology   *Visualizing Patterns*

With technology, students can explore a wider range of topics than in the past. Geometry software activities help students understand the properties of geometric figures.

5.6 Parallels in Space

Learn how to...
- recognize relationships among parallel and intersecting planes

So you can...
- identify parallel lines and planes in space
- analyze real-world examples of parallel and perpendicular lines in space

There are two possibilities for how two planes can be related: the planes may be parallel or they may intersect in a line. But what happens if there are three planes?

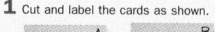

EXPLORATION
COOPERATIVE LEARNING

Investigating Planes and Their Intersections

Work with another student. You will need:
- three index cards
- scissors
- two toothpicks

1 Cut and label the cards as shown.

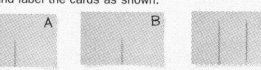

2 Hold cards *A* and *B* so that they model parallel planes. Can you position plane *C* so it is parallel to both *A* and *B*? Can plane *C* intersect one plane but not the other? Can *C* intersect both planes? If so, how are the lines of intersection related?

3 Use cards *A* and *B* to model intersecting planes. Can plane *C* intersect one plane but not the other? Can *C* intersect both planes?

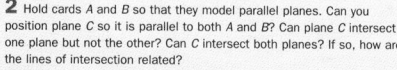

4 Can three planes form exactly one line of intersection? two lines? three lines? Explain.

5 Draw a pair of intersecting lines on one card. Place one end of a toothpick on the point of intersection and position the toothpick so it is perpendicular to both of the lines. A line that is perpendicular to both lines is perpendicular to the plane of the card.

6 Hold two toothpicks to model two different lines that are each perpendicular to the same card. Describe the relationship between the lines.

Modeling with Manipulatives *Building Understanding*

Exploring with manipulatives helps students develop an intuitive base for understanding concepts and techniques.

Real-World Experiments *Working with Data*

Portfolio projects give students opportunities to apply geometry to real-world situations.

PORTFOLIO PROJECT

Applying Solar Geometry

The sun can help to heat your home during the winter. The roof of a passive solar home allows sunlight to shine in during the winter, when the sun is low in the sky. These roofs also help keep homes cool during the summer when the sun is higher.

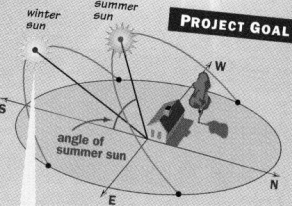

PROJECT GOAL Use the angle of the sun to plan the roof overhang for a passive solar home.

At noon on December 21, the sun's angle is 90° − latitude − 23.5°.

Analyzing Angles

The house in the diagrams shown is in Chicago, Illinois, at a **latitude of 42° north** . To find the angle of the sun at noon on the longest day of the year, June 21, use this expression:

$$90° - 42° + 23.5°$$

After June 21, the days shorten until the sun reaches its lowest point in the sky on December 21. You can use the angle of the sun to find the best angle for a roof.

The angle of the roof allows the sun's rays in during the winter.

The angle of the roof blocks the sun's rays during the summer.

1. Find the amount of overhang, *BC*, of a roof of slope 30° in the Chicago home shown. Explain your method. (*Hint:* Find *BC* in terms of *x* and solve for *x*.)

2. Examine the diagram of a solar home with a flat roof, as shown above. Find the amount of overhang, *y*, that blocks the summer sun.

Connect *geometry to the real world and to other subjects and math topics.*

Real-World Applications   *Mathematics in Context*

Throughout this course new concepts are introduced, practiced, and extended in the context of real-world applications. Seeing the wide variety of settings in which geometry is useful helps students value mathematics.

Connection   ART

How do you make a flat surface appear to have depth? Ancient Egyptians used overlapping forms to suggest depth. Chinese artists used misty spaces to separate nearby forms from distant ones. In the fifteenth and sixteenth centuries, Italian artists explored painting objects as though seen through a window with one eye shut. This method is called *linear perspective*.

Above: Egypt, 1550–1295 B.C.
Right: China, fifteenth century

Italy, around 1470

31. a. Imagine that you are in the painting. Are $\overleftrightarrow{AB}$ and $\overleftrightarrow{CD}$ parallel?
  b. Think of the painting as a flat canvas with paint on it. Are $\overleftrightarrow{AB}$ and $\overleftrightarrow{CD}$ parallel? Explain.

32. a. Imagine that you are in the painting. Name two skew lines.
  b. Think of the painting as a flat canvas. Are the lines from part (a) still skew? Why or why not?

33. Name two lines that are parallel both when you look at the painting as a flat canvas and when you imagine that you are in the painting.

34. Investigation You will need tracing paper. Carefully trace the columns and the lines of the floor and roof.
  a. Extend $\overleftrightarrow{AB}$ and $\overleftrightarrow{CD}$. Label the point of intersection *P*. Where is *P*?
  b. Extend $\overleftrightarrow{EF}$ and $\overleftrightarrow{GH}$. Where do they intersect?
  c. Open-ended Problem Make a conjecture based on your answers to parts (a) and (b). Can you find a counterexample?
  d. Point *P* is called the *vanishing point*. Where is the vanishing point in the illustration at the left?

SECTION

6.2 | Exploring Congruence

Learn how to...
* identify corresponding parts of polygons

So you can...
* match congruent polygons such as windows and window frames

The parts of a car come from many different sources. For all of the pieces to fit together, they must be exactly the right size and shape. For example, the window must match the window frame exactly.

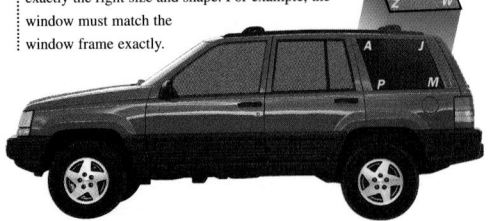

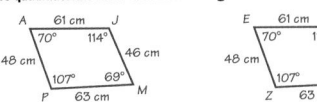

Corner *E* on the window *corresponds* to corner *A* in the car frame. Polygons *AJMP* and *ERWZ* are **congruent polygons** if each part (angle or side) of *AJMP* is congruent to the **corresponding part** of *ERWZ*. Congruent polygons have the same size and shape.

EXAMPLE 1

Are quadrilaterals *AJMP* and *ERWZ* congruent?

WATCH OUT! ▶
Corresponding angles in congruent polygons are not the same as corresponding angles of parallel lines.

SOLUTION

Use the diagram to tell which pairs of corresponding parts are congruent.

$\angle A \cong \angle E$	$\angle M \cong \angle W$	$\overline{AJ} \cong \overline{ER}$	$\overline{MP} \cong \overline{WZ}$
$\angle J \cong \angle R$	$\angle P \cong \angle Z$	$\overline{JM} \cong \overline{RW}$	$\overline{PA} \cong \overline{ZE}$

All eight parts (four angles and four sides) of quadrilateral *AJMP* are congruent to the corresponding parts of quadrilateral *ERWZ*, so quadrilateral *AJMP* ≅ quadrilateral *ERWZ*.

Interdisciplinary Problems   *Connecting Learning*

Each chapter has several clusters of exercises that focus on the connections of geometry to careers and to other subject areas, both within and outside of mathematics.

INTERVIEW Debby Tewa

Look back at the article on pages 490–492.

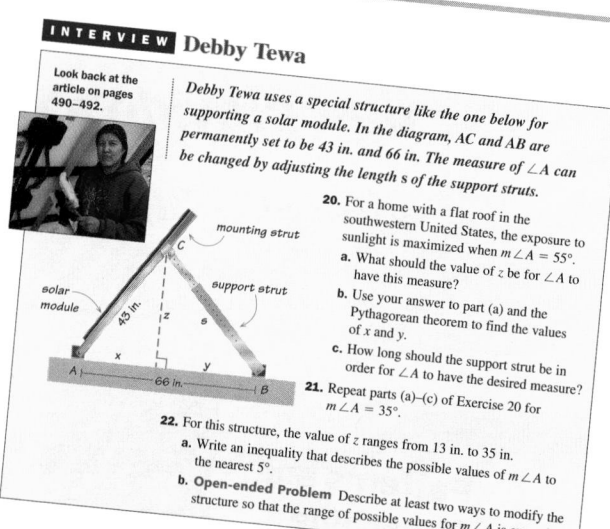

Debby Tewa uses a special structure like the one below for supporting a solar module. In the diagram, AC and AB are permanently set to be 43 in. and 66 in. The measure of $\angle A$ can be changed by adjusting the length s of the support struts.

mounting strut
solar module
45 in.
support strut
z
s
C
x
y
A — 66 in. — B

20. For a home with a flat roof in the southwestern United States, the exposure to sunlight is maximized when $m \angle A = 55°$.
 a. What should the value of z be for $\angle A$ to have this measure?
 b. Use your answer to part (a) and the Pythagorean theorem to find the values of x and y.
 c. How long should the support strut be in order for $\angle A$ to have the desired measure?

21. Repeat parts (a)–(c) of Exercise 20 for $m \angle A = 35°$.

22. For this structure, the value of z ranges from 13 in. to 35 in.
 a. Write an inequality that describes the possible values of $m \angle A$ to the nearest 5°.
 b. Open-ended Problem Describe at least two ways to modify the structure so that the range of possible values for $m \angle A$ is expanded.

Sketch a 45-45-90 triangle and a 30-60-90 triangle. Label each side length. Then use the triangles to find the exact value of each ratio.

23. sin 45° **24.** cos 45° **25.** cos 60°

26. sin 60° **27.** sin 30° **28.** cos 30°

29. AVIATION A jet takes off at a 15° angle. The jet's air speed is 300 ft/s.
 a. Write an equation that gives the distance d (in feet) that the jet has traveled through the air in terms of t, the number of seconds it has been in the air.
 b. Write equations for the horizontal distance h and the vertical distance v that the jet travels in t seconds.
 c. After 10 s, what horizontal distance has the jet traveled? What is the jet's altitude?

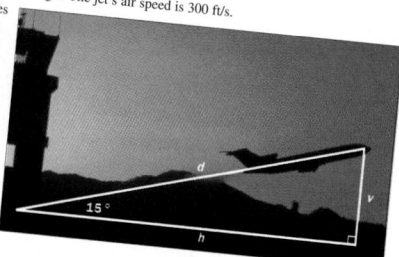
d
v
$15°$
h

10.4 Sine and Cosine Ratios **519**

SECTION

4.2 Equations of Lines

Learn how to...
• find the slope of a segment or a line
• write equations of lines in slope-intercept form

So you can...
• graph and compare equations of lines
• investigate geometric relationships using lines

Maybe you would like to go to the seashore but live too far inland. Or maybe you would like to try downhill skiing but live in a region without mountains or snow. At indoor beaches and ski resorts that overcome the limitations of geography and climate, people swim and ski all year.

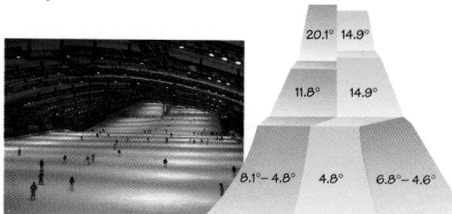

20.1° 14.9°
11.8° 14.9°
8.1°– 4.8° 4.8° 6.8°– 4.6°

The world's largest indoor ski slope is located in Funabashi, Japan. It is called the *LaLaport Skidome* and has three courses—a red or *expert* course, a yellow or *intermediate* course, and a blue or *beginners'* course. Each course consists of alternating flat and sloped regions. A diagram of the red course is shown below. All distances are rounded to the nearest meter.

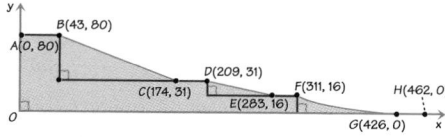

B(43, 80)
A(0, 80)
D(209, 31)
C(174, 31)
F(311, 16)
E(283, 16)
H(462, 0)
O
G(426, 0)
x
y

THINK AND COMMUNICATE

1. How many meters does the red course drop in its first sloped part, between point B and point C?

2. What is the horizontal distance from B to C?

You can express how steep the first sloped part of the red course is by using the *slope formula*.

4.2 Equations of Lines **173**

Build understanding of the concepts that provide a strong foundation for future courses and careers.

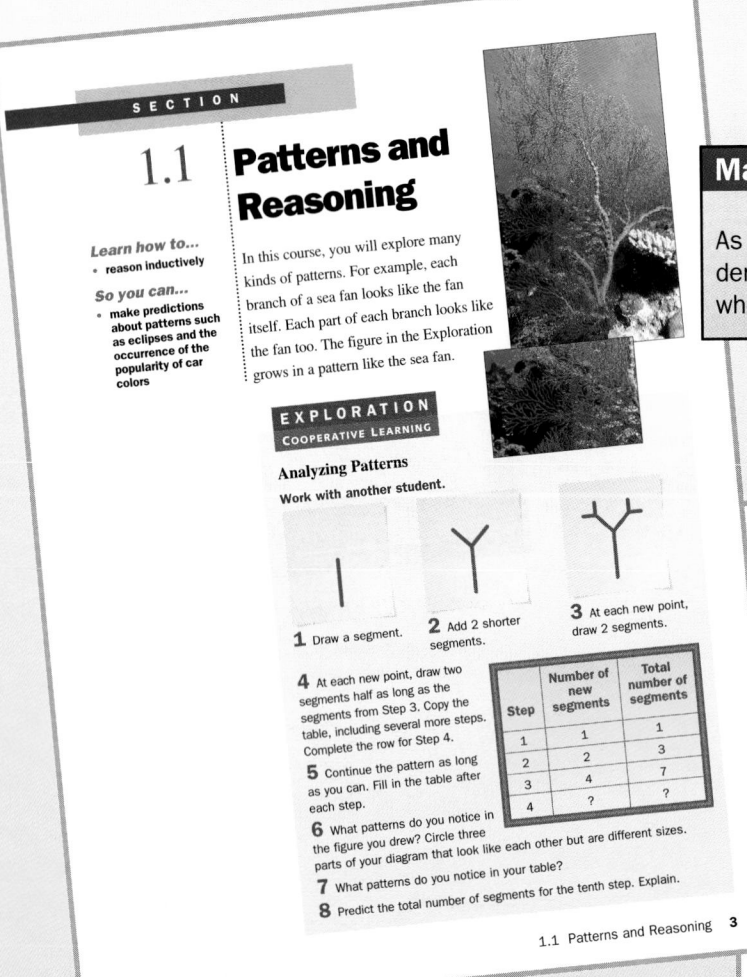

SECTION

1.1 Patterns and Reasoning

Learn how to...
• reason inductively

So you can...
• make predictions about patterns such as eclipses and the occurrence of the popularity of car colors

In this course, you will explore many kinds of patterns. For example, each branch of a sea fan looks like the fan itself. Each part of each branch looks like the fan too. The figure in the Exploration grows in a pattern like the sea fan.

EXPLORATION
COOPERATIVE LEARNING

Analyzing Patterns
Work with another student.

1 Draw a segment.

2 Add 2 shorter segments.

3 At each new point, draw 2 segments.

4 At each new point, draw two segments half as long as the segments from Step 3. Copy the table, including several more steps. Complete the row for Step 4.

5 Continue the pattern as long as you can. Fill in the table after each step.

Step	Number of new segments	Total number of segments
1	1	1
2	2	3
3	4	7
4	?	?

6 What patterns do you notice in the figure you drew? Circle three parts of your diagram that look like each other but are different sizes.

7 What patterns do you notice in your table?

8 Predict the total number of segments for the tenth step. Explain.

1.1 Patterns and Reasoning **3**

Mathematical Reasoning *A Unifying Theme*

As an introduction to reasoning and proof, students look at patterns and learn to recognize when to use inductive or deductive reasoning.

SECTION

3.4 Two-Column Proof

A proof shows that a statement is true. Often more than one correct proof can be written. Two proofs of the same theorem may present the same statements and reasons, but look very different. Compare Mary's and Carlotta's proofs of the Vertical Angles Theorem.

Learn how to...
• write a proof in two-column format

So you can...
• use and present convincing arguments in a variety of formats

Given: ∠1 and ∠3 are vertical angles.
Prove: ∠1 ≅ ∠3

$m\angle 1 + m\angle 2 = 180°$ and $m\angle 2 + m\angle 3 = 180°$, because ∠1 and ∠2 are a linear pair, and ∠2 and ∠3 are a linear pair. This means that $m\angle 1 + m\angle 2 = m\angle 2 + m\angle 3$, by the Substitution Property. Subtracting $m\angle 2$ from both sides gives $m\angle 1 = m\angle 3$, so ∠1 ≅ ∠3.

Given: ∠1 and ∠3 are vertical angles.
Prove: ∠1 ≅ ∠3

Statement	Reason
1. ∠1 and ∠3 are vertical angles.	1. Given
2. $m\angle 1 + m\angle 2 = 180°$ $m\angle 2 + m\angle 3 = 180°$	2. Angles in a linear pair are supplements.
3. $m\angle 1 + m\angle 2 = m\angle 2 + m\angle 3$	3. Substitution Property (Step 2)
4. $m\angle 1 = m\angle 3$	4. Subtraction Property
5. ∠1 ≅ ∠3	5. Definition of congruent angles

THINK AND COMMUNICATE

1. How are the two proofs different? How are they the same?

2. Mary and Carlotta proved the same theorem. Which proof do you prefer? Explain why.

130 Chapter 3 *Reasoning in Geometry*

Justification and Proof *Communicating Ideas*

Students learn to explain their thinking verbally and to justify their conclusions in writing using key steps, paragraph, two-column, and flow proofs.

Surface Area

In Chapter 2 you learned how to sketch a net for a prism by drawing the faces connected to one another. The area of a net for a three-dimensional figure is called the **surface area** of the figure, abbreviated *S.A.*

The area of the lateral faces is called the **lateral area**. The lateral faces of a prism can be arranged to form one rectangle.

The perimeter of the base is one dimension of the rectangle.

The two bases are congruent, so their areas are equal.

The height of the prism is the other dimension of the rectangle.

$$\text{Surface Area} = \text{Lateral area} + \text{Area of bases}$$
$$= (\text{Perimeter of base} \times \text{height}) + 2(\text{Area of base})$$
$$= ph + 2B$$

THINK AND COMMUNICATE

1. Imagine cutting along the height of a tube and laying it flat. What shape is the resulting figure?

2. You learned in Chapter 2 that prisms are named according to their bases. What do you think a *regular prism* is?

Surface Areas of Prisms and Cylinders

The surface area of a prism or cylinder is the sum of the lateral area and the area of the bases.

$$S.A. = ph + 2B \qquad\qquad S.A. = 2\pi rh + 2\pi r^2$$

Visualizing Mathematics   *Multiple Perspectives*

Three-dimensional objects are illustrated in a variety of ways to give students a better understanding of their properties.

Mathematical Modeling   *A Tool for Understanding*

Geometric figures are used to model real-world objects and situations so that students can draw conclusions from the properties of the figures.

SECTION

6.6 Properties of Isosceles Triangles

Learn how to...
* apply the Isosceles Triangle Theorem and its converse

So you can...
* find measures in isosceles triangles
* write proofs using isosceles triangles

In one type of rock climbing, called *top roping*, climbers tie themselves to a rope that is supported by *anchors* at the top of the climb. If a climber slips, the anchors catch the fall. For safety, climbers use at least two anchors.

The illustration below shows two anchors in a horizontal crack of a rock face. To be sure that the force on the anchors is equal, the angles formed by the anchors and the crack should be congruent.

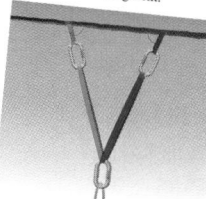

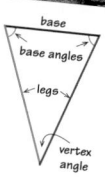

base

base angles

legs

vertex angle

isosceles triangle

The triangle formed by these anchors is an isosceles triangle. The sides opposite the congruent angles are the **legs** and the third side is the **base**. The congruent angles are **base angles** and the third angle is the **vertex angle**.

THINK AND COMMUNICATE

1. **a.** If the red anchor is longer than the blue anchor, are the base angles congruent? If not, which angle is larger?
   b. Use the triangle inequality theorems on page 280. Which theorem supports your answer to part (a)?

2. If a climber adjusts the anchors so they are the same length, do you think that the base angles will be congruent?

Assess *students' progress in ways that support learning.*

Embedded Assessment   *A Part of Learning*

Ongoing Assessment questions in each section ask students to apply concepts and explain their thinking. *Journal* writing gives students a chance to reflect on their own learning process.

ONGOING ASSESSMENT

23. Cooperative Learning Work in a group of three people. You will need graph paper and scissors. Two ways to cut a 3 in. by 3 in. square into two congruent polygons are shown. Find at least five more ways. Sketch, label, and name each differently shaped pair of congruent polygons that you cut.

You may use concave polygons.

SPIRAL REVIEW

Give the postulate, definition, or property that makes each statement about the diagram true. *(Section 3.2)*

24. $\angle B \cong \angle D$ **25.** $\overline{AC} \cong \overline{AC}$

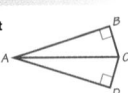

ASSESS YOUR PROGRESS

VOCABULARY

congruent polygons (p. 285)   **corresponding part** (p. 285)

Tell whether or not a triangle can be formed from the given lengths. *(Section 6.1)*

1. 1 in., 4 in., 10 in.   **2.** 2 ft, 4 ft, 2 ft   **3.** 5.4 m, 0.5 m, 5.0 m

The lengths of two sides of a triangle are given. What can you conclude about the length of the third side? *(Section 6.1)*

4. 1 in., 4 in.   **5.** 2 ft, 4 ft   **6.** 18 cm, 12 cm

7. Name all of the triangles in the diagram that appear to be congruent. *(Section 6.2)*

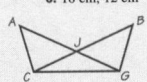

8. $\triangle DEF \cong \triangle HKL$ *(Section 6.2)*
a. Find the value of r.
b. Find the value of p.

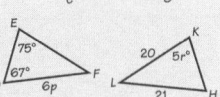

9. Journal Sue says that to tell if three numbers can be the lengths of the sides of a triangle, you only need to check if the sum of the two smaller numbers is greater than the third number. Do you agree? Explain.

CHAPTER

1

Assessment

For Questions 15–19, use the diagram at the right.
15. Give at least three names for plane *ABC*.
16. What is the intersection of plane *ABE* and $\overleftrightarrow{EC}$?
17. What is the intersection of $\overleftrightarrow{AD}$ and $\overleftrightarrow{DE}$?
18. Name a pair of skew lines.
19. Name two lines that appear to be parallel.

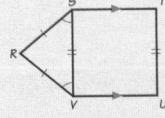

Sketch each situation.
20. Line ℓ intersects $\overleftrightarrow{AB}$ at point *B*.
21. Plane *ABC* intersects line *m* at point *B*.
22. Noncollinear points *A*, *B*, *C*, and *D* lie in plane *H* and $\overleftrightarrow{AB} \parallel \overleftrightarrow{CD}$.

SECTIONS 1.5, 1.6, *and* 1.7

Use the diagram at the right.
23. List as many conclusions as you can about the diagram.
24. Describe a conclusion that looks correct but may not be.

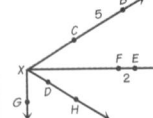

25. Point *C* is the midpoint of $\overline{XB}$ and $XB = XE$.
a. Find *XC*.   b. Find *XB*.   c. Find *XF*.
26. $XD = 3y - 1$, $DH = 3$, and $XH = 4y + 1$. Find the value of *y* and the length *XH*.
27. $\overrightarrow{XE}$ bisects $\angle BXD$, $\overrightarrow{XD}$ bisects $\angle BXG$, and $m \angle BXF = 30°$. Find $m \angle BXD$ and $m \angle BXG$.

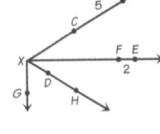

28. Open-ended Problem Draw two triangles. Measure the angles and sides of the triangles. Make a conjecture about the relationship between the shortest side and the smallest angle in any triangle. Draw another triangle and test your conjecture.

PERFORMANCE TASK

29. Design a border for wallpaper or fabric with a repeating pattern. Use as many of the following geometric relationships and objects as you can: angle bisector, parallel lines, midpoint, reflection symmetry, rotational symmetry, translational symmetry, vertex, congruent segments, congruent angles. Draw at least two repeats of your design. Label the geometric figures and relationships that you used.

Monitoring Progress   *Check for Understanding*

A variety of types of assessment questions are included throughout the course. A *Performance Task* is included in each chapter assessment.

Modeling Cavalieri's Principle

Archimedes discovered the formula for the volume of a sphere by imagining cutting a sphere, cone, and cylinder into very thin slices and comparing the slices. This method was developed further by Cavalieri, and became part of the foundation for calculus. Imagine slicing each object at the same height. How do you think the areas of the cross sections are related?

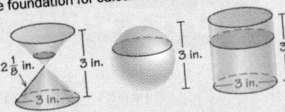

PROJECT GOAL Compare the volumes of a sphere, cylinder, and two cones by comparing cross sections of the objects.

Making the Shapes

Work with a partner to make the sphere, cylinder, and two cones with the dimensions shown above. Use modeling clay.

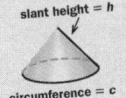 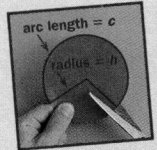 

CONE To make a mold for the cone above, cut a circle out of strong plastic and draw a sector as shown. Tape the sector's edges together.

SPHERE You can check the diameter of your sphere by using a circular cutout with a diameter that is slightly larger than 3 in.

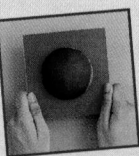 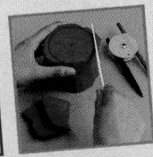

CYLINDER You might want to make your cylinder too large and then trim it carefully by using a compass and plastic knife.

Portfolio Assessment *Demonstrating Growth*

The *Portfolio Projects* provide opportunities for original student work that can be part of a mathematical portfolio. In each project, students are asked to present their results and assess their work.

Comparing Cross Sections

1. Cut each object carefully into six slices of equal thickness. Use fishing line or dental floss.

2. Combine the clay from the bottom slice of the cones and the bottom slice of the sphere to make a new disk of equal thickness.

3. Compare the new disk with the bottom slice of the cylinder. How were the cross sectional areas of the bottom slices of the objects related?

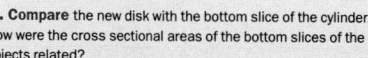

4. Repeat Steps 2 and 3 five times, comparing the cross sections of the slices at each height.

Presenting Your Project

Write a report describing your experiment. Include answers to the following questions.

- How are the areas of the cross sections of the objects related? How are the volumes of the objects related? How do you know?

- How can you use the volume formulas for a cylinder and a cone to find the volume formula for a sphere?

You may want to extend your project to include some of the ideas below:

- Look up *Cavalieri's Principle* in an encyclopedia or a mathematics dictionary. Summarize it and explain how you used it in your project.

- If *x* is the height at which you slice each object, write an expression for the area of each cross section. Show that the relationship you discovered is true for any length *x* that is less than 3 in.

Self-Assessment
What grade would you give yourself for this project? Why? If you did the experiment again, what would you do differently? Why?

Planning Pages for Every Chapter

CHAPTER

3 Reasoning in Geometry

OVERVIEW

Connecting to Prior and Future Learning

⇔ In Chapter 3, students continue their study of reasoning from Chapter 1 as they learn about inductive and deductive reasoning, converses, negations, and contrapositives.

⇔ Students also learn about postulates, definitions, and properties in this chapter. They will discover how to use these ideas to justify statements about geometric figures, which leads to a study of paragraph proof and two-column proof.

⇔ This chapter also includes an introduction to the Pythagorean theorem. Students prove this theorem using manipulatives and concepts from algebra. Students can review simplifying radicals by studying pages 700 and 701 in the **Student Resources Toolbox**.

Chapter Highlights

Interview with Mary-Jacque Mann: The use of logic to solve crimes is highlighted in this interview with Mary-Jacque Mann. Related exercises can be found on pages 128 and 133.

Explorations in Chapter 3 involve using geometry software to explore the relationship between exterior and interior angles of a triangle in Section 3.3, and using the areas of right triangles cut out of a square to prove the Pythagorean theorem in Section 3.6.

The Portfolio Project: Students use a classification system to classify a guitar as a stringed, wind, percussion, or keyboard instrument. They then develop and use a classification system for something that they are interested in.

Technology: Students use geometry software to explore the relationship between the measures of interior and exterior angles of a triangle in Section 3.3 and to discover a property of quadrilaterals in Section 3.6. A spreadsheet is used in Section 3.6 to find Pythagorean triples.

OBJECTIVES

Section	Objectives	NCTM Standards
3.1	• Use deductive reasoning to reach conclusions. • Make a convincing argument. • Recognize valid and invalid arguments.	1, 2, 3, 4, 7
3.2	• Recognize and use postulates, definitions, and properties. • Justify statements about geometric figures.	1, 2, 3, 4, 7
3.3	• Write and understand mathematical proofs. • Use mathematical reasoning to prove that a statement is always true.	1, 2, 3, 4, 7
3.4	• Write a proof in two-column format. • Use and present convincing arguments in a variety of formats.	1, 2, 3, 4, 7
3.5	• Write the converse of a conditional statement. • Recognize and use converses in logical arguments.	1, 2, 3, 4, 7
3.6	• Find the lengths of the sides of a right triangle. • Decide if a triangle is a right triangle. • Find lengths of parts of figures and real-life objects.	1, 2, 3, 4, 7
3.7	• Determine whether a triangle is acute or obtuse from the lengths of its sides. • Write inverses and contrapositives of statements. • Recognize inverses and contrapositives in logical arguments.	1, 2, 3, 4, 7

108A

OVERVIEW

The **Overview** provides a summary of connections to prior and future learning and highlights the chapter interview, explorations, portfolio project, and use of technology.

OBJECTIVES

Chapter **Objectives** gives objectives and NCTM Standards for each section.

INTEGRATION

Mathematical Connections	3.1	3.2	3.3	3.4	3.5	3.6	3.7
geometry	111–116*	117–123	124–129	130–135	136–140	141–147	148–153
algebra	112, 115	123				142–144, 146	153
data analysis, probability, discrete math							153
patterns and functions	113, 114		129				
logic and reasoning	111–116	117–123	124–129	130–135	136–140	147	149–153

Interdisciplinary Connections and Applications							
history and geography						146	
reading and language arts			127		139		
sports and recreation					138		
zoology and archaeology	115					145	
marketing	116						
personal finance	113						
architecture and design						146, 147	
cooking, navigation, measurement, clocks, carpentry, travel, auto maintenance		120–122				143	150, 151

** **Bold page numbers** indicate that a topic is used throughout the section.*

TECHNOLOGY

	opportunities for use with	
Section	Student Book	Support Material
3.1		**Technology Book:** Spreadsheet Activity 3
3.3	geometry software	**Technology Book:** Calculator Activity 4 TI-92 Activity 3
3.6	scientific calculator graphing calculator geometry software	**Technology Book:** Spreadsheet Activity 3 TI-92 Activity 3 **Geometry Inventor Activity Book:** Activity 5
3.7		**Technology Book:** TI-92 Activity 3

108B

INTEGRATION

The **Integration** chart highlights the mathematical and interdisciplinary connections and applications found throughout the chapter.

TECHNOLOGY

The **Technology** chart highlights opportunities to use technology in both the student book and support materials.

Regular Scheduling (45 min)

Section	Materials Needed	Core Assignment	Extended Assignment	Applications	Communication	Technology
				exercises that feature		
3.1		1–14, 24–29	1–29	19, 20, 22, 23	14–20, 22, 24, 25	
3.2	protractor	1–14, 22–28, 30–34, AYP*	1–34, AYP	15–21, 29	16–18, 20, 29	
3.3	geometry software or ruler, protractor	1–6, 10, 11, 15–26	1–26	7–9, 12–14	7–9, 14, 21	
3.4		1–4, 10–16, 18–23, AYP	1–23, AYP	5–9	8, 9, 18	
3.5		1–9, 11, 18–28	1–28	10, 13–17, 19	10	
3.6	scissors, ruler, calculator, graph paper, heavy paper, geometry software	**Day 1:** 1–27 odd, 31–35 **Day 2:** 2–26 even, 47–50	**Day 1:** 1–35 odd **Day 2:** 2–36 even, 37–50	29 37–40, 45	29, 31 36, 47	46
3.7	graph paper, scissors, ruler, protractor	**Day 1:** 1–14, 16–24 **Day 2:** 29–34, 36–45, AYP	**Day 1:** 1–24 **Day 2:** 25–45, AYP	15 25–28	15	
Review/ Assess		**Day 1:** 1–10 **Day 2:** 11–20 **Day 3:** Ch. 3 Test	**Day 1:** 1–10 **Day 2:** 11–20 **Day 3:** Ch. 3 Test			
Portfolio Project		Allow 2 days.	Allow 2 days.			

Yearly Pacing (with Portfolio Project)	Chapter 3 Total	Chapters 1–3 Total	Remaining	Total
	14 days	42 days	118 days	160 days

Block Scheduling (90 min)

	Day 15	Day 16	Day 17	Day 18	Day 19	Day 20	Day 21
Teach/Interact	3.1 3.2	3.3: Exploration, page 125 3.4	3.5 3.6: Exploration, page 141	Continue with 3.6 3.7: Exploration, page 148	Continue with 3.7 Review	Review Port. Proj.	Ch. 3 Test Port. Proj.
Apply/Assess	**3.1:** 1–14, 24–29 **3.2:** 1–14, 22–28, 30–34, AYP*	**3.3:** 1–6, 10, 11, 15–26 **3.4:** 1–4, 10–23, AYP	**3.5:** 1–9, 11, 18–28 **3.6:** 1–27 odd, 31–35	**3.6:** 2–26 even, 47–50 **3.7:** 1–14, 16–24	**3.7:** 29–34, 36–45, AYP **Review:** 1–10	**Review:** 11–20 **Port. Proj.**	**Ch. 3 Test Port. Proj.**

NOTE: A one-day block has been added for the Portfolio Project—timing and placement to be determined by teacher.

Yearly Pacing (with Portfolio Project)	Chapter 3 Total	Chapters 1–3 Total	Remaining	Total
	7 days	21 days	59 days	80 days

* **AYP** is Assess Your Progress.

108C

PLANNING GUIDE

The **Planning Guide** gives materials, pacing, and suggested assignments for each section, and block scheduling assignments.

Section	Practice Bank	Study Guide*	Assessment Book*	Visuals	Explorations Lab Manual	Lesson Plans	Technology Book
3.1	16	3.1		Warm-Up 3.1		3.1	Spreadsheet Act. 3
3.2	17	3.2	Test 11	Warm-Up 3.2		3.2	
3.3	18	3.3		Warm-Up 3.3	Master 10	3.3	Calculator Act. 4 TI-92 Act. 3
3.4	19	3.4	Test 12	Warm-Up 3.4	Add. Expl. 4	3.4	
3.5	20	3.5		Warm-Up 3.5		3.5	
3.6	21	3.6		Warm-Up 3.6 Folder 4	Master 2	3.6	Spreadsheet Act. 3 TI-92 Act. 3
3.7	22	3.7	Test 13	Warm-Up 3.7.	Add. Expl. 5 Masters 1, 11	3.7	TI-92 Act. 3
Review Test	23	Chapter Review	Tests 14, 15, Alternative Assessment			Review Test	

*Spanish versions of *Study Guide* and *Assessment Book* are available.

Chapter Support
- Course Guides
- Lesson Plans
- Portfolio Project Book:
  Additional Project 2:
  Using Reasoning to Make Decisions
- Preparing for College Entrance Tests
- Multi-Language Glossary
- *Test Generator* Software
- Professional Handbook
- Challenge Problems

Software Support

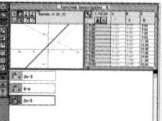

McDougal Littell Mathpack
Geometry Inventor

Internet Support

http://www.hmco.com
Next go to McDougal Littell; then the Education Center; then Secondary Math.

Books, Periodicals
Hirschorn, Daniel B. and Denisse R. Thompson. "Technology and Reasoning in Algebra and Geometry." *Mathematics Teacher* (February 1996): pp. 138–142.
McGivney, Jean M. and Thomas C. DeFranco. "Geometry Proof Writing: A Problem-Solving Approach a la Polya." *Mathematics Teacher* (October 1995): pp. 552–555.
Chazan, Daniel and Richard Houde. *How to Use Conjecturing and Microcomputers to Teach Geometry.* Reston, VA: NCTM, 1989.

Serra, Michael. *Discovering Geometry: An Inductive Approach.* Berkeley, CA: Key Curriculum Press.

Activities, Manipulatives
Miller, William A. and Linda Wagner. "Pythagorean Dissection Puzzles." *Mathematics Teacher* (April 1993): pp. 302–314.
Naraine, Bishnu. "If Pythagoras Had a Geoboard." *Mathematics Teacher* (February 1993): pp. 137–148.

Software
Jackiw, Nicholas, designer. *The Geometer's Sketchpad.* (Ver. 3.0) for Macintosh or Windows. Berkeley, CA: 1995.

Videos
Apostol, Tom M. *The Theorem of Pythagoras.* Videotape and guide. Reston, VA: NCTM, 1988.

Internet
Find Classroom Materials on the World Wide Web at the Teacher's Place:
http://forum.swarthmore.edu/classroom.html

108D

LESSON SUPPORT

The **Lesson Support** chart lists all support materials for each section.

OUTSIDE RESOURCES

Outside Resources lists books, periodicals, manipulatives and activities, software, videos, and Internet addresses.

A Teaching Plan for Every Section

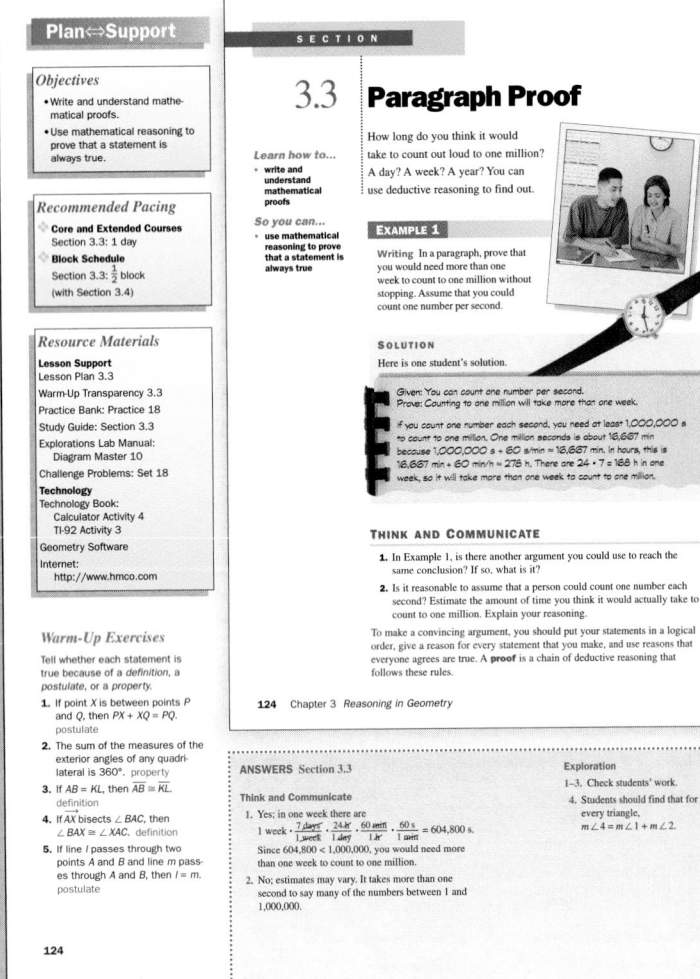

Plan⇔Support

Objectives
- Write and understand mathematical proofs.
- Use mathematical reasoning to prove that a statement is always true.

Recommended Pacing
◆ **Core and Extended Courses**
   Section 3.3: 1 day
◆ **Block Schedule**
   Section 3.3: $\frac{1}{2}$ block
   (with Section 3.4)

Resource Materials
Lesson Support
Lesson Plan 3.3
Warm-Up Transparency 3.3
Practice Bank: Practice 18
Study Guide: Section 3.3
Explorations Lab Manual:
   Diagram Master 10
Challenge Problems: Set 18
Technology
Technology Book:
   Calculator Activity 4
   TI-92 Activity 3
Geometry Software
Internet:
   http://www.hmco.com

Warm-Up Exercises
Tell whether each statement is true because of a *definition*, a *postulate*, or a *property*.
1. If point *X* is between points *P* and *Q*, then *PX* + *XQ* = *PQ*. postulate
2. The sum of the measures of the exterior angles of any quadrilateral is 360°. property
3. If *AB* = *KL*, then $\overline{AB} \cong \overline{KL}$. definition
4. If $\overrightarrow{AX}$ bisects ∠ *BAC*, then ∠ *BAX* ≅ ∠ *XAC*. definition
5. If line *l* passes through two points *A* and *B* and line *m* passes through *A* and *B*, then *l* = *m*. postulate

SECTION

3.3 Paragraph Proof

Learn how to...
- write and understand mathematical proofs

So you can...
- use mathematical reasoning to prove that a statement is always true

How long do you think it would take to count out loud to one million? A day? A week? A year? You can use deductive reasoning to find out.

EXAMPLE 1

Writing In a paragraph, prove that you would need more than one week to count to one million without stopping. Assume that you could count one number per second.

SOLUTION

Here is one student's solution.

Given: You can count one number per second.
Prove: Counting to one million will take more than one week.

If you count one number each second, you need at least 1,000,000 s to count to one million. One million seconds is about 16,667 min because 1,000,000 s ÷ 60 s/min = 16,667 min. In hours, this is 16,667 min ÷ 60 min/h = 278 h. There are 24 • 7 = 168 h in one week, so it will take more than one week to count to one million.

THINK AND COMMUNICATE

1. In Example 1, is there another argument you could use to reach the same conclusion? If so, what is it?
2. Is it reasonable to assume that a person could count one number each second? Estimate the amount of time you think it would actually take to count to one million. Explain your reasoning.

To make a convincing argument, you should put your statements in a logical order, give a reason for every statement that you make, and use reasons that everyone agrees are true. A **proof** is a chain of deductive reasoning that follows these rules.

124 Chapter 3 *Reasoning in Geometry*

ANSWERS Section 3.3

Think and Communicate
1. Yes; in one week there are
   1 week • $\frac{7 \text{ days}}{1 \text{ week}}$ • $\frac{24 \text{ hr}}{1 \text{ day}}$ • $\frac{60 \text{ min}}{1 \text{ hr}}$ • $\frac{60 \text{ s}}{1 \text{ min}}$ = 604,800 s.
   Since 604,800 < 1,000,000, you would need more than one week to count to one million.
2. No; estimates may vary. It takes more than one second to say many of the numbers between 1 and 1,000,000.

Exploration
1–3. Check students' work.
4. Students should find that for every triangle,
   m∠4 = m∠1 + m∠2.

Plan⇔Support

- Section Objectives
- Recommended Pacing
- Resource Materials
- Warm-Up Exercises

A **theorem** is a conjecture that can be proved to be true. A proof that is written in complete sentences is called a **paragraph proof**. When you write a proof, you make a series of **statements** and give the **reason** for each statement. Given information, postulates, definitions, and theorems can be used as reasons in a proof.

The **hypothesis** is the information you are **given**.

The **conclusion** is the statement you want to prove.

Given: You can count one number per second.
Prove: Counting to one million will take more than one week.

If you count one number each second, you need at least 1,000,000 s to count to one million. One million seconds is about 16,667 min because 1,000,000 s ÷ 60 s/min = 16,667 min. In hours, this is 16,667 min ÷ 60 min/h = 278 h. There are 24 • 7 = 168 h in one week, so it will take more than one week to count to one million.

For each **statement** of a step in your reasoning, include the reason the statement is true.

In the Exploration, you will learn a theorem about the exterior angles of a triangle. You will see how to prove this theorem using facts and properties that you already know.

EXPLORATION
COOPERATIVE LEARNING

Measuring Exterior Angles

Work with another student.
You will need:
- geometry software
- a ruler and protractor

1 Draw a triangle with one side extended, as in the diagram. Find the measures of ∠1, ∠2, ∠3, and ∠4.

2 Round the angle measures to the nearest degree. Record the measures in a table.

m∠1	m∠2	m∠3	m∠4	m∠1 + m∠2
26°	78°	76°	104°	104°
?	?	?	?	?
?	?	?	?	?

3 Move one vertex, or draw new triangles, to form at least five different triangles. Record the angle measures for each triangle in your table.

4 Make a conjecture about the measures of the interior and exterior angles of any triangle. Compare your conjecture with other groups.

Exploration Note

Purpose
The purpose of this Exploration is to have students discover that the measure of an exterior angle of a triangle is equal to the sum of the measures of the two interior angles that are not adjacent to it.

Materials/Preparation
Each pair of students needs geometry software or a ruler and protractor.

Procedure
Students draw a triangle and an exterior angle at one vertex. They then find the measures of the interior angles, the exterior angle, and the sum of the nonadjacent interior angles, and record the data in a table. They repeat this process for several triangles and then make a conjecture about the measures of the interior and exterior angles of any triangle.

Closure
Students should see that the measure of an exterior angle equals the sum of the measures of the interior angles that are not adjacent to it.

Explorations Lab Manual
See the Manual for more commentary on this Exploration.
Diagram Master 10

Teach⇔Interact

Additional Example 1

Suppose someone starts a rumor by telling it to 5 other people. Each of these people within one hour tells the rumor to 5 new people. Assume that the rumor continues to spread at this rate. In a paragraph, prove that 6 billion people will have heard the rumor in less than one 24-hour day. At the end of the first hour, 5 new people have heard the rumor. At the end of the second hour, $5 \times 5 = 5^2$, or 25 new people have heard it. At the end of the third hour, another $5 \times 5 \times 5 = 5^3$, or 125 people have heard the rumor. If the rumor continues to spread at this rate, at the end of the 14th hour, $5^{14} = 6,103,515,625$ new people will have heard the rumor. Since this number is greater than 6 billion, and since 14 hours is less than 24 hours, more than 6 billion people will have heard the rumor in less than one 24-hour day.

Section Note

Writing Proofs
Solving problems in mathematics, and in particular, writing proofs in geometry, always involves making a convincing argument in which each step of the argument can be justified by previously established facts. Writing paragraph proofs introduces students to the concept of proof by relying upon their past experiences in writing complete sentences to compose a paragraph. This approach is a natural introduction to showing students how to organize their thinking in a logical way.

Using Technology
Students can use the *Geometry Inventor* software from the *McDougal Littell Mathpack* to explore the measures of exterior and interior angles of a triangle.

125

Teach⇔Interact

- Additional Examples
- Closure Questions
- Notes on the student lesson, including:

  Technology
  Explorations
  Spatial Reasoning
  Writing Proofs
  Multicultural Information
  Second-Language Learners

Apply⇔Assess

- Suggested Assignment
- *Practice Bank* facsimile
- Notes on the exercises, including:

Writing Proofs
Construction Note
Problem Solving
Technology
Integrating Algebra
Assessment

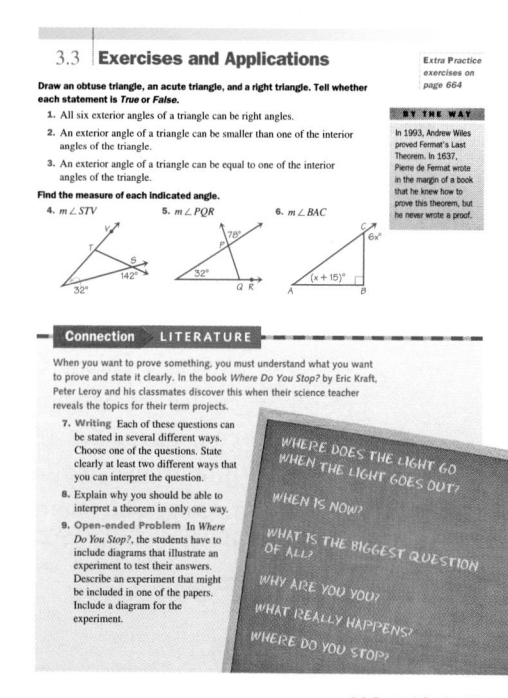

3.3 | Exercises and Applications

Extra Practice exercises on page 664

Draw an obtuse triangle, an acute triangle, and a right triangle. Tell whether each statement is *True* or *False*.

1. All six exterior angles of a triangle can be right angles.
2. An exterior angle of a triangle can be smaller than one of the interior angles of the triangle.
3. An exterior angle of a triangle can be equal to one of the interior angles of the triangle.

Find the measure of each indicated angle.

4. $m \angle STV$
5. $m \angle PQR$
6. $m \angle BAC$

BY THE WAY

In 1993, Andrew Wiles proved Fermat's Last Theorem. In 1637, Pierre de Fermat wrote in the margin of a book that he knew how to prove this theorem, but he never wrote a proof.

Connection | LITERATURE

When you want to prove something, you must understand what you want to prove and state it clearly. In the book *Where Do You Stop?* by Eric Kraft, Peter Leroy and his classmates discover this when their science teacher reveals the topics for their term projects.

7. **Writing** Each of these questions can be stated in several different ways. Choose one of the questions. State clearly at least two different ways that you can interpret the question.

8. Explain why you should be able to interpret a theorem in only one way.

9. **Open-ended Problem** In *Where Do You Stop?*, the students have to include diagrams that illustrate an experiment to test their answers. Describe an experiment that might be included in one of the papers. Include a diagram for the experiment.

> WHERE DOES THE LIGHT GO WHEN THE LIGHT GOES OUT?
> WHEN IS NOW?
> WHAT IS THE BIGGEST QUESTION OF ALL?
> WHY ARE YOU YOU?
> WHAT REALLY HAPPENS?
> WHERE DO YOU STOP?

3.3 Paragraph Proof **127**

Exercises and Applications

1–3. Check students' work.
1. False; False; False.
2. True; False; False.
3. False; False; True.
4. 70°
5. 110°
6. 36°

7. Answers may vary. An example is given. You might interpret the question "What is the biggest question of all?" as "Which question is printed in the largest size type?" or as "Which question is the most important?"

8. so that everyone can agree on what a theorem says and whether its proof is correct

9. Answers may vary. An example is given. I interpreted the question "Where do you stop?" to mean "Where do you stop when stopping for a stop sign?" I chose a stop sign at a busy intersection and observed traffic for awhile, noting how far from the intersection cars stopped. Diagrams will vary.

127

Apply⇔Assess

Suggested Assignment

◆ **Core Course**
Exs. 1–6, 10, 11, 15–26

◆ **Extended Course**
Exs. 1–26

◆ **Block Schedule**
Day 16 Exs. 1–6, 10, 11, 15–26

Exercise Notes

Historical Connection
By the Way Point out to students that Fermat's Last Theorem was the most famous unsolved problem of the last 300 years. The theorem asserts that the equation $x^n + y^n = z^n$, where n is an integer greater than 2, has no solutions that are positive integers. Many of the world's greatest mathematicians tried and failed to solve this problem since Fermat's time, until Wiles solved it in 1993.

Geometric Thinking
Exs. 1–3 Ask students to explain the connections they see between the statements in these exercises and the Exterior Angle Theorem.

Integrating Algebra
Exs. 4–6 Students should have no difficulty using the Exterior Angle Theorem to write equations that can be used to find the indicated measures.

Interdisciplinary Problems
Exs. 7–9 The use of deductive reasoning is not limited to geometry or mathematics. It is used in all academic disciplines and, in fact, in every aspect of life. Ask students to explain why they think deductive reasoning is important in both school subjects and outside of school as well.

Apply⇔Assess

Exercise Notes

Communication: Discussion
Exs. 10, 11 Discuss with students which of the figures in these exercises is easier to work with and why. Clearly, it is easier to work with the figure in Ex. 10 because it contains only two angles. However, there is a matter of notation to consider as well. The angles in Ex. 10 are named with the numbers 1 and 2, while the angles in Ex. 11 are named using three letters. It is necessary to examine the figure in Ex. 11 carefully to find the angles referred to, and this is time consuming. Some parts of Ex. 11 could be stated and answered more easily if the six smallest angles were named with numbers. (This is true for parts (c) and (d). However, parts (a) and (b) require 3-letter notation.) When students draw their own diagram for a proof, they should think about the best way to label the diagram.

Application
Exs. 12–14 These exercises can be used to reinforce the idea that deductive thinking can be used in any area of life, not just in geometry.

Second-Language Learners
Exs. 12–14 Students learning English may benefit from having the terms *species*, *alligator*, *caiman*, *iris of eye*, *lower jaw*, *ear coverlets*, and *crossbands* explained orally before they begin these exercises. For Ex. 14, you may want to allow students to write their responses in a list rather than in paragraph form.

10. Complete the proof of the theorem: All right angles are congruent.
Given: ?
Prove: $\angle 1 \cong \angle 2$
From ?, $m \angle 1 = 90°$ and $m \angle 2 = 90°$. So $m \angle 1 = m \angle 2$ by ?. Therefore ?, by the definition of congruent angles.

11. a. The measures of what angles are equal to $m \angle BGF - m \angle AGB$?
b. The measure of which angle is equal to $m \angle BGC + m \angle CGD$? Why?
c. What is the sum $m \angle CGD + m \angle DGE + m \angle EGF$? Why?
d. If $\angle EGF \cong \angle AGF$, explain why $\angle BGC \cong \angle CGD$.

INTERVIEW | Mary-Jacque Mann

Look back at the article on pages 108–110.

At the Wildlife Forensics Lab, Mary-Jacque Mann's colleagues may need to identify an animal whose skin was used to create an object. At other times, they may need to identify a whole animal. They might use a key to identify the species of an alligator or a caiman.

1. Iris of eye is dark brown, red, or orange.
   - Yes – go to 2
   - No – go to 4
2. Has fewer than 20 teeth on each side of the lower jaw.
   - Yes – Chinese Alligator
   - No – go to 3
3. Ear coverlets contrast with the lighter color of the head.
   - Yes – Schneider's Caiman
   - No – Cuvier's Caiman
4. Back is black with distinct yellow or white crossbands.
   - Yes – go to 5
   - No – one of the other Caiman species
5. Has three or more large dark spots on the sides of the jaws.
   - Yes – Black Caiman
   - No – American Alligator

12. In the key, what species of alligators are identified by the number of teeth they have?

13. Identify the animal at the left. Write a statement and reason for each step in the process of identification.

14. Write a paragraph explaining how you would know that an animal is a Cuvier's Caiman.

128 Chapter 3 *Reasoning in Geometry*

---

10. $\angle 1$ and $\angle 2$ are right angles; the definition of a right angle; the Substitution Property; $\angle 1 \cong \angle 2$

11. a. $\angle AGF$ and $\angle CGD$
b. $\angle BGD$ by the Angle Addition Postulate
c. 180°; because C, G, and F are collinear points
d. $\angle AGF \cong \angle CGD$ and $\angle EGF \cong \angle BGC$, since

vertical angles are congruent. By the definition of congruent angles and the Substitution Property, $\angle BGC \cong \angle CGD$.

12. Chinese Alligator
13. American Alligator
14. A Cuvier's Caiman has an iris that is dark brown, red, or orange; has 20 or more teeth on each side of the lower jaw; and has ear coverlets that do not contrast with the color of its head.

15. Yes; since the triangle is equilateral, two of its sides are congruent.
16. No; for example, let $m \angle H = 60°$. Then $m \angle M = 120°$ and $m \angle L = 60°$. $\angle H$ is not a supplement of $\angle L$. (Only if all three angles are right angles is $\angle H$ a supplement of $\angle L$.)
17. No; for example, let $m \angle 1 = 60°$. Then $m \angle 3 = 120°$ and $m \angle 2 = 120°$. $\angle 1$ and $\angle 2$ are not congruent angles.

18. Yes; since $m \angle U + m \angle A + m \angle R = 180°$, $m \angle A + m \angle R = 180° - 90° = 90°$.
19. B
20. a. $\angle 1$ is an exterior angle of $\triangle XYZ$ at Z (Given); $m \angle 1 = m \angle X + m \angle Y$ (Exterior Angle Theorem); $m \angle Y = m \angle 1 - m \angle X$ (Subtraction Property)

128

ANSWERS

Answers to Explorations, Think and Communicate questions, Checking Key Concept exercises, and Exercises and Applications are conveniently located at the bottom of each page.

In addition to the section side-column notes, a set of **Progress Check** questions is provided for each Assess Your Progress in the student book.

Pacing and Making Assignments

PACING CHART

A yearly Pacing Chart and daily assignments are provided for three courses—a core course, an extended course, and a block-scheduled course. The core and extended courses require 160 days, and the block-scheduled course requires 80 days. These time frames include days for using the Portfolio Projects and time for review and testing. The Pacing Chart below shows the number of days allotted for each of the three courses. Semester and trimester divisions are indicated by red and blue rules, respectively.

Chapter	1	2	3	4	5	6	7	8	9	10	11	12
Core Course	14	14	14	12	13	13	13	13	13	15	15	11
Extended Course	14	14	14	12	13	13	13	13	13	15	15	11
Block Schedule	7	7	7	$6\frac{1}{2}$	$6\frac{1}{2}$	$6\frac{1}{2}$	$6\frac{1}{2}$	$6\frac{1}{2}$	$6\frac{1}{2}$	$7\frac{1}{2}$	$7\frac{1}{2}$	$5\frac{1}{2}$

trimester   semester   trimester

Core Course

The Core Course is intended for students who enter with typical mathematical and problem-solving skills. The course covers all twelve chapters. The daily assignments provide students with substantial work with the skills and concepts presented in each lesson. The exercises assigned range from exercises that involve straightforward application of the new material to exercises involving higher-order thinking skills.

Extended Course

The Extended Course is intended for students who enter with strong mathematical and problem-solving skills and who are able to understand new concepts quickly. The course covers all twelve chapters. The daily assignments include all material in the core course plus additional exercises that focus on higher-order thinking skills.

Block-Scheduled Course

The Block-Scheduled Course is intended for schools that use longer periods, typically 90-minute blocks, for instruction. The course covers all twelve chapters. The exercises assigned range from exercises that involve straightforward application of the new material to exercises involving higher-order thinking skills. All material in the core course is included, plus some additional exercises requiring higher-order thinking skills.

Part of the Block-Scheduled Course for Chapter 3 is shown on the facing page. The entire chart for each chapter is located on the interleaved pages preceding the chapter.

The Planning Guide for each chapter is located on the interleaved pages preceding the chapter. Part of the Planning Guide for Chapter 3 is shown here.

Regular Scheduling (45 min)

Section	Materials Needed	Core Assignment	Extended Assignment	exercises that feature		
				Applications	Communication	Technology
3.1		1–14, 24–29	1–29	19, 20, 22, 23	14–20, 22, 24, 25	
3.2	protractor	1–14, 22–28, 30–34, AYP*	1–34, AYP	15–21, 29	16–18, 20, 29	
3.3	geometry software or ruler, protractor	1–6, 10, 11, 15–26	1–26	7–9, 12–14	7–9, 14, 21	
3.4		1–4, 10–16, 18–23, AYP	1–23, AYP	5–9	8, 9, 18	
3.5		1–9, 11, 18–28	1–28	10, 13–17, 19	10	

Applications

Each section contains exercises that relate the mathematics of that section to real-world applications. These exercises are usually assigned in the daily assignments and are listed in the Planning Guide under the *Applications* head.

Communication

Each section contains exercises that require students to communicate mathematically. These exercises have students discuss or write about the mathematical concepts presented in the section and are usually assigned in the daily assignments. These exercises are denoted by in-line heads in the Exercises and Applications sets and are listed in the Planning Guide under the *Communication* head.

Technology

Each chapter contains exercises that involve technology, usually graphing calculators, spreadsheets, or geometry software. Technology-based exercises are usually assigned in the daily assignments. Exercises that require technology or are especially appropriate for using technology have a logo (shown in the chart above) beside them in the textbook. These exercises are listed in the Planning Guide under the *Technology* head.

Block Scheduling (90 min)

	Day 15	Day 16	Day 17	Day 18	Day 19	Day 20	Day 21
Teach/Interact	3.1 3.2	3.3: Exploration, page 125 3.4	3.5 3.6: Exploration, page 141	Continue with 3.6 3.7: Exploration, page 148	Continue with 3.7 Review	Review Port. Proj.	Ch. 3 Test Port. Proj.
Apply/Assess	**3.1:** 1–14, 24–29 **3.2:** 1–14, 22–28, 30–34, AYP*	**3.3:** 1–6, 10, 11, 15–26 **3.4:** 1–4, 10–23, AYP	**3.5:** 1–9, 11, 18–28 **3.6:** 1–27 odd, 31–35	**3.6:** 2–26 even, 47–50 **3.7:** 1–14, 16–24	**3.7:** 29–34, 36–45, AYP **Review:** 1–10	**Review:** 11–20 **Port. Proj.**	**Ch. 3 Test Port. Proj.**

Geometry

Explorations and Applications

$$\angle BCD \cong \angle CDE$$

Authors

Douglas B. Aichele
Patrick W Hopfensperger
Miriam A. Leiva
Marguerite M. Mason
Stuart J. Murphy
Vicki J. Schell
Matthias C. Vheru

McDougal Littell
A HOUGHTON MIFFLIN COMPANY

Evanston, Illinois • Boston • Dallas

i

Authors

Douglas B. Aichele Regents Professor of Mathematics,
Oklahoma State University, Stillwater, Oklahoma

Patrick W. Hopfensperger Mathematics Teacher, Homestead High School,
Mequon, Wisconsin

Miriam A. Leiva Cone Distinguished Professor for Teaching and Professor
of Mathematics, University of North Carolina at Charlotte

Marguerite M. Mason Assistant Professor of Mathematics Education,
University of Virginia, Charlottesville, Virginia

Stuart J. Murphy Visual Learning Specialist, Evanston, Illinois

Vicki J. Schell Professor of Mathematics and Mathematics Education,
Lenoir-Rhyne College, Hickory, North Carolina

Matthias C. Vheru Mathematics Teacher, Locke High School,
Los Angeles, California

ISBN: 0-395-72285-3   123456789—VH—01 00 99 98 97

Internet Web Site: http://www.hmco.com

Manuscript Reviewers

Judy B. Basara — Curriculum Chair, St. Hubert's High School, Philadelphia, Pennsylvania

Jean Freedman — Mathematics Teacher, Miami Southridge Senior High School, Miami, Florida

Elda López — Mathematics Teacher, Jefferson Davis High School, Houston, Texas

David K. Masunaga — Mathematics Teacher, Iolani School, Honolulu, Hawaii

Irene Nordé — High School Mathematics Supervisor, Detroit Public Schools, Detroit, Michigan

Mary Ann Payne — Mathematics Teacher, Vines High School, Plano, Texas

Pamela Summers — Coordinator of Secondary Mathematics and Science, Lubbock Independent School District, Lubbock, Texas

Straight Line Editorial Development, Inc. — Editorial Consultants, San Francisco, California

Manuscript Reviewers read and reacted to draft manuscript, focusing on its effectiveness from a teaching/learning viewpoint.

Student Advisors

Amanda Belleville, Atherton High School, Louisville, KY
Bryan Brilhart, Francis Scott Key High School, Union Bridge, MD
Mayra Caldera, Estancia High School, Costa Mesa, CA
Kristy Cervenka, Lyons Township High School, Western Springs, IL
Emily Chavie, White Bear Lake Senior High School, St. Paul, MN
Michael Croakman, Sandalwood Senior High School, Jacksonville, FL
Stephanie Decker, Independence High School, Charlotte, NC
Melissa DeShazo, Truman Senior High School, Independence, MO
Erin Devine, Glen Este High School, Cincinnati, OH
Adam Dunlap, South Charleston High School, Charleston, WV
Monique Dupree, Arlington High School, Indianapolis, IN
Ryan Eaton, Durfee High School, Fall River, MA
Blair Edwards, Putnam City North High School, Oklahoma City, OK
Laetesia Ible, Dulles High School, Stafford, TX
Tanner Jacobsen, Monterey Senior High School, Lubbock, TX
Lisa Kandarapally, Wauwatosa West High School, Wauwatosa, WI
Meghan Keedy, Mandeville High School, Mandeville, LA
Erica Mariola, Deerfield Beach High School, Deerfield, FL
Billy Mower III, Gateway Regional High School, Woodbury Heights, NJ
Shayla Charelle Sanford, Lamar High School, Houston, TX
Aimee Shillito, Mountain Pointe High School, Phoenix, AZ
Brandon Slatt, Central Dauphin East Senior High School, Harrisburg, PA
Vedran Sosa, Miami Beach High School, Miami Beach, FL
Sanaz Tehrani, Homestead High School, Cupertino, CA

Contents

Patterns, Lines, and Planes

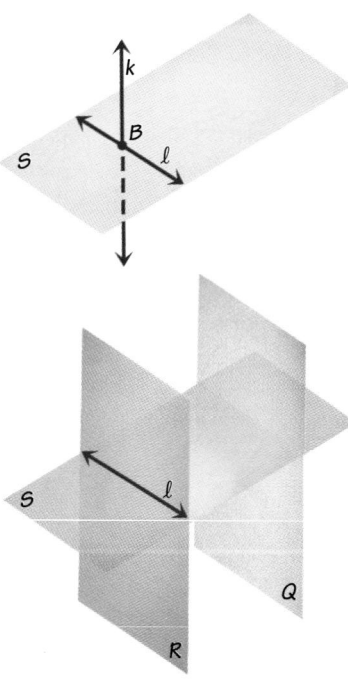

1.4 *Representing planes* 23

Interview *Latecia Leavy* xxxii

1.1 Patterns and Reasoning 3
   Exploration: *Analyzing Patterns* 3

1.2 Transformations and Symmetry 8

1.3 Making Conjectures 14
   Exploration: *Tracing Networks* 14

1.4 Modeling Points, Lines, and Planes 21
   Exploration: *Representing Points, Lines, and Planes* 21

1.5 Segments and Their Measures 28

1.6 Working with Angles 34

1.7 Bisecting Segments and Angles 40

Review   Chapter Review 50

Assessment   Ongoing Assessment 7, 13, 19, 27, 32, 39, 46
   Assess Your Progress 20, 33, 47
   Journal writing 20, 33, 47
   Portfolio Project: *Investigating Symmetry* 48
   Chapter Assessment 52

Applications

Interview:
Latecia Leavy
Tumbler
12, 17, 37

—————— Connection ——————

Astronomy	7	**Manufacturing**	31
Communication	13	**Science**	39
History	19	**Crafts**	45
Art	26		

Applying Algebra 16, 32, 41, 43, 44, 46

Additional applications: chemistry, travel, social studies, history, skiing, racing, boating

Triangles and Polygons

Interview *Virgil Lueth*		54
2.1	**Types of Angles**	57
2.2	**Classifying Triangles**	64
	Exploration: *Sorting Triangles*	64
2.3	**Types of Polygons**	72
2.4	**Angles in Polygons**	79
	Exploration: *Finding Angle Measures in Polygons*	79
2.5	**Parallelograms**	87
	Exploration: *Analyzing Parallelograms*	87
2.6	**Building Prisms**	93
Review **Chapter Review**		102
Algebra Review/Preview		106

Assessment	**Ongoing Assessment**	63, 71, 78, 86, 92, 98
	Assess Your Progress	71, 86, 99
	Journal writing	71, 86, 99
	Portfolio Project: *Building the Platonic Solids*	100
	Chapter Assessment	104

2.2 *Sorting Triangles*   64

Applications

Interview:
Virgil Lueth
Mineralogist
68, 82, 85, 98

——— Connection ———

Biology	61	**Crafts**	77
Communications	62	**Art**	83
History	70	**Optics**	91

Applying Algebra   66, 67, 68, 80, 84, 86, 88, 104

Additional applications: architecture, bicycling, sports, computer-aided design, games

Contents   **v**

CHAPTER 3

Reasoning in Geometry

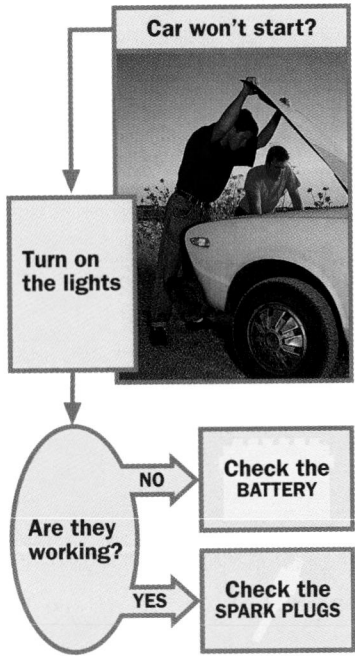

Car won't start?

Turn on the lights

Are they working?

NO → **Check the BATTERY**

YES → **Check the SPARK PLUGS**

3.7 *Auto Maintenance*   151

Interview *Mary-Jacque Mann*   108

3.1   **Inductive and Deductive Reasoning**   111

3.2   **Postulates, Definitions, and Properties**   117

3.3   **Paragraph Proof**   124
     Exploration: *Measuring Exterior Angles*   125

3.4   **Two-Column Proof**   130

3.5   **Converses of Statements**   136

3.6   **The Pythagorean Theorem**   141
     Exploration: *Proving the Pythagorean Theorem*   141

3.7   **Negations and Contrapositives**   148
     Exploration: *Analyzing Triangles*   148

Review   **Chapter Review**   156

Assessment   **Ongoing Assessment**   116, 123, 129, 134, 140, 147, 152

     Assess Your Progress   123, 135, 153

     Journal writing   123, 135, 153

     Portfolio Project: *Classifying Information*   154

     Chapter Assessment   158

     Cumulative Assessment *Ch. 1–3*   160

Applications

Interview: **Mary-Jacque Mann**
Forensic Scientist   128, 133

———— Connection ————

Zoology   115   Language   139
Navigation   121   History   146
Measurement   122   Auto Maintenance   151
Literature   127

Applying **Algebra**   112, 115, 142, 144

Additional applications: personal finance, marketing, carpentry, archeology, architecture, design, travel

Coordinates in Geometry

Interview   *Marc Hannah*   162

4.1   Finding Distances and Midpoints   165

4.2   Equations of Lines   173

4.3   Exploring Parallels and Perpendiculars   180
   Exploration: *Comparing Slopes of Lines*   180

4.4   Equations of Circles   187

4.5   Coordinates and Proof   194
   Exploration:
   Placing a Parallelogram on a Coordinate Plane   194

4.6   Coordinates in Three Dimensions   201

Review   **Chapter Review**   210

Algebra **Review/Preview**   214

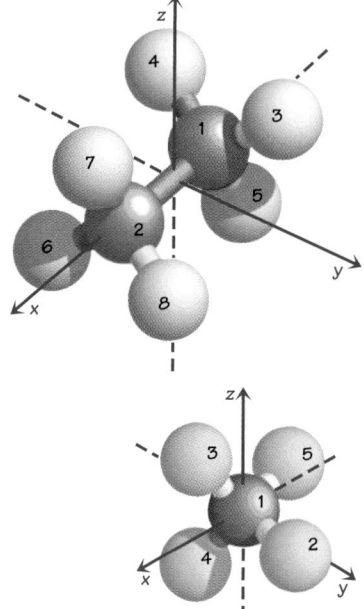

Assessment   **Ongoing Assessment**   172, 179, 186, 193, 200, 206

                    Assess Your Progress   186, 207

                    Journal writing   186, 207

                    Portfolio Project:
                    Exploring Taxicab Geometry   208

                    Chapter Assessment   212

4.6 *Ethane and methane molecules*
   201, 205

Applications

Interview: *Marc Hannah*
Computer-system Designer 170, 198, 206

———— Connection ————

Puzzles	171	**Quilt Patterns**	185
Engineering	177	**Architecture**	192
City Planning	178		

Applying Algebra 170, 177, 185, 193, 196, 200

Additional applications: sports, recreation, accessibility, city planning, carpentry, printing, chemistry

CHAPTER

5

Parallel Lines

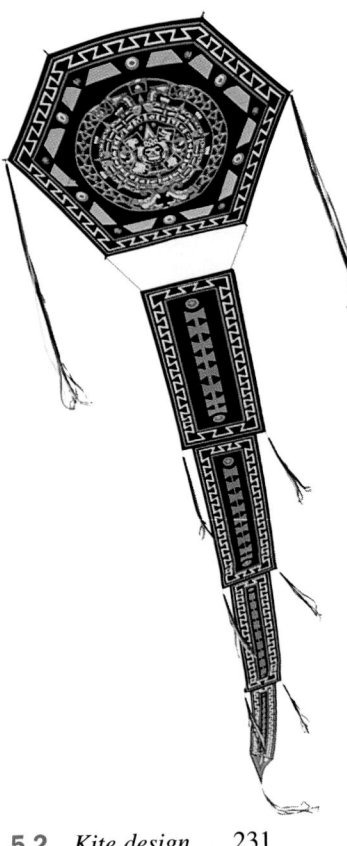

5.2 *Kite design* 231

Interview	*José Saínz*	216
5.1	**Parallel Lines and Transversals**	219
	Exploration: *Angles and Transversals*	221
5.2	**Properties of Parallel Lines**	226
5.3	**Types of Proofs**	234
5.4	**Conditions for Parallel Lines**	242
	Exploration: *Drafting Parallel Lines*	242
5.5	**Proving Theorems About Parallels**	249
5.6	**Parallels in Space**	256
	Exploration:	
	Investigating Planes and Their Intersections	256
5.7	**Constructing Parallels and Perpendiculars**	263
Review	**Chapter Review**	272

Assessment	**Ongoing Assessment**	225, 233, 241, 248, 254, 262, 269
	Assess Your Progress	233, 255, 269
	Journal writing	233, 255, 269
	Portfolio Project: *Creating Technical Drawings*	270
	Chapter Assessment	274

Applications

Interview: *José Saínz*
Kite Designer 231, 252, 262

——— Connection ———

Botany	224	**Design**	254
Architecture	232	**Electronics**	259
Engineering	240	**Technical Drawing**	261
Horticulture	246	**Paper Folding**	266
Dance	247		

Applying Algebra 222, 224, 227, 230, 232, 243, 245

Additional applications: biology, automobiles, art, design, engineering, city planning, aviation

Conjectures About Triangles

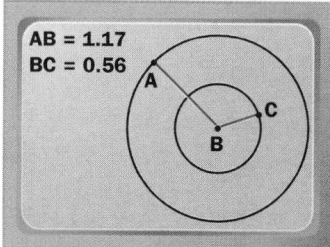

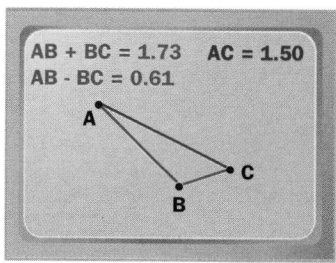

6.1 *Exploring using technology* 279

Interview *Madeleine Fleming*   276

6.1   **Triangle Inequalities**   279
    Exploration: *Comparing Sides of Triangles*   279

6.2   **Exploring Congruence**   285
    Exploration: *Investigating Overlapping Triangles*   287

6.3   **Congruent Triangles: SSS and SAS**   292

6.4   **Congruent Triangles: ASA, AAS, and HL**   299

6.5   **Applying Congruence**   306

6.6   **Properties of Isosceles Triangles**   313

6.7   **Altitudes, Medians, and Bisectors**   319

Review   **Chapter Review**   328

Algebra Review/Preview   332

Assessment   **Ongoing Assessment**   284, 291, 298, 305, 312, 318, 324

    Assess Your Progress   291, 312, 325

    Journal writing   291, 312, 325

    Portfolio Project: *Building a Mobile*   326

    Chapter Assessment   330

    Cumulative Assessment *Ch. 4–6*   334

Applications

Interview:
Madeleine Fleming
Optical Physicist
290, 295, 317

—— Connection ——

Probability   283
Interior Design   284
Rescue Safety   297
Astronomy   305

Aerial Photography   310
Physics   323

Applying Algebra   286, 289, 310, 315, 317, 322, 323, 325, 330

Additional applications: art, pottery, biology, cars, rock climbing, architecture

Quadrilaterals, Areas, and Volumes

Interview *Walt Stone*   336

7.1 **Classifying Quadrilaterals**   339
     Exploration: *Investigating Diagonals*   340

7.2 **Identifying Parallelograms**   346

7.3 **Conditions for Special Parallelograms**   353

7.4 **Areas of Triangles and Quadrilaterals**   360
     Exploration: *Discovering Area Formulas*   361

7.5 **Areas of Regular Polygons and Circles**   367

7.6 **Prisms and Cylinders**   374
     Exploration: *Comparing Volumes*   374

Review **Chapter Review**   384

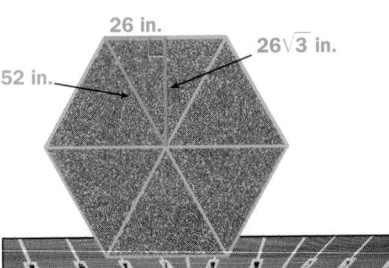

7.5 *Grass for the Silverdome*   367

Assessment   **Ongoing Assessment**   345, 352, 358,
                                                  366, 373, 381

               Assess Your Progress   359, 381

               Journal writing   359, 381

               Portfolio Project: *Designing a Cottage*   382

               Chapter Assessment   386

Applications

Interview:
Walt Stone
Alderman
366, 380

——————— Connection ———————

Engineering	344	**Biology**	371
Machinery	351	**Consumer**	
Social Studies	357	**Economics**	379
Gardening	365		

Applying Algebra 341, 351, 354, 357, 365, 380

Additional applications: auto repair, carpentry, drafting, crafts, quilting, history, clothing, sports, architecture, package design, nutrition

Using Transformations

Interview   *Terri Johnson*   388

8.1   **Using Reflections**   391
      Exploration: *Reflecting a Polygon Over a Line*   392

8.2   **Reflections with Coordinates**   398

8.3   **Translations**   405
      Exploration: *Comparing Translated Polygons*   405

8.4   **Applying Rotations**   412

8.5   **Glide Reflections**   419

8.6   **Dilations**   426
      Exploration: *Making an Enlargement*   427

Review   **Chapter Review**   436

Algebra **Review/Preview**   440

Assessment   **Ongoing Assessment**   397, 404, 411, 418, 425, 432

                 Assess Your Progress   411, 433

                 Journal writing   411, 433

                 Portfolio Project:
                 Tessellating the Plane   434

                 Chapter Assessment   438

8.4   *Hawaiian quilts*   417

Applications

Interview: ***Terri Johnson***
                Architect   395, 409, 418

——————— Connection ———————

Audio-Visual   **Hawaiian Quilts**   417
Presentation   396   **Computer**
Miniature Golf   403   **Simulation**   424
Architecture   410   **Literature**   431

Applying Algebra 395, 402

Additional applications: periscopes, optics,
computers, furniture design, fabric design,
animal tracks, manufacturing

Similar Polygons

Interview *Loy Arcenas*   442

9.1 **Properties of Similar Figures**   445

9.2 **Similar Triangles**   453

9.3 **Proportions and Similarity**   461
     Exploration:
     Finding Proportions in Triangles   461

9.4 **Areas and Volumes of Similar Figures**   468

9.5 **Geometric Probability**   475
     Exploration:
     Investigating Probability Based on Area   475

Review **Chapter Review**   484

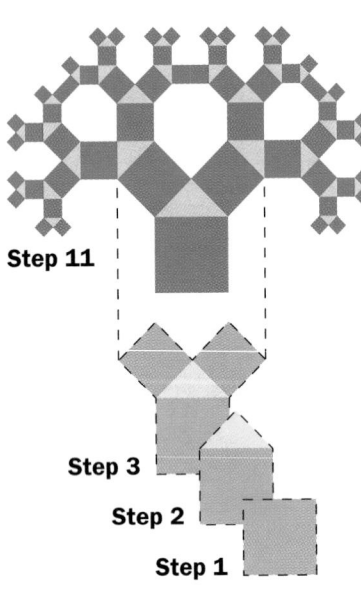

Step 11

Step 3

Step 2

Step 1

Assessment	**Ongoing Assessment**	452, 460, 466, 474, 480
	Assess Your Progress	467, 481
	Journal writing	467, 481
	Portfolio Project: *Scaling the Planets*	482
	Chapter Assessment	486
	Cumulative Assessment *Ch. 7–9*	488

9.1 *Self-similarity*   452

Applications

Interview:
Loy Arcenas
Set Designer
465, 468, 472

──────── Connection ────────

Astronomy	450	**Consumer**	
Surveying	457	**Economics**	473
Electronics	459	**Zoology**	479

Applying Algebra 448, 449, 462, 464, 466, 470, 474

Additional applications: graphic design, technical drawing, movies, geography, fractals, sculpture, balloonning, advertising

Applying Right Triangles

Interview *Debby Tewa*   490

10.1 Similar Right Triangles   493
      Exploration: *Comparing Right Triangles*   493

10.2 Special Right Triangles   500

10.3 The Tangent Ratio   507

10.4 Sine and Cosine Ratios   514
      Exploration: *Analyzing Ratios in Triangles*   514

10.5 Using Vectors   521

10.6 Areas and Trigonometry   529

10.7 Pyramids and Cones   535

Review **Chapter Review**   544

Algebra **Review/Preview**   548

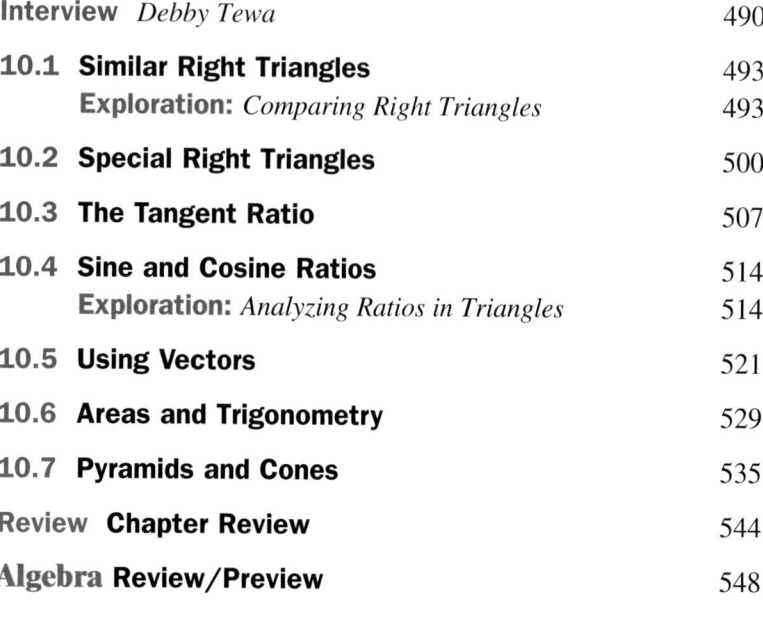

Assessment   **Ongoing Assessment**   499, 506, 513, 520,
                                              528, 534, 541

              Assess Your Progress   506, 528, 541

              Journal writing   506, 528, 541

              Portfolio Project: *Applying Solar Geometry*   542

              Chapter Assessment   546

Angles of grain piles

	wheat	corn	oat
$m \angle A$	27.0	27.5	28.0
tan A	0.5095	0.5206	0.5317

10.3 *Angles of grain piles*   507, 513

Applications

Interview:
Debby Tewa
Solar Engineer
503, 519

──────── Connection ────────

Literature	498	Geology	526
Architecture	504	History	533
Transportation	512	Architecture	539
Astronomy	518		

Applying Algebra 494, 497, 499, 505, 520, 534

Additional applications: home repair, music, sports, history, forestry, agriculture, navigation, paleontology, theater, aviation, orienteering, archaeology, packaging

Contents   **xiii**

11

Circles and Spheres

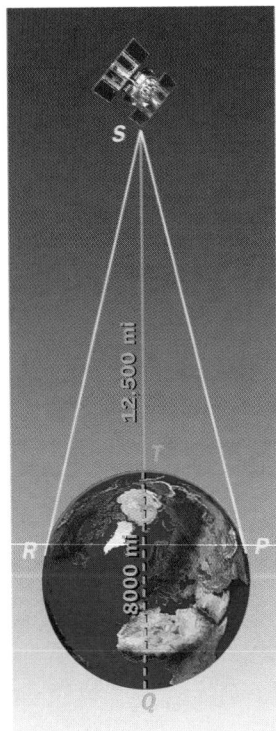

11.4 *Global positioning* 575

Interview *Ron Courson* 550

11.1 Angles and Circles 553
    Exploration: *Comparing Central and Inscribed Angles* 553

11.2 Tangents, Secants, and Chords 560

11.3 Applying Properties of Chords 567
    Exploration: *Circumscribing a Triangle* 567

11.4 Segment Lengths in Circles 573

11.5 Sectors and Arc Lengths 579

11.6 Surface Areas and Volumes of Spheres 586

11.7 Volumes of Similar Solids 592

11.8 Spherical Geometry 597
    Exploration: *Triangles on a Sphere* 597

Review Chapter Review 606

Assessment Ongoing Assessment 559, 565, 572, 578, 585, 591, 596, 603

   Assess Your Progress 566, 585, 603

   Journal writing 566, 585, 603

   Portfolio Project: *Modeling Cavalieri's Principle* 604

   Chapter Assessment 608

Applications

Interview:
Ron Courson
Athletic Trainer/
Physical Therapist
557, 589

――――― Connection ―――――

Biology	564	**Biology**	595
Highway Safety	571	**Cooking**	596
Earth Science	577	**Geography**	601
Rock Climbing	584	**Hyperbolic**	
Sports	590	**Geometry**	602

Applying Algebra 558, 562, 568, 574, 576, 587, 595

Additional applications: global positioning, sports, irrigation, catering, architecture, astronomy, physics

Coordinates for Transformations

Interview *Adriana Ocampo*		610
12.1 Dilations with Matrices		613
12.2 Translations with Matrices		619
12.3 Multiplying Matrices		626
12.4 Reflections with Matrices		634
Exploration: *Reflections Using Matrix Multiplication*		634
12.5 Rotations with Matrices		641
Exploration: *Matrices with Zeros on a Diagonal*		641
Review **Chapter Review**		652

Assessment	**Ongoing Assessment**	618, 625, 633, 640, 647
	Assess Your Progress	625, 647
	Journal writing	625, 647
	Portfolio Project: *Creating Fractals*	648
	Chapter Assessment	654
	Cumulative Assessment *Ch. 10–12*	656

12.3 *Morphing* 626

Applications

Interview: **Adriana Ocampo**
Planetary Geologist 623, 645

—————— Connection ——————

Digital Maps	617	**Art Photography**	637
Color Photocopiers	624	**Virtual Reality**	639
Special Effects	631	**Video Games**	646

Applying Algebra 630, 640

Additional applications: computers, morphing

Student Resources

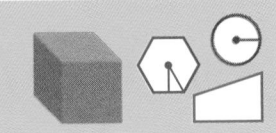

Extra Practice   659–686

Toolbox   687–711

Using Geometric Tools and Transformations   687
Operations with Variable Expressions   690
Linear Equations   694
Graphing on the Coordinate Plane   696
Quadratic Equations   700
Formulas for Geometric Figures   702
Inequalities   705
Ratio and Proportion   707
Data Analysis and Probability   708
Matrices   711

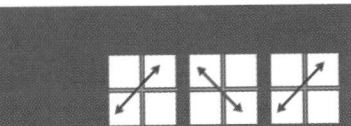

Postulates, Properties, Theorems, and Constructions   712–720

List of Postulates and Properties   712
List of Theorems   714
List of Constructions   720

Technology Handbook   721–729

Using a Graphing Calculator   721
Using a Spreadsheet   725
Using Geometry Software   726

Tables   730–733

Table of Measures   730
Table of Symbols   731
Table of Squares and Square Roots   732
Table of Trigonometric Ratios   733

Appendices   734–741

Appendix 1: A Brief History of Geometric Systems   734
Appendix 2: Truth Tables and Logic   736

Glossary   742–747

Index   748–758

Selected Answers

About the Interviews

Using Mathematics in Careers

Each chapter of this book starts with a personal interview with someone who uses mathematics in his or her life. You may be surprised by the wide range of careers that are included. These are the people you will be reading about:

- **Tumbler**
  Latecia Leavy
- **Mineralogist**
  Virgil Lueth
- **Forensic Scientist**
  Mary-Jacque Mann
- **Computer-system Designer**
  Marc Hannah
- **Kite Designer**
  José Saínz
- **Optical Physicist**
  Madeleine Fleming

- **Alderman**
  Walt Stone
- **Architect**
  Terri Johnson
- **Set Designer**
  Loy Arcenas
- **Solar Electric Technician**
  Debby Tewa
- **Athletic Trainer/Physical Therapist**
  Ron Courson
- **Planetary Geologist**
  Adriana Ocampo

Debby Tewa

After each interview, the *Explore and Connect* questions guide you in learning more about the topic being discussed. In each chapter there are *Related Examples and Exercises* that show how the mathematics you are learning is used by the person highlighted in the interview.

José Saínz

Marc Hannah

Welcome to Geometry
Explorations and Applications

GOALS OF THE COURSE

This book will help you use mathematics in your daily life and prepare you for success in future courses and careers.

In this course you will:
- Study the geometry concepts that are most important for today's students
- Apply these concepts to solve many different types of problems
- Learn how calculators and computers can help you solve problems

You will have a chance to develop your skills in:
- Reasoning and problem solving strategies
- Using geometric properties to solve real-world problems
- Communicating orally and in writing
- Studying and learning independently and as a team member

Pyramid Arena, Memphis, Tennessee

MATHEMATICAL CONTENT

This contemporary geometry course gives you a strong background in the types of mathematical reasoning and problem solving that will be important in your future.

The book emphasizes:

- Visualizing and analyzing geometric relationships in two and three dimensions
- Developing inductive and deductive reasoning skills
- Investigating connections of geometry to algebra, probability, trigonometry, and discrete mathematics

ACTIVE LEARNING

To learn geometry successfully, you need to get involved!

There will be many opportunities in this course for you to participate in:

- Explorations of mathematical concepts
- Cooperative learning activities
- Small-group and whole-class discussions

So don't sit back and be a spectator. If you join in and share your ideas, everyone will learn more.

Course Overview

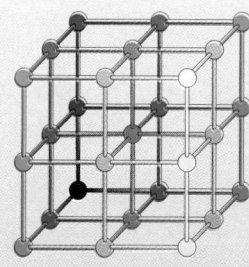

To get an overview of your course, turn to pages xx–xxxi to see some of the types of problems you will solve and topics you will explore.

- **Applications and Connections**   pp. xx–xxi
- **Explorations and Cooperative Learning**   pp. xxii–xxiii
- **Logical Reasoning and Visualization**   pp. xxiv–xxv
- **Using Technology**   pp. xxvi–xxvii
- **Integrating Math Topics**   pp. xxviii–xxix
- **Building for the Future**   pp. xxx–xxxi

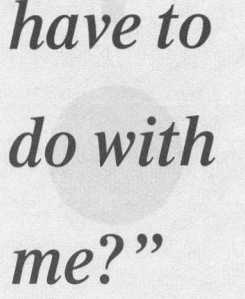

"What does geometry have to do with me?"

Applications and Connections

Geometry is about you and the world around you.

In this course you'll learn how geometry can help answer many different types of questions in daily life and in careers.

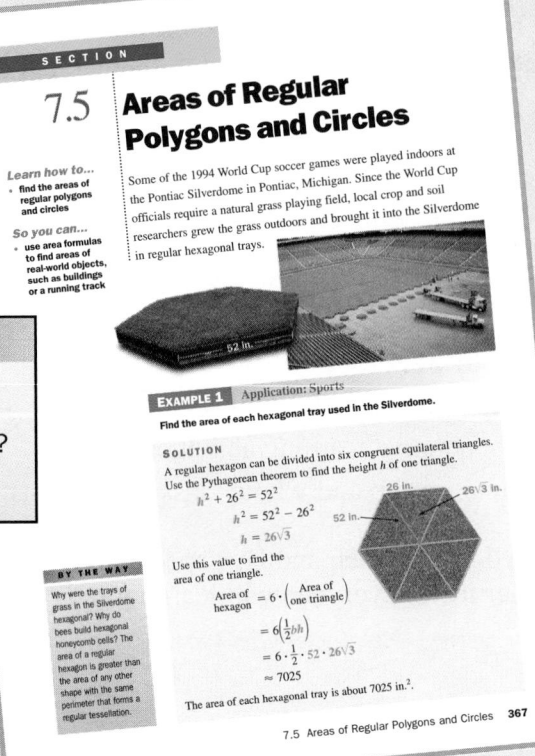

SECTION

7.5 Areas of Regular Polygons and Circles

Learn how to...
- find the areas of regular polygons and circles

So you can...
- use area formulas to find areas of real-world objects, such as buildings or a running track

Some of the 1994 World Cup soccer games were played indoors at the Pontiac Silverdome in Pontiac, Michigan. Since the World Cup officials require a natural grass playing field, local crop and soil researchers grew the grass outdoors and brought it into the Silverdome in regular hexagonal trays.

52 in.

EXAMPLE 1 Application: Sports

Find the area of each hexagonal tray used in the Silverdome.

SOLUTION
A regular hexagon can be divided into six congruent equilateral triangles. Use the Pythagorean theorem to find the height h of one triangle.

$$h^2 + 26^2 = 52^2$$
$$h^2 = 52^2 - 26^2$$
$$h = 26\sqrt{3}$$

26 in.   26√3 in.
52 in.

Use this value to find the area of one triangle.

$$\text{Area of hexagon} = 6 \cdot \left(\text{Area of one triangle}\right)$$
$$= 6\left(\tfrac{1}{2}bh\right)$$
$$= 6 \cdot \tfrac{1}{2} \cdot 52 \cdot 26\sqrt{3}$$
$$\approx 7025$$

The area of each hexagonal tray is about 7025 in.²

BY THE WAY

Why were the trays of grass in the Silverdome hexagonal? Why do bees build hexagonal honeycomb cells? The area of a regular hexagon is greater than the area of any other shape with the same perimeter that forms a regular tessellation.

7.5 Areas of Regular Polygons and Circles   **367**

Engineering

Why were hexagonal trays used to install a grass playing field indoors?

(Chapter 7, page 367)

Auto Repair

What kind of quadrilateral can help you change a tire?

(Chapter 7, page 345)

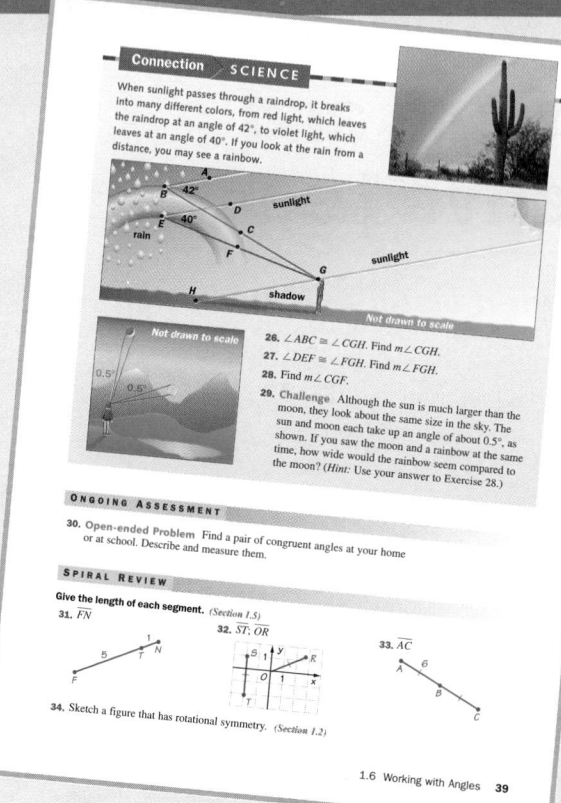

Connection SCIENCE

When sunlight passes through a raindrop, it breaks into many different colors, from red light, which leaves the raindrop at an angle of 42°, to violet light, which leaves at an angle of 40°. If you look at the rain from a distance, you may see a rainbow.

Not drawn to scale

Not drawn to scale

26. $\angle ABC \cong \angle CGH$. Find $m\angle CGH$.

27. $\angle DEF \cong \angle FGH$. Find $m\angle FGH$.

28. Find $m\angle CGF$.

29. Challenge Although the sun is much larger than the moon, they look about the same size in the sky. The sun and moon each take up an angle of about 0.5°, as shown. If you saw the moon and a rainbow at the same time, how wide would the rainbow seem compared to the moon? (*Hint:* Use your answer to Exercise 28.)

ONGOING ASSESSMENT

30. Open-ended Problem Find a pair of congruent angles at your home or at school. Describe and measure them.

SPIRAL REVIEW

Give the length of each segment. *(Section 1.5)*

31. $\overline{FN}$ **32.** $\overline{ST}$; $\overline{OR}$ **33.** $\overline{AC}$

34. Sketch a figure that has rotational symmetry. *(Section 1.2)*

1.6 Working with Angles   **39**

Connections Exercises

These clusters of exercises, which appear throughout each chapter, focus on the connections of geometry to other subjects, careers, and branches of mathematics.

Biology

What does the angle of the crest of a Stellar's jay tell you?

(Chapter 2, page 61)

Crafts

How can you fold a piece of paper to cut out a perfect five-pointed star?

(Chapter 2, page 77)

"Do we just sit back and listen?"

Explorations and Cooperative Learning

In this course you'll be an active learner.

Working individually and in groups, you'll investigate questions and then present and discuss your results. Here are some of the topics you'll explore.

Combining Congruent Triangles

What kinds of shapes can you make by combining two congruent triangles?

(Chapter 6, page 287)

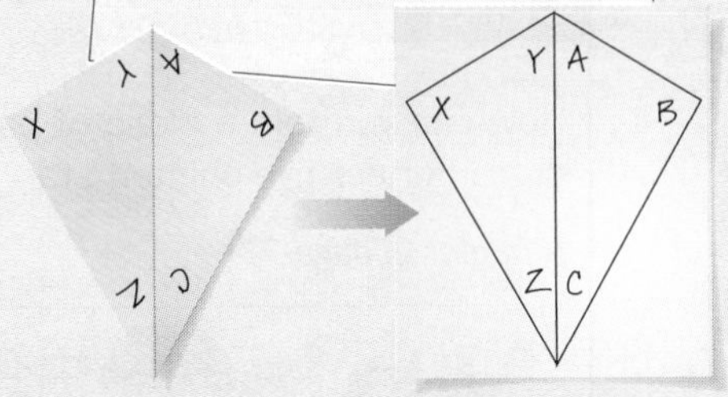

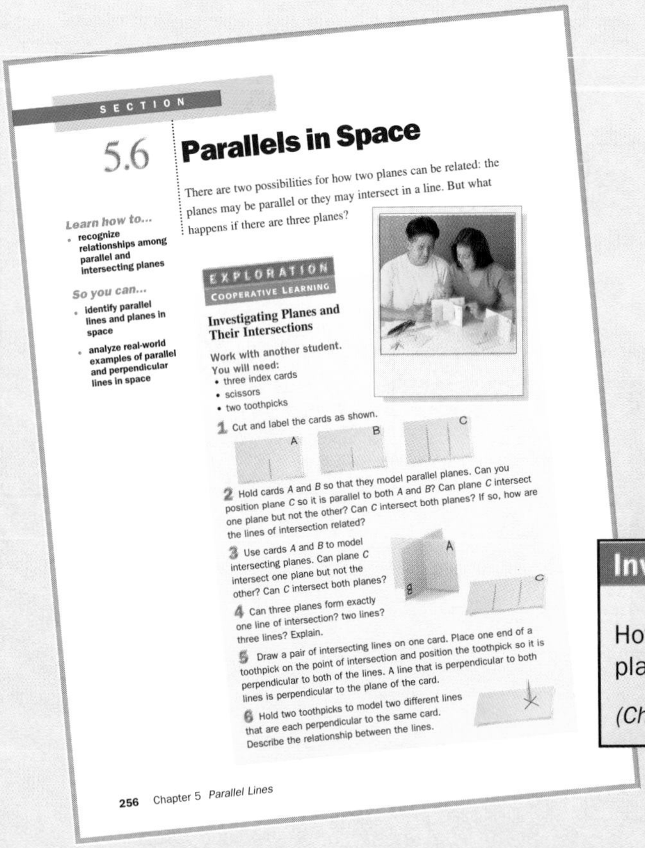

SECTION

5.6 Parallels in Space

There are two possibilities for how two planes can be related: the planes may be parallel or they may intersect in a line. But what happens if there are three planes?

Learn how to...
- recognize relationships among parallel and intersecting planes

So you can...
- identify parallel lines and planes in space
- analyze real-world examples of parallel and perpendicular lines in space

EXPLORATION
COOPERATIVE LEARNING

Investigating Planes and Their Intersections

Work with another student.
You will need:
- three index cards
- scissors
- two toothpicks

1 Cut and label the cards as shown.

2 Hold cards A and B so that they model parallel planes. Can you position plane C so it is parallel to both A and B? Can plane C intersect one plane but not the other? Can C intersect both planes? If so, how are the lines of intersection related?

3 Use cards A and B to model intersecting planes. Can plane C intersect one plane but not the other? Can C intersect both planes?

4 Can three planes form exactly one line of intersection? two lines? three lines? Explain.

5 Draw a pair of intersecting lines on one card. Place one end of a toothpick on the point of intersection and position the toothpick so it is perpendicular to both of the lines. A line that is perpendicular to both lines is perpendicular to the plane of the card.

6 Hold two toothpicks to model two different lines that are each perpendicular to the same card. Describe the relationship between the lines.

256 Chapter 5 *Parallel Lines*

Investigating Intersecting Planes

How many lines of intersection can three planes form?

(Chapter 5, page 256)

Portfolio Projects

These open-ended projects give you a chance to explore applications of the topics you have studied.

Tessellating the Plane

A tessellation is a pattern formed by repeating a shape to cover a plane without any gaps or overlapping. A brick walkway is an example of a tessellation that uses rectangles to cover the plane. A tiled wall is another example. So is a honeycomb—not all tessellations are made with quadrilaterals.

PROJECT GOAL Use transformations with different polygons to make tessellations.

Making Quadrilateral Tilings

1. Work with a partner. Cut several congruent rectangles from cardboard to represent bricks, and use transformations to make three different tessellations with them. Discuss with your partner the transformations you used.

2. Cut a quadrilateral that is not a rectangle from a piece of cardboard. Trace the quadrilateral several times to make a tessellation. Discuss the transformations that you used to make your pattern.

3. Choose any vertex on your sketch and measure the angles at that vertex. What do you notice about the sum of the measures of these angles? How might you have predicted this result without measuring?

434 Chapter 8 *Using Transformations*

Investigating Symmetry

What kind of symmetry can you find in objects around you?

(Chapter 1, page 48)

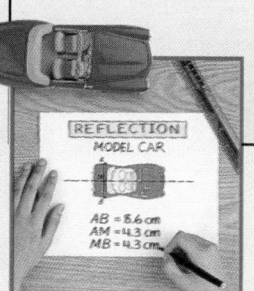 

Constructing a Perpendicular Bisector

How can you draw the perpendicular bisector of a line without measuring the angle or the length of the line?

(Chapter 5, page 265)

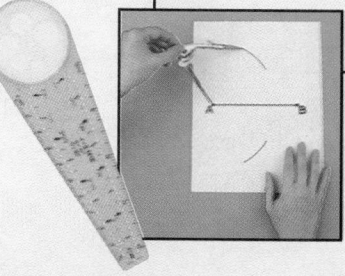 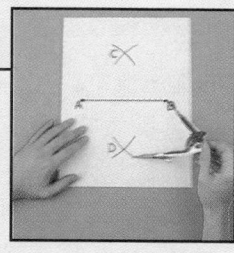 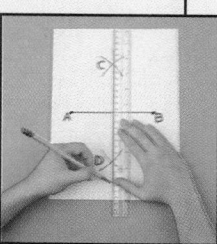

"How is geometry different from algebra?"

Logical Reasoning and Visualization

In this course you'll learn new ways to see and understand mathematics.

You will study the properties of two- and three-dimensional shapes and learn new ways to reason mathematically.

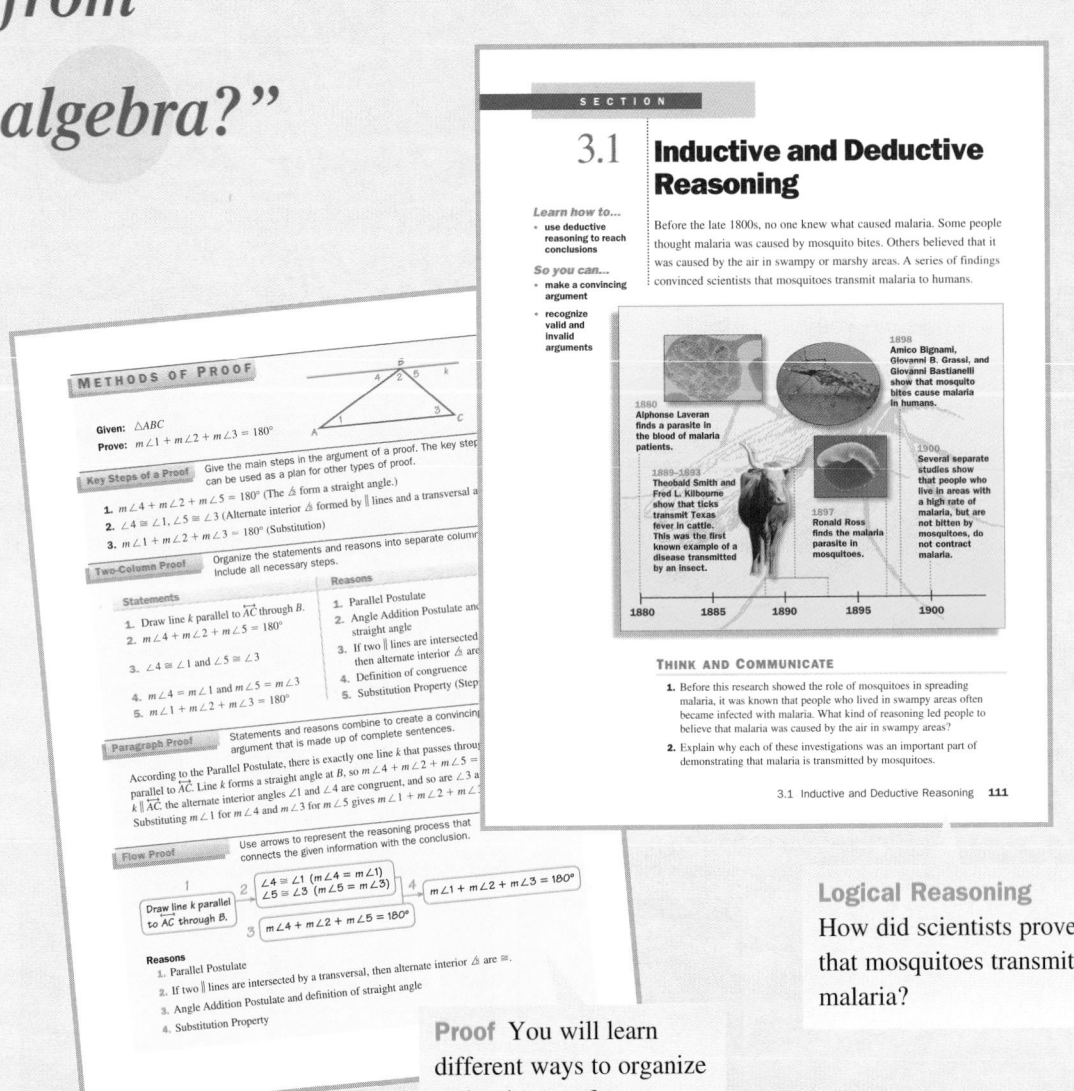

METHODS OF PROOF

Given: △ABC
Prove: $m\angle 1 + m\angle 2 + m\angle 3 = 180°$

Key Steps of a Proof   Give the main steps in the argument of a proof. The key steps can be used as a plan for other types of proof.

1. $m\angle 4 + m\angle 2 + m\angle 5 = 180°$ (The ⟂ form a straight angle.)
2. $\angle 4 \cong \angle 1, \angle 5 \cong \angle 3$ (Alternate interior ⟂ formed by ∥ lines and a transversal a
3. $m\angle 1 + m\angle 2 + m\angle 3 = 180°$ (Substitution)

Two-Column Proof   Organize the statements and reasons into separate column Include all necessary steps.

Statements	Reasons
1. Draw line *k* parallel to $\overleftrightarrow{AC}$ through *B*.	1. Parallel Postulate
2. $m\angle 4 + m\angle 2 + m\angle 5 = 180°$	2. Angle Addition Postulate and straight angle
3. $\angle 4 \cong \angle 1$ and $\angle 5 \cong \angle 3$	3. If two ∥ lines are intersected then alternate interior ⟂ are
4. $m\angle 4 = m\angle 1$ and $m\angle 5 = m\angle 3$	4. Definition of congruence
5. $m\angle 1 + m\angle 2 + m\angle 3 = 180°$	5. Substitution Property (Step

Paragraph Proof   Statements and reasons combine to create a convincing argument that is made up of complete sentences.

According to the Parallel Postulate, there is exactly one line *k* that passes throu parallel to $\overleftrightarrow{AC}$. Line *k* forms a straight angle at *B*, so $m\angle 4 + m\angle 2 + m\angle 5 =$ $k \parallel \overleftrightarrow{AC}$, the alternate interior angles $\angle 1$ and $\angle 4$ are congruent, and so are $\angle 3$ a Substituting $m\angle 1$ for $m\angle 4$ and $m\angle 3$ for $m\angle 5$ gives $m\angle 1 + m\angle 2 + m\angle$

Flow Proof   Use arrows to represent the reasoning process that connects the given information with the conclusion.

1 — Draw line *k* parallel to $\overline{AC}$ through *B*.
2 — $\angle 4 \cong \angle 1$ ($m\angle 4 = m\angle 1$) / $\angle 5 \cong \angle 3$ ($m\angle 5 = m\angle 3$)
4 — $m\angle 1 + m\angle 2 + m\angle 3 = 180°$
3 — $m\angle 4 + m\angle 2 + m\angle 5 = 180°$

Reasons
1. Parallel Postulate
2. If two ∥ lines are intersected by a transversal, then alternate interior ⟂ are ≅.
3. Angle Addition Postulate and definition of straight angle
4. Substitution Property

Proof You will learn different ways to organize and write proofs.

SECTION

3.1 Inductive and Deductive Reasoning

Learn how to...
- use deductive reasoning to reach conclusions

So you can...
- make a convincing argument
- recognize valid and invalid arguments

Before the late 1800s, no one knew what caused malaria. Some people thought malaria was caused by mosquito bites. Others believed that it was caused by the air in swampy or marshy areas. A series of findings convinced scientists that mosquitoes transmit malaria to humans.

1898 Amico Bignami, Giovanni B. Grassi, and Giovanni Bastianelli show that mosquito bites cause malaria in humans.

1880 Alphonse Laveran finds a parasite in the blood of malaria patients.

1889–1893 Theobald Smith and Fred L. Kilbourne show that ticks transmit Texas fever in cattle. This was the first known example of a disease transmitted by an insect.

1897 Ronald Ross finds the malaria parasite in mosquitoes.

1900 Several separate studies show that people who live in areas with a high rate of malaria, but are not bitten by mosquitoes, do not contract malaria.

1880   1885   1890   1895   1900

THINK AND COMMUNICATE

1. Before this research showed the role of mosquitoes in spreading malaria, it was known that people who lived in swampy areas often became infected with malaria. What kind of reasoning led people to believe that malaria was caused by the air in swampy areas?

2. Explain why each of these investigations was an important part of demonstrating that malaria is transmitted by mosquitoes.

3.1 Inductive and Deductive Reasoning   **111**

Logical Reasoning
How did scientists prove that mosquitoes transmit malaria?

Visualizing 3-Dimensional Objects

What does the surface of a three-dimensional shape look like when you unfold it?

(Chapter 2, page 95)

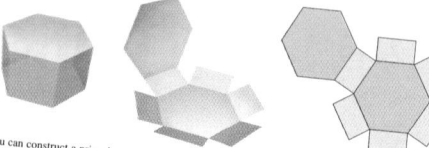

Unfolding a Prism

Imagine cutting this prism along some of its edges, then opening and unfolding it. The resulting plane figure is called a **net**.

You can construct a prism by folding up its net and taping the edges together.

EXAMPLE 2

Sketch a net for this prism.

SOLUTION

Method 1

Sketch all the lateral faces attached to one of the bases of the prism.
Sketch the other base on the opposite edge of one of the lateral faces.

Method 2

Sketch all the lateral faces attached to each other.
Place the bases at opposite ends of two of the lateral faces.

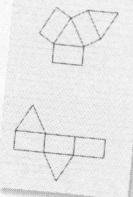

THINK AND COMMUNICATE

3. Describe another way to sketch a net for the prism above. How many different methods do you think there are?

4. Describe one possible net for a cube.

2.6 Building Prisms **95**

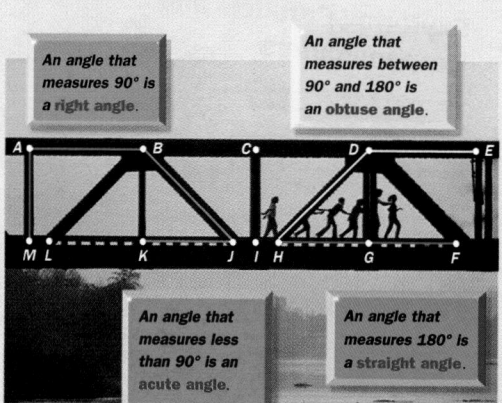

An angle that measures 90° is a right angle.

An angle that measures between 90° and 180° is an obtuse angle.

An angle that measures less than 90° is an acute angle.

An angle that measures 180° is a straight angle.

Analyzing Geometric Relationships

Geometry plays an important role in objects you see every day.

(Chapter 2, page 57)

Visual Thinking

How can you show that the measures of the angles of a triangle add up to 180°? *(Chapter 2, page 65)*

"How can I visualize the problem?"

Using Technology

Calculators and computers can help you see mathematical relationships.

In this course there are many opportunities to use technology to model problem situations, identify patterns, and find solutions.

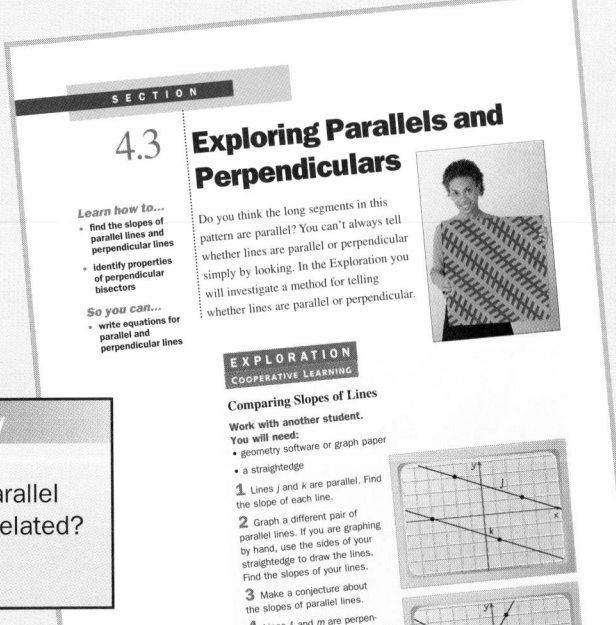

Graphing Technology

How are the slopes of parallel and perpendicular lines related?

(Chapter 4, page 180)

Matrix Operations

How are matrix operations used to change images on a computer screen?

(Chapter 12, page 614)

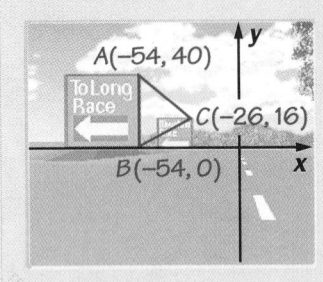

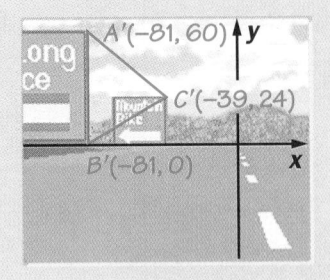

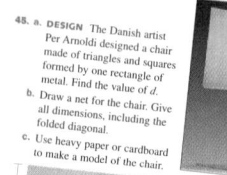

45. a. **DESIGN** The Danish artist Per Arnoldi designed a chair made of triangles and squares formed by one rectangle of metal. Find the value of d.

b. Draw a net for the chair. Give all dimensions, including the folded diagonal.

c. Use heavy paper or cardboard to make a model of the chair.

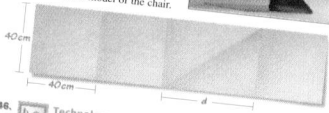
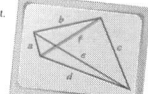

46. 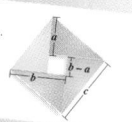 **Technology** Use geometry software.

a. Draw a quadrilateral. Draw the diagonals of the quadrilateral. Measure the length of each segment.

b. Repeat part (a) with at least five different quadrilaterals. Record the lengths in a table.

c. For what quadrilaterals will the equation $a^2 + b^2 + c^2 + d^2 = e^2 + f^2$ be true? Explain your reasoning.

ONGOING ASSESSMENT

47. a. **Visual Thinking** Explain how this diagram illustrates a proof of the Pythagorean theorem. Write the algebraic steps for the proof.

b. How is this diagram different from the one that you used in the Exploration on page 141? Explain.

SPIRAL REVIEW

Write each statement as a conditional. Circle the hypothesis and underline the conclusion. *(Section 1.3)*

48. Angles in a linear pair are supplementary angles.

49. Perpendicular lines intersect at right angles.

50. Write the converse of each statement in Exercises 48 and 49. *(Section 3.5)*

3.6 The Pythagorean Theorem **147**

Technology Exercises

In these exercises you will be using geometry software or spreadsheets to practice, apply, and extend what you have learned.

Geometry Software

How can you use a computer to investigate a relationship between parallel lines and similar triangles?

(Chapter 9, page 461)

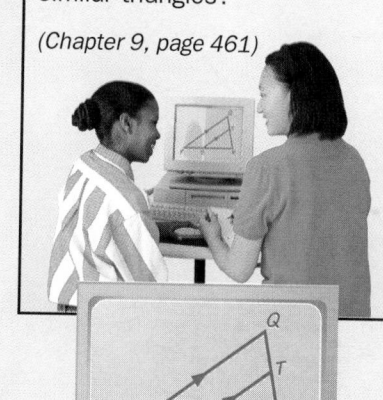

	A	B	C	D	E
		Diameter at Equator (km)	Mean distance from sun (million km)	Scale diameter (cm)	Scale distance (cm)
1					
2	Sun	1400000	0		
3	Mercury	4870	57.9		
4	Venus				
5	Earth				

Spreadsheet

What is an appropriate scale to use to make a scale model of the solar system?

(Chapter 9, page 482)

"Can I solve this problem with geometry?"

Integrating Math Topics

Sometimes you need to combine geometry with other math topics in order to find a solution.

In this course you'll see how you can solve problems by integrating geometry with algebra, probability, trigonometry, and discrete mathematics.

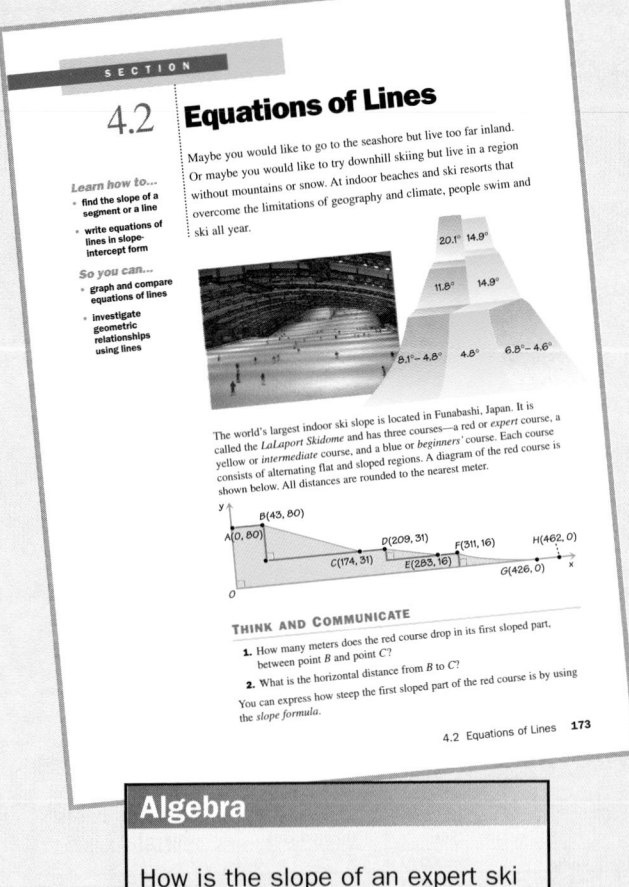

SECTION

4.2 Equations of Lines

Maybe you would like to go to the seashore but live too far inland. Or maybe you would like to try downhill skiing but live in a region without mountains or snow. At indoor beaches and ski resorts that overcome the limitations of geography and climate, people swim and ski all year.

Learn how to...
- find the slope of a segment or a line
- write equations of lines in slope-intercept form

So you can...
- graph and compare equations of lines
- investigate geometric relationships using lines

The world's largest indoor ski slope is located in Funabashi, Japan. It is called the *LaLaport Skidome* and has three courses—a red or *expert* course, a yellow or *intermediate* course, and a blue or *beginners'* course. Each course consists of alternating flat and sloped regions. A diagram of the red course is shown below. All distances are rounded to the nearest meter.

THINK AND COMMUNICATE

1. How many meters does the red course drop in its first sloped part, between point B and point C?

2. What is the horizontal distance from B to C?

You can express how steep the first sloped part of the red course is by using the *slope formula*.

4.2 Equations of Lines   **173**

Trigonometry

What is the horizontal distance a jet travels in the first ten seconds after takeoff?

(Chapter 10, page 519)

Algebra

How is the slope of an expert ski course different from the slope of a beginner course?

(Chapter 4, page 173)

Transformational Geometry

How can you model a reflection using transformations?

(Chapter 1, pages 8–9)

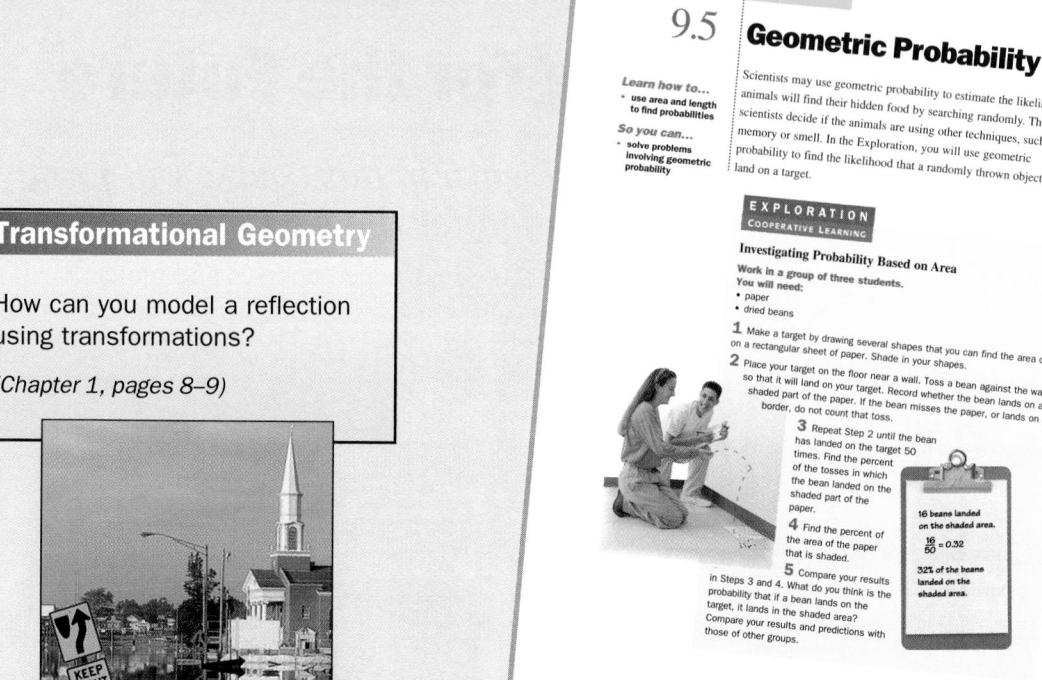

Probability

What is the probability that an object thrown at random will land on a target?

(Chapter 9, page 475)

Discrete Mathematics

How many squares remain after the third step in creating Sierpinski's carpet?

(Chapter 9, page 480)

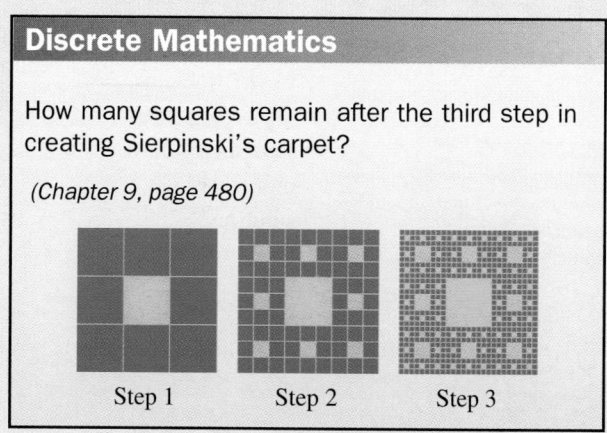

| Step 1 | Step 2 | Step 3 |

"When will I ever use this?"

Building for the Future

The skills you'll learn in this course will form a strong foundation for the future.

They'll prepare you for more advanced courses and increase your career opportunities.

Problem Solving and Communication

The exercise sets help you develop your problem solving and communication skills.

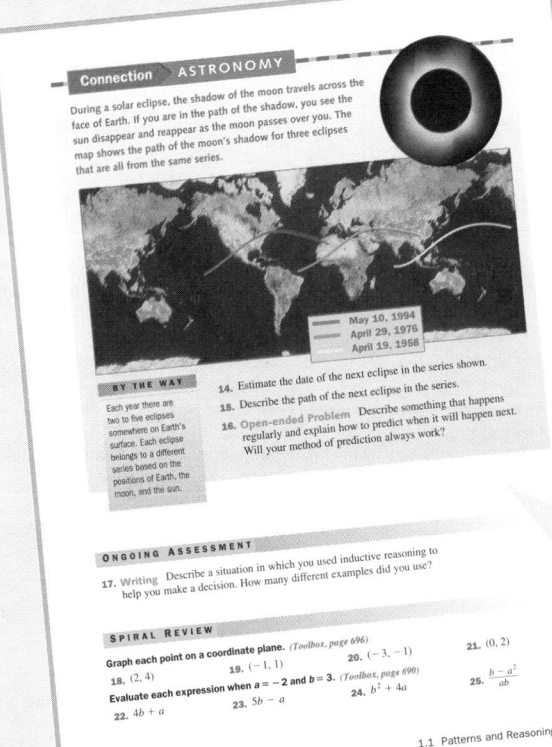

Connection ▸ ASTRONOMY

During a solar eclipse, the shadow of the moon travels across the face of Earth. If you are in the path of the shadow, you see the sun disappear and reappear as the moon passes over you. The map shows the path of the moon's shadow for three eclipses that are all from the same series.

May 10, 1994
April 29, 1976
April 19, 1958

BY THE WAY

Each year there are two to five eclipses somewhere on Earth's surface. Each eclipse belongs to a different series based on the positions of Earth, the moon, and the sun.

14. Estimate the date of the next eclipse in the series shown.
15. Describe the path of the next eclipse in the series.
16. **Open-ended Problem** Describe something that happens regularly and explain how to predict when it will happen next. Will your method of prediction always work?

ONGOING ASSESSMENT

17. **Writing** Describe a situation in which you used inductive reasoning to help you make a decision. How many different examples did you use?

SPIRAL REVIEW

Graph each point on a coordinate plane. *(Toolbox, page 696)*

18. $(2, 4)$ 19. $(-1, 1)$ 20. $(-3, -1)$ 21. $(0, 2)$

Evaluate each expression when $a = -2$ and $b = 3$. *(Toolbox, page 690)*

22. $4b + a$ 23. $5b - a$ 24. $b^2 + 4a$ 25. $\dfrac{b - a^2}{ab}$

1.1 Patterns and Reasoning   **7**

---

22. **Open-ended Problem** Give an example of a scale drawing or a scale model you have seen that is an enlargement of a real-life object or group of objects.

23. a. **TECHNICAL DRAWING** If you drew the structures in the table below so that 1 in. represents 100 ft, what would the scale be? How tall would each structure be in your drawing?

CN Tower	TMG Offices	Washington Monument	Empire State Building
Toronto, Canada	Tokyo, Japan	Washington, DC	New York City, NY
1815 ft	793 ft	555 ft	1250 ft

b. **Open-ended Problem** On a sheet of $8\frac{1}{2}$ in. by 11 in. paper, sketch and label a scale drawing of the four structures. Explain how you chose the scale for your drawing.

24. **TRANSFORMATIONS** Explain why dilations are sometimes called *similarity transformations.*

25. **Challenge** What do you think must be true for two prisms to be similar? Write a definition for similar prisms.

26. **MOVIES** For most of the movie *King Kong*, the scale models of King Kong were built to appear about 18 ft tall. When King Kong climbs the Empire State Building, the models make him appear to be 24 ft tall.

a. Make a scale drawing that shows King Kong at both 18 ft and 24 ft tall, a 6 ft tall person, and the Empire State Building (1250 ft).

b. **Writing** Why do you think the movie makers chose to make King Kong appear taller when he climbs the Empire State Building?

27. **SAT/ACT Preview** Select the pair of words that *best* expresses a relationship similar to that expressed in the pair *gallon : liquid.*

A. ruler : paper   B. week : month   C. length : width

D. degree : temperature   E. foot : meter

28. **GEOGRAPHY** The Nile, in Africa, is the world's longest river. It would be about the length of a 24 in. shoelace (61 cm) if drawn   that 1 cm represents 110 km. What is the scale? About how long is   Nile?

451

SAT/ACT Preview

These exercises help you prepare for college entrance exams.

Like many real-world problems, **Open-ended Problems** have more than one solution.

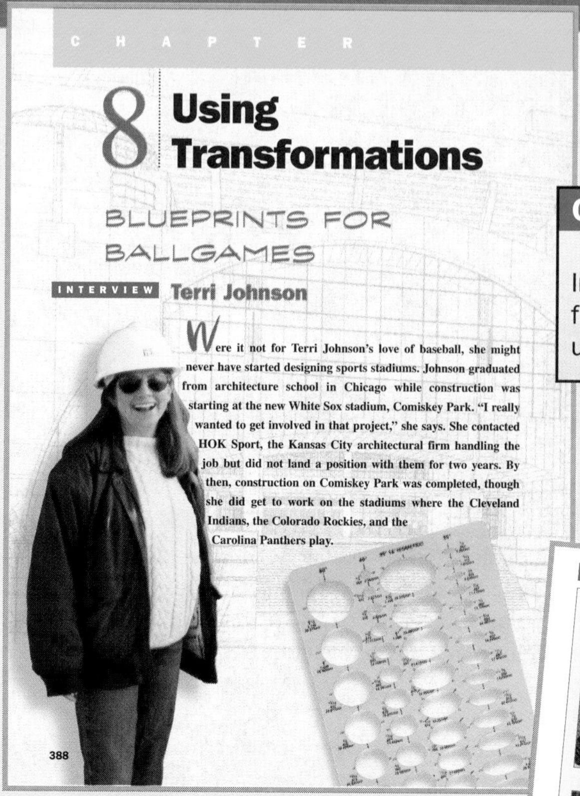
CHAPTER

8 Using Transformations

BLUEPRINTS FOR BALLGAMES

INTERVIEW Terri Johnson

Were it not for Terri Johnson's love of baseball, she might never have started designing sports stadiums. Johnson graduated from architecture school in Chicago while construction was starting at the new White Sox stadium, Comiskey Park. "I really wanted to get involved in that project," she says. She contacted HOK Sport, the Kansas City architectural firm handling the job but did not land a position with them for two years. By then, construction on Comiskey Park was completed, though she did get to work on the stadiums where the Cleveland Indians, the Colorado Rockies, and the Carolina Panthers play.

388

Terri Johnson, *Architect*
Chapter 8

Career Preview

In the Interviews and exercises, you will find problems that show how math is used in a variety of careers.

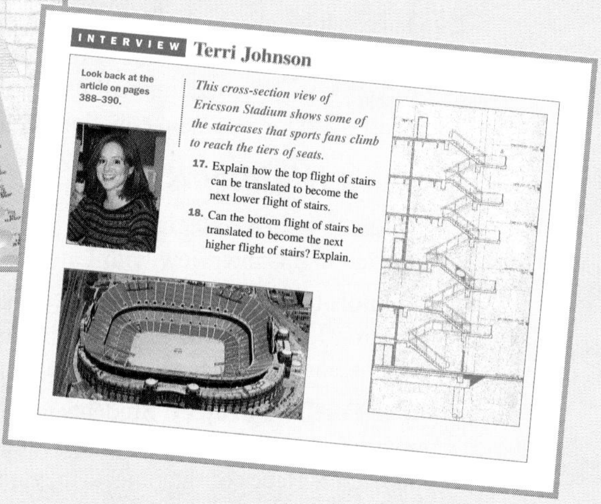
INTERVIEW Terri Johnson

Look back at the article on pages 388–390.

This cross-section view of Ericsson Stadium shows some of the staircases that sports fans climb to reach the tiers of seats.

17. Explain how the top flight of stairs can be translated to become the next lower flight of stairs.

18. Can the bottom flight of stairs be translated to become the next higher flight of stairs? Explain.

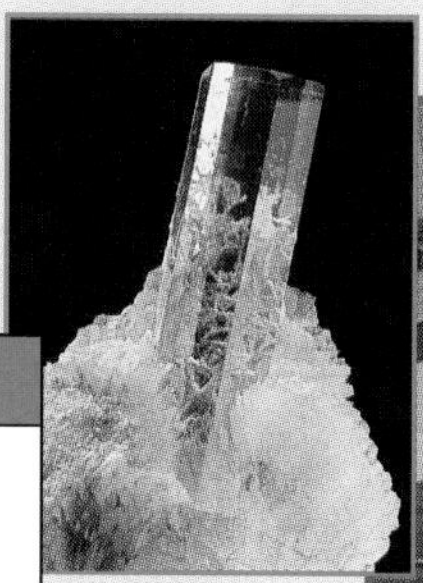

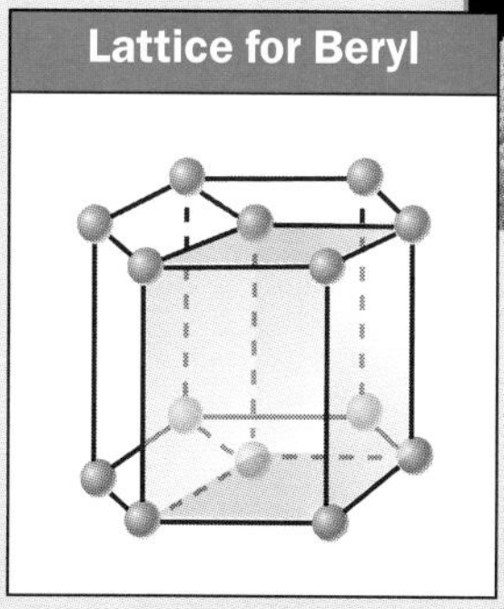
Lattice for Beryl

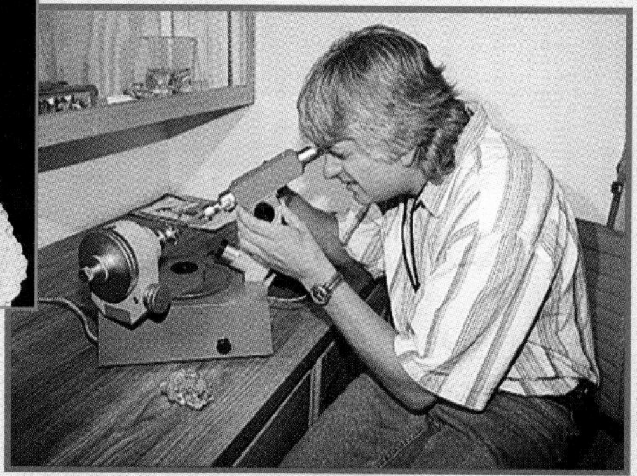

Virgil Lueth, *Mineralogist*
Chapter 2

xxxi

1 Patterns, Lines, and Planes

OVERVIEW

Connecting to Prior and Future Learning

⇔ Chapter 1 introduces students to patterns and reasoning. Students begin to use their reasoning skills as they learn to write and analyze conjectures to determine if conditional statements are true or false.

⇔ Students also begin their work with transformations and symmetry in this chapter. As with patterns and reasoning, these concepts will be important to their study of geometry.

⇔ This chapter also includes an introduction to the basic undefined terms of point, line, and plane. These basic elements are then used to introduce segments, angles, and concepts related to these figures. Students will use algebra when bisecting segments and angles, and can review solving two-step equations on page 696 in the **Student Resources Toolbox**.

Chapter Highlights

Interview with Latecia Leavy: The relationship between geometry and tumbling is highlighted in this interview with Latecia Leavy. Related exercises are on pages 12, 17, and 37.

Explorations in Chapter 1 involve analyzing patterns in Section 1.1, tracing networks in Section 1.3, and using physical objects, such as foam trays and uncooked spaghetti to model points, lines and planes in Section 1.4.

The Portfolio Project: Students identify objects that have symmetry and analyze the transformations that preserve their shapes and sizes. Students' presentation of their results include labeled sketches of the objects they measure.

 Technology: Students use geometry software to draw a cube in Section 1.4 and to explore rotations around a given point in Section 1.6.

OBJECTIVES

Section	Objectives	NCTM Standards
1.1	• Reason inductively. • Make predictions about patterns in real-life situations.	1, 2, 3, 4, 5
1.2	• Identify and perform transformations. • Describe movement and patterns in real-life situations.	1, 2, 3, 4, 5, 8
1.3	• Write and analyze conjectures. • Tell whether conditional statements are true or false.	1, 2, 3, 4, 5
1.4	• Use points, lines, and planes. • Describe and sketch relationships between points, lines, and planes in real-world objects.	1, 2, 3, 4, 7
1.5	• Represent rays and segments. • Interpret diagrams. • Find lengths and distances in various real-world situations.	1, 2, 3, 4, 7, 8
1.6	• Name and measure angles. • Describe angles in various real-world situations.	1, 2, 3, 4, 7
1.7	• Identify bisectors of segments and angles. • Find real-world distances. • Find angle measures in real-world situations.	1, 2, 3, 4, 7, 8

INTEGRATION

Mathematical Connections	1.1	1.2	1.3	1.4	1.5	1.6	1.7
geometry	**3–7***	**8–13**	**14–20**	**21–27**	**28–33**	**34–39**	**40–47**
algebra			16		32		41, 43, 44, 46
data analysis, probability, discrete math			14				
patterns and functions	**3–7**		20		33		
logic and reasoning	**3–7**	13, 18	**14–20**	26, 27	32	39	47

Interdisciplinary Connections and Applications							
history and geography			19	24			
biology and earth science						39	
chemistry and physics	6						
arts and entertainment				26			
sports and recreation				25		37	44, 46
astronomy	7						
social studies			17				
communication, travel, manufacturing, crafts		13	17		31		45

*__Bold page numbers__ indicate that a topic is used throughout the section.

TECHNOLOGY

Section	Student Book	Support Material
		opportunities for use with
1.1	graphing calculator	**Technology Book:** Spreadsheet Activity 1
1.2		**Technology Book:** Calculator Activity 1 TI-92 Activity 1 **Geometry Inventor Activity Book:** Activities 29, 30
1.3		**Technology Book:** Spreadsheet Activity 1
1.4	geometry software	**Technology Book:** Calculator Activity 2
1.6	geometry software	
1.7		**Geometry Inventor Activity Book:** Activity 2

PLANNING GUIDE

Regular Scheduling (45 min)

Section	Materials Needed	Core Assignment	Extended Assignment	exercises that feature		
				Applications	Communication	Technology
1.1	graph paper	2–5, 10, 11, 18–25	1–25	9, 12, 14, 15	8, 9, 16, 17	
1.2	graph paper	1–9, 11–17, 23–27	1–27	11–13, 18–23	1–3, 10, 14, 18, 21	
1.3	graph paper	3–14, 23–31, AYP*	1–31, AYP	1–5, 19–22	18, 29	
1.4	3 trays or stiff paper, tape, scissors, uncooked spaghetti, tracing paper, geometry software	**Day 1:** 1–19 **Day 2:** 20–30, 37–45	**Day 1:** 1–19 **Day 2:** 20–45	31–35	11–17 34c, 37	36
1.5		**Day 1:** 1–13, 18–23 **Day 2:** 24, 25, 28–34, AYP	**Day 1:** 1–23 **Day 2:** 24–34, AYP	14–16 27	17 27, 28	
1.6	protractor, geometry software	1–9, 11–18, 30–34	1, 5–8, 10–34	10, 19, 20, 26–30	1, 19	21–25
1.7	square sheet of paper	1–8, 10–22, 29–32, 34–39, AYP	1–11, 16–22, 28–39, AYP	9, 28	9b, 23, 27, 33, 34	
Review/ Assess		**Day 1:** 1–14 **Day 2:** 15–29 **Day 3:** Ch. 1 Test	**Day 1:** 1–14 **Day 2:** 15–29 **Day 3:** Ch. 1 Test			
Portfolio Project		Allow 2 days.	Allow 2 days.			

Yearly Pacing (with Portfolio Project)	Chapter 1 Total 14 days			Remaining 146 days	Total 160 days

Block Scheduling (90 min)

	Day 1	Day 2	Day 3	Day 4	Day 5	Day 6	Day 7
Teach/Interact	1.1: Exploration, page 3 1.2	1.3: Exploration, page 14 1.4: Exploration, page 21	Continue with 1.4 1.5	Continue with 1.5 1.6	1.7 Review	Review Port. Proj.	Ch. 1 Test Port. Proj.
Apply/Assess	**1.1:** 2–5, 10, 11, 18–25 **1.2:** 1–9, 11–17, 23–27	**1.3:** 3–14, 23–31, AYP* **1.4:** 1–19	**1.4:** 20–30, 37–45 **1.5:** 1–13, 18–23	**1.5:** 24, 25, 28–34, AYP **1.6:** 1–9, 11–18, 30–33	**1.7:** 1–8, 10–22, 29–32, 34–39, AYP **Review:** 1–14	**Review:** 15–28 **Port. Proj.**	**Ch. 1 Test** **Port. Proj.**

NOTE: A one-day block has been added for the Portfolio Project—timing and placement to be determined by teacher.

Yearly Pacing (with Portfolio Project)	Chapter 1 Total 7 days			Remaining 73 days	Total 80 days

* **AYP** is Assess Your Progress.

Section	Practice Bank	Study Guide*	Assessment Book*	Visuals	Explorations Lab Manual	Lesson Plans	Technology Book
1.1	1	1.1		Warm-Up 1.1	Masters 2, 6	1.1	Spreadsheet Act. 1
1.2	2	1.2		Warm-Up 1.2	Add. Expl. 1 Master 2	1.2	Calculator Act. 1 TI-92 Act. 1
1.3	3	1.3	Test 1	Warm-Up 1.3	Masters 2, 7	1.3	Spreadsheet Act. 1
1.4	4	1.4		Warm-Up 1.4 Folder 1		1.4	Calculator Act. 2
1.5	5	1.5	Test 2	Warm-Up 1.5	Master 2	1.5	
1.6	6	1.6		Warm-Up 1.6		1.6	
1.7	7	1.7	Test 3	Warm-Up 1.7		1.7	
Review Test	8	Chapter Review	Tests 4, 5, Alternative Assessment			Review Test	

*__Spanish versions__ of *Study Guide* and *Assessment Book* are available.

Chapter Support

- Course Guides
- Lesson Plans
- Portfolio Project Book: Additional Project 1: History of Geometry
- Preparing for College Entrance Tests
- Multi-Language Glossary
- *Test Generator* Software
- Professional Handbook
- Challenge Problems

Software Support

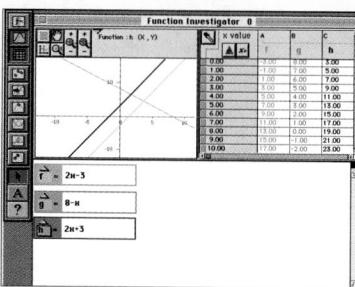

McDougal Littell Mathpack
Geometry Inventor

Internet Support

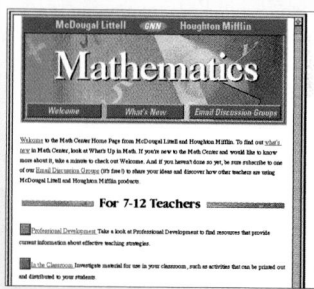

http://www.hmco.com
Next go to McDougal Littell; then the Education Center; then Secondary Math.

Books, Periodicals

Lindquist, Mary M., ed. *Learning and Teaching Geometry, K–12, 1987 Yearbook.* Reston, VA: NCTM, 1987.

Coxford, Arthur F., Linda Burks, Claudia Giamati, and Joyce Jonik. *Geometry from Multiple Perspectives: Addenda Series, Grades 9–12.* Reston, VA: NCTM, 1991.

Engelhardt, John. *Geometry in Our World.* Color slides with commentary. Reston, VA: NCTM, 1987.

Activities, Manipulatives

Kempe, A. B. *How to Draw a Straight Line.* Historic lecture from 1877. Reston, VA: NCTM, 1977.

Software

Spring Branch Software, Inc. *Tools of Mathematics: Points, Lines, and Planes.* Macintosh. Acton, MA: William K. Bradford Publishing Co.

Jackiw, Nicholas, designer. *The Geometer's Sketchpad.* (Ver. 3.0) for Macintosh or Windows. Berkeley, CA: Key Curriculum Press, 1995.

Videos

Geometry: New Tools for New Technologies. Hour One, "Snowbound: Euler Circuits"; Hour Two, "Symmetry: Rigid Motions and Patterns." Arlington, VA: COMAP.

Internet

The American Mathematical Society (AMS) and the Society for Industrial and Applied Mathematics (SIAM) maintain a Web page that deals with nonacademic careers in mathematics. Find this at:
    http://www.ams.org/careers/

Patterns, Lines, and Planes

SPRINGING Across Obstacles

INTERVIEW Latecia Leavy

Tumbling

Tumbling has existed for centuries. While its exact beginning is difficult to trace, it is known that many cultures have tumbling in their history. For example, the ancient Greeks used tumbling as a training device; some African peoples used it as an integral part of their rituals, and other cultures of the world used it in theatrical entertainment and dance.

In Germany in the early 1800s, Frederick Ludwig Jahn was attempting to train young Germans to be physically fit. Tumbling was a part of his comprehensive athletic plan. Eventually, groups of athletes were competing and a new sport was created. In this way, Jahn is seen as the inventor of the competitive sport of gymnastics. The sport came to the United States in the 1820s. In 1886, the first United States National Tumbling Championship was held.

Latecia Leavy

Latecia Leavy started performing with the Jesse White Tumbling Team when she was 12 years old. Today, she is on another tumbling team and is also a cheerleader for her high school.

As a member of the Jesse White Tumbling Team, Latecia Leavy has performed at various places such as the United Center and Soldier Field in Chicago, Springfield, Illinois, and St. Louis, Missouri. The tumbling team has also toured in Japan, Canada, Bermuda, and has been involved in making three movies and three commercials.

Latecia Leavy studies mathematics by day at a Chicago high school and then, in the late afternoon, she takes the "crash course." In a typical session with the Jesse White Tumbling Team, the 15-year-old Leavy is just a blur as she whips across the ground in a dizzying series of backward half-flips ("flip-flops") and back somersaults. Leavy and her teammates also catapult from a mini-trampoline, soaring over people, cars, and other objects, before doing a roll and safely returning to the planet Earth.

"Everybody should find something they like to do as much as I enjoy tumbling."

For Leavy, geometry is more than an academic exercise. "You have to take off from the trampoline at the right angle, so you don't go off sideways and crash into somebody," she explains. Leavy began tumbling when she was eight years old and continues to practice several hours each week. It barely seems like work to her because she enjoys tumbling so much. "Although this is not for everyone," she says, "everybody should find something they like to do as much as I enjoy tumbling."

Entertaining Teamwork

Leavy and about 60 other youths—mostly residents of Chicago housing projects—make up the tumbling team founded and coached by Jesse White. The group has traveled around the world, displaying breathtaking feats. They have appeared in movies and TV shows and entertained audiences at professional sports events. "You've got to have some experience to make this team," says White, "but members also need a good attitude, and they have to understand teamwork."

> **"You have to spring from the trampoline at a precise angle for a vertical landing."**

1

Background

Trampolines

The trampoline originated in the circus when the landing nets that protected high wire and trapeze performers were tightened. In the 1930s, George Nissen designed and produced the first modern trampoline. It was the first "bed" suspended from springs on a frame. Nissen also developed other pieces of apparatus and was very much involved in the early development of the sport of trampolining.

The small trampoline, often called the mini-trampoline, is considered safe for jumping and tumbling only if used at various angles. Trampolines on which flips and tricks may be performed are much larger and are parallel to the ground.

Second-Language Learners

Students learning English might benefit from working with a peer tutor when reading this interview. The peer tutor can define key terms and rephrase parts of the interview that contain challenging vocabulary.

Multicultural Note

Although talented women from all over the world competed in the 1996 World Gymnastics Championships, the nations that are generally considered to have the best overall women's gymnastic programs are the United States, Romania, Russia, China, and Ukraine. In the 1996 Summer Olympics, Ukrainian Lilia Podkopayeva won three medals, and Romanians Simona Amanar and Gina Gogean both went home with four. Mo Hiulan and Bi Wenjiing, representing China, each brought home a medal. Dominique Dawes became one of the first African Americans ever to win an Olympic medal in gymnastics, and she won two. Her teammate, Chinese American Amy Chow, also won two Olympic medals.

Mathematical Connection

As a tumbler, deciding what stunts or tricks to do that will please an audience is often difficult. Feats of daring, such as the *Superman* stunt, are often crowd pleasers, but other tricks and stunts can also be very well accepted by audiences. All tricks and stunts require a lot of practice, however, to ensure that they are performed safely and correctly. In Section 1.2, several other stunts and tricks that tumblers perform are explored. In this section, students are asked to analyze what types of transformations these stunts involve. In Section 1.3, several conjectures about the art of tumbling are used to investigate the *if-then* form of conditional statements. Then, in Section 1.6, students look at trampolines placed at various angles and are asked to analyze the effect of the angles.

Explore and Connect

Writing

If students have difficulty visualizing the writing question, have them complete the project first to better understand the effect of the angle of the trampoline. Also, make sure students answer the question in terms of both height and distance.

Project

You may wish to suggest that students work with a partner to complete this project. Then both students can decide how the path of the pebble varies.

Research

The library or an encyclopedia may be good sources for this research.

2

Angling for Success

In one trick, teammates form a "Pyramid," with three people on the base, two more on the second level, and another two on top. Then someone else springs from the trampoline after a running take-off and sails over the human pyramid. Tumblers have to run fast before hitting the trampoline or they won't make it over the top.

White sets the trampoline at about a 45 degree angle, as steep as it will go, to ensure maximum height and distance. The 45 degree setting is crucial for the most spectacular trick of the show, called "Superman." A tumbler springs from the trampoline and soars over 20 or more people—a feat Leavy is still training for. "It's an amazing thing to see...and do," she says.

Latecia speaks with a spectator before a performance.

1. Writing White sets the trampoline at the angle shown above so the tumbler will fly as far as possible. How do you think the tumbler's path would be different if the trampoline were horizontal? Explain your reasoning.

2. Project Make a model trampoline by taping a piece of a large balloon over the top of an open can or box. Throw a pebble at the trampoline while it is tilted at different angles. How does the path of the pebble vary? Make a poster to report your results to your class.

3. Research Some gymnasts use a *springboard* instead of a trampoline. Find out what a springboard is. Can you set it at different angles?

Mathematics
& Latecia Leavy

In this chapter, you will learn more about how ideas of geometry are related to tumbling.

Related Exercises

Section 1.2
• Exercises 11–13

Section 1.3
• Exercises 3–5

Section 1.6
• Exercise 10

1.1 Patterns and Reasoning

Learn how to...
- reason inductively

So you can...
- make predictions about patterns such as eclipses and the occurrence of the popularity of car colors

In this course, you will explore many kinds of patterns. For example, each branch of a sea fan looks like the fan itself. Each part of each branch looks like the fan too. The figure in the Exploration grows in a pattern like the sea fan.

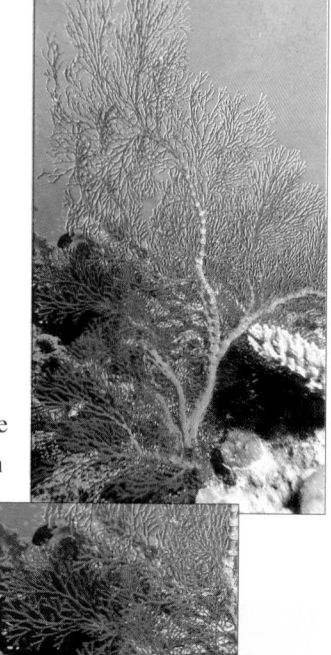

EXPLORATION
COOPERATIVE LEARNING

Analyzing Patterns

Work with another student.

1 Draw a segment.

2 Add 2 shorter segments.

3 At each new point, draw 2 segments.

4 At each new point, draw two segments half as long as the segments from Step 3. Copy the table, including several more steps. Complete the row for Step 4.

5 Continue the pattern as long as you can. Fill in the table after each step.

6 What patterns do you notice in the figure you drew? Circle three parts of your diagram that look like each other but are different sizes.

7 What patterns do you notice in your table?

8 Predict the total number of segments for the tenth step. Explain.

Step	Number of new segments	Total number of segments
1	1	1
2	2	3
3	4	7
4	?	?

Exploration Note

Purpose
The purpose of this Exploration is to have students discover that by identifying a pattern in several examples of a geometric figure, they can use the pattern to make predictions about the subsequent figures.

Materials/Preparation
No special materials are required.

Procedure
Students copy the figures shown in Steps 1–3 and continue the pattern by following the directions in Step 4. They copy and extend their table with information about the figures. Finally, they examine the figures and

the data in the table to find patterns and make predictions about subsequent figures.

Closure
Discuss the results of Steps 6–8. Discuss the idea that by analyzing several examples of a situation to determine a pattern, the pattern can then be used to make predictions about the situation.

Explorations Lab Manual
See the Manual for more commentary on this Exploration.
Diagram Master 6

For answers to the Exploration, see answers in back of book.

Plan⇔Support

Objectives
- Reason inductively.
- Make predictions about patterns in real-life situations.

Recommended Pacing
❖ **Core and Extended Courses**
Section 1.1: 1 day
❖ **Block Schedule**
Section 1.1: $\frac{1}{2}$ block
(with Section 1.2)

Resource Materials
Lesson Support
Lesson Plan 1.1
Warm-Up Transparency 1.1
Practice Bank: Practice 1
Study Guide: Section 1.1
Explorations Lab Manual: Diagram Masters 2, 6
Challenge Problems: Set 1
Technology
Technology Book: Spreadsheet Activity 1
Graphing Calculator
Internet: http://www.hmco.com

Warm-Up Exercises

What comes next in each pattern?

1. 11:00 A.M., 11:30 A.M., 12:00 noon, 12:30 P.M., ___?___ 1:00 P.M.

2.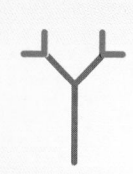

3.

4. 1 2 3 4 5, 2 3 4 5 1, 3 4 5 1 2, 4 5 1 2 3, ___?___ 5 1 2 3 4

5. (0, 0), (2, –3), (4, –6), (6, –9), ___?___ (8, –12)

Section Note

Communication: Discussion
A brief discussion of the concept of inductive reasoning is important to ensure that all students understand this fundamental idea. Discuss the meaning of the words *to generalize.* Ask students to provide some examples of using inductive reasoning to make a generalization.

About the Example

Problem Solving
Identifying patterns and then using inductive reasoning to make a prediction is a powerful problem-solving strategy. This point should be made explicit to students after discussing the Solution in the Example. Point out that if the problem were to predict the number of smaller triangles in the *twenty-fifth* figure, then using a drawing approach, as in Method 1, would not be viable. But an analysis of some drawings is necessary in order to find a formula, as in Method 2. Thus, the algebraic solution is based upon a geometric approach.

Additional Example

Predict the number of squares in the sixth figure of the pattern.

Method 1 Look for a pattern and continue it.

Draw 2 squares.
Add 4 squares.
Add 6 squares.
Add 8 squares.
Add 10 squares.
Add 12 squares.

Count the number of squares in the sixth figure. There are 42 squares in the sixth figure.
Method 2 Write a formula for the number of squares in the *n*th figure.

Figure	1	2	3	4...	n
Number of squares	2	6	12	20...	$n(n+1)$

$1\cdot 2 \quad 2\cdot 3 \quad 3\cdot 4 \quad 4\cdot 5... \quad n(n+1)$

There are $n(n+1)$ squares in the *n*th figure, so in the sixth figure there are $6 \cdot 7 = 42$ squares.

When you make a prediction based on several examples, you are using **inductive reasoning**. For example, if your eyes itch each time you play with a cat, you use inductive reasoning to generalize and decide that you are allergic to cats.

EXAMPLE

Predict the number of small triangles in the seventh figure of the pattern.

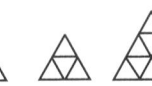

SOLUTION

Look for a pattern.

Method 1

Continue the pattern.

Count the small triangles in the seventh figure.

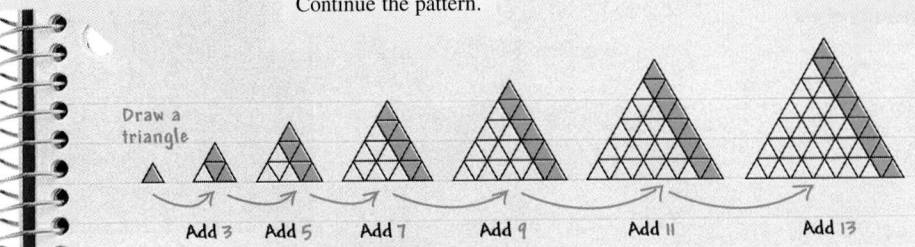

Draw a triangle

Add 3   Add 5   Add 7   Add 9   Add 11   Add 13

There are 49 small triangles in the seventh figure.

Method 2

Write a formula for the number of small triangles in the *n*th figure.

Figure	1	2	3	4	...	n
Number of small triangles	1	4	9	16	...	n^2

$$1^2 \quad 2^2 \quad 3^2 \quad 4^2 \qquad n^2$$

There are n^2 small triangles in the *n*th figure of the pattern, so in the seventh figure there are 7^2, or 49 small triangles.

The number of small triangles is the square of the figure number.

THINK AND COMMUNICATE

Describe each pattern and give the next two numbers or figures.

1. 5, 4, 3, 2, 1, 0, . . .

2.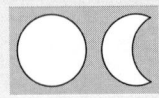

3. Describe at least two ways to continue this pattern: 1, 2, 1, ...

4   Chapter 1 *Patterns, Lines, and Planes*

4

✓ CHECKING KEY CONCEPTS

For Questions 1–4, use the shapes and table below.

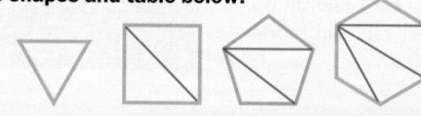

Number of sides	3	4	5	6
Number of triangles	1	2	3	4

1. Copy the table, leaving room for two more columns. What patterns do you notice?

2. Sketch the next two shapes in the sequence. Record the number of triangles in each shape.

3. Give a formula for the number of triangles in a shape with n sides.

4. Use your formula from Question 3 to find the number of triangles in a shape that has 100 sides.

1.1 Exercises and Applications

Extra Practice exercises on page 659

1. a. The table describes the number of points in the figure you drew in the Exploration on page 3. Copy the table and add three rows.

   b. What patterns do you notice in your table? What will the total number of points be for the tenth stage? Explain.

Step	Number of new points	Total number of points
1	2	2
2	2	4
3	4	8

Use inductive reasoning to find the next two numbers in each pattern.

2. $-1, -3, -5, -7, \underline{?}, \underline{?}$

3. $0, 3, 8, 15, \underline{?}, \underline{?}$

4. $1, 3, 7, 13, \underline{?}, \underline{?}$

5. $2, \dfrac{3}{2}, \dfrac{4}{3}, \underline{?}, \underline{?}$

Use inductive reasoning to sketch the next shape in each pattern.

6.      7.

8. **Open-ended Problem** Describe one way you have used inductive reasoning outside of your mathematics class. What did you predict? How much information did you base your prediction on? How accurate was your prediction?

1.1 Patterns and Reasoning **5**

Exercises and Applications

1. a. The table values for the first seven steps are given.

Step	Number of new points	Total number of points
1	2	2
2	2	4
3	4	8
4	8	16
5	16	32
6	32	64
7	64	128

b. From Step 2 on, the number of new points at step n is 2^{n-1} and the total number of points is 2^n. The total number of points at the tenth stage will be $2^{10} = 1024$.

2. $-9, -11$

3. $24, 35$

4. $21, 31$

5. $\dfrac{5}{4}, \dfrac{6}{5}$

6.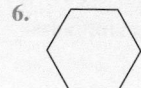

7.  ‾◁

8. Answers may vary. An example is given. I predicted my family would have spaghetti for dinner on Tuesday. I based my prediction on the fact that we have had spaghetti for dinner every Tuesday for as long as I can remember. My prediction was inaccurate. We had a power failure on Tuesday and ordered take-out pizza for dinner.

Teach⇔Interact

Checking Key Concepts

Geometric Thinking
Questions 1–4 take students through the inductive reasoning process in a step-by-step approach. First they find a pattern, then create the next two shapes to confirm their pattern, then use their pattern to find a formula, and then use the formula to make a prediction.

Closure Question

Describe how inductive reasoning is used to make a prediction based upon several examples. *The examples are analyzed to find a pattern. The pattern is then used to make a prediction.*

Apply⇔Assess

Suggested Assignment

❖ **Core Course**
Exs. 2–5, 10, 11, 18–25

❖ **Extended Course**
Exs. 1–25

❖ **Block Schedule**
Day 1 Exs. 2–5, 10, 11, 18–25

Exercise Notes

Problem Solving
Exs. 2–5 Finding the difference between successive terms will quickly reveal a pattern for each of these exercises. For Ex. 5, it may help some students to point out that the first number can be written as the fraction $\dfrac{2}{1}$.

Common Error
Ex. 8 A discussion of the last question in the exercise, *How accurate was your prediction?*, can make students aware of the fact that a prediction based on inductive reasoning may not be true. Students who assume that inductive reasoning always leads to correct predictions are making an error in reasoning. This idea is discussed in Section 1.3, where students learn that a conjecture based on inductive reasoning may be true or false.

5

Exercise Notes

Application
Ex. 9 This is a good exercise to help students realize that although the trends observed in the table may be used to make a prediction for the most popular colors in 1996, the prediction may not be accurate. For example, for some unknown reason, the color bright red may again become the most popular color in 1996.

Interdisciplinary Problems
Ex. 12 The structural formulas in this exercise are for (reading from left to right) methane, ethane, propane, and butane. They are among a large class of organic compounds known as hydrocarbons.

Research
Ex. 12 Science can provide an endless source of examples of how inductive reasoning has been used to make predictions. You may wish to ask students who have taken science courses to research some examples for presentation to the class.

Using Technology
Ex. 13 Students who know how to use a graphing calculator to find regression equations can use the calculator to discover that $y = x^3 - 3x^2 + 3x - 2$ gives y-values of –1, 0, 7, 26, and 63 for x-values of 1, 2, 3, 4, and 5, respectively. Students who use other approaches are more likely to come up with the expression $(n - 1)^3 - 1$, which is equal to $n^3 - 3n^2 + 3n - 2$. On the TI-82, press STAT 1 and enter the numbers 1, 2, 3, 4, and 5 in L₁. (These are the position numbers for the terms in the given pattern.) Then press ▶ and enter the given terms –1, 0, 7, 26, and 63 in L₂. Press STAT ▶ 5 ENTER to display the coefficients for a linear regression equation. A linear regression equation is $y = 15.4x - 27.2$. Next, press STAT ▶ 6 ENTER to get the quadratic regression equation $y = 6x^2 - 20.6x + 14.8$. For a cubic regression equation, press STAT ▶ 7 ENTER to get $y = x^3 - 3x^2 + 3x - 2$. Replace x with 1, 2, 3, 4, and 5 in each of these equations to see which equation yields the given pattern –1, 0, 7, 26, and 63. The equation that works is the cubic equation.

---

9. Writing Each year, auto makers introduce new colors for their cars. One way auto makers choose colors for new cars is to find out which colors sold well in the past. The table shows color popularity for compact cars and sports cars from 1990 through 1995.

Color	Popular Car Colors (percent of total sales for the year)					
	1990 sales	1991 sales	1992 sales	1993 sales	1994 sales	1995 sales
White	20.6%	21.0%	19.3%	15.3%	15.3%	14.4%
Bright red	22.0%	19.3%	12.4%	12.9%	11.2%	9.5%
Medium red	5.8%	7.7%	10.6%	11.4%	10.5%	11.3%
Bright blue	4.8%	7.5%	4.3%	3.8%	5.4%	2.3%
Silver	6.4%	6.8%	4.9%	5.2%	2.4%	6.3%
Green	0.4%	0.6%	6.3%	16.1%	12.8%	15.2%

a. Describe any trends that you observe in color popularity from 1990 through 1995.

b. Based on the trends, what do you think were the most popular colors in 1996? Explain.

Write a formula for the value of the *n*th term in each pattern.

10.

Term	1	2	3	4	5	6	...	n
Value	–3	–2	–1	0	1	2	...	?

11.

Term	1	2	3	4	5	6	...	n
Value	4	7	10	13	16	19	...	?

12. CHEMISTRY Give the molecular and structural formulas for the next two compounds in the alkane series. The first four compounds are shown below.

Molecular formula	CH_4	C_2H_6	C_3H_8	C_4H_{10}

Structural formula

$$H-\overset{\overset{\displaystyle H}{|}}{\underset{\underset{\displaystyle H}{|}}{C}}-H \qquad H-\overset{\overset{\displaystyle H}{|}}{\underset{\underset{\displaystyle H}{|}}{C}}-\overset{\overset{\displaystyle H}{|}}{\underset{\underset{\displaystyle H}{|}}{C}}-H$$

13. Challenge Find the next two numbers in the pattern below and explain your reasoning. –1, 0, 7, 26, 63, . . .

---

9. a. Answers may vary. An example is given. The colors that were most popular in 1990, white and bright red, became less popular, while the least popular color, green, became the most popular. Bright blue remained a fairly unpopular color, while the popularity of medium red increased and that of silver stayed about the same.

b. Answer may vary. An example is given. white, green, and medium red; Changes seem to be gradual, so the colors popular in 1995 will probably still be popular in 1996.

10. $n - 4$

11. $3n + 1$

12. Molecular formulas: C_5H_{12}; C_6H_{14}; Structural formulas:

13. 124, 215; The nth term is $(n - 1)^3 - 1$.

14. Estimates may vary; about May 20, 2012 and May 30, 2030.

During a solar eclipse, the shadow of the moon travels across the face of Earth. If you are in the path of the shadow, you see the sun disappear and reappear as the moon passes over you. The map shows the path of the moon's shadow for three eclipses that are all from the same series.

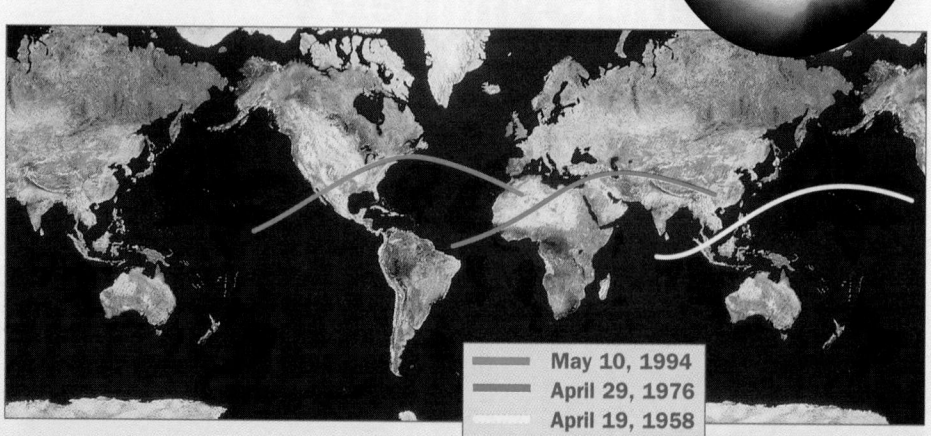

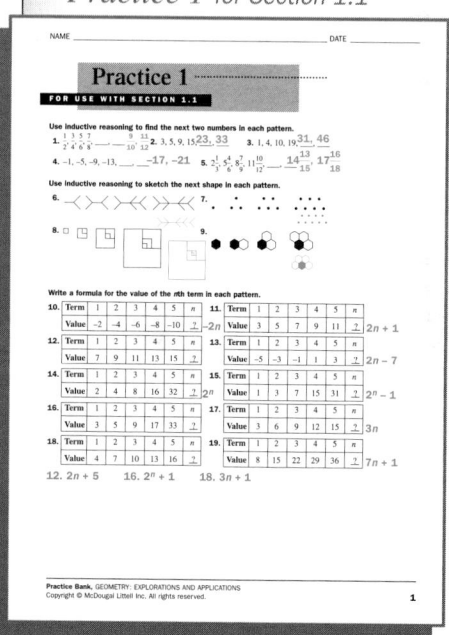

━━━	May 10, 1994
━━━	April 29, 1976
━━━	April 19, 1958

BY THE WAY

Each year there are two to five eclipses somewhere on Earth's surface. Each eclipse belongs to a different series based on the positions of Earth, the moon, and the sun.

14. Estimate the date of the next eclipse in the series shown.

15. Describe the path of the next eclipse in the series.

16. Open-ended Problem Describe something that happens regularly and explain how to predict when it will happen next. Will your method of prediction always work?

ONGOING ASSESSMENT

17. Writing Describe a situation in which you used inductive reasoning to help you make a decision. How many different examples did you use?

SPIRAL REVIEW

Graph each point on a coordinate plane. *(Toolbox, page 696)*

18. $(2, 4)$ **19.** $(-1, 1)$ **20.** $(-3, -1)$ **21.** $(0, 2)$

Evaluate each expression when $a = -2$ and $b = 3$. *(Toolbox, page 690)*

22. $4b + a$ **23.** $5b - a$ **24.** $b^2 + 4a$ **25.** $\dfrac{b - a^2}{ab}$

1.1 Patterns and Reasoning   **7**

Exercise Notes

Assessment Note
Ex. 17 This exercise should help you assess how well students understand the process of inductive reasoning. You may wish to discuss and relate the idea of *making a decision* to making a prediction.

Second-Language Learners
Ex. 17 Some students learning English may find it difficult to compose a written description of how they used inductive reasoning when making a decision. Consider offering students the alternative of orally describing a situation in which they used inductive reasoning.

Practice 1 for Section 1.1

NAME _____ DATE _____

Practice 1
FOR USE WITH SECTION 1.1

Use inductive reasoning to find the next two numbers in each pattern.

1. $\frac{1}{2}, \frac{3}{4}, \frac{5}{6}, \frac{7}{8}$ ___ ___ $\frac{9}{10}, \frac{11}{12}$ **2.** 3, 5, 9, 15, ___ ___ 23, 33   **3.** 1, 4, 10, 19, ___ ___ 31, 46

4. –1, –5, –9, –13, ___ ___ –17, –21   **5.** $2\frac{1}{5}, 5\frac{4}{6}, 8\frac{7}{9}, 11\frac{10}{12}$ ___ ___ $14\frac{13}{15}, 17\frac{16}{18}$

Use inductive reasoning to sketch the next shape in each pattern.

6. ○ ⟨⟩⟨⟩⟨⟩⟨⟨ ⟩⟩ ⟫⟫ ⟪⟪   **7.**

8. □ ⬚   **9.** ● ● ● ◐ ◯ ◯

Write a formula for the value of the nth term in each pattern.

10. Term	1	2	3	4	5	n
Value	–2	–4	–6	–8	–10	?

–2n

11. Term	1	2	3	4	5	n
Value	3	5	7	9	11	?

2n + 1

12. Term	1	2	3	4	5	n
Value	7	9	11	13	15	?

13. Term	1	2	3	4	5	n
Value	–5	–3	–1	1	3	?

2n – 7

14. Term	1	2	3	4	5	n
Value	2	4	8	16	32	?

2n

15. Term	1	2	3	4	5	n
Value	1	3	7	15	31	?

2n – 1

16. Term	1	2	3	4	5	n
Value	3	5	9	17	33	?

17. Term	1	2	3	4	5	n
Value	3	6	9	12	15	?

3n

18. Term	1	2	3	4	5	n
Value	4	7	10	13	16	?

19. Term	1	2	3	4	5	n
Value	8	15	22	29	36	?

7n + 1

12. 2n + 5   16. 2ⁿ + 1   18. 3n + 1

---

15. across the Pacific Ocean and Southeast Asia

16. Answers may vary. An example is given. Leap years occur every four years with some exceptions. To predict when one will happen next, determine the next year divisible by 4. If the next such year ends in a double zero, ignore it unless it is divisible by 400. This method of prediction always works.

17. Answers may vary. An example is given. I used inductive reasoning to predict how long it would take me to complete my math homework assignment. I considered how long I had spent on math homework every night last week and found the average.

18–21.

A coordinate plane graph showing points: (2, 4), (0, 2), (–1, 1), (–3, –1)

22. 10

23. 17

24. 1

25. $\frac{1}{6}$

Objectives

- Identify and perform transformations.
- Describe movement and patterns in real-life situations.

Recommended Pacing

❖ **Core and Extended Courses**
Section 1.2: 1 day

❖ **Block Schedule**
Section 1.2: $\frac{1}{2}$ block
(with Section 1.1)

Resource Materials

Lesson Support
Lesson Plan 1.2
Warm-Up Transparency 1.2
Practice Bank: Practice 2
Study Guide: Section 1.2
Explorations Lab Manual:
    Additional Exploration 1
    Diagram Master 2
Challenge Problems: Set 2

Technology
Technology Book:
    Calculator Activity 1
    TI-92 Activity 1
McDougal Littell Mathpack
    Geometry Inventor Activity Book:
    Activities 29, 30

Internet:
    http://www.hmco.com

Warm-Up Exercises

Give the coordinates of each point.

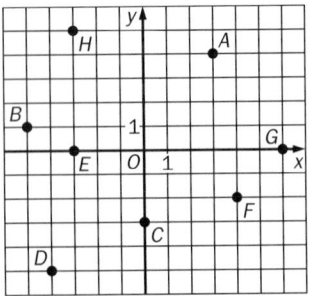

1. A (3, 4)   **2.** B (–5, 1)
3. C (0, –3)   **4.** D (–4, –5)
5. E (–3, 0)   **6.** F (4, –2)
7. G (6, 0)   **8.** H (–3, 5)

SECTION

1.2 | Transformations and Symmetry

Learn how to...
- **identify and perform transformations**

So you can...
- **describe movement and patterns, such as in tumbling and Braille letters**

In the photos, the signs are **reflected** in the water. The potter makes pottery by sculpting clay as it **rotates** on the potter's wheel. When a goose runs in a straight line across snow, it leaves **translated** sets of tracks.

Rotation, reflection, and translation are examples of **transformations**. When you transform a figure, the new figure is called the **image**. In these transformations, the image is the same size and shape as the original figure.

The reflection in the water is the **image** of the sign.

The back of the bowl is an **image** of the front.

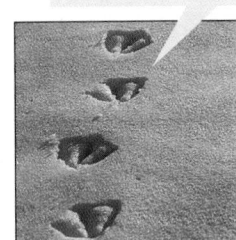

Each set of goose tracks is an **image** of the first set.

EXAMPLE 1

Perform each transformation.

a. Reflect the triangle over the red line.

b. Rotate the triangle by any amount around the green point.

c. Translate the triangle any distance to the right.

d. Rotate the triangle a complete turn around the green point.

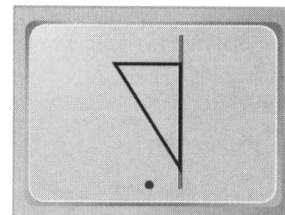

SOLUTION

a. Each point flips across the line.

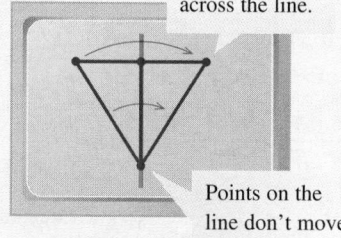

Points on the line don't move.

b.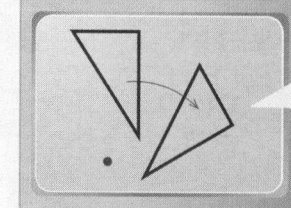

Each point rotates by the same amount.

c.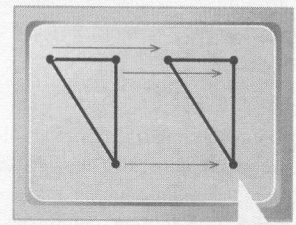

Each point moves the same distance to the right.

d.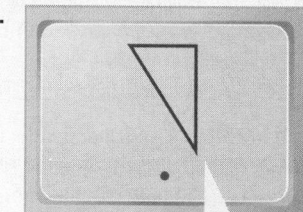

Each point returns to its original position.

THINK AND COMMUNICATE

1. How are the two transformations below alike? How are they different?

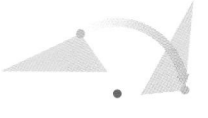 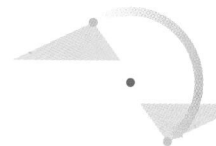

2. The second transformation is called a *half-turn*. Explain what this term means.

3. Is there more than one way to rotate a figure around a point? Explain.

4. Is there more than one way to reflect a figure over a line? Explain.

1.2 Transformations and Symmetry **9**

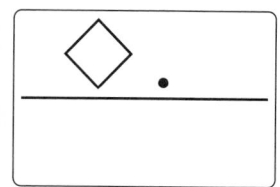
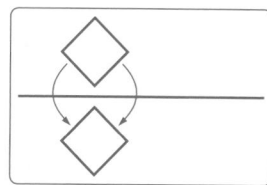
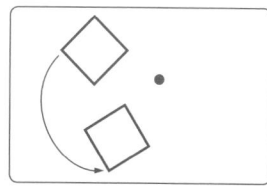
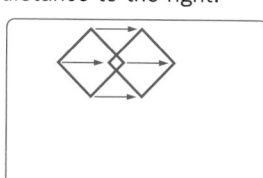
9

Additional Example 2

How do the coordinates of the points on a triangle change when you translate the triangle down?

Translate a triangle 3 units down. Each *y*-coordinate decreases by 3.

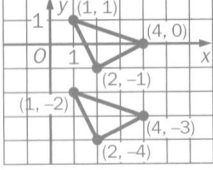

Translate a triangle 2 units down. Each *y*-coordinate decreases by 2.

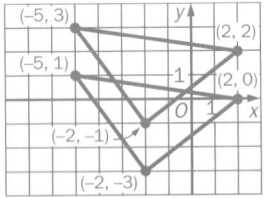

Translate a triangle 100 units down. Each *y*-coordinate decreases by 100.

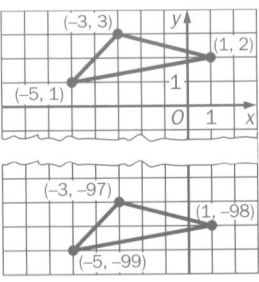

When you translate a triangle down *n* units, the *y*-coordinate of each point decreases by *n* units.

Section Note

Geometric Thinking

Many students may have an understanding of the concept of symmetry on an intuitive level. You can ascertain this by asking for some examples of things that are symmetrical. The text presents a mathematical definition of symmetry based upon the concept of transformation. Relate this view of symmetry to the examples of symmetry given by the class.

Additional Example 3

Describe each object's symmetry.

a.

The table has rotational symmetry and reflection symmetry. It looks exactly the same no matter how you turn it. Also, a plane of reflection runs through its center.

10

EXAMPLE 2

How do the coordinates of the points on a triangle change when you translate the triangle to the right? Try many different examples.

SOLUTION

Translate a triangle 3 units to the right.

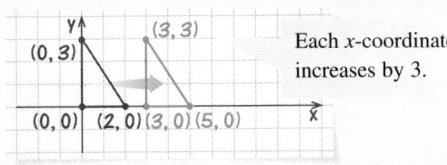

Each *x*-coordinate increases by 3.

Translate a triangle 1 unit to the right.

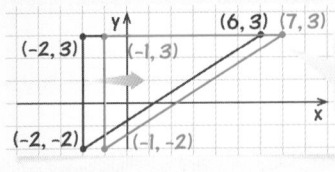

Each *x*-coordinate increases by 1.

Translate a triangle 5 units to the right.

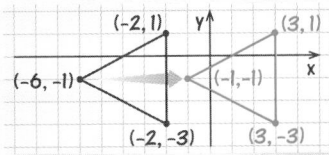

Translate a triangle 102 units to the right.

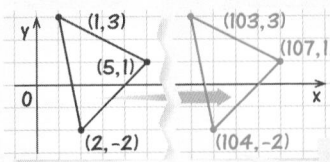

Each *x*-coordinate increases by 102.

Each *x*-coordinate increases by 5.

When you translate a triangle to the right *n* units, the *x*-coordinate of each point increases by *n* units.

If a figure and its image coincide, the figure has **symmetry**. For example, if you reflect the butterfly over the red line, the image coincides exactly with the original. The butterfly has *reflection symmetry* and the line is a **line of symmetry**. A figure may also have rotational or translational symmetry.

EXAMPLE 3

Describe the symmetry of each object.

a.  **b.**

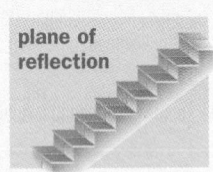

SOLUTION

a. The umbrella has rotational symmetry. It looks exactly the same after you spin it part way around the handle.

b. The staircase has both translational and reflection symmetry. If you translate the staircase up one step, it looks the same. Also, a plane of reflection runs down the center of the stairs.

plane of reflection

10   Chapter 1 *Patterns, Lines, and Planes*

Checking Key Concepts

1.

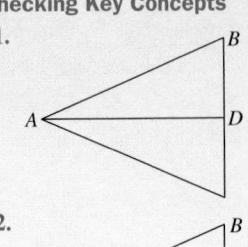

2.

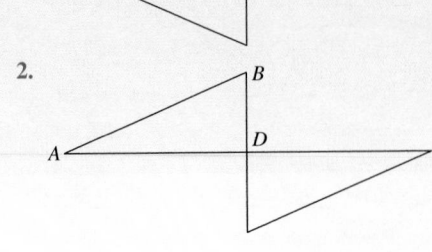

3.

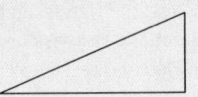
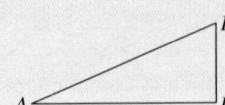

4. (7, –2), (12, 4), (10, –3)

5. reflection symmetry

6. reflection and rotational symmetry

7. translation symmetry

☑ CHECKING KEY CONCEPTS

For Questions 1–4, copy the triangle at the right.

1. Reflect the triangle over side *AD*.

2. Rotate the triangle a half-turn around point *D*.

3. Translate the triangle 2 in. to the left.

4. The *vertices*, or corner points, of a triangle are *A*(0, −2), *B*(5, 4), and *C*(3, −3). Predict the coordinates of the vertices of the image if you translate this triangle 7 units to the right.

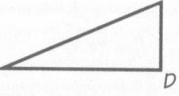

Describe the symmetry of each object.

5.

6.

7.

1.2 | Exercises and Applications

Extra Practice exercises on page 659

Open-ended Problems **Describe at least two real-world objects that have each type of symmetry.**

1. reflection symmetry
2. rotational symmetry
3. translational symmetry

Copy each diagram. Sketch the reflection of the triangle over the red line.

4.

5.

6.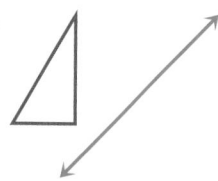

Sketch the fabric. Include the blue line and green point in your sketch. Sketch the image of each transformation.

7. Reflect the figure over the blue line.

8. Rotate the figure a half-turn around the green point.

9. Translate the figure any distance to the right.

10. **Research** Find a crossword puzzle in a newspaper. Describe the symmetry of the puzzle.

Exercises and Applications

1–3. Answers may vary. Examples are given.

1. the human body; most chairs

2. an ice cream cone; an automobile tire

3. a checkerboard; most tiled floors

4.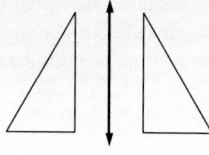

5.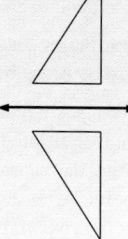

6.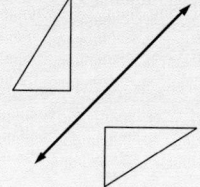

7–9. See answers in back of book.

10. Answers may vary. Most puzzles have rotational symmetry (half-turn around the center of the puzzle).

Teach⇔Interact

Additional Example 3 (continued)

b.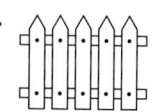

The fence has translational and reflection symmetry. If you translate it one slat to the left or right, it looks the same. Also, a plane of reflection runs through its center.

Checking Key Concepts

Using Manipulatives
When discussing questions 1–3, you may wish to use a large cardboard cutout of a triangle to show students the transformations.

Closure Question

How can you tell whether a figure has symmetry? You can check for symmetry by drawing the figure and then applying various transformations to it. If you can fold the figure along a line so that the two halves match, then the figure has reflection symmetry. If you can rotate it less than a full turn so that it coincides with itself, then it has rotational symmetry. If you can move it in one direction so that it coincides with itself, then it has translational symmetry.

Apply⇔Assess

Suggested Assignment

❖ **Core Course**
Exs. 1–9, 11–17, 23–27

❖ **Extended Course**
Exs. 1–27

❖ **Block Schedule**
Day 1 Exs. 1–9, 11–17, 23–27

Exercise Notes

 Using Technology
Exs. 1–3 Students can use Activities 29 and 30 in the *Geometry Inventor Activity Book* to explore reflection and rotational symmetry.

11

Exercise Notes

Integrating Algebra

Ex. 14 Coordinate geometry provides one of the most direct ways to integrate algebra and geometry. This exercise and Example 2 on page 10 begin this process by having students translate the vertices of a triangle up, down, left, and right and then having them summarize their results. Further work with coordinate geometry can be found in Chapters 4 and 12.

Geometric Thinking

Ex. 14 Upon doing this cooperative learning activity, students not only see how the coordinates of the vertices of a triangle change when translated, but they should also observe that the size and shape of the image triangle are the same as that of the original triangle. Since a translation, in a sense, *moves* a figure without altering its size and shape, it is often called a *rigid motion*. Rotations and reflections are also rigid motions.

Teaching Tip

Exs. 15–17 After discussing these exercises, students may question how the bicycles pictured in Ex. 15 can have translational symmetry, for if you translate a finite number of identical objects, at least one of the objects and their images do not coincide. Explain to students that translational symmetry only applies to an infinite number of identical objects, for then the objects and their images do coincide. You may wish to demonstrate this by using a tessellation composed of regular hexagons or squares on a geometric plane.

INTERVIEW Latecia Leavy

Look back at the article on pages xxxii–2.

From the opening floorwalk through the closing somersault, each Jesse White show contains many tumbling transformations and patterns.

11. Which transformation best describes the floorwalk? Why?

12. What kind of symmetry does this pose have? Explain.

13. Which transformation best describes this flip? Why?

14. Cooperative Learning Work in a group of three people. Use the results from Example 2 on page 10. One person should complete part (a), another should complete part (b), and the third should complete part (c). Work together for parts (d) and (e).

a. Graph $D(-2, 0)$, $E(0, 2)$, and $F(-5, 2)$ and connect the points to draw a triangle. How do the coordinates change when you translate the triangle up 2 units? up 3 units? up n units?

b. Graph $D(-2, 0)$, $E(0, 2)$, and $F(-5, 2)$ and connect the points to draw a triangle. How do the coordinates change when you translate the triangle down 2 units? down 3 units? down n units?

c. Graph $D(-2, 0)$, $E(0, 2)$, and $F(-5, 2)$ and connect the points to draw a triangle. How do the coordinates change when you translate the triangle to the left 2 units? to the left 3 units? to the left n units?

d. Summarize your answers from parts (a)–(c) and the results in Example 2.

e. If you translate a triangle to the right a units and then down b units, how will the coordinates change? Explain.

Describe the symmetry of each object or group of objects.

15.

16.

17.

11. translational; Each gymnast's position is an image of the first gymnast's position.

12. reflection; A plane of reflection runs down the center of the pyramid.

13. rotational; Each point of the gymnast's body rotates around some point by the same amount.

14. **a.** The y-coordinates increase by 2; increase by 3; increase by n.

b. The y-coordinates decrease by 2; decrease by 3; decrease by n.

c. The x-coordinates decrease by 2; decrease by 3; decrease by n.

d. Let n be any positive number. If a figure is translated to the right n units, the x-coordinate of each point increases by n units. If a figure is translated to the left n units, the x-coordinate of each point decreases by n units. If a figure is translated up n units, the y-coordinate of each point increases by n units. If a figure is translated down n units, the y-coordinate of each point decreases by n units.

e. Each x-coordinate will increase by a and each y-coordinate will decrease by b. A translation to the right increases the x-coordinate and a translation down decreases the y-coordinate.

Connection | COMMUNICATION

Braille is a code of raised dots that can be read by touch. It was developed by a 15-year-old blind French student named Louis Braille. He got the idea from a dot code used to send messages to soldiers at night. The Braille alphabet is based on a cell three dots high and two dots wide.

A sportswriter uses Braille while working at a game.

A B C D E F G H I J

K L M N O P Q R S T

U V X Y Z W

BY THE WAY

Louis Braille did not include the letter W in his original alphabet because it is rarely used in French. W was added to the Braille alphabet for use in English.

18. Writing Compare the first ten letters of the Braille alphabet with the second ten letters. What patterns do you see?

19. Which Braille letters are reflections of each other?

20. Which Braille letters are rotations of each other?

21. Writing There are no Braille letters that are translations of each other. Why do you think this is so?

22. Challenge Find a word that has rotational symmetry or reflection symmetry when written in Braille.

ONGOING ASSESSMENT

23. Open-ended Problem A door rotates when you open it. What other objects do you rotate? translate? What can you use to reflect an object?

SPIRAL REVIEW

Solve each equation. *(Toolbox, pages 695 and 696)*

24. $-2x = 35$ **25.** $\frac{n}{3} = -12$ **26.** $5 - 3p = -13$

27. Anna buys a book about juggling for \$9 and some juggling bags for \$1.50 each. Write a variable expression for the amount she pays, based on the number of juggling bags she buys. *(Toolbox, page 692)*

1.2 Transformations and Symmetry   13

Apply⇔Assess

Exercise Notes

Student Progress
Exs. 18–21 These exercises provide a quick check on how well students can identify patterns and recognize transformations.

Assessment Note
Ex. 23 Discuss this exercise in class. Students can make a list of objects for each type of transformation. They may wish to include their lists in their journals.

Practice 2 for Section 1.2

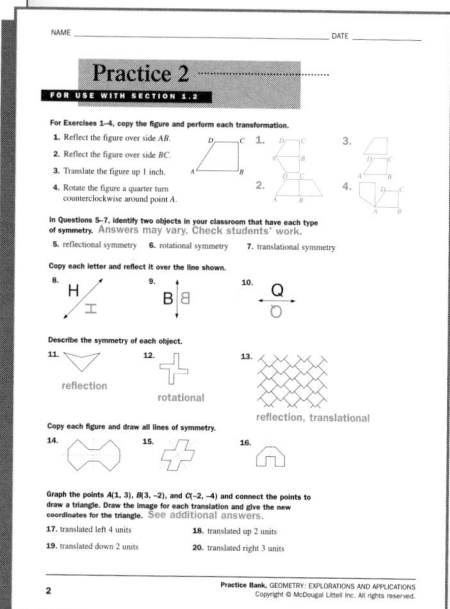

15. The bicycles have translational symmetry.

16. The door has reflection symmetry about a line down the center of the door.

17. The starfish has rotational symmetry (about its center) and reflection symmetry. It has five planes of symmetry, one running down the middle of each leg.

18. When the second ten letters are arranged in a row beneath the first ten as shown in the text, each letter can be formed from the one above it by adding a dot in the first column, third row.

19. B, C; D, F; E, I; H, J; R, W

20. B, C; E, I; R, W; F, H, D, and J

21. Letters that are are translations of each other would be indistinguishable.

22. Answers may vary. Examples are given. fad (reflection symmetry); bib (rotational symmetry)

23. Answers may vary. Examples are given. the pages of a book; a sliding door; a mirror

24. $-17\frac{1}{2}$

25. -36

26. 6

27. $1.5n + 9$, where n = the number of bags

13

Warm-Up Exercises

Tell whether each statement is *always true*, *sometimes true*, or *never true*.

1. If a whole number ends in 2, then its square ends in 4.
always true

2. If the square of a whole number ends in 4, then the number itself ends in 2. sometimes true; The number might end in 8.

3. If a whole number ends in exactly 3 zeros, then its square also ends in exactly 3 zeros. never true

4. If a figure is a circle, then it has reflection symmetry. always true

5. If a figure is a circle, then it has translational symmetry. never true

---

SECTION

1.3 Making Conjectures

Can you draw a house like this one without lifting your pencil or retracing an edge? The house is an example of a *network*, a figure made by connecting points. In the Exploration, you will discover how to tell whether a network is traceable.

Learn how to...
- write and analyze conjectures

So you can...
- tell whether conditional statements are true or false

EXPLORATION
COOPERATIVE LEARNING

Tracing Networks

Work with another student.

1 Copy each network.

vertex

edge

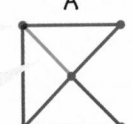 A 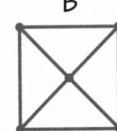 B 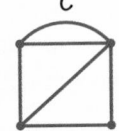 C  D  E

2 Try to trace each network without lifting your pencil or retracing an edge. Which networks are traceable?

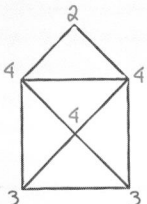

A vertex is **odd** if an odd number of edges meet at the vertex.

A vertex is **even** if an even number of edges meet at the vertex.

3 For each network, count how many edges meet at each vertex.

4 Copy and complete the table, including a row for each figure. (*Note:* Vertices is the plural of vertex.) Describe any patterns that you notice.

5 Sketch a network with four odd and zero even vertices. Is it traceable?

6 Make a conjecture about how you can tell if a network is traceable.

Network	Number of odd vertices	Traceable?
House	2	Yes
A	?	?
B	?	?
C	?	?

 Exploration Note

Purpose
The purpose of this Exploration is to have students discover a rule that can be used to tell whether a network is traceable.

Materials/Preparation
No special materials are required.

Procedure
Students copy the five networks shown in Step 1 and work together to see which networks can be traced without lifting their pencils or retracing an edge. They copy and complete a table showing the number of odd vertices of a network and if it is traceable.

They then look for a pattern that can be used to predict whether a network is traceable.

Closure
Discuss the rules that students write for Step 6. Help students understand that if a network has more than two odd vertices, it is not traceable.

Explorations Lab Manual
See the Manual for more commentary on this Exploration.
Diagram Master 7

For answers to the Exploration, see answers in back of book.

A prediction based on inductive reasoning, like the rule you wrote in the Exploration, is called a **conjecture**. Conjectures are usually *conditional statements*. A **conditional statement** can be written in the form "if P, then Q." A conditional statement may be true or false.

> If **all the vertices of a network are even**, then **it is traceable**.

> If **today is July 4**, then **it is Independence Day in the United States**.

The *if* part is called the **hypothesis** and the *then* part is called the **conclusion**. Sometimes the hypothesis comes after the conclusion.

> A network is traceable only if **all its vertices are even**.

> It is Independence Day in the United States if **today is July 4**.

EXAMPLE 1

For each conditional statement, identify the hypothesis and the conclusion.

a. Cars with underinflated tires waste gasoline.

b. Every point on the *y*-axis has an *x*-coordinate of zero.

SOLUTION

a. If a car has underinflated tires, then it wastes gasoline.

> First rewrite each statement using the words *if* and *then*.

> Hypothesis: a car has underinflated tires
> Conclusion: the car wastes gasoline

> The *if* part is the hypothesis and the *then* part is the conclusion.

b. If a point is on the y-axis, then its x-coordinate is zero.

> Hypothesis: a point is on the y-axis
> Conclusion: the x-coordinate of the point is zero

THINK AND COMMUNICATE

Tell whether each conditional statement is *True* or *False*. If it is false, explain why.

1. If you translate a triangle three units to the right, then the *x*-coordinate of each point increases by three units.

2. If it's January, then it's cold out.

3. A number is divisible by 9 if it is divisible by 3.

ANSWERS Section 1.3

Think and Communicate

1. True.

2. False; even in parts of the world where it is winter in January, it is not always cold then.

3. False; for example, 3 is divisible by 3 but not by 9.

Section Notes

Historical Connection
The concept of a network is part of a branch of mathematics called topology. Topology is the study of the properties of figures whose size and shape are changed by a transformation. Any topological property is also a geometric property, but many geometric properties are not topological properties. The origins of topology go back to the 18th century with the solution of the Königsberg bridge problem by Leonhard Euler in 1735. Topology was developed as a branch of mathematics during the early decades of the 20th century.

Topic Spiraling: Preview
Point out to students that the conditional statement *If all the vertices are even, then the network is traceable* is true. Ask students to form a new statement that exchanges the hypothesis and conclusion; that is, *If a network is traceable, then all the vertices are even.* Ask: Is this conditional true or false? Why? (False; from the Exploration on page 14, students learned that a traceable network can have two odd vertices.) This activity previews the work in Section 3.5 on converses.

Additional Example 1

For each conditional statement, identify the hypothesis and the conclusion.

a. To have a savings account at a bank, you must keep a minimum balance in the account.
Rewrite the statement using the words *if* and *then*.
If you have a savings account at a bank, then you must keep a minimum balance in the account.
Hypothesis: You have a savings account at a bank.
Conclusion: You must keep a minimum balance in the account.

b. All whole numbers greater than 99 have at least 3 digits.
Rewrite the statement using the words *if* and *then*.
If a whole number is greater than 99, then the number has at least 3 digits.
Hypothesis: A whole number is greater than 99.
Conclusion: The number has at least 3 digits.

About Example 2

Alternate Approach
When discussing this Example, point out that there are two ways to see that $n^2 - n + 41$ is not prime. One procedure is to systematically replace n with different whole numbers and check each result. Another way is to examine the expression $n^2 - n + 41$. When n is 41 or any multiple of 41, then each term of $n^2 - n + 41$ is divisible by 41. Hence, the value of $n^2 - n + 41$ is also divisible by 41 and, thus, is not prime.

Additional Example 2

Linda started with the fraction $\frac{2}{3}$ and successively added 1 to both numerator and denominator to get the fractions $\frac{3}{4}, \frac{4}{5}, \frac{5}{6}$, and so on.

She noted that $\frac{2}{3} < \frac{3}{4} < \frac{4}{5} < \frac{5}{6} < \dots$.
Linda made a conjecture that if you start with a positive fraction and successively add 1 to both numerator and denominator, you always get an increasing sequence of fractions. Is this conjecture true?
No, the conjecture is not true. Linda started with a fraction that is less than 1. But if you start with a fraction greater than 1, such as $\frac{5}{3}$, you get $\frac{6}{4}, \frac{7}{5}, \frac{8}{6}$, and so on. These fractions form a decreasing sequence.

Checking Key Concepts

Challenge
When discussing question 6, ask students why the statement must always be true. (If n is odd, then n^2 is also odd. Hence, $n^2 + n + 11$ is odd because the sum of three odd numbers is always odd. If n is even, then n^2 is even. Therefore, $n^2 + n + 11$ is the sum of two even numbers, which is even, and an odd number. The sum of an even and odd number is an odd number.)

Closure Question

What is a conjecture and how is one usually written? A conjecture is a prediction based on inductive reasoning. It is usually written as a conditional statement using the words *if* and *then*, such as *If P, then Q*.

Counterexamples

As you probably decided in Questions 1–3 on page 15, a conjecture is not true unless it is always true. If you can find one example for which the hypothesis is true but the conclusion is false, then the conjecture is false. Such an example is called a **counterexample**.

EXAMPLE 2   Connection: Algebra

Kevin used a computer to test the integers from 1 to 35 in the formula $P = n^2 - n + 41$ and see if the result is a prime number. He found that P was a prime number each time. Kevin made a conjecture:

If n is a positive integer, then $n^2 - n + 41$ is a prime number.

Is this conjecture true?

Formula for Primes?		
n	$n^2 - n + 41$	Prime?
1	41	Yes
2	43	Yes
3	47	Yes
4	53	Yes
33	1097	Yes
34	1163	Yes
35	1231	Yes
100%		

WATCH OUT!
Even though Kevin's conjecture works for 35 numbers, one counterexample is enough to show that the conjecture is false.

SOLUTION

Test 41 in the formula: $41^2 - 41 + 41 = 41^2$, and 41^2 is not prime. Since there is one value of n that doesn't work in the formula, Kevin's conjecture is false.

THINK AND COMMUNICATE

4. Write a conjecture that has a counterexample.

5. Write a conjecture that does not have a counterexample.

☑ CHECKING KEY CONCEPTS

Identify the hypothesis and the conclusion of each statement.

1. If a figure is a square, then it has four lines of symmetry.

2. The expression $(a - b)^2$ represents a positive number if a does not equal b.

3. People who live in glass houses shouldn't throw stones.

Tell whether each statement is *True* or *False*. If it is false, give a counterexample.

4. If a person has seen a doctor, then the person has a broken arm.

5. If $n = 3$, then $2n - 2 = 4$.

6. If n is a whole number, then $n^2 + n + 11$ is an odd number.

16   Chapter 1 *Patterns, Lines, and Planes*

Think and Communicate

4–5. Answers may vary. Examples are given.

4. Every odd number is prime. Counterexample: 9

5. For every even integer k, there is an integer n such that $k = 2n$.

Checking Key Concepts

1. a figure is a square; it has four lines of symmetry

2. a does not equal b; the expression $(a - b)^2$ represents a positive number

3. people live in glass houses; they shouldn't throw stones

4. False; for example, the person may be having a routine physical.

5. True.

6. True.

1.3 **Exercises and Applications**

Apply⇔Assess (right column header)

1. TRAVEL The map shows part of the city of Kyoto, Japan. The bureau of tourism recommends a walking tour to appreciate the beauty of the path along the canal. Nancy Kim has planned the path shown. Can she follow the path without retracing her steps? How do you know?

Extra Practice exercises on page 660

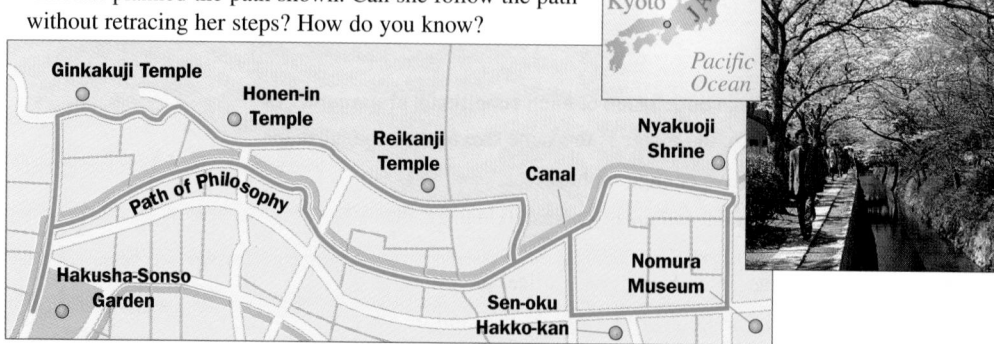

2. SOCIAL STUDIES Compare the two systems of counting marks shown below. Make a conjecture about the figures for 8 and 9 in the Japanese system.

American	I	II	III	IIII	✖	✖I	✖II
Japanese	一	T	F	下	正	正一	正T

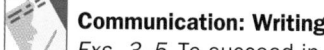

INTERVIEW Latecia Leavy

Look back at the article on pages xxxii–2.

A good tumbling performance depends on many things. The tumblers must be fast, steady, and careful. As Latecia Leavy describes the challenges of tumbling, she often uses conditional statements.

Rewrite each statement using the words *if* and *then*. Identify the hypothesis and conclusion.

3. When you're at the bottom of a human pyramid, you have to be steady.

4. You have to spring from the trampoline at a precise angle for a vertical landing.

5. Tumblers who don't run fast enough before hitting the trampoline won't make it to the other side of the pyramid.

1.3 Making Conjectures **17**

Right sidebar:

Suggested Assignment
- **Core Course** Exs. 3–14, 23–31, AYP
- **Extended Course** Exs. 1–31, AYP
- **Block Schedule** Day 2 Exs. 3–14, 23–31, AYP

Exercise Notes

Teaching Tip *Ex. 1* Some students may find it helpful to draw a new diagram that looks more like the networks shown in the Exploration on page 14.

Research *Ex. 1* You may wish to suggest that students research the Königsberg bridge problem and write a short report on it for their notebooks or journals.

Interdisciplinary Problems *Ex. 2* Ask students to conjecture how numbers greater than 9 might be written in Japanese. They can check their conjectures by consulting books on translating between English and Japanese.

Communication: Writing *Exs. 3–5* To succeed in learning geometry, students should become proficient in writing and understanding statements in if-then form. They must also understand the difference between the hypothesis of a statement and its conclusion. Review the answers to these exercises and try to identify any student who may need extra help with writing conditional statements.

Exercises and Applications

1. No; a network for her path has 3 odd vertices.

2. In the Japanese stsyem, the figure for 6 is a combination of the figures for 5 and 1. Similarly, the figure for 7 is a combination of the figures for 5 and 2. The figures for 8 and 9 are probably combinations, respectively, of the figures for 5 and 3 and 5 and 4.

American	✖ III	✖ IIII
Japanese	正 F	正 F

3. If you're at the bottom of a human pyramid, then you have to be steady. hypothesis: you're at the bottom of a human pyramid; conclusion: you have to be steady

4. If you want to make a vertical landing, then you have to spring from the trampoline at a precise angle. hypothesis: you want to make a vertical landing; conclusion: you have to spring from the trampoline at a precise angle

5. If a tumbler doesn't run fast enough before hitting the trampoline, he or she won't make it to the other side of the pyramid. hypothesis: a tumbler doesn't run fast enough; conclusion: he or she won't make it to the other side of the pyramid

17

Visual Thinking
Exs. 6–8 Ideas for helping students develop their visual learning skills are provided at strategic points throughout the side-column notes. Visual thinking skills include:

1. Observation
2. Identification
3. Recognition
4. Recall
5. Interpretation
6. Exploration
7. Correlation
8. Generalization
9. Inference
10. Perception
11. Communication
12. Self-Expression

Student Progress
Exs. 9–13 Review the answers to these exercises in class by calling upon different students to respond to each one. Ask students to state Ex. 12 as a conditional statement using if-then form. All students should be able to do these exercises correctly.

Integrating Algebra
Ex. 14 To determine whether this conditional statement is true or false, students need to remember that a solution to an equation makes the equation a true statement when the variable is replaced by the solution. Hence, this conditional is true because $2^2 - 5 \cdot 2 + 6 = 0$ is a true statement ($0 = 0$). However, 2 is not the only solution. Although some students may have studied quadratic equations in their Algebra One course and thus be able to solve the equation to find both solutions (2 and 3), others may not have. Even so, these other students can try substituting a few other numbers such as 0, 1, –2, 3 into the equation to discover that 3 is also a solution. Then you may wish to show them that $x^2 - 5x + 6 = (x - 3)(x - 2) = 0$. Therefore, $x = 3$ and $x = 2$ are the solutions.

Using Manipulatives
Exs. 15–18 In doing these exercises, students will discover that repeatedly folding a sheet of paper quickly becomes difficult. This might affect students' ability to test their conjectures in Ex. 18. Try to have available some very thin sheets of tissue paper that students can use.

Visual Thinking For each figure, write a conjecture in the form "If the figure has ? symmetry, then the hidden part of the figure looks like..." Complete the conjecture with a sketch of the hidden part.

6.  7.  8.

Identify the hypothesis and conclusion of each conditional statement.

9. If you reflect a shape, the image is the same size as the original shape.

10. I'll buy a book if I have enough money.

Tell whether each statement is *True* or *False*. If it is false, give a counterexample.

11. If a network is traceable, then all the vertices are even.

12. A person who lives in Southern California lives in Los Angeles.

13. If a and b are real numbers, then $a + b = b + a$.

14. **Challenge** If $x^2 - 5x + 6 = 0$, then $x = 2$.

Investigation For Exercises 15–18, you will need a piece of paper that is at least $8\frac{1}{2}$ in. by 11 in.

Fold the paper in half. Unfold it and notice that the crease divides the paper into two sections.

15. Refold the paper. Fold it in half again. How many sections are there when you unfold the paper?

16. Refold the paper. Fold it in half again. How many sections are there when you unfold the paper?

17. Make a conjecture about the number of sections if you fold the paper in half n times. Identify the hypothesis and conclusion of your conjecture.

18. **Writing** Test your conjecture by folding the paper two more times. Do you think your conjecture is *True* or *False*? Explain.

18   Chapter 1 *Patterns, Lines, and Planes*

6. If the figure has rotational (or reflection) symmetry, then the hidden part of the figure looks like:

7. If the figure has reflection symmetry, then the hidden part of the figure looks like:

8. If the figure has translation symmetry, then the hidden part of the figure looks like:

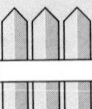

9. hypothesis: you reflect a shape; conclusion: the image is the same size as the original shape

10. hypothesis: I have enough money; conclusion: I'll buy a book

11. False; for example, the network in Figure A on page 14 is traceable, but two of its five vertices are odd.

12. False; for example, a person living in San Diego lives in Southern California but not in Los Angeles.

13. True.

14. False; x might be equal to 3.

15. 4 sections

Connection HISTORY

Natural events have had a great effect on history. For example, on the night of August 29, 1776, George Washington's army was able to escape the British because the East River was covered in a thick fog.

Identify the hypothesis and conclusion of each conditional statement. (*Hint:* Write the hypothesis and conclusion using "was not" instead of "had not been" and so on.)

19. If the East River had not been cloaked by fog, George Washington's army might have been trapped at the western end of Long Island.

20. If there had not been a drought in the Southern Great Plains in the 1930s, the area might not have been hit by the huge dust storms that drove thousands of families away.

21. If the winter of 1620 had not been unusually mild, then the Pilgrims might not have survived their first winter in Massachusetts.

22. Open-ended Problem A solar eclipse during a battle in ancient Mesopotamia frightened the armies so much that they made peace. Modern astronomers used what they know about eclipses to calculate the exact date of the battle. Use these two facts to write a conditional statement like those in Exercises 19–21. Identify your hypothesis and conclusion.

Use each hypothesis to write a true conditional statement.

23. Hypothesis: Today is January 1.

24. Hypothesis: $5x + 7 = -3x - 5$.

25. Hypothesis: A point is located on the x-axis.

26. Hypothesis: A shape has reflection symmetry.

27. Hypothesis: You translate a point up one unit on a coordinate plane.

28. SAT/ACT Preview If $A = x$ and $B = x^2$, then:

A. $A \geq B$ **B.** $A \leq B$ **C.** $A = B$ **D.** relationship cannot be determined

ONGOING ASSESSMENT

29. Open-ended Problem Write a conditional statement about yourself. Identify the hypothesis and conclusion. Is there a counterexample?

1.3 Making Conjectures   **19**

---

16. 8 sections

17. If a paper is folded in half n times, when the paper is unfolded, the creases divide the paper into 2^n sections. hypothesis: a paper is folded in half n times; conclusion: when the paper is unfolded, the creases divide the paper into 2^n sections

18. True; the number of sections doubles each time.

19. hypothesis: the East River was not cloaked by fog; conclusion: George Washington's army was trapped at the western end of Long Island

20. hypothesis: there was not a drought in the Southern Great Plains in the 1930s; conclusion: the area was not hit by the huge dust storms that drove thousands of families away

21. hypothesis: the winter of 1620 was not unusually mild; conclusion: the Pilgrims did not survive their first winter in Massachusetts

22. Answers may vary. An example is given. If modern astronomers did not know how to calculate dates of eclipses, they would not be able to determine when a battle took place in ancient Mesopotamia; hypothesis: astronomers do not know how to calculate dates of eclipse; conclusion: they are not able to determine when a battle took place in ancient Mesopotamia.

23–29. See answers in back of book.

Apply⇔Assess

Exercise Notes

Communication: Reading
Exs. 19–21 Ask students to read the hypothesis and conclusion of each conditional statement. The beginning of each statement with the word *if* is a clue to identifying each part of the conditional.

Multicultural Note
Ex. 21 Many factors contributed to the Pilgrims' surviving their first winter in Massachusetts. Perhaps the most important of these factors was the cooperation and helpfulness of the local Native American group, the Wampanoags. Three Wampanoags, in particular, Squanto, Samoset, and the chief, Massasoit, helped the Pilgrims during their first years in Massachusetts by showing them how to take advantage of the wilderness around them. Among other things, they taught the Pilgrims to cultivate corn and pumpkins and to fish. Without the help of the Wampanoags, the Pilgrims might have starved during their first winter. The Wampanoags under Massasoit maintained a peaceful relationship with the Pilgrims for forty years, giving the new settlers time to form a strong settlement.

Communication: Writing
Exs. 23–27 Most students will probably write their conditional statements so that the hypothesis comes before the conclusion. If this is the case, ask some students to restate their conditionals so that the conclusion comes before the hypothesis. Make sure, however, that in so doing, they do not interchange the hypothesis and the conclusion by switching the placement of *if* and *then*.

Assessment Note
Ex. 29 Students should use a conditional statement of a general nature. Unless the conditional is a general statement, the notion of a counterexample is not appropriate.

19

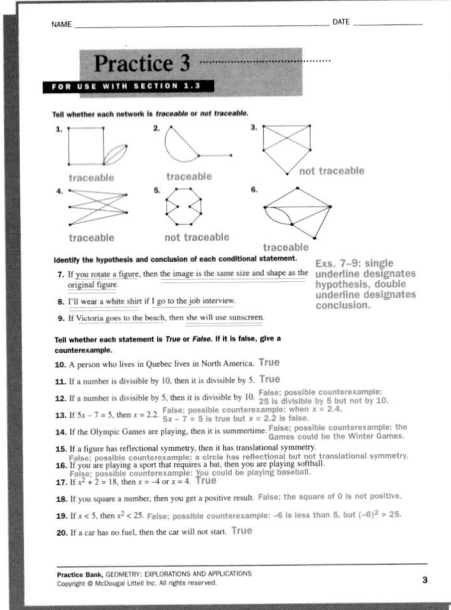
30. **Open-ended Problem** Sketch a simple figure that has reflection symmetry. Draw the line of symmetry. *(Section 1.2)*

31. Graph the equation $y = 4 + \frac{1}{3}x$. *(Toolbox, page 697)*

ASSESS YOUR PROGRESS

VOCABULARY

inductive reasoning (p. 4)	**line of symmetry** (p. 10)
reflect (p. 8)	**conjecture** (p. 15)
rotate (p. 8)	**conditional statement** (p. 15)
translate (p. 8)	**hypothesis** (p. 15)
transformation (p. 8)	**conclusion** (p. 15)
image (p. 8)	**counterexample** (p. 16)
symmetry (p. 10)	

For Exercises 1–4, use the toothpick figures below.

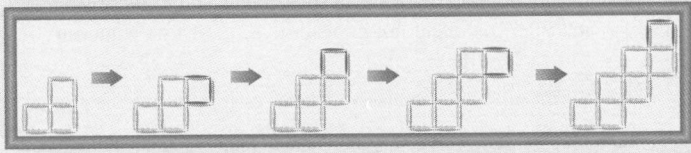

1. Copy and complete the table. What pattern do you notice? *(Section 1.1)*

Number of squares	3	4	5	6	7
Number of toothpicks	10	?	?	?	?

2. Find the number of toothpicks needed to make 10 squares. Describe how you found your answer. *(Section 1.1)*

3. Write a conjecture about the number of toothpicks needed to make n squares. Identify the hypothesis and conclusion. *(Section 1.3)*

4. Describe the symmetry of each figure. *(Section 1.2)*

5. Copy the diagram. Then sketch the image after each transformation. *(Section 1.2)*
   a. Reflect the letter over the red line.
   b. Rotate the letter around the green point by a half turn.
   c. Translate the letter two inches down.

6. Tell whether the statement "if a and b are integers, then $a \div b \leq a$" is *True* or *False*. If it is false, give a counterexample. *(Section 1.3)*

7. **Journal** Describe how you look for a pattern in a series of numbers.

20   Chapter 1 *Patterns, Lines, and Planes*

30.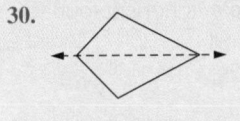

31.

Assess Your Progress

1.

Number of squares	3	4	5	6	7
Number of toothpicks	10	13	16	19	22

Each additional square requires 3 additional toothpicks.

2. 31 toothpicks; Answers may vary. An example is given. I noticed that for n squares, the number of toothpicks is $3n + 1$.

3. If you want to make a toothpick figure having n squares, you need $3n + 1$ toothpicks.

4. The first, third, and fifth figures have reflection symmetry. The second and fourth have rotational symmetry.

5. See answers in back of book.

6. False; for example, if $a = -2$ and $b = -1$, then $a \div b = 2$ and $a \div b > a$.

7. Answers may vary. An example is given. First, I consider the differences between consecutive numbers, then their sums, products, or quotients. Next, I might consider whether each number is related to its position number; for example, whether, say, the first term is 1^2, the second is 2^2, and so on.

1.4 Modeling Points, Lines, and Planes

Learn how to...
- use points, lines, and planes

So you can...
- describe and sketch relationships between points, lines, and planes in real-world objects

Figures in geometry are made of points, lines, and planes. Instead of defining *point*, *line*, and *plane*, mathematicians describe the relationship between them. A point represents a location and has no size. A line is made of many points, and a plane contains many points and lines.

EXPLORATION
COOPERATIVE LEARNING

Representing Points, Lines, and Planes

Work with another student.
You will need:
- three foam trays or pieces of stiff paper
- tape
- scissors
- several pieces of uncooked spaghetti

Make a model like the one shown.
The pieces of spaghetti represent lines and the trays represent planes.

Tape three trays together. Poke one hole in each tray. Label the holes as points *A*, *B*, and *C*.

1 On the tray that contains point *A*, draw a line through *A*. Label the line *ℓ*. Put a piece of spaghetti through the holes at *A* and *B*. Do line *ℓ* and the spaghetti have any points in common other than point *A*?

2 Draw a line on the same tray as line *ℓ* so that the two lines don't share any points, even though they continue forever. Label the new line *m*.

3 Add a line to your model that is not on the same tray as line *ℓ* and does not share any points with *ℓ*.

4 Is it possible for two different trays to contain the same line? If so, add such a line to your model and label it *s*.

5 Is it possible for a line and a plane to have exactly one point in common? exactly two points in common? no points in common? Give an example of each possibility, drawing lines on your model as necessary.

Exploration Note

Purpose
The purpose of this Exploration is to have students observe the possible relationships between two lines, between a line and a plane, and between two planes.

Materials/Preparation
Each pair of students needs three trays or pieces of stiff paper, tape, scissors, and several pieces of uncooked spaghetti.

Procedure
Students should join the trays or pieces of paper as shown in the diagram. They then follow the instructions in Steps 1–5 to explore possible relationships between various lines and planes.

Closure
Students should see that for any two points, there is one line through them; two lines intersect in a point; and that in a plane, lines can have no points or a single point in common. They should also see that two lines in different planes do not share any points, and if two planes intersect, then their intersection is a line.

Explorations Lab Manual
See the Manual for more commentary on this Exploration.

For answers to the Exploration, see following page.

Resource Materials

Lesson Support
Lesson Plan 1.4

Warm-Up Transparency 1.4

Overhead Visuals:
  Folder 1: Lines and Planes in Space

Practice Bank: Practice 4

Study Guide: Section 1.4

Challenge Problems: Set 4

Technology
Technology Book:
  Calculator Activity 2

Geometry Software

Internet:
  http://www.hmco.com

Objectives
- Use points, lines, and planes.
- Describe and sketch relationships between points, lines, and planes in real-world objects.

Recommended Pacing
❖ **Core and Extended Courses**
   Section 1.4: 2 days
❖ **Block Schedule**
   Section 1.4: 2 half-blocks (with Sections 1.3 and 1.5)

Warm-Up Exercises

How would you describe each of the following by using the term *point*, *line*, or *plane*?

1. the tip of a pen point
2. the top of a desk plane
3. the edge of a sheet of paper line
4. the floor of a room plane
5. the edge of a ruler line

Section Notes

Geometric Thinking

This initial discussion of points, lines, and planes may be confusing to students because of concepts such as a point having no size and lines and planes extending forever. Point out that in geometry, points, lines, and planes are abstract ideas that do not exist physically. The drawings in the book are simply models for points, lines, and planes. Remind students that the numbers they know from arithmetic are also purely abstractions—they have no physical reality, although the symbol for a number, such as 5, is a physical thing.

Spatial Reasoning

Discuss how the textbook uses various techniques in diagrams to suggest three dimensions. Call attention to shading and how dashed lines are used to show hidden portions of figures. Some students have excellent powers of visualization and can learn how to sketch three-dimensional figures quickly; others may need to make frequent use of concrete models to learn how to show spatial relationships clearly. You might find it helpful to call on students to sketch different configurations of points, lines, and planes on the board. Point out the good features of the sketches and discuss ideas for improving those that are not well drawn.

Communication: Reading/Writing

A number of different symbolic notations are introduced for naming points, lines, and planes, and it is essential that students be able to read and write these symbols correctly.

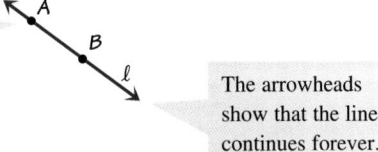

WATCH OUT!
In this book, if a phrase refers to more than one figure, the figures are not the same. That is, "two points" means "two *different* points."

For any two points, there is exactly one line through the points. You can name a line by using a single letter or by naming two points on the line. Line ℓ can also be called $\overleftrightarrow{AB}$ or $\overleftrightarrow{BA}$.

Points that are on the same line are **collinear**.

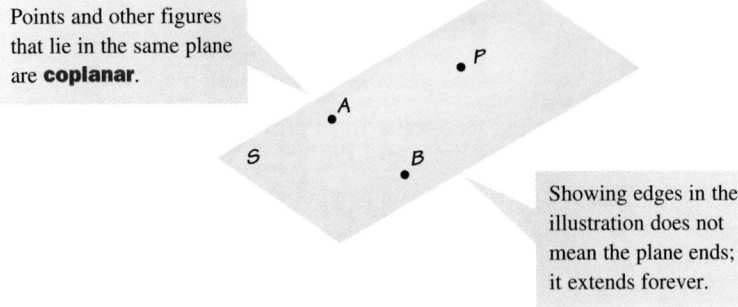

The arrowheads show that the line continues forever.

For any three noncollinear points, there is exactly one plane through the points. You can name a plane by using three noncollinear points in the plane. Plane S can also be called plane ABP or plane PAB, for example.

Points and other figures that lie in the same plane are **coplanar**.

Showing edges in the illustration does not mean the plane ends; it extends forever.

If two different figures share at least one point, then they **intersect**. The shared points are called the **intersection**. If two lines in the same plane do not intersect, then they are **parallel**. If two lines do not intersect and are not parallel, then they are **skew**.

Point P is the intersection of lines n and m.

The **red arrows** show that lines ℓ and m are parallel. You can write $\ell \parallel m$.

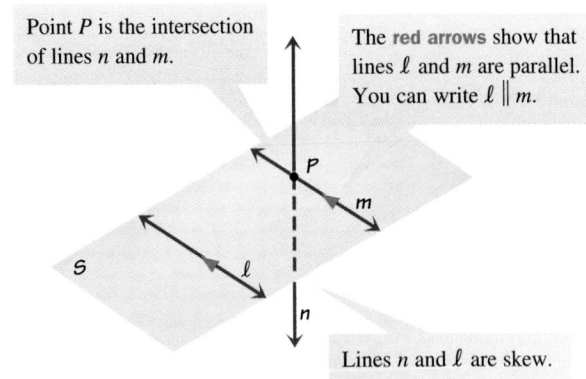

Lines n and ℓ are skew.

ANSWERS Section 1.4

Exploration

1. No.

2, 3. Check students' work.

4. Yes (where the two trays meet).

5. Yes (see art); No; Yes (see art).

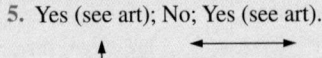
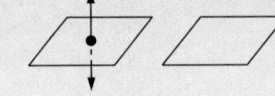

Two planes can intersect in a line. Planes that do not intersect are parallel. If a line and a plane intersect, then their intersection is a point or a line.

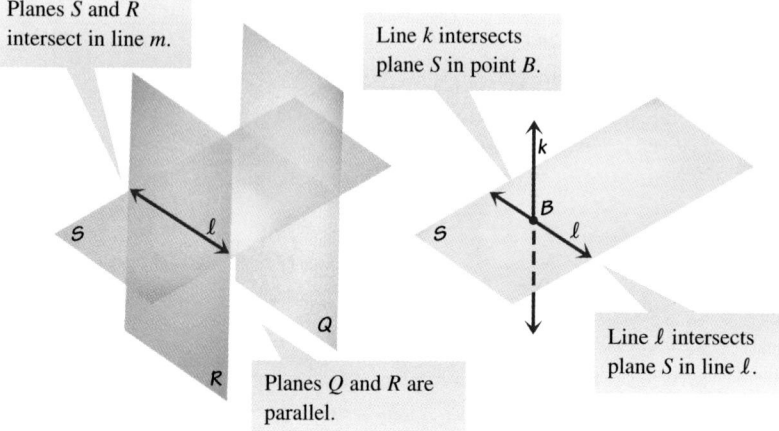

Planes S and R intersect in line m.

Line k intersects plane S in point B.

Line ℓ intersects plane S in line ℓ.

Planes Q and R are parallel.

EXAMPLE

Sketch each situation.

a. $\overleftrightarrow{HI}$ intersects plane E in point Q.

b. Line m and line n are skew.

SOLUTION

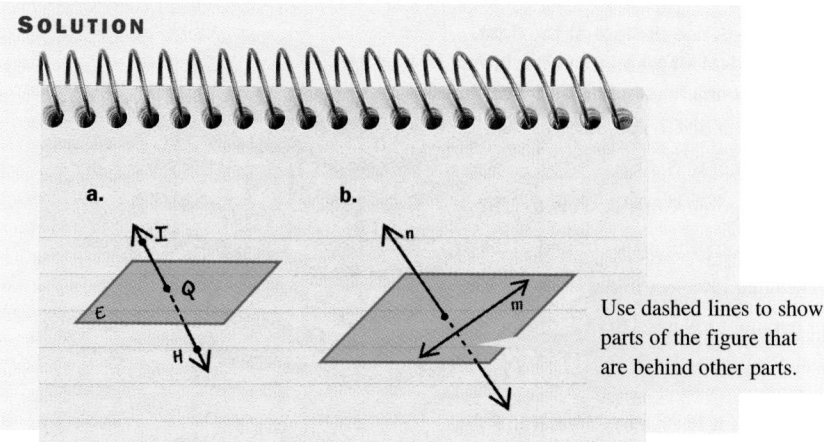

a.

b.

Use dashed lines to show parts of the figure that are behind other parts.

THINK AND COMMUNICATE

Tell whether each statement is *True* or *False*. Explain your answer.

1. If two lines intersect, then they are coplanar.

2. For any line, there are many planes that contain the line.

3. It is possible for three points to be noncoplanar.

4. It is possible for four points to be noncoplanar.

Section Notes

Geometric Thinking
The notion of a line *intersecting* a plane in a line, as line *m* intersects plane *S* in the drawing, may seem to be an unusual use of the word *intersect* to some students. These students are probably thinking of intersect in a physical sense as something passing through something else. To help clear up any misunderstanding of this term, read the sentence on page 22 that states the mathematical meaning of the term: if two different figures share *at least one point*, then they intersect.

Communication: Drawing
Suggest that students try to redraw the diagram above the Example without using shading. They should use dashed lines to indicate hidden portions of the planes. Ask students whether they prefer this technique or one that uses shading.

Additional Example

Sketch each situation.

a. Planes *J* and *K* intersect in line *m* that is skew to line *n*.

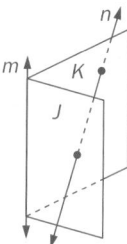

b. Two parallel lines *p* and *q* intersect plane *C* in points *T* and *U*, respectively.

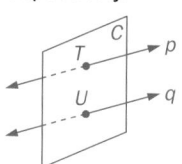

Think and Communicate

When discussing these questions, you may wish to have students draw sketches or use manipulatives to demonstrate whether each statement is true or false.

Think and Communicate

1.

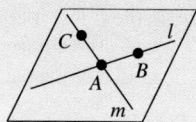

True. In the sketch, points *A*, *B*, and *C* determine a plane. Since *A* and *B* are on *l* and in the plane, line *l* is in the plane. Also, since *A* and *C* are on *m* and in the plane, line *m* is in the plane. Therefore, *l* and *m* are coplanar.

2.

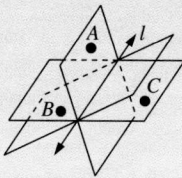

True. In the sketch, line *l*, together with each of the points not on *l*, determines a different plane. There are infinitely many such points.

3. False. If the points are collinear, they are in any plane containing the line. If they are noncollinear, exactly one plane contains them.

4.

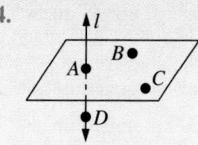

True. In the sketch, line *l* intersects plane *ABC* in point *A*. Point *D* on *l* is not in the plane.

Checking Key Concepts

Using Manipulatives
Some students may find it helpful to have a physical model to refer to when answering these questions. They may use a small cardboard box and pencils to model the relationships.

Closure Question

Describe the possible relationships between two intersecting planes and a line that is not contained in either of the planes. *The line may be parallel to or skew to the line of intersection of the two planes or it may intersect that line.*

Suggested Assignment

❖ **Core Course**
Day 1 Exs. 1–19
Day 2 Exs. 20–30, 37–45

❖ **Extended Course**
Day 1 Exs. 1–19
Day 2 Exs. 20–45

❖ **Block Schedule**
Day 2 Exs. 1–19
Day 3 Exs. 20–30, 37–45

Exercise Notes

Using Manipulatives
Exs. 1–6 Students who have difficulty using the diagram to answer these exercises may wish to construct a model for reference.

Multicultural Note
Exs. 1–6 Kites were first developed in China over 2000 years ago. Many of the early legends about kites tell of their use in military operations. During the Han dynasty (206 B.C. to A.D. 221), the Chinese military used kites with bamboo whistles attached to them to frighten their enemies. As the kites flew through the sky, air passed through the whistles. The sound made by the whistles terrified their enemies, who thought that gods were predicting their defeat. Another legend tells of General Han Hsin, who used a kite to estimate the distance between his troops and the inside of a fortress he had surrounded. He dug a tunnel based on his measurements to the inside of the fortress and took his enemy by surprise.

✓ CHECKING KEY CONCEPTS

For Questions 1–7, use the figure at the right.

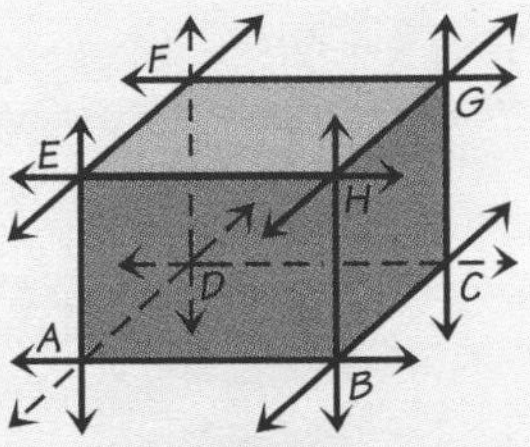

1. What is the intersection of $\overleftrightarrow{AB}$ and $\overleftrightarrow{EA}$?

2. Which lines are parallel to $\overleftrightarrow{EF}$?

3. Tell whether each group of points or lines is *collinear, coplanar but noncollinear,* or *noncoplanar.*

   a. *E, F,* and *H*   **b.** *D* and *B*   **c.** $\overleftrightarrow{AD}$ and *B*
   d. *A, D,* and *G*   **e.** *A, B, C,* and *G*   **f.** $\overleftrightarrow{AD}$ and $\overleftrightarrow{CG}$

4. Name all the lines in the diagram that are skew to $\overleftrightarrow{FD}$.

5. What is the intersection of $\overleftrightarrow{AB}$ and plane *HGC*?

6. What is the intersection of plane *EHB* and plane *GCB*?

7. Which plane is parallel to plane *EFH*?

8. Are points that are collinear also coplanar? Explain.

1.4 Exercises and Applications

Extra Practice exercises on page 660

HISTORY For Exercises 1–6, use the kite at the right. The kite was part of a survival kit for pilots in World War II. The kite string was an antenna for sending an S O S signal.

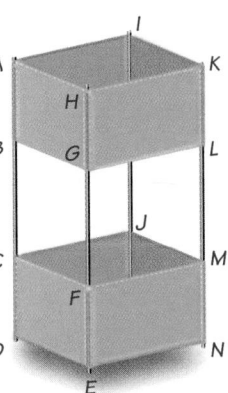

1. What is the intersection of $\overleftrightarrow{CJ}$ and $\overleftrightarrow{MJ}$?

2. Name four lines parallel to $\overleftrightarrow{BG}$.

3. Name a line through point *K* that is skew to $\overleftrightarrow{HE}$.

4. What is the intersection of plane *CDE* and plane *MNE*?

5. Name two planes that intersect in $\overleftrightarrow{AB}$.

6. Name a plane parallel to plane *ABG*.

Tell whether each statement is *True* or *False*. If the statement is false, sketch a counterexample or give a counterexample from the figure at the right.

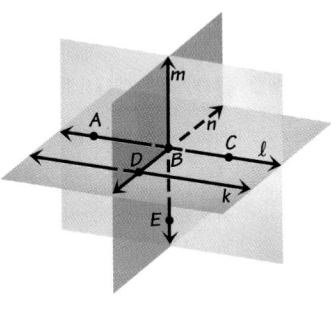

7. Two points are always collinear.

8. If two lines are not skew, then they are coplanar.

9. If three points are collinear, then there is exactly one plane through the points.

10. If three lines intersect at one point, then the lines are coplanar.

24   Chapter 1   *Patterns, Lines, and Planes*

Checking Key Concepts

1. point *A*

2. $\overleftrightarrow{AD}, \overleftrightarrow{HG},$ and $\overleftrightarrow{BC}$

3. **a.** coplanar but noncollinear
   b. collinear
   c. coplanar but noncollinear
   d. coplanar but noncollinear
   e. noncoplanar
   f. noncoplanar

4. $\overleftrightarrow{EH}, \overleftrightarrow{AB}, \overleftrightarrow{HG},$ and $\overleftrightarrow{BC}$

5. point *B*   6. $\overleftrightarrow{HB}$

7. plane *ADC*

8. Yes; they all lie in every plane that contains the line.

Exercises and Applications

1. point *J*

2. any four of $\overleftrightarrow{AH}, \overleftrightarrow{CF}, \overleftrightarrow{IK}, \overleftrightarrow{JM},$ and $\overleftrightarrow{DE}$

3. $\overleftrightarrow{IK}$ 4. $\overleftrightarrow{HE}$

5. plane *ABG*, plane *ABI*

6. plane *IKL*

7. True.   8. True.

9. False; in the figure, points *A, B,* and *C* are collinear and lie in two of the planes shown.

10. False; in the figure, $\overleftrightarrow{AB}, \overleftrightarrow{DB},$ and $\overleftrightarrow{BE}$ intersect at *B* and are not coplanar.

11. parallel

12. skew

13. intersecting

SKIING Describe the relationship between the two skis in each photo.

11.

12.

13.

Open-ended Problem Describe something in your school or classroom that fits each description.

14. parallel lines   **15.** skew lines   **16.** parallel planes   **17.** intersecting planes

18. a. What do the arrowheads on the ends of the lines *t* and *n* mean?

   b. What do the red arrows in the middle of the lines mean?

19. Which name best describes the plane?
   A. plane *n*   **B.** plane *ADC*   **C.** plane *S*

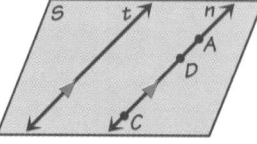

Sketch each situation.

20. Line ℓ contains points *A*, *B*, *X*, and *Y*.

21. $\overleftrightarrow{FG}$ and $\overleftrightarrow{DE}$ intersect in point *C*.

22. $s \parallel r$

23. Noncollinear points *D*, *E*, and *F* lie in plane *W*.

24. Collinear points *A*, *F*, *X*, and *Z* lie in plane *H*.

25. $\overleftrightarrow{RM}$ lies in plane *A*.

26. The intersection of $\overleftrightarrow{PQ}$ and plane *B* is point *D*.

Use the diagram at the right.

27. Tell which line is parallel to $\overleftrightarrow{BE}$.

28. Tell which plane is parallel to plane *BXE*.

29. Tell which lines are skew to $\overleftrightarrow{CF}$.

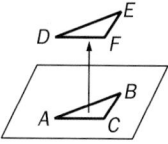

30. TRANSFORMATIONS Copy the triangle and translate it up about 2 in. Label the new triangle so that the image of *A* is *D*, the image of *B* is *E*, and the image of *C* is *F*.

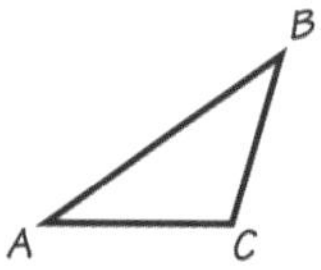

   a. Draw $\overleftrightarrow{AD}$, $\overleftrightarrow{BE}$, and $\overleftrightarrow{CF}$. Tell whether the lines are *parallel*, *intersecting*, or *skew*.

   b. Draw a simple figure and translate it in any direction. Draw lines through points on the figure and their images as in part (a). Tell whether the lines are *parallel*, *intersecting*, or *skew*.

   c. Make a conjecture based on your answers to parts (a) and (b).

1.4 Modeling Points, Lines, and Planes   **25**

Exercise Notes

Geometric Thinking
Exs. 16, 17 After students describe examples of parallel and intersecting planes in the school or classroom, ask them if it is possible for two planes to be skew. (No.) They should be able to support their answer with a logical explanation. (Because planes extend forever in each direction, two planes that are not parallel would always intersect in a line somewhere in three-dimensional space.)

Spatial Reasoning
Exs. 20–26 The first six of these exercises represent relationships that are either on a line or in a plane, and all students should be able to make a correct sketch. Ex. 26 is a simple three-dimensional relationship, but some students probably will need help with this sketch. Suggest to these students that they stick a pencil through a sheet of paper and then use this model to make a sketch.

Challenge
Ex. 30 Ask students if a translation of triangle *ABC* can result in an image that is not in the same plane as triangle *ABC*. (Yes.) Have students support their ideas by using sketches or cutout copies of the triangle. (A sample sketch is shown below.)

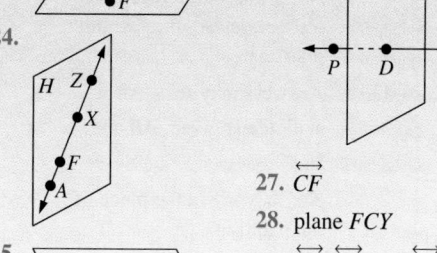

14–17. Answers may vary. Examples are given.

14. the top and bottom edges of one wall

15. the top edge of the front wall and the corner where the back wall meets a side wall

16. the floor and ceiling

17. the ceiling and a wall

18. a. The lines continue forever.
   b. The lines are parallel.

19. C

20–26. Answers may vary. Examples are given.

20.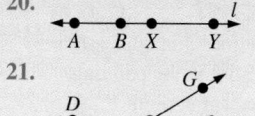

21.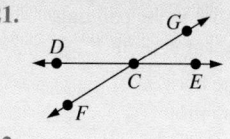

22.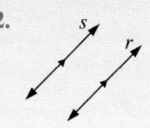

23.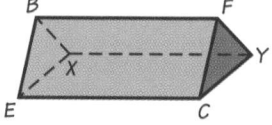

24.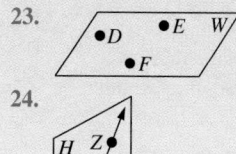

25.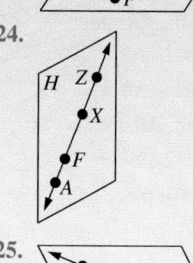

26.

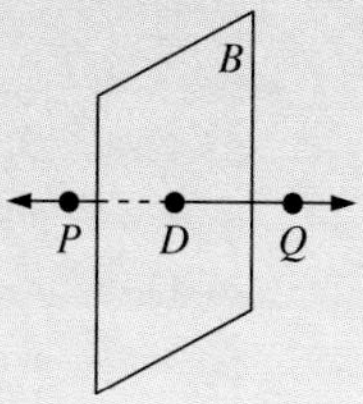

27. $\overleftrightarrow{CF}$

28. plane *FCY*

29. $\overleftrightarrow{BX}$, $\overleftrightarrow{EX}$, and $\overleftrightarrow{XY}$

30. Check students' work.
   a. parallel
   b. parallel
   c. If a figure is translated, lines that connect points on the original figure and their images are parallel.

25

Spatial Reasoning

Exs. 31–33 The concept of depth is fundamental to understanding the concept of linear perspective. Kinesthetically demonstrate depth to students by holding a pencil up in the front of the classroom and asking them to hold another pencil in front of themselves. Ask them to compare how they see the two pencils. Ask: Which pencil appears larger to you? Point out that objects farther away appear smaller than equal-sized objects that are nearer to us, and that artists make use of this fact about perception to create a sense of depth in painting on flat surfaces.

Interdisciplinary Problems

Exs. 31–34 You might consider inviting an art teacher to visit the class to discuss some of the techniques that artists use to create three-dimensional effects. For Ex. 34, you might request that the teacher bring some good examples of paintings or sketches that can help students gain a deeper understanding of vanishing points and vanishing lines.

Student Study Tip

Ex. 34 Part (d) of this exercise introduces the concept of a *vanishing point*. Suggest to students that they think of a straight railroad track on level ground that can be viewed for a great distance. Looking down the tracks, they seem to have a point in common at the horizon even though the tracks are parallel. This vanishing point is also called the *point at infinity* in non-Euclidean geometries.

Connection   ART

How do you make a flat surface appear to have depth? Ancient Egyptians used overlapping forms to suggest depth. Chinese artists used misty spaces to separate nearby forms from distant ones. In the fifteenth and sixteenth centuries, Italian artists explored painting objects as though seen through a window with one eye shut. This method is called *linear perspective*.

*Above: Egypt,
1550–1295 B.C.
Right: China, fifteenth century*

Italy, around 1470

31. **a.** Imagine that you are in the painting. Are $\overleftrightarrow{AB}$ and $\overleftrightarrow{CD}$ parallel?

    b. Think of the painting as a flat canvas with paint on it. Are $\overleftrightarrow{AB}$ and $\overleftrightarrow{CD}$ parallel? Explain.

32. **a.** Imagine that you are in the painting. Name two skew lines.

    b. Think of the painting as a flat canvas. Are the lines from part (a) still skew? Why or why not?

33. Name two lines that are parallel both when you look at the painting as a flat canvas and when you imagine that you are in the painting.

34. **Investigation** You will need tracing paper. Carefully trace the columns and the lines of the floor and roof.

    a. Extend $\overleftrightarrow{AB}$ and $\overleftrightarrow{CD}$. Label the point of intersection *P*. Where is *P*?

    b. Extend $\overleftrightarrow{EF}$ and $\overleftrightarrow{GH}$. Where do they intersect?

    c. **Open-ended Problem** Make a conjecture based on your answers to parts (a) and (b). Can you find a counterexample?

    d. Point *P* is called the *vanishing point*. Where is the vanishing point in the illustration at the left?

31. **a.** Yes.

    b. No; if the segments drawn were extended, they would intersect.

32. **a.** Answers may vary. An example is given. $\overleftrightarrow{AB}$ and $\overleftrightarrow{HD}$

    b. No; they are in the plane of the painting.

33. $\overleftrightarrow{GH}$ and $\overleftrightarrow{EF}$

34. **a.** Point *P* is at the horizon of the painting.

    b. at point *P*

    c. All lines that appear to be parallel in the painting intersect at point *P*. No.

    d. at the horizon of the illustration

35. Challenge The etching at the right is by M. C. Escher, a Dutch artist.

a. Look closely at the building. What is unusual about it?

b. Describe two lines that are skew and intersect.

c. Could you build a model of this building? Explain.

36. 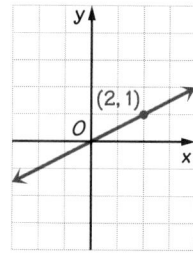 **Technology** Use geometry software and these steps to draw a cube.

a. Draw a point. A point is a *zero-dimensional* figure.

b. Copy the point. Connect the old point and the new point by drawing a segment between them. Move the new point until the segment is horizontal. A segment is a *one-dimensional* figure.

c. Copy the segment. Connect each endpoint of the segment with the same endpoint on the new segment. Move the new segment until you form a square. A square is a *two-dimensional* figure.

d. Copy the square and connect the endpoints. Move the new square until the figure looks like a cube. A cube is a *three-dimensional* figure.

ONGOING ASSESSMENT

37. Writing How would you describe a point, a line, and a plane to someone who has not taken geometry?

SPIRAL REVIEW

Tell whether each statement is *True* or *False*. If it is false, give a counterexample. *(Section 1.3)*

38. If I turn the lights off, then I can't see.

39. If $x = 5$, then $x^2 = 25$.

Give the slope of each line. *(Toolbox, page 697)*

40. $y = 4x + 7$ **41.** $y = x$ **42.** $y = -\frac{1}{2}x + 3$

For each graph, give the slope and the y-intercept. *(Toolbox, page 697)*

43. 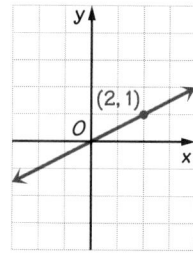   **44.** 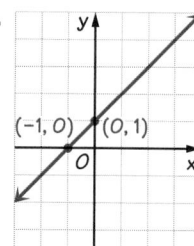   **45.**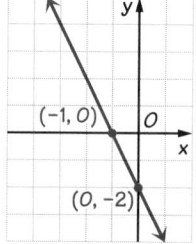

1.4 Modeling Points, Lines, and Planes **27**

Apply⇔Assess

Exercise Notes

Research
Ex. 35 Students who wish to examine M.C. Escher's art in depth should have little difficulty finding reproductions in books. Escher makes extensive use of perspective and transformations to create unusual and fascinating images.

Assessment Note
Ex. 37 You may wish to have some students read their descriptions in class. This would provide a good review of these basic ideas for the entire class.

Practice 4 for Section 1.4

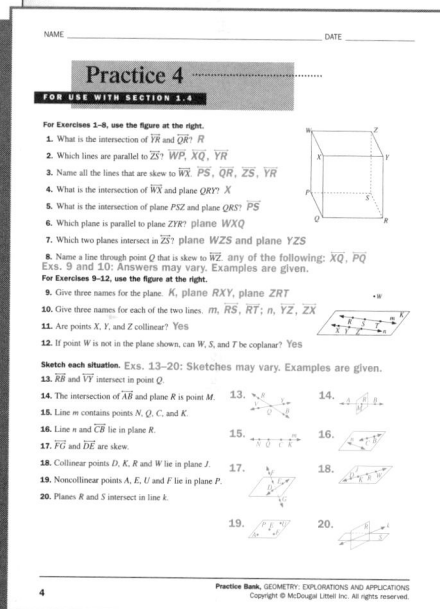

35. a. The figure is distorted. Lines that should be parallel intersect. Lines that should be skew are coplanar.

b. Answers may vary. An example is given. The column at the front left corner of the building should be skew to the edge of the back wall, but intersects it.

c. No; what appears to be a building is an optical illusion. In a real building, for example, edges representing skew lines do not intersect.

36. Check students' work.

37. Answers may vary. An example is given. Points, lines, and planes are geometric figures. A point indicates location and has no size, but can be represented in drawings by a dot. A line extends forever in opposite directions. A plane extends forever in all four directions.

38. False; a person with normal vision can see with the lights off during the day.

39. True.

40. 4

41. 1

42. $-\frac{1}{2}$

43. $y = \frac{1}{2}x$

44. $y = x + 1$

45. $y = -2x - 2$

Warm-Up Exercises

Give the coordinates of each point.

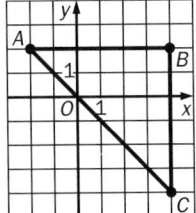

1. A (−2, 2)

2. B (4, 2)

3. C (4, −4)

Solve each equation.

4. $x + x = 9$ 4.5

5. $2x = x + 5$ 5

6. $(2k − 1) + (3k + 6) = 20$ 3

7. $(3.5n + 2) − 5 = 13$ $\frac{32}{7}$

Learn how to...

• **represent rays and segments**

• **interpret diagrams**

So you can...

• **find lengths and distances in various situations, such as manufacturing and astronomy**

1.5 Segments and Their Measures

Tunnels beneath cities and towns are carefully constructed. One way to make sure a tunnel is straight is to use a *laser alignment tool*. The laser projects a straight ray of light in the direction of the tunnel, and guides the machine that digs the tunnel.

In geometry, a **ray** is part of a line with one endpoint. You can name a ray by its endpoint and another point on the ray. Always name the endpoint first.

endpoint

You can call this ray $\overrightarrow{AB}$ or $\overrightarrow{AC}$.

A **segment** is part of a line with two endpoints. Name a segment by using its endpoints. Segments that are equal in length are **congruent segments**.

Tick marks are used to show congruence.

You can call this segment $\overline{EF}$ or $\overline{FE}$.

You can write "$\overline{DE}$ is congruent to $\overline{EF}$" by writing $\overline{DE} \cong \overline{EF}$. The length of the segment $\overline{DE}$, written DE, is the distance between D and E.

EXAMPLE 1

Give the length of each segment.

a. $\overline{QN}$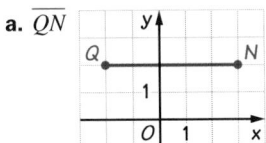

b. $\overline{EF}$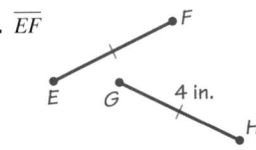

4 in.

SOLUTION

a. Count how many units long $\overline{QN}$ is. $QN = 5$.

b. The tick marks show that $\overline{EF} \cong \overline{GH}$, so $EF = GH$. $EF = 4$ in.

You might say that a store is *between* home and school even though the store is not on a straight line from home to school. But in geometry, a point is *between* two other points only if the three points are collinear.

Betweenness and Segment Addition

If $XY + YZ = XZ$, then point Y is between points X and Z on $\overline{XZ}$.

If point Y is between points X and Z, then $XY + YZ = XZ$.

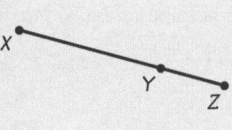

EXAMPLE 2

Find each length.
a. AC
b. CB

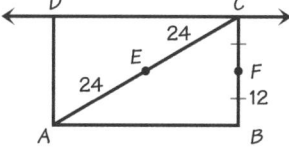

SOLUTION

a. Point E is between points A and C, so $AC = AE + EC$.

$$AC = AE + EC \qquad \text{Use Segment Addition.}$$
$$= 24 + 24$$
$$= 48$$

b. The red tick marks show that $\overline{CF} \cong \overline{FB}$, so $CF = FB$.

$$CF = 12$$

Point F is between C and B, so $CB = CF + FB$.

$$CB = CF + FB \qquad \text{Use Segment Addition.}$$
$$= 12 + 12$$
$$= 24$$

THINK AND COMMUNICATE

Use the figure in Example 2.

1. Which two segments are congruent to $\overline{AE}$? How do you know?

2. Describe how $\overrightarrow{DC}$ is different from $\overrightarrow{CD}$.

3. Make a sketch that shows that $MN = NO = 1$ in. In your sketch, how long is $\overline{MO}$? Must M, N, and O be collinear? Compare your answer with your classmates' answers.

1.5 Segments and Their Measures **29**

ANSWERS Section 1.5

Think and Communicate

1. $\overline{EC}$ and $\overline{CB}$; The figure shows that $AE = EC$, and the solution in part (b) shows that $CB = 24$.

2. $\overrightarrow{DC}$ and $\overrightarrow{CD}$ have different endpoints and extend in opposite directions.

3. Check students' work. M, N, and O may or may not be collinear. If they are, $MO = 2$. If not, $MO < 2$.

Teach⇔Interact

Section Note

Geometric Thinking
Stress the distinction between length and congruence. Discuss the fact that the use of the overhead bar in the symbol $\overline{EF}$ indicates that $\overline{EF}$ names a segment whose endpoints are E and F. When there is no bar, the symbol refers to the length of the segment. Urge students to be careful to use the symbol $\cong$ when they mean congruence and the symbol $=$ when they mean length.

Additional Example 1

Give the length of each segment.

a. $\overline{AB}$

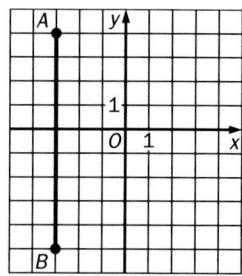

Count how many units long $\overline{AB}$ is. $AB = 9$

b. $\overline{QR}$

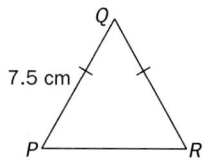

The tick marks show that $\overline{QR} \cong \overline{QP}$, so $QR = QP$. $QR = 7.5$ cm

Additional Example 2

Find each length.

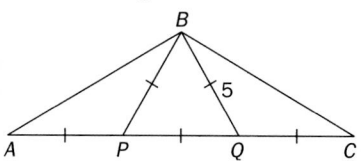

a. PQ The tick marks show that $\overline{PQ} \cong \overline{BQ}$, so $PQ = BQ = 5$.

b. AC All of the segments with tick marks are congruent, so they all have the same length. Thus, $AP = 5$, $PQ = 5$, and $QC = 5$. Point Q is between point P and point C, so $PC = PQ + QC$. $PC = 5 + 5 = 10$ Point P is between point A and point C, so $AC = AP + PC$. $AC = 5 + 10 = 15$

29

Communication: Drawing
Students need to understand what kind of information they can conclude from a diagram and what kind they cannot. Stress that a diagram can communicate certain information such as congruence or parallel segments only if the appropriate marks are shown on it. For example, it *appears* that $\overline{AD} \parallel \overline{BC}$ and $\overline{AB} \cong \overline{CD}$, but since the diagram is not marked as such, it would be incorrect to assume this information. Ask students if they can conclude that point E is halfway between points A and C. (No.) Have them explain their answer.

Visual Thinking
Ask students to sketch any triangle, to draw a line segment connecting one vertex of the triangle to the midpoint of the opposite side, and to label any point on that segment. Then, ask them to indicate what they can and cannot conclude about the elements of that triangle. This activity involves the visual skills of *identification* and *interpretation*.

Checking Key Concepts

Student Progress
All students should be able to answer questions 1–10 correctly before proceeding to work on the exercises that follow.

Closure Question

Define a ray, a segment, and congruent segments. A ray is a part of a line. It has one endpoint and continues forever in one direction. A segment is a part of a line and has two endpoints. Congruent segments are equal in length.

Diagrams

In geometry, you should be careful about what you conclude from a sketch or diagram. For example, you can conclude that all the points shown in the diagram below are coplanar, that $\overrightarrow{AB}$, $\overrightarrow{AC}$, and $\overrightarrow{AD}$ all intersect at point A, and that E is between A and C. Notice that segments and rays that are contained in parallel lines are also parallel.

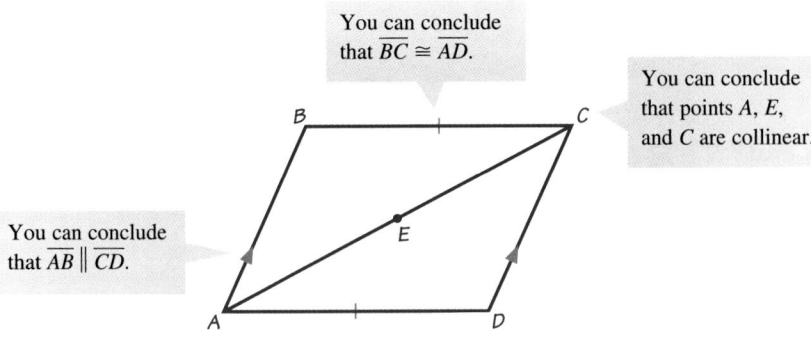

You can conclude that $\overline{BC} \cong \overline{AD}$.

You can conclude that points A, E, and C are collinear.

You can conclude that $\overrightarrow{AB} \parallel \overrightarrow{CD}$.

The diagram does *not* tell you that $\overrightarrow{AD} \parallel \overrightarrow{BC}$ or that $\overline{AB} \cong \overline{CD}$.

☑ CHECKING KEY CONCEPTS

1. Give another name for $\overline{XN}$.

2. Give another name for $\overrightarrow{QP}$.

3. Is $\overrightarrow{PQ}$ the same as $\overrightarrow{QP}$? Explain why or why not.

Give the length of each segment.

4. $\overline{WF}$ 5. $\overline{KM}$ 6. $\overline{BG}$

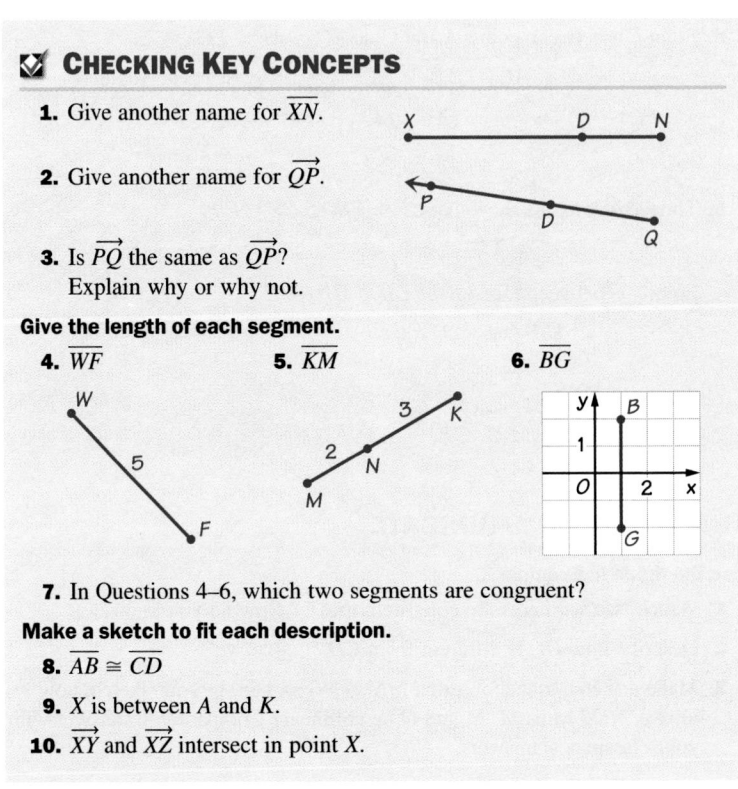

7. In Questions 4–6, which two segments are congruent?

Make a sketch to fit each description.

8. $\overline{AB} \cong \overline{CD}$

9. X is between A and K.

10. $\overrightarrow{XY}$ and $\overrightarrow{XZ}$ intersect in point X.

Checking Key Concepts

1. $\overline{NX}$

2. $\overrightarrow{QD}$

3. No; $\overrightarrow{PQ}$ and $\overrightarrow{QP}$ have different endpoints and extend in different directions.

4. 5

5. 5

6. 4

7. $\overline{WF}$ and $\overline{KM}$

8–10. Answers may vary. Examples are given.

8.
```
A   6   B
•———————•
  C  6
  •———•
       •D
```

9.
```
•  •      •
A  X      K
```

10.

1.5 | Exercises and Applications

Extra Practice
exercises on
page 660

Give the length of each segment.

1. $\overline{AB}$ **2.** $\overline{BE}$ **3.** $\overline{DH}$

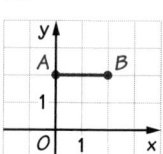 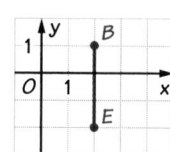 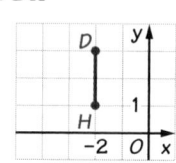

Give the length of each segment.

4. $\overline{CG}$

5. $\overline{BC}$

6. $\overline{AC}$

7. List three pairs of congruent segments in the figure.

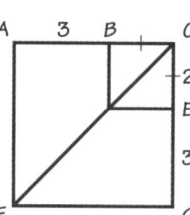

Make a sketch to fit each description.

8. $\overrightarrow{AB}$ **9.** $\overline{CD}$

10. $ST = 2$ in.   **11.** $\overline{EF} \cong \overline{GE}$

12. Point M is between points N and P.   **13.** $\overrightarrow{JK}$ intersects $\overline{LQ}$ in point R.

Connection   MANUFACTURING

When designers plan a new product, they must consider how the product will be transported or stored. Often, long narrow objects are designed to come apart into pieces.

About how long will each object be when it is assembled?

14.

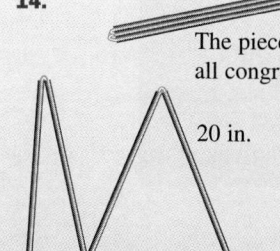

The pieces are all congruent.

20 in.

Tent poles would never fit in a car trunk at their full length. But folded up, the poles can fit into a bag with the tent.

15.

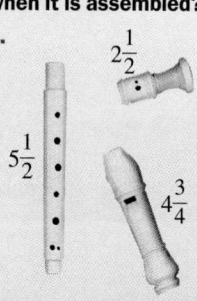

$2\frac{1}{2}$

$5\frac{1}{2}$

$4\frac{3}{4}$

A recorder comes apart so it can be cleaned and stored easily. You can store a recorder in pieces.

16.

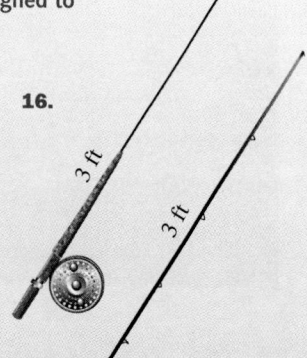

3 ft

3 ft

Portable fishing poles are popular with hikers who don't want to carry much food with them.

Apply⇔Assess

Suggested Assignment

❖ **Core Course**
  Day 1 Exs. 1–13, 18–23
  Day 2 Exs. 24, 25, 28–34, AYP
❖ **Extended Course**
  Day 1 Exs. 1–23
  Day 2 Exs. 24–34, AYP
❖ **Block Schedule**
  Day 3 Exs. 1–13, 18–23
  Day 4 Exs. 24, 25, 28–34, AYP

Exercise Notes

Common Error
Ex. 7 Some students may write = when they should write ≅. Remind these students that the equals sign can be used only when lengths (that is, numbers) are involved. When the segments themselves are involved, the congruence symbol must be used.

Teaching Tip
Exs. 8–13 You may wish to have one or more students draw the sketches on the board in order to review the meaning of the notations and the concepts involved in these exercises.

Exercises and Applications

1. 2

2. 3

3. 2

4. 5

5. 2

6. 5

7. $\overline{AB} \cong \overline{EG}; \overline{BC} \cong \overline{CE};$
   $\overline{AC} \cong \overline{CG}$

8–13. Answers may vary. Examples are given.

8.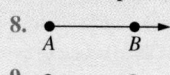

9.

10.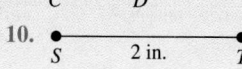

11.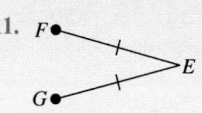

12.

13.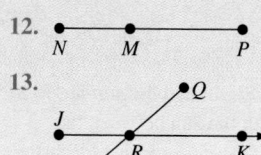

14. 80 in.

15. $12\frac{3}{4}$ in.

16. 6 ft

17. a. Writing In the picture at the right, Tyler is between Juna and Ian, Ian is between Tyler and Charelra, and Charelra is next to Nicole and Ian. Juna is wearing a light blue shirt. What is each of the other students wearing? Explain how you got your answer.

   b. Is Ian between Juna and Nicole? Explain.

For Exercises 18–23, give the letter of the best description.

18. $\overline{AB}$ A. the line through points A and B

19. $\overleftrightarrow{AB}$ B. the distance between points A and B

20. $\overrightarrow{AB}$ C. is congruent to

21. AB D. the segment connecting points A and B

22. $\cong$ E. the ray from point A through point B

23. $\parallel$ F. is parallel to

24. ALGEBRA $PM = 3x + 2$, $MN = x + 4$, and $PN = 18$. Find the values of x and of PM.

25. ALGEBRA $ST = 2y + 7$, $RT = 7y + 1$, and $RS = 9$. Find the values of y and of RT.

26. Open-ended Problem Make a sketch that has five points labeled, and includes one pair of congruent segments, one ray, and three collinear points. Name the ray and the congruent segments.

27. Challenge Earth and Saturn both orbit the sun. On average, Earth is 1.5×10^{11} m from the sun and Saturn is 1.4×10^{12} m from the sun. About how far can Saturn be from Earth? About how close can Saturn be to Earth? Explain your answer. Include a diagram.

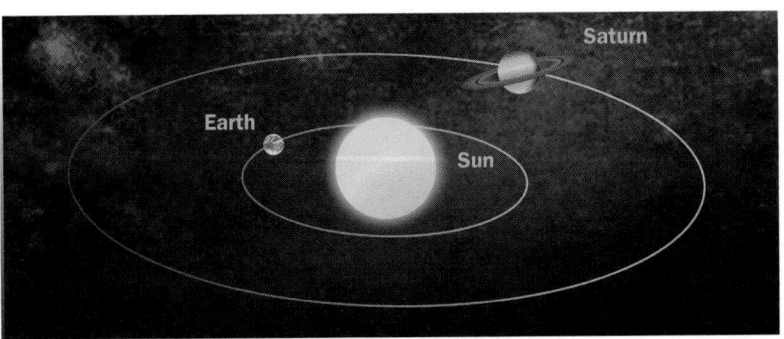

ONGOING ASSESSMENT

28. Writing What information does the diagram give you about the points and segments in the diagram? Explain how you know.

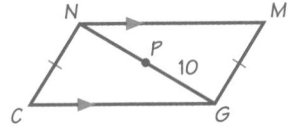

32   Chapter 1   *Patterns, Lines, and Planes*

17. **a.** Tyler is wearing a red and black shirt; Ian is wearing a striped shirt; Charles is wearing a white shirt; Nicola is wearing a black shirt. Answers may vary. An example is given. I got my answer by using the word *between* to locate the position of each student, starting with the facts that Juna is wearing a light blue shirt and that Tyler is between Juna and Ian.

   b. Yes; Ian is between Tyler and Charles, Tyler is between Juna and Ian, and Charelra is between Ian and Nicole.

18. D   19. A   20. E

21. B   22. C   23. F

24. 3; 11   25. 3; 22

26. Answers may vary. An example is given. The diagram shows $\overline{AB} \cong \overline{BD}$, $\overrightarrow{AC}$, and collinear points A, B, and C.

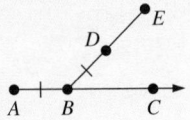

27. 1.55×10^{12} m; 1.25×10^{12} m; Saturn is farthest from Earth when Saturn, Earth, and the sun are in a line, with the sun between the planets. Saturn is closest to Earth when Saturn, Earth, and the sun are in a line, with Earth between the sun and Saturn.

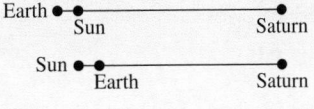

28. $\overline{NM} \parallel \overline{CG}$, which is indicated by the arrows. $\overline{NC} \cong \overline{MG}$, which is indicated by the tick marks. The length of $\overline{PG}$ is given as 10. The drawing also indicates that N, P, and G are collinear and N, M, G, C, and P are coplanar.

29–32. Answers may vary. Examples are given.

29.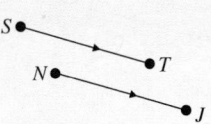

30.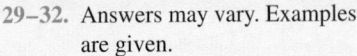

31.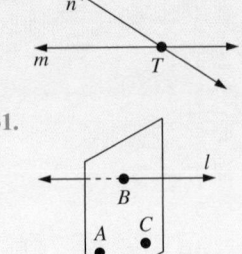

Sketch each situation. *(Section 1.4)*

29. $\overleftrightarrow{ST} \parallel \overleftrightarrow{NJ}$

30. Lines *n* and *m* intersect in point *T*.

31. Line *ℓ* intersects plane *ABC* in point *B*.

32. Points *P*, *M*, and *R* are noncollinear.

Sketch the next figure in each pattern. *(Section 1.1)*

33.

34. ∨ < ∧ > ∨ < ∧ > ∨ <

ASSESS YOUR PROGRESS

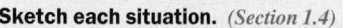

VOCABULARY

collinear (p. 22)   skew (p. 22)
coplanar (p. 22)   ray (p. 28)
intersect (p. 22)   endpoint (p. 28)
intersection (p. 22)   segment (p. 28)
parallel (p. 22)   congruent segments (p. 28)

For Exercises 1–5, use the figure at the right. *(Section 1.4)*

1. Name two lines that intersect at point *C*.

2. What is the intersection of $\overleftrightarrow{CD}$ and plane *AED*?

3. What is the intersection of $\overleftrightarrow{AE}$ and plane *BCD*?

4. Which two planes intersect in $\overleftrightarrow{AD}$?

5. Name a line that is skew to $\overleftrightarrow{AB}$.

Sketch each situation. *(Section 1.4)*

6. $n \parallel m$

7. Points *A*, *B*, and *C* are collinear.

8. $\overrightarrow{QF}$ lies in plane *Z*.

9. Line *ℓ* intersects $\overleftrightarrow{RS}$ in point *T*.

Make a sketch to fit each description. *(Section 1.5)*

10. $AB = 7$

11. Point *S* is between *T* and *U*.

12. Open-ended Problem Draw a segment on a coordinate graph and give its length.

Give the length of each segment. *(Section 1.5)*

13. $\overline{BD}$

14. $\overline{AE}$

15. $\overline{AC}$

16. Name a segment that is congruent to $\overline{AE}$. *(Section 1.5)*

17. Journal How many lines contain a particular segment? How many rays contain a particular segment? Explain your answers.

1.5 Segments and Their Measures **33**

Assess Your Progress

Review the meaning of the vocabulary terms with the class. Students should be able to sketch figures to illustrate each term.

Journal Entry
Suggest that students support their answers with diagrams.

Progress Check 1.4–1.5

See page 51.

Practice 5 for Section 1.5

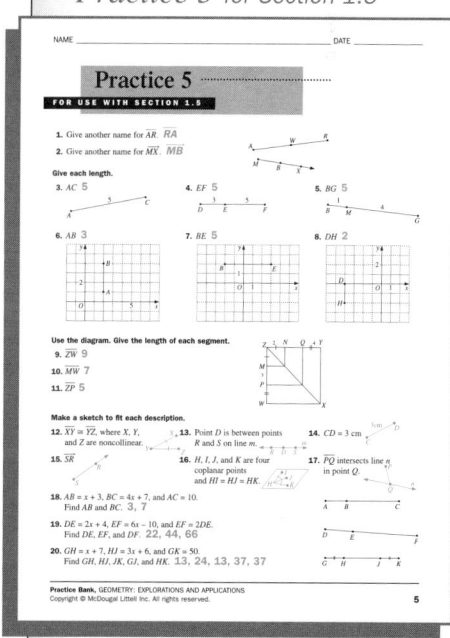

32.

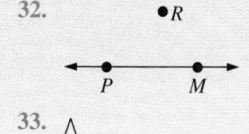

33. ∧

34.

Assess Your Progress

1. any two of $\overleftrightarrow{EC}$, $\overleftrightarrow{DC}$, and $\overleftrightarrow{BC}$

2. point *D*

3. point *A*

4. plane *AED* and plane *ACD*

5. $\overleftrightarrow{ED}$ or $\overleftrightarrow{EC}$

6–12. Answers may vary. Examples are given.

6.

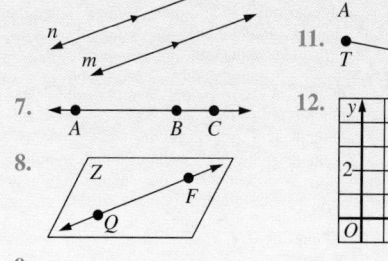

7.

8.

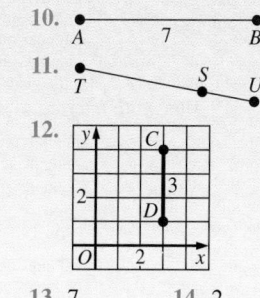

9.

10.

11.

12.

13. 7   **14.** 2

15. 4   **16.** $\overline{CE}$

17. A segment has two end-points. For any two points, there is exactly one line through the given points. So exactly one line contains a given segment. On the other hand, since there are infinitely many points on a line and any one of those can be the endpoint of a ray, there are infinitely many rays that contain a particular segment.

Objectives

- Name and measure angles.
- Describe angles in various real-world situations.

Recommended Pacing

❖ **Core and Extended Courses**
Section 1.6: 1 day

❖ **Block Schedule**
Section 1.6: $\frac{1}{2}$ block
(with Section 1.5)

Resource Materials

Lesson Support
Lesson Plan 1.6
Warm-Up Transparency 1.6
Practice Bank: Practice 6
Study Guide: Section 1.6
Challenge Problems: Set 6

Technology
Geometry Software
Internet:
   http://www.hmco.com

Warm-Up Exercises

Use the diagram.

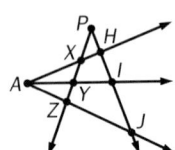

Name all the rays that have the given endpoint. Answers may vary. Examples are given.

1. P $\vec{PX}, \vec{PJ}$

2. A $\vec{AX}, \vec{AI}, \vec{AZ}$

3. X $\vec{XH}, \vec{XZ}$

4. Y $\vec{YI}, \vec{YZ}$

Name two rays that contain each segment. Answers may vary. Examples are given.

5. $\overline{XH}$ $\vec{XH}, \vec{AX}$

6. $\overline{IJ}$ $\vec{IJ}, \vec{HI}$

SECTION

1.6   Working with Angles

Learn how to...
- name and measure angles

So you can...
- describe angles in various situations, such as space flight, tumbling, and finding your latitude

In 1970, *Apollo 13* blasted into space. The astronauts hoped to set foot on the moon. Just over two days into the trip, an oxygen tank exploded and the space capsule had to return to Earth. Getting back was not easy. The capsule had to maintain its course for the correct *reentry angle*.

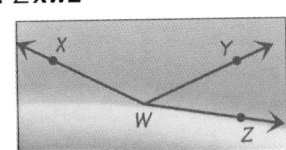

Apollo 13

5.3°

Ideal reentry path

7.7°

At an angle greater than 7.7°, the capsule would slow down too quickly.

Earth's Atmosphere

Earth

Not drawn to scale

If the capsule entered the atmosphere at an angle less than 5.3°, it would bounce off the atmosphere.

An **angle** is formed by two rays with a common endpoint. The endpoint is the **vertex** of the angle and the rays are the sides of the angle. You can name an angle by its vertex, by the vertex and a point on each ray, or by a number. When three points name an angle, the vertex is in the middle.

EXAMPLE 1

Give as many different names as you can for each angle.

a. ∠ *ABC*

b. ∠ *XWZ*

SOLUTION

a. ∠ *B*, ∠ 1, ∠ *CBA*

b. ∠ *ZWX*. You cannot use ∠ *W* as a name for ∠ *ZWX* because ∠ *W* could also refer to ∠ *XWY* and ∠ *YWZ*.

You use a protractor to measure an angle in *degrees* (°). The measure of an angle is greater than 0° and less than or equal to 180°. Write the measure of ∠TSV as m∠TSV. Two angles that have the same measure are **congruent angles**.

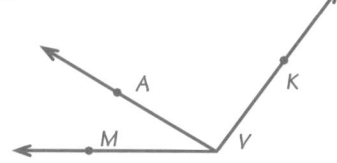

m∠TSV = 105°

105°

The red marks show that the angles are congruent.
∠TSU ≅ ∠USW.

EXAMPLE 2

Use a protractor to measure each angle, to the nearest degree.

a. ∠AVM

b. ∠AVK

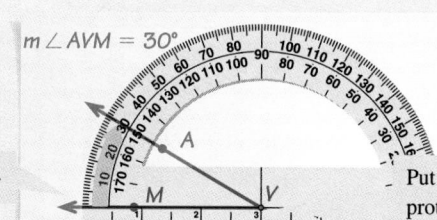

SOLUTION

a. Copy the angle first.

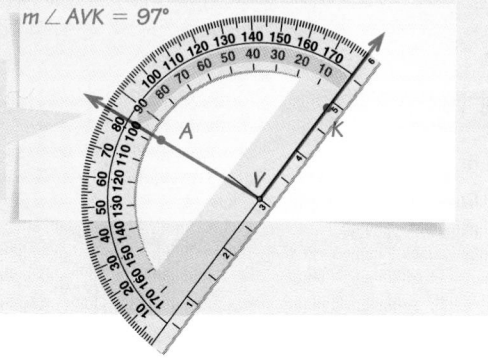

m∠AVM = 30°

Line up one ray with 0°.

Put the center of the protractor over the vertex of the angle.

b.

m∠AVK = 97°

When you copy the angle, you may need to extend rays so that they cross the scale.

THINK AND COMMUNICATE

1. Look at the numbers along the scale of the protractor. When should you use the numbers that are on the outside? When should you use the numbers that are on the inside?

2. Measure ∠MVK. How is m∠MVK related to m∠AVM and m∠AVK?

ANSWERS Section 1.6

Think and Communicate

1. when you line up one ray with the 0° on the outside scale; when you line up one ray with the 0° on the inside scale

2. m∠MVK = 129°; m∠MVK = m∠AVM + m∠AVK

Additional Example 1

Give as many different names as you can for each angle.

a. ∠PQR

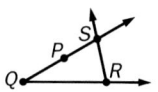

∠Q, ∠RQP, ∠SQR, ∠RQS

b. ∠AYC

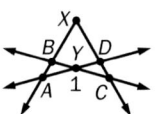

∠1, ∠CYA
You cannot use ∠Y for ∠AYC because ∠Y could also refer to ∠AYB, ∠CYD, or ∠BYD.

Additional Example 2

Use a protractor to measure each angle.

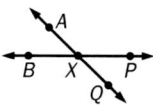

a. ∠AXP Copy the angle first. Extend the rays so that they cross the scale.

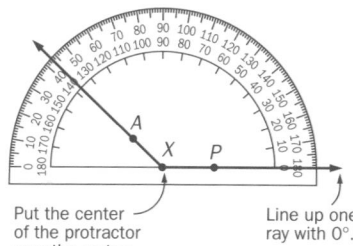

Put the center of the protractor over the vertex of the angle.

Line up one ray with 0°.

m∠AXP = 137°

b. ∠QXP

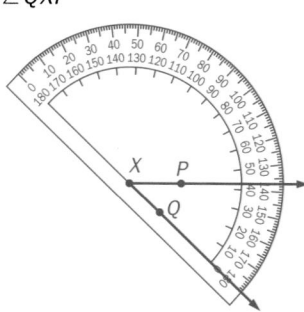

m∠QXP = 43°

Think and Communicate

Some students may tend to read the wrong scale on their protractors. Question 1 should help clear up this difficulty.

Teach⇔Interact

Section Note

Geometric Thinking

Since students have learned on page 35 that the measure of an angle can equal 180°, a logical question that may be asked is: What is the interior of an angle whose measure is 180°? Point out that for such an angle, such as $\angle ABC$ below, either side of $\overleftrightarrow{AC}$ can be considered its interior.

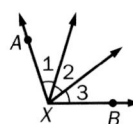

Additional Example 3

Points P, Q, and R are collinear. Find $m\angle PQS$.

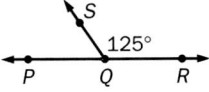

Point Q is between points P and R, so $m\angle PQR = 180°$. Use Angle Addition.

$m\angle PQS + m\angle SQR = m\angle PQR$
$m\angle PQS + 125° = 180°$
$m\angle PQS = 55°$

Additional Example 4

$m\angle AXB = 108°$
Find $m\angle 1$.

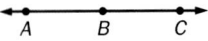

The marks mean that $\angle 1 \cong \angle 2$, $\angle 2 \cong \angle 3$, and $\angle 1 \cong \angle 3$.
So, $m\angle 1 = m\angle 2$, $m\angle 2 = m\angle 3$, and $m\angle 1 = m\angle 3$.

$m\angle 1 + m\angle 2 + m\angle 3 = m\angle AXB$
You can substitute $m\angle 1$ for $m\angle 2$ and $m\angle 3$.

$m\angle 1 + m\angle 1 + m\angle 1 = 108°$
$3m\angle 1 = 108°$
$m\angle 1 = 36°$

To reenter Earth's atmosphere safely, *Apollo 13* had to stay inside the safe reentry corridor. The inside of an angle is called the angle's *interior*.

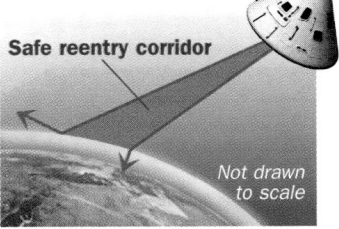

Safe reentry corridor

Not drawn to scale

Angle Addition

If D is in the interior of $\angle ABC$, then $m\angle ABD + m\angle DBC = m\angle ABC$.

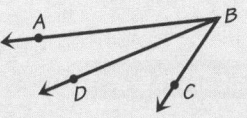

EXAMPLE 3

Find $m\angle BOC$.

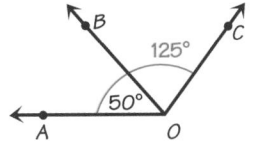

SOLUTION

Point B is in the interior of $\angle AOC$, so

Use Angle Addition.

$m\angle AOB + m\angle BOC = m\angle AOC$
$50° + m\angle BOC = 125°$
$m\angle BOC = 75°$

EXAMPLE 4

$m\angle 1 = 24°$
$m\angle 2 = 62°$
Find $m\angle PRS$.

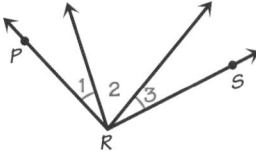

SOLUTION

The red marks mean that $\angle 1 \cong \angle 3$, so $m\angle 1 = m\angle 3$.

Substitute **24** for $m\angle 1$ and $m\angle 3$, and **62** for $m\angle 2$.

$m\angle PRS = m\angle 1 + m\angle 2 + m\angle 3$
$= 24 + 62 + 24$
$m\angle PRS = 110°$

☑ CHECKING KEY CONCEPTS

Give another name for each angle.

1. ∠1

2. ∠2

3. ∠XYZ

4. $m\angle 1 = 62°$ and $m\angle 2 = 47°$. Find $m\angle XYZ$.

Find the measure of each angle to the nearest degree.

5. ∠ABG 6. ∠CBD 7. ∠CBG

8. Name two angles that are congruent.

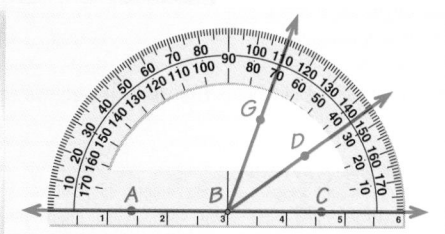

1.6 | Exercises and Applications

Extra Practice exercises on page 660

1. **Writing** Sketch three rays that share a vertex. Label the vertex *A* and label another point on each ray. Explain why you can't name any of the angles ∠*A*.

Use a protractor to find the measure of each angle, to the nearest degree.

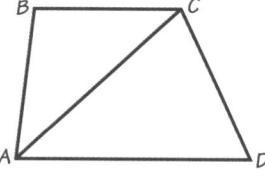

2. ∠BAD 3. ∠CAB

4. ∠CAD 5. ∠D

6. ∠B 7. ∠BCD

8. ∠BCA 9. ∠ACD

10. **TUMBLING** One of the Jesse White tumblers' most exciting tricks is called *Superman*. A tumbler springs from the trampoline and soars over 20 or more people. If the trampoline is at about a 45° angle to the floor, the tumbler will fly as far as possible. Estimate the angle of each trampoline.

a.   b.   c.   d.

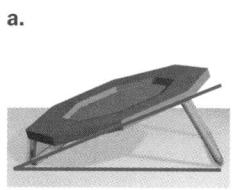 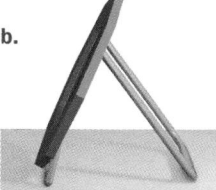  

e. Which of these trampoline positions would be best for the *Superman* stunt?

Teach⇔Interact

Checking Key Concepts

Communication: Discussion Discuss these questions in class to be sure students understand how to name and measure an angle. Use questions 5–7 to check that students can read the measure of an angle from a protractor correctly.

Closure Question

How is an angle formed and what measures can an angle have?
An angle is formed by two rays that have a common endpoint. The measure of an angle is greater than 0° and less than or equal to 180°.

Apply⇔Assess

Suggested Assignment

❖ **Core Course**
Exs. 1–9, 11–18, 30–34

❖ **Extended Course**
Exs. 1, 5–8, 10–34

❖ **Block Schedule**
Day 4 Exs. 1–9, 11–18, 30–34

Exercise Notes

Geometric Thinking
Exs. 2–9 After students have measured all eight angles, ask them if there is another method they could use to find the measure of some of these angles. (Yes, Angle Addition can be used for the angles sharing vertices *A* and *C* once you measure two of the three angles at each vertex.)

Checking Key Concepts

1. ∠WYZ or ∠ZYW

2. ∠WYX or ∠XYW

3. ∠ZYX

4. 109°

5. 110°

6. 35°

7. 70°

8. ∠CBD and ∠DBG

Exercises and Applications

1. Answers may vary. An example is given.

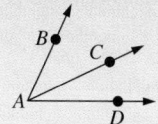

In the example, none of the angles can be named ∠A because that name could refer to ∠BAC, ∠BAD, or ∠CAD.

2–9. Answers may vary slightly.

2. 82°   3. 41°

4. 41°   5. 65°

6. 98°   7. 115°

8. 41°   9. 74°

10. Answers may vary. Examples are given.

a. 20°   b. 70°

c. 0°   d. 50°

e. d

Exercise Notes

Give the measure of each angle.

11. ∠*MNO*

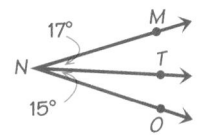

12. ∠*PQR*

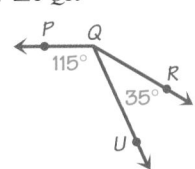

13. ∠*YZA*

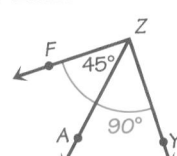

For Exercises 14–18, use the diagram at the right.

14. Find *m*∠*CEG*.

15. Find *m*∠*GED*.

16. Find *m*∠*HED*.

17. Which angle is congruent to ∠*CEG*?

18. *m*∠*HEV* = 60°. Find *m*∠*CEV*.

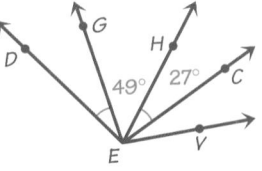

19. **Open-ended Problem** Look at the objects and surfaces around you and estimate the measures of any angles you see. Which angle measure(s) seems most common? Why do you think this is so?

20. **Research** Work with another person. Use an encyclopedia, star map, or astronomy guide to find the North Star.

   a. You can use the North Star to estimate your latitude. One of you should point one arm at the North Star and the other at the horizon. The other should estimate the angle between the arms. The measure of the angle is about the same as your latitude.

   b. Use a map or an almanac to find your latitude. How close is your estimate? What could cause your estimate to be far from the actual latitude?

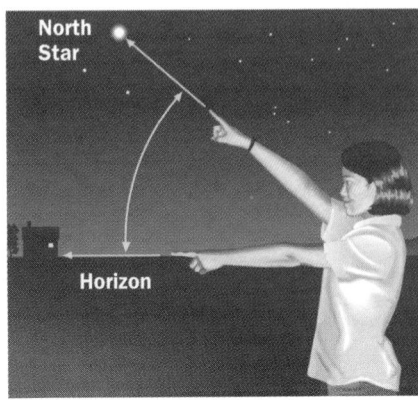

 Technology **You will need geometry software.**

Step 1 Construct $\overline{AB}$ and point *O*.

Step 2 Rotate $\overline{AB}$ around point *O* by 90°. Label the new endpoints *C* and *D*, as shown, so *C* is the image of *A* and *D* is the image of *B*.

Step 3 Construct $\overline{AO}$ and $\overline{CO}$.

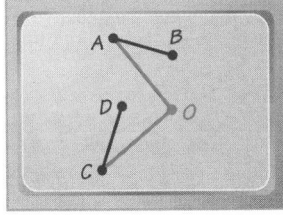

21. Measure ∠*AOC*.

22. Move $\overline{AB}$. Describe the effect on *m*∠*AOC*.

23. Move point *O* and describe the effect on *m*∠*AOC*.

24. Make a conjecture concerning the measure of ∠*DOB*.

25. Measure ∠*DOB*. Is your conjecture correct?

38 Chapter 1 *Patterns, Lines, and Planes*

11. 32°

12. 150°

13. 45°

14. 76°

15. 27°

16. 76°

17. ∠*HED*

18. 33°

19. Answers may vary. Right angles are commonly used in construction. For example, walls meet at right angles, ceilings and walls meet at right angles, and so on. So, many of the angles you see will be right angles.

20. a. Check students' work.

   b. Answers may vary. Problems with vision or with estimation, for example, could cause estimates to be far from the actual latitude.

21. 90°

22. It is unchanged.

23. It is unchanged.

24. *m*∠*DOB* = 90°

25. 90°; Yes.

26. 42°

27. 40°

28. 2°

29. 4 times as wide

Connection ▸ SCIENCE

When sunlight passes through a raindrop, it breaks into many different colors, from red light, which leaves the raindrop at an angle of 42°, to violet light, which leaves at an angle of 40°. If you look at the rain from a distance, you may see a rainbow.

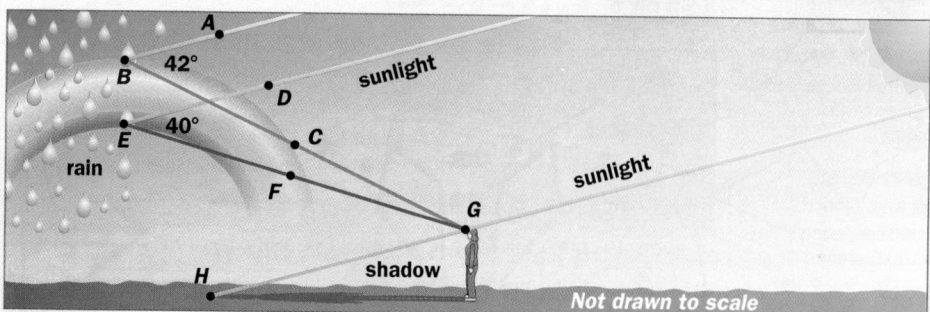

rain

Not drawn to scale

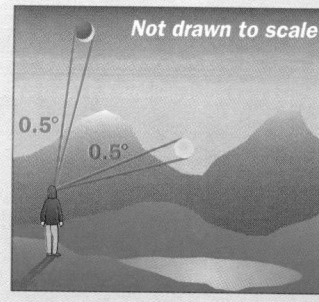

Not drawn to scale

26. $\angle ABC \cong \angle CGH$. Find $m\angle CGH$.

27. $\angle DEF \cong \angle FGH$. Find $m\angle FGH$.

28. Find $m\angle CGF$.

29. **Challenge** Although the sun is much larger than the moon, they look about the same size in the sky. The sun and moon each take up an angle of about 0.5°, as shown. If you saw the moon and a rainbow at the same time, how wide would the rainbow seem compared to the moon? (*Hint:* Use your answer to Exercise 28.)

ONGOING ASSESSMENT

30. **Open-ended Problem** Find a pair of congruent angles at your home or at school. Describe and measure them.

SPIRAL REVIEW

Give the length of each segment. (*Section 1.5*)

31. $\overline{FN}$

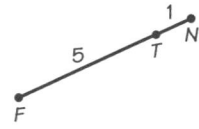

32. $\overline{ST}$; $\overline{OR}$

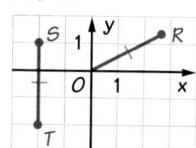

33. $\overline{AC}$

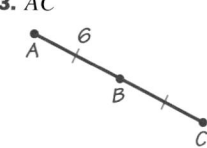

34. Sketch a figure that has rotational symmetry. (*Section 1.2*)

1.6 Working with Angles **39**

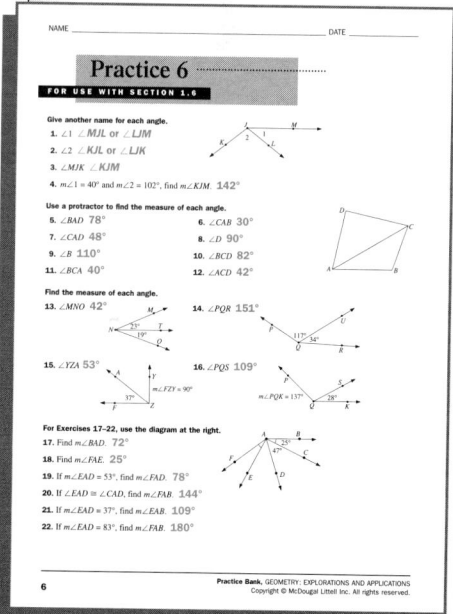
30. Answers may vary. An example is given. I have two windows in my room. The two panels of the curtains on each window are tied back at the sides. The panels make a 40° angle at the top.

34. Answers may vary. An example is given.

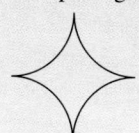

31. 6

32. 3; 3

33. 12

39

Warm-Up Exercises

1. What lines of symmetry does this figure have?

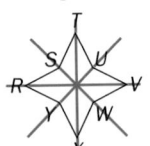

$\overleftrightarrow{TX}, \overleftrightarrow{RV}, \overleftrightarrow{SW}, \overleftrightarrow{UY}$

2. Suppose $MN = 28$ cm. Find EN.

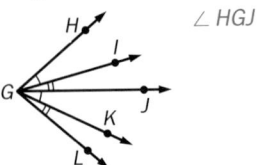

14 cm

3. Name an angle that is congruent to $\angle JGL$.

$\angle HGJ$

40

1.7 Bisecting Segments and Angles

Learn how to...

• identify bisectors of segments and angles

So you can...

• find real-world distances, such as the distance between two cities

• find angle measures, such as in boating

Origami, the art of paper folding, originated in China almost 2000 years ago. Although the word *origami* is Japanese, the art was also practiced in North Africa and Spain. The folded figures suggest not only the form of an object, but its motion or even its character. One of the most popular figures is the crane, sometimes used as a symbol of peace.

THINK AND COMMUNICATE

The first two folds for making an origami crane are shown.

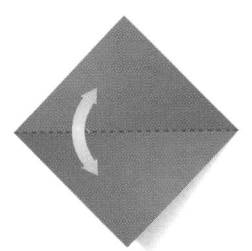

fold and unfold

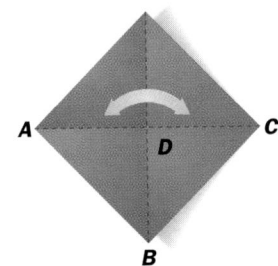

fold and unfold

1. Do you think that $\angle ABD \cong \angle CBD$? Why or why not?

2. Do you think that $\overline{AD} \cong \overline{DC}$? Why or why not?

3. What transformation is suggested by these folds?

40 Chapter 1 *Patterns, Lines, and Planes*

ANSWERS Section 1.7

Think and Communicate

1. Yes; when the paper is folded, $\angle ABD$ and $\angle CBD$ coincide.

2. Yes; when the paper is folded, $\overline{AD}$ and $\overline{DC}$ coincide.

3. reflection

When you reflect, translate, or rotate a figure, the image is congruent to the original figure. In the diagram on page 40, $\overline{DC}$ is the reflection of $\overline{AD}$ over $\overline{DB}$, and $\overline{AD} \cong \overline{DC}$. Point D is called the *midpoint* of $\overline{AC}$. The **midpoint** of a segment is the point that divides the segment into two congruent segments.

If M is the midpoint of $\overline{XY}$, then:

$XM = MY$ and $\overline{XM} \cong \overline{MY}$

$XM = \frac{1}{2}XY$ and $MY = \frac{1}{2}XY$

A **bisector of a segment** is a line, segment, ray, or plane that intersects the segment at its midpoint.

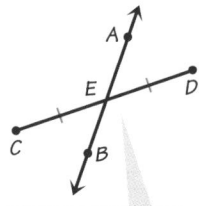

$\overleftrightarrow{AB}$ bisects $\overline{CD}$.

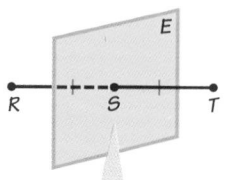

$\overline{MN}$ and $\overline{XY}$ bisect each other.

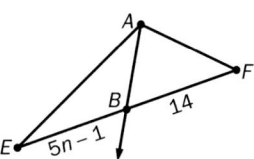

Plane E bisects $\overline{RT}$.

EXAMPLE 1   Connection: Algebra

$\overrightarrow{RS}$ bisects $\overline{PQ}$.
Find the value of n.

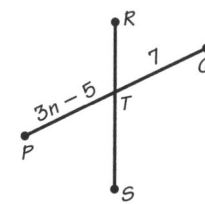

Toolbox p. 696
Solving Two-Step Equations

SOLUTION

Because $\overrightarrow{RS}$ bisects $\overline{PQ}$, T is the midpoint of $\overline{PQ}$.

$$PT = TQ$$
$$3n - 5 = 7 \qquad \text{Add 5 to both sides of the equation.}$$
$$3n = 12$$
$$n = 4 \qquad \text{Divide both sides by 3.}$$

THINK AND COMMUNICATE

4. Can a line have a midpoint? Explain why or why not.

5. In the diagram, $\overline{DE} \cong \overline{EF}$, but E is not the midpoint of $\overline{DF}$. Explain why not.

6. Explain how to find PQ in Example 1.

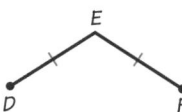

1.7 Bisecting Segments and Angles **41**

Learning Styles: Kinesthetic

Kinesthetic learners should fold a piece of paper, as shown in the Think and Communicate on page 40, to see the actual angle and segment bisectors. They can then label the points, A, B, C, and D on their paper and use it when reading the paragraph at the top of this page.

Section Note

Teaching Tip
Stress to students that only in the middle diagram do the segments bisect each other. In the diagram on the left, there is no indication that $\overline{CD}$ bisects $\overline{AB}$ since no marks on the diagram show that this is so. In general, one segment can bisect another without being bisected itself.

Additional Example 1

$\overrightarrow{AB}$ bisects $\overline{EF}$. Find the value of n.

Because $\overrightarrow{AB}$ bisects $\overline{EF}$, B is the midpoint of $\overline{EF}$.
$$EB = BF$$
$$5n - 1 = 14$$
$$5n = 15$$
$$n = 3$$

Think and Communicate

Questions 4 and 5 should help clear up two common misconceptions about midpoints, namely that lines have midpoints and that any point equidistant from two other points is their midpoint.

Think and Communicate

4. No; a line continues forever, so it cannot be divided into two congruent parts.

5. E is not on $\overline{DF}$, so it does not divide $\overline{DF}$ into two congruent segments.

6. Since $\overrightarrow{RS}$ bisects $\overline{PQ}$, $PT = TQ$ and $PQ = 2 \cdot TQ = 14$.

Additional Example 2

Find the coordinates of the midpoint of $\overline{PQ}$.

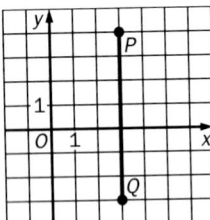

$\overline{PQ}$ is 7 units long. The midpoint of $\overline{PQ}$ divides the segment into two segments, each 3.5 units long. Point M (3, 0.5) is on $\overline{PQ}$ and is 3.5 units from both P and Q.

Section Notes

Geometric Thinking

Although the bisector of a segment can be a line, segment, ray, or plane, the bisector of an angle can only be a ray or a line. Ask students why an angle bisector cannot be a segment.

Using Technology

Students can use Activity 2 in the *Geometry Inventor Activity Book* to explore properties of angle bisectors and bisected angles.

EXAMPLE 2

Find the coordinates of the midpoint of $\overline{GH}$.

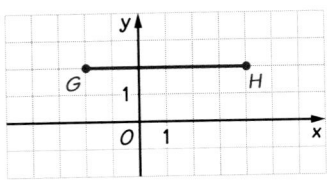

SOLUTION

$\overline{GH}$ is 6 units long. The midpoint of $\overline{GH}$ divides it into two segments, each 3 units long.

Count how far G is from H. $GH = 6$.

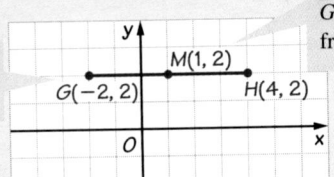

Point $M(1, 2)$ is on $\overline{GH}$ and is 3 units from both G and H.

THINK AND COMMUNICATE

7. In Example 2, explain how you could fold the paper to find the coordinates of the midpoint of $\overline{GH}$.

8. In Example 2, how is the *x*-coordinate of the midpoint related to the *x*-coordinates of the endpoints?

Angle Bisectors

If you fold one ray of an angle onto the other ray, and unfold the paper, the crease divides the angle into two congruent parts. The crease is an example of an *angle bisector*.

A **bisector of an angle** is a ray or line that divides the angle into two congruent angles.

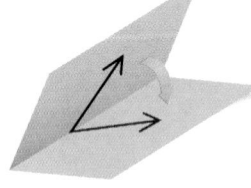

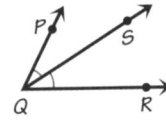

If $\overrightarrow{QS}$ bisects $\angle PQR$, then:

$$m\angle PQS = m\angle SQR \text{ and } \angle PQS \cong \angle SQR$$

$$m\angle PQS = \frac{1}{2}m\angle PQR \text{ and } m\angle SQR = \frac{1}{2}m\angle PQR$$

42   Chapter 1 *Patterns, Lines, and Planes*

Think and Communicate

7. Fold the paper so that G falls on top of H. The point where the crease meets $\overline{GH}$ is the midpoint of $\overline{GH}$.

8. It is the average of the *x*-coordinates of the endpoints.

EXAMPLE 3 **Connection: Algebra**

$\overrightarrow{SV}$ bisects $\angle RST$.
Find the value of
x and $m\angle VST$.

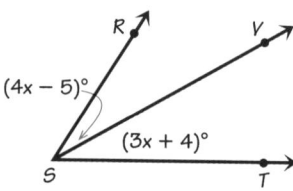

SOLUTION

Because $\overrightarrow{SV}$ bisects $\angle RST$, $m\angle RSV = m\angle VST$.

$$m\angle RSV = m\angle VST$$
$$(4x - 5)° = (3x + 4)° \qquad \text{Add 5 to both sides.}$$
$$4x = 3x + 9$$
$$x = 9 \qquad \begin{array}{l}\text{Subtract } 3x \text{ from} \\ \text{both sides.}\end{array}$$
$$m\angle VST = (3x + 4)°$$
$$= 3(9) + 4 \qquad \text{Substitute 9 for } x.$$
$$= 31°$$

So $x = 9$ and $m\angle VST = 31°$.

✓ **CHECKING KEY CONCEPTS**

The midpoint of $\overline{QR}$ is M, the midpoint of $\overline{PR}$ is Q, and $QM = 1.7$.
Find each length.

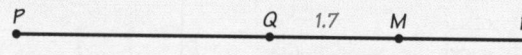

1. QR **2.** MR **3.** PR **4.** PM

$\overrightarrow{DB}$ bisects $\angle ADC$.
Find each measure.

5. $m\angle BDC$

6. $m\angle ADC$

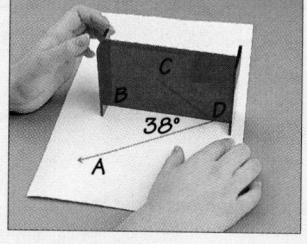

7. Sketch a segment with endpoints J and L. Show how to use paper folding to find its midpoint, K.

8. $\overrightarrow{CD}$ bisects $\angle BCE$. Find the value of x.

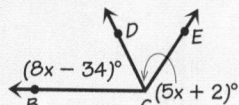

1.7 Bisecting Segments and Angles **43**

Checking Key Concepts

1. 3.4

2. 1.7

3. 6.8

4. 5.1

5. 38°

6. 76°

7. Check students' work. Fold the paper so that J falls on top of L. Unfold the paper. The point where the crease meets $\overline{JL}$ is the midpoint of $\overline{JL}$.

8. 12

About Example 3

Integrating Algebra
The algebraic approach taken in this Example provides an opportunity for students to review their equation-solving skills. After students have studied the Example, point out that since $\overrightarrow{SV}$ bisects $\angle RST$, then you could write $m\angle VST$ as $(4x - 5)°$ or $m\angle RSV$ as $(3x + 4)°$. In either case, however, this would not have made it possible to find the value of x. Ask students why.

Additional Example 3

$\overrightarrow{WQ}$ bisects $\angle PWR$. Find the value of x and $m\angle PWR$.

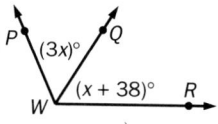

Because $\overrightarrow{WQ}$ bisects $\angle PWR$,
$m\angle PWQ = m\angle QWR$.
$$m\angle PWQ = m\angle QWR$$
$$(3x)° = (x + 38)°$$
$$2x = 38$$
$$x = 19$$
$$m\angle PWR = (3x)° + (x + 38)°$$
$$= (3 \cdot 19)° + (19 + 38)°$$
$$= 57° + 57°$$
$$= 114°$$
So, $x = 19$ and $m\angle PWR = 114°$.

Closure Question

How does the definition of an angle bisector compare to the definition of a segment bisector? An angle bisector is a ray or line that divides the angle into two congruent angles. A segment bisector is a line, segment, ray, or plane that divides a segment into two congruent segments. Thus, both types of bisectors divide their respective figures into two congruent parts, but an angle bisector can only be a ray or a line, whereas a segment bisector can be a line, segment, ray, or plane.

Extra Practice
exercises on
page 661

1.7 Exercises and Applications

In the diagram, $\overline{AB}$ and $\overline{EF}$ bisect each other. Find each length.

1. MB
2. EF
3. MF
4. AB

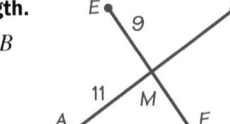

Tell whether each statement is *True* or *False*.

5. Y and W are both midpoints of $\overline{XZ}$.

6. Y is the midpoint of $\overline{XW}$.

7. If YZ = 6, then YW = 12.

8. If XW = 12, then YZ = 12.

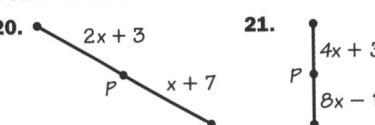

9. **RACING** The Iditarod Sled Dog Race is held each spring. Racers (called "mushers") are pulled by teams of sled dogs from Anchorage to Nome, Alaska.

   a. The route changes each year. In odd-numbered years, the city of Iditarod is roughly at the midpoint of the 1161 mi race. About how far from Nome is Iditarod?

   b. **Writing** How is the midpoint of a race like the midpoint of a segment? How is it different?

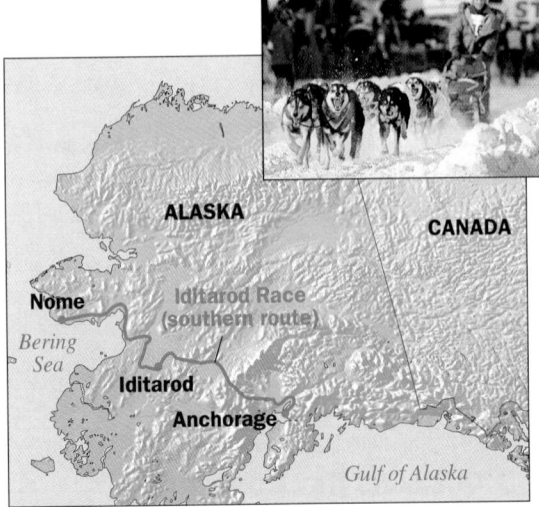

Find the coordinates of the midpoint of each segment.

10. $\overline{AB}$
11. $\overline{CD}$
12. $\overline{EF}$
13. $\overline{GH}$
14. $\overline{IJ}$
15. $\overline{OA}$

Make a sketch to illustrate each statement.

16. M is the midpoint of $\overline{SG}$.

17. A is the midpoint of $\overline{CD}$ and B is the midpoint of $\overline{CA}$.

18. $\overline{PQ}$ bisects $\overline{DE}$, but $\overline{DE}$ does not bisect $\overline{PQ}$.

19. $\overrightarrow{XY}$ bisects $\angle AXP$.

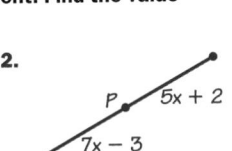

ALGEBRA In each figure, *P* is the midpoint of the segment. Find the value of each variable.

20. $2x + 3$, $x + 7$

21. $4x + 3$, $8x - 11$

22. $5x + 2$, $7x - 3$

44   Chapter 1 *Patterns, Lines, and Planes*

Exercises and Applications

1. 11
2. 18
3. 9
4. 22
5. False.
6. True.
7. False.
8. True.
9. a. about 580.5 mi

b. Answers may vary. An example is given. The midpoint of a race is the point halfway between two endpoints (the start and the finish). It divides the course into 2 approximately equal parts, but the midpoint of the race is not the midpoint of a segment unless the entire course is a segment.

10. (2, 2)
11. (−2, 3)
12. $\left(-2\frac{1}{2}, -1\right)$
13. (4, 1)
14. (1, −4)
15. (1, 0)

16–19. Answers may vary. Examples are given.

16. S M G

17. C 2 B 2 A 4 D

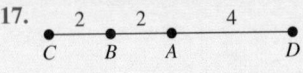

You can use origami to make a drinking cup that will actually hold water. The figures below show a square piece of paper with vertices *A*, *B*, *C*, and *D*. You will locate points *E*, *F*, *G*, and *H* by folding.

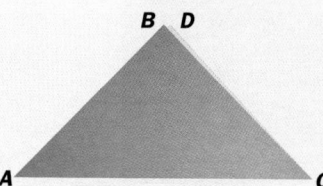

Step 1 Fold the bottom half of the paper up along $\overline{AC}$, bringing point *B* to *D*.

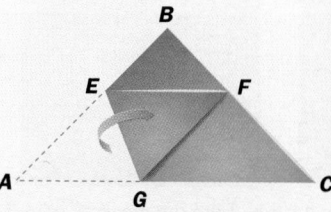

Step 2 Fold $\overline{AC}$ so point *A* meets $\overline{BC}$ at point *F* and $\overline{EF} \parallel \overline{GC}$.

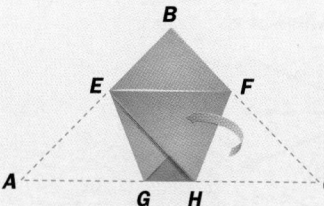

Step 3 Fold $\overline{BC}$ at *F* so point *C* meets $\overline{AB}$ at point *E*.

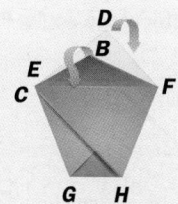

Step 4 Separate the two points, *B* and *D*, at the top. Fold the rear piece back. Fold the front piece forward and tuck it into the front pocket.

Step 5 Squeeze the sides together lightly and you have a drinking cup.

Unfold your cup. Make a sketch showing the creases you have made. Label the corners of the paper and the endpoints of each crease as shown above.

23. **Writing** Name three pairs of congruent segments and at least three pairs of congruent angles. Explain how you know that the parts are congruent.

24. **Open-ended Problem** Find four angles that are all congruent to each other. How do you know that they are congruent?

25. Do any of the creases represent angle bisectors? If so, tell which ones.

26. Do any of the creases represent segment bisectors? If so, tell which ones.

27. **Research** Find directions for another origami figure, such as a crane or a pajarita. Fold the figure. Describe any segment bisectors or angle bisectors in the completed figure.

1.7 Bisecting Segments and Angles **45**

18.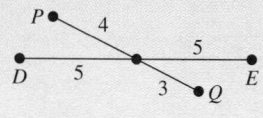

19.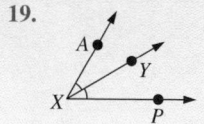

20. 4

21. $3\frac{1}{2}$

22. $2\frac{1}{2}$

23. Answers may vary. Examples are given. $\overline{AB}$ and $\overline{AD}$; $\overline{CB}$ and $\overline{CD}$; $\overline{AE}$ and $\overline{CF}$; $\angle ABC$ and $\angle ADC$; $\angle DAG$ and $\angle DCH$; $\angle AGE$ and $\angle CHF$; In each case, the segments and angles are congruent because they fit over each other exactly when the paper is folded.

24. Answers may vary. An example is given. $\angle GAE$, $\angle AEG$, $\angle HFC$, and $\angle HCF$ are all congruent because they fit over each other exactly when the paper is folded.

25. Yes; $\overline{AC}$ bisects $\angle BAD$ and $\angle BCD$, $\overline{GE}$ bisects $\angle FEA$ and $\overline{FH}$ bisects $\angle EFC$.

26. No.

27. Check students' work.

Integrating Algebra
Exs. 29–33 These exercises show how algebra can be used to model and solve problems involving angle and segment bisectors. A demonstration of the solutions to these exercises at the board will be useful to those students who need a review of the algebraic skills involved.

Problem Solving
Ex. 33 This exercise provides an example of using an equation to solve a geometric problem. In this case, the question as to whether *N* is the midpoint of $\overline{MP}$ can be answered by solving the equation $(3x + 5) + (5x + 2) = 24$. Remind students that they cannot answer this question by looking at the diagram, since there are no marks to indicate that $MN = NP$.

Cooperative Learning
Ex. 34 For best results in part (a), students should use triangles whose angles measure 90° or less. In part (b), the size of the angles is less important. However, very small angles should not be used because it is difficult to fold them accurately to get a good angle bisector.

Topic Spiraling: Preview
Ex. 34 This exercise previews some of the concepts of Section 6.7. You may wish to extend the exercise by having students check the medians and altitudes of triangles.

28. BOATING A boat traveling through water forms two waves as part of its *wake*. If the boat is traveling in a straight line and there is no wind, then the trail of the boat bisects the angle formed by the waves. The measure of the angle depends on the speed of the boat and the shape of its hull.

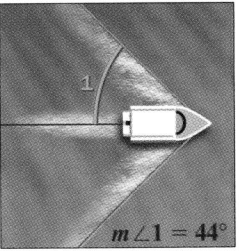
$m\angle 1 = 44°$

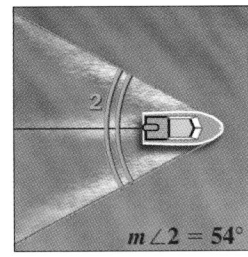
$m\angle 2 = 54°$

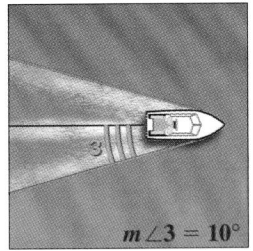
$m\angle 3 = 10°$

a. Find the measure of the angle between the waves of the wake.

b. Find the measure of the angle between the boat's trail and one wave of the wake.

c. Find the measure of the angle between the waves of the wake.

ALGEBRA In each figure, $\overline{PQ}$ is the bisector of the angle. Find the value of *x*.

29.
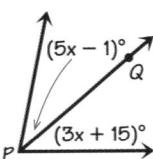

30.
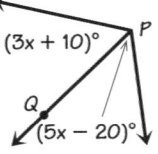

31.
22°

32. $\overrightarrow{BD}$ bisects $\angle ABE$ and $\overrightarrow{BE}$ bisects $\angle ABC$.
a. Find the value of *x*.
b. Find $m\angle DBC$.

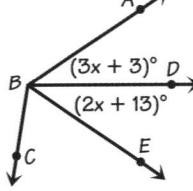

33. Challenge In the diagram, $MP = 24$. Is *N* the midpoint of $\overline{MP}$? Explain why or why not.

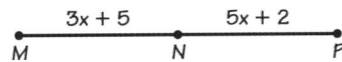

ONGOING ASSESSMENT

34. Cooperative Learning Work with another person.
a. Draw a triangle on a blank piece of paper. Fold each side of the triangle to find its midpoint. What do you notice about the three creases?
b. Draw another large triangle on a piece of paper. Fold each angle of the triangle to form its bisector. What do you notice about the three creases?

---

28. a. 88°
    b. 27°
    c. 20°
    d. Answers may vary. An example is given. I think the boat in diagram (c) is the fastest. The more rapidly the front edge of the wave is pulled forward, the narrower the angle.
29. 8

30. 15
31. 16
32. a. 10
    b. 99°
33. No; solving $(3x + 5) + (5x + 2) = 24$ gives $x = 2.125$. So, $MN = 11.375$ and $NP = 12.625$. Therefore, $MN \neq NP$, and *N* is not the midpoint of $\overline{MP}$.
34. a. They intersect in one point.
    b. They intersect in one point.

Find the measure of each angle. *(Section 1.6)*

35. ∠DFG

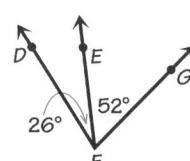

36. ∠1

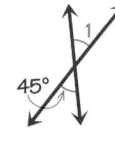

37. ∠ZXQ

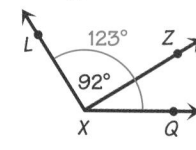

Use inductive reasoning to find the next two numbers in each pattern.
(Section 1.1)

38. 2, −3, −8, −13, ?, ?   **39.** 2, 5, 10, 17, ?, ?

ASSESS YOUR PROGRESS

VOCABULARY

angle (p. 34)   **midpoint** (p. 41)
vertex of an angle (p. 34)   **bisector of a segment** (p. 41)
congruent angles (p. 35)   **bisector of an angle** (p. 42)

Give as many names as you can for each angle. *(Section 1.6)*

1. ∠1

2. ∠EHF

Find the measure of each angle.
(Section 1.6)

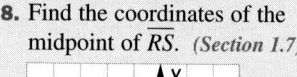

3. ∠EHG   **4.** ∠GFH   **5.** ∠EFH

6. K is the midpoint of $\overline{JL}$ and JK = 13. Find KL and JL. *(Section 1.7)*

7. $\overrightarrow{AB}$ bisects $\overline{PQ}$. Find the value of x. *(Section 1.7)*

8. Find the coordinates of the midpoint of $\overline{RS}$. *(Section 1.7)*

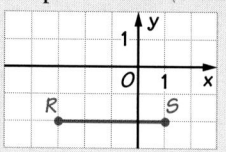

9. $\overrightarrow{NP}$ bisects ∠MNO. Find the value of a and m∠MNO. *(Section 1.7)*

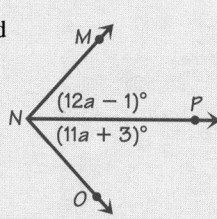

10. Journal Draw an angle dark enough so you can see it through the paper. Describe how you can fold the paper to form:
   a. the angle bisector   **b.** a congruent angle

1.7 Bisecting Segments and Angles   **47**

Assess Your Progress

Review the vocabulary terms with students. Ask them to give definitions and draw sketches for the terms.

Journal Entry
Students may wish to cut out angles, fold them, and paste them in their journals.

Progress Check 1.6–1.7

See page 51.

Practice 7 for Section 1.7

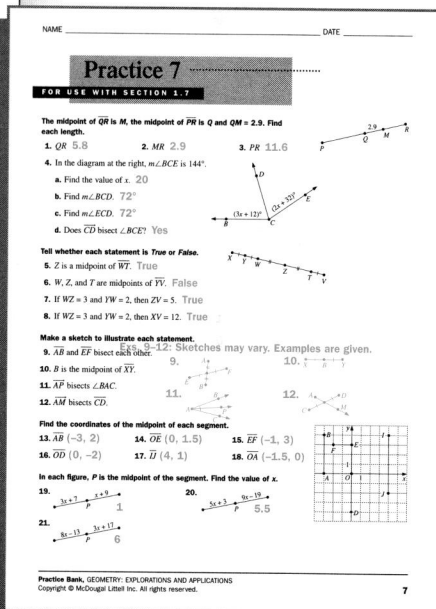

---

35. 78°

36. 45°

37. 31°

38. −18, −23

39. 26, 37

Assess Your Progress

1. ∠FGH; ∠HGF; ∠G

2. ∠FHE

3. 130°

4. 27°

5. 52°

6. 13; 26

7. 1

8. (−1, −2)

9. 4; 94°

10. a. Fold the paper so that one side of the angle falls on top of the other.

b. Fold the paper along one side of the angle and trace the other side.

Mathematical Goals

- Identify objects that have symmetry.
- Analyze transformations that preserve shapes and sizes.

Planning

Materials
- ruler
- protractor
- paper and pencil
- poster board (if necessary)

Project Teams
Students can work individually to collect their objects and make their drawings. If they decide to work in pairs, every aspect of the project should be completed as a team.

Guiding Students' Work

Before students begin, make sure they understand the terminology involved with this project. Terms such as *reflection*, *translation*, *rotation*, *symmetry*, and *transformation* should be clearly understood by every student. If students cannot find an object that exhibits a type of symmetry (especially objects in nature), have them find a picture of an object and then draw a sketch of it for the project. Also, students may need help selecting points on an object to illustrate a type of symmetry. If this is the case, have them make some rough sketches of the object and experiment with selecting different points to decide which ones are best to illustrate the symmetry of the object. This approach can also help students understand and investigate some of the questions in the project.

Second-Language Learners
Allowing students to make a poster instead of writing a report, or to give an oral presentation, is an excellent alternative for second-language learners. You may need to make sure that all students fully understand the directions given, so that they can complete the activity correctly.

Investigating Symmetry

From hummingbirds to hubcaps, from leaves to ladders, symmetry can be found all around you. Countless natural and artificial objects have symmetry.

PROJECT GOAL Identify objects that have symmetry and analyze transformations that preserve their shapes and sizes.

Reflection, Translation, and Rotation

1. Find five or more objects that exhibit symmetry. At least one of them should be something that is found in nature, and at least one should be artificial. Make sure you have examples of each type of symmetry (reflection, translation, and rotation).

2. Identify a way to transform each object so that the image coincides with the object. Measure the transformation as described below. Then sketch the object. Label the angles and distances you measured.

ROTATION Choose a point *R* on the object and locate point *S*, the image of *R* after a rotation. Find the point *O* around which the object is rotated and measure $\angle ROS$, the angle of rotation.

TRANSLATION Choose a point *T* on the object and locate point *U*, the image of *T* after a translation. Measure *TU*.

REFLECTION Choose a point *A* on the object and locate point *B*, the image of *A* after a reflection. Measure *AB* and find the midpoint *M* of $\overline{AB}$. Include the line (or plane) of symmetry in your sketch.

General Rubric for Projects
Each project can be evaluated in many possible ways. The following rubric is just one way to evaluate these open-ended projects. It is based on a 4-point scale.

4 The student fully achieves all mathematical and project goals. The presentation demonstrates clear thinking and explanation. All work is complete and correct.

3 The student substantially achieves the mathematical and project goals. The main thrust of the project and the mathematics behind it is understood, but there may be some minor misunderstanding of content, errors in computation, or weakness in presentation.

2 The student partially achieves the mathematical and project goals. A limited grasp of the main mathematical ideas or project requirements is demonstrated. Some of the work may be incomplete, misdirected, or unclear.

1 The student makes little progress toward accomplishing the goals of the project because of lack of understanding or lack of effort.

Presenting Your Results

Make a poster, write a report, or give a verbal presentation of your results. Include sketches of each object that you used. Label each sketch with the type of symmetry the object has. On your sketches, label the points that you used and the angles and distances that you measured.

You may also want to consider the following ideas:

- Do any of the objects have more than one type of symmetry? Tell how many types of symmetry each has and explain how you know.

- Choose one of the objects that has rotational symmetry. Are there other angles that can be used to rotate the object so that the image coincides with the object? Explain your answer.

- Do any of the objects that you measured have features that make them not truly symmetrical? Explain.

Extending Your Project

You can extend your project by examining some of the ideas below:

- Describe how symmetry is used by a group that you know, such as a marching band or gymnastics team. Use sketches or photographs to help you explain what types of symmetry are used.

- Ask a graphic designer, scientist, architect, or other professional how symmetry is used in his or her field.

Self-Assessment

Describe how you took the measurements of your objects. Were any of the objects particularly difficult to measure? Why? How did measuring and sketching the objects improve your understanding of symmetry?

Portfolio Project **49**

Guiding Students' Work

Rubric for Chapter Project

4 Students collect the required number of objects and accurately draw them on paper. Choices of points on the objects are made as are the measurements for the transformations on the sketches. The presentation of the results is well done, and it addresses the ideas and questions presented for the project. Students extend the project in one of the two ways listed and give an accurate and complete self-assessment of their work.

3 Students collect the required number of objects and accurately draw them on paper. Choices of points are made, but one of the transformations is not completed correctly or is missing some information regarding how it was done. Students make a presentation, but it does not completely convey all of the information about the project. Students extend the project in one of the two ways described and give a good self-assessment of their work.

2 Students collect the objects but make several mistakes in the sketches. Also, some of the information about the transformations is missing on the sketches. Students make a presentation, but it is evident they do not completely understand the mathematical ideas of the project because their presentation is incomplete or unclear. Students do not extend the project but do make an attempt at self-assessment.

1 Students collect only a few of the objects. They attempt to make sketches but do not complete them and the transformation being described is not evident. If a presentation is made, it is not complete and shows a lack of understanding of the mathematics of the project. No attempt is made to extend the project or make a self-assessment. Students should be encouraged to speak with the teacher as soon as possible to review their work and to make a new start on the project.

Progress Check 1.1–1.3

1. Draw the next figure in the pattern. *(Section 1.1)*

2. Copy the figure and reflect it over the line. *(Section 1.2)*

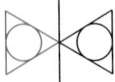

3. Identify the hypothesis and conclusion of the conditional statement. Then tell whether the conditional is *True* or *False*. If it is false, give a counter-example. *(Section 1.3)*
If $a < b$, then $a^2 < b^2$.

Hypothesis: $a < b$
Conclusion: $a^2 < b^2$
The conditional is false. Counterexamples may vary. An example is given. $-5 < 1$ is true, but it is not true that $(-5)^2 < 1^2$.

---

1 | Review

STUDY TECHNIQUE

What study techniques have you tried before? Write two brief paragraphs starting with these two phrases:

• To study for a mathematics test I usually . . .

• A study technique that has helped me in the past is . . .

VOCABULARY

inductive reasoning (p. 4)	**intersect** (p. 22)
reflect (p. 8)	**intersection** (p. 22)
rotate (p. 8)	**parallel** (p. 22)
translate (p. 8)	**skew** (p. 22)
transformation (p. 8)	**ray** (p. 28)
image (p. 8)	**endpoint** (p. 28)
symmetry (p. 10)	**segment** (p. 28)
line of symmetry (p. 10)	**congruent segments** (p. 28)
conjecture (p. 15)	**angle** (p. 34)
conditional statement (p. 15)	**vertex of an angle** (p. 34)
hypothesis (p. 15)	**congruent angles** (p. 35)
conclusion (p. 15)	**midpoint** (p. 41)
counterexample (p. 16)	**bisector of a segment** (p. 41)
collinear (p. 22)	**bisector of an angle** (p. 42)
coplanar (p. 22)	

SECTIONS 1.1, 1.2, *and* 1.3

Inductive reasoning involves identifying patterns and making a prediction based on those patterns.

     Prediction: next figure will be

Inductive reasoning may lead to a **conjecture**, which is a **conditional statement** that says if one thing is true, then something else will be true.

Consider the prime numbers: 2, 3, 5, 7, 11, 13, 17, 19, 23, . . .

You might conjecture: If **a number ends in 3**, then **the number is prime**.
                                    hypothesis   conclusion

This conjecture is false because it has a **counterexample**: 33 is not prime.

A **transformation** shifts a figure by **reflecting** it over a line or plane, **rotating** it around a given point, or **translating** it in a given direction. The new figure is called the **image**. A figure has **symmetry** if it coincides with its image after a transformation.

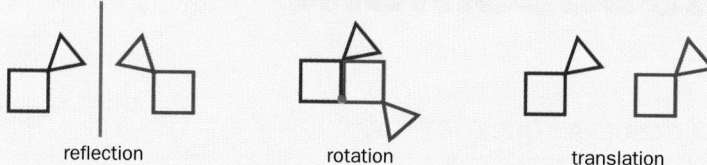

reflection   rotation   translation

SECTION 1.4

Points are **collinear** if they are on the same line.

Points or lines are **coplanar** if they lie on the same plane.

Two figures **intersect** if they have at least one point in common.

Parallel lines lie in the same plane but do not intersect.

Skew lines do not intersect and are not parallel.

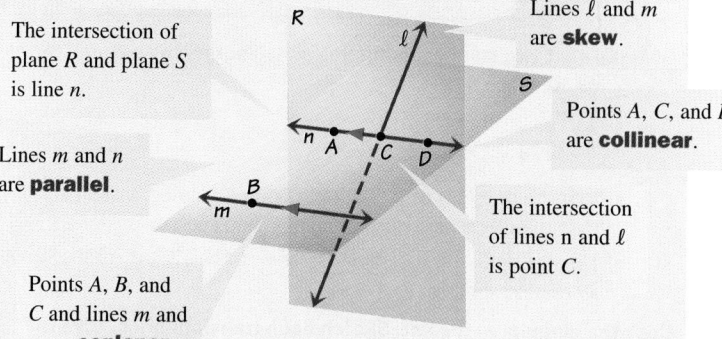

The intersection of plane R and plane S is line n.

Lines m and n are **parallel**.

Points A, B, and C and lines m and n are **coplanar**.

Lines ℓ and m are **skew**.

Points A, C, and D are **collinear**.

The intersection of lines n and ℓ is point C.

SECTIONS 1.5, 1.6, and 1.7

A **ray** is a part of a line with one endpoint.

Two rays with a common endpoint, or **vertex**, form an angle. The **bisector of an angle** is a ray that divides an angle into two congruent angles.

If $\overrightarrow{XV}$ bisects $\angle YXW$ and $m\angle YXW = 40°$, then $m\angle YXV = m\angle VXW = 20°$, and $\angle YXV \cong \angle VXW$.

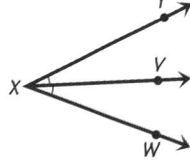

A **segment** is part of a line with two endpoints.

A **midpoint** divides a segment into two **congruent** segments. If S is the midpoint of $\overline{RT}$, then $\overline{RS} \cong \overline{ST}$ and $RS = ST$.

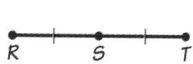

The segments are **congruent**.   The segments' lengths are **equal**.

If three points such as R, S, and T are collinear, then $RS + ST = RT$.

Review   **51**

Progress Check 1.4–1.5

Tell whether each statement is *True* or *False*. If false, sketch a counterexample. *(Section 1.4)*

1. Parallel lines are coplanar. True.

2. If points A, B, and C are collinear and A, D, and E are collinear, then A, B, C, D, and E are coplanar. True.

3. If lines l and m are skew and lines l and n are skew, then lines m and n are skew. False; counterexample:

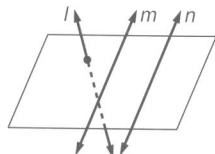

4. If point X is between points P and Q and $\overline{XP} \cong \overline{XQ}$, then $XP = \frac{1}{2}PQ$. *(Section 1.5)* True.

5. If $\overline{CD}$ is contained in $\overrightarrow{MN}$, then $\overleftrightarrow{CD}$ is contained in $\overrightarrow{MN}$. *(Section 1.5)* False; counterexample:

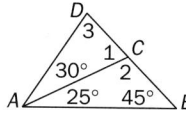

6. If $\overline{KL} \cong \overline{UW}$, $AB = 2KL$, and $EF = 2UW$, then $\overline{AB} \cong \overline{EF}$. *(Section 1.5)* True.

Progress Check 1.6–1.7

Give as many names as you can for each angle. *(Section 1.6)*

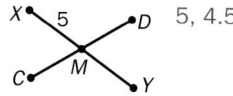

1. $\angle 3$ $\angle D$, $\angle ADC$, $\angle CDA$, $\angle ADB$, $\angle BDA$,

2. $\angle ACB$ $\angle 2$, $\angle BCA$

3. Find the measure of $\angle DAB$. *(Section 1.6)* 55°

4. M is the midpoint of $\overline{CD}$ and $\overline{XY}$, and $CD = 9$. Find MY and CM. *(Section 1.7)*

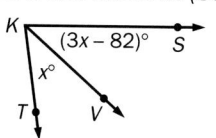

5, 4.5

5. $\overrightarrow{KV}$ bisects $\angle SKT$. Find the value of x and $m\angle SKT$. *(Section 1.7)*

$x = 41$
$m\angle SKT = 82°$

51

Chapter 1 Assessment
Form A Chapter Test

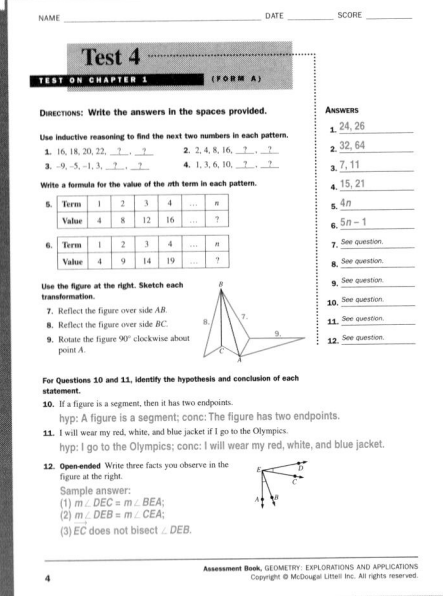

Chapter 1 Assessment
Form B Chapter Test

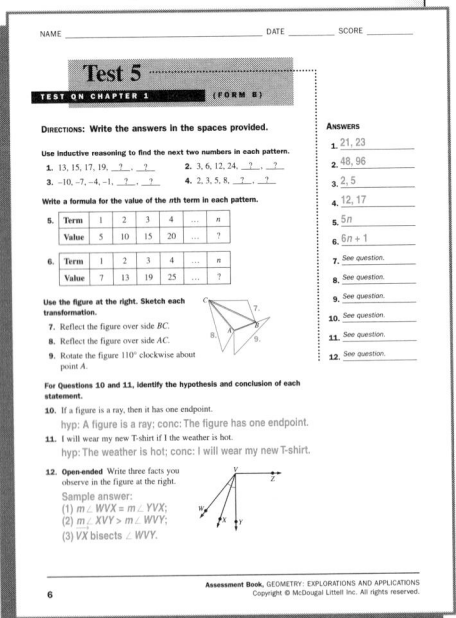

Assessment

VOCABULARY QUESTIONS

1. What kind of reasoning can help you identify patterns?

2. What are the two parts of a conditional statement?

3. After a transformation, what is the new figure called? What are three common transformations?

4. **Writing** Explain the difference between *congruent* and *equal* for two segments.

SECTIONS 1.1, 1.2, *and* 1.3

Give the next two numbers or figures and describe the pattern.

5. 4, 12, 36, 108, $\underline{?}$, $\underline{?}$

6.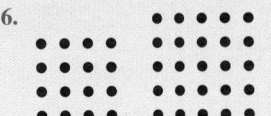

7. a. Rewrite the statement "All tall people are good basketball players" in if-then form. Identify the hypothesis and conclusion.

   b. Do you think the statement is *True* or *False*? If it is false, give a counterexample.

Copy the diagram at the left. Sketch each transformation. Give the coordinates of the vertices of the images.

8. reflect over the *x*-axis

9. translate 6 units up

10. rotate a half-turn about the origin

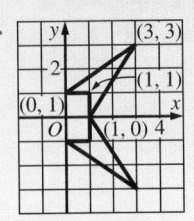

SECTION 1.4

For Exercises 11–14, tell whether each statement is *True* or *False*. If a statement is false, explain why.

11. Two intersecting lines are coplanar.

12. Three noncollinear points are always coplanar.

13. Two skew lines can be coplanar.

14. Two planes can intersect in a point.

ANSWERS Chapter 1

Assessment

1. inductive

2. the hypothesis (the "if") and the conclusion (the "then")

3. image; rotation, reflection, translation

4. *Congruent* refers to the segments, *equal* refers to their lengths. Two segments are congruent if their lengths are equal.

5. 324, 972

6.

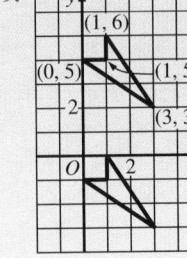

7. a. If a person is tall, then he or she is a good basketball player. hypothesis: a person is tall; conclusion: he or she is a good basketball player

   b. False; there are many counterexamples, including tall people who do not play basketball at all.

8.

9.

For Questions 15–19, use the diagram at the right.

15. Give at least three names for plane *ABC*.

16. What is the intersection of plane *ABE* and $\overleftrightarrow{EC}$?

17. What is the intersection of $\overleftrightarrow{AD}$ and $\overleftrightarrow{DE}$?

18. Name a pair of skew lines.

19. Name two lines that appear to be parallel.

Sketch each situation.

20. Line ℓ intersects $\overleftrightarrow{AB}$ at point *B*.

21. Plane *ABC* intersects line *m* at point *B*.

22. Noncollinear points *A*, *B*, *C*, and *D* lie in plane *H* and $\overleftrightarrow{AB} \parallel \overleftrightarrow{CD}$.

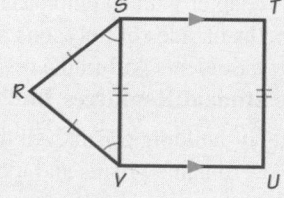

SECTIONS 1.5, 1.6, *and* 1.7

Use the diagram at the right.

23. List as many conclusions as you can about the diagram.

24. Describe a conclusion that looks correct but may not be.

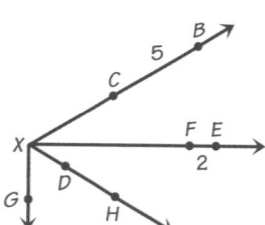

25. Point *C* is the midpoint of $\overline{XB}$ and $XB = XE$.
  a. Find *XC*. **b.** Find *XB*. **c.** Find *XF*.

26. $XD = 3y - 1$, $DH = 3$, and $XH = 4y + 1$. Find the value of *y* and the length *XH*.

27. $\overrightarrow{XE}$ bisects $\angle BXD$, $\overrightarrow{XD}$ bisects $\angle BXG$, and $m \angle BXF = 30°$. Find $m \angle BXD$ and $m \angle BXG$.

28. **Open-ended Problem** Draw two triangles. Measure the angles and sides of the triangles. Make a conjecture about the relationship between the shortest side and the smallest angle in any triangle. Draw another triangle and test your conjecture.

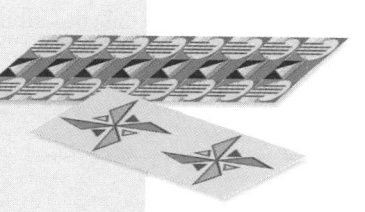

PERFORMANCE TASK

29. Design a border for wallpaper or fabric with a repeating pattern. Use as many of the following geometric relationships and objects as you can: angle bisector, parallel lines, midpoint, reflection symmetry, rotational symmetry, translational symmetry, vertex, congruent segments, congruent angles. Draw at least two repeats of your design. Label the geometric figures and relationships that you used.

Assessment **53**

Chapter 1
ALTERNATIVE ASSESSMENT

1. **Project** Cut pictures of objects from magazines that have the following characteristics. Make a well-labelled display of your pictures.
   a. 1 line of reflection symmetry
   b. 2 lines of reflection symmetry
   c. 4 lines of reflection symmetry

2. **Project** Take photographs of at least 3 objects in your neighborhood that have reflection symmetry, rotational symmetry, or translational symmetry. Describe how these symmetries are represented in your photographs.

3. **Open-ended Problem** Draw a figure that has 6 lines of reflection symmetry. What are the characteristics of a figure of this type?

4. **Performance Task** Which of the following can be bisected? Explain why or why not.
   a. ray
   b. segment
   c. line
   d. angle

5. **Performance Task** Draw a set of coordinate axes on a large piece of paper. Place your hand in the first quadrant and trace around it. Draw a reflection of your drawing over the *x*-axis. Now draw a reflection of this image over the *y*-axis. How is the final image related to the original tracing? Is there a single transformation that would have the same effect?

6. **a.** Find and describe a real-world example of each of the following: parallel lines, intersecting lines, skew lines, parallel planes, and intersecting planes.
   b. Why do you think there are no skew planes?

92

10.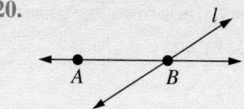

11. True.

12. True.

13. False; two lines in the same plane are either parallel or they intersect.

14. False; if two planes intersect, they intersect in a line.

15. plane *S*, plane *ABD*, plane *CBD* (The order of the letters can also be changed; for example, the plane could also be called plane *ACB*.)

16. point *E*

17. point *D*

18. Answers may vary. An example is given. $\overleftrightarrow{AB}$ and $\overleftrightarrow{CE}$

19. $\overleftrightarrow{AB}$ and $\overleftrightarrow{DC}$; $\overleftrightarrow{AD}$ and $\overleftrightarrow{BC}$

20–22. Answers may vary. Examples are given.

20.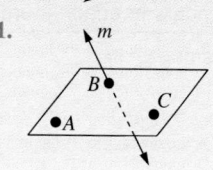

21.

22.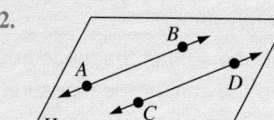

23. $\overline{RV} \cong \overline{RS}$, $\angle V \cong \angle S$, $\overline{SV} \cong \overline{TU}$, $\overline{ST} \parallel \overline{VU}$

24, 25. See answers in back of book.

26. 1; 5   **27.** 60°; 120°

28. Check students' work. The shortest side of a triangle is opposite the smallest angle.

29. Check students' work.

2 Triangles and Polygons

OVERVIEW

Connecting to Prior and Future Learning

⇔ Students begin their work in Chapter 2 by studying different types of angles. This basic concept is essential to students' understanding of geometry and mathematics in general.

⇔ Polygons are introduced in this chapter with an emphasis on triangles and parallelograms. The measures of angles and the lengths of sides of polygons are explored, as are the lengths of diagonals and lines of symmetry. Students will find a review of simplifying and evaluating expressions on page 690 in the **Student Resources Toolbox** helpful.

⇔ This chapter closes with a study of building prisms. All of the concepts presented in this chapter are used as students learn how to name prisms and identify their parts.

Chapter Highlights

Interview with Virgil Lueth: The use of mathematics in geology is highlighted in this interview with Virgil Lueth. The corresponding exercises on pages 68, 82, 85, and 98 allow students to explore this application even further.

Explorations in Chapter 2 include sorting triangles by certain rules created by students in Section 2.2, using dot paper or a geoboard and a graphing calculator or graph paper to find the measures of the angles in polygons in Section 2.4, and using notebook paper, patty paper, a ruler, and a protractor to analyze parallelograms in Section 2.5.

The Portfolio Project: In this project on Platonic solids, students discover the regular solids as they build them from each of several regular polygons. After building each Platonic solid, students make nets and use them to investigate different properties of each solid.

Technology: Students use geometry software to explore the relationships between equiangular and equilateral triangles in Section 2.2. In Section 2.4, graphing calculators are used to graph data describing the sum of the measures of the angles in polygons.

OBJECTIVES

Section	Objectives	NCTM Standards
2.1	• Identify and classify angles. • Estimate and calculate the measures of angles.	1, 2, 3, 4, 7
2.2	• Name, label, and classify triangles. • Sketch triangles and find the measures of their angles. • Use types of triangles to describe real-world objects.	1, 2, 3, 4, 7
2.3	• Describe the properties of a figure. • Identify polygons.	1, 2, 3, 4, 7
2.4	• Find the sum of the measures of the interior angles of a polygon. • Find the sum of the measures of the exterior angles of a polygon. • Find specific measures for a given polygon.	1, 2, 3, 4, 7
2.5	• Identify types of quadrilaterals. • Find the measures of angles and segments in quadrilaterals.	1, 2, 3, 4, 7
2.6	• Name prisms and identify parts of prisms. • Sketch three-dimensional objects. • Analyze nets.	1, 2, 3, 4, 7

INTEGRATION

Mathematical Connections	2.1	2.2	2.3	2.4	2.5	2.6
geometry	**57–63***	**64–71**	**72–78**	**79–86**	**87–92**	**93–99**
algebra		66–68		80, 84, 86	88	
patterns and functions			76			99
logic and reasoning	61	69	76, 78	83, 85, 86	91, 92	96, 98

Interdisciplinary Connections and Applications						
history and geography		70				
biology and earth science	61					
arts and entertainment				83		
sports and recreation		71		80		97
communications	62					
crafts, optics			77		91	

*__Bold page numbers__ indicate that a topic is used throughout the section.

TECHNOLOGY

Section	opportunities for use with	
	Student Book	**Support Material**
2.1		**Technology Book:** Spreadsheet Activity 2
2.2	geometry software	
2.3		**Technology Book:** Spreadsheet Activity 2 **Geometry Inventor Activity Book:** Activity 9
2.4	graphing calculator	**Technology Book:** Calculator Activity 3 **Geometry Inventor Activity Book:** Activity 11
2.5		**Technology Book:** TI-92 Activity 2
2.6		**Technology Book:** Spreadsheet Activity 2

Regular Scheduling (45 min)

Section	Materials Needed	Core Assignment	Extended Assignment	exercises that feature		
				Applications	Communication	Technology
2.1		1–4, 9–19, 27–35	1–35	5–8, 20–26, 33	8, 9, 26, 29, 31c	
2.2	triangles, compass, straightedge, geometry software or ruler, protractor	**Day 1:** 1–11, 14–24 **Day 2:** 25–29, 33–36, AYP*	**Day 1:** 1–24 **Day 2:** 25–36, AYP	12, 13 25–30	5, 13, 17, 18	31, 32
2.3	transparent mirror, scissors	1–10, 17–30	1–30	11–16	6, 8b, 14, 16, 26	
2.4	geoboard or dot paper, rubber bands, graphing calculator or graph paper	**Day 1:** 1–9, 12–16, 19–22 **Day 2:** 24–35, 42–45, AYP	**Day 1:** 1–23 **Day 2:** 24–45, AYP	10, 11 36–41	10, 11 37c, 41–43	
2.5	lined paper, patty paper or tracing paper, ruler, protractor, scissors	1–10, 16–30	1–30	13–15	10, 13, 16	
2.6		**Day 1:** 1–16 **Day 2:** 17–21, 24, 27, 29–33, AYP	**Day 1:** 1–16 **Day 2:** 17–33, AYP	23, 25, 26	11, 13 24, 27b, 29	
Review/ Assess		**Day 1:** 1–15 **Day 2:** 16–31 **Day 3:** Ch. 2 Test	**Day 1:** 1–15 **Day 2:** 16–31 **Day 3:** Ch. 2 Test			
Port. Proj.		Allow 2 days.	Allow 2 days.			

Yearly Pacing (with Portfolio Project)	Chapter 2 Total 14 days	Chapters 1–2 Total 28 days	Remaining 132 days	Total 160 days

Block Scheduling (90 min)

	Day 8	Day 9	Day 10	Day 11	Day 12	Day 13	Day 14
Teach/Interact	2.1 2.2: Exploration, page 64 Construction, page 69	Continue with 2.2 2.3: Construction, page 77	2.4: Exploration, page 79	2.5: Exploration, page 87 2.6	Continue with 2.6 Review	Review Port. Proj.	Ch. 2 Test Port. Proj.
Apply/Assess	**2.1:** 1–4, 9–19, 27–35 **2.2:** 1–11, 14–24	**2.2:** 25–29, 33–36, AYP* **2.3:** 1–10, 17–30	**2.4:** 1–9, 12–16, 19–22, 24–35, 42–45, AYP	**2.5:** 1–10, 16–30 **2.6:** 1–16	**2.6:** 17–21, 24, 27, 29–33, AYP **Review:** 1–15	**Review:** 16–31 **Port. Proj.**	**Ch. 2 Test Port. Proj.**

NOTE: A one-day block has been added for the Portfolio Project—timing and placement to be determined by teacher.

Yearly Pacing (with Portfolio Project)	Chapter 2 Total 7 days	Chapters 1–2 Total 14 days	Remaining 66 days	Total 80 days

*__AYP__ is Assess Your Progress.

Section	Practice Bank	Study Guide*	Assessment Book*	Visuals	Explorations Lab Manual	Lesson Plans	Technology Book
2.1	9	2.1		Warm-Up 2.1 Folder 2		2.1	Spreadsheet Act. 2
2.2	10	2.2	Test 6	Warm-Up 2.2 Folder 3	Master 8	2.2	
2.3	11	2.3		Warm-Up 2.3	Add. Expl. 2	2.3	Spreadsheet Act. 2
2.4	12	2.4	Test 7	Warm-Up 2.4	Add. Expl. 3 Masters 2, 3, 9	2.4	Calculator Act. 3
2.5	13	2.5		Warm-Up 2.5		2.5	TI-92 Act. 2
2.6	14	2.6	Test 8	Warm-Up 2.6		2.6	Spreadsheet Act. 2
Review Test	15	Chapter Review	Tests 9, 10, Alternative Assessment			Review Test	

*Spanish versions of *Study Guide* and *Assessment Book* are available.

Chapter Support

- Course Guides
- Lesson Plans
- Preparing for College Entrance Tests
- Multi-Language Glossary
- *Test Generator* Software
- Professional Handbook
- Challenge Problems

Software Support

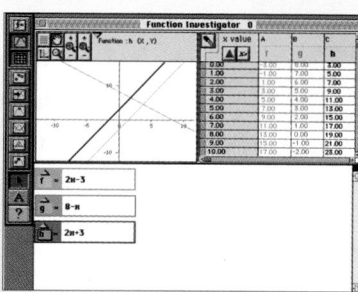

McDougal Littell Mathpack
Geometry Inventor

Internet Support

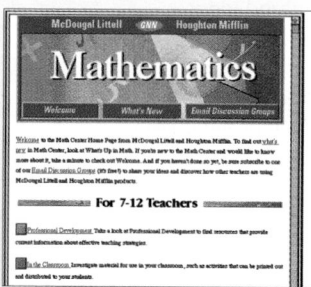

http://www.hmco.com
Next go to McDougal Littell; then the Education Center; then Secondary Math.

Books, Periodicals

Zbiek, Rose Mary. "The Pentagon Problem: Geometric Reasoning with Technology." *Mathematics Teacher* (February 1996): pp. 86–90.

Activities, Manipulatives

Miller, William A. and Robert G. Clason. "Golden Triangles, Pentagons, and Pentagrams." *Mathematics Teacher* (May 1994): pp. 338–353.

Software

Jackiw, Nicholas, designer. *The Geometer's Sketchpad.* (Ver. 3.0) for Macintosh or Windows. Berkeley, CA: Key Curriculum Press, 1995.

Laborde, Jean-Marie and Franck Bellemain, designers. *Cabri Geometry II.* Dallas, TX: Texas Instruments, 1994.

TI-92 Geometry. (Resident on the TI-92 graphing calculator.) Dallas, TX: Texas Instruments, 1995.

Videos

Geometry: New Tools for New Technologies. Hour Two, "Structures: Will It Fall Down?" Arlington, VA: COMAP.

Internet

For information about the Connected Geometry project and materials available from Janson Publications, Dedham, MA, visit the World Wide Web site at
http://www.edc.org/LTT/ConnGeo/

2 Triangles and Polygons

Exploring CRYSTALS

INTERVIEW Virgil Lueth

Classifying Crystals

In crystallography, crystals are classified according to symmetry. There are seven general crystal systems, each of which may be described in terms of imaginary axes, called *crystallographic axes*. All the axes intersect in the center of the crystal and extend through the centers of the sides, not through the edges. In this way, the axes show the directions followed by the crystal's edges.

Virgil Lueth

Virgil Lueth grew up in western Wisconsin in a small town called Spring Valley. Both of his parents were schoolteachers. He moved to Spring Valley when he was five years old, which is when his interest in crystals began. In college, he decided to major in geology and studied many specialized ideas regarding crystallography. During his studies, he began to realize that crystals are, in a sense, nature's artwork and that although they are the same in many ways, they are also different. Before he began work at his current job, Lueth was a college professor for five years. In his work today, he finds it gratifying to work with children because of their eagerness to learn. When asked what he thinks young people like most about crystals, he cannot name only one thing. For some it is their shape, for some the color, and for others it is relating the crystal to something else.

"The crystals in the rocks were always intriguing."

"**C**rystals are everywhere," explains Virgil Lueth, state mineralogist of New Mexico and crystallographer. "All gems are crystals. Rocks are made up of crystals. Crystals are the basis of everything."

Each type of crystal has a predictable shape because the angles found in the molecules determine the angles formed by the faces of the crystal. "Ever notice when you look at snowflakes that they're always six-sided, or hexagonal?" Lueth asks. "When you freeze water into ice, the water molecules line up. They always line up in the same way and repeat these six-sided patterns."

Lueth got hooked on crystals when he was very young. "I always liked rocks, and where I grew up there were a lot of little cavities in the rocks that were lined with quartz crystal. The crystals in the rocks were always intriguing to me."

54

A Multifaceted Career

After studying geology in college and working as a professor, Lueth became state mineralogist for New Mexico. What does Lueth do now? His job is anything but dull. "I identify rocks and minerals for the public free of charge. I do the same thing for my scientific colleagues. I also maintain the museum at the Bureau of Mines. Here we exhibit classic minerals from around the world and especially New Mexico," Lueth says.

> "When I'm telling people about minerals and crystallography, that's pure enjoyment for me."

"I also do geological and geochemical research aimed at finding new ore deposits," he continues. "I do field work mostly in New Mexico, and I map the rocks, using geometry to determine the orientation of rocks in three dimensions. If there's a gold deposit down there, you want to know exactly where it is in order to drill for it and dig it out." A favorite part of Lueth's job is explaining the science he loves to visitors—everyone from young children and college students to senior citizens. "When I'm telling people about minerals and crystallography, that's pure enjoyment for me," Lueth says.

55

Background

The Bureau of Mines

The Bureau of Mines is an agency of the United States Department of the Interior. Created in 1910, the bureau has a number of responsibilities relating to mining and minerals. One responsibility is to ensure that the United States has enough copper, lead, silver, zinc, and other nonfuel minerals. Another is to conduct research to develop safe and economical methods for mining, processing, using, and recycling minerals. A third is to investigate ways to decrease land damage caused by mining. The Bureau of Mines also publishes statistics on various areas.

Second-Language Learners

You may need to define the following terms before students read the interview: *crystals* (minerals with a special structure), *crystallographer* (a scientist who studies crystals), *mineralogist* (a scientist who studies minerals), *gems* (certain minerals and other materials that are valued for their beauty), *ore deposits* (places in the earth where minerals are found), *opaque* (not shiny; unable to be seen through), *metallic luster* (a shine like metals have).

Multicultural Note

Minerals found in parts of the Southwest have been used in Native American jewelry for well over a thousand years. Turquoise is the most common stone used, but azurite, malachite, jet, serpentine, lapis lazuli, and others are also popular. Mexicans introduced metalworking to the Navajos in the mid-1800s, who then passed on the craft to the Pueblo peoples. The Navajo, Hopi, and Zuni peoples have developed unique styles of working silver and gold and shaping stones to create their pieces of jewelry. Today, about 65% of the residents of the Zuni reservation in New Mexico are involved in jewelry making.

Mathematical Connection

In crystallography, classifying crystals and understanding their structure is a necessary and important part of the field. In order to classify crystals, crystallographers must understand the definitions and meanings associated with the many terms involved. Many prefixes and suffixes are used in different combinations to describe a wide variety of crystals. In Section 2.2, students look at different crystals and try to use the terminology learned in the section to describe the forms of crystals. Another way to study crystals is to consider some of the angle measures in them. In Example 3 of Section 2.4, students find the measures of interior and exterior angles of one kind of crystal. Also in the exercises of this section, students explore other angle relationships in other forms of crystals. Then, in Section 2.6, students look at the molecular structure of crystals, the shapes these molecules form, and the angle relationships created in the structure.

Explore and Connect

Project
This activity can help students understand more about the field of crystallography. It can also help them with the extension of the project at the end of the chapter, where one of their choices is to make a goniometer.

Research
Students should be able to find many books containing information about rocks and minerals in a library. Encyclopedias may also be a good source for this research.

Writing
You may want to suggest that students keep their responses to these questions in their journals. Then, when cubes are discussed, they can compare their definitions to the one given in the book.

The Geometry of Crystals

Crystallographers study the organization of atoms and molecules in crystals. To unlock this hidden geometry, they measure the crystal's flat surfaces. "We can do it by measuring the angles between faces using what's called a contact goniometer," Lueth explains. The crystallographer rests the base of the goniometer on one crystal face and the movable arm on the adjoining face. The arm points at a scale that gives the angle between the two faces.

"By compiling these angles," Lueth says, "we decide which geometric system the crystal belongs to. The angles measured correspond to specific forms which are characteristic of particular crystal systems. For example, right angles are characteristic of the cubic system."

Using these facts, crystallographers can determine the exact internal structure of the crystal and what forms are possible—key information in nearly every area of science and technology.

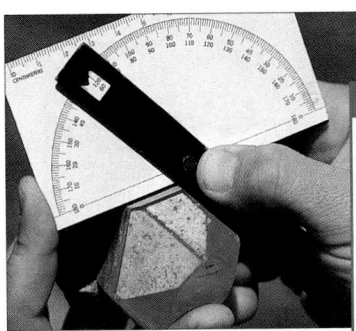

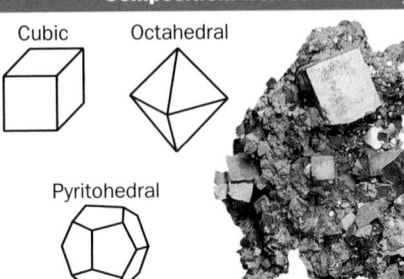

Pyrite
Composition: Iron sulfide FeS$_2$

Pyrite is the most common sulfide. Its symmetry is isometric, which means its axes are the same length and meet at right angles. Pyrite usually takes the form of a cube, though it can also be octahedral or pyritohedral.

Cubic   Octahedral   Pyritohedral

Explore and Connect

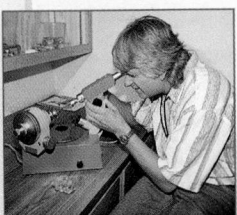

The machine above is another kind of goniometer used to measure much smaller crystals.

1. Project Find out more about goniometers. Then make your own goniometer and use it to measure the angles of everyday objects, like the roof of a birdhouse.

2. Research What types of rocks and minerals are found in your area? Choose a mineral and learn about its structure. Summarize your research in a written report or an oral presentation.

3. Writing The pyrite crystal often takes the form of a cube. What type of angles does a cube have? What shape is each face of a cube? Write a possible definition of a cube.

Mathematics & Virgil Lueth

In this chapter, you will learn more about how mathematics is related to crystallography.

Related Examples and Exercises

Section 2.2
• Exercises 12 and 13

Section 2.4
• Example 3
• Exercises 36–41

Section 2.6
• Exercises 25 and 26

2.1 Types of Angles

The beams of this bridge over the Kwai River in Thailand form many angles. When describing structures like this bridge, you will find it helps to be able to name and recognize different kinds of angles.

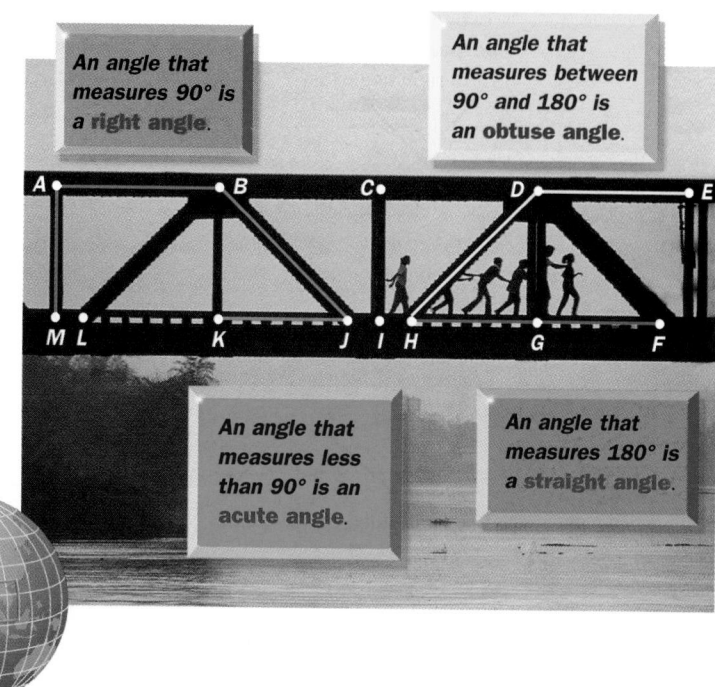

An angle that measures 90° is a right angle.

An angle that measures between 90° and 180° is an obtuse angle.

An angle that measures less than 90° is an acute angle.

An angle that measures 180° is a straight angle.

Thailand
Bangkok

Kwai River

THINK AND COMMUNICATE

Use the photograph of the bridge.

1. Find an acute angle, an obtuse angle, a right angle, and a straight angle different from those used in the definitions above.

2. In the diagram, ∠BCI and ∠ICD are both right angles. Explain why ∠BCD must be a straight angle.

3. In the diagram, ∠CDG is a right angle.
   a. Explain why ∠CDF must be an obtuse angle.
   b. Explain why ∠CDH and ∠HDG must be acute angles.

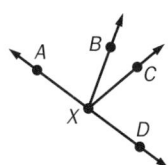

Section Notes

Teaching Tip
Ask students if an angle can be complementary or supplementary to itself. (Yes, in both cases.) What is the measure of any angle that is complementary to itself? (45°) supplementary to itself? (90°)

Geometric Thinking
Ask students if they think all angles have complements and supplements and to explain their reasoning. (If an angle measure is 90° or greater, then there is no angle measure that can be added to it to get 90°. Therefore, only acute angles have complements. With the exception of straight angles, all angles have supplements. If $m\angle A < 180°$, then any angle with measure $180° - m\angle A$ is a supplement of $\angle A$.)

Common Error
In subsequent work in this section, some students may confuse the terms *complementary* and *supplementary*. To help correct this error, review the definitions of these terms on this page. You can also provide a memory tip such as the following. Alphabetically, the first letter of *complementary* comes before the first letter of *supplementary* and the number 90 comes before 180.

Additional Example 1

a. Find the measures of a complement and a supplement of $\angle ABC$.

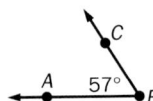

Complement: $90° - 57° = 33°$
Supplement: $180° - 57° = 123°$

b. Sketch $\angle ABC$ and its supplement adjacent to one another.
Sketch $\angle ABC$. Draw the line of which $\overrightarrow{BA}$ is a part. Let D be a point on $\overleftrightarrow{AB}$ that is not on $\overrightarrow{BA}$. Because $\angle ABD$ is a straight angle, $\angle CBD$ and $\angle ABC$ are supplements.

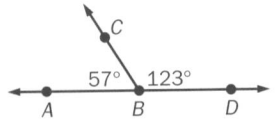

You can classify a pair of angles by looking at points and rays in common or by finding the sum of their measures.

Adjacent angles are two coplanar angles that share a vertex and a side but do not overlap.

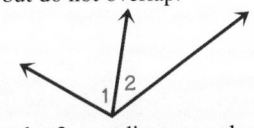

$\angle 1$ and $\angle 2$ are adjacent angles.

Two adjacent angles form a **linear pair** if their nonshared rays form a straight angle.

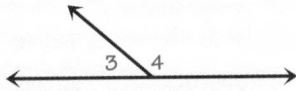

$\angle 3$ and $\angle 4$ form a linear pair.

Two lines, segments, or rays are **perpendicular** if they intersect to form right angles.

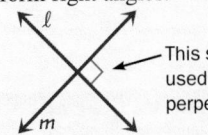

This symbol is used to indicate perpendicular lines.

The statement $\ell \perp m$ is read, "ℓ is perpendicular to m."

Vertical angles are non-adjacent, non-overlapping angles formed by two intersecting lines.

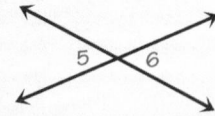

$\angle 5$ and $\angle 6$ are vertical angles.

Two angles are **complements** of each other if their measures add up to 90°.

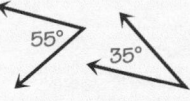

They are called *complementary angles*.

Two angles are **supplements** of each other if their measures add up to 180°.

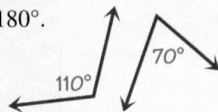

They are called *supplementary angles*.

EXAMPLE 1

a. Find the measures of a complement and a supplement of $\angle SPT$.
b. Sketch $\angle SPT$ and its complement adjacent to one another.

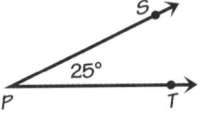

SOLUTION

a. Complement: $90° - 25° = 65°$
Supplement: $180° - 25° = 155°$

b. Sketch $\angle SPT$. Draw a ray perpendicular to $\overrightarrow{PT}$ at P. Because $\angle RPT$ is a right angle, $\angle RPS$ and $\angle SPT$ are complements.

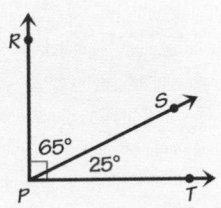

58   Chapter 2 *Triangles and Polygons*

The position of some angles can tell you how their measures are related. The needle on this fuel gauge forms a linear pair with the base of the gauge. The two angle measures change when the needle moves, but their sum is always 180°.

24° + 156° = **180°**   110° + 70° = **180°**

Linear Pair Property

The angles that form a linear pair are supplementary.

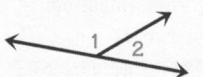

If $\angle 1$ and $\angle 2$ form a linear pair, then $m\angle 1 + m\angle 2 = 180°$.

These scissors form vertical angles. The angle between the blades stays congruent to the angle between the handles when you open and close the scissors.

Vertical Angles Property

Vertical angles are congruent.

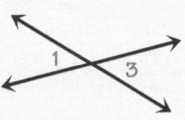

If $\angle 1$ and $\angle 3$ are vertical angles, then $\angle 1 \cong \angle 3$.

EXAMPLE 2

Find the value of each variable. Give the properties that you use.

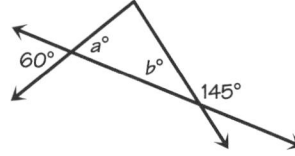

SOLUTION

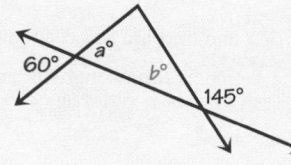

Vertical angles are congruent.

$a = 60$

Angles in a linear pair are supplementary.

$b = 180 - 145$
$b = 35$

Section Notes

Spatial Reasoning
Point out that the term *linear pair* can be used only for adjacent supplementary angles. Have students use cutouts or tracings to show that any two supplementary angles can be positioned so that they form a linear pair. Similarly, any two complementary angles can be positioned to be adjacent. In their new positions, the nonshared sides of the adjacent angles form a right angle.

Topic Spiraling: Preview
The Vertical Angles Property is presented as a property without proof; that is, students are being asked to *assume* that vertical angles are congruent. Other properties of geometric figures are also presented without proof in this chapter. This approach has the advantage of introducing students to some interesting properties on an intuitive level early in the course. These properties can then be proved when the concept of proof is introduced in the next chapter.

Geometric Thinking
Most students will find the Linear Pair Property and Vertical Angles Property intuitively clear and, thus, true from their diagrams. You may wish, however, to engage students in an informal discussion of *why* these two properties are true. They are easily derived from the definitions on page 58 and the properties of equality that students know from previous mathematics courses.

Additional Example 2

Find the value of each variable. Give the properties that you use.

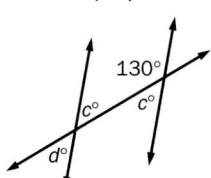

Angles in a linear pair are supplementary.
$c = 180 - 130$
$c = 50$
The angles labeled $c°$ and $d°$ are vertical angles, and vertical angles are congruent. Since $c = 50$, then $d = 50$.

Think and Communicate

Some students may point out that they can make part (c) of question 4 a true statement by using the words *supplementary* or *congruent*. While this is a correct observation, the sequence of statements for parts (a)–(c) is more interesting if students complete statement (c) with the word *congruent*. Then they have a statement that follows as a *logical conclusion* from (a) and (b). The results of questions 5 are summarized below it and generalized in the discussion about angles formed by perpendicular lines.

Section Note

Topic Spiraling: Preview/Review
The discussion in *Think and Communicate* question 5 previews the work in Chapter 3 on reasoning in geometry. Also, the two conditional statements about angles formed by perpendicular lines can be used to review the ideas of conditional statement, hypothesis, and conclusion from Chapter 1.

Additional Example 3

Tell whether each statement is *True* or *False*. Explain your reasoning.

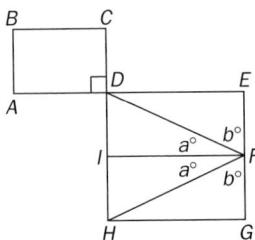

a. $\angle CDE \cong \angle IDE$

True. Since $\overline{CH} \perp \overline{AE}$, $\overline{CH}$ and $\overline{AE}$ form congruent adjacent angles. $\angle CDE$ and $\angle IDE$ are congruent angles formed by these segments.

b. $\overline{IF} \perp \overline{EG}$

True. Since $m\angle IFE = (a + b)°$ and $m\angle IFG = (a + b)°$, then $\angle IFE$ and $\angle IFG$ are congruent adjacent angles. Therefore, $\overline{IF} \perp \overline{EG}$.

THINK AND COMMUNICATE

Complete each statement.

4. a. All right angles measure $\underline{\ ?\ }°$.

   b. Angles with the same measure are $\underline{\ ?\ }$ to each other.

   c. All right angles are $\underline{\ ?\ }$ to each other.

5. Use the diagram at the right.

   a. Since $\overrightarrow{CB} \perp \overleftrightarrow{AD}$, $m\angle ACB = \underline{\ ?\ }°$.

   b. Since $\angle ACB$ and $\angle BCD$ form a linear pair, $m\angle ACB + m\angle BCD = \underline{\ ?\ }°$.

   c. Therefore, $m\angle BCD = \underline{\ ?\ }°$.

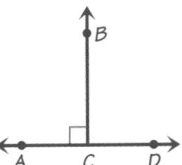

Notice that rays and segments can also be perpendicular, if they intersect to form right angles. By definition, perpendicular lines form right angles. Since all right angles are congruent, the angles formed by perpendicular lines are congruent.

Angles Formed by Perpendicular Lines

If two lines are perpendicular, then they form congruent adjacent angles.

If two lines form congruent adjacent angles, then the lines are perpendicular.

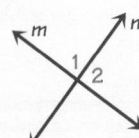

If $m \perp n$, then $\angle 1 \cong \angle 2$.
If $\angle 1 \cong \angle 2$, then $m \perp n$.

EXAMPLE 3

Tell whether each statement is *True* or *False*. Explain your reasoning.

a. $\overline{AJ} \perp \overline{IB}$

b. $\angle CGH \cong \angle HGK$

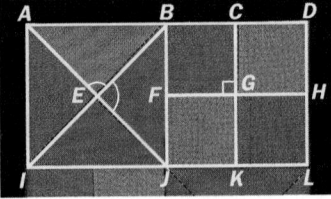

SOLUTION

a. True. Since $\angle AEB \cong \angle BEJ$ and they are adjacent angles, $\overline{AJ} \perp \overline{IB}$.

b. True. Since $\overline{CK} \perp \overline{FH}$, the segments form congruent adjacent angles.

Think and Communicate

4. a. 90

   b. congruent

   c. congruent

5. a. 90

   b. 180

   c. 90

✓ CHECKING KEY CONCEPTS

For Questions 1–3, find the measures of a complementary angle, a supplementary angle, and a vertical angle for the given angle.

1. $m \angle Z = 40°$ 2. $m \angle W = 82°$ 3. $m \angle Y = x°$

4. The measure of $\angle ABC$ is 80°. $\angle ABC$ and $\angle CBD$ form a linear pair.
   a. Sketch $\angle ABC$ and $\angle CBD$.
   b. What is the measure of $\angle CBD$? How do you know?

5. Complete the following statement with *always*, *sometimes*, or *never*.
   Lines that form supplementary vertical angles are ? perpendicular.

2.1 | Exercises and Applications

Extra Practice
*exercises on
page 661*

Complete each statement.

1. $\angle AHB$ and ? are vertical angles.

2. $\angle BJG$ and ? are congruent adjacent angles.

3. $\angle FJE$ and ? are complementary angles.

4. $\angle GJE$ and ? are supplementary angles that do not form a linear pair.

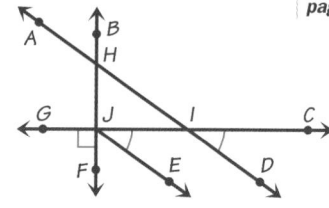

Connection ▸ BIOLOGY

angle of crest

Scientists who study birds have noticed that the angle of the crest of the Stellar's jay can indicate the mood or activity of the bird. For Exercises 5–7, match the description of the crest with the corresponding picture.

5. The angle of the crest is close to 0° during courtship.

6. While eating sunflower seeds, the angle of the crest is about 30°.

7. The crest is almost at a right angle during aggression.

8. **Writing** What type of angles can the crest of a Stellar's jay form? Explain.

A.

B.

C.

2.1 Types of Angles   **61**

Checking Key Concepts

Geometric Thinking
When discussing question 3, point out that $\angle Y$ has a complement only if $x < 90$ and a supplement only if $x < 180$. For question 5, students should be able to explain how they decided on their answer.

Closure Question

Suppose $\angle A$ is an acute angle. How many degrees difference is there between the measure of a complement of $\angle A$ and the measure of a supplement of $\angle A$? 90°

Apply⇔Assess

Suggested Assignment

❖ **Core Course**
  Exs. 1–4, 9–19, 27–35
❖ **Extended Course**
  Exs. 1–35
❖ **Block Schedule**
  Day 8 Exs. 1–4, 9–19, 27–35

Exercise Notes

Communication: Reading
Exs. 1–4 These four exercises contain six mathematical terms. Have students read aloud their answers with each statement to informally assess their understanding of these basic terms.

Interdisciplinary Problems
Exs. 5–8 These exercises illustrate a fascinating fact that the geometric concept of an angle can be used to understand the mood or activity of a Stellar's jay simply by observing the angle of its crest.

Checking Key Concepts

1. complementary: 50°; supplementary: 140°; vertical: 40°

2. complementary: 8°; supplementary: 98°; vertical: 82°

3. complementary, $x < 90°$: $90° - x°$; supplementary, $x < 180°$: $180° - x°$; vertical: $x°$

4. a.

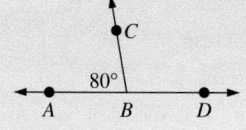

b. 100°; The sum of the measures of the angles in a linear pair is 180°.

5. always

Exercises and Applications

1. $\angle FHD$

2. $\angle BJC$ or $\angle GJF$

3. $\angle CJE$

4. $\angle CID$

5. B   6. A

7. C

8. Answers may vary. An example is given. The angles of the crest are between 0° and 90°, so they are acute.

Exercise Notes

Spatial Reasoning
Ex. 9 By applying the definition of adjacent angles to these three diagrams, students should be able to explain why the given pairs of angles are not adjacent.

Cooperative Learning
Exs. 10–19 Some students may enjoy working in small groups to make up exercises similar to these for group discussion. Students who do so should be careful to use the terminology correctly. In particular, they should keep in mind when discussing complementary angles and supplementary angles that these kinds of angles come in *pairs*. Thus, in the sample diagram below, it would be incorrect to ask whether ∠BXC, ∠CXD, and ∠DXE are complementary, since only two angles can be complementary.

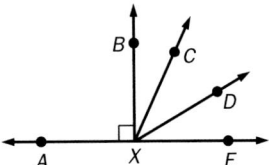

A good question would be whether the sum of the measures of ∠BXC, ∠CXD, and ∠DXE is 90°.

Application
Exs. 20–26 These exercises help students to review the names of different angles in an applied setting. Ask students what number between 1 and 7 is missing and how would it be shown.

Communication: Writing
Ex. 29 Students will probably find it easier to provide a good response to this exercise if they sketch a diagram to accompany their explanation.

---

9. Writing Explain why the angles in each pair are *not* adjacent.

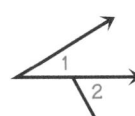

a. ∠1 and ∠2   b. ∠3 and ∠4   c. ∠ABC and ∠ABD

Find the measure of each angle in the diagram at the right. Tell which properties or definitions you use.

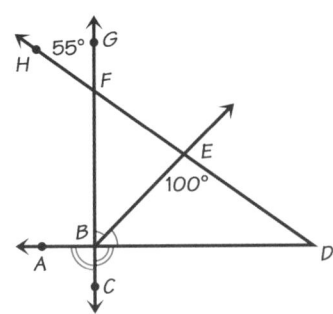

10. ∠EFB   11. ∠FEB   12. ∠ABC

13. ∠EBD   14. ∠GFE   15. ∠ABE

Use the diagram at the right. Tell whether each statement is *True* or *False*.

16. ∠FEB and ∠BED are complements.

17. ∠HFB ≅ ∠GFE

18. ∠ABG and ∠HFB are adjacent angles.

19. ∠ABC and ∠EBD are vertical angles.

Connection   COMMUNICATION

Semaphore is a signaling system that uses flags to send messages. The signaler holds a flag in each hand and positions his or her arms differently for each number or letter. For each number given in Exercises 20–25, tell what type of angle is formed by the raised arm and the side of the body.

20. 1   21. 2   22. 3   23. 5   24. 6   25. 7

26. a. **Open-ended Problem** Write instructions for signaling one of the numbers shown. Describe the position of each arm by the type of angle it makes with the side of the body.

b. **Cooperative Learning** Have a classmate follow the instructions you wrote in part (a) and guess which number you chose. If your partner finds the instructions unclear, work together to revise them.

62   Chapter 2 *Triangles and Polygons*

---

9. a. The angles do not share a vertex.

b. The angles do not share a side.

c. The angles overlap.

10. 55°; Vertical Angles Property

11. 80°; Linear Pair Property

12. 90°; If two lines form congruent adjacent angles, then the lines are perpendicular.

13. 45°; If two lines form congruent adjacent angles, then the lines are perpendicular; Linear Pair Property or Vertical Angles Property; definition of complementary angles.

14. 125°; Linear Pair Property

15. 135°; Linear Pair Property (using Ex. 13)

16. False.   17. True.

18. False.   19. False.

20. acute   21. right

22. obtuse   23. obtuse

24. right   25. acute

26. a. Answers may vary. An example is given. To signal the number "7," hold the flag in your right arm straight downward, while holding the flag in your left arm to the left side of the body at an angle of about 45° down from horizontal.

b. Check students' work.

27. 140°

28. D

29. They are complementary. Because the rays are perpendicular, they meet at a right angle. Then the sum of the measures of the adjacent angles is 90°; that is, they are complementary.

27. What is the measure of a supplement of a complement of a 50° angle?

28. SAT/ACT Preview What is the measure of a complement of a supplement of a 140° angle?
    A. 40°   **B.** 90°   **C.** 140°   **D.** 50°   **E.** 180°

29. Writing The nonshared rays of two adjacent acute angles are perpendicular. How are their measures related? Explain your reasoning.

30. a. ∠BFC and ∠CFD are complementary angles. Find m∠CFD.
    b. ∠AFB and ∠BFC are complementary angles. Find m∠AFB.
    c. Make a conjecture about the measures of two angles that are complementary to the same angle. Explain your reasoning.

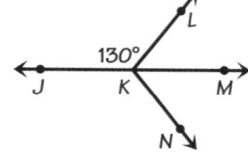

31. a. Find m∠LKM. How did you get your answer?
    b. ∠JKL and ∠MKN are supplementary angles. Find m∠MKN.
    c. Make a conjecture about the measures of two angles that are supplementary to the same angle. Explain your reasoning.

32. Challenge Show that $\overleftrightarrow{TU} \perp \overleftrightarrow{PW}$.

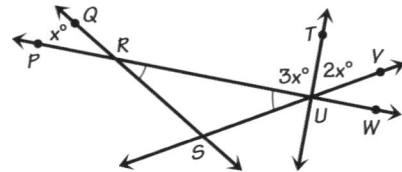

ONGOING ASSESSMENT

33. Open-ended Problem Sketch a building, a bridge, or a stained glass window. Include and label examples of acute, obtuse, right, straight, vertical, supplementary, and complementary angles, perpendicular segments, and a linear pair. Mark all right angles and congruent angles in your sketch.

SPIRAL REVIEW

34. Make a sketch of plane R intersecting plane M in $\overleftrightarrow{XY}$. *(Section 1.4)*

35. Use a protractor to measure each angle. *(Section 1.6)*
    a. ∠A
    b. ∠B
    c. ∠C

Exercise Notes

Geometric Thinking
Exs. 30, 31 These exercises help students to learn how to think and reason geometrically. Ask students how the conjecture in Ex. 31 relates to the Vertical Angles Property. This conjecture is proved as a theorem in Chapter 3.

Integrating Algebra
Ex. 32 This exercise provides an opportunity for students to use algebraic notation to demonstrate a geometric result.

Assessment Note
Ex. 33 This exercise provides a good review of all the concepts of this section. Students may create their own sketches or use copies of drawings or photographs of actual buildings, bridges, or stained glass windows.

Practice 9 for Section 2.1

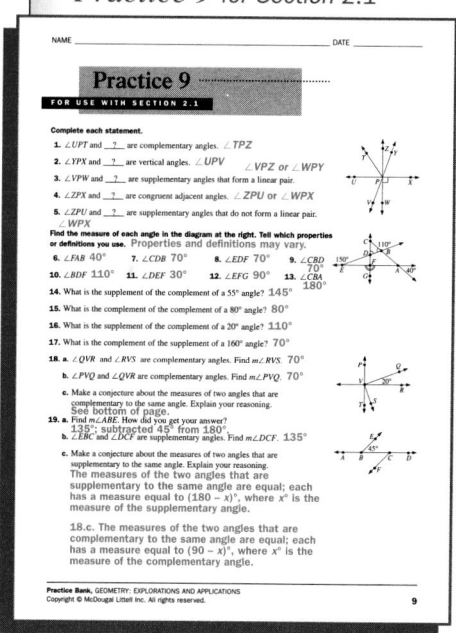

---

30. a. 50°   **b.** 50°

  c. Two angles that are complementary to the same angle are congruent. This is because if
m∠1 + m∠2 = 90° and
m∠1 + m∠3 = 90°, then
m∠2 = 90° − m∠1 and
m∠3 = 90° − m∠1, so
m∠2 = m∠3.

31. a. 50°; by the Linear Pair Property
  b. 50°

  c. Two angles that are supplementary to the same angle are congruent. This is because if
m∠1 + m∠2 = 180° and
m∠1 + m∠3 = 180°, then
m∠2 = 180° − m∠1 and
m∠3 = 180° − m∠1, so
m∠2 = m∠3.

32. By the Vertical Angles Property, ∠QRP ≅ ∠WRS, so m∠WRS = x°. Because ∠SUP ≅ ∠WRS, m∠SUP = x°. Then by combining angles and using the Linear Pair Property, x° + 3x° + 2x° = 180°, so x = 30. Then 3x° = 90°, so m∠TUP = 90° and, thus, $\overleftrightarrow{TU} \perp \overleftrightarrow{PW}$.

33. Check students' work.

34.

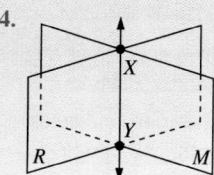

35. a. 70°
    b. 40°
    c. 70°

Objectives

- Name, label, and classify triangles.
- Sketch triangles and find the measures of their angles.
- Use types of triangles to describe real-world objects.

Recommended Pacing

❖ **Core and Extended Courses**
Section 2.2: 2 days

❖ **Block Schedule**
Section 2.2: 2 half-blocks (with Sections 2.1 and 2.3)

Resource Materials

Lesson Support
Lesson Plan 2.2

Warm-Up Transparency 2.2

Overhead Visuals:
Folder 3: Angle Sum of a Triangle

Practice Bank: Practice 10

Study Guide: Section 2.2

Explorations Lab Manual:
Diagram Master 8

Challenge Problems: Set 10

Assessment Book: Test 6

Technology
Geometry Software

Internet:
http://www.hmco.com

Warm-Up Exercises

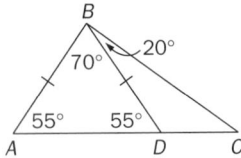

1. Name two congruent segments.
   $\overline{AB}$ and $\overline{DB}$

2. Name two congruent angles.
   $\angle BAD$ and $\angle BDA$

3. Name two perpendicular segments. $\overline{AB}$ and $\overline{BC}$

4. Name two complementary angles. $\angle ABD$ and $\angle DBC$

5. Name two different pairs of supplementary angles. $\angle ADB$ and $\angle CDB$; $\angle BAD$ and $\angle BDC$

6. Name an obtuse angle. $\angle BDC$

2.2 Classifying Triangles

A **triangle** is the figure formed by the segments whose endpoints are three noncollinear points. The shape of a triangle depends on the measures of its angles and the lengths of its sides.

Learn how to...
- **name, label, and classify triangles**
- **sketch triangles and find the measures of their angles**

So you can...
- **use types of triangles to describe objects such as crystals**

EXPLORATION
COOPERATIVE LEARNING

Sorting Triangles

Work in a group of four students. You will need:
- a set of triangles like the ones at the right

1 One student should write down a "secret rule" that a second student will use to sort the triangles.

2 The second student, the sorter, should choose triangles that fit the rule as the other students watch. At any time, another student can stop the sorter and guess the rule.

3 Repeat Steps 1 and 2 until five rules have been used. Make sure each student has a chance to write a rule, sort the triangles, and guess a rule.

4 Make a list of the rules your group used. What other rules can you think of? What rules did the other groups in your class use?

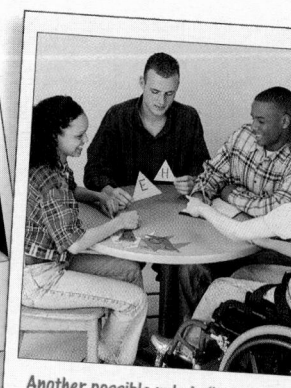

One rule might be "triangles with an obtuse angle."

Another possible rule is "triangles with three congruent sides."

Exploration Note

Purpose
The purpose of this Exploration is to have students discover five rules that can be used to sort or classify a triangle by considering its sides or its angles.

Materials/Preparation
Each group of students needs paper or cardboard cutouts of the triangles shown.

Procedure
One student in the group writes a rule for sorting the triangles. A second student uses the rule to choose those triangles that fit it. The other two students try to guess the rule

that determined how the triangles were sorted. All students participate in writing a rule, sorting the triangles, and guessing a rule.

Closure
Use a class discussion to compare the sorting rules that students discover. Students should see that angle relationships and side relationships are two ways of sorting any group of triangles.

Explorations Lab Manual
See the Manual for more commentary on this Exploration.
Diagram Master 8

The segments that form a triangle are called its **sides**. The endpoints of the sides are called **vertices** (plural of *vertex*). The triangle with vertices *A*, *B*, and *C* is called △*ABC*, read as "triangle *ABC*."

You can classify a triangle by considering its sides.

A **scalene triangle** has no congruent sides.

An **isosceles triangle** has at least two congruent sides.

An **equilateral triangle** has three congruent sides.

 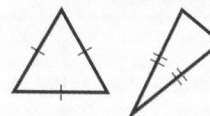 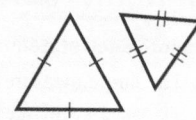

You can also classify a triangle by considering its angles.

An **equiangular triangle** has three congruent angles.

An **acute triangle** has three acute angles.

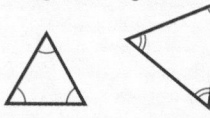 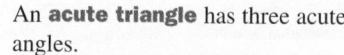  

A **right triangle** has a right angle.

An **obtuse triangle** has an obtuse angle.

THINK AND COMMUNICATE

1. Which triangles on page 64 are scalene? Which are acute? right? obtuse?

2. Are all equilateral triangles isosceles? Explain.

You may already know what the sum of the angle measures of a triangle is. Here is a quick way to recall this sum:

Cut any triangle out of a piece of paper.

Tear off two angles of the triangle.

Place the angles so that they are adjacent.

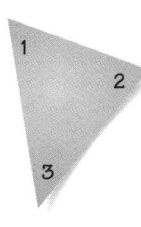 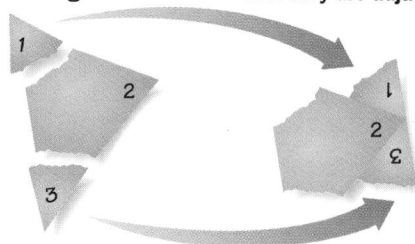

Together the three angles form a straight angle, so the sum of their measures is 180°.

ANSWERS Section 2.2

Exploration

1–4. Sample rules include "triangles with a right angle," "triangles with only acute angles," or "triangles with no congruent sides."

Think and Communicate

1. scalene: A, B, F, J, K; acute: A, F, I; right: D, G, J; obtuse: B, C, K

2. Yes; an equilateral triangle has three congruent sides, so it will always have at least two congruent sides, which is the definition of an isosceles triangle.

Section Notes

Student Progress

Many students will probably be familiar with the names of some of the triangles discussed here from their previous study of geometry. However, it is still important to discuss these ideas thoroughly so all students can remember them correctly.

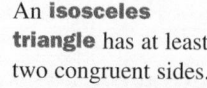 **Communication: Drawing**
After students have had an opportunity to study the various types of triangles, suggest that they draw examples of these triangles in their notebooks or journals as a study guide. Remind students to mark congruent sides and congruent angles as shown in the text.

Visual Thinking

Ask students to find triangles in magazine photographs, to highlight the triangles with markers, and to classify the triangles by their sides and angles. Encourage the class to create a wall display that demonstrates the various ways in which triangles can be classified. This activity involves the visual skills of *identification* and *interpretation*.

Think and Communicate

It is important for students to understand that all equilateral triangles are also isosceles. Students sometime think that all classification rules must separate the objects being classified into completely separate categories. This is not always the case, and there are many examples in mathematics and in everyday life where completely exclusive categories do not exist.

Section Note

Alternate Approach

Students can also measure the angles of various triangles with a protractor and add them to confirm that the sum for each triangle is 180°. The approach used on this page has the advantage of providing visual evidence that the sum is 180°.

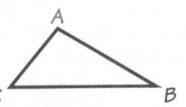

Section Note

Family Involvement
Working at home, students may
wish to involve family members in
an exploration of finding the sum of
the angle measures of a triangle.
They can use the approach of cutting any triangle out of a piece of
paper (as shown at the bottom of
page 65), or they can draw many
different types of triangles and measure the angles using a protractor.

About Example 1

Spatial Reasoning
Example 1 illustrates the use of a
technique that can help students
improve their spatial reasoning
powers, that is, to separate
overlapping figures into separately
drawn figures. Encourage students
to do this whenever they are working with overlapping figures.

Additional Example 1

In the figure below, △ABC and
△XYZ are equiangular. Find
m∠BKX.

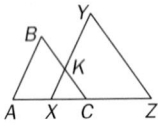

First find m∠XKC.
m∠XKC + m∠KXC + m∠KCX =
180°
Since ∠KXC is an angle of equiangular triangle XYZ and ∠KCX is an
angle of equiangular triangle ABC,
their measures are 60°. Therefore,
m∠XKC + 60° + 60° = 180°
    m∠XKC + 120° = 180°
         m∠XKC = 60°.
Because ∠XKC and ∠BKX form
a linear pair, m∠XKC + m∠BKX =
180°. Now find m∠BKX.
60° + m∠BKX = 180°
Therefore, m∠BKX = 120°.

Additional Example 2

Sketch each triangle, if possible. If
it is not possible, explain why not.

a. a triangle with two obtuse angles
   Sketch an obtuse angle.

   Choose a point Y on one side of
   the angle to form a side of the
   triangle.

The Angles of a Triangle

The sum of the angle measures of a
triangle is 180°.

In △ABC, $m\angle A + m\angle B + m\angle C = 180°$.

THINK AND COMMUNICATE

Complete each statement.

   3. The angle measures of an equiangular triangle are ? .

   4. The sum of the three angle measures of a triangle is ? .

   5. The measure of each angle of an equiangular triangle is ? .

EXAMPLE 1   Connection: Algebra

In the figure, $m\angle BDC = 10°$ and
$\angle BAD \cong \angle BDA$. △ACD is an
equiangular triangle.
Find $m\angle ABD$.

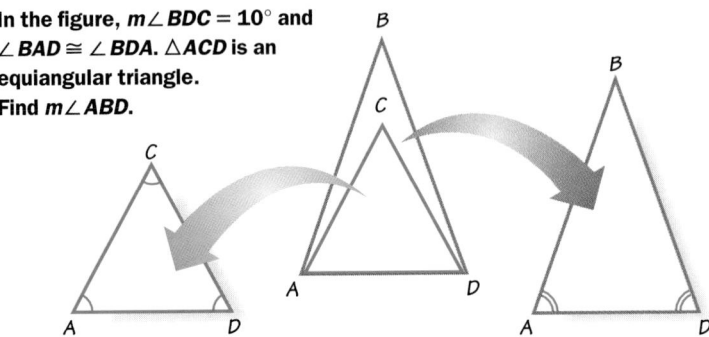

SOLUTION

First find $m\angle BDA$ and $m\angle BAD$.

Because ∠BDC and ∠CDA are adjacent:

$m\angle BDA = m\angle BDC + m\angle CDA$

$\qquad = 10° + 60°$ 　　Since △ACD is an equiangular triangle,
$\qquad\qquad\qquad\qquad m\angle DAC = m\angle ACD = m\angle CDA = 60°.$
$\qquad = 70$

Because $m\angle BAD = m\angle BDA$, $m\angle BAD = 70°$ as well.

Now find $m\angle ABD$.

$m\angle ABD = 180° - (m\angle BAD + m\angle BDA)$

$\qquad = 180 - (70 + 70)$

$\qquad = 180 - 140$

$\qquad = 40$

Therefore, $m\angle ABD = 40°$.

Toolbox p. 690
*Simplifying and
Evaluating
Expressions*

66   Chapter 2 *Triangles and Polygons*

Think and Communicate

3. equal

4. 180°

5. 60°

Notice that Triangle B on page 64 is both scalene and obtuse. Not all combinations of types are possible, however.

EXAMPLE 2

Sketch each triangle, if possible. If it is not possible, explain why not.

a. an acute isosceles triangle   **b.** an obtuse right triangle

SOLUTION

a. Sketch an acute angle.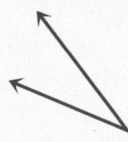

Make the sides into congruent segments.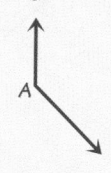

Sketch the third side. $\triangle XYZ$ is an acute isosceles triangle.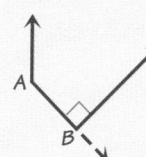

b. Sketch an obtuse angle.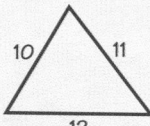

Sketch a ray perpendicular to one of its sides.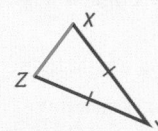

The sum of $m\angle A$ and $m\angle B$ is greater than 180°, so no such triangle can exist.

☑ CHECKING KEY CONCEPTS

Classify each triangle. Be as specific as possible.

1.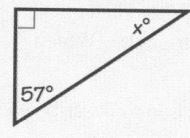

2.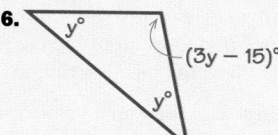

Sketch each triangle. Mark any right angles, congruent angles, and congruent sides.

3. a scalene right triangle   **4.** an isosceles obtuse triangle

ALGEBRA Find each unknown angle measure.

5.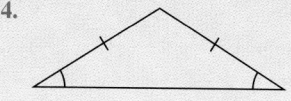

6.

Checking Key Concepts

1. scalene, acute triangle

2. isosceles, right triangle

3, 4. Answers may vary. Examples are given.

3.

4.

5. 33°

6. $y° = 39°$; $(3y - 15)° = 102°$

Additional Example 2 (continued)

Sketch an obtuse angle at point Y.

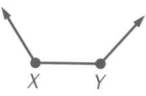

The sum of $m\angle X$ and $m\angle Y$ is greater than 180°, so no such triangle can exist.

b. an equiangular right triangle
Sketch a right angle.

Choose a point B on one side of $\angle A$.

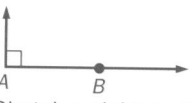

Sketch a right angle with vertex at B.

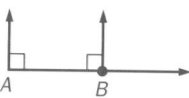

The sum of $m\angle A$ and $m\angle B$ is 180°, so no such triangle can exist.

Checking Key Concepts

Integrating Algebra
In questions 5 and 6, students must use equations to find the unknown angle measures. When discussing question 5, point out that the value of x can be found by using either of two equations ($x + 57 + 90 = 180$ or $x + 57 = 90$). When answering question 6, some students may only find the value of y. Remind these students that they need to substutute this value into $3y - 15$ to find that angle measure.

Closure Question

Describe two ways of classifying triangles and give the types of triangles for each classification. One way to classify triangles is according to the number of its congruent sides: scalene (no congruent sides), isosceles (two or more congruent sides), and equilateral (all three sides congruent). The other classification is according to its angles: acute (three acute angles), right (contains a right angle), obtuse (contains an obtuse angle), and equiangular (all three angles congruent).

Exercise Notes

Geometric Thinking
Exs. 1–7 These exercises help students learn how to think geometrically by requiring that they base their explanations and answers upon previously established facts.

Integrating Algebra
Exs. 8–11 These exercises provide an opportunity for students to use their algebraic skills to find information about angle measures. A brief review of the skills involved (combining like terms and solving simple equations) may be necessary for some students.

Interview Note: Research
Exs. 12, 13 Students can consult encyclopedias or books on minerals and crystallography to find what other forms crystals may have. Suggest that they prepare a brief report that can be shared with the class or included in a portfolio.

2.2 | Exercises and Applications

Extra Practice
exercises on page 662

Sketch each triangle, if possible. If it is not possible, explain why not.

1. an isosceles right triangle
2. a scalene equilateral triangle
3. an obtuse equiangular triangle
4. an obtuse scalene triangle

5. **Writing** A given triangle has two congruent angles. Are the two angles acute, right, or obtuse? How do you know?

6. **Open-ended Problem** Draw an obtuse triangle with two congruent angles. Label each angle with its measure.

7. **Logical Reasoning** Sketch a right triangle, △PQR, with a right angle at Q.
   a. $m\angle P + m\angle R = \underline{\ ?\ }$
   b. How are the acute angles in a right triangle related? Explain.

ALGEBRA Find each unknown angle measure.

8.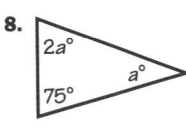

9.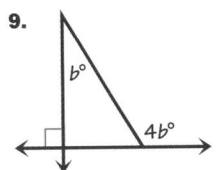

10.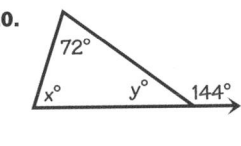

11.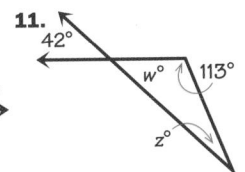

INTERVIEW Virgil Lueth

Look back at the article on pages 54–56.

The geometric shape you see when you look at a crystal is called its form. The different forms are often described in terms of their faces. Since many of the faces are triangular, identifying different types of triangles can help you learn the forms that crystals have.

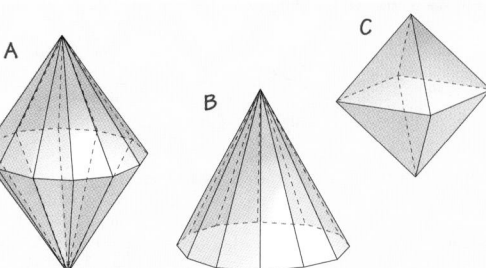
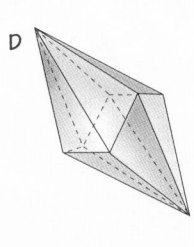

12. A *hexagonal scalenohedron* has twelve faces. When the crystal develops perfectly, all of the faces are scalene triangles. Which diagram best fits this description?

13. **Writing** Use the terms you learned in this section to describe each of the other forms shown.

Exercises and Applications

1. Answers may vary. An example is given.
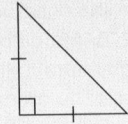

2. not possible; A scalene triangle has no congruent sides, while an equilateral triangle has three congruent sides.

3. not possible; An equiangular triangle has three angles that measure 60°, so it has only acute angles.

4. Answers may vary. An example is given.
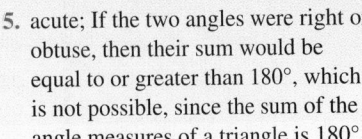

5. acute; If the two angles were right or obtuse, then their sum would be equal to or greater than 180°, which is not possible, since the sum of the angle measures of a triangle is 180°.

6. Answers may vary. An example is given.
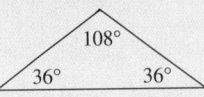

7. a. 90°
   b. They are complementary. Because the right angle measures 90°, the other two angle measures must total 180° − 90° = 90°.

8. $a° = 35°$; $2a° = 70°$
9. $b° = 30°$; $4b° = 120°$
10. $x° = 72°$; $y° = 36°$
11. $w° = 42°$; $z° = 25°$
12. D
13. Crystal A is formed from 24 congruent, acute, isosceles triangles, with 12 triangles sharing the top and bottom vertices, and 4 triangles meeting at each of 12 vertices around the middle. Crystal B is formed from 12 congruent, acute, isosceles triangles all sharing the top vertex,

CONSTRUCTION

EQUILATERAL TRIANGLE

Construct an equilateral triangle.

A geometric *construction* is a method of drawing figures using only a compass and a straightedge (a ruler with no marks). The steps below show how to construct an equilateral triangle.

1. Draw a ray. Label the endpoint *A*.

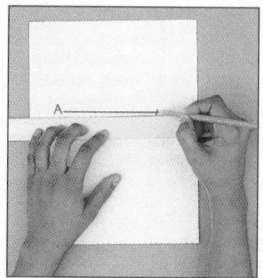

2. Place the point of the compass at *A*. Using any compass setting, draw an arc that intersects the ray at a point *B*.

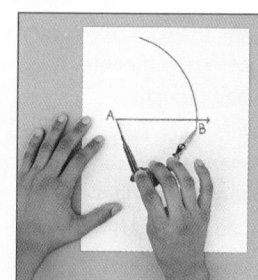

3. Without changing the setting, place the point of the compass at *B* and draw another arc. Label the intersection *C*.

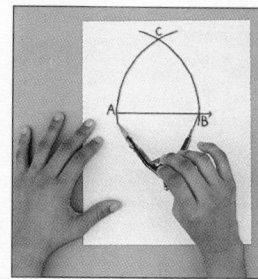

4. Draw $\overline{AC}$ and $\overline{BC}$ to form equilateral $\triangle ABC$.

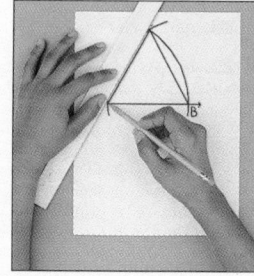

14. How do you know that $\overline{AB} \cong \overline{AC}$ and that $\overline{AB} \cong \overline{BC}$?

15. Construct an equilateral triangle.

16. Construct an equilateral triangle with side length *JK*. (*Hint:* Open the compass so that the distance from the point to the pencil is *JK*.)

17. Writing Suppose you extended the arcs in Steps 2 and 3 on both sides of $\overline{AB}$. Explain why the result is two equilateral triangles that share a side.

18. Challenge Explain how to construct an isosceles triangle by drawing one arc with the point of the compass at a given point *X*.

Complete each statement with *always*, *sometimes*, or *never*.

19. A triangle ? has two right angles.

20. A triangle ? has exactly one acute angle.

21. A triangle ? has at least two acute angles.

22. An obtuse triangle ? has exactly two acute angles.

23. An isosceles triangle ? has three congruent sides.

24. An equilateral triangle is ? isosceles.

2.2 Classifying Triangles **69**

Apply⇔Assess

Exercise Notes

Second-Language Learners
Exs. 14–18 Some students learning English may benefit from a discussion of the parts of the word *equilateral*. Explain to students that the prefix, *equi-*, means "equal," and that the base word, *lateral*, means "of the sides." Knowing what the word parts mean can help students to understand and remember the mathematical concept. You may want to expand the discussion by asking students what *equiangular* means.

Historical Connection
Exs. 14–18 Ancient Greek mathematicians viewed the drawing of constructions as having two rules: a straightedge can be used to draw a straight line through any two given distinct points, and a compass can be used to draw a circle or an arc with any point as the center and passing through a given second point. Many interesting and intricate constructions can be made following these two rules. You may wish to ask some students to research why the compass and straightedge are called *Euclidean* tools.

Construction Note
Exs. 14–18 It may take some time for students to appreciate fully an approach to drawing that uses only a compass and a straightedge. After students gain more experience doing constructions, point out that such an approach draws special attention to symmetry and other general properties of geometric figures. For Ex. 18, point out that construction of an isosceles triangle requires only one arc, whereas construction of an equilateral triangle (a more specialized isosceles triangle) requires two arcs.

Communication: Discussion
Exs. 19–24 These types of *always, sometimes,* or *never* exercises are helpful in determining students' levels of understanding of the concepts involved. If you have a discussion of these exercises, have students explain their answers in order to assess their level of understanding.

with 2 triangles meeting at each of 12 vertices around the base. Crystal C is formed from 8 congruent, equilateral, equiangular triangles, with 4 triangles meeting at each vertex.

14. Since the compass setting is not changed, and *C* lies on arcs of radius *AB* drawn from both *A* and *B*, the distance from *C* to *A* and *B* is *AB*, so $\overline{AB} \cong \overline{AC}$ and $\overline{AB} \cong \overline{BC}$.

15. Check students' work.

16. Check students' work.

17. Because the compass is not adjusted and its point is not moved, the two arcs that intersect below $\overline{AB}$ will also both have length *AB*, as did those in Ex. 14, so you will form two equilateral triangles sharing the side $\overline{AB}$.

18. Draw an arc with the point of the compass at point *X*. Because the radius of the arc is always the same, any two segments drawn from *X* to the arc and connected by a segment between these two intersections with the arc will form an isosceles triangle.

19. never   **20.** never

21. always   **22.** always

23. sometimes   **24.** always

Exercise Notes

Exercise Notes

Career Connection

Exs. 25–29 Many students are probably familiar with the term *surveyor* but may not fully understand what a surveyor does. Point out that surveyors have a long history, going back to early times in the development of the first farming communities when land boundaries were established. These early surveyors used basic geometric ideas they knew were true from experience. Today, surveyors use sophisticated tools, such as a transit, to determine angles, bearings, and levels. Surveyors are usually employed by civil engineering firms involved in building construction, road construction, or in the mapping of large tracts of land.

Historical Connection

Ex. 30 Students may find it interesting to research the Great Trigonometrical Survey of India, which was conducted in the nineteenth century. Students can find information on this famous project by consulting works on cartography and surveying, encyclopedias, and appropriate historical sources.

Geometric Thinking

Exs. 31, 32 Review with students the idea that the conjectures they make in these exercises are predictions about triangles based upon a limited number of observations. Emphasize that conjectures are not facts; they are statements that we think are true but which may turn out to be false. Point out that in Chapter 3 methods will be developed for determining whether a conjecture is true or false.

Connection HISTORY

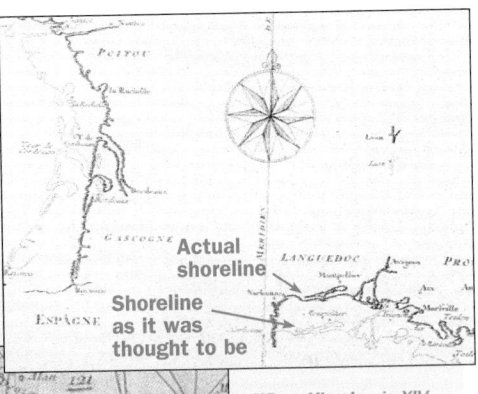

In the seventeenth and eighteenth centuries, a survey was conducted in France. The method used, *triangulation*, involved constructing a network of triangles throughout the country. By knowing some of the side lengths and angle measures of the triangles, the surveyors calculated the size of France, finding it much smaller than they had thought.

Find each angle measure in the map at the right.

25. $m \angle CAB$

26. $m \angle ABC$

27. $m \angle DCB$

28. $m \angle CED$

29. $m \angle CDE$

Actual shoreline

Shoreline as it was thought to be

When King Louis XIV saw the difference between the shoreline as they believed it to be and the actual shoreline, he remarked that the survey had cost him more land than a war.

The map at the left shows some of the triangles used for the survey.

30. **Research** Find out more about triangulation. What kind of mathematics is involved? What information do the surveyors need to determine distances? How do they measure the angles?

Technology For Exercises 31 and 32, you will need geometry software or a ruler and a protractor.

31. **a.** Use the steps on page 69 or the software to construct an equilateral triangle. Measure each angle. What are the measures?

b. Draw an equiangular triangle. Measure each side of the triangle. How are the lengths related?

c. Make a conjecture about the relationship between equiangular triangles and equilateral triangles.

32. **a.** Draw two congruent segments that share an endpoint, as shown. Connect the other two endpoints. What kind of triangle is it?

b. Measure the two angles that are opposite the congruent sides. These are called the *base angles*. What do you notice?

c. Repeat parts (a) and (b) using a different side length and a different angle measure.

d. Make a conjecture about the base angles of an isosceles triangle.

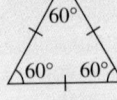

25. 55°   26. 55°

27. 67.5°   28. 67.5°

29. 72.5°

30. Check students' work.

31. **a.** 60°

b. They are equal.

c. Equilateral triangles are equiangular, and equiangular triangles are equilateral.

32. **a.** isosceles

b. They are congruent.

c. The results are the same.

d. The base angles of an isoceles triangle are congruent.

33. Answers may vary. Examples are given.

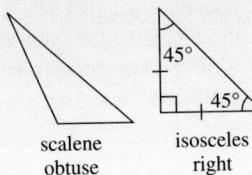

scalene obtuse

45°
45°

isosceles right

60°
60° 60°

equilateral
equiangular
acute

34. **a.** 5 segments

b. $\overline{AB} \cong \overline{ED}, \overline{BC} \cong \overline{CD}$

c. $\angle A \cong \angle E, \angle B \cong \angle D$

35. **a.** 5 segments

33. Open-ended Problem Sketch three or four triangles so that you have at least one of each type: scalene, isosceles, equilateral, equiangular, acute, right, and obtuse. Classify each triangle. Mark all congruent angles and sides, and give angle measures when possible.

SPIRAL REVIEW

For each figure, (a) state how many segments are shown, (b) list any congruent segments, and (c) list any congruent angles. *(Sections 1.5 and 1.6)*

34.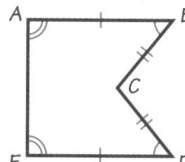

35.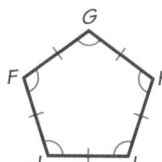

36.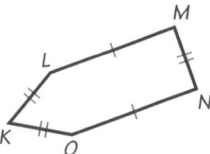

ASSESS YOUR PROGRESS

VOCABULARY

types of angles (p. 57)	**supplements** (p. 58)
adjacent angles (p. 58)	**triangle** (p. 64)
linear pair (p. 58)	**side of a triangle** (p. 65)
perpendicular (p. 58)	**vertex of a triangle** (p. 65)
vertical angles (p. 58)	**types of triangles** (p. 65)
complements (p. 58)	

Find the value of each variable. Tell which property or definition you use. *(Sections 2.1 and 2.2)*

1. a **2.** b

3. c **4.** d

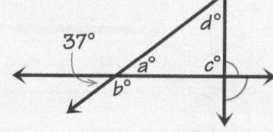

BICYCLING Imagine sitting on a bicycle. For each pedal position, (a) sketch the angle formed at your knee, (b) estimate the measure of the angle, and (c) tell what type of angle it is. *(Section 2.1)*

5. when the pedal is farthest from the ground

6. when the pedal is closest to the ground

7. when the pedal is in the middle of a stroke

8. Name and classify the four triangles shown in the diagram. Be as specific as possible. *(Section 2.2)*

9. Journal Choose ten vocabulary words from Sections 2.1 and 2.2. Describe a way to remember the definition of each.

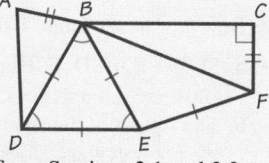

2.2 Classifying Triangles   **71**

Apply⇔Assess

Exercise Notes

Assessment Note
Ex. 33 This exercise provides a good opportunity for students to review the important definitions of this section.

Assess Your Progress

Journal Entry
Have students share their answers in class so that everyone can re-vise his or her journal entry after a first draft. Devising good methods for remembering important geomet-ric ideas will help all students suc-ceed in their study of geometry.

Progress Check 2.1–2.2

See page 102.

Practice 10 for Section 2.2

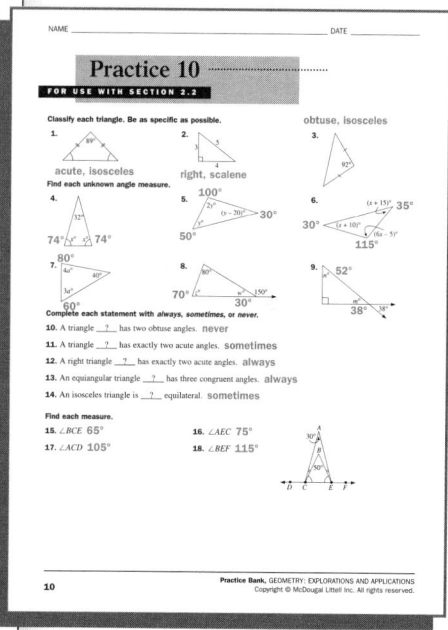

---

b. $\overline{FG} \cong \overline{GH} \cong \overline{HI} \cong \overline{IJ} \cong \overline{JF}$

c. $\angle F \cong \angle G \cong \angle H \cong \angle I \cong \angle J$

36. a. 5 segments

b. $\overline{KL} \cong \overline{KO} \cong \overline{MN}, \overline{LM} \cong \overline{ON}$

c. No congruent angles are indicated.

Assess Your Progress

1. 37; Vertical Angles Property

2. 143; Linear Pair Property

3. 90; If two lines form congruent adjacent angles, then the lines are perpendicular; Linear Pair Property or Vertical Angles Property.

4. 53°; The sum of the angle measures of a triangle is 180°.

5–9. Answers may vary. Examples are given.

5. a.   b. 90°   c. right

6. a.   b. 167°   c. obtuse

7. a.   or

pedal   pedal
at front   at back

b. 140°, 95°

c. obtuse

8. △*ABD*: scalene, acute triangle; △*BDE*: equilateral triangle (thus, equiangular and acute); △*BEF*: isosceles, obtuse triangle; △*BCF*: scalene, right triangle

9. Check students' work.

Warm-Up Exercises

Use the figure below.

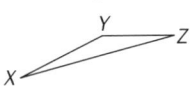

1. Are points *X*, *Y*, and *Z* collinear or *noncollinear*? noncollinear

2. Name all of the line segments in the figure. $\overline{XY}, \overline{YZ}, \overline{XZ}$

How many lines of symmetry does each figure have?

3. 3

4. 2

5. 4

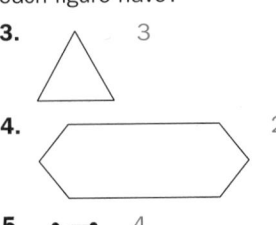

SECTION

2.3 Types of Polygons

Learn how to...
- **describe the properties of a figure**

So you can...
- **identify polygons**

The *Generalife* is a thirteenth century palace overlooking the Alhambra in Granada, Spain. The plans below show part of the Generalife gardens which contain many different geometric designs and shapes.

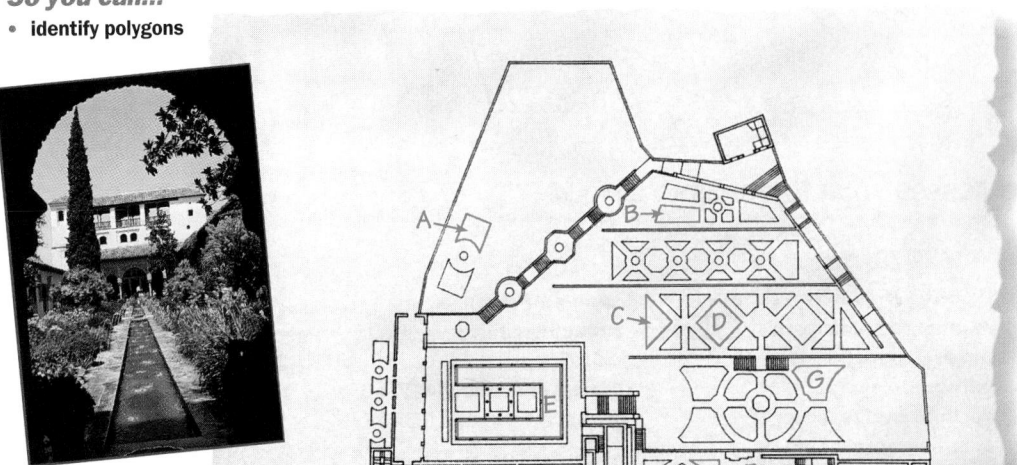

THINK AND COMMUNICATE

1. How are shapes E and H alike?

2. How are shapes E and F different from shapes B, C, D, and H?

3. **a.** Shapes B, C, D, E, F, and H are examples of *polygons*. Shapes A, G, and J are not polygons. How would you define polygon?

   b. The figures at the right are not polygons. Does this affect your definition from part (a)? If so, change it as needed.

ANSWERS Section 2.3

Think and Communicate

1. They each have eight sides.

2. Answers may vary. An example is given. Shapes E and F have "indentations," so that you can connect parts of each figure with line segments that pass outside of the figures. In shapes B, C, D, and H, any segment you can draw connecting two points of the figure stays within the figure.

3. Answers may vary. Examples are given.

   a. A polygon is a plane figure made by connecting line segments.

   b. Yes; a polygon is a plane figure that encloses space. Its sides are line segments that intersect only at their endpoints, with exactly two segments meeting at each intersection point.

A **polygon** is a closed plane figure whose sides are segments that intersect only at their endpoints. No two sides with a common endpoint are collinear.

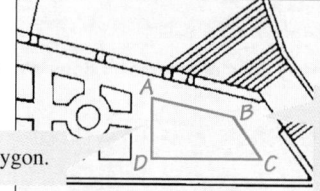

$\overline{AD}$ is a side of the polygon.

Point B is a vertex of the polygon. $\angle B$ is an angle of the polygon.

One way to classify a polygon is by the number of sides it has.

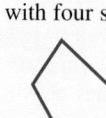

A **triangle** is a polygon with three sides.

A **quadrilateral** is a polygon with four sides.

A **pentagon** is a polygon with five sides.

A **hexagon** is a polygon with six sides.

Here are names for some other polygons:

Number of sides	7	8	9	10	n
Name	heptagon	octagon	nonagon	decagon	n-gon

A polygon can be either *concave* or *convex*. In a convex polygon, no segment can be drawn outside of the polygon to connect two vertices.

For example, on page 72, shape E is concave and shape B is convex.

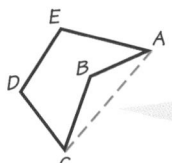

This polygon is concave because $\overline{AC}$ lies outside of the polygon.

You can describe a polygon by comparing the lengths of its sides or the measures of its angles.

If all the sides of a polygon are congruent, then it is an **equilateral** polygon.

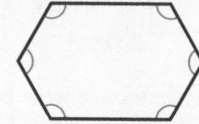

If all the angles of a polygon are congruent, then it is an **equiangular** polygon.

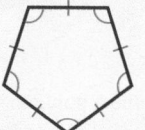

If a polygon is both equilateral and equiangular, then it is a **regular** polygon.

2.3 Types of Polygons   **73**

Section Notes

Multicultural Note
The Alhambra and the Generalife were both built in Granada, Spain by Islamic Moorish rulers. The Moors, a people from North Africa, founded Granada around A.D. 750 and ruled until 1492, when they surrendered the city to the Catholic Spanish monarchs. The Alhambra was built between the years 1248 and 1354 and served as both a palace and a fortress. (*Alhambra* is an Arabic word meaning "the red one.") Today, the Alhambra is one of the best examples of Islamic architecture in Europe. The Generalife, which stands next to the Alhambra, was built in 1318 as a summer retreat for the Moorish kings of Granada. Although little remains of the Generalife palace, the beauty of the gardens continues to draw large numbers of visitors.

 **Communication: Drawing**
An initial discussion of polygons should include drawings to illustrate the various types. In addition to the four polygons shown in the text, either you or some students can draw more examples on the board. Be sure to show examples of polygons having seven to ten sides, as identified in the table. Impress upon students the need to remember the names of the polygons having from three to ten sides.

Spatial Reasoning
When discussing the concept of a polygon as a *closed* plane figure, ask a student to demonstrate at the board what an *open* figure would look like. Also, lead students to understand that another way to define a convex polygon is to say that for any two points chosen inside the polygon, all points of the segment connecting these two points are always inside of the polygon. Use diagrams to illustrate this alternative definition.

Teaching Tip
Students have learned in the previous section that every equilateral triangle is equiangular and every equiangular triangle is equilateral. The two first diagrams at the bottom of this page show that polygons in general do not possess this property. However, as shown by the third diagram, a polygon can be *both* equilateral and equiangular, in which case it is a regular polygon.

EXAMPLE 1

About Example 1

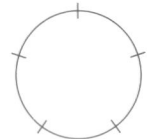
Communication: Drawing The procedure demonstrated here is important because it can be used to draw other regular polygons. Discuss different ways of locating eight equally spaced points for Step 1. Some students may suggest using a protractor. Since $\frac{360}{8} = 45$, they could also draw a number of adjacent 45° angles with their vertices at the center of the circle. The points where their sides intersect the circle give the desired equally spaced points. Another procedure would be to fold the circle to form halves, then quarters, then eighths.

Additional Example 1

How many lines of symmetry does a regular pentagon have?

Step 1 Sketch a circle. Mark five points evenly spaced around the circle.

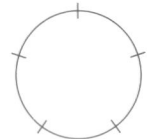

Step 2 Connect the points to form a regular pentagon.

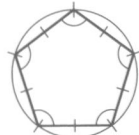

Step 3 Fold the paper or use a transparent mirror to find the lines of symmetry.

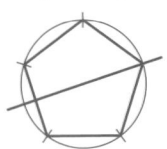

line through a vertex and the center of the circle

There are five lines of symmetry, each passing through a vertex and the center of the circle.

Think and Communicate

For question 4(a), students can use a protractor to mark off twelve 30° arcs that go around a circle. Some students may suggest using a straightedge and compass to construct an equilateral triangle. The triangle can be folded in half to obtain a 30° angle that can be used to mark off the twelve points.

How many lines of symmetry does a regular octagon have?

SOLUTION

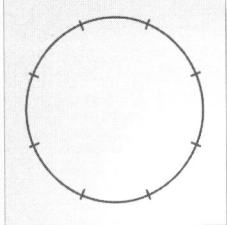

Step 1 Sketch a circle. Mark eight points evenly spaced around the circle.

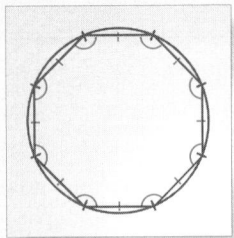

Step 2 Connect the points to form a regular octagon.

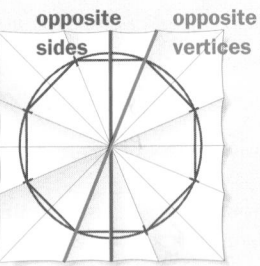

opposite sides   opposite vertices

Step 3 Fold the paper or use a transparent mirror to find the lines of symmetry.

There are four lines of symmetry that pass through **opposite vertices** and four lines of symmetry that bisect **opposite sides**. A regular octagon has 8 lines of symmetry.

THINK AND COMMUNICATE

4. **a.** Sketch a regular 12-gon. How many lines of symmetry does it have?
   b. Repeat part (a), using a different regular polygon. Make a conjecture about how many lines of symmetry a regular *n*-gon has.

5. Rachel says that all regular polygons have rotational symmetry as well as line symmetry. Do you agree or disagree? Explain.

Parts of Polygons

Two vertices of a polygon connected by a side are called *consecutive vertices*. Also, two angles that share a side are called **consecutive angles**, and two sides that share a vertex are called **consecutive sides**. A segment that connects nonconsecutive vertices is called a **diagonal**.

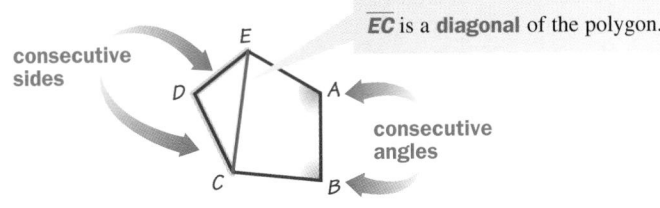

$\overline{EC}$ is a **diagonal** of the polygon.

consecutive sides

consecutive angles

To name a polygon, start at any vertex and write consecutive vertices in order. Two names for the pentagon above are *BAEDC* and *ABCDE*.

Think and Communicate

4. a.

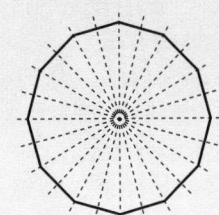

There are 12 lines of symmetry—six through opposite vertices and six that bisect opposite sides.

b.

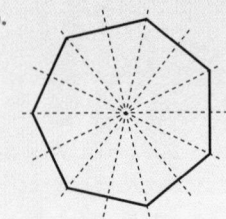

Answers may vary. An example is given. A regular heptagon has seven lines of symmetry, each connecting a vertex with the mid-

point of the opposite side. Conjecture: A regular *n*-gon has *n* lines of symmetry.

5. agree; All the angles of a regular polygon are congruent, and the sides are equal in length, so you can rotate a regular polygon until the original position of a line of symmetry is taken by the adjacent line of symmetry. Because each line of symmetry divides the regular polygon into halves that are exactly the same, the repositioned figure will appear the same. You can then rotate the polygon further until the next line of symmetry moves into the

THINK AND COMMUNICATE

Use the polygon at the right.

6. Name two consecutive angles of the polygon.

7. Name a pair of consecutive sides of the polygon.

8. Name a diagonal of the polygon.

9. Name this polygon two different ways.

10. Explain why *PRTSQ* is not a name for the polygon.

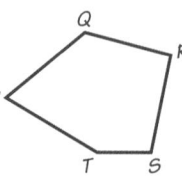

EXAMPLE 2

Sketch a hexagon that has two consecutive right angles.

SOLUTION

Sketch a segment with a right angle at each end.

Add three more sides to make a hexagon.

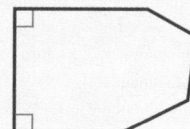

☑ CHECKING KEY CONCEPTS

Classify each polygon. Be as specific as possible. Then use the vertices to name the polygon.

1.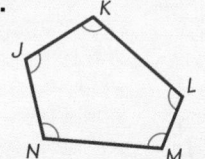

2.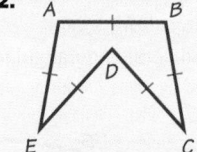

3.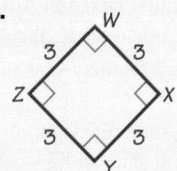

Look back at the plans for the gardens on page 72.

4. What kind of polygon is shape F?

5. Explain why shape G is not a polygon.

Tell whether each statement is *True* or *False*. Explain your reasoning.

6. Every regular polygon is equilateral.

7. Every equiangular polygon is regular.

8. A diagonal of a convex polygon can lie outside of the polygon.

9. Every equilateral polygon is convex.

Think and Communicate

You may wish to discuss with students how many ways there are to answer questions 6–9.

Additional Example 2

Sketch a quadrilateral that has exactly two nonconsecutive right angles.

Sketch a segment with the vertex of a right angle at one end and the vertex of an obtuse angle at the other end.

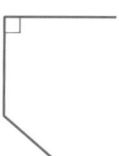

At a point on the second side of the obtuse angle, sketch another right angle. Make the new side long enough to intersect the side of the first right angle to make a quadrilateral.

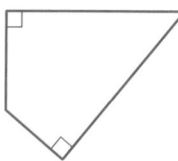

Checking Key Concepts

Geometric Thinking
For questions 6–9, suggest that students sketch a counterexample for any statement they say is false. Remind students that general statements such as these are true only if they are *always* true. However, saying that a statement is false may mean it is always false or only false in certain cases.

Closure Question

List the properties that a figure must have to be a polygon.
The figure must be a closed plane figure. It must be made up entirely of line segments that intersect only at their endpoints.

---

position of the original line of symmetry, and so on, until you return to the original position.

6. ∠P and ∠Q, ∠Q and ∠R, ∠R and ∠S, ∠S and ∠T, or ∠T and ∠P

7. $\overline{PQ}$ and $\overline{QR}$, $\overline{QR}$ and $\overline{RS}$, $\overline{RS}$ and $\overline{ST}$, $\overline{ST}$ and $\overline{TP}$, or $\overline{TP}$ and $\overline{PQ}$

8. $\overline{QT}$, $\overline{QS}$, $\overline{SP}$, $\overline{RT}$, or $\overline{RP}$

9. PQRST, QRSTP, RSTPQ, STPQR, TPQRS, PTSRQ, TSRQP, SRQPT, RQPTS, or QPTSR

10. The vertices are not listed consecutively.

Checking Key Concepts

1–3. Names may vary. Examples are given.

1. convex, equiangular pentagon; *JKLMN*

2. concave, equilateral pentagon; *ABCDE*

3. (convex) regular quadrilateral (a square); *WXYZ*

4. a concave pentagon

5. Its borders include curves, and not just line segments.

6. True; by the definition of a regular polygon.

7. False; a rectangle is an example of an equiangular polygon that is not regular.

8. False; by the definition of a convex polygon.

9. False; the equilateral polygon shown at the bottom of page 73 is not convex.

Extra Practice
exercises on
page 662

Suggested Assignment

❖ **Core Course**
Exs. 1–10, 17–30

❖ **Extended Course**
Exs. 1–30

❖ **Block Schedule**
Day 9 Exs. 1–10, 17–30

Exercise Notes

Challenge
Ex. 5 After students complete the table, ask them to make a conjecture about a formula that gives the number of diagonals for an *n*-sided polygon. (The formula is $d = \frac{n(n-3)}{2}$, where *n* is the number of sides of the polygon and *d* is the number of diagonals.)

Using Technology
Ex. 5 As an alternative way to have students explore the number of diagonals in a polygon and to discover an expression that relates the number of diagonals to the number of sides of the polygon, you may wish to assign Activity 9 in the *Geometry Inventor Activity Book*.

Communication: Writing
Ex. 6 Students should be encouraged to illustrate their explanations with supporting diagrams.

Problem Solving
Exs. 9, 10 Part (b) of Ex. 9 reminds students that vertices can be listed in a clockwise or counterclockwise order. A useful strategy to find how many names a polygon has is to systematically list them. Ask students how many different ways an *n*-gon can be named. (2*n* different ways)

2.3 | Exercises and Applications

Classify each polygon. Be as specific as possible.

1.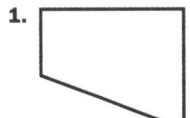
2.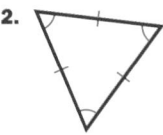
3.
4.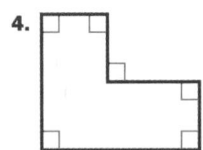

5. **Logical Reasoning** Copy and complete the table. Use any patterns you see to predict what the next column will look like.

triangle	quadrilateral	pentagon	hexagon	heptagon	octagon
		?	?	?	?
3 sides	4 sides	?	?	?	?
0 diagonals	2 diagonals	?	?	?	?

6. **Writing** Can a regular polygon be concave? Explain your reasoning.

7. **a.** Use a circle to sketch a regular nonagon.
   b. Use a circle to sketch a nonagon that is *not* regular. How is the method you used different from your method in part (a)?

8. **a.** Imagine stretching a regular hexagon *ABCDEF* so that $\overline{AB}$ and $\overline{DE}$ become longer than the other sides. Sketch the resulting figure.
   b. Is hexagon *ABCDEF* still regular? Is it still equilateral? equiangular? Explain.

9. **a.** When naming the polygon at the right, how many different vertices can you start with?
   b. If you start with vertex *G*, how many different vertices could come next?
   c. How many different names does this polygon have?

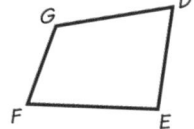

10. How many different names does each polygon have?

a.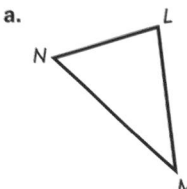
b.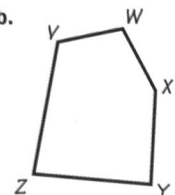
c.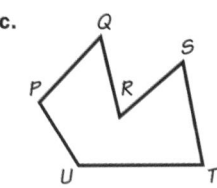

Exercises and Applications

1. convex quadrilateral

2. regular triangle (an equilateral triangle)

3. concave, equilateral 12-gon

4. concave hexagon

5. pentagon: 5 sides, 5 diagonals; hexagon: 6 sides, 9 diagonals; heptagon: 7 sides, 14 diagonals; octagon: 8 sides, 20 diagonals; Check students' work.

Prediction—nonagon: 9 sides, 27 diagonals

6. No; a concave polygon "bends in" on itself. That is, at least one of the angles has measure greater than 180°. But some of the angles must have measure less than 180°, or the figure could not close. Then a concave polygon cannot be equiangular, so it cannot be regular.

7. **a.**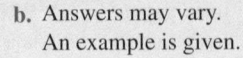

b. Answers may vary. An example is given.

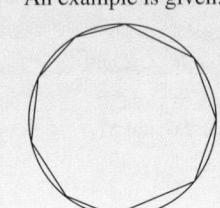

To sketch a regular nonagon, you must mark nine evenly spaced points around the circle before connecting consecutive points. For a nonagon that is not regular, you can connect any nine points consecutively around the circle.

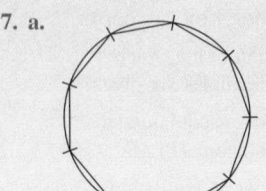

Paper-cutting is a craft found in many countries, including China, Japan, Mexico, and Poland. Use an $8\frac{1}{2}$ in. by 10 in. piece of paper for this paper-cutting activity.

Step 1 Fold the paper in half, as shown. Then crease the paper by folding and unfolding it in half horizontally and vertically.

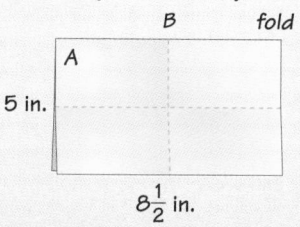

Step 2 Folding at *B*, bring corner *A* right and down until it meets the horizontal crease.

Step 3 Bring corner *A* left until the folded edges meet.

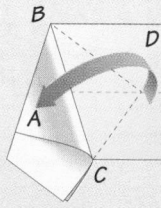

Step 4 Bring corner *D* left and fold on $\overline{BC}$.

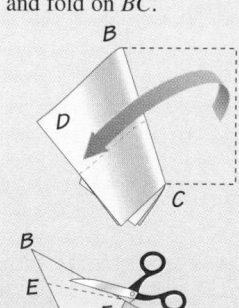

Step 5 Bring corner *D* right until $\overline{BD}$ lines up with $\overline{BC}$.

Step 6 Cut along a segment, $\overline{EF}$, so that $\triangle BEF$ is an obtuse triangle. When you unfold the paper, you should have a five-pointed star.

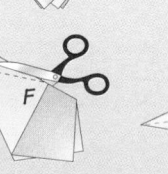
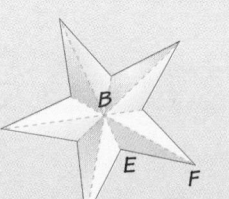

11. How many triangles in the diagram of the star have the same side lengths and angle measures as $\triangle BEF$?

12. What type of polygon is the resulting star?

13. How many lines of symmetry does the star have?

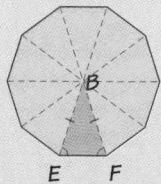

14. **a.** The polygon at the left can be made by changing the shape of $\triangle BEF$. What type of polygon is it? Is it equilateral? equiangular? Explain.

   b. What kind of triangle is $\triangle BEF$ in the polygon at the left?

   c. Describe how to cut $\triangle BEF$ in Step 6 to make the polygon.

15. Suppose you cut along $\overline{EF}$ so that $\triangle BEF$ is a right triangle, as shown at the right. What type of polygon is the resulting figure? Sketch the polygon, marking any congruent angles and sides.

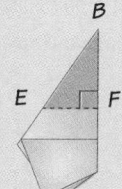

16. **Challenge** Describe how to cut $\triangle BEF$ in Step 6 to make an equilateral convex decagon that is not regular.

Exercise Notes

Using Manipulatives
Exs. 11–16 Try to set aside some time for students to do the paper-folding and cutting activities in class. If possible, provide students with clean sheets of white paper and have several pairs of scissors available for students to share.

Spatial Reasoning
Exs. 11–16 These exercises help students to develop their skills in visualizing geometric figures in both two and three dimensions. They also reinforce many basic ideas about polygons introduced in this section and in Chapter 1. For Ex. 15, you may wish to encourage students to make a conjecture about what kind of polygon they will get before they unfold the paper.

Research
Exs. 11–16 Students who enjoy this paper-cutting activity would probably be interested in origami, which involves folding paper to make figures. There are a number of excellent books that interested students may consult.

---

8. **a.** Answers may vary. An example is given.

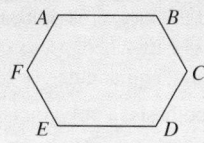

   b. No; No; Yes; because two of the sides are longer than the other four, the hexagon is not equilateral and, therefore, is not regular. It is equiangular, however, because stretching two opposite sides does not change the angle at which any of the segments meet.

9. **a.** 4 vertices
   b. 2 vertices
   c. 8 names

10. **a.** 6 names
    b. 10 names
    c. 12 names

11. 10 triangles (including $\triangle BEF$)

12. a concave, equilateral decagon

13. 5 lines of symmetry

14. **a.** a regular decagon; Yes; Yes. Each triangle has the same length base (the length of the cut), so the polygon is equilateral. Because each angle of the decagon is composed of two of the base angles of the triangles and all the base angles are congruent, the decagon is also equiangular.

    b. acute isosceles

    c. Cut it so that $BE = BF$.

15. a regular pentagon

16. Pick the point *F* at the start of the cut. Choose the other side of the cut, *E*, so that $\angle F$ is acute, $\angle E$ is acute, and $m\angle E \neq m\angle F$.

Exercise Notes

Communication: Drawing
Exs. 17–19 Students can support their answers by copying the figures and drawing all lines of symmetry.

Spatial Reasoning
Exs. 20–25 Students would benefit from seeing examples drawn by their classmates at the board. They would be able to see which figures have the same basic shape and which ones do not.

Cooperative Learning
Exs. 20–25 You may wish to arrange small groups of four students to work on these exercises. Try to form groups containing both peer tutors and students who are having difficulty learning geometry.

Practice 11 for Section 2.3

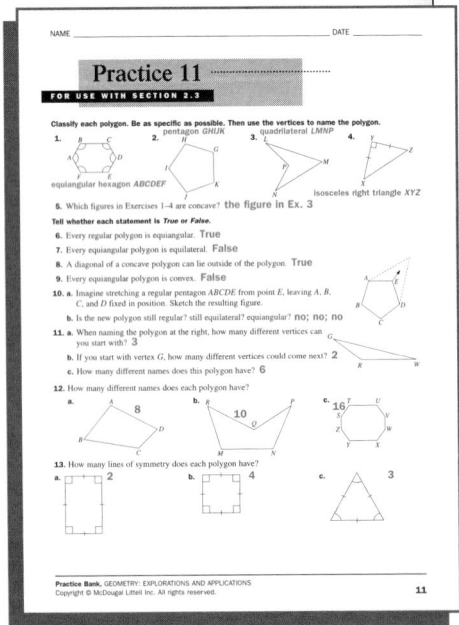

How many lines of symmetry does each polygon have?

17.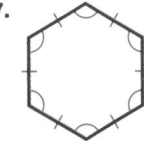

18.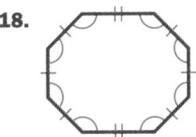

19.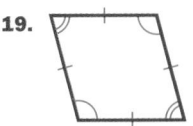

Open-ended Problems **Sketch an example of each figure.**

20. a pentagon with three congruent obtuse angles

21. a concave hexagon with one line of symmetry

22. a quadrilateral with three right angles

23. a convex hexagon with two consecutive congruent sides

24. a polygon divided into two pentagons by a diagonal

25. Try to make or sketch each polygon described in the table. Copy and complete the table by sketching each polygon that is possible.

Polygon	Triangle	Quadrilateral	Pentagon
Concave	not possible	?	?
Convex	?	?	?
Regular	?	?	?
Equilateral only	?	?	?
Equiangular only	?		?

ONGOING ASSESSMENT

26. **Writing** Describe at least two ways to classify a polygon. What parts of the polygon should you examine?

SPIRAL REVIEW

27. Identify the hypothesis and conclusion of this conditional statement: If a polygon is regular, then it is equiangular. *(Section 1.3)*

Find each unknown angle measure. *(Sections 2.1 and 2.2)*

28. 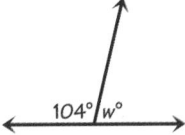 104°/w°

29. 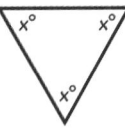 t°, t°, t°

30. 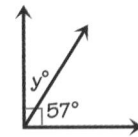 y°, 57°

17. 6 lines of symmetry
18. 2 lines of symmetry
19. 2 lines of symmetry
20–24. Answers may vary. Examples are given.

20. [figure]
21. [figure]

22. [figure]
23. [figure]
24.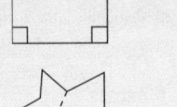

25. See answers in back of book.
26. Answers may vary. An example is given. You can classify a polygon by its number of sides, and by whether it is concave or convex. You can also classify it by the lengths of its sides and the measures of its angles, that is, by whether it is equilateral, equiangular, or regular.

27. hypothesis: a polygon is regular; conclusion: it is equiangular

28. 76°
29. 60°
30. 33°

2.4 Angles in Polygons

You can find the sum of the angle measures of any polygon just by knowing the number of sides it has. From now on in this book, unless stated otherwise, the term "polygon" refers to a convex polygon.

Learn how to...

• find the sum of the measures of the interior angles of a polygon

• find the sum of the measures of the exterior angles of a polygon

So you can...

• find specific measures for a given polygon

EXPLORATION
COOPERATIVE LEARNING

Finding Angle Measures in Polygons

Work in a group of four students.
You will need:

• dot paper or a geoboard and rubber bands

• a graphing calculator or graph paper

Repeat Steps 1 and 2 for each of the following polygons: triangle, quadrilateral, pentagon, hexagon, and heptagon.

1 Sketch the polygon on dot paper or make it on the geoboard. Draw all of its diagonals from one vertex. The diagonals should divide the polygon into triangles.

2 Make a table like the one below. How are the numbers in each column related to the numbers in the previous column?

Type of polygon	Number of sides	Number of triangles formed	Sum of angle measures of triangles	Sum of angle measures of polygon
triangle	3	1	180°	180°
?	?	?	?	?

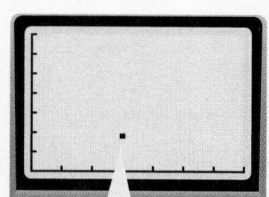

The point (3, 180) represents the sum of the measures of the angles of a triangle.

3 Graph the data as shown. Put the number of sides on the x-axis and the sum of the angle measures on the y-axis. Describe any patterns you see in the graph.

4 Extend your table to include an n-gon. Make a conjecture about the relationship between the number of sides of a polygon and the sum of its angle measures.

Exploration Note

Purpose
The purpose of this Exploration is to have students discover that the sum of the angle measures of an n-gon is $(n - 2)180°$.

Materials/Preparation
Each group needs dot paper or a geoboard and rubber bands and a graphing calculator or graph paper.

Procedure
Students use their dot paper or geoboards to make each type of convex polygon. For each polygon with more than 3 sides, students select a vertex and draw all diagonals that have that vertex as an endpoint as they complete the appropriate row of their table

in Step 2. Finally, students plot points for the data in columns 2 and 5 of their table.

Closure
Have three groups describe the graphs they obtained in Step 3. Then discuss the groups' conjectures for Step 4. Students should agree that the sum of the angle measures of an n-gon is $(n - 2)180°$.

Explorations Lab Manual
See the Manual for more commentary on this Exploration.
Diagram Masters 2, 3, 9

For answers to the Exploration, see answers in back of book.

Plan⇔Support

Objectives

• Find the sum of the measures of the interior angles of a polygon.

• Find the sum of the measures of the exterior angles of a polygon.

• Find specific measures for a given polygon.

Recommended Pacing

❖ **Core and Extended Courses**
Section 2.4: 2 days

❖ **Block Schedule**
Section 2.4: 1 block

Resource Materials

Lesson Support
Lesson Plan 2.4

Warm-Up Transparency 2.4

Practice Bank: Practice 12

Study Guide: Section 2.4

Explorations Lab Manual:
Additional Exploration 3
Diagram Masters 2, 3, 9

Challenge Problems: Set 12

Assessment Book: Test 7

Technology
Technology Book:
Calculator Activity 3

McDougal Littell Mathpack
Geometry Inventor Activity Book:
Activity 11

Graphing Calculator

Internet:
http://www.hmco.com

Warm-Up Exercises

1. Complete the equation using the diagram.

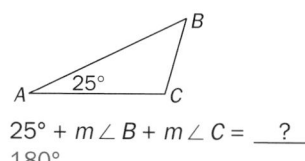

$25° + m\angle B + m\angle C =$ ___?___
180°

Find each value.

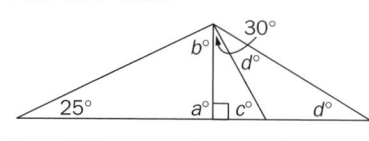

2. a 90°

3. b 55°

4. c 60°

5. d 30°

Section Note

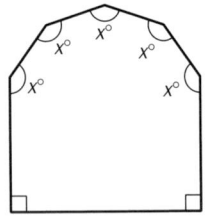

Using Technology
To make the graph in the Exploration using a TI-82, press STAT and choose 1:Edit. Use the arrow keys to move the cursor on top of L1 and press CLEAR ENTER. Repeat for L2. Then move the cursor to the first position of L1, type the first number of sides, 3, followed by ENTER. Continue with the next number of sides, 4, and so on. Then position the cursor on the first position of L2 and enter the sum of the angle measures in this list. To graph, press ZOOM and move the cursor down until it is on the option 9:ZoomStat and press ENTER.

Additional Example 1

The back end of a barn is shaped like the polygon in the diagram.

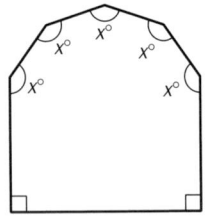

The side walls are perpendicular to the ground and the angles in the roof are congruent. Find the measure of each angle in the roof.
Step 1 Find the sum of the angle measures of a heptagon.
Sum of the angle measures
$$= (n - 2)180°$$
$$= (7 - 2)180°$$
$$= 900°$$
Step 2 Write and solve an equation.
$$90° + 90° + x° + x° + x° + x° + x° = 900°$$
Solve for x.
$$180 + 5x = 900$$
$$5x = 720$$
$$x = 144$$
Each angle in the roof has a measure of 144°.

Section Note

Using Technology
Students can explore angles of polygons and the sum of the angle measures of various polygons by using Activity 11 in the *Geometry Inventor Activity Book*.

In the Exploration you saw how to divide an *n*-gon into $(n - 2)$ triangles so that the sum of the angle measures of the triangles equals the sum of the angle measures of the polygon. This leads to the following formula.

The Angles of a Polygon

In a polygon with *n* sides, the sum of the angle measures is $(n - 2)180°$.

EXAMPLE 1   Application: Sports

In baseball and softball, home plate is a pentagon with three right angles and two congruent obtuse angles. Find the measure of each obtuse angle in home plate.

SOLUTION

Step 1 Find the sum of the angle measures of a pentagon.

Sum of the angle measures $= (n - 2)180°$
$$= (5 - 2)180$$
$$= 540$$

Step 2 Write and solve an equation.
$$90° + 90° + x° + 90° + x° = 540°$$
Solve for x.
$$270 + 2x = 540$$
$$2x = 270$$
$$x = 135$$

Each obtuse angle of home plate measures 135°.

EXAMPLE 2   Connection: Algebra

What polygon, if any, has angle measures that add up to 1080°?

SOLUTION

Use the formula above.

$$(n - 2)180° = \text{Sum of the angle measures of the polygon}$$
$$(n - 2)180 = 1080 \qquad \text{Divide both sides by 180.}$$
$$n - 2 = 6$$
$$n = 8$$

The polygon whose angle measures add up to 1080° is an octagon.

1. How could you use the graph that you made in the Exploration on page 79 to solve Example 2?

2. Can the sum of the angle measures of a polygon be 450°? Explain.

Exterior Angles of a Polygon

The angles that you looked at in the Explorations are *interior angles* of the polygons. If one side of a polygon is extended, the ray forms an *exterior angle* with an adjacent side of the polygon.

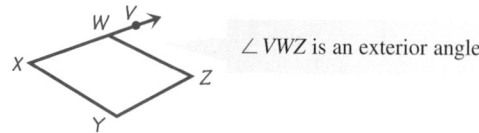

∠VWZ is an exterior angle.

At each vertex of a polygon, an interior angle and its adjacent exterior angle form a linear pair. You can use this fact to find the sum of the measures of the exterior angles.

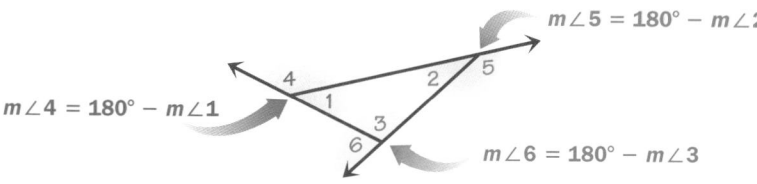

$m\angle 4 = 180° - m\angle 1$

$m\angle 5 = 180° - m\angle 2$

$m\angle 6 = 180° - m\angle 3$

$$m\angle 4 + m\angle 5 + m\angle 6 = (180° - m\angle 1) + (180° - m\angle 2) + (180° - m\angle 3)$$
$$= 3(180) - m\angle 1 - m\angle 2 - m\angle 3$$
$$= 540 - (m\angle 1 + m\angle 2 + m\angle 3)$$
$$= 540 - 180$$
$$= 360°$$

For an *n*-gon:

$$\begin{array}{c} \text{Sum of the measures} \\ \text{of } n \text{ exterior angles} \end{array} = \begin{array}{c} \text{Sum of the measures} \\ \text{of } n \text{ linear pairs} \end{array} - \begin{array}{c} \text{Sum of the measures} \\ \text{of } n \text{ interior angles} \end{array}$$
$$= n(180°) - (n - 2)180°$$
$$= 180n - 180n + 360$$
$$= 360°$$

The Exterior Angles of a Polygon

The sum of the measures of the exterior angles of any polygon is 360°.

2.4 Angles in Polygons **81**

....

ANSWERS Section 2.4

Think and Communicate

1. Extend the line until it reaches the *y*-value of 1080°. Then see if the corresponding *x*-value is an integer. If so, this is the value of *n*.

2. No; the sum of the angle measures of a polygon must be a multiple of 180°, and 450° is not divisible by 180°.

Additional Example 2

What polygon, if any, has angle measures that add up to 2000°? Use the formula for the sum of the measures of an *n*-gon.

$(n - 2)180°$ = Sum of the angle measures of the polygon

$(n - 2)180° = 2000$

$n - 2 = \dfrac{100}{9}$

$n - 2 = 11\dfrac{1}{9}$

$n = 13\dfrac{1}{9}$

Since the number of sides of a polygon must be a whole number, there is no polygon whose angle measures add up to 2000°.

Think and Communicate

For question 2, students can use the fact that $(n - 2)180 = 450$ does not have a whole-number solution. Alternatively, they could use the table or graph from the Exploration on page 79.

Section Notes

Teaching Tip
Point out to students that any time an angle is referred to as "an angle of a polygon," it is assumed that the angle is an *interior* angle. An exterior angle of a polygon is always a supplement of the adjacent interior angle.

Integrating Algebra
In the demonstration that the sum of the measures of the exterior angles of any polygon is 360°, you may wish to have students explain why the sequence of equations below the triangle is true. For example, $m\angle 4 = (180° - m\angle 1)$, so $(180° - m\angle 1)$ can be *substituted* for $m\angle 4$ (both expressions represent the same real number). Ask students also how the expression $540 - (m\angle 1 + m\angle 2 + m\angle 3)$ is arrived at from the preceding expression.

Second-Language Learners
You may want to give students learning English examples of how the terms *interior* and *exterior* are used with the parts of a house in order to clarify the use of these terms in mathematics.

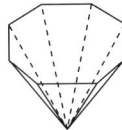

Additional Example 3

A gemstone has been cut so that the top face is in the shape of a regular octagon.

Find the measure of each interior and each exterior angle of a regular octagon.

Method 1 First find the sum of the measures of the interior angles.

$(n - 2)180° = (8 - 2)180°$
$= 1080°$

Then divide the sum by 8 to find the measure of each interior angle.

$\frac{1080}{8} = 135°$

Divide 360° by 8 to find the measure of each exterior angle.

$\frac{360°}{8} = 45°$

Method 2 Start by finding the measure of one exterior angle.

$\frac{360°}{8} = 45°$

Each interior angle is the supplement of an exterior angle.

$180° - 45° = 135°$

Closure Question

How would you find the sum of the measures of the interior measures of the angles of a polygon, and the sum of the measures of the exterior angles of a polygon?

To find the sum of the interior angles, use the formula $(n - 2)180°$, where n is the number of sides of the polygon. The sum of the measures of the exterior angles of any polygon is 360°.

THINK AND COMMUNICATE

3. Repeat the calculations in the middle of page 81 for a hexagon. Is the result different? Explain.

4. What is the measure of each exterior angle of a regular quadrilateral?

EXAMPLE 3 Interview: Virgil Lueth

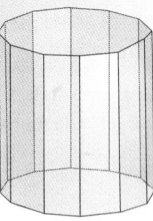

Some crystals take the form of a *dihexagonal prism*, whose cross-section is an equiangular 12-gon. Find the measure of each interior and each exterior angle of an equiangular 12-gon.

SOLUTION

Method 1

First find the sum of the measures of the interior angles.

$(n - 2)180° = (12 - 2)180°$
$= 1800°$

Then divide the sum by 12 to find the measure of each **interior angle**.

$\frac{1800°}{12} = 150°$

Divide 360° by 12 to find the measure of each **exterior angle**.

$\frac{360°}{12} = 30°$

Method 2

Start by finding the measure of one **exterior angle**.

$\frac{360°}{12} = 30°$

Each **interior angle** is the supplement of an exterior angle.

$180° - 30° = 150°$

☑ CHECKING KEY CONCEPTS

Find the sum of the measures of the interior angles of each polygon.

1. hexagon   2. 25-gon   3. $2x$-gon

What polygon, if any, has angle measures that add up to the given sum?

4. 1440°   5. 4860°   6. 2700°

7. Find $m\angle 1$.

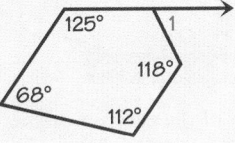

125°   1
118°
68°
112°

82   Chapter 2 *Triangles and Polygons*

Think and Communicate

3. The result is the same, since the sum of the measures of the exterior angles of any polygon is 360°. The only difference is that you have to find the sum of six exterior angles.

4. 90°

Checking Key Concepts

1. 720°

2. 4140°

3. $(2x - 2)180°$, or $(x - 1)360°$

4. decagon

5. 29-gon

6. 17-gon

7. 63°

Exercises and Applications

1. 360°   2. 900°

3. 13,140°   4. 115°

5. 135°

6. $z° = 80°$; $(z - 5)° = 75°$; $(z + 50)° = 130°$

7. interior: 108°; exterior: 72°

8. interior: $147\frac{3}{11}°$; exterior: $32\frac{8}{11}°$

9. interior: 168.75°; exterior: 11.25°

2.4 | **Exercises and Applications**

Extra Practice
exercises on page 662

Find the sum of the measures of the interior angles of each polygon.

1. quadrilateral

2. heptagon

3. 75-gon

Find the value of each variable.

4.
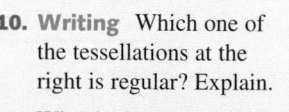

5.
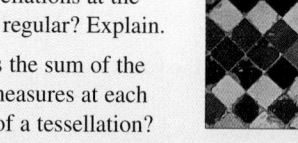

6.

Find the measure of each interior and exterior angle of the polygon.

7. regular pentagon

8. regular 11-gon

9. regular 32-gon

Connection | **ART**

A *tessellation* is a pattern of polygons that covers the plane without gaps or overlaps. Tessellations have been used by many cultures to cover walls, floors, and ceilings. A regular tessellation is formed by repeating a single regular polygon. The angle measures of a polygon determine whether it can form a regular tessellation.

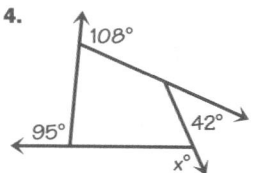
Floor tiles from Mexico

10. Writing Which one of the tessellations at the right is regular? Explain.

11. What is the sum of the angle measures at each vertex of a tessellation? Explain your reasoning.

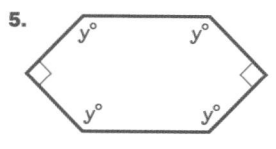
Wall tiles from Spain

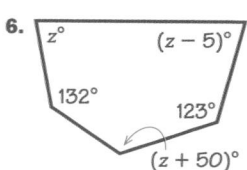
Floor tiles from Italy

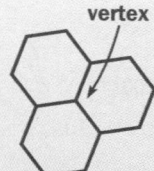
vertex

For Exercises 12–15, answer the following questions:
a. What is the measure of an interior angle of a regular *n*-gon?
b. How many regular *n*-gons can meet at a vertex without overlapping?
c. Can a regular *n*-gon form a tessellation?

12. $n = 3$

13. $n = 4$

14. $n = 5$

15. $n = 6$

16. a. How many regular heptagons can meet at a vertex without overlapping?
b. How many regular octagons can meet at a vertex without overlapping?
c. Explain why only two regular *n*-gons can meet at a vertex without overlapping for all $n \geq 7$.
d. Which regular polygons can form a regular tessellation?

2.4 Angles in Polygons **83**

---

10. The wall tiles from Spain form a regular tessellation because the pattern is made by repeating a single regular polygon, a square. The tiles from Mexico and Italy both contain nonregular polygons.

11. 360°; You can draw a circle that contains each vertex so that it intersects only the sides adjacent to each vertex. These angles combined form one rotation around a circle, so the sum of their measures is 360°.

12. a. 60°   **b.** 6
      c. Yes.

13. a. 90°   **b.** 4
      c. Yes.

14. a. 108°   **b.** 3
      c. No.

15. a. 120°   **b.** 3
      c. Yes.

16. a. 2 regular heptagons
      b. 2 regular octagons

c. For $n \geq 7$, the interior angles have measures greater than 128°. Because 128° divides into 360° only twice (and the interior angles of a regular *n*-gon are always less than 180°), only two regular *n*-gons will meet without overlapping when $n \geq 7$.

d. triangle, quadrilateral, hexagon

Apply⇔Assess

Suggested Assignment

❖ **Core Course**
 Day 1 Exs. 1–9, 12–16, 19–22
 Day 2 Exs. 24–35, 42–45, AYP

❖ **Extended Course**
 Day 1 Exs. 1–23
 Day 2 Exs. 24–45, AYP

❖ **Block Schedule**
 Day 10 Exs. 1–9, 12–16,
      19–22, 24–35, 42–45,
      AYP

Exercise Notes

Integrating Algebra
Exs. 4–6 These exercises can be completed by writing and solving an equation. Ask students to explain the reasoning they used to write their equations.

Application
Exs. 10, 11 Tessellations are commonly used to create decorative patterns in floor tiles and wallpaper. You may wish to have students consult books on architecture and design to gather more examples of tessellations.

Multicultural Note
Exs. 10, 11 Many Native American tribes have traditionally used tessellated patterns in various forms of art, especially in basketry, weaving, and sewing. Peoples as diverse as the Flatheads from Montana, the Menominees from the Great Lakes, the Miamis from Indiana and Ohio, the Senecas from New York, and the Washo from the Great Basin near Lake Tahoe have created pieces of art that incorporate tessellations. Often the tessellated pattern is one of many themes in a piece of art, and does not entirely cover it.

Problem Solving
Ex. 16 A thorough discussion of part (c) of this exercise would be helpful for many students. Ask students how they can be sure that two angles of a regular polygon can always meet without overlapping. (No matter how many angles a regular polygon has, the angle measure *a*° of an angle of the polygon is less than 180°. Since *a*° < 180°, then 2*a*° < 360°.)

Exercise Notes

Common Error

Ex. 23 Some students may incorrectly assume that any set of positive numbers with a sum of 6(180) = 1080 will work. Remind students, if necessary, that the exterior angles must have measures whose sum is 360°.

Integrating Algebra

Exs. 24–29 A review of the solutions to these exercises can help students see if they are applying their algebraic skills correctly.

Geometric Thinking

Ex. 33 Some students may approach this problem by trying to draw a polygon with the given angle measures. After several failures, they may be tempted to assert that no such polygon exists. The conclusion is correct, but the thinking that leads to it is not. Students should understand that the fact they have not been successful in drawing the polygon does not imply it cannot be drawn. They may have simply not found the right approach. The way to demonstrate conclusively that no such polygon exists is to examine the sum of the measures of the corresponding exterior angles.

Spatial Reasoning

Ex. 35 This investigation exercise illustrates how a geometric figure can be used to arrive at an important fact, namely that the sum of the measures of an *n*-gon is (*n* − 2)180°. Students can see that by examining more than one polygon they will get the same result, and that from an investigation of different polygons, a *conjecture* can be made. However, other methods will have to be used (deductive reasoning in Chapter 3) to verify that the conjecture is true *for all polygons.*

Exercises 17 and 18 show equiangular windows from the wall of a Chinese garden. Find the measure of an interior and exterior angle of each.

17.

18.

What polygon, if any, has angle measures that add up to the given sum?

19. 3600°   **20.** 3060°   **21.** 2160°   **22.** $(180y − 360)°$

23. Open-ended Problem Find a possible set of eight measures for the interior angles of an octagon that is not equiangular.

ALGEBRA In Exercises 24–26, the measures of the exterior angles of a polygon are given. Find the value of each variable.

24. 65°, 70°, 115°, *w*°   **25.** 105°, *z*°, 4*z*°   **26.** 90°, 90°, 40°, 40°, *x*°, *x*°

ALGEBRA Find the measure of each interior angle.

27. 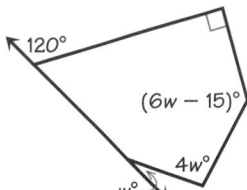   **28.** 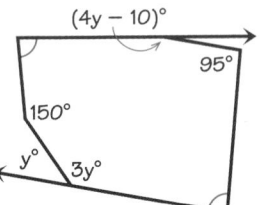   **29.**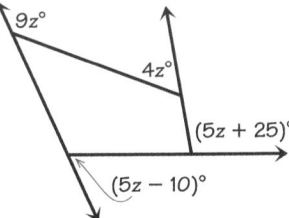

The measure of one interior angle of a regular *n*-gon is given. Find the value of *n*.

30. 144°   **31.** 162°   **32.** 171°

33. Challenge Show that the interior angles of a convex polygon cannot include four angles that measure 90°, 75°, 130°, and 60°.

34. SAT/ACT Preview Which of the following *cannot* be the sum of the measures of the interior angles of a polygon?

   A. 1260°   **B.** 360°   **C.** 1500°   **D.** 180°   **E.** 2700°

35. Investigation Draw a polygon. From a point inside the polygon, draw a segment to each vertex.

   a. How many sides does the polygon have? How many triangles are formed by the segments?

   b. What is the sum of the angle measures of the triangles?

   c. Which angles of the triangles do not share a vertex with the polygon? What is the sum of the measures of these angles?

   d. Show how these steps lead to the expression (180*n* − 360)°, or the same expression on page 80.

84   Chapter 2 *Triangles and Polygons*

---

17. interior: 120°; exterior: 60°

18. interior: 135°; exterior: 45°

19. 22-gon

20. 19-gon

21. 14-gon

22. *y*-gon

23. Answers may vary. An example is given. 139°, 138°, 137°, 136°, 134°, 133°, 132°, 131°

24. 110°

25. *z*° = 51°; 4*z*° = 204°

26. 50°

27. *w*° = 25°; 4*w*° = 100°; (6*w* − 15)° = 135°

28. *y*° = 45°; 3*y*° = 135°; (4*y* − 10)° = 170°

29. 4*z*° = 60°; 9*z*° = 135°; (5*z* − 10)° = 65°; (5*z* + 25)° = 100°

30. 10   31. 20

32. 40

33. The measures of the exterior angles corresponding to these interior angles are 90°, 105°, 50°, and 120°. But the sum of the measures of the exterior angles of a convex polygon is 360°, and 90° + 105° + 50° + 120° = 365°.

34. C

35. a. Answers may vary. An example is given.

7 sides; 7 triangles

b. 1260°

c. the angles with vertex at the interior point; 360°

d. The sum of the measures of the angles of the *n*-gon is the sum of the measures of the angles of the *n* triangles, 180*n*, minus the sum of the measures of those angles of the triangles that

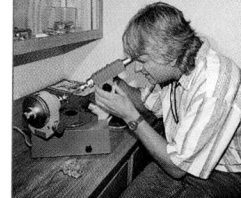

Look back at the article on pages 54–56.

Imagine two planes inside a crystal, each perpendicular to a face. The angle that these planes form when they meet inside the crystal is called the **interfacial angle.** *When it is not possible to measure this angle directly, a* **contact goniometer** *is used to measure a congruent angle outside the crystal.*

angle measured by goniometer

interfacial angle

36. What is the sum of the measures of the interior angles of quadrilateral *ABCD*?

37. Logical Reasoning Complete the statements in parts (a) and (b). Explain your answers.
  a. $m\angle ABC + m\angle ADC = \underline{\ ?\ }°$
  b. $m\angle ABC + m\angle EBC = \underline{\ ?\ }°$
  c. Use your completed statements from parts (a) and (b) to explain why $\angle ADC \cong \angle EBC$.

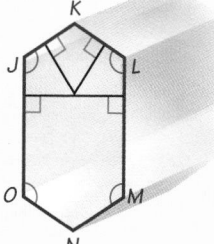

At the left is a cross section of an aragonite crystal. Use the diagram for Exercises 39–41.

38. What is the measure of the interfacial angle that the goniometer at the left is measuring?

BY THE WAY

The interfacial angles of a crystal are congruent to the angles found in a molecule of the substance. Crystallographers now use lasers to measure the angles directly from the molecules.

39. The interfacial angle of faces $\overline{JK}$ and $\overline{KL}$ is 64.8°.
  a. Find $m\angle K$.
  b. Find the measures of $\angle L$, $\angle M$, $\angle O$, and $\angle J$. How did you get your answers?
  c. What is the interfacial angle of faces $\overline{KL}$ and $\overline{LM}$?

40. Name the three pairs of faces that have an interfacial angle congruent to the interfacial angle of faces $\overline{KL}$ and $\overline{LM}$. Explain why these four angles must be congruent.

41. Writing Explain why $\angle K \cong \angle N$.

2.4 Angles in Polygons **85**

Exercise Notes

Topic Spiraling: Preview
Exs. 37, 39–41 The logical reasoning that students employ to solve these exercises provides a sound preparation and introduction to the content of Chapter 3, Reasoning in Geometry.

---

are not part of the angles of the polygon, 360. This gives 180*n* – 360° for the sum of the measures of the angles of the polygon. So, for a heptagon, for example, the sum is 180(7) – 360 = 1260 – 360 = 900°, which agrees with earlier results.

36. 360°

37. a. 180°; The sum of the measures of interior angles of a

quadrilateral is 360°, so $m\angle ABC + m\angle ADC = 360 - 90 - 90 = 180°$.

  b. 180°; The sum of the measure of the angles in a linear pair is 180°.

  c. Two angles that are supplementary to the same angle are congruent.

38. 125°

39. a. 115.2°

b. $m\angle L = 122.4°$; $m\angle M = 122.4°$; $m\angle O = 122.4°$; $m\angle J = 122.4°$; The sum of the interior angles of the pentagon containing vertices *J*, *K*, and *L* is 540°. Since two of the angles are right angles, and $m\angle K = 115.2°$, the sum of the measures of $\angle L$ and $\angle J$ is 540 – (180 + 115.2) = 244.8°. Then, since they are congruent, each measures 122.4° (and $\angle M$ and $\angle O$ are congruent to these two).

  c. 57.6°

40. $\overline{JK}$ and $\overline{JO}$, $\overline{JO}$ and $\overline{NO}$, $\overline{MN}$ and $\overline{LM}$; Each angle is the fourth interior angle of a quadrilateral whose other three angles measure 90°, 90°, and 122.4°, so the measure of each angle is 360° – (90° + 90° + 122.4°) = 57.6°.

41. The sum of the measures of the interior angles of *JKLMNO* must be 720°, so $m\angle N = 720 - 4(122.4) - 115.2 = 115.2°$. Thus, $m\angle K \cong m\angle N$ (from Ex. 39(a)).

Exercise Notes

Assessment Note

Ex. 42 This exercise can be used to assess students' understanding of the result about the sum of the angle measures of a polygon. It can also provide feedback on reasoning and writing skills.

Assess Your Progress

Journal Entry

Encourage students to use a ruler or a straightedge to draw their polygons. This will ensure that they create neat and accurate figures.

Progress Check 2.3–2.4

See page 103.

Practice 12 for Section 2.4

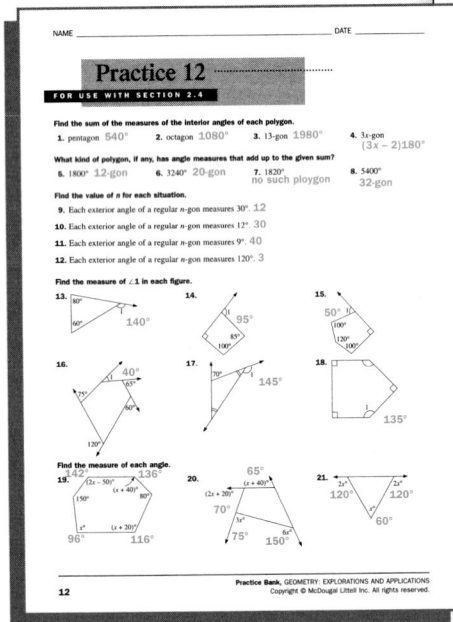

---

42. Writing Suppose you know that the angles of a regular polygon are acute. Explain how you can tell how many sides the polygon has. Can you do the same thing if you know that the angles are right? if they are obtuse? Explain why or why not.

43. Explain the difference between skew lines and parallel lines. *(Section 1.4)*

Sketch each figure. *(Section 2.3)*

44. a hexagon with three right angles

45. a concave octagon

ASSESS YOUR PROGRESS

VOCABULARY

polygon (p. 73)   equiangular (p. 73)
triangle (p. 73)   regular (p. 73)
quadrilateral (p. 73)   consecutive angles (p. 74)
pentagon (p. 73)   consecutive sides (p. 74)
hexagon (p. 73)   diagonal (p. 74)
equilateral (p. 73)

Classify each polygon. Be as specific as possible. *(Section 2.3)*

1.    2. 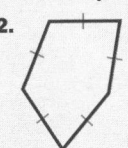   3. 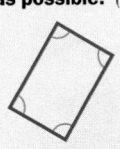   4.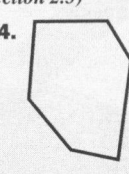

5. a. Sketch a regular polygon *ABCDEF*.

 b. How many lines of symmetry does *ABCDEF* have?

 c. Name two consecutive sides of *ABCDEF*.

 d. Name two consecutive angles of *ABCDEF*. *(Section 2.3)*

6. ALGEBRA Three interior angles of a hexagon measure 70°, 120°, and 110°. The other three angles are congruent. Find the measure of each angle. *(Section 2.4)*

7. Find the measure of each interior and exterior angle of a regular 24-gon. *(Section 2.4)*

8. Journal Make a diagram of the different polygons you learned about in Section 2.3. Give angle measures whenever possible.

---

42. If the interior angles are acute, then the exterior angles are obtuse. Since the sum of their measures must be 360°, there must be fewer than four vertices. The only polygon with fewer than four vertices is a triangle, so the regular polygon has three sides. If the interior angles are right angles, then so are the exterior angles, and 360 ÷ 90 = 4, so there are four exterior angles and, thus, four sides. If the interior angles are obtuse, then the exterior angles are acute. 360 ÷ n ≤ 90, so n ≥ 4, which has any number of possibilities.

43. Parallel lines are coplanar lines that do not intersect, while skew lines are not coplanar and also do not intersect.

44, 45. Answers may vary. Examples are given.

44.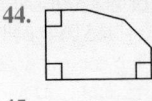

45.

Assess Your Progress

1. concave hexagon

2. convex, equilateral pentagon

3. convex, equiangular quadrilateral (a rectangle)

4. convex hexagon

5. a.

b. 6 lines of symmetry

c. $\overline{AB}$ and $\overline{BC}$, $\overline{BC}$ and $\overline{CD}$, $\overline{CD}$ and $\overline{DE}$, $\overline{DE}$ and $\overline{EF}$, $\overline{EF}$ and $\overline{AF}$ or $\overline{AF}$ and $\overline{AB}$

d. ∠A and ∠B, ∠B and ∠C, ∠C and ∠D, ∠D and ∠E, ∠E and ∠F, or ∠F and ∠A

6. 140°

7. interior: 165°; exterior: 15°

8. Check students' work.

2.5 Parallelograms

A *parallelogram* is a type of quadrilateral. In this section, you will explore the properties of parallelograms and learn about the special types of parallelograms.

Learn how to...
- **identify types of quadrilaterals**

So you can...
- **find the measures of angles and segments in quadrilaterals**

EXPLORATION
COOPERATIVE LEARNING

Analyzing Parallelograms

Work in a group of four students. You will need:
- lined notebook paper
- patty paper or tracing paper
- a ruler
- a protractor

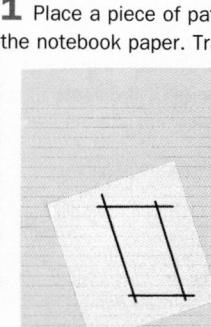

1 Place a piece of patty paper on the notebook paper. Trace two lines as shown.

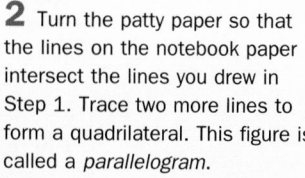

2 Turn the patty paper so that the lines on the notebook paper intersect the lines you drew in Step 1. Trace two more lines to form a quadrilateral. This figure is called a *parallelogram.*

3 Measure each side of the parallelogram. What do you notice?

4 Measure each angle. How do the measures of opposite angles compare?

5 Draw a diagonal of the parallelogram and find its midpoint. Draw the other diagonal. What do you notice?

6 Repeat Steps 1–5 at least two more times. Use different side lengths and angle measures for each parallelogram.

7 Make three conjectures about the parts of a parallelogram.

Exploration Note

Purpose
The purpose of this Exploration is to have students discover some properties of a parallelogram.

Materials/Preparation
Each group needs lined paper, patty paper or tracing paper, a ruler, and a protractor.

Procedure
Students trace parallel lines from a sheet of lined paper in order to draw several parallelograms. For each parallelogram, they measure the sides and angles, draw a diagonal and mark its midpoint, and then draw the other diagonal. Students then observe how

the measures of opposite sides and angles are related and where the diagonals intersect. They then conjecture what must be true about these parts of parallelograms.

Closure
Have students discuss their conjectures about the parts of a parallelogram. Students should observe that the opposite sides are congruent, the opposite angles are congruent, and the diagonals bisect each other.

Explorations Lab Manual
See the Manual for more commentary on this Exploration.

For answers to the Exploration, see following page.

Objectives
- Identify types of quadrilaterals.
- Find the measures of angles and segments in quadrilaterals.

Recommended Pacing
❖ **Core and Extended Courses**
Section 2.5: 1 day
❖ **Block Schedule**
Section 2.5: $\frac{1}{2}$ block
(with Section 2.6)

Resource Materials

Lesson Support
Lesson Plan 2.5
Warm-Up Transparency 2.5
Practice Bank: Practice 13
Study Guide: Section 2.5
Challenge Problems: Set 13

Technology
Technology Book:
  TI-92 Activity 2
Internet:
  http://www.hmco.com

Warm-Up Exercises

Find the value of each variable.

1. 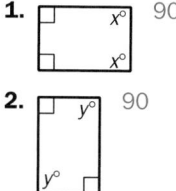 90

2. 90

3. 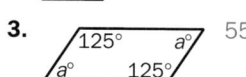 55

4. Can one pair of opposite angles of a quadrilateral both have measures of 50° while the other pair have measures of 130°?
Yes.

5. Can one pair of opposite angles of a quadrilateral both have measures of 40° while the other pair have measures of 135°?
No.

Teach⇔Interact

About Example 1

Communication: Discussion
This Example illustrates that only one angle measure in a parallelogram is needed to find the remaining angle measures and that two side lengths are needed to find the remaining side lengths. Discuss these ideas with students.

Additional Example 1

Find the value of each variable for the figure shown. In the figure, quadrilaterals *ABCD* and *AFED* are parallelograms.

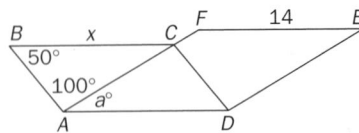

Since opposite sides of a parallelogram are congruent, *AD* = *FE* = 14. Since $\overline{AD}$ is also opposite $\overline{BC}$ in □*ABCD*, *BC* = 14, Therefore, *x* = 14.
Opposite angles of a parallelogram are congruent. So *m*∠*BAD* = *m*∠*BCD* and since *m*∠*BAD* = 100° + *a*°, then *m*∠*BCD* = 100° + *a*°. Since ∠*ABC* and ∠*CDA* are opposite angles in a parallelogram, *m*∠*CDA* = 50°. Now,
(100° + *a*°) + (100° + *a*°) + 50° + 50° = 360
300 + 2*a* = 360
2*a* = 60
a = 30
Therefore, *a* = 30.

Section Note

Geometric Thinking
Ask students to explain why a square is both a rhombus and a rectangle.

A **parallelogram** is a quadrilateral with both pairs of opposite sides parallel. The expression □*ABCD* is read "parallelogram *ABCD*." You may have discovered the following properties of parallelograms in the Exploration.

Opposite Parts of a Parallelogram

The opposite sides of a parallelogram are congruent.

The opposite angles of a parallelogram are congruent.

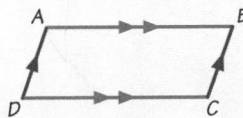

If *ABCD* is a parallelogram, then $\overline{AB} \cong \overline{DC}$ and $\overline{AD} \cong \overline{BC}$.

If *ABCD* is a parallelogram, then ∠*A* ≅ ∠*C* and ∠*B* ≅ ∠*D*.

EXAMPLE 1   Connection: Algebra

Find the value of each variable in □*PQRS*. Explain your reasoning.

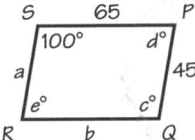

SOLUTION

Since opposite sides of a parallelogram are congruent, *SR* = *PQ* and *SP* = *RQ*. So *a* = 45 and *b* = 65.

Opposite angles of a parallelogram are congruent. So ∠*S* ≅ ∠*Q* and *c* = 100. Use the sum of the angle measures of a quadrilateral to find the other angle measures.

$c° + d° + e° + 100° = 360°$

$100 + d + d + 100 = 360$

$2d + 200 = 360$

$d = 80$

Since the opposite angles are congruent, you can substitute **100** for *c* and *d* for *e*.

Solve for *d*.

Therefore, *d* = 80 and *e* = 80.

There are three special kinds of parallelograms.

A **rectangle** is an equiangular parallelogram.

A **rhombus** is an equilateral parallelogram.

A **square** is a regular parallelogram.

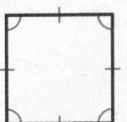

ANSWERS Section 2.5

Exploration

3. The opposite sides are congruent.

4. The opposite angles are congruent.

5. The diagonals intersect at their midpoints.

6. The results are the same.

7. Opposite sides of a parallelogram are congruent. Opposite angles of a parallelogram are congruent. The diagonals of a parallelogram bisect each other.

This diagram shows how these quadrilaterals are related.

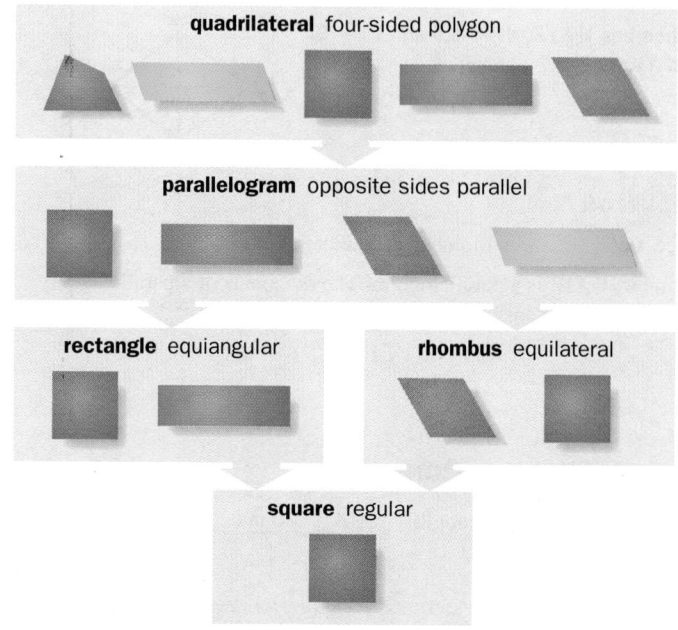

quadrilateral four-sided polygon

parallelogram opposite sides parallel

rectangle equiangular   **rhombus** equilateral

square regular

A quadrilateral may or may not have some of the properties of the quadrilaterals linked below it, but each quadrilateral in the diagram has all the properties of the quadrilaterals linked above it.

> **Diagonals of Parallelograms**
>
> The diagonals of a parallelogram bisect each other.
>
> If *ABCD* is a parallelogram, then $\overline{AC}$ bisects $\overline{BD}$ and $\overline{BD}$ bisects $\overline{AC}$.

THINK AND COMMUNICATE

1. What is the measure of each angle of a rectangle? How do you know?

2. How does the diagram above show that a parallelogram is a special type of quadrilateral?

3. When is a rectangle a quadrilateral? a parallelogram? a square?

4. When is a rhombus a quadrilateral? a parallelogram? a square?

5. Is a square a quadrilateral? a parallelogram? a rectangle? a rhombus?

6. Do the diagonals bisect each other in a rectangle? in a rhombus? in a square? Explain.

2.5 Parallelograms **89**

Think and Communicate

1. 90°; A rectangle is a quadrilateral, so the sum of its angle measures is 360°. Because it is equiangular, each of the four angles has the same measure, so 360 ÷ 4 = 90°.

2. Because the parallelogram is below the quadrilateral in the diagram, it must have the property of being a quadrilateral. It is a special type of quadrilateral because all parallelograms are quadrilaterals, but not all quadrilaterals are parallelograms.

3. always; always; when it is also equilateral

4. always; always; when it is also equiangular

5. Yes; Yes; Yes; Yes.

6. Yes; Yes; Yes; the diagonals of all parallelograms bisect each other. Because rectangles, rhombuses, and squares are all special kinds of parallelograms, their diagonals must bisect each other.

Additional Example 2

In the diagram, quadrilaterals *ABCD* and *ACXY* are squares, and *DX* = 5. Find the values of *n* and *k*.

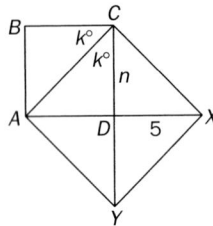

In square *ACXY*, diagonals $\overline{CY}$ and $\overline{AX}$ bisect each other. Therefore, *AD* = *DX* = 5. Since $\overline{AD}$ is a side of square *ABCD* and all sides of a square are congruent, then *n* = *CD* = *AD* = 5.
Since *ABCD* is a square,
$m\angle BCD = 90°$.
$m\angle BCD = k° + k° = 90°$
$$2k = 90$$
$$k = 45$$
Therefore, *n* = 5 and *k* = 45.

Closure Question

Describe a parallelogram and state three important properties that are true for all parallelograms.
A parallelogram is a quadrilateral in which both pairs of opposite sides are parallel. Both pairs of opposite sides of a parallelogram are congruent. Both pairs of opposite angles are congruent. The diagonals of a parallelogram bisect each other.

Apply⇔Assess

Suggested Assignment

❖ **Core Course**
Exs. 1–10, 16–30

❖ **Extended Course**
Exs. 1–30

❖ **Block Schedule**
Day 11 Exs. 1–10, 16–30

Exercise Notes

Geometric Thinking
Exs. 1–9 When reviewing the answers to these exercises, be sure students can give appropriate reasons to support their answers.

EXAMPLE 2

In rhombus *WXYZ*, *XW* = 15 and *VY* = 9. Find *XY* and *WY*.

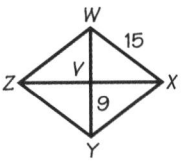

SOLUTION

Since *XW* = 15 and a rhombus is equilateral, *XY* = 15.

Rhombus *WXYZ* is a parallelogram. The diagonals of a parallelogram bisect each other, so *V* is the midpoint of $\overline{WY}$.

$$WY = 2(VY) = 2(9) = 18$$

✓ CHECKING KEY CONCEPTS

1. Find each length or angle measure of □*ABCD*.
  a. $m\angle A$
  b. $m\angle B$
  c. *CD*

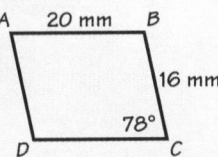

Tell whether each statement is *True* or *False*. If the statement is false, sketch a counterexample and justify your answer.

2. Every square is a rhombus.

3. Every rectangle is a square.

4. The diagonals of a rhombus bisect each other.

2.5 Exercises and Applications

Extra Practice exercises on page 663

Find each length or angle measure.

1. rectangle *UVWX*
  a. *UW*
  b. $m\angle VWX$
  c. *XW*

2. rhombus *EFGH*
  a. *HG*
  b. *GF*
  c. $m\angle F$

3. □*JKLM*
  a. $m\angle JLM$
  b. $m\angle KLM$
  c. $m\angle JMN$

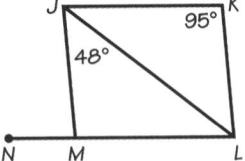

90   Chapter 2 *Triangles and Polygons*

Checking Key Concepts	Exercises and Applications	4. 8	5. 13
1. a. 78°	1. a. 15 m	6. 90°	7. 108°
b. 102°	b. 90°	8. 36°	9. 36°
c. 20 mm	c. 12 m		

Checking Key Concepts

1. a. 78°
   b. 102°
   c. 20 mm

2. True.

3. False; a rectangle is a square only if it is also equilateral.

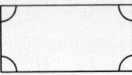

4. True.

Exercises and Applications

1. a. 15 m
   b. 90°
   c. 12 m

2. a. 14 ft
   b. 14 ft
   c. 60°

3. a. 37°
   b. 85°
   c. 85°

4. 8
5. 13
6. 90°
7. 108°
8. 36°
9. 36°

10. Both are correct. A square is a rhombus because it is an equilateral parallelogram, but it is also equiangular. A square is a rectangle because it is an equiangular parallelogram, but it is also equilateral.

11. a. Answers may vary.
    b. a rhombus

Find each length or angle measure of rectangle *JKLM*.

4. *NK*

5. *ML*

6. *m* ∠ *JML*

7. *m* ∠ *JNK*

8. *m* ∠ *KJN*

9. *m* ∠ *JKN*

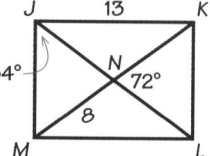

10. **Writing** Sarita says that a square is an equiangular rhombus. Bob says that a square is an equilateral rectangle. Who is correct? Explain.

Investigation **For Exercises 11 and 12, you will need scissors and a sheet of paper.**

11. **a.** Fold a sheet of paper in quarters as shown. Cut off the folded corner. What shape do you think you will have when you unfold the paper?

    b. Unfold the paper. What shape did you get?

12. Tell whether it is possible to cut the folded paper to produce each quadrilateral. If it is possible, describe how. If not, explain why not.

    a. a parallelogram that is not a rhombus

    b. a rhombus

    c. a rectangle that is not a square

    d. a square

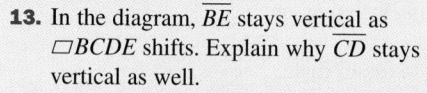

Connection ▸ OPTICS

When amateur astronomers or birdwatchers need to hold binoculars steady for a long period of time, they use special binocular mounts. This mount features a parallelogram that shifts to adjust to the height of the viewer.

13. In the diagram, $\overline{BE}$ stays vertical as ▱*BCDE* shifts. Explain why $\overline{CD}$ stays vertical as well.

14. Suppose the mount is set so that $m \angle ABE = 55°$.

    a. Find $m \angle BCD$.

    b. Find $m \angle FCB$.

    c. What is the angle of elevation when the line of sight, $\overline{GC}$, makes a 67° angle with the top piece of the mount, $\overline{AC}$?

15. **Challenge** Show that the angle of elevation of the binoculars does not change as ▱*BCDE* shifts.

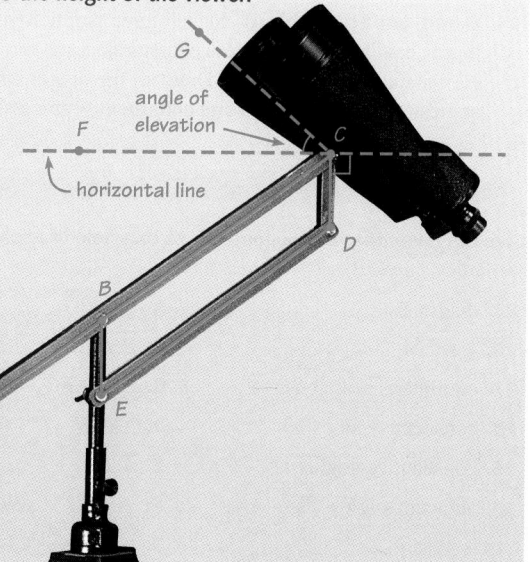

2.5 Parallelograms **91**

91

---

12. **a.** No; the figure will always be equilateral because the length of each side is the length of the cut.

    b. Yes; the figure will always be a rhombus given the directions in Ex. 11.

    c. No; the figure will always be equilateral because the length of each side is the length of the cut.

    d. Yes; cut off the corner so that it is in the shape of an isosceles triangle.

13. Answers may vary. An example is given. Because *BCDE* is a parallelogram, opposite sides must remain parallel. Because $\overline{BE}$ stays vertical, $\overline{CD}$ must remain vertical to remain parallel.

14. **a.** 55°

    b. 35°

    c. 32°

15. The angle of the binoculars with respect to the vertical segment $\overline{CD}$, or to ∠ *GCD*, is fixed by the binocular mount. Since $m \angle GCD = m \angle BCD + m \angle FCB + m \angle GCF$, $m \angle GCF = m \angle GCD - (m \angle BCD + m \angle FCB) = m \angle GCD - 90°$. Then, because $m \angle GCD$ is constant, $m \angle GCF$ (the measure of the angle of elevation) is constant.

Apply⇔Assess

Exercise Notes

Communication: Writing
Ex. 10 This exercise provides a good opportunity for students to demonstrate their understanding of the definitions on page 88 and the diagram at the top of page 89.

Second-Language Learners
Ex. 11 Students learning English may be unfamiliar with this use of the word *quarters*. Explain that a quarter of an item is equivalent to one-fourth of the whole of that item. Folding paper "in quarters" means folding paper in four equal parts. Then ask students to tell why they think the coin is called a quarter. (because it is one-fourth of one dollar)

Spatial Reasoning
Exs. 11, 12 Ask students to first try to predict the kind of figure that will result from the cut they plan to make. This requires thinking carefully about such properties as symmetry and parallelism. Each prediction can then be checked by cutting and unfolding.

Application
Exs. 13–15 Ask students if they can think of other devices that make use of properties of parallelograms in their construction and operation. (Examples: parallel rulers, pliers whose jaws remain parallel, stands for certain types of ironing boards, director's chair)

Interdisciplinary Problems
Exs. 13–15 The science of optics is a subspecialty of the science of physics. An optical instrument, such as binoculars, a telescope, a magnifying glass, or a microscope, is designed to enable a person to see objects more clearly. Critical to the development of these instruments was the creation of the lens, which forms the enlarged image.

Exercise Notes

Visual Thinking
Exs. 17–23 Ask students to sketch two or three polygons for each of the categories shown in the chart. Encourage them to mark all right angles, parallel sides, and congruent sides and angles. This activity involves the visual skills of *interpretation* and *recognition*.

Teaching Tip
Exs. 17–23 A copy of this table with the answers filled in can be placed on an overhead projector so students can review their work. Students should be able to justify their answers.

Assessment Note
Ex. 24 This exercise provides a good opportunity for students to show how well they understand the terminology involved and the conventions for marking diagrams.

Practice 13 for Section 2.5

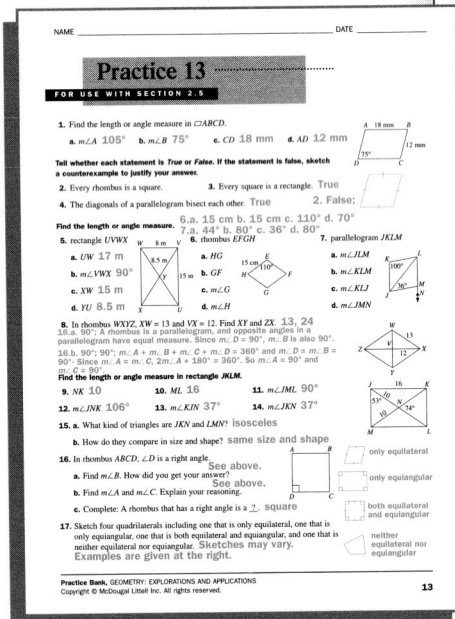

---

16. Logical Reasoning In □*ABCD*, ∠*D* is a right angle.
   a. Find *m*∠*B*. How did you get your answer?
   b. Find *m*∠*A* and *m*∠*C*. Explain your reasoning.
   c. Complete: A parallelogram that has one right angle is a ___?___ .

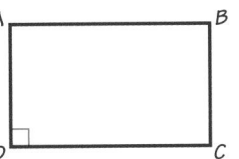

Copy the table. Use check marks to show which polygons have each property.

		quadrilateral	parallelogram	rectangle	rhombus	square
17.	Opposite sides are parallel.	?	?	?	?	?
18.	Opposite sides are congruent.	?	?	?	?	?
19.	Opposite angles are congruent.	?	?	?	?	?
20.	All angles are congruent.	?	?	?	?	?
21.	All angles are right angles.	?	?	?	?	?
22.	All sides are congruent.	?	?	?	?	?
23.	Diagonals bisect each other.	?	?	?	?	?

ONGOING ASSESSMENT

24. Open-ended Problem Sketch three parallelograms including one that is equilateral, one that is equiangular, and one that is neither equilateral nor equiangular. Draw all the diagonals. Mark all right angles, congruent angles, and congruent segments.

SPIRAL REVIEW

Give the letter that corresponds to an example of each term.
(Sections 2.1 and 2.3)

25. diagonals	**A.** $\overline{CB}, \overline{BG}$
26. vertices	**B.** $\overline{BH}, \overline{EG}$
27. equilateral polygon	**C.** *BCDEFG*
28. consecutive sides	**D.** *ABGH*
29. consecutive angles	**E.** *A, F*
30. adjacent angles	**F.** ∠*ABH*, ∠*HBG*
	G. ∠*BCD*, ∠*CDE*

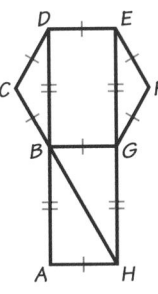

92   Chapter 2 *Triangles and Polygons*

---

16. a. 90°; Opposite angles of a parallelogram are congruent.

   b. *m*∠*A* + *m*∠*C*, since ∠1 and ∠1 are opposite angles of a parallelogram. *m*∠*A* + *m*∠*C* = 180°, since the sum of the angle measures of a quadrilateral is 360° and *m*∠*B* + *m*∠*D* = 180°. Then 2*m*∠*A* = 180°, so *m*∠*A* = *m*∠*C* = 90°.

   c. rectangle

17. parallelogram; rectangle; rhombus; square

18. parallelogram; rectangle; rhombus; square

19. parallelogram; rectangle; rhombus; square

20. rectangle; square

21. rectangle; square

22. rhombus; square

23. parallelogram; rectangle; rhombus; square

24. Answers may vary. Examples are given.

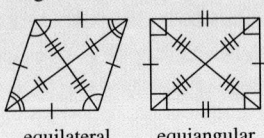

equilateral   equiangular

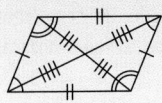

neither equilateral nor equiangular

25. B
26. E
27. C
28. A
29. G
30. F

2.6 Building Prisms

Three-dimensional objects are often designed using a process called *computer-aided design*. For example, an architect might enter the plans for this house into a computer system in order to look at different aspects of the design.

Learn how to...
* **name prisms and identify parts of prisms**

So you can...
* **sketch three-dimensional objects**
* **analyze nets**

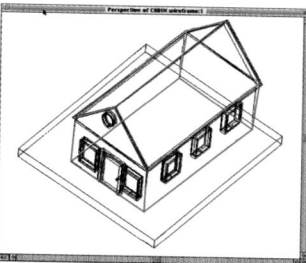

There are several ways to visualize a three-dimensional object. One way is to look through the *faces* of the object to the *edges*, as in the drawing at the left.

Another way to look at an object is from different positions. For instance, an architect might look at the front, side, and top views of the house, as shown.

Side View

Top View

Front View

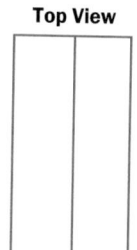

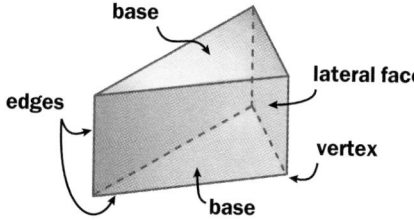

The house above and the figure at the left are *prisms*. A **prism** is a three-dimensional figure, with two congruent faces called the **bases**, that lie in parallel planes. The other faces of a prism, called the **lateral faces**, are formed by connecting the corresponding vertices of the bases. In this book, the lateral faces of prisms are rectangles. A prism's **vertices** are connected by segments called **edges**.

2.6 Building Prisms   **93**

Plan⇔Support

Objectives
* Name prisms and identify parts of prisms.
* Sketch three-dimensional objects.
* Analyze nets.

Recommended Pacing

❖ **Core and Extended Courses**
Section 2.6: 2 days

❖ **Block Schedule**
Section 2.6: 2 half-blocks
(with Section 2.5 and Chapter 2 Review)

Resource Materials

Lesson Support
Lesson Plan 2.6
Warm-Up Transparency 2.6
Practice Bank: Practice 14
Study Guide: Section 2.6
Challenge Problems: Set 14
Assessment Book: Test 8

Technology
Technology Book:
   Spreadsheet Activity 2
Internet:
   http://www.hmco.com

Warm-Up Exercises

Complete each statement.

1. Lines in the same plane that have no points in common are called ___?___ lines. parallel

2. Opposite sides of a rhombus are ___?___ and ___?___. congruent, parallel

3. A rectangle has ___?___ right angles. four

4. Opposite angles of a parallelogram are ___?___. congruent

5. A rectangle that is also a rhombus is called a ___?___. square

Teach⇔Interact

Section Note

Teaching Tip
Point out to students that any pair of opposite faces of a cube can be thought of as the bases of the cube.

Additional Example 1

Sketch each prism.

a. a hexagonal prism whose bases are not regular

Draw the   Connect
bases.   corresponding
                vertices.

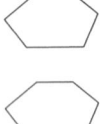 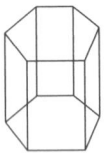

Make the edges that are hidden from view into dashed lines.

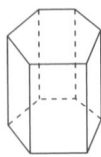

b. a prism whose bases are quadrilaterals, with one pair of sides parallel in each base

Draw the bases.

Connect corresponding vertices.

Make the edges that are hidden from view into dashed lines.

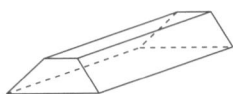

Section Note

Spatial Reasoning
Point out that if solid lines are shown for hidden edges, then the figure may appear to be two-dimensional or may give rise to an optical illusion. If lines for the hidden edges are omitted completely, then some people may view it as two-dimensional, while others may view it as solid. Those who view it as solid, however, cannot be certain that the figure is indeed a prism.

Prisms are classified by the shapes of their bases.

The bases of this prism are triangles, so it is called a triangular prism.

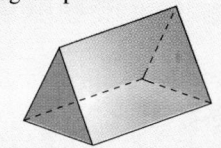

A **cube** is a rectangular prism whose faces are all squares.

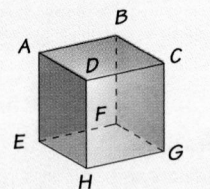

WATCH OUT! ▶

Notice that the bases of a prism do not have to be horizontal.

THINK AND COMMUNICATE

1. How many faces does the triangular prism have? How many faces does the cube have?

2. **a.** Name three different possible pairs of bases for the cube.

    b. Are there different possible pairs of bases for the triangular prism? Explain why or why not.

EXAMPLE 1

Sketch each prism.

a. a rectangular prism   **b. a pentagonal prism**

SOLUTION

Start by sketching one base. Sketch the other base either above or below the first base, or next to it. Make sure that the corresponding edges of the bases are parallel.

Draw the bases.   **Connect corresponding vertices.**   **Make the edges that are hidden from view into dashed lines.**

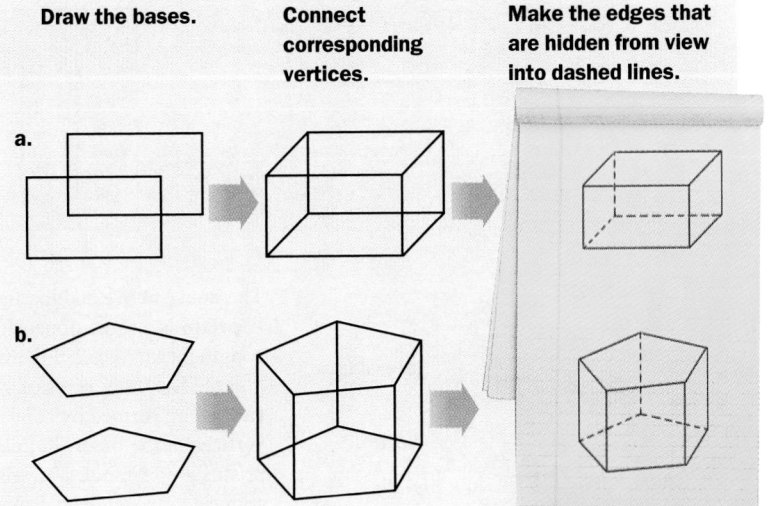

ANSWERS Section 2.6

Think and Communicate

1. 5 faces; 6 faces

2. **a.** *ABCD* and *EFGH*, *ADHE* and *BCGF*, *ABFE* and *DCGH*

    b. No; there are only two faces that lie in parallel planes.

Unfolding a Prism

Imagine cutting this prism along some of its edges, then opening and unfolding it. The resulting plane figure is called a **net**.

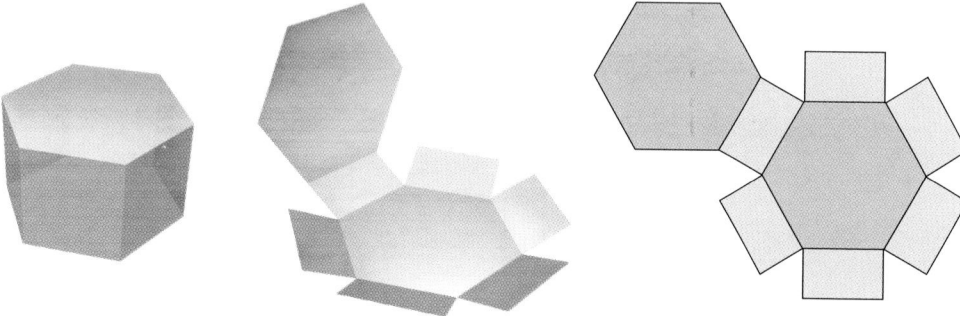

You can construct a prism by folding up its net and taping the edges together.

EXAMPLE 2

Sketch a net for this prism.

SOLUTION

Method 1

Sketch all the lateral faces attached to one of the bases of the prism.
Sketch the other base on the opposite edge of one of the lateral faces.

Method 2

Sketch all the lateral faces attached to each other.
Place the bases at opposite ends of two of the lateral faces.

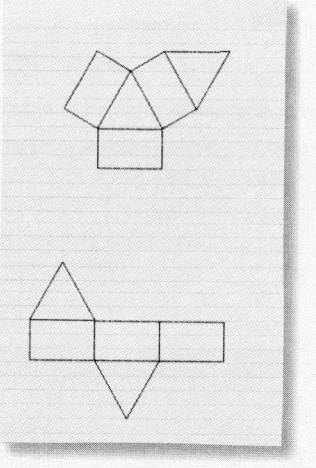

THINK AND COMMUNICATE

3. Describe another way to sketch a net for the prism above. How many different methods do you think there are?

4. Describe one possible net for a cube.

2.6 Building Prisms **95**

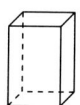
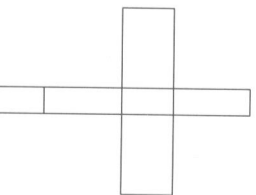
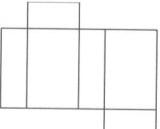
95

Checking Key Concepts

Using Manipulatives
When discussing questions 3 and 4, some students may wish to trace the figures, cut them out, and fold them to check their answers.

Closure Question

How would you describe a prism?
A prism is a three-dimensional figure with two congruent faces, called the bases, that lie in parallel planes. The other faces of a prism, called the lateral faces, are formed by connecting the corresponding vertices of the bases. A prism's vertices are connected by line segments called edges.

Apply⇔Assess

Suggested Assignment

❖ **Core Course**
Day 1 Exs. 1–16
Day 2 Exs. 17–21, 24, 27, 29–33, AYP

❖ **Extended Course**
Day 1 Exs. 1–16
Day 2 Exs. 17–33, AYP

❖ **Block Schedule**
Day 11 Exs. 1–16
Day 12 Exs. 17–21, 24, 27, 29–33, AYP

Exercise Notes

Historical Connection
Exs. 11–13 Leonard Euler was a Swiss mathematician who lived from 1707 to 1783. He made important contributions to geometry, analysis, and number theory. Euler was responsible for some mathematical notation used today. He was the first to use the letter *e* to designate the base of natural logarithms and the letter *i* for the imaginary unit $\sqrt{-1}$.

✔ CHECKING KEY CONCEPTS

1. Sketch the front, top, and side views of a rectangular prism. Then sketch the prism.

2. Use the diagram of a cube on page 94. How many edges does it have? How many vertices?

Visual Thinking Tell whether or not each figure is a net for a prism.

3.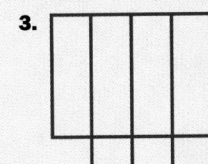

4.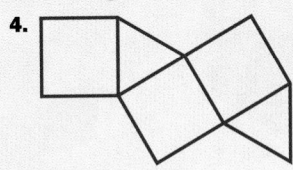

2.6 | Exercises and Applications

Extra Practice exercises on page 663

Use the prism at the right.

1. Name the two bases of the prism. What type of prism is it?

2. Sketch the front, top, and side views of the prism.

3. How many lateral faces does the prism have?

4. How many edges meet at vertex *C*? Name them.

5. Name two skew edges of the prism.

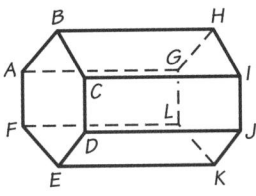

Use the prisms in this section or sketch the prisms to complete the chart.

	Type of Prism	Number of faces, F	Number of vertices, V	Number of edges, E
6.	triangular	?	?	?
7.	rectangular	?	?	?
8.	pentagonal	?	?	?
9.	hexagonal	?	?	?
10.	heptagonal	?	?	?

11. **Logical Reasoning** Describe any patterns that you see in the chart. Use these patterns to find *F*, *V*, and *E* for a prism whose base is an *n*-gon.

12. Use your answer to Exercise 11 to write an expression for *E* in terms of *F* and *V*. (This relationship is known as *Euler's formula.*)

13. Check that the relationship you found in Exercise 12 is true for another type of prism. Do you think that the relationship will work for any prism? Explain why or why not.

Checking Key Concepts

1. Answers may vary. An example is given.

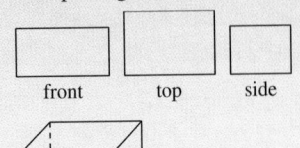

front   top   side

2. 12 edges; 8 vertices
3. No.   4. Yes.

Exercises and Applications

1. *ABCDEF, GHIJKL*; hexagonal prism

2.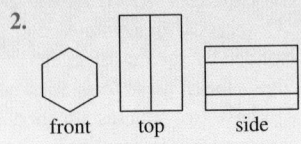

front   top   side

3. 6 lateral faces
4. 3 edges; $\overline{BC}$, $\overline{CI}$, and $\overline{CD}$
5. Answers may vary. Examples are given. $\overline{AF}$ and $\overline{CI}$, $\overline{HG}$ and $\overline{EK}$

6. 5 faces; 6 vertices; 9 edges
7. 6 faces; 8 vertices; 12 edges
8. 7 faces; 10 vertices; 15 edges
9. 8 faces; 12 vertices; 18 edges
10. 9 faces; 14 vertices; 21 edges
11. Answers may vary. An example is given. The number of faces is two more than the number of sides of a base. The number of vertices is twice the number of vertices of a base. The number of

Sketch each prism.

14. an octagonal prism

15. a prism whose bases are obtuse triangles

16. Open-ended Problem Sketch a prism that is not a cube but has two lateral faces that are squares.

Visual Thinking **Tell whether or not each figure is a net for a prism.**

17.

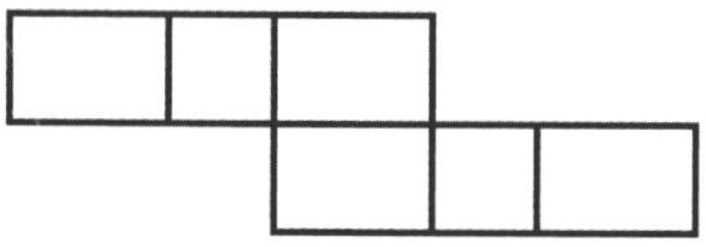

18.

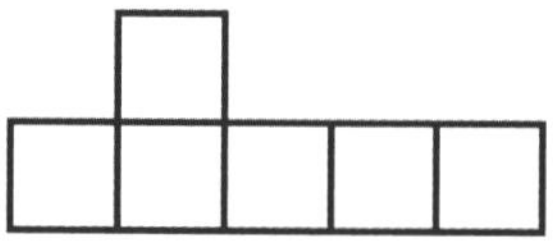

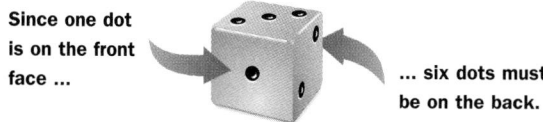

Sketch a net for each prism.

19. a rectangular prism

20. a pentagonal prism

21. a prism whose bases are nonregular hexagons

22. Research Most cardboard boxes are made from cutouts that are not actually nets. Examine an unfolded cardboard box. How is it different from a net? What are some possible reasons for the differences?

23. GAMES Many games use a die shaped like a cube, with the faces labeled 1 through 6. The sides are labeled so that when you add the numbers on any pair of opposite faces, the total is 7.

Since one dot is on the front face ...

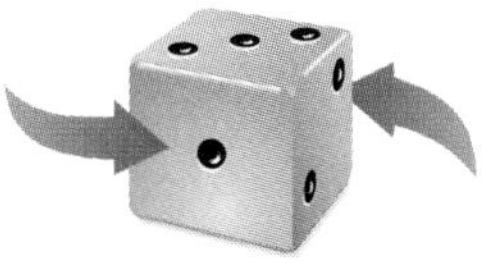

... six dots must be on the back.

a. How many dots are on the left face of this die? How many dots are on the bottom face?

b. Sketch a net for the die, writing the numbers 1 through 6 inside the appropriate squares.

24. Writing All of the prisms that you have seen in this section and will see later in this book are *right prisms.* Prisms that are not right prisms are called *oblique prisms.* Here are some examples of both types of prisms.

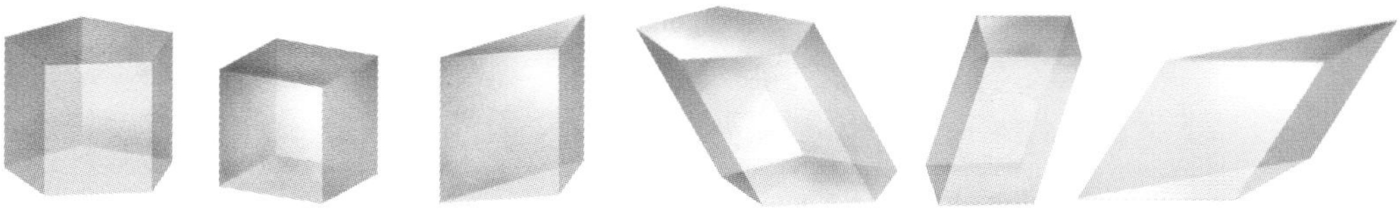

Right prisms **Oblique prisms**

Suppose that your friend is going to take a quiz in which he or she is shown a prism and must tell whether it is right or oblique. Write a note that explains what your friend should know.

Apply⇔Assess

Exercise Notes

Using Manipulatives

Exs. 14–21 Models made from cardboard or construction paper can be helpful for students who have difficulty sketching three-dimensional figures or sketching a net for a prism. You may wish to have students make some of their own models or have some commercial-made models available in class for students to handle when doing these exercises.

Communication: Writing

Ex. 24 This exercise introduces oblique prisms and offers students an opportunity to compare and contrast these prisms with the right prisms that they have just studied. You may wish to have some models of oblique prisms available for students to use or you may wish to conduct a short discussion of these types of prisms with the class.

edges is three times the number of sides of a base. For a prism whose base is an n-gon, $F = n + 2$, $V = 2n$, and $E = 3n$.

12. Because $E = 3n$ and $F + V = (n + 2) + 2n = 3n + 2$, $E = F + V - 2$.

13. Answers may vary. An example is given. For an octagonal prism, $F = 10$, $V = 16$, and $E = 24$, so $F + V - 2 = 26 - 2 = E$. This will work for any prism, because the values determined in Ex. 11 have the relationship determined in Ex. 12 for any value of n.

14–16. Answers may vary. Examples are given.

14.

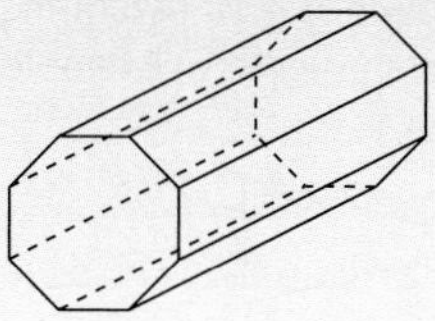

15.

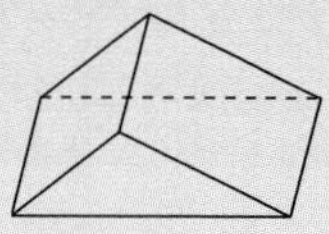

16.

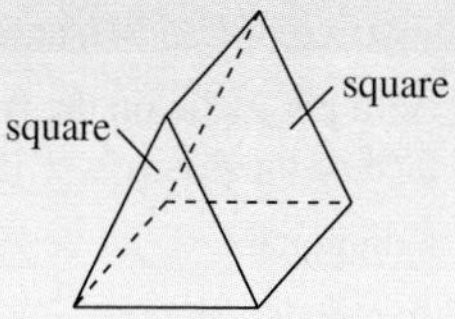

17. Yes.

18. No.

19–21. Answers may vary. Examples are given.

19.

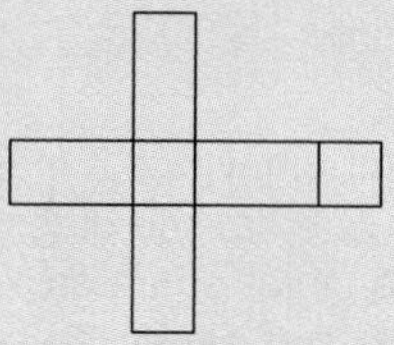

20.

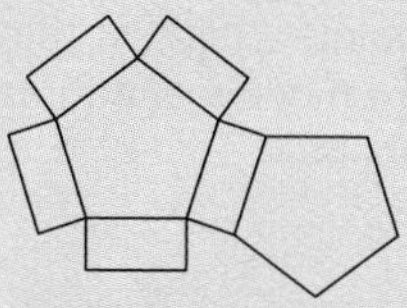

21.

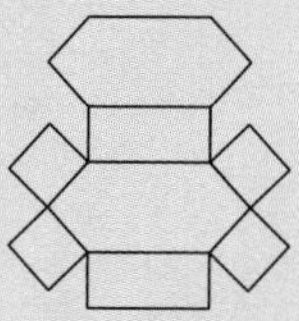

22. Answers may vary. An example is given. Unfolded boxes are usually connected so that all the lateral sides form a flattened loop. To lay them flat, you would have to make an additional cut along an edge between two lateral sides and unfold. Also, the flaps that form the base or top each form only half of that face, and they overlap so that the resulting base and top are two layers thick. These modifications help to make the finished box more rigid and stronger than a folded net would be.

23, 24. See answers in back of book.

Exercise Notes

Interview Note: Research
Exs. 25, 26 Students can consult a book on minerals and crystallography for other examples of crystals that have prismatic forms.

Interdisciplinary Problems
Exs. 25, 26 Since many objects in nature have the shapes of polygons or prisms, geometric concepts often serve as powerful models for studying and understanding ideas in physics and chemistry. An arrangement of molecules in a crystal, as shown in these exercises, can be modeled by a prism. Remind students that all matter is composed of tiny particles called molecules. The molecules of any substance, such as beryl, are identical. Molecules are not the smallest particles of matter but are composed of atoms, which, in turn, are composed of various subatomic particles. Physicists today know that even some of the subatomic particles, such as protons and neutrons, are composed of more fundamental particles called *quarks,* which, along with electrons, are considered the ultimate building blocks of all physical things.

Challenge
Ex. 27 Students may find it interesting to see whether the diagonals of a rectangular prism are equal in length and if they intersect at their midpoints. Suggest they use cardboard models and thread for the diagonals.

INTERVIEW **Virgil Lueth**

Look back at the article on pages 54–56.

In crystallography, a lattice is a set of points that shows the arrangement of molecules in a given crystal. Each point represents a molecule, and the length of each segment corresponds to the distance between the molecules. The diagrams below show two lattices that are shaped like prisms.

Lattice for Pyrite

Lattice for Beryl

Use the lattices above.

25. In the lattice for pyrite, the segments are all the same length and all meet at right angles. What type of prism do the segments form?

26. Each base of the lattice for beryl is a regular hexagon formed by three identical rhombuses. Find the measure of each angle of one of these rhombuses.

A *diagonal of a prism* connects two vertices that do not lie on the same face of the prism. Use this definition for Exercises 27 and 28.

27. **a.** Name two other diagonals of this prism.
   b. Explain why $\overline{AH}$ is not a diagonal of the prism.
   c. How many diagonals does this prism have?

28. **Challenge** How many diagonals does a hexagonal prism have?

ONGOING ASSESSMENT

29. **Cooperative Learning** Work with another student.
   a. Sketch a prism. Without showing it to the other student, describe the prism. The other student should sketch the prism, based on the description.
   b. Compare the two sketches. How similar are they? Should the description have been different? If so, revise it so it is more accurate.

---

25. a cube

26. The acute angles measure 60°, and the obtuse angles measure 120°.

27. **a.** any two of $\overline{CE}$, $\overline{BH}$, and $\overline{GA}$
   b. Both points lie on the front face of the prism.
   c. 4 diagonals

28. 18 diagonals

29. a, b. Check students' work.

30. A rhombus is a quadrilateral. It is a parallelogram, so opposite sides are congruent, opposite angles are congruent, and its diagonals bisect each other. It is also equilateral.

31. An isosceles triangle is a triangle in which at least two of the sides are congruent. An example is given.

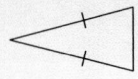

32. −2, −5   33. 25, 36

Assess Your Progress

1. **a.** 60°
   b. 120°
   c. 27

2. **a.** 90°
   b. 14
   c. 50

3. **a.** 9
   b. 45°
   c. 135°

30. List all of the properties of a rhombus that you have learned. *(Section 2.5)*

31. Define *isosceles triangle* and sketch an example. *(Section 2.2)*

Use inductive reasoning to find the next two terms in each number pattern.
(Section 1.2)

32. 10, 7, 4, 1, _?_ , _?_   **33.** 1, 4, 9, 16, _?_ , _?_

ASSESS YOUR PROGRESS

VOCABULARY

parallelogram (p. 88)	**lateral face** (p. 93)
rectangle (p. 88)	**vertex of a prism** (p. 93)
rhombus (p. 88)	**edge** (p. 93)
square (p. 88)	**cube** (p. 94)
prism (p. 93)	**net** (p. 95)
base (p. 93)	

Find each length or angle measure. *(Section 2.5)*

1. ▱*WXYZ*
  a. $m\angle X$
  b. $m\angle Y$
  c. *WX*

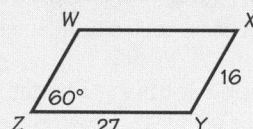

2. rectangle *ABCD*
  a. $m\angle ABC$
  b. *BC*
  c. *BD*

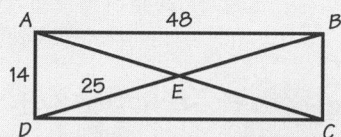

3. rhombus *LMNO*
  a. *MN*
  b. $m\angle M$
  c. $m\angle LOP$

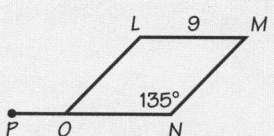

4. Sketch a hexagonal prism. Label one of its bases and one of its lateral faces. *(Section 2.6)*

5. Sketch two different nets for a rectangular prism. *(Section 2.6)*

6. Journal Sketch a rectangle, a rhombus, and a square. Draw all of the lines of symmetry for each shape. How are the lines of symmetry for the square related to the ones for the rectangle and the rhombus?

Assess Your Progress

Journal Entry
Allow students an opportunity to check their journal entries by placing the sketches for this activity on the board. Volunteers can then come to the board to draw the lines of symmetry.

Progress Check 2.5–2.6

See page 103.

Practice 14 for Section 2.6

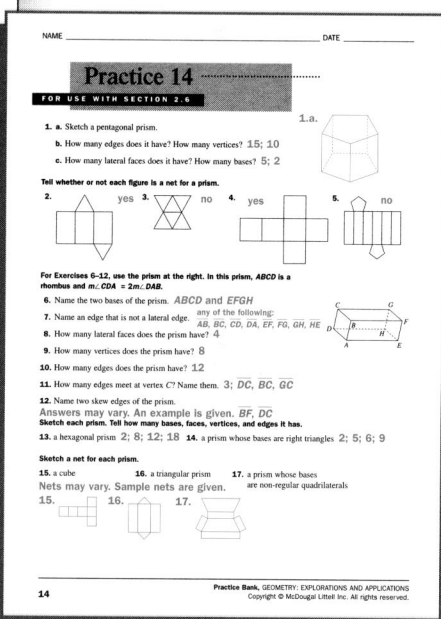

4, 5. Answers may vary. Examples are given.

4.
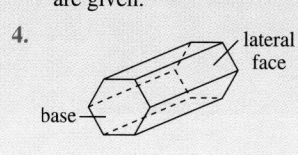

5.
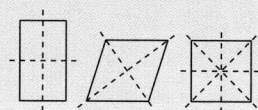

6. Answers may vary. An example is given.

The rectangle has two lines of symmetry connecting the midpoints of opposite sides. The rhombus has two lines of symmetry connecting opposite vertices. The square has four lines of symmetry—two connecting the midpoints of opposite sides and two connecting opposite vertices.

Mathematical Goals

- Draw a regular triangle, a square, a regular pentagon, and a regular hexagon.
- Construct the Platonic solids.
- Draw nets for the Platonic solids.

Planning

Materials
- paper and pencil
- protractor
- ruler
- tape
- scissors

Project Teams

Students should select a partner and complete all phases of the project together, especially the construction of the solids where four hands are sometimes necessary. It may also be helpful for students to discuss their ideas before constructing the solids.

Guiding Students' Work

Some students may have difficulty visualizing and creating three-dimensional solids. If this is the case, work with them on the first solid to get them started. Also, have students draw their triangles, squares, pentagons, and hexagons so they are large enough to work with as three-dimensional shapes. It may take a while for students to realize that the hexagon is not used in constructing the solids and that the equilateral triangle is used three times. When students have completed making their solids, they should see that the tetrahedron has 4 triangular faces, the cube has 6 square faces, the octahedron has 8 triangular faces, the dodecahedron has 12 pentagonal faces, and the icosahedron has 20 triangular faces.

Second-Language Learners

Students learning English may need an explanation of what the term *closed three-dimensional vertex* means. Make sure that students fully understand the term before they begin Step 2.

Building the Platonic Solids

A *regular solid* is made of only one type of regular polygon. Over 2000 years ago, the Greek mathematician Euclid proved that only five regular solids, called the *Platonic solids*, are possible. The subject of this activity is the last proposition in Euclid's book, *The Elements*.

PROJECT GOAL Work with a partner to discover the regular solids and investigate how to make nets for them.

Discovering the Five Solids

1. Draw a regular triangle, a square, a regular pentagon, and a regular hexagon using what you know about the angle measures of regular polygons. Cut out six copies of each. Cut out additional copies as needed.

2. Tape three triangles together so that they share a common vertex, as shown below. Can they fold up to form a closed three-dimensional vertex? If so, use another piece of tape to hold the sides together. If not, explain why they don't make a closed vertex.

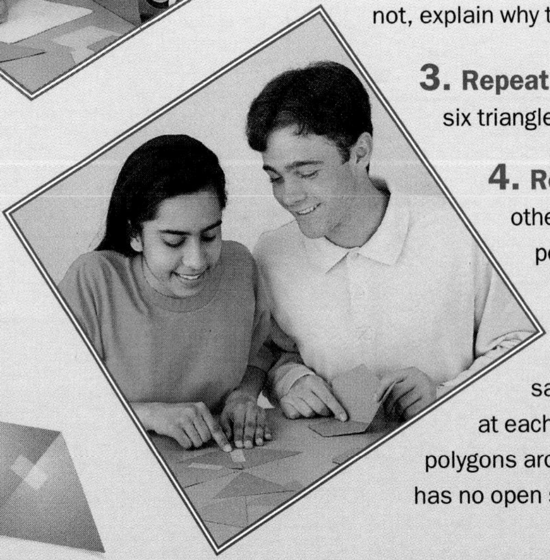

3. Repeat Step 2 for four, five, and six triangles.

4. Repeat Steps 2 and 3 for the other polygons. You will find five possible vertices.

5. Finish constructing the five solids so that the same number of polygons meet at each vertex. Continue adding polygons around each vertex until the solid has no open spaces.

Use the information at the right to identify each of your solids. Tell what kind of polygons were used for the faces of each solid, and how many polygons meet at each vertex.

tetrahedron ■	4 faces
hexahedron (cube) ■	6 faces
octahedron ■	8 faces
dodecahedron ■	12 faces
icosahedron ■	20 faces

Making Nets for the Solids

Cut or remove some of the tape from the icosahedron. Leave enough edges attached to keep it in one piece, but separate enough edges so that the pieces lie flat. Sketch the resulting figure. Use your sketch to make an accurate net for the solid from heavy paper. Do the same for the dodecahedron.

Try to make nets for the other solids without taking them apart. Use these nets to make solids out of heavy cardboard.

Extending Your Project

Continue your investigation of Platonic solids by exploring one of the following topics.

• Find all the different possible nets for the tetrahedron. Find at least three nets for each of the other solids. What makes one net easier to assemble than another?

• Count the number of vertices, faces, and edges of each solid. Use a table like the one on page 96 to record your data. What patterns do you see? Two pairs of Platonic solids are related by their vertices, faces, and edges. Find these pairs and describe the relationships.

• Use a protractor to make a contact goniometer like the kind crystallographers use (see page 56). Find the interfacial angles of each solid.

Self-Assessment

Describe the work you and your partner did for this project. What did you learn during your investigation? What advice would you give to someone who plans to do this project in the future?

Portfolio Project **101**

Progress Check 2.1–2.2

1. Find the measures of the complement and supplement of ∠ *CTB.* *(Section 2.1)*

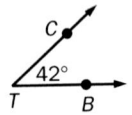

complement: 48°
supplement: 138°

Use the figure below to answer Exs. 2–4. *(Section 2.1)*

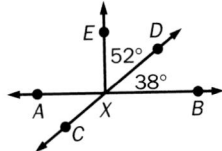

2. Name two angles that are adjacent to ∠ *EXD.* any two of the following: ∠ *CXE,* ∠ *AXE,* ∠ *DXB*

3. Name a pair of vertical angles. ∠ *AXC* and ∠ *DXB,* or ∠ *AXD* and ∠ *CXB*

4. Name two right angles. ∠ *EXB,* ∠ *AXE*

In the figure, $\overline{PQ} \perp \overline{SQ}$. Tell whether each statement is *True* or *False.* *(Section 2.2)*

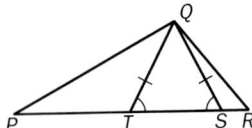

5. △ *PQR* is an acute triangle.
  False.

6. △ *TQS* is isosceles. True.

7. △ *PQT* is an obtuse triangle.
  True.

STUDY TECHNIQUE

Make a sketch or give an example for each of the vocabulary words that you learned in this chapter.

VOCABULARY

types of angles (p. 57)
adjacent angles (p. 58)
linear pair (p. 58)
perpendicular (p. 58)
vertical angles (p. 58)
complements (p. 58)
supplements (p. 58)
triangle (p. 64)
side of a triangle (p. 65)
vertex of a triangle (p. 65)
types of triangles (p. 65)
polygon (p. 73)
triangle (p. 73)
quadrilateral (p. 73)
pentagon (p. 73)
hexagon (p. 73)
equilateral (p. 73)

equiangular (p. 73)
regular (p. 73)
consecutive angles (p. 74)
consecutive sides (p. 74)
diagonal (p. 74)
parallelogram (p. 88)
rectangle (p. 88)
rhombus (p. 88)
square (p. 88)
prism (p. 93)
base (p. 93)
lateral face (p. 93)
vertex of a prism (p. 93)
edge (p. 93)
cube (p. 94)
net (p. 95)

SECTIONS 2.1 *and* 2.2

Angles are classified by their measures. An angle can be **acute, right, obtuse,** or **straight.** Pairs of angles can be classified by their positions or by their measures.

  Vertical angles are congruent. Angles that form a **linear pair** are supplementary.

  If two lines are **perpendicular,** they form congruent adjacent angles. If two lines form congruent adjacent angles, the lines are perpendicular.

  When classified by its sides, a triangle can be **scalene, isosceles,** or **equilateral.** When classified by its angles, a triangle can be **acute, obtuse, right,** or **equiangular.**

  The sum of the measures of the interior angles of any triangle is 180°.

SECTIONS 2.3 and 2.4

A **polygon** is a closed plane figure whose sides are segments that meet only at their vertices.

A polygon is named by consecutive vertices. Two names for the polygon at the right are hexagon *BCDEFA* and hexagon *DCBAFE*.

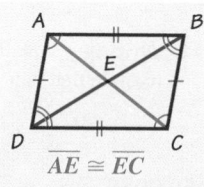

consecutive angles

A **diagonal** connects two nonconsecutive vertices.

consecutive sides

In an **equilateral** polygon, all sides are congruent.
In an **equiangular** polygon, all angles are congruent.
A **regular** polygon is both equilateral and equiangular.

Polygons are classified by the number of sides they have.

Number of sides	3	4	5	6	*n*
Name	triangle	quadrilateral	pentagon	hexagon	*n*-gon

The sum of the measures of the interior angles of an *n*-gon is $(n - 2)180°$.
The sum of the measures of the exterior angles of any polygon is 360°.

SECTIONS 2.5 and 2.6

A **parallelogram** is a quadrilateral with both pairs of opposite sides parallel. The opposite sides of a parallelogram are congruent. The opposite angles of a parallelogram are congruent. The diagonals of a parallelogram bisect each other.

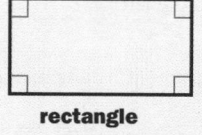

$$\overline{AE} \cong \overline{EC}$$
$$\overline{BE} \cong \overline{ED}$$

There are three special types of parallelograms.

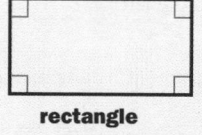 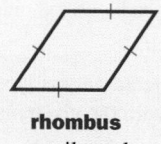 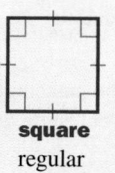

rectangle
equiangular

rhombus
equilateral

square
regular

A **prism** is a figure with two congruent parallel faces called **bases**. The **lateral faces** of the prism are rectangles. The line segments connecting the **vertices** are called **edges**.

Prisms are named by the shapes of their bases. This is a triangular prism.

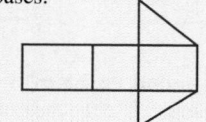

This **net** can be folded and taped to make the triangular prism.

Review **103**

1. Sketch a pentagon that is not regular and show its diagonals. *(Section 2.3)* Answers may vary. An example is given.

120°
90°

2. How many lines of symmetry does a regular 18-gon have? *(Section 2.3)* 18 lines of symmetry

3. Can a quadrilateral have four obtuse angles? Explain. *(Section 2.4)* No; the sum of the angle measures of the polygon would be greater than 360°, which is not possible.

4. Can a pentagon have exterior angles whose measures are 120°, 140°, 60°, 50°, and 20°? Explain. *(Section 2.4)* No; the sum of the given angle measures is 390°. All polygons have exterior angles whose measures have a sum of 360°.

Progress Check 2.5–2.6

In the figure below, $\overline{AF} \parallel \overline{CD}$ and quadrilateral *BCEF* is a square. Tell whether each statement or equation is *True* or *False*. Explain your reasoning. *(Section 2.5)*

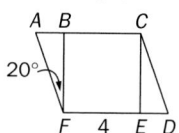

1. $\overline{AC}$ is parallel to $\overline{FD}$. True; the opposite sides $\overline{BC}$ and $\overline{FE}$ of square *BCEF* are parallel, since every square is a parallelogram.

2. $m\angle AFD = 110°$ *(Section 2.5)* True; $\angle AFB$ and $\angle BFD$ are adjacent angles and $m\angle BFD = 90°$, since *BCEF* is a square.

3. Quadrilateral *ACDF* is a parallelogram. True; $\overline{AC} \parallel \overline{FD}$ and $\overline{AF} \parallel \overline{CD}$.

4. How many vertices does a pentagonal prism have? *(Section 2.6)* 10 vertices

5. How many edges does a heptagonal prism have? *(Section 2.6)* 21 edges

Chapter 2 Assessment
Form A Chapter Test

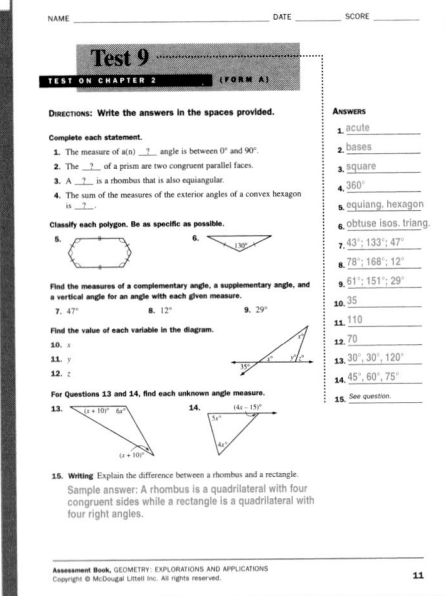

Chapter 2 Assessment
Form B Chapter Test

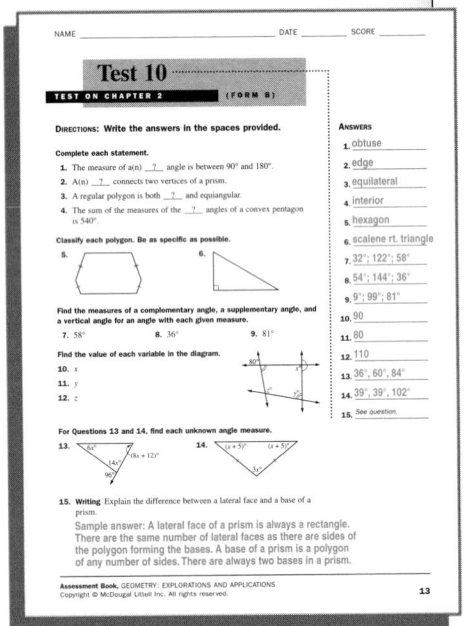

CHAPTER

2 Assessment

VOCABULARY QUESTIONS

Complete each statement.

1. The measure of a(n) ? angle is between 90° and 180°.

2. A(n) ? connects two nonconsecutive vertices of a polygon.

3. A regular quadrilateral is called a(n) ? .

4. The sum of the measures of the ? angles of an octagon is 360°.

Use the diagram at the left. Give an example of each of the following.

5. vertical angles

6. congruent adjacent angles

7. Explain the difference between a prism and a net.

Classify each polygon. Be as specific as possible.

8.   9.   10.

11. **Writing** Explain why it is impossible to sketch an equilateral triangle that is not isosceles.

SECTIONS 2.1 and 2.2

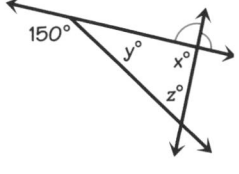

12. Find the value of each variable in the diagram at the left.

   a. x **b.** y **c.** z

Find the measures of a complementary angle, a supplementary angle, and a vertical angle for an angle with the given measure.

13. 71°   14. 39°   15. 85°

ALGEBRA Find each unknown angle measure.

16.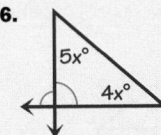

17.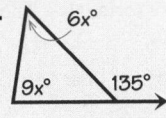

18.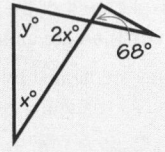

19. **Open-ended Problem** Find a possible set of three angle measures of a triangle. Use the measures to classify the triangle.

104   Chapter 2 *Triangles and Polygons*

ANSWERS Chapter 2

Assessment

1. obtuse

2. diagonal

3. square

4. exterior

5. ∠ *BEG* and ∠ *DEF*, or ∠ *BED* and ∠ *GEF*

6. ∠ *ACF* and ∠ *ACB*

7. A prism is a three-dimensional figure formed by connecting the corresponding vertices of two congruent faces that lie in parallel planes. A net is a two-dimensional representation of this three-dimensional figure formed by drawing the faces of the prism as if the prism had been cut along some of its edges (without disconnecting any of the faces), unfolded, and then laid out flat.

8. rhombus (a convex, equilateral parallelogram)

9. convex, equiangular pentagon

10. isosceles right triangle

11. An equilateral triangle has three congruent sides. This means that it always has at least two congruent sides, which is the definition of an isosceles triangle.

12. a. 90
    b. 30
    c. 60

13. complementary: 19°; supplementary: 109°; vertical: 71°

14. complementary: 51°; supplementary: 141°; vertical: 39°

15. complementary: 5°; supplementary: 95°; vertical: 85°

16. $4x° = 40°; 5x° = 50°$

17. $6x° = 54°; 9x° = 81°$

18. $x° = 34°; 2x° = 68°; y° = 78°$

19. Answers may vary. An example is given. 88°, 89°, 3°; acute

104

SECTIONS 2.3 *and* 2.4

For Questions 20–22, use the diagram at the right.

20. What type of polygon is the figure? Be as specific as possible.

21. Give three different names for the polygon.

22. Name all the diagonals with one endpoint at *A*.

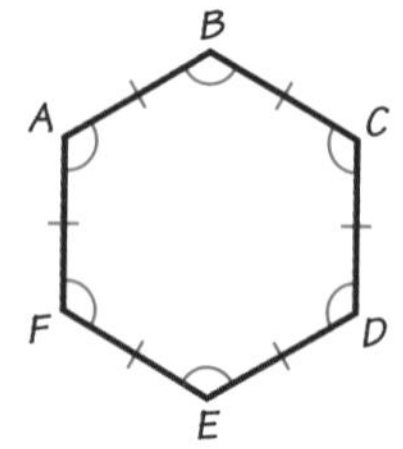

Find the value of each variable.

23.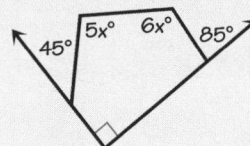

24.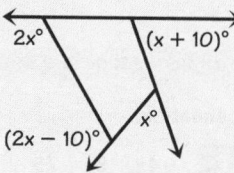

25. Each exterior angle of an equiangular *n*-gon measures 12°. Find the value of *n*.

26. Find the measure of each interior angle of a regular 24-gon.

SECTIONS 2.5 *and* 2.6

Find each length or angle measure.

27. rhombus *DEFG*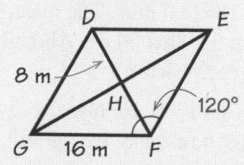

a. *FE*

b. *DF*

c. $m \angle DEF$

28. parallelogram *STUV*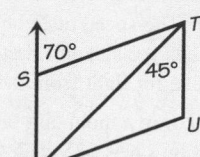

a. $m \angle U$

b. $m \angle TVU$

c. $m \angle STV$

29. Sketch a pentagonal prism.

30. a. What type of prism is shown at the right?

  b. How many lateral faces does the prism have?

  c. Sketch two different nets for the prism.

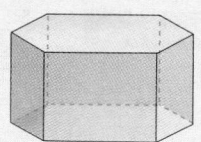

PERFORMANCE TASK

31. Make a poster that shows what you know about prisms. Identify the parts of a prism and give instructions for drawing a prism on paper. Show how a given prism can be made from different nets.

Assessment **105**

Chapter 2
ALTERNATIVE ASSESSMENT

1. **Performance Task** Make a poster that shows all the characteristics of each of the following types of triangles.
   equilateral, equiangular, isosceles, scalene, obtuse, acute, right

2. **Group Activity** Write each of the following terms on separate index cards. Then draw a diagram of each item on another index card. Shuffle the cards and lay them in rows face down. Take turns turning over pairs of cards. If you turn over a term and the matching diagram, you keep the cards and get another turn. If you turn over two cards that do not form a matching pair, turn them back over and the next person gets a turn. The game ends when all of the cards have been taken. The person with the most cards wins.

   Terms:
   triangle
   quadrilateral
   pentagon
   hexagon
   parallelogram
   rectangle
   square
   rhombus
   complementary angles
   supplementary angles
   edge
   face
   linear pair of angles
   vertical angles

3. **Open-ended Problem** Students sometimes get supplementary angles and complementary angles confused. Devise a way to help such students remember how to distinguish between these two terms. You can use a song, a poem, a picture, etc. Present your method to the class.

4. **a.** How many different ways can you name a triangle using its vertices?
   b. How many different ways can you name a quadrilateral using its vertices?

5. **Performance Task** Draw a triangle on a piece of paper and cut it out. Use paper-folding to demonstrate that the sum of the measures of the three angles of your triangle is 180°. Suppose everyone in your class used a different triangle to show that the sum of the measures of the angles of a triangle is 180°. Would this prove that the statement is always true? Explain your answer.

93

6. **Open-ended Problem** Sketch each triangle, if possible, and label its important features. If it is not possible to sketch such a triangle, explain why not.
   a. an isosceles equilateral triangle
   b. an isosceles equiangular triangle
   c. an obtuse scalene triangle
   d. an isosceles right triangle

7. **Performance Task** Triangle *ABC* is a right triangle with two congruent sides. Sketch such a triangle. What other information do you know about triangle *ABC*?

8. **Performance Task** Is the following statement true or false? Explain your answer.
   A polygon that is not regular does not have any symmetries.

20. regular hexagon (convex, equiangular, and equilateral)

21. any three of the following:
ABCDEF, BCDEFA, CDEFAB, DEFABC, EFABCD, FABCDE, AFEDCB, FEDCBA, EDCBAF, DCBAFE, CBAFED, BAFEDC

22. $\overline{AC}, \overline{AD}, \overline{AE}$

23. $5x° = 100°; 6x° = 120°$

24. $x° = 60°; 2x° = 120°;$
$(x + 10)° = 70°;$
$(2x − 10)° = 110°$

25. 30

26. 165°

27. a. 8 m
  b. 16 m
  c. 60°

28. a. 110°
  b. 25°
  c. 25°

29. Answers may vary. An example is given.

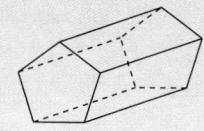

30. a. a hexagonal prism
  b. 6 lateral faces

c. Answers may vary. Examples are given.

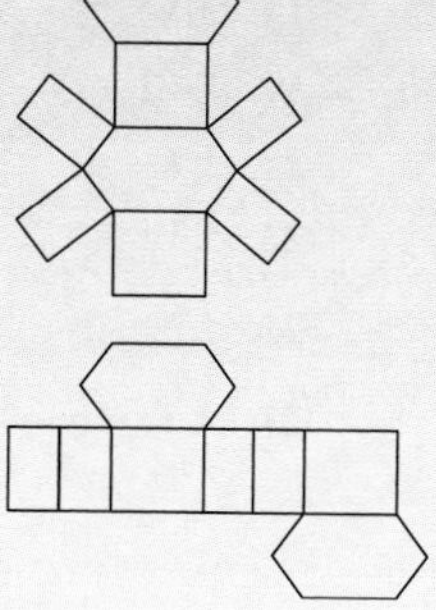

31. Check students' work.

105

Algebra
Review/Preview

These exercises review algebra topics you will use in the next chapters.

SOLVING LINEAR EQUATIONS

Solve each equation.

Toolbox p. 694
Linear Equations

EXAMPLE $-4x + 5 = 29$

$-4x + 5 - 5 = 29 - 5$ ← Subtract 5 from both sides.

$-4x = 24$

$\dfrac{-4x}{-4} = \dfrac{24}{-4}$ ← Divide both sides by -4.

$x = -6$

1. $-81 = -37 + b$ **2.** $-3y - 4 = 8$ **3.** $5z + 16 = -34$

4. $27 = 9 + 8p$ **5.** $12 - 10q = 60$ **6.** $-200 = -32 - 7r$

7. $\dfrac{w - 6}{2} = 8$ **8.** $\dfrac{1 + k}{3} = -17$ **9.** $\dfrac{5}{4}n - 4 = 11$

Toolbox p. 692
Operations with Variable Expressions

10. ASTRONOMY The speed of light is about 186,000 mi/s. The mean distance between the sun and Earth is about 93,000,000 mi. About how long does it take for light from the sun to reach Earth?

11. Mario wants to buy a mountain bike that costs $780. He has already saved $130. If he saves $25 each week, after how many weeks will he have enough money to buy the bike?

SIMPLIFYING RADICALS

Simplify each expression.

Toolbox p. 700
Simplifying Radicals

EXAMPLES

$\sqrt{48} = \sqrt{16 \cdot 3}$
$= \sqrt{16} \cdot \sqrt{3}$
$= 4\sqrt{3}$

$\sqrt{50} + \sqrt{18} = \sqrt{25 \cdot 2} + \sqrt{9 \cdot 2}$
$= \sqrt{25} \cdot \sqrt{2} + \sqrt{9} \cdot \sqrt{2}$
$= 5\sqrt{2} + 3\sqrt{2}$
$= 8\sqrt{2}$

12. $\sqrt{12}$ **13.** $\sqrt{27}$ **14.** $\sqrt{45}$

15. $\sqrt{75}$ **16.** $5\sqrt{7} - 3\sqrt{7}$ **17.** $6\sqrt{32} + 9\sqrt{32}$

18. $\sqrt{20} + \sqrt{5}$ **19.** $\sqrt{72} - \sqrt{8}$ **20.** $\sqrt{24} + 7\sqrt{3}$

ANSWERS Chapter 2

Algebra Review/Preview

1. -44 2. -4

3. -10 4. $\dfrac{9}{4}$

5. -4.8 6. 24

7. 22 8. -52

9. 12 10. about 500 s

11. 26 weeks 12. $2\sqrt{3}$

13. $3\sqrt{3}$ 14. $3\sqrt{5}$

15. $5\sqrt{3}$ 16. $2\sqrt{7}$

17. $60\sqrt{2}$ 18. $3\sqrt{5}$

19. $4\sqrt{2}$ 20. $2\sqrt{6} + 7\sqrt{3}$

21. ± 9 22. ± 12

23. ± 3 24. ± 8

25. ± 2 26. $\pm\dfrac{1}{2}$

27.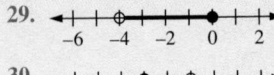

28.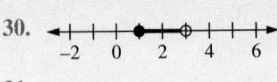

29.

30.

31.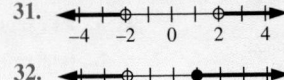

32.

33.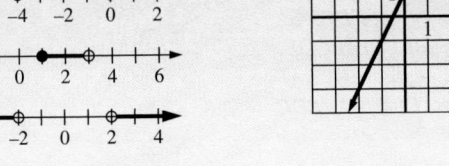

SOLVING QUADRATIC EQUATIONS

Solve each equation.

EXAMPLE

$3x^2 + 9 = 84$

$3x^2 = 75$ ← Subtract 9 from both sides.

$x^2 = 25$ ← Divide both sides by 3.

$x = \pm\sqrt{25}$ ← Take the square root of both sides.

$x = \pm 5$

Toolbox p. 701
Quadratic Equations

21. $x^2 = 81$ **22.** $134 = y^2 - 10$ **23.** $3r^2 = 27$

24. $-\frac{5}{2}t^2 = -160$ **25.** $5p^2 - 9 = 11$ **26.** $17 - 8n^2 = 15$

GRAPHING INEQUALITIES

Graph each inequality on a number line.

EXAMPLES

$x \geq 1$

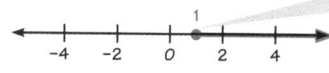

A closed circle at 1 means that $x = 1$ *does* satisfy the inequality.

$-3 < p < 2$

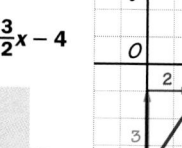

Open circles at -3 and 2 mean that $p = -3$ and $p = 2$ *do not* satisfy the inequality.

$z \leq -2$ or $z > 0$

27. $x \geq -2$ **28.** $u < 1$ **29.** $-4 < v \leq 0$

30. $1 \leq y < 3$ **31.** $k < -2$ or $k > 2$ **32.** $m < 0$ or $m \geq 3$

GRAPHING EQUATIONS OF LINES

Graph each equation.

EXAMPLE $y = \frac{3}{2}x - 4$

Step 3 Connect the two plotted points with a line.

Step 1 The y-intercept is -4. Graph the point $(0, -4)$.

Step 2 The slope is $\frac{3}{2}$. Start at $(0, -4)$. Count **3 units up** and **2 units right**. Graph this point.

33. $y = 2x + 1$ **34.** $y = 6x - 2$ **35.** $y = 4 - x$

36. $y = \frac{4}{3}x + 2$ **37.** $y = -\frac{2}{5}x - 3$ **38.** $3x + y = 5$

Toolbox p. 697
Graphing Linear Equations

Algebra Review/Preview   **107**

34.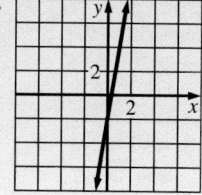

35.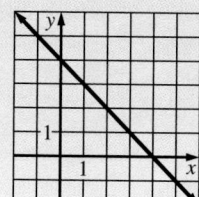

36.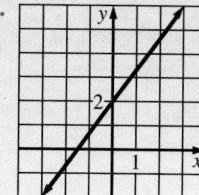

37.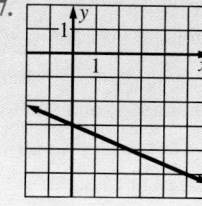

38.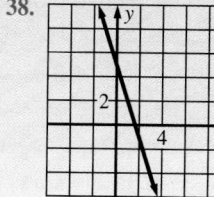

3 Reasoning in Geometry

OVERVIEW

Connecting to Prior and Future Learning

⟺ In Chapter 3, students continue their study of reasoning from Chapter 1 as they learn about inductive and deductive reasoning, converses, negations, and contrapositives.

⟺ Students also learn about postulates, definitions, and properties in this chapter. They will discover how to use these ideas to justify statements about geometric figures, which leads to a study of paragraph proof and two-column proof.

⟺ This chapter also includes an introduction to the Pythagorean theorem. Students prove this theorem using manipulatives and concepts from algebra. Students can review simplifying radicals by studying pages 700 and 701 in the **Student Resources Toolbox**.

Chapter Highlights

Interview with Mary-Jacque Mann: The use of logic to solve crimes is highlighted in this interview with Mary-Jacque Mann. Related exercises can be found on pages 128 and 133.

Explorations in Chapter 3 involve using geometry software to explore the relationship between exterior and interior angles of a triangle in Section 3.3, and using the areas of right triangles cut out of a square to prove the Pythagorean theorem in Section 3.6.

The Portfolio Project: Students use a classification system to classify a guitar as a stringed, wind, percussion, or keyboard instrument. They then develop and use a classification system for something that they are interested in.

Technology: Students use geometry software to explore the relationship between the measures of interior and exterior angles of a triangle in Section 3.3 and to discover a property of quadrilaterals in Section 3.6. A spreadsheet is used in Section 3.6 to find Pythagorean triples.

OBJECTIVES

Section	Objectives	NCTM Standards
3.1	• Use deductive reasoning to reach conclusions. • Make a convincing argument. • Recognize valid and invalid arguments.	1, 2, 3, 4, 7
3.2	• Recognize and use postulates, definitions, and properties. • Justify statements about geometric figures.	1, 2, 3, 4, 7
3.3	• Write and understand mathematical proofs. • Use mathematical reasoning to prove that a statement is always true.	1, 2, 3, 4, 7
3.4	• Write a proof in two-column format. • Use and present convincing arguments in a variety of formats.	1, 2, 3, 4, 7
3.5	• Write the converse of a conditional statement. • Recognize and use converses in logical arguments.	1, 2, 3, 4, 7
3.6	• Find the lengths of the sides of a right triangle. • Decide if a triangle is a right triangle. • Find lengths of parts of figures and real-life objects.	1, 2, 3, 4, 7
3.7	• Determine whether a triangle is acute or obtuse from the lengths of its sides. • Write inverses and contrapositives of statements. • Recognize inverses and contrapositives in logical arguments.	1, 2, 3, 4, 7

Mathematical Connections	3.1	3.2	3.3	3.4	3.5	3.6	3.7
geometry	**111–116***	**117–123**	**124–129**	**130–135**	**136–140**	**141–147**	**148–153**
algebra	112, 115	123				142–144, 146	153
data analysis, probability, discrete math							153
patterns and functions	113, 114		129				
logic and reasoning	**111–116**	**117–123**	**124–129**	**130–135**	**136–140**	147	149–153

Interdisciplinary Connections and Applications

	3.1	3.2	3.3	3.4	3.5	3.6	3.7
history and geography						146	
reading and language arts			127		139		
sports and recreation					138		
zoology and archaeology	115					145	
marketing	116						
personal finance	113						
architecture and design						146, 147	
cooking, navigation, measurement, clocks, carpentry, travel, auto maintenance		120–122				143	150, 151

*__Bold page numbers__ indicate that a topic is used throughout the section.

Section	Student Book	*opportunities for use with*
		Support Material
3.1		**Technology Book:** Spreadsheet Activity 3
3.3	geometry software	**Technology Book:** Calculator Activity 4 TI-92 Activity 3
3.6	scientific calculator graphing calculator geometry software	**Technology Book:** Spreadsheet Activity 3 TI-92 Activity 3 **Geometry Inventor Activity Book:** Activity 5
3.7		**Technology Book:** TI-92 Activity 3

PLANNING GUIDE

Regular Scheduling (45 min)

Section	Materials Needed	Core Assignment	Extended Assignment	*exercises that feature*		
				Applications	Communication	Technology
3.1		1–14, 24–29	1–29	19, 20, 22, 23	14–20, 22, 24, 25	
3.2	protractor	1–14, 22–28, 30–34, AYP*	1–34, AYP	15–21, 29	16–18, 20, 29	
3.3	geometry software or ruler, protractor	1–6, 10, 11, 15–26	1–26	7–9, 12–14	7–9, 14, 21	
3.4		1–4, 10–16, 18–23, AYP	1–23, AYP	5–9	8, 9, 18	
3.5		1–9, 11, 18–28	1–28	10, 13–17, 19	10	
3.6	scissors, ruler, calculator, graph paper, heavy paper, geometry software	**Day 1:** 1–27 odd, 31–35 **Day 2:** 2–26 even, 47–50	**Day 1:** 1–35 odd **Day 2:** 2–36 even, 37–50	29 37–40, 45	29, 31 36, 47	46
3.7	graph paper, scissors, ruler, protractor	**Day 1:** 1–14, 16–24 **Day 2:** 29–34, 36–45, AYP	**Day 1:** 1–24 **Day 2:** 25–45, AYP	15 25–28	15	
Review/ Assess		**Day 1:** 1–10 **Day 2:** 11–20 **Day 3:** Ch. 3 Test	**Day 1:** 1–10 **Day 2:** 11–20 **Day 3:** Ch. 3 Test			
Portfolio Project		Allow 2 days.	Allow 2 days.			

Yearly Pacing (with Portfolio Project)	Chapter 3 Total 14 days	Chapters 1–3 Total 42 days	Remaining 118 days	Total 160 days

Block Scheduling (90 min)

	Day 15	Day 16	Day 17	Day 18	Day 19	Day 20	Day 21
Teach/Interact	3.1 3.2	3.3: Exploration, page 125 3.4	3.5 3.6: Exploration, page 141	Continue with 3.6 3.7: Exploration, page 148	Continue with 3.7 Review	Review Port. Proj.	Ch. 3 Test Port. Proj.
Apply/Assess	**3.1:** 1–14, 24–29 **3.2:** 1–14, 22–28, 30–34, AYP*	**3.3:** 1–6, 10, 11, 15–26 **3.4:** 1–4, 10–23, AYP	**3.5:** 1–9, 11, 18–28 **3.6:** 1–27 odd, 31–35	**3.6:** 2–26 even, 47–50 **3.7:** 1–14, 16–24	**3.7:** 29–34, 36–45, AYP **Review:** 1–10	**Review:** 11–20 **Port. Proj.**	**Ch. 3 Test Port. Proj.**

NOTE: A one-day block has been added for the Portfolio Project—timing and placement to be determined by teacher.

Yearly Pacing (with Portfolio Project)	Chapter 3 Total 7 days	Chapters 1–3 Total 21 days	Remaining 59 days	Total 80 days

*****AYP** is Assess Your Progress.*

Section	Practice Bank	Study Guide*	Assessment Book*	Visuals	Explorations Lab Manual	Lesson Plans	Technology Book
3.1	16	3.1		Warm-Up 3.1		3.1	Spreadsheet Act. 3
3.2	17	3.2	Test 11	Warm-Up 3.2		3.2	
3.3	18	3.3		Warm-Up 3.3	Master 10	3.3	Calculator Act. 4 TI-92 Act. 3
3.4	19	3.4	Test 12	Warm-Up 3.4	Add. Expl. 4	3.4	
3.5	20	3.5		Warm-Up 3.5		3.5	
3.6	21	3.6		Warm-Up 3.6 Folder 4	Master 2	3.6	Spreadsheet Act. 3 TI-92 Act. 3
3.7	22	3.7	Test 13	Warm-Up 3.7	Add. Expl. 5 Masters 1, 11	3.7	TI-92 Act. 3
Review Test	23	Chapter Review	Tests 14, 15, Alternative Assessment			Review Test	

*Spanish versions of *Study Guide* and *Assessment Book* are available.

Chapter Support

- Course Guides
- Lesson Plans
- Portfolio Project Book: Additional Project 2: Using Reasoning to Make Decisions
- Preparing for College Entrance Tests
- Multi-Language Glossary
- *Test Generator* Software
- Professional Handbook
- Challenge Problems

Software Support

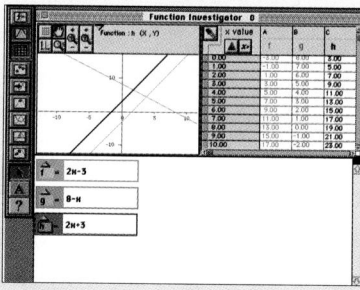

McDougal Littell Mathpack
Geometry Inventor

Internet Support

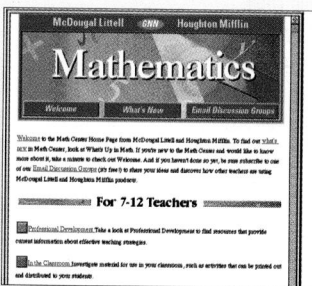

http://www.hmco.com
Next go to McDougal Littell; then the Education Center; then Secondary Math.

Books, Periodicals

Hirschorn, Daniel B. and Denisse R. Thompson. "Technology and Reasoning in Algebra and Geometry." *Mathematics Teacher* (February 1996): pp. 138–142.

McGivney, Jean M. and Thomas C. DeFranco. "Geometry Proof Writing: A Problem-Solving Approach a la Polya." *Mathematics Teacher* (October 1995): pp. 552–555.

Chazan, Daniel and Richard Houde. *How to Use Conjecturing and Microcomputers to Teach Geometry.* Reston, VA: NCTM, 1989.

Serra, Michael. *Discovering Geometry: An Inductive Approach.* Berkeley, CA: Key Curriculum Press.

Activities, Manipulatives

Miller, William A. and Linda Wagner. "Pythagorean Dissection Puzzles." *Mathematics Teacher* (April 1993): pp. 302–314.

Naraine, Bishnu. "If Pythagoras Had a Geoboard." *Mathematics Teacher* (February 1993): pp. 137–148.

Software

Jackiw, Nicholas, designer. *The Geometer's Sketchpad.* (Ver. 3.0) for Macintosh or Windows. Berkeley, CA: 1995.

Videos

Apostol, Tom M. *The Theorem of Pythagoras.* Videotape and guide. Reston, VA: NCTM, 1988.

Internet

Find Classroom Materials on the World Wide Web at the Teacher's Place:
http://forum.swarthmore.edu/classroom.html

3 Reasoning in Geometry

Interview Notes

Background

Forensic Science

Forensic science is the technique of using scientific methods to solve crimes. One of the first crime laboratories was established in 1910 by the French physician Edmond Locard in Lyon, France. In the United States, the first crime laboratory was established in Los Angeles, California. Today, there are about 250 crime laboratories in the United States.

There are various specialties associated with forensic science, such as forensic psychiatry, forensic toxicology, and forensic pathology. In order to be a forensic scientist, a person needs at least a bachelor's degree in chemistry, and courses in botany, anatomy, geology, physics, and physiology.

Mary-Jacque Mann

Before beginning work at the National Fish and Wildlife Forensics Laboratory, Mary-Jacque Mann worked at the Smithsonian Institution in Washington, D.C., and at the New Mexico State Police Crime Lab. At the National Fish and Wildlife Forensics Laboratory, Mann works in the criminalistics section, where microscopic, chemical, and instrumental techniques are used to identify and compare evidence.

Logical Crime SOLVER

INTERVIEW Mary-Jacque Mann

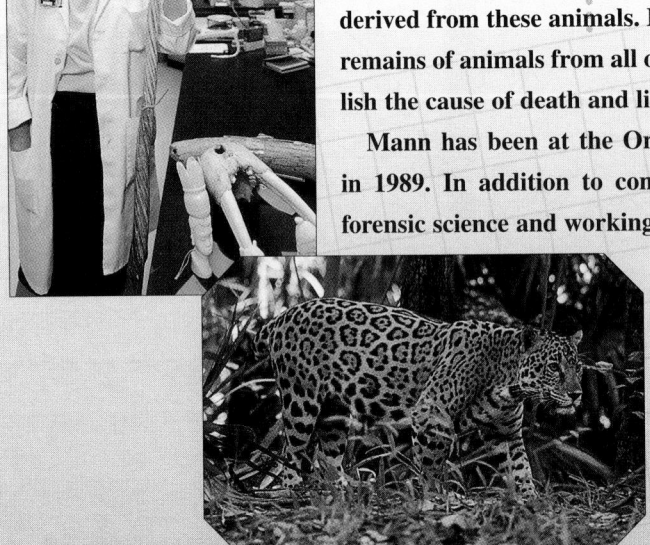

Mary-Jacque Mann uses scientific tools and logical analysis to track down criminals. As a senior forensic scientist at the National Fish and Wildlife Forensics Laboratory in Ashland, Oregon, Mann investigates crimes against wildlife—hunting protected animals and importing outlawed products derived from these animals. Mann and her colleagues test the remains of animals from all over the world. They try to establish the cause of death and link a suspect to the crime.

Mann has been at the Oregon laboratory since it opened in 1989. In addition to completing an advanced degree in forensic science and working for six years in a crime lab, she was a volunteer at the center before she joined the staff. "This is the only full-service wildlife forensics lab in the world," she says. "I'm at the best possible place to pursue the kind of work I like to do."

" ... We use logic to solve actual crimes. We're looking for a higher standard ... "

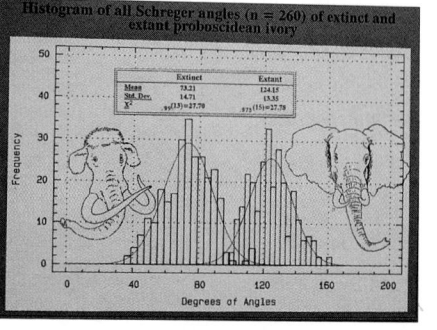

The histogram shows the results of the research on mammoth and elephant ivory.

Making a Difference

"There's a lot of variety in this job, and the analysis is challenging," she says. "I also feel we are making a difference." Staff members at the lab identify samples of wild animal hair, feathers, fur, or blood that are found at a crime scene. Mann may try to identify the gun involved in a crime. One of the first projects she worked on at the lab was finding a way to identify different kinds of ivory.

In order to protect elephants worldwide, the United States banned the importation and sale of elephant ivory. Mammoth ivory and some other materials that look like elephant ivory are still legal. The lab wanted to find a way to tell whether an object is made of elephant ivory without damaging the object. Working with Ed Espinoza, Mann found a technique to distinguish between elephant ivory products and other forms of ivory. After doing thousands of measurements, they discovered that the faint etchings known as Schreger lines found in elephant ivory form different angles from those found in mammoth ivory. (Mammoths, relatives of the modern elephant, became extinct about 10,000 years ago.)

109

Mathematical Connection

Following specific procedures and methods when collecting, analyzing, and presenting evidence is crucial to forensic science. If specific procedures are not followed, evidence may be contaminated or an incorrect analysis may result. The methods and procedures involved in forensic science follow a logical pattern. Two of these patterns are studied in the exercises of this chapter. In Section 3.3, students are introduced to a key that can be used to identify the species of an alligator or caiman. Students use the key and describe its logical reasoning. In Section 3.4, an ivory identification chart is shown. Students are asked to analyze information and classify a piece of material.

Explore and Connect

Project
There are thousands of choices for animals in this project. The library, an encyclopedia, or the Internet may be good sources for information. Students may also want to write to zoos or conservation programs for information and pictures for their poster.

Writing
If there is a crime lab in your area, students may want to call or write to answer these questions. They could also write to the Wildlife Forensics Lab for more information.

Research
You may want to suggest that students write a report on mammoths. Recent findings on mammoths can also be included.

The Ivory Angle

Mann sums up this finding in two rules of thumb:

- If the Schreger angles in an ivory object average 115° or more, then the object is made of elephant ivory.
- If the Schreger angles average 90° or less, then the object is made of mammoth ivory.

If the object is neither elephant nor mammoth ivory, a set of rules formulated by Espinoza and Mann can be used to determine whether the object is made of whale, walrus, hippopotamus, or warthog ivory. "As in other sciences, if you follow a logical progression from beginning to end, going through a specified sequence of steps, you can get a predictable result," Mann says.

With this identification technique, officials are able to enforce a law banning the importation of elephant ivory, essentially shutting down sales of illegal ivory in the United States. Mann hopes to see more cases where the combination of logic, the scientific method, and international law can be applied to protect endangered animals like the African and Asian elephant.

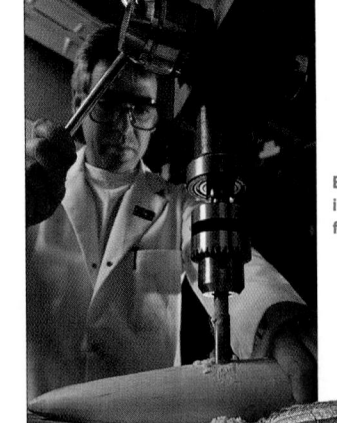

Ed Espinoza cuts ivory samples from a tusk.

Schreger lines on mammoth tusk

Schreger lines on elephant tusk

EXPLORE AND CONNECT

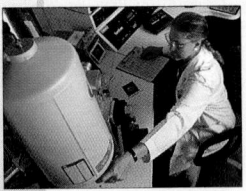

Mary-Jacque Mann uses an electron microscope to determine the source of dust found on a piece of evidence.

1. Project Make a poster about an endangered animal. Explain why the animal has become endangered. Have the measures taken to protect the animal affected local economy? Have the protection measures succeeded in helping the animal survive?

2. Writing Find out what forensic scientists do in a crime lab. How do you think the work in a regular crime lab is different from the work done at the Wildlife Forensics Lab?

3. Research Find out when the mammoths lived. In what parts of the world did they live? How did humans rely on mammoths?

Mathematics & Mary-Jacque Mann

In this chapter, you will learn more about how mathematics is related to forensic science.

Related Exercises

Section 3.3
- Exercises 12–14

Section 3.4
- Exercises 5–8

3.1 Inductive and Deductive Reasoning

Learn how to...
- use deductive reasoning to reach conclusions

So you can...
- make a convincing argument
- recognize valid and invalid arguments

Before the late 1800s, no one knew what caused malaria. Some people thought malaria was caused by mosquito bites. Others believed that it was caused by the air in swampy or marshy areas. A series of findings convinced scientists that mosquitoes transmit malaria to humans.

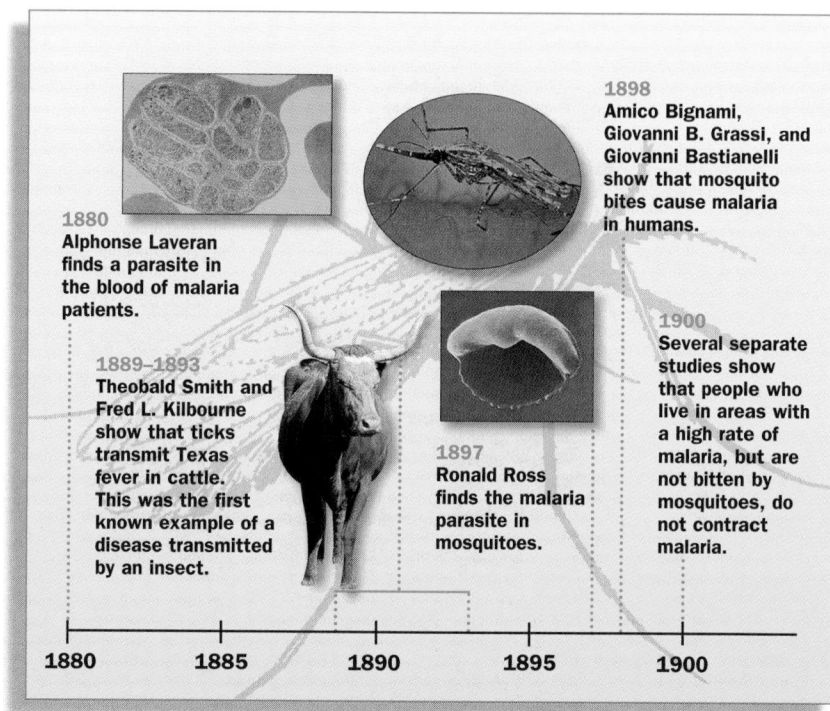

1880
Alphonse Laveran finds a parasite in the blood of malaria patients.

1889–1893
Theobald Smith and Fred L. Kilbourne show that ticks transmit Texas fever in cattle. This was the first known example of a disease transmitted by an insect.

1897
Ronald Ross finds the malaria parasite in mosquitoes.

1898
Amico Bignami, Giovanni B. Grassi, and Giovanni Bastianelli show that mosquito bites cause malaria in humans.

1900
Several separate studies show that people who live in areas with a high rate of malaria, but are not bitten by mosquitoes, do not contract malaria.

1880   1885   1890   1895   1900

THINK AND COMMUNICATE

1. Before this research showed the role of mosquitoes in spreading malaria, it was known that people who lived in swampy areas often became infected with malaria. What kind of reasoning led people to believe that malaria was caused by the air in swampy areas?

2. Explain why each of these investigations was an important part of demonstrating that malaria is transmitted by mosquitoes.

3.1 Inductive and Deductive Reasoning   **111**

ANSWERS   Section 3.1

Think and Communicate

1. inductive reasoning

2. Answers may vary. An example is given. The 1880 study found a parasite in the blood of malaria patients; subsequent investigations could focus on finding the source of the parasite. The 1889–1893 studies showed that it was possible for insects to transmit diseases to humans; subsequent investigations could focus on finding an insect that carries the malaria parasite. The 1897 study identified the malaria parasite in mosquitoes, forming a link between the disease, its victims, and mosquitoes. The 1898 study proved the connection between mosquitoes and malaria infection. The 1900 study suggested that malaria is transmitted only by mosquito bites.

Plan⇔Support

Objectives
- Use deductive reasoning to reach conclusions.
- Make a convincing argument.
- Recognize valid and invalid arguments.

Recommended Pacing
❖ **Core and Extended Courses**
  Section 3.1: 1 day
❖ **Block Schedule**
  Section 3.1: $\frac{1}{2}$ block
  (with Section 3.2)

Resource Materials
Lesson Support
Lesson Plan 3.1
Warm-Up Transparency 3.1
Practice Bank: Practice 16
Study Guide: Section 3.1
Challenge Problems: Set 16
Technology
Technology Book:
  Spreadsheet Activity 3
Internet:
  http://www.hmco.com

Warm-Up Exercises
What comes next in each pattern?

1. $\frac{2}{3}, \frac{4}{9}, \frac{8}{27}, \frac{16}{81},$? $\frac{32}{243}$

2. [box patterns] ?

3. All people who bought tickets before 7 P.M. got good seats. Ty bought his ticket at 6 P.M. Did he get a good seat? Yes.

4. If the butler did it, he would have mud on his shoes. There is no mud on his shoes, and he has only one pair of shoes. Did the butler do it? No.

5. People who eat undercooked hamburgers may get sick. Mrs. Lura ate a hamburger and did not get sick. Was the hamburger she ate undercooked? not possible to say

111

Section Note

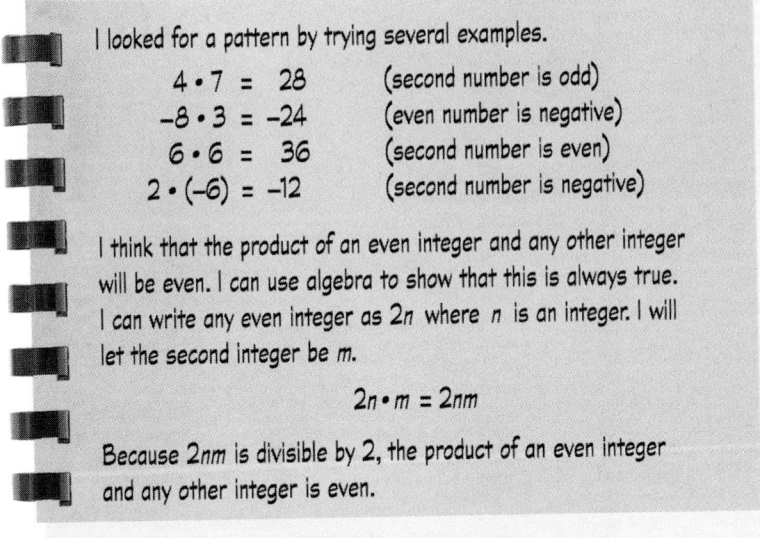

Communication: Discussion
Discuss the difference between inductive and deductive reasoning. Emphasize that deductive reasoning shows that something is *always* true. Once a mathematical fact, such as a property of a geometric figure, is established deductively, this fact is always true.

About Example 1

Topic Spiraling: Review
This Example makes use of the algebraic property that any even number has a factor of 2 and thus is divisible by 2. Review this idea with students by using four or five simple examples of even numbers and show that for any number n, the number $2n$ is even.

Additional Example 1

Does the square of any whole number ending in 5 end in 25? Write a convincing argument that explains your conclusion.
Look for a pattern by trying several examples.

$5 \cdot 5 = 25$
$15 \cdot 15 = 225$
$25 \cdot 25 = 625$
$75 \cdot 75 = 5625$

The square of any whole number ending in 5 ends in 25. Algebra shows why this is always true. Any whole number that ends in 5 can be written as $10k + 5$, where k is a whole number. The square of $10k + 5$ is $100k^2 + 100k + 25$. The number $100k^2$ is a whole number that ends in two 0's and the same is true of $100k$. The sum $100k^2 + 100k$ ends in two 0's. Then $100k^2 + 100k + 25$ is a whole number that ends in 25.

When you make conjectures based on patterns, such as noticing that people who live in swampy areas become infected with malaria, you use inductive reasoning. You use *deductive reasoning* to determine whether your conjectures are always true. **Deductive reasoning** involves using facts, definitions, and accepted properties in a logical order to reach a conclusion.

EXAMPLE 1   Connection: Algebra

Writing Is the product of an even integer and any other integer even or odd? Write a convincing argument that explains your conclusion.

SOLUTION

In order to find a pattern, you should try examples with several different types of numbers. You might want to make a table, or an organized list. Here is Ruby's solution.

> I looked for a pattern by trying several examples.
>
> $4 \cdot 7 = 28$ (second number is odd)
> $-8 \cdot 3 = -24$ (even number is negative)
> $6 \cdot 6 = 36$ (second number is even)
> $2 \cdot (-6) = -12$ (second number is negative)
>
> I think that the product of an even integer and any other integer will be even. I can use algebra to show that this is always true. I can write any even integer as $2n$ where n is an integer. I will let the second integer be m.
>
> $$2n \cdot m = 2nm$$
>
> Because $2nm$ is divisible by 2, the product of an even integer and any other integer is even.

THINK AND COMMUNICATE

3. What kind of reasoning did Ruby use when she looked for a pattern in the solution of Example 1? Using only this kind of reasoning, could she be certain of her conclusion?

4. What kind of reasoning did Ruby use to complete her solution? Which part of the argument do you find most convincing? Why?

5. Why do you think she used both kinds of reasoning?

Think and Communicate

3. inductive reasoning; No.

4. deductive reasoning; the part that uses deductive reasoning; because something justified correctly by using deductive reasoning is always true

5. Answers may vary. An example is given. She used inductive reasoning to see if she could make a conjecture about the product of an even integer and any another integer. She predicted that this product is always even. Then she used deductive reasoning to show that her conjecture is correct.

EXAMPLE 2

Tell whether each argument uses *inductive* or *deductive* reasoning.

a. In the United States, you must be at least 18 years old to vote. Jamal is 19 years old. Jamal is old enough to vote in the United States.

b.

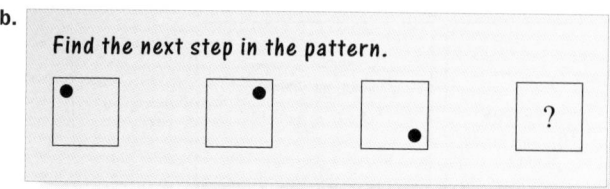

Find the next step in the pattern.

At each step, the black circle rotates clockwise to the next corner, so the next picture in the pattern is $\longrightarrow$

SOLUTION

a. This argument uses deductive reasoning. The conclusion is based on two known facts.

b. This argument uses inductive reasoning. The conclusion is based on the observation of a pattern in the given steps.

THINK AND COMMUNICATE

6. In part (a) of Example 2, you cannot conclude that Jamal can vote in the United States. Give a reason why Jamal might not be able to vote in the United States.

7. Jonah is taller than Emily and Matt is shorter than Emily. Who is taller, Jonah or Matt? Did you use *inductive* or *deductive* reasoning?

☑ CHECKING KEY CONCEPTS

Tell whether each argument uses *inductive* or *deductive* reasoning.

1. By observing many individual cases, people concluded that malaria was caused by breathing air in swampy areas.

2. All students must study Algebra 1 before studying Geometry. Mia is studying Geometry. Therefore, Mia has studied Algebra 1.

3. Any quadrilateral with four congruent angles is a rectangle. A square has four congruent angles. A square is a rectangle.

4. **PERSONAL FINANCE** Martha Greene has been studying the profits of mutual funds for the last six months. She decides to invest in the fund that has had the highest profits during this time.

BY THE WAY

Malaria continues to be a health problem in many areas of the world. Scientists have found that one very effective way to prevent the disease is to use bed nets to prevent mosquito bites during the night.

3.1 Inductive and Deductive Reasoning **113**

Think and Communicate

6. Answers may vary. An example is given. Jamal may not be a citizen of the United States.

7. Jonah; deductive

Checking Key Concepts

1. inductive reasoning

2. deductive reasoning

3. deductive reasoning

4. inductive reasoning

Think and Communicate

Use question 5 to emphasize the fact that learning geometry, as well as other kinds of mathematics, means learning how to use both inductive and deductive reasoning.

Additional Example 2

Tell whether each argument uses *inductive* or *deductive* reasoning.

a. Find the next step in the pattern.

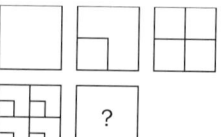

The next step in the pattern is to divide each of the large squares into four smaller squares.

This argument uses inductive reasoning. The conclusion is based on the observation of the pattern in the given steps.

b. If the pizza is overcooked, then the oven was too hot or the pizza was left in the oven more than 10 minutes. The pizza is overcooked after being taken out of the oven in 9 minutes. The oven was too hot.

This argument uses deductive reasoning. The conclusion is based on known facts.

Checking Key Concepts

Teaching Tip

In discussing these questions, ask students to explain *why* they think an argument uses inductive or deductive reasoning.

Closure Question

Explain the meaning of deductive reasoning and inductive reasoning.
Deductive reasoning uses known facts, definitions, principles, and properties in a logical order to reach a conclusion. Inductive reasoning uses specific examples to make a conjecture that something is true or false.

Extra Practice
exercises on
page 663

3.1 Exercises and Applications

Tell whether each argument uses *inductive* or *deductive* reasoning.

1. The sun has risen every day of my life, so the sun will rise tomorrow.

2.

Dilbert reprinted by permission of United Feature Syndicate, Inc.

3. The probability of being left-handed is about $\frac{1}{10}$. About 20% of the musicians that I know are left-handed. So left-handed people are more likely than right-handed people to be musicians.

4. All finches eat seeds. The Eurasian Hawfinch, Rosita's Bunting, and the Painted Bunting are all finches. These three birds all eat seeds.

Use inductive reasoning to predict the next two terms or figures in each pattern. Justify your prediction.

5.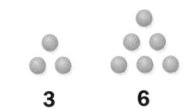

    1   3   6

6.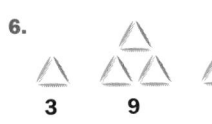

    3   9   18

7. 1, 2, 4, 8, . . .

8. 1, 8, 27, 64, 125, . . .

9. 1, 2, 6, 24, 120, . . .

10. 2, 9, 16, 23, . . .

For Exercises 11–13, select the conclusion that makes the statement true.

11. All equilateral triangles are isosceles. △*ABC* is equilateral. Therefore, △*ABC* (*is, is not,* or *may be*) isosceles.

12. Jean is a member of her school's outing club. Jean is a rock climber. So, the members of this club (*are, are not,* or *may be*) rock climbers.

13. If a triangle is isosceles, it has two congruent angles. △*ABC* has no congruent angles. Therefore, △*ABC* (*is, is not,* or *may be*) isosceles.

14. a. Give a convincing argument that the next two numbers in the sequence 1, 2, 3, . . . could be 4, 5.

    b. Give a convincing argument that the next two numbers in the sequence 1, 2, 3, . . . could be 6, 12.

    c. Give another pair of numbers that could be the next two numbers in the sequence 1, 2, 3, ... Explain your reasoning.

15. **Open-ended Problem** Describe a situation in your life when you used deductive reasoning.

16. Visual Thinking This pattern is made by placing points on a circle and drawing each segment that connects a pair of points.

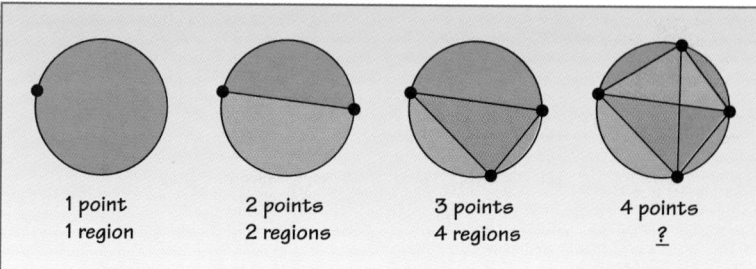

1 point	2 points	3 points	4 points
1 region	2 regions	4 regions	?

a. Complete the table at the right by sketching diagrams. Count the number of regions created.

b. **Open-ended Problem** Predict the number of regions formed by 6 points and their segments. Explain your prediction.

c. Sketch diagrams to check your prediction in part (b).

Number of points	Number of regions
2	2
3	4
4	?
5	?

17. Challenge Change the pattern in Exercise 16 so that you draw only segments connecting adjacent points. Write a formula for the number of regions as a function of the number of points. Give a convincing argument that your formula is correct.

18. ALGEBRA Suppose you multiply any two odd integers. Is the product odd or even? Explain your answer with a convincing argument. (*Hint:* An odd number can be written as $2n + 1$.)

Connection ▸ ZOOLOGY

Upside-down catfish live in water that may contain little oxygen except near the surface. Zoologist Lauren Chapman suspected that the fish swim upside down to reach this oxygen more easily. To test this theory, Chapman and colleagues studied these catfish in an aquarium. When they lowered the oxygen content of the water, the catfish swam upside down.

19. What kind of reasoning did Chapman use to test her theory? Do you think this experiment provides convincing evidence that her theory is correct? Explain why or why not.

20. The catfish eat surface plankton. Swimming upside down may allow the catfish to reach the surface plankton more easily. Describe an experiment you could use to test this theory.

Exercise Notes

Cooperative Learning
Exs. 16, 17 Students may enjoy working with a partner to do these exercises. For Ex. 17, ask for volunteers to present their arguments to the class.

Integrating Algebra
Ex. 18 Suggest that students examine some examples of products of odd numbers before writing an argument.

Interdisciplinary Problems
Exs. 19, 20 Point out that scientists do experiments to test their theories, as in these studies of catfish. Emphasize that experiments give specific information or data that can be used to support or disprove a theory. Theories are often changed or modified based upon the results of experiments.

---

b. Each term after the first is twice the previous term.

c. Answers may vary. Examples are given. 5 and 8; Each term after the second is the sum of the two previous terms.

15. Answers may vary. An example is given. I went to see my math teacher after school. I saw her coat hanging in her classroom and concluded that she had not yet left the school.

16. a. 8; 16

b. Predictions may vary. An example is given. 32; It seems that each time a point is added, the number of regions doubles.

c. 31

17. Let n be the number of points, where $n > 1$, and r be the number of regions. Then $r = n + 1$. Answers may vary. An example is given. When n points are drawn, there are n segments that can be drawn connecting adjacent points. The circle is divided into n regions bounded by part of the circle and one segment, and one region bounded by the segments, or $n + 1$ regions in all.

18. odd; Answers may vary. An example is given. Let $2n + 1$ and $2m + 1$ represent the odd integers. Their product is $(2n + 1)(2m + 1) = 4mn + 2m + 2n + 1 = 2(2mn + m + n) + 1$.

Since the product is 1 more than an even number, the product is odd.

19. inductive reasoning; An example is given. I do not think this experiment provides convincing evidence that her theory is correct. The aquarium is not the natural habitat of the catfish. Perhaps the fish are responding to the stress of confinement.

20. Answers may vary. An example is given. Set up two aquariums. Put plankton on the surface of one and none in the other. Check to see whether the fish swim upside down. If they do so in both aquariums, it is unlikely that the plankton is the reason.

Exercise Notes

Student Progress

Exs. 21, 22 Most students are probably more familiar with inductive reasoning than deductive reasoning because of their past experiences in life and in previous mathematics courses. As they progress in this course, they will begin to develop a more complete and mathematically mature understanding of deductive reasoning.

Second-Language Learners

Ex. 25 You may want to allow students learning English to share the advertisements they have chosen with a peer tutor before they complete this exercise.

Assessment Note

Ex. 25 This exercise provides a good opportunity to assess how well students understand deductive reasoning as well as providing an opportunity to explore ideas concerning valid and invalid reasoning.

Practice 16 for Section 3.1

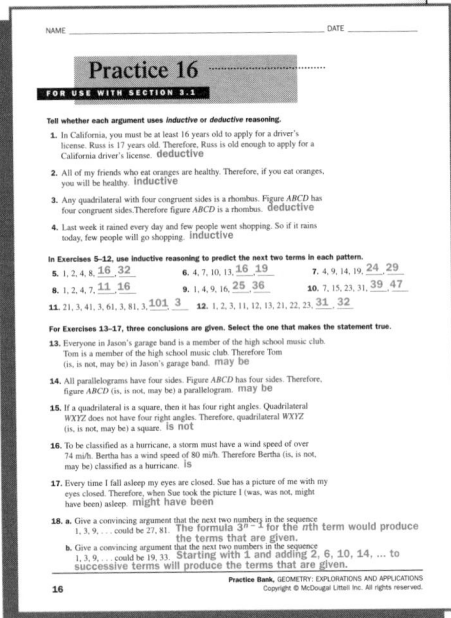

21. What type of reasoning does Rosemary use in her note to her friend? Do you think her conclusion is reasonable?

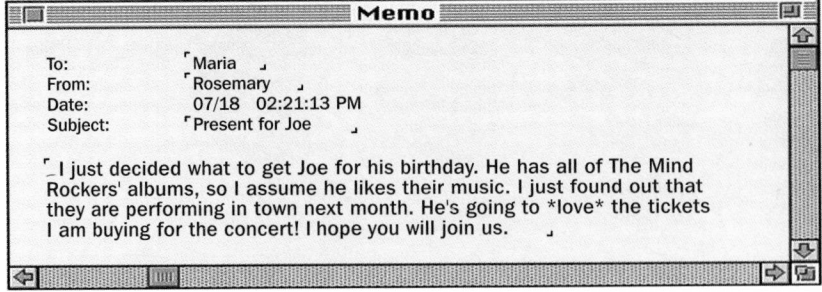

Memo

To: Maria
From: Rosemary
Date: 07/18 02:21:13 PM
Subject: Present for Joe

I just decided what to get Joe for his birthday. He has all of The Mind Rockers' albums, so I assume he likes their music. I just found out that they are performing in town next month. He's going to *love* the tickets I am buying for the concert! I hope you will join us.

MARKETING Andre would like to work as a buyer for a music store. He made an appointment to talk to Karla Myers, the buyer at a music store.

22. Karla said that she makes her decisions based on what she observes, by analyzing her mistakes, and also by trying to think ahead. Explain how she uses both inductive reasoning and deductive reasoning.

23. Open-ended Problem Suppose that you are the buyer for a music store. A new band is releasing an album. What would you consider in deciding how many CDs and cassettes to order for your store?

24. Writing Explain the difference between inductive and deductive reasoning. Why do mathematicians and scientists use both deductive reasoning and inductive reasoning in their work?

ONGOING ASSESSMENT

25. Cooperative Learning Work in a group of three people. Find six advertisements in a magazine or newspaper. Each advertisement should make a claim about a service or product. Each of you should determine what kind of reasoning is used to reach the conclusions in each advertisement. Discuss whether the claims are true or false. Choose one advertisement and work together to give a deductive argument that could be used to support the claim made in it.

SPIRAL REVIEW

26. Sketch a triangular prism. *(Section 2.6)*

Tell whether each angle is *acute*, *right*, or *obtuse*. *(Section 2.1)*

27. ∠AFC

28. ∠CFE

29. ∠AFE

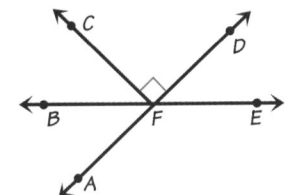

---

21. inductive reasoning; Yes.

22. When Karla makes decisions based on what she observes, she is using inductive reasoning. In analyzing her mistakes, she probably uses deductive reasoning. In trying to think ahead, she may use both inductive and deductive reasoning.

23. Answers may vary. An example is given. For a new band with no previous releases, I would see if any of the musicians were involved in other groups and find out how popular those groups were. I would consider local album sales for the type of music the band plays. I would also determine if the band had performed publicly and, if so, how well tickets had sold.

24. Answers may vary. An example is given. A person using inductive reasoning makes a conjecture based on observations. A person using deductively reasoning reaches a conclusion by using facts, definitions, and accepted principles and properties in a logical order. A conclusion reached by correct deductive reasoning must be true, while a conjecture reached by inductive reasoning may or may not be true. Inductive reasoning may provide scientists and mathematicians with conjectures that they may then prove deductively. They can use deductive reasoning to show that a statement is always true.

25. Check students' work.

26.

27. right

28. obtuse

29. obtuse

3.2 | Postulates, Definitions, and Properties

Learn how to...

- **recognize and use postulates, definitions, and properties**

So you can...

- **justify statements about geometric figures**

Members of the Continental Congress signed the Declaration of Independence in August 1776. Thomas Jefferson wrote most of this document, which declared the colonies independent of British rule. In the document, Jefferson stated the assumptions that the congress made about the rights of men.

> **"...We** hold these truths to be self-evident, that all men are created equal, that they are endowed by their Creator with certain unalienable Rights, that among these are Life, Liberty, and the pursuit of Happiness ..."

BY THE WAY

Abigail Adams urged her husband John to include rights for women in the laws being written by the Continental Congress.

THINK AND COMMUNICATE

1. What does it mean for a statement to be self-evident? According to the Declaration of Independence, what truths are self-evident?

2. Why do you think Jefferson needed to state his assumptions?

3. Jefferson used the phrase "all men are created equal." What meanings does the word *men* have in English? How does the meaning of the word *men* affect the meaning of this statement? How do you think Jefferson defined *men*?

In mathematics, a statement that is accepted without proof is called a **postulate**. Jefferson's *self-evident truths* were his postulates for the Declaration of Independence. Postulates may be used as the starting points of our convincing arguments.

3.2 Postulates, Definitions, and Properties   **117**

Plan⇔Support

Objectives

- Recognize and use postulates, definitions, and properties.
- Justify statements about geometric figures.

Recommended Pacing

❖ **Core and Extended Courses**
Section 3.2: 1 day

❖ **Block Schedule**
Section 3.2: $\frac{1}{2}$ block
(with Section 3.1)

Resource Materials

Lesson Support
Lesson Plan 3.2
Warm-Up Transparency 3.2
Practice Bank: Practice 17
Study Guide: Section 3.2
Challenge Problems: Set 17
Assessment Book: Test 11

Technology
Internet:
    http://www.hmco.com

Warm-Up Exercises

Tell whether the given statement is *always true*, *sometimes true*, or *never true*.

1. If $\overline{AB} \cong \overline{CD}$ and $\overline{CD} \cong \overline{EF}$, then $\overline{AB}$ and $\overline{EF}$ have the same length. always true

2. If ∠ 1 is obtuse, then $m \angle 1 > 90°$. always true

3. If ∠ XYZ is acute, then the supplement of ∠ XYZ is also acute. never true

4. Rhombuses are squares. sometimes true

5. Squares are rhombuses. always true

Additional Example 1

Sketch an illustration of the postulate:

If point Y is between points X and Z, then X, Y, and Z are collinear and $XY + YZ = XZ$.

Draw any three points X, Y, and Z such that Y is between X and Z. The points will be collinear and the sum of XY and YZ will equal XZ.

If point Y is not between X and Z, then X, Y, and Z are not collinear and $XY + YZ \neq XZ$.

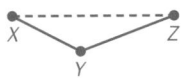

Section Notes

Geometric Thinking

In the paragraph on definitions, call attention to the words *if and only if* in the last sentence. Point out that these words provide a way of combining two true if-then statements into a single statement: If a shape is a square, then it is a regular quadrilateral, *and* if a shape is a regular quadrilateral, then it is a square. Thus, in this definition, the hypothesis and conclusion are interchangeable.

Teaching Tip

Students learn in the Watch Out! that some words used in geometry are not defined. You may wish to point out that it is not possible to define every term in geometry without being forced to use *circular* forms of definition. Students can check this fact by looking up a word in a dictionary, then looking up the words used to define it. They will discover that all the words are eventually defined in terms of each other.

Some of the ideas that you learned in Chapters 1 and 2 are postulates. For example, the statement "If point Y is between points X and Z, then X, Y, and Z are collinear and $XY + YZ = XZ$" can now be called the Segment Addition *Postulate*. A list of the postulates that you can use in convincing arguments is given on page 712.

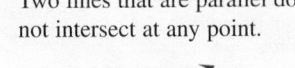

Sketch an illustration of the postulate:

If two lines intersect, they intersect in one and only one point.

SOLUTION

Draw any two lines that intersect. They will intersect at only one point.

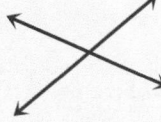

Two lines that are parallel do not intersect at any point.

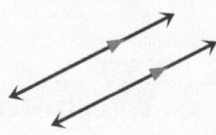

Definitions

> **WATCH OUT!** ▶
>
> Some words used in geometry are not defined. *Point, line,* and *plane* are undefined terms. In Chapter 1, these terms were described, but not defined.

Another important feature of an argument is the **definition**, or meaning, of each word that is used. In the Declaration of Independence, the definition of *men* that is used affects the meaning of Jefferson's self-evident truths. In mathematics, if a word is defined, then there is no question about its meaning.

For example, a polygon is a square *if and only if* it is a regular quadrilateral. This is an example of a *biconditional* statement.

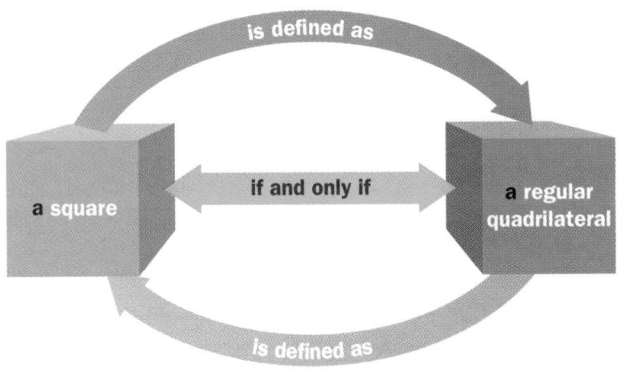

Properties

You can use properties of equality to reason about lengths of segments and measures of angles. Some of these properties are also true for congruence.

Equality and Congruence

Reflexive Property

$a = a$ $\overline{PQ} \cong \overline{PQ}; \angle 1 \cong \angle 1$

Symmetric Property

If $a = b$, then $b = a$.   If $\overline{PQ} \cong \overline{RS}$, then $\overline{RS} \cong \overline{PQ}$.

Transitive Property

If $a = b$ and $b = c$, then $a = c$.   If $\angle 1 \cong \angle 2$ and $\angle 2 \cong \angle 3$, then $\angle 1 \cong \angle 3$.

You can add or subtract the same value from both sides of an equation. In geometry, this can lead to new information about lengths and measures.

Properties of Equality

Addition Property

If $a = b$, then $a + c = b + c$.   If $PQ = RS$, then $PQ + QR = QR + RS$.

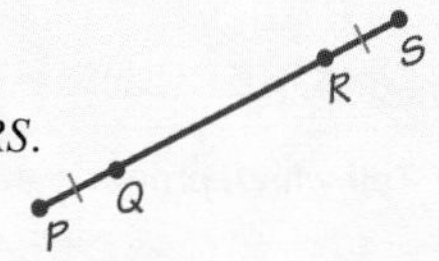

Subtraction Property

If $a = b$, then $a - c = b - c$.   If $m\angle JNL = m\angle KNM$, then $m\angle JNL - m\angle KNL = m\angle KNM - m\angle KNL$.

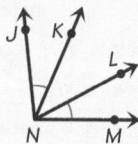

Substitution Property

If $a = b$, then a can be substituted for b (and b for a) in an expression.   If $m\angle 1 + m\angle 2 = 180°$ and $m\angle 1 = m\angle 3$, then $m\angle 3 + m\angle 2 = 180°$.

THINK AND COMMUNICATE

4. Use the diagram above. Explain how the Addition Property supports the statement "If $\overline{PQ} \cong \overline{RS}$, then $\overline{PR} \cong \overline{QS}$."

5. Use the diagram above. Explain how the Subtraction Property supports the statement "If $\angle JNL \cong \angle KNM$, then $\angle JNK \cong \angle LNM$."

Think and Communicate

4. If $\overline{PQ} \cong \overline{RS}$, then $PQ = RS$. By the Addition Property, $PQ + QR = QR + RS$. By the Segment Addition Postulate, $PQ + QR = PR$ and $QR + RS = QS$. By the Substitution Property, $PR = QS$, so $\overline{PR} \cong \overline{QS}$.

5. If $\angle JNL \cong \angle KNM$, then $m\angle JNL = m\angle KNM$. By the Subtraction Property, $m\angle JNL - m\angle KNL = m\angle KNM - m\angle KNL$. By the Angle Addition Property and the Subtraction Property, $m\angle JNK = m\angle JNL - m\angle KNL$ and $m\angle LNM = m\angle KNM - m\angle KNL$. By the Substitution Property, $m\angle JNK = m\angle LNM$, so $\angle JNK \cong \angle LNM$.

Section Notes

Teaching Tip
To help clarify and reinforce the meaning of the Reflexive Property, Symmetric Property, and Transitive Property, ask students to consider other examples of relationships that have some or all of these properties. Some examples from mathematics and from everyday situations are:

> (transitive, but neither symmetric nor reflexive)

≤ (reflexive and transitive, but not symmetric)

is a multiple of (reflexive and transitive, but not symmetric)

belongs to the same species as (reflexive, symmetric, and transitive)

is a pen pal of (symmetric, but neither reflexive nor transitive)

is a source of food for (not reflexive, symmetric, or transitive)

Student Study Tip
Besides the Addition Property and Subtraction Property of Equality, students have learned in previous courses that there is also a Multiplication Property and a Division Property. If $a = b$, then $ac = bc$ and if $a = b$ and $c \neq 0$, then $\frac{a}{c} = \frac{b}{c}$. These two properties are sometimes useful when working with measures of angles or segments.

Challenge
After discussing the Think and Communicate questions, you may wish to explore the Properties of Equality further by asking students whether this statement is true or false: If $a = b$ and $c = d$, then $a + c = b + d$. (True.) Is this statement the same as the Addition Property? (No.) How can the statement be justified on the basis of the Properties of Equality? ($a + c = a + c$ by the Reflexive Property. If $a = b$ and $c = d$, then $a + c = b + d$ by the Substitution Property.) Similar reasoning can be used to show that if $a = b$ and $c = d$, then $a - c = b - d$.

Additional Example 2

Give the postulate, definition, property, or previous statement part that makes each statement true.

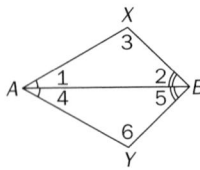

a. $m\angle 1 = m\angle 4$; $m\angle 2 = m\angle 5$
Definition of congruent angles

b. $m\angle 1 + m\angle 2 = m\angle 1 + m\angle 2$
Reflexive Property

c. $m\angle 1 + m\angle 2 = m\angle 4 + m\angle 5$
This follows from parts (a) and (b) by the Substitution Property.

d. $m\angle 1 + m\angle 2 + m\angle 3 = 180°$;
$m\angle 4 + m\angle 5 + m\angle 6 = 180°$
The sum of the angle measures of a triangle is 180°.

e. $m\angle 1 + m\angle 2 + m\angle 3 = m\angle 4 + m\angle 5 + m\angle 6$
This follows from part (d) by the Transitive Property.

f. $m\angle 3 = m\angle 6$
This follows from parts (c) and (e) by the Substitution Property and the Subtraction Property.

Checking Key Concepts

Communication: Drawing
Students should be able to answer questions 1, 3, and 4 without using drawings. However, if there are students who cannot do so, have them make drawings to visualize the situations.

Closure Question

Name three kinds of statements that can be used to justify other statements about geometric figures. postulates, definitions, and properties

120

You can use postulates, definitions, and properties as reasons in convincing arguments.

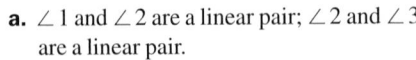

Give the postulate, definition, property, or previous statement that makes each statement true.

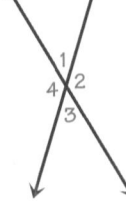

a. $\angle 1$ and $\angle 2$ are a linear pair; $\angle 2$ and $\angle 3$ are a linear pair.

b. $m\angle 1 + m\angle 2 = 180°$; $m\angle 2 + m\angle 3 = 180°$

c. $m\angle 1 + m\angle 2 = m\angle 2 + m\angle 3$

d. $m\angle 1 = m\angle 3$

SOLUTION

a. Definition of linear pair

b. Angles in a linear pair are supplementary. (Linear Pair Property)

c. Because both expressions in part (b) are equal to 180°, the Substitution Property says that they equal each other.

d. This follows from part (c) by the Subtraction Property.

☑ CHECKING KEY CONCEPTS

Tell which property makes each statement true.

1. $\angle ABC \cong \angle CBA$

2. **COOKING** If a recipe calls for $\frac{1}{2}$ cup buttermilk, you can use $\frac{1}{2}$ cup milk mixed with 1 teaspoon of lemon juice instead.

3. If $\overline{GH} \cong \overline{JK}$, then $\overline{JK} \cong \overline{GH}$.

4. If $m\angle 1 = m\angle 2$ and $m\angle 2 = m\angle 3$, then $m\angle 1 = m\angle 3$.

For each sketch, list the postulate(s) that it illustrates.

5.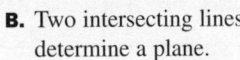

6.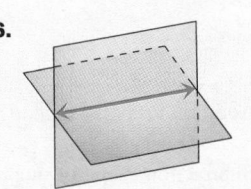

A. Through any two points there is exactly one line.

B. Two intersecting lines determine a plane.

C. Two planes can intersect in a line.

D. If A is between X and Y, then $XA + AY = XY$.

Checking Key Concepts

1. Reflexive Property

2. Substitution Property

3. Symmetric Property

4. Substitution Property

5. A

6. C

3.2 Exercises and Applications

Extra Practice exercises on page 664

Tell whether each statement is a *definition* or a *postulate*.

1. If two lines intersect, then they intersect in exactly one point.

2. Two lines in the same plane that never intersect are parallel.

3. An isosceles triangle has at least two congruent sides.

Tell which property makes each statement true.

4. $m \angle ABC = m \angle ABC$

5. If $\overline{AB} \cong \overline{CD}$ and $\overline{CD} \cong \overline{EF}$, then $\overline{AB} \cong \overline{EF}$.

6. If $m \angle A + m \angle B = 180°$ and $m \angle A = 45°$, then $45° + m \angle B = 180°$.

For Exercises 7–14, give the postulate, definition, property, or previous statement that makes each statement about the diagram true.

7. $\overline{BC} \cong \overline{BC}$

8. $CE = AC$

9. $AB = DE$

10. $\overline{CE} \cong \overline{DF}$

11. $m \angle PTR = m \angle QTS$

12. $\angle PTQ \cong \angle QTP$

13. $m \angle PTQ = m \angle RTS$

14. $\angle QTR \cong \angle QTR$

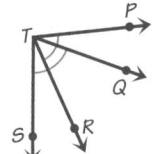

Connection ▶ NAVIGATION

Earth is divided into regions by imaginary circles called *meridians*. Because meridians are used as references for time and distance, it's important to have a common definition of the meridian that will be the "zero point." At a conference in 1884, representatives from twenty-five nations established the meridian through Greenwich, England, as the *prime meridian*, designated 0°.

15. The longitude of a location tells how many degrees east or west of the prime meridian it is. Before this agreement, many countries used different prime meridians. What information was needed in order to describe the longitude of a particular location?

16. **Research** Explain how degrees of longitude are related to degrees in an angle.

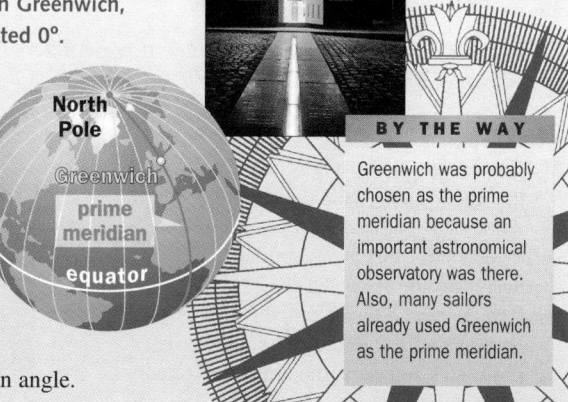

BY THE WAY

Greenwich was probably chosen as the prime meridian because an important astronomical observatory was there. Also, many sailors already used Greenwich as the prime meridian.

3.2 Postulates, Definitions, and Properties **121**

Apply⇔Assess

Suggested Assignment

❖ **Core Course**
Exs. 1–14, 22–28, 30–34, AYP

❖ **Extended Course**
Exs. 1–34, AYP

❖ **Block Schedule**
Day 15 Exs. 1–14, 22–28, 30–34, AYP

Exercise Notes

Geometric Thinking
Exs. 1–3 If some students are unsure whether a statement is a definition or a postulate, suggest that they try to rewrite the statement in the form of two conditional statements. If this can be done, then the statement is a definition.

Topic Spiraling: Preview
Exs. 7–14 These exercises prepare students for Section 3.3, where they will begin to write proofs.

Application
Exs. 15, 16 Geometry, algebra, and trigonometry are used in modern navigation. There may be some students in the class who have some basic knowledge of navigation. If so, invite them to share their knowledge with the class. It may be helpful to discuss how using latitude and longitude designations compares with using ordered pairs of numbers to locate points in the coordinate plane.

Student Study Tip
Exs. 15, 16 On a sphere such as Earth, the intersection of the sphere and a plane that passes through the center of the sphere forms a *great circle*. The shortest distance between any two points on a sphere is the length of an arc that is part of a great circle. This is why airline pilots usually fly great circle routes on long trips.

Second-Language Learners
Exs. 15, 16 You may want to pair students learning English with English proficient partners for these exercises, since they require well-developed language skills in both reading and writing.

Exercises and Applications

1. postulate

2. definition

3. definition

4. Reflexive Property

5. Transitive Property

6. Substitution Property

7. Reflexive Property

8. given, Addition Property, Segment Addition Postulate

9. given, definition of congruent segments

10. given, Addition Property, Segment Addition Postulate, definition of congruent segments

11. given, definition of congruent angles

12. Reflexive Property

13. Ex. 11, Angle Addition Postulate, Subtraction Property

14. Reflexive Property

15. the number of degrees east or west of the prime meridian and the meridian designated as the prime meridian

16. Each degree used to measure an angle is equal in magnitude to $\frac{1}{360}$ of a complete revolution. Each degree of longitude is equal to $\frac{1}{360}$ of a circle.

Research

Exs. 17–20 Science and technology rely on accurate measurements and, hence, on the instruments used to obtain them. Interested students can research the design and use of measuring instruments and present their findings to the class. They should concentrate on instruments that are used in situations where geometry plays an important role.

Visual Thinking

Ex. 21 Ask students to create sketches that demonstrate why the sundial shadow rotation is different in the southern hemisphere than in the northern hemisphere. Ask them to show the hands on a clock and tell the time from a counterclockwise point of view. This activity involves the visual skills of *recognition* and *interpretation*.

 **Application**

Ex. 29 Ask students to recall any experiences they may have had trying to fix a table that was not stable. Usually one leg of the table was slightly shorter than the other three. A three-footed table or pot is stable, but the top surface may not be parallel to the floor. Discuss why this may be true.

Multicultural Note

Ex. 29 Several scholars believe that the Chinese may have reached the Americas long before Columbus. Some theorize that the Chinese came as early as the Shang period (1766 B.C. to 1122 B.C.) to Central America. They site evidence of artistic similarities, such as jade carving. Others think that the Chinese may have arrived in Mexico or Peru during the Han dynasty (206 B.C. to A.D. 221), based on such evidence as similar cylindrical tripod vessels, lotus designs, and feline motifs.

Connection MEASUREMENT

In order for scientific researchers around the world to communicate effectively with each other, they must agree on consistent definitions and standards for their procedures. They can calibrate a process by comparing their results to a standard.

17. **Writing** An ordinary bathroom scale is a measurement instrument. Explain how you can tell if your scale is calibrated correctly.

18. Explain why the thermostat on an oven must be calibrated correctly.

19. **Open-ended Problem** Name three other appliances or instruments that require calibration.

20. A researcher has a Standard Reference Material (SRM) and a list of the exact amounts of the chemicals in the SRM. Explain how the researcher can use this information to calibrate equipment.

21. **CLOCKS** In the northern hemisphere, the shadows on a sundial rotate in the direction we call clockwise. In the southern hemisphere, shadows on a sundial rotate counterclockwise. If clocks had been invented in the southern hemisphere, what might the meanings of clockwise and counterclockwise have been?

Give the postulate, definition, property, or previous statement that makes the statement about the diagram true.

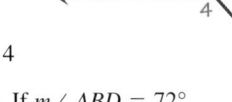

22. $\angle 1 \cong \angle 2$ 23. $\angle 2 \cong \angle 3$

24. $\angle 1 \cong \angle 3$ 25. $\angle 3 \cong \angle 4$

26. $\angle 1 \cong \angle 4$ 27. $m\angle 1 = m\angle 4$

28. **SAT/ACT Preview** Use the diagram. If $m\angle ABD = 72°$ and $m\angle CBE = 123°$, then $m\angle DBE = \underline{\ ?\ }$.

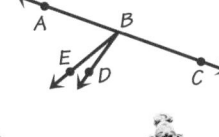

   A. 168°   B. 23°   C. 12°   D. 156°   E. 15°

29. **Challenge** Some anthropologists believe that the Chinese had contact with Central Americans before 100 A.D. Part of the evidence for this is the similarity in the designs of three-footed pots in both cultures. Explain why a three-footed pot is more stable than one with four feet. Use geometric postulates and properties to support your explanation.

Han dynasty jar, 206 B.C.–221 A.D.

Teotihuacán jar, 100–200 A.D.

---

17. Answers may vary. An example is given. You could place an object of known weight on your scale, for example, a 10 lb bag of potatoes, and see if your scale gives the correct weight.

18. Answers may vary. An example is given. Fairly exact temperature settings are important in preparing most cooked foods. For example, many cakes require an oven temperature of 350°F. An oven that heats to

400° when the thermostat is set at 350° would produce less than satisfactory results.

19. Answers may vary. Examples are given. an iron, a radio, a furnace, and the speedometer, odometer, and gas gauge on a car

20. A researcher can measure some of the chemicals present in the SRM. If the instrument does not read the exact amount

indicated on the list, the researcher can recalibrate the equipment and retest, repeating until the equipment measures the amount indicated on the list.

21. The meanings of the words would probably have been reversed.

22. Vertical angles are congruent.

23. given information

24. Exs. 22 and 23 and the Transitive Property

25. Vertical angles are congruent.

26. Exs. 24 and 25 and the Transitive Property

27. Ex. 26 and the definition of congruent angles

28. E

29. Answers may vary. An example is given. Since three points determine a plane, a

30. Open-ended Problem Write three statements about the figure at the right. Give the postulate, definition, or property that makes each statement true.

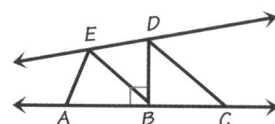

Find the measures of a complementary angle, a supplementary angle, and a vertical angle for the given angle. *(Section 2.1)*

31. $m\angle A = 63°$ **32.** $m\angle Y = 14°$ **33.** $m\angle H = 2y°$

34. Use a protractor to draw $\triangle ABC$ so that $m\angle A = 35°$, $m\angle B = 70°$, and $m\angle C = 75°$. *(Section 1.6)*

ASSESS YOUR PROGRESS

deductive reasoning (p. 112)   **definition** (p. 118)
postulate (p. 117)

Tell whether each argument uses *inductive* or *deductive* reasoning.
(Section 3.1)

1. $\triangle JKL$ is a regular triangle. A regular n-gon has n lines of symmetry. $\triangle JKL$ has three lines of symmetry.

2. $7 \cdot 3 = 21$ $5 \cdot 9 = 45$ $-3 \cdot 11 = -33$
The product of two odd numbers is an odd number.

Select the conclusion that makes the statement true. *(Section 3.1)*

3. Quadrilateral $EFGH$ is a parallelogram. It has four congruent sides. Therefore, it (*is*, *is not*, *might be*) a rhombus.

4. All of the Hawaiian Islands were formed by volcanoes. Little Diomede is not a Hawaiian Island. Therefore, Little Diomede (*was*, *was not*, *might have been*) formed by a volcano.

Give the postulate, definition, property, or previous statement that makes the statement about the diagram true. *(Section 3.2)*

5. $\overrightarrow{FC}$ bisects $\angle DFB$.

6. $\angle AFB \cong \angle EFD$

7. $m\angle AFC = m\angle EFC$

8. $\angle AFE \cong \angle DFB$

9. Journal Give an example of a postulate, a definition, and a property. Explain how the three are different. Explain how they are the same.

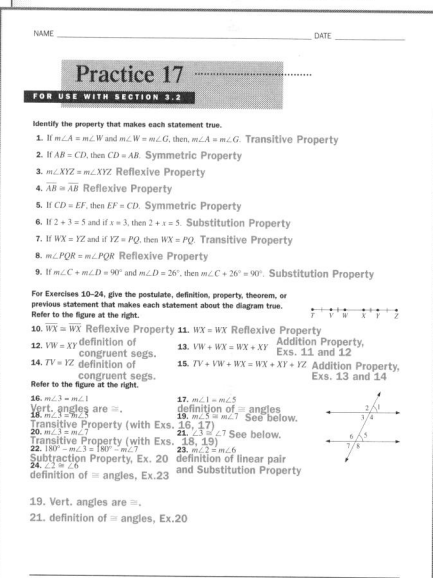

Kilauea is one of the active volcanoes in the state of Hawaii.

Exercise Notes

Assessment Note
Ex. 30 Suggest that students try to give at least one statement for each of the three kinds of reasons.

Assess Your Progress

Journal Entry
Students' responses to Ex. 9 can help you to determine their understanding of these basic types of statements.

Progress Check 3.1–3.2

See page 156.

Practice 17 for Section 3.2

NAME _____   DATE _____

Practice 17

FOR USE WITH SECTION 3.2

Identify the property that makes each statement true.

1. If $m\angle A = m\angle W$ and $m\angle W = m\angle G$, then $m\angle A = m\angle G$. Transitive Property
2. If $AB = CD$, then $CD = AB$. Symmetric Property
3. $m\angle XYZ = m\angle XYZ$ Reflexive Property
4. $\overline{AB} \cong \overline{AB}$ Reflexive Property
5. If $CD = EF$, then $EF = CD$. Symmetric Property
6. If $2 + 3 = 5$ and if $x = 3$, then $2 + x = 5$. Substitution Property
7. If $WX = YZ$ and if $YZ = PQ$, then $WX = PQ$. Transitive Property
8. $m\angle PQR = m\angle PQR$ Reflexive Property
9. If $m\angle C + m\angle D = 90°$ and $m\angle D = 26°$, then $m\angle C + 26° = 90°$. Substitution Property

For Exercises 10–24, give the postulate, definition, property, theorem, or previous statement that makes each statement about the diagram true. Refer to the figure at the right.

10. $\overline{WX} = \overline{WX}$ Reflexive Property 11. $WX = WX$ Reflexive Property
12. $VW = XY$ definition of congruent segs. 13. $VW + WX = WX + XY$ Addition Property, Exs. 11 and 12
14. $TV = YZ$ definition of congruent segs. 15. $TV + VW + WX = WX + XY + YZ$ Addition Property, Exs. 13 and 14

Refer to the figure at the right.

16. $m\angle 3 = m\angle 1$ 17. $m\angle 1 = m\angle 5$
18. $m\angle 3 = m\angle 2$ Vert. angles are ≅.   19. $m\angle 5 = m\angle 7$ definition of ≅ angles See below.
20. $m\angle 3 = m\angle 7$ Transitive Property (with Exs. 16, 17)   21. $m\angle 3 = m\angle 7$ See below.
22. $m\angle 3 = 180° - m\angle 7$ Transitive Property (with Exs. 18, 19)   23. $m\angle 2 = m\angle 6$
24. $m\angle 3 = m\angle 6$ Subtraction Property, Ex. 20 definition of ≅ angles, Ex.23   definition of linear pair and Substitution Property

19. Vert. angles are ≅.
21. definition of ≅ angles, Ex.20

---

three-footed pot is likely to balance easily. A four-footed pot might be difficult to balance, since the plane determined by three of the feet may not be the same as the plane determined by three others.

30. Answers may vary. Examples are given. (1) $AB + BC = AC$ (Segment Addition Postulate); (2) $m\angle ABE + m\angle BEA + m\angle EAB = 180°$ (The sum of the angle measures of a

triangle is 180°.); (3) $\angle ABE$ and $\angle EBC$ are a linear pair. (definition of a linear pair)

31. 27°; 117°; 63°

32. 76°; 166°; 14°

33. $(90 - 2y)°$ for $y \le 45$; $(180 - 2y)°$ for $y \le 90$; $2y°$

34.

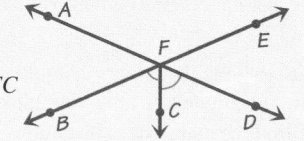

Assess Your Progress

1. deductive
2. inductive
3. is
4. might have been
5. definition of angle bisector
6. Vertical angles are congruent.
7. Ex. 6, definition of congruent angles, given, and the Addition Property

8. Vertical angles are congruent.

9. Answers may vary. An example is given. Postulate: The Angle Addition Postulate; definition: Congruent angles are angles with equal measures; property: If $x = y$, then $ax = ay$. A postulate is a basic geometric assumption, a definition is a clearly stated meaning of a word or phrase, and a property is an assumption about the real numbers. Each provides a way to reach conclusions in a logical argument.

Objectives

- Write and understand mathematical proofs.
- Use mathematical reasoning to prove that a statement is always true.

Recommended Pacing

❖ **Core and Extended Courses**
Section 3.3: 1 day

❖ **Block Schedule**
Section 3.3: $\frac{1}{2}$ block
(with Section 3.4)

Resource Materials

Lesson Support
Lesson Plan 3.3
Warm-Up Transparency 3.3
Practice Bank: Practice 18
Study Guide: Section 3.3
Explorations Lab Manual:
 Diagram Master 10
Challenge Problems: Set 18

Technology
Technology Book:
 Calculator Activity 4
 TI-92 Activity 3
Geometry Software
Internet:
 http://www.hmco.com

Warm-Up Exercises

Tell whether each statement is true because of a *definition*, a *postulate*, or a *property*.

1. If point *X* is between points *P* and *Q*, then *PX* + *XQ* = *PQ*. postulate

2. The sum of the measures of the exterior angles of any quadrilateral is 360°. property

3. If *AB* = *KL*, then $\overline{AB} \cong \overline{KL}$. definition

4. If $\overrightarrow{AX}$ bisects ∠ *BAC*, then ∠ *BAX* ≅ ∠ *XAC*. definition

5. If line *l* passes through two points *A* and *B* and line *m* passes through *A* and *B*, then *l* = *m*. postulate

124

SECTION

3.3 Paragraph Proof

Learn how to...

- **write and understand mathematical proofs**

So you can...

- **use mathematical reasoning to prove that a statement is always true**

How long do you think it would take to count out loud to one million? A day? A week? A year? You can use deductive reasoning to find out.

EXAMPLE 1

Writing In a paragraph, prove that you would need more than one week to count to one million without stopping. Assume that you could count one number per second.

SOLUTION

Here is one student's solution.

> Given: You can count one number per second.
> Prove: Counting to one million will take more than one week.
>
> If you count one number each second, you need at least 1,000,000 s to count to one million. One million seconds is about 16,667 min because 1,000,000 s ÷ 60 s/min ≈ 16,667 min. In hours, this is 16,667 min ÷ 60 min/h ≈ 278 h. There are 24 • 7 = 168 h in one week, so it will take more than one week to count to one million.

THINK AND COMMUNICATE

1. In Example 1, is there another argument you could use to reach the same conclusion? If so, what is it?

2. Is it reasonable to assume that a person could count one number each second? Estimate the amount of time you think it would actually take to count to one million. Explain your reasoning.

To make a convincing argument, you should put your statements in a logical order, give a reason for every statement that you make, and use reasons that everyone agrees are true. A **proof** is a chain of deductive reasoning that follows these rules.

124 Chapter 3 *Reasoning in Geometry*

ANSWERS Section 3.3

Think and Communicate

1. Yes; in one week there are
$$1 \text{ week} \cdot \frac{7 \text{ days}}{1 \text{ week}} \cdot \frac{24 \text{ hr}}{1 \text{ day}} \cdot \frac{60 \text{ min}}{1 \text{ hr}} \cdot \frac{60 \text{ s}}{1 \text{ min}} = 604{,}800 \text{ s.}$$
Since 604,800 < 1,000,000, you would need more than one week to count to one million.

2. No; estimates may vary. It takes more than one second to say many of the numbers between 1 and 1,000,000.

Exploration

1–3. Check students' work.

4. Students should find that for every triangle,
$$m\angle 4 = m\angle 1 + m\angle 2.$$

A **theorem** is a conjecture that can be proved to be true. A proof that is written in complete sentences is called a **paragraph proof**. When you write a proof, you make a series of **statements** and give the **reason** for each statement. Given information, postulates, definitions, and theorems can be used as reasons in a proof.

The **hypothesis** is the information you are **given**.

The **conclusion** is the statement you want to **prove**.

Given: You can count one number per second.
Prove: Counting to one million will take more than one week.

If you count one number each second, you need at least 1,000,000 s to count to one million. One million seconds is about 16,667 min because 1,000,000 s ÷ 60 s/min ≈ 16,667 min. In hours, this is 16,667 min ÷ 60 min/h ≈ 278 h. There are 24 • 7 = 168 h in one week, so it will take more than one week to count to one million.

For each **statement** of a step in your reasoning, include the **reason** the statement is true.

In the Exploration, you will learn a theorem about the exterior angles of a triangle. You will see how to prove this theorem using facts and properties that you already know.

EXPLORATION
COOPERATIVE LEARNING

Measuring Exterior Angles

Work with another student.
You will need:
- geometry software
- a ruler and protractor

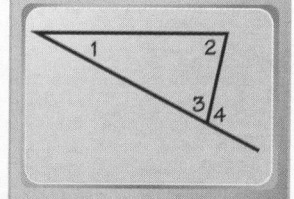

1 Draw a triangle with one side extended, as in the diagram. Find the measures of ∠1, ∠2, ∠3, and ∠4.

2 Round the angle measures to the nearest degree. Record the measures in a table.

m∠1	m∠2	m∠3	m∠4	m∠1 + m∠2
26°	78°	76°	104°	104°
?	?	?	?	?
?	?	?	?	?

3 Move one vertex, or draw new triangles, to form at least five different triangles. Record the angle measures for each triangle in your table.

4 Make a conjecture about the measures of the interior and exterior angles of any triangle. Compare your conjecture with other groups.

Exploration Note

Purpose
The purpose of this Exploration is to have students discover that the measure of an exterior angle of a triangle is equal to the sum of the measures of the two interior angles that are not adjacent to it.

Materials/Preparation
Each pair of students needs geometry software or a ruler and protractor.

Procedure
Students draw a triangle and an exterior angle at one vertex. They then find the measures of the interior angles, the exterior angle, and the sum of the nonadjacent

interior angles, and record the data in a table. They repeat this process for several triangles and then make a conjecture about the measures of the interior and exterior angles of any triangle.

Closure
Students should see that the measure of an exterior angle equals the sum of the measures of the interior angles that are not adjacent to it.

Explorations Lab Manual
See the Manual for more commentary on this Exploration.

Diagram Master 10

Teach⇔Interact

Additional Example 1

Suppose someone starts a rumor by telling it to 5 other people. Each of these people within one hour tells the rumor to 5 new people. Assume that the rumor continues to spread at this rate. In a paragraph, prove that 6 billion people will have heard the rumor in less than one 24-hour day. At the end of the first hour, 5 new people have heard the rumor. At the end of the second hour, $5 \times 5 = 5^2$, or 25 new people have heard it. At the end of the third hour, another $5 \times 5 \times 5 = 5^3$, or 125 people have heard the rumor. If the rumor continues to spread at this rate, at the end of the 14th hour, $5^{14} = 6,103,515,625$ new people will have heard the rumor. Since this number is greater than 6 billion, and since 14 hours is less than 24 hours, more than 6 billion people will have heard the rumor in less than one 24-hour day.

Section Note

Writing Proofs
Solving problems in mathematics, and in particular, writing proofs in geometry, always involves making a convincing argument in which each step of the argument can be justified by previously established facts. Writing paragraph proofs introduces students to the concept of proof by relying upon their past experiences in writing complete sentences to compose a paragraph. This approach is a natural introduction to showing students how to organize their thinking in a logical way.

Using Technology
Students can use the *Geometry Inventor* software from the *McDougal Littell Mathpack* to explore the measures of exterior and interior angles of a triangle.

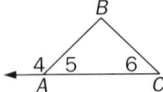

Additional Example 2

Write a paragraph proof to show that if $m\angle 4 + m\angle 6 = 180°$, then $m\angle 5 = m\angle 6$.

Step 1 Write the hypothesis and conclusion.
Given: $m\angle 4 + m\angle 6 = 180°$
Prove: $m\angle 5 = m\angle 6$

Step 2 Write the proof. Give a reason for each statement.
It is given that $m\angle 4 + m\angle 6 = 180°$. Because $\angle 4$ and $\angle 5$ form a linear pair, $m\angle 4 + m\angle 5 = 180°$. Since $m\angle 4 + m\angle 5 = 180°$, then by the Symmetric Property, $180° = m\angle 4 + m\angle 5$. Then, $m\angle 4 + m\angle 6 = 180° = m\angle 4 + m\angle 5$. By the Transitive Property, $m\angle 4 + m\angle 6 = m\angle 4 + m\angle 5$. Now, by the Reflexive Property, $m\angle 4 = m\angle 4$. Using the Subtraction Property, subtract $m\angle 4$ from both sides of the equation $m\angle 4 + m\angle 6 = m\angle 4 + m\angle 5$ to get $m\angle 6 = m\angle 5$, or $m\angle 5 = m\angle 6$.

Section Note

Writing Proofs

When students write proofs, it is important that they organize the statements in a logical order and justify each statement with a reason. Stress that the whole point of a proof is to demonstrate that the conclusion follows from the hypothesis. You may want to suggest that before writing a proof, students should first think about how they can proceed, that is, to try to develop some idea or plan for writing the proof. This plan can then be used to guide the writing of the proof itself.

Closure Question

What is a paragraph proof?
A paragraph proof is a convincing argument written in complete sentences that shows how the hypothesis of a statement leads to the conclusion. The statements that are used to get from the hypothesis to the conclusion are justified by referring to appropriate postulates, definitions, properties, and previously proved theorems.

126

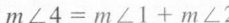

Exterior Angle Theorem

The measure of an exterior angle of a triangle is equal to the sum of the measures of the two interior angles that are not adjacent to it.

$$m\angle 4 = m\angle 1 + m\angle 2$$

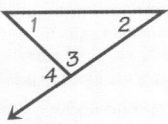

EXAMPLE 2

Write a paragraph proof of the Exterior Angle Theorem.

SOLUTION

Step 1 Write the hypothesis and conclusion.
    Given: A triangle with interior angles $\angle 1, \angle 2, \angle 3$. $\angle 4$ exterior to $\angle 3$.
    Prove: $m\angle 4 = m\angle 1 + m\angle 2$

Step 2 Sketch a diagram based on the given information.

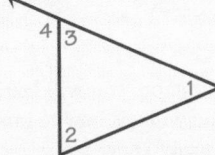

Step 3 Write the proof. Give a reason for each statement.
In any triangle, the sum of the measures of the angles is $180°$, so $m\angle 1 + m\angle 2 + m\angle 3 = 180°$. Because $\angle 4$ and $\angle 3$ are a linear pair, $m\angle 4 + m\angle 3 = 180°$. Using the Substitution Property, substitute $m\angle 4 + m\angle 3$ for $180°$ in the first equation. This gives a new equation: $m\angle 1 + m\angle 2 + m\angle 3 = m\angle 4 + m\angle 3$. Using the Subtraction Property, subtract $m\angle 3$ from both sides to get $m\angle 1 + m\angle 2 = m\angle 4$.

☑ CHECKING KEY CONCEPTS

For Questions 1–3:
a. Sketch a diagram to illustrate the statement.
b. State the hypothesis and the conclusion.
   1. If a quadrilateral is a square, then it has four right angles.
   2. If a figure is a triangular prism, then it has five faces and nine edges.
   3. If a quadrilateral is a rhombus, then its diagonals bisect each other.

Draw an obtuse triangle, an acute triangle, and a right triangle. Tell whether each statement is *True* or *False*.
   4. All six exterior angles of a triangle may be obtuse.
   5. All six exterior angles of a triangle can have different measures.

Checking Key Concepts

1. a.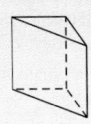

   b. hypothesis: A quadrilateral is a square. conclusion: It has four right angles.

2. a.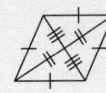

b. hypothesis: A figure is a triangular prism. conclusion: It has five faces and nine edges.

3. a. [diagram of rhombus with diagonals]

   b. hypothesis: A quadrilateral is a rhombus. conclusion: Its diagonals bisect each other.

4–5. Check students' work.

   4. True (if the triangle is an acute triangle).

   5. False; the two exterior angles at each vertex are congruent.

3.3 Exercises and Applications

Extra Practice exercises on page 664

Draw an obtuse triangle, an acute triangle, and a right triangle. Tell whether each statement is *True* or *False*.

1. All six exterior angles of a triangle can be right angles.

2. An exterior angle of a triangle can be smaller than one of the interior angles of the triangle.

3. An exterior angle of a triangle can be equal to one of the interior angles of the triangle.

BY THE WAY

In 1993, Andrew Wiles proved Fermat's Last Theorem. In 1637, Pierre de Fermat wrote in the margin of a book that he knew how to prove this theorem, but he never wrote a proof.

Find the measure of each indicated angle.

4. $m \angle STV$

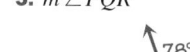

5. $m \angle PQR$

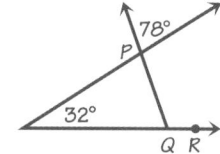

6. $m \angle BAC$

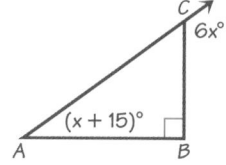

Connection — LITERATURE

When you want to prove something, you must understand what you want to prove and state it clearly. In the book *Where Do You Stop?* by Eric Kraft, Peter Leroy and his classmates discover this when their science teacher reveals the topics for their term projects.

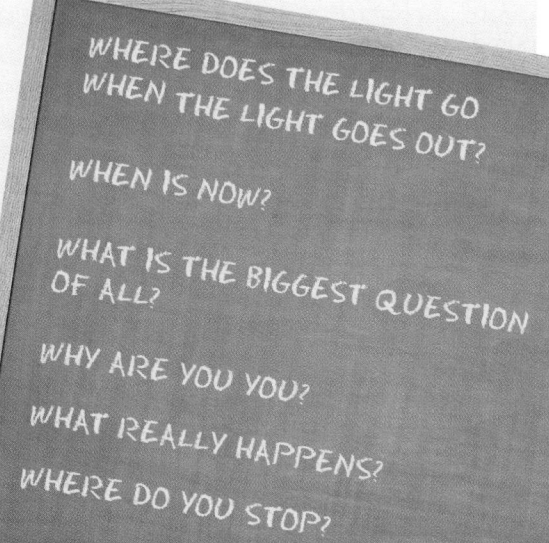

7. **Writing** Each of these questions can be stated in several different ways. Choose one of the questions. State clearly at least two different ways that you can interpret the question.

8. Explain why you should be able to interpret a theorem in only one way.

9. **Open-ended Problem** In *Where Do You Stop?*, the students have to include diagrams that illustrate an experiment to test their answers. Describe an experiment that might be included in one of the papers. Include a diagram for the experiment.

3.3 Paragraph Proof **127**

Exercise Notes

Communication: Discussion
Exs. 10, 11 Discuss with students which of the figures in these exercises is easier to work with and why. Clearly, it is easier to work with the figure in Ex. 10 because it contains only two angles. However, there is a matter of notation to consider as well. The angles in Ex. 10 are named with the numbers 1 and 2, while the angles in Ex. 11 are named using three letters. It is necessary to examine the figure in Ex. 11 carefully to find the angles referred to, and this is time consuming. Some parts of Ex. 11 could be stated and answered more easily if the six smallest angles were named with numbers. (This is true for parts (c) and (d). However, parts (a) and (b) require 3-letter notation.) When students draw their own diagram for a proof, they should think about the best way to label the diagram.

Application
Exs. 12–14 These exercises can be used to reinforce the idea that deductive thinking can be used in any area of life, not just in geometry.

Second-Language Learners
Exs. 12–14 Students learning English may benefit from having the terms *species*, *alligator*, *caiman*, *iris of eye*, *lower jaw*, *ear coverlets*, and *crossbands* explained orally before they begin these exercises. For Ex. 14, you may want to allow students to write their responses in a list rather than in paragraph form.

---

10. Complete the proof of the theorem: All right angles are congruent.

    Given: _?_
    Prove: $\angle 1 \cong \angle 2$
    From _?_ , $m \angle 1 = 90°$ and $m \angle 2 = 90°$. So $m \angle 1 = m \angle 2$ by _?_ . Therefore _?_ , by the definition of congruent angles.

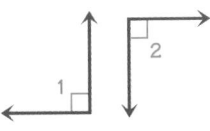

11. **a.** The measures of what angles are equal to $m \angle BGF - m \angle AGB$?
    b. The measure of which angle is equal to $m \angle BGC + m \angle CGD$? Why?
    c. What is the sum $m \angle CGD + m \angle DGE + m \angle EGF$? Why?
    d. If $\angle EGF \cong \angle AGF$, explain why $\angle BGC \cong \angle CGD$.

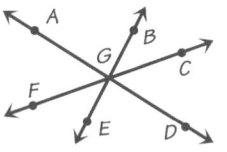

INTERVIEW Mary-Jacque Mann

Look back at the article on pages 108–110.

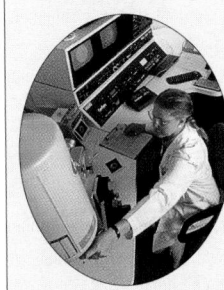

At the Wildlife Forensics Lab, Mary-Jacque Mann's colleagues may need to identify an animal whose skin was used to create an object. At other times, they may need to identify a whole animal. They might use a key to identify the species of an alligator or a caiman.

1. **Iris of eye is dark brown, red, or orange.**
   - Yes – go to 2
   - No – go to 4
2. **Has fewer than 20 teeth on each side of the lower jaw.**
   - Yes – Chinese Alligator
   - No – go to 3
3. **Ear coverlets contrast with the lighter color of the head.**
   - Yes – Schneider's Caiman
   - No – Cuvier's Caiman
4. **Back is black with distinct yellow or white crossbands.**
   - Yes – go to 5
   - No – one of the other Caiman species
5. **Has three or more large dark spots on the sides of the jaws.**
   - Yes – Black Caiman
   - No – American Alligator

12. In the key, what species of alligators are identified by the number of teeth they have?

13. Identify the animal at the left. Write a statement and reason for each step in the process of identification.

14. Write a paragraph explaining how you would know that an animal is a Cuvier's Caiman.

---

10. $\angle 1$ and $\angle 2$ are right angles; the definition of a right angle; the Substitution Property; $\angle 1 \cong \angle 2$

11. **a.** $\angle AGF$ and $\angle CGD$
    b. $\angle BGD$ by the Angle Addition Postulate
    c. 180°; because C, G, and F are collinear points
    d. $\angle AGF \cong \angle CGD$ and $\angle EGF \cong \angle BGC$, since

vertical angles are congruent. By the definition of congruent angles and the Substitution Property, $\angle BGC \cong \angle CGD$.

12. Chinese Alligator

13. American Alligator

14. A Cuvier's Caiman has an iris that is dark brown, red, or orange; has 20 or more teeth on each side of the lower jaw; and has ear coverlets that do not contrast with the color of its head.

15. Yes; since the triangle is equilateral, two of its sides are congruent.

16. No; for example, let $m \angle H = 60°$. Then $m \angle M = 120°$ and $m \angle L = 60°$. $\angle H$ is not a supplement of $\angle L$. (Only if all three angles are right angles is $\angle H$ a supplement of $\angle L$.)

17. No; for example, let $m \angle 1 = 60°$. Then $m \angle 3 = 120°$ and $m \angle 2 = 120°$. $\angle 1$ and $\angle 2$ are not congruent angles.

18. Yes; since $m \angle U + m \angle A + m \angle R = 180°$, $m \angle A + m \angle R = 180° - 90° = 90°$.

19. B

20. **a.** $\angle 1$ is an exterior angle of $\triangle XYZ$ at Z (Given); $m \angle 1 = m \angle X + m \angle Y$ (Exterior Angle Theorem); $m \angle Y = m \angle 1 - m \angle X$ (Subtraction Property)

128

For Exercises 15–18, tell whether each conclusion follows from the hypothesis. Explain why or why not.

15. $\triangle VJS$ is equilateral. Therefore, $\triangle VJS$ is isosceles.

16. $\angle H$ is a supplement of $\angle M$. $\angle M$ is a supplement of $\angle L$. Therefore, $\angle H$ is a supplement of $\angle L$.

17. $m\angle 1 + m\angle 3 = 180°$ and $\angle 2 \cong \angle 3$. Therefore, $\angle 2 \cong \angle 1$.

18. In $\triangle UAR$, $\angle U$ is a right angle. Therefore, $\angle A$ is a complement of $\angle R$.

19. SAT/ACT Preview In the figure, if $d = 97$, then $a = \underline{\ ?\ }$.

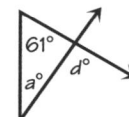

  A. 61   **B.** 36   **C.** 38
  D. 83   **E.** 45

20. a. Match each statement with the correct reason in the proof below.

  b. Challenge Arrange the statements and reasons in complete sentences to make a complete paragraph proof. Draw a diagram for the proof.

  Given: $\angle 1$ is an exterior angle of $\triangle XYZ$ at Z.
  Prove: $m\angle Y = m\angle 1 - m\angle X$

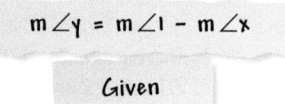

Exterior Angle Theorem	$m\angle y = m\angle 1 - m\angle x$
$\angle 1$ is an exterior angle of $\triangle XYZ$ at Z.	Given
Subtraction Property	$m\angle 1 = m\angle x + m\angle y$

ONGOING ASSESSMENT

21. Writing Write a paragraph proof of the theorem: The sum of the measures of the exterior angles of a triangle is 360°. Look at page 81 for an outline of an argument you can use for your proof. Be sure to include a diagram, given and prove statements, and reasons for your proof.

SPIRAL REVIEW

Tell which property makes each statement true. *(Section 3.2)*

22. $\angle ACE \cong \angle ACE$

23. If $\overline{AB} \cong \overline{BC}$ and $\overline{BC} \cong \overline{CD}$, then $\overline{AB} \cong \overline{CD}$.

24. If $m\angle 1 + m\angle 2 = 180°$ and $m\angle 2 = m\angle 5$, then $m\angle 1 + m\angle 5 = 180°$.

25. Describe at least two ways to continue the pattern 2, 4, 8, ... *(Section 1.1)*

26. $\angle 1$ and $\angle 2$ are vertical angles. If $m\angle 1 = 56°$, what is $m\angle 2$? How do you know? *(Section 2.1)*

Exercise Notes

Geometric Thinking
Exs. 15–20 These exercises provide an opportunity for students to think geometrically about properties of figures and to reason deductively to reach correct conclusions. Students' explanations in Exs. 15–18 can be used to assess their understanding of the concepts involved.

Assessment Note
Ex. 21 This exercise will demonstrate how well students understand the concept of proof. A well-organized proof, accompanied by an appropriate diagram, would make a good journal entry.

Practice 18 for Section 3.3

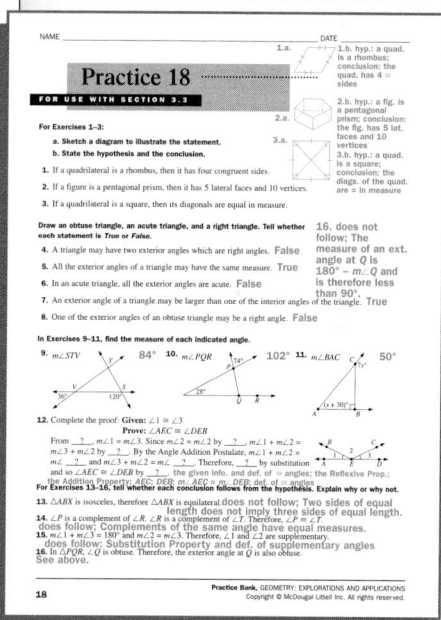

---

b.

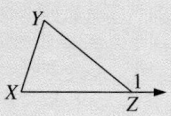

Since $\angle 1$ is an exterior angle of $\triangle XYZ$ at Z, by the Exterior Angle Theorem, $m\angle 1 = m\angle X + m\angle Y$. By the Subtraction Property, it follows that $m\angle 1 - m\angle X = m\angle Y$.

21.

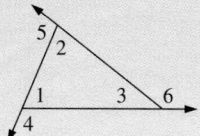

Given: $\angle 4$, $\angle 5$, and $\angle 6$ are the exterior angles of a triangle.
Prove: $m\angle 4 + m\angle 5 + m\angle 6 = 360°$.
Since $\angle 1$ and $\angle 4$, $\angle 2$ and $\angle 5$, and $\angle 3$ and $\angle 6$ are linear pairs, $m\angle 4 = 180° - m\angle 1$, $m\angle 5 = 180° - m\angle 2$, and $m\angle 6 = 180° - m\angle 3$. Then $m\angle 4 + m\angle 5 + m\angle 6 = 180° - m\angle 1 + 180° - m\angle 2 + 180 - m\angle 3 = 540° - (m\angle 1 + m\angle 2 + m\angle 3) = 540° - 180°$. (The sum of the angle measures of a triangle is 180°, Substitution Property). So $m\angle 4 + m\angle 5 + m\angle 6 = 360°$.

22. Reflexive Property

23. Transitive Property

24. Substitution Property

25. Answers may vary. Examples are given. 2, 4, 8, 16, 32, 64, … (The nth term is $2n$.); 2, 4, 8, 14, 22, 32, … (For $n > 1$, the nth term is found by adding $2n - 2$ to the previous term.)

26. 56°; Vertical angles are congruent.

129

Warm-Up Exercises

Why is each statement true?

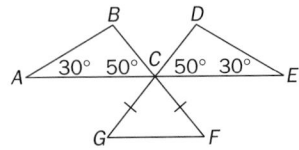

1. $\angle BCA \cong \angle DCE$
   Definition of congruent angles

2. $\angle BCD \cong \angle GCF$
   Vertical angles are congruent.

3. $m\angle BCE = m\angle BCD + 50°$
   Angle Addition Postulate

4. $\overline{GC} \cong \overline{FC}$ Given

5. $m\angle B + 30° + 50° = 180°$
   The sum of the angle measures of a triangle is 180°.

6. $m\angle D = 100°$
   The sum of the angle measures of a triangle is 180°;
   $m\angle D = 180° - 80° = 100°$.

SECTION

3.4 Two-Column Proof

Learn how to...
- write a proof in two-column format

So you can...
- use and present convincing arguments in a variety of formats

A proof shows that a statement is true. Often more than one correct proof can be written. Two proofs of the same theorem may present the same statements and reasons, but look very different. Compare Mary's and Carlotta's proofs of the Vertical Angles Theorem.

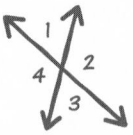

Given: $\angle 1$ and $\angle 3$ are vertical angles.
Prove: $\angle 1 \cong \angle 3$

$m\angle 1 + m\angle 2 = 180°$ and $m\angle 2 + m\angle 3 = 180°$, because $\angle 1$ and $\angle 2$ are a linear pair, and $\angle 2$ and $\angle 3$ are a linear pair. This means that $m\angle 1 + m\angle 2 = m\angle 2 + m\angle 3$, by the Substitution Property. Subtracting $m\angle 2$ from both sides gives $m\angle 1 = m\angle 3$, so $\angle 1 \cong \angle 3$.

Given: $\angle 1$ and $\angle 3$ are vertical angles.
Prove: $\angle 1 \cong \angle 3$

Statement	Reason
1. $\angle 1$ and $\angle 3$ are vertical angles.	1. Given
2. $m\angle 1 + m\angle 2 = 180°$ $m\angle 2 + m\angle 3 = 180°$	2. Angles in a linear pair are supplements.
3. $m\angle 1 + m\angle 2 = m\angle 2 + m\angle 3$	3. Substitution Property (Step 2)
4. $m\angle 1 = m\angle 3$	4. Subtraction Property
5. $\angle 1 \cong \angle 3$	5. Definition of congruent angles

THINK AND COMMUNICATE

1. How are the two proofs different? How are they the same?

2. Mary and Carlotta proved the same theorem. Which proof do you prefer? Explain why.

130   Chapter 3 *Reasoning in Geometry*

ANSWERS Section 3.4

Think and Communicate

1. Answers may vary. An example is given. One is a paragraph proof and the other is written with numbered steps and reasons. However, the proofs include the same statements and reasons.

2. Answers may vary. An example is given. I prefer Carlotta's proof because I think it is well organized and easy to follow.

Carlotta wrote a **two-column proof**. This format contains the statements and reasons of the proof arranged in two columns. Two-column proofs and paragraph proofs contain the same information, but they are organized differently.

In Chapter 2, you learned that vertical angles are congruent. You can use theorems like this one as reasons in proofs.

Vertical Angles Theorem

Vertical angles are congruent.

If $\angle 1$ and $\angle 2$ are vertical angles, then $\angle 1 \cong \angle 2$.

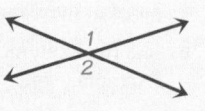

The proof of the Vertical Angles Theorem uses the fact that each of the vertical angles is supplementary to the same angle. You can prove the more general theorem that if two angles are supplementary to the same angle, the two angles are congruent.

EXAMPLE

Write a two-column proof of this theorem: If two angles are supplementary to the same angle, then the angles are congruent.

SOLUTION

Given: $\angle 1$ and $\angle 2$ are supplementary.
$\angle 1$ and $\angle 3$ are supplementary.

Prove: $\angle 2 \cong \angle 3$

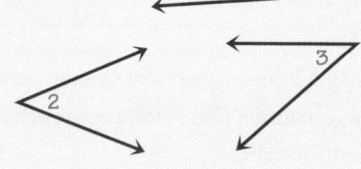

Statements	Reasons
1. $\angle 1$ and $\angle 2$ are supplementary. $\angle 1$ and $\angle 3$ are supplementary.	1. Given
2. $m\angle 1 + m\angle 2 = 180°$ $m\angle 1 + m\angle 3 = 180°$	2. Definition of supplementary angles
3. $m\angle 1 + m\angle 2 = m\angle 1 + m\angle 3$	3. Substitution Property (Step 2)
4. $m\angle 2 = m\angle 3$	4. Subtraction Property
5. $\angle 2 \cong \angle 3$	5. Definition of congruent angles

When you use the Transitive Property or the Substitution Property as a reason, list each step you used.

You will use both equality and congruence in proofs about geometric relationships. Remember that equality refers to numbers and congruence refers to figures.

Learning Styles: Visual

Although proofs in advanced mathematics are almost always written in paragraph form, the two-column format is helpful to students just beginning to learn how to write proofs. By numbering statements and reasons in a two-column proof, students learn how to organize their thinking in a logical way so as to proceed from what is given to what must be proved. Some students, however, may prefer paragraph proofs because they resemble the way they have learned how to write in their English courses.

Additional Example

Write a two-column proof of the following statement: If $\angle 1$ and $\angle 2$ are adjacent angles, $\angle 1$ and $\angle 3$ are complementary angles, and $\angle 2 \cong \angle 3$, then the nonshared sides of $\angle 1$ and $\angle 2$ are perpendicular.

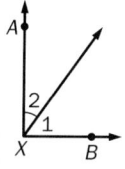
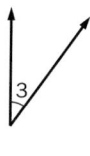

Given: $\angle 1$ and $\angle 2$ are adjacent.
$\angle 1$ and $\angle 3$ are complementary.
$\angle 2 \cong \angle 3$
Prove: $\overrightarrow{XA} \perp \overrightarrow{XB}$
Statements (Reasons)
1. $\angle 1$ and $\angle 2$ are adjacent.
   $\angle 1$ and $\angle 3$ are complementary.
   $\angle 2 \cong \angle 3$ (Given)
2. $m\angle 1 + m\angle 3 = 90°$ (Definition of complementary angles)
3. $m\angle 1 + m\angle 2 = 90°$ (Substitution Property (Steps 1 and 2))
4. $m\angle 1 + m\angle 2 = m\angle AXB$ (Definition of adjacent angles and Angle Addition Postulate)
5. $m\angle AXB = 90°$ (Transitive Property (Steps 3 and 4))
6. $\overrightarrow{XA} \perp \overrightarrow{XB}$ (Definition of perpendicular lines)

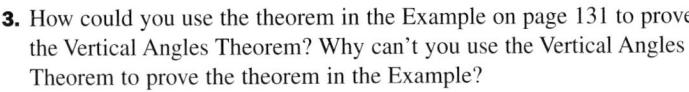

Teach⇔Interact

Checking Key Concepts

Geometric Thinking
Point out that when a result about a given figure is to be proved, it is not necessary to list all the information the figure contains in the "Given." In question 1, for example, the figure makes it clear that ∠1 and ∠2 form a linear pair and that ∠3 and ∠4 form a linear pair.

Closure Question

Describe how to write a proof in two-column format. *A two-column proof should begin with a statement of the given information, what is to be proved, and an appropriate diagram. The proof lists those statements that lead from the given information to the conclusion. These statements are numbered and listed in a logical order in the left-hand column of the proof. The right-hand column gives numbered reasons why the statements in the left-hand column are true.*

Apply⇔Assess

Suggested Assignment

❖ **Core Course**
Exs. 1–4, 10–16, 18–23, AYP

❖ **Extended Course**
Exs. 1–23, AYP

❖ **Block Schedule**
Day 16 Exs. 1–4, 10–16, 18–23, AYP

Exercise Notes

Geometric Thinking
Exs. 1–3 Diagrams to illustrate these three statements may help many students to see what is wrong with each one.

132

THINK AND COMMUNICATE

3. How could you use the theorem in the Example on page 131 to prove the Vertical Angles Theorem? Why can't you use the Vertical Angles Theorem to prove the theorem in the Example?

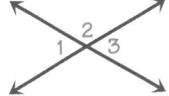

4. In the diagram at the left, ∠1 is supplementary to ∠2 and ∠2 is supplementary to ∠3, but you should not use this diagram for the proof in the Example. Explain why not.

5. In a two-column proof, the reasons are often stated very briefly. In the Example, explain how each reason supports the statement.

✓ CHECKING KEY CONCEPTS

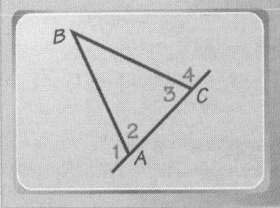

1. Copy and complete the two-column proof.
   Given: $m\angle 2 = m\angle 3$
   Prove: $m\angle 1 = m\angle 4$

Statements	Reasons
1. $m\angle 1 + m\angle 2 = 180°$ $\quad m\angle 3 + m\angle 4 = 180°$	1. Angles in a linear pair are supplements.
2. $m\angle 1 + m\angle 2 = m\angle 3 + m\angle 4$	2. _?_
3. _?_	3. Given
4. _?_	4. Subtraction Property

2. State the theorem proved in Question 1.

State the conclusion that follows from each hypothesis. Give a reason for each conclusion.

3. The sum of the measures of the interior angles of a polygon is 3060°. Therefore, the number of sides is _?_.

4. If ∠ABC ≅ ∠ABD and ∠ABD ≅ ∠CBE, then _?_.

3.4 | Exercises and Applications

Extra Practice
exercises on page 664

For Exercises 1–3, tell what is wrong with each statement. Then rewrite the statement so that it is always true.

1. Angles that are supplementary to the same angle are supplementary.

2. Angles that are congruent are vertical angles.

3. The measure of an exterior angle of a triangle is greater than the measure of each interior angle of a triangle.

132 Chapter 3 *Reasoning in Geometry*

Think and Communicate

3. Using the diagram on page 130, you could point out that ∠1 and ∠3 are both supplementary to ∠2; you could then conclude from the theorem in the Example that ∠1 ≅ ∠3. The angles in the theorem in the Example are not vertical angles.

4. The diagram shows a more specific case than the one described in the Example. Each pair of angles is not only supplementary but it is also a linear pair, so that the angles that are supposed to be proved congruent are vertical angles.

5. Step 1: Given information Step 2: Supplementary angles are two angles whose measures have the sum 180°. Step 3: $m\angle 1 + m\angle 3$ can be substituted for 180° in the first equation of Step 2. Step 4: $m\angle 1$ is subtracted from both sides of the equation in Step 3. Step 5: Congruent angles have equal measures.

Checking Key Concepts

1. Substitution Property (Step 1); $m\angle 2 = m\angle 3$; $m\angle 1 = m\angle 4$

2. If two interior angles of a triangle are congruent, then the two exterior angles with the same vertices are also congruent.

4. Copy and complete this two-column proof.

Given: $\angle 2 \cong \angle 3$

Prove: $\angle 1 \cong \angle 4$

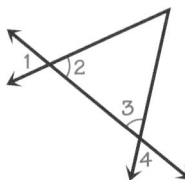

Statements	Reasons
1. _?_	1. Given
2. _?_ ; _?_	2. Vertical Angles Theorem
3. $\angle 1 \cong \angle 4$	3. _?_

INTERVIEW Mary-Jacque Mann

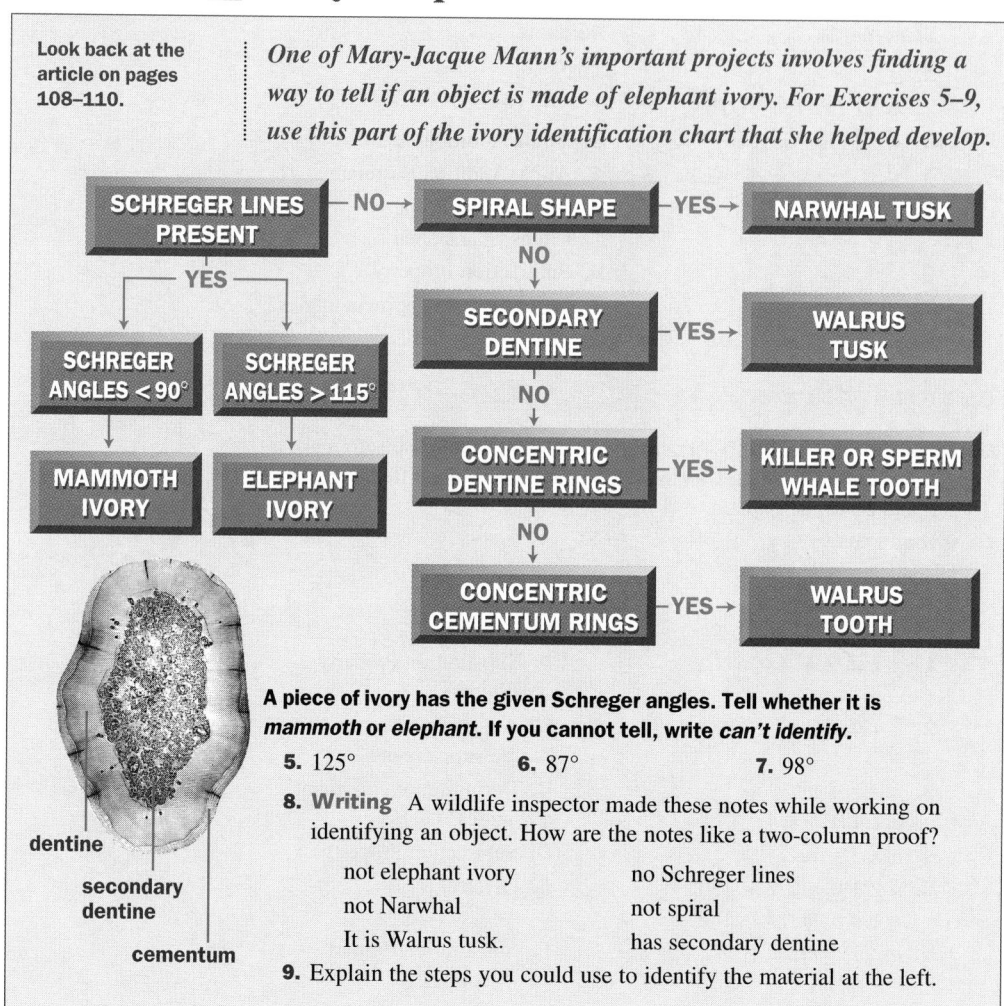

Look back at the article on pages 108–110.

One of Mary-Jacque Mann's important projects involves finding a way to tell if an object is made of elephant ivory. For Exercises 5–9, use this part of the ivory identification chart that she helped develop.

SCHREGER LINES PRESENT —NO→ SPIRAL SHAPE —YES→ NARWHAL TUSK

—YES—

SCHREGER ANGLES < 90°   SCHREGER ANGLES > 115°

NO ↓

SECONDARY DENTINE —YES→ WALRUS TUSK

MAMMOTH IVORY   ELEPHANT IVORY

NO ↓

CONCENTRIC DENTINE RINGS —YES→ KILLER OR SPERM WHALE TOOTH

NO ↓

CONCENTRIC CEMENTUM RINGS —YES→ WALRUS TOOTH

dentine
secondary dentine
cementum

A piece of ivory has the given Schreger angles. Tell whether it is *mammoth* or *elephant*. If you cannot tell, write *can't identify*.

5. 125°   **6.** 87°   **7.** 98°

8. Writing A wildlife inspector made these notes while working on identifying an object. How are the notes like a two-column proof?

not elephant ivory   no Schreger lines

not Narwhal   not spiral

It is Walrus tusk.   has secondary dentine

9. Explain the steps you could use to identify the material at the left.

3.4 Two-Column Proof **133**

Exercise Notes

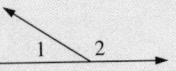

Communication: Drawing/Writing

Exs. 14–16 You may wish to suggest that after students draw a diagram for the proof of each theorem, they then rewrite each theorem as a conditional statement. This should help them to determine more easily what is given and what is to be proved.

Assessment Note

Ex. 18 Ask students to explain how they arrived at the ordering they used for the statements and reasons. Students should understand that many proofs can be written with slight variation in the ordering of the steps. This is particularly true for proofs having many steps.

Alternate Approach

Ex. 18 One approach students can use in this exercise is to work backward from the end of the proof to the beginning. The last statement in the left-hand column should be the conclusion, that is, $\angle 2 \cong \angle 4$. This statement would imply that $m\angle 2 = m\angle 4$. Students should continue to work backward in this way to put the reasons in correct order.

- - - - - - - - - - - - - - - - - -

10. 21° 11. 94°

12. 86° 13. 159°

14–16. Answers may vary. Examples are given.

14. a.
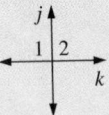

b. Given: $\angle 1$ and $\angle 2$ are a linear pair. Prove: $\angle 1$ and $\angle 2$ are supplementary.

15. a.
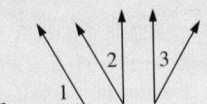

b. Given: $j \perp k$. Prove: $\angle 1 \cong \angle 2$

16. a.

b. Given: $\angle 1$ and $\angle 2$ are complementary; $\angle 1$ and $\angle 3$ are complementary. Prove: $\angle 2 \cong \angle 3$

17. $\angle PTR \cong \angle QTS$; definition of congruence; $m\angle QTS = m\angle 2 + m\angle 3$; $m\angle 1 + m\angle 2 = m\angle 2 + m\angle 3$; 2, 3, and 4; $m\angle 1 = m\angle 3$; $\angle 1 \cong \angle 3$

Use the information in the diagram to find the measure of each angle.

10. $\angle FGE$ 11. $\angle BGA$ 12. $\angle BGC$ 13. $\angle BGD$

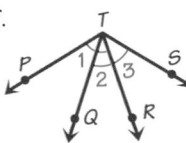

For Exercises 14–16:

a. Sketch and label a diagram for the proof of each theorem.

b. Write the Given and Prove statements for each theorem.

14. The angles in a linear pair are supplementary.

15. Two perpendicular lines form congruent adjacent angles.

16. Complements of the same angle are congruent.

17. **Challenge** Complete the proof.

Given: $\angle PTR \cong \angle QTS$

Prove: $\angle 1 \cong \angle 3$

Statements	Reasons
1. _?_	1. Given
2. $m\angle PTR = m\angle QTS$	2. _?_
3. $m\angle PTR = m\angle 1 + m\angle 2$	3. Angle Addition Postulate
4. _?_	4. Angle Addition Postulate
5. _?_	5. Substitution Property (Steps _?_, _?_, and _?_)
6. _?_	6. Subtraction Property
7. _?_	7. Definition of congruent angles

ONGOING ASSESSMENT

18. **Open-ended Problem** Arrange the statements and reasons to make a valid two-column proof. Can the steps be arranged in a different order and still be valid? Explain.

Given: $\angle 1 \cong \angle 3$

Prove: $\angle 2 \cong \angle 4$

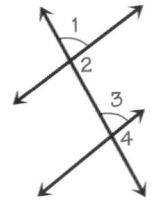

Statements	Reasons
1. $\angle 2 \cong \angle 4$	A. Substitution Property (Step(s) _?_)
2. $\angle 1 \cong \angle 3$ ($m\angle 1 = m\angle 3$)	B. Angles in a linear pair are supplements.
3. $m\angle 1 + m\angle 2 = m\angle 1 + m\angle 4$	C. Definition of congruent angles
4. $m\angle 2 = m\angle 4$	D. Subtraction Property
5. $m\angle 1 + m\angle 2 = 180°$ $m\angle 3 + m\angle 4 = 180°$	E. Given
6. $m\angle 1 + m\angle 2 = m\angle 3 + m\angle 4$	F. Substitution Property (Step(s) _?_)

18.

Statements	Reasons
1. $m\angle 1 + m\angle 2 = 180°$; $m\angle 3 + m\angle 4 = 180°$	1. Angles in a linear pair are supplementary.
2. $m\angle 1 + m\angle 2 = m\angle 3 + m\angle 4$	2. Substitution Property (Step 1)
3. $\angle 1 \cong \angle 3$ ($m\angle 1 = m\angle 3$)	3. Given (Definition of congruent angles)
4. $m\angle 1 + m\angle 2 = m\angle 1 + m\angle 4$	4. Substitution Property (Steps 2 and 3)
5. $m\angle 2 = m\angle 4$	5. Subtraction Property
6. $\angle 2 \cong \angle 4$	6. Definition of congruent angles

19. If <u>two angles are congruent</u>, then (their measures are equal.)

20. If <u>two lines are parallel to the same line</u>, then (the two lines are parallel to each other.)

21. If <u>a quadrilateral is a parallelogram</u>, then (its diagonals bisect each other.)

22.
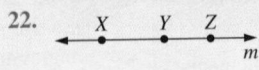

23.
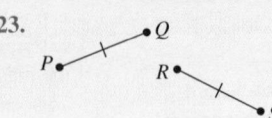

For Exercises 19–21, rewrite each statement in conditional form. Underline the hypothesis and circle the conclusion of each statement. *(Section 1.3)*

19. Congruent angles have equal measures.

20. Two lines parallel to the same line are parallel to each other.

21. The diagonals of a parallelogram bisect each other.

Sketch each situation. *(Sections 1.4 and 1.5)*

22. Points X, Y, and Z lie on line m.

23. $\overline{PQ} \cong \overline{RS}$

ASSESS YOUR PROGRESS

VOCABULARY

proof (p. 124)   **conclusion** (p. 125)
theorem (p. 125)   **statement** (p. 125)
paragraph proof (p. 125)   **reason** (p. 125)
hypothesis (p. 125)   **two-column proof** (p. 131)

1. Write a paragraph proof of this theorem: If two angles are supplementary to the same angle, then the angles are congruent. Use the Example on page 131 as a guide. *(Section 3.3)*

2. Rewrite the proof in Example 2 on page 126 as a two-column proof. *(Section 3.4)*

3. Complete this proof. *(Section 3.4)*

Given: $\angle 1 \cong \angle 3$
Prove: $\angle 2 \cong \angle 4$

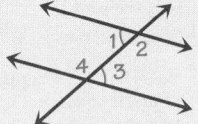

Statements	Reasons
1. $\angle 1 \cong \angle 3$	1. _?_
2. $m\angle 1 = m\angle 3$	2. _?_
3. $m\angle 1 + m\angle 2 = 180°$ $m\angle 3 + m\angle 4 = 180°$	3. _?_
4. $m\angle 1 + m\angle 2 = m\angle 3 + m\angle 4$	4. _?_
5. $m\angle 1 + m\angle 2 = m\angle 1 + m\angle 4$	5. _?_
6. $m\angle 2 = m\angle 4$	6. _?_
7. $\angle 2 \cong \angle 4$	7. _?_

4. Journal Is *a paragraph proof* or *a two-column proof* easier for you to read and understand? Which type of proof do you think is easier to write? Explain your answers.

3.4 Two-Column Proof **135**

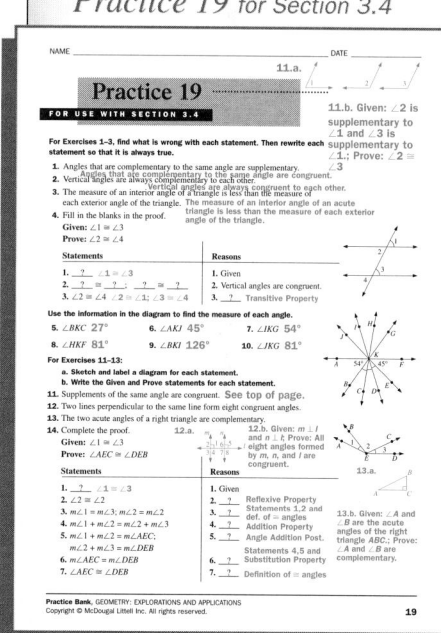

Assess Your Progress

1. Given: $\angle 1$ and $\angle 2$ are supplementary. $\angle 1$ and $\angle 3$ are supplementary.
Prove: $\angle 2 \cong \angle 3$.
By the definition of supplementary angles, $m\angle 1 + m\angle 2 = 180°$ and $m\angle 1 + m\angle 3 = 180°$. Therefore, by the Substitution Property, $m\angle 1 + m\angle 2 = m\angle 1 + m\angle 3$. Subtracting $m\angle 1$ from both sides gives $m\angle 2 = m\angle 3$, so $\angle 2 \cong \angle 3$.

2.

Statements	Reasons
1. $m\angle 1 + m\angle 2 + m\angle 3 = 180°$	1. The sum of the angle measures of a triangle is 180°.
2. $m\angle 4 + m\angle 3 = 180°$	2. Angles in a linear pair are supplementary.
3. $m\angle 1 + m\angle 2 + m\angle 3 = m\angle 4 + m\angle 3$	3. Substitution Property (Steps 1 and 2)
4. $m\angle 1 + m\angle 2 = m\angle 4$	4. Subtraction Property

3. **Reasons**

1. Given
2. Definition of congruent angles
3. Angles in a linear pair are supplementary.
4. Substitution Property (Step 3)
5. Substitution Property (Steps 2 and 4)
6. Subtraction Property
7. Definition of congruent angles

4. Answers may vary. An example is given. I find two-column proofs easier to read and understand because they are well organized and the steps are listed line by line. However, I think it is easier to write a paragraph proof because I am able to write in the same way I reason.

135

Objectives

- Write the converse of a conditional statement.
- Recognize and use converses in logical arguments.

Recommended Pacing

❖ **Core and Extended Courses**
  Section 3.5: 1 day
❖ **Block Schedule**
  Section 3.5: $\frac{1}{2}$ block
  (with Section 3.6)

Resource Materials

Lesson Support
Lesson Plan 3.5
Warm-Up Transparency 3.5
Practice Bank: Practice 20
Study Guide: Section 3.5
Challenge Problems: Set 20

Technology
Internet:
  http://www.hmco.com

Warm-Up Exercises

A, B, and C are the following statements.
A: ∠1 and ∠2 are vertical angles.
B: ∠1 and ∠2 are adjacent angles.
C: ∠1 and ∠2 are congruent.

1. Suppose a conditional statement has A as its hypothesis and C as its conclusion. Is the statement *True* or *False*? True.

2. Suppose a conditional statement has C as its hypothesis and A as its conclusion. Is the statement *True* or *False*? False.

3. Consider the conditional statement, "If A, then B." Is the statement *True* or *False*? False.

136

SECTION

3.5 Converses of Statements

Learn how to...
- write the converse of a conditional statement

So you can...
- recognize and use converses in logical arguments

It's true that if you're in Peru, then you're in South America. Is it also true that if you're in South America, then you must be in Peru? You will explore statements like this, called *converses*, in this lesson.

Angel Falls is the highest waterfall in the world.

South America's rain forests have more than 40,000 varieties of plants.

The Atacama Desert is one of the driest deserts in the world.

THINK AND COMMUNICATE

Tell whether each statement is *True* or *False*.
If it is false, give a counterexample.

1. **a.** If I am in the Atacama Desert, then I am in one of the driest places in the world.
   b. If I am in one of the driest places in the world, then I am in the Atacama Desert.

2. **a.** If I am in Paraguay, then I am not at the seashore.
   b. If I am not at the seashore, then I am in Paraguay.

3. What relationship do you see between the pairs of statements in Questions 1 and 2? If one statement in a pair is true, must the other statement also be true?

136   Chapter 3 *Reasoning in Geometry*

ANSWERS Section 3.5

Think and Communicate

1. a. True.
   b. False; I might be in another of the driest places in the world, such as the Sahara Desert.
2. a. True.
   b. False; I could be at an inland location anywhere in the world.

3. In each pair, the hypothesis of one statement is the conclusion of the other statement. No.

You can write the **converse** of a conditional statement by interchanging the **hypothesis** and **conclusion** of the original statement.

Statement

If you are at Angel Falls, then you are at the tallest waterfall in the world.

Converse

If you are at the tallest waterfall in the world, then you are at Angel Falls.

EXAMPLE

Use this theorem: If a triangle is a right triangle, then two of its angles are complementary.

a. Draw a diagram illustrating the theorem. State the hypothesis and the conclusion of the theorem.

b. Draw a diagram illustrating the converse of the theorem. State the hypothesis and the conclusion of the converse.

SOLUTION

a.

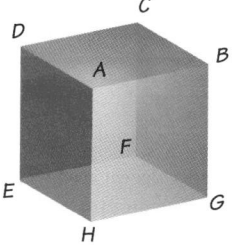

Given: $\triangle PQR$ is a right triangle, with a right angle at R.

Prove: $\angle P$ and $\angle Q$ are complementary.

b.

Given: In $\triangle PQR$, $\angle P$ and $\angle Q$ are complementary.

Prove: $\triangle PQR$ is a right triangle.

The given information does not tell you that the triangle is a right triangle.

◀ **WATCH OUT!**

Do not assume that the converse of a theorem is true. The converse of a theorem must be proved before you know that it is true.

THINK AND COMMUNICATE

Use this statement and diagram: If two lines are parallel, then they do not intersect.

4. Name two lines that are parallel. Name two lines that are skew.

5. Do skew lines ever intersect? Are skew lines ever parallel?

6. Give the converse of the statement. Is the converse true or false?

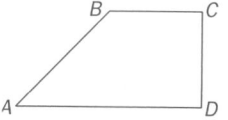

3.5 Converses of Statements **137**

Think and Communicate

4. Answers may vary. An example is given. $\overleftrightarrow{AD}$ and $\overleftrightarrow{BC}$; $\overleftrightarrow{AB}$ and $\overleftrightarrow{EH}$

5. No; No.

6. If two lines do not intersect, then they are parallel; False.

Teach⇔Interact

Section Note

Common Error
A common error that many students make in logical reasoning is to assume that the converse of a true statement is also true. This incorrect assumption is brought to students' attention in the *Watch Out!* on this page. Use the material presented on page 136 and at the top of this page to show students that the converse of a true statement can be either true or false.

Additional Example

Use this statement: If a quadrilateral is a rhombus, then it has a pair of parallel sides.

a. Draw a diagram illustrating the statement. State the hypothesis and conclusion of the statement.

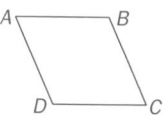

Given: Quadrilateral *ABCD* is a rhombus.
Prove: *ABCD* has a pair of parallel sides.

b. Draw a diagram illustrating the converse of the statement. State the hypothesis and conclusion of the converse.

Given: Quadrilateral *ABCD* has a pair of parallel sides.
Prove: *ABCD* is a rhombus.
The given information does not tell you that the quadrilateral has sides of equal length.

Think and Communicate

Students may need to be reminded of the definition of skew lines from Chapter 1: Skew lines are lines that do not intersect and are not parallel.

Checking Key Concepts

Communication: Writing
Be sure students understand how to write the statements in questions 5–7 as conditionals. Some students may have difficulty translating sentences into conditionals and, thus, confuse the hypothesis and conclusion of a statement. Further practice with this skill is provided in Exs. 4–9.

Second-Language Learners
Students learning English may not understand the difference between "the official language" and "an offical language," which may cause confusion when they try to create converses. Explain that "the" implies that all the official languages are listed, whereas "a" or "an" implies that there are additional official languages. In the case of Peru, Quechua and Spanish are the two official languages.

Closure Question

What is the converse of the statement: If *A* is true, then *B* is true? Is the converse of a true conditional statement *always true, sometimes true*, or *never true*? If *B* is true, then *A* is true. The converse of a true condiional statement is sometimes true.

Apply⇔Assess

Suggested Assignment

❖ **Core Course**
Exs. 1–9, 11, 18–28

❖ **Extended Course**
Exs. 1–28

❖ **Block Schedule**
Day 17 Exs. 1–9, 11, 18–28

Exercise Notes

Communication: Discussion
Exs. 1–9 Have students discuss whether the original statement and its converse are true or false. This will help students to see that the converse of a true statement is not necessarily true.

French Guiana
Suriname
Guyana
Venezuela
Colombia
Ecuador

Peru
Bolivia
Chile
Paraguay
Uruguay
Argentina

Brazil

- ▨ Dutch
- ☐ English
- ▨ French
- ▨ Portuguese
- ▨ Spanish
- ▨ Spanish, Guarani
- ▨ Spanish, Quechua, Aymara

☑ CHECKING KEY CONCEPTS

Use the map of South America at the left. Write the converse of each statement. Is the statement *True* or *False*? the converse?

1. If the official language is English, then you are in Guyana.

2. If you are in Peru, then an official language is Quechua.

3. If you are in Brazil, then the official language is Portuguese.

4. Write a true conditional statement about South America. Write the converse of your statement. Is it *True* or *False*? (You may also use the map on page 136.)

For Questions 5–7:
a. Rewrite each statement as a conditional.
b. Write the converse of each statement.

5. A square is a rectangle.

6. I will buy a car if I earn enough money.

7. The sum of the measures of the angles in a triangle is 180°.

3.5 Exercises and Applications

Extra Practice exercises on page 664

Write the converse of each statement. Tell whether the converse is *True* or *False*.

1. BASEBALL If the American League team won the first four games of the World Series, then they are the champions.

2. If the opposite sides of a quadrilateral are parallel, then the quadrilateral is a parallelogram.

3. If a rectangle is 2 ft long and 3 ft wide, then its area is 6 ft².

For Exercises 4–9:
a. Rewrite the statement as a conditional.
b. Write the converse of the statement.

4. The opposite angles of a parallelogram are congruent.

5. The supplements of congruent angles are congruent.

6. The supplement of an obtuse angle is an acute angle.

7. Isosceles triangles are not scalene.

8. The diagonals of a parallelogram bisect each other.

9. The diagonals of a square are congruent and perpendicular.

10. Research Find three examples of conditional statements in newspapers or magazines. Give the converse of each statement. Is the statement true? Is the converse true? Explain your reasoning.

Checking Key Concepts
1–7. See answers in back of book.

Exercises and Applications

1. If the American League team is the championship team, then it won the first four games of the World Series; False.

2. If a quadrilateral is a parallelogram, then the opposite sides of the quadrilateral are parallel; True.

3. If the area of a rectangle is 6 ft², then the rectangle is 2 ft long and 3 ft wide; False.

4. **a.** If a quadrilateral is a parallelogram, then the opposite angles are congruent.
b. If the opposite angles of a quadrilateral are congruent, then the quadrilateral is a parallelogram.

5. **a.** If two angles are congruent, then supplements of those angles are congruent.
b. If the supplements of two angles are congruent, then the two angles are congruent.

6–9. See answers in back of book.

10. Check students' work.

11. Writing Match each statement with the correct reason. Then rewrite the statements in the correct order to prove the theorem in the Example.

Given: In △*PQR*, ∠*R* is a right angle.

Prove: ∠*P* and ∠*Q* are complementary angles.

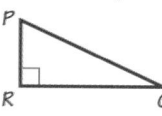

Statements	Reasons
1. $m\angle P + m\angle Q = 90°$	A. Definition of complementary angles
2. $m\angle P + m\angle Q + m\angle R = 180°$	B. Given
3. $m\angle R = 90°$	C. Definition of right angle
4. ∠*P* and ∠*Q* are complements.	D. Substitution Property (Step(s) _?_)
5. In △*PQR*, ∠*R* is a right angle.	E. Subtraction Property
6. $m\angle P + m\angle Q + 90° = 180°$	F. The sum of the angle measures of a triangle is 180°.

12. Challenge Write a proof of the converse of the theorem in Exercise 11.

Connection ⟩ LANGUAGE

Translating between languages presents several difficult problems. Words often have more than one meaning, so a translator must decide which definition is intended from the context.

For Exercises 13–16:
a. Tell whether the statement is *always*, *sometimes*, or *never* true.
b. Give the converse. Tell whether the converse is *True* or *False*.

13. If a Japanese sentence uses the word "sumu," an English translation would use the word "live."

14. If an English sentence uses the word "live," then a Japanese translation could use the word "ikiru."

Japanese	English
sumu	to live, reside
EXAMPLE: I live on Maple Lane.
ikiru	to live, be alive, breathe
EXAMPLE: I live in the twentieth century.

A. check (money)

B. check (investigate, look over, inspect)

15. If I want to sign the word "check," I would use the sign shown in diagram A.

16. If I use the sign shown in B, I want to sign the word "check."

17. Open-ended Problem Give an example of an English word that has several different meanings. Write a true conditional statement about using this word. Is the converse true?

Exercise Notes

Research
Ex. 10 Students have seen several examples of statements that are not in if-then form but that can be written in this form. Remind students to keep this in mind as they look for examples in newspapers and magazines.

Writing Proofs
Exs. 11, 12 By having students match statements and reasons in Ex. 11 and then arrange the statements in the correct order, students are given further understanding of how to construct a two-column proof. They should be able to use this understanding to write the proof for Ex. 12.

Common Error
Ex. 12 Some students may write the converse as "If ∠*P* and ∠*Q* are complementary angles, then ∠*R* is a right angle." Correct this error by pointing out that without placing the angles in the context of a triangle, the statement has no meaning. Point out that the "Given" in the proof in Ex. 11 started with the words "In △*PQR*."

Interdisciplinary Problems
Exs. 13–16 Students who are taking world language courses may be able to propose similar exercises that can be considered in a class discussion.

Challenge
Exs. 13–16 Ask students if the types of problems that present themselves when translating between Japanese and English would also occur when translating a geometry textbook from either language to the other. Focus on how terms are defined in geometry. Do mathematical words have more than one meaning?

---

11. The order of the statements may vary. An example is given.

Statements	Reasons
1. In △*PQR*, ∠*R* is a right angle.	1. Given
2. $m\angle R = 90°$	2. Definition of a right angle
3. $m\angle P + m\angle Q + m\angle R = 180°$	3. The sum of the angle measures of a triangle is 180°.
4. $m\angle P + m\angle Q + 90° = 180°$	4. Substitution Property (Steps 2 and 3)
5. $m\angle P + m\angle Q = 90°$	5. Subtraction Property
6. ∠*P* and ∠*Q* are complements.	6. Definition of complementary angles

12, 13. See answers in back of book.

14. a. sometimes true

   b. If a Japanese translation uses the word "ikiru," then an English sentence could use the word "live"; True.

15. a. sometimes true

   b. If I use the sign shown in A, then I want to sign the word "check"; True.

16. a. always true

   b. If I want to sign the word "check," then I use the sign shown in B; False.

17. Answers may vary. An example is given. The word "duck" has several meanings including "any of various wild or domesticated swimming birds of the family *Anatidae*" and "to lower the head or body." If there is a duck flying overhead, then I yell "Duck." The converse (If I yell, "Duck," then there is a duck flying overhead) is not always true.

Exercise Notes

Writing Proofs

Ex. 18 Stress the fact that the reasons used in a proof must be given information, postulates, definitions, properties, or previously proved theorems. If some students make up their own reasons, ask them to identify their reasons as one of the above mentioned statements.

Assessment Note

Ex. 19 In order to assess students' understanding of conditional statements and true and false converses, have volunteers read their responses to the class for discussion.

Practice 20 for Section 3.5

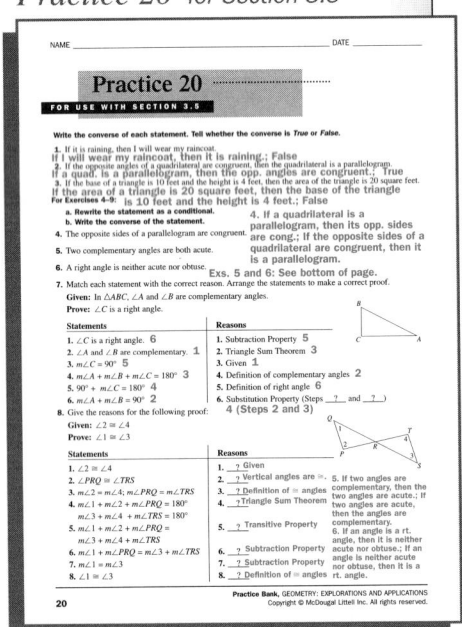

---

18. Give the reasons for the following proof:

    Given: ∠*FMR* ≅ ∠*FRM*

    ∠1 ≅ ∠3

    Prove: ∠2 ≅ ∠4

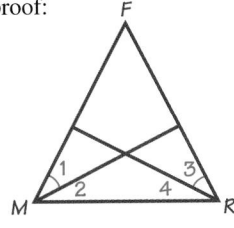

Statements	Reasons
1. ∠*FMR* ≅ ∠*FRM*; ∠1 ≅ ∠3	1. ?
2. *m*∠*FMR* = *m*∠*FRM*	2. ?
3. *m*∠*FMR* = *m*∠1 + *m*∠2; *m*∠*FRM* = *m*∠3 + *m*∠4	3. ?
4. *m*∠1 + *m*∠2 = *m*∠3 + *m*∠4	4. ?
5. *m*∠1 = *m*∠3	5. ?
6. *m*∠1 + *m*∠2 = *m*∠1 + *m*∠4	6. ?
7. *m*∠2 = *m*∠4	7. ?
8. ∠2 ≅ ∠4	8. ?

ONGOING ASSESSMENT

19. **Open-ended Problem** Give an example of a true conditional statement with a true converse that you might hear at lunch or at a sports event. Give an example of a true conditional statement with a false converse.

SPIRAL REVIEW

20. Use this theorem: If two exterior angles at different vertices of a triangle are congruent, then two interior angles of the triangle are congruent. *(Section 3.4)*

    a. Sketch and label a diagram to illustrate a proof of this theorem.

    b. Write the Given and Prove statements for this theorem.

Find each length or angle measure in rectangle *ABCD*. *(Section 2.5)*

21. *m*∠*BCD*   22. *m*∠*CED*

23. *AC*   24. *EC*

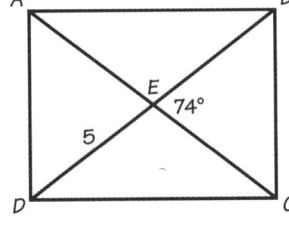

Sketch each triangle, if possible. If it is not possible, explain why not. *(Section 2.2)*

25. an acute scalene triangle

26. a right equiangular triangle

27. an isosceles obtuse triangle

28. an obtuse equilateral triangle

140   Chapter 3 *Reasoning in Geometry*

18.	Reasons
	1. Given
	2. Definition of congruent angles
	3. Angle Addition Postulate
	4. Substitution Property (Steps 2 and 3)
	5. Definition of congruent angles
	6. Substitution Property (Steps 4 and 5)
	7. Subtraction Property
	8. Definition of congruent angles

19. Answers may vary. An example is given. True statement with true converse: If I have at least $1.75, then I can buy the school lunch. (Converse: If I can buy the school lunch, then I have at least $1.75.) True statement with false converse: If I buy the school lunch, then I eat in the lunch room. (Converse: If I eat in the lunch room, then I buy the school lunch.)

20. Answers may vary. Examples are given.

    a.

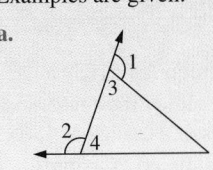

    b. Given: ∠1 ≅ ∠2
       Prove: ∠3 ≅ ∠4

21. 90°   22. 106°

23. 10   24. 5

25.

26. not possible; The measure of each angle of an equiangular triangle is 60°.

27.

28. not possible; Each angle of an equilateral triangle measures 60°.

3.6 The Pythagorean Theorem

Learn how to...

- **find the lengths of the sides of a right triangle**
- **decide if a triangle is a right triangle**

So you can...

- **find lengths of parts of figures and real-life objects**

The *Pythagorean theorem* states a special relationship among the lengths of the sides of a right triangle. This theorem is named after the Greek mathematician Pythagoras (around 560–480 B.C.), although the relationship was known even earlier by other cultures.

EXPLORATION
COOPERATIVE LEARNING

Proving the Pythagorean Theorem

Work in a group of three students.
You will need:

- paper • scissors • ruler

1 Cut out four identical right triangles. In a right triangle, each of the two shorter sides is called a **leg**. The side opposite the right angle is called the **hypotenuse**.

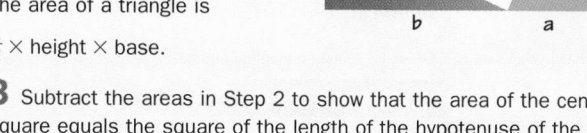

2 Arrange your triangles in a square. Measure each length and find the area of the square. Find the sum of the areas of the right triangles. The area of a triangle is $\frac{1}{2} \times$ height $\times$ base.

3 Subtract the areas in Step 2 to show that the area of the central square equals the square of the length of the hypotenuse of the right triangles.

| Area of the central square | = | Area of the large square | − | Area of the 4 right triangles |

4 Compare your results with those of other groups.

5 Use *a*, *b*, and *c* for the lengths of the sides of the triangles, as shown. Use algebra to show that in any right triangle, the sum of the squares of the lengths of the legs equals the square of the length of the hypotenuse.

Plan⇔Support

Objectives

- Find the lengths of the sides of a right triangle.
- Decide if a triangle is a right triangle.
- Find lengths of parts of figures and real-life objects.

Recommended Pacing

❖ **Core and Extended Courses**
   Section 3.6: 2 days

❖ **Block Schedule**
   Section 3.6: 2 half-blocks
   (with Sections 3.5 and 3.7)

Resource Materials

Lesson Support
Lesson Plan 3.6

Warm-Up Transparency 3.6

Overhead Visuals:
   Folder 4: Pythagorean Theorem

Practice Bank: Practice 21

Study Guide: Section 3.6

Explorations Lab Manual:
   Diagram Master 2

Challenge Problems: Set 21

Technology
Technology Book:
   Spreadsheet Activity 3
   TI-92 Activity 3

Scientific Calculator

Graphing Calculator

Geometry Software

McDougal Littell Mathpack
   Geometry Inventor Activity Book:
      Activity 5

Internet:
   http://www.hmco.com

Exploration Note

Purpose
The purpose of this Exploration is to discover that in a right triangle, the sum of the squares of the lengths of the legs equals the square of the length of the hypotenuse.

Materials/Preparation
Each group of students needs paper, scissors, and a ruler.

Procedure
Students cut out four identical right triangles and arrange them to form two squares. They then find the area of the large square and the sum of the areas of the right triangles. Next, they subtract the sum of the areas of

the triangles from the area of the large square to find the area of the central square. Groups compare their results and then use algebra to show that in any right triangle, the sum of the squares of lengths of the legs equals the square of the length of the hypotenuse.

Closure
Discuss the algebraic results to be sure students understand the Pythagorean theorem.

Explorations Lab Manual
See the Manual for more commentary on this Exploration.

For answers to the Exploration, see following page.

Warm-Up Exercises

1. Find the area of a square whose sides are 10 units long.
   100 square units

2. The square of what number is 2704? 52

Evaluate each expression.

3. $8^2 + 3^2$ 73

4. $12^2 + 5^2$ 169

5. $30^2 + 40^2$ 2500

Teach⇔Interact

Section Note

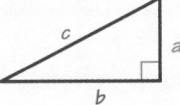

Using Technology
You may want to suggest that students use Activity 5 in the *Geometry Inventor Activity Book* to explore the relationship between the sides of a right triangle.

Additional Example 1

The legs of a right triangle are *a* and *b* units long. The hypotenuse is *c* units long. Find the unknown lengths in each right triangle.

a. $a = 12, b = 16$
$$a^2 + b^2 = c^2$$
$$12^2 + 16^2 = c^2$$
$$144 + 256 = c^2$$
$$\sqrt{400} = \sqrt{c^2}$$
$$\pm 20 = c$$
The length is positive, so $c = 20$.

b. $b = 2\sqrt{5}, c = 13$
$$a^2 + b^2 = c^2$$
$$a^2 + (2\sqrt{5})^2 = 13^2$$
$$a^2 + 20 = 169$$
$$a^2 = 149$$
$$\sqrt{a^2} = \sqrt{149}$$
$$a = \pm\sqrt{149}$$
$$a \approx \pm 12.21$$
The length is positive, so $a = \sqrt{149}$, or about 12.2.

Additional Example 2

Simplify each expression.

a. $\pm\sqrt{50}$
$$\pm\sqrt{50} = \pm\sqrt{25 \cdot 2}$$
$$= \pm\sqrt{25} \cdot \sqrt{2}$$
$$= \pm 5\sqrt{2}$$

b. $\sqrt{33}$
Since there is no perfect-square factor for 33 other than 1, $\sqrt{33}$ is already simplified.

> **The Pythagorean Theorem**
>
> In a right triangle, the sum of the squares of the lengths of the legs is equal to the square of the length of the hypotenuse.
> $$a^2 + b^2 = c^2$$

EXAMPLE 1

The legs of a right triangle are *a* and *b* units long. The hypotenuse is *c* units long. Find the unknown length for each right triangle.

a. $a = 12, b = 9$ **b.** $a = 3, c = 2\sqrt{6}$

SOLUTION

Find the square roots of both sides of the equation. The **radical** symbol, $\sqrt{\ }$, indicates the square root of a number.

a.
$$a^2 + b^2 = c^2$$
$$12^2 + 9^2 = c^2$$
$$144 + 81 = c^2$$
$$\sqrt{225} = \sqrt{c^2}$$
$$\pm 15 = c$$
The length is positive, so $c = 15$.

b.
$$a^2 + b^2 = c^2$$
$$3^2 + b^2 = (2\sqrt{6})^2$$
$$9 + b^2 = 24$$
$$b^2 = 15$$
$$\sqrt{b^2} = \sqrt{15}$$
$$b = \pm\sqrt{15}$$
$$b \approx \pm 3.87$$
The length is positive, so $b = \sqrt{15}$, or about 3.9.

In part (b) of Example 1, you can choose to give the answer as a radical or as a decimal approximation. In a real-world situation, the approximation that a calculator gives is usually accurate enough. In a mathematical situation, using radicals may make a calculation easier. Radicals in a solution are usually *simplified*.

EXAMPLE 2   Connection: Algebra

Simplify each expression.

a. $\pm\sqrt{48}$ **b.** $-\sqrt{7}$

Toolbox p. 700
Simplifying Radicals

SOLUTION

a. $\pm\sqrt{48} = \pm\sqrt{16 \cdot 3}$
$$= \pm\sqrt{16} \cdot \sqrt{3}$$
$$= \pm 4\sqrt{3}$$

b. Because 7 is a prime number, $\sqrt{7}$ is already simplified.

142   Chapter 3   *Reasoning in Geometry*

ANSWERS   Section 3.6

Exploration

1–3. Check students' work.

4. Results do not depend on the right triangle that is chosen.

5. The area of the central square is c^2, the area of the large square is $(a + b)^2$ and the area of each right triangle is $\frac{1}{2}ab$.
According to the results of Step 3,
Area of the central square =
Area of the large square −
Area of the 4 right triangles, or
$$c^2 = (a + b)^2 - 4\left(\frac{1}{2}ab\right) =$$
$$a^2 + 2ab + b^2 - 2ab = a^2 + b^2.$$
Then the square of the length of the hypotenuse is equal to the sum of the squares of the lengths of the legs.

THINK AND COMMUNICATE

Tell whether each expression is in simplified form.

1. $2\sqrt{12}$ **2.** $\sqrt{51}$ **3.** $3\sqrt{19}$ **4.** $\sqrt{27}$

5. Two sides of a right triangle are 4 and 5 units long. Explain why the third side could be either 3 or $\sqrt{41}$ units long.

The Converse of the Pythagorean Theorem

The converse of the Pythagorean theorem is true. You can use the converse to determine whether a triangle is a right triangle.

> **Converse of the Pythagorean Theorem**
>
> If a, b, and c are the lengths of the sides of a triangle, and $a^2 + b^2 = c^2$, then the triangle is a right triangle.

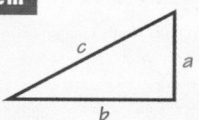

EXAMPLE 3   Application: Carpentry

Richard Mason is making a birdhouse for a Great Crested Flycatcher. The peak of the roof should be a right angle. He put the roof on the feeder and found that it does not fit correctly. What should he do to fix this problem?

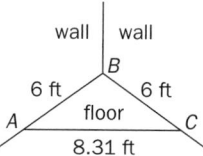

SOLUTION

Use the converse of the Pythagorean theorem to find out if the dimensions of the front of the birdhouse and the peak of the roof form a right triangle.

<table>
<tr><th>Front</th><th>Roof peak</th></tr>
<tr><td>$a^2 + b^2 \overset{?}{=} c^2$</td><td>$a^2 + b^2 \overset{?}{=} c^2$</td></tr>
<tr><td>$11.9^2 + 12.0^2 \overset{?}{=} 16.9^2$</td><td>$14.9^2 + 15^2 \overset{?}{=} 20.2^2$</td></tr>
<tr><td>$141.61 + 144 \overset{?}{=} 285.61$</td><td>$222.01 + 225 \overset{?}{=} 408.04$</td></tr>
<tr><td>$285.61 = 285.61$</td><td>$447.01 \neq 408.04$</td></tr>
<tr><td>This is a right triangle.</td><td>This is not a right triangle.</td></tr>
</table>

The angle at the top of the front is a right angle. The pieces of the roof do not meet at a right angle. Richard should change the angle of the roof to match the front of the birdhouse.

Think and Communicate

For question 5, be sure students understand that since the question does not specify that 5 is the length of the hypotenuse, two possibilities exist.

Section Note

Topic Spiraling: Review
Since the converse of the Pythagorean theorem is true, you may wish to ask students to express the theorem and its converse using *if and only if* language.

Additional Example 3

Angela Rodriguez is remodeling an old house that she recently bought. She has decided to put new tiles on the floor of the bathroom. Each tile is square and therefore has corners that are 90°. Angela wants to check that the angle of the floor corner is 90°, using the measurements shown.

Use the converse of the Pythagorean theorem to find out if $\triangle ABC$ is a right triangle.

$a^2 + b^2 \overset{?}{=} c^2$
$6^2 + 6^2 \overset{?}{=} 8.31^2$
$36 + 36 \overset{?}{=} 69.0561$
$72 \neq 69.0561$

$\triangle ABC$ is not a right triangle. The angle of the floor corner is not 90°.

Think and Communicate

1. No; $\sqrt{12} = 2\sqrt{3}$, so $2\sqrt{12} = 4\sqrt{3}$.

2. Yes; 51 has no perfect-square factor.

3. Yes; 19 is a prime number.

4. No; $\sqrt{27} = 3\sqrt{3}$.

5. The third side could be either a leg or the hypotenuse. If the hypotenuse is 5 units long, then $a = \sqrt{5^2 - 4^2} = 3$ units. If the legs are 4 and 5 units long, then $c = \sqrt{4^2 + 5^2} = \sqrt{41}$ units long.

Checking Key Concepts

Common Error
Without a diagram, as in questions 7–9, some students may incorrectly add the square of the hypotenuse to the square of one of the legs. Remind students that they must add the squares of the two smaller numbers together and then check to see if their sum equals the square of the larger number.

Closure Question

How can you tell if three given side lengths of a triangle are or are not the lengths of the sides of a right triangle. What theorem would you use to do this? Use the converse of the Pythagorean theorem. Check to see whether the sum of the squares of the two shorter sides equals the square of the longest side. If the answer is yes, then the triangle is a right triangle. If the answer is no, then the triangle is not a right triangle.

Suggested Assignment

❖ **Core Course**
Day 1 Exs. 1–27 odd, 31–35
Day 2 Exs. 2–26 even, 47–50

❖ **Extended Course**
Day 1 Exs. 1–35 odd
Day 2 Exs. 2–36 even, 37–50

❖ **Block Schedule**
Day 17 Exs. 1–27 odd, 31–35
Day 18 Exs. 2–26 even, 47–50

Exercise Notes

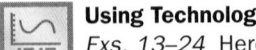 **Using Technology**
Exs. 13–24 Here is a short TI-82 program that uses the Pythagorean theorem to find the value of c, when given a and b. The equal symbol is found by pressing [2nd] [TEST]. You may wish to have students write another program that gives the value of a or b when given c and b or a.
PROGRAM:PYTHTHEO
:Input "ENTER A:", A
:Input "ENTER B:", B
:Disp "C=",√(A² + B²)

144

✓ CHECKING KEY CONCEPTS

Use a calculator to approximate each square root to the nearest hundredth. Then simplify each expression.

1. $\sqrt{49}$ **2.** $\pm\sqrt{54}$ **3.** $-\sqrt{24}$

Find each unknown length.

4.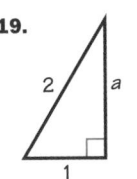

5.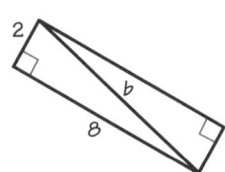

6.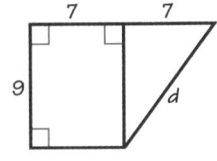

Can the given numbers be the lengths of the sides of a right triangle?

7. 20 ft, 21 ft, 29 ft   **8.** 8 m, 10 m, 12 m   **9.** 10 cm, 24 cm, 36 cm

---

3.6 | Exercises and Applications

Extra Practice exercises on page 665

ALGEBRA Use a calculator to approximate each square root to the nearest hundredth. Then simplify each expression.

1. $\sqrt{121}$ **2.** $-\sqrt{80}$ **3.** $\pm\sqrt{27}$

4. $\pm\sqrt{47}$ **5.** $-\sqrt{405}$ **6.** $\sqrt{196}$

Can the given numbers be the lengths of the sides of a right triangle?

7. 1.2 m, 1.6 m, 2 m   **8.** 8 in., 9 in., 12 in.   **9.** 5.6 cm, 2.8 cm, 6.4 cm

10. 7, 25, 24   **11.** $6\frac{2}{3}$, 4, $5\frac{1}{3}$ **12.** $\sqrt{2}$, $\sqrt{3}$, $\sqrt{5}$

The legs of a right triangle are a and b units long. The hypotenuse is c units long. Find the unknown length for each right triangle, to the nearest hundredth.

13. $a = 5, c = 12$ **14.** $a = 6, b = 8$ **15.** $b = 9, c = 15$

16. $a = 21, b = 28$ **17.** $b = 5, c = 6$ **18.** $a = 273, b = 136$

Find each unknown length.

19.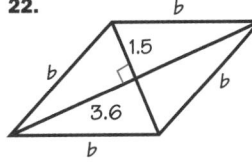

20.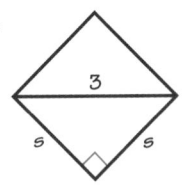

21.

22.

23.

24.

Checking Key Concepts
1. 7; 7
2. ±7.35; $\pm3\sqrt{6}$
3. -4.90; $-2\sqrt{6}$
4. 5   5. 8
6. 10.9   7. Yes.
8. No.   9. No.

Exercises and Applications
1. 11; 11
2. -8.94; $-4\sqrt{5}$

3. ±5.20; $\pm3\sqrt{3}$
4. ±6.86; $\pm\sqrt{47}$
5. -20.12; $-9\sqrt{5}$
6. 14; 14   7. Yes.
8. No.   9. No.
10. Yes.   11. Yes.
12. Yes.   13. 10.91
14. 10   15. 12
16. 35   17. 3.32
18. 305   19. $\sqrt{3} \approx 1.73$

20. $2\sqrt{17} \approx 8.25$
21. $\sqrt{130} \approx 11.40$
22. 3.9
23. $\frac{3\sqrt{2}}{2} \approx 2.12$
24. 26   25. Yes.
26. No.   27. No.
28. Answers may vary. Examples are given. 15 cm and 20 cm, 7 cm and 24 cm, 5 cm and $10\sqrt{6}$ cm

Tell whether or not each parallelogram can be a rectangle.

25.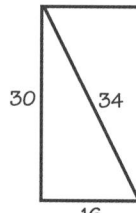

26.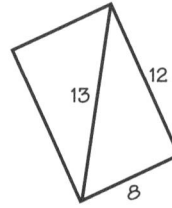

27.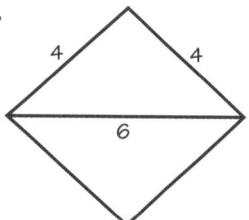

28. Open-ended Problem Each diagonal of a rectangle is 25 cm long. Find three possible pairs of lengths for the sides of the rectangle.

29. ARCHEOLOGY One of the two temples at Tikal in Guatemala has 49 steps leading up to the top of the pyramid. Each step is approximately 0.31 m deep and 0.36 m high. Estimate the height and depth of the entire flight of stairs. Estimate d, the length of the flight of stairs. Explain your reasoning.

30. a. Find the length of the diagonal, d, of the rectangular prism. (*Hint:* First find the value of e.)

b. Use the result of part (a) to find the length of the longest object that will fit inside a rectangular box with the given dimensions.

c. Find the length of the longest object that will fit inside a cube with edges that are 4 ft long.

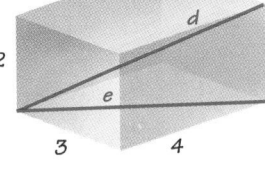

31. Writing Use rectangle $ABCD$. Find the lengths of the diagonals, $\overline{AC}$ and $\overline{BD}$. What do you notice? Do you think this will be true for all rectangles? Explain your reasoning.

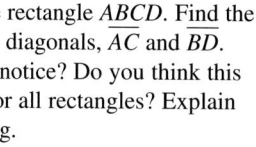

Plot each group of points on graph paper. Connect the points with segments. Find the length of each segment.

32. $A(3, 4)$, $B(6, 4)$, $C(6, 8)$ **33.** $C(2, 7)$, $D(2, -4)$, $E(10, -4)$

34. $M(-3, 9)$, $N(5, -6)$ **35.** $P(12, 7)$, $Q(5, 4)$

36. Writing Explain the difference between the Pythagorean theorem and its converse. Give an example of when to use each one.

3.6 The Pythagorean Theorem   **145**

29. 17.64 m; 15.19 m; The flight of stairs is about $\sqrt{17.64^2 + 15.19^2} \approx 23.28$ m long because the length is the hypotenuse of a right triangle with legs of 17.64 m and 15.19 m.

30. a. $d \approx 5.4$

b. about 5.4

c. about 6.9 ft

31. $AC = BD = 37$; The diagonals have the same length. For all rectangles, the lengths of the diagonals are equal, since opposite sides are congruent and all four angles are right angles.

32.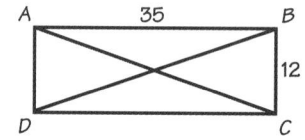

33.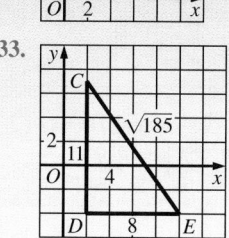

34.

35.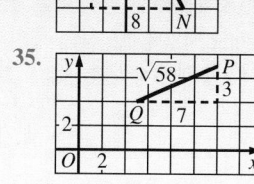

36. The Pythagorean theorem gives the relationships among the lengths of the sides of a right triangle. The converse of the Pythagorean theorem indicates under what conditions three segments form a right triangle. If the lengths of two sides of a right triangle are given, you can use the Pythagorean theorem to find the length of the third side. If you are given the lengths of all three sides of a triangle, you can use the converse of the Pythagorean theorem to determine whether the triangle is a right triangle.

Exercise Notes

Ex. 41 This exercise can also be done using the list features of the TI-82. Press STAT and choose 1:Edit. Use the arrow keys to move the cursor on top of L1 and press CLEAR ENTER. Repeat for L2 and L3. Then move the cursor to the first position of L1, type the first value for leg length, 119, followed by ENTER. Continue with the next leg length, 3367, and so on. Then position the cursor on the first position of L2 and enter the values for the length of the hypotenuse in this list. To produce a column for the length of the other leg, move the cursor on top of L3 and type √(L2² − L1²) followed by ENTER.

Historical Connection

Exs. 41–44 The Babylonian tablet mentioned in these exercises is contained in a collection at Columbia University in New York City. Two thousand years after the tablet was created, Arab mathematicians showed that all reduced Pythagorean triples can be found by using the expressions given in Ex. 42.

Problem Solving

Ex. 43 Some students may notice that part (b) does not mention that p and q cannot both be odd, but that condition is mentioned in the rule for finding a reduced Pythagorean triple. Ask why it is necessary that p and q not both be odd. (If p and q are both odd, then p^2 and q^2 will both be odd. Hence, $p^2 − q^2$ and $p^2 + q^2$ will both be even. Since $2pq$ is always even, the values of $p^2 − q^2$, $p^2 + q^2$, and $2pq$ will have a common factor whenever p and q are both odd.)

ARCHITECTURE In the blueprint, the *pitch triangle* gives the ratio of the length of the roof to the height of the roof. Sketch each triangle and give its dimensions.

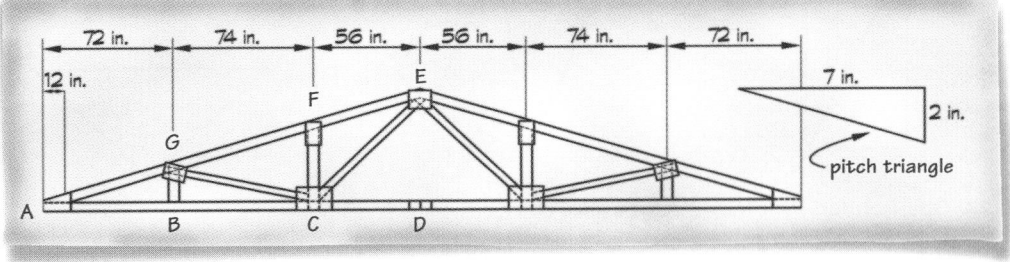

37. △ADE   **38.** △ACF   **39.** △BGC   **40.** △CED

Connection ▶ HISTORY

Three integers that can be the lengths of the sides of a right triangle are called a *Pythagorean triple*. A Babylonian tablet made between 1800 and 1650 B.C. suggests that the Babylonians knew the Pythagorean theorem.

Babylonian Empire in 1686 B.C.

41. Spreadsheets The tablet gives two of the three numbers in 15 Pythagorean triples. Some of these pairs are given in the table. Find the missing number in each triple.

leg	hypotenuse
119	169
3367	4825
4601	6649
65	97
319	481
2291	3541
481	769
4961	8161

42. One way to find a Pythagorean triple is to choose two integers p and q, then find the values of $2pq$, $p^2 − q^2$, and $p^2 + q^2$. Choose three different pairs of values for p and q. Find the values above and show that they are Pythagorean triples.

43. Challenge The three numbers in a *reduced* Pythagorean triple have no common factors except 1. You can find a reduced Pythagorean triple by using values of p and q that have no common factors and are not both odd in the expressions given in Exercise 42.

   a. Choose values of p and q that meet the conditions above. Show that they produce a reduced Pythagorean triple.

   b. Use algebra to show that a Pythagorean triple found by using values of p and q with a common factor, n, where n is a positive integer greather than 1, will not be reduced.

44. ALGEBRA Show that $(2pq)^2 + (p^2 − q^2)^2 = (p^2 + q^2)^2$ when $q = 1$.

37–40. Answers are given to the nearest hundredth of an inch.

37.

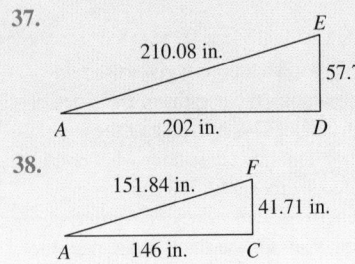

210.08 in.   57.71   202 in.   A   D   E

38.

151.84 in.   F   41.71 in.   A   146 in.   C

39. Use △ABG to find BG.

G   76.96 in.   21.14 in.   B   74 in.   C

40. Use the length of $\overline{ED}$ found in Ex. 37.

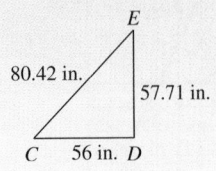

80.42 in.   57.71 in.   C   56 in.   D   E

41. 120; 3456; 4800; 72; 360; 2700; 600; 6480

42. Answers may vary. Examples are given.

p	q	$2pq$	$p^2 − q^2$	$p^2 + q^2$
9	2	36	77	85
8	3	48	55	73
7	4	56	33	65

In each case, $(2pq)^2 + (p^2 − q^2) = 4p^2q^2 + p^4 − 2p^2q^2 + q^4 = p^4 + 2p^2q^2 + q^4 = (p^2 + q^2)^2$.

43. a. Answers may vary. The examples given in the answer for Ex. 42 meet the condition stated and produce a reduced Pythagorean triple.

   b. Let $p = jn$ and $q = kn$. Then $2pq = 2jkn^2$, $p^2 − q^2 = j^2n^2 − k^2n^2 = (j^2 − k^2)n^2$, and $p^2 + q^2 = j^2n^2 + k^2n^2 = (j^2 + k^2)n^2$. Since each term of the Pythagorean triple $2pq$, $p^2 − q^2$, and $p^2 + q^2$ contains the factor n^2, the Pythagorean triple is not reduced.

45. a. DESIGN The Danish artist Per Arnoldi designed a chair made of triangles and squares formed by one rectangle of metal. Find the value of d.

b. Draw a net for the chair. Give all dimensions, including the folded diagonal.

c. Use heavy paper or cardboard to make a model of the chair.

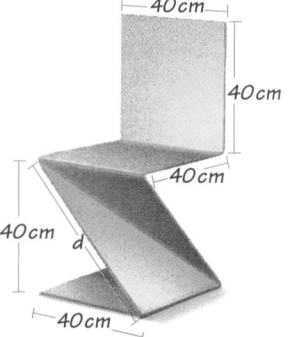

46. 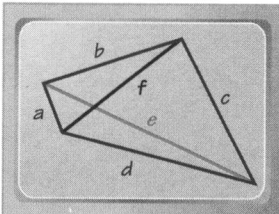 **Technology** Use geometry software.

a. Draw a quadrilateral. Draw the diagonals of the quadrilateral. Measure the length of each segment.

b. Repeat part (a) with at least five different quadrilaterals. Record the lengths in a table.

c. For what quadrilaterals will the equation $a^2 + b^2 + c^2 + d^2 = e^2 + f^2$ be true? Explain your reasoning.

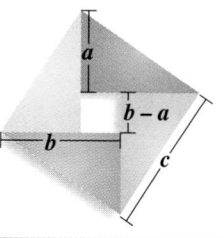

ONGOING ASSESSMENT

47. a. Visual Thinking Explain how this diagram illustrates a proof of the Pythagorean theorem. Write the algebraic steps for the proof.

b. How is this diagram different from the one that you used in the Exploration on page 141? Explain.

SPIRAL REVIEW

Write each statement as a conditional. Circle the hypothesis and underline the conclusion. *(Section 1.3)*

48. Angles in a linear pair are supplementary angles.

49. Perpendicular lines intersect at right angles.

50. Write the converse of each statement in Exercises 48 and 49. *(Section 3.5)*

3.6 The Pythagorean Theorem **147**

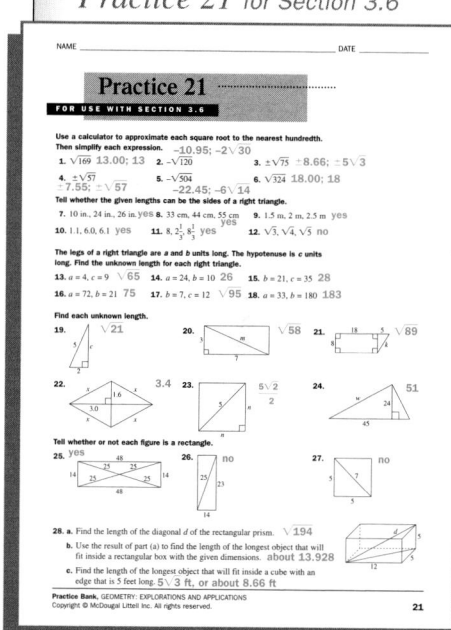
Apply⇔Assess

Exercise Notes

Research
Ex. 47 About 370 different proofs of the Pythagorean theorem have been found. Some students may enjoy researching and presenting to the class some of the more elementary proofs of the theorem.

Assessment Note
Ex. 47 In the Exploration on page 141, students worked in groups to prove the Pythagorean theorem. This exercise, which is similar to the Exploration, can be used to assess whether each student individually has understood how to prove the theorem.

Practice 21 for Section 3.6

---

44. When $q = 1$, $(2pq)^2 + (p^2 - q^2)^2 = (2p)^2 + (p^2 - 1)^2 = p^4 + 2p^2 + 1$ and $(p^2 + q^2)^2 = (p^2 + 1)^2 = p^4 + 2p^2 + 1$.

45. a. $d \approx 56.57$ cm

b.
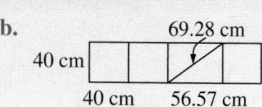

c. Check students' work.

46. a, b. Check students' work.

c. The equation is true if the quadrilateral is a rectangle. In

that case, $e = f$, $a = c$, and $b = d$, so $a^2 + b^2 + c^2 + d^2 = 2a^2 + 2b^2 = 2(a^2 + b^2)$ and $e^2 + f^2 = 2e^2$. Thus, $a^2 + b^2 = e^2$. Since $a^2 + b^2 = e^2$ holds for a rectangle, the equation is true.

47. a. Each right triangle has legs of lengths a and b and a hypotenuse of length c. The inner square has sides of length $b - a$, and the large outer square has sides of

length c. The area of the large square is equal to the areas of the four triangles plus the area of the small square;
$c^2 = 4\left(\frac{1}{2}ab\right) + (b - a)^2 = 2ab + b^2 - 2ab + a^2 = a^2 + b^2$.
Therefore, $a^2 + b^2 = c^2$.

b. Answers may vary. An example is given. In the diagram on page 141, the hypotenuses of the four right triangles formed the sides of the inner square.

In this diagram, the hypotenuses of the four right triangles form the sides of the outer square.

48. If (two angles form a linear pair,) then they are supplementary angles.

49. If (two lines are perpendicular,) then they intersect at right angles.

50. If two angles are supplementary, then they form a linear pair. (False.) If two lines form right angles, then the lines are perpendicular. (True.)

147

Objectives

- Determine whether a triangle is acute or obtuse from the lengths of its sides.
- Write inverses and contra-positives of statements.
- Recognize inverses and contra-positives in logical arguments.

Recommended Pacing

❖ **Core and Extended Courses**
Section 3.7: 2 days

❖ **Block Schedule**
Section 3.7: 2 half-blocks (with Section 3.6 and Chapter 3 Review)

Resource Materials

Lesson Support
Lesson Plan 3.7

Warm-Up Transparency 3.7

Practice Bank: Practice 22

Study Guide: Section 3.7

Explorations Lab Manual:
Additional Exploration 5
Diagram Masters 1, 11

Challenge Problems: Set 22

Assessment Book: Test 13

Technology
Technology Book:
TI-92 Activity 3

Internet:
http://www.hmco.com

Warm-Up Exercises

1. Can a triangle have a right angle and an obtuse angle? No.

2. Can a triangle have three acute angles? Yes.

3. Can a triangle have only two acute angles? Yes.

4. Is the inequality $5^2 + 12^2 > 13^2$ *True* or *False*? False.

5. Is it possible to have a square that is not a rhombus? No.

SECTION

3.7 Negations and Contrapositives

Learn how to...
- **determine whether a triangle is acute or obtuse from the lengths of its sides**
- **write inverses and contrapositives of statements**

So you can...
- **recognize inverses and contrapositives in logical arguments**

You can use the converse of the Pythagorean theorem to find out whether a triangle is a right triangle. Can the squares of the lengths of the sides tell you anything more about a triangle that is *not* a right triangle?

EXPLORATION
COOPERATIVE LEARNING

Analyzing Triangles

Work with another student.
You will need:
- graph paper
- scissors

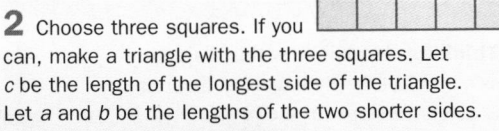

1 Cut ten squares out of graph paper. Use several different side lengths.

2 Choose three squares. If you can, make a triangle with the three squares. Let c be the length of the longest side of the triangle. Let a and b be the lengths of the two shorter sides.

Lengths of sides	$a^2 + b^2$	c^2	Type of triangle
3, 4, 5	25	25	right
4, 4, 5	32	25	acute
?	?	?	?

3 Make at least ten different triangles. Copy and complete the table.

4 What conjectures can you make about the relationship between the squares of the side lengths and the type of triangle?

5 Use your conjectures to predict whether a triangle with the given side lengths is *right*, *obtuse*, or *acute*.
  a. 11, 11, 15
  b. 11, 13, 18
  c. 5, 11, 12

Exploration Note

Purpose
The purpose of this Exploration is to discover whether a triangle is right, obtuse, or acute.

Materials/Preparation
Each group needs graph paper and scissors.

Procedure
Students cut ten squares out of graph paper, using different side lengths. They select three squares and place them to make a triangle. They should order the side lengths from shortest to longest and complete the table. Students repeat this procedure and then conjecture how the side

lengths can be used to predict whether the triangle is right, obtuse, or acute.

Closure
Students should see that if c is the length of the longest side and a and b are the lengths of the other two sides, then the triangle is a right triangle when $c^2 = a^2 + b^2$, obtuse when $c^2 > a^2 + b^2$, and acute when $c^2 < a^2 + b^2$.

Explorations Lab Manual
See the Manual for more commentary on this Exploration.

Diagram Masters 1, 11

One conjecture you might have made in the Exploration is "If $\triangle ABC$ is not a right triangle, then $a^2 + b^2 \neq c^2$." The hypothesis of this statement is the **negation** of the statement "$\triangle ABC$ is a right triangle." You can use a letter to indicate a statement. For example, the negation of "P" is "not P." You can use negations to change a conditional into a different statement.

CONDITIONAL	
If P, then Q.	If a serving of a food has less than 140 mg of sodium, then it may be labeled *low sodium*.

INVERSE	
If *not P*, then *not Q*.	If a serving of a food does not have less than 140 mg of sodium, then it may *not* be labeled *low sodium*.

CONVERSE	
If Q, then P.	If a food may be labeled *low sodium*, then it has less than 140 mg of sodium per serving.

CONTRAPOSITIVE	
If *not Q*, then *not P*.	If a food may *not* be labeled *low sodium*, then it does *not* have less than 140 mg of sodium per serving.

THINK AND COMMUNICATE

1. The conditional statement above is true. Is the inverse *True* or *False*? Is the converse? Is the contrapositive?

2. Use the statement "If $AB = CD$, then $\overline{AB} \cong \overline{CD}$." Is the inverse *True* or *False*? Is the converse? Is the contrapositive?

EXAMPLE

Rewrite this statement as a conditional. Give the inverse, converse, and contrapositive of the statement.

An equilateral triangle is an isosceles triangle.

SOLUTION

Conditional: If a triangle is equilateral, then it is isosceles.

Inverse: If a triangle is *not* equilateral, then it is *not* isosceles.

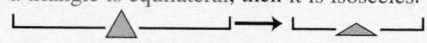

Converse: If a triangle is isosceles, then it is equilateral.

Contrapositive: If a triangle is *not* isosceles, then it is *not* equilateral.

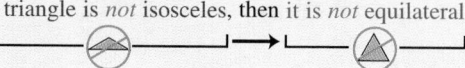

3.7 Negations and Contrapositives   **149**

ANSWERS Section 3.7

Exploration

1, 2. Check students' work.

3. See answers in back of book.

4. Let the lengths of the sides of a triangle be a, b, and c, with c the length of the longest side. Then if $a^2 + b^2 = c^2$, the triangle is a right triangle; if $a^2 + b^2 > c^2$, the triangle is an acute triangle; if $a^2 + b^2 < c^2$, the triangle is an obtuse triangle.

5. a. acute
   b. obtuse
   c. acute

Think and Communicate

1. True; True; True.
2. True; True; True.

A triangle has sides whose lengths, from least to greatest, are *k*, *m*, and *n*. Write an inequality that must be true if the triangle is acute; if the triangle is obtuse.

acute: $n^2 < k^2 + m^2$

obtuse: $n^2 > k^2 + m^2$

Apply⇔Assess

Suggested Assignment

❖ **Core Course**
Day 1 Exs. 1–14, 16–24
Day 2 Exs. 29–34, 36–45, AYP

❖ **Extended Course**
Day 1 Exs. 1–24
Day 2 Exs. 25–45, AYP

❖ **Block Schedule**
Day 18 Exs. 1–14, 16–24
Day 19 Exs. 29–34, 36–45, AYP

Exercise Notes

Common Error
Exs. 1–8 Some students may make errors because they fail to check the proper inequality. Remind these students that they should always compare the square of the length of the longest side to the sum of the squares of the lengths of the two shorter sides.

Spatial Reasoning
Exs. 1–8 To help students remember how to use side lengths to classify triangles as right, obtuse, or acute, appeal to their spatial intuition. Imagine holding two pencils so that their eraser ends are in contact, as shown on the following page.

✓ CHECKING KEY CONCEPTS

Tell whether the triangle with sides of the given lengths is *right*, *obtuse*, or *acute*.

1. 10, 14, 17   **2.** 11, 12, 17   **3.** $\sqrt{3}$, 5, 6

Tell whether each statement from the Example on page 149 is *True* or *False*. If a statement is false, give a counterexample.

4. the conditional   **5.** the inverse   **6.** the contrapositive

---

3.7 | Exercises and Applications

Extra Practice exercises on page 665

For Exercises 1–8, tell whether the triangle with sides of the given lengths is *right*, *obtuse*, or *acute*.

1. 2, 3, 4   **2.** 3, 4, 5   **3.** 4, 5, 6   **4.** 1.5, 2.0, 2.5
5. 0.6, 0.8, 0.9   **6.** 1, 1, $\sqrt{2}$ **7.** 1, $\sqrt{2}$, 2   **8.** $\sqrt{2}$, 2, $\sqrt{5}$

9. Open-ended Problem Using 3 and 7 as the lengths of two sides each time, draw an acute, a right, and an obtuse triangle. Estimate the length of the third side of each triangle, then measure the diagrams to check your estimates.

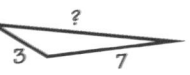

For Exercises 10–13, tell whether each statement is *True* or *False*.

10. An obtuse triangle may have sides of lengths 4, 7, and 10.

11. If the lengths of the sides of a triangle are 2 m, 7 m, and 8 m, then the triangle is an acute triangle.

12. In any triangle, the square of the length of one side is equal to the sum of the squares of the lengths of the other two sides.

13. If two sides of a triangle are 5 cm and 12 cm long, then the third side must be 13 cm long.

14. SAT/ACT Preview If A = $m \angle CED$ and B = 90°, then:
A. A < B   **B.** A > B   **C.** A = B
D. relationship cannot be determined

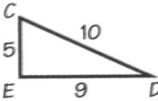

15. a. TRAVEL The highways Route 14, Route 92, and Route 175 form a triangle. Do these highways form a right triangle? Explain your reasoning.

b. Writing Interstate highways that run north-south have odd numbers. Interstate highways that run east-west have even numbers. Do these three highways follow this numbering rule? Why do you think Route 14 has an even number?

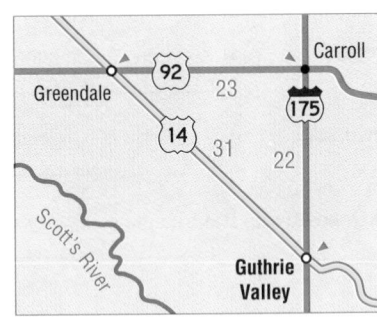

Checking Key Concepts

1. acute   2. obtuse
3. obtuse   4. True.
5. False; a triangle with exactly two congruent sides is not equilateral, yet it is isosceles.
6. True.

Exercises and Applications

1. obtuse   2. right
3. acute   4. right
5. acute   6. right
7. obtuse   8. acute

9. Drawings and estimates may vary. Check students' work.
10. True.   11. False.
12. False.   13. False.
14. A
15. a. No; $23^2 + 22^2 \neq 31^2$.
b. No. Route 92 and Route 175 do not intersect in a right angle, so it is not possible for Route 92 to run east-west, while Route 175 runs north-south. The choice

of an even number for Route 14 seems arbitrary.

16. Inverse: If it is not raining, then the streets are not wet. Contrapositive: If the streets are not wet, then it is not raining.

17. Inverse: If it does not snow, then classes are not canceled. Contrapositive: If classes are not canceled, then it does not snow.

18. Inverse: If a triangle is not acute, then it has an obtuse

angle. Contrapositive: If a triangle has an obtuse angle, then it is not acute.

19. Inverse: If a quadrilateral is not a parallelogram, then its opposite sides are not congruent. Contrapositive: If the opposite sides of a quadrilateral are not congruent, then the quadrilateral is not a parallelogram.

20. Inverse: If two lines are not perpendicular, then they do not form adjacent supplementary angles.

Write the inverse and the contrapositive of each statement.

16. If it is raining, then the streets are wet.

17. If it snows, then classes are canceled.

18. If a triangle is an acute triangle, then it has no obtuse angles.

19. The opposite sides of a parallelogram are congruent.

20. Perpendicular lines form adjacent supplementary angles.

For Exercises 21–23, write the inverse of each statement. Is the inverse *True* or *False*?

21. If two lines are perpendicular, then they form congruent adjacent angles.

22. The diagonals of a square are perpendicular.

23. Vertical angles are congruent.

24. Which of these statements has the same meaning as "The diagonals of a rhombus are perpendicular"?

   A. If $ABCD$ is a rhombus, then $\overline{AC}$ is perpendicular to $\overline{BD}$.
   B. If $\overline{AC}$ is perpendicular to $\overline{BD}$, then $ABCD$ is a rhombus.
   C. If $ABCD$ is not a rhombus, then $\overline{AC}$ is not perpendicular to $\overline{BD}$.
   D. If $\overline{AC}$ is not perpendicular to $\overline{BD}$, then $ABCD$ is not a rhombus.
   E. If $ABCD$ is a parallelogram, then $\overline{AC}$ is not perpendicular to $\overline{BD}$.
   F. If $\overline{AC}$ is not perpendicular to $\overline{BD}$, then $ABCD$ is a parallelogram.

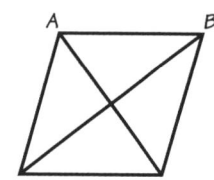

Connection ▶ AUTO MAINTENANCE

If you drive a car, chances are good that one day it will not start. The first thing to check is whether you have left the lights turned on and run down the battery. If the engine cranks when you turn the key, but the car does not start, then you can check several possible sources of the problem.

For Exercises 25–27, write the inverse and the contrapositive of the statement.

25. If the battery needs to be recharged, then the lights will dim when the key is turned.

26. If the lights are bright when the key is turned, then the spark plugs may be bad.

27. If the battery and spark plugs are working correctly, then fuel may not be getting to the engine.

28. **Open-ended Problem** Write another if-then statement based on auto maintenance. Is its converse true?

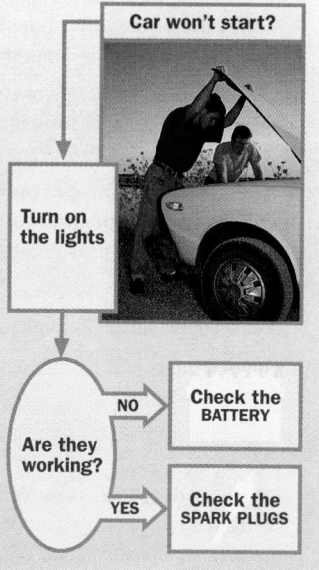

3.7 Negations and Contrapositives **151**

Exercise Notes

Spatial Reasoning (continued)

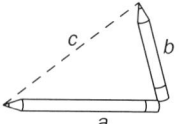

If these pencils are the two shorter sides of a triangle, then c^2 will equal $a^2 + b^2$ when the angle formed by the two shorter sides is a right angle. If the angle is decreased slightly, the triangle becomes acute and the length c will decrease slightly. When this happens, c^2 will become less than $a^2 + b^2$. If the angle between the sides of the lengths a and b is slightly greater than 90°, then the triangle will be obtuse and c^2 will be greater than $a^2 + b^2$.

Communication: Reading
Exs. 16–24 You may wish to have students read their answers to these exercises in class. When doing this, write *If P, then Q* on the board for reference along with its inverse and contrapositive. Since most of these exercises involve properties of geometric figures, a class discussion will also provide a good review of many important geometric ideas.

Career Connection
Exs. 25–28 As these exercises illustrate, trying to locate the source of a problem in a car can involve logical reasoning. Given the high technology of today's modern cars, automobile mechanics are trained to use many sophisticated types of diagnostic equipment to find problems and make repairs. Future mechanics are trained in technical high schools, community colleges, specialized automobile training schools, and by working on the job as an apprentice to an experienced mechanic. They can own and work in their own repair shops, in the service departments of automobile dealerships, or for other small business owners of repair shops.

Contrapositive: If two lines do not form adjacent supplementary angles, then they are not perpendicular.

21. If two lines are not perpendicular, then they do not form congruent adjacent angles; True.

22. If a quadrilateral is not a square, then its diagonals are not perpendicular; False.

23. If two angles are not vertical angles, then they are not congruent; False.

24. A and D

25. Inverse: If the battery does not need to be recharged, then the lights will not dim when the key is turned. Contrapositive: If the lights do not dim when the key is turned, then the battery does not need to be recharged.

26. Inverse: If the lights are not bright when the key is turned, then the spark plugs may not be bad. Contrapositive: If the spark plugs are not bad, then the lights are not bright when the key is turned.

27. Inverse: If the battery or the spark plugs are not working correctly, then fuel may be getting to the engine. Contrapositive: If fuel is getting to the engine, then the battery and spark plugs are not working correctly.

28. Answers may vary. Check students' work.

Without measuring, tell whether the side lengths of each triangle appear to be accurate. Explain your reasoning.

29.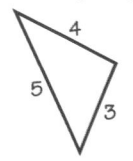

30.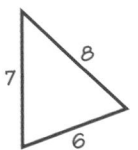

31.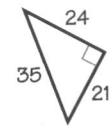

32.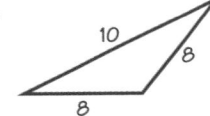

33.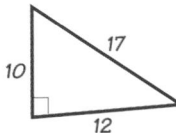

34.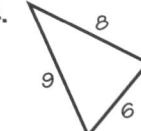

35. **Cooperative Learning** Work with another person. One of you should do part (a) and the other should do part (b). Work together on parts (c) and (d). You will need a ruler and a protractor.

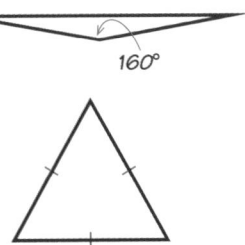

a. Use the obtuse triangle. Find the ratio of the sum of the lengths of the two shorter sides to the length of the longest side. Experiment with other obtuse triangles. Write an inequality to describe the lower limit of the ratio.

b. Use the equilateral triangle. What is the ratio of the sum of the lengths of two sides to the length of the third side?

c. **Challenge** Write an inequality to describe all the possible values of the ratio of the sum of the lengths of the two shorter sides of a triangle to the length of the longest side.

d. Did you use *inductive* or *deductive* reasoning to answer part (c)? Can you find a triangle for which your inequality does not work?

For Exercises 36–38:
a. Write the contrapositive of the statement.
b. Draw a diagram for the proof of the contrapositive.
c. Write the given and prove statements for the proof.

36. If a quadrilateral is a square, then it is a parallelogram.

37. If two angles are right angles, then they are congruent.

38. The diagonals of a parallelogram bisect each other.

ONGOING ASSESSMENT

39. **Open-ended Problem** Write two conditional statements. One should be from a real-world situation, and one should be mathematical. Write the inverse, converse, and contrapositive of each statement. Tell whether each of these statements is *True* or *False*. Give a counterexample for each false statement.

Answers:

29. Yes; the lengths are the lengths of the sides of a right triangle and the triangle appears to be a right triangle.

30. Yes; the lengths are the lengths of the sides of an acute triangle and the triangle appears to be acute.

31. No; the lengths are not the lengths of the sides of a right triangle.

32. No; the lengths are the lengths of the sides of an acute triangle and the triangle appears to be obtuse.

33. No; the lengths are not the lengths of the sides of a right triangle.

34. Yes; the lengths are the lengths of the sides of an acute triangle and the triangle appears to be acute.

35. Let a and b be the lengths of the shorter sides of a triangle and c the length of the longest side.

a. Estimates may vary; about $\frac{31}{30}$; $\frac{a+b}{c} > 1$.

b. 2

c. $1 < \frac{a+b}{c} < 2$

d. inductive; Answers will vary. The inequality applies to all triangles.

36–38. Sketches and labels may vary.

36. a. If a quadrilateral is not a parallelogram, then it is not a square.

b.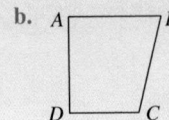

c. Given: *ABCD* is not a parallelogram.
Prove: *ABCD* is not a square.

37. a. If two angles are not congruent, then they are not both right angles.

b.

c. Given: ∠*Y* is not congruent to ∠*X*.
Prove: $m\angle Y \neq 90°$ or $m\angle X \neq 90°$.

38. a. If the diagonals of a quadrilateral do not bisect each other, then the quadrilateral is not a parallelogram.

b.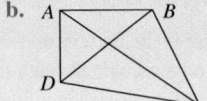

c. Given: $\overline{AC}$ and $\overline{BD}$ do not bisect each other.
Prove: *ABCD* is not a parallelogram.

39. See answers in back of book.

40. 5   41. 2

42. 4.6   43. Yes.

44. No.   45. Yes.

Assess Your Progress

1. $2\sqrt{6} \approx 4.90$

2. $6\sqrt{5} \approx 13.42$

3. $\frac{5\sqrt{5}}{2} \approx 5.59$

4. right   5. obtuse

6. obtuse   7. acute

8. acute   9. obtuse

Find the mean of each pair of numbers. (*Toolbox, page 708*)

40. 7, 3   **41.** −4, 8   **42.** 2.7, 6.5

Tell whether the given lengths can be the sides of a right triangle.
(*Section 3.6*)

43. 15, 20, 25   **44.** 6, 9, 11   **45.** 1.6, 3, 3.4

ASSESS YOUR PROGRESS

VOCABULARY

converse (p. 137)   **negation** (p. 149)

leg of a right triangle (p. 141)   **inverse** (p. 149)

hypotenuse (p. 141)   **contrapositive** (p. 149)

radical (p. 142)

Find each unknown length. (*Section 3.6*)

1. 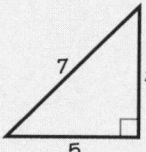   **2.** 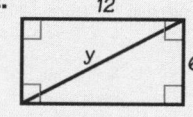   **3.**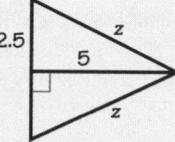

Tell whether the triangle with sides of the given lengths is *acute*, *obtuse*, or *right*. (*Sections 3.6 and 3.7*)

4. 20, 29, 21   **5.** 6, 12, 8   **6.** 3.6, 7.2, 9.8

7. 4.3, 6.5, 5.8   **8.** $\sqrt{16}, \sqrt{12}, \sqrt{7}$ **9.** $4\frac{1}{2}, 3\frac{1}{3}, 7\frac{1}{2}$

For Exercises 10 and 11, write the inverse, the converse, and the contrapositive of each statement. (*Sections 3.5 and 3.7*)

10. The diagonals of a rectangle are congruent.

11. If the exterior rays of two adjacent angles are perpendicular, then the angles are complements.

12. Use the statement: If *PQRS* is a parallelogram, then $\angle P \cong \angle R$.
   a. Write the contrapositive of the statement.
   b. Draw a diagram for a proof of the contrapositive.
   c. Write the Given and Prove statements for a proof of the contrapositive. (*Section 3.7*)

13. Journal Make a chart with examples of the four types of conditional statements that you have learned. Assume that the conditional is true. Indicate on the chart whether each of the others is *always*, *never*, or *sometimes* true.

Apply ⟺ Assess

Exercise Notes

Topic Spiraling: Review
Exs. 40–42 These exercises will help students review how to find the mean, which they will use in Chapter 4.

Progress Check 3.5–3.7

See page 157.

Practice 22 for Section 3.7

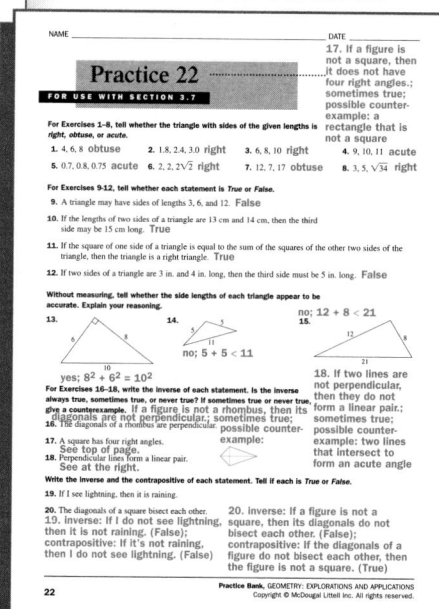

---

10. Inverse: If a quadrilateral is not a rectangle, then its diagonals are not congruent. Converse: If the diagonals of a quadrilateral are congruent, then the quadrilateral is a rectangle. Contrapositive: If the diagonals of a quadrilateral are not congruent, then the quadrilateral is not a rectangle.

11. Inverse: If the exterior rays of two adjacent angles are not perpendicular, then the angles are not complements. Converse: If two adjacent angles are complements, then the exterior rays of the angles are perpendicular. Contrapositive: If two adjacent angles are not complements, then the exterior rays of the angles are not perpendicular.

12. a. In quadrilateral *PQRS*, if $\angle P$ is not congruent to $\angle R$, then *PQRS* is not a parallelogram.

   b.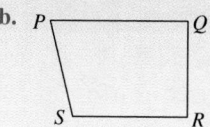

   c. Given: $\angle P$ is not congruent to $\angle R$.
     Prove: *PQRS* is not a parallelogram.

13. Answers may vary. Check students' work. If a given conditional is true, then the contrapositive is also true. The inverse and converse of the conditional are either both true or both false.

Mathematical Goals

- Make and use a classification system.
- Develop a system for classifying a group of objects.
- Draw a diagram for a classification system.

Planning

Materials
- objects collected by students
- items needed to make a display

Project Teams
You may wish to have students work in pairs or in groups of three to complete this project. When working in groups, the objects collected should be of interest to all members of the group. Group members can discuss how they want to proceed and who will record the results of their work.

Guiding Students' Work

If students have difficulty following the given classification system, have them review Exs. 12–14 on page 128 and Exs. 5–9 on page 133. If necessary, work through these exercises with students. Also, it may be difficult for some students to draw a diagram to classify their system. If students are having difficulty in this area, have them discuss their classification system first, using words to describe how it works. Then they can attempt to write down their ideas and use them to create the diagram.

Second-Language Learners
If necessary, invite volunteers to describe what purpose a classifying system serves, and then give additional examples of common classifying systems such as card catalogs or record store sections. You may also want to have students learning English look up *musical instruments* in an encyclopedia to read about the different classes of instruments (*stringed*, *wind*, *percussion*, *keyboard*). This should help them associate instruments and categories with their English names.

Classifying Information

How does a music store owner decide how to display instruments? Instruments could be grouped by the type of music performed on them or they could be grouped into types of instrument. The owner might use a classification system. We use systems like this to classify just about everything to help keep track of what we have or know.

PROJECT GOAL   Make and use a classification system for something of interest to you.

Using a Classification System

1. Classify a guitar as a *stringed*, *wind*, *percussion*, or *keyboard* instrument by using the diagram below.

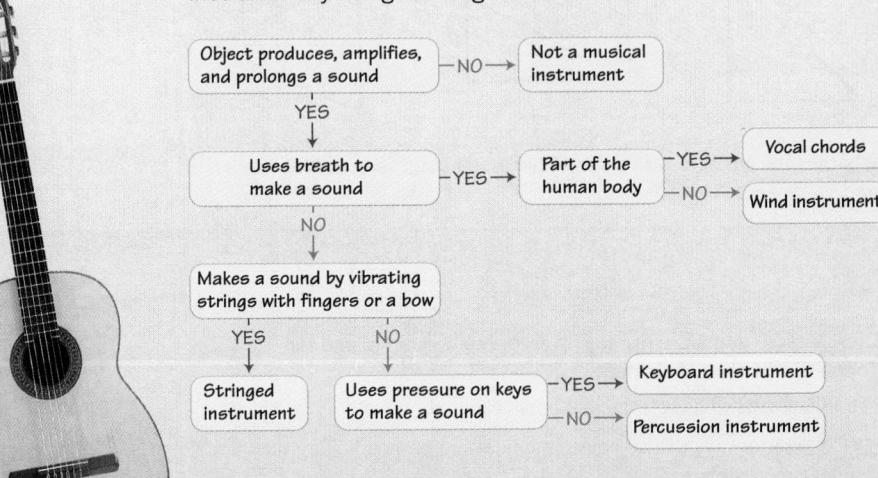

2. Choose some other instruments to classify using the diagram above. Does the diagram classify all of the instruments the way you expect them to be classified? How could you deal with instruments that fit into more than one category?

3. Discuss other ways the diagram could be arranged. Would it make sense to have more than two options in some places? What changes would you make to help classify instruments correctly?

Where Does it Belong?

Develop a system for classifying a group of objects that you collect. If you do not have a collection, choose something that you are interested in, such as types of music or Internet sites.

Draw a diagram for your classification system. Show how to classify at least five different items in the group using your system. If you need to, adjust your classification system.

Presenting Your Project

Make a display of your classification system and how it works. Explain what group of items your system classifies.

Look at some of the other displays. Analyze how well you think one other classification system works. Give a copy of your analysis to the group that created that system.

You may want to extend your project and explore one of the ideas below:

- Interview someone who works in marketing. Describe how they classify customers for various markets.

- Expand the musical instrument diagram. Add branches after the existing instrument categories.

- Research how astronomers classify stars. What type of criteria do they use to categorize a star?

Self-Assessment

What are some important points to remember when developing a classification system? What items are the most difficult to classify? Why?

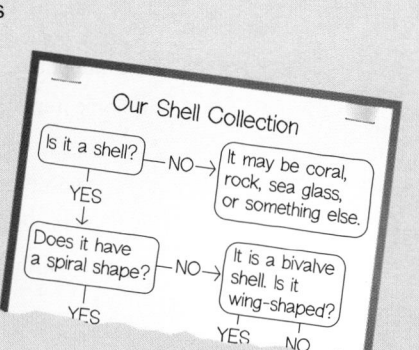

Spiral Shells

Coral Rock

Hinged Shells

Our Shell Collection

Is it a shell? —NO→ It may be coral, rock, sea glass, or something else.

↓ YES

Does it have a spiral shape? —NO→ It is a bivalve shell. Is it wing-shaped?

↓ YES

YES   NO

Guiding Students' Work

Rubric for Chapter Project

4 Students correctly classify the guitar and other instruments using the diagram. They also make good suggestions for rearranging the diagram. Students develop a system for classifying their objects and draw an accurate diagram. They also correctly classify at least five items from their system. Students make a display for their system and provide a clear explanation of what it classifies. Students make an analysis of one of the other displays and give a copy to the group that made the display. Students also extend the project in one of the ways given, and perform an insightful self-assessment.

3 Students correctly classify the guitar and other instruments using the diagram, but their ideas on rearranging the diagram are not clear. Students develop a system for classification, but there are some errors in their diagram. Five items are classified using their diagram. Students analyze another system correctly and provide the group who designed the system with a copy of their analysis. Students extend the project and complete an accurate self-assessment.

2 Students classify the guitar and other instruments but make several mistakes. Also, students do not find ways to rearrange the given diagram. Students select a group of objects to classify but do not draw a correct diagram and do not correctly classify five items. An attempt is made to analyze another system, but the analysis is incomplete. An attempt is made to extend the project, but it is also not complete. Students complete a self-assessment but do not answer all of the questions.

1 Students attempt to classify the guitar and other instruments but do it incorrectly. No effort is made to rearrange the diagram, and a diagram for their system is incomplete or not done. Students do not analyze another group's diagram. No attempt is made to extend the project or complete a self-assessment. Students should be encouraged to speak with the teacher as soon as possible to review their work and to make a new start on the project.

Chapter Support

Course Guide: Chapter 3

Lesson Plans: Chapter 3

Practice Bank:
Cumulative Practice 23

Study Guide: Chapter 3 Review

Challenge Problems: Set 23

Assessment Book:
Chapter Tests 14 and 15
Chapter 3 Alternative Assessment

Test Generator Software

Portfolio Project Book:
Additional Project 2:
Using Reasoning to Make
Decisions

Preparing for College Entrance Tests

Professional Handbook

Progress Check 3.1–3.2

1. Tell whether the argument uses *inductive* or *deductive* reasoning. *(Section 3.1)*
To be a registered nurse, it is necessary to have a knowledge of metric units of volume. Leon wants to be a registered nurse. Therefore, he should acquire a knowledge of metric units of volume. deductive reaoning

Give the postulate, definition, or property that makes each statement true. *(Section 3.2)*

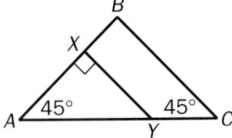

2. ∠ A is a complement of ∠ C.
Definition of complementary angles

3. $m \angle A + m \angle B + m \angle C = 180°$
The sum of the angle measures of a triangle is 180°.

4. If $m \angle A + m \angle B + m \angle C = 180°$, then $45° + m \angle B + 45° = 180°$.
Substitution Property

5. If $m \angle B + 90° = 180°$, then $m \angle B = 90°$. Subtraction Property

Progress Check 3.3–3.4

In the figure, points *A*, *X*, and *D* are collinear, and points *C*, *X*, and *B* are collinear. Also, ∠ 3 ≅ ∠ 4 and ∠ 5 ≅ ∠ 6.

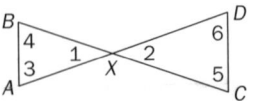

Review

STUDY TECHNIQUE

Reread each section of the chapter. Pay special attention to the section titles and the description, below each section number, of what you should learn in that section. Review any topics that you found especially difficult. Write a short summary of each section to use as a study guide.

VOCABULARY

deductive reasoning (p. 112)
postulate (p. 117)
definition (p. 118)
proof (p. 124)
theorem (p. 125)
paragraph proof (p. 125)
statement (p. 125)
reason (p. 125)
hypothesis of a proof (p. 125)

conclusion of a proof (p. 125)
two-column proof (p. 131)
converse (p. 137)
leg of a right triangle (p. 141)
hypotenuse (p. 141)
radical (p. 142)
negation (p. 149)
inverse (p. 149)
contrapositive (p. 149)

SECTIONS 3.1 *and* 3.2

You can use inductive reasoning to make conjectures based on examples or patterns. **Deductive reasoning** is used to give a convincing argument that a conjecture is true. Use given information, definitions, and accepted assumptions, **properties** and **postulates**, in a logical chain of reasoning.

Some important postulates and properties include the following:

Reflexive Property	Addition Property
Substitution Property	Subtraction Property
Symmetric Property	Segment Addition Postulate
Transitive Property	Angle Addition Postulate

SECTIONS 3.3 *and* 3.4

A **proof** is a convincing argument that can be used to show that an important conjecture, called a **theorem**, is true.

Given: $\angle 1$ and $\angle 2$ are right angles.

Prove: $\angle 1 \cong \angle 2$

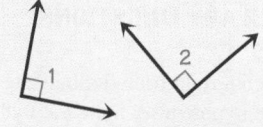

Use a diagram to represent the given information.

$\angle 1$ and $\angle 2$ are right angles, so by the definition of a right angle, $m\angle 1 = 90°$ and $m\angle 2 = 90°$. By the Substitution Property, $m\angle 1 = m\angle 2$. By the definition of congruent angles, $\angle 1 \cong \angle 2$.

In a **two-column proof**, the statements and reasons are arranged in columns.

SECTIONS 3.5, 3.6, *and* 3.7

You can use the Pythagorean theorem to find the lengths of the sides of a right triangle. The converse of the Pythagorean theorem tells you whether a triangle is a right triangle.

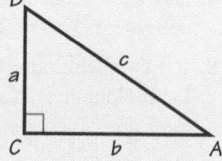

- If $\angle C$ is a right angle, then $a^2 + b^2 = c^2$.
- If $a^2 + b^2 = c^2$, then $\angle C$ is a right angle.

You can also use inequalities related to the converse of the Pythagorean theorem to tell if a triangle is acute or obtuse. Suppose the lengths of the sides of a triangle are a, b, and c, and c is the length of the longest side.

- If $a^2 + b^2 > c^2$, then the triangle is an acute triangle.
- If $a^2 + b^2 < c^2$, then the triangle is an obtuse triangle.

You can rewrite a conditional statement several different ways.

Conditional	If P, then Q.	If $x = 3$, then $x^2 = 9$.
Converse	If Q, then P.	If $x^2 = 9$, then $x = 3$.
Inverse	If *not P*, then *not Q*.	If $x \neq 3$, then $x^2 \neq 9$.
Contrapositive	If *not Q*, then *not P*.	If $x^2 \neq 9$, then $x \neq 3$.

If a conditional statement is true, then its contrapositive is also true. The converse and inverse of the statement are either both true or both false.

Progress Check 3.3–3.4 (cont.)

1. Are $\angle 1$ and $\angle 2$ vertical angles? How do you know? *(Section 3.3)* Yes; definition of vertical angles.

2. Is it true that $m\angle 3 + m\angle 4 + m\angle 1 = 180°$? How do you know? *(Section 3.3)* Yes; the sum of the angle measures of any triangle is 180°.

3. Complete the two-column proof. *(Section 3.4)*

A   B   C   D

Given: B is the midpoint of $\overline{AC}$;
       C is the midpoint of $\overline{BD}$.

Prove: $AB = CD$

Statements	Reasons
1. B is the midpoint of $\overline{AC}$; C is the midpoint $\overline{BD}$.	1. ?
2. $\overline{AB} \cong \overline{BC}, \overline{BC} \cong \overline{CD}$	2. ?
3. $\overline{AB} \cong \overline{CD}$	3. ?
4. $AB = CD$	4. ?

1. Given
2. Definition of midpoint
3. Transitive Property of congruence
4. Definition of congruent segments

Progress Check 3.5–3.7

1. Write the converse of the following statement and tell whether the converse is *True* or *False*. If *l* and *m* are parallel lines, then *l* and *m* lie in the same plane. *(Section 3.5)* If *l* and *m* are lines in the same plane, then *l* and *m* are parallel; False.

2. Alberto decided that since $5^2 + 12^2 = 13^2$, a triangle with side lengths 5, 12, and 13 is a right triangle. Did he use the Pythagorean theorem or its converse? Was he correct in what he decided? *(Sections 3.5 and 3.6)* converse; Yes.

The legs of a right triangle have lengths a and b, and the hypotenuse has length c. Find the missing side lengths. *(Section 3.6)*

3. $a = 6$, $b = 4$ $c = 2\sqrt{13}$

4. $b = \sqrt{7}$, $c = \sqrt{10}$ $a = \sqrt{3}$

5. Write the inverse and contrapositive of the following conditional statement. If $\angle 1$ and $\angle 2$ form a linear pair, then $\angle 1$ and $\angle 2$ are supplementary. *(Section 3.7)* Inverse: If $\angle 1$ and $\angle 2$ do not form a linear pair, then they are not supplementary.
Contrapositive: If $\angle 1$ and $\angle 2$ are not supplementary, then $\angle 1$ and $\angle 2$ do not form a linear pair.

Chapter 3 Assessment
Form A Chapter Test

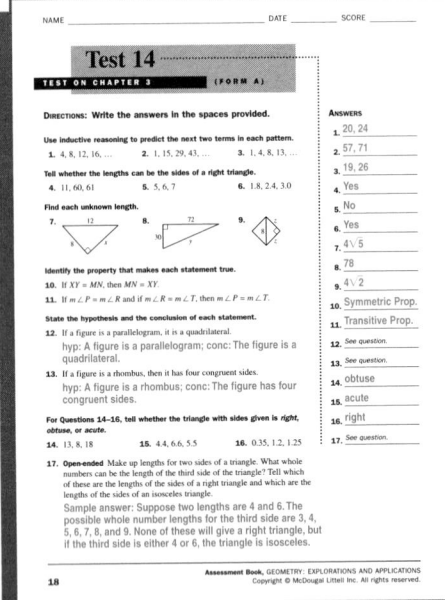

3 Assessment

VOCABULARY QUESTIONS

1. Sketch a right triangle. Indicate which sides are the legs and which is the hypotenuse. State the Pythagorean theorem in terms of your triangle.

2. Complete each statement. To show that a theorem is true, write a _?_. The given information is the _?_ and the statement that you want to prove is the _?_.

3. Write the inverse of this conditional: If point Y is between points X and Z, then $XY + YZ = XZ$.

SECTIONS 3.1 *and* 3.2

For Questions 4 and 5, select the conclusion that makes the statement true.

4. Deanna used deductive reasoning to show that two angles are congruent. The angles (*must, may,* or *cannot*) be congruent.

5. If a parallelogram is a square, then it has four right angles. Parallelogram $ABCD$ is not a square. Therefore, $ABCD$ (*has, does not have,* or *may have*) four right angles.

6. Tell whether the argument uses *inductive* or *deductive* reasoning.

$$1 = 1^2$$
$$1 + 3 = 4 = 2^2$$
$$1 + 3 + 5 = 9 = 3^2$$
$$1 + 3 + 5 + 7 = 16 = 4^2$$
$$\text{So } 1 + 3 + 5 + 7 + 9 = 5^2$$

Give the postulate, definition, property, or previous statement that makes the statement about the diagram true.

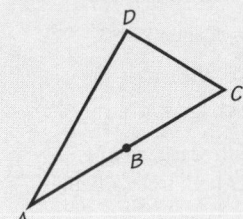

7. There is only one line that contains points A and B.

8. If $AB = BC$, then B is the midpoint of $\overline{AC}$.

9. If $\overline{AB} \cong \overline{BC}$ and $\overline{BC} \cong \overline{CD}$, then $\overline{AB} \cong \overline{CD}$.

10. $m\angle A + m\angle C + m\angle D = 180°$

Chapter 3 Assessment
Form B Chapter Test

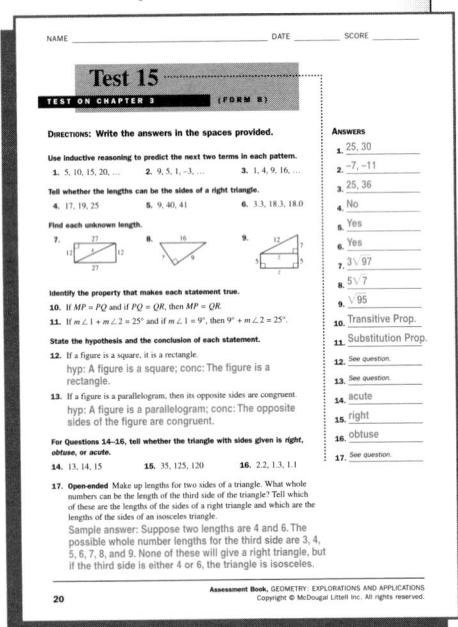

ANSWERS Chapter 3

Assessment

1. Sketches and labels may vary. An example is given. For the triangle in the sketch, the Pythagorean theorem can be stated as $a^2 + b^2 = c^2$.

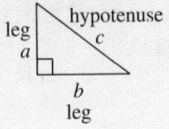

2. proof; hypothesis; conclusion

3. If point Y is not between points X and Z, then $XY + YZ \neq XZ$.

4. must

5. may have

6. inductive

7. Through any two points there is exactly one line.

8. definition of midpoint

9. Transitive Property

10. The sum of the angle measures of a triangle is 180°.

11. Given; Vertical angles are congruent; Substitution Property (Steps 1 and 2); $\angle 1 \cong \angle 4$; Substitution Property (Steps 2 and 3)

12. a. Sketches and labels may vary.

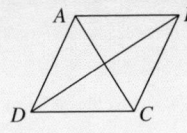

b. Given: $ABCD$ is a rhombus. Prove: $\overline{AC} \perp \overline{BD}$

13. a. Converse: If the diagonals of a quadrilateral are perpendicular, then it is a rhombus. Inverse: If a quadrilateral is not a rhombus, then its diagonals are not perpendicular. Contrapositive: If the diagonals of a quadrilateral are not perpendicular, then it is not a rhombus.

SECTIONS 3.3 and 3.4

11. Copy and complete the proof.

Given: $\angle 2 \cong \angle 3$

Prove: $\angle 1 \cong \angle 4$

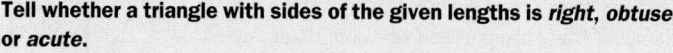

Statements	Reasons
1. $\angle 2 \cong \angle 3$	1. ?
2. $\angle 1 \cong \angle 2$, $\angle 3 \cong \angle 4$	2. ?
3. $\angle 1 \cong \angle 3$	3. ?
4. ?	4. ?

12. a. Sketch and label a diagram for the proof of the theorem: The diagonals of a rhombus are perpendicular.

b. Write the Given and Prove statements for the theorem.

SECTIONS 3.5, 3.6, and 3.7

13. Use the theorem in Question 12.

a. Write the converse, the inverse, and the contrapositive of the theorem.

b. Choose one of your statements from part (a). Draw a diagram for the proof of the statement you chose. Write the Given and Prove statements for the proof.

14. Open-ended Problem Give a conditional that is true, but whose converse is false.

15. A rectangle is 10 cm long and 8 cm wide. Find the length of each diagonal.

16. Find the length h in the diagram at the right.

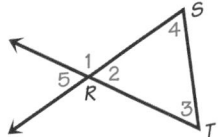

Tell whether a triangle with sides of the given lengths is *right*, *obtuse*, or *acute*.

17. 2.1, 2.8, 3.6   **18.** $\sqrt{7}$, 3, 4   **19.** 2, 9, 9

PERFORMANCE TASK

20. Use this diagram. Write at least two true conditionals and two false conditionals about the diagram. For each conditional, write the conditional, the converse, the inverse, and the contrapositive, and tell whether each is *True* or *False*. Do you notice a pattern in your results?

Chapter 3 Assessment
Form C Alternative Assessment

Chapter 3
ALTERNATIVE ASSESSMENT

1. **Performance Task** Explain the difference between inductive and deductive reasoning to a classmate.

2. **a.** Give an example of inductive reasoning.
   b. Give an example of deductive reasoning.

3. **Performance Task** Casey is trying to win a stuffed animal at the school carnival. She watches a numbered wheel spin ten times and notices that her lucky number 7 came up three times. She concludes that 7 would be a good number to play to win the animal. Do you agree or disagree? Use inductive reasoning to explain your answer.

4. **Open-ended Problem** Give an example in which inductive reasoning is not valid.

5. **Research Project** Research the topics of *criminal law* and *trials*. Explain how inductive and deductive reasoning are used in court cases.

6. **Project** Draw a cartoon to help you remember one of the following properties: Reflexive Property, Symmetric Property, Transitive Property, Addition Property, or Subtraction Property.

7. **Research Project** Interview someone who works in a medical profession. Find out some ways that inductive and deductive reasoning are used to help make decisions in the field of medicine.

8. **Group Activity** Copy a two-column proof from the textbook or your homework onto a piece of poster board. Cut the second column (reasons) out of the proof. Now cut the reasons apart and cut the step numbers off. Have the rest of your group members try to match these reasons to the statements shown in the Statement column of your proof.

Assessment Book, GEOMETRY: EXPLORATIONS AND APPLICATIONS
Copyright © McDougal Littell Inc. All rights reserved.   **95**

---

b. For any of the statements, the diagram given in the answer to Ex. 12 can be used.
Converse—Given: $\overline{AC} \perp \overline{BD}$; Prove: *ABCD* is a rhombus.
Inverse—Given: *ABCD* is not a rhombus; Prove: $\overline{AC}$ is not perpendicular to $\overline{BD}$.
Contrapositive—Given: $\overline{AC}$ is not perpendicular to $\overline{BD}$; Prove: *ABCD* is not a rhombus.

14. Answers may vary. An example is given. If a quadrilateral is a square, then it is a rectangle. The converse is false. It is not true that if a quadrilateral is a rectangle, then it is a square.

15. $2\sqrt{41} \approx 12.81$

16. $3\sqrt{3} \approx 5.20$

17. obtuse

18. right

19. acute

20. Answers may vary. One true conditional and one false conditional are given. True conditional: If $m\angle 2 = 90°$, then *RST* is a right triangle. Converse: If *RST* is a right triangle, then $m\angle 2 = 90°$. (False.) Inverse: If $m\angle 2 \neq 90°$, then *RST* is not a right triangle. (False.) Contrapositive: If *RST* is not a right triangle, then $m\angle 2 \neq 90°$. (True.)

False conditional: If $\angle 3 \cong \angle 4$, then $\angle 2 \cong \angle 5$. Converse: If $\angle 2 \cong \angle 5$, then $\angle 3 \cong \angle 4$. (False.) Inverse: If $\angle 3$ is not congruent to $\angle 4$, then $\angle 2$ is not congruent to $\angle 5$. (False.) Contrapositive: If $\angle 2$ is not congruent to $\angle 5$, then $\angle 3$ is not congruent to $\angle 4$. (False.) A statement and its contrapositive are always either both true or both false. Its converse and its inverse are always either both true or both false.

Cumulative Assessment
CHAPTERS 1–3

CHAPTER 1

For Questions 1–3, use the table below. Suppose that each person in a group of people shakes hands with every other person once.

Number of people	2	3	4	5	6	7
Number of handshakes	1	3	6	10	?	?

1. **LOGICAL REASONING** Use inductive reasoning to predict the total number of handshakes for 6 people and for 7 people.

2. Make a conjecture about the number of handshakes added when one more person joins a group of *n* people. Identify the hypothesis and conclusion of your conjecture.

3. **Writing** How could you test the conjecture that you made in Question 2? Would your method allow you to be sure that your conjecture is correct? Why or why not?

4. **Open-ended Problem** Draw a figure that has one line of symmetry.

5. **Open-ended Problem** Write a conditional statement that is false. Give a counterexample to show that the statement is false.

For Questions 6–9, use the diagram.

6. If $\overline{BD}$ bisects $\overline{AC}$, $AE = 3x + 4$, and $AC = 38$, find the value of *x*.

7. Give as many different names as you can for $\angle 1$.

8. Find the measures of $\angle ABC$ and $\angle ACF$.

9. Name an angle bisector and two congruent angles.

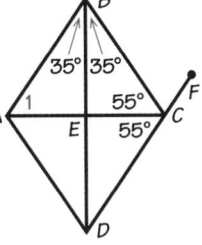

CHAPTER 2

Complete each statement with *always*, *sometimes*, or *never*.

10. Two angles formed by perpendicular lines are _?_ complementary.

11. An equilateral triangle is _?_ isosceles.

12. An isosceles triangle is _?_ an acute triangle.

13. The diagonals of a rhombus _?_ bisect each other.

ANSWERS   Chapters 1–3

Cumulative Assessment

1. 15 handshakes; 21 handshakes

2. When one more person joins a group of *n* people, *n* handshakes are added.
   Hypothesis: one more person joins a group of *n* people;
   Conclusion: *n* handshakes are added

3–5. Answers may vary. Examples are given.

3. You could use a group of people or a group of diagrams to extend the table shown on page 160. Add a third line to the table, showing the number of handshakes added for each number of people in the group, and see if *n* handshakes are added when one person joins a group of *n* people. No; you could not be sure that the conjecture is correct because a conclusion based on inductive reasoning is not necessarily true.

4.

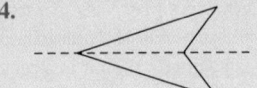

5. If a number is a prime number, then the number is an odd integer. The number 2 is a counterexample, since it is an even prime number.

6. 5

7. $\angle BAE$, $\angle BAC$, $\angle EAB$, $\angle CAB$

8. 70°; 125°

9. one of the following: $\overrightarrow{BD}$, $\angle ABD$, and $\angle CBD$, or $\overrightarrow{CA}$, $\angle BCA$, and $\angle DCA$

10. never

11. always

12. sometimes

13. always

14. Writing Suppose *ABCD* is a parallelogram. Explain how you know that ∠*A* and ∠*B* are supplementary angles.

For each figure, find the values of *x* and *y*.

15.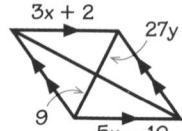

16.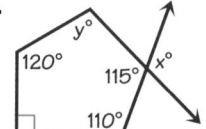

17.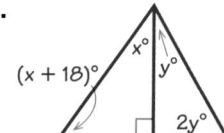

18. Sketch an equilateral pentagon and a regular quadrilateral. Describe the lines of symmetry of each figure and mark any congruent sides or angles.

19. Open-ended Problem Sketch a prism whose bases are isosceles triangles and sketch a net for the prism.

CHAPTER 3

20. Writing Write a convincing argument that uses deductive reasoning to explain why the opposite sides of a rectangle are parallel.

Give the postulate, definition, property, or theorem that makes the statement about the diagram true.

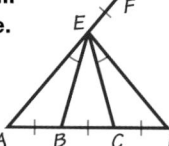

21. *m*∠*A* + *m*∠*D* = *m*∠*DEF*

22. *m*∠*AEC* = *m*∠*DEB*

23. *AB* + *BC* = *AC*

24. Draw and label a diagram for the proof of the statement: The bisectors of the angles in a linear pair form a right angle. Write the Given and Prove statements for the proof.

25. Write a paragraph proof of the theorem: The measure of each angle of an equiangular triangle is 60°.

26. Write the inverse, the converse, and the contrapositive of the following statement: The diagonals of a rhombus are perpendicular.

Find the value of *x*.

27.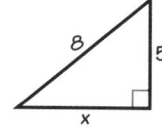

28.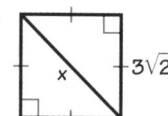

29.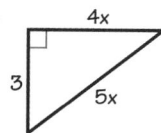

30. Open-ended Problem Use the theorem in Question 26. Draw a rhombus and label possible lengths for the sides and the diagonals of the rhombus.

14. Since *ABCD* is a parallelogram, ∠*A* ≅ ∠*C* and ∠*B* ≅ ∠*D*. The sum of the angle measures of a quadrilateral is 360°, so *m*∠*A* + *m*∠*B* + *m*∠*C* + *m*∠*D* = 360°; substituting, *m*∠*A* + *m*∠*B* + *m*∠*A* + *m*∠*B* = 360°; 2 · *m*∠*A* + 2 · *m*∠*B* = 360°; 2(*m*∠*A* + *m*∠*B*) = 360°; *m*∠*A* + *m*∠*B* = 180°; by the definition of supplementary angles, ∠*A* and ∠*B* are supplementary.

15. *x* = 6; *y* = $\frac{1}{3}$

16. *x* = 115; *y* = 105

17. *x* = 36; *y* = 30

18, 19. Answers may vary. Examples are given.

18.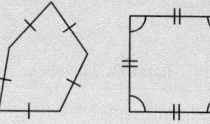

The pentagon has no lines of symmetry. The regular quadrilateral, or square, has four lines of symmetry: the lines that contain the diagonals and the lines that join the midpoints of opposite sides.

19.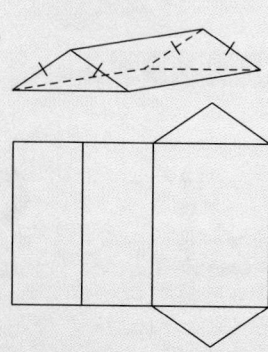

20. Every rectangle is a parallelogram, and the opposite sides of a parallelogram are parallel. Therefore, the opposite sides of a rectangle are parallel.

21. The measure of an exterior angle of a triangle is equal to the sum of the measures of the two interior angles that are not adjacent to it.

22. the Angle Addition Postulate

23. the Segment Addition Postulate

24. Answers may vary. An example is given.

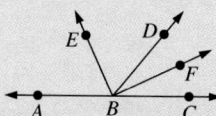

Given: ∠*ABD* and ∠*DBC* form a a linear pair. $\overrightarrow{BE}$ bisects ∠*ABD* and $\overrightarrow{BF}$ bisects ∠*DBC*.
Prove: ∠*EBF* is a right angle.

25. The sum of the angle measures of a triangle is 180°. Since the angles of an equiangular triangle are congruent, their measures are equal. Thus, the measure of each angle is $\frac{180°}{3}$ = 60°.

26. Statement: If a quadrilateral is a rhombus, then its diagonals are perpendicular.
Inverse: If a quadrilateral is not a rhombus, then its diagonals are not perpendicular.
Converse: If the diagonals of a quadrilateral are perpendicular, then the quadrilateral is a rhombus.
Contrapositive: If the diagonals of a quadrilateral are not perpendicular, then the quadrilateral is not a rhombus.

27. $\sqrt{39}$ ≈ 6.24

28. 6

29. 1

30. Answers may vary. An example is given.

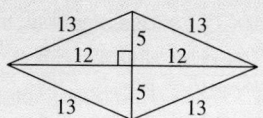

4 Coordinates in Geometry

OVERVIEW

Connecting to Prior and Future Learning

⟺ Students continue their study of coordinate graphing from Algebra 1. The chapter opens with a study of distances and midpoints.

⟺ After finding distances and midpoints, students learn about equations of lines and the properties of parallel and perpendicular lines. Many of the ideas presented in this section are a review from Algebra 1 and are essential to understanding other ideas yet to be presented in geometry.

⟺ The last half of Chapter 4 presents several more important concepts, including the equations of circles, coordinate proofs, and coordinates in three dimensions.

Chapter Highlights

Interview with Marc Hannah: This interview highlights the relationship between mathematics and computer graphics. Related exercises can be found on pages 170, 198, and 206.

Explorations in Chapter 4 involve comparing slopes of lines in Section 4.3, and placing a parallelogram on a coordinate plane in Section 4.5.

The Portfolio Project: Taxicab geometry, the non-Euclidean geometry that a taxicab and a pedestrian must obey, is the focus of this project. Students compare Euclidean geometry to taxicab geometry and calculate traveling distances along a city grid.

Technology: In Section 4.1, students use spreadsheets to explore distances and geometry software to explore properties of quadrilaterals. Graphing software can be used in Section 4.3 to explore the slopes of lines. In Section 4.5, geometry software is used to prove a property of the diagonals of a rectangle.

OBJECTIVES

Section	Objectives	NCTM Standards
4.1	• Find the distance between two points and the coordinates of the midpoint of a segment. • Determine distances in real-world situations. • Classify polygons by the lengths of their sides.	1, 2, 3, 4, 5, 8
4.2	• Find the slope of a segment or a line and write equations of lines in slope-intercept form. • Graph and compare equations of lines. • Investigate geometric relationships using lines.	1, 2, 3, 4, 5, 8
4.3	• Find the slopes of parallel lines and perpendicular lines. • Identify properties of perpendicular bisectors. • Write equations for parallel and perpendicular lines.	1, 2, 3, 4, 5, 8
4.4	• Identify circles and parts of circles. • Write equations for circles. • Describe circular shapes in the real world.	1, 2, 3, 4, 5, 8
4.5	• Place figures on coordinate axes and label their vertices. • Plan and write coordinate geometry proofs. • Use algebraic methods to verify conjectures about triangles and quadrilaterals. • Improve reasoning skills.	1, 2, 3, 4, 5, 8
4.6	• Find coordinates in three dimensions. • Use the Distance Formula and the Midpoint Formula in three dimensions. • Identify and describe relationships between geometric figures in three dimensions.	1, 2, 3, 4, 5, 8

INTEGRATION

Mathematical Connections	4.1	4.2	4.3	4.4	4.5	4.6
geometry	**165–172***	**173–179**	**180–186**	**187–193**	**194–200**	**201–207**
algebra	170			193	196, 198, 200	
logic and reasoning	171	179	182, 185	192	**194–200**	206

Interdisciplinary Connections and Applications						
chemistry and physics						204, 205
arts and entertainment	171					
sports and recreation	166	174, 175				
engineering		177				
city planning		178	184			
carpentry			184			
architecture				192		
accessibility, quilt patterns, printing		176	182, 185	191		

****Bold page numbers** indicate that a topic is used throughout the section.*

TECHNOLOGY

Section	opportunities for use with	
	Student Book	**Support Material**
4.1	scientific calculator spreadsheet software geometry software	**Technology Book:** Calculator Activity 5
4.2	scientific calculator graphing calculator	**Technology Book:** TI-92 Activity 4 **Function Investigator with Matrix Analyzer Activity Book:** Activities 3, 4, 5
4.3	graphing calculator geometry software	**Technology Book:** Spreadsheet Activity 4 **Geometry Inventor Activity Book:** Activity 15
4.4	graphing calculator	**Technology Book:** Calculator Activity 6 TI-92 Activity 4 **Geometry Inventor Activity Book:** Activity 12
4.5	scientific calculator geometry software	**Geometry Inventor Activity Book:** Activities 4, 7
4.6	graphing calculator	**Technology Book:** Spreadsheet Activity 4

Regular Scheduling (45 min)

Section	Materials Needed	Core Assignment	Extended Assignment	exercises that feature		
				Applications	Communication	Technology
4.1	graph paper, spreadsheet software, cardboard or heavy paper, geometry software	4–17, 21, 22, 35–43	1–43	18–29	1, 3, 17, 29, 31	2, 32–34
4.2	graph paper	1–7, 9–16, 23–33, 36–48	1–7 odd, 8, 9–15 odd, 17–22, 23–33 odd, 34–48	7, 17–22	8, 19b, 36	
4.3	geometry software or graph paper, straightedge	**Day 1:** 1–12, 14–22 **Day 2:** 28–31, 33–40, AYP*	**Day 1:** 1–23 **Day 2:** 24–40, AYP	13, 23 24–27	13b, 23 33	
4.4	graph paper, compass, circular object	1–9, 11–20, 23–25, 28–44	1–9 odd, 10, 11–21 odd, 22, 23, 25–44	10, 21, 22	26, 33	
4.5	graph paper, scissors, straightedge, geometry software	1–3, 8–19, 21–24, 28–37	1–8, 10–13, 16–37	4–7	7, 20, 26, 28d	27
4.6	3-D graph paper, graph paper	1–21, 30–34, AYP	1–21 odd, 22–34, AYP	23–28	22, 29c	
Review/ Assess		**Day 1:** 1–16 **Day 2:** 17–32 **Day 3:** Ch. 4 Test	**Day 1:** 1–16 **Day 2:** 17–32 **Day 3:** Ch. 4 Test			
Portfolio Project		Allow 2 days.	Allow 2 days.			

Yearly Pacing (with Portfolio Project)	Chapter 4 Total 12 days	Chapters 1–4 Total 54 days	Remaining 106 days	Total 160 days

Block Scheduling (90 min)

	Day 22	Day 23	Day 24	Day 25	Day 26	Day 27
Teach/Interact	4.1 4.2	4.3: Exploration, page 180	4.4 4.5: Exploration, page 194	4.6 Review	Review Port. Proj.	Ch. 4 Test Port. Proj.
Apply/Assess	**4.1:** 4–17, 21, 22, 35–43 **4.2:** 1–7, 9–16, 23–33, 36–48	**4.3:** 1–12, 14–22, 28–31, 33–40, AYP*	**4.4:** 1–9, 11–20, 23–25, 28–44 **4.5:** 1–3, 8–19, 21–24, 28–37	**4.6:** 1–21, 30–34, AYP **Review:** 1–16	**Review:** 17–32 **Port. Proj.**	**Ch. 4 Test Port. Proj.**

NOTE: A one-day block has been added for the Portfolio Project—timing and placement to be determined by teacher.

Yearly Pacing (with Portfolio Project)	Chapter 4 Total 6 days	Chapters 1–4 Total 27 days	Remaining 53 days	Total 80 days

*AYP is Assess Your Progress.

162C

Section	Practice Bank	Study Guide*	Assessment Book*	Visuals	Explorations Lab Manual	Lesson Plans	Technology Book
4.1	24	4.1		Warm-Up 4.1	Add. Expl. 6 Masters 1, 2	4.1	Calculator Act. 5
4.2	25	4.2		Warm-Up 4.2	Add. Expl. 7 Master 2	4.2	TI-92 Act. 4
4.3	26	4.3	Test 16	Warm-Up 4.3	Master 2	4.3	Spreadsheet Act. 4
4.4	27	4.4		Warm-Up 4.4	Master 2	4.4	Calculator Act. 6 TI-92 Act. 4
4.5	28	4.5		Warm-Up 4.5 Folder 5	Master 2	4.5	
4.6	29	4.6	Test 17	Warm-Up 4.6	Masters 2, 5	4.6	Spreadsheet Act. 5
Review Test	30	Chapter Review	Tests 18, 19, Alternative Assessment			Review Test	

*__Spanish versions__ of *Study Guide* and *Assessment Book* are available.

Chapter Support

- Course Guides
- Lesson Plans
- Preparing for College Entrance Tests
- Multi-Language Glossary
- *Test Generator* Software
- Professional Handbook
- Challenge Problems

Software Support

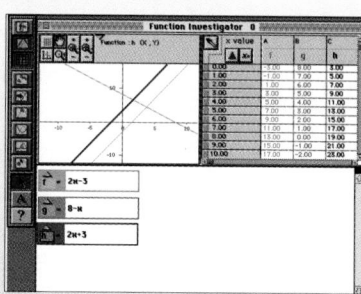

McDougal Littell Mathpack
Geometry Inventor
Function Investigator
Matrix Analyzer

Internet Support

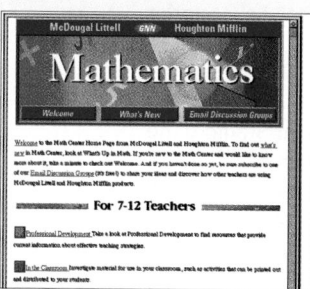

http://www.hmco.com
Next go to McDougal Littell; then the
Education Center; then Secondary Math.

OUTSIDE RESOURCES

Books, Periodicals

Eddins, Susan K., Evelyn Osman Maxwell, and Floramma Stanislaus. "Geometric Transformations, Part 1." *Mathematics Teacher* (March 1994): pp. 177–181, 187–189.

Activities, Manipulatives

Douglas, Lew. "Pentagrams and Spirals." *Mathematics Teacher* (November 1996): pp. 680–687.

Breuningsen, Chris, Bill Bower, Linda Antinone, and Elisa Breuningsen. "From Here to There." *Real-World Math with the CBL System.* Activity 1: pp. 11–16. Texas Instruments, 1995.

Software

Laborde, Jean-Marie and Franck Bellemain, designers. *Cabri Geometry II.* Dallas, TX: Texas Instruments, 1994.

TI-92 Geometry. (Resident on the TI-92 graphing calculator.) Dallas, TX: Texas Instruments, 1995.

Videos

Classroom Connect: Internet Curriculum Integration Videos. Part 1: "Integrating the Internet 101." Lancaster, PA. Order through e-mail:
    connect@classroom.net

Internet

From an Internet account, access lesson plans for K–12 mathematics, compiled by the Eisenhower Network, using the command:
    gopher enc.org

Silicon Valley

Silicon Valley is the name often used when referring to the area from San Jose northwest to Palo Alto in California. It is one of the nation's leading regions for manufacturing computers and related technology. The element silicon, used to make the chips or semiconductors used in computers, gives the region its name.

Marc Hannah

Marc Hannah was born and grew up in Chicago, Illinois. His father was an accountant and his mother was a teacher. He attended and graduated from Kenwood Academy in Chicago, and he then went on to earn a degree in electrical engineering from the Illinois Institute of Technology in 1977. During that summer, he began working for Bell Laboratories designing computer graphics. While at Bell Laboratories, he decided to enter graduate school, earning both his master's degree and his doctorate from Stanford University. At Stanford, he met James Clark, with whom he cofounded Silicon Graphics in 1981. Today, most of Hannah's work at Silicon Graphics involves designing computer chips.

CHAPTER

4 Coordinates in Geometry

Making the Virtual Look Real

INTERVIEW Marc Hannah

Marc Hannah has liked math and science ever since he was young. "I always found those courses easier and more interesting," he says. Hannah majored in electrical engineering in college and then studied computer graphics. Teaming up with university professor James Clark, Hannah began working on the "geometry engine," a computer system that performs calculations to describe objects and display them realistically on a computer screen from any perspective. This collaboration eventually led to the co-founding of Silicon Graphics, a world-leading company that makes computer workstations.

"The most important thing, in terms of choosing a career, is that you enjoy what you do."

Marc Hannah and a co-worker discuss a design for a computer chip.

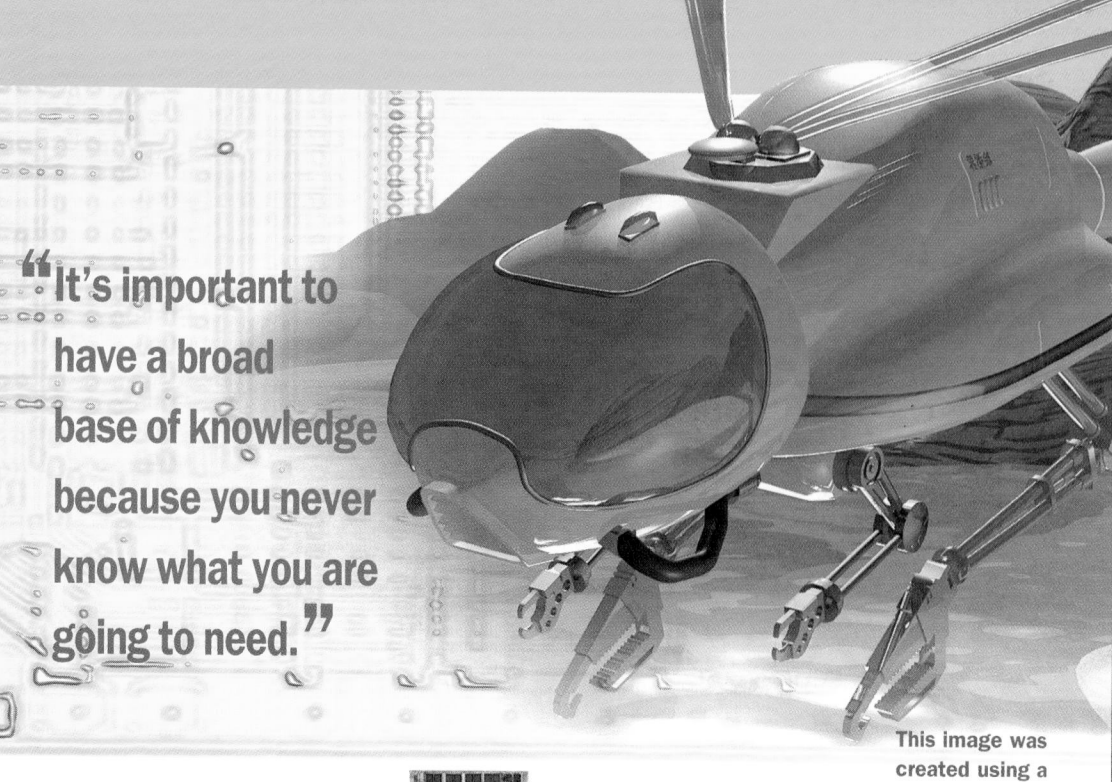

"It's important to have a broad base of knowledge because you never know what you are going to need."

This image was created using a Silicon Graphics computer.

Visualizing Data in Three Dimensions

Computer graphics workstations are used for flight simulators, industrial design, and medical scanning. They are also used to generate special effects in movies such as *Jurassic Park*. These applications all involve representing objects in digital form and presenting them convincingly on the screen.

"Whether we're trying to depict a brain in a CAT scan or the view a pilot might see flying into Chicago, it's all done by choosing coordinates in three dimensions for every point in the image," Hannah explains. "These complicated pictures are broken down into geometric shapes, such as triangles and other polygons, which are illuminated and manipulated on the screen."

Building a Faster Machine

As a Vice President and Chief Scientist for Silicon Graphics, Hannah tries to predict what computer features will be needed in the future and the best way to build them. He designs computer chips that will perform calculations quickly. "I try to figure out how our customers use our systems, what they'd like to do, and how to build the fastest machine possible."

Hannah spends a lot of time working with various design groups in the company. "We put together complex systems, which means that no single individual can do it alone," he says. "It's important to be able to work with a team of people and express your ideas clearly."

163

Interview Notes

Background

Computer Chips
The manufacturing of computer chips is very precise and complex. Chips are usually made of silicon, which in its pure form does not conduct electricity, but does if impurities are added. To make computer chips, a wafer of silicon that measures from 1 to 8 inches in diameter is used. The design for the integrated circuit is reduced to microscopic size, and these very small designs, called masks, are used as stencils to make hundreds of chips on one wafer through a series of steps that includes etching, heating, and imparting impurities into the silicon.

Second-Language Learners

Students learning English might find it interesting to learn that the verb *majored* (specialized in at college) is related to the English adjective *major*, meaning "greater in importance, size, number, or scope." You may also want to make sure that all students have some understanding of the terms *computer graphic workstations*, *flight simulators*, *industrial design*, and *medical scanning*.

Multicultural Note

Women have been important contributors to the development of computers and computer programming. Augusta Ada Byron, countess of Lovelace, is often credited with being the first computer programmer. In 1843, she described a punch-card program that would allow George Babbage's "Analytical Engine" to compute Bernoulli numbers. Admiral Grace Hopper, another pioneer in computer science, worked on several early computer models and her work was fundamental in the invention of both the modern compiler and a computer programming language call COBOL. Another important computer programmer, Evelyn Boyd Granville, one of the first African American women to get her doctorate in mathematics, designed computer programs used in both the Mercury and Apollo projects.

Mathematical Connection

There are three basic components involved with making and using computers. First, there is the hardware, which is the physical computer and its components. Second, the software, or programs, to run the computer is needed. Third, a person needs to know how to use both the hardware and software to complete the work or task desired. In Section 4.1, applications dealing with software are explored. In this section, students use variations of the Midpoint Formula in relation to computer programming. In Section 4.5, applications involving hardware are discussed. In this section, students look at various computer chips and relate them to coordinate geometry. In Section 4.6, the exercises discuss using a computer to create computer graphics. Students are involved in the three-dimensional aspects of colors in computer graphics.

Explore and Connect

Writing
Describing the regions in the diagram is a good way to help students understand how to locate points in the interior of the triangle. You may wish to point out that lines k, l, and m separate the plane into three distinct sets of dots on the screen: those inside the triangle, those on the sides of the triangle, and those outside the triangle.

Project
Drawing and shading the polygon extends the idea explored in the Writing activity. As an extension, students can find equations for the lines they drew. They can also try to write inequalities describing the interior of their polygon.

Research
There are also movies, documentaries, or instructional films that show how some of the special effects in specific movies were created. Students may also want to view these films and use them as part of their presentation.

Creating an Image

Coordinates are used in Marc Hannah's work to display images on a computer screen. To draw the interior of a triangle, the computer must color all the dots on the screen (called *pixels*) that lie inside the triangle.

"The way I define a triangle is by specifying the x- and y-coordinates of its vertices," says Hannah.

"Next I have to find out which pixels are within the boundaries of the triangle."

One way to do this is to find equations for the lines that pass through the vertices. Then the computer can calculate whether any particular pixel is above or below each of the three lines.

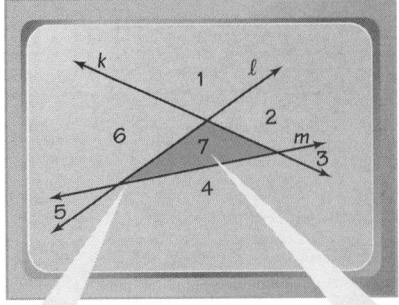

Each line passes through two vertices of the triangle.

Region 7, the interior of the triangle, is all the points that are *below* line k, *below* line l, and *above* line m.

Explore and Connect

Marc Hannah near one of his company's workstations

1. Writing In the diagram above, each region can be described by referring to its place *above* or *below* each of the three lines. Describe the positions of Regions 1–6 this way.

2. Project Draw a polygon on a coordinate plane. Show how you can describe all the points in its interior by referring to the lines that pass through its vertices.

3. Research Find a movie in which computer graphics were used to create special effects. Rent or borrow a video of it and present selections to your class. Describe how the computer-generated special effects are combined with the real actors and settings.

Mathematics & Marc Hannah

In this chapter, you will learn more about how mathematics is related to computer graphics.

Related Exercises

Section 4.1
• Exercises 18–20

Section 4.5
• Exercises 4–7

Section 4.6
• Exercises 25–28

4.1 Finding Distances and Midpoints

Plan⇔Support

Objectives
- Find the distance between two points.
- Find the coordinates of the midpoint of a segment.
- Determine distances in real-world situations.
- Classify polygons by the lengths of their sides.

Recommended Pacing
❖ **Core and Extended Courses**
  Section 4.1: 1 day
❖ **Block Schedule**
  Section 4.1: $\frac{1}{2}$ block
  (with Section 4.2)

Resource Materials
Lesson Support
Lesson Plan 4.1
Warm-Up Transparency 4.1
Practice Bank: Practice 24
Study Guide: Section 4.1
Explorations Lab Manual:
  Additional Exploration 6
  Diagram Masters 1, 2
Challenge Problems: Set 24
Technology
Technology Book:
  Calculator Activity 5
Scientific Calculator
Spreadsheet Software
Geometry Software
Internet:
  http://www.hmco.com

Learn how to...
- find the distance between two points
- find the coordinates of the midpoint of a segment

So you can...
- determine distances, such as at a swimming competition
- classify polygons by the lengths of their sides

At hand-timed swimming competitions, timers at the end of each lane start their watches when they hear the starter's signal. If the starter is closer to the swimmers than to the timers, each swimmer will hear the signal before his or her timer does. This gives each swimmer a different advantage, which must be calculated to adjust the race times.

The lanes begin 1 m from the edge. Each lane is 2.4 m wide.

Each lane has a swimmer and a timer at opposite ends.

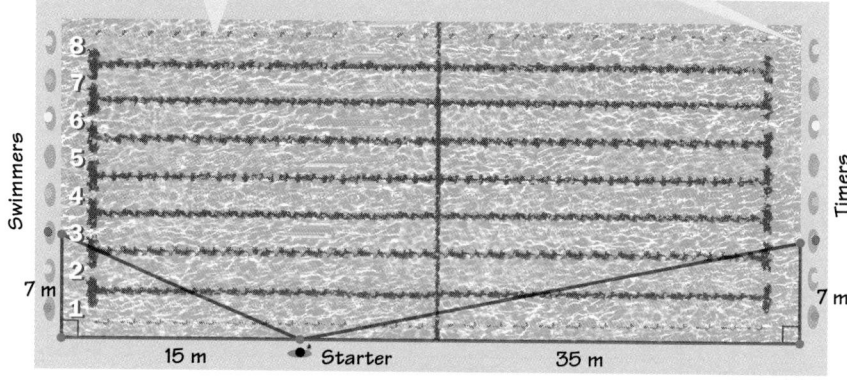

THINK AND COMMUNICATE

1. Use the Pythagorean theorem to find the distance, to the nearest hundredth of a meter, between the starter and the swimmer in lane 3.

2. Use the Pythagorean theorem to find the distance, to the nearest hundredth of a meter, between the starter and the timer in lane 3.

3. Use 346 m/s for the speed of the starting signal. How long does it take the sound, to the nearest thousandth of a second, to reach the swimmer in lane 3? to reach the timer in lane 3?

4. How much sooner, to the nearest thousandth of a second, does the swimmer in lane 3 hear the signal than the timer? How should you adjust the swimmer's race time to account for this advantage?

4.1 Finding Distances and Midpoints **165**

ANSWERS Section 4.1

Think and Communicate

1. 16.55 m
2. 35.69 m
3. 0.048 s; 0.103 s
4. 0.055 s; You should add 0.055 s to the swimmer's race time.

Warm-Up Exercises

1. Complete the Pythagorean theorem: For a right triangle with legs of lengths a and b and hypotenuse of length c, ___?___.
   $a^2 + b^2 = c^2$

Use the Pythagorean theorem to find the length of the third side of each right triangle.

2. $a = 12$, $b = 5$ 13
3. $a = 4$, $b = 7$ 8.1
4. $a = 6.3$, $c = 12.4$ 10.7
5. $b = 43.1$, $c = 157.8$ 151.8

Teaching Tip

The Distance Formula can be derived by using the Pythagorean theorem. Point out that the Distance Formula uses the coordinates of the endpoints of the legs of the triangle rather than the lengths of the legs. Students should realize that the coordinates of the vertex of the right angle are (x_2, y_1).

Second-Language Learners

Some students learning English may benefit from a discussion of the parts of the word *subscript*. Point out that the prefix *sub-* means "below" or "beneath," as in *submarine*. Add that *script* means "writing." If students are native speakers of Spanish, you may want to tell them that *script* comes from the same root as the Spanish word *escribir*, meaning "to write."

Additional Example 1

Refer to Example 1. Suppose the starter moves to $P(20, 0)$. Find the distance, to the nearest hundredth of a meter, between:

a. the starter and the swimmer in lane 6.

Place a coordinate plane with $(0, 0)$ at one corner of the pool.

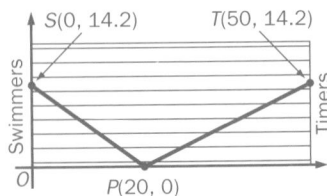

The starter is at $P(20.00, 0)$ and the swimmer in lane 6 is at $S(0, 14.20)$.

$PS =$
$\sqrt{(0 - 20.00)^2 + (14.20 - 0)^2}$
$= \sqrt{(-20.00)^2 + 14.20^2}$
$= \sqrt{400 + 201.64}$
≈ 24.53

The swimmer in lane 6 is about 24.53 m from the starter.

b. the starter and the timer in lane 6.

The starter is at $P(20.00, 0)$ and the timer in lane 6 is at $T(50, 14.20)$.

$PT =$
$\sqrt{(50.00 - 20.00)^2 + (14.20 - 0)^2}$
$= \sqrt{30.00^2 + 14.2^2}$
$= \sqrt{900 + 201.64}$
≈ 33.19

The timer in lane 6 is about 33.19 m from the starter.

To find distances on a coordinate plane, the Pythagorean theorem is expressed as the Distance Formula. *Subscripts* are used to name the coordinates of points. You say "*x* sub 1" and "*y* sub 1" for a point with coordinates (x_1, y_1).

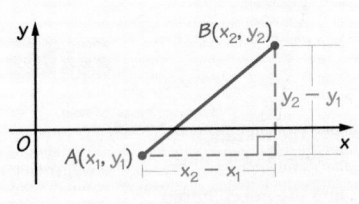

The Distance Formula

The distance between the points $A(x_1, y_1)$ and $B(x_2, y_2)$ is:

$$AB = \sqrt{(x_2 - x_1)^2 + (y_2 - y_1)^2}$$

EXAMPLE 1 Application: Sports

Find the distance, to the nearest hundredth of a meter, between:

a. the starter and the swimmer in lane 8.

b. the starter and the timer in lane 8.

SOLUTION

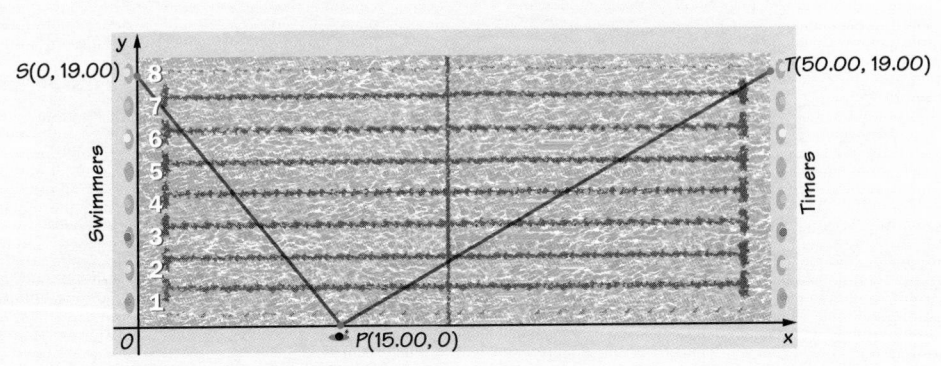

Place a coordinate grid with $(0, 0)$ at one corner of the pool.

a. The starter is at $P(15.00, 0)$ and the swimmer in lane 8 is at $S(0, 19.00)$.

$PS = \sqrt{(0 - 15.00)^2 + (19.00 - 0)^2}$
$= \sqrt{(-15.00)^2 + 19.00^2}$
$= \sqrt{225 + 361}$
≈ 24.21

The swimmer in lane 8 is about 24.21 m from the starter.

b. The starter is at $P(15.00, 0)$ and the timer in lane 8 is at $T(50.00, 19.00)$.

$PT = \sqrt{(50.00 - 15.00)^2 + (19.00 - 0)^2}$
$= \sqrt{35.00^2 + 19.00^2}$
$= \sqrt{1225 + 361}$
≈ 39.82

The timer in lane 8 is about 39.82 m from the starter.

THINK AND COMMUNICATE

5. **a.** How long does it take the starting signal to reach the swimmer in lane 8? the timer in lane 8? (Use 346 m/s for the speed of sound.)

   b. How much sooner does the swimmer in lane 8 hear the signal than the timer in lane 8? How should you adjust the swimmer's race time to account for this advantage?

6. Refer to *Think and Communicate* Questions 4 and 5. Compare the time advantages of the swimmer in lane 3 and the swimmer in lane 8.

You can also use coordinates to find midpoints easily. The coordinates of the midpoint of a segment are the means of the coordinates of the endpoints.

BY THE WAY

Differences among the times of swimmers finishing in first, second, and third places may be only hundredths or thousandths of a second.

The Midpoint Formula

The midpoint of the segment joining the points $A(x_1, y_1)$ and $B(x_2, y_2)$ has these coordinates:

$$\text{coordinates of} \atop \text{midpoint of } \overline{AB} = \left(\frac{x_1 + x_2}{2}, \frac{y_1 + y_2}{2}\right)$$

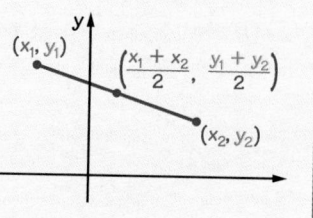

EXAMPLE 2

Use quadrilateral *OBCD*. Find the coordinates of the midpoint of $\overline{BD}$.

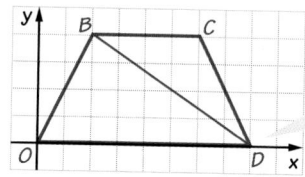

Unless marked otherwise, each grid square represents one unit.

SOLUTION

Use the Midpoint Formula with $B(2, 4)$ and $D(8, 0)$.

$$\begin{aligned}
\text{coordinates of} \atop \text{midpoint of } \overline{BD} &= \left(\frac{x_1 + x_2}{2}, \frac{y_1 + y_2}{2}\right) \\
&= \left(\frac{2 + 8}{2}, \frac{4 + 0}{2}\right) \\
&= (5, 2)
\end{aligned}$$

The coordinates of the midpoint of $\overline{BD}$ are (5, 2).

4.1 Finding Distances and Midpoints   **167**

Think and Communicate

5. **a.** 0.070 s; 0.115 s

   b. 0.045 s; You should add 0.045 s to the swimmer's race time.

6. The time advantage of the swimmer in lane 3 is 0.01 s greater than that of the swimmer in lane 8.

Section Note

Teaching Tip
It may help some students to think of the coordinates of the midpoint of a segment as the mean, or average, of the coordinates of the endpoints.

About Example 2

Geometric Thinking
Discuss the placement of the quadrilateral in the coordinate plane. Students should note that one vertex is at the origin and one of the sides is along the *x*-axis. Ask why this position was chosen. (Some of the coordinates are zeros, making the calculations in the formula easier.) Point out that when using coordinates to find properties of geometric figures, this position is commonly used.

Additional Example 2

Use quadrilateral *ORST*. Find the coordinates of the midpoint of $\overline{OS}$.

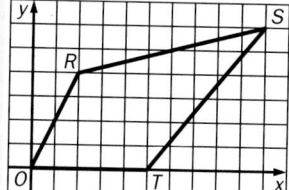

Use the Midpoint Formula with $O(0, 0)$ and $S(10, 6)$.

$$\begin{aligned}
\text{coordinates of midpoint of } \overline{OS} \\
= \left(\frac{x_1 + x_2}{2}, \frac{x_2 + y_1}{2}\right) \\
= \left(\frac{0 + 10}{2}, \frac{0 + 6}{2}\right) \\
= (5, 3)
\end{aligned}$$

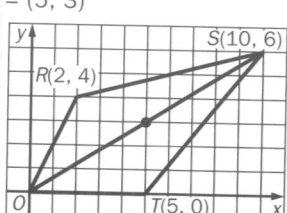

The coordinates of the midpoint of $\overline{OS}$ are (5, 3).

167

Additional Example 3

The vertices of quadrilateral *WXYZ* are *W*(–1, 4), *X*(3, 5), *Y*(4, 1), *Z*(0, 0). Give the most specific name for quadrilateral *WXYZ*.

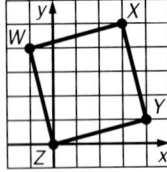

Use the Distance Formula to find the lengths of the sides of quadrilateral *WXYZ*.

$WX = \sqrt{(3 - (-1))^2 + (5 - 4)^2}$
$\quad = \sqrt{4^2 + 1^2}$
$\quad = \sqrt{17}$

$XY = \sqrt{(4 - 3)^2 + (1 - 5)^2}$
$\quad = \sqrt{1^2 + (-4)^2}$
$\quad = \sqrt{17}$

$YZ = \sqrt{(0 - 4)^2 + (0 - 1)^2}$
$\quad = \sqrt{(-4)^2 + (-1)^2}$
$\quad = \sqrt{17}$

$WZ = \sqrt{(0 - (-1))^2 + (0 - 4)^2}$
$\quad = \sqrt{1^2 + (-4)^2}$
$\quad = \sqrt{17}$

Since $WX = XY = YZ = WZ$, quadrilateral *WXYZ* is a rhombus. To determine if *WXYZ* is a square, find the length of each diagonal and use the converse of the Pythagorean theorem.

$WY = \sqrt{(4 - (-1))^2 + (1 - 4)^2}$
$\quad = \sqrt{5^2 + 3^2}$
$\quad = \sqrt{34}$

$XZ = \sqrt{(3 - 0)^2 + (5 - 0)^2}$
$\quad = \sqrt{3^2 + 5^2}$
$\quad = \sqrt{34}$

Since $(\sqrt{17})^2 + (\sqrt{17})^2 = (\sqrt{34})^2$, all four angles of the rhombus are right angles. Quadrilateral *WXYZ* is a square.

Think and Communicate

Students will most likely answer question 8 by using the converse of the Pythagorean theorem. You may wish to refer students back to this question when they study the relationship of the slopes of perpendicular lines in Section 4.3. In answering question 9, students should use the Midpoint Formula to verify that the intersection of the line of symmetry with $\overline{AB}$ is the midpoint.

You can use the Distance Formula to classify triangles and other polygons based on the lengths of their sides.

EXAMPLE 3

The vertices of △*ABC* are *A*(–3, 1), *B*(2, 6), and *C*(3, 0). Give the most specific name for △*ABC*.

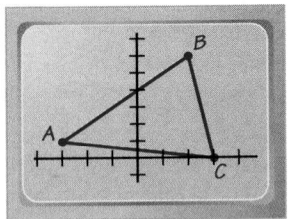

SOLUTION

Use the Distance Formula to find the lengths of the sides of △*ABC*.

$AB = \sqrt{(2 - (-3))^2 + (6 - 1)^2}$
$\quad = \sqrt{5^2 + 5^2}$
$\quad = \sqrt{50}$
$\quad = 5\sqrt{2}$

To find *AB*, use *A*(**–3**, **1**) as (x_1, y_1) and *B*(**2**, **6**) as (x_2, y_2).

$BC = \sqrt{(3 - 2)^2 + (0 - 6)^2}$
$\quad = \sqrt{1^2 + (-6)^2}$
$\quad = \sqrt{37}$

To find *BC*, use *B*(**2**, **6**) as (x_1, y_1) and *C*(**3**, **0**) as (x_2, y_2).

$AC = \sqrt{(3 - (-3))^2 + (0 - 1)^2}$
$\quad = \sqrt{6^2 + (-1)^2}$
$\quad = \sqrt{37}$

To find *AC*, use *A*(**–3**, **1**) as (x_1, y_1) and *C*(**3**, **0**) as (x_2, y_2).

Since $BC = AC$, △*ABC* is isosceles. Since $AC \neq AB$, △*ABC* is not equilateral.

THINK AND COMMUNICATE

7. Find the coordinates of the midpoint of $\overline{OC}$, the other diagonal of quadrilateral *OBCD* in Example 2.

8. How can you check whether the isosceles triangle in Example 3 is a right triangle?

9. Draw the isosceles triangle in Example 3 on graph paper and draw a line of symmetry for it. Where does the line of symmetry you drew intersect $\overline{AB}$?

Think and Communicate

7. (3, 2)

8. Use the converse of the Pythagorean theorem; $\overline{AB}$ is the longest side. Since $(AB)^2 = 50$ and $(BC)^2 + (AC)^2 = 74$, △*ABC* is not a right triangle.

9.

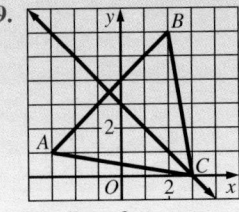

The line of symmetry intersects $\overline{AB}$ at its midpoint, $\left(-\frac{1}{2}, \frac{7}{2}\right)$.

☑ CHECKING KEY CONCEPTS

Use the diagram. Find each length.

1. *AB*   2. *CD*

3. *EF*   4. *GH*

Find the coordinates of the midpoint of each segment.

5. $\overline{AB}$ 6. $\overline{CD}$

7. $\overline{EF}$ 8. $\overline{GH}$

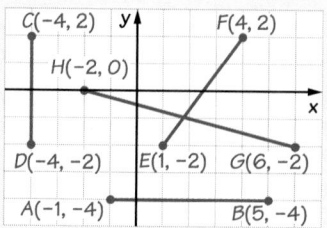

C(–4, 2) F(4, 2)
H(–2, 0)
D(–4, –2) E(1, –2) G(6, –2)
A(–1, –4) B(5, –4)

4.1 Exercises and Applications

Extra Practice exercises on page 665

For Exercises 1–3, refer to the swimmers in Example 1.

1. **Cooperative Learning** Work in a group of three people. Assume the starter is at *P*(15.00, 0).

   a. Copy and complete the swimmers' time advantage table below. Each person should complete two rows. Two rows are already completed.

   b. Discuss your results. Which swimmers have greater time advantages? Which have smaller time advantages?

Lane	Swimmer position	Timer position	PS	PT	PS/346	PT/346	Advantage
8	S(0, 19.00)	T(50.00, 19.00)	24.21	39.82	0.070	0.115	0.045
7	S(0, 16.60)	?	?	?	?	?	?
6	S(0, 14.20)	?	?	?	?	?	?
5	S(0, 11.80)	?	?	?	?	?	?
4	S(0, 9.40)	?	?	?	?	?	?
3	S(0, 7.00)	T(50.00, 7.00)	16.55	35.69	0.048	0.103	0.055
2	S(0, 4.60)	?	?	?	?	?	?
1	S(0, 2.20)	?	?	?	?	?	?

2. **Spreadsheets** Use a spreadsheet to create a table like the one above. Include cells where you can specify the coordinates of *P*, the position of the starter. How does changing the position of the starter along the edge of the pool affect the results in your table?

3. **Open-ended Problem** Describe where the starter *P* could be positioned so that no swimmer has a time advantage.

Closure Question

How is the Distance Formula related to the Pythagorean theorem?
On a coordinate plane, the lengths of the legs of a right triangle whose hypotenuse has endpoints at $A(x_1, y_1)$ and $B(x_2, y_2)$ can be expressed as $x_2 - x_1$ and $y_2 - y_1$. Substituting these two expressions into the Pythagorean theorem and solving for *AB* gives the Distance Formula.

Apply⇔Assess

Suggested Assignment

❖ **Core Course**
Exs. 4–17, 21, 22, 35–43

❖ **Extended Course**
Exs. 1–43

❖ **Block Schedule**
Day 22 Exs. 4–17, 21, 22, 35–43

Exercise Notes

Interdisciplinary Problems
Exs. 1–3 These exercises illustrate how geometry and algebra can be used with spreadsheet technology to solve a complex problem in swimming. This situation can be extended to examine other athletic events that involve a starting sound and a timer.

Cooperative Learning
Exs. 1–3 Provide sufficient class time for the groups to complete these exercises. You may want to have the group members assign the rows for Ex. 1(a) and then have students complete their rows for homework before the rest of the work is completed in class.

Using Technology
Ex. 2 This exercise requires the use of spreadsheet software. Each group should work together to design the spreadsheet before working at a computer. Students will need an absolute cell reference when using the cells that hold the coordinates of *P* in another formula. This will enable them to enter the formulas for lane 8, and then use *fill down* to complete the rest of the spreadsheet.

Checking Key Concepts
1. 6
2. 4
3. 5
4. $2\sqrt{17}$
5. (2, –4)
6. (–4, 0)
7. (2.5, 0)
8. (2, –1)

Exercises and Applications
1. See answers in back of book.
2. Answers may vary. Check students' work. If the starter is at (25, 0) or at (25, 20), no swimmer has a time advantage over another. If the starter is closer to the swimmers than to the timers, then the swimmer closest to the starter has the greatest advantage. If the starter is closer to the timers than to the swimmers, then the swimmer that is farthest away has the greatest advantage.
3. either at (25, 0) or at (25, 20)

Exercise Notes

Student Progress
Exs. 4–15 Students should be able to complete these straightforward exercises with little or no difficulty. Students who are having difficulty should refer to Example 2 (for Exs. 10–15) and Example 3 (for Exs. 4–9).

Common Error
Exs. 4–15 Students tend to make errors when using the Distance Formula and the Midpoint Formula. Some errors are computational in nature and others involve confusing the operations in the formulas. To help avoid these errors, suggest to students that they write the formula as the first step. This way they will have the correct operations to look at and also will be better able to see where they are subtracting negative numbers.

Using Manipulatives
Exs. 16, 17, 26–29 These exercises give students an opportunity to explore the concepts of distance and midpoint using physical models. In Exs. 16 and 17, students use paper folding to find the midpoint of a segment. In Exs. 26–29, various arrangements of the tangram puzzle are examined.

Interview Note: Geometric Thinking
Exs. 18–20 To help students better understand the formula used in these exercises, point out that in cases where $x_1 < x_2$ and $y_1 < y_2$, the differences between coordinates are the horizontal and vertical distances between the two points. Thus, the x-coordinate of the midpoint is the x-coordinate of the first point plus half the horizontal distance between the two points. The y-coordinate of the midpoint is the y-coordinate of the first point plus half the vertical distance between the two points. Have students investigate cases where one or both of the coordinates of the second point are smaller than the coordinates of the first point.

Find the lengths of the sides of each polygon whose vertices are given. Give the most specific name for each polygon.

4. $A(2, -2)$, $B(-1, 1)$, $C(11, 7)$

5. $D(2, -1)$, $E(4, 2)$, $F(7, 0)$

6. $G(-5, 0)$, $H(1, -3)$, $K(4, 3)$

7. $J(-3, 0)$, $K(0, 3)$, $L(2, 1)$, $M(-1, -2)$

8. $P(2, -4)$, $Q(4, 0)$, $R(2, 2)$, $S(0, 0)$

9. $W(-3, 3)$, $X(2, 2)$, $Y(3, -3)$, $Z(-2, -2)$

Find the coordinates of the midpoint of the segment with each given pair of endpoints.

10. $A(0, 0)$, $B(8, 4)$

11. $C(-5, 7)$, $D(-5, 9)$

12. $E(2, -7)$, $F(-2, 7)$

13. $G(2, 5)$, $H(5, 2)$

14. $N(1, 1)$, $J(-3, -3)$

15. $K(7, 4)$, $L(-3, 1)$

Investigation For Exercises 16 and 17, use paper folding to locate midpoints.

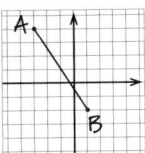 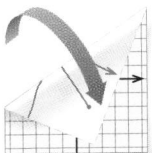 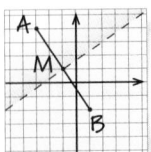

Step 1 Draw a pair of axes on graph paper. Graph any two points, A and B.

Step 2 Carefully fold A onto B. Make a crease.

Step 3 Label the intersection of the crease and $\overline{AB}$ as midpoint M.

16. Open-ended Problem Choose two points, A and B, that have integer coordinates. Fold to locate the midpoint. Label the midpoint M and give its coordinates.

17. Writing Sally says that if a point P lies on the crease shown above, then $PA = PB$. Do you agree or disagree? Use the crease you made in Exercise 16 and the Distance Formula to support your answer.

INTERVIEW Marc Hannah

Look back at the article on pages 162–164.

Computer graphics programmers sometimes use this variation of the Midpoint Formula:

$$\left(x_1 + \frac{1}{2}(x_2 - x_1), y_1 + \frac{1}{2}(y_2 - y_1)\right)$$

18. Use the formula to find the midpoint of the segment with endpoints (2, 6) and (6, 14). Does your result agree with what you expect?

19. ALGEBRA Show that the formula above is the same as the Midpoint Formula shown on page 167.

20. Check that the formulas below give the coordinates of the points that divide the segment with endpoints (0, 9) and (3, 15) into three congruent parts.

$$\left(x_1 + \frac{1}{3}(x_2 - x_1), y_1 + \frac{1}{3}(y_2 - y_1)\right) \text{ and } \left(x_1 + \frac{2}{3}(x_2 - x_1), y_1 + \frac{2}{3}(y_2 - y_1)\right)$$

---

4. $AB = 3\sqrt{2}$; $BC = 6\sqrt{5}$; $AC = 9\sqrt{2}$; scalene right triangle

5. $DE = EF = \sqrt{13}$; $DF = \sqrt{26}$; isosceles right triangle

6. $GH = HI = 3\sqrt{5}$; $GI = 3\sqrt{10}$; isosceles right triangle

7. $JK = LM = 3\sqrt{2}$; $JM = KL = 2\sqrt{2}$; rectangle

8. $PQ = PS = 2\sqrt{5}$; $QR = SR = 2\sqrt{2}$; quadrilateral (Some students may recognize that *PQRS* is a kite.)

9. $WX = XY = YZ = WZ = \sqrt{26}$; rhombus

10. (4, 2)

11. (−5, 8)

12. (0, 0)

13. (3.5, 3.5)

14. (−1, −1)

15. (2, 2.5)

16. Answers may vary. Check students' work.

17. agree; Answers may vary. Check students' work.

18. (4, 10); Yes; $\left(\frac{6+2}{2}, \frac{14+6}{2}\right) =$ (4, 10).

19. $x_1 + \frac{1}{2}(x_2 - x_1) = x_1 + \frac{x_2}{2} - \frac{x_1}{2} = \frac{x_1}{2} + \frac{x_2}{2}$; $y_1 + \frac{1}{2}(y_2 - y_1) = y_1 + \frac{y_2}{2} - \frac{y_1}{2} = \frac{y_1}{2} + \frac{y_2}{2}$

20. The formula gives the points (1, 11) and (2, 13). The distance from (0, 9) to (3, 15) is $3\sqrt{5}$. The distance from (0, 9) to (1, 11) is $\sqrt{5}$ and from (0, 9) to (2, 13) is $2\sqrt{5}$.

The Chinese puzzle game called *ch'i ch'ae pan* has seven pieces with straight edges. You may know this puzzle by the name *tangram*. The pieces can be arranged into a square and many other shapes.

Use the diagram of the tangram pieces in the form of a square.

21. Find the coordinates of all 10 points at the vertices of the tangram pieces.

22. The small square has edge length 1. Find the edge lengths of the other six pieces.

23. Use cardboard or heavy paper to make a set of tangram pieces. Use your answers to Exercise 22 to mark the lengths of the sides on your pieces.

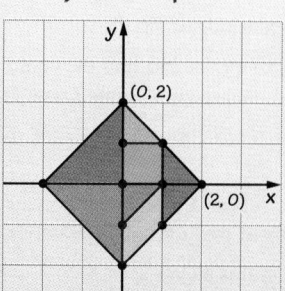

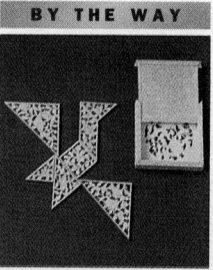
Use the diagram of the tangram pieces in the form of a running figure.

24. Each vertex marked in red on the diagram is a midpoint of a tangram edge. Find the coordinates of the unlabeled midpoints.

25. **Challenge** Find coordinates for all the vertices of the running figure.

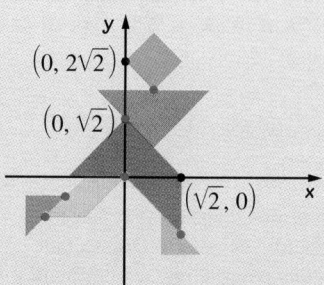

Use your tangram pieces to build each figure. Then sketch it on a coordinate plane.

26.

27.

28.

29. **Cooperative Learning** Work with another person. Create your own tangram figure and draw its outline on graph paper. Label the coordinates of some of the vertices. Exchange papers with the other person and try to build the figure you are given.

30. **Open-ended Problem** Sketch three different segments that have $(0, 0)$ as a midpoint. Write the coordinates of the endpoints of each segment. What do you notice about the coordinates?

31. **Writing** Suppose points A and B have integer coordinates. Under what circumstances will the midpoint of $\overline{AB}$ have integer coordinates?

4.1 Finding Distances and Midpoints **171**

Apply⇔Assess

Exercise Notes

Application
Exs. 21–29 For these exercises, students use the Distance Formula and the Midpoint Formula to analyze several figures formed by a tangram puzzle.

Multicultural Note
Exs. 21–29 The Chinese called the tangram the "board of wisdom" or the "seven-board of cunning," since manipulating the pieces required some skill. Although people in China were probably familiar with the game much earlier, the first known book of tangrams was not published until 1813. Tangram books spread to the West shortly after that; by 1818, they had appeared in Europe and the United States. The game of tangram can be played in many ways, the only rules being that all seven pieces must be used and they should form a two-dimensional figure.

Research
Exs. 21–29 Some students may want to research the history of the tangram puzzle and its figures. Students can use their research to create a bulletin board display.

Communication: Discussion
Exs. 26–28 After completing these exercises, ask several students to share their drawings with the class. There should be some variation in the way students placed each figure in the coordinate plane and how they choose to scale it. This should provide a basis for a discussion of the advantages and disadvantages of various choices.

Integrating Number Theory
Exs. 30, 31 In both of these exercises, students consider some properties of numbers. In Ex. 30, they discover that a segment whose midpoint is the origin has endpoints whose corresponding coordinates are opposites. In Ex. 31, they see that for a midpoint to have integer coordinates, the sums of corresponding coordinates must be even. This occurs when the corresponding coordinates are either both even or both odd.

Challenge
Ex. 30 Ask students to write an algebraic proof that if the midpoint of a segment is the origin, the corresponding coordinates of the endpoints of the segment are opposites.

---

21. $(0, 2)$, $(0, 1)$, $(1, 1)$, $(-2, 0)$, $(0, 0)$, $(1, 0)$, $(2, 0)$, $(0, -1)$, $(1, -1)$, $(0, -2)$

22. large triangles: 2, 2, and $2\sqrt{2}$; medium triangle: $\sqrt{2}$, $\sqrt{2}$, and 2; small triangles: 1, 1, and $\sqrt{2}$; parallelogram: 1, $\sqrt{2}$, 1, $\sqrt{2}$

23. Check students' work.

24–27. See answers in back of book.

28.

29. Answers may vary. Check students' work.

30. Answers may vary. Examples are given. $(9, 3)$ and $(-9, -3)$, $(2.3, -5.7)$ and $(-2.3, 5.7)$, and $\left(-\sqrt{2}, -\frac{1}{2}\right)$ and $\left(\sqrt{2}, \frac{1}{2}\right)$; In each pair, the *x*-coordinates and *y*-coordinates are opposites.

31. if the *x*-coordinates are either both even or both odd and the *y*-coordinates are either both even or both odd

Apply⇔Assess

Exercise Notes

Using Technology
Exs. 32–34 By using geometry software, students can make conjectures based on numerous examples. You may wish to extend these exercises by having students investigate what type of quadrilateral *PQRS* is when the diagonals not only have the same midpoint, but (1) have the same length, (2) are perpendicular, and (3) have the same length and are perpendicular.

Assessment Note
Ex. 35 This open-ended problem gives students an opportunity to not only practice using the Distance Formula but also to realize that the geometric properties of a figure are independent of the placement of the coordinate system. This realization will be especially useful in Section 4.5, when students write coordinate proofs.

Practice 24 for Section 4.1

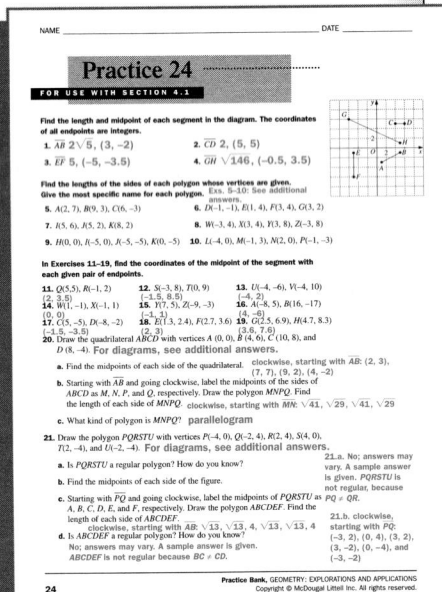

Technology For Exercises 32–34, use geometry software.

Step 1 Draw a quadrilateral that has no parallel or congruent sides and no right angles. Label the vertices *P, Q, R,* and *S*.

Step 2 Find the length of each side of *PQRS*.

Step 3 Draw the diagonals of *PQRS* and find the coordinates of their midpoints. Label the midpoints *L* and *M*.

Step 4 Move the vertices of *PQRS* until *L* and *M* are the same point.

32. Repeat the steps above several times. What do you notice about the lengths of the sides of *PQRS* after Step 4?

33. What type of quadrilateral is *PQRS* after Step 4? Write a conjecture that describes what you have discovered.

34. Investigation Draw another quadrilateral. Locate and connect the midpoints of its sides. What type of quadrilateral is formed?

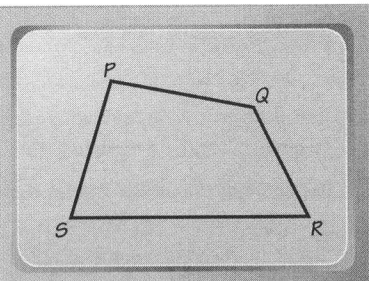

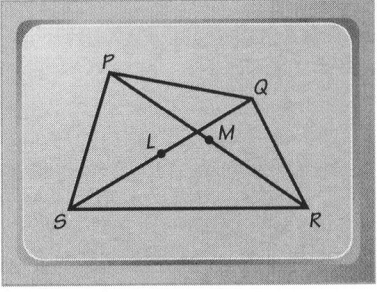

ONGOING ASSESSMENT

35. Open-ended Problem Use the triangle on the geoboard.
   a. Choose one peg of the geoboard to represent the point with coordinates (0, 0). Draw a pair of axes through this point. Find the coordinates of each vertex of the triangle.
   b. Find the lengths of the sides of the triangle.
   c. Tell whether the triangle is *equilateral, isosceles,* or *scalene.*
   d. Suppose you choose a different peg to represent the point with coordinates (0, 0). Which of the results above would change? Which would not?

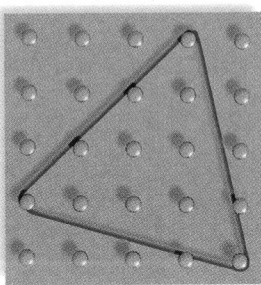

SPIRAL REVIEW

Tell whether the triangle with sides of the given lengths is *right*, *obtuse*, or *acute*. *(Section 3.7)*

36. 3, 7, $\sqrt{58}$ **37.** 2, $\sqrt{7}$, 2$\sqrt{3}$ **38.** 0.5, 1, 1

Open-ended Problems Sketch an example of each figure. *(Section 2.3)*

39. a pentagon with exactly one line of symmetry   **40.** an equilateral parallelogram

Find the slope of each line. *(Toolbox, p. 697)*

41. $y = -4x + 7$ **42.** $2y = 3x + 8$ **43.** $x + y = 5$

---

32. *PQ* = *RS* and *QR* = *PS*

33. parallelogram; If the diagonals of a quadrilateral bisect each other (or have the same midpoint), then the quadrilateral is a parallelogram.

34. parallelogram

35. a. Answers may vary. An example is given. If the vertex at the far left is chosen as the origin, then the vertices are (0, 0), (3, 3), and (4, −1).

  b. 3$\sqrt{2}$, $\sqrt{17}$, and $\sqrt{17}$

  c. isosceles

  d. The coordinates of the vertices would change, the lengths of the sides and the classification of the triangle would not.

36. right

37. obtuse

38. acute

39, 40. Answers may vary. Examples are given.

39.

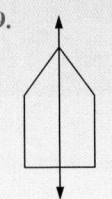

40.

41. −4

42. $\dfrac{3}{2}$

43. −1

4.2 Equations of Lines

Learn how to...
- find the slope of a segment or a line
- write equations of lines in slope-intercept form

So you can...
- graph and compare equations of lines
- investigate geometric relationships using lines

Maybe you would like to go to the seashore but live too far inland. Or maybe you would like to try downhill skiing but live in a region without mountains or snow. At indoor beaches and ski resorts that overcome the limitations of geography and climate, people swim and ski all year.

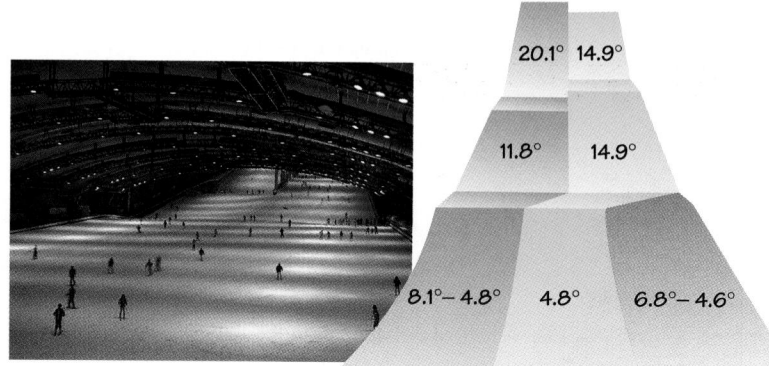

The world's largest indoor ski slope is located in Funabashi, Japan. It is called the *LaLaport Skidome* and has three courses—a red or *expert* course, a yellow or *intermediate* course, and a blue or *beginners'* course. Each course consists of alternating flat and sloped regions. A diagram of the red course is shown below. All distances are rounded to the nearest meter.

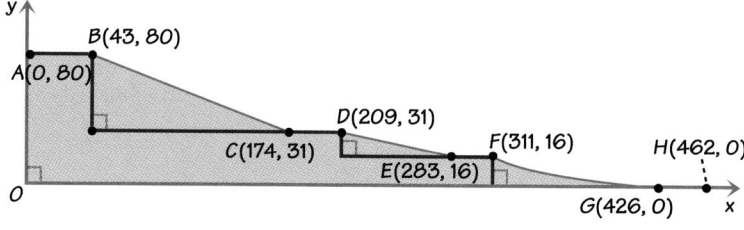

THINK AND COMMUNICATE

1. How many meters does the red course drop in its first sloped part, between point B and point C?

2. What is the horizontal distance from B to C?

You can express how steep the first sloped part of the red course is by using the *slope formula*.

4.2 Equations of Lines **173**

Plan⟺Support

Objectives
- Find the slope of a segment or a line.
- Write equations of lines in slope-intercept form.
- Graph and compare equations of lines.
- Investigate geometric relationships using lines.

Recommended Pacing
❖ **Core and Extended Courses**
Section 4.2: 1 day
❖ **Block Schedule**
Section 4.2: $\frac{1}{2}$ block
(with Section 4.1)

Resource Materials
Lesson Support
Lesson Plan 4.2
Warm-Up Transparency 4.2
Practice Bank: Practice 25
Study Guide: Section 4.2
Explorations Lab Manual:
    Additional Exploration 7
    Diagram Master 2
Challenge Problems: Set 25
Technology
Technology Book:
    TI-92 Activity 4
Scientific Calculator
Graphing Calculator
McDougal Littell Mathpack
    Function Investigator with Matrix
    Analyzer Activity Book:
        Activities 3–5
Internet:
    http://www.hmco.com

ANSWERS Section 4.2

Think and Communicate
1. 49 m
2. 131 m

Warm-Up Exercises

Solve for *b*.
1. $3 = 6 + b$ –3
2. $\frac{1}{2} = -4 + b$ $4\frac{1}{2}$

Solve for *y*.
3. $3 = \frac{y - 2}{5}$ 17
4. $y - 3 = 2x + 4$ $y = 2x + 7$
5. $m = \frac{y - b}{x - 0}$ $y = mx + b$

Think and Communicate

Questions 1 and 2 focus students' attention on the horizontal and vertical changes in a portion of the ski slope. These changes are used to introduce the concept of slope.

Section Note

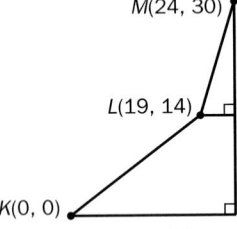 **Communication: Discussion** Discuss the material in the Slope Formula box. Point out that the Slope Formula can be used to find the slope of any nonvertical line. Ask students to use the Slope Formula to show why the slope of a horizontal line is 0 and why the slope of a vertical line is undefined.

Additional Example 1

A diagram of an indoor rock climbing course is shown below. All distances are measured in feet.

Find the slope of the segment of the course from L to M.
The coordinates of L are (19, 14). The coordinates of M are (24, 30). Use the Slope Formula. Substitute 19 for x_1, 14 for y_1, 24 for x_2, and 30 for y_2.

$$m = \frac{y_2 - y_1}{x_2 - x_1}$$
$$= \frac{30 - 14}{24 - 19}$$
$$= \frac{16}{5}$$
$$= 3.2$$

The slope of the segment of the course from L to M is 3.2.

The Slope Formula

The **slope** m of a line containing the points (x_1, y_1) and (x_2, y_2) is:

$$m = \frac{y_2 - y_1}{x_2 - x_1}$$

The slope of a **horizontal** line is **0**.

The slope of a **vertical** line is **undefined**.

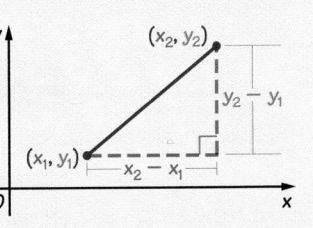

EXAMPLE 1   Application: Recreation

Find the slope of the first sloped part of the red course of the LaLaport Skidome.

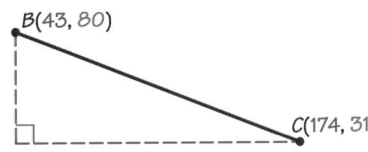

SOLUTION

The coordinates of point B are (**43, 80**). The coordinates of point C are (**174, 31**).

Use the slope formula.

$$m = \frac{y_2 - y_1}{x_2 - x_1}$$

$$= \frac{31 - 80}{174 - 43}$$

Substitute **43** for x_1, **80** for y_1, **174** for x_2, and **31** for y_2.

$$= -\frac{49}{131}$$

$$= -0.374$$

The slope of the first sloped part of the red course is about -0.37.

BY THE WAY

In the United Kingdom, where snow is rare, skiers can ski on slopes made of a synthetic plastic material.

THINK AND COMMUNICATE

3. What does the negative sign mean in the value of the slope in Example 1?

4. In Example 1, Meg used (174, 31) for (x_1, y_1) and (43, 80) for (x_2, y_2). Will this affect the value of the slope? Explain why or why not.

5. Use the diagram on page 173. Calculate the slope of the second sloped part of the red course, between points D and E.

Think and Communicate

Questions 3–5 should help students to better understand negative slopes. Students should realize that the greater the absolute value of the slope, the steeper the segment.

Think and Communicate

3. that the course slopes downward

4. about −0.20

5. No; the slope can be found by using $\frac{8-31}{43-174}$, which is equal to $-\frac{49}{131}$.

The y-coordinate of the point where a graph crosses the y-axis is called a **y-intercept**. If the y-intercept of a line is b, then $(0, b)$ is on the line. Substitute $(0, b)$ for (x_1, y_1) in the definition of slope and solve for y:

$$m = \frac{y - b}{x - 0} \;\rightarrow\; y = m(x - 0) + b$$

This leads to the *slope-intercept form* of an equation of a line.

Slope-Intercept Form of an Equation of a Line

The equation of a line with slope m and y-intercept b can be expressed in **slope-intercept form**:

$$y = mx + b$$
slope ↗   ↘ y-intercept

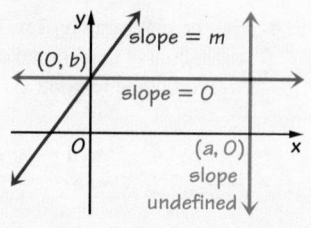

An equation of a **horizontal** line that contains $(0, b)$ is $y = b$.

An equation of a **vertical** line that contains $(a, 0)$ is $x = a$.

EXAMPLE 2   Application: Recreation

Write an equation for the line that represents the first sloped part of the red course in the LaLaport Skidome.

SOLUTION

Point C with coordinates $(174, 31)$ is on the line. From Example 1, the slope of this part of the course is about -0.37. Use the slope-intercept form. Find the value of b, the y-intercept:

$$y = mx + b$$
$$31 = (-0.37)(174) + b$$
$$b \approx 95$$

Substitute **174** for x, **31** for y, and $-\mathbf{0.37}$ for m.

An equation of the line that represents the first sloped part of the red course is $y = -0.37x + 95$.

THINK AND COMMUNICATE

6. Use the equation from Example 2 to estimate the height of a point 100 m from the beginning of the red course.

7. Suppose that the first sloped part of the red course were extended to the left until it met the y-axis. How high would the top of the course be? How does this compare with its actual height?

Think and Communicate

6. about 58 m

7. about 95 m; about 15 m higher than the actual height

Teach⇔Interact

Section Note

Using Technology
Students can use Function Investigator Activities 3, 4, and 5 in the *Function Investigator with Matrix Analyzer Activity Book* to explore the meaning of m and b in the slope-intercept form of an equation of a line, $y = mx + b$.

About Example 2

Alternate Approach
You may want to introduce the method for finding the equation of a line when given two points on it as a four-step algorithm.
Step 1: Use the Slope Formula to find the slope of the line.
Step 2: Substitute the slope and the coordinates of one of the points into the slope-intercept form.
Step 3: Solve the equation for b.
Step 4: Substitute the slope and y-intercept into the slope-intercept form to get the equation of the line.

Additional Example 2

Refer to Additional Example 1. Write an equation for the line that contains the segment of the rock climbing course from L to M.
Point L with coordinates $(19, 14)$ is on the line. From Additional Example 1, the slope of this part of the course is 3.2. Use the slope-intercept form. Substitute 3.2 for m, 19 for x, and 14 for y. Find the value of b, the y-intercept.
$y = mx + b$
$14 = (3.2)(19) + b$
$b \approx -46.8$
An equation of the line that contains the segment of the rock climbing course from L to M is $y = 3.2x - 46.8$.

Think and Communicate

After completing question 6, ask students why it would not be appropriate to use the equation from Example 2 to estimate the height of a point 300 m from the beginning of the red course. (The equation only describes the course for x-values from 43 to 174.) Question 7 leads students to locate the y-intercept they found in Example 2 on the graph.

175

Checking Key Concepts

Visual Thinking
Ask students to sketch the lines for each of questions 1–4. Encourage them to compare their sketches with those of other students and to discuss any differences. This activity involves the visual skills of *interpretation* and *communication*.

Closure Question

Describe how you write an equation for a line if you know the coordinates of two points on the line. Use the Slope Formula to find the slope of the line. Substitute the slope and the coordinates of one of the points into the slope-intercept form, and solve for *b*. Write the equation using the values for *m* and *b*.

Suggested Assignment

❖ **Core Course**
Exs. 1–7, 9–16, 23–33, 36–48

❖ **Extended Course**
Exs. 1–7 odd, 8, 9–15 odd, 17–22, 23–33 odd, 34–48

❖ **Block Schedule**
Day 22 Exs. 1–7, 9–16, 23–33, 36–48

Exercise Notes

Common Error
Exs. 1–6 Computational errors are common when finding slope. Students sometimes confuse the numerator and denominator of the formula, or subtract the *x*- and *y*-coordinates in different orders. Suggest that students graph the points to check that the slope they find makes sense. In addition to checking that the sign of the slope is correct, they should also check to see that a steep line has a slope with a large absolute value.

☑ **CHECKING KEY CONCEPTS**

1. Find the slope of the line that contains the points with coordinates (5, 1) and (7, 1).

2. Write an equation of the line that contains the point with coordinates (3, 4) and has slope 2.

3. Write an equation of the line that contains the point with coordinates (0, 2) and has slope −1.

4. Use the diagram on page 173. Suppose the LaLaport Skidome had a single course that started at a height of 80 m and continued at a constant slope to point *G*. What would be the slope of the course?

4.2 | Exercises and Applications

Extra Practice exercises on page 666

Find the slope of the line that contains each pair of points given.

1. (0, 1) and (2, 9)
2. (2, 1) and (4, −3)
3. (1, 3) and (5, −2)
4. (−4, 7) and (1, −1)
5. (3, 2) and (−3, −1)
6. (2, −5) and (−1, −5)

7. **ACCESSIBILITY** Ramps are cut into curbs to allow people in wheelchairs to enter and exit sidewalks. In the United States, ramps have sloped sides whenever the sidewalk at the top of the ramp is narrower than 4 ft. The slopes may not be greater than $\frac{1}{12}$.

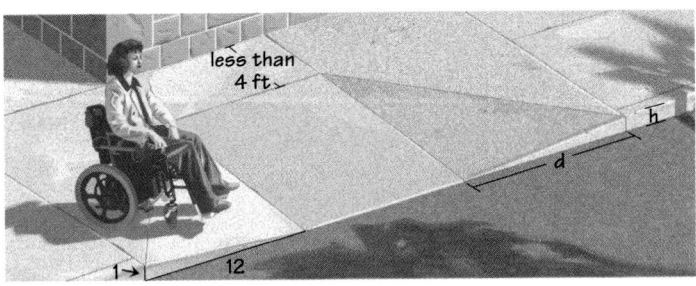

a. Suppose *h* is 3 in. Find the minimum length *d*.
b. Suppose *d* is 42 in. Find the maximum height *h*.
c. **Research** Find a wheelchair ramp at your school or another public building. If possible, find its dimensions and calculate its slope.

8. **Writing** Is $m = \frac{y_1 - y_2}{x_1 - x_2}$ another formula for the slope of a line? Explain why or why not.

Checking Key Concepts

1. 0
2. $y = 2x - 2$
3. $y = -x + 2$
4. If the course extended from $A(0, 80)$ to $G(426, 0)$, its slope would be about −0.19.

Exercises and Applications

1. 4
2. −2
3. $-\frac{5}{4}$
4. $-\frac{8}{5}$
5. $\frac{1}{2}$
6. 0

7. a. 36 in. or 3 ft
   b. 3.5 in.
   c. Check students' work.
8. Yes; $\frac{y_1 - y_2}{x_1 - x_2} \cdot \frac{-1}{-1} = \frac{y_2 - y_1}{x_2 - x_1}$.

Write an equation for each line described.

9. contains $(-3, -1)$; has slope 2

10. contains $(1, 1)$; has slope -2

11. contains $(4, -2)$; has slope $-\frac{3}{4}$

12. contains $(-3, 2)$; has slope 3.5

13. the vertical line that contains $(13, 5)$

14. the horizontal line that contains $(7, -9)$

15. contains $(-2, 1)$ and $(2, 2)$

16. contains $(-1, 3)$ and $(3, -1)$

Connection ▸ ENGINEERING

To supply cities with water when the source is a long distance away, artificial channels called *aqueducts* may be built. More than 2000 years after it was built, a Roman aqueduct still stands in southern France. It brought water from a source in Uzès to the city of Nîmes. The water traveled downhill, dropping only 17 m across a distance of 50 km. The steepest part of the aqueduct is between Uzès and the Pont du Gard, a bridge across the Gardon River.

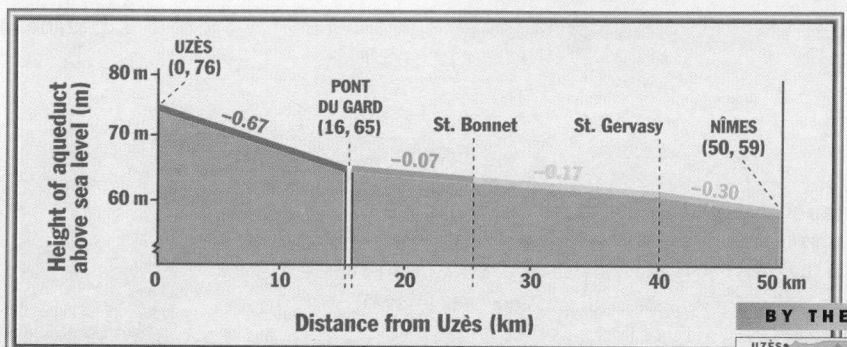

17. To approximate the slope of the steepest part of the aqueduct, what fraction could you use? (*Note:* Use "m/km" as the unit of measure for the slope.)

18. Suppose the aqueduct started at Uzès and ended at Nîmes, but had a constant slope. Write a linear equation to describe its course.

19. a. The Pont du Gard is located 16 km from Uzès. Using the equation you wrote in Exercise 18, find the height of the aqueduct at the bridge. How much lower than this is the actual bridge?

    b. **Writing** Why do you think the Romans made the first part of the aqueduct steeper than the rest?

BY THE WAY

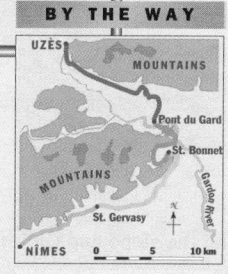

The aqueduct takes an indirect route because a mountain blocks the direct route from Uzès to Nîmes.

4.2 Equations of Lines **177**

---

9. $y = 2x + 5$

10. $y = -2x + 3$

11. $y = -0.75x + 1$

12. $y = 3.5x + 11$

13. $x = 13$

14. $y = -9$

15. $y = 0.25x + 1.5$

16. $y = -x + 2$

17. about $-\frac{19}{50}$ m/km

18. $y = -0.24x + 71$

19. a. about 67 m; about 2 m

    b. Answers may vary. An example is given. The large drop at the beginning allowed a large portion of the aqueduct to be built at a lower height, which substantially decreased the amount of materials needed to build the aqueduct. Also, the large initial drop increased the speed of the water so it could travel a greater distance.

Apply⇔Assess

Exercise Notes

Family Involvement
Ex. 7 Students may wish to involve family members in doing this exercise. After completing parts (a) and (b), they can discuss where to locate a wheelchair ramp to complete part (c).

Problem Solving
Ex. 8 A problem-solving strategy that students can use to answer this question is to try some examples and then make a conjecture. By studying different examples, students can gain insight into the problem. The given formula can be proved equivalent to the Slope Formula by multiplying the right side by $\frac{-1}{-1}$.

Student Progress
Exs. 9–16 Students should be able to write the equations for these lines quickly and accurately. If some students need additional practice, you can pair them together and have them create and solve problems similar to these. Students should check that their equations are correct by examining the slope and by substituting the point into the equation to see if it is a solution.

Interdisciplinary Problems
Exs. 17–19 In these exercises, students use the concepts of slope and linear equations to study the design of a Roman aqueduct. Some students may be interested in finding out more about Roman aqueducts. If so, ask them to share their information with the class.

Using Manipulatives
Exs. 17–19 Students may want to investigate and compare the speed of the water in the actual aqueduct with the aqueduct modeled in Ex. 18. To do this, they can build scale models of both aqueducts using cardboard. Students can then use toy race cars to represent the water and let them run down the aqueducts to see how the design affects the speed of the car.

Apply⇔Assess

Exercise Notes

Application
Exs. 20–22 In completing
these exercises, students see some
of the ways that lines and linear
equations can be used to plan and
describe a city. You may wish to
have students change the location
of the origin to the Museum of Art
and then find the new coordinates
of the three locations and an equation for the line connecting the
William Penn statue and the *Ghost*
mobile. How do these two equations compare?

Historical Connection
Exs. 20–22 Alexander Calder was
one of the most original sculptors
of the 20th century. He is known as
the first sculptor to use movement
as a major element of his work.
During a period from the 1930s to
the 1970s, Calder became
renowned for his wind-propelled
mobiles. Calder earned a degree in
mechanical engineering before
attending art school. Some of his
larger works adorn public settings
in Brussels, Montreal, New York
City, Chicago, and Mexico City.

Teaching Tip
Exs. 23–30 Remind students that
to find the slope and *y*-intercept of
a line, the equation must be in
slope-intercept form.

Using Technology
Exs. 23–30 Students can
check their answers to these problems using a graphing calculator.
They should enter the equation
on the Y= list and graph it in the
standard window. They can then
determine if the slope has the
correct sign, and use trace to find
the *y*-intercept. The equations in
Exs. 28–30 need to be solved for
y before being entered into the
calculator. Students need to adjust
the window in Ex. 25 to see the
y-intercept.

178

Connection ▸ CITY PLANNING

The sculptor Alexander Calder was born in Philadelphia, Pennsylvania.
Both his father and his father's father were also named Alexander and
were also sculptors. Artworks by three generations of Alexander Calders
are placed along the line defined by the Benjamin Franklin Parkway in
downtown Philadelphia.

Museum of Art
(−1760 m, 1232 m)

The Swann Memorial
fountain, in Logan Circle,
is by Alexander Stirling
Calder (1870–1945).

(−655 m, 458.5 m)

City Hall
(0 m, 0 m)

Looking northwest from
the fountain you can see
the Museum of Art. Inside
it is *Ghost*, a mobile by
Alexander "Sandy" Calder
(1898–1976).

**Use the map of Philadelphia. Look at the
line that passes through the William Penn
statue and the *Ghost* mobile.**

Looking southeast from the
fountain, you can see the
William Penn statue atop City
Hall. The statue is by
Alexander Milne Calder
(1846–1923).

20. Use the coordinates of the William Penn
statue and *Ghost* to find the slope of the
line that connects them.

21. What is an equation for the line?

22. Show that the Swann Memorial fountain lies on the same line.

Find the slope and *y*-intercept of each line.

23. $y = 3x + 5$

24. $y = -2x + 4$

25. $y = -5(x + 5)$

26. $y = 4$

27. $y = -\frac{2}{3}x$

28. $y + 6 = \frac{1}{2}(x - 4)$

29. $y - 1 = 4(x - 1)$

30. $3y = 6x + 9$

178 Chapter 4 *Coordinates in Geometry*

---

20. -0.7

21. $y = -0.7x$

22. $-0.7(-655) = 458.5$

23. $3; 5$

24. $-2; 4$

25. $-5; -25$

26. $0; 4$

27. $-\frac{2}{3}; 0$

28. $\frac{1}{2}; -8$

29. $4; -3$

30. $2; 3$

31. slope of $\overline{AB}$ = slope of $\overline{CD}$ = 0;
slope of $\overline{AD}$ = slope of $\overline{BC}$ = −2

32. slope of $\overline{CD}$ = slope of $\overline{EF}$ = $\frac{2}{3}$;
slope of $\overline{DE}$ = slope of $\overline{CF}$ = $-\frac{2}{3}$

33. slope of $\overline{FO}$ = $\frac{1}{4}$; slope of $\overline{OH}$ = 4;
slope of $\overline{GH}$ = $-\frac{1}{4}$; slope of $\overline{FG}$ = 2

Find the slopes of the sides of each quadrilateral.

31.

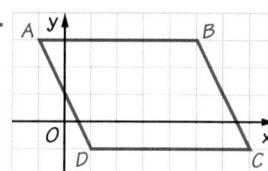

32.

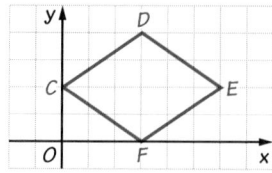

33.

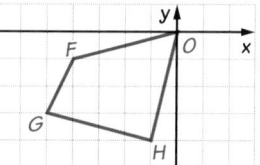

34. Open-ended Problem Draw several lines that have the same y-intercept but different slopes. What do you notice about them?

35. Challenge A line contains the points (2, 3) and (6, 1). Find the slope of the line. Then substitute the coordinates of each point in the definition of slope to create two equations for the line. Use algebra to show that the two equations are equivalent.

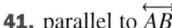

ONGOING ASSESSMENT

36. Writing A line contains the points $(a, 0)$ and $(0, b)$.

a. Find the slope of the line.

b. Write an equation for the line.

c. In parts (a) and (b) you used variables to find the slope and equation of a line. How is this different from using numeric coordinates?

SPIRAL REVIEW

Find the coordinates of the midpoint of the segment with each given pair of endpoints. *(Section 4.1)*

37. (0, 0) and (12, 6)

38. (−2, 5) and (2, −5)

39. (−7, 6) and (−7, −3)

40. (3, 5) and (6, −3)

Use the diagram at the right. Name all the lines in the diagram that fit each condition. *(Section 1.4)*

41. parallel to $\overleftrightarrow{AB}$

42. skew to $\overleftrightarrow{DH}$

43. parallel to $\overleftrightarrow{BC}$

44. skew to $\overleftrightarrow{EF}$

45. intersect $\overleftrightarrow{CG}$

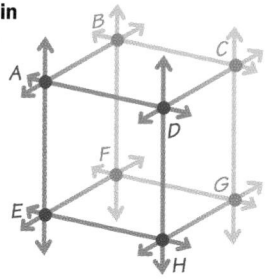

Tell whether each statement is *True* or *False*. If the statement is false, sketch a counterexample and justify your answer. *(Section 2.5)*

46. The opposite sides of a parallelogram are parallel.

47. All angles of a rhombus are right angles.

48. The diagonals of a square bisect each other.

Apply⇔Assess

Exercise Notes

Topic Spiraling: Preview

Exs. 31–33, 46–48 After completing Exs. 31–33, students may begin to see the relationship between the slopes of lines that are parallel or perpendicular. The slopes of parallel and perpendicular lines are examined in the next section. Some of the properties of polygons are explored using coordinates in Section 4.5. Exs. 46–48 involve students in beginning to think about these ideas.

Assessment Note

Ex. 36 This exercise provides a general formula for writing the equation of a line if the x- and y-intercepts are known. You may wish to have students describe the formula in words to assess their understanding of this.

Practice 25 for Section 4.2

NAME ____________ DATE ____________

Practice 25

FOR USE WITH SECTION 4.2

Find the slope of the line that contains each pair of given points.

1. (0, 3) and (3, 9) 2
2. (5, 1) and (−2, −3) $\frac{4}{7}$
3. (11, 2) and (6, −2) $\frac{4}{5}$
4. (−4, −7) and (−1, −1) 2
5. (5, 3) and (−5, −1) $\frac{2}{5}$
6. (2, 3) and (−1, 3) 0

Write an equation for each line.

7. contains (−4, −1); has slope $\frac{1}{2}$
8. contains (−1, −1); has slope −3
9. contains (5, −1); has slope $-\frac{3}{5}$
10. contains (−4, 5); has slope −2
11. the vertical line containing (−1, −2)
12. the horizontal line containing (−4, 1)

7. $y = \frac{1}{2}x + 1$
8. $y = -3x - 4$
9. $y = -\frac{3}{5}x + 2$
10. $y = -2x - 3$
11. $x = -1$
12. $y = 1$

Find the slope and y-intercept of each line.

13. $y = 2x - 5$ 2, −5
14. $y = -\frac{1}{2}x + 5$ $-\frac{1}{2}$, 5
15. $y = -\frac{3}{2}(x + 1)$ $-\frac{3}{2}$, $-\frac{3}{2}$
16. $y = -7$ 0, −7
17. $y = -\frac{4}{5}x$ $-\frac{4}{5}$, 0
18. $y - 7 = \frac{2}{3}(x - 6)$ $\frac{2}{3}$, 3

Find the slopes of the sides of each figure. The coordinates of all vertices are integers.

19. clockwise from $\overline{AB}$: $\frac{5}{2}$, −4, $-\frac{3}{4}$
20. clockwise from $\overline{EF}$: $-\frac{2}{9}$, $\frac{4}{3}$, $\frac{2}{7}$, 8
21. clockwise from $\overline{PQ}$: −1, 3, $\frac{1}{2}$, $-\frac{7}{2}$, 0

22. Find the slope of the line that contains the points with coordinates (9, 2) and (−3, 2). 0
23. Write an equation of the line that contains the point with coordinates (4, 3) and whose slope is $\frac{1}{2}$. $y = \frac{1}{2}x + 1$
24. Write an equation of the line that contains the point with coordinates (7, 0) and whose slope is −2. $y = -2x + 14$
25. Write an equation of the line that contains the point with coordinates (−7, −3) and whose slope is 0. $y = -3$

Practice Bank, GEOMETRY: EXPLORATIONS AND APPLICATIONS

25

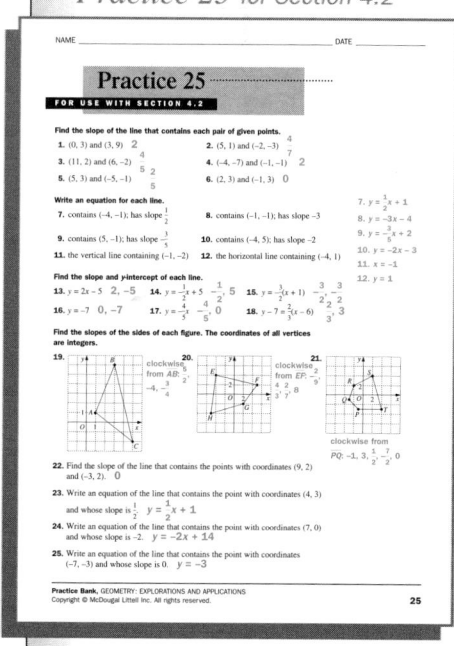

34. No two of the lines are parallel.

35. $-\frac{1}{2}$; $\frac{y-3}{x-2} = -\frac{1}{2}$ and $\frac{y-1}{x-6} = -\frac{1}{2}$; Each of these equations can be written in the slope-intercept form $y = -\frac{1}{2}x + 4$, so the equations are equivalent.

36. a. $-\frac{b}{a}$

b. $y = -\frac{b}{a}(x) + b$

c. When coordinates that include variables are used to find the slope and equation of a line, the only difference is that the slope and equation contain these variables rather than specific numbers.

37. (6, 3)

38. (0, 0)

39. (−7, 1.5)

40. (4.5, 1)

41. $\overleftrightarrow{CD}$, $\overleftrightarrow{EF}$, and $\overleftrightarrow{GH}$

42. $\overleftrightarrow{AB}$, $\overleftrightarrow{EF}$, $\overleftrightarrow{BC}$, and $\overleftrightarrow{FG}$

43. $\overleftrightarrow{AD}$, $\overleftrightarrow{EH}$, and $\overleftrightarrow{FG}$

44. $\overleftrightarrow{DH}$, $\overleftrightarrow{CG}$, $\overleftrightarrow{BC}$, and $\overleftrightarrow{AD}$

45. $\overleftrightarrow{BC}$, $\overleftrightarrow{DC}$, $\overleftrightarrow{GC}$, $\overleftrightarrow{HG}$, $\overleftrightarrow{FG}$

46. True.

47. False; if the rhombus is not a square, the angles are not right angles.

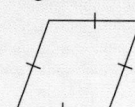

48. True.

Objectives

• Find the slopes of parallel lines and perpendicular lines.

• Identify properties of perpendicular bisectors.

• Write equations for parallel and perpendicular lines.

Recommended Pacing

❖ **Core and Extended Courses**
  Section 4.3: 2 days

❖ **Block Schedule**
  Section 4.3: 1 block

Resource Materials

Lesson Support

Lesson Plan 4.3

Warm-Up Transparency 4.3

Practice Bank: Practice 26

Study Guide: Section 4.3

Explorations Lab Manual:
  Diagram Master 2

Challenge Problems: Set 26

Assessment Book: Test 16

Technology

Technology Book:
  Spreadsheet Activity 4

Graphing Calculator

Geometry Software

McDougal Littell Mathpack
  Geometry Inventor Activity Book:
    Activity 15

Internet:
  http://www.hmco.com

Warm-Up Exercises

Find the slope of the line through each pair of points.

1. $(-3, 4)$, $(7, -1)$ $-\dfrac{1}{2}$

2. $(6, -2)$, $(9, 4)$ 2

Give the slope and y-intercept for each line.

3. $y = -3x + 6$ -3; 6

4. $y = 5(x - 2)$ 5; -10

5. Write an equation for the line through $(5, -2)$ with slope -3.
  $y = -3x + 13$

SECTION

4.3 | Exploring Parallels and Perpendiculars

Learn how to...

• find the slopes of parallel lines and perpendicular lines

• identify properties of perpendicular bisectors

So you can...

• write equations for parallel and perpendicular lines

Do you think the long segments in this pattern are parallel? You can't always tell whether lines are parallel or perpendicular simply by looking. In the Exploration you will investigate a method for telling whether lines are parallel or perpendicular.

EXPLORATION
COOPERATIVE LEARNING

Comparing Slopes of Lines

Work with another student.
You will need:
• geometry software or graph paper
• a straightedge

1 Lines j and k are parallel. Find the slope of each line.

2 Graph a different pair of parallel lines. If you are graphing by hand, use the sides of your straightedge to draw the lines. Find the slopes of your lines.

3 Make a conjecture about the slopes of parallel lines.

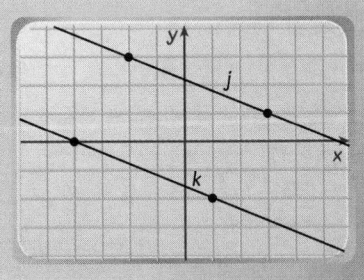

4 Lines ℓ and m are perpendicular. Find the slope of each line.

5 Graph a different pair of perpendicular lines. If you are graphing by hand, use a corner of a piece of paper to help you. Find the slopes of the lines.

6 Make a conjecture about the slopes of perpendicular lines.

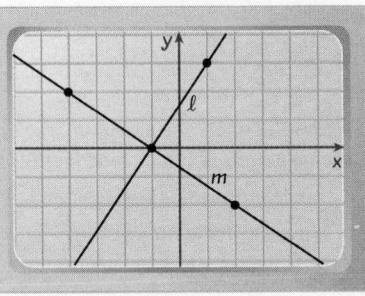

Exploration Note

Purpose
The purpose of this Exploration is to have students discover the relationship between the slopes of parallel lines and the relationship between the slopes of perpendicular lines.

Materials/Preparation
Each group of students needs geometry software or graph paper, and a straightedge.

Procedure
Students graph pairs of parallel lines, calculate their slopes, and make a conjecture

about the slopes of parallel lines. They then graph pairs of perpendicular lines, calculate their slopes, and make a conjecture about the slopes of perpendicular lines.

Closure
Students should see that the slopes of parallel lines are equal, and that the product of the slopes of perpendicular lines is –1.

Explorations Lab Manual
See the Manual for more commentary on this Exploration.

Diagram Master 2

Slopes of Parallel Lines and Perpendicular Lines

Two nonvertical lines are parallel if and only if their slopes are equal.

If $m_1 = m_2$, then $\ell_1 \parallel \ell_2$.

If $\ell_1 \parallel \ell_2$, then $m_1 = m_2$.

Two nonvertical lines are perpendicular if and only if the product of their slopes is -1.

If $m_1 \cdot m_2 = -1$, then $\ell_1 \perp \ell_2$.

If $\ell_1 \perp \ell_2$, then $m_1 \cdot m_2 = -1$.

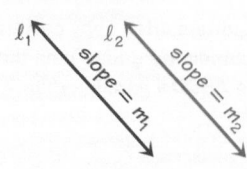

A **perpendicular bisector** of a segment is a line, ray, or segment that bisects the segment and is perpendicular to it. You can use what you know about perpendicular lines to find out if a line is a perpendicular bisector.

EXAMPLE 1

Is $\overleftrightarrow{CD}$ a perpendicular bisector of $\overline{AB}$? Explain.

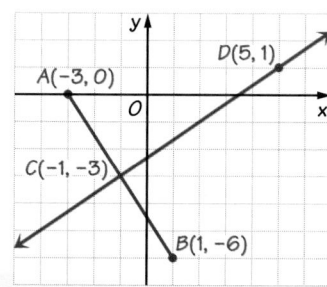

SOLUTION

The midpoint of the segment with endpoints $A(-3, 0)$ and $B(1, -6)$ is $C(-1, -3)$. So $\overleftrightarrow{CD}$ bisects $\overline{AB}$.

Next find the slopes of $\overline{AB}$ and $\overleftrightarrow{CD}$.

$$\text{slope of } \overline{AB} = \frac{-6 - 0}{1 - (-3)} \qquad \text{slope of } \overline{CD} = \frac{1 - (-3)}{5 - (-1)}$$

$$= \frac{-6}{4} \qquad\qquad\qquad = \frac{4}{6}$$

$$= -\frac{3}{2} \qquad\qquad\qquad = \frac{2}{3}$$

The product of the slopes is $-\frac{3}{2} \cdot \frac{2}{3} = -1$, so $\overleftrightarrow{CD} \perp \overline{AB}$.

Since $\overleftrightarrow{CD}$ is perpendicular to $\overline{AB}$, and $\overleftrightarrow{CD}$ bisects $\overline{AB}$, $\overleftrightarrow{CD}$ is a perpendicular bisector of $\overline{AB}$.

4.3 Exploring Parallels and Perpendiculars **181**

ANSWERS Section 4.3

Exploration

1. slope of j = slope of $k = -\frac{2}{5}$

2. Answers may vary. Check students' work. The lines have equal slopes or are both vertical.

3. The slopes of nonvertical parallel lines are equal.

4. slope of $l = -\frac{4}{3}$; slope of $m = \frac{3}{4}$

5. Answers may vary. Check students' work. The product of the slopes of the lines is -1, or one of the lines is vertical and the other horizontal.

6. The product of the slopes of two nonvertical perpendicular lines is -1.

Teach⇔Interact

Section Notes

Integrating Algebra
In this section, algebra is used to classify pairs of lines as parallel, perpendicular, or neither. The theorems stated on this page can be used to help classify triangles and quadrilaterals, as well as to identify a perpendicular bisector of a segment.

Topic Spiraling: Review
Students should recognize that the use of the words *if and only if* in the theorems about parallel and perpendicular lines means that two statements are involved, namely, the theorem as conjectured in the Exploration on page 180 and its converse.

About Example 1

Teaching Tip
Ask students how they can verify that point $C(-1, -3)$ is the midpoint of $\overline{AB}$. (They can find the average of the x-coordinates and the average of the y-coordinates of points A and B.)

Additional Example 1

In quadrilateral $QRST$, is $\overline{QS}$ the perpendicular bisector of $\overline{RT}$? Explain.

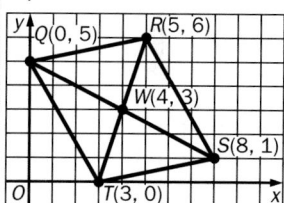

The midpoint of the segment with endpoints $R(5, 6)$ and $T(3, 0)$ is $W(4, 3)$. So $\overline{QS}$ bisects $\overline{RT}$. Next find the slopes of $\overline{QS}$ and $\overline{RT}$.

$$\text{slope of } \overline{QS} = \frac{1 - 5}{8 - 0}$$
$$= \frac{-4}{8}$$
$$= -\frac{1}{2}$$
$$\text{slope of } \overline{RT} = \frac{0 - 6}{3 - 5}$$
$$= \frac{-6}{-2}$$
$$= 3$$

The product of the slopes is $-\frac{1}{2} \cdot 3 = -\frac{3}{2}$, which is not equal to -1, so $\overline{QS}$ is not perpendicular to $\overline{RT}$. Thus, $\overline{QS}$ is not a perpendicular bisector of $\overline{RT}$.

181

Additional Example 2

Use the quilt pattern below. Is *HIJK* a rectangle?

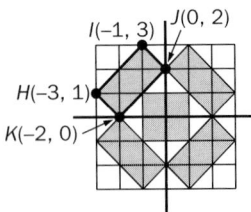

Find the slope of each side of *HIJK*.

slope of $\overline{HI} = \dfrac{3-1}{-1-(-3)} = 1$

slope of $\overline{IJ} = \dfrac{2-3}{0-(-1)} = -1$

slope of $\overline{JK} = \dfrac{0-2}{-2-0} = 1$

slope of $\overline{HK} = \dfrac{0-1}{-2-(-3)} = -1$

(slope of $\overline{HI}$) · (slope of $\overline{IJ}$) = −1, so $\overline{HI} \perp \overline{IJ}$.

(slope of $\overline{IJ}$) · (slope of $\overline{JK}$) = −1, so $\overline{IJ} \perp \overline{JK}$.

(slope of $\overline{JK}$) · (slope of $\overline{HK}$) = −1, so $\overline{JK} \perp \overline{HK}$.

(slope of $\overline{HI}$) · (slope of $\overline{HK}$) = −1, so $\overline{HI} \perp \overline{HK}$.

HIJK has four right angles, so it is a rectangle.

Additional Example 3

Write an equation of the line that passes through *W*(–2, 1) and is parallel to $y = \dfrac{2}{3}x - 2$.

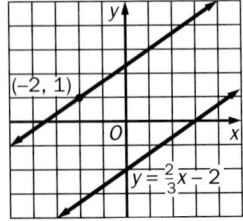

The line $y = \dfrac{2}{3}x - 2$ has slope $\dfrac{2}{3}$. Any line parallel to this line has slope $\dfrac{2}{3}$. Use the slope-intercept form to find *b*.

$y = mx + b$

$1 = \dfrac{2}{3} \cdot (-2) + b$

$b = \dfrac{7}{3}$

An equation of the line is $y = \dfrac{2}{3}x + \dfrac{7}{3}$.

EXAMPLE 2   Application: Quilt Patterns

Quilters often copy patterns onto a coordinate grid to plan their work. Is *ABCD* a parallelogram?

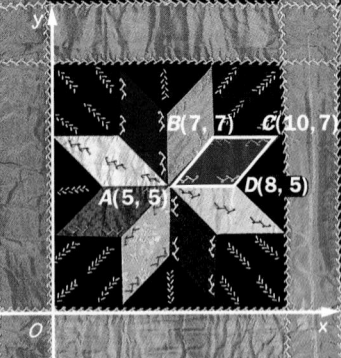

SOLUTION

Find the slope of each side of *ABCD*.

slope of $\overline{AB} = \dfrac{7-5}{7-5} = 1$

slope of $\overline{CD} = \dfrac{5-7}{8-10} = 1$

slope of $\overline{BC} = \dfrac{7-7}{10-7} = 0$ slope of $\overline{AD} = \dfrac{5-5}{8-5} = 0$

The slope of $\overline{AB}$ = the slope of $\overline{CD}$, so $\overline{AB} \parallel \overline{CD}$.

The slope of $\overline{BC}$ = the slope of $\overline{AD}$, so $\overline{BC} \parallel \overline{AD}$.

By the definition of a parallelogram, *ABCD* is a parallelogram.

THINK AND COMMUNICATE

1. State each theorem on page 181 as a conditional and its converse.

2. Why do the theorems on page 181 apply to nonvertical lines only?

3. Which formula can you use to find out if *ABCD* in Example 2 is a rhombus? Explain.

4. Explain how you know that *ABCD* in Example 2 is not a square.

EXAMPLE 3

Write an equation of the line that passes through (3, −3) and is perpendicular to the line $y = -2x + 3$.

SOLUTION

The line $y = -2x + 3$ has slope −2. Any line perpendicular to this line has slope $\dfrac{1}{2}$.

Use the slope-intercept form to find *b*.

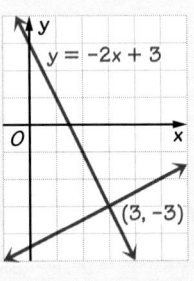

$y = mx + b$

$-3 = \dfrac{1}{2} \cdot 3 + b$ Substitute (**3**, −**3**) for (*x*, *y*) and $\dfrac{1}{2}$ for *m*.

$b = -4.5$

An equation of the line is $y = \dfrac{1}{2}x - 4.5$.

Think and Communicate

1. (1) If two nonvertical lines are parallel, then their slopes are equal. If the slopes of two non-vertical lines are equal, then the lines are parallel. (2) If two nonvertical lines are perpendicular, then the product of their slopes is −1. If the product of the slopes of two non-vertical lines is −1, then the lines are perpendicular.

2. The slope of a vertical line is undefined.

3. You can use the Distance Formula to check whether all four sides are equal in length.

4. The product of the slopes of consecutive sides is 0, not −1.

THINK AND COMMUNICATE

5. What is the slope of any line that is perpendicular to a vertical line?
6. What is the slope of any line that is parallel to a vertical line?
7. Give an equation for a line that is parallel to the line $x = 4$.
8. Give an equation for a line that is perpendicular to the line $x = 4$.

☑ CHECKING KEY CONCEPTS

Tell whether each pair of lines is *parallel, perpendicular,* or *neither.*

1. $y = -3x + 6$; $y = -\frac{1}{3}x$
2. $y = -5x - 11$; $y = \frac{1}{5}x - 11$
3. $x = 1$; $x = 5$
4. For the slope of a line that is perpendicular to the line $y = mx + b$, Emily uses $\frac{-1}{m}$. Will this method always work? Explain.

Find the slope of each line.

5. a line parallel to the line $y = 7x + 1$
6. a line perpendicular to the line $y = -2x - 3$
7. a line parallel to the x-axis
8. a line perpendicular to the x-axis

4.3 Exercises and Applications

Extra Practice *exercises on page 666*

Tell whether each pair of lines is *parallel, perpendicular,* or *neither.*

1. $y = -3$; $x = -3$
2. $y = 2x - 7$; $y = 7x - 2$
3. $y = -\frac{4}{5}x + 1$; $y = -0.8x - 5$

Find the slope of each line.

4. a line parallel to the line $y = -x + 3$
5. a line parallel to the line $y = 5$
6. a line perpendicular to the line $y = 7x - 4$
7. a line perpendicular to the line $y = -3x$
8. **SAT/ACT Preview** Which equation represents a line that is perpendicular to the line with equation $y = -\frac{1}{2}x - 7$?

   A. $y = -\frac{1}{2}x + 7$ **B.** $y = -2x - 7$ **C.** $y = 2x + 7$
   D. $y = -7x - \frac{1}{2}$ **E.** $y = \frac{1}{2}x + 7$

Use slopes to tell whether each triangle with the given vertices is a right triangle.

9. $A(-4, 1)$, $B(2, 3)$, $C(5, -6)$
10. $D(-5, 0)$, $E(0, 5)$, $F(5, 0)$
11. $G(1, 1)$, $H(4, -1)$, $J(-1, 3)$
12. $K(-4, -4)$, $L(0, 0)$, $M(-6, 6)$

Think and Communicate

5. 0
6. undefined
7. $x = a$ for any $a \neq 4$.
8. $y = b$ for any number b.

Checking Key Concepts

1. neither
2. perpendicular
3. parallel
4. No; if the line $y = mx + b$ is a horizontal line, then $m = 0$ and Emily cannot use $\frac{-1}{m}$ for the slope of the perpendicular line because $\frac{-1}{m}$ is undefined.
5. 7
6. $\frac{1}{2}$
7. 0
8. undefined

Exercises and Applications

1. perpendicular
2. neither
3. parallel
4. –1
5. 0
6. $-\frac{1}{7}$
7. $\frac{1}{3}$
8. C
9. Yes.
10. Yes.
11. No.
12. Yes.

Teach⇔Interact

Think and Communicate

Questions 5–8 have students investigate the special case of lines that are parallel and perpendicular to vertical lines. You may wish to expand this discussion by asking these questions about horizontal lines.

Closure Question

What is the relationship between the slopes of parallel lines? between the slopes of perpendicular lines? The slopes of parallel lines are equal. The product of the slopes of perpendicular lines is –1.

Apply⇔Assess

Suggested Assignment

Core Course
Day 1 Exs. 1–12, 14–22
Day 2 Exs. 28–31, 33–40, AYP

Extended Course
Day 1 Exs. 1–23
Day 2 Exs. 24–40, AYP

Block Schedule
Day 23 Exs. 1–12, 14–22, 28–31, 33–40, AYP

Exercise Notes

Using Technology
Exs. 1–7 As an alternative way to have students explore the slopes of parallel and perpendicular lines, you may wish to assign Activity 15 in the *Geometry Inventor Activity Book.*

Common Error
Exs. 4–8 Students may incorrectly give a parallel line instead of a perpendicular line, or visa versa. To help avoid these errors, have students underline the word *parallel* or *perpendicular* in the question and then write immediately *same slope* or *slopes with product –1* on their paper to help them remember which slope to use.

13. CITY PLANNING Parking lots may use different designs in different
locations.

A   B   C

   a. Which of these plans use perpendicular lines? Which use parallel lines?

   b. Writing What are some advantages and disadvantages of each
      parking plan?

Give the most specific name for each quadrilateral with the given vertices.

14. $A(-4, -2)$, $B(5, 4)$, $C(7, 1)$, $D(-2, -5)$ **15.** $E(0, -3)$, $F(1, 1)$, $G(7, 0)$, $H(5, -4)$

16. $J(-3, 2)$, $K(-1, 6)$, $L(1, 0)$, $M(-1, -4)$ **17.** $P(-2, 3)$, $Q(3, 4)$, $R(4, -1)$, $S(-1, -2)$

18. Open-ended Problem Find the vertices of a quadrilateral with
perpendicular diagonals that are neither horizontal nor vertical.

**For each set of points, tell whether $\overleftrightarrow{CD}$ is a perpendicular bisector of $\overline{AB}$.
If not, tell why not.**

19. $A(0, 0)$, $B(4, -4)$, $C(2, -2)$, $D(5, 1)$ **20.** $A(-1, 1)$, $B(3, 3)$, $C(1, 2)$, $D(3, -3)$

21. $A(-2, 4)$, $B(2, -2)$, $C(1, 1)$, $D(4, 3)$ **22.** $A(1, 3)$, $B(3, -3)$, $C(-1, -1)$, $D(5, 1)$

23. CARPENTRY A *carpenter's square* can be used to find slopes and draw
parallel and perpendicular lines. It looks like two perpendicular rulers
that are joined at the 0 point.

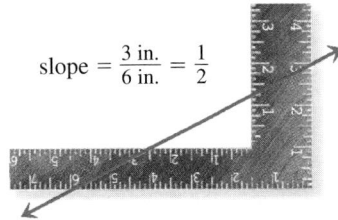

slope $= \dfrac{3 \text{ in.}}{6 \text{ in.}} = \dfrac{1}{2}$

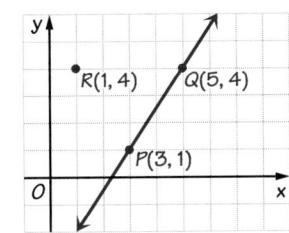

   a. Describe how to use a carpenter's square to find the slope of $\overleftrightarrow{PQ}$.
      Write an equation for $\overleftrightarrow{PQ}$.

   b. Describe how to use a carpenter's square to draw a line perpendicular
      to $\overleftrightarrow{PQ}$ that passes through R. Write an equation of the line.

   c. Describe how to use a carpenter's square and the line from part (b) to
      draw a line parallel to $\overleftrightarrow{PQ}$ that passes through R. Write an equation
      of the line.

184   Chapter 4   *Coordinates in Geometry*

---

13. **a.** perpendicular: B and C;
    parallel: A, B, and C

   b. Answers may vary. Examples are
    given. Plans A and B allow dri-
    vers (when space is available) to
    enter a space and drive through,
    so they can later leave without
    backing out. Plan C would make
    it very difficult for drivers to cut
    across lanes, improving safety.
    Plans A and C would probably
    require one-way traffic.

14. rectangle

15. quadrilateral

16. parallelogram

17. square

18. Answers may vary. An
    example is given. (0, 0),
    (5, 0), (8, 4), and (3, 4)

19. Yes.

20. No; $\overleftrightarrow{CD}$ is not perpen-
    dicular to $\overline{AB}$.

21. No; $\overleftrightarrow{CD}$ does not bisect $\overline{AB}$.

22. Yes.

23. **a.** Line up the carpenter's
    square so that P lies on the
    horizontal edge of the ruler
    and Q lies on the vertical
    edge. Then read the coordi-
    nates on each edge. The
    fraction $\dfrac{\text{horizontal coordinate}}{\text{vertical coordinate}}$
    is equal to the slope.
    $y = \dfrac{3}{2}x - \dfrac{7}{2}$

   b. Line up the carpenter's square so
    that one edge lies along $\overleftrightarrow{PQ}$ and
    the other edge passes through
    point R. Draw along the second
    edge to locate the required line.
    $y = -\dfrac{2}{3}x + \dfrac{14}{3}$

   c. Use the procedure described in
    part (b) to draw a line that is per-
    pendicular to the line drawn in
    part (b) and passes through point
    R. Since the new line and $\overleftrightarrow{PQ}$ are
    both perpendicular to the line

In order to transfer a quilt pattern, quilters may use a square that is 12 in. on a side, as shown below. By using slopes, they can check that they have copied a pattern correctly. Also, they can tell what kinds of quadrilaterals are in the pattern.

For each quadrilateral marked in the quilt patterns shown:
a. **Give the coordinates of the vertices.**
b. **Find the slopes of the sides.**
c. **Tell whether it is a *parallelogram*, a *rhombus*, or *neither*.**

24. guiding star

25. tumbling maple leaf

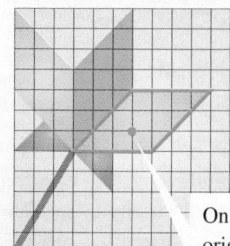

On each square, the origin is indicated by a red dot.

This Arapaho woman is preparing a quilt for the Sundance River Indian Reservation in Wyoming.

26. desert rose basket

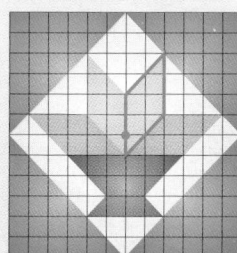

27. Arkansas traveler

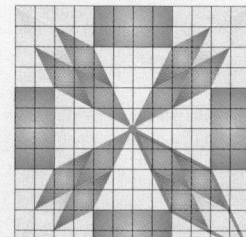

Write an equation for each line.

28. The line passes through $P(-4, 2)$ and is parallel to the line $y = -\frac{4}{3}x + 5$.

29. The line passes through $Q(3, 7)$ and is parallel to the line $y = 2$.

30. The line passes through $R(0, -2)$ and is perpendicular to the line $y = x - 4$.

31. The line passes through $S(2, 6)$ and is perpendicular to the *x*-axis.

32. **Challenge** Use the points $T(a, b)$, $U(c, d)$, $V(b, c)$, and $W(d, a)$, where $a \neq c$ and $b \neq d$.
   a. Show that $\overleftrightarrow{TU} \perp \overleftrightarrow{VW}$.
   b. Why is it necessary to state that $a \neq c$ and $b \neq d$?

4.3 Exploring Parallels and Perpendiculars   **185**

Exercise Notes

Cooperative Learning
Exs. 24–27 Students may benefit from completing these exercises in pairs. Have each student find the vertices for two of the patterns, then switch patterns and find the slopes of the sides. The partners can work together to determine the type of each quadrilateral. Each group could make a design of their own and represent it on a 12-by-12 grid.

Multicultural Note
Exs. 24–27 Many cultures practice the art of quilting, but not all use quilted materials for bedcovers and wall hangings. In Africa, peoples in the southern Saharan region traditionally used quilted garments as a military tool: warriors and their horses were outfitted with quilted armor to prevent arrows and other weapons from harming them. The Japanese quilting art called *sashiko* ("little stabs") was developed to provide people in rural Japan with warm winter clothes. When soaked in water, these quilted garments also served to protect firefighters from heat and flames. In India, Jat parents put quilts in their daughters' dowries, since the number of quilts a household owns bolsters their status in the community. Other Indian ethnic groups make lightweight quilts, often carried by travelers who use them as pillows or blankets.

 **Using Technology**
Exs. 28, 30 Students can check their solutions to these exercises by graphing both equations on a graphing calculator. For Ex. 30, students should choose a square graphing window. On the TI-82, this can be done by choosing ZSquare from the ZOOM menu. A square window ensures that perpendicular lines appear perpendicular.

Teaching Tip
Exs. 29, 31 It may help students to sketch a graph for these problems before writing an equation.

Challenge
Ex. 32 Encourage students who have difficulty beginning this exercise to start with a specific example. They can then follow the same procedure for the general case. Part (b) helps students realize the necessity of the given conditions to eliminate the cases in which either $\overleftrightarrow{TU}$ or $\overleftrightarrow{VW}$ are vertical lines.

---

drawn in part (b), these two lines are parallel. $y = \frac{3}{2}x + \frac{5}{2}$

24. a. $(0, 0)$, $(1, -2)$, $(0, -6)$, $(-1, -2)$
   b. $-2, 4, -4, 2$
   c. neither

25. a. $(-3, -1)$, $(1, -1)$, $(4, 2)$, $(0, 2)$
   b. $0, 1, 0, 1$
   c. parallelogram

26. a. $(0, -1)$, $(2, 1)$, $(2, 4)$, $(0, 2)$

b. 1, undefined, 1, undefined
   c. parallelogram

27. a. $(0, 0)$, $(2, -4)$, $(6, -6)$, $(4, -2)$
   b. $-2, -\frac{1}{2}, -2, -\frac{1}{2}$
   c. rhombus

28. $y = -\frac{4}{3}x - \frac{10}{3}$

29. $y = 7$

30. $y = -x - 2$ 31. $x = 2$

32. a. slope of $\overleftrightarrow{TU} = \frac{d-b}{c-a}$ and slope of $\overleftrightarrow{VW} = \frac{a-c}{d-b}$; The product of the slopes is $\frac{d-b}{c-a} \cdot \frac{a-c}{d-b} = \frac{a-c}{c-a} = \frac{-(c-a)}{c-a} = -1$, so the lines are perpendicular.

b. If $a = c$, then $\overleftrightarrow{TU}$ is a vertical line and its slope is undefined. Similarly, if $b = d$, $\overleftrightarrow{VW}$ is vertical and its slope is undefined. (In this case, $\overleftrightarrow{TU} \parallel \overleftrightarrow{VW}$.)

Apply⟺Assess

Exercise Notes

Assessment Note
Ex. 33 Students review the concepts of this section while discovering that a line perpendicular to one of two parallel lines is also perpendicular to the other.

Assess Your Progress

Journal Entry
You may wish to have students share their methods and examples to be certain all students understand this procedure, especially with horizontal and vertical segments.

Progress Check 4.1–4.3

See page 210.

Practice 26 for Section 4.3

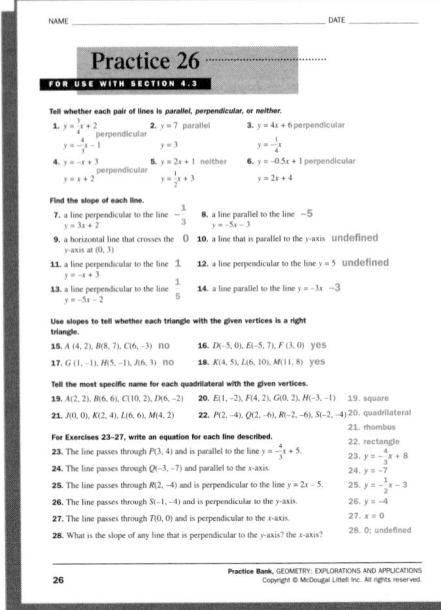

33. Cooperative Learning Work with another person.

a. Working with your partner, choose a slope. Each of you should write an equation for a line with that slope. Graph the lines on the same coordinate plane and label them *l* and *m*.

b. Write an equation of a line that is perpendicular to *l*. Graph it and label it *q*.

c. What is the relationship between lines *m* and *q*? Complete this conjecture: "If a line is perpendicular to one of two parallel lines, then _?_."

SPIRAL REVIEW

Find the slope and *y*-intercept of each line. *(Section 4.2)*

34. $y = -3x - 2$ **35.** $y = x$ **36.** $y = 9$

37. Sketch a pentagonal prism. *(Section 2.6)*

The legs of a right triangle are *a* and *b* units long. The hypotenuse is *c* units long. Find the unknown length for each right triangle. *(Section 3.6)*

38. $a = 9.5, b = 16.8$ **39.** $a = 46.8, c = 49.3$ **40.** $b = 89.9, c = 90.1$

ASSESS YOUR PROGRESS

VOCABULARY

slope (p. 174)   **slope-intercept form** (p. 175)
y-intercept (p. 175)   **perpendicular bisector** (p. 181)

1. Use $\triangle ABC$ at the right. *(Section 4.1)*

a. Find the perimeter of $\triangle ABC$.

b. Label the midpoints of the sides *M, N,* and *P*. Find their coordinates.

c. Find the perimeter of $\triangle MNP$. What do you notice?

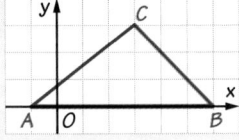

Find the slope of the line that contains the given points. *(Section 4.2)*

2. $(0, 0)$ and $(5, -3)$ **3.** $(-2, 5)$ and $(7, 5)$ **4.** $(-3, -3)$ and $(1, 7)$

Give the slope of each line. *(Section 4.3)*

5. a line parallel to the line $y = 4x + 3$

6. a line perpendicular to the line $y = -5x - 2$

7. Journal Describe how to find an equation for the perpendicular bisector of a segment if you know the coordinates of the segment's endpoints. Draw several examples, showing your method. Include at least one horizontal segment and one vertical segment.

186 Chapter 4 *Coordinates in Geometry*

---

33. Answers may vary. Examples are given.

a. The lines *l* and *m* with equations $y = -2x + 3$ and $y = -2x - 1$ are parallel lines with slope –2.

b. Line *q* with equation $y = \frac{1}{2}x + 3$ is perpendicular to line *l* at the point $(0, 3)$.

c. Lines *m* and *q* are perpendicular. the line is perpendicular to the other parallel line

34. –3; –2

35. 1; 0

36. 0; 9

37. Answers may vary. An example is given.

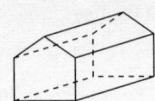

38. $c = 19.3$

39. $b = 15.5$

40. $a = 6.0$

Assess Your Progress

1. a. $12 + 3\sqrt{2}$

b. Let *M, N,* and *P* be the midpoints of $\overline{AB}, \overline{AC},$ and $\overline{BC}$, respectively. $M = (2.5, 0)$, $N = (1, 1.5), P = (4.5, 1.5)$

c. $6 + \frac{1}{3}\sqrt{2}$; The perimeter of $\triangle MNP$ is half the perimeter of $\triangle ABC$.

2. $-\frac{3}{5}$

3. 0

4. $\frac{5}{2}$

5. 4

6. $\frac{1}{5}$

7. See answers in back of book.

186

4.4 Equations of Circles

Learn how to...

- identify circles and parts of circles
- write equations for circles

So you can...

- describe circular shapes, such as architectural forms

Architects have often used domes to create impressive interior spaces for public buildings. One of the best examples is the rotunda at the University of Virginia, designed by Thomas Jefferson. A rotunda is a circular building with a domed roof. A cross section of the building shows that a circle drawn to touch the dome also touches the center of the base.

$\overline{CD}$ is a *radius*.

$\overline{AB}$ is a *diameter*.

The center of a circle is not part of the circle.

A **circle** is the set of all points in a plane that are an equal distance from a given point, the **center**, which is also in the plane. A **diameter** is a segment that passes through the center of a circle and whose endpoints lie on the circle. A **radius** is a segment whose endpoints are the center and a point on the circle. In the diagram above, the circle with center C is called $\odot C$ (read "circle C").

THINK AND COMMUNICATE

In the cross section of the University of Virginia rotunda, *AB* is about 77 ft.

1. $\overline{DF}$ is a diameter of $\odot C$ that connects the dome of the building with its base. Find the length DF.

2. Find the lengths AC, CB, DC, and CF.

3. How do you know that all the *radii* (plural of *radius*) of a circle are congruent?

4.4 Equations of Circles **187**

Plan⟺Support

Objectives

- Identify circles and parts of circles.
- Write equations for circles.
- Describe circular shapes in the real world.

Recommended Pacing

❖ **Core and Extended Courses**
Section 4.4: 1 day

❖ **Block Schedule**
Section 4.4: $\frac{1}{2}$ block
(with Section 4.5)

Resource Materials

Lesson Support
Lesson Plan 4.4
Warm-Up Transparency 4.4
Practice Bank: Practice 27
Study Guide: Section 4.4
Explorations Lab Manual:
   Diagram Master 2
Challenge Problems: Set 27

Technology
Technology Book:
   Calculator Activity 6
   TI-92 Activity 4
Graphing Calculator
McDougal Littell Mathpack
   Geometry Inventor Activity Book:
   Activity 12
Internet:
   http://www.hmco.com

Warm-Up Exercises

Find the distance between each pair of points.

1. $(3, -1)$ and $(4, 5)$ $\sqrt{37}$
2. $(-5, -3)$ and $(-3, 8)$ $5\sqrt{5}$
3. $(0, 4)$ and $(-6, -4)$ 10

Evaluate each expression for $x = 3$ and $y = -4$.

4. $(x - 3)^2 + (y + 3)^2$ 1
5. $(x + 5)^2 + (y - 6)^2$ 164

ANSWERS Section 4.4

Think and Communicate

1. about 77 ft
2. Each length is about 38.5 ft.
3. All points on a circle are the same distance from the center.

Teach⇔Interact

Section Notes

Using Technology
Students can investigate the basic properties of circles by using Activity 12 in the *Geometry Inventor Activity Book*.

Integrating Algebra
In this section, students learn that circles can be represented using an algebraic equation. The length of a radius or diameter of a circle can be determined using the Distance Formula, and the center of a circle can be determined using the Midpoint Formula.

Think and Communicate

Questions 1–3 give students experience using the vocabulary associated with circles, while they discover that all radii and diameters of the same circle are congruent.

About Example 1

Geometric Thinking
This Example can help students see the connection between the equation of a circle with center (0, 0) and the Distance Formula. Stress that the Distance Formula is being used to find the distance between the center of the circle (0, 0) and any point (x, y) on the circle. This distance is equal to the radius of the circle.

Additional Example 1

The inner wall of the Pantheon is a circle with diameter about 142 ft. Write an equation for this wall if the center of the Pantheon is given the coordinates (0, 0).
Since the diameter is 142 ft, the radius is 71 ft.
Distance from (x, y) to $(0, 0) = 71$
$\sqrt{(x-0)^2 + (y-0)^2} = 71$
$\qquad x^2 + y^2 = 71^2$
$\qquad x^2 + y^2 = 5041$

188

The Pantheon, a Roman building dating from the first century, is the inspiration for many later buildings, including the University of Virginia rotunda.

You can write an equation for a circle by placing it on a coordinate plane and using the Distance Formula.

EXAMPLE 1

Write an equation for the inner wall of the University of Virginia rotunda.

The inner wall of the rotunda is a circle with radius about 72 ft.

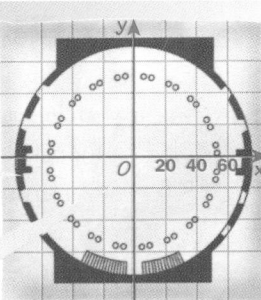

The center of the rotunda is at (0, 0).

SOLUTION

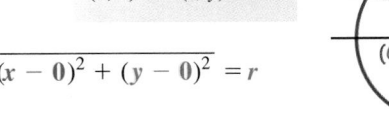

Distance from (x, y) to $(0, 0) = 72$
$$\sqrt{(x-0)^2 + (y-0)^2} = 72 \qquad r \approx 72 \text{ ft}$$
$$x^2 + y^2 = (72)^2$$
$$x^2 + y^2 = 5184$$

In general, every point (x, y) on a circle with center $(0, 0)$ and radius r is r units from the center. An equation for such a circle is:

The distance between $(0, 0)$ and (x, y) is r.

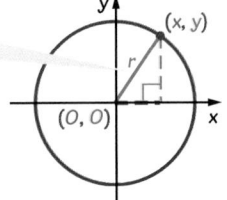

$$\sqrt{(x-0)^2 + (y-0)^2} = r$$

Squaring both sides of the above equation eliminates the square root symbol.

Equation of a Circle with Center (0, 0)

An equation of the circle with center $(0, 0)$ and radius r is:
$$x^2 + y^2 = r^2$$

THINK AND COMMUNICATE

4. The outer wall of the University of Virginia rotunda has a radius of about 77 ft. What is an equation for this circle?

5. The round skylight in the dome of the University of Virginia rotunda has a diameter of about 16 ft. What is an equation for this circle?

188 Chapter 4 *Coordinates in Geometry*

Think and Communicate

4. $x^2 + y^2 = 5929$

5. $x^2 + y^2 = 64$

The Distance Formula also leads to an equation for a circle that is not centered at (0, 0). The point (h, k) is usually used to represent the center of such a circle.

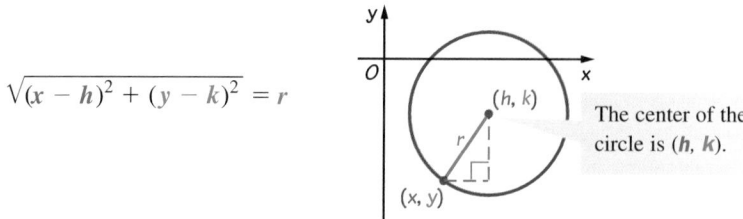

$$\sqrt{(x - h)^2 + (y - k)^2} = r$$

The center of the circle is (h, k).

Squaring both sides of the equation above eliminates the square root symbol.

Equation of a Circle with Center (h, k)

An equation of the circle with center (h, k) and radius r is:

$$(x - h)^2 + (y - k)^2 = r^2$$

EXAMPLE 2

Write an equation of the circle with center (2, 4) and radius 5.

SOLUTION

Use the equation of a circle.

$(x - h)^2 + (y - k)^2 = r^2$

$(x - 2)^2 + (y - 4)^2 = 5^2$ Substitute **(2, 4)** for (h, k) and **5** for r.

$(x - 2)^2 + (y - 4)^2 = 25$

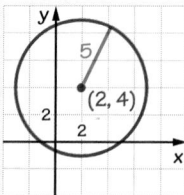

THINK AND COMMUNICATE

6. Suppose the center of the circle in Example 2 were $(-2, -4)$ instead of (2, 4). How would the equation of the circle change?

7. What happens to the equation of a circle centered at (h, k) if you use (0, 0) as the center?

8. In a plane, two or more circles that have the same center are called **concentric circles**. How are the equations for two concentric circles alike? How are they different? Explain.

4.4 Equations of Circles **189**

Think and Communicate

6. The equation would be $(x + 2)^2 + (y + 4)^2 = 25$.

7. You get the equation for a circle with center at the origin, $x^2 + y^2 = r^2$.

8. If two circles are concentric with different radii, the left sides of their equations are the same, while the right sides are different. If the common center is (h, k) and the radii are r_1 and r_2, the equations are $(x - h)^2 + (y - k)^2 = r_1^2$ and $(x - h)^2 + (y - k)^2 = r_2^2$.

Teach⇔Interact

Section Notes

Teaching Tip
You may wish to demonstrate the development of the equation of a circle with center (h, k) at the board. Use an illustration of a circle, labeled with the center, and an arbitrary point (x, y) on the circle. Students should understand that the Distance Formula is used to find the distance between the center and the arbitrary point, and that this distance is equal to the radius.

Alternate Approach
You can introduce the circle with center (h, k) by using the concept of a translation, as presented in Chapter 1. This circle is a translation of the circle with center (0, 0) right h units (left if h is negative), and up k units (down if k is negative).

Teaching Tip
Encourage students to check their equations by picking a point on the graph of a circle and substituting it into the equation.

Additional Example 2

Write an equation of the circle with center (–3, 1) and radius 9.

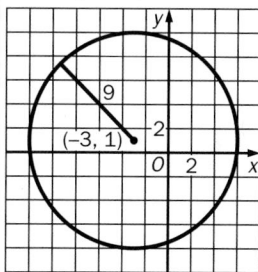

Use the equation of a circle. Substitute (–3, 1) for (h, k) and 9 for r.

$(x - h)^2 + (y - k)^2 = r^2$

$(x - (-3))^2 + (y - 1)^2 = 9^2$

$(x + 3)^2 + (y - 1)^2 = 81$

Think and Communicate

Question 6 highlights the equation of a circle when the coordinates of the center are negative numbers. In question 7, students discover that the equation for a circle with center (0, 0) is a special case of the equation for a circle with center (h, k). Students may need to draw some concentric circles and find their equations before answering question 8.

189

About Example 3

Geometric Thinking

Stress that the given equation needs to be written in the form of the general equation before the center and radius can be read. Point out that the expression in the first term is written as a double negative to achieve the required subtraction, and that the constant term is written as a square. Students should notice that in Step 2, four points were found on the circle by counting 4 units from the center in both horizontal and vertical directions.

Additional Example 3

Sketch the circle with equation $(x - 1)^2 + (y + 4)^2 = 3$.
Rewrite the equation of the circle in the form $(x - h)^2 + (y - k)^2 = r^2$.
$(x - 1)^2 + (y + 4)^2 = 3$
Rewrite $y + 4$ as $y - (-4)$ and 3 as $(\sqrt{3})^2$.
$(x - 1)^2 + (y - (-4))^2 = (\sqrt{3})^2$
The center is $(1, -4)$. The radius is $\sqrt{3} \approx 1.7$. Sketch the circle using center $(1, -4)$ and radius 1.7.
Step 1 Graph the center, $(1, -4)$.
Step 2 Graph some points that are 1.7 units from the center.
Step 3 Sketch the circle.

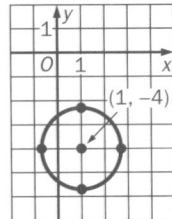

Think and Communicate

Ask students what the relationship is between the circles in questions 9 and 10 and the one in Example 3. (They are concentric circles.)

Closure Question

Suppose you are given the graph of a circle. How would you write an equation for the circle?
Determine the coordinates of the center and the radius of the circle from the graph. Substitute the values for the center (h, k) and radius r into the equation of a circle, $(x - h)^2 + (y - k)^2 = r^2$.

EXAMPLE 3

Sketch the circle with equation $(x + 2)^2 + (y - 3)^2 = 16$.

SOLUTION

Rewrite the equation of the circle in the form $(x - h)^2 + (y - k)^2 = r^2$.

$$(x + 2)^2 + (y - 3)^2 = 16$$

Rewrite $x + 2$ as $x - (-2)$.   Rewrite 16 as 4^2.

$$(x - (-2))^2 + (y - 3)^2 = 4^2$$

The center is $(-2, 3)$.   The radius is 4.

Now sketch the circle using the center $(-2, 3)$ and radius 4.

Step 1 Graph the center, $(-2, 3)$.

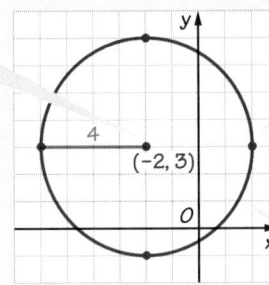

Step 2 Graph some points that are 4 units from the center.

Step 3 Sketch the circle by hand or with a compass.

THINK AND COMMUNICATE

9. What radius should you use to sketch a circle with equation $(x + 2)^2 + (y - 3)^2 = 25$?

10. What radius should you use to sketch a circle with equation $(x + 2)^2 + (y - 3)^2 = 7$?

☑ CHECKING KEY CONCEPTS

1. What is the radius of a circle whose diameter is 10?

2. Write an equation of a circle with center $(0, 0)$ and radius 7.

3. Write an equation of a circle with center $(2, -3)$ and radius 4.

4. Sketch the circle with equation $(x - 4)^2 + (y + 1)^2 = 9$.

190   Chapter 4 *Coordinates in Geometry*

Think and Communicate

9. 5

10. $\sqrt{7} \approx 2.6$

Checking Key Concepts

1. 5

2. $x^2 + y^2 = 49$

3. $(x - 2)^2 + (y + 3)^2 = 16$

4.

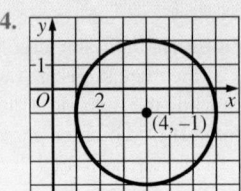

Exercises and Applications

1. $x^2 + y^2 = 9$

2. $(x - 3)^2 + (y + 3)^2 = 4$

3. $(x + 2)^2 + (y - 3)^2 = 9$

4. $x^2 + y^2 = 2.25$

5. $(x - 5)^2 + (y - 2)^2 = 49$

6. $(x - 3)^2 + (y - 4)^2 = 25$

7. $(x + 2)^2 + (y - 1)^2 = 1$

8. $(x + 6)^2 + (y + 8)^2 = 12.25$

9. $(x - 10)^2 + (y + 20)^2 = 400$

10. a. $(x - 4.5)^2 + (y - 4.5)^2 = 20.25$

  b. $\odot B$: $(x - 4.5)^2 + (y - 5.5)^2 = 10.24$; $\odot C$: $(x - 4.5)^2 + (y - 3.5)^2 = 10.24$

4.4 **Exercises and Applications**

Extra Practice
exercises on
page 666

Write an equation of each circle.

1. 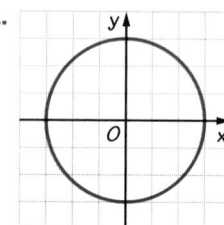   **2.** 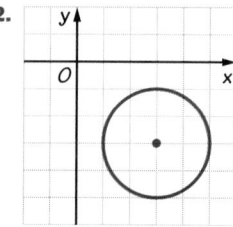   **3.**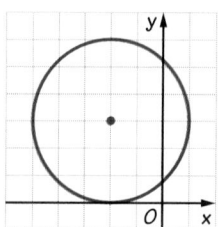

Write an equation of the circle with each given center and radius.

4. center $(0, 0)$
radius 1.5

5. center $(5, 2)$
radius 7

6. center $(3, 4)$
radius 5

7. center $(-2, 1)$
radius 1

8. center $(-6, -8)$
radius 3.5

9. center $(10, -20)$
radius 20

10. PRINTING Letters in a type design, or *typeface,* are carefully drawn, often with thick and thin areas. Typeface designers often work on a coordinate grid. In the design for the letter *O* shown below, three circles were used to draw the curves.

   a. The outside of the letter is $\odot A$, with center $(4.5, 4.5)$ and radius 4.5. Write an equation for $\odot A$.

   b. The inside of the letter is shaped using $\odot B$ and $\odot C$. Write equations for these circles.

   c. **Open-ended Problem** How do you think the curves shown in blue were drawn?

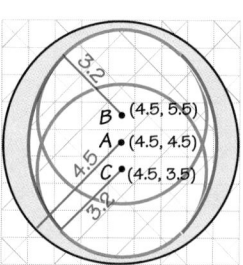

Sketch each circle. Label the coordinates of the center. Draw a radius and label it with its length.

11. $x^2 + y^2 = 64$

12. $(x - 1)^2 + (y + 5)^2 = 9$

13. $(x - 3)^2 + (y - 1)^2 = 10$

14. $(x + 5)^2 + (y - 3)^2 = 121$

15. $(x + 2)^2 + y^2 = 25$

16. $(y - 3)^2 + (x - 4)^2 = 25$

Make a sketch to fit each description.

17. $\odot A$ with diameter $\overline{CD}$ and radius $\overline{AB}$

18. two concentric circles with center Q

19. two circles that intersect in two points

20. two circles that intersect in exactly one point

4.4 Equations of Circles   **191**

Suggested Assignment

❖ **Core Course**
Exs. 1–9, 11–20, 23–25, 28–44

❖ **Extended Course**
Exs. 1–9 odd, 10, 11–21 odd, 22, 23, 25–44

❖ **Block Schedule**
Day 24 Exs. 1–9, 11–20, 23–25, 28–44

Exercise Notes

Common Error
Exs. 1–9 Students may forget to square the radius when writing an equation for a circle or they may make sign errors within the parentheses. To help avoid these errors, students should write the general form of the equation of a circle to provide a visual reference for the exponent and subtraction signs.

Application
Ex. 10 This exercise introduces students to an application of circles involving the design of a typeface. Interested students may want to find out more about how typefaces are designed.

Using Technology
Exs. 11–16 Students can use technology to check the graphs of these equations. They need to solve each equation for *y* and enter each half of the circle as a separate function. For example, the equation in Ex. 12 is entered as the two functions $y = -5 + \sqrt{9 - (x - 1)^2}$ and $y = -5 - \sqrt{9 - (x - 1)^2}$. On a graphics calculator, a square window must be used to avoid distorting the circle. On the TI-82, a square window can be made by choosing ZSquare from the ZOOM menu. The trace features of the calculator can be used to locate some of the points on the circle. These can then be compared with points on students' graphs.

Challenge
Exs. 19, 20 Have students draw the circles for these two exercises on graph paper. They can then find the equations for both circles, and verify that the coordinates of the point(s) of intersection satisfy both equations.

---

  c. Answers may vary. I think the curves were drawn using the point *A* as the center and a radius of 3.2.

11.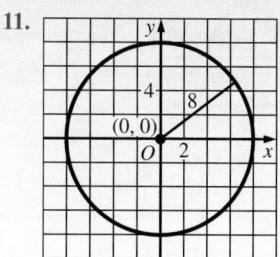

12.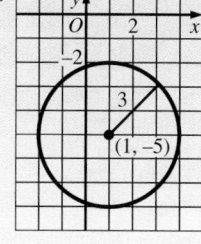

13.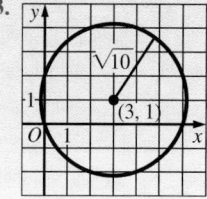

14–20. See answers in back of book.

Exercise Notes

Multicultural Note

Exs. 21, 22 The land traditionally occupied by the Inuit spans 6000 miles, from Siberia through Alaska and Canada to Greenland. The people in that area were divided into three major cultures: the Alaskan, the Central, and the Greenland. Of these groups, only the Central Inuit of Canada lived in igloos during the winter, although others made temporary snowhouses when they traveled. The Alaskan and Greenland peoples generally spent their winters in sod houses, some of which were built to last for only one winter and others for many winters. In summer, most Inuit lived in tents made from sealskin or caribou skin. Today, the approximately 140,000 Inuits in the Arctic Region live in modern buildings.

Career Connection

Exs. 21, 22 Cultural anthropologists study societies such as the Iglulik people. A cultural anthropologist might look at the architecture of the igloo cluster and see how its design is related to the social framework of the Iglulik people and the environment in which they live. The social framework might include family and other relationships, the marriage ritual, religious ritual, cultural traditions, and politics. Cultural anthropologists generally study various peoples by spending long periods of time living in their community. Because societies are composed of so many elements, cultural anthropologists may have various backgrounds that include the study of anthropology, sociology, biology, literature, political science, art, philosophy, and religion.

 Communication: Writing

Exs. 25, 26 Students use the result of Ex. 25 to help them determine the steps that should be taken to find the equation of a circle in Ex. 26. They should see that the center of a circle is always the midpoint of its diameter.

Integrating Algebra

Exs. 27–31 Students need to rewrite the given equation to determine the center and radius of the circle in Ex. 27. Part of this rewriting requires students to factor a quadratic trinomial. In Exs. 28–31, students use substitution to determine if a point lies on a circle.

Connection   ARCHITECTURE

The Inuit peoples of the Arctic use domes in the construction of their traditional homes, called *igloos.* To construct an igloo, the builder carves a circle in the snow, then arranges snow blocks cut from the interior of the circle to form a dome. Igloos are sometimes built in clusters, with arched passageways connecting the domes.

21. The Iglulik people of Hudson Bay, Canada, built this igloo cluster. Dimensions are marked in feet. Write equations for ⊙*B* and ⊙*E*.

22. Use the Distance Formula to find the distance between the center of igloo *C* and the center of igloo *E*.

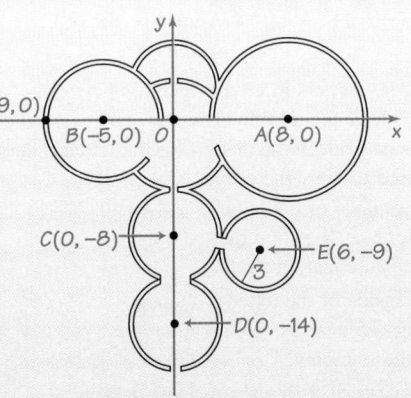

A diameter of each circle is marked. Find the coordinates of the center of the circle, the length of the radius, and an equation of each circle.

23.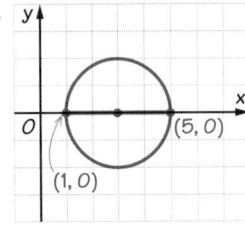

24.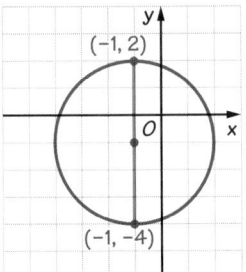

25.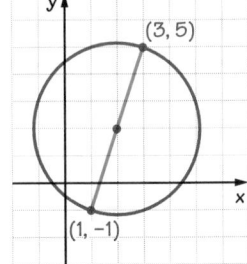

26. **Writing** Explain how to write an equation of a circle if you are given only the coordinates of the endpoints of a diameter of the circle, as in Exercise 25.

27. **Challenge** Sketch the circle with equation $x^2 - 4x + 4 + y^2 - 25 = 0$.

21. $(x + 5)^2 + y^2 = 16$; $(x - 6)^2 + (y + 9)^2 = 9$

22. $\sqrt{37} \approx 6.1$ ft

23. $(3, 0)$; 2; $(x - 3)^2 + y^2 = 4$

24. $(-1, -1)$; 3; $(x + 1)^2 + (y + 1)^2 = 9$

25. $(2, 2)$; $\sqrt{10}$; $(x - 2)^2 + (y - 2)^2 = 10$

26. Find the coordinates of the midpoint of the diameter. This point is the center (h, k) of the circle. Use the Distance Formula to find the distance r between point (h, k) and one of the given points. This is the radius r of the circle. Substitute for (h, k) and r in the general equation for a circle to obtain the required equation. (To find r, you could also use the Distance Formula to find the length of the diameter and divide by 2.)

27. The given equation is equivalent to $(x - 2)^2 + y^2 = 5^2$.

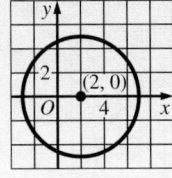

28. No; $9^2 + 0^2 \neq 18$.

29. Yes; $(2\sqrt{2})^2 + 2^2 = 12$.

ALGEBRA Tell whether the given point is on the circle with the given equation. Explain your answer.

28. $(9, 0)$; $x^2 + y^2 = 18$

29. $(2\sqrt{2}, 2)$; $x^2 + y^2 = 12$

30. $(3, 5)$; $(x - 3)^2 + (y + 1)^2 = 36$

31. $(7, 5\sqrt{3})$; $(x - 2)^2 + y^2 = 100$

32. SAT/ACT Preview Which point does *not* lie on the circle with equation $(x - 2)^2 + (y + 5)^2 = 36$?

A. $(2, 1)$ **B.** $(8, -5)$ **C.** $(-4, -5)$ **D.** $(2, -5)$ **E.** $(2, -11)$

ONGOING ASSESSMENT

33. Cooperative Learning Work with another person. You will each need a circular object and graph paper.

   a. Each of you should draw a set of axes on graph paper. Without showing the other person your object, place it on your paper and trace around it with a pencil.

   b. On a different piece of paper, write an equation of the circle you traced.

   c. Exchange equations with the other person. Sketch the graph of the equation you receive.

   d. Compare your sketch with the circular object the other person traced. Explain any differences.

SPIRAL REVIEW

Give the most specific name for each quadrilateral with the given vertices.
(Section 4.3)

34. $A(0, 0)$, $B(5, 0)$, $C(7, 4)$, $D(2, 4)$

35. $W(7, 0)$, $X(0, 3)$, $Y(-7, 0)$, $Z(0, -3)$

36. $P(1, 0)$, $Q(0, 2)$, $R(4, 4)$, $S(5, 2)$

37. $J(-1, -1)$, $K(-1, 2)$, $L(2, 2)$, $M(2, -1)$

In the diagram, $\overline{AB}$ and $\overline{CD}$ bisect each other at point E. Find each length.
(Section 1.7)

38. EB

39. CD

40. AB

41. CE

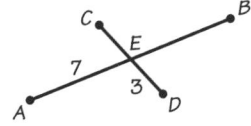

Tell whether each statement is *True* or *False*. If a statement is false, sketch a counterexample and justify your answer. *(Section 2.5)*

42. Every rhombus is a square.

43. Every rhombus is a parallelogram.

44. Every equiangular parallelogram is a square.

4.4 Equations of Circles   **193**

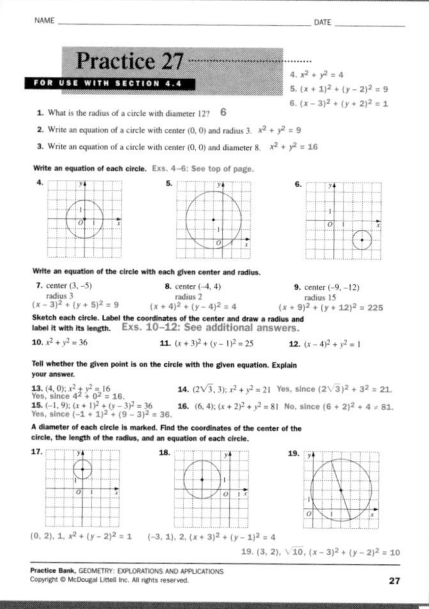
---

30. Yes; $(3 - 3)^2 + (5 + 1)^2 = 36$.

31. Yes; $(7 - 2)^2 + (5\sqrt{3})^2 = 100$.

32. D

33. Answers may vary. Check students' work.

34. parallelogram

35. rhombus

36. rectangle

37. square

38. 7

39. 6

40. 14

41. 3

42. False; a rhombus that is not a rectangle is not a square.

43. True.

44. False; some rectangles are not equilateral.

4.5 Coordinates and Proof

Variables and equations can help you study geometric figures and prove theorems about them. The first step is to give coordinates for points on the figure. The Exploration will help you become familiar with this process.

Objectives

- Place figures on coordinate axes and label their vertices.
- Plan and write coordinate geometry proofs.
- Use algebraic methods to verify conjectures about triangles and quadrilaterals.
- Improve reasoning skills.

Recommended Pacing

❖ **Core and Extended Courses**
 Section 4.5: 1 day
❖ **Block Schedule**
 Section 4.5: $\frac{1}{2}$ block
 (with Section 4.4)

Resource Materials

Lesson Support
Lesson Plan 4.5

Warm-Up Transparency 4.5

Overhead Visuals:
 Folder 5: Methods of Proof

Practice Bank: Practice 28

Study Guide: Section 4.5

Explorations Lab Manual:
 Diagram Master 2

Challenge Problems: Set 28

Technology
Scientific Calculator

Geometry Software

McDougal Littell Mathpack
 Geometry Inventor Activity Book:
 Activities 4, 7

Internet:
 http://www.hmco.com

Learn how to...

- place figures on coordinate axes and label their vertices
- plan and write coordinate geometry proofs

So you can...

- use algebraic methods to verify conjectures about triangles and quadrilaterals
- improve your reasoning skills

EXPLORATION
COOPERATIVE LEARNING

Placing a Parallelogram on a Coordinate Plane

Work in a group of four students.
You will need: graph paper, scissors, straightedge

1 On graph paper, each student should draw a quadrilateral with vertices $E(0, 0)$, $F(6, 8)$, $G(11, 8)$, and $H(5, 0)$. Cut out your quadrilateral and label the vertices on both sides. Color it if you wish.

2 On a separate piece of graph paper, draw coordinate axes. Place the cut-out polygon on the coordinate plane so that all four vertices have integer coordinates. Each student should use a different placement.

3 Trace your cut-out polygon onto your graph paper. Label the vertices E, F, G, and H and write their coordinates.

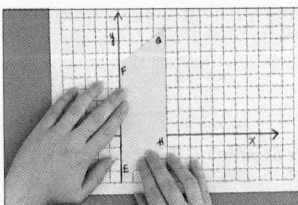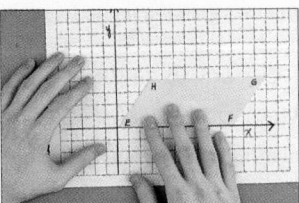

4 Use the Slope Formula to show that $EFGH$ is a parallelogram.

5 Use the Midpoint Formula to show that the midpoint of $\overline{EG}$ is the same as the midpoint of $\overline{FH}$. Which theorem from Section 2.5 does this illustrate?

6 Compare your answers with those of others in your group. How are your answers different? How are they the same?

7 Does the position of a parallelogram on a coordinate plane affect whether or not the diagonals bisect each other? Explain.

Warm-Up Exercises

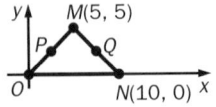

1. Find OM and MN. $5\sqrt{2}$; $5\sqrt{2}$
2. Find the slope of $\overline{OM}$ and $\overline{MN}$. 1; –1
3. Find the coordinates of the midpoints of $\overline{OM}$ and $\overline{MN}$. $P(2.5, 2.5)$; $Q(7.5, 2.5)$
4. Find PQ and ON. What do you notice? 5; 10; PQ is half of ON.

Exploration Note

Purpose
The purpose of this Exploration is to have students discover that properties of geometric figures are not affected by the placement of the figure on a coordinate plane.

Materials/Preparation
Each group of students needs graph paper, scissors, and a straightedge.

Procedure
Each student places a congruent quadrilateral in a different location on the coordinate plane so that all four vertices have integer coordinates. Students then use these coordinates to show that the quadrilateral is a

parallelogram, and that its diagonals bisect each other.

Closure
Discuss the fact that although the coordinates for the midpoints and the slopes of the sides of the quadrilateral are different for each placement, the basic properties that the quadrilateral is a parallelogram and that the diagonals bisect each other are not affected by the placement of the quadrilateral on a coordinate plane.

Explorations Lab Manual
See the Manual for more commentary on this Exploration. Diagram Master 2

In the Exploration, you showed that the diagonals of a given parallelogram bisect each other. To prove that this is true for *all* parallelograms, you must label a parallelogram in a general way, using variables to represent coordinates. Use only as many variables as necessary.

EXAMPLE 1

Prove that the diagonals of a parallelogram bisect each other.

SOLUTION

Plan ahead: Use variables for the coordinates of the vertices. Use the Midpoint Formula to show that the midpoint of $\overline{EG}$ is the same as the midpoint of $\overline{FH}$.

Proof

You need only three variables to label points *E, F, G,* and *H.* Use the definition of a parallelogram: the slopes of opposite sides are equal.

Step 1 Place the parallelogram with vertex *E* at (**0, 0**) and one side along the *x*-axis.

Step 3 Use (**b, c**) for point *H.*

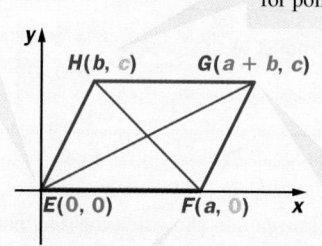

Step 4 Point *G* must have coordinates (**a + b, c**), so that the slopes of $\overline{EH}$ and $\overline{FG}$ will be equal.

Step 2 Use (**a, 0**) for point *F* because it is on the *x*-axis.

Now use the Midpoint Formula to find the midpoints of diagonals $\overline{EG}$ and $\overline{FH}$.

$$\text{midpoint of } \overline{EG} = \left(\frac{x_1 + x_2}{2}, \frac{y_1 + y_2}{2}\right)$$

$$= \left(\frac{0 + (a + b)}{2}, \frac{0 + c}{2}\right)$$

$$= \left(\frac{a + b}{2}, \frac{c}{2}\right)$$

$$\text{midpoint of } \overline{FH} = \left(\frac{x_1 + x_2}{2}, \frac{y_1 + y_2}{2}\right)$$

$$= \left(\frac{a + b}{2}, \frac{0 + c}{2}\right)$$

$$= \left(\frac{a + b}{2}, \frac{c}{2}\right)$$

The midpoints of the diagonals are the same point because they have identical coordinates. Therefore, the diagonals bisect each other.

4.5 Coordinates and Proof **195**

About Example 1

Writing Proofs
Example 1 introduces students to the idea of a coordinate geometry proof. When discussing the proof, stress that the placement of the figure with vertex *E* at the origin and $\overline{EF}$ along the *x*-axis minimizes the number of variables needed and, in so doing, simplifies the algebra involved in finding the midpoints of $\overline{EG}$ and $\overline{FH}$.

Additional Example 1

Prove that any point on the perpendicular bisector of a segment is equidistant from the endpoints of the segment.
Plan ahead: Use variables for the coordinates of the endpoints of the segment and for the point on the perpendicular bisector. Use the Distance Formula to show that AC and BC are equal.
Proof: You need only two variables to label points *A*, *B*, and *C*. Use what you already know about perpendicular bisectors: they divide the segment into two equal parts and they are perpendicular to the segment.
Step 1 Place the segment on the *x*-axis with one vertex at (–*a*, 0) and the other vertex at (*a*, 0).
Step 2 As the perpendicular bisector divides the segment and is perpendicular to it, it must be on the *y*-axis.
Step 3 Use *C*(0, *b*) for a point on the perpendicular bisector of $\overline{AB}$ because it lies on the *y*-axis.

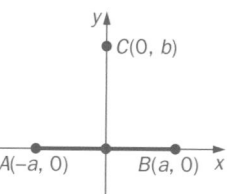

Now use the Distance Formula to find the distance from point *C* on the perpendicular bisector to each of the endpoints *A* and *B* of the segment.

$$AC = \sqrt{(x_2 - x_1)^2 + (y_2 - y_1)^2}$$
$$= \sqrt{(0 - (-a))^2 + (b - 0)^2}$$
$$= \sqrt{a^2 + b^2}$$
$$BC = \sqrt{(x_2 - x_1)^2 + (y_2 - y_1)^2}$$
$$= \sqrt{(0 - a)^2 + (b - 0)^2}$$
$$= \sqrt{a^2 + b^2}$$

The lengths of AC and BC are equal. Therefore, point *C* is equidistant from the endpoints *A* and *B*.

195

Think and Communicate

You may wish to extend question 2 by having students prove the result for a rhombus. Ask students what else they can prove about the diagonals of a rhombus. (See Additional Example 2.)

Additional Example 2

Prove that the diagonals of a rhombus are perpendicular.

Plan ahead: Place a rhombus on a coordinate plane. Find the slope of each diagonal of the rhombus. Show that the product of the slopes of the diagonals is –1.

Proof: If the length of each side of the rhombus is c, then point X has coordinates $(c, 0)$.

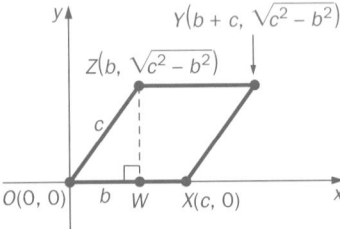

Since $\overline{OZ}$ is a side of the rhombus, it has length c, and if b is the x-coordinate of Z, $OW = b$. The Pythagorean theorem can be used to find the length of $\overline{ZW}$.

$$ZW^2 + b^2 = c^2$$
$$ZW^2 = c^2 - b^2$$
$$ZW = \sqrt{c^2 - b^2}$$

The y-coordinates of points Z and Y are $\sqrt{c^2 - b^2}$. Because $OXYZ$ is a parallelogram, the x-coordinate of point Y is $b + c$.

Use the Slope Formula to find the slope of each diagonal of the rhombus.

$$\text{slope of } \overline{OY} = \frac{y_2 - y_1}{x_2 - x_1}$$
$$= \frac{\sqrt{c^2 - b^2} - 0}{b + c - 0}$$
$$= \frac{\sqrt{c^2 - b^2}}{b + c}$$

$$\text{slope of } \overline{XZ} = \frac{y_2 - y_1}{x_2 - x_1}$$
$$= \frac{0 - \sqrt{c^2 - b^2}}{c - b}$$
$$= -\frac{\sqrt{c^2 - b^2}}{c - b}$$

The product of the slopes is
$$\left(\frac{\sqrt{c^2 - b^2}}{b + c}\right)\left(-\frac{\sqrt{c^2 - b^2}}{c - b}\right)$$
$$= -\frac{c^2 - b^2}{c^2 - b^2} = -1.$$

The product of the slopes is –1, so the diagonals are perpendicular. Therefore, the diagonals of any rhombus are perpendicular.

196

THINK AND COMMUNICATE

1. Show that the slopes of $\overline{EF}$ and $\overline{HG}$ are equal in the parallelogram in Example 1. Then show that the slopes of $\overline{EH}$ and $\overline{FG}$ are equal.

2. Vivian says that the result of Example 1 is true for a rhombus. Explain her reasoning.

The proof in Example 1 is a **coordinate geometry proof** that uses the Midpoint Formula. In Example 2, the Distance Formula and the Slope Formula are used in another proof.

EXAMPLE 2   Connection: Algebra

Prove that the diagonals of a square are congruent and perpendicular.

SOLUTION

Plan ahead: Place a square on a coordinate plane. Use the Distance Formula to show that the diagonals are congruent. Use the Slope Formula to show that the product of the slopes of the diagonals is − 1.

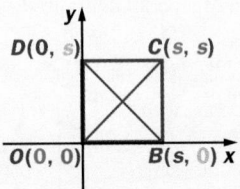

Proof

If the square has sides of length s, then it can be placed on a coordinate plane as shown. B and D are s units from O. C is s units from B and D.

Use the Distance Formula to show that the diagonals are congruent.

$$OC = \sqrt{(x_2 - x_1)^2 + (y_2 - y_1)^2} \qquad BD = \sqrt{(x_2 - x_1)^2 + (y_2 - y_1)^2}$$
$$= \sqrt{(s - 0)^2 + (s - 0)^2} \qquad\qquad = \sqrt{(0 - s)^2 + (s - 0)^2}$$
$$= \sqrt{s^2 + s^2} \qquad\qquad\qquad = \sqrt{(-s)^2 + s^2}$$
$$= \sqrt{2s^2} \qquad\qquad\qquad\qquad = \sqrt{2s^2}$$
$$= s\sqrt{2} \qquad\qquad\qquad\qquad = s\sqrt{2}$$

Because $OC = BD$, the *diagonals are congruent.*

Use the Slope Formula to show that the diagonals are perpendicular.

$$\text{slope of } \overline{OC} = \frac{y_2 - y_1}{x_2 - x_1} \qquad\qquad \text{slope of } \overline{BD} = \frac{y_2 - y_1}{x_2 - x_1}$$
$$= \frac{s - 0}{s - 0} \qquad\qquad\qquad\qquad = \frac{s - 0}{0 - s}$$
$$= 1 \qquad\qquad\qquad\qquad\qquad = -1$$

The product of the slopes is − 1, so the *diagonals are perpendicular.*

Therefore, the diagonals of any square are congruent and perpendicular.

196   Chapter 4   *Coordinates in Geometry*

Think and Communicate

1. slope of $\overline{EF} = \frac{0 - 0}{a - 0} = 0$;

   slope of $\overline{HG} = \frac{c - c}{(a + b) - b} = 0$

2. Yes; every rhombus is a parallelogram.

3. Adam used the facts that the sides of a square are congruent and that adjacent sides are perpendicular. Clint used the facts that the diagonals are

congruent, perpendicular, and bisect each other.

4. Yes; his diagram makes no assumptions about the diagonals of a square, so it can be used to prove the theorem in Example 2.

5. No; his diagram assumes that the diagonals are congruent and perpendicular. He cannot use his diagram to prove these properties.

6. Yes; every square is a parallelogram.

7. No; some rectangles are not squares.

THINK AND COMMUNICATE

3. Adam and Clint each place squares on a coordinate plane, as shown. What properties of a square did Adam use in placing his square? What properties did Clint use?

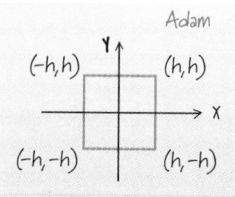

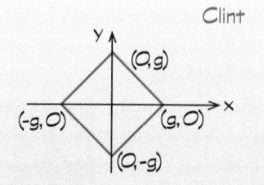

4. Can Adam use his diagram to prove the theorem in Example 2? Explain.

5. Can Clint use his diagram to prove the theorem in Example 2? Explain.

6. Example 1 shows that the diagonals of a parallelogram bisect each other. A square is a parallelogram. Can you conclude that the diagonals of a square bisect each other? Explain.

7. Example 2 shows that the diagonals of a square are congruent and perpendicular. A square is a rectangle. Can you conclude that the diagonals of a rectangle are congruent and perpendicular? Explain.

☑ CHECKING KEY CONCEPTS

1. Three vertices of a rectangle are $(0, 0)$, $(b, 0)$, and $(0, h)$. Sketch the rectangle and find the coordinates of the fourth vertex.

2. Show how to use the coordinates a, b, and c to label the vertices of a triangle on a coordinate plane.

3. Describe the steps needed to prove that the diagonals of a rectangle are congruent.

4.5 Exercises and Applications

Extra Practice exercises on page 667

Find the missing coordinates without using any new variables.

1. rectangle

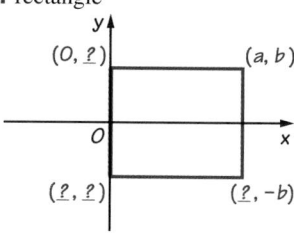

2. parallelogram

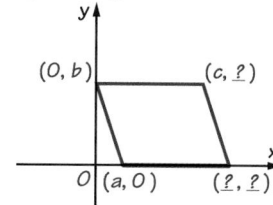

3. isosceles triangle

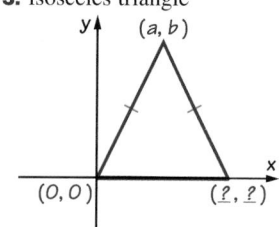

4.5 Coordinates and Proof **197**

Think and Communicate

Questions 3–5 reinforce the idea that there is more than one way to place a figure in the coordinate plane when writing a coordinate geometry proof. Students should realize, however, that the properties used to place the figure and label the coordinates cannot be those that need to be proved. Discuss questions 6 and 7. It is important that students understand the difference between these two situations and why certain conclusions can be made or not made.

Checking Key Concepts

Visual Thinking
For question 3, ask students to create a diagram that demonstrates that the diagonals of a rectangle are congruent. Encourage them to label congruent sides, angles, and line segments. Ask them to present their sketches to the class and to discuss the related steps of proof. This activity involves the visual skills of *correlation* and *communication*.

Closure Question

When placing a polygon on a coordinate plane to write a proof, how many variables should be used? How can you place an isosceles right triangle in the coordinate plane in such a way that only one variable is used? Use only as many variables as necessary. Let a be the length of the legs of the triangle. Then the vertices of the triangle could be $(0, a)$, $(0, 0)$, and $(a, 0)$.

Apply⇔Assess

Checking Key Concepts

1. Check students' work. The fourth vertex is (b, h).

2. Answers may vary. An example is given.

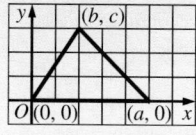

3. Answers may vary. An example is given. Place a rectangle on a coordinate plane. Use the Distance Formula to show that the diagonals are congruent.

Exercises and Applications

1. $(0, b)$, $(0, -b)$, $(a, -b)$

2. (c, b), $(a + c, 0)$

3. $(2a, 0)$

Suggested Assignment

❖ **Core Course**
Exs. 1–3, 8–19, 21–24, 28–37

❖ **Extended Course**
Exs. 1–8, 10–13, 16–37

❖ **Block Schedule**
Day 24 Exs. 1–3, 8–19, 21–24, 28–37

Geometric Thinking
Exs. 1–3 Students should be able to use a property of the figure to explain how they determined the coordinates of the vertices.

Interview Note: Application
Exs. 4–7 These exercises introduce students to the way computer engineers use coordinate systems to define the placement of components on a computer chip.

Topic Spiraling: Review
Ex. 7 This exercise uses the concept of a translation. Refer students to Section 1.1 if they need to review this type of transformation.

Problem Solving
Exs. 8–15 The problems in Exs. 8–10 are specific examples of the result proved in Ex. 11. Similarly, the problems in Exs. 12–14 are specific examples of the result proved in Ex. 15. Students should model their proofs of the general results on the results for the specific examples. This approach is often helpful when students are not certain how to start writing a general proof.

Writing Proofs
Exs. 11, 15, 24 Remind students to follow the format used in Example 1 and Example 2 when writing these coordinate geometry proofs. First, they should write a *Plan ahead* statement, including the steps to be used. Then, they should place the figure in the coordinate plane and assign coordinates based on the definition of the figure or on previously proved facts. Students should then write the proof, generally using one or more of these formulas: Slope Formula, Distance Formula, Midpoint Formula.

Journal Entry
Exs. 11, 15, 16, 19, 24, 25 Suggest that students keep a list of the results proved in these exercises. They can also include a plan for each proof and a diagram.

198

INTERVIEW Marc Hannah

Look back at the article on pages 162–164.

On computer chips, like this one that Marc Hannah designed, components may be used in several locations. In designing a chip, coordinates define the placements of components. A grid has been placed on the chip.

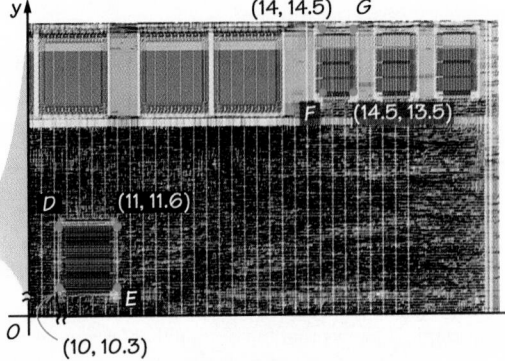

4. The three Random Access Memory (RAM) components are congruent rectangles. Supply the missing coordinates for points *A*, *B*, and *C* in the diagram below.

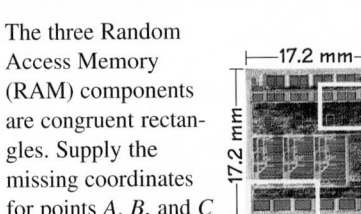

5. The three register files are rectangles. Supply the missing coordinates for points *D* and *E* in the diagram above.

6. The three Address Units are rectangles. Supply the missing coordinates for points *F* and *G* in the diagram above.

7. TRANSFORMATIONS What type of transformation moves each of the components to a congruent component? Explain.

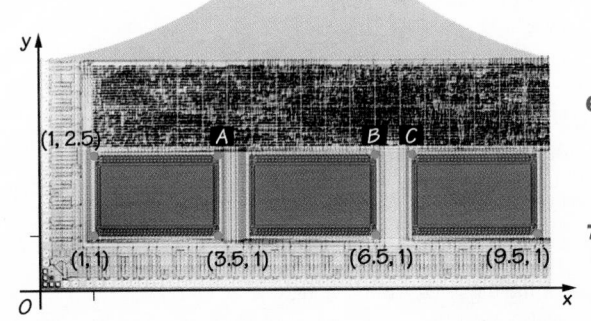

For each triangle *ABO*, find the midpoint *M* of the hypotenuse. Then find the lengths *MA*, *MB*, and *MO*.

8.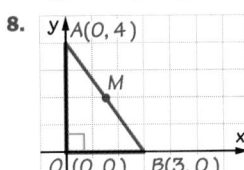

9.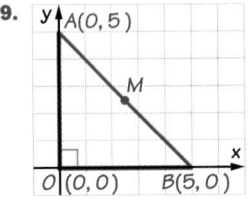

10.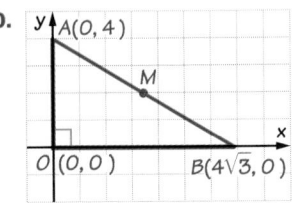

11. Draw a right triangle like the ones in Exercises 8–10, using variables for coordinates. Use your figure to prove that the midpoint of the hypotenuse of a right triangle is the same distance from all three vertices of the triangle.

198   Chapter 4 *Coordinates in Geometry*

---

4. $A(3, 2)$; $B(5.5, 2)$; $C(8, 2)$

5. $D(10, 11.6)$; $E(11, 10.6)$

6. $F(14, 13.5)$; $G(14.5, 14.5)$

7. translation; Each component is the image of a congruent component under a horizontal translation.

8. $(1.5, 2)$; $MA = MB = MO = 2.5$

9. $(2.5, 2.5)$; $MA = MB = MO = \frac{5}{2}\sqrt{2}$

10. $(2\sqrt{3}, 2)$; $MA = MB = MO = 4$

11. Answers may vary. An example is given.

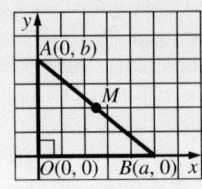

Using the Midpoint Formula, $M = \left(\frac{a}{2}, \frac{b}{2}\right)$.
By the Distance Formula, $MA = MB = MO = \frac{1}{2}\sqrt{a^2 + b^2}$.

12–14. Let *M* and *N* be the midpoints of the sides of the triangle.

12. $M = (1, 2)$ and $N = (4, 2)$. The slope of $\overline{MN} = 0 =$ slope of the base, so $\overline{MN}$ is parallel to the base.
$MN = 3 = \frac{1}{2} \times$ length of base

13. $M = (2, 2)$ and $N = (4.5, 2)$. The slope of $\overline{MN} = 0 =$ slope of the base, so $\overline{MN}$ is parallel to the base.
$MN = 2.5 = \frac{1}{2} \times$ length of base

The midpoints of two sides of a triangle are endpoints of a segment. Show that the segment is parallel to the third side and half as long.

12.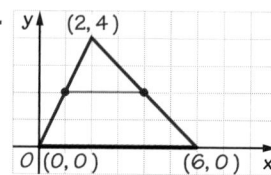

13.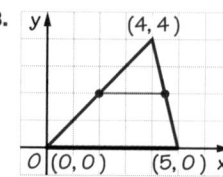

14.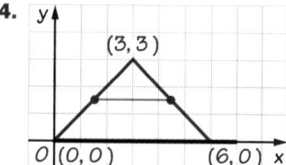

15. Draw a triangle like the ones in Exercises 12–14, using variables for coordinates. Use your figure to prove the result suggested by those exercises.

16. Use the diagram of rectangle *PQRS*.
 a. Find the missing coordinates.
 b. Find the lengths of the diagonals.
 c. Describe your results in part (b) as a theorem.

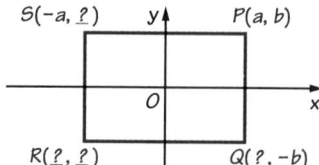

Use the diagram of quadrilateral *OKLM*. The midpoints of the sides are *E*, *F*, *G*, and *H*.

17. Find the coordinates of *E*, *F*, *G*, and *H*.

18. Find the midpoints of $\overline{EG}$ and $\overline{FH}$.

19. Describe your results from Exercise 18 as a theorem.

20. Writing Why do you think the coordinates 2*a*, 2*b*, 2*c*, 2*d*, and 2*e* are used in the diagram instead of *a*, *b*, *c*, *d*, and *e*?

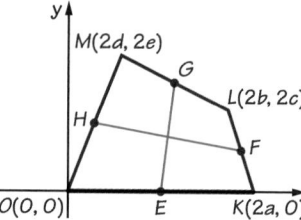

Open-ended Problems For each type of quadrilateral, draw several examples on graph paper. Find the midpoints of the sides and join them to form a quadrilateral. Tell what kind of quadrilateral is formed.

21. rectangle   **22.** rhombus   **23.** parallelogram

24. Prove that the midpoints of the sides of any quadrilateral form a parallelogram. Use a diagram similar to the one used in Exercises 17–20.

25. Challenge Use the coordinates given in the diagram below. Complete the proof.
 Given: The diagonals of parallelogram *OXYZ* are perpendicular.
 Prove: *OXYZ* is a rhombus.
 a. Find the slope of each diagonal.
 b. The diagonals are perpendicular, so the product of their slopes is −1. Use this fact and your results from part (a) to complete:
 $$\frac{?}{?} \cdot \frac{?}{?} = -1$$
 c. Use algebra to solve the equation in part (b) for *p*.
 d. Show that all sides of *OXYZ* have length *p*. Therefore, *OXYZ* is a rhombus.

(Diagram: Z(q, r), Y(p + q, r), O(0, 0), X(p, 0))

4.5 Coordinates and Proof   **199**

Apply⟺Assess

Exercise Notes

 Using Technology
Exs. 12–14 You may wish to suggest that students use Activity 4 in the *Geometry Inventor Activity Book* to show that the segment connecting the midpoints of two sides of a triangle is one half the length of the third side of the triangle.

Geometric Thinking
Ex. 20 Students should note the approach taken in this exercise to make the calculation of the midpoints easier. Whenever a proof involves the calculation of midpoints, students should name the coordinates with variable expressions of the form 2*x*. This technique can be used in Ex. 24 as well.

 Communication: Drawing
Exs. 21–24 Ask students to share their drawings for Exs. 21–23. You might have students make their drawings on overhead transparencies so they can be shown easily to the class. Students could also include some examples of quadrilaterals that are not parallelograms. The result proved in Ex. 24 may be surprising to some students since it is not obvious that this would be true for all quadrilaterals.

Challenge
Ex. 25 The algebra required in this exercise is fairly complex. It involves multiplying binomials and solving a simple quadratic equation for one variable. Note that this exercise is the converse of the result proved in Additional Example 2 in the side column of page 196.

---

14. *M* = (1.5, 1.5) and *N* = (4.5, 1.5). The slope of $\overline{MN}$ = 0 = slope of the base, so $\overline{MN}$ is parallel to the base. $MN = 3 = \frac{1}{2} \times$ length of base

15. Answers may vary. An example is given.

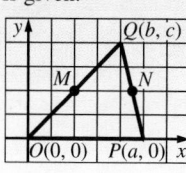

Using the Midpoint Formula, $M = \left(\frac{b}{2}, \frac{c}{2}\right)$ and $N = \left(\frac{a+b}{2}, \frac{c}{2}\right)$. The slope of $\overline{MN}$ = 0 = slope of $\overline{OP}$, so $\overline{MN} \parallel \overline{OP}$. By the Distance Formula, $MN = \frac{a}{2} = \frac{1}{2} \times OP$. This proves that the segment that joins the midpoints of two sides of a triangle is parallel to the third side and half as long as the third side.

16. a. $Q(a, -b)$, $S(-a, b)$, $R(-a, -b)$
 b. $PR = QS = 2\sqrt{a^2 + b^2}$
 c. The diagonals of a rectangle are congruent.

17. $E(a, 0)$, $F(a + b, c)$, $G(b + d, c + e)$, $H(d, e)$

18. $\left(\frac{a+b+d}{2}, \frac{c+e}{2}\right)$ is the midpoint of $\overline{EG}$ and of $\overline{FH}$.

19. The segments joining the midpoints of the opposite sides of a quadrilateral bisect each other.

20. If 2*a*, 2*b*, 2*c*, 2*d*, and 2*e* are used, the coordinates of *E*, *F*, *G*, and *H* do not involve fractions.

21–23. Check students' work.

21. rhombus

22. rectangle

23. parallelogram

24, 25. See answers in back of book.

Exercise Notes

Cooperative Learning

Ex. 26 This cooperative learning activity gives students additional practice using the characteristics of a figure to assign coordinates using the minimum number of variables.

Integrating Algebra

Ex. 27 Part (b) of this exercise requires students to solve a system of equations in two variables.

Assessment Note

Ex. 28 Students who can answer all four parts of this exercise correctly will have demonstrated a firm understanding of the mathematics in this section. For others, a discussion of the answers may help to clear up any misunderstandings and difficulties encountered in this section.

Practice 28 for Section 4.5

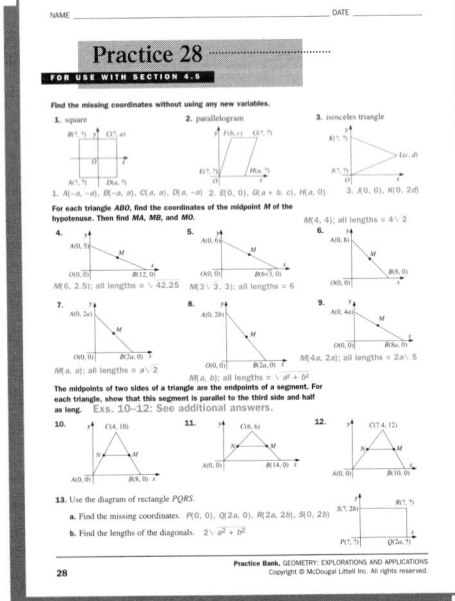

26. Cooperative Learning Work with another person. Each of you should draw a polygon on a coordinate plane, labeling some coordinates with variables but leaving other coordinates unlabeled. Write what kind of polygon it is. Exchange papers with your partner and supply the missing coordinates, using as few new variables as possible.

27. 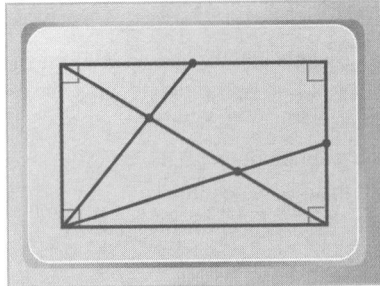 **Technology** Use geometry software to draw segments from one vertex of a rectangle to the midpoints of two other sides, as shown. Draw the diagonal of the rectangle that crosses these segments.

a. Show that the segments divide the diagonal into three equal parts.

b. **ALGEBRA** Prove the result from part (a) using a coordinate geometry proof. (*Hint:* Write equations for the lines that contain the segments and diagonal. Solve systems of equations to find the intersection points.)

ONGOING ASSESSMENT

28. Use the figure at the right. Assume $a > b$.

a. Prove that the diagonals of *PQRS* are congruent.

b. Label the midpoints of the diagonals *M* and *N* and find their coordinates.

c. Show that $MN = \frac{1}{2}(PS - QR)$.

d. **Writing** Write a paragraph that summarizes the main steps of your proof.

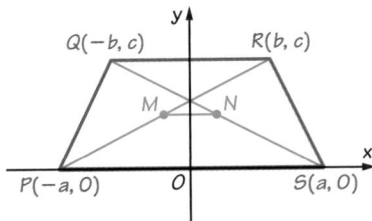

SPIRAL REVIEW

Write an equation of the circle with each given center and radius.

(Section 4.4)

29. center $(7, 0)$
radius 6

30. center $(2, -5)$
radius 5

31. center $(-1, 4)$
radius 7

Find the sum of the measures of the interior angles of each polygon.

(Section 2.4)

32. pentagon

33. hexagon

34. 12-gon

Find each unknown length. *(Section 3.6)*

35.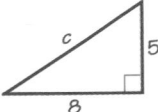

36.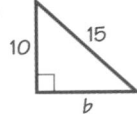

37.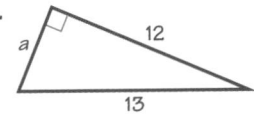

200 Chapter 4 *Coordinates in Geometry*

26. Check students' work.

27. a. Check students' work.

b. Answers may vary. An example is given.

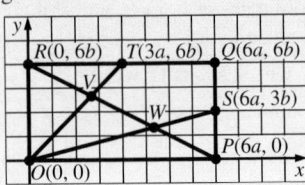

$\overleftrightarrow{OT}$ has equation $y = \frac{2b}{a}x$, $\overleftrightarrow{OS}$ has equation $y = \frac{b}{2a}x$, and $\overleftrightarrow{PR}$ has equation $y = -\frac{b}{a}x + 6b$. $\overleftrightarrow{OT}$ and $\overleftrightarrow{PR}$ intersect at the point *V* at which $\frac{2b}{a}x = -\frac{b}{a}x + 6b$; $x = 2a$ and $y = -\frac{b}{a}(2a) + 6b = 4b$. *V* is the point $(2a, 4b)$. $\overleftrightarrow{OS}$ and $\overleftrightarrow{PR}$ intersect at the point *W* at which $\frac{b}{2a}x = -\frac{b}{a}x + 6b$; $x = 4a$ and $y = -\frac{b}{a}(4a) + 6b = 2b$. *W* is the point $(4a, 2b)$. Using the Distance Formula, $PR = 6\sqrt{a^2 + b^2}$ and $RV = VW = WP = 2\sqrt{a^2 + b^2}$. Therefore, $\overline{OT}$ and $\overline{OS}$ divide $\overline{PR}$ into three congruent parts.

28. See answers in back of book.

29. $(x - 7)^2 + y^2 = 36$

30. $(x - 2)^2 + (y + 5)^2 = 25$

31. $(x + 1)^2 + (y - 4)^2 = 49$

32. 540°

33. 720°

34. 1800°

35. $\sqrt{89} \approx 9.43$

36. $5\sqrt{5} \approx 11.18$

37. 5

4.6 Coordinates in Three Dimensions

Learn how to...

- **find coordinates in three dimensions**
- **use the Distance Formula and the Midpoint Formula in three dimensions**

So you can...

- **identify and describe relationships between geometric figures in three dimensions**

The element carbon is the basis of all life on Earth and is one of the key components of coal, oil, and natural gas. Atoms of carbon combine with other atoms to form molecules that have a three-dimensional structure.

The molecules of the natural gas *ethane* contain two carbon atoms and six hydrogen atoms. To describe the positions of the atoms within the molecule, you can use a **three-dimensional coordinate system**. The table shows the coordinates of the center of each atom.

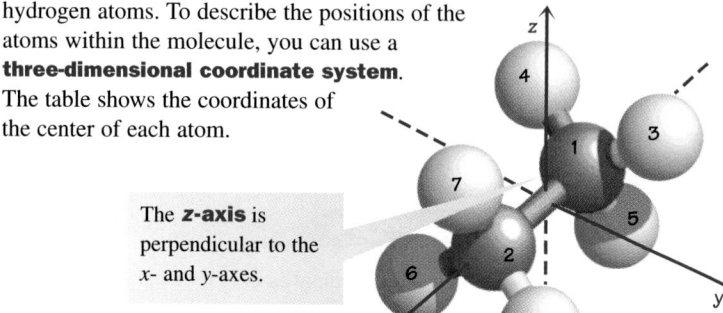

The **z-axis** is perpendicular to the *x*- and *y*-axes.

Atom	Type	x	y	z
1	carbon	−0.76	0.00	0.00
2	carbon	0.76	0.00	0.00
3	hydrogen	−1.12	0.89	0.51
4	hydrogen	−1.12	−0.89	0.51
5	hydrogen	−1.12	0.00	−1.03
6	hydrogen	1.12	−0.89	−0.51
7	hydrogen	1.12	0.00	1.03
8	hydrogen	1.12	0.89	−0.51

Coordinates are given to the nearest hundredth of an angstrom (Å). 10^{10} Å = 1 meter.

THINK AND COMMUNICATE

1. Name three atoms that have the same *x*-coordinate.

2. a. What do you notice about the coordinates of the two carbon atoms?

   b. What point is halfway between the centers of these two atoms?

4.6 Coordinates in Three Dimensions **201**

ANSWERS Section 4.6

Think and Communicate

1. atoms 3, 4, and 5, or atoms 6, 7, and 8

2. a. The *x*-coordinates are opposites, and the *y*-coordinates and *z*-coordinates are 0.

   b. the origin

Plan⇔Support

Objectives

- Find coordinates in three dimensions.
- Use the Distance Formula and the Midpoint Formula in three dimensions.
- Identify and describe relationships between geometric figures in three dimensions.

Recommended Pacing

❖ **Core and Extended Courses**
Section 4.6: 1 day

❖ **Block Schedule**
Section 4.6: $\frac{1}{2}$ block (with Chapter 4 Review)

Resource Materials

Lesson Support
Lesson Plan 4.6

Warm-Up Transparency 4.6

Practice Bank: Practice 29

Study Guide: Section 4.6

Explorations Lab Manual: Diagram Masters 2, 5

Challenge Problems: Set 29

Assessment Book: Test 17

Technology
Technology Book: Spreadsheet Activity 4

Graphing Calculator

Internet: http://www.hmco.com

Warm-Up Exercises

Find the distance between each pair of points.

1. (−5, 2) and (−1, 4) $2\sqrt{5}$

2. (1, −3) and (7, 5) 10

Find the midpoint of the segment joining each pair of points.

3. (4, 1) and (0, −3) (2, −1)

4. (−3, −1) and (9, 4) $\left(3, \frac{3}{2}\right)$

5. A rectangle has vertices (4, −3), (4, 4), and (8, 4). What are the coordinates of the fourth vertex? (8, −3)

Learning Styles: Kinesthetic

Students with a kinesthetic learning style may benefit from building models of the figures in this section. A model of the ethane molecule can be made from gum drops and toothpicks. Coordinate axes can be made from straws held together at the origin with clay. Units can be marked on the straws. Students can then make either rectangular or triangular prisms and place them on the coordinate axes.

Section Note

Spatial Reasoning

When drawing three axes, the x-axis should appear to come out of the paper, and extend beyond it. To achieve this result, the actual angle between the positive x-axis and positive y-axis should be about 140°. The tick marks on the x-axis are more closely spaced than those on the other axes to make figures appear correct. Remind students that in a three-dimensional drawing, the parts that are hidden by other parts are represented by dashed lines. Discuss the points in the coordinate planes. Note that the three coordinate planes intersect at the origin, whose coordinates are (0, 0, 0). Ask students which coordinate is zero in each of the three coordinate planes.

Additional Example 1

Find the coordinates of vertices D and F of the triangular prism.

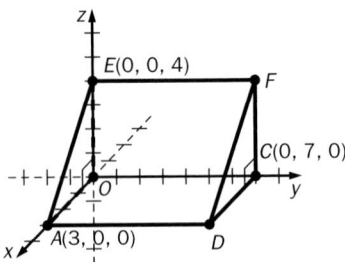

D has the same x-coordinate as A and the same y-coordinate as C. Its z-coordinate is 0. D has coordinates (3, 7, 0). F has the same y-coordinate as C and the same z-coordinate as E. Its x-coordinate is 0. F has coordinates (0, 7, 4).

202

The position of each point in space is given by an **ordered triple (x, y, z)**. For example, in the graph below, P has coordinates (**2, 3, 4**). To locate P, start at the origin. Move 2 units along the x-axis, 3 units parallel to the y-axis, and 4 units parallel to the z-axis.

The axes intersect at the *origin*, the point (0, 0, 0).

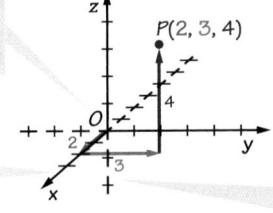

The **x-axis** appears to extend out from the plane of the paper.

The dashed part of each axis represents the negative direction.

Each pair of axes determines a *coordinate plane*. At least one coordinate of every point that lies in a coordinate plane is 0.

Every point in the **yz-plane** has x-coordinate 0.

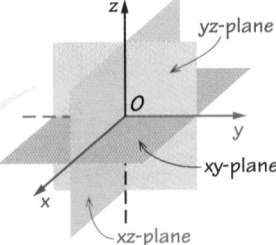

EXAMPLE 1

Find the coordinates of vertices *D* and *E* of the rectangular prism.

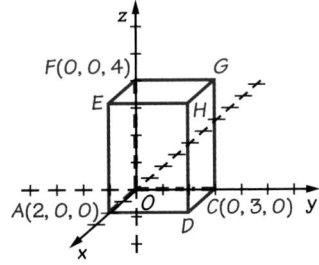

SOLUTION

D has the same x-coordinate as *A* and the same y-coordinate as *C*. Its z-coordinate is 0. *D* has coordinates (2, 3, 0).

E has the same x-coordinate as *A* and the same z-coordinate as *F*. Its y-coordinate is 0. *E* has coordinates (2, 0, 4).

THINK AND COMMUNICATE

3. In Example 1, what are the coordinates of *O*? of *G*? of *H*?

4. What is the z-coordinate of every point in the xy-plane?

202 Chapter 4 *Coordinates in Geometry*

Think and Communicate

3. (0, 0, 0); (0, 3, 4); (2, 3, 4)

4. 0

In two dimensions, each coordinate of the midpoint of a segment is the mean of the corresponding coordinates of the endpoints. This is true in three dimensions as well.

The Midpoint Formula in Three Dimensions

The midpoint of the segment that joins any two points $A(x_1, y_1, z_1)$ and $B(x_2, y_2, z_2)$ has these coordinates:

$$\text{coordinates of the} \atop \text{midpoint of } \overline{AB} = \left(\frac{x_1 + x_2}{2}, \frac{y_1 + y_2}{2}, \frac{z_1 + z_2}{2} \right)$$

EXAMPLE 2

The coordinates of *E* are (2, 0, 2), and the coordinates of *C* are (−1, 3, −2). Find the coordinates of the midpoint of $\overline{EC}$.

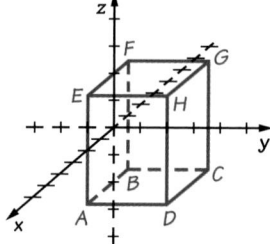

SOLUTION

Find the mean of each pair of corresponding coordinates:

$$\text{coordinates of} \atop \text{midpoint of } \overline{EC} = \left(\frac{2 + (-1)}{2}, \frac{0 + 3}{2}, \frac{2 + (-2)}{2} \right)$$

$$= \left(\frac{1}{2}, \frac{3}{2}, \frac{0}{2} \right)$$

$$= \left(\frac{1}{2}, 1\frac{1}{2}, 0 \right)$$

The distance between two points in three dimensions can be found using a formula similar to the Distance Formula for two dimensions.

The Distance Formula in Three Dimensions

The distance AB between two points in space $A(x_1, y_1, z_1)$ and $B(x_2, y_2, z_2)$ is:

$$AB = \sqrt{(x_2 - x_1)^2 + (y_2 - y_1)^2 + (z_2 - z_1)^2}$$

Section Note

Research
Students may see the transition from two dimensions to three dimensions reflected in the Midpoint Formula and the Distance Formula, and wonder if these ideas can be extended to higher dimensions. (Yes.) Interested students may want to explore how the concept of four dimensions can be interpreted mathematically.

About Example 2

Spatial Reasoning
With the coordinates of points *E* and *C* given, the coordinates of all the other vertices of the prism can be determined. You may wish to have students find the coordinates of vertices *A, B, D, F, G,* and *H,* and the midpoints of $\overline{FD}$, $\overline{GA}$, and $\overline{HB}$. They will see that all four segments intersect at the midpoint.

Additional Example 2

The coordinates of *P* are (3, 3, 4), and the coordinates of *I* are (0, −2, −1). Find the coordinates of the midpoint of $\overline{PI}$.

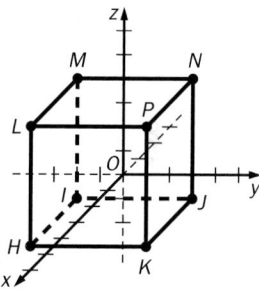

Find the mean of each pair of corresponding coordinates:
coordinates of the midpoint of $\overline{PI}$

$$= \left(\frac{3 + 0}{2}, \frac{3 + (-2)}{2}, \frac{4 + (-1)}{2} \right)$$

$$= \left(\frac{3}{2}, \frac{1}{2}, \frac{3}{2} \right)$$

$$= \left(1\frac{1}{2}, \frac{1}{2}, 1\frac{1}{2} \right)$$

Section Note

Alternate Approach
The Distance Formula in Three Dimensions is derived in Ex. 29 on page 206. You may want to use the method of this exercise with a particular example, such as finding the distance between points *E* and *C* from Example 2, before discussing the formula with the class.

Additional Example 3

Use the data for the ethane molecule shown on page 201. Find the distance between the centers of atoms 5 and 8 to the nearest hundredth angstrom.

Atom 5 has coordinates $(-1.12, 0, -1.03)$, and atom 8 has coordinates $(1.12, 0.89, -0.51)$.

$d =$

$\sqrt{(1.12-(-1.12))^2 + (0.89-0)^2 + (-0.51-(-1.03))^2}$

$= \sqrt{2.24^2 + 0.89^2 + 0.52^2}$

$= \sqrt{6.0801}$

≈ 2.47

The distance between atoms 1 and 3 is about 2.47 Å.

Checking Key Concepts

Using Manipulatives
Most students should be able to complete these questions. Students who have difficulty with questions 1–3 may need to use a physical model. You could make a model of this figure using a triangular block and two index cards. Tape the cards at a 90° angle; these represent the positive halves of the xz- and yz-planes. Place the cards upright on a desk, with the desktop representing the positive xy-plane. The block can then be placed in the appropriate position. Although the scale of the block may not be the same as the figure, you can ask students questions such as "Which two points have the same x-coordinates?" to get them to find the coordinates for points P, L, and M in the figure.

Closure Question

Compare the coordinates of points, the Midpoint Formula, and the Distance Formula for two and three dimensions. In two dimensions, each point has two coordinates, while in three dimensions, each point has three coordinates. The Midpoint Formulas for each dimension are found by averaging corresponding coordinates. The Distance Formulas are similar, except that in three dimensions, the square of the difference of the z-coordinates of the two points is added under the radical sign.

EXAMPLE 3   Application: Chemistry

Use the data for the ethane molecule shown on page 201. Find the distance between the centers of atoms 1 and 3 to the nearest hundredth of an angstrom.

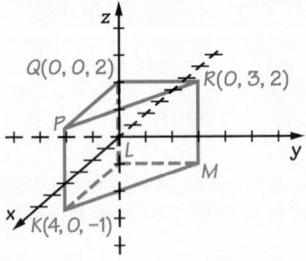

SOLUTION

Atom 1 has coordinates $(-0.76, 0, 0)$ and atom 3 has coordinates $(-1.12, 0.89, 0.51)$.

$$d = \sqrt{(-1.12 - (-0.76))^2 + (0.89 - 0)^2 + (0.51 - 0)^2}$$
$$= \sqrt{(-0.36)^2 + (0.89)^2 + (0.51)^2}$$
$$= \sqrt{1.1818}$$
$$\approx 1.09$$

The distance between atoms 1 and 3 is about 1.09 Å.

✓ CHECKING KEY CONCEPTS

Use the triangular prism. Give the coordinates of each vertex.

1. P
2. L
3. M

Find the coordinates of the midpoint of the segment with the given endpoints.

4. $(7, 0, 6)$ and $(5, -8, 1)$ 5. $(2, 6, -4)$ and $(-4, 2, -4)$

6. $\left(1, \frac{1}{2}, -2\right)$ and $\left(-1, \frac{1}{2}, 3\right)$ 7. $(0, 0, 0)$ and $(4, -6, -1)$

Find the length of the segment with the given endpoints.

8. $(2, 4, 3)$ and $(5, 0, -3)$ 9. $(1, 9, -3)$ and $(0, 12, -5)$

204   Chapter 4 *Coordinates in Geometry*

Checking Key Concepts

1. $(4, 0, 2)$

2. $(0, 0, -1)$

3. $(0, 3, -1)$

4. $\left(6, -4, \frac{7}{2}\right)$

5. $(-1, 4, -4)$

6. $\left(0, \frac{1}{2}, \frac{1}{2}\right)$

7. $\left(2, -3, -\frac{1}{2}\right)$

8. $\sqrt{61} \approx 7.81$

9. $\sqrt{14} \approx 3.74$

4.6 | Exercises and Applications

*Extra Practice
exercises on
page 667*

For Exercises 1–9, use the rectangular prism.
Give the coordinates of each vertex.

1. *P*
2. *R*
3. *S*
4. *T*
5. *O*
6. *V*

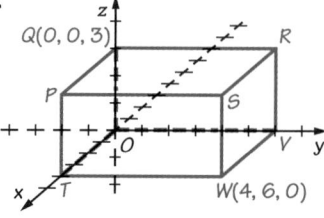

Tell which coordinate plane is parallel to each face of the prism.

7. *PQRS*
8. *SRVW*
9. *PSWT*

Tell which coordinate plane contains each point.

10. $A(2, 4, 0)$
11. $B(1, 0, 2)$
12. $C(0, -1, 4)$

Find the coordinates of the midpoint of the segment with the given endpoints.

13. $(0, 0, 6)$ and $(10, -4, 3)$
14. $(1, 3, -4)$ and $(8, 0, 4)$
15. $(6, -3.5, -9)$ and $(2, 2.5, 7)$

Find the length of the segment with the given endpoints.

16. $(1, 0, 2)$ and $(-2, 3, 5)$
17. $(1, 0, 0.5)$ and $(5, -2, 1.5)$
18. $(15, 3, -7)$ and $(10, -9, 10)$

**The coordinates of the vertices of a triangle are given. Tell whether the
triangle is *isosceles*, *equilateral*, or *scalene*.**

19. $R(-2, 4, -3)$, $S(0, 1, 3)$, $T(-2, -3, -3)$

20. $A(5, 3, 4)$, $B(8, 8, 11)$, $C(4, -1, 8)$

21. $X(-1, -3, -1)$, $Y(2, 3, 6)$, $Z(3, -1, 6)$

22. **Writing** Use your answers to Exercises 19–21. Are any of the triangles
right triangles? Explain how you know.

CHEMISTRY The table gives the coordinates of the atoms of a methane
molecule to the nearest hundredth of an angstrom.

Atom	Type	*x*	*y*	*z*
1	carbon	−0.26	−0.36	0.00
2	hydrogen	0.26	0.73	0.00
3	hydrogen	0.77	−0.73	0.89
4	hydrogen	0.77	−0.73	−0.89
5	hydrogen	−0.77	−0.73	0.00

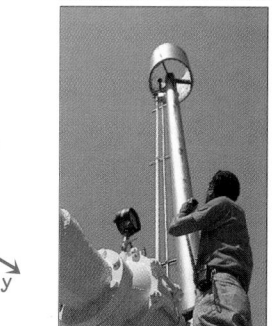

23. **Open-ended Problem** Show that the triangle formed by the centers of
the carbon atom and any two hydrogen atoms is isosceles.

24. **Open-ended Problem** Show that the triangle formed by any three
hydrogen atoms is equilateral.

Methane gas can be
captured at landfills. It
is then converted into
usable energy.

Apply⇔Assess

Suggested Assignment

❖ **Core Course**
Exs. 1–21, 30–34, AYP

❖ **Extended Course**
Exs. 1–21 odd, 22–34, AYP

❖ **Block Schedule**
Day 25 Exs. 1–21, 30–34, AYP

Exercise Notes

Spatial Reasoning
Exs. 1–6 You may wish to point out
to students that the coordinates of
a rectangular prism can be determined whenever the coordinates of
two vertices are given that are
diagonally opposite one another.

Using Technology
Exs. 13–18 You may wish
to challenge students to write a
short program for a graphing calculator that can find the midpoint and
length of a segment in space when
given its endpoints. Below is a
sample program written for a TI-82.

```
PROGRAM:SPACE
:Input "X1", A
:Input "X2", B
:Input "Y1", C
:Input "Y2", D
:Input "Z1", E
:Input "Z2", F
:√((B−A)^2+(D−C)^2+(F−E)^2)→G
:Disp "MIDPOINT IS (",(A+B)/2,",",
(C+D)/2,",",(E+F)/2,")"
:DISP "DISTANCE IS",G
```

Common Error
Exs. 19–21, 23, 24 Some students may make computational
errors with these exercises, especially with Exs. 23 and 24. Remind
students to pay careful attention
to negative signs and to their
computations with decimals.

Communication: Writing
Ex. 22 Some students may
not be able to answer this exercise
because they are attempting to
find slopes of perpendicular lines
and there is no slope formula in
three dimensions. Ask these students what other methods exist to
determine if a triangle is a right
triangle.

Exercises and Applications

1. $(4, 0, 3)$
2. $(0, 6, 3)$
3. $(4, 6, 3)$
4. $(4, 0, 0)$
5. $(0, 0, 0)$
6. $(0, 6, 0)$
7. the *xy*-plane
8. the *xz*-plane
9. the *yz*-plane
10. the *xy*-plane
11. the *xz*-plane
12. the *yz*-plane
13. $(5, -2, 4.5)$
14. $(4.5, 1.5, 0)$
15. $(4, -0.5, -1)$
16. $3\sqrt{3} \approx 5.20$
17. $\sqrt{21} \approx 4.58$
18. $\sqrt{458} \approx 21.40$
19. isosceles
20. scalene
21. scalene
22. No; none of the triangles have
side lengths that form a
Pythagorean triple.
23. Answers may vary. An example is given. The distance
between atoms 1 and 3 and
atoms 1 and 4 are both about
1.41 Å, so the triangle formed
by the centers of atoms 1, 3,
and 4 is isosceles.
24. Answers may vary. An example is given. The distances
between atoms 2 and 3, atoms
3 and 4, and atoms 2 and 4 are
all about 1.78 Å, so the triangle formed by the centers of
atoms 2, 3, and 4 is equilateral.

Exercise Notes

Interview Note: Application
Exs. 25–28 These exercises implicitly introduce students to the idea that a three-dimensional coordinate system can be used in other ways than to locate a point in space. The coordinates in this cube represent the intensity of color in a mixture rather than the position of a point in space.

Teaching Tip
Exs. 25–28 The color cube in these exercises can be interpreted using probability. For example, ask students how many combinations of colors can be made by mixing red, blue, and yellow in intensities of 0%, 50%, and 100%. The intensity of any one color is independent of the others. There are three intensities for each of the three primary colors, so there are 3 · 3 · 3 = 27 different combinations. Point out that there are 27 dots on the cube, one to represent each of these combinations. Ask students to determine the number of color combinations if intensities of 25% and 75% are also used. (125)

Challenge
Ex. 29 This problem can be used to lead students to the Distance Formula for Three Dimensions. By using the Pythagorean theorem twice, students can find the distance from point P to point Q, and in so doing, arrive at PQ in terms of its x-, y-, and z-coordinates.

Assessment Note
Ex. 30 Students who have difficulty with this problem should review *Think and Communicate* question 1 on page 201. Using the model of the ethane molecule, students should realize that if one of the coordinates is constant, the points lie in a plane. The plane is perpendicular to the axis for which the coordinates are constant.

INTERVIEW Marc Hannah

Look back at the article on pages 162–164.

The colors you see when you look at a color computer screen are all created by combining the three primary colors of light—red, blue, and green—in different proportions. The intensity of the red, blue, and green light can vary from 0% to 100%. The colors created this way can be displayed in a color cube.

25. What are the coordinates of the three vertices of the color cube that are not labeled? What colors are at those locations?

26. Open-ended Problem Choose two edges of the color cube. Find the coordinates of the midpoint of each edge that you chose. What color is at each midpoint?

27. What are the coordinates of the point at the center of the color cube? What color is at this location?

28. Open-ended Problem Find the coordinates of three points on the color cube that form an equilateral triangle.

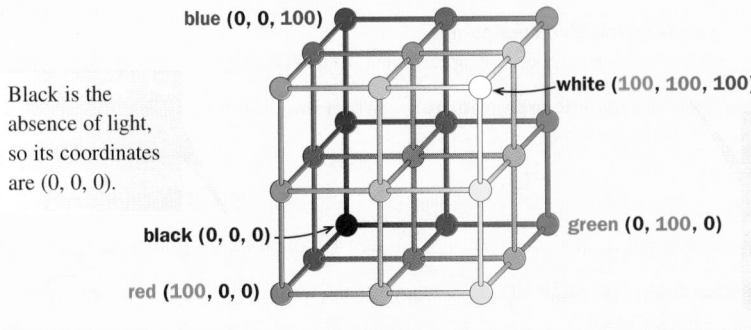

Black is the absence of light, so its coordinates are (0, 0, 0).

blue (0, 0, 100)

white (100, 100, 100)

black (0, 0, 0)

red (100, 0, 0)

green (0, 100, 0)

White is made by mixing the three light primaries at 100% intensity, so its coordinates are (100, 100, 100).

29. Challenge Given two points P and Q in space, a rectangular prism can be constructed so that P and Q are the endpoints of a diagonal of the prism. Each of the edges is parallel to an axis.

a. Use the Pythagorean theorem to find the length of the dashed diagonal.

b. Use the Pythagorean theorem to find PQ.

c. Writing Explain how finding the length of the diagonal of a prism is related to the Distance Formula in three dimensions.

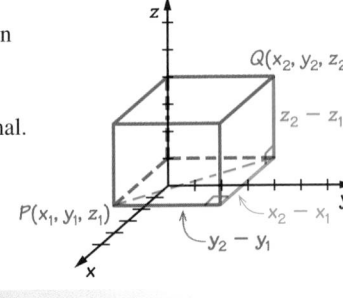

ONGOING ASSESSMENT

30. Open-ended Problem Choose any number. Imagine all the points in a three-dimensional coordinate system that have that number for a z-coordinate. What figure do you think they form? Describe the figure and sketch it. Make a conjecture about points that have the same z-coordinate.

206 Chapter 4 *Coordinates in Geometry*

---

25. (100, 100, 0) represents a shade of yellow that is equal parts red and green; (0, 100, 100) represents a shade of blue-green that is equal parts green and blue; (100, 0, 100) represents a shade of magenta that is equal parts red and blue.

26. Answers may vary. An example is given. The midpoint of (0, 0, 0) and (100, 0, 0) is the point (50, 0, 0), which represents a dark shade of red (a red-black mix). The midpoint

of (0, 0, 100) and (100, 100, 100) is the point (50, 50, 100), which corresponds to light blue (a blue-white mix).

27. (50, 50, 50); a shade of gray that is equal parts red, green, and blue

28. Answers may vary. An example is given. (100, 0, 0), (0, 100, 0), and (0, 0, 100)

29. a. The segment marked $x_2 - x_1$ has endpoints (x_1, y_2, z_1) and (x_2, y_2, z_1), so its length is $x_2 - x_1$. The segment marked $y_2 - y_1$ has endpoints (x_1, y_1, z_1) and (x_1, y_2, z_1), so its length is $y_2 - y_1$. The segment marked $z_2 - z_1$ has endpoints (x_2, y_2, z_1) and (x_2, y_2, z_2), so its length is $z_2 - z_1$.

b. $\sqrt{(x_2 - x_1)^2 + (y_2 - y_1)^2}$

c. $\sqrt{(x_2 - x_1)^2 + (y_2 - y_1)^2 + (z_2 - z_1)^2}$

d. The formula for the distance between two points $P(x_1, y_1, z_1)$ and $Q(x_2, y_2, z_2)$ can be derived by using the Pythagorean theorem to determine the length of the diagonal $\overline{PQ}$ of the prism determined by points P and Q.

30. The set of all points with z-coordinate n is a plane that is parallel to the xy-plane and |n| units from the xy-plane. For example, the set of all points with z-coordinate

SPIRAL REVIEW

Write the converse of each statement. *(Section 3.5)*

31. If a triangle is equilateral, then all its sides are congruent.

32. If a quadrilateral is a parallelogram, then its opposite sides are congruent.

Sketch each situation. *(Section 1.4)*

33. $\overleftrightarrow{AB} \parallel \overleftrightarrow{CD}$

34. $\overleftrightarrow{AB}$ intersects plane S at point N.

ASSESS YOUR PROGRESS

VOCABULARY

circle (p. 187)
center of a circle (p. 187)
diameter (p. 187)
radius, radii (p. 187)
concentric circles (p. 189)
coordinate geometry proof (p. 196)
three-dimensional coordinate system (p. 201)
***z*-axis** (p. 201)
ordered triple (*x, y, z*) (p. 202)

Sketch each circle. Give the coordinates of the center. Draw a radius and label it with its length. *(Section 4.4)*

1. $x^2 + (y - 3)^2 = 4$

2. $(x - 4)^2 + y^2 = 16$

Answer Exercises 3–6 to complete the proof. Use the coordinates given in the diagram. *(Section 4.5)*

Given: The vertices of isosceles $\triangle ABC$ are $A(-k, 0)$, $B(0, 3k)$, and $C(k, 0)$; D and E are the midpoints of $\overline{AB}$ and $\overline{CB}$.

Prove: $\overline{AE} \perp \overline{CD}$

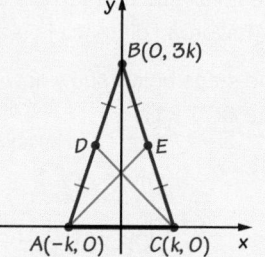

3. Find the coordinates of D and E.

4. Find the slopes of $\overline{AE}$ and $\overline{CD}$.

5. Find the product of the slopes of $\overline{AE}$ and $\overline{CD}$.

6. Explain why you can conclude that $\overline{AE} \perp \overline{CD}$.

The coordinates of the vertices of a triangle are given. Tell whether the triangle is *isosceles, equilateral,* or *scalene.* *(Section 4.6)*

7. $D(5, 0, 0)$, $E(0, 5, 0)$, $F(0, 0, 5)$

8. $U(0, 7, 1)$, $V(2, 5, 3)$, $W(4, 3, 1)$

9. Journal Create a coordinate geometry problem that requires the Distance Formula, the Midpoint Formula, or the Slope Formula. Your problem can be an exercise about a figure or a proof of a conjecture. Write your problem and a complete solution. Discuss any difficulties you had in creating and solving your problem.

Apply⇔Assess

Assess Your Progress

Journal Entry

This journal activity allows students to consolidate the skills they have learned by creating an original problem of their own. After reading students' problems, you may wish to share the most interesting ones with the class.

Progress Check 4.4–4.6

See page 211.

Practice 29 for Section 4.6

NAME ______ DATE ______

Practice 29

FOR USE WITH SECTION 4.6

Use the triangular prism. Given the coordinates of each vertex.

1. P (6, 0, 5)
2. L (0, 0, −2)
3. M (0, 4, −2)

Use the rectangular prism. Give the coordinates of each vertex.

4. P (6, 0, 4)
5. R (0, 8, 4)
6. S (6, 8, 4)
7. T (6, 0, 0)
8. O (0, 0, 0)
9. V (0, 8, 0)

Tell which coordinate plane contains each face of the rectangular prism.

10. $QRVO$ *yz*-plane
11. $TOVW$ *xy*-plane
12. $QOTP$ *xz*-plane

Tell which coordinate plane contains each point.

13. $N(2, 0, 2)$ *xz*-plane
14. $M(3, 4, 0)$ *xy*-plane
15. $L(0, 3, 3)$ *yz*-plane

Find the coordinates of the midpoint of the segment with the given endpoints.

16. (6, 8, 0) and (8, −2, 2) (7, 3, 1)
17. (7, 5, −1) and (9, −5, −3) (8, 0, −2)
18. (2.5, 4, −3.5) and (−2.5, −7, 1.5)
19. (3.4, 2, 0) and (1.8, 4.4, −3) (2.6, 3.2, −1.5)
20. (8, 2, 4) and (4, 6, 10) (6, 4, 7)
21. (11, 5, −7) and (7, −3, 5) (9, 1, −1)

18. (0, −1.5, −1)

Find the length of the segment with the given endpoints.

22. (6, 7, 0) and (8, −2, 2) $\sqrt{89}$
23. (2, 3, −0.5) and (5, −1, 1.5) $\sqrt{29}$
24. (3, 2, 4) and (0, 5, −3) $\sqrt{67}$
25. (2, 7, −1) and (1, 13, −3) $\sqrt{41}$
26. (12, −5, 6) and (9, 4, −3) $3\sqrt{19}$
27. (5, 6, 7) and (8, 9, 10) $3\sqrt{3}$

The coordinates of the vertices of a triangle are given. Tell whether the triangle is *isosceles, equilateral,* or *scalene.*

28. $A(-1, 3, 2)$, $B(4, 0, 3)$, $C(1, 5, 2)$ isosceles
29. $D(2, -4, 7)$, $E(-1, 1, 7)$, $F(8, 0, -2)$ scalene

Practice Bank, GEOMETRY: EXPLORATIONS AND APPLICATIONS

29

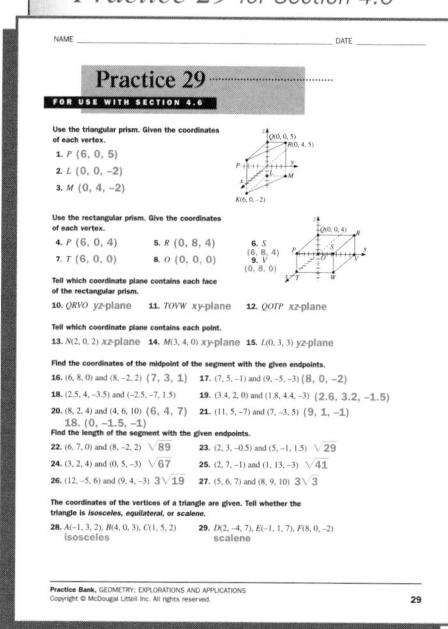

3 is a plane that is 3 units above the *xy*-plane as shown.

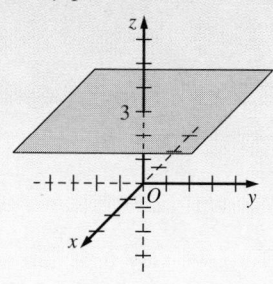

31. If all the sides of a triangle are congruent, then the triangle is equilateral.

32. If the opposite sides of a quadrilateral are congruent, then the quadrilateral is a parallelogram.

33, 34. Answers may vary. Examples are given.

33.

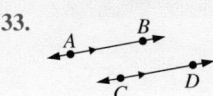

34.

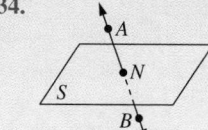

Assess Your Progress

1.

2.

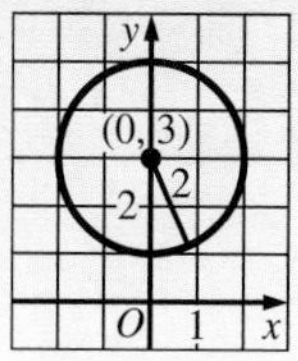

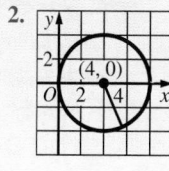

3. $D\left(-\frac{k}{2}, \frac{3k}{2}\right)$; $E\left(\frac{k}{2}, \frac{3k}{2}\right)$

4. slope of $\overline{AE} = 1$; slope of $\overline{CD} = -1$

5. −1

6. If the product of the slopes of two lines is −1, then the lines are perpendicular.

7. equilateral

8. isosceles

9. See answers in back of book.

- Understand the concept of taxicab distance.
- Compare Euclidean geometry to taxicab geometry.
- Find taxicab distances on city maps.
- Find Euclidean distances.
- Make a map of a city.

Planning

Materials
- graph paper and pencil
- ruler
- a map of a city (optional)
- poster board

Project Teams
Students can work individually or in pairs to complete the project. If students work in pairs, both students should be involved in completing all phases of the project.

Guiding Students' Work

If students are using the map of an actual city, have them select a portion of the map that contains mostly square blocks of equal length. Then let each block represent one unit on their grid. To find the taxicab "circle," have students draw all the paths a person can take on the grid. The horizontal and vertical lines connecting the endpoints are the circle. When students are trying to calculate the value of π in taxicab geometry, remind them that the circumference of a circle is actually its perimeter. The perimeter of their taxicab circle can be found by adding the distances around the edge. Then have students find the ratio $\frac{\text{perimeter}}{\text{diameter}}$ to find the value of π. Students should understand that the value of π in the taxicab circle will be greater than the value of π in an actual circle.

Second-Language Learners
To serve as a reminder to all students learning English, ask a student to show on the board a line that is *horizontal*, one that is *vertical*, and one that is at a *diagonal*.

PORTFOLIO PROJECT

Exploring Taxicab Geometry

You probably think that the shortest distance from *A* to *B* is along a straight line. This is true in the Euclidean geometry of the coordinate plane. What happens if you're in a city at point *A* and you want to go to point *B*? You can't go straight through buildings! You have to go along the streets. **Taxicab geometry** is the non-Euclidean geometry that a taxicab and a pedestrian must obey.

New York City

| PROJECT GOAL | Compare Euclidean geometry to taxicab geometry and calculate traveling distances along a city grid. |

Getting from Here to There

In taxicab geometry, there is more than one "shortest path" from *A* to *B*.

The rules in taxicab geometry are simple: You may move horizontally and vertically but never diagonally. The diagram below shows several ways of moving from *A* to *B*. Each path covers 10 blocks.

The **taxicab distance** between $A(x_1, y_1)$ and $B(x_2, y_2)$ is the sum of the horizontal and vertical distances along a shortest path from *A* to *B*. Pythagorean distance is less than or equal to taxicab distance.

The steps below show how to find the taxicab distance between $A(1, 3)$ and $B(7, -1)$.

Taxicab Distance
$$AB = |x_2 - x_1| + |y_2 - y_1|$$

Absolute value is used because distances are always positive.

$$AB = |x_2 - x_1| + |y_2 - y_1|$$
$$= |7 - 1| + |-1 - 3|$$
$$= |6| + |-4|$$
$$= 6 + 4$$
$$= 10$$

Pythagorean Distance
$$AB = \sqrt{(x_2 - x_1)^2 + (y_2 - y_1)^2}$$

Making a Taxicab Map

1. Make a map showing the grid of a real or invented city. Choose an origin. Plot two points at street intersections representing a school and a park entrance. The points should not lie on a horizontal or a vertical line. Label their coordinates.

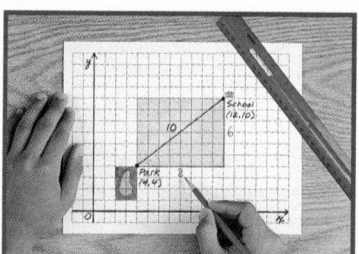

> The **points between A and B** form a rectangular region.

> The **points equidistant from A and B** do not always form a straight line.

2. Find the Euclidean and taxicab distances between the school and the park. How do they compare? Draw the taxicab region that represents all points between the school and the park.

3. Choose a location for an apartment that is the same taxicab distance from the school as it is from the park. Find the coordinates of several possible locations for the apartment.

Making a Presentation

Suppose you walk your dog a round-trip distance of 10 blocks each day. Show on your map every place you can reach. This is the interior of a taxicab "circle" centered at the apartment with a radius of 5 blocks. Present your labeled and completed map to the class.

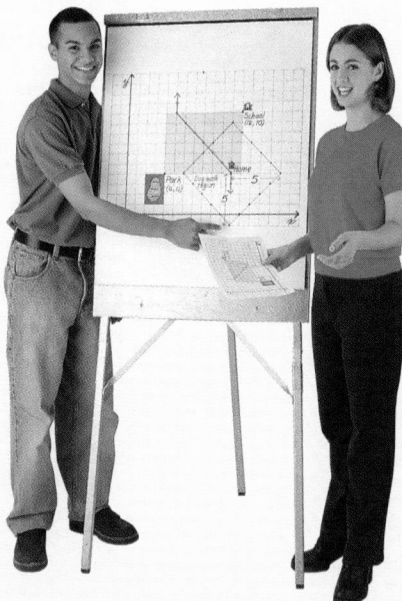

Extending Your Project

Find the taxicab circumference and taxicab diameter of the dog-walk "circle". If π is the ratio of circumference to diameter, what is the value of π in taxicab geometry?

Self-Assessment

What advice would you give to someone who plans to do the project? What are some other problems that you might think differently about after working with taxicab geometry?

Portfolio Project **209**

Project Notes

Guiding Students' Work

Rubric for Chapter Project

4 Students make a well-drawn map of a city and correctly find the distances between their school and the park entrance. The taxicab region is also correctly drawn and the given selections for the apartment location are accurate. Students make an accurate drawing of their taxicab circle and present their work to the class in a clear, accurate, and well-thought out way. Students also extend the project correctly and make an insightful self-assessment of their work.

3 Students make a map of a city and find the distances between their school and the park entrance, but there is a computational error in their work. The taxicab region is drawn on the map and the given selections for the apartment are correct. Students correctly draw their taxicab circle and make a presentation of their map to the class, but it is evident that some of the mathematical ideas involved in the project are not completely understood. Students provide a good self-assessment for the project.

2 Students make a map of a city and find the distances between their school and the park entrance but make an error or two in computing the distances. The taxicab region is drawn on the map, but one or two of the given selections for the apartment are not correct. Students draw a taxicab circle but do not complete the extension of the project. Also, the self-assessment is brief and lacking in detail.

1 Students' maps of the city are poorly drawn and they do not find all of the distances. Students make a poor presentation of their work and no attempt is made at extending the project or at self-assessment. Students should be encouraged to speak with the teacher as soon as possible to review their work and to make a new start on the project.

Chapter Support

Course Guide: Chapter 4

Lesson Plans: Chapter 4

Practice Bank:
  Cumulative Practice 30

Study Guide: Chapter 4 Review

Challenge Problems: Set 30

Assessment Book:
  Chapter Tests 18 and 19
  Chapter 4 Alternative Assessment

Test Generator Software

Preparing for College Entrance Tests

Professional Handbook

Progress Check 4.1–4.3

Segment $\overline{DE}$ has endpoints $D(-3, 1)$ and $E(5, -7)$.

1. Find the coordinates of the midpoint M of $\overline{DE}$. *(Section 4.1)* $M(1, -3)$

2. Find DM and ME and verify that they are equal. *(Section 4.1)* $DM = ME = 4\sqrt{2}$

3. Find the slope of $\overline{DE}$. *(Section 4.2)* -1

4. Find an equation of the line through M and $F(6, 2)$. *(Section 4.2)* $y = x - 4$

5. Is $\overline{FM}$ the perpendicular bisector of $\overline{DE}$? Explain. *(Section 4.3)* Yes. The slope of $\overline{FM}$ is 1, so $\overline{FM}$ and $\overline{DE}$ are perpendicular, and $\overline{FM}$ intersects $\overline{DE}$ at its midpoint.

6. Quadrilateral $LMNP$ has coordinates $L(-5, -2)$, $M(-2, 4)$, $N(2, 7)$, $P(-1, 1)$. Is quadrilateral $LMNP$ a parallelogram? Explain. *(Section 4.3)* Yes.

   slope of $\overline{LM} = \dfrac{4 - (-2)}{-2 - (-5)} = 2$;

   slope of $\overline{NP} = \dfrac{1 - 7}{-1 - 2} = 2$;

   slope of $\overline{MN} = \dfrac{7 - 4}{2 - (-2)} = \dfrac{3}{4}$;

   slope of $\overline{LP} = \dfrac{1 - (-2)}{-1 - (-5)} = \dfrac{3}{4}$;

   Since $\overline{LM} \parallel \overline{NP}$ and $\overline{MN} \parallel \overline{LP}$, $LMNP$ is a parallelogram.

CHAPTER

4 Review

STUDY TECHNIQUE

One way to prepare for a test is to write one. Write a test for this chapter, then exchange it with a classmate. After taking the tests, correct and discuss them. Review any topics that gave you difficulty.

VOCABULARY

slope (p. 174)

y-intercept (p. 175)

slope-intercept form (p. 175)

perpendicular bisector (p. 181)

circle (p. 187)

center of a circle (p. 187)

diameter (p. 187)

radius, radii (p. 187)

concentric circles (p. 189)

coordinate geometry
  proof (p. 196)

three-dimensional
  coordinate system (p. 201)

z-axis (p. 201)

ordered triple (x, y, z) (p. 202)

SECTIONS 4.1, 4.2, *and* 4.3

Use the *Distance Formula* to find the length of a segment.
$$AB = \sqrt{(x_2 - x_1)^2 + (y_2 - y_1)^2}$$

Use the *Midpoint Formula* to find the midpoint of a segment.
$$\text{midpoint of } \overline{AB} = \left(\frac{x_1 + x_2}{2}, \frac{y_1 + y_2}{2} \right)$$

An equation of $\overleftrightarrow{AB}$ is in **slope-intercept form** $y = mx + b$, where m is the slope and b is the y-intercept.

Use the slope formula to find the **slope** m of $\overline{AB}$.
$$m = \frac{y_2 - y_1}{x_2 - x_1}$$

A line parallel to $\overleftrightarrow{AB}$ has the same slope but a different y-intercept. The slope of a line perpendicular to $\overleftrightarrow{AB}$ is the negative reciprocal of the slope of $\overleftrightarrow{AB}$.

SECTIONS 4.4, 4.5, *and* 4.6

A **circle** is the set of all points in a plane that are the same distance from a fixed point, the **center** of the circle. The distance from the center of a circle to any point on the circle is the **radius**.

An equation of $\odot O$ with center at the origin $(0, 0)$ and radius $r = 3$ is:

$$x^2 + y^2 = 3^2 \qquad \text{Use } x^2 + y^2 = r^2.$$

An equation of $\odot P$ with center $(h, k) = (4, 3)$ and radius $r = 2$ is:

$$(x - 4)^2 + (y - 3)^2 = 4$$

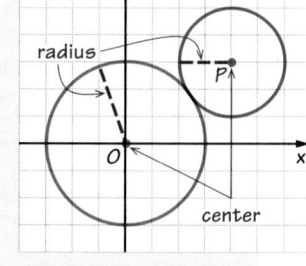

$$\text{Use } (x - h)^2 + (y - k)^2 = r^2.$$

Start a **coordinate geometry proof** by placing the figure on a coordinate plane. Use the properties of the figure to label its coordinates using as few variables as possible. Then use the Distance Formula, the Midpoint Formula, and the properties of slopes of parallel and perpendicular lines to prove theorems.

For example, in quadrilateral $RSTU$, you can use the Distance Formula to show $RT = \sqrt{(a + b)^2 + c^2} = SU$, so you can prove that the diagonals of $RSTU$ are congruent.

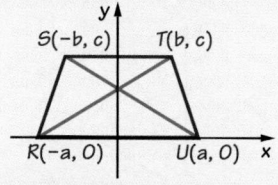

To locate a point in space, three coordinates are needed. Point P has x-coordinate 2, y-coordinate 0, and z-coordinate -3. It lies in the xz-plane. The coordinates of point Q are given by the ordered triple $(5, 4, 6)$.

Find the midpoint of $\overline{PQ}$ by using the Midpoint Formula.

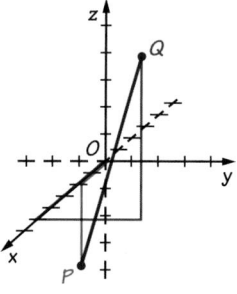

$$\text{midpoint of } \overline{PQ} = \left(\frac{x_1 + x_2}{2}, \frac{y_1 + y_2}{2}, \frac{z_1 + z_2}{2} \right)$$

$$= \left(\frac{2 + 5}{2}, \frac{0 + 4}{2}, \frac{-3 + 6}{2} \right)$$

$$= \left(\frac{7}{2}, 2, \frac{3}{2} \right)$$

Find the length of $\overline{PQ}$ by using the Distance Formula.

$$PQ = \sqrt{(x_2 - x_1)^2 + (y_2 - y_1)^2 + (z_2 - z_1)^2}$$

$$= \sqrt{(5 - 2)^2 + (4 - 0)^2 + (-3 - 6)^2}$$

$$= \sqrt{3^2 + 4^2 + (-9)^2}$$

$$\approx 10.29$$

Review **211**

Progress Check 4.4–4.6

1. Sketch the circle $(x - 1)^2 + (y + 2)^2 = 16$. Give the coordinates of the center, draw a radius, and label it with its length. *(Section 4.4)*

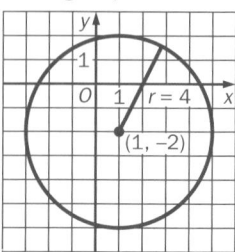

2. Write an equation for the circle with center $(-3, 2)$ and radius 5. *(Section 4.4)*

$(x + 3)^2 + (y - 2)^2 = 25$

3. Find the missing coordinates for the isosceles triangle without using any new variables. *(Section 4.5)*

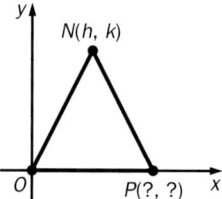

Point P has coordinates $(2h, 0)$.

4. Describe how you would use the figure above to show that in an isosceles triangle, the segment joining the vertex to the midpoint of the base is perpendicular to the base. *(Section 4.5)* Find the coordinates of the midpoint of $\overline{OP}$. Label this point M. Show that the slope of $\overline{MN}$ is undefined, which means $\overline{MN}$ is vertical. Since $\overline{MN}$ intersects the x-axis (and $\overline{OP}$) at M and is vertical, it is perpendicular to the x-axis (and $\overline{OP}$).

5. Find the length and the midpoint of the segment joining points $(4, -5, 1)$ and $(-7, 0, 2)$ *(Section 4.6)*

$7\sqrt{3}; \left(-\frac{3}{2}, -\frac{5}{2}, \frac{3}{2} \right)$

Chapter 4 Assessment
Form A Chapter Test

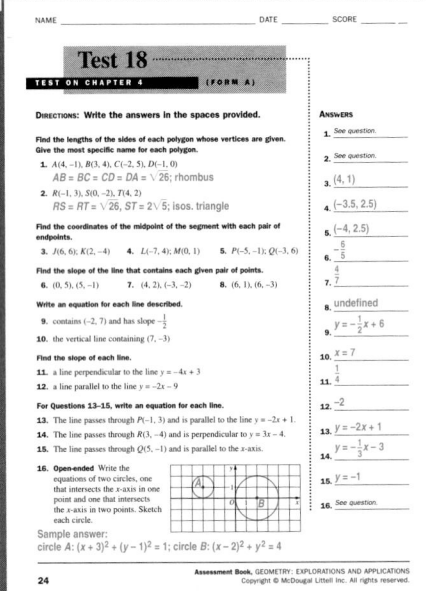

Chapter 4 Assessment
Form B Chapter Test

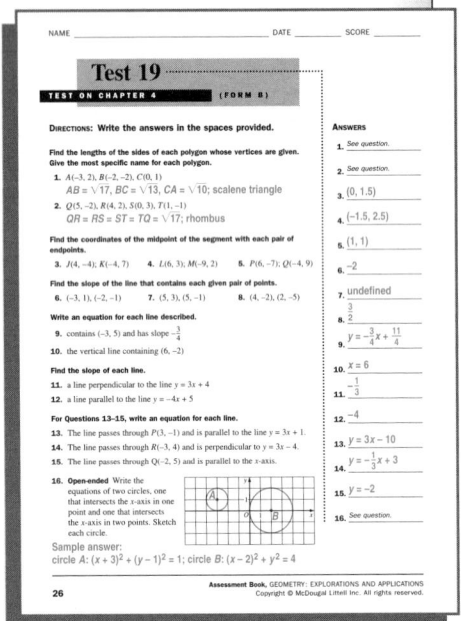

4 Assessment

VOCABULARY QUESTIONS

For each question, match a word to the given value.

The line with equation $y = -x + 1.8$

1. -1 **2.** 1.8

The circle with equation $x^2 + (y + 7)^2 = 100$

3. $(0, -7)$ **4.** 10   **5.** 20

    A. diameter
    B. slope
    C. center
    D. radius
    E. y-intercept

SECTIONS 4.1, 4.2, and 4.3

For Exercises 6 and 7, (a) find the lengths of each side of the polygon, and (b) give the most specific name for the polygon.

6.
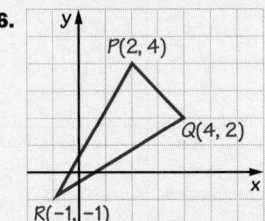

7.
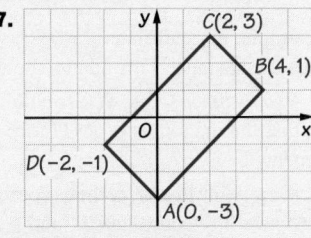

Find the slope of the line that contains each pair of points.

8. $(3, 7)$ and $(9, 4)$ **9.** $(-2, 9)$ and $(0, 1)$ **10.** $(0, 5)$ and $(0, 12)$

Give the slope and y-intercept of each line.

11. $y = 3x - 12$ **12.** $y = \frac{3}{4}x + 1$ **13.** $y = 4 - x$

For Exercises 14–17, write an equation for each line.

14. contains $(7, 3)$; has slope 2   **15.** contains $(0, 0)$; has slope $-\frac{3}{2}$

16. the vertical line that contains $(1, -1)$

17. contains $(9, -1)$; has slope 0

18. The vertices of $\triangle JKL$ are $J(3, 4)$, $K(4, 1)$, and $L(-2, -1)$. Use slopes to tell if $\triangle JKL$ is a right triangle.

19. Writing Suppose you know the coordinates of points P and M, and M is the midpoint of $\overline{PQ}$. Explain how to find the coordinates of Q.

ANSWERS Chapter 4

Assessment

1. B   **2.** E   **3.** C

4. D   **5.** A

6. a. $PR = \sqrt{34}$; $QR = \sqrt{34}$; $PQ = 2\sqrt{2}$

  b. isosceles triangle

7. a. $AB = 4\sqrt{2}$; $CD = 4\sqrt{2}$; $BC = 2\sqrt{2}$; $AD = 2\sqrt{2}$

  b. rectangle

8. $-\frac{1}{2}$

9. -4

10. undefined

11. 3; -12

12. $\frac{3}{4}$; 1

13. -1; 4

14. $y = 2x - 11$

15. $y = -\frac{3}{2}x$

16. $x = 1$

17. $y = -1$

18. The slope of $\overline{JK}$ is -3 and the slope of $\overline{KL}$ is $\frac{1}{3}$. Since $(-3)\left(\frac{1}{3}\right) = -1$, $\overline{JK} \perp \overline{KL}$, so $\triangle JKL$ is a right triangle.

19. Let $P = (x_1, y_1)$, $M = (x_2, y_2)$, and $Q = (x_3, y_3)$. Since M is the midpoint of PQ, $x_2 = \frac{x_1 + x_3}{2}$ and $y_2 = \frac{y_1 + y_3}{2}$. Then $x_3 = 2x_2 - x_1$ and $y_3 = 2y_2 - y_1$.

20. $x^2 + y^2 = 9$

21. $x^2 + y^2 = 5$

22. $(x - 4)^2 + (y - 2)^2 = 81$

23. $(x - 6)^2 + (y + 2)^2 = 1.25$

24. a.
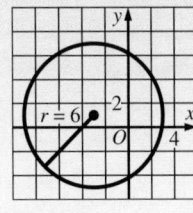

  b. $(-3, 1)$

  c. See part (a).

Write an equation of the circle with the given center and radius.

20. center $(0, 0)$ and radius 3

21. center $(0, 0)$ and radius $\sqrt{5}$

22. center $(4, 2)$ and radius 9

23. center $(6, -2)$ and radius 2.5

24. a. Sketch the circle with equation $(x + 3)^2 + (y - 1)^2 = 36$.

b. Give the coordinates of the center.

c. Draw a radius and label it with its length.

Answer Exercises 25–27 to complete the proof.

Given: $\triangle PQR$ is isosceles.
S, T, and O are the midpoints
of $\overline{PQ}$, $\overline{QR}$, and $\overline{PR}$.

Prove: $\triangle STO$ is isosceles.

25. Find the coordinates of point R
without using any new variables.

26. Find the coordinates of points S,
T, and O.

27. a. Find the lengths TS, TO, and SO.

b. Compare the lengths. Explain why $\triangle STO$ is isosceles.

28. Writing Give an example of an ordered triple, and explain how to
locate that point in a three-dimensional coordinate system.

For Exercises 29 and 30, use points $M(-3, 5, 0)$ and $N(1, 7, -6)$.

29. Find the coordinates of the midpoint of $\overline{MN}$.

30. Find the length of $\overline{MN}$.

31. Open-ended Problem Sketch a rectangular prism that has point
$(1, 3, 4)$ as a vertex. Label all the vertices of the prism.

PERFORMANCE TASK

32. Open-ended Problem Choose one of the following quadrilaterals:
parallelogram, rectangle, rhombus, or square.

a. Draw two examples of the quadrilateral on graph paper. One
example should have vertices that are labeled with numbers as
coordinates and the other should have vertices that are labeled with
variables.

b. State a fact about the quadrilateral and use coordinate geometry to
prove the fact.

Assessment **213**

Chapter 4 Assessment
Form C Alternative Assessment

Chapter 4

ALTERNATIVE ASSESSMENT

1. Performance Task The vertices of $\triangle ABC$ are $A(2, 9)$, $B(2, 1)$, and $C(8, 1)$.
Use the Distance Formula and the Converse of the Pythagorean theorem to
prove that $\triangle ABC$ is a right triangle.

2. a. The endpoints of $\overline{EF}$ are $E(2, 14)$ and $F(7, 2)$. Graph $\overline{EF}$ on a coordinate
plane. Draw a right triangle DEF that has $\overline{EF}$ as its hypotenuse. Find the
lengths of $\overline{DE}$ and $\overline{DF}$ from your graph. Use these leg lengths and the
Pythagorean theorem to find the length of $\overline{EF}$.

b. Find the length of $\overline{EF}$ using the Distance Formula.

c. Discuss the similarities in the two methods for finding the length of $\overline{EF}$.

3. Project Write a program for the graphics calculator that finds the distance
between two points whose coordinates are given.

4. Project Write a program for the graphics calculator that finds the midpoint
of a segment given the coordinates of its two endpoints.

5. Open-ended Problem Write the equations of the lines described.

a. two lines that are parallel

b. two lines that are perpendicular

6. Performance Task Draw a rectangle with vertices whose coordinates are
$(-1, 2)$, $(-1, -4)$, $(3, -4)$, and $(3, 2)$.

a. Find the perimeter and area of the rectangle.

b. Multiply the coordinates of the vertices of the rectangle by 3.

c. Draw the rectangle whose vertices are those you found in part b. Find
the perimeter and area of this new rectangle.

d. Compare the perimeter of the rectangle in part c to the perimeter of the
original rectangle.

e. Compare the area of the rectangle in part c to the area of the original
rectangle.

Assessment Book, GEOMETRY: EXPLORATIONS AND APPLICATIONS
Copyright © McDougal Littell Inc. All rights reserved.

96

---

25. $(-2a, 0)$

26. $S(a, b)$; $T(-a, b)$; $O(0, 0)$

27. a. $TS = 2a$; $TO = SO = \sqrt{a^2 + b^2}$

b. $TO = SO$; $\triangle STO$ is isosceles by
definition.

28. Answers may vary. An example is
given. To locate $(1, 2, 3)$, start at the
origin. Move 1 unit along the positive
x-axis, 2 units parallel to the y-axis,
and 3 units parallel to the z-axis.

29. $(-1, 6, -3)$

30. $MN = 2\sqrt{14}$

31, 32. Answers may vary. Examples are
given.

31.

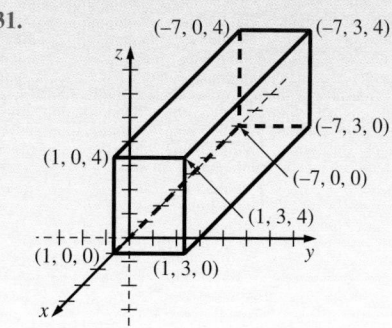

32.

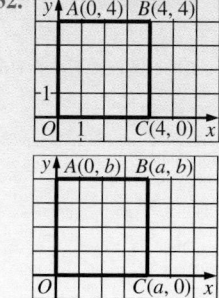

The diagonals of a square
are perpendicular.

Statements (Reasons)

1. slope of $\overline{AC} = \dfrac{b - 0}{0 - b} = -1$; slope of

$\overline{OB} = \dfrac{b - 0}{b - 0} = 1$ (Def. of slope)

2. slope of $\overline{AC} \cdot$ slope of $\overline{OB} =$

$-1 \cdot 1 = -1$ (Substitution)

3. $\overline{AC} \perp \overline{OB}$ (Two nonvertical lines are
perpendicular if and only if the
product of their slopes is -1.

Algebra
Review/Preview

These exercises review algebra topics you will use in the next chapters.

Toolbox p. 695
Linear Equations

SOLVING LINEAR EQUATIONS

Solve each equation.

EXAMPLE
$$4(x + 3) = -x + 22$$
$$4x + 12 = -x + 22 \quad \longleftarrow \text{Use the distributive property.}$$
$$5x + 12 = 22 \quad \longleftarrow \text{Add } x \text{ to both sides.}$$
$$5x = 10 \quad \longleftarrow \text{Subtract 12 from both sides.}$$
$$x = 2 \quad \longleftarrow \text{Divide both sides by 5.}$$

1. $7x - 1 = 4x + 5$ **2.** $5(2u - 1) = 4(u - 1)$

3. $8s + 18 = 6(2s + 3)$ **4.** $2(r + 1) = 4r + 1$

Solve each equation.

EXAMPLE
$$\frac{a}{2} + \frac{a}{3} = 10$$
$$6\left(\frac{a}{2} + \frac{a}{3}\right) = 6 \cdot 10 \quad \longleftarrow \begin{array}{l} \text{Multiply both sides by the least} \\ \text{common denominator, } 6. \end{array}$$
$$\frac{6a}{2} + \frac{6a}{3} = 60$$
$$3a + 2a = 60$$
$$5a = 60$$
$$a = 12 \quad \longleftarrow \text{Divide both sides by 5.}$$

5. $\dfrac{x}{2} + \dfrac{1}{7} = 1$ **6.** $\dfrac{y}{3} + \dfrac{y}{5} = 16$ **7.** $\dfrac{1}{12}w - \dfrac{1}{3}w = 2$

8. $\dfrac{5}{6}c + \dfrac{2}{15} = \dfrac{7}{10}c$ **9.** $\dfrac{t}{2} = \dfrac{3}{4} - \dfrac{6t}{24}$ **10.** $\dfrac{n + 3}{3} - \dfrac{n}{4} = \dfrac{n}{3}$

Solve each equation.

EXAMPLE
$$-1.6 + 3.68y = 4y$$
$$100(-1.6 + 3.68y) = 100(4y) \quad \longleftarrow \begin{array}{l} \text{Multiply both sides by } \mathbf{100} \text{ to} \\ \text{eliminate the decimals.} \end{array}$$
$$-160 + 368y = 400y$$
$$-160 = 32y \quad \longleftarrow \text{Subtract } 368y \text{ from both sides.}$$
$$-5 = y \quad \longleftarrow \text{Divide both sides by 32.}$$

11. $1.3x + 7 = 10.9$ **12.** $8 - 0.7p = 2.18p + 3.5$

13. $2.2y - 12.8 = 6.6$ **14.** $3.12u - 5 = 4.37u$

ANSWERS Chapter 4

Algebra Review/Preview

1. 2
2. $\dfrac{1}{6}$
3. 0
4. $\dfrac{1}{2}$
5. $\dfrac{12}{7}$
6. 30
7. –8
8. –1
9. 1
10. 4
11. 3
12. 1.5625
13. $\dfrac{97}{11}$
14. –4

SOLVING LINEAR SYSTEMS

Solve each system of equations.

EXAMPLE

$x + y = 3$

$5x - 2y = -13$

Solve one equation for *y*.

$y = 3 - x$

Substitute $3 - x$ for *y* in the other equation and solve for *x*.

$5x - 2(3 - x) = -13$

$5x - 6 + 2x = -13$

$7x - 6 = -13$

$7x = -7$

$x = -1$

To find *y*, substitute -1 for *x* in either of the original equations.

$x + y = 3$

$-1 + y = 3$

$y = 4$

The solution (*x*, *y*) is (−1, 4).

Toolbox p. 694
*Evaluating
Equations for
Given Values*

15. $y = 2x$
$5x - y = 30$

16. $x + y = 15$
$4x + 3y = 38$

17. $2a + 3b = 0$
$a - 6b = -5$

18. $2p - q = 17$
$3p + 4q = -13$

19. $-r + 4s = -8$
$2r - 8s = 6$

20. $\frac{v}{2} = 2 - u$
$6u + 3v = 12$

21. Tickets to an amusement park cost $20 for adults and $12 for children. On one day, 4250 tickets were sold, and the total cost was $63,400. How many adults and how many children visited the park that day?

SOLVING LINEAR INEQUALITIES

Solve each inequality. Graph the solution on a number line.

EXAMPLES

$4x - 7 \geq 1$

$4x \geq 8$

$\frac{4x}{4} \geq \frac{8}{4}$

$x \geq 2$

$-7 - 3m < 5$

$-3m < 12$

$\frac{-3m}{-3} > \frac{12}{-3}$

$m > -4$

When multiplying or dividing by a negative number, reverse the direction of the inequality.

Toolbox p. 705
*Solving
Inequalities*

22. $x - 2 \geq 3$

23. $\frac{2}{3}z < 12$

24. $-\frac{3}{5}y \geq 6$

25. $12p + 3 \geq 15$

26. $-8d + 19 < -29$

27. $5(1 - k) \leq 4(3 - k)$

Algebra Review/Preview   **215**

---

15. (10, 20)

16. (−7, 22)

17. $\left(-1, \frac{2}{3}\right)$

18. (5, −7)

19. no solution

20. Values that satisfy one equation will always satisfy the other; there are infinitely many solutions.

21. 1550 adults and 2700 children

22. $x \geq 5$

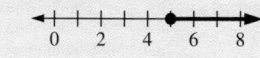

23. $z < 18$

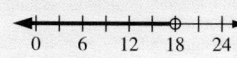

24. $y \leq -10$

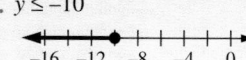

25. $p \geq 1$

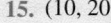

26. $d > 6$

27. $k \geq -7$

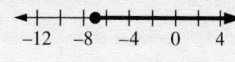

5 Parallel Lines

OVERVIEW

Connecting to Prior and Future Learning

⟺ Students study the relationships among the different angles formed by parallel lines and a transversal. They then use these relationships to find the measures of these angles.

⟺ Flow proofs are introduced in this chapter and students can now choose from a variety of types of proofs. Students also learn how to draw auxiliary lines to aid in proving statements. Students have an opportunity to use their knowledge of proofs as they prove theorems about parallels in Section 5.5.

⟺ The study of parallel lines continues as students learn about conditions for parallel lines and parallel lines in space. In Section 5.7, students learn how to construct parallel lines and perpendicular lines. All of these concepts are important to students' future work in geometry.

Chapter Highlights

Interview with José Saínz: This interview highlights the use of mathematics in designing kites. Related exercises are on pages 231, 252, and 262.

Explorations can be found throughout Chapter 5. The exploration in Section 5.1 has students explore angles and transversals. In Section 5.4, students use a straightedge or a ruler to draw parallel lines as an architect, engineer, or other designer would, and in Section 5.6, students investigate planes in space and their intersections.

Portfolio Project: Creating technical drawings by making orthographic projections and isometric drawings of objects is the focus of this project.

 Technology: In Section 5.7, students use geometry software to discover a different way to construct the bisector of a segment.

OBJECTIVES

Section	Objectives	NCTM Standards
5.1	• Identify pairs of angles formed by transversals and lines. • Find the measures of angles formed by transversals and lines. • Analyze real-world examples of intersecting lines.	1, 2, 3, 4, 5, 7
5.2	• Find the measures of alternate interior angles and same-side interior angles. • Identify trapezoids. • Prove statements about special angles. • Find congruent angles in real-world objects.	1, 2, 3, 4, 5, 7
5.3	• Write a flow proof. • Introduce auxiliary lines in a diagram. • Choose from a variety of methods for proving statements.	1, 2, 3, 4, 7
5.4	• Apply the converses of theorems about parallel lines and transversals. • Use facts about angles to prove that two lines are parallel.	1, 2, 3, 4, 5, 7
5.5	• Apply conditions to prove that lines are parallel. • Analyze real-world objects.	1, 2, 3, 4, 5, 7
5.6	• Recognize relationships among parallel and intersecting planes. • Identify parallel lines and planes in space. • Analyze real-world examples of parallel and perpendicular lines in space.	1, 2, 3, 4, 7
5.7	• Complete compass-and-straightedge constructions. • Construct lines parallel or perpendicular to given lines.	1, 2, 3, 4, 7

Mathematical Connections	5.1	5.2	5.3	5.4	5.5	5.6	5.7
geometry	**219–225***	**226–233**	**234–241**	**242–248**	**249–255**	**256–262**	**263–269**
algebra	222–224	227, 230, 232, 233	241	243, 245	252, 255	262	
logic and reasoning	221, 225	232	**234–241**	244, 247, 248	**249–255**	261, 262	268

Interdisciplinary Connections and Applications							
arts and entertainment				247			268
botany and horticulture	224			246			
architecture		232					
engineering			240				
city planning					251, 253		
electronics						259	
aviation						260	
automobiles, design, technical drawing, paper folding	223				254	261	266

*__Bold page numbers__ *indicate that a topic is used throughout the section.*

TECHNOLOGY

Section	opportunities for use with	
	Student Book	**Support Material**
5.1	geometry software	
5.2		**Geometry Inventor Activity Book:** Activity 1
5.3		**Geometry Inventor Activity Book:** Activity 10
5.4	graphing calculator	
5.5		**Technology Book:** Calculator Activity 7
5.7	geometry software	**Technology Book:** TI-92 Activity 5

Regular Scheduling (45 min)

Section	Materials Needed	Core Assignment	Extended Assignment	exercises that feature		
				Applications	Communication	Technology
5.1	lined paper, ruler, protractor	1–10, 17–26, 28–30, 32–35	1–35	11–16, 27	4, 13, 15, 31, 32b	
5.2	ruler, protractor	**Day 1:** 1–17, 22, 23 **Day 2:** 24–28, 30–37, AYP*	**Day 1:** 1–23 **Day 2:** 24–37, AYP	18–21 28, 29	20 24–27, 35a	
5.3		1–4, 6, 7, 12–17, 19–22	1–22	8–11	5, 11a, 19	
5.4	straightedge or ruler, index card, scissors	1–11, 15, 16, 21–27	1–27	12–14, 18–20	12, 14, 20, 23, 24	
5.5		1–3, 11–17, 23–29, AYP	1–29, AYP	4–10, 19–21	7, 9, 10, 13, 15–20, 22, 24	
5.6	index cards, scissors, toothpicks, protractor, ruler	1–7, 10–14, 18, 19, 24–28, 31–34	1–34	8, 9, 15–17, 20–23, 29, 30	4, 8–10, 14b, 19, 20, 23, 29	
5.7	straightedge, compass	1–7, 12–14, 16, 18, 23–26, AYP	1–26, AYP	8–11, 19–22	7, 9–11, 14, 16, 17, 21, 23	15
Review/ Assess		**Day 1:** 1–11 **Day 2:** 12–22 **Day 3:** Ch. 5 Test	**Day 1:** 1–11 **Day 2:** 12–22 **Day 3:** Ch. 5 Test			
Port. Proj.		Allow 2 days.	Allow 2 days.			

Yearly Pacing (with Portfolio Project)	**Chapter 5 Total** 13 days	**Chapters 1–5 Total** 67 days	**Remaining** 93 days	**Total** 160 days

Block Scheduling (90 min)

	Day 28	Day 29	Day 30	Day 31	Day 32	Day 33	Day 34
Teach/Interact	5.1: Exploration, page 221 5.2	Continue with 5.2 5.3	5.4: Exploration, page 242 5.5	5.6: Exploration, page 256 5.7: Constructions: pages 263, 264, 265, 267	Review Port. Proj.	Review Port. Proj.	Ch. 5 Test 6.1: Exploration, page 279
Apply/Assess	**5.1:** 1–10, 17–26, 28–30, 32–35 **5.2:** 1–17, 22, 23	**5.2:** 24–28, 30–37, AYP* **5.3:** 1–4, 6, 7, 12–17, 19–22	**5.4:** 1–11, 15, 16, 21–27 **5.5:** 1–3, 11–17, 23–29, AYP	**5.6:** 1–7, 10–14, 18, 19, 24–28, 31–34 **5.7:** 1–7, 12–14, 16, 18, 23–26, AYP	**Review:** 1–11 **Port. Proj.**	**Review:** 12–22 **Port. Proj.**	**Ch. 5 Test** **6.1:** 1–20

NOTE: A one-day block has been added for the Portfolio Project—timing and placement to be determined by teacher.

Yearly Pacing (with Portfolio Project)	**Chapter 5 Total** $6\frac{1}{2}$ days	**Chapters 1–5 Total** $33\frac{1}{2}$ days	**Remaining** $46\frac{1}{2}$ days	**Total** 80 days

*__AYP__ is Assess Your Progress.

Section	Practice Bank	Study Guide*	Assessment Book*	Visuals	Explorations Lab Manual	Lesson Plans	Technology Book
5.1	31	5.1		Warm-Up 5.1	Master 12	5.1	
5.2	32	5.2	Test 20	Warm-Up 5.2		5.2	
5.3	33	5.3		Warm-Up 5.3		5.3	
5.4	34	5.4		Warm-Up 5.4	Add. Expl. 8	5.4	
5.5	35	5.5	Test 21	Warm-Up 5.5		5.5	Calculator Act. 7
5.6	36	5.6		Warm-Up 5.6 Folder 6		5.6	
5.7	37	5.7	Test 22	Warm-Up 5.7	Master 2	5.7	TI-92 Act. 5
Review Test	38	Chapter Review	Tests 23, 24, Alternative Assessment			Review Test	

*Spanish versions of *Study Guide* and *Assessment Book* are available.

Chapter Support

- Course Guides
- Lesson Plans
- Preparing for College Entrance Tests
- Multi-Language Glossary
- *Test Generator* Software
- Professional Handbook
- Challenge Problems

Software Support

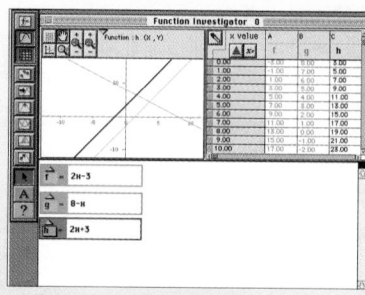

McDougal Littell Mathpack
Geometry Inventor

Internet Support

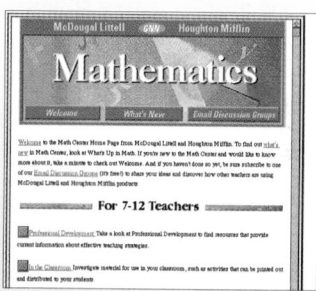

http://www.hmco.com
Next go to McDougal Littell; then the
Education Center; then Secondary Math.

Books, Periodicals

Woodward, Ernest and Thomas Ray Hamel. "The Use of Dot Paper in Geometry Lessons." *Mathematics Teacher* (October 1993): pp. 558–561.

Activities, Manipulatives

Serra, Michael. *Patty Paper Geometry.* Activities to explore constructions. Berkeley, CA: Key Curriculum Press.

Olson, Alton T. *Mathematics Through Paper Folding.* Reston, VA: NCTM, 1975.

Johnson, Donovan A. *Paper Folding for the Mathematics Class.* Reston, VA: NCTM, 1957.

Software

Jackiw, Nicholas, designer. *The Geometer's Sketchpad.* (Ver. 3.0) for Macintosh or Windows. Berkeley, CA: Key Curriculum Press, 1995.

Laborde, Jean-Marie and Franck Bellemain, designers. *Cabri Geometry II.* Dallas, TX: Texas Instruments, 1994.

TI-92 Geometry. (Resident on the TI-92 graphing calculator.) Dallas, TX: Texas Instruments, 1995.

Internet

Investigate science and mathematics resources (no graphics) at:
http://www-hpcc.astro.washington.edu/scied/science.html

Background

Kites

Kites have existed for more than 2000 years. Though many people think of kites in recreational terms, their history gives another view. In China (where the kite probably originated) during the Han dynasty, the military attached bamboo pipes to kites. When flown over the enemy, the wind going through the pipes made a whistling sound that caused the enemy to panic and flee.

Kites have been used in construction, meteorology, and aviation. In 1847, a kite helped pull a cable across the Niagara River as part of the construction of the river's first bridge. Benjamin Franklin conducted his famous electricity experiment with a kite. The Wright Brothers used kites to study ideas about wing warping, and their results helped them make the first airplane in 1903.

José Saínz

José Saínz immigrated to the United States from Mexico. He earned a university degree in drafting and currently works for San Diego Gas and Electric designing electrical systems. To make his Aztec calendar kite, which he calls "Azteca," Saínz projected an image of the calendar onto his garage wall. He then traced it and used the tracing as a pattern for the kite. Saínz created his own color scheme since the calendar itself has no color. "Azteca" won three major awards in the 1992 American Kite Association's Annual Convention. Saínz also entered a number of smaller kites in other categories. All of these kites won prizes as well.

5 Parallel Lines

Watching the Colors Come Alive

INTERVIEW **José Saínz**

José Saínz had no idea what he was getting into when a friend at work asked him to check out a dual-line stunt kite during lunch. But he was so taken with the stunt kite, which could perform tricks in the air, that by the end of the week he had bought one of his own. That was in 1989, and his life hasn't been the same since.

Now one of the top kite designers in the country, the Mexican-born Saínz spends two to three hours every night in his home workshop in San Diego, planning and building his latest creations. It's a hobby that takes him to kite exhibitions and contests all over the world, in which he often captures the top prizes. "Ironically, the friend who introduced me to all this dropped out of kiting soon afterwards," he says. "But I got stuck."

It's All in the Design

Saínz uses a variety of geometric shapes in his kite designs. His first kite, a bird design called "Ave," was square. His award-winning Aztec Calendar kite is a giant hexagon with a tail that is 55 ft long. Other designs are rectangular, triangular, hexagonal, or diamond-shaped.

In competition, kites are judged for aesthetics, structural integrity, craftsmanship, and, most importantly, flight characteristics. "The kite should be stable in the air and fly at a good angle, the higher the better," Saínz explains. The angle at which a conventional single-line kite will fly (and therefore the height of the kite), depends on its shape and the wind speed. "If you have two kites of the exact same shape, they will fly at the same angle and their lines will be parallel," he adds. The two kiters can stand very close together, yet the lines will not get tangled.

> **"The kite should be stable in the air and fly at a good angle, the higher the better."**

217

Background

The Aztec Calendar

The Aztec calendar, derived from the Mayan calendar and very similar to it, is a dating system that combines a ritual cycle of 260 days with the solar cycle of 365 days. The ritual cycle was divided into 13 periods of 20 days, and the solar cycle was divided into 18 periods of 20 days with an additional 5 days. Every 52 years, the two calendars returned to the same positions. This day was celebrated as the Binding Up of the Years, or the New Fire Ceremony.

Second-Language Learners

The first two sentences contain three colloquialisms that may need to be clarified. First, explain that *José Saínz had no idea what he was getting into* means that, at first, José Saínz did not realize that what he was beginning would become a big part of his life. Next, explain that to *check out* a dual-line stunt kite means to look at and try out the kite. Finally, mention that when it says that Saínz was *taken by* the stunt kite, it means that the kite impressed him a great deal, and he liked it very much.

Multicultural Note

The Aztecs built an impressive empire in central Mexico during the fifteenth and sixteenth centuries. According to Aztec mythology, the Aztecs moved south from Aztlan to central Mexico in search of a new homeland. This myth is in part substantiated by anthropological evidence: the Aztecs do seem to have originated in the southwestern United States or northern Mexico, and their language, Nahuatl, is related to the languages spoken by several tribes in that region. By 1400, the Aztecs were beginning to dominate central Mexico, and they continued to expand their empire for more than one hundred years. Their capital at Tenochtitlan may have had a population of as many as 100,000 people. Today, more than one million Mexican citizens are descendants of the Aztecs and retain the Nahuatl language.

Mathematical Connection

While a kite is defined as a geometric shape, kites used for flying can be almost any geometric shape. Because kites are geometric, inherent in them are many geometric properties. Some of these geometric properties are studied in this chapter. In Section 5.2, properties of parallel lines that exist in a regular hexagonal kite are explored. In this section, students use the Aztec Calendar kite to study properties of parallel lines. In Section 5.5, properties of parallel and perpendicular lines are studied. In this section, students look at cross-supports and vertical supports used in the construction of the Aztec Warrior kite. In Section 5.6, dual-line stunt kites are used to investigate ideas about parallels in space.

Explore and Connect

Writing

It may be easier for students to visualize this question if they use some type of object such as pencils or toothpicks to illustrate what the question is asking.

Research

Students may find it helpful to investigate what types of kites are available for practical use. Then they may want to select either the flat kite, the box kite, the delta kite, or the parafoil kite for their research.

Project

Before designing their kites, students may wish to study the construction of kites in order to determine what purpose each part serves in keeping the kite in the air. For example, a kite will not fly without a tail. Students may also want to make a full-sized model of their kite to see if it will actually fly.

Synchronized Precision Maneuvers

Dual-line stunt kites are guided by two lines that run parallel from the kiter's hands to the *bridle point*, where they attach to the kite. One of the most exciting events in an exhibition takes place when members of a four-person team stand side by side, each flying a dual-line, controllable kite, performing synchronized precision maneuvers in

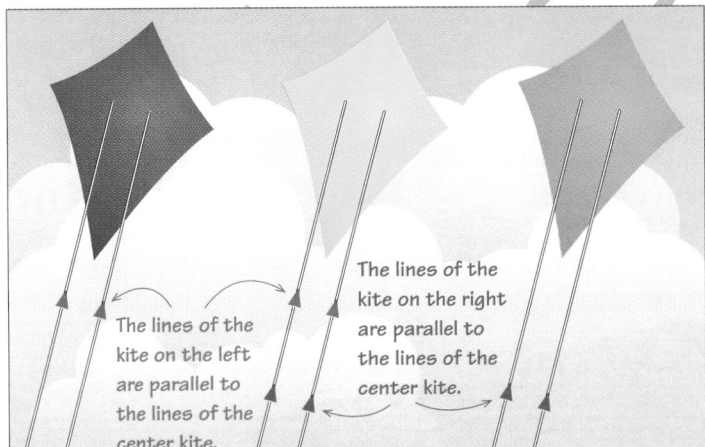

The lines of the kite on the left are parallel to the lines of the center kite.

The lines of the kite on the right are parallel to the lines of the center kite.

the sky. "It's like an air show," Saínz explains. "All these kites are doing flips and loops at the same time which, of course, requires maintaining parallel lines."

In events where a three-person team stands side by side, the teammates on the left and right try to keep their lines parallel with the lines of the center person. As a result, the left and right teammates' lines are parallel to each other. This observation leads to the *Dual Parallels Theorem*, which you will learn more about in this chapter.

You can use the Dual Parallels Theorem to conclude that the lines of the left and right kites are parallel.

Explore and Connect

José Saínz with a
***rokkaku* (Japanese for "six-sided") kite.**

1. Writing In the team formation shown above, if the lines of the right-end person were not parallel to the lines of the center person, how would the lines of the right-end person be related to the lines of the left-end person?

2. Research Kites are used for practical as well as recreational uses. Write a short paragraph about one practical use for a kite. Explain the function of the kite and describe its structure.

3. Project Design a kite. It must have a balanced design in order to fly. First make a sketch of your kite. Then use paper, straws, string, and other materials to make the kite.

Mathematics
& José Saínz

In this chapter, you will learn more about how mathematics is related to kite design.

Related Exercises

Section 5.2
• Exercises 18–21

Section 5.5
• Exercises 4–6

Section 5.6
• Exercises 29 and 30

5.1

Parallel Lines and Transversals

Learn how to...

• **identify pairs of angles formed by transversals and lines**

So you can...

• **find the measures of these angles**

• **analyze real-world examples of intersecting lines**

What produces the oxygen that you breathe? Plants feeding themselves! The leaves of a plant produce food by combining water with carbon dioxide that people exhale. In the process, plants make the oxygen that people and animals depend on. The veins of a leaf are pipelines that make it all possible by carrying water and food through the leaf. In many leaves, the veins resemble intersecting lines.

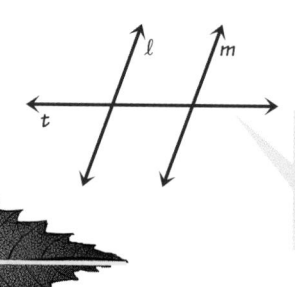

A **transversal** is a line that intersects two or more other lines in the same plane at different points.

A transversal forms several angles with the two lines it intersects. Special names are given to pairs of these angles.

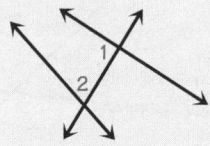

Same-side interior angles lie on the same side of a transversal between the two lines that it intersects.

∠1 and ∠2 are same-side interior angles.

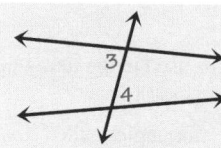

Alternate interior angles lie on opposite sides of a transversal between the two lines that it intersects.

∠3 and ∠4 are alternate interior angles.

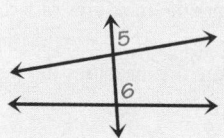

Corresponding angles lie on the same side of a transversal, in corresponding positions with respect to the two lines that it intersects.

∠5 and ∠6 are corresponding angles.

5.1 Parallel Lines and Transversals **219**

Plan⇔Support

Objectives

• Identify pairs of angles formed by transversals and lines.

• Find the measures of angles formed by transversals and lines.

• Analyze real-world examples of intersecting lines.

Recommended Pacing

❖ **Core and Extended Courses**
Section 5.1: 1 day

❖ **Block Schedule**
Section 5.1: $\frac{1}{2}$ block
(with Section 5.2)

Resource Materials

Lesson Support
Lesson Plan 5.1

Warm-Up Transparency 5.1

Practice Bank: Practice 31

Study Guide: Section 5.1

Explorations Lab Manual:
Diagram Master 12

Challenge Problems: Set 31

Technology
Geometry Software

Internet:
http://www.hmco.com

Warm-Up Exercises

Use the diagram.

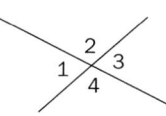

1. Name two angles that form a linear pair. ∠1 and ∠2; ∠2 and ∠3; ∠3 and ∠4; ∠1 and ∠4

2. Name two pairs of vertical angles. ∠1 and ∠3; ∠2 and ∠4

Complete.

3. Angles that form a linear pair are __?__. supplementary

4. Vertical angles are __?__. congruent

5. Supplementary angles have measures that add to __?__. 180°

Learning Styles: Verbal/Visual

To help students remember the terms in this section, encourage those with verbal learning styles to think about the common meanings of the words *transverse*, *interior*, *alternate*, and *corresponding*, and how they relate to the word *transversal* and the names of the various angles created by a transversal. Students with visual learning styles should draw illustrations for each term and then use them to remember the definitions.

Additional Example 1

Name all pairs of same-side interior angles, alternate interior angles, and corresponding angles in this figure.

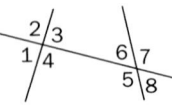

Same-side interior angles: ∠3 and ∠6, ∠4 and∠5
Alternate interior angles: ∠3 and ∠5, ∠4 and ∠6
Corresponding angles: ∠2 and ∠6, ∠3 and ∠7, ∠1 and ∠5, ∠4 and ∠8

Think and Communicate

In answering questions 2 and 3, students develop word associations that will help them remember the terms in this section. Question 4 is a prelude to the Exploration in which students discover that the corresponding angles formed when two parallel lines are intersected by a transversal are congruent. Point out however, that it is not known whether the lines in this diagram are parallel because they are not marked.

EXAMPLE 1

Name all pairs of same-side interior angles, alternate interior angles, and corresponding angles in this figure.

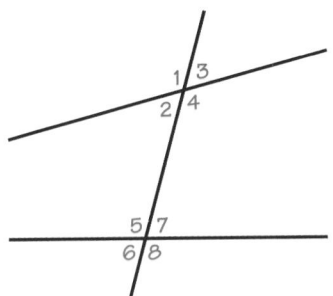

SOLUTION

Same-side interior angles	Alternate interior angles	Corresponding angles
∠2 and ∠5	∠2 and ∠7	∠1 and ∠5
∠4 and ∠7	∠4 and ∠5	∠2 and ∠6
		∠3 and ∠7
		∠4 and ∠8

THINK AND COMMUNICATE

Use the diagram.

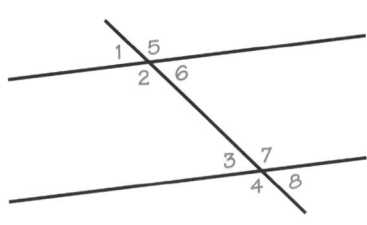

1. Name a pair of same-side interior angles and a pair of alternate interior angles.

2. Why is the word *interior* used in the terms *same-side interior angles* and *alternate interior angles*?

3. Why are same-side interior angles called *same-side*? Why are alternate interior angles called *alternate*?

4. a. Estimate and compare the measures of two corresponding angles in the diagram above.

   b. Estimate and compare the measures of two corresponding angles from Example 1.

ANSWERS   Section 5.1

Think and Communicate

1. ∠2 and ∠3 or ∠6 and ∠7; ∠2 and ∠7 or ∠3 and ∠6

2. *Interior* means "inside" and the angles are inside the space bounded by the two lines.

3. Same-side interior angles are located on the same side of the transversal. Alternate interior angles are on opposite, or alternate, sides of the transversal.

4. a. It appears that the measures of two corresponding angles are equal.

   b. It appears that the measures of two corresponding angles are equal.

Exploration

1. The lines are parallel because the lines on the paper are parallel; they lie in the same plane and do not intersect.

2, 3. Check students' work.

4. The measures of the angles in both pairs should be the same. If two parallel lines are intersected by a transversal, then corresponding angles are congruent.

5. Both sums should be 180°. If two parallel lines are intersected by a transversal, then same-side interior angles are supplementary.

6. The measures of the angles in both pairs should be the same. If two parallel lines are intersected by a transversal, then alternate interior angles are congruent.

7. Each of the conjectures in Steps 4–6 is true for for any two parallel lines intersected by a transversal.

8. Answers may vary. An example is given. No, I do not think the conjectures would still be true because the diagram in Example 1 suggests that they are not true.

EXPLORATION
COOPERATIVE LEARNING

Angles and Transversals

Work in a group of three students.
You will need:
- lined paper or geometry software
- a ruler or straightedge
- a protractor

1 Working individually, each member of your group should draw two parallel lines, using a ruler and paper or geometry software. Discuss how you know that the lines you drew are parallel.

2 Draw a transversal through the lines and number the angles formed.

3 Measure each angle in your diagram and record the results in a table.

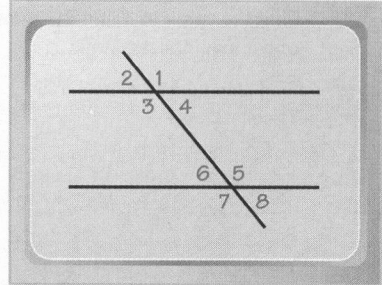

Angle	Measure
1	
2	
3	
4	
5	
6	
7	

4 Compare the measures of two corresponding angles. What do you notice? Compare the measures of a different pair of corresponding angles. Make a conjecture about corresponding angles.

5 Compare the measures of two same-side interior angles. What do you notice? Compare the measures of a different pair of same-side interior angles, and make a conjecture about same-side interior angles.

6 Compare the measures of two alternate interior angles. What do you notice? Measure a different pair of alternate interior angles. Make a conjecture about alternate interior angles.

7 Discuss the conjectures you made in Steps 4–6 with the other members of your group. Are your conjectures true for their diagrams?

8 If the lines you drew in Step 1 were *not* parallel, do you think your conjectures would still be true? Explain why or why not.

Exploration Note

Purpose
The purpose of this Exploration is to have students discover the relationships between pairs of angles formed by two parallel lines intersected by a transversal.

Materials/Preparation
Each group needs lined notebook paper or geometry software, a ruler or straightedge, and a protractor.

Procedure
Each student draws two parallel lines with a transversal, measures all the angles, and compares the measures of pairs of corresponding angles, same-side interior angles,

and alternate interior angles. They then make and compare conjectures and discuss what happens when the lines are not parallel.

Closure
Students should understand that when a transversal intersects a pair of parallel lines, corresponding angles are congruent, same-side interior angles are supplementary, and alternate interior angles are congruent.

Explorations Lab Manual
See the Manual for more commentary on this Exploration.
Diagram Master 12

For answers to the Exploration, see facing page.

Section Notes

Geometric Thinking
In the Exploration, students use inductive reasoning to make conjectures about the relationship between corresponding angles, same-side interior angles, and alternate interior angles. In Step 7, as students in each group compare their results, they have more evidence to support their conjectures. However, remind students that inductive reasoning is used to make a conjecture, but deductive reasoning is needed to *prove* that a conjecture is true for every case. Some true conjectures can be accepted as postulates, and this is the case with corresponding angles being congruent.

Topic Spiraling: Preview
In addition to the Corresponding Angles Postulate, the Exploration previews the Alternate Interior Angles Theorem and the Same-Side Interior Angles Theorem. These theorems are presented in the next section.

Spatial Reasoning
Students should understand that parallel lines do not need to be horizontal for the properties they have discovered in the Exploration to be true. Have students draw parallel lines that are oriented in different directions, draw a transversal, number the angles, and then describe the relationships between the angles based on their discoveries in the Exploration.

About Example 2

Multicultural Note

In the early days of mathematical development, algebra and geometry were used separately. Chinese mathematicians were the first to integrate the two together. In the third century A.D., mathematician Liu Hui wrote the *Sea Island Mathematical Manual*, so named because the first problem in it relates to the surveying of an island in the sea. This book contains the first known examples of using algebraic equations to describe geometrical shapes. The idea of combining algebra and geometry spread to the Middle East in the middle of the ninth century but did not appear in European mathematics until A.D. 1220.

Topic Spiraling: Review

This Example uses the Linear Pair Property and the Vertical Angles Property that were presented in Section 2.1. You may want to review these properties prior to discussing this Example.

 Communication: Drawing
Remind students that the arrows on $\overleftrightarrow{AB}$ and $\overleftrightarrow{CD}$ indicate that they are parallel. Lines cannot be assumed parallel unless these arrows are present or unless it is given that they are parallel.

Additional Example 2

Find the value of *x* and the measure of each numbered angle.

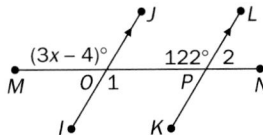

∠ *MOJ* ≅ ∠ *MPL* because they are corresponding angles formed by a transversal and two parallel lines.
$(3x - 4)° = 122°$
$\qquad 3x = 126$
$\qquad x = 42$
∠ 1 ≅ ∠ *MOJ* because they are vertical angles.
$m∠1 = m∠MOJ = 122°$
∠ 2 and ∠ *MPL* are supplementary because they form a linear pair.
$m∠2 = 180° − m∠MPL$
$\qquad = 180 − 122$
$\qquad = 58°$
So $x = 42$, $m∠1 = 122°$, $m∠2 = 58°$.

222

In the Exploration, you probably discovered several facts about the angles formed when parallel lines are intersected by a transversal. This postulate describes one of them.

> **Corresponding Angles Postulate**
>
> If two parallel lines are intersected by a transversal, then corresponding angles are congruent.
>
> If $k \parallel \ell$, then $∠1 ≅ ∠2$.

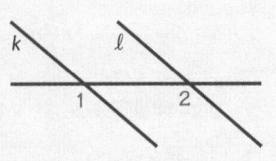

EXAMPLE 2   Connection: Algebra

Find the value of *x* and the measure of each numbered angle.

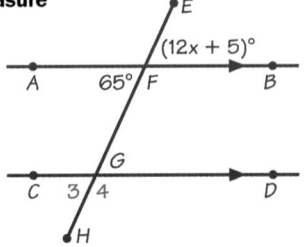

SOLUTION

$∠EFB ≅ ∠AFG$ because they are vertical angles.

$(12x + 5)° = 65°$
$\qquad 12x = 60$
$\qquad x = 5$

$∠3 ≅ ∠AFG$ because they are corresponding angles formed by a transversal and two parallel lines.

$m∠3 = m∠AFG$
$\qquad = 65°$

$∠4$ and $∠3$ are supplementary because they form a linear pair.

$m∠4 = 180° − m∠3$
$\qquad = 180 − 65$
$\qquad = 115°$

So $x = 5$, $m∠3 = 65°$, and $m∠4 = 115°$.

Checking Key Concepts

1. *j*

2. $∠3$ and $∠5$, $∠4$ and $∠6$

3. $∠3$ and $∠6$, $∠4$ and $∠5$

4. Any two of the following:
   $∠1$ and $∠5$, $∠3$ and $∠7$, $∠2$ and $∠6$

5. a. 60°

   b. 60°

   c. 58

☑ CHECKING KEY CONCEPTS

Use the diagram.

1. Which line is the transversal?

2. Name two pairs of same-side interior angles.

3. Name two pairs of alternate interior angles.

4. Name two pairs of corresponding angles.

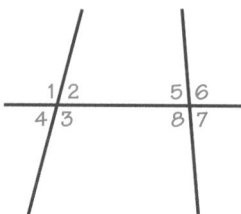

5. Given that $m \angle 1 = 120°$, find:

   a. $m \angle 3$ **b.** $m \angle 6$ **c.** the value of x

5.1 | Exercises and Applications

Extra Practice
exercises on
page 667

Use the diagram for Exercises 1–4.

1. Name all pairs of same-side interior angles.

2. Name all pairs of alternate interior angles.

3. Name all pairs of corresponding angles.

4. **Writing** Find two pairs of angles that could be called *same-side exterior angles*, and explain why that term could be used.

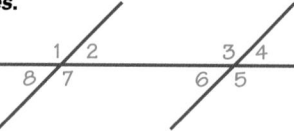

Classify each pair of angles as *corresponding angles*, *alternate interior angles*, or *same-side interior angles*.

5. $\angle 5$ and $\angle 7$ 6. $\angle 7$ and $\angle 3$

7. $\angle 3$ and $\angle 1$ 8. $\angle 6$ and $\angle 2$

9. $\angle 2$ and $\angle 3$ 10. $\angle 8$ and $\angle 6$

AUTOMOBILES The windshield wipers of this car move in such a way that they are always parallel to each other.

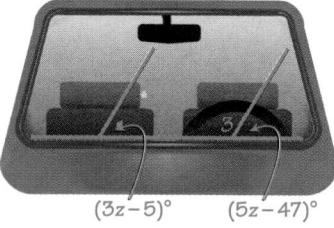

(3z−5)°   (5z−47)°

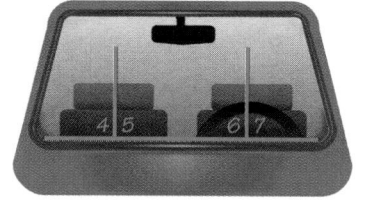

11. **ALGEBRA** Find the value of z and $m \angle 3$.

12. If you know that $\angle 4 \cong \angle 7$, what can you say about the windshield wipers? Explain.

5.1 Parallel Lines and Transversals   **223**

Interdisciplinary Problems
Exs. 13–16 These exercises provide students with an example of a connection between the science of botany and geometry.

Career Connection
Exs. 13–16 Botany is a subfield of biology and is concerned with the study of plant life. Among the topics botanists may examine are plant classification, plant form and structure, plant life processes, plant interaction with the environment, plant diseases, plant genetics, and plant fossils. Some botanists concentrate on a particular type of plant. Botanists need to have a college degree in biology and often take courses in physical science and mathematics as well. Many botanists have advanced degrees in their field and work at universities doing research, or they may work in the private sector for firms that specialize in areas such as agriculture or the environment.

Using Manipulatives
Exs. 13–16 Ask students to bring in beech tree leaves and other leaves so that the vein patterns can be studied. Ask questions such as: Is the angle between the parallel lines and the transversal the same on each beech tree leaf? What leaf patterns are on leaves from different trees? Do any of them contain parallel lines and transversals?

Student Progress
Exs. 19–26 Students should be able to complete these exercises with little or no difficulty. Students who have trouble with them but not with Exs. 5–10, need to learn to focus on the part of the drawing that is relevant to the question. Have these students identify the two lines and the transversal for each exercise. Each of the two angles has a side that is part of the transversal. They may want to reproduce the part of the drawing containing these two lines and the transversal.

Application
Ex. 27 This open-ended exercise guides students to discover how the mathematics of lines and transversals applies to their everyday lives. Ask students to share their sketches with the class, or you might have them display their work on a bulletin board.

Connection   BOTANY

Beech trees are found throughout the Northern Hemisphere. The secondary veins of the European beech, which all intersect the central vein, are often parallel to each other.

13. Is k a transversal of m and ℓ? Explain why or why not.

14. Given that $m\angle 1 = 142°$, find $m\angle 2$ and $m\angle 3$.

15. If you know $m\angle 2$, can you use the Corresponding Angles Postulate to find $m\angle 4$? Explain why or why not.

16. Research Find out about another type of leaf that has veins like this one. What makes this pattern of veins efficient?

BY THE WAY

Many insects, known as leaf insects, are naturally camouflaged because their wings mimic the veined pattern of a leaf.

ALGEBRA In Exercises 17 and 18, find each value specified.

17. Find the values of x, y, and z.

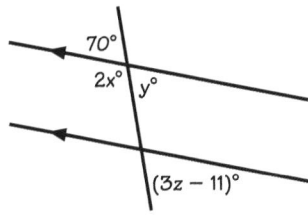

18. Find the value of w and $m\angle 1$.

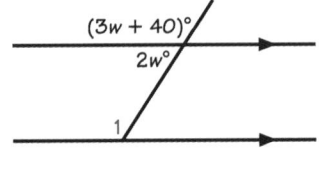

Classify each pair of angles as *corresponding angles, alternate interior angles, same-side interior angles,* or *none of these.*

19. $\angle 2$ and $\angle 3$

20. $\angle 7$ and $\angle 12$

21. $\angle 3$ and $\angle 16$

22. $\angle 10$ and $\angle 2$

23. $\angle 4$ and $\angle 10$

24. $\angle 6$ and $\angle 3$

25. $\angle 7$ and $\angle 11$

26. $\angle 14$ and $\angle 16$

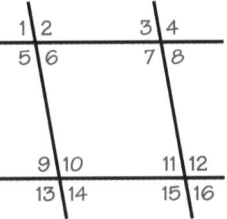

27. Open-ended Problem Sketch an example of lines intersected by a transversal in a piece of furniture, in the architecture of your school building, or in some other object. Label a pair of corresponding angles and a pair of same-side interior angles.

---

13. No; k does not intersect l.

14. 38°; 142°

15. No; lines j and k are not parallel, so the postulate cannot be used.

16. Answers may vary. Research should include the fact that veins carry water and food from the stem and distribute it throughout the leaf.

17. $x = 55$; $y = 70$; $z = 27$

18. $w = 28$; $m\angle 1 = 124$

19. same-side interior angles

20. alternate interior angles

21. none of these

22. corresponding angles

23. none of these

24. alternate interior angles

25. same-side interior angles

26. corresponding angles

27. Check students' work.

28. Copy and complete the proof.

Given: $m \parallel n; j \parallel k$
Prove: $\angle 1 \cong \angle 4$

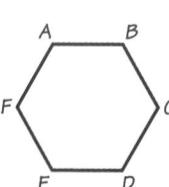

Statements	Reasons
1. $m \parallel n, j \parallel k$	1. _?_
2. $\angle 1 \cong$ _?_	2. If two $\parallel$ lines are intersected by a transversal, then corresponding angles are $\cong$.
3. $\angle 2 \cong \angle 3$	3. _?_
4. $\angle 3 \cong \angle 4$	4. _?_
5. _?_	5. Transitive Property (Steps 2, 3, and 4)

Use this regular hexagon for Exercises 29–31.

29. Name the three pairs of parallel segments in the hexagon.

30. Visual Thinking If $\overline{AB}$, $\overline{BC}$, and $\overline{ED}$ are extended into lines, which one is the transversal?

31. Challenge If all six sides of the hexagon are extended into lines, which of the lines are transversals of $\overleftrightarrow{AB}$ and $\overleftrightarrow{BC}$? Explain.

ONGOING ASSESSMENT

32. Open-ended Problem Sketch two pairs of parallel lines that intersect to form a parallelogram.
   a. On your diagram, label a pair of same-side interior angles, a pair of alternate interior angles, and a pair of corresponding angles.
   b. Choose one of the pairs from part (a) in which the two angles are congruent. Explain how you know that they are congruent.

SPIRAL REVIEW

33. The endpoints of $\overline{AB}$ are $A(3, -7, 4)$ and $B(2, 0, 6)$. Find the midpoint and length of $\overline{AB}$. *(Section 4.6)*

Tell whether each statement is *True* or *False*. If the statement is false, sketch a counterexample. *(Section 2.5)*

34. Every rectangle is a parallelogram.

35. If two sides of a quadrilateral are parallel, then it is a parallelogram.

Exercise Notes

Challenge
Ex. 31 Students should realize that they need to eliminate the segments that are parallel to $\overline{AB}$ and $\overline{BC}$.

Assessment Note
Ex. 32 You might have several students work this problem at the board. They can then explain their reasoning to help others assess their own work.

Practice 31 for Section 5.1

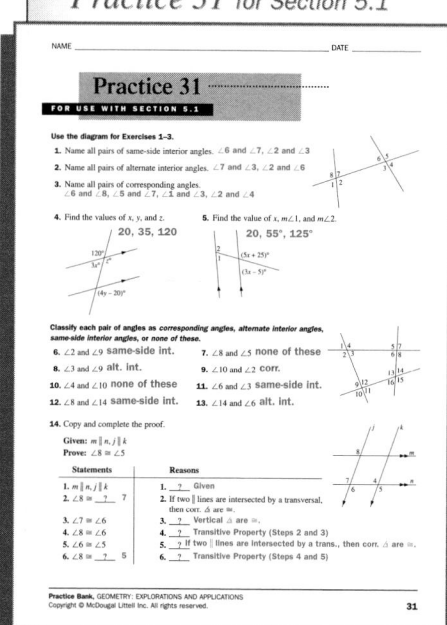

28. (1) Given; (2) $\angle 2$; (3) Vertical $\angle$s are $\cong$; (4) If two $\parallel$ lines are intersected by a transversal, then corresponding $\angle$s are $\cong$; (5) $\angle 1 \cong \angle 4$.

29. $\overline{AB}$ and $\overline{DE}$; $\overline{BC}$ and $\overline{EF}$; $\overline{CD}$ and $\overline{FA}$

30. $\overleftrightarrow{BC}$

31. $\overleftrightarrow{AF}$ and $\overleftrightarrow{CD}$, since these lines would intersect both $\overleftrightarrow{AB}$ and $\overleftrightarrow{BC}$

32. Answers may vary. For example, consider the figure in Ex. 28.

a. Let $\angle 5$ be the interior angle to j and k that is formed by the intersection of m and k. $\angle 1$ and $\angle 5$ are same-side interior angles. $\angle 1$ and $\angle 3$ are alternate interior angles and $\angle 1$ and $\angle 2$ are corresponding angles.

b. $\angle 1 \cong \angle 2$ by the Corresponding Angles Postulate. $\angle 2$ and $\angle 3$ are vertical angles, so $\angle 2 \cong \angle 3$. Therefore, alternate interior angles $\angle 1$ and $\angle 3$ are congruent by the Transitive Property.

33. $(2.5, -3.5, 5)$; $3\sqrt{6}$

34. True.

35. False; answers may vary. An example is given.

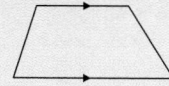

- Find the measures of alternate interior angles and same-side interior angles.
- Identify trapezoids.
- Prove statements about special angles.
- Find congruent angles in real-world objects.

Recommended Pacing

❖ **Core and Extended Courses**
Section 5.2: 2 days

❖ **Block Schedule**
Section 5.2: 2 half-blocks (with Sections 5.1 and 5.3)

Resource Materials

Lesson Support
Lesson Plan 5.2
Warm-Up Transparency 5.2
Practice Bank: Practice 32
Study Guide: Section 5.2
Challenge Problems: Set 32
Assessment Book: Test 20

Technology
McDougal Littell Mathpack
   Geometry Inventor Activity Book:
      Activity 1

Internet:
   http://www.hmco.com

Warm-Up Exercises

Find the measure of each angle.

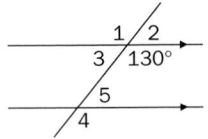

1. ∠1 130°
2. ∠2 50°
3. ∠3 50°
4. ∠4 130°
5. ∠5 50°

SECTION

5.2 Properties of Parallel Lines

Learn how to...
- **find the measures of alternate interior angles and same-side interior angles**
- **identify trapezoids**

So you can...
- **prove statements about special angles**
- **find congruent angles in kites, windows, and other real-world objects**

Cloth embroidered by the Shoowa people of central Zaire is admired for the beauty and complexity of its geometric designs. In the piece shown, parallel lines appear to be woven, suggesting basketwork.

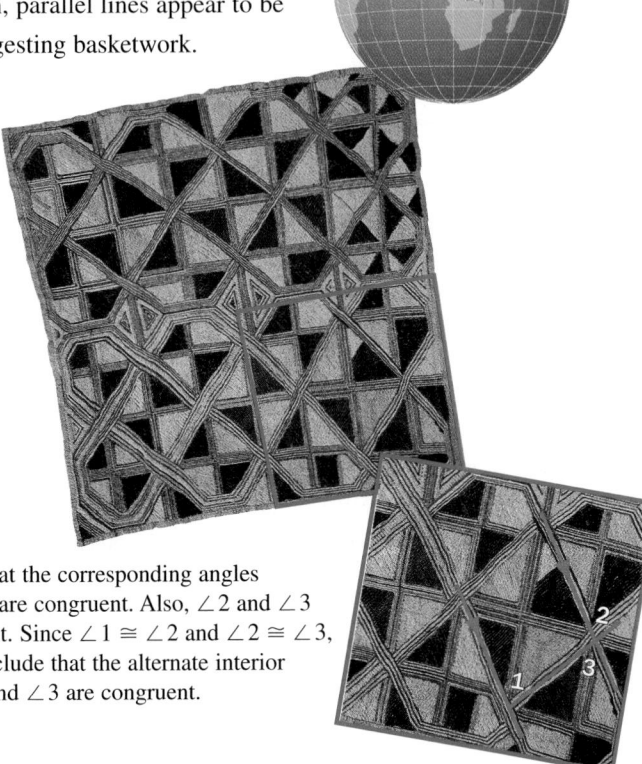

You know that the corresponding angles ∠1 and ∠2 are congruent. Also, ∠2 and ∠3 are congruent. Since ∠1 ≅ ∠2 and ∠2 ≅ ∠3, you can conclude that the alternate interior angles ∠1 and ∠3 are congruent.

THINK AND COMMUNICATE

Use the section of Shoowa cloth above.

1. How do you know that ∠2 ≅ ∠3?

2. If ∠1 ≅ ∠2 and ∠2 ≅ ∠3, how can you justify the conclusion that ∠1 ≅ ∠3?

226   Chapter 5 *Parallel Lines*

Alternate Interior Angles Theorem

If two parallel lines are intersected by a transversal, then alternate interior angles are congruent.

If $k \parallel \ell$, then $\angle 1 \cong \angle 3$.

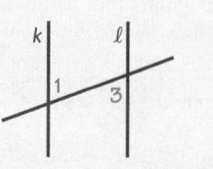

EXAMPLE 1   Connection: Algebra

Find the value of y.

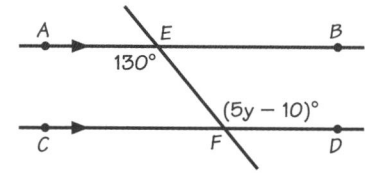

SOLUTION

Because $\angle EFD$ and $\angle AEF$ are alternate interior angles formed by a transversal and two parallel lines, they are congruent.

$$m \angle EFD = m \angle AEF$$
$$(5y - 10)° = 130°$$
$$5y = 140 \qquad \text{Solve for } y.$$
$$y = 28$$

THINK AND COMMUNICATE

Use the diagram in Example 1.

3. What is $m \angle EFD$?

4. What is $m \angle EFC$? How do you know?

5. What do you notice about $m \angle AEF$ and $m \angle EFC$?

Same-Side Interior Angles Theorem

If two parallel lines are intersected by a transversal, then same-side interior angles are supplementary.

If $j \parallel k$, then $m \angle 4 + m \angle 5 = 180°$.

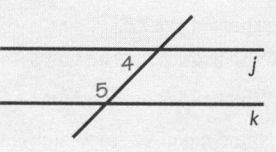

5.2 Properties of Parallel Lines   **227**

Section Notes

Teaching Tip
Students first saw examples of the relationships between alternate interior angles and same-side interior angles in the Exploration on page 221. You may wish to have students refer back to the conjectures they made at that time.

Visual Thinking
Encourage students to sketch a variety of diagrams that explain the two theorems presented on this page. Ask them to present their diagrams to the class. This activity involves the visual skills of *recognition* and *communication*.

Using Technology
Students can use Activity 1 in the *Geometry Inventor Activity Book* to explore properties of parallel lines.

Common Error
Some students tend to apply the two theorems on this page and the Corresponding Angles Postulate even when two nonparallel lines are intersected by a transversal. Stress to students that for any of these statements to apply, the lines intersected by the transversal must be parallel.

Think and Communicate

After discussing the Alternate Interior Angles Theorem, students should be able to use their responses to questions 1 and 2 to help them write a paragraph proof of this theorem. This proof can then be used in Ex. 33 on page 232.

Additional Example 1

Find the value of z.

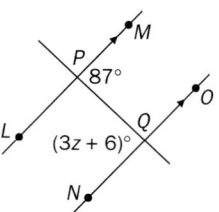

Because $\angle MPQ$ and $\angle NQP$ are alternate interior angles formed by a transversal and two parallel lines, they are congruent.
$$m \angle NQP = m \angle MPQ$$
$$(3z + 6)° = 87°$$
$$3z = 81$$
$$z = 27$$

Think and Communicate

3. 130°

4. 50°; $\angle EFC$ and $\angle EFD$ form a linear pair.

5. The sum of the measures is 180°.

227

Section Notes

Writing Proofs
You may wish to have students use the key steps for a proof of the Same-Side Interior Angles Theorem to write a paragraph proof or two-column proof for this theorem. When discussing this proof, have students read aloud all of the symbols to be certain they understand what each symbol represents.

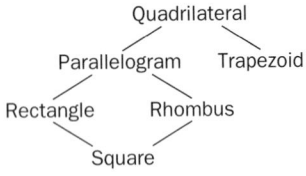 **Communication: Discussion**
Ask students to discuss the meaning of the phrase *exactly one* in the definition of trapezoid. They should realize that a trapezoid cannot have two pairs of parallel sides. Thus, trapezoids are not parallelograms. Some students may prefer to define a trapezoid as a quadrilateral with one pair of parallel sides and one pair of nonparallel sides.

Geometric Thinking
After defining trapezoid, ask students to draw a diagram that shows the hierarchy of quadrilaterals, such as the one below.

```
              Quadrilateral
             /            \
    Parallelogram        Trapezoid
      /        \
 Rectangle    Rhombus
       \        /
         Square
```

Think and Communicate

Ask students to identify the parallel lines and transversal they are using before answering question 8. It may help some students to extend the parallel lines shown in the trapezoid.

You already know how to write proofs of theorems in two different ways: two-column proof and paragraph proof. Sometimes, you may want to be less formal and write down only the *key steps of a proof*. The key steps can also serve as a plan for a more formal proof.

Here are the key steps of a proof of the Same-Side Interior Angles Theorem.

Given: $j \parallel k$
Prove: $m \angle 4 + m \angle 5 = 180°$

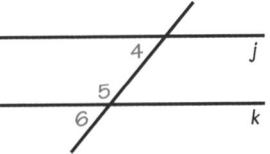

1. $m \angle 6 + m \angle 5 = 180°$ (The ⦞ form a linear pair.)
2. $\angle 4 \cong \angle 6$ (Corresponding ⦞ formed by $\parallel$ lines and a transversal are $\cong$.)
3. $m \angle 4 + m \angle 5 = 180°$ (Substitution)

A quadrilateral with exactly one pair of parallel sides is called a **trapezoid**. Each of the quadrilaterals highlighted in red below is a trapezoid.

The two parallel sides of a trapezoid are called the **bases**.

The two non-parallel sides are called the **legs**.

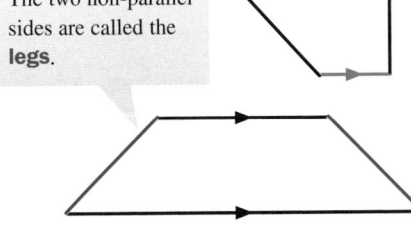

THINK AND COMMUNICATE

Use trapezoid *WXYZ*.

6. Which sides are the bases?
7. Which sides are the legs?
8. Name a pair of same-side interior angles.

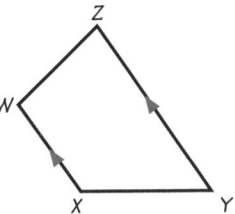

Think and Communicate
6. $\overline{WX}$ and $\overline{YZ}$
7. $\overline{WZ}$ and $\overline{XY}$
8. $\angle W$ and $\angle Z$; $\angle X$ and $\angle Y$

You can prove statements about trapezoids using the postulates and theorems that you know about parallel lines.

EXAMPLE 2

In trapezoid *ABCD*, $\overline{AB} \parallel \overline{DC}$. Write a paragraph proof showing that $m\angle A + m\angle D = 180°$.

SOLUTION

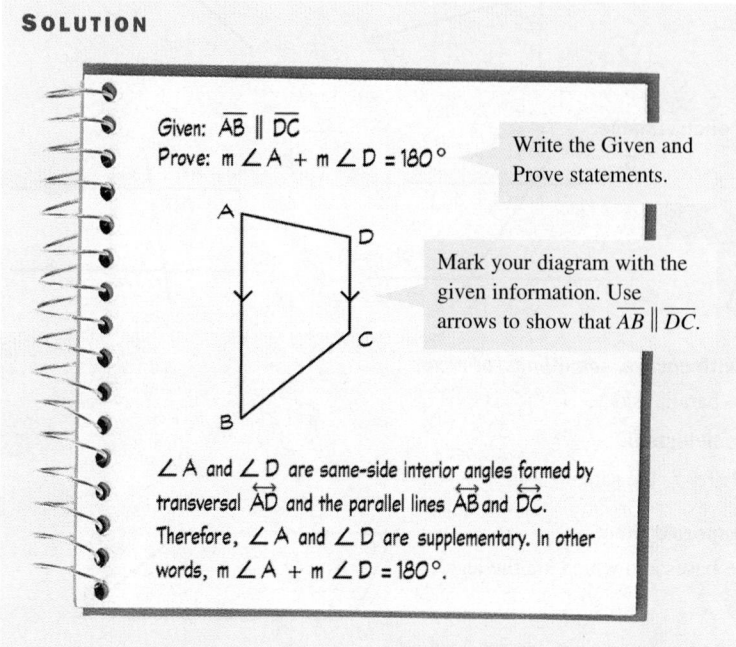

Given: $\overline{AB} \parallel \overline{DC}$
Prove: $m\angle A + m\angle D = 180°$

Write the Given and Prove statements.

Mark your diagram with the given information. Use arrows to show that $\overline{AB} \parallel \overline{DC}$.

$\angle A$ and $\angle D$ are same-side interior angles formed by transversal $\overleftrightarrow{AD}$ and the parallel lines $\overleftrightarrow{AB}$ and $\overleftrightarrow{DC}$. Therefore, $\angle A$ and $\angle D$ are supplementary. In other words, $m\angle A + m\angle D = 180°$.

☑ CHECKING KEY CONCEPTS

Find the values indicated.

1. Find the value of z and $m\angle 1$.

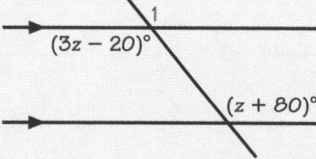

2. Find the value of x and $m\angle 2$.

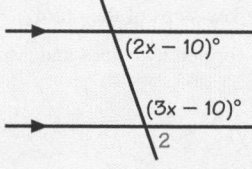

3. Sketch a trapezoid in which one base is twice as long as the other base.

Teach⇔Interact

Additional Example 2

In trapezoid *WXYZ*, $\overline{WX} \parallel \overline{ZY}$. $\overline{XZ}$ is a diagonal of the trapezoid. Write a paragraph proof showing that $m\angle WXZ = m\angle YZX$.
Write the Given and Prove statements.
Given: $\overline{WX} \parallel \overline{ZY}$
Prove: $m\angle WXZ = m\angle YZX$
Mark your diagram with the given information. Use arrows to show that $\overline{WX} \parallel \overline{ZY}$.

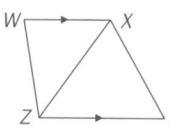

$\angle WXZ$ and $\angle YZX$ are alternate interior angles formed by transversal $\overleftrightarrow{ZX}$ and the parallel lines $\overleftrightarrow{WX}$ and $\overleftrightarrow{ZY}$. Therefore, $\angle WXZ$ and $\angle YZX$ are congruent. In other words, $m\angle WXZ = m\angle YZX$.

Checking Key Concepts

Geometric Thinking
Check that students' drawings for question 3 contain one pair of parallel sides and one pair of nonparallel sides.

Closure Question

Describe the postulate and theorems that you have learned about the angles formed when a traversal intersects a pair of parallel lines.
The Corresponding Angles Postulate states that the corresponding angles are congruent. The Alternate Interior Angles Theorem states that the alternate interior angles are congruent. The Same-Side Interior Angles Theorem states that same-side interior angles are supplementary.

Checking Key Concepts

1. $z = 50$; $m\angle 1 = 130°$

2. $x = 40$; $m\angle 2 = 70°$

3. Answers may vary. An example is given.

Extra Practice
exercises on
page 668

Suggested Assignment

❖ **Core Course**
  Day 1 Exs. 1–17, 22, 23
  Day 2 Exs. 24–28, 30–37, AYP
❖ **Extended Course**
  Day 1 Exs. 1–23
  Day 2 Exs. 24–37, AYP
❖ **Block Schedule**
  Day 28 Exs. 1–17, 22, 23
  Day 29 Exs. 24–28, 30–37,
          AYP

Exercise Notes

Student Progress
Exs. 1–6 Students should be able to complete these exercises with little or no difficulty. Those who have trouble should be encouraged to review the postulates and theorems used in the first two sections of this chapter and to write them out for reference.

Integrating Algebra
Exs. 4–6, 31, 32, 36, 37 These exercises require students to write and solve an algebraic equation. After solving the equation, students need to check to see if additional work needs to be done to complete the exercise.

Geometric Thinking
Exs. 7–9 These exercises can help to sharpen students' understanding of the definition of a trapezoid. Students should refer to the definition of a trapezoid on page 228 if they are uncertain about their answers.

Cooperative Learning
Exs. 14–17 These exercises are appropriate for small group work. Each student should try to draw the trapezoid. Group members can then compare their results.

5.2 Exercises and Applications

Find the measure of each numbered angle.

1.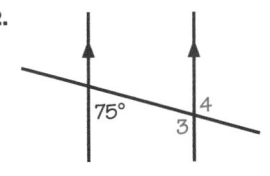

2.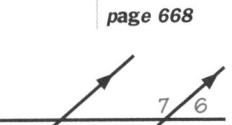

3.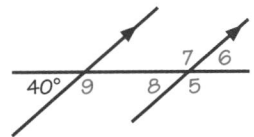

ALGEBRA Find the value of each variable.

4.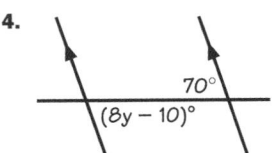

5.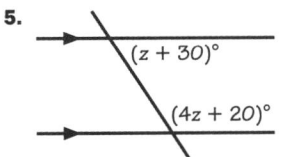

6.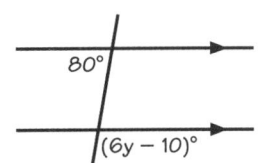

Complete each statement with *always*, *sometimes*, or *never*.

7. A trapezoid ? has two parallel sides.

8. A trapezoid is ? a parallelogram.

9. The legs of a trapezoid are ? the same length.

For Exercises 10–12, use trapezoid *JKLM*.

10. Tell which sides are the bases and which are the legs.

11. Find $m \angle M$.

12. Can you find $m \angle K$ using the Same-Side Interior Angles Theorem? If so, find it. If not, explain why not.

13. **a.** Sketch a copy of the diagram and mark it with the given information.

   Given: $\overleftrightarrow{WX} \parallel \overleftrightarrow{YZ}$
   Prove: $\angle 1 \cong \angle 2$

   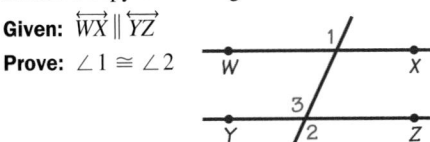

   b. Copy and complete these key steps of the proof.

   1. ? (Corresponding ∠s formed by ∥ lines and a transversal are ≅.)

   2. $\angle 3 \cong \angle 2$ (?)

   3. ? (Transitive Property)

Sketch each trapezoid, if possible. If it is not possible, explain why not.

14. Each of the bases is longer than each leg.
15. The trapezoid has two right angles.
16. Each of the bases is shorter than each leg.
17. The trapezoid has three acute angles.

230   Chapter 5 *Parallel Lines*

Exercises and Applications

1. $m \angle 1 = m \angle 2 = 125°$

2. $m \angle 3 = m \angle 4 = 105°$

3. $m \angle 5 = m \angle 7 = m \angle 9 = 140°$;
   $m \angle 6 = m \angle 8 = 40°$

4. 10   5. 26

6. $18\frac{1}{3}$ 7. always

8. never   9. sometimes

10. bases: $\overline{LK}$ and $\overline{MJ}$; legs: $\overline{LM}$ and $\overline{KJ}$

11. 115°

12. No; $\overline{LM}$ is not parallel to $\overline{KJ}$. If you knew $m \angle J$, you could use the Same-Side Interior Angles Theorem to find $m \angle K$.

13. **a.**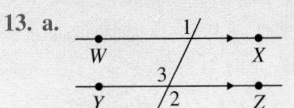

   b. (1) $\angle 1 \cong \angle 3$; (2) Vertical ∠s are ≅; (3) $\angle 1 \cong \angle 2$.

14–16. Answers may vary. Examples are given.

14.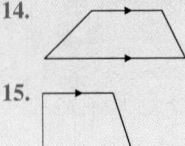

15.

16.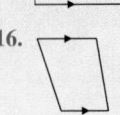

17. not possible; Answers may vary. For example, Example 2 on page 229 showed that the

angles of a trapezoid consist of two pairs of supplementary angles. Two acute angles cannot be supplementary, so a trapezoid cannot have three acute angles.

18. $\overleftrightarrow{PQ}$ and $\overleftrightarrow{ST}$

19. Answers may vary. An example is given. $\angle UPS$ and $\angle RSP$

20. Answers may vary. An example is given. Each acute angle with vertex at O has measure

Look back at the article on pages 216–218.

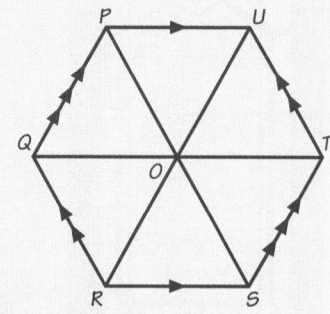

Without its 55 ft long tail, the Aztec Calendar kite designed by José Saínz is a regular hexagon. The cloth is supported by six hollow sticks that are joined at the center of the kite and extend to the six vertices of the hexagon.

BY THE WAY

The Aztec Calendar kite won first prize in its class, the Grand Champion Award, and the People's Choice Award at the American Kitefliers Association's annual convention.

18. $\overleftrightarrow{PS}$ is a transversal of $\overleftrightarrow{UP}$ and $\overleftrightarrow{SR}$. Name another pair of parallel lines that have $\overleftrightarrow{PS}$ for a transversal.

19. Open-ended Problem Name a pair of angles on the kite that the Alternate Interior Angles Theorem guarantees are congruent.

20. Writing Explain why all of the acute angles are congruent in the diagram of the kite above.

21. Copy the diagram of the kite and extend $\overline{QR}$ and $\overline{UT}$ into lines. Label a point on each new ray. Name two angles that are supplementary by the Same-Side Interior Angles Theorem.

For Exercises 22 and 23, use the diagram.

22. SAT/ACT Preview Which angles are supplementary to $\angle 1$?
   A. $\angle 2$ and $\angle 4$ only
   B. $\angle 4$, $\angle 5$, and $\angle 8$
   C. $\angle 2$, $\angle 5$, and $\angle 6$
   D. $\angle 2$, $\angle 4$, $\angle 6$, and $\angle 8$

23. Tell which angles are congruent to $\angle 7$.

Cooperative Learning For Exercises 24–27, work with another person.

24. Each of you should use a ruler to draw a large diagram of two lines that are not parallel. Draw three non-parallel transversals of the lines and label three pairs of same-side interior angles *on the same side of the transversals*, as shown.

25. Measure the angles. Find $m\angle 1 + m\angle 2$ and $m\angle 3 + m\angle 4$.

26. Predict the value of $m\angle 5 + m\angle 6$. Then check your prediction by measuring.

27. Writing Compare your results with your partner's results. What can you conclude?

$\dfrac{360°}{6} = 60°$. $\overleftrightarrow{PS}, \overleftrightarrow{QT},$ and $\overleftrightarrow{RU}$ are lines of symmetry for the regular hexagon, so each of these lines bisects two interior angles of the hexagon. Since the measure of each interior angle of the regular hexagon is 120°, each acute angle with vertex at a vertex of the hexagon measures 60°. All the acute angles in the diagram are congruent, since each angle measures 60°.

21. Answers may vary. An example is given. $\angle XUR$ and $\angle QRU$

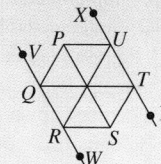

22. D

23. $\angle 2, \angle 4,$ and $\angle 6$

24. Check students' work.

25–27. Answers may vary. Examples are given.

25. The sum of the measures is the same for both pairs of angles.

26. The sum of the measures is the same as in Ex. 25.

27. If two lines are intersected by transversals, then the sum of the measures of any two same-side interior angles on the same sides of the transversals is always the same.

Exercise Notes

Interview Note: Research
Exs. 18–21 After viewing this kite, some students may wish to find out more about the Aztec Calendar and how it is represented here.

Second-Language Learners
Exs. 18–21 The different parts of the interview with José Saínz contain vocabulary that is specific to kites. You may want to begin a list of words and definitions with *kite* and *tail,* As you continue the chapter, add *flaps* and *supports* (page 252) and *dual-time, stunt, flips, loops,* and *stack* (page 262).

Using Manipulatives
Ex. 21 Some students may have difficulty focusing on a part of a diagram. To help with this, have them place a clear transparency sheet over the illustration. They can then trace over and extend $\overline{QR}$ and $\overline{UT}$ into lines and draw either $\overline{QT}$ or $\overline{RU}$ for a transversal. The transparency can be removed and the supplementary angles marked. The transparency can be replaced on the diagram to name the angles. This technique can be used whenever students need to focus their attention on part of a complicated drawing.

Communication: Writing
Ex. 27 This exercise gives students practice writing the results of an investigation. Each person should discuss and compare his or her result with their partner, and then write their own conclusions. You may wish to have the partners exchange papers and critique each other's writing.

Exercise Notes

232

This stained glass window was designed by Frank Lloyd Wright for the Bradley house in Kankakee, Illinois. In the part of the window shown, $\overleftrightarrow{AC} \parallel \overleftrightarrow{HD}$ and $\overleftrightarrow{CE} \parallel \overleftrightarrow{BF}$.

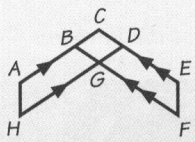

28. Use the Corresponding Angles Postulate and the theorems that you learned in this lesson.

   a. Name four angles that are supplementary to ∠*CBG*.

   b. Name two angles that are congruent to ∠*CBG*.

29. Challenge Name four trapezoids in the part of the window shown above.

30. a. In parallelogram *WXYZ*, $m \angle X = 60°$. Sketch *WXYZ* and find $m \angle W$ and $m \angle Y$.

   b. Sketch a diagram and write the key steps of a proof of this theorem: Consecutive angles of a parallelogram are supplementary.

ALGEBRA Find the measure of each numbered angle.

31.
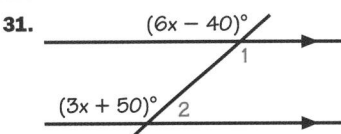

32.
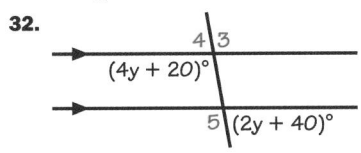

33. Write a two-column proof of the Alternate Interior Angles Theorem.

34. Each sentence of this paragraph proof has something wrong with it. Rewrite the proof correctly.

   Given: $j \parallel k$; $m \perp j$

   Prove: $m \perp k$

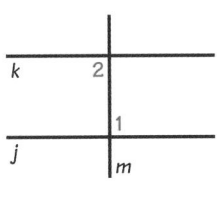

Since *j* ll *k* , ∠1 is a right angle. Because ∠1 and ∠2 are corresponding angles formed by two parallel lines and a transversal, ∠1 ≅ ∠2. Therefore, ∠2 is an acute angle. So by the definition of perpendicular lines, *m* ⊥ *j*.

---

28. a. ∠*ABG*, ∠*BCD*, ∠*BGD*, ∠*GDE*

   b. ∠*HGB* and ∠*CDG*

29. *ABGH*, *ACDH*, *DEFG*, *BCEF*

30. a.
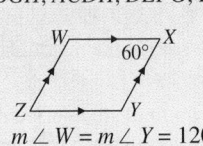

   $m \angle W = m \angle Y = 120°$

   b. Check students' work. Answers may vary. An example is given.
   Given: *ABCD* is a ▱.
   Prove: ∠*A* and ∠*B* are supplementary.

Since *ABCD* is a ▱, $\overleftrightarrow{AB} \parallel \overleftrightarrow{CD}$ and $\overleftrightarrow{AD} \parallel \overleftrightarrow{BC}$. Therefore, ∠*A* and ∠*B* are supplementary. (If two ∥ lines are intersected by a transversal, then same-side interior ⦞ are supplementary.) Similarly, it can be shown that the following pairs of ⦞ are supplementary: ∠*B* and ∠*C*, ∠*C* and ∠*D*, ∠*D* and ∠*A*.

31. $m \angle 1 = 140°$; $m \angle 2 = 40°$

32. $m \angle 3 = 100°$; $m \angle 4 = 80°$; $m \angle 5 = 100°$

33. Answers may vary. An example is given.
Given: $k \parallel l$
Prove: ∠1 ≅ ∠3

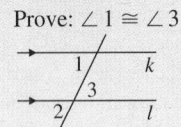

Statements (Reasons)

1. $k \parallel l$ (Given)
2. ∠1 ≅ ∠2 (If two ∥ lines are intersected by a transversal, then corresponding ⦞ are ≅.)

35. Open-ended Problem Sketch two parallel lines and two transversals that form a trapezoid. Number all of the angles formed.

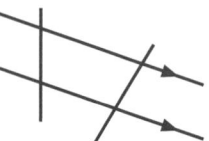

  a. List three pairs of supplementary angles. Explain how you know that they are supplementary.

  b. Use your results from part (a) to write a theorem about the angles of a trapezoid.

SPIRAL REVIEW

ALGEBRA **Find each unknown angle measure.** *(Section 2.2)*

36.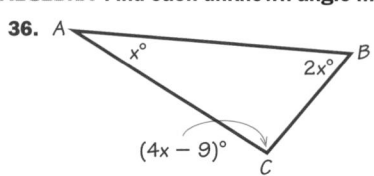

37.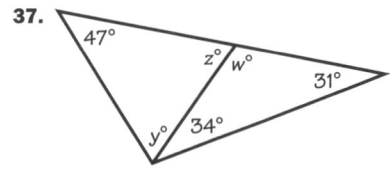

ASSESS YOUR PROGRESS

VOCABULARY

transversal (p. 219)
same-side interior angles (p. 219)
alternate interior angles (p. 219)
corresponding angles (p. 219)

trapezoid (p. 228)
bases of a trapezoid (p. 228)
legs of a trapezoid (p. 228)

Use the diagram. *(Section 5.1)*

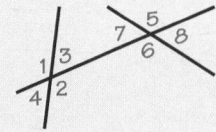

1. Name all pairs of same-side interior angles, alternate interior angles, and corresponding angles.

ALGEBRA **Find the value of each variable.** *(Sections 5.1 and 5.2)*

2.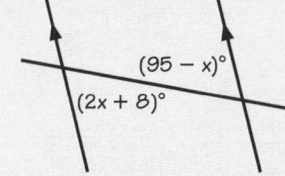

3.

$(6z - 119)°$

$\frac{1}{3}y°$

$y°$

$(3z - 5)°$

4. Sketch and label a trapezoid. Tell which sides are the bases and which are the legs. *(Section 5.2)*

5. Journal Sketch two parallel lines and a transversal and label the angles formed. Name all of the angles that are supplementary to one of the angles, and explain how you know they are supplementary.

5.2 Properties of Parallel Lines **233**

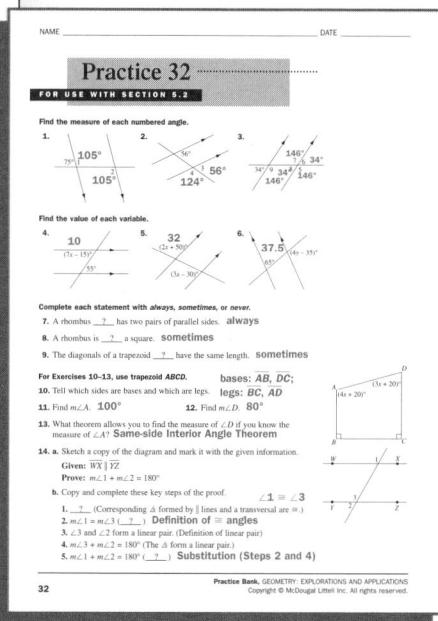
---

3. ∠2 ≅ ∠3 (Vertical ⦞ are ≅.)
4. ∠1 ≅ ∠3 (Transitive Property (Steps 2 and 3))

34. Since $m \perp j$, ∠1 is a right ∠. Because ∠1 and ∠2 are alternate interior ⦞ formed by two ∥ lines and a transversal, ∠1 ≅ ∠2. Therefore, ∠2 is a right ∠. So, by the definition of ⊥ lines, $m \perp k$.

35. Answers may vary. Examples are given.

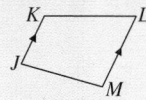

a. ∠1 and ∠2 (a linear pair); ∠6 and ∠7 (same-side interior angles); ∠1 and ∠4 (∠1 ≅ ∠3 since they are corresponding angles; ∠3 and ∠4 are supplementary since they are a linear pair; by the Substitution Property, $m\angle 1 + m\angle 4 = 180°$).

b. The angles of a trapezoid that include a non-parallel side are supplementary.

36. $m\angle A = 27°$; $m\angle B = 54°$; $m\angle C = 99°$

37. $w° = 115°$; $z° = 65°$; $y° = 68°$

Assess Your Progress

1. same-side interior angles: ∠3 and ∠7, ∠2 and ∠6; alternate interior angles: ∠3 and ∠6, ∠2 and ∠7; corresponding angles: ∠1 and ∠7, ∠3 and ∠5, ∠4 and ∠6, ∠2 and ∠8

2. 29

3. $y = 135$; $z = 38$

4. Answers may vary. For the example shown, $\overline{JK}$ and $\overline{ML}$ are the bases and $\overline{KL}$ and $\overline{JM}$ are the legs.

5. See answers in back of book.

233

Objectives

- Write a flow proof.
- Introduce auxiliary lines in a diagram.
- Choose from a variety of methods for proving statements.

Recommended Pacing

❖ **Core and Extended Courses**
Section 5.3: 1 day

❖ **Block Schedule**
Section 5.3: $\frac{1}{2}$ block
(with Section 5.2)

Resource Materials

Lesson Support
Lesson Plan 5.3
Warm-Up Transparency 5.3
Practice Bank: Practice 33
Study Guide: Section 5.3
Challenge Problems: Set 33

Technology
McDougal Littell Mathpack
   Geometry Inventor Activity Book:
      Activity 10

Internet:
   http://www.hmco.com

Warm-Up Exercises

The measures of two angles in a triangle are given. Find the measure of the third angle.

1. 32°, 68° 80°

2. 150°, 21° 9°

3. 60°, 60° 60°

Find each unknown angle measure.

4. 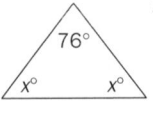 52°; 52°

5. 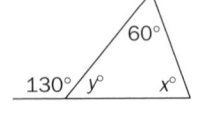 x = 70, y = 50

SECTION

5.3 Types of Proofs

Learn how to...
- **write a flow proof**
- **introduce auxiliary lines in a diagram**

So you can...
- **choose from a variety of methods for proving statements**

Have you ever gotten stuck while trying to prove something? There are many different strategies you can use. One proof format that may help you to organize your ideas is called **flow proof**. As you can see in the proof below, this method displays the relationships between the statements in a proof.

Given: $n \parallel p$
Prove: $m\angle 1 + m\angle 2 = 180°$

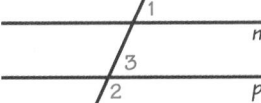

One way to start a flow proof is to write the given information on the left side of your paper and the statement that will be proved on the right side. You need to show how these statements can be logically connected.

① n ∥ p $m\angle 1 + m\angle 2 = 180°$

Reasons
1. Given

You can try out different ideas with a flow proof. For example, even if you are not sure how to show that $\angle 1$ is supplementary to $\angle 2$, you might think it is related to the fact that $\angle 3$ is supplementary to $\angle 2$.

① n ∥ p

② $m\angle 3 + m\angle 2 = 180°$ →? $m\angle 1 + m\angle 2 = 180°$

Reasons
1. Given
2. Angles in a linear pair are supplements.

If you can substitute $m\angle 1$ for $m\angle 3$, the proof will be complete. You can, because $\angle 1$ and $\angle 3$ are congruent and therefore $m\angle 1 = m\angle 3$.

The complete flow proof shows how the given information leads logically to the conclusion.

① n ∥ p ③ $\angle 1 \cong \angle 3$ $(m\angle 1 = m\angle 3)$ ④

② $m\angle 3 + m\angle 2 = 180°$ → $m\angle 1 + m\angle 2 = 180°$

Reasons
1. Given
2. Angles in a linear pair are supplements.
3. If two ∥ lines are intersected by a transversal, then corresponding ∠s are ≅.
4. Substitution Property

234 Chapter 5 *Parallel Lines*

Using a line or segment that is not part of the given information can sometimes help you prove a statement. An *auxiliary line* is a line that is added to a diagram to help complete a proof.

These two postulates are often used to justify including auxiliary lines in diagrams. You can use them in proofs about parallels and perpendiculars.

Parallel Postulate

Through a point not on a given line, there is exactly one line parallel to the given line.

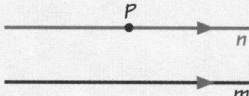

Through P, there is exactly one line n that is parallel to m.

Perpendicular Postulate

Through a point not on a given line, there is exactly one line perpendicular to the given line.

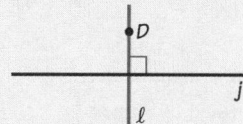

Through D, there is exactly one line ℓ that is perpendicular to j.

THINK AND COMMUNICATE

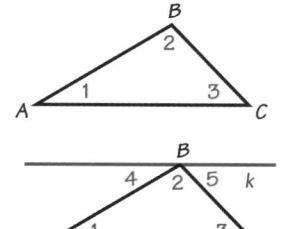

1. Does either of the theorems in Section 5.2 apply to this diagram? Why or why not?

2. In order to apply the theorems in Section 5.2 to the second diagram, what do you need to know about lines k and $\overleftrightarrow{AC}$?

3. What can you conclude about the angles in the diagram if $k \parallel \overleftrightarrow{AC}$?

In Questions 1–3 above, you saw how to introduce an auxiliary line to help prove that the sum of the angles of a triangle is 180°. On the next two pages, this familiar result is proved as the *Triangle Sum Theorem*. The proof is shown in four of the formats that you have learned.

ANSWERS Section 5.3

Think and Communicate

1. No; the diagram does not contain a pair of parallel lines.
2. that they are parallel
3. The sum of the measures of the angles is 180°. ($m\angle 4 + m\angle 2 + m\angle 5 = 180°$; $m\angle 4 = m\angle 1$ and $m\angle 5 = m\angle 3$ so, by the Substitution Property, $m\angle 1 + m\angle 2 + m\angle 3 = 180°$.)

Teach⇔Interact

Learning Styles: Visual

Visual learners may find the flow proof format the most comfortable type of proof to understand and write. Later in this section, students see how a flow proof is written in diagram form. The diagram can help students organize their thinking and understand how each step leads to the next. It also allows similar ideas to be grouped together. Some students may find it helpful to make a flow proof prior to writing a two-column or paragraph proof.

Section Notes

Writing Proofs

The flow proof format can be helpful for students who have difficulty knowing where to start a proof. Instead of focusing on the conclusion, they can start with the given information and write down what follows logically from that. They continue to proceed step by step in this way, while checking to see if they are getting closer to the conclusion. If this does not work, they can go back and try a different first step. Or, if students do not see a logical conclusion from the given information, they can start with the Prove and work backward. Sometimes a flow proof can be started in the middle, if there is a conclusion that can be drawn from the diagram. Then, the writer needs to work backward to the Given and forward to the Prove.

Using Manipulatives

If a student has difficulty ordering the steps in a flow proof, suggest that he or she write each step on an index card. The cards can then be arranged into the proper order to write the proof. This approach can also be helpful for students who start a proof with a step in the middle, or work backward from the conclusion.

Think and Communicate

Questions 1–3 introduce students to some of the ideas used in the proof of the Triangle Sum Theorem on the next page, including the need for introducing an auxiliary line.

Section Notes

Communication: Discussion
The four methods of proof given on these two pages should be discussed thoroughly. Write the name of each type of proof on the board and have the class brainstorm to make a list of the similarities and differences of the different methods. Students should understand that each method of proof uses the same logical steps, although the Key Steps of a Proof method omits the less important ones.

Teaching Tip
Point out that the initial figure for the Triangle Sum Theorem simply consists of a triangle with vertices labeled *A*, *B*, and *C*. The auxiliary line *k* is drawn in the first step using the Parallel Postulate.

Writing Proofs
Many students have difficulties writing proofs for two major reasons: either they fail to see the chain of reasoning needed to connect the given information with the conclusion or they are unsure of how to write out the steps and what level of detail to include. Having more than one method of writing a proof provides students with a good opportunity to be successful in writing proofs because they now can choose a method they understand and feel comfortable with. This should allow students more time to concentrate on learning how to establish the logical chain of reasoning that is the essence of a proof.

METHODS OF PROOF

In each of these proofs of the Triangle Sum Theorem, the same chain of reasoning is used to connect the given information with the conclusion. You can choose which method of proof to use based on how you want to organize the information and how many details you want to include.

Given: $\triangle ABC$
Prove: $m\angle 1 + m\angle 2 + m\angle 3 = 180°$

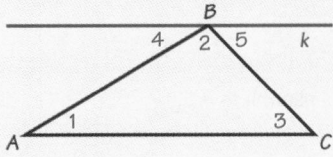

Key Steps of a Proof

Give the main steps in the argument of a proof. The key steps can be used as a plan for any of the other types of proof.

1. $m\angle 4 + m\angle 2 + m\angle 5 = 180°$ (The $\angle$s form a straight angle.)
2. $\angle 4 \cong \angle 1$, $\angle 5 \cong \angle 3$ (Alternate interior $\angle$s formed by $\parallel$ lines and a transversal are $\cong$.)
3. $m\angle 1 + m\angle 2 + m\angle 3 = 180°$ (Substitution)

Two-Column Proof

Organize the statements and reasons in separate columns. Include all necessary steps.

Statements	Reasons
1. Draw line *k* parallel to $\overleftrightarrow{AC}$ through *B*.	1. Parallel Postulate
2. $m\angle 4 + m\angle 2 + m\angle 5 = 180°$	2. Angle Addition Postulate and definition of a straight angle
3. $\angle 4 \cong \angle 1$ and $\angle 5 \cong \angle 3$	3. If two $\parallel$ lines are intersected by a transversal, then alternate interior angles are $\cong$.
4. $m\angle 4 = m\angle 1$ and $m\angle 5 = m\angle 3$	4. Definition of congruent angles
5. $m\angle 1 + m\angle 2 + m\angle 3 = 180°$	5. Substitution Property (Steps 2 and 4)

Paragraph Proof

Statements and reasons combine to create a convincing argument that is made up of complete sentences.

According to the Parallel Postulate, there is exactly one line k that passes through B and is parallel to $\overleftrightarrow{AC}$. Line k forms a straight angle at B, so $m\angle 4 + m\angle 2 + m\angle 5 = 180°$. Because $k \parallel \overleftrightarrow{AC}$, the alternate interior angles $\angle 1$ and $\angle 4$ are congruent. For the same reason, $\angle 3$ and $\angle 5$ are congruent. Substituting $m\angle 1$ for $m\angle 4$ and $m\angle 3$ for $m\angle 5$ gives $m\angle 1 + m\angle 2 + m\angle 3 = 180°$.

Flow Proof

Use arrows to represent the reasoning process that connects the given information with the conclusion.

1 Draw line k parallel to $\overleftrightarrow{AC}$ through B.

2 $\angle 4 \cong \angle 1$ $(m\angle 4 = m\angle 1)$
   $\angle 5 \cong \angle 3$ $(m\angle 5 = m\angle 3)$

3 $m\angle 4 + m\angle 2 + m\angle 5 = 180°$

4 $m\angle 1 + m\angle 2 + m\angle 3 = 180°$

Reasons

1. Parallel Postulate

2. If two $\parallel$ lines are intersected by a transversal, then alternate interior $\angle s$ are $\cong$.

3. Angle Addition Postulate and definition of a straight angle

4. Substitution Property

THINK AND COMMUNICATE

4. Which of the methods of proof do you think is easiest to understand? Explain your choice.

5. Which method do you think is the most organized? Explain your choice.

6. Which of the methods of proof do you think shows the chain of logical reasoning the best? Explain your choice.

7. What other method of proof have you learned? How is it different from these methods? How is it like them?

5.3 Types of Proofs **237**

Teach⇔Interact

Section Note

Communication: Drawing
Help students to see how they can use boxes and arrows to organize a flow proof. Point out that related ideas can be contained in a single box. Writing a flow proof gives students a visual representation of the reasoning process.

Think and Communicate

These questions direct students' attention to the qualities of each type of proof. Students' responses to questions 1 and 2 may differ depending on individual learning styles and preferences. Ask several volunteers to share their responses with the class to expose all students to the variety of responses. After discussing these questions, you may wish to suggest that students write a journal entry to summarize each method of proof.

Section Notes

Second-Language Learners
You may want to allow students learning English to discuss the *Think and Communicate* questions in their first language. Allowing students to ask questions and share ideas in their first language places the focus on understanding the mathematical concepts involved rather than on the level of proficiency in English.

Visual Thinking
After completing the *Think and Communicate* questions, you may wish to have students work in small groups to create their own process diagrams of two different methods of proof for the same situation. Encourage them to use their own words and diagram style. Have each group present its diagrams to the class and explain which of the two is easiest to use. This activity involves the visual skills of *exploration* and *communication*.

237

About the Example

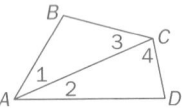 **Using Technology**
Students can use Activity 10 in the *Geometry Inventor Activity Book* to discover relationships between the angles in various types of quadrilaterals.

Additional Example

Prove the statement in the Example that the sum of the measures of the angles of a quadrilateral is 360° using the key steps of a proof, a two-column proof, and a paragraph proof.

Given: *ABCD* is a quadrilateral.
Prove: $m\angle A + m\angle B + m\angle C + m\angle D = 360°$

Key Steps of a Proof

1. Draw $\overline{AC}$. (Through any two points, there is exactly one line.)
2. $m\angle 1 + m\angle B + m\angle 3 + m\angle 2 + m\angle 4 + m\angle D = 360°$ (The sum of the angle measures of each triangle is 180°.)
3. $m\angle A + m\angle B + m\angle C + m\angle D = 360°$. (Substitution)

Two-Column Proof

Statements (Reasons)

1. Draw $\overline{AC}$. (Through any two points, there is exactly one line.)
2. $m\angle 1 + m\angle B + m\angle 3 = 180°$; $m\angle 2 + m\angle 4 + m\angle D = 180°$ (The sum of the angle measures of a triangle is 180°.)
3. $m\angle 1 + m\angle 2 + m\angle B + m\angle 3 + m\angle 4 + m\angle D = 360°$ (Addition Property (Step 2))
4. $m\angle 1 + m\angle 2 = m\angle A$; $m\angle 3 + m\angle 4 = m\angle C$ (Angle Addition Postulate)
5. $m\angle A + m\angle B + m\angle C + m\angle D = 360°$ (Substitution Property (Steps 3 and 4))

Paragraph Proof

Since through any two points there is exactly one line, draw $\overline{AC}$. Since the sum of the angle measures of a triangle is 180°, $m\angle 1 + m\angle B + m\angle 3 = 180°$ and $m\angle 2 + m\angle 4 + m\angle D = 180°$. Then $m\angle 1 + m\angle B + m\angle 3 + m\angle 2 + m\angle 4 + m\angle D = 360°$ by the Addition Postulate. Because $m\angle 1 + m\angle 2 = m\angle A$ and $m\angle 3 + m\angle 4 = m\angle C$ by the Angle Addition Postulate, substituting these measures gives $m\angle A + m\angle B + m\angle C + m\angle D = 360°$.

238

Write a flow proof of this statement:

The sum of the measures of the angles of a quadrilateral is 360°.

SOLUTION

Plan Ahead: Sketch and label a diagram. To prove the statement, draw an auxiliary segment that splits the quadrilateral into two triangles.

Given: *ABCD* is a quadrilateral.
Prove: $m\angle A + m\angle B + m\angle C + m\angle D = 360°$

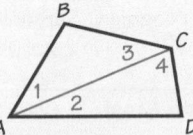

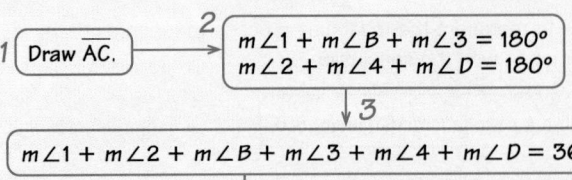

Reasons

1. Through any two points there is exactly one line.
2. The sum of the angle measures of a triangle is 180°.
3. Addition Property
4. Angle Addition Postulate
5. Substitution Property

☑ CHECKING KEY CONCEPTS

Questions 1 and 2 refer to this piece of a flow proof.

1. What statement should be in the blank in the proof labeled "2"?

2. **Open-ended Problem** Sketch a triangle that might accompany this flow proof.

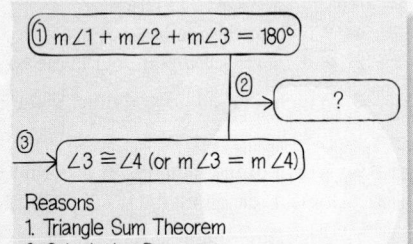

Reasons
1. Triangle Sum Theorem
2. Substitution Property

Checking Key Concepts

1. $m\angle 1 + m\angle 2 + m\angle 4 = 180°$

2. Answers may vary. An example is given.

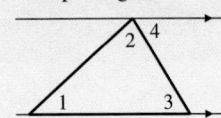

Exercises and Applications

1. (1) $\angle A$ and $\angle B$ are complements;
   (3) $m\angle A + m\angle B + m\angle C = 180°$;
   (4) 180°; Substitution Property (Steps 2 and 3),
   (5) $m\angle C = 90°$; Subtraction Property.

2. *m*; *n*; $\angle 2$; $j \parallel k$; $\angle 2$; they are corresponding angles; $\angle 1 \cong \angle 3$

3. Answers may vary. An example is given. I might have started with the statement to be proved and work backward to the given statement.

4. **Statements (Reasons)**
   1. $n \parallel p$ (Given)
   2. $m\angle 3 + m\angle 2 = 180°$ (⊿ in a linear pair are supplements.)
   3. $m\angle 1 = m\angle 3$ (If two ∥ lines are intersected by a transversal, then corresponding ⊿ are ≅; Def. of ≅ ⊿)
   4. $m\angle 1 + m\angle 2 = 180°$ (Substitution Property (Steps 2 and 3))

5.3 | **Exercises and Applications**

Extra Practice exercises on page 668

For Exercises 1 and 2, copy and complete each proof.

1. Given: In $\triangle ABC$, $\angle A$ and $\angle B$ are complements.

  Prove: $m\angle C = 90°$

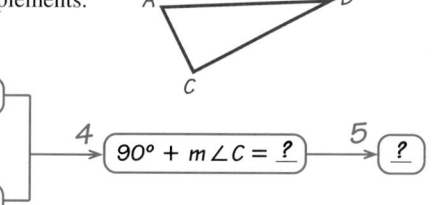

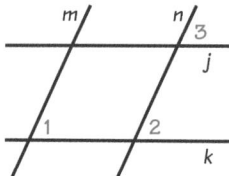

Reasons

1. Given

2. Definition of complementary angles

3. The sum of the angle measures of a triangle is 180°.

4. ?

5. ?

2. Given: $j \parallel k$; $m \parallel n$

  Prove: $\angle 1 \cong \angle 3$

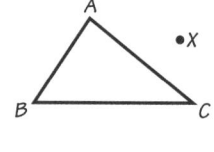

Since ? ∥ ?, it follows that $\angle 1 \cong$? because they are corresponding angles. Since ?, it follows that ? $\cong \angle 3$ because ?. By the Transitive Property, ?.

3. Open-ended Problem One advantage of using the flow proof method is that you can start at the beginning, middle, or end of the proof. How else might you have started writing the flow proof on page 234?

4. Rewrite the flow proof on page 234 as a two-column proof.

5. Writing In the piece of a proof to the right, Ashley introduces two auxiliary lines. Write a note to Ashley that explains what is wrong with her reasoning in each of the two sentences shown.

Draw the line through A that is perpendicular to $\overline{BC}$ at its midpoint. Then draw the line through A and X that is parallel to $\overline{BC}$.

A _corollary_ of a theorem is a statement that can be easily proved using the theorem. Explain how each corollary of the Triangle Sum Theorem follows from the theorem.

6. Each angle of an equiangular triangle measures 60°.

7. In a triangle, there can be at most one obtuse or right angle.

5. Answers may vary. An example is given. Dear Ashley, The Perpendicular Postulate guarantees that there is a line through A that is perpendicular to $\overline{BC}$. There is also a line through A and the midpoint of $\overline{AB}$ (since through any two points there is exactly one line). You cannot assume that these two lines are the same line. Similarly, you cannot

assume that the line through A and X and the line through A that is perpendicular to $\overline{AB}$ are the same line. I hope this helps you understand more about auxiliary lines.

6. Since all three angles of an equiangular triangle are congruent, the measure of each is $\frac{1}{3}$ of 180°, or 60°.

7. If a triangle contained more than one angle with measure 90° or greater, then the sum of the measures of the three angles would be greater than 180°. Therefore, if one angle is right or obtuse, the other two angles must be acute in order for the angle sum to be exactly 180°.

Exercise Notes

Connection   ENGINEERING

Spanish architect and engineer Santiago Calatrava designed the Alamillo Bridge for Exposicion Universal de 1992, in Seville, Spain. It is one of eight new bridges over the Guadalquivir River that were built for the Expo. The cables of the bridge are all parallel.

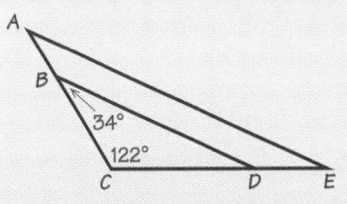

8. Sketch a copy of the diagram of the bridge at the left, including the two labeled angle measures. Find $m\angle CAE$. What postulate or theorem did you use?

9. Find $m\angle AEC$. What theorem or postulate did you use?

10. Find $m\angle BDC$. What theorem or postulate did you use?

BY THE WAY

The 142 m tall pylon in the Alamillo Bridge is made of steel filled with concrete. It is heavy enough to counterbalance the deck of the bridge, so no cables are needed on the other side of the pylon.

11. Use this theorem: If two angles of one triangle are congruent to two angles of another triangle, then the third angles are congruent.
   a. Does the theorem apply to the triangles that you explored in Exercises 8–10? Explain why or why not.
   b. Write a two-column proof of the theorem using the diagrams below.

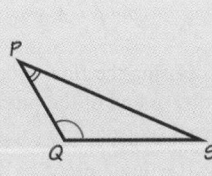
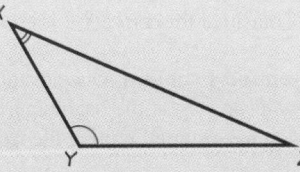

Use the method of proof indicated. First copy the diagram and mark it with the given information.

12. Flow proof
   Given: $\overleftrightarrow{HV} \parallel \overleftrightarrow{TU}$; $\angle 6 \cong \angle 7$
   Prove: $\angle 7 \cong \angle 8$

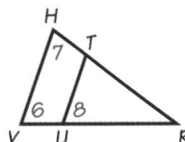

13. Key steps of a proof
   Given: $\angle 1 \cong \angle 5$
   Prove: $\angle 2 \cong \angle 4$

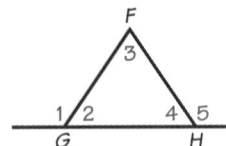

---

8. Check students' work. 34°; the Corresponding Angles Postulate

9. 24°; the Triangle Sum Theorem

10. 24°; the Corresponding Angles Postulate or the Triangle Sum Theorem

11. a. Yes; both △*BCD* and △*ACE* contain ∠*C*. Since $\overline{BD} \parallel \overline{AE}$, ∠*A* ≅ ∠*DBC*. The theorem tells you that the third pair of angles, ∠*E* and ∠*BDC*, must be congruent.

b. Given: ∠*P* ≅ ∠*X*; ∠*Q* ≅ ∠*Y*
Prove: ∠*S* ≅ ∠*Z*
Statements (Reasons)
1. ∠*P* ≅ ∠*X*; ∠*Q* ≅ ∠*Y* (Given)
2. $m\angle P = m\angle X$; $m\angle Q = m\angle Y$ (Def. of ≅ ∠)
3. $m\angle P + m\angle Q + m\angle S = 180°$; $m\angle X + m\angle Y + m\angle Z = 180°$ (The sum of the ∠ measures of a △ is 180°.)
4. $m\angle P + m\angle Q + m\angle S = m\angle X + m\angle Y + m\angle Z$ (Substitution Property (Step 3))
5. $m\angle S = m\angle Z$ (Subtraction Property)
6. ∠*S* ≅ ∠*Z* (Def. of ≅ ∠)

12. See answers in back of book.

13. Check students' work.
1. $m\angle 1 + m\angle 2 = 180°$; $m\angle 4 + m\angle 5 = 180°$ (The ∠ form a linear pair.)
2. $m\angle 1 + m\angle 2 = m\angle 4 + m\angle 5$ (Substitution)
3. $m\angle 1 = m\angle 5$ (Given)
4. ∠2 ≅ ∠4 (Subtraction Property)

14. See answers in back of book.

15. **Statements (Reasons)**
1. ∠2 ≅ ∠4 (Given)
2. ∠5 ≅ ∠6 (Vertical ∠ are ≅.)
3. ∠1 ≅ ∠3 (Ex. 11: If two ∠ of one △ are ≅ to two ∠ of another △, then the third ∠ are ≅.)

Use the method of proof indicated. First copy the diagram and mark it with the given information.

14. Flow proof

Given: $\overline{YU} \parallel \overline{ST}$; $\angle YXS \cong \angle UXT$

Prove: $\angle S \cong \angle T$

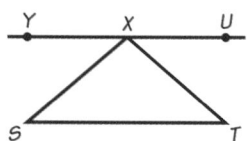

15. Two-column proof

Given: $\angle 2 \cong \angle 4$

Prove: $\angle 1 \cong \angle 3$

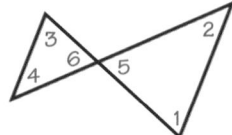

16. Paragraph proof

Given: *JKLM* is a parallelogram.

Prove: $\angle 7 \cong \angle 9$

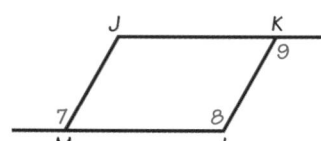

17. Use any method.

Given: $\overline{AB} \parallel \overline{CD}$; $\overline{BC} \parallel \overline{DE}$

Prove: $\angle ABC \cong \angle CDE$

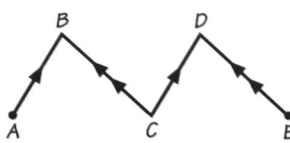

18. Challenge Sketch and label a diagram and write a two-column proof.

Given: In trapezoid *PQRS*, $\overline{PQ} \parallel \overline{RS}$ and $\angle P \cong \angle Q$.

Prove: $\angle R \cong \angle S$

ONGOING ASSESSMENT

19. Open-ended Problem Choose one of the proofs that you did in Exercises 12–18, and rewrite it using a different method of proof. Explain which method you like better and why.

SPIRAL REVIEW

20. ALGEBRA Find the measures of $\angle 1$ and $\angle 2$.
(Section 5.2)

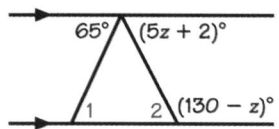

Write the converse of each statement. Tell whether the converse is *True* or *False*. *(Section 3.5)*

21. If a triangle is equilateral, then the triangle is isosceles.

22. If a bird has wings, then the bird can fly.

Apply⇔Assess

Exercise Notes

Teaching Tip

Exs. 14, 16, 17 Proofs involving parallel lines frequently use the Corresponding Angles Postulate, the Alternate Interior Angles Theorem, or the Same-Side Interior Angles Theorem. Students can extend the parallel lines in figures, such as the ones in these exercises, to help them identify where the theorems can be used.

Topic Spiraling: Review

Exs. 21, 22 These exercises provide a review of the converse of a statement. The converses of the Corresponding Angles Postulate, the Alternate Interior Angles Theorem, and the Same-Side Interior Angles Theorem are discussed in the next section.

Practice 33 for Section 5.3

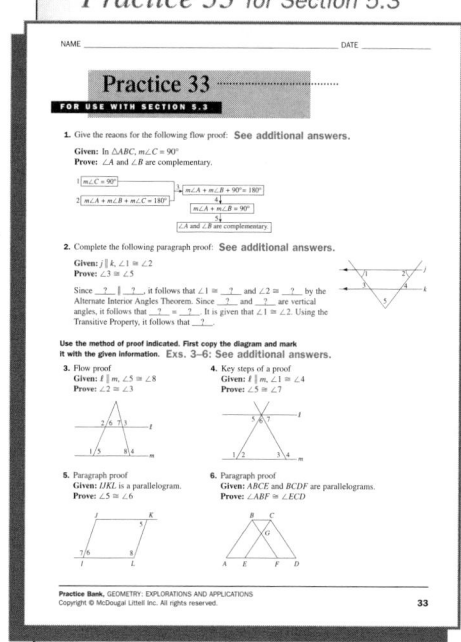

- -

16. Since *JKLM* is a ▱, $\overline{JM} \parallel \overline{KL}$ and $\overline{JK} \parallel \overline{ML}$. Therefore, corresponding ∠s 7 and 8 are ≅. Also, alternate interior ∠s 8 and 9 are ≅. By the Transitive Property, $\angle 7 \cong \angle 9$.

17. Methods of proof may vary. An example is given. Since $\overline{AB} \parallel \overline{CD}$, $\angle ABC \cong \angle BCD$. (If two ∥ lines are intersected by a transversal, then alternate interior ∠s are ≅.) Similarly, since $\overline{BC} \parallel \overline{DE}$, $\angle BCD \cong \angle CDE$. By the Transitive Property, $\angle ABC \cong \angle CDE$.

18. Methods of proof may vary. An example is given.

Statements (Reasons)

1. $\overline{PQ} \parallel \overline{RS}$ (Given)
2. $\angle P$ and $\angle S$ are supplementary; $\angle Q$ and $\angle R$ are supplementary. (If two ∥ lines are intersected by a transversal, then same-side interior ∠s are supplementary.)
3. $m\angle P + m\angle S = 180°$; $m\angle Q + m\angle R = 180°$ (Def. of supplementary ∠s)
4. $m\angle P + m\angle S = m\angle Q + m\angle R$ (Substitution Property (Step 3))
5. $\angle P \cong \angle Q$ or $m\angle P = m\angle Q$ (Given; Def. of ≅ ∠s)
6. $m\angle S = m\angle R$ (Subtraction Property)
7. $\angle R \cong \angle S$ (Def. of ≅ ∠s)

19. Answers may vary. Check students' work.

20. $m\angle 1 = 65°$; $m\angle 2 = 62°$

21. If a triangle is isosceles, then the triangle is equilateral; False.

22. If a bird can fly, then the bird has wings; True.

Objectives

- Apply the converses of theorems about parallel lines and transversals.
- Use facts about angles to prove that two lines are parallel.

Recommended Pacing

❖ **Core and Extended Courses**
Section 5.4: 1 day

❖ **Block Schedule**
Section 5.4: $\frac{1}{2}$ block
(with Section 5.5)

Resource Materials

Lesson Support

Lesson Plan 5.4

Warm-Up Transparency 5.4

Practice Bank: Practice 34

Study Guide: Section 5.4

Explorations Lab Manual:
Additional Exploration 8

Challenge Problems: Set 34

Technology

Graphing Calculator

Internet:
http://www.hmco.com

Warm-Up Exercises

Write the converse of each statement.

1. If it rains today, we will stay inside. If we stay inside, it is raining today.

2. If a quadrilateral is a square, then it is a rhombus. If a quadrilateral is a rhombus, then it is a square.

3. If today is Friday, then pizza is served for lunch. If pizza is served for lunch, then today is Friday.

4. If a quadrilateral has four right angles, it is a rectangle. If a quadrilateral is a rectangle, it has four right angles.

5.4 Conditions for Parallel Lines

Learn how to...

- apply the converses of theorems about parallel lines and transversals

So you can...

- use facts about angles to prove that two lines are parallel

Architects, engineers, and other designers may use drafting tools called T-squares and triangles to make precise technical drawings. In the Exploration, you will make your own triangle and use it to explore lines and transversals.

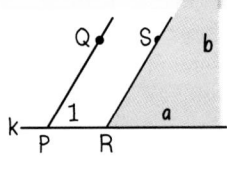

EXPLORATION
COOPERATIVE LEARNING

Drafting Parallel Lines

Work with another student.
You will need:
- a straightedge or ruler
- an index card
- scissors

1 Using a straightedge, draw one diagonal of the index card. Cut along the diagonal to form two right triangles, one for each student. Label the two triangles differently, as shown.

2 Each of you should use a straightedge to draw a line k. Place your triangle on your paper so leg a lies along k. Trace the hypotenuse to draw $\overline{PQ}$. Label $\angle 1$, the acute angle formed by k and $\overline{PQ}$.

3 Slide the triangle along line k to a new position. Trace the hypotenuse to draw $\overline{RS}$. Label $\angle 2$, the acute angle formed by k and $\overline{RS}$.

4 Use k as a transversal. What is the name for the pair of angles you labeled?

5 How do lines $\overleftrightarrow{PQ}$ and $\overleftrightarrow{RS}$ appear to be related? Compare your results with your partner's results. Make a conjecture based on your results.

 Exploration Note

Purpose
The purpose of this Exploration is for students to discover that if two lines are intersected by a transversal and corresponding angles are congruent, then the lines are parallel.

Materials/Preparation
Each group of students needs a straightedge or ruler and an index card.

Procedure
Students cut an index card on the diagonal to form two right triangles. They use their triangles to draw two lines that intersect a transversal so that the corresponding angles are congruent.

Closure
Students should understand that when a transversal intersects a pair of lines so that the corresponding angles are congruent, then the lines are parallel.

Explorations Lab Manual
See the Manual for more commentary on this Exploration.

Using the triangles in the Exploration, you probably discovered the following postulate. Notice that it is the converse of a postulate that you learned earlier.

Converse of the Corresponding Angles Postulate

If two lines are intersected by a transversal and corresponding angles are congruent, then the lines are parallel.

If $\angle 1 \cong \angle 2$, then $k \parallel l$.

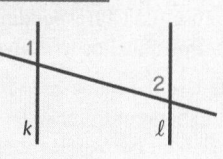

EXAMPLE 1   Connection: Algebra

Find the value of x that will allow you to prove that $\overleftrightarrow{AB} \parallel \overleftrightarrow{CD}$.

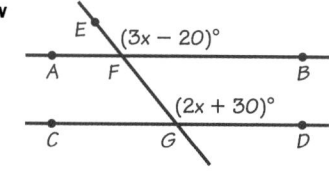

SOLUTION

$\angle EFB$ and $\angle FGD$ are corresponding angles. The postulate above states that if $\angle EFB \cong \angle FGD$, then $\overleftrightarrow{AB} \parallel \overleftrightarrow{CD}$.

$$m \angle EFB = m \angle FGD$$
$$(3x - 20)° = (2x + 30)°$$

Subtract $2x$ from both sides.

$$3x - 2x - 20 = 30$$
$$x - 20 = 30$$
$$x = 50$$

If $x = 50$, then $m \angle EFB = 130°$ and $m \angle FGD = 130°$. So, $\overleftrightarrow{AB}$ will be parallel to $\overleftrightarrow{CD}$ if $x = 50$.

In the Exploration you saw that the converse of the Corresponding Angles Postulate is true. In the exercises, you will show that the converse of the Alternate Interior Angles Theorem is also true.

Converse of the Alternate Interior Angles Theorem

If two lines are intersected by a transversal and alternate interior angles are congruent, then the lines are parallel.

If $\angle 3 \cong \angle 4$, then $m \parallel n$.

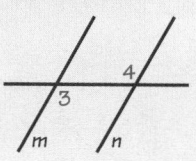

5.4 Conditions for Parallel Lines   **243**

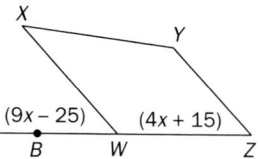

Think and Communicate

Question 1 helps students distinguish between situations in which the Corresponding Angles Postulate is needed and those in which its converse is needed. Similar distinctions can be made for the Alternate Interior Angles Theorem and the Same-Side Interior Angles Theorem. In each case, students need to use the original theorem or postulate if they know that the lines are *parallel*, and they want to determine information about the angles. The converses of these theorems are used when information is given about the angles, and students want to show that the lines are parallel. Questions 2 and 3 lead into the Converse of the Same-Side Interior Angles Theorem.

About Example 2

Alternate Approach
The Converse of the Same-Side Interior Angles Theorem can also be proven using the Converse to the Alternate Interior Angles Theorem. You may want to challenge students to write such a proof.

Additional Example 2

Prove that if two sides of a quadrilateral are parallel, and one pair of opposite angles are congruent, then the quadrilateral is a parallelogram.
Given: $\overline{AD} \parallel \overline{BC}$, $m\angle 1 = m\angle 3$
Prove: *ABCD* is a parallelogram.

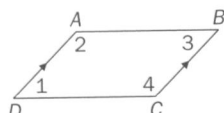

Statements (Reasons)
1. $\overline{AD} \parallel \overline{BC}$ (Given)
2. $m\angle 2 + m\angle 3 = 180°$ (If parallel lines are intersected by a transversal, same-side interior angles are supplementary.)
3. $m\angle 1 = m\angle 3$ (Given)
4. $m\angle 1 + m\angle 2 = 180°$ (Substitution Property (Steps 2 and 3))
5. $\overline{AB} \parallel \overline{DC}$ (If two lines are intersected by a transversal and same-side interior angles are supplementary, then the lines are parallel.)
6. *ABCD* is a parallelogram. (Def. of parallelogram (Steps 1 and 5))

THINK AND COMMUNICATE

1. Suppose you know that $\overline{AB} \parallel \overline{CD}$ and you want to prove that $\angle BAC \cong \angle DCE$. Should you use the Corresponding Angles Postulate, or its converse?

2. Use the Same-Side Interior Angles Theorem to make a conjecture about two angles formed by $\overleftrightarrow{CD}$, $\overleftrightarrow{EF}$, and $\overleftrightarrow{AE}$.

3. What is the converse of your answer to Question 2?

Converse of the Same-Side Interior Angles Theorem

If two lines are intersected by a transversal and same-side interior angles are supplementary, then the lines are parallel.

If $m\angle 1 + m\angle 2 = 180°$, then $j \parallel k$.

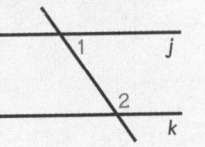

EXAMPLE 2

Prove the Converse of the Same-Side Interior Angles Theorem.

SOLUTION

Given: $m\angle 1 + m\angle 2 = 180°$
Prove: $j \parallel k$

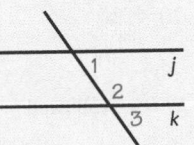

Statements	Reasons
1. $m\angle 1 + m\angle 2 = 180°$	1. Given
2. $m\angle 3 + m\angle 2 = 180°$	2. The ∠s in a linear pair are supplementary.
3. $m\angle 1 + m\angle 2 = m\angle 3 + m\angle 2$	3. Transitive Property (Steps 1 and 2)
4. $m\angle 1 = m\angle 3$	4. Subtraction Property
5. $j \parallel k$	5. If two lines are intersected by a transversal and corresponding ∠s are ≅, then the lines are ∥.

244   Chapter 5   *Parallel Lines*

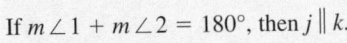

Think and Communicate

1. the Corresponding Angles Postulate

2. If $\overleftrightarrow{CD} \parallel \overleftrightarrow{EF}$, then $\angle DCE$ is supplementary to $\angle CEF$.

3. If $\angle DCE$ is supplementary to $\angle CEF$, then $\overleftrightarrow{CD} \parallel \overleftrightarrow{EF}$.

Checking Key Concepts

1. If two lines are intersected by a transversal and same-side interior angles are supplementary, then the lines are parallel.

2. If two lines are intersected by a transversal and alternate interior angles are congruent, then the lines are parallel.

☑ CHECKING KEY CONCEPTS

For each diagram, state the theorem or postulate that justifies why $\ell \parallel m$.

1.

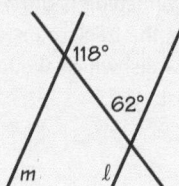

2.

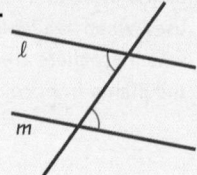

5.4 | Exercises and Applications

Extra Practice exercises on page 669

ALGEBRA For each diagram, find the value of the variable that will allow you to prove that two lines are parallel.

1.

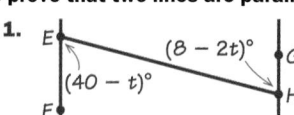

2.

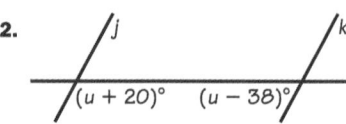

3.

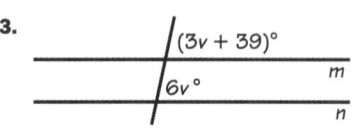

4.

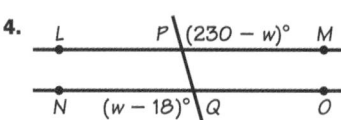

For each pair of lines indicated, tell whether they are *parallel* or *not parallel*. Explain your reasoning.

5. $\overleftrightarrow{RW}$ and $\overleftrightarrow{SV}$

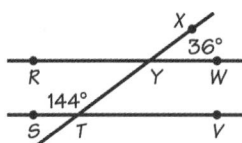

6. $\overleftrightarrow{EF}$ and $\overleftrightarrow{GH}$

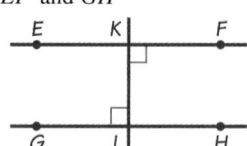

7. c and d

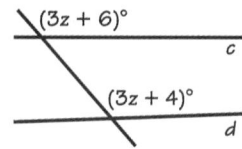

8. Sketch and label a diagram and write the key steps of a proof of the Converse of the Alternate Interior Angles Theorem.

For Exercises 9–11, use the diagram.

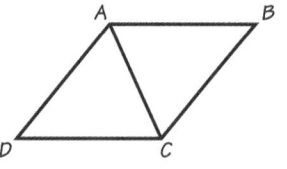

9. If $\angle DAC \cong \angle ACB$, which segments are parallel?

10. If $\angle DCA \cong \angle BAC$, which segments are parallel?

11. If $m\angle DCB + m\angle ABC = 180°$, which segments are parallel?

Exercises and Applications

1. −32

2. 99

3. 13

4. 124

5. parallel; Answers may vary. An example is given. $m\angle YTV = 36°$, so $\angle XYW$ and $\angle YTV$ are congruent, and since corresponding angles are congruent, the lines are parallel.

6. parallel; $\angle FKH$ and $\angle KLG$ are congruent, and since alternate interior angles are congruent, the lines are parallel.

7. not parallel; The labeled corresponding angles cannot be congruent.

8. Answers may vary. An example is given.

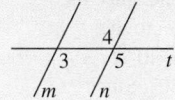

1. $\angle 3 \cong \angle 4$ (Given)

2. $\angle 4 \cong \angle 5$ (Vertical ∠s are ≅.)

3. $\angle 3 \cong \angle 5$ (Transitive Property)

4. $m \parallel n$ (Converse of the Corresponding Angles Postulate)

9. $\overline{AD}$ and $\overline{BC}$

10. $\overline{AB}$ and $\overline{DC}$

11. $\overline{AB}$ and $\overline{DC}$

Teach⇔Interact

Closure Question

When would you use the Converse of the Alternate Interior Angles Theorem or the Converse of the Same-Side Interior Angles Theorem. You would use either theorem to show that a pair of lines is parallel. The Converse of the Alternate Interior Angles Theorem would be used if the alternate interior angles are congruent. The Converse of the Same-Side Interior Angles Theorem would be used if the same-side interior angles are supplementary.

Apply⇔Assess

Suggested Assignment

❖ **Core Course**
  Exs. 1–11, 15, 16, 21–27
❖ **Extended Course**
  Exs. 1–27
❖ **Block Schedule**
  Day 30 Exs. 1–11, 15, 16, 21–27

Exercise Notes

Common Error
Exs. 1–4 Some students may not understand when to make the two expressions equal or when to add them and make the sum equal to 180°. A visual check of the drawing can help students determine which angles look congruent and which look supplementary.

 **Using Technology**
Exs. 1–4, 7 Students can check their solutions to these exercises using the graphing features of a TI-82 calculator by entering two equations on the Y= list, one for each side of the equation that they need to solve. For instance, in Ex. 2, students should enter Y1 = X + 20 + X − 38 and Y2 = 180. Students can use the answer they found algebraically to help set the viewing window. The solution to the equation should be within the range of x-values set. After graphing, the intersection can be found by choosing 5:intersect from the CALCULATE menu. The x-coordinate of the intersection is the correct answer, which is 99.

Exercise Notes

Visual Thinking

Exs. 12–14 Ask students to work in teams to create their own designs for an *espalier*. Encourage students to be creative. Ideas might include representing a school mascot or outlining the name of a popular sports team. Ask the teams to present their designs to the class and to discuss the relationships between parallel lines, transversals, and angles. Display the designs for an Espalier Show that demonstrates the conditions of parallel lines. This activity involves the visual skills of *correlation* and *self-expression*.

Career Connection

Exs. 12–14 Horticulture is a branch of agriculture that deals with plants grown in gardens. Horticultural scientists are college-educated. They study other sciences as well to help them develop methods to increase the quality and yield of these plants. They are also concerned with protecting plants from insects and disease. Horticulturists may be employed in the nursery industry, which specializes in growing trees and shrubs; the plant-growing industry, which grows vegetable and flower plants; and the seed industry, which grows plants for seeds.

Multicultural Note

Exs. 12–14 There are several different ways in which people train plants to grow. The art of *bonsai* requires that plants, generally trees or shrubs, be trained to reflect age and growth but also to maintain a miniature size. Bonsai are grown in trays or dishes, usually to a height between two inches and three feet, and are created for aesthetic purposes. Most bonsai are trained to be asymmetrical in shape. Although many ancient peoples grew trees in pots, they did so only for mobility and other practical reasons. The Chinese were the first people to plant trees in pots for aesthetic value, creating the art of bonsai. In the 6th century A.D., Buddhist monks traveling from China to Japan probably brought the art of bonsai with them. Modern bonsai began in the late 19th and early 20th centuries in Japan, with the introduction of new techniques and styles. Today, there are bonsai enthusiasts throughout the world.

246

Connection **HORTICULTURE**

An *espalier* is a tree or shrub that is trained to grow in a flat plane, often in a regular pattern. Espalier gardening is often used when available space is limited. Also, the trellises on which espaliers are grown are often near walls, which shelters the plants from cold temperatures and frost.

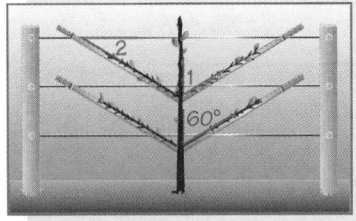

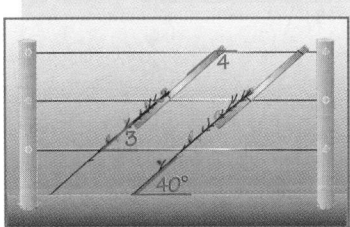

12. Open-ended Problem On this espalier trellis, the lowest wire is placed at a 90° angle to the vertical endposts. Describe how to position the remaining wires to make them parallel to the lowest one.

13. In the *palmette* style, shown in the middle diagram, all of the permanent branches on each side of the main trunk are parallel. Wooden stakes tied to the trellis wires train branches to grow in the desired direction.

   a. With what measure of ∠1 should the stake on the right be tied to grow the upper right branch parallel to the lower right branch?

   b. For this espalier to be symmetrical about the main trunk, with what measure of ∠2 should the stake on the left be tied?

14. In the *single oblique* cordon style, single trunks (cordons) are grown in parallel rows.

   a. If the right cordon is planted at an angle of 40° with the horizontal, describe how the other cordon should be positioned.

   b. Find $m\angle 3$ and $m\angle 4$. Explain your reasoning.

15. a. Copy the diagram and mark it with the given information.

   b. Open-ended Problem Write a paragraph proof that uses either the Converse of the Corresponding Angles Postulate or the Converse of the Alternate Interior Angles Theorem.

   c. Write the theorem that you proved in part (b).

Given: ∠JKF ≅ ∠GLM
Prove: $\overleftrightarrow{EF} \parallel \overleftrightarrow{GH}$

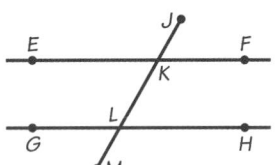

246 Chapter 5 *Parallel Lines*

---

12. Answers may vary. An example is given. Place each of the three wires so they make 90° angles with the left post. Then, by the Converse of the Corresponding Angles Postulate, the wires will be parallel to the lowest one.

13. a. 60°

   b. 30°

14. a. at an angle of 40° from the horizontal, so that the corresponding angles formed by the ground (the transversal) and the cordons (the lines) will be congruent

   b. $m\angle 3 = 40°$; Since the cordons are parallel, the angle formed by the second cordon and the ground is 40°, and since the wires are parallel to the ground, $m\angle 3 = 40°$. $m\angle 4 = 180° - 40° = 140°$; Since ∠3 and ∠4 are same-side interior angles formed by two parallel lines and a transversal, the angles are supplementary.

15. a. Check students' work.

   b. Proofs may vary. An example is given. ∠JKF ≅ ∠GLM (Given). Also, ∠GLM ≅ ∠KLH because vertical ∠s are ≅. Thus, ∠JKF ≅ ∠KLH by the Transitive Property. Then $\overleftrightarrow{EF} \parallel \overleftrightarrow{GH}$ because if two lines are intersected by a transversal

16. Copy and complete this flow proof.

Given: $\angle 3$ and $\angle 4$ are supplementary angles.
Prove: $h \parallel \ell$

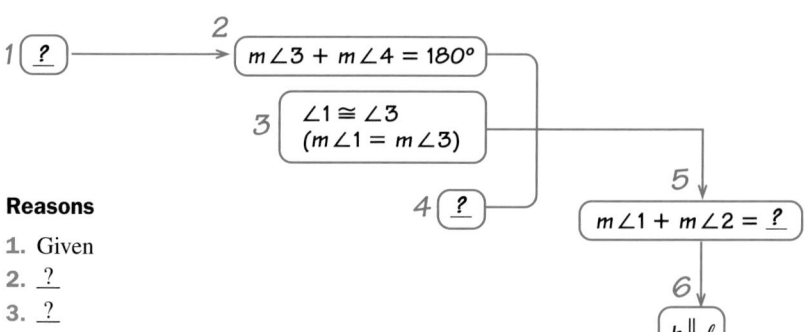

1 ⬜ **?** ──2──> $m\angle 3 + m\angle 4 = 180°$

3 ⬜ $\angle 1 \cong \angle 3$ $(m\angle 1 = m\angle 3)$

4 ⬜ **?**

5 ⬇ $m\angle 1 + m\angle 2 = $ **?**

6 ⬇ $h \parallel \ell$

Reasons

1. Given
2. **?**
3. **?**
4. Vertical angles are congruent.
5. Substitution Property (Steps **?** , **?** , and **?**)
6. **?**

17. Challenge Find the value of each variable that will allow you to prove that *ABCD* is a parallelogram. Then find the measure of each angle of the parallelogram.

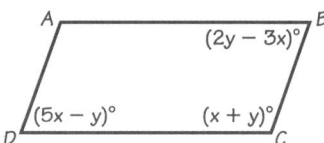

$(2y - 3x)°$
$(5x - y)°$
$(x + y)°$

Connection ▶ **DANCE**

In dance, precise positioning of the arms and legs is crucial to give a pose its visual impact. In this move, each dancer must be sure that his leg is at the correct angle, so that it is parallel to the legs of the other dancers.

18. Given: $a \parallel b$; $\angle 1 \cong \angle 4$
Prove: $c \parallel d$

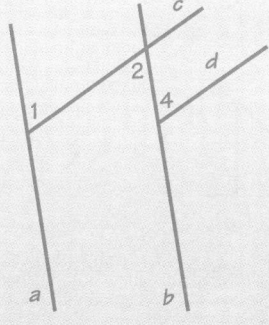

19. Visual Thinking Carefully redraw the diagram so that $\angle 1 \cong \angle 4$, but lines *a* and *b* are *not* parallel.

20. If the given information in Exercise 18 did not include "$a \parallel b$," would you still be able to prove that $c \parallel d$? Explain why or why not.

Apply⇔Assess

Exercise Notes

Topic Spiraling: Review
Ex. 13 Part (b) of this exercise requires students to use the Triangle Sum Theorem.

Cooperative Learning
Ex. 15 You may wish to have students work with a partner on this exercise. Each person would be responsible for writing the proof using one of the two converses. Partners can then compare and contrast each method.

Integrating Algebra
Ex. 17 This challenge question integrates algebra and geometry. To answer the question, students must write and solve a system of equations in two variables. If students are having difficulty with this exercise, you may wish to review methods for solving systems of equations.

Interdisciplinary Problems
Exs. 18–20 Students are given an introduction to the relationship between geometry and dance in these exercises. Interested students may want to look for other ways in which these two disciplines are connected. Have them look for books on dance that contain photographs of dancers to note the geometric structures of the various arrangements. A well-done project in this area would make an interesting bulletin board display.

Geometric Thinking
Ex. 20 Students must determine whether a given piece of information is vital to the conclusion of the proof. A Given that is vital to a conclusion is often called a *necessary condition*. A necessary condition is one that must be present for the conclusion to be reached.

---

and corresponding angles are congruent, the lines are parallel.

c. If two lines are intersected by a transversal and alternate exterior angles are congruent, then the lines are parallel.

16. (1) $\angle 3$ and $\angle 4$ are supplementary $\angle$s; (2) Def. of supplementary $\angle$s; (3) Vertical $\angle$s are $\cong$; (4) $\angle 2 \cong \angle 4$; (5) 180°; 2; 3; 4; (6) If two lines are intersected by a transversal and same-side

interior $\angle$s are supplementary, then the lines are $\parallel$.

17. $x = 30$; $y = 80$; $\angle A = 110°$, $\angle B = 70°$, $\angle C = 110°$, $\angle D = 70°$

18. Since $a \parallel b$, $\angle 1 \cong \angle 2$. (If two $\parallel$ lines are intersected by a transversal, then alternate interior $\angle$s are $\cong$.) It is given that $\angle 1 \cong \angle 4$, so $\angle 2 \cong \angle 4$ by the Transitive Property. Then $c \parallel d$ because if two lines are intersected by a transversal and

alternate interior $\angle$s are $\cong$, then the lines are $\parallel$.

19. Answers may vary. An example is given.

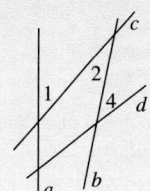

20. No; as the drawing in Ex. 18 shows, it is possible for $\angle 1$ and $\angle 4$ to be congruent with *c* not parallel to *d*.

Exercise Notes

Writing Proofs

Ex. 21 This exercise focuses on the importance of the order of statements in a proof. When writing proofs, it is common for students to have some steps out of order. Doing exercises of this type should help provide a remedy for this.

Communication: Writing

Ex. 23 This exercise allows students to show that they understand the concepts of this section without having to write a formal proof.

Assessment Note

Ex. 24 This open-ended problem will help students assess their understanding of the material in this section and how it is related to the Exploration on page 242.

Practice 34 for Section 5.4

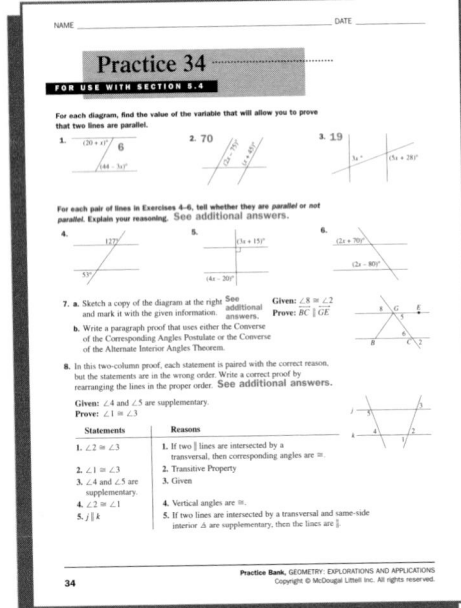

21. In this two-column proof, each statement is paired with the correct reason, but the statements are in the wrong order. Write a correct proof by rearranging the lines in the proper order.

Given: $\angle 1 \cong \angle 3$
Prove: $\angle 4$ and $\angle 5$ are supplements.

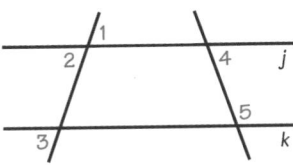

Statements	Reasons
1. $j \parallel k$	**1.** If two lines are intersected by a transversal and corresponding ∡ are ≅, then the lines are ∥.
2. $\angle 2 \cong \angle 3$	**2.** Transitive Property
3. $\angle 2 \cong \angle 1$	**3.** Vertical angles are congruent.
4. $\angle 4$ and $\angle 5$ are supplements.	**4.** If two ∥ lines are intersected by a transversal, then same-side interior ∡ are supplementary.
5. $\angle 1 \cong \angle 3$	**5.** Given

22. Open-ended Problem Sketch three lines m, n, and p, and a transversal that intersects them. Label all the angles formed. For each pair of lines, write two statements that could be used to prove that the lines are parallel.

23. Writing From this diagram, Ted concludes that $x = 60$. Explain how you know that he is wrong.

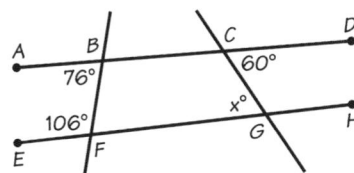

24. Open-ended Problem Explain how you can use the Converse of the Alternate Interior Angles Theorem and the triangle that you used in the Exploration on page 242 to draw parallel lines.

25. Write a flow proof. *(Section 5.3)*

Given: $\overleftrightarrow{LM} \parallel \overleftrightarrow{QW}$
$m \angle LMW + m \angle Q = 90°$

Prove: $\triangle WQD$ is a right triangle.

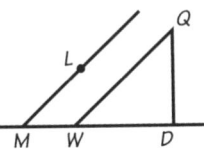

Write an equation for each line described. *(Section 4.3)*

26. The line passes through $P(-2, 3)$ and is parallel to $y = x - 4$.

27. The line passes through $R(0, -6)$ and is perpendicular to $y = -3x + 7$.

248   Chapter 5 *Parallel Lines*

---

21. Statements (Reasons)
  1. $\angle 1 \cong \angle 3$ (Given)
  2. $\angle 2 \cong \angle 1$ (Vertical ∡ are ≅.)
  3. $\angle 2 \cong \angle 3$ (Transitive Property)
  4. $j \parallel k$ (If two lines are intersected by a transversal and corresponding ∡ are ≅, then the lines are ∥.)
  5. $\angle 4$ and $\angle 5$ are supplements. (If two ∥ lines are intersected by a transversal, then same-side interior ∡ are supplementary.)

22. Answers may vary. Examples are given.

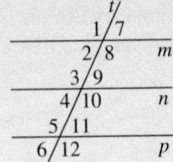

m and n: $\angle 1 \cong \angle 3$ or $\angle 8$ and $\angle 9$ are supplementary.
m and p: $\angle 2 \cong \angle 11$ or $\angle 1 \cong \angle 5$.
n and p: $\angle 4$ and $\angle 5$ are supplementary or $\angle 8 \cong \angle 11$.

23. If Ted is right, $\overleftrightarrow{AD}$ and $\overleftrightarrow{EH}$ are parallel lines. However, since $\angle ABF$ and $\angle BFE$ are not supplementary angles, the lines are not parallel.

24. Follow Steps 1 and 2 of the Exploration. Then rotate your triangle 180° so that leg a lies along line k but is on the opposite side of the line. Trace the hypotenuse to draw and label a segment $\overline{RS}$. Label $\angle 2$, the

acute angle formed by k and $\overline{RS}$. Since $\angle 1$ and $\angle 2$ are congruent alternate interior angles, $\overleftrightarrow{PQ}$ and $\overleftrightarrow{RS}$ are parallel lines.

25. See answers in back of book.

26. $y = x + 5$

27. $y = \frac{1}{3}x - 6$

5.5 Proving Theorems About Parallels

Learn how to...

• apply conditions to prove that lines are parallel

So you can...

• analyze real-world objects such as kites and desk lamps

Objectives

• Apply conditions to prove that lines are parallel.

• Analyze real-world objects.

Recommended Pacing

❖ **Core and Extended Courses**
Section 5.5: 1 day

❖ **Block Schedule**
Section 5.5: $\frac{1}{2}$ block
(with Section 5.4)

Resource Materials

Lesson Support
Lesson Plan 5.5
Warm-Up Transparency 5.5
Practice Bank: Practice 35
Study Guide: Section 5.5
Challenge Problems: Set 35
Assessment Book: Test 21

Technology
Technology Book:
Calculator Activity 7
Internet:
http://www.hmco.com

For passenger safety, railway engineers need to make sure that the rails of the track are always the same distance apart. To measure the distance, they use a *gauge bar* that fits between the rails. The gauge bar must be perpendicular to both rails so that each measurement is accurate.

If the gauge bar fits snugly, then it is perpendicular to both rails.

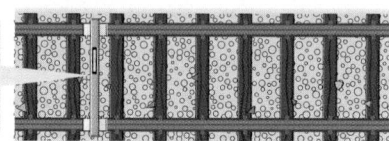

The following theorem guarantees that the rails are parallel if the gauge bar fits snugly.

Dual Perpendiculars Theorem

In a plane, if two lines are both perpendicular to a third line, then the two lines are parallel.

If *m*, *n*, and *t* are coplanar, $m \perp t$ and $n \perp t$, then $m \parallel n$.

The *distance between two parallel lines* is the length of a segment that connects the lines and is perpendicular to both of them.

The uniform distance between the rails of a track is called the *gauge*.

The **distance from a point to a line** is the length of the perpendicular segment from the point to the line. The distance from *Q* to *k* is 5 units.

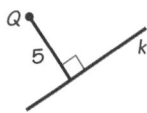

Warm-Up Exercises

What would you need to know about the given angles to prove that lines *l* and *k* are parallel.

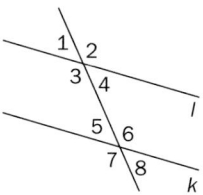

1. ∠3 and ∠5 These angles must be supplementary.

2. ∠1 and ∠5 These angles must be congruent.

3. ∠3 and ∠6 These angles must be congruent.

4. ∠4 and ∠8 These angles must be congruent.

5. Let $m \angle 4 = x + 5$ and $m \angle 6 = 3x + 3$. Find the value of *x* that would allow you to prove that *l* and *k* are parallel.   43

Section Notes

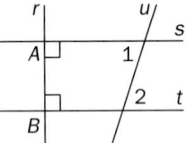

Communication: Reading
There is a great deal of information on the first page of this section. You might use it as an exercise in critical reading. Have students read the page to themselves in class and take notes. Then ask different students to read one of the notes they wrote and explain why they thought it was important. They should point out the Dual Perpendiculars Theorem, as well as definitions for the distance between two parallel lines, and the distance from a point to a line.

Geometric Thinking
Students understand intuitively that the distance from a point to a line, as defined, is the *shortest* distance from the point to the line. Ask them to explain mathematically why it must be the shortest distance. They can use the Pythagorean theorem to do this.

Additional Example 1

In the diagram below, line *r* is perpendicular to both lines *s* and *t*. Write a paragraph proof to show that ∠1 is congruent to ∠2.

Given: $r \perp s, r \perp t$
Prove: $\angle 1 \cong \angle 2$
Since *s* and *t* are perpendicular to *r*, by the Dual Perpendiculars Theorem, $s \parallel t$. Since *u* is a transversal of parallel lines *s* and *t*, the alternate interior angles, ∠1 and ∠2, are congruent.

Think and Communicate

These questions help students to understand that while slight variations for the drawings in a proof are possible, so long as the given information is represented correctly, the proof will have the same conclusion.

250

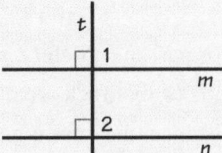

Write a paragraph proof of the Dual Perpendiculars Theorem.

SOLUTION

Write the given information and the statement you want to prove, and sketch a diagram showing the given information.

Given: Lines *m*, *n*, and *t* are coplanar.
$m \perp t; n \perp t$
Prove: $m \parallel n$

Using your diagram as a guide, write a proof of the theorem:

Since *m* and *n* are perpendicular to *t*, ∠1 and ∠2 are both right angles and are congruent. Since the transversal *t* forms congruent corresponding angles with the two lines *m* and *n*, $m \parallel n$.

Suppose you know that the first rung of a ladder is parallel to the second rung, the second rung is parallel to the third rung, and so on. The Dual Parallels Theorem will allow you to show that any two rungs of the ladder are parallel.

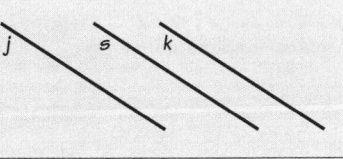

Dual Parallels Theorem

If two lines are both parallel to a third line, then the two lines are parallel.

If $j \parallel s$ and $k \parallel s$, then $j \parallel k$.

In the exercises, you will explain how to prove the Dual Parallels Theorem for lines that are coplanar.

THINK AND COMMUNICATE

1. If you used this diagram instead of the one shown in Example 1, would the proof for Example 1 be any different? Explain why or why not.

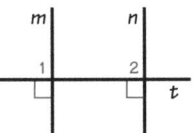

2. Are there other ways you could draw a diagram of three lines *j*, *k*, and *s* for the Dual Parallels Theorem? If there are, describe them.

ANSWERS   Section 5.5

Think and Communicate

1. No, the proof would be the same because corresponding angles formed by lines *m* and *n* and transversal *t* are still congruent.

2. Yes; answers may vary. Examples are given. Line *j* could lie "between" the other two lines, or line *k* could lie "between" the other two lines. Also, the lines could be drawn so that they do not all lie in one plane.

EXAMPLE 2 Application: City Planning

In this map of part of Chicago, North Avenue is parallel to Chicago Avenue, and Orleans Street is perpendicular to both Division Street and Chicago Avenue. What can you conclude about ∠1 and ∠2?

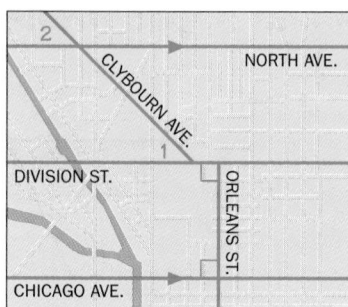

SOLUTION

Division is parallel to Chicago, because they are both perpendicular to Orleans.

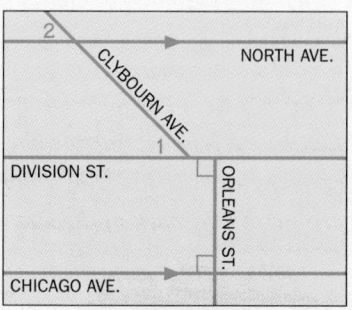

BY THE WAY

The plan of Chicago is based on the points of the compass. This method was used in Indian, Chinese, Aztec, and African cultures as long ago as 2000 B.C.

Because North is parallel to Chicago and Division is parallel to Chicago, the Dual Parallels Theorem allows you to conclude that North is parallel to Division.

∠1 ≅ ∠2, because they are corresponding angles formed by a transversal and two parallel lines.

☑ CHECKING KEY CONCEPTS

For Questions 1–3, use the diagram.

1. What can you conclude using the Dual Perpendiculars Theorem?

2. What do you know about $\overleftrightarrow{AB}$ and $\overleftrightarrow{FD}$? Explain how you know.

3. What can you conclude from your answers to Questions 1 and 2? Explain your reasoning.

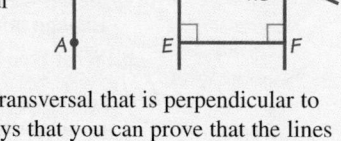

4. Sketch two coplanar lines and a transversal that is perpendicular to both of them. Describe all the ways that you can prove that the lines are parallel.

5.5 Proving Theorems About Parallels **251**

Checking Key Concepts

1. $\overleftrightarrow{CE} \parallel \overleftrightarrow{FD}$

2. $\overleftrightarrow{AB} \parallel \overleftrightarrow{FD}$ because if two lines are intersected by a transversal and same-side interior angles are supplementary, then the lines are parallel.

3. $\overleftrightarrow{CE} \parallel \overleftrightarrow{AB}$; If two lines are both parallel to a third line, then the lines are parallel.

4.

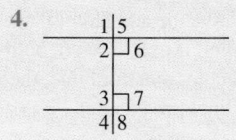

All the numbered angles are right angles. The lines can be proved parallel by showing corresponding angles 1 and 3, 2 and 4, 5 and 7, or 6 and 8 are congruent; by showing alternate interior angles 2 and 7 or 3 and 6 are congruent; by showing that same-side interior angles 2 and 3 or 6 and 7 are supplementary; or by using the Dual Perpendiculars Theorem.

Teach↔Interact

About Example 2

Common Error
Some students may want to conclude directly from the map that Division Street and North Avenue are parallel. Caution them that they cannot assume this is true unless it is stated as given or marked on the drawing itself. In this case, the Dual Perpendiculars Theorem is used to show that Division is parallel to Chicago, and then the given information along with the Dual Parallels Theorem is used to show that North Avenue is parallel to Division Street.

Additional Example 2

In this city map, Top Street is perpendicular to both Front Street and Back Street. ∠2 is congruent to ∠3. What can you conclude about ∠1 and ∠2?

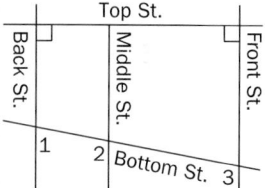

Because Back Street and Front Street are both perpendicular to Top Street, they are parallel by the Dual Perpendiculars Theorem. Also, because ∠2 and ∠3 are congruent corresponding angles formed by a transversal, Middle Street and Front Street are parallel. Because Middle is parallel to Front and Back is parallel to Front, the Dual Parallels Theorem allows you to conclude that Middle is parallel to Back. ∠1 and ∠2 are supplementary, because they are same-side interior angles formed by a transversal and two parallel lines.

Closure Question

Describe the Dual Perpendiculars Theorem and the Dual Parallels Theorem. The Dual Perpendiculars Theorem says that in a plane, if two lines are perpendicular to a third line, then the two lines are parallel. The Dual Parallels Theorem states that if two lines are both parallel to a third line, then the two lines are parallel.

251

Suggested Assignment

❖ **Core Course**
Exs. 1–3, 11–17, 23–29, AYP

❖ **Extended Course**
Exs. 1–29, AYP

❖ **Block Schedule**
Day 30 Exs. 1–3, 11–17, 23–29, AYP

Exercise Notes

Interview Note: Application
Exs. 4–6 These exercises demonstrate how the geometric concepts of perpendicularity and parallelism can be incorporated into the design and construction of a kite.

Historical Connection
Ex. 4–6 The Aztec people were the dominant power in 15th century Mexico. They controlled a vast empire that stretched across central Mexico from the Atlantic to the Pacific. The Aztec people developed a sophisticated culture and built a number of large and beautiful cities.

5.5 | Exercises and Applications

Extra Practice exercises on page 669

Use the diagram for Exercises 1–3.

1. What do you know about $\overleftrightarrow{LT}$ and $\overleftrightarrow{NU}$? Explain how you know.

2. Use your answer to Exercise 1. What can you conclude using the Dual Parallels Theorem?

3. **ALGEBRA** Find the value of x and $m\angle LPQ$.

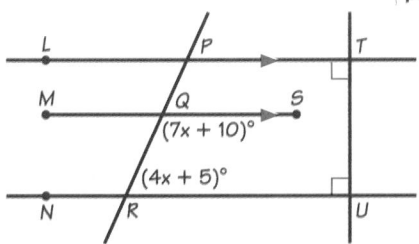

INTERVIEW **José Saínz**

Look back at the article on pages 216–218.

José Saínz has found inspiration for many of his designs in his own heritage. His Aztec Warrior kite displays the warrior in a black rectangle, surrounded by colorful flaps. There are eight supports across the back of the kite, and six vertical supports.

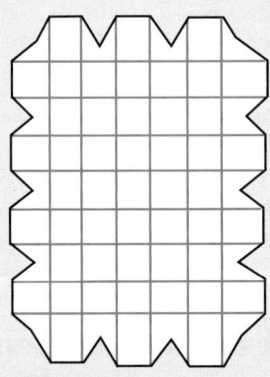

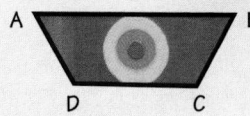

4. If all of the cross-supports are perpendicular to one vertical support, what do you know about all of the cross-supports? Explain.

5. In quadrilateral $ABCD$, $m\angle ABC = 60°$ and $m\angle BCD = 120°$.
   a. What do you know about $\overline{AB}$ and $\overline{CD}$? Explain how you know.
   b. What type of quadrilateral is $ABCD$?

6. If you know that $\overline{AB}$ is parallel to one of the cross-supports, what can you conclude about $\overline{CD}$? Explain.

Exercises and Applications

1. They are parallel; in a plane, if two lines are both perpendicular to a third line, then the two lines are parallel.

2. $\overleftrightarrow{MS} \parallel \overleftrightarrow{NU}$

3. $x = 15$; $m\angle LPQ = 65°$

4. All the cross-supports must be parallel to each other. Since all the cross-supports are perpendicular to the same line, they are all parallel by the Dual Perpendiculars Theorem.

5. a. They are parallel; if two lines are intersected by a transversal and same-side interior angles are supplementary, then the lines are parallel.
   b. a trapezoid

6. $\overline{CD}$ is parallel to that cross-support; if two lines are both parallel to a third line, then the two lines are parallel.

7. Webster Avenue and Grant Place are parallel; if two lines are both perpendicular to a third line, then the two lines are parallel.

8. They are parallel.

9. Webster Avenue and Belden Avenue are parallel; if two lines are both parallel to a third line, then the two lines are parallel.

10. $\angle 3$ and $\angle 5$; Answers may vary. Examples are given. Since Belden Avenue and Webster Avenue are parallel (Ex. 9), alternate interior angles 1 and 3 are congruent, and corresponding angles 1 and 5 are congruent.

11. $4\sqrt{6}$

12. $2\sqrt{3}$

13. No; by the Perpendicular Postulate, through a point not on a given line, there is exactly one line perpendicular to the given line.

14. A

CITY PLANNING **For Exercises 7–10, use this street map.**

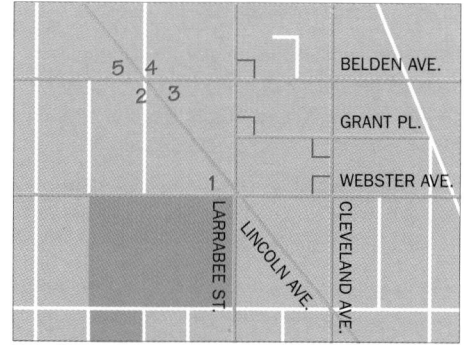

7. What can you conclude from the fact that Cleveland Avenue is perpendicular to both Webster Avenue and Grant Place? Explain your reasoning.

8. What does the Dual Perpendiculars Theorem tell you about Grant Place and Belden Avenue?

9. Based on your answers to Exercises 7 and 8, what do you know? Explain your reasoning.

10. Name each of the numbered angles that is congruent to ∠1, and explain how you know.

11. Find the distance from P to $\overleftrightarrow{AB}$.

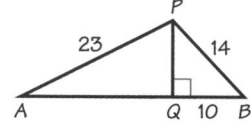

12. Find the distance from O to $\overleftrightarrow{EF}$.

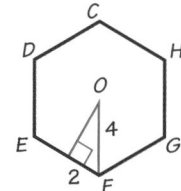

13. **Writing** The definition of the *distance from a point to a line* refers to the length of the perpendicular segment from the point to the line. Is it possible that there might be more than one such segment? Explain why or why not.

14. **SAT/ACT Preview** Which is greater, a or b?

$a = LM$

$b = $ the distance from L to $\overleftrightarrow{MN}$

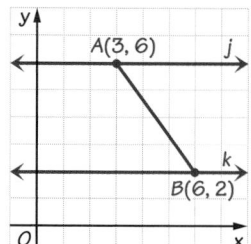

    A. a is greater.   **B.** b is greater.   **C.** a and b are equal.

15. **Open-ended Problem** Describe all the different ways that you can use to prove that two lines are parallel.

16. **Visual Thinking** Explain why the words *in a plane* are used in the statement of the Dual Perpendiculars Theorem. Include a sketch in your answer.

17. Carmela thinks that the distance between lines j and k is 5 units.

    a. Writing Do you agree or disagree with Carmela? Explain your answer.

    b. Find the distance between j and k.

18. **Writing** Explain why you can find the distance between two lines only if the lines are parallel.

Exercise Notes

Family Involvement
Ex. 7 Many cities in the United States are laid out in a way that streets and avenues are perpendicular to one another. Thus, streets are parallel and avenues are parallel. Students can look for maps of U.S. cities that display this pattern (the Borough of Manhattan in New York City is a good example) and involve family members in some of the ideas of this section.

Student Progress
Exs. 7–10 Students should be able to complete these exercises successfully. Encourage those that have difficulty to follow the reasoning of Example 2.

Spatial Reasoning
Exs. 11, 12 Students need to realize that these exercises ask for the distance from a point to a line. They should copy and mark the diagrams to make this more apparent by highlighting the point and highlighting and extending the line.

Topic Spiraling: Review
Exs. 12, 13 These exercises require the use of two concepts from previous sections. Students need to use the Pythagorean theorem for Ex. 12. In Ex. 13, they need to use the Perpendicular Postulate.

Journal Entry
Ex. 15 Students may want to save their responses to this exercise in their journals so they can refer to them when completing proofs that involve parallel lines.

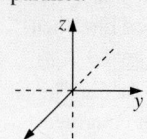
Communication: Drawing
Ex. 18 Encourage students to draw examples to support their answer for this exercise. This should reinforce the fact that it is impossible to draw a line that is perpendicular to both the given lines if they are not parallel.

---

15. Answers may vary. Two lines can be proved to be parallel using the Converse of the Corresponding Angles Postulate, the Converse of the Alternate Interior Angles Theorem, the Converse of the Same-Side Interior Angles Theorem, the Dual Perpendiculars Theorem, or the Dual Parallels Theorem.

16. The theorem is not true if the phrase *in a plane* is omitted.

Sketches may vary. The example shows the axes of a three-dimensional coordinate system. The x-axis and the y-axis are both perpendicular to the z-axis, but they are not parallel.

17. **a.** disagree; AB is not the distance from j to k because $\overline{AB}$ is not perpendicular to both j and k.

    b. 4 units

18. Answers may vary. An example is given. The distance between two lines is only defined for two parallel lines. If the lines are not parallel, then there can be no segment that is perpendicular to both lines.

Exercise Notes

🌐 Application
Exs. 19–21 These exercises can provide students with an insight into how geometry is involved in the design of a desk lamp. You may wish to bring a lamp with this type of adjustable arm into class to help demonstrate some of the concepts involved in these exercises.

Cooperative Learning
Exs. 19–21 This group of exercises can be done in small groups of three students. Having lamps of the type described here available for students to experiment with will aid in the formulation and discussion of their answers.

Challenge
Ex. 22 Encourage students to use graph paper to help solve this problem. By creating a diagram, students should get a better idea of the method necessary to find the distance.

Assessment Note
Ex. 24 Students' explanations should be in the form of a plan to write a proof of the Dual Parallels Theorem.

This desk lamp is designed with an adjustable arm so that different parts of the desk can be lit. The two quadrilaterals in the arm are flexible, making it possible to adjust the position of the lamp.

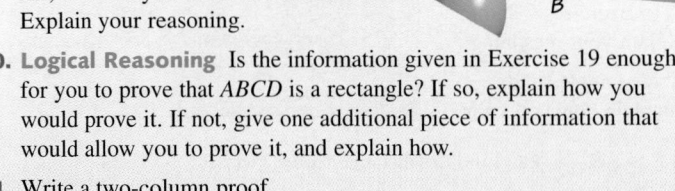

19. If $\overline{AD}$ and $\overline{BC}$ are both perpendicular to $\overline{AB}$, what do you know about $\overline{AD}$ and $\overline{BC}$? Explain your reasoning.

20. Logical Reasoning Is the information given in Exercise 19 enough for you to prove that *ABCD* is a rectangle? If so, explain how you would prove it. If not, give one additional piece of information that would allow you to prove it, and explain how.

21. Write a two-column proof.
Given: $\overline{FG} \parallel \overline{EH}$; $\angle FEH \cong \angle HGF$
Prove: *EFGH* is a parallelogram.

22. Challenge Find the exact distance from the point $P(6, 4)$ to the line $y = -\frac{1}{3}x - 4$. Describe your method.

23. Copy and complete this flow proof. First copy the diagram and mark it with the given information.
Given: $\angle 1 \cong \angle 2$; $\angle 3 \cong \angle 4$
Prove: $n \parallel p$

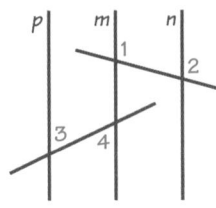

Reasons
1. ?
2. Given
3. ?
4. If two lines are intersected by a transversal and alternate interior angles are $\cong$, then the lines are $\parallel$.
5. ?

1 $\boxed{\angle 1 \cong \angle 2}$ —3→ $\boxed{n \parallel m}$ —5→ $\boxed{?}$
2 $\boxed{?}$ —4→ $\boxed{?}$

ONGOING ASSESSMENT

24. Writing Use the diagram. Explain how drawing the transversal, *t*, allows you to prove the Dual Parallels Theorem for coplanar lines.

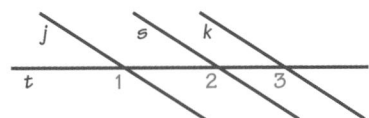

---

19. $\overline{AD} \parallel \overline{BC}$; In a plane, if two lines are both perpendicular to a third line, then the two lines are parallel.

20. No. Answers may vary. An example is given. If you also knew that $\overline{AD} \perp \overline{DC}$, then you could use the Dual Perpendiculars Theorem again to conclude that $\overline{AB} \parallel \overline{CD}$. Then you would know that *ABCD* is a parallelogram. Since its opposite angles are congruent, all four angles must be right angles. Thus, *ABCD* is an equiangular parallelogram, that is, a rectangle.

21. Proofs may vary. An example is given.
Statements (Reasons)
1. $\overline{FG} \parallel \overline{EH}$ (Given)
2. $\angle EFG$ and $\angle FEH$ are supplementary. (If two $\parallel$ lines are intersected by a transversal, then same-side interior $\angle$ are supplementary.)
3. $m\angle EFG + m\angle FEH = 180°$ (Def. of supplementary $\angle$)
4. $\angle FEH \cong \angle HGF$ or $m\angle FEH = m\angle HGF$ (Given; Def. of $\cong \angle$)
5. $m\angle EFG + m\angle HGF = 180°$ (Substitution Property (Steps 3 and 4))

6. $\overline{EF} \parallel \overline{GH}$ (If two lines are intersected by a transversal and same-side interior $\angle$ are supplementary, the lines are $\parallel$.)
7. *EFGH* is a $\square$. (Def. of $\square$)

22. $3\sqrt{10}$; The distance is the length of the perpendicular segment from P to y. The segment has slope 3 and, so, equation $y = 3x - 14$. I solved $y = -\frac{1}{3}x - 4$ and $y = 3x - 14$ simultaneously to find the intersection of the lines, $Q(3, -5)$. $PQ = 3\sqrt{10}$

23. Check students' work.
(1) Given; (2) $\angle 3 \cong \angle 4$; (3) If two lines are intersected by a transversal and corresponding $\angle$ are $\cong$, then the lines are $\parallel$; (4) $p \parallel m$; (5) $n \parallel p$ (If two lines are both $\parallel$ to a third line, then the two lines are $\parallel$.)

24. When you draw transversal *t*, two pairs of congruent corresponding angles are formed, and these can be used to show that lines *j* and *k* are parallel. If two parallel lines are intersected by a transversal, then corresponding angles are congruent. So since $j \parallel s$, $\angle 1 \cong \angle 2$, and since $k \parallel s$, then $\angle 2 \cong \angle 3$. Therefore, $\angle 1 \cong \angle 3$ by the Transitive Property, and so $j \parallel k$. (If two lines are intersected by a transversal and corresponding angles are congruent, then the lines are parallel.)

25–27. Answers may vary. Examples are given.

Sketch each situation. *(Section 1.4)*

25. Line *j* and line *k* are skew.

26. $\overline{PQ}$ lies in plane *W*.

27. The intersection of $\overleftrightarrow{RS}$ and plane *X* is point *T*.

Write an equation for each line described. *(Section 4.3)*

28. The line passes through $M(2, -5)$ and is parallel to the line $y = 3x - 18$.

29. The line passes through $N(3, 0)$ and is perpendicular to the line $y = 7x$.

ASSESS YOUR PROGRESS

VOCABULARY

flow proof (p. 234)

distance from a point to a line (p. 249)

Write a proof, using any method. *(Sections 5.3–5.5)*

1. Given: $m \angle J + m \angle L = m \angle K$
Prove: $\triangle JKL$ is a right triangle.

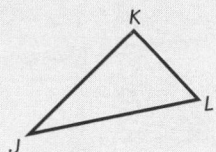

2. Given: $\angle 1 \cong \angle 2$; $\overleftrightarrow{DC} \parallel \overleftrightarrow{BA}$
Prove: $\overleftrightarrow{FE} \parallel \overleftrightarrow{BA}$

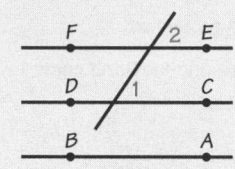

Tell which segments are parallel. Explain your reasoning. *(Section 5.4)*

3.

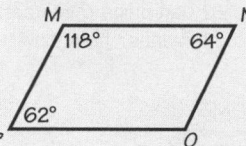

4.

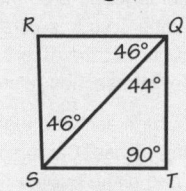

ALGEBRA For each diagram, find the value of the variable that will allow you to prove that two lines are parallel. *(Section 5.4)*

5.

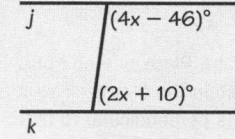

6.

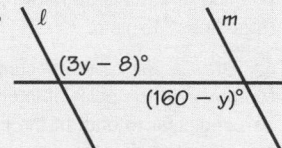

7. Journal This chapter includes four different methods of proof: *paragraph proof, two-column proof, flow proof,* and *key steps of a proof.* Describe the advantages and disadvantages of each method.

5.5 Proving Theorems About Parallels **255**

Apply⇔Assess

Assess Your Progress

Journal Entry
Students should refer primarily to Section 5.3 when writing this entry. They can also look at the proofs in the Examples in other sections and refer back to Sections 3.3 and 3.4 as well. Ask several students to read their entries so that each student can assess his or her own work.

Progress Check 5.3–5.5

See page 272.

Practice 35 for Section 5.5

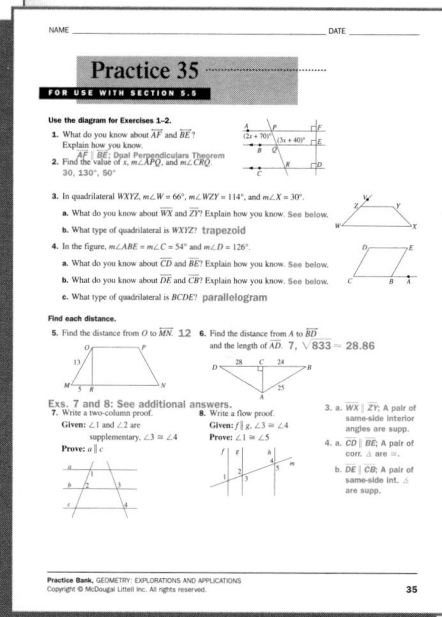

25.

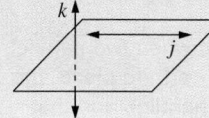

26.

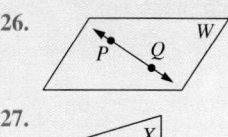

27.

28. $y = 3x - 11$

29. $y = -\frac{1}{7}x + \frac{3}{7}$

Assess Your Progress

1. Since the sum of the angle measures of a triangle is 180°, $m \angle J + m \angle K + m \angle L = 180°$. It is given that $m \angle J + m \angle L = m \angle K$, so by the Substitution Property, $m \angle K + m \angle K = 180°$. $2 \cdot m \angle K = 180°$, and so $m \angle K = 90°$. $\angle K$ is a right

angle by the definition of right angle and *JKL* is a right triangle by the definition of right triangle.

2. Statements (Reasons)
1. $\angle 1 \cong \angle 2$ (Given)
2. $\overleftrightarrow{FE} \parallel \overleftrightarrow{DC}$ (If two lines are intersected by a transversal and corresponding ≰ are ≅, then the lines are ∥.)
3. $\overleftrightarrow{DC} \parallel \overleftrightarrow{BA}$ (Given)
4. $\overleftrightarrow{FE} \parallel \overleftrightarrow{BA}$ (If two lines are both ∥ to a third line, then the two lines are ∥.)

3. $\overline{MN} \cong \overline{PO}$; If two lines are intersected by a transversal and same-side interior angles are supplementary, then the lines are parallel.

4. $\overline{RQ} \parallel \overline{ST}$; Answers may vary. An example is given. $\angle RQT$ and $\angle T$ are both right angles. In a plane, if two lines are both perpendicular to a third line, then the lines are parallel.

5. 36   **6.** 42

7. See answers in back of book.

255

Parallels in Space

Objectives

- Recognize relationships among parallel and intersecting planes.
- Identify parallel lines and planes in space.
- Analyze real-world examples of parallel and perpendicular lines in space.

Recommended Pacing

❖ **Core and Extended Courses**
Section 5.6: 1 day

❖ **Block Schedule**
Section 5.6: $\frac{1}{2}$ block
(with Section 5.7)

Resource Materials

Lesson Support

Lesson Plan 5.6

Warm-Up Transparency 5.6

Overhead Visuals:
    Folder 6: Points, Lines, and
        Planes

Practice Bank: Practice 36

Study Guide: Section 5.6

Challenge Problems: Set 36

Technology
Internet:
    http://www.hmco.com

Warm-Up Exercises

Tell whether each statement is True or False.

1. Two lines are coplanar if they intersect.   True.

2. Two planes can intersect in a line or a point.   False.

3. A line and a plane can intersect in a point or a line.   True.

4. Two planes can be parallel.   True.

5. For any line, there are exactly two planes that contain the line.   False.

Learn how to...

- recognize relationships among parallel and intersecting planes

So you can...

- identify parallel lines and planes in space

- analyze real-world examples of parallel and perpendicular lines in space

There are two possibilities for how two planes can be related: the planes may be parallel or they may intersect in a line. But what happens if there are three planes?

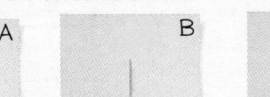

EXPLORATION
COOPERATIVE LEARNING

Investigating Planes and Their Intersections

Work with another student. You will need:
- three index cards
- scissors
- two toothpicks

1 Cut and label the cards as shown.

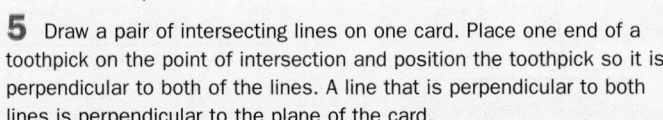

2 Hold cards *A* and *B* so that they model parallel planes. Can you position plane *C* so it is parallel to both *A* and *B*? Can plane *C* intersect one plane but not the other? Can *C* intersect both planes? If so, how are the lines of intersection related?

3 Use cards *A* and *B* to model intersecting planes. Can plane *C* intersect one plane but not the other? Can *C* intersect both planes?

4 Can three planes form exactly one line of intersection? two lines? three lines? Explain.

5 Draw a pair of intersecting lines on one card. Place one end of a toothpick on the point of intersection and position the toothpick so it is perpendicular to both of the lines. A line that is perpendicular to both lines is perpendicular to the plane of the card.

6 Hold two toothpicks to model two different lines that are each perpendicular to the same card. Describe the relationship between the lines.

In the Exploration, you may have discovered this property of intersecting planes. The proof is not included.

Intersecting Planes Theorem

If two parallel planes are intersected by a third plane, then the lines of intersection are parallel.

If plane $X \parallel$ plane Y, then $l \parallel n$.

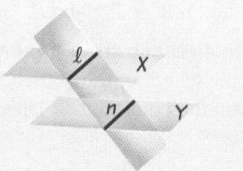

EXAMPLE 1

In this file box, faces *ABCD* and *EFGH* lie in parallel planes. How are edges $\overline{EF}$ and $\overline{AB}$ related?

 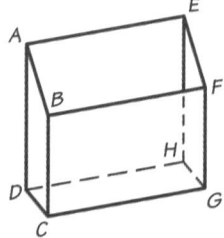

SOLUTION

$\overleftrightarrow{EF}$ and $\overleftrightarrow{AB}$ are the lines of intersection formed when the parallel planes *ABCD* and *EFGH* are intersected by plane *EFBA*. By the Intersecting Planes Theorem, the two lines are parallel. So $\overline{EF} \parallel \overline{AB}$.

THINK AND COMMUNICATE

Use the file box shown in Example 1.

1. What does the Intersecting Planes Theorem allow you to conclude about $\overline{HG}$ and $\overline{DC}$? Explain.

2. Noah believes he can use the Intersecting Planes Theorem to show that $\overline{AB} \parallel \overline{DC}$. Explain why this theorem *cannot* be used in this case.

You have learned that two lines parallel to a third line are parallel to each other. This theorem, which is not proved, shows that planes are related in the same way.

Parallel Planes Theorem

If two planes are both parallel to a third plane, then the two planes are parallel.

If plane $P \parallel$ plane Z and plane $Q \parallel$ plane Z, then plane $P \parallel$ plane Q.

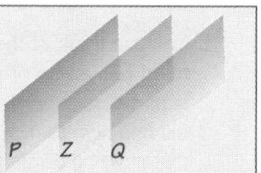

5.6 Parallels in Space **257**

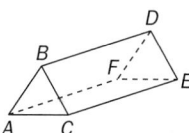

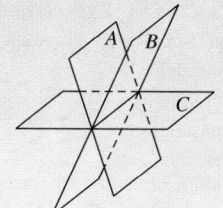 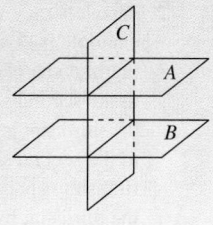 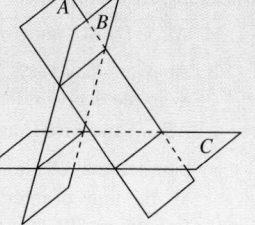

Additional Example 2

Sketch and label a diagram for this theorem. Then write the Given and Prove statements.

If a line is perpendicular to two other lines at the same point, then it is perpendicular to the plane containing those two lines.

Sketch and label a diagram showing a plane containing two lines that intersect in a point. Draw a third line that intersects both of these lines at their point of intersection, and is perpendicular to both lines. Use dashed lines for the parts of the line that are hidden from view. Use the labels from your sketch to write the Given and Prove statements.

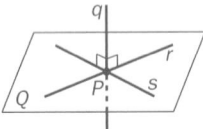

Given: Lines r, q, and s intersect at point P, $r \perp q$, $s \perp q$.
Prove: plane $Q \perp$ line q

Closure Question

Describe the different ways that three planes can be related.

Three planes can be parallel; they can intersect in a single point or a single line; each pair of planes can intersect in a different line; if two of the planes are parallel, the three planes can intersect in two parallel lines.

A **line is perpendicular to a plane** if it intersects the plane and is perpendicular to every line in the plane that passes through the point of intersection. In this figure, line $m \perp$ plane P.

The **distance from a point to a plane** is the length of the perpendicular segment from the point to the plane.

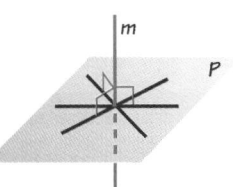

EXAMPLE 2

Sketch and label a diagram for this theorem. Then write the Given and Prove statements.

If two lines are both perpendicular to the same plane, then the lines are parallel.

SOLUTION

Sketch and label a diagram showing a plane and two lines perpendicular to the plane.

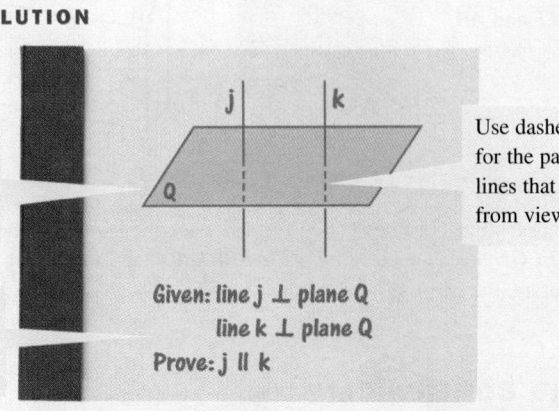

Use dashed lines for the parts of the lines that are hidden from view.

Use the labels in your sketch to write the Given and Prove statements.

Given: line $j \perp$ plane Q
line $k \perp$ plane Q
Prove: $j \parallel k$

☑ CHECKING KEY CONCEPTS

1. Use the cube at the right.
  a. $\overline{AB}$ is perpendicular to plane *EAD*. Name three segments that are perpendicular to $\overline{AB}$.
  b. Explain how to show that $\overline{AB} \parallel \overline{HG}$.

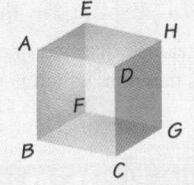

2. You can think of the slats of a vertical window blind as a group of planes. When the blind is partly open, each plane is parallel to the one next to it. Explain how you know that any two of the planes are parallel.

258 Chapter 5 *Parallel Lines*

Checking Key Concepts

1. **a.** $\overline{AE}$, $\overline{AD}$, $\overline{BC}$, and $\overline{BF}$

  b. Answers may vary. An example is given. Planes *ABCD* and *EFGH* are parallel and $\overleftrightarrow{AB}$ and $\overleftrightarrow{HG}$ are the lines of intersection when *ABCD* and *EFGH* are intersected by plane *ABGH*. By the Intersecting Planes Theorem, $\overline{AB} \parallel \overline{HG}$.

2. Answers may vary. An example is given. Since the first slat is parallel to the second slat, and the third slat is parallel to the second slat, the first slat is parallel to the third slat by the Parallel Planes Theorem. Similar reasoning can be used to show that any two of the planes are parallel.

Exercises and Applications

1–3. Answers may vary. Examples are given.

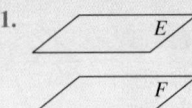

5.6 | Exercises and Applications

Extra Practice
exercises on
page 670

Sketch each situation.

1. Plane $E \parallel$ plane F.

2. Plane A and plane B intersect in line p.

3. Lines j and k lie in plane Z and intersect at point D; line m is perpendicular to plane Z and also passes through D.

4. **Open-ended Problem** Tell which of the theorems and definitions that you learned in this section are illustrated by the planes formed by the triangles in the roof and the lines formed by the cross-supports. Explain your choices.

In this diagram, plane $X \parallel$ plane Y.

5. a. How are $\overline{AB}$ and $\overline{FC}$ related? Explain how you know.

   b. What type of quadrilateral is $ABDF$?

6. Suppose $m \angle ABD = 65°$. Find $m \angle FDB$.

7. Name two angles that are congruent to $\angle ABD$.

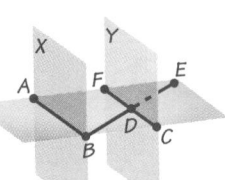

Connection   ELECTRONICS

A *capacitor* is a basic component of televisions, radios, electronic toys, computers, and almost all electronic devices. Many old radios use an adjustable capacitor to tune in different stations.

8. What do you need to know about the front upper plate and the front lower plate in order to show that any upper plate and any lower plate lie in parallel planes? Explain your reasoning.

The planes of all of the plates in the upper group are parallel.

The planes of all of the plates in the lower group are parallel.

Tuning knob attaches here

9. **Writing** When the tuning knob of the radio is turned, the top group rotates to fit between the plates of the bottom group. Explain how the Parallel Planes Theorem shows that none of the plates from the upper group will touch any of the plates from the lower group.

3.

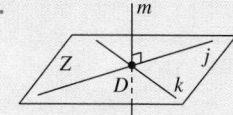

4. (1) Intersecting Planes Theorem; The two opposite planes formed by the triangles in the roof are parallel, so the top edges of the roof are parallel. (2) A line is perpendicular to a plane if it intersects the plane and is perpendicular to

every line in the plane that passes through the point of intersection; Each cross-support is perpendicular to the planes formed by the two triangles since they are perpendicular to every line in the plane that passes through the point of intersection.

5. a. They are parallel; if two parallel planes are intersected by a third plane, the lines of intersection are parallel.

b. a trapezoid

6. 115°

7. $\angle FDE$ and $\angle BDC$

8. The front upper plate and front lower plate need to be parallel.

9. Since all of the plates in the upper group are parallel, all of the plates in the lower group are parallel, and the front upper plate is parallel to the front lower plate; all of the plates, both

lower and upper, must be parallel to each other by the Parallel Planes Theorem and therefore will not touch each other.

Suggested Assignment

❖ **Core Course**
Exs. 1–7, 10–14, 18, 19, 24–28, 31–34

❖ **Extended Course**
Exs. 1–34

❖ **Block Schedule**
Day 31 Exs. 1–7, 10–14, 18, 19, 24–28, 31–34

Exercise Notes

Spatial Reasoning
Exs. 1–3, 12–14, 17 These exercises provide an opportunity for students to develop their spatial reasoning powers by drawing planes and lines. Remind students to use dashed lines for parts of a figure that are hidden from view. Encourage those students who are having difficulty to use the figures in this section as a model. Some students may also want to build models using index cards and toothpicks before making their sketches.

Application
Exs. 4, 15–17 These exercises allow students to see how the geometry of planes and lines can be applied to building construction and aviation.

Interdisciplinary Problems
Exs. 8, 9 These problems combine geometry and electronics to study a radio capacitor. You might ask some students with an interest in this area to find out more about capacitors and how they are used in electronic devices. Ask these students to share their findings with the class.

Exercise Notes

Teaching Tip

Exs. 10, 11 Students can copy these diagrams and then shade or color the parallel faces to help them distinguish the parallel lines formed by the intersection of the parallel faces with a third face.

Journal Entry

Exs. 12, 13, 16 Encourage students to write a journal entry that describes and illustrates the three theorems in these exercises. They can also include the other theorems from this section. This entry can be used for quick reference when needed.

Historical Connection

Exs. 15–17 Orville and Wilbur Wright designed the Kitty Hawk, which on December 17, 1903 in the town of Kitty Hawk, on the Outer Banks of North Carolina, became the first heavier-than-air, self-propelled aircraft to fly. By 1907, the Wright brothers had developed more sophisticated airplanes that were capable of carrying two people as far as 125 miles.

Using Manipulatives

Exs. 18, 19 Some students may have difficulty visualizing $\overleftrightarrow{AD}$ being perpendicular to plane *CDE*. Suggest that they use pencils or pens to model the segments and the fact that $\overleftrightarrow{AD} \perp$ plane *CDE*.

In the diagram at the right, faces *MPTQ* and *NOSR* lie in parallel planes.

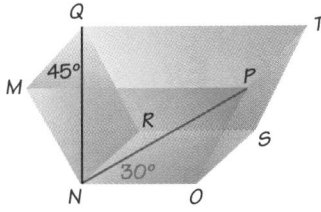

10. How are edges $\overline{OS}$ and $\overline{PT}$ related? Explain your reasoning.

11. Find $m\angle QNR$ and $m\angle MPN$, if possible. If it is not possible, explain why not.

For Exercises 12 and 13, sketch and label a diagram for the theorem. Then write the Given and Prove statements.

12. If a plane is perpendicular to one of two parallel lines, then the plane is also perpendicular to the other line.

13. If two planes are both perpendicular to the same line, then the planes are parallel.

14. **a.** Sketch a cube and label all eight vertices.

   b. Writing Use your diagram and the theorem in Exercise 13 to explain why two opposite faces of a cube lie in parallel planes.

AVIATION For Exercises 15–17, use this photograph of a Wright brothers' biplane. Although airplane wings are curved in order to produce lift, these two wings can be represented by parallel planes.

15. Each of the struts (vertical supports connecting the wings) is perpendicular to the plane of the upper wing. Explain how you know that all of the struts are parallel.

16. **a.** How do the struts appear to be related to the lower wing?

   b. Complete: If a line is perpendicular to one of two parallel planes, then ?.

17. **a.** Sketch and label a diagram for the theorem that you completed in Exercise 16(b).

   b. Write the Given and Prove statements for the theorem.

In this diagram, $\overleftrightarrow{AD} \perp$ plane *CDE*.

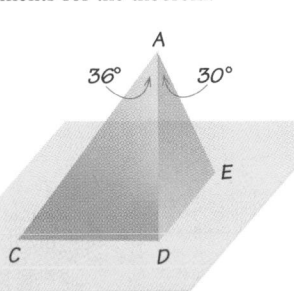

18. **a.** Find $m\angle ADC$. Explain your reasoning.

   b. Find $m\angle ACD$.

19. **a.** Find $m\angle AED$.

   b. Is it possible that $\overleftrightarrow{AE} \perp$ plane *CDE*? Explain your reasoning.

---

10. $\overline{OS} \parallel \overline{PT}$; If two parallel planes are intersected by a third plane, then the lines of intersection are parallel.

11. $m\angle QNR = 45°$; $m\angle MPN = 30°$

12, 13. Answers may vary. Examples are given.

12.

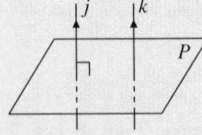

   Given: line *j* ∥ line *k*;
       plane *P* ⊥ line *j*
   Prove: plane *P* ⊥ line *k*

13.

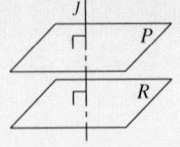

   Given: plane *P* ⊥ line *j*;
       plane *R* ⊥ line *j*
   Prove: plane *P* ∥ plane *R*

14. **a.** Labels may vary. An example is given.

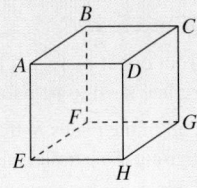

   b. Answers may vary. An example is given. The planes that contain faces *AEHD* and *BFGC* are parallel because both are perpendicular to the line $\overleftrightarrow{AB}$.

15. By the theorem in Example 2 on p. 258, the struts must all be parallel.

16. **a.** The struts appear to be perpendicular to the plane of the lower wing.

   b. the line is perpendicular to the other plane also

17. **a.** Answers may vary. An example is given.

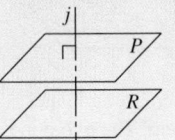

   b. Given: plane *P* ∥ plane *R*;
       line *j* ⊥ plane *P*
   Prove: line *j* ⊥ plane *R*

260

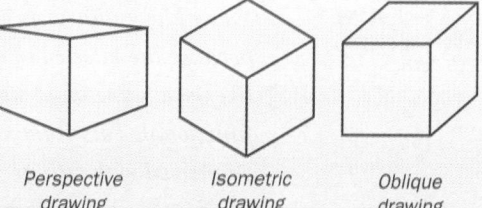

There are many ways to represent a three-dimensional object on a flat piece of paper. This cube is drawn using three different methods. *Oblique* drawing is the method you learned in Section 2.6.

Perspective drawing   *Isometric drawing*   *Oblique drawing*

20. Writing Are segments that are parallel in a cube represented by parallel segments in each drawing? Are segments that are congruent in a cube represented by congruent segments in each drawing? Write a comparison of the three methods based on your answers to these questions.

21. Follow the instructions at the right to make an isometric drawing of a prism that is 3 cm high with a square base that measures 5 cm by 5 cm.

22. Make an isometric drawing of a rectangular prism with edges of lengths 4 cm, 6 cm, and 8 cm.

23. Research Look in magazines, books, or newspapers to find a sketch or drawing of a building. Are any of the methods above used in the sketch? If so, why do you think that method was used?

Step 1 Draw three rays from a common endpoint. One should be vertical and the other two should be 30° from a horizontal line.

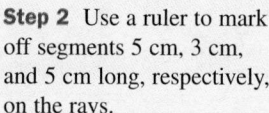

Step 2 Use a ruler to mark off segments 5 cm, 3 cm, and 5 cm long, respectively, on the rays.

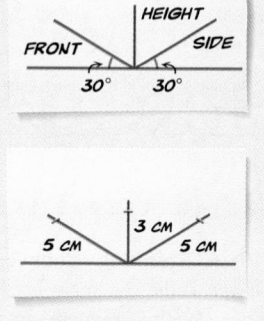

Step 3 Finish the diagram by drawing parallel and congruent segments as shown.

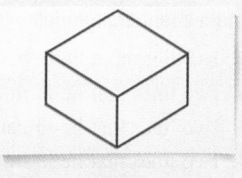

24. Challenge Sketch three planes that intersect in exactly one point.

In this rectangular prism, $\overleftrightarrow{AB} \perp$ plane *BCDE*.

25. SAT/ACT Preview Which of the green segments is (are) perpendicular to $\overleftrightarrow{AB}$?
A. $\overline{BE}$ only
B. $\overline{BC}$ only
C. $\overline{BE}$ and $\overline{BC}$ only
D. $\overline{BE}$, $\overline{BC}$, and $\overline{BD}$

26. a. Find the distance from *F* to plane *BCD*.
   b. Find the distance from *A* to plane *BCD*.

27. Find the distance between points *D* and *B*.

28. Find the distance from *D* to plane *EFA*.

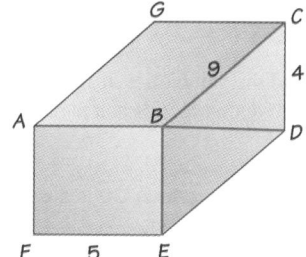

5.6 Parallels in Space **261**

---

18. a. 90°; Since $\overleftrightarrow{AD} \perp$ plane *CDE*, $\overleftrightarrow{AD} \perp \overleftrightarrow{DC}$; perpendicular lines intersect to form right angles, so $m\angle ADC = 90°$.
   b. 54°

19. a. 60°
   b. No; since $m\angle AED \neq 90°$, $\overleftrightarrow{AE}$ is not perpendicular to $\overleftrightarrow{ED}$, so $\overleftrightarrow{AE}$ cannot be perpendicular to plane *CDE*.

20. Answers may vary. An example is given. A perspective drawing represents an object as it would appear in a photograph. A portion that is closer to the front appears larger than a portion that is farther away. Segments that are parallel are not drawn to be parallel, and segments that are congruent are not drawn to be congruent. In an isometric drawing, segments that are parallel are drawn to be parallel, and segments that are congruent are drawn to be congruent. In an oblique drawing, segments that are parallel are drawn to be parallel, but segments that are congruent are not drawn to be congruent in order to make the figure appear as it would in real life.

21. Check students' work. Sketches should look like the one shown in Step 3.

22.

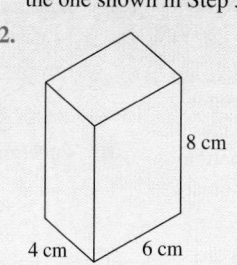
8 cm
4 cm   6 cm

23. Answers may vary.

24. Answers may vary. An example is given.

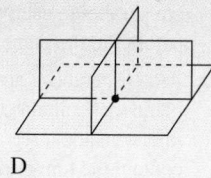

25. D
26. a. 5   **b.** 5
27. $\sqrt{97}$
28. 9

Exercise Notes

Spatial Reasoning
Exs. 20–23 These exercises can help students to develop their spatial reasoning abilities. A class discussion of Ex. 20 would help to ensure that all students understand the key differences of the three drawing methods.

Communication: Writing
Ex. 20 After students write a comparison of the three drawing methods, ask them to share their answers with the class. This will not only allow you to assess how well students understand these methods, but will also help other students to improve their understanding.

Challenge
Ex. 24 Students have difficulty visualizing three planes intersecting in a single point. You may wish to point out that the walls of the classroom are planes that intersect in a line and then ask students how the floor or ceiling intersects with this line.

Topic Spiraling: Review
Ex. 27 The solution to this exercise requires using the Pythagorean theorem.

Apply⟺Assess

Exercise Notes

Interview Note: Geometric Thinking
Exs. 29, 30 These exercises allow students to see how the geometry of planes in space relates to the design of stunt kites.

Assessment Note
Ex. 31 This exercise provides a good opportunity to assess students' understanding of the material in this section. After completing the exercise, have students work in small groups to compare and discuss their responses.

Practice 36 for Section 5.6

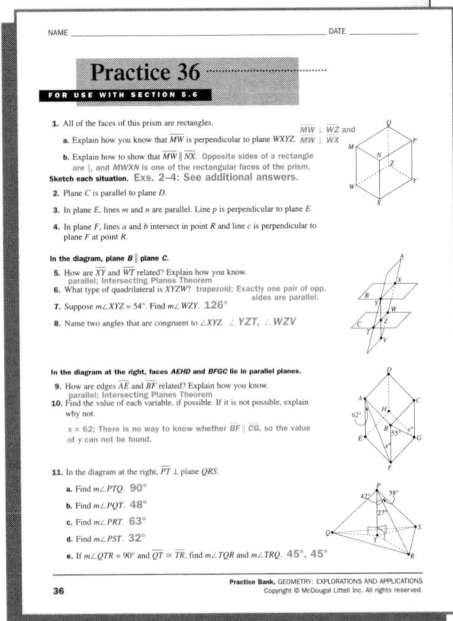

INTERVIEW José Saínz

Look back at the article on pages 216–218.

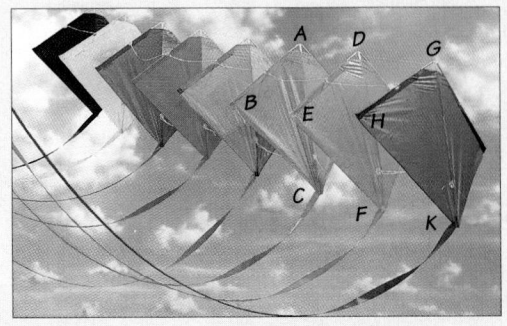

Many kite enthusiasts enjoy flying dual-line stunt kites because the kites can perform flips and loops. Kite designers like José Saínz build stunt kites to be attractive as well as functional. This stunt kite is a stack of eight kites that are connected with string.

29. Plane *ABC* is parallel to plane *DEF*, and plane *DEF* is parallel to plane *GHK*. What does the Parallel Planes Theorem allow you to conclude? Explain your reasoning.

30. Edges $\overline{AB}$, $\overline{DE}$, and $\overline{GH}$ are coplanar. Explain what you can conclude using the Intersecting Planes Theorem.

ONGOING ASSESSMENT

31. **Visual Thinking** Complete each statement with *always*, *sometimes*, or *never*. If the statement is *always* true, sketch an example. If it is *never* true, sketch a counterexample. If it is *sometimes* true, sketch an example and a counterexample.

   a. Two lines that lie in parallel planes are _?_ parallel.
   b. Two lines that lie in intersecting planes are _?_ parallel.
   c. Two lines that lie in parallel planes _?_ intersect.
   d. Two lines that lie in intersecting planes _?_ intersect.
   e. Two lines that intersect are _?_ coplanar.
   f. Two lines that are parallel are _?_ coplanar.

SPIRAL REVIEW

32. **ALGEBRA** Find the value of each variable that will allow you to prove that *EFGH* is a parallelogram. *(Section 5.4)*

$(2x - 7)°$ $(3x + 7)°$
$(90 - 5y)°$

For each set of points, tell whether $\overleftrightarrow{CD}$ is a perpendicular bisector of $\overline{AB}$. If not, tell why not. *(Section 4.3)*

33. $A(-5, 1)$, $B(3, 5)$, $C(-2, 5)$, $D(0, 1)$
34. $A(-5, -1)$, $B(1, -9)$, $C(-2, -5)$, $D(3, -2)$

---

29. Plane *ABC* is parallel to plane *GHK*. The planes of the top and bottom kites are parallel to each other, since each is parallel to the plane of the middle kite.

30. $\overline{AB}$, $\overline{DE}$, and $\overline{GH}$ are all parallel. The plane that contains the three edges intersects the three parallel planes that contain the top, middle, and bottom kites, so the lines of intersection are parallel.

31. Sketches may vary. Examples are given.

a. sometimes
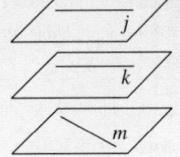

b. sometimes
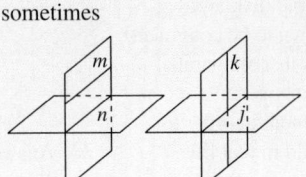

c. never
d. sometimes

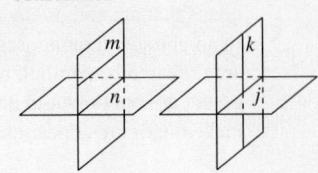

e. always
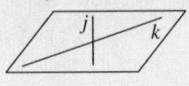

f. always
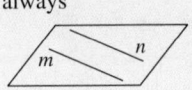

32. x = 36; y = −5
33. Yes.
34. No; $\overleftrightarrow{CD}$ is not perpendicular to $\overline{AB}$, nor does $\overleftrightarrow{CD}$ intersect $\overline{AB}$ at its midpoint.

5.7 Constructing Parallels and Perpendiculars

Learn how to...
- complete compass-and-straightedge constructions

So you can...
- construct lines parallel or perpendicular to given lines

Measuring tools such as rulers and protractors can be used to draw precise geometric figures. Geometric figures can also be *constructed* using a compass and a straightedge (a ruler with no marks). A compass is used to construct circles and *arcs*, unbroken parts of circles. A straightedge is used to construct lines, rays, and segments.

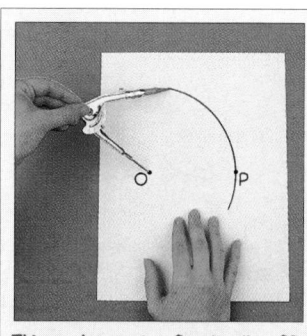

This arc has center *O* and radius *OP*.

CONSTRUCTION

CONGRUENT ANGLES

Given an angle, construct an angle congruent to it.

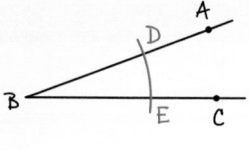
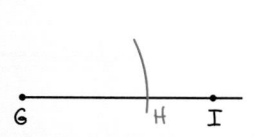

1. Using any radius and center *B*, draw an arc that intersects $\overrightarrow{BA}$ at *D* and $\overrightarrow{BC}$ at *E*. Draw a ray $\overrightarrow{GI}$. Using the same radius, draw an arc with center *G*. Label the intersection *H*.

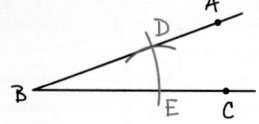
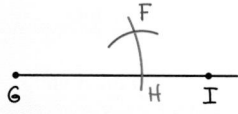

2. Open the compass to distance *DE*. Draw an arc with this radius and center *H* that intersects the arc from Step 1. Label the intersection *F*.

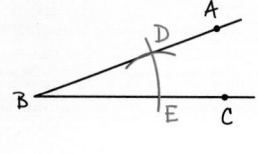
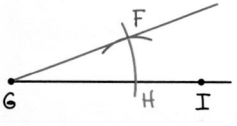

3. Use your straightedge to draw $\overrightarrow{GF}$.
$\angle FGH \cong \angle ABC$

5.7 Constructing Parallels and Perpendiculars **263**

Plan⇔Support

Objectives
- Complete compass-and-straight-edge constructions.
- Construct lines parallel or perpendicular to given lines.

Recommended Pacing
❖ **Core and Extended Courses**
  Section 5.7: 1 day
❖ **Block Schedule**
  Section 5.7: $\frac{1}{2}$ block
  (with Section 5.6)

Resource Materials
Lesson Support
Lesson Plan 5.7
Warm-Up Transparency 5.7
Practice Bank: Practice 37
Study Guide: Section 5.7
Explorations Lab Manual:
  Diagram Master 2
Challenge Problem: Set 37
Assessment Book: Test 22
Technology
Technology Book:
  TI-92 Activity 5
Geometry Software
Internet:
  http://www.hmco.com

Warm-Up Exercises

Complete each statement.

1. When parallel lines are intersected by a transversal, corresponding angles are __?__.
congruent

2. If a pair of __?__ planes are intersected by a third plane, then the intersections are parallel lines. parallel

3. If two lines are perpendicular to a plane, then they are __?__.
parallel

4. A perpendicular bisector intersects a segment at its __?__.
midpoint

Learning Styles: Verbal

While compass-and-straightedge constructions are an excellent way for students to gain insight into geometric ideas, they can be difficult for students with a verbal learning style. These students may want to write directions for each construction. They can then use one index card for each construction, writing the name on one side and the directions with an illustration on the other side.

Section Notes

Communication: Reading
Have students read the first paragraph of this section in class. Discuss the difference between a drawing and a construction. Make certain that students understand that when doing a construction, they are not allowed to measure. In fact, point out that straightedges and compasses have no measuring scales on them.

Construction Note

In the construction on page 263, point out to students that $\angle ABC$ can be any given angle. To stress this point, and to extend the construction, have students draw a right angle and an obtuse angle and then construct a congruent angle for each of these angles. For the construction on this page, if students are having trouble constructing $\angle CJD$, suggest that they turn their papers so that $\overleftrightarrow{AJ}$ forms the horizontal part of the angle.

Additional Example

Given $\angle 1$, construct a pair of vertical angles that are congruent to angle $\angle 1$.

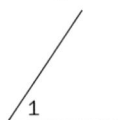

Step 1 Draw a line $\overleftrightarrow{PQ}$ and label a point R that is between points P and Q. Using the Congruent Angles Construction, construct $\angle QRS$, with side $\overrightarrow{RQ}$, that is congruent to $\angle 1$.

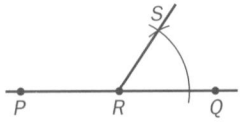

264

EXAMPLE

Given $\angle 1$ and $\angle 2$, construct an angle whose measure is equal to $m \angle 1 + m \angle 2$.

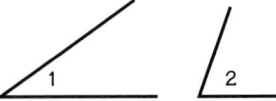

SOLUTION

Step 1 Draw a ray $\overrightarrow{QP}$. Using the Congruent Angles Construction, construct an angle with side $\overrightarrow{QP}$ that is congruent to $\angle 1$. Label it $\angle LQM$.

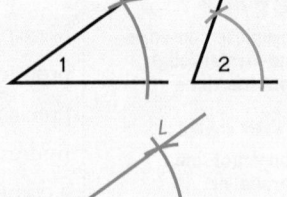

Step 2 Use the Congruent Angles Construction to construct an angle that is adjacent to $\angle LQM$ and congruent to $\angle 2$. Label it $\angle JQK$.

$$m \angle JQM = m \angle LQM + m \angle JQK$$
$$= m \angle 1 + m \angle 2$$

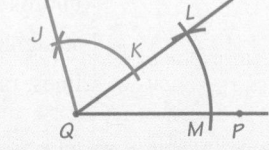

In the Example, the Congruent Angles Construction was used as part of a different construction. This is also true in the following construction.

CONSTRUCTION

PARALLEL LINES

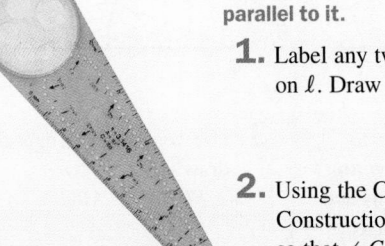

Given a line, construct a line parallel to it.

1. Label any two points A and B on ℓ. Draw $\overleftrightarrow{AJ}$.

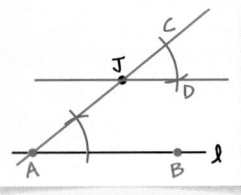

2. Using the Congruent Angles Construction, construct $\angle CJD$ so that $\angle CJD$ and $\angle JAB$ are congruent corresponding angles. $\overleftrightarrow{JD} \parallel \ell$.

CONSTRUCTION

PERPENDICULAR BISECTOR

Given a segment, construct its perpendicular bisector.

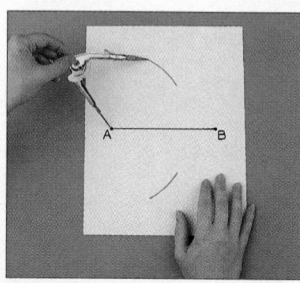 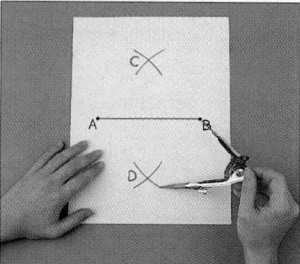 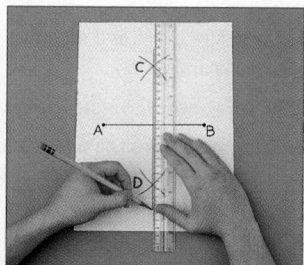

1. Using any radius greater than $\frac{1}{2}AB$, draw **two arcs with center** A, one on each side of $\overline{AB}$.

2. Using the same radius as in Step 1, draw **two arcs with center** B that intersect the arcs you drew in Step 1. Label the points C and D.

3. Draw $\overleftrightarrow{CD}$. $\overleftrightarrow{CD}$ is the perpendicular bisector of $\overline{AB}$.

THINK AND COMMUNICATE

1. In the Parallel Lines Construction, $\overleftrightarrow{JD}$ is constructed so that it is parallel to ℓ. Are there any other lines through J that are parallel to ℓ? Explain why or why not.

2. In the Perpendicular Bisector Construction, are there any other lines that contain the midpoint of $\overline{AB}$ and are perpendicular to $\overline{AB}$? Explain.

☑ CHECKING KEY CONCEPTS

Draw a figure like the one shown. Then complete the construction.

1. Construct an angle with side $\overrightarrow{TU}$ that is congruent to $\angle 1$.

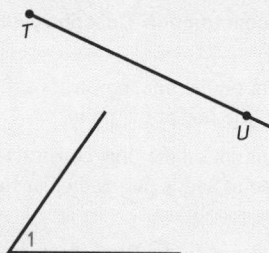

2. Construct the line through R that is parallel to m.

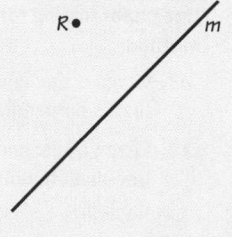

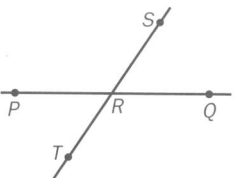

BY THE WAY

Constructions can be done using geometry software, paperfolding, some special graphing calculators, or a straightedge and compass, like the ones in this lesson.

3. Draw any segment $\overline{GH}$. Construct its perpendicular bisector.

5.7 Constructing Parallels and Perpendiculars **265**

ANSWERS Section 5.7

Think and Communicate

1. No; the Parallel Postulate states that through a point not on a given line, there is exactly one line parallel to the given line.

2. No; the definition of a perpendicular bisector of a segment states that it is a line, ray, or segment that bisects the segment and is perpendicular to it.

Checking Key Concepts

1–3. Check students' work. Students should use the indicated method.

1. the Parallel Lines Construction

2. the Perpendicular Bisector Construction

3. the Congruent Angles Construction

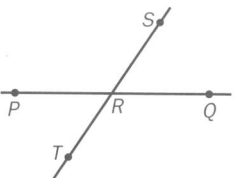

Additional Example (continued)

Step 2 Construct $\overrightarrow{RT}$ by extending $\overrightarrow{RS}$. $\angle PRT$ is congruent to $\angle QRS$ because they are vertical angles. $\angle PRT$ and $\angle QRS$ are vertical angles that are congruent to $\angle 1$.

Section Note

Construction Note

In the Perpendicular Bisector Construction, ask students why the radius must be greater than $\frac{1}{2}AB$. (If it were less, the two arcs would not intersect.) This construction is based on the theorem that any point on the perpendicular bisector of a segment is equidistant from the endpoints of the segment.

Think and Communicate

These questions help students to understand that if the steps of the construction are followed, there is only one possible way to construct a line parallel to a given line through a given point, or to construct a perpendicular bisector of a given segment.

Checking Key Concepts

Student Progress

These questions test students' ability to make the three constructions presented in this section. All students should be able to do these constructions correctly. If any student is having difficulty with a particular construction, refer him or her to the appropriate construction directions.

Closure Question

Describe how to do each construction presented in this section. For congruent angles, see page 263. For parallel lines, see page 264. For perpendicular bisector, see this page.

265

266

5.7 Exercises and Applications

Extra Practice
exercises on
page 670

1. Draw any obtuse ∠ *RST*. Construct an angle congruent to ∠ *RST*.

2. Draw any line *k* and a point *Q* not on *k*. Construct the line through *Q* that is parallel to *k*.

3. Draw a vertical segment and construct its perpendicular bisector.

Draw two angles like the ones shown. In Exercises 4–6, construct an angle with the indicated measure.

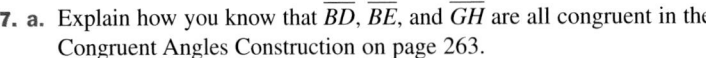

4. $(x + y)°$

5. $(x - y)°$

6. $(180 - 2y)°$

7. **a.** Explain how you know that $\overline{BD}$, $\overline{BE}$, and $\overline{GH}$ are all congruent in the Congruent Angles Construction on page 263.

   b. Writing Suppose you are given a segment, $\overline{XY}$. Explain how you could construct a segment congruent to $\overline{XY}$.

Connection ▶ PAPER FOLDING

Many constructions that can be done with a compass and straightedge can also be done using paper folding. Use paper you can see through. Every time you fold the paper, crease it and draw a dashed line along the crease.

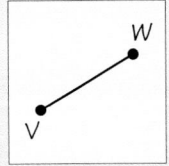 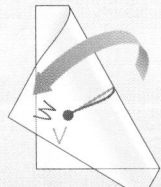 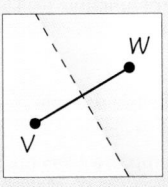

Paper folding can be used to make complex paper sculptures, such as this squid.

8. To construct the perpendicular bisector of $\overline{VW}$, fold the paper so that *W* meets *V*. The crease is the perpendicular bisector of $\overline{VW}$. How do you know that the crease is perpendicular to $\overline{VW}$? How can you tell that the crease intersects $\overline{VW}$ at its midpoint?

Use paper folding for each construction. Describe your method.

9. Given a line and a point on the line, construct a line that is perpendicular to the line at the point.

10. Given a line and a point not on the line, construct a line that contains the point and is perpendicular to the line.

11. Given a line and a point not on the line, construct a line that contains the point and is parallel to the line.

Suggested Assignment

❖ **Core Course**
Exs. 1–7, 12–14, 16, 18, 23–26, AYP

❖ **Extended Course**
Exs. 1–26, AYP

❖ **Block Schedule**
Day 31 Exs. 1–7, 12–14, 16, 18, 23–26, AYP

Exercise Notes

Student Progress

Exs. 1–6 Students generally enjoy doing constructions and should be successful with these basic ones. If you assign these exercises for class work, you can observe students' progress and provide assistance as needed.

Construction Note

Ex. 7 Part (a) of this exercise focuses students' attention on the fact that when you make two arcs with the same radius, the distances from each center to each arc are equal. This fact is then used to construct congruent segments in part (b).

Using Manipulatives

Exs. 9–11 These exercises give students an opportunity to do constructions using paper folding. These constructions make use of symmetry over the fold line. Figures that are symmetric over the fold line have congruent corresponding angles and segments.

Cooperative Learning

Exs. 9–11 You may wish to have students complete these exercises in small groups. Each group member should have a piece of paper and fold it to achieve the desired results. Students can compare their methods and help one another if any difficulties arise. Each member of the group should be able to do all of the constructions at the completion of the activity.

Exercises and Applications

1–4. Check students' work. Students should use the indicated method.

1. the Congruent Angles Construction

2. the Parallel Lines Construction

3. the Perpendicular Bisector Construction

4. the method illustrated in the Example on p. 264

5. Constructions may vary. In the example, $m∠PQR = x°$, $m∠SQR = y°$, and $m∠PQS = (x - y)°$.

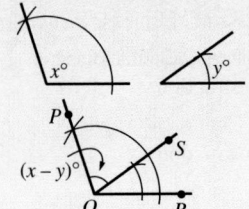

6. Constructions may vary. In the example, $m∠VWX = y° = m∠XWY$, and $m∠YWZ = (180 - 2y)°$.

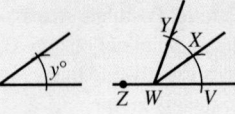

7. **a.** The same radius was used to draw the arcs used to locate points *E*, *D*, and *H*.

   b. Draw any ray, say $\overrightarrow{AB}$. Open the compass to the distance *XY*. Draw an arc with center *A* and radius *XY*

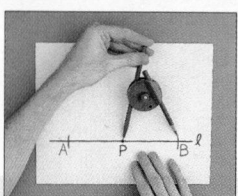

PERPENDICULAR LINES

In the construction on page 265, you learned how to construct the perpendicular bisector of a segment. Here are two related constructions. Only the first step of each construction is shown.

Given a line ℓ and a point P on ℓ, construct the line through P that is perpendicular to ℓ.

1. Using any radius, draw two arcs with center P that intersect ℓ at points A and B.

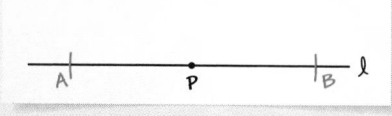

2. ?

Given a line g and a point Q not on g, construct the line through Q that is perpendicular to g.

1. Using any radius larger than the distance from Q to g, draw two arcs with center Q that intersect g at points M and N.

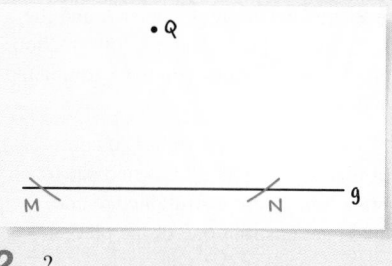

2. ?

12. a. Complete the construction on the left above by completing the missing step. (*Hint: P* is the midpoint of $\overline{AB}$.)

  b. Draw any line j and a point X on j. Construct the line through X that is perpendicular to j.

13. a. Complete the construction on the right above.

  b. Draw any line k and a point H not on k. Construct the line through H that is perpendicular to k.

14. Writing Describe a method for constructing parallel lines that is based on the Dual Perpendiculars Theorem. Use your method to construct two parallel lines. Do you prefer your method or the one on page 264?

15. **Technology** Here is a different method for the construction on the right above.

Step 1 Label any two points X and Y on g. Draw a circle with center X and radius XQ.

Step 2 Draw a circle with center Y and radius YQ. Label Z, the other point of intersection of the two circles.

Step 3 Draw $\overleftrightarrow{QZ}$.

  a. Repeat Exercise 13(b) using this method.

  b. Do you prefer this method or the method above? Explain.

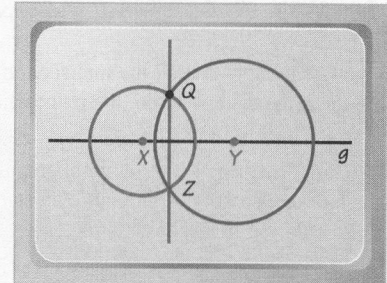

Exercise Notes

 Communication: Discussion *Exs. 12, 13* Prior to assigning these exercises, you may wish to discuss the beginning of each construction. Ask students how these constructions are different from the one for a perpendicular bisector.

Alternate Approach *Exs. 14, 15, 16* In Exs. 14 and 16, students develop alternate methods for constructing a parallel line that are based on the Dual Perpendiculars Theorem and the Converse of the Alternate Interior Angles Theorem. Ex. 15 leads students through an alternative method for constructing a perpendicular to a line through a given point not on the line.

Second-Language Learners *Exs. 14–17, 21* Since these exercises require a written response, you may want to consider assigning a peer tutor to each student learning English. Students can use their tutors as a resource if they encounter problems while writing their responses.

Spatial Reasoning *Ex. 15* The illustration uses full circles rather than arcs for the construction. Some students may wish to use this approach as they do their constructions.

Label the intersection point C. Draw $\overleftrightarrow{CP}$. $\overleftrightarrow{CP} \perp \overline{AB}$

  b. Check students' work.

13. a. Using M and N as centers and a radius greater than $\frac{1}{2}MN$, draw arcs that intersect below line g at a point R. Draw $\overleftrightarrow{QR}$. $\overleftrightarrow{QR} \perp g$

  b. Check students' work.

14. Answers may vary. An example is given. Choose a point P on a line l. Use the construction described in Ex. 12 to construct the line through P that is perpendicular to l. Choose a point Q on the perpendicular line and use the same construction to construct the line m perpendicular to $\overleftrightarrow{PQ}$ at point Q; $m \parallel l$. Check students' work. Preferences may vary.

15. a. Check students' work.

  b. Preferences may vary.

that intersects $\overrightarrow{AB}$ at point C. $\overline{AC} \cong \overline{XY}$

8. The crease forms congruent adjacent supplementary angles, so the crease is perpendicular to $\overline{VW}$. Since W was positioned to meet V, the crease divides $\overline{VW}$ into two congruent parts. The point of intersection is the midpoint of $\overline{VW}$.

9–11. Answers may vary. Examples are given.

9. Given a line t and a point P on t, fold the paper so that the crease goes through point P and the parts of line t on either side of P fold over on each other. The crease is perpendicular to t at P.

10. Given a line m and a point Q that is not on m, fold the paper so that the crease goes through point Q, and one part of line m is on top of the other part of line m. The crease is perpendicular to m and contains point Q.

11. Given a line m and a point Q that is not on m, fold the paper as described in Ex. 10 to construct the perpendicular line, n, from Q to m. Then fold the paper as described in Ex. 9 to construct the perpendicular line, t, to n at point Q. Since n is perpendicular to both m and t, $m \parallel t$.

12. a. Using any radius that is greater than $\frac{1}{2}AB$, draw two arcs on the same side of line l, one with center A and the other with center B.

Family Involvement
Exs. 17–22 Students may wish to involve family members in these construction exercises. Family members who have taken geometry may be a helpful reference with these more involved constructions.

Construction Note
Ex. 18 Ask students what figure they would get by drawing segments between consecutive points on the circle shown. (a regular hexagon)

Teaching Tip
Exs. 19–22 Before students begin these exercises, you may wish to have them develop a method for constructing a square using a compass and straightedge.

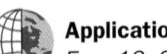

Application
Exs. 19–22 These exercises demonstrate how some common patterns in art can be created using compass-and-straightedge constructions.

16. a. the Converse of the Corresponding Angles Postulate; Since ∠ *CJD* and ∠ *JAB* are congruent corresponding angles formed by two lines and a transversal, the two lines are parallel.

   b. Answers may vary. An example is given. Refer to the diagram on page 264. Do Step 1, as shown. Then use the Congruent Angles Construction to construct ∠ *EJA* so that ∠ *EJA* and ∠ *JAB* are congruent alternate interior angles. $\overleftrightarrow{EJ} \parallel l$

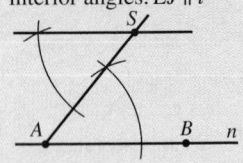

17. Answers may vary. Examples are given. Draw a line *l* and choose point *P* on *l*. Use the method of Ex. 12 to construct a line through *P* that is perpendicular to *l*. With any convenient radius, draw congruent arcs with center *P* intersecting both sides of one of the right angles formed. Label the intersection points *Q* and *R*. Draw △ *PQR*, which is an isosceles right triangle. *m* ∠ *PQR* = *m* ∠ *PRQ* = 45°. To construct a 135° angle, use the method of the Example on p. 264 to construct an angle whose measure is *m* ∠ *QPR* + *m* ∠ *PQR*.

16. a. In the Parallel Lines Construction on page 264, which postulate or theorem guarantees that $\overleftrightarrow{JD} \parallel \ell$? Explain.

   b. Describe a method for constructing parallel lines that uses the Converse of the Alternate Interior Angles Theorem. Then draw any line *n* and a point *S* not on *n*. Use your method to construct the line through *S* parallel to *n*.

17. **Challenge** Construct a 45° angle and a 135° angle. Explain why your methods work.

18. Follow these steps to construct a daisy pattern using only a compass.

   Step 1 Draw a circle with any radius and choose a point *A* on the circle. Using the same radius, draw an arc with center *A* that intersects the circle at *B* and *F*.

   Step 2 Draw an arc with center *B* and the same radius, and label the new intersection point *C*. Continue until you have completed the pattern.

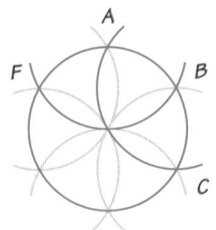

19. **ART** The pattern at the right is made up of nine square pieces. Given a square, describe how to use a straightedge and compass to construct one of the pieces.

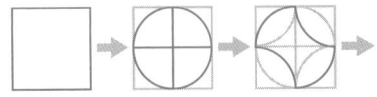

ART Many star patterns are based on a regular octagon.

20. Follow these directions to make a regular octagon. Start with a square. Draw the diagonals and locate their intersection. Using the intersection as the center, draw a circle that touches the square in four points. The eight intersection points shown are the vertices of a regular octagon.

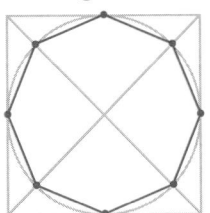

21. Explain how you know that all the intersection points in the diagram above are the same distance from the point where the diagonals intersect.

22. **Open-ended Problem** Starting with a regular octagon like the one you made in Exercise 20, design your own star pattern by connecting vertices and coloring in regions.

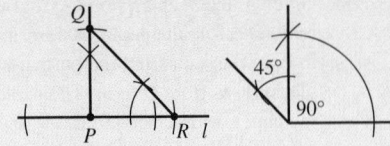

18. Check students' work.

19. Answers may vary. An example is given. Use the Perpendicular Bisector Construction to construct the perpendicular bisector of two adjacent sides of the square. Set your compass so its radius is half the length of a side of the square.

Using the upper left vertex and the lower right vertex of the square as centers, draw arcs connecting two midpoints of the sides, as shown. Shade the quarter-circles in the upper left and lower right corners of the square, and also shade the upper right and lower left corners of the square.

20. Check students' work.

21. The intersection points are all on a circle with center the point where the diagonals intersect.

22. Check students' work.

23. a. Check students' work. Students should use the Congruent Angles Construction and the Perpendicular Bisector Construction.

   b. Check students' work.

23. Cooperative Learning Work with another person. Draw an angle and a segment on a piece of paper. Then exchange papers with your partner.

  a. Construct an angle congruent to the angle you received and construct the perpendicular bisector of the segment you received.

  b. Exchange papers again. Check your partner's work by using a protractor and a ruler to measure the angles and segments.

SPIRAL REVIEW

Graph each system of inequalities. *(Toolbox, page 706)*

24. $x \le 4$
$\quad y \ge -2$

25. $y > x$
$\quad y > 0$

26. $x > 5$
$\quad x > -2$

ASSESS YOUR PROGRESS

VOCABULARY

line perpendicular to a plane (p. 258)
distance from a point to a plane (p. 258)

In the diagram, plane $P \parallel$ plane R
and plane $R \parallel$ plane Q. *(Section 5.6)*

  1. What does the Parallel Planes Theorem allow you to conclude?

  2. How are $\overline{MN}$ and $\overline{KL}$ related? Explain your reasoning.

  3. Suppose $m \angle NMF = 58°$. Find $m \angle MFL$ and $m \angle GHF$.

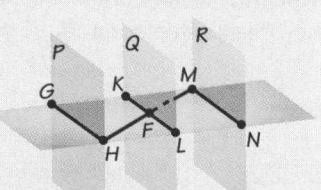

For Exercises 4–6, draw a figure like the one shown. Then complete the construction. *(Section 5.7)*

  4. Construct an angle congruent to $\angle L$.

  5. Construct the perpendicular bisector of $\overline{JL}$.

  6. Construct the line through P that is parallel to $\overleftrightarrow{JL}$.

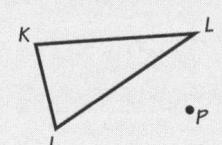

  7. Journal Compare the Parallel Planes Theorem to the Dual Parallels Theorem. Then compare the theorem in Example 2 on page 258 to the Dual Perpendiculars Theorem. How are the theorems alike? How are they different?

5.7 Constructing Parallels and Perpendiculars **269**

Exercise Notes

Assessment Note
Ex. 23 For this exercise, students work with a partner to assess their understanding of the material in this section. You might also have them construct a line parallel to the given segment.

Assess Your Progress

Journal Entry
This entry has students contrast some theorems about lines in a plane with some similar theorems about planes in space. Contrasting these two situations and noting the similarities and differences should help students gain insight into both.

Progress Check 5.6–5.7

See page 273.

Practice 37 for Section 5.7

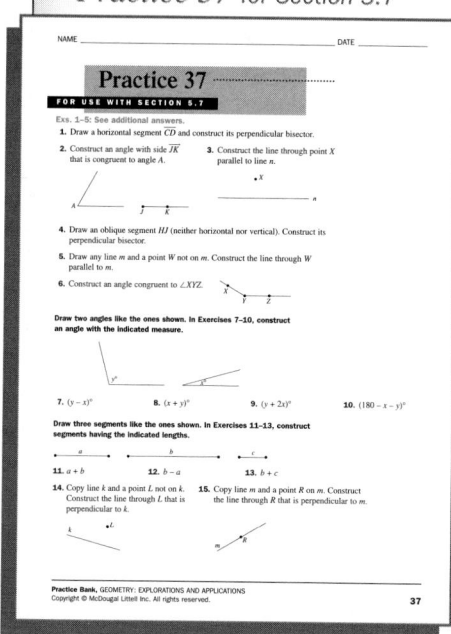

24.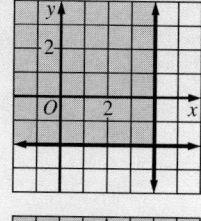

25.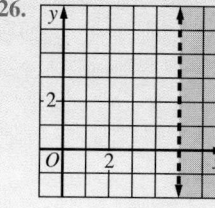

26.

third plane, the lines of intersection are parallel.

  3. 122°; 58°

  4–6. Checks students' work. Students should use the indicated method.

  4. the Congruent Angles Construction

Assess Your Progress

  1. plane $P \parallel$ plane Q

  2. They are parallel; if two parallel planes are intersected by a

  5. the Perpendicular Bisector Construction

  6. the Parallel Lines Construction

7. Answers may vary. An example is given. The Parallel Planes Theorem and the Dual Parallels Theorem are alike in that both allow you to conclude that two figures must be parallel if each of the figures is parallel to a third figure. The difference is that the Parallel Planes Theorem involves planes, whereas the Dual Parallels Theorem involves lines. The theorem in Example 2 on p. 258 is like the Dual Perpendiculars Theorem in that both allow you to conclude that two lines are parallel if the lines are both perpendicular to the same figure. The difference is that the theorem in the example involves lines perpendicular to a plane, whereas the Dual Perpendiculars Theorem involves lines perpendicular to a line. Also, the Duals Perpendiculars Theorem is restricted to figures all in one plane, and the theorem in the example involves figures that do not all lie in one plane.

- Understand the concepts of orthographic projection and isometric drawing.
- Make orthographic projections and isometric drawings of objects.

Planning

Materials
- paper and pencil
- ruler or tape measure
- isometric dot paper
- poster board

Project Teams
Students can work individually or in pairs to complete this project. If students work in pairs, each student should select two objects and make an orthographic projection and an isometric drawing of each object.

Guiding Students' Work

Some students may have difficulty picturing the different views of their object to make the orthographic projection. Suggest to these students that they physically look at their object from each view by moving either the object or themselves. For the isometric drawings, again have students look at their objects from the position in which they are drawing them. If students are still having difficulty, tell them to concentrate on one line in the drawing at a time.

Second-Language Learners

This project uses language that may be difficult for students learning English. You might want to define for them the following terms: *technical drawings*, *product design engineers*, *manufacture*, *industry*, *consumers*. Also, make sure that students understand the definitions and descriptions of isometric drawings, isometric axes, and isometric lines, since understanding these terms is necessary to complete the project.

PORTFOLIO PROJECT

Creating Technical Drawings

Did you know that many of the things you use every day began as technical drawings? Product design engineers use technical drawing to design and manufacture products for industry and for consumers.

PROJECT GOAL   Make orthographic projections and isometric drawings of objects.

Visualizing Three-Dimensional Objects

Orthographic projection and *isometric drawing* are two ways that are often used to represent three-dimensional objects.

ORTHOGRAPHIC PROJECTION
This orthographic projection shows three perpendicular views of an object. Each view shows a single side of the object.

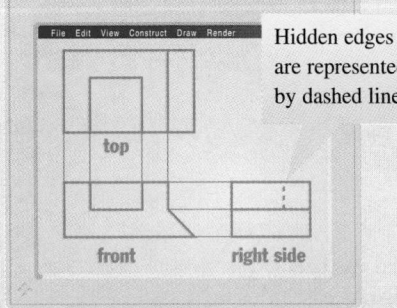

Hidden edges are represented by dashed lines.

Each isometric line is drawn parallel to an isometric axis.

isometric axes

Some lines are not isometric.

ISOMETRIC DRAWING This isometric drawing shows three sides of the object in one diagram. At one corner of the object, three edges meet at 90° angles. These edges are drawn on the *isometric axes*, which meet at 120°. A line on the object that is parallel to one of

270   Chapter 5 *Parallel Lines*

Making the Drawings

1. Choose two objects at your school or where you live that have simple shapes and have most of their edges on isometric lines. Use objects that are already made, such as a bookshelf or table, or create your own, by stacking simple pieces, such as wooden blocks.

2. Make an orthographic projection and an isometric drawing of each object as shown. Measurements along isometric lines should be approximately proportional to the actual distances they represent.

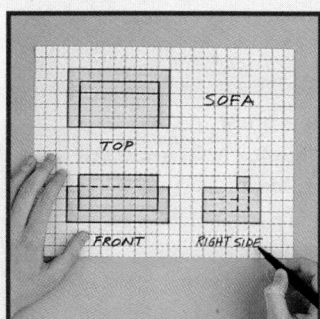 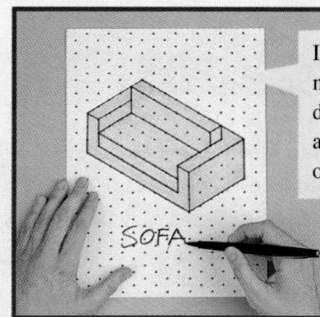

Isometric dot paper makes it easy to draw isometric lines and measure distances on them.

Displaying Your Drawings

Make a poster displaying the drawings you made of each object. Include a brief description or a photograph of each object. You can extend your project by exploring one of the following ideas.

- Draw an orthographic projection of a different object. Exchange drawings with another person and make isometric drawings from each others' orthographic projections.

- Use a computer and CAD (Computer-Aided Design) software to make orthographic projections and isometric drawings.

Self-Assessment

What did you find challenging about each type of drawing? Which type
of drawing was easier for you to make? Explain your answers.

Guiding Students' Work

Rubric for Chapter Project

4 Students make good selections of objects and correctly make the orthographic projection and the isometric drawing. Students make a creative poster and include an accurate description or photograph of each object. Students extend the project in one of the two ways listed and write their ideas clearly in the self-assessment of the project. The students' work is well organized, accurate, and demonstrates a thorough understanding of the mathematics of the project.

3 Students make good selections of objects and make the orthographic projection and the isometric drawing, but one or two mistakes are made in the drawings. Students create a well-designed poster and include descriptions or photographs of each object. Students extend the project in one of the two ways listed but make some mistakes or show minimal effort in doing so. Students also complete a self-assessment of the project. It is evident from the students' work that most of the mathematical ideas of the project are understood.

2 Students make selections of objects and attempt to draw both orthographic projections and isometric drawings but make several mistakes in their work. Students create a display, but it does not contain a brief description or a photograph of each object. If students make an attempt at extending the project, it is incomplete. From the students' work, it is clear that the mathematical ideas of the project are not fully understood. Students provide a self-assessment, but it does not completely answer the questions in the project.

1 Students make selections of objects but do not complete the orthographic projection or the isometric drawing. If students put together a poster of their work, it is incomplete and poorly done. No attempt is made to extend the project or to provide a self-assessment of the work. Students should be encouraged to speak with the teacher as soon as possible to review their work and to make a new start on the project.

Progress Check 5.1–5.2

Use the diagram.

1. List all four pairs of same-side interior angles in figure QRST. (Section 5.1) $\angle Q$ and $\angle R$, $\angle R$ and $\angle S$, $\angle S$ and $\angle T$, $\angle T$ and $\angle Q$

2. Which of the pairs of angles in figure QRST are supplementary? (Section 5.1) $\angle R$ and $\angle S$, $\angle T$ and $\angle Q$

3. Name the bases and legs of trapezoid QRST. (Section 5.2) bases: $\overline{QR}$ and $\overline{TS}$; legs: $\overline{QT}$, $\overline{RS}$

Find the measure of each numbered angle. (Section 5.2)

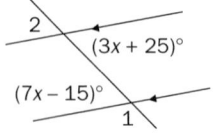

4. $\angle 1$ 125°

5. $\angle 2$ 55°

Progress Check 5.3–5.5

1. List four types of proof. (Section 5.3) paragraph proof, two-column proof, flow proof, and key steps of a proof

5 Review

STUDY TECHNIQUE

Without looking at the book, make a list of all the important postulates and theorems in this chapter. Make a sketch to illustrate each one. Then check each section to see if you missed any postulates or theorems.

VOCABULARY

transversal (p. 219)

same-side interior angles (p. 219)

alternate interior angles (p. 219)

corresponding angles (p. 219)

trapezoid (p. 228)

bases of a trapezoid (p. 228)

legs of a trapezoid (p. 228)

flow proof (p. 234)

distance from a point to a line (p. 249)

line perpendicular to a plane (p. 258)

distance from a point to a plane (p. 258)

SECTIONS 5.1 and 5.2

If two parallel lines are intersected by a **transversal**, then:

- corresponding angles are congruent.
  Example: If $r \parallel s$, then $\angle 1 \cong \angle 2$.

- Same-side interior angles are supplementary.
  Example: If $r \parallel s$, then $m\angle 2 + m\angle 3 = 180°$.

- Alternate interior angles are congruent.
  Example: If $r \parallel s$, then $\angle 3 \cong \angle 4$.

A **trapezoid** is a quadrilateral with exactly one pair of parallel sides called **bases**. The other two sides are the **legs**.

SECTIONS 5.3, 5.4, and 5.5

When you write a proof, you can choose from several different formats.

A *paragraph proof* combines the most important statements and reasons into sentences that form a convincing argument.

A *two-column proof* includes all the necessary steps with the statements and reasons listed in two separate columns.

A *flow proof* uses arrows to connect numbered statements in a logical order.

A *key-steps proof* includes the most important parts of a proof in an informal format.

Two lines must be parallel if they are intersected by a transversal and:

- corresponding angles are congruent.
  Example: If $\angle 5 \cong \angle 6$, then $j \parallel k$.
- alternate interior angles are congruent.
  Example: If $\angle 5 \cong \angle 7$, then $j \parallel k$.
- same-side interior angles are supplementary.
  Example: If $m\angle 7 + m\angle 8 = 180°$, then $j \parallel k$.

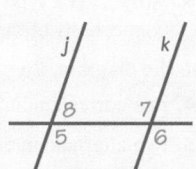

Also, two lines must be parallel if:

- they are coplanar and both are perpendicular to a third line.
- both lines are parallel to a third line.

The **distance from a point to a line** is length of the perpendicular segment from the point to the line.

SECTIONS 5.6 and 5.7

Two planes can either intersect or be parallel.

A **line is perpendicular to a plane** if the line intersects the plane and is perpendicular to every line in the plane that contains the point of intersection.

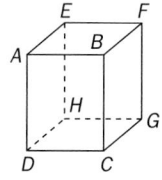

Line $\ell \perp$ plane S. The distance from point A to plane S is 5 units.

The **distance from a point to a plane** is the length of the perpendicular segment from the point to the plane.

A straightedge and a compass are used to make geometric diagrams called *constructions*. You can construct parallel and perpendicular lines, congruent angles, and perpendicular bisectors.

Use the diagram.

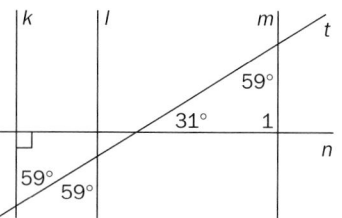

2. Explain how you know that $\angle 1$ is a right angle. *(Section 5.3)*
The sum of the measures of the angles of a triangle is 180°.

3. What can you say about lines k and l? Explain. *(Section 5.4)*
They are parallel because the alternate interior angles formed by transversal t are congruent.

4. What can you say about lines k and m? Explain. *(Section 5.5)*
They are parallel because they are both perpendicular to n.

5. Why is l parallel to m? *(Section 5.5)* Since k is parallel to both l and m, they are parallel to each other.

Progress Check 5.6–5.7

In the figure at the right, faces *AEHD* and *BFGC* lie in parallel planes.

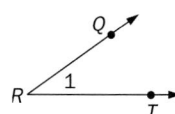

1. What can you say about $\overline{AE}$ and $\overline{BF}$? *(Section 5.6)* They are parallel.

2. What else would have to be known to say that $\overline{FG}$ and $\overline{BC}$ are parallel? *(Section 5.6)* Faces *ABCD* and *EFGH* lie in parallel planes.

3. Describe how to find the distance between a point and a plane. *(Section 5.6)* The distance from a point to a plane is the length of the perpendicular segment from the point to the plane.

Use the diagram.

4. Construct an angle congruent to $\angle 1$. *(Section 5.7)* Check students' work.

5. Construct the perpendicular bisector of $\overline{RT}$. *(Section 5.7)* Check students' work.

Chapter 5 Assessment
Form A Chapter Test

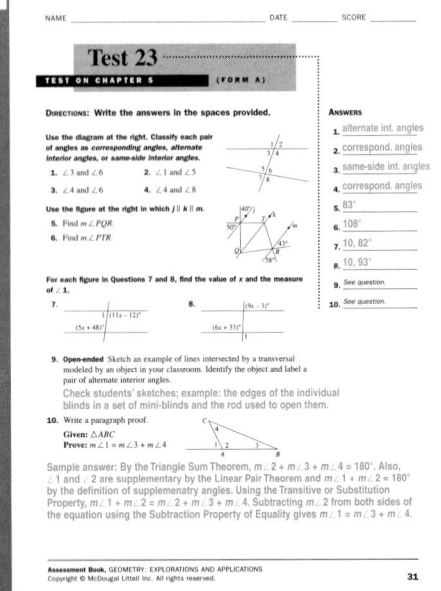

Chapter 5 Assessment
Form B Chapter Test

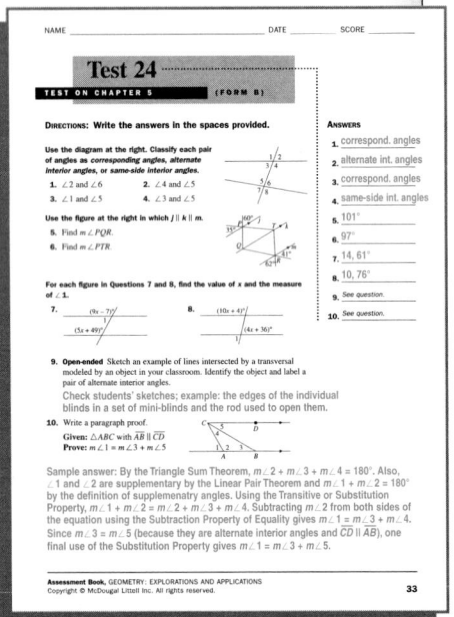

CHAPTER

5 Assessment

VOCABULARY REVIEW

Complete each statement.

1. A quadrilateral with exactly two parallel sides is called a(n) _?_ and the parallel sides are called the _?_ .

2. A(n) _?_ is a line that is coplanar with two other lines and intersects those lines.

3. A(n) _?_ is a type of proof that uses arrows to show the logical connections of the statements.

Use the diagram. Give an example of each of the following.

4. two corresponding angles

5. two alternate interior angles

6. two same-side interior angles

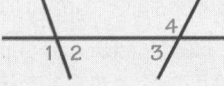

7. **Writing** Compare the definitions for the distance from a point to a line and the distance from a point to a plane. How are they alike? How are they different?

SECTIONS 5.1 *and* 5.2

ALGEBRA Find the value of each variable.

8.

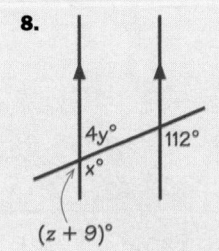

9.

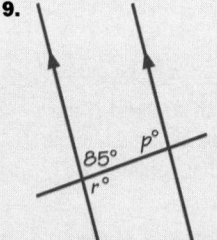

10.

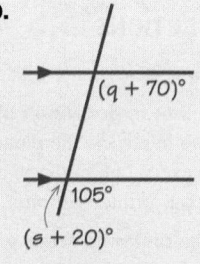

11. Sketch trapezoid *WXYZ* with bases $\overline{WX}$ and $\overline{YZ}$.

    a. What must be true about $\angle X$ and $\angle Y$? Justify your conclusion.

    b. Can you make any conclusions about $\angle Y$ and $\angle Z$? Explain why or why not.

274 Chapter 5 *Parallel Lines*

ANSWERS Chapter 5

Assessment

1. trapezoid; bases

2. transversal

3. flow proof

4. $\angle 1$ and $\angle 3$

5. $\angle 2$ and $\angle 4$

6. $\angle 2$ and $\angle 3$

7. Answers may vary. An example is given. Both are defined as the length of a perpendicular segment. There are no significant differences.

8. $z = 59$; $y = 17$; $x = 112$

9. $p = r = 95°$

10. $s = 55$; $q = 35$

11. **a.** $\angle X$ and $\angle Y$ are supplementary since $\overline{WX} \parallel \overline{YZ}$; if two parallel lines are intersected by a transversal, then same-side interior angles are supplementary.

    b. Yes; $\angle Y$ and $\angle Z$ are not supplementary. If they were, $\overline{WZ}$ and $\overline{XY}$ would have to be parallel and *WXYZ* would not be a trapezoid. (They may, however, be congruent.)

SECTIONS 5.3, 5.4, *and* 5.5

Use the diagram and this theorem: If two parallel lines are cut by a transversal, then the bisectors of two corresponding angles are parallel.

12. Write the Given and Prove statements.

13. Write a flow proof of the theorem.

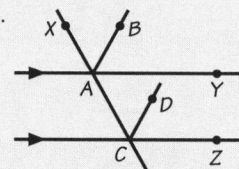

Use the diagram and the given information. State the postulate or theorem that would allow you to conclude that line ℓ and line m are parallel.

14. $\angle 1 \cong \angle 2$

15. $m\angle 3 = 90°$ and $m\angle 4 = 90°$

16. $\angle 2 \cong \angle 6$

17. $m\angle 3 + m\angle 5 = 180°$

18. $\ell \parallel n$ and $m \parallel n$

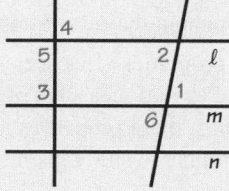

SECTIONS 5.6 *and* 5.7

19. In the diagram at the right, planes P and Q are parallel and lines j and k are parallel. Explain why lines k and ℓ must also be parallel.

20. Writing Suppose plane X is parallel to plane Y, and plane X is parallel to plane Z. Are planes Y and Z *always, sometimes*, or *never* intersecting planes? Explain.

21. Draw any line m and a point C not on m. Construct a line that is parallel to m and passes through point C.

PERFORMANCE TASK

22. Write a report that describes how each group of figures could intersect. Include sketches to illustrate each possibility.

- a line and a plane
- a line and two parallel planes
- two parallel lines and a plane
- a line and two intersecting planes

Assessment **275**

---

12. Given: $\overleftrightarrow{AY} \parallel \overleftrightarrow{CZ}$; $\overrightarrow{AB}$ bisects $\angle XAY$; $\overleftrightarrow{CD}$ bisects $\angle ACZ$.
Prove: $\overleftrightarrow{AB} \parallel \overleftrightarrow{CD}$

13. See answers in back of book.

14. If two lines are intersected by a transversal and alternate interior angles are congruent, then the lines are parallel.

15. In a plane, two lines perpendicular to the same line are parallel.

16. If two lines are intersected by a transversal and corresponding angles are congruent, then the lines are parallel.

17. If two lines are intersected by a transversal and same-side interior angles are supplementary, then the lines are parallel.

18. If two lines are both parallel to a third line, then they are parallel to each other.

19. j is parallel to l; thus, l must be parallel to k.

20. never; If two planes are both parallel to a third plane, then the two planes are parallel.

21. Check students' work.

22. Check students' work. Given a line and a plane, the line may be in the plane, parallel to the plane, or intersect it in one point. Given a line and two parallel planes, the line may intersect each plane in one point, it may be parallel to

both planes, or it may be in one of the planes. Given two parallel lines and a plane, the lines may be parallel to the plane, each may intersect the plane in a point, or the lines may be in the plane. Given a line and two intersecting planes, the line may be parallel to both planes, it may be parallel to one and intersect the other in a point, intersect each plane in one point, be in one of the planes, or in both (that is, be the intersection).

6 Conjectures about Triangles

OVERVIEW

Connecting to Prior and Future Learning

⇔ Students' study of triangles continues as they learn about triangle inequalities and congruent triangles. The postulates and theorems used to prove triangles congruent are the main focus of this chapter and will be used throughout the rest of this course.

⇔ Students also learn about special properties of isosceles triangles and the Isosceles Triangle Theorem. These concepts are used to find measures of angles and to write proofs. As with the congruence postulates and theorems, these properties and this theorem are important and students will use them as they continue to study geometry.

⇔ This chapter also introduces medians and altitudes of triangles. Students learn how to identify and draw these segments. They also explore the properties of these segments in triangles and study real-world applications of these properties.

Chapter Highlights

Interview with Madeleine Fleming: Geometric relationships are an important part of Madeleine Fleming's job as an optical physics specialist. Related exercises can be found on pages 290, 295, and 317.

Explorations focus on using geometry software to compare sides of triangles and using paper and scissors to investigate overlapping triangles.

The Portfolio Project: Pairs of students use properties of triangles to create a mobile. Students finish their work on this project by writing a brief description of their mobile.

Technology: The use of technology in Chapter 6 includes using geometry software in Section 6.1 to compare the lengths of the sides of a triangle and using the sine button on a calculator to find a distance.

OBJECTIVES

Section	Objectives	NCTM Standards
6.1	• Apply the Triangle Inequality Theorems. • Determine if a triangle can be formed from three given lengths. • Use the lengths of two sides of a triangle to describe the length of the third side.	1, 2, 3, 4, 5, 7
6.2	• Identify corresponding parts of polygons. • Match congruent polygons to real-world objects.	1, 2, 3, 4, 5, 7
6.3	• Use the SSS and SAS Postulates. • Prove triangles are congruent without proving all corresponding parts are congruent.	1, 2, 3, 4, 7
6.4	• Use the ASA Postulate, the AAS Theorem, and the HL Theorem. • Prove that triangles are congruent. • Use triangulation to find the positions of distant objects.	1, 2, 3, 4, 7
6.5	• Use corresponding parts of congruent triangles. • Write proofs that use congruent triangles to prove other statements. • Measure distances indirectly.	1, 2, 3, 4, 7
6.6	• Apply the Isosceles Triangle Theorem and its converse. • Find measures in isosceles triangles and write proofs using isosceles triangles.	1, 2, 3, 4, 5, 7
6.7	• Identify and draw medians and altitudes of triangles. • Explore the properties of medians and altitudes in triangles.	1, 2, 3, 4, 5, 7

Mathematical Connections	6.1	6.2	6.3	6.4	6.5	6.6	6.7
geometry	**279–284***	**285–291**	**292–298**	**299–305**	**306–312**	**313–318**	**319–325**
algebra		286, 289			310	315, 317, 318	322, 323, 325
data analysis, probability, discrete math	283						
logic and reasoning	282	289, 290	293, 294	300, 304	307–311	314, 316, 318	320, 322, 324

Interdisciplinary Connections and Applications

	6.1	6.2	6.3	6.4	6.5	6.6	6.7
biology and earth science		289					
chemistry and physics							323
arts and entertainment	280						
sports and recreation						316	
interior design	284						
astronomy				305			
architecture						316	
pottery, optics, rescue safety, automobiles, aerial photography	282		295, 297	303	310		

***Bold page numbers** indicate that a topic is used throughout the section.*

		opportunities for use with	
Section	**Student Book**	**Support Material**	
6.1	graphing calculator geometry software	**Technology Book:** Spreadsheet Activity 5 **Geometry Inventor Activity Book:** Activity 3	
6.3	geometry software	**Technology Book:** Calculator Activity 8 TI-92 Activity 6	
6.4	scientific calculator geometry software		
6.6	graphing calculator		
6.7	geometry software	**Technology Book:** Calculator Activity 9 **Geometry Inventor Activity Book:** Activity 6	

Regular Scheduling (45 min)

Section	Materials Needed	Core Assignment	Extended Assignment	exercises that feature		
				Applications	Communication	Technology
6.1	geometry software, ruler, uncooked spaghetti, small bag	**Day 1:** 1–20 **Day 2:** 21–24, 27, 28, 30–32, 39–44	**Day 1:** 1–20 **Day 2:** 21–44	26–28, 34–38	25, 26, 29–33	
6.2	scissors, graph paper	1–9, 12–15, 22–25, AYP*	1–25, AYP	17–21	11, 16, 23	
6.3	compass, straightedge, ruler, protractor, geometry software	1, 2, 4–11, 15–30	1–30	3, 13, 14	12–14, 18, 22b, 24	
6.4	ruler, protractor, compass, straightedge, calculator	1–10, 12–17, 20–26	1–26	11, 18, 19	17c, 18	19
6.5	compass, straightedge	1–9, 12–15, 17–25, AYP	1–25, AYP	10, 11	10, 14, 15	
6.6		2–5, 7–17, 19–21, 23–28	1–7, 9, 11, 13–28	1, 5, 6, 16–18	4, 6, 17, 18, 23	
6.7	scissors, compass, straightedge, cardboard, geometry software	1–7, 11–16, 18–21, AYP	1–21, AYP	8–10	10, 13–15	
Review/ Assess		**Day 1:** 1–12 **Day 2:** 13–24 **Day 3:** Ch. 6 Test	**Day 1:** 1–12 **Day 2:** 13–24 **Day 3:** Ch. 6 Test			
Port. Proj.		Allow 2 days.	Allow 2 days.			

Yearly Pacing (with Portfolio Project)	Chapter 6 Total 13 days	Chapters 1–6 Total 80 days	Remaining 80 days	Total 160 days

Block Scheduling (90 min)

	Day 34	Day 35	Day 36	Day 37	Day 38	Day 39	Day 40
Teach/Interact	Ch. 5 Test 6.1: Exploration, page 279	Continue with 6.1 6.2: Exploration, page 287	6.3: Construction, page 292 6.4	6.5: Construction, page 311 6.6	6.7 Review	Review Port. Proj.	Ch. 6 Test Port. Proj.
Apply/Assess	**Ch. 5 Test** **6.1:** 1–20	**6.1:** 21–24, 27, 28, 30–32, 39–44 **6.2:** 1–9, 12–15, 22–25, AYP*	**6.3:** 1, 2, 4–11, 15–30 **6.4:** 1–10, 12–17, 20–26	**6.5:** 1–9, 12–15, 17–25, AYP **6.6:** 2–5, 7–17, 19–21, 23–28	**6.7:** 1–7, 11–16, 18–21, AYP **Review:** 1–12	**Review:** 13–24 **Port. Proj.**	**Ch. 6 Test Port. Proj.**

NOTE: A one-day block has been added for the Portfolio Project—timing and placement to be determined by teacher.

Yearly Pacing (with Portfolio Project)	Chapter 6 Total $6\frac{1}{2}$ days	Chapters 1–6 Total 40 days	Remaining 40 days	Total 80 days

*AYP is Assess Your Progress.

Section	Practice Bank	Study Guide*	Assessment Book*	Visuals	Explorations Lab Manual	Lesson Plans	Technology Book
6.1	39	6.1		Warm-Up 6.1 Folder 7	Master 1	6.1	Spreadsheet Act. 5
6.2	40	6.2	Test 25	Warm-Up 6.2	Master 2	6.2	
6.3	41	6.3		Warm-Up 6.3	Master 2	6.3	Calculator Act. 8 TI-92 Act. 6
6.4	42	6.4		Warm-Up 6.4		6.4	
6.5	43	6.5	Test 26	Warm-Up 6.5		6.5	
6.6	44	6.6		Warm-Up 6.6	Add. Expl. 9	6.6	
6.7	45	6.7	Test 27	Warm-Up 6.7 Folder 8		6.7	Calculator Act. 9
Review Test	46	Chapter Review	Tests 28–30, Alternative Assessment			Review Test	

*__Spanish versions__ of _Study Guide_ and _Assessment Book_ are available.

Chapter Support

- Course Guides
- Lesson Plans
- Portfolio Project Book:
  Additional Project 3:
  Triangles in Architecture
- Preparing for College Entrance Tests
- Multi-Language Glossary
- _Test Generator_ Software
- Professional Handbook
- Challenge Problems

Software Support

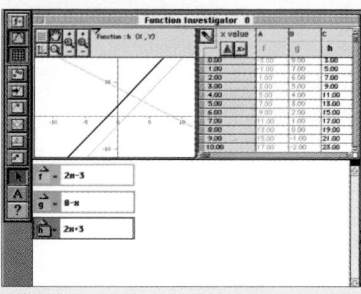

McDougal Littell Mathpack
Geometry Inventor

Internet Support

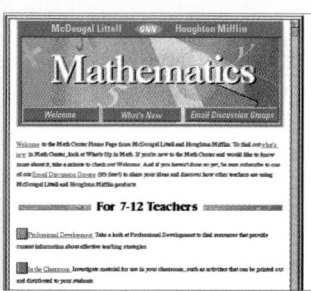

http://www.hmco.com
Next go to McDougal Littell; then the
Education Center; then Secondary Math.

Books, Periodicals

Bradell, Joseph L. "Helping Students Write Paragraph Proofs in Geometry." _Mathematics Teacher_ (October 1994): pp. 498–502.

Godbold, A. Landy, Jr. "Geometry, Iteration, and Finance." _Mathematics Teacher_ (November 1996): pp. 646–651.

Hirschhorn, Daniel B. "Why is the SSA Triangle-Congruence Theorem Not Included in Textbooks?" _Mathematics Teacher_ (May 1990): pp. 358–361.

Activities, Manipulatives

Exploring Geometry. Blackline masters for use with _The Geometer's Sketchpad._ "Construction: Angle Bisectors": p. 71. Berkeley, CA: Key Curriculum Press.

Software

Jackiw, Nicholas, designer. _The Geometer's Sketchpad._ (Ver. 3.0) for Macintosh or Windows. Berkeley, CA: Key Curriculum Press, 1995.

Laborde, Jean-Marie and Franck Bellemain, designers. _Cabri Geometry II._ Dallas, TX: Texas Instruments, 1994.

Internet

The U.S. Dept. of Education's On-Line Library includes a Technology Resource Guide at:
   http://www.ed.gov/
and also at:
   gopher ed.gov

Background

Optics

Optics is the branch of physics that studies light. There are two major branches of optics: physical optics and geometrical optics. Physicists who study physical optics are concerned with the nature of light and how it is emitted from different sources or bodies. They investigate the physical processes of light and how it is transmitted from place to place. Physicists who study geometrical optics are concerned with how light travels and how it is affected by different materials. There are numerous applications for the properties of light. For example, the discovery and development of microscopes, telescopes, eyeglasses, and mirrors all involve understanding how light behaves.

Madeleine Fleming

Madeleine Fleming is originally from New York State and works at the 3M Center in Minneapolis, Minnesota. She earned her doctorate in optics from the University of Rochester. At 3M, Fleming works on microreplication, which is the process of covering the surfaces of large sheets of plastic, rubber, and other moldable materials with microscopic ridges, pyramids, or other raised features. In addition to the road signs and light pipes, the microreplication process is also used in other products, such as a computer mouse pad that gives users pinpoint control of their cursors, and sandpaper that can be used to polish such things as golf clubs, hip implants, and other metal items. Fleming is part of 3M's Women in Science program.

6 Conjectures About Triangles

Shedding LIGHT on the Matter

INTERVIEW **Madeleine Fleming**

Madeleine Fleming uses geometry every day in her job as an optical physics specialist at the 3M Optics Technology Center in Minnesota. Fleming designs surfaces covered with microscopic structures—tiny bumps, ridges, indentations, and furrows—that bend and reflect light. "In designing these *microstructures*, I need to understand geometric relationships," Fleming, who has a Ph.D. in optics, says. "In particular, I need to find out what happens to a ray of light when it bumps into or passes through one of these microstructures."

"Math can help us understand the world around us, but we can also use it to build the world around us."

Illuminating Math

Fleming's job is not just about determining the precise shape, angle, and position of microstructures. It also involves people. As part of the Visiting Wizards and TECH programs, Fleming and other 3M scientists travel to elementary, junior high, and high schools, showing students how math and science are applied in the real world. Her advice for getting others interested in math is simple: "Look to the applications!"

> **"Math is interesting in its own right, but what is even more interesting is what you can do with it."**

Some people may not appreciate math for the sake of math, Fleming says, but they can appreciate what it can do and how it can be applied. "The reason I use math every day is because it's practical."

For example, geometry was used to design the microstructures on these light pipes. The pipe is smooth on the inside and grooved on the outside. This design causes light to travel long distances down the pipe, while allowing some light to leak out along the way. A similar surface is used in laptop computer screens to direct the screen's light toward the user.

277

Background

Heavy Traffic

In the United States, there are about 4,000,000 miles of streets, roads, and highways, and the estimated distance traveled by U.S. drivers is about 2 trillion miles per year. This heavy volume of traffic has led to a need for the development of devices that manage traffic flow in an efficient way. One such device is the detector, which is a magnet or switch placed in or on the road. When a car passes over a detector, the traffic light changes to let the car go through the intersection. Detectors can be used at intersections where it is beneficial to have the traffic itself dictate when the traffic lights change.

Second-Language Learners

You might want to clarify for students that the title, *Shedding Light on the Matter*, is a play on words. Explain that to *shed light* on something is a common English idiom that means "to make understandable or more obvious." Point out that the idiom is doubly appropriate in this interview, since the text "sheds light" on the work done by Madeleine Fleming, and she "sheds light" on different "matters" by designing light-reflective surfaces.

Multicultural Note

The first road signs were created and used by the ancient Romans, who paved 53,000 miles of road. To control the flow of traffic, the Romans invented stop signs, one-way streets, two-lane highways, and several other devices. Two thousand years later, the invention of the automobile required refinement in traffic control devices. Concerned with the increasing numbers of accidents, African American inventor Garrett Morgan came up with the idea of using electric light signals to indicate which cars should stop and which should go at an intersection. In 1923, he patented the first traffic light.

Mathematical Connection

In the process of microreplication, all kinds of triangles and triangular grooves are used to create different angles of reflection. Once the best angle measures are found for a particular project, the type of figure that best reflects the light is found and replicated, or reproduced, on the surface thousands or sometimes millions of times. In this way, the surface created contains a large number of structures that are congruent. Students investigate some of the surfaces created using microreplication in the exercises of this chapter. In Sections 6.2 and 6.6, students explore the surface of reflective film used for road signs and computer screens. In Section 6.2, students look at congruent polygons in the film, while in Section 6.6, they use the theorems and properties discussed in the chapter to explain how triangles in the reflective film are congruent and isosceles. In Section 6.3, students explore the surface of a light pipe and explain how two triangles that make up the surface can be said to be congruent.

Explore and Connect

Writing
A brief discussion of students' responses to this question can help them appreciate the value that mathematics and science bring to practical applications.

Research
You may wish to suggest that students also use a dictionary to look up the words *reflection* and *incidence*, then relate those meanings to the ones found in a physics textbook or encyclopedia.

Project
As an alternative to making a collage, students may also make a collection of objects that contain triangles. When students put together their display, have them organize the objects by the types of triangles they contain.

Knowing All the Angles

Street and highway signs, coated with a thin plastic film made by 3M, can be found all over the world. The back of the clear film contains about 7000 tiny pyramids per square inch. The base of each pyramid is an isosceles triangle. All these triangles are *congruent*, meaning that they are the same size and shape.

The light from the headlights of a car passes through the film and reflects inside the pyramids on the back. The pyramids are designed so that the light reflects back to the driver, no matter where the car is. "In determining the angles, we work out the geometry, figuring out where the sign is, where the headlights are, and where the driver is," Fleming explains.

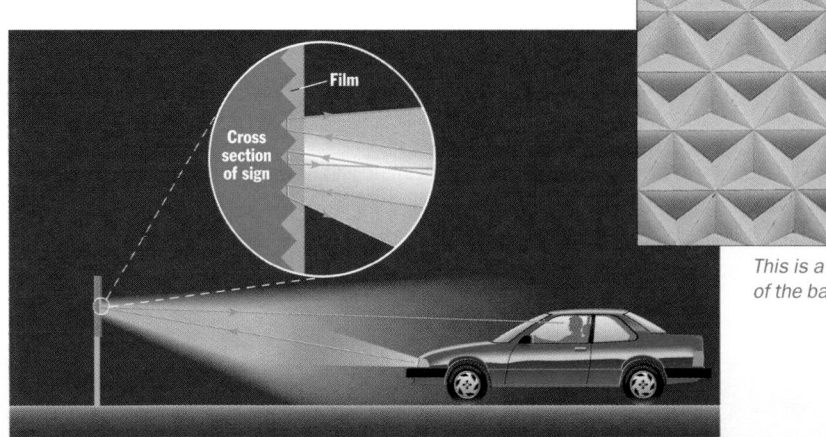

This is a photomicrograph of the back of the film.

Explore and Connect

Madeleine Fleming explains how microstructures are used to make thin lenses.

1. Writing Why do you think it is important for street signs to reflect light from car headlights back to the driver?

2. Research Using a physics textbook or an encyclopedia, look up the meanings of *angle of reflection* and *angle of incidence*. What do these terms mean? How are the angles related?

3. Project Make a collage showing many different triangles, including some congruent triangles. You could cut out illustrations from newspapers and magazines, or draw triangles you see around you. Indicate congruent triangles and any isosceles triangles.

Mathematics & Madeleine Fleming

In this chapter, you will learn more about how mathematics is related to optical microstructures.

Related Exercises

Section 6.2
• Exercises 18–21

Section 6.3
• Exercise 3

Section 6.6
• Exercises 16–18

278 Chapter 6 *Conjectures About Triangles*

6.1 Triangle Inequalities

Learn how to...
- apply the Triangle Inequality Theorems

So you can...
- determine if a triangle can be formed from three given lengths
- use the lengths of two sides of a triangle to describe the length of the third side

Can a triangle have sides of lengths 3 in., 4 in., and 5 in.? How about 3 in., 4 in., and 10 in.? If you know the lengths of two sides of a triangle, are there any limits on how long the third side can be? In the Exploration you will investigate the lengths of sides in triangles.

EXPLORATION
COOPERATIVE LEARNING

Comparing Sides of Triangles

Work with another student.
You will need:
- geometry software

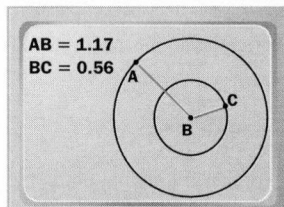

1 Construct two circles with the same center. Label the center *B*. Construct and label point *A* on the larger circle and point *C* on the smaller circle.

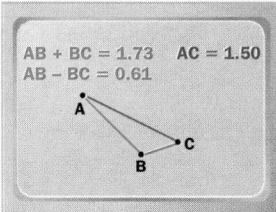

2 Construct and measure $\overline{AB}$ and $\overline{BC}$. Hide the circles so only $\angle ABC$ is showing.

3 Calculate $AB + BC$ and $AB - BC$.

4 Using a different type or color of line, construct $\overline{AC}$. Measure AC.

5 Compare *AC* with *AB + BC* as you move points *A* and *C* to different positions on the circles. Compare *AC* with *AB − BC*. What do you notice?

6 When *AC* is at its largest value, do $\overline{AB}$, $\overline{BC}$, and $\overline{AC}$ form a triangle? Explain why or why not. Do the segments form a triangle when *AC* is at its smallest value?

7 If segments $\overline{AB}$, $\overline{BC}$, and $\overline{AC}$ form a triangle, write two inequalities to describe *AC* in terms of *AB* and *BC*.

 Exploration Note

Purpose
The purpose of this Exploration is to have students discover how the lengths of the sides of a triangle are related.

Materials/Preparation
Each group needs geometry software.

Procedure
Students graph two concentric circles and draw a radius for each. The ends of the radii are connected to form a triangle. Students then calculate the sum and difference of the radii and compare these to the length of the third side. The points on the circle are moved and the length of the third side is compared

to the sum and difference of the lengths of the other two sides. Although the circles are hidden, students should realize that *AB* and *BC* remain constant because they are on the circles.

Closure
Students should see that the length of one side of a triangle is less than the sum of the lengths of the other two sides and greater than the difference of the lengths of the other two sides.

Explorations Lab Manual
See the Manual for more commentary on this Exploration.

For answers to the Exploration, see following page.

Plan⇔Support

Objectives
- Apply the Triangle Inequality Theorems.
- Determine if a triangle can be formed from three given lengths.
- Use the lengths of two sides of a triangle to describe the length of the third side.

Recommended Pacing
❖ **Core and Extended Courses**
Section 6.1: 2 days

❖ **Block Schedule**
Section 6.1: 2 half-blocks (with Chapter 5 Test and Section 6.2)

Resource Materials

Lesson Support
Lesson Plan 6.1

Warm-Up Transparency 6.1

Overhead Visuals:
    Folder 7: Triangle Inequality

Practice Bank: Practice 39

Study Guide: Section 6.1

Explorations Lab Manual:
    Diagram Master 1

Challenge Problems: Set 39

Technology
Technology Book:
    Spreadsheet Activity 5

Graphing Calculator

Geometry Software

McDougal Littell Mathpack
    Geometry Inventor Activity Book:
    Activity 3

Internet:
    http://www.hmco.com

Warm-Up Exercises

Complete with > or <.
1. 14 + 15 __?__ 28   >
2. 16 − 12 __?__ 15   <
3. 9 + 2 __?__ 17 − 3   >
Solve each inequality.
4. *x* + 3 < 5   *x* < 2
5. *y* − 6 > 2   *y* > 8

279

Alternate Approach

If geometry software is not available, the Exploration can be done using a hinged object. The object must be able to open to 180°. Place the hinge flat on a sheet of paper. Use the ends of the hinge to locate points *A* and *C*, and the joint of the hinge for point *B*. Join the three points to form a triangle. The hinge can be opened or closed to produce various triangles for which *AB* and *BC* are constant.

About Example 1

Teaching Tip

Recommend that students use a systematic approach when writing the inequalities in this type of problem. Point out that in each case, the numbers appear in the first inequality in the same order as they were originally given. Then, the numbers can be rotated in a pattern to form each of the other inequalities.

Additional Example 1

A gardener wants to make a triangular garden. He has 23 feet of fence with which to form the perimeter of the garden. Can the gardener make a garden with the given side lengths?

a. 3 ft, 5 ft, 15 ft

Use the first Triangle Inequality Theorem.
$3 + 5 \not> 15$
$5 + 15 > 3$ ✓
$15 + 3 > 5$ ✓
A triangular garden cannot be made using these side lengths.

b. 8 ft, 7 ft, 8 ft

$8 + 7 > 8$ ✓
$7 + 8 > 8$ ✓
$8 + 8 > 7$ ✓
A triangular garden can be made using these side lengths.

c. 3 ft, 10 ft, 10 ft

$3 + 10 > 10$ ✓
$10 + 10 > 3$ ✓
$10 + 3 > 10$ ✓
A triangular garden can be made using these side lengths.

In the Exploration, you probably noticed that *AC* is less than $AB + BC$ regardless of the shape of $\triangle ABC$. You may also have noticed how the length of $\overline{AC}$ varies with the measure of $\angle B$. For example, if $\angle B$, which is opposite $\overline{AC}$, is the largest angle of the triangle, then $\overline{AC}$ is the longest side.

> **Triangle Inequality Theorems**
>
> The sum of the lengths of any two sides of a triangle is greater than the length of the third side.
>
>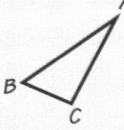
>
> In $\triangle ABC$, $AB + BC > AC$, $AB + AC > BC$, and $AC + BC > AB$.
>
> One side of a triangle is longer than a second side if and only if the angle opposite the first side is larger than the angle opposite the second side.
>
> In $\triangle ABC$, $AB > BC$ if and only if $m\angle C > m\angle A$.

THINK AND COMMUNICATE

1. In the Exploration, you probably discovered that $AC > AB - BC$. How is this inequality related to the first Triangle Inequality Theorem?

2. In $\triangle MHP$, $MH = 4.7$, $HP = 6.3$, $m\angle P < m\angle H$, and $m\angle H < m\angle M$. What can you conclude about the length *MP*?

EXAMPLE 1   **Application: Art**

A *stretcher* is the wooden frame that holds the canvas for a painting. The pieces of wood that form the stretcher are called *strips*. Can an artist make a stretcher from strips of the given lengths?

a. 9 in., 18 in., 7 in.   **b.** 10 in., 5 in., 15 in.   **c.** 14 in., 11 in., 8 in.

SOLUTION

Use the first Triangle Inequality Theorem.

a. $9 + 18 > 7$ ✔
$18 + 7 > 9$ ✔
$7 + 9 \not> 18$
These strips cannot form a triangle.

b. $10 + 5 \not> 15$
$5 + 15 > 10$ ✔
$15 + 10 > 5$ ✔
These strips cannot form a triangle.

c. $8 + 11 > 14$ ✔
$11 + 14 > 8$ ✔
$14 + 8 > 11$ ✔
These strips form a triangle.

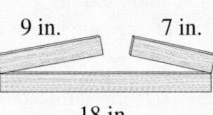

9 in.   7 in.   18 in.   10 in.   5 in.   15 in.

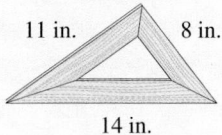

11 in.   8 in.   14 in.

ANSWERS Section 6.1

Exploration

Steps 1–4. Answers may vary. Check students' work.

5. Students should find that $AC < AB + BC$ and that $AC > AB - BC$ for every choice of points *A* and *C*.

6. No; *A*, *B*, and *C* are collinear. No; *C* is between *A* and *B*.

7. $AC < AB + BC$ and $AC > AB - BC$

Think and Communicate

1. It is equivalent to the inequality $AC + BC > AB$.

2. $4.7 < MP < 6.3$

EXAMPLE 2

In △PQR, PR = 15 ft and RQ = 12 ft.

a. What do you know about the length PQ?
b. Sketch three possible shapes of △PQR.

SOLUTION

a. Use the first Triangle Inequality Theorem.

$PQ + PR > RQ$	$PQ + RQ > PR$	$PR + RQ > PQ$
$PQ + 15 > 12$	$PQ + 12 > 15$	$15 + 12 > PQ$
$PQ > -3$	$PQ > 3$	$PQ < 27$

$\overline{PQ}$ is longer than 3 ft and shorter than 27 ft.

b. Three possible shapes of the triangle are shown.

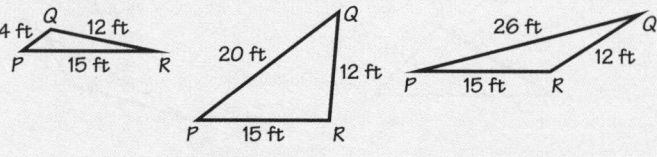

THINK AND COMMUNICATE

3. a. In Example 2, why can't $\overline{PQ}$ be one foot long?

    b. Maureen thinks that the solution to Example 2 should be "$\overline{PQ}$ is longer than -3 ft and shorter than 27 ft." What is wrong with Maureen's statement?

4. a. In Example 2, why can't $\overline{PQ}$ be three feet long?

    b. Why can't $\overline{PQ}$ be 27 ft long?

5. Write an inequality to describe the possible lengths of $\overline{PQ}$ in Example 2.

☑ **CHECKING KEY CONCEPTS**

Can a triangle be formed from sides of the given lengths?

  1. 5 ft, 5 ft, 5 ft   **2.** 15 in., 10 in., 12 in.

  3. 10 m, 10 m, 19.9 m   **4.** 5 cm, 13 cm, 6 cm

The lengths of two sides of a triangle are given. What can you conclude about the length of the third side?

  5. 13 ft, 19 ft   **6.** 9.9 cm, 10.1 cm   **7.** 8 m, 8 m

  8. In △ABC, $m \angle C > m \angle B$ and $m \angle B > m \angle A$.
    What do you know about the
    length of $\overline{AB}$?

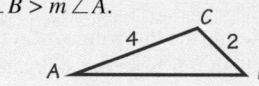

Think and Communicate

3. a. By the first Triangle Inequality, the length of $\overline{PQ}$ must be greater than 3 ft.

    b. Part (a) of the solution gives three inequalities, all of which must be true. One of these is $PQ > 3$.

4. a. By the first Triangle Inequality, the length of $\overline{PQ}$ must be greater than 3 ft.

    b. By the first Triangle Inequality, the length of $\overline{PQ}$ must be less than 27 ft.

5. $3 < PQ < 27$

Checking Key Concepts

1. Yes.

2. Yes.

3. Yes.

4. No.

5. It is between 6 ft long and 32 ft long.

6. It is between 0.2 cm long and 20.0 cm long.

7. It is less than 16 m long, and like every length, greater than 0 m long.

8. It is greater than 4 and less than 6.

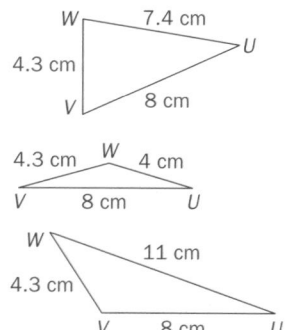

Suggested Assignment

❖ **Core Course**
Day 1 Exs. 1–20
Day 2 Exs. 21–24, 27, 28,
30–32, 39–44

❖ **Extended Course**
Day 1 Exs. 1–20
Day 2 Exs. 21–44

❖ **Block Schedule**
Day 34 Exs. 1–20
Day 35 Exs. 21–24, 27, 28,
30–32, 39–44

Exercise Notes

Using Technology
Exs. 1–5 Students who have experience writing programs for a graphing calculator can be challenged to write one that can be used to perform these exercises. When given three numbers, the program should determine if the numbers can be the lengths of the sides of a triangle.

Problem Solving
Exs. 1–12 Before beginning these exercises, you may wish to suggest that students write a procedure that can be used to solve each type of problem: one procedure when the three sides are given, and one when two of the three sides are given.

Common Error
Exs. 13–18 Some students may make the error of assuming information about a figure based upon its appearance. Remind these students that they cannot assume that a particular side or angle of a triangle is the largest or smallest from the appearance of the figure itself. This information must be specified.

Geometric Thinking
Ex. 20 Remind students that to demonstrate that a statement is true requires a proof, while a statement can be shown to be false by giving a single counterexample.

Spatial Reasoning
Exs. 21–24 Students should graph these points on a coordinate plane. They should then realize that the only time three given points do not form a triangle is when they are collinear.

6.1 | Exercises and Applications

Extra Practice exercises on page 670

Tell whether or not a triangle can be formed from sides of the given lengths.

1. 2 in., 7 in., 8 in.
2. 33 cm, 12 cm, 21 cm
3. 13 ft, 14 ft, 15 ft

4. 17 m, 23 m, 7 m
5. 55 ft, 34 ft, 21 ft
6. $x, x, 3x$

The lengths of two sides of a triangle are given. What can you conclude about the length of the third side?

7. 4 ft, 3 ft
8. 7 in., 1 in.
9. 9 cm, 12 cm

10. 10.7 m, 21.8 m
11. $7\frac{1}{4}$ in., $6\frac{3}{4}$ in.
12. $x, x + 4$

What can you conclude about the length of $\overline{AB}$?

13.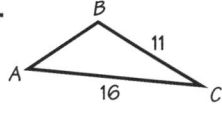

14. $m\angle C > m\angle A$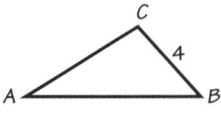

15. $m\angle A < m\angle C$ and $m\angle C < m\angle B$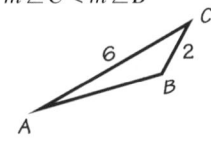

16. $m\angle C < m\angle B$ and $m\angle B < m\angle A$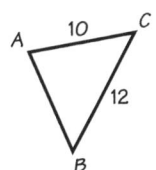

17. $m\angle B < m\angle A$ and $m\angle A < m\angle C$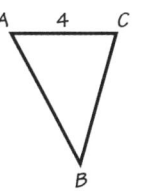

18. $m\angle B > m\angle A$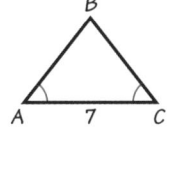

19. SAT/ACT Preview In △MNP, if $m\angle 1 < m\angle 2$, then: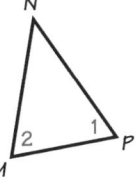
A. $MN > NP$
B. $MN < NP$
C. $MN = NP$
D. Relationship cannot be determined.

20. Logical Reasoning Is this statement *True* or *False*? If it is false, give a counterexample.
> The sum of the measures of two angles of a triangle is always greater than the measure of the third angle.

Can the three points be the vertices of a triangle? If not, explain why not.

21. $D(1, 6), E(1, 9), F(14, 8)$
22. $G(7, -3), H(0, -3), I(-3, -3)$

23. $J(1, 1), K(5, 5), L(-14, -14)$
24. $M(0, 0), N(3, 2), P(-231, 782)$

25. The university is 1 mi from the plaza, and the plaza is $\frac{1}{4}$ mi from the market place. What do you know about the distance from the university to the market place? Explain.

26. POTTERY In her studio, Tammy wants the throwing wheel to be 3 ft from the sink, and the table to be both 5 ft from the sink and 4 ft from the wheel. Is this arrangement possible? Explain why or why not.

Exercises and Applications

1. Yes.
2. No.
3. Yes.
4. Yes.
5. No.
6. No.
7. It is between 1 ft long and 7 ft long.
8. It is between 6 in. long and 8 in. long.
9. It is between 3 cm long and 21 cm long.
10. It is between 11.1 m long and 32.5 m long.

11. It is between $\frac{1}{2}$ in. long and 14 in. long.
12. It is between 4 units long and $2x + 4$ units long.
13. $5 < AB < 27$
14. $AB > 4$
15. $4 < AB < 6$
16. $2 < AB < 10$
17. $AB > 4$
18. $3.5 < AB < 7$
19. B

20. False. For example, the sum of the measures of the acute angles of a right triangle is equal to the measure of the right angle. It is also possible for the sum of the measures of two angles to be less than the measure of the third angle. Consider an isosceles triangle with a 100° vertex angle and 40° base angles.

21–24. Answers may vary. Examples are given.

Connection · PROBABILITY

In Exercises 27–29, you will estimate the experimental probability of forming a triangle from three segments of random lengths.

> **Toolbox p. 710**
> *Probability*

Cooperative Learning Work in a group of four people. You will need a ruler, 16 whole pieces of uncooked spaghetti, and a small bag.

Step 1 Write the lengths "1 cm," "2 cm," . . . , "24 cm" on twenty-four small pieces of paper and put them in the bag.

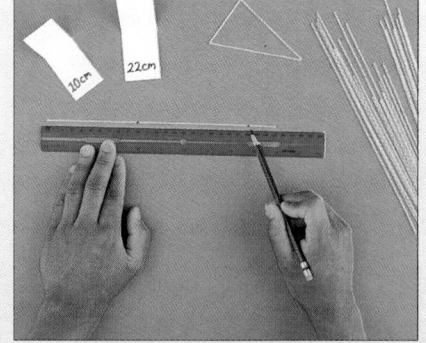

Step 2 Shake the bag to mix the pieces of paper. Decide where to break a piece of spaghetti by drawing two numbers from the bag. Measure both distances from the same end of the spaghetti and break the spaghetti at each location. Put the numbers back in the bag.

27. Each person should do Step 2 three times. Record the lengths of the spaghetti pieces and whether or not they form a triangle.

28. Share your results from Exercise 27. Use the formula

Probability = $\dfrac{\text{Total number of successes}}{\text{Total number of tries}}$ to estimate the probability of forming a triangle from a piece of broken spaghetti.

29. Writing Explain why you should use the pieces of paper to decide where to break the spaghetti instead of breaking the spaghetti wherever you like.

Cooperative Learning For Exercises 30–32, work in a group of three people. For Exercise 30, one of you should do part (a), another should do part (b), and the third should do part (c). Work together on Exercises 31 and 32. You will need a ruler.

30. Try to draw a quadrilateral with sides of the given lengths. Is it possible? Explain why or why not.

a. 2 in., 3 in., 4 in., and 9 in.   **b.** 2 in., 3 in., 4 in., and 12 in.   **c.** 2 in., 3 in., 4 in., and 8 in.

31. Writing Compare your answers to parts (a), (b), and (c) in Exercise 30. Make a conjecture about the relationship between the lengths of the sides of a quadrilateral.

32. Three sides of a quadrilateral are 3 cm, 3 cm, and 3 cm long. Describe all the possible lengths of the fourth side.

33. Challenge Is it possible for the perimeter of △*WBF* to be 24? Explain why or why not.

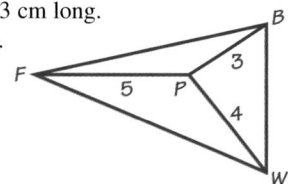

6.1 Triangle Inequalities   **283**

---

21. Yes; the three points are not collinear.

22. No; all three points lie on the horizontal line $y = -3$.

23. No; all three points lie on the line $y = x$.

24. Yes; the three points are not collinear.

25. The distance from the university to the market place is more than $\frac{3}{4}$ mi and less than $1\frac{1}{4}$ mi, by the first Triangle Inequality.

26. Yes; the three positions will be the vertices of a right triangle, since $3^2 + 4^2 = 5^2$.

27. See answers in back of book.

28. Answers may vary. For example, using the data in Ex. 27, the probability of forming a triangle is $\frac{3}{12} = 0.25$.

29. It would be difficult for someone to break the spaghetti into pieces of random lengths.

30. a. No. If the first three segments are drawn end to end, without any angle between them, their total length is 9 in. This means that the length of the fourth side of the quadrilateral must be less than 9 in.

  b. No. See explanation given for part (a).

  c. Yes. See diagram.

  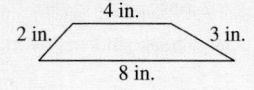

31. The length of the longest side of a quadrilateral is less than the sum of the lengths of the other three sides.

32. The length of the fourth side must be greater than 0 cm and less than 9 cm.

33. No; by the first Triangle Inequality, $BF < 8$, $FW < 9$, and $BW < 7$. Therefore, the perimeter of △*WBF* is less than 23.

Apply⇔Assess

Exercise Notes

Integrating Probability
Exs. 27–29 In these exercises, students calculate the experimental probability that three segments of random lengths form a triangle. Remind students that experimental probability is found by conducting a number of trials of an experiment, and then finding the ratio of the total number of successes to the total number of trials.

Second-Language Learners
Exs. 27–29 It may be helpful to second-language learners to discuss the concept of probability in a group or with the class. They should understand that *probability* is a topic in mathematics in which the chance of an event happening is expressed precisely with numbers. During the discussion, guide them to recognize *probable* and *probably* as related words that are used to talk about the chance, or likelihood, of something happening.

Cooperative Learning
Exs. 27–29 These exercises allow students to share their ideas and help to strengthen their understanding of the Triangle Inequality Theorems. Allow sufficient classroom time for students to complete the activities. Then ask one or more of the groups to report their findings to the class and discuss the results.

Exercise Notes

Application
Exs. 34–38 In these exercises, students see how the Triangle Inequality Theorems can be applied to design a kitchen.

Communication: Drawing
Ex. 38 This open-ended problem is a good opportunity for students who enjoy design to demonstrate what they have learned in this section. You may wish to display some of the best sketches on the bulletin board.

Assessment Note
Ex. 39 Having students create their own example is an excellent way to assess their understanding of the objectives of this section. Ask students to share their examples with the class and to describe how they thought of them.

Practice 39 for Section 6.1

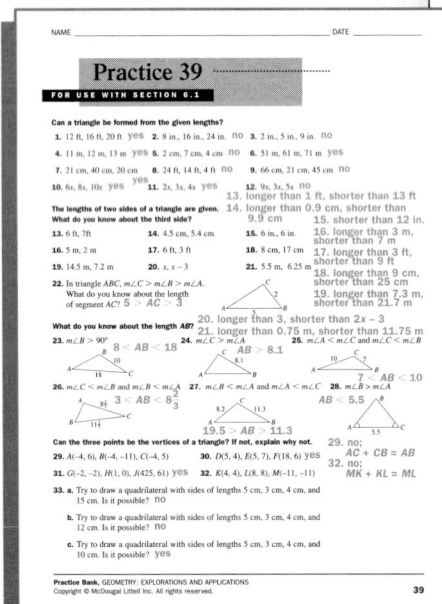

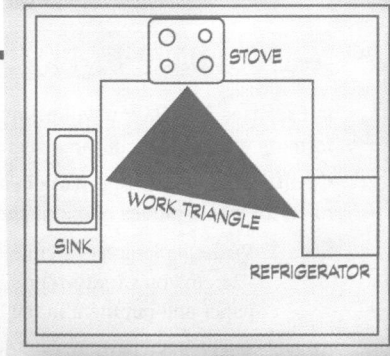

Connection   INTERIOR DESIGN

When designing a kitchen, architects look at the *work triangle* to see if the kitchen will be easy to use. The sides of the work triangle are the paths between the sink, the stove, and the refrigerator. The perimeter of a work triangle should be less than 22 ft and more than 12 ft.

34. What is the range of possible lengths for any side of the work triangle? How did you get your answer?

35. In addition to the guideline above, each side of the work triangle should be between 3.5 ft and 7 ft long. Suppose two sides of the triangle are each 3.5 ft long. What is the range of possible lengths for the third side?

36. Ideally, a work triangle should be equilateral. Give the side lengths of four work triangles that meet this condition as well as the conditions given above.

37. Research Measure the work triangle in a home kitchen. How well does it meet these guidelines?

38. Open-ended Problem Design a kitchen that meets the guidelines described above. Sketch the kitchen, including doorways and dimensions.

ONGOING ASSESSMENT

39. a. Open-ended Problem Give an example of three measurements that *cannot* be the lengths of the sides of a triangle.

   b. Give an example of three measurements that can be the lengths of the sides of a scalene triangle, $\triangle ABC$.

   c. In your triangle from part (b), which angle has the largest measure? the smallest measure?

SPIRAL REVIEW

Sketch each situation. *(Section 1.4)*

40. Points G, S, and N are collinear.   **41.** Line ℓ intersects plane R in point T.

Find the slope and y-intercept of each line. *(Section 4.2)*

42. $y = -3x + 7$ **43.** $y = 4$ **44.** $y = -2x$

34. Any side must have a positive length that is less than 22 ft. Answers may vary.

35. The third side must be more than 5 ft long and be less than 7 ft long.

36. Answers may vary. For example, all three sides could be 5 ft long, 5.5 ft long, 6 ft long, or 7 ft long.

37, 38. Check students' work.

39. Answers may vary. Examples are given.

   a. 8 mm, 10 mm, 20 mm

   b. $AB = 8$ mm, $BC = 10$ mm, and $AC = 14$ mm

   c. $\angle B$; $\angle C$

40.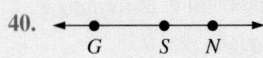

41.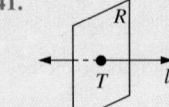

42. -3; 7

43. 0; 4

44. -2; none

6.2 Exploring Congruence

Plan⇔Support

Objectives

- Identify corresponding parts of polygons.
- Match congruent polygons to real-world objects.

Recommended Pacing

❖ **Core and Extended Courses**
Section 6.2: 1 day

❖ **Block Schedule**
Section 6.2: $\frac{1}{2}$ block
(with Section 6.1)

Resource Materials

Lesson Support
Lesson Plan 6.2
Warm-Up Transparency 6.2
Practice Bank: Practice 40
Study Guide: Section 6.2
Explorations Lab Manual:
    Diagram Master 2
Challenge Problems: Set 40
Assessment Book: Test 25

Technology
Internet:
    http://www.hmco.com

Learn how to...

- **identify corresponding parts of polygons**

So you can...

- **match congruent polygons such as windows and window frames**

The parts of a car come from many different sources. For all of the pieces to fit together, they must be exactly the right size and shape. For example, the window must match the window frame exactly.

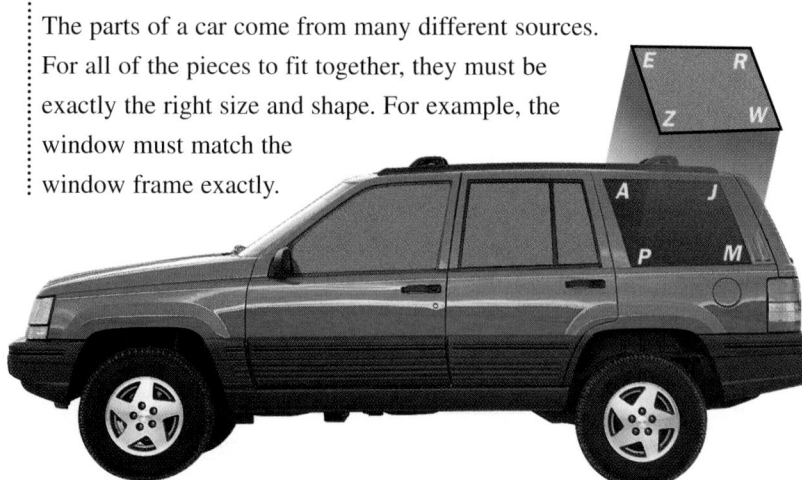

Corner *E* on the window *corresponds* to corner *A* in the car frame. Polygons *AJMP* and *ERWZ* are **congruent polygons** if each part (angle or side) of *AJMP* is congruent to the **corresponding part** of *ERWZ*. Congruent polygons have the same size and shape.

EXAMPLE 1

Are quadrilaterals *AJMP* and *ERWZ* congruent?

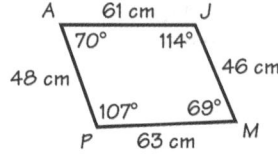
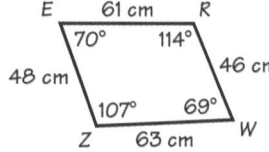

SOLUTION

Use the diagram to tell which pairs of corresponding parts are congruent.

$\angle A \cong \angle E$	$\angle M \cong \angle W$	$\overline{AJ} \cong \overline{ER}$	$\overline{MP} \cong \overline{WZ}$
$\angle J \cong \angle R$	$\angle P \cong \angle Z$	$\overline{JM} \cong \overline{RW}$	$\overline{PA} \cong \overline{ZE}$

All eight parts (four angles and four sides) of quadrilateral *AJMP* are congruent to the corresponding parts of quadrilateral *ERWZ*, so quadrilateral *AJMP* ≅ quadrilateral *ERWZ*.

WATCH OUT! ▶

Corresponding angles in congruent polygons are not the same as corresponding angles of parallel lines.

6.2 Exploring Congruence **285**

Warm-Up Exercises

Use quadrilateral *RDWZ*.

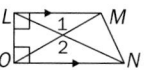

1. Identify the congruent sides.
$\overline{RD} \cong \overline{ZW}, \overline{RZ} \cong \overline{DW}$

2. Identify the congruent angles.
$\angle R \cong \angle W, \angle Z \cong \angle D$

Why are each pair of angles congruent in the figure below?

3. $\angle 1$ and $\angle 2$
Vertical angles are congruent.

4. $\angle MLO$ and $\angle LON$
All right angles are congruent.

5. $\angle LMO$ and $\angle MON$
When parallel lines are intersected by a transversal, alternate interior angles are congruent.

Teaching Tip
When discussing the *Watch Out!* on page 285, explain that the term *corresponding* means the two angles (or sides) are in the same relative position. In the case of parallel lines, corresponding angles are in the same position relative to the transversal and the two lines. In the case of polygons, corresponding sides or angles are in the same position relative to the other sides and angles. Point out that congruent polygons have the same size and same shape.

Additional Example 1

Are pentagons *STUVW* and *EFGHI* congruent?

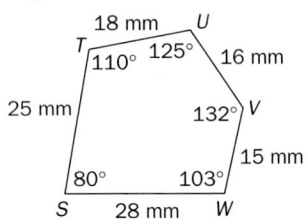

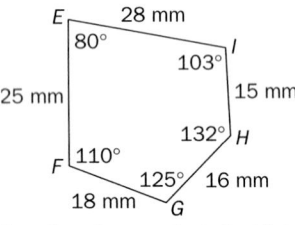

Use the diagram to tell which pairs of corresponding parts are congruent.

$\angle S \cong \angle E$	$\overline{ST} \cong \overline{EF}$
$\angle T \cong \angle F$	$\overline{TU} \cong \overline{FG}$
$\angle U \cong \angle G$	$\overline{UV} \cong \overline{GH}$
$\angle V \cong \angle H$	$\overline{VW} \cong \overline{HI}$
$\angle W \cong \angle I$	$\overline{WS} \cong \overline{IE}$

All ten parts (five angles and five sides) of pentagon *STUVW* are congruent to the corresponding parts of pentagon *EFGHI*.
STUVW ≅ *EFGHI*

Think and Communicate

It is essential for students to understand that when identifying congruent polygons, the vertices of one polygon must be listed in the same order as the corresponding vertices in the other polygon. In this way, the congruent parts of the polygons can be determined directly from the names. Questions 1–3 reinforce this idea.

When you identify congruent polygons, list the vertices of one polygon in the same order as the corresponding vertices of the other polygon. For example, quadrilateral *MPAJ* ≅ quadrilateral *WZER* indicates that *M* corresponds to *W*, *P* corresponds to *Z*, *A* corresponds to *E*, and *J* corresponds to *R*.

THINK AND COMMUNICATE

1. List all of the corresponding parts of quadrilaterals *JKLM* and *PRQS* if *JKLM* ≅ *QSPR*.

2. How else can you write *JKLM* ≅ *QSPR* so that it remains true?

3. If △*XYZ* ≅ △*UVW* is $\overline{ZX} \cong \overline{WU}$? Explain how you know.

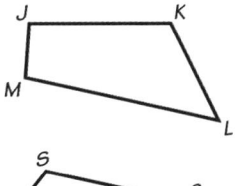

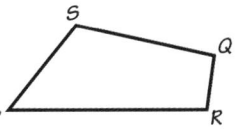

EXAMPLE 2 Connection: Algebra

polygon **MNOPQ** ≅ polygon **VUTSR**

a. Find the value of *x*.
b. Find the value of *d*.

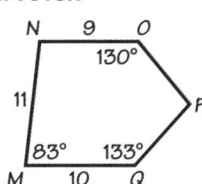

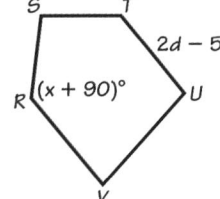

SOLUTION

Since *MNOPQ* ≅ *VUTSR*, the corresponding parts of *MNOPQ* and *VUTSR* are congruent.

a. $\angle R \cong \angle Q$ ∠*R* corresponds to ∠*Q*.

$$m\angle R = m\angle Q$$
$$(x + 90)° = 133°$$
$$x = 133 - 90$$
$$x = 43$$

b. $\overline{TU} \cong \overline{ON}$ $\overline{TU}$ corresponds to $\overline{ON}$.

$$TU = ON$$
$$2d - 5 = 9$$
$$2d = 14$$
$$d = 7$$

ANSWERS Section 6.2

Think and Communicate

1. $\overline{JK}$ and $\overline{QS}$, $\overline{KL}$ and $\overline{SP}$, $\overline{LM}$ and $\overline{PR}$, $\overline{JM}$ and $\overline{QR}$, ∠*J* and ∠*Q*, ∠*K* and ∠*S*, ∠*L* and ∠*P*, and ∠*M* and ∠*R*

2. Answers may vary. Examples are given. *KJML* ≅ *SQRP*, *JMLK* ≅ *QRPS*, and *LMJK* ≅ *PRQS*

3. Yes; vertex *Z* corresponds to vertex *W* and vertex *X* corresponds to vertex *U*, so $\overline{ZX}$ corresponds to $\overline{WU}$. Therefore, the segments are congruent.

Investigating Overlapping Triangles

Work with another student.
You will need:
• scissors

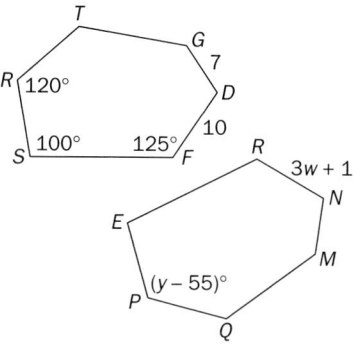

1 Fold in one corner of a piece of paper to form a triangle. Trace the triangle on the paper. What type of triangle is it?

2 Cut along the traced edges and along the fold. Explain why the two triangles are congruent.

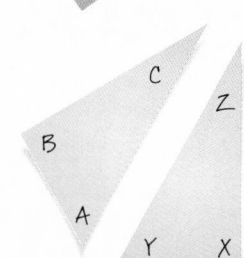

3 Label the vertices of each triangle as shown. Label each vertex on both sides of the paper. Complete: △BCA ≅ _?_ .

4 Use your two triangles to form each shape below. After you form the shape, sketch it, label the vertices, and name the congruent triangles.

Example:

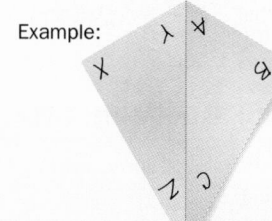

△BCA ≅ △XZY

5 What other shapes can you form with △ABC and △YXZ? Sketch and label at least five new shapes and name the congruent triangles in each one.

Exploration Note

Purpose
The purpose of this Exploration is to have students discover how two congruent triangles can be combined to form other shapes and, thus, to prepare students to find congruent triangles in diagrams associated with proofs.

Materials/Preparation
Each group of students needs a pair of scissors.

Procedure
Students use reflection to make a pair of congruent right triangles. They then arrange them to form different shapes and name the congruent triangles in each one.

Closure
Ask various groups to draw the five new shapes they formed in Step 5 on the board and to list the congruent triangles in each one. Discuss the results with the class.

Explorations Lab Manual
See the Manual for more commentary on this Exploration.

For answers to the Exploration, see answers in back of book.

Teach⇔Interact

Additional Example 2

hexagon *RTGDFS* ≅ hexagon *PQMNRE*

a. Find the value of *y*.
Since *RTGDFS* ≅ *PQMNRE*, the corresponding parts of *RTGDFS* and *PQMNRE* are congruent.
$$\angle P \cong \angle R$$
$$m\angle P = m\angle R$$
$$(y - 55)° = 120°$$
$$y = 120 + 55$$
$$y = 175$$

b. Find the value of *w*.
$$\overline{NR} \cong \overline{DF}$$
$$NR = DF$$
$$3w + 1 = 10$$
$$3w = 9$$
$$w = 3$$

Section Notes

Spatial Reasoning
Some students may have difficulty finding the corresponding parts of congruent triangles when the two figures overlap. To help students see the congruence, they can use a clear transparency sheet to trace over one triangle and label its vertices. The transparency can then be placed directly over the other triangle in such a way that they match. The correspondence can then be seen clearly. This method can be used for the shapes students make in the Exploration and for many of the exercises.

Family Involvement
You may wish to suggest that students do the Exploration with family members. After completing Step 5, they can then point out that designs found on linens, wallpaper, pottery, and clothing often contain congruent triangles. Students can then enlist family members to look for some examples at home. They can sketch the design, label the vertices, and name the congruent figures.

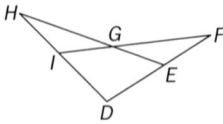

Additional Example 3

Name all of the triangles that appear to be congruent.

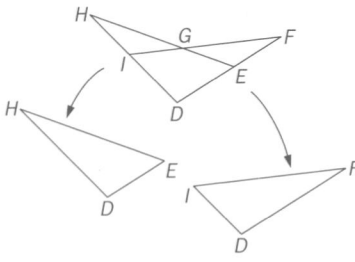

There are two triangles that appear to be congruent and do not overlap: △ HIG ≅ △ FEG.
There are two triangles that appear to be congruent and do overlap. Imagine sliding the figure apart.

△ HED ≅ △ FID
It appears that △ HIG ≅ △ FEG and △ HED ≅ △ FID.

Checking Key Concepts

Student Progress
You may wish to have students complete these questions individually and then check their work with a partner. Students should have a thorough understanding of the solutions to these problems before moving on to the exercises.

Closure Question

How can you use the names of congruent polygons to determine the congruent sides? The sides that are named using the first two letters in the name of each of the polygons are congruent. The sides named by the second and third letters in the name of each polygon are also congruent. Continue pairing the letters from the names of the polygons in order to find the names of other congruent sides. After using the last two letters, use the last and first letter of the name of each polygon to determine the final pair of congruent sides.

EXAMPLE 3

Name all of the triangles that appear to be congruent.

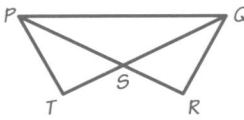

SOLUTION

There are two triangles that appear to be congruent and do not overlap:

$$\triangle PST \cong \triangle QSR$$

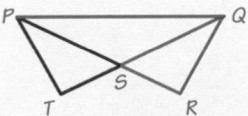

There are two triangles that appear to be congruent and do overlap, just like the triangles in the Exploration. Imagine sliding the figure apart:

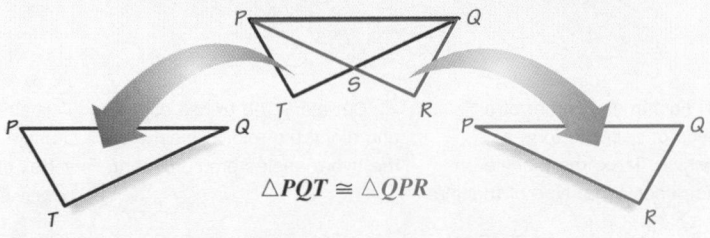

$$\triangle PQT \cong \triangle QPR$$

It appears that △PST ≅ △QSR and △PQT ≅ △QPR.

THINK AND COMMUNICATE

4. List all six pairs of corresponding parts for △PQT and △QPR in Example 3.

5. a. What part do △PQT and △QPR share?
   b. Which property tells you that this part is congruent to itself?

✓ CHECKING KEY CONCEPTS

1. If △GHJ ≅ △BCA, then △HGJ ≅ ? .

2. Name all of the triangles that appear to be congruent.

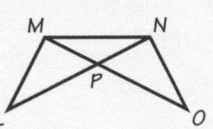

3. △UQR ≅ △VSW
   a. Find the value of y.
   b. Find the value of t.

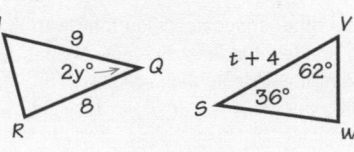

Think and Communicate

4. $\overline{PQ}$ and $\overline{QP}$, $\overline{QT}$ and $\overline{PR}$, $\overline{PT}$ and $\overline{QR}$, ∠PQT and ∠QPR, ∠T and ∠R, ∠QPT and ∠PQR

5. a. $\overline{PQ}$
   b. the Reflexive Property

Checking Key Concepts

1. △CBA

2. △LMP ≅ △ONP and △LMN ≅ △ONM

3. a. 18
   b. 5

Exercises and Applications

1. Yes; △FME ≅ △ELF.

2. No.

3. Yes; △WXY ≅ △WVU.

4. Yes; △CDH ≅ △GFH and △ECF ≅ △EGD.

5. $\overline{BC}$

6. m ∠ M = 48°

7. WC = 12

8. 2

6.2 Exercises and Applications

Extra Practice
exercises on
page 671

Are any polygons in the figure congruent? If so, name the congruent polygons.

1.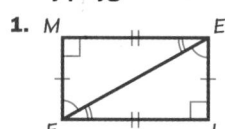

2.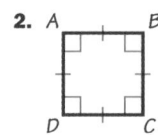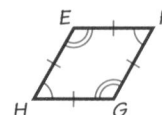

3.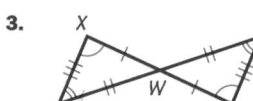

4.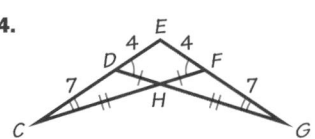

**For Exercises 5–7, △MPQ ≅ △WBC.
Complete each statement.**

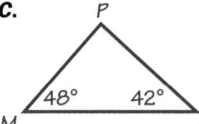 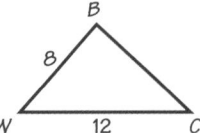

5. $\overline{PQ} \cong$?

6. $m \angle W =$?

7. $MQ =$?

ALGEBRA For Exercises 8–10, △WUN ≅ △TRE.

8. Find the value of x.

9. Find the value of y.

10. **Challenge** Find the value of z.

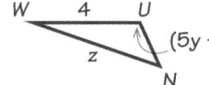 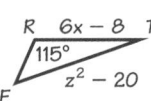

11. **Open-ended Problem** Draw two congruent rectangles on a
coordinate grid. Explain how you know that they are congruent.

**For Exercises 12–14, tell which triangle appears to be congruent to the
given triangle.**

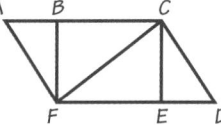

12. △ABF ≅ ?   13. △FBC ≅ ?   14. △AFC ≅ ?

15. Sketch a scalene obtuse triangle, △KXM. Sketch △NBE so that
△KXM ≅ △NBE.

   a. Is △XMK ≅ △BNE?

   b. Is △KMX ≅ △NEB?

16. **Writing** How many different ways can you write △JKL ≅ △MNP?
Explain how you found your answer.

17. **BIOLOGY** The Clara Satin moth, which lives in
Australia and Tasmania, belongs to the family
geometridae. The wings of a Clara Satin
moth are covered by a geometric pattern.
Sketch the moth using only straight lines.
Label each vertex and name at least six pairs
of polygons that appear to be congruent.

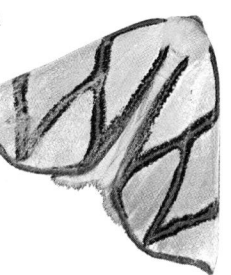

6.2 Exploring Congruence   **289**

---

9. 21   10. 5

11. Answers may vary. An
example is given.

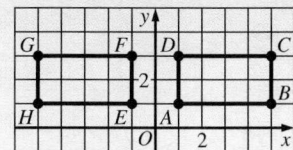

$ABCD \cong HEFG$ because
corresponding angles and sides
are congruent.

12. △DEC

13. △CEF   14. △DCF

15. Check student's work.

   a. No.

   b. Yes.

16. 6 ways; Answers may vary; for
example, students might use a
systematic list, the Multiplication Counting Principle, or the
permutation $_3P_3 = 6$.

17. Answers may vary. An
example is given.

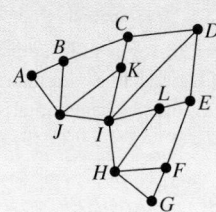

△ABJ ≅ △FGH;
△CDI ≅ △EDI;
△JKI ≅ △HLI;
quad BCKJ ≅ quad FELH;
quad ACIJ ≅ quad EGHI;
pent ACDIJ ≅ pent GEDIH

Suggested Assignment

❖ **Core Course**
Exs. 1–9, 12–15, 22–25, AYP

❖ **Extended Course**
Exs. 1–25, AYP

❖ **Block Schedule**
Day 35 Exs. 1–9, 12–15,
22–25, AYP

Exercise Notes

Topic Spiraling: Review
Exs. 1, 3, 4 You may want to
review the Reflexive Property and
Vertical Angles Property before
assigning these exercises.

Student Progress
Exs. 5–7 Students should be able
to complete problems similar to
these when given the names
without the corresponding figures.

Integrating Algebra
Ex. 10 This exercise requires that
students solve a quadratic equation, which can be done by using
either factoring or the quadratic formula. You may wish to discuss with
students why the negative solution
to the equation can be disregarded
in this situation.

Communication: Drawing

Ex. 15 Ask students to
suggest ideas on how to draw a
figure congruent to a given figure.
Some may measure all the angles
and sides, while others may trace
the figure or use a cutout of the
figure to trace.

Problem Solving
Ex. 16 Problems of this type can
be solved by using a pattern to
make a systematic list. For example, students can make a list by
rotating the first letter of each triangle name to the last position.

Exercise Notes

Second-Language Learners

Exs. 18–21 To help students learning English with the vocabulary contained in the different parts of the interview with Madeleine Fleming, you may want to begin a list of words and their definitions with *reflective film* and *street signs*. As you continue the chapter, add *optical surfaces*, *light pipe*, *mold*, and *ridges* (page 295).

Geometric Thinking

Ex. 21 Students should realize that when congruent pairs of polygons are combined together, the result is another congruent polygon.

Writing Proofs

Ex. 22 Students should begin this proof and all others by marking the diagram with the given information. As they begin to write a proof, they should also have a plan in mind. You may wish to ask students to write a plan for this proof. They need to remember that to show two polygons are congruent, they must show that all the corresponding angles and all the corresponding sides are congruent. Point out to students how each piece of given information is used. When students are writing their own proofs, they should review the given information to see that they have used it all.

INTERVIEW ## Madeleine Fleming

Look back at the article on pages 276–278.

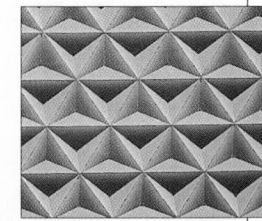

Madeleine Fleming designs surfaces made up of millions of tiny congruent figures. For example, the reflective film at the right contains about 7000 pyramids in each square inch. The film is used to make street signs.

Open-ended Problems For Exercises 18–21, complete each statement.

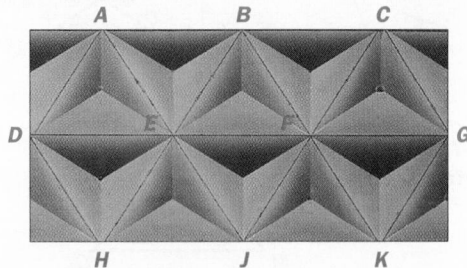

18. It appears that $\triangle AEB \cong$? .

19. It appears that quadrilateral $ABFJ \cong$? .

20. It appears that pentagon $AJKGC \cong$? .

21. Name three other pairs of polygons that appear to be congruent.

22. Copy and complete the proof.

Given: $\overline{AB} \cong \overline{DC}$
$\overline{AD}$ and $\overline{BC}$ bisect each other.
$\overleftrightarrow{AB} \parallel \overleftrightarrow{CD}$

Prove: $\triangle ABE \cong \triangle$?

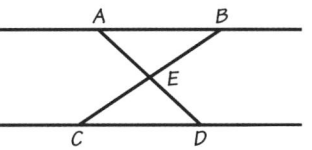

Statements	Reasons
1. $\overline{AB} \cong \overline{DC}$	**1.** Given
2. $\overline{AD}$ and $\overline{BC}$ bisect each other.	**2.** ?
3. $\overline{AE} \cong$?	**3.** Definition of segment bisector
4. $\overline{BE} \cong \overline{CE}$	**4.** ?
5. $\overleftrightarrow{AB} \parallel \overleftrightarrow{CD}$	**5.** ?
6. $\angle BAE \cong \angle CDE$	**6.** ?
7. $\angle$? $\cong \angle DCE$	**7.** If two ∥ lines are intersected by a transversal, then alternate interior angles are ≅.
8. $\angle$? $\cong \angle$?	**8.** Vertical angles are congruent.
9. ? $\cong \triangle DCE$	**9.** Definition of congruent polygons

---

18–21. Answers may vary. Examples are given.

18. $\triangle EJF$

19. quad $BCGK$

20. pent $HBCGK$

21. quad $ABFE \cong$ quad $EFKJ$; quad $CGKF \cong$ quad $BFJE$; trap $ACFE \cong$ trap $HKFE$

22. Prove: $\triangle ABE \cong \triangle DCE$; (2) Given; (3) $\overline{DE}$; (4) Def. of segment bisector; (5) Given; (6) If two ∥ lines are intersected by a transversal, then alternate interior ∠s are ≅; (7) ABE; (8) AEB, DEC; (9) $\triangle ABE$

23. Answers may vary. Examples are given.

$ABEF \cong DEBC$

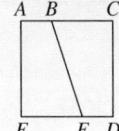

$ABGHEF \cong DEHGBC$

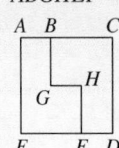

$ABEF \cong CBED$

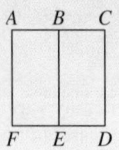

$ABGHJIEF \cong DEIJHGBC$

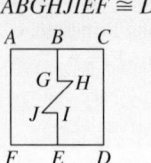

$AEHFGCD \cong CGFHEAB$

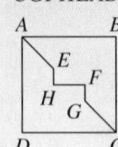

$ABGHJIEF \cong DEIJHGBC$

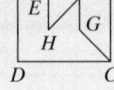

$AEHFGCD \cong CGFHEAB$

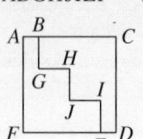

ONGOING ASSESSMENT

23. **Cooperative Learning** Work in a group of three people. You will need graph paper and scissors. Two ways to cut a 3 in. by 3 in. square into two congruent polygons are shown. Find at least five more ways. Sketch, label, and name each differently shaped pair of congruent polygons that you cut.

△ABC ≅ △FED

You may use concave polygons.

SPIRAL REVIEW

Give the postulate, definition, or property that makes each statement about the diagram true. *(Section 3.2)*

24. ∠B ≅ ∠D

25. $\overline{AC} \cong \overline{AC}$

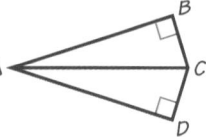

ASSESS YOUR PROGRESS

VOCABULARY

congruent polygons (p. 285)   **corresponding part** (p. 285)

Tell whether or not a triangle can be formed from the given lengths. *(Section 6.1)*

1. 1 in., 4 in., 10 in.   2. 2 ft, 4 ft, 2 ft   3. 5.4 m, 0.5 m, 5.0 m

The lengths of two sides of a triangle are given. What can you conclude about the length of the third side? *(Section 6.1)*

4. 1 in., 4 in.   5. 2 ft, 4 ft   6. 18 cm, 12 cm

7. Name all of the triangles in the diagram that appear to be congruent. *(Section 6.2)*

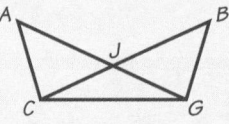

8. △DEF ≅ △HKL *(Section 6.2)*
   a. Find the value of r.
   b. Find the value of p.

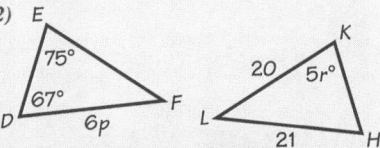

9. **Journal** Sue says that to tell if three numbers can be the lengths of the sides of a triangle, you only need to check if the sum of the two smaller numbers is greater than the third number. Do you agree? Explain.

6.2 Exploring Congruence **291**

Apply⇔Assess

Exercise Notes

Topic Spiraling: Preview
Exs. 24, 25 In the next section, students will use properties such as the ones used in these exercises to prove two triangles are congruent.

Assess Your Progress

Journal Entry
In this journal entry, students develop a method for checking if three lengths can form a triangle. Ask several students to read their responses to ensure that all students understand the logic involved.

Progress Check 6.1–6.2

See page 328.

Practice 40 for Section 6.2

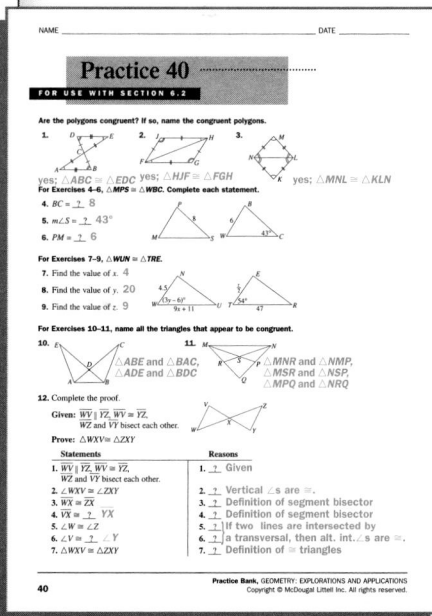

24. All right angles are congruent.

25. Reflexive Property

Assess Your Progress

1. No.

2. No.

3. Yes.

4. It is greater than 3 in. and less than 5 in.

5. It is greater than 2 ft and less than 6 ft.

6. It is greater than 6 cm and less than 30 cm.

7. △AJC ≅ △BJG and △ACG ≅ △BGC

8. a. 15
   b. 3.5

9. Sue is correct. Suppose the lengths of the sides of a triangle are x, y, and z, and that z is the longest length. Since z > x, z + y > x. Since z > y, z + x > y. Therefore, the only inequality that must be checked is whether x + y is greater than z, that is, whether the sum of the lengths of the two shortest sides is greater than the length of the longest side. If so, the three sides can form a triangle.

291

Objectives

- Use the SSS and SAS Postulates.
- Prove that triangles are congruent without proving that all six corresponding parts are congruent.

Recommended Pacing

❖ **Core and Extended Courses**
Section 6.3: 1 day

❖ **Block Schedule**
Section 6.3: $\frac{1}{2}$ block
(with Section 6.4)

Resource Materials

Lesson Support
Lesson Plan 6.3

Warm-Up Transparency 6.3

Practice Bank: Practice 41

Study Guide: Section 6.3

Explorations Lab Manual:
  Diagram Master 2

Challenge Problems: Set 41

Technology
Technology Book:
  Calculator Activity 8
  TI-92 Activity 6

Geometry Software

Internet:
  http://www.hmco.com

Warm-Up Exercises

1. If △ ADF ≅ △ BKI, which angle is congruent to ∠ D?   ∠ K

2. If △ ADF ≅ △ BKI, which side is congruent to $\overline{DF}$? $\overline{KI}$

3. If △ ADF ≅ △ BKI, which side is congruent to $\overline{IB}$? $\overline{FA}$

4. Name the triangles that appear congruent in the figure.
△ SUR ≅ △ TUV, △ RST ≅ △ VTS

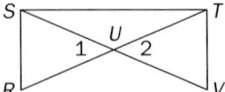

5. In the figure above, how do you know that ∠ 1 ≅ ∠ 2?
Vertical angles are congruent.

SECTION

6.3

Congruent Triangles: SSS and SAS

You probably know some good shortcuts for getting from one place to another. Sometimes shortcuts are better than longer routes. In this section, you will discover two good shortcuts for showing that triangles are congruent.

CONSTRUCTION

CONSTRUCTING A TRIANGLE FROM THREE SIDES

Given △**ABC**, construct a triangle congruent to it by copying only the sides.

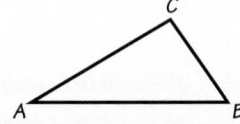

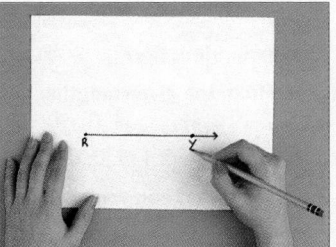

1. Draw a ray. Label it $\overrightarrow{RY}$.

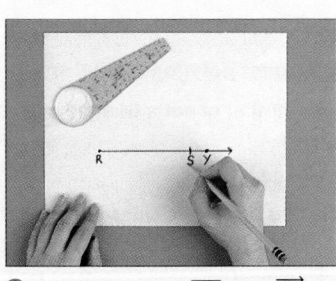

2. Copy segment $\overline{AB}$ onto $\overrightarrow{RY}$ and label the segment $\overline{RS}$.

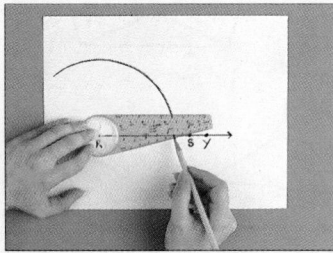

3. Open your compass to radius *AC*. Use this radius to draw a long arc with center *R*.

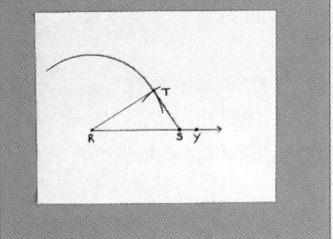

4. Using radius *BC* and center *S*, draw an arc that intersects the arc from Step 3. Label the intersection *T*. Draw △*RST*.

THINK AND COMMUNICATE

Use the construction on page 292.

1. Explain why corresponding sides of △*ABC* and △*RST* are congruent.

2. Measure the corresponding angles. Are the triangles congruent?

3. Use the construction to copy △*ABC* in as many ways as possible. For example, draw the arcs below $\overrightarrow{RY}$, or draw $\overrightarrow{RY}$ diagonally.

4. Is there any way to construct △*RST* so that $\overline{RS} \cong \overline{AB}$, $\overline{RT} \cong \overline{AC}$, and $\overline{ST} \cong \overline{BC}$, but △*RST* and △*ABC* are not congruent? Explain.

As you saw in the construction, triangles with congruent corresponding sides are congruent. This idea is helpful when you want to prove that two triangles are congruent.

Side-Side-Side Postulate (SSS Postulate)

If three sides of a triangle are congruent to three sides of another triangle, then the triangles are congruent.

If $\overline{WN} \cong \overline{FR}$, $\overline{NA} \cong \overline{RO}$, and $\overline{AW} \cong \overline{OF}$, then △*AWN* ≅ △*OFR*.

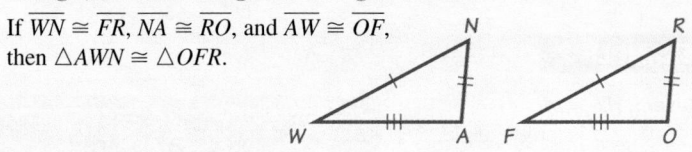

EXAMPLE 1

Given: $\overline{AB} \cong \overline{CB}$; $\overline{AD} \cong \overline{CD}$

Prove: △*ABD* ≅ △*CBD*

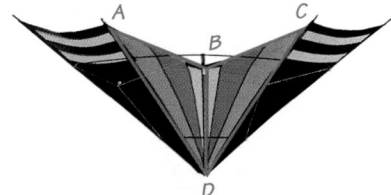

SOLUTION

Plan Ahead: Copy the diagram and mark the parts that you know are congruent. Use the fact that $\overline{BD} \cong \overline{BD}$ and the SSS Postulate.

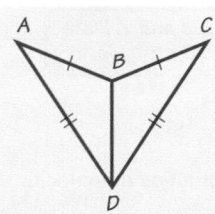

Statements	Reasons
1. $\overline{AB} \cong \overline{CB}$; $\overline{AD} \cong \overline{CD}$	1. Given
2. $\overline{BD} \cong \overline{BD}$	2. Reflexive Property
3. △*ABD* ≅ △*CBD*	3. SSS Postulate

6.3 Congruent Triangles: SSS and SAS **293**

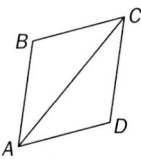

Section Note

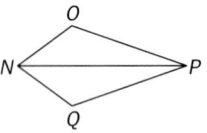

Communication: Reading
Have students read the first paragraph at the top of this page. Ask a volunteer to explain what is meant by an *included* angle. Have students draw a triangle in their notebooks or journals and illustrate the definition of an included angle. Students have the opportunity in Exs. 15–19 on page 297 to construct congruent triangles using two sides and the included angle.

About Example 2

Teaching Tip
Discuss the organization of the flow proof of this Example. Point out that the three parts immediately preceding the conclusion are the two pairs of congruent sides and the congruent included angles.

Additional Example 2

Given: $\overline{NO} \cong \overline{NQ}$
$\overline{NP}$ bisects $\angle ONQ$.
Prove: $\triangle NOP \cong \triangle NQP$

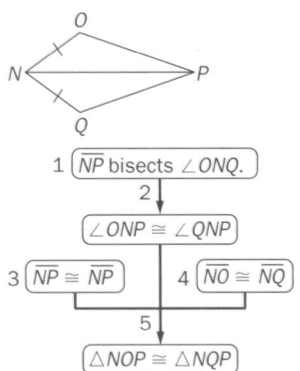

Plan Ahead: Use the definition of angle bisector and the Reflexive Property.

Reasons
1. Given
2. Def. of angle bisector
3. Reflexive Property
4. Given
5. SAS Postulate

The Side-Angle-Side Postulate

In $\triangle STU$, $\angle S$ is *included* between sides $\overline{ST}$ and $\overline{SU}$. In the Exercises, you will construct congruent triangles by copying two sides and the included angle. No matter how you do the construction, the resulting triangle is always the same size and shape. This leads to another postulate.

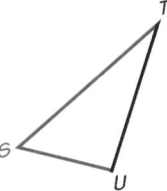

Side-Angle-Side Postulate (SAS Postulate)

If two sides and the included angle of one triangle are congruent to two sides and the included angle of another triangle, then the triangles are congruent.

If $\overline{XY} \cong \overline{UV}$, $\angle X \cong \angle U$, and $\overline{XZ} \cong \overline{UW}$, then $\triangle XYZ \cong \triangle UVW$.

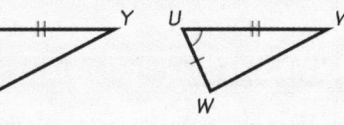

EXAMPLE 2

Given: $\overline{HJ} \cong \overline{ML}$
$\angle J$ is a right angle.
$\angle L$ is a right angle.
K is the midpoint of $\overline{JL}$.
Prove: $\triangle HJK \cong \triangle MLK$

SOLUTION

Plan Ahead: Use the definition of midpoint and what you know about right angles.

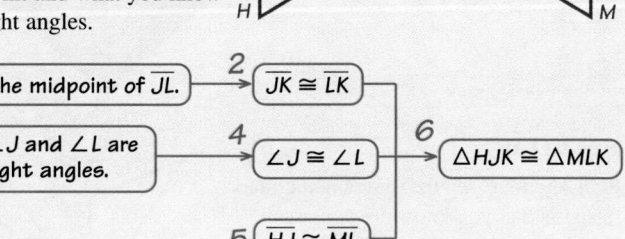

Reasons

1. Given
2. Definition of midpoint
3. Given
4. All right angles are congruent.
5. Given
6. SAS Postulate

✓ CHECKING KEY CONCEPTS

Decide whether or not you can prove that the triangles are congruent. If you can, tell which postulate you would use.

1.    **2.** 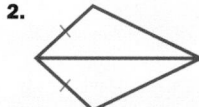   **3.**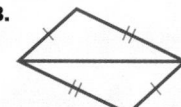

For Questions 4 and 5, use the information in the diagram.

4. What other information do you need to know in order to use the SSS Postulate to prove that $\triangle PQR \cong \triangle TQR$?

5. What other information do you need to know in order to use the SAS Postulate to prove that $\triangle PQR \cong \triangle TSR$?

6. How are the SSS Postulate and the SAS Postulate alike? How are they different?

7. Given: $\overline{XY} \cong \overline{ZW}; \overline{XY} \parallel \overline{ZW}$
   Prove: $\triangle XYZ \cong \triangle ZWX$

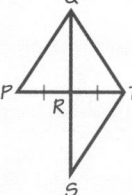

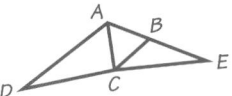

6.3 | Exercises and Applications

Extra Practice exercises on page 671

1. Which angle is included between $\overline{AC}$ and $\overline{BC}$?

2. Open-ended Problem Name two sides and their included angle in $\triangle ACD$.

3. OPTICS Madeleine Fleming designs optical surfaces such as the surface of this light pipe. A mold is used to make tiny parallel ridges on the outside of the pipe. The ridges and mold fit together as shown. Which postulate can you use to explain why $\triangle ABC \cong \triangle DEF$?

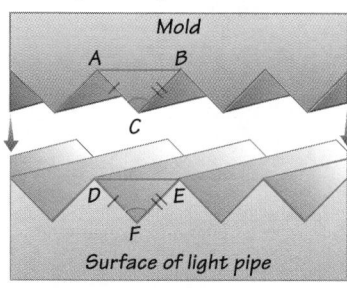

Mold
Surface of light pipe

Checking Key Concepts

Geometric Thinking
In questions 1–3, students determine whether there is sufficient information to prove that the triangles are congruent. This is an important skill. In some cases, the necessary information can be deduced from properties, postulates, and theorems such as the Reflexive Property and the Vertical Angles Theorem. When writing the proof in question 7, students should examine the given information and determine which triangle postulate they will used prior to writing the proof.

Closure Question

Describe the SSS Postulate and SAS Postulate. How do these postulates help you prove two triangles are congruent? The SSS Postulate states that if three sides of a triangle are congruent to three sides of another triangle, then the triangles are congruent. The SAS Postulate states that if two sides and the included angle of one triangle are congruent to two sides and the included angle of another triangle, then the triangles are congruent. These postulates allow you to prove two triangles are congruent without showing that all the corresponding angles and corresponding sides of the triangles are congruent.

Apply⇔Assess

Suggested Assignment

❖ **Core Course**
Exs. 1, 2, 4–11, 15–30
❖ **Extended Course**
Exs. 1–30
❖ **Block Schedule**
Day 36 Exs. 1, 2, 4–11, 15–30

Exercise Notes

 **Application**
Ex. 3 In this exercise, students learn how congruent triangles are used in the design and construction of optical surfaces.

Checking Key Concepts

1. Yes; SAS.

2. No.

3. Yes; SSS.

4. $\overline{PQ} \cong \overline{TQ}$

5. $\overline{RQ} \cong \overline{RS}$

6. Answers may vary. An example is given. Each postulate uses three pairs of congruent corresponding parts to prove that two triangles are congruent. The SSS

Postulate uses three pairs of sides, while the SAS Postulate uses two pairs of sides and one pair of included angles.

7. **Statements (Reasons)**
 1. $\overline{XY} \cong \overline{ZW}; \overline{XY} \parallel \overline{ZW}$ (Given)
 2. $\angle WZX \cong \angle YXZ$ (If two $\parallel$ lines are intersected by a transversal, then alternate interior $\angle$ are $\cong$.)
 3. $\overline{XZ} \cong \overline{ZX}$ (Reflexive Property)
 4. $\triangle XYZ \cong \triangle ZWX$ (SAS Postulate)

Exercises and Applications

1. $\angle ACB$

2. Answers may vary. An example is given. $\angle D$ is included between $\overline{DA}$ and $\overline{DC}$.

3. SAS Postulate

Exercise Notes

Teaching Tip

Exs. 4–9 In both triangle postulates, three pieces of information are needed to conclude that the triangles are congruent. Some students may want to number the congruent parts when examining figures such as these. For example, when they identify two sides as congruent, they could put a number 1 next to each side. Then when they see a side congruent to itself by the Reflexive Property, they can put a 2 by that side. The next congruence they discover can be numbered 3. They then need to check that the three congruences are either all sides or two sides and the included angle. For the triangles that do not have sufficient information, ask what else is needed to prove them congruent, and which postulate would be used to do so.

Writing Proofs

Ex. 11 Encourage students to examine the proof in Example 1 if they need help writing the proof for this exercise. Students should look at the given information, mark the diagram, and decide which triangle postulate to use before writing the proof. Suggest to students that they keep a list of other theorems, properties, and postulates that are frequently used in triangle-congruence proofs. In so doing, they can refer to this list for ideas.

Cooperative Learning

Ex. 12 This is a short cooperative learning exercise that can be completed in the first 10 minutes or last 10 minutes of a class period. You may want to discuss the method for drawing a quadrilateral prior to dividing the class into groups.

Common Error

Ex. 12 Some students try to extend the triangle congruence postulates to quadrilaterals and other polygons. Successful completion of this exercise should help students begin to realize that these postulates apply only to triangles.

296

Tell whether you can prove that the triangles are congruent. If you can, name the congruent triangles and tell which postulate you can use.

4.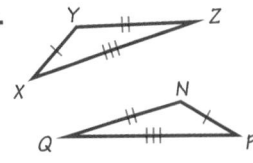

5.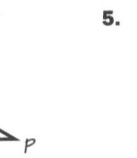

6.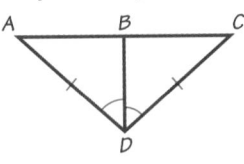

7. $\overline{FH} \cong \overline{GK}$ and $\overline{KL} \cong \overline{HE}$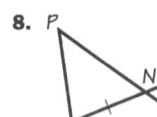

8.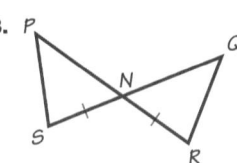

9.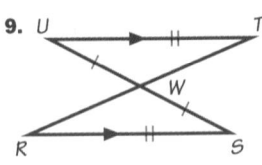

10. Copy and complete the flow proof.
   Given: Y is the midpoint of $\overline{XZ}$.
   $\overline{XW} \cong \overline{ZW}$
   Prove: $\triangle XWY \cong \triangle ZWY$

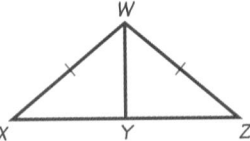

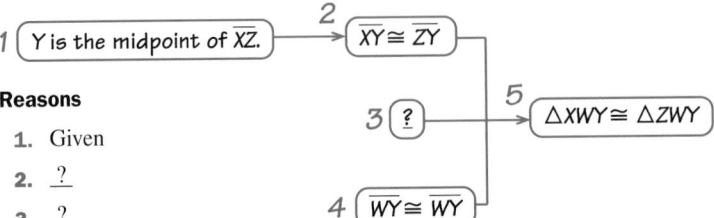

Reasons
1. Given
2. ?
3. ?
4. ?
5. ?

11. Write a two-column proof.
   Given: $\overline{AB} \cong \overline{CB}$; $\overline{BE} \cong \overline{BD}$
   Prove: $\triangle ABE \cong \triangle CBD$

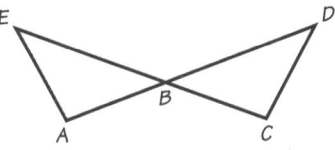

12. **Cooperative Learning** Work in a group of four people. You will need a ruler, a compass, and a protractor.

   a. Each member in your group should draw quadrilateral *GHJK* with *GH* = 4 cm, *HJ* = 6 cm, *JK* = 8 cm, and *KG* = 10 cm.

   b. Compare your quadrilateral with other quadrilaterals drawn in your group. Are all of the quadrilaterals congruent?

   c. Do you think there is an SSSS postulate for proving quadrilaterals congruent? Explain why or why not.

296   Chapter 6 *Conjectures About Triangles*

---

4. Yes; $\triangle XYZ \cong \triangle PNQ$; SSS Postulate.

5. Yes; $\triangle ABD \cong \triangle CBD$; SAS Postulate.

6. No.

7. No.

8. No.

9. No.

10. (2) Def. of midpoint
   (3) $XW \cong ZW$, Given
   (4) Reflexive Property
   (5) SSS Postulate

11. **Statements (Reasons)**
   1. $\overline{AB} \cong \overline{CB}$; $\overline{BE} \cong \overline{BD}$ (Given)
   2. $\angle ABE \cong \angle CBD$ (Vertical ∠ are ≅.)
   3. $\triangle ABE \cong \triangle CBD$ (SAS Postulate)

12. a. Check students' work.

   b. The quadrilaterals probably will not all be congruent.

   c. No; the quadrilaterals drawn in any group should provide counterexamples.

When rescuers enter a partially collapsed building, they must reinforce damaged doorways for safety. If the doorway will not be used during the rescue, diagonal braces can be used.

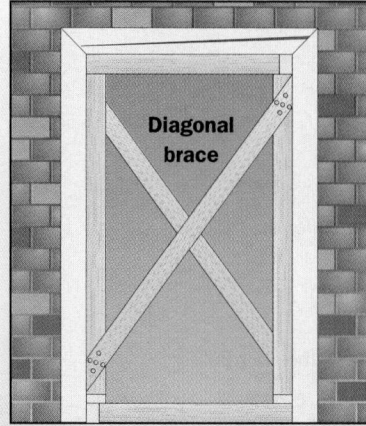

Diagonal brace

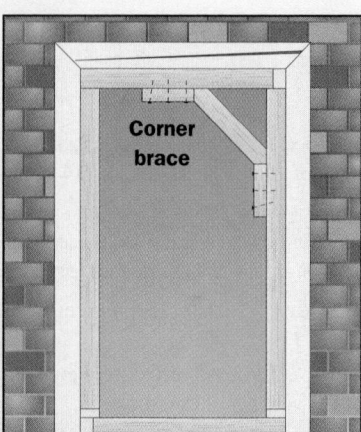

Corner brace

13. a. Writing A quadrilateral might change shape under pressure, but a triangle will not. Use the SSS Postulate to explain why this is true.

 b. Explain why diagonal braces make a doorway more stable.

14. Challenge If the doorway must not be blocked, rescuers can strengthen it with a corner brace instead of a diagonal brace. Explain why a corner brace makes the doorway more stable.

Investigation For Exercises 15–19, you will need geometry software, or a compass and straightedge. Draw a triangle and label the vertices *A*, *B*, and *C*. You will construct copies of △*ABC* by copying $\overline{AB}$, ∠*A*, and $\overline{AC}$.

15. Make a copy of △*ABC* by first copying $\overline{AB}$, then ∠*A*, then $\overline{AC}$.

16. Make a copy of △*ABC* by first copying ∠*A*, then $\overline{AB}$, then $\overline{AC}$.

17. Make a copy of △*ABC* by first copying $\overline{AC}$, then ∠*A*, then $\overline{AB}$.

18. Writing How are the triangles you drew in Exercises 15–17 different? How are they alike?

19. Is there a way to construct a copy of △*ABC* so that two of the sides are congruent to $\overline{AB}$ and $\overline{AC}$ and the included angle is congruent to ∠*A*, but the triangle is not congruent to △*ABC*? If so, give an example.

20. SAT/ACT Preview For △*GHJ* and △*PQR*, $\overline{GH} \cong \overline{PQ}$, $\overline{HJ} \cong \overline{QR}$, and ∠*H* ≅ ∠*Q*. △*HGJ* ≅ __?__

 A. △*PQR*   **B.** △*QRP*   **C.** △*RQP*   **D.** △*QPR*   **E.** cannot be determined

6.3 Congruent Triangles: SSS and SAS **297**

Apply⇔Assess

Exercise Notes

Using Manipulatives
Exs. 13, 14 You can use physical models to help students understand the difference between a triangle with given side lengths and a quadrilateral with given side lengths. To make the quadrilateral, cut out 4 thin strips of wood of various lengths and join them together using nuts and bolts. Similarly, construct a triangle using three strips of wood. A toy construction set can also be used to make these models. Leave the bolts loose enough to turn. After completing Ex. 13, give one student the triangle model and another the quadrilateral model. Ask them to try and move the model to make a different shape. The sides of the triangle model cannot move; hence, there is only one possible triangle with these three sides. The sides of the quadrilateral model, however, can move to form an infinite number of quadrilaterals. Thus, the side lengths do not determine a unique quadrilateral. These models will also help students understand Ex. 14.

Research
Exs. 13, 14, 21 You might have students look for other types of construction to see how triangles are used to ensure that buildings are stable. Roof rafters on a typical house are a good example.

Construction Note
Exs. 15–19 Prior to assigning these exercises, you may wish to review the method for constructing an angle congruent to a given angle.

Teaching Tip
Exs. 15–19 These exercises provide evidence to support the SAS postulate. Students should feel confident upon completing these exercises that only one triangle can be constructed when given the lengths of two sides and the measure of the included angle.

Alternate Approach
Exs. 15–19 If you wish, you can lead the class through this investigation using geometry software and an overhead screen. Ask students for input at each stage of the investigation.

13. a. According to the SSS Postulate, the shape of a triangle with sides of given lengths is determined. No other triangle has sides of the same length and a different shape.

 b. Diagonal braces create triangles that will not change shape under pressure.

14. A corner brace forms a triangle at one corner, making that corner more stable yet allowing passage through the doorway. The increased stability of one side of the doorway helps make the entire doorway more stable.

15–17. Check students' work.

18. The triangles were constructed differently and are oriented differently; each is congruent to △*ABC* and to each other.

19. No.

20. D

Apply⇔Assess

Exercise Notes

Integrating Algebra
Ex. 22 This exercise demonstrates
how the Distance Formula can be
combined with the SSS Postulate
to prove two triangles on a coordinate plane are congruent.

Topic Spiraling: Review/Preview
Ex. 23 Reflections and symmetry
were first studied in Section 1.2.
Here, students see that a reflected
image is congruent to its preimage.
This idea will be explored further in
Section 8.2.

Assessment Note
Ex. 24 This exercise provides students with an opportunity to consolidate their understanding of how
to prove that two triangles are congruent. Ask several volunteers to
read their responses to the class
so that all students can assess
their own work.

Practice 41 for Section 6.3

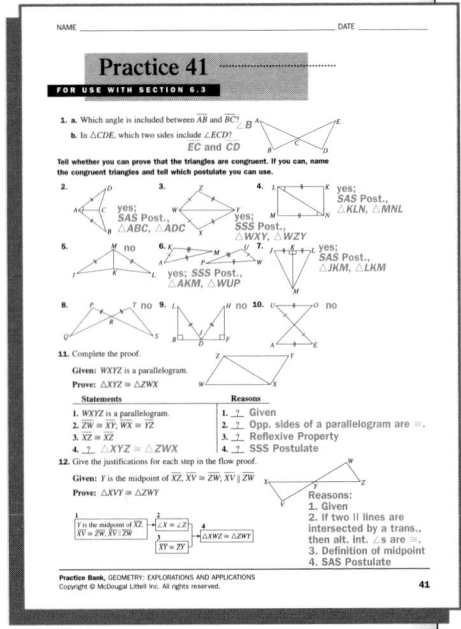

---

21. Write a proof using any format.
    Given: $\overline{PQ} \cong \overline{RQ}$; $\overline{PS} \cong \overline{RS}$
    Prove: $\triangle PQS \cong \triangle RQS$

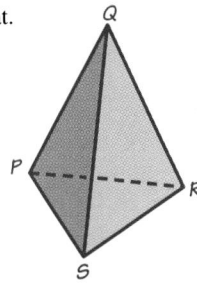

22. The vertices of $\triangle ABC$ are $A(-3, 1)$, $B(-8, 5)$, and $C(-1, 8)$. The
    vertices of $\triangle DEF$ are $D(0, 1)$, $E(4, 6)$, and $F(7, -1)$.
    a. Graph $\triangle ABC$ and $\triangle DEF$.
    b. Explain how to use the Distance Formula to tell whether the triangles
       are congruent. Which postulate must you use?

23. **TRANSFORMATIONS** The vertices of $\triangle ABC$ are $A(1, 7)$, $B(6, 3)$,
    and $C(3, 1)$.
    a. Graph $\triangle ABC$ and reflect it over the x-axis. What are the coordinates
       of the vertices of the image?
    b. Reflect $\triangle ABC$ over the y-axis. What are the coordinates of the
       vertices of the image?
    c. Are the two images of $\triangle ABC$ congruent? Explain why or why not.

ONGOING ASSESSMENT

24. **Writing** Describe three ways to prove that two triangles are congruent.
    Choose one of the ways and give an example.

SPIRAL REVIEW

Name all of the triangles that appear to be congruent. *(Section 6.2)*

25.
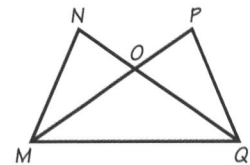

26.
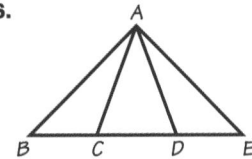

27. The floor plan for a building has a scale of 1 in. = 10 ft. Write this as a
    ratio in lowest terms. *(Toolbox, page 707)*

Find each unknown angle measure. *(Section 2.2)*

28.
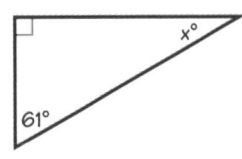

29.
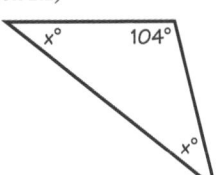

30.
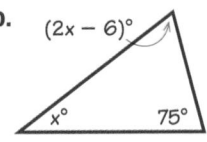

298 Chapter 6 *Conjectures About Triangles*

---

21. Answers may vary. An example is
    given. It is given that $\overline{PQ} \cong \overline{RQ}$ and
    that $\overline{PS} \cong \overline{RS}$. By the Reflexive
    Property, $\overline{QS} \cong \overline{QS}$. Then, by the
    SSS Postulate, $\triangle PQS \cong \triangle RQS$.

22. a.
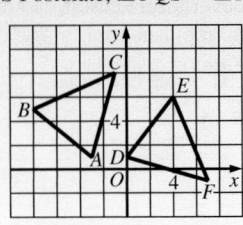

b. Use the Distance
   Formula to find the
   length of each side of
   the two triangles.
   Check whether all
   three sides of one triangle are congruent to
   the three sides of the
   other triangle. If so,
   the triangles are con-
   gruent by the SSS
   Postulate.

23. a.
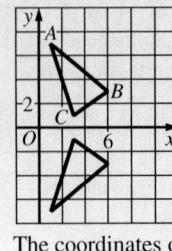

The coordinates of
the vertices of the
image of $\triangle ABC$
are $(1, -7)$, $(6, -3)$,
and $(3, -1)$.

b.
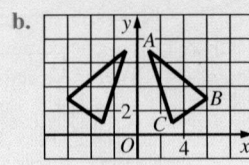

The coordinates of the
vertices of the image
of $\triangle ABC$ are $(-1, 7)$,
$(-6, 3)$, and $(-3, 1)$.

c. Yes; the lengths of the
   sides of each image of
   $\triangle ABC$ are the same as

the lengths of the
sides of $\triangle ABC$, so
by the SSS Postu-
late, each image is
congruent to $\triangle ABC$.

24–30. See answers in
back of book.

298

6.4

Congruent Triangles: ASA, AAS, and HL

Learn how to...
- **use the ASA Postulate, the AAS Theorem, and the HL Theorem**

So you can...
- **prove that triangles are congruent**
- **find the positions of distant objects, such as stars and islands**

In 1576, King Frederik II of Denmark gave the astronomer Tycho Brahe an island to encourage him to remain in Denmark. Brahe used *triangulation* to map his island and find its exact location in relation to the rest of Denmark.

$\overline{AB}$ is *included* between $\angle ABP$ and $\angle BAP$.

1.5 mi

Suppose you are on an island and want to find the distance from the island to a known point, P, on shore. Choose two points, A and B, on the island and measure $\overline{AB}$, $\angle PAB$, and $\angle ABP$. You can use these measures to find BP.

THINK AND COMMUNICATE

1. Make a scale drawing of $\triangle ABP$: Use a ruler to draw a baseline 1.5 in. long. Then use a protractor to draw the two angles, extending the rays until they intersect at point P. Measure the distance from B to P in your drawing.

2. Compare your triangle with those of your classmates. Is the length BP about the same in each triangle? Are all of the triangles congruent?

3. How can you use your scale drawing to estimate the actual distance from point B on the island to point P on shore?

6.4 Congruent Triangles: ASA, AAS, and HL   **299**

Plan⇔Support

Objectives
- Use the ASA Postulate, the AAS Theorem, and the HL Theorem.
- Prove that triangles are congruent.
- Use triangulation to find the positions of distant objects.

Recommended Pacing
❖ **Core and Extended Courses**
Section 6.4: 1 day
❖ **Block Schedule**
Section 6.4: $\frac{1}{2}$ block
(with Section 6.3)

Resource Materials
Lesson Support
Lesson Plan 6.4
Warm-Up Transparency 6.4
Practice Bank: Practice 42
Study Guide: Section 6.4
Challenge Problems: Set 42
Technology
Scientific Calculator
Geometry Software
Internet:
http://www.hmco.com

Warm-Up Exercises

The measures of two angles of a triangle are given. Find the measure of the third angle.

1. 23°, 54° 103°

2. 119°, 51° 10°

3. A right triangle has legs of lengths 3 and 6. Find the length of the hypotenuse. $3\sqrt{5}$

4. The length of a leg of a right triangle is 12 and the length of its hypotenuse is 13. What is the length of the other leg? 5

5. $\overrightarrow{CD}$ bisects $\angle ECF$. If $m\angle DCF = 29°$, find $m\angle ECF$. 58°

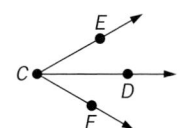

ANSWERS Section 6.4

Think and Communicate

1. Check students' work. $BP \approx 6$ in.

2. If the figures are drawn correctly, BP should be about the same in all the drawings; Yes.

3. Since 1.5 in. was used to represent 1.5 mi, 6 in. in the scale drawing represents an actual distance of 6 mi.

Integrating Measurement

In this section, students learn a method of indirect measurement called *triangulation*. This method is based on the fact that when one triangle has two angles and the included side congruent to two angles and the included side of another triangle, the triangles are congruent.

Historical Connection

Tycho Brahe is famous for his work in astronomy. He is credited with collecting the most accurate astronomical data prior to the invention of the telescope. After Brahe's death, the astronomer Johannes Kepler used the data to discover the laws of planetary motion.

Think and Communicate

After completing question 2, students should discover that all their triangles are congruent. This leads to the formal presentation of the ASA Postulate.

Additional Example 1

Given: $\overleftrightarrow{DE} \parallel \overleftrightarrow{GH}$
$\overline{GE}$ bisects $\overline{DH}$.
Prove: $\triangle DFE \cong \triangle HFG$

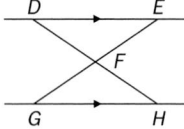

Plan Ahead: Use the definition of segment bisector to show that $\overline{DF} \cong \overline{HF}$. Use the fact $\overleftrightarrow{DE} \parallel \overleftrightarrow{GH}$ to show that $\angle EDF \cong \angle GHF$. Use the fact that vertical angles are congruent to show that $\angle DFE \cong \angle HFG$.

Statements (Reasons)

1. $\overline{GE}$ bisects $\overline{DH}$. (Given)
2. $\overline{DF} \cong \overline{HF}$ (Def. of segment bisector)
3. $\overleftrightarrow{DE} \parallel \overleftrightarrow{GH}$ (Given)
4. $\angle EDF \cong \angle GHF$ (If two parallel lines are intersected by a transversal, then alternate interior angles are congruent.)
5. $\angle DFE \cong \angle HFG$ (Vertical angles are congruent.)
6. $\triangle DFE \cong \triangle HFG$ (ASA Postulate)

The Angle-Side-Angle Postulate

When you and your classmates used triangulation to find the distance *BP*, your triangles should have all been congruent. The method of triangulation leads to another shortcut for proving that triangles are congruent.

Angle-Side-Angle Postulate (ASA Postulate)

If two angles and the included side of one triangle are congruent to two angles and the included side of another triangle, then the triangles are congruent.

If $\angle A \cong \angle L$,
$\overline{AC} \cong \overline{LN}$,
and $\angle C \cong \angle N$,
then $\triangle ABC \cong \triangle LMN$.

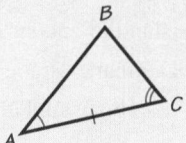 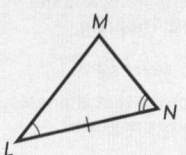

EXAMPLE 1

Given: $\overrightarrow{HK}$ bisects $\angle JHL$.
$\overrightarrow{KH}$ bisects $\angle JKL$.

Prove: $\triangle JHK \cong \triangle LHK$

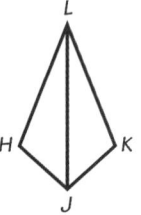

SOLUTION

Plan Ahead: Use the definition of angle bisector to show that $\angle JHK \cong \angle LHK$ and $\angle JKH \cong \angle LKH$. Use the fact that $\overline{HK}$ is a side of both triangles.

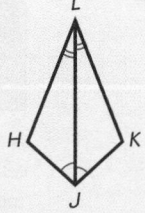

Statements	Reasons
1. $\overrightarrow{HK}$ bisects $\angle JHL$.	1. Given
2. $\angle JHK \cong \angle LHK$	2. Definition of angle bisector
3. $\overline{HK} \cong \overline{HK}$	3. Reflexive Property
4. $\overrightarrow{KH}$ bisects $\angle JKL$.	4. Given
5. $\angle JKH \cong \angle LKH$	5. Definition of angle bisector
6. $\triangle JHK \cong \triangle LHK$	6. ASA Postulate

So far you have learned three shortcuts for proving that two triangles are congruent. SSS, SAS, and ASA are all postulates. You can use these postulates to prove two other shortcuts, which are theorems.

THINK AND COMMUNICATE

4. Find the measures of $\angle A$ and $\angle D$.

5. If two angles of a triangle are congruent to two angles of another triangle, then what can you tell about the third angles?

6. Explain how you know that $\triangle ABC \cong \triangle DEF$.

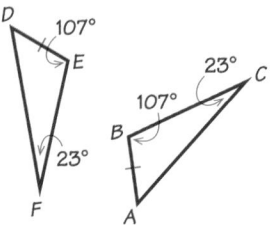

Angle-Angle-Side Theorem (AAS Theorem)

If two angles and a non-included side of one triangle are congruent to the corresponding parts of another triangle, then the triangles are congruent.

If $\angle Q \cong \angle X$, $\angle R \cong \angle Y$, and $\overline{RS} \cong \overline{YZ}$, then $\triangle QRS \cong \triangle XYZ$.

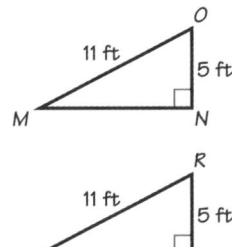

In the triangles at the right, you can use the Pythagorean theorem to see that $MN = 4\sqrt{6}$ and $QP = 4\sqrt{6}$. So $\triangle MNO \cong \triangle QPR$ by the SSS Postulate or the SAS Postulate.

In general, if two pairs of corresponding sides in a right triangle are congruent, you can show that the third pair of corresponding sides is congruent. This leads to a shortcut that works only for right triangles.

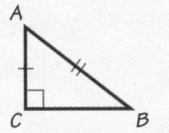
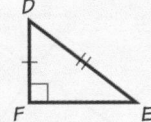

Hypotenuse-Leg Theorem (HL Theorem)

If the hypotenuse and a leg of one right triangle are congruent to the corresponding parts of another right triangle, then the triangles are congruent.

If $\triangle ABC$ and $\triangle DEF$ are right triangles, $\overline{AB} \cong \overline{DE}$, and $\overline{AC} \cong \overline{DF}$, then $\triangle ABC \cong \triangle DEF$.

6.4 Congruent Triangles: ASA, AAS, and HL **301**

Teach⇔Interact

Think and Communicate

Questions 4–6 lead students through the reasoning that can be used to prove the AAS Theorem.

Section Notes

Journal Entry
After discussing the AAS Theorem and the HL Theorem, ask students to write a journal entry that lists all the triangle congruence postulates and theorems they have learned. This list can be used as a reference when completing the exercises.

Common Error
Some students may try to prove two triangles congruent by using AAA or SSA. Point out that two triangles having three pairs of congruent angles are not necessarily congruent. Similarly, two triangles with two pairs of sides and the non-included angles congruent are not necessarily congruent. This second case is explored more thoroughly in Ex. 17 on page 304.

Challenge
You may wish to challenge students students to write a proof for the AAS Theorem or the HL Theorem. For the AAS Theorem, they should use the Triangle Sum Theorem to show that the remaining angles are also congruent and then use the ASA Postulate to prove the triangles are congruent. For the HL Theorem, they should use the Pythagorean theorem to show that the remaining legs are also congruent and then use the SAS Postulate to prove the triangles are congruent. You may wish to ask a student to write these proofs on an overhead transparency and explain them to the class.

Think and Communicate

4. $m \angle A = m \angle D = 50°$

5. The third angles are also congruent.

6. $\angle D \cong \angle A$, $\overline{DE} \cong \overline{AB}$, and $\angle E \cong \angle B$, so $\triangle ABC \cong \triangle DEF$ by the ASA Postulate.

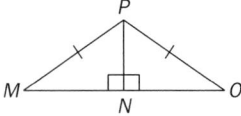

About Example 2

Teaching Tip
Point out to students that the HL Theorem can only be used with right triangles and that it is necessary in the proof to state that the triangles are right triangles.

Additional Example 2

Given: $\overline{PN} \perp \overline{MO}$
$\overline{PM} \cong \overline{PO}$
Prove: $\triangle MNP \cong \triangle ONP$

Plan Ahead: Because $\triangle MNP$ and $\triangle ONP$ are right triangles, use the HL Theorem. $\overline{PN}$ is a leg of both triangles.

Statement (Reasons)
1. $\overline{PN} \perp \overline{MO}$ (Given)
2. $\angle MNP$ and $\angle ONP$ are right angles. (Def. of perpendicular)
3. $\triangle MNP$ and $\triangle NOP$ are right triangles. (Def. of right triangle)
4. $\overline{PM} \cong \overline{PO}$ (Given)
5. $\overline{PN} \cong \overline{PN}$ (Reflexive Property)
6. $\triangle MNP \cong \triangle ONP$ (HL Theorem)

Checking Key Concepts

Geometric Thinking
After students complete these questions, have them discuss their responses with a partner. In question 4, make sure students understand that there is no AAA postulate or theorem.

Closure Question

List and state in words the postulate and theorems you learned in this section to prove that two triangles are congruent. Angle-Side-Angle Postulate: If two angles and the included side of one triangle are congruent to two angles and the included side of another triangle, then the triangles are congruent. Angle-Angle-Side Theorem: If two angles and a non-included side of one triangle are congruent to the corresponding parts of another triangle, then the triangles are congruent. Hypotenuse-Leg Theorem: If the hypotenuse and a leg of one right triangle are congruent to the corresponding parts of another triangle, then the triangles are congruent.

302

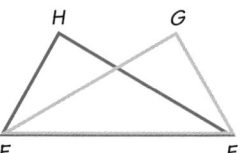

Given: $\overline{EH} \cong \overline{FG}$
$\angle H$ and $\angle G$ are right angles.
Prove: $\triangle EHF \cong \triangle FGE$

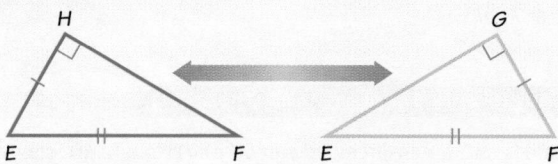

SOLUTION

Plan Ahead: Sketch the triangles separately. Because $\triangle EHF$ and $\triangle FGE$ are right triangles, use the HL Theorem. $\overline{EF}$ is the hypotenuse of both triangles.

Statements	Reasons
1. $\angle H$ and $\angle G$ are right angles.	1. Given
2. $\triangle EHF$ and $\triangle FGE$ are right triangles.	2. Definition of right triangle
3. $\overline{EH} \cong \overline{FG}$	3. Given
4. $\overline{EF} \cong \overline{FE}$	4. Reflexive Property
5. $\triangle EHF \cong \triangle FGE$	5. HL Theorem

☑ CHECKING KEY CONCEPTS

1. Draw and label a triangle $\triangle ABC$. Name two angles and their included side. Then name the same two angles and a non-included side.

Tell which method(s) you can use to prove the triangles congruent. If no method can be used, write *none*.

2.  3. 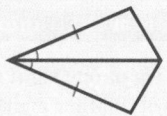 4.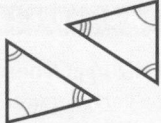

5.  6. 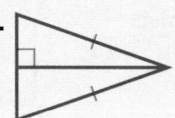 7.

Checking Key Concepts

1. Check students' work. In any $\triangle ABC$, $\overline{AB}$ is the side included between $\angle A$ and $\angle B$. The non-included sides for $\angle A$ and $\angle B$ are $\overline{AC}$ and $\overline{BC}$.
2. AAS Theorem, ASA Postulate
3. SAS Postulate
4. none
5. ASA Postulate, AAS Theorem
6. HL Theorem, SSS Postulate
7. none

6.4 | Exercises and Applications

Extra Practice exercises on page 672

Tell which method(s) you can use to prove that the triangles are congruent. If no method can be used, write *none*.

1.
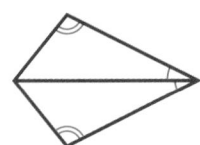

2.
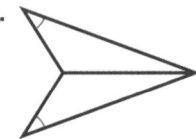

3.
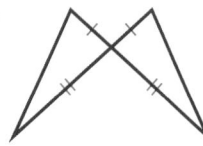

4.
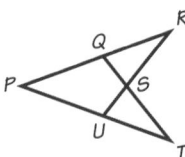

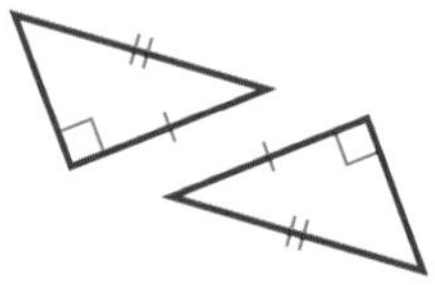

5.

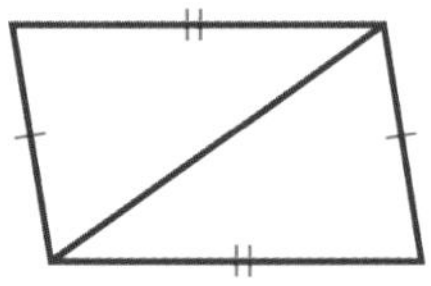

6.
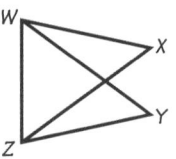

For each exercise, name a pair of overlapping congruent triangles. Tell which method you can use to prove the triangles congruent.

7. **Given:** $\overline{RU} \cong \overline{TQ}$
   $\angle R \cong \angle T$

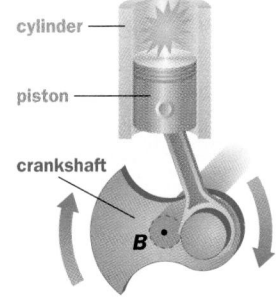

8. **Given:** $\overline{AD} \cong \overline{BC}$
   $\overline{AB} \perp \overline{BC}$
   $\overline{AB} \perp \overline{DA}$

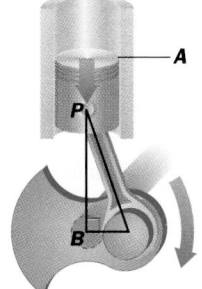

9. **Given:** $\angle X \cong \angle Y$
   $\angle WZX \cong \angle ZWY$

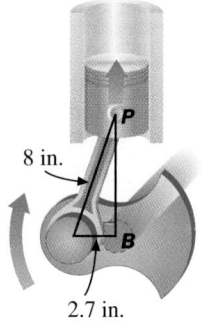

10. **Open-ended Problem** Choose one of the diagrams in Exercises 7–9. Plan a proof for the exercise that you chose.

11. **AUTOMOBILES** In a car engine, exploding gas causes pistons to move up and down in a cylinder. The pistons are attached to the crankshaft, which rotates around point *B* as shown below.

    a. When the shaft is at the position shown in blue, the piston is at *A*. Where is the piston in relation to *A* when the crankshaft is at the position shown in green? Which postulate or theorem justifies your answer?

    b. How close can point *P* get to point *B*? How far can it get from point *B*? Which postulate or theorem justifies your answer?

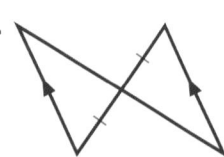

cylinder

piston

crankshaft

A

P

P

8 in.

2.7 in.

B

B

B

6.4 Congruent Triangles: ASA, AAS, and HL   **303**

Exercises and Applications

1. AAS Theorem, ASA Postulate

2. none

3. SAS Postulate

4. HL Theorem, SSS Postulate

5. SSS Postulate

6. ASA Postulate, AAS Theorem

7. $\triangle PRU \cong \triangle PTQ$; AAS Theorem, ASA Postulate

8. $\triangle ABC \cong \triangle BAD$; SAS Postulate, SSS Postulate

9. $\triangle XWZ \cong \triangle YZW$; AAS Theorem, ASA Postulate

10. Answers may vary. One proof for Ex. 7 would include the given information and the fact that $\angle P \cong \angle P$. One proof for Ex. 8 would include the given information and the fact that $\overline{AB} \cong \overline{BA}$. One proof for Ex. 9 would include the given information and the fact that $\overline{WZ} \cong \overline{ZW}$.

11. a. The piston is again at point *A*; HL Theorem.

    b. about 4.8 in.; about 10.2 in.; Pythagorean theorem

Suggested Assignment

❖ **Core Course**
   Exs. 1–10, 12–17, 20–26

❖ **Extended Course**
   Exs. 1–26

❖ **Block Schedule**
   Day 36 Exs. 1–10, 12–17, 20–26

Exercise Notes

Journal Entry
Exs. 1–6 If students have created a list of triangle congruence postulates and theorems as suggested in the note on page 301, they will find it useful when doing these exercises.

Teaching Tip
Exs. 7–9, 15 Remind students that it is often helpful to separate overlapping figures such as the ones in these exercises. One way to do this is to trace the figures using a transparent sheet of tissue paper.

Cooperative Learning
Exs. 7–10 For Ex. 10, you might ask students to work with a partner to plan a proof for each of Exs. 7–9. The partners should discuss each plan together, and then take turns writing the plans.

Application
Ex. 11 In this exercise, students see an application of congruent triangles to the position of the crankshaft and piston in a car engine.

Exercise Notes

Writing Proofs
Exs. 12, 13, 15, 16 Students can choose any of the proof formats for these exercises. You may wish to have them chose at least two different types.

Communication: Drawing
Ex. 14 Students may find it helpful to draw a sketch of each description before answering the question.

Geometric Thinking
Ex. 17 This exercise leads students through the construction of a triangle that has two sides and a non-included angle congruent to those in a given triangle. It should be obvious that in this case, two triangles are possible, only one of which is congruent to the given triangle. This investigation clearly shows that there can be no SSA postulate or theorem.

Communication: Writing
Ex. 17 This is a good opportunity for students to demonstrate their understanding of the triangle postulates and theorems.

Interdisciplinary Problems
Exs. 18, 19 These exercises illustrate a method used by astronomers to determine the distances to stars.

Multicultural Note
Exs. 18, 19 Many native peoples throughout the Americas made skillful astronomical observations. In the Southwest, Hopi sunwatchers decided what day was the last feasible time to plant, based on precise observations of the summer solstice. A Mayan document contains astronomical tables for the years A.D. 755 to 788 that predict eclipses and the movement of Venus to a high level of accuracy. At Monte Albán, an ancient Zapotec site in Mexico, one of the buildings may have served as an observatory, since it is aligned with the positions of certain heavenly objects at specific times of the year. In the foothills of the Colombian Andes, a Kogi temple has four upright stakes that mark the solstices.

12. Given: ∠*A* is a right angle.
 ∠*C* is a right angle.
 $\overrightarrow{BD}$ bisects ∠*ABC*.
 Prove: △*ADB* ≅ △*CDB*

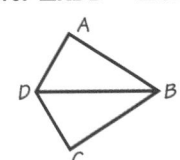

13. Given: *JKLM* is a rectangle.
 $\overline{JP} \cong \overline{KP}$
 Prove: △*JMP* ≅ △*KLP*

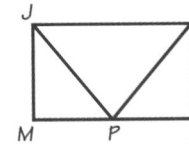

14. Visual Thinking Tell which postulate or theorem about congruent triangles justifies each statement. Explain your choice.

 a. If the legs of one right triangle are congruent to the legs of another right triangle, then the triangles are congruent.

 b. If a leg and the adjacent acute angle of one right triangle are congruent to the corresponding leg and acute angle of another triangle, then the triangles are congruent.

15. Given: $\overline{PQ} \parallel \overline{VS}$
 $\overline{QU} \parallel \overline{ST}$
 $\overline{PQ} \cong \overline{VS}$
 Prove: △*PQU* ≅ △*VST*

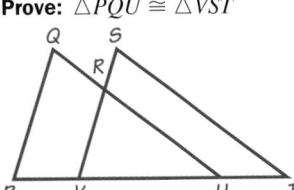

16. Given: $\overline{NH} \perp$ plane *EHG*
 $\overline{GN} \cong \overline{EN}$
 Prove: △*GNH* ≅ △*ENH*

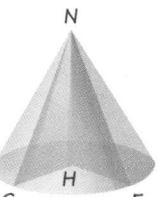

17. Investigation You will need geometry software or a compass and straightedge. Draw an acute or obtuse triangle and label the vertices *A*, *B*, and *C*. Follow Steps 1–4.

 Step 1 Copy ∠*A*. Label the new angle ∠*XDY* and construct $\overrightarrow{DX}$.

 Step 2 Copy $\overline{AB}$ on $\overrightarrow{DY}$. Label the new segment $\overline{DE}$.

 Step 3 Construct a **circle** with center *E* and radius *BC*.

 Step 4 Construct $\overline{EF}$ so that *EF* = *BC* and *F* is on $\overleftrightarrow{DX}$.

 a. How many different ways are there to do Step 4? Do all of the ways result in a triangle that is congruent to △*ABC*? Explain.

 b. Draw a right triangle with a right angle at *C* and hypotenuse $\overline{AB}$. Repeat Steps 1–4 and part (a).

 c. Writing Explain why there is an HL Theorem but not an SSA Theorem.

12. Since ∠*A* and ∠*C* are right angles, ∠*A* ≅ ∠*C*. Since $\overrightarrow{BD}$ bisects ∠*ABC*, ∠*ABD* ≅ ∠*CBD*. By the Reflexive Property, $\overline{BD} \cong \overline{BD}$. Therefore, △*ADB* ≅ △*CDB* by the AAS Theorem.

13. Answers may vary. An example is given. Since *JKLM* is a rectangle, ∠*M* and ∠*L* are right angles. Therefore, △*JMP* and △*KLP* are right triangles. A rectangle is a parallelogram, and the opposite sides of a parallelogram are congruent, so $\overline{JM} \cong \overline{KL}$. $\overline{JP} \cong \overline{KP}$ (given), so △*JMP* ≅ △*KLP* by the HL Theorem.

14. Answers may vary. Examples are given.

 a. SAS Postulate; the included angle is the right angle.

 b. ASA Postulate; a leg of a right triangle is included between the right angle and the adjacent acute angle.

15. Since $\overline{PQ} \parallel \overline{VS}$, ∠*P* ≅ ∠*SVT*. (If two ∥ lines are intersected by a transversal, then corresponding ∠s are ≅.) Since $\overline{QU} \parallel \overline{ST}$, ∠*QUP* ≅ ∠*T*. (If two ∥ lines are intersected by a transversal, then corresponding ∠s are ≅.) $\overline{PQ} \cong \overline{VS}$ (Given), so △*PQU* ≅ △*VST* by the AAS Theorem.

16, 17. See answers in back of book.

Astronomers can find the distance to a nearby star by noticing how it moves against the background of more distant stars.

18. Challenge Earth is on opposite sides of the sun at different times of the year as shown at the right. Explain how you can tell that astronomers have enough information to find the distance to star *S*. Use at least one postulate or theorem in your answer.

19. 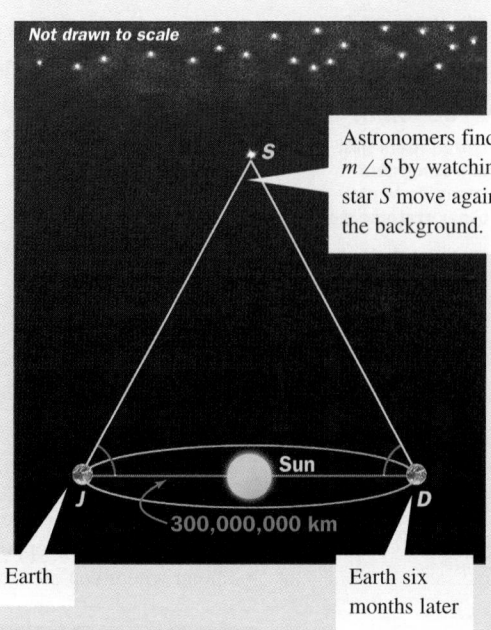 **Technology** You can use the *sine* button on a calculator and the formula

$$SD = \frac{\sin{(m \angle J)}}{\sin{(m \angle S)}} \cdot JD$$

to find the distance to star *S*. For *Alpha Centauri*, the nearest star to Earth other than the sun, $m \angle S \approx 0.0004°$.

a. Find $m \angle J$ for *Alpha Centauri*.

b. About how far away is *Alpha Centauri*?

BY THE WAY

The Yokut tribe of California used astronomy to time the Mourning Ceremony with the setting of the evening star.

Not drawn to scale

Astronomers find $m \angle S$ by watching star *S* move against the background.

Sun

300,000,000 km

Earth

Earth six months later

ONGOING ASSESSMENT

20. a. Open-ended Problem List all the methods that you know for proving that two triangles are congruent.

b. Choose two methods from part (a). For each method, sketch and label a pair of triangles and mark the information needed to use that method to prove that the triangles are congruent.

SPIRAL REVIEW

For Exercises 21–23, $\triangle ABC \cong \triangle JLK$. **Complete each statement.** *(Section 6.2)*

21. $\overline{AB} \cong \underline{?}$ **22.** $\angle C \cong \underline{?}$ **23.** $\overline{KJ} \cong \underline{?}$

Find the measure of each interior and exterior angle of the polygon. *(Section 2.4)*

24. regular triangle **25.** regular 7-gon **26.** equiangular 24-gon

6.4 Congruent Triangles: ASA, AAS, and HL **305**

Apply⇔Assess

Exercise Notes

Integrating Trigonometry
Ex. 19 Students may not be familiar with the concept of sine. Explain to students that given a right triangle, the sine of either acute angle is the ratio of the side opposite the angle to the hypotenuse. Explain also that the sine of a given angle is always the same. Diagrams may help to clarify these ideas. Trigonometry is introduced in Chapter 10.

Assessment Note
Ex. 20 This exercise is an excellent way for students to consolidate what they have learned about triangle congruence. You may wish to have students check their responses with a partner.

Practice 42 for Section 6.4

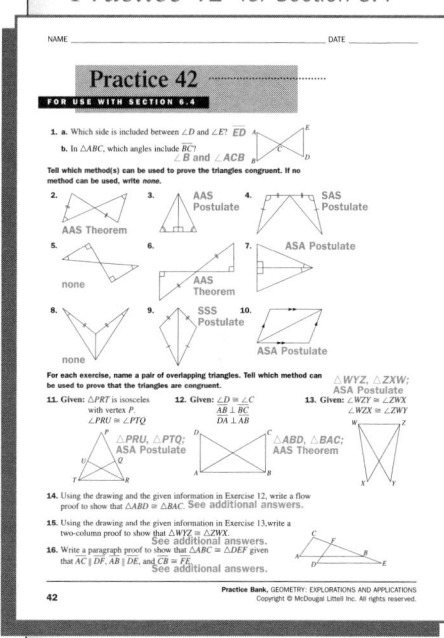

---

18. Answers may vary. An example is given. Since $\angle J \cong \angle D$, $\triangle SJD$ is an isosceles triangle. Once $m \angle S$ is found, $m \angle J$ and $m \angle D$ can be calculated. $(m \angle J = m \angle D = \frac{1}{2}(180° - m \angle S))$ Then a scale drawing can be made using JD, $m \angle D$, and $m \angle J$. By the ASA Postulate, any triangle with two angles and an included side congruent to the given sides and angle will be congruent to $\triangle SJD$. The distance to the star can be calculated by measuring the appropriate distance on the scale drawing and multiplying by the scale factor.

19. a. 89.9998°

b. about 4.3×10^{13} km

20. a. the definition of congruent triangles, the SSS Postulate, the SAS Postulate, the ASA Postulate, the AAS Theorem, and the HL Theorem

b. Answers may vary. An example is given. To prove that $\triangle ABC \cong \triangle DEF$ by the SSS Postulate, you must show that $\overline{AB} \cong \overline{DE}$, $\overline{BC} \cong \overline{EF}$, and $\overline{AC} \cong \overline{DF}$. To prove that $\triangle ABC \cong \triangle DEF$ by the ASA Postulate, you could show that $\angle A \cong \angle D$, $\overline{AB} \cong \overline{DE}$, and that $\angle B \cong \angle E$.

21. $\overline{JL}$

22. $\angle K$

23. $\overline{CA}$

24. 60°; 120°

25. $128\frac{4}{7}°$; $51\frac{3}{7}°$

26. 165°; 15°

Warm-Up Exercises

In the figure, $\overline{AC} \cong \overline{EC}$. What additional information is needed to prove that $\triangle ACB \cong \triangle ECD$ using each of the following.

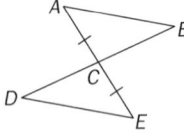

1. ASA Postulate $\angle A \cong \angle E$
2. SAS Postulate $\overline{CB} \cong \overline{CD}$
3. AAS Theorem $\angle D \cong \angle B$
4. HL Theorem $\angle ACB$ and $\angle DCE$ are right angles, and $\overline{AB} \cong \overline{ED}$.

6.5 | Applying Congruence

Learn how to...
- **use corresponding parts of congruent triangles**

So you can...
- **write proofs that use congruent triangles to prove other statements**
- **measure distances indirectly**

Finding the distance between two points can be as simple as pacing it off or using a yardstick or another measuring device. But if the two points are on opposite sides of a busy highway or a stream, you may need to find another method. Congruent triangles can help!

Here is one way to find the distance across a stream:

Choose a point on the other side of the stream directly opposite the point where you are standing.

Put on a visor or a cap with a visor. Look at the point you chose and adjust the visor until it is in line with your eye and the point.

Turn right or left. Note the point on the ground that is now in line with your eye and the visor tip.

Pace off or measure the distance from your feet to the point.

THINK AND COMMUNICATE

1. Explain why $\triangle EFD$ and $\triangle EFG$ are congruent right triangles.

2. Which segment in $\triangle EFG$ is the same length as the distance across the stream?

306 Chapter 6 *Conjectures About Triangles*

ANSWERS Section 6.5

Think and Communicate

1. Answers may vary. An example is given. $\overline{EF} \cong \overline{EF}$ by the Reflexive Property. $\angle EFD$ and $\angle EFG$ are both right angles (assuming that you are standing up straight and the ground is level) so $\angle EFD \cong \angle EFG$. $\angle FED \cong \angle FEG$ because both were formed by

your eye and the visor tip.
Thus, $\triangle EFD \cong \triangle EFG$ by the ASA Postulate.

2. $\overline{FG}$

The definition of congruent polygons tells you that if two triangles are congruent, then all pairs of corresponding parts are congruent. You can use this idea to prove that the distance across the stream is the same as the distance along the bank.

EXAMPLE 1

Given: $\angle EFD$ and $\angle EFG$
are right angles.
$\angle DEF \cong \angle GEF$
Prove: $\overline{DF} \cong \overline{GF}$

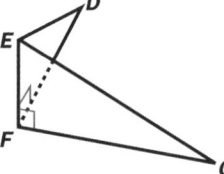

SOLUTION

Plan Ahead: Prove that $\triangle EFD \cong \triangle EFG$. Then $\overline{DF} \cong \overline{GF}$ because they are corresponding parts of congruent triangles.

Statements	Reasons
1. $\angle EFD$ and $\angle EFG$ are right angles.	1. Given
2. $\angle EFD \cong \angle EFG$	2. All right angles are congruent.
3. $\overline{EF} \cong \overline{EF}$	3. Reflexive Property
4. $\angle DEF \cong \angle GEF$	4. Given
5. $\triangle EFD \cong \triangle EFG$	5. ASA Postulate
6. $\overline{DF} \cong \overline{GF}$	6. Definition of congruent triangles

When applying the definition of congruent polygons to triangles in a proof, write *definition of congruent triangles*.

Perpendicular Bisectors

If point A is the same distance from point B as it is from point C, then A is *equidistant* from B and C. You will use this idea to learn more about perpendicular bisectors.

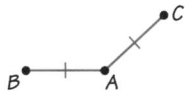

THINK AND COMMUNICATE

$\overleftrightarrow{PR}$ **is the perpendicular bisector of** $\overline{QS}$.

3. Is R equidistant from Q and S? Explain why or why not.

4. Which theorem or postulate tells you that $\triangle PQR \cong \triangle PSR$?

5. Explain why P is equidistant from Q and S.

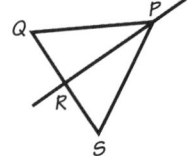

Think and Communicate

3. Yes; $\overleftrightarrow{PR}$ bisects $\overline{QS}$, that is, intersects it at its midpoint. So R is the midpoint of $\overline{QS}$ and $QR = RS$.

4. SAS Postulate

5. $\overline{PQ} \cong \overline{PS}$ by the definition of congruent triangles, so $PQ = PS$.

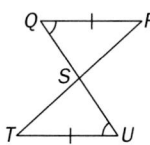

Think and Communicate

Questions 3–5 on page 307 lead students through the reasoning for the proof of the Perpendicular Bisector Theorem. Students may want to refer to their responses to these questions when they complete the proof in Ex. 7 on page 310.

Section Note

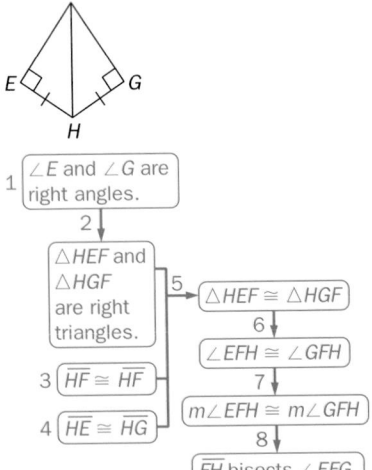

Communication: Discussion Discuss with students the paragraph on this page which identifies the various types of statements that can be proved by using corresponding parts of congruent triangles. Discuss how corresponding parts can be used in each case. For example, to prove that a point is a midpoint, show that the two segments formed by the point are corresponding sides of congruent triangles.

Additional Example 2

Given: $\overline{HE} \cong \overline{HG}$
$\angle E$ and $\angle G$ are right angles.
Prove: $\overline{FH}$ bisects $\angle EFG$.

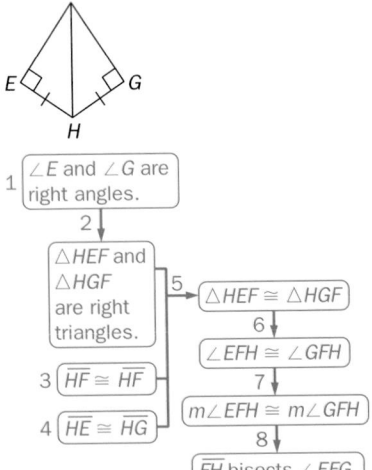

Reasons
1. Given
2. Def. of right triangle
3. Reflexive Property
4. Given
5. HL Theorem
6. Def. of congruent triangles
7. Def. of congruence
8. Def. of angle bisector

The questions on the previous page lead to the Perpendicular Bisector Theorem. You will prove this theorem in the exercises.

The Perpendicular Bisector Theorem

If a point is on the perpendicular bisector of a segment, then the point is equidistant from the endpoints of the segment.

Any point on ℓ is equidistant from A and B.

You can use corresponding parts of congruent triangles to prove statements about midpoints, angle bisectors, segment bisectors, parallel segments, isosceles triangles, and other figures.

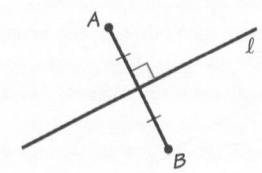
EXAMPLE 2

Given: $\overline{JK} \cong \overline{LM}$; $\overline{JM} \cong \overline{LK}$
Prove: $\overline{JM} \parallel \overline{LK}$

SOLUTION

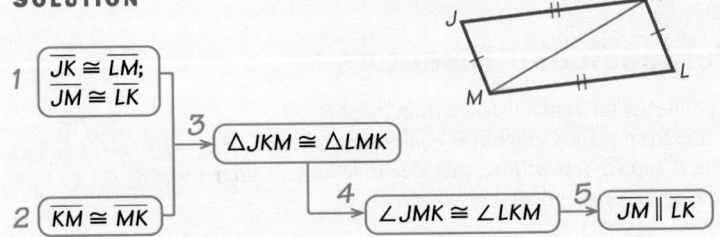

Reasons
1. Given
2. Reflexive Property
3. SSS Postulate
4. Definition of congruent triangles
5. If two lines are intersected by a transversal and alternate interior angles are $\cong$, then the lines are $\parallel$.

308   Chapter 6   *Conjectures About Triangles*

Checking Key Concepts

1. $\triangle DGH$ and $\triangle EGF$
2. $\triangle DEH$ and $\triangle EDF$
3. $\triangle DEF$ and $\triangle EDH$
4. $\triangle XWY$ and $\triangle ZYW$ or $\triangle WVX$ and $\triangle YVZ$
5. $\triangle XVY$ and $\triangle ZVW$ or $\triangle WVX$ and $\triangle YVZ$
6. $\triangle VYX$ and $\triangle VWX$
7. P, R 8. 4
9. isosceles

✓ CHECKING KEY CONCEPTS

Name a pair of triangles that must be congruent in order for you to use the definition of congruent triangles to prove each statement.

1. $\overline{GH} \cong \overline{GF}$

2. $\overline{HE} \cong \overline{FD}$

3. $\angle EDF \cong \angle DEH$

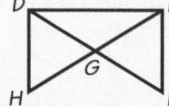

4. $\angle XWY \cong \angle ZYW$

5. $\overline{VX} \cong \overline{VZ}$

6. $\angle WYX \cong \angle YWX$

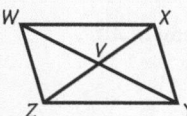

$\overleftrightarrow{QS}$ **is the perpendicular bisector of** $\overline{PR}$.

7. Complete: T is equidistant from points $\underline{?}$ and $\underline{?}$.

8. Find the length QR.

9. What type of triangle is $\triangle PSR$?

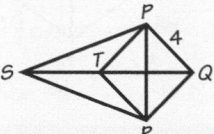

6.5 Exercises and Applications

Extra Practice exercises on page 672

Name a pair of triangles that must be congruent in order for you to use the definition of congruent triangles to prove each statement.

1. $\overline{BE} \cong \overline{CE}$

2. $\angle ABE \cong \angle DCE$

3. $\overline{AC} \cong \overline{DB}$

4. $\angle BCA \cong \angle CBD$

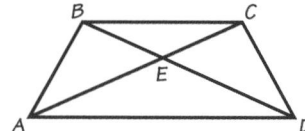

5. **Given:** $\overline{PS} \parallel \overline{QR}$
   $\angle P \cong \angle R$

   Prove: $\overline{PQ} \cong \overline{RS}$

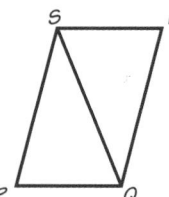

6. **Given:** $\angle ABC$ and $\angle DEF$ are right angles.
   $\overline{AB} \cong \overline{DE}$
   $\overline{BC} \cong \overline{EF}$

   Prove: $\angle BAC \cong \angle EDF$

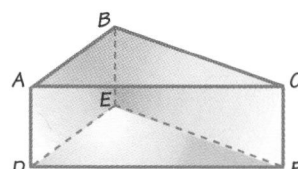

6.5 Applying Congruence   **309**

Checking Key Concepts

Writing Proofs
Questions 1–6 can lead to a discussion of the proof-writing strategy of working backward. If students were required to prove statements such as these, it might help them to start with the conclusion, look for the congruent triangles, and then look at the Given to see if they can prove these triangles are congruent. Questions 7–9 give students some ideas of how the Perpendicular Bisector Theorem can be used.

Closure Question

Describe a method for proving that a part of one triangle is congruent to a part of another triangle. Use information that is given or can be found by using other theorems or postulates to prove that the two triangles are congruent. This can be done by using either SSS, SAS, ASA, AAS, or HL. Then, because the triangles are congruent, their corresponding parts are congruent.

Apply⇔Assess

Suggested Assignment

❖ **Core Course**
   Exs. 1–9, 12–15, 17–25, AYP
❖ **Extended Course**
   Exs. 1–25, AYP
❖ **Block Schedule**
   Day 37 Exs. 1–9, 12–15, 17–25, AYP

Exercise Notes

Student Progress
Exs. 5, 6 Students should be able to write both of these proofs. Remind students having difficulty that they should mark the diagrams with the given information and write a plan before beginning. Also, encourage students to use Example 1 and Example 2 in this section as models for their proofs.

Exercises and Applications

1. $\triangle ABE$ and $\triangle DCE$

2. $\triangle ABD$ and $\triangle DCA$ or $\triangle ABE$ and $\triangle DCE$

3. $\triangle ABC$ and $\triangle DCB$ or $\triangle ABD$ and $\triangle DCA$

4. $\triangle ABC$ and $\triangle DCB$

5. Statements (Reasons)
1. $\overline{PS} \parallel \overline{QR}$ (Given)
2. $\angle PSQ \cong \angle RQS$ (If two $\parallel$ lines are intersected by a transversal, then alternate interior $\angle$s are $\cong$.)
3. $\angle P \cong \angle R$ (Given)
4. $\overline{QS} \cong \overline{QS}$ (Reflexive Property)
5. $\triangle PSQ \cong \triangle RQS$ (AAS Theorem)
6. $\overline{PQ} \cong \overline{RS}$ (Def. of $\cong \triangle$s)

6. Statements (Reasons)
1. $\angle ABC$ and $\angle DEF$ are right $\angle$s. (Given)
2. $\angle ABC \cong \angle DEF$ (All right $\angle$s are $\cong$.)
3. $\overline{AB} \cong \overline{DE}$ and $\overline{BC} \cong \overline{EF}$ (Given)
4. $\triangle ABC \cong \triangle DEF$ (SAS Postulate)
5. $\angle BAC \cong \angle EDF$ (Def. of $\cong \triangle$s)

Integrating Algebra

Exs. 8, 9 In these exercises, variable expressions are used to represent the lengths of two segments. Students must use geometric properties to write an equation and then solve the equation to find the value of x or y.

Application

Exs. 10, 11 These exercises expose students to one of the ways congruent triangles can be used to solve problems encountered by aerial surveyors.

Career Connection

Exs. 10, 11 A surveyor is a person who determines the dimensions or layout of a particular area, most often land areas, boundaries of construction sites, and property lines. Triangulation is one of the methods surveyors use to determine great distances. Topographic surveyors determine the shape of the ground surface, and most commonly use aerial photography to produce maps. Hydrographic surveyors make maps of the bottoms of large bodies of water. Land surveyors mark boundaries for property lines. Construction surveyors set boundaries to control the construction of buildings, bridges, highways, and other projects. Surveyors are usually employed by federal or local governments, or by civil engineering firms, construction firms, or small surveying businesses. Surveying courses are generally found under engineering or construction disciplines, although some colleges offer an undergraduate degree in surveying. Surveying jobs generally require several years of practical work experience as well as some formal education.

310

7. Copy and complete the key steps of this proof of the Perpendicular Bisector Theorem.

   Given: Line ℓ is the perpendicular bisector of $\overline{AB}$.

   Prove: For any point P on ℓ, $PA = PB$.

   1. Draw $\overline{PA}$ and $\underline{\ ?\ }$. (Through any two points there is $\underline{\ ?\ }$ line.)

   2. $\ell \perp \overline{AB}$ and C is the midpoint of $\overline{AB}$. (Definition of $\underline{\ ?\ }$)

   3. $\triangle APC \cong \triangle BPC$ $(\underline{\ ?\ })$

   4. $\underline{\ ?\ }$ $(\underline{\ ?\ })$

ALGEBRA In each diagram, ℓ is the perpendicular bisector of $\overline{AB}$.

8. Find the value of x.

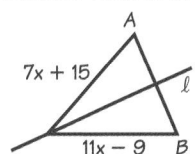

9. Find the value of y.

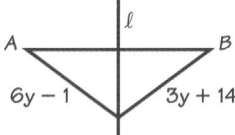

Connection ▸ AERIAL PHOTOGRAPHY

Aerial surveyors use airplanes to survey large areas of land. They fly back and forth over the area, taking pictures with a camera pointed straight down. If the area being photographed is not flat, the surveyor may need to adjust the plane's altitude during the flight.

10. **Writing** Suppose the surveyor takes all of the pictures from the same altitude above sea level. Will photograph 1 show as much of the ground as photograph 2? Explain why or why not.

11. To be sure that all of the photographs cover the same amount of ground (and have the same scale), the pilot adjusts the plane's altitude so that the camera is always the same height above the ground.

    Given: $\overline{MB} \cong \overline{NQ}$

    $\angle AMB \cong \angle BMC \cong \angle PNQ \cong \angle QNR$

    $\angle ABM \cong \angle PQN$

    Prove: $\overline{AC} \cong \overline{PR}$

    (*Hint:* Use congruent triangles to show that $\overline{AB} \cong \overline{PQ}$. Then use congruent triangles to show that $\overline{BC} \cong \overline{QR}$.)

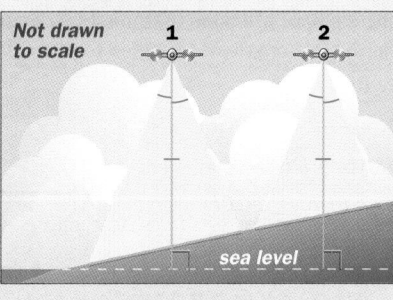

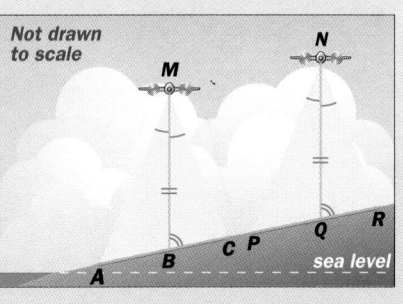

310   Chapter 6 *Conjectures About Triangles*

---

7. (1) $\overline{PB}$; exactly one; (2) perpendicular bisector; (3) SAS Postulate; (4) $PA = PB$; Definition of congruent triangles

8. 6   9. 5

10. Photograph 1 will show more ground than photograph 2 because the airplane is farther from the ground at position 1 than at position 2.

11. It is given that $\angle AMB \cong \angle PNQ$, $\angle ABM \cong \angle PQN$, and $\overline{MB} \cong \overline{NQ}$.

Therefore, $\triangle ABM \cong \triangle PQN$ by the ASA Postulate. $\overline{AB} \cong \overline{PQ}$ by the definition of congruent triangles. It is given that $\angle BMC \cong \angle QNR$, $\angle MBC \cong \angle NQR$, and $\overline{MB} \cong \overline{NQ}$. Therefore, $\triangle BMC \cong \triangle QNR$ by the ASA Postulate. $\overline{BC} \cong \overline{QR}$ by the definition of congruent triangles. By substitution, $AB + BC = PQ + QR$. Since $AB + BC = AC$ and $PQ + QR = PR$, $\overline{AC} \cong \overline{PR}$ by substitution. Therefore, $\overline{AC} \cong \overline{PR}$.

12. a. Check students' work. $\overline{ZX} \cong \overline{ZY}$ because the same radius was used to draw the arcs in Step 2.

    b. $\triangle BXZ \cong \triangle BYZ$ by the SSS Postulate. $\angle XBZ \cong \angle YBZ$ by the definition of congruent triangles. Then by the definition of an angle bisector, $\overrightarrow{BZ}$ bisects $\angle ABC$.

13. Check students' work.

14. Both constructions involve drawing arcs or pairs of arcs to determine points. The construction of the perpendicular bisector involves constructing a line, so two points must be constructed. Two pairs of intersecting arcs, all with the same radius are drawn. The second construction of the bisector of an angle involves constructing a ray. The endpoint of the ray is the vertex of the angle, so only one point must be constructed. A single arc and one pair of intersecting arcs must be drawn.

ANGLE BISECTOR

Given an angle, construct its bisector.

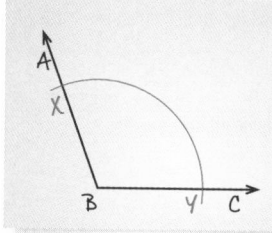 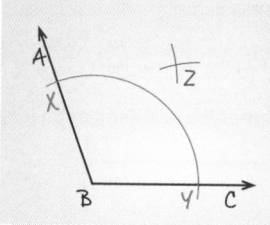 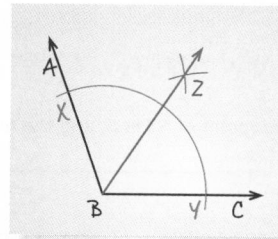

1. Using center *B* and any radius, draw **an arc** that intersects $\overrightarrow{BA}$ and $\overrightarrow{BC}$. Label the points of intersection *X* and *Y*.

2. Draw **an arc** with center *X* and a convenient radius. Draw **an arc** with center *Y* and the same radius. Label the intersection of the arcs *Z*.

3. Draw $\overrightarrow{BZ}$. $\overrightarrow{BZ}$ bisects ∠*ABC*.

12. a. Sketch the figure in Step 3. Include $\overline{ZX}$ and $\overline{ZY}$. How do you know that $\overline{ZX} \cong \overline{ZY}$?

    b. Explain how you know that $\overrightarrow{BZ}$ bisects ∠*ABC*.

13. Open-ended Problem Draw an obtuse angle and construct its bisector.

14. Writing Compare the constructions for an angle bisector and a segment bisector. How are the constructions alike? How are they different?

15. Writing State the converse of the Perpendicular Bisector Theorem. Do you think that it is true? Explain why or why not.

16. Challenge Copy the figure. Use constructions to locate a point that is equidistant from *X*, *Y*, and *Z*.

•*Y*

X•

•*Z*

17. Given: $\overline{WQ} \cong \overline{YQ}$; $\overline{XQ} \cong \overline{ZQ}$
    Prove: $\overline{WX} \parallel \overline{YZ}$

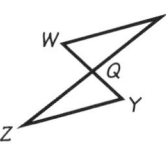

18. Given: $\overleftrightarrow{SQ}$ is the perpendicular bisector of $\overline{PR}$.
    Prove: ∠*SPT* ≅ ∠*SRT*

19. SAT/ACT Preview In △*RST* and △*ABC*, $\overline{RS} \cong \overline{AB}$, $\overline{ST} \cong \overline{BC}$, and $\overline{TR} \cong \overline{CA}$. Which angle is congruent to ∠*T*?

A. ∠*R*   **B.** ∠*A*   **C.** ∠*B*   **D.** ∠*C*   **E.** cannot be determined

6.5 Applying Congruence   **311**

15. If a point is equidistant from the endpoints of a segment, then the point is on the perpendicular bisector of the segment. Yes; answers may vary. For example, suppose a point *P* is equidistant from the end points of $\overline{AB}$. Then *PA* = *PB*. Draw the line through *P* and perpendicular to $\overline{AB}$, intersecting $\overline{AB}$ at *Q*. (Through a point not on a given line, there is exactly one line perpendicular to a given line.) $\overline{PQ} \cong \overline{PQ}$, so by the HL Theorem,

△*APQ* ≅ △*BPQ*. Therefore, by the definition of congruent triangles, $\overline{AQ} \cong \overline{BQ}$, and so point *Q* is the midpoint of $\overline{AB}$. $\overleftrightarrow{PQ} \perp \overline{AB}$ and bisects $\overline{AB}$, so $\overleftrightarrow{PQ}$ is the perpendicular bisector of $\overline{AB}$.

16. Check students' work. The point that is equidistant from *X*, *Y*, and *Z* is the intersection of the perpendicular bisectors of $\overline{XY}$, $\overline{YZ}$, and $\overline{XZ}$.

17. Statements (Reasons)
1. ∠*WQX* ≅ ∠*YQZ* (Vertical ∠s are ≅.)
2. $\overline{WQ} \cong \overline{YQ}$ and $\overline{XQ} \cong \overline{ZQ}$ (Given)
3. △*WQX* ≅ △*YQZ* (SAS Postulate)
4. ∠*W* ≅ ∠*Y* (Def. of ≅ △s)
5. $\overline{WX} \parallel \overline{YZ}$ (If two lines are intersected by a transversal and alternate interior ∠s are ≅, the lines are ∥.)

18. $\overleftrightarrow{SQ}$ is the perpendicular bisector of $\overline{PR}$ (given), so *SP* = *SR* and *TP* = *TR* by the Perpendicular Bisector Theorem. *ST* = *ST* by the Reflexive Property, so △*STP* ≅ △*STR* (SSS Postulate). Then ∠*SPT* ≅ ∠*SRT* by the definition of congruent triangles.

19. D

Assess Your Progress

If students have not already done so, you may wish to suggest that they make a list of all the triangle congruence postulates and theorems. They could also include other statements that are used frequently in these types of proofs, such as the Reflexive Property and the properties about angles formed by parallel lines and transversals.

Journal Entry
Students may have different approaches to this problem. Ask several students to read their descriptions to the class so that others can hear the variety of responses possible.

Progress Check 6.3–6.5

See page 329.

Practice 43 *for Section 6.5*

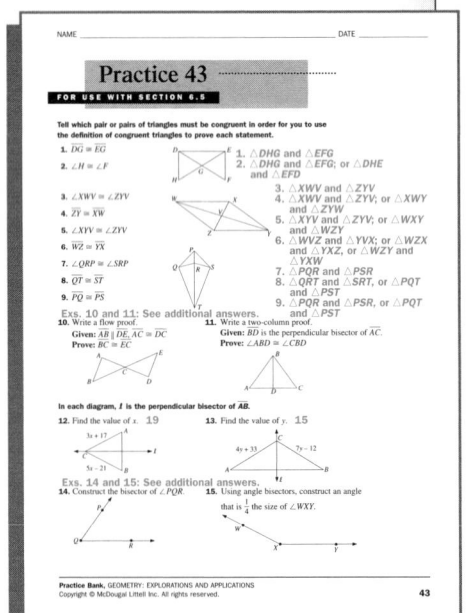

ONGOING ASSESSMENT

20. **Open-ended Problem** In $\triangle PQR$, $\overline{PQ} \cong \overline{PR}$ and S is the midpoint of $\overline{QR}$. Sketch the figure and make a list of everything that you can prove about it.

SPIRAL REVIEW

The midpoint of $\overline{RT}$ is S, and the midpoint of $\overline{RU}$ is T. Find each length. *(Section 1.7)*

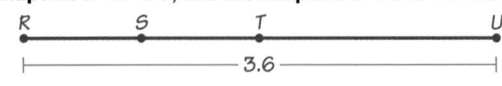

21. RT	22. ST	23. SU

$\overrightarrow{NP}$ bisects $\angle MNO$. Find each measure. *(Section 1.7)*

24. $m \angle MNP$

25. $m \angle MNO$

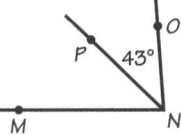

ASSESS YOUR PROGRESS

Tell whether you can prove that the triangles are congruent. If you can, tell which method(s) you can use. *(Sections 6.3 and 6.4)*

1.	2.	3.

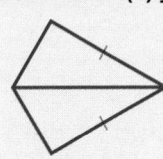 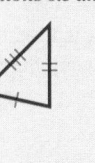 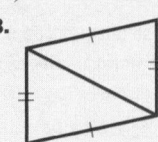

4.	5.	6.

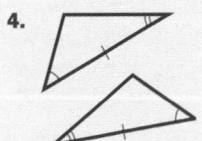 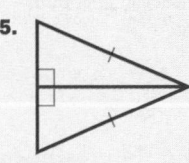 

Name a pair of triangles that must be congruent in order for you to use the definition of congruent triangles to prove each statement. *(Section 6.5)*

7. $\angle YZV \cong \angle XWV$ 8. $\overline{ZX} \cong \overline{WY}$

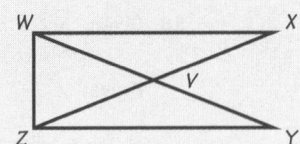

9. Suppose $\overleftrightarrow{AB}$ is the perpendicular bisector of $\overline{CD}$. Sketch the figure using only points A, B, C, and D. Include two pairs of congruent segments and label them. *(Section 6.5)*

10. **Journal** Describe how to use congruent triangles to find the distance between two points at opposite ends of a lake. Include a sketch in your explanation.

312 Chapter 6 *Conjectures About Triangles*

---

20. Answers may vary. An example is given.

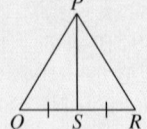

$\triangle PQR$ is an isosceles triangle; $\overline{QS} \cong \overline{RS}$; $\overline{PS} \cong \overline{PS}$; $\triangle PQS \cong \triangle PRS$; $\angle Q \cong \angle R$, $\angle PSQ \cong \angle PSR$, and $\angle QPS \cong \angle RPS$; $\overrightarrow{PS}$ bisects $\angle QPR$; $\overline{PS} \perp \overline{QR}$; $\overleftrightarrow{PS}$ is the perpendicular bisector of $\overline{QR}$.

21. 1.8

22. 0.9

23. 2.7

25. 43°

26. 86°

Assess Your Progress

1. No.

2. Yes; SSS Postulate or SAS Postulate.

3. Yes; SSS Postulate.

4. Yes; ASA Postulate or AAS Theorem.

5. Yes; HL Theorem or SSS Postulate.

6. Yes; ASA Postulate or AAS Theorem.

7. $\triangle YVZ$ and $\triangle XVW$

8. $\triangle XWZ$ and $\triangle YZW$

9. Answers may vary. An example is given.

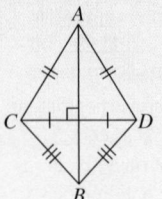

10. Check students' work.

6.6 Properties of Isosceles Triangles

Plan⇔Support

Learn how to...

• apply the Isosceles Triangle Theorem and its converse

So you can...

• find measures in isosceles triangles

• write proofs using isosceles triangles

In one type of rock climbing, called *top roping*, climbers tie themselves to a rope that is supported by *anchors* at the top of the climb. If a climber slips, the anchors catch the fall. For safety, climbers use at least two anchors.

The illustration below shows two anchors in a horizontal crack of a rock face. To be sure that the force on the anchors is equal, the angles formed by the anchors and the crack should be congruent.

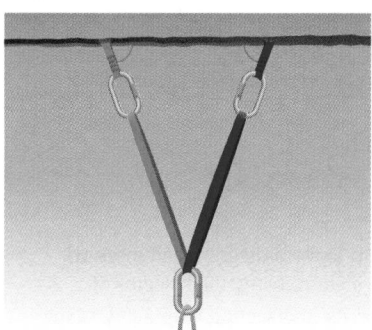

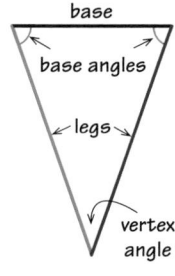

base

base angles

legs

vertex angle

isosceles triangle

The triangle formed by these anchors is an isosceles triangle. The sides opposite the congruent angles are the **legs** and the third side is the **base**. The congruent angles are **base angles** and the third angle is the **vertex angle**.

THINK AND COMMUNICATE

1. a. If the red anchor is longer than the blue anchor, are the base angles congruent? If not, which angle is larger?

b. Use the triangle inequality theorems on page 280. Which theorem supports your answer to part (a)?

2. If a climber adjusts the anchors so they are the same length, do you think that the base angles will be congruent?

6.6 Properties of Isosceles Triangles **313**

Objectives

• Apply the Isosceles Triangle Theorem and its converse.

• Find measures in isosceles triangles.

• Write proofs using isosceles triangles.

Recommended Pacing

❖ **Core and Extended Courses**
Section 6.6: 1 day

❖ **Block Schedule**
Section 6.6: $\frac{1}{2}$ block
(with Section 6.5)

Resource Materials

Lesson Support
Lesson Plan 6.6

Warm-Up Transparency 6.6

Practice Bank: Practice 44

Study Guide: Section 6.6

Explorations Lab Manual: Additional Exploration 9

Challenge Problems: Set 44

Technology
Graphing Calculator

Internet:
http://www.hmco.com

Warm-Up Exercises

Complete with < or >.

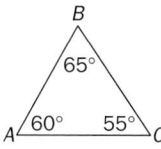

B

65°

60° 55°

A   *C*

1. AB _?_ BC <

2. BC _?_ AC <

Solve each equation.

3. $4x - 6 = 3x + 9$ 15

4. $5x - 11 = x - 3$ 2

5. $9x = 12x - 4$ $\frac{4}{3}$

ANSWERS Section 6.6

Think and Communicate

1. a. No; the angle made by the blue anchor.

   b. One side of a triangle is longer than a second side if and only if the angle opposite the first side is larger than the angle opposite the second side.

2. Yes.

Think and Communicate

Questions 1 and 2 help students understand the Isosceles Triangle Theorem and its converse by recalling the second Triangle Inequality Theorem. Because this theorem is a biconditional, it can be used to support both the Isosceles Triangle Theorem and its converse.

Section Notes

 Communication: Writing
Ask students to write the Isosceles Triangle Theorem and its converse as a biconditional statement. (Two sides of a triangle are congruent if and only if the angles opposite the sides are congruent.)

Geometric Thinking

It is often possible to prove a theorem using different approaches. In Example 1, the Isosceles Triangle Theorem is proved by dividing an isosceles triangle into congruent triangles. In Ex. 7 on page 316, the theorem is proved by showing that the triangle is congruent to its mirror image. To further point out the use of different approaches, you may wish to have students share their proofs of the Converse of the Isosceles Triangle Theorem in Ex. 21 on page 318.

Additional Example 1

Given: $\overline{AB} \cong \overline{CB}$ and $\overline{AE} \cong \overline{CD}$
Prove: $\angle 1 \cong \angle 2$

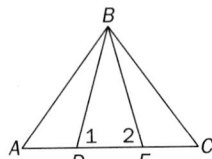

Plan Ahead: Use the Isosceles Triangle Theorem to prove that $\angle A \cong \angle C$. Then prove that $\triangle ABE \cong \triangle CBD$. Use the Isosceles Triangle Theorem again to prove that $\angle 1 \cong \angle 2$.

Statements (Reasons)
1. $\overline{AB} \cong \overline{CB}$ (Given)
2. $\angle A \cong \angle C$ (Isosceles Triangle Theorem)
3. $\overline{AE} \cong \overline{CD}$ (Given)
4. $\triangle ABE \cong \triangle CBD$ (SAS Postulate)
5. $\overline{BE} \cong \overline{BD}$ (Def. of congruent triangles)
6. $\angle 1 \cong \angle 2$ (Isosceles Triangle Theorem)

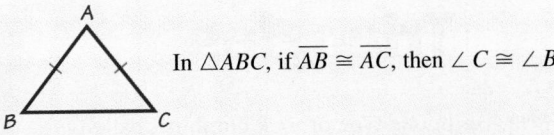

The Isosceles Triangle Theorem

If two sides of a triangle are congruent, then the angles opposite the sides are congruent.

In $\triangle ABC$, if $\overline{AB} \cong \overline{AC}$, then $\angle C \cong \angle B$.

The Converse of the Isosceles Triangle Theorem

If two angles of a triangle are congruent, then the sides opposite the angles are congruent.

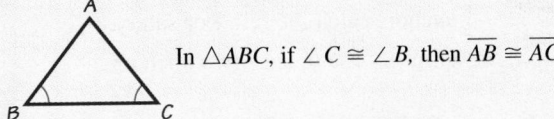

In $\triangle ABC$, if $\angle C \cong \angle B$, then $\overline{AB} \cong \overline{AC}$.

Example 1 shows one way to prove the Isosceles Triangle Theorem. You will prove the Converse of the Isosceles Triangle Theorem in the exercises.

EXAMPLE 1

Prove the Isosceles Triangle Theorem.

SOLUTION

Given: In $\triangle XYZ$, $\overline{XZ} \cong \overline{YZ}$.
Prove: $\angle X \cong \angle Y$

Plan Ahead: Draw a diagram and mark the given information. Draw the bisector of $\angle Z$ and let P be its intersection with $\overline{XY}$. Use the SAS Postulate to prove that $\triangle XZP \cong \triangle YZP$. Therefore, $\angle X \cong \angle Y$.

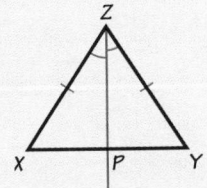

Statements	Reasons
1. $\overline{XZ} \cong \overline{YZ}$	1. Given
2. Draw $\overrightarrow{ZP}$, the angle bisector of $\angle Z$.	2. Angle Bisector Construction (p. 311)
3. $\angle XZP \cong \angle YZP$	3. Definition of angle bisector
4. $\overline{ZP} \cong \overline{ZP}$	4. Reflexive Property
5. $\triangle XZP \cong \triangle YZP$	5. SAS Postulate
6. $\angle X \cong \angle Y$	6. Definition of congruent triangles

EXAMPLE 2   **Connection: Algebra**

Find $m\angle D$.

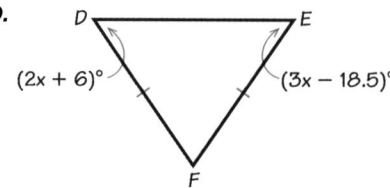

SOLUTION

$\overline{DF} \cong \overline{EF}$ so, by the Isosceles Triangle Theorem, $\angle D \cong \angle E$.

$$m\angle D = m\angle E$$
$$(2x + 6)° = (3x - 18.5)° \qquad \text{◀ First find } x.$$
$$6 = 3x - 18.5 - 2x$$
$$6 + 18.5 = x$$
$$24.5 = x$$
$$m\angle D = (2x + 6)° \qquad \text{◀ Now use } x \text{ to}$$
$$\qquad\qquad\qquad\qquad \text{find } m\angle D.$$
$$= 2(24.5) + 6$$
$$= 49 + 6$$
$$= 55°$$

Check

$$m\angle E = m\angle D$$
$$3x - 18.5 = 55$$
$$3(24.5) - 18.5 \stackrel{?}{=} 55$$
$$73.5 - 18.5 \stackrel{?}{=} 55$$
$$55 = 55 ✔$$

☑ **CHECKING KEY CONCEPTS**

Complete.

1. $AB = \underline{?}$

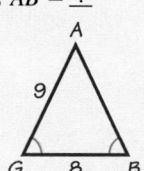

2. $m\angle Z = \underline{?}$

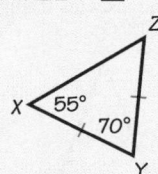

ALGEBRA   **Find each length or angle measure.**

3. $m\angle D = \underline{?}$

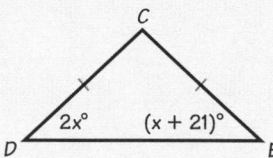

4. $NP = \underline{?}$

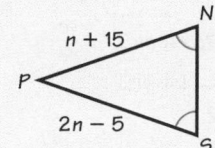

6.6   Properties of Isosceles Triangles   **315**

Checking Key Concepts

1. 9

2. 55°

3. 42°

4. 35

Teach⇔Interact

About Example 2

Teaching Tip
Make sure students see that the expression for $\angle E$ is used to check the solution to this problem. This expression should also be equal to 55 because $\angle D \cong \angle E$.

Additional Example 2

Find LM.

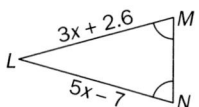

$\angle M \cong \angle N$ so, by the Converse of the Isosceles Triangle Theorem, $\overline{LM} \cong \overline{LN}$. First find x.

$$LM = LN$$
$$3x + 2.6 = 5x - 7$$
$$2.6 = 5x - 7 - 3x$$
$$7 + 2.6 = 2x$$
$$9.6 = 2x$$
$$4.8 = x$$

Now use x to find LM.

$$LM = 3x + 2.6$$
$$= 3(4.8) + 2.6$$
$$= 14.4 + 2.6$$
$$= 17$$

Check

$$LN = LM$$
$$5x - 7 = 17$$
$$5(4.8) - 7 \stackrel{?}{=} 17$$
$$24 - 7 \stackrel{?}{=} 17$$
$$17 = 17 ✓$$

Checking Key Concepts

Student Progress
Students should be able to do questions 1 and 2 successfully. Students who have difficulty with questions 3 and 4 may need to review the procedures for solving linear equations that contain variable expressions on both sides of the equals sign.

Closure Question

Describe the Isosceles Triangle Theorem and its converse. The Isosceles Triangle Theorem states that if two sides of a triangle are congruent, the angles opposite the sides are also congruent. Its converse states that if two angles of a triangle are congruent, the sides opposite the congruent angles are congruent.

315

Suggested Assignment

❖ **Core Course**
Exs. 2–5, 7–17, 19–21, 23–28

❖ **Extended Course**
Exs. 1–7, 9, 11, 13–28

❖ **Block Schedule**
Day 37 Exs. 2–5, 7–17, 19–21, 23–28

Exercise Notes

Application
Ex. 1 This exercise is a practical application of the Converse of the Isosceles Triangle Theorem. Some students may need to refer back to the opening situation on page 313.

Spatial Reasoning
Exs. 2–4 For these exercises, students need to be able to focus on specific components in the drawing. If they have difficulty doing this, suggest that they copy the drawing and mark the congruent parts.

Multicultural Note
Exs. 5, 6 The roofs of houses can be built from diverse materials and take many different forms. All roofs function to cover an area, but some serve additional purposes. The Dogon people of Mali build flat roofs above their kitchens, with a ladder leading from the kitchen to the top of the roof. During the hot season, the roof doubles as a sleeping area. In Togo, the conical shape of the straw roofs helps to prevent the torrential rain from rotting the straw and leaking into the house. A traditional Norwegian house has an organic roof, made of bark covered by dirt, roots, and grass. The roots and grass become intertwined as they grow and renew themselves year after year, keeping the house insulated and saving its owners the work of regularly replacing the roof. In Central Asia, the traditional housing unit of many nomads is a round shelter called the *yurt*. The roof of a yurt has a hole in the middle to provide sunlight and to serve as a chimney. When a yurt is pitched facing south, the sunlight that enters through the hole acts as a clock. For similar reasons, many Native American peoples also left holes in the center of their roofs.

6.6 | Exercises and Applications

Extra Practice exercises on page 672

1. **ROCK CLIMBING** A rock climber sets up two anchors in a horizontal crack of a rock face. One anchor is two feet long. How long must the other anchor be for the angles formed by the anchors and the crack to be equal?

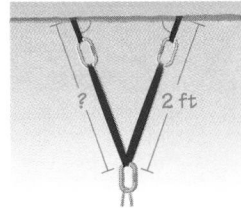

For Exercises 2–4, use the diagram.

2. $\angle 1 \cong \angle 2$. Name two congruent segments.

3. $\overline{KP} \cong \overline{PD}$. Name two congruent angles.

4. $\triangle KPD$ and $\triangle SPR$ are both isosceles.
   a. Are $\triangle KPD$ and $\triangle SPR$ congruent? Explain.
   b. Are $\triangle KPS$ and $\triangle DPR$ congruent? Explain.

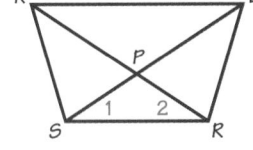

ARCHITECTURE For Exercises 5 and 6, use the photographs of roofs from around the world.

Lithuania

Japan

United States

Germany

5. Which roofs appear to form isosceles triangles?

6. **Open-ended Problem** Give some possible reasons why isosceles triangles are so common in roof building. When might it make more sense to use a scalene triangle instead of an isosceles triangle?

7. Copy and complete this alternate proof of the Isosceles Triangle Theorem.
   Given: In $\triangle ABC$, $\overline{AB} \cong \overline{AC}$.
   Prove: $\angle B \cong \angle C$

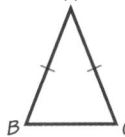

   Plan Ahead: Imagine picking $\triangle ABC$ up and flipping it over so that $\overline{AB}$ lands on $\overline{AC}$. Use the SAS Postulate to show that $\triangle BAC \cong \triangle CAB$.

Statements	Reasons
1. ?	1. Given
2. $\angle A \cong \angle A$	2. ?
3. $\triangle BAC \cong \triangle CAB$	3. ?
4. ?	4. ?

Exercises and Applications

1. 2 ft

2. $\overline{PS}$ and $\overline{PR}$

3. $\angle PKD$ and $\angle PDK$

4. a. No; corresponding sides are not congruent.
   b. Yes; $\overline{PK} \cong \overline{PD}$ and $\overline{PS} \cong \overline{PR}$. Vertical angles $\angle KPS$ and $\angle DPR$ are congruent, so $\triangle KPS \cong \triangle DPR$ (SAS Postulate).

5. the roofs in Lithuania, Japan, and Germany

6. Answers may vary. An example is given. Roofs are often build to be symmetric, both for aesthetic reasons and for practical reasons, such as equally distributing snow accumulation. Symmetry requires the use of isosceles triangles. It might make more sense to use a scalene triangle when a design element, such as a solar panel or a distinctive chimney, imposes special restrictions on the roof.

7. (1) $\overline{AB} \cong \overline{AC}$; (2) Reflexive Property; (3) SAS Postulate; (4) $\angle B \cong \angle C$ and Def. of $\cong$ $\triangle$s

ALGEBRA For Exercises 8–13, find each length or angle measure.

8. AB
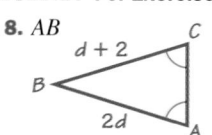

9. PR
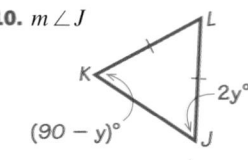

10. $m \angle J$
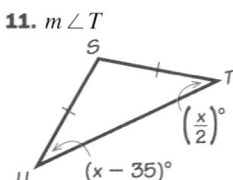

11. $m \angle T$
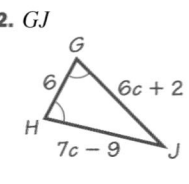

12. GJ
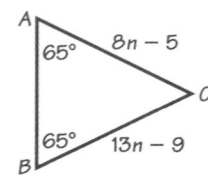

13. ZW
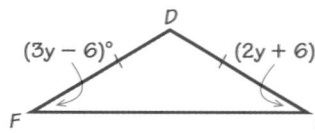

14. Find the value of n and the length of $\overline{AC}$.

15. Find the value of y and $m \angle D$.

Look back at the article on pages 276–278.

Madeleine Fleming designs optical surfaces like the reflective film described on page 290. The back of the film has thousands of tiny pyramids on it. One of the pyramids is shown below.

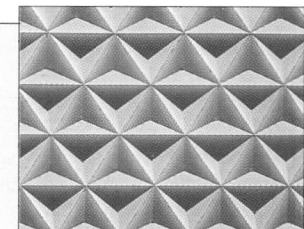

16. Prove that $\triangle ABC \cong \triangle DBC$.

17. **Writing** Explain how you can conclude that $\triangle ABD$ is isosceles.

18. **Writing** Name another isosceles triangle in the pyramid. Explain how you know that it is isosceles.

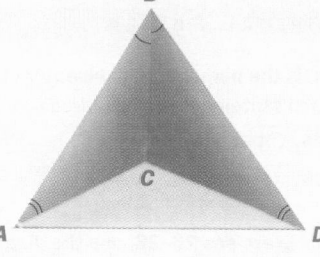

BY THE WAY

In many states and other countries, license plates coated with a retroreflective film have a security mark. The mark can be seen only from a 30° angle above the plate.

8. 4
9. 2
10. 60°
11. 35°
12. 68
13. 5.5
14. 0.8; 1.4
15. 12; 120°

16. $\angle A \cong \angle D$, $\angle ABC \cong \angle DBC$, and $\overline{BC} \cong \overline{BC}$, so $\triangle ABC \cong \triangle DBC$ by the AAS Theorem.

17. From Ex. 16, $\triangle ABC \cong \triangle DBC$, so $\overline{AB} \cong \overline{DB}$ by the definition of congruent triangles. Since $\triangle ABD$ has two congruent sides, it is isosceles.

18. $\triangle ACD$; from Ex. 16, $\triangle ABC \cong \triangle DBC$, so $\overline{AC} \cong \overline{DC}$ by the definition of congruent triangles. Since $\triangle ACD$ has two congruent sides, it is isosceles.

Apply⇔Assess

Exercise Notes

Visual Thinking
Ex. 6 Ask students to use photos or sketches to demonstrate the reasons why isosceles triangles are often used in roof building, and to show situations where it would be more appropriate to use scalene triangles. This activity involves the visual skills of *recognition* and *generalization*.

Second-Language Learners
Ex. 7 Students learning English may benefit from a kinesthetic demonstration of a triangle flipping over onto itself.

Integrating Algebra
Exs. 8–15 In these exercises, the lengths of two sides or the measures of two angles of a triangle are represented using algebraic expressions. The Isosceles Triangle Theorem, or its converse, must be used to write an equation.

Common Error
Exs. 8–15 Some students may give the value of the variable as the answer to these exercises. Point out that the variable must be used to find the indicated measure.

Using Technology
Exs. 8–15 The solutions to the equations in these exercises can also be found or checked with a graphing calculator. For example, to use a TI-82 for Ex. 8, press $\boxed{Y=}$ and enter the equations $Y_1 = X + 2$ and $Y_2 = 2X$. Then choose 6:ZStandard from the ZOOM menu. Press $\boxed{2nd}$ $\boxed{CALC}$ and choose 5:intersect. Press $\boxed{ENTER}$ three times and the calculator will display the x- and y- coordinates of the intersection. The solution to the equation $x + 2 = 2x$ is the x-coordinate of the intersection.

Problem Solving
Ex. 13 In this exercise, students need to realize that this is an equiangular triangle and, thus, an equilateral triangle.

Challenge
Ex. 13 After completing this exercise, you may wish to ask students to write a proof that an equiangular triangle is also an equilateral triangle.

Exercise Notes

Topic Spiraling: Review

Ex. 19 This exercise provides students with an opportunity to review the concepts of reflection symmetry and a line of symmetry from Section 1.2, and to see how these ideas apply to isosceles triangles.

Writing Proofs

Ex. 21 You may wish to have students compare and contrast their proof of the Converse of the Isosceles Triangle Theorem with the proof of the Isosceles Triangle Theorem given in Example 1 on page 314.

Topic Spiraling: Preview

Exs. 24–28 These exercises preview some of the concepts about triangles that are examined in the next section.

Practice 44 for Section 6.6

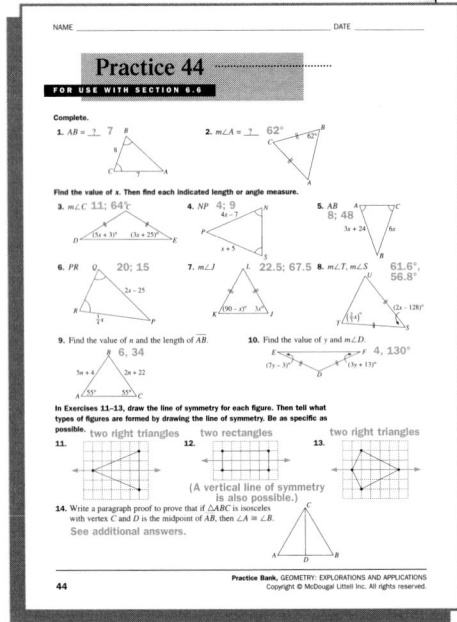

---

19. **TRANSFORMATIONS** In △PQR, ∠Q ≅ ∠R. Explain why △PQR has reflection symmetry and identify the line of symmetry.

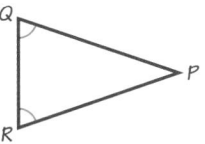

20. **Given:** $\overline{AB} \cong \overline{BC}$; $\overline{DE} \cong \overline{EC}$
    Prove: ∠A ≅ ∠D

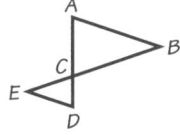

21. Prove the converse of the Isosceles Triangle Theorem.
    Given: In △XYZ, ∠X ≅ ∠Y.
    Prove: $\overline{XZ} \cong \overline{YZ}$

    Plan Ahead: Draw a diagram and mark the given information. Draw the bisector of ∠Z and label its intersection with $\overline{XY}$ point P. Prove that the triangles are congruent. Then use the definition of congruent triangles.

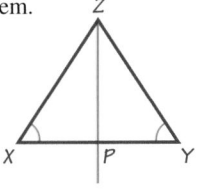

22. **Challenge** In △YSW, $\overline{SY} \cong \overline{SW}$, $\overrightarrow{WR}$ bisects ∠SWY, and $\overrightarrow{YT}$ bisects ∠SYW. Name three pairs of congruent triangles. Justify your answers.

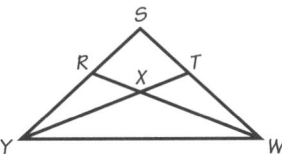

ONGOING ASSESSMENT

23. **Writing** Describe how scissors are like an isosceles triangle. Which part of the scissors are the legs of the triangle? the vertex angle? Why wouldn't it make sense for scissors to be like scalene triangles?

SPIRAL REVIEW

$\overleftrightarrow{PC}$ is the perpendicular bisector of $\overline{AB}$. Tell why each statement is true. *(Section 6.5)*

24. $\overline{CA} \cong \overline{CB}$

25. $\overline{AP} \cong \overline{BP}$

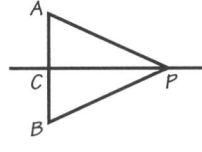

For Exercises 26–28, use the diagram at the right.

26. List all of the angle bisectors shown in the diagram. *(Section 1.7)*

27. List all of the bisectors of $\overline{AB}$. *(Section 1.7)*

28. List all of the pairs of perpendicular lines. *(Section 2.1)*

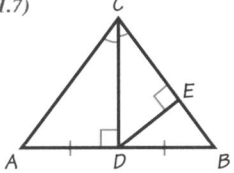

318 Chapter 6 *Conjectures About Triangles*

---

19. As shown in Ex. 7 on page 316, if you could pick up △ PQR and flip it over so that $\overrightarrow{PQ}$ lands on $\overrightarrow{PR}$, the resulting triangle would be congruent to △ PQR. "Flipping the triangle over" is equivalent to reflecting △ PQR over the perpendicular bisector of $\overline{QR}$, which is the line of symmetry. (This line also bisects ∠P.)

20. **Statements (Reasons)**
    1. $\overline{AB} \cong \overline{BC}$; $\overline{DE} \cong \overline{EC}$ (Given)
    2. ∠A ≅ ∠ACB; ∠D ≅ ∠DCE (If two sides of a △ are ≅, then the ∠s opposite those sides are ≅.)
    3. ∠ACB ≅ ∠DCE (Vertical ∠s are ≅.)
    4. ∠A ≅ ∠D (Transitive Property (Steps 2 and 3))

21. Draw the bisector of ∠ XZY. (Every angle has a bisector.) Let P be the intersection of the bisector and $\overline{XY}$. ∠XZP ≅ ∠YZP by the definition of angle bisector. ∠X ≅ ∠Y (given) and $\overline{ZP} \cong \overline{ZP}$ (Reflexive Property). △XPZ ≅ △YPZ (AAS Theorem), and so $\overline{XZ} \cong \overline{YZ}$ by the definition of congruent triangles.

22, 23. See answers in back of book.

24. A line that bisects a segment intersects it at its midpoint.

25. If a point is on the perpendicular bisector of a segment, then the point is equidistant from the endpoints of the segment.

26. $\overrightarrow{CD}$ and $\overrightarrow{ED}$

27. $\overline{DC}$ and $\overline{DE}$

28. $\overleftrightarrow{AB}$ and $\overleftrightarrow{DC}$, $\overleftrightarrow{BC}$ and $\overleftrightarrow{DE}$

6.7

Altitudes, Medians, and Bisectors

Learn how to...
- identify and draw medians and altitudes of triangles

So you can...
- explore the properties of these segments in triangles

There are many different kinds of tents, but almost all of them use poles to support the roof. To set up this tent, you put the bottom of the pole in the middle of the bottom edge of the tent. This design ensures that the pole will be perpendicular to the ground.

Both sides of the roof are the same length.

The bottom of the pole is halfway between the corners of the tent.

The tent pole is both a *median* and an *altitude* of the triangular front of the tent. A **median of a triangle** is a segment from a vertex to the midpoint of the opposite side. An **altitude of a triangle** is a perpendicular segment from a vertex to the line that contains the opposite side.

EXAMPLE 1

For each triangle, *M* is the midpoint of $\overline{AB}$. Copy the triangle and draw the median and altitude from vertex *C*.

a.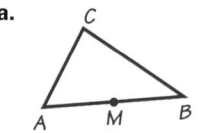

b.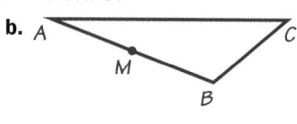

Notice that an altitude can lie outside of the triangle.

SOLUTION

a.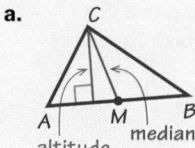

b.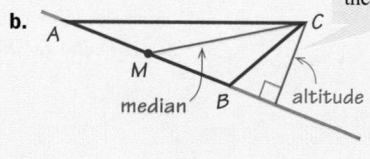

Plan⇔Support

Objectives
- Identify and draw medians and altitudes of triangles.
- Explore the properties of medians and altitudes in triangles.

Recommended Pacing
❖ **Core and Extended Courses**
Section 6.7: 1 day

❖ **Block Schedule**
Section 6.7: $\frac{1}{2}$ block
(with Review)

Resource Materials

Lesson Support
Lesson Plan 6.7

Warm-Up Transparency 6.7

Overhead Visuals:
   Folder 8: Concurrent Lines

Practice Bank: Practice 45

Study Guide: Section 6.7

Challenge Problems: Set 45

Assessment Book: Test 27

Technology
Technology Book:
   Calculator Activity 9

Geometry Software

McDougal Littell Mathpack
   Geometry Inventor Activity Book:
      Activity 6

Internet:
   http://www.hmco.com

Warm-Up Exercises

In the figure, $QT = 4$, $\overline{QS}$ bisects $\angle RQT$, $\overline{WU}$ is the perpendicular bisector of $\overline{QT}$, and $\overline{QR} \cong \overline{QT}$.

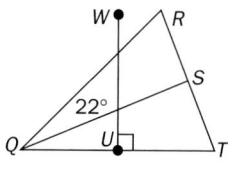

1. $m\angle SQT = \underline{\ ?\ }$ 22°

2. $m\angle RQT = \underline{\ ?\ }$ 44°

3. $m\angle R = \underline{\ ?\ }$ 68°

4. $QU = \underline{\ ?\ }$ 2

5. $UT = \underline{\ ?\ }$ 2

Visual learners may have an easier time remembering the definitions on page 319 if they draw a sketch and label the important parts.

Additional Example 1

Refer to Example 1. Draw P, the midpoint of $\overline{AC}$, and the median and altitude from vertex B.

a.

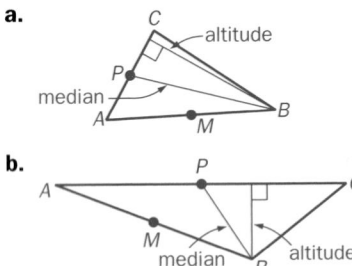

b.

Section Note

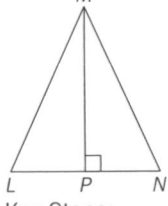

Using Technology
You may wish to have students use geometry software to explore the relationships between the altitude, median, and angle bisectors in scalene, isosceles, and equilateral triangles.

Additional Example 2

In $\triangle LMN$, $\overline{MP}$ is an altitude and a median. Prove $\triangle LMN$ is isosceles.
Given: $\overline{MP}$ is an altitude of $\triangle LMN$.
$\overline{MP}$ is a median of $\triangle LMN$.
Prove: $\triangle LMN$ is isosceles.

Key Steps:
1. $\overline{MP} \perp \overline{LN}$ (Def. of altitude)
2. P is the midpoint of $\overline{LN}$. (Def. of median)
3. $\overline{MP}$ is the perpendicular bisector of $\overline{LN}$. (Def. of perpendicular bisector)
4. $\overline{MP} \cong \overline{MP}$ (Reflexive Property)
5. $\triangle LPM \cong \triangle NPM$ (SAS Postulate, with $\overline{LP} \cong \overline{NP}$ and $\angle LPN \cong \angle NPM$)
6. $\overline{LM} \cong \overline{MN}$ (Def. of congruent triangles)
7. $\triangle LMN$ is isosceles. (Def. of isosceles triangle)

Segments in Isosceles Triangles

The altitude, the median, and the angle bisector from the vertex angle of an isosceles triangle to its base are all the same segment. They are also the same as the perpendicular bisector of the base of the triangle. You can prove these relationships using the definition of congruent triangles.

EXAMPLE 2

In isosceles $\triangle DBC$, $\overline{DB} \cong \overline{DC}$. Prove that the bisector of $\angle D$ is also an altitude of $\triangle DBC$.

SOLUTION

Given: In $\triangle DBC$, $\overline{DB} \cong \overline{DC}$;
$\overrightarrow{DA}$ bisects $\angle BDC$.
Prove: $\overline{DA}$ is an altitude of $\triangle DBC$.

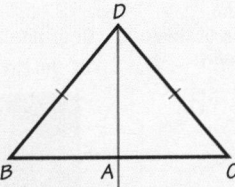

Key steps:
1. $\triangle BDA \cong \triangle CDA$ (SAS Postulate, with $\overline{DA} \cong \overline{DA}$ and $\overline{DB} \cong \overline{DC}$)
2. $\angle DAB \cong \angle DAC$ (Definition of congruent triangles)
3. $\overline{DA} \perp \overline{BC}$ ($\angle DAB$ and $\angle DAC$ are congruent adjacent angles.)
4. $\overline{DA}$ is an altitude of $\triangle DBC$. (Definition of altitude)

THINK AND COMMUNICATE

1. In Example 2, how would the proof be different if you wanted to prove that $\overline{DA}$ is the perpendicular bisector of $\overline{BC}$?

2. In a scalene triangle, are the angle bisectors also altitudes? Are the angle bisectors also medians? Explain your answers or sketch some examples.

3. Sketch a triangle in which the altitude, the median, and the angle bisector from one vertex are three different segments. Can you sketch a triangle in which they are just two different segments? Explain.

4. In $\triangle ACD$, $\angle ADC$ is a right angle and $\overline{AD} \cong \overline{CD}$. $\overline{DB}$ is an altitude of the triangle. How could you prove that $\triangle ABD$ and $\triangle CBD$ are both isosceles?

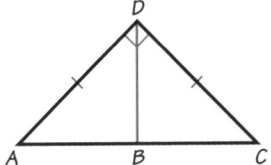

ANSWERS Section 6.7

Think and Communicate

1. You would add a step after Step 3 in which you use the definition of congruent triangles to prove that $\overline{AB} \cong \overline{AC}$, so that A is the midpoint of $\overline{BC}$. The final step would then be that $\overline{DA}$ is the perpendicular bisector of $\overline{BC}$.

2. No; No. Answers may vary. An example is given. In $\triangle PQR$, $\overrightarrow{QS}$ bisects $\angle Q$ and $\overline{QT}$ is the median from vertex Q.

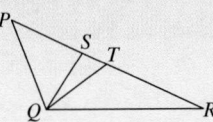

3. Answers may vary. An example is given. In $\triangle DEF$, $\overline{DA}$ is the altitude from vertex D, $\overrightarrow{DB}$ is the bisector of $\angle D$, and $\overline{DM}$ is the median from vertex D.

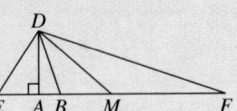

The altitude, median, and angle bisector cannot be just two different segments. For example, if a segment is both an altitude and a median, then it is also a perpendicular bisector. Then the triangle is isosceles and the segment is also an angle bisector.

EXAMPLE 3

In this roof truss, $\overline{WX} \cong \overline{WY}$, $\overrightarrow{WV}$ bisects $\angle XWY$, and $\overline{UT}$ is a median of $\triangle UVX$. Find each length or measure.

a. $m \angle WVX$

b. XT

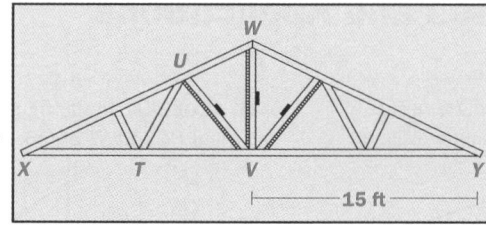

Trusses are made up of beams in triangular patterns. They have the strength and rigidity to span large distances with small amounts of material.

SOLUTION

a. Since $\overrightarrow{WV}$ bisects the vertex angle of an isosceles triangle, it is also the perpendicular bisector of the base.
$$WV \perp XY$$
$$m \angle WVX = 90°$$

b. Since $\overrightarrow{WV}$ is the perpendicular bisector of the base,
$$XV = VY$$
$$= 15$$

Because $\overline{UT}$ is a median of $\triangle UVX$, the midpoint of $\overline{XV}$ is T.

$$XT = \frac{1}{2}(XV) \quad \longleftarrow \text{Definition of midpoint}$$

$$= \frac{1}{2}(15)$$

$$= 7.5$$

☑ CHECKING KEY CONCEPTS

In $\triangle DEF$, M is the midpoint of $\overline{DE}$ and $\angle DQE$ is a right angle.

1. Name an altitude of $\triangle DEF$.

2. Name a median of $\triangle DEF$.

3. Could $\overrightarrow{EQ}$ be an angle bisector of $\triangle DEF$? Explain.

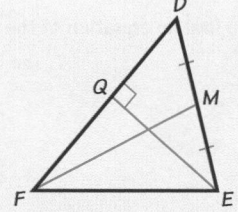

4. Sketch an isosceles triangle with an angle bisector, a median, and an altitude from the same vertex that are not the same segment.

5. In $\triangle PQR$, $\overline{PQ} \cong \overline{QR}$, $\overline{PS}$ is a median, and $PQ = 22$ cm. Find the length of $\overline{SR}$.

6.7 Altitudes, Medians, and Bisectors **321**

Additional Example 3

In $\triangle ACE$, $m \angle AEC = 65°$, $\overline{AC} \cong \overline{AE}$, $\overline{AD}$ is an altitude, and $\overline{BE}$ bisects $\angle AEC$. Find each measure or length.

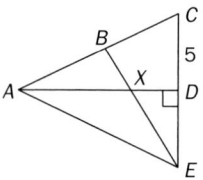

a. $m \angle DXE$

$m \angle AEC = 65°$ and $\overline{BE}$ bisects $\angle AEC$, so
$$m \angle XED = \frac{1}{2}(m \angle AEC)$$
$$= \frac{1}{2}(65°) = 32.5°.$$
Since $\overline{AD}$ is an altitude, $m \angle EDX = 90°$. By the Triangle Sum Theorem, $m \angle DXE + m \angle EDX + m \angle XED = 180°$.
$$m \angle DXE + 90° + 32.5° = 180°$$
$$m \angle DXE = 57.5°$$

b. CE

Since $\overline{AC} \cong \overline{AE}$, $\triangle ACE$ is an isosceles triangle, and thus the altitude from the vertex angle is also a median. By the definition of median, D is the midpoint of $\overline{CE}$. Thus,
$$CE = 2(CD)$$
$$= 2(5)$$
$$= 10$$

Checking Key Concepts

Teaching Tip
Question 4 ties together the concepts of this section. Since students' triangles may vary, you may wish to ask various students to draw their triangles on the board. Be certain to have at least one right isosceles triangle and one obtuse isosceles triangle so that students may compare these cases.

Closure Question

How are a median and an altitude alike? How are they different?
Both a median and an altitude are segments that join a vertex of a triangle with a point on the line that contains the side opposite the vertex. For a median, that point is the midpoint of the side. For an altitude, the point is chosen so that the altitude is perpendicular to the line that contains the opposite side.

---

4. Answers may vary. An example is given. Since $\overline{AD} \cong \overline{CD}$, $\angle A \cong \angle C$, and since $\angle D$ is a right angle, $m \angle A = m \angle C = 45°$. $\overline{DB}$ is an altitude, so $\angle ABD$ and $\angle CBD$ are both right angles and $\triangle ABD$ and $\triangle CBD$ are right triangles. $\overline{DB} \cong \overline{DB}$, so $\triangle ABD \cong \triangle CBD$ (HL Theorem). Then $m \angle ADB = m \angle CDB = 45°$ (definition of congruent $\triangle$s).

Because $\angle A$ and $\angle ADB$ both measure 45°, $\triangle ABD$ is isosceles. Because $\angle C$ and $\angle CBD$ both measure 45°, $\triangle CBD$ is isosceles.

Checking Key Concepts
1. $\overline{EQ}$ 2. $\overline{FM}$
3. Yes; if $\overline{ED} \cong \overline{EF}$, then $\overline{EQ}$ is both an altitude from E and the bisector of $\angle DEF$.

4. Answers may vary. An example is given.

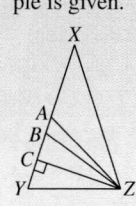

5. 11 cm

321

Suggested Assignment

❖ **Core Course**
Exs. 1–7, 11–16, 18–21, AYP

❖ **Extended Course**
Exs. 1–21, AYP

❖ **Block Schedule**
Day 38 Exs. 1–7, 11–16, 18–21, AYP

Exercise Notes

Integrating Algebra
Exs. 2, 11, 12 In these exercises, variable expressions are used to represent side lengths or angle measures. Students must use the geometric properties learned in this section to write an expression or an equation.

Writing Proofs
Exs. 3–5 The proofs in these exercises all involve isosceles triangles and the fact that the median and altitude from a vertex angle are the same as the bisector of that angle. These proofs can be done by using the method of Example 2 on page 320.

Second-Language Learners
Ex. 5 To make sure that all students learning English know how to work with the conditional statements in Ex. 5, you may want to ask each of them to create a conditional statement using the sentence frame *If it rains tomorrow, then _____*.

Topic Spiraling: Review
Exs. 6, 7 These exercises reinforce the concepts of the section and review concepts from coordinate geometry and algebra, including slope, the equation of a line, and the Midpoint Formula.

6.7 Exercises and Applications

Extra Practice exercises on page 672

Find each length or measure.

1. In $\triangle JKL$, $\overline{JK} \cong \overline{JL}$, and $\overline{JM}$ is both a median and an angle bisector.
  a. $m\angle KMJ$
  b. KL

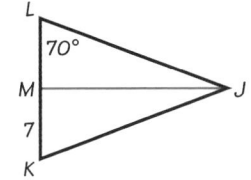

2. ALGEBRA In the figure, $\overline{PQ} \cong \overline{PR}$, and $\overline{PS}$ and $\overline{ST}$ are medians.
  a. QT
  b. QR

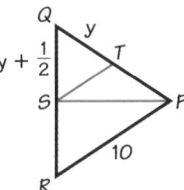

3. a. Write the converse of the theorem in Example 2 on page 320.
  b. Write a plan and prove the statement that you wrote in part (a).

4. a. Copy and complete the proof.

Given: In $\triangle ABC$, $\overline{AC} \cong \overline{BC}$.
$\overline{CD}$ is the median to $\overline{AB}$.
Prove: $\overline{CD}$ is the altitude to $\overline{AB}$.

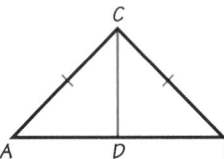

Plan Ahead: Use the SSS Postulate to show that $\triangle ACD \cong \triangle BCD$. Then $\angle ADC$ and $\angle BDC$ are right angles, so $\overline{CD}$ is the altitude to $\overline{AB}$.

b. Write the theorem that you proved in part (a) as an if-then statement.

5. Open-ended Problem $\triangle XYZ$ is an isosceles triangle with legs $\overline{XY}$ and $\overline{XZ}$, and W is on $\overline{YZ}$. Choose one of the following statements and write a proof for it.

If $\overline{XW}$ is the median to $\overline{YZ}$, then it is the perpendicular bisector of $\overline{YZ}$.

If $\overrightarrow{XW}$ bisects $\angle YXZ$, then it is the perpendicular bisector of $\overline{YZ}$.

If $\overline{XW}$ is the altitude to $\overline{YZ}$, then it is the median to $\overline{YZ}$.

If $\overline{XW}$ is the perpendicular bisector of $\overline{YZ}$, then it is an altitude of $\triangle XYZ$.

Sketch each segment or ray. Then find an equation of the line that contains it.

6. the median from P to $\overline{ON}$

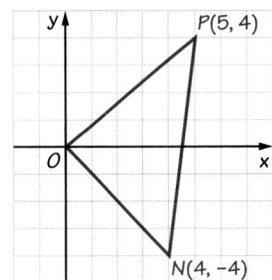

7. the bisector of $\angle P$

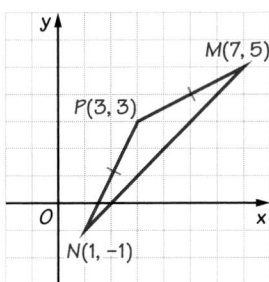

322   Chapter 6 *Conjectures About Triangles*

Exercises and Applications

1. a. 90°   **b.** 14

2. a. 5   **b.** 11

3. a. In an isosceles triangle, the altitude to the base of the triangle is also the bisector of the vertex angle.

  b. Answers may vary. An example is given. Plan of Proof: Use the diagram on page 320. Use the HL Theorem to prove that $\triangle DAB \cong \triangle DAC$. Then $\angle ADB \cong \angle ADC$, so $\overrightarrow{DA}$ bisects $\angle BDC$. Check students' proofs.

4. a. Statements (Reasons)

1. $\overline{AC} \cong \overline{BC}$ and $\overline{CD}$ is the median to $\overline{AB}$. (Given)
2. $\overline{AD} \cong \overline{BD}$ (Def. of median)
3. $\overline{CD} \cong \overline{CD}$ (Reflexive Property)
4. $\triangle ACD \cong \triangle BCD$ (SSS Postulate)
5. $\angle CDA \cong \angle CDB$ (Def. of $\cong \triangle$s)
6. $\overline{CD} \perp \overline{AB}$ (If two lines form $\cong$ adjacent $\angle$s, then the lines are $\perp$.)
7. $\overline{CD}$ is the altitude to $\overline{AB}$. (Def. of altitude)

b. If $\triangle ABC$ is an isosceles triangle and $\overline{CD}$ is the median to $\overline{AB}$, then $\overline{CD}$ is the altitude to $\overline{AB}$.

5. Check students' work. Plans of proof are given. Statement 1: Use the SSS Postulate to prove that $\triangle XWY \cong \triangle XWZ$. Then $\angle XWY$ and $\angle XWZ$ are right angles, and since W is the midpoint of $\overline{YZ}$, $\overline{XW}$ is the perpendicular bisector of $\overline{YZ}$. Statement 2: Use the SAS Postulate to prove that $\triangle XWY \cong \triangle XWZ$. Then $\angle XWY$ and $\angle XWZ$ are right angles, and W is the midpoint of $\overline{YZ}$. Then $\overline{XW}$ is the perpendicular bisector of

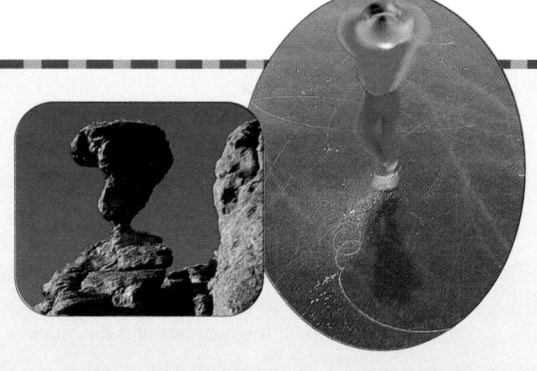

Connection ▷ PHYSICS

Every object has a *center of gravity*. You can balance an object on a pedestal by putting the pedestal directly under the center of gravity. A freely spinning object spins around its center of gravity.

Investigation For Exercises 8 and 9, you will need scissors, a compass, a straightedge, and cardboard.

8. a. Draw a large acute scalene triangle on the cardboard. Cut out the triangle and balance it on the tip of your finger or a pencil. Label the center of gravity.

b. Construct the perpendicular bisector of each side of the triangle to locate its midpoint. (Tape the triangle to a large sheet of paper if necessary.) Carefully draw each median of the triangle. The intersection of the medians is the triangle's *centroid*. Is the centroid the same as the center of gravity?

9. Draw and cut out several cardboard parallelograms. Find the center of gravity for each parallelogram. What intersecting lines can you draw to find the center of gravity of a parallelogram?

10. Writing Sarah wants to make a wind chime. She will hang three metal rods of equal weight from the vertices of a metal triangle. Describe how Sarah can use geometry to find the point at which she can attach a wire so that the triangle will balance.

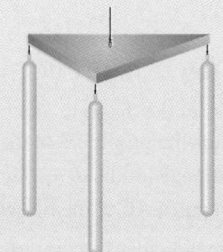

BY THE WAY

Some pagodas in China had more than one hundred wind bells attached to their corners. The sound was believed to drive away evil.

ALGEBRA Find the value of *x*.

11. $\overline{KL}$ is an altitude of $\triangle HJK$.

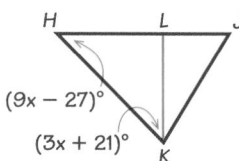

12. $\overline{PO}$ is the perpendicular bisector of $\overline{MN}$.

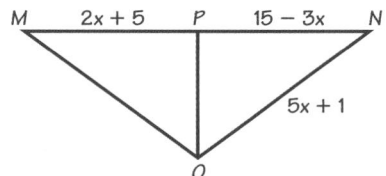

6.7 Altitudes, Medians, and Bisectors   **323**

$\overline{YZ}$. Statement 3: Use the HL Theorem to prove that $\triangle XWY \cong \triangle XWZ$. Then W is the midpoint of $\overline{YZ}$, so $\overline{XW}$ is the median to $\overline{YZ}$. Statement 4: If $\overline{XW}$ is the perpendicular bisector of $\overline{YZ}$, then by the definition of perpendicular bisector, $\overline{XW} \perp \overline{YZ}$, so $\overline{XW}$ is an altitude of $\triangle XYZ$.

6. Check students' work.
$y = 2x - 6$

7. Check students' work.
$y = -x + 6$

8. a. Check students' work.
b. Check students' work. Yes.

9. the lines that contain the diagonals of the parallelogram

10. Answers may vary. An example is given. Sarah needs to locate the midpoint of each side of the triangle. She should draw the three medians of the

triangle. The intersection of the medians is the center of gravity of the triangle.

11. 8

12. 2

323

Apply⇔Assess

Exercise Notes

Interdisciplinary Problems
Exs. 8–10 In these exercises, students see that a special point in a triangle, called the *centroid*, has a physical application. The centroid is the intersection of the medians in a triangle. This point is also the center of gravity of the triangle, or the place where the triangle would balance.

Using Manipulatives
Exs. 8–10 In these exercises, students use cardboard cutouts of triangles and parallelograms to experiment with the center of gravity. In Ex. 8(b), students need to realize that they are constructing the perpendicular bisector only to find the midpoint of each segment. They then need to draw the median from the midpoint to the vertex.

 #### Using Technology
Exs. 8–10 Students can use Activity 6 in the *Geometry Inventor Activity Book* to explore the medians of a triangle, including the centroid.

Cooperative Learning
Exs. 8–10, 13–15 These groups of exercises would make a worthwhile cooperative learning activity. The groups could work together on Exs. 8 and 9 and then discuss Ex. 10. After the discussion, each student should write a response to Ex. 10. After working together on Exs. 13–15, ask students to write an individual paper defining the incenter, centroid, and orthocenter, and describing the properties they have discovered about each point.

Exercise Notes

Geometric Thinking
Ex. 13 Students should realize that they can use the Perpendicular Bisector Theorem to show that the incenter of the triangle is equidistant from each of the three vertices.

Challenge
Ex. 17 Students may have difficulty finding the error in this proof. If so, point out to students that they should not assume that the perpendicular bisector of $\overline{AC}$ contains the point *B*. They first need to prove that the perpendicular bisector contains the point *B*, using the converse of the Perpendicular Bisector Theorem. This theorem was discussed in Ex. 15 on page 311.

Assessment Note
Ex. 18 This exercise gives students an opportunity to compare and contrast the different triangle segments presented in this section. A review of the answers in class would allow students to assess their understanding of the concepts involved.

Topic Spiraling: Preview
Exs. 19–21 Quadrilaterals are the topic of the next chapter. In these exercises, students can use what they have learned about medians, altitudes, and angle bisectors to observe certain facts about the sides, angles, and diagonals of quadrilaterals.

Investigation For Exercises 13–15, you will need geometry software or a compass and straightedge.

13. **a.** Draw a triangle and label it △*XYZ*. Construct the bisectors of the angles and label their intersection *P*. The point *P* is the *incenter* of the triangle. Can the incenter of a triangle ever be outside the triangle? Explain why or why not.

  b. Construct the perpendicular segment from *P* to any side of △*XYZ*. Use the length of the segment as the radius to draw a circle with center *P*. What do you notice? Write a conjecture about the relationship between the incenter and the sides of a triangle.

14. **a.** Draw a triangle and label it △*ABC*. Construct the three medians of the triangle and label them as shown. The intersection, *G*, of the medians is the *centroid* of the triangle. Can the centroid of a triangle ever be outside the triangle? Explain why or why not.

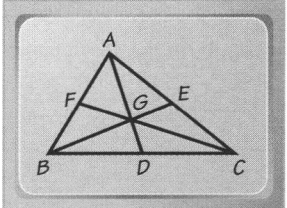

  b. Measure the parts of each median and find $\frac{AG}{GD}$, $\frac{CG}{GF}$, and $\frac{BG}{GE}$. What do you notice?

15. Draw a triangle and label it △*PQR*. Construct the three altitudes of the triangle and label their intersection *C*. *C* is the *orthocenter* of △*PQR*. Can the orthocenter of a triangle ever be outside the triangle? Explain.

16. **SAT/ACT Preview** For the diagram at the right, which of the following statements must be true?

  I. $\overline{QS}$ is a median of △*PQR*.
  II. $\overline{QS}$ is an altitude of △*PQR*.
  III. $\overline{QS}$ is the bisector of ∠*Q*.

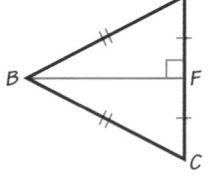

  A. I only   **B.** I and II only   **C.** III only   **D.** I, II, and III   **E.** none of the above

17. **Challenge** Describe the errors in the diagram and plan for this proof. Make an appropriate diagram for the proof and mark what you know.

  Given: △*ABC* is an isosceles triangle with $\overline{AB} \cong \overline{BC}$.

  Prove: The perpendicular bisector of $\overline{AC}$ is the bisector of ∠*B*.

  Plan Ahead: Let $\overleftrightarrow{BF}$ be the perpendicular bisector of $\overline{AC}$. Use the HL Theorem to prove that △*ABF* ≅ △*CBF*. Then ∠*ABF* ≅ ∠*CBF* because of the definition of congruent polygons. So $\overleftrightarrow{BF}$ is the bisector of ∠*ABC*.

ONGOING ASSESSMENT

18. **Visual Thinking** Imagine a scalene triangle, △*ABC*. What do the segments in each group have in common?
  a. the perpendicular bisector of $\overline{AB}$ and the altitude to $\overline{AB}$
  b. the median to $\overline{AB}$ and the perpendicular bisector of $\overline{AB}$
  c. the median to $\overline{AB}$, the altitude to $\overline{AB}$, and the angle bisector of ∠*C*

---

13. **a.** Check students' work. No; each angle bisector lies inside the angle it bisects, that is, inside the triangle in which it is drawn.

  b. Check students' work. The circle touches each side of the triangle in just one point; the incenter of a triangle is equidistant from its sides.

14. **a.** Check students' work. No; each median joins a vertex

of the triangle to the midpoint of the opposite side, so each median lies entirely inside the triangle in which it is drawn.

  b. Each ratio is approximately equal to $\frac{2}{1}$, or 2.

15. Check students' work. Yes; if a triangle is obtuse, then the lines containing the altitudes intersect outside the triangle.

16. D

17. Answers may vary. An example is given. The diagram assumes that the perpendicular bisector of $\overline{AC}$ contains point *B*. In a correct diagram, *F* would be the midpoint of $\overline{AC}$; you would then prove that $\overline{BF} \perp \overline{AC}$ and that ∠*ABF* ≅ ∠*CBF*, so that $\overline{BF}$ is both the perpendicular bisector of $\overline{AC}$ and the bisector of ∠*ABC*. Alternatively, a correct diagram could show $\overline{BF}$, the

Find each length or angle measure. *(Section 2.5)*

19. rhombus *ABCD*
  a. $m \angle B$
  b. *AD*

20. ▱*LMNO*
  a. *LO*
  b. $m \angle M$

21. ▱*WXYZ*
  a. *QY*
  b. *XQ*

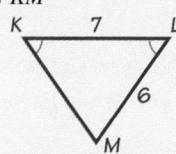 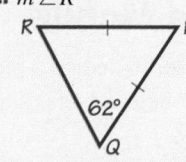 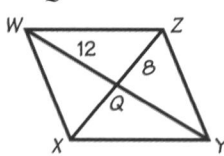

ASSESS YOUR PROGRESS

VOCABULARY

legs of an isosceles
  triangle (p. 313)
base of an isosceles
  triangle (p. 313)
base angles of an
  isosceles triangle (p. 313)

vertex angle of an isosceles
  triangle (p. 313)
median of a triangle (p. 319)
altitude of a triangle (p. 319)

Find each length or angle measure. *(Section 6.6)*

1. *KM*
2. $m \angle R$

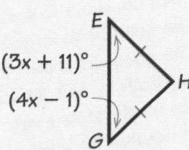

In the diagram, line $n \perp \overleftrightarrow{SH}$, $\overline{JS} \perp \overleftrightarrow{SH}$, and $\overline{GT} \cong \overline{TH}$. *(Section 6.7)*

3. Name a perpendicular bisector.

4. Name an altitude of $\triangle GHJ$.

ALGEBRA **Find each length or angle measure.** *(Sections 6.6 and 6.7)*

5. $m \angle E$

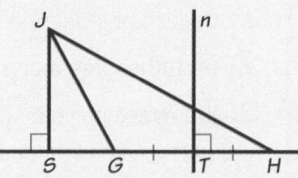

6. $\overline{BD}$ is a median of $\triangle ABC$. Find *AD*.

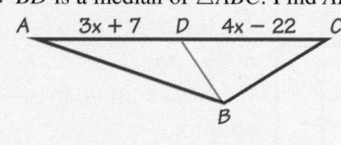

7. Journal Write a summary of what you know about angles and segments in isosceles triangles. How are scalene triangles different from isosceles triangles?

6.7 Altitudes, Medians, and Bisectors **325**

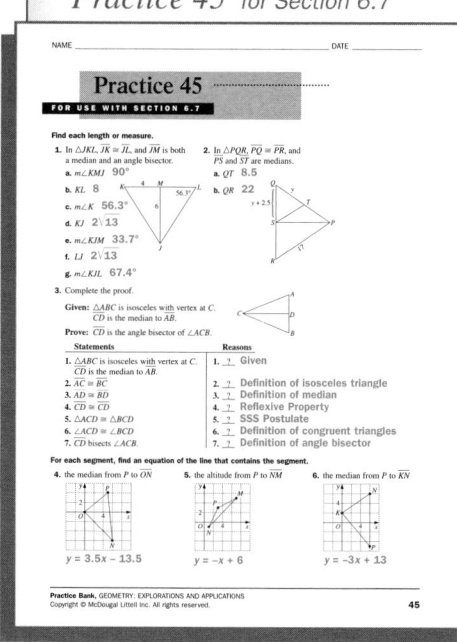
---

perpendicular segment drawn from point *B* to $\overline{AC}$; you would then prove that *F* is the midpoint of $\overline{AC}$ and that $\angle ABF \cong \angle CBF$, so that $\overline{BF}$ is both the perpendicular bisector of $\overline{AC}$ and the bisector of $\angle ABC$.

18. a. Both are perpendicular to $\overline{AB}$.
  b. Both contain the midpoint of $\overline{AB}$.
  c. All three contain point *C* and intersect $\overleftrightarrow{AB}$.

19. a. 130°
  b. 22
20. a. 12
  b. 92°
21. a. 12
  b. 8

Assess Your Progress

1. 6
2. 62°
3. line *n*
4. $\overline{JS}$
5. 47°
6. 94

7. Answers may vary. An example is given. In an isosceles triangle, the base angles are congruent. Also, the segment that joins the midpoint of the base to the opposite vertex is the median to the base, the bisector of the vertex angle, the altitude to the base, and the perpendicular bisector of the base. In a scalene triangle, the median to a side, the altitude to that side, the perpendicular bisector of that side, and the angle bisector that intersects that side would be four different segments.

Mathematical Goals

- Design a mobile that illustrates properties of triangles studied in this chapter.
- Create a mobile that contains at least one pair of congruent triangles and at least one isosceles triangle.
- In the mobile, illustrate angle bisectors, altitudes, medians, and perpendicular bisectors.

Planning

Materials

- cardboard, poster board, or strong paper
- string, yarn, thread, or a similar material
- wire, sticks, or something sturdy
- tape and scissors
- ruler and protractor

Project Teams

Students can select a partner and work together to complete the project. Each pair of students should plan, design, and construct the mobile as a team.

Guiding Students' Work

Encourage students to think of inventive ways to make their mobiles. Suggest that they start by thinking of ways they can illustrate congruent triangles or isosceles triangles in other polygons. Once students have designed their mobile and are beginning to put it together, much of the balancing will be trial and error. This can be frustrating for some students. Encourage them not to give up.

Second-Language Learners

Point out to students that in addition to naming a certain kind of structure, the word *mobile* can be an adjective meaning "able to be moved." It is related to the English words *move* and *movement*. If students are native speakers of Spanish, you might want to point out that the word comes from the same Latin root as the Spanish words *movil*, *mover*, and *movimiento*.

PORTFOLIO PROJECT

Building a Mobile

How can you make a sculpture that looks like it's moving? Sculptors use curves, lines, and texture to create a feeling of movement. Some sculptures actually move. Delicately balanced sculptures that move in the breeze are called *mobiles*.

Antennae with Red and Blue Dots
by Alexander Calder

PROJECT GOAL Create a mobile that illustrates many of the properties of triangles that you have learned in this chapter.

Design the Mobile

Work with a partner to design a mobile. Decide how you will meet each requirement below. Make a sketch of what you would like your mobile to look like.

1. Create at least one pair of congruent triangles.

2. Include at least one isosceles triangle.

3. Illustrate angle bisectors, altitudes, medians, and perpendicular bisectors.

4. Balance at least one triangle at its *centroid*. (See page 323.)

5. Illustrate at least one of the properties or theorems you have learned in this chapter.

We'll use straws to illustrate the first triangle inequality. The green straws don't form a triangle because the sum of the lengths of the shorter straws is less than the length of the longest straw. The pink straws don't form a triangle because

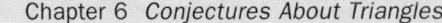

pink   green   orange

Making it Balance

Materials: You will need something to make triangles from, such as cardboard; something to hang the triangles with, such as string; and something sturdy to support the triangles, such as wire or sticks. Tape may also be helpful.

Balance your mobile so it moves in a gentle breeze or when you touch it. It will take some patience to make your mobile balance. Don't give up! If your mobile has more than one level, balance it from the bottom up.

Move the string until you find the point where it balances.

Balance the top last.

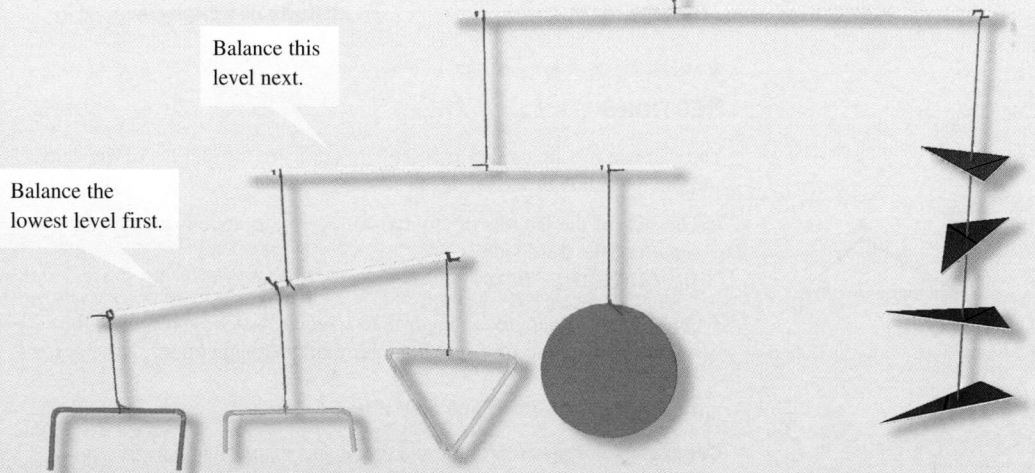

Balance this level next.

Balance the lowest level first.

Presenting Your Project

Write a brief description of your mobile. Include an explanation of how your mobile meets each of the requirements on page 326.

Self-Assessment
Compare your completed mobile with your original design and ideas. If you were to make another mobile, what would you do differently? What tips would you give to someone who wants to build a mobile? What do you like best about your mobile?

Project Notes

Guiding Students' Work

Rubric for Chapter Project

4 Students design and construct a mobile that meets all of the requirements listed for the project. Students may also decide to illustrate additional properties or theorems from the chapter in their mobile. Students balance their mobile and write a clear and concise description of how it was designed and how it illustrates the listed requirements. Students give a self-assessment of the project and answer the questions asked.

3 Students design and construct a mobile that meets most of the requirements listed for the project. The mobile balances and students write a description of how their mobile was designed. However, the description does not explain how all of the requirements were met. Students complete a self-assessment of the project and answer all of the questions asked.

2 Students design a mobile and attempt to construct it, but the mobile does not meet all of the requirements listed for the project and the construction is not adequate because the mobile does not balance. Students write a description of the mobile, but it is not complete. A self-assessment is attempted, but it does not answer all of the questions asked.

1 Students begin a design of a mobile but do not complete its construction. Few of the requirements listed for the project are met and no description written. Students do not complete a self-assessment of the project. Students should be encouraged to speak with the teacher as soon as possible to review their work and to make a new start on the project.

Chapter Support

Course Guide: Chapter 6

Lesson Plans: Chapter 6

Practice Bank:
  Cumulative Practice 46

Study Guide: Chapter 6 Review

Challenge Problems: Set 46

Assessment Book:
  Chapter Tests 28 and 29
  Cumulative Test 30
  Chapter 6 Alternative Assessment

Test Generator Software

Portfolio Project Book:
  Additional Project 3:
    Triangles in Architecture

Preparing for College Entrance Tests

Professional Handbook

Progress Check 6.1–6.2

Can a triangle be formed from segments having the given lengths?
(Section 6.1)

1. 5 cm, 9 cm, 3 cm No.

2. 1.1 in., 1.2 in., 2.2 in. Yes.

3. In $\triangle RST$, $RS = 5$, $RT = 7$, and $\angle R$ is the largest angle. What do you know about ST? *(Section 6.1)* $7 < ST < 12$

4. Name the triangles that appear to be congruent. *(Section 6.2)*

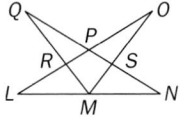

$\triangle QRP \cong \triangle OSP$,
$\triangle LRM \cong \triangle NSM$,
$\triangle LOM \cong \triangle NQM$

5. $\triangle WSA \cong \triangle ABW$. Find the value of t. *(Section 6.2)* 8

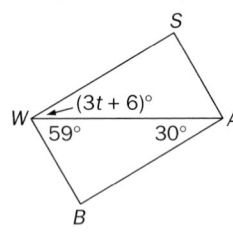

STUDY TECHNIQUE

Work in a group of three students to write a brief summary of Section 6.1. Then each member should write a summary of the major concepts in two more sections of this chapter. Share all of the summaries with your group.

VOCABULARY

congruent polygons (p. 285)

corresponding part (p. 285)

legs of an isosceles triangle (p. 313)

base of an isosceles triangle (p. 313)

base angles of an isosceles triangle (p. 313)

vertex angle of an isosceles triangle (p. 313)

median of a triangle (p. 319)

altitude of a triangle (p. 319)

SECTIONS 6.1 *and* 6.2

You can use two inequality theorems to compare the lengths of the sides and the measures of the angles of a triangle.

- The sum of the lengths of any two sides of a triangle is greater than the length of the third side.
  In $\triangle ABC$, $AB + BC > AC$, $AB + AC > BC$, and $BC + AC > AB$.

- One side of a triangle is longer than a second side if and only if the angle opposite the first side is larger than the angle opposite the second side.
  In $\triangle ABC$, $AB > BC$ if and only if $m \angle C > m \angle A$.

Congruent polygons have the same size and shape. Two polygons are congruent if and only if their **corresponding angles** and **corresponding sides** are congruent. Be sure to list corresponding vertices in the same order.

polygon $ABCD \cong$ polygon $PQRS$

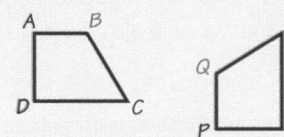

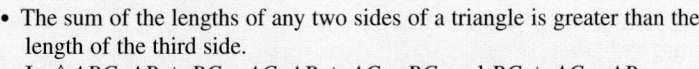
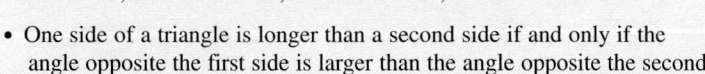

$\overline{AB} \cong \overline{PQ}$ $\angle A \cong \angle P$
$\overline{BC} \cong \overline{QR}$ $\angle B \cong \angle Q$
$\overline{CD} \cong \overline{RS}$ $\angle C \cong \angle R$
$\overline{DA} \cong \overline{SP}$ $\angle D \cong \angle S$

$DCBA \cong SRQP$
$CDAB \cong RSPQ$

SECTIONS 6.3 and 6.4

To prove that two triangles are congruent, you can use the definition of congruent triangles or you can use a shortcut.

SSS Postulate	SAS Postulate	ASA Postulate
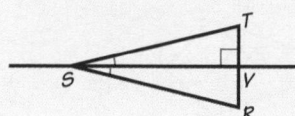		
three sides	2 sides and included angle	2 angles and included side
AAS Theorem	HL Theorem	
two angles and non-included side	hypotenuse and leg	There is no SSA shortcut!

For each shortcut, if the indicated corresponding parts are congruent, then the triangles are congruent.

SECTION 6.5

You can use congruent triangles and perpendicular bisectors to prove that parts of triangles are congruent.

If you prove $\triangle RSV \cong \triangle TSV$, then you can prove that $\angle R \cong \angle T$ and $\overline{RS} \cong \overline{TS}$.

If $\overleftrightarrow{AB}$ is the perpendicular bisector of $\overline{CD}$, then $\overline{CB} \cong \overline{DB}$ and $\overline{CA} \cong \overline{DA}$.

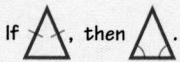

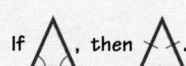

SECTIONS 6.6 and 6.7

In an isosceles triangle, the base angles are congruent. Conversely, if two angles of a triangle are congruent, then the sides opposite those angles are congruent.

If ⟍, then ⟍.   If ⟍, then ⟍.

In an isosceles triangle, the bisector of the vertex angle is also a median, an altitude, and the perpendicular bisector of the base.

Review   **329**

329

Progress Check 6.3–6.5

In the figure, $\overline{SV}$ is the perpendicular bisector of $\overline{RT}$. Tell which postulate or theorem you can use to prove that each pair of triangles is congruent. *(Sections 6.3, 6.4, and 6.5)*

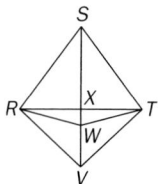

1. $\triangle RSX$ and $\triangle TXS$ SAS

2. $\triangle RXW$ and $\triangle TXW$ SSS, SAS, or HL

3. $\triangle RWV$ and $\triangle TWV$ SSS

Tell which pair of triangles must be congruent in order to prove each statement. *(Section 6.5)*

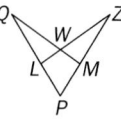

4. $\overline{LW} \cong \overline{MW}$ $\triangle QLW$ and $\triangle ZMW$

5. $\angle QMP \cong \angle ZLP$ $\triangle QMP$ and $\triangle ZLP$

Progress Check 6.6–6.7

Find each length or angle measure. *(Section 6.6)*

1. $\angle G$ 82°

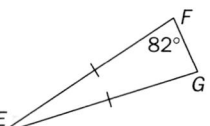

2. LK 12

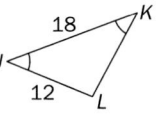

Use the figure below for Exs. 3–5. *(Section 6.7)*

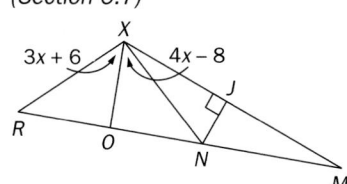

3. Name a triangle whose altitude is drawn. $\triangle XNM$ or $\triangle NXJ$

4. What would have to be true for $\overline{XN}$ to be the median of $\triangle XOM$? $\overline{ON} \cong \overline{NM}$

5. $\overline{OX}$ bisects $\angle RXN$. Find $m\angle OXN$. 48°

329

Chapter 6 Assessment
Form A Chapter Test

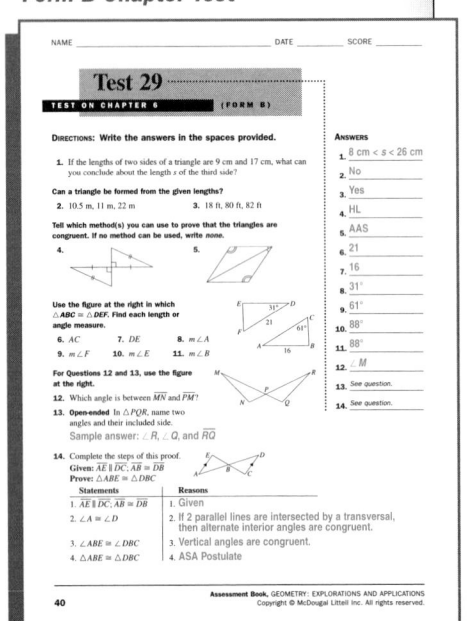

Chapter 6 Assessment
Form B Chapter Test

6 Assessment

VOCABULARY QUESTIONS

Match each part of the figure with at least one description.

1. $\angle G$ **A.** vertex angle of an isosceles triangle

2. $\overline{HK}$ **B.** base angle of an isosceles triangle

3. $\overline{HJ}$ **C.** median of a triangle

4. $\overline{GH}$ **D.** altitude of a triangle

5. $\angle F$ **E.** leg of an isosceles triangle

             F. perpendicular bisector

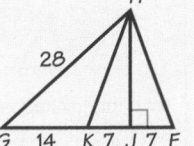

SECTIONS 6.1 *and* 6.2

6. Can you form a triangle using the lengths 4.8 ft, 1.3 ft, and 3.4 ft?

7. The lengths of two sides of a triangle are 6 cm and 8 cm. What can you conclude about the length of the third side of the triangle?

8. In $\triangle ABC$, $m\angle A = 50°$, $m\angle B = 30°$, and $m\angle C = 100°$. Identify the longest and shortest sides of $\triangle ABC$.

9. Name all the triangles that appear to be congruent in the diagram at the right.

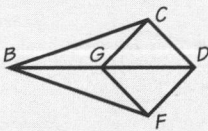

For Exercises 10 and 11, polygon *MRKV* $\cong$ polygon *SBWD*.

10. Complete: polygon $WDSB \cong \underline{\ ?\ }$

11. **ALGEBRA** Find the values of x, y, and z.

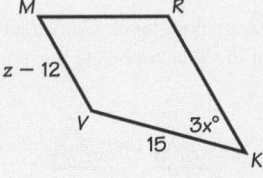 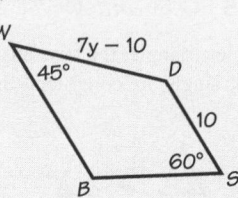

ANSWERS   **Chapter 6**

Assessment

1. A
2. C, E
3. C, D, F
4. E
5. B
6. No.
7. It is between 14 cm long and 2 cm long.
8. $\overline{AB}$; $\overline{AC}$
9. $\triangle GCD$, $\triangle GFD$; $\triangle DCG$, $\triangle DFG$; $\triangle BCD$, $\triangle BFD$; $\triangle BCG$, $\triangle BFG$

10. polygon $KVMR$
11. $x = 15$; $y = \frac{25}{7}$; $z = 22$
12. $\triangle XTY \cong \triangle ZTW$; SAS Postulate
13. $\triangle TUV \cong \triangle TWV$; SSS Postulate
14. none
15. $\triangle SPR \cong \triangle QRP$; ASA Postulate or AAS Theorem
16. none
17. $\triangle EKF \cong \triangle GHF$; ASA Postulate or AAS Theorem

18. Answers may vary. An example is given. Draw a square, $ABCD$ and diagonals $\overline{AC}$ and $\overline{BD}$. Given: $ABCD$ is a square. Prove: $\triangle ADC \cong \triangle BCD$
19. **Statements (Reasons)**
  1. Line m is the perpendicular bisector of $\overline{CD}$. (Given)
  2. $\overline{FC} \cong \overline{FD}$ (If a point is on the perpendicular bisector of a segment, then the point is equidistant from the endpoints of the segment.)

Name two congruent triangles. Tell which method(s) can be used to prove that the triangles are congruent. If no triangles are congruent, write *none*.

12.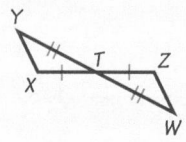

13.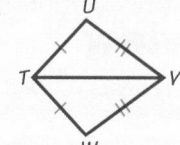

14.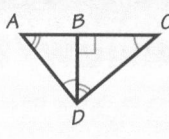

15.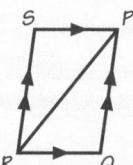

16.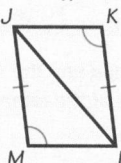

17.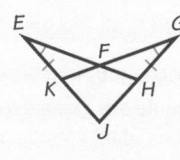

18. Open-ended Problem Sketch and label a pair of overlapping triangles that you could prove congruent using the SAS Postulate. List the given and prove statements for your diagram.

19. Given: Line *m* is the perpendicular bisector of $\overline{CD}$.

Prove: $\angle C \cong \angle D$

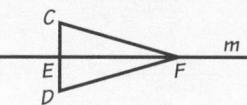

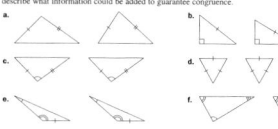

SECTIONS 6.6 *and* 6.7

20. In $\triangle QRS$, $\overline{RQ} \cong \overline{RS}$. Find the value of *n*.

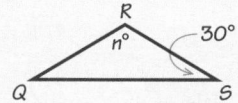

In $\triangle JKL$, $\overline{JM} \perp \overline{KL}$ and $\overline{JN} \cong \overline{NL}$.

21. a. Name a median of $\triangle JKL$.

b. Name an altitude of $\triangle JKL$.

22. If $\overline{KJ} \cong \overline{KL}$, what can you conclude about $\overline{KN}$?

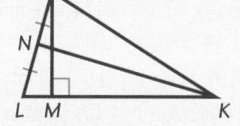

23. Writing Can the lines that contain the altitudes of a triangle intersect at a point on the triangle? Explain why or why not.

PERFORMANCE TASK

24. Open-ended Problem Write a proof: If $\overline{XP}$ is both a median and an altitude of $\triangle XYZ$, then $\triangle XYZ$ is an isosceles triangle.

....

3. $\angle C \cong \angle D$ (If two sides of a triangle are $\cong$, then the $\angle$s opposite the sides are $\cong$.)

20. 120°

21. a. $\overline{KN}$

b. $\overline{JM}$

22. $\overline{KN}$ is the perpendicular bisector of $\overline{JL}$.

23. Yes; the altitudes of a right triangle intersect at the vertex of the right angle.

24. Given: $\overline{XP}$ is a median of $\triangle XYZ$; $\overline{XP}$ is an altitude of $\triangle XYZ$.

Prove: $\triangle XYZ$ is an isosceles triangle.

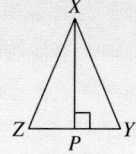

Statements (Reasons)

1. $\overline{XP}$ is a median of $\triangle XYZ$. (Given)
2. *P* is the midpoint of $\overline{YZ}$. (Def. of median)
3. $\overline{PY} \cong \overline{PZ}$ (Def. of midpoint)
4. $\overline{XP}$ is an altitude of $\triangle XYZ$. (Given)
5. $\overline{XP} \perp \overline{YZ}$ (Def. of altitude)
6. $\angle XPY \cong \angle XPZ$ (Two lines are perpendicular if and only if they form congruent adjacent angles.)
7. $\overline{XP} \cong \overline{XP}$ (Reflexive Property)
8. $\triangle XPY \cong \triangle XPZ$ (SAS Postulate)
9. $\overline{XY} \cong \overline{XZ}$ (Def. of congruent triangles)
10. $\triangle XYZ$ is an isosceles triangle. (Def. of isosceles triangle)

Algebra
Review/Preview

These exercises review algebra topics you will use in the next chapters.

USING FORMULAS

In each of Exercises 1–6, a formula and the values of some variables in the formula are given. Find the value of the remaining variable. Assume that the values of all variables are positive.

Toolbox p. 694
Evaluating Equations for Given Values

EXAMPLE $V = \pi r^2 h$; $V = 500$ and $h = 7$

Substitute 500 for V and 7 for h in the formula $V = \pi r^2 h$. Then solve for r.

$$500 = \pi r^2(7)$$

$$\frac{500}{7\pi} = r^2 \quad \longleftarrow \text{ Divide both sides by } 7\pi.$$

$$\frac{500}{7(3.14)} \approx r^2 \quad \longleftarrow \text{ Use } 3.14 \text{ for } \pi.$$

$$22.75 \approx r^2$$

$$\sqrt{22.75} \approx r \quad \longleftarrow \text{ Take the positive square root.}$$

$$4.77 \approx r$$

1. $V = \pi r^2 h$; $r = 2$ and $h = 3$ **2.** $V = \pi r^2 h$; $V = 72$ and $r = 6$

3. $A = \pi r^2$; $A = 10.6$ **4.** $d = rt$; $d = 150$ and $t = 3.5$

5. $P = 2l + 2w$; $P = 20$ and $l = 4$ **6.** $\frac{1}{f_1} + \frac{1}{f_2} = \frac{1}{f}$; $f_1 = 8$ and $f_2 = 24$

7. PHYSICS After t seconds, the distance d (feet) traveled by an object moving in a straight line and accelerating at a constant rate is $d = vt + \frac{1}{2}at^2$, where v is the initial speed (ft/s) and a is the acceleration (ft/s^2). How far does an object whose initial speed is 20 ft/s and whose acceleration is 4 ft/s^2 travel in 10 s?

Solve each equation for the specified variable.

Toolbox p. 702
Working with Formulas

EXAMPLE $y = mx + b$ for m

$$y - b = mx \quad \longleftarrow \text{ Subtract } b \text{ from both sides.}$$

$$\frac{y - b}{x} = m \quad \longleftarrow \text{ Divide both sides by } x.$$

8. $C = 2\pi r$ for r **9.** $A = lw$ for l

10. $A = \frac{1}{2}bh$ for h **11.** $P = 2l + 2w$ for w

12. $d = vt + \frac{1}{2}at^2$ for v **13.** $V = \frac{4}{3}\pi r^3$ for r

---

ANSWERS   Chapter 6

Algebra Review/Preview

1. 37.68
2. 0.64
3. 1.84
4. 42.86
5. 6
6. 6
7. 400 ft

8. $r = \frac{C}{\pi}$

9. $l = \frac{A}{w}$

10. $h = \frac{2A}{b}$

11. $w = \frac{1}{2}P - l$

12. $v = \frac{2d - at^2}{t}$

13. $r = \sqrt[3]{\frac{3V}{4\pi}}$

Find each ratio as a fraction in lowest terms.

EXAMPLE   6 days; 3 weeks

$$\frac{6 \text{ days}}{3 \text{ weeks}} = \frac{6 \text{ days}}{21 \text{ days}}$$

Express both quantities in the same units.

$$= \frac{2}{7}$$

Toolbox p. 707
Creating a Ratio

14. 20 lbs; 12 oz   **15.** 30 min; 4 h   **16.** 6 ft; 14 in.

17. 10 km; 800 m   **18.** $4\frac{2}{3}$ kg; 120 g   **19.** 6000 s; $3\frac{1}{2}$ h

Solve each proportion.

EXAMPLE $\frac{5}{4} = \frac{x}{12}$

$$5 \cdot 12 = 4x$$
$$60 = 4x$$
$$15 = x$$

If $\frac{a}{b} = \frac{c}{d}$, then $ad = bc$.

Toolbox p. 708
Solving Proportions

20. $\frac{9}{y} = \frac{3}{4}$ **21.** $\frac{a}{8} = \frac{5}{12}$ **22.** $\frac{4}{11} = \frac{14}{b}$ **23.** $\frac{15v}{64} = \frac{45}{32}$

SOLVING LINEAR SYSTEMS

Solve each system of equations.

EXAMPLE $4x - 5y = 23$
$$3x + 10y = 31$$

Multiply both sides of the first equation by 2 so the y terms are opposites.

$$2(4x - 5y) = 2(23) \longrightarrow 8x - 10y = 46$$
$$3x + 10y = 31 \longrightarrow \underline{+ (3x + 10y) = + (31)}$$
$$11x = 77$$
$$x = 7$$

Add to eliminate y. Then solve for x.

To find y, substitute 7 for x in either of the original equations.

$$4(7) - 5y = 23$$
$$-5y = -5$$
$$y = 1$$

Subtract 28 from both sides.

The solution (x, y) is $(7, 1)$.

24. $x + 2y = -7$ **25.** $2u + 5v = 14$ **26.** $18r - 5s = 17$
        $3x - 8y = 7$ $6u + 7v = 10$ $6r + 10s = -6$

27. $2a + 3b = 1$ **28.** $7c - 2d = 10$ **29.** $6m + 8n = -3$
        $4a + 6b = 5$ $-14c + 4d = -20$ $9m + 6n = -7$

Algebra Review/Preview   **333**

---

14. $\frac{80}{3}$

15. $\frac{1}{8}$

16. $\frac{36}{7}$

17. $\frac{25}{2}$

18. $\frac{350}{9}$

19. $\frac{10}{21}$

20. 12

21. $3\frac{1}{3}$

22. $38\frac{1}{2}$

23. 6

24. $(-3, -2)$

25. $(-3, 4)$

26. $\left(\frac{2}{3}, -1\right)$

27. no solution

28. Values that satisfy one equation will always satisfy the other; there are infinitely many solutions.

29. $\left(-1\frac{1}{18}, \frac{5}{12}\right)$

Cumulative Assessment
CHAPTERS 4–6

CHAPTER 4

For Questions 1–5, use the points $P(3, -5)$ and $Q(-2, 5)$.

1. Find the length of $\overline{PQ}$ and the coordinates of the midpoint of $\overline{PQ}$.

2. **ALGEBRA** Find the slope of the line that passes through P and Q.

3. Find an equation of the line that contains $(8, 0)$ and is perpendicular to $\overleftrightarrow{PQ}$.

4. **Open-ended Problem** Find the coordinates of a point R so that $\triangle PQR$ is an isosceles triangle. Use the Distance Formula to show that the triangle is isosceles.

5. Write an equation of a circle whose center is P and contains point Q.

Sketch each circle. Label the coordinates of the center. Draw a radius and label it with its length.

6. $(x - 3)^2 + y^2 = 9$ 7. $x^2 + y^2 = 25$ 8. $x^2 + (y + 3)^2 = 2$

Answer Questions 9–11 to complete the proof. Use the coordinates given in the diagram.

Given: $\overleftrightarrow{RS}$ is the perpendicular bisector of $\overline{AB}$.

Prove: $RA = RB$

9. **Writing** Explain how you know that the origin is the midpoint of $\overline{AB}$.

10. **Writing** Explain why $\overleftrightarrow{RS}$ must be the y-axis.

11. Find the x-coordinate of point R. Then show that $RA = RB$.

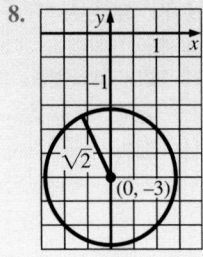

CHAPTER 5

For Questions 12–14, use the diagram at the right.

12. Suppose $m\angle 1 = 70°$ and $m\angle 2 = (9x - 2)°$. Find the value of x.

13. Suppose $m\angle 3 = (6y - 22)°$ and $m\angle 5 = (2y + 2)°$. Find the value of y.

14. **Writing** What can you conclude about the quadrilateral formed by the four lines in the diagram? Explain how you reached your conclusions.

 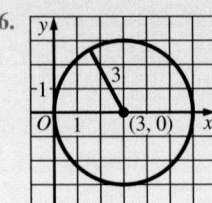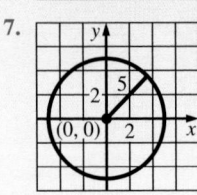

15. Write a flow proof.

Given: Lines n, p, and t are coplanar.
$n \parallel p$; $t \perp n$

Prove: $t \perp p$

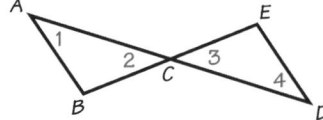

16. Suppose plane P is parallel to plane Q. State a conclusion based on this information and each additional statement.

a. Plane R intersects planes P and Q.

b. Plane R is parallel to plane Q.

17. Using the diagram, write the key steps of a proof.

Given: $\angle 1 \cong \angle 2$
$\angle 3 \cong \angle 4$

Prove: $\overleftrightarrow{AB} \parallel \overleftrightarrow{DE}$

CHAPTER 6

18. Writing Explain why a triangle can be formed with the lengths 5 in., 7 in., and 3 in. Sketch the triangle.

19. $\triangle AMR \cong \triangle LZJ$. Complete each statement.

a. $\overline{ZJ} \cong \underline{\ ?\ }$ **b.** $\angle M \cong \underline{\ ?\ }$ **c.** $\triangle ZLJ \cong \underline{\ ?\ }$

Tell which method(s) you can use to prove that the triangles are congruent. If no method can be used, write *none*.

20. 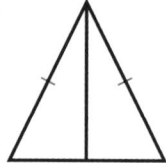   **21.** 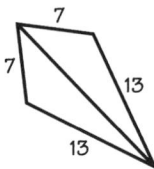   **22.**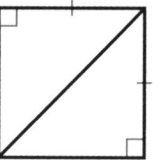

23. Given: $\angle 1 \cong \angle 2$
$\angle 3 \cong \angle 4$

Prove: $\overline{PQ} \cong \overline{RS}$

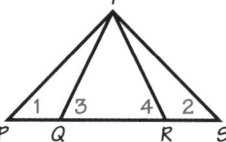

Complete each statement with *always*, *sometimes*, or *never*.

24. An altitude of a triangle is $\underline{\ ?\ }$ a side of the triangle.

25. A median of a scalene triangle is $\underline{\ ?\ }$ an angle bisector of the triangle.

26. If an altitude of a triangle is also a median, then the triangle is $\underline{\ ?\ }$ isosceles.

Cumulative Assessment **335**

18. $5 + 7 > 3$, $5 + 3 > 7$, and $3 + 7 > 5$; The lengths satisfy the first Triangle Inequality, so a triangle can be formed.

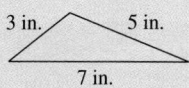

19. a. $\overline{MR}$

b. $\angle Z$

c. $\triangle MAR$

20. none

21. SSS Postulate

22. HL Theorem

23. Answers may vary. An example is given. Since $\angle 1 \cong \angle 2$, $\overline{PT} \cong \overline{ST}$ (converse of the Isosceles Triangle Theorem). $\angle 3 \cong \angle 4$ (given). $\triangle PTR \cong \triangle STQ$ (AAS Theorem), so $\overline{PR} \cong \overline{SQ}$ (def. of congruent triangles). Therefore, $\overline{PQ} \cong \overline{RS}$ (Segment Addition Postulate and Subtraction Property).

24. sometimes

25. never

26. always

15.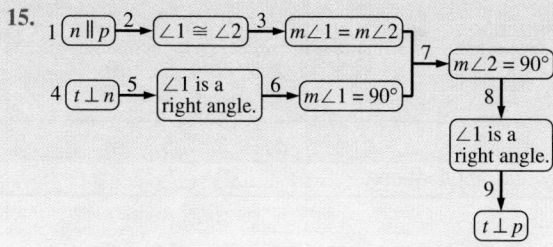

Reasons: (1) Given; (2) If two $\parallel$ lines are intersected by a transversal, then corresponding $\angle$s are $\cong$; (3) Def. of congruent angles; (4) Given; (5) Def. of perpendicular lines; (6) Def. of right angle; (7) Substitution Property; (8) Def. of right angle; (9) Def. of perpendicular lines

16. a. The lines of intersection are parallel.

b. Plane R is parallel to plane P.

17. $\angle 1 \cong \angle 2$ (given), $\angle 2 \cong \angle 3$ (vertical angles), and $\angle 3 \cong \angle 4$ (given). Thus, $\angle 1 \cong \angle 4$ (Transitive Property), and so $\overleftrightarrow{AB} \parallel \overleftrightarrow{DE}$ (alternate interior angles are congruent).

335

7 Quadrilaterals, Areas, and Volumes

OVERVIEW

Connecting to Prior and Future Learning

⟺ Students begin this chapter by classifying quadrilaterals and by determining whether a quadrilateral is a parallelogram. Students then use properties of special parallelograms to classify them as rectangles or rhombuses.

⟺ The study of area begins with an exploration that leads students to discover area formulas. Students then find areas of regular polygons and circles. Students may benefit from a review of finding perimeter, area, and volume by using pages 703 and 704 in the **Student Resources Toolbox**.

⟺ This chapter also includes a study of prisms and cylinders. Students learn how to find the volume, lateral area, and surface area. These basic mathematical concepts will be used again in this and future mathematics courses.

Chapter Highlights

Interview with Walt Stone: Creating access to buildings for people with disabilities is emphasized in this interview with Walt Stone. Related exercises are on pages 366 and 380.

Explorations in Chapter 7 involve investigating diagonals of a rhombus in Section 7.1. In Section 7.4, students discover area formulas for different polygons. In Section 7.6, students compare the volumes of prisms.

The Portfolio Project: The use of area and volume to design a building is the focus of the project in Chapter 7. Students use given information on minimum room areas to make a detailed floor plan of a cottage. They complete their work by finding the volume of their cottage, which can be used to choose a heating system for the building.

Technology: Students use geometry software in Section 7.2 to create a quadrilateral with given properties and then use the software to check that the quadrilateral is a parallelogram. They also use geometry software and a graphing calculator in Section 7.5 to explore the area and perimeter of polygons as the number of sides increases.

OBJECTIVES

Section	Objectives	NCTM Standards
7.1	• Identify quadrilaterals by using their properties. • Recognize the relationships between the diagonals of special quadrilaterals. • Use properties of quadrilaterals in real-world applications.	1, 2, 3, 4, 5, 7, 8
7.2	• Tell whether a quadrilateral is a parallelogram. • Write indirect proofs. • Reach conclusions about special quadrilaterals in real-world and geometric situations.	1, 2, 3, 4, 5, 7, 8
7.3	• Classify parallelograms using their properties. • Prove that figures are special types of parallelograms.	1, 2, 3, 4, 5, 7, 8
7.4	• Find the areas of triangles and quadrilaterals. • Use the formulas for finding the areas of triangles and quadrilaterals in real-world applications.	1, 2, 3, 4, 5, 7, 8
7.5	• Find the areas of regular polygons and circles. • Use area formulas of regular polygons and circles to find areas of real-world objects.	1, 2, 3, 4, 5, 7
7.6	• Analyze the parts of prisms and cylinders. • Find the volumes, lateral areas, and surface areas of prisms and cylinders.	1, 2, 3, 4, 5, 7

Mathematical Connections	7.1	7.2	7.3	7.4	7.5	7.6
geometry	**339–345***	**346–352**	**353–359**	**360–366**	**367–373**	**374–381**
algebra	341, 344	351	354, 357	363–365	372	380
logic and reasoning	343, 345	346, 348, 350, 351	356–358	365, 366	373	380

Interdisciplinary Connections and Applications						
history and geography				364, 366		
biology and earth science					371	
sports and recreation					367, 370	
engineering and machinery	344	351				
drafting and architecture		352			373	
package design and consumer economics						377, 379
crafts and quilting			356, 357			
auto repair, carpentry, social studies, clothing, gardening, nutrition	345	346	357	364, 365		378, 379

__Bold page numbers__ indicate that a topic is used throughout the section.

TECHNOLOGY

Section	opportunities for use with	
	Student Book	**Support Material**
7.1	scientific calculator	**Technology Book:** TI-92 Activity 7
7.2	geometry software	**Geometry Inventor Activity Book:** Activity 8
7.4	scientific calculator	**Technology Book:** Spreadsheet Activity 6 **Geometry Inventor Activity Book:** Activities 6, 7
7.5	graphing calculator geometry software	**Technology Book:** Calculator Activity 10 Spreadsheet Activity 6 **Geometry Inventor Activity Book:** Activity 12

336B

Regular Scheduling (45 min)

Section	Materials Needed	Core Assignment	Extended Assignment	exercises that feature		
				Applications	Communication	Technology
7.1	drinking straws, string, protractor, graph paper	1–19, 26–28, 30–33	1–33	21–24	4, 21, 23, 30	
7.2	geometry software	1–6, 12, 14, 16–19, 21–24, 27, 29–38	1–4, 7–17, 20–38	25, 26, 28	7, 12–14, 28	8–11
7.3	graph paper, straightedge, compass, ruler, protractor	**Day 1:** 1–10, 14–19 **Day 2:** 20–22, 26–30, 32–38, AYP*	**Day 1:** 1–19 **Day 2:** 20–38, AYP	11, 13 23–25	25, 28, 33	
7.4	rectangular sheet of paper, scissors, graph paper	**Day 1:** 1–6, 9–20 **Day 2:** 21–27, 35–41	**Day 1:** 1–20 **Day 2:** 21–41	7, 8 28–34	9 27b, 29	
7.5	straightedge, compass	1–9, 15–23, 28–34	1–3, 7–34	11–13, 25–27	13, 14c, 19, 25, 27, 28	14
7.6	rectangular sheet of paper, tape, popcorn or beans	1–6, 8–11, 21–27, AYP	1–27, AYP	7, 12–18	15, 19	
Review/ Assess		**Day 1:** 1–13 **Day 2:** 14–27 **Day 3:** Ch. 7 Test	**Day 1:** 1–13 **Day 2:** 14–27 **Day 3:** Ch. 7 Test			
Port. Proj.		Allow 2 days.	Allow 2 days.			

Yearly Pacing (with Portfolio Project)	Chapter 7 Total 13 days	Chapters 1–7 Total 93 days	Remaining 67 days	Total 160 days

Block Scheduling (90 min)

	Day 41	Day 42	Day 43	Day 44	Day 45	Day 46	Day 47
Teach/Interact	7.1: Exploration, page 340 7.2: Construction, page 350	7.3: Construction, page 358	7.4: Exploration, page 361	7.5: Construction, page 372 7.6: Exploration, page 374	Review Port. Proj.	Review Port. Proj.	Ch. 7 Test 8.1: Exploration, page 392
Apply/Assess	**7.1:** 1–19, 26–28, 30–33 **7.2:** 1–6, 12, 14, 16–19, 21–24, 27, 29–38	**7.3:** 1–10, 14–22, 26–30, 32–38, AYP*	**7.4:** 1–6, 9–27, 35–41	**7.5:** 1–9, 15–23, 28–34 **7.6:** 1–6, 8–11, 21–27, AYP	**Review:** 1–13 **Port. Proj.**	**Review:** 14–27 **Port. Proj.**	**Ch. 7 Test** **8.1:** 1–9, 14–17, 25–36

NOTE: A one-day block has been added for the Portfolio Project—timing and placement to be determined by teacher.

Yearly Pacing (with Portfolio Project)	Chapter 7 Total $6\frac{1}{2}$ days	Chapters 1–7 Total $46\frac{1}{2}$ days	Remaining $33\frac{1}{2}$ days	Total 80 days

*__AYP__ is Assess Your Progress.

Section	Practice Bank	Study Guide*	Assessment Book*	Visuals	Explorations Lab Manual	Lesson Plans	Technology Book
7.1	47	7.1		Warm-Up 7.1	Master 2	7.1	
7.2	48	7.2		Warm-Up 7.2		7.2	TI-92 Act. 7
7.3	49	7.3	Test 31	Warm-Up 7.3	Master 2	7.3	
7.4	50	7.4		Warm-Up 7.4 Folder 9	Add. Expl. 10 Master 2	7.4	Spreadsheet Act. 6
7.5	51	7.5		Warm-Up 7.5	Add. Expl. 11	7.5	Calculator Act. 10 Spreadsheet Act. 6
7.6	52	7.6	Test 32	Warm-Up 7.6		7.6	
Review Test	53	Chapter Review	Tests 33, 34, Alternative Assessment			Review Test	

*__Spanish versions__ of *Study Guide* and *Assessment Book* are available.

Chapter Support

- Course Guides
- Lesson Plans
- Portfolio Project Book:
  Additional Project 4:
    Designing a Bridge
- Preparing for College Entrance Tests
- Multi-Language Glossary
- *Test Generator* Software
- Professional Handbook
- Challenge Problems

Software Support

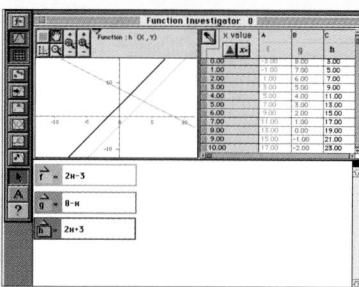

McDougal Littell Mathpack
Geometry Inventor

Internet Support

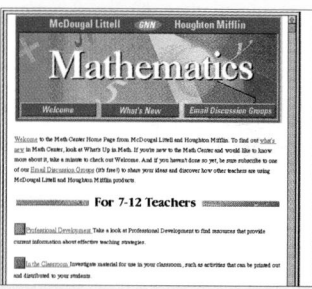

http://www.hmco.com
Next go to McDougal Littell; then the
Education Center; then Secondary Math.

OUTSIDE RESOURCES

Books, Periodicals

Wills, Herbert, III. *Leonardo's Dessert: No Pi*. Reston, VA: NCTM, 1985.

Oliver, Bernard M. "Heron's Remarkable Triangle Area Formula." *Mathematics Teacher* (February 1993): pp. 161–163.

Housinger, Margaret M. "Trap a Surprise in an Isosceles Trapezoid." *Mathematics Teacher* (January 1996): pp. 12–14.

Activities, Manipulatives

Zbiek, Rose Mary. "Multiple Connections." *Mathematics Teacher* (November 1996): pp. 628–634.

Hopley, Ronald B. "Nested Platonic Solids: A Class Project in Solid Geometry." *Mathematics Teacher* (May 1994): pp. 312–318.

Breuningsen, Chris, Bill Bower, Linda Antinone, and Elisa Breuningsen. "Meet You at the Intersection." *Real-World Math with the CBL System*. Activity 4: pp. 27–32. Texas Instruments, 1995.

Software

Jackiw, Nicholas, designer. *The Geometer's Sketchpad*. (Ver. 3.0) for Macintosh or Windows. Berkeley, CA: Key Curriculum Press, 1995.

Laborde, Jean-Marie and Franck Bellemain, designers. *Cabri Geometry II*. Dallas, TX: Texas Instruments, 1994.

TI-92 Geometry. (Resident on the TI-92 graphing calculator.) Dallas, TX: Texas Instruments, 1995.

Videos

Apostol, Tom M. *The Story of Pi*. Reston, VA: NCTM, 1989.

7 Quadrilaterals, Areas, Volumes

Alderman

An alderman is an elected official who helps in governing a city. The term *alderman* was first used in the United States during colonial days when the colonies were setting up local governments. In cities that are divided into wards, only members of a ward can vote in the election of their alderman. A city council member or commissioner, on the other hand, is elected city-wide. Both aldermen and city council members are part of the legislative branch of city government and are involved with making the laws for the city.

Walt Stone

Born in Waukegan, Illinois and raised in California, Walt Stone moved to southwest Missouri when he was 23 years old. In addition to being an alderman, Stone served in the U.S. Army and has been honorably discharged. His hobbies include cutting coin jewelry, wheeling his racing chair seven to ten miles a day, and gardening. As an alderman, Stone accomplished many things that not only assist persons with disabilities, but also able-bodied individuals as well. For example, the elderly benefit from the easier access to streets and buildings, as do people who are pushing strollers. His motivation for the job came from wanting to do something for his community.

ACCESS *for Everyone*

INTERVIEW Walt Stone

Walt Stone's civic and professional work of the past several years can be summed up in one word: "access." A former alderman in Branson, Missouri, and current chairperson of the town's Advisory Council on Disabilities, Stone is dedicated to a simple proposition. "My main goal is to make Branson accessible to everyone, so that everybody who lives here, or comes here, can enjoy the town," he says.

Branson, a southern Missouri town with some 4500 residents, draws about six million visitors each year because of its picturesque Ozarks setting and its growing reputation as a country music center. Thanks to the efforts of Stone and others, Branson is also becoming an easier place for people with disabilities to get around.

> " I never expected to use much math in my job, but now I'm finding that I use it just about every day. "

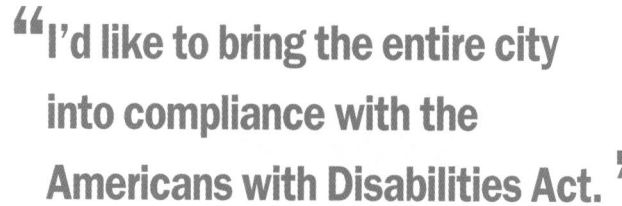

"I'd like to bring the entire city into compliance with the Americans with Disabilities Act. "

A Crusade for Ramps

Stone has worked with the local public works department and state highway department to make 400 "curb cuts" (for both wheelchairs and baby strollers) along a four-mile stretch of road that runs through the town's central business district. Of the 34 country music theaters in Branson, about 95 percent are now wheelchair-accessible, according to Stone. "At the depot for the scenic railway, we put in wheelchair ramps from the street so you can roll right onto the train," he says. People in wheelchairs can also freely roam the town's lakefront, as well as a 62-acre park and playground. "I'd like to bring the entire city into compliance with the Americans with Disabilities Act," says Stone.

Personal Motivation

Stone himself has had to rely on a wheelchair since injuring his spine in a motorcycle accident 25 years ago. Although this experience helped motivate his latest campaign, it has not made the task any easier. He's had no formal training for his job and instead has had to figure out on his own how to implement the Americans with Disabilities Act—a highly technical, three-inch thick document. That effort, in turn, has required a lot of measurements and calculations—a surprising development for someone who, until recently, considered himself "math illiterate."

337

Background

The Ozarks

The Ozark region, which includes the Ozark Mountains, the Ozark National Forest, and other areas of flatlands, covers about 40,000 square miles in Illinois, Missouri, Arkansas, and Oklahoma. The mountains rise from 1500 feet to 2300 feet above sea level, their highest point being the Boston Mountains of Arkansas. There are numerous natural resources found and used in the Ozark region, including timber, limestone, lead, and iron. Cattle are also raised for both dairy products and beef, and apples, grapes, peaches and wheat are grown. With all of these features, however, its beauty is still considered to be its greatest feature. The region contains many breathtaking sites and a varying terrain, from steep hills with dense forests to deep valleys containing swift-running streams and river gorges.

Second-Language Learners

You may need to define the following terms for students learning English: *summed up* (summarized), *civic* (of a city), *picturesque* (attractive or striking, worthy of being photographed or painted), *federal standards* (regulations made by the U.S. government).

Multicultural Note

One of the many inventors who have used their talents to create devices that improve life for people with disabilities is Martine Kempf. When she was twenty-three, Kempf invented a speech-recognition control system to help people with disabilities manage certain tasks without help from others. She was inspired to develop this system by her father, who had polio and was confined to a wheelchair. Katalavox, as the voice control system is now called, is used in power wheelchairs and allows people to maneuver without the help of an attendant. People with disabilities can also use Kempf's voice control system to turn lights on and off, to control a television, to answer and dial a phone, and so on. The system can be adapted to any language.

Mathematical Connection

Part of the purpose of the Americans with Disabilities Act (ADA) was to allow equal access to public buildings for all citizens. There are many ways to make buildings accessible to everyone, from installing ramps and elevators so that each floor of the building can be reached to installing parking places that are large enough for individuals with a disability to get in and out of their vehicles. Two of the tasks are explored in this chapter. In Section 7.5, students use area formulas to solve problems regarding handicapped parking spaces. In this section, students look at the problem in terms of finding the total area for a specific number of parking spaces, and also in terms of creating as many parking spaces as possible in a given amount of area. In Section 7.6, students investigate the construction of ramps. In this section, students calculate the dimensions of a ramp and then use the dimensions to find the volume of cement necessary to construct the ramp.

Explore and Connect

Writing

Students may need to refer to Section 2.6 to review the information about prisms. Students may also find it helpful to use a three-dimensional object that is the shape of the ramp to answer these questions.

Research

Students may also want to ask their contact person what has been implemented in the past as well as what is currently being implemented to make their city more accessible. As an extension to the research, students may wish to conduct an in-depth interview with their contact person and write an article for the school newspaper.

Project

Before making a net for a ramp, students should read page 95 in Chapter 2. They can also refer to any entries on nets they may have recorded in their journals.

Meeting the Standards

"I never expected to use math much in my job, but now I'm finding that I use it just about every day," Stone says. For instance, Stone finds the number of parking spaces that have to be set aside for drivers or passengers with disabilities. The area of these spaces must meet federal standards. Wheelchair ramps should also meet federal standards. The ramps cannot exceed a specified angle; if they're too steep, most people would not be able to make their way up. "The ramps are made of concrete and the volume of concrete that has to be poured can be determined in advance, depending on the length and height of the ramp," Stone explains.

You can find the volume of a prism-shaped ramp using the following formula:

$$V = Bh$$

where the volume equals the product of the area of the base and the height of the prism.

Base h

■ Explore and Connect

Walt Stone plans a curb cut in a Branson sidewalk.

1. Writing   Look at the prism above. What type of prism is it? How would you find the area of the base? Does it matter where you measure the height of the prism? Explain.

2. Research   Contact the person who is responsible for implementing the Americans with Disabilities Act in your city. Find out what the city's implementation goals for the next five years are. How do these goals compare to the goals for Branson, Missouri?

3. Project   Make a net for a ramp. Look back at Chapter 2 where you first learned how to make a net for a prism. Use the methods in Chapter 2 to make your net.

Mathematics & Walt Stone

In this chapter, you will learn more about how mathematics is related to access for people with disabilities.

Related Exercises

Section 7.4
• Exercises 31–33

Section 7.6
• Exercises 16–18

7.1 Classifying Quadrilaterals

Learn how to...

• classify quadrilaterals by using their properties

• recognize the relationships between the diagonals of special quadrilaterals

So you can...

• use properties of quadrilaterals in real-world applications, such as engineering and auto repair

You deal with classifications and properties every day. For instance, since apples are fruits and fruits have seeds, you know that apples have seeds. The same kind of reasoning can be used in geometry.

In Chapter 5 you learned about trapezoids. Another special quadrilateral is the *kite*. A **kite** is a convex quadrilateral that has two pairs of congruent sides, but no pair of opposite sides is congruent.

      kites   **nonkites**

Many quadrilaterals belong to more than one group. For example, if a given quadrilateral is a rhombus, then it is also a parallelogram. This diagram shows how quadrilaterals can be divided into specific groups.

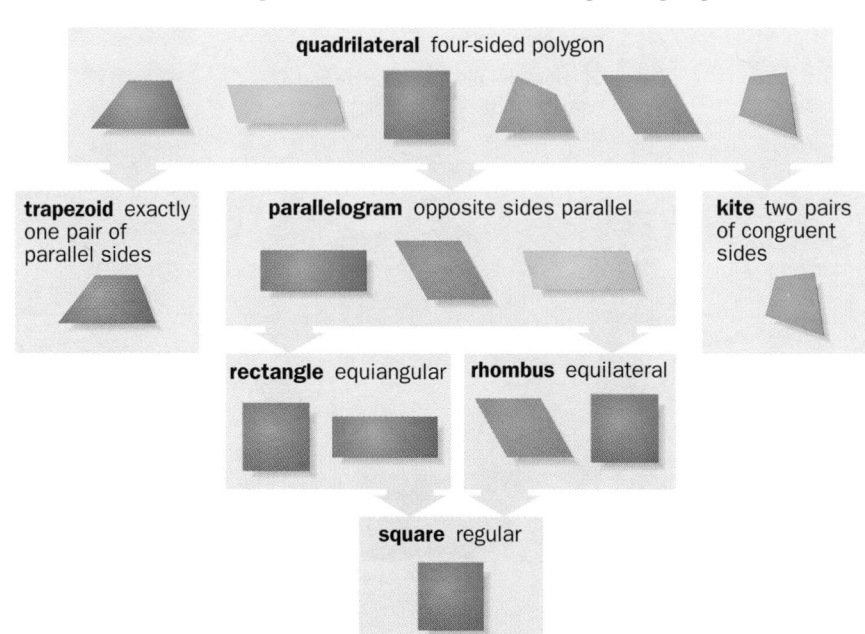

quadrilateral four-sided polygon

trapezoid exactly one pair of parallel sides

parallelogram opposite sides parallel

kite two pairs of congruent sides

rectangle equiangular

rhombus equilateral

square regular

Objectives

• Identify quadrilaterals by using their properties.

• Recognize the relationships between the diagonals of special quadrilaterals.

• Use properties of quadrilaterals in real-world applications.

Recommended Pacing

❖ **Core and Extended Courses**
Section 7.1: 1 day

❖ **Block Schedule**
Section 7.1: $\frac{1}{2}$ block
(with Section 7.2)

Resource Materials

Lesson Support
Lesson Plan 7.1
Warm-Up Transparency 7.1
Practice Bank: Practice 47
Study Guide: Section 7.1
Explorations Lab Manual:
   Diagram Master 2
Challenge Problems: Set 47
Technology
Scientific Calculator
Internet:
   http://www.hmco.com

Warm-Up Exercises

Tell whether each statement is *True* or *False*.

1. A square is a regular parallelogram. True.

2. A rhombus is an equilateral parallelogram. True.

3. A parallelogram is a trapezoid. False.

4. A rectangle is an equiangular parallelogram. True.

5. If two lines are perpendicular, the product of their slopes is 1. False.

Section Notes

Second-Language Learners
To help students learning English understand and remember the characteristics of each quadrilateral, encourage them to make a chart. Each column can list a category, such as *polygon*, *quadrilateral*, and *parallelogram*. The entry for each quadrilateral should have a check mark in each column that is a category within which that quadrilateral fits.

Teaching Tip
This may be the first time that some students have seen the definition for a kite. Stress that the congruent sides must be adjacent, and that all four sides cannot be congruent.

 **Communication: Drawing**
Students should be encouraged to study the diagram on page 339 carefully. They should see that each quadrilateral has all the properties of any quadrilateral above it to which it is linked. Thus, a rectangle is also a parallelogram. Stress that the quadrilaterals must (a) *be linked* and (b) be linked *above* for them to have the properties.

Think and Communicate

Questions 2–4 essentially provide a proof of the third property listed in the box on page 341. You may wish to have students write a formal proof of this property after completing these questions.

THINK AND COMMUNICATE

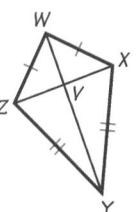

1. How many lines of symmetry does a kite have?
2. Use the fact that $\triangle WZY \cong \triangle WXY$ to show that $\triangle ZWV \cong \triangle XWV$.
3. Explain why $\overline{ZV} \cong \overline{XV}$ and $\angle ZVW \cong \angle XVW$.
4. Tell whether this statement is *True* or *False*: In kite $WXYZ$, $\overline{WY}$ is the perpendicular bisector of $\overline{ZX}$.

When you use the diagram on page 339, it is helpful to remember that a quadrilateral has all the characteristics of all the groups linked above it.

EXPLORATION
COOPERATIVE LEARNING

Investigating Diagonals

Work with another student.
You will need:
- four drinking straws
- string
- a protractor

1 Thread the string once through each straw and a second time through the first straw. You should have a movable rhombus.

2 Hold the rhombus steady and use the ends of the string to form its diagonals.

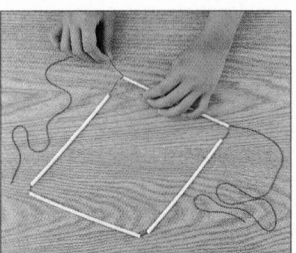

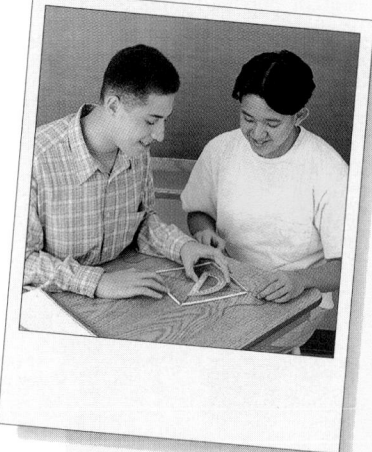

3 Measure the angles that the diagonals form. What do you notice?

4 Measure the adjacent angles at each vertex. How are the measures related?

5 Repeat Steps 2–4 twice more. Adjust the rhombus so that the straws meet at different angles each time.

6 Make two conjectures about the diagonals of a rhombus.

 Exploration Note

Purpose
The purpose of this Exploration is to have students discover that the diagonals of a rhombus are perpendicular and that the adjacent angles at each vertex have equal measures.

Materials/Preparation
Each group needs four drinking straws, a protractor, and a piece of string that is about 8 times as long as one of the straws.

Procedure
Students thread the string through the straws to form a movable rhombus. They thread the string once more through the

first straw. The ends of the string are then used as the diagonals. Students move the rhombus into different positions, and each time measure the angles that are formed by the diagonals.

Closure
Students should see that the diagonals of a rhombus are perpendicular and that each diagonal bisects the angles at the vertices of the rhombus.

Explorations Lab Manual
See the Manual for more commentary on this Exploration.

In the Exploration, you discovered the relationship between the diagonals of a rhombus. The diagonals of rectangles and kites also have special relationships. You may recall one of these theorems from Chapter 4.

Diagonals of Special Quadrilaterals

The diagonals of a rhombus are perpendicular.

If *ABCD* is a rhombus, then $\overline{AC} \perp \overline{BD}$.

The diagonals of a rectangle are congruent.

If *EFGH* is a rectangle, then $\overline{EG} \cong \overline{FH}$.

Exactly one diagonal of a kite is a line of symmetry for the kite and the perpendicular bisector of the other diagonal.

If *JKLM* is a kite, then $\triangle JKL \cong \triangle JML$, $\overline{MK} \perp \overline{JL}$, and $\overline{MN} \cong \overline{KN}$.

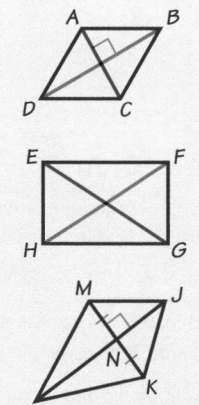

EXAMPLE 1   **Connection: Algebra**

Write a coordinate proof to show that the diagonals of any rectangle are congruent.

SOLUTION

Step 1 To show that two segments in the coordinate plane are congruent, you need to use the Distance Formula. To make the calculations easier, place as many vertices as possible on the axes. One possible choice of vertices is shown.

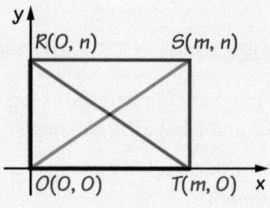

Step 2 Find *RT* and *SO* in rectangle *RSTO*.

$$RT = \sqrt{(0 - m)^2 + (n - 0)^2}$$

Remember that $(-m)^2 = m^2$.

$$= \sqrt{m^2 + n^2}$$

$$SO = \sqrt{(m - 0)^2 + (n - 0)^2}$$

$$= \sqrt{m^2 + n^2}$$

Therefore, $RT = SO$. The diagonals of any rectangle are congruent.

7.1 Classifying Quadrilaterals   **341**

ANSWERS   Section 7.1

Think and Communicate

1. one line of symmetry

2. $\triangle WZY \cong \triangle WXY$, so $\angle ZWV \cong \angle XWV$ by the definition of congruent triangles. $\overline{WZ} \cong \overline{WX}$ (given) and $\overline{WV} \cong \overline{WV}$ (Reflexive Property), so $\triangle ZWV \cong \triangle XWV$ by the SAS Postulate.

3. definition of congruent triangles

4. True.

Exploration

1–2. Check students' work.

3. They are right angles.

4. They are equal.

5. Check students' work.

6. The diagonals of a rhombus are perpendicular. Each diagonal bisects two angles of the rhombus.

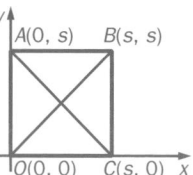

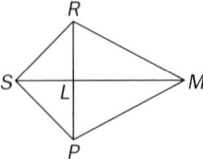

Additional Example 2

The figure below is a kite with $\overline{SR} \cong \overline{SP}$. Given that $PR = 9$ units and $SR = 6$ units, find SL.

R, S, L, M, P figure

Since $SRMP$ is a kite, its line of symmetry $\overline{SM}$ bisects its other diagonal. Thus,

$LR = \frac{1}{2}(PR) = \frac{1}{2}(9) = 4.5$.

The diagonals of a kite are also perpendicular; thus, $\triangle SRL$ is a right triangle. Use the Pythagorean theorem to find SL. Substitute 4.5 for LR and 6 for SR.
$SL^2 + LR^2 = SR^2$
$SL^2 + 4.5^2 = 6^2$
$\qquad SL^2 = 36 - 20.25$
$\qquad SL^2 = 15.75$
$\qquad\;\; SL \approx 3.97$
SL is about 4 units long.

Checking Key Concepts

Geometric Thinking
After completing question 2, challenge students to make a conjecture about the diagonals of a rhombus based on this question and the first property in the box on page 341. (The diagonals of a rhombus are perpendicular bisectors of each other.)

Closure Question

Can you draw a circle whose center is the intersection of the diagonals of a rectangle and contains all four of the rectangle's vertices? Explain. Yes, the diagonals of a rectangle are congruent. Also, since a rectangle is a parallelogram, its diagonals bisect each other. Thus, the four segments formed by the intersecting diagonals are congruent. A circle with radius equal to the length of one of these segments and whose center is at the intersection of the diagonals will contain all four vertices of the rectangle.

342

EXAMPLE 2

The square in this Samoan cloth is divided by its diagonals. If _MN_ = 10 units, find _JK_.

SOLUTION

Since a square is a parallelogram, its diagonals bisect each other. A square is also a rectangle, so its diagonals are congruent.

$$JN = KN = LN = MN = 10 \text{ units}$$

Because a square is a rhombus, the diagonals are perpendicular. Therefore, $\triangle JNK$ is a right triangle with hypotenuse $\overline{JK}$.

Use the Pythagorean theorem to find JK.

$$JK^2 = JN^2 + KN^2$$
$$JK^2 = 10^2 + 10^2 \qquad \text{Substitute 10 for } JN \text{ and } KN.$$
$$JK^2 = 200$$
$$JK \approx 14.14$$

The side of the square, $\overline{JK}$, is about 14 units long.

☑ CHECKING KEY CONCEPTS

1. A kite has side lengths of 2 and 8. Sketch what the kite might look like and label all four side lengths.

2. Show that each diagonal of a rhombus is a line of symmetry for the rhombus.

Find each length.

3. rhombus $JKLM$
   a. MN
   b. MK

4. rectangle $DEFG$
   a. DF
   b. GH

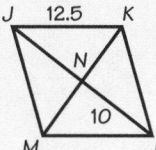

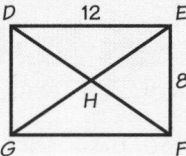

342 Chapter 7 _Quadrilaterals, Areas, and Volumes_

Checking Key Concepts

1. Answers may vary. An example is given.

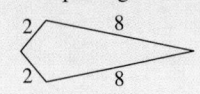

2. Answers may vary. An example is given.

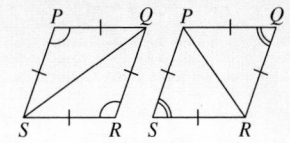

If $PQRS$ is a rhombus, then $PQRS$ is a parallelogram, and $\angle P \cong \angle R$ and $\angle Q \cong \angle S$. Then by the SAS Postulate, $\triangle PSQ \cong \triangle RSQ$ and $\triangle QPR \cong \triangle SPR$. That is, each diagonal is a line of symmetry for the rhombus.

3. a. 7.5
   b. 15

4. a. $4\sqrt{13} \approx 14.4$
   b. $2\sqrt{13} \approx 7.2$

Exercises and Applications

1. $AD = 4; CD = 5$

2. $JK = 3; ML = 7$

3. a. $\overline{WX} \cong \overline{WZ}$ because they are radii of the same circle. Similarly, $\overline{YX} \cong \overline{YZ}$.

   b. Draw two circles with centers A and C and different radii that intersect in two points, B and D. Draw $\overline{AB}, \overline{BC}, \overline{CD},$ and $\overline{AD}$. $ABCD$ has two pairs of congruent sides, but opposite sides are

7.1 Exercises and Applications

Find the missing side lengths of each kite.

1.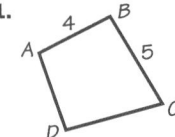

2.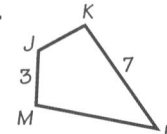

Extra Practice exercises on page 672

3. **a.** Show that *WXYZ* at the right is a kite.
   b. CONSTRUCTION Describe a way to construct a kite by drawing two circles with different radii. Use your answer from part (a) to justify your method.

4. **Writing** The segments that join the midpoints of consecutive sides of a rectangle form a rhombus. The segments that join the midpoints of consecutive sides of a rhombus form a rectangle. What type of quadrilateral is formed by the segments that join the midpoints of consecutive sides of a square? Use the diagram on page 339 to justify your answer.

5. **Visual Thinking** In the diagram below, $\overline{BH}$ and $\overline{DF}$ divide rectangle *ACJG* into four congruent rectangles. Use a theorem you learned in this section to show that *DBFH* is equilateral.

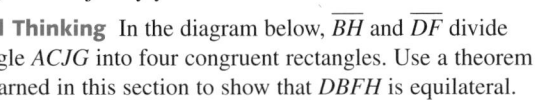

Find each length.

6. rectangle *ABCD*
   a. *BC*
   b. *AB*

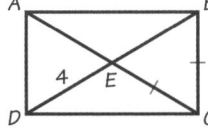

7. rhombus *PQRS*
   a. *SR*
   b. *TQ*

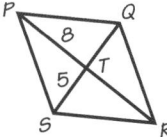

8. square *WXYZ*
   a. *WY*
   b. *XY*

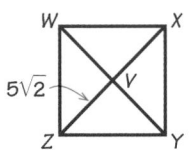

9. kite *FGHJ*
   a. *KF*
   b. *KJ*

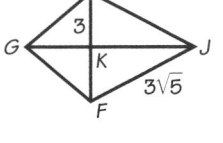

10. **Logical Reasoning** A *right trapezoid* is a trapezoid with a right angle. An *isosceles trapezoid* is a trapezoid whose legs are congruent. Copy the diagram of types of quadrilaterals shown on page 339. Include the right trapezoid and the isosceles trapezoid and explain how you decided where to place them.

Right trapezoid   **Isosceles trapezoid**

7.1 Classifying Quadrilaterals   **343**

---

not congruent. Two adjacent sides are radii of circle *A* and two are radii of circle *C* and *A* and *C* have different radii. *ABCD* is a kite.

4. a square; A square is both a rectangle and a rhombus so the segments that join the midpoints of consecutive sides is both a rhombus and a rectangle.

5. If two rectangles are congruent, then their diagonals are congruent. (Given rectangle

PQRS ≅ rectangle *TUVW*, if you draw diagonals $\overline{PR}$ and $\overline{TV}$, △*RQP* ≅ △*VUT* by the SAS Postulate.) Then since the diagonals of any rectangle are congruent, $\overline{BD}$, $\overline{BF}$, $\overline{HD}$, and $\overline{HF}$ are congruent and *DBFH* is equilateral and, therefore, a rhombus.

6. **a.** 4   **b.** $4\sqrt{3} \approx 6.9$

7. **a.** $\sqrt{89} \approx 9.4$ **b.** 5

8. **a.** $10\sqrt{2} \approx 14.1$ **b.** 10

9. **a.** 3   **b.** 6

10. The isosceles trapezoid and the right trapezoid should be placed separately under the trapezoid. Neither satisfies the requirements for any other special quadrilateral. The two are separate cases, however, because no isosceles trapezoid can be a right trapezoid, and no right trapezoid can be an isosceles trapezoid.

Suggested Assignment

❖ **Core Course**
   Exs. 1–19, 26–28, 30–33

❖ **Extended Course**
   Exs. 1–33

❖ **Block Schedule**
   Day 41 Exs. 1–19, 26–28, 30–33

Exercise Notes

Construction Note
Ex. 3 You may wish to suggest that students organize their description in part (b) in a series of steps. Students can exchange papers and follow the descriptions to see if they do construct a kite.

Communication: Writing
Ex. 4 This exercise gives students an opportunity to demonstrate their understanding of the relationships that exist between a rectangle and a square and a rhombus and a square. Students' explanations should demonstrate they understand that any figure that is both a rectangle and a rhombus must be a square.

Student Progress
Exs. 6–9 Students should be able to complete these exercises with little or no difficulty by using what they have learned about the diagonals of quadrilaterals and the Pythagorean theorem. Some students may need a brief review of squaring and adding radical expressions for Exs. 8 and 9.

Geometric Thinking
Ex. 10 This exercise introduces two special types of trapezoids. You may wish to have students make some conjectures about properties of these trapezoids and then prove these conjectures. (See Ex. 12 on page 344 for one such conjecture.)

343

344

---

11. Write the key steps of a proof showing that the diagonals of a rhombus are perpendicular.

12. **ALGEBRA** Write a coordinate proof to show that the diagonals of an isosceles trapezoid are congruent.

13. Use what you know about the diagonals of a rectangle to prove that for any right triangle, the midpoint of the hypotenuse is equidistant from all three vertices of the triangle.

> **Given:** $\triangle ABC$ is a right triangle with a right angle at B.
> D is the midpoint of $\overline{AC}$.
>
> **Prove:** $AD = BD = CD$

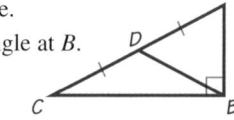

In the diagram at the right, *BDGF* is a rectangle and *FDHJ* is a kite. For Exercises 14–19, find each length.

14. *FD*   15. *FG*   16. *BE*

17. *ED*   18. *GH*   19. *JH*

20. What type of parallelogram is *ADEB*? Explain.

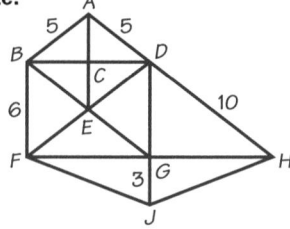

Connection   ENGINEERING

Nearly 2000 years ago, the Persians began building irrigation systems called *qanats*. To measure the slope of a *qanat*, they used a level made of a square wooden frame with a weight at one corner. When the square was hung by the corner opposite the weight, the stick that connected the other two corners was horizontal.

21. **Writing** Why do you think the weight was necessary for the level to work? Why did the frame have to be hung from the corner opposite the weight?

22. Explain why the stick at diagonal $\overline{DB}$ is always horizontal.

23. Does the frame have to be square? What other shape(s) could be used? Explain your reasoning.

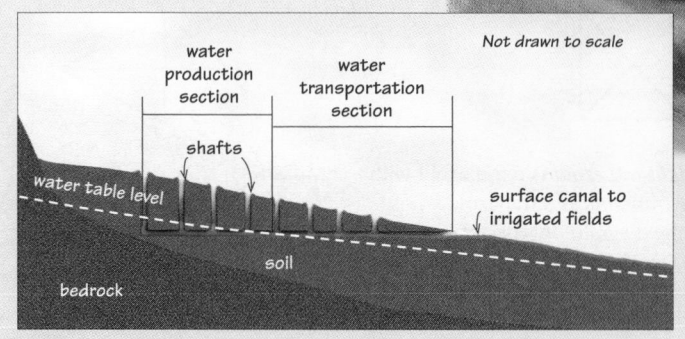

---

11. Answers may vary. An example is given.
Given: *ABCD* is a rhombus.
Prove: $\overline{AC} \perp \overline{BD}$

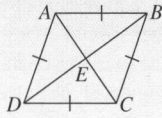

ABCD is a rhombus, so *ABCD* is a parallelogram. Then $\overline{EA} \cong \overline{EC}$ since the diagonals of a parallelogram bisect each other.

$\triangle ABE \cong \triangle CBE$ by the SSS Postulate. $\triangle AEB \cong \triangle CEB$, so $\overline{AC}$ and $\overline{BD}$ form congruent adjacent angles; therefore, $\overline{AC} \perp \overline{BD}$.

12. Given: isosceles trapezoid
$OPQR$
Prove: $\overline{PR} \cong \overline{OQ}$

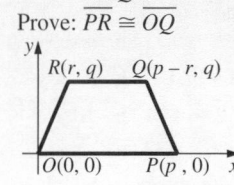

$PR = \sqrt{(p - r)^2 + (0 - q)^2} = \sqrt{(p - r)^2 + q^2}$; $OQ = \sqrt{((p - r) - 0)^2 + (q - 0)^2} = \sqrt{(p - r)^2 + q^2}$; $PR = OQ$, so by the definition of congruent segments, $\overline{PR} \cong \overline{OQ}$.

13. Construct a line through A parallel to $\overline{BC}$ and a line through C parallel to $\overline{AB}$. (Through a point not on a line, there is exactly one line parallel to the given line.) Label E, the point where the constructed lines intersect.

24. AUTO REPAIR The jack at the right is used to lift cars off the ground. As the handle is turned, vertices *X* and *Z* move together so that $\overline{ZX}$ stays parallel to the ground.

   a. What type of parallelogram is *WXYZ*?

   b. What do you know about the angle $\overline{WY}$ makes with the ground? Explain.

   c. Research Diagonal $\overline{WY}$ represents the direction of *force* that raises the car off the ground. Find out why the jack works best when the direction of the force is perpendicular to the ground.

25. Open-ended Problem Sketch a rhombus, a rectangle, or a kite. Sketch the diagonals of the figure. Mark all right angles and use the Pythagorean theorem to give possible lengths for each segment.

26. Open-ended Problem Sketch a rectangle *ABCD*. Draw $\overline{AC}$ and $\overline{BD}$ and label the point where they intersect *E*. Measure $\angle BAC$ and use this measure to find the measures of all the angles in your diagram.

27. Open-ended Problem Sketch a rhombus *EFGH*. Draw $\overline{EG}$ and $\overline{FH}$. Measure $\angle GEF$ and use this measure to find the measures of all the angles in your diagram.

28. SAT/ACT Preview In quadrilateral *WXYZ*, $\overline{WX} \cong \overline{XY}$ and $\overline{WZ} \cong \overline{ZY}$. What type of quadrilateral is *WXYZ*?

   I. kite   **II.** rhombus   **III.** square

   A. I only   **B.** I or II   **C.** II or III   **D.** I or III   **E.** I, II, or III

29. Challenge In rectangle *JKLM*, $\overline{JL}$ and $\overline{KM}$ intersect at *N*, and $\overline{JM} \cong \overline{NL}$. Sketch rectangle *JKLM* and prove that $m \angle LJK + m \angle LMK = m \angle KLN$.

ONGOING ASSESSMENT

30. Writing The diagonals of any convex quadrilateral form four non-overlapping triangles. For which quadrilaterals are these triangles right? isosceles? congruent? Explain.

SPIRAL REVIEW

Find each length or angle measure. *(Sections 6.6 and 6.7)*

31. a. *AC*
   b. $m \angle B$

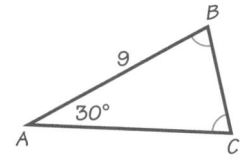

32. a. $m \angle R$
   b. $m \angle Q$

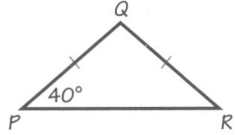

33. a. $m \angle KML$
   b. $m \angle LKM$

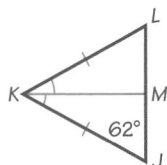

7.1 Classifying Quadrilaterals   **345**

Apply⇔Assess

Exercise Notes

Communication: Reading
Exs. 26, 27 These exercises provide an excellent opportunity for students to read a description and then use it to produce a drawing of their own. Students should label the vertices of their figures clockwise in the same order as they appear in the name of the quadrilateral.

Assessment Note
Ex. 30 Students having difficulty with this exercise should refer to the diagram on page 339 and the box on page 341.

Practice 47 for Section 7.1

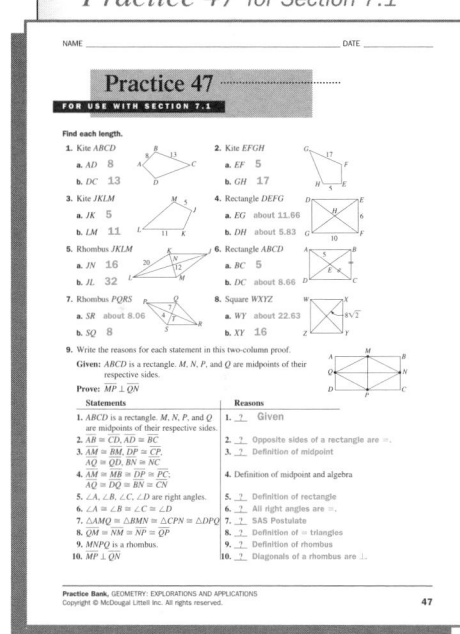

ABCE is a rectangle. Since *ABCE* is a parallelogram, its diagonals bisect each other. Since *ABCE* is a rectangle, its diagonals are congruent. Then *AD = BD = CD*.

14. 10   **15.** 8

16. 5   **17.** 5

18. 8   **19.** $\sqrt{73} \approx 8.5$

20. rhombus; *ADEB* is an equilateral parallelogram.

21. The weight allowed the force of gravity to pull down on one diagonal of the square, making it vertical, forcing the other diagonal, which is perpendicular to the first, to be horizontal. Hanging the square from the corner opposite the weight ensures the diagonal is vertical, making the other diagonal horizontal.

22. It is perpendicular to the other diagonal, which is vertical.

23. No; for example, the frame could be a nonsquare quadrilateral with perpendicular diagonals, such as a rhombus or kite.

24. a. rhombus

   b. It is a right angle; $\overline{ZX}$ is parallel to the ground and $\overline{WY}$ is perpendicular to $\overline{ZX}$ because *WXYZ* is a rhombus.

   c. It is the vertical component of the force that raises the car; if the direction is perpendicular to the ground, the entire force is vertical.

25. Answers may vary. Examples of a rhombus, rectangle, and kite are given.

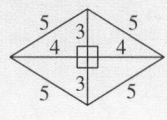
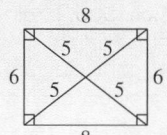
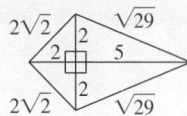

26–33. See answers in back of book.

345

Objectives

• Tell whether a quadrilateral is a parallelogram.

• Write indirect proofs.

• Reach conclusions about special quadrilaterals in real-world and geometric situations.

Recommended Pacing

❖ **Core and Extended Courses**
Section 7.2: 1 day

❖ **Block Schedule**
Section 7.2: $\frac{1}{2}$ block
(with Section 7.1)

Resource Materials

Lesson Support
Lesson Plan 7.2

Warm-Up Transparency 7.2

Practice Bank: Practice 48

Study Guide: Section 7.2

Challenge Problems: Set 48

Technology
Technology Book:
  TI-92 Activity 7

Geometry Software

McDougal Littell Mathpack
  Geometry Inventor Activity Book:
  Activity 8

Internet:
  http://www.hmco.com

Warm-Up Exercises

Describe the properties of the diagonals of each type of quadrilateral.

1. kite One diagonal is a line of symmetry and is the perpendicular bisector of the other diagonal.

2. rhombus The diagonals are perpendicular and they bisect each other.

3. trapezoid no special properties

4. parallelogram The diagonals bisect each other.

5. square The two diagonals are congruent. Each diagonal is the perpendicular bisector of the other.

346

SECTION

7.2 | Identifying Parallelograms

Learn how to...
• tell whether a quadrilateral is a parallelogram
• write indirect proofs

So you can...
• reach conclusions about special quadrilaterals in real-world and geometric situations

Suppose you want to make two segments of a structure parallel. One way is to make them the opposite sides of a parallelogram. But how do you make a parallelogram?

EXAMPLE 1 **Application: Carpentry**

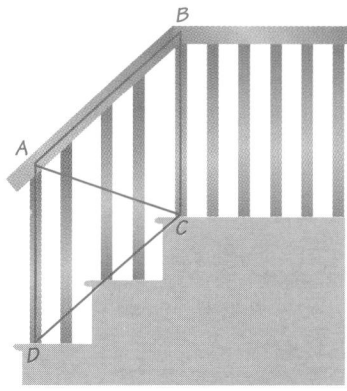

Susan Leonard is building a railing for the front steps of her house. Show that if the end posts are the same length and both perpendicular to the ground, the hand railing will be parallel to the segment that connects the top and bottom steps.

Given: $\overline{AD} \parallel \overline{BC}$; $\overline{AD} \cong \overline{BC}$
Prove: $ABCD$ is a parallelogram.

SOLUTION

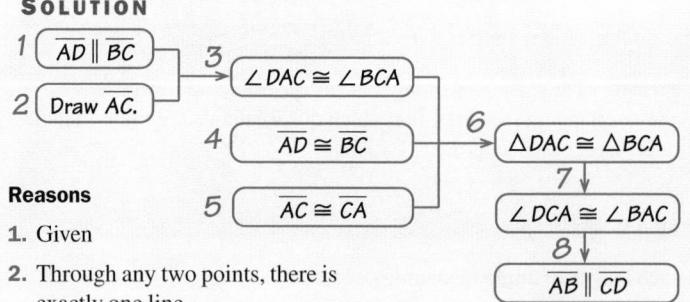

Reasons

1. Given

2. Through any two points, there is exactly one line.

3. If $\parallel$ lines are intersected by a transversal, then alternate interior ∠s are ≅.

4. Given

5. Reflexive Property

6. SAS Postulate

7. Definition of congruent triangles

8. If two lines form ≅ alternate interior ∠s, then the lines are $\parallel$.

9. Definition of parallelogram

In Chapter 2 you learned several properties of parallelograms. If a quadrilateral has one of these properties, you can prove that it is a parallelogram.

Ways to Prove that a Quadrilateral is a Parallelogram

Show that both pairs of opposite sides are parallel. (Definition)

If $\overline{AB} \parallel \overline{CD}$ and $\overline{AD} \parallel \overline{BC}$, then $ABCD$ is a parallelogram.

Show that both pairs of opposite sides are congruent.

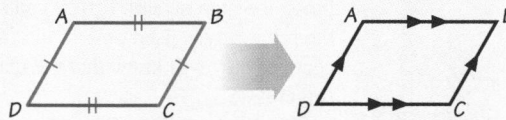

If $\overline{AB} \cong \overline{CD}$ and $\overline{AD} \cong \overline{BC}$, then $ABCD$ is a parallelogram.

Show that both pairs of opposite angles are congruent.

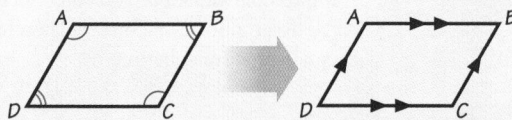

If $\angle A \cong \angle C$ and $\angle B \cong \angle D$, then $ABCD$ is a parallelogram.

Show that one pair of opposite sides is both parallel and congruent.

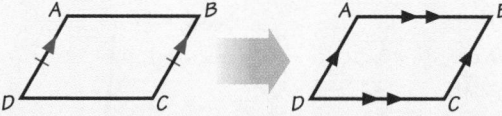

If $\overline{AD} \parallel \overline{BC}$ and $\overline{AD} \cong \overline{BC}$, then $ABCD$ is a parallelogram.

Show that the diagonals bisect each other.

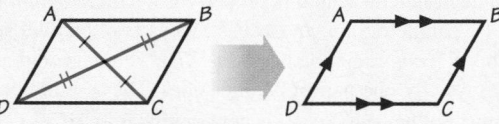

If $\overline{AC}$ bisects $\overline{BD}$ and $\overline{BD}$ bisects $\overline{AC}$, then $ABCD$ is a parallelogram.

THINK AND COMMUNICATE

1. Which method above can you use to show that an equilateral quadrilateral must be a parallelogram?

2. Which method above can you use to show that an equiangular quadrilateral must be a parallelogram?

3. Use one of the methods above to show that any two congruent triangles can be arranged to form a parallelogram.

Additional Example 1

Refer to Example 1 on page 346. Susan's friend proposes that she build the same railing using the following method. Cut two pieces of wood whose lengths are equal to the desired height of the railing. Cut two other pieces with lengths equal to that of the steps (DC in the figure). Nail the congruent pieces so that they are opposite each other. Show that this method forms a parallelogram.

Given: $\overline{AD} \cong \overline{BC}$; $\overline{AB} \cong \overline{DC}$
Prove: $ABCD$ is a parallelogram.

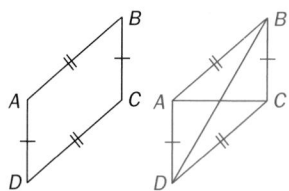

Statements (Reasons)

1. $\overline{AD} \cong \overline{BC}$, $\overline{AB} \cong \overline{DC}$ (Given)
2. Draw $\overline{BD}$. (A segment can be drawn between any two points.)
3. $\overline{BD} \cong \overline{BD}$ (Reflexive Property)
4. $\triangle ABD \cong \triangle CDB$ (SSS Postulate)
5. $\angle ABD \cong \angle CDB$ (Def. of congruent triangles)
6. $\overline{AB} \parallel \overline{DC}$ (If two lines form $\cong$ alternate interior ⊿, then the lines are $\parallel$.)
7. Draw $\overline{AC}$. (A segment can be drawn between any two points.)
8. $\overline{AC} \cong \overline{AC}$ (Reflective Property)
9. $\triangle ABC \cong \triangle CDA$ (SSS Postulate)
10. $\angle ACB \cong \angle CAD$ (Def. of congruent triangles)
11. $\overline{AD} \parallel \overline{BC}$ (If two lines form $\cong$ alternate interior ⊿, then the lines are $\parallel$.)
12. $ABCD$ is a parallelogram. (Def. of parallelogram)

Section Note

Geometric Thinking
The chart on this page presents the different ways to prove that a quadrilateral is a parallelogram. Review with students what the hypothesis is for each method. Stress that when proving that a quadrilateral is a parallelogram, the given information should be examined to see which hypothesis is met or can be met using the given information.

ANSWERS Section 7.2

Think and Communicate

1. Show that both pairs of opposite sides are congruent.

2. Show that both pairs of opposite angles are congruent.

3. Answers may vary. An example is given. Arrange the two triangles with one pair of congruent sides together in reverse order.

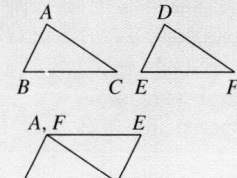

$\triangle ABC \cong \triangle DEF$; since both pairs of opposite sides of the resulting quadrilateral are congruent, the quadrilateral is a parallelogram.

Section Notes

Writing Proofs

Discuss the concept of an indirect proof thoroughly in class. Students should understand that this type of proof is based upon the idea that facts in mathematics, or in real life, cannot contradict one another. If this happens, then one of the so-called "facts" is false and the other, therefore, must be true.

Visual Thinking

Ask students to create a diagram of an indirect proof of something from their own lives. *Those gym shoes are mine because they are not..., I was at the game because I was not...*, and so on. Encourage them to present their diagrams to the class and to explain how indirect proof is different from other types of proof. This activity involves the visual skills of *exploration* and *inference*.

About Example 2

Teaching Tip

When reading through this proof with students, ask them to identify the assumption and the contradiction it implies.

Additional Example 2

Write an indirect proof that the line of symmetry in a kite cannot be bisected by the other diagonal.

Given: In kite *BCDE*, $\overline{CD} \cong \overline{ED}$, $\overline{BC} \cong \overline{BE}$, and the line of symmetry is $\overline{BD}$.

Prove: $\overline{BX} \not\cong \overline{DX}$

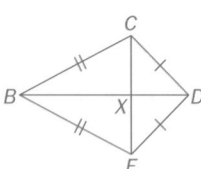

Suppose $\overline{BX} \cong \overline{DX}$. Since the diagonals of a kite are perpendicular, $\angle BXC$ and $\angle CXD$ are right angles. Also, $\overline{XC} \cong \overline{XC}$ by the Reflexive Property, so that $\triangle BXC \cong \triangle DXC$ by SAS. By the definition of congruent triangles, $\overline{BC} \cong \overline{CD}$. By substitution, all four sides of kite *BCDE* are congruent. This contradicts the definition of kite, which states that it has two distinct pairs of congruent sides. So $\overline{BX} \not\cong \overline{DX}$.

Indirect Proof

All of the types of proof you have learned in this book are *direct* proofs. These proofs consist of statements that follow directly from each other, leading from the given information to the conclusion. To use *indirect reasoning*, you need to show that all other cases are impossible, so the conclusion must be true.

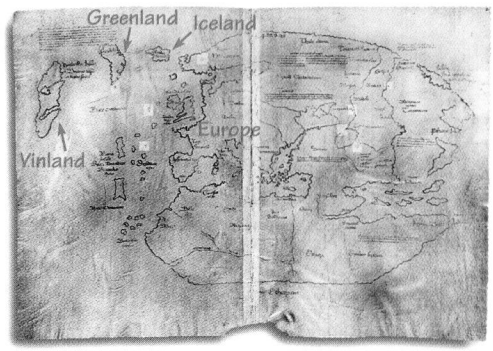

Greenland Iceland
Europe
Vinland

For example, this map found in 1965 implies that the Norse landed in America before Columbus. Some scholars have tried to prove the map invalid. If they can show that the ink is a type used only in the twentieth century, they will know that the map is a forgery.

To write an **indirect proof**, start by assuming that the conclusion is false. When logic leads you to a contradiction or an impossible situation, you can conclude that the assumption is incorrect. Therefore, the conclusion must be true.

EXAMPLE 2

Write an indirect proof to show that the parallel sides of a trapezoid cannot be congruent.

SOLUTION

Given: In *JKLM*, $\overline{JK} \parallel \overline{ML}$.
Prove: $\overline{JK} \not\cong \overline{ML}$

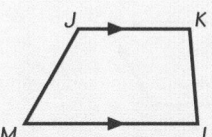

Suppose $\overline{JK} \cong \overline{ML}$. Then quadrilateral *JKLM* has a pair of opposite sides that are both parallel and congruent. The third theorem on page 347 states that a quadrilateral with this property is a parallelogram. By the definition of a parallelogram, *JKLM* has two pairs of parallel sides. However, this is a contradiction because *JKLM* is a trapezoid and therefore has exactly one pair of parallel sides. The assumption that $\overline{JK} \cong \overline{ML}$ must not be true. So $\overline{JK}$ is not congruent to $\overline{ML}$.

THINK AND COMMUNICATE

4. Kiyana has a vase of roses on her desk where she does her homework each night. Give a convincing argument that she is not allergic to roses.

5. Suppose you want to write an indirect proof to show that a given angle is acute. What three cases should you prove are impossible?

6. Sara wrote an indirect proof to show that a given quadrilateral is a trapezoid by proving that it is not a parallelogram or a kite. What is wrong with her reasoning?

Think and Communicate

4. If Kiyana were allergic to roses, then sitting at her desk every night would cause an allergic reaction, making her very uncomfortable. She would eventually either discard the roses or stop sitting at her desk. So Kiyana is probably not allergic to roses.

5. that $\angle A$ is right, obtuse, or straight

6. He did not show that all other possibilities are false. For example, the quadrilateral could be one with no special properties.

Checking Key Concepts

1. *VWXY* is a parallelogram because $\overline{XY}$ and $\overline{WV}$ are both parallel and congruent.

   a. 14   b. 6

   c. 10.3

2. *PQRS* is a parallelogram because its diagonals, $\overline{PR}$ and $\overline{SQ}$, bisect each other.

   a. 17   b. 122°   c. 58°

3. Answers may vary. An example is given. Let *PQRS* be a parallelogram and point *T* be the intersection of the diagonals. Since the diagonals bisect each other and vertical angles are congruent, $\triangle PTQ \cong \triangle RTS$ and $\triangle PTS \cong \triangle RTQ$ (SAS Postulate). Then $\overline{PQ} \cong \overline{SR}$ and $\overline{PS} \cong \overline{QR}$, so *PQRS* is a parallelogram.

✓ CHECKING KEY CONCEPTS

Show that each quadrilateral is a parallelogram. Then find each length or measure.

1. **a.** VX
   b. YZ
   c. WX

2. **a.** QR
   b. $m \angle SRQ$
   c. $m \angle PSR$

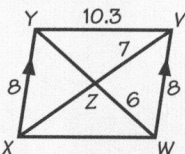

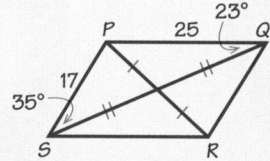

3. Write the key steps of a proof showing that if the diagonals of a quadrilateral bisect each other, then the quadrilateral is a parallelogram.

7.2 Exercises and Applications

Extra Practice exercises on page 673

Show that each quadrilateral is a parallelogram. Then find each length or measure.

1. **a.** $m \angle DAB$
   b. BC
   c. DC

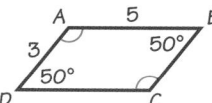

2. **a.** EJ
   b. HF
   c. HG

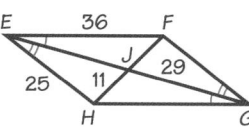

Sketch each quadrilateral. Mark the given information. If the quadrilateral must be a parallelogram, explain why.

3. $ABCD$ with $\overline{AB} \cong \overline{CD}$ and $\overline{BC} \cong \overline{AD}$

4. $WXYZ$ with $\overline{WX} \parallel \overline{YZ}$ and $\overline{XY} \cong \overline{WZ}$

5. $FGHI$ with $\angle F \cong \angle I$ and $\angle G \cong \angle H$

6. $JKLM$ with $\angle J \cong \angle L$ and $\angle K \cong \angle M$

7. **Open-ended Problem** Draw a parallelogram. Describe the method you used and how you knew the result would be a parallelogram.

📈 **Technology** Use geometry software to create a quadrilateral with the given property. Then use the slope feature to check that the quadrilateral is a parallelogram.

8. Opposite sides are congruent.

9. Opposite angles are congruent.

10. The diagonals bisect each other.

11. One pair of sides is congruent and parallel.

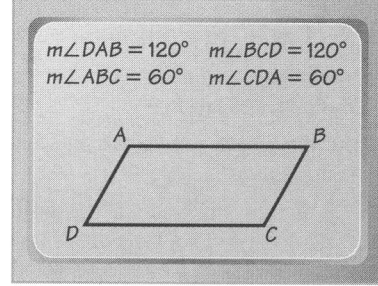

$m \angle DAB = 120°$ $m \angle BCD = 120°$
$m \angle ABC = 60°$ $m \angle CDA = 60°$

Exercises and Applications

1. $ABCD$ is a parallogram because both pairs of opposite angles are congruent.
   a. 130°   **b.** 3
   c. 5

2. $EFGH$ is a parallelogram because both pairs of opposite sides are parallel since alternate interior angles are congruent.

a. 29   **b.** 22
c. 36

3.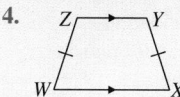
   $ABCD$ must be a parallelogram because both pairs of opposite sides are congruent.

4.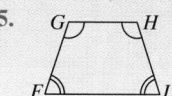

5.

$WXYZ$ may or may not be a parallelogram. In the example, $WXYZ$ is not a parallelogram.

$FGHI$ may or may not be a parallelogram. In the example, $FGHI$ is not a parallelogram.

6, 7. See answers in back of book.

8–11. Check students' work.

Teach⇔Interact

Think and Communicate

Questions 5 and 6 help students to understand that the initial assumption, together with the conclusion, must include all possibilities. If this is not the case, then the indirect proof may not be valid.

Checking Key Concepts

Teaching Tip
Ask students to give their reasons in questions 1 and 2 without referring to the chart on page 347.

Closure Question

Describe five ways to prove a quadrilateral is a parallelogram. Show both pairs of opposite sides are parallel. Show both pairs of opposite sides are congruent. Show both pairs of opposite angles are congruent. Show one pair of opposite sides is both parallel and congruent. Show the diagonals bisect each other.

Apply⇔Assess

Suggested Assignment

❖ **Core Course**
Exs. 1–6, 12, 14, 16–19, 21–24, 27, 29–38

❖ **Extended Course**
Exs. 1–4, 7–17, 20–38

❖ **Block Schedule**
Day 41 Exs. 1–6, 12, 14, 16–19, 21–24, 27, 29–38

Exercise Notes

Topic Spiraling: Review
Ex. 1 Students should recall that the sum of the measures of the interior angles in a quadrilateral is 360°.

📝 **Communication: Drawing**
Exs. 3–6 These exercises give students practice sketching a quadrilateral from a written description, a skill useful when creating diagrams for proofs.

📝 **Communication: Discussion**
Ex. 7 Students would benefit from a discussion of their answers to this exercise by seeing methods other than their own.

349

Exercise Notes

Construction Note
Exs. 12–15 Prior to beginning these exercises, students should practice Steps 1–5 to construct a parallelogram with a compass and straightedge.

Cooperative Learning
Exs. 12–15 These exercises are appropriate for small group work. Students can take turns doing the constructions in Steps 1–5 and answering the questions in Exs. 12–15.

Geometric Thinking
Exs. 12–15 Unlike the constructions done in the previous chapter to demonstrate the triangle congruence theorems, these constructions can produce different parallelograms, rhombuses, and isosceles trapezoids from the same given side lengths. You may wish to have students start with the same side lengths for Ex. 15 and see how many different trapezoids are produced. Students should observe that all rectangles with two given adjacent side lengths are congruent, and all squares with a given side length are congruent.

CONSTRUCTION

PARALLELOGRAM

Given two segments, construct a parallelogram with sides congruent to the segments.

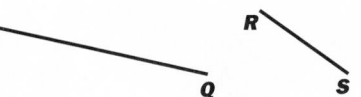

1. Construct $\overline{AB}$ congruent to $\overline{PQ}$.

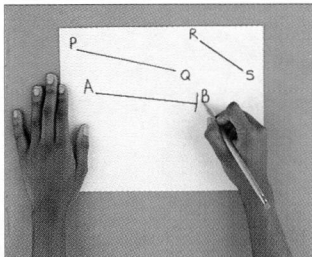

2. Construct $\overline{BC}$ congruent to $\overline{RS}$.

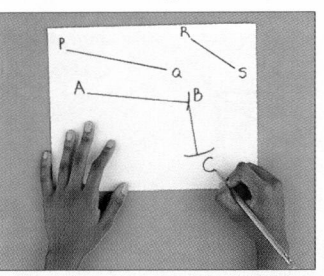

3. Draw an arc with center A and radius BC.

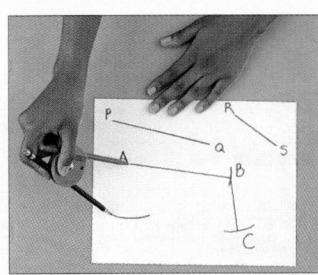

4. Draw an arc with center C and radius AB. The arc should intersect the arc you drew in Step 3.

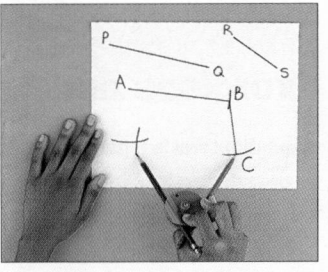

5. Label the point where the arcs intersect D. Draw $\overline{AD}$ and $\overline{CD}$. $ABCD$ is a parallelogram.

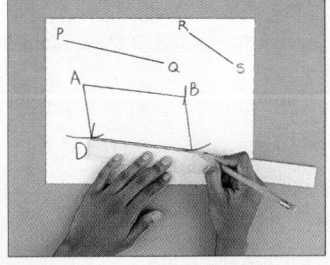

12. Writing Explain why $ABCD$ must be a parallelogram.

13. Open-ended Problem Describe two different ways to check that $ABCD$ is a parallelogram.

14. Writing How can you use the construction shown above to construct a rhombus?

15. Challenge How can you use this construction to construct an isosceles trapezoid, given the length of one base and the length of the congruent sides? (*Hint:* How would you locate D differently?)

350 Chapter 7 *Quadrilaterals, Areas, and Volumes*

---

12. Both pairs of opposite sides are congruent.

13. Answers may vary. Examples are given. Measure the angles to show that both pairs of opposite angles are congruent. Measure the segments of each diagonal to show that the diagonals bisect each other.

14. Start with congruent segments $\overline{PQ}$ and $\overline{RS}$.

15. Suppose the base has length PQ and each leg has length RS. Construct parallelogram $ABCD$ as described. Draw an arc with center B and radius BC intersecting $\overline{DC}$ in a point other than C. Label the intersection point E and draw $\overline{BE}$. $ABED$ is an isosceles trapezoid.

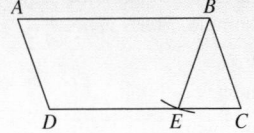

16–19. Answers may vary. Examples are given.

16.

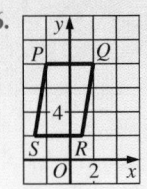

$PQ = SR = 4$; $PS = QR = \sqrt{37}$; $PQRS$ is a parallelogram because both pairs of opposite sides are congruent.

17.

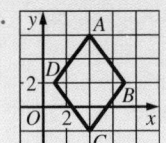

$(4, 2)$ is the midpoint of both $\overline{AC}$ and $\overline{DB}$. $ABCD$ is a parallelogram because its diagonals bisect each other.

ALGEBRA Graph the points. Show that they are the vertices of a parallelogram.

16. $P(-2, 8), Q(2, 8), R(1, 2), S(-3, 2)$ **17.** $A(4, 6), B(7, 2), C(4, -2), D(1, 2)$

18. $K(4, 7), L(9, 6), M(6, 3), N(1, 4)$ **19.** $F(0, 4), G(3, 5), H(2, 0), I(-1, -1)$

20. Open-ended Problem Give the coordinates of four points that are the vertices of a nonrectangular parallelogram. Tell which theorem you used to choose the points.

21. Write the key steps of a proof showing that if the opposite angles of a quadrilateral are congruent, then the quadrilateral is a parallelogram.

22. Write an indirect proof to show that $\triangle AEB \not\cong \triangle CEB$ in kite $ABCD$.

Given: kite $ABCD$

Prove: $\triangle AEB \not\cong \triangle CEB$

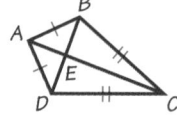

23. Write an indirect proof to show that each side of a kite is congruent to exactly one other side of the kite.

24. Write an indirect proof to show that if a trapezoid has a right angle, then it has exactly two right angles. (*Hint:* You will need to consider more than one case.)

Connection   MACHINERY

The people at the right are photographing an airplane. To get an aerial view, they are standing on a *scissors lift*. Eight congruent metal pieces that bisect each other, $\overline{AE}$, $\overline{BD}$, $\overline{DH}$, $\overline{EG}$, $\overline{GL}$, $\overline{HK}$, $\overline{KP}$, and $\overline{LN}$ move together to raise and lower the platform of the lift.

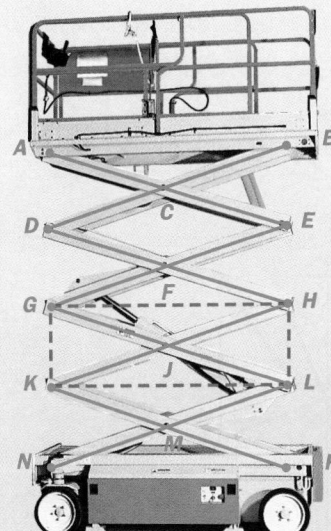

25. a. What type of quadrilaterals are *CEFD*, *FHJG*, and *JLMK*? How do you know?

    b. Use what you know about vertical angles to show that these three quadrilaterals remain congruent even while the lift is in motion.

26. You can also think of the metal pieces as the diagonals of quadrilaterals. For instance, $\overline{GL}$ and $\overline{HK}$ are the diagonals of quadrilateral *GKLH*.

    a. How do you know that *ADEB*, *DGHE*, *GKLH*, and *KNPL* are parallelograms?

    b. When the lift rises, $\overline{NL}$ and $\overline{PK}$ move at the same rate, forcing $\overline{KL}$ to stay horizontal. Show that the platform of the lift always stays horizontal.

Exercise Notes

Integrating Algebra
Exs. 16–19 For these exercises, students can chose to use the methods from the chart on page 347 to prove that each quadrilateral is a parallelogram. Depending on the method chosen, the Distance Formula, Slope Formula, and/or the Midpoint Formula must be used. After students complete the exercises, ask them why they chose a particular method for each problem.

Common Error
Ex. 24 Some students may think that it is sufficient to show that a trapezoid with a right angle cannot have three right angles. This leaves out the possibilities of 0, 1, or 4 right angles. The assumption for the indirect proof should be equivalent to: Suppose that the trapezoid with a right angle has fewer than two right angles or more than two right angles.

Application
Exs. 25, 26 In these exercises, students see that the design of the piece of machinery called a *scissors lift* is based on the properties of the diagonals of rectangles. Students may wish to create a model of a scissors lift using cardboard and paper fasteners to help in doing these exercises.

3. $m\angle J + m\angle L + m\angle K + m\angle M = 360°$ (The sum of the $\angle$ measures of a quad. is $360°$.)

4. $2m\angle J + 2m\angle K = 360°$
$2m\angle L + 2m\angle M = 360°$
(Addition Property)

5. $m\angle J + m\angle K = 180°$
$m\angle L + m\angle M = 180°$ (Algebra)

6. $\overline{JK} \parallel \overline{ML}$ and $\overline{KL} \parallel \overline{JM}$ (If two lines are intersected by a transversal and the same-side interior $\angle$s are supplementary, then the lines are $\parallel$.)

7. *JKLM* is a parallelogram. (Def. of parallelogram)

18.

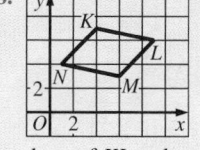

slope of KL = slope of $MN = -\frac{1}{5}$; slope of $\overline{KN}$ = slope of $\overline{LM}$ = 1; *KLMN* is a parallelogram by definition.

19.

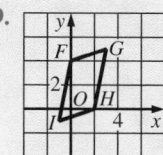

slope of $\overline{FG}$ = slope of $\overline{IH} = \frac{1}{3}$; $FG = IH = \sqrt{10}$; *FGHI* is a parallelogram because one pair of sides, $\overline{FG}$ and $\overline{IH}$, are both congruent and parallel.

20. Answers may vary. An example is given. The quadrilateral with vertices $A(0, 0)$, $B(7, 0)$, $D(3, 4)$, and $C(10, 4)$ is a parallelogram. If one pair of sides of a quadrilateral is both congruent and parallel, the quadrilateral is a parallelogram.

21.

Given: *JKLM*; $\angle J \cong \angle L$, $\angle K \cong \angle M$
Prove: *JKLM* is a parallelogram.
Statements (Reasons)
1. $\angle J \cong \angle L$ and $\angle K \cong \angle M$ (Given)
2. $m\angle J = m\angle L$ and $m\angle K = m\angle M$ (Def. of $\cong$)

22. Suppose that $\triangle AEB \cong \triangle CEB$. Then $\overline{AB} \cong \overline{CB}$ by the definition of congruent triangles. $\overline{AB} \cong \overline{AD}$, so $\overline{CB} \cong \overline{AD}$. But this is a contradiction. Opposite sides of a kite are not congruent. Therefore, $\triangle AEB \not\cong \triangle CEB$.

23–26. See answers in back of book.

351

Apply⇔Assess

Exercise Notes

Geometric Thinking

Ex. 27 This exercise shows students the relationship among the coordinates of the vertices of a parallelogram when it is placed with one vertex at the origin and one side along the *x*-axis. This positioning is most useful when doing coordinate proofs with parallelograms.

Assessment Note

Ex. 29 This assessment exercise requires students to recall all the properties used to show that a quadrilateral is a parallelogram, thus providing an excellent opportunity to assess students' knowledge of these properties.

Practice 48 for Section 7.2

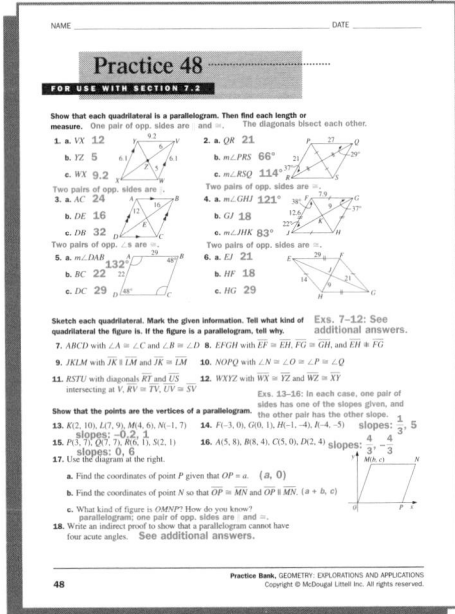

---

27. Use the diagram at the right.

    a. Find the coordinates of point *P* given that $OP = c$.

    b. Find the *y*-coordinate of point *N* so that $\overline{MN} \parallel \overline{OP}$.

    c. Find the *x*-coordinate of point *N* so that $\overline{MN} \cong \overline{OP}$.

    d. Which theorem on page 347 ensures that *MNPO* is a parallelogram? Explain your reasoning.

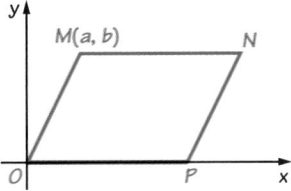

28. DRAFTING Drafting tables are designed so that they can tilt at different angles. You adjust the table shown by joining its legs at different holes. How would you fasten the legs together if you wanted the tabletop to be parallel to the floor? Explain your reasoning.

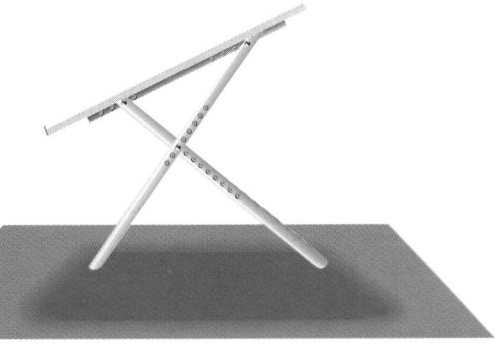

ONGOING ASSESSMENT

29. Open-ended Problem Sketch and label a quadrilateral *WXYZ*. Include both diagonals. Give four different sets of properties that can be used to prove that *WXYZ* is a parallelogram.

SPIRAL REVIEW

Find each length. *(Section 7.1)*

30. *JL* in rhombus *JKLM*

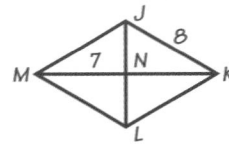

31. *RS* in kite *PQRS*

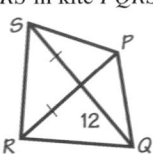

32. *BD* in rectangle *ABCD*

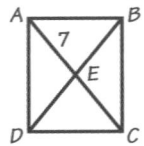

Find the length and midpoint of each segment in the diagram. *(Section 4.1)*

33. $\overline{AB}$ **34.** $\overline{CF}$ **35.** $\overline{DC}$

36. $\overline{HJ}$ **37.** $\overline{EF}$ **38.** $\overline{GB}$

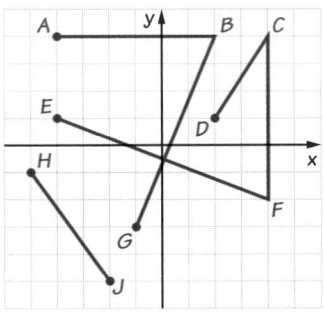

---

27. a. $(c, 0)$

    b. *b*

    c. $a + c$

    d. If one pair of sides of a quadrilateral is both parallel and congruent, then the quadrilateral is a parallelogram. $\overline{MN}$ and $\overline{OP}$ are both parallel and congruent.

28. The legs should be positioned so that they bisect each other. Then the quadrilateral whose vertices are the tops and bottoms of the four legs will be a parallelogram and the top of the table will be parallel to the floor.

29. Check students' work. Properties may include any four of the following: $\overline{JK} \parallel \overline{ML}$ and $\overline{JM} \parallel \overline{KL}$; $\overline{JK} \cong \overline{ML}$ and

$\overline{JM} \cong \overline{KL}$; $\overline{JK} \parallel \overline{ML}$ and $\overline{JK} \cong \overline{ML}$ or $\overline{JM} \parallel \overline{KL}$ and $\overline{JM} \cong \overline{KL}$; $\angle KJM \cong \angle KLM$ and $\angle JML \cong \angle JKL$; $\overline{JL}$ and $\overline{MK}$ bisect each other.

30. $2\sqrt{15} \approx 7.7$

31. $12\sqrt{2} \approx 17.0$

32. 14

33. 6; $(-1, 4)$

34. 6; $(4, 1)$

35. $\sqrt{13} \approx 3.6; \left(3, 2\frac{1}{2}\right)$

36. 5; $\left(-3\frac{1}{2}, -3\right)$

37. $\sqrt{73} \approx 8.5; \left(0, -\frac{1}{2}\right)$

38. $\sqrt{58} \approx 7.6; \left(\frac{1}{2}, \frac{1}{2}\right)$

7.3 Conditions for Special Parallelograms

Learn how to...

- classify parallelograms using their properties

So you can...

- prove that figures are special types of parallelograms

Many chairs and tables have rectangular frames. Since the lengths of the pieces that make up the frames stay the same over time, the frames remain parallelograms. Under the pressure of use, however, the angles may change.

One way to check if a frame is rectangular is to measure its diagonals. If the diagonals are congruent, the frame must be a rectangle.

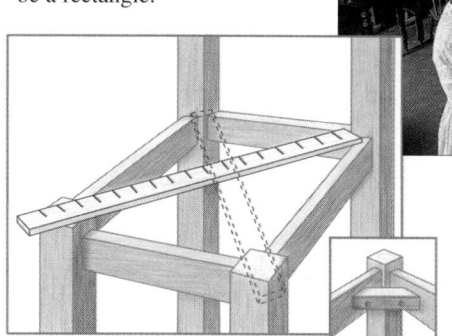

You can keep a frame rectangular by securing one corner at a right angle. When the parallelogram frame has a right angle, it is a rectangle.

Ways to Prove that a Parallelogram is a Rectangle

If the diagonals of a parallelogram are congruent, then it is a rectangle.

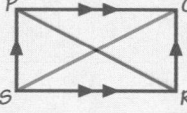  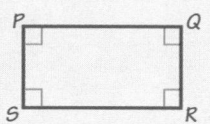

If $\overline{PR} \cong \overline{QS}$ in $\square PQRS$, then $PQRS$ is a rectangle.

If one angle of a parallelogram is a right angle, then it is a rectangle.

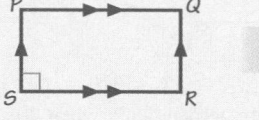  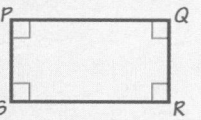

If one angle of $\square PQRS$ is a right angle, then $PQRS$ is a rectangle.

7.3 Conditions for Special Parallelograms **353**

Objectives

- Classify parallelograms using their properties.
- Prove that figures are special types of parallelograms.

Recommended Pacing

❖ **Core and Extended Courses**
Section 7.3: 2 days

❖ **Block Schedule**
Section 7.3: 1 block

Resource Materials

Lesson Support
Lesson Plan 7.3

Warm-Up Transparency 7.3

Practice Bank: Practice 49

Study Guide: Section 7.3

Explorations Lab Manual:
Diagram Master 2

Challenge Problems: Set 49

Assessment Book: Test 31

Technology
Internet:
http://www.hmco.com

Warm-Up Exercises

Complete.

1. If the diagonals of a quadrilateral bisect each other, the quadrilateral is a ___?___ . parallelogram

2. If both pairs of ___?___ sides in a quadrilateral are congruent, the quadrilateral is a parallelogram. opposite

3. If one pair of opposites sides in a quadrilateral is ___?___ and ___?___ , the quadrilateral is a parallelogram. parallel, congruent

4. If both pairs of opposite ___?___ in a quadrilateral are congruent, the quadrilateral is a parallelogram. angles or sides

Teach⇔Interact

Think and Communicate

Question 1 helps focus students' attention on the fact that they must know that the quadrilateral is a parallelogram before applying either of the theorems on page 353.

About Example 1

Geometric Thinking

Point out to students that a figure can be placed with the diagonals on the coordinate axes only if it is known that the diagonals are perpendicular. Example 1 uses the fact that the diagonals of a parallelogram bisect each other to center the figure at the origin and name the vertices.

Additional Example 1

Joining the midpoints of the sides of a rectangle forms a parallelogram. Write a coordinate proof to show that the figure formed by joining these midpoints is a rhombus.

Step 1 Place the rectangle on the coordinate axes with one vertex at the origin and two sides along the x- and y-axes.

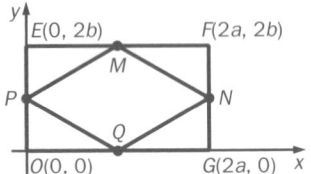

Step 2 Use the Midpoint Formula to find the coordinates of each midpoint.

midpoint of $\overline{EF} = M\left(\dfrac{0+2a}{2}, \dfrac{2b+2b}{2}\right)$

$= M(a, 2b)$

midpoint of $\overline{FG} = N\left(\dfrac{2a+2a}{2}, \dfrac{2b+0}{2}\right)$

$= N(2a, b)$

midpoint of $\overline{GO} = Q\left(\dfrac{0+0}{2}, \dfrac{2a+0}{2}\right)$

$= Q(a, 0)$

midpoint of $\overline{OE} = P\left(\dfrac{0+2b}{2}, \dfrac{0+0}{2}\right)$

$= P(0, b)$

Step 3 Use the Slope Formula to show that the diagonals of *PMNQ* are perpendicular.

slope of $\overline{PN} = \dfrac{b-b}{2a-0} = 0$

slope of $\overline{MQ} = \dfrac{2b-0}{a-a}$ is undefined.

$\overline{PN}$ is horizontal and $\overline{MQ}$ is vertical; hence, they are perpendicular. Since *PMNQ* is a parallelogram, it is also a rhombus by the result proved in Example 1.

354

THINK AND COMMUNICATE

1. In quadrilateral *JKLM*, $\overline{JL} \cong \overline{KM}$. Explain why the first theorem on page 353 cannot be used to show that *JKLM* is a rectangle.

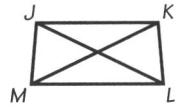

2. Use a property of parallelograms to explain why a parallelogram with one right angle must be a rectangle.

There are also ways to tell whether a parallelogram is a rhombus. The example below shows how to do this by looking at the diagonals of the parallelogram.

EXAMPLE 1   **Connection: Algebra**

Write a coordinate proof to show that if the diagonals of any parallelogram are perpendicular, then it is a rhombus.

SOLUTION

Step 1 Place the diagonals along the x- and y-axes, since they are perpendicular. To make sure the quadrilateral is a parallelogram, choose the vertices so that the origin is the midpoint of both diagonals. For instance, $K(a, 0)$ and $M(-a, 0)$ are both a units from $(0, 0)$.

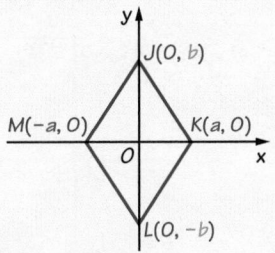

Step 2 Use the Distance Formula to show that $\square JKLM$ is equilateral.

$$JK = \sqrt{(0-a)^2 + (b-0)^2}$$

$$= \sqrt{a^2 + b^2}$$

$$KL = \sqrt{(a-0)^2 + (0-(-b))^2}$$

$$= \sqrt{a^2 + b^2}$$

> Remember that $0 - (-b) = 0 + b$.

$$ML = \sqrt{(-a-0)^2 + (0-(-b))^2}$$

$$= \sqrt{a^2 + b^2}$$

$$JM = \sqrt{(0-(-a))^2 + (b-0)^2}$$

$$= \sqrt{a^2 + b^2}$$

By substitution, $JK = KL = ML = JM$. Therefore, if the diagonals of a parallelogram are perpendicular, then it is a rhombus.

354   Chapter 7   *Quadrilaterals, Areas, and Volumes*

ANSWERS Section 7.3

Think and Communicate

1. *JKLM* is not necessarily a parallelogram.

2. Since opposite angles of a parallelogram are congruent, the angle opposite the given right angle must also be a right angle. The other two angles must also be right angles because they are both supplementary and congruent. (If two parallel lines are intersected by a transversal, then same-side interior angles are supplementary. Both pairs of opposite angles of a parallelogram are congruent.)

Ways to Prove that a Parallelogram is a Rhombus

If the diagonals of a parallelogram are perpendicular, then it is a rhombus.

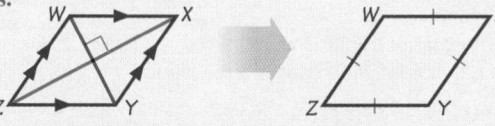

If $\overline{WY} \perp \overline{XZ}$ in $\square WXYZ$, then $WXYZ$ is a rhombus.

If two consecutive sides of a parallelogram are congruent, then it is a rhombus.

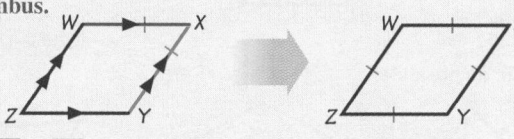

If $\overline{WX} \cong \overline{XY}$ in $\square WXYZ$, then $WXYZ$ is a rhombus.

EXAMPLE 2

Classify quadrilateral *ABCD*. Be as specific as possible.

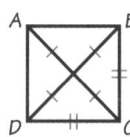

SOLUTION

Since $\overline{AC}$ and $\overline{BD}$ bisect each other, $ABCD$ is a parallelogram.

The diagram also shows that $\overline{AC} \cong \overline{BD}$. So $\square ABCD$ is a rectangle because its diagonals are congruent.

Since $\overline{BC}$ and $\overline{CD}$ are congruent consecutive sides, $\square ABCD$ is a rhombus.

Since $\square ABCD$ is both a rectangle and a rhombus, it is a square.

☑ CHECKING KEY CONCEPTS

Classify each quadrilateral. Be as specific as possible.

1.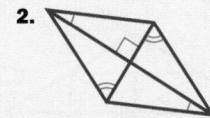
2.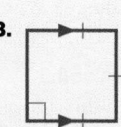
3.

4. Write a flow proof to show that if two consecutive sides of a parallelogram are congruent, then it is a rhombus.

7.3 Conditions for Special Parallelograms **355**

Checking Key Concepts

1. parallelogram

2. rhombus

3. square

4. Given: $\square DEFG$ with $\overline{DE} \cong \overline{EF}$
   Prove: $DEFG$ is a rhombus.

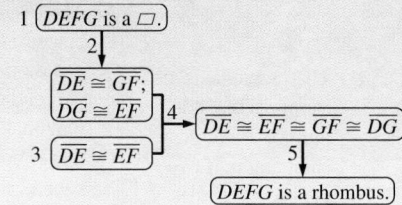

Reasons

1. Given
2. Def. of parallelogram
3. Given
4. Transitive Property
5. Def. of rhombus

Teach⇔Interact

About Example 2

Geometric Thinking
As with the theorems on page 353, remind students that they must first show or be given that the quadrilateral is a parallelogram before applying theorems on this page.

Additional Example 2

Classify quadrilateral *LMNO*. Be as specific as possible.

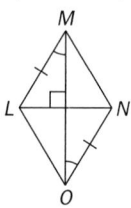

Since $\angle LMO \cong \angle NOM$, $\overline{LM} \parallel \overline{ON}$. Quadrilateral *LMNO* is a parallelogram because it has a pair of opposite sides, $\overline{LM}$ and $\overline{ON}$, that are parallel and congruent. Since $\overline{LN} \perp \overline{MO}$, $\square LMNO$ is also a rhombus.

Checking Key Concepts

Teaching Tip
For questions 1–3, make sure that students first justify that each of these figures is a parallelogram.

Visual Thinking
For question 4, ask students to explain their flow proofs to the class. This activity involves the visual skills of *interpretation* and *communication*.

Closure Question

Describe how you can further classify a quadrilateral that you know is a parallelogram. Check to see if any angles are right angles or if the diagonals are congruent. If either of these conditions is met, the quadrilateral is a rectangle. Check to see if the diagonals are perpendicular or if two consecutive sides are congruent. Satisfying either of these conditions indicates that the quadrilateral is a rhombus. If the quadrilateral is both a rhombus and a rectangle, then it is also a square.

355

Exercise Notes

Student Progress
Exs. 2–9 Students should be able to classify each of these figures. They should also be able to write a justification for their classification similar to the one given in the Solution for Example 2 on page 355. Students who have difficulty should be encouraged to keep a list of the methods that can be used to prove a quadrilateral is a parallelogram, rectangle, or rhombus, and to use the list when necessary.

Common Error
Exs. 2–9 You may need to remind some students to show that the quadrilateral is first a parallelogram.

 Communication: Drawing
Ex. 10 This open-ended problem gives students an opportunity to see visually that before applying the theorems of this section, one must show or be given that the quadrilateral is a parallelogram. Without that restriction, as this exercise shows, other figures, namely, kites, are a possibility.

 Application
Exs. 11, 13 In Ex. 11, students see how the theorems of this section can be used to justify why the paper-folding technique produces a square. For Ex. 13, students analyze a quilting pattern by using what they have learned to determine that a quadrilateral is a square.

Alternate Approach
Ex. 12 You may wish to ask students to write a coordinate proof to show that the figure formed by the midpoints of the sides of a rhombus is a rectangle.

7.3 | **Exercises and Applications**

Extra Practice exercises on page 673

1. **Logical Reasoning** Explain why you *cannot* use the theorems you learned in this section to classify each quadrilateral below as a rectangle or a rhombus.

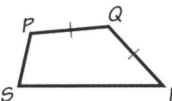 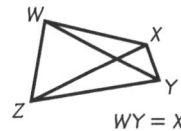 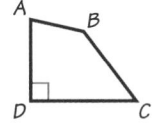 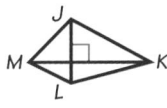

$WY = XZ$

Classify each quadrilateral. Be as specific as possible.

2. 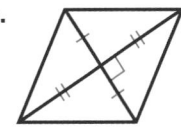 **3.** 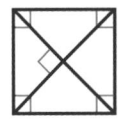 **4.** 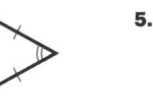 **5.**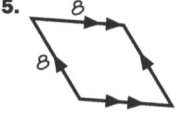

6. 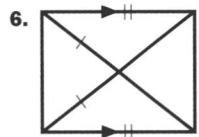 **7.** 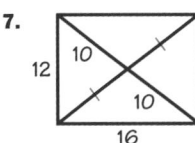 **8.** 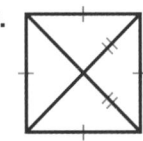 **9.**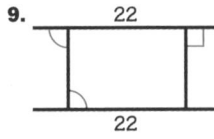

10. **Open-ended Problem** Sketch an example of a nonsquare quadrilateral whose diagonals are both congruent and perpendicular.

11. **CRAFTS** Two common crafts in Japan are paper-folding (origami) and paper-cutting (kirigami). Since both crafts usually require a square sheet of paper, it is helpful to know how to cut a nonsquare rectangular sheet of paper into a square.

Fold the paper so that a short edge lines up with a long edge.

Then cut along the edge of the two-layer triangle.

Why is the resulting rectangle a square? Which theorem does this process illustrate?

12. **Visual Thinking** Use the diagram at the right and a theorem you learned in this section to explain why the segments that join the midpoints of consecutive sides of a rhombus form a rectangle.

Exercises and Applications

1. The quadrilaterals are not necessarily parallelograms.

2. rhombus 　3. square

4. rhombus 　5. rhombus

6. rectangle 　7. rectangle

8. square 　9. rectangle

10. Answers may vary. One example is a kite with congruent diagonals. Another example is shown, in which the quadrilateral has no other special properties.

11. The rectangle is a rhombus that contains at least one right angle; if two consecutive sides of a parallelogram are congruent, then it is a rhombus.

12. See answers in back of book.

13. Answers may vary. An example is given. The resulting figure is a parallelogram because the diagonals bisect each other. It is a rhombus because the diagonals are perpendicular. It is a rectangle because the diagonals are congruent. A parallelogram that is both a rectangle and a rhombus is a square.

14–19. Answers may vary. Numerical examples should follow the given form, where each variable represents a positive number.

13. **QUILTING** Many quilts feature squares that are made up of triangles. Explain why placing four congruent isosceles right triangles as shown at the right always results in a square.

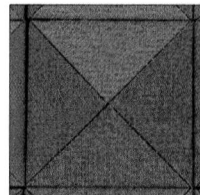

ALGEBRA Give the possible coordinates of the vertices of each quadrilateral so that the diagonals lie on the *x*- and *y*-axes.

14. a kite

15. a nonsquare rhombus

16. a square

17. a rectangle

18. a trapezoid

19. a parallelogram

20. Prove that if two consecutive angles of a parallelogram are congruent, then the parallelogram is a rectangle.

   Given: In $\square WXYZ$, $\angle W \cong \angle X$.
   Prove: $\square WXYZ$ is a rectangle.

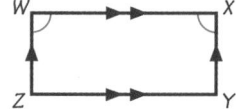

21. **ALGEBRA** Write a coordinate proof to show that if exactly one diagonal of a quadrilateral is the perpendicular bisector of the other diagonal, then the quadrilateral is a kite.

22. Write the key steps of a proof showing that if the diagonals of a parallelogram are congruent, then the parallelogram is a rectangle.

Connection   SOCIAL STUDIES

Although traditional houses of the Kpelle people of Liberia are circular, modern Kpelle structures are rectangular. To determine the space that a building will occupy, the Kpelle sometimes place two long wooden sticks across each other as the diagonals of the rectangle.

23. How must the lengths of the sticks be related in order to produce a rectangular structure?

LIBERIA

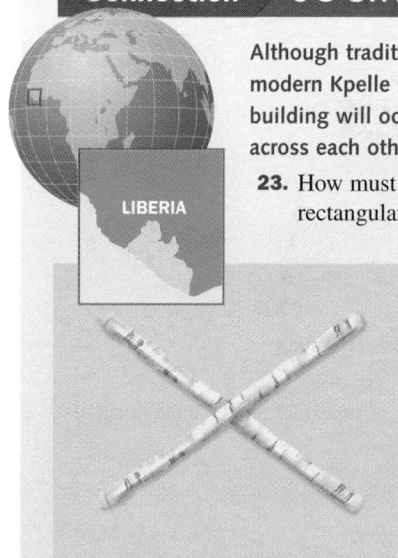

24. If the sticks are the appropriate lengths, does it matter how they are placed? Why or why not?

25. **Writing** A square building requires fewer materials than a nonsquare rectangular one that covers the same area. How should the sticks be placed for a square house? Explain your reasoning.

7.3 Conditions for Special Parallelograms   **357**

---

14. $(0, a)$, $(0, b)$, $(-c, 0)$, $(0, c)$
    $(a \neq b)$

15. $(0, a)$, $(0, -a)$, $(-b, 0)$, $(b, 0)$
    $(a \neq b)$

16. $(0, a)$, $(0, -a)$, $(-a, 0)$, $(a, 0)$

17. If the diagonals of a rectangle are perpendicular, the rectangle must be a square. See Ex. 16.

18. $(-a, 0)$, $(0, b)$, $(b, 0)$, $(0, -a)$

19. If the diagonals of a parallelogram are perpendicular, the parallelogram must be a rhombus. See Exs. 15 and 16. (The rhombus may or may not be a square.)

20–22. See answers in back of book.

23. They must be equal.

24. Yes; they must bisect each other so the resulting quadrilateral will be a parallelogram.

25. The sticks should be placed so that they bisect each other and are perpendicular. If the sticks bisect each other, they will be diagonals of a parallelogram (and, therefore, a rectangle). If they are perpendicular, they are the diagonals of a rhombus. A rectangular rhombus is a square.

Apply⇔Assess

Exercise Notes

Integrating Algebra
Exs. 14–19 In these exercises, students try to place quadrilaterals in the coordinate plane so that the diagonals are formed by the axes. Ask students to state the theorems that allow the figures to be arranged as described.

Cooperative Learning
Exs. 14–19 You may wish to have students work on these exercises in small groups. Group members can then share ideas on how to place the figures in the coordinate plane.

Writing Proofs
Exs. 20–22 For Ex. 20, you may want to suggest that students write either a two-column proof or flow proof. Also, after completing the coordinate proof for Ex. 21, students may want to write a two-column or flow proof for this theorem. To do this, students need to use the Reflexive Property and the HL Theorem to show that the triangles formed by the diagonals are two pairs of congruent triangles. Then they can use the definition of congruent triangles to show that two pairs of adjacent sides of the quadrilateral are congruent. Finally, they can show that the other pairs of triangles are not congruent and, thus, the other pairs of adjacent sides are not congruent.

Interdisciplinary Problems
Exs. 23, 24 These problems give students an insight into how mathematics can be used in a social studies context to analyze the building methods used by the Kpelle people of Liberia.

Multicultural Note
Exs. 23–25 The Kpelle people live in the two West African nations of Liberia and Guinea. In Liberia, they form the largest single ethnic group. In Guinea, the Kpelle comprise a smaller proportion of the population. Many African peoples whose traditional homes are round, like the Kpelle, are building more and more rectangular houses. This change is attributed to several factors, including Western influence and the rectangular configuration and rigidity of modern building materials, such as corrugated tin roofing material and cement blocks.

Exercise Notes

Construction Note
Exs. 26–30 Students should do the construction at the top of this page prior to doing these exercises. In Ex. 30, students need to realize that $\overline{JL}$ is the diagonal of the rectangle. To use the circle method, $\overline{JL}$ must be one of the diameters. The perpendicular bisector of $\overline{RS}$ needs to be constructed to find the midpoint of $\overline{RS}$ and, hence, the center of the circle. Once the circle is drawn, any other diameter can be chosen to make the rectangle.

Challenge
Ex. 31 This problem requires that students write and use algebraic expressions involving one or two variables. The values of the variables are not unique. In fact, any values of x and y that satisfy the equation $x + y = 90$ will form a rectangle. This restriction is a result of the fact that the angles labeled $2y°$ and $2x°$ form a straight angle. Since it is not known whether the figure is a parallelogram, students must show that all four angles are right angles.

Student Study Tip
Ex. 32 Encourage students to make a sketch of this figure on graph paper. The solution can then be seen clearly, and using the Distance Formula, Midpoint Formula, or Slope Formula can be avoided. These formulas would have to be used, however, if a proof of the result were desired.

Assessment Note
Ex. 33 One method for drawing a square can be found in Ex. 25. The construction described in Ex. 30 can also be adapted to produce a square. Students should realize, however, that these two methods are essentially the same: each one uses two diagonals of equal length that are the perpendicular bisectors of each other. To find a different method, students need to focus on the sides and angles of the figure rather than the diagonals.

—CONSTRUCTION—

RECTANGLE

1. Construct ⊙E.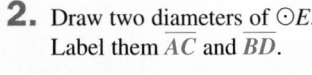

2. Draw two diameters of ⊙E. Label them $\overline{AC}$ and $\overline{BD}$.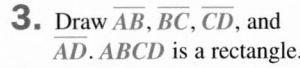

3. Draw $\overline{AB}, \overline{BC}, \overline{CD}$, and $\overline{AD}$. *ABCD* is a rectangle.

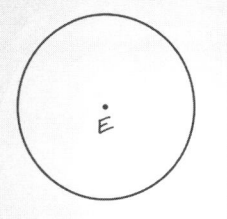
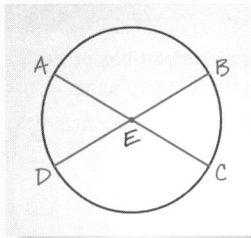
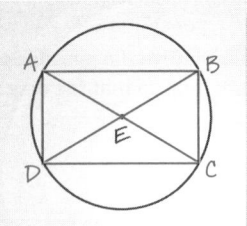

26. How do you know that quadrilateral *ABCD* is a parallelogram?

27. What guarantees that ▱*ABCD* will be a rectangle?

28. Writing The diagonals of two rectangles are congruent, but the rectangles are not congruent. Describe how to construct the rectangles.

29. In Section 5.7, you learned how to construct the perpendicular bisector of a given segment.
   a. Construct two segments, $\overline{WY}$ and $\overline{XZ}$, that are perpendicular bisectors of each other.
   b. Draw $\overline{WX}, \overline{XY}, \overline{YZ}$, and $\overline{WZ}$. What type of quadrilateral is *WXYZ*? How do you know?

30. Construct a rectangle *JKLM* so that $\overline{JL} \cong \overline{RS}$.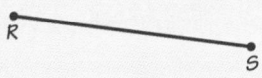

31. Challenge Prove that quadrilateral *ABCD* is a rectangle.

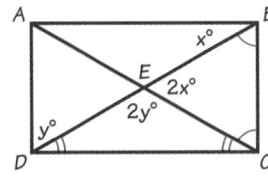

32. SAT/ACT Preview The coordinates of the vertices of a quadrilateral are $(1, 4), (3, 0), (1, -2)$, and $(-3, 0)$. The quadrilateral is:

   A. a rhombus   **B.** a kite   **C.** a square   **D.** a trapezoid   **E.** none of these

ONGOING ASSESSMENT

33. Open-ended Problem Using a ruler and a protractor, use two different methods to draw a square. Describe your methods and prove that they produce squares.

---

26. *EA, EB, EC*, and *ED* are all radii of the circle, so they are congruent and the diagonals of *ABCD* bisect each other.

27. $\overline{AC} \cong \overline{DB}$ (They are both diameters of the circle.)

28. Construct the first rectangle as described in Steps 1–3. To construct the second, use the same radius for circle *E*, but choose diameters that intersect at a different angle than the diameters in your first rectangle.

29. a.
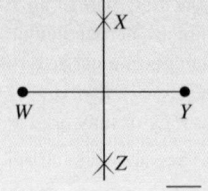
Draw a segment $\overline{WY}$. When you construct its perpendicular bisector, let *X* and *Y* be the points where the two pairs of arcs intersect. $\overline{WY}$ and $\overline{XZ}$ are perpendicular bisectors of each other.

(You may choose different points *X* and *Y* by drawing any two arcs with the same radius and center at the midpoint of $\overline{WY}$.)

b. Check students' work; rhombus; the diagonals bisect each other, so *WXYZ* is a parallelogram. The diagonals are perpendicular, so *WXYZ* is a rhombus.

30–33. See answers in back of book.

Find each length or angle measure. *(Section 6.6)*

34. *JK*

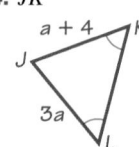

35. *m* ∠ *Z*

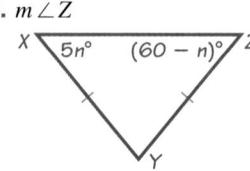

36. *m* ∠ *E*

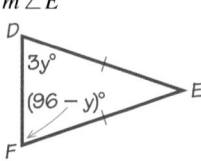

Write an equation for each line described. *(Section 4.2)*

37. contains (6, 2); has slope −3

38. contains (0, −1); has slope 5

ASSESS YOUR PROGRESS

VOCABULARY

kite (p. 339)

indirect proof (p. 348)

Find each length. *(Section 7.1)*

1. rectangle *WXYZ*
  a. *VY*
  b. *XZ*
  c. *ZY*

2. rhombus *JKLM*
  a. *JK*
  b. *NK*
  c. *MK*

3. kite *PQRS*
  a. *TQ*
  b. *TR*
  c. *PQ*

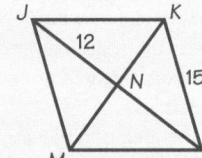

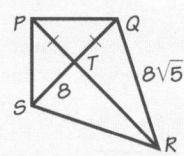

4. Write an indirect proof to show that
∠ *W* ≇ ∠ *Y* in kite *WXYZ*. *(Section 7.2)*

Given: *WXYZ* is a kite.

Prove: ∠ *W* ≇ ∠ *Y*

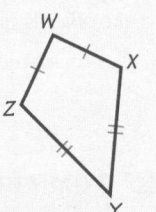

Classify each quadrilateral. Be as specific as possible. Give reasons for your answers. *(Sections 7.2 and 7.3)*

5.
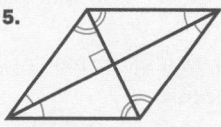

6.
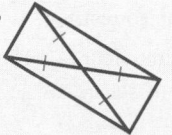

7.
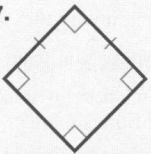

8. Journal List all the ways that you have learned to identify each special type of quadrilateral shown on page 339. Which methods do you think are the easiest to remember and to use?

7.3 Conditions for Special Parallelograms **359**

Assess Your Progress

Journal Entry
This journal entry can help students summarize what they have learned in the first half of this chapter. They should realize that all given information must be considered when determining which of the methods to use.

Progress Check 7.1–7.3

See page 384.

Practice 49 for Section 7.3

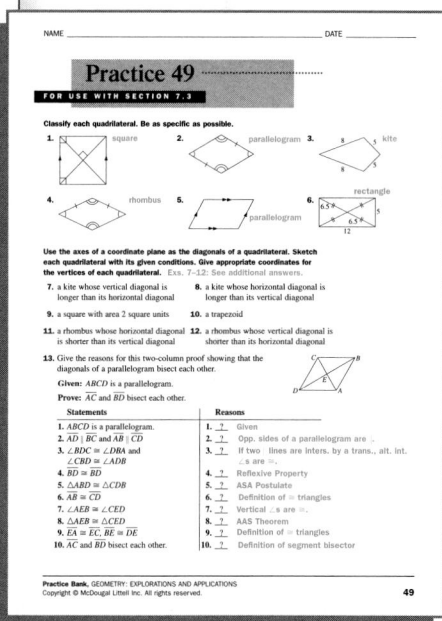

---

34. 6
35. 50°
36. 72°
37. *y* = −3*x* + 20
38. *y* = 5*x* − 1

Assess Your Progress

1. a. 5   b. 10
   c. 2√21 ≈ 9.2
2. a. 15   b. 9
   c. 18
3. a. 8   b. 16
   c. 8√2 ≈ 11.3

4. Suppose that ∠ *W* ≅ ∠ *Y*. Since it is given that $\overline{WZ} ≅ \overline{WX}$ and $\overline{YZ} ≅ \overline{YX}$, and $\overline{WY} ≅ \overline{WY}$ by the Reflexive Property, △ *WZY* ≅ △ *WXY* by the SSS Postulate. Then ∠ *Z* ≅ ∠ *X*. However, if ∠ *W* ≅ ∠ *Y*, then *WXYZ* is a parallelogram since both pairs of opposite angles are congruent. This is a contradiction, since opposite sides of a kite are not parallel. The assumption that ∠ *W* ≅ ∠ *Y* must not be true. So ∠ *W* ≇ ∠ *Y*.

5–7. Answers may vary. Examples are given.

5. rhombus; Both pairs of opposite sides are parallel, so the quadrilateral is a parallelogram by definition; it is a rhombus because the diagonals are perpendicular.

6. rectangle; The diagonals bisect each other, so the quadrilateral is a parallelogram; it is a rectangle because the diagonals are congruent.

7. square; It is given that the quadrilateral is equiangular. Then both pairs of opposite angles are congruent and the quadrilateral is an equiangular parallelogram, that is, a rectangle. Since two consecutive sides are congruent, it is a rhombus. A rectangular rhombus is a square.

8. See answers in back of book.

359

Objectives

- Find the areas of triangles and quadrilaterals.
- Use the formulas for finding the areas of triangles and quadrilaterals in real-world applications.

Recommended Pacing

❖ **Core and Extended Courses**
Section 7.4: 2 days

❖ **Block Schedule**
Section 7.4: 1 block

Resource Materials

Lesson Support
Lesson Plan 7.4

Warm-Up Transparency 7.4

Overhead Visuals:
  Folder 9: Areas of Simple
    Polygons

Practice Bank: Practice 50

Study Guide: Section 7.4

Explorations Lab Manual:
  Additional Exploration 10
  Diagram Master 2

Challenge Problems: Set 50

Technology
Technology Book:
  Spreadsheet Activity 6

McDougall Littell Mathpack:
  Geometry Inventor Activity Book:
    Activities 6, 7

Internet:
  http://www.hmco.com

Warm-Up Exercises

1. Find the side length of an equilateral triangle with perimeter 15. **5**

2. Find the side length of a square with perimeter 36. **9**

3. Find length of the legs of an isosceles triangle with base 12 and perimeter 28. **8**

4. Find the lengths of the other sides of a parallelogram with one side 4 and perimeter 40. **4, 16, and 16**

5. Find the height of an equilateral triangle whose sides are 5 units long. **about 4.3 units**

360

7.4 Areas of Triangles and Quadrilaterals

Learn how to...
- find the areas of triangles and quadrilaterals

So you can...
- use these formulas in real-world applications, such as tiling a floor

It takes $15\frac{1}{2}$ rows to fill the length of the room.

A row of 12 tiles stretches across the width of the room.

Paul Bornstein's porch is 12 ft wide and $15\frac{1}{2}$ ft long. He wants to cover the floor with 1 ft by 1 ft square tiles. How many tiles does he need? Does it matter what shape the room is? Is there a quick way to figure it out?

Paul needs 12 rows of $15\frac{1}{2}$ tiles. So $12 \times 15\frac{1}{2} = 186$ tiles will cover the full *area* of the floor of his porch.

The **area** of a two-dimensional figure is the number of *square units* enclosed within the boundary of the figure. A square unit is the space enclosed by a 1 unit by 1 unit square. In this situation, the area of each tile is one square foot, written as 1 ft².

To find the area of a rectangle, multiply the lengths of two adjacent sides. One length is considered to be the **base** and the other is the **height**.

Area of a rectangle = **base** × **height**

$$A = bh$$

THINK AND COMMUNICATE

1. Praktisha says that the base of Paul's floor is 12 ft and the height is $15\frac{1}{2}$ ft. Danielle thinks that the base is $15\frac{1}{2}$ ft and the height is 12 ft. Explain why they are both correct.

2. Suppose you want to tile the floor of a 12 ft by 12 ft square room.
   a. What is the height of the square? What is the base of the square?
   b. Find the area of the square.

3. Explain how to find the area of the figure at the right. What units will the area be measured in?

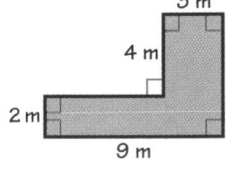

Toolbox p. 704
Finding Area and Volume

ANSWERS Section 7.4

Think and Communicate

1. Either length can be considered the base.

2. **a.** 12 ft; 12 ft
   b. 144 ft²

3. Divide the figure into rectangles. An example is given.

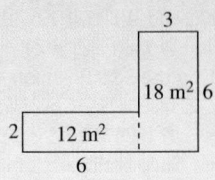

The area will be measured in square meters (m²).

EXPLORATION
COOPERATIVE LEARNING

Discovering Area Formulas

Work in a group of three students.
You will need:
- a rectangular sheet of paper
- scissors

1 Cut the paper into two noncongruent rectangles. Label the sides as shown.

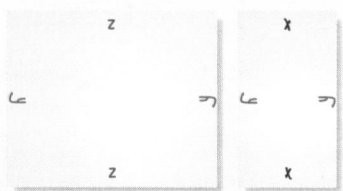

2 Cut along one diagonal of one rectangle to make two congruent right triangles, as shown.

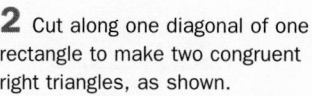

3 How do the base and height (the lengths of the legs) of one triangle compare to the base and height of the rectangle it was cut from? How does the area of one triangle compare with the area of that rectangle? Describe the area of one triangle in terms of its base and height.

4 Use all three pieces to make a rectangle.

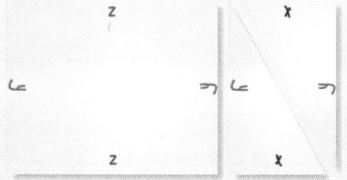

> The rectangle's **base** is $x + z$ and its **height** is y. So the area is $A = (x + z)y$.

5 Use all three pieces to make a nonrectangular parallelogram whose *height*, the distance between two of its parallel sides, is y. Write an expression for the base. Does the parallelogram have the same area as the rectangle you made in Step 4? Describe the area of the parallelogram in terms of its base and height.

6 Use all three pieces to make a trapezoid whose height is y. Write an expression for each base and the sum of the bases. Does the trapezoid have the same area as the rectangle you made in Step 4? Describe the area of the trapezoid in terms of its height and the sum of its bases.

Section Notes

Common Error

When finding the area of a parallelogram or trapezoid, some students may mistakenly use the length of the side as the height, since this was the case for a rectangle. Refer these students to the figures used in the Exploration to remind them that the height is the perpendicular distance from the base to a vertex opposite the base.

Teaching Tip

Although the formula for the area of a triangle was found using a right triangle in the Exploration, point out to students that the formula $A = \frac{1}{2}bh$ applies to any triangle. In fact, you may wish to have students draw two nonright congruent triangles, label a different base in each triangle, measure the base and height using this base, and then find the area. This should convince students that the formula works for any triangle and that they can use any side of a triangle for the base.

Additional Example 1

Find the area of an isosceles trapezoid with perimeter 48 in. if the lengths of the bases are 14 in. and 18 in.

Subtract the lengths of the bases from the perimeter.

$48 - 14 - 18 = 16$

Divide by 2 to find the lengths of the equal sides.

$\frac{16}{2} = 8$

Sketch the trapezoid.

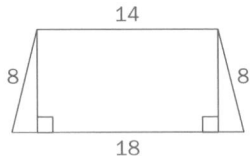

The quadrilateral in the trapezoid is a rectangle; thus, the length of each horizontal side is 14. To find the base of each of the two congruent right triangles, subtract 14 from 18 and divide by 2 to get 2 in. Use the Pythagorean theorem to find the height of the trapezoid.

$h^2 = 8^2 - 2^2 = \sqrt{60}$

Use the formula for the area of a trapezoid. Substitute 14 for b_1, 18 for b_2, and $\sqrt{60}$ for h.

In the area formulas given below, the *height* is the distance from the base to the highest point of the figure.

Area Formulas

The area of a rectangle is the product of the base and the height.

$A = bh$

The area of a square is the square of the length of one side.

$A = s^2$

The area of a triangle is half the product of the base and the height.

$A = \frac{1}{2}bh$

The area of a parallelogram is the product of the base and the height.

$A = bh$

The area of a trapezoid is the product of the height and the mean of the bases.

$A = \frac{1}{2}(b_1 + b_2)h$

EXAMPLE 1

Toolbox p. 703
Finding Perimeter

Find the area of an equilateral triangle with perimeter 18 cm.

SOLUTION

The **base** of an equilateral triangle with perimeter 18 cm is **6 cm**. Sketch the triangle and use the Pythagorean theorem to find its height.

Use the formula given above.

$3^2 + h^2 = 6^2$
$h^2 = 6^2 - 3^2$
$h = 3\sqrt{3}$

$A = \frac{1}{2}bh$

$= \frac{1}{2}(6)(3\sqrt{3})$ Substitute **6** for *b* and $3\sqrt{3}$ for *h*.

≈ 15.6

The area of the triangle is about 15.6 cm².

362   Chapter 7   *Quadrilaterals, Areas, and Volumes*

Exploration

1, 2. Check students' work.

3. The base and height of the triangle are the base and height of the rectangle. The area of one triangle is half the area of the rectangle. The area of the triangle is half the product of its base and its height.

4. Check students' work.

5.
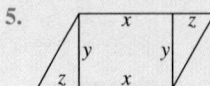

$x + z$; Yes; the area of the parallelogram is the product of its base and its height.

6.
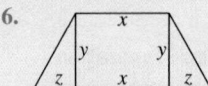

x and $x + 2z$; $2x + 2z$; Yes; the area of the trapezoid is half the product of its height and the sum of the lengths of its bases.

EXAMPLE 2 **Connection: Algebra**

Find the area of the trapezoid.

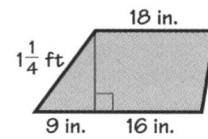

SOLUTION

Use the Pythagorean theorem to find the height h of the trapezoid.

$$9^2 + h^2 = 15^2$$

$$h^2 = 15^2 - 9^2$$

> Make sure the measurements have the same units.
>
> $1\frac{1}{4}$ ft = 15 in.

$$h = \sqrt{225 - 81}$$

$$h = 12$$

Use the formula for the area of a trapezoid.

$$A = \frac{1}{2}(b_1 + b_2)h$$

$$= \frac{1}{2}[18 + (9 + 16)](12)$$

> Use the Segment Addition Postulate to find the length of the lower base.

$$= \frac{1}{2}(43)(12)$$

$$= 258$$

The area of the trapezoid is 258 in.^{2}.

☑ **CHECKING KEY CONCEPTS**

Find the area of each polygon.

1.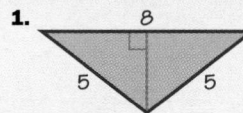

2.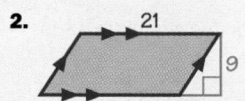

3.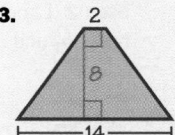

4.

5. a rectangle with side lengths 26 ft and 9 ft

6. a square with side length 19 cm

7. a right triangle with one leg of length 14 in. and hypotenuse 30 in.

7.4 Areas of Triangles and Quadrilaterals **363**

Checking Key Concepts

1. 12

2. 189

3. 64

4. 24

5. 234 ft^2

6. 361 cm^2

7. $56\sqrt{11} \approx 185.7$ in.^{2}

Teach⇔Interact

Additional Example 1 (continued)

$$A = \frac{1}{2}(b_1 + b_2)h$$

$$= \frac{1}{2}(14 + 18)(\sqrt{60})$$

$$= 16\sqrt{60}$$

$$\approx 123.9$$

The area of the trapezoid is about 123.9 in.^{2}

About Example 2

Teaching Tip

Make sure students understand that one of the measurements in this problem has to be converted. In all area formulas, the lengths must be given in the same units.

Additional Example 2

Find the area of the parallelogram.

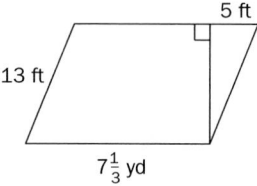

Use the Pythagorean theorem to find the height h of the parallelogram. Because the opposite sides are equal, the hypotenuse of the triangle is 13 ft.

$$5^2 + h^2 = 13^2$$

$$h^2 = 13^2 - 5^2$$

$$h^2 = 144$$

$$h = 12 \text{ ft}$$

Use the formula for the area of a parallelogram. Make sure the measurements have the same units.

$7\frac{1}{3}$ yd = 22 ft

$$A = bh$$

$$= (22)(12)$$

$$= 264$$

The area of the parallelogram is 264 ft^2.

Checking Key Concepts

Communication: Writing
Encourage students to first write the formula for the area of each polygon. This should help them remember the formulas as well as avoid errors.

Closure Question

What are the formulas for the area of a rectangle, a square, a triangle, a parallelogram, and a trapezoid? The formulas are given on page 362.

363

Exercise Notes

Student Study Tip

Exs. 1–6 Encourage students to first determine the type of figure they are working with and to write down the formula for its area. They can then determine which variables they have values for and which ones they need to find. In Ex. 6, the figure is a kite. As students do not have a formula for the area of a kite, they need to find the area of each of the four triangles.

Historical Connection

Ex. 7 Heron, commonly referred to as Hero of Alexandria, made many discoveries in geometry and physics. Some of his work on the reflection of light by mirrors was fundamental to the development of modern optics and dynamics. He also was an inventor and is said to have constructed a steam engine.

Common Error

Ex. 8 After calculating the area in square feet, some students may try to convert to square yards by dividing this value by 3. Remind these students that there are 9 square feet in 1 square yard. Student can see this by drawing a model of a square yard, dividing each side into 3 feet, and seeing that there are 9 squares formed. To avoid this type of error, students should convert all measurements before calculating the area.

Using Manipulatives

Exs. 14–16 In this group of exercises, students use paper folding and cutting to develop an alternate formula for the area of a trapezoid in terms of its height and midline.

 Using Technology
Exs. 14–16 Students can explore the properties of the midline of a trapezoid by using Activity 7 in the *Geometry Inventor Activity Book*.

7.4 | Exercises and Applications

Extra Practice
exercises on
page 674

Find the area of each polygon.

1.
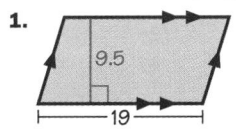

2.
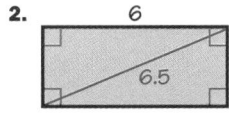

3.
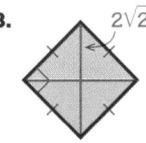

4.
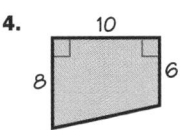

5.
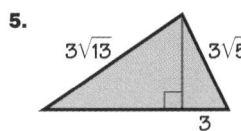

6.
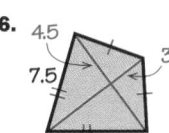

7. **HISTORY** Heron, a Greek mathematician who lived over 2000 years ago, developed the following formula to find the area of a triangle with side lengths a, b, and c. In this formula, $s = \frac{1}{2}(a + b + c)$.

$$A = \sqrt{s(s - a)(s - b)(s - c)}$$

Use Heron's Formula to find the area of the triangle in Example 1.

8. **CLOTHING** A typical Chilean poncho is rectangular, 10 ft long by 6 ft wide. How many square yards of wool cloth are needed for a poncho?

9. **Writing** The sides of two parallelograms are congruent. Must the areas of the parallelograms be the same? Explain.

Find the area of each polygon.

10. an isosceles triangle with base 10 cm and perimeter 34 cm

11. a trapezoid with height 2 m and bases that add up to 18 m

12. a square with perimeter 22 ft

13. a rectangle with side length 5 in. and diagonal 13 in.

Investigation The *midline* of a trapezoid is the segment that is parallel to the bases and bisects the legs. Cut a trapezoid out of paper and fold to find its midline.

14. Follow the steps at the right. Show that the pieces form a parallelogram in Step 3.

15. **ALGEBRA** Find an expression for the length of the midline of the trapezoid in terms of its bases.

16. Write a formula for the area of a trapezoid in terms of the length of its midline, m, and its height, h.

Step 1 Label the midline m.

Step 2 Label the bases b_1 and b_2.

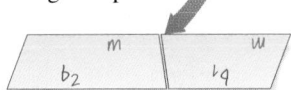

Step 3 Then cut along the midline and arrange the pieces as shown.

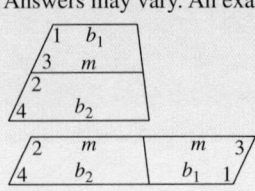

Exercises and Applications

1. 180.5
2. 15
3. 16
4. 70
5. 36
6. 40.5

7. For the triangle in Example 1, $a = b = c = 6$ and $s = 9$.
$A = \sqrt{9(9 - 6)(9 - 6)(9 - 6)} = \sqrt{243} = 9\sqrt{3} \approx 15.6$ cm^2

8. $6\frac{2}{3}$ yd^2

9. No; the heights may not be the same. Consider, for example, a square and a rhombus that is not a square whose sides are the same length as the square.

10. $5\sqrt{119} \approx 54.5$ cm^2

11. 18

12. 30.25 ft^2

13. 60 in.^{2}

14. Answers may vary. An example is given.

The bases and the midline of the trapezoid are all parallel, so $\angle 1 \cong \angle 2$ and $\angle 3 \cong \angle 4$. Then both pairs of opposite angles of the quadrilateral formed in Step 3 are congruent and the quadrilateral is a parallelogram.

ALGEBRA The coordinates of the vertices of a polygon are given. Sketch the polygon and find its area.

17. $S(0, 6)$, $T(0, 9)$, $U(7, 9)$, $V(7, 6)$

18. $D(1, 1)$, $E(1, 5)$, $F(5, 5)$, $G(5, 1)$

19. $A(2, 1)$, $B(4, 8)$, $C(10, 1)$

20. $W(-2, 4)$, $X(0, 4)$, $Y(-3, 0)$, $Z(-5, 0)$

21. $J(0, 3)$, $K(5, 3)$, $L(8, 6)$, $M(3, 6)$

22. $P(-2, 9)$, $Q(-2, -1)$, $R(-5, 5)$, $S(-5, 7)$

23. Sketch a rectangle and a parallelogram that have the same base and the same height. Explain how the formula for the area of a rectangle leads to the formula for the area of a nonrectangular parallelogram.

24. Sketch a parallelogram and one of its diagonals. Explain how the formula for the area of a parallelogram leads to the formula for the area of a triangle.

25. ALGEBRA Write a coordinate proof to show that the area of a rhombus is half the product of the lengths of the diagonals.

26. Open-ended Problem Sketch a trapezoid with area 24 cm². Label all the dimensions of the trapezoid in your sketch.

27. Logical Reasoning The base of $\triangle ABD$ is 24 and the height is 8.

   a. Find the areas of $\triangle ABC$ and $\triangle ACD$.

   b. Writing Describe how to divide a triangle into two or more smaller triangles whose areas are equal. Tell why your method works.

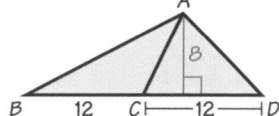

Connection   GARDENING

Many communities set aside land for gardens. The ground is usually divided into plots whose areas are equal, but often the plots are different shapes. The plans below show one layout for the Jesse Frey Community Garden in San Jose, California.

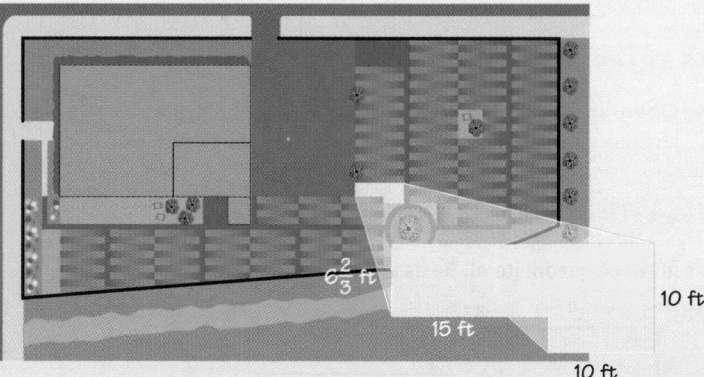

28. A standard plot is a 4 ft by 25 ft rectangle. Find the area of one plot.

29. Writing Describe a way to divide the shaded region into plots with the same area as a 4 ft by 25 ft plot.

30. Open-ended Problem Do you think the shape of a garden plot is important? Why or why not?

---

15. $m = \frac{1}{2}(b_1 + b_2)$

16. $A = mh$

17. rectangle; 21

18. square; 16

19. triangle; 28

20. parallelogram; 8

21. parallelogram; 15

22. trapezoid; 18

23. If a rectangle and a parallelogram that is not a rectangle

have the same base and height, they have the same area. To see why, picture cutting the parallelogram along the perpendicular segment and positioning the resulting triangle as shown.

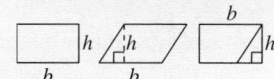

24.

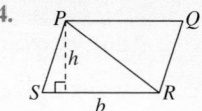

$\overline{PR}$ divides parallelogram $PQRS$ into two congruent triangles, $\triangle PSR$ and $\triangle RQP$. The area of the parallelogram is bh, so the area of each triangle is $\frac{1}{2}bh$.

25–27. See answers in back of book.

28. 100 ft²

29. Divide the plot into two rectangular plots, one 10 ft by 10 ft, the other 6 ft 8 in. by 15 ft.

30. Answers may vary. Examples are given. Yes. I think the shape of a garden plot is important because it can affect how the plants are watered with an irrigation system. It can also affect the placement of plants in a sunny or shaded area. Tending a garden that is long and narrow may be easier than one that is square.

Exercise Notes

Cooperative Learning
Exs. 17–22 You may wish to have students work in groups of three on these exercises. Each student can perform a different task. One student can sketch the polygon, another can find the lengths using the Distance Formula if necessary, and the third can use the information in the appropriate formula to find the area. Students should rotate roles for each exercise.

Communication: Drawing
Ex. 26 This exercise gives students an opportunity to experiment with the concept of the area of a trapezoid by using a drawing. Have students share their trapezoids with the class. Ask students how they determined the dimensions of their trapezoids.

Application
Exs. 28–30 These exercises introduce students to some considerations that must be made when a large area of land is to be divided into plots with equal areas. Students should come to see the variety of shapes that are possible for a particular area. For Ex. 30, students should consider the implications for gardening a plot of a particular shape. For example, you can ask them to consider the shape or shapes that would be easiest to water with a sprinkler.

Teaching Tip
Exs. 28–30 It may help some students to make a model of the garden on graph paper, with each square representing one square foot.

Apply⇔Assess

Exercise Notes

Interview Note: Application
Exs. 31–33 In this set of exercises, students see how considerations of area affect the design of handicapped parking.

Research
Ex. 34 Interested students may want to find out more about Babylonian mathematics. One area of research could be the way the Babylonians represented numbers.

Assessment Note
Ex. 35 This exercise offers students a creative way to demonstrate their understanding of the area formulas in this section. Ask several volunteers to explain how they determined the dimensions for their figures. You may wish to mention that the rectangle, trapezoid, and parallelogram could be made using the techniques of the Exploration on page 361.

Practice 50 for Section 7.4

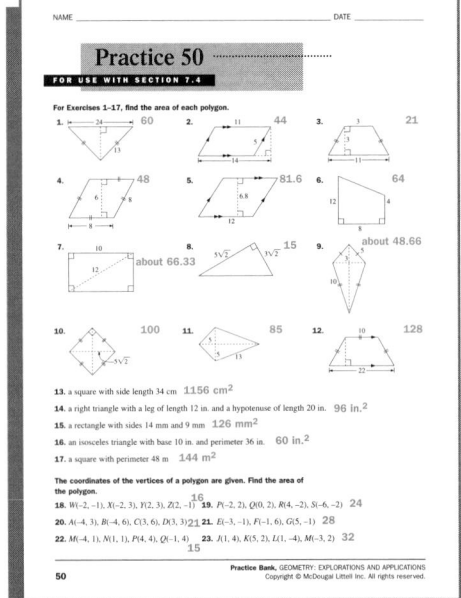

INTERVIEW **Walt Stone**

Look back at the article on pages 336–338.

One of Walt Stone's many accomplishments was to increase the number of parking places for people with disabilities in Branson. To plan this, he had to calculate the area that the spaces would need.

31. By regulation, each rectangular space for people with disabilities must be 8 ft wide and 20 ft long. What is the area of one space?

32. There must be 5 ft between spaces. Find the minimum area covered by four spaces that are next to each other.

33. How many parking spaces will fit in an area that is 35 ft wide?

34. HISTORY The following formula for the area of a trapezoid was used in calculations on an ancient Babylonian tablet:

$$A = \frac{b_1 + b_2}{2} \cdot \frac{s_1 + s_2}{2}$$

In the formula, s_1 and s_2 are the lengths of the nonparallel sides.

a. How is the formula on the tablet different from the correct formula?

b. For which types of quadrilaterals can this formula be used?

c. Challenge For which trapezoids does this formula give an answer that is approximately correct? Does it *overestimate* or *underestimate* the area? Explain.

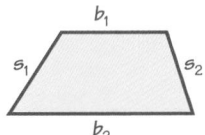

ONGOING ASSESSMENT

35. Open-ended Problem Sketch a triangle, a rectangle, a trapezoid, and a nonrectangular parallelogram that all have the same area.

SPIRAL REVIEW

Classify each quadrilateral. Be as specific as possible. *(Sections 7.2 and 7.3)*

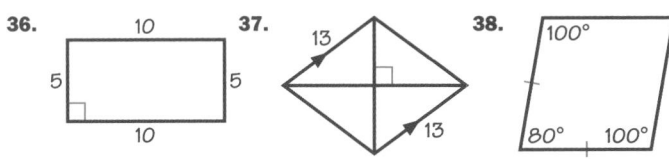

Sketch each figure. Tell how many lines of symmetry it has. *(Section 2.3)*

39. regular pentagon   **40.** regular 7-gon   **41.** regular 16-gon

366 Chapter 7 *Quadrilaterals, Areas, and Volumes*

---

31. 160 ft² **32.** 940 ft²

33. 3 parking spaces

34. a. The formula on the tablet can be obtained by substituting $\frac{s_1 + s_2}{2}$ for the height of the trapezoid in the correct formula.

b. rectangles and squares

c. trapezoids that are nearly rectangular; overestimates (s_1 will be nearly equal to s_2, so $\frac{s_1 + s_2}{2} \approx s_1$ and $s_1 > h$.)

35. Answers may vary. Examples are given.

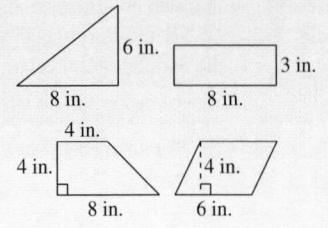

36. rectangle

37. rhombus

38. rhombus

39.

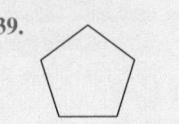

5 lines of symmetry

40.

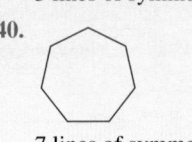

7 lines of symmetry

41.

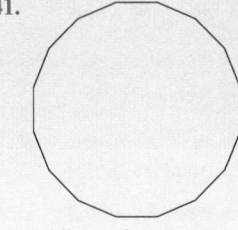

16 lines of symmetry

7.5

Areas of Regular Polygons and Circles

Learn how to...

• find the areas of regular polygons and circles

So you can...

• use area formulas to find areas of real-world objects, such as buildings or a running track

Some of the 1994 World Cup soccer games were played indoors at the Pontiac Silverdome in Pontiac, Michigan. Since the World Cup officials require a natural grass playing field, local crop and soil researchers grew the grass outdoors and brought it into the Silverdome in regular hexagonal trays.

52 in.

EXAMPLE 1 Application: Sports

Find the area of each hexagonal tray used in the Silverdome.

SOLUTION

A regular hexagon can be divided into six congruent equilateral triangles. Use the Pythagorean theorem to find the height h of one triangle.

$$h^2 + 26^2 = 52^2$$
$$h^2 = 52^2 - 26^2$$
$$h = 26\sqrt{3}$$

Use this value to find the area of one triangle.

$$\text{Area of hexagon} = 6 \cdot \left(\text{Area of one triangle} \right)$$
$$= 6\left(\frac{1}{2}bh\right)$$
$$= 6 \cdot \frac{1}{2} \cdot 52 \cdot 26\sqrt{3}$$
$$\approx 7025$$

26 in.

26√3 in.

52 in.

The area of each hexagonal tray is about 7025 in.^{2}.

BY THE WAY

Why were the trays of grass in the Silverdome hexagonal? Why do bees build hexagonal honeycomb cells? The area of a regular hexagon is greater than the area of any other shape with the same perimeter that forms a regular tessellation.

7.5 Areas of Regular Polygons and Circles **367**

Plan⇔Support

Objectives

• Find the areas of regular polygons and circles.

• Use area formulas to find areas of real-world objects.

Recommended Pacing

❖ **Core and Extended Courses**
Section 7.5: 1 day

❖ **Block Schedule**
Section 7.5: $\frac{1}{2}$ block
(with Section 7.6)

Resource Materials

Lesson Support
Lesson Plan 7.5

Warm-Up Transparency 7.5

Practice Bank: Practice 51

Study Guide: Section 7.5

Explorations Lab Manual:
    Additional Exploration 11

Challenge Problems: Set 51

Technology
Technology Book:
    Calculator Activity 10
    Spreadsheet Activity 6

Graphing Calculator

Geometry Software

McDougal Littell Mathpack
    Geometry Inventor Activity Book:
    Activity 12

Internet:
    http://www.hmco.com

Warm-Up Exercises

Is the statement *True* or *False*?

1. A hexagon has 6 sides. True.

2. A heptagon has 9 sides. False.

3. A regular polygon is both equiangular and equilateral. True.

4. A rhombus is a regular quadrilateral. False.

5. The sum of the measures of the interior angles of a pentagon is 360°. False.

Additional Example 1

Find the area of the regular heptagon.

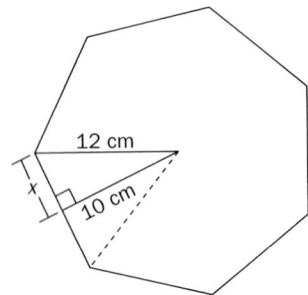

A regular heptagon can be divided into seven congruent isosceles triangles. Use the Pythagorean theorem to find the length of half the base of one triangle.

$x^2 + 10^2 = 12^2$
$\quad x^2 = 12^2 - 10^2$
$\quad\ x = 2\sqrt{11}$

The base is $2x = 2(2\sqrt{11}) = 4\sqrt{11}$. Use this value to find the area of one triangle.

Area of heptagon
$= 7 \cdot (\text{Area of one triangle})$
$= 7\left(\frac{1}{2}bh\right)$
$= 7 \cdot \frac{1}{2} \cdot 4\sqrt{11} \cdot 10$
≈ 464

The area of the heptagon is about 464 cm².

Additional Example 2

The apothem of a regular octagon is 4.8 and the distance from the center to each vertex is 6. Find the area of the octagon. Sketch the octagon. Then use the Pythagorean theorem to find the length of half of each side of the octagon.

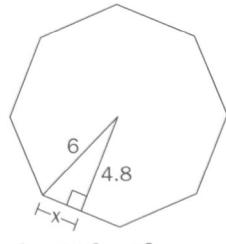

$x^2 + 4.8^2 = 6^2$
$\quad x^2 = 6^2 - 4.8^2$
$\quad\ x \approx 3.6$

Each side of the octagon has length $2x = 7.2$, so its perimeter is $8 \cdot 7.2$. Use the formula for the area of a regular polygon.

$A = \frac{1}{2}ap$
$\ \ = \frac{1}{2}(4.8)(8 \cdot 7.2)$
$\ \ = 138.24$

The area of the octagon is 138.24 square units.

In Example 1 you saw how to divide a hexagon into six congruent triangles. Any regular *n*-gon can be divided into *n* congruent triangles by drawing a segment from the center to each vertex.

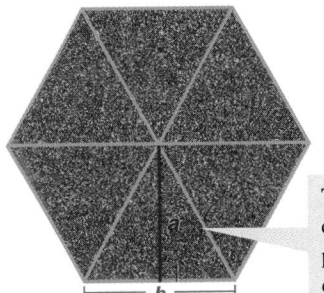

Area of a regular *n*-gon = $n \times$ Area of one triangle
$= n \times \frac{1}{2}bh$
$= \frac{1}{2}h(nb)$
$= \frac{1}{2}ap$

The distance from the center of a regular polygon to a side is called the **apothem, a**.

The product of *n* and the side length *b* is the **perimeter, p**.

Area of a Regular Polygon

The area of any regular polygon is half the product of the apothem and the perimeter.

$$A = \frac{1}{2}ap$$

EXAMPLE 2

The length of a side of a regular pentagon is 12 and the distance from the center to each vertex is 10.2. Find the area of the pentagon.

SOLUTION

Sketch the pentagon. Then use the Pythagorean theorem to find the apothem.

$6^2 + a^2 = (10.2)^2$
$\quad a^2 = (10.2)^2 - 6^2$
$\quad\ a = \sqrt{68.04}$

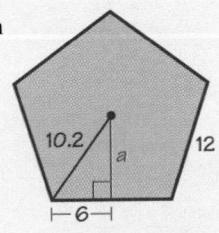

Then use the formula.

$A = \frac{1}{2}ap$
$\ \ = \frac{1}{2}\left(\sqrt{68.04}\right)(5 \cdot 12)$
$\ \ \approx 247.5$

Multiply the number of sides by the side length to find the perimeter.

The area of the pentagon is about 248.

THINK AND COMMUNICATE

1. You can divide any regular *n*-gon into *n* congruent triangles. Why are the triangles equilateral in a regular hexagon? What type of triangles do you think they are in other regular polygons? Explain your reasoning.

2. In Example 1, you found the apothem of a regular hexagon by using the length of one side. Can you find the apothem of *any* regular polygon using only the side length? Why or why not?

3. The length of each side of a square is *s*. Find expressions for the perimeter and apothem in terms of *s*. Then show that $A = \frac{1}{2}ap = s^2$ for a square.

Area of a Circle

The pattern below shows regular polygons *inscribed* in circles. The radius of a regular polygon is the radius of the circle that the polygon is inscribed in. As the sequence continues and the number of sides of the polygon increases, the length of the apothem approaches the length of the radius. The perimeter and area of the polygon approach the perimeter and area of the circle.

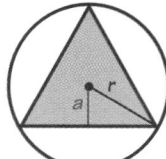 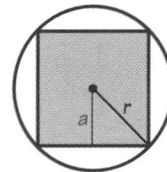 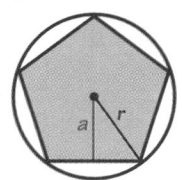 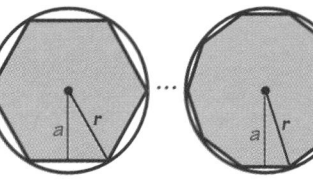

The perimeter of a circle is called the circle's *circumference*. The ratio $\frac{circumference}{diameter}$ is a constant value for all circles. It is an irrational number denoted by the Greek letter π ("pi") and is approximately 3.14, or $\frac{22}{7}$. This leads to the formula $C = 2\pi r$.

You can think of the **radius** of a circle as the **apothem** of a regular polygon, and the **circumference** as the **perimeter**:

$$\text{Area of a regular polygon} = \frac{1}{2}ap$$

$$\text{Area of a circle} = \frac{1}{2}r(2\pi r)$$

$$= \pi r^2$$

Area and Circumference of a Circle

The area of a circle with radius *r* is πr^2.

The circumference of a circle with radius *r* is $2\pi r$.

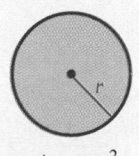

$$A = \pi r^2$$
$$C = 2\pi r$$

7.5 Areas of Regular Polygons and Circles   **369**

ANSWERS   Section 7.5

Think and Communicate

1. The measure of each interior angle of a regular hexagon is 120°. Two angles of each triangle are 60° angles, so the third angle is also a 60° angle. An equiangular triangle is equilateral. In other regular polygons, the triangles are isosceles. For any regular polygon, the distance from the center to each vertex is the same.

2. No; for other regular polygons, the triangles are not equilateral.

3. $4s; \frac{s}{2}; A = \frac{1}{2}ap = \frac{1}{2}(4s)\left(\frac{s}{2}\right) = s^2$

Section Notes

Communication: Discussion
Define the term *inscribed* as meaning that the polygon has all its vertices on the circle. Students should understand that the phrase *the apothem approaches the length of the radius* means that as the number of sides of the polygons increase, the length of the apothem gets closer and closer to the length of the radius of the circle. Define the term *circumference* as the perimeter of a circle and stress that the ratio of the circumference to diameter is always equal to π. Point out how this leads to the formula for circumference by replacing the diameter by two times the radius, 2*r*, and solving for *C*. Discuss how the area formula of a circle can be thought of as a special case of the area formula for a regular polygon with the apothem replaced by radius and the perimeter replaced by circumference.

Integrating Algebra
The introduction of the concept of the area of a circle uses sequences. Point out to students that there are several sequences that can be considered. First, there is the sequence of figures. Notice that there are several figures missing before the last figure given. Then there is the sequence of apothem lengths, the sequence of perimeters of the polygons, and the sequence of areas of the polygons. Ask students to describe what happens to each of these sequences as the number of sides increases.

Using Manipulatives
You may wish to try the following activity to help students gain insight into the value of π and the circumference of a circle. Give students different sized cylinders (various food cans work well), a piece of string, and a ruler. Have them wrap the string around the cylinder, mark where it first overlaps, and measure the string with the ruler to get the circumference of the cylinder. Then they should measure the diameter of the cylinder and calculate the ratio $\frac{circumference}{diameter}$. By comparing their ratios, students will see that regardless of the size of the cylinder, the ratio of the circumference of a circle to its diameter is always equal to π.

Additional Example 3

This figure is a square inscribed in a circle. Find the area of the shaded region.

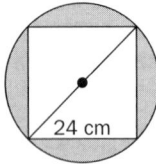

24 cm

The shaded region is a circle with a square removed. Use the Pythagorean theorem to find the length of the diagonal, d, of the square.
$d^2 = 24^2 + 24^2$
$d = 24\sqrt{2}$
The radius of the circle is half the length of the diagonal of the square, or $12\sqrt{2}$.
Area of shaded region
= Area of circle − Area of square
$= \pi r^2 - s^2$
$= \pi(12\sqrt{2})^2 - 24^2$
≈ 1233
The area of the shaded region is about 1233 cm².

Checking Key Concepts

Geometric Thinking
Questions 3 and 6 may be difficult for some students. In question 3, students are given the apothem and must use it to find the side length. After using the Pythagorean theorem, they need to double the result (which is half the side length) to get the side length. In question 6, students need to use the circumference formula to find the value of r, and then use it in the area formula.

Closure Question

What is the apothem of a regular polygon? How is it used to find the area of the polygon? The apothem is the distance from the center of the polygon to one of its sides. To find the area of the polygon, multiply one half the apothem by the perimeter of the polygon.

EXAMPLE 3   Application: Sports

A high school running track surrounds a field that needs to be watered. Find the area enclosed by the track.

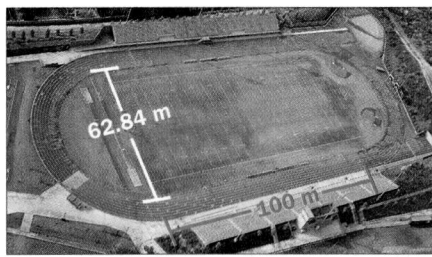

62.84 m

100 m

SOLUTION

The region is a rectangle with a semicircle at each end.

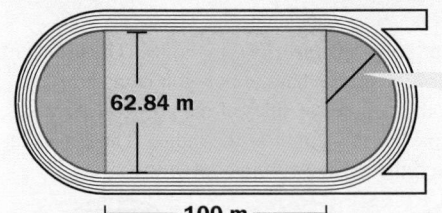

62.84 m

100 m

The diameter of each semicircle is 62.84 m, so the **radius** is **31.42 m**.

$$\text{Area enclosed by the track} = \left(\begin{array}{c}\text{Area of}\\\text{rectangle}\end{array}\right) + 2\left(\begin{array}{c}\text{Area of one}\\\text{semicircle}\end{array}\right)$$

$$= (bh) + 2\left(\frac{1}{2}\pi r^2\right)$$

The area of a semicircle is half the area of a circle.

$$= (100 \cdot 62.84) + \pi(31.42)^2$$

$$\approx 9385.43$$

The area enclosed by the track is about 9390 m².

☑ **CHECKING KEY CONCEPTS**

Find the area of each regular polygon or circle.

1.
3√2
6

2.
12

3.
16.1
13

4.
9

5.
13

6.
$c = 8\pi$

Checking Key Concepts

1. 36
2. $216\sqrt{3} \approx 374.1$
3. $65\sqrt{90.21} \approx 617.54$
4. $81\pi \approx 254$
5. $\frac{169\pi}{4} \approx 133$
6. $16\pi \approx 50$

7.5 | Exercises and Applications

Extra Practice exercises on page 674

Find the area of each regular polygon.

1.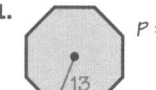
$P = 80$
13

2.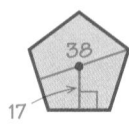
38
17

3.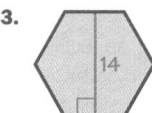
14

Find the area of each circle.

4.
$\frac{1}{2}$

5.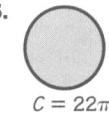
$C = 22\pi$

6.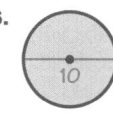
10

Find the area of each shaded region.

7.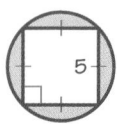
5

8.
8

9.
3 1

10. Open-ended Problem Sketch two concentric circles. Give an example of what their radii could be if the area enclosed between the circles is 12π.

Connection ▶ BIOLOGY

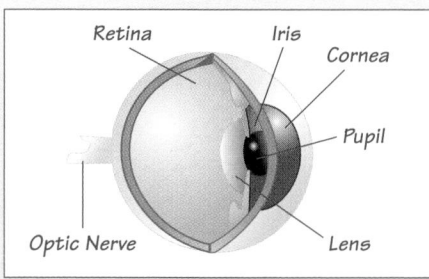
Retina   Iris   Cornea   Pupil   Optic Nerve   Lens

The *pupil* is the opening through which light enters the eye. The *iris*, the colored part of the eye, opens and closes around the pupil to change the amount of light that enters the eye.

Entering a dark room

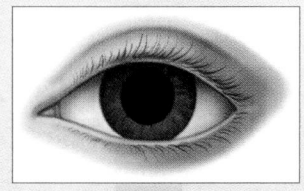

For Exercises 11 and 12, find the area of the pupil.

11. When you enter a dark room, your iris opens until the pupil is about 8 mm in diameter.

12. In a well-lit room, your iris closes in around the pupil, giving it a diameter of about 4 mm.

13. Writing Look back at Exercises 11 and 12. What is the ratio of the diameters of the pupils? What is the ratio of the areas? Is this what you expected? Explain.

Entering a well-lit room

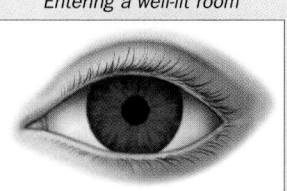

Exercises and Applications

1. 480
2. $510\sqrt{2} \approx 721$
3. $98\sqrt{3} \approx 170$
4. $\frac{\pi}{4} \approx 0.8$
5. $121\pi \approx 380$
6. $25\pi \approx 79$
7. $\frac{25\pi}{2} - 25 \approx 14$
8. $16\pi - 24\sqrt{3} \approx 9$

9. $7\pi \approx 22$
10.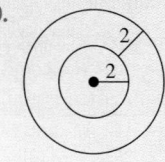
2
2

If the radii are both integers, they are 2 and 4.

11. $16\pi \approx 50$ mm^2
12. $4\pi \approx 13$ mm^2

13. $2:1$; $4:1$; Answers may vary. An example is given. Yes; if $\frac{r_1}{r_2} = \frac{2}{1}$, then $r_1 = 2r_2$ and $\frac{\pi(2r_2)^2}{\pi r_2^2} = \frac{4\pi r_2^2}{\pi r_2^2} = \frac{4}{1}$.

Apply⇔Assess

Suggested Assignment

❖ **Core Course**
Exs. 1–9, 15–23, 28–34

❖ **Extended Course**
Exs. 1–3, 7–34

❖ **Block Schedule**
Day 44 Exs. 1–9, 15–23, 28–34

Exercise Notes

Problem Solving
Exs. 1–9 These exercises provide a variety of problems that students can solve by applying the concepts and area formulas they learned in this section. The problems are structured in a way that before an area formula can be used, the values of the variables in the formula must first be determined. For example, in Ex. 1, students need to use the Pythagorean theorem to first find the apothem of the polygon. In Ex. 3, students need to use the fact that a hexagon can be divided into 6 equilateral triangles. In Ex. 5, students need to use the circumference formula to find the radius and then substitute the radius in the area formula.

Interdisciplinary Problems
Exs. 11–13 In these exercises, students see how the area formula for a circle can be used to find the area of the pupil of an eye in a dark room and in a well-lit room.

Topic Spiraling: Preview
Ex. 13 This exercise previews the work on areas of similar figures in Section 9.4. Students should realize that when the diameter of the pupil is halved, the area is quartered. Ask students what happens when the diameter is divided by 3, or by 5, and why. (The area is divided by 9 and 25, respectively. This is a result of the fact that the area of a circle varies directly with the square of the diameter.) Have students predict what will happen if the diameter is multiplied by 7. (The area is multiplied by 49.)

Exercise Notes

Integrating Functions

Ex. 14 In this exercise, students examine the perimeter and area of a polygon inscribed in a circle of radius 1 as a function of the number of its sides. They graph these functions and use them to verify that the perimeters approach 2π and the areas approach π. To help students see this, you might also have them graph the equations $y = 2\pi$ and $y = \pi$.

Using Technology

Ex. 14 To create the graphs for part (b) of this exercise on a TI-82, press [STAT] [ENTER] and clear each of the first three lists by using the arrow keys to position the cursor over the list name and pressing [CLEAR] [ENTER]. Position the cursor in the first row under L1 and enter the number of sides in this column. Similarly, enter the corresponding perimeters and areas in L2 and L3, respectively. Press [2nd] [STATPLOT], choose 1:Plot1 and highlight On, the scatter plot picture, L1, and L2 by positioning the cursor over each of these items and pressing [ENTER]. Press [2nd] [STATPLOT] again and choose 2:Plot2 and highlight On, the scatter plot picture, L1, and L3. To see the graph, choose 9:ZoomStat from the ZOOM menu.

Geometric Thinking

Ex. 14 After students complete this exercise, ask them what would happen to the areas and perimeters if the radius of the circle were changed to three. They should realize that the perimeters approach the circumference of the circle, 6π, and the areas approach the area of the circle, 9π.

Construction Note

Exs. 15–19 This construction is based on the fact that the distance from the center to a vertex in a regular hexagon is equal to the length of a side. The distance from the center to a vertex is also equal to the radius of the circle in this construction. Students need to realize these facts to make their construction for Ex. 19.

---

14.  **Technology** Work with two other people. You will need geometry software and a graphing calculator.

   a. Cooperative Learning Each student should construct two different regular polygons, each inscribed in a circle whose radius is 1. Find the perimeter and area of each polygon.

   b. ALGEBRA Make two graphs of your group's data with the number of sides of the polygon on the *x*-axis. For the first graph, put the perimeter of the polygon on the *y*-axis. For the second graph, put the area of the polygon on the *y*-axis. Sketch the two graphs.

   c. Writing Look at your graphs. How do the perimeters and areas change as the number of sides of the polygons increases? Suppose you inscribe a 100-gon in a circle of radius 1. What do you think the perimeter and area would be? Explain.

CONSTRUCTION

REGULAR HEXAGON

1. Construct a circle.

2. Using the same radius, place the point of the compass at any point *A* on the circle and draw an arc. Label the intersection *B*.

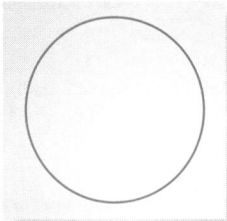

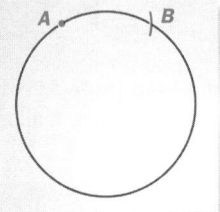

3. Place the point of the compass at *B* and draw an arc that intersects the circle at point *C*. Repeat this process to find *D*, *E*, and *F*.

4. Connect the consecutive points around the circle. *ABCDEF* is a regular hexagon.

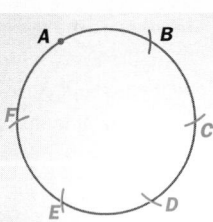

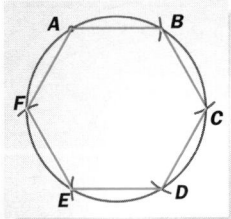

15. How do you know that hexagon *ABCDEF* is equilateral?

16. Explain why hexagon *ABCDEF* is equiangular. (*Hint:* What type of triangles are formed when you connect the opposite vertices of hexagon *ABCDEF*?)

17. How can you change the last step to construct an equilateral triangle instead of a regular hexagon? Explain your reasoning.

18. Construct a regular hexagon.

19. Writing Construct a regular hexagon with side length 3 cm. Write a description of your method. Include enough information so that someone else can follow your steps.

---

14. See answers in back of book.

15. Each side was constructed using the same compass radius.

16. Draw all the diagonals of the hexagon. Six triangles are formed. Since the circle and all sides of the hexagon were constructed using the same compass radius, each triangle is equilateral. The measure of each angle of one of the triangles is 60°, so the measure of each angle of the

hexagon is 120° and the hexagon is equiangular.

17. Connect only vertices *A*, *C*, and *E* or only vertices *B*, *D*, and *F*. To show that △*ACE* is equilateral, draw $\overline{AC}$, $\overline{AE}$, and $\overline{EC}$. Since *ABCDEF* is a regular hexagon, △*FEA*, △*BAC*, and △*DCE* are congruent by the SAS Postulate, and $\overline{AC} \cong \overline{AE} \cong \overline{EC}$.

18. Check students' work.

19. Set the radius of the compass to 3 cm. Follow Steps 1–4. The resulting regular hexagon will have side lengths of 3 cm.

20–23. Answers are given to the nearest tenth.

20. $2\sqrt{3} \approx 3.5$

21. about 44.7

22. $12\sqrt{3} \approx 20.8$

23. 359.0

24. $50\sqrt{3} \approx 86.6$

Find the area of each shaded region. Each outer polygon is regular.

20.
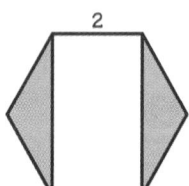
2

21.
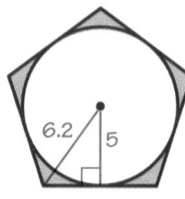
6.2 5

22.
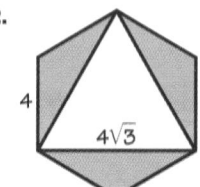
4
$4\sqrt{3}$

23.

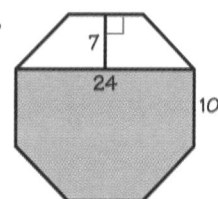
7
24
10

24. **Challenge** Find the area of the shaded region.

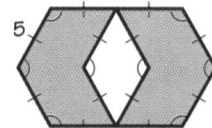
5

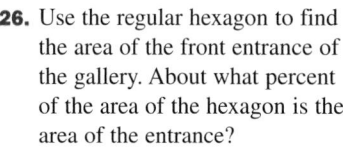

ARCHITECTURE The front entrance of the Clore Gallery, part of the Tate Gallery in London, is shown at the right. The sloped sides are the same length and make 120° angles with each other and with the vertical sides. Use this information for Exercises 25–27.

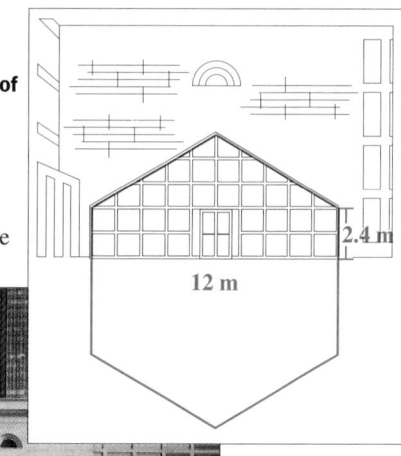
2.4 m
12 m

25. **Writing** You can think of the shape of the front entrance as part of a regular hexagon, as shown. Why is this possible?

26. Use the regular hexagon to find the area of the front entrance of the gallery. About what percent of the area of the hexagon is the area of the entrance?

27. Describe another method you can use to find the area of the front entrance. Which method do you think is easier? Explain.

ONGOING ASSESSMENT

28. **Writing** Explain how you can find the area of a regular *n*-gon by finding the area of *n* isosceles triangles.

SPIRAL REVIEW

Find the area of each polygon. *(Section 7.4)*

29. a square with perimeter 28 yd

30. a rectangle with base 5 mm and perimeter 18 mm

Sketch a net for each prism. *(Section 2.6)*

31. an octagonal prism

32. a cube

33. a triangular prism

34. a heptagonal prism

7.5 Areas of Regular Polygons and Circles **373**

Apply⟺Assess

Exercise Notes

Cooperative Learning
Exs. 20–23 You may wish to have students work with a partner on these exercises.

Application
Exs. 25–27 In these exercises, the geometry of a regular hexagon is used to analyze the building design of part of the Tate Gallery in London.

Topic Spiraling: Preview
Exs. 31–34 The area of a net for a prism is the same as the surface area of the prism. The surface area of prisms and cylinders is discussed in the next section.

Practice 51 for Section 7.5

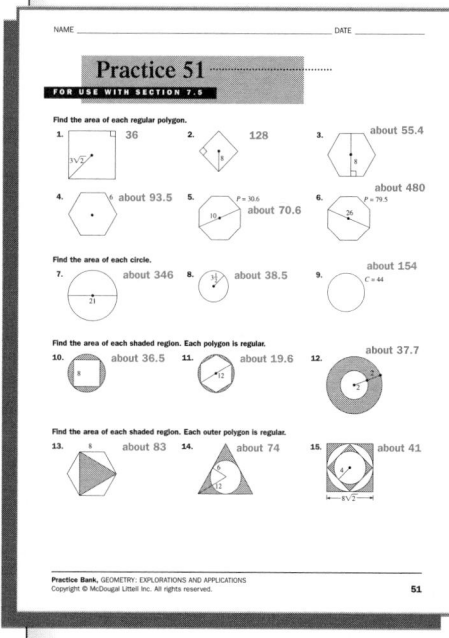

---

25. Three of the angles are 120° angles and the sides of the roof are congruent. The shape is then part of a regular hexagon whose sides are the same length as the sides of the roof.

26. about 599 ft

27. The front entrance can be divided into an isosceles triangle and a rectangle. To find the height of the triangle, draw the six diagonals of the hexagon. Answers may vary. The method in Ex. 26 is simpler.

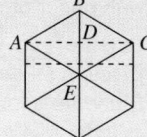

△*AEB* is an equilateral triangle; *AD* = $12\sqrt{3}$, so *BD* = 12. The area of the triangle is $\frac{1}{2}(12\sqrt{3})(24)$. The total area is $\frac{1}{2}(12\sqrt{3})(24) + (24\sqrt{3})(8) \approx 582$ ft².

28. If you draw segments from the center of an *n*-gon to each vertex, they divide the polygon into *n* congruent isosceles triangles. The area of the polygon is the sum of the areas of the triangles.

29. 49   30. 20

31–34. Answers may vary. Examples are given.

31.
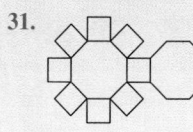

32.
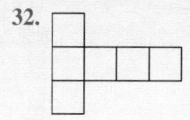

33.
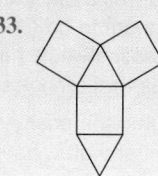

34.
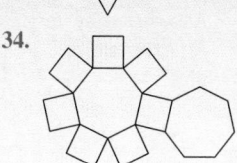

373

7.6 | Prisms and Cylinders

Objectives

- Analyze the parts of prisms and cylinders.
- Find the volumes, lateral areas, and surface areas of prisms and cylinders.

Recommended Pacing

❖ **Core and Extended Courses**
Section 7.6: 1 day

❖ **Block Schedule**
Section 7.6: $\frac{1}{2}$ block
(with Section 7.5)

Resource Materials

Lesson Support
Lesson Plan 7.6
Warm-Up Transparency 7.6
Practice Bank: Practice 52
Study Guide: Section 7.6
Challenge Problems: Set 52
Assessment Book: Test 32

Technology
Scientific Calculator
Internet:
   http://www.hmco.com

Warm-Up Exercises

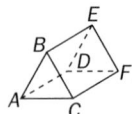

1. Name the bases in the prism.
   $\triangle ABC$ and $\triangle DEF$

2. Name the lateral faces. quadrilaterals *ABED*, *BCFE*, and *ACFD*

3. What type of prism is it?
   triangular prism

4. What would have to be true for the prism to be a regular prism?
   The bases would have to be equilateral triangles.

5. Assume the prism is a regular prism and draw its net.
   Answers may vary.
   An example is
   given.

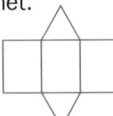

Learn how to...

- **analyze the parts of prisms and cylinders**

So you can...

- **find the volumes, lateral areas, and surface areas of prisms and cylinders**

Suppose two tubes of different sizes are made from identical pieces of paper. Do they both hold the same amount? You will answer this question in the Exploration.

EXPLORATION
COOPERATIVE LEARNING

Comparing Volumes

**Work with another student.
You will need:**

- rectangular pieces of paper that are the same size
- tape
- popcorn or dried beans

1 Tape the two long sides of one piece of paper together to form a tube.

2 Tape together the two short sides of another piece of paper, identical to the one used in Step 1, to form a shorter, wider tube.

3 Stand the long tube on its end and fill it with popcorn or beans.

4 Place the wide tube around the thin tube, as shown. Slowly raise the thin tube so that the contents fill the wide tube. Which tube holds more? How can you tell?

5 Repeat Steps 1–4 using paper of different dimensions. Do you get the same result?

6 Do you think the *radius* of a tube or the *height* of a tube has more effect on the amount it holds? Explain your reasoning.

Exploration Note

Purpose
The purpose of the Exploration is to have students discover that the radius of a cylinder has more effect on its volume than does the height.

Materials/Preparation
Each group needs rectangular pieces of paper that are the same size, tape, and popcorn or dried beans.

Procedure
Students tape together the long sides of one sheet of paper and the short sides of another to form two different sized tubes.

They fill the taller tube with popcorn or beans, and then transfer them to the shorter tube. This procedure is repeated with paper of different dimensions.

Closure
Discuss students' responses to Step 6. Students should agree that the radius of a cylinder has more effect on its volume than does the height.

Explorations Lab Manual
See the Manual for more commentary on this Exploration.

The **volume** of a solid is the number of *cubic units* needed to fill the solid. If the edges of a cube are one unit long, then its volume is one cubic unit. To find the volume of a prism, you multiply the number of cubes that "cover" or "make up" the base by the number of layers of cubes that fill the prism.

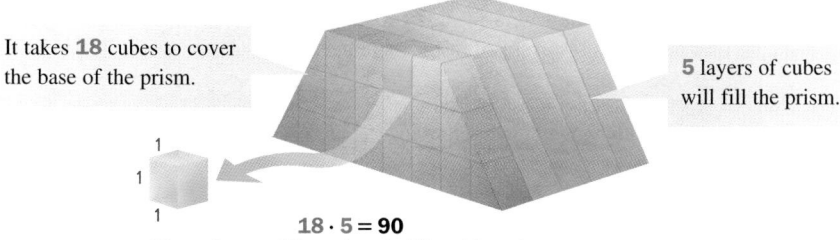

It takes **18** cubes to cover the base of the prism.

5 layers of cubes will fill the prism.

$18 \cdot 5 = 90$

The volume of the prism is 90 cubic units.

The same reasoning can be applied to *cylinders*. A **cylinder** is a space figure whose bases are circles of the same size.

base

height

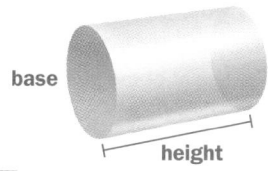

◀ **WATCH OUT!**

The height of a figure is not always vertical, and the base is not always the bottom.

Volumes of Prisms and Cylinders

The volume of a prism or a cylinder is the product of the height and the area of the base.

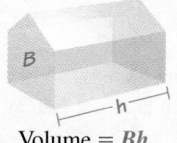

Volume $= Bh$

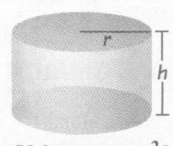

Volume $= \pi r^2 h$

EXAMPLE 1

Find the volume of this prism.

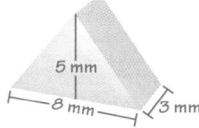

5 mm

8 mm

3 mm

SOLUTION

Multiply the area of the triangular base by the height of the prism.

$$\text{Volume} = Bh$$
$$= \left(\frac{1}{2} \cdot 8 \cdot 5\right) \cdot 3$$

Use the formula for the area of a triangle.

$$= 20 \cdot 3$$
$$= 60$$

The volume of the prism is 60 mm^3.

7.6 Prisms and Cylinders **375**

ANSWERS Section 7.6

Exploration

1–3. Check students' work.

4. the wide tube; The amount of popcorn that filled the thin tube does not fill the wide tube.

5. Yes.

6. Answers may vary. An example is given. I think the radius has more effect. The tube with the greater radius had greater volume even though it was much shorter than the tube with the smaller radius.

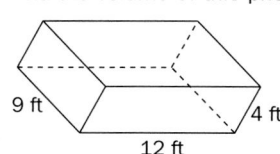

Using Manipulatives

You may wish to give students the opportunity to construct their own nets for the triangular prism shown on this page. They can then cut the nets out and fold them into the prism. Students can use color markers to mark the edges. Then, when they open up the prism, they can see how the sides match up. This should help students understand the formula for the surface area of a prism. Similarly, have students tape the ends of a rectangular piece of paper into a cylinder. Point out how the height of the cylinder is the length of the paper and the circumference of the cylinder is the width of the paper. This should help them understand the formula for the lateral surface area of a cylinder.

Teaching Tip

Point out the similarities in the formulas for the surface areas of prisms and cylinders. The perimeter is similar to the circumference, and the area of the base for a cylinder is πr^2. Also, remind students that surface area is measured in square units.

Second-Language Learners

You may want to point out to students learning English that *lateral* is an English adjective meaning "of the side." If students are Spanish-speakers, you might want to point out the similarity between the word *lateral* and the Spanish words *lateral* ("lateral") and *lado* ("side").

Surface Area

In Chapter 2 you learned how to sketch a net for a prism by drawing the faces connected to one another. The area of a net for a three-dimensional figure is called the **surface area** of the figure, abbreviated *S.A.*

The area of the lateral faces is called the **lateral area**. The lateral faces of a prism can be arranged to form one rectangle.

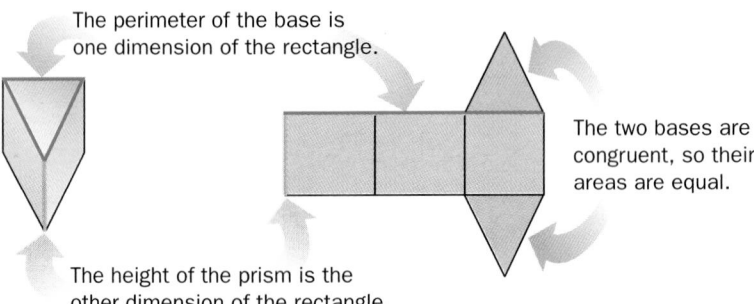

The perimeter of the base is one dimension of the rectangle.

The two bases are congruent, so their areas are equal.

The height of the prism is the other dimension of the rectangle.

$$\text{Surface Area} = \text{Lateral area} + \text{Area of bases}$$
$$= (\text{Perimeter of base} \times \text{height}) + 2(\text{Area of base})$$
$$= ph + 2B$$

THINK AND COMMUNICATE

1. Imagine cutting along the height of a tube and laying it flat. What shape is the resulting figure?

2. You learned in Chapter 2 that prisms are named according to their bases. What do you think a *regular prism* is?

Surface Areas of Prisms and Cylinders

The surface area of a prism or cylinder is the sum of the lateral area and the area of the bases.

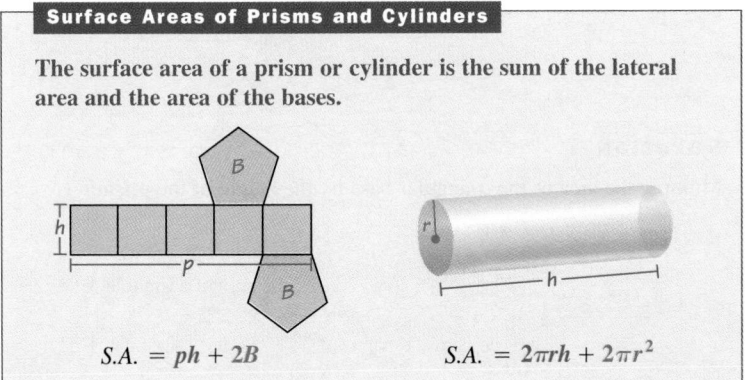

$$S.A. = ph + 2B \qquad S.A. = 2\pi rh + 2\pi r^2$$

Think and Communicate

1. rectangle

2. a prism whose bases are congruent regular polygons

EXAMPLE 2

Find the volume and surface area of the cylinder that the net folds into.

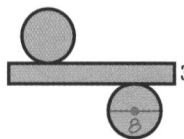

SOLUTION

The diameter of the base is 8, so the radius is 4. Use this value in the formulas for volume and surface area.

$$\text{Volume} = \pi r^2 h \qquad S.A. = 2\pi rh + 2\pi r^2$$
$$= \pi(4)^2 \cdot 3 \qquad\qquad = (8\pi \cdot 3) + 2\pi(4)^2$$
$$= 48\pi \qquad\qquad\qquad = 56\pi$$
$$\approx 150.8 \qquad\qquad\quad \approx 175.9$$

The volume of the cylinder is approximately 150.8.
The surface area of the cylinder is approximately 175.9.

EXAMPLE 3   Application: Package Design

The gift box at the right is a regular hexagonal prism. How much cardboard was used to make the box?

SOLUTION

Use the Pythagorean theorem to find the apothem of the hexagonal base.

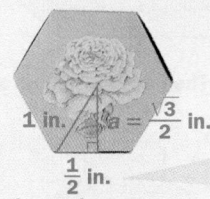

The length of each side is 1 in., so the length of half of the side is $\frac{1}{2}$ in.

Substitute this value into the formula for surface area.

$$S.A. = ph + 2B$$
$$= \left[(6 \cdot 1) \cdot \frac{21}{4}\right] + 2\left[\frac{1}{2} \cdot \frac{\sqrt{3}}{2} \cdot (6 \cdot 1)\right]$$
$$= \frac{126}{4} + 3\sqrt{3}$$
$$\approx 36.7$$

To find the area of the base, substitute $\frac{\sqrt{3}}{2}$ for *a* and $6 \cdot 1$ for *p* in the formula $A = \frac{1}{2}ap$.

About 36.7 in.² of cardboard was used to make the box.

7.6   Prisms and Cylinders   **377**

Additional Example 2

Find the volume and surface area of the regular prism that the net folds into.

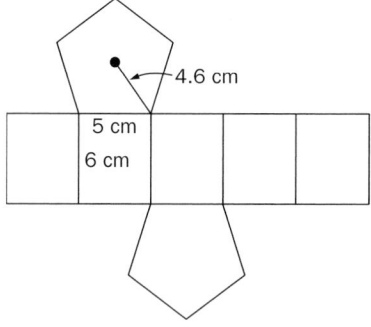

Use the Pythagorean theorem to find the apothem of the pentagonal base.
$$a^2 = 4.6^2 - 2.5^2$$
$$a \approx 3.9$$
Use this value in the formulas for volume and surface area.
Volume = Bh
$$= \left(\frac{1}{2} \cdot a \cdot p\right) \cdot h$$
$$\approx \left[\frac{1}{2} \cdot 3.9 \cdot (5 \cdot 5)\right] \cdot 6$$
$$\approx 292.5$$
S.A.
$$= ph + 2B$$
$$= ph + 2\left(\frac{1}{2} \cdot a \cdot p\right)$$
$$\approx [(5 \cdot 5) \cdot 6] + 2\left[\frac{1}{2} \cdot 3.9 \cdot (5 \cdot 5)\right]$$
$$\approx 247.5$$
The volume of the prism is approximately 292.5 cm³. The surface area of the prism is approximately 247.5 cm².

Additional Example 3

The mailing tube below can be used to mail posters or architectural plans. How much cardboard was needed to make the tube?

$3\frac{3}{4}$ in. $32\frac{1}{2}$ in.

The diameter of the base is $3\frac{3}{4}$ in., so the radius is $\frac{15}{8}$ in. Substitute this value into the formula for surface area.
$$S.A. = 2\pi rh + 2\pi r^2$$
$$= \left(2\pi \cdot \frac{15}{8} \cdot \frac{65}{2}\right) + 2\pi\left(\frac{15}{8}\right)^2$$
$$= \frac{4125}{32}\pi$$
$$\approx 404.97$$
It takes about 404.97 in.² of cardboard to produce one mailing tube.

377

Checking Key Concepts

Teaching Tip
You may need to remind students that a regular prism is one in which the bases are regular polygons. The lateral faces are not necessarily regular. For question 3, to find the area of the base, students need to use the Pythagorean theorem and the fact that a hexagon can be divided into six equilateral triangles. In question 4, students need to use the formula for the circumference of a circle to find the radius.

Closure Question

Describe how to find the volumes and surface areas of a prism and a cylinder. To find the volume of a prism, multiply the area of the base by the height of the prism. If the base is regular, take half of the apothem and multiply it by the perimeter of the base to find the area of the base. To find the surface area of a prism, first find the product of the perimeter of the base and the height of the prism and then add it to two times the area of the base. To find the volume of a cylinder, multiply the area of the base by the height of the cylinder. To find the surface area of a cylinder, first find the lateral area of the cylinder and then add it to two times the area of the base.

Apply⇔Assess

Suggested Assignment

❖ **Core Course**
Exs. 1–6, 8–11, 21–27, AYP

❖ **Extended Course**
Exs. 1–27, AYP

❖ **Block Schedule**
Day 44 Exs. 1–6, 8–11, 21–27, AYP

Exercise Notes

Student Study Tip
Exs. 1–6 You may wish to suggest to students that prior to beginning these exercises, they write the four formulas for volume and surface area in their journals for reference.

378

✓ **CHECKING KEY CONCEPTS**

Find the volume and surface area of each regular prism or cylinder.

1.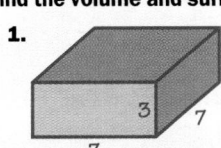

2.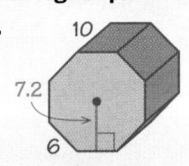

3.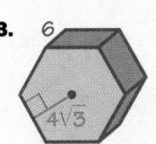

4.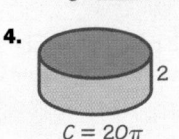

5. Sketch the figure that the net folds into. Then find its volume and surface area.

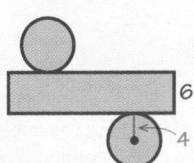

7.6 Exercises and Applications

Extra Practice exercises on page 674

Find the volume and surface area of each prism or cylinder.

1.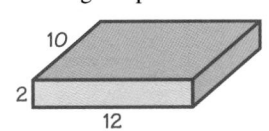

2. rectangular prism
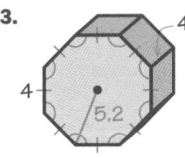

3.

Sketch the figure that each net folds into. Then find its volume and surface area.

4.

5.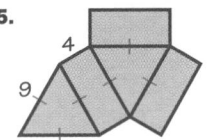

6.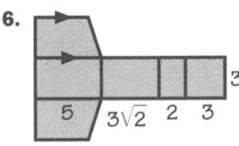

7. NUTRITION According to the package label, each 1 in.³ serving of a certain brand of cheese has 110 calories. Each package contains a block of cheese that is 1 in. by $2\frac{1}{4}$ in. by 6 in.

a. How many servings are in each package?

b. How many calories are in a $\frac{1}{4}$ in. slice off the smallest end of the block?

Checking Key Concepts

1. 147; 182　　2. 1728; 825.6
3. $576\sqrt{3} \approx 997.7$;
$192\sqrt{3} + 288 \approx 620.6$
4. $200\pi \approx 628.3$; $240\pi \approx 754.0$
5.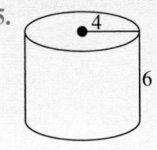

$96\pi \approx 301.6$; $80\pi \approx 251.3$

Exercises and Applications

1. $\frac{891\pi}{4} \approx 699.8$; $\frac{279\pi}{2} \approx 438.3$
2. 240; 328　　3. 307.2; 281.6
4.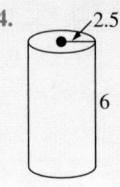

$\frac{75\pi}{2} \approx 117.8$; $\frac{85\pi}{2} \approx 133.5$

5.
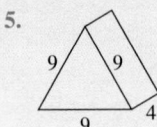

$81\sqrt{3} \approx 140.3$;
$\frac{81\sqrt{3} + 216}{2} \approx 178.1$

6.
$3\sqrt{2}$
2　3
5　3

Find the volume of each space figure described.

8. Each edge of a cube is 3 ft long.

9. The radius of a cylinder is 3 m and the lateral area is 15π m^2.

10. The perimeter of each base of a regular hexagonal prism is 24 cm and the area of one lateral face is 40 cm^2.

11. A triangular prism is 10 in. high. The perimeter of each base is 12 in. and the area of each base is 6 in.^{2}

12. **CARPENTRY** David Taylor plans to reshingle the sides of the house shown at the right. One bundle of shingles covers 25 ft^2.

a. Sketch the faces of the prism that represent the walls of the house. Are they all lateral faces? Explain why or why not.

b. Estimate how many bundles of shingles David will need.

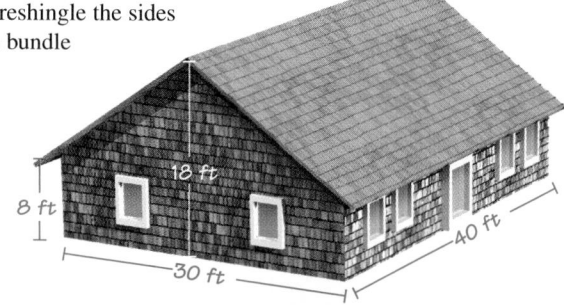

Connection   CONSUMER ECONOMICS

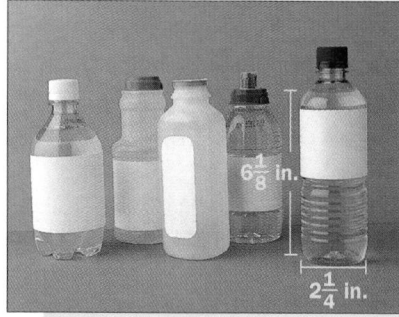

Many beverages are sold in cylindrical bottles. However, some manufacturers use other shapes to make their products more practical or appealing.

13. a. A cylindrical 400 mL bottle has a diameter of $2\frac{1}{4}$ in. and a height of $6\frac{1}{8}$ in. Find the volume of the bottle in cubic inches.

b. Use the conversion factor $\dfrac{1 \text{ mL}}{0.061 \text{ in.}^3}$ to check your answer to part (a).

14. One brand of spring water comes in a bottle in the shape of a regular triangular prism. The bottle just fits in a cup holder in a car. The cup holder is circular with diameter $3\frac{1}{2}$ in.

a. The height of an equilateral triangle is one and one-half times the radius of the triangle. Find the length of each side of the base of the triangular bottle.

b. How tall does the bottle have to be to hold 400 mL of water? How do you know?

15. a. **Open-ended Problem** Estimate the amount of plastic used to make each of the bottles described in Exercises 13 and 14. Explain your reasoning.

b. **Open-ended Problem** Which bottle do you think costs less to make? Which do you think is easier to store? Overall, which design do you think is more practical and why?

7.6 Prisms and Cylinders   **379**

Apply⇔Assess

Exercise Notes

Application
Exs. 7, 12 These exercises provide students with examples of how the concepts of volume and surface area can be applied to solve problems about nutrition and carpentry.

Communication: Reading
Exs. 8–11 These exercises require students to find the volume of an object that is described in words. Encourage students to sketch a diagram of each figure and to reread the description in order to label the parts of the sketch. In Exs. 9 and 10, students have to use the formula for the lateral surface area to find the height of the object.

Interdisciplinary Problems
Exs. 13–15 These exercises expose students to some of the considerations that are important to consumer economists in the design of beverage containers. Students first consider two differently shaped bottles. They then use the surface area formulas to compare the amount of plastic needed to produce the bottles. In Ex. 15(b), students also consider other factors that can affect the design of the bottles.

Cooperative Learning
Exs. 13–15 Students may benefit from working with a partner on these exercises. This will allow the pair to exchange ideas and check each other's work. After working together and discussing all the questions, each student should write his or her own answers.

Career Connection
Exs. 13–15 Economists study the effects of varying the production, distribution, or consumption of goods and services. A college degree with a major in economics is generally the minimum educational requirement to become an economist. Many economists, however, have completed graduate study at the masters or doctoral level. Economists can find work in government agencies or private firms. Large investment banks, corporations, money management firms, brokerage houses, economic forecasting firms, and other businesses involved in the financial markets often employ highly trained economists to help manage their operations and investment decisions.

---

7. a. 13.5 servings

b. about 62 calories

8. 27 ft^3 9. $\dfrac{45\pi}{2} \approx 70.7$ m^3

10. $240\sqrt{3} \approx 415.7$ cm^3

11. 60 in.^{3}

12. a. Each sketch represents two walls.

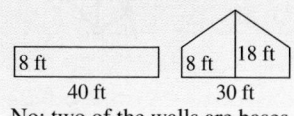

No; two of the walls are bases of the prism.

b. about 57 bundles (No allowance was made for the area of the windows and doors, which will not be covered.)

13. a. $\dfrac{3969\pi}{512} \approx 24.4$ in.^{3}

b. $\dfrac{1 \text{ mL}}{0.061 \text{ in.}^3} = \dfrac{x \text{ mL}}{24.4 \text{ in.}^3}$; $x = 400$ mL

14. a. $\dfrac{7\sqrt{3}}{4} \approx 3.0$ in.

b. The area of a base is about $\dfrac{147\sqrt{3}}{64} \approx 4$ in.^{2}. From Ex. 13 (a), a bottle that holds 400 mL has volume about 24.4 in.^{3}. Then the height of the bottle must be about $\dfrac{24.4}{4} \approx 6.1$ in.

15. See answers in back of book.

379

Interview Note: Application
Exs. 16, 17 In these exercises, students see how the formula for the volume of a prism can be used to determine the amount of cement needed for the construction of a wheelchair ramp.

Communication: Writing
Ex. 19 This exercise requires students to study the diagrams of cylinders in this section and to use what they discover about them to describe a method for drawing cylinders. The use of a cylindrical object would help students to visualize these diagrams. As an extension, you may wish to have students use their method to sketch a cylinder.

Teaching Tip
Ex. 20 There is not a unique solution to this problem. It may help students to pick a value for one of the dimensions, and then use the formula for surface area to find the other dimension.

Integrating Algebra
Exs. 21–23 These exercises provide students with an excellent review of the methods for representing points and finding distances in a three-dimensional coordinate system. The Distance Formula for points in space needs to be used to find some of the distances in these problems.

INTERVIEW Walt Stone

Look back at the article on pages 336–338.

Legislators like Walt Stone make sure that wheelchair ramps are built to meet certain requirements. If they are too steep or too narrow, they can be dangerous. Some ramps are wooden, but they can also be made of cement. The shape of a cement ramp is usually a prism.

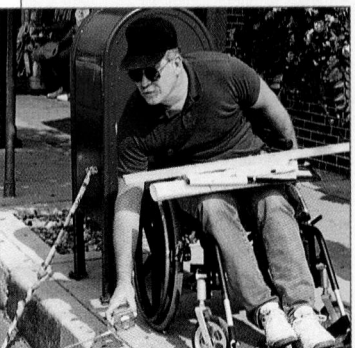

16. A ramp must be at least 3 ft wide and extend at least 12 in. horizontally for every inch it rises vertically.

  a. How long must a ramp be if it reaches a step that is $2\frac{1}{4}$ ft above the ground?

  b. Find the minimum amount of cement needed to make a ramp up to a step that is $2\frac{1}{4}$ ft high.

17. Challenge Find the amount of cement needed to make the ramp shown.

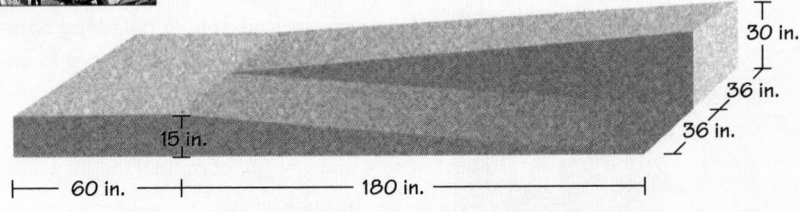

18. Research Measure the dimensions of a cement ramp in your school, library, or neighborhood and calculate the amount of cement needed to build it.

19. Writing Look at the diagrams of cylinders in this section. How would you sketch a cylinder? How is it different from sketching a prism?

20. Open-ended Problem Sketch a cylinder with surface area 72π cm². Give all dimensions of the cylinder and find its volume.

ALGEBRA The vertices of a prism are given. Sketch the prism and find its volume.

21. $T(3, -1, 1)$, $V(5, 2, 1)$, $W(7, -1, 1)$, $X(3, -1, 5)$, $Y(5, 2, 5)$, $Z(7, -1, 5)$

22. $J(0, 0, -2)$, $K(0, -3, -2)$, $L(11, -3, -2)$, $M(11, 0, -2)$, $P(0, 0, 3)$, $Q(0, -3, 3)$, $R(11, -3, 3)$, $S(11, 0, 3)$

23. $A(-2, 5, 0)$, $B(0, 9, 0)$, $C(6, 9, 0)$, $D(4, 5, 0)$, $E(-2, 5, -8)$, $F(0, 9, -8)$, $G(6, 9, -8)$, $H(4, 5, -8)$

16. a. about 27 ft   b. 91.13 ft³
17. 150 ft³
18. Answers may vary.
19. Answers may vary. An example is given. To sketch a cylinder, I would draw two ovals to represent the bases and connect them to form the cylinder. Drawing a cylinder requires drawing only two bases and two segments. Drawing a prism requires drawing two bases and as many segments as there are sides of the base.
20. Answers may vary. Check students' work. One suitable cylinder has radius 1.24 cm and height 8 cm. Its volume is about 38.6 cm³.

21. V = 21
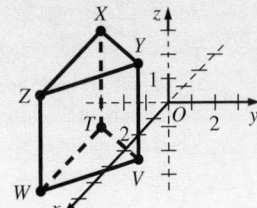

22. V = 165
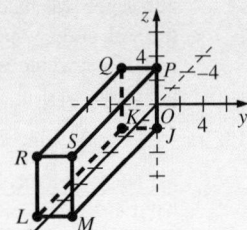

23. V = 192
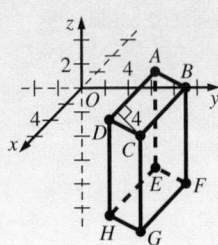

24. Open-ended Problem Sketch a prism and a cylinder that have approximately the same volume. Give the dimensions and surface area of both objects.

SPIRAL REVIEW

Copy each diagram. Sketch the reflection of the triangle over the red line.
(Section 1.2)

25.

26.

27.

ASSESS YOUR PROGRESS

VOCABULARY

area (p. 360)   **cylinder** (p. 375)
apothem (p. 368)   **surface area** (p. 376)
volume (p. 375)   **lateral area** (p. 376)

Find the area of each figure. *(Sections 7.4 and 7.5)*

1.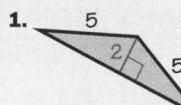

2.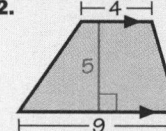

3.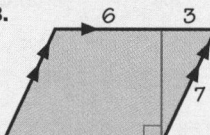

4.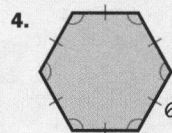

5. 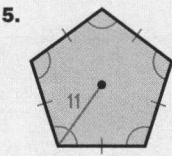 $P = 65$

6. 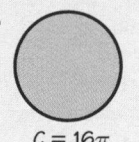 $C = 16\pi$

Find the volume and surface area of each prism or cylinder. *(Section 7.6)*

7.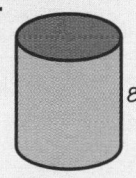

8. rectangular prism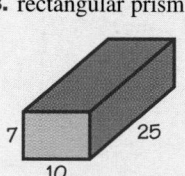

9.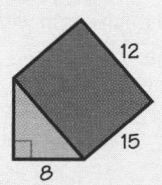

10. Journal List all of the area and volume formulas you learned in this chapter and sketch a diagram to illustrate each one. Show how some formulas can be used to find the others.

7.6 Prisms and Cylinders **381**

Apply⇔Assess

Exercise Notes

Topic Spiraling: Review/Preview
Exs. 25–27 These exercises review reflection over a line. Reflections and other transformations will be examined in detail in the next chapter.

Assess Your Progress

Journal Entry
This journal entry can help students review the formulas learned in this chapter. Students should include formulas for the areas of triangles, quadrilaterals, regular polygons, and circles, and the volumes and surface areas of prisms and cylinders.

Progress Check 7.4–7.6

See page 385.

Practice 52 for Section 7.6

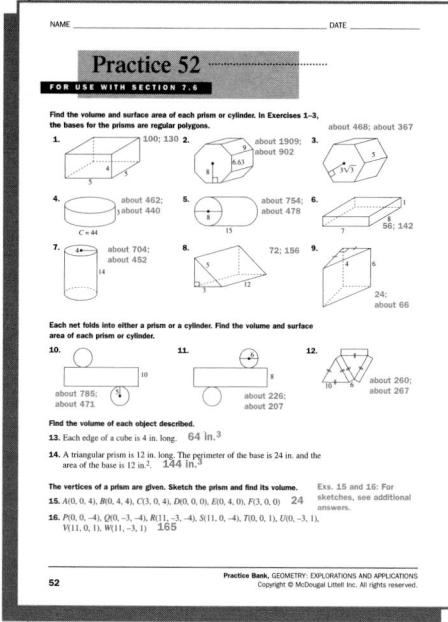

---

24. Answers may vary.
Examples are given.

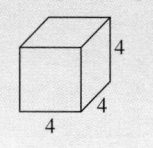 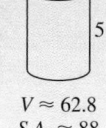

$V = 64$ $V \approx 62.8$
$S.A. = 96$ $S.A. \approx 88$

25.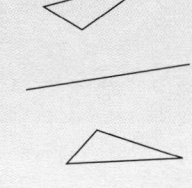

26.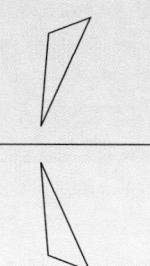

27.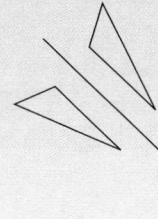

Assess Your Progress

1. $2\sqrt{21} \approx 9.2$ 2. 32.5

3. $18\sqrt{10} \approx 56.9$ 4. $54\sqrt{3} \approx 93.5$

5. $\dfrac{65\sqrt{78.75}}{2} \approx 288.4$

6. $64\pi \approx 201.1$

7. $128\pi \approx 402.1; 96\pi \approx 301.6$

8. $1750; 990$

9. $240\sqrt{5} \approx 536.7; 92\sqrt{5} + 300 \approx 505.7$

10. See answers in back of book.

381

Mathematical Goals

- Design a floor plan using specified guidelines.
- Find the area of a pentagon.
- Find the volume of a pentagonal prism.

Planning

Materials
- pencil and paper
- ruler
- graph paper or dot paper
- calculator
- tape measure (if necessary)

Project Teams
Students form groups of three and work together to complete the project. When ideas are discussed and decisions are made, all team members should be involved. You may wish to suggest that each student make a drawing of a detailed floor plan and then use the best one for the group's final proposal.

Guiding Students' Work

At the beginning of the project, students can work together to decide on three different dimensions of the cottage, then each student can draw one of the rough sketches. If possible, show students an example of a floor plan for a house or other building so they can see how to mark doors and windows. You may also want to explain how this is done before students begin the project. For example, students need to indicate on their floor plan which way each door opens. Students should use a ruler when drawing the plan and graph or dot paper. If students are placing appliances and furniture in the room, it may be helpful to have them cut scale models of each piece out of paper. These can then be moved about to decide where they will be located.

Second-Language Learners
Encourage students learning English to look up in a dictionary the words with which they are unfamiliar. Possibilities include *cottage*, *constraints*, *restrictions*, and *guidelines*.

PORTFOLIO PROJECT

Designing a Cottage

An architect who designs a building must work within certain constraints, such as the area the building will cover. Within these restrictions, there is a lot of room for creativity.

PROJECT GOAL — Design a floor plan for a one-story cottage, using specified guidelines.

Understanding the Specifications

You and two other students will design a cottage. The floor plan will be a rectangle whose area is between 550 ft^2 and 650 ft^2. The cottage should have two entrance doors and at least four rooms, as described in the table. Any hallways you include should be between three and four feet wide.

Minimum Room Areas	
Living Room	300 ft^2
Bedroom	90 ft^2
Kitchen	65 ft^2
Bathroom	50 ft^2

Arranging the Floor Plan

1. Determine at least three possibilities for the length and width of the rectangular cottage. For each possibility, make a rough sketch showing the locations of the rooms.

2. Make a detailed cottage floor plan, using one of your sketches. Draw the rectangle on graph paper or dot paper, then determine the exact dimensions and placement of the rooms. Label each room with its name, dimensions, and area.

Indicate the scale of your floor plan.

3. Show the locations of the windows and doors. You may also want to show appliances and furniture.

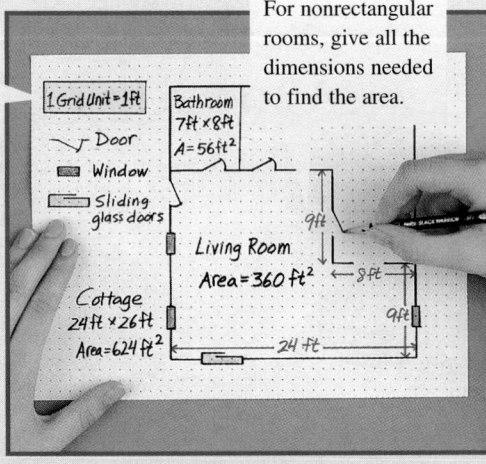

For nonrectangular rooms, give all the dimensions needed to find the area.

The Third Dimension

Knowing the volume of a building is important for choosing a heating system. The completed cottage will be shaped like a pentagonal prism. Sketch one of the bases, labeling all the lengths you need to find its area. Find the area of the pentagon. Then find the volume of the house.

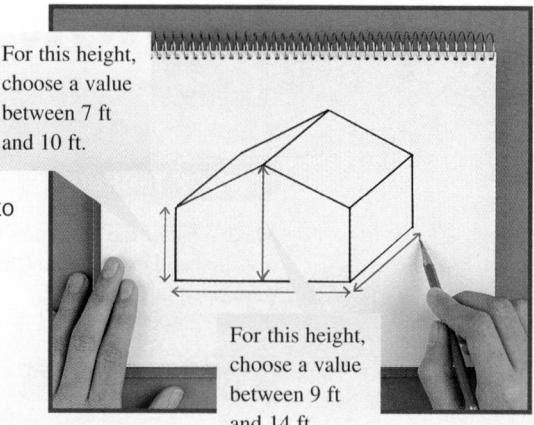

For this height, choose a value between 7 ft and 10 ft.

For this height, choose a value between 9 ft and 14 ft.

Submitting Your Plan

Create a proposal describing the cottage to its future owners. Include the detailed floor plan and explain the features of your design. Also explain how you calculated the volume of the cottage. You can extend the project by exploring one of the following ideas:

- Use a computer with drawing or CAD (Computer-Aided Design) software to help you create your plan.

- Take measurements of rooms where you live or at your school. How do they compare with those in the cottage you designed?

- How would the floor plan be different if it were based on a shape other than a rectangle? Design a floor plan that uses other shapes that you learned about in this chapter.

- Talk to an architect in your area. What other factors affect the design of a house?

Self-Assessment

How did your group decide which of the rough sketches to use for the final floor plan? If you did not have to follow the specifications given in this project, how would you change your design? Explain your reasoning.

Portfolio Project **383**

Project Notes

Guiding Students' Work

Rubric for Chapter Project

4 Students draw at least three rough sketches for their cottage and then design a detailed floor plan. Their final floor plan shows exact dimensions. Windows, doors, furniture, and appliances are clearly labeled. Students make a three-dimensional sketch of the cottage and correctly calculate its volume. Students create a detailed proposal that describes the cottage for the future owners, and extend the project in one of the four ways listed. A self-assessment is completed that addresses all of the questions asked.

3 Students draw three rough sketches and a detailed floor plan for their cottage. Their final floor plan shows exact dimensions, but some of them are incorrect. Students make a three-dimensional sketch of the cottage and calculate its volume. A proposal describing the cottage is created and explains the features of the cottage; however, it lacks some detail. Students extend the project in one of the four ways listed. Students also complete a self-assessment of the project and address the questions asked.

2 Students draw rough sketches and a detailed floor plan for the cottage, but the sketches are not complete and the detailed plan is poorly done. Students do not label the dimensions of the room, but windows and doors are shown. A three-dimensional sketch is not drawn and the volume is calculated incorrectly. Students attempt to create a proposal describing their cottage, but it is incomplete. Students do not extend the project, and if a self-assessment is done, it answers only some of the questions asked.

1 Students try to make a detailed floor plan, but it is incomplete and does not show dimensions, windows, or doors. The volume of the cottage is not calculated. Students do not create a proposal to describe their cottage and do not extend the project or complete a self-assessment. Students should be encouraged to speak with the teacher as soon as possible to review their work and to make a new start on the project.

Progress Check 7.1–7.3

In the figure, *UVWZ* is a square and *OVPW* is a kite. Find each length. *(Section 7.1)*

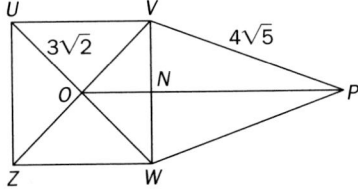

1. $\overline{VZ}$ $6\sqrt{2}$
2. $\overline{VW}$ 6
3. $\overline{VN}$ 3
4. $\overline{NP}$ $\sqrt{71}$

5. Write an indirect proof that a trapezoid cannot have three right angles. *(Section 7.2)*
Suppose that a trapezoid can have three right angles. Then, because the sum of the measures of the interior angles of a quadrilateral is 360°, the trapezoid has four right angles. Thus, it must be a rectangle. A rectangle has two pairs of parallel sides and, hence, cannot be a trapezoid. Therefore, a trapezoid cannot have three right angles.

7 | Review

STUDY TECHNIQUE

Make a list of the important ideas you learned in Chapter 2 and then revisited and extended in Chapter 7. Also include any new ideas that are developed in Chapter 7.

VOCABULARY

kite (p. 339)
indirect proof (p. 348)
area (p. 360)
apothem (p. 368)

volume (p. 375)
cylinder (p. 375)
surface area (p. 376)
lateral area (p. 376)

SECTIONS | 7.1, 7.2, *and* 7.3

The chart on page 339 describes how quadrilaterals can be classified by their properties. The chart on page 347 shows how to prove that a quadrilateral is a parallelogram.

Diagrams of Quadrilaterals

The diagonals of a rhombus are perpendicular.

The diagonals of a rectangle are congruent.

One diagonal of a kite is a line of symmetry for the kite and the perpendicular bisector of the other diagonal.

The charts on pages 353 and 355 show how to prove that a parallelogram is a rectangle or a rhombus.

An **indirect proof** shows that the conclusion must be true if all other possibilities are false. For example, to prove that two angles of a triangle cannot be obtuse:

Assume that $\angle A$ and $\angle B$ of $\triangle ABC$ are both obtuse. Then $m\angle A > 90°$ and $m\angle B > 90°$, and the sum of the angles of the triangle is greater than $180° + m\angle C$. But the sum of the measures of the angles of a triangle must be 180°. Therefore two angles of a triangle cannot be obtuse.

384   Chapter 7 *Quadrilaterals, Areas, and Volumes*

SECTIONS 7.4, 7.5, *and* 7.6

The area of a two-dimensional figure is the number of square units enclosed within the boundary of the figure. The chart on page 362 gives the area formulas for several polygons.

The area of a circle with radius r is $A = \pi r^2$ and its circumference is $C = 2\pi r$. If the radius of a circle is 4 cm, then the area is 16π cm^2 and the circumference is 8π cm.

The area A of a regular polygon is given by $A = \frac{1}{2}ap$, where a is the **apothem** of the polygon and p is the perimeter.

For example, each side of a regular heptagon is 5 in. and the distance from the center to each vertex is 5.7 in. To find the area:

Step 1 Sketch the polygon, and use the Pythagorean theorem to find the apothem.

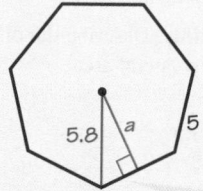

$$2.5^2 + a^2 = 5.8^2$$
$$a^2 = 5.8^2 - 2.5^2$$
$$a \approx 5.2$$

The length of one leg of the triangle is half the length of a side of the heptagon.

Step 2 Use the formula for the area of a regular polygon.

$$A = \frac{1}{2}ap$$
$$\approx \frac{1}{2}(5.2)(7 \cdot 5)$$
$$\approx 74.8$$

The area of the heptagon is about 74.8 in.^{2}.

The **volume** of a solid is the number of cubic units needed to fill the solid. The volume of a prism or a **cylinder** is the product of its height and the area of its base, $V = Bh$. The **surface area** of a figure is the area of its net. The **lateral area** of a figure is the sum of the areas of all faces of the figure except its bases.

Volume	Surface Area
$V = Bh$ $= (\pi \cdot 49)(3)$ $= 147\pi$ ≈ 461.58 m^3	$S.A. = 2\pi rh + 2\pi r^2$ $= (2\pi \cdot 7)(3) + 2\pi \cdot 7^2$ $= 140\pi$ ≈ 439.60 m^2
$V = Bh$ $= (4 \cdot 8)(2)$ $= 64$ ft^3	$S.A. = ph + 2B$ $= (2 \cdot 4 + 2 \cdot 8)(2) + (2)(32)$ $= 112$ ft^2

Review **385**

Classify each quadrilateral. Be as specific as possible. Give reasons for your answers. *(Sections 7.2 and 7.3)*

6.

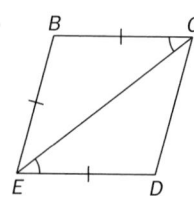

BCDE is a parallelogram because $\overline{BC}$ and $\overline{ED}$ are parallel and congruent. It is also a rhombus because consecutive sides $\overline{BE}$ and $\overline{BC}$ are congruent.

7.

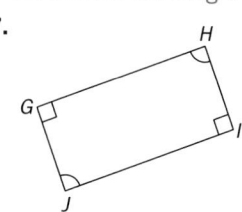

GHIJ is a parallelogram because it has two pairs of opposite angles that are congruent. It is a rectangle because at least one of the angles is a right angle.

Progress Check 7.4–7.6

Find the area of each figure. *(Sections 7.4 and 7.5)*

1.

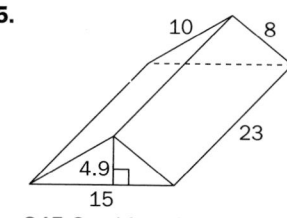

128 square units

2. a rectangle with width 6 and perimeter 36 72 square units

3. a regular hexagon with side length 10 $150\sqrt{3}$ square units

4. a circle whose circumference is 36 about 103.1 square units

Find the volume and surface area of each figure. *(Section 7.6)*

5.

845.3 cubic units; 832.5 square units

6. a cylinder with length 3.5 and diameter 12 about 395.8 cubic units; about 358.1 square units

385

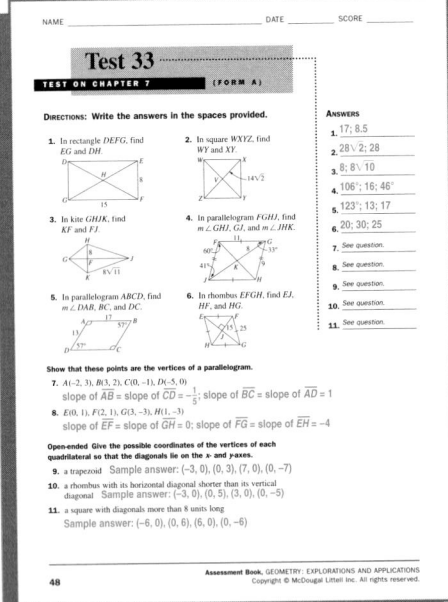
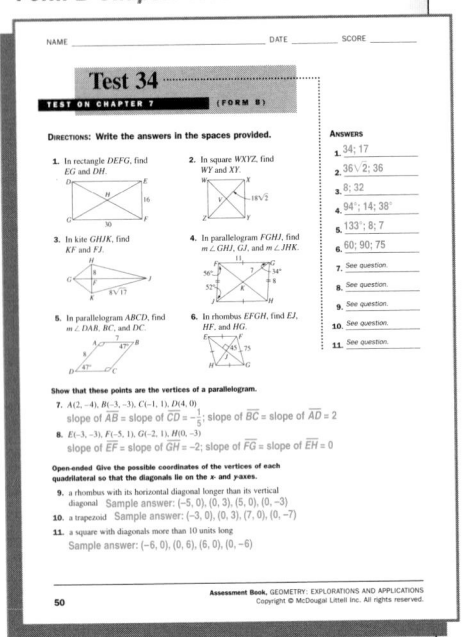
VOCABULARY QUESTIONS

1. **Open-ended Problem** Sketch a kite. Label the lengths of two noncongruent sides.

2. Complete: To write an ? proof, start by assuming that the conclusion is false.

3. Complete: The volume of a cylinder is equal to the area of the ? times the height.

4. **Writing** Describe the difference between the surface area of a prism and its lateral area.

SECTIONS 7.1, 7.2, and 7.3

Find each length or angle measure.

5. kite *ABCD*
   a. *AD*
   b. $m \angle ADC$

6. rectangle *WXYZ*
   a. *WY*
   b. *YZ*

7. square *LMNP*
   a. *LN*
   b. *MN*

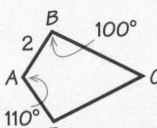
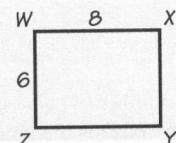
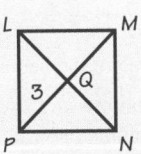

Sketch each quadrilateral. Mark the given information. If the quadrilateral is a parallelogram, explain why.

8. *JKLM* with $\angle J \cong \angle L$ and $\angle K \cong \angle M$

9. *ABCD* with $\angle A \cong \angle D$ and $\angle B \cong \angle C$

10. *WXYZ* with $\overline{WZ} \cong \overline{XY}$ and $\overline{WX} \cong \overline{YZ}$

11. Write an indirect proof to show that a kite cannot be a parallelogram.

Classify each quadrilateral. Be as specific as possible.

12.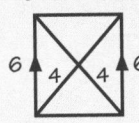

13.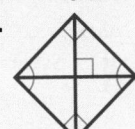

14.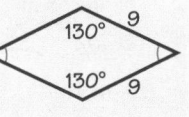

386 Chapter 7 *Quadrilaterals, Areas, and Volumes*

ANSWERS   Chapter 7

Assessment

1. Answers may vary. An example is given.

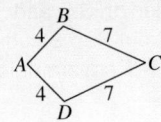

2. indirect

3. base

4. The lateral area is the area of the lateral faces (the sides that are not bases). The surface area is the lateral area plus the area of the bases.

5. a. 2   b. 100°
6. a. 10   b. 8
7. a. 6   b. $3\sqrt{2}$
8.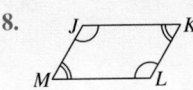

If both pairs of opposite angles of a quadrilateral are congruent, the quadrilateral is a parallelogram.

9.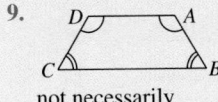

not necessarily

10.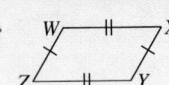

If both pairs of opposite sides of a quadrilateral are congruent, the quadrilateral is a parallelogram.

11. Answers may vary. An example is given. Suppose that a kite is a parallelogram. Then both pairs of opposite sides are congruent. But opposite sides of a kite are not congruent, so a kite is not a parallelogram.

12. rectangle

13. square

14. rhombus

15. Given: $\overline{AB} \parallel \overline{DC}$
$\overline{AB} \cong \overline{DC}$

Prove: $\triangle AED \cong \triangle CEB$

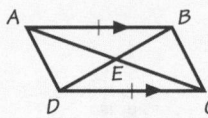

SECTIONS 7.4, 7.5, *and* 7.6

Find the area of each polygon.

16.
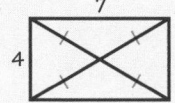

17.
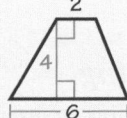

18.
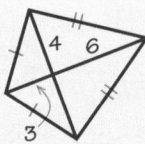

Find the area of each shaded region. Each polygon is a regular polygon.

19.
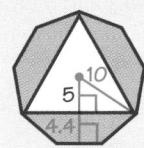

20.
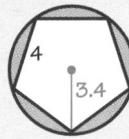

21.
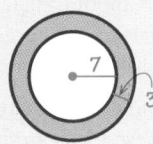

22. HISTORY In the third century B.C., an ice pit, or refrigerator, was built for the first Ch'in Emperor Shih Huang Ti. It was made of terra cotta rings about 5 ft 8 in. in diameter and 3 ft high, and extended 43 ft below the surface.

a. Find the lateral surface area of one terra cotta ring.

b. Find the volume of a cylindrical ice pit 43 ft deep and with diameter 5 ft 8 in.

23. The area of the base of a cylinder is 7 cm^2 and its height is 3 cm.

a. Find its volume.   **b.** Find its surface area.

24. A hexagonal prism is 8 in. high and the edges of a base of the prism are each 4 in. long.

a. Find its volume.   **b.** Find its lateral area.

PERFORMANCE TASK

25. Design a stained glass window using triangles and quadrilaterals. How do you know which quadrilaterals are rectangles, rhombuses, trapezoids, parallelograms, or squares? How would you find the area of each color of glass in your design? If you know the weight of 1 in.^{3} of glass, how can you find the total weight of the glass in the window?

Chapter 7 Assessment
Form C Alternative Assessment

Chapter 7
ALTERNATIVE ASSESSMENT

1. Group Activity Form groups of four. Draw each of the following quadrilaterals on a separate index card.

Sort the cards into stacks, combining the cards whose quadrilaterals have similar characteristics. If needed, use a ruler and a protractor to help. Write a brief description of the common characteristics for each stack. Repeat the activity for a different set of common characteristics.

2. Project Draw a quadrilateral on a coordinate plane. Write the coordinates of the vertices on a separate sheet of paper. Exchange your set of coordinates with a partner, who is to use the Slope Formula and the Distance Formula to determine what type of quadrilateral has these coordinates for its vertices.

3. Open-ended Problem Create a poster that displays all of the important characteristics of each quadrilateral.
   a. square
   b. rectangle
   c. parallelogram
   d. rhombus
   e. trapezoid
   f. kite

4. Project Take photographs or find pictures of bridges, construction sites, buildings, etc. where triangles are used for support. Include some three-dimensional triangular supports. Make a booklet of your pictures. Why do you think triangles are often used for supports?

Assessment Book, GEOMETRY: EXPLORATIONS AND APPLICATIONS
Copyright © McDougal Littell Inc. All rights reserved.   **99**

5. Project A fruit juice company would like to reduce its production costs for one of their juice containers. Suppose the company president has hired you to find a way to reduce the amount of aluminum used in their 12 ounce juice cans. The cans currently used each hold 355 cm^3 (ml) of juice. Your job is to find the can that uses the least amount of aluminum and still has a volume of 355 cm^3. Write a proposal to convince the company's Board of Directors that your can is the best choice for meeting their needs. In your proposal, discuss issues such as consumer appeal, ease of use, environmental conservation, etc. Construct a prototype of your can.

6. a. Draw a rectangle with an area of 36 cm^2. Label its dimensions.
   b. Draw a triangle with an area of 36 cm^2. Label its dimensions.
   c. Draw a rectangle with a perimeter of 36 cm. Label its dimensions.
   d. Draw a triangle with a perimeter of 36 cm. Label its dimensions.
   e. Is your answer to each of parts a–d unique? Explain why or why not.

7. Open-ended Problem Draw a container with an irregular shape. Calculate the surface area and the volume of your container. Describe how you got your results.

100   **Assessment Book**, GEOMETRY: EXPLORATIONS AND APPLICATIONS
Copyright © McDougal Littell Inc. All rights reserved.

---

15. Statements (Reasons)
1. $\overline{AB} \parallel \overline{DC}$; $\overline{AB} \cong \overline{DC}$ (Given)
2. *ABCD* is a parallelogram. (If one pair of opposite sides of a quadrilateral is both parallel and congruent, the quadrilateral is a parallelogram.)
3. $\overline{AC}$ and $\overline{BD}$ bisect each other. (The diagonals of a parallelogram bisect each other.)
4. $\overline{EA} \cong \overline{EC}$; $\overline{EB} \cong \overline{ED}$ (Def. of bisector);
5. $\angle AED \cong \angle CEB$ (Vertical ⩜ are ≅.)
6. $\triangle AED \cong \triangle CEB$ (SAS Postulate)

16. 28

17. 16

18. 36

19–24. Answers are given to the nearest tenth.

19. 158.7

20. 8.8

21. 160.2

22. a. 51.8 ft^2
   b. 1021.7 ft^3

23. a. 21 cm^3
   b. 42.3 cm^2

24. a. 332.6 in.^{3}
   b. 192 in.^{2}

25. Answers may vary. To find the weight, determine the volume in cubic inches and multiply by the weight of 1 in.^{3} of glass.

8 Using Transformations

OVERVIEW

Connecting to Prior and Future Learning

⇔ In Chapter 8, students continue their study of reflections, translations, and rotations. The chapter opens with a study of the use of coordinates to describe reflections.

⇔ In Section 8.3, students learn about translations. They explore the orientations and measures of corresponding sides of translated figures, and use coordinates to describe the translations. Translations are combined with reflections in Section 8.5 to produce glide reflections.

⇔ Rotations, similar figures, and dilations are discussed in the second half of the chapter. Many real-world applications of these ideas are included.

Chapter Highlights

Interview with Terri Johnson: The relationship between architecture and geometry is highlighted in this interview. Related exercises are on pages 395, 409, and 418.

Explorations can be found throughout Chapter 8. In Section 8.1, students reflect a polygon over a line. In Section 8.3, students compare translated polygons. Students enlarge a triangle in Section 8.6.

The Portfolio Project: Students use transformations with different polygons to make tessellations. Students summarize their work by writing a report describing their tessellations, including sketches of four of their tessellations.

 Technology: Students use geometry software to explore the results of reflecting a quadrilateral over a line in Section 8.1, to compare translated polygons in Section 8.3, to compare reflections and rotations in Section 8.4, and to investigate dilations in Section 8.6. Students write a program for a graphing calculator to find the coordinates of points reflected over the x-axis, the y-axis, and the line $y = x$.

OBJECTIVES

Section	Objectives	NCTM Standards
8.1	• Draw the reflection of a polygon and a line of reflection for a figure and its image. • Use reflections to solve real-world problems.	1, 2, 3, 4, 7
8.2	• Find the coordinates of a polygon reflected over the x-axis, the y-axis, or the line $y = x$. • Describe how coordinates change after reflections. • Explore situations involving reflections that can be modeled on a coordinate plane.	1, 2, 3, 4, 5, 8
8.3	• Find the coordinates of the image of a figure after a translation. • Describe a translation based on an original figure and its image. • Describe patterns that involve translation.	1, 2, 3, 4, 5, 8
8.4	• Rotate a figure around a center of rotation. • Find the coordinates of the vertices of a polygon that has been rotated around the origin. • Describe rotations of real-world objects.	1, 2, 3, 4, 5, 8
8.5	• Combine a translation and a reflection to form a glide reflection. • Find the image of a figure after a glide reflection. • Identify, describe, and create patterns that use glide reflections.	1, 2, 3, 4, 5, 8
8.6	• Compare lengths in a figure and its image after a dilation. • Find the scale factor and center of a dilation. • Make reductions and enlargements of figures. • Describe the relationship between a figure and its image after a dilation.	1, 2, 3, 4, 5, 8

INTEGRATION

Mathematical Connections	8.1	8.2	8.3	8.4	8.5	8.6
geometry	**391–397***	**398–404**	**405–411**	**412–418**	**419–425**	**426–433**
algebra	395	402				
patterns and functions			408, 409		419, 422–424	
logic and reasoning	397	403	410	418	424	430, 432

Interdisciplinary Connections and Applications						
reading and language arts						431
sports and recreation		403				
periscopes and optics		402				
computers			408		424	
architecture			410			
furniture design and fabric design				416	423	
audio-visual presentation, Hawaiian quilts, animal tracks, manufacturing	396			417	421, 425	430

****Bold page numbers** *indicate that a topic is used throughout the section.*

TECHNOLOGY

Section	opportunities for use with	
	Student Book	Support Material
8.1	geometry software	**Geometry Inventor Activity Book:** Activities 16–19, 29
8.2	graphing calculator	**Geometry Inventor Activity Book:** Activities 16–18 **Function Investigator Activity Book:** Activities 6, 7
8.3	geometry software	**Geometry Inventor Activity Book:** Activities 22, 26, 27
8.4	geometry software	**Geometry Inventor Activity Book:** Activities 20, 21, 26, 27, 30
8.5		**Technology Book:** Spreadsheet Activity 7 TI-92 Activity 8 **Geometry Inventor Activity Book:** Activities 25, 27
8.6	geometry software	**Technology Book:** Calculator Activity 11 **Geometry Inventor Activity Book:** Activities 23, 24, 28 **Function Investigator Activity Book:** Activity 8

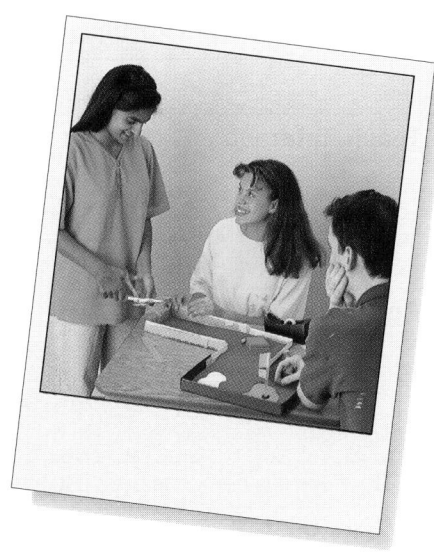

Regular Scheduling (45 min)

Section	Materials Needed	Core Assignment	Extended Assignment	exercises that feature		
				Applications	Communication	Technology
8.1	patty, tracing, or wax paper, protractor, ruler, graph paper, geometry software	1–9, 14–17, 25–36	1–36	10–13, 18–22	13, 24	23
8.2	graph paper, graphing calculator, compass, straightedge	1–10, 12, 14–18, 22–32	1–32	11, 13–19	15, 21, 22	20
8.3	geometry software or patty paper, ruler, graph paper	**Day 1:** 1–11, 13–15 **Day 2:** 17–21, 25–29, AYP*	**Day 1:** 1–16 **Day 2:** 17–29, AYP	12 17–21	16c 25d	16
8.4	graph paper, polar graph paper, geometry software	**Day 1:** 1–14, 16–18 **Day 2:** 19–32, 34–43	**Day 1:** 1–18 **Day 2:** 19–43	15 19–23, 29–32	18 31, 32, 34	24
8.5	graph paper	1–4, 6–12, 15–23, 25–36	1–36	10–13, 16–22, 25–28	16, 19	
8.6	ruler, protractor, graph paper, geometry software	1–7, 10–16, 18, 20–25, AYP	1–25, AYP	8, 10–13	9, 20d, 21	17
Review/ Assess		**Day 1:** 1–9 **Day 2:** 10–18 **Day 3:** Ch. 8 Test	**Day 1:** 1–9 **Day 2:** 10–18 **Day 3:** Ch. 8 Test			
Port. Proj.		Allow 2 days.	Allow 2 days.			

Yearly Pacing (with Portfolio Project)	Chapter 8 Total 13 days	Chapters 1–8 Total 106 days	Remaining 54 days	Total 160 days

Block Scheduling (90 min)

	Day 47	Day 48	Day 49	Day 50	Day 51	Day 52	Day 53
Teach/Interact	Ch. 7 Test 8.1: Exploration, page 392	8.2 8.3: Exploration, page 405	Continue with 8.3 8.4	Continue with 8.4 8.5	8.6: Exploration, page 427 Review	Review Port. Proj.	Ch. 8 Test Port. Proj.
Apply/Assess	**Ch. 7 Test** **8.1:** 1–9, 14–17, 25–36	**8.2:** 1–10, 12, 14–18, 22–32 **8.3:** 1–11, 13–15	**8.3:** 17–21, 25–29, AYP* **8.4:** 1–14, 16–18	**8.4:** 19–32, 34–43 **8.5:** 1–4, 6–12, 15–23, 25–36	**8.6:** 1–7, 10–16, 18, 20–25, AYP **Review:** 1–9	**Review:** 10–18 **Port. Proj.**	**Ch. 8 Test** **Port. Proj.**

NOTE: A one-day block has been added for the Portfolio Project—timing and placement to be determined by teacher.

Yearly Pacing (with Portfolio Project)	Chapter 8 Total $6\frac{1}{2}$ days	Chapters 1–8 Total 53 days	Remaining 27 days	Total 80 days

*__AYP__ is Assess Your Progress.

LESSON SUPPORT

Section	Practice Bank	Study Guide*	Assessment Book*	Visuals	Explorations Lab Manual	Lesson Plans	Technology Book
8.1	54	8.1		Warm-Up 8.1		8.1	
8.2	55	8.2		Warm-Up 8.2	Add. Expl. 12 Masters 1, 2	8.2	
8.3	56	8.3	Test 35	Warm-Up 8.3	Masters 1, 2	8.3	
8.4	57	8.4		Warm-Up 8.4 Folder 10	Master 2	8.4	
8.5	58	8.5		Warm-Up 8.5	Masters 1, 2	8.5	Spreadsheet Act. 7 TI-92 Act. 8
8.6	59	8.6	Test 36	Warm-Up 8.6	Add. Expl. 13 Master 2	8.6	Calculator Act. 11
Review Test	60	Chapter Review	Tests 37, 38, Alternative Assessment			Review Test	

*Spanish versions of *Study Guide* and *Assessment Book* are available.

Chapter Support

- Course Guides
- Lesson Plans
- Preparing for College Entrance Tests
- Multi-Language Glossary
- *Test Generator* Software
- Professional Handbook
- Challenge Problems

Software Support

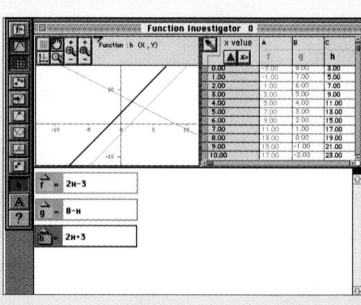

McDougal Littell Mathpack
Geometry Inventor
Function Investigator

Internet Support

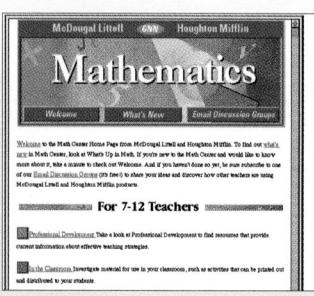

http://www.hmco.com
Next go to McDougal Littell; then the
Education Center; then Secondary Math.

OUTSIDE RESOURCES

Books, Periodicals

Eddins, Susan K., Evelyn Osman Maxwell, and Floramma Stanislaus. "Geometric Transformations, Part 1." *Mathematics Teacher* (March 1994): pp. 177–181, 187–189.

Activities, Manipulatives

Smith, Scott. *Geometry Connections: Tessellations.* Allows students to explore transformations to create drawings. Acton, MA: William K. Bradford Publishing Co.

Software

deLange, Jan. *Gliding.* Macintosh. Pleasantville, NY: Wings for Learning/Sunburst.

Edwards, Lois and Kevin Lee. *TesselMania!* Macintosh, Windows, or CD-ROM versions. Berkeley, CA: Key Curriculum Press.

Jackiw, Nicholas, designer. *The Geometer's Sketchpad.* (Ver. 3.0) for Macintosh or Windows. Berkeley, CA: Key Curriculum Press, 1995.

Videos

Donald in Mathmagic Land. Burbank, CA: Walt Disney Co., 1959. Distributed by Buena Vista Home Video, Burbank, CA.

Internet

Search for mathematics-related software, teaching materials, and other gophers through:

gopher archives.math.utk.edu

Ericsson Stadium

Ericsson Stadium was constructed in Charlotte, North Carolina for the Carolina Panthers, a National Football League team. Its total cost was $187 million and its construction was not publicly funded. Opened in June 1996, the surface of the stadium is grass and the seating capacity is 72,350 people. Inside, the stadium contains 430 concession sites for food, drink, and memorabilia, a 24 ft by 32 ft color replay video board, a 17 ft by 32 ft animation board, and a 10 ft by 50 ft scoreboard. Outside, the main entrances are flanked by six 18 ft statues of fierce-looking black panthers with neon-green eyes. Surrounding the stadium are spacious grass lawns containing as many as 150 oak trees.

Terri Johnson

Terri Johnson grew up in Wheeling, Illinois, a northwest suburb of Chicago. After graduation from high school, Johnson attended Iowa State University to study architecture. She transferred to the University of Illinois in Chicago after three years at Iowa State and continued to study architecture, graduating with a Bachelor or Architecture degree in June 1990. While studying at UIC, Johnson worked part-time at an architectural firm designing houses. In April 1992, she moved to Kansas City to work for HOK Sport.

CHAPTER

8 Using Transformations

BLUEPRINTS FOR BALLGAMES

INTERVIEW Terri Johnson

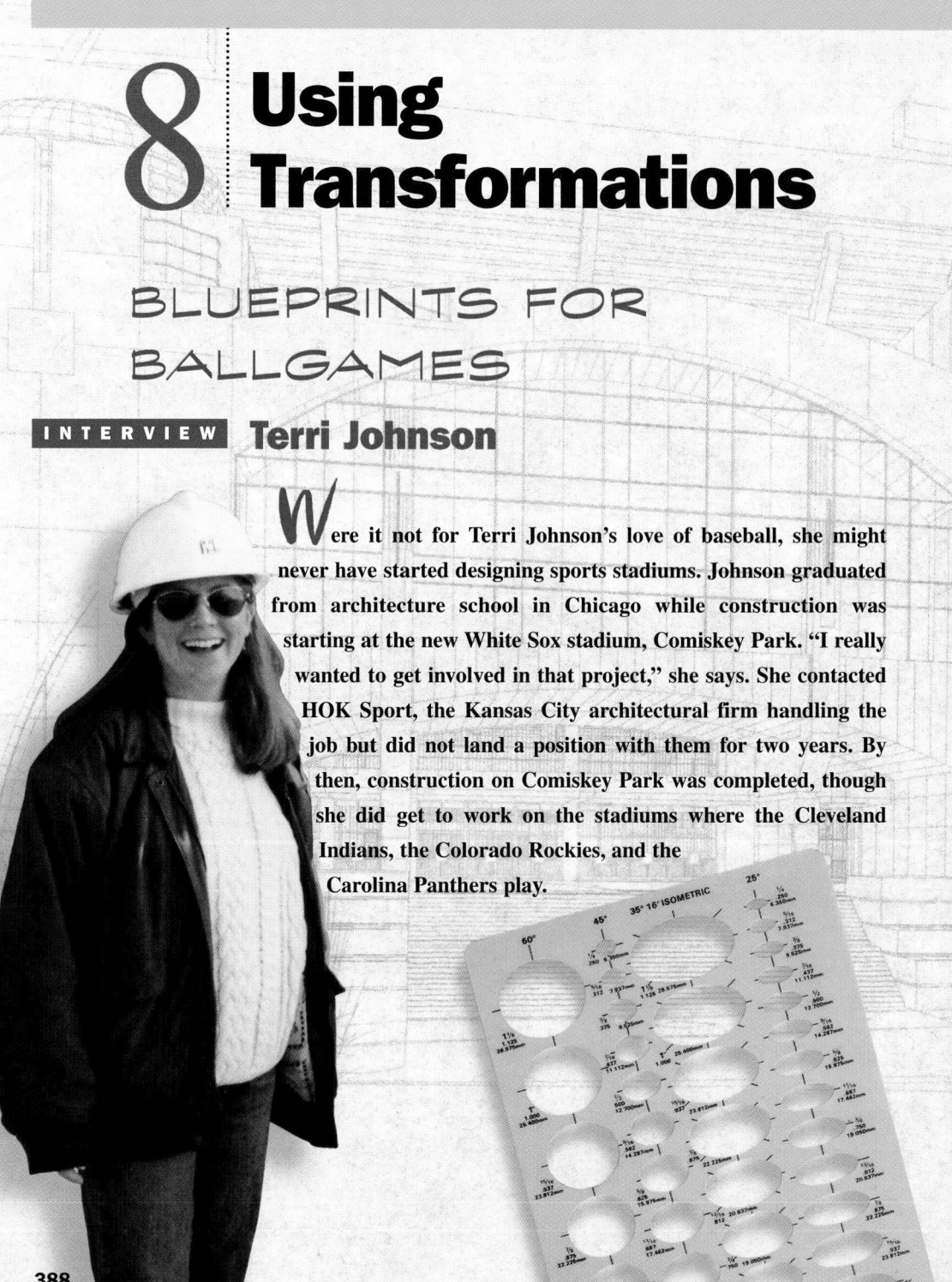

Were it not for Terri Johnson's love of baseball, she might never have started designing sports stadiums. Johnson graduated from architecture school in Chicago while construction was starting at the new White Sox stadium, Comiskey Park. "I really wanted to get involved in that project," she says. She contacted HOK Sport, the Kansas City architectural firm handling the job but did not land a position with them for two years. By then, construction on Comiskey Park was completed, though she did get to work on the stadiums where the Cleveland Indians, the Colorado Rockies, and the Carolina Panthers play.

388

> " GEOMETRY IS A PRACTICAL TOOL FOR AN ARCHITECT. IT'S SOMETHING WE USE ALMOST EVERY DAY. "

Architecture

The word *architecture* often refers to the design of buildings or the buildings themselves. To become an architect, a person needs to complete a long program of study and pass certain examinations. First, in high school, students should take classes in art, mechanical drawing, history, social studies, mathematics, and computer science. Upon completing high school, students need to obtain an undergraduate degree in architecture from an accredited college or university. Then, students must pass a design test, work as an intern with a licensed architect, engineer, or contractor, and lastly, take a licensing examination. The American Institute of Architects (AIA) based in Washington, D.C. publishes literature on architecture and can provide information about architectural schools.

A Grid Plan

Every project begins with a "grid geometry plan," an overall scheme that shows how the building will be laid out and supported structurally. Next come more advanced plans that contain the details contractors need to start building a stadium or sports arena. "We get very specific in our construction drawings, down to the team logos in the locker room carpet of Chicago's United Center." Johnson notes.

Building elements such as stairways, doors, windows, seats, and aisles are spelled out precisely in the final drawings submitted by HOK Sport. A computer is a crucial tool throughout this process, used in all designs, presentations, and supporting documents. "A knowledge of geometry and math is also essential for figuring out how three-dimensional spaces work," Johnson says.

Reflecting on Architecture

Symmetry is a basic element used in stadium designs. The entrance to Ericsson Stadium, for example, is symmetrical: If you draw a vertical line through the middle of the building's front facade, the left half and right half are mirror images of each other.

"We use mirror images to design concession stands as well," Johnson explains. "One or two are designed and then reflected over the stadium's lines of symmetry. This cuts down on design time and makes the stadium easier to build."

Second-Language Learners

Consider allowing students learning English to read the text of this interview with a peer tutor or aide who can define difficult vocabulary terms and describe the different parts of a stadium (*locker room, aisle, front facade, concession stand*).

Multicultural Note

Baseball was invented in the eastern United States in the middle of the nineteenth century. Although some credit Abner Doubleday with inventing baseball, most historical evidence suggests that it gradually developed from rounders, a game played in England as early as the 1600s. Shortly after baseball became popular in the United States, it spread to other nations; international students, sailors, and other travelers carried baseball to Japan, Cuba, the Dominican Republic, Mexico, Puerto Rico, Venezuela, Panama, Nicaragua, and various other places. Today, many talented players from these nations play for professional teams in their home countries; others come to the United States to play professional baseball.

389

Mathematical Connection

Transformations—reflections, translations, rotations, and dilations—can be seen in many natural and artificial objects. Almost all varieties of flowers, trees, and seeds contain some type of transformation. Man-made objects such as clothing, automobiles, machines, and buildings often exhibit a wide variety of transformations. In this chapter, students look at how transformations were used to help architects plan and build Ericsson Stadium. In Section 8.1, students study reflections and how to use them to construct the entrance to the stadium. In Section 8.3, students see how translations can be applied to the construction of the stairs of the stadium. Then in Section 8.4, students apply rotations to the design of the seating arrangements in the stadium.

Explore and Connect

Writing

After students have written their explanations, have them compare their work with another student who selected a different quadrant to reflect.

Research

Students may use magazines or newspapers to find pictures of buildings for examples of symmetry, or they can actually visit the buildings to look for symmetry.

Project

Have students plan their model before actually constructing it by making a scale drawing. When making their drawings, students should use reflection and rotational symmetry.

Clever Repetition

The Ericsson Stadium is divided into four quadrants. After designing Quadrant A, the architects can generate Quadrants B and D by reflecting Quadrant A over the horizontal and vertical axes shown in the diagram. Quadrant C is created by reflecting Quadrant A over both axes, one after the other. "Once you do that, the entire building is composed," Johnson notes.

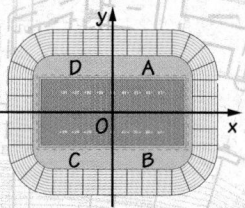

Building elements, such as a section of stairway, can be reproduced on paper via a process called *translation*, which involves reflecting a set of stairs twice over two parallel lines. The result is that a new stairway, identical to the first, appears in another position in the drawing. This process enables architects to use the same set of stairs in their designs and then translate it to various levels and locations in the building.

> "WE HAVE TO TURN OUR DESIGNS INTO BUILDINGS. THAT DEPENDS ON GEOMETRY."

Explore and Connect

Terri Johnson uses an architect's scale to measure a drawing.

1. Writing Choose Quadrant B, C, or D in the stadium above. Explain how you can generate the entire stadium plan by reflecting your quadrant.

2. Research Find examples of symmetry in the architecture of other building types, such as schools, theaters, houses, or office buildings. Describe the symmetry you find in each building.

3. Project Build a model of a house or an apartment building that features reflection or rotational symmetry.

Mathematics & Terri Johnson

In this chapter, you will learn more about how mathematics is related to architecture.

Related Exercises

Section 8.1
• Exercises 10–13

Section 8.3
• Exercises 17, 18

Section 8.4
• Exercises 29–32

8.1

Using Reflections

A canoe designer takes great care in shaping the *hull,* the part that touches the water, so that the canoe will be both stable and fast. Because canoes have reflection symmetry, a designer needs to draw only one half of the hull in detail. The other half can be drawn by reflecting the design over the line of symmetry, which is also called the **line of reflection**.

Top view

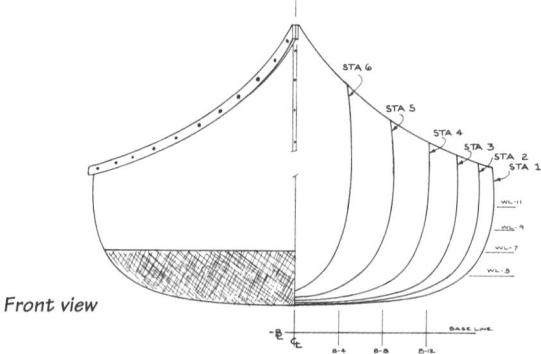

Front view

A **reflection** is a transformation in which each point of a figure has an **image** that is the same distance from the line of reflection as the original point. A point and its image lie on opposite sides of the line of reflection unless the point is on the line of reflection itself.

BY THE WAY

Boat builders like Rollin Thurlow, who designed this canoe, call the line of reflection the *centerline* of the boat.

THINK AND COMMUNICATE

1. In the front view of the canoe design above, is the line of reflection horizontal or vertical?

2. Copy the diagram of the front view of the canoe. Mark a point that is *not* on the line of reflection. Describe where the image of that point lies on the diagram.

3. Mark a point on the canoe design that lies on the line of reflection. Describe where the image of that point lies.

ANSWERS Section 8.1

Think and Communicate

1. vertical

2, 3. Check students' work.

2. The image of a point not on the line of reflection is on the opposite side of the line of reflection. The point and its image are the same distance from the line.

3. The image of a point on the line of reflection is the point itself.

Learning Styles: Verbal

Students who are verbal learners may need some extra practice working with reflections. To identify lines of reflection, such as the one in the canoe, have them trace the drawing and fold the paper until they find a fold line for which the images on either side match. This fold line is a line of symmetry, or a line of reflection.

Section Notes

Geometric Thinking

The definition of a *reflection* given on page 391 is an intuitive introduction to this concept. In the Exploration, students discover that a reflection of a point *A* over line *m* is a point *A'*, such that *m* is the perpendicular bisector of $\overline{AA'}$.

Using Technology

Students can examine the concept of a reflection of a point and a line segment by using Activities 16 and 17 in the *Geometry Inventor Activity Book*.

Common Error

Students may confuse the concepts of line of symmetry and line of reflection. Point out that every line of symmetry of a figure is also a line of reflection, but that the converse is not true; that is, every line of reflection is not a line of symmetry.

Think and Communicate

In questions 2 and 3, students contrast the reflections of points that do not lie on the line of reflection with those that do.

Section Note

Second-Language Learners

Make sure that second-language learners understand what is meant by a transformation that *preserves congruence* and *changes orientation*. You might want to hold a mirror to a clock to demonstrate both: first, the shape and size of the clock is the same in the reflection (preserving congruency), and second, the numbers go from 1 to 12 in the opposite direction on the reflection than on the real clock (changing orientation).

EXPLORATION
COOPERATIVE LEARNING

Reflecting a Polygon Over a Line

Work in a group of four students.
You will need:
- patty paper, tracing paper, or wax paper
- a ruler and a protractor

1 Fold your paper to make a crease. Label it line *m*.

2 Draw △*ABC* on one side of line *m*. Reflect points *A*, *B*, and *C* over line *m* by folding.

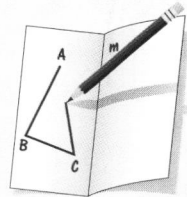

3 Mark the image points *A'*, *B'*, and *C'*. Draw △*A'B'C'*.

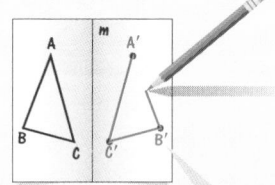

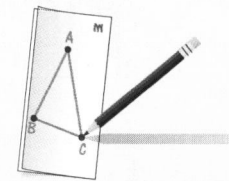

4 Draw $\overline{AA'}$, $\overline{BB'}$, and $\overline{CC'}$. Label the points where these segments intersect line *m* as *P*, *Q*, and *R*, respectively.

5 Compare the lengths *AP* and *PA'*. Compare *BQ* and *QB'*. Compare *CR* and *RC'*. What do you notice?

6 Measure ∠*APR*, ∠*BQP*, and ∠*CRQ*. What do you notice?

7 How is line *m* related to $\overline{AA'}$, $\overline{BB'}$, and $\overline{CC'}$?

8 Compare *AB* and *A'B'*. Compare *BC* and *B'C'*. Compare *CA* and *C'A'*. What do you know about △*ABC* and △*A'B'C'*?

> The image of point *B* is often called point *B'* (read "*B* prime").

In the Exploration, if you look at the vertices of △*ABC* in alphabetical order, you see that they are arranged in a *counterclockwise orientation*. The vertices of △*A'B'C'* are arranged in a *clockwise orientation*.

counterclockwise orientation	clockwise orientation
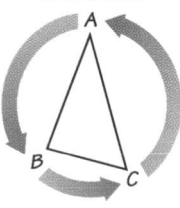	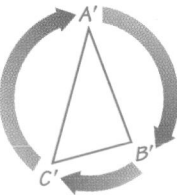

Exploration Note

Purpose
The purpose of this Exploration is to have students discover that a line of reflection is the perpendicular bisector of the segment connecting a point and its image and that the image of a figure is congruent to the original figure.

Materials/Preparation
Each group needs patty paper, tracing paper, or wax paper and a ruler and protractor.

Procedure
Students fold the paper to make a line. They draw a triangle on one side and use folding to find the image. They then compare the lengths of the segments connecting a point

on the figure and its image point to the same point on the line of reflection. They also measure the angles that these segments make with the line of reflection.

Closure
Discuss students' responses to Steps 5–8 to bring out the fact that a reflection of a point *A* over line *m* is a point *A'*, such that *m* is the perpendicular bisector of $\overline{AA'}$. Students should also understand that the image triangle *A'B'C'* is congruent to the original triangle *ABC*.

Explorations Lab Manual
See the Manual for more commentary on this Exploration.

Properties of Reflections

A reflection is a transformation that preserves congruence.

A reflection changes orientation.

The line of reflection is the perpendicular bisector of every segment that connects a point and its image.

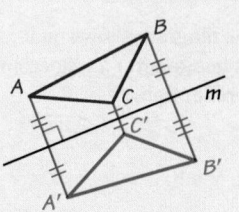

About Example 1

Spatial Reasoning
Some students may need detailed instructions for drawing the image of a point after a reflection. Tell these students to draw a line perpendicular to the line of reflection that contains the point they want to reflect. To find its exact location, they should measure the distance between the point being reflected and the line of reflection. The image point is the same distance from the line of reflection as measured along the perpendicular.

EXAMPLE 1

Copy the diagram and draw the image of △*XYZ* after reflection over line *k*.

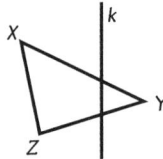

SOLUTION

Reflect each vertex of △*XYZ*. Label the image vertices *X′*, *Y′*, and *Z′*.

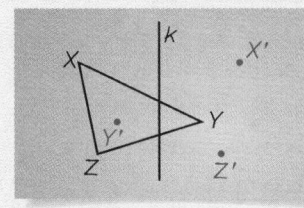

Connect the image vertices to draw the image △*X′Y′Z′*.

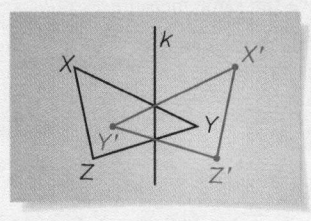

Additional Example 1

Copy the diagram and draw the image of △ *LMN* after reflection over line *j*.

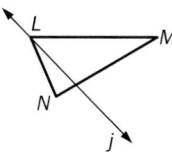

Reflect each vertex of △ *LMN*. Since *L* is on the reflection line, it is its own image. Label the image vertices *L′*, *M′*, and *N′*.

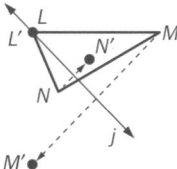

Connect the image vertices to draw image △ *L′M′N′*.

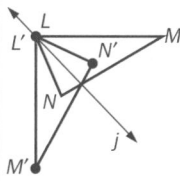

THINK AND COMMUNICATE

4. In Example 1, is the orientation of △*XYZ* counterclockwise or clockwise?

5. What is the orientation of △*X′Y′Z′*?

6. Line *k* is the perpendicular bisector of $\overline{XX'}$. Name two other segments that have line *k* as a perpendicular bisector.

Think and Communicate

These questions help students understand two of the properties of reflection that are presented at the top of this page. You may wish to ask students how they could show that △ *XYZ* is congruent to △ *X′Y′Z′*. (They can measure the lengths of the sides of both triangles and compare them.)

8.1 Using Reflections   **393**

• •

Exploration

1–4. Check students' work.

5. $AP = PA'$; $BQ = QB'$; $CR = RC'$; Each pair of lengths is equal.

6. ∠ *APR*, ∠ *BQP*, and ∠ *CRQ* are all right angles.

7. Line *m* is the perpendicular bisector of $\overline{AA'}$, $\overline{BB'}$, and $\overline{CC'}$.

8. $AB = A'B'$; $BC = B'C'$; $CA = C'A'$; △ *ABC* ≅ △ *A′B′C′*

Think and Communicate

4. clockwise

5. counterclockwise

6. $\overline{YY'}$; $\overline{ZZ'}$

About Example 2

Teaching Tip
Students should understand that any point and its image can be used to locate the line of reflection.

Additional Example 2

The diagram shows quadrilateral *RSTU* and its image after a reflection. Draw the line of reflection.

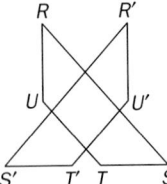

Draw a segment connecting vertex *R* and its image *R'*. The line of reflection is the perpendicular bisector of this segment, line *k*.

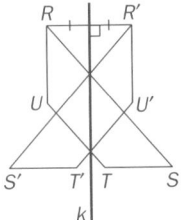

Checking Key Concepts

Student Progress
These questions will help you assess whether students understand the concept of reflection over a line and the properties of reflection. Students who have difficulty with any of these questions should reread pages 391–393 and study Examples 1 and 2 carefully.

Closure Question

What is a reflection and what properties does a reflection have?
A reflection is a transformation in which each point of a figure has an image point that is the same distance from the line of reflection as the original point. The line of reflection is the perpendicular bisector of the segment connecting the original point and its image. The properties of reflections are stated at the top of page 393.

EXAMPLE 2

The diagram shows quadrilateral *JKLM* and its image after a reflection. Draw the line of reflection.

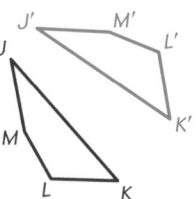

SOLUTION

Draw a segment connecting vertex *K* and its image *K'*. The line of reflection is the perpendicular bisector of this segment, line *n*.

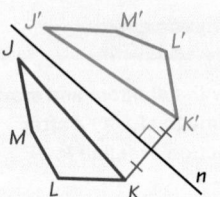

☑ CHECKING KEY CONCEPTS

In the diagram, *P'Q'R'S'* is the image of quadrilateral *PQRS* after a reflection over line *j*.

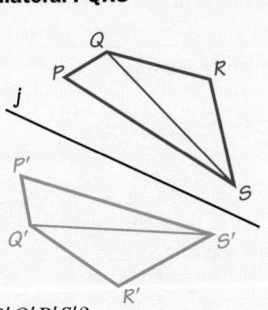

1. What is the image of $\overline{PQ}$? of $\overline{RS}$?
2. Name a triangle congruent to $\triangle QRS$.
3. Name a triangle congruent to $\triangle SPQ$.
4. Name two angles in the diagram that have the same measure.
5. What is the image of *P'Q'R'S'* after a reflection over line *j*?
6. What are the orientations of *PQRS* and *P'Q'R'S'*?

8.1 **Exercises and Applications**

Extra Practice exercises on page 675

Copy each diagram. Sketch the image of the polygon after a reflection over the given line. Label the image polygon.

1.

2.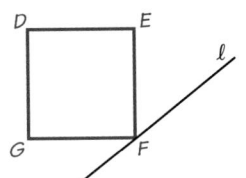

3.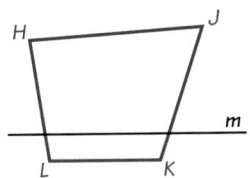

394   Chapter 8 *Using Transformations*

Checking Key Concepts

1. $\overline{P'Q'}$; $\overline{R'S'}$

2. $\triangle Q'R'S'$

3. $\triangle S'P'Q'$

4. $\angle P$ and $\angle P'$; $\angle R$ and $\angle R'$; $\angle PQS$ and $\angle P'Q'S'$; $\angle RQS$ and $\angle R'Q'S'$; $\angle PSQ$ and $\angle P'S'Q'$; $\angle RSQ$ and $\angle R'S'Q'$

5. quadrilateral *PQRS*

6. clockwise; counterclockwise

Exercises and Applications

1.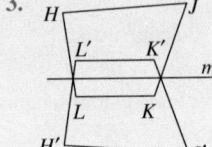

2. (see figure)

3.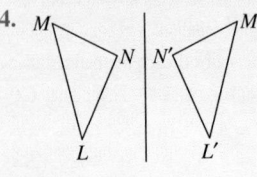

4. (see figure)

Each diagram shows a polygon and its image after a reflection. Copy each diagram and draw the line of reflection.

4.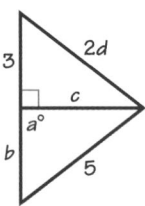

5.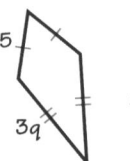

6.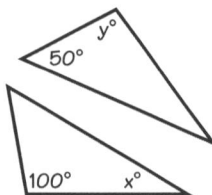

ALGEBRA **Each diagram shows a polygon and its image after a reflection. Find the value of each variable.**

7.

8.

9.

INTERVIEW **Terri Johnson**

Look back at the article on pages 388–390.

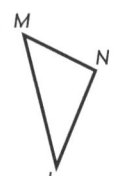

Terri Johnson reviews a presentation of the stadium.

The diagram shows the entrance to Ericsson Stadium, which Terri Johnson helped develop from the design.

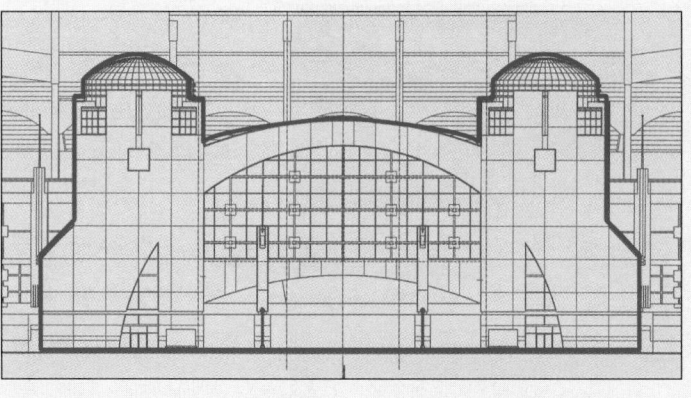

10. Trace the items outlined in green on a piece of paper.

11. Fold your paper to find the line of symmetry for the stadium entrance.

12. Open-ended Problem Choose any shape on one side of the entrance. Find its image on the other side.

13. Writing The towers on each side of the entrance are chambers for stairwells. What parts of one tower have reflection symmetry? What parts do not?

8.1 Using Reflections **395**

5.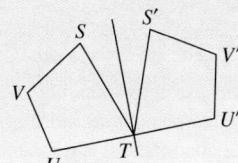

6.

7. $a = 90$; $b = 3$; $c = 4$; $d = 2\frac{1}{2}$

8. $p = 5$; $q = 3$

9. $x = 30$; $y = 100$

10, 11.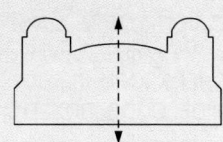

12–13. Answers may vary. Examples are given.

12. The image of the square window on the left tower is the square window on the right tower.

13. the windows in the upper part of each tower; the curved windows at the bottom of each tower

395

Exercise Notes

Student Progress

Exs. 14–17 Most students should be able to complete these exercises successfully. Suggest that students list all the letters to make sure that they have not omitted any.

Application

Exs. 18–22 This set of exercises introduces students to a interesting application of the concept of reflection to an audio-visual presentation. Students will need to spend some time reading the description and analyzing the illustration before answering the questions.

Interdisciplinary Problems

Exs. 18–22 This application involves concepts from physics concerning the properties of reflected light. The law of reflection states that for light reflecting off any smooth surface, the angle of incidence is equal to the angle of reflection. The angle of incidence is the angle between the incoming light and the surface. The angle of reflection is the angle between the surface and the reflected light.

Family Involvement

Exs. 18–22 Students whose families own a slide projector and a large enough mirror may wish to actually set up the situation pictured in the diagram with the help of family members. Students and their family members could then use this setup to discuss and answer these exercises.

For Exercises 14–17, use block letters, as shown.

14. If you write the word OBOE in block letters, as shown, it has a horizontal line of symmetry. What other letters have this kind of symmetry? Use any of these letters to write a word.

15. If you write the word AIM in block letters, as shown, it has a vertical line of symmetry. What other letters have this kind of symmetry? Use any of these letters to write a word.

16. What letters have both horizontal and vertical reflection symmetry?

17. What letters have neither horizontal nor vertical reflection symmetry?

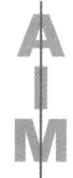

Connection — AUDIO-VISUAL PRESENTATION

Rear projection is a way to set up a slide projector so that its picture shines on the back of a screen rather than on the front. Many stages or auditoriums are not large enough to put a projector directly behind the screen, in the position marked *virtual projector* in the diagram. Instead, a mirror is used to reflect the image from the *actual projector* to the screen.

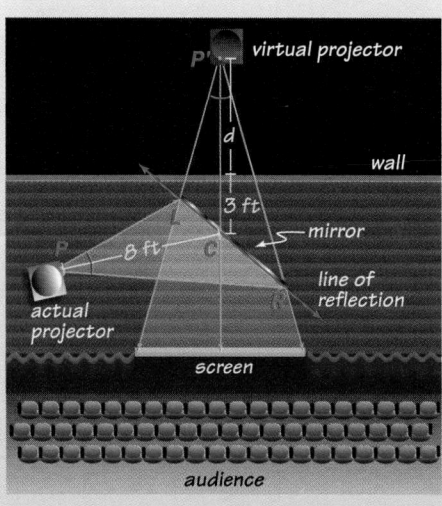

18. What is the image of *P* after the reflection in the mirror?

19. What are the images of *L*, *C*, and *R*, points that lie on the mirror?

20. Name three pairs of congruent triangles shown in the diagram.

21. Explain why $PC = P'C$.

22. Using the distances in the diagram, estimate *d*, the extra stage depth that would be needed if the mirror method were not used.

BY THE WAY

When using rear projection, the operator must put slides into the projector backwards so they will be displayed correctly after they are reflected in the mirror.

396   Chapter 8 *Using Transformations*

14. C, D, H, I, K, X; Answers may vary. Examples are given. CHECK, COOKIE, BOOK, HIDE, CODE, BOXED

15. H, O, T, U, V, W, X, Y; Answers may vary. Examples are given.

```
Y  W  M  M  Y  W
O  A  A  Y  A  H
U  X  T  T  M  I
T     H  H     M
H
```

16. H, I, O, X

17. F, G, J, L, N, P, Q, R, S, Z

18. *P'*

19. *L*, *C*, and *R* (Each point is its own image.)

20. △*PCL* and △*P'CL*; △*PLR* and △*P'LR*; △*PCR* and △*P'CR*

21. *C* is on the perpendicular bisector of $\overline{PP'}$.

22. 5 ft

23. a, b. Check students' work.

 c. Because corresponding angles and corresponding sides of the quadrilateral and its image are congruent, the quadrilaterals are, by definition, congruent.

23. **Technology** Use geometry software. Draw a quadrilateral *ABCD* and a line of reflection.

a. Reflect *ABCD* over the line of reflection. Label its vertices.

b. Measure the corresponding segments and angles of *ABCD* and its image.

c. Explain how you can use the definition of congruent polygons and the results of part (b) to conclude that *ABCD* and its image are congruent.

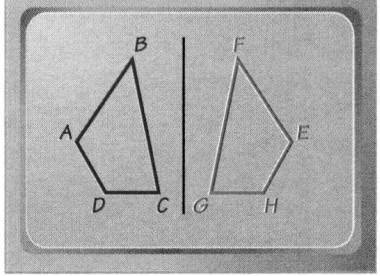

24. Challenge Suppose $\triangle ABC \cong \triangle A'B'C'$ and the midpoints of $\overline{AA'}$, $\overline{BB'}$, and $\overline{CC'}$ lie on a line. Is the line a line of reflection? Make a sketch and explain your answer.

ONGOING ASSESSMENT

25. Open-ended Problem Make an interesting design that uses a polygon and several reflections. Label the vertices of the polygon and its images and name some congruent polygons.

SPIRAL REVIEW

Find the area of the shaded region in each regular polygon or circle.
(Section 7.5)

26.
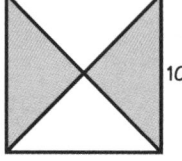
10

27.
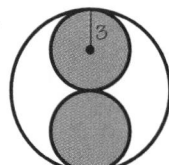
3

28.
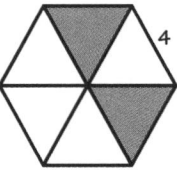
4

Tell which triangle appears to be congruent to the given triangle. *(Section 6.2)*

29. $\triangle ABF \cong$?

30. $\triangle FGH \cong$?

31. $\triangle JME \cong$?

32. $\triangle CEH \cong$?

Write an equation for each line described. *(Section 4.2)*

33. contains (3, 4); has slope 2

34. contains (−3, 4); has slope −2

35. the horizontal line that contains (8, 7)

36. the vertical line that contains (8, 7)

Apply⇔Assess

Exercise Notes

Using Technology
Ex. 23 This exercise is similar to the Exploration on page 392 except that geometry software is used to reflect the figure and to measure the angles and sides. Discuss part (c) to assess students' understanding of how they know that the figures are congruent.

Assessment Note
Ex. 25 This open-ended problem gives students an opportunity to show what they have learned about reflection. Students should label the points of reflection using prime notation. They should also label the lines of reflection.

Practice 54 for Section 8.1

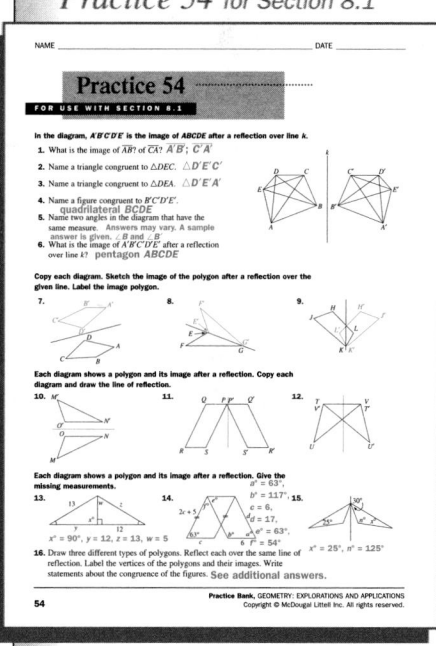

---

24. not necessarily; $\triangle ABC$ and $\triangle A'B'C'$ in the diagram are congruent. The midpoints of $\overline{AA'}$, $\overline{BB'}$, and $\overline{CC'}$ all lie on the y-axis. The y-axis is not a line of reflection for the figures because it is not the perpendicular bisector of any of the segments.

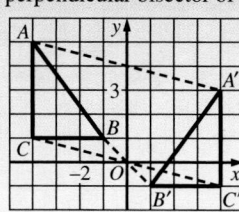

25. Check students' work.

26. 50

27. $18\pi \approx 56.5$

28. $8\sqrt{3} \approx 13.9$

29. $\triangle DCE$

30. $\triangle EJH$

31. $\triangle GLF$

32. $\triangle BFH$

33. $y = 2x - 2$

34. $y = -2x - 2$

35. $y = 7$

36. $x = 8$

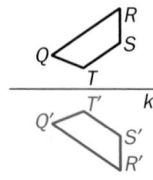
398

SECTION

8.2 Reflections with Coordinates

Learn how to...

- **find the coordinates of a polygon reflected over the *x*-axis, the *y*-axis, or the line *y* = *x***

- **describe how coordinates change after reflections**

So you can...

- **explore situations involving reflections that can be modeled on a coordinate plane**

When a polygon is placed on a coordinate plane, you can describe a reflection of the polygon by comparing the coordinates of the polygon and the coordinates of image vertices. Notice what happens to the coordinates of the triangle shown below as it is reflected over the *y*-axis.

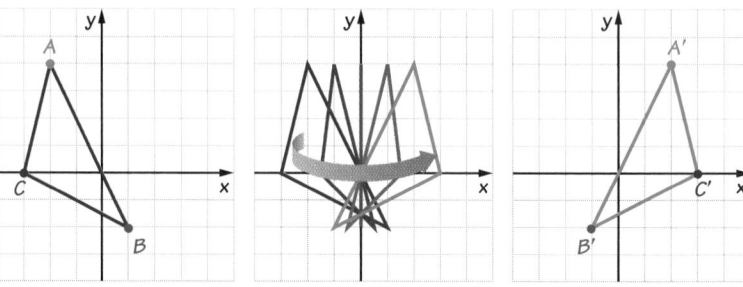

$$A(-2, 4) \rightarrow A'(2, 4)$$
$$B(1, -2) \rightarrow B'(-1, -2)$$
$$C(-3, 0) \rightarrow C'(3, 0)$$

The arrow symbol means "goes to."

THINK AND COMMUNICATE

1. When a point is reflected over the *y*-axis, which of its coordinates changes? How does it change?

2. Choose a point on △*ABC* that is not a vertex. Estimate its coordinates. Find the coordinates of its image after a reflection over the *y*-axis.

3. Which points on △*ABC* do not change position when they are reflected over the *y*-axis? Estimate the coordinates of these points.

ANSWERS Section 8.2

Think and Communicate

1. the *x*-coordinate; The *x*-coordinate of the image is the opposite of the *x*-coordinate of the original point.

2. Answers may vary. An example is given.
   (−1, 2); (1, 2)

3. points on the *y*-axis; Estimates may vary. The actual coordinates are (0, 0) and $\left(0, -1\frac{1}{2}\right)$.

EXAMPLE 1

Reflect quadrilateral **WXYZ** over the *x*-axis. Give the coordinates of the vertices of the image.

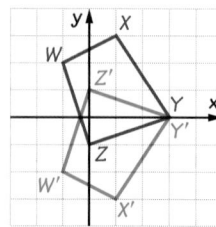

SOLUTION

When a point is reflected over the *x*-axis, only the *y*-coordinate changes.

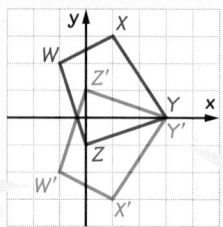

W is 2 units *above* the *x*-axis, so **W'** must be 2 units *below* the *x*-axis.

Points that are on the line of reflection do not change their position.

$$W(-1, 2) \rightarrow W'(-1, -2)$$
$$X(1, 3) \rightarrow X'(1, -3)$$
$$Y(3, 0) \rightarrow Y'(3, 0)$$
$$Z(0, -1) \rightarrow Z'(0, 1)$$

Notice in the Solution to Example 1 that each segment joining a point and its image is vertical. If the reflection were over the *y*-axis, such segments would be horizontal.

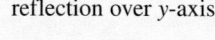
Reflecting Over the Coordinate Axes

When a point (a, b) is reflected over a coordinate axis, one coordinate changes to its opposite and the other coordinate does not change.

reflection over *x*-axis

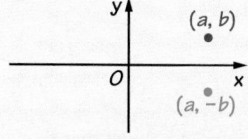

$(a, b) \rightarrow (a, -b)$

reflection over *y*-axis

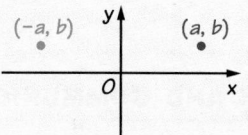

$(a, b) \rightarrow (-a, b)$

8.2 Reflections with Coordinates **399**

Teach⇔Interact

Section Note

Alternate Approach
You may wish to have students work in small groups to discover what happens to the coordinates of a triangle reflected over the *y*-axis. Each group needs graph paper and a ruler. Each student in a group should draw a triangle on a coordinate plane with integer coordinates at the vertices, reflect the triangle over the *y*-axis, and find the coordinates of the vertices of the image. Then have students discuss the pattern they observe. This procedure can be repeated using reflection over the *x*-axis.

Think and Communicate

These questions have students consider how the coordinates of points are changed when they are reflected over the *y*-axis.

Additional Example 1

Reflect quadrilateral *DEFG* over the *x*-axis. Give the coordinates of the vertices of the image.

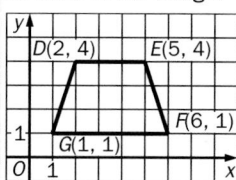

When a point is reflected over the *x*-axis, only the *y*-coordinate changes. *D* is 4 units *above* the *x*-axis, so *D'* is 4 units *below* the *x*-axis.

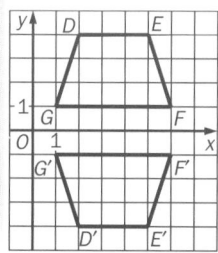

$D(2, 4) \rightarrow D'(2, -4)$
$E(5, 4) \rightarrow E'(5, -4)$
$F(6, 1) \rightarrow F'(6, -1)$
$G(1, 1) \rightarrow G'(1, -1)$

399

Section Notes

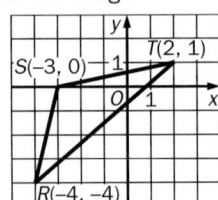

Communication: Discussion
Discuss reflecting over the line $y = x$. Ask students to give the coordinates of some points on line $y = x$. Ask what happens if the point being reflected is on the line $y = x$? (The x- and y-coordinates of points on the line are the same, so when you reflect the point and switch coordinates, the point remains on the line.) You may wish to extend this discussion to include reflecting over the line $y = -x$.

Visual Thinking
Ask students to sketch a shape of their own choosing on a coordinate plane so that it crosses the line $y = x$. Then, ask them to sketch the reflection of that shape over the line. This activity involves the visual skills of *exploration* and *generalization*.

About Example 2

Geometric Thinking
Ask students what happens to a point on the x-axis when it is reflected over the line $y = x$. (It becomes a point on the y-axis.)

Additional Example 2

Reflect △ RST over the line $y = x$. Give the coordinates of the vertices of the image.

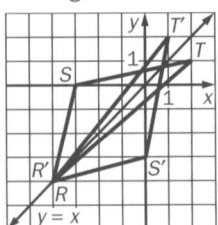

Draw the line $y = x$. Reverse the coordinates of each point to find its image.

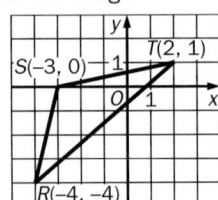

$R(-4, -4) \rightarrow R'(-4, -4)$
$S(-3, 0) \rightarrow S'(0, -3)$
$T(2, 1) \rightarrow T'(1, 2)$

You can reflect points over lines other than the coordinate axes, but it is not usually easy to write a simple rule showing what happens to the coordinates of a point (a, b). One exception is reflection over the line $y = x$. For reflections over this line, all you have to do is switch the coordinates to find the image of a point.

Reflecting Over the Line $y = x$

When a point (a, b) is reflected over the line $y = x$, the coordinates are reversed.

reflection over the line $y = x$

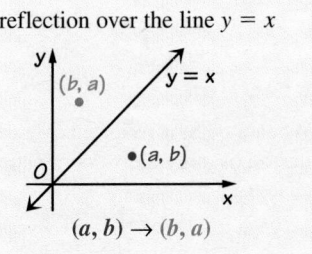

$(a, b) \rightarrow (b, a)$

EXAMPLE 2

Reflect △ *JKL* over the line $y = x$. Give the coordinates of the vertices of the image.

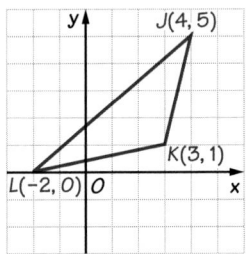

SOLUTION

Draw the line $y = x$. Reverse the coordinates of each point to find its image.

$J(4, 5) \rightarrow J'(5, 4)$
$K(3, 1) \rightarrow K'(1, 3)$
$L(-2, 0) \rightarrow L'(0, -2)$

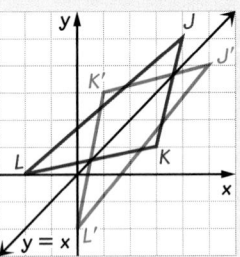

THINK AND COMMUNICATE

4. Use the coordinates of the points in Example 2 to check that the slopes of $\overline{JJ'}$, $\overline{KK'}$, and $\overline{LL'}$ are all -1. Explain why this is so.

5. Find the coordinates of the midpoint of $\overline{KK'}$. Explain how you know that the point is on the line $y = x$.

Think and Communicate

4. slope of $\overline{JJ'} = \dfrac{4-5}{5-4} = -1$;

slope of $\overline{KK'} = \dfrac{3-1}{1-3} = -1$;

slope of $\overline{LL'} = \dfrac{-2-0}{0-(-2)} = -1$;

Each segment is perpendicular to the line $y = x$, which has slope 1. (Two nonvertical lines are perpendicular if and only if the product of their slopes is -1.)

5. (2, 2); The line $y = x$ is the perpendicular bisector of $\overline{KK'}$, so it

must contain the midpoint of $\overline{KK'}$; also, (2, 2) is a solution of $y = x$.

Checking Key Concepts

1.

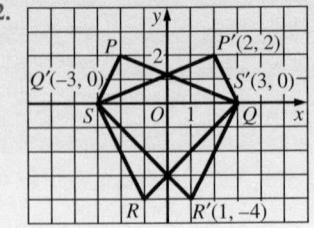

2.

☑ CHECKING KEY CONCEPTS

Copy the diagram and reflect *PQRS* over each
line. Give the coordinates of the vertices
of the image.

1. the *x*-axis

2. the *y*-axis

3. the line *y* = *x*

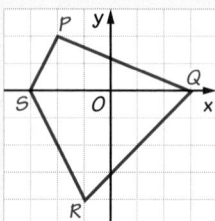

8.2 | **Exercises and Applications**

Extra Practice
exercises on
page 675

Copy each polygon and reflect it over the *y*-axis. Give the coordinates of the
vertices of the image.

1.
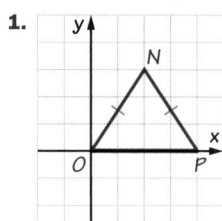

2.
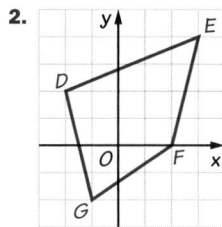

3.
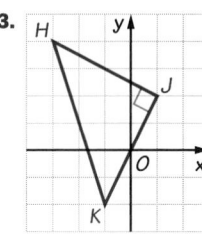

Copy each polygon and reflect it over the *x*-axis. Give the coordinates of the
vertices of the image.

4.
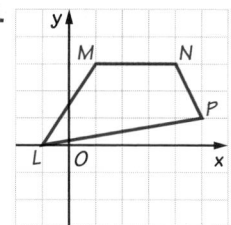

5.
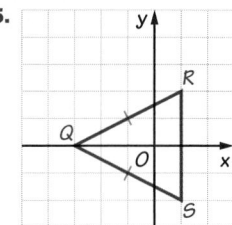

6.
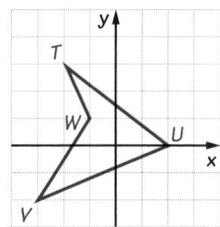

Copy each polygon and reflect it over the line *y* = *x*. Give the coordinates of
the vertices of the image.

7.
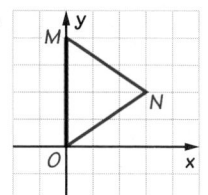

8.
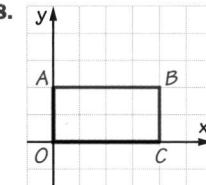

9.
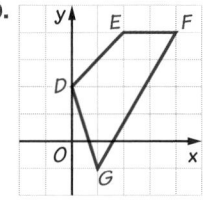

8.2 Reflections with Coordinates **401**

3.
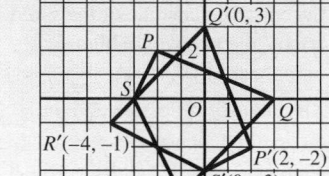

Exercises and Applications

1. *N*′(−2, 3); *O*′(0, 0); *P*′(−4, 0)

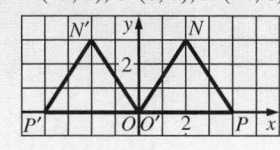

2. *D*′(2, 2); *E*′(−3, 4); *F*′(−2, 0);
G′(1, −2)

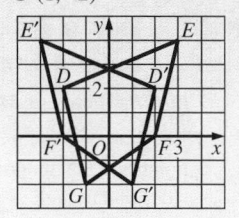

3–9. See answers in back of book.

Teach⇔Interact

Checking Key Concepts

Teaching Tip
It may be helpful for students to
make a separate graph for each of
the three reflections. Encourage
them to check their work by comparing the coordinates of the image
with the original coordinates to see
that $(a, b) \rightarrow (a, -b)$, $(a, b) \rightarrow$
$(-a, b)$, and $(a, b) \rightarrow (b, a)$ when
reflected over the *x*-axis, the *y*-axis,
and the line *y* = *x*, respectively.

Closure Question

Describe what happens to the coordinates of the vertices of a polygon
when the polygon is reflected over
the *x*-axis, the *y*-axis, and the line
y = *x*. When reflected over the
x-axis, the *x*-coordinates of the
vertices remain the same and the
y-coordinates are opposites. When
reflected over the *y*-axis, the
y-coordinates remain the same and
the *x*-coordinates are opposites.
When reflected over the line *y* = *x*,
the coordinates are reversed.

Apply⇔Assess

Suggested Assignment

❖ **Core Course**
Exs. 1–10, 12, 14–18, 22–32

❖ **Extended Course**
Exs. 1–32

❖ **Block Schedule**
Day 48 Exs. 1–10, 12, 14–18,
22–32

Exercise Notes

Common Error
Exs. 1–6 Some students may confuse the coordinate that becomes
the opposite in the image point.
Stress that the coordinate that
changes is the one that is not the
same as the axis of reflection. In
other words, to reflect a point over
the *x*-axis, change the *y*-coordinate
and to reflect over the *y*-axis,
change the *x*-coordinate. Also,
encourage students to check the
image figure by examining it visually to see that the reflection is over
the indicated axis.

Exercise Notes

Integrating Algebra

Ex. 10 For this exercise, students need to use the Midpoint Formula to verify that when a figure is reflected over the line $y = x$, this line bisects the segment that joins each point to its image.

Interdisciplinary Problems

Exs. 11, 13 These two exercises involve the reflection of light by a mirror. The study of light and its reflection is a part of the science of physics. Interested students may want to find out more about light reflection by consulting a physics book.

Using Manipulatives

Ex. 11 If possible, bring a periscope to class for students to use. You may be able to borrow one from the science department. Or, you may wish to ask some students to make a periscope using two small mirrors and cardboard.

Topic Spiraling: Review

Ex. 13 To complete part (a) of this exercise, students need to recall the theorem that the segment joining the midpoints of any two sides of a triangle is half as long as the third side of the triangle. Students first encountered this theorem in Exs. 12–15 on page 199.

Application

Exs. 14–19 Students should find this application interesting. Before beginning to work the exercises, they need to read the description and study the diagram. For each bounce of the ball, students should draw the point that was aimed at, the point where the ball actually went, and the reflection line. It would be a good idea to use a different color marker for each shot. Students will have the opportunity to design their own miniature golf hole in Ex. 22.

10. ALGEBRA Choose one of the diagrams for Exercises 7–9 on the previous page. Find the midpoints of the segments that join each vertex to its image. Show that each midpoint lies on the line $y = x$.

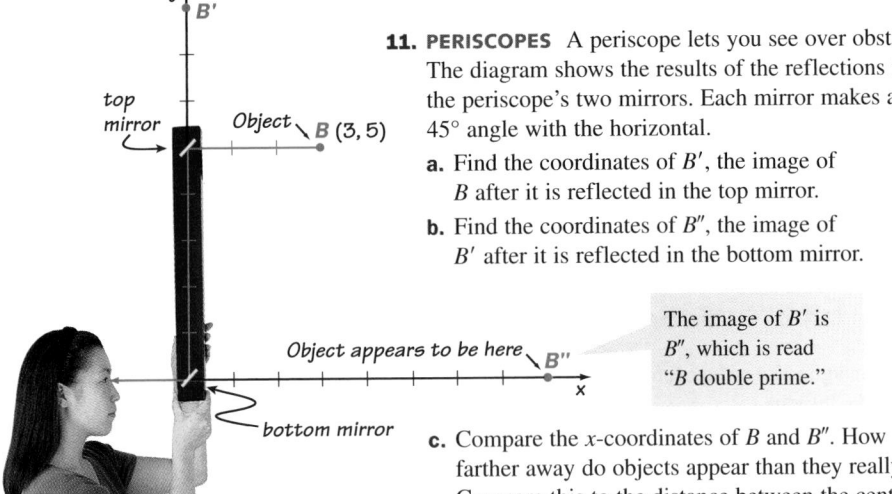

11. PERISCOPES A periscope lets you see over obstacles. The diagram shows the results of the reflections in the periscope's two mirrors. Each mirror makes a 45° angle with the horizontal.

   a. Find the coordinates of B', the image of B after it is reflected in the top mirror.

   b. Find the coordinates of B'', the image of B' after it is reflected in the bottom mirror.

> The image of B' is B'', which is read "B double prime."

   c. Compare the x-coordinates of B and B''. How much farther away do objects appear than they really are? Compare this to the distance between the centers of the mirrors.

12. Investigation Draw a polygon on a coordinate plane and label the coordinates of its vertices.

   a. Reflect the polygon over the line $y = -x$ and find the coordinates of the vertices of the image.

   b. Complete the rule that describes this transformation:

   reflection over the line $y = -x$
   $$(a, b) \to (\underline{\ ?\ }, \underline{\ ?\ })$$

13. OPTICS A person with an eye at E looks at points M and N on a mirror and sees the top and bottom of his face at T' and B', as shown.

   a. Use the information in the diagram and what you know about reflections and the midpoints of two sides of a triangle to show that $MN = \frac{1}{2}TB$. (*Hint:* See Exercises 12–15 on page 199.)

   b. What does your answer to part (a) suggest is a good size for a hand-held mirror?

   c. Open-ended Problem Draw and label a sketch that shows how you can see a reflection of your entire body in a mirror that is half your height.

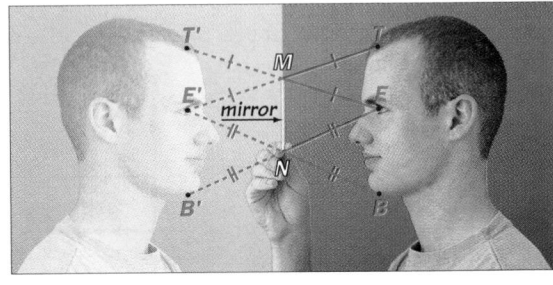

402 Chapter 8 *Using Transformations*

---

10. (7) $\overline{MM'}$: (2, 2); $\overline{NN'}$: $\left(\frac{5}{2}, \frac{5}{2}\right)$; Since $O' = O$, there is no segment; (8) $\overline{AA'}$: (1, 1); $\overline{BB'}$: (3, 3); $\overline{CC'}$: (2, 2); Since $O' = O$, there is no segment; (9) $\overline{DD'}$: (1, 1), $\overline{EE'}$: (3, 3), $\overline{GG'}$: (0, 0); Since $F' = F$, there is no segment. All of these points are on the line $y = x$ because the coordinates are equal.

11. a. (0, 8)   **b.** (8, 0)

   c. The x-coordinate of B'' is 5 more than the x-coordinate of B; 5 units; the distances are the same.

12. a. Check students' work.

   b. $-b$; $-a$

13. a. In $\triangle E'TB$, M is the midpoint of $\overline{E'T}$ and N is the midpoint of $\overline{E'B}$. Therefore, $MN = \frac{1}{2}TB$, since the segment whose endpoints are the midpoints of two sides of a triangle is parallel to the third side and half as long.

   b. slightly longer than half the length of an average face, perhaps 6 in. long

   c. Answers may vary. An example is given.

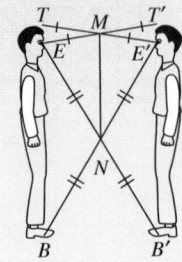

14. If a ball were aimed directly at the cup from any one of the given points, the ball would hit a bumpboard.

15. The angle of reflection and the angle of incidence are congruent, so $\angle QPA \cong \angle OPH$. But vertical angles QPA and OPJ are also congruent, so $\angle HPO \cong \angle JPO$. So if Alisa aims the ball at point J, the ball will hit point H, the reflection of J in the x-axis.

In miniature golf, when a ball with no spin hits a bumpboard, its *angle of reflection* is congruent to its *angle of incidence.* In the diagram below, adapted from a real course,

$$m\angle QPA = m\angle OPH.$$

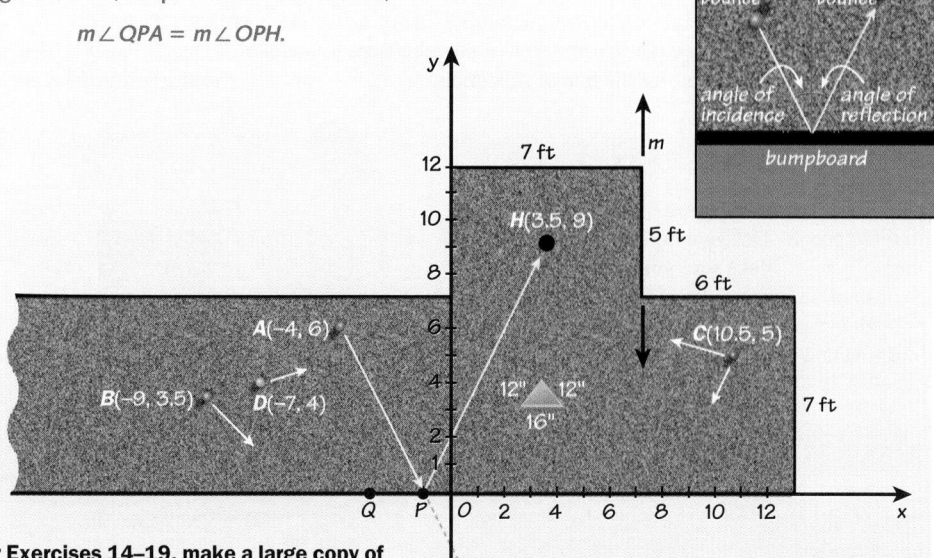

Exercise Notes

Cooperative Learning

Exs. 14–19 This set of exercises is appropriate for small group work. Divide the class into groups of four students each. The groups should discuss each exercise and then individual students can help record the answers. One student can begin by reading each exercise; another student can be responsible for the drawing; a third student can write the coordinates of the points; and the fourth student can be responsible for writing and reporting the group's findings.

Multicultural Note

Exs. 14–19 Miniature golf is usually played for recreation, but golfing on traditional courses can be a career. Professional golf players are of many ethnicities and nationalities. Lee Trevino, a Mexican American from Texas, taught himself to play golf as a young boy and went on to win both the U.S. and British Opens. Fijian Vijay Singh won the 1993 Rookie of the Year award for his impressive golf skills. Masashi Ozaki and Isao Aoki are two of the many Japanese golfers who have made a name for themselves in American golf. In 1996, Tiger Woods became the first young African American to play on the PGA (Professional Golf Association) Tour; he also won the Las Vegas Invitational, only his fifth professional event. Women's golf has comparable diversity: the legendary Mexican American golfer Nancy Lopez became the eleventh inductee to the Women's Golf Hall of Fame in 1987; South Korean Pearl Sinn won the 1988 U.S. Women's Amateur Championship; and Ayako Okamoto, Japan's most popular female golfer, has won seventeen professional events.

For Exercises 14–19, make a large copy of the diagram on graph paper.

14. Explain why a player at $A(-4, 6)$, $B(-9, 3.5)$, $C(10.5, 5)$, or $D(-7, 4)$ cannot aim directly at the cup at $H(3.5, 9)$.

15. **Writing** Alisa, the player at A, wants to reach H by bouncing a ball off the bumpboard along the *x*-axis. Explain why aiming at J (the image of H after reflection over the *x*-axis) is a good idea.

16. Basha, the player at B, aims at J. Draw the path of the ball. What are the coordinates of the point where the ball hits the bumpboard?

17. Carlos, at C, aims at J. Draw the path of the ball. What are the coordinates of the point where the ball hits the bumpboard?

18. Suppose Carlos decides instead to bounce the ball against the bumpboard along the *y*-axis. Reflect H across the *y*-axis. Label the image K and give its coordinates. Draw the path of the ball. What are the coordinates of the point where the ball hits the bumpboard?

19. **Challenge** Show how Dianne, the player at D, can bounce a ball off the bumpboard that lies along line m so that it reaches H.

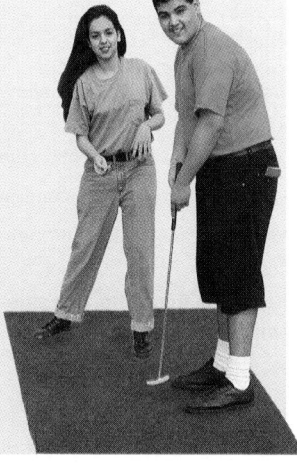

Challenge

Ex. 19 If students have difficulty with this problem, you may want to tell them that it involves more than one reflection.

8.2 Reflections with Coordinates   **403**

16. $(-5.5, 0)$

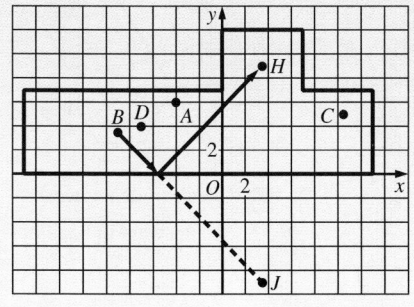

17. $(8, 0)$

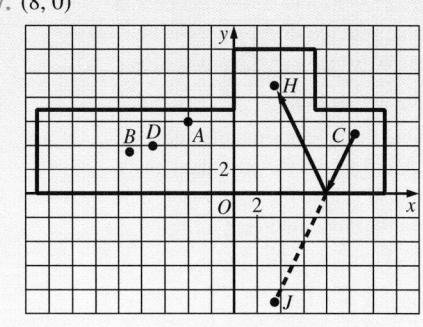

18, 19. See answers in back of book.

Exercise Notes

Geometric Thinking

Ex. 21 To solve this problem, students need to use what they have learned in this section in a new context. Encourage them to use a specific example to help determine the steps to take. They should make a drawing to help find the reflection line. To write the equation of the line, students need to recall the procedure for writing an equation of a line when given two points on it.

Assessment Note

Ex. 22 At the completion of the group work, you may wish to have each student write a summary of the process the group went through to find the path of a hole in one. This will allow you to assess individual understanding of the concepts involved.

Practice 55 for Section 8.2

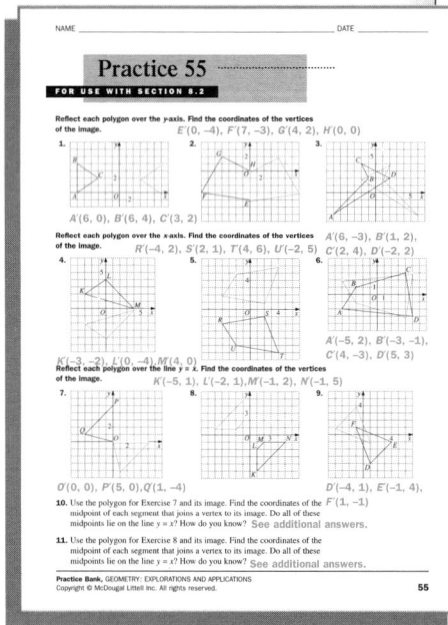

---

20. 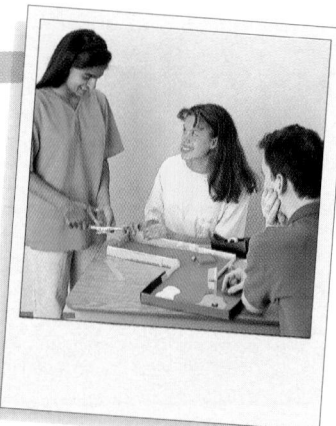 **Technology** Write a program for your graphing calculator that asks for the coordinates of a point and gives the coordinates of the images of that point after reflection in the *x*-axis, the *y*-axis, and the line $y = x$.

21. Writing Suppose you know the coordinates of the vertices of a polygon and its image after a reflection, but the line of reflection is not the *x*-axis, the *y*-axis, or the line $y = x$. Describe the steps you could take to find an equation for the line of reflection.

ONGOING ASSESSMENT

22. Cooperative Learning Work in a group of three people. Design a hole for a miniature golf course. Make sure your design has obstacles so that a player must bounce a ball against the bumpboards or other surfaces. Show how it is possible to get a hole in one.

SPIRAL REVIEW

Each diagram shows a polygon and its image after a reflection. Copy each diagram and draw the line of reflection. *(Section 8.1)*

23. 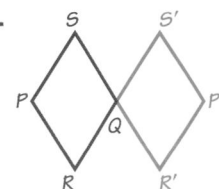　　**24.** 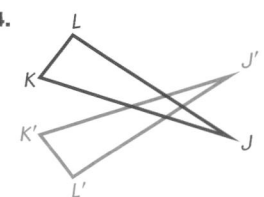　　**25.**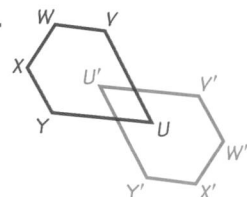

Draw a figure like the one shown. Then complete the construction. *(Section 5.7)*

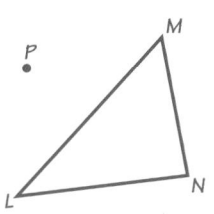

26. Construct the perpendicular bisector of $\overline{MN}$.

27. Construct an angle congruent to $\angle N$.

28. Construct the line through *P* that is parallel to $\overleftrightarrow{LM}$.

Sketch the front, top, and side view of a prism with the given base. Then sketch the prism. *(Section 2.6)*

29. triangular　　　　**30.** square

31. rectangular　　　　**32.** hexagonal

---

20. :Disp "x"
:Input A
:Disp "y"
:Input B
:–1 • A → C:–1 • B → D
:Disp "over x: x = ",A,"y = ",D
:Disp "over y: x = ",C,"y = ",B
:Disp "over y = x: x = ",B,"y = ",A
:Return

21. Answers may vary. An example is given. The line of reflection is the perpendicular bisector of the line joining any vertex to its image. I would choose a vertex. I would find the slope, *m*, of the line containing the vertex and its image. The slope of the perpendicular bisector is $-\frac{1}{m}$. I would use the Midpoint Formula to find the midpoint of the segment joining the vertex and its image. Finally, I would substitute the coordinates of the midpoint in the slope-intercept form $y = -\frac{1}{m}x + b$ and solve for *b* to find the equation of the line of reflection.

22. Check students' work.

23.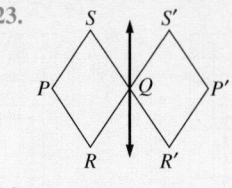

24.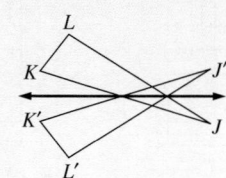

25.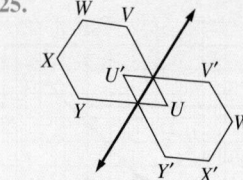

26–28. Check students' work.

29–32. See answers in back of book.

8.3 Translations

When you stand between two parallel mirrors, you see many reflections of yourself. Half of them face you and half face the other way. This is a way to see the back of your head! In the Exploration, you will see what happens when you transform a figure by reflecting it twice across parallel lines.

Learn how to...
• find the coordinates of the image of a figure after a translation
• describe a translation based on an original figure and its image

So you can...
• describe patterns that involve translation, such as architectural decoration

EXPLORATION
COOPERATIVE LEARNING

Comparing Translated Polygons

Work in a group of four students.
You will need:
• geometry software or patty paper
• a ruler

1 Draw points *P* and *Q* on a horizontal line and draw vertical lines through them. Label the lines *m* and *n*. Draw △*ABC* to the left of line *m*.

2 Reflect △*ABC* over line *m* and label the image △*A′B′C′*. Reflect △*A′B′C′* over line *n* and label the image △*A″B″C″*.

> The image of *A′* is *A″*, which is read "*A* double prime."

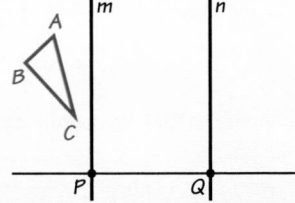 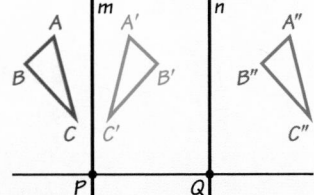

3 What are the orientations of △*ABC* and △*A″B″C″*?

4 Measure *AA″*, *BB″*, *CC″*, and *PQ*. What do you notice?

5 Change the distance between lines *m* and *n* and see what happens to the distances *AA″*, *BB″*, and *CC″*. What do you notice?

Exploration Note

Purpose
The purpose of this Exploration is to have students discover that the result of two reflections across a pair of parallel lines is the same as a translation.

Materials/Preparation
Each group of students needs geometry software or patty paper and a ruler.

Procedure
Students draw a pair of vertical parallel lines that intersect a horizontal line. A triangle is drawn to the left of the left-most vertical line. This triangle is then reflected across the vertical line, and its image is reflected across the other vertical line. Students then examine the distances between the vertices of the original triangle and its image after both reflections.

Closure
Discuss the results of Steps 3–5 so students understand that the two reflections create a translation and that the shift or distance of the translation is twice the distance between the parallel lines.

Explorations Lab Manual
See the Manual for more commentary on this Exploration.

For answers to the Exploration, see following page.

Plan⟺Support

Objectives
• Find the coordinates of the image of a figure after a translation.
• Describe a translation based on an original figure and its image.
• Describe patterns that involve translation.

Recommended Pacing
❖ **Core and Extended Courses**
Section 8.3: 2 days
❖ **Block Schedule**
Section 8.3: 2 half-blocks (with Sections 8.2 and 8.4)

Resource Materials
Lesson Support
Lesson Plan 8.3
Warm-Up Transparency 8.3
Practice Bank: Practice 56
Study Guide: Section 8.3
Explorations Lab Manual: Diagram Masters 1, 2
Challenge Problems: Set 56
Assessment Book: Test 35
Technology
Geometry Software
McDougal Littell Mathpack Geometry Inventor Activity Book: Activities 22, 26, 27
Internet: http://www.hmco.com

Warm-Up Exercises

The vertices of △ *DEG* are *D*(−3, 1), *E*(4, −6), and *G*(−1, −1). Give the vertices of the image △ *D′E′G′*.

1. reflection over the *y*-axis
D′(3, 1), *E′*(−4, −6), *G′*(1, −1)

2. reflection over the line *y* = *x*
D′(1, −3), *E′*(−6, 4), *G′*(−1, −1)

3. reflection over the *x*-axis
D′(−3, −1), *E′*(4, 6), *G′*(−1, 1)

4. Point *H*(4, −5) is reflected and its image point is *H′*(4, 5). Name the line of reflection.
the *x*-axis

5. Point *T*(5, 3) is reflected and its image point is *T′*(−3, −5). Name the line of reflection. *y* = −*x*

Section Notes

Geometric Thinking
Students should see intuitively that a translation preserves congruence and orientation. You may wish to point out also that since a translation can be thought of as a combination of two reflections, each of which preserves congruence, that a translation must also preserve congruence.

Communication: Discussion
You may wish to give students some examples of translations in coordinate notation and have them describe the translation in words. Discuss the fact that if h is positive, the shift is to the right; if h is negative, the shift is left. Similarly, if k is positive, the shift is up; if k is negative, the shift is down.

Second-Language Learners
To help students learning English keep track of the differences between the kinds of transformations they will learn about in this chapter, encourage them to make a journal entry for each type (reflection, translation, glide reflection, and dilation). Their entry can include a bilingual definition of the term, several examples of the type of transformation, and a description of the ways in which the transformation is different from the others.

Additional Example 1

Sketch the image of $\triangle LOP$ after the translation $(a, b) \rightarrow (a + 3, b - 2)$.

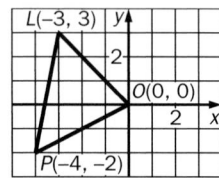

Substitute and simplify to find the coordinates of the vertices of the image.

$L(-3, 3) \rightarrow (-3 + 3, 3 - 2)$
$\quad = L'(0, 1)$
$O(0, 0) \rightarrow (0 + 3, 0 - 2)$
$\quad = O'(3, -2)$
$P(-4, -2) \rightarrow (-4 + 3, -2 - 2)$
$\quad = P'(-1, -4)$

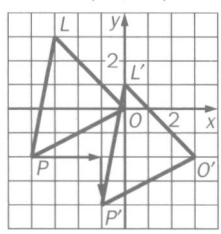

If you repeat the Exploration using nonvertical lines, you can compare the coordinates of the vertices of $\triangle ABC$ and $\triangle A''B''C''$ and see that each vertex is shifted the same distance in the same direction.

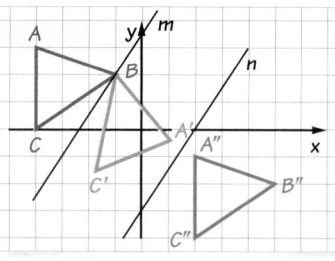

$A(-4, 3) \rightarrow A''(2, -1)$
$B(-1, 2) \rightarrow B''(5, -2)$
$C(-4, 0) \rightarrow C''(2, -4)$

In the diagram above, for example, each vertex is shifted 6 units to the right and 4 units down. This kind of transformation is called a *translation*.

Properties of Translations

A **translation** is a transformation that can be described in coordinate notation this way:

$$(a, b) \rightarrow (a + h, b + k)$$

Every point shifts h units horizontally and k units vertically.

A translation preserves congruence.

A translation preserves orientation.

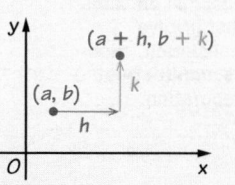

EXAMPLE 1

Sketch the image of $\triangle XYZ$ after the translation $(a, b) \rightarrow (a - 3, b + 4)$.

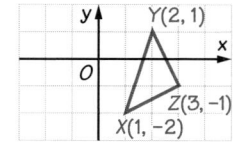

SOLUTION

Substitute and simplify to find the coordinates of the vertices of the image.

$X(1, -2) \rightarrow (1 - 3, -2 + 4) = X'(-2, 2)$
$Y(2, 1) \rightarrow \quad (2 - 3, 1 + 4) = Y'(-1, 5)$
$Z(3, -1) \rightarrow (3 - 3, -1 + 4) = Z'(0, 3)$

Plot the vertices of the image triangle and connect them.

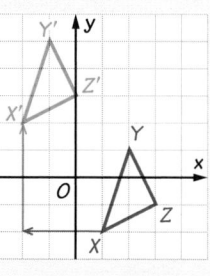

406   Chapter 8 *Using Transformations*

ANSWERS   Section 8.3

Exploration

1–2. Check students' work.

3. counterclockwise

4. Each distance is twice the distance between lines m and n.

5. Each distance is still twice the distance between lines m and n.

Think and Communicate

1. Every point is shifted the same distance in the same direction.

2. Yes; $(a, b) \rightarrow (a - 3.5, b + 7)$

3. Yes. Answers may vary. Examples are given.
$(a, b) \rightarrow (a + 6, b)$ (Each point is shifted to the right 6 units.);
$(a, b) \rightarrow (a, b - 13)$ (Each point is shifted down 13 units.)

Checking Key Concepts

1.

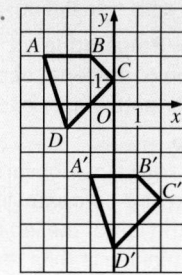

In Chapter 1 you saw that a design has *translational symmetry* if you can translate it so that the design and its image coincide.

EXAMPLE 2

Describe how to translate the pattern so that it matches itself.

SOLUTION

Step 1 Find the coordinates of any point in the design and a corresponding point in a nearby image.

$$P(3.5, 7) \qquad P'(7, 0)$$

Step 2 Subtract the coordinates to find the horizontal change h and vertical change k.

$$h = 7 - 3.5 \qquad k = 0 - 7$$
$$= 3.5 \qquad\qquad = -7$$

Step 3 Use the values of h and k to describe the translation using coordinate notation:

$$(a, b) \rightarrow (a + 3.5, b - 7)$$

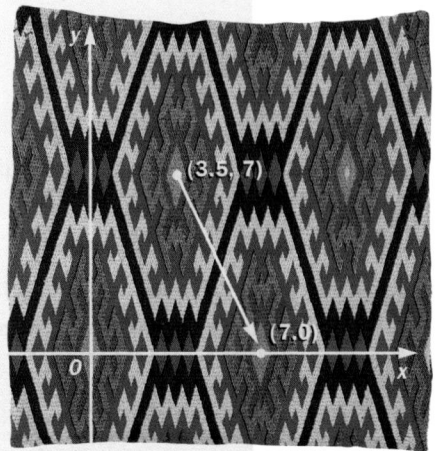

THINK AND COMMUNICATE

1. In Example 2, why is it sufficient to pick a single point and its image?

2. If you translate the design the same distance as in the Solution above but in the opposite direction, does the design exactly match itself? Describe that translation using coordinate notation.

3. Are there other translations that move the design onto itself? Give one or two examples and a general description of these translations.

☑ CHECKING KEY CONCEPTS

Copy the diagram and draw the image of *ABCD* after each translation.

1. $(a, b) \rightarrow (a + 2, b - 5)$
2. $(a, b) \rightarrow (a - 3, b)$
3. $(a, b) \rightarrow (a, b + 4)$
4. $(a, b) \rightarrow (a - 2, b + 5)$

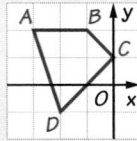

Describe each translation using coordinate notation.

5. Every point moves to the left 3 units and up 4 units.
6. Every point moves to the right 5 units.

8.3 Translations   **407**

About Example 2

Geometric Thinking

The Solution to this Example is not unique. There are an infinite number of translations that would move the pattern on top of itself. For example, a different translation would be found if students used the same point A, but chose as A′ the corresponding point on the image horizontally to the right of the original figure. Students explore these ideas in *Think and Communicate* questions 2 and 3.

Additional Example 2

Describe how to translate the pattern so that it matches itself.

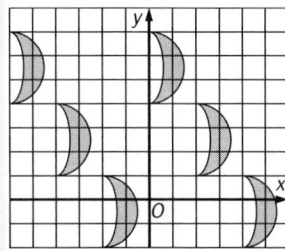

Step 1 Find the coordinate of any point in the design and a corresponding point in a nearby image, for example, $B(2, 4)$ and $B'(-2, 1)$.

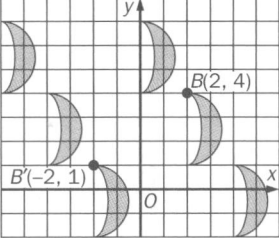

Step 2 Subtract the coordinates to find the horizontal change h and vertical change k.

$h = -2 - 2 = -4$
$k = 1 - 4 = -3$

Step 3 Use the values of h and k to describe the translation using coordinate notation:

$(a, b) \rightarrow (a - 4, b - 3)$

Closure Question

How would you describe a translation of a point in coordinate notation? What properties of a figure does a translation preserve?

A translation is a shift that can be described in coordinate notation as $(a, b) \rightarrow (a + h, b + k)$. Every point moves h units horizontally and k units vertically. A translation preserves congruence and orientation.

2.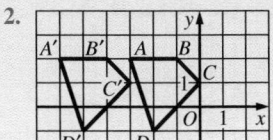

3.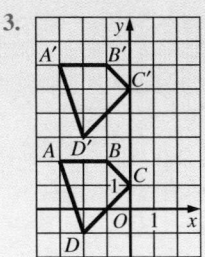

4.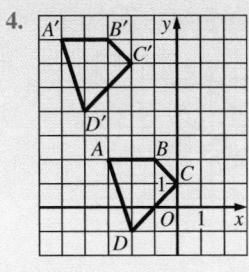

5. $(a, b) \rightarrow (a - 3, b + 4)$

6. $(a, b) \rightarrow (a + 5, b)$

Extra Practice
exercises on
page 676

Suggested Assignment

❖ **Core Course**
  Day 1 Exs. 1–11, 13–15
  Day 2 Exs. 17–21, 25–29, AYP
❖ **Extended Course**
  Day 1 Exs. 1–16
  Day 2 Exs. 17–29, AYP
❖ **Block Schedule**
  Day 48 Exs. 1–11, 13–15
  Day 49 Exs. 17–21, 25–29, AYP

Exercise Notes

Student Progress
Exs. 1–8 Students should have little or no difficulty with these exercises. Answers to Exs. 1–4 can be checked visually to see that the image is congruent with the original figure. For Exs. 5–8, you may wish to have students use the coordinates of an actual point and graph its image for each translation. In so doing, students can verify their use of coordinate notation.

Common Error
Exs. 1–8 Some students may translate a figure the correct distance but in the wrong direction. Ask these students to think of a coodinate plane. Positive numbers (+) are to the right or up; negative numbers (–) are to the left or down. The movement under a translation corresponds to this.

Geometric Thinking
Exs. 9–11 The translations for these polygons can be found by inspecting the graphs and counting the vertical and horizontal shifts. Students should also know how to find the translation using the method of Example 2. This method is necessary when the coordinates are not integers, or when they are variables.

Application
Ex. 12 This exercise illustrates how the concept of translation is used with a computer to generate background patterns.

8.3 Exercises and Applications

Copy *EFGH* and draw its image after each translation.

1. $(a, b) \rightarrow (a - 1, b + 3)$
2. $(a, b) \rightarrow (a, b + 7)$
3. $(a, b) \rightarrow (a + 5, b - 2)$
4. $(a, b) \rightarrow (a - 4, b)$

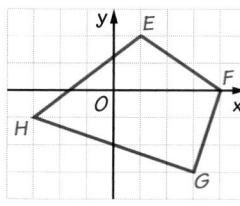

Describe each translation using coordinate notation.

5. Every point moves to the left 7 units and up 7 units.
6. Every point moves to the left 6 units.
7. Every point moves up 4 units.
8. Every point moves to the right 2 units and down 1 unit.

The image of each polygon after a translation is shown in red. Describe each translation using words or coordinate notation.

9.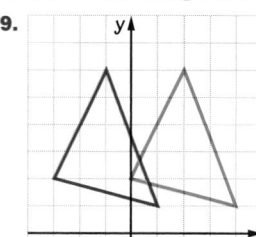

10.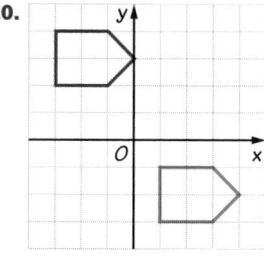

11.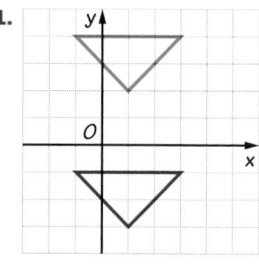

12. **COMPUTERS** On some computers you can design a square 8 pixels wide by 8 pixels high that is translated across the screen to create a background pattern. (A *pixel* is the smallest dot on a computer screen.) Square A is at the upper left corner of the computer screen. The translation

$$(a, b) \rightarrow (a + 8, b)$$

moves square A onto square B.

  a. Describe the translations that move square A onto squares C, D, E, and F.

  b. Open-ended Problem Create your own 8 pixel by 8 pixel pattern on graph paper. Translate it several times to show the background pattern it generates.

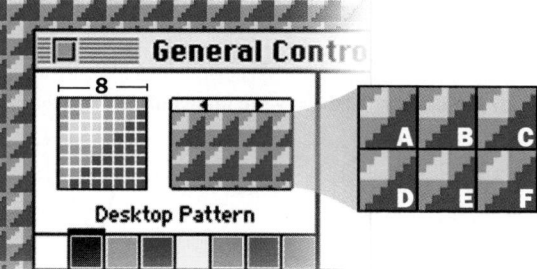

13. **SAT/ACT Preview** The distance between a point and its image after the translation $(a, b) \rightarrow (a + 7, b - 5)$ is:

  A. 7   **B.** 5   **C.** 12   **D.** $\sqrt{74}$ **E.** $\sqrt{24}$

408   Chapter 8 *Using Transformations*

Exercises and Applications

1.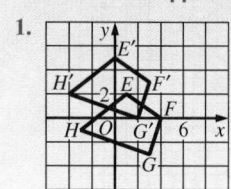

2.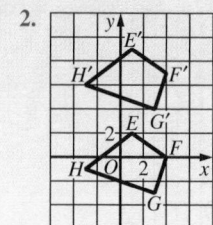

3.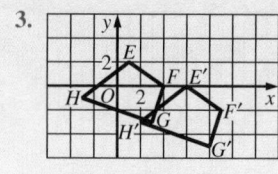

4.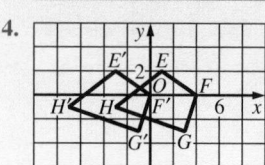

Open-ended Problems Each pattern has translational symmetry. Describe how to translate the pattern so that it matches itself.

14.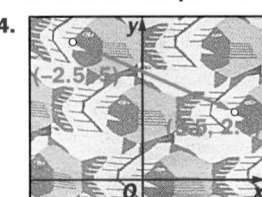

15.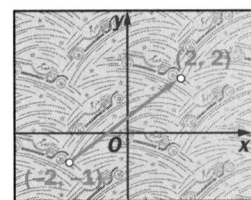

16.  **Technology** Use geometry software to repeat the steps of the Exploration on page 405: Draw parallel lines m and n and $\triangle ABC$. Reflect the triangle over line m and then reflect the result over line n.

 a. Measure AA'', BB'', and CC''.

 b. Change the shape of $\triangle ABC$ by moving its vertices. What happens to AA'', BB'', and CC''?

 c. **Writing** Change the position of $\triangle ABC$ by moving it around on the coordinate plane. Explain what happens to $\triangle A''B''C''$. Does it matter which side of line m or n $\triangle ABC$ is on?

INTERVIEW Terri Johnson

Look back at the article on pages 388–390.

This cross-section view of Ericsson Stadium shows some of the staircases that sports fans climb to reach the tiers of seats.

 17. Explain how the top flight of stairs can be translated to become the next lower flight of stairs.

 18. Can the bottom flight of stairs be translated to become the next higher flight of stairs? Explain.

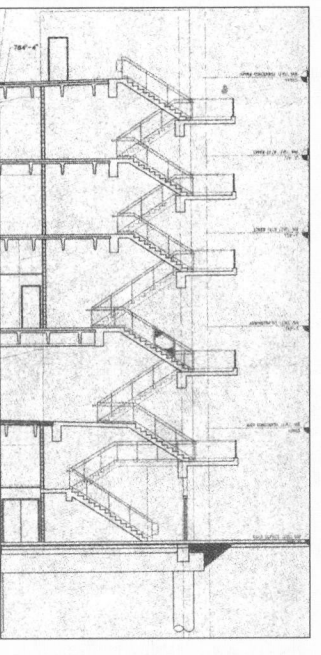

Apply⇔Assess

Exercise Notes

Teaching Tip
Exs. 14, 15 Students should be encouraged to follow the three-step method outlined for problems such as these in Example 2 on page 407.

 Using Technology
Ex. 16 In this exercise, students repeat the steps of the Exploration on page 405 by using geometry software. The software, however, enables them to go beyond what was done previously to examine the results for many different triangles, including those located anywhere in the plane.

Cooperative Learning
Ex. 16 This exercise can be done with small groups of students sharing one computer. This will allow students to test ideas and theories on each other. Have students take turns drawing the triangles and finding the measurements.

Interview Note: Application
Exs. 17, 18 In these exercises, students see how the concept of translation can be used in architectural design.

5. $(a, b) \rightarrow (a - 7, b + 7)$

6. $(a, b) \rightarrow (a - 6, b)$

7. $(a, b) \rightarrow (a, b + 4)$

8. $(a, b) \rightarrow (a + 2, b - 1)$

9. Every point moves to the right 3 units; $(a, b) \rightarrow (a + 3, b)$.

10. Every point moves to the right 4 units and down 5 units; $(a, b) \rightarrow (a + 4, b - 5)$.

11. Every point moves up 5 units; $(a, b) \rightarrow (a, b + 5)$.

12. a. $(a, b) \rightarrow (a + 16, b)$; $(a, b) \rightarrow (a, b - 8)$; $(a, b) \rightarrow (a + 8, b - 8)$; $(a, b) \rightarrow (a + 16, b - 8)$

 b. Check students' work.

13. D

14–15. Answers may vary. Examples are given.

14. Translate every point to the right 6 units and down 2.5 units.

15. Translate every point to the right 4 units and up 3 units.

16. a. Each distance is twice the distance between lines m and n.

 b. Each distance is still twice the distance between lines m and n.

 c. $\triangle A''B''C''$ moves so that the distance between a vertex and its image is twice the distance between lines m and n; No.

17. The top flight can be translated down by a distance equal to the distance between the floors.

18. No; the bottom flight of stairs does not have the same slope as the next higher flight.

Exercise Notes

Historical Connection
Ex. 19–21 One of the disciples of Louis Sullivan was the famous 20th century architect Frank Lloyd Wright.

Interdisciplinary Problems
Exs. 19–21, 23 In Exs. 19–21, students see how the design of building ornamentation involves the use of reflection. In Ex. 23, students relate the idea of a translation to describe certain musical forms.

Communication: Drawing
Ex. 22 You may wish to expand this exercise by asking students to make a sketch of the pattern from the clothing on graph paper. They could then find a translation that would copy the pattern onto itself. Students could also draw in any reflection lines they see. Ask several students to display their work and explain the transformations to the class.

Communication: Listening
Ex. 23 If possible, bring a tape of a fugue or canon to class and have students listen to it. After hearing it a few times, students can discuss the translations in the music.

Connection   **ARCHITECTURE**

Louis H. Sullivan (1856–1924) was an influential architect who designed some of the earliest skyscrapers in Chicago, Illinois. Sullivan's buildings often had features inspired by plant forms, which he called "organic ornamentation."

The photograph shows terra cotta ornamentation on the Van Allen & Son store in Clinton, Iowa. Reflection lines have been added.

19. Name two reflection lines that can be used to translate Block A onto Block C.

20. Name two reflection lines that can be used to translate Block A onto Block B.

21. Open-ended Problem Show that there is more than one correct answer to Exercises 19 and 20.

22. Open-ended Problem Look through some clothing catalogs or newspapers to find three examples of patterns that have translational symmetry. Cut them out and attach them to a piece of paper to show in class. Do any of the patterns also have reflection symmetry?

23. Research A *round*, such as "Row, Row, Row Your Boat," is a song that includes a melody that repeats later in the piece. In classical music, this idea is also used in *canons* and *fugues*. Find some sheet music that displays these forms and explain how the concept of "translation" can be used to describe them.

24. Challenge In the Exploration, you saw that a double reflection over two *parallel* lines is a translation. What kind of transformation happens if you use a double reflection over two *intersecting* lines? Make a sketch that illustrates your answer.

410   Chapter 8 *Using Transformations*

19–20. Answers may vary. Examples are given.

19. lines *q* and *s*

20. lines *p* and *q*

21. In addition to the examples given above, Block A can be translated onto Block C by using lines *p* and *r* or lines *r* and *s* as lines of reflection. Block A can be translated onto Block C by using lines *r* and *s* or lines *q* and *r* as lines of reflection.

22. Answers may vary.

23. Answers may vary. When you look at the sheet music for a fugue or canon, you will see parts of the melody that are repeated and can be thought of as translations along the staff. Other parts may be transposed to a higher or lower pitch and can be thought of as translations up or down the staff as well as along it.

24. rotation; Sketches may vary. In the example shown, $\triangle ABC$ is reflected over the *y*-axis and then the *x*-axis. $\triangle A''B''C''$ can also be produced by rotating $\triangle ABC$ 180° around the origin.

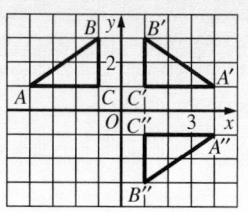

25. a, b. Answers may vary. An example is given. $\triangle ABC$ has vertices $A(0, 1)$, $B(2, 4)$, and $C(5, 5)$.

a. $A'(3, 6)$; $B'(5, 9)$; $C'(8, 10)$

b. $A''(-1, 8)$; $B''(1, 11)$; $C''(4, 12)$

c. $(a, b) \rightarrow (a - 1, b + 7)$

d. $A'(-4, 3)$; $B'(-2, 6)$; $C'(1, 7)$; $A''(-1, 8)$; $B''(1, 11)$; $C''(4, 12)$; $(a, b) \rightarrow (a - 1, b + 7)$; The order does not matter because of the commutative and associative properties of addition. For example, for any values

25. a. On graph paper, draw a triangle and label it △*ABC*. Translate △*ABC* using Translation 1. Find the coordinates of the vertices of △*A'B'C'*.

Translation 1: $(a, b) \rightarrow (a + 3, b + 5)$

b. Translate △*A'B'C'* using Translation 2. Find the coordinates of the vertices of △*A''B''C''*.

Translation 2: $(a, b) \rightarrow (a - 4, b + 2)$

c. Use coordinate notation to describe a single translation that moves △*ABC* directly to △*A''B''C''*.

d. Writing Repeat parts (a)–(c) using Translation 2 first, followed by Translation 1. Explain why the order in which you combine two translations does not affect the results.

SPIRAL REVIEW

Tell whether the triangle with sides of the given lengths is *right*, *obtuse*, or *acute*. *(Section 3.7)*

26. 3, 5, 6   **27.** 2, 7, 7   **28.** 12, 16, 20

29. Draw a pentagon and use a protractor to measure each angle, to the nearest degree. *(Section 1.6)*

ASSESS YOUR PROGRESS

VOCABULARY

line of reflection (p. 391)   **image** (p. 391)
reflection (p. 391)   **translation** (p. 406)

Use the diagram for Exercises 1–4. *K'L'M'N'* is the image of *KLMN*.
(Section 8.1)

1. What is the image of $\overline{LM}$? of $\overline{KL}$?

2. Name a triangle congruent to △*KMN*.

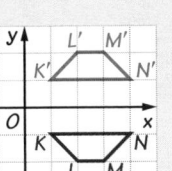

Copy the polygon *KLMN*. Reflect it over each line.
Give the coordinates of the vertices of the image.
(Section 8.2)

3. the line $y = x$ **4.** the *y*-axis

Describe each translation using coordinate notation. *(Section 8.3)*

5. Every point moves to the right 5 units and down 7 units.

6. Every point moves to the left 3 units and up 4 units.

7. Journal Describe how the number of reflections affects the orientation of a figure that is reflected more than once. Make a sketch that illustrates your conclusions.

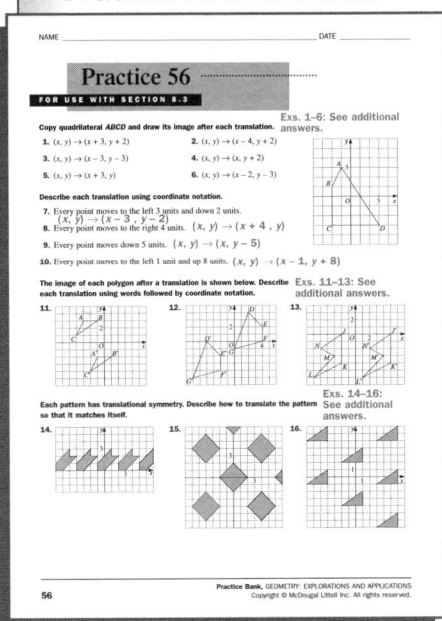

---

of *h* and *k*, $(a + h) + k = a + (h + k) = a + (k + h) = (a + k) + h$.

26. obtuse   **27.** acute

28. right

29. Answers may vary. An example is given.

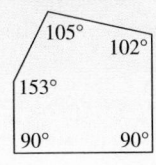

105°
102°
153°
90°   90°

Assess Your Progress

1. *L'M'*; *K'L'*

2. △*K'M'N'*

3.

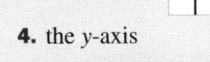

4.

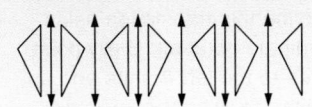

5. $(a, b) \rightarrow (a + 5, b - 7)$

6. $(a, b) \rightarrow (a - 3, b + 4)$

7. Each reflection reverses the orientation of a figure. Then an even number of reflections results in a figure with the same orientation as the original figure, while an odd number of reflections results in a figure with orientation opposite that of the original figure. This can be observed in the sketch.

Objectives

- Rotate a figure around a center of rotation.
- Find the coordinates of the vertices of a polygon that has been rotated 90°, 180°, or 270° around the origin.
- Describe rotations of real-world objects.

Recommended Pacing

❖ **Core and Extended Courses**
Section 8.4: 2 days

❖ **Block Schedule**
Section 8.4: 2 half-blocks
(with Sections 8.3 and 8.5)

Resource Materials

Lesson Support

Lesson Plan 8.4

Warm-Up Transparency 8.4

Overhead Visuals:
Folder 10: Slopes and Rotations

Practice Bank: Practice 57

Study Guide: Section 8.4

Explorations Lab Manual:
Diagram Master 2

Challenge Problems: Set 57

Technology

Geometry Software

McDougal Littell Mathpack
Geometry Inventor Activity Book:
Activities 20, 21, 26, 27, 30

Internet:
http://www.hmco.com

8.4 Applying Rotations

When sails are attached to a windmill, the tilted frames catch the wind and turn around a fixed point.

Learn how to...

- rotate a figure around a center of rotation
- find the coordinates of the vertices of a polygon that has been rotated 90°, 180°, or 270° around the origin

So you can...

- describe rotations, such as those found in patterns on quilts

A transformation in which every point moves along a circular path around a fixed point is called a **rotation**. The fixed point is called the **center of rotation**. Segments drawn from a point and its image to the center of rotation always form the same angle, the **angle of rotation**.

In this book, rotations are measured counterclockwise.

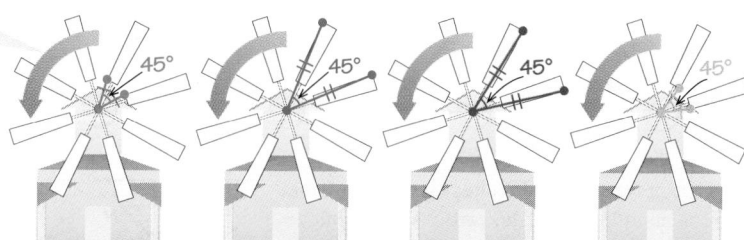

BY THE WAY

In the Netherlands, windmill owners positioned their sails to convey information. Stopping them just before the uppermost sail was vertical meant a celebration of good news.

Properties of Rotations

A rotation is a transformation that preserves congruence.

A rotation preserves orientation.

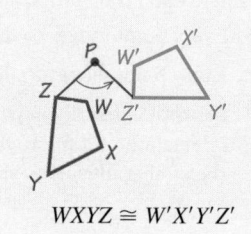

$WXYZ \cong W'X'Y'Z'$

Warm-Up Exercises

Write coordinates for the image of $A(4, -2)$ for each transformation.

1. reflection over the *y*-axis
$A'(-4, 2)$

2. the translation $(a, b) \rightarrow (a - 2, b + 1)$ $A'(2, -1)$

3. the reflection $(a, b) \rightarrow (b, a)$
$A'(-2, 4)$

4. Name the line of reflection in Ex. 3 above. $y = x$

5. Describe this translation using coordinate notation: Every point moves to the left 4 units and up 3 units. $(a, b) \rightarrow (a - 4, b + 3)$

You can use a protractor and either a ruler or a compass to help you find the images of points and figures after a rotation.

EXAMPLE 1

Draw the image of the given polygon after a 110° rotation with center P.

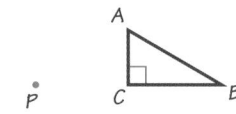

SOLUTION

Step 1 Draw a segment from vertex *A* to the center of rotation point *P*.

Step 2 Measure a 110° angle counterclockwise. Draw a ray.

Step 3 Use a ruler or compass to locate *A′* along the ray so that *PA′* = *PA*.

Step 4 Repeat Steps 1–3 for each vertex. Connect the vertices to form the image polygon.

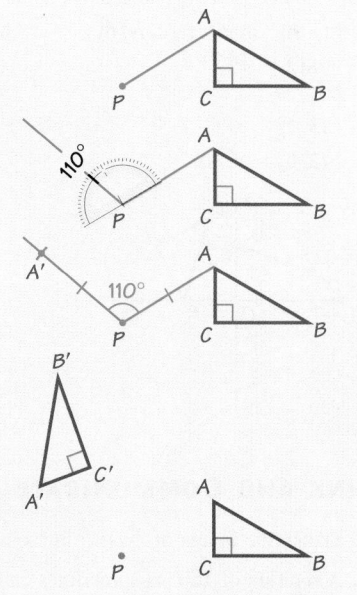

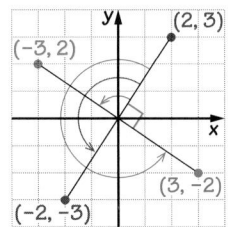

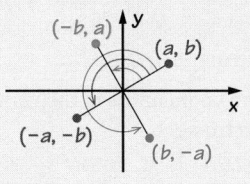

If you rotate a point 90°, 180°, and 270° around the origin on a coordinate plane, you will notice some patterns in the coordinates of the images.

In the diagram at the right, the point (2, 3) has been rotated three times.

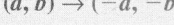

| Rotating a Point 90°, 180°, or 270° Around the Origin |

When a point (a, b) is rotated counterclockwise around the origin:

90° rotation: $(a, b) \rightarrow (-b, a)$

180° rotation: $(a, b) \rightarrow (-a, -b)$

270° rotation: $(a, b) \rightarrow (b, -a)$

8.4 Applying Rotations   **413**

Teach ⇔ Interact

Additional Example 1

Draw the image of the given polygon after a 85° rotation with center *R*.

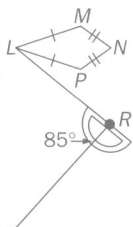

Step 1 Draw a segment from vertex *L* to the center of rotation.
Step 2 Measure an 85° angle counterclockwise. Draw a ray.

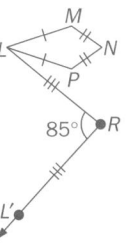

Step 3 Use a ruler or compass to locate *L′* along the ray so that *RL′* = *RL*.

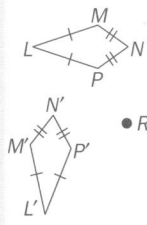

Step 4 Repeat Steps 1–3 for each vertex. Connect the vertices to form the image polygon.

Section Note

Alternate Approach

You may wish to have students discover the relationships between the coordinates of points and their images after rotations of 90°, 180°, and 270°. This can be accomplished by having students work in small groups. Students can begin by drawing a triangle in the first quadrant and then, using the technique of Example 1, they can draw its image after each rotation with the center at the origin. Each rotation can be done on a separate sheet of paper. Have students find the coordinates of the vertices of the image triangle and compare them with those of the the original triangle.

413

Additional Example 2

Sketch the image of the given polygon after each rotation around the origin.

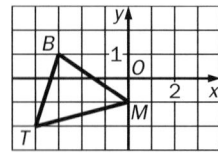

a. 90°

Substitute to find the coordinates of the vertices of each image. In a 90° rotation around the origin, $(a, b) \rightarrow (-b, a)$.
$B(-3, 1) \rightarrow B'(-1, -3)$
$M(0, -1) \rightarrow M'(1, 0)$
$T(-4, -2) \rightarrow T'(2, -4)$

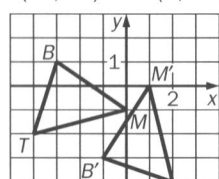

b. 270°

In a 270° rotation around the origin, $(a, b) \rightarrow (b, -a)$.
$B(-3, 1) \rightarrow B'(1, 3)$
$M(0, -1) \rightarrow M'(-1, 0)$
$T(-4, -2) \rightarrow T'(-2, 4)$

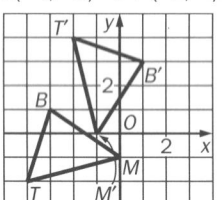

Checking Key Concepts

Geometric Thinking

These questions can help you assess students' skills at determining a rotation visually and algebraically. You may need to remind students that the rotations are counterclockwise around the origin.

Closure Question

How is a rotation defined and what properties does it have? A rotation is a transformation in which every point moves along a circular path around a fixed point, called the center of rotation. Segments drawn from a point and its image to the center of rotation always form the same angle, called the angle of rotation. A rotation preserves congruence and orientation.

EXAMPLE 2

Sketch the image of the given polygon after each rotation around the origin.

a. 90°   **b.** 180°

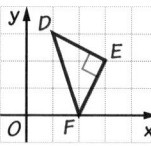

SOLUTION

Substitute to find the coordinates of the vertices of each image.

a. In a 90° rotation around the origin, $(a, b) \rightarrow (-b, a)$.
$D(1, 3) \rightarrow D'(-3, 1)$
$E(3, 2) \rightarrow E'(-2, 3)$
$F(2, 0) \rightarrow F'(0, 2)$

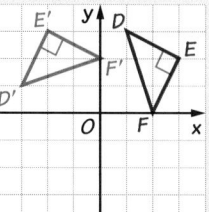

b. In a 180° rotation around the origin, $(a, b) \rightarrow (-a, -b)$.
$D(1, 3) \rightarrow D'(-1, -3)$
$E(3, 2) \rightarrow E'(-3, -2)$
$F(2, 0) \rightarrow F'(-2, 0)$

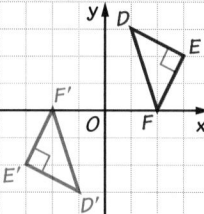

THINK AND COMMUNICATE

1. Sketch the image of △DEF in Example 2 after a 270° rotation.

2. A rotation of 180° is sometimes called a **half-turn**. What could a 90° rotation be called?

3. The center of a 180° rotation is the midpoint of every segment that connects a point and its image. Explain why this is true.

☑ CHECKING KEY CONCEPTS

Name the image of each triangle after the rotation around the origin.

1. Rotate △ABC 90°.

2. Rotate △DEF 180°.

3. Rotate △JKL 270°.

4. Rotate △GHI 180°.

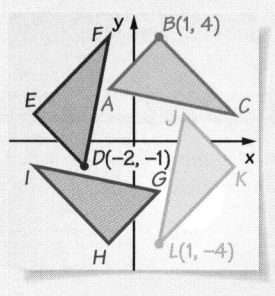

Give the coordinates of the vertices of each triangle.

5. △DEF   6. △GHI   7. △JKL

ANSWERS   Section 8.4

Think and Communicate

1.

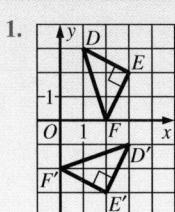

2. a quarter-turn

3. Segments drawn from a point and from its image to the center of rotation always form the angle of rotation. If the angle of rotation is 180°, a point and its image are the endpoints of a segment and the center of rotation is its midpoint.

Checking Key Concepts

1. △DEF   2. △JKL

3. △GHI   4. △ABC

5. $D(-2, -1)$; $E(-4, 1)$; $F(-1, 4)$

6. $G(1, -2)$; $H(-1, -4)$; $I(-4, -1)$

7. $J(2, 1)$; $K(4, -1)$; $L(1, -4)$

Exercises and Applications

1–4. Estimates may vary.

1. about 90°

2. between 0° and about 135°

Exercises and Applications

Extra Practice
exercises on
page 676

Estimate the angle of rotation involved in each situation.

1. turning a doorknob

2. bending your arm at the elbow

3. looking over your shoulder

4. opening a jar of peanut butter

Copy each diagram. Draw the image of the polygon after a rotation with the given measure and center P.

5. rotate 45°

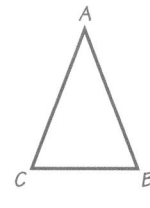

6. rotate 120°

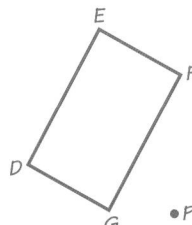

7. rotate 70°

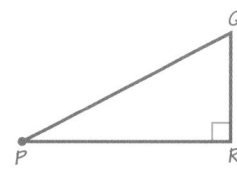

Copy each diagram and sketch the image of each polygon after a 90° rotation around the origin.

8.

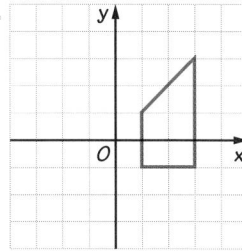

9.

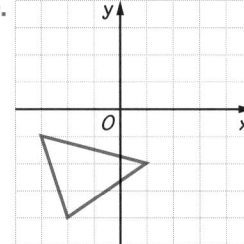

10.

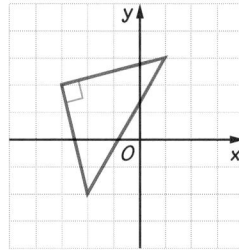

A figure has *rotational symmetry* if it looks the same after a rotation that is less than 360°. Does each object have rotational symmetry? If so, name each angle of rotation that results in an image that matches the original figure.

11.

12.

13.

14. **Open-ended Problem** Find an object in your home or at school that has rotational symmetry. Make a sketch and describe the symmetry.

8.4 Applying Rotations **415**

Suggested Assignment

❖ **Core Course**
  Day 1 Exs. 1–14, 16–18
  Day 2 Exs. 19–32, 34–43

❖ **Extended Course**
  Day 1 Exs. 1–18
  Day 2 Exs. 19–43

❖ **Block Schedule**
  Day 49 Exs. 1–14, 16–18
  Day 50 Exs. 19–32, 34–43

Exercise Notes

Spatial Reasoning
Exs. 1–4 Students should be able to visualize angles of rotation between 0° and 360°. It may help some students to identify angles of rotation with circular movement. An angle of 90° represents a quarter-turn around a circle. A movement of 180° is halfway around a circle. A 270° angle is three-quarters of the way around and 360° is a full turn around. Other angles can be placed in between these four in order to estimate their measure. Remind students that a positive angle of rotation is in a counterclockwise direction.

Second-Language Learners
Exs. 1–4 Encourage students to pantomine each action listed to ensure that second-language learners understand what is meant by each phrase.

Communication: Drawing
Exs. 8–10 The angle lines that are used to draw a rotation are not part of the rotation itself. They should be drawn lightly in pencil and can be erased after the rotation is complete.

Problem Solving
Exs. 12, 13 Students may have some difficulty with these exercises. In Ex. 12, students should focus on the waffles, not on the entire waffle iron. When they do, they will see that the waffles have rotational symmetry with angles of 72°, 144°, 216°, and 288°. In Ex. 13, students may focus on the outer ring of flowers and say that the figure has rotational symmetry with angles of 40°, 80°, 120°, 160°, 200°, 240°, 280°, and 320°. This is not so, however, since these angles do not apply to the central flower.

---

3. about 90°

4. about 360°

5.

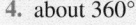

6.

7.

8–14. See answers in back of book.

415

Exercise Notes

Application
Ex. 15 In this exercise, students see how the concept of rotation can be used to understand the design and construction of objects, in this case, an end table. In part (b), the drawer is rotated –10°. Tell students that a negative angle of rotation has the same magnitude as a positive angle, but it is in the clockwise direction. Thus, an angle of –10° is a 10 degree rotation in the clockwise direction.

Communication: Discussion
Exs. 16, 17 Provide students with a sheet of polar graph paper and discuss its features prior to assigning these exercises. The concentric circles give the distance a point is from the center. The angle measures give the angle the point is from the right half of the horizontal line. Angles between two points on the paper can be determined by subtraction. Thus, the angle of rotation between a point on the 40° line and one on the 70° line is 70 – 40 = 30°.

Geometric Thinking
Ex. 18 The use of polar graph paper introduces students to the idea that the location of a point can be described by giving its distance from the center and the angle of rotation from the right-hand horizontal line. The description of locations in this manner is called the polar coordinate system. In polar coordinates, the point *R* in Ex. 18 is represented by *R*(5, 40), which means that *R* is 5 units from the center and 40° from the right-hand horizontal line. Students should understand that any rotation can be accomplished easily in polar coordinates. The first coordinate, the radius, remains the same. The second coordinate is increased by the angle of rotation.

15. FURNITURE DESIGN Furniture designer Marc Desplaines of San Francisco built this end table with drawers that are rotated in opposite directions. Two legs support each drawer.

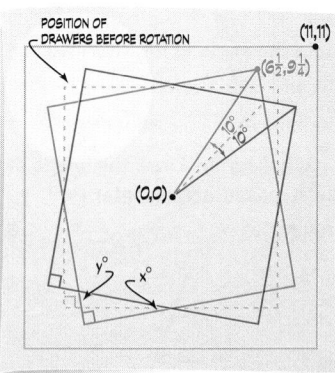

In the top view above, measurements are given to the nearest quarter inch.

a. The lower square drawer, shown in red, is rotated 10° away from the gray position. One corner is at $\left(6\frac{1}{2}, 9\frac{1}{4}\right)$. Find the coordinates of the other three corners.

b. The upper square drawer, shown in green, is rotated $-10°$ away from the gray position. Use what you know about reflections to find the coordinates of its corners.

c. Because the drawers are rotated, you can see a triangular corner of the lower drawer. Find the values of *x* and *y* in the triangle.

Copy each polygon on polar graph paper. Draw its image after the given rotation.

16. 50° counterclockwise

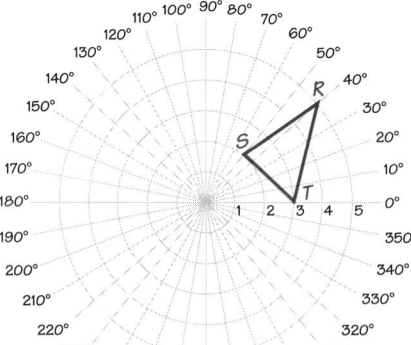

17. 120° counterclockwise

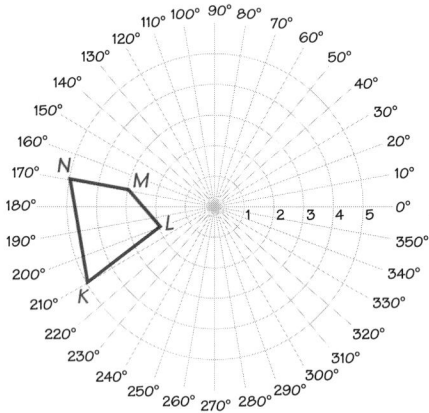

18. Writing The *polar coordinates* of △*RST* in Exercise 16 are *R*(5, 40°), *S*(2, 50°), and *T*(3, 0°).

a. Write the polar coordinates of △*R′S′T′*, the image triangle you drew in Exercise 16.

b. Describe how the polar coordinates of △*RST* and △*R′S′T′* are related to each other.

416 Chapter 8 *Using Transformations*

---

15. See answers in back of book.

16.

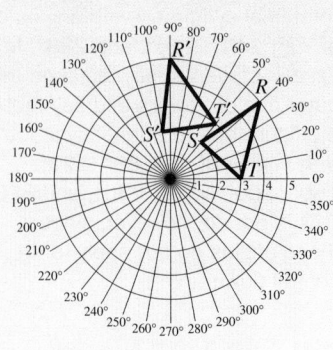

17.

18. a. *R′*(5, 90°); *S′*(2, 100°); *T′*(3, 50°)

b. Each point and its image have the same first coordinate. The second coordinate of an image point is 50° greater than the second coordinate of the corresponding vertex of △ *RST*.

19–22. The center of each rotation is the center of the square.

19. 90° rotation

20. 180° rotation

21. no rotation

22. 270° rotation

23. a–c. Answers may vary. Students may choose any two lines that intersect at an angle of 45°. Reflecting Triangle 1 over one of the lines and reflecting its image over the other has the same effect as rotating Triangle 1 either 90° or –90° around the center of the square.

24. a. △*A″B″C″* has the same orientation as △*ABC* and is the image of △*ABC*

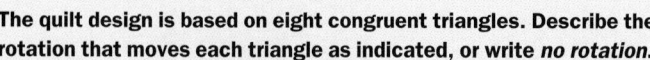

Connection HAWAIIAN QUILTS

Elizabeth Akana, who lives on the island of Oahu, is an expert in the art of Hawaiian quilting. These quilts are made by folding a square piece of cloth in half three times to form a triangle, and then cutting along the fold lines. When the design is unfolded, it has four lines of symmetry, as shown. The design is then sewn onto other cloth using a pattern of stitches that ripples outward.

The quilt design is based on eight congruent triangles. Describe the rotation that moves each triangle as indicated, or write *no rotation*.

19. Triangle 1 onto Triangle 3 **20.** Triangle 2 onto Triangle 6

21. Triangle 7 onto Triangle 8 **22.** Triangle 4 onto Triangle 2

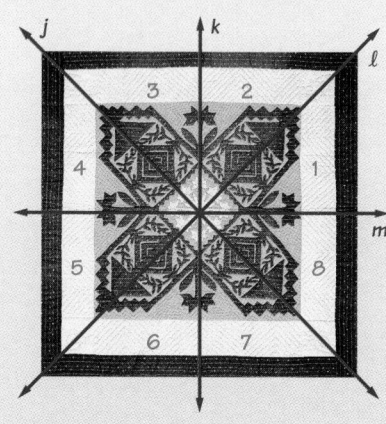

23. Investigation Choose two of the lines of symmetry on the quilt diagram that intersect at an angle of 45°.

　a. What happens to Triangle 1 if it is reflected over one line and the image is reflected over the other line?

　b. Describe the single rotation that has the same effect as the reflections in part (a).

　c. Repeat parts (a) and (b) using two other lines that intersect at an angle of 45°. What do you notice?

24.  **Technology** Use geometry software. Draw two intersecting lines and △ABC. Reflect △ABC over one line, then reflect the image, △A′B′C′, over the other line.

　a. How is the image △A″B″C″ related to the original △ABC?

　b. Change the angle at which the lines intersect. How is the measure of the angle related to the angle of the rotation you see? Write a conjecture.

Sketch the image of $\overline{AB}$ after a 180° rotation around the origin.

25. 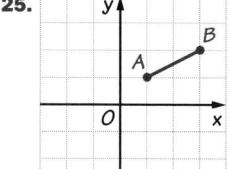 **26.** 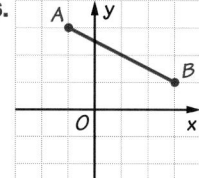 **27.**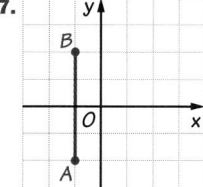

28. Choose one of sketches you made in Exercises 25–27. Use slopes to show that quadrilateral ABA′B′ is a parallelogram.

8.4 Applying Rotations **417**

---

under a rotation around the point where the lines intersect.

b. The measure of the angle of rotation is twice that of the angle at which the lines intersect.

25.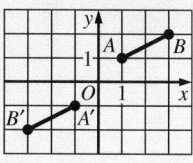

26.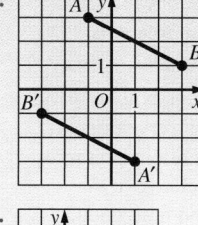

27.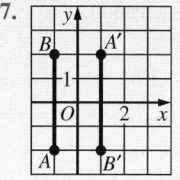

28. (25) Slope of $\overline{AB}$ = slope of $\overline{A'B'} = \frac{1}{2}$; slope of $\overline{AB'}$ = slope of $\overline{A'B} = \frac{3}{4}$. Then ABA′B′ is a parallelogram by definition. (26) Slope of $\overline{AB}$ = slope of $\overline{A'B'} = -\frac{1}{2}$; slope of $\overline{AB'}$ = slope of $\overline{A'B} = 2$. Then ABA′B′ is a parallelogram by definition. (Note that ABA′B′ is a rectangle, since adjacent sides are perpendicular.) (27) $\overline{AB}$ and $\overline{A'B'}$ are both vertical, so they are parallel. $\overline{AB'}$ and $\overline{A'B}$ are both horizontal, so they are parallel. Then ABA′B′ is a parallelogram by definition. (Note that ABA′B′ is a rectangle, since adjacent sides are perpendicular.)

Exercise Notes

🌐 Application
　Exs. 19–22 In this application, students see how the concept of rotation can be used to analyze a quilt pattern.

Using Manipulatives
Exs. 19–22 You may wish to have students simulate the procedure used to make this quilt pattern. Give each student a large square sheet of paper and a pair of scissors. They should fold the paper as described in the introduction to these exercises and cut out a design. They can then unfold the design and see the rotational symmetry. Ask students if they can fold the paper in such a way as to get a different rotational symmetry.

Research
Exs. 19–22 Quilting has been important to many cultures. Some students may be interested in finding books about quilting to look for different patterns. They can copy a pattern and look for rotational symmetry.

Multicultural Note
Exs. 19–22 Hawaiian quilts are well known for their intricate and beautiful "snowflake patterns," so called because the patterns are cut from folded cloth, much like paper snowflakes are cut from folded paper. Quilting was first introduced in Hawaii by American women around the middle of the 19th century. It quickly gained popularity among Hawaiian women, who drew from their experience designing traditional *tapa* cloths to invent patterns unlike those made by mainland quilters. Contemporary Hawaiian quilt designs represent a broad spectrum of themes, including nature, history, and personal experiences.

Geometric Thinking
Exs. 23, 24 In these exercises, students discover that a series of two reflections over intersecting lines results in a rotation. They should also see that the magnitude of the rotation is twice the smaller angle between the two lines. You may want to discuss the similarity of this result to the one students discovered in the Exploration on page 405 about two reflections over parallel lines.

417

Exercise Notes

Interview Note: Geometric Thinking
Exs. 29–32 In this set of exercises, students examine the design of the seating sections for Ericsson Stadium to look for rotations. This leads them to discover that a double reflection in perpendicular lines is the same as a rotation of 180° around the intersection point of the lines.

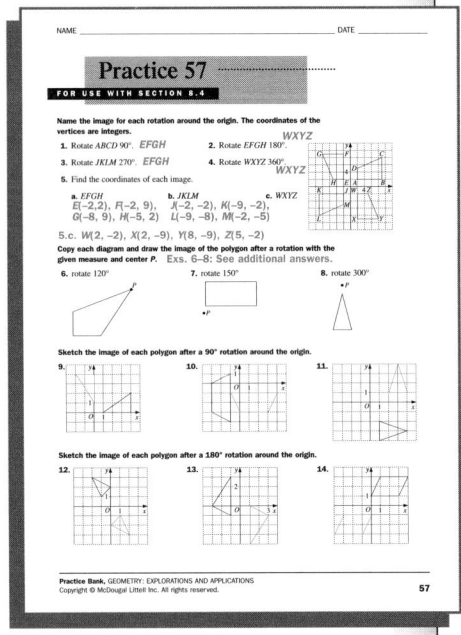

Communication: Writing
Ex. 34 This exercise gives students an opportunity to explore relationships involving reflections, rotations, and translations. Before writing their responses, students may need to try some examples.

Practice 57 for Section 8.4

NAME _____ DATE _____

Practice 57

FOR USE WITH SECTION 8.4

Name the image for each rotation around the origin. The coordinates of the vertices are integers.

1. Rotate *ABCD* 90°. *EFGH*   2. Rotate *EFGH* 180°. *WXYZ*
3. Rotate *JKLM* 270°. *EFGH*   4. Rotate *WXYZ* 360°. *WXYZ*
5. Find the coordinates of each image.
   a. *EFGH*   b. *JKLM*   c. *WXYZ*
   E(−2,2), *F*(−2, 9),   *J*(−2, −2), *K*(−9, −2),
   G(−8, 9), *H*(−5, 2)   *L*(−9, −8), *M*(−2, −5)
5.c. *W*(2, −2), *X*(2, −9), *Y*(8, −9), *Z*(5, −2)
Copy each diagram and draw the image of the polygon after a rotation with the given measure and center *P*. Exs. 6–8: See additional answers.
6. rotate 120°   7. rotate 150°   8. rotate 300°
Sketch the image of each polygon after a 90° rotation around the origin.
9.   10.   11.
Sketch the image of each polygon after a 180° rotation around the origin.
12.   13.   14.

Practice Bank, GEOMETRY: EXPLORATIONS AND APPLICATIONS
Copyright © McDougal Littell Inc. All rights reserved.   57

INTERVIEW Terri Johnson

Look back at the article on pages 388–390.

The diagram shows a simplified drawing of the Ericsson Stadium seating sections. After designing one quadrant of seating, the architects reflect the design over horizontal and vertical axes.

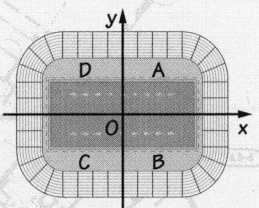

29. What rotation moves quadrant A to quadrant C?

30. What rotation moves quadrant B to quadrant D?

31. Does a 90° rotation move quadrant D to quadrant C? Explain.

32. Explain how your answers to Exercises 29 and 30 are related to the results you found in Exercise 24.

33. **Challenge** Suppose you are given only a figure and its image after a rotation. How can you use perpendicular bisectors to find the center of the rotation? Make a sketch and justify your method.

ONGOING ASSESSMENT

34. a. **Writing** If you reflect a polygon over a line, can you rotate the image back onto the original polygon? Explain why or why not.

    b. **Writing** If you translate a polygon, can you rotate the image back onto the original polygon? Explain why or why not.

SPIRAL REVIEW

Describe each translation using coordinate notation. *(Section 8.3)*

35. Every point moves to the left 3 units and up 6 units.

36. Every point moves to the right 2 units and down 6 units.

Name a pair of triangles that must be congruent in order to prove each statement. *(Section 6.5)*

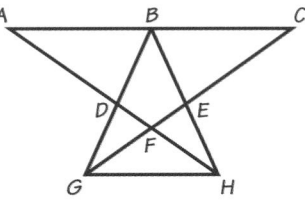

37. $\overline{FD} \cong \overline{FE}$ 38. $\angle BDF \cong \angle BEF$

39. $\overline{AH} \cong \overline{CG}$ 40. $\angle ABG \cong \angle CBH$

Find the coordinates of the midpoint of the segment with each given pair of endpoints. *(Section 4.1)*

41. $A(5, 2)$, $B(5, -2)$ 42. $C(-1, 3)$, $D(5, -3)$ 43. $E(4, 0)$, $F(-4, 3)$

418 Chapter 8 *Using Transformations*

---

29. a 180° rotation around the origin

30. a 180° rotation around the origin

31. Yes; a 90° rotation of quadrant *D* has the same effect as a reflection of quadrant *D* over the *x*-axis.

32. Each rotation has the same effect as reflection in one axis, followed by reflection in the other. The axes intersect at a 90° angle, so the measure of the angle of rotation is 2(90°) = 180°.

33. A point and its image after a rotation are equidistant from the center of the rotation. Then the center is on the perpendicular bisector of each segment determined by a point and its image. You can find the center of rotation by drawing two or more such segments and constructing their perpendicular bisectors. The intersection is the center of the rotation. Check students' work.

34. a. No; reflections reverse orientation, rotations preserve it.

    b. No; although both transformations preserve orientation, translations preserve position relative to the vertical and horizontal. Rotations do not.

35. $(a, b) \rightarrow (a - 3, b + 6)$
36. $(a, b) \rightarrow (a + 2, b - 6)$

37. △*DFG* and △*EFH*
38. △*BDH* and △*BEG*
39. △*ABH* and △*CBG*
40. △*ABD* and △*CBE*
41. (5, 0)
42. (2, 0)
43. $\left(0, 1\frac{1}{2}\right)$

8.5 Glide Reflections

Learn how to...

• combine a translation and a reflection to form a glide reflection

• find the image of a figure after a glide reflection

So you can...

• identify, describe, and create patterns that use glide reflections

Patterns like the ones below are used to decorate pottery, fabrics, and walls in many cultures. They are called *border* or *frieze* patterns. You can see that reflections, translations, and rotations have been used to create these repeating patterns.

Pattern A: Nazca

Pattern B: Oklahoma

Pattern C: Yurok

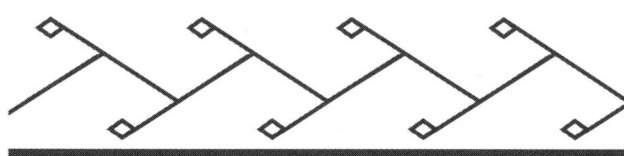

Pattern D: Navaho

BY THE WAY

The Nazcan people inhabited a desert coast south of what is now Lima, Peru. They made giant drawings using white rocks, which can be seen from the air.

THINK AND COMMUNICATE

1. Which of the patterns above use reflections?

2. Which of the patterns use translations?

3. Which of the patterns use rotations?

4. Describe how you could make Pattern D by using a translation followed by a reflection.

8.5 Glide Reflections **419**

ANSWERS Section 8.5

Think and Communicate

1. Patterns A, B, and C

2. Patterns A, B, C, and D

3. Pattern C

4. Pattern D can be generated by translating the pattern half the distance between the diamonds along the top, then reflecting over the horizontal centerline of the pattern. Pattern D can also be generated by a simple translation whose length is the distance between the diamonds along the top.

Plan⇔Support

Objectives

• Combine a translation and a reflection to form a glide reflection.

• Find the image of a figure after a glide reflection.

• Identify, describe, and create patterns that use glide reflections.

Recommended Pacing

❖ **Core and Extended Courses**
Section 8.5: 1 day

❖ **Block Schedule**
Section 8.5: $\frac{1}{2}$ block
(with Section 8.4)

Resource Materials

Lesson Support
Lesson Plan 8.5

Warm-Up Transparency 8.5

Practice Bank: Practice 58

Study Guide: Section 8.5

Explorations Lab Manual:
   Diagram Masters 1, 2

Challenge Problems: Set 58

Technology
Technology Book:
   Spreadsheet Activity 7
   TI-92 Activity 8

McDougal Littell Mathpack
   Geometry Inventor Activity Book:
   Activities 25, 27

Internet:
   http://www.hmco.com

Warm-Up Exercises

Write the coordinate rule for each transformation.

1. a reflection over the *y*-axis
$(a, b) \rightarrow (-a, b)$

2. a rotation of 90°
$(a, b) \rightarrow (-b, a)$

3. a translation *h* units horizontally and *k* units vertically
$(a, b) \rightarrow (a + h, b + k)$

4. a rotation of 270°
$(a, b) \rightarrow (b, -a)$

5. a reflection over the line *y = x*
$(a, b) \rightarrow (b, a)$

Learning Styles: Verbal

Identifying the transformation in boarder designs, such as the ones on page 419, may be difficult for students who are verbal learners. Have them use small pieces of clear transparency film and copy what they think is the smallest portion of the design that can be used to create the entire design. They can then try reflecting and translating the transparency to see if it creates the design.

Section Note

Using Technology
Students can use Activity 25 in the *Geometry Inventor Activity Book* to investigate glide reflections and their properties.

Think and Communicate

After completing question 6, ask students to describe the differences they see between the first two patterns, which combine translations and reflections, with the last two that contain glide reflections. (The first two patterns contain both translations and reflections, whereas a glide reflection is a single transformation that is both a translation and reflection. In a glide reflection, there is no way to fold the paper so that the pattern lies on itself.)

Additional Example 1

Sketch the image of the triangle with the given vertices after a glide reflection using the given translation and reflection.
$T(0, -5)$, $U(2, -2)$, $V(6, -1)$
translation: $(a, b) \rightarrow (a + 4, b + 4)$
reflection: over the line $y = x$
The translation moves each point to the right 4 units and up 4 units. The reflection switches the coordinates.

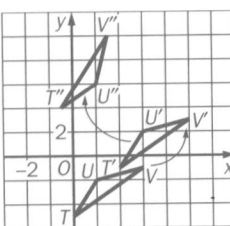

The coordinates change as shown.
translation   reflection
$T(0, -5) \rightarrow T'(4, -1) \rightarrow T''(-1, 4)$
$U(2, -2) \rightarrow U'(6, 2) \rightarrow U''(2, 6)$
$V(6, -1) \rightarrow V'(10, 3) \rightarrow V''(3, 10)$

420

A translation followed by a reflection over a line parallel to the translation is called a **glide reflection**.

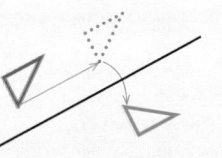

Properties of Glide Reflections

The line of reflection in a glide reflection is *parallel* to the direction of the translation.

A glide reflection preserves congruence.

A glide reflection reverses orientation.

THINK AND COMMUNICATE

5. What does the word "glide" refer to in the phrase *glide reflection*?

6. Which pattern(s) on page 419 can be described using glide reflections?

EXAMPLE 1

Sketch the image of the triangle with the given vertices after a glide reflection using the given translation and reflection.
$A(-5, 1)$, $B(-4, 4)$, $C(1, 2)$
translation: $(a, b) \rightarrow (a + 6, b)$
reflection: over the x-axis

SOLUTION

Plot $\triangle ABC$. Then translate it and reflect it.

The **translation** moves each point to the right 6 units.

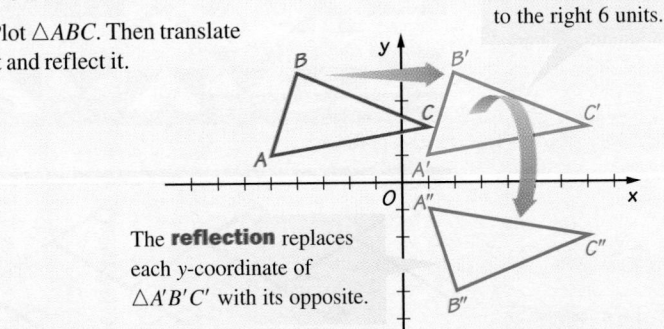

The **reflection** replaces each y-coordinate of $\triangle A'B'C'$ with its opposite.

The coordinates change as shown.

	translation		reflection
$A(-5, 1)$	$\rightarrow$	$A'(1, 1)$	$\rightarrow$ $A''(1, -1)$
$B(-4, 4)$	$\rightarrow$	$B'(2, 4)$	$\rightarrow$ $B''(2, -4)$
$C(1, 2)$	$\rightarrow$	$C'(7, 2)$	$\rightarrow$ $C''(7, -2)$

add 6 to the x-coordinate   change sign of y-coordinate

....

Think and Communicate

5. the translation

6. Patterns C and D

EXAMPLE 2   Application: Animal Tracks

The tracks made by the feet of a walking black bear form a glide reflection. Copy the tracks and describe a translation and reflection that combine to create the glide reflection.

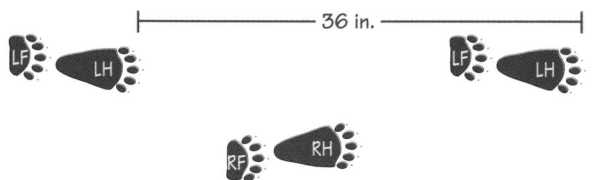

LEGEND
LF → left foreleg
LH → left hind leg
RF → right foreleg
RH → right hind leg

SOLUTION

The distance between footprints made by the same foot, called the *stride*, is about 36 in. The translation needed for the glide reflection is half this amount, about 18 in.

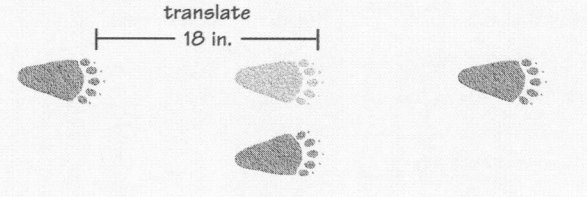

The line of reflection runs along the middle of the tracks, as shown in the diagram below. The line of reflection is parallel to the direction of the translation.

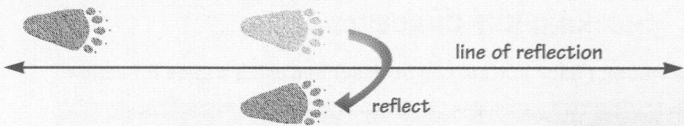

THINK AND COMMUNICATE

7. Is the stride of the hind leg the same as the stride of the foreleg for a black bear?

8. Explain why the translation needed for the glide reflection is half the bear's stride.

8.5 Glide Reflections   **421**

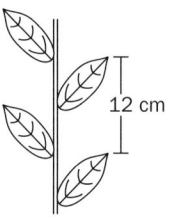
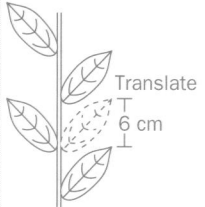

Think and Communicate

7. Yes.

8. If the bear is walking normally, the feet on one side hit the ground when the feet on the other side are in the middle of their stride.

Communication: Discussion
Examine and discuss the summary of the transformations at the top of this page. Students should understand how each transformation affects the congruence and orientation of a figure.

Journal Entry
You may wish to suggest that students summarize the material at the top of this page in their journals. Ask them to include the coordinate rules for each type of transformation when possible.

Multicultural Note
The art from many traditional cultures contains examples of reflections, translations, and rotations. In Africa, these repetitive patterns can be found on Kikuyu wooden shields, Kuba raffia cloths, Benin bronzes, Yoruba adire cloths, and Asante brass weights, to name a few. Native American designs also contain many examples of reflection, translation, and rotation. Navajo woven blankets, Crow saddle bags, Cheyenne work bags, Iroquois knife sheaths, Pomo baskets, and many other artistic items from Native American peoples show repetitive patterns.

Checking Key Concepts

Geometric Thinking
Students should be able to identify the translation and line of reflection in the designs for question 1 and 2. In questions 3 and 4, students should verify that the line of reflection is parallel to the direction of the translation.

Closure Question

Does every combination of a translation and a reflection result in a glide reflection? Explain. No, in a glide reflection, the line of reflection must be parallel to the direction of the translation.

Transformations that Preserve Congruence

This chart summarizes the four transformations described so far in this chapter. All these transformations preserve congruence.

These transformations reverse orientation.

Reflection (Sections 8.1 and 8.2)

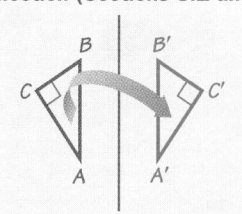

Glide Reflection (Section 8.5)

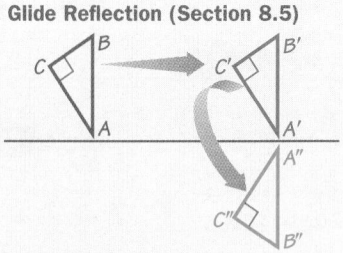

These transformations preserve orientation.

Translation (Section 8.3)

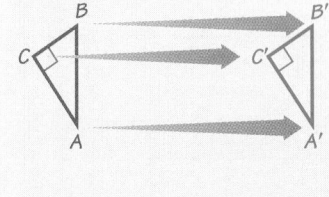

Rotation (Section 8.4)

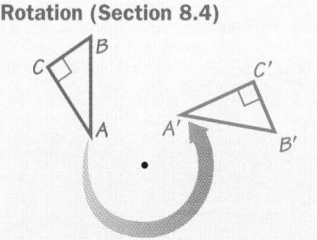

☑ CHECKING KEY CONCEPTS

Tell whether each pattern can be described using a glide reflection.

1.

2.

Sketch the image of the triangle with the given vertices after a glide reflection using the given translation and reflection.

3. $A(0, 0)$, $B(1, 3)$, $C(4, 4)$
translation: $(a, b) \rightarrow (a, b - 4)$
reflection: over the y-axis

4. $D(0, 0)$, $E(2, 3)$, $F(5, 0)$
translation: $(a, b) \rightarrow (a + 2, b + 2)$
reflection: over the line $y = x$

Checking Key Concepts

1. Yes.

2. No; the figures in each pattern block have the same orientation.

3.

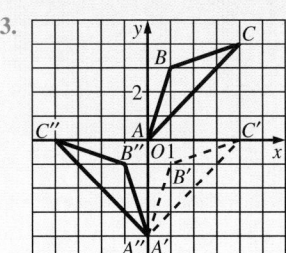

4.

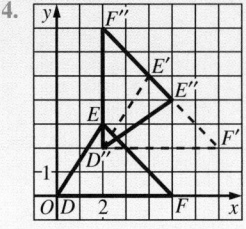

Exercises and Applications

1.

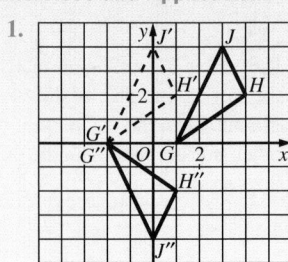

Extra Practice exercises on page 676

Sketch the image of the triangle with the given vertices after a glide reflection using the given translation and reflection.

1.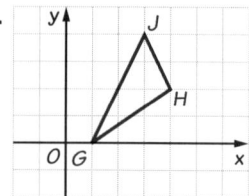

translation: $(a, b) \rightarrow (a - 3, b)$
reflection: over the x-axis

2.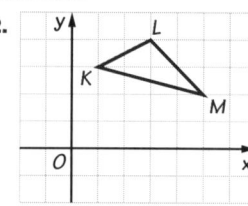

translation: $(a, b) \rightarrow (a - 5, b - 5)$
reflection: over the line $y = x$

3. $P(-3, -2), Q(0, 0), R(2, -4)$
translation: $(a, b) \rightarrow (a, b + 1)$
reflection: over the y-axis

4. $X(-2, 1), Y(0, 4), Z(3, 5)$
translation: $(a, b) \rightarrow (a + 4, b - 4)$
reflection: over the line $y = -x$

5. Challenge Choose one of the glide reflections from Exercises 1–4. Describe a rule for the transformation using coordinate notation:

$$(a, b) \rightarrow (\ ?\ , \ ?\ )$$

Draw a sketch of each situation, showing how a glide reflection is involved.

6. rungs on a telephone pole

7. snow-shoe or ski tracks

8. wallpaper or stencil pattern

9. closed zipper

FABRIC DESIGN These patterns are taken from men's ties. Tell whether each pattern uses a glide reflection. If it does, make a sketch, showing the direction and amount of translation, and draw the line of reflection.

10.

11.

12.

13. Research Find other examples of border patterns like the ones on page 419. Describe the transformations that are used to create them.

14. Open-ended Problem Create a pattern that uses a glide reflection.

15. SAT/ACT Preview Which of the following are preserved by a glide reflection?

I. length   II. orientation   III. angle measure

A. I only   **B.** I and II   **C.** II and III

D. I and III   **E.** I, II, and III

Suggested Assignment

❖ **Core Course**
Exs. 1–4, 6–12, 15–23, 25–36

❖ **Extended Course**
Exs. 1–36

❖ **Block Schedule**
Day 50 Exs. 1–4, 6–12, 15–23, 25–36

Exercise Notes

Student Progress
Exs. 1–4 Students should now feel comfortable with the coordinate notation for transformations, and they should know the notations for reflections over the x-axis, the y-axis, and the line $y = x$.

Challenge
Ex. 5 In completing this exercise, students learn that it is possible to describe a glide reflection using a single rule.

Second-Language Learners
Exs. 6–9 Invite students learning English to ask for assistance if they do not understand some of the words in these exercises.

Application
Exs. 6–12, 25–29 Glide reflections are found in a large variety of real-world settings. These exercises include examples from fabric design and animal tracks.

Career Connection
Exs. 10–12 A textile designer is a person who creates designs that decorate fabrics. The fabrics can then used to make clothing or linens such as sheets and towels. Textile designers make extensive use of reflections, translations, rotation, and glide reflections. Most textile designers have at least a bachelor's degree with a major in art or design.

Communication: Drawing
Ex. 14 Ask students to share their work with the class. Point out how those patterns whose initial figure is on one side of the line of reflection differ from those whose initial figure includes the line of reflection. Ask students to make other comparisons between the different patterns.

2–4. See answers in back of book.

5. (1): $(a, b) \rightarrow (a - 3, -b)$;
(2): $(a, b) \rightarrow (b - 5, a - 5)$;
(3): $(a, b) \rightarrow (-a, b + 1)$;
(4): $(a, b) \rightarrow (4 - b, -a - 4)$

6–9. Answers may vary. Examples are given.

6.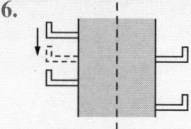

7.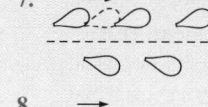

8.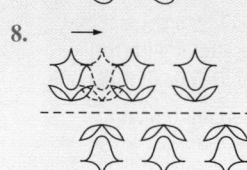

9.

10. Yes.

11. Yes.

12. No.

13, 14. Check students' work.

15. D

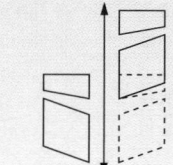

Exercise Notes

Interdisciplinary Problems
Exs. 16–22 The concept of glide reflection is used in these exercises to help students understand a computer simulation. You may want to read and discuss the introduction to these problems in class before assigning them to students to make certain that students understand how each generation is created.

Cooperative Learning
Exs. 16–22 Students will need sufficient time to complete these exercises. You may want to consider having students work in their groups for this entire set of exercises.

Writing Proofs
Ex. 24 This exercise requires students to write a coordinate proof about a property of glide reflections. Ask a student who thinks he or she has a valid proof to present it at the board.

A computer game called "Life," developed by John Conway, reveals how complex behavior can arise from simple rules. The game uses a grid of square cells, some filled and others empty.

The number of filled neighbors of each cell determines whether the cell will be filled or empty in the next "generation," according to these rules:
- An empty cell becomes filled if it has exactly three filled neighbors.
- A filled cell stays filled if it has two or three filled neighbors.

Five generations of an initial population called a "glide-reflection spaceship" are shown.

Every cell has 8 neighbors. The blue cell has three filled neighbors.

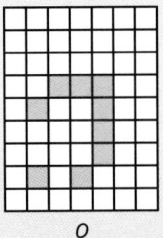 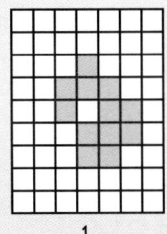 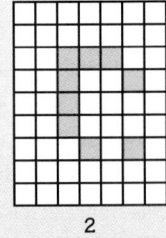 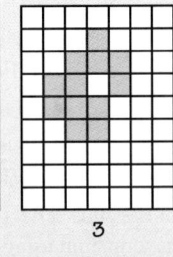 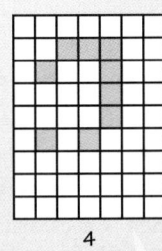

   0   *1*   *2*   *3*   *4*

16. **Cooperative Learning** Work in a group of four people. Each person should choose one of the generations 1–4. Verify that the generation you choose is correctly "born" from the previous one.

17. Which generation is a glide reflection of generation 0? of generation 1?

18. What transformation moves generation 0 to generation 4?

19. **Writing** What happens to this population if the game continues? Explain why you think this population is called a "spaceship."

When counting cell neighbors, imagine the grid extends forever in all directions.

Use graph paper to plot several generations of each initial population using the rules of the game of Life. Describe the glide reflections that appear.

20. 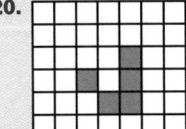   21. 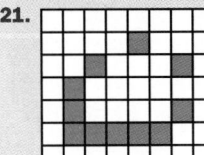   22.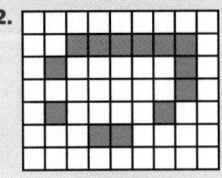

23. **Visual Thinking** In Section 8.3 you saw that a translation can be created by combining two reflections in parallel lines. Show how to create a glide reflection by combining three reflections. How are the lines related?

24. **Challenge** On a coordinate plane, draw a point and its image under a glide reflection that uses the *x*-axis as the line of reflection.
    a. What do you notice about the midpoint of the segment that joins the point and its image?
    b. Make a conjecture and prove it using coordinates.
    c. Can you extend your conjecture to other kinds of glide reflections?

---

16. Check students' work.

17. generation 2; generation 3

18. the translation $(a, b) \rightarrow (a, b + 2)$

19. Generations 0–4 are repeated in order over and over. Answers may vary. An example is given. I think the population is called a spaceship because if you look at the figures in generations 0, 4, 8, and so on, they have the same shape but appear to be moving through space.

20–22. See answers in back of book.

23. A glide reflection is a translation followed by a reflection over a line parallel to the translation. Then, if *j* and *k* are parallel lines and line *m* is perpendicular to both *j* and *k*, reflection over *j*, then *k*, then *m* has the same effect as a glide reflection.

24. Answers may vary. An example is given.

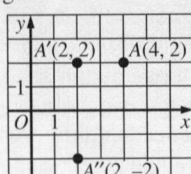

a. The midpoint of $\overline{AA''}$, $(2, 0)$, is on the line of reflection, the *x*-axis.

b. If the line of reflection for a glide reflection is the *x*-axis, then the midpoint of the segment determined by any point

and its image is on the *x*-axis. Let a glide reflection consist of the translation $(a, b) \rightarrow (a + k, b)$ followed by reflection over the *x*-axis. (Since the reflection must be parallel to the translation, the translation must be horizontal.) Any point $P(x, y)$ is translated to $P'(x + k, y)$, then reflected to $P''(x + k, -y)$. The midpoint of PP'' is $\left(\frac{2x + 2k}{2}, \frac{y + (-y)}{2}\right) = (x + k, 0)$, which is on the *x*-axis.

c. Yes; the conjecture is true for all glide reflections. That is, the midpoint of the segment determined by any point and its image after a glide reflection is on the line of reflection.

25. translation length: 0.6 cm (6 mm)

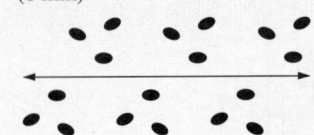

26. translation length: 6 in.

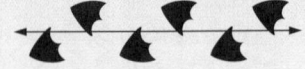

27. translation length: 11 in.

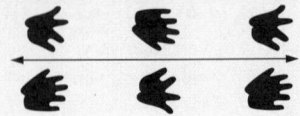

ANIMAL TRACKS Each set of animal tracks can be described using a glide reflection. Copy the tracks, find the length of the translation, and draw the line of reflection needed.

25. ant (6 legs)

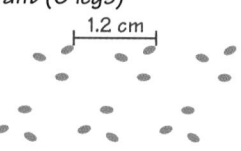
1.2 cm

26. herring gull (2 legs)

12 in.

27. raccoon (4 legs)

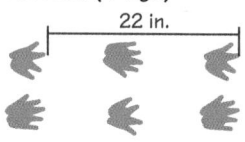
22 in.

28. scorpion (8 legs)

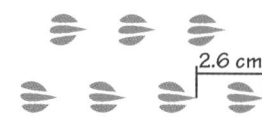
2.6 cm

29. Investigation Measure your own stride and draw a scale diagram of your footprints, showing how a glide reflection can describe them.

ONGOING ASSESSMENT

30. Open-ended Problem Experiment to find out if the order in which you perform the translation and reflection affects the final image of a glide reflection. Are the results different if you reflect first, then translate?

SPIRAL REVIEW

Copy each diagram and sketch the image of each polygon after a 180° rotation around the origin. *(Section 8.4)*

31.

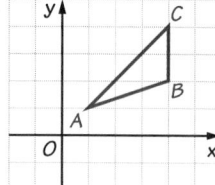

32.

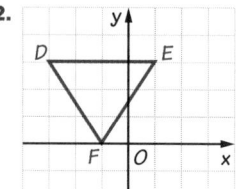

Open-ended Problems Describe at least two objects where you live or at school that have each type of symmetry. *(Section 1.2)*

33. reflection symmetry   **34.** rotational symmetry

The midpoints of two sides of a triangle are endpoints of a segment. Show that the segment is parallel to the third side and half as long. *(Section 4.5)*

35.

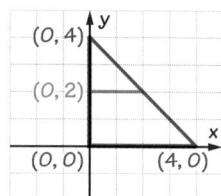

36.

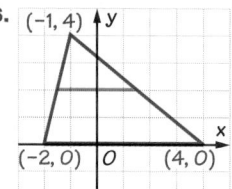

8.5 Glide Reflections   **425**

Exercise Notes

Teaching Tip
Ex. 29 Students can walk in loose soil or sand to create footprints. These prints can then be measured to find the stride length.

Assessment Note
Ex. 30 Students should try several examples before reaching their conclusion. Ask students if they can support their conclusion from examining the changes in the coordinates.

Practice 58 for Section 8.5

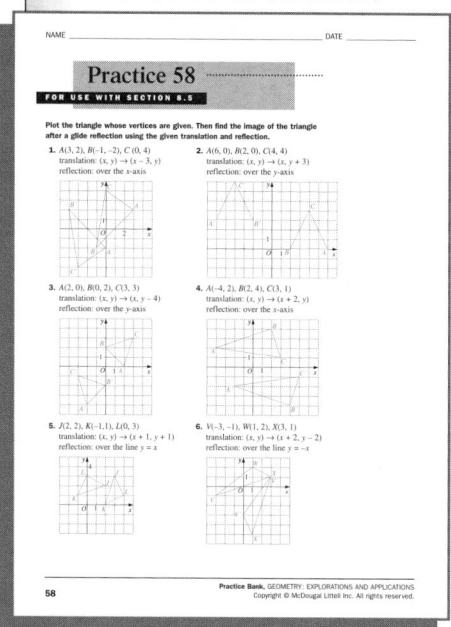

---

28. translation length: 1.3 cm

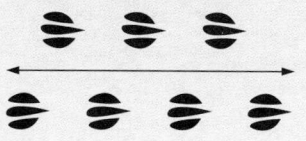

29. Check students' work.

30. No; the order does not affect the final image.

31.

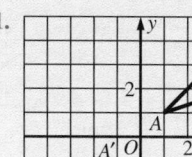

32.

D ... E, F' O ... x, E' ... D'

33–34. Answers may vary. Examples are given.

33. a rectangular table top; a window, a chair

34. a car steering wheel, a circular table top, a square table top

35. The third side is horizontal and is 4 units long. The other midpoint of the segment is (2, 2), so its length is 2, which is half the length of the third side. The segment is horizontal, so it is parallel to the third side.

36. The third side is horizontal and is 6 units long. The endpoints of the segment are $\left(-\frac{3}{2}, 2\right)$ and $\left(\frac{3}{2}, 2\right)$. Its length is 3, which is half the length of the third side. The segment is horizontal, so it is parallel to the third side.

425

Objectives

- Compare lengths in a figure and its image after a dilation.
- Find the scale factor and center of a dilation.
- Make reductions and enlargements of figures.
- Describe the relationship between a figure and its image after a dilation.

Recommended Pacing

❖ **Core and Extended Courses**
Section 8.6: 1 day

❖ **Block Schedule**
Section 8.6: $\frac{1}{2}$ block
(with Chapter 8 Review)

Resource Materials

Lesson Support
Lesson Plan 8.6
Warm-Up Transparency 8.6
Practice Bank: Practice 59
Study Guide: Section 8.6
Explorations Lab Manual:
    Additional Exploration 13
    Diagram Master 2
Challenge Problems: Set 59
Assessment Book: Test 36

Technology
Technology Book:
    Calculator Activity 11
Geometry Software
McDougal Littell Mathpack
    Geometry Inventor Activity Book:
        Activities 23, 24, 28
    Function Investigator Activity
        Book: Activity 8
Internet:
    http://www.hmco.com

SECTION

8.6 | Dilations

Many photocopiers make reduced or enlarged copies as well as same-size copies. A youth organization used this built-in feature to make a small version of their logo for a notebook cover and a larger version for their T-shirts.

Learn how to...

- **compare lengths in a figure and its image after a dilation**
- **find the scale factor and center of a dilation**

So you can...

- **make reductions and enlargements of figures**
- **describe the relationship between a figure and its image after a dilation**

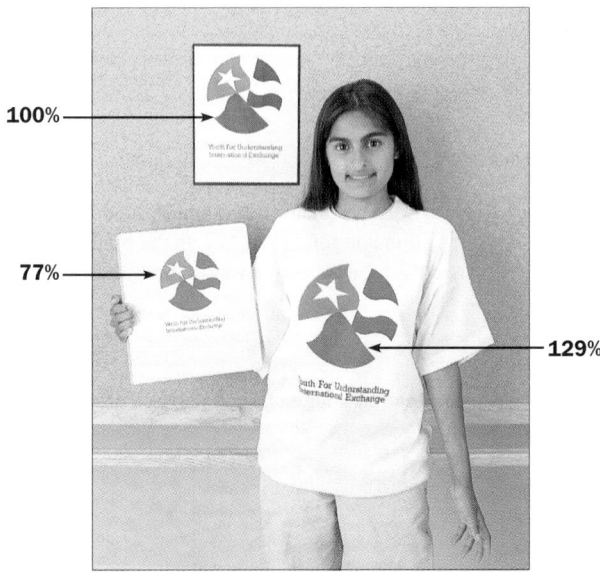

100%

77%

129%

A 77% reduction was used to make the smaller logo. Every segment in this logo is 0.77 times as long as the corresponding segment in the original logo. The number 0.77 is called the *scale factor* for this reduction. A scale factor can be a percent, a decimal, or a fraction.

THINK AND COMMUNICATE

1. A 129% enlargement was used to make the larger logo. Explain how the lengths of two corresponding segments in the enlarged figure and the original logo are related.

2. What scale factor would you use to produce a same-size copy?

3. The photocopier has 77% and 64% settings. How could you combine them to make an image that is a 50% reduction (half size)?

426   Chapter 8 *Using Transformations*

Warm-Up Exercises

Find the length of $\overline{RY}$ if $PT = 12$.

1. $\overline{RY}$ is three times the length of $\overline{PT}$. 36
2. $\overline{RY}$ is half as long as $\overline{PT}$. 6
3. $\overline{RY}$ is $\frac{3}{2}$ the length of $\overline{PT}$. 18
4. $\overline{RY}$ is 0.45 the length of $\overline{PT}$. 5.4
5. $\overline{RY}$ is 75% as long as $\overline{PT}$. 9

ANSWERS   Section 8.6

Think and Communicate

1. Each segment in the enlargement is 1.29 times as long as the corresponding segment in the original logo.

2. 100% (or 1)

3. Make a copy at one setting, then copy the image at the second setting. Since 0.77(0.64) = 0.4928, the effect is approximately equal to a 50% reduction. (Either setting may be used first, since 0.77(0.64) = 0.64(0.77).)

Exploration

1–2. Check students' work.

3. Corresponding angles are congruent.

4. Each length in $\triangle A'B'C'$ is twice the corresponding length in $\triangle ABC$.

5. The resulting triangle is congruent to $\triangle A'B'C'$, but its position is different.

EXPLORATION
COOPERATIVE LEARNING

Making an Enlargement

Work with another student.
You will need:

- a ruler

- a protractor

1 Draw any △*ABC*. Choose any point *O* outside the triangle. Draw rays from *O* through each vertex of the triangle.

2 Use your ruler to find the length *OA*. Double this number and use it to place *A'* on $\overrightarrow{OA}$ so that *OA'* = 2 • *OA*. Use the same process to find *B'* and *C'*. Draw △*A'B'C'*.

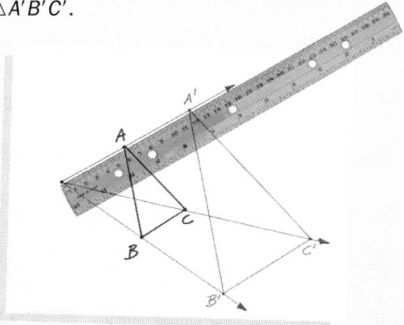

3 Use a protractor to compare the corresponding angles in your two triangles. What do you notice?

4 Use a ruler to compare the corresponding lengths in your two triangles. What do you notice?

5 Choose another point *P*, inside △*ABC*. Repeat Steps 1–4 using *P* instead of *O*. How is your image triangle different? How is it the same?

The enlargements in the Exploration, like the photocopy of the youth organization logo, do *not* preserve congruence, because corresponding lengths are not the same. The transformations in the Exploration are examples of *dilations*.

Exploration Note

Purpose
The purpose of this Exploration is to have students learn how to draw a dilation and to discover the properties of a dilation.

Materials/Preparation
Each group needs a ruler and a protractor.

Procedure
Students follow Steps 1 and 2 to create a triangle that is the image of another triangle after a dilation whose center is outside the triangle. The angles and sides of the two tri-

angles are measured and compared. They then repeat the procedure with a center of dilation inside the triangle.

Closure
After completing Steps 1–5, students should understand that after a dilation, the side lengths of a figure change, but the angle measures remain the same.

Explorations Lab Manual
See the Manual for more commentary on this Exploration.

Teach⇔Interact

Section Notes

Communication: Discussion
Read and discuss the introduction to this section on page 426. Stress that a scale factor is a number that is used to change the size of a figure. This is done by multiplying the length of any segment of the figure by the scale factor to produce the length of the corresponding segment in the image.

Research
After reading the *By the Way* note on page 426, have students find the dimensions of tabloid-sized, legal-sized, and letter-sized paper. They should then verify that the reductions and enlargement described in this note produce the correct size image.

Using Technology
Dilating triangles and other polygons is explored in detail in Activities 23 and 24, respectively, in the *Geometry Inventor Activity Book*.

Topic Spiraling: Preview
A figure and its image after a dilation are similar figures. Similar figures are studied in detail in Chapter 9.

Section Note

Communication: Discussion
Have students read the summary box on the properties of dilations and discuss these properties by using an illustration of the figure in the box at the board. Stress that the scale factor is a positive number.

Additional Example 1

Draw the image of *ABCD* after a dilation with center *O* and scale factor $\frac{3}{2}$.

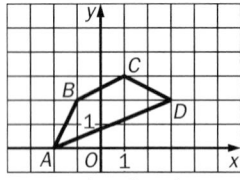

Draw $\overrightarrow{OA}$, $\overrightarrow{OB}$, $\overrightarrow{OC}$, and $\overrightarrow{OD}$.

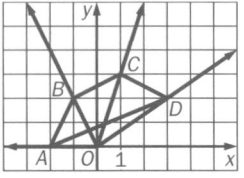

The scale factor is $\frac{3}{2}$.

Place *A'* on $\overrightarrow{OA}$ so that $OA' = \frac{3}{2}OA$.

Place *B'* on $\overrightarrow{OB}$ so that $OB' = \frac{3}{2}OB$.

Place *C'* on $\overrightarrow{OC}$ so that $OC' = \frac{3}{2}OC$.

Place *D'* on $\overrightarrow{OD}$ so that $OD' = \frac{3}{2}OD$.

Draw *A'B'C'D'*.

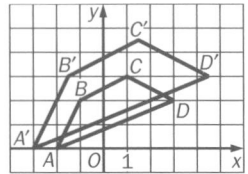

Think and Communicate

Question 5 focuses students' attention on the fact that the ratio of corresponding segments of the image and original figure after a dilation is equal to the scale factor. Make sure students understand that in the ratio for a scale factor, the numerator is the length of a segment in the image and the denominator is the length of a segment in the original figure.

Properties of Dilations

A **dilation** with **center** *O* and positive **scale factor** *k* is a transformation in which every point *P* has an image *P'* placed on $\overrightarrow{OP}$ so that

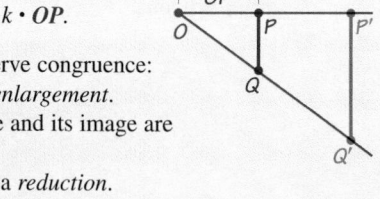

$$\frac{OP'}{OP} = k \quad \text{and} \quad OP' = k \cdot OP.$$

Dilations do not usually preserve congruence:
    If $k > 1$, the dilation is an *enlargement*.
    If $k = 1$, the original figure and its image are
       congruent.
    If $0 < k < 1$, the dilation is a *reduction*.

Dilations preserve angle measures.

The image of *any* segment after a dilation is always *k* times as long as the original segment. In the diagram above, for example, $P'Q' = k \cdot PQ$.

EXAMPLE 1

Draw the image of *JKLM* after a dilation with center *O* and scale factor $\frac{1}{2}$.

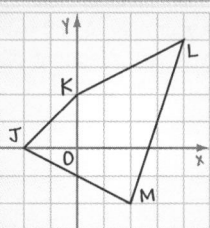

SOLUTION

Copy *JKLM* on graph paper.
Draw $\overrightarrow{OJ}$, $\overrightarrow{OK}$, $\overrightarrow{OL}$, and $\overrightarrow{OM}$.

The scale factor is $\frac{1}{2}$.

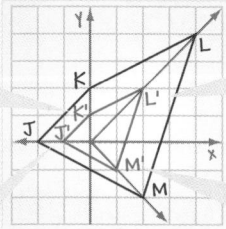

Place *K'* on $\overrightarrow{OK}$ so that $OK' = \frac{1}{2}OK$.

Place *J'* on $\overrightarrow{OJ}$ so that $OJ' = \frac{1}{2}OJ$.

Place *L'* on $\overrightarrow{OL}$ so that $OL' = \frac{1}{2}OL$.

Place *M'* on $\overrightarrow{OM}$ so that $OM' = \frac{1}{2}OM$.

THINK AND COMMUNICATE

4. Is the image in Example 1 an *enlargement* or a *reduction*?

5. Use the Distance Formula to find the lengths of two corresponding segments of *JKLM* and *J'K'L'M'*. What is their ratio?

Think and Communicate

4. reduction

5. $JK = 2\sqrt{2}$, $J'K' = \sqrt{2}$, $\frac{J'K'}{JK} = \frac{1}{2}$;

$KL = 2\sqrt{5}$, $K'L' = \sqrt{5}$,

$\frac{K'L'}{KL} = \frac{1}{2}$; $LM = 2\sqrt{10}$, $L'M' =$

$\sqrt{10}$, $\frac{L'M'}{LM} = \frac{1}{2}$; $JM = 2\sqrt{5}$,

$J'M' = \sqrt{5}$, $\frac{J'M'}{JM} = \frac{1}{2}$

EXAMPLE 2

△*P′Q′R′* is the image of △*PQR* after a dilation. Find the center and the scale factor of the dilation.

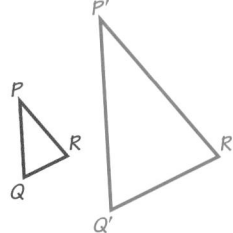

SOLUTION

Draw lines through each pair of corresponding vertices. Measure $\overline{OP'}$ and $\overline{OP}$.

$OP' = 30$ mm

The scale factor is the ratio $\frac{OP'}{OP}$.

$OP = 12$ mm

$\frac{OP'}{OP} = \frac{30}{12} = 2.5$

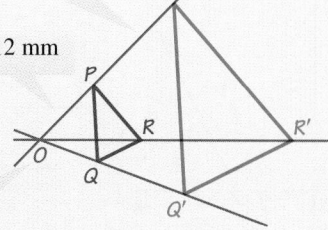

The intersection of the lines is the center of the dilation. Label it *O*.

The dilation has center *O* and scale factor 2.5.

> ◀ **WATCH OUT!**
> The ratio $\frac{PP'}{OP}$ is *not* the scale factor of the dilation.

THINK AND COMMUNICATE

6. In Example 2, what is the ratio $\frac{OQ'}{OQ}$? What is the ratio $\frac{OR'}{OR}$?

7. Make a conjecture about the perimeters of △*PQR* and △*P′Q′R′*.

☑ CHECKING KEY CONCEPTS

Draw the image of △*ABC* after each dilation.

1. center *C* and scale factor 2

2. center *A* and scale factor $\frac{1}{3}$

3. Let *M* and *N* be the midpoints of $\overline{AB}$ and $\overline{BC}$, respectively. Find the center and scale factor of the dilation that reduces △*ABC* to △*MBN*.

4. Explain how a dilation is different from the other transformations that you have learned about in this chapter.

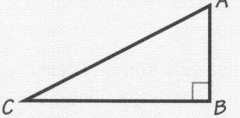

Think and Communicate

6. 2.5; 2.5

7. The ratio of the perimeter of △*P′Q′R′* to the perimeter of △*PQR* is 2.5.

Checking Key Concepts

1.

2.

3. *B*; $\frac{1}{2}$

4. A dilation does not preserve congruence. All the other transformations I have learned about in this chapter do preserve congruence.

Additional Example 2

GH′M′K′ is the image of *GHMK* after a dilation. Find the center and the scale factor of the dilation.

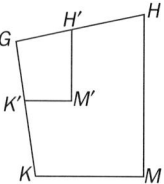

Draw lines through each pair of corresponding vertices.

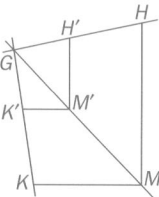

Measure *GM′* and *GM*. The scale factor is the ratio $\frac{GM'}{GM}$.

GM′ = 11 mm and *GM* = 25 mm.

$\frac{GM'}{GM} = \frac{11}{25} = 0.44$

The intersection of the lines is the center of the dilation. In this case, it is the vertex *G*. The dilation has center *G* and scale factor 0.44.

Think and Communicate

Question 6 shows students that they could have used any two corresponding parts from the triangle and its image to determine the scale factor in Example 2. In question 7, students extend what they have learned about dilation to consider the relative size of the perimeters of the figure and its image.

Checking Key Concepts

Geometric Thinking
For question 1, the center of dilation is also a vertex. Discuss how the image is different from one where the center is outside the figure or inside the figure.

Closure Question

Describe how a figure and its image after a dilation are related.
If the scale factor, *k*, is greater than 1, the image is an enlargement. If *k* = 1, the original figure and its image are congruent. If *k* is less than 1, the image is a reduction.

8.6 | **Exercises and Applications**

Extra Practice
exercises on
page 677

Draw the image of each figure after a dilation with center *O* and the given scale factor.

1. scale factor 3

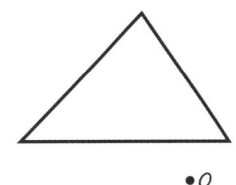

2. scale factor 1.5

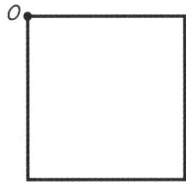

3. scale factor 0.75

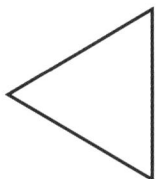

4. scale factor 3

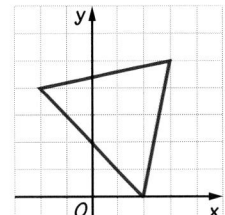

5. scale factor $\frac{1}{3}$

6. scale factor 4

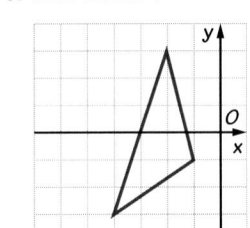

7. a. Use the coordinates of the vertices of the figure in Exercise 4. Copy and complete the table.

b. Complete this conjecture based on your table: The image of (a, b) after a dilation with center *O* and scale factor *k* has coordinates (_?_ , _?_).

c. Test your conjecture by comparing the coordinates of points and their images in Exercises 5 and 6.

Figure	Scale	Image
(_?_ , _?_)	3	(_?_ , _?_)
(_?_ , _?_)	3	(_?_ , _?_)
(_?_ , _?_)	3	(_?_ , _?_)

8. MANUFACTURING European paper has dimensions in the ratio $1 : \sqrt{2}$. For example, letter-size paper, called A4, is 210 mm × 297 mm. Each size can be enlarged to the next size by using a scale factor of $\sqrt{2}$.

a. The next paper size larger than A4 is called A3. What are the dimensions of A3 paper?

b. Challenge What reduction percent would you use to photocopy one size of European paper onto the next *smaller* size?

9. Writing Elena wants to compile some of her family's favorite recipes in a notebook. The recipes are on 3 in. × 5 in. cards. She wants to enlarge them in a photocopier to fit onto 8.5 × 11 in. letter paper. Choose an appropriate enlargement she can use. Explain your choice.

Exercises and Applications

1.

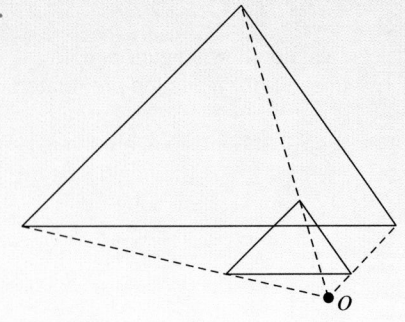

2.

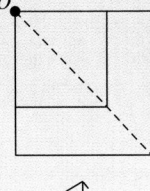

3.

4.

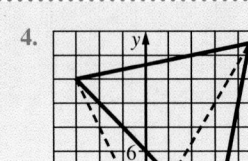

5.

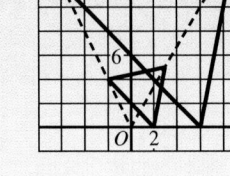
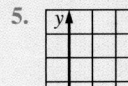
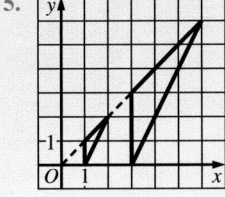

6.

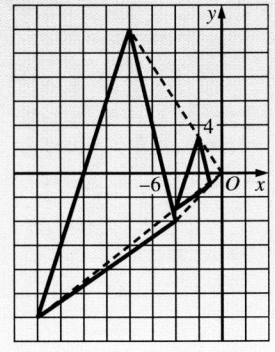

Before the invention of photocopiers, people could use *pantographs* to make enlargements. The title character in Mark Twain's *Pudd'nhead Wilson* uses one to make large drawings of fingerprints to display as evidence at a courtroom trial.

*H*e made fine and accurate reproductions of a number of his [fingerprint] "records," and then enlarged them on a scale of ten to one with his pantograph. [He] made each individual line of the bewildering maze of whorls or curves or loops . . . stand out bold and black by reinforcing it with ink [so] they resembled the markings of a block of wood that has been sawed across the grain, and the dullest eye could detect at a glance, and at a distance of many feet, that no two of the patterns were alike.

In the pantograph shown, *ABCD* is a parallelogram with joints that move so that *O, D,* and *E* always lie on a line. Point *O* is fixed to the drawing board.

To enlarge a figure, you put a stylus (a pointed object) at *D* and a pencil at *E*. If you guide the stylus at *D* so it traces your drawing, the pencil at *E* will draw an enlargement.

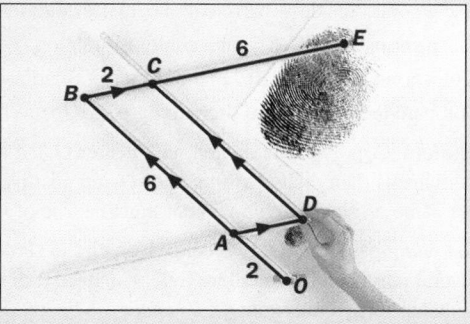

10. Find the value of each ratio.

   a. $\dfrac{OB}{OA}$ **b.** $\dfrac{BE}{AD}$ **c.** $\dfrac{OE}{OD}$

11. $\triangle OBE$ is the image of $\triangle OAD$ after a dilation. What is the center of the dilation? What is the scale factor?

12. What kind of dilation results if you put the stylus at *E* and a pencil at *D* and trace a drawing with the stylus?

13. Sketch the dimensions of a pantograph that Pudd'nhead Wilson could have used to enlarge his fingerprint collection by a scale factor of 10.

Exercise Notes

Interdisciplinary Problems
Exs. 10–13 In these exercises, students use what they have learned about dilation and scale factor to understand a paragraph of Mark Twain's novel *Pudd'nhead Wilson*. You might ask if anyone has read the novel and, if so, could he or she provide a brief description of the main plot.

Communication: Reading
Exs. 10–13 Before assigning these exercises, you may want to have volunteers read the opening paragraph and the description of the pantograph. Students need to work carefully through this description to understand how the pantograph works.

Second-Language Learners
Exs. 10–13 As students learning English read and respond to the literature passage, suggest that they look up any unfamiliar terms in a dictionary. If the language is too difficult, students can ask a peer tutor to rephrase or paraphrase the text.

Geometric Thinking
Ex. 12 After completing this exercise, you may wish to ask students to investigate the relationship between the scale factor of a dilation which transforms a figure into an image, and the scale factor of a dilation which transforms the image back into the original figure. (They are reciprocals.) This question is explored in Ex. 19.

9. Answers may vary. An example is given. A 220% enlargement is the largest that will allow the entire contents of the card to be copied onto the sheet of paper.

10. a. 4

   b. 4

   c. 4

11. O; 4

12. reduction

13.

7. a.	Figure	Scale	Image
	(2, 0)	3	(6, 0)
	(3, 5)	3	(9, 15)
	(−2, 4)	3	(−6, 12)

   b. (ka, kb)

c.	Figure	Scale (Ex. 5)	Image	Figure	Scale	Image (Ex. 6)
	(3, 0)	$\frac{1}{3}$	(1, 0)	(−1, −1)	4	(−4, −4))
	(3, 3)	$\frac{1}{3}$	(1, 1)	(−2, 3)	4	(−8, 12)
	(6, 6)	$\frac{1}{3}$	(2, 2)	(−4, −3)	4	(−16, −12)

8. a. $210\sqrt{2}$ mm $\times$ $297\sqrt{2}$ mm $\approx$ 297 mm $\times$ 420 mm

   b. $\dfrac{\sqrt{2}}{2} \approx 0.71$ or 71%

431

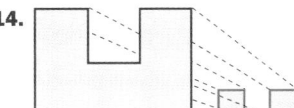

Exercise Notes

Common Error

Exs. 14–16 Some students may use the incorrect ratio when determining the scale factor. Students should check that their scale factor makes sense. It should be greater than 1 if the dilation is an enlargement and less than 1 if the dilation is a reduction.

Problem Solving

Ex. 20 After completing this exercise, ask students to use what they have learned to develop a method for dilating any triangle by a given scale factor *k*. They should first find the perpendicular bisectors of each side of the triangle. Their intersection is the center of dilation. Students can then proceed as follows. Using this center, they can draw a circle that contains the vertices of the triangle. Multiply the radius of this circle by the scale factor to find the radius of a circle that contains the vertices of the image. Draw this circle with the same center. Draw a ray from the center of the circle through each vertex of the original triangle. Connect the points where the second circle intersects the rays. This is the desired image.

Assessment Note

Ex. 21 After constructing the figure and its image, students may want to copy them using tracing paper and give the copy to their partners. This will ensure that the lines and center are not visible.

Copy each figure. The red figure is the image of the blue figure after a dilation. Find the center and the scale factor of the dilation.

14. 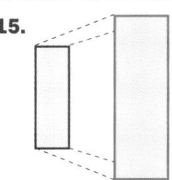 15. 16.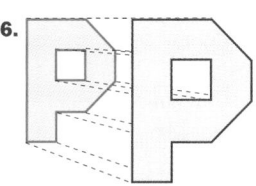

17. 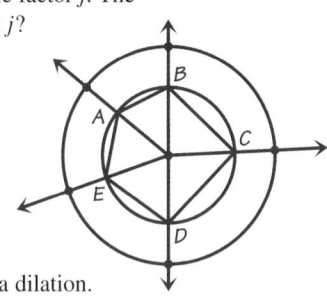 **Technology** Use geometry software to draw a polygon and two points.

   a. Dilate the polygon using one point as the center of dilation. Then dilate the image using the other point as a new center and a different scale factor.

   b. Your final polygon in part (a) and the original polygon are related by a single dilation. Find its scale factor.

   c. What do you notice about the location of the center in relation to the other two centers of dilation?

18. **Open-ended Problem** Draw any △*ABC* on graph paper. For each center of dilation, draw the image of △*ABC* after a dilation of scale factor 2.

   a. Use a point outside △*ABC* as the center of dilation.

   b. Use a vertex of △*ABC* as the center of dilation.

   c. Use a point inside △*ABC* as the center of dilation.

   d. Compare the images you drew in parts (a)–(c). What do you notice?

19. **Challenge** A polygon is dilated with center *O* and positive scale factor *k*. Its image is then dilated with center *O* and positive scale factor *j*. The image is the original polygon. What must be true of *k* and *j*?

20. **a.** Use a compass to draw two concentric circles. Plot several points on the smaller circle. Connect your points to form a polygon.

   b. Draw a ray from the center of the circle through each vertex.

   c. Connect the points where the circle intersects your rays. What do you notice about this polygon?

   d. **Writing** Explain why your diagram is an example of a dilation.

ONGOING ASSESSMENT

21. **Cooperative Learning** Work with another person. Use a ruler. Each person should draw a figure on plain paper and its image after a dilation. Trace the figure and its image without tracing the center or your construction lines. Exchange papers and find the center and scale factor your partner used.

---

14. center: *P*; scale factor: $\frac{1}{2}$

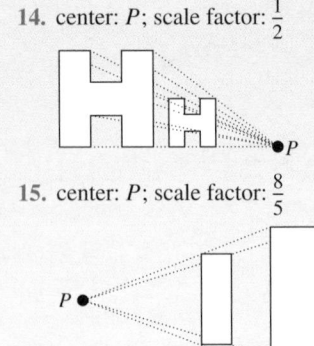

15. center: *P*; scale factor: $\frac{8}{5}$

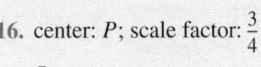

16. center: *P*; scale factor: $\frac{3}{4}$

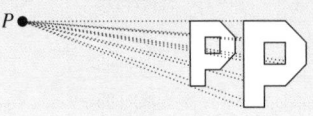

17. **a.** Check students' work.

   b. The scale factor of the single dilation is the product of the two scale factors.

   c. The center of the single dilation is on the line determined by the other two centers. If the first dilation has center *P* and scale factor *k* and the second has center *Q* and scale factor *j*, the single dilation has scale factor *jk* and center *R* on $\overline{PQ}$ such that $\frac{PR}{PQ} = \frac{j-1}{jk-1}$.

18. See answers in back of book.

19. $k = \frac{1}{j}$

20. **a, b.** Check students' work.

   c. It has the same shape as the original polygon, but corresponding sides are not congruent. Corresponding sides are, however, parallel.

   d. Let *O* be the center of the circles and *k* the ratio of the radii of the two circles. Every point *P* on the original polygon has an image *P′* on $\overrightarrow{OP}$ so that $\frac{OP'}{OP} = k$.

Complete each construction. *(Section 5.7)*

22. Draw an angle ∠ *KLM*. Then construct an angle congruent to it.

23. Draw any $\overleftrightarrow{AB}$ and a point *R* not on $\overleftrightarrow{AB}$. Construct the line through *R* that is perpendicular to $\overleftrightarrow{AB}$.

Find the lengths of the sides of each polygon whose vertices are given.
(Section 4.1)

24. *P*(0, 0), *Q*(5, 0), *R*(2, 3)   **25.** *X*(1, 1), *Y*(3, 3), *Z*(4, 0)

ASSESS YOUR PROGRESS

VOCABULARY

rotation (p. 412)   **glide reflection** (p. 420)
center of rotation (p. 412)   **dilation** (p. 428)
angle of rotation (p. 412)   **center of a dilation** (p. 428)
half-turn (p. 414)   **scale factor** (p. 428)

Sketch the images of each point after 90° and 180° rotations around the origin. *(Section 8.4)*

1. *X*(−2, 3)   **2.** *Y*(5, 0)

3. *Z*(3, −6)   **4.** *W*(0, −2)

Sketch the image of the triangle with the given vertices after a glide reflection using the given translation and reflection. *(Section 8.5)*

5. *A*(0, 0), *B*(5, 1), *C*(2, 6)   **6.** *D*(−3, 2), *E*(0, −1), *F*(3, 4)
translation: (*a*, *b*) → (*a* − 3, *b*)   translation: (*a*, *b*) → (*a*, *b* − 2)
reflection: over the *x*-axis   reflection: over the *y*-axis

Draw the image of each figure after a dilation with center O and the given scale factor. *(Section 8.6)*

7. scale factor 2   **8.** scale factor $\frac{1}{4}$

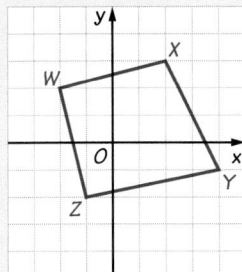   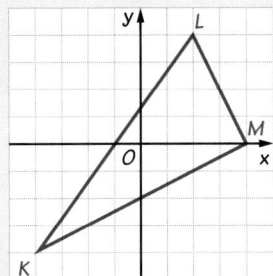

9. Journal You know how to describe *every* translation using coordinate notation. Write a description of the types of reflections, rotations, and dilations that you can describe using coordinate notation.

8.6 Dilations   **433**

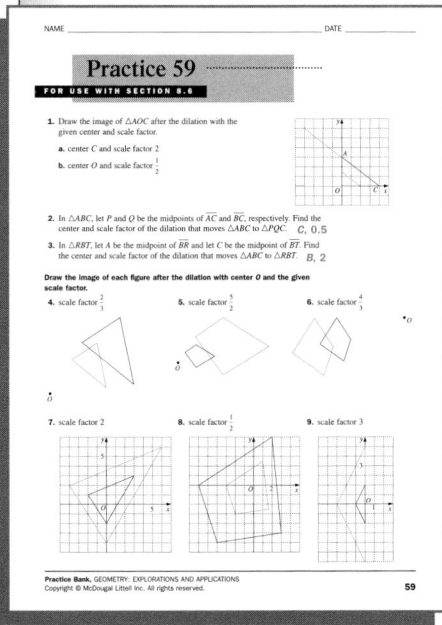

21–23. Check students' work.

24. *PQ* = 5; *PR* = $\sqrt{13}$ ≈ 3.6; *QR* = 3$\sqrt{2}$ ≈ 4.2

25. *XY* = 2$\sqrt{2}$ ≈ 2.8; *XZ* = *YZ* = $\sqrt{10}$ ≈ 3.2

Assess Your Progress

1–4. The image of a point *P* after a 90° rotation around the origin is labeled *P′*. The image after a 180° rotation around the origin is labeled *P″*.

1.
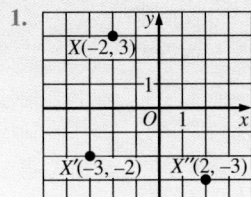

2.
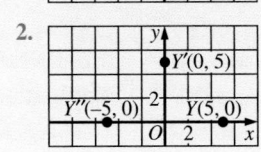

3.
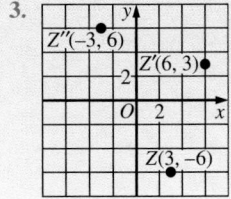

4.
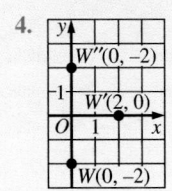

5.
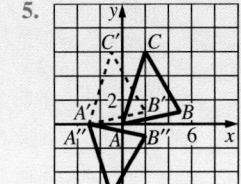

6.
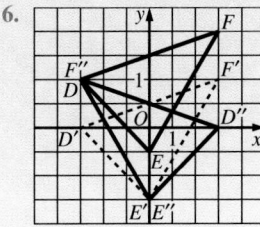

7–9. See answers in back of book.

Mathematical Goals

- Use transformations with triangles, rectangles, and other quadrilaterals to make tessellations.
- Create patterns that tessellate.
- Discover features of tessellating patterns.

Planning

Materials
- pencil and paper
- scissors
- cardboard
- tape
- protractor

Project Teams
Suggest that students select their own partner and work together to complete the project. In this way, students can combine their ideas to come up with the different patterns for the project.

Guiding Students' Work

There are many patterns that can be created for this project. Encourage students to use various sizes of rectangles. When students explore tessellations with other polygons, have them complete at least six transformations or combinations of transformations. Then ask them if any of the resulting tessellations can be described using transformations in more than one way. This is a good way for students to see that rotations and translations are actually combinations of two reflections.

Second-Language Learners
Make sure that all students learning English work with English proficient partners for this project, since it contains vocabulary and instructions that may be confusing to some students.

Tessellating the Plane

A tessellation is a pattern formed by repeating a shape to cover a plane without any gaps or overlapping. A brick walkway is an example of a tessellation that uses rectangles to cover the plane. A tiled wall is another example. So is a honeycomb—not all tessellations are made with quadrilaterals.

PROJECT GOAL Use transformations with different polygons to make tessellations.

Making Quadrilateral Tilings

1. Work with a partner. Cut several congruent rectangles from cardboard to represent bricks, and use transformations to make three different tessellations with them. Discuss with your partner the transformations you used.

2. Cut a quadrilateral that is *not* a rectangle from a piece of cardboard. Trace the quadrilateral several times to make a tessellation. Discuss the transformations that you used to make your pattern.

3. Choose any vertex on your sketch and measure the angles at that vertex. What do you notice about the sum of the measures of these angles? How might you have predicted this result without measuring?

Exploring with Other Polygons

Cut two identical rectangles from cardboard.

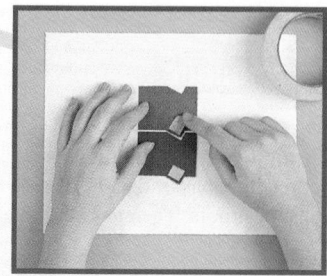

6. Trace the new shape many times to make a tessellation.

4. Place the two rectangles together and cut a piece from the bottom edges of both.

5. Tape a piece to each shape so the piece fills the missing space in the other shape.

Use this method or geometry software to produce three other shapes that will form tessellations when transformations are used. You can start with triangles or other polygons that tessellate. Experiment with reflections, translations, rotations, or combinations of these transformations.

Writing a Report

Write a report that describes your tessellations with quadrilaterals and other polygons.

- Describe the kinds of polygons that can be used to make tessellations and the transformations used with each kind.

- Include sketches or printouts of four of the tessellations that you describe in your report.

You may want to extend your report by including some tessellations that use combinations of two or more polygons to cover the plane.

Self-Assessment

Describe how you and your partner worked together in this project. How did your partner help you understand something about transformations and tessellations? How did you help your partner?

Portfolio Project   **435**

Project Notes

Guiding Students' Work

Rubric for Chapter Project

4 Students use transformations to complete the tessellations described in the project. The patterns students make are creative and accurately drawn. Students write a report describing their tessellations, how polygons can be used to make tessellations, and include four of the tessellations they describe. The report is well organized and clearly written. Students also extend the project using combinations of polygons that tessellate to cover the plane. Students also complete a self-assessment of the project.

3 Students use transformations to complete most of the tessellations for the project. A report is written and is complete, but the sketches of the tessellations are not always well done. Students extend the project by including one tessellation that uses combinations of two or more polygons to cover the plane. Students complete a self-assessment of the project but do not answer all of the questions.

2 Students use transformations to complete part of the project; however, some of the drawings and examples of tessellations are missing. Students write a report about their project, but it is also incomplete and shows a minimal effort. Some evidence of experimentation with transformations is shown in the report along with some sketches of tessellations, but students do not extend the project or complete a self-assessment.

1 Students show little or no effort to make the tessellations described in the project. The tessellations that are made are incomplete and do not show how transformations were used to develop the pattern. Students do not write a report, extend the project, or complete a self-assessment of their work. Students should be encouraged to speak with the teacher as soon as possible to review their work and to make a new start on the project.

Progress Check 8.1–8.3

Given quadrilateral *PROS*.

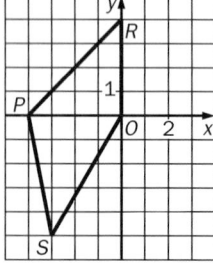

1. Draw quadrilateral *P′R′O′S′*, the reflection of *PROS* over the *y*-axis. *(Section 8.1)*

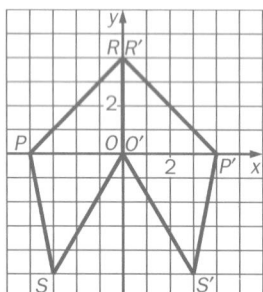

2. Name a segment in *P′R′O′S′* that is congruent to $\overline{SO}$. *(Section 8.1)* $\overline{S'O'}$

3. Give the coordinates of the vertices of *P′R′O′S′*. *(Section 8.2)* *P′*(4, 0), *R′*(0, 4), *O′*(0, 0), *S′*(3, −5)

4. Suppose quadrilateral *PROS* is reflected over the line *y = x*. Give the coordinates of the image points. *(Section 8.2)* *P′*(0, −4), *R′*(4, 0), *O′*(0, 0), *S′*(−5, −3)

5. Sketch the image of *PROS* after the translation (*a*, *b*) → (*a* + 1, *b* − 2). *(Section 8.3)*

CHAPTER

8 Review

STUDY TECHNIQUE

Write at least six questions about the chapter. Focus on the concepts that you had the most difficulty learning. Three should be short answer questions about specific situations or details. Three should be more involved questions, justifications, or proofs. Then answer the questions.

VOCABULARY

line of reflection (p. 391)

reflection (p. 391)

image (p. 391)

translation (p. 406)

rotation (p. 412)

center of rotation (p. 412)

angle of rotation (p. 412)

half-turn (p. 414)

glide reflection (p. 420)

dilation (p. 428)

center of a dilation (p. 428)

scale factor (p. 428)

SECTIONS 8.1 *and* 8.2

A *reflection* is a transformation that preserves congruence but reverses orientation. After a **reflection**, each point of a figure and its **image** lie equidistant from the **line of reflection**, but on opposite sides of it.

The line of reflection is the perpendicular bisector of every segment that connects a point and its image.

When reflected over the **y-axis**, the *x*-coordinate becomes its opposite. (*a*, *b*) → (−*a*, *b*)

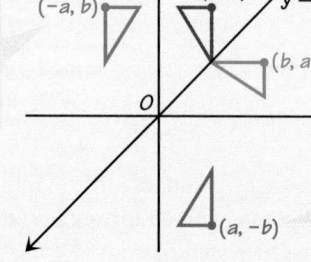

When reflected over the line *y = x*, the coordinates switch places. (*a*, *b*) → (*b*, *a*)

When reflected over the **x-axis**, the *y*-coordinate becomes its opposite. (*a*, *b*) → (*a*, −*b*)

Points that lie on the line of reflection do not change their position.

SECTIONS 8.3 *and* 8.4

A **translation** is a shift that can be described as $(a, b) \rightarrow (a + h, b + k)$. Every point moves h units horizontally and k units vertically. A translation preserves congruence and orientation.

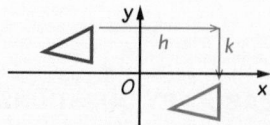

In a **rotation**, every point moves along a circular path around a fixed point called the **center of rotation**. The angle formed by any point, the center of rotation, and the image of the point is called the **angle of rotation**.

A rotation preserves congruence and orientation. When a point is rotated counterclockwise around the origin:

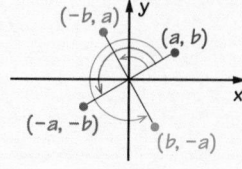

90°: $(a, b) \rightarrow (-b, a)$
180°: $(a, b) \rightarrow (-a, -b)$
270°: $(a, b) \rightarrow (b, -a)$

SECTIONS 8.5 *and* 8.6

A **glide reflection** is the result of a translation followed by a reflection over a line parallel to the translation. A glide reflection preserves congruence but reverses orientation.

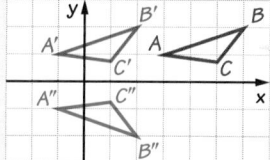

Translated to the left 4 units and then reflected over the x-axis

A **dilation** is a transformation that may enlarge or reduce a figure. In a dilation with **center** O, for every point A and its image A', there is a positive **scale factor** k such that $\frac{OA'}{OA} = k$, and $OA' = k \cdot OA$.

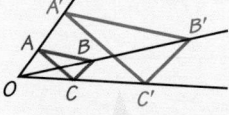 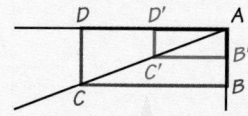

The center of a dilation is at the intersection of lines that connect points and their images.

If $k > 1$, the dilation is an *enlargement*.

If $k < 1$, the dilation is a *reduction*.

The image of any segment after a dilation is k times as long as the original segment. Dilations do not preserve congruence unless $k = 1$.

Review   **437**

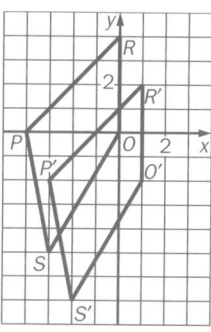

6. Describe this translation using coordinate notation. Every point moves 3 units to the left and 4 units down. *(Section 8.3)*
$(a, b) \rightarrow (a - 3, b - 4)$

Progress Check 8.4–8.6

Given $\triangle ABC$ with vertices $A(-6, 0)$, $B(0, -2)$, and $C(-3, -3)$. Draw its image after each transformation.

1. rotation of 90° around the origin *(Section 8.4)*

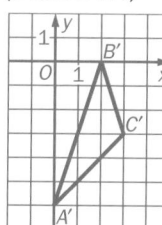

2. rotation of 270° around the origin *(Section 8.4)*

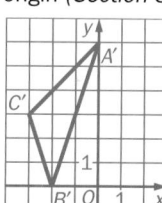

3. a glide reflection with translation $(a, b) \rightarrow (a, b + 2)$ and reflection over the y-axis *(Section 8.5)*

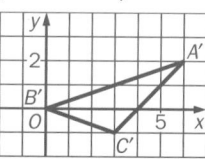

4. a dilation with center at the origin and scale factor $\frac{1}{3}$ *(Section 8.6)*

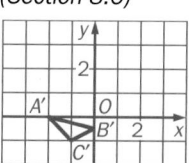

5. What center and scale factor would be needed to transform the image from Ex. 4 to the original figure? *(Section 8.6)*
center $(0, 0)$ and scale factor 3

437

Chapter 8 Assessment
Form A Chapter Test

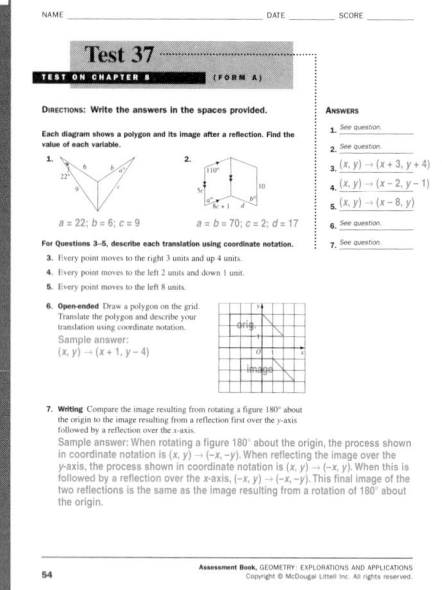

NAME _____ DATE _____ SCORE _____

Test 37
TEST ON CHAPTER 8 **(FORM A)**

DIRECTIONS: Write the answers in the spaces provided.

Each diagram shows a polygon and its image after a reflection. Find the value of each variable.

1.   2.

$a = 22; b = 6; c = 9$ $a = b = 70; c = 2; d = 17$

For Questions 3–5, describe each translation using coordinate notation.

3. Every point moves to the right 3 units and up 4 units.
4. Every point moves to the left 2 units and down 1 unit.
5. Every point moves to the left 8 units.

6. **Open-ended** Draw a polygon on the grid. Translate the polygon and describe your translation using coordinate notation.
Sample answer: $(x, y) \rightarrow (x + 1, y - 4)$

7. **Writing** Compare the image resulting from rotating a figure 180° about the origin to the image resulting from a reflection first over the y-axis followed by a reflection over the x-axis.
Sample answer: When rotating a figure 180° about the origin, the process shown in coordinate notation is $(x, y) \rightarrow (-x, -y)$. When reflecting the image over the y-axis, the process shown in coordinate notation is $(x, y) \rightarrow (-x, y)$. When this is followed by a reflection over the x-axis, $(-x, y) \rightarrow (-x, -y)$. This final image of the two reflections is the same as the image resulting from a rotation of 180° about the origin.

ANSWERS
1. See question.
2. See question.
3. $(x, y) \rightarrow (x + 3, y + 4)$
4. $(x, y) \rightarrow (x - 2, y - 1)$
5. $(x, y) \rightarrow (x - 8, y)$
6. See question.
7. See question.

54   Assessment Book, GEOMETRY: EXPLORATIONS AND APPLICATIONS
Copyright © McDougal Littell Inc. All rights reserved.

Chapter 8 Assessment
Form B Chapter Test

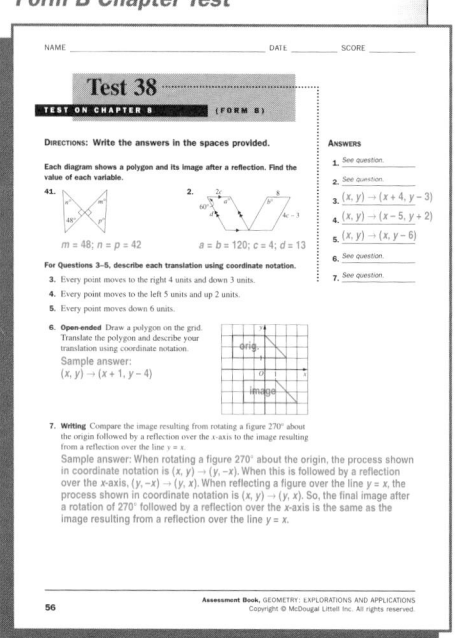

NAME _____ DATE _____ SCORE _____

Test 38
TEST ON CHAPTER 8 **(FORM B)**

DIRECTIONS: Write the answers in the spaces provided.

Each diagram shows a polygon and its image after a reflection. Find the value of each variable.

41.   2.

$m = 48; n = p = 42$ $a = b = 120; c = 4; d = 13$

For Questions 3–5, describe each translation using coordinate notation.

3. Every point moves to the right 4 units and down 3 units.
4. Every point moves to the left 5 units and up 2 units.
5. Every point moves down 6 units.

6. **Open-ended** Draw a polygon on the grid. Translate the polygon and describe your translation using coordinate notation.
Sample answer: $(x, y) \rightarrow (x + 1, y - 4)$

7. **Writing** Compare the image resulting from rotating a figure 270° about the origin followed by a reflection over the x-axis to the image resulting from a reflection over the line y = x.
Sample answer: When rotating a figure 270° about the origin, the process shown in coordinate notation is $(x, y) \rightarrow (y, -x)$. When this is followed by a reflection over the x-axis, $(y, -x) \rightarrow (y, x)$. When reflecting a figure over the line y = x, the process shown in coordinate notation is $(x, y) \rightarrow (y, x)$. So, the final image after a rotation of 270° followed by a reflection over the x-axis is the same as the image resulting from a reflection over the line y = x.

ANSWERS
1. See question.
2. See question.
3. $(x, y) \rightarrow (x + 4, y - 3)$
4. $(x, y) \rightarrow (x - 5, y + 2)$
5. $(x, y) \rightarrow (x, y - 6)$
6. See question.
7. See question.

56   Assessment Book, GEOMETRY: EXPLORATIONS AND APPLICATIONS
Copyright © McDougal Littell Inc. All rights reserved.

VOCABULARY QUESTIONS

Copy the diagram. In each space, write the name of the transformation indicated by the arrow of the same color.

1. ?

2. ?

3. ?

4. ?

SECTIONS 8.1 *and* 8.2

Copy each diagram. Sketch the image of each polygon after a reflection over the given line. Label the image polygon.

5.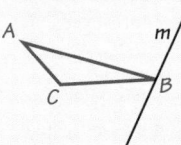

6.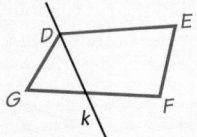

7. The diagram shows a triangle and its image after a reflection. Draw the line of reflection.

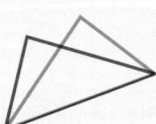

8. Reflect *MNPQ* over the line $y = x$. Give the coordinates of the vertices of the image.

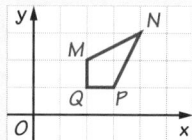

9. On a golf course, a player at $A(2, 5)$ wants to reach the cup at $C(10, 3)$ by bouncing off the bumpboard at a point B along the x-axis.

   a. Make a sketch to represent the shot. Draw the path of the ball.

   b. What are the coordinates of point B on the bumpboard?

438   Chapter 8 *Using Transformations*

ANSWERS   **Chapter 8**

Assessment

1–4.

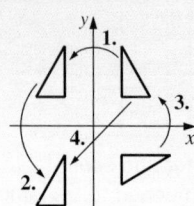

1. reflection   2. translation
3. rotation   4. glide reflection

5.

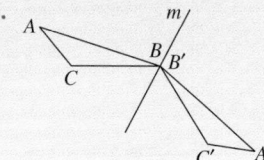

6.

7.

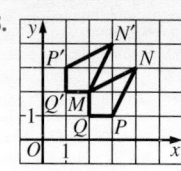

8.

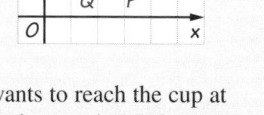

$M'(2, 2); N'(3, 4); P'(1, 3); Q'(1, 2)$

9. a.

b. $(7, 0)$

10. $(a, b) \rightarrow (a + 3, b - 1)$

11. $(a, b) \rightarrow (a - 1, b - 1)$

SECTIONS 8.3 *and* 8.4

The image of each polygon after a translation is shown in red. Describe each translation using coordinate notation.

10.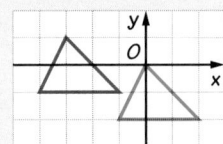

11.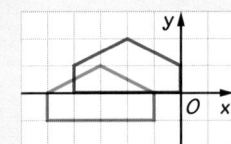

12. $\triangle ABC$ has vertices $A(-2, 1)$, $B(1, 2)$, $C(2, 4)$. Sketch its image after a 180° rotation around the origin.

13. Open-ended Problem Draw a triangle and a point P outside it. Rotate the triangle 90° counterclockwise around P.

SECTION 8.5 *and* 8.6

14. For the following glide reflection, find the distance of the translation and draw the line of reflection.

15. Sketch the image of the triangle with vertices $D(2, 2)$, $E(-1, 1)$, $F(3, -2)$ after a glide reflection using the translation: $(a, b) \to (a + 2, b - 1)$ and reflection over the line $y = -x$. Find the vertices of the image.

Draw the image of each triangle after a dilation with center *P* and the given scale factor.

16. scale factor $\frac{1}{2}$

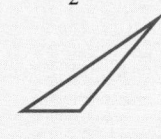

17. scale factor $\frac{3}{2}$

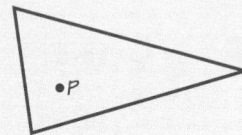

PERFORMANCE TASK

18. On graph paper, prepare two designs for a wallpaper company. Use a simple design element, such as a curved leaf. Make one design that uses rotations and reflections. Make another that uses glide reflections. Describe the transformations using coordinate notation if possible.

Assessment **439**

12.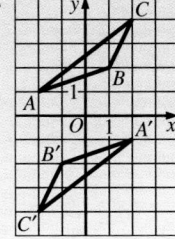

13. Answers may vary. An example is given.

14. 6.2 cm

15.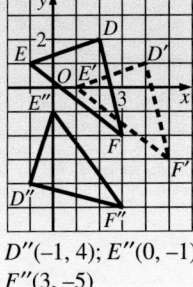

$D''(-1, 4)$; $E''(0, -1)$; $F''(3, -5)$

16.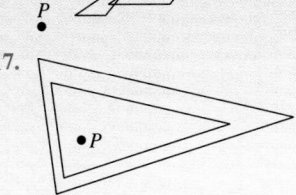

17.

18. Answers may vary. Check students' work.

439

Algebra
Review/Preview

These exercises review algebra topics you will use in the next chapters.

Simplify each expression. If it is not possible, write *cannot be simplified*.

Toolbox p. 690
Simplifying and Evaluating Expressions

EXAMPLES

$$x + \frac{1}{2}x^2 - \frac{5}{6}x^2 + 3x$$

$$= \left(\frac{1}{2}x^2 - \frac{5}{6}x^2\right) + (1x + 3x) \quad \longleftarrow \text{ Group the like terms.}$$

$$= \left(\frac{1}{2} - \frac{5}{6}\right)x^2 + (1 + 3)x \quad \longleftarrow \text{ Use the distributive property.}$$

$$= \left(\frac{3}{6} - \frac{5}{6}\right)x^2 + 4x \quad \longleftarrow \text{ Use the least common denominator (LCD).}$$

$$= -\frac{2}{6}x^2 + 4x$$

$$= -\frac{1}{3}x^2 + 4x \quad \longleftarrow \text{ Simplify the answer.}$$

$4y^2 + 5y$ ***cannot be simplified*** because $4y^2$ and $5y$ are not like terms.

1. $2l + 2w$ **2.** $12y^2 - y^2$ **3.** $7a + 4b - ab$

4. $\pi r^2 + 4\pi r^2$ **5.** $n^2 - 3n + 6n^2$ **6.** $a^2 - a - (a - 4a^2)$

7. $\frac{1}{3}x^2 + \frac{3}{4}x^2$ **8.** $\frac{1}{5}k^2 + \frac{1}{2}k + \frac{3}{10}k^2$ **9.** $x^2 + 7.4x^2 - 9.8x^2$

Simplify.

Toolbox p. 700
Simplifying Radicals

EXAMPLES

$$\sqrt{96} = \sqrt{16 \cdot 6} \quad \longleftarrow \text{ 16 is the largest perfect square factor of 96.}$$

$$= \sqrt{16} \cdot \sqrt{6} \quad \longleftarrow \sqrt{ab} = \sqrt{a} \cdot \sqrt{b}$$

$$= 4\sqrt{6} \quad \longleftarrow \sqrt{16} = \sqrt{4^2} = 4$$

$$(2\sqrt{5})^2 = 2\sqrt{5} \cdot 2\sqrt{5}$$

$$= 2 \cdot 2 \cdot \sqrt{5} \cdot \sqrt{5}$$

$$= 4 \cdot 5 \quad \longleftarrow \sqrt{5} \cdot \sqrt{5} = \sqrt{25} = 5$$

$$= 20$$

$$\sqrt{\frac{8}{81}} = \frac{\sqrt{8}}{\sqrt{81}} = \frac{\sqrt{4 \cdot 2}}{\sqrt{9 \cdot 9}} = \frac{2\sqrt{2}}{9}$$

ANSWERS   Chapter 8

Algebra Review/Preview

1. cannot be simplified
2. $11y^2$
3. cannot be simplified
4. $5\pi r^2$
5. $7n^2 - 3n$
6. $5a^2 - 2a$
7. $\frac{13}{12}x^2$
8. $\frac{1}{2}k^2 + \frac{1}{2}k$
9. $-1.4x^2$
10. $10\sqrt{3}$
11. $7\sqrt{6}$
12. $6\sqrt{5}$
13. 37
14. $\frac{15}{2}$
15. $\frac{3\sqrt{5}}{4}$
16. $\frac{1}{18}$
17. 11
18. 192
19. 105
20. $25\sqrt{6}$

10. $\sqrt{300}$ **11.** $\sqrt{294}$ **12.** $\sqrt{180}$ **13.** $\sqrt{37^2}$

14. $\sqrt{\dfrac{15}{4}}$ **15.** $\sqrt{\dfrac{45}{16}}$ **16.** $\sqrt{\dfrac{1}{324}}$ **17.** $\left(\sqrt{11}\right)^2$

18. $\left(8\sqrt{3}\right)^2$ **19.** $7\sqrt{225}$ **20.** $5\sqrt{10}\cdot\sqrt{15}$ **21.** $2\sqrt{3}\cdot 6\sqrt{6}$

SOLVING QUADRATIC EQUATIONS

Solve each quadratic equation.

EXAMPLES

$$x^2 = 4^2 + \left(4\sqrt{5}\right)^2 \qquad 5^2 + u^2 = 13^2$$
$$x^2 = 16 + 80 \qquad\qquad 25 + u^2 = 169$$
$$x^2 = 96 \qquad\qquad\qquad u^2 = 144$$
$$x = \pm\sqrt{96} \qquad\qquad u = \pm\sqrt{144}$$
$$x = \pm 4\sqrt{6} \qquad\qquad u = 12 \text{ or } u = -12$$
$$x = 4\sqrt{6} \text{ or } x = -4\sqrt{6}$$

Toolbox p. 701
Solving Simple Quadratic Equations

22. $x^2 = 9^2 + 12^2$ **23.** $y^2 = 5^2 + \left(5\sqrt{3}\right)^2$ **24.** $z^2 + 4^2 = 8^2$

25. $a\cdot 4a = 12\cdot 18$ **26.** $s^2 = \left(\sqrt{2}\right)^2 + \sqrt{4}$ **27.** $k^2 = 3^2 + 2^2$

28. $x^2 + 9^2 = 13^2$ **29.** $b\cdot 5b = 15\cdot 7$ **30.** $2\sqrt{6^2} + 6\sqrt{2^2} = c^2$

31. $x^2 + x^2 = 6^2$ **32.** $x^2 + \left(9\sqrt{3}\right)^2 = 4x^2$ **33.** $x^2 + \left(x\sqrt{3}\right)^2 = 8^2$

34. Refer to Exercise 32. Suppose x, $9\sqrt{3}$, and $2x$ represent the lengths of the sides of a right triangle. Find the value of x and the length of the hypotenuse of the triangle.

SOLVING MORE PROPORTIONS

Solve each proportion.

EXAMPLES

$$\frac{a}{a+6} = \frac{3}{5} \qquad\qquad 12:x = x:2 \leftarrow \frac{12}{x} = \frac{x}{2}$$
$$5a = 3(a+6) \qquad\qquad x\cdot x = 12\cdot 2$$
$$5a = 3a + 18 \qquad\qquad x^2 = 24$$
$$2a = 18 \qquad\qquad\qquad x = \pm\sqrt{24}$$
$$a = 9 \qquad\qquad\qquad = \pm 2\sqrt{6}$$

Toolbox p. 707
Ratio and Proportion

35. $\dfrac{x}{21-x} = \dfrac{16}{12}$ **36.** $5x:2x = 2x:16$ **37.** $\dfrac{1.8}{0.4} = \dfrac{y+2}{5}$

38. $\dfrac{12}{r} = \dfrac{9}{15}$ **39.** $\dfrac{t+6}{t} = \dfrac{7t}{5t}$ **40.** $3^2:5^2 = 225:A$

41. $3:z = z:5$ **42.** $\dfrac{A}{4\pi} = \dfrac{3^2}{(1.2)^2}$ **43.** $\dfrac{63}{x} = \dfrac{x}{7}$

44. $n:8 = 10:n$ **45.** $\dfrac{4}{x^2} = \dfrac{1}{9}$ **46.** $\dfrac{25}{4x} = \dfrac{x}{36}$

Algebra Review/Preview **441**

---

21. $36\sqrt{2}$
22. ± 1.5
23. ± 10
24. $\pm 4\sqrt{3}$
25. $\pm 3\sqrt{6}$
26. ± 2
27. $\pm\sqrt{13}$
28. $\pm 2\sqrt{22}$
29. $\pm\sqrt{21}$

30. $\pm 2\sqrt{6}$
31. $\pm 3\sqrt{2}$
32. ± 9
33. 4
34. 9; 18
35. 12
36. 20
37. 20.5
38. 20

39. 15
40. 625
41. $\pm\sqrt{15}$
42. 25π
43. ± 21
44. $\pm 4\sqrt{5}$
45. ± 6
46. ± 15

9 Similar Polygons

OVERVIEW

Connecting to Prior and Future Learning

⇔ Students continue their work with polygons by studying the properties of similar figures, with a special emphasis on similar triangles in Sections 9.2 and 9.3. The concept of proportions is used throughout the chapter, so students may find it helpful to study page 708 in the **Student Resources Toolbox** which presents a review of solving proportions.

⇔ Students find areas and volumes of similar figures in Section 9.4. This section builds on the concepts presented in the first part of the chapter, and students continue to use ratios and proportions in their work.

⇔ This chapter also includes an introduction to geometric probability. Students use their knowledge of finding the area of a figure as they study this topic. All of the concepts presented in this chapter are building blocks for future studies in mathematics.

Chapter Highlights

Interview with Loy Arcenas: Mathematics is used by set designers when designing sets for plays. This fact is emphasized in the interview with Loy Arcenas, a set designer. Students have an opportunity to study this relationship in more detail in the related exercises found on pages 465, 468, and 472.

Explorations in Chapter 9 ask students to use geometry software, or patty paper and a ruler, to find proportions in triangles, and paper and dried beans to investigate probability based on area.

The Portfolio Project: Students use the properties of similar figures and spreadsheets to create a model of our solar system. Students choose a scale and then create scale models of the sun and each planet in the solar system. Students summarize their work by presenting their model and data to the class.

Technology: Students use spreadsheets to explore scale factors in Section 9.1. In Section 9.2, they use geometry software to explore properties of similar quadrilaterals. Graphing software or a graphing calculator is used in Section 9.4 to explore similar cubes.

OBJECTIVES

Section	Objectives	NCTM Standards
9.1	• Use ratios and proportions to find measures of similar figures. • Identify similar figures. • Use scale models to find the dimensions of an object being modeled.	1, 2, 3, 4, 5, 7
9.2	• Use special rules to identify similar triangles. • Write proofs involving similar triangles.	1, 2, 3, 4, 5, 7
9.3	• Write many different proportions using the sides of similar figures. • Find lengths and measures in similar polygons.	1, 2, 3, 4, 5, 7
9.4	• Compare measures of similar figures. • Find perimeters, areas, and volumes of similar figures.	1, 2, 3, 4, 5, 7
9.5	• Use area and length to find probabilities. • Solve problems involving geometric probability.	1, 2, 3, 4, 5, 9, 12

Mathematical Connections	9.1	9.2	9.3	9.4	9.5
geometry	**445–452***	**453–460**	**461–467**	**468–474**	**475–481**
algebra	448, 449		462, 464, 466	470, 474	
data analysis, probability, discrete math	450			472, 474	**475–481**
patterns and functions	452				480
logic and reasoning	450, 451	453, 458, 460	463, 466	472	478

Interdisciplinary Connections and Applications					
history and geography	451				
sports and recreation					480
graphic design and technical drawing	447, 451				
astronomy	450				
consumer economics and advertising				473	477, 480
electronics		459			
surveying		457			
movies, sculpture, set design, zoology	451, 452			472	479

*__Bold page numbers__ indicate that a topic is used throughout the section.

Section	opportunities for use with	
	Student Book	Support Material
9.1	graphing calculator spreadsheet software	**Technology Book:** Calculator Activity 12
9.2	geometry software	**Technology Book:** Spreadsheet Activity 8
9.3	geometry software	**Technology Book:** Spreadsheet Activity 8 **Function Investigator Activity Book:** Activity 23
9.4	graphing calculator geometry software	**Technology Book:** TI-92 Activity 9 **Geometry Inventor Activity Book:** Activity 4
9.5	scientific calculator	**Probability Constructor Activity Book:** Activities 7, 10, 14, 15

Regular Scheduling (45 min)

Section	Materials Needed	Core Assignment	Extended Assignment	exercises that feature		
				Applications	Communication	Technology
9.1	ruler, spreadsheet software, graph paper	**Day 1:** 1–18 **Day 2:** 19, 23, 24, 26–28, 31–37	**Day 1:** 1–18 **Day 2:** 19–37	19–21, 23, 26, 28, 29, 31	23, 24, 26b, 29	20, 21
9.2	geometry software or ruler, protractor	1–8, 12, 14–17, 22–31	1–31	9–11, 18–21	9, 11, 22b, 23, 27b	24–26
9.3	geometry software or patty paper, ruler, compass	**Day 1:** 1–9, 14–21 **Day 2:** 23–31, AYP*	**Day 1:** 1–21 **Day 2:** 22–31, AYP	10–13	10, 17 23	
9.4	geometry software or graphing calculator, graph paper	**Day 1:** 1–13, 15, 18 **Day 2:** 20–22, 24–26, 28–39	**Day 1:** 1–19 **Day 2:** 20–39	17 20–23	16d, 17 22c, 26b, 27	16
9.5	dried beans	1–6, 8–11, 19–27, AYP	1–27, AYP	12–18, 20	13, 21	
Review/ Assess		**Day 1:** 1–8 **Day 2:** 9–17 **Day 3:** Ch. 9 Test	**Day 1:** 1–8 **Day 2:** 9–17 **Day 3:** Ch. 9 Test			
Portfolio Project		Allow 2 days.	Allow 2 days.			

Yearly Pacing (with Portfolio Project)	Chapter 9 Total 13 days	Chapters 1–9 Total 119 days	Remaining 41 days	Total 160 days

Block Scheduling (90 min)

	Day 54	Day 55	Day 56	Day 57	Day 58	Day 59	Day 60
Teach/Interact	9.1	9.2 9.3: Exploration, page 461	Continue with 9.3 9.4	Continue with 9.4 9.5: Exploration, page 475	Review Port. Proj.	Review Port. Proj.	Ch. 9 Test 10.1: Exploration, page 493
Apply/Assess	**9.1:** 1–19, 23, 24, 26–28, 31–37	**9.2:** 1–8, 12, 14–17, 22–31 **9.3:** 1–9, 14–21	**9.3:** 23–31, AYP* **9.4:** 1–13, 15, 18	**9.4:** 20–22, 24–26, 28–39 **9.5:** 1–6, 8–11, 19–27, AYP	**Review:** 1–8 **Port. Proj.**	**Review:** 9–17 **Port. Proj.**	**Ch. 9 Test** 10.1: 1–8, 10–19, 26–32

NOTE: A one-day block has been added for the Portfolio Project—timing and placement to be determined by teacher.

Yearly Pacing (with Portfolio Project)	Chapter 9 Total $6\frac{1}{2}$ days	Chapters 1–9 Total $59\frac{1}{2}$ days	Remaining $20\frac{1}{2}$ days	Total 80 days

*__AYP__ is Assess Your Progress.

Section	Practice Bank	Study Guide*	Assessment Book*	Visuals	Explorations Lab Manual	Lesson Plans	Technology Book
9.1	61	9.1		Warm-Up 9.1	Master 2	9.1	Calculator Act. 12
9.2	62	9.2		Warm-Up 9.2 Folder 11		9.2	Spreadsheet Act. 8
9.3	63	9.3	Test 39	Warm-Up 9.3 Folder 11	Master 2	9.3	Spreadsheet Act. 8
9.4	64	9.4		Warm-Up 9.4 Folder 11	Master 1	9.4	TI-92 Act. 9
9.5	65	9.5	Test 40	Warm-Up 9.5	Add. Expl. 14	9.5	
Review Test	66	Chapter Review	Tests 41, 42, Alternative Assessment			Review Test	

*Spanish versions of *Study Guide* and *Assessment Book* are available.

Chapter Support

- Course Guides
- Lesson Plans
- Portfolio Project Book:
  Additional Project 5:
    Creating Special Effects
- Preparing for College Entrance Tests
- Multi-Language Glossary
- *Test Generator* Software
- Professional Handbook
- Challenge Problems

Software Support

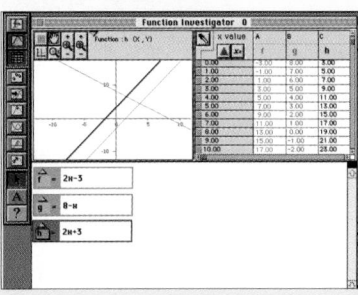

McDougal Littell Mathpack
Geometry Inventor
Function Investigator
Probability Constructor

Internet Support

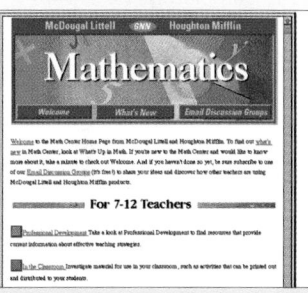

http://www.hmco.com
Next go to McDougal Littell; then the
Education Center; then Secondary Math.

Books, Periodicals

Sandefur, James T. "Using Similarity to Find Length and Area." *Mathematics Teacher* (May 1994): pp. 319–325.

Haruta, Mako E., Mark Flaherty, Jean McGivney, and Raymond J. McGivney. "Coin Tossing." *Mathematics Teacher* (November 1996): pp. 642–645.

Software

Department of Mathematics and Computer Science of the North Carolina School of Science and Mathematics. *Geometric Probability.* Materials and software. IBM. Reston, VA: NCTM, 1988.

Jackiw, Nicholas, designer. *The Geometer's Sketchpad.* (Ver. 3.0) for Macintosh or Windows. Berkeley, CA: Key Curriculum Press, 1995.

Laborde, Jean-Marie and Franck Bellemain, designers. *Cabri Geometry II.* Dallas, TX: Texas Instruments, 1994.

TI-92 Geometry. (Resident on the TI-92 graphing calculator.) Dallas, TX: Texas Instruments, 1995.

Videos

Math TV: *Geometry.* Includes fractals, Fibonacci series, symmetry, and non-Euclidean geometry. Arlington, VA: COMAP.

Apostol, Tom M. *Similarity.* Reston, VA: NCTM, 1989.

Internet

Investigate Internet Math Projects at:
    http://forum.swarthmore.edu/
    projects.html

Background

The Stage

When deciding on how to make a set for a play, the set designer must know the type of stage that will be used. In theater, there are four basic types of stage. The first is the *proscenium stage* or picture frame stage. This type of stage often contains a curtain that is used to hide or reveal scenery and allows for elaborate three-dimensional sets to be constructed. The second type of stage is the *open stage,* which allows the audience to view the play from three sides. Large pieces of scenery are used only at the back of the stage so the audience's view is not obstructed. The third type of stage is the *theater-in-the-round*, which allows the audience to completely surround the stage and view the play from all four sides. Scene changes for this stage are made either in darkness or in view of the audience. The fourth stage type is called *flexible theater*, which allows the stage and audience to be rearranged for each individual production.

Loy Arcenas

Loy Arcenas came to New York City in 1978 as part of the Third World Theatre Arts Studies program at La Mama ETC. Initially, Arcenas had difficulty finding work, but by the early 1980s, he was working steadily. Since then, he has designed sets throughout the United States. In 1993, Arcenas won the Obie Award for Sustained Achievement. To date, he has designed sets in Off-Off Broadway, Off-Broadway, Broadway, regional theater, opera, and television. His favorite, though, is regional theater.

CHAPTER

9 | Similar Polygons

Setting the Scene

INTERVIEW Loy Arcenas

oy Arcenas planned to become a doctor before his career took a dramatic turn. After completing premedical studies at the University of the Philippines, he went to London for training at the Drama Studio and a design course at the English National Opera. He tried both acting and set design. Now, Arcenas is one of the most sought-after set designers in the United States.

When designing a set, Arcenas begins by reading the play. While reading, he has "small ideas that may seem to have nothing to do with one another." Then he finds what these ideas have in common and uses them to sketch the set for the play.

"Much of theater design is based on illusion; there are many tricks involved."

Sculptures that Move

Arcenas relies on careful calculations to guarantee that the pieces in his set will fit together when assembled in the shop. But his job involves a lot more than details. It's an artistic endeavor: the sets are movable sculptures that set the mood for the play. In Shakespeare's comedy *All's Well That Ends Well*, Arcenas used soft blue tones to create a dream-like quality. His set for Shakespeare's history *Henry IV* was dominated by large, sculpted metallic pieces that suggested wealth and power.

The biggest challenge is establishing a mood. The props, backdrops, and lights he chooses can affect the actors and audience emotionally. "It's really about creating an illusion," he says. When he does the job right, the set makes the action on stage seem real to the actors and audience.

Scenes from *All's Well That Ends Well*

"Set design is really about the relation of the actors to the space around them."

Next, he builds a *scale model* out of cardboard and other inexpensive materials. This model helps him understand and improve his plan for the stage. The models always include miniature people so that Arcenas can check the proportions. "Otherwise, we might end up with a well-proportioned room with a door that's just five feet tall," he says. Using the model, he also checks that the audience will be able to see the whole stage.

Scene from *Henry IV*

443

Background

Konstantin Stanislavski

Konstantin Stanislavski (1863–1938) was the stage name of Konstantin Alekseyev, a Russian actor who developed the *Method* style of acting. This type of acting was the first to require actors to call upon past experiences that might be relevant to their role. This method completely changed how theater was performed. Before Stanislavski developed his method, acting was often artificial. To train actors to perform using his method, Stanislavski founded the Moscow Art Theater with Vladimir Nemirovich-Danchenko in 1898. This theater became one of the most influential theaters of the 1900s.

Second-Language Learners

Students learning English may not recognize the play on words in the first sentence. If necessary, explain that Loy Arcenas's career took a *dramatic turn* in more than one way. First, taking a *dramatic turn* generally means making a big change—the dramatic turn in Arcenas's career was from medicine to set design. Second, Arcenas's career took a *dramatic turn* because it turned toward drama.

Multicultural Note

Theater takes many different forms around the world. In Jamaican *jonkonnu*, the dancers dress up in colorful costumes as a variety of folk figures. Jonkonnu, which reflects Jamaica's mixed African and European heritage, features a "Break Out," in which the dancers simultaneously perform individual dances. Vietnamese water puppetry was invented by rice farmers who were working in flooded fields. Performers stand up to their chests in water, which hides the rods and strings used to control the puppets, provides sound effects, and serves as a stage. Indonesian puppetmasters perform with shadow puppets. The flat puppets are cut from leather, and the play's action takes place behind a lit screen.

Mathematical Connection

Set designers are like architects; they plan, design, and create a structure for a specific use. Many of the tasks involved with set design are also involved with architecture. For example, the set must make efficient use of space and be pleasing to the people who are using it. Also, blueprints are drawn and scale models are built before completing the final structure. The exercises and examples in this chapter explore the use of architectural design techniques in the creation of theatrical sets. In Section 9.3, students investigate problems that relate to the construction of the set for *Once On This Island*. In this section, students use proportions to see how scale was used in both the models and blueprints made for this set. Then, in Section 9.4, students investigate areas and volumes of similar figures by using another stage set model.

Explore and Connect

Research
Contacting a local theater group may provide leads to people who would be willing to give an interview. If the interviewee has worked at different types of theaters, students can ask what the advantages and disadvantages are of each type of theater.

Project
If theatrical plays are performed in your area, you may want to encourage students to see a live performance and analyze the use of the scenery in it.

Writing
Students may want to work in pairs or groups of three to brainstorm a list of ideas for the activities they could see.

On the Scale of Things

Arcenas uses his scale models to check that all the pieces of the set will fit together. The pieces in the model are all *similar* to the pieces of the stage set. The lengths all change by the same factor, and the angles do not change at all. The *scale* of the set tells how the lengths will change between the model and the set. He often uses a 1:48 scale, so $\frac{1}{4}$ in. on his model corresponds to 1 ft in the actual set. Arcenas is helped in this process by his facility with math and his knack for visualizing objects in three dimensions. "Initially, all this has to be orchestrated in your mind," he says. "Then you can use the model to help you think about the real thing."

Scale Dimensions for Door *All's Well That Ends Well*			
	Scale	Height (in.)	Width (in.)
Model	1:48	$\frac{13}{16}$	$2\frac{1}{2}$
Blueprint	1:24	$1\frac{5}{8}$	5
Stage Set	1:1	39	120

Explore and Connect

Arcenas paints a piece for a scale model set of *Henry IV*.

1. Research Interview someone who participates in theater in your community. Ask how the available space at the theater affects the number and types of sets that the theater can use. Report your findings.

2. Project Borrow a video of a stage play. Watch how the set is used to provide a setting for each scene. Are some pieces of the set used for several different locations? Analyze the use of the set in a poster or report.

3. Writing Imagine that you are in the studio audience for a taping of a television show. Describe some of the activities that you would be able to see that the television audience does not see.

Mathematics
& Loy Arcenas

In this chapter, you will learn more about how mathematics is related to set design and scale models.

Related Exercises

Section 9.3
• Exercises 10–13

Section 9.4
• page 468
• Exercise 17

9.1

Properties of Similar Figures

Learn how to...
- use ratios and proportions to find measures of similar figures
- identify similar figures

So you can...
- use scale models to find the dimensions of an object being modeled, for example

When working on a new sign, a designer will usually use *scale drawings* instead of full-size drawings. The person who builds the sign will use these drawings to make sure that the sign looks exactly as planned.

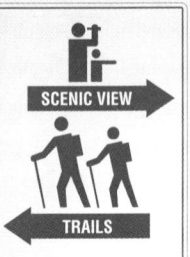

Matilda Johnson
State Park Signage
Scale 1:4

Read 1:4 as "1 to 4."

Lengths in the scale drawing are different from corresponding lengths on the actual sign, but all angles in the drawing are congruent to corresponding angles on the sign. You can use the *scale* of the drawing to find the size of a shape in the sign. The scale of the drawing is 1:4. Therefore, an object that is 1 in. tall on the drawing will be 4 in. tall on the sign.

THINK AND COMMUNICATE

1. If the designer wants the letters to be 3 in. tall on the sign, how tall should they be in the drawing?

2. Blueprints are scale drawings. A blueprint is labeled "1 cm to 5 m." Is the scale 1 to 5? Explain why or why not.

9.1 Properties of Similar Figures **445**

ANSWERS Section 9.1

Think and Communicate

1. $\frac{3}{4}$ in.

2. No; the scale is 1 to 500. A 1 cm object on the drawing will be 5 m or 500 cm in the actual figure.

Plan⇔Support

Objectives
- Use ratios and proportions to find measures of similar figures.
- Identify similar figures.
- Use scale models to find the dimensions of an object being modeled.

Recommended Pacing
❖ **Core and Extended Courses**
   Section 9.1: 2 days
❖ **Block Schedule**
   Section 9.1: 1 block

Resource Materials
Lesson Support
Lesson Plan 9.1
Warm-Up Transparency 9.1
Practice Bank: Practice 61
Study Guide: Section 9.1
Explorations Lab Manual:
   Diagram Master 2
Challenge Problems: Set 61
Technology
Technology Book:
   Calculator Activity 12
Graphing Calculator
Spreadsheet Software
Internet:
   http://www.hmco.com

Warm-Up Exercises

1. Write the ratio 1:3 in fraction form. $\frac{1}{3}$

The ratio of math to English classes is 3:5.

2. How many English classes are there if there are 27 math classes?   45

3. How many math classes are there if there are 35 English classes?   21

Find the value of *x*.

4. $\frac{x}{12} = \frac{3}{64}$ $\frac{9}{16}$ or 0.5625

5. $\frac{15}{x} = \frac{5}{7}$ 21

Teach⇔Interact

Toolbox p. 708
Proportions

Section Notes

Integrating Algebra
The study of similar figures requires the use of ratios and proportions. A brief review of these concepts may be appropriate for some students.

Second-Language Learners
Students learning English may find it helpful to discuss the nonmathematical uses of the word *similar*. If necessary, explain to students that two things are similar when they are alike but not identical.

Additional Example 1

Tell whether the shapes are similar.

a.

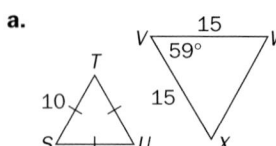

Since △*STU* is equilateral, it is also equiangular. Thus, each angle measures 60°. Since $m\angle XVW = 59°$, the corresponding angles are not congruent. The triangles are not similar.

b.

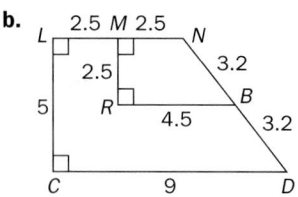

Step 1 Since all right angles are congruent, $\angle L \cong \angle RMN \cong \angle BRM \cong \angle C$. $\angle N$ is congruent to $\angle N$ by the Reflexive Property. The sum of the measures of the interior angles in a quadrilateral is 360°. Thus, $m\angle D = 360° - (90° + 90° + m\angle N)$ and $m\angle RBN = 360° - (90° + 90° + m\angle N)$. By substitution, $m\angle D = m\angle RBN$. Therefore, $\angle D \cong \angle RBN$, and all the angles are congruent.
Step 2 Check the ratios of corresponding sides.
$$\frac{LC}{MR} = \frac{5}{2.5} = 2$$
$$\frac{LM}{MN} = \frac{2.5 + 2.5}{2.5} = \frac{5}{2.5} = 2$$
$$\frac{ND}{NB} = \frac{3.2 + 3.2}{3.2} = \frac{6.4}{3.2} = 2$$
$$\frac{CD}{RB} = \frac{9}{4.5} = 2$$
$$\frac{LC}{MR} = \frac{LN}{MN} = \frac{ND}{NB} = \frac{CD}{RB}$$
Because corresponding angles are congruent and corresponding sides are in proportion, *LNDC ~ MNBR*.

446

A **proportion** is an equation that shows that two ratios are equal. The proportions $\frac{a}{b} = \frac{c}{d}$ and $a:b = c:d$ can be read as "*a* is to *b* as *c* is to *d*."

Similar Figures

Two figures are **similar** if these two conditions are true:

1. Corresponding angles are congruent.

2. The lengths of corresponding sides are in proportion.

The symbol ~ means "is similar to."

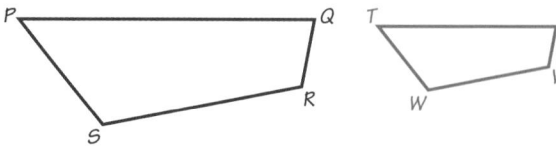

PQRS ~ TUVW

When you name similar figures, be sure to name corresponding vertices in the same order. For example, if *PQRS ~ TUVW*, then:

1. $\angle P \cong \angle T$, $\angle Q \cong \angle U$, $\angle R \cong \angle V$, and $\angle S \cong \angle W$

2. $\frac{PQ}{TU} = \frac{QR}{UV} = \frac{RS}{VW} = \frac{SP}{WT}$

WATCH OUT! ▶

When you write a proportion, the ratios must compare *corresponding* lengths.

EXAMPLE 1

Tell whether the shapes in each pair are similar.

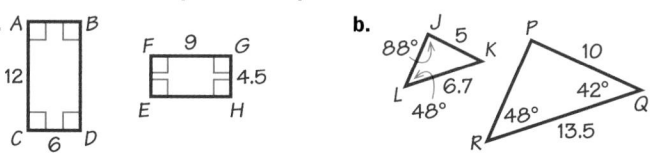

SOLUTION

a. Step 1 The angles are all right angles, so corresponding angles are congruent.

Step 2 Check the ratios of corresponding sides.

$$\frac{AB}{EF} \overset{?}{=} \frac{BD}{FG}$$

$$\frac{6}{4.5} \overset{?}{=} \frac{12}{9}$$

$$6 \cdot 9 \overset{?}{=} 4.5 \cdot 12$$

$$54 = 54 \checkmark$$

You need to check only one proportion, because opposite sides of a rectangle are congruent.

Find the cross products.

Because corresponding angles are congruent and corresponding sides are in proportion, *ABDC ~ EFGH*.

446 Chapter 9 *Similar Polygons*

b. Step 1 Are the angles congruent?

Find $m \angle K$ and $m \angle P$.

$$m \angle K = 180° - (88° + 48°) = 44°$$

$$m \angle P = 180° - (42° + 48°) = 90°$$

Corresponding angles are not congruent. The triangles are not similar.

EXAMPLE 2 Application: Graphic Design

The hiking symbol in the scale drawing on page 445 is 3 in. wide and 2.5 in. tall. The symbol will be 10 in. tall on the sign. Find the width of the symbol on the sign.

SOLUTION

Set up a proportion. Let w = width of symbol in inches.

$$\frac{\text{height of symbol in drawing}}{\text{height of symbol on sign}} = \frac{\text{width of symbol in drawing}}{\text{width of symbol on sign}}$$

$$\frac{2.5}{10} = \frac{3}{w}$$

$$2.5w = 10 \cdot 3$$

$$w = \frac{30}{2.5}$$

$$w = 12$$

The hiking symbol will be 12 in. wide on the sign.

> **Check:**
> Use the scale of the drawing. One unit in the drawing is equal to 4 units in the sign, so $3 \cdot 4 = 12.$ ✔

THINK AND COMMUNICATE

3. The scale of the drawing to the park sign is 1:4. What is the scale of the park sign to the drawing of the sign?

4. $\triangle ABC \sim \triangle PRQ$. Is $\frac{AB}{PR} = \frac{RQ}{BC}$? Explain why or why not. Give a proportion that is true.

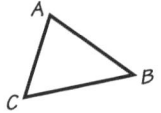
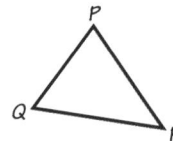

5. If two polygons are congruent, must they be similar?

6. If two figures are similar but not congruent, how are they alike? How are they different?

7. If the lengths of the hypotenuses of two right triangles are in the ratio of 1:1, must the triangles be similar? Explain your answer using a sketch.

Think and Communicate

3. 4:1

4. No; the ratios do not compare corresponding measures. Answers may vary. An example is given. $\frac{AB}{PR} = \frac{BC}{RQ}$

5. Yes.

6. Corresponding angles are congruent; corresponding sides are not equal in measure.

7. not necessarily; Right triangles may have congruent hypotenuses without being similar. Sketches may vary. An example is given.

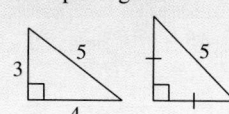

447

Additional Example 3

Tell whether $\triangle SRV$ is similar to $\triangle RVT$.

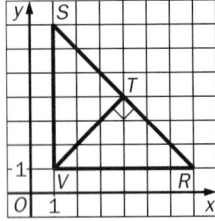

For this Example, it is best to work with the sides first.

Step 1 Are the sides in proportion? Use the Distance Formula or count to find SV and RV.

$SV = RV = 6$.

Use the Pythagorean theorem to find SR in $\triangle SRV$.

$$SR = \sqrt{SV^2 + VR^2}$$
$$= \sqrt{6^2 + 6^2}$$
$$= \sqrt{36 \cdot 2}$$
$$= 6\sqrt{2}$$

Since $\triangle SRV$ is an isosceles triangle, the altitude through the vertex is also a median. Thus,

$$TR = \frac{1}{2}SR = \frac{1}{2} \cdot 6\sqrt{2} = 3\sqrt{2}.$$

Use the Pythagorean theorem to find VT in $\triangle RVT$.

$$VT = \sqrt{VR^2 - TR^2}$$
$$= \sqrt{6^2 - (3\sqrt{2})^2}$$
$$= \sqrt{36 - 18}$$
$$= \sqrt{18}$$
$$= \sqrt{9 \cdot 2}$$
$$= 3\sqrt{2}$$

Check the proportion:
Because both triangles are isosceles, you need to check only the proportion of one pair of corresponding legs and the bases.

$$\frac{\text{leg of } \triangle RVT}{\text{leg of } \triangle SRV} = \frac{\text{base of } \triangle RVT}{\text{base of } \triangle SRV}$$

$$\frac{VT}{RV} \stackrel{?}{=} \frac{RV}{SR}$$

$$\frac{3\sqrt{2}}{6} \stackrel{?}{=} \frac{6}{6\sqrt{2}}$$

$$3\sqrt{2} \cdot 6\sqrt{2} \stackrel{?}{=} 6 \cdot 6$$

$$36 = 36 \checkmark$$

All sides are in proportion.

Step 2 Are the angles congruent?
Because $\angle SVR$ and $\angle RTV$ are right angles, $\angle SVR \cong \angle RTV$. It was shown in Step 1 that the triangles are both isosceles. Thus, the base angles of the triangles are congruent: $\angle S \cong \angle R$ and $\angle RVT \cong \angle R$. All angles are congruent and all sides are in proportion. By the definition of similar figures, $\triangle SRV \sim \triangle RVT$.

EXAMPLE 3   Connection: Algebra

Tell whether the two triangles are similar.

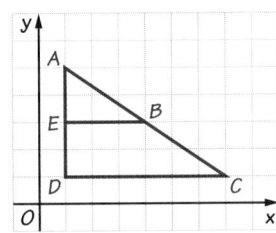

SOLUTION

Step 1 Are the angles congruent?

$\angle A \cong \angle A$, by the Reflexive Property.
Because $\angle ADC$ and $\angle AEB$ are right angles, $\angle ADC \cong \angle AEB$.
$\overline{EB} \parallel \overline{DC}$, so $\angle ABE \cong \angle ACD$.
Corresponding angles of the triangle are congruent.

Step 2 Are the sides in proportion?

Use the Distance Formula or count to find AE, AD, EB, and DC.

$AE = 2$, $AD = 4$, $EB = 3$, and $DC = 6$.

Use the Pythagorean theorem to find AC and AB.

$$AC^2 = AD^2 + DC^2 \qquad\qquad AB^2 = AE^2 + EB^2$$
$$AC = \sqrt{AD^2 + DC^2} \qquad\quad AB = \sqrt{AE^2 + EB^2}$$
$$= \sqrt{4^2 + 6^2} \qquad\qquad\quad = \sqrt{2^2 + 3^2}$$
$$= \sqrt{52} \qquad\qquad\qquad\quad = \sqrt{13}$$
$$= \sqrt{4 \cdot 13} = 2\sqrt{13}$$

$$AC = 2\sqrt{13} \text{ and } AB = \sqrt{13}$$

Check the proportion:

$$\frac{\text{short leg of } \triangle ABE}{\text{short leg of } \triangle ACD} = \frac{\text{long leg of } \triangle ABE}{\text{long leg of } \triangle ACD} = \frac{\text{hypotenuse of } \triangle ABE}{\text{hypotenuse of } \triangle ACD}$$

$$\frac{AE}{AD} \stackrel{?}{=} \frac{EB}{DC} \stackrel{?}{=} \frac{AB}{AC}$$

$$\frac{2}{4} \stackrel{?}{=} \frac{3}{6} \stackrel{?}{=} \frac{\sqrt{13}}{2\sqrt{13}}$$

$$\frac{1}{2} = \frac{1}{2} = \frac{1}{2} \checkmark$$

All angles are congruent and all sides are in proportion. By the definition of similar figures, $\triangle ABE \sim \triangle ACD$.

Checking Key Concepts

1. Yes; corresponding angles are congruent. (The measure of the third angle of the larger triangle is $180° - (112° + 30°) = 38°$.) Lengths of corresponding sides are in proportion. $\frac{12}{8} = \frac{15}{10} = \frac{22.5}{15} = \frac{3}{2}$

2. No; lengths of corresponding sides are not in proportion. $\frac{3}{5} \neq \frac{5}{8}$

3. $\triangle TUV \sim \triangle CAB$; $\frac{TU}{CA} = \frac{UV}{AB} = \frac{TV}{CB}$

4. $ABCDE \sim TSRQP$; $\frac{AB}{TS} = \frac{BC}{SR} = \frac{CD}{RQ} = \frac{DE}{QP} = \frac{AE}{TP}$

5. $\frac{1}{4}$

6. Answers may vary. An example is given.

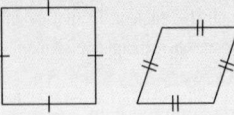

☑ CHECKING KEY CONCEPTS

Tell whether the polygons in each pair are similar. Explain your reasoning.

1.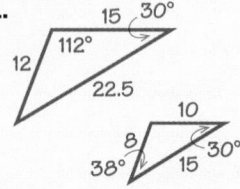

2.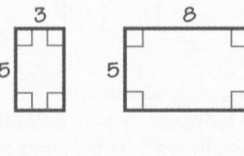

Name each pair of similar figures. Give two proportions that are true.

3.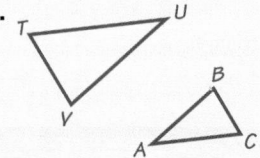

4.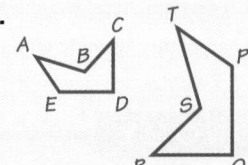

5. If the ratio of one pair of corresponding sides of similar triangles is $\frac{1}{4}$, what are the ratios of other pairs of corresponding sides?

6. Open-ended Problem Sketch two quadrilaterals that are not similar but whose corresponding sides are in proportion.

9.1 Exercises and Applications

Extra Practice
exercises on
page 677

ALGEBRA Solve each proportion.

1. $\frac{2}{4} = \frac{x}{16}$

2. $\frac{2}{5} = \frac{9}{y}$

3. $\frac{4}{z} = \frac{6}{z+1}$

4. $2:16 = y:20$

5. $2:x = x:8$

6. $z:z+2 = 30:42$

Tell whether the polygons in each pair are similar. Explain your reasoning.

7.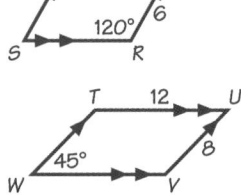

8.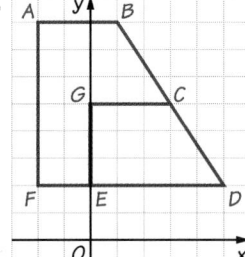

9.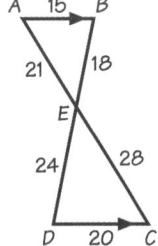

9.1 Properties of Similar Figures **449**

Exercises and Applications

1. 8
2. $22\frac{1}{2}$
3. 2
4. $2\frac{1}{2}$
5. 4
6. 5

7. No; since *PQRS* is a parallel-ogram, $m \angle P = 120°$ and $m \angle S = m \angle Q = 60°$. Similarly, since *TUVW* is a parallelogram, $m \angle U = 45°$ and $m \angle T = m \angle V = 135°$. Then no two angles of the figures are congruent.

8. No; lengths of corresponding sides are not in proportion. For example, $\frac{AB}{GC} = 1$, but $\frac{AF}{GE} = 2$.

9. Yes; $\angle AEB \cong \angle CED$ since vertical angles are congruent. Since $\overleftrightarrow{AB} \parallel \overleftrightarrow{DC}$, $\angle B \cong \angle D$ and $\angle A \cong \angle C$. So corresponding angles are congruent. Also, $\frac{AE}{CE} = \frac{EB}{ED} = \frac{AB}{CD} = \frac{3}{4}$.

Teach⇔Interact

Checking Key Concepts

Geometric Thinking
Question 5 leads students to an alternative statement of the second condition for similar figures given in the summary box on page 446, namely, the ratio of corresponding sides is constant. Question 6 helps students to understand why two conditions must be true in order to say that two figures are similar.

Closure Question

How are similar figures like con-gruent figures and how are they different? Similar figures, like congruent figures, have congruent corresponding angles. Unlike con-gruent figures, however, the corre-sponding sides of similar figures do not need to be congruent. Instead, the lengths of corresponding sides of similar figures must be in proportion.

Apply⇔Assess

Suggested Assignment

❖ **Core Course**
  Day 1 Exs. 1–18
  Day 2 Exs. 19, 23, 24, 26–28, 31–37
❖ **Extended Course**
  Day 1 Exs. 1–18
  Day 2 Exs. 19–37
❖ **Block Schedule**
  Day 54 Exs. 1–19, 23, 24, 26–28, 31–37

Exercise Notes

Topic Spiraling: Review
Exs. 1–6 These exercises provide students with an opportunity to review methods for solving propor-tions. Remind students that the cross products of a proportion are equal. That is, if $\frac{a}{b} = \frac{c}{d}$, then $ad = bc$.

Student Study Tip
Exs. 7–9 Students may have to use previously learned theorems or postulates to show that various angles have equal measures (or are congruent). You may wish to sug-gest that students keep a list of theorems and postulates they can use for this purpose.

449

Exercise Notes

Communication: Drawing
Exs. 13–18 Encourage students to draw their sketches before answering each exercise.

Using Technology
Exs. 19–21 These exercises can also be done using the list features of a TI-82 graphing calculator. First, go to the home screen and enter the scale factor as a variable by pressing 0.011 ÷ 1400000 STO▶ ALPHA [A] ENTER on the home screen. Then press STAT ENTER and clear the first four lists by positioning the cursor over the name of each list and pressing CLEAR ENTER . Enter the data for the mean distance from the sun and the diameter in L1 and L2. To find the mean distance from the planetarium to each planet in the scale model in kilometers, place the cursor on L3, and press 2nd [L1]∗ ALPHA [A] ENTER . Repeat this procedure to enter the diameter of each planet in the scale model in kilometers in L4. This time multiply A by the values in L2.

Multicultural Note
Exs. 19–21 Many cultures' traditional beliefs about the structure of the universe were influenced by their mythologies as well as by observation of nature. The Enuma Elish, the ancient Babylonian creation myth, states that the world began as liquid chaos, out of which grew Anu, the heavens, and Nudimmut, the Earth. Ancient Egyptians believed Earth was the center of the universe, and the sun traveled through the sky. One Hindu belief held that the universe consisted of three layers: the flat Earth, the rainy and windy atmosphere, and the sunny, fiery heavens. Both ancient Greek and Chinese scholars, however, were forming ideas about the structure of the universe based on scientific observations. In the fourth century B.C., the Greek Heraclidis Ponticus suggested that Venus and Mercury were satellites of the sun and Aristarchus argued that the sun was in fact larger than Earth. By the third century B.C., Chinese scholars had determined that Earth was in steady motion and some believed that the seasons were the result of Earth's movements.

The polygons in each pair are similar. Find the value of each variable.

10.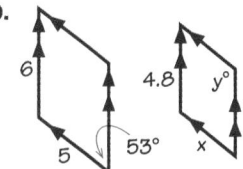

11.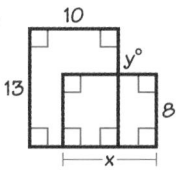

12.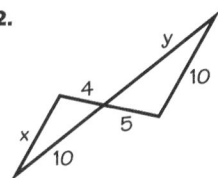

Logical Reasoning Tell whether the polygons in each pair are *always*, *sometimes*, or *never* similar. If a pair is always or sometimes similar, sketch an example. If a pair is never or sometimes similar, sketch a counterexample.

13. two squares

14. two rhombuses

15. two rectangles

16. two equilateral triangles

17. two isosceles trapezoids

18. an obtuse triangle and an acute triangle

Connection   ASTRONOMY

Toolbox p. 709
Scientific Notation

The Lakeview Museum Community Solar System in Peoria, Illinois, is the largest scale model of the solar system in the world. The sizes of and distances between the sun and planets are represented on the same scale. The sun for the model is the Lakeview Museum's planetarium, which is 11 m in diameter.

19. The diameter of the sun is about 1,400,000 km. Find the scale of the model. Remember that the units of the scale must be the same.

 Spreadsheets For Exercises 20 and 21, use a spreadsheet and the scale that you found in Exercise 19.

20. Use a formula to find the mean distance from the planetarium to each planet in the scale model.

21. Find the diameter of each planet in the scale model.

	A	B	C
1	Planet	Mean distance from sun (km)	Diameter (km)
2	Mercury	5.79×10^7	4870
3	Venus	1.08×10^8	12100
4	Earth	1.50×10^8	12800
5	Mars	2.29×10^8	6770
6	Jupiter	7.79×10^8	140000
7	Saturn	1.43×10^9	117000
8	Uranus	2.87×10^9	47000
9	Neptune	4.50×10^9	45500
10	Pluto	5.91×10^9	11400

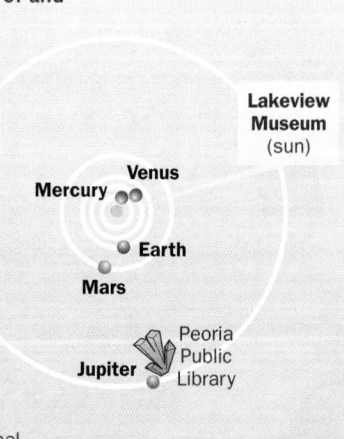

450   Chapter 9 *Similar Polygons*

---

10. $x = 4$; $y = 127$

11. $x = 10.4$; $y = 90$

12. $x = 8$; $y = 12.5$

13. always

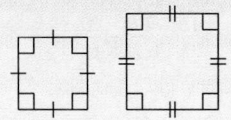

14. sometimes

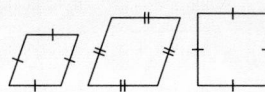

15. sometimes

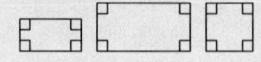

16. always

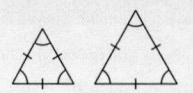

17. sometimes

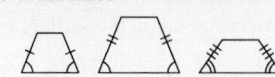

18. never

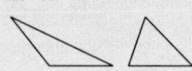

19. 11 : 1,400,000,000

20. Distances are given to the nearest meter. Mercury: 455 m; Venus: 849 m; Earth: 1179 m; Mars: 1799 m; Jupiter: 6121 m; Saturn: 11,212 m; Uranus: 22,558 m; Neptune: 35,326 m; Pluto: 46,459 m

22. Open-ended Problem Give an example of a scale drawing or a scale model you have seen that is an enlargement of a real-life object or group of objects.

23. a. TECHNICAL DRAWING If you drew the structures in the table below so that 1 in. represents 100 ft, what would the scale be? How tall would each structure be in your drawing?

CN Tower	TMG Offices	Washington Monument	Empire State Building
Toronto, Canada	Tokyo, Japan	Washington, DC	New York City, NY
1815 ft	793 ft	555 ft	1250 ft

b. Open-ended Problem On a sheet of $8\frac{1}{2}$ in. by 11 in. paper, sketch and label a scale drawing of the four structures. Explain how you chose the scale for your drawing.

24. TRANSFORMATIONS Explain why dilations are sometimes called *similarity transformations*.

25. Challenge What do you think must be true for two prisms to be similar? Write a definition for similar prisms.

26. MOVIES For most of the movie *King Kong*, the scale models of King Kong were built to appear about 18 ft tall. When King Kong climbs the Empire State Building, the models make him appear to be 24 ft tall.

a. Make a scale drawing that shows King Kong at both 18 ft and 24 ft tall, a 6 ft tall person, and the Empire State Building (1250 ft).

b. **Writing** Why do you think the movie makers chose to make King Kong appear taller when he climbs the Empire State Building?

27. SAT/ACT Preview Select the pair of words that *best* expresses a relationship similar to that expressed in the pair *gallon : liquid*.

A. ruler : paper   **B.** week : month   **C.** length : width

D. degree : temperature   **E.** foot : meter

28. GEOGRAPHY The Nile, in Africa, is the world's longest river. It would be about the length of a 24 in. shoelace (61 cm) if drawn so that 1 cm represents 110 km. What is the scale? About how long is the Nile?

9.1 Properties of Similar Figures   **451**

Exercise Notes

 Communication: Discussion
Ex. 22 Ask students to share some of the drawings and models they have seen. Can they determine the scale if it was not given? Did any students see drawings or models that were not to scale? This means that the drawing or model and the real object were not similar figures. Figures on billboards and public statues are examples of models not drawn to scale. Ask students if they can think of other examples of a scale drawing or a scale model that is an enlargement of a real-life object.

Cooperative Learning
Ex. 23 You may wish to have students do part (b) with a partner. Students can then discuss what a reasonable scale may be and each student can then draw two of the figures.

Application
Exs. 23, 26, 28 These exercises contain applications of similar figures to technical drawing, movie making, and geography.

Geometric Thinking
Ex. 24 This exercise helps students to understand the connection between similarity and dilation. Students should realize that any image formed by a dilation is similar to the original figure.

Challenge
Ex. 25 You may wish to extend this exercise by having students attempt to draw sketches of similar prisms.

---

21. Distances are given to the nearest millimeter. Mercury: 38 mm; Venus: 95 mm; Earth: 100 mm; Mars: 53 mm; Jupiter: 1097 mm; Saturn: 916 mm; Uranus: 369 mm; Neptune: 357 mm; Pluto: 90 mm

22. Answers may vary. Examples are given. a drawing of an insect or a biological model of a human ear

23. a. 1 : 1200; CN Tower: 18.15 in.; TMG Offices: 7.93 in.; Washington Monument: 5.55 in.; Empire State Building: 12.5 in.

b. Answers may vary. Check students' work.

24. A figure and its image under a dilation are similar.

25. Answers may vary. An example is given. Similar prisms have the same shape but not necessarily the same size. Two prisms are *similar* if their bases are similar and all corresponding lengths are in proportion.

26. a. Check students' work.

b. because the Empire State Building is so tall

27. D

28. 1 : 11,000,000; about 6710 km

Exercise Notes

Research

Exs. 29, 30 The study of fractals is part of a new branch of mathematics called Chaos Theory. Interested students may want to do some research to find out more about fractals.

Assessment Note

Ex. 32 This exercise provides an excellent opportunity for students to demonstrate their knowledge of similar figures and their properties. Some students may consider adding their work on this exercise to their portfolios.

Topic Spiraling: Review/Preview

Exs. 33–35 These exercises review methods of proving triangles congruent. Methods of proving triangles similar will be presented in Section 9.2.

Practice 61 for Section 9.1

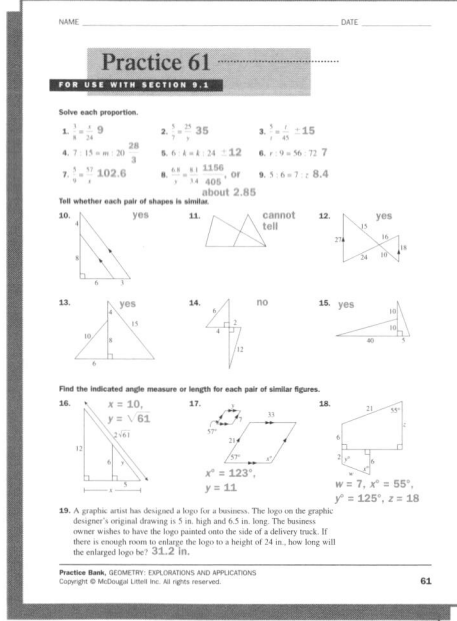

29. FRACTALS The pattern at the right is made by adding two squares or an isosceles right triangle at each step. When two squares connect to form a rectangle, you stop adding shapes to that part of the pattern.

a. Writing In the diagram at the right, all of the triangles are similar and all of the squares are similar. Are all of the groups of one triangle with three surrounding squares similar? Explain why or why not.

b. Trace or sketch the pattern. Find two groups of more than four shapes in the pattern that are similar but not congruent.

30. Open-ended Problem Use the steps in Exercise 29 to make a design. Instead of using an isosceles right triangle in Step 2, use a different kind of triangle. (For example, you could use a right triangle with legs in a ratio of 1 : 2, or an isosceles obtuse triangle with a 100° vertex angle.) Draw at least the first five steps of the pattern.

31. SCULPTURE The Crazy Horse Monument, based on a marble sculpture by Korczak Ziolkowski, will be 563 ft high. Lengths for the sculpture in the mountain are taken from the 16 ft tall scale model of the sculpture. What is the scale of the model? Explain how workers can use the scale model to determine lengths on the mountain.

BY THE WAY

The Crazy Horse Monument is being carved in South Dakota. Crazy Horse was a leader of the Oglala Sioux. He is remembered for his courage and wisdom.

ONGOING ASSESSMENT

32. Open-ended Problem Make a scale drawing of a diagram in this book. On your drawing, give the scale that you used. Show that two of the shapes in your drawing are similar to the shapes in the original drawing. Give the page number of the original diagram.

SPIRAL REVIEW

Tell which method(s) you can use to prove that the triangles are congruent. If no method can be used, write *none*. (*Section 6.4*)

33. 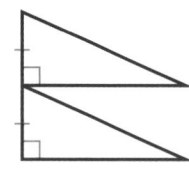   **34.** 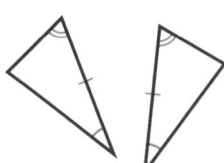   **35.**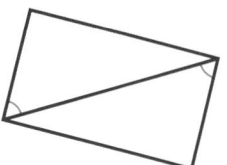

Open-ended Problems Draw a quadrilateral *ABCD* on the coordinate plane. Draw the image of *ABCD* after each dilation. (*Section 8.6*)

36. center *A* and scale factor 4   **37.** center *C* and scale factor $\frac{1}{2}$

29. a. Yes; corresponding angles are congruent and lengths of corresponding sides are in proportion.

b. Answers may vary. In the example, the two groups are circled.

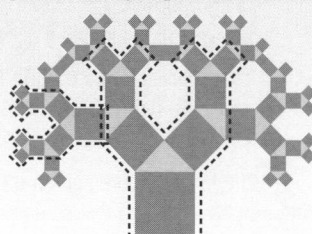

30. Answers may vary. In the example, the triangle in Step 2 is a right triangle. The ratio of the lengths of the legs is 1:2.

31. $1 : 35\frac{3}{16}$; Workers can multiply any length on the model by $35\frac{3}{16}$ to find the appropriate length on the monument.

32. Check students' work.

33. none

34. ASA

35. none

36, 37. Check students' work.

9.2

Similar Triangles

You know several shortcuts for proving that two triangles are congruent without checking each angle and side. In this lesson you will learn some shortcuts for proving that triangles are similar.

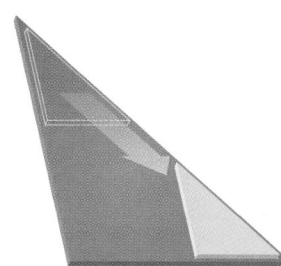

When two triangles have two pairs of congruent angles, the third pair must be congruent, too. If you experiment with two triangles in which all three pairs of angles are congruent, you will find that the lengths of their sides are in proportion.

Angle-Angle (AA) Similarity Postulate

If two angles of a triangle are congruent to two angles of another triangle, then the triangles are similar.

EXAMPLE 1

Given: $\overline{QT} \parallel \overline{RS}$
Prove: $\triangle PRS \sim \triangle PQT$

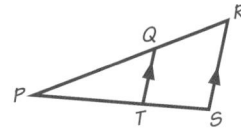

SOLUTION

Given: $\overline{QT} \parallel \overline{RS}$
Prove: $\triangle PRS \sim \triangle PQT$

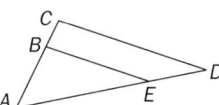

Reasons

1. Given

2. If two $\parallel$ lines are intersected by a transversal, then corresponding angles are $\cong$.

3. Reflexive Property

4. AA Similarity Postulate

9.2 Similar Triangles **453**

Plan⇔Support

Objectives

*   Use special rules to identify similar triangles.
*   Write proofs involving similar triangles.

Recommended Pacing

❖ **Core and Extended Courses**
Section 9.2: 1 day

❖ **Block Schedule**
Section 9.2: $\frac{1}{2}$ block
(with Section 9.3)

Resource Materials

Lesson Support
Lesson Plan 9.2

Warm-Up Transparency 9.2

Overhead Visuals:
 Folder 11: Similar Polygons in
  Space

Practice Bank: Practice 62

Study Guide: Section 9.2

Challenge Problems: Set 62

Technology
Technology Book:
 Spreadsheet Activity 8

Geometry Software

Internet:
 http://www.hmco.com

Warm-Up Exercises

1. Name the triangles that appear similar. $\triangle ACD \sim \triangle ABE$

Assume the triangles named in Ex. 1 are similar.

2. Write two true proportions.
$\frac{AB}{AC} = \frac{AE}{AD}$, $\frac{AE}{AD} = \frac{BE}{CD}$, and $\frac{AB}{AC} = \frac{BE}{CD}$

3. If $AB = 6$, $AE = 8$, and $AC = 7$, find AD. $9\frac{1}{3}$

4. If $CD = 12$, $AD = 15$, and $BE = 10$, find AE. 12.5

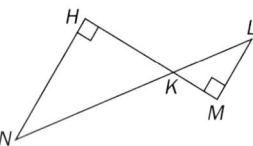
Additional Example 1

Given: △ *NHK* and △ *LMK* are right triangles.
Prove: △ *NHK* ~ △ *LMK*

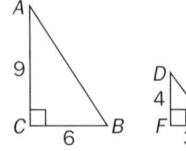

Given: △ *NHK* and △ *LMK* are right triangles.

Prove: △ *NHK* ~ △ *LMK*
Statements (Reasons)
1. △ *NHK* and △ *LMK* are right triangles. (Given)
2. ∠ *NHK* ≅ ∠ *LMK* (All right ∠ are ≅.)
3. ∠ *NKH* and ∠ *MKL* are vertical angles. (Def. of vertical ∠)
4. ∠ *NKH* ≅ ∠ *MKL* (Vertical ∠ are ≅.)
5. △ *NHK* ~ △ *LMK* (AA Similarity Postulate)

Section Note

Second-Language Learners
To help students learning English remember and differentiate between the theorems and postulates about triangle congruency and those about triangle similarity, encourage them to start a two-column list. One column can show and describe all the theorems and postulates that prove congruency (SSS, SAS, ASA, AAS, HL) and the other can show the theorems and postulates that prove similarity (AA Similarity, SAS Similarity, SSS Similarity, Midpoint Similarity, Triangle Proportionality).

Additional Example 2

Is it possible to prove that the triangles in each pair are similar? Explain why or why not.

a.

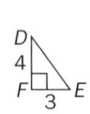

No. Although ∠ *ACB* and ∠ *DFE* are right angles and thus congruent, the sides including these angles are not in proportion.

$$\frac{CB}{FE} = \frac{6}{3} = 2$$

$$\frac{AC}{DF} = \frac{9}{4}$$

THINK AND COMMUNICATE

1. Describe another way to prove that △*PRS* ~ △*PQT* in Example 1.

2. Sketch two isosceles right triangles. Are they similar? Why?

As with congruent triangles, there are several ways to prove that two triangles are similar. The proofs of these theorems have been omitted.

Side-Angle-Side (SAS) Similarity Theorem

If an angle of one triangle is congruent to an angle of another triangle, and the sides including these angles are in proportion, then the triangles are similar.

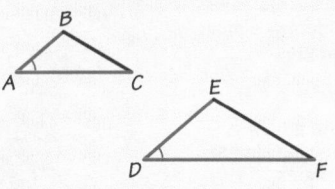

If ∠*A* ≅ ∠*D* and $\frac{AB}{DE} = \frac{AC}{DF}$, then △*ABC* ~ △*DEF*.

Side-Side-Side (SSS) Similarity Theorem

If all corresponding sides of two triangles are in proportion, then the triangles are similar.

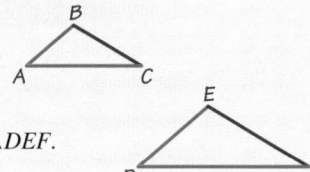

If $\frac{AB}{DE} = \frac{BC}{EF} = \frac{AC}{DF}$, then △*ABC* ~ △*DEF*.

EXAMPLE 2

Is it possible to prove that the triangles in each pair are similar? Explain why or why not.

a.
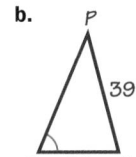

b.

SOLUTION

a. Yes. ∠*AED* ≅ ∠*CEB* because they are vertical angles. ∠*A* ≅ ∠*C* because they are alternate interior angles formed by two parallel lines and a transversal. So △*AED* ~ △*CEB* by the AA Similarity Postulate.

b. No. The congruent angles are not included between the given sides.

ANSWERS Section 9.2

Think and Communicate

1. Use the same method but two different pairs of congruent angles, either ∠ *P* ≅ ∠ *P* and ∠ *PTQ* ≅ ∠ *PSR* or ∠ *PQT* ≅ ∠ *PRS* and ∠ *PTQ* ≅ ∠ *PSR*.

2. Any two isosceles right triangles are similar; the two right angles are congruent and all four acute angles are congruent. Then the triangles are similar by the AA Similarity Postulate.

EXAMPLE 3

Tell whether the triangles in each pair are similar. Explain your reasoning.

a.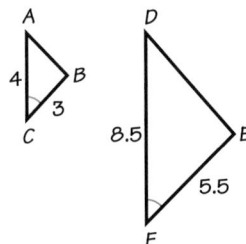

b.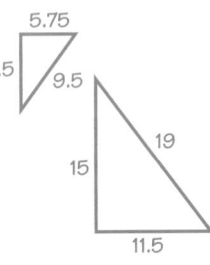

Fenda Gandega paints a wall of her family's home in Djajibinni, Mauritania.

SOLUTION

a. You know that $\angle C \cong \angle F$. Check whether the sides that include the angles are in proportion.

$$\frac{DF}{AC} \overset{?}{=} \frac{EF}{BC}$$

$$\frac{8.5}{4} \overset{?}{=} \frac{5.5}{3} \qquad \text{Check whether the fractions are equal.}$$

$$2.125 \neq 1.8\overline{3}$$

The sides are not in proportion, so the triangles are not similar.

b. Are the corresponding sides in proportion?

$$\frac{5.75}{11.5} \overset{?}{=} \frac{7.5}{15} \overset{?}{=} \frac{9.5}{19}$$

$$\frac{1}{2} = \frac{1}{2} = \frac{1}{2} ✔$$

By the SSS Similarity Theorem, the triangles are similar.

THINK AND COMMUNICATE

3. Explain why $\triangle ABC \sim \triangle LJK$. Find the length JK.

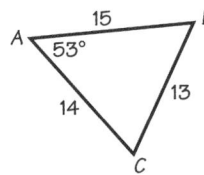 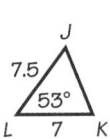

4. How are the SSS and SAS Similarity Theorems like the SSS and SAS Postulates for congruent triangles? How are they different?

5. Two isosceles triangles each have a vertex angle of 50°. Must the triangles be similar? Explain your reasoning.

Think and Communicate

3. SAS Similarity Theorem; $JK = 6.5$

4. Answers may vary. Both SSS statements relate the lengths of the sides of one triangle to the lengths of the sides of a second triangle. The congruence postulate requires that corresponding sides be congruent; the similarity theorem requires

that the lengths of corresponding sides be in proportion. Both SAS statements relate congruent angles in two triangles and the sides including those angles. The congruence postulate requires that corresponding sides be congruent; the similarity theorem requires that the lengths of corresponding sides be in proportion.

5. Yes; in each triangle, since the measure of the vertex angle is 50°, the measure of each base angle is $\frac{1}{2}(180° - 50°) = 65°$. Then the triangles are similar by the AA Similarity Postulate.

Additional Example 2 (continued)

b.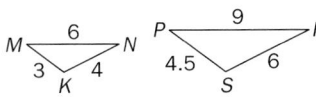

Yes. $\frac{MK}{PS} = \frac{KN}{SR} = \frac{NM}{RP} = \frac{2}{3}$

All corresponding sides are in proportion. So $\triangle MKN \sim \triangle PSR$ by the SSS Similarity Theorem.

Additional Example 3

Tell whether the triangles are similar. Explain your reasoning.

a. $\triangle AEC$ and $\triangle FDE$

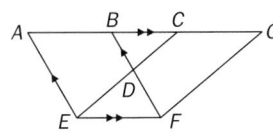

$\angle AEC \cong \angle FDE$ because they are alternate interior angles formed by two parallel lines intersected by transversal $\overline{EC}$. For the same reason, $\angle ECA \cong \angle DEF$. By the AA Similarity Postulate, the triangles are similar.

b. $\triangle LMN$ and $\triangle PRN$

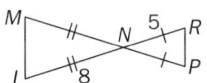

$\angle LNM$ and $\angle RNP$ are vertical angles and, hence, congruent. Check whether the sides that include the angles are in proportion. Use the fact that these are isosceles triangles to find MN and PN.

$$\frac{PN}{LN} \overset{?}{=} \frac{RN}{MN}$$

$$\frac{5}{8} = \frac{5}{8} ✓$$

The sides are in proportion, so the triangles are similar by the SAS Similarity Theorem.

Think and Communicate

Question 4 asks students to compare the triangle similarity theorems with the triangle congruence postulates. By thinking about the similarities and differences between these theorems and postulates, students can strengthen their understanding of how to apply them correctly.

Checking Key Concepts

Geometric Thinking

These questions provide a comprehensive review of the concepts in this section. You might have students check their work with a partner. This should help students identify any misunderstandings before beginning the exercises.

Visual Thinking

Encourage students to search for magazine photographs that show similar triangles. Ask them to use markers to highlight the triangles and to demonstrate how the SAS Similarity Theorem and the SSS Similarity Theorem can be used to prove that the triangles are similar. This activity involves the visual skills of *generalization* and *communication*.

Closure Question

Describe three ways to show that two triangles are similar. Show that two pairs of corresponding angles are congruent. Show that one pair of corresponding angles are congruent and that the sides including these angles are in proportion. Show that all corresponding sides are in proportion.

Suggested Assignment

❖ **Core Course**
Exs. 1–8, 12, 14–17, 22–31

❖ **Extended Course**
Exs. 1–31

❖ **Block Schedule**
Day 55 Exs. 1–8, 12, 14–17, 22–31

Exercise Notes

Student Progress

Exs. 1–6 Students should be able to complete Exs. 1–3 without much difficulty. For Exs. 4–6, you may wish to suggest that students redraw the similar triangles with the same orientation to simplify seeing corresponding parts.

456

☑ CHECKING KEY CONCEPTS

Explain why the triangles in each pair are similar.

1.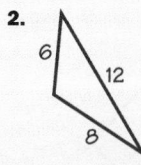

2.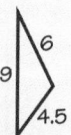

Is it possible to prove that the triangles in each pair are similar? Explain why or why not.

3.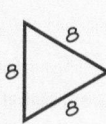

4.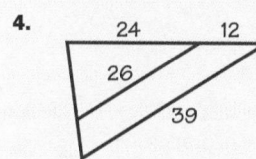

5. If two angles of a triangle are congruent to two angles of another triangle, what can you conclude about the triangles?

6. Complete: You can use the SAS Similarity Theorem to show that two triangles are similar if you know that: _?_ .

7. The length of each side of △ABC is one third the length of the corresponding side in △QRS. What can you conclude about the triangles? Why?

9.2 | Exercises and Applications

Extra Practice
exercises on
page 678

Is it possible to prove that the triangles in each pair are similar? Explain why or why not.

1.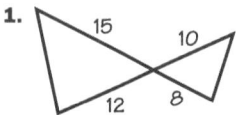

2.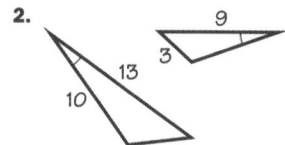

3.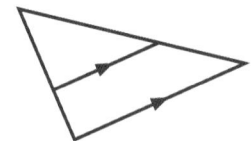

Choose two triangles in each diagram. Explain why they are similar.

4.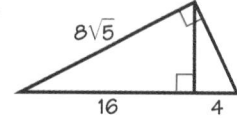

5.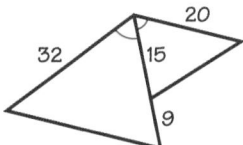

6.

Checking Key Concepts

1. by the SAS Similarity Theorem; Vertical angles are congruent and $\frac{7}{10} = \frac{14}{20}$.

2. by the SSS Similarity Theorem; $\frac{4.5}{6} = \frac{6}{8} = \frac{9}{12} = \frac{3}{4}$

3. Yes; since the ratio of the lengths of any two corresponding sides is $\frac{5}{8}$, the triangles are similar by the SSS Similarity Theorem.

4. No; the angles that are known to be congruent are not included by the sides known to be in proportion.

5. They are similar.

6. An angle of one triangle is congruent to an angle of the other triangle, and the sides including these angles are in proportion.

7. They are similar by the SSS Similarity Theorem.

Exercises and Applications

1. Yes; the triangles are similar by the SAS Similarity Theorem since the two vertical angles are congruent and the sides including the angles are in proportion.

2. No; the lengths of all three pairs of sides are not given; nor do we have the lengths of the sides including the congruent angles.

7. Given: $\overline{BC} \parallel \overline{ED}$; $\overline{AB} \parallel \overline{DC}$
   Prove: $\triangle ABC \sim \triangle CDE$

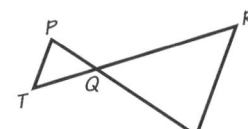

8. Given: $\overline{PT} \parallel \overline{SR}$
   Prove: $\triangle QTP \sim \triangle QRS$

Connection   SURVEYING

Surveyors use modern instruments to measure distances and heights accurately. You can estimate heights and distances using only a tape measure and your knowledge of similar triangles.

9. Writing Carol D'Anjow measured the length of her shadow and the length of the shadow of a tree.

  a. Use the measurements in the diagram to find the height of the tree.

  b. Explain why you must measure both shadows at about the same time for this method to work. Explain why this method may not work for a tree in the forest or a building in the city.

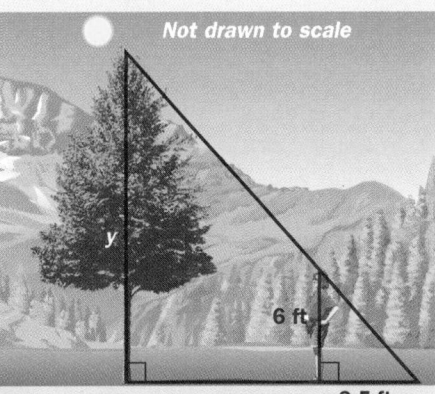

Not drawn to scale

6 ft

3.5 ft

24 ft

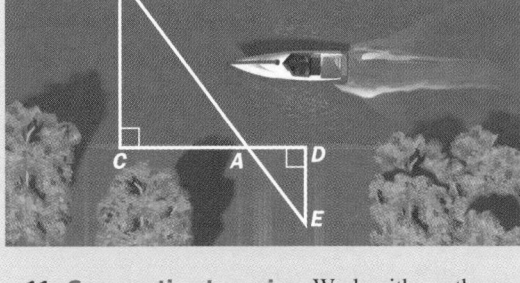

Not drawn to scale

10. a. Explain why $\triangle ABC \sim \triangle AED$ in the diagram at the left.

  b. Clarice Yee wants to estimate the width of a river. She measures three distances, and records that $AC = 5$ m, $AD = 2$ m, and $DE = 8.3$ m. Draw a diagram showing these measurements. How wide is the river?

11. Cooperative Learning Work with another person. Choose one of the methods used in Exercises 9 and 10. Use this method to estimate a height or distance in your neighborhood or at school. Tell which method you chose and explain how you estimated the height or distance.

---

3. Yes; it can be shown the triangles are similar by the AA Similarity Postulate. Because the two lines are parallel, it can be shown that two angles of one triangle are congruent to two angles of the other triangle.

4. by the SAS Similarity Theorem since $\frac{8\sqrt{5}}{20} = \frac{16}{8\sqrt{5}}$ and the included angles are congruent, or by the SSS Similarity Theorem (First use the Pythagorean

theorem to find the lengths of the remaining sides.)

5. by the SAS Similarity Theorem; The sides including the congruent angles are in proportion; $\frac{20}{32} = \frac{15}{24} = \frac{5}{8}$.

6. by the AA Similarity Theorem; The two pairs of parallel lines can be used to show that two angles of one triangle are congruent to two angles of the other.

7. Since $\overline{BC} \parallel \overline{ED}$ and $\overline{AB} \parallel \overline{DC}$, $\angle BCA \cong \angle DEC$ and $\angle A \cong \angle ECD$. (If two $\parallel$ lines are intersected by a transversal, then alternate interior $\angle$s are $\cong$.) Then $\triangle ABC \sim \triangle CDE$ by the AA Similarity Postulate.

8–10. See answers in back of book.

11. Check students' work.

Exercise Notes

Writing Proofs
Exs. 7, 8 Students should copy and mark the diagrams and think about a plan for each proof. They can then choose the type of proof format they wish to use.

Application
Exs. 9–11 These exercises give students an opportunity to explore two different methods for making indirect measurements using similar triangles. The first method compares the lengths of shadows to determine the height of a tree. The second method determines the distance across a river by using one triangle across the river and a similar triangle on the bank. In Ex. 11, students have the opportunity to test one of these methods.

Multicultural Note
Exs. 9–11 Direct and indirect evidence suggest that the ancient Babylonians, Chinese, and Egyptians all surveyed land to some extent. All three cultures taxed people based on the amount of land in their possession (requiring measurement of distance and area); created irrigation systems (requiring leveling instruments); and planned buildings (requiring right angles). The Egyptians, Chinese, and Babylonians probably used many surveying instruments such as marked cords to measure distances, plumb bobs to check verticality, and squares to determine right angles. In later periods, other groups of people refined surveying technology and instruments. The Indians and Muslims made important contributions to surveying methods before and during the Middle Ages, with such things as the Indian circles method to establish the meridian line, vertical staffs to measure heights by shadows, and highly developed astrolabes for estimating heights and other measurements.

Cooperative Learning
Ex. 11 You may want to discuss Exs. 9 and 10 in class before having students complete this activity. Have groups share their experiences with the class. Students should answer questions such as: What types of problems did you run into? What did you consider when choosing the method used?

Exercise Notes

Teaching Tip
Exs. 12, 13 Remind students to look for similar figures other than just triangles.

Communication: Writing
Ex. 14 You may wish to ask students to write the theorem they proved in this exercise.

Writing Proofs
Exs. 14, 16, 17 To continue to help students develop their proof-writing skills, ask for three volunteers to present their proofs for these exercises at the board. Students can review these sample proofs and compare them to their own. You may wish to ask students who used different methods for any of these exercises to share their methods with the class.

Problem Solving
Ex. 15 This exercise helps to broaden students' problem-solving skills by developing the idea that when similar right triangles are involved in a problem, it may not be necessary to use the Pythagorean theorem to find the length of a side. Instead, ratios of corresponding sides and proportions can be used.

---

12. **Open-ended Problem** In the diagram, $ABCD \sim EFGH$. Name three other pairs of similar figures in the diagram.

13. **Challenge** Use the diagram for Exercise 12. Name all the similar figures shown in the diagram.

14. Copy and complete the proof.

Given: $\angle A \cong \angle P$
$\overline{AB} \cong \overline{AC}$
$\overline{PR} \cong \overline{PQ}$

Prove: $\triangle ABC \sim \triangle PQR$

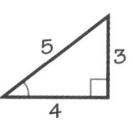

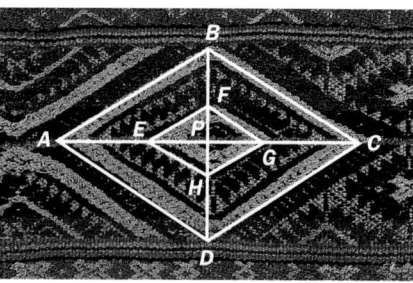

Indian tapestry from Peru

Statements	Reasons
1. _?_	1. Given
2. $\overline{AB} \cong \overline{AC}$; $\overline{PQ} \cong \overline{PR}$ ($AB = AC$; $PQ = PR$)	2. _?_
3. $\dfrac{AB}{PQ} = \dfrac{AC}{PR}$	3. Algebra
4. $\triangle ABC \sim \triangle PQR$	4. _?_

15. **a. Logical Reasoning** Describe how to find the value of c without using the Pythagorean theorem.

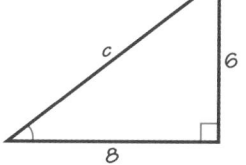

b. Find each missing length without using the Pythagorean theorem.

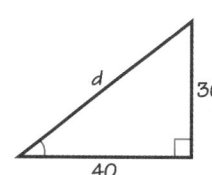 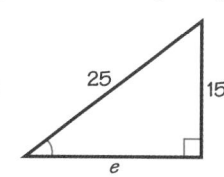 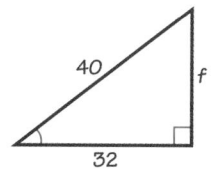

16. **Given:** $\dfrac{XZ}{PR} = \dfrac{YZ}{QR}$
$\angle Z$ and $\angle R$ are right angles.

Prove: $\triangle XYZ \sim \triangle PQR$

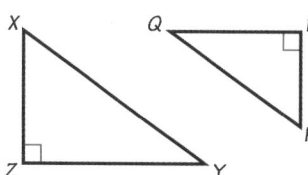

17. **Given:** $\overline{AB} \perp \overline{AE}$
$\overline{ED} \perp \overline{AE}$

Prove: $\triangle ABC \sim \triangle EDC$

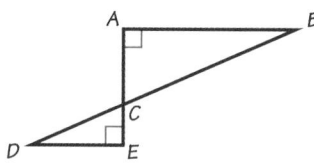

....

12. Answers may vary. Examples are given. $\triangle PEH$ and $\triangle PAD$, $\triangle PGF$ and $\triangle PCB$; $\triangle PHG$ and $\triangle PDC$

13. The following are similar: $\triangle PEH$, $\triangle PAD$, $\triangle PGF$, and $\triangle PCB$; $\triangle PEF$, $\triangle PAB$, $\triangle PGH$, and $\triangle PCD$; $\triangle ABC$, $\triangle CDA$, $\triangle EFG$, and $\triangle GHE$; $\triangle ABD$, $\triangle EFH$, $\triangle CDB$, and $\triangle GHF$; $AEHD$ and $CGFB$; $ABFE$ and $CDHG$.

14. (1) $\angle A \cong \angle P$; (2) Given (def. of congruence); (4) SAS Similarity Theorem

15. **a.** Since the right angles are congruent and the sides including the right angles are in proportion $\left(\dfrac{3}{6} = \dfrac{4}{8} = \dfrac{1}{2}\right)$, the triangles are similar. The ratio of the lengths of corresponding sides is $\dfrac{1}{2}$, so $\dfrac{5}{c} = \dfrac{1}{2}$ and $c = 10$.

b. $d = 50$; $e = 20$; $f = 24$

16. **Statements (Reasons)**
1. $\dfrac{XZ}{PR} = \dfrac{YZ}{QR}$; $\angle Z$ and $\angle R$ are right angles. (Given)
2. $\angle Z \cong \angle R$ (All right $\angle$s are $\cong$.)
3. $\triangle XYZ \cong \triangle PQR$ (SAS Similarity Theorem)

17. Answers may vary. An example is given.
Statements (Reasons)
1. $\overline{AB} \perp \overline{AE}$ and $\overline{ED} \perp \overline{AE}$. (Given)
2. $\overline{AB} \parallel \overline{ED}$ (In a plane, two lines $\perp$ to the same line are $\parallel$.)
3. $\angle A \cong \angle E$ and $\angle B \cong \angle D$. (If two $\parallel$ lines are intersected by a transversal, alternate interior $\angle$s are $\cong$.)
4. $\triangle ABC \sim \triangle EDC$ (AA Similarity Postulate)

18. 5
19. 6
20. 8

The total *resistance* for a circuit depends on the resistance of each *resistor* in the circuit. Let R = the total resistance for the circuit, R_1 = the resistance of one part of the circuit, and R_2 = the resistance of a second part of the circuit. You can use the formula $\frac{1}{R} = \frac{1}{R_1} + \frac{1}{R_2}$ to find the total resistance for the circuit.
Follow these steps to find the value of R for a circuit with the given values of R_1 and R_2.

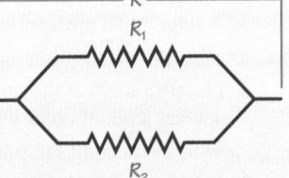

Step 1 Plot R_1 and R_2.

Step 3 The intersection of $\overline{R_1R_2}$ and the R-axis shows the total resistance.

Step 2 Draw $\overline{R_1R_2}$.

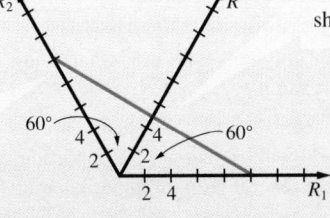

18. $R_1 = 10$; $R_2 = 10$ **19.** $R_1 = 15$; $R_2 = 10$ **20.** $R_1 = 24$; $R_2 = 12$

21. Challenge You can use the diagram below to explain why the graphs you made in Exercises 18–20 give the correct resistance for the circuits. Follow these steps to prove that in the diagram $\frac{1}{R} = \frac{1}{R_1} + \frac{1}{R_2}$. Let $x = R_1$, $y = R_2$, and $z = R$.

a. Extend $\overrightarrow{BA}$ through E so that $\overline{EC} \parallel \overline{AD}$. Explain why $\triangle AEC$ is equilateral. Find the length EC.

b. Explain why $\triangle ECB \sim \triangle ADB$.

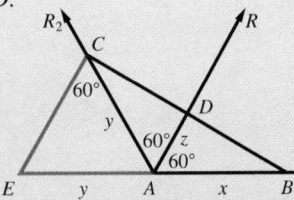

c. Use the proportion $\frac{EB}{AB} = \frac{EC}{AD}$ to show that $xy = xz + yz$.

d. Divide both sides of the equation in part (c) by xyz to show that $\frac{1}{z} = \frac{1}{y} + \frac{1}{x}$. Explain why the proof is now complete.

22. a. Each side of regular pentagon $ABCDE$ is x units long, and each side of regular pentagon $PQRST$ is y units long. Explain why $ABCDE \sim PQRST$.

b. Writing Are all regular polygons with the same number of sides similar? Explain.

9.2 Similar Triangles **459**

---

21. a. $m\angle E = m\angle DAB = 60°$ because $\overline{EC} \parallel \overline{AD}$ and, therefore, corresponding angles are congruent. $m\angle CAE = 60°$ because the sum of the measures of the angles of a triangle is $180°$. $\triangle AEC$ is equiangular, therefore, it is equilateral. $EC = y$

b. $\angle E \cong \angle DAB$ and $\angle B \cong \angle B$, so $\triangle ECB \sim \triangle ADB$ by the AA Similarity Postulate.

c. $\frac{EB}{AB} = \frac{EC}{AD} = \frac{y+x}{x} = \frac{y}{z}$; $(y+x)z = xy$; $xz + yz = xy$

d. $\frac{xz+yz}{xyz} = \frac{xy}{xyz}$; $\frac{1}{y} + \frac{1}{x} = \frac{1}{z}$, so $\frac{1}{z} = \frac{1}{y} + \frac{1}{x}$.

22. a. Since both pentagons are regular, the angles are all congruent. The lengths of corresponding sides are in proportion since the ratio of the lengths of any two sides is $\frac{x}{y}$.

b. For any integer n ($n \geq 3$), any two regular n-gons are similar because all the angles are congruent and the lengths of corresponding sides are in proportion. If the lengths of each side of one n-gon is j and the length of each side of the other n-gon is k, the ratio of the lengths of any two sides is $\frac{j}{k}$.

Exercise Notes

Communication: Reading
Exs. 18–21 The introduction to these exercises contains a fair amount of information and an example that students need to understand in order to do the problems. In addition, students must be able to follow the steps to use the graph to solve the resistance equation. Most students would benefit from having this material read aloud and discussed in class.

Integrating Algebra
Exs. 18–21 These exercises provide a nice example of how a geometric figure can help solve an algebraic equation.

Research
Exs. 18–21 Since many of the ideas in these exercises may be unfamiliar to students, you may wish to ask those students interested in science to research electrical circuits, including resistors and circuit diagrams. These students could then present their information to the class the day that the problems are assigned. With the help of a science teacher, they might even be able to build a small circuit to bring in as an example.

Career Connection
Exs. 18–21 The field of electronics offers a wide array of career opportunities. Electronic components are found in an increasing number of products including cars, microwave ovens, computers, and home security systems. There are a great many job opportunities to design, manufacture, or service these products. These career opportunities all require some technical expertise and various levels of advanced education. As a special assignment, you can ask each student to research one career in electronics. Ask them to report on the training required and the job opportunities available.

Geometric Thinking
Ex. 22 In this exercise, students prove a result for regular pentagons, and then generalize it to all regular polygons. You may wish to have students state this result formally.

459

Exercise Notes

 Using Technology
Exs. 24–26 Students use geometry software to experiment with quadrilaterals to see if the triangle similarity properties presented in this section are valid for four-sided figures. Students then generalize their results for polygons in general.

Assessment Note
Ex. 27 A class discussion of students' counterexamples and rewritten conjectures will help all students to clarify their understanding of many of the concepts in this section.

23. **Writing** Which similarity relationship is most closely related to the ASA Theorem for congruent triangles? Explain your choice.

 Technology For Exercises 24–26, use geometry software or a ruler and a protractor.

24. **a.** If all the side lengths of two quadrilaterals are in proportion, must the quadrilaterals be similar? Draw several pairs of quadrilaterals with sides in proportion to illustrate your answer.

   b. Is your conclusion in part (a) true for polygons with more than four sides? Give an example or a counterexample.

25. **a.** If all of the corresponding angles of two quadrilaterals are congruent, must the quadrilaterals be similar? Draw several pairs of quadrilaterals with congruent angles to illustrate your answer.

   b. Is your conclusion in part (a) true for polygons with more than four sides? Give an example or a counterexample.

26. If two quadrilaterals have two pairs of corresponding angles that are congruent and these angles are included by sides that are in proportion, must the quadrilaterals be similar? Draw several pairs of quadrilaterals to illustrate your answer.

ONGOING ASSESSMENT

27. **a. Open-ended Problem** A student made this conjecture based on the proof in Exercise 14. Show that this conjecture is false by finding a counterexample.

   b. Writing Rewrite the conjecture to make it a true statement. Explain why your statement is always true.

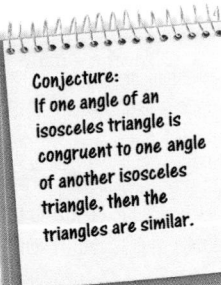
Conjecture:
If one angle of an isosceles triangle is congruent to one angle of another isosceles triangle, then the triangles are similar.

Practice 62 for Section 9.2

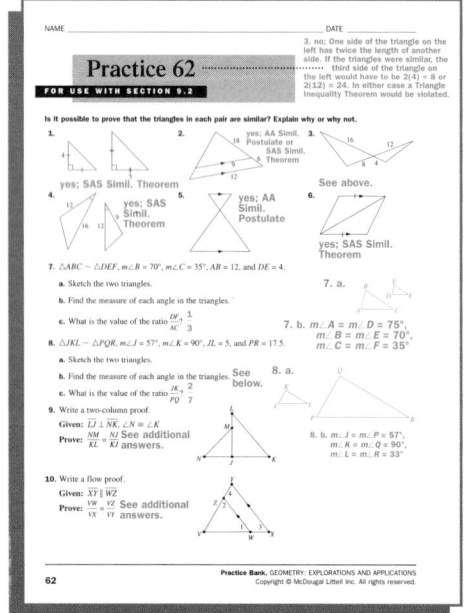

SPIRAL REVIEW

28. Write a two-column proof. *(Section 6.3)*
   Given: $\overline{AD} \cong \overline{DC}$; $\angle ADB \cong \angle BDC$
   Prove: $\triangle ABD \cong \triangle CBD$

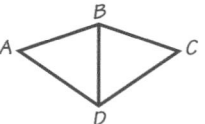

Tell whether the polygons in each pair are similar. Explain your reasoning. *(Section 9.1)*

29.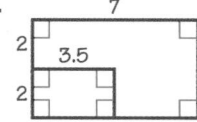

30.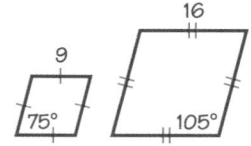

31.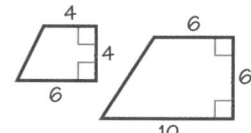

460   Chapter 9   *Similar Polygons*

---

23. The AA Similarity Postulate relates two angles of one triangle to two angles of another, just as the ASA Congruence Theorem does. In the similarity case, there is no requirement for the lengths of the included sides.

24–26. Sketches may vary. Examples are given.

24. **a.** not necessarily
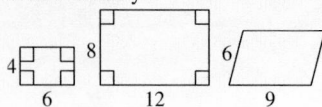

b. Yes.
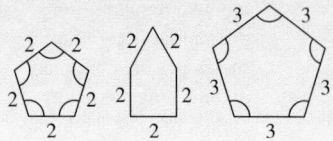

25. **a.** not necessarily
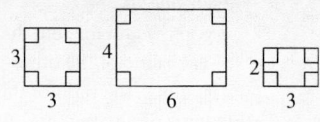

b. Yes.
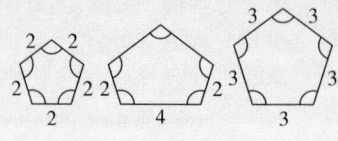

26. not necessarily
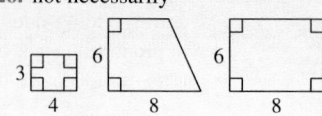

27, 28. See answers in back of book.

29. Yes; corresponding angles are congruent and the lengths of corresponding sides are in proportion.

30. Yes; corresponding angles are congruent and the lengths of corresponding sides are in proportion.

31. No; $\frac{4}{6} \neq \frac{6}{10}$.

9.3 Proportions and Similarity

Learn how to...
- write many different proportions using the sides of similar figures

So you can...
- find lengths and measures in similar polygons

You know that in similar figures the lengths of the sides are in proportion. You can rewrite these proportions in several ways to help you learn more about similar figures. In the Exploration, you will investigate one way that parallel lines and similar figures are related.

EXPLORATION
COOPERATIVE LEARNING

Finding Proportions in Triangles

Work with another student.
You will need:
- geometry software, or patty paper and a ruler

1 Draw a scalene triangle, $\triangle PQR$. Construct $\overline{ST}$ so that $\overline{ST} \parallel \overline{PQ}$ and $\overline{ST}$, $\overline{RT}$, and $\overline{RS}$ form a smaller triangle, $\triangle RST$. Find the lengths of the sides of the triangles.

2 Find $\frac{PS}{SR}$ and $\frac{QT}{TR}$. Record your results.

3 Explain why $\triangle RPQ \sim \triangle RST$.

4 Each of you should repeat Steps 1 and 2 with three different triangles. Make a conjecture about these triangles based on your results.

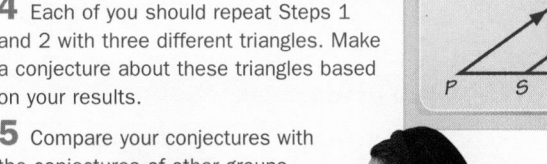

5 Compare your conjectures with the conjectures of other groups.

For answers to the Exploration, see following page.

 Exploration Note

Purpose
The purpose of this Exploration is to have students discover the Triangle Proportionality Theorem, which is stated on page 462.

Materials/Preparation
Each group of students needs geometry software, or patty paper and a ruler.

Procedure
Students draw a scalene triangle with a segment that intersects two sides of the triangle and is parallel to the third side. Students measure the parts of the two divided sides and examine the ratios of these parts. They also explain why the two triangles are

similar. This procedure is repeated three times with different triangles, and a conjecture is made about the triangles.

Closure
After comparing their conjectures, students should realize that if a segment is parallel to one side of a triangle and intersects the other two sides, then it divides those sides proportionally.

Explorations Lab Manual
See the Manual for more commentary on this Exploration.

Plan⇔Support

Objectives
- Write many different proportions using the sides of similar figures.
- Find lengths and measures in similar polygons.

Recommended Pacing
❖ **Core and Extended Courses**
Section 9.3: 2 days

❖ **Block Schedule**
Section 9.3: 2 half-blocks (with Sections 9.2 and 9.4)

Resource Materials

Lesson Support
Lesson Plan 9.3

Warm-Up Transparency 9.3

Overhead Visuals:
 Folder 11: Similar Polygons in Space

Practice Bank: Practice 63

Study Guide: Section 9.3

Explorations Lab Manual: Diagram Master 2

Challenge Problems: Set 63

Assessment Book: Test 39

Technology
Technology Book:
 Spreadsheet Activity 8

Geometry Software

McDougal Littell Mathpack
 Function Investigator Activity Book: Activity 23

Internet:
 http://www.hmco.com

Warm-Up Exercises

Write a reason to justify each statement.

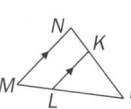

1. $\angle MNP \cong \angle LKP$ If two $\parallel$ lines are intersected by a transversal, then corresponding $\angle$s are $\cong$.

2. $\angle P \cong \angle P$ Reflexive Property

3. $\triangle MNP \sim \triangle LKP$ AA Similarity Postulate

4. $\frac{PK}{PN} = \frac{PL}{PM}$ Def. of similarity

5. If $MN = 15$, $PM = 21$, and $PL = 14$, find LK.   10

Section Notes

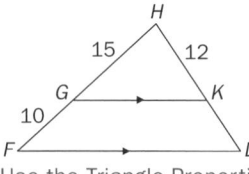

Communication: Reading
Read and discuss the Triangle Proportionality Theorem and the Midsegment Theorem. In the diagram for the Triangle Proportionality Theorem, ask students why triangle *PST* is similar to triangle *PQR*. ($\angle PST \cong \angle PQR$ and $\angle PTS \cong \angle PQR$. By the AA Similarity Postulate, $\triangle PST \sim \triangle PQR$.) Ask also: Is the proportion $\frac{PS}{PQ} = \frac{PT}{PR}$ true, and if so, why? (Yes; definition of similar figures.) The same types of questions can be asked for the diagram in the Midsegment Theorem. These questions can prepare students to understand the first four steps of the proof of the Triangle Proportionality Theorem given on page 463.

Geometric Thinking

The Midsegment Theorem was proved previously by using a coordinate geometry proof. Point out to students that this theorem can be proved by using the Triangle Proportionality Theorem.

Additional Example 1

Find the length *KL*.

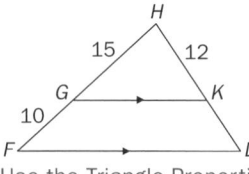

Use the Triangle Proportionality Theorem.

$$\frac{HG}{GF} = \frac{HK}{KL}$$

$$\frac{15}{10} = \frac{12}{KL}$$

Find the cross products.

$$15 \cdot KL = 10 \cdot 12$$

$$KL = \frac{120}{15}$$

$$= 8$$

In the Exploration, you saw that a line that intersects two sides of a triangle and is parallel to the third side forms a triangle that is similar to the original one. The proof of this theorem is on page 463.

Triangle Proportionality Theorem

If a segment is parallel to one side of a triangle and intersects the other two sides, then it divides those sides proportionally.

If $\overline{ST} \parallel \overline{QR}$, then $\frac{PS}{SQ} = \frac{PT}{TR}$.

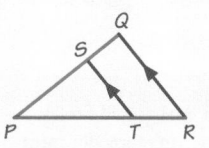

The *midsegment* joins the midpoints of two segments. The Midsegment Theorem is a special case of the Triangle Proportionality Theorem.

Midsegment Theorem

If the midpoints of two sides of a triangle are joined by a segment, then the segment is parallel to the third side of the triangle and its length is half the length of the third side of the triangle.

If D is the midpoint of $\overline{AB}$ and E is the midpoint of $\overline{BC}$, then $\overline{DE} \parallel \overline{AC}$ and $DE = \frac{1}{2}AC$.

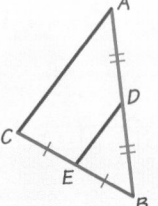

EXAMPLE 1   Connection: Algebra

Find the length *JN*.

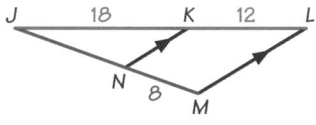

SOLUTION

Use the Triangle Proportionality Theorem.

$$\frac{JN}{NM} = \frac{JK}{KL}$$

$$\frac{JN}{8} = \frac{18}{12} \qquad \text{Find the cross products.}$$

$$12 \cdot JN = 8 \cdot 18$$

$$JN = \frac{144}{12}$$

$$JN = 12$$

Check

$$\frac{12}{8} \stackrel{?}{=} \frac{18}{12}$$

$$\frac{3}{2} = \frac{3}{2} \checkmark$$

ANSWERS Section 9.3

Exploration

1. Check students' work.

2. For all triangles produced according to the directions in Step 1, $\frac{PS}{SR} = \frac{QT}{TR}$.

3. Answers may vary. An example is given. Since $\overline{ST} \parallel \overline{PQ}$, $\angle Q \cong \angle RTS$ and $\angle P \cong \angle TSR$.

(If two $\parallel$ lines are intersected by a transversal, corresponding $\angle$s are $\cong$.) So $\triangle RPQ \sim \triangle RST$ by the AA Similarity Theorem.

4. Answers may vary. An example is given. A segment that is parallel to one side of a triangle and intersects the other two sides forms a triangle similar to the original triangle.

5. Check students' work.

Proportions

Two equations are *equivalent* if one equation can be changed into the other using algebra. For example, $\frac{a}{b} = \frac{c}{d}$ is equivalent to $ad = bc$ because you can get the second equation by multiplying both sides of the first equation by bd.

Properties of Proportions

All of the proportions below are equivalent to each other.

$$\frac{a}{b} = \frac{c}{d} \qquad \frac{b}{a} = \frac{d}{c} \qquad \frac{a}{c} = \frac{b}{d} \qquad \frac{c}{a} = \frac{d}{b}$$

These proportions are also equivalent to the ones above.

$$\frac{a}{a+b} = \frac{c}{c+d} \qquad \frac{b}{a+b} = \frac{d}{c+d} \qquad \frac{a+b}{a} = \frac{c+d}{c}$$

THINK AND COMMUNICATE

In the diagram, $\overline{EF} \parallel \overline{JG}$. Complete each proportion.

1. $\dfrac{y}{x} = \underline{\ ?\ }$

2. $\dfrac{x}{x+y} = \underline{\ ?\ }$

3. $\dfrac{5}{x} = \underline{\ ?\ }$

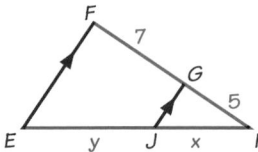

When you use equivalent proportions in proofs, give *A property of proportions* as the reason. Equivalent proportions can be used to prove the Triangle Proportionality Theorem.

Given: $\overline{ST} \parallel \overline{QR}$

Prove: $\dfrac{PS}{SQ} = \dfrac{PT}{TR}$

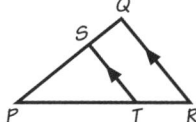

Statements	Reasons
1. $\overline{ST} \parallel \overline{QR}$	1. Given
2. $\angle PST \cong \angle PQR$; $\angle PTS \cong \angle PRQ$	2. If two $\parallel$ lines are intersected by a transversal, then corresponding angles are $\cong$.
3. $\triangle PST \sim \triangle PQR$	3. AA Similarity Postulate
4. $\dfrac{PS}{PQ} = \dfrac{PT}{PR}$	4. Definition of similar figures
5. $PQ = PS + SQ$; $PR = PT + TR$	5. Segment Addition Postulate
6. $\dfrac{PS}{PS + SQ} = \dfrac{PT}{PT + TR}$	6. Substitution Property (Steps 4 and 5)
7. $\dfrac{PS}{SQ} = \dfrac{PT}{TR}$	7. A property of proportions

9.3 Proportions and Similarity **463**

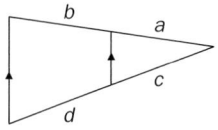

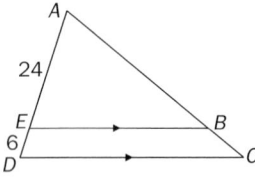

Additional Example 2

Find each value.

A

24

E
6
D

B

C

a. $\dfrac{AB}{BC}$

Use the Triangle Proportionality
Theorem. $\dfrac{AB}{BC} = \dfrac{AE}{ED} = \dfrac{24}{6} = \dfrac{4}{1}$

b. $\dfrac{AB}{AB+BC}$ $\dfrac{AB}{AB+BC} = \dfrac{AE}{AE+ED} = $
$\dfrac{24}{24+6} = \dfrac{24}{30} = \dfrac{4}{5}$

Closure Question

A segment is parallel to one side of
a triangle and intersects the other
two sides. Describe any properties
involved.   The segment divides the
two sides it intersects proportionally. If the segment joins the midpoints of the two sides it intersects,
then it is half the length of the third
side.

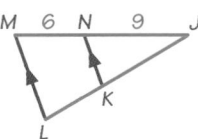

Apply⇔Assess

Suggested Assignment

❖ **Core Course**
  Day 1 Exs. 1–9, 14–21
  Day 2 Exs. 23–31, AYP

❖ **Extended Course**
  Day 1 Exs. 1–21
  Day 2 Exs. 22–31, AYP

❖ **Block Schedule**
  Day 55 Exs. 1–9, 14–21
  Day 56 Exs. 23–31, AYP

Exercise Notes

Common Error
Ex. 2 Some students may use the
Triangle Proportionality Theorem
directly and write the proportion
$\dfrac{y}{12} = \dfrac{5}{4}$ rather than the correct

proportion $\dfrac{y}{12} = \dfrac{5}{9}$. Remind students
that if they are finding the length of
an *entire* side of a triangle, they
must use the *entire* side of the
similar triangle.

464

EXAMPLE 2

Find each value.

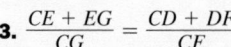

a. $\dfrac{LK}{KJ}$ **b.** $\dfrac{KN}{LM}$

SOLUTION

Use the Triangle Proportionality Theorem.

a. $\dfrac{LK}{KJ} = \dfrac{MN}{NJ} = \dfrac{6}{9} = \dfrac{2}{3}$ **b.** $\dfrac{KN}{LM} = \dfrac{JN}{JM} = \dfrac{9}{15} = \dfrac{3}{5}$

☑ CHECKING KEY CONCEPTS

Tell whether each proportion is *True* or *False*.

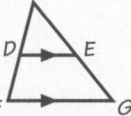

1. $\dfrac{CD}{DF} = \dfrac{CE}{EG}$ **2.** $\dfrac{DE}{FG} = \dfrac{CG}{CE}$

3. $\dfrac{CE+EG}{CG} = \dfrac{CD+DF}{CF}$ **4.** $\dfrac{CE}{DF} = \dfrac{CD}{EG}$

Find the value of each ratio or length.

5. a. $\dfrac{GH}{HD}$ **6. a.** $\dfrac{PT}{TS}$

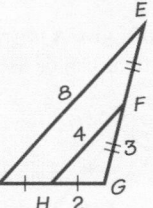

b. EF **b.** $\dfrac{TQ}{SR}$

c. DG **c.** $\dfrac{PS}{PT}$

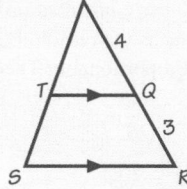

9.3 ⋮ **Exercises and Applications**

Extra Practice
exercises on
page 678

Find the value of each variable.

1.

5

4

3   x

2.

5   y

4   12

3.

z

10

4   18

ALGEBRA Complete each equation so it is equivalent to the proportion $\dfrac{x}{y} = \dfrac{3}{4}$.

4. $\dfrac{y}{x} = \underline{\ ?\ }$ **5.** $\dfrac{4}{y} = \underline{\ ?\ }$ **6.** $\dfrac{4}{4+y} = \underline{\ ?\ }$

464   Chapter 9   *Similar Polygons*

Checking Key Concepts

1. True.   2. False.

3. True.   4. False.

5. a. 1   6. a. $\dfrac{4}{3}$

   b. 3   b. $\dfrac{4}{7}$

   c. 4   c. $\dfrac{7}{4}$

Exercises and Applications

1. $x = 3\dfrac{3}{4}$

2. $y = 6\dfrac{2}{3}$

3. $z = 2\dfrac{2}{9}$

4. $\dfrac{4}{3}$

5. $\dfrac{3}{x}$

6. $\dfrac{3}{3+x}$

7. 2 cm; 2.25 cm; 1.5 cm

8. 1 cm; 1.125 cm; 0.75 cm

For Exercises 7 and 8, use the diagram at the right. $AB = 3$ cm, $CA = 4$ cm, and $CB = 4.5$ cm.

7. D is the midpoint of $\overline{AC}$ and E is the midpoint of $\overline{CB}$. Find the lengths CD, CE, and DE.

8. F is one fourth the distance from C to A, and G is one fourth the distance from C to B. Find the lengths CF, CG, and FG.

9. a. Sketch $\triangle KLM$ so that $\overline{EF}$ connects the midpoints of $\overline{KM}$ and $\overline{KL}$, $\overline{CD}$ connects the midpoints of $\overline{KE}$ and $\overline{KF}$, and $\overline{AB}$ connects the midpoints of $\overline{KC}$ and $\overline{KD}$.

   b. $AB = 4$ cm. Find the lengths CD, EF, and ML.

   c. $KA = 3$ cm and $KB = 6$ cm. Find the lengths EM, FL, and KF.

INTERVIEW Loy Arcenas

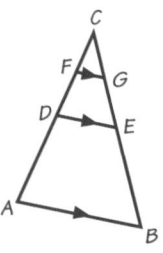

Look back at the article on pages 442–444.

Loy Arcenas gives both a scale model and blueprints to the builders of a stage set he has designed. The blueprints and model shown here are for a production of the musical Once On This Island, *by Lynn Ahrens and Stephen Flaherty.*

10. **Writing** Loy Arcenas always includes a model of a person in his scale models. Explain how this can help him understand whether he has made the pieces of the set a reasonable size.

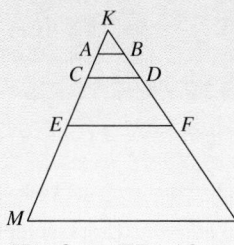

11. On the blueprint for *Tree Layer #2*, Arcenas gives the final dimensions as 34 ft 6 in. wide and 12 ft 10 in. tall. The drawing is $17\frac{1}{4}$ in. by $6\frac{7}{16}$ in. Find the scale of the drawing.

12. The scale of the model stage is 1:48. If Arcenas made *Tree Layer #2* for his scale model, what would its dimensions be? The final dimensions are given in Exercise 11.

13. **Open-ended Problem** In the drawing for a *Typical Flower*, Arcenas does not give a scale. He tells the builder to choose a size for the 25 flowers based on the tree layers. What is a reasonable size for a *Typical Flower*? Explain your reasoning.

Exercise Notes

Problem Solving
Ex. 8 In this exercise, students must translate the phrase "*F* is one fourth the distance from *C* to *A*" into a ratio. It may help some students to use an example. If $CF = 1$, then $CA = 4$ and $\frac{CF}{CA} = \frac{1}{4}$. Similarly, if $CF = 3$, then $CA = 12$ and $\frac{CF}{CA} = \frac{1}{4}$. Thus, the above phrase can be translated as "the ratio of *CF* to *CA* is 1 : 4."

Geometric Thinking
Ex. 9 To solve this exercise, students must apply the Midsegment Theorem repeatedly, each time finding the side length of a larger triangle.

Interview Note: Application
Exs. 10–13 A scale model and a set of blueprints that are used to build the stage set for a musical are examined in these exercises. Students gain experience working between the scale of the model and the scale of the blueprint. They see how using a scale model of a person can aid the designer in making the pieces of the set a reasonable size.

---

9. a. Answers may vary. An example is given.

   K, A, B, C, D, E, F, M, L

   b. $CD = 8$ cm; $EF = 16$ cm; $ML = 32$ cm

   c. $EM = 12$ cm; $FL = 24$ cm; $KF = 24$ cm

10. He can tell how the set will appear in relation to the actors.

11. 1 : 24

12. $8\frac{5}{8}$ in. tall and $3\frac{5}{24}$ in. wide

13. Answers may vary. An example is given. a square with sides about $\sqrt{2}$, or about 1.4 ft; *Tree Layer #2* is 34 ft 6 in. by 12 ft 10 in. and has an area of about 47 ft². If the flowers are supposed to cover about half of this layer, then they should have a total area of about 24 ft² or about 2 ft² each. If they are roughly square, then they should have a side of about $\sqrt{2}$ or 1.4 ft.

Exercise Notes

Geometric Thinking

Exs. 14–16 You may wish to ask students how this theorem is related to the Triangle Proportionality Theorem. To demonstrate the relationship visually, students can extend the transversals $\overleftrightarrow{AC}$ and $\overleftrightarrow{FD}$ to form a triangle.

Integrating Algebra

Exs. 18–21 In Exs. 18–20, students use algebra to show that certain proportions are equivalent. Students should see that the problem in Ex. 21 can be solved by assigning a variable to represent the height of the tree. The proportion $\frac{5}{x} = \frac{4}{15}$ can then be solved to find the height of the tree.

Challenge

Ex. 22 Some students may need a hint that they can use the fact that the opposite sides of a parallelogram are congruent.

Cooperative Learning

Ex. 23 This activity provides students with an opportunity to explore other triangle proportions. After completing part (c), students should compare their summaries with those of other groups to check for accurate conclusions.

This theorem is related to the Triangle Proportionality Theorem: Parallel lines divide transversals into segments that are in proportion. Use this theorem to complete each proportion.

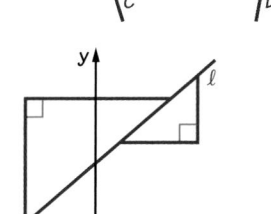

14. $\frac{FE}{ED} = \underline{\ ?\ }$ **15.** $\frac{CB}{DE} = \underline{\ ?\ }$ **16.** $\frac{AC}{AB} = \underline{\ ?\ }$

17. a. Investigation On a coordinate plane, draw a line, ℓ, that is not parallel to an axis. Draw three different right triangles, each with hypotenuse on ℓ and both legs parallel to the axes. Explain why your three triangles are similar.

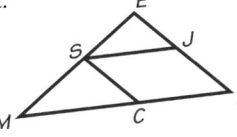

   b. Draw a line j that is parallel to ℓ. Draw a right triangle with hypotenuse on j and legs parallel to the axes. Is this triangle similar to the ones you drew in part (a)? Explain your reasoning.

ALGEBRA Use algebra to show that each statement is true.

18. The proportion $\frac{a}{b} = \frac{c}{d}$ is equivalent to $\frac{a+b}{b} = \frac{c+d}{d}$.

19. The proportion $\frac{a}{b} = \frac{c}{d}$ is equivalent to $\frac{b}{a} = \frac{d}{c}$.

20. The proportion $\frac{a}{b} = \frac{c}{d}$ is equivalent to $\frac{a}{a-b} = \frac{c}{c-d}$.

21. SAT/ACT Preview A 5 ft tall sign has a 4 ft shadow. At the same time, a nearby tree has a 15 ft shadow. How tall is the tree?

   A. $20\frac{3}{4}$ ft   **B.** $16\frac{1}{2}$ ft   **C.** $18\frac{3}{4}$ ft   **D.** 12 ft   **E.** $20\frac{1}{2}$ ft

22. Challenge Write a proof in any format.

   Given: $\overline{JS} \parallel \overline{KM}$; $\overline{SC} \parallel \overline{EK}$

   Prove: $\frac{EJ}{SC} = \frac{JS}{CM}$

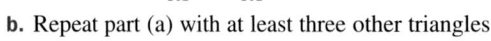

ONGOING ASSESSMENT

23. Cooperative Learning Work with another person. Use geometry drawing software or a compass and ruler.

   a. Draw $\triangle RST$. Draw the bisector of $\angle T$. Label the point where the bisector intersects $\overline{RS}$ as point Y. Find the lengths of the segments and the ratios $\frac{ST}{RT}$ and $\frac{SY}{RY}$.

   b. Repeat part (a) with at least three other triangles.

   c. Writing Write a summary of your results.

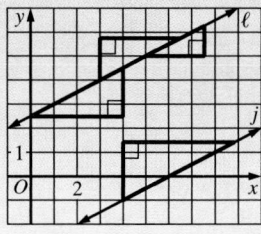

---

14. $\frac{AB}{BC}$

15. $\frac{AB}{FE}$, $\frac{AC}{FD}$

16. $\frac{FD}{FE}$

17. Answers may vary. An example is given.

   a. In each pair of triangles, each leg of one triangle is parallel to a leg of the other. The parallel lines can be used to show that two angles of one triangle are congruent to two angles of the other. Then the triangles are similar by the AA Similarity Postulate. (Alternatively, the parallel lines can be used to show that one angle of one triangle is congruent to one angle of the other. Then, since one angle of each triangle is a right angle, the triangles are similar by the AA Similarity Postulate.)

   b. Yes; the reasoning is similar to that described in part (a). However, additional parallel lines (horizontals and verticals) may be needed to show the triangles are similar.

18–20. Answers may vary. Examples are given.

18. Since $\frac{a}{b} = \frac{c}{d}$, $\frac{a}{b} + 1 = \frac{c}{d} + 1$. Then $\frac{a}{b} + \frac{b}{b} = \frac{c}{d} + \frac{d}{d}$ and $\frac{a+b}{b} = \frac{c+d}{d}$.

19. If $\frac{a}{b} = \frac{c}{d}$, then $1 \div \frac{a}{b} = 1 \div \frac{c}{d}$ or $\frac{b}{a} = \frac{d}{c}$.

20. $\frac{a}{b} = \frac{c}{d}$ and $\frac{a}{a-b} = \frac{c}{c-d}$ are both equivalent to $ad = bc$. (Cross multiply $\frac{a}{a-b} = \frac{c}{c-d}$: $ac - ad = ac - bc$; $-ad = -bc$; $ad = bc$)

24. An isosceles right triangle with sides of lengths 1, 1, and $\sqrt{2}$ is dilated using a scale factor of 3. What is the perimeter of the image? What is the area of the image? *(Section 8.6)*

Find the area of each polygon. *(Section 7.4)*

25. a rectangle whose sides are 13 ft and 8 ft long

26. a triangle whose base is 5 m and whose height is 6 m

27. a rectangle with a side of length 12 in. and a diagonal 13 in.

Express each probability as a decimal between 0 and 1, inclusive.
(Toolbox, page 710)

28. a 100% chance   **29.** 1 chance in 1000   **30.** a 3 in 8 chance   **31.** a 4% chance

ASSESS YOUR PROGRESS

VOCABULARY

proportion (p. 446)   **similar** (p. 446)

Solve each proportion. *(Section 9.1)*

1. $\dfrac{4}{10} = \dfrac{x}{25}$ **2.** $3:y = 15:21$ **3.** $z:6 = 36:27$

Tell whether the polygons in each pair are similar. *(Sections 9.1 and 9.2)*

4. 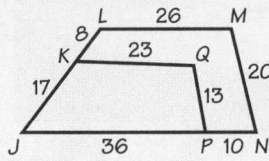   **5.**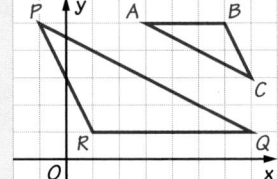

The triangles in each pair are similar. Find the value of each variable.
(Section 9.3)

6. 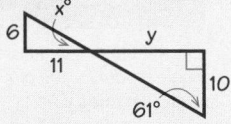   **7.**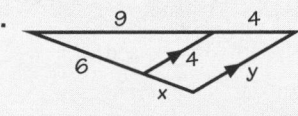

Complete each equation so it is equivalent to the proportion $\dfrac{x}{y} = \dfrac{2}{5}$.
(Section 9.3)

8. $\dfrac{y}{x} = \underline{\ ?\ }$ **9.** $\dfrac{2}{x} = \underline{\ ?\ }$ **10.** $\dfrac{5+y}{y} = \underline{\ ?\ }$

11. Journal The symbol for congruence, $\cong$, is a combination of the symbol for equality, $=$, and the symbol for similarity, $\sim$. Do you think this makes sense? Explain why or why not.

9.3 Proportions and Similarity   **467**

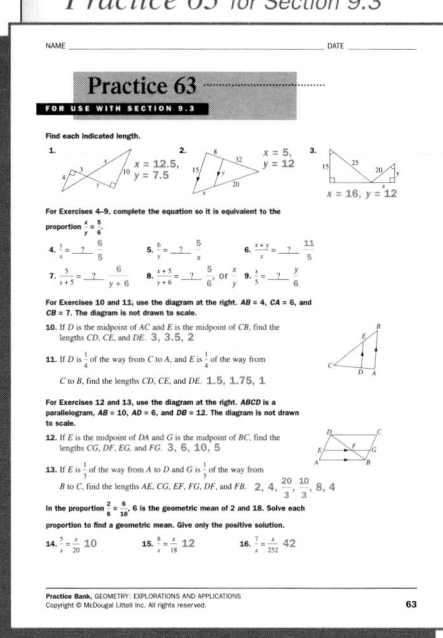

---

21. C

22. Answers may vary. An example is given.

  Statements (Reasons)

  1. $\overline{JS} \parallel \overline{KM}$; $\overline{SC} \parallel \overline{EK}$ (Given)
  2. $\angle E \cong \angle MSC$; $\angle ESJ \cong \angle M$
  (If two $\parallel$ lines are intersected by a transversal, then corresponding $\angle$s are $\cong$.)
  3. $\triangle MSC \sim \triangle SEJ$ (AA Similarity Postulate)
  4. $\dfrac{EJ}{SC} = \dfrac{JS}{CM}$ (Def. of $\sim$ $\triangle$s)

23. a. $\dfrac{ST}{RT} = \dfrac{SY}{RY}$

  b. Check students' work.

  c. The bisector of an angle of a triangle divides the side opposite the angle into segments proportional to the other two sides.

24. $3(2 + \sqrt{2})$; $4\dfrac{1}{2}$

25. 104 ft^2 **26.** 15 m^2

27. 60 in.^{2} **28.** 1

29. 0.001   **30.** 0.375

31. 0.04

Assess Your Progress

1. 10   **2.** 4.2

3. 8   **4.** No.

5. Yes.

6. $x = 29$; $y = 18\dfrac{1}{3}$

7. $x = 2\dfrac{2}{3}$; $y = 5\dfrac{7}{9}$

8. $\dfrac{5}{2}$ **9.** $\dfrac{5}{y}$

10. $\dfrac{2+x}{x}$

11. Answers may vary. An example is given. I think this makes sense because I think of congruence as a special sort of similarity. Two figures are congruent if they are similar, and the ratio of the lengths of any two corresponding sides is 1. That is, the lengths of corresponding sides are equal.

Objectives

• Compare measures of similar figures.

• Find perimeters, areas, and volumes of similar figures.

Recommended Pacing

❖ **Core and Extended Courses**
Section 9.4: 2 days

❖ **Block Schedule**
Section 9.4: 2 half-blocks (with Sections 9.3 and 9.5)

Resource Materials

Lesson Support

Lesson Plan 9.4

Warm-Up Transparency 9.4

Overhead Visuals:
Folder 11: Similar Polygons in Space

Practice Bank: Practice 64

Study Guide: Section 9.4

Explorations Lab Manual:
Diagram Master 1

Challenge Problems: Set 64

Technology
Technology Book:
TI-92 Activity 9

Graphing Calculator

Graphing Software

McDougal Littell Mathpack
Geometry Inventor Activity Book:
Activity 4

Internet:
http://www.hmco.com

Warm-Up Exercises

Find the area and the perimeter of each figure.

1.
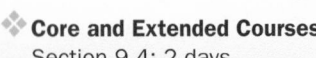
42.5; 37

2. a right triangle with legs of lengths 4 and 12
24; about 28.65

3. a rectangle with sides of length 2.8 and 3.6   10.08; 12.8

4. Find the area and circumference of a circle with radius 6.   36π; 12π

468

SECTION

9.4 | Areas and Volumes of Similar Figures

Learn how to...

• compare measures of similar figures

So you can...

• find perimeters, areas, and volumes of similar figures

When Loy Arcenas designs a stage set, he has to consider the amount of space available for the set. The set builder uses Arcenas' model and scale drawings to estimate the amount of material needed to build the set. Relationships between similar figures can help the builder make decisions about area and materials.

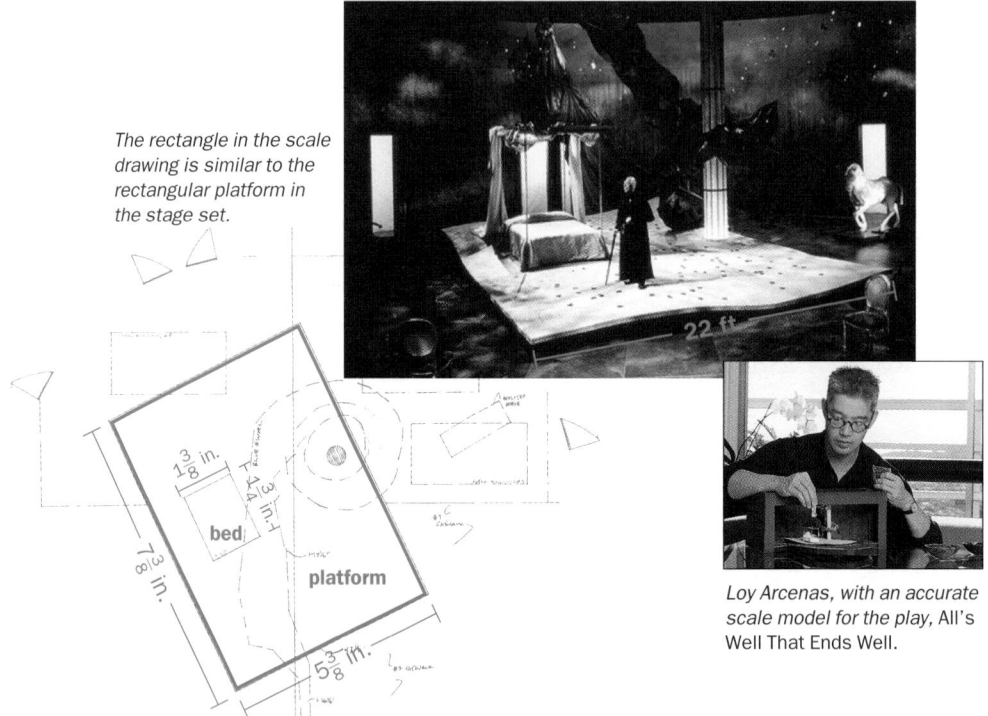

The rectangle in the scale drawing is similar to the rectangular platform in the stage set.

Loy Arcenas, with an accurate scale model for the play, All's Well That Ends Well.

THINK AND COMMUNICATE

1. What is the ratio of a side length in the scale drawing to the corresponding side length in the stage set?

2. Find the perimeters of the two rectangles. What is the ratio of the perimeters?

3. What is the ratio of the areas of the two rectangles?

Examples of linear measures of a figure are lengths of its sides, medians, diameters, altitudes, perimeter, and circumference. In similar figures, all ratios of a linear measure of one figure to the corresponding measure in the other are equal. As you saw on page 468, the areas of similar figures are related by a different ratio.

If two similar two-dimensional figures have corresponding lengths whose ratio is $a:b$, then:

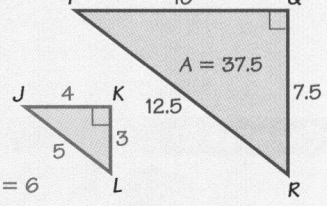

- The ratio of all corresponding linear measures is $a:b$.

$$\frac{JK}{PQ} = \frac{KL}{QR} = \frac{JL}{PR} = \frac{2}{5}$$

- The ratio of their perimeters is $a:b$.

$$\frac{\text{Perimeter of } \triangle JKL}{\text{Perimeter of } \triangle PQR} = \frac{2}{5}$$

- The ratio of their areas is $a^2:b^2$.

$$\frac{\text{Area of } \triangle JKL}{\text{Area of } \triangle PQR} = \frac{2^2}{5^2} = \frac{4}{25}$$

EXAMPLE 1

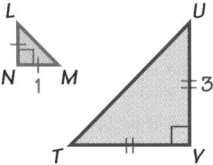

a. Explain why $\triangle LMN \sim \triangle TUV$.

b. Find the ratio of the lengths of the corresponding sides.

c. Find the ratio of the perimeters.

d. Find the ratio of the areas.

SOLUTION

a. $\frac{LN}{TV} = \frac{MN}{UV}$ and $\angle N \cong \angle V$, so $\triangle LMN \sim \triangle TUV$ by the SAS Similarity Theorem.

b. The triangles are similar and $MN:UV = 1:3$, so $LN:TV = 1:3$ and $LM:TU = 1:3$.

c. The ratio of the perimeters is also $1:3$.

d. The ratio of the areas of $\triangle LMN$ and $\triangle TUV$ is $1^2:3^2 = 1:9$.

THINK AND COMMUNICATE

4. All squares are similar. Do you think that all cubes are similar? Explain your reasoning.

5. Are all circles similar? Explain your reasoning.

6. Two right rectangular prisms have similar bases. If the prisms are similar, what other facts about the prisms must be true?

9.4 Areas and Volumes of Similar Figures   **469**

Think and Communicate

4. Answers may vary. An example is given. Yes; all ratios of a linear measure of one square or cube to a corresponding measure of any other square or cube are equal.

5. Yes; all circles have the same shape.

6. All ratios of a linear measure of one prism to a corresponding measure of any other prism must be equal.

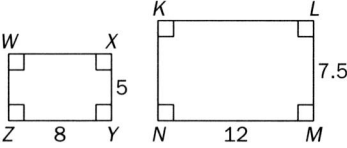

Teach⇔Interact

About Example 2

Geometric Thinking
Students should understand that the similarity relationships for surface areas and volumes allow them to find the surface area or volume of one figure if they know the surface area or volume of a similar figure. This can be done without directly using the formulas for volume or surface area.

Additional Example 2

These similar prisms have bases that are regular pentagons. The surface area of the larger prism is 680 in.^{2} and its volume is 1400 in.^{3}. Find the surface area and volume of the smaller prism.

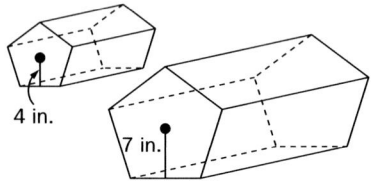

4 in.

7 in.

Find the ratio of a linear measure.
$$\frac{\text{apothem of smaller prism}}{\text{apothem of larger prism}} = \frac{4}{7}$$

The ratio of the surface areas is $\frac{a^2}{b^2}$.

Let S = the surface area of the smaller prism.

$$\frac{\text{S.A. of smaller prism}}{\text{S.A. of larger prism}} = \frac{S}{680} = \frac{4^2}{7^2}$$

$$49S = 16 \cdot 680$$
$$S = \frac{10,880}{49}$$
$$S \approx 222$$

The surface area of the smaller prism is about 222 in.^{2}

The ratio of the volumes is $\frac{a^3}{b^3}$.

Let V = the volume of the smaller prism.

$$\frac{\text{Volume of smaller prism}}{\text{Volume of larger prism}} = \frac{V}{1400} = \frac{4^3}{7^3}$$

$$343V = 64 \cdot 1400$$
$$V = \frac{89,600}{343}$$
$$V \approx 261$$

The volume of the smaller prism is about 261 in.^{3}

Areas and Volumes of Similar Figures

If two similar three-dimensional figures have corresponding lengths whose ratio is $a:b$, then:

- The ratio of all corresponding linear measures is $a:b$.
  $$\frac{AB}{PQ} = \frac{DE}{ST} = \frac{3}{4}$$

- The ratio of their surface areas (S.A.) is $a^2:b^2$.
  $$\frac{\text{S.A. of I}}{\text{S.A. of II}} = \frac{3^2}{4^2} = \frac{9}{16}$$

- The ratio of their volumes (V) is $a^3:b^3$.
  $$\frac{V \text{ of I}}{V \text{ of II}} = \frac{3^3}{4^3} = \frac{27}{64}$$

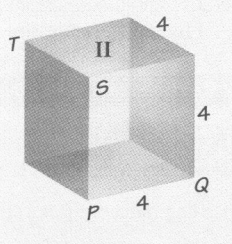

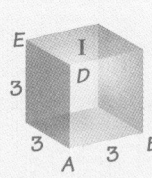

EXAMPLE 2   Connection: Algebra

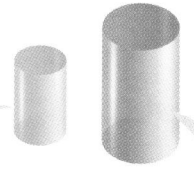

radius = 2 cm
surface area = 32π cm^2
volume = 24π cm^3

radius = 3 cm

These cylinders are similar. Find the surface area and volume of the larger cylinder.

SOLUTION

Find the ratio of a linear measure: $\frac{a}{b} = \frac{\text{radius of smaller cylinder}}{\text{radius of larger cylinder}} = \frac{2}{3}$.

Let S = the surface area of the larger cylinder.

$$\frac{\text{S.A. of smaller cylinder}}{\text{S.A. of larger cylinder}} = \frac{2^2}{3^2} = \frac{32\pi}{S}$$

The ratio of the surface areas is $\frac{a^2}{b^2}$.

$$4S = 9 \cdot 32\pi$$
$$S = \frac{288\pi}{4}$$
$$S = 72\pi$$

The surface area of the larger cylinder is 72π cm^2.
Let V = the volume of the larger cylinder.

$$\frac{\text{Volume of smaller cylinder}}{\text{Volume of larger cylinder}} = \frac{2^3}{3^3} = \frac{24\pi}{V}$$

The ratio of the volumes is $\frac{a^3}{b^3}$.

$$8V = 27 \cdot 24\pi$$
$$V = \frac{648\pi}{8}$$
$$V = 81\pi$$

The volume of the larger cylinder is 81π cm^3.

470   Chapter 9 *Similar Polygons*

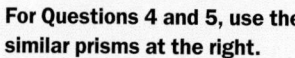
✓ CHECKING KEY CONCEPTS

For Questions 1–3, $\triangle PQR \sim \triangle JKL$.

1. Find the ratio $\frac{QR}{KL}$.

2. Find the ratio of the areas.

3. Find the ratio of the perimeters.

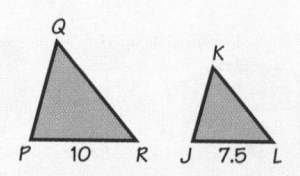

For Questions 4 and 5, use the similar prisms at the right.

4. Find the surface area of II.

5. Find the volume of I.

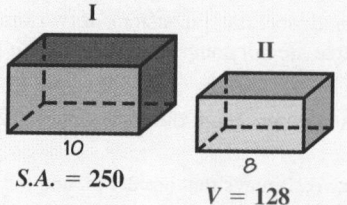

9.4 | **Exercises and Applications**

Extra Practice exercises on page 679

The figures in each pair are similar. Find each missing value.

1. $P = 25.2$ $P = \underline{?}$
   $A = \underline{?}$ $A = \underline{?}$

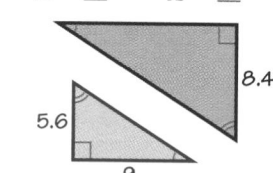

2. $P = 28$ $P = \underline{?}$
   $A = \underline{?}$ $A = 6$

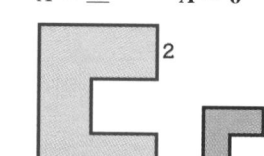

3. $P = 24$ $P = 36$
   $A = 28$ $A = \underline{?}$
   $EF = \underline{?}$

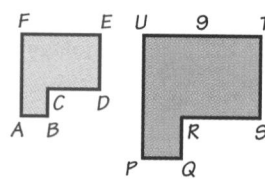

4. $C = \underline{?}$ $C = 10\pi$
   $A = 9\pi$ $r = \underline{?}$
               $A = 25\pi$

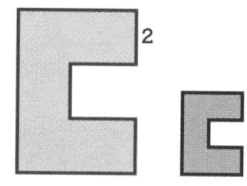

5. Area of $\triangle ABE = 24$
   Area of $\triangle ACD = 54$
   $DC = \underline{?}$

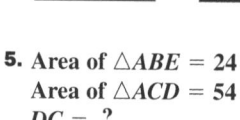

6. $P = 32$ $P = \underline{?}$
   $A = \underline{?}$ $A = 108$

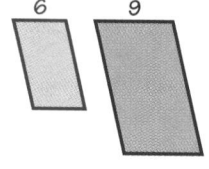

7. $\triangle ABC \sim \triangle PQR$ and $AB:PQ = 3:5$.

   a. What is the ratio of the areas of the triangles?

   b. The lengths of the sides of $\triangle ABC$ are 6, 12, and 12. Find the length of the altitude to the shortest side of $\triangle ABC$. What is the corresponding length in $\triangle PQR$?

9.4 Areas and Volumes of Similar Figures   **471**

Teach⟺Interact

Checking Key Concepts

Geometric Thinking
After completing questions 4 and 5, ask students to share the proportions they used to solve each problem. Some students may have used different but equivalent proportions, and students should see that there are different proportions that can be used to solve these types of problems.

Closure Question
What are some examples of linear measures? How can you use linear measures to determine the relationship between the areas, surface areas, or volumes of similar figures? Some examples of linear measures are the lengths of sides, diagonals, medians, diameters, apothems, altitudes, perimeters, and circumferences of figures. In similar figures, if the ratio of corresponding linear measures is $a:b$, then the ratio of areas or surface areas is $a^2:b^2$, and the ratio of volumes is $a^3:b^3$.

Apply⟺Assess

Suggested Assignment

❖ **Core Course**
  Day 1 Exs. 1–13, 15, 18
  Day 2 Exs. 20–22, 24–26, 28–39

❖ **Extended Course**
  Day 1 Exs. 1–19
  Day 2 Exs. 20–39

❖ **Block Schedule**
  Day 56 Exs. 1–13, 15, 18
  Day 57 Exs. 20–22, 24–26, 28–39

Exercise Notes

Problem Solving
Exs. 1–6 You may want to ask students to write an algorithm for solving problems such as these. They could model the algorithm on the Solution to Example 2. Students should see that the first step is to find the ratio of a linear measure.

 Communication: Drawing
Ex. 7 Encourage students to sketch the figures in this exercise to help them answer the questions.

Checking Key Concepts

1. $\frac{4}{3}$

2. $\frac{16}{9}$

3. $\frac{4}{3}$

4. 160

5. 250

Exercises and Applications

1. small triangle: $A = 25.2$;
   large triangle: $P = 37.8$;
   $A \approx 56.7$

2. $A = 24$; $P = 14$

3. $EF = 6$; $A = 63$

4. $C = 6\pi$; $r = 5$

5. 9

6. $A = 48$; $P = 48$

7. a. $3:5$; $9:25$

   b. $3\sqrt{15}$; $5\sqrt{15}$

Geometric Thinking

Exs. 8, 9 Many students may now realize that any two regular polygons with the same number of sides or any two circles are similar. This does not apply, however, to prisms and cylinders. For prisms whose bases are regular polygons to be similar, the ratio of the side lengths must be equal to the ratio of heights. Similarly, for cylinders, the ratio of radii must be equal to the ratio of heights.

Integrating Algebra

Ex. 15 You may wish to show students how these results can be proved algebraically. For example, let $h_1 =$ the height of the smaller triangle, and $b_1 =$ the base of the smaller triangle. Then the area of the triangle is $A_1 = \frac{1}{2}b_1 h_1$. When the side lengths are doubled, the larger triangle has base $2b_1$. The height of the new triangle is $2h_1$ because the two triangles are similar and all linear measures are in proportion. The area of the larger triangle is $A = \frac{1}{2}(2b_1)(2h_1) = 4\left(\frac{1}{2}b_1 h_1\right) = 4A_1$.

Using Technology

Ex. 16 To use a TI-82 for this exercise, press [STAT] and clear the first three lists by positioning the cursor over the name of the list and pressing [CLEAR] [ENTER]. Enter the edge lengths as L1. Position the cursor over L2 and type in the expression for the surface area, 6L1² and press [ENTER]. Similarly, create a volume column by entering its formula for L3. To graph the surface area as a function of edge length, press [2nd] [STATPLOT] [ENTER]. Turn Plot1 on, choose the scatter plot picture, L1 for Xlist, and L2 for Ylist. Press [ZOOM] and choose 9:ZoomStat. To graph the volume as a function of edge length, repeat the procedure above to change Ylist to L3.

Geometric Thinking

Ex. 18 This exercise helps students make a connection between dilations and similar figures. Students should understand that a figure and its image under dilation are always similar. The scale factor of the dilation is the ratio of the corresponding linear measures of the image to the original figure.

Tell whether the solids in each pair are *similar*, *not similar*, or if there is *not enough information to determine*.

8.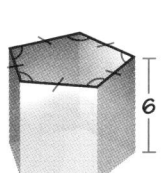

9.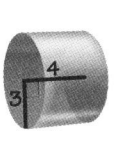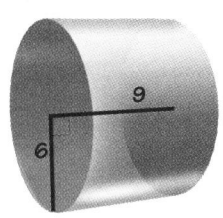

10. **Visual Thinking** Sketch and label the dimensions of two similar three-dimensional figures that are not congruent. How do you know that they are similar?

For Exercises 11–14, sketch the figures. Mark the given information on your sketch.

11. Two equilateral triangles have altitudes that are 2.4 in. and 5.2 in. long. What is the ratio of corresponding side lengths? of the perimeters? of the areas?

12. Two regular octagons have sides that are 8 in. and 9 in. long, respectively. What is the ratio of two corresponding diagonals?

13. The volumes of two cubes are in the ratio 8:125. What is the ratio of the lengths of their sides? of their surface areas?

14. **Challenge** The sides of a triangle are 12 cm, 16 cm, and 20 cm long. How long are the sides of a similar triangle whose area is 120 cm²?

15. If the lengths of the sides of a triangle are doubled, how does this change the length of the medians? the perimeter? the area of the triangle?

16. **Technology** Use graphing software or a graphing calculator.

   a. Sketch six cubes with different edge lengths. Record the edge length and find the surface area and volume of each cube. Use a spreadsheet or record your data in a table.

   b. Graph the surface area as a function of edge length. Describe the shape of the graph.

   c. Graph the volume as a function of edge length. Describe the shape of the graph.

   d. **Writing** What do the graphs in parts (b) and (c) tell you about how surface area and volume change as the edge length changes?

Cubes		
A	**B**	**C**
Edge length (cm)	Surface area (cm^2)	Volume (cm^3)
2	24	8
3		
0.5		

17. **SET DESIGN** Look at the scale drawing of the stage set for *All's Well That Ends Well* on page 468. What are the dimensions of the bed in the stage set? Explain your reasoning.

---

8. similar

9. not similar

10. Answers may vary. An example is given. All ratios of a linear measure of the smaller cube to a corresponding measure of the larger cube are equal.

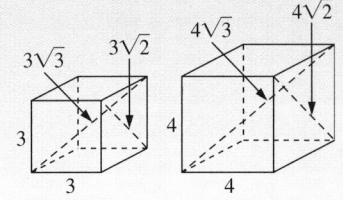

11–14. Check students' work.

11. $\frac{13}{6}, \frac{13}{6}, \frac{169}{368}$

12. $\frac{8}{9}$

13. 2:5; 4:25

14. 15 cm; 20 cm; 25 cm

15. doubled; doubled; quadrupled

16. See answers in back of book.

17. 84 in. by 66 in.; The scale of the drawing is 1:48 and the dimensions of the bed in the drawing are $1\frac{3}{4}$ in. by $1\frac{3}{8}$ in.

18. a. $169x$

   b. $\frac{w}{169}$

19. $\frac{9}{16}$

20. a. $4\frac{7}{12}$ in.

   b. 12 oz

18. a. The sides of quadrilateral *ABCD* are dilated by a scale factor of 13. If the area of *ABCD* is *x*, find the area of the image in terms of *x*.

   b. If the area of the image is *w*, find the area of *ABCD* in terms of *w*.

19. Challenge The volumes of two similar cylinders are 135π cm^3 and 320π cm^3. Find the ratio of their surface areas.

Connection   CONSUMER ECONOMICS

Manufacturers often sell the same item in different size packages so that you can purchase a *single serving* or an *economy size* package of the same product. The packages often appear to be similar so that consumers will recognize them as the same product.

8 oz   No. 2

20. a. An *8 oz* can has diameter 3 in. and height 4 in. A *No. 2* can is similar to an *8 oz* can and has a diameter of $3\frac{7}{16}$ in. Find the height of a *No. 2* can.

   b. How many ounces does a *No. 2* can hold? Round your answer to the nearest ounce. (*Hint:* Remember that an ounce is a measure of volume.)

21. a. Is the large box of cereal similar to the medium size box? Is the large box similar to the single serving box? Explain.

   b. The rectangular front faces of all three boxes are similar. Explain why the manufacturer might want the faces to be similar even if the boxes are not similar.

22. a. Find the volume of each of the three boxes of cereal.

   b. If the large box costs $3.29, give an appropriate price for each of the other two boxes.

   c. Open-ended Problem Often a product in a larger size package costs less per unit than a smaller size package. Give a reason why this is the case and write a convincing argument for your idea.

23. Research Find another product that comes in two different size containers. Are the containers similar? Sketch the containers and give the dimensions of each container. Why might you decide to buy the smaller container? the larger container?

7.2 cm
10.3 cm
4.2 cm

15.8 cm
22.5 cm
4.7 cm

19.4 cm
27.5 cm
5.5 cm

9.4 Areas and Volumes of Similar Figures   **473**

---

21. a. No; the ratios of the lengths of corresponding sides are not proportional.

   b. The front faces of the boxes are what you see when you look at the shelf. If they are similar, you might make assumptions about the volumes of the boxes that are not necessarily true.

22. a. 311.47 cm^3; 1670.85 cm^3; 2934.25 cm^3

   b. small box: $.35 medium box: $1.87

   c. Answers may vary. An example is given. The part of the unit cost related to packaging is relatively less for larger packages, because the packaging costs are a function of surface area.

23. Answers may vary. Check students' work. A shopper might decide on a smaller container to avoid waste or because storage space is limited. He or she might choose the larger container based on need or economic reasons.

Exercise Notes

Application
Exs. 20–23 Students should enjoy this practical application of similarity to consumer economics. Point out how helpful the concept of similarity is when determining the number of ounces in the #2 can (part (b) of Ex. 20). If the number of ounces was determined by using the formula for the volume of a cylinder, the result would be in cubic inches. This would then have to be converted to ounces. By using the proportions derived from the similarity of the two cans, students can avoid converting units. Make sure they understand that regardless of the units use for volumes, the ratio of volumes is constant.

Research
Ex. 23 You can extend this activity by having students compare the costs of the two sizes as well. If the containers are similar, are the costs in proportion as well? One good example is pizza. Is the ratio of the cost of a small pizza to a large pizza the same as the ratio of the diameter of the small pizza to the large pizza? If not, which pizza is a better deal?

Multicultural Note
By the Way (page 474) Many cultures have children's stories about people who are unusually small and the adventures they have had. Japan's Issunboshi is one such character. Issunboshi (translated as "one-inch boy") is so small that he eats only one grain of rice and a tiny fish for a meal, and uses a needle for a sword. The tale of Issunboshi varies somewhat from version to version, but generally follows this outline: Issunboshi travels to Kyoto to offer his services to a great lord, who gives him the job of protecting his beautiful daughter. One day, as the lord's daughter and Issunboshi are returning home after visiting a temple, two *oni* (evil beings) attack the lord's daughter; Issunboshi manages to fight them off despite his small size. The oni leave their magic mallet behind, which grants Issunboshi's wish to become a full-sized man. Impressed by Issunboshi's courage and dedication, the lord gives Issunboshi permission to marry his daughter.

Exercise Notes

Using Manipulatives

Ex. 26 This exercise allows students to further explore the concept of similarity for three-dimensional figures using paper models. Students should realize that it is possible for two nonsimilar figures to have the same surface area, and that having the same surface area does not necessarily mean that the figures have the same volume.

Assessment Note

Ex. 28 This exercise provides an opportunity for students to recall and use the Midsegment Theorem and the SSS Similarity Theorem.

Practice 64 for Section 9.4

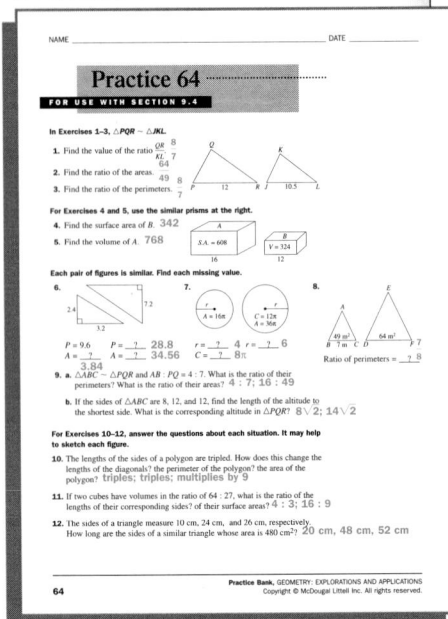

---

24. TRANSFORMATIONS If the scale factor of a dilation is 0.25, what is the ratio of the perimeter of the original polygon to the perimeter of its image? What is the ratio of the areas?

25. ALGEBRA Let *s* = the length of a side of a square. Prove using algebra that if the square is dilated by a scale factor of *k*, the area is changed by a scale factor of k^2.

26. a. Investigation Use two $8\frac{1}{2}$ in. by 11 in. sheets of paper and roll each to form an open-ended cylinder. One cylinder has height $8\frac{1}{2}$ in., and the other has height 11 in. Are the two cylinders similar?

b. Explain why the open-ended cylinders have the same lateral surface area. Do they have the same volume? Explain your reasoning.

27. Writing Many folktales and children's stories describe what it would be like to be very small. Write a paragraph about a one-inch tall person. Describe three ordinary objects that this person could use and what each could be used for.

ONGOING ASSESSMENT

28. Visual Thinking On graph paper, draw a triangle so that its vertices are on grid points.

a. Find the midpoint of each side of the triangle. Connect the midpoints to create four new triangles.

b. Explain why the four new triangles are all congruent to each other and why each one is similar to the original triangle.

c. Give two different reasons why the area of each of the four new triangles is one fourth the area of the original triangle.

SPIRAL REVIEW

29. *ABCD* is a trapezoid with $\overline{AB} \parallel \overline{CD}$ and diagonals intersecting at *E*. Prove: $\triangle ABE \sim \triangle CDE$. *(Section 9.2)*

30. In $\triangle JKL$, $\overline{ST} \parallel \overline{LK}$, *JS* = 6, *SL* = 3, and *ST* = 5. Find the length *LK*. *(Section 9.3)*

31. Miguel studied a population of 43 sea lions. The following year, 32 of these sea lions returned to the area. What is the probability that any given sea lion returned? *(Toolbox, page 710)*

Write each fraction as a percent.

32. $\frac{1}{2}$ **33.** $\frac{8}{7}$ **34.** $\frac{3}{5}$ **35.** $\frac{15}{4}$

Write each percent as a fraction in lowest terms.

36. 18%   **37.** 0.5%   **38.** 210%   **39.** 95%

---

24. $\frac{4}{1}; \frac{16}{1}$

25. The length of a side of the image square is *kx*. Its area is $(kx)^2 = k^2x^2$.

26. a. No; the radius of the taller cylinder is $\frac{8\frac{1}{2}}{2\pi}$. The radius of the shorter is $\frac{11}{2\pi}$. Then the ratio of the radius of the taller cylinder to the shorter is $\frac{8\frac{1}{2}}{11}$, while the ratio of the heights is $\frac{11}{8\frac{1}{2}}$.

b. The lateral surface area of each cylinder is the area of the sheet of paper, which is the same. The volumes are not equal. The volume of the taller cylinder is $\pi\left(\frac{8\frac{1}{2}}{2\pi}\right)^2 (11) \approx 63$ in.^{3}. The volume of the shorter cylinder is $\pi\left(\frac{11}{2\pi}\right)^2 \left(8\frac{1}{2}\right) \approx 82$ in.^{3}.

27. Answers may vary. Examples are given. A matchbox could be used as a bed. A cotton square (the kind used to apply medicine or remove makeup) could be used as a quilt. The tops from small jars or bottles could be used as tables or chairs. A soup bowl could be used as a wading pool.

28–30. See answers in back of book.

31. $\frac{32}{43} \approx 74\%$

32. 50%

33. $114\frac{2}{7}\% \approx 114.3\%$

34. 60%

35. 375%

36. $\frac{9}{50}$

37. $\frac{1}{200}$

38. $\frac{21}{10}$

39. $\frac{19}{20}$

9.5 Geometric Probability

Learn how to...
• use area and length to find probabilities

So you can...
• solve problems involving geometric probability

Scientists may use geometric probability to estimate the likelihood that animals will find their hidden food by searching randomly. This helps scientists decide if the animals are using other techniques, such as memory or smell. In the Exploration, you will use geometric probability to find the likelihood that a randomly thrown object will land on a target.

EXPLORATION
COOPERATIVE LEARNING

Investigating Probability Based on Area

Work in a group of three students.
You will need:
• paper
• dried beans

1 Make a target by drawing several shapes that you can find the area of on a rectangular sheet of paper. Shade in your shapes.

2 Place your target on the floor near a wall. Toss a bean against the wall so that it will land on your target. Record whether the bean lands on a shaded part of the paper. If the bean misses the paper, or lands on a border, do not count that toss.

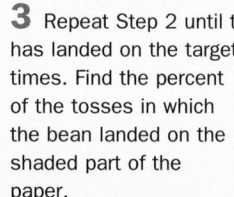

3 Repeat Step 2 until the bean has landed on the target 50 times. Find the percent of the tosses in which the bean landed on the shaded part of the paper.

4 Find the percent of the area of the paper that is shaded.

5 Compare your results in Steps 3 and 4. What do you think is the probability that if a bean lands on the target, it lands in the shaded area? Compare your results and predictions with those of other groups.

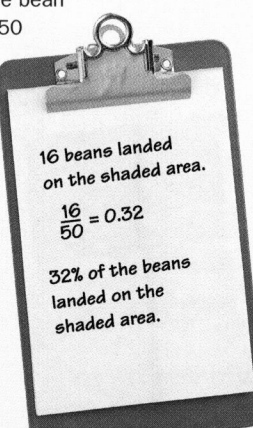

16 beans landed on the shaded area.

$\frac{16}{50} = 0.32$

32% of the beans landed on the shaded area.

Exploration Note

Purpose
The purpose of this Exploration is for students to discover how areas can be used to determine probabilities.

Materials/Preparation
Each group needs paper and dried beans.

Procedure
Students draw and shade several shapes for which they can find the areas on their paper. They then use this paper as a target and toss a bean until it lands on the paper 50 times. Students then calculate the percent of tosses that land on shaded areas and compare this percent with the percent

of the paper that is shaded. After this comparison, students make a conjecture about the probability of a bean landing in a shaded area if it lands on the paper.

Closure
Students should realize that the probability of a bean landing on the shaded area of the target is equal to the ratio of the shaded area to total area of the target.

Explorations Lab Manual
See the Manual for more commentary on this Exploration.

For answers to the Exploration, see following page.

Plan⇔Support

Objectives
• Use area and length to find probabilities.
• Solve problems involving geometric probability.

Recommended Pacing
❖ **Core and Extended Courses**
  Section 9.5: 1 day
❖ **Block Schedule**
  Section 9.5: $\frac{1}{2}$ block
  (with Section 9.4)

Resource Materials
Lesson Support
Lesson Plan 9.5
Warm-Up Transparency 9.5
Practice Bank: Practice 65
Study Guide: Section 9.5
Explorations Lab Manual: Additional Exploration 14
Challenge Problems: Set 65
Assessment Book: Test 40
Technology
Scientific Calculator
McDougal Littell Mathpack Probability Constructor Activity Book: Activities 7, 10, 14, 15
Internet: http://www.hmco.com

Warm-Up Exercises

A circular spinner is divided into 8 sections. Three are red, two are yellow, and one each is black, blue, or green. Find the probability of each outcome.

1. The spinner lands on green. $\frac{1}{8}$

2. The spinner lands on red. $\frac{3}{8}$

3. The spinner does not land on yellow. $\frac{3}{4}$

4. The spinner lands on red or black. $\frac{1}{2}$

5. The spinner lands on white. 0

Section Note

Integrating Probability

The concept of geometric probability provides a good example of how two seemingly unrelated branches of mathematics can be used to solve problems. Geometric probability also involves work with ratios, decimals, and percents.

About Example 1

Spatial Reasoning

Encourage students to look for and use equal areas, rather than calculate each area separately. In part (b), for example, rather than calculate the area of each triangle, students should realize that each triangle has equal area, and that half of them are shaded.

Additional Example 1

For each rectangular target, find the probability that a bean tossed at random will land on the shaded area.

a.

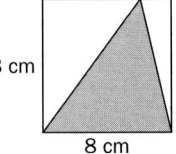

8 cm

8 cm

Find the shaded area and the total area. Because this rectangle is a square, both the base and the height of the triangle must be 8 cm.

Shaded area $= \frac{1}{2} \cdot 8 \cdot 8 = 32$ cm^2

Total area $= 8 \cdot 8 = 64$ cm^2

Probability of hitting a shaded

area $= \dfrac{\text{Shaded area}}{\text{Total area}}$

$= \dfrac{32}{64}$

$= \dfrac{1}{2}$

The probability of hitting the shaded area is $\frac{1}{2}$.

Toolbox p. 710
Probability

Probability is a ratio that tells how likely an event is to happen. For example, if a teacher randomly chooses 5 students from a class of 20 students:

$$\text{Probability of being chosen} = \frac{\text{Number of students chosen}}{\text{Number of students in class}} = \frac{5}{20}$$

The probability can also be written as $\frac{1}{4}$, or 0.25, or 25%.

In **geometric probability**, the probability of an event is given by ratios that compare lengths, areas, perimeters, or other measures. In the Exploration, you investigated the experimental probability of hitting a target with a randomly tossed object. If you cannot count the number of possible outcomes, you can use areas to find the theoretical probability.

EXAMPLE 1

For each rectangular target, find the probability that a bean tossed at random will land on the shaded area.

a.
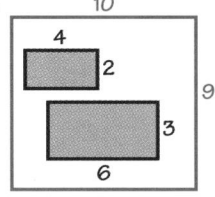

b.
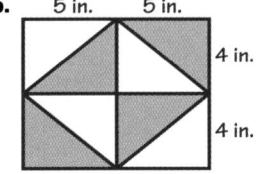

SOLUTION

a. Find the shaded area and the total area.

$$\text{Shaded area} = 4 \cdot 2 + 6 \cdot 3 = 26 \text{ in.}^2$$
$$\text{Total area} = 10 \cdot 9 = 90 \text{ in.}^2$$

$$\text{Probability of hitting a shaded area} = \frac{\text{Shaded area}}{\text{Total area}}$$

$$= \frac{26}{90}$$

$$= \frac{13}{45}$$

The probability of hitting the shaded area is $\frac{13}{45}$, or about 0.29.

b. The rectangle is divided into 8 congruent triangles. Four of the triangles are shaded.

$$\text{Probability of hitting a shaded area} = \frac{\text{Number of shaded triangles}}{\text{Total number of triangles}}$$

$$= \frac{4}{8}$$

$$= \frac{1}{2}$$

The probability of hitting the shaded area is $\frac{1}{2}$.

ANSWERS Section 9.5

Exploration

1–4. Check students' work.

5. about $\dfrac{\text{Shaded area}}{\text{Area of target}}$

THINK AND COMMUNICATE

1. Give each probability in Example 1 as a percent.

2. In part (a) of Example 1, what is the probability that the bean will land on the unshaded part of the target?

The probabilities in the Exploration and Example 1 use areas. You can also find the probability of an event using lengths. For example, if you chose a point P on $\overline{AB}$ at random, the probability that P is on $\overline{AC} = \dfrac{\text{length of } \overline{AC}}{\text{length of } \overline{AB}}$.

EXAMPLE 2   Application: Consumer Economics

A store uses cash register tape that has a red star printed on it every 200 cm. If the register tape for your receipt has a red star on it, you will receive 10% off your next purchase at the store. Joel received 25 cm of cash register tape with his purchase. What is the probability that he received a red star?

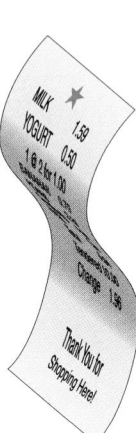

SOLUTION

To find the probability, set up a ratio.

$$\text{Probability of receiving a red star} = \frac{\text{Length of receipt}}{\text{Distance between red stars}}$$

$$= \frac{25}{200}$$

$$= 0.125$$

The probability that Joel received a red star is 12.5%.

THINK AND COMMUNICATE

3. In Example 2, at least how long would Joel's receipt need to be to guarantee that he receives a red star?

4. If the red stars were 150 cm apart, would Joel be *more* or *less* likely to receive a red star than he is in Example 2?

5. If an event has a probability of $\frac{1}{2}$, how often would you expect it to happen? Will it always occur this often?

6. If an event has a probability of 1, how often would you expect it to happen? If an event has a probability of 0, how often would you expect it to happen?

9.5 Geometric Probability **477**

 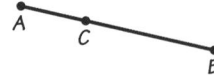
477

Checking Key Concepts

Geometric Thinking
These questions cover the concepts of this section in depth. You may wish to ask volunteers to explain their responses to the class so all students can assess their under-standing of the ideas involved. Students should be able to justify why the shaded triangle in question 2 is congruent to the other small tri-angles. (Since each side of the shaded triangle is formed by joining the midpoints of two sides of the larger triangle, its length is half the length of the side of the larger tri-angle. By the SSS Postulate, all of the triangles are congruent.)

Closure Question

Describe how to use area or length to calculate probability. Find the area or length of the desired out-come and the total area or length. The probability of the desired out-come is the ratio of the area or length of the desired outcome to the total area or length.

Suggested Assignment

❖ **Core Course**
Exs. 1–6, 8–11, 19–27, AYP

❖ **Extended Course**
Exs. 1–27, AYP

❖ **Block Schedule**
Day 57 Exs. 1–6, 8–11, 19–27, AYP

Exercise Notes

Student Progress
Exs. 1–4 Students should be able to calculate geometric probabilities when given drawings such as these that involve the formulas for finding areas of rectangles, triangles, and circles.

☑ CHECKING KEY CONCEPTS

For Questions 1 and 2, find the probability of hitting the shaded area of the target with a bean tossed at random that hits the target.

1.   **2.**

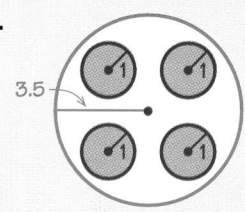   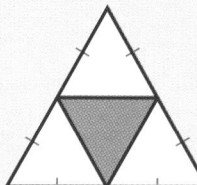

3. If the area of a target is *z* square units and the area of the shaded area on the target is *w* square units, what is the probability of a bean tossed at random landing on the shaded area of the target?

4. A store uses register tape that has a green star printed every 75 cm. If the register tape that your receipt is printed on has a green star, you will receive a prize. If your receipt is 9 cm long, what is the probability that you will receive a prize?

5. What is the probability that a point selected at random on $\overline{AB}$ is closer to *B* than to *A*?

9.5 | **Exercises and Applications**

Extra Practice exercises on page 679

For Exercises 1–4, find the probability of hitting the shaded area of the target with a bean tossed at random that hits the target.

1.   **2.**   **3.**   **4.**

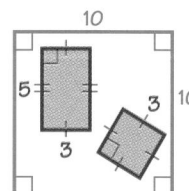 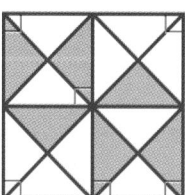 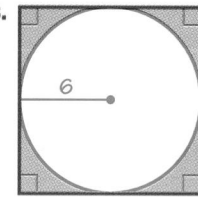 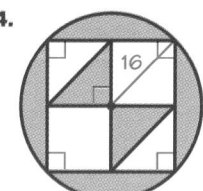

5. The midpoint of $\overline{AB}$ is *M*. What is the probability that a point selected at random on the segment is closer to *M* than it is to *A* or to *B*?

6. a. Visual Thinking In △*ABC*, $\overline{DE}$ connects the midpoints of $\overline{AC}$ and $\overline{BC}$. What is the ratio of the area of △*CDE* to the area of △*ABC*?

   b. If △*ABC* is a dart board, what is the probability of a dart thrown at random that hits the target landing on △*CDE*? on *ABED*?

7. Open-ended Problem Design a target for which the probability of a bean tossed at random landing on the shaded area is $\frac{1}{3}$.

478   Chapter 9   *Similar Polygons*

Checking Key Concepts

1. $\frac{16}{49}$

2. $\frac{1}{4}$

3. $\frac{w}{z}$

4. $\frac{3}{25}$

5. $\frac{1}{2}$

Exercises and Applications

1. $\frac{6}{25}$

2. $\frac{3}{8}$

3. $\frac{4-\pi}{4} \approx 21.5\%$

4. $\frac{256\pi - 384}{256\pi} \approx 52.3\%$

5. $\frac{1}{2}$

6. a. $\frac{1}{4}$

   b. $\frac{1}{4}, \frac{3}{4}$

7. Answers may vary. An example is given.

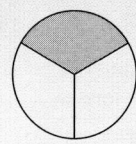

If you choose a random point on a side of the polygon, what is the probability that it is on the given segment?

8. $\overline{AB}$

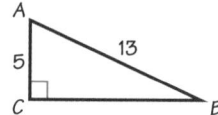

9. $\overline{QR}$

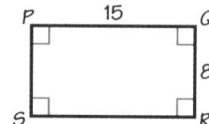

10. a. In $\triangle ABC$, $\overline{XY} \parallel \overline{AB}$, $CA = 3 \cdot CX$, and $CB = 3 \cdot CY$. What is the ratio of the area of $\triangle XYC$ to the area of $\triangle ABC$?

    b. If $\triangle ABC$ is a dart board, what is the probability of a dart thrown at random hitting $ABYX$?

11. a. A square dart board is 3 ft wide. It has three concentric circles on it. The radii of the circles are 15 in., 10 in., and 5 in., respectively. Darts are thrown and hit the target at random. What is the probability of hitting the area inside the smallest circle?

    b. What is the probability of hitting the area inside the largest circle but not inside the 10 in. circle?

Connection ⟩ ZOOLOGY

In the wild, many animals collect and hide food when it is plentiful and retrieve it when food sources are scarce. Research suggests that the animals may use landmarks to remember the location of hidden food.

In one study, a researcher used a rectangular site 116 cm by 238 cm. Landmarks such as rocks and bare twigs were added and chipmunks were allowed to bury seeds in the site.

12. Suppose a chipmunk buried 30 caches of seeds in this rectangular site. The area in which a chipmunk must dig to locate a cache is about 4.5 cm². What is the probability the chipmunk would discover one of its caches by choosing four places to dig at random?

13. **Writing** In one trial, a chipmunk located three caches in four digs, for a success rate of 75%. Do you think it is likely that the chipmunk searched at random? Explain your reasoning.

14. **Open-ended Problem** The research suggests that chipmunks use landmarks to find hidden food. What other methods might they use?

BY THE WAY

Yellow pine chipmunks are common in the western United States. They collect and hide seeds in late summer and early fall, and they return later to eat the seeds or move them to underground nests for use during the winter.

9.5 Geometric Probability   **479**

---

8. $\frac{13}{30} \approx 0.43$

9. $\frac{4}{23}$

10. a. $\frac{1}{9}$

    b. $\frac{8}{9}$

11. a. $\frac{25\pi}{1296} \approx 6\%$

    b. $\frac{125\pi}{1296} \approx 30\%$

12. $\frac{135}{27,608} \approx 0.5\%$

13. No; as demonstrated in Ex. 12, the probability that a chipmunk would discover even a single cache by searching at random is extremely low.

14. Answers may vary. Examples are given. Chipmunks may be able to smell the seeds or may recognize disturbances in a site that indicate digging.

479

Exercise Notes

Visual Thinking
Ex. 15 Ask students to create diagrams that demonstrate the geometric probability factors of this exercise. Encourage them to discuss their diagrams with the class. This activity involves the visual skills of *recognition* and *communication*.

 Application
Exs. 15, 20 These exercises introduce students to applications of probability in ballooning and advertising. Ex. 15 uses geometric probability. Ex. 20 is a straightforward probability question that can be done without using geometry.

Communication: Discussion
Exs. 16–18 Discuss the steps to this group of problems before assigning them. Suggest that students look for a pattern to find the number of squares remaining after Step 3 or Step 4. This will help them to understand the expression for the number of squares in Step *n*. You may also wish to discuss with students why Sierpinski's carpet is self-similar. (See Ex. 29 on page 452.)

Challenge
Exs. 16–18 You may wish to challenge students by having them repeat these exercises using an equilateral triangle with area 1 square unit. Have students divide this equilateral triangle into four congruent equilateral triangles by connecting the midpoints of the sides. Remove the middle triangle. Repeat this step for the remaining triangles. Ask students how many triangles remain at each step.

Assessment Note
Ex. 21 As a means of assessment, students can exchange their designs with a partner and check to see that it satisfies the conditions of the problem. Students can then discuss what they found in each other's work.

480

---

15. BALLOONING In one event at the Albuquerque International Balloon Fiesta, balloon pilots try to hit a target on the ground with a small bag of bird seed or other small object dropped from the balloon. The area of the target is 196 ft^2, and the object must land within 200 ft of the center of the target to be scored in the competition.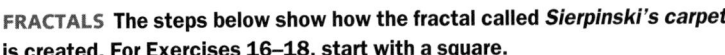

   a. A bag of bird seed dropped at random lands within 200 ft of the center of the target. What is the probability that it hits the target?

   b. Marilyn Rubin's bag of bird seed lands 5 ft from the center of the target. What is the probability that another pilot's randomly dropped bag will land closer to the target than her bag? Do you think this is related to the actual probability that Marilyn will win the competition? Explain.

FRACTALS The steps below show how the fractal called *Sierpinski's carpet* is created. For Exercises 16–18, start with a square.

Step 1 Divide the square into 9 equal squares, as shown. Remove the middle square.

Step 2 Repeat Step 1 in each of the remaining 8 squares.

Step 3 Repeat Step 1 in each of the remaining 64 squares.

Step *n* Repeat Step 1 in each of the remaining $8^{(n-1)}$ squares.

16. The area of the original square is 1 square unit. What is the total area remaining after Step 2?

17. After Step 2, what is the probability that a point chosen at random in the original square will be on one of the remaining squares?

18. a. Challenge After Step 3, how many squares will remain? Explain how you know.

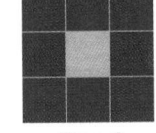 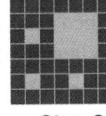 

   Step 1   **Step 2**   **Step 3**

   b. What is the probability that a point chosen at random in the original square will be on one of the remaining squares?

19. SAT/ACT Preview A box contains 10 blue marbles and 5 red marbles. What is the probability that a randomly chosen marble will be red?

  A. $\frac{1}{4}$ **B.** $\frac{1}{3}$ **C.** $\frac{2}{3}$ **D.** $\frac{1}{2}$ **E.** $\frac{3}{4}$

20. ADVERTISING During a one-hour television show, there are eight minutes of commercials. If you turn on the television at a random time during a one-hour show, what is the probability that you will turn on the television during a commercial?

ONGOING ASSESSMENT

21. Open-ended Problem A bean is tossed at random and hits a target. Design a target that meets each condition. Explain your reasoning.

   a. The probability that the bean lands on the shaded part is $\frac{1}{2}$.

   b. The probability that the bean lands on the shaded part is 0.6.

---

15. a. about 0.16%

   b. about 0.4%; No; the bags are not being dropped at random. They are aimed and are much more likely than randomly thrown objects to land near the target.

16. $\frac{64}{81}$

17. $\frac{64}{81} \approx 79\%$

18. a. 512 squares; $8^{(4-1)} = 8^3 = 512$

   b. $\frac{512}{729} \approx 70\%$

19. B

20. $\frac{2}{15} \approx 0.13$

21. Answers may vary. Examples are given.

   a.

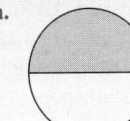

The half-circles are congruent; probability = $\frac{\text{Shaded area}}{\text{Total area}} = \frac{1}{2}$.

b.

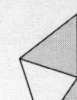

The five triangles are congruent; probability = $\frac{\text{Shaded area}}{\text{Total area}} = \frac{3}{5} = 0.6$.

Complete each equation so that it is equivalent to the proportion $\frac{x}{y} = \frac{5}{8}$.
(Section 9.3)

22. $\frac{y}{x} = \frac{?}{}$ **23.** $\frac{8+5}{8} = \frac{?}{}$ **24.** $\frac{8}{y} = \frac{?}{}$

Tell whether a triangle with sides of the given lengths is *right*, *obtuse*, or
acute. *(Section 3.7)*

25. 9, 12, 15   **26.** 5, 16, 20   **27.** 3, 5, 7

ASSESS YOUR PROGRESS

VOCABULARY

geometric probability (p. 476)

The figures in each pair are similar. Find each missing value. *(Section 9.4)*

1. $A = \frac{?}{}$ $A = \frac{?}{}$ **2.** $P = \frac{?}{}$ $P = 540$
    $P = \frac{?}{}$ $P = 21$ $A = 4422.6$ $A = \frac{?}{}$

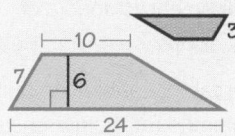
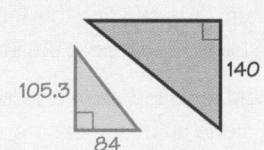

3. The ratio of the volumes of two similar rectangular prisms is $\frac{27}{125}$.
What is the ratio of the lengths of any two corresponding sides?
(Section 9.4)

**Find the probability of hitting the shaded area of the target with a dart
thrown at random that hits the target.** *(Section 9.5)*

4. 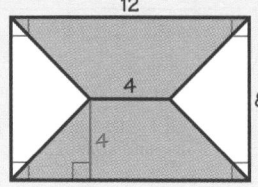   **5.**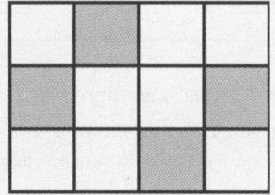

6. On $\overline{AB}$, what is the probability of a
point selected at random being within
2 cm of point G? *(Section 9.5)*

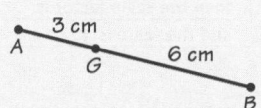

7. Journal Give an example of a situation in which you could use
geometric probability to find the probability of an event. Explain why
you could not count all of the possible outcomes in this situation.

9.5 Geometric Probability   **481**

Exercise Notes

Topic Spiraling: Review/Preview
Exs. 22–24 This review of work
with proportions will be useful in
Section 10.1 when students study
the geometric mean.

Assess Your Progress

Journal Entry
A brief review in class of some
students' examples will help others
to assess the accuracy of these
situations as well as their own.

Progress Check 9.4–9.5

See page 485.

Practice 65 for Section 9.5

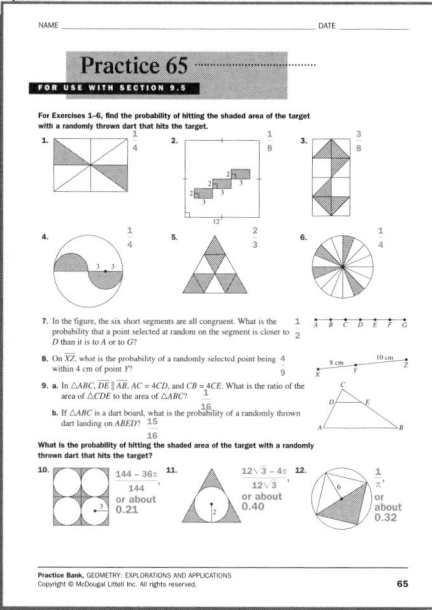

---

22. $\frac{8}{5}$

23. $\frac{x+y}{y}$

24. $\frac{5}{x}$

25. right

26. obtuse

27. obtuse

Assess Your Progress

1. large trapezoid: $A = 102$, $P = 49$; small trapezoid: $A = 18\frac{36}{49}$

2. $P = 324$; $A = 12{,}285$

3. $\frac{3}{5}$

4. $\frac{2}{3}$

5. $\frac{4}{9}$

6. $\frac{4}{9}$

7. Answers may vary. An example is given. I could use geometric probability to find the probability of winning a game where a coin is tossed onto a grid and a prize is won if the coin lands in a certain area. I cannot count all of the possible outcomes; there are infinitely many.

Scaling the Planets

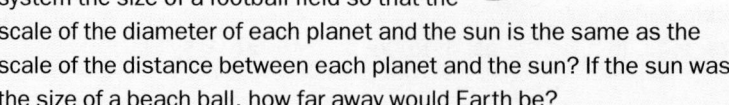

Could you make a scale model of the solar system the size of a football field so that the scale of the diameter of each planet and the sun is the same as the scale of the distance between each planet and the sun? If the sun was the size of a beach ball, how far away would Earth be?

PROJECT GOAL Create a model of our solar system.

Organizing the Data

1. Find the diameter of the sun and each of the nine planets in our solar system. You can find this information in an encyclopedia or almanac. Be sure to record the source of your data.

2. Organize your data in a spreadsheet or table. Label everything clearly. Include units of measure. Include columns for the dimensions of the scale model and the actual dimensions of the solar system.

	A	B	C	D	E
1		Diameter at Equator (km)	Mean distance from sun (million km)	Scale diameter (cm)	Scale distance (cm)
2	Sun	1400000	0		
3	Mercury	4870	57.9		
4	Venus				
5	Earth				

3. Choose a convenient size for Earth's scale diameter. Use this size to calculate the scale for the model. Your scale must use the same units for Earth's actual diameter and its scale diameter.

21	If the diameter of Earth =	3 cm
22	then the scale factor is	2.35E−09
23	and the scale is 1:	425000000

This is another expression for $2.35 \cdot 10^{-9}$.

Project Notes

Mathematical Goals

- Collect and organize data in a spreadsheet.
- Choose an appropriate scale for the solar system model.
- Create scale models of the sun and planets.

Planning

Materials
- tape measure
- materials for making planetary models

Project Teams
Students can work alone or in groups of two or three to complete the project. If students work in groups, make sure they work together, and everyone contributes equally to the project.

Guiding Students' Work

When students are selecting a scale, they may need to make several calculations before deciding upon a scale to use for their model. Creating three-dimensional models may be difficult and time consuming for some students who are working alone. However, group members can share in this activity to create a realistic model of the solar system. Suggest that all members of a group participate in presenting the group's report.

Second-Language Learners
Students may want to look up the word *scale* in a dictionary to find out what meaning this word has in various contexts.

Making the Model

1. Use the scale you found to calculate the size of the sun and each planet in a model. If the sizes of the models will be unreasonable, choose a new scale for the diameters.

2. Choose a scale to show the distances of the planets from the sun. Can you use the same scale that you chose for the diameters?

3. Create scale models of the sun and each planet in the solar system. Your models may be two- or three-dimensional. Place these models at their scale distance from the sun.

Presenting Your Results

Present your model and data to the class. In your report, you should include:

- the source of your data

- your spreadsheet with an explanation of each formula you used

- a discussion of why you chose the scale(s) that you used for your model

You may want to extend your report by investigating some of these topics:

- Find information about the planets and photos of them to include in your report. Use the Internet, magazines, or other sources. Be sure to include the source(s) you use.

- Find the scale distance between our sun and another star. Include it in your model, if possible.

- Discuss the size of scale model of the solar system if the diameter of each planet and the distance between each planet and the sun are at the same scale. See the Connection to Astronomy on page 450 for one possibility.

Self-Assessment

Explain how you chose the scale you used for your model. How did making this model change your understanding of the size of the solar system?

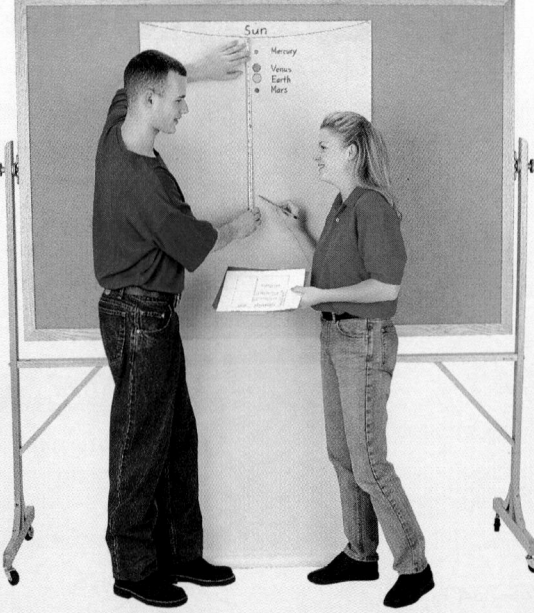

483

Guiding Students' Work

Rubric for Chapter Project

4 Students find correct diameters for the sun and planets and organize them in a spreadsheet or table. The scale chosen is convenient. The actual scale model is neatly done, and the locations and sizes of the planets are correct. Students write a detailed report about their project that includes all necessary information. The presentation made to the class is clear and well organized. Students extend the project in one of the ways listed and answer the self-assessment questions.

3 Students find correct diameters for the sun and planets and organize them in a spreadsheet or table. The scale chosen is convenient. Students construct a scale model but make some mistakes in the location of the planets. Students write a report, but from their presentation to the class, it is clear that they do not have a complete understanding of the mathematics of the project. Students extend the project in one of the ways listed and complete a self-assessment.

2 Students find the diameters of the sun and planets and organize them in a spreadsheet or table. Students do not select a convenient scale for the model and this is evident in their actual model. In the model, many planets are not correctly spaced and distances are not calculated correctly. Students write a report, but it is incomplete. From the presentation, it is clear the students do not have an understanding of the mathematics of the project. Students do not extend the project but complete a self-assessment.

1 Students do not find all of the diameters for the sun and planets nor do they correctly calculate the distances for the planets they have. Students do not select a scale, and if an actual model is made, the planets are placed at random distances. If students prepare a report, it is incomplete and not well done. Students make no effort to extend the project or complete a self-assessment. Students should be encouraged to speak with the teacher as soon as possible to review their work and to make a new start on the project.

Chapter Support

Course Guide: Chapter 9

Lesson Plans: Chapter 9

Practice Bank:
   Cumulative Practice 66

Study Guide: Chapter 9 Review

Challenge Problems: Set 66

Assessment Book:
   Chapter Tests 41 and 42
   Chapter 9 Alternative Assessment

Test Generator Software

Portfolio Project Book:
   Additional Project 5:
      Creating Special Effects

Preparing for College Entrance Tests

Professional Handbook

Progress Check 9.1–9.3

1. Solve $\frac{3}{33} = \frac{9}{y}$. *(Section 9.1)*   99

Tell how you would show that the polygons are similar. *(Sections 9.1 and 9.2)*

2. △ *RTU* and △ *RUS*

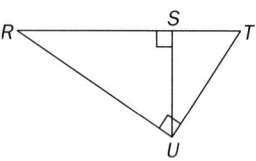

Use the AA Similarity Postulate because both triangles contain ∠ *R* and a right angle.

3.

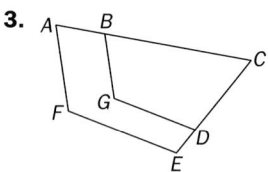

Show that all the corresponding sides are in proportion and all the corresponding angles are congruent.

Find the values of the variables for each condition. *(Sections 9.2 and 9.3)*

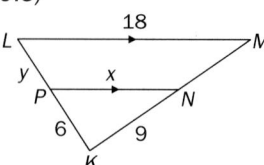

4. *KM* = 27   *x* = 6; *y* = 12

5. *P* is the midpoint of $\overline{LK}$ and *N* is the midpoint of $\overline{MK}$.   *x* = 9; *y* = 6

CHAPTER

9 | Review

STUDY TECHNIQUE

Describe any exercises or ideas in this chapter that you found difficult but now understand. How did you resolve your difficulties? Now describe some things you still don't understand. Can you use some of the methods you used before to help you resolve these difficulties?

VOCABULARY

proportion (p. 446)   **geometric probability** (p. 476)

similar (p. 446)

This means the pentagons are **similar**.

SECTIONS | 9.1, 9.2, *and* 9.3

Two figures are **similar** if corresponding angles are congruent and corresponding sides are in proportion.

Pentagon *ABCDE* ~ pentagon *PQRST*.

Corresponding sides are in the ratio 3 : 1.

$$\frac{AB}{PQ} = \frac{6}{2} = 3$$

To find the lengths of other sides, set up and solve proportions:

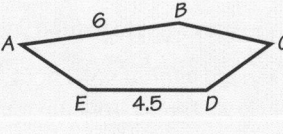

$$\frac{3}{1} = \frac{4.5}{ST} \qquad \frac{3}{1} = \frac{BC}{1.2}$$

$$3ST = 4.5 \qquad (3)(1.2) = BC$$

$$ST = 1.5 \qquad BC = 3.6$$

To show that two triangles are similar, you can use:

• AA Similarity Postulate

• SAS Similarity Theorem

• SSS Similarity Theorem

The Triangle Proportionality Theorem and the Midsegment Theorem use similar triangles to give shortcuts to find lengths.

You can use properties of proportions to write proportions that are equivalent to a given proportion.

484   Chapter 9 *Similar Polygons*

SECTION 9.4

If two figures are similar and the ratio of the lengths of two corresponding sides is $a:b$, then:

the ratio of the perimeters or any pair of corresponding lengths is $a:b$.

Ratio of perimeters: $\frac{5}{9}$

the ratio of the areas is $a^2:b^2$.

Ratio of areas: $\frac{25}{81}$

the ratio of the volumes is $a^3:b^3$.

Ratio of volumes: $\frac{125}{729}$

The rectangular prisms below are similar, so **rectangle ABCD ~ rectangle STUW**.

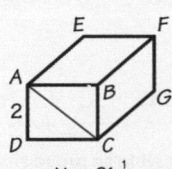

$\frac{AD}{SW} = \frac{2}{3}$, so $\frac{AC}{SU} = \frac{2}{3}$ and $AC = 3\frac{1}{3}$

The surface area of the larger prism is 108. To find the surface area, A, of the smaller prism:

$$\frac{A}{108} = \frac{2^2}{3^2}$$

$$9A = 4 \cdot 108$$

$$A = 48$$

The surface area of the smaller prism is 48.

The volume of the smaller prism is $21\frac{1}{3}$. To find the volume, V, of the larger prism:

$$\frac{21\frac{1}{3}}{V} = \frac{2^3}{3^3}$$

$$21\frac{1}{3} \cdot 27 = 8V$$

$$V = 72$$

The volume of the larger prism is 72.

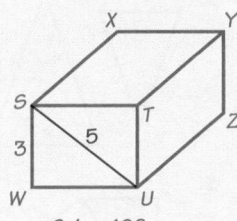

$V = 21\frac{1}{3}$

S.A. = 108

SECTION 9.5

In geometric probability, you use ratios of areas, lengths, and other measures to find the probability of an event.

The probability of hitting the shaded area of the target with a bean tossed at random that hits the target is $\frac{\text{shaded area}}{\text{total area}} = \frac{10}{25} = \frac{2}{5}$.

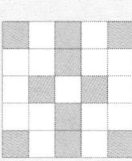

1. Find the ratio of the circumferences and the ratio of the areas for the two circles. *(Section 9.4)*

$\frac{3}{5}$; $\frac{9}{25}$

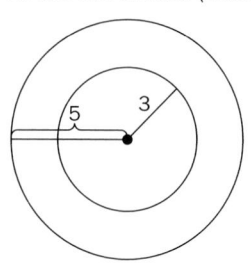

2. A triangular prism has height 12 cm and volume 263 cm³. Find the volume of a similar prism with height 5 cm. *(Section 9.4)* about 19 cm³

3. The ratio of the volumes of two rectangular solids is $\frac{27}{343}$. Find the ratio of their surface areas. *(Section 9.4)* $\frac{9}{49}$

4. Find the probability of hitting the shaded area of the target with a bean thrown at random that hits the target. *(Section 9.5)*

$\frac{1}{2}$

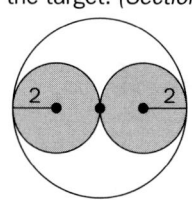

5. A highway is 156 miles long. What is the probability that if an accident occurs somewhere on the highway, it occurs within 18 miles of the police barracks that is located 75 miles from the start of the highway. *(Section 9.5)* about 0.23

Chapter 9 Assessment
Form A Chapter Test

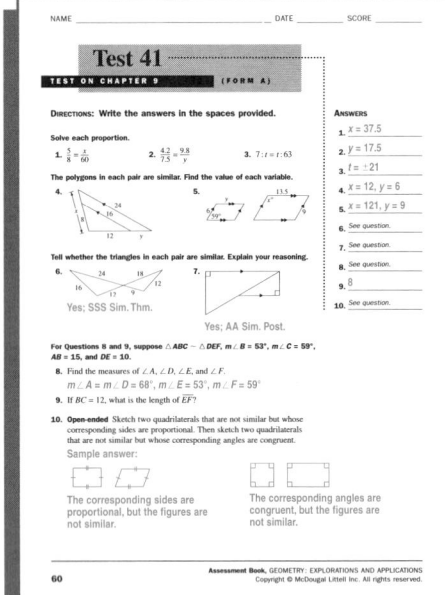

Chapter 9 Assessment
Form B Chapter Test

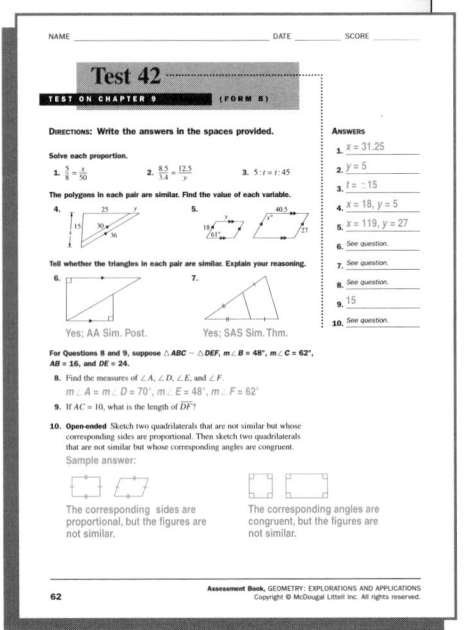

9 | Assessment

VOCABULARY QUESTIONS

1. **Writing** Explain the difference between *congruent* and *similar*. Can two similar figures be congruent? Explain.

2. **Open-ended Problem** Write two equivalent ratios. Then write the ratios as a proportion.

SECTIONS 9.1, 9.2, *and* 9.3

Is it possible to prove that the triangles in each pair are similar? If so, tell which postulate or theorem you would use.

3.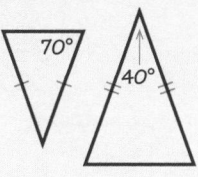

4.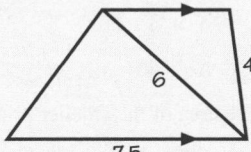

5.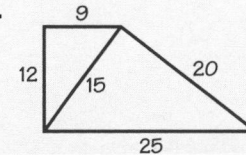

6.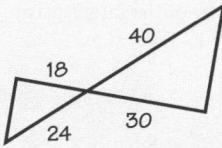

Find the value of each variable.

7.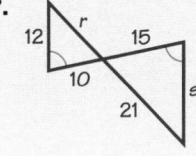

8.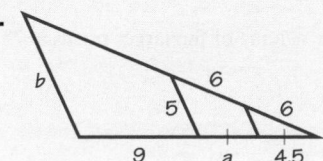

9.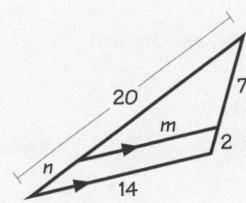

10.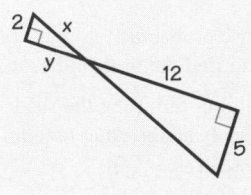

ANSWERS Chapter 9

Assessment

1. Answers may vary. An example is given. Congruent figures have the same shape and the same size. Similar figures have the same shape but not necessarily the same size. Two similar figures are congruent if the ratio of the lengths of corresponding sides is 1, that is, if corresponding sides are congruent.

2. Answers may vary. An example is given. $6:8$ and $3:4$ or $\dfrac{6}{8} = \dfrac{3}{4}$

3. Yes; AA Postulate or SAS Similarity Theorem.

4. No.

5. Yes; SSS Similarity Theorem.

6. Yes; SAS Similarity Theorem.

7. $r = 14$, $s = 18$

8. $a = 4.5$, $b = 10$

9. $n = \dfrac{40}{9}$; $m = \dfrac{98}{9}$

10. $x = 5.2$; $y = 4.8$

SECTION 9.4

The figures in each pair are similar. Find each missing value.

11. $P = 50$ $P = 20$

12. $A = 7.2 \text{ m}^2$ $A = \underline{\ ?\ }$

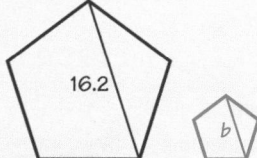

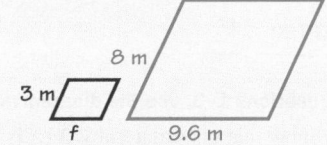

13. $V = \underline{\ ?\ }$

$V = 682.5 \text{ in.}^3$

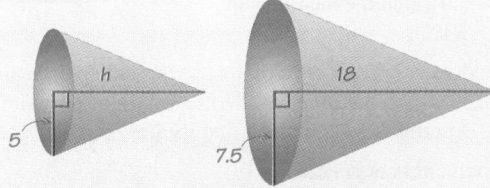

14. The heights of two similar cylinders are in the ratio $2:3$.

  a. The area of the base of the larger cylinder is 54 cm^2. What is the area of the base of the smaller cylinder?

  b. The volume of the smaller cylinder is 72 cm^3. What is the volume of the larger cylinder? Explain your reasoning.

SECTION 9.5

15. If you spin the spinner, what is the probability that it will land on 1?

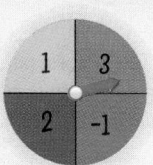

16. If you toss a bean at random on the target, what is the probability it will land on the shaded area?

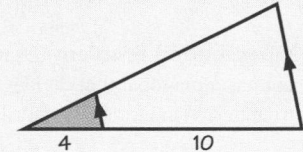

PERFORMANCE TASK

17. Open-ended Problem Research the prices of pizzas, cakes, ice cream, or other food that is available in similar shapes but different sizes. Calculate and compare the prices for the area or volume. What did you discover? Why do you think there are these differences?

Chapter 9
ALTERNATIVE ASSESSMENT

1. **Project** Find a drawing or a photograph of two objects that appear to be out of proportion. Use a ruler and your knowledge of proportions to confirm that the figures are not realistic.

2. **Research Project** Find out how scale models are used in moviemaking to reduce expenses. Write a report of your findings.

3. **Project** There were actually two gorilla models used in the movie *King Kong*, each a different size. For the scenes where King Kong was on Skull Island, the model was 18 feet tall. In the New York City scenes, a 24 foot tall model was used so that King Kong still seemed large compared to the Empire State Building. An average adult male gorilla is about 6 feet tall and weighs about 450 pounds.
   a. Use this information to predict the weight of the two different versions of King Kong. (*Hint:* If the height of an animal is increased by a factor of n, the weight of the animal is increased by a factor of n^3.)
   b. In a later movie, *King Kong vs. Godzilla*, King Kong grew again in order to match the 100-meter tall Godzilla. Give the height in feet and predict the weight in pounds of this new King Kong.

4. **Performance Task** Design a geometric probability experiment using dried beans and centimeter grid paper to measure the area of a large leaf. Conduct the experiment several times. Use transparent centimeter grid paper to find the area of your leaf. Write a report about your procedure and your results. What factors could affect the accuracy of your experiment?

5. **Group Activity** Conduct the following geometric probability experiment to predict the amount of Earth's surface that is covered by water. Toss a globe beach ball around the group. Each time a member of the group catches the ball, record whether their right thumb is located on land or water. (*Note:* When a thumb is covering both land and water, majority rules: if more land than water is covered, record the result as "land"; if more water than land is covered, record "water." If the amounts are too close to judge, do not count that trial of the experiment.)
   a. Keep a tally for 25 tosses. Use the results to calculate the percent of the surface that is covered by water.
   b. Repeat the experiment for 50 tosses.
   c. We know that approximately 75% of Earth's surface is covered with water. Did increasing the number of tosses affect the accuracy of your prediction?

6. **Open-ended Problem** Draw a line with a nonzero slope on a piece of centimeter graph paper.
   a. Give the coordinates of two points on the line. Use a centimeter ruler to measure the rise and run of the portion of the line between the two points. Use the measurements to state the slope of the line.
   b. Give the coordinates of two other points on the line. Use a centimeter ruler to measure the rise and run of the portion of the line between these two points. Use the measurements to state the slope of the line. Compare this slope with the slope found in part a. Use what you know about similar triangles to explain the results of this comparison.

---

11. 6.48

12. $A = 51.2 \text{ m}^2, f = 3.6 \text{ m}$

13. $V \approx 202.2 \text{ cm}^3, h = 12 \text{ cm}$

14. a. 24 cm^2

   b. $243 \text{ cm}^3; \dfrac{2^3}{3^3} = \dfrac{72}{x}$

15. $\dfrac{1}{4}$

16. $\dfrac{4}{49}$

17. Answers may vary.

Cumulative Assessment
CHAPTERS 7–9

CHAPTER 7

For Questions 1–3, use the diagram at the right.

1. Show that quadrilateral *WXYZ* is a parallelogram.

2. Find the perimeter and the area of *WXYZ*.

3. If $m \angle WZY = 60°$, find the measures of $\angle WXY$ and $\angle XYZ$.

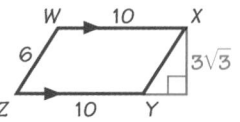

4. **Writing** Quadrilateral *ABCD* is a rhombus whose diagonals are 10 cm and 24 cm long. Sketch the rhombus. Write everything you know about the rhombus. Describe how you could find its area.

Find the area of each circle or polygon.

5.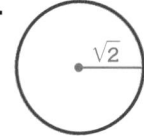

6.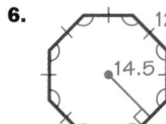

7.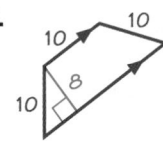

8. a regular hexagon whose perimeter is 60 in.

Find the volume and surface area of each cylinder or prism.

9. a cylinder with radius 12 and height 20

10. the triangular prism shown at the right

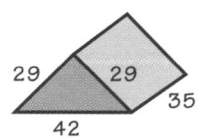

11. **Open-ended Problem** Draw a quadrilateral that is not a parallelogram, but has a diagonal that divides the quadrilateral into two congruent triangles. What kind of quadrilateral must your sketch be? Explain.

CHAPTER 8

The vertices of △*DEF* are *D*(− 4, 2), *E*(3, − 1), and *F*(0, − 2). For Questions 12–16, find the coordinates of the vertices of △*DEF* after each transformation described.

12. after a reflection over the given line:

    a. the *y*-axis

    b. the line $y = x$

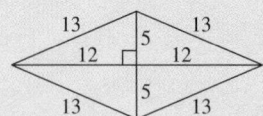
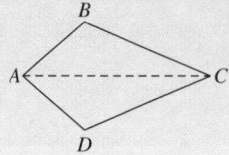

13. after a rotation around the origin the given amount:

    a. 270°   **b.** 180°

14. after a dilation with scale factor $\frac{1}{2}$ with the given center:

    a. (0, 0)   **b.** $D(-4, 2)$

15. after a glide reflection with translation $(a, b) \rightarrow (a - 2, b)$ and reflection over the x-axis

16. Which of the transformations described in Questions 12–15 preserve congruence? Which preserve orientation?

17. Writing Suppose you reflect a figure over two parallel lines. How is this like reflecting the figure over two intersecting lines? How is it different?

18. Open-ended Problem Draw any polygon. Then draw its image after an enlargement with center at one vertex. Find the area of the original figure and of the image. Make a conjecture about the relationship between the scale factor of the dilation and the ratio of the areas.

CHAPTER 9

19. Writing Use the diagram. Write four true statements about the diagram. Explain your reasoning.

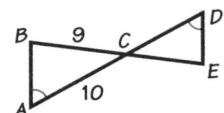

Is it possible to prove that the triangles are similar? Explain why or why not.

20. 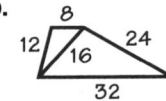   **21.** 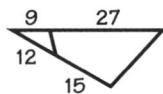   **22.**

23. The ratio of the areas of two regular pentagons is $9:25$. Find the ratio of their perimeters.

In the figure at the right, M, N, and P are the midpoints of the sides of △ABC.

24. Find the perimeter of $\triangle MNP$.

25. The area of $\triangle ABC$ is 150 square units. Find the area of $\triangle MNP$.

26. The height of a rectangular prism is 12 in. and its surface area is 552 in.^{2}. The height of a similar prism is 18 in. Find the surface area of the larger prism and find the ratio of their volumes.

27. GAMES This target is an equilateral triangle inscribed in a circle. A dart is thrown at random and hits the target. What is the probability of hitting the shaded area?

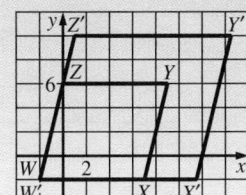

---

17. Answers may vary. An example is given. When you reflect a figure over two parallel lines, the combined effect is a translation; when you reflect a figure over two intersecting lines, the combined effect is a rotation. Both types of transformations preserve orientation and congruence.

18. Answers may vary. An example is given.

Parallelogram $WXYZ$ has area $5(4) = 20$, and parallelogram $W'X'Y'Z'$ has area $7.5(6) = 45$. The scale factor of the enlargement is 1.5, and the ratio of the areas is $\frac{45}{20} = 2.25 = (1.5)^2$. I predict that the ratio of the areas of the enlarged figure and the original figure is always equal to the square of the scale factor of the dilation.

19. Answers may vary. Examples are given.

    $\angle ACB \cong \angle DCE$ (vertical angles), $\triangle ABC \sim \triangle EDC$ (AA Postulate), $\angle B \cong \angle E$; $\frac{AB}{DE} = \frac{BC}{DC} = \frac{AC}{EC}$ (def. of similar triangles)

20. Yes; SSS Similarity Theorem.

21. Yes; SAS Similarity Theorem.

22. No; not enough information is given.

23. $3:5$ **24.** 30

25. 37.5 square units   **26.** 1242 in.^{2}; $8:27$

27. about 0.59

10 Applying Right Triangles

OVERVIEW

Connecting to Prior and Future Learning

⟺ The work on similar figures in Chapter 9 is extended in the first section of this chapter to similar right triangles. As students work with right triangles, they will be using skills acquired in algebra for working with radical expressions. Students can review simplifying radicals on page 700 in the **Student Resources Toolbox**.

⟺ In Section 10.2, students study 45-45-90 and 30-60-90 triangles. Students then study tangent, sine, and cosine in the next two sections, and learn how to use these ratios to solve problems.

⟺ The chapter closes with a presentation of vectors, the relationship between trigonometry and area, and finding the surface area and volume of pyramids and cones.

Chapter Highlights

Interview with Debby Tewa: The relationship between mathematics and solar energy is highlighted in this interview. Related exercises are on pages 503 and 519.

Explorations in Chapter 10 use paper, a ruler, and scissors to compare right triangles, and geometry software or a ruler and protractor to analyze ratios in triangles.

The Portfolio Project: Using the angle of the sun to plan the roof overhang for a passive solar home is the focus of this project. Students analyze popular types of roofs and then use the mathematical results of their project to design a house.

 Technology: Students use a scientific calculator throughout this chapter to find the values of trigonometric functions and their inverses, and they use geometry software in Section 10.4 to explore ratios in triangles.

OBJECTIVES

Section	Objectives	NCTM Standards
10.1	• Recognize relationships among the triangles formed by the altitude to the hypotenuse of a right triangle. • Find the geometric mean of two numbers. • Use similar right triangles to estimate lengths.	1, 2, 3, 4, 5, 7
10.2	• Find relationships among side lengths in 45-45-90 triangles and in 30-60-90 triangles. • Find lengths and distances in real-world problems involving special right triangles, squares, and equilateral triangles.	1, 2, 3, 4, 5, 7
10.3	• Find the tangent of an acute angle. • Use tangents to find lengths and angle measures in right triangles. • Solve real-world problems involving angles and lengths in right triangles.	1, 2, 3, 4, 5, 9
10.4	• Find the sine and cosine of an acute angle. • Find the measure of an acute angle whose sine or cosine is given. • Find the lengths of the sides of right triangles. • Solve real-world problems using sines and cosines.	1, 2, 3, 4, 5, 9
10.5	• Identify vector quantities and express them in component form. • Use vectors and vector sums to analyze real-world situations.	1, 2, 3, 4, 5, 12
10.6	• Use trigonometry to find areas of polygons and volumes of prisms. • Find areas of real-world objects using trigonometry.	1, 2, 3, 4, 5, 9
10.7	• Apply formulas for surface areas and volumes of pyramids and cones. • Analyze real-world pyramids and cones.	1, 2, 3, 4, 5, 7

INTEGRATION

Mathematical Connections	10.1	10.2	10.3	10.4	10.5	10.6	10.7
geometry	**493–499***	**500–506**	**507–513**	**514–520**	**521–528**	**529–534**	**535–541**
algebra	494, 497, 499	505	511	520		534	
patterns and functions	499						
logic and reasoning	498, 499	505, 506	513	520	527	534	540

Interdisciplinary Connections and Applications							
history and geography		502				529, 533	
reading and language arts	498						
arts and entertainment	499			516			
sports and recreation	500				526		
architecture		504					539, 540
navigation			510		523, 527		
aviation			510	519	527		
home repair, forestry, agriculture, paleontology, transportation, astronomy, geology, archaeology, packaging	496	505	508, 511–513	518	526		536, 537

__Bold page numbers__ indicate that a topic is used throughout the section.

TECHNOLOGY

Section	opportunities for use with	
	Student Book	**Support Material**
10.1	scientific calculator	**Technology Book:** Spreadsheet Activity 9 **Geometry Inventor Activity Book:** Activity 5
10.2	scientific calculator	**Technology Book:** Spreadsheet Activity 9
10.3	scientific calculator	
10.4	scientific calculator graphing calculator geometry software	
10.5	scientific calculator	**Technology Book:** TI-92 Activity 10
10.6	scientific calculator graphing calculator	**Technology Book:** Calculator Activity 13
10.7	scientific calculator	**Technology Book:** Spreadsheet Activity 9

Regular Scheduling (45 min)

Section	Materials Needed	Core Assignment	Extended Assignment	exercises that feature		
				Applications	Communication	Technology
10.1	ruler, scissors, straws, tape	1–8, 10–19, 26–32	1, 2, 3–15 odd, 16–32	20–23, 25	25	
10.2	compass, straightedge	**Day 1:** 1–7, 9–17 **Day 2:** 19–27, 29–35, AYP*	**Day 1:** 1–18 **Day 2:** 19–35, AYP	9–18 26	9 25, 26, 29	
10.3	scientific calculator, protractor	1–16, 18–21, 25–27, 29–33	1–33	17, 22–24	17c, 22, 24b, 29	
10.4	geometry software or ruler, protractor, scientific calculator	**Day 1:** 1–17 **Day 2:** 20–34, 36–43	**Day 1:** 1–17 **Day 2:** 18–43	18–22, 29, 30	22b, 30, 34	
10.5	graph paper	**Day 1:** 1–20 **Day 2:** 22–25, 27, 30–37, AYP	**Day 1:** 1–20 **Day 2:** 21–37, AYP	15–18 27–29	15c 26, 30	
10.6	ruler, protractor	1–11, 13–17, 22–25, 27–31	1–31	18–21	14, 21, 27	
10.7	beans or popcorn, ruler, scissors	1–9, 14–16, 18–28, AYP	1–28, AYP	11–14	10, 14c, 17–20, 22	
Review/ Assess		**Day 1:** 1–15 **Day 2:** 16–30 **Day 3:** Ch. 10 Test	**Day 1:** 1–15 **Day 2:** 16–30 **Day 3:** Ch. 10 Test			
Port. Proj.		Allow 2 days.	Allow 2 days.			

Yearly Pacing (with Portfolio Project)	Chapter 10 Total 15 days	Chapters 1–10 Total 134 days	Remaining 26 days	Total 160 days

Block Scheduling (90 min)

	Day 60	Day 61	Day 62	Day 63	Day 64	Day 65	Day 66	Day 67
Teach/Interact	Ch. 9 Test 10.1: Exploration, page 493	10.2: Construction, pages 504, 505	10.3 10.4: Exploration, page 514	Continue with 10.4 10.5	Continue with 10.5 10.6	10.7 Review	Review Port. Proj.	Ch. 10 Test Port. Proj.
Apply/Assess	**Ch. 9 Test** **10.1:** 1–8, 10–19, 26–32	**10.2:** 1–7, 9–17, 19–27, 29–35, AYP*	**10.3:** 1–16, 18–21, 25–27, 29–33 **10.4:** 1–17	**10.4:** 20–34, 36–43 **10.5:** 1–20	**10.5:** 22–25, 27, 30–37, AYP **10.6:** 1–11, 13–17, 22–25, 27–31	**10.7:** 1–9, 14–16, 18–28, AYP **Review:** 1–15	**Review:** 16–30 **Port. Proj.**	**Ch. 10 Test** **Port. Proj.**

NOTE: A one-day block has been added for the Portfolio Project—timing and placement to be determined by teacher.

Yearly Pacing (with Portfolio Project)	Chapter 10 Total $7\frac{1}{2}$ days	Chapters 1–10 Total 67 days	Remaining 13 days	Total 80 days

*_**AYP** is Assess Your Progress._

Section	Practice Bank	Study Guide*	Assessment Book*	Visuals	Explorations Lab Manual	Lesson Plans	Technology Book
10.1	67	10.1		Warm-Up 10.1 Folder 12		10.1	Spreadsheet Act. 9
10.2	68	10.2	Test 43	Warm-Up 10.2		10.2	Spreadsheet Act. 9
10.3	69	10.3		Warm-Up 10.3		10.3	
10.4	70	10.4		Warm-Up 10.4	Add. Expl. 15 Master 13	10.4	
10.5	71	10.5	Test 44	Warm-Up 10.5	Masters 1, 2	10.5	TI-92 Act. 10
10.6	72	10.6		Warm-Up 10.6		10.6	Calculator Act. 13
10.7	73	10.7	Test 45	Warm-Up 10.7	Add. Expl. 16 Master 2	10.7	Spreadsheet Act. 9
Review Test	74	Chapter Review	Tests 46, 47, Alternative Assessment			Review Test	

*__Spanish versions__ of *Study Guide* and *Assessment Book* are available.

Chapter Support

- Course Guides
- Lesson Plans
- Preparing for College Entrance Tests
- Multi-Language Glossary
- *Test Generator* Software
- Professional Handbook
- Challenge Problems

Software Support

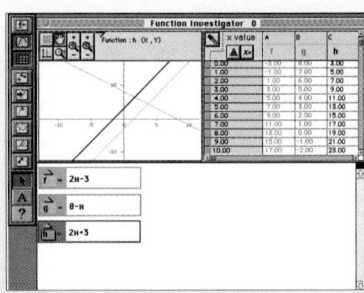

McDougal Littell Mathpack
Geometry Inventor

Internet Support

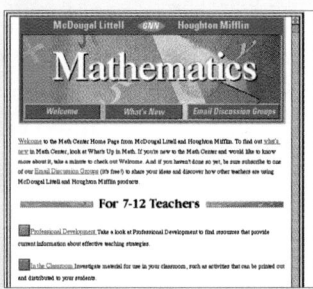

http://www.hmco.com
Next go to McDougal Littell; then the
Education Center; then Secondary Math.

OUTSIDE RESOURCES

Books, Periodicals

Nowlin, Donald. "Practical Geometry Problems: The Case of the Ritzville Pyramids." *Mathematics Teacher* (March 1993): pp. 198–200.

Stanton, Robert O. "An Unlawful Use of the Law of Sines." *Mathematics Teacher* (February 1993): pp. 164, 165.

Vonder Embse, Charles and Arne Engebretsen. "Using Interactive Geometry Software for Right-Angle Trigonometry." *Mathematics Teacher* (October 1996): pp. 602–605.

Software

deLange, Jan. *Gliding*. Macintosh. Pleasantville, NY: Wings for Learning/Sunburst, 1992.

Jackiw, Nicholas, designer. *The Geometer's Sketchpad.* (Ver. 3.0) for Macintosh or Windows. Berkeley, CA: Key Curriculum Press, 1995.

Laborde, Jean-Marie and Franck Bellemain, designers. *Cabri Geometry II.* Dallas, TX: Texas Instruments, 1994.

TI-92 Geometry. (Resident on the TI-92 graphing calculator.) Dallas, TX: Texas Instruments, 1995.

Videos

Apostol, Tom M. *Sines and Cosines*, Parts 1 and 2. Reston, VA: NCTM.

Internet

A Web site for "Multimedia Math" is located at:

http://www.sa.ua.edu/m3func.html

Background

Pueblo Dwellings

Traditional Hopi houses, or pueblos, are made of sandstone and adobe (or stucco). Flat slabs of sandstone, a soft red or yellow colored rock, were put together with adobe, a kind of mud mortar. Using this method, a four- to five-story apartment-type structure was built. Heavy wood beams were lined across the walls to form a flat ceiling or roof, then poles, branches, leaves, and grass were used to fill it in. A mud mortar was then used to complete the roof or ceiling. The rooms themselves did not contain many doors or windows, and ladders were often used to go up to another room. Often, these units were quite extensive and provided shelter for an entire town. In the Hopi tradition, women owned the houses.

The outer rooms of the buildings were often used for living areas since they received more sunlight. Inner rooms were used for storage or to live during the colder seasons. Cooking or pottery making was often done on the balconies or terraces.

Debby Tewa

Debby Tewa's original intentions were to become a physical education teacher. After completing high school in California, she attended Northern Arizona University. Currently, she is installing solar-electric systems in native Hopi and Navajo homes. For her own home, Tewa has also installed a solar shower. Her shower consists of a water-filled plastic bag set out in the sun. The sunlight heats the water, and Tewa pours this into an insulated tank. A battery-operated pump then moves the water through a shower head.

10 Applying Right Triangles

Technology and Tradition

INTERVIEW Debby Tewa

You might say Debby Tewa lives in two very different worlds. Like many of her customers, she lives in a traditional sandstone house in Hotevilla, a Hopi village in Arizona. As an electrician, she has mastered the latest in high technology.

Tewa works every day to bring these two worlds together, providing electricity for Hopi, Navajo, and other native American people who would otherwise have none. "I'm fortunate to be in the position to explain the components of these systems to older people in the Hopi community," she says.

Since 1988, when it was founded by the Hopi Foundation, the Solar Electric Enterprise has equipped about 300 homes with photovoltaic panels—rooftop solar panels that convert sunlight into electricity. The electricity, in turn, can be used to run lights, power tools, and kitchen appliances.

"**Solar energy teaches people to conserve energy and to become more self-sufficient.**"

Background

Solar Energy Development
In ancient times, the Greeks and Chinese knew how to use the sun's rays to start fires. Today, we use the sun's energy to heat water and homes, and for electricity. Much is being done today to make better and more efficient use of solar energy. The branch of chemistry called photochemistry involves the development of technological uses of solar energy. Chemists are trying to convert sunlight into energy in various ways, such as using sunlight to produce fuels like hydrogen gas and methanol, or developing more efficient ways to convert sunlight into electricity. Photochemists are also trying to imitate the process of photosynthesis with artificially created molecules. In this way, the sun's energy could be converted into the chemical energy of food, which could then be converted into coal or petroleum through geological processes.

"I never planned this career. It just happened."

Clean Energy

Photovoltaic (or solar) panels produce "clean energy," Tewa says. By using them, the damage to the environment associated with conventional energy sources and the need for power lines is reduced. Solar energy, she adds, is "appropriate not just for Hopi villages, but for everybody. It teaches people to conserve energy and to become more self-sufficient."

Tewa has toured throughout the United States, lecturing about the advantages of solar energy. She has also traveled to Ecuador, where she helped people with their solar installations. It's an exciting, and somewhat surprising, turn of events for her. Tewa never would have become an electrician, or a solar energy expert, if she had not been recruited for trade school shortly after leaving college. "I never planned this career," she explains. "It just happened." And she and her customers are glad that it did.

Second-Language Learners

Some students learning English might find it helpful to know that the word *solar* is the adjectival form of *sun*. Tell native speakers of Spanish that the word *solar* comes from the same Latin word as the Spanish word *sol* (sun).

Multicultural Note

The Hopi are a Native American people from the Southwest, often grouped with other peoples under the more general term Pueblo. The name Hopi is a shortened form of the Hopi word *hopituh*, meaning "peaceful ones." Traditional Hopi culture places a high value on respect for nature and other people, and the Hopi have been peaceful throughout much of their history. Like other Pueblo peoples, the Hopi build their homes in villages on or near mesas. These homes traditionally had no doors or windows and were entered from the top, but most contemporary Hopi homes show some modern adaptations. Today, about 9000 Hopi live on or near three mesas that make up the Hopi Reservation in northeastern Arizona.

491

Mathematical Connection

Photovoltaic cells, or solar cells, are used to capture energy from the sun. They consist of thin slices of semiconductor materials pressed between two plates of glass. When the sun shines on a photovoltaic cell, electric current flows from one side of the cell to the other. The shape of the solar cells and the angle in which they are placed are two important factors in how much electrical energy is created. In Section 10.2, students explore the shape of the photovoltaic cells and use the properties of special right triangles to calculate lengths and areas. In Section 10.4, students investigate how the solar panels are installed. They use the trigonometric ratios of sine and cosine to calculate the lengths of supporting structures for installing solar energy cells, as well as finding possible angle measures when the solar panels are tilted.

Explore and Connect

Writing
After students have had an opportunity to answer these questions in writing, ask some students to read their answers in class. Ask the other students if they agree with the explanation of the reasoning given.

Research
Students may want to use the library, Internet, or an encyclopedia as possible sources for this research.

Project
Suggest to students that they contact their local town hall or municipal government to find out if anyone may be able to tell them about a building with solar panels or answer the questions for this project. If not, they may be able to refer students to someone else who can answer the questions.

The Geometry of Solar Energy

Solar panels should be placed at a certain angle to maximize exposure to sunlight throughout the year. The angle depends on location. For example, in the southwestern United States where Debby Tewa lives, the solar panels are typically mounted at a 55° angle. Most Hopi roofs are essentially flat. The panels, roof, and support structures form a triangle. "If you know the length of the panel you need, you can use trigonometry to determine the size of the mounting materials," Tewa says.

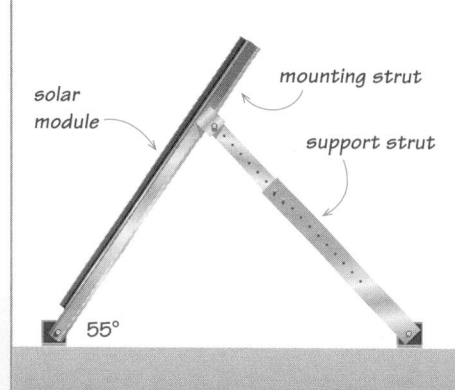

" ...you can use trigonometry to determine the size of the materials. "

Explore and Connect

Debby Tewa demonstrates how she adjusts the solar module to the correct angle.

1. Writing What do you know about any two right triangles with 55° angles? How could Debby Tewa use this relationship in her calculations? Explain your reasoning.

2. Research Find out more about solar energy. List some of the environmental and economic benefits of using solar energy.

3. Project Visit a building that has solar panels and talk with the people who take care of them. At what angle are the panels positioned? Do the solar panels operate differently depending on the weather or the season? How does using solar energy affect the electric bills for the building? Report your findings to your class.

Mathematics & Debby Tewa

In this chapter, you will learn more about how mathematics is related to solar energy.

Related Exercises

Section 10.2
• Exercises 9–13

Section 10.4
• Exercises 20–22

10.1 | **Similar Right Triangles**

Learn how to...

• **recognize relationships among the triangles formed by the altitude to the hypotenuse of a right triangle**

• **find the geometric mean of two numbers**

So you can...

• **estimate lengths, such as the height of a house**

When you draw the altitude to the hypotenuse of a right triangle, you create two smaller right triangles. How are the smaller triangles related to the original triangle? You will find out in the Exploration.

EXPLORATION
COOPERATIVE LEARNING

Comparing Right Triangles

Work with another student.
You will need:

• a rectangular sheet of paper

• a ruler   • scissors

1 Draw a diagonal of the paper to create two congruent right triangles. Fold the paper to find the altitude to the hypotenuse of one of the triangles.

2 Along the fold, draw the segment that corresponds to $\overline{DE}$ below. Mark the three right angles as shown.

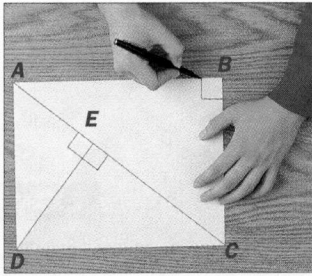

3 Explain why $\angle CAB \cong \angle DCE$ and $\angle ACB \cong \angle DAE$. Mark the triangles to indicate that these pairs of angles are congruent. What can you conclude about $\triangle CAB$ and $\triangle DCE$? about $\triangle CAB$ and $\triangle ADE$? How do you know?

4 Explain why $\angle CDE \cong \angle DAE$. Mark the triangles to indicate that these angles are congruent. What can you conclude about $\triangle DCE$ and $\triangle ADE$?

5 Cut along $\overline{AC}$ and $\overline{DE}$. Arrange the triangles so that three congruent acute angles match up, as shown at the left. How does this arrangement support the angle relationships in Steps 3 and 4?

Exploration Note

Purpose
The purpose of this Exploration is to have students discover the relationships that exist among triangles when the altitude to the hypotenuse of a right triangle is drawn.

Materials/Preparation
Each pair of students needs a rectangular sheet of paper, a ruler, and scissors.

Procedure
Students draw a diagonal of the paper to divide it into two congruent right triangles. The altitude to the hypotenuse of one of the triangles is drawn. The angles of the three

triangles are examined to establish the fact that all three triangles are similar.

Closure
Discuss the results of Steps 3–5 so students understand that when the altitude is drawn to the hypotenuse of a right triangle, the two triangles formed are similar to the original triangle and to each other.

Explorations Lab Manual
See the Manual for more commentary on this Exploration.

For answers to the Exploration, see following page.

Objectives

• Recognize relationships among the triangles formed by the altitude to the hypotenuse of a right triangle.

• Find the geometric mean of two numbers.

• Use similar right triangles to estimate lengths.

Recommended Pacing

❖ **Core and Extended Courses**
Section 10.1: 1 day

❖ **Block Schedule**
Section 10.1: $\frac{1}{2}$ block
(with Chapter 9 Test)

Resource Materials

Lesson Support
Lesson Plan 10.1

Warm-Up Transparency 10.1

Overhead Visuals:
Folder 12: Similar Right Triangles

Practice Bank: Practice 67

Study Guide: Section 10.1

Challenge Problems: Set 67

Technology
Technology Book:
Spreadsheet Activity 9

Scientific Calculator

McDougal Littell Mathpack
Geometry Inventor Activity Book:
Activity 5

Internet:
http://www.hmco.com

Warm-Up Exercises

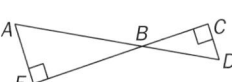

1. Are the triangles similar?
Yes; AA Similarity Postulate.

2. Which angle in $\triangle AEB$ corresponds to $\angle BDC$ in $\triangle DCB$?
$\angle BAE$

3. Which side in $\triangle AEB$ corresponds to $\overline{DB}$ in $\triangle DCB$? $\overline{AB}$

Complete each ratio.

4. $\frac{EB}{CB} = \frac{AE}{?}$ DC **5.** $\frac{AE}{?} = \frac{DC}{DB}$ AB

493

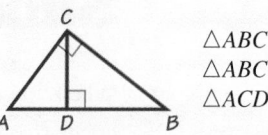

Learning Styles: Verbal

Students with verbal learning styles may have difficulty visualizing the three triangles shown with the Similar Right Triangles Theorem. Encourage these students to separate the three triangles as illustrated on page 495. Separating the triangles should help them to see the corresponding sides and angles in the triangles. This method is used in Step 5 of the Exploration when the triangles are cut apart.

Think and Communicate

The proportions used in question 2 all involve geometric means. After studying Example 1, you may wish to have students look back at this question and make some conjectures about their results. This would be a good way to introduce the two Geometric Mean Theorems on page 495.

Section Note

Teaching Tip
You may want to remind students that the values p and s in the proportion $\frac{p}{q} = \frac{r}{s}$ are called the *extremes*.

About Example 1

Teaching Tip
You may need to remind students that they can use cross products to solve a proportion. Thus, if $\frac{a}{b} = \frac{c}{d}$, then $ad = bc$.

Additional Example 1

Find the geometric mean of 3 and 15.
Let x = the geometric mean of 3 and 15. Write and solve a proportion. The geometric mean is always positive.

$$\frac{3}{x} = \frac{x}{15}$$
$$45 = x^2$$
$$\sqrt{45} = x$$
$$3\sqrt{5} = x$$

The geometric mean of 3 and 15 is $3\sqrt{5}$.

In the Exploration, you discovered the following result.

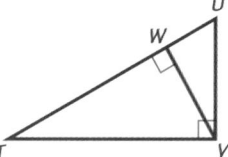

Similar Right Triangles Theorem

If the altitude is drawn to the hypotenuse of a right triangle, then the two triangles formed are similar to the original triangle and to each other.

$\triangle ABC \sim \triangle ACD$
$\triangle ABC \sim \triangle CBD$
$\triangle ACD \sim \triangle CBD$

THINK AND COMMUNICATE

1. a. Which angle in $\triangle TUV$ corresponds to $\angle TVW$ in $\triangle TVW$?

b. Identify the three pairs of similar triangles in the diagram.

c. Which side in $\triangle TUV$ corresponds to $\overline{UW}$ in $\triangle VUW$?

2. Use corresponding parts of similar triangles to complete each statement.

a. $\dfrac{TU}{VU} = \dfrac{VU}{?}$ **b.** $\dfrac{TU}{TV} = \dfrac{?}{TW}$ **c.** $\dfrac{UW}{?} = \dfrac{VW}{TW}$

The Geometric Mean

In the proportion $\frac{p}{q} = \frac{r}{s}$, the values q and r are called the *means*. If a, b, and x are positive numbers and $\frac{a}{x} = \frac{x}{b}$, then x is called the **geometric mean** of a and b.

EXAMPLE 1 **Connection: Algebra**

Find the geometric mean of 2 and 18.

SOLUTION

Let x = the geometric mean of 2 and 18. Write and solve a proportion.

$$\frac{2}{x} = \frac{x}{18}$$
$$36 = x^2$$
$$\sqrt{36} = x$$ The geometric mean is always positive.
$$6 = x$$

The geometric mean of 2 and 18 is 6.

When you draw the altitude to the hypotenuse of a right triangle, you can use the geometric mean to describe the relationships among the lengths of the segments formed. The diagrams below illustrate these relationships.

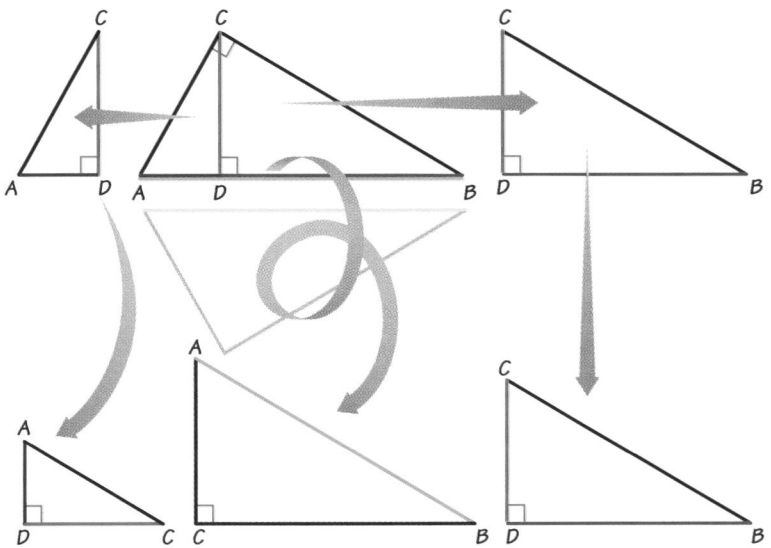

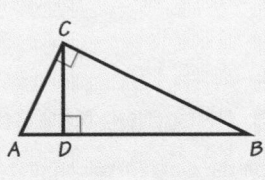

Since $\triangle ACD \sim \triangle CBD$, you can write $\dfrac{AD}{CD} = \dfrac{CD}{BD}$. Therefore, CD is the geometric mean of AD and BD.

Since $\triangle ABC \sim \triangle ACD$, you can write $\dfrac{AB}{AC} = \dfrac{AC}{AD}$. Therefore, AC is the geometric mean of AB and AD.

Since $\triangle ABC \sim \triangle CBD$, you can write $\dfrac{AB}{CB} = \dfrac{CB}{DB}$. Therefore, CB is the geometric mean of AB and DB.

Notice that each value in the proportions above is a length of a segment in $\triangle ABC$: either the altitude, a leg, the hypotenuse, or a segment of the hypotenuse.

Geometric Mean Theorems

If the altitude is drawn to the hypotenuse of a right triangle, then the length of the altitude is the geometric mean of the lengths of the segments of the hypotenuse.

If the altitude is drawn to the hypotenuse of a right triangle, then the length of each leg is the geometric mean of the lengths of the hypotenuse and the segment of the hypotenuse adjacent to that leg.

$$\frac{AD}{CD} = \frac{CD}{BD}$$

$$\frac{AB}{AC} = \frac{AC}{AD} \quad \text{and} \quad \frac{AB}{CB} = \frac{CB}{DB}$$

10.1 Similar Right Triangles **495**

Section Notes

Using Manipulatives

You may wish to have each student make a model for the diagram at the top of this page. Students should trace and cut out the three triangles on the second row of the diagram. They can then place the two smaller triangles on top of the larger one and label the vertices as they are in the diagram. The labels should be placed inside the vertices. Students should then follow the arrows to move each piece of the triangle. This activity will help students visualize the manipulations illustrated in the diagram.

Communication: Reading
When students read the middle paragraph of this page, they should verify the proportion given for each similarity. Students should understand that the triangles are similar because of the Similar Right Triangles Theorem.

Communication: Drawing
The following three drawings can help students remember the Geometric Mean Theorems. The geometric mean is labeled x and the extremes are labeled a and b.

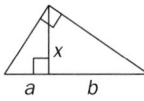

$$\frac{a}{x} = \frac{x}{b}$$

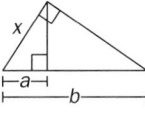

$$\frac{a}{x} = \frac{x}{b}$$

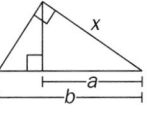

$$\frac{a}{x} = \frac{x}{b}$$

About Example 2

Alternate Approach

In problems involving a geometric mean, suggest that students label the geometric mean as x in the diagram, and the extremes as a and b. Then they can write the proportion as $\dfrac{a}{x} = \dfrac{x}{b}$.

Additional Example 2

Jacob plans to install shutters on his house. He wants to estimate the height of the top edge of the second floor window to determine how long a ladder is needed. He uses the same method as Aida in Example 2. Use the diagram to find h, the height of the top of the second floor window.

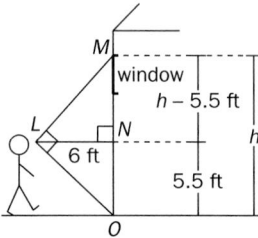

Use the Geometric Mean Theorem that includes the length of the altitude.

$$\frac{MN}{LN} = \frac{LN}{ON}$$

$$\frac{h - 5.5}{6} = \frac{6}{5.5}$$

$$5.5(h - 5.5) = 6^2$$

$$5.5h - 30.25 = 36$$

$$5.5h = 66.25$$

$$h \approx 12.05$$

The height of the top of the second floor window is about 12 ft.

Closure Question

State the Similar Right Triangles Theorem and describe the Geometric Mean Theorem that involves the shorter leg of a right triangle.

If the altitude is drawn to the hypotenuse of a right triangle, then the two triangles formed are similar to the original triangle and to each other. If the altitude is drawn to the hypotenuse of a right triangle, then the length of the shorter leg is the geometric mean of the length of the hypotenuse and the shorter segment of the hypotenuse.

EXAMPLE 2 **Application: Home Repair**

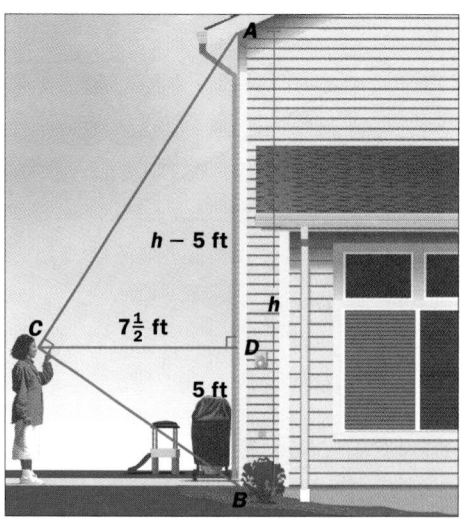

Aida Ramos wants to buy a ladder that reaches the roof of her house. To find how tall the ladder should be, she estimates the height of the edge of the roof above the ground as follows:

Step 1 Aida holds a notebook near her eye. She backs away from the house until the edge of the roof and the base of the house are in line with the notebook's edges, as shown in the diagram.

Step 2 Aida measures her distance from the house.

Use the diagram to find h, the height of the edge of the roof.

SOLUTION

Use the Geometric Mean Theorem that includes the length of the altitude.

$$\frac{AD}{CD} = \frac{CD}{BD}$$

> BD is the distance from Aida's eyes to the ground, about 5 ft.

$$\frac{h - 5}{7.5} = \frac{7.5}{5}$$

$$5(h - 5) = (7.5)^2 \qquad \text{Use the Distributive Property.}$$

$$5h - 25 = 56.25$$

$$5h = 81.25$$

$$h = 16.25$$

The height of the edge of the roof is about 16 ft.

✓ CHECKING KEY CONCEPTS

Use the diagram to complete each statement.

1. $\angle GEF \cong \angle\ \underline{\ ?\ }$
2. $\triangle EFG \sim \triangle\,\underline{\ ?\ } \sim \triangle\,\underline{\ ?\ }$
3. $\dfrac{FE}{FG} = \dfrac{FG}{?}$
4. GH is the geometric mean of $\underline{\ ?\ }$ and $\underline{\ ?\ }$.
5. If $EH = 9$ and $EF = 25$, then $EG = \underline{\ ?\ }$.

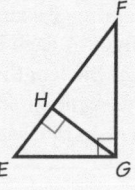

Checking Key Concepts

1. $\angle HGF$
2. EGH; GFH
3. FH
4. EH; HF
5. 15

Exercises and Applications

1. $\triangle JKL,\ \triangle JMK,\ \triangle KML$
2. $\triangle RTS,\ \triangle RVT,\ \triangle SVT$
3. 15
4. $4\sqrt{3}$
5. $6\sqrt{5}$
6. $\dfrac{2}{3}$
7. 1
8. 4

9. a. Answers may vary. An example is given. 4 and 25
   b. No; 10 is the geometric mean of any two positive numbers whose product is 100.
10. 4
11. 18
12. 6.5
13. $x = 5\sqrt{5}$; $y = 10\sqrt{5}$

10.1 Exercises and Applications

Extra Practice exercises on page 680

Identify the similar triangles in each diagram.

1.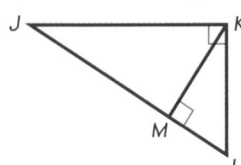

2.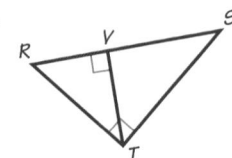

ALGEBRA Find the geometric mean of the given numbers.

3. 45 and 5   **4.** 8 and 6   **5.** 12 and 15

6. 4 and $\frac{1}{9}$ **7.** $\frac{3}{7}$ and $\frac{7}{3}$ **8.** 0.2 and 80

Toolbox p. 700
Simplifying Radicals

9. a. Open-ended Problem Find two numbers whose geometric mean is 10.

  b. Writing Are the numbers you found in part (a) the *only* two numbers whose geometric mean is 10? Explain.

Find the value of each variable.

10.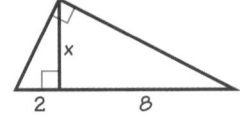

11.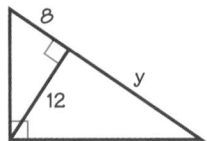

12.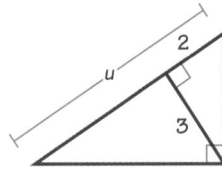

13.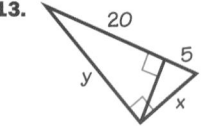

14.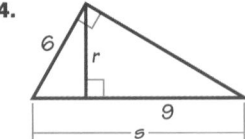

15.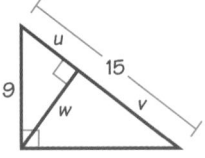

16. Draw a 3-4-5 right triangle and the altitude to the hypotenuse. Use the Geometric Mean Theorems to find the length of the altitude and the lengths of the segments of the hypotenuse.

For Exercises 17 and 18, prove the Geometric Mean Theorems on page 495. Use the diagram at the right for both exercises.

17. Given: In $\triangle ABC$, $\angle BCA$ is a right angle.
  $\overline{CD}$ is the altitude to the hypotenuse.

  Prove: $\dfrac{e}{d} = \dfrac{d}{f}$

18. Given: In $\triangle ABC$, $\angle BCA$ is a right angle.
  $\overline{CD}$ is the altitude to the hypotenuse.

  Prove: $\dfrac{c}{a} = \dfrac{a}{e}$ and $\dfrac{c}{b} = \dfrac{b}{f}$

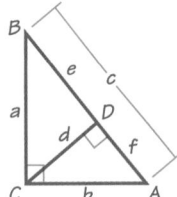

---

14. $r = 3\sqrt{3}$; $s = 12$

15. $u = 5.4$; $v = 9.6$; $w = 7.2$

16.
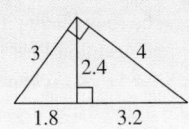

17. $\triangle ACD \sim \triangle CBD$. (If the altitude is drawn to the hypotenuse of a right triangle, then the two triangles formed are similar to the original triangle and to each other.) Then, since corresponding sides of similar triangles are in proportion,
$\dfrac{BD}{CD} = \dfrac{CD}{AD}$, or $\dfrac{e}{d} = \dfrac{d}{f}$.

18. $\triangle ABC \sim \triangle ACD$ and $\triangle ABC \sim \triangle CBD$. (If the altitude is drawn to the hypotenuse of a right triangle, then the two triangles formed are similar to the original triangle and to each other.) Then, since corresponding sides of similar triangles are in proportion,
$\dfrac{BA}{BC} = \dfrac{BC}{BD}$ and $\dfrac{BA}{CA} = \dfrac{CA}{DA}$, or
$\dfrac{c}{a} = \dfrac{a}{e}$ and $\dfrac{c}{b} = \dfrac{b}{f}$.

Suggested Assignment

❖ **Core Course**
  Exs. 1–8, 10–19, 26–32

❖ **Extended Course**
  Exs. 1, 2, 3–15 odd, 16–32

❖ **Block Schedule**
  Day 60 Exs. 1–8, 10–19, 26–32

Exercise Notes

Problem Solving
Exs. 3–8 After students complete these exercises, ask them if they can find a short cut for determining the geometric mean. They may see that when *a* and *b* are positive numbers, the geometric mean of *a* and *b* is $\sqrt{ab}$.

Common Error
Exs. 10–15 For exercises such as these, some students may use the incorrect geometric mean, or they may use an incorrect length for the geometric mean. To help students avoid these errors, have them first decide whether they should use a geometric mean involving a leg or the altitude. Then, they need to identify the two lengths that are the extremes in the proportion. Remind students that when using the altitude, the two extremes are the lengths of the segments of the hypotenuse. When using a leg, the two extremes are the lengths of the hypotenuse and the part of the hypotenuse adjacent to the leg. In some problems, students may need to write a new variable expression for the part of the triangle they need. For example, in Ex. 12, the proportion should involve the lengths of the two segments of the hypotenuse, so the longer segment must be represented by $u - 2$.

Writing Proofs
Exs. 17, 18 For each of these proofs, students need to show that two of the triangles are similar. They can refer to page 495 to help them with this task. For both proofs, students may wish to redraw the figure and label only the variables involved in the proof. Some students may prefer to write two separate proofs for Ex. 18, one for each of the proportions.

498

---

19. One proof of the Pythagorean theorem uses the Geometric Mean Theorem that includes the legs of the triangle. Copy and complete the proof.

Given: In $\triangle ABC$, $\angle ACB$ is a right angle.
$\overline{CD}$ is the altitude to the hypotenuse.

Prove: $a^2 + b^2 = c^2$

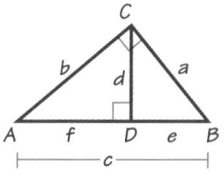

Statements	Reasons
1. In $\triangle ABC$, $\angle ACB$ is a right angle. $\overline{CD}$ is the altitude to the hypotenuse.	**1.** Given
2. $\frac{c}{a} = \frac{a}{e}$, $\frac{c}{b} = \frac{b}{f}$	**2.** _?_
3. _?_ $= ce$; _?_ $= cf$	**3.** _?_
4. _?_ $+$ _?_ $= ce + cf$	**4.** _?_
5. _?_ $+$ _?_ $= c(e + f)$	**5.** _?_
6. $e + f =$ _?_	**6.** _?_
7. $a^2 + b^2 = c^2$	**7.** _?_

Connection ▸ LITERATURE

The MYSTERIOUS ISLAND
BY JULES VERNE

Pictures by N. C. WYETH

NEW YORK
CHARLES SCRIBNER'S SONS

In Jules Verne's *The Mysterious Island*, Cyrus Harding, one of several travelers stranded on an island, uses similar right triangles to estimate the height of a cliff.

THE MYSTERIOUS ISLAND

Cyrus Harding had provided himself with a straight stick, twelve feet long. . . . Having reached a spot . . . nearly five hundred feet from the cliff, which rose perpendicularly, Harding thrust the pole two feet into the sand . . . perpendicularly. . . .

That done, he retired the necessary distance, when, lying in the sand, his eye glanced at the same time at the top of the pole and the crest of the cliff. He carefully marked the place with a little stick. . . .

The first distance was fifteen feet between the stick and the place where the pole was thrust into the sand.

The second distance between the stick and the bottom of the cliff was five hundred feet.

20. Visual Thinking Draw and label a diagram that models the situation described in the passage.

21. Identify the similar triangles in the diagram you drew in Exercise 20.

22. Find the height of the cliff.

23. Open-ended Problem What other distances in your diagram can you find? What theorem(s) or postulate(s) would you need to use to calculate these distances?

---

19. (2) If an altitude is drawn to the hypotenuse of a right triangle, then the length of the altitude is the geometric mean of the lengths of the hypotenuse and the segment of the hypotenuse adjacent to that leg. (3) a^2; b^2; a property of proportions (4) $a^2 + b^2$; Addition Property (5) $a^2 + b^2$; Distributive Property (6) c; Segment Addition Postulate (7) Substitution Property

20–23. Answers are based on the sketch given for Ex. 20 (which is not drawn to scale.) S represents the location of the stick, P the bottom of the pole, Q the top of the pole, X the bottom of the cliff, and Y the top of the cliff.

20.

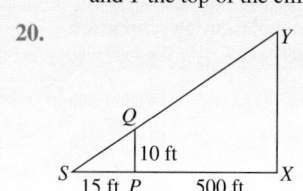

21. $\triangle SPQ \sim \triangle SXY$

22. about 343 ft

23. SX can be found using the Segment Addition Postulate; $SX = 515$ ft. SQ and SY can be found using the Pythagorean theorem; $SQ = 5\sqrt{13} \approx 18$ ft; $SY \approx 619$ ft. QY can be found using the Segment Addition Postulate; $QY \approx 601$ ft.

24. ALGEBRA Let a and b be positive numbers. The *arithmetic mean* of
a and b is $\dfrac{a+b}{2}$.

  a. Show that the geometric mean of a and b is $\sqrt{ab}$.

  b. Challenge Show that the arithmetic mean of a and b is greater than
    or equal to the geometric mean.

  c. When are the arithmetic and geometric means of a and b equal?

25. MUSIC In a *panpipe* based on a
chromatic scale, the length of the
nth pipe is the geometric mean of
the lengths of the $(n-1)$th pipe
and the $(n+1)$th pipe, where
$n = 2, 3, 4, \ldots, 12$. For example,
the length of the second pipe is the
geometric mean of the lengths of
the first and third pipes.

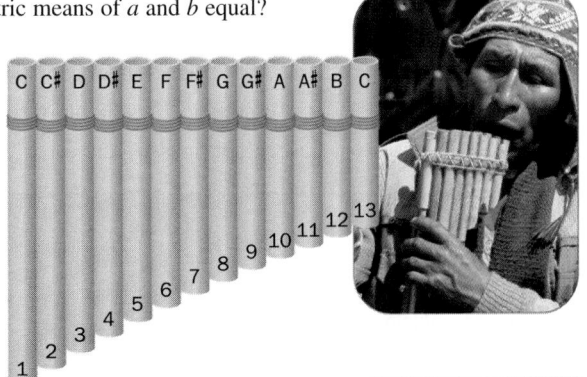

  a. Copy and complete the table.
    Round your answers to the
    nearest tenth of a centimeter.

Number of pipe	1	2	3	4	5	6	7	8	9	10	11	12	13
Length (cm)	16.4	15.5	?	?	?	?	?	?	?	?	?	?	?

  b. Cooperative Learning Work with a partner to make a panpipe
    using 13 plastic straws, a ruler, scissors, and masking tape. Cut the
    straws so that they have the lengths you found in part (a). Place the
    straws on a flat surface from longest to shortest, and align the top
    edges. Use the masking tape to attach the straws together.

BY THE WAY

A panpipe consists of
several pipes of
different lengths
attached together. It is
played by blowing across
the tops of the pipes.
Longer pipes produce
lower-pitched notes than
shorter pipes.

ONGOING ASSESSMENT

26. Open-ended Problem Choose values for RU
and SU in the diagram. Use the values you chose
and the Geometric Mean Theorems to find RT, TU,
and ST. Use the Pythagorean theorem to check
your answers.

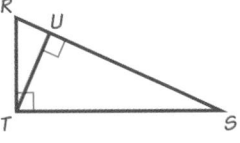

SPIRAL REVIEW

**Find the coordinates of the midpoint of the segment with each given pair of
endpoints.** *(Section 4.1)*

27. $A(0, 0)$, $B(6, -2)$ **28.** $C(-7, 1)$, $D(11, -5)$ **29.** $E(3, -9)$, $F(4, 8)$

**The legs of a right triangle are a and b units long. The hypotenuse is c units
long. Find the unknown length for each right triangle to the nearest
hundredth.** *(Section 3.6)*

30. $a = 6$, $b = 8$ **31.** $a = 5$, $c = 11$ **32.** $b = 48$, $c = 50$

10.1 Similar Right Triangles   **499**

Practice 67 for Section 10.1

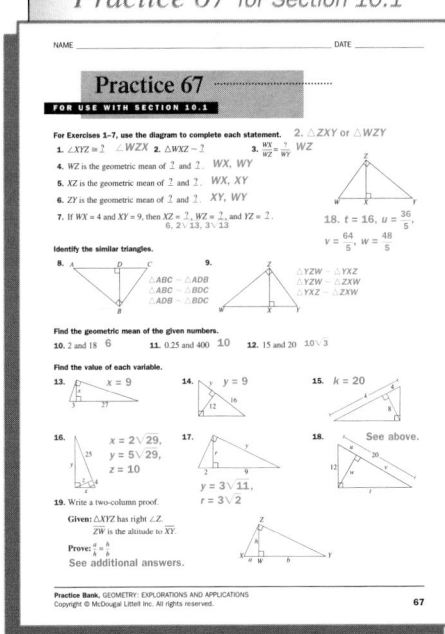

---

24. a. The geometric mean, x, of a and b is the
number such that $\dfrac{a}{x} = \dfrac{x}{b}$. Find the cross
products: $x^2 = ab$, so $x = \sqrt{ab}$.

  b. Answers may vary. An example is given.
    Suppose that $\dfrac{a+b}{2} < \sqrt{ab}$.

$$\frac{a+b}{2} < \sqrt{ab}$$
$$a + b < 2\sqrt{ab}$$
$$a^2 + 2ab + b^2 < 4ab$$
$$a^2 - 2ab + b^2 < 0$$
$$(a-b)^2 < 0$$

This is a contradiction, since the square of any number
is positive. Then the assumption that $\dfrac{a+b}{2} < \sqrt{ab}$ must
be incorrect. Then $\dfrac{a+b}{2} \geq \sqrt{ab}$.

  c. when a and b are equal

25. See answers in back of book.

26. Answers may vary. For any values of RU and SU,
$RT = \sqrt{RU(RU + SU)}$, $TU = \sqrt{RU \cdot SU}$, and
$ST = \sqrt{SU(RU + SU)}$. For example, if $RU = 2$ and $SU = 8$,
then $RT = 2\sqrt{5}$, $TU = 4$, and $ST = 4\sqrt{5}$. $RU^2 + UT^2 =$
$2^2 + 4^2 = 20 = RT^2$; $TU^2 + SU^2 = 4^2 + 8^2 = 80 = ST^2$.

27. $(3, -1)$

28. $(2, -2)$

29. $\left(\dfrac{7}{2}, -\dfrac{1}{2}\right)$

30. $c = 10$

31. $b = 4\sqrt{6} \approx 9.80$

32. $a = 14$

499

Objectives

- Find relationships among side lengths in 45-45-90 triangles and 30-60-90 triangles.
- Find lengths and distances in real-world problems involving special right triangles, squares, and equilateral triangles.

Recommended Pacing

❖ **Core and Extended Courses**
Section 10.2: 2 days

❖ **Block Schedule**
Section 10.2: 1 block

Resource Materials

Lesson Support
Lesson Plan 10.2

Warm-Up Transparency 10.2

Practice Bank: Practice 68

Study Guide: Section 10.2

Challenge Problems: Set 68

Assessment Book: Test 43

Technology
Technology Book:
    Spreadsheet Activity 9

Scientific Calculator

Internet:
    http://www.hmco.com

Warm-Up Exercises

Find the measure of each angle.

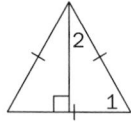

1. ∠1   60°

2. ∠2   30°

Simplify each square root.

3. $\sqrt{18}$ $3\sqrt{2}$

4. $\sqrt{20}$ $2\sqrt{5}$

5. $\sqrt{108}$ $6\sqrt{3}$

10.2 Special Right Triangles

If you've ever been to a baseball game, you know that a runner at first base will often try to "steal" second. To prevent the runner from doing so, sometimes the catcher throws the ball from home plate to second base. How far does the catcher need to throw the ball?

Learn how to...

- find relationships among side lengths in 45-45-90 triangles and in 30-60-90 triangles

So you can...

- easily find lengths and distances in real-world problems involving special triangles, squares, and equilateral triangles

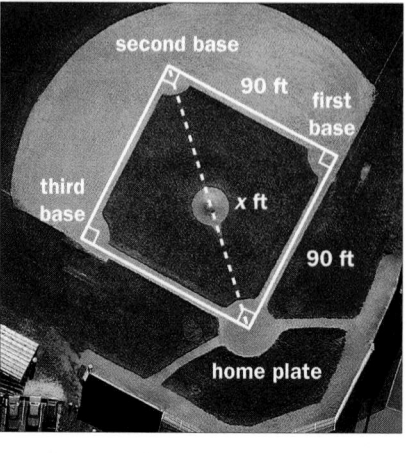

An official baseball diamond is a 90 ft by 90 ft square. The three bases and home plate mark each corner of the square.

Home plate, first base, and second base form an isosceles right triangle. Isosceles right triangles are also called **45-45-90 triangles**. You can use the Pythagorean theorem to find the relationships among the lengths of the sides of a 45-45-90 triangle.

EXAMPLE 1   **Application: Sports**

Find the distance from home plate to second base in a baseball diamond.

SOLUTION

Use the Pythagorean theorem.

$$x^2 = 90^2 + 90^2$$
$$x^2 = 2 \cdot 90^2$$
$$x = \sqrt{2 \cdot 90^2}$$
$$x = 90\sqrt{2}$$

> The square root of the product of two numbers is the product of the square roots of the numbers.

The distance from home plate to second base is $90\sqrt{2}$ ft, or about 127.28 ft.

500   Chapter 10 *Applying Right Triangles*

45-45-90 Triangles

In a 45-45-90 triangle, the length of the hypotenuse is $\sqrt{2}$ times the length of one leg.

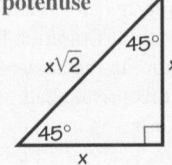

An isosceles right triangle often appears in real-world problems because it is half of a square, as in Example 1. Likewise, it is helpful to know the dimensions of a **30-60-90 triangle**, which is half of an equilateral triangle.

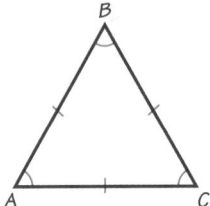

An equilateral triangle is also equiangular.
$m\angle A = m\angle B = m\angle C = 60°$

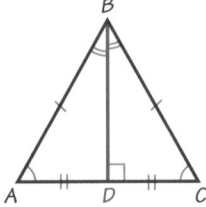

$m\angle ABD = m\angle DBC = 30°$
$AD = CD$

$\triangle DBC$ is a 30-60-90 triangle.

THINK AND COMMUNICATE

1. Explain why $m\angle DBC = 30°$.

2. Explain why all right triangles formed by drawing one altitude of an equilateral triangle are similar.

3. Describe the relationship between the lengths DC and BC. Explain your reasoning.

4. Let $DC = 4$.
   a. Use your answer to Question 3 to find BC.
   b. Use the Pythagorean theorem to find BD.

5. Suppose $DC = k$. Find BC and BD in terms of k.

30-60-90 Triangles

In a 30-60-90 triangle, the length of the hypotenuse is twice the length of the shorter leg, and the length of the longer leg is $\sqrt{3}$ times the length of the shorter leg.

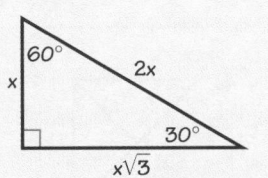

◀ **WATCH OUT!**

The shorter leg of a 30-60-90 triangle is opposite the 30° angle, and the longer leg is opposite the 60° angle.

10.2 Special Right Triangles   **501**

ANSWERS   Section 10.2

Think and Communicate

1. $\overrightarrow{BD}$ bisects $\angle ABC$; since $m\angle ABC = 60°$, $m\angle DBC = 30°$.

2. by the AA Similarity Postulate (Each triangle includes a right angle and a 60° angle.)

3. $DC = \frac{1}{2}BC$; $DC = \frac{1}{2}AC$ and $AC = BC$.

4. a. 8
   b. $4\sqrt{3}$

5. $2k$; $k\sqrt{3}$

Teach⇔Interact

Section Note

Alternate Approach
You may wish to have students explore the relationships between the lengths of the sides of a 45-45-90 triangle by using various approaches. One approach would involve doing pencil and paper calculations (similar to Example 1) for several triangles in which the lengths of the legs are integers. Another approach would be to have students use geometry software to draw various 45-45-90 triangles and then find the ratio of the length of the hypotenuse to the length of the legs. A third approach would be to use a spreadsheet or the list features of a graphing calculator. The first column would contain the various side lengths; the second column, the length of the hypotenuse, calculated using the method of Example 1; and the third column would contain the ratio of the length of the hypotenuse to the length of the legs. Similar approaches can be used for a 30-60-90 triangle.

Additional Example 1

An official softball diamond is a 60 ft by 60 ft square. Find the distance from home plate to second base in a softball diamond.
Use the Pythagorean theorem.
$x^2 = 60^2 + 60^2$
$x^2 = 2 \cdot 60^2$
$x = \sqrt{2 \cdot 60^2}$
$x = 60\sqrt{2}$
The distance from home plate to second base is $60\sqrt{2}$ ft, or about 84.85 ft.

Section Note

Student Study Tip
To help students remember the relationships in a 30-60-90 triangle, stress the fact that this triangle is one-half of an equilateral triangle. By drawing the other half of a 30-60-90 triangle, students should be able to see which sides are x and $2x$. The hint from the *Watch Out!* note at the bottom of this page can also be used to check that the two legs have the correct lengths.

501

Additional Example 2

A park has a wading pool in the shape of a regular hexagon. The length of a segment from the center to each vertex is 5 m. Find the apothem of the hexagon.

Since a regular hexagon can be divided into six equilateral triangles, it can be divided into twelve congruent 30-60-90 triangles. In each 30-60-90 triangle, the hypotenuse is $2x$. This expression equals 5 m. Write an equation and solve for x.

$$2x = 5$$
$$x = \frac{5}{2}$$

The length of the leg of the triangle opposite the 60° angle is $\sqrt{3}$ times the value of x. So the apothem of the hexagon is $\frac{5\sqrt{3}}{2}$ m, or about 4.33 m.

Checking Key Concepts

Teaching Tip
In question 4, students should use the 45-45-90 relationships of the smaller triangles to see that the hypotenuse of the larger triangle is 14.

Closure Question

In a 45-45-90 triangle, if the length of one leg is x, what is the length of the other leg and the length of the hypotenuse? In a 30-60-90 triangle, suppose you know the length of the shorter leg. Describe how to find the length of the longer leg and the length of the hypotenuse. In the 45-45-90 triangle, the length of the other leg is x, and the length of the hypotenuse is $x\sqrt{2}$. In the 30-60-90 triangle, the length of the longer leg is $\sqrt{3}$ times the length of the shorter leg. The length of the hypotenuse is twice the length of the shorter leg.

EXAMPLE 2   Application: History

The Conimbriga archeological site is located south of Coimbra, Portugal. One structure in the ancient Roman village is this private bath in the shape of a regular hexagon. The apothem of the hexagon is about 2 m. Find the length of each side of the bath.

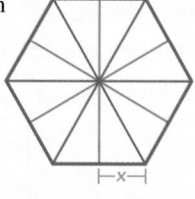

SOLUTION

Since a regular hexagon can be divided into six congruent equilateral triangles, it can be divided into twelve congruent 30-60-90 triangles.

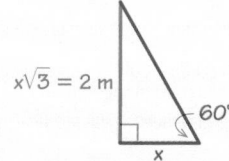

Let x = the length of the short leg in each 30-60-90 triangle. Then the side opposite the 60° angle is $x\sqrt{3}$, which equals 2 m. Write an equation and solve for x.

Divide both sides of the equation by $\sqrt{3}$.

$$x\sqrt{3} = 2$$
$$x = \frac{2}{\sqrt{3}}$$

The length of each side of the hexagon is twice the value of x. So the side length of the hexagon is $\frac{4}{\sqrt{3}}$ m, or about 2.31 m.

✓ CHECKING KEY CONCEPTS

Find the missing side lengths of each triangle.

1.   2.   3.

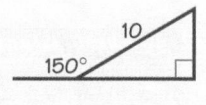

4.   5.   6.

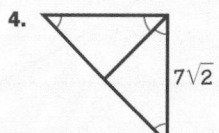
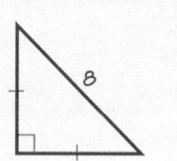
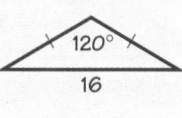

502   Chapter 10 *Applying Right Triangles*

Checking Key Concepts

1. $3; 3\sqrt{2}$

2. $6\sqrt{3}, 12$

3. $5; 5\sqrt{3}$

4. $7\sqrt{2}; 14$

5. $4\sqrt{2}; 4\sqrt{2}$

6. $\frac{16}{\sqrt{3}}; \frac{16}{\sqrt{3}}$

Exercises and Applications

1. 10   2. $5\sqrt{3}$

3. 4   4. $15\sqrt{2}$

5. $n = \frac{5}{\sqrt{3}}; m = \frac{10}{\sqrt{3}}$

6. $x = 45; y = 13\sqrt{2}$

Extra Practice
exercises on
page 680

Find the exact value of each variable.

1.
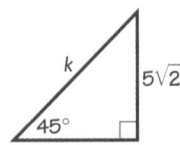

2.
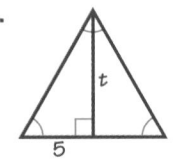

3.
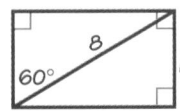

4.
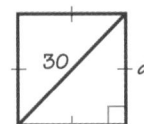

5.
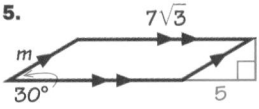

6.
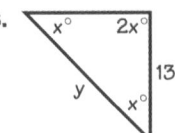

7. **Open-ended Problem** Choose one of the figures in Exercises 1–6 and find its area. Explain your method.

8. **Logical Reasoning** Suppose a 30-60-90 triangle with leg lengths a and b has the same area as a 45-45-90 triangle with leg length x. Prove that x is the geometric mean of a and b.

INTERVIEW Debby Tewa

Look back at the article on pages 490–492.

One of the solar modules that Debby Tewa uses contains solar cells that are approximately equiangular octagons with the dimensions shown. You can think of the octagon as a square with its corners removed.

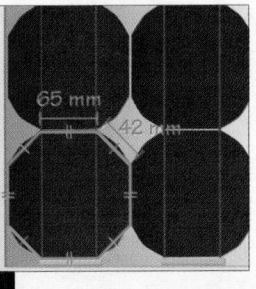

9. **Writing** What type of triangles must the corners be if the octagon is equiangular? Explain your reasoning.

10. Use your answer to Exercise 9 to find the side length of the square.

11. Find the area of the square.

12. Find the area of each triangular corner.

13. Use your answers to Exercises 11 and 12 to estimate the area of each solar cell.

10.2 Special Right Triangles **503**

Exercise Notes

Student Progress
Exs. 1–6 These exercises should be a good indicator as to how well students understand the 30-60-90 and 45-45-90 triangle relationships. Students having difficulty with these exercises should refer back to page 501.

Geometric Thinking
Ex. 8 This nonroutine problem gives students an opportunity to think about the relationships between the side lengths of 30-60-90 triangles and 45-45-90 triangles that have the same area. You may wish to ask students to compare the perimeters of the two triangles. Do they think that, in general, one of the triangles will have a larger perimeter than the other triangle with equal area? (Yes, the 30-60-90 triangle will have a larger perimeter.)

Interview Note: Application
Exs. 11, 12 In these exercises, 45-45-90 triangles can be used to help find the area of the octagonal solar cell.

Suggested Assignment

❖ **Core Course**
 Day 1 Exs. 1–7, 9–17
 Day 2 Exs. 19–27, 29–35, AYP

❖ **Extended Course**
 Day 1 Exs. 1–18
 Day 2 Exs. 19–35, AYP

❖ **Block Schedule**
 Day 61 Exs. 1–7, 9–17, 19–27, 29–35, AYP

---

7. (1) $A = \frac{1}{2}bh = \frac{1}{2}(5\sqrt{2})(5\sqrt{2}) = 25$

 (2) $A = \frac{1}{2}bh = \frac{1}{2}(10)(5\sqrt{3}) = 25\sqrt{3}$

 (3) $A = bh = 4\sqrt{3}(4) = 16\sqrt{3}$

 (4) $A = d^2 = (15\sqrt{2})^2 = 450$

 (5) $A = \frac{1}{2}h(b_1 + b_2) =$

 $\frac{1}{2}\left(\frac{5\sqrt{3}}{3}\right)(7\sqrt{3} + 5 + 7\sqrt{3}) =$

 $\frac{25\sqrt{3} + 210}{6}$

 (6) $A = \frac{1}{2}bh = \frac{1}{2}(13)(13) = \frac{169}{2}$

8. Area of 30-60-90 triangle = $\frac{1}{2}ab$;

 Area of 45-45-90 triangle = $\frac{1}{2}x^2$;

 If the areas are equal, then $\frac{1}{2}ab = \frac{1}{2}x^2$.

 This is equivalent to $ab = x^2$, or $\frac{a}{x} = \frac{x}{b}$.

9. 45-45-90; The measure of each angle of an equiangular octagon is 135°.

10. about 115.4 mm

11. about 13,317.16 mm²

12. about 571.22 mm²

13. 11,032 mm² to the nearest square millimeter

Connection   ARCHITECTURE

Throughout history, various cultures have used arches in their
buildings and structures. The *Gothic arch* is formed by two arcs that
intersect at a vertex. Some Gothic arches are based on the shape of
an equilateral triangle.

14. CONSTRUCTION Follow these steps to draw the shape of a Gothic
arch. Tell which three points of your arch determine the equilateral
triangle, and how you know it is equilateral.

1. Draw $\overline{AB}$.

2. Draw an arc with center
    A and radius *AB*.

3. Draw an arc with center
    B and radius *AB*.

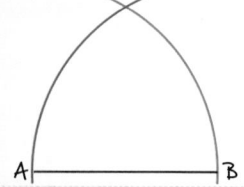

15. Suppose the width of an arch
where it starts to curve is
12 in. What is the height *h*
of the curved part of the
arch? How do you know?

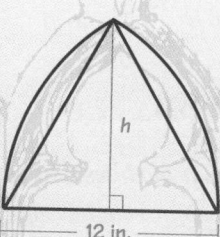

Find the total height of each arch.

16.

35 in.

30 in.

Regency Display Cabinet

17.

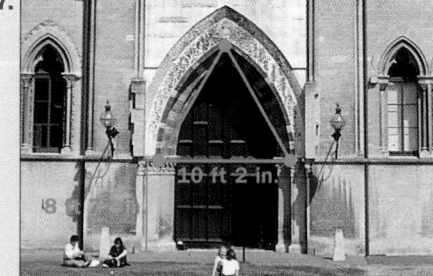

10 ft 2 in.

Oxford College, England

18. Research Find a Gothic arch in your neighborhood or a picture of
one that is based on an equilateral triangle. Use the steps in Exercise 14
to draw the arch. Find out the width of the arch and use what you know
about special triangles to calculate the height of the arch.

14. Check students' work. The
equilateral triangle is determined by *A*, *B*, and the point
where the two arcs you drew
intersect. The two arcs were
drawn with radius *AB*.

15. $6\sqrt{3} \approx 10.4$ in.; The height is
the length of an altitude of an
equilateral triangle. The altitude determines two 30-60-90
triangles with shorter leg 6 in.
long. The altitude is the longer
leg of each triangle.

16. about 60.3 in.

17. about 18 ft 2 in.

18. Check students' work.

19. $x = 45; y = 12$

20. $j = 6 + 6\sqrt{3}; k = 6\sqrt{2}$

21. $h = 2\sqrt{2}; f = \sqrt{6}; g = \sqrt{3}$

22. $n = 15; a = 8; b = 4 + 4\sqrt{3}$

23. C

24. Draw a segment, $\overline{AB}$. Construct a
perpendicular to $\overline{AB}$ at *A*. Place the
compass point at *A* and draw an arc
with radius *AB* intersecting the
perpendicular at a point *C*.
Draw $\overline{BC}$. $\triangle ABC$ is a right triangle
and $AB = AC$, so $\angle C \cong \angle B$. (If
two sides of a triangle are congru-
ent, the angles opposite the sides
are congruent.) Since $\angle C$ and $\angle B$
are the acute angles of a right triangle and are congruent, $m \angle C =$
$m \angle B = 45°$.

ALGEBRA Find the value of each variable.

19.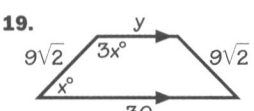

20.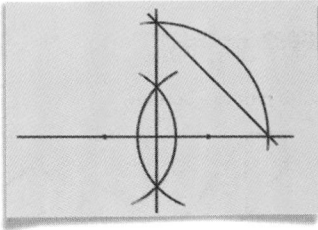

21.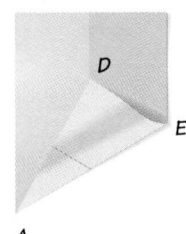

22.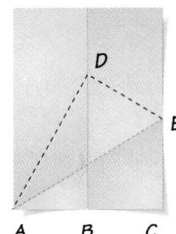

23. **SAT/ACT Preview** The distance from the center of a square to one of its vertices is 4 cm. What is the area of the square?

A. 16 cm^2 B. $4\sqrt{2}$ cm^2 C. 32 cm^2 D. $16\sqrt{2}$ cm^2 E. 8 cm^2

24. CONSTRUCTION Use the diagram to describe how to construct a 45-45-90 triangle.

25. **Investigation** Although it is not possible to trisect every angle with a compass and a straightedge, you can trisect any given angle by folding paper. To trisect a 90° angle, use a rectangular sheet of paper.

Crease the paper by folding it in half and opening it again.

Folding at *A*, bring *C* up to meet the first crease at *D*.

Unfold the paper.
$m \angle EAC = 30°$

  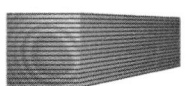

a. Let $AC = x$. Find AB and AD in terms of x. Explain your reasoning.

b. What type of triangle is $\triangle ADB$? How do you know?

c. What relationship does $\overrightarrow{AE}$ have to $\angle DAC$? Explain.

d. Use your answers to parts (a)–(c) to write a paragraph proof showing that $m \angle EAC = 30°$.

26. **FORESTRY** Ed Cushman wants to cut a log into usable boards. To use the log most efficiently, he cuts the cylindrical log into a prism with square bases. Then he cuts the prism into boards. If the diameter of the log is 18 in., how wide can the boards be? Explain.

10.2 Special Right Triangles **505**

Exercise Notes

Problem Solving
Exs. 19–22 In doing these exercises, students need to use the properties of special triangles more than once. For each figure, they should first look for a situation that involves only one variable.

Spatial Reasoning
Exs. 19–22 Some students may find these exercises easier to do if they separate the triangles and orient them as the triangles are oriented in the statements of the special triangle properties on page 501.

Integrating Algebra
Exs. 19–22 To find the value of each variable in there exercises, students need to write and solve several equations involving the properties of special triangles.

Construction Note
Ex. 24 Encourage students to analyze and perform the construction pictured in the diagram. You may wish to have students do this in pairs or groups to stimulate discussion on how the construction works.

Application
Ex. 26 In completing this exercise, students learn how the properties of the 45-45-90 triangle can be used to determine the size of boards that can be made from a tree of a given diameter.

---

25. a. $AB = \frac{1}{2}x$ because $\overline{AC}$ was folded in half; $AD = x$ because *D* was chosen so that $AD = AC$.

b. 30-60-90; $AD = x$ and $AB = \frac{1}{2}x$, so by the Pythagorean theorem, $DB = \frac{\sqrt{3}}{2}x$. Then $\triangle ADB$ is a 30-60-90 triangle because it is similar to any 30-60-90 triangle by the SSS Similarity Theorem.

c. $\overrightarrow{AE}$ bisects $\angle DAC$ as $\triangle EAD$ and $\triangle EAC$ are congruent.

d. Given: $\triangle ADB$ obtained by paperfolding as described.
Prove: $m \angle EAC = 30°$
$\overline{AD} \cong \overline{AC}$ and $\overline{DB}$ bisects $\overline{AC}$, so $AB = \frac{1}{2}AD$. By the Pythagorean theorem, $DB = \frac{AB\sqrt{3}}{2}$, so $\triangle ADB$ is a 30-60-90 triangle with $m \angle DAC = 60°$. From the way the paper was folded, $\triangle EAD \cong \triangle EAC$. Thus, by the definition of congruent triangles, $\angle EAD \cong \angle EAC$. But $m \angle EAD + m \angle EAC = m \angle DAC$. So $m \angle EAC = \frac{1}{2}m \angle DAC = \frac{1}{2}(60°) = 30°$.

26. about 12.7 in.; The diameter of the log is the length of a diagonal of the square base of the prism. The width of a board is the length of a side of the square. The diagonal divides the square into two 45-45-90 triangles. If the length of the hypotenuse is 18 in., the length of each leg is $\frac{18}{\sqrt{2}}$, or about 12.7 in.

505

Exercise Notes

Assessment Note

Ex. 29 You may wish to ask two or more students to read their answers to the class. A presentation at the board by the students would enable them to illustrate their responses with appropriate diagrams.

Assess Your Progress

Journal Entry

This activity provides an excellent review of the major concepts of Sections 10.1 and 10.2. Students should check their diagrams against those in the text before putting them in their journals.

Progress Check 10.1–10.2

See page 544.

Practice 68 for Section 10.2

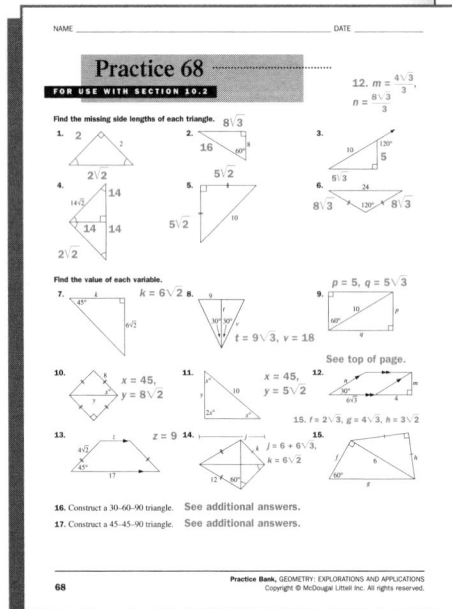

27. Look back at page 501. Use Questions 1–5 to write a formal proof of the 30-60-90 triangle theorem.

28. Challenge Write an equation for a line that makes a 60° angle with the *x*-axis. How did you get your answer?

ONGOING ASSESSMENT

29. Writing How can you find the area of a square if you know the length of its diagonal? How can you find the area of an equilateral triangle if you know the length of one side?

SPIRAL REVIEW

Explain why the triangles in each pair are similar. *(Section 9.2)*

30.    **31.** 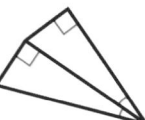   **32.**

Find the slope of the line that contains each pair of points given. *(Section 4.2)*

33. (0, 0) and (4, 7)   **34.** (−1, 0) and (0, 1)   **35.** (2, −3) and (−4, 5)

ASSESS YOUR PROGRESS

VOCABULARY

geometric mean (p. 494)   **30-60-90 triangle** (p. 501)
45-45-90 triangle (p. 500)

Find the geometric mean of the given numbers. *(Section 10.1)*

1. 3 and 27   **2.** 5 and 8   **3.** 21 and 35

Find the value of each variable. *(Section 10.1)*

4.   **5.**   **6.**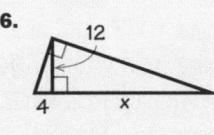

Find the value of each variable. *(Section 10.2)*

7. 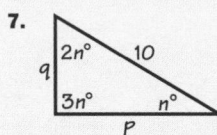   **8.** 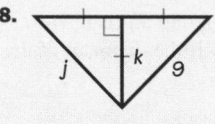   **9.**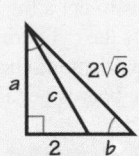

10. Journal Draw and label several triangles to illustrate the main ideas you learned in Sections 10.1 and 10.2.

---

27. Given: Equilateral $\triangle ABC$;
$\overrightarrow{BD}$ bisects $\angle ABC$.

Prove: $AD = \frac{x}{2}$ and $BD = \frac{\sqrt{3}}{2}x$.

$m\angle ABD = 30°$ (Def. of angle bisector); $\overline{BD} \perp \overline{AC}$ and $\overline{AD} = \overline{DC}$ (the bisector of the vertex angle of an isosceles triangle is the perpendicular bisector of the base). Therefore, $\triangle BAD$ is a 30-60-90 triangle. Since $AB = AC$, $AD = \frac{1}{2}x$. By the Pythagorean theorem, $x^2 = \left(\frac{1}{2}x\right)^2 + (BD)^2$.

So $(BD)^2 = x^2 - \frac{1}{4}x^2 = \frac{3}{4}x^2$.

Thus, $BD = \frac{\sqrt{3}}{2}x$.

28. $y = \sqrt{3}x$;
Let *m* be the line through the origin that makes a 60° angle with the *x*-axis. The hypotenuse of a triangle with vertices (0, 0), (1, 0), and (1, $\sqrt{3}$) is on line *m*. Line *m* contains (0, 0) and (1, $\sqrt{3}$). The slope of *m* is $\sqrt{3}$ and the *y*-intercept is 0.

29. See answers in back of book.

30. SAS Similarity Theorem

31. AA Similarity Postulate

32. AA Similarity Postulate

33. $\frac{7}{4}$

34. 1

35. $-\frac{4}{3}$

Assess Your Progress

1. 9   **2.** $2\sqrt{10}$

3. $7\sqrt{15}$

4. $c = 7$; $d = 7\sqrt{5}$

5. $t = \frac{11}{\sqrt{2}}$; $s = \frac{22}{\sqrt{2}}$

6. 36

7. $n = 30$; $q = 5$; $p = 5\sqrt{3}$

8. $j = 9$; $k = \frac{9}{\sqrt{2}}$

9. $a = 2\sqrt{3}$; $b = 2\sqrt{3} - 2$; $c = 4$

10. See answers in back of book.

506

10.3

The Tangent Ratio

Learn how to...

- find the tangent of an acute angle

- use tangents to find lengths and angle measures in right triangles

So you can...

- solve real-world problems involving angles and lengths in right triangles

If you travel through farming areas, you may see cone-shaped piles of grain. The angle that the grain makes with the ground is called the *angle of repose*. The measure of this angle depends on the type of grain. For example, the angle of repose for wheat measures 27°.

The angle of repose determines the *coefficient of friction of grain on grain.* This coefficient, which measures how easily the kernels of grain slide against each other, is equal to the *tangent* of the angle of repose.

In a right triangle, the **tangent** of an acute angle is the ratio of the length of the leg *opposite the angle* to the length of the leg *adjacent to the angle*.

Write the tangent of angle *A* as "tan *A*."

$$\tan A = \frac{\text{opposite}}{\text{adjacent}} = \frac{BC}{AC}$$

EXAMPLE 1

Use the diagram to find tan *P* and tan *Q*.

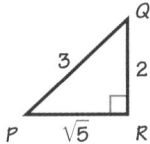

SOLUTION

Use the definition of tangent.

$$\tan P = \frac{\text{length of leg opposite } \angle P}{\text{length of leg adjacent to } \angle P} = \frac{2}{\sqrt{5}}$$

$$\tan Q = \frac{\text{length of leg opposite } \angle Q}{\text{length of leg adjacent to } \angle Q} = \frac{\sqrt{5}}{2}$$

10.3 The Tangent Ratio **507**

Plan⇔Support

Objectives

- Find the tangent of an acute angle.

- Use tangents to find lengths and angle measures in right triangles.

- Solve real-world problems involving angles and lengths in right triangles.

Recommended Pacing

❖ **Core and Extended Courses**
Section 10.3: 1 day

❖ **Block Schedule**
Section 10.3: $\frac{1}{2}$ block
(with Section 10.4)

Resource Materials

Lesson Support
Lesson Plan 10.3

Warm-Up Transparency 10.3

Practice Bank: Practice 69

Study Guide: Section 10.3

Challenge Problems: Set 69

Technology
Scientific Calculator

Internet:
http://www.hmco.com

Warm-Up Exercises

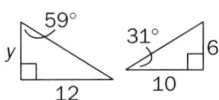

1. Are the two triangles above similar? Explain. Yes; since the sum of the angle measures of a triangle is 180°, both triangles have angles of 31° and 59°. Thus, they are similar by the AA Similarity Postulate.

2. Find the value of *y*.   7.2

3. Find the length of the hypotenuse of each triangle. about 14; about 11.7

What is the decimal value of each expression?

4. $\frac{6}{\sqrt{2}}$ about 4.24

5. $\frac{3}{\sqrt{5}}$ about 1.34

Section Notes

Integrating Trigonometry

This section introduces students to trigonometry by using a right triangle to define the tangent of an angle. In more advanced mathematics courses, students will learn that the tangent can be defined as a function whose domain is the set of all angles.

Teaching Tip

Make sure students understand the meanings of the terms *leg opposite* and *leg adjacent*. Point out that the legs designated as opposite and adjacent can change because they depend on which acute angle is being considered.

Additional Example 1

Use the diagram to find tan *L* and tan *N*.

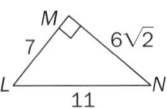

Use the definition of tangent.

$$\tan L = \frac{\text{length of leg opposite } \angle L}{\text{length of leg adjacent to } \angle L}$$

$$= \frac{6\sqrt{2}}{7}$$

$$\tan N = \frac{\text{length of leg opposite } \angle N}{\text{length of leg adjacent to } \angle N}$$

$$= \frac{7}{6\sqrt{2}}$$

Think and Communicate

When discussing question 2, make sure students understand that any two right triangles containing a given acute angle are similar. Thus, the tangent of that angle is the same regardless of the right triangle that is used to find its value.

THINK AND COMMUNICATE

1. Look back at the solution to Example 1. How are $m \angle P$ and $m \angle Q$ related? How are tan *P* and tan *Q* related? Explain.

2. a. What can you conclude about $\triangle XWV$ and $\triangle XYZ$?

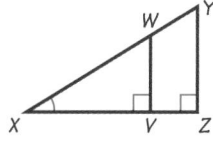

   b. Use your answer from part (a) to explain why the tangent of a given angle measure is constant.

3. a. For each triangle shown, measure $\angle A$ and find tan *A*.

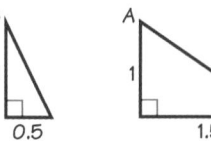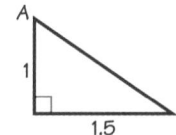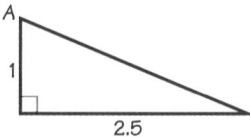

   b. What happens to tan *A* as $m \angle A$ approaches 90°?

   c. Is there a limit to how high the tangent values can be? Explain.

The tangent of an angle in a right triangle depends only on the measure of the angle and not on the size of the triangle. In other words, an expression like tan 27° has a fixed value that is independent of any particular right triangle.

You can approximate the tangent of an angle by measuring. However, many calculators have a tangent key that gives more precise values.

Table p. 733
Trigonometric Ratios

EXAMPLE 2   Application: Agriculture

Suppose the radius of the base of a pile of wheat is 30 ft. Find the height *h* of the pile.

SOLUTION

$$\tan A = \frac{\text{opposite}}{\text{adjacent}}$$

$$\tan 27° = \frac{h}{30}$$

$$30 \tan 27 = h$$

$$30(0.5095) \approx h$$

$$15.3 \approx h$$

```
tan 27
        .5095254495
Ans*30
        15.28576348
```

The height of the pile of wheat is about 15.3 ft.

ANSWERS   Section 10.3

Think and Communicate

1. $\angle P$ and $\angle Q$ are the acute angles of a right triangle, so $\angle P$ and $\angle Q$ are complementary; $\tan P = \dfrac{1}{\tan Q}$; $\tan P = \dfrac{QR}{PR} = \dfrac{1}{\frac{PR}{QR}} = \dfrac{1}{\tan Q}$.

2. a. $\triangle XWV \sim \triangle XYZ$

   b. Any two right triangles that contain a given acute angle are similar. The ratio of the lengths of two corresponding sides is always the same, so the tangent of the given angle is constant.

3. a. Estimates of angle measures may vary. 27°, $\frac{1}{2}$; 56°, 1.5; 68°, 2.5

   b. tan *A* increases.

   c. No; picture the triangle shown in part (a) as $m \angle A$ approaches 90°. The length of the leg opposite $\angle A$ increases without limit, so tan *A* increases without limit, as well.

As well as being able to give tangents of angles, most scientific calculators can determine the measure of an angle given its tangent. If $\angle A$ is an acute angle and $\tan A = x$, then $\tan^{-1} x = m\angle A$. The expression $\tan^{-1} x$ is read as "the inverse tangent of x."

EXAMPLE 3

Find $m\angle U$ in $\triangle TUV$.

SOLUTION

Use a scientific calculator to find the measure of an angle whose tangent is $\frac{15}{8}$.

$$\tan U = \frac{15}{8}$$

$$m\angle U = \tan^{-1} 1.875$$

$$m\angle U \approx 61.9°$$

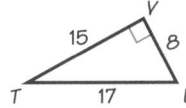

tan⁻¹(15/8)
          61.92751306

The measure of $\angle U$ is about 61.9°.

Angles of Elevation and Depression

Suppose a person in a boat sees the top of a lighthouse in the distance. The angle that the person's line of sight makes with the horizontal is called the **angle of elevation**.

Suppose the lighthouse operator spots the boat from the top of the lighthouse at the same time. The angle that the operator's line of sight makes with the horizontal is called the **angle of depression**. You can use parallel lines and alternate interior angles to show that these two angles are congruent.

angle of depression

Additional Example 2

Suppose that the angle of repose for a certain type of grain is 32°. Find the radius r of the base of a pile of this grain if the height is 18 ft.

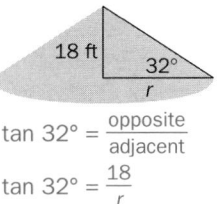

$$\tan 32° = \frac{\text{opposite}}{\text{adjacent}}$$

$$\tan 32° = \frac{18}{r}$$

$$r \tan 32 = 18$$

$$r = \frac{18}{\tan 32}$$

$$r \approx \frac{18}{0.6249}$$

$$r \approx 28.8$$

The radius of the base of the pile of grain is about 28.8 ft.

Section Note

Using Technology
You may need to remind students to put their calculators in degree mode to find tangents and inverse tangents when working with angles in degrees.

Additional Example 3

Find $m\angle T$ in $\triangle RST$.

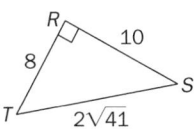

Use a scientific calculator to find the measure of an angle whose tangent is $\frac{10}{8}$.

$$\tan T = \frac{10}{8} = \frac{5}{4}$$

$$m\angle T = \tan^{-1} 1.25$$

$$m\angle T \approx 51.3°$$

The measure of $\angle T$ is about 51.3°.

Section Note

Second-Language Learners
In order to make the concepts of elevation and depression clearer for second-language learners, discuss the meanings of the two words outside of a mathematical context. If necessary, define *elevation* as "lifting up" and *depression* as "pressing down." Ask students how knowing these definitions can help them understand and remember the mathematical concepts.

Section Note

Visual Thinking

Ask students to find photographs in magazines and newspapers that can be used to demonstrate angles of elevation and depression. Encourage them to use markers to highlight these angles on the photos, and to estimate the size of the angle and the length of one leg. Then, ask them to determine the other measures. This activity involves the visual skills of *interpretation* and *exploration*.

Additional Example 4

The angle of depression from a hot air balloon to its landing target is 12°. If the balloon is 175 ft high, find its distance measured along the ground from the target.

Let d = the distance measured along the ground from the balloon to the target. Draw and label a diagram that models the situation. The angle of elevation is congruent to the angle of depression.

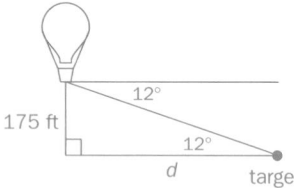

Write an equation involving d and the angle of elevation.

$$\tan 12° = \frac{h}{d}$$
$$\tan 12 = \frac{175}{d}$$
$$d \tan 12 = 175$$
$$d = \frac{175}{\tan 12}$$
$$d \approx 823$$

The distance measured along the ground from the balloon to the target is about 823 ft.

Closure Question

Describe how to find the measures of the acute angles of a right triangle if the lengths of the legs are known.

Choose one of the acute angles and identify the leg opposite and leg adjacent to this angle. Find the ratio $\frac{\text{length of leg opposite}}{\text{length of leg adjacent to}}$. Use the $\tan^{-1}$ button on a scientific calculator to find the angle having this tangent ratio. This is one of the acute angles. To find the other acute angle, subtract the measure of the first acute angle from 90°.

EXAMPLE 4   Application: Navigation

Suppose the angle of elevation from a ship to the top of a lighthouse is 6°. The lighthouse is 60 ft tall and the cliff is 250 ft high. Find the distance from the ship to the base of the cliff.

SOLUTION

Let h = the distance from the base of the cliff to the top of the lighthouse.
Let d = the distance from the ship to the base of the cliff.
Draw and label a diagram that models the situation.

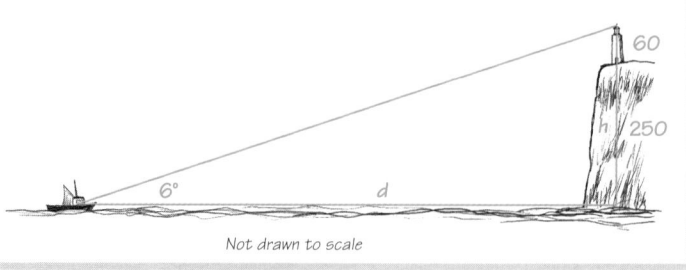

Not drawn to scale

Write an equation involving h and d:

$$\tan 6° = \frac{h}{d} \qquad \text{Substitute } 250 + 60 = 310 \text{ for } h.$$

$$\tan 6 = \frac{310}{d} \qquad \text{Multiply both sides by } d.$$

$$d \tan 6 = 310 \qquad \text{Divide both sides by } \tan 6.$$

$$d = \frac{310}{\tan 6}$$

$$d \approx 2949$$

The distance from the ship to the base of the cliff is about 2950 ft.

BY THE WAY

During the third century B.C., the Egyptians constructed the Pharos of Alexandria, the tallest lighthouse ever built. This lighthouse was over 400 ft high and guided ships for about 1500 years.

☑ **CHECKING KEY CONCEPTS**

For Questions 1 and 2, use △JKL.

1. Find tan J and tan K.

2. Find $m \angle J$ and $m \angle K$.

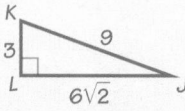

3. AVIATION The angle of depression from an airplane flying at an altitude of 5000 ft to the closer end of a runway is 9°.

  a. Find the horizontal distance from the airplane to the runway.

  b. Suppose the runway is 7000 ft long. Find the angle of elevation from the farther end of the runway to the airplane.

510   Chapter 10 *Applying Right Triangles*

Checking Key Concepts

1. $\frac{1}{2\sqrt{2}}$; $2\sqrt{2}$

2. about 19.5°; about 70.5°

3. **a.** about 31,569 ft

  b. about 7.4°

Exercises and Applications

1. $\frac{3}{4}$; $\frac{4}{3}$

2. $\frac{\sqrt{17}}{8}$; $\frac{8}{\sqrt{17}}$

3. $\frac{5}{4}$; $\frac{4}{5}$

4. 0.8391

5. 2.4751

6. 0.0998

7. 572.9572

8. **a.** Answers may vary. Examples are given.

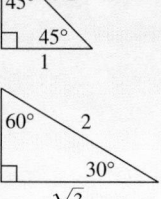

10.3 Exercises and Applications

Extra Practice
exercises on
page 681

For Exercises 1–3, find tan A and tan B.

1.
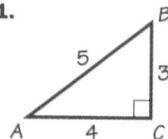

2.
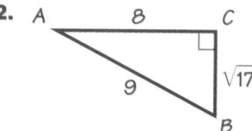

3.
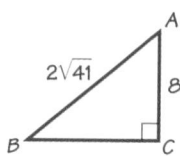

Find the value of each ratio. Round your answers to four decimal places.

4. $\tan 40°$

5. $\tan 68°$

6. $\tan 5.7°$

7. $\tan 89.9°$

8. a. **Open-ended Problem** Sketch a 45-45-90 triangle and a 30-60-90 triangle. Label each side length.

   b. Use your diagrams from part (a) to find the exact values of tan 30°, tan 45°, and tan 60°.

Find the measure of the acute angle that satisfies the given equation. Round your answers to the nearest tenth of a degree.

9. $\tan A = \frac{2}{7}$

10. $\tan B = \frac{19}{5}$

11. $\tan R = 1.268$

12. $\tan Y = 0.4779$

Find the value of each variable. Round your answers to the nearest tenth.

13.
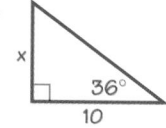

14.
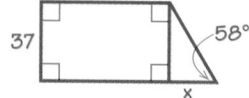

15.
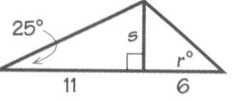

16. **Open-ended Problem** Draw a right triangle. Measure the legs and use $\tan^{-1}$ to find the measure of each acute angle. Use a protractor to check your answers.

17. **PALEONTOLOGY** The diagram at the right illustrates several characteristics of a two-legged dinosaur's *gait* (way of walking). In the diagram, AC is called the *pace length*, BD is called the *height at the hip*, and $\angle ABC$ is called the *angle of gait*. One study suggests that the normal pace length for a two-legged dinosaur was about 0.65 times the dinosaur's height at the hip.

   a. **ALGEBRA** Let $BD = x$. Find AC, AD, and DC in terms of x.

   b. Use your answers to part (a) to find $m\angle ABD$ and $m\angle CBD$.

   c. **Writing** What was the angle of gait for two-legged dinosaurs? Explain why it did not depend on the size of the dinosaur.

10.3 The Tangent Ratio **511**

---

b. $\tan 30° = \frac{1}{\sqrt{3}}$; $\tan 45° = 1$;

   $\tan 60° = \sqrt{3}$

9. 15.9°

10. 75.3°

11. 51.7°

12. 25.5°

13. 7.3

14. 23.1

15. $s = 5.1$; $r = 40.4°$

16. Answers may vary. An example is given. Angle measures are given to the nearest degree.

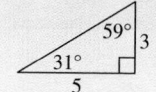

17. a. $AC = 0.65x$; $AD = DC = 0.325x$

    b. $m\angle ABD = m\angle CBD \approx 18°$

    c. about 36°; As shown in part (b), the angle of gait was 36°, no matter what the value of x (the height at the hip) was.

511

Exercise Notes

18. MEASUREMENT Two buildings are 75 ft apart. Carolyn, who is standing on the roof of the shorter building, measures the angle of elevation from her eyes to the top of the taller building to be 28°. She measures the angle of depression from her eyes to the bottom of the taller building to be 53°. Find the height of each building, given that Carolyn's eyes are 5 ft above the roof of the shorter building.

Find the measure of each acute angle. Round your answers to the nearest tenth of a degree.

19.
R, P, 12, 37, Q

20.
B, A, E, C, 12.6, 9.1, D

21.
X, $n\sqrt{5}$, $n\sqrt{5}$, Y, $2n$, Z

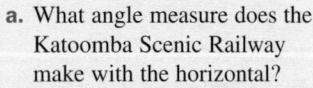
Connection   **TRANSPORTATION**

A railway's steepness is measured by its *gradient*. The gradient is the ratio of the railway's rise to its run and is usually expressed as a percent. That is, a railway with a gradient of *g*% rises *g* ft vertically for every 100 ft traveled horizontally.

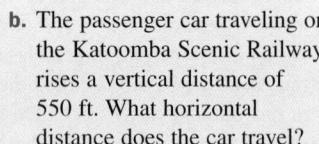

22. Writing What is the relationship between a railway's gradient expressed as a percent and the angle that the railway makes with the horizontal? Explain.

23. The steepest railway in the world is the Katoomba Scenic Railway in the Blue Mountains of New South Wales, Australia. A single passenger car is pulled by two steel cables up the length of the railway. The gradient of the railway is 122%.

g ft
100 ft
gradient = g%

   a. What angle measure does the Katoomba Scenic Railway make with the horizontal?

   b. The passenger car traveling on the Katoomba Scenic Railway rises a vertical distance of 550 ft. What horizontal distance does the car travel?

   c. The length of the Katoomba Scenic Railway is 1020 ft. Find the vertical distance that the passenger car travels when it goes from the bottom of the railway to the top. (*Hint:* Use the diagram at the right, where *A* is the angle you found in part (a). Express *y* in terms of *x*, and use the Pythagorean theorem. You will learn a simpler way to solve this problem in Section 10.4.)

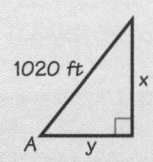
1020 ft, x, A, y

18. The shorter building is about 94.5 ft tall; the taller building is about 139.4 ft tall.

19. $m\angle P = 18.9°$; $m\angle Q = 71.1°$

20. $m\angle ACD = m\angle CAD = m\angle BAC = m\angle BCA = 35.8°$; $m\angle BDC = m\angle ADB = m\angle ABD = m\angle CBD = 54.2°$

21. $m\angle Y = m\angle Z = 63.4°$; $m\angle X = 53.2°$

22. The gradient expressed as a percent is equal to the tangent of the angle. The tangent of the angle is equal to $\frac{g}{100}$.

23. Answers may vary slightly due to rounding.
   a. about 50.7°
   b. about 450.2 ft
   c. about 789.3 ft

24. AGRICULTURE The table shows the angle of repose for wheat, corn, and oat. Recall that the tangent of the angle of repose is the coefficient of friction of grain on grain.

  a. Copy and complete the table.

  b. **Writing** What happens to the coefficient of friction of grain on grain as the angle of repose increases? Which grain in the table do you think slides most easily? Explain your reasoning.

Angles of grain piles			
	wheat	corn	oat
$m \angle A$	27.0	27.5	28.0
tan A	?	?	?

25. A tree casts a shadow 25 ft long. The angle of elevation from the tip of the shadow to the top of the tree is 64°. Find the height of the tree.

26. Prove that if $\angle A$ and $\angle B$ are complements, then $(\tan A)(\tan B) = 1$.

27. Prove that if $0 < m \angle A < 45°$, then $\tan A < 1$.

28. Challenge A glider is approaching a large river. The angle of depression from the glider to the near side of the river is 14°. The angle of depression from the glider to the far side of the river is 10°. The altitude of the glider is 3000 ft. Find the width w of the river.

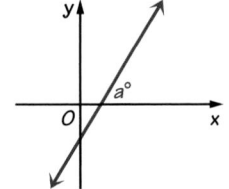

Not drawn to scale

3000 ft

x — river — w

ONGOING ASSESSMENT

29. Writing For any line whose slope is positive, let $a°$ be the measure of the angle that the line makes with the x-axis. (Measure this angle counterclockwise from the x-axis to the line, so that the angle is always acute.) Describe how a is related to the slope of the line.

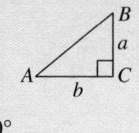

SPIRAL REVIEW

30. The hypotenuse of a 30-60-90 triangle is 8 cm long. What are the lengths of the triangle's legs? *(Section 10.2)*

Can a triangle be formed from sides of the given lengths? *(Section 6.1)*

31. 4, 10, 12   **32.** 6, 6, 8   **33.** 1, 3, 5

10.3 The Tangent Ratio   **513**

Practice 69 for Section 10.3

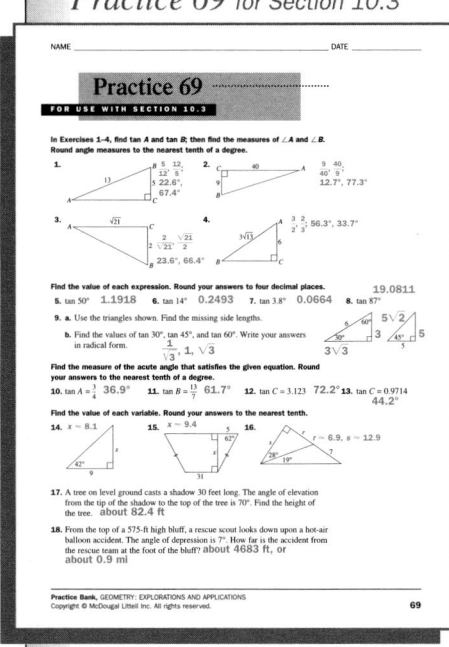

---

24. a. Answers may vary slightly due to rounding.

	wheat	corn	oat
$m \angle A$	27.0	27.5	28.0
tan A	0.5095	0.5206	0.5317

  b. It increases. I think that the lower the coefficient of friction, the more easily the grain slides. Therefore, I think that, of the grains in the table, wheat slides most easily.

25. about 51.3 ft

26. In $\triangle ABC$, $\angle A$ and $\angle B$ are complementary, since the sum of the measures of the acute angles of a right triangle is 90°. By definition, $\tan A = \frac{a}{b}$ and $\tan B = \frac{b}{a}$. Therefore,

$(\tan A)(\tan B) = \frac{a}{b} \cdot \frac{b}{a} = 1$.

27. In $\triangle ABC$, $0 < m \angle A < 45°$. Since $m \angle A + m \angle B = 90°$, $45° < m \angle B < 90°$. This means that $b > a$ because the side opposite the larger angle is longer than the side opposite the smaller angle. Since $\tan A = \frac{a}{b}$ and $\frac{a}{b} < 1$, $\tan A < 1$.

28. about 4981.5 ft

29. The slope of the line is the tangent of the angle.

30. 4 cm and $4\sqrt{3}$ cm

31. Yes.

32. Yes.

33. No.

513

Objectives

- Find the sine and cosine of an acute angle.
- Find the measure of an acute angle whose sine or cosine is given.
- Find the lengths of the sides of right triangles.
- Solve real-world problems using sines and cosines.

Recommended Pacing

❖ **Core and Extended Courses**
Section 10.4: 2 days

❖ **Block Schedule**
Section 10.4: 2 half-blocks
(with Sections 10.3 and 10.5)

Resource Materials

Lesson Support
Lesson Plan 10.4
Warm-Up Transparency 10.4
Practice Bank: Practice 70
Study Guide: Section 10.4
Explorations Lab Manual:
    Additional Exploration 15
    Diagram Master 13
Challenge Problems: Set 70

Technology
Scientific Calculator
Graphing Calculator
Geometry Software
Internet:
    http://www.hmco.com

Warm-Up Exercises

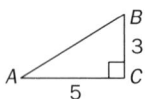

Use the diagram to find each value.

1. $\tan A$ $\frac{3}{5}$ **2.** $\tan B$ $\frac{5}{3}$

3. $m\angle A$ about 31.0°

4. A girl is flying a kite with an angle of elevation of 25°. A shadow cast directly below the kite is 40 ft from the girl. If the girl is holding the string 3 ft above the ground, how high is the kite?   about 21.7 ft

---

SECTION

10.4 | Sine and Cosine Ratios

The tangent ratio you learned about in the last section is part of an important branch of mathematics called *trigonometry*. The word "trigonometry" comes from Greek words meaning "triangle measurement," although trigonometry probably originated in ancient Egypt and Mesopotamia. In this section, you will learn about two other trigonometric ratios.

Learn how to...

- **find the sine and cosine of an acute angle**
- **find the measure of an acute angle whose sine or cosine is given**

So you can...

- **find the lengths of the sides of right triangles**
- solve problems in engineering, for example

EXPLORATION
COOPERATIVE LEARNING

Analyzing Ratios in Triangles

Work with another student.
You will need:

- geometry software or a ruler and protractor

1 For each angle measure in the table below, complete these steps:

- One student should draw right △ABC so that ∠A has the given measure.

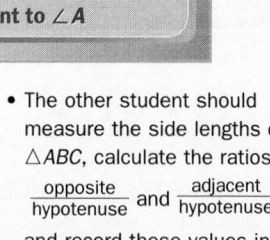

$m\angle A$	opposite / hypotenuse	adjacent / hypotenuse
20°	?	?
40°	?	?
60°	?	?
80°	?	?

- The other student should measure the side lengths of △ABC, calculate the ratios $\frac{\text{opposite}}{\text{hypotenuse}}$ and $\frac{\text{adjacent}}{\text{hypotenuse}}$, and record these values in the table. Round each value to the nearest hundredth.

2 As $m\angle A$ increases, what happens to each ratio?

3 What value do you think each ratio approaches as $m\angle A$ approaches 0°? as $m\angle A$ approaches 90°? Explain your answers.

4 For what value of $m\angle A$ do you think the two ratios will be equal? Draw a right triangle with this angle measure and find out if you are correct.

Exploration Note

Purpose
The purpose of this Exploration is to have students investigate the values of the sine and cosine ratios for various acute angles.

Materials/Preparation
Each group of students needs geometry software or a ruler and protractor.

Procedure
Students draw right triangles that contain various acute angles and find the ratio of the side opposite the angle to the hypotenuse, and the ratio of the side adjacent to the angle and the hypotenuse. They consider what happens to the values of these ratios as the angle measures increase and

as the angle measures approach 0. They also consider what value of $m\angle A$ will make the two ratios equal.

Closure
Call upon various students to answer the questions in Steps 2–4. Students should understand that as the angle measures increase, the sine ratio approaches 1 and the cosine ratio approaches 0. The opposite effect is seen as the angle measures approach 0°.

Explorations Lab Manual
See the Manual for more commentary on this Exploration.
Diagram Master 13

In a right triangle, the **sine** of an acute angle is the ratio of the length of the leg opposite the angle to the length of the hypotenuse. The **cosine** of an acute angle is the ratio of the length of the leg adjacent to the angle to the length of the hypotenuse. Sines and cosines of acute angles are always between 0 and 1.

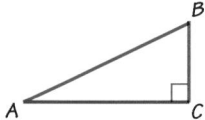

$$\sin A = \frac{\text{opposite}}{\text{hypotenuse}} = \frac{BC}{AB}$$ — Write the sine of $\angle A$ as "sin A."

$$\cos A = \frac{\text{adjacent}}{\text{hypotenuse}} = \frac{AC}{AB}$$ — Write the cosine of $\angle A$ as "cos A."

EXAMPLE 1

In $\triangle JKL$, find sin J and cos J.

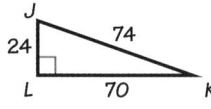

SOLUTION

$$\sin J = \frac{\text{opposite}}{\text{hypotenuse}} = \frac{70}{74} \approx 0.9459$$

$$\cos J = \frac{\text{adjacent}}{\text{hypotenuse}} = \frac{24}{74} \approx 0.3243$$

In Section 10.3, you used a scientific calculator to approximate tangents of angles. Most calculators also have keys for approximating sines and cosines.

EXAMPLE 2

Find the values of x and y in the triangle shown.

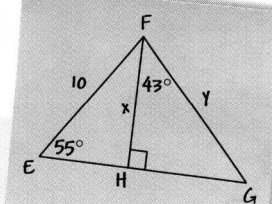

SOLUTION

Step 1 Use a sine ratio in $\triangle EFH$ to find the value of x.

$$\sin 55° = \frac{x}{10}$$

$$10 \sin 55 = x$$

$$10(0.8192) \approx x$$

$$8.19 \approx x$$

Use a scientific calculator to find the value of sin 55°.

Step 2 Use a cosine ratio in $\triangle FGH$ to find the value of y.

$$\cos 43° = \frac{x}{y}$$

$$\cos 43 \approx \frac{8.19}{y}$$ — Multiply both sides by y.

$$y \cos 43 \approx 8.19$$

$$y \approx \frac{8.19}{\cos 43}$$ — Use a scientific calculator to find the value of cos 43°.

$$y \approx \frac{8.19}{0.7314}$$

$$y \approx 11.20$$

10.4 Sine and Cosine Ratios **515**

ANSWERS Section 10.4

Exploration

1. Answers may vary slightly due to rounding.

$m \angle A$	$\dfrac{\text{opposite}}{\text{hypotenuse}}$	$\dfrac{\text{adjacent}}{\text{hypotenuse}}$
20°	0.34	0.94
40°	0.64	0.77
60°	0.87	0.50
80°	0.98	0.17

2. The ratio $\dfrac{\text{opposite}}{\text{hypotenuse}}$ increases; the ratio $\dfrac{\text{adjacent}}{\text{hypotenuse}}$ decreases.

3. As $m \angle A$ approaches 0°, the length of the leg opposite $\angle A$ approaches 0, so $\dfrac{\text{opposite}}{\text{hypotenuse}}$ approaches 0. The length of the adjacent leg approaches the length of the hypotenuse, so $\dfrac{\text{adjacent}}{\text{hypotenuse}}$ approaches 1. As $m \angle A$ approaches 90°, the length of the leg opposite $\angle A$ approaches the length of the hypotenuse, so $\dfrac{\text{opposite}}{\text{hypotenuse}}$ approaches 1. The length of the adjacent leg approaches 0, so $\dfrac{\text{adjacent}}{\text{hypotenuse}}$ approaches 0.

4. 45°

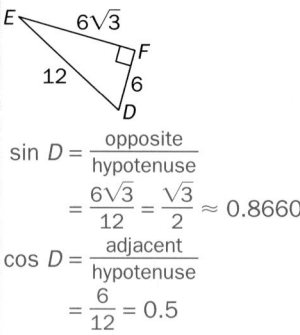
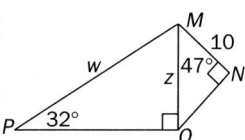

Section Note

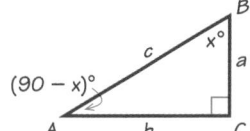

Using Technology
Students can check their answer to *Think and Communicate* question 2 by using a graphing calculator. Have students enter the four functions $y = \sin x$, $y = \sin (90 - x)$, $y = \cos x$, and $y = \cos (90 - x)$ as Y1, Y2, Y3, and Y4 on the Y= list. Then set the viewing window with the values Xmin = 0, Xmax = 90, Xscl = 10, Ymin = –2, Ymin = 2, and Yscl = 1. Press GRAPH and then press TRACE when the graphs appear. The small number in the upper-right corner of the screen tells you which function the cursor is following. Use the right and left arrow keys to move the cursor to a place where the graphs are separate. Then use the up and down arrow keys to change the number of the graph that is being traced. Using this method, students can see that the graphs of Y1 and Y4 coincide and the graphs of Y2 and Y3 coincide. Thus, $\sin x = \cos (90 - x)$ and $\cos x = \sin (90 - x)$.

Additional Example 3

A stage is constructed so that it is 42.5 ft from the back of the stage area to the front as measured horizontally. The stage itself is 43 ft long. At what angle does the stage slant? Is the stage acceptable?

Use a cosine ratio to find the measure of the angle that the stage makes with the horizontal.

$$\cos x° = \frac{42.5}{43}$$

$$x = \cos^{-1}\left(\frac{42.5}{43}\right)$$

Use a scientific calculator to evaluate $\cos^{-1}\left(\frac{42.5}{43}\right)$.

$$x \approx 8.75°$$

Since the angle measure is greater than 5°, the stage is too steep for the actors.

THINK AND COMMUNICATE

1. a. In Example 2, what is $m \angle EFH$? What is $m \angle FGH$?

   b. Find the value of x in Example 2 using the cosine of $\angle EFH$. Then find the value of y using the sine of $\angle FGH$. Check that your values of x and y match those in Example 2.

2. Use the diagram at the left to find $\sin x°$, $\cos (90 - x)°$, $\cos x°$, and $\sin (90 - x)°$ in terms of a, b, and c. How are the sine and cosine of an angle related to the sine and cosine of its complement? Explain.

In Section 10.3, you found the measure of an angle given its tangent. You can also find the measure of an angle given its sine or cosine. If $\angle A$ is an acute angle and $\sin A = x$, then $\sin^{-1} x = m \angle A$. Also, if $\angle B$ is an acute angle and $\cos B = y$, then $\cos^{-1} y = m \angle B$. These expressions are read as "inverse sine of x" and "inverse cosine of y."

EXAMPLE 3 **Application: Theater**

A *raked stage* is a stage that is slanted toward the audience, like a ramp. In general, a raked stage that slants much more than 5° is too steep for the actors to move around on easily. Suppose a theater company builds a raked stage on a stage 29 ft deep. If the raked stage is supported at the back by a post that is 4 ft long and perpendicular to it, at what angle does the stage slant? Is the stage acceptable?

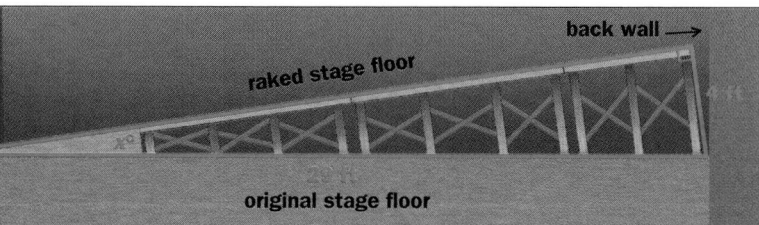

SOLUTION

Use a sine ratio to find the measure of the angle that the stage makes with the horizontal.

$$\sin x° = \frac{4}{29}$$

$$x = \sin^{-1}\left(\frac{4}{29}\right)$$

$$\approx \sin^{-1} 0.1379 \qquad \text{Use a scientific calculator to find } \sin^{-1} 0.1379.$$

$$\approx 7.93°$$

Since the angle measure is greater than 5°, the stage is probably too steep for the actors.

516   Chapter 10 *Applying Right Triangles*

Think and Communicate

1. a. 35°; 47°

   b. 8.19; 11.20

2. $\sin x° = \dfrac{b}{c}$; $\cos (90 - x)° = \dfrac{b}{c}$;

$\cos x° = \dfrac{a}{c}$; $\sin (90 - x)° = \dfrac{a}{c}$;

The sine of an angle is the cosine of its complement. The cosine of an angle is the sine of its complement. In a right triangle with acute angles A and B, $\angle A$ and $\angle B$ are complements and the leg opposite $\angle A$ is adjacent to $\angle B$, while the leg adjacent to $\angle A$ is opposite $\angle B$.

Here is a summary of the three trigonometric ratios that you can use to solve problems.

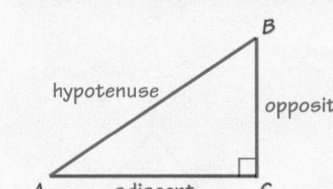

Trigonometric Ratios for Right Triangles

$$\tan A = \frac{\text{opposite}}{\text{adjacent}} = \frac{BC}{AC}$$

$$\sin A = \frac{\text{opposite}}{\text{hypotenuse}} = \frac{BC}{AB}$$

$$\cos A = \frac{\text{adjacent}}{\text{hypotenuse}} = \frac{AC}{AB}$$

☑ CHECKING KEY CONCEPTS

For Questions 1 and 2, use △LMN.

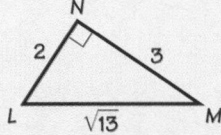

1. Find sin L and cos L. Use sin L to find $m \angle L$ to the nearest degree.

2. Find sin M and cos M. Use cos M to find $m \angle M$ to the nearest degree.

3. Latricia is flying a kite on level ground. Her hands are 4 ft above the ground. The angle between the string and the horizontal measures 50°, and the string is 300 ft long. How high above the ground is the kite? Round your answer to the nearest foot.

4. Find the values of x and y in △PQR.

10.4 | **Exercises and Applications**

Extra Practice exercises on page 681

For Exercises 1–3, find sin A, cos A, sin B, and cos B.

1.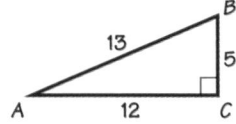

2.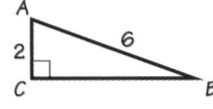

3.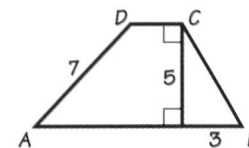

Find the value of each ratio. Round your answers to four decimal places.

4. sin 59°   5. sin 3.6°   6. cos 14°   7. cos 70.2°

Find the measure of an acute angle that satisfies the given equation. Round your answers to the nearest tenth of a degree.

8. $\sin A = \frac{3}{8}$ 9. $\sin M = 0.7874$ 10. $\cos B = 0.1096$ 11. $\cos Y = \frac{24}{25}$

10.4 Sine and Cosine Ratios   **517**

517

Student Progress

Exs. 12–17 Students who are having difficulty with these exercises should first identify the sides of each triangle and their relation to the known or desired length or angle measure. This should enable them to choose the correct trigonometric ratio.

Problem Solving

Exs. 14, 17 In both of these exercises, students need to recognize that the triangles are isosceles. They should then draw the altitude to the base of each triangle, and use trigonometric ratios to find the answer. Students should understand that trigonometric ratios cannot be used with the original triangles because they are not right triangles.

Interdisciplinary Problems

Exs. 18, 19 These exercises apply the sine ratio to understanding the effect that the atmosphere has on the brightness of a star. All of the trigonometric functions have extensive applications to astronomy and to other sciences, especially physics.

Communication: Reading

Exs. 18, 19 Before beginning these two exercises, students need to read and interpret the opening paragraph and the figure shown. Most students would benefit from having this material read aloud and discussed in class prior to attempting to do the exercises.

Second-Language Learners

Exs. 18, 19 Second-language learners might need guidance in understanding the scientific language in the connection to astronomy. Have students look up in a dictionary any unfamiliar words, such as *horizon, overhead, atmosphere,* or *intensity.*

Find the value of each variable.

12.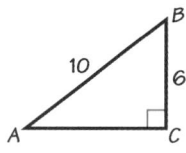

13.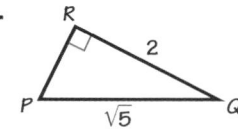

14.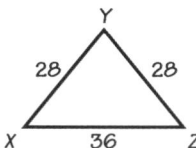

Find the measure of each acute angle. Round your answers to the nearest tenth of a degree.

15.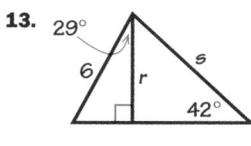

16.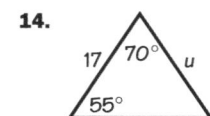

17.

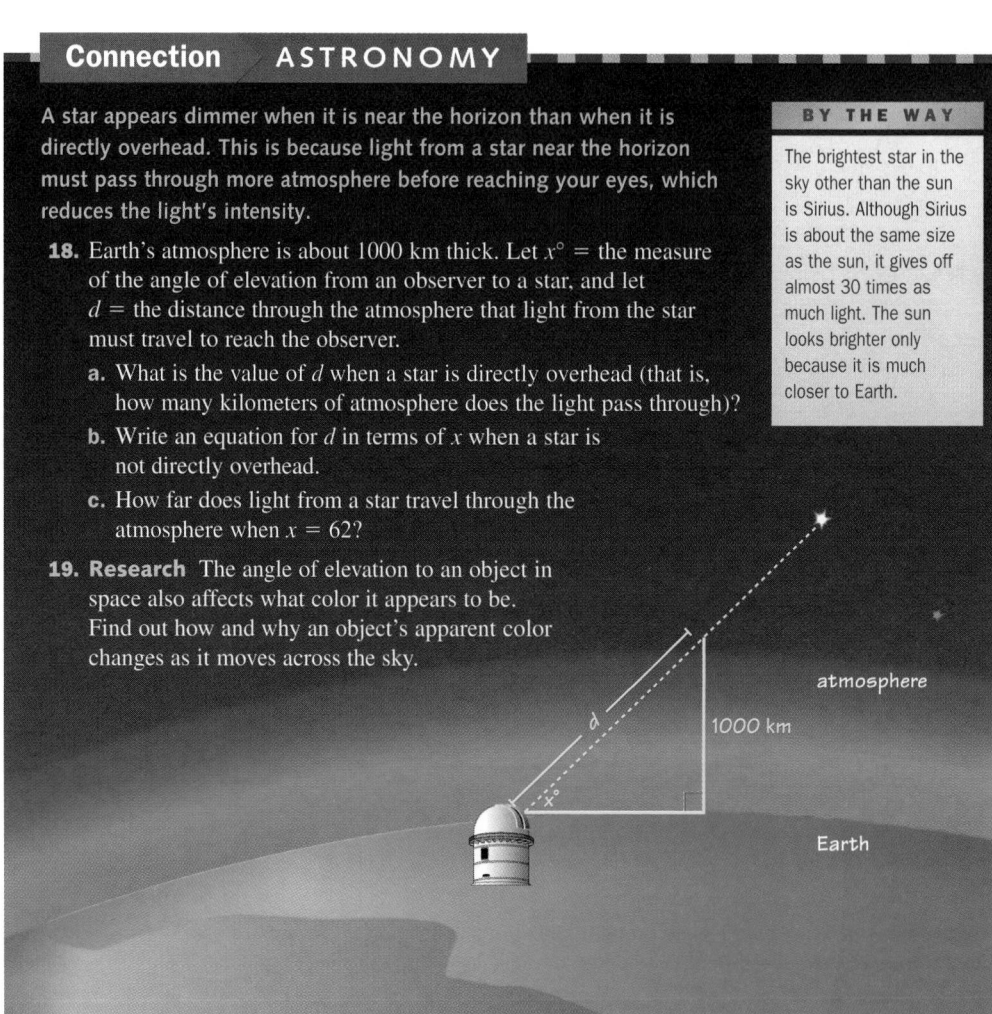

Connection   ASTRONOMY

A star appears dimmer when it is near the horizon than when it is directly overhead. This is because light from a star near the horizon must pass through more atmosphere before reaching your eyes, which reduces the light's intensity.

18. Earth's atmosphere is about 1000 km thick. Let $x°$ = the measure of the angle of elevation from an observer to a star, and let d = the distance through the atmosphere that light from the star must travel to reach the observer.

   a. What is the value of d when a star is directly overhead (that is, how many kilometers of atmosphere does the light pass through)?

   b. Write an equation for d in terms of x when a star is not directly overhead.

   c. How far does light from a star travel through the atmosphere when $x = 62$?

19. Research The angle of elevation to an object in space also affects what color it appears to be. Find out how and why an object's apparent color changes as it moves across the sky.

BY THE WAY

The brightest star in the sky other than the sun is Sirius. Although Sirius is about the same size as the sun, it gives off almost 30 times as much light. The sun looks brighter only because it is much closer to Earth.

12–14. Answers are given to two decimal places.

12. $x = 19.28$; $y = 22.98$

13. $r = 5.25$; $s = 7.85$

14. $u = 17.00$; $v = 19.50$

15. $m\angle A = 36.9°$; $m\angle B = 53.1°$

16. $m\angle P = 63.4°$; $m\angle Q = 26.6°$

17. $m\angle X = m\angle Z = 50.0°$; $m\angle Y = 80°$

18. a. 1

   b. $d = \dfrac{1}{\sin x°}$

   c. about 1.133 kilometers

19. Answers may vary. As an object moves through space, its light is reflected and absorbed to different degrees by Earth's atmosphere. Interstellar dust also affects various wavelengths differently.

20. a. about 35.2 in.

   b. $x \approx 24.7$ in.; $y \approx 41.3$ in.

   c. about 54.3 in.

21. a. about 24.7 in.

   b. $x \approx 35.2$ in.; $y \approx 30.8$ in.

   c. about 39.5 in.

INTERVIEW Debby Tewa

Look back at the article on pages 490–492.

Debby Tewa uses a special structure like the one below for supporting a solar module. In the diagram, AC and AB are permanently set to be 43 in. and 66 in. The measure of ∠A can be changed by adjusting the length s of the support struts.

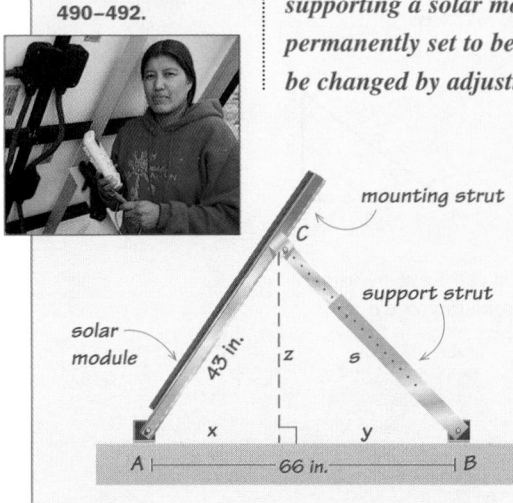

mounting strut

support strut

solar module

43 in.

z s

x y

A   66 in.   B

C

20. For a home with a flat roof in the southwestern United States, the exposure to sunlight is maximized when $m\angle A = 55°$.

 a. What should the value of z be for $\angle A$ to have this measure?

 b. Use your answer to part (a) and the Pythagorean theorem to find the values of x and y.

 c. How long should the support strut be in order for $\angle A$ to have the desired measure?

21. Repeat parts (a)–(c) of Exercise 20 for $m\angle A = 35°$.

22. For this structure, the value of z ranges from 13 in. to 35 in.

 a. Write an inequality that describes the possible values of $m\angle A$ to the nearest 5°.

 b. **Open-ended Problem** Describe at least two ways to modify the structure so that the range of possible values for $m\angle A$ is expanded.

Sketch a 45-45-90 triangle and a 30-60-90 triangle. Label each side length. Then use the triangles to find the exact value of each ratio.

23. $\sin 45°$ **24.** $\cos 45°$ **25.** $\cos 60°$

26. $\sin 60°$ **27.** $\sin 30°$ **28.** $\cos 30°$

29. AVIATION A jet takes off at a 15° angle. The jet's air speed is 300 ft/s.

 a. Write an equation that gives the distance d (in feet) that the jet has traveled through the air in terms of t, the number of seconds it has been in the air.

 b. Write equations for the horizontal distance h and the vertical distance v that the jet travels in t seconds.

 c. After 10 s, what horizontal distance has the jet traveled? What is the jet's altitude?

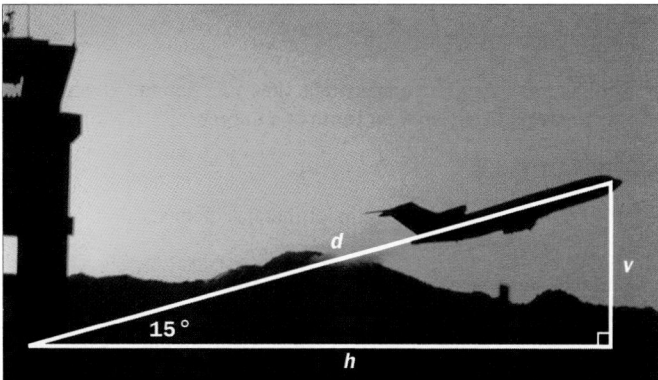

d

v

15°

h

10.4 Sine and Cosine Ratios   **519**

22. a. $15° \le m\angle A \le 55°$

 b. Answers may vary. Examples are given. Adapt the mounting strut so that the support struts may attach to it at different points. Hinge the connection between the support strut and the mounting strut so that the distance AB can vary.

23–28. Check students' work.

23. $\dfrac{1}{\sqrt{2}}$ **24.** $\dfrac{1}{\sqrt{2}}$

25. $\dfrac{1}{2}$ **26.** $\dfrac{\sqrt{3}}{2}$

27. $\dfrac{1}{2}$ **28.** $\dfrac{\sqrt{3}}{2}$

29. a. $d = 300t$

 b. $h = d\cos 15° = 300t\cos 15° \approx 289.78t$;
 $v = d\sin 15° = 300t\sin 15° \approx 77.65t$

 c. about 2897.8 ft; about 776.5 ft

Practice 70 for Section 10.4

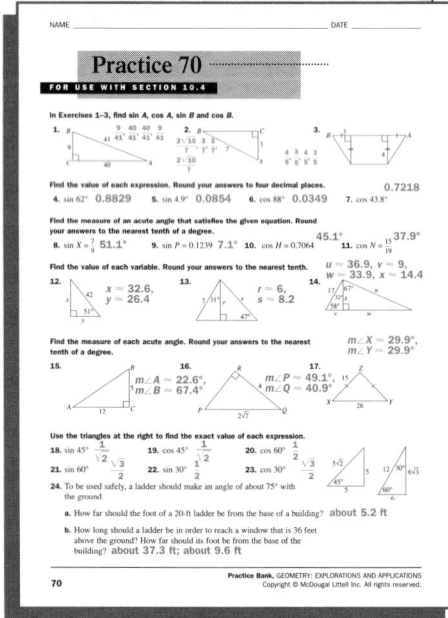

30. Writing Explain how to decide which trigonometric ratio (sine, cosine, or tangent) is best for solving a particular problem.

31. SAT/ACT Preview Suppose that $0° < m \angle X < 45°$. If A = sin X and B = cos X, then:

   A. A > B   **B.** B > A   **C.** A = B

   D. relationship cannot be determined

For Exercises 32 and 33, use △ABC at the right.

32. Prove that each equation is true.

   a. sin A = cos B   **b.** $\tan A = \dfrac{\sin A}{\cos A}$

33. ALGEBRA Prove that $(\sin A)^2 + (\cos A)^2 = 1$.
(*Hint:* Use the Pythagorean theorem.)

34. Cooperative Learning Write a problem that you can solve by using a sine or cosine ratio. Have a classmate solve the problem you wrote.

35. Challenge A weight is suspended from a string attached to two vertical poles, as shown. The heights of the poles are h_1 ft and h_2 ft, the distance between the poles is d ft, and the length of the string is ℓ ft. According to the laws of physics, the weight will come to rest at a position such that $\angle ABC$ and $\angle DBE$ have equal measures. Let $n° =$ the measure of each of these angles. Show that $\sin n° = \dfrac{d}{\ell}$. (This means that $m \angle ABE$ does not depend on the heights of the poles.)

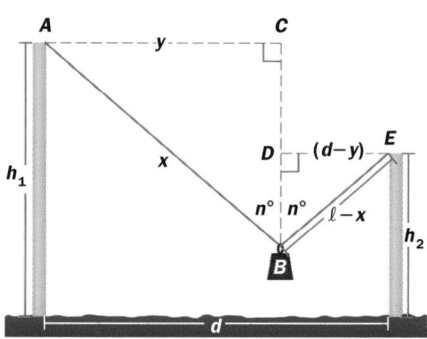

ONGOING ASSESSMENT

36. Open-ended Problem Draw a right triangle. Use a ruler and a protractor to find the length of the hypotenuse and the measure of an acute angle. Use sine and cosine ratios to find the lengths of the triangle's legs. Check your answers by measuring the legs directly.

SPIRAL REVIEW

Find the measure of an acute angle that satisfies the given equation. Round your answers to the nearest tenth of a degree. (*Section 10.3*)

37. tan A = 3   **38.** $\tan R = \dfrac{5}{11}$ **39.** tan X = 1.437

40. Find the volume of a cylinder that has diameter 6 in. and height 8 in. (*Section 7.6*)

Find the coordinates of the midpoint of the segment with each given pair of endpoints. (*Section 4.1*)

41. (0, 0) and (3, 4)   **42.** (−1, 2) and (5, −6)   **43.** (7, 1) and (−3, −5)

520   Chapter 10 *Applying Right Triangles*

---

30. Answers may vary. An example is given. It is usually best to use the ratio that involves the least calculation. Suppose you know the lengths of any two sides of a right triangle and need to find the measure of one of the acute angles. By first using the Pythagorean theorem, you can use any of the three ratios. It would be best to use the two known lengths to avoid calculation errors. If the lengths of two legs are known, use a tangent ratio. If the length of one leg and the hypote-nuse are known, use a sine or cosine ratio. Similarly, given the measure of one acute angle and the lengths of two sides, it would be best to use the ratio involv-ing the known lengths.

31. B

32. a. By definiion, $\sin A = \dfrac{a}{c}$; $\cos B = \dfrac{a}{c}$.

b. By definition, $\tan A = \dfrac{a}{b}$; $\dfrac{\sin A}{\cos A} = \dfrac{\frac{a}{c}}{\frac{b}{c}} = \dfrac{a}{b}$.

33. $(\sin A)^2 + (\cos A)^2 = \left(\dfrac{a}{c}\right)^2 + \left(\dfrac{b}{c}\right)^2 = \dfrac{a^2 + b^2}{c^2} = 1$
(By the Pythagorean theorem, $a^2 + b^2 = c^2$.)

34. Answers may vary. Check students' work.

35. From △ABC, $\sin n° = \dfrac{y}{x}$. From △DBE, $\sin n° = \dfrac{d-y}{l-x}$. Then $\dfrac{y}{x} = \dfrac{d-y}{l-x}$. By a property of proportions, $yl - yx = xd - xy$, so $yl = dx$. Then $\dfrac{y}{x} = \dfrac{d}{l}$, or $\sin n° = \dfrac{d}{l}$.

36. See answers in back of book.

37. 71.6°   **38.** 24.4°   **39.** 55.2°

40. $72\pi \approx 226.2$ in.³   **41.** $\left(\dfrac{3}{2}, 2\right)$

42. (2, −2)   **43.** (2, −2)

10.5 Using Vectors

Learn how to...
- **identify vector quantities and express them in component form**

So you can...
- **use vectors and vector sums to analyze real-world situations**

Orienteering events take place on an established course in a wilderness area. Participants use a map and a magnetic compass to travel between checkpoints.

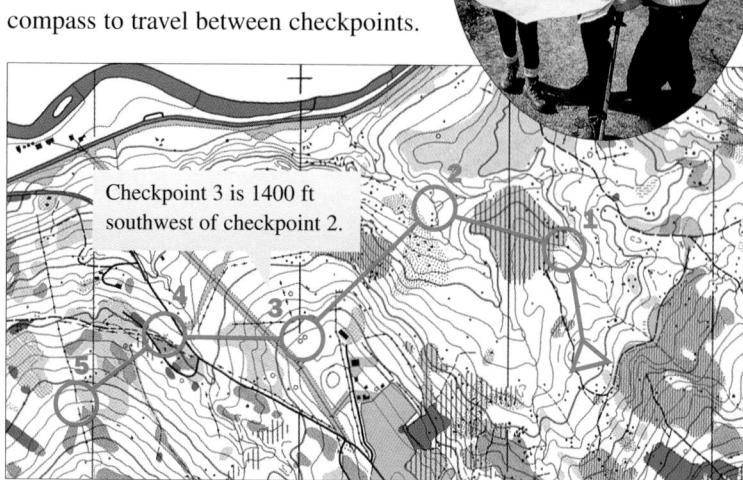

Checkpoint 3 is 1400 ft southwest of checkpoint 2.

In order to plan a course from checkpoint 2 to checkpoint 3, an orienteer needs to know both the direction (southwest) and the distance (1400 ft). This quantity, 1400 ft southwest, is an example of a *vector*.

A **vector** is a quantity that has both *magnitude* (size) and *direction*. A vector is represented by an arrow drawn between two points. For example, the *initial point* of the vector at the right is A, and the *terminal point* is B. The name of the vector is $\overrightarrow{AB}$, read as "vector AB."

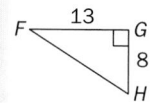

WATCH OUT! ▶

The symbol for a vector looks like the symbol for a ray, but they are *not* the same thing.

By drawing a vector on the coordinate plane, you can easily find its horizontal and vertical components.

The **horizontal component** of $\overrightarrow{CD}$ is 3.

The **vertical component** of $\overrightarrow{CD}$ is 4.

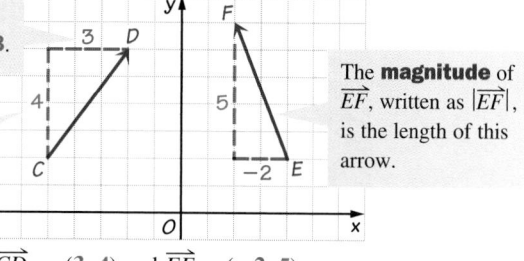

The **magnitude** of $\overrightarrow{EF}$, written as $|\overrightarrow{EF}|$, is the length of this arrow.

In component form, $\overrightarrow{CD} = (3, 4)$ and $\overrightarrow{EF} = (-2, 5)$.

10.5 Using Vectors **521**

Plan⇔Support

Objectives

- Identify vector quantities and express them in component form.
- Use vectors and vector sums to analyze real-world situations.

Recommended Pacing

❖ **Core and Extended Courses**
Section 10.5: 2 days

❖ **Block Schedule**
Section 10.5: 2 half-blocks (with Sections 10.4 and 10.6)

Resource Materials

Lesson Support
Lesson Plan 10.5
Warm-Up Transparency 10.5
Practice Bank: Practice 71
Study Guide: Section 10.5
Explorations Lab Manual: Diagram Masters 1, 2
Challenge Problems: Set 71
Assessment Book: Test 44

Technology
Technology Book: TI-92 Activity 10
Scientific Calculator
Internet: http://www.hmco.com

Warm-Up Exercises

Find each value.

Find each value.
(right triangle with F at top-left, G at top-right, H at bottom-right; $FG = 13$, $GH = 8$)

1. FH about 15.3

2. $m\angle F$ about 31.6°

3. $m\angle H$ about 58.4°

If the figure were placed on a coordinate plane so that point F is at $(1, 10)$, find the coordinates of points G and H.

4. G $(14, 10)$

5. H $(14, 2)$

521

Learning Styles:
Kinesthetic

One way to help kinesthetic learners understand vectors is to provide them with a geoboard and rubber bands. Tell them to mark one place on the rubber band to represent the tip of the vector. Students can then practice modeling vectors and giving their components. They can also use the rubber bands to model scalar multiplication and vector addition.

Section Note

Teaching Tip
Students should understand the following: (1) a vector has a specific magnitude and direction; (2) the coordinate form of a vector does not represent a point in the plane; (3) the notation for the magnitude of a vector is the same as that used for absolute value.

About Example 1

Geometric Thinking
Make sure students understand the note in the Solution of this Example about using 3 instead of –3.

Additional Example 1

Express $\overrightarrow{GH}$ in component form. Find $|\overrightarrow{GH}|$ and the value of w.

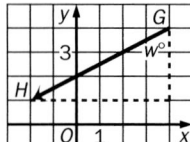

In component form, $\overrightarrow{GH} = (-6, -3)$.

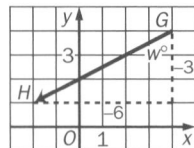

To find $|\overrightarrow{GH}|$, use the Pythagorean theorem. Use positive numbers for the length of each component.

$$|\overrightarrow{GH}| = \sqrt{6^2 + 3^2}$$
$$= \sqrt{45}$$
$$\approx 6.7$$

Use the tangent ratio to find the value of w.

$$\tan w° = \frac{6}{3} = 2$$
$$w = \tan^{-1}(2)$$
$$w \approx 63.4$$

If you know the components of a vector, you can use the Pythagorean theorem to find its magnitude. You can use a trigonometric ratio to find the angle that describes its direction.

EXAMPLE 1

Express $\overrightarrow{PQ}$ in component form. Find $|\overrightarrow{PQ}|$ and the value of z.

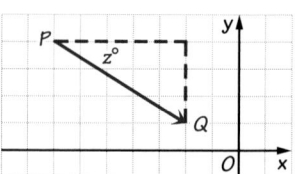

SOLUTION

In component form, $\overrightarrow{PQ} = (5, -3)$.

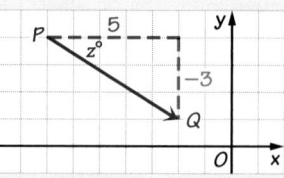

To find $|\overrightarrow{PQ}|$, use the Pythagorean theorem:

$$|\overrightarrow{PQ}| = \sqrt{5^2 + 3^2}$$ ← Use 3, *not* –3, → $$\tan z° = \frac{3}{5}$$
$$= \sqrt{34}$$ because the $$z = \tan^{-1}\left(\frac{3}{5}\right)$$
$$\approx 5.8$$ length of a $$z \approx 31.0$$
                  segment is
                  always positive.

Use the tangent ratio to find the value of z:

The magnitude and direction of a vector are important, but its location is not. Vectors $\overrightarrow{GH}$ and $\overrightarrow{JK}$ are **equal vectors** because they have the same magnitude and direction.

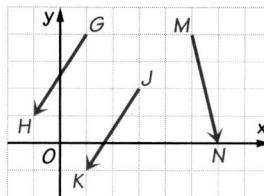

THINK AND COMMUNICATE

Use the graph above.

1. Express each of the three vectors in component form. What do you notice about the components of equal vectors?

2. Give the components of a vector $\overrightarrow{OP}$ that has the same magnitude as $\overrightarrow{MN}$ but a different direction.

3. Give the components of a vector $\overrightarrow{QR}$ that has the same direction as $\overrightarrow{MN}$ but a different magnitude.

ANSWERS   Section 10.5

Think and Communicate

1. $\overrightarrow{GH} = (-2, -3)$; $\overrightarrow{JK} = (-2, -3)$; $\overrightarrow{MN} = (1, -4)$; The horizontal components are equal, as are the vertical components.

2. Answers may vary. An example is given. $(-1, 4)$

3. Answers may vary. An example is given. $(2, -8)$

Multiplying a Vector by a Number

Suppose the vector $\overrightarrow{AB}$ represents the path taken by a plane traveling for 1 hour at a constant velocity. The path of a plane traveling for 2 hours with the same speed and direction as the first plane is represented by $2\overrightarrow{AB}$. The process of multiplying a vector by a real number is called **scalar multiplication**.

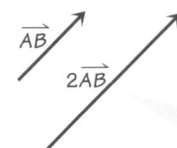

$2\overrightarrow{AB}$ has the same direction as $\overrightarrow{AB}$ and twice the magnitude.

Scalar Multiplication

When a vector is multiplied by a real number k, the length of the vector is multiplied by $|k|$.

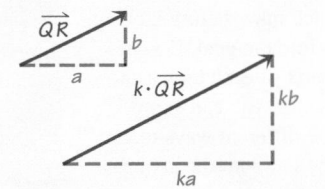

If $\overrightarrow{QR} = (a, b)$, then $k \cdot \overrightarrow{QR} = (ka, kb)$.

EXAMPLE 2   Application: Navigation

A ship travels in a straight path to a location 20 mi west and 15 mi north of its initial point.

a. Graph a vector $\overrightarrow{ST}$ that represents the ship's path. How far did the ship travel?

b. Express $3\overrightarrow{ST}$ in component form. What is the magnitude of $3\overrightarrow{ST}$?

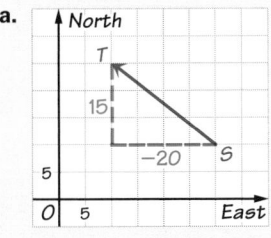

SOLUTION

a.

The distance traveled is the magnitude of the vector.

$$\overrightarrow{ST} = (-20, 15)$$

$$|\overrightarrow{ST}| = \sqrt{20^2 + 15^2}$$

$$= 25$$

The ship traveled 25 mi.

b. $3\overrightarrow{ST} = (3(-20), 3(15))$

$= (-60, 45)$

$|3\overrightarrow{ST}| = |3| \cdot |\overrightarrow{ST}|$

$= 3 \cdot 25$

$= 75$

In component form, $3\overrightarrow{ST} = (-60, 45)$. The magnitude of $3\overrightarrow{ST}$ is 75 mi.

10.5 Using Vectors   **523**

Section Note

 Communication: Discussion
You may wish to point out that since the scalar k is a real number, it can be negative. Then for $k < 0$, the direction of the vector is reversed. You can use algebra to show students that when a vector is multiplied by k, its length is multiplied by $|k|$.

$|k\overrightarrow{QR}| = \sqrt{(ka)^2 + (kb)^2} = \sqrt{k^2a^2 + k^2b^2} = \sqrt{k^2(a^2 + b^2)} = |k|\sqrt{a^2 + b^2} = |k||\overrightarrow{QR}|$

About Example 2

Geometric Thinking
Students can check the Solution to this Example by finding $|3\overrightarrow{ST}|$, using the Pythagorean theorem, and verifying that $|3\overrightarrow{ST}|$ is equal to $3|\overrightarrow{ST}|$.

$|3\overrightarrow{ST}| = \sqrt{60^2 + 45^2} = 75 = 3|\overrightarrow{ST}|$

Additional Example 2

A ship travels in a straight path to a location 25 mi south and 10 mi east of where it began.

a. Graph a vector $\overrightarrow{BP}$ that represents the ship's path. How far did the ship travel?

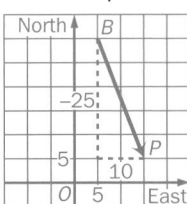

The distance traveled is the magnitude of the vector.

$\overrightarrow{BP} = (10, -25)$

$|\overrightarrow{BP}| = \sqrt{10^2 + 25^2}$

≈ 27

The ship traveled about 27 mi.

b. Express $-2\overrightarrow{BP}$ in component form. What is the magnitude of $-2\overrightarrow{BP}$?

$-2\overrightarrow{BP} = (-2(10), -2(-25))$

$= (-20, 50)$

$|-2\overrightarrow{BP}| = |-2| \cdot |\overrightarrow{BP}|$

$\approx 2 \cdot 27$

≈ 54

In component form, $-2\overrightarrow{BP} = (-20, 50)$. The magnitude of $-2\overrightarrow{BP}$ is about 54 mi.

Think and Communicate

Think and Communicate

These questions focus students' attention on the effect of multiplying a vector by a negative scalar.

Section Notes

Spatial Reasoning
When using a diagram to add vectors, students need to remember to draw the vectors in the sum *tip* to *tail*. That is, the tip of the next vector in the sum should join the tail of the previous vector. The resulting vector is drawn from the tail of the first vector in the sum to the tip of the last vector in the sum.

Visual Thinking
Encourage students to use road maps, navigational maps, trail guides, sketches of playing fields or other locational diagrams to demonstrate the components of a vector and its magnitude. Ask them to show how the vector would be expressed in component form and to determine which parts they would need to estimate to find the values of all the other variables. This activity involves the visual skills of *interpretation* and *exploration*.

About Example 3

Geometric Thinking
Students can check the Solution by using the algebraic definition of vector addition.

Additional Example 3

$\overrightarrow{TR} = (4, -2)$ and $\overrightarrow{RM} = (3, 9)$. Use a graph to find $\overrightarrow{TR} + \overrightarrow{RM}$. Write the resulting vector in component form.
Draw $\overrightarrow{TR}$. Then draw $\overrightarrow{RM}$ beginning at the terminal point of $\overrightarrow{TR}$. The sum is the vector from the initial point of $\overrightarrow{TR}$ to the terminal point of $\overrightarrow{RM}$.

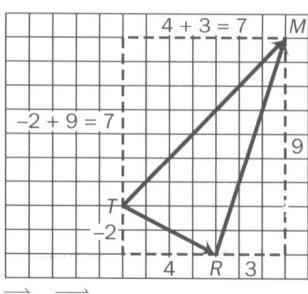

$\overrightarrow{TR} + \overrightarrow{RM} = (7, 7)$

524

THINK AND COMMUNICATE

Use the vector $\overrightarrow{OP} = (3, -2)$.

4. Express $-1\overrightarrow{OP}$ in component form.

5. Compare the magnitudes of $\overrightarrow{OP}$ and $-1\overrightarrow{OP}$.

6. Compare the directions of $\overrightarrow{OP}$ and $-1\overrightarrow{OP}$.

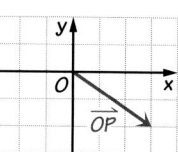

Adding Vectors

Soccer players sometimes have to pass the ball to each other before kicking it into the goal. If you think of each kick as a vector, you can express the different ways to get the ball into the goal as a vector sum.

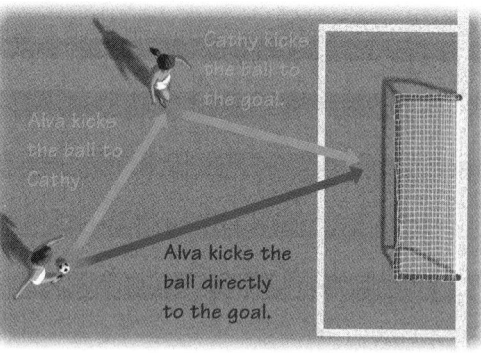

$\overrightarrow{AC} + \overrightarrow{CG} = \overrightarrow{AG}$

Vector Addition

If $\overrightarrow{GH} = (a, b)$ and $\overrightarrow{HK} = (c, d)$, then $\overrightarrow{GH} + \overrightarrow{HK} = (a + c, b + d)$.

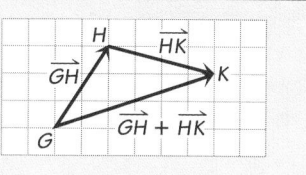

EXAMPLE 3

$\overrightarrow{KL} = (-9, -2)$ and $\overrightarrow{LM} = (3, -3)$. Use a graph to find $\overrightarrow{KL} + \overrightarrow{LM}$. Write the resulting vector in component form.

SOLUTION

Draw $\overrightarrow{KL}$. Then draw $\overrightarrow{LM}$ beginning at the terminal point of $\overrightarrow{KL}$.

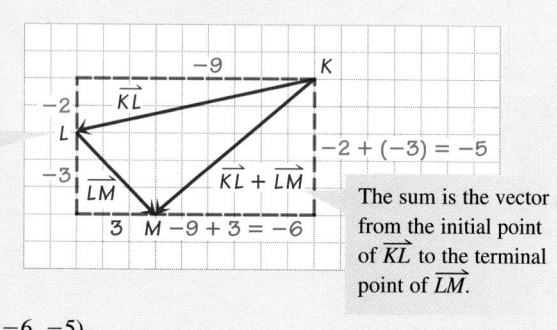

$-2 + (-3) = -5$

The sum is the vector from the initial point of $\overrightarrow{KL}$ to the terminal point of $\overrightarrow{LM}$.

$\overrightarrow{KL} + \overrightarrow{LM} = (-6, -5)$

Think and Communicate

4. $(-3, 2)$

5. $|\overrightarrow{OP}| = |-1\overrightarrow{OP}| = \sqrt{13}$

6. The directions are opposite.

Checking Key Concepts

1–3. Answers may vary.

1. $|\overrightarrow{AB}| = 5$

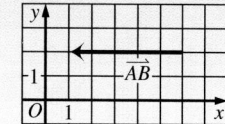

2. $|\overrightarrow{CD}| = 3\sqrt{2} \approx 4.2$

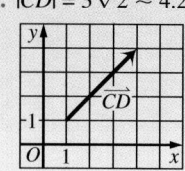

3. $|\overrightarrow{EF}| = 3\sqrt{5} \approx 6.7$

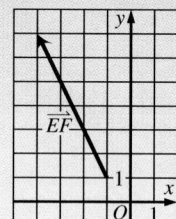

4, 5. Answers are given to the nearest tenth.

4. 26.6   5. 56.3

6–8. See answers in back of book.

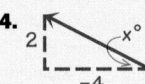

CHECKING KEY CONCEPTS

Graph each vector and find its magnitude.

1. $\overrightarrow{AB} = (-5, 0)$ **2.** $\overrightarrow{CD} = (3, 3)$ **3.** $\overrightarrow{EF} = (-3, 6)$

Find the value of each variable.

4. 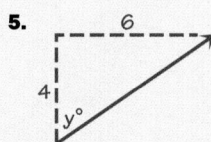   **5.**

6. In component form, $\overrightarrow{JK} = (3, 7)$. Graph $2\overrightarrow{JK}$ and express $2\overrightarrow{JK}$ in component form.

7. $\overrightarrow{WX} = (-12, 7)$ and $\overrightarrow{YZ} = (5, -21)$. Use these components to find $\overrightarrow{WX} + 4\overrightarrow{YZ}$.

8. $\overrightarrow{NP} = (1, 3)$ and $\overrightarrow{PQ} = (-5, 1)$. Use a graph to find $\overrightarrow{NP} + \overrightarrow{PQ}$ and write the resulting vector in component form.

10.5 │ Exercises and Applications

Extra Practice
exercises on
page 682

Express each vector in component form and find the value of each variable.

1. 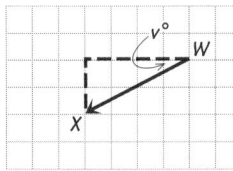   **2.** 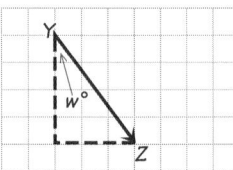   **3.**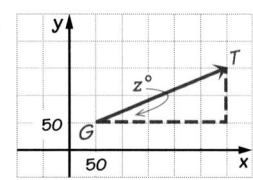

Graph each vector and find its magnitude to the nearest tenth.

4. $\overrightarrow{HJ} = (10, 4)$ **5.** $\overrightarrow{KL} = (-3, 8)$ **6.** $\overrightarrow{MN} = (-4, -4)$ **7.** $\overrightarrow{PQ} = (7, -5)$

8. Graph each scalar multiple of $\overrightarrow{OQ}$ and find its magnitude to the nearest tenth.

  a. $2\overrightarrow{OQ}$

  b. $\frac{1}{2}\overrightarrow{OQ}$

  c. $-3\overrightarrow{OQ}$

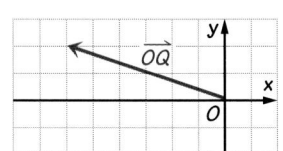

In component form, $\overrightarrow{AB} = (-3, 3)$, $\overrightarrow{CD} = (5, 4)$, and $\overrightarrow{EF} = (-2, 6)$. For Exercises 9–14, use the components to find each vector sum.

9. $\overrightarrow{AB} + \overrightarrow{CD}$ **10.** $\overrightarrow{AB} + \overrightarrow{EF}$ **11.** $\overrightarrow{CD} + \overrightarrow{EF}$

12. $2\overrightarrow{AB} + \overrightarrow{CD}$ **13.** $\overrightarrow{AB} + 2\overrightarrow{CD}$ **14.** $-2\overrightarrow{CD} + \overrightarrow{EF}$

Exercises and Applications

1–3. Values of the variables are given to the nearest tenth.

1. $(-4, -2)$; 26.6

2. $(3, -4)$; 36.9

3. $(250, 100)$; 21.8

4. $|\overrightarrow{HJ}| = 2\sqrt{29} \approx 10.8$ 5. $|\overrightarrow{KL}| = \sqrt{73} \approx 8.5$

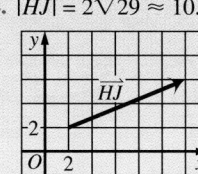   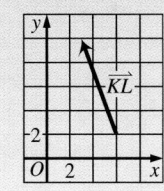

6–8. See answers in back of book.

9. $(2, 7)$

10. $(-5, 9)$

11. $(3, 10)$

12. $(-1, 10)$

13. $(7, 11)$

14. $(-12, -2)$

Checking Key Concepts

Geometric Thinking
Students should understand that the vectors in questions 1–3 can be drawn with any point as the initial point. You can use questions 6–8 to discuss the advantages and disadvantages of the graphic and algebraic methods for finding the resulting vector under scalar multiplication or vector addition.

Closure Question

How is the vector (4, –3) different from the point (4, –3)?   The point (4, –3) is a location in the coordinate plane that is 4 units right of the origin and 3 units below the origin. The vector is not a point but a quantity that has both magnitude and direction. The components describe the difference between the initial point and terminal point of the vector. In this case, the terminal point is 4 units horizontally from the initial point and 3 units down from the initial point.

Apply⇔Assess

Suggested Assignment

❖ **Core Course**
  Day 1 Exs. 1–20
  Day 2 Exs. 22–25, 27, 30–37, AYP

❖ **Extended Course**
  Day 1 Exs. 1–20
  Day 2 Exs. 21–37, AYP

❖ **Block Schedule**
  Day 63 Exs. 1–20
  Day 64 Exs. 22–25, 27, 30–37, AYP

Exercise Notes

Common Error
Exs. 1–7 In Ex. 1–3, students need to pay attention to the direction of the arrow when determining the sign of the components. Also, in Exs. 1–7, some students may forget to take the absolute value of the components before using them in a tangent ratio or in the Pythagorean theorem. Students who make these types of errors should review Example 1 on page 522.

Exercise Notes

ORIENTEERING Luis must pass through either checkpoint *Q* or checkpoint *R* on his way from checkpoint *P* to checkpoint *S*. Use the map below for Exercises 15 and 16.

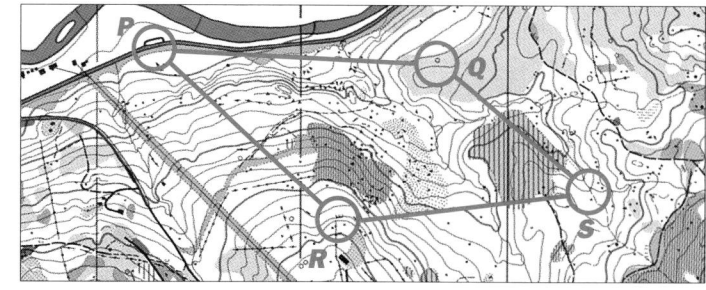

15. The vectors on the map have these components (in feet):
$\vec{PQ} = (2200, -100)$ $\vec{QS} = (1400, -1000)$ $\vec{PR} = (1500, -1300)$ $\vec{RS} = (2100, 200)$
   a. Express $\vec{PQ} + \vec{QS}$ in component form.
   b. Express $\vec{PR} + \vec{RS}$ in component form.
   c. Compare your answers to parts (a) and (b). Explain how these answers can be the result of two different vector sums.

16. a. Find $|\vec{PQ}|$ and $|\vec{QS}|$. How far will Luis travel if he passes through checkpoint *Q*?
   b. Find $|\vec{PR}|$ and $|\vec{RS}|$. How far will Luis travel if he passes through checkpoint *R*?
   c. Compare your answers to parts (a) and (b). Which route is shorter?
   d. **Open-ended Problem** What other factors, besides distance, do you think an orienteer might consider when choosing a path?

Connection   GEOLOGY

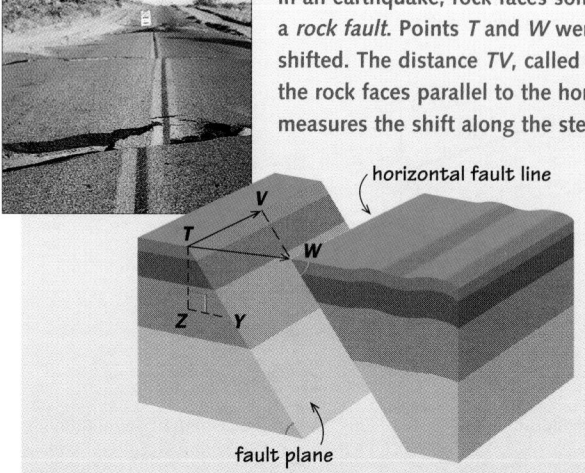

In an earthquake, rock faces sometimes move past each other along a *rock fault*. Points *T* and *W* were the same point before the rocks shifted. The distance *TV*, called the *strike slip*, measures the shift of the rock faces parallel to the horizontal fault line. The *dip-slip*, *VW*, measures the shift along the steepest line of the fault plane.

horizontal fault line

fault plane

17. If the strike-slip is 72 ft and the dip-slip is 56 ft, find the magnitude of the total shift, $\vec{TW}$, to the nearest foot.

18. If *TY* = 56 ft and *TZ* = 43 ft, find $m \angle TYZ$, the angle at which the fault plane is inclined to the horizontal.

---

15. a. (3600, −1100)
   b. (3600, −1100)
   c. Both represent paths from checkpoint *P* to *S*; any path from checkpoint *P* to *S* can be represented by a vector with initial checkpoint *P* and terminal checkpoint *S*.

16. a–c. Answers are given to the nearest foot.
   a. 2202 ft; 1720 ft; 3922 ft
   b. 1985 ft; 2110 ft; 4095 ft

   c. the route through checkpoint *Q*
   d. Answers may vary. Examples are given. the difficulty of the terrain or the view

17. 91 ft   18. about 50.2°

19. a. $\vec{AB} + \vec{CD} = (9, 0)$

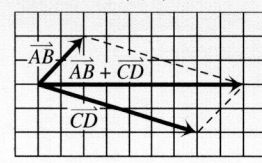

b. $\vec{AB} + \vec{CD} = (9, 0)$

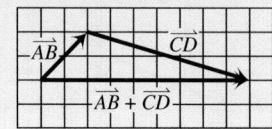

c. In the parallelogram method, the length of the side opposite $\vec{CD}$ equals $|\vec{CD}|$. Since $\vec{AB}$ is the same in both diagrams, the diagonal of the parallelogram has the same

length as the third side of the triangle in the second diagram.

20. Answers may vary. An example is given.

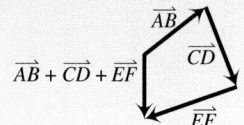

$\vec{AB} + \vec{CD} + \vec{EF}$

19. You can use the *parallelogram method* to add two vectors. Draw both vectors with the same initial point, as shown. Then draw two segments to complete a parallelogram. The sum of the vectors lies along a diagonal of the parallelogram and has the same initial point as the original vectors.

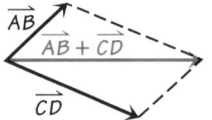

   a. In component form, $\overrightarrow{AB} = (2, 2)$ and $\overrightarrow{CD} = (7, -2)$. Use the parallelogram method to find $\overrightarrow{AB} + \overrightarrow{CD}$.

   b. Find $\overrightarrow{AB} + \overrightarrow{CD}$ using the method used in Example 3 on page 524.

   c. Prove that the methods you used in parts (a) and (b) both yield the same result.

20. **Visual Thinking** Suppose you want to add three vectors. Sketch what this might look like.

21. **Open-ended Problem** Give an example of two different vector sums that yield the same result.

22. Suppose A and B are points on the coordinate plane. Prove that $\overrightarrow{BA} = -1\overrightarrow{AB}$.

The magnitude and direction of a vector are labeled on each diagram. Find each unknown length. Then express the vector in component form.

23.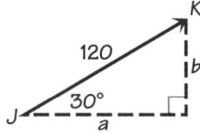

24.

25.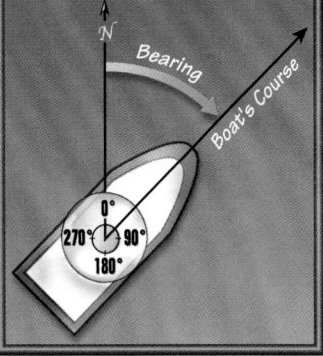

26. **Writing** Given two points W and Z, explain how the ray $\overrightarrow{WZ}$ and the vector $\overrightarrow{WZ}$ are different.

27. **AVIATION** In the diagram at the right, $\overrightarrow{VL}$ is an airplane's *velocity vector*. The horizontal component, g, is called the *ground speed*, and the vertical component, r, is the *rate of climb*.

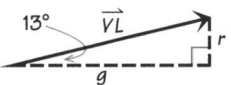

   a. Suppose the airplane takes off at an angle of 13° to the horizontal, with an air speed of 225 mi/h. Find the airplane's ground speed in miles per hour.

   b. Find the rate of climb in miles per hour for the airplane in part (a).

NAVIGATION The *bearing* of a ship or airplane gives its direction of travel. The bearing is the angle, measured clockwise from north, that the velocity vector makes with a vector that points due north.

28. A ship's velocity vector is $\overrightarrow{TU} = (10, 12)$. Draw $\overrightarrow{TU}$ and find its bearing.

29. **Challenge** In miles per hour, an airplane's velocity vector is $(-180, -130)$. Find the plane's bearing and speed.

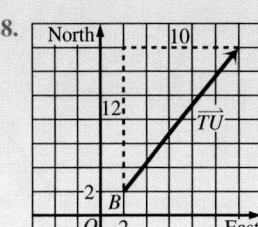

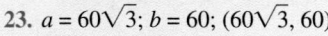

21. Answers may vary. An example is given.
$(6, -2) + (-4, 1) = (2, -1)$ and $(3, 4) + (-1, -5) = (2, -1)$

22. Let $\overrightarrow{BA} = (x, y)$. Then $\overrightarrow{AB} = (-x, -y)$. $-1(-x, -y) = (x, y) = \overrightarrow{BA}$. Therefore, $\overrightarrow{BA} = -1\overrightarrow{AB}$.

23. $a = 60\sqrt{3}$; $b = 60$; $(60\sqrt{3}, 60)$

24. $c \approx 7.2$; $d \approx 15.4$; $(-7.2, 15.4)$

25. $e = f = 2\sqrt{2}$; $(-2\sqrt{2}, -2\sqrt{2})$

26. The ray $\overrightarrow{WZ}$ has a fixed location in the coordinate plane; it has an endpoint at W and passes through Z. It is infinitely long. The vector $\overrightarrow{WZ}$ has two endpoints, is not infinitely long, and has no fixed location in the coordinate plane.

27. a. about 219.2 mi/h
   b. about 50.6 mi/h

28.

$\tan B = \dfrac{10}{12} = 0.8333;$

$m\angle B = 39.8°$

29. about 234.2°; about 222.0 mi/h

Apply⟺Assess

Exercise Notes

Alternate Approach
Ex. 19 This exercise presents the parallelogram method for adding vectors. After completing the exercise, you may wish to discuss with students why this method is called the parallelogram method and what properties of a parallelogram allow the method to work.

Communication: Writing
Ex. 26 Students often confuse rays and vectors since the rotations and visual representations are similar. Completing this exercise should help students to clarify and separate these two concepts.

Application
Exs. 27-29 These exercises show students that a vector can represent a quantity that is not a distance. The vector $\overrightarrow{VL}$, for example, represents velocity, with the magnitude of the vector representing the speed, and the direction of the vector representing the direction of the motion.

Problem Solving
Exs. 28, 29 Students need to find the angle formed by the vector and its components in each of these exercises. They then use this information to find the bearing. Stress that the bearing is measured clockwise from north.

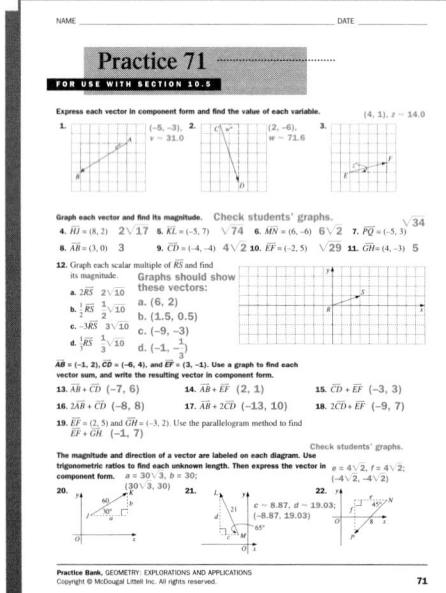
ONGOING ASSESSMENT

30. **Cooperative Learning** On paper, sketch vectors showing several paths in your classroom. For example, you might show the path from your desk to another desk, or from your teacher's desk to the door. Have another person use the vectors to move around the classroom.

SPIRAL REVIEW

Find the measure of the acute angle that satisfies each equation. Round your answers to the nearest tenth of a degree. *(Sections 10.3 and 10.4)*

31. $\sin D = \dfrac{5}{9}$ 32. $\tan Z = 3.461$ 33. $\cos G = 0.1096$

34. $\sin R = \dfrac{\sqrt{3}}{2}$ 35. $\tan J = 1$ 36. $\cos B = \dfrac{3}{5}$

37. What is the formula for the volume of a prism? *(Section 7.6)*

ASSESS YOUR PROGRESS

VOCABULARY

tangent of an angle (p. 507)   vector (p. 521)
angle of elevation (p. 509)   magnitude of a vector (p. 521)
angle of depression (p. 509)   equal vectors (p. 522)
sine (p. 515)   scalar multiplication (p. 523)
cosine (p. 515)

Find the measure of the acute angle that satisfies each equation. Round your answers to the nearest tenth of a degree. *(Sections 10.3 and 10.4)*

1. $\tan X = 13.73$ 2. $\sin A = 0.2$ 3. $\cos K = 0.7501$

Find the value of each variable. Round your answers to the nearest tenth. *(Sections 10.3 and 10.4)*

4. 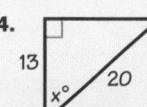   5. 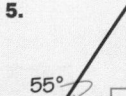   6.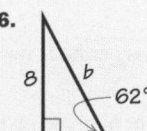

Graph each vector and find its magnitude. *(Section 10.5)*

7. $\overrightarrow{PQ} = (-7, 3)$ 8. $\overrightarrow{RS} = (25, 40)$

In component form, $\overrightarrow{AB} = (-4, -5)$, $\overrightarrow{CD} = (7, -2)$, and $\overrightarrow{EF} = (3, 1)$. Graph each vector and express it in component form. *(Section 10.5)*

9. $2\overrightarrow{AB}$ 10. $\overrightarrow{AB} + \overrightarrow{CD}$ 11. $-3\overrightarrow{EF} + \overrightarrow{CD}$

12. **Journal** Sketch a right triangle. Label the length of the hypotenuse and the length of one leg. Explain how to find the measures of both acute angles of the triangle.

528   Chapter 10 *Applying Right Triangles*

---

30. Check students' work.

31. 33.7°   32. 73.9°   33. 83.7°

34. 60.0°   35. 45.0°   36. 53.1°

37. $V = Bh$, where B is the area of a base and h is the height of the prism.

Assess Your Progress

1. 85.8°   2. 11.5°

3. 41.4°   4. 49.5

5. 8.6   6. 9.1

7. $|\overrightarrow{PQ}| = \sqrt{58} \approx 7.6$

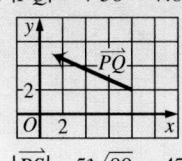

8. $|\overrightarrow{RS}| = 5\sqrt{89} \approx 47.2$

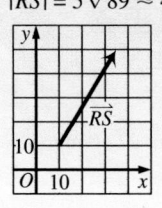

9. $(-8, -10)$

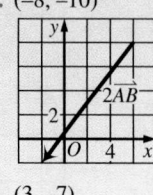

10. $(3, -7)$

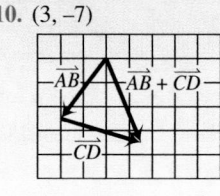

11. $(10, -1)$

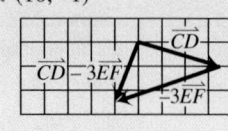

12. Answers may vary. An example is given.

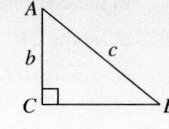

To find $m \angle A$, first find $\dfrac{b}{c}$.

Since $\cos A = \dfrac{b}{c}$, $m \angle A = \cos^{-1} \dfrac{b}{c}$ and $m \angle B = 90° - m \angle A$. Alternatively, to find $m \angle B$, first find $\dfrac{b}{c}$. Since $\sin B = \dfrac{b}{c}$, $m \angle B = \sin^{-1} \dfrac{b}{c}$ and $m \angle A = 90° - m \angle B$.

528

10.6 Areas and Trigonometry

Learn how to...
- **use trigonometry to find areas of polygons and volumes of prisms**

So you can...
- **find areas of real-world objects such as barns**

When you think of a barn, you probably think of a rectangular building. In the early part of the twentieth century, however, barns of many different shapes were built in parts of the United States. A barn in the shape of a regular 12-gon was built in northern Indiana in 1912. The diameter of the barn is 56 ft.

EXAMPLE 1   Application: History

Find the area of the floor of the barn described above.

SOLUTION

The 12-gon can be divided into 12 congruent isosceles triangles. The vertex angle of each triangle measures 30°, so each base angle measures 75°.

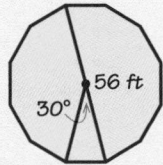

Use trigonometry to find the height and base of each triangle.

$$\sin 75° = \frac{a}{28}$$

$$28 \sin 75 = a$$

$$27.0 \approx a$$

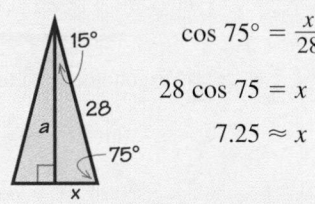

$$\cos 75° = \frac{x}{28}$$

$$28 \cos 75 = x$$

$$7.25 \approx x$$

The apothem of the 12-gon is about 27 ft and each side length is about $14\frac{1}{2}$ ft. Use these values in the formula for the area of a regular polygon.

$$A = \frac{1}{2}ap$$

$$\approx \frac{1}{2}(27)(174)$$

The perimeter of the barn is about $12 \cdot 14\frac{1}{2} = 174$.

$$\approx 2349$$

The area of the floor of the barn is about 2350 ft².

Objectives
- Use trigonometry to find areas of polygons and volumes of prisms.
- Find areas of real-world objects using trigonometry.

Recommended Pacing
❖ **Core and Extended Courses**
  Section 10.6: 1 day
❖ **Block Schedule**
  Section 10.6: $\frac{1}{2}$ block
  (with Section 10.5)

Resource Materials

Lesson Support
Lesson Plan 10.6
Warm-Up Transparency 10.6
Practice Bank: Practice 72
Study Guide: Section 10.6
Challenge Problems: Set 72

Technology
Technology Book:
  Calculator Activity 13
Scientific Calculator
Graphing Calculator
Internet:
  http://www.hmco.com

Warm-Up Exercises

Find the area of each figure.
1. a regular hexagon with apothem $4\sqrt{3}$ and side length 8 $96\sqrt{3}$, or about 166.3
2. a parallelogram with height 3 and base 11   33
3. a trapezoid with height 6 and bases of length 4 and 9   39

Find the value of each variable.
4.   7.5

5.   3.5

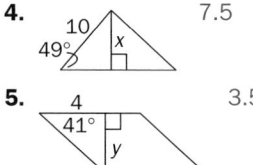

Additional Example 1

A barn is in the shape of a regular octagon with a 56 ft diameter. Find the area of the barn floor.   The octagon can be divided into 8 congruent isosceles triangles. The vertex angle of each triangle measures 45°, so each base angle measures 67.5°.

Use trigonometry to find the height and base of each triangle.

$$\sin 67.5° = \frac{a}{28}$$
$$28 \sin 67.5 = a$$
$$26.0 \approx a$$
$$\cos 67.5° = \frac{x}{28}$$
$$28 \cos 67.5 = x$$
$$10.7 \approx x$$

The height of the triangle, which is also the apothem of the polygon, is about 26 ft, and the length of one side of the barn is about 21.4 ft. The perimeter of the barn is $8 \cdot 21.4 \approx 171$ ft. Use these values in the formula for the area of a regular polygon.

$$A = \frac{1}{2}ap \approx \frac{1}{2}(26)(171)$$
$$\approx 2223$$

The area of the floor of the barn is about 2223 ft².

Additional Example 2

Find the area of this parallelogram.

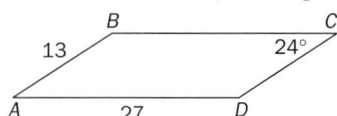

As the opposite angles of a parallelogram are congruent, $m\angle A = 24°$. Draw an altitude from B to $\overline{AD}$.

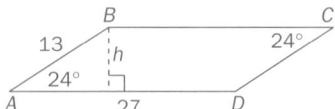

Use trigonometry to find h.

$$\sin 24° = \frac{h}{13}$$
$$13 \sin 24 = h$$
$$5.29 \approx h$$

Use this value in the formula for the area of a parallelogram.
$$A = bh \approx (27)(5.29)$$
$$\approx 142.8$$

The area of the parallelogram is approximately 143.

THINK AND COMMUNICATE

Look back at Example 1.

1. Explain how dividing the floor into triangles is an important step in the solution.

2. Describe two different ways to find the angle measures of each isosceles triangle in the barn.

As Example 1 shows, sometimes you can find the area of a figure, even if the measures that you need are not given. You can often use trigonometry to find these missing dimensions.

EXAMPLE 2

Find the area of this rhombus.

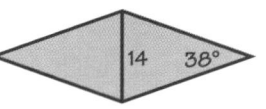

SOLUTION

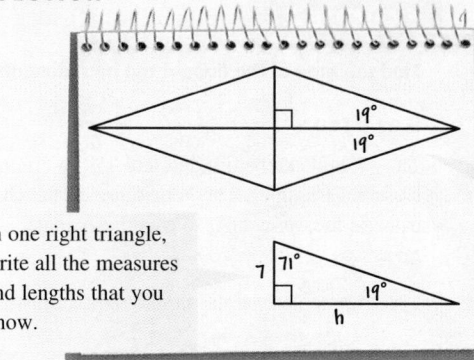

In one right triangle, write all the measures and lengths that you know.

Create four congruent right triangles by drawing the other diagonal.

Use trigonometry to find the value of h.

$$\tan 71° = \frac{h}{7}$$
$$7 \tan 71 = h$$
$$20.33 \approx h$$

The area of the rhombus is four times the area of one right triangle.

$$\text{Area of rhombus} = 4 \cdot \frac{1}{2}bh$$
$$\approx 4 \cdot \frac{1}{2}(7)(20.33)$$
$$\approx 284.6$$

Substitute **7** for b and **20.33** for h.

The area of the rhombus is approximately 285.

ANSWERS   Section 10.6

Answers may vary slightly due to rounding.

Think and Communicate

1. Dividing into triangles produces isosceles and right triangles, which allows you to use trigonometric ratios.

2. (1) The 12-gon is regular, so it is equiangular. The sum of the angle measures is $(12 - 2)(180°)$, so the measure of each angle is $\frac{1800°}{12} = 150°$. Then the measure of each base angle of an isosceles triangle is $\frac{150°}{2} = 75°$ and the measure of each vertex angle is $180° - 2(75°) = 30°$. (2) The sum of the measures of the vertex angles is 360°, so the measure of each is $\frac{360°}{12} = 30°$. Then the measure of each base angle is $\frac{1}{2}(180° - 30°) = 75°$.

EXAMPLE 3

Each base of this prism is an isosceles trapezoid. Find the volume of the prism.

5 in.

48°

8 in.

13 in.

SOLUTION

Step 1 Find the height of the trapezoid.

$$\tan 48° = \frac{h}{4}$$

$$4 \tan 48 = h$$

$$4.44 \approx h$$

5

h

48°

4   5   4

These segments create two congruent right triangles.

Then use the formula for the area of a trapezoid.

$$\text{Area of base} = \frac{1}{2}(b_1 + b_2)h$$

$$\approx \frac{1}{2}(5 + 13)(4.44)$$

$$\approx 40$$

Step 2 Find the volume of the prism.

$$\text{Volume} = (\text{Area of base}) \times (\text{Height of prism})$$

$$\approx 40 \cdot 8$$

$$\approx 320$$

The volume of the prism is about 320 in.^{3}.

☑ CHECKING KEY CONCEPTS

Find the area of each polygon.

1.
12
59°

2.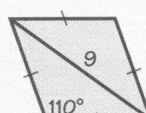
9
110°

3.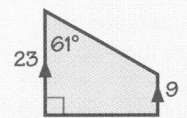
23 61°
9

4. a regular decagon with side length 10 cm

5. a regular pentagon with apothem 8 in.

6. an isosceles triangle with a vertex angle of 40° and leg length 3 ft

10.6 Areas and Trigonometry **531**

Additional Example 3

Each base of this prism is a regular pentagon. Find the volume of the prism.

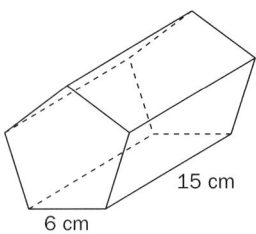

15 cm

6 cm

The pentagon can be divided into 5 congruent triangles with vertex angle 72° and base angles 54°.

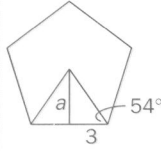

a   54°

3

Step 1 Find the apothem.

$$\tan 54° = \frac{a}{3}$$

$$3 \tan 54 = a$$

$$4.1 \approx a$$

Step 2 The perimeter of the base is $5 \cdot 6 = 30$. Use the formula for the area of a regular polygon.

$$A = \frac{1}{2} ap$$

$$\approx \frac{1}{2}(4.1)(30)$$

$$\approx 61.5$$

Step 3 Find the volume of the prism.

Volume = (Area of base) × (Height of prism)

$$\approx (61.5)(15)$$

$$\approx 922.5$$

The volume of the prism is about 922.5 cm^3.

Section Note

Visual Thinking
Ask students to find a room, building, outdoor area, or other space with an unusual shape. Encourage them to create a sketch of the perimeter of the floor of that space, and to measure or estimate as many sides or angles as necessary to find the area of the floor using trigonometry. Ask them to discuss their sketches with the class. This activity involves the visual skills of *interpretation* and *generalization*.

Checking Key Concepts

1–6. Answers are given to the nearest tenth.

1. 43.2

2. 30.3

3. 404.8

4. 770.0 cm^2

5. 232.0 in.^{2}

6. 2.8 ft^2

Without using trigonometry, what information must you know about a regular polygon to find its area? How does trigonometry help you find the area? *Without using trigo-nometry, you must know two of these three lengths: the apothem, the distance from the center to a vertex, or the length of a side. Using trigonometry, you need to know only one of the lengths mentioned above because you always know the angles of the congruent triangles that form the regular polygon.*

Apply⇔Assess

Suggested Assignment

❖ **Core Course**
   Exs. 1–11, 13–17, 22–25, 27–31
❖ **Extended Course**
   Exs. 1–31
❖ **Block Schedule**
   Day 64 Exs. 1–11, 13–17, 22–25, 27–31

Exercise Notes

Problem Solving
Exs. 1–6 When solving problems such as these, students can follow a systematic approach. First, they should identify the type of figure given and write the formula for its area. Then, they should look to see what information is given and what they need to know. Next, students should find the missing information using a trigonometric ratio. Finally, they substitute the information into the area formula to find the area of the polygon.

Student Study Tip
Exs. 7–10 After students sketch each figure, they should highlight one of the congruent triangles in the figure. If students are unsure how to proceed, they should use Example 1 on page 529 as a model.

 Communication: Drawing
   Ex. 14 This exercise demonstrates a method used to find the area of composite figures, namely, dividing the original figure into other figures whose areas can be found and then adding those areas. You may wish to put some composite figures on the board and have students find their areas.

10.6 **Exercises and Applications**

Extra Practice exercises on page 682

Find the area of each polygon.

1.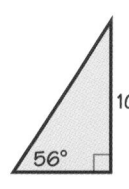

2.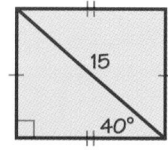

3.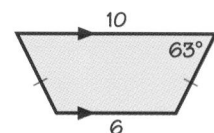

4.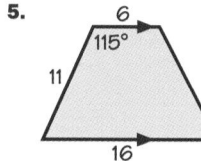

5.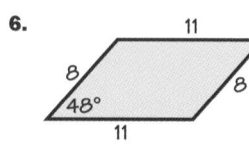

6.

For Exercises 7–10, sketch each figure described. Then find its area.

7. The apothem of a regular pentagon is 12 cm.

8. A regular 9-gon has radius 5 mm.

9. A parallelogram has 10 m and 50 m side lengths and a 25° angle.

10. The angle measures of a trapezoid are 90°, 90°, 105°, and 75°. The bases are 8 yd and 9 yd long.

11. In the diagram at the right, $m \angle ABC = 115°$.
   a. Find BD and AD.
   b. Use BD and a trigonometric ratio to find DC.
   c. Use your answers to parts (a) and (b) to find the area of $\triangle ABC$.

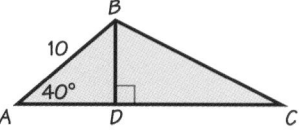

12. Open-ended Problem Draw any $\triangle ABC$. Measure $\angle A$, $\angle B$, and $\overline{AB}$. Then repeat parts (a)–(c) of Exercise 11 to find the area of $\triangle ABC$.

13. Investigation You will need a ruler and a protractor. Copy this graph.
   a. Choose any point on the line and draw a perpendicular segment from the point to the *x*-axis. Measure the length of the segment.
   b. Use trigonometry and the length you found in part (a) to find the area of the triangle whose sides are the line, the *x*-axis, and the perpendicular segment.
   c. Graph another line that has a positive slope. Measure the angle that the line makes with the *x*-axis and repeat parts (a) and (c).

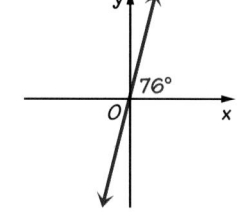

14. Writing Describe two different ways to find the area of this trapezoid. Include diagrams for each method.

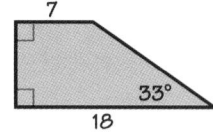

Exercises and Applications

Answers are given to the nearest tenth.

1. 33.5   2. 110.4
3. 31.2   4. 308.0
5. 110.0   6. 64.9

7–10. Check students' work.

7. 522 cm²   8. 71.9 mm²
9. 210 m²   10. 31.7 yd²

11. a. 6.4; 7.7

b. 13.7
c. 68.5

12. Check students' work.

13. a. Answers may vary. An example is given.

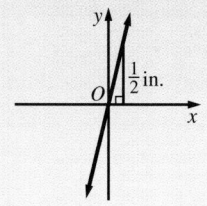

b. about 0.025 in.²
c. Answers may vary.

14. First, use $\tan 33° = \dfrac{h}{11}$ to determine h ($h \approx 7.1$). Then find the area of the trapezoid using the formula $A = \dfrac{1}{2}h(b_1 + b_2) \approx 88.75$.

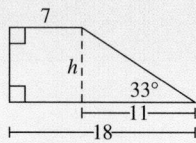

Find the volume of each prism.

15.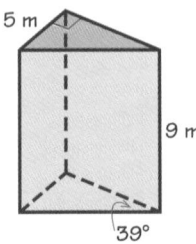
5 m
9 m
39°

16.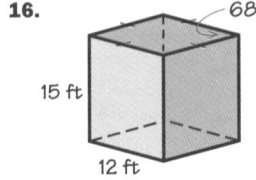
68°
15 ft
12 ft

17.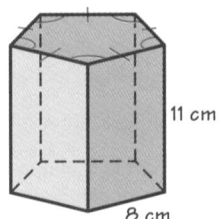
11 cm
8 cm

Connection ▷ HISTORY

In 1792, George Washington designed a special barn for his new wheat treading process. Horses walked on the wheat on the second level of the barn. Their hooves separated the straw from the grain, which fell through grooves in the floor to the first level. In 1996, this barn, a regular 16-gon, was reconstructed in Mount Vernon, Virginia.

George Washington's original plans

The reconstructed barn

18. Each side of the barn is 10 ft 3 in. Find the area of the barn floor.

19. At the center of each level of the barn is a grain storage room in the shape of a regular octagon. The wheat from the field was stored in the upper level storage room until the horses were brought in to walk on it. Each side of the room is about 10 ft 5 in. Find the area of the storage room floor.

20. The horses walked around the outside of the storage room. Use your answers to Exercises 18 and 19 to find the area of the floor where they walked on the wheat.

21. **Writing** Compare the area of the storage room floor to the area of the floor where the horses walked. Why might Washington have been interested in comparing these areas?

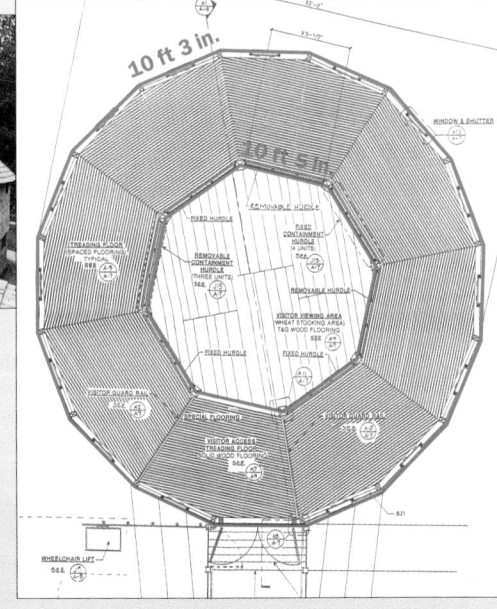
10 ft 3 in.
10 ft 5 in.

The modern plans for the reconstructed barn

10.6 Areas and Trigonometry **533**

Exercise Notes

Using Technology
Ex. 13 Students can check the areas they found in this exercise by using a TI-82 graphing calculator. The calculator will give an approximate result. Recall that the slope of the given line is equal to tan 76°, or about 4 (see Ex. 29, page 513). The equation of the line is thus approximated by $y = 4x$. Enter this equation on the Y= list. Set the window values to Xmin = −9.4, Xmax = 9.4, Xscl = 1, Ymin = −20, Ymax = 20, Yscl = 5. Press GRAPH and then TRACE when the graph appears. The window values chosen cause the x-coordinates of the points traced to be convenient values. Have students trace to the point they used on their graph. Note that the absolute value of the y-coordinate of the point should be approximately equal to the length measured by the student. Press 2nd [CALC] and choose 7:∫f(x)dx. If the point chosen was to the left of the y-axis, press ENTER, trace to the origin, and press ENTER again. The absolute value of the number given is the area of the shaded triangle. If the point chosen is to the right of y-axis, after choosing option 7 from the CALC menu, move the cursor to the origin and press ENTER, then move the cursor to the point and press ENTER again.

Geometric Thinking
Exs. 15–17 Students can begin these exercises by identifying the base of the prism. They should then draw the base and use trigonometry to find its area.

Interdisciplinary Problems
Exs. 18–21 In these exercises, students use what they have learned about finding the areas of regular polygons to analyze a barn designed by George Washington.

Research
Exs. 18–21 Ask interested students to find out how the straw and grain are separated by modern wheat farmers.

Or, find the area of the square and the right triangle shown in the figure below and add to find the area of the trapezoid.

$A \approx 7 \cdot 7 + \frac{1}{2}(7.1)(11) = 88.05$

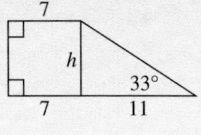
7
h
7   11
33°

15. 138.9 m³
16. 2002.7 ft³
17. 1211.2 cm³
18. 2112.7 ft²
19. 524.3 ft²
20. 1588.4 ft²

21. The area of the storage room floor is about one fourth the area of the floor where the horses walked. Answers may vary. An example is given. I think it would be important to know whether there was enough wheat in the storage area to cover the walking area to a reasonable depth. Knowing that, Washington could determine when the horses should be brought in to walk on the wheat.

Exercise Notes

Integrating Algebra
Ex. 22 In this exercise, students use algebra, geometry, and trigonometry to write an alternate formula for finding the area of a triangle. Ask students to describe a situation in which this formula would be useful. (when you know two side lengths and the measure of the included angle of the triangle)

Assessment Note
Ex. 27 This exercise provides a good opportunity to have students assess each other's work. Assign each student a partner. They should exchange papers, read each other's summary, and then describe any changes that can be made to improve the summary.

Practice 72 for Section 10.6

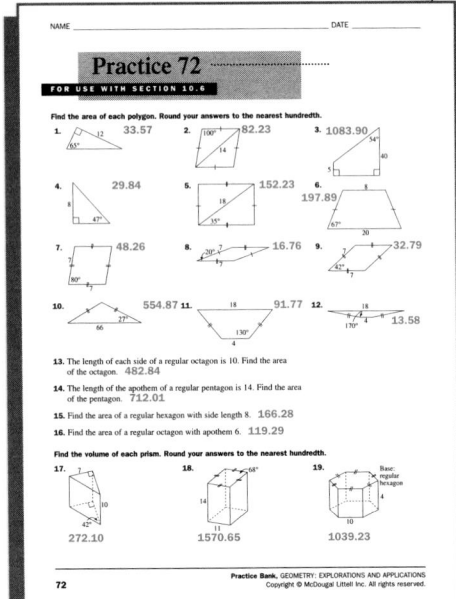

22. ALGEBRA In this section, you learned how to use trigonometry to find the areas of right triangles. You can also use trigonometry to find the area of a non-right triangle. Use △ABC at the right.

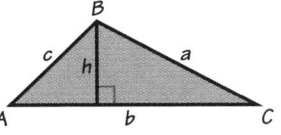

a. Complete: $\sin A = \frac{?}{?}$.

b. Solve the equation that you wrote in part (a) for h. Use the result to write an expression for the area of △ABC.

c. The formula that you found in part (b) can be used to find the area of a triangle when two side lengths and the measure of the included angle are known. Use this formula to find the area of the triangle below.

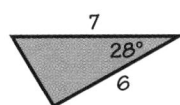

23. Prove that the area of △XYZ with a right angle at Z is $A = \frac{1}{2}XZ^2 \tan X$.

24. Prove that the area of an isosceles triangle with a base angle that measures $x°$ and a base of length b is $A = \frac{1}{4}b^2 \tan x°$.

25. SAT/ACT Preview $ABCD$ is a square, X is the midpoint of $\overline{AD}$, and Y is the midpoint of $\overline{CD}$. Which regions have equal areas?

A. △ABD and △DYB
B. △BDC and △ABX
C. △BDY and △BCY
D. Trapezoid ABYD and quadrilateral BXDY
E. None of these

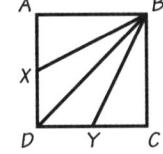

26. Challenge Find the surface area of this prism.

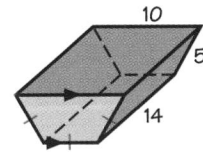

ONGOING ASSESSMENT

27. Writing Write a summary of how to use trigonometry to find areas and volumes. Include sketches and examples if necessary.

SPIRAL REVIEW

Find the magnitude of each vector. *(Section 10.5)*

28. $\overrightarrow{AB} = (3, -4)$ **29.** $\overrightarrow{HJ} = (1, 1)$ **30.** $\overrightarrow{MP} = (-7, -11)$

31. The radius of a cylinder is 12 cm, and the height is 3 cm. Find the volume of the cylinder. *(Section 7.6)*

534   Chapter 10 *Applying Right Triangles*

---

22. a. $\frac{h}{c}$

  b. $h = c \sin A$; Area $= \frac{1}{2}bc \sin A$

  c. about 9.9

23. The area of △XYZ is

$$A = \frac{1}{2}(XZ)(YZ).$$

Since $\tan X = \frac{YZ}{XZ}$,

$YZ = XZ \cdot \tan X$. By substitution, $A = \frac{1}{2}(XZ)^2 \tan X$.

24. The area of △ABC is $A = \frac{b}{2} \cdot BD$.

$\tan x° = \frac{BD}{b}$, so

$BD = \frac{b}{2} \tan x°$. By substitution, $A = \frac{1}{4}b^2 \tan x°$.

25. E

26. $350 + \frac{75\sqrt{3}}{2} \approx 415$

27. Answers may vary. An example is given. You can use trigonometric ratios to find unknown lengths in right triangles. In polygons that are not right triangles, you can first draw right triangles, then use trigonometric ratios to find unknown lengths.

28. 5

29. $\sqrt{2} \approx 1.4$

30. $\sqrt{170} \approx 13.0$

31. $432\pi \approx 1357.2$ cm^2

10.7 Pyramids and Cones

Learn how to...

- apply formulas for surface areas and volumes of pyramids and cones

So you can...

- analyze real-world pyramids and cones, such as paper cups

Many major sports and musical events in Memphis, Tennessee, take place at the Pyramid Arena, shown below. Built from 1989 to 1991, the Arena was modeled after the Great Pyramids in Egypt.

In geometry, a **pyramid** is a three-dimensional figure with one **base** that is a polygon. The other faces of a pyramid, called the **lateral faces**, are triangles that connect the base to the **vertex**.

The Pyramid Arena is a *regular pyramid*. In a **regular pyramid**, the base is a regular polygon and the lateral faces are congruent isosceles triangles.

The **height**, *h*, of a regular pyramid is the distance from the vertex to the center of the base.

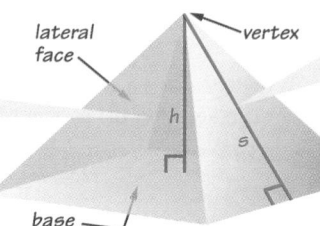

The **slant height**, *s*, of a regular pyramid is the altitude of a lateral face.

You can use a net to find the formula for the surface area of a pyramid.

$$\text{Surface area} = \text{Lateral area} + \text{Area of base}$$

$$= n\left(\frac{1}{2}xs\right) + B$$

$$= \frac{1}{2}(nx)s + B$$

$$\text{S.A.} = \frac{1}{2}ps + B$$

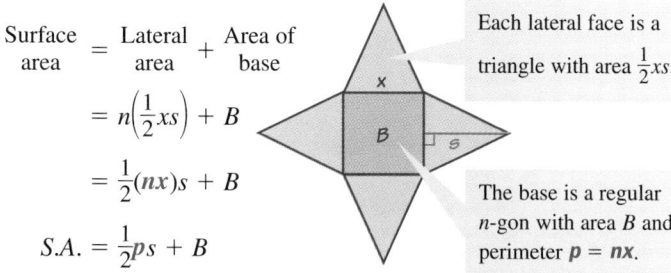

Each lateral face is a triangle with area $\frac{1}{2}xs$.

The base is a regular *n*-gon with area *B* and perimeter $p = nx$.

This formula can be used to find the surface area of any regular pyramid.

10.7 Pyramids and Cones **535**

Plan ⇔ Support

Objectives

- Apply formulas for surface areas and volumes of pyramids and cones.
- Analyze real-world pyramids and cones.

Recommended Pacing

❖ **Core and Extended Courses**
Section 10.7: 1 day

❖ **Block Schedule**
Section 10.7: $\frac{1}{2}$ block
(with Chapter 10 Review)

Resource Materials

Lesson Support
Lesson Plan 10.7
Warm-Up Transparency 10.7
Practice Bank: Practice 73
Study Guide: Section 10.7
Explorations Lab Manual:
   Additional Exploration 16
   Diagram Master 2
Challenge Problems: Set 73
Assessment Book: Test 45

Technology
Technology Book:
   Spreadsheet Activity 9
Scientific Calculator
Internet:
   http://www.hmco.com

Warm-Up Exercises

Find the area of each triangle.

1. 32° 10   about 31.2

2. 9 4   about 16.1

Find the volume and surface area of each figure.

3. a cylinder with radius 13 and height 10   about 5309; about 10,047

4. a square prism with base sides of length 5 and height 12 300; 290

Additional Example 1

An architect designed a clock tower in the shape of a regular triangular pyramid.

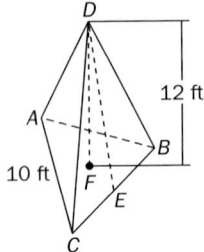

The height of the pyramid is 12 ft and the length of each side of the base is 10 ft. The distance from the center of the base to the side of the base is about 2.9 ft. Find the volume and surface area of the pyramid.

To find the volume, first find the area of the base.

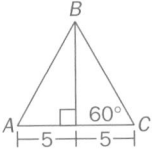

The base is an equilateral triangle, so its altitude forms a 30-60-90 triangle. Its altitude is $5\sqrt{3}$, or about 8.7 ft.

$$\text{Area of the base} \approx \frac{1}{2}(10)(8.7)$$
$$\approx 43.5 \text{ ft}^2$$

$$V = \frac{1}{3}Bh$$
$$\approx \frac{1}{3}(43.5)(12)$$
$$\approx 174$$

The volume of the pyramid is about 174 ft³.

To find the surface area, sketch right △DFE and use the Pythagorean theorem to find s, the slant height of the pyramid.

$$s = \sqrt{DF^2 + FE^2}$$
$$\approx \sqrt{12^2 + 2.9^2}$$
$$\approx 12.3$$

Substitute 12.3 for s in the formula for surface area.

$$S.A. = \frac{1}{2}ps + B$$
$$\approx \frac{1}{2}(3 \cdot 10)(12.3) + 43.5$$
$$\approx 228$$

The surface area of the pyramid is about 228 ft².

Volume and Surface Area of Pyramids

The volume of a pyramid is one third the product of the height and the area of the base.

$$V = \frac{1}{3}Bh$$

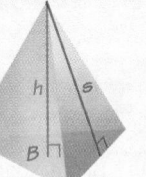

The surface area of a regular pyramid is the sum of the lateral area and the area of the base.

$$S.A. = \frac{1}{2}ps + B$$

EXAMPLE 1   Application: Archeology

When the pyramid of Khafre was built in Giza, Egypt, it was about 471 ft high. Its base is a square, $707\frac{3}{4}$ ft on a side. Find the original volume and surface area of the pyramid.

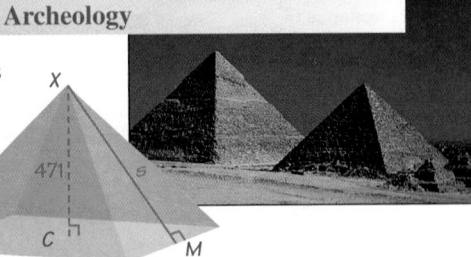

SOLUTION

Find the volume:

$$V = \frac{1}{3}Bh$$

The area of the square base is $B = (707.75)^2$.

$$= \frac{1}{3}(707.75)^2(471)$$
$$\approx 78,643,000$$

To find the surface area, first find s, the slant height of the pyramid.

$$s = \sqrt{XC^2 + CM^2}$$
$$= \sqrt{471^2 + (353.875)^2}$$
$$\approx 589$$

Substitute 589 for s in the formula for surface area.

$$S.A. = \frac{1}{2}ps + B$$

$$CM = \frac{1}{2}(707.75)$$
$$= 353.875$$

$$\approx \frac{1}{2}(4 \cdot 707.75)(589) + (707.75)^2$$
$$\approx 1,335,000$$

The original volume of the pyramid was about 78,643,000 ft³. The original surface area was about 1,335,000 ft².

Like prisms, pyramids are classified by the shapes of their bases. The Pyramid Arena and the Giza Pyramids are *square pyramids*. You can sketch a square pyramid by following the directions below.

Step 1 Draw the base and locate its center.

Step 2 Draw a vertical segment from the center of the base.

Step 3 Join the vertices of the base to the endpoint of the segment. Make the hidden edges into dashed lines.

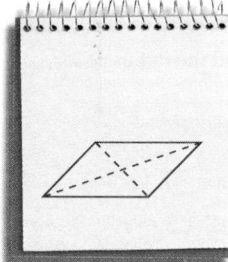 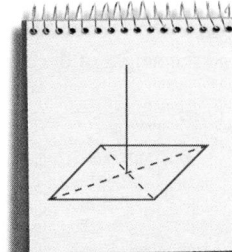 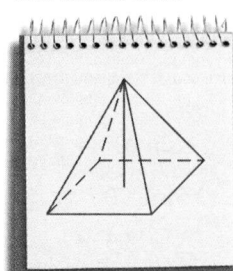

Cones

A **cone** is a three-dimensional figure with one circular base and a vertex. In this book, you will learn about *right cones*. The vertex of a right cone is directly above the center of the base.

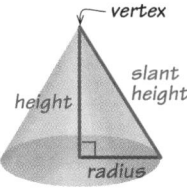

THINK AND COMMUNICATE

1. Describe how a right cone is like a regular pyramid.

2. How would you find the perimeter and area of the base of a cone?

You can use the formulas for volume and surface area of pyramids to find the formulas for volume and surface area of cones.

Volume and Surface Area of Cones

The volume of a cone is one third the product of the height and the area of the base.

$$V = \frac{1}{3}Bh$$

$$= \frac{1}{3}\pi r^2 h$$

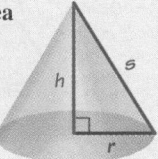

The surface area of a cone is the sum of the lateral area and the area of the base.

$$S.A. = \frac{1}{2}ps + B$$

$$= \frac{1}{2}(2\pi r)s + \pi r^2$$

$$= \pi rs + \pi r^2$$

10.7 Pyramids and Cones **537**

EXAMPLE 2 Application: Packaging

Additional Example 2

A tin bird feeder is made in the shape of a cone with the dimensions shown. How much bird feed, in cubic centimeters and in liters, can the feeder hold? How much tin is required to make the bird feeder?

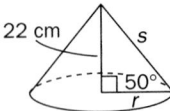

First use trigonometry to find the slant height of the cone and the radius of its base.

$$\sin 50° = \frac{22}{s}$$

$$\frac{22}{\sin 50°} = s$$

$$28.7 \approx s$$

$$\tan 50° = \frac{22}{r}$$

$$\frac{22}{\tan 50°} \approx r$$

$$18.5 \text{ cm} \approx r$$

Use these values for s and r to find the cone's volume and surface area.

$$V = \frac{1}{3}Bh$$

$$= \frac{1}{3}\pi r^2 h$$

$$= \frac{1}{3}\pi (18.5)^2(22)$$

$$\approx 7881$$

The bird feeder can hold about 7881 cm³ or 7.9 liters of bird feed.

$$S.A. = \frac{1}{2}ps + B$$

$$= \pi rs + \pi r^2$$

$$= \pi(18.5)(28.7) + \pi(18.5)^2$$

$$\approx 2741.85$$

At least 2741.85 cm² of tin is required to make the bird feeder.

Closure Question

How are the volumes of a pyramid and a cone related to the volumes of a prism and a cylinder with the same bases and the same heights? Describe how to find the surface area of a regular pyramid and of a cone. *The volume of a pyramid is one third the volume of a prism with the same base and height. The volume of a cone is one third the volume of a cylinder with the same base and height. The surface area of a regular pyramid and of a cone is the sum of the lateral area and the area of the base.*

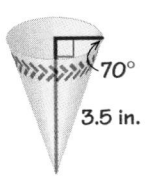

One type of paper cup is cone-shaped, with the dimensions shown. How much water, in cubic inches, can the cup hold? How much paper is required to make the cup?

SOLUTION

First use trigonometry to find the height of the cup and the radius of its base.

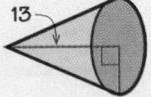

$$\sin 70° = \frac{h}{3.5}$$

$$(3.5)\sin 70 = h$$

$$3.3 \approx h$$

$$\cos 70° = \frac{r}{3.5}$$

$$(3.5)\cos 70 = r$$

$$1.2 \approx r$$

Substitute these values for h and r into the formulas for volume and surface area.

$$V = \frac{1}{3}\pi r^2 h$$

$$= \frac{1}{3}\pi(1.2)^2(3.3)$$

$$\approx 5.0$$

$$S.A. = \pi r s$$

$$= \pi(1.2)(3.5)$$

$$\approx 17.7$$

The cup can hold about 5.0 in.³ of water. At least 17.7 in.² of paper is required to make the cup.

☑ CHECKING KEY CONCEPTS

1. The volume of this cone is 117π. Find the radius of the base and the slant height.

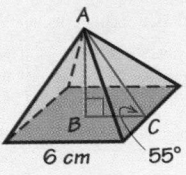

Find the volume and surface area of the right cone or regular pyramid.

2.

6 ft

2 ft

3.

6 cm   55°

Checking Key Concepts

1. 5.2; 14.0
2. 25.1 ft³; 48.1 ft²
3. 51.4 cm³; 98.8 cm²

Exercises and Application

1. 37.7; 75.4
2. 720.0; 531.7
3. 314.2; 282.7
4. 378.2; 333.8
5. 593.8; 497.1
6. 51.8; 92.3

7–9. Answers may vary. Examples are given.

7.

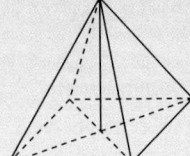

10.7 | Exercises and Applications

Extra Practice exercises on page 682

Find the volume and surface area of each right cone or regular pyramid.

1.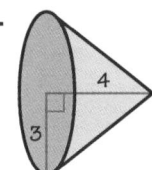

2.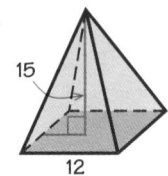

3.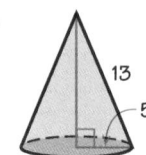

4.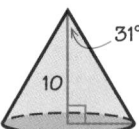

5.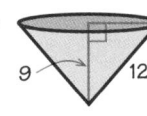

6.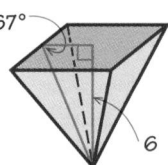

Use the steps shown on page 537 to sketch each pyramid.

7. a square pyramid   **8.** a triangular pyramid   **9.** a pentagonal pyramid

10. Writing Write directions that a friend could follow to sketch a cone.

Connection ARCHITECTURE

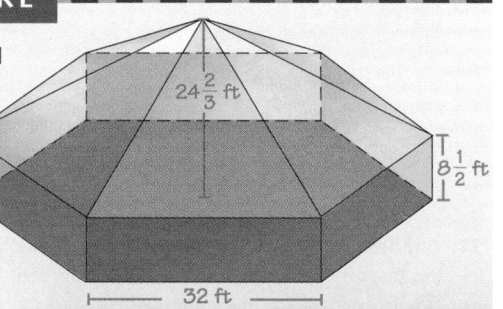

Burt Rutan, an airplane designer, has an unusual home in the Mojave Desert. The top part of his house is a regular hexagonal pyramid. The bottom part is a regular hexagonal prism.

11. a. Find the lateral area of the pyramid.

  b. How much paint is needed to paint the pyramid's outer surface? (One gallon covers about 400 ft².)

12. a. Find the area of the hexagonal base.

  b. Find the volume of the pyramid.

  c. Find the volume of the prism. Then find the volume of the entire house.

13. Open-ended Problem Design a house using pyramids and cones. Sketch the house and label its dimensions. Find the volume and the external surface area of the house.

BY THE WAY

Even the furniture in Burt Rutan's house is unusual. The pool table is a parallelogram with two 60° angles and two 120° angles.

10.7 Pyramids and Cones   **539**

Apply⇔Assess

Suggested Assignment

❖ **Core Course**
  Exs. 1–9, 14–16, 18–28, AYP

❖ **Extended Course**
  Exs. 1–28, AYP

❖ **Block Schedule**
  Day 65 Exs. 1–9, 14–16, 18–28, AYP

Exercise Notes

Topic Spiraling: Review
Exs. 1–6 Some of these exercises require the use of trigonometry. Others require using the Pythagorean theorem. Discuss how students can identify when either of these is required.

Communication: Writing
Ex. 10 Students should first study the steps given on page 537 for sketching a pyramid. After considering how a cone and a pyramid are different, students can alter the steps to draw a cone.

Application
Exs. 11–14 These problems provide students with some examples of how architects have incorporated pyramids and prisms into their building designs. Students also learn why knowing the surface area of the building is important when buying paint. Ask students why knowing the volume of a building may be important. (One reason is to know how much air needs to be heated or cooled during cold or warm weather.)

Historical Connection
Exs. 11–14 In 1984, U.S. architect I.M. Pei designed an expansion for the Louvre, one of the world's great art museums. His design included a glass pyramid which functions as the entrance to the museum.

8.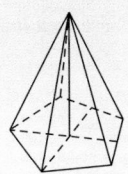

9.

10. Answers may vary. An example is given. First draw an oval to represent the circular base of the cone. Indicate the center and draw a diameter. Draw a segment perpendicular to the diameter and choose a point on the perpendicular to be the vertex. Draw a segment from the vertex to each endpoint of the diameter. Make the hidden part of the base dashed.

11. a. 3562 ft²

  b. 8.9 gallons

12. a. 2660 ft²

  b. 21,870 ft³

  c. 22,610 ft³; 44,480 ft³

13. Check students' work.

Exercise Notes

Geometric Thinking

Ex. 14 In comparing the volume of a cone to the volume of a 16-sided regular pyramid, students see that the volume of the cone can be used to approximate the volume of the pyramid. Ask students why they would want to use this approximation. (The volume of the cone is easier to calculate than the volume of the pyramid.) The technique of approximating a smooth figure like a cone, circle, or curve with a many-sided figure, or vice versa, is one that is used often in mathematics. In calculus, the area of rectangles of various heights are used to approximate the area under a curve.

Integrating Algebra

Exs. 15, 16 In these exercises, students must use the formula for the information they are given, and then solve for the unknown variable.

Cooperative Learning

Exs. 17–19 This is an excellent hands-on activity that can help students see the relationship between a cone and a cylinder that have the same base and height. Partners should work together on Exs. 17 and 18. You may want to suggest that they first discuss Ex. 19 and then independently write their responses.

Challenge

Exs. 17–19 Students should use their paper models to verify that the heights and the bases of the two figures are the same. In Ex. 19, they should use algebra and geometry to demonstrate these facts. Students can start by finding the circumference of each base. Then they need to recognize that the radius of the semicircle is the slant height of the cone.

14. **ARCHITECTURE** The Misumi Ferry Terminal in Japan is a 16-sided regular pyramid, 25 m high. The base of the building is 34 m in diameter.

   a. Find the volume of the terminal.

   b. Find the volume of a cone with diameter 34 m and height 25 m.

   c. **Logical Reasoning** Compare your answers to parts (a) and (b). When does it make sense to estimate the volume of a pyramid by finding the volume of a cone? Explain your reasoning.

15. The volume of a right cone is 144π and the area of its base is 36π. Find the radius, height, and slant height of the cone.

16. A pyramid has a height of 15 ft and a volume of 450 ft^3. What is the area of the base of the pyramid?

Cooperative Learning **Work with another person. You will need heavy paper and dried beans or popcorn kernels. You will compare the volumes of a cone and a cylinder that have the same height and congruent bases.**

Step 1 Cut a rectangle and a semicircle out of paper using the dimensions shown. Tape the edges to make a cylinder and a cone.

Step 2 Fill the cone with beans. Pour the beans from the cone into the cylinder. Repeat until the cylinder is full.

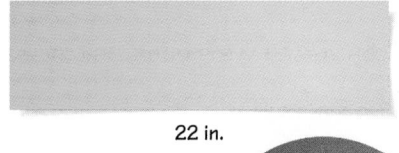

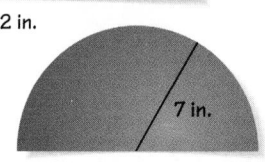

6.1 in.

22 in.

7 in.

17. **Challenge** Show that the cylinder and cone have the same height and approximately congruent bases.

18. What is the ratio of the volume of the cone to the volume of the cylinder? Explain your reasoning.

19. **Writing** Compare the formula for the volume of a cone with the formula for the volume of a cylinder. Is the relationship between the formulas supported by your results of this activity? Explain.

20. Suppose the base of a cone has radius 10 cm and the base angle of the cone is 65°. If the vertex is removed by cutting through the cone along a plane parallel to the base, the resulting figure is called a *truncated cone*.

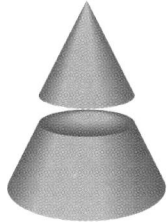

   a. Find the volume of the original cone.

   b. The removed vertex is also a cone. What is the base angle of this cone? Find the volume of the vertex if its base has a radius of 5 cm.

   c. Find the volume of the truncated cone.

---

14. a. 7372 m^3 b. 7566 m^3

   c. The volume of the cone and the pyramid are close. Answers may vary. If you need only a rough estimate of the volume of a pyramid and you know the diameter of its base, you can always estimate the volume by finding the volume of a cone having a base with the same diameter. If the base of the pyramid is a regular *n*-sided polygon, the volume of the cone approximates the volume of the pyramid more closely as *n* increases. The area of the base of the pyramid gets closer to the area of the base of the cone.

15. $r = 6$; $h = 12$; $s = 6\sqrt{5} \approx 13.4$

16. 90 ft^2

17. The two bases have the same radius if their circumferences are the same.

circumferemce of base of cone =
$\frac{1}{2}$(circumference of circle with radius 7) =
$\frac{1}{2}(2\pi \cdot 7) \approx 22.0$ in. This is the same as the circumferemce of the base of the cylinder. The height of the cylinder is 6.1 in. Since the circumference of the base of the cone is 22 in., the radius of the base of the cone is $\frac{22}{2\pi} \approx 3.5$ in. The slant height of the cone is 7 in., so by the Pythagorean theorem, its height is $\sqrt{7^2 - 3.5^2} \approx 6.1$ in.

18. 1:3; The cone was filled with beans three times in order to fill the cylinder.

19. Yes; the volume of a cone with base area B and height h is $\frac{1}{3}Bh$. The volume of a cylinder with base area B and height h is Bh.

20. a. 2246 cm^3

   b. 65°; 281 cm^3

   c. 1965 cm^3

21. The volume formulas you learned can also be applied to *oblique* pyramids and cones. Find the volume of each figure.

a.

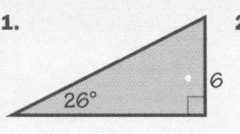

b.

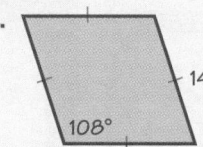

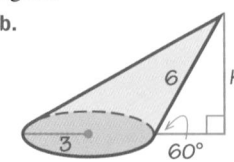

ONGOING ASSESSMENT

22. Cooperative Learning Work with another person. Make a chart showing all of the surface area and volume formulas you have learned so far. Sketch a figure for each, labeling all of its dimensions.

SPIRAL REVIEW

Graph each vector and find its magnitude. *(Section 10.5)*

23. $\overrightarrow{AB} = (0, -7)$ **24.** $\overrightarrow{CD} = (-5, 12)$ **25.** $\overrightarrow{EF} = (550, -400)$

Find the geometric mean of the given numbers. *(Section 10.1)*

26. 4 and 16   **27.** 10 and 1000   **28.** 30 and 60

ASSESS YOUR PROGRESS

VOCABULARY

pyramid (p. 535)	**regular pyramid** (p. 535)
base of a pyramid (p. 535)	**height of a pyramid** (p. 535)
lateral face of a pyramid (p. 535)	**slant height** (p. 535)
vertex of a pyramid (p. 535)	**cone** (p. 537)

Find the area of each polygon. *(Section 10.6)*

1.

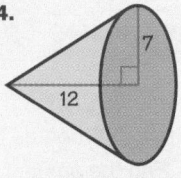

2.

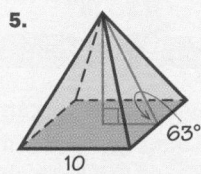

3.

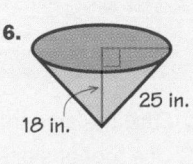

Find the volume and surface area of each right cone or regular pyramid. *(Section 10.7)*

4.

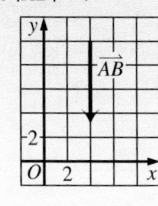

5.

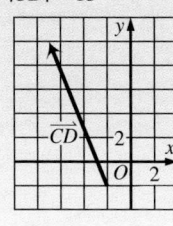

6.

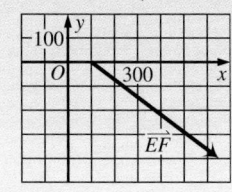

7. Journal Compare the formulas for volume and surface area of pyramids to the formulas for volume and surface area of prisms.

10.7 Pyramids and Cones **541**

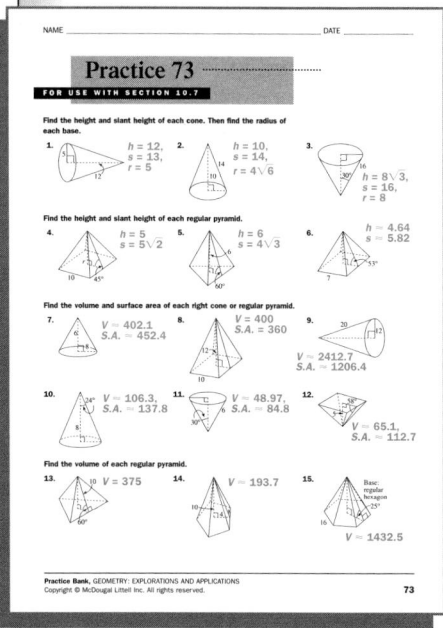
---

21. a. 8.0
    b. 49.0

22. See answers in back of book.

23. $|\overrightarrow{AB}| = 7$

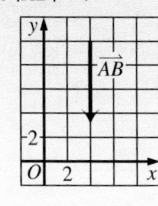

24. $|\overrightarrow{CD}| = 13$

25. $|\overrightarrow{EF}| = \sqrt{462,500} \approx 680.0$

26. 8
27. 100
28. $30\sqrt{2} \approx 42.4$

Assess Your Progress

1. 36.9
2. 186.2
3. 344.4
4. 615.8; 459.6
5. 326.7; 320.0
6. 5641.5; 2299.0
7. Answers may vary. Both volume formulas depend on the area of the base

of the figure and its height. Given a pyramid and a prism with congruent bases and the same height, the volume of the pyramid is one third that of the prism. A pyramid has only one base and sides that are triangles. A prism has two bases and sides that are rectangles. That accounts for the differences in the formulas for surface area. The surface area of a pyramid with base area B, perimeter p, and slant height s is $\frac{1}{2}ps + B$. The surface area of a prism with base area B, perimeter p, and height h is $ph + 2B$.

Mathematical Goals

- Use a formula to calculate the angle of the sun.
- Use trigonometry to design a solar roof.

Planning

Materials
- paper and pencil
- ruler
- protractor
- drawing supplies (if necessary)

Project Teams
Students can work in pairs to complete the project. Each partner should do all of the calculations independently and then partners should compare results with one another.

Guiding Students' Work

If students are having difficulty with the initial steps for the project, have them label angles and lengths in the figure. Remind them that the chapter is about applying right triangles, and ask them if they can find a right triangle in the drawing. Then have them copy and label the right triangle. When drawing their designs, students may need to make one or two rough drafts before drawing their final versions.

Second-Language Learners
You may need to explain the meaning of *greenhouse effect* to students learning English. You can describe it as "trapping sunlight, so as to hold its warmth." You might also want to mention that the term *greenhouse effect* is often used to describe the phenomenon in which Earth traps the sun's heat.

PORTFOLIO PROJECT

Applying Solar Geometry

The sun can help to heat your home during the winter. The roof of a *passive solar* home allows sunlight to shine in during the winter, when the sun is low in the sky. These roofs also help keep homes cool during the summer when the sun is higher.

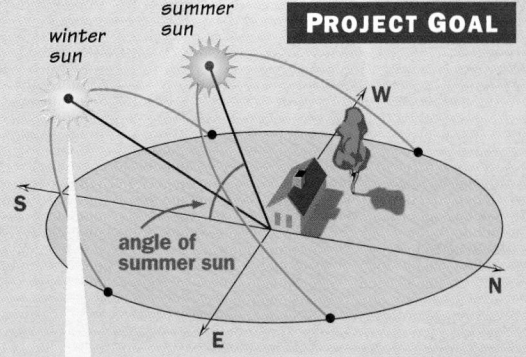

PROJECT GOAL Use the angle of the sun to plan the roof overhang for a passive solar home.

At noon on December 21, the sun's angle is 90° − **latitude** − 23.5°.

Analyzing Angles

The house in the diagrams shown is in Chicago, Illinois, at a **latitude of 42°** north . To find the angle of the sun at noon on the longest day of the year, June 21, use this expression:

$$90° - 42° + 23.5°$$

After June 21, the days shorten until the sun reaches its lowest point in the sky on December 21. You can use the angle of the sun to find the best angle for a roof.

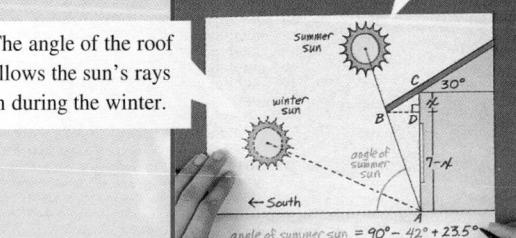

The angle of the roof blocks the sun's rays during the summer.

The angle of the roof allows the sun's rays in during the winter.

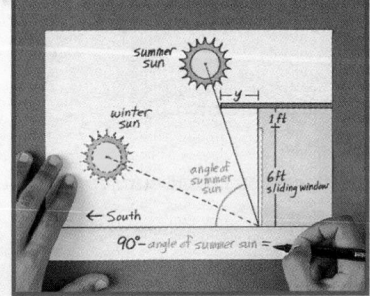

1. **Find** the amount of overhang, *BC*, of a roof of slope 30° in the Chicago home shown. Explain your method. (*Hint:* Find *BC* in terms of *x* and solve for *x*.)

2. **Examine** the diagram of a solar home with a flat roof, as shown above. Find the amount of overhang, *y*, that blocks the summer sun.

You can get ideas for your design by looking at homes in your neighborhood.

Designing Your Roof

3. Look up the latitude of your town. Calculate the angle of the sun at noon on the longest and shortest days of the year.

4. Use the angle that you found in Step 3 to design your sloping solar roof. Use trigonometry to find the amount of overhang needed in the summer. Include the dimensions of windows or doors affected by the overhang.

Presenting Your Results

Use diagrams, photographs, or models to illustrate your results. Describe your design and be sure to include these points:

- Describe how a passive solar roof performs different functions at different times of the year.

- Explain the differences between overhangs of flat and sloping roofs. Which type do you prefer, and why?

- How do weather conditions, location, and the direction the house faces affect its solar efficiency?

Extending Your Project

Here are some ideas for extending your project:

- Is there only one correct angle for the roof of a passive solar house? Explain why or why not.

- Why is 23.5° used to calculate the sun's angle? (*Hint:* Find the definition of *ecliptic*.)

- Are there other ways to heat a house using solar energy?

- If possible, talk to an architect to see if your plan is realistic.

Self-Assessment

Did your design for a solar home turn out the way you expected? How did you use trigonometry in your design? Would you consider using some type of passive solar heating if you were building a house? Why or why not?

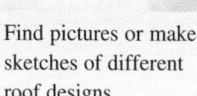

Find pictures or make sketches of different roof designs.

Portfolio Project **543**

Progress Check 10.1–10.2

1. Find the geometric mean of 10
  and 45. *(Section 10.1)* $15\sqrt{2}$

Find the value of each variable.
(Sections 10.1 and 10.2)

2.
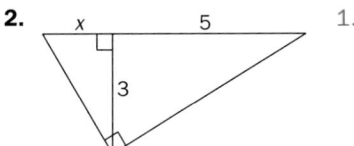

3.
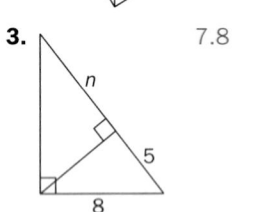
7.8

4.
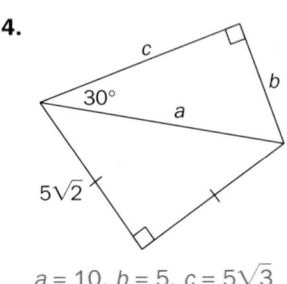

  $a = 10,\ b = 5,\ c = 5\sqrt{3}$

5. Find the apothem of a regular
  hexagon with side length 6.
  (Section 10.2) $3\sqrt{3}$

CHAPTER

10 | Review

STUDY TECHNIQUE

List three techniques you have used in the past to study for a test. Exchange lists with another student and use at least two of the suggestions to study this chapter in preparation for the test.

VOCABULARY

geometric mean (p. 494)	**equal vectors** (p. 522)
45-45-90 triangle (p. 500)	**scalar multiplication** (p. 523)
30-60-90 triangle (p. 501)	**pyramid** (p. 535)
tangent of an angle (p. 507)	**base of a pyramid** (p. 535)
angle of elevation (p. 509)	**lateral face of a pyramid** (p. 535)
angle of depression (p. 509)	**vertex of a pyramid** (p. 535)
sine (p. 515)	**regular pyramid** (p. 535)
cosine (p. 515)	**height of a pyramid** (p. 535)
vector (p. 521)	**slant height** (p. 535)
magnitude of a vector (p. 521)	**cone** (p. 537)

SECTIONS 10.1 *and* 10.2

You can find the positive number x that is the **geometric mean** of two positive numbers a and b by solving the proportion $\frac{a}{x} = \frac{x}{b}$.

When the altitude is drawn to the hypotenuse of a right triangle:

- the two triangles formed are similar to the original triangle and to each other.

- the length of the altitude is the geometric mean of the lengths of the segments of the hypotenuse.

- the length of each leg is the geometric mean of the lengths of the hypotenuse and the segment of the hypotenuse adjacent to that leg.

These diagrams show the relationships between the side lengths for a **45-45-90 triangle** and for a **30-60-90 triangle**.

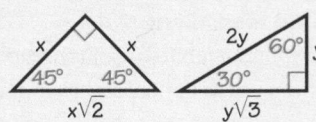

SECTIONS 10.3, 10.4, and 10.5

You can find side lengths and angle measures of a right triangle by using these *trigonometric ratios*:

$$\sin A = \frac{a}{c} \qquad \cos A = \frac{b}{c} \qquad \tan A = \frac{a}{b}$$

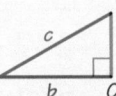

A vector is a quantity with both magnitude and direction. The **magnitude** of $\overrightarrow{RS}$ is the length RS. In the diagram below, $\overrightarrow{RS}$ has *horizontal component* -4 and *vertical component* 3.

You can use the Pythagorean theorem to find the magnitude of a vector and trigonometry to find the angle that a vector makes with the horizontal. Vectors with the same magnitude and direction are **equal vectors**.

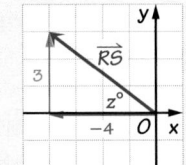

To multiply a vector by a real number k, multiply each component by k. This is called **scalar multiplication**. Multiplying a vector by k multiplies the length of the vector by $|k|$. To add two vectors, add their corresponding components.

SECTIONS 10.6 and 10.7

Trigonometry can be used to find the area, surface area, or volume of some figures. For example:

$$\sin 50° = \frac{h}{8} \qquad A = bh$$
$$h \approx 10.4 \qquad \approx 10(10.4)$$
$$= 104$$

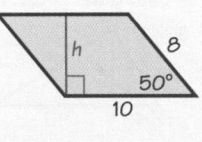

A **pyramid** is a figure with one **base** that is a polygon. In a **regular pyramid**, the base is a regular polygon and the **lateral faces** are congruent isosceles triangles.

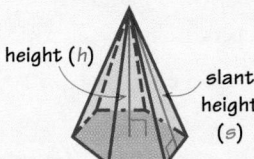

$$V = \frac{1}{3}Bh$$
$$S.A. = \frac{1}{2}ps + B$$

The figure at the right is a **cone**. In a *right cone*, the vertex lies directly above the center of the base.

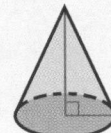

$$V = \frac{1}{3}\pi r^2 h$$
$$S.A. = \pi rs + \pi r^2$$

Review **545**

Find the value of each variable. Round answers to the nearest tenth. *(Sections 10.3 and 10.4)*

1.

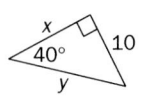

$x \approx 11.9$; $y \approx 15.6$

2.

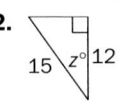

36.9

3. Graph $\overrightarrow{PR} = (-3, 6)$ and find its magnitude. *(Section 10.5)*

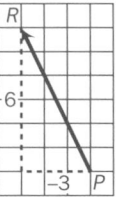

$3\sqrt{5}$, or about 6.7

If $\overrightarrow{EG} = (1, -5)$ and $\overrightarrow{GH} = (-7, -2)$, find the component form of each vector. *(Section 10.5)*

4. $2\overrightarrow{GH}$ $(-14, -4)$

5. $\overrightarrow{EG} + \overrightarrow{GH}$ $(-6, -7)$

Progress Check 10.6–10.7

Find the area of each figure. Round answers to the nearest tenth. *(Section 10.6)*

1. a regular pentagon with side length 9 cm   132.0 cm²

2.

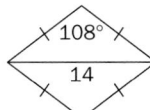

about 71.4

3. Find the volume of the prism below. *(Section 10.6)*

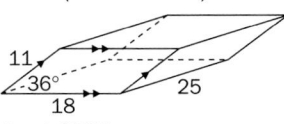

about 2925

Find the volume and surface area of each figure. *(Section 10.7)*

4. a cone with height 12 in. and radius 3 in.   about 113.1 in.³; about 144.9 in.²

5. a square pyramid with height 16 m whose base has a side length of 10 m   about 533.3 m³; about 435.2 m²

Chapter 10 Assessment
Form A Chapter Test

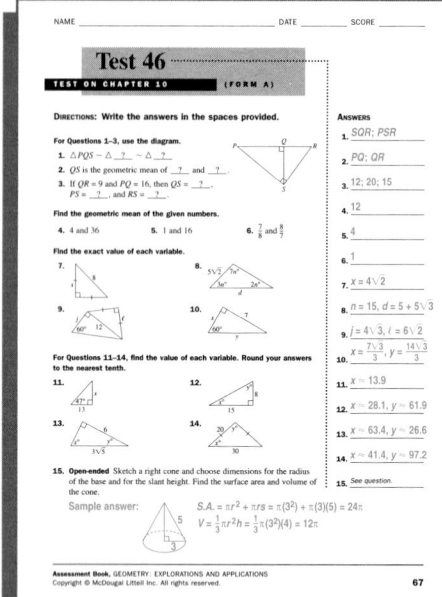

Chapter 10 Assessment
Form B Chapter Test

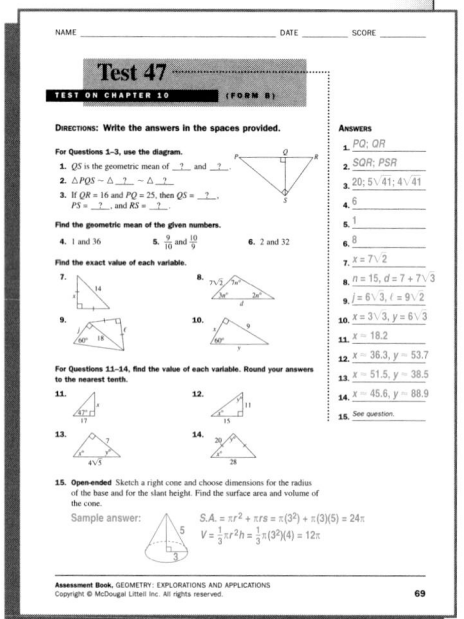

VOCABULARY QUESTIONS

1. Sketch a regular pyramid and a right cone. On each figure, label the base, vertex, height, and slant height.

2. Refer to the right triangle shown. Find the sine, cosine, and tangent of $\angle R$ and $\angle S$ in terms of RS, ST, and RT.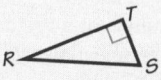

3. Explain how to find the geometric mean of two positive numbers.

4. A train is traveling at a speed of 50 mi/h. Can you represent this situation using a vector? Explain why or why not.

5. Explain the difference between *angle of elevation* and *angle of depression*.

SECTIONS 10.1 *and* 10.2

For Questions 6–11, find the value of each variable.

6.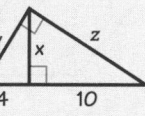

7.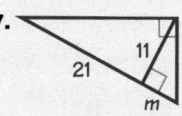

8.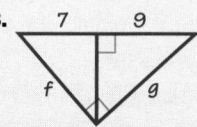

9.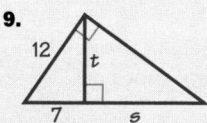

10.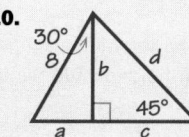

11.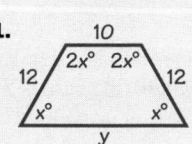

12. Find the geometric mean of the given numbers.

    a. $\dfrac{5}{6}$ and $\dfrac{1}{2}$ **b.** 1 and 1000   **c.** $\sqrt{2}$ and $32\sqrt{2}$

Find the area of each polygon.

13.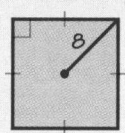

14.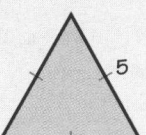

15.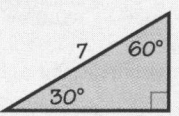

ANSWERS Chapter 10

Assessment

1. Answers may vary. Examples are given.

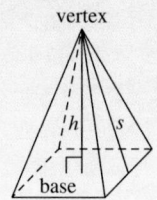 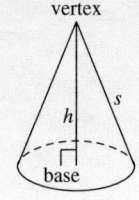

2. $\sin R = \dfrac{TS}{RS}$; $\cos R = \dfrac{RT}{RS}$; $\tan R = \dfrac{TS}{RT}$; $\sin S = \dfrac{RT}{RS}$; $\cos S = \dfrac{ST}{RS}$; $\tan S = \dfrac{RT}{TS}$

3. Find the square root of their product.

4. No; a vector has both magnitude and direction.

5. An angle of elevation is formed with the horizontal and the line of sight to a point above the viewing point. An angle of depression is formed with the horizontal and the line of sight to a point below the viewing point.

6. $x = 2\sqrt{10}$; $y = 2\sqrt{14}$; $z = 2\sqrt{35}$

7. $m = \dfrac{121}{21}$; $r = \sqrt{\dfrac{68{,}002}{441}}$

8. $f = 4\sqrt{7}$; $g = 12$

9. $s = \dfrac{95}{7}$; $t = \sqrt{95}$

10. $a = 4$; $b = 4\sqrt{3}$; $c = 4\sqrt{3}$; $d = 4\sqrt{6}$

11. $x = 30$; $y = 6\sqrt{3}$

12. **a.** $\sqrt{\dfrac{5}{12}}$

    b. $10\sqrt{10}$

    c. 8

13. 128

14. $\dfrac{25}{4}\sqrt{3}$

15. $\dfrac{49}{4}\sqrt{3}$

16. 0.6101

17. 0.7923

18. 0.7701

19. about 1.4 ft

20. about 32.5 ft tall

SECTIONS 10.3, 10.4, and 10.5

Find the value of each ratio. Round your answers to four decimal places.

16. sin 37.6° **17.** cos 37.6° **18.** tan 37.6°

19. A wheelchair ramp that extends 20 ft forms a 4° angle with the level ground. About how high above the ground is the top of the ramp?

20. MEASUREMENT When Susan Karp stands 10 ft from a building, the angle of elevation to the top of the building is about 70°. If Susan's eyes are 5 ft above the ground, about how tall is the building?

21. In $\triangle ABC$, $AB = BC = 12$ and $AC = 18$. Find $m \angle A$, $m \angle B$, and $m \angle C$.

In Questions 22 and 23, $\vec{AB} = (-15, -8)$ and $\vec{CD} = (9, -3)$.

22. Graph $\vec{AB}$ and $\frac{1}{3}\vec{CD}$ and find the magnitude of each vector.

23. Use the components to find the sum $\vec{AB} + \vec{CD}$.

24. Writing Suppose $\vec{PQ} = (a, b)$. Explain how to find the magnitude of $\vec{PQ}$ and the angle $\vec{PQ}$ makes with the horizontal.

SECTIONS 10.6 and 10.7

25. The perimeter of a regular 7-gon is 63. Find the area of the 7-gon.

26. The perimeter of a rhombus is 80 cm, and one of its angles measures 146°. Find the area of the rhombus.

Find the volume and the surface area of the prism, the cone, and the regular pyramid.

27.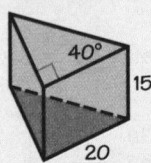

28.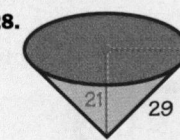

29.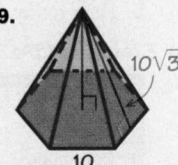

30. Open-ended Problem Sketch a square pyramid and choose dimensions for the base and for the slant height. Find the surface area and the volume of the pyramid.

PERFORMANCE TASK

31. Suppose you know the value of one trigonometric ratio of an acute angle. Describe how to find the values of the other two trigonometric ratios of that angle.

21. $m \angle B \approx 97.2°$, $m \angle A = m \angle C \approx 41.4°$

22.

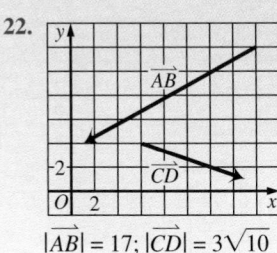

$|\vec{AB}| = 17$; $|\vec{CD}| = 3\sqrt{10}$

23. $(-6, -11)$

24. The magnitude of $\vec{PQ}$ is $\sqrt{a^2 + b^2}$. The tangent of the angle that $\vec{PQ}$ makes with the horizontal is $\frac{b}{a}$, so the angle is $\tan^{-1}\frac{b}{a}$.

25–29. Answers may vary slightly due to rounding. Answers are given to the nearest tenth.

25. 296.1

26. 225.38 cm²

27. 1279.5; 2520

28. 3078.8; 8796.5

29. 562.9; 1299.0

30. Check students' work.

31. Answers may vary. An example is given. Sketch $\triangle ABC$ with right angle C, and side lengths a, b, and c. Label the triangle according to the information you have. For example, suppose you know that $\sin A = 0.8829$. Let $a = 0.8829$ and $c = 1$. Use the Pythagorean theorem to find b. You will have sufficient information to find both $\cos A$ and $\tan A$.

Algebra
Review/Preview

These exercises review algebra topics you will use in the next chapters.

SIMPLIFYING MORE EXPRESSIONS

Simplify each expression. If it is not possible, write *cannot be simplified*.

Toolbox p. 690

Operations with Variable Expressions

EXAMPLES $2x^3 + 4x^2 - 3x^3$

$$= 2x^3 - 3x^3 + 4x^2 \qquad \text{Group the like terms.}$$

$$= -x^3 + 4x^2$$

$2\pi r^2 + 2\pi rh$ **cannot be simplified** because the terms have different variable parts.

1. $y + y^3 - y + y^3$ **2.** $2x^2 + 4xh$ **3.** $9\pi r^2 - 4\pi r^2$

4. $\pi r^2 h + \frac{1}{3}\pi r^2 h$ **5.** $\frac{1}{2}s^2 + 2s^2$ **6.** $\frac{4}{3}\pi r^3 + \frac{1}{2}\pi r^3$

7. $5n + 15n^2 + n^3$ **8.** $(2z)^3 + (3z)^2 + z^3$ **9.** $(x^2 - x) - (x + x^3)$

EXPLORING MATRICES

Toolbox p. 711

Matrices

EXAMPLES The matrix $\begin{bmatrix} -1 & 6 & 5 \\ -3 & -3 & 0 \end{bmatrix}$ has two rows and three columns. Its dimensions are 2 × 3.

10. a. How many rows and how many columns does a 5 × 2 matrix have?

  b. How many elements does a 5 × 2 matrix have?

For Exercises 11–13, use the table and matrix of statistics for hockey teams.

Team	W	L	T
Nordiques	30	13	5
Penguins	29	16	3
Bruins	27	18	3
Sabres	22	19	7
Whalers	19	24	5
Canadiens	18	23	7
Senators	9	34	5

$$M = \begin{bmatrix} 30 & 13 & 5 \\ 29 & 16 & 3 \\ 27 & 18 & 3 \\ 22 & 19 & 7 \\ 19 & 24 & 5 \\ 18 & 23 & 7 \\ 9 & 34 & 5 \end{bmatrix}$$

11. Give the dimensions of matrix M.

12. How many wins did the Bruins have?

13. How many losses did the Canadiens have?

ANSWERS   Chapter 10

Algebra Review/Preview

1. $2y^3$

2. cannot be simplified

3. $5\pi r^2$

4. $\frac{4}{3}\pi r^2 h$

5. $\frac{5}{2}s^2$

6. $\frac{11}{6}\pi r^3$

7. cannot be simplified

8. $9z^3 + 9z^2$

9. $-x^3 + x^2 - 2x$

10. a. 5 rows, 2 columns

    b. 10 elements

11. 7 × 3

12. 27 wins

13. 23 losses

14. 1 × 2; 3 × 1; 3 × 2; 3 × 1; 3 × 2

15. a. not possible

    b. $\begin{bmatrix} 5 \\ -9 \\ 8 \end{bmatrix}$

    c. not possible

    d. $\begin{bmatrix} 6 & -2 \\ 3 & -4 \\ 0 & 4 \end{bmatrix}$

Simplify each expression if possible.

$\begin{bmatrix} 4 & 1 \\ -3 & 2 \end{bmatrix} + \begin{bmatrix} 1 & 0 \\ 3 & -5 \end{bmatrix} = \begin{bmatrix} 4+1 & 1+0 \\ -3+3 & 2+(-5) \end{bmatrix} = \begin{bmatrix} 5 & 1 \\ 0 & -3 \end{bmatrix}$

$\begin{bmatrix} 4 & 1 \\ -3 & 2 \end{bmatrix} + \begin{bmatrix} -1 & 6 & 5 \\ -3 & -3 & 0 \end{bmatrix}$ **cannot be evaluated because the matrices have different dimensions.**

Use matrices *A*, *B*, *C*, *D*, and *E* for Questions 14–17.

$A = \begin{bmatrix} 1 & -1 \end{bmatrix} \quad B = \begin{bmatrix} 7 \\ -6 \\ 0 \end{bmatrix} \quad C = \begin{bmatrix} 6 & -9 \\ 0 & -2 \\ 1 & 4 \end{bmatrix} \quad D = \begin{bmatrix} -2 \\ -3 \\ 8 \end{bmatrix} \quad E = \begin{bmatrix} 0 & 7 \\ 3 & -2 \\ -1 & 0 \end{bmatrix}$

14. Give the dimensions of each matrix.

15. Add each pair of matrices, if possible. If the matrices cannot be added, write *not possible*.

    a. $A + C$ **b.** $B + D$ **c.** $B + E$ **d.** $C + E$

16. Subtract each pair of matrices, if possible. If the matrices cannot be subtracted, write *not possible*.

    a. $B - D$ **b.** $A - E$ **c.** $C - E$ **d.** $E - C$

17. Writing Compare your answers to parts (c) and (d) of Question 16. Does the order in which you subtract matrices affect the result? Do you think that the order in which you add two matrices affects the result? Explain why or why not.

USING FORMULAS

Solve each formula or equation for the indicated variable.

Toolbox p. 702
Working with Formulas

EXAMPLES

Solve $x^2 + y^2 = r^2$ **for *y*.**

$$y^2 = r^2 - x^2$$

$$y = \pm\sqrt{r^2 - x^2}$$

Solve $\frac{4}{3}\pi r^3 = 36\pi$ **for *r*.**

$$\pi r^3 = \left(\frac{3}{4}\right)36\pi$$

$$r^3 = 27 \qquad 3 \cdot 3 \cdot 3 = 27$$

$$r = 3$$

Multiply both sides by $\frac{3}{4}$, the inverse of $\frac{4}{3}$.

18. Solve $a^2 + b^2 = (2a)^2$ for *b*.

19. Solve $x^2 - y^2 = r^2$ for *y*.

20. Solve $\frac{8}{3}\pi r^2 = 384\pi$ for *r*.

21. Solve $\frac{24}{21} = \frac{x^3}{7}$ for *x*.

22. Solve $\frac{x}{360} \cdot 2\pi \cdot 15 = \frac{9\pi}{4}$ for *x*.

23. Solve $x^2 = 4(4 + 5)$ for *x*.

24. Solve $S = 4\pi r^2$ for *r*.

25. Solve $s^3 = 125$ for *s*.

26. Solve $A = \frac{1}{2}ap$ for *p*.

27. Solve $V = \frac{1}{3}\pi r^2 h$ for *h*.

Algebra Review/Preview   **549**

---

16. a. $\begin{bmatrix} 9 \\ -3 \\ -8 \end{bmatrix}$

    b. not possible

    c. $\begin{bmatrix} 6 & -16 \\ -3 & 0 \\ 2 & 4 \end{bmatrix}$

    d. $\begin{bmatrix} -6 & 16 \\ 3 & 0 \\ -2 & -4 \end{bmatrix}$

17. Yes; No; matrix addition and subtraction are based on addition and subtraction of real numbers. Addition is commutative, subtraction is not.

18. $\pm a\sqrt{3}$

19. $\pm\sqrt{x^2 - r^2}$

20. $12\sqrt{2}$

21. 2

22. 27

23. ± 6

24. $\frac{1}{2}\sqrt{\frac{S}{\pi}}$

25. 5

26. $\frac{2A}{a}$

27. $\frac{3V}{\pi r^2}$

11 Circles and Spheres

OVERVIEW

Connecting to Prior and Future Learning

⇔ The first part of Chapter 11 presents circles and the various angles, arcs, sectors, rays, lines, and segments associated with them. Students learn how the measures of many of these parts are related. Students use variable expressions as they complete this work with circles. They can review operations with variable expressions on page 690 in the **Student Resources Toolbox**.

⇔ The study of circles is expanded in Section 11.6 to a study of surface areas and volumes of spheres. Students continue to study volume in Section 11.7.

⇔ The chapter closes with a presentation of spherical geometry. Throughout the chapter, applications are presented for all of the topics involving circles and spheres.

Chapter Highlights

Interview with Ron Courson: The relationship between mathematics and treating sports injuries is discussed in this interview with Ron Courson, director of sports medicine at the University of Georgia. Exercises that relate to this interview can be found on pages 557 and 589.

Explorations are found in several sections of Chapter 11. In Section 11.1, students compare central and inscribed angles of a circle. Section 11.3 opens with an exploration of circumscribed triangles. In Section 11.8, students explore triangles on a sphere.

The Portfolio Project: This project guides students through an exploration of the relationships among the areas of the cross sections of a sphere, cylinder, and two cones. These relationships are then used to compare the volumes of the solids.

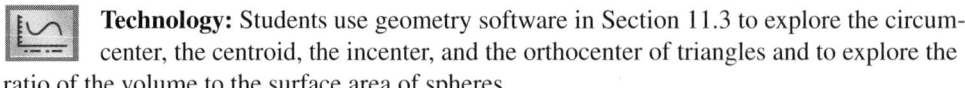 **Technology:** Students use geometry software in Section 11.3 to explore the circumcenter, the centroid, the incenter, and the orthocenter of triangles and to explore the ratio of the volume to the surface area of spheres.

OBJECTIVES

Section	Objectives	NCTM Standards
11.1	• Identify types of arcs and angles in a circle. • Find the measures of arcs and angles.	1, 2, 3, 4, 7
11.2	• Find arc and angle measures when segments intersect circles. • Solve real-world problems involving segments that intersect circles.	1, 2, 3, 4, 5, 7, 9
11.3	• Identify relationships among arcs and chords of circles. • Solve real-world problems involving arcs and chords of circles.	1, 2, 3, 4, 5, 7
11.4	• Find lengths of segments formed by chords, tangents, and secants. • Solve real-world problems involving lengths of segments in circles.	1, 2, 3, 4, 5, 7
11.5	• Find lengths of arcs and areas of sectors. • Find lengths and areas in real-world situations.	1, 2, 3, 4, 5, 7
11.6	• Calculate the surface area and volume of a sphere. • Compare spheres of different sizes that represent real-world objects.	1, 2, 3, 4, 5, 7
11.7	• Compare volumes and surface areas of similar solids. • Estimate measures of real-world objects.	1, 2, 3, 4, 5, 7
11.8	• Represent lines and shapes on a sphere. • Explore geometry on real-world spheres.	1, 2, 3, 4, 7, 14

Mathematical Connections	11.1	11.2	11.3	11.4	11.5	11.6	11.7	11.8
geometry	**553–559***	**560–566**	**567–572**	**573–578**	**579–585**	**586–591**	**592–596**	**597–603**
algebra	558	562, 563, 566	568, 570	574–576		587	595	
data analysis, probability, discrete math	559						592, 593, 595	
logic and reasoning	558	563	569–572	576, 577	583, 584	591	596	603
Interdisciplinary Connections and Applications								
history and geography								601
biology and earth science		564		577			595	
chemistry and physics						591		
sports and recreation		565			583, 584	587, 590	595	600
global positioning		560		575, 577				
architecture						589	594	
navigation, highway safety, physical therapy, irrigation, catering, fashion, automobiles, astronomy, cooking	559		571, 572		581–583	589	594, 596	

__Bold page numbers__ indicate that a topic is used throughout the section.

Section	opportunities for use with	
	Student Book	**Support Material**
11.1	geometry software	**Geometry Inventor Activity Book:** Activity 13
11.2	scientific calculator	**Geometry Inventor Activity Book:** Activity 10
11.3	scientific calculator geometry software	
11.4	scientific calculator	**Technology Book:** TI-92 Activity 11 **Geometry Inventor Activity Book:** Activity 14
11.5	scientific calculator	**Technology Book:** Spreadsheet Activity 10
11.6	graphing calculator	**Technology Book:** Spreadsheet Activity 10
11.7	scientific calculator	**Technology Book:** Calculator Activity 14

Regular Scheduling (45 min)

Section	Materials Needed	Core Assignment	Extended Assignment	Applications	Communication	Technology
				exercises that feature		
11.1	geometry software, compass, straight-edge, protractor	1–9, 13–20, 22, 27–31	1–9 odd, 10–17, 19–31	11, 12, 24–26	12, 25	
11.2		1–16, 24–27, AYP*	1–11 odd, 12–27, AYP	18–22	24	
11.3	straightedge, patty paper, compass, geometry software	1–6, 8, 9, 12, 17–26	1–26	13–16	15, 19	10
11.4		1–21, 25, 27–33	1–17 odd, 18–33	19, 23, 24	22, 23b, 24d, 25, 26b	
11.5		**Day 1:** 1–6, 8–16 **Day 2:** 20, 22–24, 26, 29–32, AYP	**Day 1:** 1–16 **Day 2:** 17–32, AYP	7 18, 19, 21–24	20a, 21c, 24, 27, 29	
11.6	graphing calculator	1–9, 13–18, 20, 21, 23, 29, 32–38	1–5, 7, 8, 10–38	10–12, 16–26, 29	28, 30, 31b, 32	27
11.7	scientific calculator	1–5, 7–12, 16–18	1–18	1, 5, 9–15	12, 15, 16	
11.8	sphere, strips of paper, tape, ruler, protractor	**Day 1:** 1–13 **Day 2:** 14–17, 23–28, AYP	**Day 1:** 1–13 **Day 2:** 14–28, AYP	11–13 14–21	1–3, 12–14, 22	
Review/ Assess		**Day 1:** 1–12 **Day 2:** 13–25 **Day 3:** Ch. 11 Test	**Day 1:** 1–12 **Day 2:** 13–25 **Day 3:** Ch. 11 Test			
Port. Proj.		Allow 2 days.	Allow 2 days.			

Yearly Pacing (with Portfolio Project)	Chapter 11 Total 15 days	Chapters 1–11 Total 149 days	Remaining 11 days	Total 160 days

Block Scheduling (90 min)

	Day 68	Day 69	Day 70	Day 71	Day 72	Day 73	Day 74	Day 75
Teach/Interact	11.1: Exploration, page 553 11.2	11.3: Exploration, page 567 11.4	11.5	11.6 11.7	11.8: Exploration, page 597	Review Port. Proj.	Review Port. Proj.	Ch. 11 Test 12.1
Apply/Assess	**11.1:** 1–9, 13–20, 22, 27–31 **11.2:** 1–16, 24–27, AYP*	**11.3:** 1–6, 8, 9, 12, 17–26 **11.4:** 1–21, 25, 27–33	**11.5:** 1–6, 8–16, 20, 22–24, 26, 29–32, AYP	**11.6:** 1–9, 13–18, 20, 21, 23, 29, 32–38 **11.7:** 1–5, 7–12, 16–18	**11.8:** 1–17, 23–28, AYP	**Review:** 1–12 **Port. Proj.**	**Review:** 13–25 **Port. Proj.**	**Ch. 11 Test** **12.1:** 1–20, 29, 32–39

NOTE: A one-day block has been added for the Portfolio Project—timing and placement to be determined by teacher.

Yearly Pacing (with Portfolio Project)	Chapter 11 Total $7\frac{1}{2}$ days	Chapters 1–11 Total $74\frac{1}{2}$ days	Remaining $5\frac{1}{2}$ days	Total 80 days

*__AYP__ is Assess Your Progress.

Section	Practice Bank	Study Guide*	Assessment Book*	Visuals	Explorations Lab Manual	Lesson Plans	Technology Book
11.1	75	11.1		Warm-Up 11.1 Folder 13	Master 14	11.1	
11.2	76	11.2	Test 48	Warm-Up 11.2		11.2	
11.3	77	11.3		Warm-Up 11.3		11.3	
11.4	78	11.4		Warm-Up 11.4		11.4	TI-92 Act. 11
11.5	79	11.5	Test 49	Warm-Up 11.5	Add. Expl. 17	11.5	Spreadsheet Act. 10
11.6	80	11.6		Warm-Up 11.6		11.6	Spreadsheet Act. 10
11.7	81	11.7		Warm-Up 11.7		11.7	Calculator Act. 14
11.8	82	11.8	Test 50	Warm-Up 11.8		11.8	
Review Test	83	Chapter Review	Tests 51, 52, Alternative Assessment			Review Test	

*__Spanish versions__ of *Study Guide* and *Assessment Book* are available.

Chapter Support

- Course Guides
- Lesson Plans
- Preparing for College Entrance Tests
- Multi-Language Glossary
- *Test Generator* Software
- Professional Handbook
- Challenge Problems

Software Support

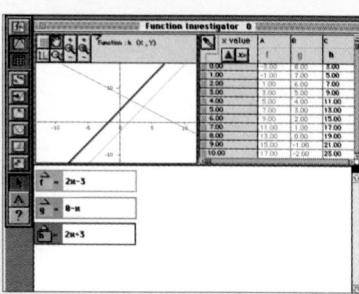

McDougal Littell Mathpack
Geometry Inventor

Internet Support

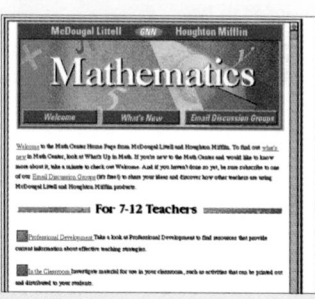

http://www.hmco.com
Next go to McDougal Littell; then the
Education Center; then Secondary Math.

Books, Periodicals

King, James. *Geometry Through the Circle with the Geometer's Sketchpad.* Blackline activity masters and sample activity disks. Berkeley, CA: Key Curriculum Press.

Lamb, John F., Jr. "Two Egyptian Construction Tools." *Mathematics Teacher* (February 1993): pp. 166, 167.

Activities, Manipulatives

Lenart Sphere Construction Materials. Activity masters, transparent sphere, and spherical construction devices. Berkeley, CA: Key Curriculum Press.

Breuningsen, Chris, Bill Bower, Linda Antinone, and Elisa Breuningsen. "Swinging Ellipses." *Real-World Math with the CBL System.* Activity 17: pp. 91–94. Texas Instruments, 1995.

Software

Smith, Scott. *Geometry Connections: Solids.* MS-DOS, 512K minimum. Acton, MA: William K. Bradford Publishing Co.

Jackiw, Nicholas, designer. *The Geometer's Sketchpad.* (Ver. 3.0) for Macintosh or Windows. Berkeley, CA: Key Curriculum Press, 1995.

Laborde, Jean-Marie and Franck Bellemain, designers. *Cabri Geometry II.* Dallas, TX: Texas Instruments, 1994.

TI-92 Geometry. (Resident on the TI-92 graphing calculator.) Dallas, TX: Texas Instruments, 1995.

Internet

Investigate "More Than Anyone Wants to Know About Pi" at:

http://uts.cc.utexas.edc/~joeting/ math/pitime.html

11 | Circles and Spheres

Treating SP**O**RTS *Injuries*

INTERVIEW Ron Courson

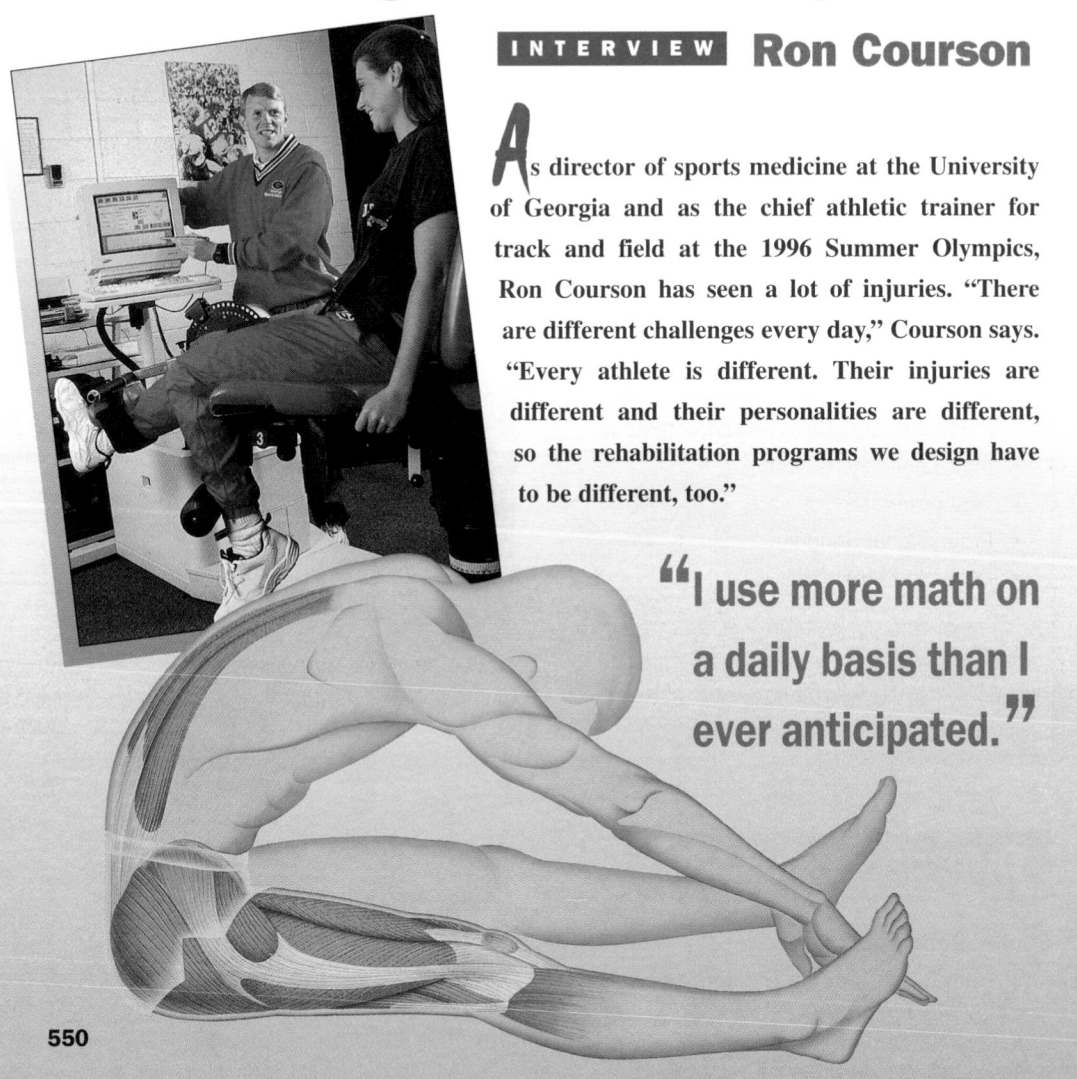

As director of sports medicine at the University of Georgia and as the chief athletic trainer for track and field at the 1996 Summer Olympics, Ron Courson has seen a lot of injuries. "There are different challenges every day," Courson says. "Every athlete is different. Their injuries are different and their personalities are different, so the rehabilitation programs we design have to be different, too."

"I use more math on a daily basis than I ever anticipated."

550

Yet all injuries have some things in common. "Almost everything we do in rehabilitation relates to the *range of motion* of joints like the shoulder, hip, or knee," Courson notes. A joint's range of motion is the arc through which it can move. "One of the first things we do in our evaluation is to measure these arcs and find out whether the athlete has a normal range of motion," Courson explains. He then recommends exercises, such as leg extensions or leg curls, which strengthen the injured part while keeping it within an appropriate arc of motion.

In addition to limiting range of motion, injuries can affect an athlete's sense of balance. To improve balance, Courson sometimes uses a technique developed in Switzerland. The athlete lies on top of a ball and tries to balance as shown. The size of the ball affects how difficult it is to balance. This technique also increases muscle strength.

Educating Athletes

Education, Courson stresses, is an important part of the rehabilitation process. "Even if we see an athlete every day for two hours, the athlete is on his or her own 22 hours a day," he says. "So athletes really need to understand the nature of their injuries and why, for instance, the knee hurts so much when it's extended through the last 45°. We show them a model so they can see all the stresses on the joint and understand the physics and geometry of the situation."

551

Background

Joints in the Body
A joint is a place where two or more bones meet. There are two types of joints: fixed and movable. Fixed joints are sometimes not seen as joints because they do not allow for a large range of movement between the bones. Examples are the joints in the skull that are fixed to protect the brain. Movable joints contain three main categories: hinge joints, pivot joints, and ball-and-socket joints. The joints at the knee and fingers are examples of hinge joints, the movement of the head from side to side is a pivot joint, and the hip and shoulder are examples of ball-and-socket joints.

Second-Language Learners

You might want to invite each second-language learner to look up one or two terms from this interview in a dictionary, and to explain the meaning of those terms to other students learning English.

Multicultural Note

Over 11,000 athletes representing 197 nations competed in the 1996 Summer Olympics. These Olympics were filled with exciting moments for people in many nations. The small nation of Burundi sent its first representative ever—runner Venuste Niyongabo, who won a gold medal in the men's 5000 meter race. Tonga won its first Olympic medal when super heavyweight boxer Paea Wolfgramm took the silver medal in his event. In the men's marathon, Josia Thugwane edged out South Korea's Lee Bong-ju, becoming the first black South African to win a gold medal in the history of the Games.

Mathematical Connection

Physical therapists apply and use a wide variety of mathematical ideas in many areas of their work. First, in assessing a person's injuries, it is often necessary for a physical therapist to define a person's range of motion and compare it to a normal range of motion. Then, when designing a rehabilitation program, a physical therapist can use many devices and techniques that involve the use of mathematics. In Section 11.1, students investigate the range of motion of a shoulder and an elbow. In this section, students calculate angle measures for shoulder flexion, shoulder extension, and elbow flexion. In Sections 11.3 and 11.6, different devices used in the rehabilitation process are explored. In Section 11.3, students investigate angle relationships in a flexibility device and how they relate to its construction. In Section 11.6, students look at spherical rehabilitation devices and calculate their volumes and surface areas.

Explore and Connect

Project
The explanation of how the range of motion was measured should be written as a brief summary or report for the project.

Writing
If some students have not had an injury, suggest they report on a family member, relative, or friend who may have.

Research
The library, Internet, or an encyclopedia may be good sources for this research project. Also, students may want to visit a tai chi instructor to ask him or her to explain the exercise.

Restricted Zones

For an injury of the *rotator cuff*, which is common to swimmers and baseball pitchers, there may be a painful arc between 130° and 150°. Courson does tests to determine precisely where that painful zone lies. He then recommends exercises that break the shoulder motion down into smaller arcs. The exercises avoid the restricted zone, preventing further injury to the athlete.

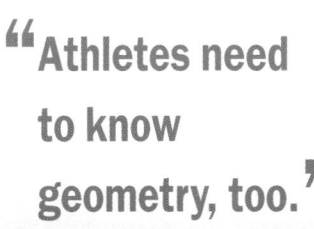

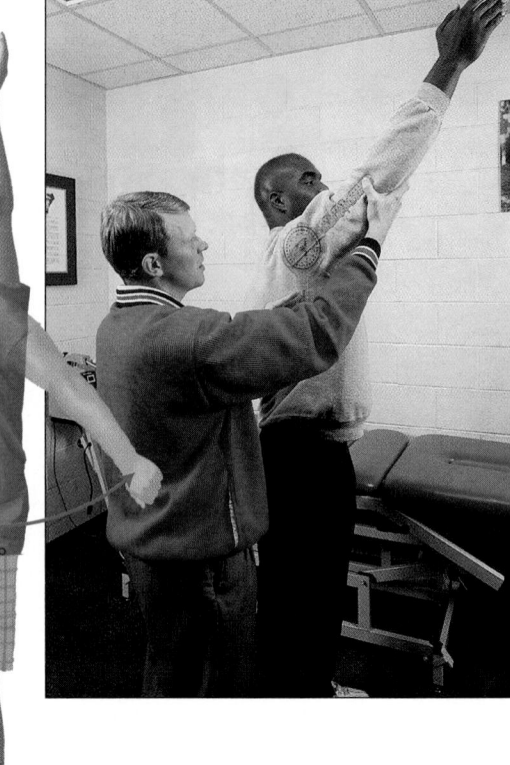

"**Athletes need to know geometry, too.**"

EXPLORE AND CONNECT

Ron Courson

1. Project Choose one joint of the body and carefully measure the range of motion for at least ten people. Be sure no one moves the joint in a way that hurts. Present your results on a poster. Include a sketch of the joint and an explanation of how you measured the range of motion.

2. Writing Describe an injury you have had and how your range of motion was affected.

3. Research One popular exercise is a modified form of *tai chi*, a traditional Chinese series of movements. Learn what some of the movements in *tai chi* are and why practicing them can be beneficial.

Mathematics & Ron Courson

In this chapter, you will learn more about how mathematics is related to athletic training and physical therapy.

Related Exercises

Section 11.1
• **Exercises 11 and 12**

Section 11.3
• **Exercise 16**

Section 11.6
• **Exercises 10–12**

11.1 Angles and Circles

The angles formed by lines that intersect circles have special properties. In the diagram, ∠*ACB* is a **central angle** because the vertex *C* is the center of a circle. ∠*ADB* is an **inscribed angle** because the vertex *D* is on the circle and the sides contain *chords* of the circle. A **chord** of a circle is a segment whose endpoints lie on the circle.

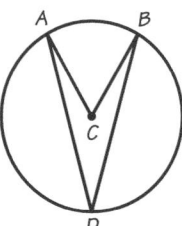

Learn how to...
- identify types of arcs and angles in a circle

So you can...
- find the measures of arcs and angles, such as range of motion

EXPLORATION
COOPERATIVE LEARNING

Comparing Central and Inscribed Angles

Work in a group of three students.
You will need:
- geometry software or a compass and a protractor

m∠ADB	m∠ACB
60°	120°

1 Choose any point on a circle *C*. From this point, draw two chords to form an acute inscribed angle. Label the angle ∠*ADB*. Find *m*∠*ADB* and record it in a table.

2 Draw ∠*ACB*. Find *m*∠*ACB* and record it in the table.

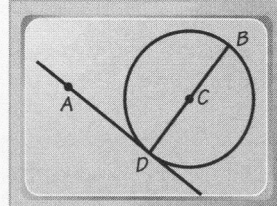

3 Repeat Steps 1 and 2 several times. Choose a different location for *D* each time, but don't change the locations of *A* and *B*. Record your results.

4 Make a conjecture based on your results in Step 3.

5 Repeat Steps 1 and 2 several times, moving only *A* to form each new acute ∠*ADB*. Record your results and make a conjecture.

6 A *tangent line* intersects a circle in only one point. Draw diameter $\overline{DB}$ and tangent $\overleftrightarrow{AD}$. Use *m*∠*ADB* and your results from Step 5 to make a conjecture about the measure of any angle formed by a tangent and a diameter. Explain your reasoning.

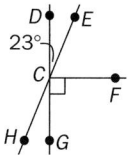

Exploration Note

Purpose
The purpose of this Exploration is to investigate the relationships between inscribed angles, central angles, and the angle formed by a tangent and the diameter that intersects the circle at the point of tangency.

Materials/Preparation
Each group of students needs geometry software or a compass and protractor.

Procedure
Students draw and measure examples of central angles and inscribed angles that intercept the same arc. They record their results and make a conjecture. They then draw a tangent to the circle and the diameter that intersects the point of tangency and make a conjecture about this angle.

Closure
Students should have discovered that the measure of an inscribed angle is half the measure of the central angle that intercepts the same arc, and that the measure of an angle formed by a tangent and a diameter is 90°.

Explorations Lab Manual
See the Manual for more commentary on this Exploration.   Diagram Master 14

For answers to the Exploration, see following page.

Plan⇔Support

Objectives
- Identify types of arcs and angles in a circle.
- Find the measures of arcs and angles.

Recommended Pacing
❖ **Core and Extended Courses**
  Section 11.1: 1 day
❖ **Block Schedule**
  Section 11.1: $\frac{1}{2}$ block
  (with Section 11.2)

Resource Materials
Lesson Support
Lesson Plan 11.1
Warm-Up Transparency 11.1
Overhead Visuals:
  Folder 13: Circles and Angles
Practice Bank: Practice 75
Study Guide: Section 11.1
Explorations Lab Manual:
  Diagram Master 14
Challenge Problems: Set 75
Technology
Geometry Software
McDougal Littell Mathpack
  Geometry Inventor Activity Book:
  Activity 13
Internet:
  http://www.hmco.com

Warm-Up Exercises
Find the measure of each angle.

1. ∠*DCG*   180°
2. ∠*DCF*   90°
3. ∠*ECF*   67°
4. ∠*ECG*   157°
5. ∠*FCH*   113°

Students with a visual learning style should use the various diagrams to help them remember the theorems and postulate in this section. Students with a verbal learning style can focus on the written statements.

Section Notes

Communication: Reading This section contains many new vocabulary terms. You may wish to suggest that students read through the section and write definitions for each new boldface term.

Using Technology Students can explore angles and circles and discover the relationship between an inscribed angle and central angle that intercept the same arc by using Activity 13 in the *Geometry Inventor Activity Book*.

Second-Language Learners Encourage students learning English to begin a list, with bilingual definitions or descriptions, of the many new terms, theorems, and postulates that are introduced in this chapter. Students can begin with the terms on page 553, including *central angle*, *inscribed angle*, and *chord*.

Geometric Thinking Point out that the Tangent Theorem is a biconditional statement. Some students may need to review the meaning of a biconditional. Ask students how they would write two conditional statements that are equivalent to the biconditional.

Think and Communicate

Question 2 focuses students' attention on the fact that any diameter of a circle produces two semicircles.

Section Note

Teaching Tip When discussing the Arc Addition Postulate, point out the similarity of this postulate to the Angle Addition Postulate and the Segment Addition Postulate.

WATCH OUT! ▶
A tangent to a circle is different from a tangent of an angle.

In Step 6 of the Exploration, $\overleftrightarrow{AD}$ is **tangent** to the circle because it is in the same plane as the circle but intersects the circle in only one point. The intersection is called the *point of tangency*. $\overrightarrow{AD}$ and $\overline{AD}$ are also tangents. You may have discovered the following result about tangents in the Exploration.

Tangent Theorem

A line, a ray, or a segment is tangent to a circle if and only if it is in the same plane as the circle and is perpendicular to a radius of the circle at the point of intersection.

$\overleftrightarrow{AC}$ is tangent to $\odot O$ if and only if $\overleftrightarrow{AC} \perp \overleftrightarrow{OB}$ at B.

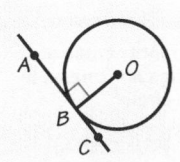

An **arc** is an unbroken part of a circle. In the circle below, points A and B divide the circle into the *minor arc $\overarc{AB}$* and the *minor arc $\overarc{ADB}$*. The arc $\overarc{BCD}$ is an example of a **semicircle**. Notice that three points are used to name a major arc or a semicircle.

The **minor arc $\overarc{AB}$** consists of point A, point B, and all the points on the circle in the interior of $\angle AEB$.

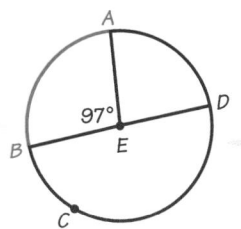

The **major arc $\overarc{ADB}$** consists of point A, point B, and all the points of the circle not on $\overarc{AB}$.

THINK AND COMMUNICATE

1. In the diagram above, $\overarc{BCD}$ is a semicircle. Name another semicircle.

2. Tony says that the endpoints of any diameter of a circle are also the endpoints of a semicircle. Do you agree? Explain.

The **measure** of a minor arc is equal to the measure of the central angle that intercepts it. The measure of a major arc is 360° minus the measure of its corresponding minor arc. For example, in $\odot E$ above, $m\overarc{AB} = 97°$ and $m\overarc{ADB} = 360° - 97° = 263°$. The measure of a semicircle is 180°. A complete circle has measure 360°.

Arc Addition Postulate

In a circle, the measure of the arc formed by two arcs that have exactly one point in common is the sum of the measures of the two arcs.

$m\overarc{AC} = m\overarc{AB} + m\overarc{BC}$

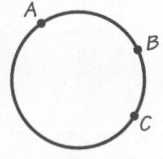

ANSWERS Section 11.1

Answers may vary slightly due to rounding.

Exploration

1–5. Check students' work.

4, 5. The measure of an inscribed angle *ADB* of a circle is half the measure of central angle *ACB*.

6. The measure of an angle formed by a tangent of a circle and a diameter is a right angle. Since $\angle DCB$ is a straight angle, it seems reasonable that

$$m\angle ADB = \frac{1}{2} \cdot m\angle DCB = \frac{1}{2} \cdot 180° = 90°.$$

Think and Communicate

1. $\overarc{BAD}$

2. Yes. Answers may vary. An example is given. In the figure, $\overarc{ADB}$ is a semicircle. Its endpoints are B and D, and $\overline{BD}$ is a diameter.

EXAMPLE 1

Find the measure of each arc.

a. $\overset{\frown}{PQR}$

b. $\overset{\frown}{PR}$

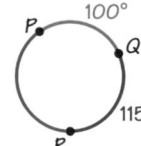

SOLUTION

a. $\overset{\frown}{PQR}$ is formed by $\overset{\frown}{PQ}$ and $\overset{\frown}{QR}$. Use the Arc Addition Postulate.

$$m\,\overset{\frown}{PQR} = m\,\overset{\frown}{PQ} + m\,\overset{\frown}{QR}$$
$$= 100° + 115°$$
$$= 215°$$

So $m\,\overset{\frown}{PQR} = 215°$.

b. $\overset{\frown}{PR}$ is the minor arc with the same endpoints as the major arc $\overset{\frown}{PQR}$.

$$m\,\overset{\frown}{PR} = 360° - m\,\overset{\frown}{PQR}$$
$$= 360° - 215°$$
$$= 145°$$

So $m\,\overset{\frown}{PR} = 145°$.

Two circles are *congruent* if their radii are equal. In the same circle or in congruent circles, arcs whose measures are equal are called **congruent arcs**. In the first circle below, $\angle C$ intercepts $\overset{\frown}{AB}$ because A and B lie on the sides of $\angle C$ and the rest of $\overset{\frown}{AB}$ is in the interior of $\angle C$.

The theorems below, which you may have discovered in the Exploration, describe some relationships between congruent arcs and the angles that intercept them. You will prove these theorems in the exercises.

Inscribed Angle Theorems

The measure of an inscribed angle is equal to half the measure of the intercepted arc.

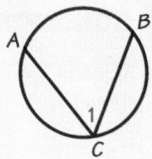

$m \angle 1 = \frac{1}{2} m\,\overset{\frown}{AB}$

In a circle or in congruent circles, inscribed angles that intercept the same arc or congruent arcs are congruent.

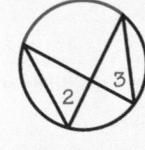

$\angle 2 \cong \angle 3$

The measure of an angle formed by a tangent and a chord is equal to half the measure of the intercepted arc.

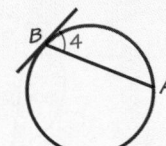

$m \angle 4 = \frac{1}{2} m\,\overset{\frown}{AB}$

Additional Example 1

Find the measure of each arc.

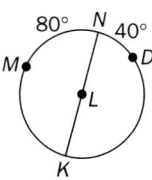

a. $\overset{\frown}{KM}$

By the Arc Addition Postulate, $m\,\overset{\frown}{KM} + m\,\overset{\frown}{MN} = m\,\overset{\frown}{KMN}$. The measure of a semicircle is 180°.

$$m\,\overset{\frown}{KM} + 80° = 180°$$
$$m\,\overset{\frown}{KM} = 180° - 80°$$
$$= 100°$$

So $m\,\overset{\frown}{KM} = 100°$.

b. $\overset{\frown}{KMD}$

$\overset{\frown}{KMD}$ is formed by $\overset{\frown}{KN}$ and $\overset{\frown}{ND}$. Use the Arc Addition Postulate.

$$m\,\overset{\frown}{KMD} = m\,\overset{\frown}{KN} + m\,\overset{\frown}{ND}$$
$$= 180° + 40°$$
$$= 220°$$

So $m\,\overset{\frown}{KMD} = 120°$.

Section Notes

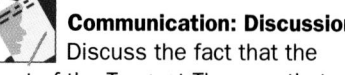 **Communication: Discussion**
Discuss the fact that the part of the Tangent Theorem that states a tangent and the radius to the point of intersection are perpendicular is a special case of the third Inscribed Angle Theorem. When the chord is a diameter, it intercepts a semicircle.

Teaching Tip
The proofs of the first and second Inscribed Angle Theorems are done as Exs. 20 and 23 on pages 558 and 559.

Teach⇔Interact

About Example 2

Geometric Thinking
Ask students why $m \angle QPT = \frac{1}{2} m \widehat{QP}$. They should see that since $\angle QPT$ is the angle formed by a chord and a tangent, its measure is half the measure of the intercepted arc. This is the third Inscribed Angle Theorem.

Additional Example 2

Find the measure of $\angle ADB$.

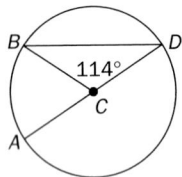

$\angle ACB$ and $\angle BCD$ form a linear pair and thus are supplementary.
$m \angle ACB = 180° - 114° = 66°$
$\angle ACB$ is the central angle that intercepts $\widehat{AB}$, so $m \widehat{AB} = m \angle ACB$.
Therefore, $m \widehat{AB} = 66°$.

$$m \angle ADB = \frac{1}{2} m \widehat{AB}$$
$$= \frac{1}{2} \cdot 66°$$
$$= 33°$$
So $m \angle ADB = 33°$.

Think and Communicate

For question 3, students should realize that the measure of any inscribed angle that intercepts a semicircle is 90°, and that the measure of any central angle that intercepts a semicircle is 180°.

Checking Key Concepts

Student Progress
Questions 1–5 provide a check of students' understanding of many of the basic terms introduced in this section. All students should be able to identify by an example the segment, angles, or line given.

556

EXAMPLE 2

Find the measure of $\angle QPT$.

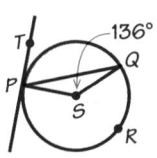

SOLUTION

$\angle QSP$ is the central angle that intercepts $\widehat{QP}$, so $m \widehat{QP} = m \angle QSP$.
Therefore, $m \widehat{QP} = 136°$.

$$m \angle QPT = \frac{1}{2} m \widehat{QP}$$
$$= \frac{1}{2} \cdot 136°$$
$$= 68°$$

So $m \angle QPT = 68°$.

BY THE WAY

On the surface of Earth, a rainbow appears as a partial or complete semicircle. A rainbow is actually a circle, but we see only part of it.

THINK AND COMMUNICATE

Use the diagram at the right.

3. **a.** What is the measure of $\angle ABC$?
   b. What is the measure of any angle that intercepts a semicircle? Explain how you know.

4. Sketch an inscribed angle that intercepts a major arc. What type of angle is it? Explain how you know.

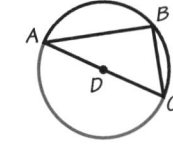

☑ CHECKING KEY CONCEPTS

Give an example of each figure in ⊙B.

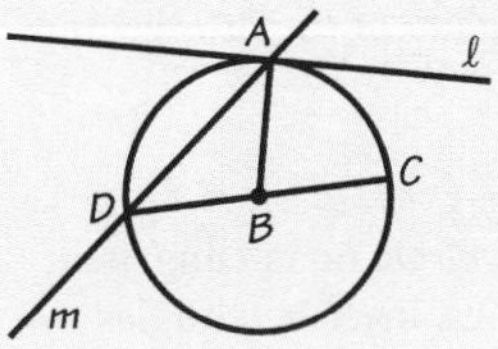

1. a chord
2. a central angle
3. an inscribed angle
4. a tangent to ⊙B
5. a minor arc and a major arc that have the same endpoints

Find each unknown measure.

6. $m \widehat{TUS} = \underline{?}$
7. $m \angle QPR = \underline{?}$
8. $m \widehat{DE} = \underline{?}$

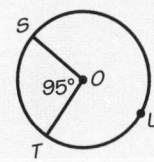
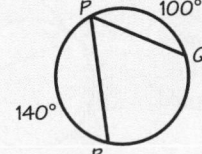
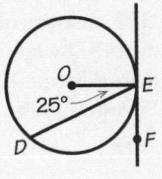

556 Chapter 11 *Circles and Spheres*

Think and Communicate

3. **a.** 90°
   b. 90°; the measure of an inscribed angle is equal to half the measure of the intercepted arc; the measure of a semicircle is 180°.

4.

obtuse angle; The measure of a major arc is greater than the measure of a semicircle, which has a measure of 180°.

Checking Key Concepts

1. $\overline{AD}$ or $\overline{DC}$
2. $\angle ABD$, $\angle ABC$, or $\angle DBC$
3. $\angle ADC$
4. line l
5. $\widehat{AC}$ and $\widehat{CDA}$, or $\widehat{AD}$ and $\widehat{ACD}$
6. 265°
7. 60°
8. 130°

11.1 Exercises and Applications

Extra Practice
exercises on
page 683

Find each measure.

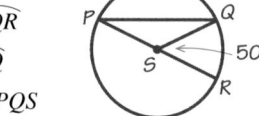

1. $m \overset{\frown}{QR}$ 2. $m \overset{\frown}{PQR}$

3. $m \angle QPR$ 4. $m \overset{\frown}{PQ}$

5. $m \overset{\frown}{QPR}$ 6. $m \angle PQS$

Copy each diagram. Find the measure of each red arc and each blue arc.

7.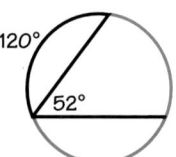
120°
52°

8.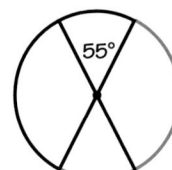
55°

9.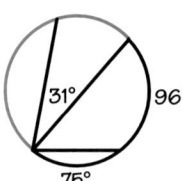
31° 96°
75°

10. **Open-ended Problem** Draw a circle. Locate and label three points on the circle so that two of the points form a 180° arc. Name at least five arcs and give their measures.

INTERVIEW **Ron Courson**

Look back at the article on pages 550–552.

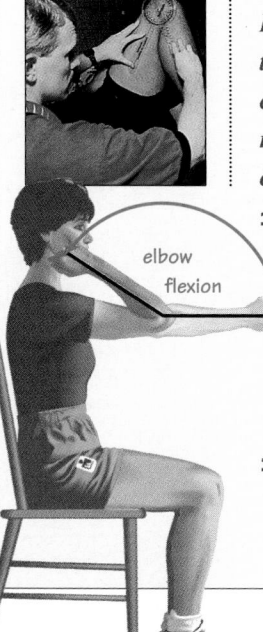
elbow flexion

A physical therapist may measure range of motion to make a diagnosis or treatment plan. Range of motion is the arc through which a part of the body can move. The diagram shows maximum range of motion for certain movements of the shoulder.

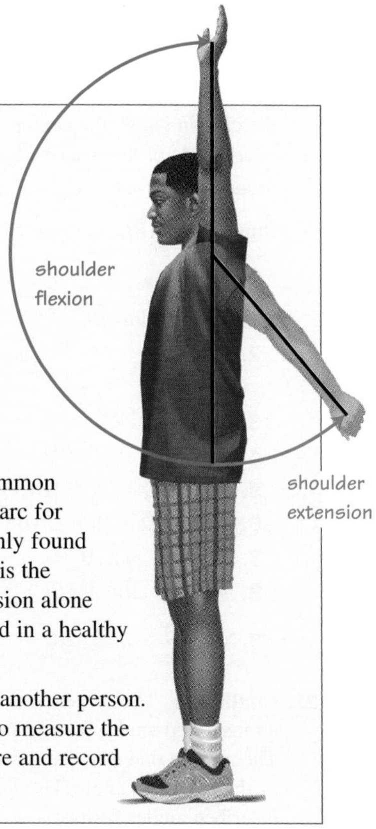

shoulder flexion

shoulder extension

11. A total arc of about 225° for shoulder flexion and extension combined is common in healthy adults. The measure of the arc for shoulder flexion alone that is commonly found in healthy adults is about 180°. What is the measure of the arc for shoulder extension alone that Ron Courson would expect to find in a healthy adult?

12. **Cooperative Learning** Work with another person. Each of you should use the diagram to measure the other person's elbow flexion. Compare and record your results.

11.1 Angles and Circles **557**

Exercises and Applications

1. 50°
2. 180°
3. 25°
4. 130°
5. 310°
6. 25°
7. 136°; 104°
8. 125°; 55°
9. 62°; 127°

10. Answers may vary. An example is given.

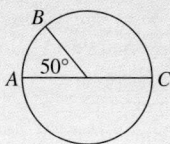
B
50°
A   C

$m \overset{\frown}{AB} = 50°$; $m \overset{\frown}{BC} = 130°$;
$m \overset{\frown}{ABC} = 180°$; $m \overset{\frown}{BCA} = 310°$;
$m \overset{\frown}{BAC} = 230°$

11. 45°

12. Answers may vary. An example is given. between 120° and 130°

Closure Question

Describe the difference between an inscribed angle and a central angle and how their measures are related if they intercept the same arc. An inscribed angle has a vertex that is a point on the circle and its sides are chords of the circle. The vertex of a central angle is the center of a circle and its sides are radii of the circle. The measure of an inscribed angle is half the measure of a central angle that intercepts the same arc.

Apply⇔Assess

Suggested Assignment

❖ **Core Course**
Exs. 1–9, 13–20, 22, 27–31
❖ **Extended Course**
Exs. 1–9 odd, 10–17, 19–31
❖ **Block Schedule**
Day 68 Exs. 1–9, 13–20, 22, 27–31

Exercise Notes

Teaching Tip
Exs. 1–9 If students have difficulty with any of these exercises, suggest that they begin by writing on copies of the diagrams the measures of the arcs or angles that can be found directly by using a central angle or an inscribed angle.

Common Error
Exs. 1–9 Some students may forget that the measure of an inscribed angle is half the measure of its intercepted arc. Refer these students to page 555 and have them review the Inscribed Angle Theorems.

Interview Note: Application
Exs. 11, 12 Students learn how an arc measure is used to describe the range of motion of the shoulder and the elbow.

Integrating Statistics
Ex. 12 You may wish to extend this activity by having students measure the elbow flexion of 20 people. They can then find the mean, median, and mode of the data and discuss which statistics best represent the data.

Exercise Notes

Integrating Algebra

Exs. 13–15, 17–19 In Exs. 13–15, angle or arc measures are represented by algebraic expressions. Students can use the geometric properties of arcs and angles in a circle to write equations that can be solved using algebra. In Exs. 17–19, variables are used to represent angles. The value of the variables can be found using geometric properties.

Communication: Drawing

Exs. 17–19 Point out to students that if a point is marked in the circle, then it is the center. Students should not assume a chord is a diameter unless it goes through a marked point or dot in the circle. For example, in Ex. 17, the chords that go through the dot are diameters. In Ex. 19, however, students cannot assume that either chord is a diameter because there is no dot showing the center.

Writing Proofs

Exs. 20, 21 These two exercises comprise the proof of the first Inscribed Angle Theorem. The proof is in three parts or cases, which is a common technique for proving results in geometry. For this proof, the center of the circle is either on a side of the inscribed angle, inside the inscribed angle, or outside the inscribed angle. Once all cases are proved, the theorem is proved.

Multicultural Note

Exs. 24–26 Ocean navigation has been a fundamental part of life in the Pacific Islands since they were first inhabited by voyagers from southern Asia. Thousands of years ago, Pacific Islanders were able to navigate their canoe-like sailing vessels across vast expanses of the Pacific Ocean without the aid of instruments or charts. To stay on course and to find land, Pacific Island navigators relied on natural guides, such as the position of the sun and stars, the flight pattern of migrating birds, and the size and shape of ocean swells. In recent times, Pacific Island peoples have begun to rely more on modern instruments, but many still know and use traditional navigational methods.

ALGEBRA Find the value of each variable and the measure of each labeled arc or angle.

13.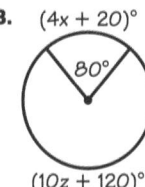
$(4x + 20)°$
$80°$
$(10z + 120)°$

14.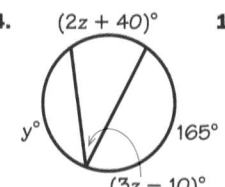
$(2z + 40)°$
$y°$
$165°$
$(3z − 10)°$

15.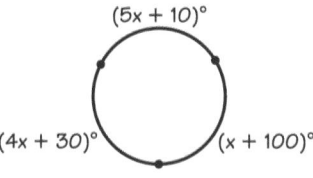
$(5x + 10)°$
$(4x + 30)°$
$(x + 100)°$

16. SAT/ACT Preview Points P, Q, and R lie on the same circle. If $A = m\,\widehat{PR} + m\,\widehat{PQR}$, and B = 360°, then:

A. $A > B$ **B.** $B > A$ **C.** $A = B$ **D.** relationship cannot be determined

ALGEBRA Find the value of each variable.

17.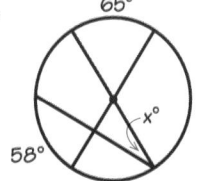
$65°$
$x°$
$58°$

18.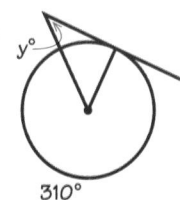
$y°$
$310°$

19.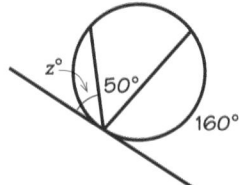
$z°$
$50°$
$160°$

20. Copy and complete the proof of the first Inscribed Angle Theorem on page 555 for the case in which the center of the circle is *on* a side of the inscribed angle.

Given: $\angle ABD$ is inscribed in $\odot C$.

Prove: $m\angle ABD = \frac{1}{2}m\widehat{AD}$

Statements	Reasons
1. Draw radius $\overline{AC}$.	**1.** ?
2. ?	**2.** All radii of a circle are congruent.
3. $\triangle ABC$ is isosceles.	**3.** ?
4. $m\angle ABC = m\angle BAC$	**4.** ?
5. $m\angle ACD = m\angle ABC + m\angle BAC$	**5.** ?
6. $m\angle ACD = 2(m\angle ABC)$	**6.** ?
7. $m\widehat{AD} = m\angle ACD$	**7.** ?
8. $m\widehat{AD} = 2(m\angle ABC)$	**8.** ?
9. $\frac{1}{2}m\widehat{AD} = m\angle ABC$	**9.** Algebra

21. Challenge The center of the circle may be *outside* or *in the interior* of an inscribed angle. Write a paragraph proof of the first Inscribed Angle Theorem in one of these cases. (*Hint:* Draw a diameter from the vertex of the inscribed angle. Use your results from Exercise 20 for the two inscribed angles formed.)

558 Chapter 11 *Circles and Spheres*

---

13. $x = 15$; $z = 16$;
$(4x + 20)° = 80°$;
$(10z + 120)° = 280°$

14. $z = 15$; $y = 125$;
$(2z + 40)° = 70°$;
$(3z − 10)° = 35°$

15. $x = 22$;
$(x + 100)° = 122°$;
$(5x + 10)° = 120°$;
$(4x + 30)° = 118°$

16. C   **17.** 28.5

18. 40   **19.** 50

20. (1) For any two points, there is exactly one line through the points.
(2) $\overline{CA} \cong \overline{CB}$
(3) Def. of an isosceles triangle
(4) Isosceles Triangle Theorem
(5) The measure of an exterior angle of a triangle is equal to the sum of the measures of the two interior angles that are not adjacent to it.
(6) Substitution (Steps 4 and 5)
(7) Def. of measure of a minor arc
(8) Substitution (Steps 6 and 7)

21. See answers in back of book.

22. Draw $\overline{SQ}$. By the Arc Addition Postulate, $m\,\widehat{SPQ} + m\,\widehat{SRQ} = 360°$. By the first Inscribed Angle Theorem, $m\angle P = \frac{1}{2}m\,\widehat{SRQ}$ and $m\angle R = \frac{1}{2}m\,\widehat{SPQ}$. Then $m\angle P + m\angle R = \frac{1}{2}m\,\widehat{SRQ} + \frac{1}{2}m\,\widehat{SPQ} = 180°$. By definition, $\angle P$ and $\angle R$ are supplementary. Similarly, $\angle S$ and $\angle Q$ are supplementary.

22. If the vertices of a polygon lie on a circle, then the polygon is *inscribed* in the circle. Prove that the opposite angles of an inscribed quadrilateral are supplementary.

   Given: Quadrilateral *PQRS* is inscribed in a circle.

   Prove: $m\angle P + m\angle R = 180°$

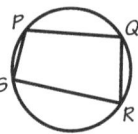

23. Prove the second Inscribed Angle Theorem on page 555.

NAVIGATION Suppose you are on a boat in the ocean and want to find your location, *T,* on the map at the right. From the boat you can see the two lighthouses, *M* and *N,* and the buoy *B.* Using a navigator's compass, you find that $m\angle MTB = 36°$ and $m\angle BTN = 56°$. Trace the shoreline, including *M, B,* and *N.* Use your diagram for Exercises 24–26. You will need a protractor, a compass, and a straightedge.

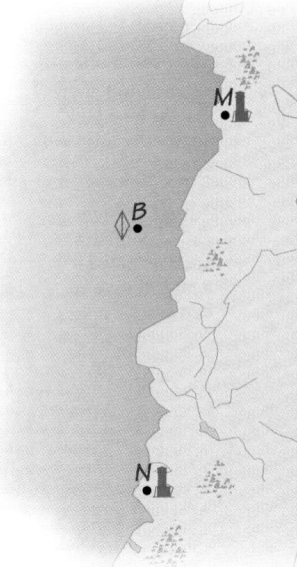

24. Draw $\triangle BMC$ so that $m\angle B = m\angle M = 90° - m\angle MTB$ and *C* is in the water. Construct $\odot C$ with radius *CM.* Explain why $m\angle C = 2m\angle MTB$.

25. **a. Writing** Circle *C* in Exercise 24 is called a *circle of position.* Use the first and second Inscribed Angle Theorems and $m\angle C$ from Exercise 24 to explain why your boat may be anywhere on $\odot C$.

   b. Could the boat not be on $\odot C$? Explain your reasoning.

26. Use points *B* and *N* and $m\angle BTN$ to draw another circle of position for your boat. Label the center of the circle *D.*

   a. How can you use $\odot D$ and $\odot C$ from Exercise 24 to find your exact position?

   b. Label your position *T* and measure $\angle MTB$ and $\angle BTN$ to check your results.

ONGOING ASSESSMENT

27. **Open-ended Problem** Copy the diagram. Without measuring, label the measures of as many of the arcs and angles in the diagram as you can.

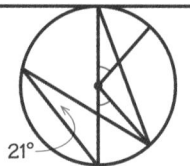

SPIRAL REVIEW

Find the volume and surface area of each right cone or regular pyramid. *(Section 10.7)*

28.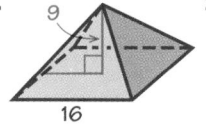

29.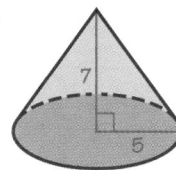

30.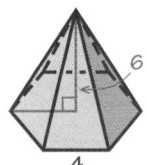

31. If you choose a point at random on $\overline{AB}$, what is the probability that the point is on $\overline{AD}$? *(Section 9.5)*

Exercise Notes

Application
Exs. 24–26 These exercises introduce students to a way in which circles and their associated angles can be used in navigation. You may wish to suggest that students complete these exercises with a partner.

Assessment Note
Ex. 27 A discussion of this exercise can help you assess students' understanding of the theorems in this section.

Practice 75 for Section 11.1

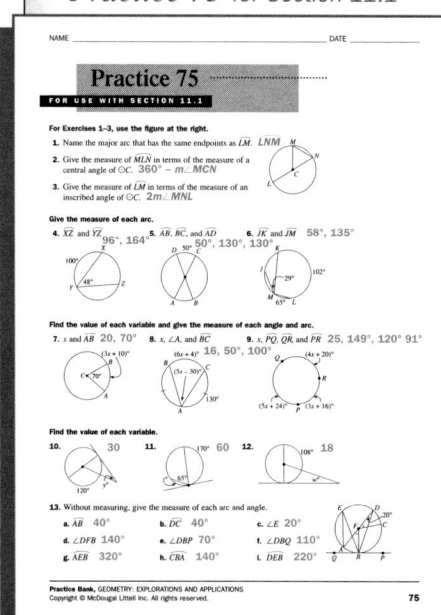

---

23. **Given:** $\odot P \cong \odot Q; \widehat{AB} \cong \widehat{CD}$
   Prove: $\angle 1 \cong \angle 2$

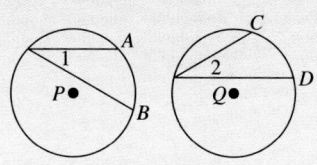

Statements (Reasons)

1. $m\angle 1 = \frac{1}{2}m\,\widehat{AB}; m\angle 2 = \frac{1}{2}m\,\widehat{CD}$

   (The measure of an inscribed angle is equal to half the measure of the intercepted arc.)

2. $\widehat{AB} \cong \widehat{CD}$ or $m\,\widehat{AB} = m\,\widehat{CD}$
   (Given; def. of congruence)

3. $\frac{1}{2}m\,\widehat{AB} = \frac{1}{2}m\,\widehat{CD}$
   (Multiplication Property)

4. $m\angle 1 = m\angle 2$ or $\angle 1 \cong \angle 2$
   (Transitive Property (Steps 1, 2, and 3); def. of congruence)

(This same proof with minor variations can be used for inscribed angles in the same circle that intercept the same arc, or inscribed angles in the same circle that intercept congruent arcs.)

24.

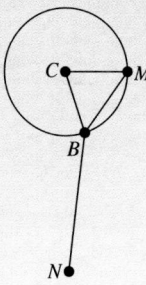

$m\angle B = m\angle M = 90° - 36° = 54°;$
$m\angle C = 180° - (m\angle B + m\angle M) =$
$180° - (54° + 54°) = 72° =$
$2m\angle MTB$

25–27. See answers in back of book.

28–30. Answers are given to the nearest tenth.

28. 768.0; 641.3

29. 183.3; 213.7

30. 83.1; 128.1

31. $\frac{23}{36} \approx 63.9\%$

559

Objectives

- Find arc and angle measures when segments intersect circles.
- Solve real-world problems involving segments that intersect circles.

Recommended Pacing

❖ **Core and Extended Courses**
Section 11.2: 1 day

❖ **Block Schedule**
Section 11.2: $\frac{1}{2}$ block
(with Section 11.1)

Resource Materials

Lesson Support
Lesson Plan 11.2

Warm-Up Transparency 11.2

Practice Bank: Practice 76

Study Guide: Section 11.2

Challenge Problems: Set 76

Assessment Book: Test 48

Technology
Scientific Calculator

McDougal Littell Mathpack
  Geometry Inventor Activity Book:
    Activity 10

Internet:
  http://www.hmco.com

Warm-Up Exercises

Find each measure to the nearest tenth.

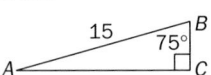

1. CB 3.9

2. AC 14.5

3. Find $m\overset{\frown}{GF}$.

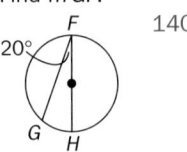

140°

Solve each equation.

4. $\dfrac{3x + 12}{2} = 39$ 22

5. $(4x + 17) + (10x - 6) = 130$
   8.5

560

SECTION

11.2 | Tangents, Secants, and Chords

Learn how to...

- find arc and angle measures when segments intersect circles

So you can...

- solve problems such as how far a migrating bird can see

Hikers can use the Global Positioning System (GPS) to keep from getting lost. A hand-held GPS receiver can calculate its exact location by comparing broadcasts from at least four satellites. Scientists chose an altitude for the satellites that minimizes the number of satellites needed to cover the globe completely.

Not drawn to scale

EXAMPLE 1   Application: Global Positioning

Find $m\overset{\frown}{RTP}$ and use it to describe how far around Earth the satellite's signal can reach.

SOLUTION

$$m\overset{\frown}{RTP} = m\angle ROP$$

$$= m\angle ROS + m\angle POS$$

$$= 2\, m\angle ROS$$

$\triangle RSO \cong \triangle PSO$ by the HL Theorem so $\angle ROS \cong \angle POS$.

Use trigonometry to find $m\angle ROS$.

$$\cos\angle ROS = \frac{RO}{OS}$$

$$= \frac{4000}{4000 + 12{,}500}$$

$$\approx 0.2424$$

$$m\angle ROS \approx 76°$$

$$m\overset{\frown}{RTP} \approx 2 \cdot 76$$

$$\approx 152°$$

The measure of a full circle is 360°, so the satellite's signal reaches about $\frac{152°}{360°}$, or about 42% of the way around Earth along $\odot O$.

---

560   Chapter 11   *Circles and Spheres*

In Example 1, trigonometry and congruent triangles were used to find the measure of an arc. You can use the rules below to find the measures of arcs formed by tangents or *secants*. A **secant** is a line, a ray, or a segment that contains a chord.

Angles Formed by Tangents and Secants

The measure of an angle formed by the intersection of two tangents, two secants, or a secant and a tangent, at a point outside a circle, is half the difference of the measures of the intercepted arcs.

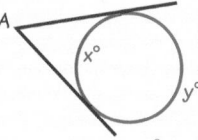 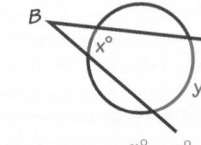 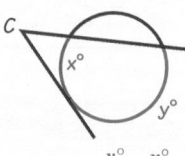

$$m \angle A = \frac{y° - x°}{2} \qquad m \angle B = \frac{y° - x°}{2} \qquad m \angle C = \frac{y° - x°}{2}$$

THINK AND COMMUNICATE

1. Use the solution to Example 1 and the theorem above to find $m \angle S$ in the diagram on page 560.

2. If the GPS satellites were closer to Earth, more satellites would be needed for the system. Explain why.

BY THE WAY

GPS is used by many people, including pilots, weather forecasters, and farmers. From any point on Earth, at least five GPS satellites are always in view.

EXAMPLE 2

Find $m \angle GJM$ and $m \widehat{GM}$.

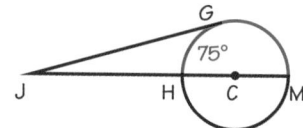

SOLUTION

Find $m \widehat{GM}$ first.

$$m \widehat{GM} + 75° = 180°$$

$\overline{HM}$ is a diameter of the circle, so $m \widehat{HGM} = 180°$.

$$m \widehat{GM} = 105°$$

Find $m \angle GJM$. Use the theorem above.

$$m \angle GJM = \frac{m \widehat{GM} - 75°}{2}$$
$$= \frac{105 - 75}{2}$$
$$= 15°$$

Therefore, $m \angle GJM = 15°$ and $m \widehat{GM} = 105°$.

11.2 Tangents, Secants, and Chords **561**

ANSWERS Section 11.2

Think and Communicate

1. 28°

2. The closer to Earth the satellite is, the smaller the arc intercepted by the tangents, and the smaller the percentage of Earth's circumference reached by the signal.

Teach⇔Interact

Additional Example 1

Scientists want a satellite's signal to reach 35% of the way around Earth along circle C. To do this, $m \widehat{AE}$ must equal 126°. How far from Earth's surface should this satellite be?

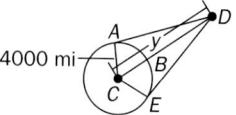

Let y = the distance from the center of Earth to the satellite. $\triangle ADC \cong \triangle EDC$ by HL Theorem, so $m \widehat{AB} = \frac{1}{2} m \widehat{AE} = \frac{1}{2} \cdot 126° = 63°$. Therefore, $m \angle ACD = 63°$. Use trigonometry to find y.

$$\cos 63° = \frac{4000}{y}$$
$$0.4540 \approx \frac{4000}{y}$$
$$y \approx \frac{4000}{0.4540}$$
$$\approx 8810$$

The distance from the center of Earth to the satellite is about 8810 miles. The radius of Earth is about 4000 miles, so that the distance of the satellite from the surface of Earth is about 8810 – 4000, or about 4810 miles.

Additional Example 2

Find $m \widehat{TP}$ and $m \widehat{SR}$.

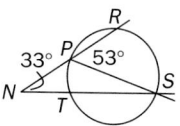

Find $m \widehat{SR}$ first. $\angle SPR$ is an inscribed angle, so

$$m \angle SPR = \frac{1}{2} m \widehat{SR}$$
$$53° = \frac{1}{2} m \widehat{SR}$$
$$106° = m \widehat{SR}$$

Find $m \widehat{TP}$. Use the theorem about angles formed by tangents and secants.

$$m \angle N = \frac{m \widehat{SR} - m \widehat{TP}}{2}$$
$$33° = \frac{106° - m \widehat{TP}}{2}$$
$$2 \cdot 33 = 106 - m \widehat{TP}$$
$$66 = 106 - m \widehat{TP}$$
$$40° = m \widehat{TP}$$

Therefore, $m \widehat{TP} = 40°$ and $m \widehat{SR} = 106°$.

Section Notes

Visual Thinking
Ask students to draw a circle and then create a diagram that shows angles formed by secants, tangents, and intersecting chords. Ask them to explain the relationship of each angle to relevant arcs of the circle. This activity involves the visual skills of *recognition* and *exploration*.

Student Study Tip
Encourage students to make a list of the properties they have learned about lines intersecting circles. The list should include the Tangent Theorem, the three Inscribed Angle Theorems, and the properties about arcs formed by tangents and secants and arcs formed by intersecting chords. The list should include a sketch for each property.

Additional Example 3

Find the value of *x* and the measure of each labeled arc or angle.

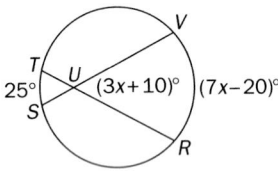

Use the theorem about arcs formed by intersecting chords.
$$\frac{(7x - 20)° + 25°}{2} = (3x + 10)°$$
$$(7x - 20) + 25 = 2(3x + 10)$$
$$7x + 5 = 6x + 20$$
$$x + 5 = 20$$
$$x = 15$$
So $(3x + 10)° = 55°$ and $(7x - 20)° = 85°$.

Closure Question

Two segments can intersect on a circle, outside a circle, or inside a circle. Describe what you know about the intercepted arcs in each situation. *If the segments intersect on a circle, they form an inscribed angle whose measure is half that of the intercepted arc. If the segments intersect outside the circle, the measure of the acute angle formed by the lines is equal to half of the difference of the measures of the intercepted arcs. If the segments intersect in the interior of the circle, the measure of an angle formed by the two lines is equal to half the sum of the measures of the intercepted arcs.*

562

You know how to find angle and arc measures if two lines intersect either *on* a circle or *outside* it. You can also find angle and arc measures if two chords intersect *inside* a circle.

Arcs Formed by Intersecting Chords

The measure of an angle formed by two chords is equal to half the sum of the measures of the intercepted arcs.

$$m \angle 1 = m \angle 2 = \frac{x° + y°}{2}$$

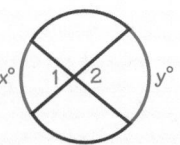

EXAMPLE 3   Connection: Algebra

Find the value of *z* and the measure of each labeled arc.

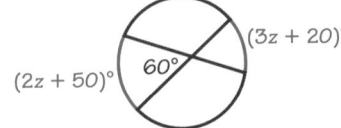

SOLUTION

Use the theorem above.
$$\frac{(3z + 20)° + (2z + 50)°}{2} = 60°$$
$$\frac{5z + 70}{2} = 60 \qquad \text{Multiply both sides of the equation by 2.}$$
$$5z + 70 = 120$$
$$5z = 50$$
$$z = 10$$

So $(3z + 20)° = 50°$ and $(2z + 50)° = 70°$.

☑ CHECKING KEY CONCEPTS

ALGEBRA Find the values of *x* and *y*.

1.

2.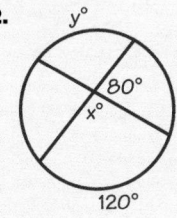

3.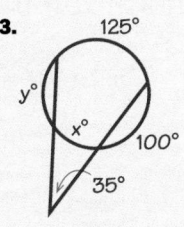

562   Chapter 11 *Circles and Spheres*

Checking Key Concepts

1. $x = 40$; $y = 60$

2. $x = 100$; $y = 80$

3. $x = 55$; $y = 80$

Exercises and Applications

1–5. Answers may vary. Examples are given.

1. $\overline{DA}$

2. $\angle ADF$

3. $\angle BFE$

4. $\angle AGB$

5. 35

6. 70

7. $y = 90$; $z = 20$

11.2 Exercises and Applications

Extra Practice
exercises on
page 683

Give an example of each figure in ⊙O.

1. a secant

2. an angle formed by two secants

3. an inscribed angle

4. an angle formed by two chords

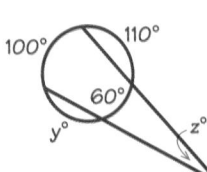

ALGEBRA Find the value of each variable.

5.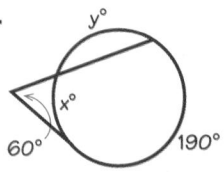

6.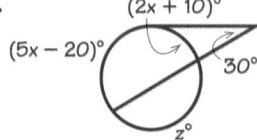

7.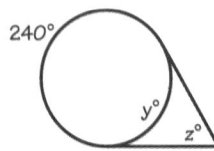

8.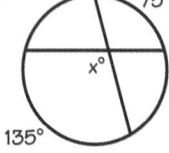

9.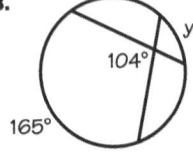

10.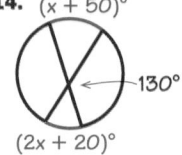

11. **Logical Reasoning** Arrange the statements below and give reasons to write a proof of the second case of the theorem on page 561.

Given: Secants $\overline{AP}$ and $\overline{CP}$ intersect as shown.

Prove: $m\angle 1 = \dfrac{m\widehat{AC} - m\widehat{BD}}{2}$

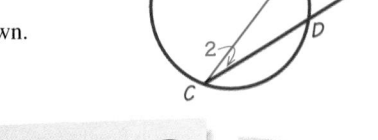

$m\angle 1 = \frac{1}{2}m\widehat{AC} - \frac{1}{2}m\widehat{BD}$

$m\angle 1 = \dfrac{m\widehat{AC} - m\widehat{BD}}{2}$

Secants $\overline{AP}$ and $\overline{CP}$ intersect as shown.

$m\angle 3 = \frac{1}{2}m\widehat{AC};\ m\angle 2 = \frac{1}{2}m\widehat{BD}$

$m\angle 1 = m\angle 3 - m\angle 2$

$m\angle 3 = m\angle 2 + m\angle 1$

Draw chord $\overline{BC}$.

$m\angle 1 = \frac{1}{2}\left(m\widehat{AC} - m\widehat{BD}\right)$

ALGEBRA Copy each diagram. Find the measure of each labeled angle or arc.

12.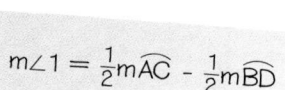

13.

14.

11.2 Tangents, Secants, and Chords **563**

---

8. $x = 70$; $y = 100$

9. $x = 30$; $z = 160$

10. $y = 120$; $z = 60$

11. **Statements (Reasons)**

1. Secants $\overline{AP}$ and $\overline{CP}$ intersect as shown. (Given)

2. Draw chord $\overline{BC}$. (For any two points, there is exactly one line through the two points.)

3. $m\angle 3 = \frac{1}{2}m\widehat{AC}$; $m\angle 2 = \frac{1}{2}m\widehat{BD}$ (The measure of an inscribed angle is half the measure of the intercepted arc.)

4. $m\angle 3 = m\angle 2 + m\angle 1$ (The measure of an exterior angle of a triangle is equal to the sum of the measures of the two interior angles that are not adjacent to it.)

5. $m\angle 1 = m\angle 3 - m\angle 2$ (Subtraction Property)

6. $m\angle 1 = \frac{1}{2}m\widehat{AC} - \frac{1}{2}m\widehat{BD}$ (Substitution Property (Steps 3 and 5))

7. $m\angle 1 = \frac{1}{2}(m\widehat{AC} - m\widehat{BD})$ (Distributive Property)

8. $m\angle 1 = \dfrac{m\widehat{AC} - m\widehat{BD}}{2}$ (Algebra)

12. $105°$

13. $43°$

14. $(x + 50°) = 60°$; $(2x + 20)° = 40°$

563

Apply⟺Assess

Suggested Assignment

❖ **Core Course**
Exs. 1–16, 24–27, AYP

❖ **Extended Course**
Exs. 1–11 odd, 12–27, AYP

❖ **Block Schedule**
Day 68 Exs. 1–16, 24–27, AYP

Exercise Notes

Student Progress
Exs. 1–4 These exercises assess whether students understand the vocabulary of this section. Students should be able to answer these exercises correctly before proceeding to the subsequent exercises.

Geometric Thinking
Exs. 5–10, 12–14 After completing these exercises, you may wish to ask students to list the properties they used. Besides those learned in this section, the list should include the fact that the measure of a circle is 360°, and that vertical angles are congruent.

Common Error
Exs. 8–10, 12–14 The kinds of errors students tend to make with these types of exercises are more algebraic than geometric. Students who have incorrect answers should check to see that they have written the correct equation and that they have solved it correctly.

Writing Proofs
Ex. 11 You may wish to ask for two or three volunteers to write their proofs on the board for examination and discussion by the class. This activity would provide a good review of the reasons used.

Exercise Notes

Writing Proofs

Exs. 15, 16 Students may use the proof format of their choice for these proofs. They should think about a plan before beginning the proof. Students may get useful ideas by studying Example 1 on page 560 and Ex. 11 on page 563.

Student Progress

Ex. 17 This open-ended problem gives students an opportunity to show what they have learned in this section. You may also want to suggest that they create a problem like those in Exs. 12–14. Have students exchange their papers with a partner and solve the problems to see that one of the variables has a value of 35.

Interdisciplinary Problems

Exs. 18–20 If any student has difficulty with these exercises, point out that the method used to solve Exs. 18 and 19 is similar to that of Example 1 on page 560. Students can refer to the Solution for that Example for help.

Career Connection

Exs. 18–20 People who study birds are called ornithologists. Ornithology is a branch of biology that encompasses all facets of bird life, including bird anatomy, behavior, ecology, evolution, wildlife management, and veterinary medicine.

In Exercises 15 and 16, you will prove two cases of the theorem on page 561.

15. Given: Secant $\overline{CP}$ and tangent $\overline{AP}$

Prove: $m \angle P = \dfrac{m\widehat{AC} - m\widehat{AB}}{2}$

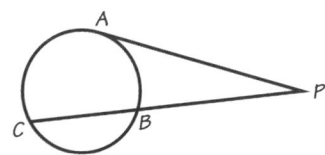

16. Given: Tangents $\overline{AP}$ and $\overline{BP}$

Prove: $m \angle P = \dfrac{m\widehat{ACB} - m\widehat{AB}}{2}$

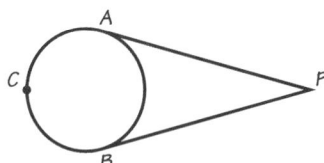

17. Open-ended Problem Create a problem like the ones in Exercises 5–10. The value of one of the variables should be 35.

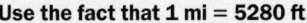

Connection   BIOLOGY

Many factors affect the altitude at which birds fly while migrating between their summer and winter habitats. The amount of land that a bird can see while flying depends on its altitude.

Use the fact that 1 mi = 5280 ft.

18. During migration, a bald eagle flies at an altitude between 100 ft and 1000 ft.
   a. What is the measure of the arc of Earth that an eagle can see when flying at 100 ft? at 1000 ft?
   b. What percent of the circumference of Earth can an eagle see at each height?
   c. The circumference of Earth is about 25,000 mi. Use your answer to part (b) to determine the length of the arc the eagle can see at each height. Explain what these numbers represent.

19. Whooping cranes usually fly at an altitude between 900 ft and 4500 ft during migration. How much of the circumference of Earth can the crane see at each of these altitudes? Give your answer as a percent and as a length. (*Hint:* Look at Exercise 18 for the steps to answer this question.)

20. **Research** Find out why whooping cranes migrate at a higher altitude than bald eagles do.

100 ft

4000 mi

Not drawn to scale

564   Chapter 11   *Circles and Spheres*

---

15. Statements (Reasons)
1. Draw chord $\overline{AB}$. (For any two points, there is exactly one line through the two points.)
2. $m \angle PAB = \frac{1}{2} m\widehat{AB}$ (The measure of an angle formed by a tangent and a chord is equal to half the measure of the intercepted arc.)
3. $m \angle ABC = \frac{1}{2} m\widehat{AC}$ (The measure of an inscribed angle is half the measure of the intercepted arc.)
4. $m \angle ABC = m \angle PAB + m \angle P$ (The measure of an exterior angle of a triangle is equal to the sum of the measures of the two interior angles that are not adjacent to it.)
5. $m \angle P = m \angle ABC - m \angle PAB$ (Subtraction Property)
6. $m \angle P = \frac{1}{2} m\widehat{AC} - \frac{1}{2} m\widehat{AB}$ (Substitution Property (Steps 2, 3, and 5))
7. $m \angle P = \dfrac{m\widehat{AC} - m\widehat{AB}}{2}$ (Distributive Property)

16. Statements (Reasons)
1. Draw chord $\overline{PC}$ intersecting the circle at D. (For any two points, there is exactly one line through the two points.)
2. $m \angle APC = \dfrac{m\widehat{AC} - m\widehat{AD}}{2}$; $m \angle BPC = \dfrac{m\widehat{BC} - m\widehat{BD}}{2}$ (Ex. 15)
3. $m \angle APC + m \angle BPC =$
$\dfrac{m\widehat{AC} - m\widehat{AD}}{2} + \dfrac{m\widehat{BC} - m\widehat{BD}}{2} =$
$\dfrac{(m\widehat{AC} + m\widehat{BC}) - (m\widehat{AD} + m\widehat{BD})}{2}$
(Addition Property; algebra)

4. $m\widehat{ACB} = m\widehat{AC} + m\widehat{BC}$;
$m\widehat{AB} = m\widehat{AD} + m\widehat{BD}$
(Arc Addition Postulate)
5. $m \angle P = m \angle APC + m \angle BPC$ (Angle Addition Postulate)
6. $m \angle P = \dfrac{m\widehat{ACB} - m\widehat{AB}}{2}$ (Substitution (Steps 3, 4, and 5))

17. Answers may vary. Check students' work.

18. a. about 0.35°; about 1.12°
   b. about 0.097%; about 0.31%
   c. about 24 mi; about 78 mi; The numbers represent how much of the circumference of Earth that the eagle can see from a given altitude.

19. about 0.29% ≈ 73 mi; about 0.66% ≈ 165 mi

SPORTS In the game of soccer, the goalkeeper tries to prevent an opponent from kicking the ball into the net and scoring a goal.

21. **a. Writing** The goalkeeper may try to stay on the line that bisects the angle formed by the opponent and the goal posts as shown below. Explain why this will help the goalkeeper prevent a goal.

   b. The goalkeeper will also approach an opponent who is trying to score. Explain how this helps the goalkeeper prevent a goal.

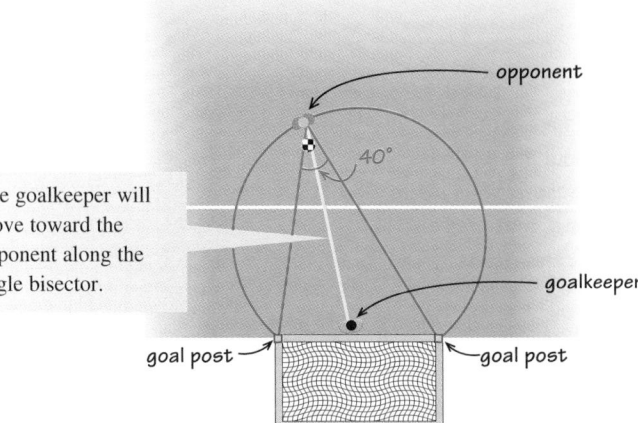

The goalkeeper will move toward the opponent along the angle bisector.

22. **a.** Explain why the angle formed by the goal posts and the opponent is 40° for every point on the circle in the diagram.

   b. Open-ended Problem An approaching opponent wants to make the angle 40° or larger. Sketch some examples and explain why a wider angle provides a better opportunity for making a goal. What is the best position for the opponent to attempt a goal? Explain your reasoning.

23. **a. Challenge** Find the value of *x*.

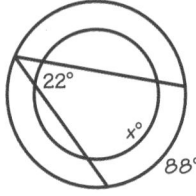

   b. Write an equation for *c* in terms of *a* and *b*.

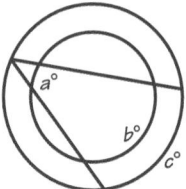

ONGOING ASSESSMENT

24. **Writing** Write a note to a student who missed class for this lesson. Explain what information is needed to find each arc or angle measure: $m\,\widehat{AB}$, $m\angle C$, $m\angle CPD$.

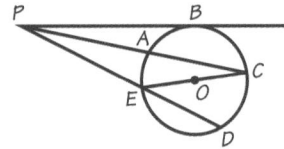

11.2 Tangents, Secants, and Chords **565**

Apply⟺Assess

Assess Your Progress

Journal Entry
A brief discussion of this exercise will help students to informally assess their understanding of the theorems involved and how they are related.

Progress Check 11.1–11.2

See page 606.

Practice 76 for Section 11.2

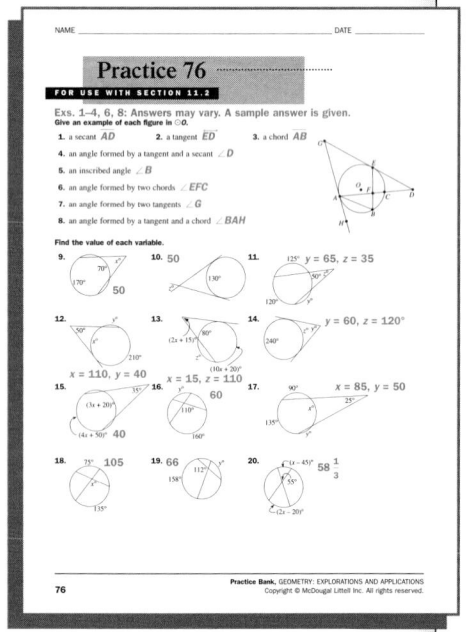

25. Describe this translation using coordinate notation: Every point moves to the left 5 units and up 2 units. *(Section 8.3)*

Sketch each figure. *(Section 11.1)*

26. a circle and a line that is tangent to it

27. a circle and two congruent inscribed angles

ASSESS YOUR PROGRESS

VOCABULARY

central angle (p. 553)   **minor arc** (p. 554)
inscribed angle (p. 553)   **major arc** (p. 554)
chord (p. 553)   **measure of an arc** (p. 554)
tangent to a circle (p. 554)   **congruent arcs** (p. 555)
arc (p. 554)   **secant** (p. 561)
semicircle (p. 554)

Find each unknown arc or angle measure. *(Section 11.1)*

1. $m\widehat{ABC} = \underline{?}$ **2.** $m\angle PQR = \underline{?}$ **3.** $m\angle SMT = \underline{?}$

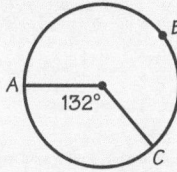 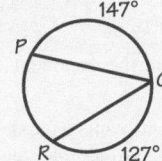 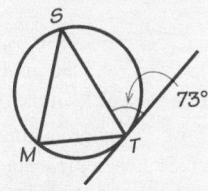

4. Each vertex of equilateral $\triangle ABC$ lies on $\odot M$. Explain why $m\widehat{AB} = m\widehat{BC} = m\widehat{AC}$.

ALGEBRA **Find the measure of each labeled arc.** *(Section 11.2)*

5. 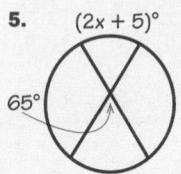   **6.** 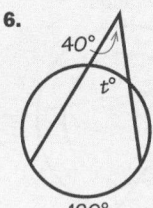   **7.**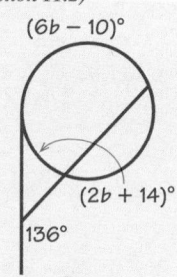

8. Journal Compare the theorem that describes how to find the measure of an inscribed angle and the theorem that describes how to find the measure of an angle formed by two secants. Are these theorems related? Explain your reasoning.

566   Chapter 11 *Circles and Spheres*

---

25. $(a, b) \rightarrow (a - 5, b + 2)$

26, 27. Answers may vary. Examples are given.

26.

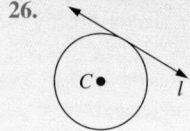

27.

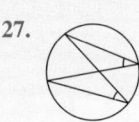

566

Assess Your Progress

1. 228°

2. 43°

3. 73°

4. Each angle of the triangle is an inscribed angle with measure 60°. Then each of the arcs has measure 120°. (The measure of an inscribed angle is equal to half the measure of the intercepted arc.)

5. $3x° = 75°$; $(2x + 5)° = 55°$

6. $t° = 40°$

7. $(6b - 10)° = 158°$; $(2b + 14)° = 70°$

8. Yes; every chord is part of a secant. The first Inscribed Angle Measure Theorem is a special case of the theorem for finding the measure of an angle formed by two secants. In this case, the secants intersect at a point on the circle. If you think of two secants intersecting at a point

P outside the circle and then imagine P moving closer and closer to the circle, you can imagine that one of the two intercepted arcs gets smaller and smaller. If P were on the circle, the measure of the smaller arc would be zero, and both theorems would give the same value for $m\angle P$.

11.3

Applying Properties of Chords

Learn how to...
* identify relationships among arcs and chords of circles

So you can...
* solve problems involving arcs and chords of circles

What is the "center" of a triangle? One way you could define the "center" is the point that is an equal distance from each vertex. This point is called the *circumcenter*. It is the center of the circle that passes through the vertices of the triangle. In the Exploration, you will *circumscribe a circle about a triangle.*

Circumscribing a Triangle

Work with another student. You will need:
* patty paper
* compass
* straightedge

1 Draw any △*DEF* on a piece of patty paper.

2 Carefully fold the paper to form the perpendicular bisector of each side of the triangle.

3 The three perpendicular bisectors should meet at a point. Label this point *C*.

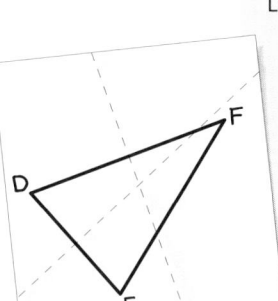

4 Draw a circle with *C* as the center and *CD* as the radius. Describe your results. How do you know that all three vertices are the same distance from *C*?

5 Compare your results from Step 4 with the results from other groups. Make a conjecture.

6 Check whether your conjecture in Step 5 is true for other triangles. Be sure to try right, obtuse, and acute triangles.

Exploration Note

Purpose
The purpose of this Exploration is for students to learn that the perpendicular bisectors of the sides of a triangle intersect in a point which is the center of the circle that can be circumscribed about the triangle.

Materials/Preparation
Each pair of students needs patty paper, a compass, and a straightedge.

Procedure
Students draw a triangle on patty paper and fold it to form the perpendicular bisectors of each side of the triangle. They use the point of intersection of the three perpendicular

bisectors to draw a circle whose radius is equal to the distance from the center to one of the vertices.

Closure
Discuss the results of Steps 5 and 6 so students understand that the circle intersects all three vertices of the triangle. This result leads to the discussion of the Chord Bisector Theorems on page 568.

Explorations Lab Manual
See the Manual for more commentary on this Exploration.

For answers to the Exploration, see following page.

Plan⟺Support

Objectives
* Identify relationships among arcs and chords of circles.
* Solve real-world problems involving arcs and chords of circles.

Recommended Pacing
❖ **Core and Extended Courses**
  Section 11.3: 1 day
❖ **Block Schedule**
  Section 11.3: $\frac{1}{2}$ block
  (with Section 11.4)

Resource Materials
Lesson Support
Lesson Plan 11.3
Warm-Up Transparency 11.3
Practice Bank: Practice 77
Study Guide: Section 11.3
Challenge Problems: Set 77
Technology
Scientific Calculator
Geometry Software
Internet:
  http://www.hmco.com

Warm-Up Exercises

Find each value.

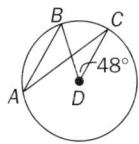

1. $m\overarc{BC}$ 48°

2. $m\angle BAC$ 24°

$\overline{LM}$ is the perpendicular bisector of $\overline{RV}$.

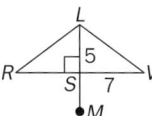

3. Find *RV*.   14

4. Is △*RSL* ≅ △*VSL*? Explain.
   Yes; by the SAS Postulate.

5. Find *LV*.   about 8.6

Teach⇔Interact

Section Notes

Second-Language Learners

To help students learning English understand and remember the concept of circumscribing a circle about a triangle, you may want to discuss the different parts of the word *circumscribe*. Point out that the prefix *circum-* means "around," and comes from the same root as *circle*. The root *-scribe* means "write." If students are native speakers of Spanish, point out that *circum-* comes from the same root as *circulo* (circle) and that *-scribe* comes from the same root as *escribir* (to write).

Teaching Tip

To help students understand how the Exploration relates to the Chord Bisector Theorems, have them copy the triangle used in the Exploration together with the circumscribed circle. They should include only one of the perpendicular bisectors in their copy and mark the center of the circle. Point out that this perpendicular bisector goes through the center of the circle and that the side of the triangle it bisects is a chord of the circle. When students erase the other two sides of the triangle, they get a diagram equivalent to the one at the top of this page.

Additional Example 1

Find $m\,\widehat{HE}$ and $m\,\widehat{EF}$.

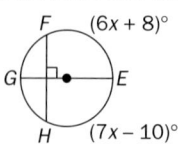

F $(6x + 8)°$
G E
H $(7x - 10)°$

Because $\overline{GE}$ is a diameter,
$m\,\widehat{HG} + m\,\widehat{HE} = 180°$ and
$m\,\widehat{GF} + m\,\widehat{FE} = 180°$.
Using substitution,
$m\,\widehat{HG} + m\,\widehat{HE} = m\,\widehat{GF} + m\,\widehat{FE}$.
Because $\overline{GE}$ is a diameter that is perpendicular to $\overline{FH}$, it bisects $\widehat{HGF}$.
$m\,\widehat{HG} = m\,\widehat{GF}$
Subtracting these equal measures from the equation above,
$m\,\widehat{HE} = m\,\widehat{FE}$
$(7x - 10)° = (6x + 8)°$
$x = 18$.
Substitute 18 for x to find $m\,\widehat{HE}$.
$m\,\widehat{HE} = (7x - 10)°$
$= 7 \cdot 18 - 10$
$= 116$
So $m\,\widehat{HE} = 116°$. Because
$m\,\widehat{HE} = m\,\widehat{FE}$, $m\,\widehat{FE} = 116°$ also.

You saw in the Exploration that each side of a triangle is a chord of the circumscribed circle, and the perpendicular bisector of each chord contains the center of the circle. The arc *corresponding to* the chord is also bisected.

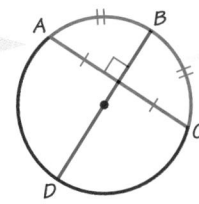

$\widehat{AC}$ **corresponds to** $\overline{AC}$.

$\widehat{AB} \cong \widehat{BC}$, so
$\overline{BD}$ bisects $\widehat{AC}$.

Chord Bisector Theorems

The perpendicular bisector of a chord passes through the center of the circle.

If $\overline{AC} \perp \overline{BD}$ and $\overline{AE} \cong \overline{CE}$, then $\overline{BD}$ contains the center of $\odot P$.

A diameter that is perpendicular to a chord bisects the chord and its corresponding arc.

If $\overline{AC} \perp \overline{BD}$ and $\overline{BD}$ contains the center of $\odot P$, then $\overline{AE} \cong \overline{CE}$ and $\widehat{AB} \cong \widehat{CB}$.

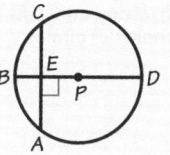

EXAMPLE 1   **Connection: Algebra**

Find $m\,\widehat{AB}$ and $m\,\widehat{BC}$.

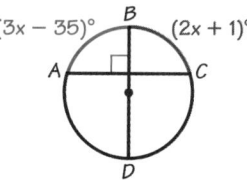

$(3x - 35)°$ B $(2x + 1)°$
A C
D

SOLUTION

Because $\overline{BD}$ is a diameter that is perpendicular to $\overline{AC}$, it bisects $\widehat{ABC}$.

$$m\,\widehat{AB} = m\,\widehat{BC}$$
$$(3x - 35)° = (2x + 1)°$$
$$x = 36$$

Substitute 36 for x to find $m\,\widehat{AB}$.

$$m\,\widehat{AB} = (3x - 35)°$$
$$= 3 \cdot 36 - 35$$
$$= 73$$

So $m\,\widehat{AB} = 73°$. Because $m\,\widehat{BC} = m\,\widehat{AB}$, $m\,\widehat{BC} = 73°$ also.

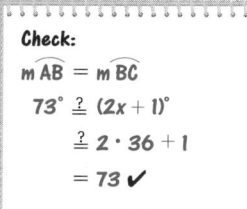

Check:
$m\,\widehat{AB} = m\,\widehat{BC}$
$73° \overset{?}{=} (2x + 1)°$
$\overset{?}{=} 2 \cdot 36 + 1$
$= 73$ ✔

ANSWERS   Section 11.3

Exploration

1–3. Check students' work.

4. All three vertices are on circle C. C is on the perpendicular bisector of each of the three segments whose endpoints are vertices of the triangle. If a point lies on the perpendicular bisector of a segment, then the point is equidistant from the endpoints of the segment. Then C is equidistant from the three vertices.

5. All groups should have the same results. The intersection of the perpendicular bisectors of the sides of a triangle is equidistant from the vertices of the triangle.

6. The conjecture given in the answer to Step 5 is true for all triangles.

You can use the second Chord Bisector Theorem to prove that two congruent chords of a circle are an equal distance from the center of the circle.

EXAMPLE 2

Write the key steps of a proof of the theorem:

In a circle, two congruent chords are equidistant from the center of the circle.

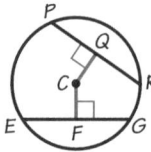

SOLUTION

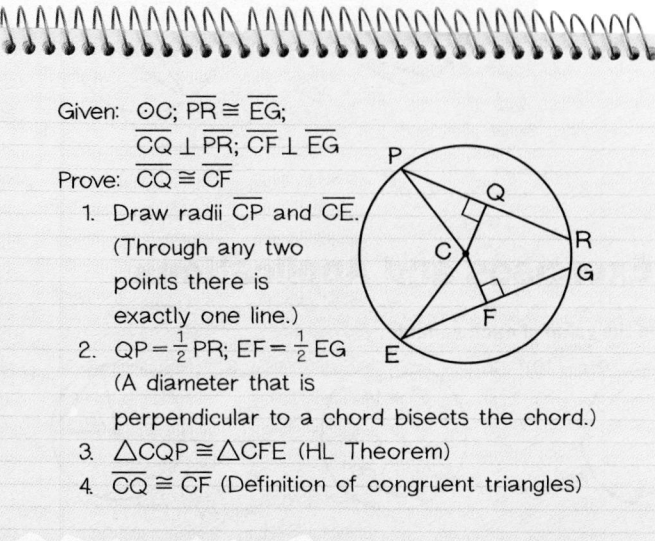

Given: ⊙C; $\overline{PR} \cong \overline{EG}$;
$\overline{CQ} \perp \overline{PR}$; $\overline{CF} \perp \overline{EG}$
Prove: $\overline{CQ} \cong \overline{CF}$

1. Draw radii $\overline{CP}$ and $\overline{CE}$.
   (Through any two points there is exactly one line.)
2. $QP = \frac{1}{2} PR$; $EF = \frac{1}{2} EG$
   (A diameter that is perpendicular to a chord bisects the chord.)
3. $\triangle CQP \cong \triangle CFE$ (HL Theorem)
4. $\overline{CQ} \cong \overline{CF}$ (Definition of congruent triangles)

The converse of the theorem in Example 2 is also true. A proof of the converse is in the exercises.

THINK AND COMMUNICATE

1. Suppose you want to prove that two congruent chords of congruent circles are equidistant from the center of each circle. Explain how the proof would be different from the proof in Example 2.

2. State the converse of the theorem in Example 2.

3. **a.** In the diagram, $\overset{\frown}{QR} \cong \overset{\frown}{RS}$. Explain how you know that $\overline{QR} \cong \overline{RS}$.
   b. If two chords of the same circle are congruent, are their corresponding arcs also congruent? Explain why or why not.

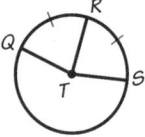

As you may have discovered in Question 3, in the same circle or congruent circles, congruent chords have congruent corresponding arcs. The converse is also true. You will prove the converse in the exercises.

11.3 Applying Properties of Chords **569**

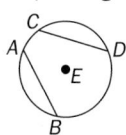
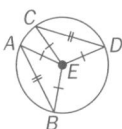
569

Closure Question

Describe two ways to show that two chords are congruent. *Show that their corresponding arcs are congruent or show that they are equidistant from the center of the circle.*

Apply⇔Assess

Suggested Assignment

❖ **Core Course**
Exs. 1–6, 8, 9, 12, 17–26

❖ **Extended Course**
Exs. 1–26

❖ **Block Schedule**
Day 69 Exs. 1–6, 8, 9, 12, 17–26

Exercise Notes

Integrating Algebra
Exs. 1–6 Variables or variable expressions are used to represent measures in these exercises. For Exs. 1–3, students need to write and solve algebraic equations.

Problem Solving
Exs. 4, 5 These exercises illustrate a common strategy that students should remember when solving similar problems. In both exercises, students know the length of the radius. They can draw another radius to form a right triangle, and use the Pythagorean theorem to find the desired length.

 Communication: Drawing
Ex. 7 This open-ended problem gives students an opportunity to use drawings to better understand the concept of a circumscribed circle. Students can use patty paper for this exercise and follow the steps of the Exploration on page 567.

Writing Proofs
Exs. 8, 9, 12 Students can choose the proof format they wish to use for these exercises. Suggest that they use at least two different types.

✓ CHECKING KEY CONCEPTS

Find each length or measure.

1. $m\widehat{BC}$ 2. AD
3. $m\widehat{CFA}$ 4. BF

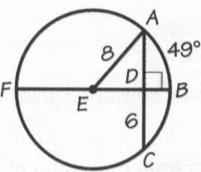

Find the value of each variable.

5. 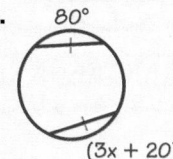   6. 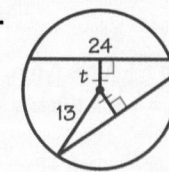   7.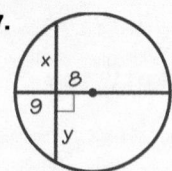

11.3 Exercises and Applications

Extra Practice exercises on page 684

ALGEBRA Find the value of each variable.

1. 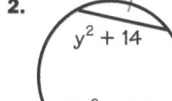   2. 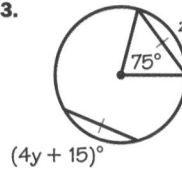   3.

4.   5.   6.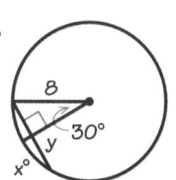

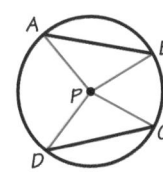

7. **Open-ended Problem** Draw a large right triangle, a large acute triangle, and a large obtuse triangle. Circumscribe a circle about each of the three triangles you drew.

8. Write a complete proof of the theorem in Example 2.

9. Write a proof of the theorem: In a circle, congruent arcs have congruent chords.

   Given: $\odot P$, $\widehat{AB} \cong \widehat{CD}$
   Prove: $\overline{AB} \cong \overline{CD}$

Checking Key Concepts

1. 49°   2. 6

3. 262°   4. 16

5. $x = 20$ 6. $t = 5$

7. $x = y = 15$

Exercises and Applications

1. $t = 18$

2. $y = 8$

3. $z = 75$; $y = 15$

4. $t = 6\sqrt{2} \approx 8.5$

5. $t = 21$; $x = 42$

6. $x = 30$; $y = 4$

7. Answers may vary. Examples are given.

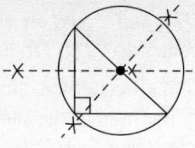

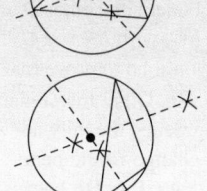

8, 9. See answers in back of book.

10. 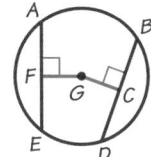 **Technology** Use geometry software or patty paper and a straightedge. Draw any △XYZ. Find the circumcenter, the centroid, the incenter, and the orthocenter of the triangle. (See Exercises 13–15 on page 324 for definitions of *centroid, incenter,* and *orthocenter.*) Are the four points collinear? Do you think this will be true for any triangle? Explain.

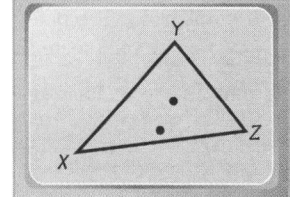

11. a. Challenge A circle is circumscribed about a square. The radius of the circle is *r*. Write an expression for the perimeter of the square in terms of *r*.

   b. Write an expression for the area of the square in part (a).

12. Write a proof of the theorem: In a circle, chords equidistant from the center are congruent.

   Given: $\overline{GC} \perp \overline{BD}$
   $\overline{GF} \perp \overline{AE}$
   $\overline{GF} \cong \overline{GC}$
   Prove: $\overline{AE} \cong \overline{BD}$

Connection ▸ **HIGHWAY SAFETY**

An accident investigator may need to determine how fast a car was moving before it was involved in an accident. If the tracks left by a car are an arc of a circle, the radius of the circle can be used to estimate the speed of the car.

13. Sheila Martinez is investigating an accident. She chooses two points *A* and *B* on the tire marks and measures *AB*. Then she finds the midpoint, *C*, of $\overline{AB}$ and measures *CD*, as shown in the diagram. Use her measurements and the Pythagorean theorem to find the radius, *r*, of the circle. Which of the theorems on page 568 did you use?

14. For a level road, the formula $S \approx 3.86\sqrt{fr}$ gives an estimate of the car's speed in miles per hour, where *f* = the *coefficient of friction* and *r* = the radius of the circle in feet. The coefficient of friction, *f*, is a measure of how slippery the road is. If *f* = 0.8 and *r* = 90 ft, find the speed of the car to the nearest mile per hour.

15. Writing Use the formula in Exercise 14. Will the marks from a fast-moving car have a *larger* or *smaller* radius than those from a slow-moving car? Explain your reasoning.

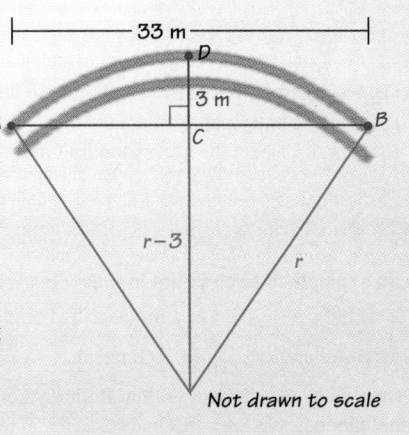

33 m

3 m

A   C   B

r−3

r

Not drawn to scale

11.3 Applying Properties of Chords   **571**

Apply⟺Assess

Exercise Notes

 Using Technology
Ex. 10 In this exercise, students use geometry software to discover an interesting result concerning the circumcenter, centroid, incenter, and orthocenter of a triangle. You may wish to have students work in pairs at a computer and discuss their results before each writes his or her own response.

Challenge
Ex. 11 Students may approach this problem in several different ways. One way is to use the properties of a 45-45-90 triangle. Another way is to use the Pythagorean theorem. Ask volunteers to explain their solutions to the class so that others can see the different approaches.

Teaching Tip
Ex. 12 You may need to point out to students that additional radii need to be drawn to complete this proof.

 Application
Exs. 13–16 These exercises provide students with examples of how the geometry of arcs and chords can be applied to estimate the speed of a car involved in an accident and to understand the construction and function of a piece of apparatus used in physical therapy. In both situations, students learn how to determine the radius of a circle by examining one of its arcs.

Communication: Writing
Ex. 15 Students should try several examples to help them understand the relationship between the speed of the car and the radius of a circle. They should then be able to explain the relationship by analyzing the equation for speed given in Ex. 14.

---

10. Answers may vary. Check students' work. Students may notice that the centroid, circumcenter, and orthocenter always appear to be collinear but the incenter is not, generally, on the same line. (This line is called *Euler's line.*) For equilateral triangles, the four points are the same point.

11. a. $\dfrac{8r}{\sqrt{2}} \approx 5.66r$

   b. $2r^2$

12. Given: $\overline{GC} \perp \overline{BD}$; $\overline{GF} \perp \overline{AE}$;
   $\overline{GF} \cong \overline{GC}$
   Prove: $\overline{AE} \cong \overline{BD}$
   Statements (Reasons)
   1. Draw $\overline{GA}$, and $\overline{GB}$. (For any two points, there is exactly one line through the points.)
   2. $\overline{GA} \cong \overline{GB}$ (Def. of radius)
   3. $\overline{GC} \perp \overline{BD}$; $\overline{GF} \perp \overline{AE}$ (Given)
   4. $\angle AFG$ and $\angle BCG$ are right angles; △AFG and △BCG are right triangles. (Def. of perpendicular lines and right triangles)

5. $\overline{GF} \cong \overline{GC}$ (Given)
6. △AFG ≅ △BCG (HL Theorem)
7. $\overline{AF} \cong \overline{BC}$ or AF = BC (Def. of congruent triangles and congruent segments)
8. 2AF = 2BC (Multiplication Property)
9. AE = 2AF; BD = 2BC (A diameter that is perpendicular to a chord bisects the chord and its corresponding arc.)
10. AE = BD or $\overline{AE} \cong \overline{BD}$ (Substitution (Steps 8 and 9); def. of congruence)

13. about 47 m; The perpendicular bisector of a chord passes through the center of the circle.

14. 33 mi/h

15. If $S \approx 3.86\sqrt{fr}$, then $r \approx \dfrac{S^2}{(3.86)^2 f} \approx \dfrac{S^2}{14.90f}$. For a constant value of *f*, as *S* increases, so does *r*. Then the marks from a fast-moving car will have a larger radius than those from a slow-moving car.

Exercise Notes

Cooperative Learning

Ex. 16 This multifaceted application is appropriate for small group work. Students can work together to solve the problem and then write their solutions independently. In doing this exercise, students use trigonometry and the Tangent Theorem.

Teaching Tip

Exs. 17, 18 These two exercises combine to prove the second Chord Bisector Theorem on page 568.

Assessment Note

Ex. 19 After students complete this exercise, you may wish to have them exchange their papers with a partner. Each partner should follow the directions to see if the desired result can be produced.

Practice 77 for Section 11.3

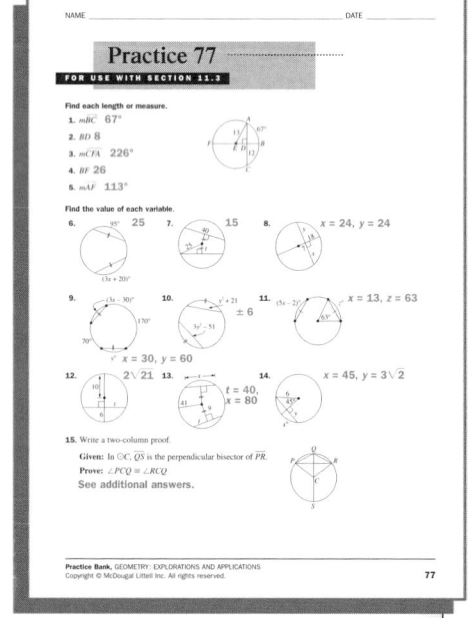

16. PHYSICAL THERAPY A physical therapist may have a client use a tilt board as a way to improve balance and flexibility. A cross section of a tilt board shows that the bottom is an arc of a circle.

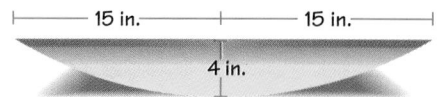

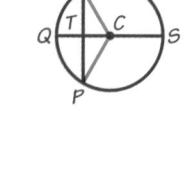

a. Imagine that the cross section is part of a circle. Use the Pythagorean theorem to find the radius of the circle to the nearest inch.

b. Use trigonometry to find the measure of the central angle that intercepts the arc of the tilt board.

c. What is the maximum angle that the flat surface of the tilt board can make with the ground while a person is standing on it? Explain your reasoning.

17. Copy the flow proof of the theorem: A diameter that is perpendicular to a chord bisects the chord. Give the reason for each step.

Given: $\odot C$; $\overline{PR} \perp \overline{QS}$

Prove: $\overline{PT} \cong \overline{RT}$

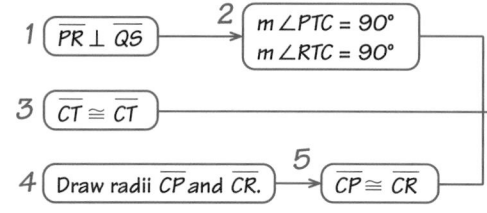

18. Write a proof of the theorem: A diameter that is perpendicular to a chord bisects the arc corresponding to the chord. (*Hint:* The proof is similar to the one in Exercise 17.)

ONGOING ASSESSMENT

19. Open-ended Problem Explain how you could use geometry software or patty paper to demonstrate that a diameter perpendicular to a chord bisects the corresponding arc.

SPIRAL REVIEW

Give an example of each figure in $\odot E$. (*Sections 11.1 and 11.2*)

20. a secant   **21.** an inscribed angle

22. a central angle   **23.** a tangent

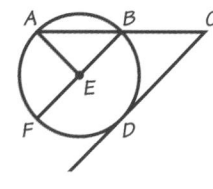

Find the value of each expression. Round your answers to four decimal places. (*Sections 10.3 and 10.4*)

24. sin 36°   **25.** cos 67°   **26.** tan 15°

572   Chapter 11 *Circles and Spheres*

---

16. a. 30 in.   **b.** 60°

  c. 30°; The maximum angle occurs when the flat surface is positioned so that the floor is tangent to the circle at one endpoint of the board. The measure of an angle formed by a tangent and a chord is half the measure of the intercepted arc (60°).

17. (1) Given
    (2) Def. of perpendicular lines

(3) Reflexive Property
(4) For any two points, there is exactly one line through the points.
(5) Def. of radius
(6) HL Theorem
(7) Def. of congruent triangles

18. Refer to the figure shown for Ex. 17 on page 572.
    Given: $\odot C$; $\overline{PR} \perp \overline{QS}$
    Prove: $\overparen{PQ} \cong \overparen{QR}$
    Draw radii $\overline{CP}$ and $\overline{CR}$. (For any two points, there is exactly one

line through the points.) $\overline{CP}$ and $\overline{CR}$ are congruent because they are radii of the same circle. Since $\overline{PR} \perp \overline{QS}$, $\triangle PCT$ and $\triangle RCT$ are right triangles. $\overline{CT}$ is congruent to itself, so $\triangle PCT \cong \triangle RCT$ by the HL Theorem. Corresponding angles PCT and RCT are congruent. Then, since $m \overparen{PQ} = m \angle PCT$ and $m \overparen{QR} = m \angle RCT$, $\overparen{PQ}$ and $\overparen{QR}$ are congruent.

19. See answers in back of book.
20. $\overline{CA}$; $\overline{BF}$; $\overline{AB}$
21. $\angle ABE$; $\angle BAE$
22. $\angle AEF$; $\angle AEB$
23. $\overrightarrow{CD}$
24. 0.5878
25. 0.3907
26. –2.1445

572

11.4 Segment Lengths in Circles

Learn how to...
- find lengths of segments formed by chords, tangents, and secants

So you can...
- solve problems such as explaining why Earth appears to be flat

In earlier sections, you learned about the measures of angles and arcs formed by chords, tangents, and secants of circles. The lengths of the segments formed by chords, tangents, and secants of a circle are also related.

THINK AND COMMUNICATE

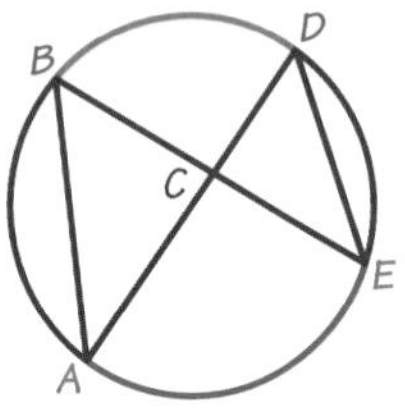

1. Explain why $\angle B \cong \angle D$ and $\angle A \cong \angle E$.

2. Explain why $\triangle ABC \sim \triangle EDC$.

3. Complete this proportion:

$$\frac{AC}{EC} = \frac{?}{DC}$$

4. a. Complete this equation:
$$AC \cdot DC = \underline{?} \cdot EC.$$

   b. Will the equation in part (a) be true for any pair of chords that intersect inside a circle? Explain.

The conclusion that you reached by answering the questions above is one of several theorems about the relationships between the lengths of segments that intersect circles.

Lengths of Chords and Secants

If $\overline{AD}$ and $\overline{BE}$ are chords of a circle and $\overline{AD}$ intersects $\overline{BE}$ at C, then $AC \cdot DC = BC \cdot EC$.

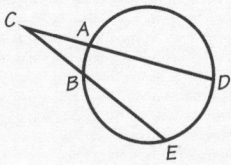

If points A, B, D, and E lie on a circle and $\overrightarrow{DA}$ and $\overrightarrow{EB}$ intersect outside the circle at C, then $CA \cdot CD = CB \cdot CE$.

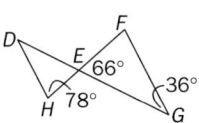

11.4 Segment Lengths in Circles **573**

Plan⇔Support

Objectives
- Find lengths of segments formed by chords, tangents, and secants.
- Solve real-world problems involving lengths of segments in circles.

Recommended Pacing
❖ **Core and Extended Courses**
Section 11.4: 1 day
❖ **Block Schedule**
Section 11.4: $\frac{1}{2}$ block
(with Section 11.3)

Resource Materials
Lesson Support
Lesson Plan 11.4
Warm-Up Transparency 11.4
Practice Bank: Practice 78
Study Guide: Section 11.4
Challenge Problems: Set 78
Technology
Technology Book:
    TI-92 Activity 11
Scientific Calculator
McDougal Littell Mathpack
    Geometry Inventor Activity Book:
    Activity 14
Internet:
    http://www.hmco.com

Warm-Up Exercises

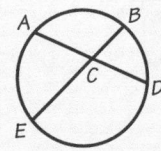

1. Is $\triangle HDE$ similar to $\triangle FGE$? Justify your answer. Yes; by the AA Similarity Postulate.

Use the diagram above to complete each proportion.

2. $\frac{DH}{ED} = \frac{?}{EG}$ *GF*

3. $\frac{HE}{FE} = \frac{ED}{?}$ *EG*

Solve.

4. $\frac{12}{35} = \frac{x}{140}$ 48

5. $14(x - 3) = 9(2x + 5)$ −21.75

ANSWERS Section 11.4

Think and Communicate

1. In a circle or in congruent circles, inscribed angles that intercept the same arc or congruent arcs are congruent.

2. AA Similarity Theorem

3. *BC*

4. a. *BC*

   b. Yes; intersecting chords will always form a pair of similar triangles for the reasons described in questions 1 and 2.

574

Think and Communicate

These questions lead to the discovery of the first theorem on page 573. Students should refer to their responses when writing the proof of this theorem in Ex. 20 on page 577.

Section Note

Teaching Tip

Encourage students to write the theorems of this section using simplified verbal phases. For example, the first theorem on page 573 can be written as: The product of the segments of one chord equals the product of the segments of the other. The second theorem can be summarized as: Outside segment times whole segment equals outside segment times whole segment.

Additional Example 1

In the figure below, $LN = 14.2$ and $PN = 13.5$. Find MN and ON.

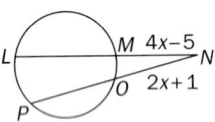

Points M, O, P, and L lie on a circle and $\overrightarrow{LM}$ and $\overrightarrow{PO}$ intersect outside the circle at N. Use the second theorem on page 573. Then use the distributive property and solve for x.

$$MN \cdot NL = ON \cdot PN$$
$$(4x - 5)14.2 = (2x + 1)13.5$$
$$56.8x - 71 = 27x + 13.5$$
$$29.8x = 84.5$$
$$x \approx 2.8$$

To find MN and ON, substitute 2.8 for x.

$MN = 4x - 5$
$\quad \approx 4 \cdot 2.8 - 5$
$\quad \approx 6.2$
$ON = 2x + 1$
$\quad \approx 2 \cdot 2.8 + 1$
$\quad \approx 6.6$
$MN \approx 6.2$ and $ON \approx 6.6$.

About Example 2

Teaching Tip

If necessary, remind students that the Global Positioning System (GPS) was described on page 560.

Lengths of Tangents and Secants

If points A, B, and D lie on a circle and $\overrightarrow{DA}$ and tangent $\overrightarrow{BC}$ intersect at C, then $AC \cdot CD = BC^2$.

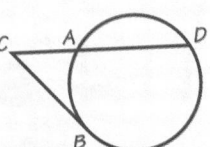

If $\overline{CA}$ and $\overline{CB}$ are tangent to a circle at A and B respectively, then $AC = BC$.

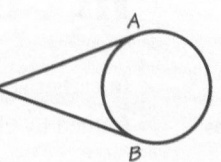

You will prove some of these theorems in the exercises. You can also use them to solve algebra problems like the one in Example 1.

EXAMPLE 1   **Connection: Algebra**

Find *RT* and *ST*.

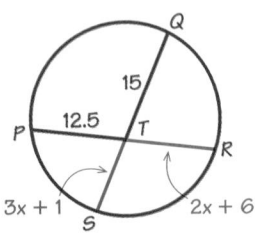

SOLUTION

$\overline{QS}$ and $\overline{PR}$ are chords that intersect inside the circle at T. Use the first theorem on page 573.

$$QT \cdot ST = PT \cdot RT$$

Use the distributive property. Then solve for x.

$$15(3x + 1) = 12.5(2x + 6)$$
$$45x + 15 = 25x + 75$$
$$20x = 60$$
$$x = 3$$

To find RT and ST, substitute 3 for x.

$$RT = 2x + 6 \qquad ST = 3x + 1$$
$$= 2 \cdot 3 + 6 \qquad = 3 \cdot 3 + 1$$
$$= 12 \qquad = 10$$

$RT = 12$ and $ST = 10$.

Toolbox p. 690
Operations with Variable Expressions

EXAMPLE 2 Application: Global Positioning

When the GPS system was designed, engineers wanted to minimize both the number of satellites and the distance that the transmitters needed to send signals. If the satellites are at an altitude of 12,500 mi, how far do the transmitters need to be able to send signals?

SOLUTION

Points R and P are the farthest points on Earth that the satellite's signal can reach without being blocked by Earth, so
$PS = RS =$ the farthest that the transmitters need to send signals.

The diameter of Earth is about 8000 mi, so $SQ \approx 8000 + 12{,}500 = 20{,}500$. Use the formula for segment lengths formed by a tangent and a secant.

$$RS^2 = TS \cdot SQ$$
$$RS^2 = 12{,}500 \cdot 20{,}500$$
$$RS^2 = 256{,}250{,}000$$
$$RS = \sqrt{256{,}250{,}000}$$
$$RS \approx 16{,}000$$

The transmitters on the GPS satellites need to be able to transmit signals about 16,000 mi.

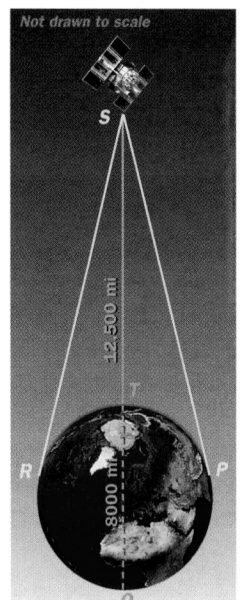
Not drawn to scale

☑ **CHECKING KEY CONCEPTS**

1. Write three true statements about the lengths of the segments in the diagram.

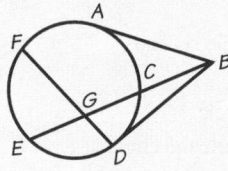

ALGEBRA Find each unknown length.

2.

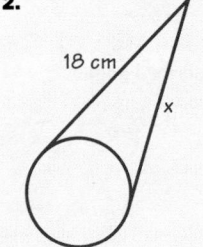

18 cm

x

3.

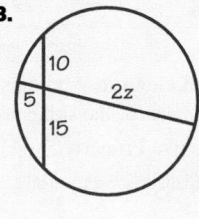

10
5
2z
15

4.

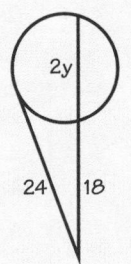

2y

24   18

11.4 Segment Lengths in Circles **575**

Checking Key Concepts

1. $BC \cdot BE = BA^2$; $EG \cdot GC = FG \cdot GD$; $BA = BD$; $BC \cdot BE = BD^2$

2. $x = 18$ cm

3. $2z = 30$

4. $2y = 14$

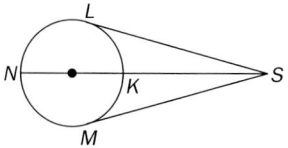

Teach⇔Interact

Additional Example 2

Suppose that another satellite is at an altitude of 9250 miles. How far do the transmitters on this satellite need to be able to send signals?

Points L and M are the farthest points on Earth that the satellite's signal can reach without being blocked by Earth, so $LS = MS =$ the farthest that the transmitters need to send signals. The diameter of Earth is about 8000 mi, so $NS \approx$ 8000 + 9250 = 17,250. Use the formula for segment lengths formed by a tangent and a secant.
$LS^2 = KS \cdot SN$
$LS^2 = 9250 \cdot 17{,}250$
$LS^2 = 159{,}562{,}500$
$LS = \sqrt{159{,}562{,}500}$
$LS \approx 12{,}600$
The transmitters on this satellite need to be able to transmit signals about 12,600 mi.

Checking Key Concepts

Student Progress
Students should complete these four problems successfully before attempting to do any of the exercises. In question 4, students need to use the expression $2y + 18$ to represent the length of the entire secant.

Closure Question

Describe how to find the length of a tangent segment using the theorems of this section.   Find the length of another tangent segment that intersects the first one outside the circle. The lengths of both segments from their intersection to the circle are equal. Find a secant that intersects the tangent outside the circle. Find the length of the part of the secant from the point of intersection to the circle, and the length of the secant from the point of intersection to the far side of the circle. The length of the tangent from the point of intersection with the secant to the circle is the square root of the product of the two secant segments.

575

Extra Practice
exercises on
page 684

Suggested Assignment

❖ **Core Course**
Exs. 1–21, 25, 27–33

❖ **Extended Course**
Exs. 1–17 odd, 18–33

❖ **Block Schedule**
Day 69 Exs. 1–21, 25, 27–33

Exercise Notes

Common Error

Exs. 1–16 Some students may choose an incorrect formula or use the wrong lengths. Encourage these students to remember each theorem using a simplified verbal phrase instead of memorizing the formulas using the letter forms given in the book. (Memorizing formulas may make it more difficult to apply a formula to a situation where the points are labeled differently.) For example, the property involving a secant and a tangent can be memorized as: tangent squared equals outside secant times whole secant.

Integrating Algebra

Exs. 7–10 For these exercises, students should examine the expression and note the lengths that are used. They then need to choose an appropriate theorem from this section to write an expression involving the lengths. Properties of algebra can be used to transform the written expression into the one given in the exercise.

Writing Proofs

Exs. 17, 20, 21 By writing the proofs for these theorems, students can develop a better understanding of their meaning and also see that they depend upon properties of similar triangles. Students can use the *Think and Communicate* questions on page 573 as well as the incomplete proof in Ex. 17 to get ideas for the proofs in Exs. 20 and 21. In Ex. 21, students need to draw a segment from *B* to *D*, which does not go through the center of the circle. They can then use the fact that ∠ *CBA* ≅ ∠ *CDB* because they intercept the same arc.

11.4 Exercises and Applications

For Exercises 1–10, use the diagram at the right. For Exercises 1–6, tell whether each statement is *True* or *False*.

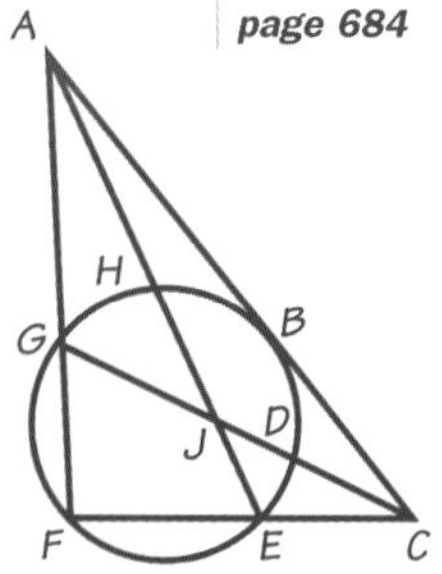

1. $AB^2 = AG \cdot AF$

2. $AF^2 = AB^2$

3. $CD \cdot DG = CE \cdot EF$

4. $JH \cdot JE = JG \cdot JD$

5. $AB^2 = CB^2$

6. $AH \cdot HJ = AG \cdot AF$

ALGEBRA Explain why each statement is true.

7. $\dfrac{CD}{CE} = \dfrac{CF}{CG}$

8. $CD^2 + CD \cdot DG = CE^2 + CE \cdot EF$

9. $CB = \dfrac{CE \cdot CF}{CB}$

10. $\dfrac{JG}{JH} = \dfrac{EJ}{JD}$

Copy the diagram and find each unknown segment length.

11.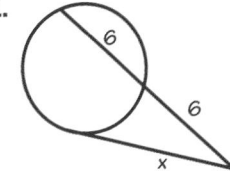

12.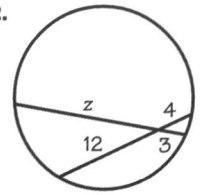

13.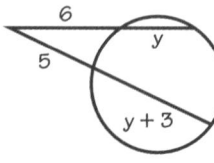

14.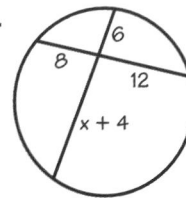

15.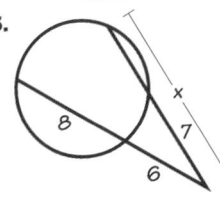

16.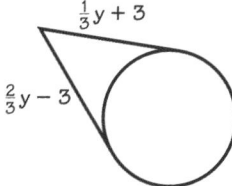

17. Copy and complete the proof of the second theorem on page 573.

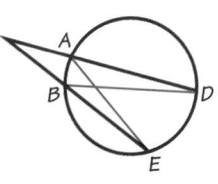

Given: *A*, *B*, *D*, and *E* lie on a circle.
$\overrightarrow{DA}$ and $\overrightarrow{EB}$ intersect outside the circle at *C*.

Prove: $CA \cdot CD = CB \cdot CE$

Statements	Reasons
1. Draw segments $\overline{AE}$ and $\overline{BD}$.	1. ?
2. ∠*ADB* ≅ ?	2. In a circle or in ≅ circles, inscribed angles that intercept the same arc are ≅.
3. ∠*C* ≅ ?	3. Reflexive Property
4. △*DBC* ~ △*EAC*	4. AA Similarity Postulate
5. $\dfrac{DC}{EC} = \dfrac{BC}{AC}$	5. ?
6. ? · ? = $CB \cdot CE$	6. ?

Exercises and Applications

1. True. 2. False.

3. False. 4. True.

5. False. 6. False.

7. $CD \cdot CG = CE \cdot CF$, so $\dfrac{CD \cdot CG}{CE \cdot CG} = \dfrac{CE \cdot CF}{CE \cdot CG}$ and $\dfrac{CD}{CE} = \dfrac{CF}{CG}$.

8. $CD \cdot CG = CE \cdot CF$, so $CD(CD + DG) = CE(CE + EF)$. Then, by the Distributive Property, $CD^2 + CD \cdot DG = CE^2 + CE \cdot CF$.

9. $CB^2 = CE \cdot CF$, so $\dfrac{CB^2}{CB} = \dfrac{CE \cdot CF}{CB}$, or $CB = \dfrac{CE \cdot CF}{CB}$.

10. $JG \cdot JD = JH \cdot EJ$, so $\dfrac{JG \cdot JD}{JD \cdot JH} = \dfrac{JH \cdot EJ}{JD \cdot JH}$ and $\dfrac{JG}{JH} = \dfrac{EJ}{JD}$.

11. $x = 6\sqrt{2} \approx 8.5$

12. $z = 16$

13. $y = 4$; $y + 3 = 7$

14. $x + 4 = 16$

15. $x = 12$

16. $\frac{1}{3}y + 3 = \frac{2}{3}y - 3 = 9$

17. (1) For any two points, there is exactly one line through the points; (2) *AEB*; (3) ∠ *C*; (5) Def. of similar figures; (6) $CA \cdot CD$; a property of proportions

18. D

19. about 26,100 mi

18. SAT/ACT Preview In the diagram, $\overline{AB}$ is tangent to the circle. $BC = \underline{\ ?\ }$

A. 20   **B.** 22   **C.** -22 **D.** 16   **E.** 18

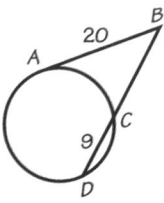

19. GLOBAL POSITIONING Use the diagram in Example 2 on page 575. One way to reduce the number of satellites needed is to place the satellites in a high orbit. A typical high orbit for a satellite is 22,400 mi. If a GPS satellite were placed in orbit at this altitude, how far would the transmitter need to send signals?

In Exercises 20 and 21, you will prove two of the theorems in this section.

20. Given: $\overline{AD}$ and $\overline{BE}$ are chords of a circle.
     $\overline{AD}$ intersects $\overline{BE}$ at C.
     Prove: $AC \cdot DC = BC \cdot EC$

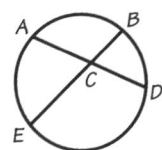

21. Given: Points A, B, and D lie on a circle.
     $\overrightarrow{DA}$ and tangent $\overline{BC}$ intersect at C.
     Prove: $AC \cdot CD = BC^2$

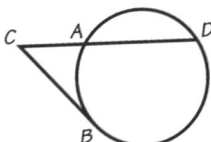

22. Writing Look at the theorems on page 574. Explain why the second theorem is a special case of the first theorem.

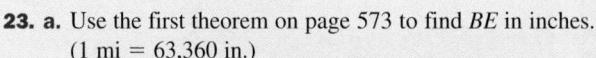

Connection   EARTH SCIENCE

For many centuries, people believed that the world is flat, but it is really round. Why does the world look flat? If you stand at C and look a mile away to A, the ground between you and A is higher than $\overline{AC}$, as shown.

23. a. Use the first theorem on page 573 to find BE in inches.
(1 mi = 63,360 in.)

     b. Writing If you are standing at point C looking toward point A, do you think that you will notice the difference in height at point B caused by the curvature of Earth? Explain your reasoning.

24. a. Challenge Use trigonometry to find $m\,\widehat{AB}$ and $m\,\widehat{AC}$.

     b. What percent of the entire measure of a circle is $m\,\widehat{AC}$?

     c. The circumference of Earth is about 25,000 mi. Use your answer to part (b) to find the length of $\widehat{AC}$.

     d. Writing Explain the meaning of the length you found in part (c).

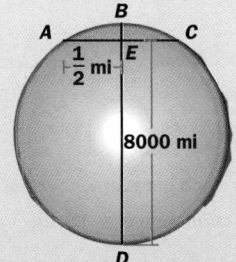

Not drawn to scale

Exercise Notes

Problem Solving

Ex. 18 This exercise can be used to discuss the elimination and guess-and-check strategy, which is often the fastest way to answer a multiple-choice question. First, you eliminate the answers that do not make sense. In this case, C is eliminated because a length cannot be negative. Also eliminate A and B because the part of the secant outside the circle must be shorter than the tangent from the same point. From the remaining choices, guess one and check it for an intersecting secant and tangent. If it checks, you have the answer. Otherwise, the remaining choice is the answer. Alternately, the answer can be found by using algebra. Then you need to use a variable to represent BC and solve the quadratic equation $x^2 + 9x - 400 = 0$.

Interdisciplinary Problems

Exs. 23, 24 In completing these exercises, students see how the geometry of chords and trigonometry can be used to explain why Earth appears flat to an observer standing on the ground. In addition, trigonometry is used to find the measure of an arc of Earth. The ratio of this arc to the whole circumference is then used to find the arc length. Arc length is one of the topics of the next section.

Historical Connection

Exs. 23, 24 The ancient Greeks Pythagoras and Aristotle knew that Earth was round and believed it to be a perfect sphere, although it is actually slightly flattened at the poles. The first accurate measurement of the circumference of Earth was made by Eratosthenes of Cyrene in the third century B.C. His estimate of the circumference of Earth was accurate to within about 15%.

---

20. Statements (Reasons)
1. Draw $\overline{AE}$ and $\overline{BD}$. (For any two points, there is exactly one line through the points.)
2. $\angle A \cong \angle B$; $\angle E \cong \angle D$ (In a circle or in congruent circles, angles that intercept the same arc or congruent arcs are congruent.)
3. $\triangle ACE \sim \triangle BCD$ (AA Similarity Theorem)
4. $\dfrac{AC}{BC} = \dfrac{CE}{CD}$ (Def. of similar figures)
5. $AC \cdot DC = BC \cdot EC$ (a property of proportions)

21. Statements (Reasons)
1. Draw $\overline{AB}$ and $\overline{BD}$. (For any two points, there is exactly one line through the points.)
2. $\angle C \cong \angle C$ (Reflexive Property)
3. $m\angle CBA = \frac{1}{2}m\,\widehat{BA}$ (The measure of an angle formed by a tangent and a chord is half the measure of the intercepted arc.)
4. $m\angle D = \frac{1}{2}m\,\widehat{BA}$ (The measure of an inscribed angle is equal to half the measure of the intercepted arc.)
5. $m\angle CBA = m\angle D$ or $\angle CBA \cong \angle D$. (Substitution Property (Steps 3 and 4))
6. $\triangle ABC \sim \triangle BCD$ (AA Similarity Theorem)
7. $\dfrac{AC}{CB} = \dfrac{CB}{CD}$ (Def. of similar figures)
8. $AC \cdot CD = CB^2$ (a property of proportions)

22. See answers in back of book.

23. a. 1.98 in.

     b. No; the height (1.98 in.) is so small compared to the distance (1 mi) that it would be unnoticeable.

24. a. $m\,\widehat{AB} \approx 0.0072°$; $m\,\widehat{AC} \approx 0.0144°$

     b. about 0.004%

     c. about 1 mi

     d. The arc and the chord determined by A and C are approximately the same length.

Exercise Notes

Communication: Writing
Ex. 25 This exercise gives students an opportunity to check and enhance their understanding of the properties of secants and tangents given on pages 573 and 574. It should also enhance students' realization that they cannot infer certain information merely by looking at a diagram unless that information is clearly marked or given.

Topic Spiraling: Preview
Exs. 28–33 These exercises preview the next section on arc lengths and the area of sectors. The formulas for finding arc lengths and areas of sectors are related to those for finding the circumference of a circle and the area of a circle.

Practice 78 for Section 11.4

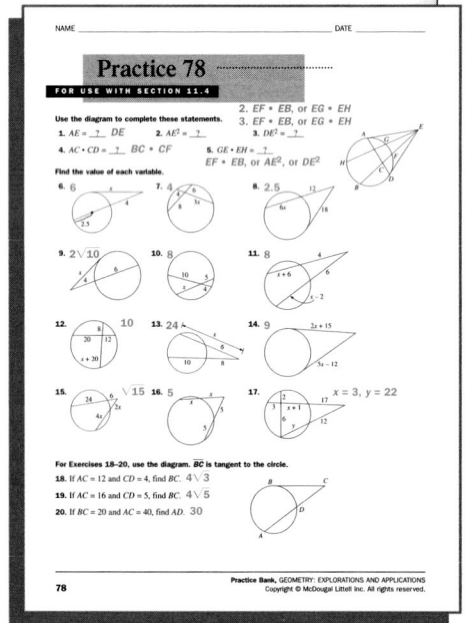

---

25. Writing Is ⊙*P* intersected by *two secants* or by *a secant and a tangent*? Explain how you know.

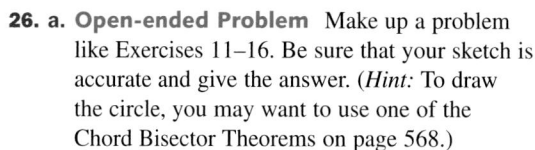

26. a. Open-ended Problem Make up a problem like Exercises 11–16. Be sure that your sketch is accurate and give the answer. (*Hint:* To draw the circle, you may want to use one of the Chord Bisector Theorems on page 568.)

   b. Describe how you drew the sketch to make it accurate.

ONGOING ASSESSMENT

27. Visual Thinking Mary used geometry software to draw a circle with two lines that contained chords of the circle. She moved the endpoints of the chords around the circle to create the pictures below. Use her diagrams to explain how the four theorems in this lesson are related.

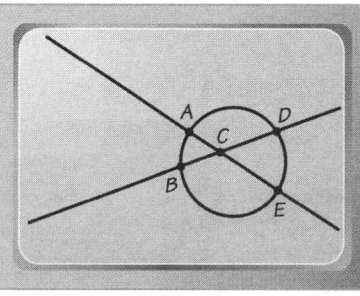
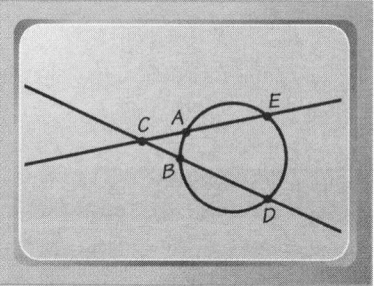

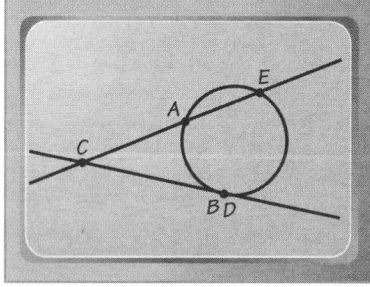
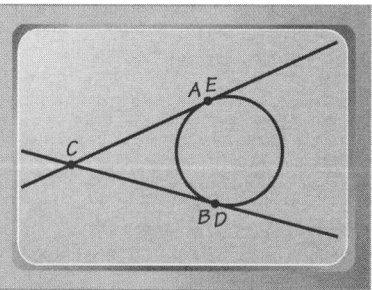

SPIRAL REVIEW

Find the area of the circle with each measure. (*Section 7.5*)

28. radius = 7 in.   **29.** circumference = 18 cm   **30.** diameter = 12 ft

Find each length or measure. (*Section 11.3*)

31. $m\widehat{AB}$

32. *AC*

33. $m\widehat{ADC}$

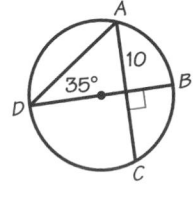

---

25. The lengths of the segments indicated are appropriate to two secants, not a secant and a tangent. If the segment that appears to be a tangent were actually a tangent, the following equation would be true: $14 \cdot 30 = 20^2$. But $14 \cdot 30 = 420$ and $20^2 = 400$, so the segment that appears to be a tangent is actually a secant. The length of the chord is 1.

26. Check students' work.

27. If the lines are secants that intersect inside the circle, they form two intersecting chords and $AC \cdot CE = BC \cdot CD$. If the lines are moved so that both remain secants but intersect outside the circle, the relationship stated above still holds, even though *A* is now between *C* and *E*, and *B* is now between *C* and *D*. (Previously *C* was between *A* and *E* and between *B* and *D*.) If $\overleftrightarrow{BD}$ is moved again so that it becomes a tangent, *B* and *D* coincide and $BD = 0$, which makes the relationship above $AC \cdot CE = BC(BC + CD) = BC(BC + 0) = BC^2$. If $\overleftrightarrow{CE}$ is moved similarly, $AC \cdot CE$ becomes AC^2.

28. 49π in.$^2 \approx 153.9$ in.^{2}

29. $\dfrac{81}{\pi}$ cm$^2 \approx 25.8$ cm^2

30. 36π ft$^2 \approx 113.1$ ft^2

31. $70°$

32. 20

33. $220°$

11.5 | Sectors and Arc Lengths

Learn how to...
• find lengths of arcs and areas of sectors

So you can...
• find lengths and areas in real-world situations, such as rock climbing or irrigation

Suppose you stop at a pizza shop for lunch. The shop sells pizza by the slice. You can afford two slices from a small pizza or one slice from a large pizza. Which will satisfy your hunger better?

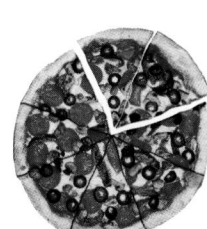 
Diameter = 10 in.

Diameter = 14 in.

EXAMPLE 1

What is the area of one slice of the large pizza?

SOLUTION

The large pizza is cut into eight pieces, so each piece is $\frac{1}{8}$ of the whole pizza. The area of each slice is $\frac{1}{8}$ of the area of the whole pizza.

$$\text{area of slice} = \frac{1}{8} \cdot \text{area of large pizza}$$

$$= \frac{1}{8} \cdot \pi(7)^2 \qquad \text{The area of a circle is } \pi r^2. \text{ The diameter of the large pizza is 14 in., so its radius } r \text{ is 7 in.}$$

$$= \frac{49}{8}\pi$$

The area of a slice of the large pizza is $\frac{49}{8}\pi$ in.^{2}, or about 19.2 in.^{2}

THINK AND COMMUNICATE

1. What is the total area of two slices from the small pizza?

2. Which lunch has the greater area (which will satisfy your hunger better), *one slice of the large pizza* or *two slices of the small pizza*?

11.5 Sectors and Arc Lengths **579**

Objectives
• Find lengths of arcs and areas of sectors.
• Find lengths and areas in real-world situations.

Recommended Pacing
❖ **Core and Extended Courses**
  Section 11.5: 2 days
❖ **Block Schedule**
  Section 11.5: 1 block

Resource Materials
Lesson Support
Lesson Plan 11.5
Warm-Up Transparency 11.5
Practice Bank: Practice 79
Study Guide: Section 11.5
Explorations Lab Manual:
  Additional Exploration 17
Challenge Problems: Set 79
Assessment Book: Test 49
Technology
Technology Book:
  Spreadsheet Activity 10
Scientific Calculator
Internet:
  http://www.hmco.com

Warm-Up Exercises

Find the area of each figure.

1. a circle with radius 5 25π, or about 78.5
2. a circle with diameter 22 121π, or about 380.1
3. a semicircle with radius 13 $\frac{169}{2}\pi$, or about 265.5
4. Find the circumference of a circle with radius 6. 12π, or about 37.7
5. Find the circumference of a circle whose area is 144π. 24π, or about 75.4

ANSWERS   Section 11.5

Think and Communicate

1. $\frac{50}{8}\pi$ in.$^2 \approx 19.6$ in.^{2}

2. two slices of the smaller pizza; The area of two slices of the smaller pieces is greater than the area of one slice of the larger pizza. Since the thicknesses are probably about the same, two slices of the smaller pieces will satisfy your hunger better that one slice of the larger pizza.

Teach⇔Interact

Additional Example 1

Yolanda wants to make a cake for her sister's birthday. She decides that each person should get a piece that has an area of about 5 in.² on top. If there are 20 guests, will a 12 in. diameter cake be large enough?

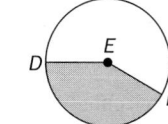

diameter = 12 in.

If there are 20 guests, each one can have $\frac{1}{20}$ of the cake. The area of the top of each piece is $\frac{1}{20}$ of the area of the top of the whole cake. The area of a circle is πr^2. The diameter of the cake is 12 in., so its radius r is 6 in.

area of a piece

$= \frac{1}{20} \cdot$ area of whole cake

$= \frac{1}{20} \cdot \pi(6)^2$

$= \frac{36}{20}\pi$

$= \frac{9}{5}\pi$

The area of the top of each slice is $\frac{9}{5}\pi$ in.², or about 5.7 in.². These pieces are larger than the ones Yolanda wants to serve, so a 12 in. diameter cake will be large enough.

Additional Example 2

The area of the shaded sector is $\frac{135}{4}\pi$ in.² and $m\angle DEF = 150°$.

Find the circumference of the circle.

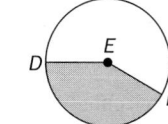

Because $m\widehat{DF} = m\angle DEF$, $m\widehat{DF} = 150°$. Use the formula for the area of a sector to find the radius of the circle.

Area of sector $= \frac{x}{360} \cdot \pi r^2$

$\frac{135}{4}\pi = \frac{150}{360} \cdot \pi r^2$

$\frac{360}{150} \cdot \frac{135}{4} = r^2$

$81 = r^2$

$9 = r$

Use this value in the formula for the circumference of a circle to find the circumference.

$C = 2\pi r$

$= 2\pi \cdot 9$

$= 18\pi$, or about 56.5 in.

580

Each pizza slice represents a geometric figure called a *sector*. A **sector** is a region of a circle that is bounded by two radii and an arc of the circle. The area of the sector depends on the measure of the arc and the radius of the circle.

Area of a Sector

If x represents the measure of the arc of a sector and r is the radius of the circle, then

Area of the sector $= \frac{x}{360} \cdot \pi r^2$.

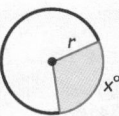

EXAMPLE 2

The area of the shaded sector is 8π cm² and $BC = 6$ cm. Find $m\angle ABC$.

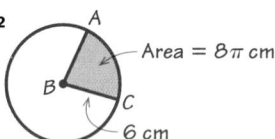

Area = 8π cm²

6 cm

SOLUTION

Use the formula for the area of a sector.

Area of sector $= \frac{x}{360} \cdot \pi r^2$

$8\pi = \frac{m\widehat{AC}}{360} \cdot \pi(6)^2$ Solve for $m\widehat{AC}$.

$8 = \frac{m\widehat{AC}}{360} \cdot 36$

$80 = m\widehat{AC}$

Because $m\angle ABC = m\widehat{AC}$, $m\angle ABC = 80°$.

The **length of an arc** is the distance from one endpoint to the other along the circle. Just as the area of a sector is a fraction of the area of the entire circle, the length of an arc is a fraction of the circle's circumference. The fraction depends on the arc's measure.

Length of an Arc

If x represents the measure of an arc and r is the radius of the circle, then

Length of the arc $= \frac{x}{360} \cdot 2\pi r$.

580   Chapter 11   *Circles and Spheres*

EXAMPLE 3 Application: Irrigation

A pivot irrigation system waters a sector of a field by spraying water from a moving arm. If the moving arm is 1800 ft long and traces an arc with measure 240°, how far does the end of the arm travel?

SOLUTION

$$\text{Length of the arc} = \frac{x}{360} \cdot 2\pi r$$

$$= \frac{240}{360} \cdot (2 \cdot \pi \cdot 1800)$$

$$= \frac{240}{360} \cdot 3600\pi$$

$$= 2400\pi$$

$$\approx 2400(3.14) \qquad \text{Use } \pi \approx 3.14.$$

$$\approx 7536 \text{ ft}$$

The end of the arm travels about 7500 ft.

BY THE WAY

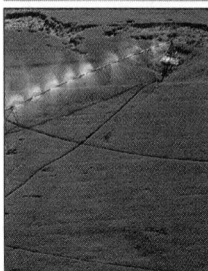

In circular irrigation systems, a long arm of sprinklers pivots around a central point. An arm that is $\frac{1}{4}$ mi long can complete a full revolution in half a day.

THINK AND COMMUNICATE

3. Describe two methods you could use to find the length of the minor arc in the situation in Example 3. What is that length?

4. Which is longer, a 50° arc in a circle of radius 7 cm, or a 50° arc in a circle of radius 12 cm? Explain.

✓ CHECKING KEY CONCEPTS

Find the area of each shaded sector and the length of each red arc. Give your answers to the nearest whole number.

1.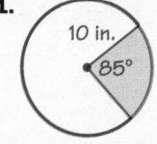
10 in.
85°

2.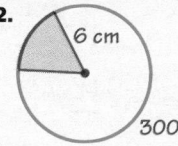
6 cm
300°

3.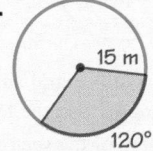
15 m
120°

4. Find the measure of an arc with length 2π ft if the radius of the circle is 4 ft.

5. The measure of an arc is 45° and the arc's length is 8 in. What is the circumference of the circle?

Think and Communicate

3. Use the formula $\frac{120}{360} \cdot 2\pi r$ or find the circumference and subtract the distance found in Example 3 from it. The length you find may vary due to rounding; it is about 3800 ft.

4. the arc in the circle of radius 12 cm;
   Since $7 < 12$, $\frac{50}{360} \cdot (2 \cdot \pi \cdot 7) <$ $\frac{50}{360} \cdot (2 \cdot \pi \cdot 12)$.

Checking Key Concepts

1. 74 in.²; 15 in.

2. 19 cm²; 31 cm

3. 236 m²; 63 m

4. 90°

5. 64 in.

Teach⟺Interact

Section Note

Multicultural Note
Successful methods of irrigation have contributed to the prosperity of many cultures, including the Incas. To support their population, the Incas created irrigation canals that traveled through the Andes and brought water to agricultural terraces carved into the mountainsides. Dr. Ann Kendall, an archeologist, studied the ancient Incan irrigation system and determined that it was more successful than contemporary irrigation methods in the same area. In an effort to revitalize farming in the region, she has spent almost 20 years working to restore parts of the original canal system that irrigated the Incan empire.

Additional Example 3

An irrigation system uses a moving arm that is 1200 ft long. The end of the arm must trace an arc that is 6300 ft long. Through what angle should the arm rotate?

Length of the arc

$$= \frac{x}{360} \cdot 2\pi r$$

$$6300 = \frac{x}{360} \cdot (2 \cdot \pi \cdot 1200)$$

$$6300 \approx \frac{x}{360} \cdot (2400 \cdot 3.14)$$

$$6300 \approx \frac{x}{360} \cdot 7536$$

$$360 \cdot \frac{6300}{7536} \approx x$$

$$301 \approx x$$

The angle of rotation of the arm is equal to the measure of the arc. The arm should rotate through an angle of about 301°.

Think and Communicate

Question 4 can be used to discuss the difference between the *measure* of an arc and the *length* of an arc. The measure of an arc is based on the central angle that intercepts the arc and is independent of the radius of the circle. The length of an arc, however, is based on the circumference of a circle and is dependent on the radius of the circle.

Closure Question

What are the formulas for the area of a sector of a circle and for the length of an arc of a circle?

Area of sector = $\frac{x}{360} \cdot \pi r^2$;

Length of arc = $\frac{x}{360} \cdot 2\pi r$

Apply⟺Assess

Suggested Assignment

❖ **Core Course**
Day 1 Exs. 1–6, 8–16
Day 2 Exs. 20, 22–24, 26, 29–32, AYP

❖ **Extended Course**
Day 1 Exs. 1–16
Day 2 Exs. 17–32, AYP

❖ **Block Schedule**
Day 70 Exs. 1–6, 8–16, 20, 22–24, 26, 29–32, AYP

Exercise Notes

Student Progress
Exs. 1–6, 8–13 Students should be able to complete these exercises correctly since they require a straightforward use of the formulas for finding the area of a sector or the length of an arc. Students having difficulty should check to see that they are using the correct formula and that their calculations do not contain errors.

Integrating Algebra
Exs. 14, 15 Students must use algebra to solve for *x* in the formulas on page 580. They can use Example 2 as a model.

 Communication: Drawing
Exs. 14–16 Encourage students to draw figures to help them organize the given information in these exercises.

Geometric Thinking
Ex. 17 This problem gives students an opportunity to explore the relationship between the radius and central angle of a sector with a given area. This should provide students with more insight into how the radius, central angle, and area of a sector are related.

11.5 Exercises and Applications

Find the area of each shaded sector to the nearest whole number.

1.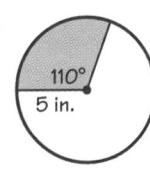
110°
5 in.

2.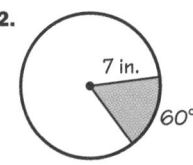
7 in.
60°

3.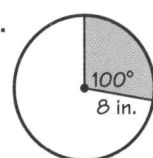
100°
8 in.

4.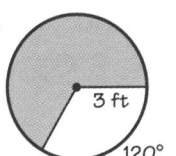
25 m

5.
3 ft
120°

6.
12 cm
58°

7. **CATERING** Wedding cakes are often constructed in *tiers*. For example, the bottom tier may have a 12 in. diameter, the middle tier a 9 in. diameter, and the top tier a 6 in. diameter. Suppose one third of the 12 in. tier is left at the end of the party.
a. If the remaining cake is a sector, what is the central angle?
b. How many wedges with arc length 2.5 in. can be cut from this remaining cake?

Find the length of the red arc in each circle. Give your answers to the nearest whole number.

8.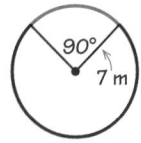
90°
7 m

9.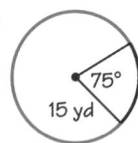
10 cm
45°

10.
75°
15 yd

11.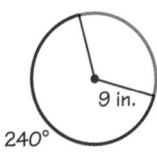
9 in.
240°

12.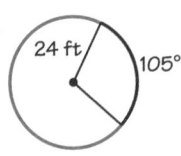
24 ft
105°

13. 110°
11 cm
130°

14. In ⊙*O*, the area of the sector with central angle ∠*AOB* is 30π cm^2 and radius *OA* is 6 cm. Find *m* $\overarc{AB}$ and the length of $\overarc{AB}$.

15. In ⊙*O*, *m* $\overarc{RS}$ = 45°. If the area of the sector with central angle ∠*ROS* is 20 in.^{2}, find the length of $\overarc{RS}$.

16. The radius of a semicircle is 5 ft. What is the length of the semicircle?

17. **Open-ended Problem** What are some possible values for the radius and central angle of a sector that has an area of 36π cm^2?

Exercises and Applications
1. 24 in.^{2}
2. 26 in.^{2}
3. 56 in.^{2}
4. 245 m^2
5. 19 ft^2
6. 146 cm^2
7. a. 120°
   b. 5 wedges
8. 11 m

9. 8 cm
10. 75 yd
11. 19 in.
12. 107 ft
13. 23 cm
14. 300°; 10π cm ≈ 31.4 cm
15. 5.6 in.
16. 5π ft ≈ 15.7 ft
17. Answers may vary. Examples are given. 12 cm, 90°; 18 cm, 40°; 36 cm, 10°

18. a. about 14 in.
    b. about 44 cm
    b. about 27 in.
19. a. 12.2 in.
    b. 73.9°
20. a. The slant height is the length of the hypotenuse of a right triangle whose side lengths are the radius and the height of the cone; l ≈ 5.4 in.; $C = 2\pi(2) = 4\pi$ ≈ 12.6 in.

18. FASHION Fans have been used in many cultures for many reasons. Folding fans can be used either fully or partly opened. For each fan, estimate the length of the arc at the edge of the fan.

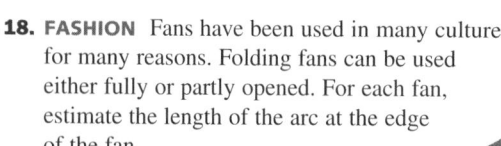

a.  ── 9 in. ──

b.  ── 21 cm ──

c. 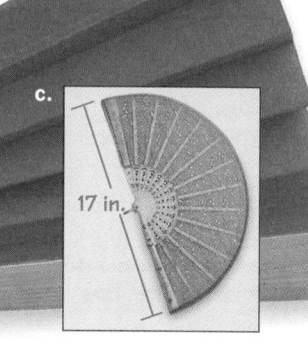 17 in.

19. AUTOMOBILES The diameter of the steering wheel of a car is 15.5 in.

  a. If you turn the steering wheel through an arc of 90° without taking your hands from the wheel, how far do your hands move?

  b. If you move your hands 10 in., what angle does the steering wheel pass through?

20. Visual Thinking Suppose you split open and flatten an ice cream cone with radius 2 in. and height 5 in.

  a. Explain how to use the radius and height of the cone to find its slant height, ℓ. Find the circumference of the base of the cone.

  b. What geometric figure does the flattened cone represent? Make a sketch of the figure. Label any dimensions you know.

  c. How can you use the slant height and circumference of the base of the cone to find the area of the figure that you sketched in part (b)?

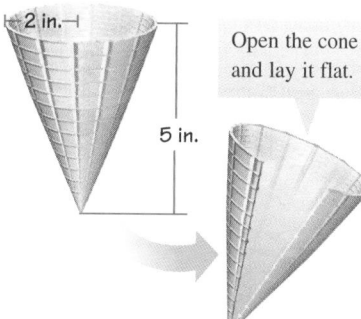
── 2 in. ──
Open the cone and lay it flat.
5 in.

21. TRACK AND FIELD On a track like the one on page 370, the racers' starting positions are staggered so the racer in the inside lane travels the same distance over the course of the race as the racer in the outside lane.

  a. For the outside lane, suppose the radius of the curve at each end of the track is 36.5 m. How far does the racer in the outside lane travel around the 180° curve?

  b. What is the radius of the curve for the inside lane if the track has 8 lanes, each 1.22 m wide? How far does the racer in the inside lane travel around the curve?

  c. Writing In some races the participants all race in the inside lane and must pass on the outside. Why does it make more sense to pass on a straight section of the track than on a curve?

11.5 Sectors and Arc Lengths **583**

---

b. a sector of a circle

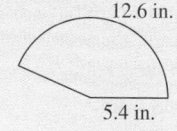
12.6 in.
5.4 in.

c. The slant height is the radius of the sector; the circumference of the base of the cone is the length of the arc. Let $x°$ be the measure of the arc of the sector. Then

$12.6 = \frac{x}{360} \cdot (2 \cdot \pi \cdot 5.4)$;
$x \approx 133.7°$. Then the area is about 34 in.².

21. Assume that the radius of each lane is measured along the center of the lane.

  a. $\frac{73\pi}{2}$ m ≈ 114.7 m

  b. 27.96 m; about 87.8 m

c. A racer travels in an arc to pass; the arc will be shorter if the other racer is moving in a straight line than if he or she is moving on a curve.

Apply⇔Assess

Exercise Notes

Research
Ex. 18 Some students may be interested in finding out about the use of fans in different cultures. Have them look for special designs and customs that are unique to that culture.

 Application
Exs. 18, 19, 21–24 These exercises introduce students to some of the many applications of sectors and arc length. The applications are drawn from fashion, automotive design, racing, and rock climbing.

Topic Spiraling: Review
Ex. 20 Students use what they have learned about arc length to review and expand upon their understanding of the lateral surface area of a cone.

Using Manipulatives
Ex. 20 It would be helpful for some students to use a physical model for this exercise. A cone can be made by cutting out and taping together a paper sector that has the dimensions given in the answer to part (b).

Interdisciplinary Problems
Ex. 21 In completing this exercise, students will understand why a staggered start is used in track and field, and how the starting positions can be determined. Part (c) gives students an opportunity to further reflect upon how the concept of arc length can play a role in racing strategy.

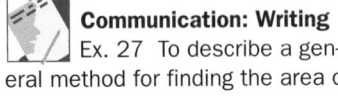
Connection   ROCK CLIMBING

Safety is important to experienced rock climbers. In the illustrations at the right, the lead climber has attached the rope to the rock face at convenient intervals. The rope attachments, along with the second climber, provide protection if the lead climber slips from the rock.

22. If the rope is attached at points A and B and the climber at point C slips, she will swing through an angle of about 150° around point B. About how far will she swing? (*Note:* Assume that the rope does not stretch.)

23. If the rope is attached at point A only and the climber at C slips, she will swing about 150° around point A. About how far will she swing?

24. **Writing** Why is it important for rock climbers to attach rope to the rock face often?

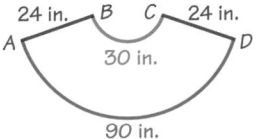

25. **Challenge** In the figure, $\overset{\frown}{AD}$ and $\overset{\frown}{BC}$ are arcs of concentric circles. $\overline{AB}$ and $\overline{CD}$ lie on radii of the larger circle. What is the area of the figure?

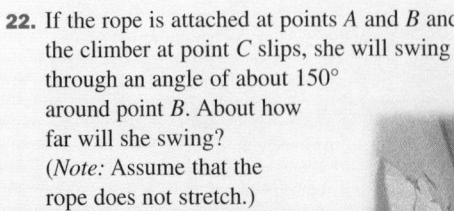
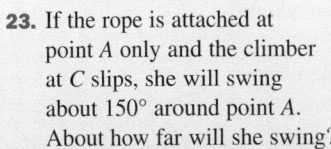
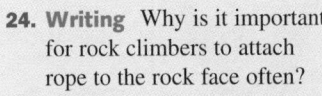

A *segment* of a circle is the portion of a sector outside of the triangle formed by the sector's two radii and a chord.

26. Find the area of each segment.

   a.
   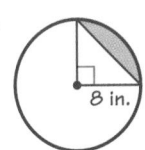

   b.
   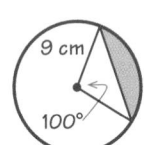

   c.
   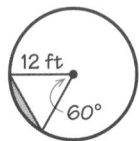

27. **Writing** Describe a general method for finding the area of a segment of a circle.

28. In part (b) of Exercise 26, what is the perimeter of the sector bounded by the major arc?

584   Chapter 11 *Circles and Spheres*

29. Cooperative Learning Work in a group of at least six people. One person should stand in the middle of a circle of people standing shoulder to shoulder.

  a. The person in the center should approximate a central angle by holding out his or her arms. How many people in the circle make up the arc intercepted by this central angle?

  b. Reduce the radius of the circle by moving closer to the student in the center. You may need to change the number of people in the circle. How many students are in the intercepted arc now?

  c. Writing Make some sketches of the results of this activity. Describe how the radius of a circle affects the lengths of arcs on the circle.

SPIRAL REVIEW

Find the surface area and volume of each figure. *(Sections 7.6 and 10.4)*

30.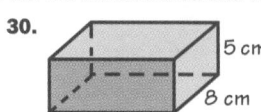
5 cm
8 cm
12 cm

31.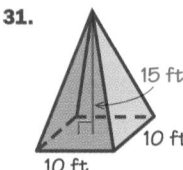
15 ft
10 ft
10 ft

32.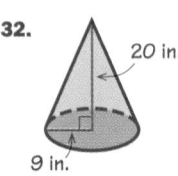
20 in.
9 in.

ASSESS YOUR PROGRESS

VOCABULARY

sector (p. 580)   **length of an arc** (p. 580)

ALGEBRA **Find the value of each variable.** *(Sections 11.3 and 11.4)*

1.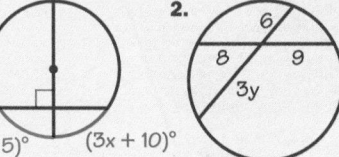
$(4x - 5)°$ $(3x + 10)°$

2.
6
8   9
3y

3.
2
5z   4

4. Find the area of the shaded sector and the length of the red arc. Give your answers to the nearest whole number. *(Section 11.5)*

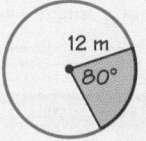
12 m
80°

5. Journal Suppose you know the area of a sector of a circle. Describe what else you need to know to find the length of the arc of the sector.

11.5 Sectors and Arc Lengths **585**

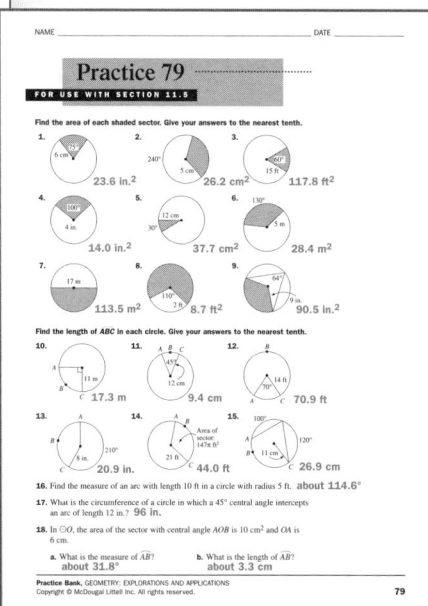

---

30. 392 cm²; 480 cm³

31. about 416 ft²; 500 ft³

32. about 874.6 in.²; about 1696.5 in.³

Assess Your Progress

1. 15

2. 4

3. 1.2

4. 101 m²; 59 m

5. the radius of the circle or the measure of the arc

Warm-Up Exercises

Name a segment that fits each description.

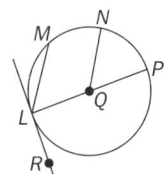

1. a diameter $\overline{LP}$
2. a radius $\overline{QN}, \overline{QP},$ or $\overline{QL}$
3. a secant $\overline{LP}$ or $\overline{LM}$
4. a tangent $\overline{LR}$

Find the volume of each figure.

5. a cone with radius 4 and height 4   about 67 cubic units
6. a cylinder with radius 4 and height 4   about 201 cubic units

---

SECTION

11.6 Surface Areas and Volumes of Spheres

Learn how to...
- calculate the surface area and volume of a sphere

So you can...
- compare spheres of different sizes, such as sports balls or planets

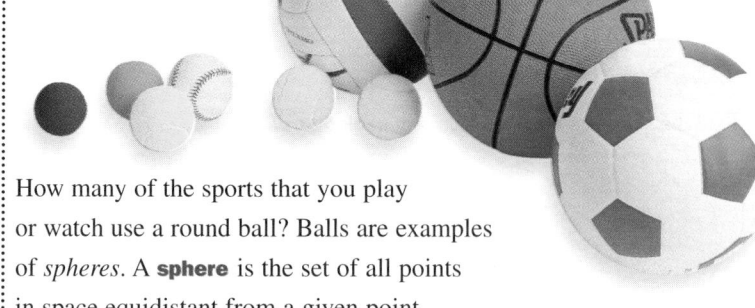

How many of the sports that you play or watch use a round ball? Balls are examples of *spheres*. A **sphere** is the set of all points in space equidistant from a given point.

Many of the terms used with spheres are the same as those used with circles.

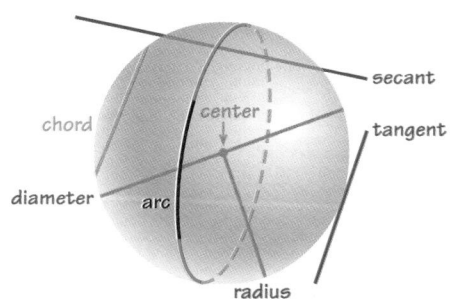

In the third century B.C., Archimedes discovered formulas for the surface area and volume of a sphere by imagining cutting a sphere, cone, and cylinder into very thin slices and comparing the slices.

Surface Area and Volume of a Sphere

The surface area of a sphere with radius r is $4\pi r^2$.

The volume of a sphere with radius r is $V = \frac{4}{3}\pi r^3$.

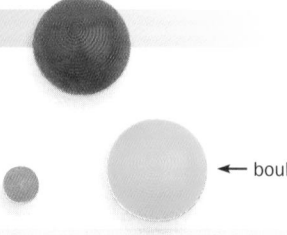

EXAMPLE 1 Application: Sports

The radius of a *boule* is 5.5 cm.
Find the surface area and volume
of a boule.

 ← boule

SOLUTION

Use **5.5 cm** for *r* in the formulas for surface area and volume of a sphere.

$$\text{Surface Area} = 4\pi r^2 \qquad\qquad \text{Volume} = \frac{4}{3}\pi r^3$$

$$= 4\pi (5.5)^2 \qquad\qquad\qquad = \frac{4}{3}\pi (5.5)^3$$

$$\approx 380 \text{ cm}^2 \qquad\qquad\qquad \approx 697 \text{ cm}^3$$

The surface area of a boule is about 380 cm² and its volume is
about 700 cm³.

THINK AND COMMUNICATE

The length of a side of a cube is *d* and the radius of a sphere is $\frac{d}{2}$.

1. Which do you think is greater, the volume of the cube or the volume of
   the sphere? Why? Check your answer by writing an expression for the
   volume of each and comparing the expressions.

2. Which do you think is greater, the surface area of the cube or the
   surface area of the sphere? Why? Check your answer.

EXAMPLE 2 Connection: Algebra

The surface area of a sphere is 24 cm². Find the radius of the sphere.

SOLUTION

$$A = 4\pi r^2$$ Use the formula for the
surface area of a sphere.

$$24 = 4\pi r^2$$

$$\frac{6}{\pi} = r^2$$ To solve for *r*, find the
square root of $\frac{6}{\pi}$.

$$\sqrt{\frac{6}{\pi}} = r$$

$$1.38 \approx r$$

The radius of the sphere is about 1.4 cm.

Teach⇔Interact

Section Notes

Communication: Discussion
Ask students how chords,
diameters, secants, tangents, and
radii of spheres are similar to those
of circles. How are they different?
Ask students to describe each of
these terms as they relate to
spheres.

Historical Connection
Archimedes proved that the volume
of a sphere is two thirds the volume
of a circumscribed cylinder. This
result leads to the formula for volume given on page 586. For example, if *r* is the radius of the sphere
and the cylinder, then the height
of the cylinder is 2*r*. The volume of
the cylinder is $\pi r^2(2r) = 2\pi r^3$. Two
thirds of this expression gives the
formula for the volume of a sphere.
Archimedes must have considered
this result one of his most important achievements because he
requested that a figure of a sphere
and circumscribed cylinder be
inscribed on his tomb.

Additional Example 1

The game of table tennis uses a
ball with radius 1.9 cm. Find the
surface area and volume of a table
tennis ball.
Use 1.9 cm for *r* in the formulas for
surface area and volume of a
sphere.
Surface Area = $4\pi r^2$
$$= 4\pi(1.9)^2$$
$$\approx 45 \text{ cm}^2$$
Volume = $\frac{4}{3}\pi r^3$
$$= \frac{4}{3}\pi(1.9)^3$$
$$\approx 29 \text{ cm}^3$$
The surface area of a table tennis
ball is about 45 cm² and its volume
is about 29 cm³.

Additional Example 2

The volume of a sphere is
$\frac{32}{3}\pi$ in.³.
Find the diameter of the sphere.
Use the formula for the volume of a
sphere and solve for *r*.
$$V = \frac{4}{3}\pi r^3$$
$$\frac{32}{3}\pi = \frac{4}{3}\pi r^3$$
$$8 = r^3$$
$$2 = r$$
The diameter of the sphere is twice
the radius, or 4 in.

ANSWERS Section 11.6

Think and Communicate

1. Answers may vary. The volume
   of the cube is greater than that
   of the sphere. The volume of
   the cube is d^3; the volume of
   the sphere is $\frac{4}{3} \cdot \pi \left(\frac{d}{2}\right)^3 =$
   $\frac{1}{6} \cdot \pi d^3 \approx \frac{1}{2}d^3$.

2. Answers may vary. The surface area of the cube is greater
   than that of the sphere. The
   surface area of the cube is $6d^2$;
   the surface area of the cube is
   $4\pi \cdot \left(\frac{d}{2}\right)^2 = \pi d^2$.

What are the formulas for finding the surface area and volume of a sphere of radius *r*? The surface area of a sphere with radius *r* is $4\pi r^2$. The volume is $V = \frac{4}{3}\pi r^3$.

Apply⇔Assess

Suggested Assignment

❖ **Core Course**
Exs. 1–9, 13–18, 20, 21, 23, 29, 32–38

❖ **Extended Course**
Exs. 1–5, 7, 8, 10–38

❖ **Block Schedule**
Day 71 Exs. 1–9, 13–18, 20, 21, 23, 29, 32–38

Exercise Notes

Multicultural Note
By the Way Many Native American tribes in the eastern part of North America—including the Cherokee, Choctaw, Creek, Iroquois, Menominee, Ojibwe, Potawatomi, Yuchi, and others—played variations of lacrosse long before the arrival of Europeans. Lacrosse was an important cultural and spiritual tradition: challenging another tribe or village to a game of lacrosse could resolve a conflict that might otherwise end in war, and was sometimes thought to bring such positive results as healing the sick or changing the weather. In some ways, Native American lacrosse was very different from contemporary lacrosse: the playing field could be up to two miles long, there were no sidelines, and some games had as many as 60 players for each team on the field at a time. Lacrosse was first adopted and regulated by European Canadians around the middle of the 19th century.

Common Error
Exs. 6–9 Students may use the diameter instead of the radius in the formulas for volume and surface area. To help students avoid this error, encourage them to write down the value of the radius in the form *r = k* before substituting it into the formula.

☑ **CHECKING KEY CONCEPTS**

For Questions 1–6, name a point, a segment, or a line that fits each description.

1. a diameter
2. a chord
3. a secant
4. a tangent
5. a radius
6. the center of the sphere
7. What is the volume of the sphere?
8. What is the surface area of the sphere?

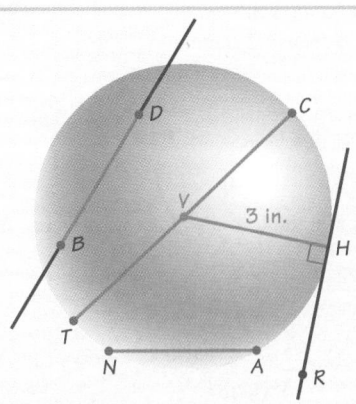

9. *Jai alai* is a fast game that originated in northern Spain. The surface area of the ball (*pelota*) is about 13 in.² What is the radius?

11.6 Exercises and Applications

Extra Practice exercises on page 685

For Exercises 1–5, match each segment or line with the best description.

1. $\overline{VC}$ **A.** a secant
2. $\overline{HG}$ **B.** a radius
3. $\overline{DC}$ **C.** a diameter
4. $\overleftrightarrow{EF}$ **D.** a tangent
5. $\overrightarrow{AB}$ **E.** a chord

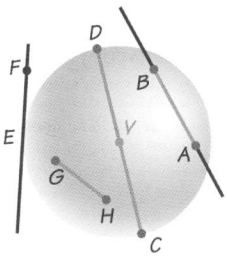

Give the surface area and the volume of each sphere to the nearest whole number.

6.
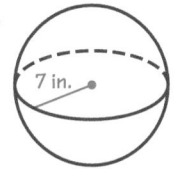
7 in.

7.
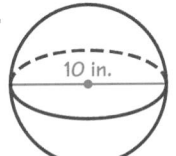
10 in.

8.
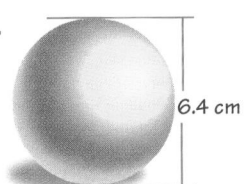
1.7 in.
golf ball

9.
6.4 cm
lacrosse ball

BY THE WAY

Lacrosse players use a stick with a net pocket on the end to throw or scoop the ball into the opposing team's goal. Lacrosse was invented by several North American peoples.

Checking Key Concepts

1. $\overline{CT}$
2. $\overline{AN}$, $\overline{BD}$, or $\overline{CT}$
3. $\overleftrightarrow{BD}$
4. $\overleftrightarrow{HR}$
5. $\overline{VT}$, $\overline{VC}$, or $\overline{VH}$
6. *V*
7. about 113 in.³
8. about 113 in.²
9. about 1 in.

Exercises and Applications

1. B
2. E
3. C
4. D
5. A
6. 616 in.²; 1437 in.³
7. 314 in.²; 524 in.³
8. 9 in.²; 3 in.³
9. 129 cm²; 137 cm³

INTERVIEW Ron Courson

Look back at the article on pages 550–552.

Different types of balls are used by athletic trainers and physical therapists like Ron Courson to improve patients' strength, balance, and proprioception.

10. Patients squeeze small putty balls to strengthen their hands and forearms. The diameter of a ball is about 2.5 in. About how much putty is in the ball?

11. When an athlete is injured, his or her sense of exactly where the injured part is or what it is doing may be thrown off. This sense is called *proprioception*. To improve proprioception for an injured arm, the physical therapist and injured athlete may play catch with a ball like the one at the left. What is the surface area of the ball?

12. Injuries may also impair an athlete's sense of balance. To improve balance, the physical therapist may ask the patient to balance on a large ball. The diameter of the ball is 80 in. What is the volume of the ball?

13. The surface area of a sphere is 35 ft^2. What is the radius?

14. The surface area of a sphere is 16 m^2. What is the diameter?

15. The volume of a sphere is 36π ft^3. What is the radius?

16. ARCHITECTURE A basic igloo is a shell of snow in the shape of half of a sphere. The size of an igloo depends on how it will be used.

  a. The diameter of a small igloo is 6 ft. Find the volume of the igloo, disregarding the entrance.

  b. If the diameter of the igloo is doubled, how is the volume affected? Disregard the entrance.

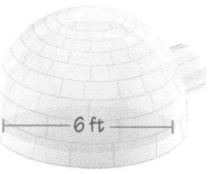

├— 6 ft —┤

ASTRONOMY **For Exercises 17–19, assume that the planets are spheres.**

17. a. The diameter of Earth is about 8000 mi. Find the surface area of Earth.

  b. About 29% of Earth's surface is land. Find the area of the land.

18. The diameter of Mars is about 4200 mi. What is the surface area of Mars?

19. Research In the solar system, which planet's surface area is the greatest? What is the surface area?

11.6 Surface Areas and Volumes of Spheres **589**

Exercise Notes

Student Study Tip
Exs. 6–9 Ask students to compare and contrast the formulas for the volume of a cone, the volume of a cylinder, and the volume of a sphere. Thinking about these three formulas as a group should help students remember them correctly.

Interview Note: Application
Exs. 10–12 Athletic trainers and physical therapists use a variety of balls with their athletes and patients. These exercises give students some examples of how the geometry of the sphere can be used to learn more about these various balls.

Problem Solving
Exs. 13–15 In these exercises, students need to use the surface area formuala or the volume formula and then solve for the radius. Some students may need to refer to Example 2 when doing these exercises.

Interdisciplinary Problems
Exs. 16–19, 29 These exercises provide students with examples of how the concepts of volume and surface area can provide insight into other disciplines. In Ex. 16, students examine the volume of an igloo. In Exs. 17–19, students compare the surface areas of different planets. In Ex. 29, students examine the physics of soap bubbles.

Integrating Algebra
Ex. 16 In part (b) of this exercise, students can use examples and inductive reasoning to reach a conclusion. Alternately, the result can be proved algebraically as follows. Let r = the radius of the small igloo. The volume of the small igloo is $\frac{1}{2}\left(\frac{4}{3}\right)\pi r^3 = \frac{2}{3}\pi r^3$. If the diameter is doubled, the radius is also doubled. The volume of the large igloo is thus $\frac{1}{2}\left(\frac{4}{3}\right)\pi (2r)^3 = 8\left(\frac{2}{3}\pi r^3\right)$, which is 8 times the volume of the small igloo. Thus, when the diameter of an igloo is doubled, its volume is multiplied by 8.

---

10. about 8.2 in.^{3} of putty

11. about 2206 in.^{2}

12. about 268,083 in.^{3}

13–15. Answers are given to the nearest tenth.

13. 1.7 ft

14. 2.3 m

15. 3.0 ft

16. a. about 113 ft^3

  b. It is multiplied by 8.

17. a. about 201,062,000 mi^2

  b. about 58,308,000 mi^2

18. about 55,418,000 mi^2

19. Jupiter; about 24.7 billion mi^2

Research

Exs. 20–23 You may wish to point out that the official basketball mentioned in these exercises is the men's basketball. The official women's basketball is smaller. Have interested students find the diameter of the women's basketball, and answer these questions using that ball.

Second-Language Learners

Exs. 20–23 As they read the text and the exercises, students learning English can use context clues, sentence sense, and the drawings to guess the meaning of any word they do not know. If necessary, suggest that they refer to a dictionary for help.

 Using Technology

Ex. 27 This exercise requires the use of a graphing calculator or spreadsheet software. On a TI-82, students can use the list features of the calculator to perform the calculations. Enter the radii as L1. Move the cursor on top of L2, and enter $(4/3)*\pi*L1^3$ followed by ENTER. L2 now has the volumes of the spheres whose radii are in L1. Similarly, enter the surface areas in L3. For L4, enter the formula L2/L3 to get the ratios of volume to surface area. To graph this ratio as a function of the radius, press 2nd [STATPLOT]. Choose Plot1, then highlight On, the scatter plot picture, L1 for Xlist, and L4 for Ylist. Choose 9:ZoomStat from the ZOOM menu to see the graph.

Integrating Functions

Ex. 27 In part (c) of this exercise, students look at a graph of the ratio $\frac{V}{S.A.}$ as a function of the radius. They should notice that the graph is linear. Challenge students to find the equation of this line. You may also want to ask them to show the same result algebraically.

$\left(\frac{V}{S.A.} = \frac{\frac{4}{3}\pi r^2}{4\pi r^2} = \frac{r}{3} = \frac{1}{3}r.\right)$ For any sphere, the ratio of its volume to surface area is given by the formula $y = \frac{1}{3}r$.

Connection ▶ **SPORTS**

An official basketball is an inflated sphere with a leather, rubber, or synthetic casing. The ball weighs 20 oz to 22 oz and the circumference is about 30 in. Basketball players throw the ball through a cast-iron rim. The diameter of the rim is 18 in.

20. Find the radius of a basketball.

21. If a player dunks the basketball right through the center of the rim, how much clearance is there between the rim and the basketball?

22. The casing of a basketball is attached with an adhesive, so no extra material is needed at the seams. How much leather is needed to cover a basketball?

23. What is the volume of a basketball?

A volleyball is an inflated sphere with a leather, rubber, or synthetic casing. It is smaller than a basketball and weighs 9 oz to 10 oz. The circumference is between 25 in. and 27 in.

24. What is the radius of a volleyball if the circumference is 25 in.? 27 in.? Use your answers to describe the range of radii for volleyballs.

25. Describe the range of surface areas for volleyballs.

26. Describe the range of volumes for volleyballs.

27. a. 📈 **Technology** The radii of five spheres are 1 ft, 2 ft, 3 ft, 4 ft, and 5 ft. Find the volume and the surface area of each sphere.

b. Find the ratio of volume to surface area $\left(\frac{V}{S.A.}\right)$ for each sphere in part (a).

c. Use a graphing calculator to graph $\frac{V}{S.A.}$ as a function of the radius. What do you notice?

28. Open-ended Problem Find an object shaped like a sphere at school or where you live. Estimate the surface area and volume of the object. Explain how you made your estimate.

---

20. about 4.75 in.

21. about 4.25 in.

22. about 284 in.^{2}

23. about 449 in.^{3}

24. from about 4.0 in. to about 4.3 in.

25. from about 201.1 in.^{2} to about 232.4 in.^{2}

26. from about 268.1 in.^{3} to about 333.0 in.^{3}

27. a, b. Answers are given to the nearest hundredth.

radius (ft)	V (ft^3)	S.A. (ft^2)	$\frac{V}{S.A.}$
1	4.19	12.56	0.33
2	33.51	50.27	0.67
3	113.10	113.10	1
4	268.08	201.06	1.33
5	523.60	314.16	1.67

c.

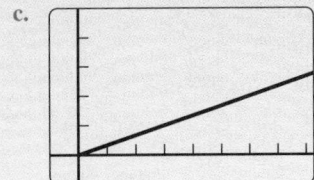

The function is linear. $\left(\frac{V}{S.A.} = \frac{1}{3}r\right)$

28. Answers may vary.

29. a. 8:1   **b.** 4:1

c. The thickness is divided by 4.

29. PHYSICS A soap bubble 4 cm in diameter is blown out until its diameter is 8 cm.

    a. What is the ratio of the new volume to the old volume?

    b. What is the ratio of the new surface area to the old surface area?

    c. If the amount of soap film doesn't change, the product of the film's thickness and its surface area is always constant. How does the thickness of the soap film change as the diameter increases from 4 cm to 8 cm?

30. a. Visual Thinking Describe the intersection of a sphere and a plane through the center of the sphere.

    b. Can the intersection of a sphere and a plane be a point? Explain.

31. Challenge The volume of a cylinder is the same as the volume of a sphere. The radius of the sphere is 1 in.

    a. Give three possibilities for the dimensions of the cylinder.

    b. Writing Is the surface area of the cylinder *always*, *sometimes*, or *never* greater than the surface area of the sphere? Explain.

ONGOING ASSESSMENT

32. Visual Thinking A *hemisphere* is half of a sphere.

    a. Which do you think is greater, the volume of the hemisphere or the volume of the cone? Why? Check your answer by finding the volume of each.

    b. Which do you think is greater, the surface area of the hemisphere or the surface area of the cone? Why? Check your answer.

hemisphere

SPIRAL REVIEW

Find the area of each shaded sector. Give your answer to the nearest whole number. *(Section 11.5)*

33.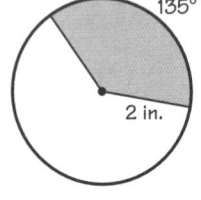

135°

2 in.

34.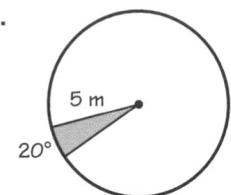

5 m

20°

35.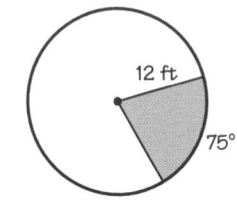

12 ft

75°

For Exercises 36–38, △ABC ~ △DEC. *(Section 9.4)*

36. Find the ratio $\frac{BC}{EC}$.

37. Find the ratio of the perimeters of △ABC and △DEC.

38. Find the ratio of the areas of △ABC and △DEC.

A

D

7 | 5

B E C

11.6 Surface Areas and Volumes of Spheres   **591**

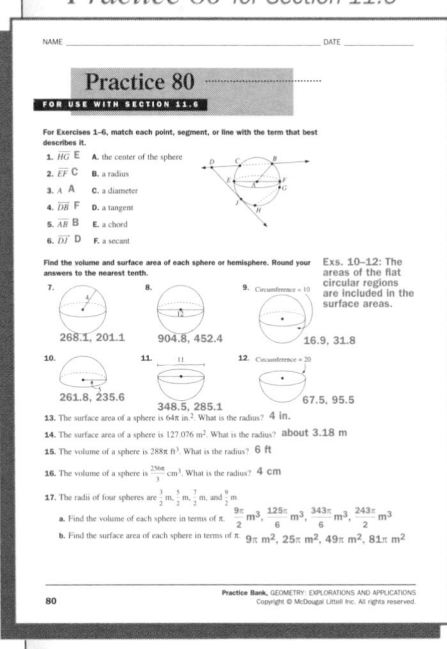
---

30. a. The intersection is a circle in the plane; the center of the circle is the center of the sphere.

    b. Yes. Answers may vary. An example is given. Let $\overline{AB}$ be a diameter of sphere *C* and let plane *P* be perpendicular to $\overline{AB}$ at *A*. *A* is the intersection of plane *P* and sphere *C*. (Picture a basketball sitting on a gym floor.)

31. a. Answers may vary. Examples are given. $r = 1$ in., $h = 1\frac{1}{3}$ in.; $r = 2$ in., $h = \frac{1}{3}$ in.; $r = \frac{1}{2}$ in., $h = 5\frac{1}{3}$ in.

    b. always; Answers may vary. An example is given. The cylinder and the sphere have the same volume and the radius of the sphere is 1 in., so $\pi r^2 h = \frac{4}{3}\pi$. Then $h = \frac{4}{3r}$. The surface area of the cylinder is $2\pi r^2 + 2\pi r h = 2\pi r^2 + \frac{8\pi}{3r}$. If you graph this function on a graphing calculator, you will see

that the minimum value is about 14.4, which is greater than 4π, the surface area of the sphere.

32. a. Answers may vary. The volumes are equal. The volume of the hemisphere is $\frac{1}{2}\left(\frac{4}{3}\pi r^3\right) = \frac{2}{3}\pi r^3$. The volume of the cone is $\frac{1}{3}(\pi r^2)(2r) = \frac{2}{3}\pi r^3$.

    b. Answers may vary. The surface area of the hemisphere is $\pi r^2 + \frac{1}{2}(4\pi r^2) = 3\pi r^2$. The slant height of the cone is $r\sqrt{5}$, so its surface area is $\pi r^2 + \sqrt{5}\pi r^2 = (\sqrt{5} + 1)\pi r^2 \approx 3.2\pi r^2$.

33. about 5 in.^{2}

34. about 4 m^2

35. about 94 ft^2

36. $\frac{7}{5}$ **37.** $\frac{7}{5}$ **38.** $\frac{49}{25}$

Objectives

- Compare volumes and surface areas of similar solids.
- Estimate measures of real-world objects.

Recommended Pacing

❖ **Core and Extended Courses**
  Section 11.7: 1 day

❖ **Block Schedule**
  Section 11.7: $\frac{1}{2}$ block
  (with Section 11.6)

Resource Materials

Lesson Support
Lesson Plan 11.7
Warm-Up Transparency 11.7
Practice Bank: Practice 81
Study Guide: Section 11.7
Challenge Problems: Set 81

Technology
Technology Book:
  Calculator Activity 14
Scientific Calculator
Internet:
  http://www.hmco.com

Warm-Up Exercises

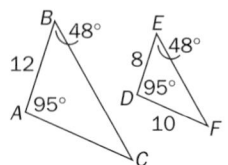

1. Are triangles *ABC* and *DEF* similar? Why?   Yes; AA Similarity Postulate.

Use the triangles above to find each value.

2. $\frac{BC}{EF}$ 1.5   **3.** *AC*   15

4. If the ratio of corresponding linear measures of two similar figures is $a:b$, what is the ratio of their areas? $a^2:b^2$

5. The ratio of the radii of two circles is $4:9$. If the smaller circle has an area of about 45 cm², what is the area of the larger circle?   about 228 cm²

11.7 Volumes of Similar Solids

Learn how to...
- compare volumes and surface areas of similar solids

So you can...
- estimate measures of objects such as weather balloons, scale models, food, and salamanders

Weather balloons are often used to send instruments high into the atmosphere. Data from the instruments help meteorologists predict the weather. There are several different sizes of weather balloons. Larger balloons can lift more equipment than smaller balloons.

EXAMPLE 1   **Connection: Data Analysis**

The table shows how much weight can be lifted by two balloons of different sizes.

Diameter of balloon	Weight balloon can lift
8 ft	17 lb
16 ft	137 lb

a. Assume both balloons are spheres. Find the volume of each balloon.

b. Compare the ratio of the diameters to the ratio of the volumes.

SOLUTION

a. Use the formula $V = \frac{4}{3}\pi r^3$ to find the volume of each balloon.

Smaller balloon: $V = \frac{4}{3}\pi(4)^3 \approx 268 \text{ ft}^3$

Larger balloon: $V = \frac{4}{3}\pi(8)^3 \approx 2145 \text{ ft}^3$

*The diameter of the smaller balloon is **8**, so **r = 4**.*

b. Find the ratio of the diameters.   Find the ratio of the volumes.

$$\frac{8}{16} = \frac{1}{2} \qquad \frac{\frac{4}{3}\pi(4)^3}{\frac{4}{3}\pi(8)^3} = \frac{(4)^3}{(8)^3} = \left(\frac{1}{2}\right)^3$$

The ratio of the volumes is the cube of the ratio of the diameters.

---

592   Chapter 11   *Circles and Spheres*

THINK AND COMMUNICATE

1. For each balloon in Example 1, find the ratio of the volume to the weight that the balloon can lift. What do you notice?

2. How can you estimate the weight that a 12 ft diameter balloon can lift?

Recall from Section 9.4 that if the ratio of corresponding lengths of two similar three-dimensional figures is $a:b$, then the ratio of their surface areas is $a^2:b^2$ and the ratio of their volumes is $a^3:b^3$. You can use these ratios to find unknown measures of similar three-dimensional figures.

EXAMPLE 2 | Connection: Estimation

A souvenir model of the Statue of Liberty has a volume of **26 mL**, excluding the pedestal. The scale of the model is about 1:200. Estimate the volume of the actual Statue of Liberty, excluding the pedestal. Assume that the model and the real statue are similar.

SOLUTION

Since the scale of the model is about 1:200, the ratio of corresponding lengths of the statue and the model is about $\frac{200}{1}$. So the ratio of their volumes is about $\left(\frac{200}{1}\right)^3$.

$$\frac{\text{Volume of statue}}{\text{Volume of model}} \approx \left(\frac{200}{1}\right)^3$$

$$\text{Volume of statue} \approx \text{Volume of model} \cdot \left(\frac{200}{1}\right)^3$$

$$\approx 26 \text{ mL} \cdot \left(\frac{200}{1}\right)^3$$

$$\approx 208{,}000{,}000 \text{ mL}$$

$$\approx 208{,}000 \text{ L} \qquad 1 \text{ L} = 1000 \text{ mL}$$

The volume of the Statue of Liberty is about 208,000 L.

BY THE WAY

You can find the volume of an object by putting it under water and measuring how much water it displaces.

THINK AND COMMUNICATE

3. The right arm of the Statue of Liberty is 42 ft long. A large model of the statue has a right arm that is $\frac{1}{2}$ ft long.
   a. What is the scale of the model?
   b. What is the ratio of the surface areas of the model and the statue?

4. The ratio of the areas of the bases of two similar cones is 4:9. Find the ratio of each measure of the cones.
   a. the radii   b. the heights   c. the surface areas   d. the volumes

---

ANSWERS   Section 11.7

Think and Communicate

1. For each balloon, the ratio is about 16:1.

2. Find the volume and divide by 16. The volume is about 905 ft, so the balloon can lift about 57 lb.

3. a. 1:84
   b. 1:7056

4. a. 2:3
   b. 2:3
   c. 4:9
   d. 8:27

Teach⇔Interact

Additional Example 1

The table shows how much juice was produced by two oranges of different sizes.

Diameter of Orange	Amount of Juice
6.6 cm	2.7 oz
8.8 cm	4.8 oz

a. Assume both oranges are spheres. Find the volume of each orange.

   Use the formula $V = \frac{4}{3}\pi r^3$ to find the volume of each orange. The diameter of the smaller orange is 6.6, so $r = 3.3$.

   Smaller orange: $V = \frac{4}{3}\pi(3.3)^3 \approx 151 \text{ cm}^3$

   Larger orange: $V = \frac{4}{3}\pi(4.4)^3 \approx 357 \text{ cm}^3$

b. Compare the ratio of the diameters to the ratio of the volumes.

   Find the ratio of the diameters.
   $$\frac{6.6}{8.8} = \frac{3}{4}$$
   Find the ratio of the volumes.
   $$\frac{\frac{4}{3}\pi(3.3)^3}{\frac{4}{3}\pi(4.4)^3} = \frac{(3.3)^3}{(4.4)^3} = \left(\frac{3}{4}\right)^3$$
   The ratio of the volumes is the cube of the ratio of the diameters.

Additional Example 2

A scale model of a new building has a volume of 8.2 ft³. The scale of the model is 1:50. Estimate the volume of the actual building. Assume that the model and the real building are similar.   Since the scale of the model is 1:50, the ratio of corresponding lengths of the building and the model is $\frac{50}{1}$. So the ratio of their volumes is $\left(\frac{50}{1}\right)^3$.

$$\frac{\text{Volume of building}}{\text{Volume of model}} = \left(\frac{50}{1}\right)^3$$

$$\text{Volume of building} \approx 8.2 \text{ ft}^3 \cdot \left(\frac{50}{1}\right)^3$$

$$\approx 1{,}025{,}000 \text{ ft}^3$$

Checking Key Concepts

Student Progress

A brief review of the answers to these questions will enable students to check their understanding of the relationships among volumes, surface areas, and other measures of similar figures.

Closure Question

How are the volumes of similar figures related?   The ratio of the volumes of similar figures is equal to the cube of the ratio of any pair of corresponding linear measures.

Apply⇔Assess

Suggested Assignment

❖ **Core Course**
Exs. 1–5, 7–12, 16–18

❖ **Extended Course**
Exs. 1–18

❖ **Block Schedule**
Day 71 Exs. 1–5, 7–12, 16–18

Exercise Notes

Interdisciplinary Problems

Exs. 1, 5, 9–12 In these exercises, students use what they have learned about similar solids to solve problems from astronomy, sports, and biology.

Communication: Discussion
Ex. 4 Ask students how they arrived at their solution for part (d). Some students may have used the similarity of the two figures. Others may have found the altitude to vertex *D* and used the formula for the area of the triangle. Point out that these two approaches should yield the same result. Ask students why one method may be preferred over the other.

☑ **CHECKING KEY CONCEPTS**

ARCHITECTURE Architects use scale models to design structures. A model of a shopping mall has the scale 1:100.

1. The model has 30 ft² of floor space. How many square feet of floor space will the shopping mall have?

2. One of the stores in the mall will have a back room with 1000 ft³ of space for storage. How much space for storage is there in the back room in the model?

The height of a cylinder is 4 in. and the height of a similar cylinder is 9 in. Find the ratio of each measure of the cylinders.

3. the radii of the bases   4. the areas of the bases

5. the surface areas   6. the volumes

Extra Practice
exercises on
page 685

11.7 Exercises and Applications

1. **ASTRONOMY** You can think of Earth and the moon as spheres. The diameter of Earth is about 12,800 km and the diameter of the moon is about 3500 km. Find the ratio of each measure.

    a. the lengths of their equators   **b.** their surface areas   **c.** their volumes

2. The cones below are similar. Find each measurement.

    a. surface area ≈ 280 cm²   **b.** surface area ≈ <u>?</u>
       $V \approx$ <u>?</u> $V \approx 1$ L

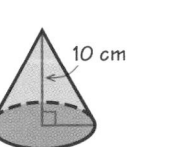
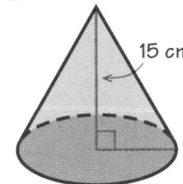

3. The heights of two similar mugs are 3.5 in. and 4 in. If the larger mug holds 12 oz, find the capacity of the smaller mug.

4. △*ABC* ~ △*EDF*.

    a. Find the lengths *AB*, *BC*, and *CA*.

    b. What is the area of △*ABC*?

    c. Find *FE*. What is the ratio of the lengths of corresponding sides of the two triangles?

    d. What is the area of △*EDF*? Explain how you found your answer.

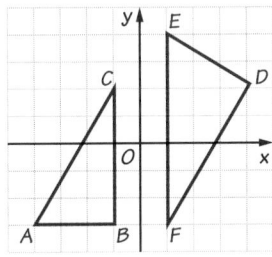

Checking Key Concepts

1. 300,000 ft²   2. 0.001 ft³

3. 4:9   4. 16:81

5. 16:81   6. 64:729

Exercises and Applications

1. a. 3.7:1

   b. 13.4:1

   c. 49:1

2. a. 0.3 L

   b. 630 cm²

3. about 8 oz

4. a. 3; 5; $\sqrt{34} \approx 5.8$

   b. 7.5

   c. 7; about 5.8:7

   d. about 10.9; Answers may vary. An example is given. I used the ratio of the areas.
   $$\frac{5.8^2}{7^2} = \frac{7.5}{A}; A \approx 10.9$$

5. about 37 oz

6. Answers may vary.

7. a. $a:b$; $a^3:b^3$

   b. smaller solid: $\frac{2}{3}\pi a^3$;
   larger solid: $\frac{2}{3}\pi b^3$

   c. $\frac{\frac{2}{3}\pi a^3}{\frac{2}{3}\pi b^3} = \frac{a^3}{b^3}$; Yes.

8. C

9. a. 2:1

   b, c. Answers may vary.

   b. about 1.26:1

   c. about 1.59:1

5. **SPORTS** A sports trophy is a bronze cup 10 in. tall. The winners of the event for which the trophy is awarded are given replicas of it that are 6 in. tall. If the replica holds 8 oz, find the capacity of the trophy.

6. **Open-ended Problem** Describe two objects in your school that are roughly similar. Estimate the ratio of corresponding lengths, the ratio of the areas, and the ratio of the volumes of the two objects.

7. **a. ALGEBRA** What is the ratio of any two corresponding lengths of the two cones? What should the ratio of the volumes be?
   b. Write an expression for the volume of each solid.
   c. Write and simplify an expression for the ratio of the volumes. Does your answer agree with your answer to part (a)?

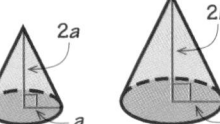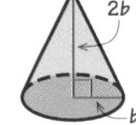

8. **SAT/ACT Preview** The volumes of two similar cones are 8π and 27π. What is the ratio of the lateral areas of the cones?

   A. $\frac{2}{3}$ **B.** $\frac{\sqrt{8}}{\sqrt{27}}$ **C.** $\frac{4}{9}$ **D.** $\frac{8}{27}$ **E.** $\frac{8^2}{27^2}$

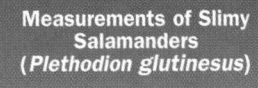

Connection ▷ **BIOLOGY**

Salamanders that don't have lungs must get all of the oxygen they need by breathing through the skin. In 1961, Juliusz Czopek published a study of the respiratory surfaces of various species of lungless salamanders. Salamanders from the same species, but of different sizes, are roughly similar.

9. The ratio of the volumes of two salamanders is the same as the ratio of their masses. For parts (a)–(d), use the data for the largest salamander and the smallest salamander in the table.
   a. What is the ratio of the volumes of the two salamanders?
   b. Use your calculator and the formula below to find the ratio of the lengths of the two salamanders.

   ratio of lengths = (ratio of volumes)$^{1/3}$

   c. What is the ratio of the surface areas of the two salamanders?
   d. Estimate the surface area of the smallest salamander.

10. Estimate the surface area of each salamander. Copy and complete the table.

11. How does the ratio of surface area to volume change as size varies?

12. **Writing** Explain why a lungless salamander the size of a horse would not be able to get enough oxygen for all of its cells.

Measurements of Slimy Salamanders (*Plethodion glutinesus*)	
Mass (g)	Surface Area (mm²)
3.40	?
3.85	?
3.96	?
5.50	?
6.80	2980

11.7 Volumes of Similar Solids **595**

Apply⇔Assess

Exercise Notes

Integrating Algebra
Ex. 7 For this exercise, students need to simplify a rational expression that involves exponents.

 Using Technology
Ex. 9 You may need to remind students that to evaluate an expression such as $4^{1/3}$ on a calculator, they must use parentheses around the exponent.

Integrating Statistics
Exs. 9–12 In these exercises, students use what they have learned about the volume and surface area of similar figures to analyze and complete data comparing the mass and surface area of salamanders.

Cooperative Learning
Exs. 9–12 You may wish to have students work together in groups of three to do these exercises. They could work as a group to find the surface area of the smallest salamander, and then work independently to find the surface area of one of the other salamanders. The group could then reconvene to check their results.

Communication: Writing
Ex. 12 For this exercise, students need to interpret the mathematical findings of Ex. 11 in a biological context. Reflecting upon the meaning of a mathematical result is essential when working in an applied situation.

---

d. about 1870 mm²

10. Estimates may vary.

Mass (g)	Surface Area (mm²)
3.40	1870
3.85	2040
3.96	2080
5.50	2600
6.80	2980

11. As the size increases, the ratio of surface area to volume decreases.

12. Answers may vary. An example is given. For the smallest and largest salamanders in the table, while the mass increases by a factor of 2, the surface area increases by a factor of only 1.59. Consider a lungless salamander weighing 1000 lb. If you use the ratio 2 : 1.59 (which would be a significant overestimate), the salamander would have a surface area of about 257,000 mm². The largest salamander in the table has about 438 mm² of skin surface per gram of mass compared to about 0.8 mm² for the horse-sized salamander.

Exercise Notes

Application

Exs. 13–15 These exercises help students to learn a valuable skill, namely, how to adjust a recipe to account for the use of an ingredient that is similar to the one required in the cookbook.

Assessment Note

Ex. 16 By listing the formulas asked for in this exercise and comparing them, students have an opportunity to strengthen their understanding of the formulas. Students may wish to put this list and their conclusions in their notebooks or journals for future reference.

Practice 81 for Section 11.7

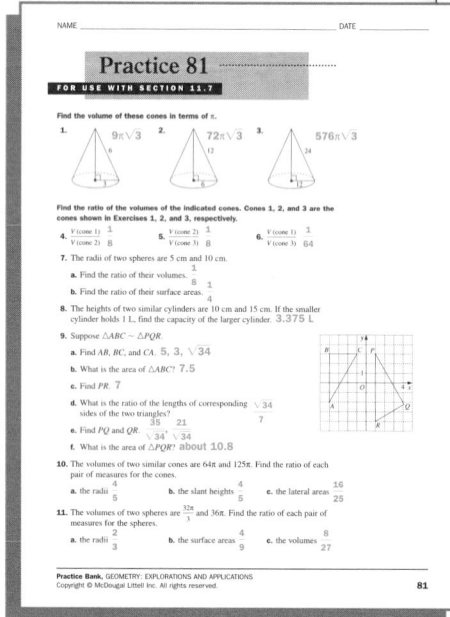

Connection   COOKING

Cooks often modify recipes to fit their needs. The recipe describes how to make a stuffed eggplant that is 10 in. long. Suppose you want to make six individual portions by stuffing six small eggplants that are each 5 in. long.

13. Assume that the stuffing from the recipe would exactly fill a 10 in. long eggplant. If you don't modify the stuffing recipe, will there be *too much* or *too little* stuffing to fill six small eggplants?

14. What fraction of the original stuffing recipe should you make to fill the six small eggplants nearly exactly?

15. **Challenge** A recipe for apple pie calls for 8 large apples. According to the cookbook, the diameter of a large apple is about four inches. If the only apples you have are about three inches in diameter, how many should you use so the pie will contain the same volume of apples? Explain how you found your answer.

ONGOING ASSESSMENT

16. **Writing** List the surface area and volume formulas for cubes, cylinders, cones, and spheres. What patterns do you notice? How are the surface area formulas alike? How are the volume formulas alike?

SPIRAL REVIEW

17. While writing a proof, Farzana introduced an auxiliary line. Which postulate or theorem can she use to justify drawing the line? *(Section 5.3)*

18. The surface area of a soccer ball is about 240 in.^{2} *(Section 11.6)*
    a. What is the radius?
    b. What is the volume?

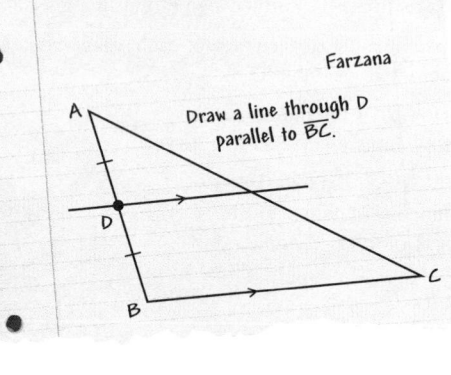

---

13. too much

14. $\frac{3}{4}$ of the original recipe

15. 19 small apples; The ratio of the diameters is $4:3$, so the ratio of the volumes is $64:27$. The volume of a large apple is $\frac{64}{27}$ that of a small apple. The volume of 8 large apples is the same as $8\left(\frac{64}{27}\right)$ or 19 small apples.

16.

Figure	Surface Area	Volume
cube (side length s)	$6s^2$	s^3
cylinder (radius r, height h)	$2\pi r^2 + 2\pi rh$	$\pi r^2 h$
cone (radius r, height h, perimeter p, slant height s)	$\pi rs + \pi r^2$	$\frac{1}{3}\pi r^2 h$
sphere (radius r)	$4\pi r^2$	$\frac{4}{3}\pi r^3$

Answers may vary. An example is given. The greatest exponent in all the surface area formulas is 2; the greatest exponent in all the volume formulas is 3. This makes sense since area involves measurements in two dimensions and volume involves measurements in three dimensions. The volume formulas for the cylinder, cone, and sphere are all quite similar. If you think of the radius of a sphere as its height, the formulas vary only by coefficient.

17. the Parallel Postulate (Through a point not on a line, there is exactly one line parallel to the given line.)

18. a. about 4.4 in.
    b. about 357 in.^{3}

11.8 Spherical Geometry

Because Earth is like a sphere, the shortest way to fly from Seattle, Washington, to Zurich, Switzerland, is to fly over Iceland! Geometry on a sphere is often surprising. You can use a basketball or a globe to explore the nature of *spherical geometry*.

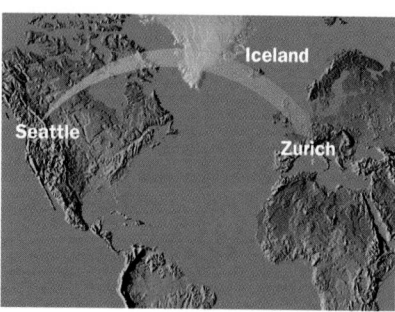

Learn how to...
- represent lines and shapes on a sphere

So you can...
- explore geometry on a sphere such as Earth

EXPLORATION
COOPERATIVE LEARNING

Triangles on a Sphere

Work in a group of two or three students.
You will need:
- a basketball, globe, or other sphere-shaped object
- five long strips of paper
- tape
- a protractor

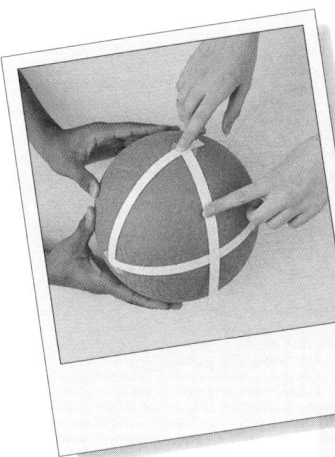

1 Tape one end of a strip of paper to the sphere. Run your finger along the strip to fit it to the sphere. Tape the strip at the other end.

2 Repeat Step 1 for two more strips to form a triangle on the sphere. Label the vertices of the triangle A, B, and C.

3 To measure ∠B, hold one of the remaining strips of paper along $\overline{AB}$ and the other along $\overline{BC}$. Tape the two strips together at B. Lay these strips on a flat surface and measure the angle. Repeat to measure ∠A and ∠C.

4 What is the sum of the measures of the angles of △ABC?

5 Form at least three other triangles of different sizes on the sphere. Find the sum of the measures of the interior angles for each triangle. What do you notice?

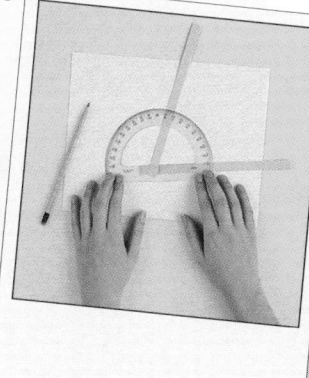

Exploration Note

Purpose
The purpose of this Exploration is to have students investigate triangles on a sphere in order to discover that the sum of the measures of the interior angles is greater than 180°.

Materials/Preparation
Each group of students needs a basketball, globe, or other sphere-shaped object, five long strips of paper, tape, and a protractor.

Procedure
Students tape three of the strips of paper on a sphere to form a triangle. The other two strips are used to form one of the angles, taped at the vertex, then laid flat to be measured. The sum of the three angles is then calculated. This process is repeated at least three other times.

Closure
Students should discover that the sum of the measures of a triangle formed on a sphere is greater than 180°.

Explorations Lab Manual
See the Manual for more commentary on this Exploration.

For answers to the Exploration, see following page.

Plan⇔Support

Objectives
- Represent lines and shapes on a sphere.
- Explore geometry on real-world spheres.

Recommended Pacing
❖ **Core and Extended Courses**
Section 11.8: 2 days
❖ **Block Schedule**
Section 11.8: 1 block

Resource Materials
Lesson Support
Lesson Plan 11.8
Warm-Up Transparency 11.8
Practice Bank: Practice 82
Study Guide: Section 11.8
Challenge Problems: Set 82
Assessment Book: Test 50
Technology
Internet:
http://www.hmco.com

Warm-Up Exercises

Complete each theorem or postulate.

1. Through any two points in a plane, there is exactly __?__.
one line

2. In a plane, if two lines are both perpendicular to a third line, then the two lines are __?__.
parallel

3. Through any point not on a given line, there is exactly one line __?__. parallel or perpendicular to the given line

4. The measures of the interior angles of a triangle is __?__.
180°

5. What is the intersection of a sphere and a plane that contains the center of the sphere?
a circle

Teach⇔Interact

Section Notes

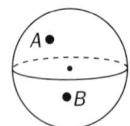

Communication: Discussion
Point out to students that all the theorems they have learned thus far are a part of Euclidean geometry, and are based on a particular set of postulates and definitions. Remind students that postulates are statements that are assumed to be true. An important idea that students should be made aware of as they study this section is that by using different postulates and different definitions, different geometries can be developed. Spherical geometry is one of these *non-Euclidean* geometries.

Student Study Tip
Encourage students to create a two-column chart with the heads *Euclidean geometry* and *Spherical geometry*. In each column, they should write the comparative rules of the two. For example, under Euclidean geometry they can write *the sum of the measures of the interior angles of a triangle always equals 180 degrees*, and under Spherical geometry, *the sum of the measures of the interior angles on a triangle is always greater than 180 degrees.*

Spatial Reasoning
If students can share some tennis balls and rubber bands, they can model the figures drawn in Examples 1 and 2, and *Think and Communicate* question 4. This should better help them to visualize the drawings given in the book.

Additional Example 1

The illustration shows two points in spherical geometry. Sketch a line through the two points.

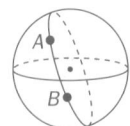

Sketch a great circle that goes through both points.

The geometry you have learned so far in this book is called *Euclidean geometry*. As you saw in the Exploration, geometry on a ball is different from Euclidean geometry. For example, on a ball, the sum of the measures of the interior angles of a triangle is always greater than 180°. Geometry on a ball is an example of *spherical geometry*.

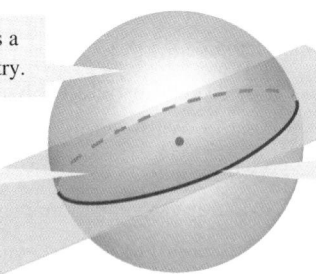

The surface of a sphere is a plane in spherical geometry.

Great circles are the lines in spherical geometry.

A **great circle** is the intersection of a sphere with a plane containing the center of the sphere.

EXAMPLE 1

The illustration shows a line ℓ in spherical geometry. Sketch another line that is perpendicular to ℓ.

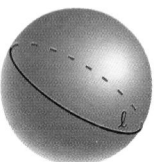

SOLUTION

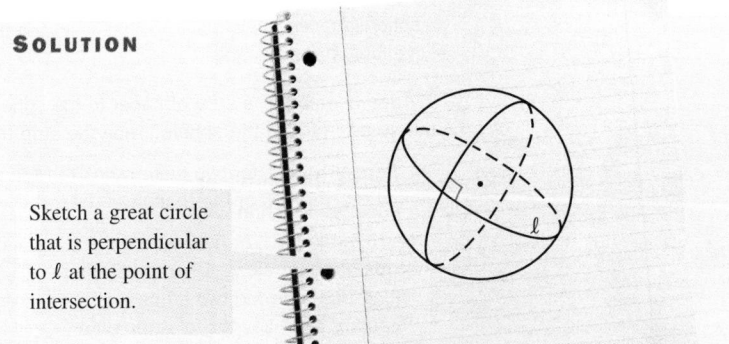

Sketch a great circle that is perpendicular to ℓ at the point of intersection.

THINK AND COMMUNICATE

The illustration shows a rubber band around a ball.

1. Is the rubber band a great circle of the sphere?

2. Imagine running your finger along the rubber band. Does the band follow the surface of the sphere in a straight line? Does it ever curve to the right or the left?

ANSWERS Section 11.8

Exploration

1–3. Check students' work.

4. Answers may vary. The sum of the measures of the interior angles is greater than 180°.

5. The sum of the measures of the interior angles is greater than 180°.

Think and Communicate

1. Yes.

2. Yes; No.

EXAMPLE 2

Show that this theorem from Euclidean geometry is *not* true for spherical geometry.

In a plane, if two lines are both perpendicular to a third line, then the two lines are parallel.

SOLUTION

The theorem is not true for spherical geometry because there is a counterexample. Imagine a globe and two different great circles that pass through both poles. Both lines are perpendicular to the equator, but they are not parallel to each other.

The surface of the globe is a plane in spherical geometry.

THINK AND COMMUNICATE

3. Explain how you know that the two lines in Example 2 that are perpendicular to the equator are not parallel to each other.

4. Amanda tried to justify the theorem in Example 2 by drawing this sketch. What mistake did Amanda make?

5. In Chapter 5, you learned Euclid's Parallel Postulate:

   Through a point not on a given line, there is exactly one line parallel to the given line.

   Is this postulate true in spherical geometry? Explain.

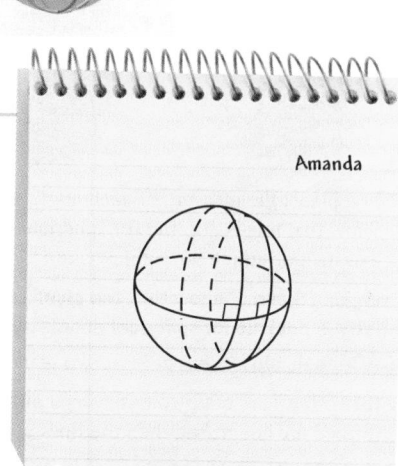

Amanda

☑ CHECKING KEY CONCEPTS

For Questions 1 and 2, draw a sphere and sketch each situation in spherical geometry. If it is not possible, explain why.

1. △*ABC* is equiangular.

2. The measure of each angle of a triangle is 60°.

Describe each situation in spherical geometry.

3.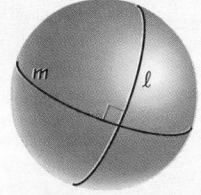

4.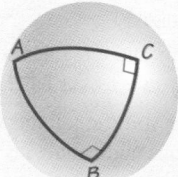

5. In Euclidean geometry, from a point not on a line there is exactly one line that is perpendicular to the given line. Do you think this is true on a sphere? Explain.

11.8 Spherical Geometry   **599**

Think and Communicate

Question 2 helps students to understand why a great circle is considered a line. Ask students to consider walking a straight line on Earth. What they consider to be a straight line on Earth is really a circle or a great circle.

Additional Example 2

Show that this theorem from Euclidean geometry is not true for spherical geometry.
Through any two points in a plane there is exactly one line.
The theorem is not true for spherical geometry because there is a counterexample. The surface of the globe is a plane in spherical geometry. Imagine the poles of a globe. An infinite number of great circles can be drawn through these two points.

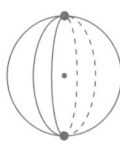

Think and Communicate

Question 4 has students analyze a common mistake in spherical geometry. Remind students that all lines in spherical geometry must be great circles. That is, their center is also the center of the sphere.

Checking Key Concepts

Teaching Tip
Some students may feel uncomfortable with spherical geometry at first. It is important to allow students to work these problems independently or with a partner and then discuss them thoroughly in class. For question 5, remind students that they need to find only one counterexample for the theorem to be false.

Think and Communicate

3. They intersect at the poles.

4. She did not draw two perpendicular lines. (One of the circles is not a great circle.)

5. No; on a sphere, through a point not on a line, there is *no* line parallel to the given line. (Look at the figure for question 1 on page 598. Imagine trying to wrap another rubber band around the ball along a great circle without having the two rubber bands cross. It is not possible.)

Checking Key Concepts

1.

2. not possible; The sum of the measures of the angles would be 180°. In spherical geometry, the sum of the measures of the interior angles of a triangle is always greater than 180°.

3. perpendicular lines *m* and *l*

4. Answers may vary. Examples are given. two lines through *A* perpendicular to the line through *B* and *C*; △*ABC*, in which two of the angles are right angles

5. No; the figure for Ex. 4 provides a counterexample. In fact, through a point not on a line, there are infinitely many lines perpendicular to the given line.

Closure Question

In Euclidean geometry, two lines can intersect in one point, an infinite number of points (coincident lines), or no points (parallel lines). What are the possibilities for the intersection of two lines in spherical geometry?   In spherical geometry, two lines can intersect in two points or an infinite number of points if they are coincident.

Extra Practice
exercises on
page 685

11.8 Exercises and Applications

1. **Writing** When the concept of a line is extended to a sphere, the result is a great circle of the sphere. Describe how you would extend the concept of *segment* to a sphere.

2. In Euclidean geometry, lines go on forever and have no ends. Do lines in spherical geometry have ends? Explain your reasoning.

3. Can you draw a ray on a sphere? Explain your reasoning.

Describe each figure. Is it possible to sketch the figure on a Euclidean plane? Explain your reasoning.

4. 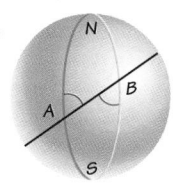   5. 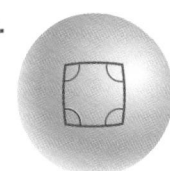   6.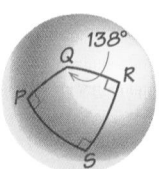

7. In a plane, the measure of each angle of an equiangular triangle is 60°. Show that this is not true for spherical triangles by sketching a triangle that includes three 90° angles.

In Exercises 8 and 9, show that the given statement from Euclidean geometry is *not* true for spherical geometry.

8. A triangle always has at least two acute angles.

9. If two lines are cut by a transversal and the corresponding angles formed are congruent, then the lines are parallel.

10. **Investigation** You will need a spherical object, three long strips of paper, tape, and a ruler. Use the strips of paper to make a triangle on the sphere. Mark the vertices of the triangle on the strips of paper, and untape the strips to measure the sides. Repeat several times. Are the Triangle Inequality Theorems on page 280 true on a sphere?

SPORTS The soccer ball at the right is a sphere covered by regular pentagons and hexagons. Two hexagons and one pentagon fit around each vertex without overlapping or leaving any gaps. On the soccer ball, as on a plane, the sum of the angles at each vertex is 360°.

11. **a.** On a plane, what is the measure of each interior angle of a regular hexagon? of a regular pentagon?
    b. On a plane, can two regular hexagons and a regular pentagon fit perfectly around a vertex as they do on a soccer ball? Explain.

12. **Open-ended Problem** Explain how you can tell that the measure of the interior angles of regular hexagons and pentagons are not the same on a sphere as on a plane.

13. The measure of each interior angle of a hexagon on a soccer ball is about 124.3°. What is the measure of each interior angle of a pentagon? Explain your reasoning.

 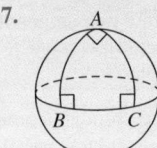

Any point on Earth can be located by its latitude and longitude measured in degrees (°) and minutes ('). One degree equals sixty minutes.

Chicago is **north** of the equator and $m\angle CTE = 41°50'$. The *latitude* of Chicago is 41°50' **north**.

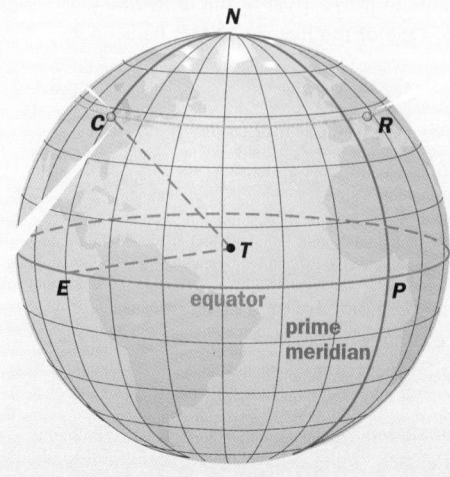

The latitude and longitude of Rome are 41°50' north, 12°40' east.

Chicago is **west** of the prime meridian, and $m\angle CNP = 87°40'$. The *longitude* of Chicago is 87°40' **west**. *Note:* $\angle CNP$ is on the surface of Earth.

14. a. The orange line on the globe above is called a *line of latitude* because all of the points on it have the same latitude. Are lines of latitude the same as lines in spherical geometry? Explain.

   b. The prime meridian is a *line of longitude*. Does the prime meridian line lie on a great circle of Earth? Explain your reasoning.

The shortest distance between two points on the globe is the length of the great-circle arc between them. This distance can be found using a *terrestrial triangle* whose third vertex is *N*, the North Pole.

15. Find the measure of $\angle CNR$ in spherical triangle *NCR*. (*Hint:* Let *P* represent a point on the prime meridian. $m\angle CNR = m\angle CNP + m\angle PNR$.)

16. $\triangle NCR$ is an isosceles triangle with $m\angle NCR = m\angle NRC \approx 51°20'$. Use your result from Exercise 15 to show that the sum of the measures of the angles of $\triangle NCR$ is greater than 180°.

17. $m\angle CTR = m\widehat{CR} \approx 70°00'$

   a. Convert the arc measure to minutes.

   b. A *nautical mile* is the length of a one-minute arc of a great circle on Earth. Find the distance from Chicago to Rome in nautical miles.

   c. A nautical mile is about 1.15 miles. Find the distance from Chicago to Rome in miles.

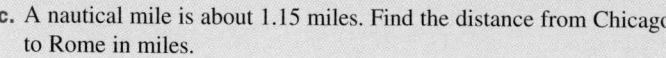

11.8 Spherical Geometry **601**

Exercise Notes

Interdisciplinary Problems
Exs. 14–17 This is an important group of exercises that not only helps students understand spherical geometry better, but also helps them to understand the concepts of latitude and longitude lines on a globe and the meaning of a nautical mile. You may need to take some time to discuss the definitions of latitude and longitude, and to give students some examples of angles measured in degrees and minutes.

Using Manipulatives
Exs. 14–17 If possible, have several globes available for students to use with these exercises. To help them become acclimated to using the globe, have students locate some of the countries they are studying in their social studies class and then estimate their location using latitude and longitude.

Integrating Measurement
Ex. 17 This exercise introduces students to the unit of measurement called a *nautical mile*. Students first convert arc measure to nautical miles, and then convert nautical miles to miles.

---

lines and the third side as the transversal, corresponding right angles are formed, but the lines are not parallel.

10. Answers may vary. For triangles in which the measure of each angle is less than 180°, the Triangle Inequality Theorems are true. If angles greater than 180° are allowed, the Triangle Inequality Theorems are not true.

11. a. 120°; 108°

   b. No; the sum of the measures of the angles at each vertex would be 348°, which would leave a gap at each vertex.

12. Answers may vary. An example is given. If they were, the figures would not fit together without gaps, since the sum of the measures of the angles at each vertex would be 348°.

13. about 111.4°; Since the sum of the measures of the angles at

each vertex on a soccer ball is 360°, the measure of each interior angle of a pentagon is about $360° - 2(124.3°) = 111.4°$.

14. a. No; the only latitude line that is a line in spherical geometry is the equator. The others are not great circles.

   b. Yes; all longitude lines are great circles.

15. 100°20'

16. $m\angle CNR + m\angle NCR + m\angle NRC \approx 100°20' + 2(51°20') = 203°0' > 180°$

17. a. 4200'

   b. about 4200 nautical miles

   c. about 4830 mi

Communication: Reading
Exs. 18–21 Have students read and then discuss the material at the top of this page before they begin these exercises. You may wish to ask student volunteers to read the material aloud in class. Provide an opportunity for questions to help ensure understanding of these new and different ideas. Point out that these geometries use all four of Euclid's postulates but not the fifth, which is altered in each geometry.

Second-Language Learners
Exs. 18–21 Students learning English will probably need some special attention to help them understand the vocabulary and ideas used in these exercises. You may wish to have these students work in groups with a peer tutor who understands this material and can explain it to them.

Problem Solving
Exs. 18–21 In doing these exercises, students should gain a greater understanding of how useful a model can be when working with abstract concepts such as hyperbolic geometry. In Exs. 18 and 19, students verify the parallel postulate for hyperbolic geometry using the model. In Ex. 20, students use the model to see that given a pair of lines that are perpendicular, it is possible to find a line that is parallel to both of them. Stress that lines are parallel if they do not intersect.

Research
Ex. 21 All of the mathematicians mentioned here were pioneers in the development of non-Euclidean geometry. Bolyai, Lobachevsky, and Gauss made important contributions to hyperbolic geometry. Riemann was instrumental in the discovery of spherical geometry.

Connection ▸ HYPERBOLIC GEOMETRY

Working in Alexandria, Egypt, in the third century B.C., Euclid tried to prove geometric concepts using a small number of postulates. Using only five postulates, he was able to prove most of the theorems that you have learned in this book. One of the five postulates he used is the *Parallel Postulate:*

Euclid: Through a point not on a given line, there is **exactly one** line parallel to the given line.

In the eighteenth century, mathematicians began to explore two different parallel postulates:

Spherical geometry: Through a point not on a given line, there is **no** line parallel to the given line.

Hyperbolic geometry: Through a point not on a given line, there is **more than one** line parallel to the given line.

Felix Klein, a German mathematician, developed a model to make it easier to explore hyperbolic geometry. In his model, the *hyperbolic plane* consists of only the interior points of a Euclidean circle. A *hyperbolic line* is the part of a Euclidean line contained in the circle.

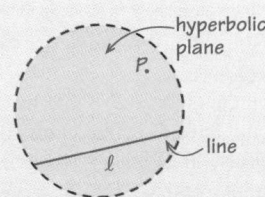

For Exercises 18–20, use Klein's model of the hyperbolic plane.

18. How do you know that two points determine a single line?

19. The figure above shows line ℓ and a point P not on line ℓ. Copy the figure and sketch several different lines through P that are parallel to (do not intersect) line ℓ.

20. In this model, angle measures and distances are not, in general, measured in the standard Euclidean manner. However, if line m is a diameter of the Euclidean circle, then another line k is perpendicular to m if k and m are perpendicular in the Euclidean sense. This figure shows two perpendicular lines. Copy the figure and sketch a third line parallel to both m and k. Which postulate of Euclidean geometry does your sketch contradict?

21. **Research** Write a report about one of the following mathematicians: G. F. Bernhard Riemann (1826–1866), Janos Bolyai (1802–1860), Nikolai Ivanovich Lobachevsky (1792–1856), or Carl Friedrich Gauss (1777–1855).

602 Chapter 11 *Circles and Spheres*

---

18. The two points determine a unique Euclidean line; the hyperbolic line is the part of that Euclidean line in the interior of the circle.

19, 20. Answers may vary. Examples are given.

19.

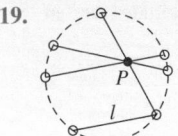

20.

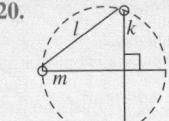

Line l is parallel to both m and k. This contradicts the Parallel Postulate. (Consider the point where lines m and k intersect. That point is not on l, but there are two lines through the point parallel to l.) It also contradicts the theorem, "If two lines are both parallel to a

third line, then the two lines are parallel." (Lines m and k are both parallel to l, but they are not parallel to each other.)

21. Answers may vary.

22. Both postulates are true on a sphere if all spheres have the same diameter. They are not true if the triangles are on spheres with different diameters. Explanations may vary. The truth of both postulates can be demonstrated using the method in the

22. Challenge Are the SSS and SAS Postulates true on a sphere? What if the two triangles are on spheres with different diameters? Explain your reasoning.

ONGOING ASSESSMENT

23. Open-ended Problem Choose a postulate or theorem from Euclidean geometry. Is it true on a sphere? Explain. Include a sketch with your explanation.

SPIRAL REVIEW

The heights of two similar pyramids are 10 cm and 7 cm. Find the ratio of each measure of the pyramids. *(Section 11.7)*

24. the areas of the bases   **25.** the lateral areas   **26.** the volumes

27. The vertices of $\triangle ABC$ are $A(-2, -1)$, $B(0, 3)$, and $C(5, -3)$. $\triangle ABC$ is reflected over the x-axis. Give the coordinates of the vertices of the image. *(Section 8.2)*

28. If $\overline{PQ}$ and $\overline{RQ}$ are tangent to $\odot C$ at P and R respectively, then: *(Section 11.4)*

    A. $PQ > RQ$ **B.** $RQ > PQ$ **C.** $PQ = RQ$

    D. relationship cannot be determined

ASSESS YOUR PROGRESS

VOCABULARY

sphere (p. 586)   **great circle** (p. 598)

1. Sketch a sphere. Draw and label a diameter, a secant, and a tangent of the sphere. *(Section 11.6)*

Find the surface area and volume of each sphere. *(Section 11.6)*

2.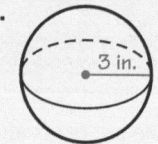

3.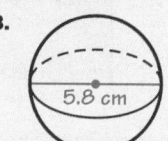

4. The slant heights of two similar cones are 5 ft and 12 ft. Find the ratio of each measure of the cones. *(Section 11.7)*

    a. the lateral areas   **b.** the volumes

5. In spherical geometry, is the sum of the measures of the interior angles of a quadrilateral 360°? Explain your reasoning. *(Section 11.8)*

6. Journal Some people think that spherical geometry is exciting and others find it unbelievable. Describe your reaction to learning that the rules of geometry change if you change the original postulates.

BY THE WAY

Another field of geometry is *projective geometry*, the study of the properties involved in transferring a three-dimensional object to a two-dimensional plane. Mathematicians in this field, such as Ruth Moufang (1905–1907), combine both algebraic and geometric concepts in their work.

11.8 Spherical Geometry **603**

Exercise Notes

Challenge
Ex. 22 Students may find it helpful to use the method of the Exploration on page 597 to model and measure spherical triangles.

Topic Spiraling: Preview
Ex. 27 In Chapter 12, students learn how a reflection can be represented by using matrices.

Assess Your Progress

Journal Entry
Finding out that there are geometries other than the one they are familiar with can be astonishing to students. This entry allows them to express some of their ideas and feelings about this significant fact.

Progress Check 11.6–11.8

See page 607.

Practice 82 for Section 11.8

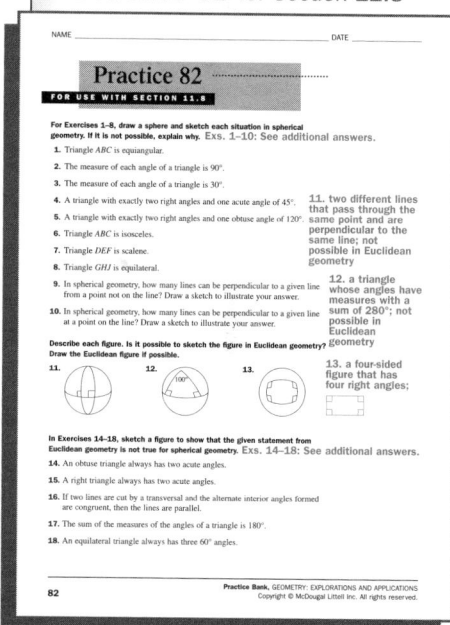

Exploration on page 597, by making a triangle and moving it to various positions on the sphere to represent the triangle on other congruent spheres. The failure of the postulates for spheres of different sizes can also be demonstrated using two spherical objects of different sizes.

23. Answers may vary. An example is given. For any two points, there is exactly one line

through the points. This postulate is not true on a sphere. Think of a globe, the poles, and lines of longitude. All the lines of longitude contain both poles. Check students' work.

24. 100:49

25. 100:49

26. 1000:343

27. $A'(-2, 1)$; $B'(0, -3)$; $C'(5, 3)$

28. C

Assess Your Progress

1. Answers may vary. An example is given.

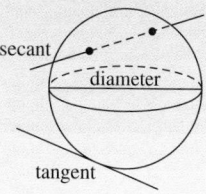

2. about 113 in.^{2}; about 113 in.^{3}

3. about 105.7 cm^2; about 102.2 cm^3

4. a. 25:144

    b. 125:1728

5. No; if you draw a diagonal, you divide the quadrilateral into two triangles. The sum of the measures of the interior angles of the each triangle is greater than 180°. Then the sum of the measures of the interior angles of the quadrilateral is greater than 360°.

6. Answers may vary.

- Make a sphere, cylinder, and two cones of a specified size.

- Find the cross sectional areas of a sphere, cylinder, and two cones.

- Compare the volumes of a sphere, cylinder, and two cones by comparing their cross sectional areas.

Planning

Materials
- modeling clay
- strong plastic
- scissors
- compass
- plastic knife
- tape
- fishing line or floss

Project Teams
Students select a partner and divide the tasks for the project evenly. Each student should mold at least one of the shapes and cut the cross sections at least once. Students should work together to write the report.

Guiding Students' Work

Constructing the cone is probably the most difficult part of the project for many students. You may need to work with the class explaining the measurements. Students should begin with a circle that has a radius of $2\frac{1}{8}$ in., or a diameter of $4\frac{1}{4}$ in. Students cut along the radius to the center and fold the circle until the diameter of the base of the cone is 3 in. To make the cylinder, students may find it easier to create a tube with a diameter of 3 in. out of a sturdy but bendable plastic and use it to cut through a piece of clay.

Second-Language Learners
You may want to give the following definitions for terms contained in this section: *slices* (thin pieces cut from a larger amount), *calculus* (a branch of mathematics), *cross sections* (a section formed by a plane cutting through an object).

Modeling Cavalieri's Principle

Archimedes discovered the formula for the volume of a sphere by imagining cutting a sphere, cone, and cylinder into very thin slices and comparing the slices. This method was developed further by Cavalieri, and became part of the foundation for calculus. Imagine slicing each object at the same height. How do you think the areas of the cross sections are related?

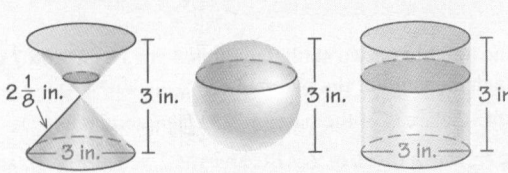

| **PROJECT GOAL** | Compare the volumes of a sphere, cylinder, and two cones by comparing cross sections of the objects. |

Making the Shapes

Work with a partner to make the sphere, cylinder, and two cones with the dimensions shown above. Use modeling clay.

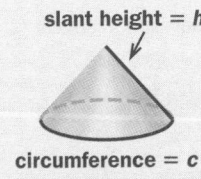
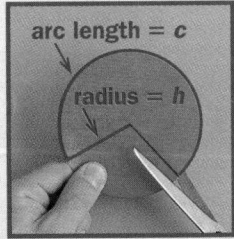

CONE To make a mold for the cone above, cut a circle out of strong plastic and draw a sector as shown. Tape the sector's edges together.

SPHERE You can check the diameter of your sphere by using a circular cutout with a diameter that is slightly larger than 3 in.

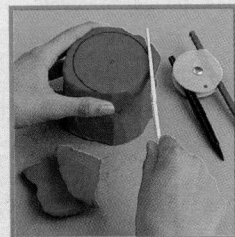

CYLINDER You might want to make your cylinder too large and then trim it carefully by using a compass and plastic knife.

Comparing Cross Sections

1. Cut each object carefully into six slices of equal thickness. Use fishing line or dental floss.

2. Combine the clay from the bottom slice of the cones and the bottom slice of the sphere to make a new disk of equal thickness.

3. Compare the new disk with the bottom slice of the cylinder. How were the cross sectional areas of the bottom slices of the objects related?

4. Repeat Steps 2 and 3 five times, comparing the cross sections of the slices at each height.

Presenting Your Project

Write a report describing your experiment. Include answers to the following questions.

- How are the areas of the cross sections of the objects related? How are the volumes of the objects related? How do you know?

- How can you use the volume formulas for a cylinder and a cone to find the volume formula for a sphere?

You may want to extend your project to include some of the ideas below:

- Look up *Cavalieri's Principle* in an encyclopedia or a mathematics dictionary. Summarize it and explain how you used it in your project.

- If x is the height at which you slice each object, write an expression for the area of each cross section. Show that the relationship you discovered is true for any length x that is less than 3 in.

Self-Assessment

What grade would you give yourself for this project? Why? If you did the experiment again, what would you do differently? Why?

Project Notes

Guiding Students' Work

Rubric for Chapter Project

4 Students perform all tasks involved with the project correctly. They also repeat the comparison five times. Students write a clear and well-organized report describing their work and answer all of the questions correctly. It is evident that the mathematics of the project is clearly understood. Students extend the project in one of the two ways described and complete an insightful self-assessment.

3 Students perform the tasks involved with the project correctly, but they do not repeat the comparison five times. Students write a report about their work and explain most of their ideas well. For the most part, students understand the concepts involved but do not understand some of the relationships of the volumes. Students extend the project in one of the two ways listed and provide a self-assessment.

2 Students attempt the project but complete the comparison only once. Students try to write a report describing their work, but because they do not understand the ideas involved, the report is incomplete and contains many errors. Students do not extend the project but attempt a self-assessment.

1 Students attempt to create the solids but do not complete the project correctly. No report is written, and students do not extend the project. No self-assessment is made. Students should be encouraged to speak with the teacher as soon as possible to review their work and to make a new start on the project.

Progress Check 11.1–11.2

Use the figure to find each value.
(Section 11.1)

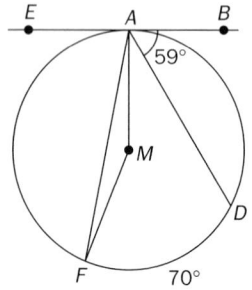

1. $m \angle EAM$ 90°
2. $m \overarc{AD}$ 118°
3. $m \angle FAD$ 35°
4. $m \angle AMF$ 172°

Use the figure to find each value.
(Section 11.2)

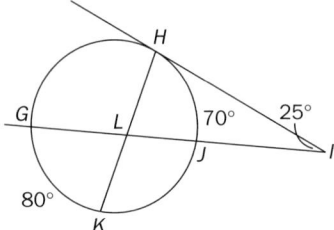

5. $m \angle GLK$ 75°
6. $m \overarc{GH}$ 120°

---

11 | Review

STUDY TECHNIQUE

Make a study sheet. Include definitions, equations, diagrams, and hints that you want to remember. Review your sheet twice a day for a week.

VOCABULARY

central angle (p. 553)
inscribed angle (p. 553)
chord (p. 553)
tangent to a circle (p. 554)
arc (p. 554)
minor arc (p. 554)
major arc (p. 554)
semicircle (p. 554)

measure of an arc (p. 554)
congruent arcs (p. 555)
secant (p. 561)
sector (p. 580)
length of an arc (p. 580)
sphere (p. 586)
great circle (p. 598)

SECTIONS 11.1 *and* 11.2

When both rays of an angle intersect a circle, the formula for finding the measure of the angle depends on the location of the angle's vertex.

Vertex at the center of the circle

The angle measure equals the measure of the intercepted arc.

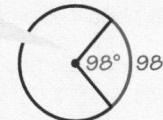

Vertex on the circle

The angle measure equals half the measure of the intercepted arc.

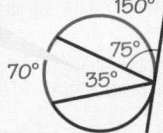

Vertex outside the circle

The angle measure equals half the *difference* of the measures of the intercepted arcs.

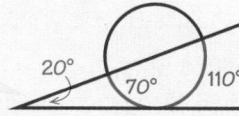

Vertex inside the circle

The angle measure equals half the *sum* of the measures of the intercepted arc.

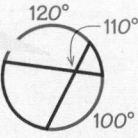

Inscribed angles that intercept the same arc or congruent arcs are congruent. A **tangent** to a circle is perpendicular to a radius at the point of tangency.

SECTIONS 11.3 and 11.4

The perpendicular bisector of a chord passes through the center of the circle. A diameter that is perpendicular to a chord bisects the chord and its corresponding arc.

You can find the lengths of some segments that intersect a circle:

Two chords
$AC \cdot CD = BC \cdot CE$

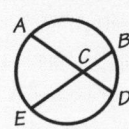

Two secants
$AC \cdot CD = BC \cdot CE$

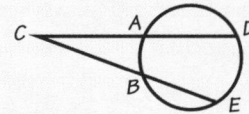

A tangent and a secant
$AC \cdot CD = BC^2$

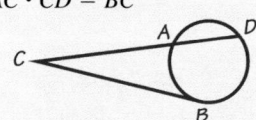

Two tangents
$AC = BC$

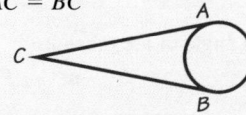

In the same circle or congruent circles, congruent chords are an equal distance from the center of the circle. Their corresponding arcs are congruent.

SECTIONS 11.5, 11.6, and 11.7

In a circle:

Area of a sector = **fraction of circle • area of circle**

$$= \frac{x}{360} \cdot \pi r^2$$

Length of an arc = **fraction of circle • circumference of circle**

$$= \frac{x}{360} \cdot 2\pi r$$

For a sphere with radius r:

Surface area $= 4\pi r^2$ Volume $= \frac{4}{3}\pi r^3$

For two similar solids, if the ratio of two corresponding lengths is $a:b$, then:
ratio of corresponding areas $= a^2:b^2$ ratio of the volumes $= a^3:b^3$

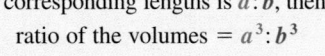

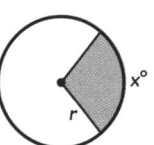

SECTION 11.8

In spherical geometry, a line is represented by a **great circle** of a sphere. The sum of the measures of the interior angles of a triangle is more than 180°, and there are no parallel lines. As a result, many of the theorems of Euclidean geometry are not true for spherical geometry.

Review   **607**

Find each value. *(Section 11.3)*

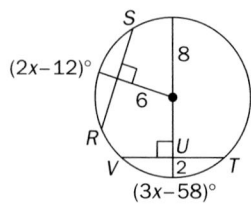

1. *UT* $2\sqrt{7}$, or about 5.3
2. $m\overset{\frown}{RS}$ 80°
3. Find the unknown length. *(Section 11.4)*

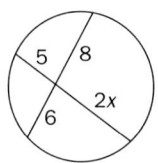

   $2x = 9.6$
4. Find the arc length of a sector of a circle with radius 18 cm and area 108π cm. *(Section 11.5)*   12π, or about 37.7 cm

Progress Check 11.6–11.8

1. Find the surface area and volume of a sphere with diameter 10 ft. *(Section 11.6)* about 314 ft²; about 524 ft³
2. Find the radius of a sphere whose surface area is 96 m². *(Section 11.6)*   about 2.8 m
3. A model of a train station has a scale of 1:100. Estimate the volume of the train station if the volume of the model is 0.25 m³. *(Section 11.7)*   250,000 m³
4. The altitudes of the bases of two similar prisms are 3 in. and 8 in., respectively. Find the ratio of their surface areas. *(Section 11.7)* $\frac{9}{64}$
5. In spherical geometry, is the following statement *True* or *False*? Explain. *(Section 11.8)*

   One line determines a plane.

   True; a line in spherical geometry is a great circle. There is only one plane (a sphere in spherical geometry) that contains this great circle.

Chapter 11 Assessment
Form A Chapter Test

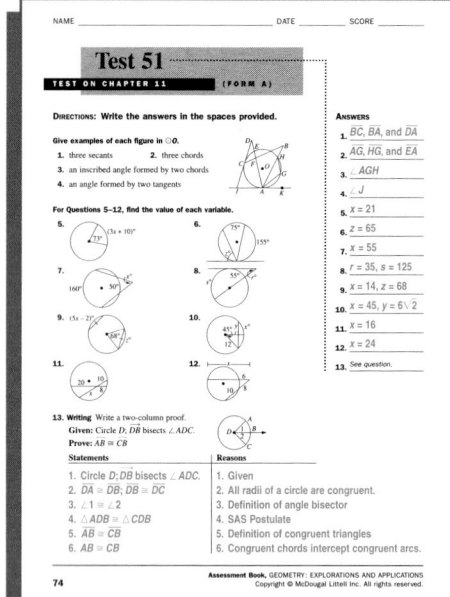

Chapter 11 Assessment
Form B Chapter Test

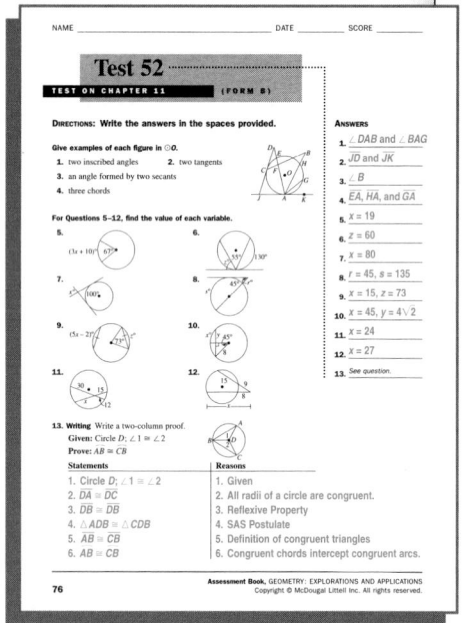

11 | Assessment

VOCABULARY REVIEW

Define each term and sketch an example.

1. congruent arcs **2.** major arc **3.** inscribed angle

4. secant **5.** tangent of a circle **6.** great circle

SECTIONS 11.1 and 11.2

Find each angle or arc measure.

7. $m \angle A$

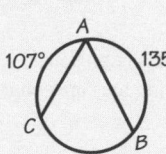

8. $m \widehat{EF}$

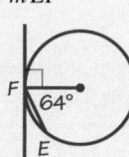

9. $m \widehat{MP}$

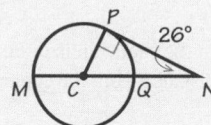

10. Given: $\widehat{AB} \cong \widehat{CD}$
Prove: $\overline{AD} \parallel \overline{BC}$

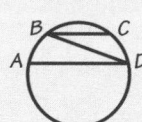

SECTIONS 11.3 and 11.4

Find the length of each labeled segment.

11.

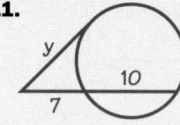

12.

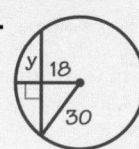

13.

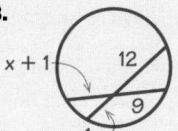

14. Given: $\widehat{MN} \cong \widehat{RS}$
Prove: $\triangle MNP \cong \triangle RSP$

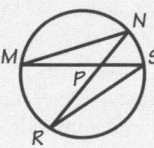

608   Chapter 11 *Circles and Spheres*

Assessment

1. arcs in congruent circles whose measures are equal

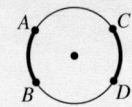

2. two points on a circle and all points of the circle not on the minor arc between the two given points

3. an angle whose vertex is on a circle and whose sides contain chords of the circle

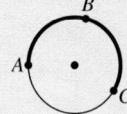

4. a line, ray, or segment that contains a chord

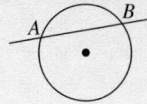

5. a line in the same plane as a circle that intersects the circle in only one point

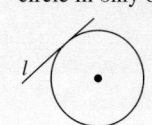

6. the intersection of a sphere with a plane containing the center of the sphere

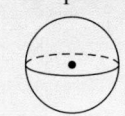

7. 59°

8. 52°

9. 116°

10. Statements (Reasons)
1. $\widehat{AB} \cong \widehat{CD}$ (Given)
2. $\angle ADB \cong \angle CBD$ (In a circle or congruent circles, angles that intercept the same arc or congruent arcs are congruent.)
3. $\overline{AD} \parallel \overline{BC}$ (If two lines are intersected by a transversal and alternate interior angles are congruent, then the lines are parallel.)

11. $\sqrt{119} \approx 10.9$ **12.** 24

13. $x + 1 = 8$; $x - 1 = 6$

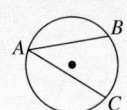

SECTIONS 11.5, 11.6, *and* 11.7

Find the area of each shaded sector and the length of each red arc.

15.
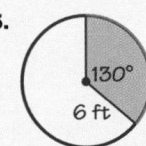
130°
6 ft

16.
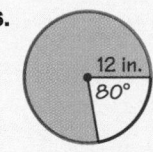
12 in.
80°

17.
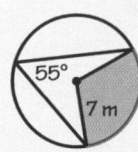
55°
7 m

Find the surface area and the volume of each sphere.

18.
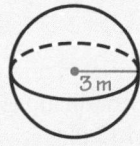
3 m

19.

6.5 cm

20. The radii of a sphere and a cone are both 7 cm. If their volumes are equal, what is the height of the cone?

21. The heights of two similar cylinders are 6 in. and 10 in. Find the ratio of each measure of the cylinders.
   a. the radii of the bases
   b. the areas of the bases
   c. the volumes

SECTION 11.8

22. How is the sum of the measures of the interior angles of a triangle different in spherical geometry than in Euclidean geometry?

23. Open-ended Problem Give an example of a statement from Euclidean geometry that is not true for spherical geometry. Explain why it is not true in spherical geometry.

24. Writing Is it possible to sketch a parallelogram on a sphere? Explain why or why not.

PERFORMANCE TASK

25. Make a list summarizing the theorems you know about angles and segments in circles. Choose two of the theorems. Do you think that they are also true for spheres? Explain your reasoning.

Chapter 11 Assessment
Form C Alternative Assessment

Chapter 11
ALTERNATIVE ASSESSMENT

1. Open-ended Problem Two local pizza restaurants are competing for business. Pizza Heaven is selling a 16-inch pizza for $12. Poppa's Pizza Palace is charging $12 for two 10-inch pizzas. Which pizza restaurant is offering the better deal? Create an ad campaign for the restaurant that is offering the better deal. Use clear, simple information to convince the customer that yours is the better deal.

2. Project Collect 5 to 10 lids of various sizes. Use a centimeter measuring tape to find the radius and circumference of each lid. Make a scatterplot of the data, plotting the radius on the horizontal axis and the circumference on the vertical axis. Use a graphics calculator with a linear regression feature to find the equation of the line of best fit for the data in the scatterplot. Compare your equation to the formula for finding the circumference of a circle. Discuss why the equation and the formula are not identical.

3. Project Collect 5 to 10 lids of various sizes. Use a centimeter ruler to find the radius of each lid and use centimeter grid paper to approximate the area of each lid. Make a scatterplot of the data, plotting the radius on the horizontal axis and the area on the vertical axis. Use a graphics calculator with a quadratic regression feature to find the equation of the line of best fit for the data in the scatterplot. Compare your equation to the formula for finding the area of a circle. Discuss why the equation and the formula are not identical.

4. Performance Task Discuss the validity of each of the following statements. Use examples in your discussion.
   a. If you know the perimeter of a rectangle, you can find its area.
   b. If you know the circumference of a circle, you can find its area.

5. Performance Task When a record spins on a turntable, it takes the same amount of time for all points on the record to complete one revolution. Suppose that a point A on the record is closer to the center of the record than some point B. Therefore, the circumference of the circle passing through point A is less than the circumference of the circle passing through point B. Explain how point B travels a greater distance in the same amount of time than does point A?

Assessment Book, GEOMETRY: EXPLORATIONS AND APPLICATIONS
Copyright © McDougal Littell Inc. All rights reserved.   **105**

6. Group Activity For this activity, draw a circle with a diameter of 7 inches on poster board. A designated group leader will call out a term from the following list: central angle, chord, diameter, inscribed angle, radius, secant, sector, semicircle, and tangent. The other members of the group then use pipe cleaners to model the term on the circle. The group leader checks the model for accuracy before calling out another term from the list. The activity continues until all of the terms have been modeled.

7. Open-ended Problem Two lines can be parallel, two planes can be parallel, and a line and a plane can be parallel. Do you think two circles can be parallel? Explain why or why not. (*Hint:* Consider slicing through a globe at two of its lines of latitude.)

106   Assessment Book, GEOMETRY: EXPLORATIONS AND APPLICATIONS
                                Copyright © McDougal Littell Inc. All rights reserved.

14. Statements (Reasons)
1. ∠*MNP* ≅ ∠*RSP*; ∠*NMP* ≅ ∠*SRP* (In a circle or congruent circles, angles that intercept the same arc or congruent arcs are congruent.)
2. $\widehat{MN}$ ≅ $\widehat{RS}$ (Given)
3. $\overline{MN}$ ≅ $\overline{RS}$ (Congruent arcs have congruent corresponding chords.)
4. △*MNP* ≅ △*RSP* (ASA Postulate)

15–19. Answers are given to the nearest tenth.

15. 40.8 ft²; 13.6 ft

16. 351.9 in.²; 58.6 in.

17. 47.0 m²; 13.4 m

18. 113.1 m²; 113.1 m³

19. 132.7 cm²; 143.8 cm³

20. 28 cm

21. a. 3 : 5
     b. 9 : 25

c. 27 : 125

22. In spherical geometry, the sum of the measures of the interior angles of a triangle is greater than 180°. In Euclidean geometry, the sum of the measures of the interior angles of a triangle is equal to 180°.

23. Answers may vary. An example is given. In Euclidean geometry, in a polygon with *n* sides, the sum of the measures of the interior angles is $(n-2)180°$. This is not true in spherical geometry. An obvious counterexample is $n = 3$.

24. No; the lines on a sphere are great circles. Since all great circles intersect in two places, there are no parallel lines in spherical geometry. Hence, there can be no parallelograms.

25. Check students' work.

12 Coordinates for Transformations

OVERVIEW

Connecting to Prior and Future Learning

⇔ The study of matrices from Algebra 1 is continued in this chapter as students learn how matrices can be used to describe transformations. The chapter opens with a study of dilations with matrices in Section 12.1, and moves to translations with matrices in Section 12.2.

⇔ A discussion of multiplying matrices in Section 12.3 is essential for students to understand the concepts presented at the end of the chapter. In this section, students learn how to transform points and polygons by using matrix multiplication and that the product is the image matrix.

⇔ Students use multiplication of matrices in Sections 12.4 and 12.5 to study reflections and rotations, respectively. Students learn how to reflect points and polygons over the *x*- and *y*-axes and other lines, and how to rotate points and polygons around the origin. These concepts provide a firm foundation for future studies in mathematics.

Chapter Highlights

Interview with Adriana Ocampo: This interview emphasizes the role mathematics plays in space travel. The application of mathematics discussed in this interview is explored in more depth in related exercises found on pages 623 and 645.

Explorations in Chapter 12 explore reflections in Section 12.4 and rotations in Section 12.5 using matrix multiplication. Students use graph paper to draw triangles and their images produced by multiplying the triangle matrix by a transformation matrix.

The Portfolio Project: Analyzing and generating fractals on a graphing calculator is the focus of this project. Students begin by analyzing the Sierpinski Triangle and finding values for the transformations used to create the triangle. They then use a program and a graphing calculator to display the triangle and to generate other fractals. Students describe the steps they used to find the transformations necessary to create their own fractals in a written report.

 Technology: Students use a calculator or geometry software to work with matrices throughout this chapter.

OBJECTIVES

Section	Objectives	NCTM Standards
12.1	• Represent points and polygons using matrices. • Multiply a matrix by a number. • Use matrices to describe dilations.	1, 2, 3, 4, 5, 8, 12
12.2	• Represent translations using matrices. • Add matrices. • Translate points and polygons using matrix addition.	1, 2, 3, 4, 5, 8, 12
12.3	• Multiply matrices. • Transform points and polygons using matrix multiplication. • Use matrix multiplication to analyze special effects.	1, 2, 3, 4, 5, 8, 12
12.4	• Represent reflections using matrices. • Reflect points and polygons using matrices. • Analyze real-world images using reflections.	1, 2, 3, 4, 5, 8, 12
12.5	• Represent rotations using matrices. • Rotate points and polygons around the origin.	1, 2, 3, 4, 5, 8, 12

INTEGRATION

Mathematical Connections	12.1	12.2	12.3	12.4	12.5
geometry	**613–618***	**619–625**	**626–633**	**634–640**	**641–647**
algebra			630	640	645
data analysis, probability, discrete math	614–618	**619–625**	**626–633**	**634–640**	**641–647**
logic and reasoning	618	623	630	639	645, 646

Interdisciplinary Connections and Applications					
digital maps	617				
color photocopies		624			
special effects			631		
art photography				637	
virtual reality				639	
video games					646

***Bold page numbers** indicate that a topic is used throughout the section.

TECHNOLOGY

Section	Student Book	opportunities for use with Support Material
12.1	graphing calculator	**Technology Book:** TI-92 Activity 12
12.2		**Technology Book:** TI-92 Activity 12
12.3	graphing calculator	**Technology Book:** TI-92 Activity 12 **Matrix Analyzer Activity Book:** Activity 1
12.4	graphing calculator	**Technology Book:** TI-92 Activity 12
12.5	graphing calculator	**Technology Book:** Calculator Activity 15 Spreadsheet Activity 11 TI-92 Activity 12

PLANNING GUIDE

Regular Scheduling (45 min)

Section	Materials Needed	Core Assignment	Extended Assignment	exercises that feature		
				Applications	Communication	Technology
12.1	graph paper, graphing calculator	1–20, 29, 32–39	1–13 odd, 14–39	21–28	27, 33	30
12.2	graph paper	1–19, 28–31, AYP*	1–13 odd, 14–31, AYP	20–27	14, 20c, 29	
12.3	graph paper, graphing calculator	1–10, 12–14, 16–23	1–23	12, 13	9b, 14d, 19	12, 15
12.4	graph paper	**Day 1:** 1–14, 16, 17 **Day 2:** 18–21, 25–41	**Day 1:** 1–17 **Day 2:** 18–41	11–15 25–31	21b, 31, 35	
12.5	graph paper, graphing calculator	1–7, 10, 12, 15–21, AYP	1–21, AYP	8, 9, 12–14	13c	13
Review/ Assess		**Day 1:** 1–13 **Day 2:** 14–28 **Day 3:** Ch. 12 Test	**Day 1:** 1–13 **Day 2:** 14–28 **Day 3:** Ch. 12 Test			
Portfolio Project		Allow 2 days.	Allow 2 days.			

Yearly Pacing (with Portfolio Project)	Chapter 12 Total 11 days	Chapters 1–12 Total 160 days	Remaining 0 days	Total 160 days

Block Scheduling (90 min)

	Day 75	Day 76	Day 77	Day 78	Day 79	Day 80
Teach/Interact	Ch. 11 Test 12.1	12.2 12.3	12.4: Exploration, page 634	12.5: Exploration, page 641 Review	Review Port. Proj.	Ch. 12 Test Port. Proj.
Apply/Assess	**Ch. 11 Test** **12.1:** 1–20, 29, 32–39	**12.2:** 1–19, 28–31, AYP* **12.3:** 1–10, 12–14, 16–23	**12.4:** 1–14, 16–21, 25–41	**12.5:** 1–7, 10, 12, 15–21, AYP **Review:** 1–13	**Review:** 14–28 **Port. Proj.**	**Ch. 12 Test** **Port. Proj.**

NOTE: A one-day block has been added for the Portfolio Project—timing and placement to be determined by teacher.

Yearly Pacing (with Portfolio Project)	Chapter 12 Total $5\frac{1}{2}$ days	Chapters 1–12 Total 80 days	Remaining 0 days	Total 80 days

*__AYP__ is Assess Your Progress.

Section	Practice Bank	Study Guide*	Assessment Book*	Visuals	Explorations Lab Manual	Lesson Plans	Technology Book
12.1	84	12.1		Warm-Up 12.1 Folder 14	Master 2	12.1	TI-92 Act. 12
12.2	85	12.2	Test 53	Warm-Up 12.2 Folder 14	Master 2	12.2	TI-92 Act. 12
12.3	86	12.3		Warm-Up 12.3	Master 2	12.3	TI-92 Act. 12
12.4	87	12.4		Warm-Up 12.4 Folder 14	Add. Expl. 18 Masters 1, 2	12.4	TI-92 Act. 12
12.5	88	12.5	Test 54	Warm-Up 12.5 Folder 14	Masters 1, 2	12.5	Calculator Act. 15 Spreadsheet Act. 11 TI-92 Act. 12
Review Test	89	Chapter Review	Tests 55–58, Alternative Assessment			Review Test	

*Spanish versions of *Study Guide* and *Assessment Book* are available.

Chapter Support

- Course Guides
- Lesson Plans
- Preparing for College Entrance Tests
- Multi-Language Glossary
- *Test Generator* Software
- Professional Handbook
- Challenge Problems

Software Support

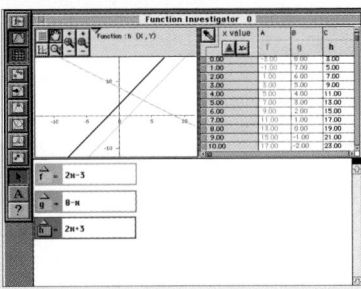

McDougal Littell Mathpack
Matrix Analyzer

Internet Support

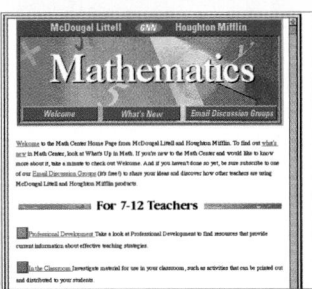

http://www.hmco.com
Next go to McDougal Littell; then the
Education Center; then Secondary Math.

Books, Periodicals

Eddins, Susan K., Evelyn Osman Maxwell, and Floramma Stanislaus. "Geometric Transformations, Part 2." *Mathematics Teacher* (April 1994): pp. 258–261, 268–270.

Lockwood, James R. and Garth E. Runion. *Deductive Systems: Finite and Non-Euclidean Geometries.* Reston, VA: NCTM, 1978.

Software

Department of Mathematics and Computer Science of the North Carolina School of Science and Mathematics. *Matrices.* Materials and software. IBM. Reston, VA: NCTM, 1988.

Laborde, Jean-Marie and Franck Bellemain, designers. *Cabri Geometry II.* Dallas, TX: Texas Instruments, 1994.

TI-92 Geometry. (Resident on the TI-92 graphing calculator.) Dallas, TX: Texas Instruments, 1995.

Videos

Donald in Mathmagic Land. Burbank, CA: Walt Disney Co., 1959. Distributed by Buena Vista Home Video, Burbank, CA.

Internet

For discussions among mathematics teachers, subscribe to the NCTM list by sending e-mail to:
 majordomo@forum.swarthmore.edu
In the body of the message, type:
 subscribe nctm-l [firstname, lastname]

CHAPTER

12 Coordinates for Transformations

Plotting EARTH and SKY

INTERVIEW Adriana Ocampo

*I*n 1969, at age fourteen, Adriana Ocampo saw how dreams can come true when she saw humans walk on the moon for the first time. "I immediately realized that I wanted to be part of that," she recalls. "It's always been my dream to get involved in space research and space missions." Now a planetary geologist at the Jet Propulsion Laboratory (JPL) in Pasadena, California, Ocampo is lucky enough to be fulfilling her dreams.

Adriana Ocampo stands among scale models of spacecraft.

"Space exploration is part of human destiny, and science makes it possible."

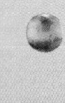

610

"Studying the processes that occur on other planets helps us better understand Earth."

An Extended Project

Born in Colombia and raised in Argentina, Ocampo came to the United States in 1970 and went to high school in South Pasadena. In a Caltech/JPL program for high school students, she learned about climatology and environmental monitoring while using sophisticated equipment. She continued working at the laboratory as a college undergraduate and as a graduate student, and has been there ever since. "This high school project extended into a career," she says.

From Earth to the Far Planets

Geology is the study of the processes that change the surface of Earth and its interior. Planetary geology, Ocampo's speciality, involves looking at the same processes on other planets and making comparisons. "By understanding how these processes occur on other planets, we gain a better understanding of our own planet," she explains.

Ocampo has worked on the Viking mission to Mars; the Voyager mission to Jupiter, Saturn, Neptune, and Uranus; and the Galileo mission to Jupiter. Her research also includes the exploration of a giant crater in the Yucatán peninsula in Mexico.

In space flight missions such as Galileo, Ocampo works chiefly as a planetary geologist. Her work relies heavily on tools developed using math and geometry.

Back to School

Ocampo participates in mentoring programs to encourage high school girls to get the training they need to pursue careers in science. "I think it's everyone's responsibility to give back a little of what they received," she says. "I understand how important it is to follow your dreams."

The Galileo mission examined Jupiter's atmosphere and some of its moons.

611

Mathematical Connection

The largest planet in our solar system, Jupiter, has a radius more than 11 times that of Earth. Also, its mass is 318 times that of Earth. While many people think of planets as having a solid structure, Jupiter is actually composed mostly of the gases hydrogen and helium. In fact, scientists are not sure Jupiter has any solid surface at all. Around Jupiter are sixteen moons. The four largest were discovered by the Italian scientist Galileo in 1610. They are Lo, the innermost satellite, Europa, Ganymede, and Callisto. All four of these moons were investigated using the Galileo spacecraft, and some of the information about them and Jupiter are explored in this chapter. In Section 12.2, students use matrices in a three-dimensional coordinate system to convert positions relating to the spacecraft. In Section 12.5, students use matrices to determine the position of the spacecraft at different times.

Explore and Connect

Project
The Internet or magazine articles would provide good sources of information on the spacecrafts and their missions. Pictures and images can also be obtained from these sources.

Writing
Students may want to work with a partner to discuss ways in which a spacecraft travels through space and what things could happen to it.

Research
Many students are fascinated about space and space travel and may enjoy presenting the results of their research to the class. Encourage these students to organize their results into a brief verbal report.

The Galileo spacecraft was launched from the space shuttle Atlantis. Black and gold fabric protects it from the heat of the sun and the cold of interplanetary space.

Galileo used the gravity fields of Venus and Earth to boost itself toward Jupiter. When it was 60,000 miles above Venus, its Near Infrared Mapping Spectrometer (NIMS) gathered data about radiant heat, shown in this false-color image.

Space Travel with Matrices

"We use matrices to define the coordinate system of the stars that we navigate the spacecraft by," Ocampo explains. Other matrices and coordinate systems define the position of the spacecraft relative to Earth. There's even a matrix to describe the orientation of the spacecraft. That's important to know so that cameras and sensors are pointed in the right direction. Navigation specialists at NASA/JPL use matrices to describe the translations and rotations that link these coordinate systems together. These calculations help keep the spacecraft on the correct trajectory.

Explore and Connect

Remote-sensing techniques for obtaining data, developed to study other planets, are now used to study our own.

1. Project Find out about the trajectory of and the planets visited by one of these spacecraft: Viking, Voyager, or Galileo. Present a report about some of the results of the mission, including some images.

2. Writing Imagine a spacecraft traveling through space. What transformations might it undergo and why? Explain.

3. Research Find out how far the large planets in the solar system are from Earth and how long it takes for spacecraft to reach them. Explain how the orbits of the planets influence the timing of the launches of spacecraft.

Mathematics & Adriana Ocampo

In this chapter, you will learn more about how mathematics relates to space science.

Related Exercises

Section 12.2
• Exercises 20 and 21

Section 12.5
• Exercises 8 and 9

612 Chapter 12 *Coordinates for Transformations*

12.1 Dilations with Matrices

In computer games, such as driving simulators, the illusion of depth is created by enlarging objects on the screen as a player nears them. The screens shown are from a bike riding simulator in which users feel they are moving down a road because objects along the roadside get bigger.

Programmers can use *matrices* to store the coordinates of objects they want to display. Then they transform the coordinates using dilations to draw larger objects.

A **matrix** is a rectangular arrangement of numbers, called **elements**, in **rows** and **columns**. *Matrices* is the plural of *matrix*. Matrices are enclosed in square brackets.

The number of rows and number of columns in a matrix are its **dimensions**. A matrix that has m rows and n columns is called an $m \times n$ (read "m by n") matrix.

You can represent a point in the coordinate plane with a **point matrix**. List the coordinates vertically. The x-coordinate is at the top. The y-coordinate is at the bottom.

$$
\begin{array}{c}
P \\
2 \text{ rows} \rightarrow \begin{bmatrix} 3 \\ 2 \end{bmatrix} \begin{array}{l} \leftarrow x\text{-coordinate} \\ \leftarrow y\text{-coordinate} \end{array} \\
\uparrow \\
\text{1 column}
\end{array}
$$

This is a 2×1 matrix.

You can represent a triangle with a **polygon matrix**. Each column represents a vertex. List the vertices in consecutive order, just as you would for a polygon.

$$
\begin{array}{c}
A\ B\ C \\
2 \text{ rows} \rightarrow \begin{bmatrix} 1 & 9 & 8 \\ 2 & 0 & 6 \end{bmatrix} \begin{array}{l} \leftarrow x\text{-coordinates} \\ \leftarrow y\text{-coordinates} \end{array} \\
\uparrow\ \uparrow\ \uparrow \\
\text{3 columns}
\end{array}
$$

This is a 2×3 matrix.

12.1 Dilations with Matrices **613**

Section Notes

Integrating Discrete Mathematics
Matrices and their operations (matrix algebra) are considered to be a branch of discrete mathematics. In this section, matrices are used to represent points and polygons in the plane. A dilation of the polygon centered about the origin is represented by scalar multiplication of the polygon matrix.

Teaching Tip
Explain that the use of matrices to represent polygons is just one of many ways matrices are used in mathematics. Some students may have already used matrices to solve systems of linear equations in algebra. Point out that while polygonal matrices always have two rows, matrices in general can have any number of rows. Also, some students may find it helpful to review their notes from Chapter 8 to refresh their understanding of dilations and other transformations they will use in this chapter, including translations, rotations, and reflections.

Communication: Reading
There are many concepts presented in the first two pages of this section. To help students learn these concepts, you can have them read the material on their own and take notes. Then ask volunteers to describe the various parts of a matrix, how to represent a point and a polygon using a matrix, and how to use scalar multiplication of matrices to represent dilations.

Student Study Tip
Point out how similar the concept of matrix scalar multiplication is to vector scalar multiplication. In fact, vectors can be represented by matrices.

Additional Example 1

Write a polygon matrix that represents hexagon *DEFGHI*. What are the dimensions of the matrix?

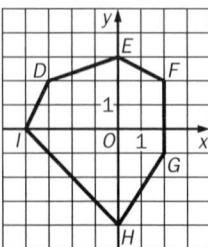

EXAMPLE 1

Write a polygon matrix that represents pentagon *VWXYZ*. What are the dimensions of the matrix?

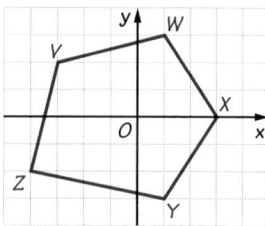

SOLUTION

Write the coordinates of the vertices in consecutive order.

$$\begin{array}{ccccc} V & W & X & Y & Z \end{array}$$
$$\begin{bmatrix} -3 & 1 & 3 & 1 & -4 \\ 2 & 3 & 0 & -3 & -2 \end{bmatrix}$$

The columns are written from left to right in the order used to name the polygon.

The dimensions of the matrix are 2×5.

THINK AND COMMUNICATE

1. Write a matrix for pentagon *ZYXWV* in Example 1. How is it different from the matrix in the solution above?

2. What are the dimensions of a matrix that represents a quadrilateral in the coordinate plane?

To dilate a polygon using $(0, 0)$ as the center of the dilation and a scale factor of 1.5, multiply each coordinate of the polygon by 1.5. Multiplying a matrix by a real number is called **scalar multiplication**.

In this chapter, light blue is used to represent transformations.

$$1.5 \cdot \begin{array}{ccc} A & B & C \end{array}$$
$$1.5 \cdot \begin{bmatrix} -54 & -54 & -26 \\ 40 & 0 & 16 \end{bmatrix} = \begin{bmatrix} -81 & -81 & -39 \\ 60 & 0 & 24 \end{bmatrix}$$

scale factor   **polygon matrix**   image matrix

The **image matrix** represents the image polygon.

Each element of the image matrix is 1.5 times the value of the corresponding element in the original polygon matrix. The original triangle and its image are shown on the screens below.

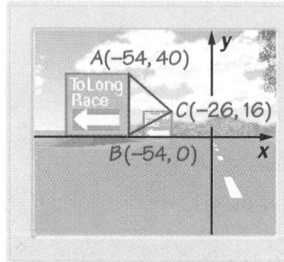 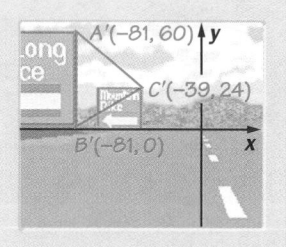

ANSWERS Section 12.1

Think and Communicate

1. $\begin{bmatrix} -4 & 1 & 3 & 1 & -3 \\ -2 & -3 & 0 & 3 & 2 \end{bmatrix}$;

The numbers appear in reverse order in each row.

2. 2×4

To multiply a matrix by a number k, multiply each element of the matrix by k:

$$k \begin{bmatrix} a & b & c \\ d & e & f \end{bmatrix} = \begin{bmatrix} ka & kb & kc \\ kd & ke & kf \end{bmatrix}$$

EXAMPLE 2

The matrix $\begin{bmatrix} -1 & 2 & 2 & 0 \\ 2 & 2 & 0 & -1 \end{bmatrix}$ represents quadrilateral **PQRS**.

a. Find the product: $3\begin{bmatrix} -1 & 2 & 2 & 0 \\ 2 & 2 & 0 & -1 \end{bmatrix}$.

b. Graph the polygons represented by $\begin{bmatrix} -1 & 2 & 2 & 0 \\ 2 & 2 & 0 & -1 \end{bmatrix}$ and $3\begin{bmatrix} -1 & 2 & 2 & 0 \\ 2 & 2 & 0 & -1 \end{bmatrix}$. Label them *PQRS* and *P'Q'R'S'*.

c. Describe the transformation that changes *PQRS* to *P'Q'R'S'*.

SOLUTION

a.
$$3\begin{matrix} P & Q & R & S \\ \begin{bmatrix} -1 & 2 & 2 & 0 \\ 2 & 2 & 0 & -1 \end{bmatrix} \end{matrix} = \begin{bmatrix} 3(-1) & 3(2) & 3(2) & 3(0) \\ 3(2) & 3(2) & 3(0) & 3(-1) \end{bmatrix}$$

$$= \begin{matrix} P' & Q' & R' & S' \\ \begin{bmatrix} -3 & 6 & 6 & 0 \\ 6 & 6 & 0 & -3 \end{bmatrix} \end{matrix}$$

b. Graph the original polygon and its image using the same pair of axes.

c. The transformation is a dilation with center at $(0, 0)$ and scale factor 3.

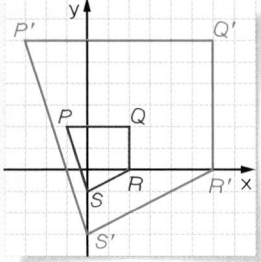

THINK AND COMMUNICATE

3. Find $\frac{1}{3}\begin{bmatrix} -3 & 6 & 6 & 0 \\ 6 & 6 & 0 & -3 \end{bmatrix}$. Explain what this product represents.

4. After a matrix is multiplied by a number, are the dimensions of the image matrix different from the dimensions of the original polygon matrix? Explain why or why not.

5. Suppose you multiply a polygon matrix by **1**. Describe the image matrix and the image polygon.

6. Suppose you multiply a polygon matrix by **0**. Describe the image matrix. What do you think the graph of the image matrix looks like?

12.1 Dilations with Matrices **615**

Think and Communicate

3. $\begin{bmatrix} -1 & 2 & 2 & 0 \\ 2 & 2 & 0 & -1 \end{bmatrix}$;

This takes *P'Q'R'S'* in Example 2 to *PQRS*. It is a dilation with center $(0, 0)$ and scale factor $\frac{1}{2}$.

4. No, multiplying a matrix by a number does not change the dimensions of the product matrix because no new rows or columns are added or subtracted.

5. The image matrix is identical to the original matrix, and the image polygon is identical to the original polygon.

6. a matrix with all elements 0; a point

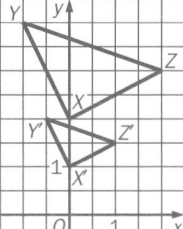

Checking Key Concepts

Student Progress
Students having difficulty with these exercises should refer to Examples 1 and 2.

Closure Question

Describe how to represent the dilation, with center (0, 0), of a polygon using scalar multiplication of a matrix.   First write the coordinates of the vertices of the polygon as a polygon matrix with the x-coordinates of the points in the first row and the corresponding y-coordinates in the second row. The points should be in the same order as they are named on the polygon. Use scalar multiplication to multiply the matrix by the scale factor of the dilation. The image matrix gives the coordinates of the image polygon.

Apply⇔Assess

Suggested Assignment

❖ **Core Course**
Exs. 1–20, 29, 32–39

❖ **Extended Course**
Exs. 1–13 odd, 14–39

❖ **Block Schedule**
Day 75 Exs. 1–20, 29, 32–39

Exercise Notes

Common Error
Ex. 13 Some students may forget to mention the center of dilation. Remind students to do this.

Integrating Algebra
Exs. 14–17 Students can solve these exercises using algebraic equations that involve the scalar and one pair of corresponding elements from each of the matrices. For example, with Ex. 14, $k \cdot 9 = 3$, so $k = \frac{1}{3}$. This can be checked using another pair of corresponding elements, such as $\frac{1}{3} \cdot 21 = 7$. After writing the matrices for the original figure and its image for Exs. 16 and 17, a linear equation can again be used to determine the scale factor.

☑ **CHECKING KEY CONCEPTS**

1. How many elements are in a 2×6 matrix? What kind of polygon could such a matrix represent?

2. Write a polygon matrix that represents parallelogram *DEFG*.

3. Multiply the matrix for *DEFG* by a scale factor of 1.5. Graph the image and label it $D'E'F'G'$.

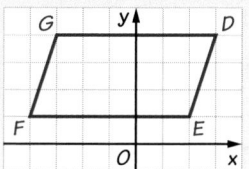

12.1 | **Exercises and Applications**

Extra Practice
exercises on page 685

Use the graph. Write a matrix that represents each figure.

1. *M*

2. *P*

3. $\overline{KN}$

4. △*MNK*

5. △*NPK*

6. *MNPK*

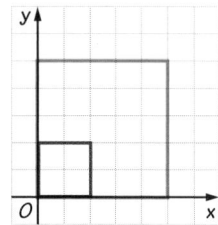

Find each product.

7. $5\begin{bmatrix} 3 \\ -1 \end{bmatrix}$

8. $4\begin{bmatrix} 1 & -3 \\ -2 & 5 \end{bmatrix}$

9. $3\begin{bmatrix} 0 & 3 & -2 \\ 0 & -1 & -4 \end{bmatrix}$

10. $2.5\begin{bmatrix} 1 & 0 & -2 & 3 \\ 1 & 5 & 0 & -4 \end{bmatrix}$

11. $1\begin{bmatrix} 10 & 5 & 15 \\ 5 & 5 & 10 \end{bmatrix}$

12. $\frac{2}{3}\begin{bmatrix} 9 & -12 \\ 0 & 3 \end{bmatrix}$

13. Describe the transformation represented by each product in Exercises 7–12.

Copy and complete each scalar multiplication problem.

14. $\underline{?}\begin{bmatrix} 9 & -3 \\ 21 & 0 \end{bmatrix} = \begin{bmatrix} 3 & -1 \\ 7 & 0 \end{bmatrix}$

15. $4\begin{bmatrix} 2 & ? & 0 \\ ? & 4 & ? \end{bmatrix} = \begin{bmatrix} 8 & 12 & ? \\ -4 & ? & 1 \end{bmatrix}$

The red polygon in each diagram is the image of the blue polygon after a dilation. Write a product to represent each dilation.

16.

17.

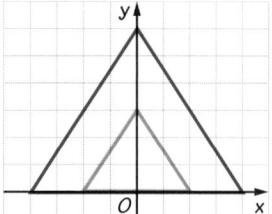

Checking Key Concepts

1. 12; a hexagon

2. $\begin{bmatrix} 3 & 2 & -4 & -3 \\ 4 & 1 & 1 & 4 \end{bmatrix}$

3. $\begin{bmatrix} 4.5 & 3 & -6 & -4.5 \\ 6 & 1.5 & 1.5 & 6 \end{bmatrix}$

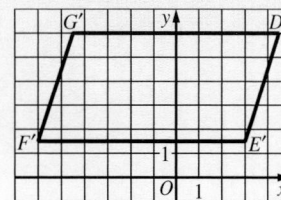

Exercises and Applications

1. $\begin{bmatrix} -3 \\ 2 \end{bmatrix}$

2. $\begin{bmatrix} 3 \\ 0 \end{bmatrix}$

3. $\begin{bmatrix} -2 & 0 \\ -2 & 4 \end{bmatrix}$

4. $\begin{bmatrix} -3 & 0 & -2 \\ 2 & 4 & -2 \end{bmatrix}$

5. $\begin{bmatrix} 0 & 3 & -2 \\ 4 & 0 & -2 \end{bmatrix}$

6. $\begin{bmatrix} -3 & 0 & 3 & -2 \\ 2 & 4 & 0 & -2 \end{bmatrix}$

7. $\begin{bmatrix} 15 \\ -5 \end{bmatrix}$

8. $\begin{bmatrix} 4 & -12 \\ -8 & 20 \end{bmatrix}$

9. $\begin{bmatrix} 0 & 9 & -6 \\ 0 & -3 & -12 \end{bmatrix}$

10. $\begin{bmatrix} 2.5 & 0 & -5 & 7.5 \\ 2.5 & 12.5 & 0 & -10 \end{bmatrix}$

11. $\begin{bmatrix} 10 & 5 & 15 \\ 5 & 5 & 10 \end{bmatrix}$

12. $\begin{bmatrix} 6 & -8 \\ 0 & 2 \end{bmatrix}$

13–17. See answers in back of book.

For each scale factor and polygon given, write a product to represent the dilation. Then graph each polygon and its image.

18. scale factor 3

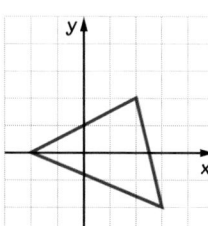

19. scale factor 2

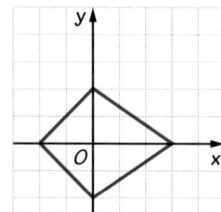

20. scale factor $\frac{3}{4}$

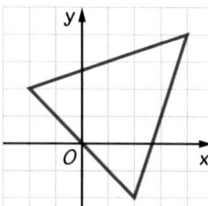

Connection ▶ DIGITAL MAPS

Digital maps available on the Internet can show a location you are interested in and then let you zoom in or out to see more or less detail. These maps show the area around Chinatown in San Francisco. From Map B, the user can zoom out to see Map A, or zoom in to see Map C. The origin is labeled X on each map.

For Exercises 21–23, the coordinates of each landmark on Map B are given. Write a point matrix for each landmark.

21. Cable Car Museum $(-58, 30)$

22. Chinatown Gate $(20, -36)$

23. Transamerica Pyramid $(54, 34)$

24. Write a triangle matrix that represents the three landmarks on Map B.

25. A dilation with scale factor $\frac{1}{2}$ reduces Map B onto Map A. Use scalar multiplication to find a triangle matrix that represents the coordinates of the landmarks on Map A.

26. A dilation with scale factor 2 enlarges Map B onto Map C. Use scalar multiplication to find a triangle matrix that represents the coordinates of the landmarks on Map C.

27. Writing Describe the dilation that enlarges Map A directly onto Map C. Explain your reasoning.

28. What point does not change its location during all of the dilations described above?

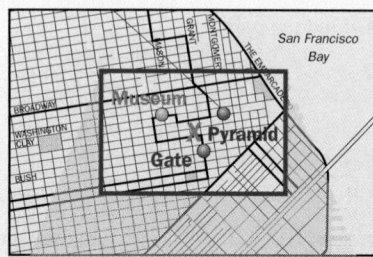

Map A

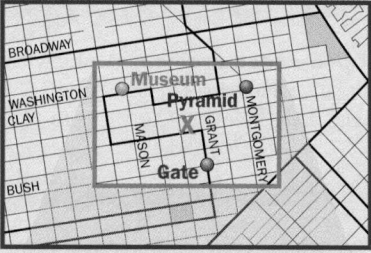

Map B

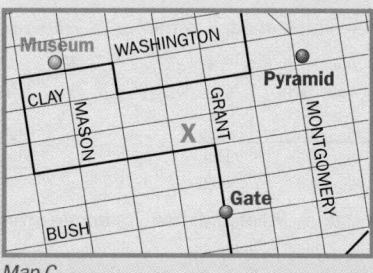

Map C

12.1 Dilations with Matrices **617**

18. $3\begin{bmatrix} -2 & 2 & 3 \\ 0 & 2 & -2 \end{bmatrix}$

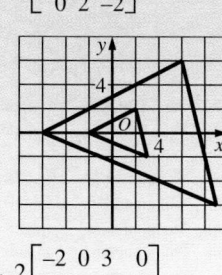

19. $2\begin{bmatrix} -2 & 0 & 3 & 0 \\ 0 & 2 & 0 & -2 \end{bmatrix}$

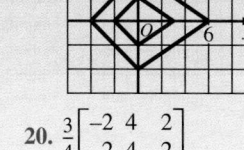

20. $\frac{3}{4}\begin{bmatrix} -2 & 4 & 2 \\ 2 & 4 & -2 \end{bmatrix}$

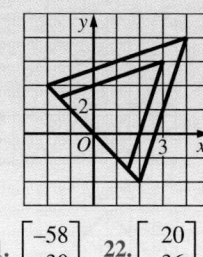

21. $\begin{bmatrix} -58 \\ 30 \end{bmatrix}$ **22.** $\begin{bmatrix} 20 \\ -36 \end{bmatrix}$

23. $\begin{bmatrix} 54 \\ 34 \end{bmatrix}$

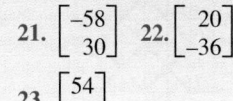

24. $\begin{bmatrix} -58 & 20 & 54 \\ 30 & -36 & 34 \end{bmatrix}$

25. $\begin{bmatrix} -29 & 10 & 27 \\ 15 & -18 & 17 \end{bmatrix}$

26. $\begin{bmatrix} -116 & 40 & 108 \\ 60 & -72 & 68 \end{bmatrix}$

27. a dilation about the "You are here" "x" by scale factor 4

28. the "x" at "You are here"

617

Exercise Notes

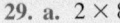
Using Technology
Ex. 30 To enter matrices on either a TI-81 or TI-82 calculator, press MATRX and choose 1:[A] from the EDIT menu. Enter the dimensions of the matrix, and then enter the elements of the matrix beginning with the one in the top left corner and moving right across the first row. After entering the first row, enter the left-most element of the second row and move right across this row. Press 2nd [QUIT] to return to the home screen. To find 5A on the TI-81, type 5 2nd [A] ENTER. On the TI-82, type 5 MATRX and press ENTER twice.

Assessment Note
Ex. 32 This exercise allows students to investigate the case of a dilation with a negative scale factor.

Practice 84 for Section 12.1

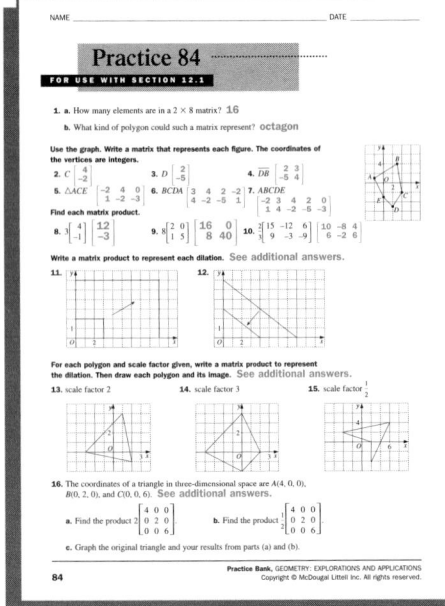

29. Suppose a matrix that represents an octagon in the coordinate plane is multiplied by a number. Give the dimensions of the image matrix.

30. 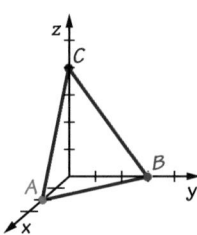 **Technology** Use a calculator or software with matrix capabilities. Matrices are named with capital letters. (See the *Technology Handbook*, pages 723–725, for more information about matrix calculations.)

    a. Store $\begin{bmatrix} 2 & 7 & 4 & -1 \\ 1 & 4 & 9 & 6 \end{bmatrix}$ as matrix A.

    b. Predict the elements in the matrices for $5A$, $3A$, and $0A$.

    c. Check your predictions using the calculator.

    [A]
    [[2 7 4 -1]
     [1 4 9 6]]
    5*[A]

31. **Challenge** A triangle in a three-dimensional coordinate system is represented by a 3×3 matrix. The matrices below represent $\triangle ABC$.

    a. Find the product: $2 \begin{bmatrix} 2 & 0 & 0 \\ 0 & 3 & 0 \\ 0 & 0 & 4 \end{bmatrix}$.

    b. Find the product: $\frac{1}{2}\begin{bmatrix} 2 & 0 & 0 \\ 0 & 3 & 0 \\ 0 & 0 & 4 \end{bmatrix}$.

    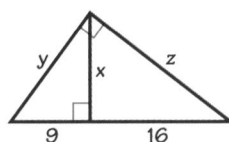

    c. Graph your results from parts (a) and (b). Describe the transformations.

ONGOING ASSESSMENT

32. **Open-ended Problem** Multiply any triangle matrix by several negative numbers. Graph each image triangle. Explain what a dilation with a *negative scale factor* looks like.

SPIRAL REVIEW

33. The volumes of two similar bottles are 27 oz and 8 oz. Find the ratio of their heights. *(Section 11.7)*

Find the value of each variable. *(Section 10.1)*

34.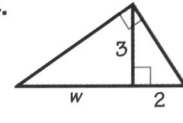

35.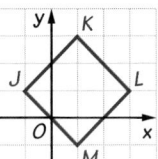

Copy square *JKLM* and draw its image after each translation. *(Section 8.3)*

36. $(a, b) \rightarrow (a + 2, b)$

37. $(a, b) \rightarrow (a, b + 2)$

38. $(a, b) \rightarrow (a - 3, b + 1)$

39. $(a, b) \rightarrow (a + 1, b - 3)$

618 Chapter 12 *Coordinates for Transformations*

---

29. a. 2×8

    b. 2×8

30. a. Check students' work.

    b. $5A = \begin{bmatrix} 10 & 35 & 20 & -5 \\ 5 & 20 & 45 & 30 \end{bmatrix}$;

    $3A = \begin{bmatrix} 6 & 21 & 12 & -3 \\ 3 & 12 & 27 & 18 \end{bmatrix}$;

    $0A = \begin{bmatrix} 0 & 0 & 0 & 0 \\ 0 & 0 & 0 & 0 \end{bmatrix}$

    c. Check students' work.

31. See answers in back of book.

32. Check students' work. When you dilate a polygon with a negative scale factor, the polygon is dilated about center (0, 0) by the absolute value of the scale factor, and rotated 180° about center (0, 0) (the effect of the negative sign).

33. $\sqrt[3]{\frac{9}{4}}$

34. $w = \frac{9}{2}$

35. $x = 12$; $y = 15$; $z = 20$

36.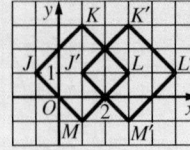

37.

38.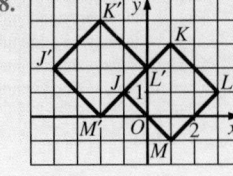

39.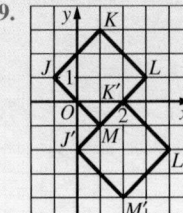

12.2 Translations with Matrices

Learn how to...
- **represent translations using matrices**
- **add matrices**

So you can...
- **translate points and polygons using matrix addition**

Dilations are not the only transformations you can represent using matrices. You can also express translations with matrices. In Chapter 8, you described a translation by describing the change in the coordinates of a point. A translation to the right 4 units and up 7 units, for example, shifts $(3, -2)$ to $(7, 5)$.

Add **4** to the *x*-coordinate.

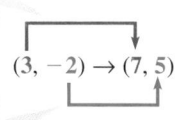

$(3, -2) \rightarrow (7, 5)$

Add **7** to the *y*-coordinate.

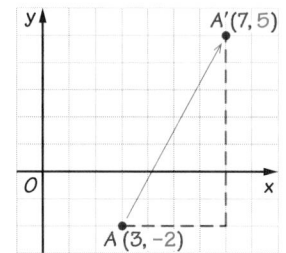

To use matrices to describe this translation, you must use **matrix addition**, the operation that adds corresponding elements of two matrices.

To add two matrices, they must have the same dimensions.

Add corresponding elements of the matrices.

In this chapter, transformations are written on the left.

$$\underbrace{\begin{bmatrix} 4 \\ 7 \end{bmatrix}}_{\substack{\text{translation} \\ \text{matrix}}} + \underbrace{\begin{bmatrix} 3 \\ -2 \end{bmatrix}}_{\substack{\text{point} \\ \text{matrix}}} = \overset{A}{\begin{bmatrix} 4 + 3 \\ 7 + (-2) \end{bmatrix}} = \overset{A'}{\underbrace{\begin{bmatrix} 7 \\ 5 \end{bmatrix}}_{\substack{\text{image} \\ \text{matrix}}}}$$

THINK AND COMMUNICATE

1. Write a translation matrix that shifts a point to the left 5 units and up 3 units. Find the image of point $A(3, -2)$ after this translation.

2. Write a translation matrix that shifts a point to the right 2 units. Find the image of point $A(3, -2)$ after this translation.

3. Write a translation matrix that shifts a point to the right *h* units and up *k* units. Find the image of point $A(3, -2)$ after this translation.

12.2 Translations with Matrices **619**

ANSWERS Section 12.2

Think and Communicate

1. $\begin{bmatrix} -5 \\ 3 \end{bmatrix}$; $(-2, 1)$

2. $\begin{bmatrix} 2 \\ 0 \end{bmatrix}$; $(5, -2)$

3. $\begin{bmatrix} h \\ k \end{bmatrix}$; $(3 + h, -2 + k)$

Plan⇔Support

Objectives
- Represent translations using matrices.
- Add matrices.
- Translate points and polygons using matrix addition.

Recommended Pacing
❖ **Core and Extended Courses**
Section 12.2: 1 day
❖ **Block Schedule**
Section 12.2: $\frac{1}{2}$ block
(with Section 12.3)

Resource Materials
Lesson Support
Lesson Plan 12.2
Warm-Up Transparency 12.2
Overhead Visuals:
    Folder 14: Matrices and
                Transformations
Practice Bank: Practice 85
Study Guide: Section 12.2
Explorations Lab Manual:
    Diagram Master 2
Challenge Problems: Set 85
Assessment Book: Test 53
Technology
Technology Book:
    TI-92 Activity 12
Internet:
    http://www.hmco.com

Warm-Up Exercises

Give the coordinates of P', the image of $P(4, -1)$, after each translation.

1. a shift to the right 3 units
   $P'(7, -1)$

2. a shift to the left 1 unit and up 2 units $P'(3, 1)$

3. a shift left 4 units and down 3 units $P'(0, -4)$

Describe the translation that transforms P to P'.

4. $P(3, 5)$, $P'(8, 2)$
   right 5 units and down 3 units

5. $P(4, -6)$, $P'(-1, -3)$
   left 5 units and up 3 units

Section Notes

Second-Language Learners
Since knowing what the word *shift* means is fundamental to understanding this section, you might want to give students a kinesthetic demonstration of a point or a figure shifting position on a coordinate plane. Review also the meanings of *up*, *down*, *left*, and *right*, since directional opposites can be difficult for second-language learners.

 Communication: Discussion
When discussing matrix addition, point out the similarity between matrix addition and vector addition. In each case, the corresponding elements are added. Stress that matrices must have the same dimensions to be added.

Additional Example 1

The matrix $\begin{bmatrix} -3 & 0 & 1 & -1 \\ 2 & 4 & 1 & -2 \end{bmatrix}$ represents quadrilateral *RSTU*.

a. Write a translation matrix that shifts *RSTU* to the right 2 units and down 1 unit. Because the polygon matrix has 4 columns, the translation matrix must also have 4 columns. Because the shift is to the right 2 units and down 1 unit, the translation matrix is $\begin{bmatrix} 2 & 2 & 2 & 2 \\ -1 & -1 & -1 & -1 \end{bmatrix}$.

b. Find the image matrix representing *R'S'T'U'*.

$\begin{bmatrix} 2 & 2 & 2 & 2 \\ -1 & -1 & -1 & -1 \end{bmatrix} +$
translation matrix

$\begin{matrix} R & S & T & U \\ \begin{bmatrix} -3 & 0 & 1 & -1 \\ 2 & 4 & 1 & -2 \end{bmatrix} \end{matrix} =$
polygon matrix

$\begin{matrix} R' & S' & T' & U' \\ \begin{bmatrix} -1 & 2 & 3 & 1 \\ 1 & 3 & 0 & -3 \end{bmatrix} \end{matrix}$
image matrix

c. Graph *RSTU* and its image.

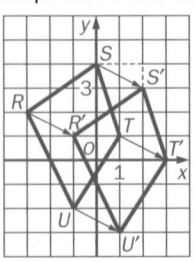

When you use matrices to translate a polygon, the dimensions of the translation matrix must match the dimensions of the polygon matrix. Each column of the translation matrix is the same as the column you would use to translate a point.

EXAMPLE 1

The matrix $\begin{bmatrix} -1 & 2 & 4 \\ 2 & 3 & 0 \end{bmatrix}$ represents △*ABC*.

a. Write a translation matrix that shifts △*ABC* to the left 3 units and down 5 units.

b. Find the image matrix representing △*A'B'C'*.

c. Graph △*ABC* and its image.

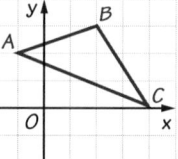

SOLUTION

a. Because the polygon matrix has 3 columns, the translation matrix must also have 3 columns. Because the shift is to the left 3 units and down 5 units, the translation matrix is:

The negative signs indicate movement *to the left* and *down*. $\begin{bmatrix} -3 & -3 & -3 \\ -5 & -5 & -5 \end{bmatrix}$

b.
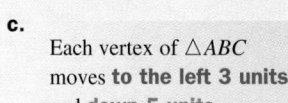
$\begin{matrix} & A\ B\ C & & A'\ B'\ C' \\ \begin{bmatrix} -3 & -3 & -3 \\ -5 & -5 & -5 \end{bmatrix} & + \begin{bmatrix} -1 & 2 & 4 \\ 2 & 3 & 0 \end{bmatrix} & = & \begin{bmatrix} -4 & -1 & 1 \\ -3 & -2 & -5 \end{bmatrix} \end{matrix}$

translation matrix   polygon matrix   image matrix

c.

Each vertex of △*ABC* moves **to the left 3 units** and **down 5 units**.

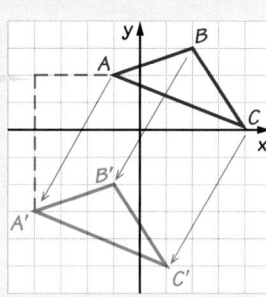

THINK AND COMMUNICATE

4. To translate △*A'B'C'* to △*ABC*, what translation matrix would you use?

5. To shift a pentagon using the same translation as in Example 1, what translation matrix would you use?

Think and Communicate

4. $\begin{bmatrix} 3 & 3 & 3 \\ 5 & 5 & 5 \end{bmatrix}$

5. $\begin{bmatrix} -3 & -3 & -3 & -3 & -3 \\ -5 & -5 & -5 & -5 & -5 \end{bmatrix}$

You can use **matrix subtraction** to find an original polygon matrix if you are given an image matrix and a translation matrix. The matrices must have the same dimensions. Subtract the corresponding elements of the matrices.

EXAMPLE 2

The matrix $\begin{bmatrix} 3 & 4 & 6 \\ -2 & 0 & -4 \end{bmatrix}$ represents $\triangle D'E'F'$, the image of $\triangle DEF$ after a translation to the right 5 units and down 3 units. Find the polygon matrix for $\triangle DEF$.

SOLUTION

Write the translation matrix and the image matrix.

The translation is to the right 5 units and down 3 units.

$$\underset{\substack{\text{translation}\\\text{matrix}}}{\begin{bmatrix} 5 & 5 & 5 \\ -3 & -3 & -3 \end{bmatrix}} + \underset{\substack{\text{polygon}\\\text{matrix}}}{\overset{D \quad E \quad F}{\begin{bmatrix} ? & ? & ? \\ ? & ? & ? \end{bmatrix}}} = \underset{\substack{\text{image}\\\text{matrix}}}{\overset{D' \quad E' \quad F'}{\begin{bmatrix} 3 & 4 & 6 \\ -2 & 0 & -4 \end{bmatrix}}}$$

$$\underset{\substack{\text{polygon}\\\text{matrix}}}{\begin{bmatrix} ? & ? & ? \\ ? & ? & ? \end{bmatrix}} = \underset{\substack{\text{image}\\\text{matrix}}}{\begin{bmatrix} 3 & 4 & 6 \\ -2 & 0 & -4 \end{bmatrix}} - \underset{\substack{\text{translation}\\\text{matrix}}}{\begin{bmatrix} 5 & 5 & 5 \\ -3 & -3 & -3 \end{bmatrix}}$$

Subtract the translation matrix from both sides of the equation.

$$= \begin{bmatrix} 3-5 & 4-5 & 6-5 \\ -2-(-3) & 0-(-3) & -4-(-3) \end{bmatrix}$$

$$= \begin{bmatrix} -2 & -1 & 1 \\ 1 & 3 & -1 \end{bmatrix}$$

$\triangle DEF$ has vertices $D(-2, 1)$, $E(-1, 3)$, and $F(1, -1)$.

☑ CHECKING KEY CONCEPTS

Add the matrices.

1. $\begin{bmatrix} 4 \\ 1 \end{bmatrix} + \begin{bmatrix} 0 \\ 3 \end{bmatrix}$

2. $\begin{bmatrix} -1 & -1 & -1 \\ 5 & 5 & 5 \end{bmatrix} + \begin{bmatrix} 0 & 3 & 7 \\ 3 & -4 & -2 \end{bmatrix}$

3. $\begin{bmatrix} 3 & 3 \\ -2 & -2 \end{bmatrix} + \begin{bmatrix} 1 & 4 \\ 5 & 2 \end{bmatrix}$

4. $\begin{bmatrix} 0 & 0 & 0 & 0 \\ 4 & 4 & 4 & 4 \end{bmatrix} + \begin{bmatrix} -2 & 3 & 2 & -5 \\ 0 & -1 & 6 & 5 \end{bmatrix}$

5. Suppose each matrix sum in Questions 1–4 represents a translation of a figure. Describe the figures and the translations.

6. Describe how to subtract matrices.

Checking Key Concepts

1. $\begin{bmatrix} 4 \\ 4 \end{bmatrix}$

2. $\begin{bmatrix} -1 & 2 & 6 \\ 8 & 1 & 3 \end{bmatrix}$

3. $\begin{bmatrix} 4 & 7 \\ 3 & 0 \end{bmatrix}$

4. $\begin{bmatrix} -2 & 3 & 2 & -5 \\ 4 & 3 & 10 & 9 \end{bmatrix}$

5. (1) The point (0, 3) is translated to the right 4 units and up 1 unit, to image point (4, 4).
(2) The triangle with vertices (0, 3), (3, −4), and (7, −2) is translated to the left 1 unit and up 5 units. The image triangle has vertices (−1, 8), (2, 1), and (6, 3).
(3) The line segment connecting (1, 5) and (4, 2) is translated to the right 3 units and down 2 units. The image segment has endpoints (4, 3) and (7, 0).

(4) A quadrilateral with vertices (−2, 0), (3, −1), (2, 6), and (−5, 5) is translated up 4 units. The image quadrilateral has vertices (−2, 4), (3, 3), (2, 10), and (−5, 9).

6. Matrices must have the same dimensions to be subtracted. Subtract each element of the second matrix from each element of the first matrix.

Teach⇔Interact

Additional Example 2

The matrix $\begin{bmatrix} 2 & -2 & 3 & -1 \\ 1 & 2 & 2 & 4 \end{bmatrix}$ represents quadrilateral *KLMN* after a translation to the left 3 units and up 2 units. Find the polygon matrix of quadrilateral *KLMN*.

Write the translation matrix and the image matrix. The translation is to the left 3 units and up 2 units.

$\begin{bmatrix} -3 & -3 & -3 & -3 \\ 2 & 2 & 2 & 2 \end{bmatrix} +$
translation matrix

$\overset{K \quad L \quad M \quad N}{\begin{bmatrix} ? & ? & ? & ? \\ ? & ? & ? & ? \end{bmatrix}} =$
polygon matrix

$\overset{K' \quad L' \quad M' \quad N'}{\begin{bmatrix} 2 & -2 & 3 & -1 \\ 1 & 2 & 2 & 4 \end{bmatrix}}$
image matrix

Subtract the translation matrix from both sides of the equation.

$\begin{bmatrix} ? & ? & ? & ? \\ ? & ? & ? & ? \end{bmatrix} =$
polygon matrix

$\begin{bmatrix} 2 & -2 & 3 & -1 \\ 1 & 2 & 2 & 4 \end{bmatrix} -$
image matrix

$\begin{bmatrix} -3 & -3 & -3 & -3 \\ 2 & 2 & 2 & 2 \end{bmatrix} =$
translation matrix

$\begin{bmatrix} 2-(-3) & -2-(-3) & 3-(-3) & -1-(-3) \\ 1-2 & 2-2 & 2-2 & 4-2 \end{bmatrix} =$

$\begin{bmatrix} 5 & 1 & 6 & 2 \\ -1 & 0 & 0 & 2 \end{bmatrix}$

Quadrilateral *KLMN* has vertices (5, −1), (1, 0), (6, 0), and (2, 2).

Checking Key Concepts

Visual Thinking
Ask students to sketch on coordinate planes the figures and the translations of those figures that are represented in the matrix sums in questions 1–4. This activity involves the visual skills of *correlation* and *exploration*.

Addition and subtraction of matrices are used to solve problems involving translations. Under what circumstances would you use matrix addition or matrix subtraction? You would use matrix addition when you know the polygon matrix and the translation matrix, and you want to find the image matrix. Matrix subtraction is used when you know the image matrix and the translation matrix, and you want to determine the polygon matrix .

Apply⇔Assess

Suggested Assignment

❖ **Core Course**
Exs. 1–19, 28–31, AYP

❖ **Extended Course**
Exs. 1–13 odd, 14–31, AYP

❖ **Block Schedule**
Day 76 Exs. 1–19, 28–31, AYP

Exercise Notes

Student Progress
Exs. 1–9 Students should be able to complete these exercises on their own. Students having difficulty should review the description of matrix addition on page 619 and matrix subtraction on page 621.

Communication: Writing
Ex. 14 This exercises gives students an opportunity to investigate whether the addition of matrices is commutative. Encourage students to try some examples. Point out that the convention of writing the translation matrix on the left makes it easier to tell which matrix is the translation matrix.

12.2 | Exercises and Applications

Extra Practice
exercises on
page 686

Add or subract the matrices.

1. $\begin{bmatrix} -3 \\ 5 \end{bmatrix} + \begin{bmatrix} 5 \\ -1 \end{bmatrix}$

2. $\begin{bmatrix} 1 & 2 & -1 \\ -3 & 4 & 6 \end{bmatrix} - \begin{bmatrix} 3 & 3 & 3 \\ 2 & 2 & 2 \end{bmatrix}$

3. $\begin{bmatrix} 1 & 1 \\ 0 & 0 \end{bmatrix} + \begin{bmatrix} -1 & 3 \\ 3 & -1 \end{bmatrix}$

4. $\begin{bmatrix} 2 & 2 & 2 & 2 \\ -4 & -4 & -4 & -4 \end{bmatrix} + \begin{bmatrix} 0 & 3 & 0 & -5 \\ -2 & 2 & 3 & 1 \end{bmatrix}$

5. Suppose each matrix sum or difference in Exercises 1–4 represents a translation of a figure. Describe the figures and the translations.

Write the matrix you would use to translate each figure as indicated.

6. a point; to the right 3 units and up 4 units

7. a segment; to the left 2 units

8. a triangle; down 5 units

9. a quadrilateral; to the right 6 units and down 1 unit

For Exercises 10–13, use the method shown in Example 1.

10. The matrix $\begin{bmatrix} 3 & 5 \\ 1 & 1 \end{bmatrix}$ represents $\overline{XY}$.

    a. Write a matrix that translates $\overline{XY}$ to the right 3 units and up 1 unit.

    b. Find the image matrix that represents $\overline{X'Y'}$.

    c. Graph $\overline{XY}$ and its image.

11. The matrix $\begin{bmatrix} -4 & -8 & 0 \\ -3 & 2 & 3 \end{bmatrix}$ represents $\triangle ABC$.

    a. Write a matrix that translates $\triangle ABC$ to the left 4 units and up 4 units.

    b. Find the image matrix that represents $\triangle A'B'C'$.

    c. Graph $\triangle ABC$ and its image.

12.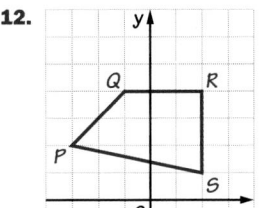

    a. Write a matrix that represents quadrilateral *PQRS*.

    b. Write a matrix that translates *PQRS* to the right 5 units and down 3 units.

    c. Find the image matrix that represents *P'Q'R'S'*. Graph the image.

13.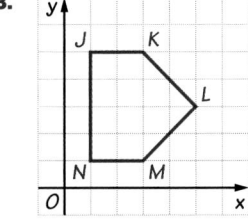

    a. Write a matrix that represents pentagon *JKLMN*.

    b. Write a matrix that translates *JKLMN* to the left 4 units and down 8 units.

    c. Find the image matrix that represents *J'K'L'M'N'*. Graph the image.

14. Writing Does it matter whether you write a translation matrix on the left side or the right side of a polygon matrix before you add the matrices to find an image matrix? Explain why or why not.

622   Chapter 12 *Coordinates for Transformations*

Exercises and Applications

1. $\begin{bmatrix} 2 \\ 4 \end{bmatrix}$

2. $\begin{bmatrix} 4 & 5 & 2 \\ -1 & 6 & 8 \end{bmatrix}$

3. $\begin{bmatrix} 0 & 4 \\ 3 & -1 \end{bmatrix}$

4. $\begin{bmatrix} 2 & 5 & 2 & -3 \\ -6 & -2 & -1 & -3 \end{bmatrix}$

5. (1) The point $(5, -1)$ is translated to the left 3 units and up 5 units.

(2) A triangle with vertices $(1, -3)$, $(2, 4)$, and $(-1, 6)$ is translated to the right 3 units and up 2 units.
(3) A line segment with endpoints $(-1, 3)$ and $(3, -1)$ is translated to the right 1 unit.
(4) A quadrilateral with vertices $(0, -2)$, $(3, 2)$, $(0, 3)$, and $(-5, 1)$ is translated to the right 2 units and down 4 units.

6. $\begin{bmatrix} 3 \\ 4 \end{bmatrix}$ **7.** $\begin{bmatrix} -2 & -2 \\ 0 & 0 \end{bmatrix}$

8. $\begin{bmatrix} 0 & 0 & 0 \\ -5 & -5 & -5 \end{bmatrix}$

9. $\begin{bmatrix} 6 & 6 & 6 & 6 \\ -1 & -1 & -1 & -1 \end{bmatrix}$

10. a. $\begin{bmatrix} 3 & 3 \\ 1 & 1 \end{bmatrix}$ **b.** $\begin{bmatrix} 6 & 8 \\ 2 & 2 \end{bmatrix}$

c.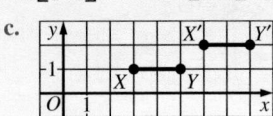

11, 12. See answers in back of book.

The red polygon in each diagram is the image of the blue polygon after a translation. Write each translation as a matrix sum.

15.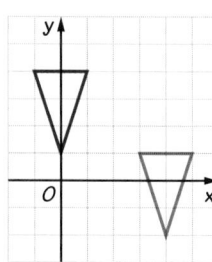

16.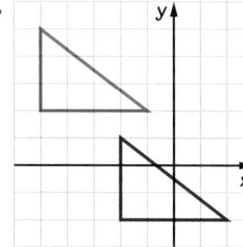

17.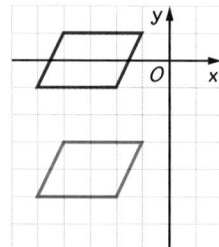

For Exercises 18 and 19, use the method shown in Example 2.

18. The matrix $\begin{bmatrix} 3 & 6 \\ 4 & 3 \end{bmatrix}$ represents $\overline{J'K'}$, the image of $\overline{JK}$ after a translation to the right 2 units and up 4 units. Find the matrix for $\overline{JK}$.

19. The matrix $\begin{bmatrix} -2 & 0 & 6 & 4 \\ 0 & -3 & 0 & 3 \end{bmatrix}$ represents $W'X'Y'Z'$, the image of quadrilateral $WXYZ$ after a translation down 6 units. Find the matrix for $WXYZ$.

INTERVIEW Adriana Ocampo

Look back at the article on pages 610–612.

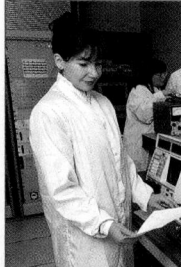

Ganymede, Europa, and Callisto are three of Jupiter's moons. To help the Galileo spacecraft take pictures of and get data from the moons, scientists such as Adriana Ocampo must know the moons' locations in relation to the spacecraft.

20. In a three-dimensional coordinate system whose origin is at the center of Jupiter, the positions of the three moons at a given time are in the matrix on the right below. The translation matrix converts the positions into a coordinate system whose origin is at the spacecraft.

10×10^5 represents 1,000,000 km.

x-coordinate → y-coordinate → z-coordinate →
$$\begin{bmatrix} -7.4 \times 10^5 & -7.4 \times 10^5 & -7.4 \times 10^5 \\ -2.7 \times 10^5 & -2.7 \times 10^5 & -2.7 \times 10^5 \\ -1.4 \times 10^5 & -1.4 \times 10^5 & -1.4 \times 10^5 \end{bmatrix} + \begin{matrix} G \quad\quad E \quad\quad C \\ \begin{bmatrix} 10 \times 10^5 & 6.4 \times 10^5 & 18 \times 10^5 \\ 3.1 \times 10^5 & 1.6 \times 10^5 & 2.0 \times 10^5 \\ 2.1 \times 10^5 & 1.4 \times 10^5 & 4.9 \times 10^5 \end{bmatrix} \end{matrix}$$
translation matrix

a. Add the matrices to find a matrix for $G'E'C'$, the moons' positions in a coordinate system whose origin is at the spacecraft.

b. Use the distance formula in three dimensions to find EC and $E'C'$.

c. **Writing** What does each of the distances you found in part (b) represent? Compare the two values and explain your results.

21. **Challenge** Write a matrix for the positions of Callisto, Jupiter, and the spacecraft in a coordinate system whose origin is at Europa.

12.2 Translations with Matrices **623**

Apply⇔Assess

Exercise Notes

Problem Solving
Exs. 15–17 These exercises require students to find the polygon matrix, translation matrix, and image matrix using a figure. You might suggest that students write a series of steps to be used to solve problems such as these. Students may find the solution to Example 1 helpful in this regard. After writing the matrix equation using the figure, students should perform the matrix addition to check their answers.

Integrating Algebra
Exs. 18, 19 When discussing these exercises, you may wish to show students how they are using algebra to solve a matrix equation and find an unknown matrix. For example, let P = the polygon matrix, T = the translation matrix, and M = the image matrix. Then, $T + P = M$. In these exercises, T and M are known, and P is unknown. The matrix equation is solved for P by subtracting T from both sides: $P = M - T$.

Interview Note
Exs. 20, 21 These exercises illustrate how matrices can be used to represent the location of a planet and other objects in space. Students learn that a translation matrix can be used to convert a planet's location from one coordinate system to another.

Spatial Reasoning
Exs. 20, 21 In previous work, students have seen that translations of objects in a fixed coordinate system move the objects to a new location (that is, their images are in a new location). In these exercises, the translation describes the differences in the object's location relative to two different coordinate systems.

13. a. $\begin{bmatrix} 1 & 3 & 5 & 3 & 1 \\ 5 & 5 & 3 & 1 & 1 \end{bmatrix}$

b. $\begin{bmatrix} -4 & -4 & -4 & -4 & -4 \\ -8 & -8 & -8 & -8 & -8 \end{bmatrix}$

c. $\begin{bmatrix} -3 & -1 & 1 & -1 & -3 \\ -3 & -3 & -5 & -7 & -7 \end{bmatrix}$

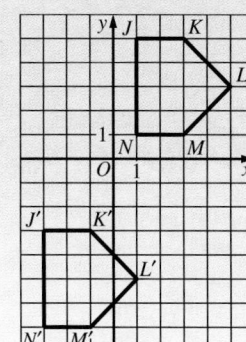

14. No, because you add the matrix elements, and addition is commutative.

15. $\begin{bmatrix} 4 & 4 & 4 \\ -3 & -3 & -3 \end{bmatrix} + \begin{bmatrix} -1 & 1 & 0 \\ 4 & 4 & 1 \end{bmatrix} = \begin{bmatrix} 3 & 5 & 4 \\ 1 & 1 & -2 \end{bmatrix}$

16. $\begin{bmatrix} -3 & -3 & -3 \\ 4 & 4 & 4 \end{bmatrix} + \begin{bmatrix} -2 & -2 & 2 \\ -2 & 1 & -2 \end{bmatrix} = \begin{bmatrix} -5 & -5 & -1 \\ 2 & 5 & 2 \end{bmatrix}$

17. $\begin{bmatrix} 0 & 0 & 0 & 0 \\ -4 & -4 & -4 & -4 \end{bmatrix} + \begin{bmatrix} -5 & -4 & -1 & -2 \\ -1 & 1 & 1 & -1 \end{bmatrix} =$ $\begin{bmatrix} -5 & -4 & -1 & -2 \\ -5 & -3 & -3 & -5 \end{bmatrix}$

18. $\begin{bmatrix} 1 & 4 \\ 0 & -1 \end{bmatrix}$

19. $\begin{bmatrix} -2 & 0 & 6 & 4 \\ 6 & 3 & 6 & 9 \end{bmatrix}$

20. See answers in back of book.

21. $\begin{bmatrix} 3.6 \times 10^5 & 0 & 11.6 \times 10^5 \\ 1.5 \times 10^5 & 0 & 0.4 \times 10^5 \\ 0.7 \times 10^5 & 0 & 3.5 \times 10^5 \end{bmatrix}$

Application
Exs. 22–27 In this series of exercises, students learn how a photocopier can use translations to center an image on the page. The information about the dimensions of the photograph and paper must be used to determine the coordinates of the vertices.

Communication: Drawing
Ex. 27 Encourage students to make some scale drawings to illustrate their response to this exercise. They can use the illustrations at the top of this page as examples.

Cooperative Learning
Ex. 28 This activity is appropriate for small group work. Each person in the group can use a different translation matrix. Then, the group can compare their results.

Integrating Functions
Ex. 28 A translation by a particular translation matrix, or multiplication by a particular scalar can be thought of as a function. The domain of these functions is all 2 × 4 matrices. The range is also the set of all 2 × 4 matrices. In this exercise, students compose these two functions and discover that the composition of the two functions is not a commutative operation. In general, the composition of functions is not a commutative operation.

Assessment Note
Ex. 29 This activity gives students an opportunity to practice matrix addition and writing translation matrices. At the same time, the group investigates a situation in which successive translations are made.

Connection COLOR PHOTOCOPIERS

Because a color photocopier scans a photograph and stores it in digital form, the copier can shift the image to a different location on the page that it prints out. Some photocopiers use this feature to center items on the output page.

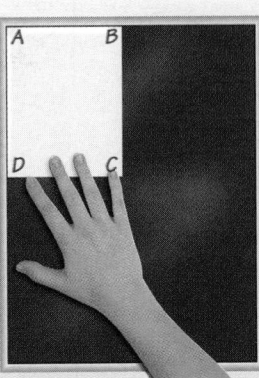
what you see

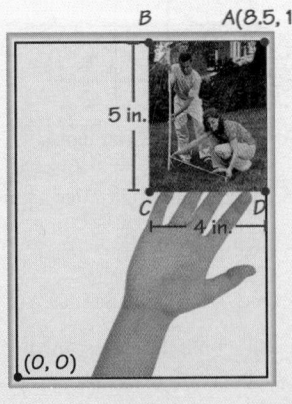
what the scanner sees

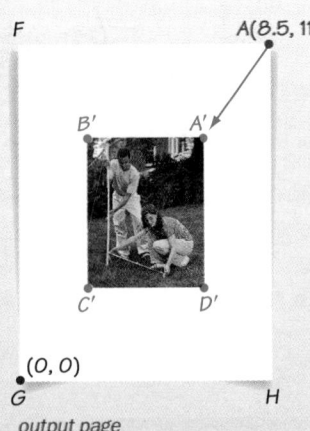
output page

22. Suppose you want to photocopy a 4 in. by 5 in. color photograph onto 8.5 in. by 11 in. paper. You place the photograph against the upper left corner of the glass plate, as shown above. Write a matrix that represents photograph *ABCD* as viewed by the scanner.

23. Explain why (4.25, 5.5) is the center of paper *AFGH*.

24. Explain why (6.5, 8.5) is the center of photograph *ABCD*.

25. To center the image, the photocopier must shift *ABCD* to *A'B'C'D'*. What translation matrix could the photocopier use?

26. Write a matrix that represents *A'B'C'D'*.

27. **Open-ended Problem** Choose dimensions different from 4 in. by 5 in. for a photograph. Describe the translation matrix the photocopier could use to center the photograph on a piece of 8.5 in. by 11 in. paper.

28. **Investigation** Use square *OPQR* shown at the right.
    a. Write a matrix that represents *OPQR*.
    b. Translate *OPQR* using a translation matrix of your choice. Then dilate the result using center (0, 0) and scale factor 2. Graph the image of *OPQR* after these two transformations.
    c. Dilate *OPQR* using center (0, 0) and scale factor 2. Then translate the result using the matrix you used in part (b). Graph the image of *OPQR* after these two transformations.
    d. Compare your answers to parts (b) and (c). Does the order in which the two transformations are applied affect the final outcome?

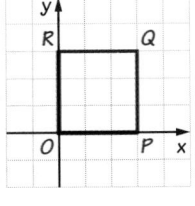

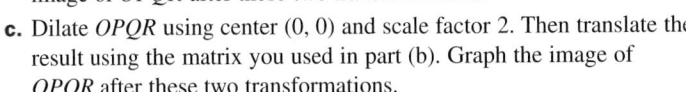
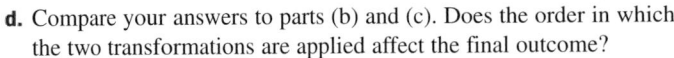

22. $\begin{bmatrix} 8.5 & 4.5 & 4.5 & 8.5 \\ 11 & 11 & 6 & 6 \end{bmatrix}$

23. 4.25 is midway between 0 and 8.5; 5.5 is midway between 0 and 11.

24. *x*-coordinate: edge (8.5) minus half photo width (2) is 6.5. *y*-coordinate: edge (11) minus half photo height (2.5) is 8.5.

25. $\begin{bmatrix} -2.25 & -2.25 & -2.25 & -2.25 \\ -3 & -3 & -3 & -3 \end{bmatrix}$

26. $\begin{bmatrix} 6.25 & 2.25 & 2.25 & 6.25 \\ 8 & 8 & 3 & 3 \end{bmatrix}$

27. a 7 in. by 7 in. photo translation matrix:
$\begin{bmatrix} -0.75 & -0.75 & -0.75 & -0.75 \\ -2 & -2 & -2 & -2 \end{bmatrix}$
In general, find the translation matrix that moves the center of the photo the scanner sees to the center of the output page.

28. a. $\begin{bmatrix} 0 & 3 & 3 & 0 \\ 0 & 0 & 3 & 3 \end{bmatrix}$

    b, c. Check students' work.
    d. Yes.

29. a, b. Check students' work.
    c. −1 times the sum of the translation matrices; This makes sense, since adding the 4 matrices is the same as adding the sum of the 4 matrices.

29. Cooperative Learning Work in a group of four people. Each person should use a different 2 × 4 translation matrix.

 a. Each person should draw a quadrilateral on a different piece of graph paper. Write a matrix for the quadrilateral. Apply your translation matrix to the quadrilateral matrix and graph the result. Label the image "1."

 b. Pass your papers in a circle. Each time you receive a paper, apply your translation matrix to the latest quadrilateral on it. Number the new quadrilateral in sequence. Continue until you get your original paper.

 c. Find a translation matrix that will take quadrilateral 4 to the original quadrilateral on your paper. Compare your answer with others in your group. What do you notice? Can you explain the results?

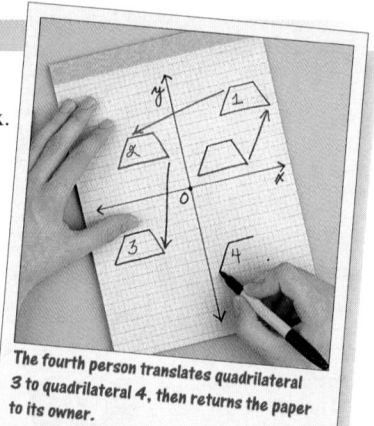

The fourth person translates quadrilateral 3 to quadrilateral 4, then returns the paper to its owner.

SPIRAL REVIEW

30. Find the area of a rectangle with a side of length 12 cm and a diagonal of length 20 cm. *(Section 7.4)*

31. Find the area of a sector of a circle with a central angle of 84° and radius 2. *(Section 11.5)*

ASSESS YOUR PROGRESS

VOCABULARY

matrix (p. 613)	**point matrix** (p. 613)
elements (p. 613)	**polygon matrix** (p. 613)
rows (p. 613)	**scalar multiplication** (p. 614)
columns (p. 613)	**matrix addition** (p. 619)
dimensions of a matrix (p. 613)	**matrix subtraction** (p. 621)

Find each product. *(Section 12.1)*

1. $5\begin{bmatrix} 0 & 4 \\ 5 & 3 \end{bmatrix}$ **2.** $1.5\begin{bmatrix} 1 & 3 & 4 \\ 0 & 2 & 3 \end{bmatrix}$

3. $0.5\begin{bmatrix} 2 & 4 & 3 & 1 \\ 0 & 4 & 6 & -2 \end{bmatrix}$ **4.** $10\begin{bmatrix} 0 & 5 & 1 \\ 0 & -2 & -3 \end{bmatrix}$

Write the matrix you would use to translate each figure. *(Section 12.2)*

5. a pentagon; to the left 1 unit and up 6 units

6. a parallelogram; to the right 5 units

7. Journal Suppose you dilate a quadrilateral using (0, 0) as the center. If the quadrilateral has one vertex in each quadrant, will the image also have one vertex in each quadrant? Explain using matrices.

12.2 Translations with Matrices **625**

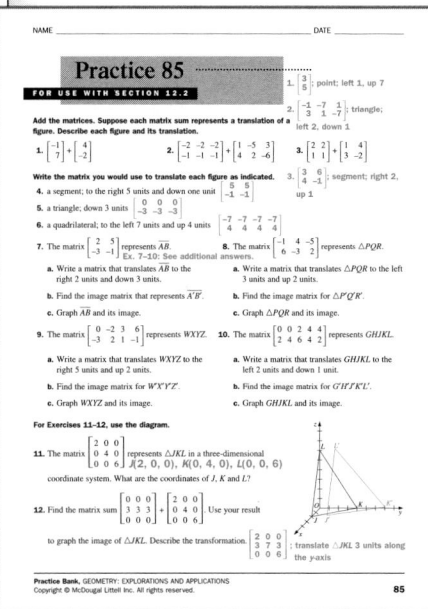
30. 192 cm²

31. about 2.93

Assess Your Progress

1. $\begin{bmatrix} 0 & 20 \\ 25 & 15 \end{bmatrix}$

2. $\begin{bmatrix} 1.5 & 4.5 & 6 \\ 0 & 3 & 4.5 \end{bmatrix}$

3. $\begin{bmatrix} 1 & 2 & 1.5 & 0.5 \\ 0 & 2 & 3 & -1 \end{bmatrix}$

4. $\begin{bmatrix} 0 & 50 & 10 \\ 0 & -20 & -30 \end{bmatrix}$

5. $\begin{bmatrix} -1 & -1 & -1 & -1 & -1 \\ 6 & 6 & 6 & 6 & 6 \end{bmatrix}$

6. $\begin{bmatrix} 5 & 5 & 5 & 5 \\ 0 & 0 & 0 & 0 \end{bmatrix}$

7. Yes. The vertices move out along the radii of a circle centered at (0, 0). With matrices: start with $\begin{bmatrix} a & -c & -e & g \\ b & d & -f & -h \end{bmatrix}$ and

multiply by scalar j, where a, $-h$, and j are positive real numbers.

$\begin{bmatrix} ja & -jc & -je & jg \\ jb & jd & -jf & -jh \end{bmatrix}$

Each image vertex is in the same quadrant as the original vertex.

Objectives

- Multiply matrices.
- Transform points and polygons using matrix multiplication.
- Use matrix multiplication to analyze special effects.

Recommended Pacing

❖ **Core and Extended Courses**
Section 12.3: 1 day

❖ **Block Schedule**
Section 12.3: $\frac{1}{2}$ block
(with Section 12.2)

Resource Materials

Lesson Support
Lesson Plan 12.3

Warm-Up Transparency 12.3

Practice Bank: Practice 86

Study Guide: Section 12.3

Explorations Lab Manual:
Diagram Master 2

Challenge Problems: Set 86

Technology
Technology Book:
TI-92 Activity 12

Graphing Calculator

McDougal Littell Mathpack
Matrix Analyzer Activity Book:
Activity 1

Internet:
http://www.hmco.com

Warm-Up Exercises

1. The following matrix sum translates △ ABC. Find the sum.

$$\begin{bmatrix} 3 & 3 & 3 \\ -2 & -2 & -2 \end{bmatrix} + \begin{bmatrix} 2 & -1 & 5 \\ -1 & 2 & 0 \end{bmatrix}$$

$$\begin{bmatrix} 5 & 2 & 8 \\ -3 & 0 & -2 \end{bmatrix}$$

2. Give the vertices of △ ABC that are represented in the polygon matrix above.
$A(2, -1)$, $B(-1, 2)$, $C(5, 0)$

3. Describe the transformation represented in the transformation matrix above.   a translation right 3 units and down 2 units

4. Multiply the polygon matrix for △ ABC by 3.
$\begin{bmatrix} 6 & -3 & 15 \\ -3 & 6 & 0 \end{bmatrix}$

12.3 Multiplying Matrices

Learn how to...
- **multiply matrices**

So you can...
- **transform points and polygons using matrix multiplication**
- **analyze special effects such as morphing**

Many movies and music videos use a special effect called *morphing*, in which one picture is gradually transformed into another. Some morphing techniques use *matrix multiplication* to transform images.

You can use **matrix multiplication** to multiply a point matrix or a polygon matrix by a transformation matrix. To transform the point $P(3, 5)$ using the transformation matrix $\begin{bmatrix} 2 & 0 \\ -7 & 4 \end{bmatrix}$, multiply the transformation matrix and the point matrix. The product is the image matrix.

Multiply elements of row 1 and **row 2** of the transformation matrix by elements of the **point matrix**.

Add the products in each row.

$$\begin{bmatrix} 2 & 0 \\ -7 & 4 \end{bmatrix} \cdot \begin{matrix} P \\ \begin{bmatrix} 3 \\ 5 \end{bmatrix} \end{matrix} = \begin{bmatrix} 2 \cdot 3 + 0 \cdot 5 \\ -7 \cdot 3 + 4 \cdot 5 \end{bmatrix}$$

trans-   point   image matrix
formation matrix
matrix

$$= \begin{matrix} P' \\ \begin{bmatrix} 6 \\ -1 \end{bmatrix} \end{matrix}$$

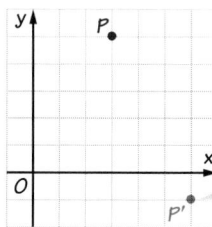

$P'(6, -1)$ is the image of the point $P(3, 5)$.

626   Chapter 12 *Coordinates for Transformations*

EXAMPLE 1

Use the transformation matrix
$\begin{bmatrix} -3 & 4 \\ -1 & 2 \end{bmatrix}$ to transform $\triangle ABC$.
Graph $\triangle ABC$ and its image.

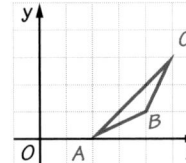

SOLUTION

Set up a matrix product with the transformation matrix on the left and the polygon matrix on the right.

Use a matrix to represent $\triangle ABC$.

To find the value in **row 1, column 3**, use **row 1** of the left matrix and **column 3** of the right matrix.

$$\begin{matrix} & A & B & C \\ \begin{bmatrix} -3 & 4 \\ -1 & 2 \end{bmatrix} & \cdot & \begin{bmatrix} 2 & 4 & 5 \\ 0 & 1 & 3 \end{bmatrix} \end{matrix} = \begin{bmatrix} -3(2) + 4(0) & -3(4) + 4(1) & -3(5) + 4(3) \\ -1(2) + 2(0) & -1(4) + 2(1) & -1(5) + 2(3) \end{bmatrix}$$

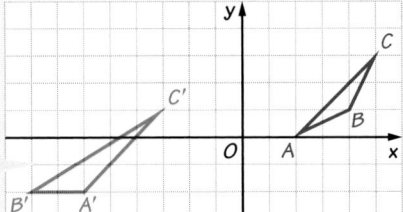

trans-
formation
matrix
**polygon
matrix**
image
matrix

$$= \begin{matrix} A' & B' & C' \\ \begin{bmatrix} -6 & -8 & -3 \\ -2 & -2 & 1 \end{bmatrix} \end{matrix}$$

$C'(-3, 1)$ is the image of $C(5, 3)$.

$\triangle A'B'C'$ is the image of $\triangle ABC$.

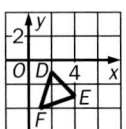

2 elements in each row

$$\begin{bmatrix} -3 & 4 \\ -1 & 2 \end{bmatrix} \begin{bmatrix} 2 & 4 & 5 \\ 0 & 1 & 3 \end{bmatrix} \leftarrow \text{2 elements in each column}$$

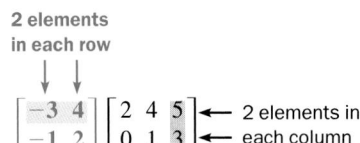

Notice that in the matrix product shown above, the number of elements in one **row** of the left matrix is equal to the number of elements in one **column** of the right matrix. This is true for all matrix multiplication.

12.3 Multiplying Matrices   **627**

Learning Styles: Kinesthetic

Students can use their fingers to help them work through a matrix multiplication problem. Use the index finger of the left hand to trace each row of the left-hand matrix. At the same time, use the right hand to trace the column that is being multiplied by the row. For example, to find the first element in the matrix product in Example 1, trace along the first row of the transformation matrix and the first column of the polygon matrix. Repeat using the first row of the transformation matrix and the second column of the polygon matrix to get the entry in the first row and second column of the product matrix. Continue until all the entries are found.

Additional Example 1

Use the transformation matrix
$\begin{bmatrix} 2 & 1 \\ 0 & -3 \end{bmatrix}$ to transform $\triangle DEF$. Graph
$\triangle DEF$ and its image.

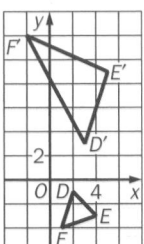

Set up a matrix product with the transformation matrix on the left and the polygon matrix on the right.

$$\begin{matrix} & & D & E & F \\ \begin{bmatrix} 2 & 1 \\ 0 & -3 \end{bmatrix} & \cdot & \begin{bmatrix} 2 & 4 & 1 \\ -1 & -3 & -4 \end{bmatrix} \end{matrix} =$$

translation
matrix
polygon
matrix

$$\begin{bmatrix} 2(2)+1(-1) & 2(4)+1(-3) & 2(1)+1(-4) \\ 0(2)+(-3)(-1) & 0(4)+(-3)(-3) & 0(1)+(-3)(-4) \end{bmatrix}$$

image matrix

$$= \begin{matrix} D' & E' & F' \\ \begin{bmatrix} 3 & 5 & -2 \\ 3 & 9 & 12 \end{bmatrix} \end{matrix}$$

$\triangle D'E'F'$ is the image of $\triangle DEF$.

627

Section Note

Teaching Tip

Unlike scalar multiplication and matrix addition, matrix multiplication may seem awkward and cumbersome at first to students, and they will need sufficient time to practice this skill. Also, you may wish to demonstrate that matrix multiplication, unlike matrix addition, is not commutative.

Think and Communicate

For question 3, students should verify that the number of columns in the left matrix is equal to the number of rows in the right matrix of the product. This is true for all matrix multiplication. If the dimensions of the left matrix are $a \times b$, and the dimensions of the right matrix are $c \times d$, then b must equal c for the matrices to be multiplied. Also, the product matrix has dimensions $a \times d$.

Additional Example 2

The matrix $\begin{bmatrix} 0 & 2 & 4 & 2 \\ 1 & 2 & 1 & 0 \end{bmatrix}$ represents rhombus *LMNP*. Use the transformation matrix $\begin{bmatrix} 1 & 0 \\ 4 & 3 \end{bmatrix}$ to transform *LMNP*. Graph *LMNP* and its image.

Method 1 Use paper and pencil. Put the transformation matrix on the left and the polygon matrix on the right.

$$\begin{bmatrix} 1 & 0 \\ 4 & 3 \end{bmatrix} \cdot \overset{L\ M\ N\ P}{\begin{bmatrix} 0 & 2 & 4 & 2 \\ 1 & 2 & 1 & 0 \end{bmatrix}} =$$

$$\begin{bmatrix} 1(0)+0(1) & 1(2)+0(2) & 1(4)+0(1) & 1(2)+0(0) \\ 4(0)+3(1) & 4(2)+3(2) & 4(4)+3(1) & 4(2)+3(0) \end{bmatrix}$$

$$= \overset{L'\ M'\ N'\ P'}{\begin{bmatrix} 0 & 2 & 4 & 2 \\ 3 & 14 & 19 & 8 \end{bmatrix}}$$

Choose an appropriate scale for the graph.

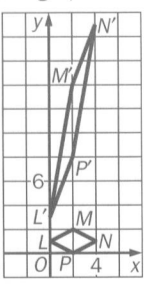

628

THINK AND COMMUNICATE

Use this matrix equation.

$$\underset{A}{\begin{bmatrix} 1 & 3 \\ 6 & -4 \end{bmatrix}} \cdot \underset{B}{\begin{bmatrix} -4 & 2 & 5 & 0 & -3 \\ 6 & 4 & -1 & -2 & 0 \end{bmatrix}} = \underset{C}{\begin{bmatrix} 14 & 14 & 2 & \underline{?} & -3 \\ -48 & -4 & 34 & 8 & -18 \end{bmatrix}}$$

1. Tell which matrix represents the:
   a. original polygon   **b.** image polygon   **c.** transformation

2. How many vertices does the original polygon have? Explain how you know.

3. Give the dimensions of matrices *A*, *B*, and *C*.

4. Explain which elements from matrix *A* and matrix *B* were used to find the value 34 in matrix *C*.

5. What value should go in the blank in row 1, column 4 of matrix *C*? Describe how to find this value.

EXAMPLE 2

The matrix $\begin{bmatrix} 0 & 4 & 4 & 0 \\ 0 & 0 & 6 & 6 \end{bmatrix}$ represents rectangle *ORST*. Use the transformation matrix $\begin{bmatrix} 2 & 3 \\ 0 & 1 \end{bmatrix}$ to transform *ORST*. Graph *ORST* and its image.

SOLUTION

Method 1

Use paper and pencil.

$$\begin{bmatrix} 2 & 3 \\ 0 & 1 \end{bmatrix} \overset{O\ R\ S\ T}{\begin{bmatrix} 0 & 4 & 4 & 0 \\ 0 & 0 & 6 & 6 \end{bmatrix}}$$

Put the **transformation matrix** on the left and the polygon matrix on the right.

$$= \begin{bmatrix} 2(0) + 3(0) & 2(4) + 3(0) & 2(4) + 3(6) & 2(0) + 3(6) \\ 0(0) + 1(0) & 0(4) + 1(0) & 0(4) + 1(6) & 0(0) + 1(6) \end{bmatrix}$$

$$= \overset{O\ R'\ S'\ T'}{\begin{bmatrix} 0 & 8 & 26 & 18 \\ 0 & 0 & 6 & 6 \end{bmatrix}}$$

Choose an appropriate scale for the graph.

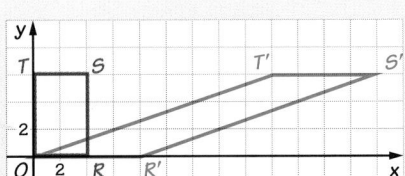

628   Chapter 12   *Coordinates for Transformations*

SOLUTION

Method 2

Use a graphing calculator or software with matrix capabilities.

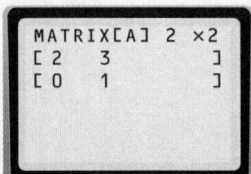

Step 1 Enter the transformation matrix as matrix *A*.

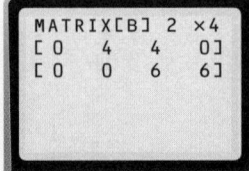

Step 2 Enter the polygon matrix as matrix *B*.

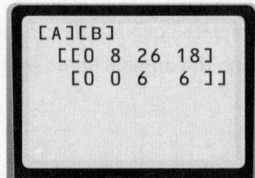

Step 3 Multiply *A* and *B* to find the image matrix.

The image matrix is $\begin{bmatrix} O & R' & S' & T' \\ 0 & 8 & 26 & 18 \\ 0 & 0 & 6 & 6 \end{bmatrix}$.

Graph *ORST* and *OR'S'T'* as shown in Method 1.

THINK AND COMMUNICATE

6. Can either the transformation in Example 1 or the transformation in Example 2 be described as a dilation, translation, reflection, or rotation? Explain why or why not.

7. In Example 2, which point in the polygon is its own image?

For more information about matrix calculations, see the *Technology Handbook*, pp. 723–725.

☑ CHECKING KEY CONCEPTS

1. Complete this matrix product.

$$\begin{bmatrix} 6 & -1 \\ 4 & 7 \end{bmatrix} \begin{bmatrix} 3 & -4 & 10 \\ 2 & 0 & -5 \end{bmatrix} = \begin{bmatrix} 16 & ? & 65 \\ ? & -16 & ? \end{bmatrix}$$

2. a. Write the polygon matrix for quadrilateral *ABCD*.

b. Use the transformation matrix $\begin{bmatrix} 3 & 1 \\ 0 & -1 \end{bmatrix}$ to transform *ABCD*.

c. Graph *ABCD* and *A'B'C'D'* on the same pair of axes.

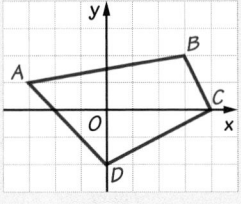

3. In order to solve Example 2, Cameron set up this matrix product. Explain what is wrong with his approach.

$$\begin{bmatrix} O & R & S & T \\ 0 & 4 & 4 & 0 \\ 0 & 0 & 6 & 6 \end{bmatrix} \begin{bmatrix} 2 & 3 \\ 0 & 1 \end{bmatrix}$$

Think and Communicate

6. No. Dilation, translation, reflection, and rotation all preserve the shape of the original figure; these two transformations did not.

7. *O*(0, 0)

Checking Key Concepts

1. $\begin{bmatrix} 16 & -24 & 65 \\ 26 & -16 & 5 \end{bmatrix}$

2. a. $\begin{bmatrix} -3 & 3 & 4 & 0 \\ 1 & 2 & 0 & -2 \end{bmatrix}$

b. $\begin{bmatrix} -8 & 11 & 12 & -2 \\ -1 & -2 & 0 & 2 \end{bmatrix}$

c.

3. The transformation matrix should be first, so the matrices cannot be multiplied in the order given.

Teach⟷Interact

Additional Example 2 (continued)

Method 2 Use a graphing calculator or software with matrix capabilities.
Step 1 Enter the transformation matrix as matrix *A*.

```
MATRIX[A] 2 X 2
[1      0        ]
[4      3        ]
```

Step 2 Enter the polygon matrix as matrix *B*.

```
MATRIX[B] 2 X 4
[0    2    4    2]
[1    2    1    0]
```

Step 3 Multiply *A* and *B* to find the image matrix.

```
[A][B]
  [[0    2    4    2]
   [3   14   19   8]]
```

The image matrix is
$\begin{bmatrix} L' & M' & N' & P' \\ 0 & 2 & 4 & 2 \\ 3 & 14 & 19 & 8 \end{bmatrix}$.
Graph *LMNP* and *L'M'N'P'* as shown in Method 1.

Think and Communicate

To answer question 6, students may need to review the concepts of reflection and rotation. In the next two sections, they will see that both reflections and rotations can be done by using matrices. After answering question 7, ask students what else remains constant under the translation. They should see that the *y*-coordinates of all the points remain the same. Point out that this is shown by the fact that the second row of the matrix was unchanged.

Closure Question

In order to multiply two matrices, what condition must exist for each matrix?   The number of elements in one row of the left matrix must equal the number of elements in one column of the right matrix.

629

Suggested Assignment

❖ **Core Course**
   Exs. 1–10, 12–14, 16–23

❖ **Extended Course**
   Exs. 1–23

❖ **Block Schedule**
   Day 76 Exs. 1–10, 12–14,
   16–23

Exercise Notes

Student Progress
Exs. 1–7 Students should know how to complete these exercises with and without the aid of a graphing calculator. You may want to suggest that they multiply the matrices using pencil and paper first, and then check their work using a calculator. Some students may find the pencil and paper method faster for Exs. 1 and 2.

Geometric Thinking
Exs. 8, 9 After identifying the identity matrix in Ex. 8, students investigate related matrices such as $\begin{bmatrix} n & 0 \\ 0 & n \end{bmatrix}$ in Ex. 9. Students compare the graphs of a triangle and its image after a transformation by one of these matrices to see that it produces a scale change with scale factor *n*. Point out that $\begin{bmatrix} n & 0 \\ 0 & n \end{bmatrix} = n \cdot \begin{bmatrix} 1 & 0 \\ 0 & 1 \end{bmatrix}$. Thus, multiplication by $\begin{bmatrix} n & 0 \\ 0 & n \end{bmatrix}$ can be thought of as multiplication by the identity matrix followed by scalar multiplication by *n*.

Integrating Algebra
Exs. 10, 11 To find the values of the variables in these two exercises, students need to use the procedure for matrix multiplication to write algebraic equations. The equations are then solved to find the unknown matrix. This method allows you to find the original point when you know the image point and the transformation matrix.

12.3 | Exercises and Applications

Extra Practice
exercises on
page 686

For Exercises 1–4:

a. Complete each matrix product.

b. Graph the original figure and its image.

$$
\text{1.} \quad \begin{bmatrix} 5 & 2 \\ 1 & 0 \end{bmatrix} \begin{bmatrix} Q \\ 1 \\ 3 \end{bmatrix}
\qquad\qquad
\text{2.} \quad \begin{bmatrix} 1 & -4 \\ 2 & -3 \end{bmatrix} \begin{bmatrix} A & B & C \\ 6 & 4 & 5 \\ 1 & 3 & 0 \end{bmatrix}
$$

$$
\text{3.} \quad \begin{bmatrix} -1 & 0 \\ 1 & -2 \end{bmatrix} \begin{bmatrix} D & E & F & G & H \\ -1 & -3 & 1 & 5 & 4 \\ -4 & -1 & 2 & 1 & -3 \end{bmatrix}
\qquad
\text{4.} \quad \begin{bmatrix} 3 & 0.5 \\ 2.4 & 1 \end{bmatrix} \begin{bmatrix} K & L \\ -1 & 3 \\ 2 & 8 \end{bmatrix}
$$

For each transformation matrix and polygon given, use the matrix to transform the polygon. Then graph the polygon and its image.

5. $\begin{bmatrix} 2 & -1 \\ 1 & -2 \end{bmatrix}$
6. $\begin{bmatrix} -3 & 1 \\ 0 & -4 \end{bmatrix}$
7. $\begin{bmatrix} 1 & 0 \\ 2 & -3.5 \end{bmatrix}$

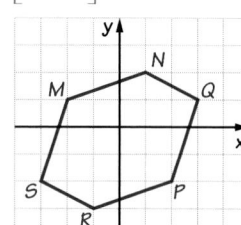
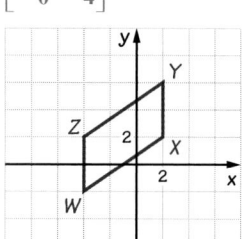
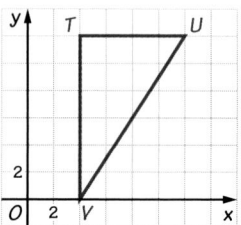

8. a. **Open-ended Problem** Write a matrix for any polygon, and use the matrix $\begin{bmatrix} 1 & 0 \\ 0 & 1 \end{bmatrix}$ to transform your polygon.

   b. Why do you think $\begin{bmatrix} 1 & 0 \\ 0 & 1 \end{bmatrix}$ is called the *identity matrix*?

9. **Cooperative Learning** Work in a group of three people. Each group member should choose a different one of these transformation matrices:

   $$\begin{bmatrix} 2 & 0 \\ 0 & 2 \end{bmatrix} \qquad \begin{bmatrix} 3 & 0 \\ 0 & 3 \end{bmatrix} \qquad \begin{bmatrix} 4 & 0 \\ 0 & 4 \end{bmatrix}$$

   a. The matrix $\begin{bmatrix} 2 & 4 & -4 \\ 3 & -3 & 0 \end{bmatrix}$ represents $\triangle ABC$. Use your transformation matrix to transform $\triangle ABC$. Graph $\triangle ABC$ and its image, $\triangle A'B'C'$.

   b. **Writing** As a group, compare your graphs from part (a). What is the geometric effect of the transformation $\begin{bmatrix} n & 0 \\ 0 & n \end{bmatrix}$? Explain.

10. **ALGEBRA** Find the values of *m* and *n*.

    $$\begin{bmatrix} 3 & 0 \\ 0 & 5 \end{bmatrix} \begin{bmatrix} m \\ n \end{bmatrix} = \begin{bmatrix} 12 \\ -35 \end{bmatrix}$$

11. **Challenge** Find the values of *p* and *q*.

    $$\begin{bmatrix} 1 & -4 \\ 2 & 7 \end{bmatrix} \begin{bmatrix} p \\ q \end{bmatrix} = \begin{bmatrix} 10 \\ 5 \end{bmatrix}$$

630 Chapter 12 *Coordinates for Transformations*

Exercises and Applications

1. a. $\begin{bmatrix} 11 \\ 1 \end{bmatrix}$

   b.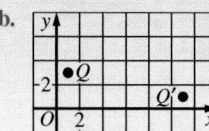

2. a. $\begin{bmatrix} 2 & -8 & 5 \\ 9 & -1 & 10 \end{bmatrix}$

   b.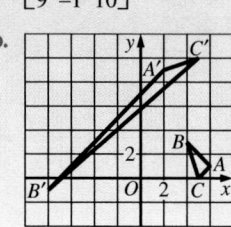

3. a. $\begin{bmatrix} 1 & 3 & -1 & -5 & -4 \\ 7 & -1 & -3 & 3 & 10 \end{bmatrix}$

   b.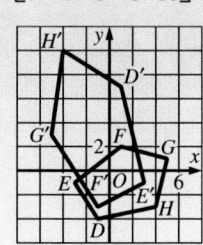

Connection ▷ SPECIAL EFFECTS

Morphing is the gradual transformation of one picture into another. In order to make a morph, a special effects artist selects pairs of *key points*, one in the starting picture and one in the final picture, which correspond to each other.

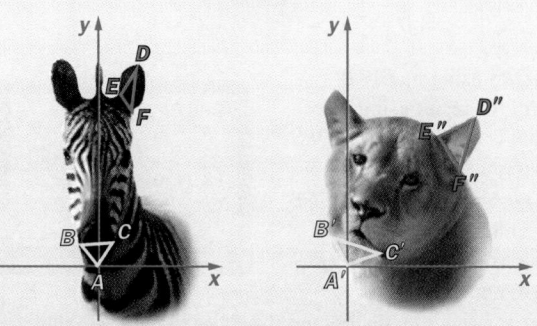

 Technology In the morph above, the transformation matrix

$$\begin{bmatrix} 1.435 & 0.473 \\ -0.647 & 0.888 \end{bmatrix}$$ is used to transform the key points $A(0, 0)$,

$B(-39, 38)$, and $C(41, 49)$ into their images A', B', and C'.

12. Use matrix multiplication to find the coordinates of A', B', and C'. Round coordinate values to the nearest integer.

The key points $D(94, 456)$, $E(47, 390)$, and $F(77, 345)$ are transformed into corresponding key points D'', E'', and F'' by a combination of matrix multiplication and matrix addition.

13. a. Multiply the polygon matrix for $\triangle DEF$ by the transformation

matrix $\begin{bmatrix} 1.55 & 0.078 \\ -0.844 & 1.237 \end{bmatrix}$ to find the matrix that represents $\triangle D'E'F'$.

b. Add the matrix $\begin{bmatrix} 109 & 109 & 109 \\ -157 & -157 & -157 \end{bmatrix}$ to the matrix that represents

$\triangle D'E'F'$ to find a matrix for the final image, $\triangle D''E''F''$.

c. Describe the transformation that is caused by the matrix addition you performed in part (b).

12.3 Multiplying Matrices **631**

---

4. a. $\begin{bmatrix} -2 & 13 \\ -0.4 & 15.2 \end{bmatrix}$

b.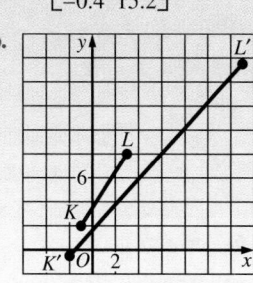

5.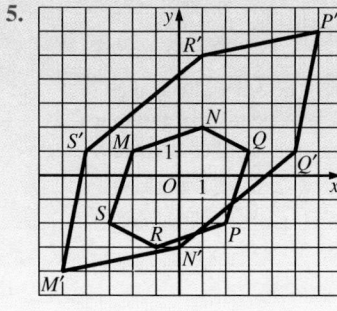

6–9. See answers in back of book.

10. $m = 4; n = -7$

11. $p = 6, q = -1$

12. $A'(0, 0)$, $B'(-38, 59)$, $C'(82, 17)$

13. a. $\begin{bmatrix} 181 & 103 & 146 \\ 485 & 443 & 362 \end{bmatrix}$

b. $\begin{bmatrix} 290 & 212 & 255 \\ 328 & 286 & 205 \end{bmatrix}$

c. a translation 109 units right and 157 units down

Exercise Notes

Application
Exs. 12, 13 Students should find this application of transformations to a technique called *morphing* interesting. Most students probably have seen examples of morphing in motion pictures or videos.

Visual Thinking
Exs. 12, 13 Ask students to find two photographs that represent images they would like to morph from one into the other. Encourage them to explain how they could use matrix multiplication to make this happen. This activity involves the visual skills of *interpretation* and *communication*.

Multicultural Note
Exs. 12, 13 Tigers, whose original habitat covered large parts of Asia, are endangered animals. Of the eight subspecies of tigers, three are already extinct (the Javan tiger became extinct as recently as the 1980s). Wildlife experts believe that there are only twenty to fifty South China tigers left in the wild. The other four subspecies, the Siberian tiger, the Indochinese tiger, the Sumatran tiger, and the Bengal tiger, are also threatened with extinction. Many nations in which these tigers are found have been working to protect them from extinction, but the task is difficult. Poaching and habitat destruction continue to be serious problems almost everywhere. Malaysia has made some impressive efforts to control poaching, and since 1976, its population of Indochinese tigers has risen from about 300 to somewhere between 520 and 650. In India, where there are thought to be about 3000 tigers, the national park system is working to keep tiger habitats intact and to educate park neighbors about the environment. Thailand has created one of the largest contiguous tiger habitats in Southeast Asia, made up of eight interconnected forest areas. Despite these efforts, there are still fewer than 6000 tigers left in the wild.

Communication: Discussion
Ex. 14 After completing this exercise, ask students to compare the result of the combined transformation with the results they got using the identity matrix in Ex. 8. They should see that the combination of transformation was equivalent to the transformation achieved by the identity matrix. In both cases, the figure and the image were identical. Have students multiply the two transformation matrices from Ex. 14. The result is the identity matrix. Have them verify that this is true regardless of the order in which the two matrices are multiplied. Two matrices whose product, in either order, is the identity matrix are called *inverse matrices*. Students will learn in later courses that inverse matrices are useful for solving matrix equations.

Integrating Functions
Ex. 15 Transformation using a particular transformation matrix can be thought of as a function. The domain of the functions in this exercise is the set of all 2×3 matrices. The range is also the set of all 2×3 matrices. Students compose two of these functions and discover that this composition is not a commutative operation.

Spatial Reasoning
Ex. 16 In this exercise, students extend the work of this section to three dimensions. They determine the image of a triangle in three-dimensional space under a particular transformation. Both the original figure and the image are then graphed.

14. Open-ended Problem Sketch any quadrilateral *KLMN* on a coordinate grid.

  a. Write a polygon matrix for *KLMN*.

  b. Using the transformation matrix $\begin{bmatrix} 3 & -2 \\ -14 & 10 \end{bmatrix}$, transform *KLMN* to find a matrix that represents *K′L′M′N′*.

  c. Multiply the matrix that represents *K′L′M′N′* by the transformation matrix $\begin{bmatrix} 5 & 1 \\ 7 & 1.5 \end{bmatrix}$ to find a matrix for the image, *K″L″M″N″*.

  d. Writing Compare the coordinates of *K″L″M″N″* to those of *KLMN*. What can you conclude about the effects of the two transformation matrices in parts (b) and (c)?

15. **Technology** Enter the polygon matrix *A* and the transformation matrices *B* and *C*, shown below, into your graphing calculator.

$$A = \begin{matrix} Q & R & S \\ \begin{bmatrix} 0 & -2 & 1 \\ 3 & 1 & -1 \end{bmatrix} \end{matrix} \qquad B = \begin{bmatrix} -1 & -3 \\ 2 & 4 \end{bmatrix} \qquad C = \begin{bmatrix} 0 & 2 \\ 1 & -4 \end{bmatrix}$$

  a. Transform △*QRS* using matrix *B*. Then transform the result using matrix *C*. What is the image of △*QRS* after these two transformations?

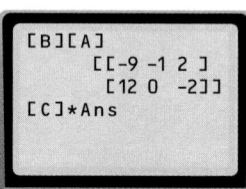

  b. Transform △*QRS* using matrix *C* first. Then transform the result using matrix *B*. What is the image of △*QRS* after these two transformations?

  c. Compare your answers to parts (a) and (b). Does the order in which the transformations are applied affect the final image? Explain your answer.

16. In the partially completed matrix product shown below, a triangle is being transformed in a three-dimensional coordinate system.

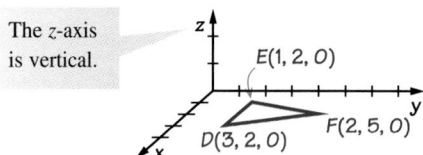

The z-axis is vertical.

E(1, 2, 0)   D(3, 2, 0)   F(2, 5, 0)

$$\begin{bmatrix} 0 & 0 & 1 \\ 1 & 0 & 0 \\ 0 & 1 & 0 \end{bmatrix} \begin{matrix} D & E & F \\ \begin{bmatrix} 3 & 1 & 2 \\ 2 & 2 & 5 \\ 0 & 0 & 0 \end{bmatrix} \end{matrix} = \begin{matrix} D' & E' & F' \\ \begin{bmatrix} 0 & ? & 0 \\ ? & 1 & 2 \\ 2 & ? & 5 \end{bmatrix} \end{matrix}$$

$1 \cdot 2 + 0 \cdot 5 + 0 \cdot 0 = 2$

  a. Write the three-dimensional coordinates of point *F* and its image, *F′*.

  b. Complete the matrix product.

  c. Graph △*DEF* and its image, △*D′E′F′*.

---

14. a, b. Answers may vary.

  c. *K″L″M″N″* will match *KLMN*.

  d. The transformation $\begin{bmatrix} 5 & 1 \\ 7 & 1.5 \end{bmatrix}$ undoes the transformation $\begin{bmatrix} 3 & -2 \\ -14 & 10 \end{bmatrix}$, and vice versa.

15. a. $\begin{bmatrix} 24 & 0 & -4 \\ -57 & -1 & 10 \end{bmatrix}$

  b. $\begin{bmatrix} 30 & 16 & -13 \\ -36 & -20 & 16 \end{bmatrix}$

  c. The order affects the image; matrix multiplication is not commutative.

16. a. *F*(2, 5, 0); *F′*(0, 2, 5)

  b. $\begin{bmatrix} 0 & 0 & 0 \\ 3 & 1 & 2 \\ 2 & 2 & 5 \end{bmatrix}$

c.

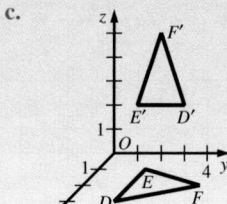

17. B

18. a, b. Answers may vary.

  c. All points are transformed onto the *y*-axis.

17. SAT/ACT Preview Use matrices A and B. Which one of these products and sums can be done?

$$A = \begin{bmatrix} 2 & -1 & 3 \\ 0 & 4 & 9 \end{bmatrix} \quad B = \begin{bmatrix} 5 & 2 \\ -1 & 4 \end{bmatrix}$$

A. AB **B.** BA **C.** $3 + A$ **D.** $A + B$

18. Open-ended Problem Complete this matrix with any two numbers to create a transformation matrix: $\begin{bmatrix} 0 & 0 \\ ? & ? \end{bmatrix}$

   a. Use matrix multiplication and your matrix to transform point $R(3, 2)$. Graph R and its image.

   b. Use your matrix to transform $\triangle JKL$, which is represented by the matrix $\begin{bmatrix} 4 & 2 & 2 \\ 0 & 3 & -3 \end{bmatrix}$. Graph $\triangle JKL$ and its image.

   c. Describe the geometric effect of your transformation matrix.

ONGOING ASSESSMENT

19. Cooperative Learning Work in a group of four people. Each group member should choose a different one of these transformation matrices:

$$\begin{bmatrix} 1 & 1 \\ 0 & 1 \end{bmatrix} \quad \begin{bmatrix} 1 & 2 \\ 0 & 1 \end{bmatrix} \quad \begin{bmatrix} 1 & 3 \\ 0 & 1 \end{bmatrix} \quad \begin{bmatrix} 1 & 4 \\ 0 & 1 \end{bmatrix}$$

   a. The matrix $\begin{bmatrix} 0 & 0 & 3 & 3 \\ 0 & 3 & 3 & 0 \end{bmatrix}$ represents square $ABCD$. Use your transformation matrix to find the image of $ABCD$.

   b. On one pair of axes, graph $ABCD$ and the image that you found, $A'B'C'D'$.

   c. Find the area of $ABCD$ and the area of $A'B'C'D'$.

   d. Compare your answers to parts (b) and (c) with others in your group. Describe how the four transformation matrices affect square $ABCD$ and its area.

SPIRAL REVIEW

Add the matrices. *(Section 12.2)*

20. $\begin{bmatrix} 1 & 1 & 1 \\ 3 & 3 & 3 \end{bmatrix} + \begin{bmatrix} 5 & 3 & -3 \\ 0 & -1 & 1 \end{bmatrix}$ **21.** $\begin{bmatrix} 1 & 3 & 4 \\ 0 & 2 & 3 \end{bmatrix} + \begin{bmatrix} 1 & 3 & 4 \\ 0 & 2 & 3 \end{bmatrix}$

22. List four different ways to prove that a quadrilateral is a parallelogram. *(Section 7.2)*

23. Rectangle $ABCD$ has vertices $A(-2, 1)$, $B(4, 1)$, $C(4, 3)$, and $D(-2, 3)$. Reflect the rectangle over each line and find the coordinates of the vertices of the image. *(Section 8.2)*

   a. the x-axis   **b.** the y-axis   **c.** the line $y = x$

12.3 Multiplying Matrices   **633**

Apply⟺Assess

Exercise Notes

Challenge
Ex. 18 After students complete this exercise, ask them to explain what is happening using the definition of matrix multiplication. Have them write their findings in their journals and ask several volunteers to read their entries to the class.

Assessment Note
Ex. 19 In this cooperative learning activity, each student is responsible for completing the work with a different transformation. Students then compare answers and write a description of their results. You may wish to extend the activity by having students make a conjecture about transformation matrices of the form $\begin{bmatrix} 1 & n \\ 0 & 1 \end{bmatrix}$.

Practice 86 for Section 12.3

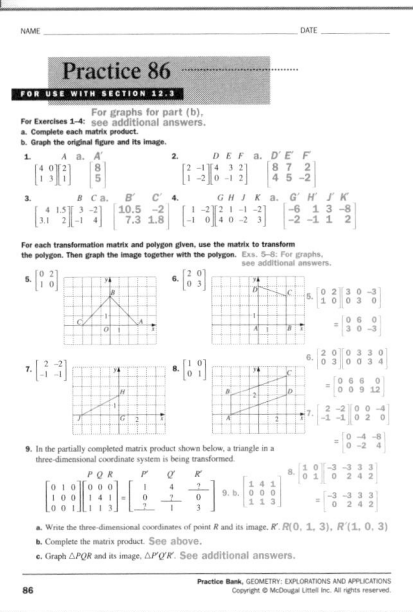

---

19. a. $\begin{bmatrix} 1 & 1 \\ 0 & 1 \end{bmatrix}\begin{bmatrix} 0 & 0 & 3 & 3 \\ 0 & 3 & 3 & 0 \end{bmatrix} = \begin{bmatrix} 0 & 3 & 6 & 3 \\ 0 & 3 & 3 & 0 \end{bmatrix}$;

$\begin{bmatrix} 1 & 2 \\ 0 & 1 \end{bmatrix}\begin{bmatrix} 0 & 0 & 3 & 3 \\ 0 & 3 & 3 & 0 \end{bmatrix} = \begin{bmatrix} 0 & 6 & 9 & 3 \\ 0 & 3 & 3 & 0 \end{bmatrix}$;

$\begin{bmatrix} 1 & 3 \\ 0 & 1 \end{bmatrix}\begin{bmatrix} 0 & 0 & 3 & 3 \\ 0 & 3 & 3 & 0 \end{bmatrix} = \begin{bmatrix} 0 & 9 & 12 & 3 \\ 0 & 3 & 3 & 0 \end{bmatrix}$;

$\begin{bmatrix} 1 & 4 \\ 0 & 1 \end{bmatrix}\begin{bmatrix} 0 & 0 & 3 & 3 \\ 0 & 3 & 3 & 0 \end{bmatrix} = \begin{bmatrix} 0 & 12 & 15 & 3 \\ 0 & 3 & 3 & 0 \end{bmatrix}$

b. Check students' work.

c. All areas are 9 square units.

d. They transform it into a parallelogram with base 3 and height 3. The area of the figure is not affected by any of the four transformations.

20. $\begin{bmatrix} 6 & 4 & -2 \\ 3 & 2 & 8 \end{bmatrix}$

21. $\begin{bmatrix} 2 & 6 & 8 \\ 0 & 4 & 6 \end{bmatrix}$

22. Both pairs of opposite sides are parallel. Both pairs of opposite sides are congruent. One pair of opposite sides are congruent and parallel. Diagonals bisect each other.

23. a. $\begin{bmatrix} -2 & 4 & 4 & -2 \\ -1 & -1 & -3 & -3 \end{bmatrix}$

b. $\begin{bmatrix} 2 & -4 & -4 & 2 \\ 1 & 1 & 3 & 3 \end{bmatrix}$

c. $\begin{bmatrix} 1 & 1 & 3 & 3 \\ -2 & 4 & 4 & -2 \end{bmatrix}$

Objectives

- Represent reflections using matrices.
- Reflect points and polygons using matrices.
- Analyze real-world images using reflections.

Recommended Pacing

❖ **Core and Extended Courses**
Section 12.4: 2 days

❖ **Block Schedule**
Section 12.4: 1 block

Resource Materials

Lesson Support
Lesson Plan 12.4

Warm-Up Transparency 12.4

Overhead Visuals:
  Folder 14: Matrices and
           Transformations

Practice Bank: Practice 87

Study Guide: Section 12.4

Explorations Lab Manual:
  Additional Exploration 18
  Diagram Masters 1, 2

Challenge Problems: Set 87

Technology
Technology Book:
  TI-92 Activity 12

Graphing Calculator

Internet:
  http://www.hmco.com

Warm-Up Exercises

Graph the point $P(3, 5)$. Reflect the point over each line below. Then write the coordinates of the image P'.

1. the x-axis $P'(3, -5)$

2. the y-axis $P'(-3, 5)$

3. the line $y = x$ $P'(5, 3)$

4. the line $y = -x$ $P'(-5, -3)$

SECTION

12.4 Reflections with Matrices

Learn how to...

- **represent reflections using matrices**

So you can...

- **reflect points and polygons using matrices**

- **analyze images such as those produced with computers**

Computer software programs for manipulating photographs allow you to reflect parts of a photograph across a line. In the images below, each half of the photograph on the left is flipped across a vertical line of reflection to create a new portrait. Because human faces are not perfectly symmetrical, the portraits look quite different.

Left   *Right*   *Left*   *Left*   *Right*   *Right*

EXPLORATION
COOPERATIVE LEARNING

Reflections Using Matrix Multiplication

Work in a group of four students.
You will need:
- graph paper

1 Each member of the group should draw a different △ABC and label the coordinates of its vertices. Each of you should write a polygon matrix that represents your triangle.

2 Multiply your triangle matrix by each transformation matrix below, putting the transformation matrix on the left. Graph the image of △ABC after each multiplication.

$$\begin{bmatrix} 1 & 0 \\ 0 & -1 \end{bmatrix} \quad \begin{bmatrix} -1 & 0 \\ 0 & 1 \end{bmatrix} \quad \begin{bmatrix} 0 & 1 \\ 1 & 0 \end{bmatrix} \quad \begin{bmatrix} 0 & -1 \\ -1 & 0 \end{bmatrix}$$

3 Compare your results. Describe the effect each transformation matrix has on a triangle.

 Exploration Note

Purpose
The purpose of this Exploration is to have students discover how reflections can be represented by using matrices.

Materials/Preparation
Each group needs graph paper.

Procedure
Each member of the group draws a different triangle and writes a matrix for it. They then multiply the triangle matrix by four different transformation matrices and graph each image.

Closure
Students should realize that each transformation matrix produces a specific reflection. These reflections are summarized on page 635.

Explorations Lab Manual
See the Manual for more commentary on this Exploration.
Diagram Master 1

The table below summarizes the reflection matrices and the effect each one has on a polygon or a point.

Summary of Matrices for Reflections

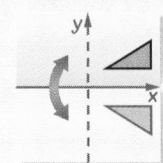

reflection over *x*-axis
$$\begin{bmatrix} 1 & 0 \\ 0 & -1 \end{bmatrix}$$

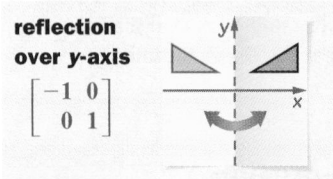

reflection over *y*-axis
$$\begin{bmatrix} -1 & 0 \\ 0 & 1 \end{bmatrix}$$

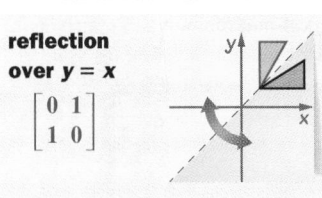

reflection over *y* = *x*
$$\begin{bmatrix} 0 & 1 \\ 1 & 0 \end{bmatrix}$$

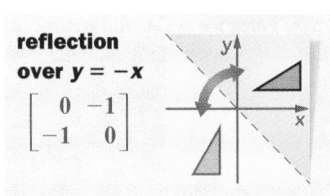

reflection over *y* = −*x*
$$\begin{bmatrix} 0 & -1 \\ -1 & 0 \end{bmatrix}$$

THINK AND COMMUNICATE

1. How is the matrix for reflection over the *x*-axis like the matrix for reflection over the *y*-axis? How is it different?

2. How is the matrix for reflection over the line *y* = *x* like the matrix for reflection over the line *y* = −*x*? How is it different?

EXAMPLE 1

Use matrix multiplication to find the image of △*ABC* after reflection over the *y*-axis. Graph the image.

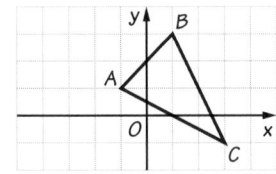

SOLUTION

Refer to the table of matrices above to find the matrix that represents reflection over the *y*-axis.

$$\begin{array}{cc} & A \quad B \quad C \\ \begin{bmatrix} -1 & 0 \\ 0 & 1 \end{bmatrix} & \begin{bmatrix} -1 & 1 & 3 \\ 1 & 3 & -1 \end{bmatrix} \end{array} = \begin{bmatrix} -1(-1)+0(1) & -1(1)+0(3) & -1(3)+0(-1) \\ 0(-1)+1(1) & 0(1)+1(3) & 0(3)+1(-1) \end{bmatrix}$$

reflection over *y*-axis   polygon matrix

$$= \begin{array}{c} A' \quad B' \quad C' \\ \begin{bmatrix} 1 & -1 & -3 \\ 1 & 3 & -1 \end{bmatrix} \\ \text{image matrix} \end{array}$$

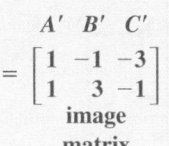
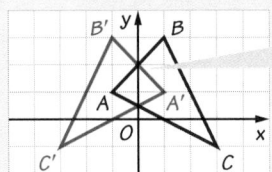

Graph the image, △*A'B'C'*

12.4   Reflections with Matrices   **635**

. . . .

Teach⇔Interact

Think and Communicate

Questions 1 and 2 help students identify the distinguishing characteristics of each matrix. Understanding these characteristics can help them to remember the matrices.

About Example 1

Geometric Thinking
After completing this Example, review the fact that when a point is reflected over the *y*-axis, its *x*-coordinate is changed to the opposite value while the *y*-coordinate remains the same. Ask students how this can be seen in the relationship between the polygon matrix and the image matrix. They should see that the numbers in the first row of the image matrix are the opposites of the numbers in the first row of the polygon matrix. The second rows of each matrix are the same.

Additional Example 1

Use matrix multiplication to find the image of △ *KLM* after reflection over the line *y* = *x*. Graph the image.

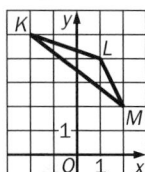

Refer to the table of matrices on this page to find the matrix that represents reflection over the line *y* = *x*.

$$\begin{array}{cc} & K \quad L \quad M \\ \begin{bmatrix} 0 & 1 \\ 1 & 0 \end{bmatrix} & \begin{bmatrix} -2 & 1 & 2 \\ 5 & 4 & 2 \end{bmatrix} \end{array} =$$

reflection over *y* = *x*   polygon matrix

$$\begin{array}{ccc} K' & L' & M' \\ \begin{bmatrix} 0(-2)+1(5) & 0(1)+1(4) & 0(2)+1(2) \\ 1(-2)+0(5) & 1(1)+0(4) & 1(2)+0(2) \end{bmatrix} \end{array}$$

$$= \begin{array}{c} K' \quad L' \quad M' \\ \begin{bmatrix} 5 & 4 & 2 \\ -2 & 1 & 2 \end{bmatrix} \\ \text{image matrix} \end{array}$$

Graph the image, △ *K'L'M'*.

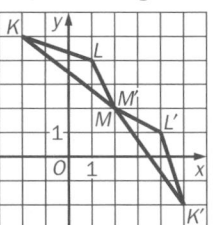

Additional Example 2

Use matrix multiplication to find the image of quadrilateral *RSTU* after reflection over the *x*-axis. Graph the image.

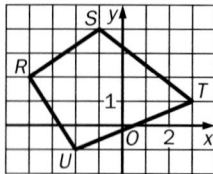

If you enter the reflection matrix as Matrix *A* and the polygon matrix as Matrix *B* in a graphing calculator, it can do the multiplication for you.

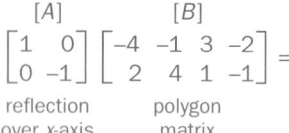

reflection   polygon
over *x*-axis   matrix

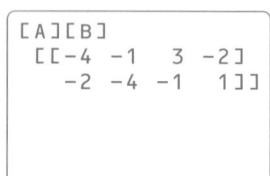

Graph the image, *R'S'T'U'*.

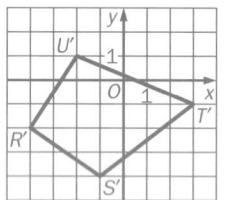

Checking Key Concepts

Teaching Tip
For questions 1 and 2, students can use a graph to determine the type of reflection matrix used.

Closure Question

What are some of the distinguishing characteristics of a reflection matrix?   A reflection matrix consists of two zeros on one of the diagonals, and either 1's or –1's or a combination of 1's and –1's on the other diagonal.

EXAMPLE 2

Use matrix multiplication to find the image of quadrilateral **DEFG** after reflection over the line $y = -x$. Graph the image.

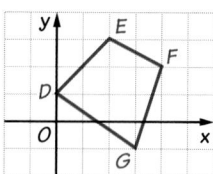

SOLUTION

If you enter the reflection matrix as Matrix *A* and the polygon matrix as Matrix *B* in a graphing calculator, it can do the multiplication for you.

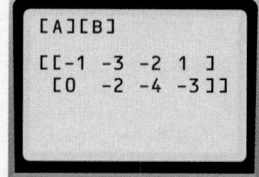

reflection   polygon
over $y = -x$ matrix

Graph the image, *D'E'F'G'*.

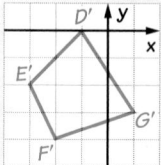

☑ CHECKING KEY CONCEPTS

1. $\begin{bmatrix} 3 \\ 5 \end{bmatrix}$ is the result after the point matrix $\begin{bmatrix} 3 \\ -5 \end{bmatrix}$ is multiplied by a reflection matrix. What reflection matrix was used?

2. $\begin{bmatrix} 1 \\ 2 \end{bmatrix}$ is the result after the point matrix $\begin{bmatrix} 2 \\ 1 \end{bmatrix}$ is multiplied by a reflection matrix. What reflection matrix was used?

Use matrix multiplication to find the image of trapezoid *JKLM* after reflection over each line. Then graph the image. Use the table on page 635 for the transformation matrices.

3. the *x*-axis

4. the *y*-axis

5. the line $y = x$

6. the line $y = -x$

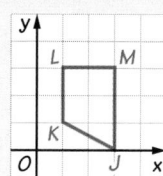

Checking Key Concepts

1. $\begin{bmatrix} 1 & 0 \\ 0 & -1 \end{bmatrix}$

2. $\begin{bmatrix} 0 & 1 \\ 1 & 0 \end{bmatrix}$

3. $\begin{bmatrix} 3 & 1 & 1 & 3 \\ 0 & -1 & -3 & -3 \end{bmatrix}$

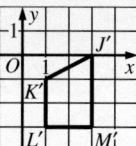

4. $\begin{bmatrix} -3 & -1 & -1 & -3 \\ 0 & 1 & 3 & 3 \end{bmatrix}$

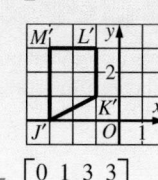

5. $\begin{bmatrix} 0 & 1 & 3 & 3 \\ 3 & 1 & 1 & 3 \end{bmatrix}$

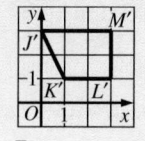

6. $\begin{bmatrix} 0 & -1 & -3 & -3 \\ -3 & -1 & -1 & -3 \end{bmatrix}$

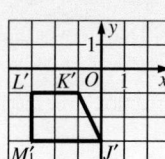

12.4 | Exercises and Applications

Extra Practice exercises on page 686

Complete each matrix product to find the image of (7, −2) after a reflection, and tell which line is the line of reflection.

1. $\begin{bmatrix} 1 & 0 \\ 0 & -1 \end{bmatrix} \begin{bmatrix} 7 \\ -2 \end{bmatrix}$

2. $\begin{bmatrix} -1 & 0 \\ 0 & 1 \end{bmatrix} \begin{bmatrix} 7 \\ -2 \end{bmatrix}$

3. $\begin{bmatrix} 0 & 1 \\ 1 & 0 \end{bmatrix} \begin{bmatrix} 7 \\ -2 \end{bmatrix}$

4. $\begin{bmatrix} 0 & -1 \\ -1 & 0 \end{bmatrix} \begin{bmatrix} 7 \\ -2 \end{bmatrix}$

5. $\begin{bmatrix} -4 \\ 2 \end{bmatrix}$ is the result after $\begin{bmatrix} 4 \\ 2 \end{bmatrix}$ is multiplied by a reflection matrix.
   What reflection matrix was used?

6. $\begin{bmatrix} -7 \\ 3 \end{bmatrix}$ is the result after $\begin{bmatrix} -3 \\ 7 \end{bmatrix}$ is multiplied by a reflection matrix.
   What reflection matrix was used?

Use matrix multiplication to find the image of trapezoid *AEUO* after reflection over each line. Then graph the image. Use the table on page 635 for the transformation matrices.

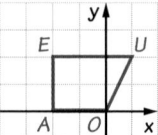

7. the *x*-axis
8. the *y*-axis
9. the line *y* = *x*
10. the line *y* = −*x*

Connection ▶ ART PHOTOGRAPHY

Marika Barnett takes photographs of buildings and then reflects them over horizontal and vertical axes using a computer. In her "Done with Mirrors" series, each composition presents the same image four ways, as if reflected in two perpendicular mirrors.

Boston Architectural Center

BY THE WAY

Hungarian-born Marika Barnett is a software engineer who enjoys creating kaleidoscopic collages using her own photographs.

Use matrix multiplication to find the image of each point after reflection over the *x*-axis and over the *y*-axis.

11. *A*(24, 10)
12. *B*(15, −5)
13. *C*(−15, −27)
14. *D*(−23, 0)

15. **Open-ended Problem** Choose one of the points in Exercises 11–14. Reflect this point over the *x*-axis and then reflect the result over the *y*-axis. Is the image the same as if you reflect the point over the *y*-axis first and then reflect the result over the *x*-axis? Explain.

Exercises and Applications

1. $\begin{bmatrix} 7 \\ 2 \end{bmatrix}$; the *x*-axis;

2. $\begin{bmatrix} -7 \\ -2 \end{bmatrix}$; the *y*-axis

3. $\begin{bmatrix} -2 \\ 7 \end{bmatrix}$; line *y* = *x*

4. $\begin{bmatrix} 2 \\ -7 \end{bmatrix}$; line *y* = −*x*

5. $\begin{bmatrix} -1 & 0 \\ 0 & 1 \end{bmatrix}$

6. $\begin{bmatrix} 0 & -1 \\ -1 & 0 \end{bmatrix}$

7. $\begin{bmatrix} -2 & -2 & 1 & 0 \\ 0 & -2 & -2 & 0 \end{bmatrix}$

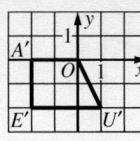

8, 9. See answers in back of book.

10. $\begin{bmatrix} 0 & -2 & -2 & 0 \\ 0 & 2 & -1 & 0 \end{bmatrix}$

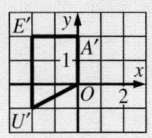

11. *A*′(24, −10); *A*′(−24, 10)
12. *B*′(15, 5); *B*′(−15, −5)
13. *C*′(−15, 27); *C*′(15, −27)
14. *D*′(−23, 0); *D*′(23, 0)
15. Yes, it is the same as rotating by 180°.

Apply⇔Assess

Exercise Notes

Student Progress

Exs. 16–19 Students should be able to complete these exercises on their own. If some students cannot arrive at the correct answers, try grouping them with another student who can serve as a peer tutor.

Topic Spiraling: Preview

Exs. 20, 21 In these exercises, students discover that a combination of reflections can result in a rotation. Rotations are the topic of the next section. Students should see that if they multiply the two reflection matrices, the result is the matrix for the rotation. Rotation matrices are discussed on page 642.

Communication: Discussion

Exs. 22–24 After completing this set of exercises, ask students to look for the pattern between the reflection matrix and the type of reflection it produces. They should see that all three of the reflection matrices have zero in all places but the diagonal. Two of the diagonal numbers are 1, while the third is –1. The –1 appears in the row whose number gives the coordinate missing from the plane of reflection. For example, in Ex. 22, the –1 is in the third row, thus, the third coordinate, *z*, is not in the name of the plane of reflection. The reflection is over the *xy*-plane. Ask students to see how this is similar to the reflection matrices for reflection over the *x*- and *y*-axes in a two-dimensional coordinate system.

Write a polygon matrix for each polygon. Then use matrix mutiplication to find the image after reflection over the given line. Use the table on page 635 for the transformation matrices.

16. the *x*-axis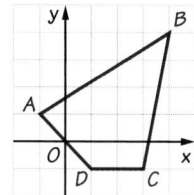

17. the *y*-axis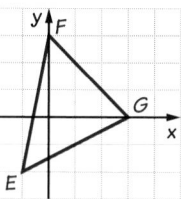

18. the line *y* = *x*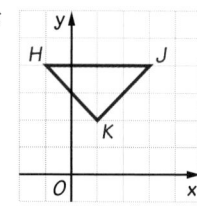

19. the line *y* = −*x*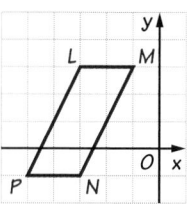

For Exercises 20 and 21, use trapezoid *PQRS*, at the right.

20. a. Write a polygon matrix for *PQRS*.

 b. Use matrix multiplication to reflect *PQRS* over the line *y* = *x*. Label the image *P'Q'R'S'*.

 c. Use matrix multiplication to reflect *P'Q'R'S'* over the *y*-axis. Label the image *P"Q"R"S"*.

 d. Describe how *P"Q"R"S"* is related to *PQRS*.

21. a. Repeat Exercise 20 but reflect *PQRS* over the *y*-axis first, then reflect the result over the line *y* = *x*. How is the image related to *PQRS*?

 b. Writing Compare your results in part (a) with those in Exercise 20. Does the order in which the reflections are applied affect the image?

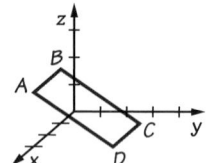

Challenge Each matrix product reflects rectangle *ABCD* in a three-dimensional coordinate system. Complete each product, graph the image, and describe the reflection.

22.
$$\begin{bmatrix} 1 & 0 & 0 \\ 0 & 1 & 0 \\ 0 & 0 & -1 \end{bmatrix} \begin{bmatrix} A & B & C & D \\ 3 & 1 & 1 & 3 \\ 0 & 0 & 3 & 3 \\ 2 & 2 & 0 & 0 \end{bmatrix} = \begin{bmatrix} A' & B' & C' & D' \\ 3 & 1 & 1 & ? \\ 0 & 0 & ? & ? \\ -2 & ? & ? & ? \end{bmatrix}$$

23.
$$\begin{bmatrix} 1 & 0 & 0 \\ 0 & -1 & 0 \\ 0 & 0 & 1 \end{bmatrix} \begin{bmatrix} A & B & C & D \\ 3 & 1 & 1 & 3 \\ 0 & 0 & 3 & 3 \\ 2 & 2 & 0 & 0 \end{bmatrix} = \begin{bmatrix} A' & B' & C' & D' \\ 3 & ? & ? & 3 \\ 0 & ? & ? & -3 \\ 2 & ? & ? & 0 \end{bmatrix}$$

24.
$$\begin{bmatrix} -1 & 0 & 0 \\ 0 & 1 & 0 \\ 0 & 0 & 1 \end{bmatrix} \begin{bmatrix} A & B & C & D \\ 3 & 1 & 1 & 3 \\ 0 & 0 & 3 & 3 \\ 2 & 2 & 0 & 0 \end{bmatrix} = \begin{bmatrix} A' & B' & C' & D' \\ ? & ? & ? & -3 \\ ? & ? & 3 & 3 \\ ? & 2 & 0 & 0 \end{bmatrix}$$

638 Chapter 12 *Coordinates for Transformations*

16. $\begin{bmatrix} -1 & 4 & 3 & 1 \\ 1 & 4 & -1 & -1 \end{bmatrix}$; $\begin{bmatrix} -1 & 4 & 3 & 1 \\ -1 & -4 & 1 & 1 \end{bmatrix}$

17. $\begin{bmatrix} -1 & 0 & 3 \\ -2 & 3 & 0 \end{bmatrix}$; $\begin{bmatrix} 1 & 0 & -3 \\ -2 & 3 & 0 \end{bmatrix}$

18. $\begin{bmatrix} -1 & 3 & 1 \\ 4 & 4 & 2 \end{bmatrix}$; $\begin{bmatrix} 4 & 4 & 2 \\ -1 & 3 & 1 \end{bmatrix}$

19. $\begin{bmatrix} -3 & -1 & -3 & -5 \\ 3 & 3 & -1 & -1 \end{bmatrix}$; $\begin{bmatrix} -3 & -3 & 1 & 1 \\ 3 & 1 & 3 & 5 \end{bmatrix}$

20. a. $\begin{bmatrix} 2 & 2 & 5 & 5 \\ 0 & 2 & 3 & 0 \end{bmatrix}$ **b.** $\begin{bmatrix} 0 & 2 & 3 & 0 \\ 2 & 2 & 5 & 5 \end{bmatrix}$

 c. $\begin{bmatrix} 0 & -2 & -3 & 0 \\ 2 & 2 & 5 & 5 \end{bmatrix}$

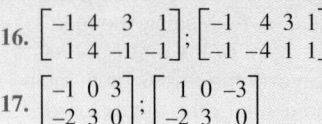

 d. *P"Q"R"S"* is the rotation of *PQRS* around (0, 0) by 90°.

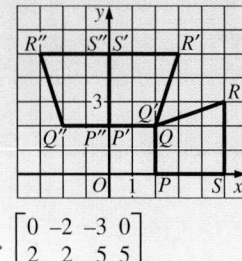

21. a. (a) $\begin{bmatrix} 2 & 2 & 5 & 5 \\ 0 & 2 & 3 & 0 \end{bmatrix}$

 (b) $\begin{bmatrix} -2 & -2 & -5 & -5 \\ 0 & 2 & 3 & 0 \end{bmatrix}$

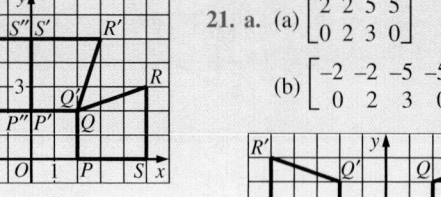

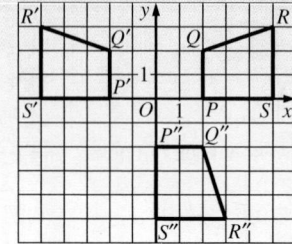

 (c) $\begin{bmatrix} 0 & 2 & 3 & 0 \\ -2 & -2 & -5 & -5 \end{bmatrix}$

 (d) *P"Q"R"S"* is the rotation of *PQRS* around (0, 0) by 270°.

 d. The order of reflection matters.

Jack is a computer-generated three-dimensional human figure based on measurements of actual people. Jack is used as a model to predict how humans might behave in a given environment, such as in a space shuttle, a submarine, or a tractor.

Unlike real people, Jack has perfect symmetry. One half of his body is reflected to create the other half. For each point on Jack's face, use matrix multiplication to find the corresponding point on the other side.

25. $A(237, 572)$ **26.** $B(167, 224)$

27. $C(-365, 374)$ **28.** $D(-433, 1004)$

The top half of Jack's eyeball can be reflected to create the lower half.

29. Write a matrix for hexagon $TUVWXY$.

30. Use matrix multiplication to find a hexagon for the eyeball's lower half.

31. Writing Do you think Jack's eyeball has any other symmetry? Explain.

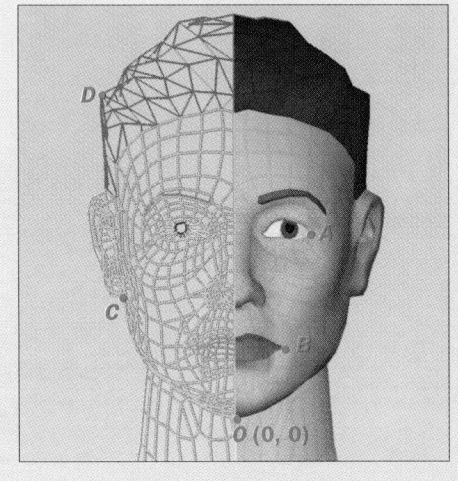

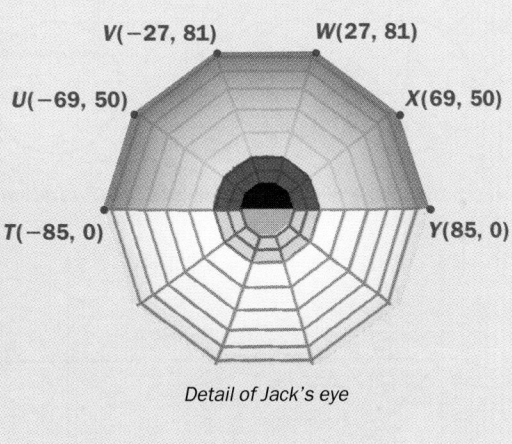

Detail of Jack's eye

Apply⇔Assess

Exercise Notes

Historical Connection
Exs. 25–31 Computer scientists have been developing virtual reality since the early 1970s. Early systems simulated three-dimensional reality using features such as colors, textures, spatial orientation, and light. In newer virtual reality systems, the user wears headgear and gloves. Sensors in this apparatus are used to incorporate movements of the user into the virtual reality system.

Communication: Writing
Ex. 31 This writing exercise gives students an opportunity to think more deeply about reflections. They should notice the rotational symmetry in the picture. Ask students which symmetry would be the most useful to the computer programmer. Point out that if one of the triangles was drawn, the other 9 could be found using rotations of 36°, 72°, 108°, 144°, 180°, 216°, 252°, 288°, 324°, and 360°.

12.4 Reflections with Matrices **639**

---

22. $\begin{bmatrix} 3 & 1 & 1 & 3 \\ 0 & 0 & 3 & 3 \\ -2 & -2 & 0 & 0 \end{bmatrix}$;

reflection over the *xy*-plane;

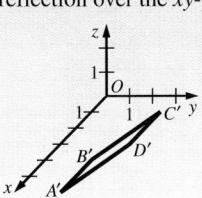

23. $\begin{bmatrix} 3 & 1 & 1 & 3 \\ 0 & 0 & -3 & -3 \\ 2 & 2 & 0 & 0 \end{bmatrix}$;

reflection over the *xz*-plane;

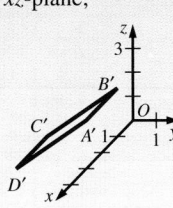

24. $\begin{bmatrix} -3 & -1 & -1 & -3 \\ 0 & 0 & 3 & 3 \\ 2 & 2 & 0 & 0 \end{bmatrix}$;

reflection over the *yz*-plane;

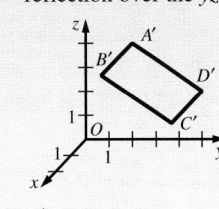

25. $A'(-237, 572)$

26. $B'(-167, 224)$

27. $C'(365, 374)$ **28.** $D'(443, 1004)$

29. $\begin{bmatrix} -85 & -69 & -27 & 27 & 69 & 85 \\ 0 & 50 & 81 & 81 & 50 & 0 \end{bmatrix}$

30. hexagon $T'U'V'W'X'Y'$ with coordinates given in this matrix:

$\begin{bmatrix} -85 & -69 & -27 & 27 & 69 & 85 \\ 0 & -50 & -81 & -81 & -50 & 0 \end{bmatrix}$

31. Yes, it has reflection symmetry using any of the diagonals as the line of reflection. It also has rotational symmetry with the center at the center of the eye and rotation by any multiple of 36°.

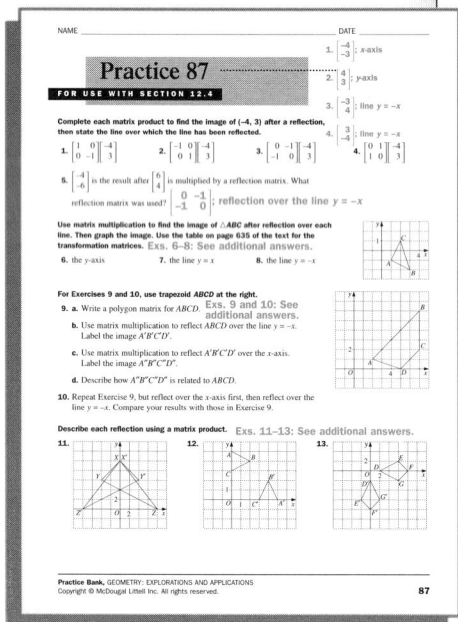
Describe each reflection using a matrix product. Include the transformation matrix, polygon matrix, and image matrix. Use the table on page 635 for the transformation matrices.

32.
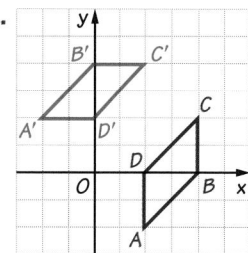

33.
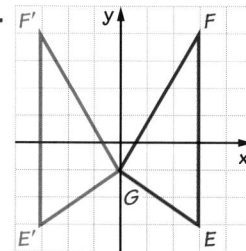

34.
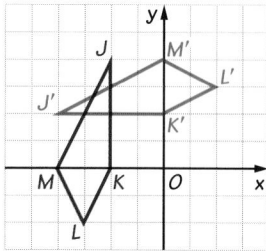

ONGOING ASSESSMENT

35. Writing Graph any segment. Use each of the matrices in the table on page 635 to reflect your segment. Compare the slope of each image segment with the slope of the original segment. Summarize your results. Write a conjecture about how different reflections affect slope.

SPIRAL REVIEW

For each transformation matrix and polygon given, use the matrix to transform the polygon. Then graph the polygon and its image. *(Section 12.3)*

36. $\begin{bmatrix} 2 & 1 \\ 1 & 2 \end{bmatrix}$
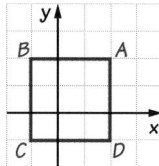

37. $\begin{bmatrix} -2 & 0 \\ 0 & -2 \end{bmatrix}$
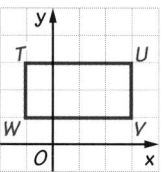

ALGEBRA Find the value of each variable. *(Section 11.4)*

38.
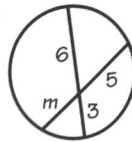

39.
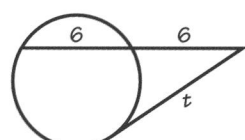

Sketch the image of each polygon after a 90° rotation around the origin. *(Section 8.4)*

40.
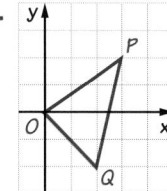

41.
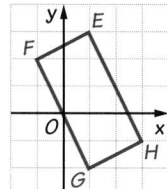

640   Chapter 12 *Coordinates for Transformations*

---

32. $\begin{bmatrix} 0 & 1 \\ 1 & 0 \end{bmatrix}\begin{bmatrix} 2 & 4 & 4 & 2 \\ -2 & 0 & 2 & 0 \end{bmatrix} = \begin{bmatrix} -2 & 0 & 2 & 0 \\ 2 & 4 & 4 & 2 \end{bmatrix}$

33. $\begin{bmatrix} -1 & 0 \\ 0 & 1 \end{bmatrix}\begin{bmatrix} 3 & 3 & 0 \\ -3 & 4 & -1 \end{bmatrix} = \begin{bmatrix} -3 & -3 & 0 \\ -3 & 4 & -1 \end{bmatrix}$

34. $\begin{bmatrix} 0 & -1 \\ -1 & 0 \end{bmatrix}\begin{bmatrix} -2 & -2 & -3 & -4 \\ 4 & 0 & -2 & 0 \end{bmatrix} =$
$\begin{bmatrix} -4 & 0 & 2 & 0 \\ 2 & 2 & 3 & 4 \end{bmatrix}$

35. over *x*-axis: $b \to -b$;
over *y*-axis: $b \to -b$;
over $y = x$: $b \to \frac{1}{b}$;
over $y = -x$: $b \to \frac{1}{b}$

When a segment is reflected over the *x*-axis or *y*-axis, the slope of the image is the opposite of the original slope. When a segment is reflected over $y = x$ or $y = -x$, the slope of the image is the inverse of the original slope.

36. See answers in back of book.

37. $\begin{bmatrix} 2 & -6 & -6 & 2 \\ -6 & -6 & -2 & -2 \end{bmatrix}$

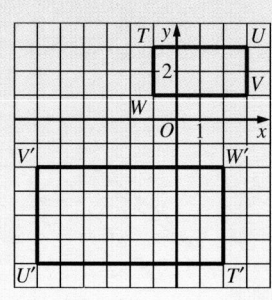

38. $m = 3.6$
39. $t = 6$

40.
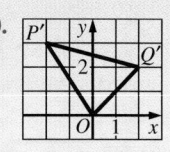

41.
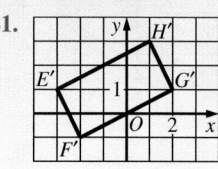

12.5 | **Rotations with Matrices**

Learn how to...
- **represent rotations using matrices**

So you can...
- **rotate points and polygons around the origin**

If a matrix has the same number of rows and columns, it has two *diagonals*: one from upper left to lower right, and one from lower left to upper right.

$$\begin{bmatrix} 3 & 7 \\ -5 & 10 \end{bmatrix}$$

$$\begin{bmatrix} 6 & 0 & 9 \\ -4 & 11 & 3 \\ 1 & -2 & 0 \end{bmatrix}$$

There are two elements on each diagonal of a 2 × 2 matrix.

There are three elements on each diagonal of a 3 × 3 matrix.

EXPLORATION
COOPERATIVE LEARNING

Matrices with Zeros on a Diagonal

Work in a group of four students. You will need:
- graph paper

1 As a group, make a list of every possible 2 × 2 matrix for which both elements on one diagonal are 0 and each element on the other diagonal is either 1 or −1.

$$\begin{bmatrix} ? & 0 \\ 0 & ? \end{bmatrix} \quad \text{or} \quad \begin{bmatrix} 0 & ? \\ ? & 0 \end{bmatrix}$$

Each of the missing elements is either 1 or −1.

2 Divide the list of matrices among the members of your group.

Multiply each matrix that you are assigned by $\begin{bmatrix} 3 & 4 & 7 \\ 1 & 3 & 1 \end{bmatrix}$, which represents △*ABC*.

3 For each matrix product in Step 2, graph △*ABC* and its image. Describe each transformation.

4 Make a chart that shows your group's results.

 Exploration Note

Purpose
The purpose of this Exploration is to have students discover the matrices for rotation.

Materials/Preparation
Each group needs graph paper.

Procedure
Students write down all the different 2 × 2 matrices for which both elements on one diagonal are 0 and each element on the other diagonal is either 1 or –1. They divide these four matrices among group members, use them to transform a given triangle

matrix, and graph the triangle and its image. These graphs are examined to find the type of transformation produced by each matrix.

Closure
Ask one group to present their results to the class. Then use the summary at the top of page 642 to generalize the results found by the groups in Step 4.

Explorations Lab Manual
See the Manual for more commentary on this Exploration.
Diagram Master 1

Section Notes

Geometric Thinking

After discussing the Summary of Matrices for Rotations, you may wish to point out that all of the rotations can be achieved by a combination of two reflections. In Exs. 20 and 21 from Section 12.4, students saw that a combination of a reflection over $y = x$ followed by a reflection over the y-axis produced a 90° rotation, and that if you reverse the order of the reflections, a 270° rotation is achieved. Some students may wish to experiment with other combinations of reflections. They should discover that a 180° reflection can be achieved by reflection over both the x- and y-axes in either order. A 360° rotation can be produced by applying any one of the reflections two times. This way of looking at rotations as a series of two reflections can also be seen in the matrices. If you multiply the matrices for the two reflections that make up a particular rotation, you get the matrix for that rotation.

Teaching Tip

Be sure to point out the *Watch Out!* note on this page. Students are not required to name the center of rotation in this section.

Additional Example 1

The matrix $\begin{bmatrix} 2 & 6 & 4 & 0 \\ 0 & 0 & -3 & -3 \end{bmatrix}$ represents parallelogram *KLMN*. Use matrix multiplication to rotate it 90°. Graph *KLMN* and its image. Find the matrix for a 90° rotation in the table on this page, and write it to the left of the polygon matrix.

$$\begin{matrix} & K & L & M & N \end{matrix}$$
$$\underbrace{\begin{bmatrix} 0 & -1 \\ 1 & 0 \end{bmatrix}}_{\substack{\text{matrix for} \\ \text{90° rotation}}} \begin{bmatrix} 2 & 6 & 4 & 0 \\ 0 & 0 & -3 & -3 \end{bmatrix} =$$

$$\begin{matrix} K' & L' & M' & N' \end{matrix}$$
$$\begin{bmatrix} 0 & 0 & 3 & 3 \\ 2 & 6 & 4 & 0 \end{bmatrix}$$

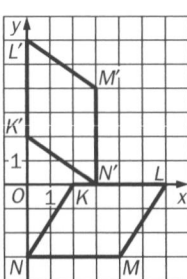

K'L'M'N' is the image of *KLMN* after a 90° rotation.

Each of the transformation matrices that you used in the Exploration is either a reflection matrix or a rotation matrix. The reflection matrices were shown on page 635; the rotation matrices are shown in the table below.

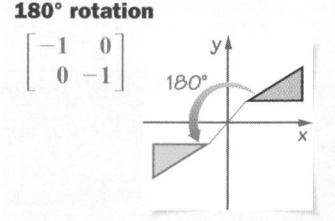

In this section, all rotations are around the origin and are measured in degrees counterclockwise.

Summary of Matrices for Rotations

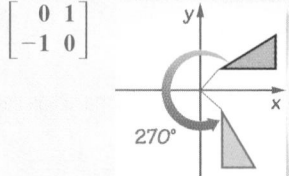

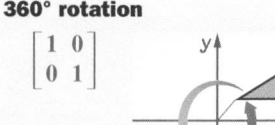

90° rotation $\begin{bmatrix} 0 & -1 \\ 1 & 0 \end{bmatrix}$

180° rotation $\begin{bmatrix} -1 & 0 \\ 0 & -1 \end{bmatrix}$

270° rotation $\begin{bmatrix} 0 & 1 \\ -1 & 0 \end{bmatrix}$

360° rotation $\begin{bmatrix} 1 & 0 \\ 0 & 1 \end{bmatrix}$

EXAMPLE 1

The matrix $\begin{bmatrix} -2 & -5 & -5 & -2 \\ 1 & 1 & 3 & 3 \end{bmatrix}$ represents rectangle *EFGH*. Use matrix multiplication to rotate it 270°. Graph *EFGH* and its image.

SOLUTION

Find the matrix for a 270° rotation in the table above, and write it to the left of the polygon matrix. Use paper and pencil or a graphing calculator to find the matrix product.

$$\begin{matrix} & E & F & G & H & & E' & F' & G' & H' \end{matrix}$$
$$\begin{bmatrix} 0 & 1 \\ -1 & 0 \end{bmatrix} \begin{bmatrix} -2 & -5 & -5 & -2 \\ 1 & 1 & 3 & 3 \end{bmatrix} = \begin{bmatrix} 1 & 1 & 3 & 3 \\ 2 & 5 & 5 & 2 \end{bmatrix}$$

matrix for 270° rotation

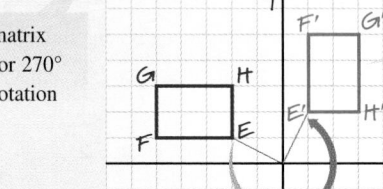

E'F'G'H' is the image of *EFGH* after a 270° rotation.

Combining Rotations and Reflections

In the Exploration, you found some transformation matrices for reflections and some for rotations. As shown below, you can apply one type of transformation to a polygon, then apply a different type of transformation to the result.

EXAMPLE 2

Use the transformation matrix $\begin{bmatrix} -1 & 0 \\ 0 & -1 \end{bmatrix}$ to transform $\triangle QRS$. Then transform the result using $\begin{bmatrix} 1 & 0 \\ 0 & -1 \end{bmatrix}$. What is the combined effect of these two transformations?

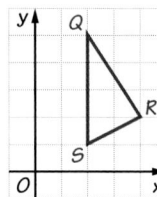

SOLUTION

Use the first matrix to transform $\triangle QRS$.

$$\begin{bmatrix} -1 & 0 \\ 0 & -1 \end{bmatrix} \overset{Q\ R\ S}{\begin{bmatrix} 2 & 4 & 2 \\ 5 & 2 & 1 \end{bmatrix}} = \overset{Q'\ R'\ S'}{\begin{bmatrix} -2 & -4 & -2 \\ -5 & -2 & -1 \end{bmatrix}}$$

Then use the second matrix to transform $\triangle Q'R'S'$.

$$\begin{bmatrix} 1 & 0 \\ 0 & -1 \end{bmatrix} \overset{Q'\ R'\ S'}{\begin{bmatrix} -2 & -4 & -2 \\ -5 & -2 & -1 \end{bmatrix}} = \overset{Q''\ R''\ S''}{\begin{bmatrix} -2 & -4 & -2 \\ 5 & 2 & 1 \end{bmatrix}}$$

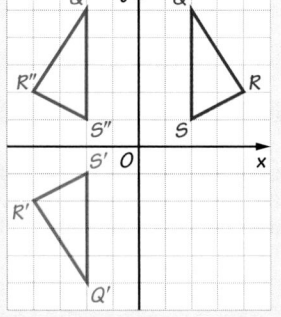

The combined effect of the two transformations is equivalent to a reflection over the y-axis.

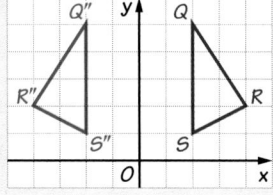

THINK AND COMMUNICATE

Use the solution to Example 2.

1. What transformation is represented by the matrix $\begin{bmatrix} -1 & 0 \\ 0 & -1 \end{bmatrix}$?

2. What transformation is represented by the matrix $\begin{bmatrix} 1 & 0 \\ 0 & -1 \end{bmatrix}$?

12.5 Rotations with Matrices   **643**

Additional Example 2

Use the transformation matrix $\begin{bmatrix} 0 & 1 \\ 1 & 0 \end{bmatrix}$ to transform $\triangle ABC$.

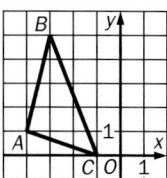

Then transform the result using $\begin{bmatrix} 0 & -1 \\ 1 & 0 \end{bmatrix}$. What is the combined effect of these two transformations? Use the first matrix to transform $\triangle ABC$.

$$\begin{bmatrix} 0 & 1 \\ 1 & 0 \end{bmatrix} \overset{A\quad B\quad C}{\begin{bmatrix} -4 & -3 & -1 \\ 1 & 5 & 0 \end{bmatrix}} = \overset{A'\ B'\ C'}{\begin{bmatrix} 1 & 5 & 0 \\ -4 & -3 & -1 \end{bmatrix}}$$

Then use the second matrix to transform $\triangle A'B'C'$.

$$\begin{bmatrix} 0 & -1 \\ 1 & 0 \end{bmatrix} \overset{A'\quad B'\quad C'}{\begin{bmatrix} 1 & 5 & 0 \\ -4 & -3 & -1 \end{bmatrix}} = \overset{A''\ B''\ C''}{\begin{bmatrix} 4 & 3 & 1 \\ 1 & 5 & 0 \end{bmatrix}}$$

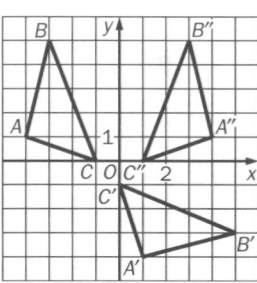

The combined effect of the two transformations is equivalent to a reflection over the y-axis.

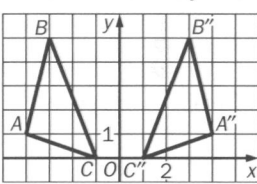

Think and Communicate

These questions ask students to determine the type of transformation produced by each matrix in Example 2. This can be done by examining the figures, or by looking for the matrices on pages 635 or 642.

Checking Key Concepts

Student Progress
These two exercises are very similar to Example 1 and Example 2. Students having difficulty answering these questions should refer back to these examples. To help students assess their progress before attempting to do the exercises, ask two volunteers to present their solutions of these problems to the class.

Closure Question

What are the possible transformations that can be achieved by multiplying a polygon matrix by a 2 × 2 matrix for which both elements on one diagonal are 0 and each element on the other diagonal is either 1 or –1? The transformation may be a reflection over the x- or y-axis, the line $y = x$, or the line $y = -x$. Or, it may be a rotation of 90°, 180°, 270°, or 360°.

Apply⇔Assess

Suggested Assignment

❖ **Core Course**
Exs. 1–7, 10, 12, 15–21, AYP

❖ **Extended Course**
Exs. 1–21, AYP

❖ **Block Schedule**
Day 78 Exs. 1–7, 10, 12, 15–21, AYP

Exercise Notes

Student Progress
Exs. 1–6 Students should be able to complete these exercises successfully. A review of the answers would provide students with an opportunity to check their work before proceeding with the remaining exercises.

Common Error
Exs. 1–6 Because the four rotation matrices are so similar, students often use the wrong one. Encourage them to check each graph to see that the correct type of rotation was achieved.

✓ CHECKING KEY CONCEPTS

1. The matrix $\begin{bmatrix} 5 & 5 & 3 & 3 \\ 1 & -2 & -2 & 3 \end{bmatrix}$ represents quadrilateral *ABDE*.

   a. Use the matrix $\begin{bmatrix} 0 & -1 \\ 1 & 0 \end{bmatrix}$ to rotate *ABDE*.

   b. Graph *ABDE* and its image. Describe the rotation.

2. The matrix $\begin{bmatrix} -3 & -5 & -2 \\ 1 & 0 & -3 \end{bmatrix}$ represents △*CGS*.

   a. Use the matrix $\begin{bmatrix} 0 & -1 \\ 1 & 0 \end{bmatrix}$ to transform △*CGS*. Then transform the result using the matrix $\begin{bmatrix} -1 & 0 \\ 0 & -1 \end{bmatrix}$.

   b. Graph △*CGS*, △*C'G'S'*, and △*C"G"S"*. What is the combined effect of the two transformations?

12.5 Exercises and Applications

Extra Practice exercises on page 686

Use matrix multiplication to find the image of each polygon after the given rotation. Then graph the polygon and its image. Use the table on page 642 for the transformation matrices.

1. 90° rotation

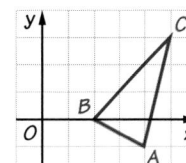

2. 270° rotation

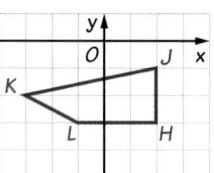

3. 360° rotation

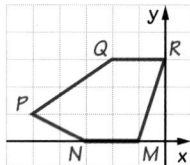

4. 270° rotation

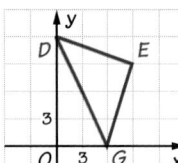

5. 180° rotation

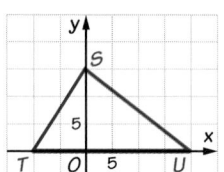

6. 90° rotation

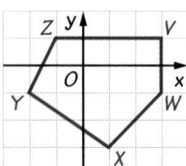

7. Open-ended Problem Choose any two rotation matrices.

   a. Use each of your rotation matrices to transform *F*(2, 7) and *H*(3, 6).

   b. For each of your rotation matrices, complete the following:
   $(a, b) \rightarrow (\underline{\ ?\ }, \underline{\ ?\ })$.

644 Chapter 12 *Coordinates for Transformations*

Checking Key Concepts

1. a. $\begin{bmatrix} -1 & 2 & 2 & -3 \\ 5 & 5 & 3 & 3 \end{bmatrix}$

   b.
   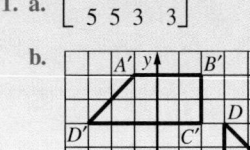
   90° rotation

2. a. $\begin{bmatrix} -1 & 0 & 3 \\ -3 & -5 & 2 \end{bmatrix}$; $\begin{bmatrix} 1 & 0 & -3 \\ 3 & 5 & 2 \end{bmatrix}$

   b.
   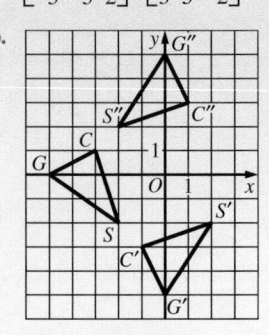
   270° rotation

Exercises and Applications

1. $\begin{bmatrix} 1 & 0 & -3 \\ 4 & 2 & 5 \end{bmatrix}$

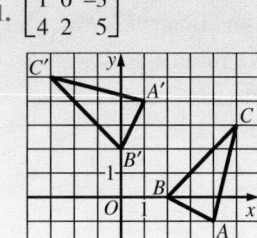

2–6. See answers in back of book.

Look back at the article on pages 610–612.

While traveling through space, the Galileo spacecraft spins around a central axis. In the spacecraft's frame of reference, it appears as if the spacecraft is stationary at the origin and the universe is rotating around it.

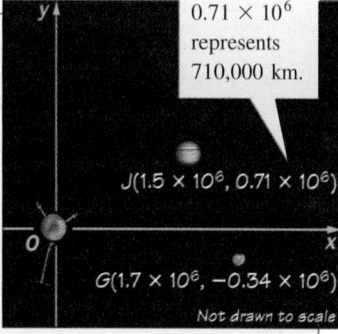

0.71×10^6 represents 710,000 km.

$J(1.5 \times 10^6, 0.71 \times 10^6)$

$G(1.7 \times 10^6, -0.34 \times 10^6)$

Not drawn to scale

8. At time 12:00:00, Ganymede (one of Jupiter's moons) and Jupiter are located at points G and J, respectively.

   a. Write point matrices for G and J. Multiply the transformation matrix $\begin{bmatrix} 0 & -1 \\ 1 & 0 \end{bmatrix}$ by these point matrices to find G' and J', Ganymede's and Jupiter's positions five seconds later, at 12:00:05.

   b. Multiply the transformation matrix $\begin{bmatrix} 0 & -1 \\ 1 & 0 \end{bmatrix}$ by your results from part (a) to find G'' and J'', the locations at 12:00:10.

   c. On one pair of axes, graph Ganymede's and Jupiter's locations at 12:00:00, 12:00:05, and 12:00:10.

   d. Based on your answers to parts (a)–(c), about how much time does it take the spacecraft Galileo to make one full (360°) turn?

9. Open-ended Problem Why might a Galileo mission scientist such as Adriana Ocampo need to know the position of Ganymede in relation to the spacecraft at a particular time?

10. The matrix $\begin{bmatrix} 5 & 5 & 3 \\ 0 & -4 & -4 \end{bmatrix}$ represents $\triangle PQR$.

   a. Use the matrix $\begin{bmatrix} -1 & 0 \\ 0 & 1 \end{bmatrix}$ to transform $\triangle PQR$.

   b. Use the matrix $\begin{bmatrix} -1 & 0 \\ 0 & -1 \end{bmatrix}$ to transform the result from part (a).

   c. Graph $\triangle PQR$, $\triangle P'Q'R'$, and $\triangle P''Q''R''$. What is the combined effect of the two transformations in parts (a) and (b)?

11. Challenge The product of two transformation matrices is also a transformation matrix.

   a. Complete the matrix product $\begin{bmatrix} -1 & 0 \\ 0 & -1 \end{bmatrix} \begin{bmatrix} 0 & -1 \\ 1 & 0 \end{bmatrix} = \begin{bmatrix} ? & ? \\ ? & ? \end{bmatrix}$.

   180°   90°

   Use the table on page 642 to find the rotation caused by the product.

   b. Complete the product $\begin{bmatrix} 0 & 1 \\ -1 & 0 \end{bmatrix} \begin{bmatrix} ? & ? \\ ? & ? \end{bmatrix} = \begin{bmatrix} -1 & 0 \\ 0 & -1 \end{bmatrix}$.

BY THE WAY

Ganymede was discovered by Italian scientist Galileo Galilei (1564–1642), after whom the spacecraft Galileo is named.

12.5 Rotations with Matrices   **645**

7. a. 90° rotation:
$F(2, 7) \to (-7, 2)$,
$H(3, 6) \to (-6, 3)$;
180° rotation:
$F(2, 7) \to (-2, -7)$,
$H(3, 6) \to (-3, -6)$;
270° rotation:
$F(2, 7) \to (7, -2)$,
$H(3, 6) \to (6, -3)$;
360° rotation:
$F(2, 7) \to (2, 7)$,
$H(3, 6) \to (3, 6)$

b. 90° rotation:
$(a, b) \to (-b, a)$;
180° rotation:
$(a, b) \to (-a, -b)$;
270° rotation:
$(a, b) \to (b, -a)$;
360° rotation:
$(a, b) \to (a, b)$

8. a. $G: \begin{bmatrix} 1.7 \times 10^6 \\ -0.34 \times 10^6 \end{bmatrix}$;

$J: \begin{bmatrix} 1.5 \times 10^6 \\ 0.71 \times 10^6 \end{bmatrix}$;

$G': \begin{bmatrix} 0.34 \times 10^6 \\ 1.7 \times 10^6 \end{bmatrix}; J': \begin{bmatrix} -0.71 \times 10^6 \\ 1.5 \times 10^6 \end{bmatrix}$

b. $G'': \begin{bmatrix} -1.7 \times 10^6 \\ 0.34 \times 10^6 \end{bmatrix}; J'': \begin{bmatrix} -1.5 \times 10^6 \\ -0.71 \times 10^6 \end{bmatrix}$;

c.

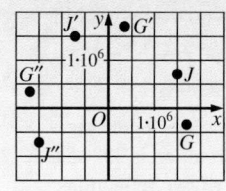

d. 20 s

9–11. See answers in back of book.

Exercise Notes

Application

Exs. 12–14 These exercises introduce students to applications of rotations in video games. Some students may have had the experience of using a video game that has this rotational effect.

Using Technology

Ex. 13 In part (b) of this exercise, students multiply two matrices to find the point matrix for *P'*. They then multiply the result by the same rotation matrix. The most efficient way to complete this problem using a graphing calculator is o enter the rotation matrix and the matrix for *P* as matrix *A* and matrix *B*, respectively. Multiply *A* and *B*, and then store the result in matrix *C*. To do this on a TI-82, after entering matrix *A* and matrix *B*, go to the home screen. Type [A][B] and press [ENTER], then type [STO►] [C]. This stores the product of matrix *A* and matrix *B* in matrix *C*. To find the point matrix for *P''*, multiply matrix *C* by matrix *A*.

Research

Ex. 14 Interested students will enjoy this research activity. You may want to have students find a computer game that uses virtual reality and demonstrate it to the class. Also, ask them to find other areas where virtual reality simulations are used. The results of this research could make an interesting bulletin board display.

Assessment Note

Ex. 15 This exercise can be used to assess students' skill in using a series of two rotation matrices. At the same time, students can discover that the order in which the rotation matrices are applied does not affect the final image. This is not the case for a combination of reflections or a combination of a reflection and a rotation.

Connection ▸ VIDEO GAMES

In *virtual reality simulators*, such as some computer games, players can feel as if they are traveling through a three-dimensional world and interacting with virtual people. When a player turns, the world appears to rotate, as shown. The new coordinates of objects are found using rotation matrices.

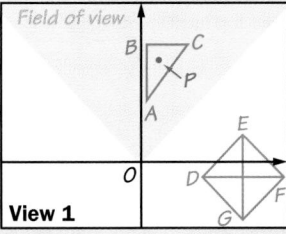 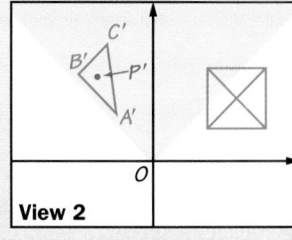 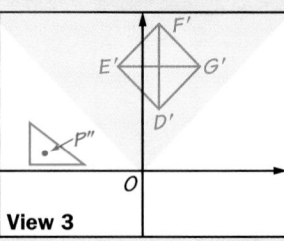

View 1   **View 2**   **View 3**

In each view, the player is at (0, 0). In View 1, the coordinates of platform *ABC*, virtual person *P*, and base *DEFG* of the pyramid are given by these matrices:

$$\begin{array}{ccc} A & B & C \end{array} \quad\quad P \quad\quad \begin{array}{cccc} D & E & F & G \end{array}$$

$$\begin{bmatrix} 2 & 2 & 19 \\ 24 & 46 & 46 \end{bmatrix} \quad \begin{bmatrix} 7 \\ 40 \end{bmatrix} \quad \begin{bmatrix} 24 & 41 & 58 & 41 \\ -6 & 11 & -6 & -23 \end{bmatrix}$$

12. a. Multiply the matrix representing *DEFG* by the rotation matrix $\begin{bmatrix} 0 & -1 \\ 1 & 0 \end{bmatrix}$ to find a matrix for *D'E'F'G'*, the pyramid in View 3.

     b. Describe the rotation caused by the transformation matrix in part (a).

13.   **Technology** Use a graphing calculator or computer software with matrix capabilities.

     a. Multiply the matrix that represents *ABC* by the rotation matrix

     $\begin{bmatrix} 0.707 & -0.707 \\ 0.707 & 0.707 \end{bmatrix}$ to find a matrix for *A'B'C'*, the platform in View 2.

     b. Multiply the point matrix for *P* by the rotation matrix $\begin{bmatrix} 0.707 & -0.707 \\ 0.707 & 0.707 \end{bmatrix}$

     to find the coordinates of *P'*. Then multiply the result by the same rotation matrix to find the coordinates of *P''*, the person in View 3.

     c. Writing What is the combined effect of the two rotations in part (b)? Find the degree measure of each rotation and explain your reasoning.

14. Research Find out more about how virtual reality is used. Present your results in a report or visual display.

12. a. $\begin{bmatrix} 6 & -11 & 6 & 23 \\ 24 & 41 & 58 & 41 \end{bmatrix}$

     b. rotation by 90°

13. a. $\begin{bmatrix} -15.554 & -31.108 & -19.089 \\ 18.382 & 33.936 & 45.955 \end{bmatrix}$

     b. *P'*(−23.331, 33.229); *P''*(−39.988, 6.998)

     c. a 90° rotation; The degree measure of each rotation is 45° so a combination of two reflections has a measure of 2(45) = 90°.

14. Answers may vary. Check students' work.

15. a, b. Check students' work.

     c. No. Rotating by *x*° and then by *y*° is rotation by $x° + y° = y° + x°$.

16. 0.7002   **17.** 0.9659

18. 0.8660   **19.** $x = 7\sqrt{2}$

20. *x* = 5   **21.** *x* = 20

Assess Your Progress

1. *x* = 2; *y* = 12

2. *g* = −2; *h* = −4

3. $\begin{bmatrix} 1 & 3 & 1 & -2 \\ 2 & 1 & -2 & 1 \end{bmatrix}; \begin{bmatrix} 2 & 1 & -2 & 1 \\ 1 & 3 & 1 & -2 \end{bmatrix}$

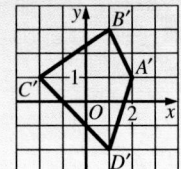

15. Open-ended Problem Write a matrix for any quadrilateral *TUVW*.

   a. Use one of the rotation matrices from page 642 to transform *TUVW*. Then transform the result using a different rotation matrix.

   b. Transform *TUVW* using the same two rotation matrices from part (a), but reverse the order in which you apply them.

   c. Do you think that the order in which the two rotations are applied has any impact on the final image? Explain your answer.

SPIRAL REVIEW

Find the value of each ratio. Round your answers to four decimal places.
(Sections 10.3 and 10.4)

16. tan 35°　　　　　**17.** sin 75°　　　　　**18.** cos 30°

Solve each proportion to find a geometric mean. *(Sections 9.1 and 10.1)*

19. $\dfrac{7}{x} = \dfrac{x}{14}$　　　　**20.** $\dfrac{1}{x} = \dfrac{x}{25}$　　　　**21.** $\dfrac{10}{x} = \dfrac{x}{40}$

ASSESS YOUR PROGRESS

VOCABULARY

matrix multiplication (p. 626)

Find the values of the variables in each matrix product. *(Section 12.3)*

1. $\begin{bmatrix} 4 & 1 \\ 3 & 2 \end{bmatrix} \begin{bmatrix} x \\ 3 \end{bmatrix} = \begin{bmatrix} 11 \\ y \end{bmatrix}$

2. $\begin{bmatrix} 1 & -1 \\ 3 & 5 \end{bmatrix} \begin{bmatrix} 2 \\ g \end{bmatrix} = \begin{bmatrix} 4 \\ h \end{bmatrix}$

Use matrix multiplication to find the image of each polygon after reflection over the given line. Then graph the image. Use the table on page 635 for the transformation matrices. *(Section 12.4)*

3. the line $y = x$

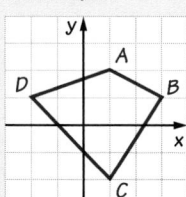

4. the line $y = -x$

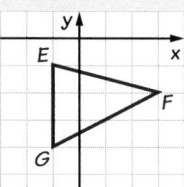

Use matrix multiplication to find the image of △*EFG* from Question 4 after the given rotation. Then graph △*EFG* and its image. Use the table on page 642 for the transformation matrices. *(Section 12.5)*

5. 90° rotation　　　　　　　**6.** 270° rotation

7. Journal Write a letter to a student who will take this geometry course next year. Describe some of the projects or problems you found interesting. Include advice about how to do well in the course.

Assess Your Progress

Students should be able to multiply two matrices using pencil and paper or a graphing calculator. In addition, they should be able to use the matrices of reflection and the matrices of rotation.

Journal Entry

This entry provides students with the opportunity to reflect on the work they have done this year. Completing this entry will help provide a sense of closure to the course.

Progress Check 12.3–12.5

See page 653.

Practice 88 for Section 12.5

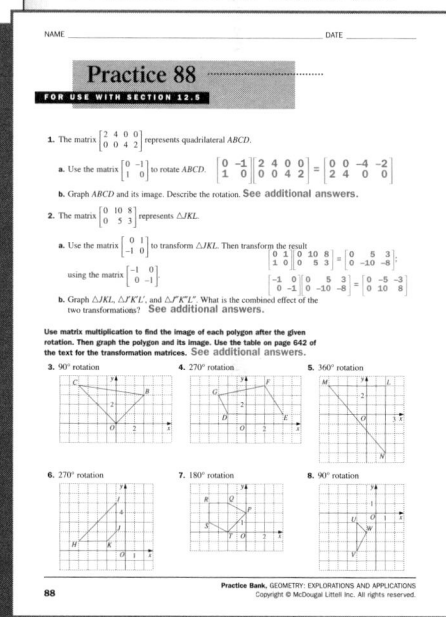

---

4. $\begin{bmatrix} -1 & 3 & -1 \\ -1 & -2 & -4 \end{bmatrix}$; $\begin{bmatrix} 1 & 2 & 4 \\ 1 & -3 & 1 \end{bmatrix}$

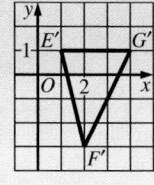

5. $\begin{bmatrix} 1 & 2 & 4 \\ -1 & 3 & -1 \end{bmatrix}$

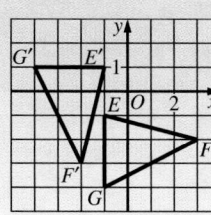

6. $\begin{bmatrix} -1 & -2 & -4 \\ 1 & -3 & 1 \end{bmatrix}$

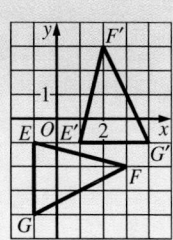

7. Answers may vary. Check students' work.

Mathematical Goals

- Analyze patterns in the Sierpinski Triangle and in other transformations.

- Find values in a transformation used to create the Sierpinski Triangle.

- Enter and execute a program on a TI-82 graphing calculator to display the Sierpinski Triangle.

- Find values in a transformation used to create other fractals and display the fractal using a TI-82 graphing calculator.

Planning

Materials

- TI-82 graphing calculator

- paper and pencil

Project Teams

Students work with a partner to complete the project. The partners should work closely together to discuss and perform each aspect of the project.

Guiding Students' Work

This project is an excellent example of using transformation matrices to create a fractal. If students understand what happens to points in certain types of transformations, they can use matrices to help draw the images. When students are finding the values in the matrices, encourage them to work step by step. Also, students may want to reread the project a few times to develop an understanding of the process used to solve for the variables.

Second-Language Learners

You might want to encourage students to determine the meaning of the term *self-similar* by looking at the word parts and at the picture of the Sierpinski Triangle. Elicit from them that *self-similar* describes a figure made up of smaller figures that look exactly like the original figure.

Creating Fractals

Like many fractals, this *Sierpinski Triangle* is *self-similar*. It is made up of smaller pieces which are geometrically similar to the whole fractal. Each of these pieces is made up of even smaller pieces that are also similar to the whole.

PROJECT GOAL Work with another student to analyze fractals and generate them on your graphing calculator.

Analyzing the Sierpinski Triangle

The Sierpinski Triangle contains three half-size copies of itself. Each copy is the image of the whole triangle after a transformation.

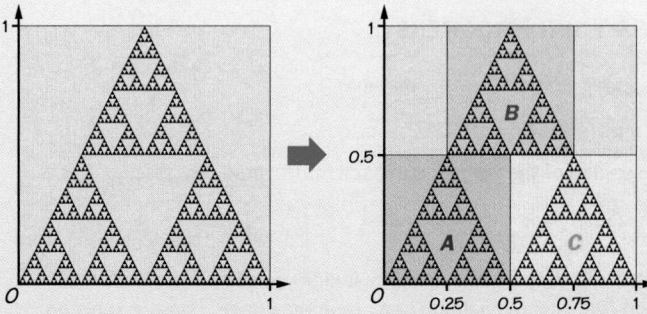

Transformation *A* is a dilation with center $(0, 0)$ and scale factor 0.5.

Transformation *B* is a dilation followed by a translation to the right 0.25 units and up 0.5 units.

This *similarity diagram* shows how the three transformations that describe the Sierpinski Triangle affect the unit square.

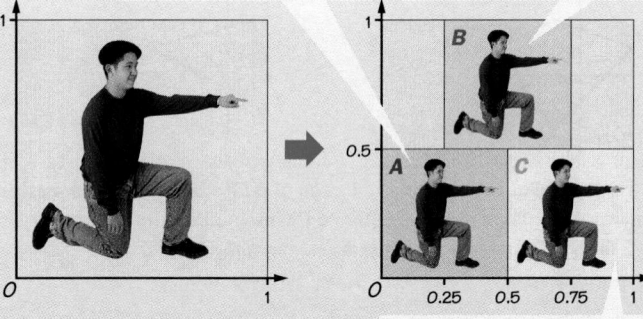

Transformation *C* is a dilation followed by a translation to the right 0.5 units.

Finding Values for the Transformations

Each transformation used to create the Sierpinski Triangle can be represented by two matrix operations. For point $P(a, b)$:

$$\overset{P}{\begin{bmatrix} p & q \\ s & t \end{bmatrix} \begin{bmatrix} a \\ b \end{bmatrix}} + \begin{bmatrix} r \\ u \end{bmatrix} = \overset{P'}{\begin{bmatrix} ? \\ ? \end{bmatrix}}$$

For each of the three transformations, you can use algebra to find the values of p, q, r, s, t, and u.

To find the values for transformation B, use the similarity diagram to see how transformation B affects the points (0, 0), (0, 1), and (1, 0).

The image of (0, 0) is **(0.25, 0.5)**.

The image of (0, 1) is **(0.25, 1)**.

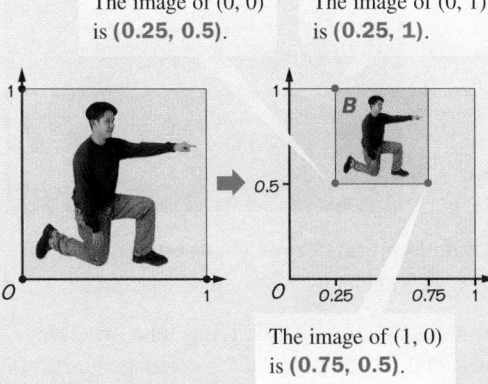

The image of (1, 0) is **(0.75, 0.5)**.

Step 1 Use the image of (0, 0) to solve for r and u.

$$(0, 0) \rightarrow (0.25, 0.5)$$

$$\begin{bmatrix} p & q \\ s & t \end{bmatrix} \begin{bmatrix} 0 \\ 0 \end{bmatrix} + \begin{bmatrix} r \\ u \end{bmatrix} = \begin{bmatrix} 0.25 \\ 0.5 \end{bmatrix}$$

$$\begin{bmatrix} p \cdot 0 + q \cdot 0 \\ s \cdot 0 + t \cdot 0 \end{bmatrix} + \begin{bmatrix} r \\ u \end{bmatrix} = \begin{bmatrix} 0.25 \\ 0.5 \end{bmatrix}$$

$$\begin{bmatrix} 0 + r \\ 0 + u \end{bmatrix} = \begin{bmatrix} 0.25 \\ 0.5 \end{bmatrix}$$

So $r = 0.25$ and $u = 0.5$.

Step 2 Use the images of (0, 1) and (1, 0) to solve for p, q, s, and t. Substitute the values of r and u that you found in Step 1.

$$(0, 1) \rightarrow (0.25, 1)$$

$$\begin{bmatrix} p & q \\ s & t \end{bmatrix} \begin{bmatrix} 0 \\ 1 \end{bmatrix} + \begin{bmatrix} r \\ u \end{bmatrix} = \begin{bmatrix} 0.25 \\ 1 \end{bmatrix}$$

$$\begin{bmatrix} p \cdot 0 + q \cdot 1 \\ s \cdot 0 + t \cdot 1 \end{bmatrix} + \begin{bmatrix} \mathbf{0.25} \\ \mathbf{0.5} \end{bmatrix} = \begin{bmatrix} 0.25 \\ 1 \end{bmatrix}$$

$$\begin{bmatrix} q + 0.25 \\ t + 0.5 \end{bmatrix} = \begin{bmatrix} 0.25 \\ 1 \end{bmatrix}$$

So $q = 0$ and $t = 0.5$.

$$(1, 0) \rightarrow (0.75, 0.5)$$

$$\begin{bmatrix} p & q \\ s & t \end{bmatrix} \begin{bmatrix} 1 \\ 0 \end{bmatrix} + \begin{bmatrix} r \\ u \end{bmatrix} = \begin{bmatrix} 0.75 \\ 0.5 \end{bmatrix}$$

$$\begin{bmatrix} p \cdot 1 + q \cdot 0 \\ s \cdot 1 + t \cdot 0 \end{bmatrix} + \begin{bmatrix} \mathbf{0.25} \\ \mathbf{0.5} \end{bmatrix} = \begin{bmatrix} 0.75 \\ 0.5 \end{bmatrix}$$

$$\begin{bmatrix} p + 0.25 \\ s + 0.5 \end{bmatrix} = \begin{bmatrix} 0.75 \\ 0.5 \end{bmatrix}$$

So $p = 0.5$ and $s = 0$.

Use the same method to find the values of p, q, r, s, t, and u for transformation A and for transformation C.

Portfolio Project **649**

Project Notes

Guiding Students' Work

Using Technology

On page 650, students use a TI-82 graphing calculator to display the Sierpinski Triangle. Refer students to the Appendices at the back of the TI-82 calculator manual if they have difficulty entering the program since the Appendices tell in which menu each command function or variable can be found. Alternatively, you may wish to enter the program yourself (or ask a student to do it), check that it works, and then distribute the program to students by linking the calculators.

The program given on page 650 will not work on the TI-81 because it requires 5 matrices, and the TI-81 has only 3. The program for the TI-83 is identical to the program for the TI-82, with one minor difference. The dim function on the TI-82 does not take parentheses; the dim function on the TI-83 does. So the line that reads {3,1}→dim[D] on the TI-82 becomes {3,1}→dim([D]) on the TI-83.

The Catalog feature on the TI-83 and TI-92 makes programming easier, since the user no longer has to hunt through menus to find each command or function. The program for the TI-92 is different than the one for the TI-82 and is given on the next page. This program can be named FRACTAL. Also, the matrices containing the values for the transformations must be stored in the calculator as variables a, b, and c. These matrices should be in the same folder as the program itself. After the program and matrices are entered, run the program by typing FRACTAL() at the home screen.

Program for the TI-92

```
:fractal()
:Prgm
:setMode("Graph","Function")
:PlotsOff
:setGraph("Axes","Off")
:FnOff
:ClrDraw
:[[0][0][1]]→d
:0→k
:0→xmin
:1→xmax
:0→ymin
:1→ymax
:ZoomSqr

:While k<2000
:int(rand()*3+1)→r
:If r=1
:a*d→e
:If r=2
:b*d→e
:If r=3
:c*d→e
:e[1,1]→d[1,1]
:e[2,1]→d[2,1]
:PtOn e[1,1],e[2,1]
:k+1→k
:EndWhile
:EndPrgm
```

```
PROGRAM: FRACTAL
:PlotsOff
:AxesOff
:FnOff
:ClrDraw
:{3,1} →dim [D]
:[[0][0][1]]→[D]
:0→K
:0→Xmin
:1→Xmax
:0→Ymin
:1→Ymax
:ZSquare
:While (K<2000)
:int (rand*3+1)→R
:If R=1
:[A]*[D]→[E]
:If R=2
:[B]*[D]→[E]
:If R=3
:[C]*[D]→[E]
:[E](1,1)→[D](1,1)
:[E](2,1)→[D](2,1)
:Pt-On([E](1,1),[E](2,1))
:K+1→K
:End
```

Displaying the Fractal

Using the program at the left and a TI-82 graphing calculator, you can display the Sierpinski Triangle using the three transformations that you found.

1. Enter the program into your graphing calculator directly, or download it from another calculator or computer. See your graphing calculator manual for more information on entering or transmitting programs.

2. Store the values you found for transformation *B* into matrix *B* on your graphing calculator. Enter the six values into a 2 × 3 matrix in this order:

$$\begin{bmatrix} p & q & r \\ s & t & u \end{bmatrix}$$

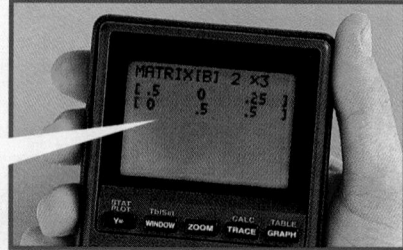

For transformation *B*, store this matrix in your graphing calculator.

3. Store the values you found for transformation *A* and transformation *C* into matrices *A* and *C*, respectively.

4. Execute the program to display the fractal. It may take several minutes for the program to finish. If you entered the correct values into the matrices, your graphing calculator screen should look like this:

Sierpinski Triangle

Matrix *A*	Matrix *B*	Matrix *C*
$\begin{bmatrix} 0.5 & 0 & 0 \\ 0 & 0.5 & 0 \end{bmatrix}$	$\begin{bmatrix} 0.5 & 0 & 0.25 \\ 0 & 0.5 & 0.5 \end{bmatrix}$	$\begin{bmatrix} 0.5 & 0 & 0.5 \\ 0 & 0.5 & 0 \end{bmatrix}$

Generating Other Fractals

Choose one of the fractals shown below. Use the corresponding similarity diagram to find the values of *p*, *q*, *r*, *s*, *t*, and *u* for each transformation. Store these values in your graphing calculator as matrices *A*, *B*, and *C*. Then run the program to display the fractal.

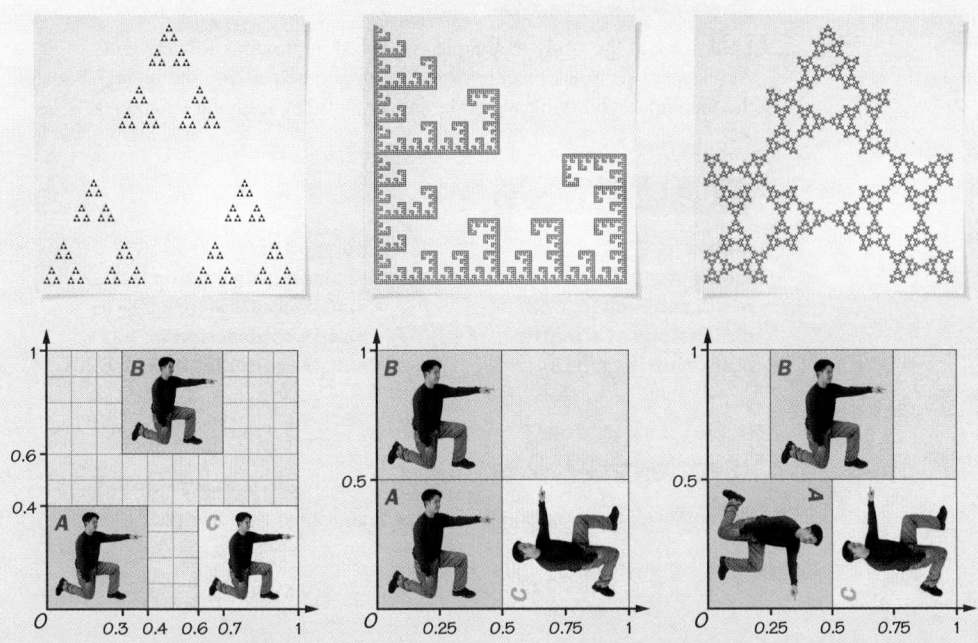

Writing a Report

In your report, give the values that you found for transformations *A* and *C* in the Sierpinski Triangle, and show the steps that you used to find them. Give the values for all three transformations in one of the fractals above, and show the steps that you used to find them. Describe how the image on your graphing calculator screen compares to the picture of the fractal above. If possible, include printouts of your calculator screen. To extend the project, draw a new similarity diagram and use your graphing calculator to display the fractal it describes.

Self-Assessment

In this project, you used some techniques that were not emphasized in the chapter. What was the most difficult part of the project? Explain.

Portfolio Project   **651**

Project Notes

Guiding Students' Work

Rubric for Chapter Project

4 Students enter the program in their graphing calculators and display the Sierpinski Triangle on their calculator screen. Students also find correct values for the six variables for one of the given transformations. Students write a complete report that contains all the information asked for; it is evident from the report that students understand the concepts discussed in the project. Students extend the project and complete a self-assessment.

3 Students enter the program in their graphing calculators and display the Sierpinski Triangle on their calculator screen. Students also make an effort to find values for the six variables for one of the given transformations in the project, but one or two miscalculations prevent them from displaying the fractal. Students write a report describing their work, but it is incomplete. Students attempt to extend the project and provide a self-assessment.

2 Students enter the program in their calculators but cannot display the Sierpinski Triangle because of errors in entering the program. Students do not complete the calculations to find the values of the six variables listed. As a result, no other fractals are generated. If a report is written, it is incomplete and does not contain most of the work requested for the project. Students do not extend the project or provide a self-assessment.

1 Students try to enter the computer program but do not succeed in doing so. Students do not attempt to complete any other part of the project. Students should be encouraged to speak with the teacher as soon as possible to review their work and to make a new start on the project.

	Matrix *A*	Matrix *B*	Matrix *C*
Disconnected Triangles	$\begin{bmatrix} 0.4 & 0 & 0 \\ 0 & 0.4 & 0 \end{bmatrix}$	$\begin{bmatrix} 0.4 & 0 & 0.3 \\ 0 & 0.4 & 0.6 \end{bmatrix}$	$\begin{bmatrix} 0.4 & 0 & 0.6 \\ 0 & 0.4 & 0 \end{bmatrix}$
Rectangular L's	$\begin{bmatrix} 0.5 & 0 & 0 \\ 0 & 0.5 & 0 \end{bmatrix}$	$\begin{bmatrix} 0.5 & 0 & 0 \\ 0 & 0.5 & 0.5 \end{bmatrix}$	$\begin{bmatrix} 0 & -0.5 & 1 \\ 0.5 & 0 & 0 \end{bmatrix}$
Star Pattern	$\begin{bmatrix} 0 & 0.5 & 0 \\ -0.5 & 0 & 0.5 \end{bmatrix}$	$\begin{bmatrix} 0.5 & 0 & 0.25 \\ 0 & 0.5 & 0.5 \end{bmatrix}$	$\begin{bmatrix} 0 & -0.5 & 1 \\ 0.5 & 0 & 0 \end{bmatrix}$

Progress Check 12.1–12.2

$\triangle RST$ has coordinates $R(3, -1)$, $S(4, 2)$, and $T(-2, -4)$.

1. Write a matrix for $\triangle RST$.
   (Section 12.1) $\begin{bmatrix} 3 & 4 & -2 \\ -1 & 2 & -4 \end{bmatrix}$

2. Multiply the matrix for $\triangle RST$ by 3. *(Section 12.1)*
   $\begin{bmatrix} 9 & 12 & -6 \\ -3 & 6 & -12 \end{bmatrix}$

3. Describe the transformation that takes $\triangle RST$ to the triangle whose vertices are given in Ex. 2. *(Section 12.1)* a dilation with center (0, 0) and scale factor 3

4. Write a matrix that translates $\triangle RST$ left 3 units and down 2 units. *(Section 12.2)*
   $\begin{bmatrix} -3 & -3 & -3 \\ -2 & -2 & -2 \end{bmatrix}$

5. Find the coordinates of $\triangle R'S'T'$, the image of $\triangle RST$ after the translations described in Ex. 4. *(Section 12.2)* $\begin{bmatrix} 0 & 1 & -5 \\ -3 & 0 & -6 \end{bmatrix}$

Review

STUDY TECHNIQUE

Look back at the study techniques you used in previous Review and Assessment sections. Decide which technique helped you the most. Choose the technique you think would be most helpful in studying Chapter 12.

VOCABULARY

matrix (p. 613)
elements (p. 613)
rows, columns (p. 613)
dimensions of a matrix (p. 613)
point matrix (p. 613)

polygon matrix (p. 613)
scalar multiplication (p. 614)
matrix addition (p. 619)
matrix subtraction (p. 621)
matrix multiplication (p. 626)

SECTIONS 12.1 *and* 12.2

A point, segment, or polygon can be represented by a **matrix**.

$A: \begin{bmatrix} -2 \\ -2 \end{bmatrix}$ $\overline{AB}: \begin{bmatrix} -2 & 0 \\ -2 & 2 \end{bmatrix}$ $ABCDE: \begin{bmatrix} -2 & 0 & 3 & 3 & 1 \\ -2 & 2 & 2 & 0 & -2 \end{bmatrix}$ ← *x*-coordinates
← *y*-coordinates

You can use **scalar multiplication** to dilate a polygon. The center of the dilation is (0, 0).

scale factor

$$\frac{1}{2} \begin{bmatrix} -2 & 0 & 3 & 3 & 1 \\ -2 & 2 & 2 & 0 & -2 \end{bmatrix} = \begin{bmatrix} -1 & 0 & \frac{3}{2} & \frac{3}{2} & \frac{1}{2} \\ -1 & 1 & 1 & 0 & -1 \end{bmatrix}$$

$\begin{matrix} A & B & C & D & E \end{matrix}$ $\begin{matrix} A' & B' & C' & D' & E' \end{matrix}$

polygon matrix   image matrix

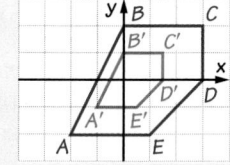

You can translate a polygon using **matrix addition**. The top row of the translation matrix translates horizontally, and the bottom row translates vertically. To add matrices, add corresponding elements.

$$\begin{bmatrix} -1 & -1 & -1 & -1 & -1 \\ 2 & 2 & 2 & 2 & 2 \end{bmatrix} + \begin{bmatrix} -2 & 0 & 3 & 3 & 1 \\ -2 & 2 & 2 & 0 & -2 \end{bmatrix} = \begin{bmatrix} -3 & -1 & 2 & 2 & 0 \\ 0 & 4 & 4 & 2 & 0 \end{bmatrix}$$

$\qquad\qquad\qquad\qquad\qquad\begin{matrix} A & B & C & D & E \end{matrix}\qquad\qquad\begin{matrix} A' & B' & C' & D' & E' \end{matrix}$

translation matrix   **polygon matrix**   image matrix

SECTIONS 12.3, 12.4, *and* 12.5

You can use **matrix multiplication** to transform polygons. For example,

the matrix $\begin{bmatrix} 1 & 0 \\ 0.5 & 1 \end{bmatrix}$ can transform a square into a parallelogram.

To multiply matrices, multiply the rows in the left matrix by the **columns** in the right matrix.

$$\begin{bmatrix} 1 & 0 \\ 0.5 & 1 \end{bmatrix} \quad \overset{O\ K\ L\ M}{\begin{bmatrix} 0 & 0 & 4 & 4 \\ 0 & 4 & 4 & 0 \end{bmatrix}} = \overset{O\ K'\ L'\ M'}{\begin{bmatrix} 0 & 0 & 4 & 4 \\ 0 & 4 & 6 & 2 \end{bmatrix}}$$

trans-   **polygon**   image
formation   **matrix**   matrix
matrix

$$0.5 \cdot 4 + 1 \cdot 0 = 2$$

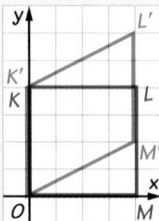

You can use matrix multiplication to reflect a polygon over certain lines, or to rotate a polygon around the origin. Use the appropriate transformation matrix from the tables below.

Matrix	Reflection
$\begin{bmatrix} 1 & 0 \\ 0 & -1 \end{bmatrix}$	reflection over x-axis
$\begin{bmatrix} -1 & 0 \\ 0 & 1 \end{bmatrix}$	reflection over y-axis
$\begin{bmatrix} 0 & 1 \\ 1 & 0 \end{bmatrix}$	reflection over y = x
$\begin{bmatrix} 0 & -1 \\ -1 & 0 \end{bmatrix}$	reflection over y = −x

Matrix	Rotation
$\begin{bmatrix} 0 & -1 \\ 1 & 0 \end{bmatrix}$	90° rotation
$\begin{bmatrix} -1 & 0 \\ 0 & -1 \end{bmatrix}$	180° rotation
$\begin{bmatrix} 0 & 1 \\ -1 & 0 \end{bmatrix}$	270° rotation
$\begin{bmatrix} 1 & 0 \\ 0 & 1 \end{bmatrix}$	360° rotation

These rotations are measured in degrees counterclockwise.

For example, to rotate $\triangle XYZ$ 90° around the origin, multiply by the appropriate transformation matrix.

$$\begin{bmatrix} 0 & -1 \\ 1 & 0 \end{bmatrix} \quad \overset{X\ Y\ Z}{\begin{bmatrix} 4 & 4 & 0 \\ 1 & 3 & 4 \end{bmatrix}} = \overset{X'\ Y'\ Z'}{\begin{bmatrix} -1 & -3 & -4 \\ 4 & 4 & 0 \end{bmatrix}}$$

90°   **polygon**   image
rotation   **matrix**   matrix

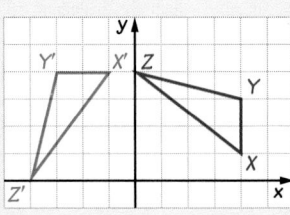

Progress Check 12.3–12.5

1. Find the values of the variables in this matrix product. *(Section 12.3)*

$$\begin{bmatrix} 3 & 2 \\ -1 & 4 \end{bmatrix} \begin{bmatrix} 5 \\ m \end{bmatrix} = \begin{bmatrix} k \\ -9 \end{bmatrix}$$

$m = -1,\ k = 13$

Use matrix multiplication to find the image of $\triangle LMN$ after each transformation. Then graph the image. *(Sections 12.4 and 12.5)*

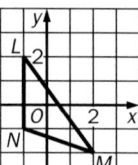

2. reflection over the x-axis

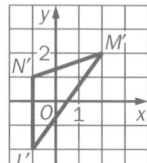

3. 180° rotation

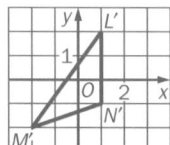

Write a matrix equation to represent each transformation of $\triangle LMN$. *(Sections 12.4 and 12.5)*

4. reflection over the line $y = -x$

$$\begin{bmatrix} 0 & -1 \\ -1 & 0 \end{bmatrix} \begin{bmatrix} -1 & 2 & -1 \\ 2 & -2 & -1 \end{bmatrix} =$$

$$\begin{bmatrix} -2 & 2 & 1 \\ 1 & -2 & 1 \end{bmatrix}$$

5. 270° rotation

$$\begin{bmatrix} 0 & 1 \\ -1 & 0 \end{bmatrix} \begin{bmatrix} -1 & 2 & -1 \\ 2 & -2 & -1 \end{bmatrix} =$$

$$\begin{bmatrix} 2 & -2 & 1 \\ 1 & -2 & 1 \end{bmatrix}$$

Chapter 12 Assessment
Form A Chapter Test

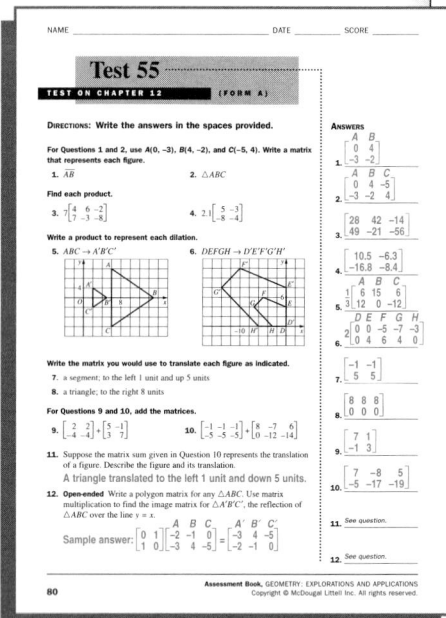

Chapter 12 Assessment
Form B Chapter Test

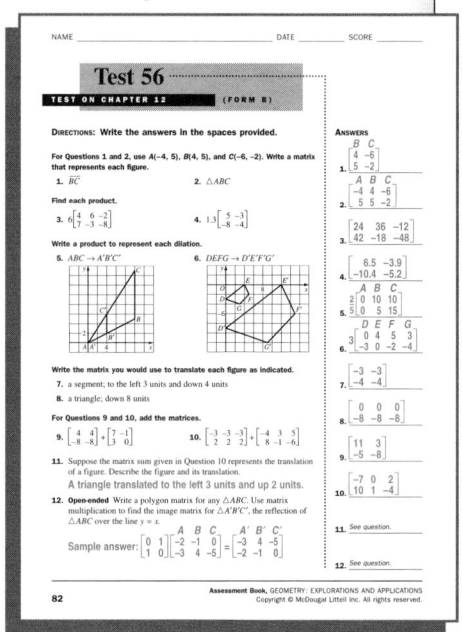

12 | Assessment

VOCABULARY QUESTIONS

1. Give an example of a matrix and state its dimensions.

2. Give an example of scalar multiplication.

3. Give an example of matrix addition.

SECTIONS 12.1 *and* 12.2

Write a matrix that represents each figure.

4. B

5. E

6. $\overline{BC}$

7. $\overline{DE}$

8. $\triangle ABC$

9. pentagon $ABCDE$

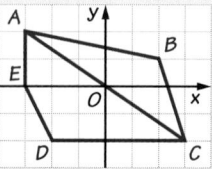

Find each product.

10. $\dfrac{5}{3}\begin{bmatrix} -6 & 3 & 15 \\ 0 & 9 & -3 \end{bmatrix}$

11. $8\begin{bmatrix} 5 & -2 \\ 3 & 9 \end{bmatrix}$

12. $0\begin{bmatrix} 1 & 8 & -5 \\ 1 & 4 & -9 \end{bmatrix}$

Add or subtract the matrices.

13. $\begin{bmatrix} -2 \\ -5 \end{bmatrix} - \begin{bmatrix} 13 \\ -4 \end{bmatrix}$

14. $\begin{bmatrix} -4 & -4 & -4 \\ 0 & 0 & 0 \end{bmatrix} + \begin{bmatrix} 1 & 2 & 3 \\ 1 & -2 & 2 \end{bmatrix}$

15. The matrix $\begin{bmatrix} -2 & 1 & 3 & 1 \\ 0 & 2 & 0 & -2 \end{bmatrix}$ represents quadrilateral $A'B'C'D'$, the image of quadrilateral $ABCD$ after a translation to the right 5 units and down 2 units. Find the matrix for $ABCD$.

The red polygon in each diagram is the image of the blue polygon after a translation. Write each translation as a matrix sum.

16.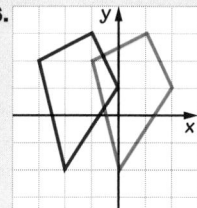

17.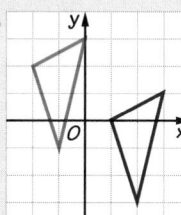

18.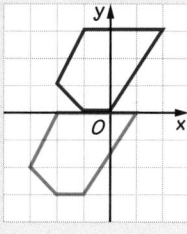

654 Chapter 12 *Coordinates for Transformations*

ANSWERS   Chapter 12

Assessment

1. $\begin{bmatrix} 3 & 2 & 1 \\ 0 & 5 & 1 \end{bmatrix}$; 2×3

2. $3 \cdot \begin{bmatrix} 1 & 6 \\ 5 & 2 \end{bmatrix} = \begin{bmatrix} 3 & 18 \\ 15 & 6 \end{bmatrix}$

3. $\begin{bmatrix} 1 & 1 \\ 2 & 3 \end{bmatrix} + \begin{bmatrix} 2 & 1 \\ 0 & 1 \end{bmatrix} = \begin{bmatrix} 3 & 2 \\ 2 & 4 \end{bmatrix}$

4. $\begin{bmatrix} 2 \\ 1 \end{bmatrix}$

5. $\begin{bmatrix} -3 \\ 0 \end{bmatrix}$

6. $\begin{bmatrix} 2 & 3 \\ 1 & -2 \end{bmatrix}$

7. $\begin{bmatrix} -2 & -3 \\ -2 & 0 \end{bmatrix}$

8. $\begin{bmatrix} -3 & 2 & 3 \\ 2 & 1 & -2 \end{bmatrix}$

9. $\begin{bmatrix} -3 & 2 & 3 & -2 & -3 \\ 2 & 1 & -2 & -2 & 0 \end{bmatrix}$

10. $\begin{bmatrix} -10 & 5 & 25 \\ 0 & 15 & -5 \end{bmatrix}$

11. $\begin{bmatrix} 40 & -16 \\ 24 & 72 \end{bmatrix}$

12. $\begin{bmatrix} 0 & 0 & 0 \\ 0 & 0 & 0 \end{bmatrix}$

13. $\begin{bmatrix} 11 \\ -9 \end{bmatrix}$

14. $\begin{bmatrix} -3 & -2 & -1 \\ 1 & -2 & 2 \end{bmatrix}$

15. $\begin{bmatrix} -7 & -4 & -2 & -4 \\ 2 & 4 & 2 & 0 \end{bmatrix}$

16. $\begin{bmatrix} 2 & 2 & 2 & 2 \\ 0 & 0 & 0 & 0 \end{bmatrix} + \begin{bmatrix} -1 & 0 & -2 & -3 \\ 3 & 1 & -2 & 2 \end{bmatrix}$

17. $\begin{bmatrix} -3 & -3 & -3 \\ 2 & 2 & 2 \end{bmatrix} + \begin{bmatrix} 3 & 2 & 1 \\ 1 & -3 & 0 \end{bmatrix}$

18. $\begin{bmatrix} -1 & -1 & -1 & -1 & -1 \\ -3 & -3 & -3 & -3 & -3 \end{bmatrix} +$
$\begin{bmatrix} 0 & -1 & -1 & -1 & 2 \\ 0 & 0 & 1 & 3 & 3 \end{bmatrix}$

19. $\begin{bmatrix} 3 & 18 \\ -9 & -38 \end{bmatrix}$

20. $\begin{bmatrix} 5 & 0 & -5 & -1 & 3 \\ -20 & 8 & 12 & -20 & -20 \end{bmatrix}$

21. $\begin{bmatrix} -3 & -2 & 2 & 1 \\ -8 & 0 & 7 & -2 \end{bmatrix}$

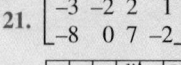

22. a. $\begin{bmatrix} -1 & 4 \\ -2 & 1 \end{bmatrix}$

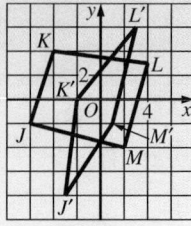

654

Complete each matrix product.

19. $\begin{bmatrix} 0 & -3 \\ -2 & 5 \end{bmatrix}\begin{bmatrix} 2 & 4 \\ -1 & -6 \end{bmatrix}$ **20.** $\begin{bmatrix} -1 & 0 \\ 4 & 4 \end{bmatrix}\begin{bmatrix} -5 & 0 & 5 & 1 & -3 \\ 0 & 2 & -2 & -6 & -2 \end{bmatrix}$

21. The matrix $\begin{bmatrix} -6 & -4 & 4 & 2 \\ -2 & 4 & 3 & -4 \end{bmatrix}$ represents quadrilateral *JKLM*. Use the

matrix $\begin{bmatrix} 0.5 & 0 \\ 1 & 1 \end{bmatrix}$ to transform *JKLM*. Graph *JKLM* and its image.

22. Use the diagram at the right.

   a. Write a matrix for $\overline{AB}$.

   b. Multiply the matrix for $\overline{AB}$ by the

   transformation matrix $\begin{bmatrix} 0 & 1 \\ 1 & 0 \end{bmatrix}$ to find

   a matrix for $\overline{A'B'}$.

   c. Graph $\overline{AB}$ and its image. Describe the transformation caused by the matrix in part (b).

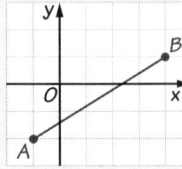

23. Open-ended Problem Sketch a polygon on graph paper.

   a. Create a design by reflecting or rotating the polygon more than once.

   b. Writing Explain how a computer could draw your design using the coordinates of the polygon's vertices and transformations.

Use each matrix given to transform quadrilateral *PRST*. Then graph the image of *PRST* and describe the transformation.

24. $\begin{bmatrix} 1 & 0 \\ 0 & -1 \end{bmatrix}$ **25.** $\begin{bmatrix} 0 & -1 \\ -1 & 0 \end{bmatrix}$

26. $\begin{bmatrix} -1 & 0 \\ 0 & -1 \end{bmatrix}$ **27.** $\begin{bmatrix} 1 & 0 \\ 0 & 1 \end{bmatrix}$

28. $\begin{bmatrix} 0 & -1 \\ 1 & 0 \end{bmatrix}$ **29.** $\begin{bmatrix} 0 & 1 \\ -1 & 0 \end{bmatrix}$

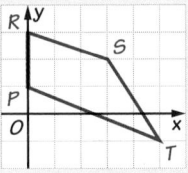

PERFORMANCE TASK

30. Open-ended Problem Write a matrix to represent a polygon. Use matrices to dilate, translate, "morph," reflect, and rotate your polygon. Make a poster displaying your results. Include matrix equations and graphs for all of the transformations.

Assessment   **655**

Chapter 12 Assessment
Form C Alternative Assessment

Chapter 12
ALTERNATIVE ASSESSMENT

1. **Performance Task** Draw a triangle in the first quadrant of a coordinate plane. Using its vertices, write a matrix to represent the triangle. Describe a matrix operation that will triple the lengths of the sides of the triangle. Does this matrix operation also triple the area of the triangle? Prove your answer.

2. **Open-ended Problem** Draw a triangle in the first quadrant of a coordinate plane. Using its vertices, write a matrix to represent the triangle. Describe a procedure involving matrix operations for dilating the triangle.

3. **Open-ended Problem** Draw a triangle in the first quadrant of a coordinate plane. Using its vertices, write a matrix to represent the triangle. Describe a procedure involving matrix operations for reflecting the triangle.

4. **Open-ended Problem** Draw a triangle in the first quadrant of a coordinate plane. Using its vertices, write a matrix to represent the triangle. Describe a procedure involving matrix operations for rotating the triangle.

5. **Open-ended Problem** Draw a triangle in the first quadrant of a coordinate plane. Using its vertices, write a matrix to represent the triangle. Describe a procedure involving matrix operations for translating the triangle.

6. **Performance Task** Draw an isosceles triangle in the first quadrant of a coordinate plane.

   a. Using its vertices, write a matrix to represent the triangle. Multiply the matrix by $\begin{bmatrix} 1 & 0 \\ 0 & 1 \end{bmatrix}$ and describe the new matrix. If the triangle modeled by the new matrix is graphed, how will it compare to the original triangle?

   b. If one or both of the 1s in the matrix $\begin{bmatrix} 1 & 0 \\ 0 & 1 \end{bmatrix}$ is negative, how would the results of part a change?

Assessment Book, GEOMETRY: EXPLORATIONS AND APPLICATIONS
Copyright © McDougal Littell Inc. All rights reserved.   107

---

b. $\begin{bmatrix} -2 & 1 \\ -1 & 4 \end{bmatrix}$

c. a reflection over $y = x$

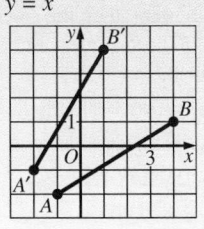

23. Check students' work.

24. $\begin{bmatrix} 0 & 0 & 3 & 5 \\ -1 & -3 & -2 & 1 \end{bmatrix}$

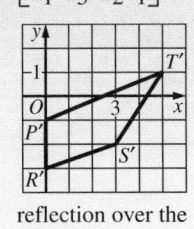

reflection over the *x*-axis

25. $\begin{bmatrix} -1 & -3 & -2 & 1 \\ 0 & 0 & -3 & -5 \end{bmatrix}$

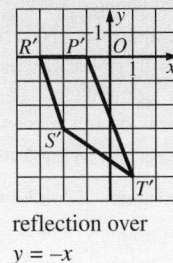

reflection over $y = -x$

26. $\begin{bmatrix} 0 & 0 & -3 & -5 \\ -1 & -3 & -2 & 1 \end{bmatrix}$

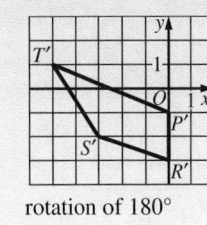

rotation of 180° about the origin

27. $\begin{bmatrix} 0 & 0 & 3 & 5 \\ 1 & 3 & 2 & -1 \end{bmatrix}$

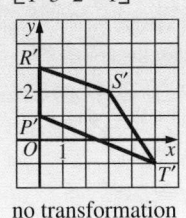

no transformation

28. See answers in back of book.

29. $\begin{bmatrix} 1 & 3 & 2 & -1 \\ 0 & 0 & -3 & -5 \end{bmatrix}$

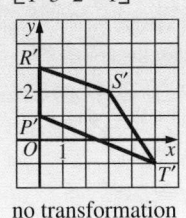

rotation of 270° about the origin

30. Answers may vary.

655

Cumulative Assessment

CHAPTERS $10-12$

CHAPTER 10

For Questions 1–3, use the diagram. Complete each statement.

1. $\triangle JKL \sim \triangle \underline{?} \sim \triangle \underline{?}$.

2. If $LM = 9$ and $MJ = 24$, then $KL = \underline{?}$.

3. If $JK = 15$ and $KL = 9$, then $m\angle J \approx \underline{?}$ and $m\angle L \approx \underline{?}$.

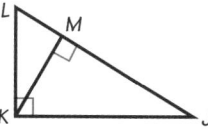

4. **Writing** Explain why any triangle with side lengths in the ratio $1:\sqrt{3}:2$ must be a 30-60-90 triangle. (*Hint:* Use the Converse of the Pythagorean Theorem and trigonometry.)

5. The perimeter of a rhombus is 80 cm, and one diagonal is 14 cm long.
   a. Find the measures of the four angles of the rhombus. Give your answers to the nearest tenth of a degree.
   b. Find the area of the rhombus.

6. If $\overrightarrow{AB} = (-10, -8)$, graph $-\frac{1}{2}\overrightarrow{AB}$ and express it in component form.

7. **Open-ended Problem** Give two vectors whose sum is $(0, 0)$. Make a conjecture about the components and the magnitudes of any two vectors whose sum is $(0, 0)$.

8. Find the area of a regular decagon whose sides are length 8 in. long.

Find the surface area and volume of the prism, right cone, and regular pyramid.

9.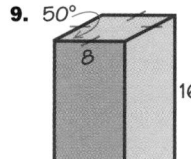

10.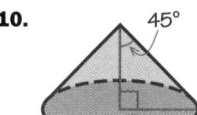

11.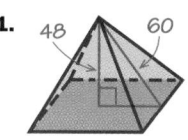

CHAPTER 11

12. **Writing** In $\odot O$, $m\overarc{AB} = 100°$ and $m\overarc{BC} = 100°$. What type of triangle is $\triangle ABC$? Find the measure of each angle of the triangle. Explain your reasoning.

Cumulative Assessment

1. *JMK, KML*

2. $3\sqrt{33} \approx 17.2$

3. $31°; 59°$

4. Let the sides have lengths x, y, and z. Then $\frac{y}{x} = \frac{\sqrt{3}}{1}$, so $y = \sqrt{3}x$. Also, $\frac{z}{x} = \frac{2}{1}$, so $z = 2x$. $x^2 + y^2 = x^2 + (\sqrt{3}x)^2 = 4x^2 = z^2$, so the triangle is a right triangle by the converse of the Pythagorean theorem. If the acute angle opposite the side of length x is $\angle A$, then $\sin A = \frac{x}{2x} = 0.5$; $\angle A = \sin^{-1}(0.5) = 30°$. Thus, the other acute angle has measure $90° - 30° = 60°$. The triangle is a 30-60-90 triangle.

5. a. $41.0°, 139.0°, 41.0°, 139.0°$
   b. about 262 cm^2

6.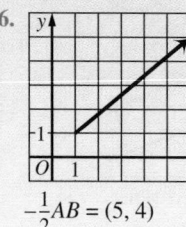
   $-\frac{1}{2}AB = (5, 4)$

7. Answers may vary. An example is given. $(-4, 3)$ and $(4, -3)$; My conjecture is that the horizontal components of two such vectors are opposites, as are the vertical components. The magnitudes of two such vectors are equal.

8. about 492 in.^{2}

9. about 610; about 784

10. about 372 in.^{2}; about 359 in.^{3}

11. 13,824; 82,944

12. isosceles; $50°, 50°,$ and $80°$; The measure of an inscribed angle is half the intercepted arc. If two angles of a triangle are congruent, then the sides opposite those angles are congruent.

13. $x = 90; y = 90$

14. $a = 2\sqrt{2} \approx 2.83$; $b = 4 - 2\sqrt{2} \approx 1.17$

15. $t = 7\sqrt{2} \approx 9.90$

656

ALGEBRA Find the value of each variable.

13.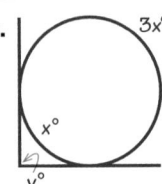

14.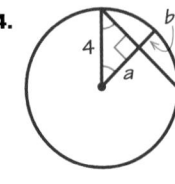

15.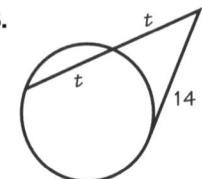

16. Open-ended Problem Two chords of a circle intersect to form a 105° angle. Find a pair of possible measures for the intercepted arcs.

17. In $\odot O$, $m\widehat{RST} = 200°$ and $OR = 18$. Find the lengths of $\widehat{RT}$ and of $\widehat{RST}$. Also find the area of the sector with central angle $\angle ROT$.

18. SPORTS The diameter of a tennis ball is 2.6 in. and the radius of a baseball is 1.4 in. What is the ratio of their volumes? of their surface areas?

19. Writing Explain why the following statement from Euclidean geometry is not true for spherical geometry: If two lines intersect, then they intersect in exactly one point.

CHAPTER 12

20. Open-ended Problem Draw any $\triangle ABC$ and label the coordinates of its vertices. Write a matrix that represents $\triangle ABC$.

    a. Choose any scale factor. Write a matrix to represent the image of $\triangle ABC$ after a dilation with your scale factor and center at $(0, 0)$.

    b. Choose any translation. Write a matrix to represent the image of $\triangle ABC$ after your translation.

The vertices of quadrilateral *PQRS* are *P*(4, 0), *Q*(−1, 3), *R*(−3, −1), and *S*(0, −2).

21. Write the polygon matrix for quadrilateral $PQRS$.

22. Multiply the polygon matrix for $PQRS$ by the transformation matrix $\begin{bmatrix} 1 & 1 \\ 1 & 1 \end{bmatrix}$. Graph $PQRS$ and its image matrix. Describe the effect of the transformation matrix.

23. Use matrix multiplication to find the image of $PQRS$ after a rotation of 90°. Graph $PQRS$ and its image. Use the table on page 642.

24. Use matrix multiplication to find the image of $PQRS$ after reflection over the line $y = -x$. Use the table on page 635.

25. Writing Name at least two types of transformations that *cannot* be represented by multiplying a polygon matrix by a 2×2 transformation matrix. Explain your reasoning.

Cumulative Assessment   **657**

16. Answers may vary. An example is given. 80° and 130°

17. $16\pi \approx 50.3$; $20\pi \approx 62.8$; $144\pi \approx 452$

18. $2197 : 2744$; $169 : 196$

19. On a sphere, a line is a great circle. Two great circles intersect in two points.

20. Answers may vary. An example is given, using points $P(5, 1)$, $Q(2, -3)$, and $R(-4, 6)$ as the vertices of the triangle. The matrix representing the triangle is $\begin{bmatrix} 5 & 2 & -4 \\ 1 & -3 & 6 \end{bmatrix}$.

  a. Using a scale factor of 3, the resulting matrix is $\begin{bmatrix} 15 & 6 & -12 \\ 3 & -9 & 18 \end{bmatrix}$.

  b. Using the translation matrix $\begin{bmatrix} 1 & 1 & 1 \\ -5 & -5 & -5 \end{bmatrix}$, the resulting matrix is $\begin{bmatrix} 6 & 3 & -3 \\ -4 & -8 & 1 \end{bmatrix}$.

21. $\begin{bmatrix} 4 & -1 & -3 & 0 \\ 0 & 3 & -1 & -2 \end{bmatrix}$

22. $\begin{bmatrix} 4 & 2 & -4 & -2 \\ 4 & 2 & -4 & -2 \end{bmatrix}$

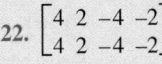

The transformation matrix transforms quadrilateral $PQRS$ into four collinear points, that is, points that lie on the line $y = x$.

23. $\begin{bmatrix} 0 & -3 & 1 & 2 \\ 4 & -1 & -3 & 0 \end{bmatrix}$

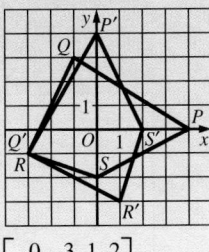

24. $\begin{bmatrix} 0 & -3 & 1 & 2 \\ -4 & 1 & 3 & 0 \end{bmatrix}$

25. translations and dilations; Translations require matrix addition and dilations require multiplication by a scalar.

657

Contents of Student Resources

Extra Practice 659–686

Toolbox 687–711

Using Geometric Tools and Transformations 687
Operations with Variable Expressions 690
Linear Equations 694
Graphing on the Coordinate Plane 696
Quadratic Equations 700
Formulas for Geometric Figures 702
Inequalities 705
Ratio and Proportion 707
Data Analysis and Probability 708
Matrices 711

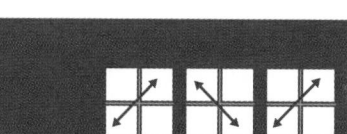

Postulates, Properties, Theorems, and Constructions 712–720

List of Postulates and Properties 712
List of Theorems 714
List of Constructions 720

Technology Handbook 721–729

Using a Graphing Calculator 721
Using a Spreadsheet 725
Using Geometry Software 726

Tables 730–733

Table of Measures 730
Table of Symbols 731
Table of Squares and Square Roots 732
Table of Trigonometric Ratios 733

Appendices 734–741

Appendix 1: A Brief History of Geometric Systems 734
Appendix 2: Truth Tables and Logic 736

Glossary 742–747

Index 748–758

Selected Answers

Extra Practice

CHAPTER 1

Use inductive reasoning to find the next two numbers in each pattern. *Section 1.1*

1. $-2, -5, -8, -11, \underline{\ ?\ }, \underline{\ ?\ }$

2. $0, 2, 5, 9, \underline{\ ?\ }, \underline{\ ?\ }$

3. $0, 1, 4, 9, \underline{\ ?\ }, \underline{\ ?\ }$

4. $1, \frac{1}{2}, \frac{2}{3}, \frac{3}{4}, \underline{\ ?\ }, \underline{\ ?\ }$

5. $7, -14, 28, -56, \underline{\ ?\ }, \underline{\ ?\ }$

6. $2, 7, 12, 17, \underline{\ ?\ }, \underline{\ ?\ }$

Use inductive reasoning to sketch the next shape in each pattern. *Section 1.1*

7.

8.
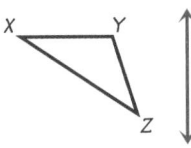

Write a formula for the value of the *n*th term in each pattern. *Section 1.1*

9.

Term	1	2	3	4	5	6	...	n
Value	2	1	0	1	2	3	...	?

10.

Term	1	2	3	4	5	6	...	n
Value	2	5	8	11	14	17	...	?

Perform each transformation. *Section 1.2*

11. Reflect the triangle over the line.

12. Rotate the triangle around point C by a half-turn.

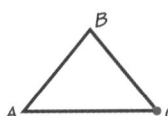

13. Translate the triangle any distance to the right.

14. Reflect the triangle over the line.

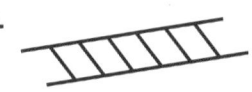

Describe the symmetry of each figure. *Section 1.2*

15.
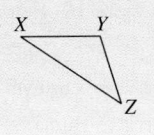

16.
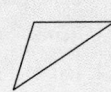

17.
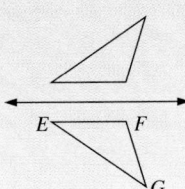

 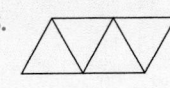
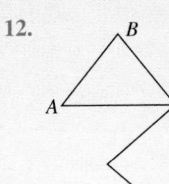
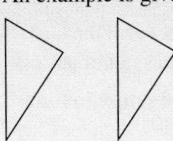

Rewrite each statement using _if_ and _then_. *Section 1.3*

18. Every square is a rectangle.

19. My car runs out of gas when I forget to fill the tank.

20. Two lines are perpendicular if they meet to form right angles.

Identify the hypothesis and the conclusion of each conditional statement. *Section 1.3*

21. $2x - 5 = 15$ if $x = 10$.   **22.** If the weather gets cold, then birds fly south.

Tell whether each statement is _True_ or _False_. If it is false, give a counterexample.
Section 1.3

23. If it is 6:00 P.M. in California, then it is 9:00 P.M. in Maryland.

24. If a banana is ripe, then it is yellow.

25. If Carmine is in Arizona, then he is in Tucson.

**Use the diagram at the right, showing plane _ABC_
parallel to plane _FGH_.** *Section 1.4*

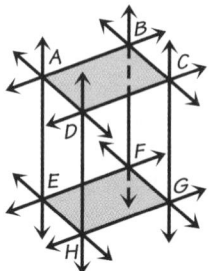

26. What is the intersection of $\overleftrightarrow{AB}$ and $\overleftrightarrow{FB}$?

27. What is the intersection of $\overleftrightarrow{BF}$ and $\overleftrightarrow{GF}$?

28. Name three lines parallel to $\overleftrightarrow{DH}$.

29. Name a line through point H that is skew to $\overleftrightarrow{FG}$.

30. What is the intersection of planes ABC and CGF?

31. Name two planes that intersect in $\overleftrightarrow{AE}$.

Give the length of each segment. *Section 1.5*

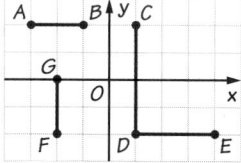

32. $\overline{AB}$ **33.** $\overline{CD}$

34. $\overline{DE}$ **35.** $\overline{FG}$

Give the length of each segment. *Section 1.5*

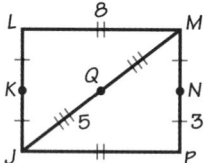

36. $\overline{JP}$ **37.** $\overline{QM}$

38. $\overline{MN}$ **39.** $\overline{JM}$

40. $\overline{MP}$ **41.** $\overline{LJ}$

Find the measure of each angle. *Section 1.6*

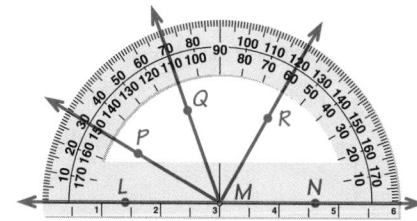

42. $\angle LMP$ **43.** $\angle LMQ$

44. $\angle PMQ$ **45.** $\angle QMN$

46. $\angle PMN$ **47.** $\angle PMR$

660   Extra Practice

---

18. If a figure is a square, then it is a rectangle.

19. If I forget to fill the tank, then my car runs out of gas.

20. If two lines meet to form right angles, then they are perpendicular.

21. hypothesis: the weather gets cold
conclusion: birds fly south

22. hypothesis: $x = 10$
conclusion: $2x - 5 = 15$

23. True.

24. False; a ripe banana is brown.

25. False; Carmine is in Huma.

26. B

27. F

28. $\overleftrightarrow{AE}, \overleftrightarrow{BF}, \overleftrightarrow{CG}$

29. $\overleftrightarrow{DH}$

30. $\overleftrightarrow{BC}$

31. AEH and AEF

32. 2   **33.** 4

34. 3   **35.** 2

36. 8   **37.** 5

38. 3   **39.** 10

40. 6   **41.** 6

42. 30°   **43.** 70°

44. 40°   **45.** 110°

46. 150°   **47.** 90°

Tell whether each statement is *True* or *False*. *Section 1.7*

48. *C* is the midpoint of $\overline{AD}$.

49. *C* is the midpoint of $\overline{BD}$.

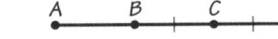

50. If $BC = 3$, then $CD = 6$.

51. If $BD = 10$, then $BC = 5$.

52. If $BC = 10$, then $AD = 30$.

In the diagram, $\overline{PQ}$ and $\overline{RS}$ bisect each other. *Section 1.7*

53. If $RT = 3$, find the length of $\overline{RS}$.

54. If $PT = 7$, find the length of $\overline{TQ}$.

55. If $PT = 3x + 1$ and $TQ = 4x - 1$, find the length of $\overline{PQ}$.

56. If $TR = 2x + 2$ and $TS = 3x - 3$, find the length of $\overline{RT}$.

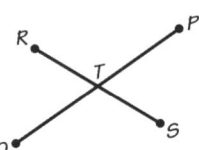

Find the midpoint of each segment. *Section 1.7*

57. $\overline{AB}$

58. $\overline{CD}$

59. $\overline{EF}$

60. $\overline{GH}$

61. $\overline{JK}$

62. $\overline{LM}$

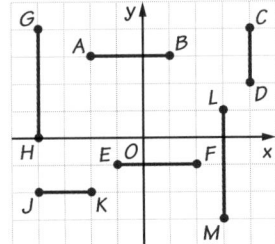

CHAPTER 2

Complete each statement about the diagram at the right. *Section 2.1*

1. $\angle EFG$ and $\underline{\ ?\ }$ are vertical angles.

2. $\angle EGF$ and $\underline{\ ?\ }$ are congruent adjacent angles.

3. $\angle FGJ$ and $\underline{\ ?\ }$ are complementary angles.

4. $\angle KGJ$ and $\underline{\ ?\ }$ are supplementary angles.

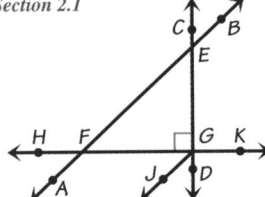

Find the measure of each angle in the diagram at the right. *Section 2.1*

5. $\angle CGH$

6. $\angle HCD$

7. $\angle HBC$

8. $\angle GCB$

9. $\angle GHC$

10. $\angle GCF$

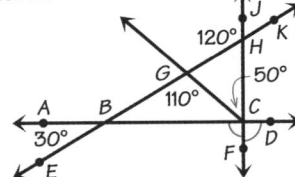

Extra Practice **661**

Find each unknown angle measure. *Section 2.2*

11.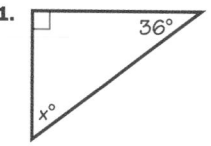

12.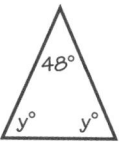

13.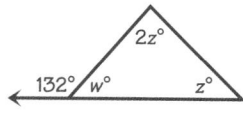

14.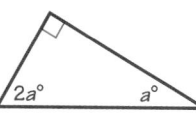

15.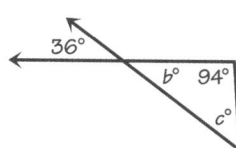

16.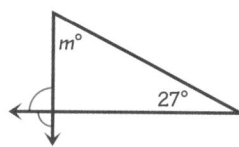

17.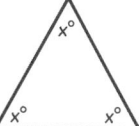

18.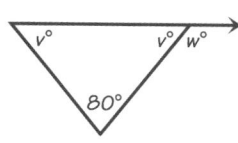

19.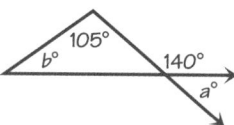

Classify each polygon. Be as specific as possible. *Section 2.3*

20.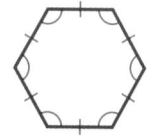

21.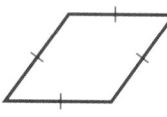

22.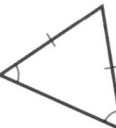

23.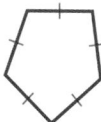

24.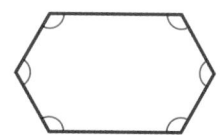

25.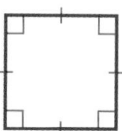

Find the sum of the measures of the interior angles of each polygon. *Section 2.4*

26. hexagon

27. 9-gon

28. 22-gon

Find each unknown angle measure. *Section 2.4*

29.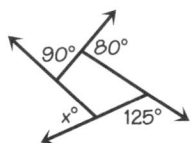

30.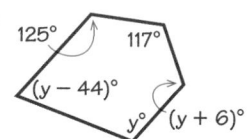

31.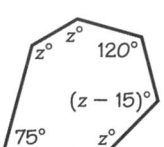

Find the measure of each interior and exterior angle of the polygon. *Section 2.4*

32. regular octagon

33. regular 10-gon

34. regular 28-gon

662 Extra Practice

11. $x = 54$

12. $y = 66$

13. $w = 48$, $z = 44$, $2z = 88$

14. $a = 30$, $2a = 60$

15. $b = 36$, $c = 50$

16. $m = 63$

17. $x = 60$

18. $v = 50$, $w = 130$

19. $a = 40$, $b = 35$

20. regular hexagon

21. equilateral quadrilateral

22. isosceles triangle

23. equilateral pentagon

24. equiangular hexagon

25. regular quadrilateral or square

26. 720°

27. 1260°

28. 3600°

29. $x° = 65°$

30. $y° = 112°$, $(y + 6)° = 118°$; $(y - 44)° = 68°$

31. $z° = 135°$; $(z - 15)° = 120°$

32. 135°; 45°

33. 144°; 36°

34. about 167°; about 13°

35. a. 8 cm

　　b. 8 cm

　　c. 125°

662

Find each length or angle measure. *Section 2.5*

35. rhombus *ABCD*
   a. *AB*
   b. *AD*
   c. *m∠C*

36. rectangle *KLMN*
   a. *NL*
   b. *ML*
   c. *m∠KLM*

37. parallelogram *PQRS*
   a. *m∠SQR*
   b. *m∠SPQ*
   c. *m∠SPT*
   d. *m∠QSR*

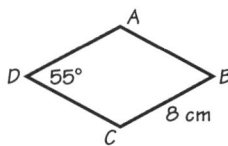 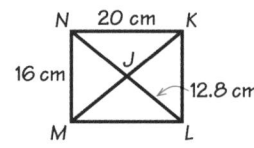 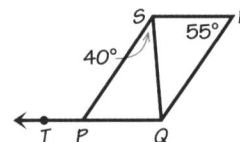

Find each length or angle measure of rectangle *ABCD*. *Section 2.5*

38. *AB*
39. *AE*
40. *m∠ADE*
41. *m∠ADC*
42. *m∠AEB*
43. *m∠EAB*

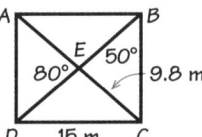

Use the prism at the right. *Section 2.6*

44. Name the two bases of the prism.

45. What type of prism is it?

46. How many lateral faces does the prism have?

47. How many edges meet at vertex *A*?

48. How many edges does the prism have?

49. How many vertices does the prism have?

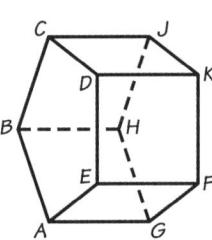

CHAPTER 3

Tell whether each argument uses *inductive* or *deductive* reasoning. *Section 3.1*

1. Rainbows only appear after a rain. It just stopped raining, so there will be a rainbow.

2. Vertical angles are congruent. ∠3 and ∠4 are vertical angles, so ∠3 ≅ ∠4.

3. All squares have four right angles. Polygon *ABCD* is a square, so *ABCD* has four right angles.

4. African violets need special light conditions. My violets are not growing well, so I don't have the correct light conditions.

Use inductive reasoning to predict the next number in each pattern. Justify your prediction. *Section 3.1*

5. 5, 8, 11, 14, _?_

6. 20, 27, 36, 47, _?_

7. $3, 1, \frac{1}{3}, \frac{1}{9}, $ _?_

---

36. a. 25.6 cm
   b. 20 cm
   c. 90°
37. a. 40°
   b. 55°
   c. 125°
   d. 85°
38. 15 m
39. 9.8 m
40. 50°

41. 90°
42. 100°
43. 40°
44. pentagon *ABCDE* and pentagon *FGHJK*
45. pentagonal prism
46. 5
47. 3
48. 15
49. 10

Chapter 3
1. inductive
2. deductive
3. deductive
4. inductive
5. 17; Each term is 3 more than the previous term.
6. 60; Each term is 7, 9, 11, 13, … more than the previous term.
7. $\frac{1}{27}$; Each term is $\frac{1}{3}$ of the previous term.

Identify the property that makes each statement true. *Section 3.2*

8. If $AB = CD$, then $CD = AB$.

9. $m\angle WXY = m\angle WXY$

10. If $XY = CD$ and $CD = 2$, then $XY = 2$.

For Exercises 11–16, give the postulate, definition, property, or previous statement that makes the statement about the diagram true. *Section 3.2*

11. $PQ = ST$ **12.** $2PQ = 2ST$

13. $QR = RS$ **14.** $QR = QR$

15. $PR = RT$ **16.** $3QR = 3RS$

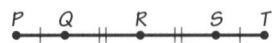

Draw an obtuse triangle, an acute triangle, and a right triangle. Tell whether each statement is *True* or *False*. *Section 3.3*

17. All six exterior angles of an obtuse triangle are obtuse.

18. Both exterior angles of one angle of a triangle may be obtuse.

Find the measure of each indicated angle. *Section 3.4*

19. $m\angle NPQ$ **20.** $m\angle BCD$ **21.** $m\angle WYZ$

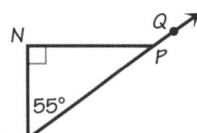   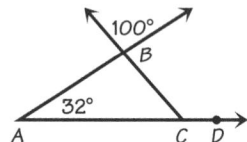   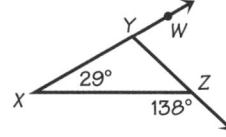

Rewrite each statement so that it is always true. *Section 3.4*

22. Angles that are complementary to the same angle are complementary.

23. Supplementary angles are angles in a linear pair.

24. Complete the proof. *Section 3.4*

Given: $m\angle 2 = m\angle 3$

Prove: $m\angle 1 = m\angle 4$

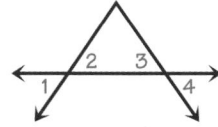

Statements	Reasons
1. $m\angle 2 = m\angle 3$	**1.** _?_
2. $m\angle 2 = m\angle 1$; $m\angle 3 = m\angle 4$	**2.** _?_
3. _?_	**3.** Substitution Property (Step(s) _?_ and _?_)

Write the converse of each statement. Tell whether the converse is *True* or *False*. *Section 3.5*

25. If a wildflower is goldenrod, then it is yellow.

26. If a woman lives in Massachusetts, then she lives in Boston.

27. If it rains, I will take my umbrella.

---

8. Symmetric Property

9. Reflexive Property

10. Transitive Property or Substitution Property

11. Definition of congruent segments

12. Addition Property

13. Definition of congruent segments

14. Reflexive Property

15. Segment Addition Postulate; Addition Property

16. Addition Property

17. False.

18. False.

19. 145°

20. 132°

21. 71°

22. Angles that are complementary to the same angle are congruent.

23. Two angles that form a linear pair are supplementary.

24. (1) Given; (2) Def. of vertical angles; (3) $m\angle 1 = m\angle 4$; 1 and 2

25. If a wildflower is yellow, then it is goldenrod. False.

26. If a woman lives in Boston, then she lives in Massachusetts. True.

27. If I take my umbrella, it will rain. False.

a. **Rewrite the statement as a conditional.**   b. **Write the converse of the statement.**

28. The opposite sides of a rectangle are congruent.

29. The complements of congruent angles are congruent.

30. The diagonals of a rhombus are perpendicular.

Use a calculator to approximate each square root to the nearest hundredth. Then simplify each expression. *Section 3.6*

31. $\sqrt{20}$ **32.** $-\sqrt{144}$ **33.** $\pm\sqrt{32}$

34. $\pm\sqrt{10}$ **35.** $-\sqrt{150}$ **36.** $\sqrt{256}$

Tell whether the given lengths can be the sides of a right triangle. *Section 3.6*

37. 1.5 m, 2 m, 2.5 m   **38.** 6 cm, 8 cm, 12 cm   **39.** 5 in., 12 in., 14 in.

40. $\sqrt{5}, \sqrt{3}, \sqrt{8}$ **41.** 5 ft, 3 ft, 8 ft   **42.** 5.5 cm, 4 cm, 9.5 cm

The legs of a right triangle are *a* and *b* units long. The hypotenuse is *c* units long. Find the unknown length for each right triangle, to the nearest hundredth. *Section 3.6*

43. $a = 12, c = 13$ **44.** $a = 24, b = 7$ **45.** $b = 12, c = 20$

46. $a = 8, c = 12$ **47.** $a = 119, b = 120$ **48.** $a = 4, b = 5$

Tell whether the triangle with sides of the given lengths is *right*, *obtuse*, or *acute*. *Section 3.7*

49. 5, 9, 12   **50.** 13, 15, 17   **51.** 7, 24, 25

52. 7, 24, 26   **53.** 10, 24, 26   **54.** 25, 25, 30

Write the inverse and the contrapositive of each statement. *Section 3.7*

55. If Jo studies hard, she will do well on the test.

56. If the piano is moved, then it will need to be tuned.

57. If Bill was born in France, then he can speak French.

CHAPTER 4

Find the length of the segment with the given endpoints. *Section 4.1*

1. $A(2, -2)$ and $B(-3, -2)$ **2.** $A(5, 8)$ and $C(10, 4)$ **3.** $D(-6, 2)$ and $B(1, -4)$

4. $F(2, 8)$ and $G(-7, 8)$ **5.** $M(-3, 4)$ and $N(3, 1)$ **6.** $R(4, -6)$ and $S(0, 3)$

Find the coordinates of the midpoint of the segment with each given pair of endpoints. *Section 4.1*

7. (2, 3) and (6, 5)   **8.** (5, -2) and (7, -4)   **9.** (-6, 2) and (-2, 8)

10. (-1, 5) and (3, -1)   **11.** (8, 2) and (2, -4)   **12.** (11, -7) and (15, 8)

55. Inverse: If Jo does not study hard, then she will not do well on the test.
Contrapositive: If Jo does not do well on the test, then she does not study hard.

56. Inverse: If the piano is not moved, then it will not need to be tuned.
Contrapositive: If the piano does not need to be tuned, then it is not moved.

57. Inverse: If Bill was not born in France, then he cannot speak French.
Contrapositive: If Bill cannot speak French, then he was not born in France.

Chapter 4

1. 5
2. $\sqrt{41} \approx 6.40$
3. $\sqrt{85} \approx 9.22$
4. 9
5. $3\sqrt{5} \approx 6.71$
6. $\sqrt{97} \approx 9.85$
7. (4, 4)
8. (6, -3)
9. (-4, 5)
10. (1, 2)
11. (5, -1)
12. $\left(13, \frac{1}{2}\right)$

28. a. If a figure is a rectangle, then its opposite sides are congruent.
    b. If the opposite sides of a figure are congruent, then the figure is a rectangle.

29. a. If two angles are congruent, then their complements are congruent.
    b. If the complements of two angles are congruent, then the angles are congruent.

30. a. If a figure is a rhombus, then its diagonals are perpendicular.
    b. If the diagonals of a figure are perpendicular, then the figure is a rhombus.

31. 4.47; $2\sqrt{5}$
32. -12; -12
33. $\pm 5.66; \pm 4\sqrt{2}$
34. $\pm 3.16; \pm\sqrt{10}$
35. -12.25; $-5\sqrt{6}$

36. 16; 16
37. Yes.
38. No.
39. No.
40. Yes.
41. No.
42. No.
43. $b = 5$
44. $c = 25$
45. $a = 16$
46. $b \approx 8.94$
47. $c = 169$
48. $c \approx 6.40$
49. obtuse
50. acute
51. right
52. obtuse
53. right
54. acute

13. $AB = 5$, $BC = 5$, $AC = 8$;
isosceles triangle

14. $AB = 8$, $BC = 5$, $CD = 8$, $DA = 5$;
parallelogram

15. $AB = 2\sqrt{5} \approx 4.47$, $BC = 5$,
$CD = 2\sqrt{5} \approx 4.47$, $DA = 5$;
parallelogram

16. 1

17. –2

18. $\frac{1}{4}$

19. –2

20. $\frac{6}{5}$

21. 3

22. 5

23. 1

24. $\frac{5}{4}$

25. $y = 2x + 1$

26. $y = -3x + 7$

27. $y = \frac{1}{4}x - \frac{11}{4}$

28. $y = -\frac{2}{3}x + 10$

29. $x = 5$

30. $y = -2$

31. $y = \frac{7}{2}x - 20$

32. $y = 6x - 60$

33. perpendicular

34. perpendicular

35. parallel

36. 2

37. 0

38. $\frac{1}{2}$

39. undefined

40. $-\frac{1}{5}$

41. 3

42. No.

43. Yes.

44. Yes.

45. No.

46. $x^2 + y^2 = 36$

47. $(x - 2)^2 + (y - 5)^2 = 16$

48. $(x - 2)^2 + (y + 3)^2 = 1$

49. $(x + 2)^2 + (y + 5)^2 = 20.25$

50.
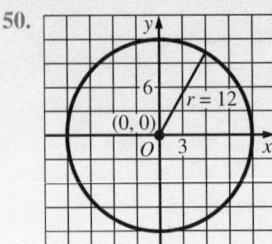

Find the lengths of each of the sides of each polygon whose vertices are given. Give the most specific name for each polygon. *Section 4.1*

13. $A(0, 4)$, $B(3, 0)$, $C(0, -4)$

14. $A(0, 3)$, $B(8, 3)$, $C(5, -1)$, $D(-3, -1)$

15. $A(-1, -4)$, $B(-3, 0)$, $C(0, 4)$, $D(2, 0)$

Find the slope of the line that contains each pair of given points. *Section 4.2*

16. $(2, 3)$ and $(5, 6)$ **17.** $(5, -2)$ and $(6, -4)$ **18.** $(-6, 2)$ and $(-2, 3)$

19. $(-1, 5)$ and $(2, -1)$ **20.** $(8, 2)$ and $(3, -4)$ **21.** $(10, -7)$ and $(15, 8)$

22. $(3, -3)$ and $(5, 7)$ **23.** $(-4, -4)$ and $(3, 3)$ **24.** $(-7, -9)$ and $(1, 1)$

Write an equation for each line. *Section 4.2*

25. contains $(1, 3)$; has slope 2   **26.** contains $(4, -5)$; has slope -3

27. contains $(3, -2)$; has slope $\frac{1}{4}$ **28.** contains $(9, 4)$; has slope $-\frac{2}{3}$

29. the vertical line that contains $(5, -2)$ **30.** the horizontal line that contains $(-4, -2)$

31. contains $(6, 1)$; has slope $\frac{7}{2}$ **32.** contains $(8, -12)$; has slope 6

Tell whether each pair of lines is *parallel, perpendicular,* or *neither*. *Section 4.3*

33. $x = 2$ **34.** $y = \frac{1}{3}x + 2$ **35.** $y = 2$
$\quad\;\; y = 3$ $\quad\;\; y = -3x - 2$ $\quad\;\; y = -\frac{1}{2}$

Find the slope of each line. *Section 4.3*

36. a line parallel to the line $y = 2x - 1$ **37.** a line parallel to the line $y = -2$

38. a line perpendicular to the line $y = -2x + 4$ **39.** a line perpendicular to the line $y = -2$

40. a line perpendicular to the line $y = 5x$ **41.** a line parallel to the line $3x - y = 1$

Use slopes to tell whether each triangle with the given vertices is a right triangle. *Section 4.3*

42. $A(0, 4)$, $B(3, 0)$, $C(0, -4)$ **43.** $C(1, 4)$, $D(3, -1)$, $E(6, 6)$

44. $E(2, 2)$, $F(8, 5)$, $G(6, 9)$ **45.** $J(-2, -5)$, $K(2, -2)$, $L(-2, 1)$

Write an equation of the circle with each given center and radius. *Section 4.4*

46. center $(0, 0)$, radius 6   **47.** center $(2, 5)$, radius 4

48. center $(2, -3)$, radius 1   **49.** center $(-2, -5)$, radius 4.5

Sketch each circle. Label the coordinates of the center and draw a radius and label it with its length. *Section 4.4*

50. $x^2 + y^2 = 144$ **51.** $(x - 2)^2 + (y - 1)^2 = 16$

52. $(x + 1)^2 + (y + 4)^2 = 12$ **53.** $(x - 5)^2 + (y + 6)^2 = 169$

51.
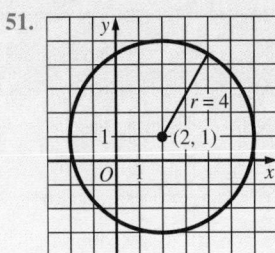

52.
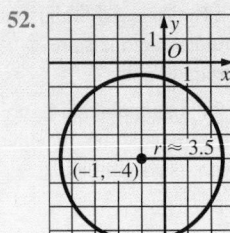

53.
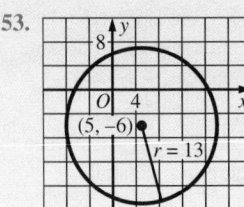

54. Tell whether the point (3, 4) is on the circle with equation $x^2 + y^2 = 5$. *Section 4.4*

Find the missing coordinates without using any new variables. *Section 4.5*

55. parallelogram

56. rectangle

57. isosceles triangle

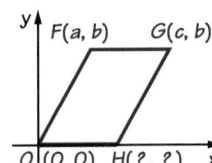

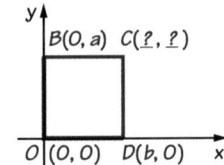

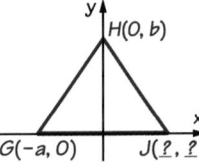

For each △ABO, find the coordinates of the midpoint M of the hypotenuse. Then find MA, MB, and MO. *Section 4.5*

58.

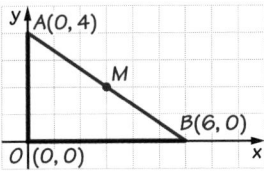

59.

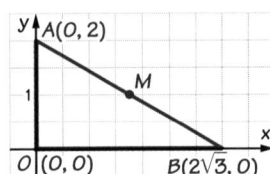

Find the coordinates of the midpoint of the segment with the given endpoints.
Section 4.6

60. (3, 4, −1) and (5, 2, 1)

61. (−2, −3, 6) and (−4, 3, 6)

62. (1, 0.5, 5) and (5, 1.5, −4)

Find the length of the segment with the given endpoints. *Section 4.6*

63. (1, 4, −1) and (5, 1, −1)

64. (3, 4, −1) and (5, 2, 1)

65. (13, 0, 1) and (5, 2, 11)

CHAPTER 5

Classify each pair of angles as *corresponding angles*, *alternate interior angles*, or *same-side interior angles*. *Section 5.1*

1. ∠3 and ∠6

2. ∠3 and ∠7

3. ∠3 and ∠5

4. ∠1 and ∠5

5. ∠5 and ∠4

6. ∠2 and ∠6

7. ∠4 and ∠11

8. ∠6 and ∠14

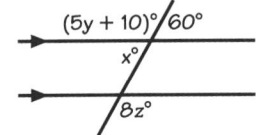

Find the values specified in each diagram. *Section 5.1*

9. Find the values of *x*, *y*, and *z*.

10. Find the value of *w* and *m*∠1.

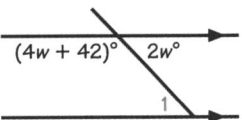

- -

54. No.

55. $H(c - a, 0)$

56. $C(b, a)$

57. $J(a, 0)$

58. $M(3, 2); MA = \sqrt{13} \approx 3.61,$
$MB = \sqrt{13} \approx 3.61,$
$MO = \sqrt{13} \approx 3.61$

59. $M(\sqrt{3}, 1); MA = 2, MB = 2,$
$MO = 2$

60. (4, 3, 0)

61. (−3, 0, 6)

62. $\left(3, 1, \frac{1}{2}\right)$

63. 5

64. $2\sqrt{3} \approx 3.46$

65. $2\sqrt{42} \approx 12.96$

Chapter 5

1. alternate interior angles

2. corresponding angles

3. same-side interior angles

4. corresponding angles

5. alternate interior angles

6. corresponding angles

7. same-side interior angles

8. corresponding angles

9. $x = 60, y = 22, z = 15$

10. $w = 23, m \angle 1 = 46°$

Find the measure of each numbered angle. *Section 5.2*

11.

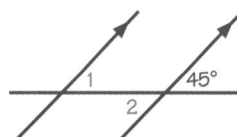

12.

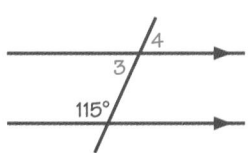

13.

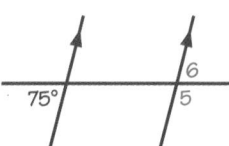

14.

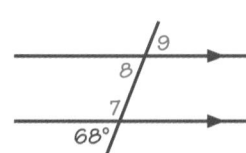

Use trapezoid KLMN. *Section 5.2*

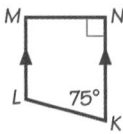

15. Tell which sides are the bases.

16. Which sides are the legs?

17. Find the measure of $\angle L$.

18. Find the measure of $\angle M$.

For Exercises 19–21, copy and complete each proof. *Section 5.3*

19. Given: $\overline{AB} \parallel \overline{CD}$

   Prove: $\angle 1 \cong \angle 3$

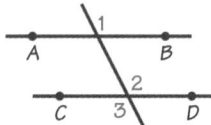

1. $\angle 1 \cong \angle 2$ (If two $\parallel$ lines are intersected by a transversal, then _?_ are $\cong$.)
2. _?_ (Vertical angles are congruent.)
3. Substitution _?_

20. Given: $j \parallel k$; $m \parallel n$

   Prove: $\angle 1 \cong \angle 4$

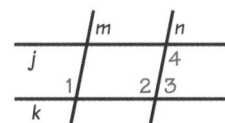

Statements	Reasons
1. $m \parallel n$	1. Given
2. $\angle 1 \cong \angle 2$	2. If two $\parallel$ lines are intersected by a transversal, then _?_ angles are congruent.
3. $j \parallel k$	3. _?_
4. $\angle 2 \cong \angle 4$	4. If two $\parallel$ lines are intersected by a transversal, then _?_ angles are congruent.
5. $\angle 1 \cong \angle 4$	5. _?_

Extra Practice

---

11. $m \angle 1 = 45°$, $m \angle 2 = 45°$

12. $m \angle 3 = 65°$, $m \angle 4 = 65°$

13. $m \angle 5 = 105°$, $m \angle 6 = 75°$

14. $m \angle 7 = 112°$, $m \angle 8 = 68°$, $m \angle 9 = 68°$

15. $\overline{LM}$ and $\overline{KN}$

16. $\overline{MN}$ and $\overline{LK}$

17. 105°

18. 90°

19. a. corresponding angles

   b. $\angle 2 \cong \angle 3$

   c. $\angle 1 \cong \angle 3$

20. (2) corresponding angles; (3) Given; (4) alternate interior angles; (5) Transitive Property

21. Given: $\overline{MN} \parallel \overline{BC}$; $\angle B \cong \angle C$

Prove: $\angle MKB \cong \angle NKC$

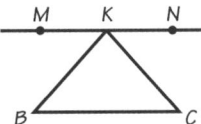

Because $\overline{MN} \parallel \overline{BC}$, $\angle MKB \cong \angle B$ and $\angle NKC \cong \angle C$, since they are
_?_ angles. Because $\angle B \cong \angle C$, it must be true that $\angle MKB \cong \angle NKC$
by the _?_ Property.

**For each diagram, find the value of the variable that will allow you to prove that
two lines are parallel.** *Section 5.4*

22.

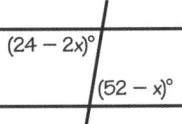

23.

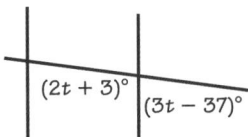

24.

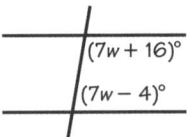

**For each pair of lines shown, tell whether they are *parallel* or *not parallel* and
explain your reasoning.** *Section 5.4*

25. ℓ and m

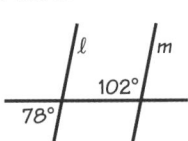

26. c and d

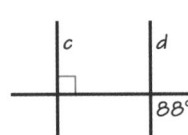

27. p and q

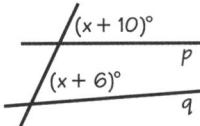

Use the diagram at the right. *Section 5.5*

28. What do you know about $\overline{PM}$ and $\overline{RS}$?

29. Using the Dual Parallels Theorem you can tell
that $\overline{AB} \parallel$ _?_ .

30. Find the value of z.

31. Find the measure of $\angle RTX$.

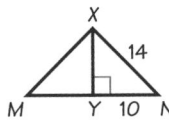

Find each distance. *Section 5.5*

32. Find the distance from X to $\overline{MN}$.

33. Find the distance from A to $\overline{PQ}$.

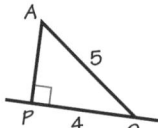

21. alternate interior; Substitution
22. $x = -28$
23. $t = 40$
24. $w = 12$
25. parallel; $78° + 102° = 180°$
26. not parallel; $90° + 88° \neq 180°$
27. not parallel; $x + 10 \neq x + 6$
28. $\overline{PM} \parallel \overline{RS}$
29. $\overline{PM}$
30. $z = 13$

31. $66°$
32. $4\sqrt{6} \approx 9.8$
33. 3

In the diagram, plane **X** ∥ plane **Z**. *Section 5.6*

34. How are $\overline{CD}$ and $\overline{FG}$ related?

35. Suppose $m\angle CDE = 76°$. Find $m\angle FED$.

36. Name two angles that are congruent to $\angle CDE$.

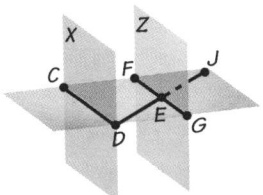

Draw two angles like the ones shown. For Exercises 37–40, construct an angle with the indicated measure. *Section 5.7*

37. $2w°$

38. $(w + v)°$

39. $(180 + v)°$

40. $(2v - w)°$

CHAPTER 6

Tell whether or not a triangle can be formed from sides of the given lengths. *Section 6.1*

1. 3 ft, 8 ft, 10 ft

2. 22 cm, 11 cm, 11 cm

3. 8 in., 9 in., 7 in.

4. 25 m, 35 m, 11 m

5. $x, 3x, 4x$

6. 16 ft, 9 ft, 25 ft

7. 5 ft, 6 ft, 4 ft

8. 2 in., 16 in., 12 in.

9. 8 cm, 14 cm, 25 cm

The lengths of two sides of a triangle are given. What can you conclude about the length of the third side? *Section 6.1*

10. 3.8 m, 14.5 m

11. $4\frac{1}{2}$ in., $3\frac{1}{4}$ in.

12. $y + 2, y$

13. 5 ft, 6 ft

14. 2 in., 16 in.

15. 8 cm, 14 cm

16. 3 ft, 8 ft

17. 20 cm, 11 cm

18. 9 in., 7 in.

What can you conclude about the length AB in each triangle? *Section 6.1*

19.

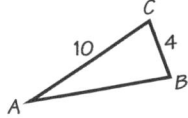

20. $m\angle B < m\angle C$

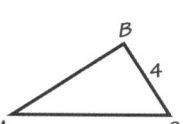

21. $m\angle A < m\angle B$

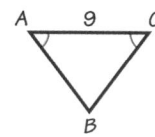

22. $m\angle B < m\angle A$ and $m\angle C < m\angle B$

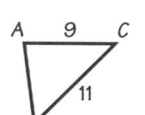

23. $m\angle C < m\angle A$

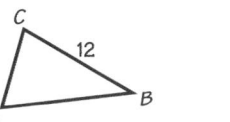

24. $m\angle B < m\angle C$ and $m\angle C < m\angle A$

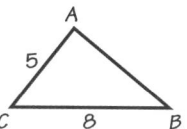

670 Extra Practice

34. $\overline{CD} \parallel \overline{FG}$

35. 104°

36. $\angle FEJ$ and $\angle DEG$

37.

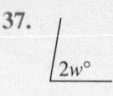

38.

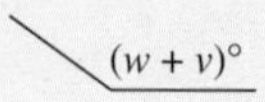

$(w + v)°$

39. $(180 + v)°$

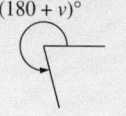

40.

$(2v - w)°$

Chapter 6

1. Yes.

2. No.

3. Yes.

4. Yes.

5. No.

6. No.

7. Yes.

8. No.

9. No.

10. It is between 10.7 m long and 18.3 m long.

11. It is between $1\frac{1}{4}$ in. long and $7\frac{3}{4}$ in. long.

12. It is between 2 units long and $2y + 2$ units long.

13. It between 1 ft long and 11 ft long.

14. It is between 14 in. long and 18 in. long.

15. It is between 6 cm long and 22 cm long.

16. It is between 5 ft long and 11 ft long.

Use the diagram at the right. *Section 6.1*

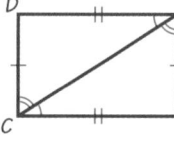

25. △DEC ≅ ?

26. ∠CDE ≅ ?

27. ∠DCE ≅ ?

28. ∠DEC ≅ ?

29. $\overline{DC}$ ≅ ?

30. $\overline{CF}$ ≅ ?

Use the diagram at the right. △**ABC** ≅ △**WXY.** *Section 6.2*

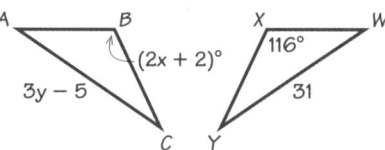

31. △CBA ≅ ?

32. ∠A ≅ ?

33. ∠C ≅ ?

34. $\overline{AC}$ ≅ ?

35. Find the value of *x*.

36. Find the value of *y*.

Decide whether or not you can prove that the triangles are congruent. If you can, tell which postulate you would use. *Section 6.3*

37.

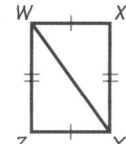

38.

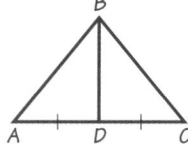

39.

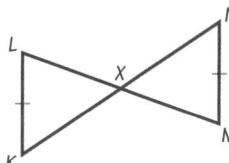

40.

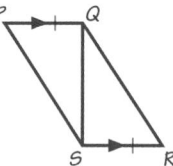

41.

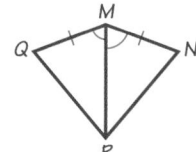

42.

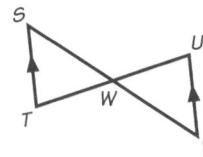

43. Copy and complete the flow proof. *Section 6.3*

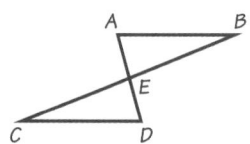

   Given: *E* is the midpoint of $\overline{BC}$.
   E is the midpoint of $\overline{AD}$.

   Prove: △ABE ≅ △DCE

Reasons

1. Given

2. Definition of ?

3. ?

4. ?

5. ?

6. ?

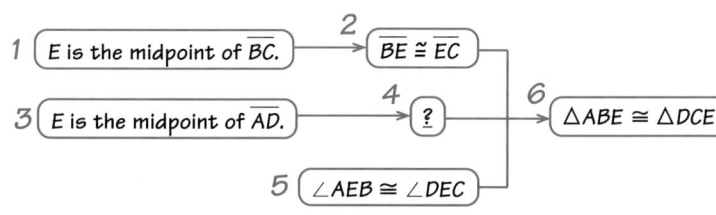

17. It is between 9 cm long and 31 cm long.

18. It is between 2 in. long and 16 in. long.

19. 6 < AB < 14

20. AB > AC

21. AB = CB

22. AB < 11

23. AB < 12

24. 8 < AB < 5

25. △FCE

26. ∠EFC

27. ∠FEC

28. ∠FCE

29. $\overline{FE}$

30. $\overline{ED}$

31. △YXW

32. ∠W

33. ∠Y

34. $\overline{WY}$

35. x = 57

36. y = 12

37. SSS Postulate

38. No.

39. No.

40. SAS Postulate

41. SAS Postulate

42. No.

43. (1) Given; (2) midpoint; (3) Given; (4) $\overline{AE} \cong \overline{ED}$; Def. of midpoint; (5) Vertical angles are ≅; (6) SAS Postulate

Tell which method you can use to prove that the triangles are congruent. If no method can be used, write *none*. *Section 6.4*

44.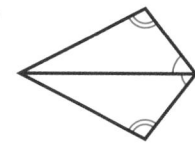

45.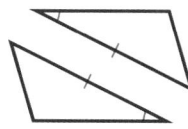

46.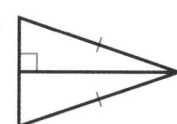

47.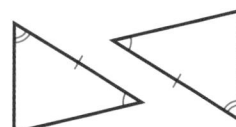

48.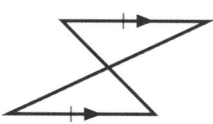

49.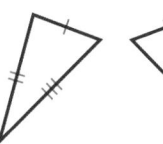

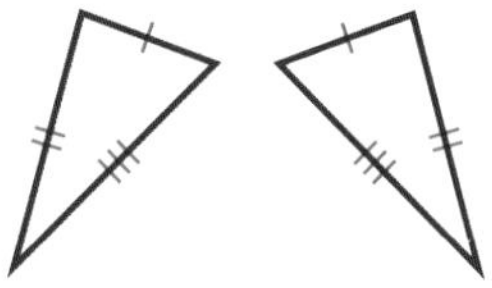

Tell which pair of triangles must be congruent in order to prove each statement. *Section 6.5*

50. $\overline{CE} \cong \overline{DE}$

51. $\overline{AD} \cong \overline{BC}$

52. $\angle ABC \cong \angle BAD$

53. $\angle ACB \cong \angle BDA$

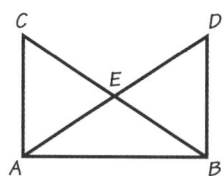

Use the diagram at the right. *Section 6.6*

54. $\angle 2 \cong \angle 3$. Name two congruent segments.

55. $\angle 7 \cong \angle 6$. Name two congruent segments.

56. $\overline{PT} \cong \overline{QT}$. Name two congruent angles.

57. $\overline{ST} \cong \overline{EC}$. Name two congruent angles.

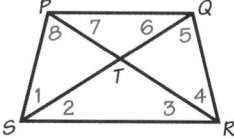

In $\triangle ABC$, $AB = AC$, $\overline{AD}$ and $\overline{CE}$ are medians and $AC = 16$. Find each length or angle measure. *Section 6.7*

58. BA

59. $m\angle BDA$

60. BE

61. EA

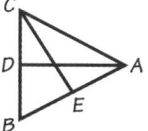

CHAPTER 7

Find each length. *Section 7.1*

1. Rectangle $PQRS$
  a. PT
  b. QT
  c. SR

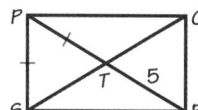

2. Kite $ABCD$
  a. DE
  b. BD
  c. BC

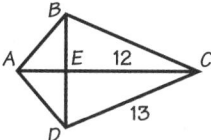

44. AAS Theorem
45. none
46. HL Theorem
47. ASA Postulate
48. AAS Theorem or ASA Postulate
49. SSS Postulate
50. $\triangle CEA \cong \triangle DEB$
51. $\triangle BAD \cong \triangle ABC$
52. $\triangle ABC \cong \triangle BAD$
53. $\triangle ACB \cong \triangle BDA$ or $\triangle ACE \cong \triangle BDE$

54. $\overline{ST} \cong \overline{RT}$ 55. $\overline{PT} \cong \overline{QT}$
56. $\angle 7 \cong \angle 6$ 57. $\angle 2 \cong \angle 3$
58. 16   59. 90°
60. 8   61. 8

Chapter 7

1. a. 5   b. 5
   c. 5
2. a. 5   b. 10
   c. 13

3. a. 6   b. 12
   c. 10
4. a. 6
   b. $3\sqrt{2} \approx 4.24$
   c. $6\sqrt{2} \approx 8.49$
5. a. 12   b. 13
   c. 20   d. 11
   e. 20
6. Paragraph proof: Since $ABCD$ is a rectangle, $\angle ADC$ and

3. Rhombus *EFGH*

a. *FJ*

b. *FH*

c. *FG*

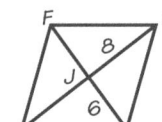

4. Square *KLMN*

a. *KN*

b. *PK*

c. *LN*

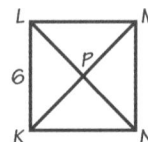

5. *ABCD* is a rectangle and *ABGH* is a parallelogram. Find each length.

a. *BC*

b. *AH*

c. *DB*

d. *HC*

e. *AC*

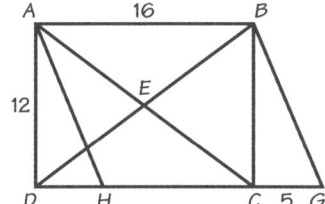

6. Use the diagram for Exercise 5 above.

Given: *ABCD* is a rectangle; *ABGH* is a parallelogram.

Prove: △*ADH* ≅ △*BCG*

Show that each quadrilateral is a parallelogram. Then find each length or measure. *Section 7.2*

7. a. *m∠NQP*

b. *NP*

c. *QN*

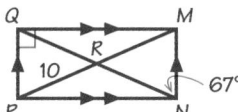

8. a. *m∠ABC*

b. *m∠ADC*

c. *AD*

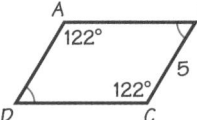

9. a. *m∠GHJ*

b. *m∠FGJ*

c. *JH*

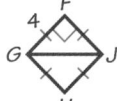

10. a. *SR*

b. *TP*

c. *QS*

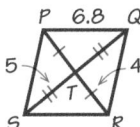

Classify each quadrilateral. Be as specific as possible. *Section 7.3*

11.

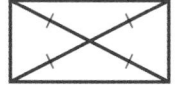

12.

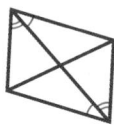

13.

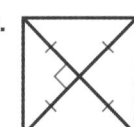

14.

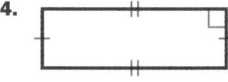

15.

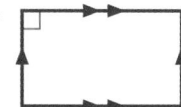

16.

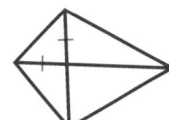

∠*BCD* are right angles. So $\overline{BC} \perp \overline{DG}$, and ∠*BCG* is also a right angle. Then ∠*ADC* ≅ ∠*BCG* since all right angles are congruent. Opposite sides of a parallelogram are congruent, so $\overline{AD} \cong \overline{BC}$ and $\overline{AH} \cong \overline{BG}$. Using the HL Theorem, △*ADH* ≅ △*BCG*.

7. Both pairs of opposite sides are parallel. (Def. of parallelogram)

a. 67°

b. 10

c. 20

8. Both pairs of opposite angles are congruent.

a. 58°   b. 58°

c. 5

9. Both pairs of opposite sides are congruent.

a. 90°   b. 45°

c. 4

10. The diagonals bisect each other.

a. 6.8

b. 4

c. 10

11. rectangle

12. parallelogram

13. square

14. rectangle

15. rectangle

16. quadrilateral

Find the area of each polygon. *Section 7.4*

17.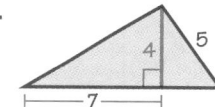

18.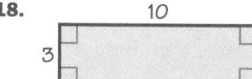

19.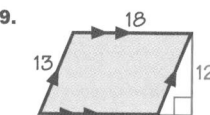

20.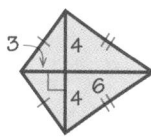

21.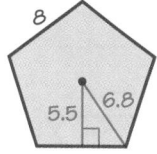

22.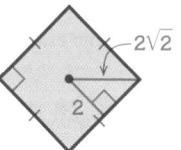

Find the area of each polygon. *Section 7.4*

23. a square with diagonal 22 cm

24. a rectangle with side 12 ft and diagonal 20 ft

25. an equilateral triangle with perimeter of 24 in.

Find the area of each regular polygon. *Section 7.5*

26.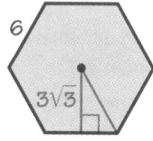

27.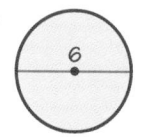

28.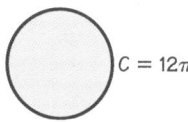

Find the area of each circle. *Section 7.5*

29.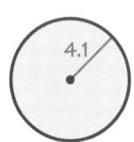

30.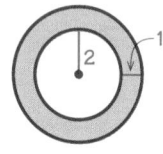

31.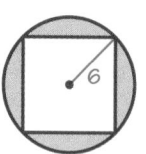

Find the area of each shaded region. *Section 7.5*

32.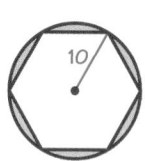

33.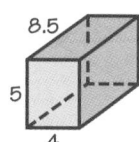

34.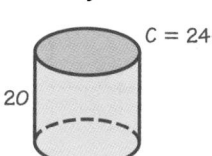

Find the volume and surface area of each prism or cylinder. *Section 7.6*

35.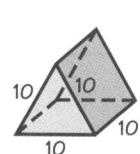

36.

37.

17. 20

18. 30

19. 216

20. 36

21. 36

22. 16

23. 242 cm^2

24. 192 ft^2

25. $16\sqrt{3} \approx 27.7$ in.^{2}

26. 110

27. 16

28. $54\sqrt{3} \approx 93.53$

29. $9\pi \approx 28.27$

30. $30\pi \approx 113.10$

31. $16.81\pi \approx 52.81$

32. $5\pi \approx 15.71$

33. $36\pi - 72 \approx 41.10$

34. $100\pi - 150\sqrt{3} \approx 54.35$

35. $V = 170$; $S.A. = 193$

36. $V = 2880\pi \approx 9047.79$;
$S.A. = 72\pi + 480\pi \approx 1734.16$

37. $V = 250\sqrt{3} \approx 433.01$;
$S.A. = 300 + 50\sqrt{3} \approx 386.60$

38. 1953.125 cm^3

39. 96 m^3

40. $200\pi \approx 628.32$ in.^{3}

Find the volume of each three-dimensional figure described. *Section 7.6*

38. Each edge of a cube is 12.5 cm long.

39. A pentagonal prism is 8 m long. The area of the base is 12 m².

40. The radius of a cylinder is 5 in. The height of the cylinder is 8 in.

CHAPTER 8

Copy each diagram. Sketch the image of the polygon after a reflection over the given line. Label the image polygon. *Section 8.1*

1.

2.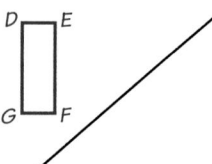

3.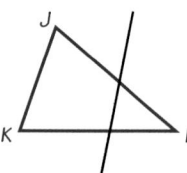

Each diagram shows a polygon and its image after a reflection. Copy each diagram and draw the line of reflection. *Section 8.1*

4.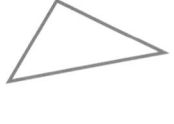

5.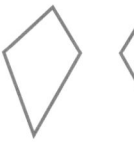

6.

Copy each polygon and reflect it over the *y*-axis. Give the coordinates of the vertices of the image. *Section 8.2*

7.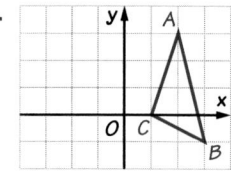

8.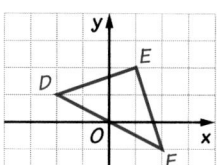

9.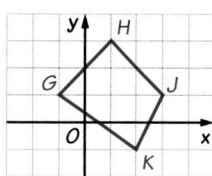

Copy each polygon and reflect it over the *x*-axis. Give the coordinates of the vertices of the image. *Section 8.2*

10.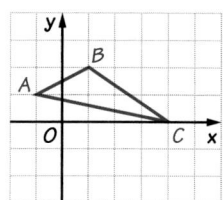

11.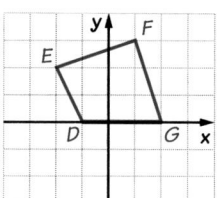

12.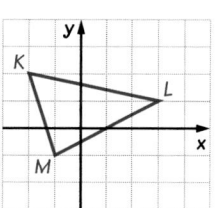

Extra Practice **675**

4.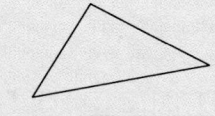

5.

6.

7. $A'(-2, 3)$, $B'(-3, -1)$, $C'(-1, 0)$

8. $D'(2, 1)$, $E'(-1, 2)$, $F'(-2, -1)$

9. $G'(1, 1)$, $H'(-1, 3)$, $J'(-3, 1)$, $K'(-2, -1)$

10. $A'(-1, -1)$, $B'(1, -2)$, $C'(4, 0)$

11. $D'(-1, 0)$, $E'(-2, -2)$, $F'(1, -3)$, $G'(2, 0)$

12. $K'(-2, -2)$, $L'(3, -1)$, $M'(-1, 1)$

Chapter 8

1.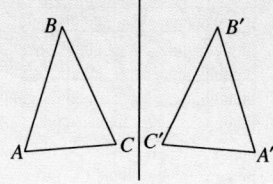

2.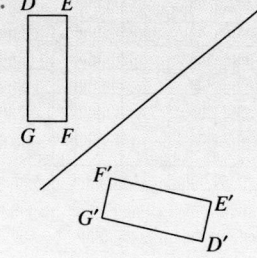

3.

675

13. $P'(2, 0)$, $Q'(3, 2)$, $R'(0, 1)$

14. $S'(1, -1)$, $T'(3, 1)$, $U'(1, 3)$,
    $V'(-1, 1)$

15. $W'(-1, -1)$, $X'(2, -1)$, $Y'(3, 1)$,
    $Z'(2, 2)$

16.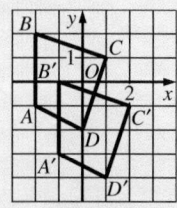

17.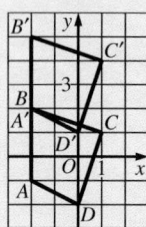

18.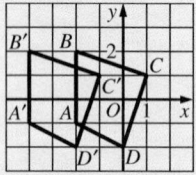

19.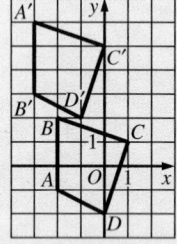

20. $(a, b) \rightarrow (a - 5, b - 3)$

21. $(a, b) \rightarrow (a + 2, b + 1)$

22. $(a, b) \rightarrow (a - 9, b)$

23.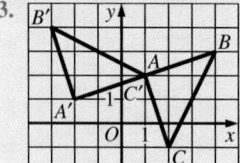

24.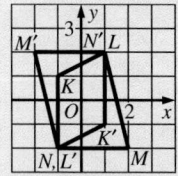

25.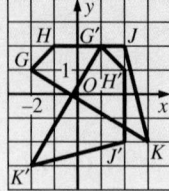

26.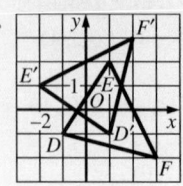

Copy each polygon and reflect it over the line $y = x$. Give the coordinates of the vertices of the image. *Section 8.2*

13.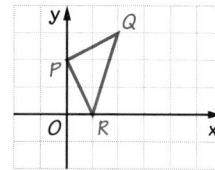

14.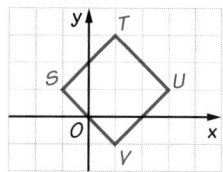

15.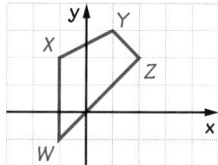

Copy ABCD and draw its image after each translation. *Section 8.3*

16. $(a, b) \rightarrow (a + 1, b - 2)$

17. $(a, b) \rightarrow (a, b + 3)$

18. $(a, b) \rightarrow (a - 2, b)$

19. $(a, b) \rightarrow (a - 1, b + 4)$

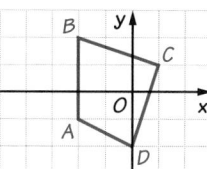

Describe each translation using coordinate notation. *Section 8.3*

20. Every point moves 5 units to the left and 3 units down.

21. Every point moves 2 units to the right and 1 unit up.

22. Every point moves 9 units to the left.

Copy each diagram and sketch the image of each polygon after the given rotation around the origin. *Section 8.4*

23. rotate 90°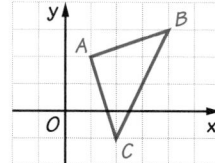

24. rotate 180°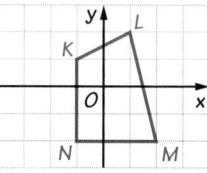

25. rotate 270°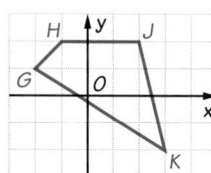

26. rotate 90°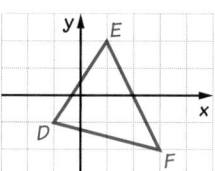

Sketch the image of the triangle with the given vertices after a glide reflection using the given translation and reflection. *Section 8.5*

27. $A(2, 0)$, $B(5, 2)$, $C(4, 4)$; translation: $(a, b) \rightarrow (a - 2, b)$; reflection: over the x-axis

28. $A(2, 6)$, $B(3, 5)$, $C(1, 4)$; translation: $(a, b) \rightarrow (a + 2, b - 1)$; reflection: over the y-axis

29. $A(-1, 0)$, $B(3, 2)$, $C(2, 4)$; translation: $(a, b) \rightarrow (a, b + 3)$; reflection: over the line $y = x$

30. $A(-3, 3)$, $B(1, 3)$, $C(-1, 0)$; translation: $(a, b) \rightarrow (a - 2, b)$; reflection: over the line $y = -x$

676 Extra Practice

27.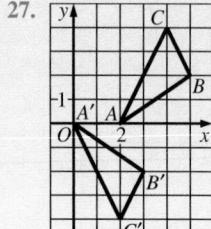

28.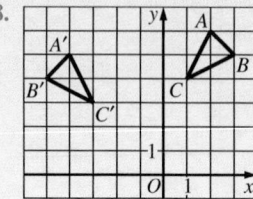

29.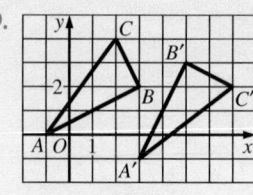

30.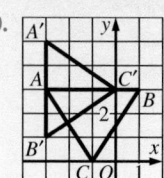

Draw the image of each polygon after a dilation with center *O* and the given scale factor. *Section 8.6*

31. scale factor 2

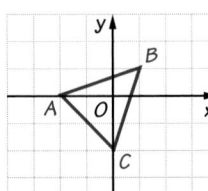

32. scale factor $\frac{1}{2}$

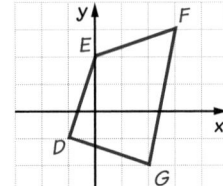

33. scale factor $\frac{2}{3}$

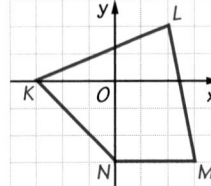

CHAPTER 9

ALGEBRA Solve each proportion. *Section 9.1*

1. $\frac{3}{5} = \frac{x}{20}$

2. $\frac{3}{4} = \frac{y}{10}$

3. $\frac{5}{z} = \frac{z}{9}$

4. $3:1 = y:20$

5. $3:x = x:27$

6. $z:5 = 12:30$

Tell whether the polygons in each pair are similar. Explain your reasoning. *Section 9.1*

7.

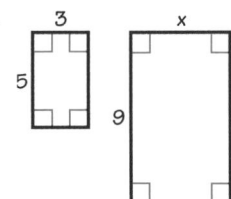

8.

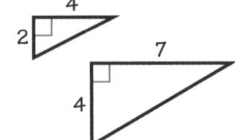

9.

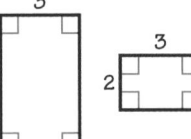

The polygons in each pair are similiar. Find the value of each variable. *Section 9.1*

10.

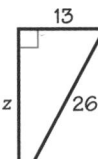

11.

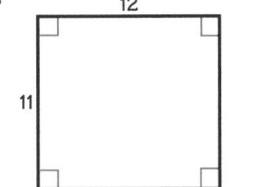
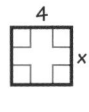

12.

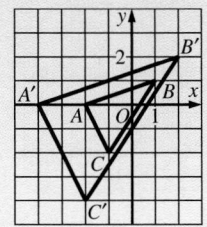

13.

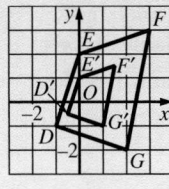
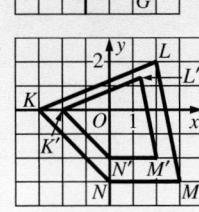

31.

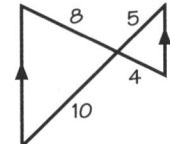

32.

33.

Chapter 9

1. $x = 12$
2. $y = 7.5$
3. $z = 3\sqrt{5}$
4. $y = 60$
5. $x = 9$
6. $z = 2$
7. Yes; corresponding sides are proportional.

8. No; corresponding sides are not proportional.
9. No; corresponding sides are not proportional.
10. $x = 5.4$
11. $x = 8.25$, $y = 2.75$
12. $w = 10$, $z = 23.4$
13. $x = 3\frac{2}{3}$

Is it possible to prove that the triangles in each pair are similar? Explain why or why not. *Section 9.2*

14.

15.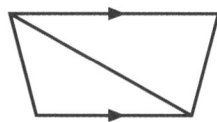

16.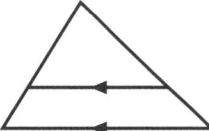

17.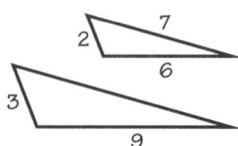

Explain why the triangles in each pair are similar. *Section 9.2*

18.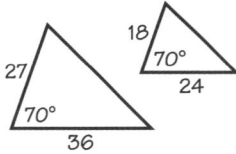

19.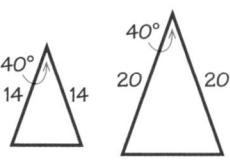

20.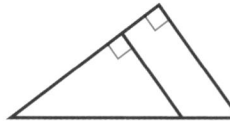

21.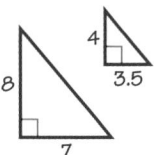

22.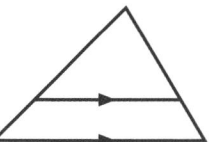

23.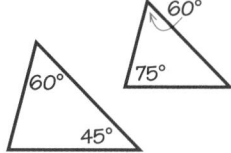

Find the value of each variable. *Section 9.3*

24.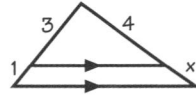

25.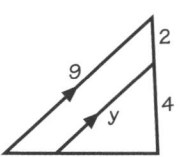

26.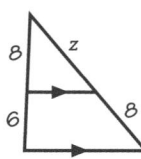

27.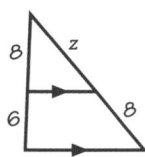

28.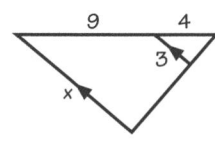

29.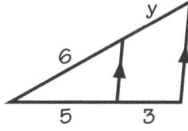

ALGEBRA Complete the equation so that it is equivalent to the proportion $\frac{x}{y} = \frac{4}{5}$. *Section 9.3*

30. $\frac{y}{x} = \underline{\ ?\ }$

31. $\frac{x}{4} = \underline{\ ?\ }$

32. $\underline{\ ?\ } = \frac{y+5}{5}$

678 Extra Practice

---

14. Yes; AA Similarity Postulate.

15. No; only one pair of corresponding angles are congruent.

16. Yes; AA Similarity Postulate.

17. No; only two pairs of corresponding sides are in proportion.

18. SAS Similarity Theorem

19. SAS Similarity Theorem

20. AA Similarity Postulate

21. SAS Similarity Theorem or SSS Similarity Theorem

22. AA Similarity Postulate

23. AA Similarity Postulate

24. $x = 1\frac{1}{3}$

25. $y = 6$

26. $z = 10\frac{2}{3}$

27. $w = 12.5$

28. $x = 9.75$

29. $y = 9.6$

30. $\frac{5}{4}$

31. $\frac{y}{5}$

32. $\frac{x+4}{4}$

The figures in each pair are similar. Find each missing value. *Section 9.4*

33. $A = 5$ $A = \underline{?}$

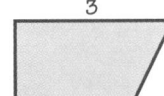 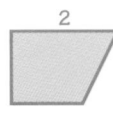

34. $P = 24$ $P = \underline{?}$
$A = \underline{?}$ $A = 54$

35. $C = 6\pi$ $C = \underline{?}$

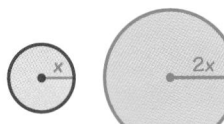

36. Area of $\triangle DEC = 16$
Area of $\triangle ABC = \underline{?}$

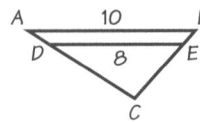

37. $P = \underline{?}$ $P = 36$

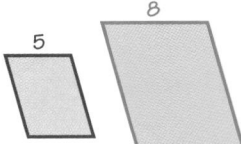

38. $A = \underline{?}$ $A = 200$

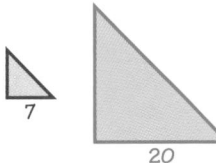

Find the probability of hitting the shaded area of the target with a bean that is tossed at random that hits the target. *Section 9.5*

39.
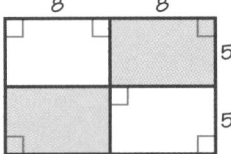

40.
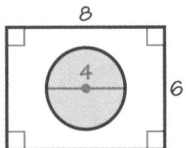

41.
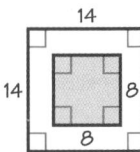

If you choose a random point on a side of the polygon, what is the probability that it is on the given segment? *Section 9.5*

42. $\overline{AB}$
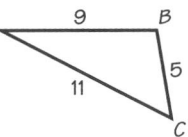

43. $\overline{PQ}$
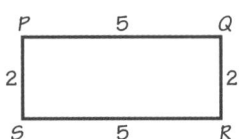

44. $\overline{JK}$
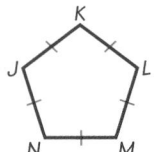

45. $\overline{CD}$
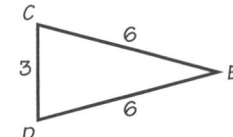

Extra Practice   **679**

---

33. $2\dfrac{2}{9}$

34. 36; 24

35. $12\pi \approx 37.70$

36. 25

37. 22.5

38. 24.5

39. $\dfrac{1}{2} = 0.5$

40. $\dfrac{\pi}{12} \approx 0.26$

41. $\dfrac{16}{49} \approx 0.33$

42. $\dfrac{9}{25} = 0.36$

43. $\dfrac{5}{14} \approx 0.36$

44. $\dfrac{1}{5} = 0.2$

45. $\dfrac{3}{15} \approx 0.2$

CHAPTER 10

Use the diagram to complete each statement. *Section 10.1*

1. $\angle QPR \cong \angle\ \underline{\ ?\ }$

2. $\triangle PQR \sim \triangle\ \underline{\ ?\ } \sim \triangle\ \underline{\ ?\ }$

3. $\dfrac{PQ}{QR} = \dfrac{PS}{\ ?\ }$

4. QS is the geometric mean of $\underline{\ ?\ }$ and $\underline{\ ?\ }$.

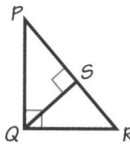

ALGEBRA   Find the geometric mean of the given numbers. *Section 10.1*

5. 3 and 27

6. 8 and 3

7. 9 and $\dfrac{1}{4}$

8. 23.04 and 34.81

9. $\dfrac{2}{5}$ and $\dfrac{5}{2}$

10. 0.5 and 50

Find the value of each variable. *Section 10.1*

11.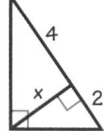

12.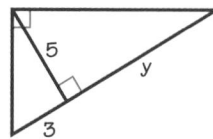

13.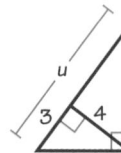

14.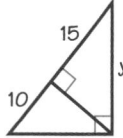

15.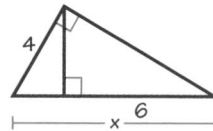

16.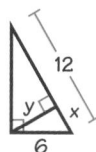

Find the value of each variable. *Section 10.2*

17.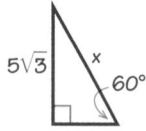

18.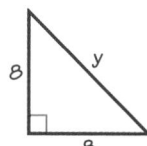

19.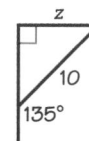

20.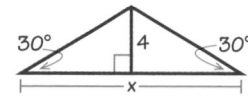

21.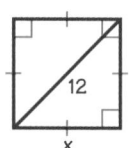

22.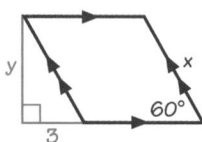

23.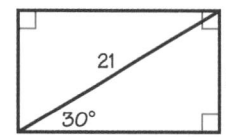

24.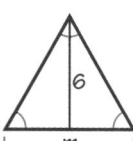

25.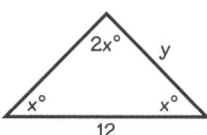

680   Extra Practice

Chapter 10

1. *SQR*

2. *PSQ, QSR*

3. *SQ*

4. *PS, RS*

5. 9

6. $2\sqrt{6} \approx 4.9$

7. $1\dfrac{1}{2}$

8. 28.32

9. 1

10. 5

11. $x = 2\sqrt{2} \approx 2.8$

12. $y = 8\dfrac{1}{3}$

13. $u = 8\dfrac{1}{3}$

14. $y = 5\sqrt{15} \approx 19.4$

15. $x = 8$

16. $x = 3, y = 3\sqrt{3} \approx 5.2$

17. $x = 10$

18. $y = 8\sqrt{2} \approx 11.3$

19. $z = 5\sqrt{2} \approx 7.1$

20. $x = 8\sqrt{3} \approx 13.9$

21. $x = 6\sqrt{2} \approx 8.5$

22. $x = 6, y = 3\sqrt{3} \approx 5.2$

23. $p = 10.5$

24. $m = \sqrt{43} \approx 6.9$

25. $x = 45, y = 6\sqrt{2} \approx 8.5$

For Exercises 26–28, find tan *A* and tan *B*. *Section 10.3*

26.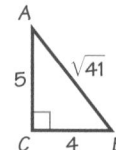

27.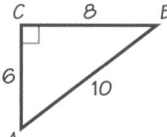

28.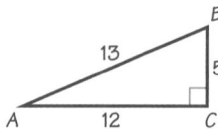

Find the value of each expression. Round your answers to four decimal places.
Section 10.3

29. $\tan 28°$ **30.** $\tan 12.5°$ **31.** $\tan 85°$

Find the value of each variable. Round your answers to the nearest tenth.
Section 10.3

32.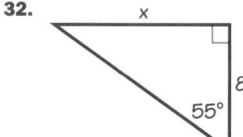

33.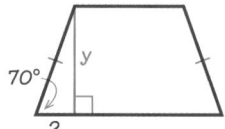

34.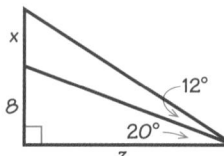

For Exercises 35–37, find sin *A*, cos *A*, sin *B*, and cos *B*. *Section 10.4*

35.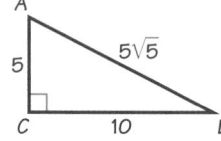

36.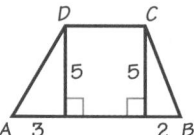

37.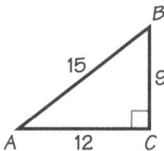

Find the value of each expression. Round your answers to four decimal places.
Section 10.4

38. $\sin 48°$ **39.** $\sin 8.5°$ **40.** $\cos 68.3°$ **41.** $\cos 12°$

Find the measure of an acute angle that satisfies the given equation. Round your answers to the nearest tenth of a degree. *Section 10.4*

42. $\sin A = \dfrac{9}{13}$ **43.** $\cos B = 0.2554$ **44.** $\sin M = \dfrac{10}{17}$ **45.** $\cos W = 0.9797$

46. $\cos D = 0.0175$ **47.** $\sin V = \dfrac{10}{13}$ **48.** $\sin C = 0.500$ **49.** $\cos Z = \dfrac{5}{9}$

Find the value of each variable. *Section 10.4*

50.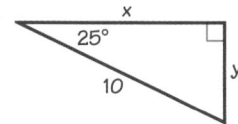

51.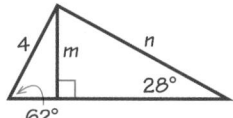

52.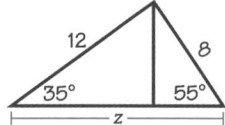

Extra Practice   **681**

---

26. $\tan A = \dfrac{4}{5}$, $\tan B = \dfrac{5}{4}$

27. $\tan A = \dfrac{4}{3}$, $\tan B = \dfrac{3}{4}$

28. $\tan A = \dfrac{5}{12}$, $\tan B = \dfrac{12}{5}$

29. 0.5317

30. 0.2217

31. 11.4301

32. $x \approx 11.4$

33. $y \approx 5.5$

34. $x \approx 5.7$, $z \approx 22.0$

35. $\sin A = \dfrac{2\sqrt{5}}{5}$, $\cos A = \dfrac{\sqrt{5}}{5}$,

     $\sin B = \dfrac{\sqrt{5}}{5}$, $\cos B = \dfrac{2\sqrt{5}}{5}$

36. $\sin A = \dfrac{5\sqrt{34}}{34}$, $\cos A = \dfrac{3\sqrt{34}}{34}$,

     $\sin B = \dfrac{5\sqrt{29}}{29}$, $\cos B = \dfrac{2\sqrt{29}}{29}$

37. $\sin A = \dfrac{3}{5}$, $\cos A = \dfrac{4}{5}$,

     $\sin B = \dfrac{4}{5}$, $\cos B = \dfrac{3}{5}$

38. 0.7431

39. 0.1478   40. 0.3697

41. 0.9781   42. 43.8°

43. 75.2°   44. 36.0°

45. 11.6°   46. 89.0°

47. 50.3°

48. 30°

49. 56.3°

50. $x \approx 9.06$, $y \approx 4.23$

51. $m \approx 3.53$, $n \approx 7.52$

52. $z \approx 14.4$

681

Find the value of each variable. Round your answers to the nearest tenth of a degree. *Section 10.4*

53.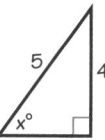

54.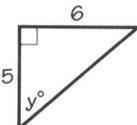

55.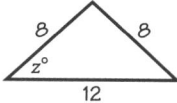

Express each vector in component form and find the value of each variable. *Section 10.5*

56.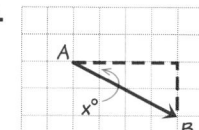

57.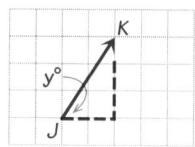

58.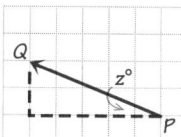

Graph each vector and find its magnitude. *Section 10.5*

59. $\overrightarrow{HJ} = (4, 9)$ 60. $\overrightarrow{JK} = (-2, 7)$ 61. $\overrightarrow{MN} = (20, -3)$ 62. $\overrightarrow{PR} = (-3, -3)$

$\overrightarrow{AB} = (2, -4)$ **and** $\overrightarrow{CD} = (-3, 7)$. **Use a graph to find each vector sum. Write the resulting vector in component form.** *Section 10.5*

63. $\overrightarrow{AB} + \overrightarrow{CD}$ 64. $2\overrightarrow{AB} + \overrightarrow{CD}$ 65. $\overrightarrow{AB} + 2\overrightarrow{CD}$

Find the area of each polygon. *Section 10.6*

66.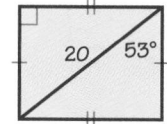

67.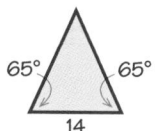

68.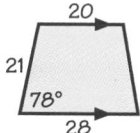

Find the volume of each prism. *Section 10.6*

69.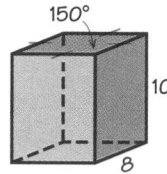

70.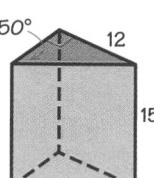

71.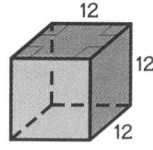

Find the volume and surface area of each right cone or regular pyramid. *Section 10.7*

72.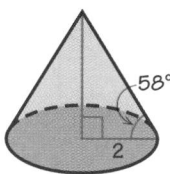

73.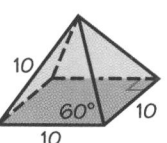

682 Extra Practice

---

53. $x \approx 53.1$

54. $y \approx 50.2$

55. $z \approx 41.4$

56. $(4, -2); x \approx 26.6$

57. $(2, 3); y \approx 56.3$

58. $(-5, 2); z \approx 21.8$

59.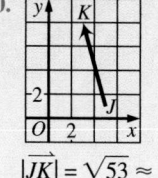
$|\overrightarrow{HJ}| = \sqrt{97} \approx 9.8$

60.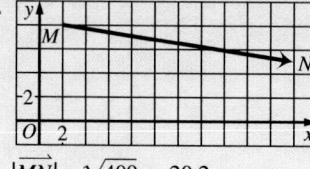
$|\overrightarrow{JK}| = \sqrt{53} \approx 7.3$

61. $|\overrightarrow{MN}| = \sqrt{409} \approx 20.2$

62.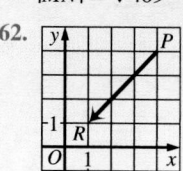
$|\overrightarrow{PR}| = 3\sqrt{2} \approx 4.2$

63. $(-1, 3)$ 64. $(1, -1)$

65. $(-4, 10)$ 66. about 192

67. about 105

68. about 493

69. 640

70. about 906

71. 1728

72. $V \approx 4.3\pi$; $S.A. \approx 11.5\pi$

73. $V = \dfrac{500\sqrt{2}}{3} \approx 235.7$;
$S.A. = 100\sqrt{3} + 100 \approx 273.2$

682

CHAPTER 11

Find each measure. *Section 11.1*

1. $m\overarc{ABC} = \underline{?}$

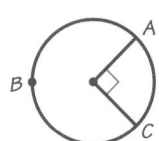

2. $m\angle WXY = \underline{?}$

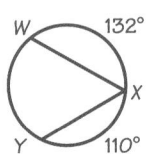

3. $m\overarc{AB} = \underline{?}$

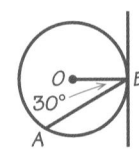

4. $m\overarc{BC} = \underline{?}$

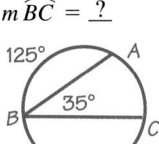

5. $m\overarc{ACB} = \underline{?}$

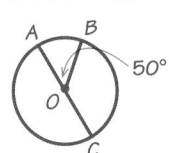

6. $m\overarc{AC} = \underline{?}$

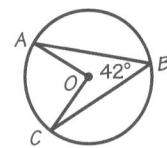

7. $m\overarc{BAC} = \underline{?}$

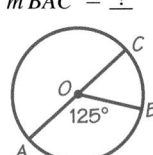

8. $m\angle AOC = \underline{?}$

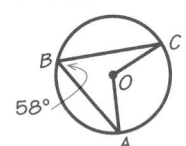

9. $m\overarc{ABC} = \underline{?}$

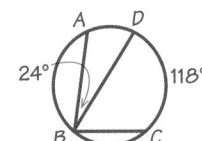

Find the value of each variable. *Section 11.2*

10.

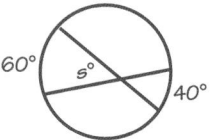

11.

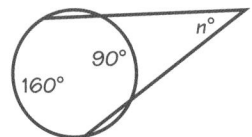

12.

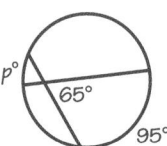

13.

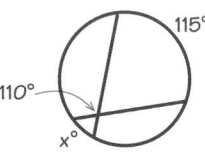

14.

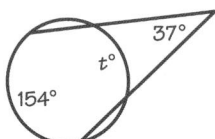

15.

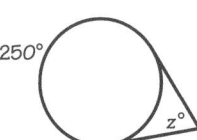

16.

17.

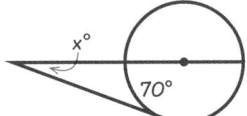

18.

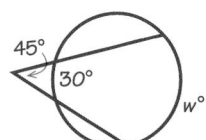

Chapter 11

1. 270°
2. 242°
3. 120°
4. 165°
5. 310°
6. 84°
7. 305°
8. 116°
9. 218°

10. $s = 50$
11. $n = 35$
12. $p = 35$
13. $x = 25$
14. $t = 58.5$
15. $z = 70$
16. $y = 60$
17. $x = 20$
18. $w = 120$

ALGEBRA   **Find the value of each variable.** *Section 11.3*

19.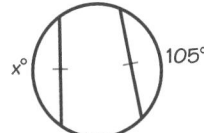

20.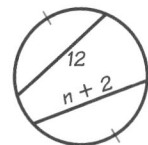

21.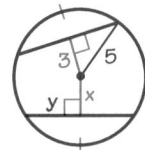

22.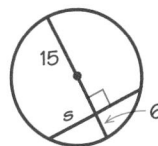

23.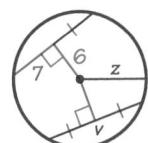

24.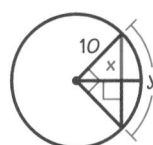

25.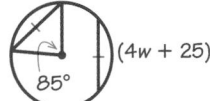

26.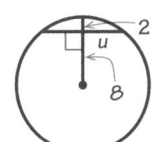

27.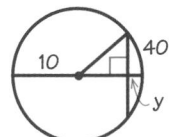

Find each unknown length. *Section 11.4*

28.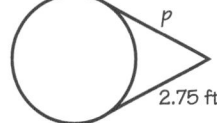

29.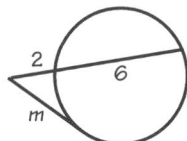

30.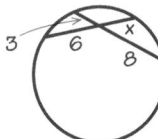

31.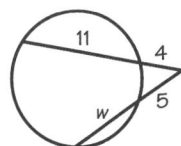

32.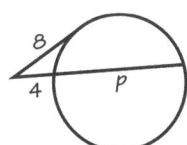

33.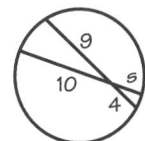

Find the area of each shaded sector to the nearest tenth. *Section 11.5*

34.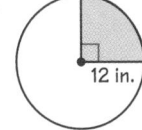

35.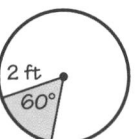

36.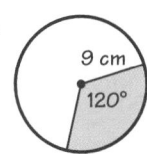

37.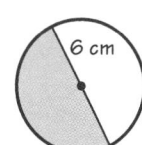

38.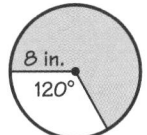

39.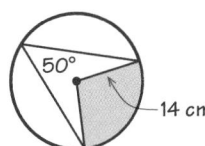

684   Extra Practice

19. $x = 105$

20. $n = 10$

21. $x = 3, y = 4$

22. $s = 12$

23. $x = 3, y = 4$

24. $x = 5\sqrt{2}, y = 90$

25. $w = 15$

26. $u = 6$

27. $y \approx 2.34$

28. $p = 2.75$

29. $m = 4$

30. $x = 4$

31. $w = 7$

32. $p = 12$

33. $s = 3.6$

34. 36π in.$^2 \approx 113.1$ in.^{2}

35. $\frac{2}{3}\pi$ ft$^2 \approx 2.1$ ft^2

36. 27π cm$^2 \approx 84.8$ cm^2

37. 18π cm$^2 \approx 56.5$ cm^2

38. $42\frac{2}{3}\pi$ in.$^2 \approx 134.0$ in.^{2}

39. $54\frac{4}{9}$ cm$^2 \approx 54.4$ cm^2

40. 10π m ≈ 31.4 m

41. $3\frac{1}{9}\pi$ in. ≈ 9.8 in.

42. $13\frac{1}{3}\pi$ m ≈ 41.9 m

43. S.A. = 576π in.$^2 \approx 1809.6$ in.^{2};
    V = 2304π in.$^3 \approx 7238.2$ in.^{3}

Find the length of the red arc in each circle. Give your answers to the nearest tenth. *Section 11.5*

40.

120°
15 m

41.
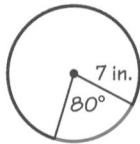
7 in.
80°

42.

120°
10 m

Give the surface area and the volume of each sphere to the nearest tenth. *Section 11.6*

43.

12 in.

44.
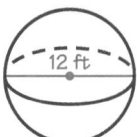
12 ft

45.

8 m

For the similar figures, find each missing volume. *Section 11.7*

46. $V = 4500$ in.^{3} $V = \underline{\ ?\ }$

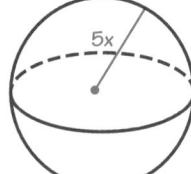
5x

2x

47. $V = \underline{\ ?\ }$ $V = \frac{16\pi}{3}$ ft^3

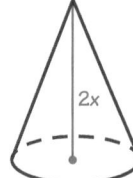
2x

x

Tell whether each statement is *true* or *false*. *Section 11.8*

48. On a sphere, parallel lines intersected by a transversal form corresponding angles that are congruent.

49. On a sphere, an equilateral triangle has angles that measure 60°.

50. On a sphere, the sum of the interior angles of a quadrilateral is 360°.

CHAPTER 12

Use the graph. Write a matrix that represents each figure. *Section 12.1*

1. A
2. B
3. AC
4. $\triangle ABC$

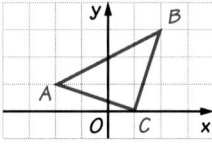

Find each product. *Section 12.1*

5. $2\begin{bmatrix} 4 \\ -3 \end{bmatrix}$

6. $1.5\begin{bmatrix} 4 & -12 \\ 6 & 14 \end{bmatrix}$

7. $\frac{1}{2}\begin{bmatrix} 8 & 6 & -6 \\ -4 & 2 & 10 \end{bmatrix}$

8. $7\begin{bmatrix} 2 & -1 \\ -3 & -2 \end{bmatrix}$

9. $2.5\begin{bmatrix} 8 & -16 \\ 6 & 12 \end{bmatrix}$

10. $\frac{2}{3}\begin{bmatrix} 9 & 6 & -6 \\ -3 & 3 & 12 \end{bmatrix}$

---

44. *S.A.* $= 144\pi$ ft$^2 \approx 452.4$ ft^2;
    $V = 288\pi$ ft$^3 \approx 904.8$ ft^3

45. *S.A.* $= 256\pi$ m$^2 \approx 804.2$ m^2;
    $V = 682\frac{2}{3}\pi$ m$^3 \approx 244.7$ m^3

46. 288π in.^{3}

47. $\frac{128\pi}{3}$ ft^3

48. True.

49. False.

50. False.

Chapter 12

1. $\begin{bmatrix} -2 \\ 1 \end{bmatrix}$

2. $\begin{bmatrix} 2 \\ 3 \end{bmatrix}$

3. $\begin{bmatrix} -2 & 1 \\ 1 & 0 \end{bmatrix}$

4. $\begin{bmatrix} -2 & 2 & 1 \\ 1 & 3 & 0 \end{bmatrix}$

5. $\begin{bmatrix} 8 \\ -6 \end{bmatrix}$

6. $\begin{bmatrix} 6 & -18 \\ 9 & 21 \end{bmatrix}$

7. $\begin{bmatrix} 4 & 3 & -3 \\ -2 & 1 & 5 \end{bmatrix}$

8. $\begin{bmatrix} 14 & -7 \\ -21 & -14 \end{bmatrix}$

9. $\begin{bmatrix} 20 & -40 \\ 15 & 30 \end{bmatrix}$

10. $\begin{bmatrix} 6 & 4 & -4 \\ -2 & 2 & 8 \end{bmatrix}$

Add the matrices. *Section 12.2*

11. $\begin{bmatrix} 4 \\ -1 \end{bmatrix} + \begin{bmatrix} -7 \\ 10 \end{bmatrix}$

12. $\begin{bmatrix} -1 & -1 \\ 6 & 6 \end{bmatrix} + \begin{bmatrix} 4 & -2 \\ -2 & -5 \end{bmatrix}$

13. $\begin{bmatrix} 2 & 2 & 2 \\ -3 & -3 & -3 \end{bmatrix} + \begin{bmatrix} 0 & 2 & 3 \\ -1 & 3 & 2 \end{bmatrix}$

Write the matrix you would use to translate each figure as indicated. *Section 12.2*

14. a point; to the right 2 units and down 1 unit

15. a segment; to the left 3 units

16. a triangle; to the right 2 units and up 1 unit

17. a quadrilateral; down 4 units

Complete each matrix product. *Section 12.3*

18. $\begin{bmatrix} 3 & -1 \\ 5 & 0 \end{bmatrix}\begin{bmatrix} 2 \\ -2 \end{bmatrix}$

19. $\begin{bmatrix} 1 & 0 \\ 0 & 1 \end{bmatrix}\begin{bmatrix} -4 & -1 \\ 10 & 3 \end{bmatrix}$

20. $\begin{bmatrix} 2 & -1 \\ 1 & -2 \end{bmatrix}\begin{bmatrix} 1 & 3 & 2 \\ 2 & 1 & -1 \end{bmatrix}$

21. $\begin{bmatrix} 0 & 0.5 \\ 1.2 & 1 \end{bmatrix}\begin{bmatrix} 2 & 4 & 8 \\ 0 & -1 & 3 \end{bmatrix}$

For each transformation matrix and polygon given, use the matrix to transform the polygon. Then graph the polygon and its image. *Section 12.3*

22. $\begin{bmatrix} -1 & 2 \\ 2 & -1 \end{bmatrix}$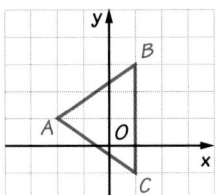

23. $\begin{bmatrix} -3 & 1 \\ 0 & 4 \end{bmatrix}$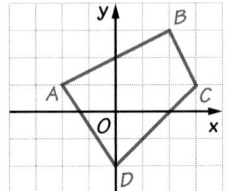

Complete each matrix product to find the image of (2, –1) after a reflection, then state the line over which the point has been reflected. *Section 12.4*

24. $\begin{bmatrix} 1 & 0 \\ 0 & -1 \end{bmatrix}\begin{bmatrix} 2 \\ -1 \end{bmatrix}$

25. $\begin{bmatrix} 0 & 1 \\ 1 & 0 \end{bmatrix}\begin{bmatrix} 2 \\ -1 \end{bmatrix}$

26. $\begin{bmatrix} -1 & 0 \\ 0 & 1 \end{bmatrix}\begin{bmatrix} 2 \\ -1 \end{bmatrix}$

27. $\begin{bmatrix} 0 & -1 \\ -1 & 0 \end{bmatrix}\begin{bmatrix} 2 \\ -1 \end{bmatrix}$

28. The matrix $\begin{bmatrix} 2 & 4 & 5 \\ 0 & -1 & 3 \end{bmatrix}$ represents $\triangle ABC$.

   a. Use the matrix $\begin{bmatrix} 0 & -1 \\ 1 & 0 \end{bmatrix}$ to transform $\triangle ABC$.

   b. Use the matrix $\begin{bmatrix} -1 & 0 \\ 0 & -1 \end{bmatrix}$ to transform the result from part (a).

   c. Graph $\triangle ABC$, $\triangle A'B'C'$, and $\triangle A''B''C''$. Describe the result of each transformation. What is the combined effect of the two transformations in parts (a) and (b)?

686 Extra Practice

---

11. $\begin{bmatrix} -3 \\ 9 \end{bmatrix}$

12. $\begin{bmatrix} 3 & -3 \\ 4 & 1 \end{bmatrix}$

13. $\begin{bmatrix} 2 & 4 & 5 \\ -4 & 0 & -1 \end{bmatrix}$

14. $\begin{bmatrix} 2 \\ -1 \end{bmatrix}$

15. $\begin{bmatrix} -3 & -3 \\ 0 & 0 \end{bmatrix}$

16. $\begin{bmatrix} 2 & 2 & 2 \\ 1 & 1 & 1 \end{bmatrix}$

17. $\begin{bmatrix} 0 & 0 & 0 & 0 \\ -4 & -4 & -4 & -4 \end{bmatrix}$

18. $\begin{bmatrix} 8 \\ 10 \end{bmatrix}$

19. $\begin{bmatrix} -4 & -1 \\ 10 & 3 \end{bmatrix}$

20. $\begin{bmatrix} 0 & 5 & 5 \\ -3 & 1 & 4 \end{bmatrix}$

21. $\begin{bmatrix} 0 & -0.5 & 1.5 \\ 2.4 & 3.8 & 12.6 \end{bmatrix}$

22.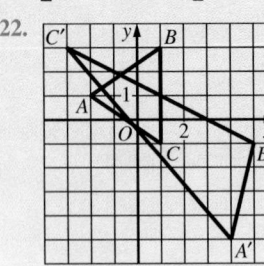

23.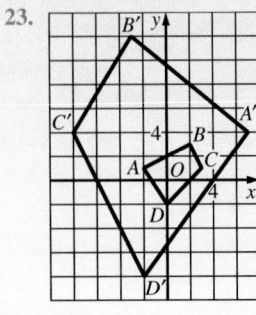

24. (2, 1)

25. (–1, 2)

26. (–2, –1)

27. (1, –2)

28. a. $\begin{bmatrix} 0 & 1 & -3 \\ 2 & 4 & 5 \end{bmatrix}$

   b. $\begin{bmatrix} 0 & -1 & 3 \\ -2 & -4 & -5 \end{bmatrix}$

   c.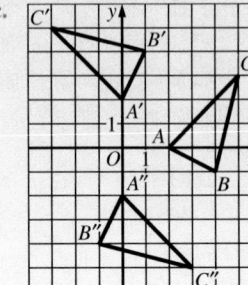

   $\triangle A'B'C'$ is the rotation of $\triangle ABC$ 90° about the origin; $\triangle A''B''C''$ is the rotation of $\triangle A'B'C'$ 180° about the origin; $\triangle A''B''C''$ is the rotation of $\triangle ABC$ 270° about the origin.

Toolbox

USING GEOMETRIC TOOLS AND TRANSFORMATIONS

Using a Protractor

A *protractor* is used to measure angles between 0° and 180°. To measure an angle, place the center mark of the protractor over the vertex of the angle and the 0° mark on one side of the angle. The numbers on one curve of the protractor measure angles counterclockwise. The numbers on the other curve of the protractor measure angles clockwise.

EXAMPLE

Use a protractor to measure the angles.

a.

b.

SOLUTION

a.
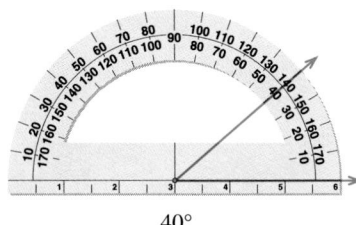

40°

b.
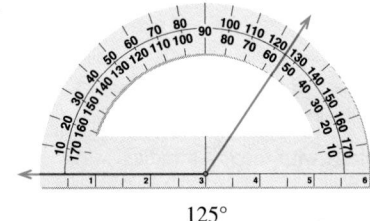

125°

PRACTICE

Use a protractor to measure the angles.

1.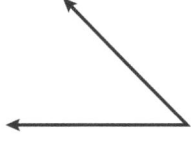

2.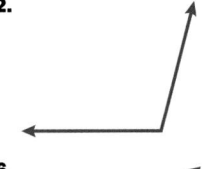

3.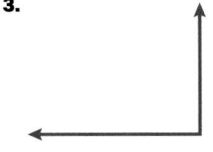

4.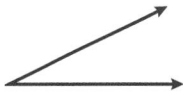

5.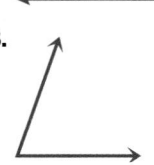

6.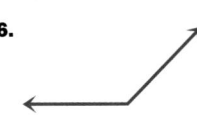

7.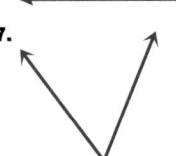

8.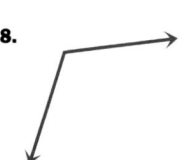

Toolbox **687**

ANSWERS Toolbox

Using a Protractor

1. 45°
2. 105°
3. 90°
4. 25°
5. 70°
6. 135°
7. 60°
8. 115°

Using a Compass

A *compass* is used to draw a circle of a certain radius. Use a ruler to mark the center of the circle and a point one radius away from the center. Place the point of the compass on the center, and adjust the compass so that the pencil is on the point one radius away. Rotate the pencil about the center to draw the circle.

EXAMPLE

Draw a circle with the given radius.

a. 1.5 in.

b. 3 cm

SOLUTION

a.

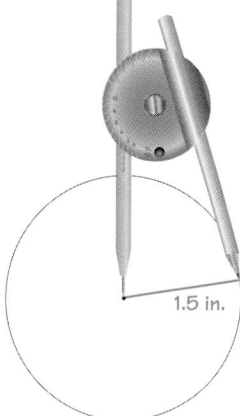

1.5 in.

b.

You can also use the measurements marked on the compass.

3 cm

PRACTICE

Draw a circle with the given radius.

1. 2 in.

2. 0.75 in.

3. $\frac{1}{2}$ in.

4. $1\frac{1}{4}$ in.

5. 6 cm

6. 3 cm

7. 4 cm

8. 2.5 cm

Transformations

A *transformation* of a geometric figure is a change in the position or size of the figure. The result of a transformation is called an *image*.

EXAMPLE

a. Translate line segment $\overline{AB}$ down 2 units and to the left 1 unit.

b. Reflect $\overline{AB}$ over the *y*-axis.

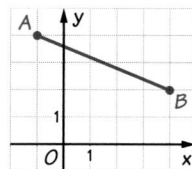

Using a Compass

1–8. Check students' work.

SOLUTION

a.

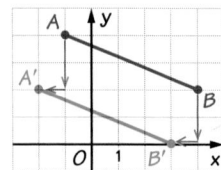

Move each point **down** two units and **left** one unit.

b.

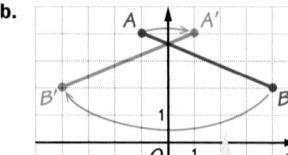

Reflect each point over the *y*-axis.

PRACTICE

Copy the diagram of △*DEF*. Sketch the image of each transformation.

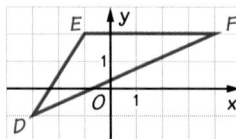

1. Translate 2 units up.

2. Reflect over the *y*-axis.

3. Translate 3 units left.

4. Reflect over the *x*-axis.

5. Translate 1 unit down and 4 units right.

6. Translate 1 unit up, then reflect over the *y*-axis.

Recognizing Symmetry

A geometric figure has symmetry if the figure can be drawn by rotating or reflecting one part.

A flower has *rotational symmetry*.

The red petal is rotated and copied to draw the other petals.

A cap has *reflection symmetry*.

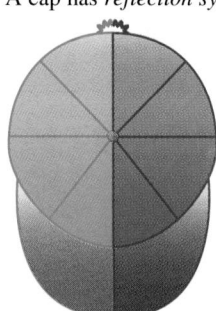

The red side of the cap is reflected to draw the other side.

EXAMPLE

State whether the figure has symmetry. If it does, tell whether it is *rotational* or *reflection symmetry*.

a.

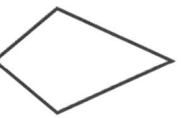

b.

c.

d.

SOLUTION

a. reflection

b. reflection and rotational

c. no symmetry

d. rotational and reflection

Toolbox   **689**

Transformations

1.

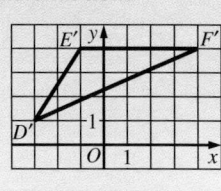

2.

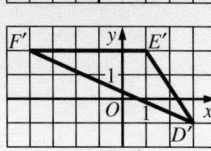

3.

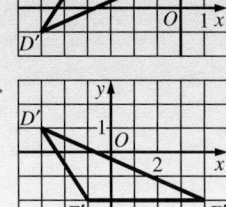

4.

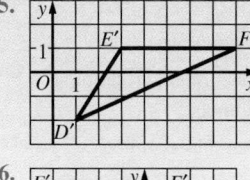

5.

6.

State whether the figure has symmetry. If it does, tell whether it is *rotational* or *reflection symmetry*.

1. 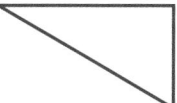 2. 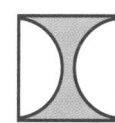 3.  4.

5. Draw an object with reflection symmetry.

6. Draw an object with rotational symmetry.

7. Draw an object that has no symmetry.

OPERATIONS WITH VARIABLE EXPRESSIONS

Simplifying and Evaluating Expressions

When you simplify a numerical expression, you must use the *order of operations*, a set of rules that guarantees that an expression has just one value.

> **Order of Operations**
>
> 1. First simplify expressions inside parentheses or other grouping symbols.
> 2. Then evaluate powers.
> 3. Next, do multiplications and divisions in order from left to right.
> 4. Last, do additions and subtractions in order from left to right.

Also, remember the following rules:
$$a - (-b) = a + b \qquad (-1)(-a) = a$$

EXAMPLE 1

Simplify each expression.

a. $-5(-2 - (-3))$

b. $\dfrac{3^3 + 2^2 - (2 - 4)}{12 \div 2 + (1 \cdot 6) + 3}$

SOLUTION

a. $5(-2 - (-3)) = -5(2 + 3)$

$$= -5(1)$$

$$= -5$$

b. $\dfrac{3^3 + 2^2 - (2 - 4)}{12 \div 2 + (1 \cdot 6) + 3} = \dfrac{27 + 4 - (-2)}{12 \div 2 + 6 + 3}$

$$= \dfrac{27 + 4 + 2}{6 + 6 + 3}$$

$$= \dfrac{33}{15} = \dfrac{11}{5}$$

The fraction bar acts like parentheses. Simplify the numerator and the denominator, then simplify the fraction.

Recognizing Symmetry

1. no symmetry

2. reflection and rotational symmetry

3. rotational symmetry

4. reflection symmetry

5–7. Answers may vary. Examples are given.

5.

6.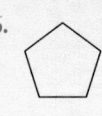

7.

A *variable expression* is an expression formed using variables, numbers, and operation symbols. For a variable expression like $x^2 + 2x^2 - 5x + 1$, the parts that are joined by plus signs and minus signs are called *terms*. Terms with the same variable parts, such as x^2 and $2x^2$, are called *like terms*.

To write an expression in simplest form, find an equivalent expression that has no parentheses and has all like terms combined. In simplest form, the expression $x^2 + 2x^2 - 5x + 1$ becomes $3x^2 - 5x + 1$.

EXAMPLE 2

Simplify each expression.

a. $-3(-t + t^2 - 5t)$

b. $\frac{1}{2}(2x + 4) + 7x - 8$

SOLUTION

a. $-3(-t + t^2 - 5t) = -3(t^2 - 6t)$ Combine like terms.

$\qquad = -3(t^2) - 3(-6t)$ Use the distributive property.

$\qquad = -3t^2 - (-18t)$

$\qquad = -3t^2 + 18t$

b. $\frac{1}{2}(2x + 4) + 7x - 8 = \frac{1}{2}(2x) + \frac{1}{2}(4) + 7x - 8$

$\qquad = x + 2 + 7x - 8$

$\qquad = 8x - 6$

To *evaluate* a variable expression, substitute a value for each variable and simplify the resulting expression using the *order of operations*.

EXAMPLE 3

Evaluate each expression when $a = -4$ and $b = 6$.

a. $(a + b)^3 - a^2 + 2b$

b. $3(2a + b) - \frac{2}{3}b(7 - b)$

SOLUTION

a. $(a + b)^3 - a^2 + 2b = (-4 + 6)^3 - (-4)^2 + 2(6)$ Substitute -4 for a and 6 for b.

$\qquad = 2^3 - (-4)^2 + 2(6)$ Use the order of operations.

$\qquad = 8 - (16) + 2(6)$

$\qquad = 8 - 16 + 12 = 4$

b. $3(2a + b) - \frac{2}{3}b(7 - b) = 3(2(-4) + 6) - \frac{2}{3} \cdot 6(7 - 6)$

$\qquad = 3(-8 + 6) - \frac{2}{3} \cdot 6(1)$

$\qquad = 3(-2) - \frac{2}{3}(6)$

$\qquad = -6 - 4 = -10$

Simplify each expression.

1. $-3(-2+5)+1$ **2.** $13+11\cdot 4-8$ **3.** $(-2-5)\cdot(3\cdot 8)$ **4.** $4\cdot 2-(3-9)$

5. $42\div(14\div 2)-15$ **6.** $(18-4^2)\cdot(-3^2)$ **7.** $12-18+15\div 3$ **8.** $-7(-3)+8\cdot 4$

9. $(63\div 3)\div(15-8)$ **10.** $\dfrac{20}{5^3\div 5-5\cdot 3}$ **11.** $\dfrac{1+3+8-2}{-15+3-3\cdot 2}$ **12.** $8\cdot 2-\dfrac{(2+2)3}{3^2-5}$

Simplify each expression.

13. $8x+2-5x-3$ **14.** $4x^2+x^3-2x^3-3x^3$ **15.** $-x^3+5x(x^2-2)$

16. $17x+3(7-2x)$ **17.** $12(t+3)+t(t-3)$ **18.** $-3x(-x-5-2x)$

19. $x(x^2+1)-x^3$ **20.** $13m-12m(5-2)$ **21.** $5x^3+7(x^2+x+1)$

22. $\dfrac{3x+5-2}{3}$ **23.** $\dfrac{3}{2}(-4x+8)-x$ **24.** $\dfrac{14x}{2}(x-2)+3(-x)$

Evaluate each expression when $t=-7$.

25. $t+2(t-1)$ **26.** $3t^2-12t+8$ **27.** $8t$ **28.** $5(t+4)^2+(2t+1)$

29. $\dfrac{1}{2}(t+13)-t$ **30.** $11-2t$ **31.** $3t-t^2+(t+2)^2$ **32.** $5(t+9)^3+\dfrac{2t-1}{t+4}$

Evaluate each expression when $x=3$ and $y=-1$.

33. $2(x+2y)$ **34.** $(x+y)(x-y)$ **35.** $12(13y+4x)-1$ **36.** $(2x+3y)^3+2x^2$

37. x^2+y^2+12-y **38.** $x+y+3x-y$ **39.** $\dfrac{2(8x-y)}{x-7y}$ **40.** $\dfrac{2x-8y}{x+y}+(x+1)^2$

Translating Phrases into Variable Expressions

Often a mathematics problem is presented in words. The words must be translated into a variable expression to solve the problem.

EXAMPLE

Rewrite each statement as a variable expression.

a. Oranges cost $1 per dozen.

b. Each player brings three markers.

c. A science class splits into lab groups with five students in each group.

d. The total cost depends on the number of tickets, at $22 each, and the number of cars parked, at $10 each.

SOLUTION

a. You pay 1 dollar for every 12 oranges, so the cost of x oranges is $\dfrac{x}{12}$.

b. The number of markers is three times the number of players, or $3p$.

c. The number of students in the class is five times the number of lab groups, or $5g$.

d. The cost for t tickets and c cars is $22t+10c$.

Simplifying and Evaluating Expressions

1. -8

2. 49

3. -168

4. 14

5. -9

6. -18

7. -1

8. 53

9. 3

10. 2

11. $-\dfrac{5}{9}$

12. 13

13. $3x-1$

14. $4x^2-4x^3$

15. $4x^3-10x$

16. $11x+21$

17. $t^2+9t+36$

18. $9x^2+15x$

19. x

20. $-23m$

21. $5x^3+7x^2+7x+7$

22. $x+1$

23. $-7x+12$

24. $7x^2-17x$

25. -23

26. 239

27. -56

28. 32

29. 10

30. 25

31. -45

32. 45

33. 2

34. 8

35. -13

36. 45

37. 23

38. 12

39. 5

40. 23

PRACTICE

Rewrite each statement as a variable expression.

1. A child must pick up three toys for each year of his or her age.

2. Lunch costs $4.25 for each main dish and $.75 for each side dish. Drinks are $1 each.

3. Lionel wants to pour orange juice evenly into 8 glasses.

4. Monica charges $20 an hour for lost data retrieval.

5. The hurricane is moving northeast at 45 miles per hour.

6. The number of tiles used in each mosaic are 44 yellow tiles, 125 white tiles, 90 blue tiles, 78 red tiles, and 13 green tiles.

Algebraic Properties

The following properties are useful when you are calculating with numbers or simplifying algebraic expressions.

Property	Commutative $a + b = b + a$ $ab = ba$	Associative $(a + b) + c = a + (b + c)$ $(ab)c = a(bc)$	Distributive $a(b + c) = ab + ac$
Summary	You can add or multiply numbers in any order without changing the result. $5 + 2 = 2 + 5$ $5 \cdot 2 = 2 \cdot 5$	When you add or multiply three or more numbers, you can regroup the numbers without changing the result. $(-4 + 5) + 3 = -4 + (5 + 3)$ $(-4 \cdot 5) \cdot 3 = -4 \cdot (5 \cdot 3)$	When a sum is multiplied by a number, you can distribute the multiplication to each of the numbers being added. $8(4 + 1) = 8 \cdot 4 + 8 \cdot 1$

EXAMPLE

Tell what property is used in the following sequence of steps:
$-3 + 16 = 16 - 3 = 13.$

SOLUTION

The order of the factors has been changed. The commutative property is used.

PRACTICE

Tell what property is used in each lettered step.

1. a. $-4(x - 1) + 6 = 6 - 4(x - 1)$
   b. $= 6 - 4x + 4$
      $= 10x - 4$

2. a. $(3x)(5y) = (3x \cdot 5)y$
   b. $= (3 \cdot 5x)y$
   c. $= (3 \cdot 5)xy$
      $= 15xy$

3. a. $4(2 + a) + 7 = 4(2) + 4(a) + 7$
   b. $= 4a + 4(2) + 7$
      $= 4a + 15$

Toolbox   **693**

Translating Phrases into Variable Expressions

1. A child y years old must pick up $3y$ toys.

2. The cost of m main dishes, s side dishes, and d drinks is $4.25m + 0.75s + d$.

3. If there is d amount of orange juice, each glass has $\frac{d}{8}$ amount.

4. Monica will charge $20h$ for working h hours.

5. The hurricane will move $45h$ miles in h hours.

6. m mosaics will require $44m$ yellow tiles, $125m$ white tiles, $90m$ blue tiles, $78m$ red tiles, and $13m$ green tiles.

Algebraic Properties

1. a. commutative
   b. distributive

2. a. associative
   b. commutative
   c. associative

3. a. distributive
   b. commutative

LINEAR EQUATIONS

Evaluating Equations for Given Values

To evaluate an equation for a given value, substitute the value for the variable.

EXAMPLE

For $y = 3x + 2$, find y when $x = -5$.

SOLUTION

$$y = 3x + 2$$
$$= 3(-5) + 2$$
$$= -15 + 2$$
$$= -13$$

PRACTICE

Find y when $x = 2$.

1. $y = x - 6$
2. $y = 3x - 5$
3. $y = 8 - 4x$
4. $y = 4(2x - 1)$
5. $y = x^2 + 7$
6. $y = 3x^2 + 8x - 14$
7. $y = \dfrac{-x + 1}{4}$
8. $y = \dfrac{3}{2}(x + 2)^2$

Translating Sentences into Equations

Often a mathematics problem is presented in words. The words must be translated into an equation to solve the problem.

EXAMPLE

Rewrite the following sentences as equations.

a. What is the cost of a bunch of bananas if bananas sell for $.70 per pound?

b. How many books will Maura read if she reads 2 books a week?

c. How far will Zachary travel if he drives at an average speed of 45 miles per hour?

d. How can you measure a distance in yards with a ruler that is only 1 ft long?

SOLUTION

a. The cost c in dollars equals 0.70 times the weight w of the bananas in pounds.
$c = 0.70w$

b. The total number of books b equals 2 times the number of weeks w.
$b = 2w$

c. The distance d in miles equals 45 times the time t in hours.
$d = 45t$

d. The distance y in yards equals 3 times the number of feet f.
$y = 3f$

Evaluating Equations for Given Values

1. $y = -4$
2. $y = 1$
3. $y = 0$
4. $y = 12$
5. $y = 11$
6. $y = 14$
7. $y = -\dfrac{1}{4}$
8. $y = 24$

Rewrite the following sentences as equations.

1. How much will Manuel pay for soda if soda costs $.80 per bottle?

2. A potter makes seventeen pieces a week. How many pieces are produced over a period of time?

3. How fast does Elaine run if she covers a distance in 15 seconds?

4. How much will Laura pay for using a computer program, if the cost is $1 for using the program plus $.05 per minute used?

5. What is the total weight of a container of food from the deli, if the container weighs 0.01 lb?

6. If a chemical reaction requires two hydrogen atoms for every oxygen atom, how much oxygen is needed?

Solving One-Step Equations

Any value of a variable that makes an equation true is called a *solution* of the equation. To find a solution, you need to get the variable alone on one side of the equation. You can do this by adding or subtracting the same value from each side of the equation.

EXAMPLE 1

Solve each equation.

a. $x + 8 = 0$ b. $y - 5 = 11$

SOLUTION

a. $x + 8 = 0$
   $x + 8 - 8 = 0 - 8$ Subtract 8 from each side.
   $x = -8$

b. $y - 5 = 11$
   $y - 5 + 5 = 11 + 5$ Add 5 to each side.
        $y = 16$

You can also isolate the variable on one side by multiplying or dividing each side of the equation by the same value.

EXAMPLE 2

Solve each equation.

a. $-3t = 21$ b. $\frac{3}{4}w = 12$

SOLUTION

a. $-3t = 21$
   $\frac{-3t}{-3} = \frac{21}{-3}$ Divide each side by –3.
   $t = -7$

b. $\frac{3}{4}w = 12$
   $\frac{4}{3} \cdot \frac{3}{4}w = \frac{4}{3} \cdot 12$ Multiply each side by $\frac{4}{3}$, the inverse of $\frac{3}{4}$.
        $w = 16$

Toolbox   **695**

Translating Sentences into Equations

1. The cost c of soda in dollars equals 0.80 times the number of bottles b. $c = 0.80b$

2. The total number of pieces p is 17 times the number of weeks w. $p = 17w$

3. Elaine's speed s in feet per second is the distance run d in feet divided by 15 s. $s = \frac{d}{15}$

4. The cost of using the program c in dollars is 1 plus 0.05 times the number of minutes the program is used, m. $c = 1 + 0.05m$

5. The total weight w of a container of food from the deli is the weight of the food f plus 0.01 lb. $w = f + 0.01$

6. The number n of oxygen atoms required is half the number of hydrogen atoms, h. $n = \frac{h}{2}$

Solve each equation.

1. $y + 2 = 7$ **2.** $x - 5 = 8$ **3.** $y + 4 = 13$

4. $3 + t = -1$ **5.** $3x = 8$ **6.** $\frac{1}{3}z = 15$

7. $t \div 8 = 4$ **8.** $-1 + x = -2$ **9.** $\frac{6}{5}y = 6$

10. $8w = 9$ **11.** $2 = \frac{1}{2}y$ **12.** $2x = -6$

Solving Two-Step Equations

Sometimes it takes more than one step to solve an equation. You may need to add or subtract, and then multiply or divide.

EXAMPLE

Solve each equation.

a. $6x + 4 = 16$ **b.** $-0.2x - 8.1 = 0.3$

SOLUTION

Add **8.1** to each side.

a.
$$6x + 4 = 16$$
$$6x + 4 - 4 = 16 - 4 \quad \text{Subtract 4 from each side.}$$
$$6x = 12$$
$$\frac{6x}{6} = \frac{12}{6} \quad \text{Divide each side by 6.}$$
$$x = 2$$

b.
$$-0.2x - 8.1 = 0.3$$
$$-0.2x - 8.1 + 8.1 = 0.3 + 8.1$$
$$-0.2x = 8.4$$
$$\frac{-0.2x}{-0.2} = \frac{8.4}{-0.2} \quad \text{Divide each side by -0.2.}$$
$$x = -42$$

PRACTICE

Solve each equation.

1. $4x + 5 = 7$ **2.** $-x + 3 = 2$ **3.** $8 \div t = 2$ **4.** $12x + 8 = -6$

5. $\frac{1}{2}y + 2 = 3$ **6.** $-\frac{5}{3}x + 3 = -12$ **7.** $\frac{1}{x} = \frac{1}{9}$ **8.** $12 = 4 - 2y$

9. $1.2x + 3 = -4.7$ **10.** $18 = -5 + 3x$ **11.** $9.90 = 3.35 + 1.2t$ **12.** $4 - 5y = 1.8$

GRAPHING ON THE COORDINATE PLANE

Graphing Points

A *coordinate plane* consists of a horizontal *x-axis* and a vertical *y-axis* that intersect at a point called the *origin*, labeled *O*. The axes divide the coordinate plane into four *quadrants* as shown at the top of page 697.

Solving One-Step Equations

1. $y = 5$
2. $x = 13$
3. $y = 9$
4. $t = -4$
5. $x = \frac{8}{3}$
6. $z = 45$
7. $t = 32$
8. $x = -1$
9. $y = 5$
10. $w = \frac{9}{8}$
11. $y = 4$
12. $x = -3$

Solving Two-Step Equations

1. $x = \frac{1}{2}$
2. $x = 1$
3. $t = 4$
4. $x = -\frac{7}{6}$
5. $y = 2$
6. $x = 9$
7. $x = 9$
8. $y = -4$
9. $x = -6.416$
10. $x = \frac{23}{3} = 7.6$
11. $t = 5.4583$
12. $y = 0.44$

Each point in a coordinate plane is associated with an *ordered pair* (a, b) of real numbers. The first number, a, is the *x-coordinate*. The second number, b, is the *y-coordinate*.

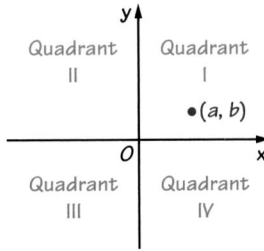

EXAMPLE

Graph the points $A(-4, -2)$ and $B(0, 3)$ in a coordinate plane. Name the quadrant, if any, in which each point lies.

SOLUTION

To graph $A(-4, -2)$, start at the origin and move **left** 4 units and **down** 2 units. The point is in Quadrant III.

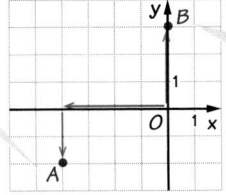

To graph $B(0, 3)$, start at the origin and move horizontally 0 units and **up** 3 units. The point is not in any quadrant.

PRACTICE

For Exercises 1–8, graph each point in the same coordinate plane. Name the quadrant, if any, in which each point lies.

1. $A(0, 1)$ **2.** $B(-3, 6)$ **3.** $C(2, -2)$ **4.** $D(7, 0)$

5. $E(0, 0)$ **6.** $F(-2, -2)$ **7.** $G(-3, 0)$ **8.** $H(4, 3)$

Give the coordinates of a point that satisfies each condition.

9. The point is in the fourth quadrant.

10. The point is on the negative *x*-axis.

Graphing Linear Equations

A *linear equation* is an equation whose graph is a line. One way to write a linear equation is $y = mx + b$. The slope is represented by the value m, and the *y-intercept* is represented by the value b.

The slope of a line is the vertical change divided by the horizontal change. The slope of a line through points (x_1, y_1) and (x_2, y_2) is:

$$m = \frac{y_2 - y_1}{x_2 - x_1}$$

Graphing Points

1–8.

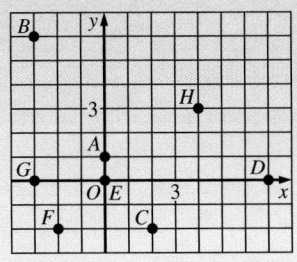

1. no quadrant

2. Quadrant II

3. Quadrant IV

4. no quadrant

5. no quadrant

6. Quadrant III

7. no quadrant

8. Quadrant I

9, 10. Answers may vary.
     Examples are given.

9. $(2, -1)$

10. $(-2, 0)$

EXAMPLE 1

Graph the equation.

a. $y = 2x - 3$

b. $y = -\dfrac{2}{3}x + 1$

SOLUTION

Graph the y-intercept. Use the slope to graph one or two more points.

a.

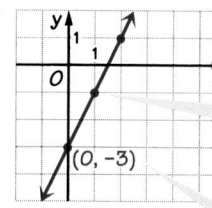

The slope is $\dfrac{2}{1}$; go up 2 and right 1.

The y-intercept is -3.

b.

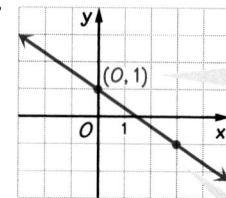

The y-intercept is 1.

The slope is $-\dfrac{2}{3}$; go down 2 and right 3.

You can find the slope of a line by putting an equation into slope-intercept form and then reading m from the equation. You can determine the slope of a graph by choosing two points and finding the vertical change and the horizontal change between them.

EXAMPLE 2

Find the slope of each line.

a. $y = 4x + 2$

b. $y = 3 - \dfrac{1}{2}x$

c. $x + y = 1$

SOLUTION

Rewrite the equations in slope-intercept form.

Subtract x from each side.

a. $y = 4x + 2$

slope: 4

b. $y = 3 - \dfrac{1}{2}x$

$y = -\dfrac{1}{2}x + 3$

slope: $-\dfrac{1}{2}$

c. $x + y = 1$

$x + y - x = 1 - x$

$y = -x + 1$

slope: -1

EXAMPLE 3

Find the slope of the line.

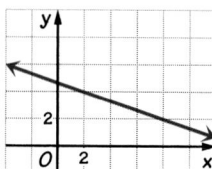

SOLUTION

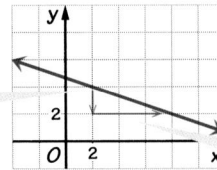

The **vertical** change is –2.

The **horizontal change** is 6.

The slope is $\frac{-2}{6}$, or $-\frac{1}{3}$.

You can find the *y*-intercept of a line by putting an equation into slope-intercept form and then reading *b* from the equation. You can determine the *y*-intercept of a graph by inspection.

EXAMPLE 4

Find the *y*-intercept of each line.

a. $y = -x + 2$ **b.** $y = 4 - 5x$ **c.** $3x + y = 0$

SOLUTION

Rewrite the equations in slope-intercept form.

a. $y = -x + 2$
y-intercept: 2

b. $y = 4 - 5x$
$y = -5x + 4$
y-intercept: 4

c. $3x + y = 0$
$y = -3x$
y-intercept: 0

EXAMPLE 5

Find the *y*-intercept of the line in Example 3.

SOLUTION

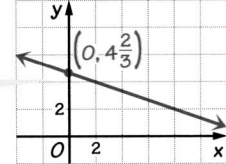

The *y*-intercept is $4\frac{2}{3}$.

PRACTICE

Find the slope and *y*-intercept of each line.

1. $y = 3x$ **2.** $y = -2x$ **3.** $y = -x + 2$ **4.** $y = -4$

5. $y = -5x - 1$ **6.** $y = -3 + 8x$ **7.** $y - x = 2$ **8.** $y = 122 + 13.8x$

9. $y = x - 3 + 2x$ **10.** $y = -\frac{5}{4}x - \frac{3}{2}$ **11.** $y = \frac{1}{2}x + \frac{1}{2}$ **12.** $y + \frac{2}{3}x - 3 = 0$

13–24. Graph each of the equations in Exercises 1–12.

Toolbox   **699**

16.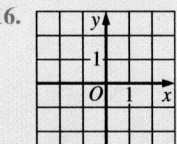

17.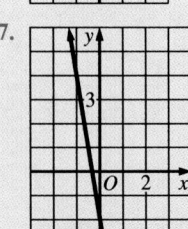

18.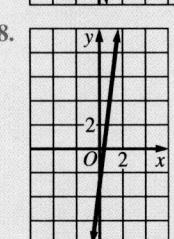

19.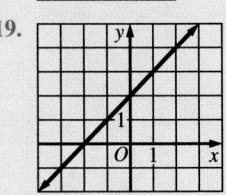

20.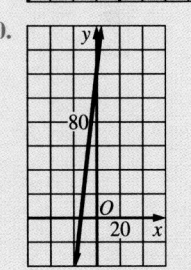

21.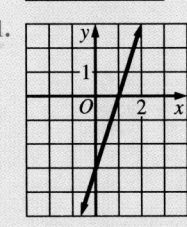

22.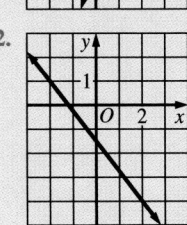

23.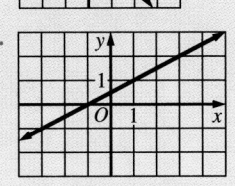

24.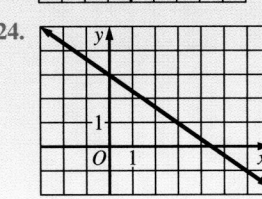

Graphing Linear Equations

1. slope 3; *y*-intercept 0

2. slope –2; *y*-intercept 0

3. slope –1; *y*-intercept 2

4. slope 0; *y*-intercept –4

5. slope –5; *y*-intercept –1

6. slope 8; *y*-intercept –3

7. slope 1; *y*-intercept 2

8. slope 13.8; *y*-intercept 122

9. slope 3; *y*-intercept –3

10. slope $-\frac{5}{4}$; *y*-intercept $-\frac{3}{2}$

11. slope $\frac{1}{2}$; *y*-intercept $\frac{1}{2}$

12. slope $-\frac{2}{3}$; *y*-intercept 3

13.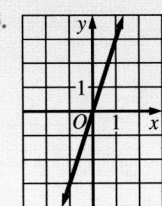

14.

15.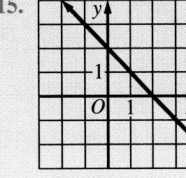

For each graph in Exercises 25–33, find the slope and the *y*-intercept.

25.
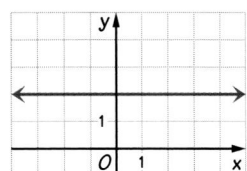

26.
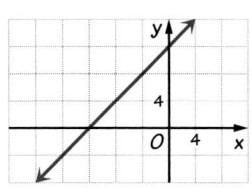

27.
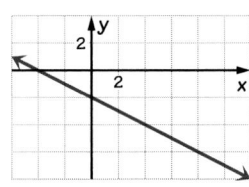

28.
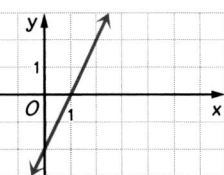

29.
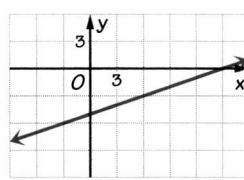

30.
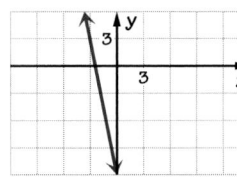

31.
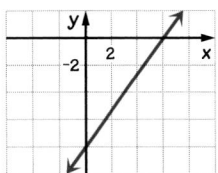

32.
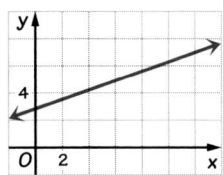

33.
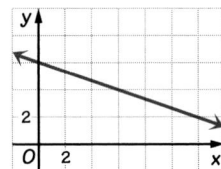

QUADRATIC EQUATIONS

Simplifying Radicals

An expression written using the symbol $\sqrt{}$ is in *radical form*. An expression in radical form is in simplest form when the expression under the radical sign is in simplest form and:

1. there is no integer under the radical sign with a perfect square factor,

2. there are no fractions under the radical sign, and

3. there are no radicals in the denominator.

Properties of Square Roots

For all nonnegative numbers a and b:

$$\sqrt{a^2} = a$$

$$\sqrt{ab} = \sqrt{a} \cdot \sqrt{b}$$

$$\sqrt{\frac{a}{b}} = \frac{\sqrt{a}}{\sqrt{b}}, b \neq 0$$

$$\left(\sqrt{a}\right)^2 = a$$

For example:

$$\sqrt{36} = 6$$

$$\sqrt{9 \cdot 5} = \sqrt{9} \cdot \sqrt{5} = 3\sqrt{5}$$

$$\sqrt{\frac{2}{9}} = \frac{\sqrt{2}}{\sqrt{9}} = \frac{\sqrt{2}}{3}$$

$$\left(\sqrt{16}\right)^2 = (4)^2 = 16$$

---

25. slope 0; *y*-intercept 2

26. slope 1; *y*-intercept 12

27. slope $-\frac{1}{2}$; *y*-intercept –2

28. slope 2; *y*-intercept –2

29. slope $\frac{1}{3}$; *y*-intercept –5

30. slope –5; *y*-intercept –12

31. slope $\frac{4}{3}$; *y*-intercept –8

32. slope $\frac{1}{3}$; *y*-intercept 3

33. slope $-\frac{1}{3}$; *y*-intercept 6

Simplify each expression.

a. $\sqrt{\dfrac{27}{4}}$

b. $8\sqrt{12} - \sqrt{3}$

SOLUTION

a. $\sqrt{\dfrac{27}{4}} = \dfrac{\sqrt{27}}{\sqrt{4}}$

$\quad = \dfrac{\sqrt{9 \cdot 39}}{\sqrt{4}}$

$\quad = \dfrac{\sqrt{3^2 \cdot 3}}{\sqrt{2^2}}$

$\quad = \dfrac{3\sqrt{3}}{2}$

b. $8\sqrt{12} - \sqrt{3} = 8\sqrt{4 \cdot 3} - \sqrt{3}$

$\qquad\qquad\qquad = 8\sqrt{2^2} \cdot \sqrt{3} - \sqrt{3}$

$\qquad\qquad\qquad = 8 \cdot 2 \cdot \sqrt{3} - \sqrt{3}$

$\qquad\qquad\qquad = 16\sqrt{3} - \sqrt{3}$

$\qquad\qquad\qquad = (16 - 1)\sqrt{3}$

$\qquad\qquad\qquad = 15\sqrt{3}$

> Simplify each term in the expression.

> Use the distributive property.

PRACTICE

Simplify each expression.

1. $\sqrt{16}$ **2.** $\sqrt{75}$ **3.** $\sqrt{2}$ **4.** $\sqrt{9 \cdot 4}$

5. $\sqrt{\dfrac{8}{9}}$ **6.** $\sqrt{\dfrac{1}{64}}$ **7.** $\sqrt{\dfrac{15}{4}}$ **8.** $\sqrt{\dfrac{48}{100}}$

9. $\sqrt{8} + \sqrt{18}$ **10.** $\sqrt{125}$ **11.** $-\sqrt{45} - \sqrt{20}$ **12.** $\sqrt{9} - \sqrt{81}$

13. $\left(\sqrt{49}\right)^2$ **14.** $\sqrt{92 - 11}$ **15.** $\sqrt{\dfrac{2 \cdot 5}{40}}$ **16.** $\dfrac{\sqrt{54}}{6}$

Solving Simple Quadratic Equations

You can solve some simple equations using radicals.

EXAMPLE 1

Solve.

a. $x^2 = 16$ **b.** $x^2 = 15$ **c.** $2x^2 + 1 = 9$

SOLUTION

Rewrite the equation, if necessary, so that it is in the form $x^2 = a$, where a is positive. Find the square root of both sides.

a. $x^2 = 16$

$\quad x = \pm\sqrt{16}$

$\quad x = 4 \text{ or } x = -4$

b. $x^2 = 15$

$\quad x = \sqrt{15}$

$\quad x = \sqrt{15} \text{ or } x = -\sqrt{15}$

c. $2x^2 + 1 = 9$

$\quad 2x^2 = 8$

$\quad x^2 = 4$

$\quad x = \pm 2$

$\quad x = 2 \text{ or } x = -2$

Simplifying Radicals

1. 4
2. $5\sqrt{3}$
3. $\sqrt{2}$
4. 6
5. $\dfrac{2\sqrt{2}}{3}$
6. $\dfrac{1}{8}$
7. $\dfrac{\sqrt{15}}{2}$
8. $\dfrac{2\sqrt{3}}{5}$
9. $5\sqrt{2}$
10. $5\sqrt{5}$
11. $-5\sqrt{5}$
12. -6
13. 49
14. 9
15. $\dfrac{1}{2}$
16. $\dfrac{\sqrt{6}}{2}$

Solve.

1. $x^2 = 25$ **2.** $x^2 = 50$ **3.** $2x^2 = 12$ **4.** $3x^2 = 108$

5. $x^2 + 5 = 6$ **6.** $-x^2 + 2 = -14$ **7.** $2x^2 - 8 = 12$ **8.** $2x^2 - 1 = -1$

9. $2x^2 = x^2 + 4$ **10.** $3x^2 = 36 - x^2$ **11.** $x^2 = 2^2 + 5^2$ **12.** $3x^2 - 2 = 16 + x^2$

FORMULAS FOR GEOMETRIC FIGURES

Working with Formulas

A *formula* is a statement of a relationship between two or more quantities.
Some common formulas include

$$\text{speed} = \frac{\text{distance}}{\text{time}}, \text{ circumference} = 2\pi \cdot \text{radius, and distance} = \frac{1}{2} \cdot \text{acceleration} \cdot (\text{time})^2.$$

When you work with formulas, you may need to rewrite the formula to isolate a
variable on one side of an equation. To use a formula, substitute a number for each
known variable and solve.

EXAMPLE 1

Rewrite the formula $s = \dfrac{d}{t}$ to find an equation for (a) d and (b) t.

SOLUTION

a. $s = \dfrac{d}{t}$ To solve for d, multiply each side by t.

$st = d$

$d = st$ Write the equation with d on the left. This is called an equation for d in terms of s and t.

b. $s = \dfrac{d}{t}$ To get t out of the denominator, multiply by t.

$st = d$

$t = \dfrac{d}{s}$ To isolate t, divide each side by s.

EXAMPLE 2

Use the formula $s = \dfrac{d}{t}$, where s is speed in mi/h, d is distance in miles, and

t is time in hours, to find each value.

a. Find s when $d = 30$ and $t = 4$.   **b.** Find t when $s = 40$ and $d = 20$.

SOLUTION

a. $s = \dfrac{d}{t}$

$s = \dfrac{30}{4}$ Put in the values you know.

$s = \dfrac{15}{2}$, or 7.5 mi/h   Include units in your answer.

b. $t = \dfrac{d}{s}$ Use the formula for t from Example 1.

$t = \dfrac{20}{40}$

$t = 0.5$ h

702   Student Resources

Solving Simple Quadratic Equations

1. $x = 5$ or $x = -5$

2. $x = 5\sqrt{2}$ or $x = -5\sqrt{2}$

3. $x = \sqrt{6}$ or $x = -\sqrt{6}$

4. $x = 6$ or $x = -6$

5. $x = 1$ or $x = -1$

6. $x = 4$ or $x = -4$

7. $x = \sqrt{10}$ or $x = -\sqrt{10}$

8. $x = 0$

9. $x = 2$ or $x = -2$

10. $x = 3$ or $x = -3$

11. $x = \sqrt{29}$ or $x = -\sqrt{29}$

12. $x = 3$ or $x = -3$

Use the formula $s = \frac{d}{t}$, where s is speed in mi/h, d is distance in miles,

and t is time in hours, to find each value.

1. Find t when $d = 4$ and $s = 12$.   **2.** Find d when $s = 50$ and $t = 0.5$.

3. Find d when $s = 35$ and $t = 1$.   **4.** Find s when $d = 100$ and $t = 1.5$.

Use the formula $A = \pi r^2$, where A is the area in cm² and r is the radius in cm,
to find each value.

5. Find A when $r = 0.4$.   **6.** Find r when $A = 9\pi$.

7. Use the formula $a = 4\frac{bc}{d}$.

    a. Write an equation for c.

    b. Find a when $b = 2$ in., $c = 12$ in., and $d = 6$ in.

Finding Perimeter

The *perimeter* of a geometric figure is the distance around the figure, or the
sum of the length of the edges.

EXAMPLE

Find the perimeter of each figure.

a.

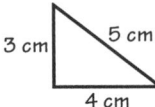

b.

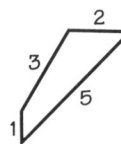

SOLUTION

a. Add together the lengths of the edges: $3 + 4 + 5 = 12$ cm.

b. $3 + 2 + 5 + 1 = 11$

PRACTICE

Find the perimeter of each figure.

1.

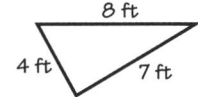

2.

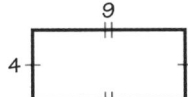

3.

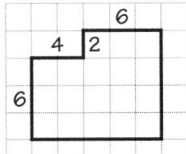

4. A square whose sides are 4 in. long.

5. A rectangle whose sides are 2 ft and 5 ft long.

6. A polygon whose sides are 3, 7, 2, 1, and 8 units long.

Working with Formulas

1. $t = \frac{1}{3}$ h

2. $d = 25$ mi

3. $d = 66\frac{2}{3}$ mi/h

4. $s = 25$ mi

5. $A = 0.16\pi$

6. $r = 3$

7. a. $c = \frac{ad}{4b}$

   b. $a = 16$ in.

Finding Perimeter

1. 19 ft

2. 26

3. 36

4. 16 in.

5. 14 ft

6. 21 units

Finding Circumference

The perimeter of a circle, called its *circumference*, is the distance around the circle. The formula for circumference is $C = 2\pi r$, where r is the *radius*. Circumference and radius are both in units of length.

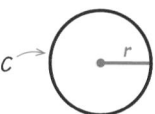

EXAMPLE

Find the circumference of a circle with radius 4 in.

SOLUTION

$$C = 2\pi r = 2\pi(4) = 8\pi \text{ in.}$$

PRACTICE

Find the circumference of a circle with each radius.

1. 8 in.   **2.** 2 cm   **3.** 1 in.   **4.** 12 in.

5. 2   **6.** 14 m   **7.** 30   **8.** 4.5

Find the radius of a circle with each circumference.

9. 18π in.   **10.** 34π **11.** 1π **12.** 2 mi

Finding Area and Volume

The *area* of a figure is a measure of the number of square units of space its surface covers; for a rectangle this is length times width, $A = \ell w$. The *volume* of a figure is the number of cubic units of space it occupies. For a rectangular prism this is length times width times height, $V = \ell w h$.

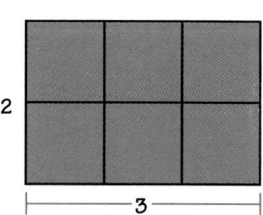

$A = 2 \cdot 3 = 6$ square units

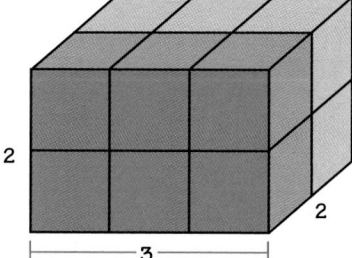

$V = 2 \cdot 3 \cdot 2 = 12$ cubic units

EXAMPLE

Find the area of a rectangle that is 2 in. long and 4 in. wide.

SOLUTION

$$A = \ell w = (2)(4) = 8 \text{ in.}^2$$

Finding Circumference

1. 16π in.
2. 4π cm
3. 2π in.
4. 24π in.
5. 4π
6. 28π m
7. 60π
8. 9π
9. 9 in.

10. 17
11. $\frac{1}{2}$
12. $\frac{1}{\pi}$ mi

Find the volume of a rectangular prism with the given dimensions.

9. $\ell = 1, w = 2, h = 5$ **10.** $\ell = 3, w = 3, h = 3$ **11.** $\ell = 10, w = 10, h = 14$

12. $\ell = 3, w = 1, h = 2$ **13.** $\ell = 4, w = 4, h = 6$ **14.** $\ell = 0.5, w = 0.5, h = 0.5$

For a rectangle, find the value of the given variable.

15. ℓ, when $A = 12$ and $w = 6$ **16.** w, when $A = 15$ and $\ell = 4$

17. w, when $A = 3$ and $\ell = 9$ **18.** A, when $\ell = 5$ and $w = 10$

INEQUALITIES

Solving Inequalities

You solve a *linear inequality* such as $2n + 1 \leq 7$ in much the same way as a linear equation. One important difference is this: When you multiply or divide by a negative number, you must reverse the direction of the inequality sign.

EXAMPLE

Solve each inequality. Graph the solution on a number line.

a. $2n + 1 \leq 7$ **b.** $-2x > 10 + 3x$

SOLUTION

a. $2n + 1 \leq 7$

$2n + 1 - 1 \leq 7 - 1$ ← Subtract **1** from each side.

$2n \leq 6$

$\dfrac{2n}{2} \leq \dfrac{6}{2}$ ← Divide each side by **2**.

$n \leq 3$

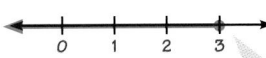

The closed circle shows that 3 *is* a solution.

b. $-2x > 10 + 3x$

$-2x - 3x > 10 + 3x - 3x$ ← Subtract **3x** from each side.

$-5x > 10$

$\dfrac{-5x}{-5} < \dfrac{10}{-5}$ ← Divide each side by **−5**, reversing the inequality sign.

$x < -2$

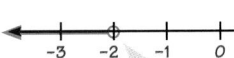

The open circle shows that −2 *is not* a solution.

PRACTICE

Solve each inequality. Graph on a number line.

1. $\dfrac{1}{2}y - 12 < -4$ **2.** $\dfrac{5}{3}n + 2 > -\dfrac{2}{3}n + 5$ **3.** $8n \leq 18 - n$ **4.** $1 - n > 1 - 2n$

5. $-3x + 2 \geq -x$ **6.** $12x < 15 - 2x$ **7.** $\dfrac{1}{4}y + 3 < 1$ **8.** $-4n - 3 \geq -2n - 1$

Toolbox **705**

Finding Area and Volume

1. 12
2. 2
3. 168
4. 3.5
5. 2.25
6. 60
7. 4
8. 20
9. 10
10. 27
11. 1400
12. 6
13. 96
14. 0.125
15. $l = 2$
16. $w = 3.75$
17. $w = \dfrac{1}{3}$
18. $A = 50$

Solving Inequalities

1. $y < 16$

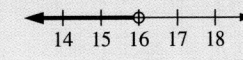

2. $n > \dfrac{9}{7}$

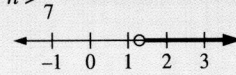

3. $n \leq 2$

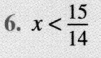

4. $n > 0$

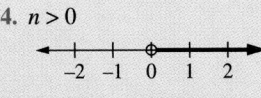

5. $x \leq 1$

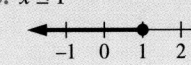

6. $x < \dfrac{15}{14}$

7. $y < -8$

8. $n \leq -1$

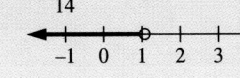

Graphing Systems of Equations and Inequalities

A *system of linear equations* is two or more linear equations that state relationships between the same variables. The solution of a system of equations is the point where the lines intersect.

A *system of linear inequalities* is two or more inequalities that state relationships between the same variables. The solution of a system of inequalities is the area included in all the graphs.

EXAMPLE 1

Graph the system $y = 2x + 1$ and $x + y = -2$. Use the graph to find the solution of the system.

SOLUTION

Put both equations into slope-intercept form.

$y = 2x + 1$

$x + y = -2$

$x + y - x = -2 - x$

$y = -x - 2$

Graph the equations.

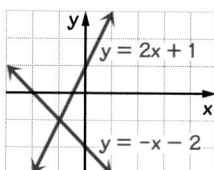

The solution is $(-1, -1)$. You can check this by substituting $(-1, -1)$ into the two equations.

EXAMPLE 2

Graph the system $y \leq \frac{1}{2}x + 1$ and $y + 3 > 0$.

SOLUTION

Put both inequalities into slope-intercept form.

$y \leq \frac{1}{2}x + 1$

$y + 3 > 0$

$y + 3 - 3 > 0 - 3$

$y > -3$

Graph the area above $y = -3$.

Graph the inequalities.

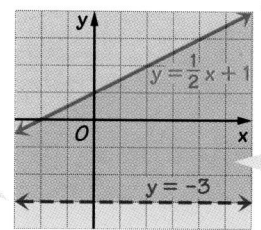

Graph the area below and including $y = \frac{1}{2}x + 1$

The solution is the area shaded in purple.

The solutions of the system $y \leq \frac{1}{2}x + 1$ and $y + 3 > 0$ are all points that are above the line $y = -3$ and are below or on the line $y = \frac{1}{2}x + 1$.

PRACTICE

Graph each system, and find the solution of the system.

1. $y = \frac{3}{2}x - 6$

$y = -\frac{1}{2}x + 6$

2. $y = -2$

$y = -\frac{1}{2}x$

3. $y - 2 = \frac{2}{5}x$

$x + 3y + 5 = 0$

4. $2x + y = 1$

$x + y = -1$

Graphing Systems of Equations and Inequalities

1.

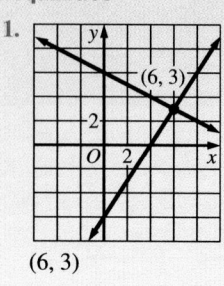

(6, 3)

2.

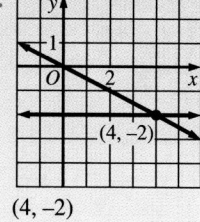

(4, −2)

3.

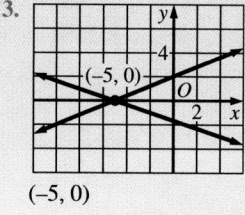

(−5, 0)

4.

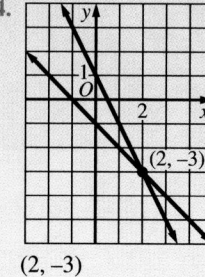

(2, −3)

Graph each system. Label the solution of the system on your graph.

5. $y > x - 2$
$y < \frac{1}{2}x + 2$

6. $y < \frac{2}{3}x + 2$
$y > \frac{2}{3}x$

7. $y \leq \frac{1}{4}x - 2$
$y - \frac{1}{2}x + 2 \leq 0$

8. $-x + 3y < 18$
$x < 3$

RATIO AND PROPORTION

Creating a Ratio

You can use *ratios* to compare two numbers using division. A ratio can be written in three ways: *a* to *b*, *a*:*b*, and $\frac{a}{b}$. Ratios are often used to give *rates*, ratios that compare two different quantities. A *unit rate* is a rate per one unit of a given quantity.

EXAMPLE 1

Write each ratio in lowest terms.

a. 10 to 15

b. $\frac{45 \text{ s}}{1 \text{ min}}$

SOLUTION

a. Write as a fraction in lowest terms:

$\frac{10}{15} = \frac{2}{3}$

b. Convert minutes to seconds:

$\frac{45 \text{ s}}{1 \text{ min}} = \frac{45 \text{ s}}{60 \text{ s}} = \frac{3}{4}$

EXAMPLE 2

What is the unit rate if you read 8 articles in 2 hours?

SOLUTION

Write the comparison as a fraction: $\frac{\text{articles}}{\text{hours}} \rightarrow \frac{8}{2} = 4$ articles per hour

PRACTICE

Write each ratio in lowest terms.

1. $12:3$

2. 18 to 40

3. $\frac{20}{100}$

4. $\frac{3}{7}$

5. 20 min to 1 h

6. 500 lb to 2 tons

7. 3 ft : 15 in.

8. 1 cm : 1 m

9. 6 months to three years

Write each unit rate.

10. 20 mi in 6 h

11. $40 in 1 h

12. $15,000 in 2 years

13. 45 h in 7 days

14. 13 cats in 4 hours

15. 120 mi in 5 days

16. 7200 cycles in 60 s

17. 18 in. in 72 min

18. 240 apples in 8 baskets

Toolbox **707**

8.
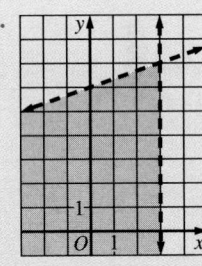

Creating a Ratio

1. $\frac{4}{1}$

2. $\frac{9}{20}$

3. $\frac{1}{5}$

4. $\frac{3}{7}$

5. $\frac{1}{3}$

6. $\frac{1}{8}$

7. $\frac{12}{5}$

8. $\frac{1}{100}$

9. $\frac{1}{6}$

10. about 3.3 miles per hour

11. $40 per hour

12. $7500 per year

13. about 6.4 hours per day

14. 3.25 cats per hour

15. 24 miles per day

16. 120 cycles per second

17. 0.25 inches per minute

18. 30 apples per basket

5.
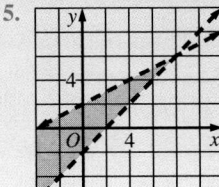

6.
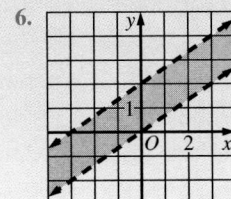

7.
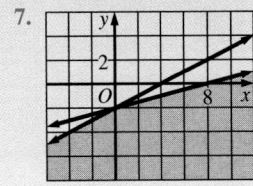

Solving Proportions

To solve a proportion, you must isolate the variable on one side.

Solve $\frac{3}{4} = \frac{x}{8}$.

SOLUTION

$$\frac{3}{4} = \frac{x}{8}$$

$$8 \cdot \frac{3}{4} = 8 \cdot \frac{x}{8} \qquad \text{Multiply each side by 8.}$$

$$6 = x$$

PRACTICE

Solve.

1. $\frac{x}{5} = \frac{4}{10}$

2. $\frac{1}{2} = \frac{x}{14}$

3. $\frac{3}{n} = 9$

4. $\frac{y}{8} = \frac{13}{4}$

5. $\frac{4}{9} = \frac{12}{x}$

6. $\frac{x}{100} = \frac{12}{60}$

7. $\frac{4}{15} = \frac{m}{3}$

8. $\frac{2}{7} = \frac{10}{x}$

DATA ANALYSIS AND PROBABILITY

Finding Averages

There are two types of averages often taken of a data set. The *mean* is the sum of all the data divided by the number of data items. The *median* is the middle number when you put the data in order from smallest to largest. If the number of data items is even, the median is the mean of the two middle numbers.

EXAMPLE

For the data set 2, 4, 8, 7, 1, 6, find each value.

a. the mean

b. the median

SOLUTION

a. Find the value: $\dfrac{\text{sum of the data}}{\text{number of data items}}$

$$\frac{2 + 4 + 8 + 7 + 1 + 6}{6} = \frac{28}{6}$$

$$= \frac{14}{3}, \text{ or } 4\frac{2}{3}$$

The mean of the data set is $4\frac{2}{3}$.

b. Put the data in order: 1, 2, 4, 6, 7, 8

The two middle numbers are 4 and 6.

Find the mean of the two middle numbers:

$$\frac{4 + 6}{2} = \frac{10}{2} = 5$$

The median of the data set is 5.

Solving Proportions

1. $x = 2$

2. $x = 7$

3. $n = \frac{1}{3}$

4. $y = 26$

5. $x = 27$

6. $x = 20$

7. $m = \frac{4}{5}$

8. $x = 35$

PRACTICE

Find the mean and the median of each data set.

1. 5, 7, 7, 10   **2.** 18, 20, 19, 20, 20   **3.** 1, 1, 3, 5, 7, 8, 8, 8   **4.** 12, 14, 11, 10

5. 14, 20, 23, 18, 7, 3   **6.** 1, 0, 3, 4, 3, 2, 0   **7.** 8, -5, 3, -4, -6, -2 **8.** 25, 22, 18, 17, 13

Scientific Notation

Very large and very small numbers are often expressed using scientific notation. A number is expressed in scientific notation if it is in the form $a \times 10^n$, where $1 \le a < 10$ and n is an integer.

EXAMPLE 1

Write each number as a decimal.

a. 2.8×10^{-4}

b. 9.21×10^6

SOLUTION

a. $2.8 \times 10^{-4} = 0.00028$

Move the decimal point 4 places to the left.

b. $9.21 \times 10^6 = 9,210,000$

Move the decimal point 6 places to the right.

EXAMPLE 2

Write each number in scientific notation.

a. 15,400,000,000

b. 0.007

SOLUTION

a. $15,400,000,000 = 1.54 \times 10^{10}$

Factoring out 10^{10} moves the decimal point 10 places to the left.

b. $0.007 = 7 \times 10^{-3}$

Factoring out 10^{-3} moves the decimal point 3 places to the right.

PRACTICE

Write each number as a decimal.

1. 1.8×10^{-5} **2.** 4.55×10^{-11} **3.** 5.2×10^9 **4.** 3.812×10^2

5. 9.84×10^1 **6.** 5.7×10^{-8} **7.** 3.9×10^{-1} **8.** 4.20×10^4

Write each number in scientific notation.

9. 0.00088   **10.** 0.12   **11.** 1,245,000   **12.** 0.0131

13. 0.00000000006   **14.** 45,000   **15.** 158,000,000   **16.** 230,000

17. 6200   **18.** 0.005   **19.** 0.0000000000011   **20.** 3,486,510

Toolbox   **709**

Finding Averages

1. mean: 7.25; median: 7
2. mean: 19.4; median: 20
3. mean: 5.125; median: 6
4. mean: 11.75; median: 11.5
5. mean: 14.16; median: 16
6. mean: 1.857142; median: 2
7. mean: –1; median: –3
8. mean: 19; median: 18

Scientific Notation

1. 0.000018
2. 0.0000000000455
3. 5,200,000,000
4. 381.2
5. 98.4
6. 0.000000057
7. 0.39
8. 42,000
9. 8.8×10^{-4}
10. 1.2×10^{-1}
11. 1.245×10^6
12. 1.31×10^{-2}
13. 6×10^{-11}
14. 4.5×10^4
15. 1.58×10^8
16. 2.3×10^5
17. 6.2×10^3
18. 5×10^{-3}
19. 1.1×10^{-12}
20. 3.48651×10^6

Probability

Probability is a ratio that measures how likely an event is to happen. *Experimental probability* is the ratio of the number of times an event actually occurs to the number of times an experiment is done. *Theoretical probability* is the ratio of the number of favorable outcomes to the total number of possible outcomes.

EXAMPLE 1

Karen makes 18 out of 30 free throws. What is the experimental probability that she will make a free throw?

SOLUTION

$$\frac{\text{number made}}{\text{number attempted}} = \frac{18}{30} = \frac{3}{5}, \text{ or } 0.60$$

A probability is usually written as a number between 0 and 1.

EXAMPLE 2

A 6-sided die is rolled. What is the theoretical probability of rolling a 1 or a 2?

SOLUTION

probability of rolling a 1 or a 2 = probability of a 1 + probability of a 2

$$= \frac{1}{6} + \frac{1}{6} = \frac{1}{3} \approx 0.33$$

PRACTICE

Find each experimental probability.

1. What is the experimental probability that Mark stands on bare floor at a random point in the room, if carpet covers $\frac{1}{4}$ of the floor?

2. José got a hit in 20 out of 58 at-bats. What is the experimental probability of getting a hit?

3. A coin is flipped and lands heads up on 2 out of 5 flips. What is the experimental probability that the coin will land heads up?

4. You have found an empty seat on the train on 8 out of 14 days. What is the experimental probability of finding an empty seat?

5. If it rains on 12 out of 20 days, what is the experimental probability that you need to bring a raincoat to school on any given day?

Find each theoretical probability.

6. Getting a head on the flip of a fair coin.

7. Drawing any 4 red cards from a standard deck of 52 cards.

8. Rolling a number greater than 3 on a 6-sided die.

9. Rolling a 7 on a 6-sided die.

10. Rolling a 10, 11, or 12 on a 20-sided die.

Probability

1. $\frac{3}{4}$, or 0.75

2. $\frac{10}{29} \approx 0.34$

3. $\frac{2}{5}$, or 0.4

4. $\frac{4}{7} \approx 0.57$

5. $\frac{3}{5}$, or 0.6

6. $\frac{1}{2}$, or 0.5

7. $\frac{1}{13} \approx 0.08$

8. $\frac{1}{2}$, or 0.5

9. $\frac{3}{20}$, or 0.15

10. 0

MATRICES

A *matrix* is a group of numbers arranged in rows and columns. If a matrix has *m* rows and *n* columns, we say its *dimensions* are *m* by *n*, or $m \times n$. If two matrices have the same dimensions, you can add or subtract them by adding or subtracting corresponding elements.

EXAMPLE 1

Evaluate each matrix expression.

a. $\begin{bmatrix} 1 & 2 \\ 3 & 4 \end{bmatrix} + \begin{bmatrix} 5 & -6 \\ 9 & 0 \end{bmatrix}$

b. $\begin{bmatrix} 3 & -2 \\ 0 & 1 \end{bmatrix} - \begin{bmatrix} -4 & 8 \\ -7 & 5 \end{bmatrix}$

SOLUTION

a. $\begin{bmatrix} 1 & 2 \\ 3 & 4 \end{bmatrix} + \begin{bmatrix} 5 & -6 \\ 9 & 0 \end{bmatrix} = \begin{bmatrix} 1+5 & 2-6 \\ 3+9 & 4+0 \end{bmatrix} = \begin{bmatrix} 6 & -4 \\ 12 & 4 \end{bmatrix}$

b. $\begin{bmatrix} 3 & -2 \\ 0 & 1 \end{bmatrix} - \begin{bmatrix} -4 & 8 \\ -7 & 5 \end{bmatrix} = \begin{bmatrix} 3-(-4) & -2-8 \\ 0-(-7) & 1-5 \end{bmatrix} = \begin{bmatrix} 7 & -10 \\ 7 & -4 \end{bmatrix}$

You can multiply a matrix by a number, or *scalar*. Multiply each element of the matrix by the scalar.

EXAMPLE 2

Evaluate $3\begin{bmatrix} 5 \\ -4 \end{bmatrix}$.

SOLUTION

$3\begin{bmatrix} 5 \\ -4 \end{bmatrix} = \begin{bmatrix} 3 \cdot 5 \\ 3 \cdot (-4) \end{bmatrix} = \begin{bmatrix} 15 \\ -12 \end{bmatrix}$

PRACTICE

Evaluate each matrix expression.

1. $\begin{bmatrix} 1 & 0 \\ 0 & 1 \end{bmatrix} + \begin{bmatrix} 5 & 5 \\ 2 & 12 \end{bmatrix}$

2. $\begin{bmatrix} 10 \\ 4 \end{bmatrix} + \begin{bmatrix} -3 \\ 6 \end{bmatrix}$

3. $\begin{bmatrix} 3 & -3 \\ 2 & 1 \end{bmatrix} - \begin{bmatrix} 1 & 1 \\ 0 & 0 \end{bmatrix}$

4. $\begin{bmatrix} -7 \\ -2 \\ -5 \end{bmatrix} + \begin{bmatrix} 8 \\ -3 \\ 0 \end{bmatrix}$

5. $\begin{bmatrix} 8 & 8 \\ 8 & 8 \end{bmatrix} + \begin{bmatrix} 2 & -1 \\ 0 & 8 \end{bmatrix}$

6. $\begin{bmatrix} 5 & 2 & 0 \\ -1 & -2 & 3 \end{bmatrix} - \begin{bmatrix} 4 & -9 & 6 \\ -3 & 0 & 7 \end{bmatrix}$

7. $2\begin{bmatrix} 1 & 0 \\ 0 & 1 \end{bmatrix}$

8. $\frac{1}{2}\begin{bmatrix} -8 & 3 \\ 6 & -4 \end{bmatrix}$

9. $8\begin{bmatrix} 9 \\ -5 \end{bmatrix}$

10. $3\begin{bmatrix} 1 & 4 & -2 \\ -4 & -1 & 1 \end{bmatrix}$

11. $-3\begin{bmatrix} 1 & 0 \\ 0 & 1 \end{bmatrix}$

12. $2\begin{bmatrix} -3 & 0 & 5 \\ 8 & -2 & -1 \\ 9 & 7 & 0 \end{bmatrix}$

13. $-\frac{5}{3}\begin{bmatrix} 5 & 3 \\ 3 & -3 \\ -1 & 6 \end{bmatrix}$

14. $\frac{3}{10}\begin{bmatrix} 15 \\ 10 \\ -40 \end{bmatrix}$

Matrices

1. $\begin{bmatrix} 6 & 5 \\ 2 & 13 \end{bmatrix}$

2. $\begin{bmatrix} 7 \\ 10 \end{bmatrix}$

3. $\begin{bmatrix} 2 & -4 \\ 2 & 1 \end{bmatrix}$

4. $\begin{bmatrix} 1 \\ -5 \\ -5 \end{bmatrix}$

5. $\begin{bmatrix} 10 & 7 \\ 8 & 16 \end{bmatrix}$

6. $\begin{bmatrix} 1 & 11 & -6 \\ 2 & -2 & -4 \end{bmatrix}$

7. $\begin{bmatrix} 2 & 0 \\ 0 & 2 \end{bmatrix}$

8. $\begin{bmatrix} -4 & \frac{3}{2} \\ 3 & -2 \end{bmatrix}$

9. $\begin{bmatrix} 72 \\ 40 \end{bmatrix}$

10. $\begin{bmatrix} 3 & 12 & -6 \\ -12 & -3 & 3 \end{bmatrix}$

11. $\begin{bmatrix} -3 & 0 \\ 0 & -3 \end{bmatrix}$

12. $\begin{bmatrix} -6 & 0 & 10 \\ 16 & -4 & -2 \\ 18 & 14 & 0 \end{bmatrix}$

13. $\begin{bmatrix} -\frac{25}{3} & -5 \\ -5 & 5 \\ \frac{5}{3} & -10 \end{bmatrix}$

14. $\begin{bmatrix} \frac{9}{2} \\ 3 \\ -12 \end{bmatrix}$

Properties and Theorems Demonstrated in Examples, Questions, and Exercises

Chapter 1

- If two lines intersect, their intersection is a point. **pp. 21, 22**
- Parallel lines lie in the same plane, but do not intersect. **pp. 21, 22**
- Skew lines lie in different planes and do not intersect. **pp. 21, 22**
- The perpendicular bisectors of the sides of a triangle are concurrent. The angle bisectors of a triangle are concurrent. **p. 46, Ex. 34**

Chapter 2

- All right angles are congruent. **p. 60, Ques. 4**
- If the nonshared rays of two adjacent acute angles are perpendicular, the angles are complementary. **p. 63, Ex. 29**
- Two angles that are complementary to the same angle are congruent. **p. 63, Ex. 30**
- Two angles that are supplementary to the same angle are congruent. **p. 63, Ex. 31**
- The measure of each angle of an equiangular triangle is $60°$. **p. 66, Ques. 3–5**
- The acute angles of a right triangle are complementary. **p. 68, Ex. 7**
- An equilateral triangle is also equiangular. An equiangular triangle is also equilateral. **p. 70, Ex. 31**
- The base angles of an isosceles triangle are congruent. **p. 70, Ex. 32**
- A regular n-gon has n lines of symmetry. **p. 74, Ques. 4**
- A regular n-gon has rotational symmetry. **p. 74, Ques. 5**
- The measure of each angle of a rectangle is $90°$. **p. 89, Ques. 1**
- A parallelogram that has one right angle is a rectangle. **p. 92, Ex. 16**
- Euler's formula ($F + V - E = 2$) **p. 96, Ex. 12**

Chapter 3

- The product of an even integer and any other integer is even. **p. 112, Example 1**
- The sum of the measures of the exterior angles of a triangle is $360°$. **p. 129, Ex. 21**
- Two angles that are supplementary to congruent angles are congruent. **p. 132, Ques. 1, 2**
- The angles in a linear pair are supplementary. **p. 134, Ex. 14**
- Two perpendicular lines form congruent adjacent angles. **p. 134, Ex. 15**
- Complements of the same angle are congruent. **p. 134, Ex. 16**
- If two angles of a triangle are complementary, the triangle is a right triangle. **p. 139, Ex. 12**
- If two exterior angles at different vertices of a triangle are congruent, then the two interior angles at the same vertices are congruent. **p. 140, Ex. 20**
- The diagonals of a rectangle are congruent. **p. 145, Ex. 31**

The sum of the squares of the lengths of the sides of a rectangle is equal to the sum of the squares of the lengths of the diagonals. **p. 147, Ex. 46**

Let a, b, and c be the lengths of the sides of a triangle with c greater than both a and b. If $a^2 + b^2 > c^2$, then the triangle is acute. If $a^2 + b^2 = c^2$, then the triangle is a right triangle. If $a^2 + b^2 < c^2$, then the triangle is obtuse. **p. 148**

Chapter 4

The quadrilateral formed by connecting the midpoints of the sides of a quadrilateral is a parallelogram. **p. 172, Ex. 34**

If a line is perpendicular to one of two parallel lines, then it is perpendicular to the other. **p. 186, Ex. 33**

The diagonals of a square are congruent and perpendicular. **p. 196**

The midpoint of the hypotenuse of a right triangle is the same distance from all three vertices of the triangle. **p. 198, Ex. 11**

The segments that join the midpoints of the opposite sides of a quadrilateral bisect each other. **p. 199, Ex. 19**

If the diagonals of a parallelogram are perpendicular, then the parallelogram is a rhombus. **p. 199, Ex. 25**

When segments are drawn from a vertex of a rectangle to the midpoints of two other sides, the diagonal that crosses the segments is divided into three congruent parts. **p. 200, Ex. 27**

The diagonals of an isosceles trapezoid are congruent. The length of the segment joining the midpoints of the diagonals is half the difference of the lengths of the bases. **p. 200, Ex. 28**

Chapter 5

The angles of a trapezoid consist of two pairs of supplementary angles. **p. 229, Example 2**

When two nonparallel lines are intersected by transversals, the sum of the measures of the same-side interior angles on the same side of the transversals is a constant. **p. 231, Exs. 24–27**

Consecutive angles of a parallelogram are supplementary. **p. 232, Ex. 30**

If two parallel lines are intersected by a transversal, same-side exterior angles are supplementary. **p. 234**

The sum of the measures of the angles of a quadrilateral is $360°$. **p. 238, Example**

In a triangle, there can be at most one obtuse or one right angle. **p. 239, Ex. 7**

If two angles of one triangle are congruent to two angles of another triangle, then the third angles are congruent. **p. 240, Ex. 11**

If one pair of base angles of a trapezoid is congruent, then the other pair of base angles is congruent. **p. 241, Ex. 18**

If two lines are intersected by a transversal and a pair of alternate exterior angles is congruent, then the lines are parallel. **p. 246, Ex. 15**

If two lines are intersected by a transversal and a pair of same-side exterior angles is supplementary, then the lines are parallel. **p. 247, Ex. 16**

If a plane is perpendicular to one of two parallel lines, then the plane is perpendicular to the other line. **p. 260, Ex. 12**

If two planes are both perpendicular to the same line, then the planes are parallel. **p. 260, Ex. 13**

If a line is perpendicular to one of two parallel planes, then the line is perpendicular to the other plane. **p. 260, Ex. 16**

Chapter 6

- The length of the longest side of a quadrilateral is less than the sum of the lengths of the other three sides. **p. 283, Ex. 31**

- If the legs of one right triangle are congruent to the legs of another right triangle, then the triangles are congruent. **p. 304, Ex. 14**

- If a leg and adjacent acute angle of one right triangle are congruent to the corresponding leg and acute angle of another right triangle, then the triangles are congruent. **p. 304, Ex. 14**

- If a point is equidistant from the endpoints of a segment, then the point is on the perpendicular bisector of the segment. **p. 311, Ex. 15**

- The bisector of the vertex angle of an isosceles triangle is also the altitude to the base. **p. 320, Example 2**

- The bisector of the vertex angle of an isosceles triangle is also the perpendicular bisector of the base. **p. 320, Ques. 1**

- The altitude to the base of an isosceles triangle is also the bisector of the vertex angle. **p. 322, Ex. 3**

- The median to the base of an isosceles triangle is also the altitude to the base. **p. 322, Ex. 4**

- The median to the base of an isosceles triangle is also the perpendicular bisector of the base. **p. 322, Ex. 5**

- The bisector of the vertex angle of an isosceles triangle is also the perpendicular bisector of the base. **p. 322, Ex. 5**

- The altitude to the base of an isosceles triangle is also the median to the base. **p. 322, Ex. 5**

- The perpendicular bisector of the base of an isosceles triangle is also the altitude to the base. **p. 322, Ex. 5**

- The medians of a triangle intersect in a point. This point is called the centroid. **p. 323, Ex. 8**

- The bisectors of the angles of a triangle intersect in a point. This point is called the incenter. The incenter is equidistant from the sides of the triangle. **p. 324, Ex. 13**

- The medians of a triangle intersect in a point that is two-thirds of the distance from each vertex to the midpoint of the opposite side. **p. 324, Ex. 14**

- The lines that contain the altitudes intersect in a point. This point is called the orthocenter. **p. 324, Ex. 15**

Chapter 7

- The segments that join the midpoints of the sides of a rectangle form a rhombus. The segments that join the midpoints of the sides of a rhombus form a square. The segments that join the midpoints of the sides of a square form a square. **p. 343, Ex. 4**

- The diagonals of an isosceles trapezoid are congruent. **p. 344, Ex. 12**

- The parallel sides of a trapezoid cannot be congruent. **p. 348**

- If two consecutive angles of a parallelogram are congruent, then the parallelogram is a rectangle. **p. 357, Ex. 20**

- If exactly one diagonal of a quadrilateral is the perpendicular bisector of the other diagonal, then the quadrilateral is a kite. **p. 357, Ex. 21**

- The area of a rhombus is half the product of the lengths of the diagonals. **p. 365, Ex. 25**

- A regular hexagon can be divided into six congruent equilateral triangles. **p. 367, Example 1**

- Any regular n-gon can be divided into n congruent triangles. **p. 369, Ques. 1**

The height of an equilateral triangle is 1.5 times the radius of the triangle. p. 379, Ex. 14(a)

Chapter 8

When a point (a, b) is reflected over the line $y = -x$, the image is point $(-b, -a)$. p. 402, Ex. 12

A translation glides all points through twice the distance from the first line of reflection to second line of reflection. p. 405

A double reflection over two intersecting lines is a rotation. The measure of the angle of rotation is twice the measure of the angle formed by the intersecting lines. p. 410, Ex. 24

A glide reflection can be created by combining three reflections. p. 424, Ex. 23

The midpoint of a segment that joins a point and its image after a glide reflection lies on the line of reflection. p. 424, Ex. 24

The image of (a, b) after a dilation with center $(0, 0)$ and scale factor k has coordinates (ka, kb). p. 430, Ex. 7

Chapter 9

If a line is parallel to one side of a triangle and intersects the other two sides, then the triangle formed is similar to the original triangle. p. 461

Parallel lines divide transversals into segments that are in proportion. p. 466, Exs. 14–16

If a ray bisects an angle of a triangle, then it divides the opposite side into segments proportional to the other two sides. p. 466, Ex. 23

Chapter 10

$\sin x° = \cos (90 - x)°$ and $\cos x° = \sin (90 - x)°$ p. 516, Ques. 2

$(\sin A)^2 + (\cos A)^2 = 1$ p. 520, Ex. 33

$\sin x° = \cos (90 - x)°$ and $\tan A = \dfrac{\sin A}{\cos A}$ p. 520, Ex. 34

The area of a triangle is equal to $\dfrac{1}{2}bc \sin A$. p. 534, Ex. 22

Chapter 11

The opposite angles of an inscribed quadrilateral are supplementary. p. 559, Ex. 22

The perpendicular bisectors of the sides of a triangle intersect in a point. This point is called the circumcenter. The circumcenter is equidistant from the vertices of the triangle. p. 567

In a circle, two congruent chords are equidistant from the center of the circle. p. 569, Example 2

In a circle or in congruent circles, congruent chords have congruent arcs. p. 569, Ques. 3

In a circle, congruent arcs have congruent chords. p. 570, Ex. 9

In a circle or in congruent circles, chords equidistant from the center are congruent. p. 571, Ex. 12

Chapter 12

The image of a polygon after a dilation with scale factor k and center $(0, 0)$ is found by multiplying the polygon matrix by k. pp. 614–615

The image of a polygon after a translation is found by adding the translation matrix and the polygon matrix. p. 620

Postulates and Properties

Patterns, Lines, and Planes

▨ For any two points, there is exactly one line through the points. **p. 22**

▨ For any three noncollinear points, there is exactly one plane through the points. **p. 22**

▨ If two planes intersect, their intersection is a line. **p. 23**

▨ If a line and a plane intersect, then their intersection is a point or a line. **p. 23**

▨ If two lines intersect, then they are coplanar. **p. 23**

▨ **Segment Addition Postulate**   Point Y is between points X and Z if and only if $XY + YZ = XZ$. **p. 29**

▨ **Angle Addition Postulate**   If D is in the interior of $\angle ABC$, then $m\angle ABD + m\angle DBC = m\angle ABC$. **p. 36**

Reasoning in Geometry

An example of each property is given.

▨ **Reflexive Property** $\overline{PQ} \cong \overline{PQ}$. **p. 119**

▨ **Symmetric Property**   If $\overline{PQ} \cong \overline{RS}$, then $\overline{RS} \cong \overline{PQ}$. **p. 119**

▨ **Transitive Property**   If $\angle 1 \cong \angle 2$ and $\angle 2 \cong \angle 3$, then $\angle 1 \cong \angle 3$. **p. 119**

▨ **Addition Property**   If $PQ = RS$, then $PQ + QR = QR + RS$. **p. 119**

▨ **Subtraction Property**   If $m\angle JNL = m\angle KNM$, then $m\angle JNL - m\angle KNL = m\angle KNM - m\angle KNL$. **p. 119**

▨ **Substitution Property**   If $m\angle 1 + m\angle 2 = 180°$ and $m\angle 1 = m\angle 3$, then $m\angle 3 + m\angle 2 = 180°$. **p. 119**

Parallel Lines

▨ **Corresponding Angles Postulate**   If two parallel lines are intersected by a transversal, then corresponding angles are congruent. **p. 222**

▨ **Parallel Postulate**   Through a point not on a given line, there is exactly one line parallel to the given line. **p. 235**

▨ **Perpendicular Postulate**   Through a point not on a given line, there is exactly one line perpendicular to the given line. **p. 235**

Converse of the Corresponding Angles Postulate If two lines are intersected by a transversal and corresponding angles are congruent, then the lines are parallel. **p. 243**

Conjectures About Triangles

SSS Postulate If three sides of a triangle are congruent to three sides of another triangle, then the triangles are congruent. **p. 293**

SAS Postulate If two sides and the included angle of one triangle are congruent to two sides and the included angle of another triangle, then the triangles are congruent. **p. 294**

ASA Postulate If two angles and the included side of one triangle are congruent to two angles and the included side of another triangle, then the triangles are congruent. **p. 300**

Using Transformations

A reflection preserves congruence. **p. 393**

A reflection changes orientation. **p. 393**

The line of reflection is the perpendicular bisector of every segment that connects a point and its image. **p. 393**

A translation preserves congruence and orientation. **p. 406**

A rotation preserves congruence and orientation. **p. 412**

The line of reflection in a glide reflection is parallel to the direction of the translation. **p. 420**

A glide reflection preserves congruence. **p. 420**

A glide reflection changes orientation. **p. 420**

A dilation does not usually preserve congruence. **p. 428**

A dilation preserves angle measures. **p. 428**

Similar Polygons

AA Similarity Postulate If two angles of a triangle are congruent to two angles of another triangle, then the two triangles are similar. **p. 453**

Circles and Spheres

Arc Addition Postulate In a circle, the measure of the arc formed by two arcs that have exactly one point in common is the sum of the measures of the two arcs. **p. 554**

Theorems

Triangles and Polygons

- The angles that form a linear pair are supplementary. **p. 59**
- Vertical angles are congruent. **pp. 59, 131**
- Two lines are perpendicular if and only if they form congruent adjacent angles. **p. 60**
- **The Triangle Sum Theorem**   The sum of the angle measures of a triangle is 180°. **pp. 66, 236–237**
- In a polygon with n sides, the sum of the angle measures is $(n - 2)180°$. **p. 80**
- The sum of the measures of the exterior angles of any polygon is 360°. **p. 81**
- The opposite sides of a parallelogram are congruent. **p. 88**
- The opposite angles of a parallelogram are congruent. **p. 88**
- The diagonals of a parallelogram bisect each other. **p. 89**

Reasoning in Geometry

- The measure of an exterior angle of a triangle is equal to the sum of the measures of the two interior angles not adjacent to it. **p. 126**
- **The Pythagorean Theorem**   In a right triangle, the square of the length of the hypotenuse is equal to the sum of the squares of the lengths of the legs. **p. 142**
- **Converse of the Pythagorean Theorem**   If a, b, and c are the lengths of the sides of a triangle and $a^2 + b^2 = c^2$, then the triangle is a right triangle. **p. 143**

Coordinates in Geometry

- **The Distance Formula**   The distance between the points $A(x_1, y_1)$ and $B(x_2, y_2)$ is $AB = \sqrt{(x_2 - x_1)^2 + (y_2 - y_1)^2}$. **p. 166**
- **The Midpoint Formula**   The midpoint of the segment joining the points $A(x_1, y_1)$ and $B(x_2, y_2)$ has the coordinates $\left(\dfrac{x_1 + x_2}{2}, \dfrac{y_1 + y_2}{2} \right)$. **p. 167**
- **The Slope Formula**   The slope m of the line containing the points (x_1, y_1) and (x_2, y_2) is $m = \dfrac{y_2 - y_1}{x_2 - x_1}$. **p. 174**

The equation of a line with slope m and y-intercept b is $y = mx + b$. **p. 175**

Two nonvertical lines are parallel if and only if their slopes are equal. **p. 181**

Two nonvertical lines are perpendicular if and only if the product of their slopes is –1. **p. 181**

An equation of the circle with center $(0, 0)$ and radius r is $x^2 + y^2 = r^2$. **p. 188**

An equation of the circle with center (h, k) and radius r is $(x - h)^2 + (y - k)^2 = r^2$. **p. 189**

The midpoint of the segment that joins any two points $A(x_1, y_1, z_1)$ and $B(x_2, y_2, z_2)$ has the coordinates $\left(\dfrac{x_1 + x_2}{2}, \dfrac{y_1 + y_2}{2}, \dfrac{z_1 + z_2}{2} \right)$. **p. 203**

The distance AB between two points in space $A(x_1, y_1, z_1)$ and $B(x_2, y_2, z_2)$ is $\sqrt{(x_2 - x_1)^2 + (y_2 - y_1)^2 + (z_2 - z_1)^2}$. **p. 203**

Parallel Lines

If two parallel lines are intersected by a transversal, then alternate interior angles are congruent. **p. 227**

If two parallel lines are intersected by a transversal, then same-side interior angles are supplementary. **p. 227**

If two lines are intersected by a transversal and alternate interior angles are congruent, then the lines are parallel. **p. 243**

If two lines are intersected by a transversal and same-side interior angles are supplementary, then the lines are parallel. **p. 244**

In a plane, if two lines are both perpendicular to a third line, then the two lines are parallel. **p. 249**

If two lines are both parallel to a third line, then the two lines are parallel. **p. 250**

If two parallel planes are intersected by a third plane, then the lines of intersection are parallel. **p. 257**

If two planes are both parallel to a third plane, then the two planes are parallel. **p. 257**

Conjectures About Triangles

The sum of the lengths of any two sides of a triangle is greater than the length of the third side. **p. 280**

One side of a triangle is longer than a second side if and only if the angle opposite the first side is larger than the angle opposite the second side. **p. 280**

Theorems **715**

AAS Theorem   If two angles and a non-included side of one triangle are congruent to the corresponding parts of another triangle, then the triangles are congruent. **p. 301**

HL Theorem   If the hypotenuse and a leg of one right triangle are congruent to the corresponding parts of another right triangle, then the triangles are congruent. **p. 301**

If a point is on the perpendicular bisector of a segment, then the point is equidistant from the endpoints of the segment. **p. 308**

The Isosceles Triangle Theorem   If two sides of a triangle are congruent, then the angles opposite the sides are congruent. **p. 314**

The Converse of the Isosceles Triangle Theorem   If two angles of a triangle are congruent, then the sides opposite the angles are congruent. **p. 314**

Quadrilaterals, Areas, and Volumes

The diagonals of a rhombus are perpendicular. **p. 341**

The diagonals of a rectangle are congruent. **p. 341**

Exactly one diagonal of a kite is a line of symmetry for the kite and the perpendicular bisector of the other diagonal. **p. 341**

If both pairs of opposite sides of a quadrilateral are congruent, then the quadrilateral is a parallelogram. **p. 347**

If both pairs of opposite angles of a quadrilateral are congruent, then the quadrilateral is a parallelogram. **p. 347**

If one pair of opposite sides of a quadrilateral is both parallel and congruent, then the quadrilateral is a parallelogram. **p. 347**

If the diagonals of a quadrilateral bisect each other, then the quadrilateral is a parallelogram. **p. 347**

If the diagonals of a parallelogram are congruent, then the parallelogram is a rectangle. **p. 353**

If one angle of a parallelogram is a right angle, then the parallelogram is a rectangle. **p. 353**

If the diagonals of a parallelogram are perpendicular, then the parallelogram is a rhombus. **p. 355**

If two consecutive sides of a parallelogram are congruent, then the parallelogram is a rhombus. **p. 355**

The area of a rectangle is the product of the base and the height. ($A = bh$) **p. 362**

The area of a square is the square of the length of one side. ($A = s^2$) **p. 362**

The area of a triangle is half the product of the base and the height. ($A = \frac{1}{2}bh$) **p. 362**

The area of a parallelogram is the product of the base and the height.
$(A = bh)$ **p. 362**

The area of a trapezoid is the product of the height and the mean of the bases. $(A = \frac{1}{2}(b_1 + b_2)h)$ **p. 362**

The area of any regular polygon is half the product of the apothem and the perimeter. $(A = \frac{1}{2}ap)$ **p. 368**

The circumference of a circle with radius r is $2\pi r$. $(C = 2\pi r)$ **p. 369**

The area of a circle with radius r is πr^2. $(A = \pi r^2)$ **p. 369**

The volume of a prism is the product of the height and the area of the base. $(V = Bh)$ **p. 375**

The volume of a cylinder is the product of the height and the area of the base. $(V = \pi r^2 h)$ **p. 375**

The surface area of a prism is the sum of the lateral area and the areas of the bases. $(S.A. = ph + 2B)$ **p. 376**

The surface area of a cylinder is the sum of the lateral area and the areas of the bases. $(S.A. = 2\pi rh + 2\pi r^2)$ **p. 376**

Similar Polygons

SAS Similarity Theorem   If an angle of one triangle is congruent to an angle of another triangle, and the sides including these angles are in proportion, then the triangles are similar. **p. 454**

SSS Similarity Theorem   If all corresponding sides of two triangles are in proportion, then the triangles are similar. **p. 454**

If a segment is parallel to one side of a triangle and intersects the other two sides, then it divides those sides proportionally. **p. 462**

If the midpoints of two sides of a triangle are joined by a segment, then the segment is parallel to the third side of the triangle and its length is half the length of the third side of the triangle. **p. 462**

If two similar two-dimensional figures have corresponding lengths whose ratio is $a:b$, then:

the ratio of all corresponding linear measures is $a:b$ **p. 469**

the ratio of their perimeters is $a:b$ **p. 469**

the ratio of their areas is $a^2:b^2$ **p. 469**

If two similar three-dimensional figures have corresponding lengths whose ratio is $a:b$, then:

the ratio of all corresponding linear measures is $a:b$ **p. 470**

the ratio of their surface areas (S.A.) is $a^2:b^2$ **p. 470**

the ratio of their volumes (V) is $a^3:b^3$ **p. 470**

Applying Right Triangles

▨ If the altitude is drawn to the hypotenuse of a right triangle, then the two triangles formed are similar to the original triangle and to each other. **p. 494**

▨ If the altitude is drawn to the hypotenuse of a right triangle, then the length of the altitude is the geometric mean of the lengths of the segments of the hypotenuse. **p. 495**

▨ If the altitude is drawn to the hypotenuse of a right triangle, then the length of each leg is the geometric mean of the lengths of the hypotenuse and the segment of the hypotenuse adjacent to that leg. **p. 495**

▨ In a 45-45-90 triangle, the hypotenuse is $\sqrt{2}$ times the length of each leg. **p. 501**

▨ In a 30-60-90 triangle, the length of the hypotenuse is twice the length of the shorter leg and the length of the longer leg is $\sqrt{3}$ times the length of the shorter leg. **p. 501**

▨ The volume of a pyramid is one third the product of the height and the area of the base. ($V = \frac{1}{3}Bh$) **p. 536**

▨ The surface area of a regular pyramid is the sum of the lateral area and the area of the base. ($S.A. = \frac{1}{2}ps + B$) **p. 536**

▨ The volume of a cone is one third the product of the height and the area of the base. ($V = \frac{1}{3}\pi r^2 h$) **p. 537**

▨ The surface area of a cone is the sum of the lateral area and the area of the base. ($S.A. = \pi rs + \pi r^2$). **p. 537**

Circles and Spheres

▨ A line, ray, or segment is a tangent of a circle if and only if it is in the same plane as the circle and is perpendicular to a radius of the circle at the point of intersection. **p. 554**

▨ The measure of an inscribed angle is equal to half the measure of the intercepted arc. **p. 555**

▨ In a circle or in congruent circles, inscribed angles that intercept the same arc or congruent arcs are congruent. **p. 555**

▨ The measure of an angle formed by a tangent and a chord is equal to half the measure of the intercepted arc. **p. 555**

▨ The measure of an angle formed by the intersection of two tangents, two secants, or a secant and a tangent, at a point outside a circle, is half the difference of the measures of the intercepted arcs. **p. 561**

The measure of an angle formed by two chords is equal to half the sum of the measures of the intercepted arcs. **p. 562**

The perpendicular bisector of a chord passes through the center of the circle. **p. 568**

A diameter that is perpendicular to a chord bisects the chord and its corresponding arc. **p. 568**

If two chords intersect in the interior of a circle, then the product of the lengths of the segments of one chord is equal to the product of the lengths of the segments of the other chord. **p. 573**

If two secant segments are drawn to a circle from an exterior point, then the product of the lengths of one secant segment and its external segment is equal to the product of the lengths of the other secant segment and its external segment. **p. 573**

If a tangent segment and a secant segment are drawn to a circle from an exterior point, then the product of the lengths of the secant segment and its external segment is equal to the square of the length of the tangent segment. **p. 574**

If two tangent segments are drawn to a circle from an exterior point, then the lengths of the tangent segments are equal. **p. 574**

If x represents the measure of the arc of a sector and r is the radius of the circle, then the area of the sector is $\dfrac{x}{360} \cdot \pi r^2$. **p. 580**

If x represents the measure of an arc and r is the radius of the circle, then the length of the arc is $\dfrac{x}{360} \cdot 2\pi r$. **p. 580**

The surface area of a sphere with radius r is $4\pi r^2$. ($S.A. = 4\pi r^2$) **p. 586**

The volume of a sphere with radius r is $\dfrac{4}{3}\pi r^3$. ($V = \dfrac{4}{3}\pi r^3$) **p. 586**

Constructions

Triangles and Polgons

■ Construct an equilateral triangle. **p. 69**

Parallel Lines

■ Given an angle, construct an angle congruent to the given angle. **p. 263**

■ Given two separate angles, construct an angle whose measure is equal to the sum of the measures of the given angles. **p. 264**

■ Given a line, construct a line parallel to the given line. **p. 264**

■ Given a segment, construct its perpendicular bisector. **p. 265**

■ Given a line and a point on the line, construct a line perpendicular to the given line and through the given point. **p. 267**

■ Given a line and a point not on the line, construct a line perpendicular to the given line and through the given point. **p. 267**

Conjectures About Triangles

■ Given three sides of a triangle, construct the triangle. **p. 292**

■ Given two sides and their included angle, construct a triangle. **p. 297**

■ Given an angle, construct its bisector. **p. 311**

Quadrilaterals, Areas, and Volumes

■ Construct a kite. **p. 343**

■ Given two segments, construct a parallelogram with sides congruent to the segments. **p. 350**

■ Construct a rectangle. **p. 358**

■ Construct a regular hexagon. **p. 372**

Applying Right Triangles

■ Construct a 45-45-90 triangle. **p. 505**

Circles and Spheres

■ Given a triangle, circumscribe a circle about the triangle. **p. 567**

Technology Handbook

This handbook introduces features of graphing calculators, spreadsheets, and geometry software that you can use with material in this book.

USING A GRAPHING CALCULATOR

This section discusses features common to most graphing calculators. Check your calculator's instruction manual for any details not provided here.

PERFORMING CALCULATIONS

The Keyboard

Look closely at your calculator's keyboard. Notice that most keys serve more than one purpose. Each key is labeled with its primary purpose, and labels for any secondary purposes appear somewhere near the key. You may need to press **2nd**, **SHIFT**, or **ALPHA** to use a key for a secondary purpose.

On the TI-82, for example, the x^2 key can be used as follows:

- Press x^2 to square a number.

- Press **2nd** and then x^2 to take a square root.

- Press **ALPHA** and then x^2 to get the letter I.

The Home Screen

Your calculator has a *home screen* where you can do calculations. You can usually enter a calculation on a graphing calculator just as you would write it on a piece of paper.

Your calculator may recognize implied multiplication and have built-in constants like pi.

Don't confuse the subtraction key, **—**, and the negation key, **(−)**.

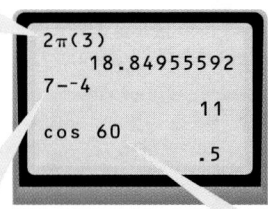

```
2π(3)
        18.84955592
7−⁻4
                  11
cos 60
               .5
```

When doing trigonometric calculations, be sure your calculator is in the degree mode.

Use your calculator to find the value of each expression.

1. $-3 - 9$ **2.** $5(6.37)$ **3.** $\sqrt{11.56}$ **4.** $\sqrt{5^2 + 9^2}$

5. 8^4 **6.** $\frac{4}{3}\pi(7^3)$ **7.** $\tan 61°$ **8.** $\sin^{-1}\left(\frac{18}{23}\right)$

DISPLAYING GRAPHS

Making a Scatter Plot

Most graphing calculators will make a scatter plot for a set of data pairs. For example, suppose you want to make a scatter plot of the data in the table.

x = length of an edge of a cube (in.)	1	2	3	4	5	6
y = surface area of the cube (in.^{2})	6	24	54	96	150	216

On the TI-82, you first need to enter the data into *lists*. Follow these steps:

Enter the lengths in list L1. Enter the surface areas in list L2.

Tell the calculator to display a scatter plot using the data in lists L1 and L2.

Press GRAPH to display the scatter plot. Adjust the viewing window as needed (see below).

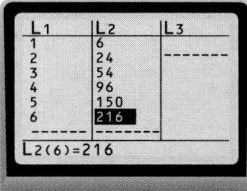 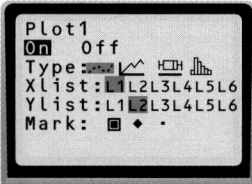 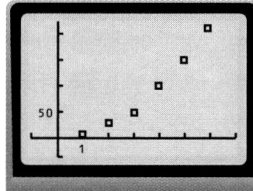

Adjusting the Viewing Window

When making a scatter plot, think of the calculator screen as a *viewing window* that lets you look at part of the coordinate plane. On the TI-82, you can adjust the viewing window by pressing WINDOW and entering appropriate values for the window variables. The values given below define a good viewing window for the scatter plot of the cube data (see above).

The interval $-1 \leq x \leq 7$ will be shown on the *x*-axis.

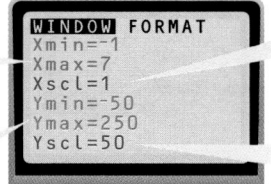

The interval $-50 \leq y \leq 250$ will be shown on the *y*-axis.

Tick marks will be **1** unit apart on the *x*-axis.

Tick marks will be **50** units apart on the *y*-axis.

....

ANSWERS   Technology Handbook

Calculator Practice

1. –12

2. 31.85

3. 3.4

4. about 10.3

5. 4096

6. about 1436.8

7. 1.804

8. 51.5°

9. The table shows the numbers of diagonals that can be drawn for polygons with different numbers of sides. Make a scatter plot of the data in the table. Adjust the viewing window so that you can see all the data points. Describe the shape of the scatter plot.

x = number of sides	3	4	5	6	7	8	9
y = number of diagonals	0	2	5	9	14	20	27

USING MATRICES

Entering a Matrix

Many graphing calculators will let you enter and perform operations on matrices. To enter a matrix on the TI-82, follow these steps:

Press MATRX, and select EDIT. Choose a name for the matrix, such as [A].

Set the dimensions of the matrix, such as 4 × 3. A matrix full of zeros appears.

Replace the zeros with the desired matrix elements.

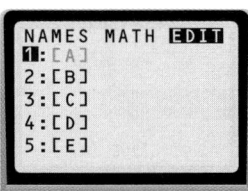
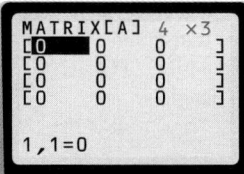
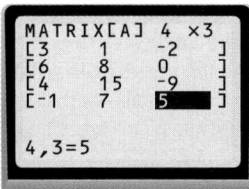

Performing Scalar Multiplication

You can use a graphing calculator to perform scalar multiplication. For example, suppose you want to find the image matrix for a dilation of a triangle, where the coordinates of the triangle's vertices are $(-1, 0)$, $(3, 4)$, and $(6, -5)$, the center of the dilation is $(0, 0)$, and the scale factor is 2. Follow these steps:

Create a 2 × 3 polygon matrix [A]. Enter the coordinates of each vertex of the triangle into a column of the matrix.

Multiply [A] by the scale factor 2 to find the image matrix for the dilation.

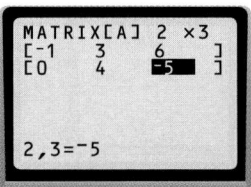
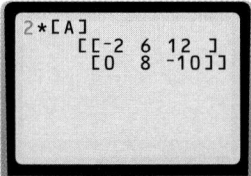

Technology Handbook **723**

Calculator Practice

9. The shape of the scatter plot is similar to half of a parabola.

Adding and Subtracting Matrices

You can use a graphing calculator to add matrices that have the same dimensions. For example, suppose you want to find the image matrix for the translation of a triangle 2 units right and 4 units down. The coordinates of the triangle's vertices are $(-3, 1)$, $(0, 5)$, and $(7, 0)$. Follow these steps:

Create a 2 × 3 translation matrix [A] that shifts a triangle 2 units right and 4 units down.

Create a 2 × 3 polygon matrix [B] containing the coordinates of the triangle's vertices.

Add [A] and [B] to find the image matrix for the translation.

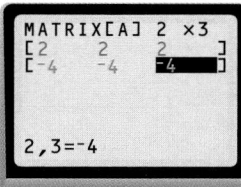

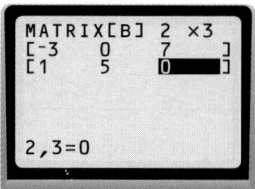

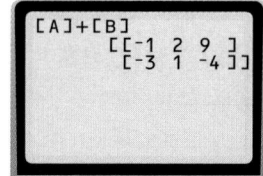

You can also use a graphing calculator to subtract matrices. For example, suppose a triangle has been translated 5 units right and 3 units up, and the coordinates of the image's vertices are $(-3, 4)$, $(2, 1)$, and $(4, 8)$. To find the coordinates of the original triangle's vertices, first create the translation matrix [C] and the image matrix [D], and then find [D] − [C]. The screen at the right shows that the original triangle's vertices are at $(-8, 1)$, $(-3, -2)$, and $(-1, 5)$.

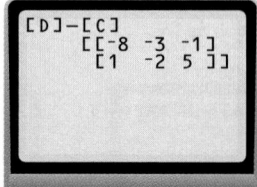

Finding the Product of Two Matrices

You can use a graphing calculator to multiply two matrices when the number of columns in the first matrix is the same as the number of rows in the second matrix. For example, suppose you want to find the image matrix for the reflection of a triangle over the y-axis. The coordinates of the triangle's vertices are $(-5, 4)$, $(-3, -3)$, and $(-2, -6)$. Follow these steps:

Create a 2 × 2 matrix [A] that reflects polygons over the y-axis. (See page 635 for a list of reflection matrices.)

Create a 2 × 3 polygon matrix [B] containing the coordinates of the triangle's vertices.

Multiply [A] and [B] to find the image matrix for the reflection.

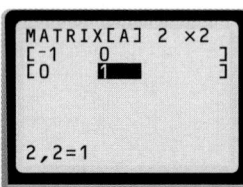

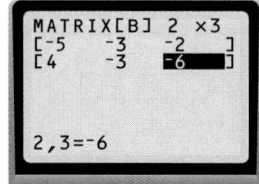

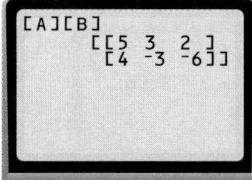

The coordinates of the vertices of a triangle are (1, 3), (4, 7), and (9, 2).
Find the image matrix for each transformation of the triangle.

10. a translation 4 units left and 1 unit up

11. a dilation with center (0, 0) and scale factor 5

12. a reflection over the line $y = -x$

13. a 90° rotation counterclockwise around (0, 0)

USING A SPREADSHEET

You can use a computer spreadsheet to perform repetitive calculations quickly and easily. For example, the spreadsheet below has been set up to calculate the surface areas of balls used in different sports. (In the spreadsheet, r is the radius of a ball and A is the ball's surface area.)

A spreadsheet is made up of cells named by a column letter and a row number, such as A3 or B4. You can enter a label, a number, or a formula into a cell.

For example, row 2 of the spreadsheet at the left contains the label "Golf ball" in cell A2, the number 2.1 in cell B2, and the formula "=4*PI()*B2^2" in cell C2. This formula tells the computer to multiply the square of the number in cell B2 by 4π and store the result in cell C2.

You don't have to type every formula in column C. Instead, you can type just the first formula and then use the *copy* or *fill down* command to generate the others.

row numbers   column letters

Ball Data

	A	B	C
1	Type of ball	r (cm)	A (cm^2)
2	Golf ball	2.1	=4*PI()*B2^2
3	Tennis ball	3.3	=4*PI()*B3^2
4	Softball	4.8	=4*PI()*B4^2
5	Volleyball	10.5	=4*PI()*B5^2
6	Basketball	11.9	=4*PI()*B6^2

The computer replaces all formulas in the spreadsheet with calculated values, as shown at the right. For example, you can see that the surface area of a basketball is about 1780 cm².

Ball Data

	A	B	C
1	Type of ball	r (cm)	A (cm^2)
2	Golf ball	2.1	55.42
3	Tennis ball	3.3	136.85
4	Softball	4.8	289.53
5	Volleyball	10.5	1385.44
6	Basketball	11.9	1779.52

14. Use a spreadsheet to find the volumes of the balls listed in the spreadsheets above.

15. For any integer n greater than 1, the set of numbers $2n$, $n^2 - 1$, and $n^2 + 1$ is a Pythagorean triple (a set of integers that can be the lengths of the sides of a right triangle). Use a spreadsheet to generate Pythagorean triples for $n = 2, 3, \ldots, 10$.

Technology Handbook   **725**

Calculator Practice

10. $\begin{bmatrix} -3 & 0 & 5 \\ 4 & 8 & 3 \end{bmatrix}$

11. $\begin{bmatrix} 5 & 20 & 45 \\ 15 & 35 & 10 \end{bmatrix}$

12. $\begin{bmatrix} -3 & -7 & -2 \\ -1 & -4 & -9 \end{bmatrix}$

13. $\begin{bmatrix} -3 & -7 & -2 \\ 1 & 4 & 9 \end{bmatrix}$

Spreadsheet Practice

14, 15. See answers in back of book.

USING GEOMETRY SOFTWARE

This section discusses features common to most geometry software. Check your software's instruction manual for any details not provided here.

CONSTRUCTING GEOMETRIC OBJECTS

Constructing Points

You can construct a point either by moving the cursor to where you want the point to be and clicking the mouse button, or by entering the point's coordinates. Points are given one-letter names, starting with *A*. Several more complicated constructions involving points are described below.

You can define a point to be on an object. (The point can't be moved off the object.)

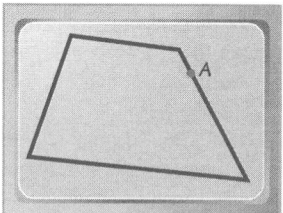

You can construct the point or points where two objects intersect.

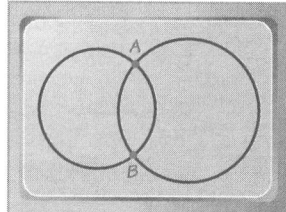

You can construct one or more points that subdivide a segment into equal parts.

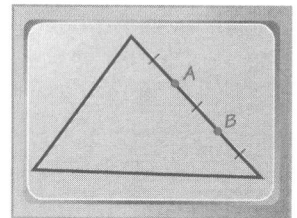

Constructing Lines, Segments, Rays, and Vectors

With most software, you can draw a line, segment, ray, or vector "freehand" by clicking the mouse button at some location on the screen and dragging away from that location to form the object. You can also construct a line by entering its equation, and a vector by specifying its tail and components. Several other constructions are described below.

You can construct a segment joining two named points, such as *B* and *D* below.

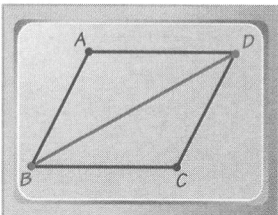

You can construct a ray that bisects an angle, such as ∠ *A* below.

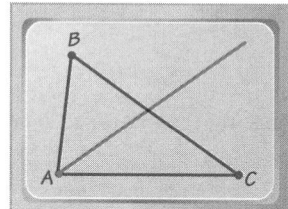

You can construct a line parallel or perpendicular to a given line or segment and through a given point.

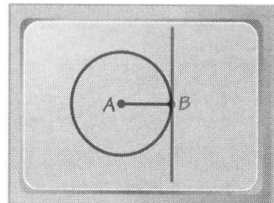

Constructing Polygons

With most software, you can construct and modify a polygon as follows:

Think of the polygon as a group of connected segments, and construct each segment individually.

The last segment should end at the vertex where you started the polygon construction.

You can change the shape of the polygon by dragging its sides and vertices.

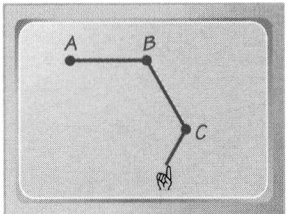 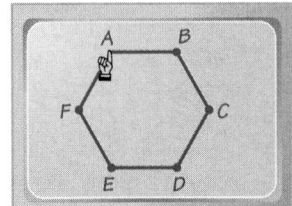 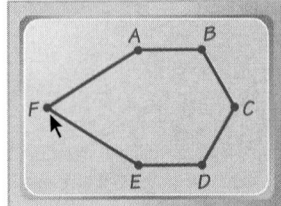

Many software programs also have tools for constructing special kinds of polygons, such as right and isosceles triangles, trapezoids, parallelograms, and regular polygons. Some programs will let you change the shape of a polygon in such a way that its perimeter or area remains constant.

Constructing Circles

Most software lets you construct a circle in one of three ways.

- You can draw the circle "freehand." This is usually done by moving the cursor to where you want the circle's center to be, clicking the mouse button, and dragging away from the center to form the circle.

- You can enter the name of a point to use for the circle's center and a number to use for the radius. For example, you can tell the computer to construct a circle whose center is point A and whose radius is 3.

- You can enter an equation for the circle.

Many software programs will also perform more complicated constructions involving circles. Two such constructions are shown below.

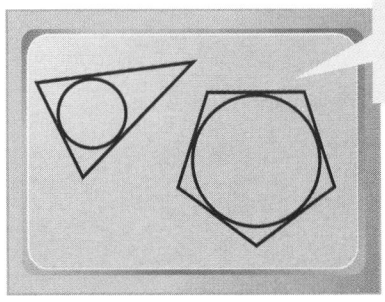

You can inscribe a circle in a triangle or regular polygon.

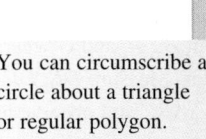

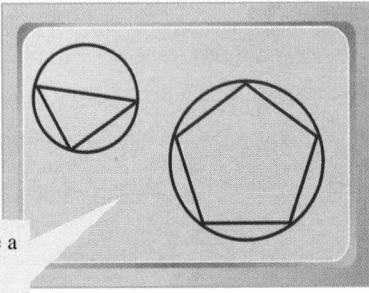

You can circumscribe a circle about a triangle or regular polygon.

Technology Handbook **727**

16. Construct a line and a point *A* not on the line. Construct a second line parallel to the first line and passing through point *A*.

17. Construct a pentagon. Then construct each diagonal of the pentagon.

18. Construct a point *A* and a circle with center *A* and radius 4. Construct a point *B* on the circle, the radius joining *A* and *B*, and a line tangent to the circle at *B*.

FINDING LENGTHS AND ANGLE MEASURES

You can use most software to find and display lengths of segments and measures of angles. The length and angle-measure displays are automatically updated if you change the corresponding segments and angles. This is useful when you want to construct segments and angles having specified lengths and measures.

For example, suppose you want to construct a triangle such that two of its sides are 10 cm and 5 cm long and have an included angle measuring 40°. Follow these steps:

Construct any triangle *ABC*. (See "Constructing Polygons" on page 727.) Define three measurement displays for ∠*BAC*, $\overline{AB}$, and $\overline{AC}$.

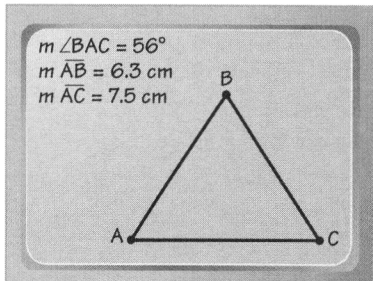

Drag vertex *B* up and to the right until *m*∠*BAC* = 40° and *AB* = 10 cm. Then drag vertex *C* to the left until *AC* = 5 cm.

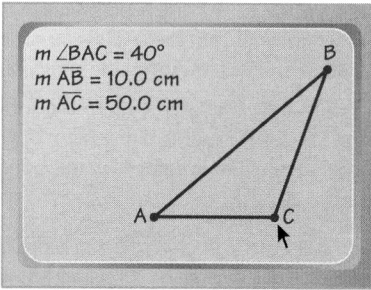

19. Construct a rhombus having a 115° angle and sides that are 8 cm long.

20. **a.** Construct a triangle *ABC* and the bisectors of the angles of the triangle. Construct the point *D* where the bisectors intersect.

    b. For each side of the triangle, construct a line perpendicular to the side through point *D*. Construct points *E*, *F*, and *G* where the perpendicular lines intersect the triangle's sides. Measure $\overline{DE}$, $\overline{DF}$, and $\overline{DG}$.

    c. Drag the sides and vertices of △*ABC* and observe how *DE*, *DF*, and *DG* change. Based on your observations, make a conjecture about the angle bisectors of a triangle.

728 Student Resources

Software Practice

16.
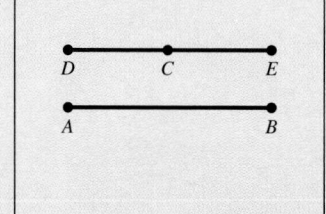

17.
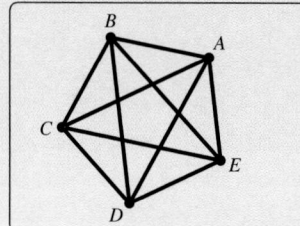

18.
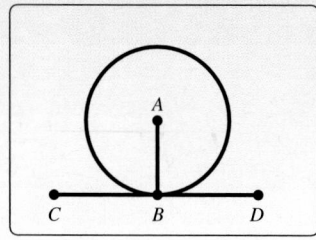

19.
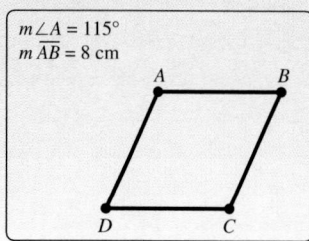

20. a, b.
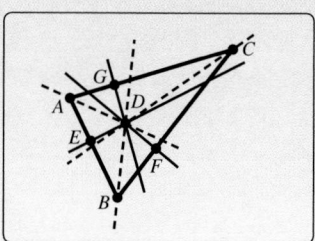

c. The point at which the three angle bisectors of a triangle intersect is equidistant from the sides of the triangle.

PERFORMING TRANSFORMATIONS

Most software will let you perform the transformations described below.
(In the computer screens shown, the original object is blue and the image
after the transformation is red.)

You can reflect an object over a line.

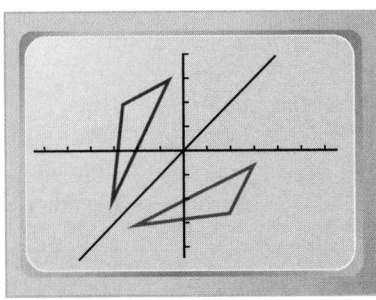

You can rotate an object around a point.

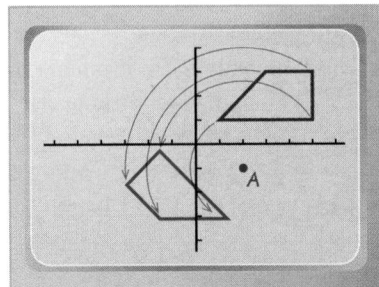

You can translate an object in the direction of an existing vector. The object is translated a distance equal to the vector's magnitude.

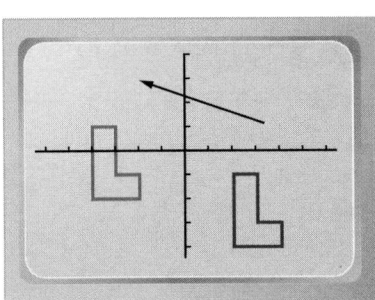

You can dilate an object using a specified center of dilation and scale factor.

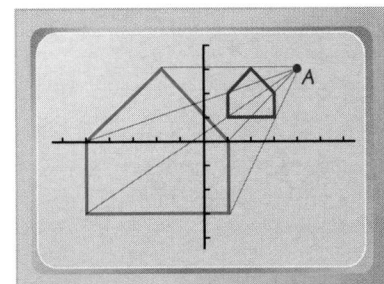

SOFTWARE PRACTICE

**Let *ABCD* be a quadrilateral with vertices at (1, 1), (2, 6), (4, 5), and (5, 2).
Construct the image of the quadrilateral after each transformation.**

21. a reflection over the line $y = -3$

22. a 200° rotation counterclockwise around $(-1, 0)$

23. a translation 5 units left and 6 units down (*Hint:* First construct a vector
whose component form is $(-5, -6)$.)

24. a dilation with center $(-2, 4)$ and scale factor 1.5

25. a glide reflection where the translation is given by $(a, b) \rightarrow (a, b - 3)$
and the reflection is over the *y*-axis

Technology Handbook **729**

21.

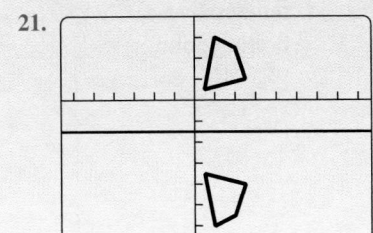

22.

23.

24.

25.

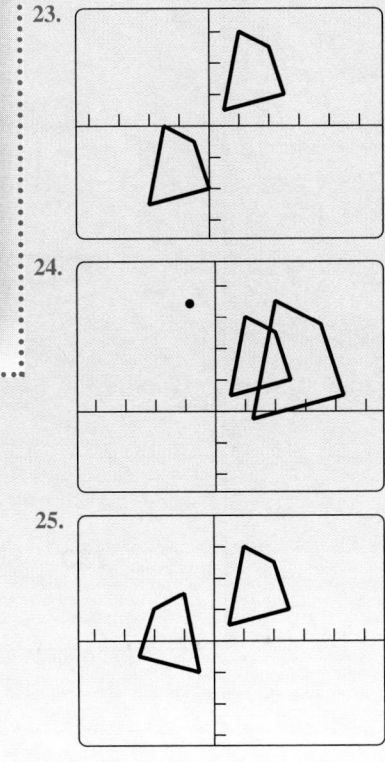

Table of Measures

Time

60 seconds (s) = 1 minute (min)
60 minutes = 1 hour (h)
24 hours = 1 day
7 days = 1 week
4 weeks (approx.) = 1 month

$\left.\begin{array}{r}365 \text{ days} \\ 52 \text{ weeks (approx.)} \\ 12 \text{ months}\end{array}\right\} = 1 \text{ year}$

10 years = 1 decade
100 years = 1 century

Metric

Length

10 millimeters (mm) = 1 centimeter (cm)

$\left.\begin{array}{r}100 \text{ cm} \\ 1000 \text{ mm}\end{array}\right\} = 1 \text{ meter (m)}$

1000 m = 1 kilometer (km)

Area

$\begin{array}{r}100 \text{ square millimeters} \\ (\text{mm}^2)\end{array} = \begin{array}{l}1 \text{ square centimeter} \\ (\text{cm}^2)\end{array}$

10,000 cm^2 = 1 square meter (m^2)
10,000 m^2 = 1 hectare (ha)

Volume

$\begin{array}{r}1000 \text{ cubic millimeters} \\ (\text{mm}^3)\end{array} = \begin{array}{l}1 \text{ cubic centimeter} \\ (\text{cm}^3)\end{array}$

1,000,000 cm^3 = 1 cubic meter (m^3)

Liquid Capacity

1000 milliliters (mL) = 1 liter (L)
1000 L = 1 kiloliter (kL)

Mass

1000 milligrams (mg) = 1 gram (g)
1000 g = 1 kilogram (kg)
1000 kg = 1 metric ton (t)

Temperature — Degrees Celsius (°C)

0°C = freezing point of water
37°C = normal body temperature
100°C = boiling point of water

United States Customary

Length

12 inches (in.) = 1 foot (ft)

$\left.\begin{array}{r}36 \text{ in.} \\ 3 \text{ ft}\end{array}\right\} = 1 \text{ yard (yd)}$

$\left.\begin{array}{r}5280 \text{ ft} \\ 1760 \text{ yd}\end{array}\right\} = 1 \text{ mile (mi)}$

Area

144 square inches (in.^{2}) = 1 square foot (ft^2)
9 ft^2 = 1 square yard (yd^2)

$\left.\begin{array}{r}43{,}560 \text{ ft}^2 \\ 4840 \text{ yd}^2\end{array}\right\} = 1 \text{ acre (A)}$

Volume

1728 cubic inches (in.^{3}) = 1 cubic foot (ft^3)
27 ft^3 = 1 cubic yard (yd^3)

Liquid Capacity

8 fluid ounces (fl oz) = 1 cup (c)
2 c = 1 pint (pt)
2 pt = 1 quart (qt)
4 qt = 1 gallon (gal)

Weight

16 ounces (oz) = 1 pound (lb)
2000 lb = 1 ton (t)

Temperature — Degrees Fahrenheit (°F)

32°F = freezing point of water
98.6°F = normal body temperature
212°F = boiling point of water

Table of Symbols

Symbol		Page
n^2	n to the 2nd power	4
$-a$	the opposite of a	5
...	and so on	6
(a, b)	ordered pair	7
$=$	is equal to	13
$\geq$	is greater than or equal to	19
$\leq$	is less than or equal to	19
$\overleftrightarrow{AB}$	line AB	22
$\parallel$	is parallel to	22
$\overrightarrow{AB}$	ray AB	28
$\overline{AB}$	segment AB	28
AB	the length of segment AB	28
$\cong$	is congruent to	28
$\circ$	degree(s)	34
$\angle ABC$	angle ABC	34
$m\angle ABC$	the measure of angle ABC	35
$\perp$	is perpendicular to	58
$\triangle ABC$	triangle ABC	65
n-gon	polygon with n sides	73
$\square ABCD$	parallelogram $ABCD$	88
$\cdot$	multiplication, times ($\times$)	112
$\sqrt{a}$	the nonnegative square root of a	142
$\pm$	plus or minus	142
$\approx$	is approximately equal to	142
$\stackrel{?}{=}$	is this statement true?	143
$\neq$	is not equal to	143
x_1	x sub 1	166

Symbol		Page
m	slope	174
$\odot C$	circle C	187
(x, y, z)	ordered triple	202
$|x|$	the absolute value of x	208
$\angle s$	angles	228
$>$	is greater than	280
$<$	is less than	280
$\not\cong$	is not congruent to	348
π	pi; an irrational number, about 3.14	369
A'	A prime	392
$\rightarrow$	goes to	398
A''	A double prime	405
$a:b$	the ratio of a to b	430
$\sim$	is similar to	446
$\tan A$	the tangent of angle A	507
$\tan^{-1} A$	the inverse tangent of angle A	509
$\sin A$	the sine of angle A	515
$\cos A$	the cosine of angle A	515
$\sin^{-1} A$	the inverse sine of angle A	516
$\cos^{-1} A$	the inverse cosine of angle A	516
$\overrightarrow{AB}$	vector AB	521
$|\overrightarrow{AB}|$	the magnitude of vector AB	523
$\overparen{AB}$	arc AB	554
$m\overparen{AB}$	the measure of arc AB	554
$\begin{bmatrix} 1 & 0 \\ 0 & 1 \end{bmatrix}$	matrix	613
a^{-n}	$\frac{1}{a^n}, a \neq 0$	709

Table of Squares and Square Roots

No.	Square	Sq. Root	No.	Square	Sq. Root	No.	Square	Sq. Root
1	1	1.000	51	2,601	7.141	101	10,201	10.050
2	4	1.414	52	2,704	7.211	102	10,404	10.100
3	9	1.732	53	2,809	7.280	103	10,609	10.149
4	16	2.000	54	2,916	7.348	104	10,816	10.198
5	25	2.236	55	3,025	7.416	105	11,025	10.247
6	36	2.449	56	3,136	7.483	106	11,236	10.296
7	49	2.646	57	3,249	7.550	107	11,449	10.344
8	64	2.828	58	3,364	7.616	108	11,664	10.392
9	81	3.000	59	3,481	7.681	109	11,881	10.440
10	100	3.162	60	3,600	7.746	110	12,100	10.488
11	121	3.317	61	3,721	7.810	111	12,321	10.536
12	144	3.464	62	3,844	7.874	112	12,544	10.583
13	169	3.606	63	3,969	7.937	113	12,769	10.630
14	196	3.742	64	4,096	8.000	114	12,996	10.677
15	225	3.873	65	4,225	8.062	115	13,225	10.724
16	256	4.000	66	4,356	8.124	116	13,456	10.770
17	289	4.123	67	4,489	8.185	117	13,689	10.817
18	324	4.243	68	4,624	8.246	118	13,924	10.863
19	361	4.359	69	4,761	8.307	119	14,161	10.909
20	400	4.472	70	4,900	8.367	120	14,400	10.954
21	441	4.583	71	5,041	8.426	121	14,641	11.000
22	484	4.690	72	5,184	8.485	122	14,884	11.045
23	529	4.796	73	5,329	8.544	123	15,129	11.091
24	576	4.899	74	5,476	8.602	124	15,376	11.136
25	625	5.000	75	5,625	8.660	125	15,625	11.180
26	676	5.099	76	5,776	8.718	126	15,876	11.225
27	729	5.196	77	5,929	8.775	127	16,129	11.269
28	784	5.292	78	6,084	8.832	128	16,384	11.314
29	841	5.385	79	6,241	8.888	129	16,641	11.358
30	900	5.477	80	6,400	8.944	130	16,900	11.402
31	961	5.568	81	6,561	9.000	131	17,161	11.446
32	1,024	5.657	82	6,724	9.055	132	17,424	11.489
33	1,089	5.745	83	6,889	9.110	133	17,689	11.533
34	1,156	5.831	84	7,056	9.165	134	17,956	11.576
35	1,225	5.916	85	7,225	9.220	135	18,225	11.619
36	1,296	6.000	86	7,396	9.274	136	18,496	11.662
37	1,369	6.083	87	7,569	9.327	137	18,769	11.705
38	1,444	6.164	88	7,744	9.381	138	19,044	11.747
39	1,521	6.245	89	7,921	9.434	139	19,321	11.790
40	1,600	6.325	90	8,100	9.487	140	19,600	11.832
41	1,681	6.403	91	8,281	9.539	141	19,881	11.874
42	1,764	6.481	92	8,464	9.592	142	20,164	11.916
43	1,849	6.557	93	8,649	9.644	143	20,449	11.958
44	1,936	6.633	94	8,836	9.695	144	20,736	12.000
45	2,025	6.708	95	9,025	9.747	145	21,025	12.042
46	2,116	6.782	96	9,216	9.798	146	21,316	12.083
47	2,209	6.856	97	9,409	9.849	147	21,609	12.124
48	2,304	6.928	98	9,604	9.899	148	21,904	12.166
49	2,401	7.000	99	9,801	9.950	149	22,201	12.207
50	2,500	7.071	100	10,000	10.000	150	22,500	12.247

Table of Trigonometric Ratios

Angle	Sine	Cosine	Tangent	Angle	Sine	Cosine	Tangent
1°	.0175	.9998	.0175	46°	.7193	.6947	1.0355
2°	.0349	.9994	.0349	47°	.7314	.6820	1.0724
3°	.0523	.9986	.0524	48°	.7431	.6691	1.1106
4°	.0698	.9976	.0699	49°	.7547	.6561	1.1504
5°	.0872	.9962	.0875	50°	.7660	.6428	1.1918
6°	.1045	.9945	.1051	51°	.7771	.6293	1.2349
7°	.1219	.9925	.1228	52°	.7880	.6157	1.2799
8°	.1392	.9903	.1405	53°	.7986	.6018	1.3270
9°	.1564	.9877	.1584	54°	.8090	.5878	1.3764
10°	.1736	.9848	.1763	55°	.8192	.5736	1.4281
11°	.1908	.9816	.1944	56°	.8290	.5592	1.4826
12°	.2079	.9781	.2126	57°	.8387	.5446	1.5399
13°	.2250	.9744	.2309	58°	.8480	.5299	1.6003
14°	.2419	.9703	.2493	59°	.8572	.5150	1.6643
15°	.2588	.9659	.2679	60°	.8660	.5000	1.7321
16°	.2756	.9613	.2867	61°	.8746	.4848	1.8040
17°	.2924	.9563	.3057	62°	.8829	.4695	1.8807
18°	.3090	.9511	.3249	63°	.8910	.4540	1.9626
19°	.3256	.9455	.3443	64°	.8988	.4384	2.0503
20°	.3420	.9397	.3640	65°	.9063	.4226	2.1445
21°	.3584	.9336	.3839	66°	.9135	.4067	2.2460
22°	.3746	.9272	.4040	67°	.9205	.3907	2.3559
23°	.3907	.9205	.4245	68°	.9272	.3746	2.4751
24°	.4067	.9135	.4452	69°	.9336	.3584	2.6051
25°	.4226	.9063	.4663	70°	.9397	.3420	2.7475
26°	.4384	.8988	.4877	71°	.9455	.3256	2.9042
27°	.4540	.8910	.5095	72°	.9511	.3090	3.0777
28°	.4695	.8829	.5317	73°	.9563	.2924	3.2709
29°	.4848	.8746	.5543	74°	.9613	.2756	3.4874
30°	.5000	.8660	.5774	75°	.9659	.2588	3.7321
31°	.5150	.8572	.6009	76°	.9703	.2419	4.0108
32°	.5299	.8480	.6249	77°	.9744	.2250	4.3315
33°	.5446	.8387	.6494	78°	.9781	.2079	4.7046
34°	.5592	.8290	.6745	79°	.9816	.1908	5.1446
35°	.5736	.8192	.7002	80°	.9848	.1736	5.6713
36°	.5878	.8090	.7265	81°	.9877	.1564	6.3138
37°	.6018	.7986	.7536	82°	.9903	.1392	7.1154
38°	.6157	.7880	.7813	83°	.9925	.1219	8.1443
39°	.6293	.7771	.8098	84°	.9945	.1045	9.5144
40°	.6428	.7660	.8391	85°	.9962	.0872	11.4301
41°	.6561	.7547	.8693	86°	.9976	.0698	14.3007
42°	.6691	.7431	.9004	87°	.9986	.0523	19.0811
43°	.6820	.7314	.9325	88°	.9994	.0349	28.6363
44°	.6947	.7193	.9657	89°	.9998	.0175	57.2900
45°	.7071	.7071	1.0000				

1

A Brief History of Geometric Systems

Geometry is primarily the study of the properties of space and figures in space. Like most of mathematics, geometry developed over thousands of years. It is the product of contributions made by many individuals and civilizations.

The Practical Origins of Geometry

The first peoples known to have written about geometry were the ancient Egyptians and Babylonians (about 4000 B.C. to 300 B.C.). They developed accurate methods for finding areas and volumes of many plane figures and solids. The Babylonians knew the Pythagorean theorem more than 1000 years before the Greek mathematician Pythagoras. Scholars believe that geometric accomplishments similar to those of the Egyptians and Babylonians also took place in ancient India and China in the same period.

Both the Egyptians and Babylonians viewed geometry as a tool for solving real-world problems, such as constructing buildings and canals and finding land areas. Their approach to geometry was based on inductive reasoning, generalizing from trial-and-error and everyday experience.

Euclid's Deductive Approach

The Greeks were responsible for two major changes in geometry. The first was from applications to the study of abstract geometric figures. The second, more important change was from inductive methods toward deductive methods. Both changes are in the work of the Greek geometer Euclid.

Euclid wrote what is probably the most important book in the history of deductive geometry, *Elements*, around 300 B.C. Euclid began by defining common geometric objects, such as points and segments, and then stated a small number of seemingly obvious geometric truths called *postulates*. For example, one of the postulates states that it is always possible to draw a segment joining two points. Euclid then used deductive reasoning to prove geometric results called *theorems* from his postulates and definitions.

Mathematical Systems

The postulate-theorem approach of geometry as a *mathematical system* that Euclid used is still vital to mathematics today. Research in the foundations of mathematics began when mathematicians looked at the relationships among postulates and appropriate ways of reasoning from them.

Non-Euclidean Geometry

One of the postulates in Euclid's *Elements*—called the *parallel postulate*—is usually stated: In a plane, given a line ℓ and a point P not on ℓ, there is exactly one line parallel to ℓ through P.

Mathematicians after Euclid were uncomfortable with the parallel postulate because it didn't seem as obvious as the other postulates. The postulate could not be verified by everyday experience because it is not possible to construct two parallel lines that extend infinitely in both directions and confirm that the lines never meet.

Many mathematicians tried to show that the parallel postulate was not needed or to replace it. In the early 1800s, Carl Friedrich Gauss of Germany, Nikolai Lobachevsky of Russia, and Janos Bolyai of Hungary independently discovered *non-Euclidean geometry*.

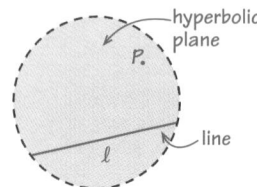

These three mathematicians replaced Euclid's parallel postulate with the following postulate: In a plane, given a line ℓ and a point P not on ℓ, there are *infinitely* many lines parallel to ℓ through P. You can visualize the postulate by defining a plane to be the interior of a circle and a line to be a chord of the circle, excluding the chord's endpoints. Then given a point P in this new plane and a line ℓ not containing P, there are in fact an infinite number of lines through P that are parallel to (do not intersect) line ℓ.

The non-Euclidean geometry formulated by Gauss, Lobachevsky, and Bolyai is called *hyperbolic geometry*. Another non-Euclidean geometry known as *spherical geometry* was derived by the German mathematician G. F. Bernhard Riemann. In spherical geometry, given any line ℓ and point P not on ℓ, there is *no* line parallel to ℓ through P.

Which type of geometry, Euclidean or non-Euclidean, is the best model for the real world? In some fields, such as engineering and surveying, Euclidean geometry still works best as a representation of the world around us. On the other hand, navigation uses spherical geometry, and Albert Einstein used non-Euclidean geometry to describe gravity in his general theory of relativity. Different non-Euclidean models are essential to physics and other sciences.

The following pages discuss particular non-Euclidean geometries:
• Taxicab Geometry (pp. 208–209) • Spherical Geometry (pp. 597–603)

RESEARCH QUESTIONS

1. **Writing** Some mathematicians who made significant contributions to geometry are Thales, Hypatia, Hipparchus, Maria Gaetana Agnesi, René Descartes, Mohammad Abu'l-Wafa, and Felix Klein. Write a report on the life and mathematical work of one of these people.

2. **Writing** An important topic not discussed in this history is projective geometry. Write a general overview of projective geometry.

ANSWERS   Appendix 1

Research

1, 2. Check students' work.

2   Truth Tables and Logic

Learn how to...

- **represent statements using logical symbols for *and, or, not,* and *implies***

- **make a truth table for a logical statement or argument**

So you can...

- **tell whether two statements are logically equivalent**

- **recognize a tautology or a contradiction**

- **tell whether a logical argument is valid**

Does the argument below seem logical to you?

> *If ABCD is a rhombus, then it is a parallelogram.*
>
> *If ABCD is a rectangle, then it is a parallelogram.*
>
> *Therefore, if ABCD is a rhombus, it is a rectangle.*

If something seems wrong to you about that reasoning, how could you convince someone it is wrong? In this appendix, you will learn one way to show whether an argument is *logically valid*.

To study logical arguments, statements are often represented by letters, such as *p, q,* and *r*.

> *p*: "Gene Kelly is in the movie *Singing in the Rain*."
> *q*: "It never rains in southern California."
> *r*: "The rain in Spain stays mainly in the plain."

If a statement is known to be true or false, it can be assigned a **truth value** of T or F. Statement *p* is true, so its truth value is T. Statement *q* is false, so its truth value is F. If you lack information about the weather and geography of Spain, then you will not be able to assign statement *r* a truth value.

To decide the truth value of a compound statement, you examine the truth values of its components.

BY THE WAY

The logical symbol $\vee$ is an abbreviation of the Latin word "vel," which means "or."

A **conjunction** joins two statements with the word "and." The symbol $\wedge$ stands for "and."

$$p \wedge q$$

"Gene Kelly is in the movie *Singing in the Rain* **and** it never rains in southern California."

This conjunction is false, because the second statement is false.

A **disjunction** joins two statements with the word "or." The symbol $\vee$ stands for "or."

$$p \vee q$$

"Gene Kelly is in the movie *Singing in the Rain* **or** it never rains in southern California."

This disjunction is true, because at least one of the statements is true.

THINK AND COMMUNICATE

Use statements *p, q,* and *r* above. Express each statement in words.

1. $p \wedge r$ **2.** $q \vee r$

3. Suppose the truth value of statement *r* is T. What are the truth values of the statements in Questions 1 and 2?

....

ANSWERS   Appendix 2

Think and Communicate

1. Gene Kelly is in the movie *Singing in the Rain* and the rain in Spain falls mainly on the plain.

2. It never rains in Southern California or the rain in Spain falls mainly on the plain.

3. T, T

A **truth table** shows how the truth value of an expression depends on the truth values of its components. The truth tables below have four rows to allow for all combinations of truth values for statements p and q.

Truth table for conjunction		
p	q	$p \wedge q$
T	T	**T**
T	F	**F**
F	T	**F**
F	F	**F**

Truth table for disjunction		
p	q	$p \vee q$
T	T	**T**
T	F	**T**
F	T	**T**
F	F	**F**

A conjunction is true only when both of its components are true.

A disjunction is false only when both of its components are false.

This book uses "or" in an *inclusive* sense. The statement $p \vee q$ means "p is true and q is false, q is true and p is false, or both p and q are true."

Two other truth tables that are useful in logic are the table for *negation* and the table for a *conditional* statement. A curly symbol ($\sim$) is used for negation. An implication arrow ($\rightarrow$) is used for a conditional.

Truth table for negation	
p	$\sim p$
T	F
F	T

Truth table for a conditional		
p	q	$p \rightarrow q$
T	T	**T**
T	F	**F**
F	T	**T**
F	F	**T**

The **negation** of the statement p is the statement $\sim p$, which means "not p" or "It is not the case that statement p is true." A statement and its negation have opposite truth values.

You can express the meaning of the conditional statement $p \rightarrow q$ in many ways: "If p, then q," "p implies q," and "q follows from p." In other words, it never happens that p occurs without q also occurring.

Notice that by convention, a conditional $p \rightarrow q$ is considered true if its hypothesis p is false. For example, the statement "If the moon is made of green cheese, then $2 + 2 = 5$" is *true*.

THINK AND COMMUNICATE

Let p stand for "$\triangle ABC$ is an isosceles triangle," and q stand for "$\triangle ABC$ is an equilateral triangle." Write each sentence using symbolic notation.

4. $\triangle ABC$ is an equilateral triangle or it is not an equilateral triangle.

5. If $\triangle ABC$ is an equilateral triangle, then it is an isosceles triangle.

Think and Communicate

4. $q \vee \sim q$

5. $q \rightarrow p$

EXAMPLE 1

EXAMPLE 1

Make a truth table for $\sim p \wedge \sim q$.

SOLUTION

Step 1 Make a table with all combinations of T and F for p and q.

Step 2 Add a column for $\sim p$ using truth values for the negation of p. Add a column for $\sim q$.

Step 3 Use the truth table for conjunction to combine the columns for $\sim p$ and $\sim q$.

Truth table for $\sim p \wedge \sim q$				
p	q	$\sim p$	$\sim q$	$\sim p \wedge \sim q$
T	T	F	F	**F**
T	F	F	T	**F**
F	T	T	F	**F**
F	F	T	T	**T**

THINK AND COMMUNICATE

6. Make a truth table for $q \rightarrow p$. Is it different from the truth table for $p \rightarrow q$? Explain.

7. Make a truth table for $\sim q \wedge \sim p$. Is it different from the truth table for $\sim p \wedge \sim q$? Explain.

If two logical statements have truth tables with identical final columns, the statements are called **logically equivalent**. In the following example, parentheses are used for grouping. As in algebra, you should evaluate statements in parentheses first.

EXAMPLE 2

Tell whether $\sim(p \vee q)$ and $\sim p \wedge \sim q$ are logically equivalent.

SOLUTION

Build a truth table for $\sim(p \vee q)$ and compare it with the truth table for $\sim p \wedge \sim q$.

Truth table for $\sim(p \vee q)$					Truth table for $\sim p \wedge \sim q$				
p	q	$(p \vee q)$	$\sim(p \wedge q)$		p	q	$\sim p$	$\sim q$	$\sim p \wedge \sim q$
T	T	T	**F**		T	T	F	F	**F**
T	F	T	**F**		T	F	F	T	**F**
F	T	T	**F**		F	T	T	F	**F**
F	F	F	**T**		F	F	T	T	**T**

The statements are logically equivalent because the final columns of the truth tables are identical.

Think and Communicate

6. Truth table for $q \rightarrow p$

p	q	$q \rightarrow p$
T	T	T
T	F	T
F	T	F
F	F	T

Yes; the statement $q \rightarrow p$ is false when p is false and q is true.

7. Truth table for $\sim q \wedge \sim p$

p	q	$\sim p$	$\sim q$	$\sim q \wedge \sim p$
T	T	F	F	F
T	F	F	T	F
F	T	T	F	F
F	F	T	T	T

No; the order of statements in a conjuction does not affect the truth table.

8. Describe the statements $\sim(p \lor q)$ and $\sim p \land \sim q$ in words. Explain why they mean the same thing.

9. Tell whether $\sim(p \lor q)$ and $\sim p \lor q$ are logically equivalent.

If a statement has all F's in the final column of its truth table, it is called a **contradiction**. If a statement has all T's in the final column of its truth table, it is called a **tautology**. Two simple examples are shown.

Truth table for $p \land \sim p$		
p	$\sim p$	$p \land \sim p$
T	F	**F**
F	T	**F**

Truth table for $p \lor \sim p$		
p	$\sim p$	$p \lor \sim p$
T	F	**T**
F	T	**T**

a contradiction   a tautology

A tautology is always true, regardless of the truth values of its component statements. When a tautology involves an implication, you can say that the argument it represents is **logically valid**.

In geometry and other forms of mathematics, you often use this reasoning:

If p is **true**, and p implies q, then q is true.

$$[p \quad \land \quad (p \to q)] \quad \to \quad q$$

You can show that this reasoning, called *modus ponens*, is logically valid by making a truth table and showing that it has all T's in its final column.

EXAMPLE 3

Show that $[p \land (p \to q)] \to q$ is a logically valid argument.

SOLUTION

Make a truth table. You can add intermediate columns to help you. Use columns 1 and 3 to calculate column 4. Repeating column 2 as column 5 may help you calculate column 6.

1	2	3	4	5	6
p	q	$p \to q$	$p \land (p \to q)$	q	$[p \land (p \to q)] \to q$
T	T	T	T	T	**T**
T	F	F	F	F	**T**
F	T	T	F	T	**T**
F	F	T	F	F	**T**

Because the last column contains only T's, the argument is logically valid.

Think and Communicate

8. The statement $\sim(p \lor q)$ is the negation of the disjunction "p or q." The statement $\sim p \land \sim q$ is the conjunction of "not p" and "not q." They mean the same thing because they are logically equivalent, as shown in the solution of Example 2. Both $\sim(p \lor q)$ and $\sim p \land \sim q$ mean "neither p nor q."

9. No; the statements are not logically equivalent because the final column in their truth tables are not identical.

Exercises and Applications

1. I win and you win.

2. I win or you win.

3. I do not win and you do not win.

4. If you win, then they do not win.

5. If you win or they win, then I do not win.

6. It is not the case that both you and I win.

7. I win or both you and they win.

8. It is not the case that you or they win.

9. $q \to r$

10. $s \to (p \wedge q)$

11. $\sim p \to \sim(s \vee r)$

12.
p	$\sim p$	$p \vee \sim p$
T	F	T
F	T	T

tautology

13.
p	$\sim p$	$p \wedge \sim p$
T	F	F
F	T	F

contradiction

14.
p	p	$p \to p$
T	T	T
F	F	T

tautology

15.
p	$\sim p$	$\sim(\sim p)$
T	F	T
F	T	F

16.
p	q	$\sim q$	$p \vee \sim q$
T	T	F	T
T	F	T	T
F	T	F	F
F	F	T	T

17.
p	q	$\sim p$	$\sim p \vee q$
T	T	F	T
T	F	F	F
F	T	T	T
F	F	T	T

18.
p	q	$\sim q$	$p \wedge \sim q$
T	T	F	F
T	F	T	T
F	T	F	F
F	F	T	F

19.
p	q	$\sim p$	$\sim p \wedge q$
T	T	F	F
T	F	F	F
F	T	T	T
F	F	T	F

20. **Truth table for $\sim(p \wedge q)$**

p	q	$p \wedge q$	$\sim(p \wedge q)$
T	T	T	F
T	F	F	T
F	T	F	T
F	F	F	T

Truth table for $\sim p \vee \sim q$

p	q	$\sim p$	$\sim q$	$\sim p \vee \sim q$
T	T	F	F	F
T	F	F	T	T
F	T	T	F	T
F	F	T	T	T

Exercises and Applications

Using the statements below, express each symbolic statement in words.

    p: "I win." q: "You win." r: "They win."

1. $p \wedge q$ **2.** $p \vee q$ **3.** $\sim p \wedge \sim q$ **4.** $q \to \sim r$

5. $(q \vee r) \to \sim p$ **6.** $\sim(p \wedge q)$ **7.** $p \vee (q \wedge r)$ **8.** $\sim(q \vee r)$

Let p stand for "$ABCD$ has two pairs of parallel sides," q stand for "$ABCD$ has four right angles," r stand for "$ABCD$ is a rectangle," and s stand for "$ABCD$ is a square." Express each sentence in symbolic notation, using parentheses if necessary.

9. If $ABCD$ has four right angles, then it is a rectangle.

10. If $ABCD$ is a square, then it has two pairs of parallel sides and four right angles.

11. If $ABCD$ does not have two pairs of parallel sides, then it is not a square or a rectangle.

Make a truth table for each symbolic statement. Identify any tautologies or contradictions.

12. $p \vee \sim p$ **13.** $p \wedge \sim p$ **14.** $p \to p$ **15.** $\sim(\sim p)$

16. $p \vee \sim q$ **17.** $\sim p \vee q$ **18.** $p \wedge \sim q$ **19.** $\sim p \wedge q$

In Exercises 20 and 21, use De Morgan's rules:

    The statement $\sim(p \vee q)$ is logically equivalent to $\sim p \wedge \sim q$.

    The statement $\sim(p \wedge q)$ is logically equivalent to $\sim p \vee \sim q$.

20. The first rule was proved in Example 2. Use truth tables to prove the second.

21. Writing Express De Morgan's rules in words.

22. Investigation On page 149, these four statements are summarized:

Conditional	**Inverse**
If p, then q.	If not p, then not q.

Converse	**Contrapositive**
If p, then q.	If not q, then not p.

  a. Write each statement using symbolic notation.

  b. Make a truth table for each statement.

  c. Tell which statements are logically equivalent.

23. The statement "p if and only if q" is called a *biconditional*. In symbols, it can be expressed as $(p \to q) \wedge (q \to p)$. Make a truth table for the biconditional.

The statements are logically equivalent because the final columns of their truth tables are identical.

21. (1) "It is not the case that p or q is true" has the same meaning as "It is not the case that p is true, and it is not the case that q is true."
(2) "It is not the case that both p and q are true" has the same meaning as "It is not the case that p is true or it is not the case that q is true."

22. **a.** conditional: $p \to q$
    converse: $q \to p$
    inverse: $\sim p \to \sim q$
    contrapositive: $\sim q \to \sim p$

  b. Truth table for conditional

p	q	$p \to q$
T	T	T
T	F	F
F	T	T
F	F	T

Truth table for converse

p	q	$q \to p$
T	T	T
T	F	T
F	T	F
F	F	T

Truth table for inverse

p	q	$\sim p$	$\sim q$	$\sim p \to \sim q$
T	T	F	F	T
T	F	F	T	T
F	T	T	F	F
F	F	T	T	T

Tell whether each statement is a *contradiction*, a *tautology*, or *neither*. Explain your answer.

24. $(p \wedge q) \to p$

25. $(p \wedge q) \wedge (\sim p \vee \sim q)$

26. $(p \vee q) \to \sim q$

27. $[(p \vee q) \wedge \sim p] \to q$

28. Open-ended Problem Write a symbolic statement that represents a tautology or a contradiction. Show its truth table.

29. The following form of reasoning is known as *modus tollens*:

> **If p implies q, and q is not true, then p is not true.**
>
> $[(p \to q) \wedge \sim q] \to \sim p$

Make a truth table to show that *modus tollens* is a logically valid form of reasoning.

30. The following form of reasoning is known as *sorites*:

> **If p implies q, and q implies r, then p implies r.**
>
> $[(p \to q) \wedge (q \to r)] \to (p \to r)$

Complete the truth table to show that *sorites* is a logically valid form of reasoning.

1	2	3	4	5	6	7	8
p	q	r	$p \to q$	$q \to r$	$(p \to q) \wedge (q \to r)$	$p \to r$	$[(p \to q) \wedge (p \to r)] \to (p \to r)$
T	T	T	T	T	T	?	?
T	T	F	T	F	F	?	?
T	F	T	F	T	?	?	?
T	F	F	F	T	?	?	?
F	T	T	T	?	?	?	?
F	T	F	T	?	?	?	?
F	F	T	T	?	?	?	?
F	F	F	T	?	?	?	?

31. Challenge You can extend the *sorites* form of reasoning. Make a truth table to show that $[(p \to q) \wedge (q \to r) \wedge (r \to s)] \to (p \to s)$. How many rows does your truth table require?

32. The word *sorites* is derived from a Greek root meaning "heap." Explain why this is appropriate.

33. Let r stand for "$ABCD$ is a rhombus," p stand for "$ABCD$ is a parallelogram," and t stand for "$ABCD$ is a rectangle." The argument at the top of page 736 can be phrased this way:

> **If r implies p, and t implies p, then r implies t.**
>
> $[(r \to p) \wedge (t \to p)] \to (r \to t)$

Make an eight-row truth table and use it to show that the above argument is *not* logically valid.

34. Research Find out how to use truth tables to model the behavior of electricity in circuits that have switches placed *in parallel* or *in series*.

Appendix 2: Truth Tables and Logic   **741**

···

Truth table for contrapositive

p	q	$\sim q$	$\sim p$	$\sim q \to \sim p$
T	T	F	F	T
T	F	T	F	F
F	T	F	T	T
F	F	T	T	T

c. The conditional and the contrapositive are logically equivalent. The converse and the inverse are logically equivalent.

23.

p	q	$p \to q$	$q \to p$	$(p \to q) \wedge (q \to p)$
T	T	T	T	T
T	F	F	T	F
F	T	T	F	F
F	F	T	T	T

24. a tautology, since there are all T's in the final column of the truth table:

p	q	$p \wedge q$	$(p \wedge q) \to p$
T	T	T	T
T	F	F	T
F	T	F	T
F	F	F	T

25–27. See answers in back of book.

28. Answers may vary. For example, the statement $\sim(p \vee q) \to (\sim p \wedge \sim q)$ is a tautology. The statement $(p \wedge q) \wedge (\sim p \wedge \sim q)$ is a contradiction.

29–32. See answers in back of book.

33. Answers may vary. An example is given. Relate circuits placed in series to truth tables by labeling a closed switch T and an open switch F and using the truth table for $p \wedge q$. Relate circuits placed in parallel to truth tables by labeling a current that will flow in a circuit T and current that will not flow F and use the truth table for $p \vee q$.

Glossary

acute angle (p. 57) An angle that measures between 0° and 90°.

acute triangle (p. 65) A triangle with three acute angles.

adjacent angles (p. 58) Two coplanar angles that share a vertex and a side but do not overlap.

alternate interior angles (p. 219) Two angles that lie on opposite sides of a transversal between the two lines that the transversal intersects.

altitude of a triangle (p. 319) A perpendicular segment from a vertex of a triangle to the line that contains the opposite side.

angle (p. 34) A figure formed by two rays (called *sides*) with a common endpoint (called the *vertex*).

angle of depression (p. 509) When a point is viewed from a higher point, the angle that the person's line of sight makes with the horizontal.

angle of elevation (p. 509) When a point is viewed from a lower point, the angle that the person's line of sight makes with the horizontal.

apothem (p. 368) The distance from the center of a regular polygon to a side.

arc (p. 554) An unbroken part of a circle.

auxiliary line (p. 235) A line, ray, or segment added to a diagram to help complete a proof.

base angles of an isosceles triangle (p. 313) The angles opposite the two congruent sides. The third angle is the *vertex angle*.

biconditional (p. 118) A statement that contains the words "if and only if." This single statement is equivalent to writing both "If p, then q" and its converse "If q, then p."

bisector of a segment (p. 41) A line, segment, ray, or plane that intersects the segment at its midpoint.

bisector of an angle (p. 42) A line or ray that divides the angle into two congruent angles.

central angle of a circle (p. 553) An angle whose vertex is the center of a circle.

chord (p. 553) A segment whose endpoints lie on a circle.

circle (p. 187) The set of all points in a plane that are an equal distance (the *radius*) from a given point (the *center*), which is also in the plane.

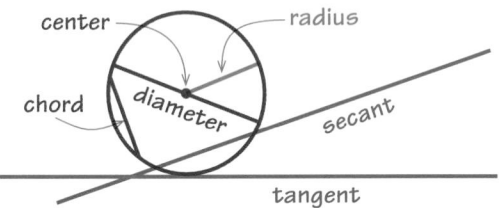

circumference of a circle (p. 369) The perimeter of a circle.

circumscribed circle (p. 369) A circle is *circumscribed* about a polygon when each vertex of the polygon lies on the circle. The polygon is *inscribed* in the circle.

circumscribed circle   inscribed circle
inscribed polygon   circumscribed polygon

circumscribed polygon (p. 324, Ex. 13) A polygon is *circumscribed* about a circle when each of the sides of the polygon is tangent to the circle. The circle is *inscribed* in the polygon.

collinear points (p. 22) Points on the same line.

complementary angles, complements (p. 58) Two angles whose measures add up to 90°.

concentric circles (p. 189) Two or more circles that lie in the same plane and have the same center.

conditional statement (p. 15) A statement that can be written in the form "If p, then q." Statement p is the *hypothesis* and statement q is the *conclusion*.

cone (p. 537) A three-dimensional figure with one circular base and a vertex. In a right cone, the *vertex* is directly above the center of the base. In the right cone shown, h is the height, ℓ is the slant height, and r is the radius.

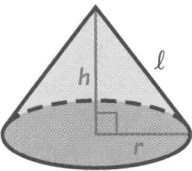 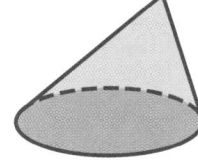

right cone   oblique cone

congruent angles (p. 35) Two angles that have the same measure.

congruent arcs (p. 555) Arcs, in the same circle or in congruent circles, that have equal measures.

congruent polygons (p. 285) Polygons that have congruent corresponding sides and angles.

conjecture (p. 15) Something believed to be true but not yet proved.

consecutive angles (p. 74) In a polygon, two angles that share a side.

consecutive sides (p. 74) In a polygon, two sides that share a vertex.

contrapositive (p. 149) The contrapositive of the conditional statement "If p, then q" is the statement "If not q, then not p."

converse (p. 137) The converse of the conditional statement interchanges the hypothesis and conclusion. "If p, then q" becomes "If q, then p."

convex polygon (p. 73) A polygon in which no segment that connects two vertices can be drawn outside the polygon.

coordinate plane (pp. 696–697) A grid formed by two axes that intersect at the *origin*. The axes divide the coordinate plane into four *quadrants*.

coplanar (p. 22) Points and other figures that lie in the same plane.

corollary of a theorem (p. 239) A statement that can be easily proved using the theorem.

corresponding angles (p. 219) Two angles that lie on the same side of a transversal, in corresponding positions with respect to the two lines that the transversal intersects.

corresponding parts (pp. 285, 445) A side (or angle) of a polygon that is matched up with a side (or angle) of a congruent or similar polygon.

cosine of an acute angle, cosine ratio (p. 515) In a right triangle, the ratio of the length of the leg adjacent to the angle to the length of the hypotenuse.

counterexample (p. 16) An example for which the hypothesis of a conditional statement is true, but the conclusion is false. A counterexample shows that the conditional statement is not always true.

cube (p. 94) A rectangular prism whose faces are all squares.

cylinder (p. 375) A space figure whose *bases* are circles of the same size. The right cylinder shown has height h and radius r.

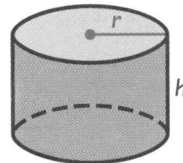 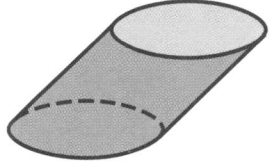

right cylinder   oblique cylinder

deductive reasoning (p. 112) Using facts, definitions, and accepted properties in a logical order to reach a conclusion or to show that a conjecture is always true.

diagonal of a polygon (p. 74) In a polygon, a segment that connects nonconsecutive vertices.

diameter of a circle (p. 187) A segment through the center of the circle with endpoints on the circle; also the length of such a segment. *See* circle.

dilation (p. 428) A transformation with center O and positive *scale factor* k in which every point P has an image P' on $\overrightarrow{OP}$ so that $OP' = k \cdot OP$. If $k > 1$, the dilation is an *enlargement*. If $k < 1$, the dilation is a *reduction*.

distance (pp. 166, 203, 249, 258) The length of the segment that connects two points, or the length of the perpendicular segment from a point to the object.

endpoint of a ray (p. 28) The point where a ray begins. A ray is named by the endpoint and another point on the ray. *See* ray.

endpoints of a segment (p. 28) The points where a segment begins and ends. These points are used to name the segment.

equiangular polygon (pp. 65, 73) A polygon whose angles are all congruent.

equidistant (pp. 209, 307) At the same distance.

equilateral polygon (pp. 65, 73) A polygon whose sides are all congruent.

Euclidean geometry (p. 598) A geometric system based on the postulates of Euclid.

exterior angle of a polygon (pp. 81, 125) An angle formed when one side of the polygon is extended. The angle is adjacent to an *interior angle* of the polygon.

geometric mean (p. 494) If a, b, and x are positive numbers and $\frac{a}{x} = \frac{x}{b}$, then x is the geometric mean of a and b.

geometric probability (p. 476) The probability of an event is given by a ratio that compares lengths, areas, perimeters, or other measures.

glide reflection (p. 420) A translation followed by a reflection over a line parallel to the translation.

great circle (p. 598) The intersection of a sphere with a plane containing the center of the sphere.

half-turn (pp. 8, 414) A rotation of 180°.

height *See* cone, cylinder, prism, *and* pyramid.

hexagon (p. 73) A polygon with six sides.

hyperbolic geometry (p. 602) A non-Euclidean geometric system based on the postulate that through a point not on a line, there is more than one line parallel to the given line.

hypotenuse (p. 141) The side opposite the right angle in a right triangle. The other two sides are the *legs*.

hypothesis in a proof (p. 125) The given information.

image (pp. 8, 391) The figure resulting from applying a transformation to a figure.

image matrix (p. 614) A matrix whose elements represent the coordinates of the image of a transformed polygon.

included angle (p. 294) An angle of a polygon whose vertex is the shared point of two sides of the polygon.

included side (p. 300) A side of a polygon whose endpoints are the vertices of two angles of the polygon.

inductive reasoning (p. 4) A type of reasoning in which a prediction or conclusion is based on an observed pattern.

inscribed angle (p. 553) An angle whose vertex is on a circle and whose sides are chords of the circle.

inscribed circle (p. 324, Ex. 13) *See* circumscribed polygon.

inscribed polygon (p. 369) *See* circumscribed circle.

intercept (p. 175) The distance from the origin to the point where a line crosses one of the axes.

interior angle of a polygon (p. 81) See exterior angle of a polygon.

inverse (p. 149) The inverse of the conditional statement "If p, then q" is the statement "If not p, then not q."

isosceles triangle (p. 65) A triangle with at least two congruent sides. The two congruent sides are the *legs*. The third side is the *base*. The angles opposite the legs are the *base angles*, and the third angle is the *vertex angle*.

kite (p. 339) A quadrilateral that has two pairs of consecutive congruent sides, but whose opposite sides are not congruent.

lateral area (pp. 376, 536, 537) For a prism and a pyramid, the area of the lateral faces. For a cylinder and a cone, the area of the curved lateral surface.

length of an arc (p. 580) Along a circle, the distance between the endpoints of an arc.

line of reflection (p. 391) The line over which a figure is flipped, or reflected.

line of symmetry (p. 10) The line over which a figure is flipped, resulting in a figure that coincides exactly with the original figure.

line perpendicular to a plane (p. 258) A line that intersects a plane and is perpendicular to every line in the plane that passes through the point of intersection.

linear pair (p. 58) Two adjacent angles form a linear pair if their nonshared rays form a straight angle.

magnitude of a vector (p. 521) The size of the vector; also the length of the arrow that represents the vector.

major arc (p. 554) Two points on a circle and all the points of the circle not on the minor arc between the two given points. A major arc measures greater than 180° and is named by three points, its endpoints, and a third point on the major arc.

matrix (matrices) (p. 613) A rectangular arrangement of numbers, called *elements*, in rows and columns. The matrix below has 2 rows and 3 columns, or *dimension* 2 × 3.

$$\begin{bmatrix} 1 & 0 & -2 \\ 5 & -3 & 4 \end{bmatrix}$$

measure of an angle (p. 35) A number between 0° and 180° that is paired with the angle.

measure of an arc (p. 554) The measure of a minor arc is equal to the measure of the central angle that intercepts it. The measure of a major arc is equal to 360° minus the measure of the corresponding minor arc. The measure of a semicircle is 180°. The measure of a complete circle is 360°.

median of a triangle (p. 319) A segment from a vertex to the midpoint of the opposite side.

midline of a trapezoid (p. 364, Ex. 16) The segment that is parallel to the bases of a trapezoid and connects the midpoints of the legs.

midpoint (p. 41) The point on a segment that divides it into two congruent segments.

minor arc (p. 554) Two points on a circle and all the points of the circle in the interior of the central angle between the two endpoints. A minor arc measures less than 180° and is named by its two endpoints.

n-gon (p. 73) A polygon with *n* sides.

negation (p. 149) The negation of a statement *p* is the statement "not *p*," or "*p* is not true."

net (p. 95) A plane figure that can be folded to form a three-dimensional figure, such as a prism.

non-Euclidean geometry (pp. 208, 597, 602) A geometric system not based on the postulates of Euclid.

obtuse angle (p. 57) An angle that measures between 90° and 180°.

obtuse triangle (p. 65) A triangle with an obtuse angle.

octagon (p. 73) A polygon with eight sides.

ordered triple (p. 202) The coordinates (x, y, z) that give the position of a point in three dimensions.

orientation in a transformation (p. 392) The order and direction in which the vertices of a given figure are arranged.

origin (pp. 696–697) On a coordinate plane, the point where the axes intersect.

parallel lines (p. 22) Lines in the same plane that do not intersect.

parallel planes (p. 23) Planes that do not intersect.

parallelogram (p. 88) A quadrilateral with both pairs of opposite sides parallel.

pentagon (p. 73) A polygon with five sides.

perpendicular (p. 58) Two lines, segments, rays, or planes that intersect to form right angles.

perpendicular bisector of a segment (p. 181) A line, ray, or segment that bisects the segment and is perpendicular to it.

point matrix (p. 613) A matrix that represents a point.

point of tangency (p. 554) The point where a tangent intersects a circle.

polar coordinates (p. 416) A system of coordinates based on distance from the origin and an angle of rotation.

polygon (p. 73) A closed plane figure whose sides are segments that intersect only at their endpoints, with each segment intersecting exactly two other segments.

polygon matrix (p. 613) A matrix that represents a polygon. Each column represents the coordinates of a vertex.

postulate (p. 117) A mathematical statement that is accepted without proof.

prism (pp. 93, 375) A three-dimensional figure with two congruent faces, called *bases*, that lie in parallel planes. The other faces, called *lateral faces*, are rectangles that connect corresponding vertices of the bases. A prism's vertices are connected by segments called *edges*. In the prisms below, *h* is the height.

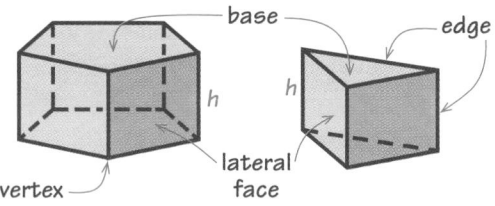

proof (p. 124) A convincing argument that can be used to show that a conjecture is true.

proportion (p. 446) An equation that shows that two ratios are equal.

pyramid (p. 535) A three-dimensional figure with one *base* that is a polygon. The other faces, called *lateral faces*, are triangles that connect the base to the *vertex*. In the right pyramid shown, *h* is the height, ℓ is the slant height, *a* is the apothem, and *r* is the radius.

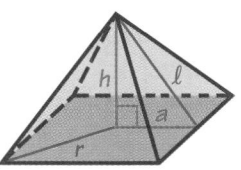
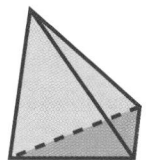

right pyramid   oblique pyramid

quadrilateral (p. 73) A polygon with four sides.

radius of a circle (p. 187) A segment whose endpoints are the center of the circle and a point on the circle; also the length of such a segment. *See* circle.

ray (p. 28) A part of a line that starts at a point, the endpoint, and continues forever in one direction.

rectangle (p. 88) An equiangular parallelogram.

rectangular prism (p. 94) A prism whose bases and faces are rectangles.

reflection (pp. 8, 391) A transformation that flips a figure over a line.

regular polygon (p. 73) A polygon that is both equilateral and equiangular.

regular pyramid (p. 535) A pyramid with a base that is a regular polygon and with lateral faces that are congruent isosceles triangles.

rhombus (p. 88) An equilateral parallelogram.

right angle (p. 57) An angle that measures 90°.

right triangle (p. 65) A triangle with a right angle.

rotation (pp. 8, 412) A transformation in which every point moves along a circular path around a fixed point called the *center of rotation*.

same-side interior angles (p. 219) Two angles that lie on the same side of a transversal between the two lines that the transversal intersects.

scalar multiplication of a matrix (p. 614) The process of multiplying each element of a matrix by a real number.

scalar multiplication of a vector (p. 523) The process of multiplying each component of a vector by a real number.

scalene triangle (p. 65) A triangle with no congruent sides.

secant of a circle (p. 561) A line, ray, or segment that contains a chord. *See* circle.

sector of a circle (p. 580) A region of a circle that is bounded by two radii and an arc of the circle.

segment (p. 28) A part of a line with two endpoints.

semicircle (p. 554) Either of the two arcs of a circle intersected by a diameter of the circle. A semicircle is named by three points, its endpoints and a third point on the semicircle.

similar (p. 446) Two figures are similar if corresponding angles are congruent and the lengths of corresponding sides are in proportion.

sine of an acute angle, sine ratio (p. 515) In a right triangle, the ratio of the length of the leg opposite the angle to the length of the hypotenuse.

skew lines (p. 22) Two lines that do not intersect and are not parallel.

slope (pp. 174, 697) The measure of the steepness of a line, or vertical change divided by horizontal change.

slope-intercept form (p. 175) The equation of a line in the form $y = mx + b$, where m is the slope of the line and b is the y-intercept.

sphere (p. 586) The set of all points in space equidistant from a given point.

spherical geometry (pp. 597–598, 602) A geometric system based on the statement that in a plane, through a point not on a given line, there is no line parallel to a given line. The surface of a sphere is a plane in spherical geometry, and a line is a *great circle*.

square (p. 88) A regular parallelogram.

straight angle (p. 57) An angle that measures 180°.

supplementary angles, supplements (p. 58) Two angles whose measures add up to 180°.

surface area (p. 376) The area of a net for a three-dimensional figure.

symmetry (p. 10) A figure has symmetry if the figure and its image coincide after a transformation.

tangent of an acute angle, tangent ratio (p. 507) In a right triangle, the ratio of the length of the leg opposite the angle to the length of the leg adjacent to the angle.

tangent to a circle (p. 554) A line in the plane of the circle that intersects the circle in only one point. *See* circle.

tangent to a sphere (p. 586) A line or plane that intersects the sphere in only one point.

taxicab geometry (p. 208) A non-Euclidean geometry based on travel on a rectangular grid.

tessellation (p. 83) A pattern of polygons that covers a plane without gaps or overlaps.

theorem (p. 125) A conjecture that can be proved to be true.

three-dimensional coordinate system (p. 201) A system in which the position of each point in space is given using three coordinate axes. The z-axis is perpendicular to the x-axis and the y-axis.

transformation (p. 8) A change made to the size or position of a figure.

translation (pp. 8, 406) A transformation that slides each point of a figure the same distance in the same direction.

transversal (p. 219) A line that intersects two or more other lines in the same plane at different points.

trapezoid (p. 228) A quadrilateral with exactly one pair of parallel sides, called the *bases*. The two non-parallel sides are called the *legs*.

triangle (pp. 64, 73) A polygon with three sides.

vector (p. 521) A quantity that has both magnitude, or size, and direction.

vertex (vertices) of a polygon (pp. 65, 73) An endpoint of a side of a polygon.

vertical angles (p. 58) Non-adjacent, non-overlapping angles formed by two intersecting lines.

y-intercept (p. 175) The y-coordinate of a point where a graph crosses the y-axis.

z-axis (p. 201) *See* three-dimensional coordinate system.

Index

Absolute value, 208–209
ACT/SAT Preview *See* SAT/ACT
 Preview.
Algebra
 applications and connections, 4, 16,
  41, 43, 80, 81, 88, 112, 141,
  142, 196, 222, 227, 243, 286,
  315, 341, 354, 363, 448, 462,
  470, 494, 500, 562, 568, 574,
  587
 Exercises, 7 (Exs. 22–25), 13
  (Exs. 24–27), 27 (Exs. 40–45),
  32, 44, 46, 67, 68, 84, 86, 104,
  115, 146, 170, 179 (Ex. 35),
  185 (Ex. 32), 193, 199 (Ex. 25),
  200, 223, 224, 230, 232, 241,
  245, 247 (Ex. 17), 252, 255,
  262, 274, 289, 310, 315, 317,
  322, 323, 325, 330, 334, 351,
  357, 364, 365, 372, 380, 395,
  402, 449, 464, 466, 474, 497,
  499, 505, 520, 534, 558, 562,
  563, 566, 570, 575, 576, 585,
  595, 616 (Exs. 14, 15), 618
  (Exs. 34, 35), 630, 640, 657
 geometric mean, 494–496
 proportions, 445–446, 494–496,
  707–708
 radical expressions, 104, 142, 166,
  168, 170 (Ex. 19), 188–189,
  193 (Exs. 28–31), 196, 200,
  203–204, 341, 354, 362, 363,
  364, 367–368, 448, 494, 500,
  522, 536, 575, 587, 700–702
 See also Equation(s) *and*
  Formula(s).
Algebra Review, 106–107, 214–215,
  332–333, 440–441, 548–549,
  690–695
Alternate exterior angles,
  230 (Ex. 13), 246 (Ex. 15)
Alternate interior angles, 219
Alternative assessment *See* Journal,
  Performance Task assessment,
  Portfolio Projects, *and* Self-
  Assessment.
Altitude
 of solids *See* Height.
 of a triangle, 319–320, 324 (Ex. 15)
And (conjunction), 736
Angle(s), 34–36, 57–60
 acute, 57, 138
 addition of, 36
 adjacent, 58, 60 (Question 4)
 alternate interior, 219–221, 226–227,
  243
 base, of an isosceles triangle,
  313–315
 bisector of, 42, 311, 324 (Ex. 13)
 central, 553–555
 in circles, 553–556
 classifying, 57–59, 219–220, 553

complementary, 58, 63 (Exs. 29, 30)
 congruent, 35, 263
 consecutive, 74
 corresponding, of congruent
  polygons, 285–288
 corresponding, and parallel lines,
  219–222, 226, 228, 243
 corresponding, of similar figures,
  445–448
 of depression, 509
 of elevation, 91, 509–510
 exterior, 81–82, 125–126,
  129 (Ex. 20)
 of incidence, 403
 included, 294
 inscribed, 553–555
 interior, 81, 125–126
 measure of, 35
 obtuse, 57, 138
 of a parallelogram, 87–88, 353
 perpendicular lines and, 60, 134
 of a polygon, 73–74, 79–82
 of a quadrilateral, 238
 of reflection, 403
 of repose, 507
 right, 57, 128 (Ex. 10)
 of rotation, 412
 same-side exterior, 223 (Ex. 4)
 same-side interior, 219–221,
  227–229, 244
 sides of, 34
 spherical, 601
 straight, 57
 supplementary, 58, 131,
  132 (Question 2), 138
 of a triangle, 65–67, 125–126,
  235–237, 240 (Ex. 11)
 trisection of, 505 (Ex. 25)
 vertex of, 34
 vertical, 58–59, 130–131
Angle Addition Postulate, 36, 712
**Angle-Angle (AA) Similarity
  Postulate,** 453, 713
Angle-Angle-Side (AAS) Theorem,
  301, 715
Angle bisector(s), 42, 311
 of a triangle, 320–321, 324 (Ex. 13),
  466 (Ex. 23), 571 (Ex. 10)
Angle-Side-Angle (ASA) Postulate,
  300, 713
Angle sum
 of a polygon, 79–80
 of a triangle, 65–66, 236–237
Apothem, 368
Applications
 accessibility, 176
 advertising, 480
 agriculture, 508, 513
 animal tracks, 421, 425
 archeology, 145, 536
 architecture, 146, 316, 373,
  382–383, 536, 540, 589, 594

art, 268, 280
 astronomy, 106, 589, 594
 automobiles, 223, 303, 583
 auto repair, 345
 aviation, 260, 510, 519, 527
 ballooning, 480
 baseball, 138
 bicycling, 71
 biology, 289
 boating, 46
 border patterns, 419
 carpentry, 143, 184, 346, 379
 catering, 582
 chemistry, 6, 204, 205
 city planning, 184, 251, 253
 clocks, 122
 clothing, 364
 computers, 408
 cooking, 120
 crafts, 356
 design, 147
 drafting, 352
 fabric design, 423
 fashion, 583
 forestry, 505
 fractals, 3, 452, 480
 furniture design, 416
 games, 97, 489
 geography, 451
 global positioning, 560, 575, 577
 government, 547
 graphic design, 447
 history, 24, 364, 366, 387, 502, 529
 home repair, 496, 519
 irrigation, 581
 manufacturing, 430, 538
 marketing, 116
 music, 499
 navigation, 510, 523, 527, 559
 nutrition, 378
 optics, 295, 402
 orienteering, 526
 package design, 377, 379, 473
 paleontology, 511
 periscope, 402
 personal finance, 113
 physical therapy, 572
 physics, 332, 591
 pottery, 282
 printing, 191
 quilts, 182, 357
 racing, 44
 recreation, 174, 175
 rock climbing, 313, 316
 sculpture, 452
 set design, 468, 472
 skiing, 25
 social studies, 17
 sports, 80, 166, 367, 370, 500, 524,
  565, 587, 595, 600, 657
 technical drawing, 451
 theater, 516

track and field, 583
travel, 17, 150
tumbling, 1–2, 37
See also By the Way, Career
   Interview, Connections,
   Historical notes, History of
   mathematics, Multicultural
   connections, Portfolio Projects,
   Research, *and* Science.
Arc Addition Postulate, 555, 713
Arc(s) of a circle, 263, 554–556
   addition of, 554–555
   classifying, 554, 555
   congruent, 555
   corresponding to chord, 568
   intercepted, 555, 561–562
   length of, 580–581
   major and minor, 554
   measure of, 554
   semicircle, 554
Arc of a sphere, 586
Area(s), 360–363
   of a circle, 369–370
   of composite figures, 476, 478
      (Exs. 1–4), 480 (Exs. 16–18),
      481 (Exs. 4, 5)
   lateral, 376–377, 535–538
   maximum, 369
   of a parallelogram, 361–362,
      365 (Ex. 23)
   of a rectangle, 360–362, 704–705
   of a regular polygon, 367–369,
      529
   of a rhombus, 365 (Ex. 25), 530
   of a sector, 579–580
   of a segment of a circle, 584
   of a square, 360, 362,
      369 (Question 3)
   and transformations, 633 (Ex. 19)
   of a trapezoid, 361–363,
      364 (Ex. 16), 366 (Ex. 34), 531
   of a triangle, 361–362, 364 (Ex. 7),
      365 (Ex. 24), 534 (Ex. 22)
   and trigonometry, 529–531
   of similar figures, 468–470
   surface, 376–377, 534 (Ex. 25),
      535–538, 586–587, 593
   using geometry software, 728
Argument *See* Convincing argument,
   Proof, *and* Reasoning.
Assessment
   Assess Your Progress, 20, 33, 47, 71,
      86, 99, 123, 135, 153, 186, 207,
      233, 255, 269, 291, 312, 325,
      359, 381, 411, 433, 467, 481,
      506, 528, 541, 566, 585, 603,
      625, 647
   Chapter Assessment, 52–53,
      104–105, 158–159, 212–213,
      274–275, 330–331, 386–387,
      438–439, 486–487, 546–547,
      608–609, 654–655
   Cumulative Assessment, 160–161,
      334–335, 488–489, 654–655
   Journal, 20, 33, 47, 71, 86, 99, 123,
      135, 153, 186, 207, 233, 255,
      269, 291, 312, 325, 359, 381,
      411, 433, 467, 481, 506, 528,
      541, 566, 585, 603, 625, 647

Ongoing Assessment, *in every
   lesson, for example*, 32, 92, 147,
   172, 200, 241, 269, 291, 358,
   418, 460, 499, 541, 578, 618
Performance Task, 53, 105, 159, 213,
   275, 331, 387, 439, 487, 547,
   609, 655
Portfolio Projects, 48–49, 100–101,
   154–155, 208–209, 270–271,
   326–327, 382–383, 434–435,
   482–483, 542–543, 604–605,
   648–651
Self-Assessment, 49, 101, 155, 209,
   271, 327, 383, 435, 483, 543,
   605, 651
Auxiliary line, 235, 596 (Ex. 17)
Axiomatic system, 117–118, 712–713,
   734–735
Axis (axes)
   of a coordinate plane, 170, 696–697
   of a three-dimensional coordinate
      system, 201–202

Base(s)
   of a cone, 537
   of a cylinder, 375
   of an isosceles triangle, 313
   of a parallelogram, rectangle, and
      square, 362
   of a prism, 93, 375–377
   of a pyramid, 535
   of a trapezoid, 228, 362
   of a triangle, 362
Base angles
   of an isosceles triangle, 313
   of a trapezoid, 228
Bearing, 527
Between, 29
Biconditional statement, 118
Bisector
   of an angle, 42, 311, 324 (Ex. 13)
   of a segment, 41
      perpendicular, 181, 265, 307–308,
      311 (Ex. 15)
By the Way, *throughout the book, for
   example*, 7, 13, 85, 115, 127,
   174, 188, 240, 246, 308, 317,
   367, 369, 396, 426, 452, 479,
   529, 539, 568, 602, 613, 639

Calculator
   graphing, 79, 372, 404, 472, 590,
      618, 629, 632, 636, 646
   scientific, 142, 144 (Exs. 1–6),
      305 (Ex. 19), 508–510,
      514–516, 595 (Ex. 9)
   See also Technology Handbook.
Calculus, underpinnings of, 369,
   604–605
Career Interview
   Arcenas, Loy, set designer, 442–444,
      465, 468, 472 (Ex. 17)
   Courson, Ron, physical therapist and
      athletic trainer, 550–552, 557,
      572 (Ex. 16), 589

Fleming, Madeleine, optical
   physicist, 276–278, 290, 295,
   317
Hannah, Marc, computer system
   designer, 162–164, 170, 198,
   206
Johnson, Terri, architect, 388–390,
   395, 409, 418
Leavy, Latecia, tumbler, 1–2, 12, 17,
   37
Lueth, Virgil, mineralogist, 54–56,
   68, 82, 85, 98
Mann, Mary-Jacque, forensic
   scientist, 108–110, 128, 133
Ocampo, Adriana, planetary
   geologist, 610–612, 623, 645
Saínz, José, kite designer, 216–218,
   231, 252, 262
Stone, Walt, alderman, 336–338,
   366, 380
Tewa, Debby, solar electric
   technician, 490–492, 503, 519
Carpenter's square, 184
Cavalieri's principle, 604–605
Center
   of a circle, 187–189
   of dilation, 428, 614–615
   of gravity, 323
   of rotation, 412
   of a sphere, 586
Central angle, 553–555
Centroid, 323 (Ex. 8), 324 (Ex. 14)
Challenge exercises, *throughout the
   book, for example*, 13, 63, 91,
   115, 185, 241, 283, 358, 397,
   418, 459, 513, 520, 591, 645
Chapter Review *See* Review.
Chapter Tests *See* Assessment.
Checking Key Concepts, *in every
   lesson, for example*, 11, 67, 120,
   138, 197, 238, 295, 370, 414,
   422, 449, 510, 538, 570, 629
Chord(s) of a circle, 553
   arc corresponding to, 568
   bisector of, 418 (Ex. 33), 568
   congruent, 568–569
   intersecting, 562, 573–574
   properties of, 567–569
Chord of a sphere, 586
Circle(s), 187
   angles of, 553–556
   arcs of, 263, 554–556, 560–562,
      580–581
   area of, 369–370
   center of, 187–189
   central angle, 553–555
   chords of, 553, 562, 567–569,
      573–574
   circumference of, 369, 704
   circumscribed, 369, 372 (Ex. 14),
      567
   concentric, 189, 191 (Ex. 18)
   diameter of, 187, 192 (Exs. 23–25)
   equations of, 188–190
   great, 598
   inscribed, 324 (Ex. 13)
   inscribed angle, 553–556
   radius of, 187
   secant of, 561, 573–574

sector of, 579–580
segment of, 584
segment lengths in, 573–575
tangent to, 554–555, 560–561, 574–575
Circumcenter, 311 (Ex. 16), 567, 571 (Ex. 10)
Circumference, 369, 704
Circumscribed circle, 369, 372 (Ex. 14), 567
Circumscribed polygon, 324 (Ex. 13)
Classifying
arcs, 554, 555
information, 154–155
quadrilaterals, 88–89, 138, 339–342
triangles, 64–65, 148, 168, 172 (Ex. 35)
Coefficient of friction, 507, 571
Collaborative learning *See* Cooperative learning, Explorations, *and* Portfolio Projects.
College entrance exams, preparing for *See* SAT/ACT Preview.
Communication
discussion, *in every lesson exposition as* Think and Communicate *questions, for example,* 29, 40, 57, 72, 74, 130, 132, 143, 174, 188, 235, 244, 265, 288, 306, 313, 340, 347, 369, 414, 419, 454, 468, 501, 522, 530, 561, 569, 581, 599, 615, 628
making a poster, display, or video, 2, 49, 53, 101, 105, 110, 146, 155, 209, 271, 278, 327, 387, 390, 444, 483, 492, 543, 552, 651
making a presentation or writing a report, 49, 56, 110, 164, 209, 275, 327, 383, 435, 444, 483, 602, 605, 612, 646, 651
writing and journal, *See* Journal *and* Writing.
See also Performance Task assessment, Portfolio Projects, *and* Research.
Compass, 69, 190, 263, 688
Complementary angles, 58, 63 (Exs. 29, 30), 137
Components of vector, 521–524
Composition of transformations, 405–406, 409, 411, 417 (Exs. 23, 24), 418 (Ex. 32), 419–422, 424 (Ex. 23), 425 (Ex. 30), 432 (Ex. 17), 624 (Ex. 28), 632 (Ex. 15), 637 (Ex. 15), 638 (Exs. 20, 21), 643–647
Computer
computer-aided design, 93, 271, 383
digital maps, 617
games, 424, 613–614, 646
geometry software, 9, 27, 38, 70, 125, 132, 147, 164, 168, 172, 180, 200, 221, 267, 279, 297, 304, 324, 349, 372, 397, 405, 409, 417, 432, 435, 460, 461, 466, 514, 553, 571, 572 (Ex. 19), 578, 726–727

graphing software, 168, 180, 472, 721–723
matrix software, 629, 646, 723–725
patterns and, 408
photographs, 634
simulation, 424
spreadsheets, 16, 146, 169, 450, 472, 482, 725
system designer, 162–164, 170, 198, 206
virtual reality, 639, 646
See also Calculator *and* Technology.
Conclusion
of a conditional statement, 15, 137
of a proof, 125–126, 132
Conditional statement, 15, 17–19, 135, 137, 149, 152 (Ex. 39)
Cone, 537–538
base of, 537
height of, 537
oblique, 541 (Ex. 18)
radius of, 537
right, 537
slant height of, 537
surface area of, 537–538
vertex of, 537
volume of, 537–538
Congruence
of angles, 35, 119
applying, 306–308
of arcs, 554
of circles, 554
determining, 28–30, 35, 60, 69, 292–294, 296 (Ex. 12), 299–302, 307
of polygons, 285–288
properties of, 60, 285–286, 307
of right angles, 60 (Question 4), 114 (Ex. 13), 128 (Ex. 10), 134 (Ex. 15)
of segments, 28, 119
and transformations, 41, 298, 393, 406, 412, 420, 422
of triangles, applying, 306–308
of triangles, methods of proof, 121, 292–295, 299–302, 304 (Ex. 14)
Conjecture, 15
making a, 3–4, 14–20, 25 (Ex. 30), 26 (Ex. 34), 38 (Ex. 24), 63 (Exs. 30, 31), 70 (Exs. 31, 32), 79, 87, 125, 148, 172 (Ex. 33), 180, 186 (Ex. 33), 206 (Ex. 30), 221, 242, 244 (Question 2), 283 (Ex. 31), 324 (Ex. 13), 340, 417 (Ex. 24), 424 (Ex. 24), 429 (Question 7), 430 (Ex. 7), 460 (Exs. 27, 28), 461, 553, 567, 640 (Ex. 36)
Conjunction (and), 736
Connections
among mathematics topics
algebra, 4, 16, 41, 43, 80, 81, 88, 112, 141, 142, 196, 222, 227, 243, 286, 315, 341, 354, 363, 448, 462, 470, 494, 500, 562, 568, 574, 587
data analysis, 592
measurement, 122, 379
probability, 283

to cross-curricular subjects
art, 26, 27, 83, 602
astronomy, 7, 305, 450, 518
biology, 61, 371, 564, 595
earth science, 577
geography, 44, 57, 70, 136, 138
history, 19, 70, 146, 533
language, 139
literature, 127, 431, 498
physics, 323
science, 39 *See also* Science.
social studies, 357
sports, 524, 587, 588, 590
to real-world applications
aerial photography, 310
architecture, 192, 232, 410, 504, 539
art photography, 637
audio-visual presentation, 396
auto maintenance, 151
botany, 224
city planning, 178
communications, 13, 62
computer games, 613
computer simulation, 424
consumer economics, 379, 473
cooking, 596
crafts, 40, 45, 77
dance, 247
design, 254
digital maps, 617
electronics, 259
engineering, 177, 240
gardening, 44, 57, 70, 136, 138, 365
geology, 526
Hawaiian quilts, 417
highway safety, 571
horticulture, 246
hydraulics, 344
interior design, 284
machinery, 351
manufacturing, 31, 285
meteorology, 592
miniature golf, 403
navigation, 121, 597, 601
nutrition, 149
optics, 91
paper folding, 266
photocopiers, 624
puzzles, 171
quilt patterns, 185
rescue safety, 297
rock climbing, 584
special effects, 631
surveying, 457
technical drawing, 261
transportation, 512
video games, 646
virtual reality, 639, 646
zoology, 115, 479
See also Applications, By the Way, Career Interview, Historical notes, History of mathematics, Multicultural connections, Portfolio Projects, Research, *and* Science.
Constructions, 69, 263, 720
angle bisector, 311

centroid of a triangle, 323 (Ex. 8), 324 (Ex. 14)
circle that circumscribes a given triangle, 567
congruent angles, 263
equilateral triangle, 69, 372 (Ex. 17)
45-45-90 triangle, 505 (Ex. 24)
Gothic arch, 504
incenter of a triangle, 324 (Ex. 13)
isosceles trapezoid, 350 (Ex. 15)
kite, 343
orthocenter of a triangle, 324 (Ex. 15)
parallel lines, 264
parallelogram, 350
perpendicular bisector, 265
perpendicular lines, 267
rectangle, 358
regular hexagon, 372
sum of angles, 264
triangle given three sides, 292
triangle given two sides and the included angle, 297
using geometry software, 726–729
Contradiction, 739
Contrapositive, 149, 151
Converse, 137, 149, 243–244
of Pythagorean theorem, 143, 148
Convex polygon, 73
Convincing argument, 112, 117–118, 120, 124, 237
writing a, 114 (Ex. 14), 115 (Exs. 17, 18), 116 (Ex. 25)
Cooperative learning
Exercises, 12, 46, 62, 98, 169, 171, 186, 193, 200, 231, 269, 283, 291, 296, 372, 404, 424, 432, 457, 466, 499, 528, 540, 541, 557, 585, 625, 630, 633
Explorations, 3, 14, 21, 64, 79, 87, 125, 141, 148, 180, 194, 221, 242, 256, 279, 287, 340, 361, 374, 392, 405, 427, 461, 475, 493, 514, 553, 567, 597, 634, 641
Portfolio Projects, 48–49, 100–101, 154–155, 208–209, 270–271, 326–327, 382–383, 434–435, 482–483, 542–543, 604–605, 648–651
Coordinate geometry
distances and midpoints, 165–168
equations of circles, 187–190
equations of lines, 173–175
parallel and perpendicular lines, 180–183
polar, 416
proof, 194–197, 198 (Ex. 11), 199 (Exs. 15, 24, 25), 200 (Exs. 27, 28), 207 (Questions 3–6), 213 (Questions 25–27, 32), 341, 344 (Ex. 12), 354, 357 (Ex. 21), 365 (Ex. 25), 424 (Ex. 24)
in three dimensions, 201–204, 380, 618 (Ex. 31), 632 (Ex. 16), 638 (Exs. 22–24)
See also Matrix *and* Transformations.
Coordinate plane, 197, 202, 696–700

Corollary, 239
Corresponding angles, 219
Corresponding Angles Postulate, 222, 712
Corresponding parts, of congruent figures, 285–288, 306–308
Cosine ratio, 514–517, 529, 538, 560, 733
Counterexample, 16
finding a, 18 (Exs. 11–14), 24 (Exs. 7–10), 27 (Exs. 38, 39), 90 (Questions 2–4), 152 (Ex. 39), 179 (Exs. 45–47), 193 (Exs. 42–45), 225 (Exs. 34, 35), 262 (Ex. 31), 282 (Ex. 20), 460 (Exs. 24, 25, 27)
Critical thinking *See* Reasoning, Study techniques, *and* Visual Thinking.
Cross-curriculum connections *See* Applications *and* Connections.
Cross sections, 206 (Ex. 28), 379 (Ex. 14), 540 (Ex. 18), 572 (Ex. 16), 604–605
Cube, 94, 101, 145 (Ex. 30)
Cubic units, 375, 730
Cultural diversity *See* Career Interview *and* Multicultural connections.
Cumulative Assessment, 160–161, 334–335, 488–489, 654–655
Customary units, 730
Cylinder, 375–377
base of, 375
height of, 375
lateral area of, 376
radius of, 374
surface area of, 376–377
volume of, 375

Data analysis, 3, 4, 6 (Ex. 9), 16, 76 (Ex. 5), 79, 96 (Ex. 11), 111, 124, 155, 169 (Exs. 1–3), 201 (Questions 1, 2), 221, 372 (Ex. 14), 450, 451, 472, 482–483, 508 (Question 3), 514, 590 (Ex. 27), 592, 595, 708–709
See also Matrix, Patterns, Portfolio Projects, *and* Technology Handbook.
Decagon, 73
Deductive reasoning, 112–113, 734–735
See also Reasoning.
Definition, 118
Degree measure, 35–36
Diagonals
of a kite, 341
of a parallelogram, 89, 138, 194–195, 347, 353–355
of a polygon, 74, 76 (Ex. 5)
of a prism, 98 (Exs. 27, 28), 145 (Ex. 30), 206 (Ex. 29)
of a rectangle, 199 (Ex. 16), 341
of a rhombus, 340–341
of special quadrilaterals, 340–342
of a square, 138, 196, 342

Diagrams
drawing, 23, 67, 94, 118, 155, 242, 258, 261, 537, 539 (Exs. 7–9)
interpreting, 22, 28, 30, 35, 58, 220
isometric, 270–271
nets, 95, 101, 147 (Ex. 45), 376–378, 535
oblique, and perspective, 261
scale drawing, 299, 445, 468
technical, 261, 270–271, 451
Diameter
of a circle, 187, 192 (Exs. 23–25)
perpendicular to chord, 568
of a sphere, 586
Dilation, 426–429, 451 (Ex. 24), 473 (Ex. 18); 474 (Ex. 24)
center of, 428
with matrices, 613–615
properties of, 428
scale factor of, 428, 614–615
Dimensions of a matrix, 613
Direction of vector, 521–522
Discrete mathematics
cellular automata (game of Life), 424
classifying and sorting, 64, 128 (Exs. 12–14) 133 (Exs. 5–9), 151 (Exs. 25–28), 154–155, 339–342
Euler's formula, 96
fractals, 3, 452, 480
geometric probability, 475–477
Heron's formula, 364 (Ex. 7)
networks, 14
Platonic solids, 100–101
writing a program, 404
See also Matrix.
Discussion *See* Communication.
Disjunction (or), 736
Distance
between two parallel lines, 249
between two points, 28, 166, 203
from a point to a line, 249
from a point to a plane, 258
taxicab, 208–209
Distance formula
on a coordinate plane, 166, 168, 188–189, 192 (Ex. 22), 196, 208, 714
in three dimensions, 203–204
Distributive property, 496, 574
Dodecahedron, 101
Drawings *See* Diagrams.

Edge
of a network, 14
of a prism, 93
Element of a matrix, 613
Elevation, angle of, 91, 509–510
Endpoint
of an arc, 554
of a ray, 28
of a segment, 28
Enlargement, 427, 428
Enrichment *See* Challenge exercises *and* Portfolio Projects.
Equation(s)
of circles, 188–190

equivalent, 463
of lines, 173–175, 505 (Ex. 23),
  697–700
matrix, 627
quadratic, 142, 188–189, 342, 494,
  500, 575, 587, 700–702
solving, 41, 43, 80, 88, 142, 222,
  227, 243, 286, 315, 332–333,
  342, 447, 462, 470, 494, 500,
  502, 508, 510, 515, 529, 562,
  568, 574, 580, 587
systems of, 333
vector, 524
See also Algebra.
Equidistant, 209, 307
Equivalent statements, 738
Estimation, 7 (Ex. 14), 32 (Ex. 27),
  37 (Ex. 10), 38 (Ex. 20), 124,
  145 (Ex. 29), 220 (Question 4),
  283 (Ex. 28), 299 (Question 3),
  306, 366 (Ex. 34), 379 (Ex. 12),
  396 (Ex. 22), 415 (Exs. 1–4),
  457, 496, 498, 540 (Ex. 14),
  590 (Ex. 28), 593
Euclidean geometry, comparison with
  non-Euclidean, 598, 734–735
Euler's formula, 96
Explorations
  analyzing parallelograms, 87
  analyzing patterns, 3
  analyzing ratios in triangles, 514
  analyzing triangles, 148
  angles and transversals, 221
  circumscribing a triangle, 567
  comparing central and inscribed
    angles, 553
  comparing right triangles, 493
  comparing sides of triangles, 279
  comparing slopes of lines, 180
  comparing translated polygons, 405
  comparing volumes, 374
  discovering area formulas, 361
  drafting parallel lines, 242
  finding angle measures in polygons,
    79
  finding proportions in triangles, 461
  investigating diagonals, 340
  investigating overlapping triangles,
    287
  investigating planes and their
    intersections, 256
  investigating probability based on
    area, 475
  making an enlargement, 427
  matrices with zeros on a diagonal,
    641
  measuring exterior angles, 125
  placing a parallelogram on a
    coordinate plane, 194
  proving the Pythagorean theorem,
    141
  reflecting a polygon over a line, 392
  reflections using matrix
    multiplication, 634
  representing points, lines, and planes,
    21
  sorting triangles, 64
  tracing networks, 14
  triangles on a sphere, 597

See also Cooperative learning,
  Investigation exercises, *and*
  Portfolio Projects.
Extensions, to Portfolio Projects, 49,
  101, 155, 209, 271, 383, 435,
  483, 542
Exterior angles
  alternate, 230 (Ex. 13), 246 (Ex. 15)
  of a polygon, 81, 125
  same-side, 223 (Ex. 4)
Extra practice, 659–686

Faces, lateral
  of a prism, 93
  of a pyramid, 535
Family involvement *See* Cooperative
  learning, Exercises.
Flow proof, 234, 237, 238, 239 (Exs. 1,
  3), 240 (Ex. 12), 241 (Ex. 14),
  247 (Ex. 16), 248 (Ex. 25),
  254 (Ex. 23), 294, 296 (Ex. 10),
  308, 346, 355 (Question 4), 453,
  572 (Ex. 17)
Formula(s)
  arc length, 580
  area, 362, 368, 369, 580, 704–705
  distance
    on a coordinate plane, 166, 168,
      188–189, 196, 208, 714
    in three dimensions, 203–204
  Euler's, 96
  Heron's, 364 (Ex. 7)
  law of sines, 305 (Ex. 19)
  midpoint
    on a coordinate plane, 167, 170
      (Ex. 19), 195, 198–199, 714
    in three dimensions, 203
  probability, 283 (Ex. 28)
  Pythagorean theorem, 141–142, 145
    (Ex. 36), 148, 166, 206 (Ex. 29)
  resistance of a circuit, 459
  slope, 174, 176, 194–195, 714
  speed of a car, 571, 702
  sum of the measures of the interior
    angles of a polygon, 80
  surface area, 376, 536, 537, 586
  volume, 375, 536, 537, 586, 704–705
  writing, 4, 702
Fractals, 3, 452, 480, 648–651
Function, 472 (Ex. 16), 590 (Ex. 27)

Geometric probability, 475–477
Geometric systems
  Euclidean and non-Euclidean,
    734–735
  hyperbolic, 602, 735
  spherical, 597–603, 735
  taxicab, 208–209, 735
Geometry software, 9, 27, 38, 70, 125,
  132, 147, 164, 168, 172, 180,
  200, 221, 267, 279, 297, 304,
  324, 349, 372, 397, 405, 409,
  417, 432, 435, 460, 461, 466,
  514, 553, 571, 572 (Ex. 19),
  578, 726–727

Glide reflection, 419–422
Glossary, 742–747
Gradient, 512
Graph(s)
  with calculator, 79, 372, 404, 472,
    590, 618, 629, 632, 636, 646,
    721–725
  with computer, 168, 180, 472
  on coordinate plane, 10, 28, 42,
    166–169, 173–176, 180–183,
    188–190, 194–197, 253,
    398–401, 405–407, 413–414,
    420, 428, 448, 521–524, 594,
    613–616, 619–621, 626–629,
    634–636, 641–644, 696–700,
    706
  polar coordinate, 416
  in three dimensions, 201–204
  time line, 111
  of vectors, 521–524
Graphing calculator, 79, 372, 404,
  472, 590, 618, 629, 632, 636,
  646, 721–725
  See also Technology.
Group activities *See* Cooperative
  learning, Explorations, *and*
  Portfolio Projects.

Half-turn, 9, 414
Hands-on activities *See* Constructions,
  Explorations, Investigation
  exercises, Manipulatives, *and*
  Portfolio Projects.
Height
  of a cone, 537
  of a cylinder, 374–375
  of a parallelogram, rectangle, square,
    trapezoid, and triangle, 362
  of a prism, 375
  of a pyramid, 145 (Ex. 29), 535
  slant, 535, 537
Hemisphere, 591 (Ex. 32)
Heron's formula, 364 (Ex. 7)
Hexagon, 73, 367, 372, 502
Hexagonal pyramid, 537
Hexahedron, 101
Historical notes, 13, 19, 24, 26, 34,
  40, 70, 72, 100, 111, 117, 121,
  122 (Ex. 29), 177, 187, 188,
  280, 299, 344, 348, 419, 502,
  510, 529, 533, 536, 577, 645
History of mathematics, 70, 100, 127,
  141, 146, 364, 366, 369, 514,
  586, 602
  Abu'l-Wafa, Mohammad, 735
  Agnesi, Maria Gaetana, 735
  Al-Kashi, 369
  Archimedes, 586
  Bolyai, Janos, 602, 735
  Cayley, Arthur, 620
  Clarke, Edith, 459
  David, Florence Nightingale, 476
  Descartes, René, 735
  Escher, M. C., 27, 602
  Euclid, 100, 599, 602, 734–735
  Euler, Leonard, 96
  Fermat, Pierre de, 127

Galilei, Galileo, 645
Gauss, Carl Friedrich, 602, 735
Heron, 364
Hipparchus, 735
Hypatia, 735
Klein, Felix, 602, 735
Lobachevsky, Nikolai Ivanovich, 602, 735
Moufang, Ruth, 603
Pythagoras, 141
Riemann, G. F., 602, 735
Thales, 735
Wiles, Andrew, 127
Hyperbolic geometry, 602
Hypotenuse, 141, 144 (Exs. 13–18)
   midpoint of, 198 (Ex. 11), 344 (Ex. 13)
Hypotenuse-Leg (HL) Theorem, 301, 716
Hypothesis
   of a conditional, 15
   of a proof, 125–126, 132

Icosahedron, 101
If and only if, 118
If-then statement, 15, 17–19, 135, 137, 149, 152 (Ex. 39)
Image, in a transformation, 8, 391
Incenter, 324 (Ex. 13), 571 (Ex. 10)
Included angle, 294
Included side, 300
Indirect measurement, 306, 312 (Question 10), 457, 496, 498
Indirect proof, 348, 351 (Exs. 22–24), 359 (Question 4), 386 (Question 9)
Inductive reasoning, 4, 112–113, 734
   See also Conjecture *and* Reasoning.
Inequalities
   linear, 705–707
   for triangles, 279–281, 600 (Ex. 10)
Inscribed angle, 553–555
Inscribed circle, 324 (Ex. 13)
Inscribed polygon, 369, 559 (Ex. 22), 567
Intercept, 175, 179 (Ex. 34), 697–699
Interior angles, 81
   alternate, 219
   same-side, 219–221, 227–229, 244
Intersecting lines, 22, 120
Intersection, of geometric figures, 22–23, 256–258, 275 (Ex. 22), 591 (Ex. 30)
Interview *See* Career Interview.
Inverse of a conditional, 149, 151
Investigation exercises, 18, 26, 84, 91, 170, 172, 297, 304, 323, 324, 364, 402, 417, 425, 466, 474, 505, 532, 600, 624, 740
Isometric drawing, 261, 270–271
Isosceles trapezoid, 200 (Ex. 28), 343 (Ex. 10), 344 (Ex. 12)
Isosceles triangle, 65, 70 (Ex. 32)
   Isosceles Triangle Theorem, 314, 716
   parts of, 313
   properties of, 138, 313–315, 320–321, 322 (Exs. 3–5)

Journal, 20, 33, 47, 71, 86, 99, 123, 135, 153, 186, 207, 233, 255, 269, 291, 312, 325, 359, 381, 411, 433, 467, 481, 506, 528, 541, 566, 585, 603, 625, 647
   See also Writing.
Justification *See* Convincing argument, Logical reasoning, Proof, *and* Reasoning.

Key steps of a proof, 228, 230 (Ex. 13), 232 (Ex. 30), 236, 240 (Ex. 13), 245 (Ex. 8), 310 (Ex. 7), 320, 344 (Ex. 11), 349 (Question 3), 351 (Ex. 21), 357 (Ex. 22), 569
Kite, 339–341, 357 (Ex. 21)

Lateral area, 376–377, 535–538
Lateral face
   of a prism, 93
   of a pyramid, 535
Latitude, 601
Law of sines, 305 (Ex. 19)
Legs
   of an isosceles triangle, 313
   of a right triangle, 141
   of a trapezoid, 228
Length(s)
   of an arc, 580–581
   proportional, 446, 461–464
   of a segment, 28
   using geometry software, 728
Limit, concept of, 152, 369, 508 (Question 3), 514
Line(s), 21–22, 118
   auxiliary, 235, 254 (Ex. 24), 596 (Ex. 17)
   concurrent, 46 (Ex. 34), 311 (Ex. 16), 323 (Ex. 8), 324 (Exs. 13–15), 571 (Ex. 10)
   contain two points, 21, 22, 120
   coplanar, 22
   distance to a point, 249
   equations of, 173–175
   horizontal, 174–175
   intersecting, 22, 120
   parallel, 22, 135, 137, 181, 219–223, 226–229, 234–237, 242–245, 249–251, 256–258, 263–265, 461–464, 466, 602
      constructing, 264
      slopes of, 180–182
   perpendicular, 58, 60, 181–182, 186 (Ex. 33), 235, 249–250, 267
      constructing, 264
      slopes of, 180–182
   perpendicular, to plane, 258, 260
   proving parallel, 242–244, 249–251, 257–258
   of reflection, 391–394
      in a glide reflection, 420–422
   skew, 22, 137

slopes of, 173–174, 180–182, 640 (Ex. 36), 697–700
   of symmetry, 10, 74, 318 (Ex. 19), 341, 342 (Question 2), 391, 396
   vertical, 174–175
Linear equations *See* Equation(s).
Linear inequalities, 705–707
Linear pair, 58–59, 126, 130
Literature
   The Mysterious Island, by Jules Verne, 498
   Pudd'nhead Wilson, by Mark Twain, 431
   Where Do You Stop?, by Eric Kraft, 127
Locus, concept of, 209
Logic *See* Proof *and* Reasoning.
Logical reasoning exercises, 68, 76, 85, 92, 160, 282, 343, 356, 365, 458, 736–741
   See also Reasoning.
Longitude, 121, 601

Magnitude of vector, 521–524
Manipulatives
   Exercises, 18, 43, 45, 77, 91, 170, 171, 172, 184, 266, 283, 291, 323, 356, 364, 431, 474 (Ex. 26), 499 (Ex. 25), 505, 540, 583, 600 (Ex. 10)
   in concept development, 40, 42, 59, 65, 376, 535, 598
   in Explorations and Projects, 21, 64, 79, 100–101, 141, 148, 194, 242, 256, 287, 340, 361, 374, 392, 475, 493, 567, 597
   See also Constructions *and* Scale model.
Mathematical systems, 734–735
Mathematicians *See* History of mathematics.
Matrix (matrices), 613, 711
   addition, 619, 711
   columns of, 613
   diagonal of, 641
   dilations with, 613–615
   dimensions of, 613, 711
   element of, 613
   identity, 630
   image, 614
   multiplication, 626–629
   point, 613
   polygon, 613–615
   reflections with, 634–636
   rotations with, 641–643
   rows of, 613
   scalar multiplication of, 614–615, 711
   subtraction, 621, 711
   translations with, 619–621
Mean
   arithmetic, 499 (Ex. 24), 708
   geometric, 494–496
Measure(s)
   of an angle, 35–36
   of angles of polygons, 79–82

Index   **753**

of an arc, 554–555
of a segment, 28
table of, 730
See also Distance formula.
Median, 319–321, 323 (Ex. 8),
324 (Ex. 14)
Metric units, 730
Mid-chapter review *See* Assessment,
Assess Your Progress.
Midline of a trapezoid, 364
Midpoint formula
on a coordinate plane, 167, 170
(Ex. 19), 194–195, 198–199,
714
in three dimensions, 203
Midpoint of a segment, 41, 181
Model, scale, 443–444, 450, 468,
482–483, 593
Modeling, mathematical, 3–5, 17
(Ex. 1), 21, 34, 38 (Ex. 20), 39,
46 (Ex. 28), 61, 79, 85, 91, 98,
121, 122, 128, 133, 136, 146,
147, 162–164, 165, 171, 173,
176, 182, 185, 188, 191, 201,
206, 231, 240, 299, 305, 379,
396, 402, 403, 444, 445, 447,
450–452, 457, 459, 465, 468,
473, 475–480, 496, 502, 510,
516, 523, 524, 538, 559, 560,
565, 571, 575, 584, 595,
597–602, 617, 639, 645
Morphing, with matrices, 626–629,
631
Multicultural connections, 13, 17, 26,
27, 40, 45, 57, 70, 72, 77, 83,
84, 100, 121, 122, 136–138,
139, 145, 146, 147, 171, 173,
174, 177, 185, 188, 192, 226,
231, 240, 251, 252, 268, 299,
305, 316, 342, 344, 348, 356,
357, 364, 369, 373, 417, 419,
452, 455, 458, 474, 502, 510,
536, 540, 587, 588, 602
See also Career Interview.
Multiple methods, 4, 82, 95, 130, 198
(Ex. 11), 236–237, 267
(Exs. 13, 15), 293 (Question 3),
297 (Exs. 15–17), 316 (Ex. 7),
344 (Ex. 13), 373 (Ex. 27),
454 (Question 1), 474 (Ex. 28),
527 (Ex. 19), 534 (Ex. 23),
572 (Ex. 19), 628–629
See also Open-ended Problems.
Multiple representations, 6 (Ex. 12),
17 (Ex. 2), 79, 95, 146, 194,
201, 361, 376, 406, 413, 444,
445, 450, 465, 468, 482–483,
495, 521, 602, 614, 632, 635,
642

Negation, 149, 737
Net, 95, 101, 147, 376–378, 535
Network, 14
***n*-gon,** 73, 368
Non-Euclidean geometry
comparison with Euclidean
geometry, 734–735

hyperbolic, 602, 735
spherical, 597–603, 735
taxicab, 208–209, 735
Number line, 705

Oblique drawing, 261
Octagon, 73
Octahedron, 101
Ongoing assessment *See* Assessment.
Open-ended Problems
Exercises, *throughout the book, for
example,* 26, 53, 84, 97, 134,
145, 184, 205, 241, 303, 334,
345, 356, 410, 425, 452, 465,
511, 557, 564, 565, 633, 647
Chapter Assessment, 53, 104, 159,
213, 331, 439, 486, 487, 547,
609, 655
Or (disjunction), inclusive, 736, 737
Ordered pair, 697
Ordered triple, 202
Orientation, in a transformation,
392–393, 406, 412, 420, 422
Origin, 696–697
Origins of geometry, Babylonian,
Egyptian, and Greek, 734–735
Orthocenter, 324 (Ex. 15), 571
(Ex. 10)
Orthographic projection, 270–271

Pantograph, 431
Paper cutting, 77 (Exs. 11–16), 91
(Exs. 11, 12), 141, 148, 194,
242, 256, 287, 356 (Ex. 11),
361, 434–435, 493
Paper folding, 18 (Exs. 15–18),
42, 45 (Exs. 23–27), 74, 77
(Exs. 11–16), 91 (Exs. 11, 12),
100–101, 170 (Exs. 16, 17), 266
(Exs. 8–11), 287, 356 (Ex. 11),
364 (Exs. 14–16), 392, 493,
505 (Ex. 25), 567
Paragraph proof, 124–126,
128 (Ex. 10), 129 (Exs. 20, 21),
135 (Question 1), 229, 232
(Ex. 34), 237, 239 (Exs. 2, 5),
241 (Ex. 16), 246 (Ex. 15), 250,
505 (Ex. 25), 558 (Ex. 21)
Parallelogram, 88–89
angle(s) of, 87–88, 232 (Ex. 30), 353
area of, 361–362, 365 (Ex. 23)
conditions for, 138, 182, 346–348,
353–355
construction of, 194, 350
diagonals of, 89, 135, 194–195,
199 (Ex. 25), 347, 353–355
properties of, 88–89, 194–195,
232 (Ex. 30)
special, 88–90, 339–342
See also Rectangle, Rhombus, *and*
Square.
Parallel Postulate, 235, 599, 602, 712,
735
Patterns
frieze, 419

identifying, 3–7, 13 (Ex. 18), 20,
76 (Ex. 5), 79, 96 (Ex. 11),
101, 112–115, 125, 146, 149,
171, 182, 184, 185, 198, 369,
372 (Ex. 14), 424 (Ex. 19),
430 (Ex. 7), 514, 596 (Ex. 16)
See also Conjecture.
in tessellations, 83, 434–435
and transformations, 419
Pentagon, 73, 368
Performance Task assessment, 53,
105, 159, 213, 275, 331, 387,
439, 487, 547, 609, 655
Perimeter, 368, 703
Perpendicular bisector, 170 (Ex. 17),
181, 184 (Exs. 19–22), 265,
307–308, 311 (Ex. 15)
Perpendicular lines, 58, 60, 180–182,
186 (Ex. 33), 196, 235,
249–250, 258, 260, 267
Perpendicular Postulate, 235, 712
Perspective drawing, 26, 261
Pi (π), 369
Plan for proof *See* Proof.
Plane(s), 21, 23, 118
coordinate, 202
intersecting, 23, 120–121, 256–257
line perpendicular to a, 258
models of, 21–23, 256–257
parallel, 23, 256–257
Platonic solids, 100–101
Point(s), 21–22, 118
collinear, 22
on a coordinate plane, 696–697
coplanar, 21–22
distance between, 28, 166, 203
matrix, 613
in space, 201–204
of tangency, 554
Polar coordinates, 416
Polygon(s), 73–74
angles of, 73–74, 79–82, 132
circumscribed, 324 (Ex. 13)
concave, 73
congruent, 285–288
convex, 73
diagonals of, 74, 76 (Ex. 5)
equiangular, 73
equilateral, 73
inscribed, 369
regular, 73–74, 100, 367–369, 529
sides of, 73–74, 170 (Exs. 4–9)
similar, 446–448
similar, perimeters and areas of,
468–470
sum of measures of angles, 79–80
vertices of, 73–74
Portfolio Projects
applying solar geometry, 542–543
building a mobile, 326–327
building the platonic solids, 100–101
classifying information, 154–155
creating fractals, 648–651
creating technical drawings, 270–271
designing a cottage, 382–383
exploring taxicab geometry, 208–209
investigating symmetry, 48–49
modeling Cavalieri's principle,
604–605

scaling the planets, 482–483
tessellating the plane, 434–435
Postulate, 117–118, 712–713, 734–735
Practice, extra, 659–686
Prism, 93–96
  base of, 93, 375
  diagonal of, 98 (Exs. 27, 28),
    145 (Ex. 30), 206 (Ex. 29)
  height of, 374–375
  oblique, 97 (Ex. 24)
  rectangular, 94
  regular, 126, 145, 376
  right, 97 (Ex. 24)
  surface area of, 376–377,
    534 (Ex. 27)
  volume of, 375, 531, 704–705
Probability, 283, 710
  geometric, 475–477
Problem-solving strategies
  building a model, 2, 21, 100–101,
    141, 147 (Ex. 45), 194, 256,
    287, 323 (Exs. 8–10), 326–327,
    340, 361, 364 (Exs. 14–16),
    374, 390, 475, 493, 540
    (Ex. 17), 583 (Ex. 20)
  choosing a method of proof,
    236–237, 241 (Ex. 19),
    255 (Ex. 7), 305 (Ex. 20)
  looking for a pattern, 3, 4, 10, 76
    (Ex. 5), 79, 96 (Exs. 6–13), 112
  making predictions, 3–5, 14–15, 38
    (Exs. 21–25), 63 (Exs. 30, 31),
    70 (Exs. 31, 32), 76 (Ex. 5), 79,
    87, 125, 148, 150 (Ex. 9), 172
    (Exs. 32–34), 180, 221, 242,
    283 (Exs. 30–32), 296 (Ex. 12),
    324 (Exs. 13, 14), 340, 430
    (Ex. 7), 461, 514, 553, 567
  making a table, 3, 4, 14, 76 (Ex. 5),
    79, 96 (Exs. 6–13), 125, 169
    (Ex. 1), 221, 499 (Ex. 25), 514,
    553
  using algebra, 4, 79,
    170 (Exs. 18–20), 181
    See also Algebra.
  using a diagram or drawing, 30, 67,
    74, 79, 87, 180, 242, 299, 392,
    427, 553, 567
  using an equation, formula, or
    proportion, 66, 80, 82, 88,
    166–168, 174, 446–448, 462,
    464, 469–470, 502, 529–531,
    536, 538, 561–562, 568,
    574–575, 579–581, 587,
    592–593
  using technology, 16, 169 (Ex. 2),
    172 (Exs. 32–34), 180, 279, 304
    (Ex. 17), 405, 460 (Exs. 24–26),
    461, 480 (Exs. 16–18), 514,
    553, 571 (Ex. 10), 578 (Ex. 27)
Projects, introductory, 2, 56, 110,
  164, 218, 278, 338, 390, 444,
  492, 552, 612
Projects, portfolio *See* Portfolio
  Projects.
Proof, 124
  coordinate geometry, 194–197,
    198 (Ex. 11), 199 (Exs. 15,
    24, 25), 200 (Exs. 27, 28),

207 (Questions 3–6), 213
    (Questions 25–27, 32), 341,
    344 (Ex. 12), 354, 357 (Ex. 21),
    365 (Ex. 25), 424 (Ex. 24)
  flow, 234, 237, 238, 239 (Exs. 1, 3),
    240 (Ex. 12), 241 (Ex. 14),
    247 (Ex. 16), 248 (Ex. 25),
    254 (Ex. 23), 294, 296 (Ex. 10),
    308, 346, 355 (Question 4), 453,
    572 (Ex. 17)
  indirect, 348, 351 (Exs. 22–24), 359
    (Question 4), 386 (Question 9)
  key steps of, 228, 230 (Ex. 13),
    232 (Ex. 30), 236, 240 (Ex. 13),
    245 (Ex. 8), 310 (Ex. 7), 320,
    344 (Ex. 11), 349 (Question 3),
    351 (Ex. 21), 357 (Ex. 22), 569
  paragraph, 124–126, 128 (Ex. 10),
    129 (Exs. 20, 21), 135
    (Question 1), 229, 232 (Ex. 34),
    237, 239 (Exs. 2, 5), 241
    (Ex. 16), 246 (Ex. 15), 250,
    505 (Ex. 25), 558 (Ex. 21)
  plan for, 238, 293, 294, 300, 302,
    303 (Ex. 10), 307, 314,
    316 (Ex. 7), 318 (Ex. 21),
    322 (Exs. 3, 4), 324 (Ex. 17)
  of Pythagorean theorem, 141,
    147 (Ex. 47), 498 (Ex. 19)
  two-column, 130–132, 133 (Ex. 4),
    134 (Exs. 17, 18), 135
    (Questions 2, 3), 139 (Ex. 11),
    140 (Ex. 18), 159 (Question 11),
    225 (Ex. 28), 232 (Ex. 33), 236,
    239 (Ex. 4), 240 (Ex. 11), 241
    (Exs. 15, 18), 244, 248 (Ex. 21),
    254 (Ex. 21), 293, 296 (Ex. 11),
    300, 302, 307, 314, 316 (Ex. 7),
    458 (Ex. 14), 463, 498 (Ex. 19),
    558 (Ex. 20), 563 (Ex. 11),
    576 (Ex. 17)
  types of, 234–238
  unspecified format, 139 (Ex. 12),
    241 (Ex. 17), 247 (Ex. 18),
    255 (Questions 1, 2), 295
    (Question 7), 298 (Ex. 21),
    304 (Exs. 12, 13, 15, 16), 309
    (Exs. 5, 6), 310 (Ex. 11), 311
    (Exs. 17, 18), 318 (Exs. 20,
    21), 322 (Exs. 3–5), 331
    (Questions 19, 24), 344
    (Ex. 13), 345 (Ex. 29), 357
    (Ex. 20), 358 (Ex. 31), 387
    (Question 13), 457 (Exs. 7, 8),
    458 (Exs. 16, 17), 466 (Ex. 22),
    497 (Exs. 17, 18), 559 (Exs. 22,
    23), 564 (Exs. 15, 16), 570
    (Exs. 8, 9), 571 (Ex. 12), 572
    (Ex. 18), 577 (Exs. 20, 21),
    608 (Questions 10, 14)
  See also Reasoning.
Property(ies)
  of congruence, 119
  distributive, 496, 574, 691, 693
  of equality, 119
  linear pair, 59
  list of, 712–713
  of parallelograms, 88–89
  of proportions, 463

of transformations, 393, 406, 412,
    420, 422
  of vertical angles, 59, 130–131
Proportion, 333, 446, 463, 494, 708
Proportional lengths, 461–464
Protractor, 35, 101, 123, 125, 152, 687
Pyramid, 535–537
  base of, 535
  height of, 535
  lateral face of, 535
  oblique, 541 (Ex. 18)
  regular, 535
  slant height of, 535
  surface area of, 535–536
  volume of, 536
Pythagorean theorem, 141–142,
    145 (Ex. 36), 147–148, 166,
    206 (Ex. 29), 301, 342,
    362–363, 367, 458 (Ex. 15),
    498 (Ex. 19), 522, 536, 714
  converse of, 143, 714
  proofs of, 141, 147 (Ex. 47),
    498 (Ex. 19)
Pythagorean triples, 146, 458 (Ex. 15)

Quadratic equations, 142, 342, 494,
    500, 575, 587, 700–702
Quadrilateral(s), 73, 167,
    184 (Exs. 14–18)
  angles of, 238
  classifying, 88–89, 138, 339–342
  exploring, 87, 91 (Exs. 11, 12),
    147 (Ex. 46), 199 (Exs. 17–23),
    200 (Ex. 27), 283 (Exs. 30, 31),
    296 (Ex. 12), 323 (Ex. 9), 349
    (Exs. 8–11), 460 (Exs. 24–26)
  planning proofs about, 347, 353, 355
  See also Kite, Parallelogram,
    Rectangle, Rhombus, Square,
    and Trapezoid.

Radical expressions, 104, 142, 166,
    168, 170 (Ex. 19), 188–189,
    196 (Exs. 28–31), 200,
    203–204, 341, 354, 362, 363,
    364, 367–368, 448, 494, 500,
    522, 536, 575, 587, 700–702
Radius (radii)
  of a circle, 187
  of a cone, 537
  of a cylinder, 375
  of a regular polygon, 369
  of a sphere, 586
Ratio(s), 707
  of lengths in similar figures, 446–449
  of perimeters, areas, and volumes in
    similar figures, 468–470,
    592–593
  trigonometric, 305 (Ex. 19),
    507–510, 514–517, 522,
    529–531, 538, 733
Ray(s), 28
Reason, in a proof, 125
Reasoning
  biconditional, 118

conclusion, 15
conditional statement, 15
conjecture, 15
contrapositive, 149
converse, 137, 149, 243–244
deductive, 112–113, 124, 734–735
De Morgan's rules, 740
errors in, 16, 239 (Ex. 5),
    248 (Ex. 23), 281 (Question 3),
    324 (Ex. 17), 348 (Question 6),
    460 (Ex. 27), 599 (Question 4),
    629 (Question 3)
hypothesis, 15
if-and-only-if, 118
inductive, 4, 112–113, 734
inverse, 149
modus ponens, 739
modus tollens, 741
negation, 149, 737
sorites, 741
truth tables, 736–741
See also Conjecture, Convincing
    argument, *and* Proof.
Rectangle, 88–89, 339
angles of, 89 (Question 1)
area of, 138, 360–362, 704–705
conditions for a, 89 (Question 1),
    92 (Ex. 16), 145 (Exs. 25–27),
    353, 357 (Exs. 20, 22)
construction of, 358
diagonals of, 199 (Ex. 16), 341
Reduction, 428
Reflection, 8–10, 41, 298 (Ex. 23),
    318 (Ex. 19), 391–394
and coordinates, 398–400
with matrices, 634–636
properties of, 393
rotation and, 643
and symmetry, 8–10, 48–49,
    318 (Ex. 19), 391, 395–396
Regular polygon, 73–74, 529
apothem of, 368
area of, 367–369
perimeter of, 368
Regular pyramid, 535
Remote interior angles, 126
Research
Exercises, 11, 38, 45, 70, 97, 121,
    138, 176, 224, 261, 284, 345,
    380, 410, 423, 473, 504, 518,
    564, 589, 602, 646, 741
introductory questions, 2, 56, 110,
    164, 218, 278, 338, 390, 444,
    492, 552, 612
Portfolio Projects, 49, 155, 271, 383,
    435, 483, 543, 605
Review
Algebra Review/Preview, 106–107,
    214–215, 332–333, 440–441,
    548–549, 687–711
Chapter Review, 50–51, 102–103,
    156–157, 210–211, 272–273,
    328–329, 384–385, 436–437,
    484–485, 544–545, 606–607,
    652–653
Checking Key Concepts, *in every
    lesson, for example*, 11, 67, 120,
    138, 197, 238, 295, 370, 414,
    422, 449, 510, 538, 570, 629

Cumulative Assessment, 160–161,
    334–335, 488–489, 654–655
Extra Practice, 659–686
Spiral Review, *in every lesson, for
    example*, 20, 27, 78, 92, 129,
    147, 172, 200, 262, 284, 305,
    352, 366, 418, 425, 452, 481,
    499, 520, 572, 578, 625, 640
Toolbox, 687–711
Rhombus, 88–90, 339–341
area of, 365 (Ex. 25), 530
conditions for a, 199 (Ex. 25),
    354–355
diagonals of, 126, 340–341
Right angle, 57, 128 (Ex. 10)
Right triangle(s), 65, 127
angles of, 68 (Ex. 7), 137,
    139 (Ex. 11)
congruent, 301
hypotenuse of, 141–142,
    198 (Ex. 11), 344 (Ex. 13)
legs of, 141
Pythagorean theorem, 141–142,
    149, 166, 206 (Ex. 29), 301,
    342, 362–363, 367, 458,
    498 (Ex. 19), 522, 536
Pythagorean theorem, converse of,
    143, 148
similar, 493–496
special (30-60-90 and 45-45-90),
    500–502
trigonometric ratios in, 507–510,
    514–517, 529–531, 538
Rotation, 8–10, 38 (Exs. 21–25), 41,
    412–414
angle of, 412
center of, 412
with matrices, 641–643
properties of, 412
reflection and, 643
and symmetry, 8–10, 48–49, 415

Same-side exterior angles, 223 (Ex. 4)
Same-side interior angles, 219–221,
    227–229, 244
SAT/ACT Preview, 19, 63, 84, 122,
    129, 150, 183, 193, 231, 253,
    261, 282, 297, 311, 324, 345,
    358, 408, 423, 451, 466, 480,
    506, 520, 534, 558, 577, 595,
    633
Scalar multiplication
of a matrix, 614
of a vector, 523
Scale drawing, 299, 455, 468
Scale factor, of a dilation, 428,
    614–615
Scale model, 443–444, 450, 468,
    482–483, 593
Scalene triangle, 65, 138
Science
Exercises and applications, 6, 7, 32,
    38, 39, 61, 68, 85, 98, 115, 122,
    128, 133, 205, 224, 289, 305,
    323, 371, 450, 459, 479, 511,
    518, 526, 564, 577, 589, 591,
    595, 623, 645

notes, 3, 7, 35, 54–56, 85, 111, 113,
    114, 115, 128, 201, 219, 224,
    367, 479, 518, 560, 592, 645
Scientific calculator *See* Calculator,
    scientific.
Scientific notation, 482, 709
Secant(s) of a circle, 561
and arc measures, 561
and segment lengths, 573–574
Secant of a sphere, 586
Sector of a circle, 579–580
Segment(s)
addition of, 29
bisector of, 41
of a circle, 584
congruent, 28
endpoints of, 28
length of, 28, 166
of a line, 28
midpoint of, 41, 167
perpendicular bisector of,
    170 (Ex. 17), 181, 265,
    307–308, 311 (Ex. 15)
Segment Addition Postulate, 29, 118,
    712
Self-Assessment, in Portfolio Projects,
    49, 101, 155, 209, 271, 327,
    383, 435, 483, 543, 605, 651
Semicircle, 554
Sequence, 369
Side(s)
adjacent to an angle, 507
of an angle, 34
consecutive, 74
corresponding, of congruent
    polygons, 285–288
corresponding, of similar polygons,
    446, 448
included, 300
opposite an angle, 507
of a polygon, 73–74
of a triangle, 65
Side-Angle-Side (SAS) Postulate, 294,
    713
**Side-Angle-Side (SAS) Similarity
    Theorem,** 454, 717
Side-Side-Side (SSS) Postulate, 293,
    713
**Side-Side-Side (SSS) Similarity
    Theorem,** 454, 717
Similarity, 445–449, 453–456,
    461–464, 468–471
and areas, 468–470
of polygons, 445–448
and proportions, 461–464
of regular polygons, 459 (Ex. 22)
of right triangles, 493–496
of solids, 470, 592–593
of triangles, 453–455, 461
and volumes, 469–470, 592–593
Simulation, 424
Sine ratio, 514–517, 529, 534 (Ex. 22),
    538, 733
Sines, law of, 305 (Ex. 19)
Skew lines, 22
Skills, maintaining *See* Review.
Slant height
of a cone, 537
of a regular pyramid, 535

Slope(s), 173–174, 640 (Ex. 36), 697–700
  of parallel and perpendicular lines, 180–182
Slope formula, 174, 176, 194–195, 697–699, 714
Slope-intercept form, 175
Software, geometry *See* Technology, computer.
Solid(s)
  Platonic, 100–101
  regular, 100
  similar, 469–470, 592–593
  See also Cone, Cube, Cylinder, Prism, Pyramid, Sphere, *and* Three-dimensional geometry.
Space *See* Three-dimensional geometry.
Spatial reasoning *See* Manipulatives, Solid(s), Three-dimensional geometry, *and* Visual thinking.
Sphere, 586
  great circle of, 598
  parts of, 586
  surface area of, 586–587
  volume of, 586–587
Spherical geometry, 597–603
Spiral Review, *in every lesson, for example,* 20, 27, 78, 92, 129, 147, 172, 200, 262, 284, 305, 352, 366, 418, 425, 452, 481, 499, 520, 572, 578, 625, 640
Spreadsheets, 16, 146, 169, 450, 472, 482, 725
Square, 88–89, 126, 138, 196, 339, 341, 355
  area of, 360, 362, 369 (Question 3)
Square roots, table of, 732
  See also Radical expressions.
Square units, 360, 730
Standard position, for a polygon on a coordinate plane, 194–197
Statements *See* Conditional statement, Contrapositive, Converse, Inverse of a conditional, Property(ies), *and* Theorem(s).
Statistics, mean and median, 708–709
  See also Data analysis.
Straightedge, 69, 263
Study techniques, 50, 102, 156, 214, 272, 328, 384, 436, 484, 544, 606, 652
Substitution *See* Property(ies).
Summaries
  area and volume, 362, 704–705
  identifying parallelograms, 347
  identifying rectangles, 353
  identifying rhombuses, 355
  list of constructions, 720
  list of postulates, 712–713
  list of related if-then statements, 149
  list of theorems, 714–719
  matrices for reflections, 635
  matrices for rotations, 642
  methods of proof, 236–237
  quadrilaterals, 339
  transformations that preserve congruence, 406, 412, 420, 422
  trigonometric ratios, 517

Sum of measures of angles
  of a polygon, 79–80
  of a triangle, 65–66, 236–237
Supplementary angles, 58, 131, 132 (Question 2), 138
Surface area(s)
  of a cone, 537–538
  of a cylinder, 376–377
  of a prism, 376–377, 534 (Ex. 27)
  of a pyramid, 535–536
  of similar solids, 470, 593
  of a sphere, 586–587
Symbols, table of, 731
Symmetry
  investigating, 48–49
  line of, 10, 74, 318 (Ex. 19), 341, 342 (Question 2), 391, 396, 639
  reflection, 10, 12, 318 (Ex. 19), 391–396, 689–690
  rotational, 8–10, 12, 415, 689–690
  translational, 10, 407, 409, 410
Systems, geometric, 734–735
Systems of linear equations, 200 (Ex. 27), 333, 706–707

Table
  of measures, 730
  of squares and square roots, 732
  of symbols, 731
  of trigonometric ratios, 733
Tangent ratio, 507–510, 522, 530, 531, 733
Tangent to a circle, 554
  and arc measures, 554–555, 560–561
  and segment lengths, 574–575
Tangent to a sphere, 586
Tangram, 171
Tautology, 739
Taxicab geometry, 208–209
Technology
  calculator
    graphing, 79, 372, 404, 472, 590, 618, 629, 632, 636, 646, 721–725
    scientific, 142, 144 (Exs. 1–6), 305 (Ex. 19), 508–509, 515–516, 595 (Ex. 9)
  computer
    computer-aided design, 93, 271, 383
    geometry software, 9, 27, 38, 70, 125, 132, 147, 164, 168, 172, 180, 200, 221, 267, 279, 297, 304, 324, 349, 372, 397, 405, 409, 417, 432, 435, 460, 461, 466, 514, 553, 571, 572 (Ex. 19), 578, 726–729
    matrix software, 629, 646, 723–725
    patterns in a, 38, 70, 408
    simulation, 424
    spreadsheets, 16, 146, 169, 450, 472, 482, 725
  systems designer, 162–164, 170, 198, 206
  See also Computer.

Technology Handbook, 721–729
Terms, undefined, 118
  See also Point(s), Line(s), *and* Plane(s).
Tessellation, 83, 367, 434–435
Tests *See* Assessment.
Tetrahedron, 101
Theorem(s), 125, 714–719, 734–735
  about area and volume, 716–717
  about circles and spheres, 718–719
  about coordinate geometry, 714–715
  about parallel and perpendicular lines, 181, 227, 243, 249, 250, 257, 258, 715
  about quadrilaterals, 716–717
  about similar polygons, 717
  about triangles, 714, 715, 718
  corollary of, 239
  writing proofs of, 125–126, 130–132
  See also Proof *and* Reasoning.
Think and Communicate *See* Communication, discussion.
Thinking skills *See* Reasoning, Study techniques, *and* Visual thinking.
Three-dimensional geometry
  coordinate system, 201–202
  distance formula, 203–204
  midpoint formula, 203
  parallel planes, 256–258
  Platonic solids, 100–101
  similar solids, 469–470, 592–593
  surface area, 376–377, 470, 535–538, 586–587
  volume, 375, 469–470, 531, 536–538, 586–587, 592–593, 704–705
  See also Cone, Cube, Cylinder, Prism, Pyramid, *and* Sphere.
Toolbox, examples and practice of previously learned mathematical skills, 687–711
Transformations, 8–13, 25 (Ex. 30), 388–433
  applications, 25, 198, 298, 318, 451, 473 (Ex. 18), 474
  composition of, 405–406, 409, 411, 417 (Exs. 23, 24), 418 (Ex. 32), 419–422, 424 (Ex. 23), 425 (Ex. 30), 432 (Ex. 17), 624 (Ex. 28), 632 (Ex. 15), 637 (Ex. 15), 638 (Exs. 20, 21), 643–647
  and congruence, 41, 298, 393, 406, 412, 420, 422
  exploring, 12 (Ex. 14), 392, 402 (Ex. 12), 405, 409 (Ex. 16), 410 (Ex. 24), 411 (Ex. 25), 417 (Exs. 23, 24), 424 (Exs. 23, 24), 425 (Ex. 30), 427, 430 (Ex. 7), 618 (Ex. 32), 624, 625, 630 (Exs. 8, 9), 632 (Exs. 14, 15), 633 (Ex. 19), 634, 637 (Ex. 15), 638 (Exs. 20, 21), 640 (Ex. 36), 641, 643–646
  with matrices, 614–615, 619–621, 626–629, 631, 634–636, 641–643
  orientation in, 392–393, 406, 412, 420, 422

properties of, 393, 406, 412, 420, 422
shear, 628
and symmetry, 8–10, 318, 391, 396,
    407, 409, 410, 415
using geometry software, 729
See also Dilation, Glide reflection,
    Matrix, Morphing, Reflection,
    Rotation, *and* Translation.
Translation, 8–10, 12 (Ex. 14),
    25 (Ex. 30), 41, 405–407
   with matrices, 619–621
   properties of, 406, 422
   and symmetry, 8–10, 48–49, 407,
     409, 410
Transversal, 219
Trapezoid, 228–229, 233 (Ex. 35),
    339, 348
   area of, 361–363, 364 (Ex. 16),
     366 (Ex. 34), 531
   bases of, 228
   isosceles, 200 (Ex. 28), 343 (Ex. 10),
     344 (Ex. 12)
   legs of, 228
   midline of, 364
   right, 343
Triangle(s), 64, 73
   acute, 65, 127, 148
   altitudes of, 319–320, 324 (Ex. 15),
     571 (Ex. 10)
   angle bisectors of, 320–321,
     324 (Ex. 13), 466 (Ex. 23),
     571 (Ex. 10)
   angles of, 65–66, 125–126, 235–237,
     240 (Ex. 11), 280
   area of, 141, 361–362, 364 (Ex. 7),
     365 (Ex. 24), 534 (Ex. 22)
   centroid of, 323 (Ex. 8), 324 (Ex. 14)
   circumcenter of, 311 (Ex. 16), 567,
     571 (Ex. 10)
   classifying, 64–65, 148, 168,
     172 (Ex. 35)
   congruent, 240 (Ex. 11), 292–295,
     299–302, 306–308
   constructing, 69, 292,
     297 (Exs. 15–19)
   equiangular, 65, 66 (Questions 3–5),
     70 (Ex. 31), 239 (Ex. 6)
   equilateral, 65, 69, 70 (Ex. 31),
     129, 152 (Ex. 35), 168, 362,
     372 (Ex. 17)
   exploring, 46 (Ex. 34), 64,
     70 (Exs. 31, 32), 141, 148, 152
     (Ex. 35), 198–199 (Exs. 8–15),
     297, 304, 323, 324,
     466 (Ex. 23), 571 (Ex. 10)
   exterior angles of, 125–126,
     129 (Ex. 21)
   45-45-90, 500–501
   incenter of, 324 (Ex. 13),
     571 (Ex. 10)
   inequalities for, 279–281,
     600 (Ex. 10)
   inscribed, 567
   isosceles, 65, 70 (Ex. 32),
     114 (Ex. 11), 121, 129, 138,
     168, 313–315, 320–321,
     322 (Exs. 3–5)
   medians of, 319–320, 323 (Ex. 8),
     324 (Ex. 14)

obtuse, 65, 127, 148, 152 (Ex. 35)
orthocenter of, 324 (Ex. 15),
    571 (Ex. 10)
perpendicular bisectors of sides,
    311 (Ex. 16), 320, 567
right, 65, 68 (Ex. 7), 127, 137,
    139 (Ex. 11), 141–143, 148,
    149, 198 (Ex. 11), 301,
    344 (Ex. 13), 493–531
    See also Right triangle(s).
scalene, 65, 138
segment joining midpoints of two
    sides, 199 (Exs. 12–15), 462
sides of, 65, 148, 279–281
similar, 453–455, 461
spherical, 597–601
sum of measures of angles, 65–66,
    236–237
terrestrial, 601
30-60-90, 501–502
vertices of, 65, 183
Triangle Sum Theorem, 66, 236–237,
    714
Triangulation, 299
Trigonometry
   applications of, 522, 529–531, 538,
     560, 572 (Ex. 16), 577 (Ex. 24)
   and area, 529–531
   law of sines, 305
   sine and cosine ratios, 514–517, 733
   summary of trigonometric ratios, 517
   table of trigonometric ratios, 733
   tangent ratio, 507–510, 733
Truth table, 737
Truth value, 736
Two-column proof, 130–132, 133
     (Ex. 4), 134 (Exs. 17, 18), 135
     (Question 2, 3), 139 (Ex. 11),
     140 (Ex. 18), 159 (Question 11),
     225 (Ex. 28), 232 (Ex. 33), 236,
     239 (Ex. 4), 240 (Ex. 11), 241
     (Exs. 15, 18), 244, 248 (Ex. 21),
     254 (Ex. 21), 293, 296 (Ex. 11),
     300, 302, 307, 314, 316 (Ex. 7),
     458 (Ex. 14), 463, 498 (Ex. 19),
     558 (Ex. 20), 563 (Ex. 11),
     576 (Ex. 17)

Undefined terms, 118
Units of measurement, 730

Valid argument, 739
Vanishing point, 26
Vector(s), 521–524
   addition of, 524, 527 (Exs. 19, 20)
   equal, 522
   scalar multiplication of, 523
Vertex (vertices)
   of an angle, 34
   of a cone, 537
   consecutive, 74
   of a network, 14
   of a polygon, 73
   of a prism, 93
   of a pyramid, 535

of a triangle, 65
Vertex angle, of an isosceles triangle,
    313
Vertical angles, 58–59, 130–131
Visual thinking, 18, 91 (Exs. 11, 12),
    96, 97, 115, 147, 225, 247, 253,
    262, 304, 324, 343, 356, 424,
    472, 474, 478, 498, 505, 527,
    578, 583, 591
   See also Manipulatives.
Vocabulary, 20, 33, 47, 50, 52, 71, 86,
    99, 102, 104, 123, 135, 153,
    156, 158, 186, 207, 210, 212,
    233, 255, 269, 272, 274, 291,
    325, 328, 330, 359, 381, 384,
    386, 411, 433, 436, 438, 467,
    481, 484, 486, 506, 528, 541,
    544, 546, 566, 585, 603, 606,
    608, 625, 647, 652, 654,
    742–747
Volume(s), 375
   of a cone, 537–538
   of a cylinder, 375
   of a prism, 375, 531, 704–705
   of a pyramid, 536
   of similar solids, 469–470, 592–593
   of a sphere, 586–587

Writing
   Chapter Assessment, 52, 104, 212,
     213, 274, 275, 331, 334–335,
     386, 486, 547, 609, 655
   Examples and Exercises, *throughout
     the book, for example,* 6, 15, 69,
     86, 97, 116, 129, 145, 176, 192,
     239, 261, 267, 289, 304, 311,
     364, 372, 403, 418, 424, 457,
     465, 466, 506, 511, 512, 559,
     565, 571, 640, 656
   introductory questions, 2, 56, 110,
     164, 218, 278, 338, 390, 444,
     492, 552, 612
   Journal, 20, 33, 47, 71, 86, 99, 123,
     135, 153, 186, 207, 233, 255,
     269, 291, 312, 325, 359, 381,
     411, 433, 467, 481, 506, 528,
     541, 566, 585, 603, 625, 647
   Portfolio Projects, 49, 101, 155, 209,
     271, 327, 383, 434, 482, 543,
     651
   See also Research.

***x*-axis,** 696
***x*-coordinate,** 697

***y*-axis,** 696
***y*-coordinate,** 697
***y*-intercept,** 175, 179 (Ex. 34)

***z*-axis,** 201
***z*-coordinate,** 201–202

Credits

277–278 3M Traffic Control Materials Division(background); 278 3M Traffic Control Materials Division(t); 280 courtesy Ellsworth Kelly, photo by Eric Pollitzer; 284 ©Jose Luis Banus-March/FPG International Corp.; 285 1992 Cindy Lewis Photography; 289 F. Collet/Ardea London Limited; 290 3M Traffic Controls Materials Division(tl, bl); 293 The Stock Market/Jeffry W. Myers; 294 Robert Frerck/Odyssey Productions/Chicago; 299 Department of Special Collections, University Research Library, UCLA(tr); 306 Superstock; 313 Glen Allison/Tony Stone Images; 316 Wysocki/Explorer/Photo Researchers, Inc.(far l); ©John Dominis/Index Stock Photography(ml); Omni-Photo Communications(mr); copyright Horst Schäfer/Peter Arnold, Inc.(far r); 317 3M Traffic Control Materials Division(tr, b); 323 William H. Mullins, The National Audubon Society Collection/Photo Researchers, Inc.(tl); Superstock(tr); 326 Tate Gallery, London/Art Resource, NY(t); 336 courtesy Walt Stone; 337 Eric R. Berndt/Unicorn Stock Photo(l, m); Andre Jenny/Unicorn Stock Photo(r); 338 courtesy Walt Stone; 342 Cameramann International; 345 Tony Freeman/PhotoEdit(t); 348 Beinecke Rare Book & Manuscript Library; 351 courtesy JLG Industries(l); Linc Cornell/Stock, Boston(r); 353 Charles Gupton/Tony Stone Images; 357 Barnaby's Picture Library(b); 360 Gary Russ/The Image Bank; 364 Julio Donoso/Contact Press Images; 365 Earl Dotter; 366 courtesy Walt Stone(l); Aneal Vohra/Unicorn Stock Photo(r); 367 Dr. John Rogers, III, and Mr. John Stier, Michigan State University, Department of Crop and Soil Sciences(l); courtesy Michigan State University, photo by Ashley Photography, Bloomfield Hills, MI(r); 370 Tony Freeman/PhotoEdit; 373 Richard Bryant/Arcaid; 380 courtesy Walt Stone; 388 Hellmuth, Obata + Kassabaum, Inc.(l); 388–389 Hellmuth, Obata + Kassabaum, Inc.(background); 389 Hellmuth, Obata + Kassabaum, Inc.(t); 389 Phil Matt(b); 390 Hellmuth, Obata + Kassabaum, Inc.; 391 drawing and portrait courtesy of Rollin Thurlow, Northwoods Canoe Company; 395 Hellmuth, Obata + Kassabaum, Inc.; 403 adapted from a design by Putt-Putt® Golf Courses of America, Inc.(t); Robert A. Daemmrich/Stock, Boston(b); 407 ©Wood River Gallery 1995; 409 Robert Frerck/Odyssey Productions/Chicago(tl); ©Wood River Gallery 1995(tr); Hellmuth, Obata + Kassabaum, Inc.(ml, br); ©Carolinas Stadium Corp. Photography by APS, Inc.(bl); 410 Cervin Robinson(t); Adam Woolfitt/Woodfin Camp and Associates(b); 412 Donald C. Johnson/The Stock Market; 415 Courtesy of Williams-Sonoma, Inc./Paul Berg Photography(m); courtesy of Antoine Proulx/Ted Dillard Photography; 417 courtesy Elizabeth A. Akana, portrait by Rita Ariyoshi(t, b); 418 Hellmuth, Obata + Kassabaum, Inc.; 419 Adapted with permission from Handbook of Regular Patterns, by Peter S. Stevens ©1981 MIT Press; 421 Animals Animals © Henry Ausloos; 422 Adapted with permission from Handbook of Regular Patterns, by Peter S. Stevens ©1981 MIT Press ; 434 Uniphoto Picture Agency/Pete Winkel(t); 442 Altman Stage Lighting, Co., Inc.(tl, tr); courtesy Loy Arcenas(background); 443 T. Charles Erickson(tl); Hartford Stage's production of Shakespeare's All's Well That Ends Well, directed by Mark Lamos and featuring Susan Appel, Kate Reid, John McDonough, and Curt Hostetter. Photo by Jennifer Lester(ml); Carol Pratt(br); 444 courtesy of Loy Arcenas(t); 450 courtesy of the Lakeview Museum; 451 Sean O'Neill/International Stock Photo(t far l); Brian Lovell/The Picture Cube(tml); Bill Ross/Westlight(tmr); Rafael Macia/Photo Researchers, Inc.(t far r); RKO (courtesy The Kobal Collection)(b); 452 Superstock; 455 Margaret Courtney-Clarke; 458 Charles Leavitt/The Picture Cube; 459 IEEE Center for the History of Electrical Engineering; 465 courtesy Loy Arcenas(tr); drawings courtesy of Loy Arcenas; 468 T. Charles Erickson(t); courtesy Loy Arcenas(bl); 479 Tom & Pat Leeson, The National Audubon Society Collection/Photo Researchers, Inc.; 480 Ferne Saltzman/Albuquerque International Balloon Fiesta; 482 NASA/JPL(t); NASA(bl); 483 NASA(m, b); 490–492 Owen Seumptewa/Native Shadows; 498 Reprinted with the permission of Atheneum Books for Young Readers, an imprint of Simon & Schuster Children's Publishing Division from THE MYSTERIOUS ISLAND by Jules Verne, illustrated by N.C. Wyeth. Copyright 1918 Charles Scribner's Sons; copyright renewed 1946 Charles Scribner's Sons.(l); 499 Mireille Vautier/Woodfin Camp and Associates; 500 Alex MacLean/Landslides; 502 Brian Brake, Science Source/Photo Researchers, Inc.; 503 Owen Seumptewa/Native Shadows; 504 Christie's Images(l); Oxford University Museum(r); 507 Tom & Pat Leeson/Photo Researchers, Inc.; 508 Tom & Pat Leeson/Photo Researchers, Inc.; 509 Superstock; 510 Bildarchiv Foto Marburg/Art Resource, NY; 511 Field Museum/Photo Researchers, Inc.; 512 courtesy Katoomba Scenic Railroad; 516 Roueche Photography; 519 Owen Seumptewa/Native Shadows(t); Hank Delespinasse/The Image Bank(b); 521 Cambridge Sports Union(tl); Val Corbett/Tony Stone Images(tr); 523 Jeff Greenberg/Photo Researchers, Inc.; 526 Cambridge Sports Union(t); Nourok Jonathon/Tony Stone Images(b); 527 ©Dan McCoy/Rainbow; 529 John T. Hanou; 533 George Washington's Mount Vernon Estate & Gardens; 535 Andre Jenny/International Stock Photo(t); 536 ©Jean Kugler/FPG International

Corp.; 539 John B. Carnett/Los Angeles Times Syndicate International; 540 Botond Bognar(t); 542–543 Superstock(b); 543 John V.A.F. Neal/Photo Researchers, Inc.; 550 David Young-Wolff/PhotoEdit(t); Leonard D. Dank/Custom Medical Stock Photo(b); 550–551 1995 Allsport USA/Michel Hans/Vandystadt/All rights reserved(t); 552 courtesy University of Georgia Athletic Association(bl); 555 ©Edmund Nagele/FPG International Corp.; 557 courtesy University of Georgia Athletic Association(ml); 560 Copyright Tom Van Sant/Geosphere Project, Santa Monica/Science Source/Photo Researchers, Inc.; 564 Tom & Pat Leeson/Photo Researchers, Inc.; 565 David Young-Wolff/PhotoEdit; 571 David R. Frazier Photolibrary; 575 Copyright Tom Van Sant/Geosphere Project, Santa Monica/Science Source/Photo Researchers, Inc.(b); 577 Bibliothèque Nationale, Paris; 579 Leo de Wys/Henryk Kaiser; 581 Shattil & Rozinski/Tom Stack & Associates(t); Alex MacLean/Landslides(b); 583 Robert Daemmrich/Tony Stone Images(b); 584 James Balog/Tony Stone Images(t); 587 Robert Fried/Stock, Boston(r); 588 CMCD, Inc. 1995(l); Lawrence Migdale(r); 590 David Young-Wolff/PhotoEdit(bl); 592 Stephen Frisch/Stock, Boston; 593 A&L Sinibaldi/Tony Stone Images(r); 595 Illustration from A FIELD GUIDE TO WESTERN REPTILES AND AMPHIBIANS, 2/e. Copyright © 1985 by Robert C. Stebbins. Reprinted by permission of Houghton Mifflin Company. All rights reserved.; 602 1996 M.C. Escher/Cordon Art-Baarn-Holland. All Rights Reserved.; 609 David Young-Wolff/PhotoEdit; 610 NASA(br); 611 NASA(b); 612 NASA(tl, m); 613–614 Tectrix Fitness Equipment; 620 Trinity College Library, Cambridge University; 626 Arni Katz/Dreamtime Systems; 631 Arni Katz/Dreamtime Systems; 637 Marika Barnett(l); Boston Globe/Wendy Maeda(r); 639 ©1996 Jack®. B. Ting, S. Sheridan, N. Badler, HMS Center, University of Pennsylvania(tl); ©1996 S. Sheridan, N. Badler, HMS Center, University of Pennsylvania(tr, bl, br) Model courtesy John Deere and Company(tr). Model courtesy Viewpoint DataLabs(bl); 645 NASA; 734 George Arthur Plimpton Collection, Rare Book and Manuscript Library, Columbia University.

Assignment Photography

Gary Choppé/Creative Image Studio xv(b), xvii(m), 216(b), 217(b), 252(l), 610(bl), 612, 623, 645.
Eduardo Fuss i(background), xxxi(b), v(b), 54–55, 56, 68, 85, 98(l).
Brent Jones iv, 1, 2, 12(tl, tm), 17(bl, br), 37(b).
Lawrence Migdale xii, 442(br), 444(b), 465(tl, b), 468(br).
Steven Ferry/P&F Communications 550(tl), 551(b), 552(tr), 589.
RMIP/Richard Haynes v(t), xix, xxii, xxiii(tl, bl, bm, br), xxvii(bl, br), 14, 21, 32, 40(l), 43, 48(bl & inset, bm, br & inset), 49(t), 60, 64, 69, 79, 87, 94, 100, 101, 117(t), 124, 127, 130, 148, 155, 180, 193, 194, 209, 216(t), 221, 242, 256, 259(b)(capacitor courtesy of the Bellingham Antique Radio Museum), 263, 264, 265, 267, 271, 279, 283, 287, 291, 292, 296, 308, 319, 323(b), 326(b), 327, 340, 350, 357(t), 361, 374, 377, 379, 382, 383, 388(r), 392, 397, 402, 404, 405, 415(l), 423, 426(logo courtesy of Youth For Understanding International Exchange), 427, 430, 431, 434(bl, bm, br), 435, 461, 466, 475, 482(br), 483(t), 491, 493, 514, 540(b), 542, 543(inset), 552(br), 567, 582, 583(t background, tl, tm, tr), 586, 587(l), 590(t, br), 593(l), 596, 597(r, l), 600, 604, 605, 624, 625, 633, 634, 641, 648, 649, 650, 651.
John Kalka/3M ix, 276, 277(r), 278(b), 290(tr), 295(br), 317(tl).
Jim Yuskavitch 108(t), 109(t).

Illustrations

Steve Cowden 38, 62, 139, 176, 445, 457, 473, 496, 524, 552, 557, 572, 584
Hannah Bonner 111(background), 115(b), 474
Piotr Kaczmereck 34, 36
Doug Stevens 278

Selected Answers

CHAPTER 1

Page 5 Checking Key Concepts

1. The number of triangles is 2 less than the number of sides. **3.** $n - 2, n \geq 3$

Pages 5–7 Exercises and Applications

3. 24, 35 **5.** $\frac{5}{4}, \frac{6}{5}$ **11.** $3n + 1$

19, 21.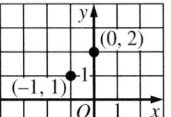
23. 17
25. $\frac{1}{6}$

Page 11 Checking Key Concepts

1.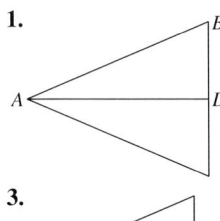

3.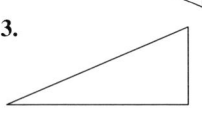

5. reflection symmetry **7.** translation symmetry

Pages 11–13 Exercises and Applications

5. **7.** **9.**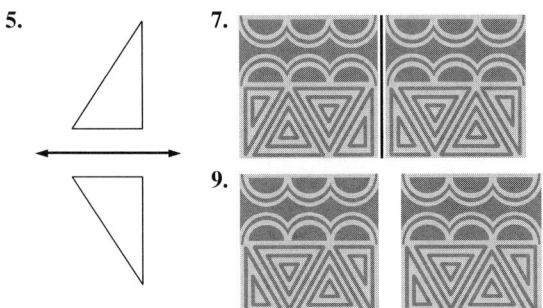

11. translational; Each gymnast's position is an image of the first gymnast's position. **13.** rotational; Each point of the gymnast's body rotates around some point by the same amount. **15.** The bicycles have translational symmetry. **17.** The starfish has rotational symmetry (about its center) and reflection symmetry. It has five planes of symmetry, one running down the middle of each leg. **25.** -36
27. $1.5n + 9$, where $n = $ the number of bags

Page 16 Checking Key Concepts

1. a figure is a square; it has four lines of symmetry
3. people live in glass houses; they shouldn't throw stones
5. True.

Pages 17–20 Exercises and Applications

3. If you're at the bottom of a human pyramid, then you have to be steady. hypothesis: you're at the bottom of a human pyramid; conclusion: you have to be steady **5.** If a tumbler doesn't run fast enough before hitting the trampoline, he or she won't make it to the other side of the pyramid. hypothesis: a tumbler doesn't run fast enough; conclusion: he or she won't make it to the other side of the pyramid **7.** If the figure has reflection symmetry, then the hidden part of the figure looks like: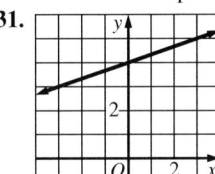

9. hypothesis: you reflect a shape; conclusion: the image is the same size as the original shape **11.** False; for example, the network in Figure A on page 14 is traceable, but two of its five vertices are odd. **13.** True. **23, 25, 27.** Answers may vary. Examples are given. **23.** If today is January 1, then tomorrow is January 2. **25.** If a point is located on the x-axis, then the y-coordinate of the point is 0. **27.** If you translate a point up one unit on a coordinate plane, then the y-coordinate of the point increases by 1.

31.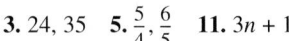

Page 20 Assess Your Progress

1.

Number of squares	3	4	5	6	7
Number of toothpicks	10	13	16	19	22

Each additional square requires 3 additional toothpicks.
2. 31 toothpicks; Answers may vary. An example is given. I noticed that for n squares, the number of toothpicks is $3n + 1$. **3.** If you want to make a toothpick figure having n squares, you need $3n + 1$ toothpicks. **4.** The first, third, and fifth figures have reflection symmetry. The second and fourth have rotational symmetry.

5. a. **b.**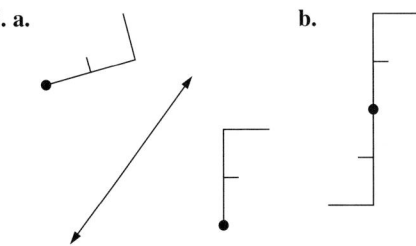

c.

6. False; for example, if $a = -2$ and $b = -1$, then $a \div b = 2$ and $a \div b > a$.

29, 31, 33. Answers may vary. Examples are given.

29.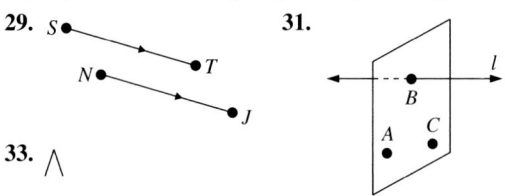

31.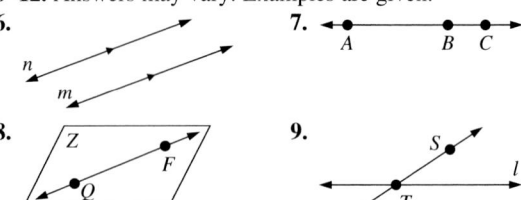

33. $\wedge$

Page 33   Assess Your Progress

1. any two of $\overleftrightarrow{EC}$, $\overleftrightarrow{DC}$, and $\overleftrightarrow{BC}$ **2.** point D **3.** point A
4. plane AED and plane ACD **5.** $\overleftrightarrow{ED}$ or $\overleftrightarrow{EC}$
6–12. Answers may vary. Examples are given.

6.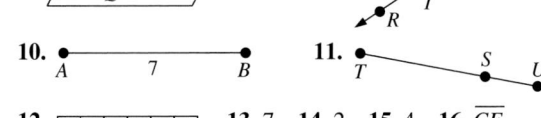

7.

8.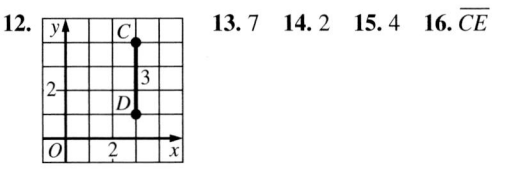

9.

10.

11.

12.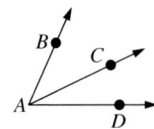

13. 7   **14.** 2   **15.** 4   **16.** $\overline{CE}$

Page 24   Checking Key Concepts

1. point A **3. a.** coplanar but noncollinear   **b.** collinear
c. coplanar but noncollinear   **d.** coplanar but noncollinear
e. noncoplanar   **f.** noncoplanar   **5.** point B **7.** plane ADC

Pages 24–27   Exercises and Applications

1. point J **3.** $\overleftrightarrow{IK}$ **5.** plane ABG, plane ABI **7.** True.
9. False; in the figure, points A, B, and C are collinear and lie in two of the planes shown.   **11.** parallel
13. intersecting   **19.** C   **21, 23, 25.** Answers may vary. Examples are given.

21.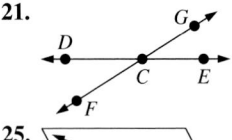

23.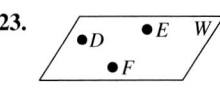

25.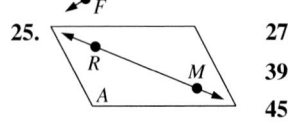

27. $\overleftrightarrow{CF}$ **29.** $\overleftrightarrow{BX}$, $\overleftrightarrow{EX}$, and $\overleftrightarrow{XY}$
39. True.   **41.** 1   **43.** $y = \frac{1}{2}x$
45. $y = -2x - 2$

Page 30   Checking Key Concepts

1. $\overline{NX}$ **3.** No; $\overrightarrow{PQ}$ and $\overrightarrow{QP}$ have different endpoints and extend in different directions.   **5.** 5   **7.** $\overline{WF}$ and $\overline{KM}$
9. Answers may vary. An example is given.

Page 30–33   Exercises and Applications

1. 2   **3.** 2   **5.** 2   **7.** $\overline{AB} \cong \overline{EG}$; $\overline{BC} \cong \overline{CE}$; $\overline{AC} \cong \overline{CG}$
9, 11, 13. Answers may vary. Examples are given.

9. 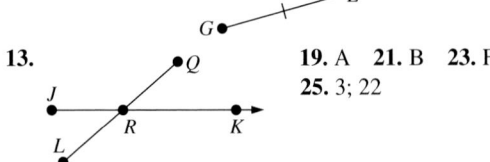   **11.**

13.   **19.** A   **21.** B   **23.** F
25. 3; 22

Page 37   Checking Key Concepts

1. $\angle WYZ$ or $\angle ZYW$ **3.** $\angle ZYX$ **5.** 110°   **7.** 70°

Pages 37–39   Exercises and Applications

1. Answers may vary. An example is given.

In the example, none of the angles can be named $\angle A$ because that name could refer to $\angle BAC$, $\angle BAD$, or $\angle CAD$.

3, 5, 7, 9. Answers may vary slightly.   **3.** 41°   **5.** 65°
7. 115°   **9.** 74°   **11.** 32°   **13.** 45°   **15.** 27°   **17.** $\angle HED$
31. 6   **33.** 12

Page 43   Checking Key Concepts

1. 3.4   **3.** 6.8   **5.** 38°   **7.** Fold the paper so that J falls on top of L. Unfold the paper. The point where the crease meets $\overline{JL}$ is the midpoint of $\overline{JL}$.

Pages 44–47   Exercises and Applications

1. 11   **3.** 9   **5.** False.   **7.** False.   **11.** $(-2, 3)$ **13.** $(4, 1)$
15. $(1, 0)$ **17, 19.** Answers may vary. Examples are given.

17.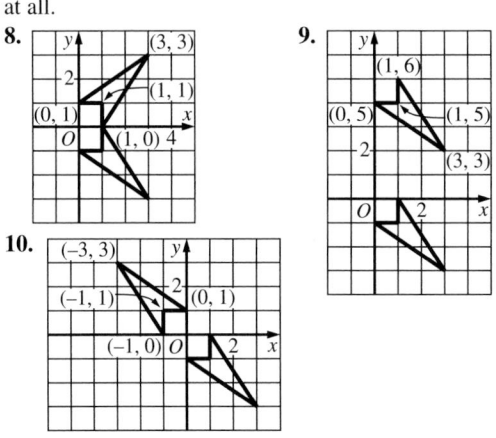

C —2— B —2— A —4— D

19.

21. $3\frac{1}{2}$ **29.** 8 **31.** 16 **35.** 78°
37. 31° **39.** 26, 37

Page 47 Assess Your Progress

1. $\angle FGH$; $\angle HGF$; $\angle G$ **2.** $\angle FHE$ **3.** 130° **4.** 27°
5. 52° **6.** 13; 26 **7.** 1 **8.** (−1, −2) **9.** 4; 94°

Pages 52–53 Chapter 1 Assessment

1. inductive **2.** the hypothesis (the "if") and the conclusion
(the "then") **3.** image; rotation, reflection, translation
5. 324, 972 **6.**

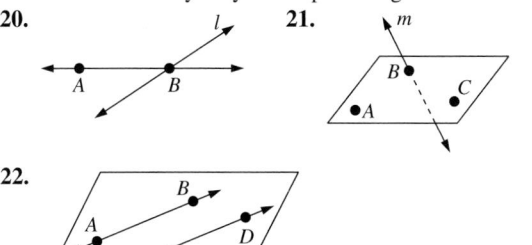

7. a. If a person is tall, then he or she is a good basketball
player. hypothesis: a person is tall; conclusion: he or she is a
good basketball player **b.** False; there are many counterexamples, including tall people who do not play basketball
at all.

8. **9.**

10.

11. True. **12.** True. **13.** False; two lines in the same plane
are either parallel or they intersect. **14.** False; if two planes
intersect, they intersect in a line. **15.** plane S, plane ABD,
plane CBD (The order of the letters can also be changed; for
example, the plane could also be called plane ACB.)
16. point E **17.** point D **18.** Answers may vary. An
example is given. $\overleftrightarrow{AB}$ and $\overleftrightarrow{CE}$ **19.** $\overleftrightarrow{AB}$ and $\overleftrightarrow{DC}$; $\overleftrightarrow{AD}$ and $\overleftrightarrow{BC}$
20–22. Answers may vary. Examples are given.
20. **21.**

22.

23. $\overline{RV} \cong \overline{RS}$, $\angle V \cong \angle S$, $\overline{SV} \cong \overline{TU}$, $\overline{ST} \parallel \overline{VU}$ **24.** Answers
may vary. Examples are given. $\overline{ST} \cong \overline{VU}$, $\overline{SV} \parallel \overline{TU}$; Angles
S, T, U, and V are all congruent to one another. **25. a.** 5
b. 10 **c.** 8 **26.** 1; 5 **27.** 60°; 120°

CHAPTER 2

Page 61 Checking Key Concepts

1. complementary: 50°; supplementary: 140°; vertical: 40°
3. complementary, $x < 90°$: $90° − x°$; supplementary,
$x < 180°$: $180° − x°$; vertical: $x°$ **5.** always

Pages 61–63 Exercises and Applications

1. $\angle FHD$ **3.** $\angle CJE$ **11.** 80°; Linear Pair Property
13. 45°; If two lines form congruent adjacent angles, then
the lines are perpendicular; Linear Pair Property or Vertical
Angles Property; definition of complementary angles.
15. 135°; Linear Pair Property (using Ex. 13) **17.** True.
19. False. **27.** 140° **31. a.** 50°; by the Linear Pair
Property **b.** 50° **c.** Two angles that are supplementary to
the same angle are congruent. This is because if $m\angle 1 +$
$m\angle 2 = 180°$ and $m\angle 1 + m\angle 3 = 180°$, then $m\angle 2 =$
$180° − m\angle 1$ and $m\angle 3 = 180° − m\angle 1$, so $m\angle 2 = m\angle 3$.
35. a. 70° **b.** 40° **c.** 70°

Page 67 Checking Key Concepts

1. scalene, acute triangle **3.** Answers may vary. An
example is given. **5.** 33°

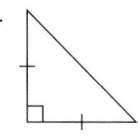

Pages 68–71 Exercises and Applications

1. Answers may vary. An example is given.

3. not possible; An equiangular triangle has three angles that
measure 60°, so it has only acute angles. **7. a.** 90°
b. They are complementary. Because the right angle
measures 90°, the other two angle measures must total
$180° − 90° = 90°$. **9.** $b° = 30°$; $4b° = 120°$ **11.** $w° = 42°$;
$z° = 25°$ **19.** never **21.** always **23.** sometimes **25.** 55°
27. 67.5° **29.** 72.5° **35. a.** 5 segments **b.** $\overline{FG} \cong \overline{GH} \cong$
$\overline{HI} \cong \overline{IJ} \cong \overline{JF}$ **c.** $\angle F \cong \angle G \cong \angle H \cong \angle I \cong \angle J$

Page 71 Assess Your Progress

1. 37; Vertical Angles Property **2.** 143; Linear Pair
Property **3.** 90; If two lines form congruent adjacent
angles, then the lines are perpendicular; Linear Pair Property
or Vertical Angles Property. **4.** 53°; The sum of the angle
measures of a triangle is 180°.

5–9. Answers may vary. Examples are given.

5. a. 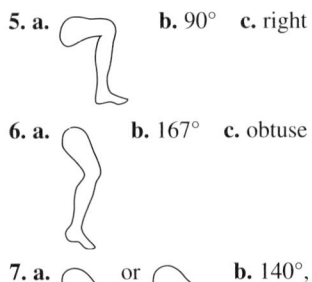 **b.** 90° **c.** right

6. a. **b.** 167° **c.** obtuse

7. a. 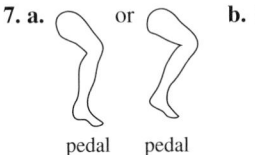 or **b.** 140°, 95° **c.** obtuse

pedal   pedal
at front at back

8. △*ABD*: scalene, acute triangle; △*BDE*: equilateral triangle (thus, equiangular and acute); △*BEF*: isosceles, obtuse triangle; △*BCF*: scalene, right triangle

Page 75 Checking Key Concepts

1, 3. Names may vary. Examples are given. **1.** convex, equiangular pentagon; *JKLMN* **3.** (convex) regular quadrilateral (a square); *WXYZ* **5.** Its borders include curves, and not just line segments. **7.** False; a rectangle is an example of an equiangular polygon that is not regular. **9.** False; the equilateral polygon shown at the bottom of page 73 is not convex.

Pages 76–78 Exercises and Applications

1. convex quadrilateral **3.** concave, equilateral 12-gon **5.** pentagon: 5 sides, 5 diagonals; hexagon: 6 sides, 9 diagonals; heptagon: 7 sides, 14 diagonals; octagon: 8 sides, 20 diagonals; Prediction—nonagon: 9 sides, 27 diagonals

7. a. 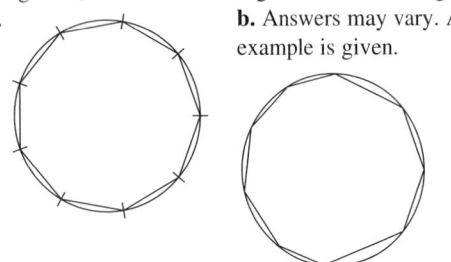 **b.** Answers may vary. An example is given.

To sketch a regular nonagon, you must mark nine evenly spaced points around the circle before connecting consecutive points. For a nonagon that is not regular, you can connect any nine points consecutively around the circle. **9. a.** 4 vertices **b.** 2 vertices **c.** 8 names **17.** 6 lines of symmetry **19.** 2 lines of symmetry **21, 23.** Answers may vary. Examples are given.

21. 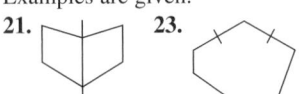 **23.**

25. Sketches may vary. Examples are given.

Polygon	Triangle	Quadrilateral	Pentagon
Concave	not possible		
Convex			
Regular			
Equilateral only	not possible		
Equiangular only	not possible		

27. hypothesis: a polygon is regular; conclusion: it is equiangular **29.** 60°

Page 82 Checking Key Concepts

1. 720° **3.** $(2x - 2)180°$, or $(x - 1)360°$ **5.** 29-gon **7.** 63°

Pages 83–86 Exercises and Applications

1. 360° **3.** 13,140° **5.** 135° **7.** interior: 108°; exterior: 72° **9.** interior: 168.75°; exterior: 11.25° **13. a.** 90° **b.** 4 **c.** Yes. **15. a.** 120° **b.** 3 **c.** Yes. **19.** 22-gon **21.** 14-gon **25.** $z° = 51°$; $4z° = 204°$ **27.** $w° = 25°$; $4w° = 100°$; $(6w - 15)° = 135°$ **29.** $4z° = 60°$; $9z° = 135°$; $(5z - 10)° = 65°$; $(5z + 25)° = 100°$ **31.** 20

35. a. Answers may vary. An example is given.

 7 sides; 7 triangles
b. 1260°
c. the angles with vertex at the interior point; 360°

d. The sum of the measures of the angles of the *n*-gon is the sum of the measures of the angles of the *n* triangles, 180*n*, minus the sum of the measures of those angles of the triangles that are not part of the angles of the polygon, 360. This gives $180n - 360°$ for the sum of the measures of the angles of the polygon. So, for a heptagon, for example, the sum is $180(7) - 360 = 1260 - 360 = 900°$, which agrees with earlier results. **43.** Parallel lines are coplanar lines that do not intersect, while skew lines are not coplanar and also do not intersect. **45.** Answers may vary. An example is given.

Page 86 Assess Your Progress

1. concave hexagon **2.** convex, equilateral pentagon **3.** convex, equiangular quadrilateral (a rectangle) **4.** convex hexagon

5. a. 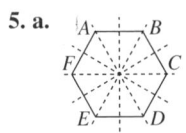 **b.** 6 lines of symmetry
c. $\overline{AB}$ and $\overline{BC}$, $\overline{BC}$ and $\overline{CD}$, $\overline{CD}$ and $\overline{DE}$, $\overline{DE}$ and $\overline{EF}$, $\overline{EF}$ and $\overline{AF}$ or $\overline{AF}$ and $\overline{AB}$

d. $\angle A$ and $\angle B$, $\angle B$ and $\angle C$, $\angle C$ and $\angle D$, $\angle D$ and $\angle E$, $\angle E$ and $\angle F$, or $\angle F$ and $\angle A$ **6.** 140° **7.** interior: 165°; exterior: 15°

Page 90 Checking Key Concepts

1. a. 78° **b.** 102° **c.** 20 mm **3.** False; a rectangle is a square only if it is also equilateral.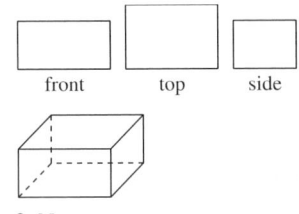

Pages 90–92 Exercises and Applications

1. a. 15 m **b.** 90° **c.** 12 m **3. a.** 37° **b.** 85° **c.** 85° **5.** 13 **7.** 108° **9.** 36° **17.** parallelogram; rectangle; rhombus; square **19.** parallelogram; rectangle; rhombus; square **21.** rectangle; square **23.** parallelogram; rectangle; rhombus; square **25.** B **27.** C **29.** G

Page 96 Checking Key Concepts

1. Answers may vary. An example is given.

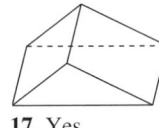

front   top   side

3. No.

Pages 96–99 Exercises and Applications

1. *ABCDEF, GHIJKL*; hexagonal prism **3.** 6 lateral faces **5.** Answers may vary. Examples are given. $\overline{AF}$ and $\overline{CI}$, $\overline{HG}$ and $\overline{EK}$ **7.** 6 faces; 8 vertices; 12 edges **9.** 8 faces; 12 vertices; 18 edges **11.** Answers may vary. An example is given. The number of faces is two more than the number of sides of a base. The number of vertices is twice the number of vertices of a base. The number of edges is three times the number of sides of a base. For a prism whose base is an *n*-gon, $F = n + 2$, $V = 2n$, and $E = 3n$. **13.** Answers may vary. An example is given. For an octagonal prism, $F = 10$, $V = 16$, and $E = 24$, so $F + V - 2 = 26 - 2 = E$. This will work for any prism, because the values determined in Ex. 11 have the relationship determined in Ex. 12 for any value of *n*. **15.** Answers may vary. An example is given.

17. Yes.

19, 21. Answers may vary. Examples are given.
19.   **21.**

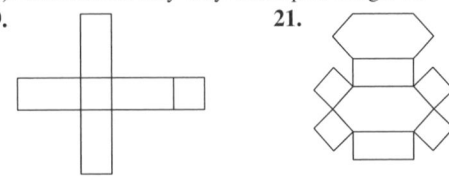

27. a. any two of $\overline{CE}$, $\overline{BH}$, and $\overline{GA}$ **b.** Both points lie on the front face of the prism. **c.** 4 diagonals **31.** An isosceles triangle is a triangle in which at least two of the sides are congruent. An example is given.   **33.** 25, 36

Page 99 Assess Your Progress

1. a. 60° **b.** 120° **c.** 27 **2. a.** 90° **b.** 14 **c.** 50 **3. a.** 9 **b.** 45° **c.** 135° **4, 5.** Answers may vary. Examples are given.

4.   **5.**

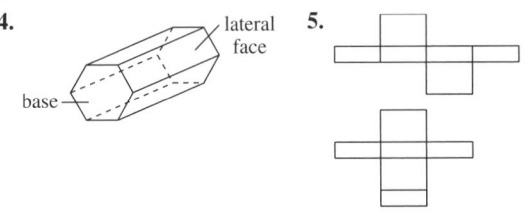

Pages 104–105 Chapter 2 Assessment

1. obtuse **2.** diagonal **3.** square **4.** exterior **5.** $\angle BEG$ and $\angle DEF$, or $\angle BED$ and $\angle GEF$ **6.** $\angle ACF$ and $\angle ACB$ **7.** A prism is a three-dimensional figure formed by connecting the corresponding vertices of two congruent faces that lie in parallel planes. A net is a two-dimensional representation of this three-dimensional figure formed by drawing the faces of the prism as if the prism had been cut along some of its edges (without disconnecting any of the faces), unfolded, and then laid out flat. **8.** rhombus (a convex, equilateral parallelogram) **9.** convex, equiangular pentagon **10.** isosceles right triangle **12. a.** 90 **b.** 30 **c.** 60 **13.** complementary: 19°; supplementary: 109°; vertical: 71° **14.** complementary: 51°; supplementary: 141°; vertical: 39° **15.** complementary: 5°; supplementary: 95°; vertical: 85° **16.** $4x° = 40°$; $5x° = 50°$ **17.** $6x° = 54°$; $9x° = 81°$ **18.** $x° = 34°$; $2x° = 68°$; $y° = 78°$ **20.** regular hexagon (convex, equiangular, and equilateral) **21.** any three of the following: *ABCDEF, BCDEFA, CDEFAB, DEFABC, EFABCD, FABCDE, AFEDCB, FEDCBA, EDCBAF, DCBAFE, CBAFED, BAFEDC* **22.** $\overline{AC}$, $\overline{AD}$, $\overline{AE}$ **23.** $5x° = 100°$; $6x° = 120°$ **24.** $x° = 60°$; $2x° = 120°$; $(x + 10)° = 70°$; $(2x - 10)° = 110°$ **25.** 30 **26.** 165° **27. a.** 8 m **b.** 16 m **c.** 60° **28. a.** 110° **b.** 25° **c.** 25° **29.** Answers may vary. An example is given.

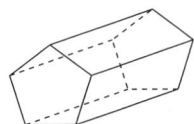

Selected Answers   **SA5**

30. a. a hexagonal prism   **b.** 6 lateral faces   **c.** Answers may vary. Examples are given.

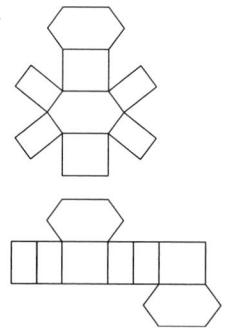

Pages 106–107   Algebra Review/Preview

1. –44   **3.** –10   **5.** –4.8   **7.** 22   **9.** 12   **11.** 26 weeks
13. $3\sqrt{3}$ **15.** $5\sqrt{3}$ **17.** $60\sqrt{2}$ **19.** $4\sqrt{2}$ **21.** ±9
23. ±3 **25.** ±2

27.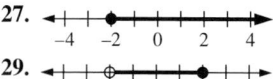

29.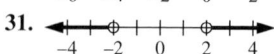

31.

33. 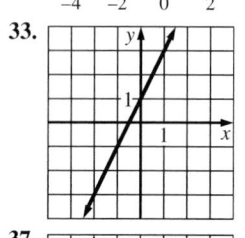   **35.**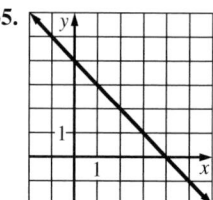

37.

CHAPTER 3

Page 113   Checking Key Concepts

1. inductive reasoning   **3.** deductive reasoning

Pages 114–116   Exercises and Applications

1. inductive reasoning   **3.** inductive reasoning
5, 7, 9. Justifications may vary. Examples are given.
5. 10 and 15; The number of dots in each row after the first is one more than the number of dots in the previous row.
7. 16, 32; Each term after the first is obtained by doubling the previous term.   **9.** 720, 5040; For $n > 1$, the nth term is found by multiplying the previous term by n.   **11.** is
13. is not   **27.** right   **29.** obtuse

Page 120   Checking Key Concepts

1. Reflexive Property   **3.** Symmetric Property   **5.** A

Pages 121–123   Exercises and Applications

1. postulate   **3.** definition   **5.** Transitive Property
7. Reflexive Property   **9.** given, definition of congruent segments   **11.** given, definition of congruent angles
13. Ex. 11, Angle Addition Postulate, Subtraction Property
23. given information   **25.** Vertical angles are congruent.
27. Ex. 26 and the definition of congruent angles   **31.** 27°; 117°; 63°   **33.** $(90 - 2y)°$ for $y \leq 45$; $(180 - 2y)°$ for $y \leq 90$; $2y°$

Page 123   Assess Your Progress

1. deductive   **2.** inductive   **3.** is   **4.** might have been
5. definition of angle bisector   **6.** Vertical angles are congruent.   **7.** Ex. 6, definition of congruent angles, given, and the Addition Property   **8.** Vertical angles are congruent.

Page 126   Checking Key Concepts

1. a. 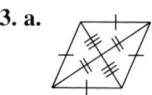   **b.** hypothesis: A quadrilateral is a square. conclusion: It has four right angles.

3. a.   **b.** hypothesis: A quadrilateral is a rhombus. conclusion: Its diagonals bisect each other.

5. False; the two exterior angles at each vertex are congruent.

Pages 127–129   Exercises and Applications

1. False; False; False.   **3.** False; False; True.   **5.** 110°
11. a. $\angle AGF$ and $\angle CGD$ **b.** $\angle BGD$ by the Angle Addition Postulate   **c.** 180°; because C, G, and F are collinear points   **d.** $\angle AGF \cong \angle CGD$ and $\angle EGF \cong \angle BGC$, since vertical angles are congruent. By the definition of congruent angles and the Substitution Property, $\angle BGC \cong \angle CGD$.   **15.** Yes; since the triangle is equilateral, two of its sides are congruent.   **17.** No; for example, let $m\angle 1 = 60°$. Then $m\angle 3 = 120°$ and $m\angle 2 = 120°$. $\angle 1$ and $\angle 2$ are not congruent angles.   **19.** B   **23.** Transitive Property   **25.** Answers may vary. Examples are given. 2, 4, 8, 16, 32, 64, … (The nth term is $2n$.); 2, 4, 8, 14, 22, 32, … (For $n > 1$, the nth term is found by adding $2n - 2$ to the previous term.)

Page 132   Checking Key Concepts

1. Substitution Property (Step 1); $m\angle 2 = m\angle 3$; $m\angle 1 = m\angle 4$ **3.** 19; If $(n - 2)180 = 3060$, where n is the number of sides, then $n = 19$.

Pages 132–135   Exercises and Applications

1. The statement is incorrect. Angles that are supplementary to the same angle are congruent.   **3.** The measure of an exterior angle of a triangle is not necessarily greater than the measure of the adjacent interior angle. Since the measure of an exterior angle of a triangle is equal to the sum of the measures of the two nonadjacent interior angles and each angle measure is positive, the measure of an exterior angle is greater than the measure of each nonadjacent interior angle. **11.** 94° **13.** 159° **15.** Answers may vary. An example is given.

a.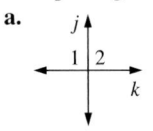
b. Given: $j \perp k$. Prove: $\angle 1 \cong \angle 2$
19. If <u>two angles are congruent</u>, then (their measures are equal.)
21. If <u>a quadrilateral is a parallelogram</u>, then (its diagonals bisect each other.)

23.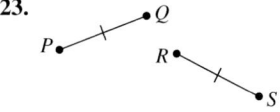

Page 135   Assess Your Progress

1. Given: $\angle 1$ and $\angle 2$ are supplementary.
$\angle 1$ and $\angle 3$ are supplementary.
Prove: $\angle 2 \cong \angle 3$.
By the definition of supplementary angles, $m\angle 1 + m\angle 2 = 180°$ and $m\angle 1 + m\angle 3 = 180°$. Therefore, by the Substitution Property, $m\angle 1 + m\angle 2 = m\angle 1 + m\angle 3$. Subtracting $m\angle 1$ from both sides gives $m\angle 2 = m\angle 3$, so $\angle 2 \cong \angle 3$.

2.

Statements	Reasons
1. $m\angle 1 + m\angle 2 + m\angle 3 = 180°$	1. The sum of the angle measures of a triangle is 180°.
2. $m\angle 4 + m\angle 3 = 180°$	2. Angles in a linear pair are supplementary.
3. $m\angle 1 + m\angle 2 + m\angle 3 = m\angle 4 + m\angle 3$	3. Substitution Property (Steps 1 and 2)
4. $m\angle 1 + m\angle 2 = m\angle 4$	4. Subtraction Property

3. Reasons

1. Given
2. Definition of congruent angles
3. Angles in a linear pair are supplementary.
4. Substitution Property (Step 3)
5. Substitution Property (Steps 2 and 4)
6. Subtraction Property
7. Definition of congruent angles

Page 138   Checking Key Concepts

1. If you are in Guyana, then the official language is English; True; True. **3.** If the official language is Portuguese, then you are in Brazil; True; True. **5. a.** If a quadrilateral is a square, then it is a rectangle. **b.** If a quadrilateral is a rectangle, then it is a square. **7. a.** If a figure is a triangle, then the sum of the angle measures is 180°. **b.** If the sum of the angle measures of a figure is 180°, then the figure is a triangle.

Pages 138–140   Exercises and Applications

1. If the American League team is the championship team, then it won the first four games of the World Series; False. **3.** If the area of a rectangle is 6 ft², then the rectangle is 2 ft long and 3 ft wide; False.

5. a. If two angles are congruent, then supplements of those angles are congruent. **b.** If the supplements of two angles are congruent, then the two angles are congruent. **7. a.** If a triangle is isosceles, then it is not scalene. **b.** If a triangle is not scalene, then it is isosceles. **9. a.** If a quadrilateral is a square, then its diagonals are congruent and perpendicular. **b.** If the diagonals of a quadrilateral are congruent and perpendicular, then the quadrilateral is a square. **21.** 90°
23. 10 **25.** 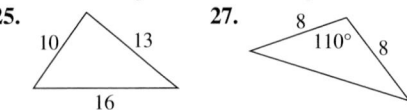 **27.**

Page 144   Checking Key Concepts

1. 7; 7 **3.** –4.90; $-2\sqrt{6}$ **5.** 8 **7.** Yes. **9.** No.

Pages 144–147   Exercises and Applications

1. 11; 11 **3.** ± 5.20; $\pm 3\sqrt{3}$ **5.** –20.12; $-9\sqrt{5}$ **7.** Yes.
9. No. **11.** Yes. **13.** 10.91 **15.** 12 **17.** 3.32
19. $\sqrt{3} \approx 1.73$ **21.** $\sqrt{130} \approx 11.40$ **23.** $\dfrac{3\sqrt{2}}{2} \approx 2.12$
25. Yes. **27.** No.

33. 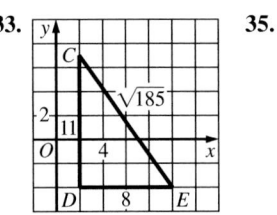 **35.**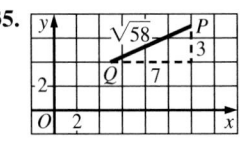

49. If (two lines are perpendicular,) then <u>they intersect at right angles.</u>

Page 150   Checking Key Concepts

1. acute **3.** obtuse **5.** False; a triangle with exactly two congruent sides is not equilateral, yet it is isosceles.

Pages 150–153   Exercises and Applications

1. obtuse **3.** acute **5.** acute **7.** obtuse **11.** False.
13. False. **17.** Inverse: If it does not snow, then classes are not canceled. Contrapositive: If classes are not canceled, then it does not snow. **19.** Inverse: If a quadrilateral is not a parallelogram, then its opposite sides are not congruent. Contrapositive: If the opposite sides of a quadrilateral are not congruent, then the quadrilateral is not a parallelogram.
21. If two lines are not perpendicular, then they do not form congruent adjacent angles; True. **23.** If two angles are not vertical angles, then they are not congruent; False.
29. Yes; the lengths are the lengths of the sides of a right triangle and the triangle appears to be a right triangle.
31. No; the lengths are not the lengths of the sides of a right triangle. **33.** No; the lengths are not the lengths of the sides of a right triangle. **37.** Sketches and labels may vary. **a.** If two angles are not congruent, then they are not both right angles.

b.

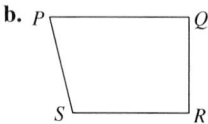

c. Given: $\angle Y$ is not congruent to $\angle X$.
Prove: $m \angle Y \neq 90°$ or $m \angle X \neq 90°$.
41. 2 **43.** Yes. **45.** Yes.

Page 153 Assess Your Progress

1. $2\sqrt{6} \approx 4.90$ **2.** $6\sqrt{5} \approx 13.42$ **3.** $\frac{5\sqrt{5}}{2} \approx 5.59$
4. right **5.** obtuse **6.** obtuse **7.** acute **8.** acute
9. obtuse **10.** Inverse: If a quadrilateral is not a rectangle, then its diagonals are not congruent. Converse: If the diagonals of a quadrilateral are congruent, then the quadrilateral is a rectangle. Contrapositive: If the diagonals of a quadrilateral are not congruent, then the quadrilateral is not a rectangle.
11. Inverse: If the exterior rays of two adjacent angles are not perpendicular, then the angles are not complements. Converse: If two adjacent angles are complements, then the exterior rays of the angles are perpendicular. Contrapositive: If two adjacent angles are not complements, then the exterior rays of the angles are not perpendicular.

12. a. In quadrilateral $PQRS$, if $\angle P$ is not congruent to $\angle R$, then $PQRS$ is not a parallelogram.

b. **c.** Given: $\angle P$ is not congruent to $\angle R$.
Prove: $PQRS$ is not a parallelogram.

Pages 158–159 Chapter 3 Assessment

1. Sketches and labels may vary. An example is given. For the triangle in the sketch, the Pythagorean theorem can be stated as $a^2 + b^2 = c^2$.

leg a / hypotenuse c / b leg

2. proof; hypothesis; conclusion **3.** If point Y is not between points X and Z, then $XY + YZ \neq XZ$. **4.** must
5. may have **6.** inductive **7.** Through any two points there is exactly one line. **8.** definition of midpoint
9. Transitive Property **10.** The sum of the angle measures of a triangle is 180°. **11.** Given; Vertical angles are congruent; Substitution Property (Steps 1 and 2); $\angle 1 \cong \angle 4$; Substitution Property (Steps 2 and 3)
12. a. Sketches and labels may vary.

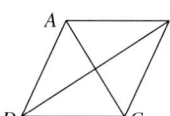 **b.** Given: $ABCD$ is a rhombus.
Prove: $\overline{AC} \perp \overline{BD}$

13. a. Converse: If the diagonals of a quadrilateral are perpendicular, then it is a rhombus. Inverse: If a quadrilateral is not a rhombus, then its diagonals are not perpendicular. Contrapositive: If the diagonals of a quadrilateral are not perpendicular, then it is not a rhombus.

b. For any of the statements, the diagram given in the answer to Ex. 12 can be used. Converse—Given: $\overline{AC} \perp \overline{BD}$; Prove: $ABCD$ is a rhombus. Inverse—Given: $ABCD$ is not a rhombus; Prove: $\overline{AC}$ is not perpendicular to $\overline{BD}$. Contrapositive—Given: $\overline{AC}$ is not perpendicular to $\overline{BD}$; Prove: $ABCD$ is not a rhombus. **15.** $2\sqrt{41} \approx 12.81$
16. $3\sqrt{3} \approx 5.20$ **17.** obtuse **18.** right **19.** acute

CHAPTERS 1–3

Pages 160–161 Cumulative Assessment

1. 15 handshakes; 21 handshakes **7.** $\angle BAE$, $\angle BAC$, $\angle EAB$, $\angle CAB$ **9.** one of the following: $\overrightarrow{BD}$, $\angle ABD$, and $\angle CBD$, or CA, $\angle BCA$, and $\angle DCA$ **11.** always
13. always **15.** $x = 6$; $y = \frac{1}{3}$ **17.** $x = 36$; $y = 30$

21. The measure of an exterior angle of a triangle is equal to the sum of the measures of the two interior angles that are not adjacent to it. **23.** the Segment Addition Postulate
25. The sum of the angle measures of a triangle is 180°. Since the angles of an equiangular triangle are congruent, their measures are equal. Thus, the measure of each angle is $\frac{180°}{3} = 60°$. **27.** $\sqrt{39} \approx 6.24$ **29.** 1

CHAPTER 4

Page 169 Checking Key Concepts

1. 6 **3.** 5 **5.** $(2, -4)$ **7.** $(2.5, 0)$

Pages 169–172 Exercises and Applications

5. $DE = EF = \sqrt{13}$; $DF = \sqrt{26}$; isosceles right triangle
7. $JK = LM = 3\sqrt{2}$; $JM = KL = 2\sqrt{2}$; rectangle
9. $WX = XY = YZ = WZ = \sqrt{26}$; rhombus **11.** $(-5, 8)$
13. $(3.5, 3.5)$ **15.** $(2, 2.5)$ **21.** $(0, 2)$, $(0, 1)$, $(1, 1)$, $(-2, 0)$, $(0, 0)$, $(1, 0)$, $(2, 0)$, $(0, -1)$, $(1, -1)$, $(0, -2)$ **37.** obtuse
39. Answers may vary. An example is given. **41.** -4 **43.** -1

Page 176 Checking Key Concepts

1. 0 **3.** $y = -x + 2$

Pages 176–179 Exercises and Applications

1. 4 **3.** $-\frac{5}{4}$ **5.** $\frac{1}{2}$ **7. a.** 36 in. or 3 ft **b.** 3.5 in.
9. $y = 2x + 5$ **11.** $y = -0.75x + 1$ **13.** $x = 13$
15. $y = 0.25x + 1.5$ **23.** 3; 5 **25.** -5; -25 **27.** $-\frac{2}{3}$; 0
29. 4; -3 **31.** slope of $\overline{AB}$ = slope of $\overline{CD}$ = 0; slope of $\overline{AD}$ = slope of $\overline{BC}$ = -2 **33.** slope of $\overline{FO}$ = $\frac{1}{4}$; slope of $\overline{OH}$ = 4; slope of $\overline{GH}$ = $-\frac{1}{4}$; slope of $\overline{FG}$ = 2

37. (6, 3) **39.** (–7, 1.5) **41.** $\overleftrightarrow{CD}$, $\overleftrightarrow{EF}$, and $\overleftrightarrow{GH}$ **43.** $\overleftrightarrow{AD}$, $\overleftrightarrow{EH}$, and $\overleftrightarrow{FG}$ **45.** $\overleftrightarrow{BC}$, $\overleftrightarrow{DC}$, $\overleftrightarrow{GC}$, $\overleftrightarrow{HG}$, $\overleftrightarrow{FG}$ **47.** False; if the rhombus is not a square, the angles are not right angles.

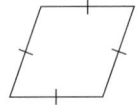

Page 183 Checking Key Concepts

1. neither **3.** parallel **5.** 7 **7.** 0

Pages 183–186 Exercises and Applications

1. perpendicular **3.** parallel **5.** 0 **7.** $\frac{1}{3}$ **9.** Yes. **11.** No.
15. quadrilateral **17.** square **19.** Yes. **21.** No; $\overleftrightarrow{CD}$ does not bisect $\overline{AB}$. **29.** $y = 7$ **31.** $x = 2$ **35.** 1; 0
37. Answers may vary. An example is given. **39.** $b = 15.5$

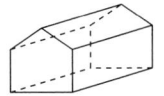

Page 186 Assess Your Progress

1. a. $12 + 3\sqrt{2}$ **b.** Let M, N, and P be the midpoints of $\overline{AB}$, $\overline{AC}$, and $\overline{BC}$, respectively. $M = (2.5, 0)$, $N = (1, 1.5)$, $P = (4.5, 1.5)$ **c.** $6 + \frac{1}{3}\sqrt{2}$; The perimeter of $\triangle MNP$ is half the perimeter of $\triangle ABC$. **2.** $-\frac{3}{5}$ **3.** 0 **4.** $\frac{5}{2}$ **5.** 4 **6.** $\frac{1}{5}$

Page 190 Checking Key Concepts

1. 5 **3.** $(x - 2)^2 + (y + 3)^2 = 16$

Pages 191–193 Exercises and Applications

1. $x^2 + y^2 = 9$ **3.** $(x + 2)^2 + (y - 3)^2 = 9$
5. $(x - 5)^2 + (y - 2)^2 = 49$ **7.** $(x + 2)^2 + (y - 1)^2 = 1$
9. $(x - 10)^2 + (y + 20)^2 = 400$

11. 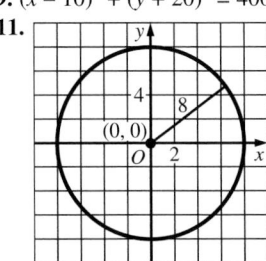 **13.**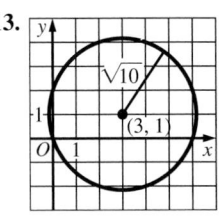

15.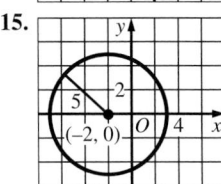

17, 19. Answers may vary. Examples are given.

17. 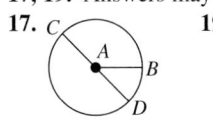 **19.**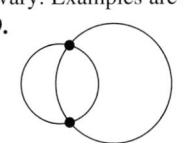

23. (3, 0); 2; $(x - 3)^2 + y^2 = 4$ **25.** (2, 2); $\sqrt{10}$;
$(x - 2)^2 + (y - 2)^2 = 10$ **29.** Yes; $(2\sqrt{2})^2 + 2^2 = 12$.
31. Yes; $(7 - 2)^2 + (5\sqrt{3})^2 = 100$. **35.** rhombus
37. square **39.** 6 **41.** 3 **43.** True.

Page 197 Checking Key Concepts

1. Answers may vary. The fourth vertex is (b, h).
3. Answers may vary. An example is given. Place a rectangle on a coordinate plane. Use the Distance Formula to show that the diagonals are congruent.

Pages 197–200 Exercises and Applications

1. $(0, b)$, $(0, -b)$, $(a, -b)$ **3.** $(2a, 0)$ **9.** (2.5, 2.5);
$MA = MB = MO = \frac{5}{2}\sqrt{2}$

11. Answers may vary. An example is given.

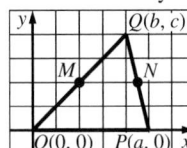

Using the Midpoint Formula, $M = \left(\frac{a}{2}, \frac{b}{2}\right)$.
By the Distance Formula, $MA = MB = MO = \frac{1}{2}\sqrt{a^2 + b^2}$.

13. Let M and N be the midpoints of the sides of the triangle. $MZ = (2, 2)$ and $N = (4.5, 2)$. The slope of $\overline{MN} = 0 =$ slope of the base, so $\overline{MN}$ is parallel to the base. $MN = 2.5 = \frac{1}{2} \times$ length of base

15. Answers may vary. An example is given.

Using the Midpoint Formula, $M = \left(\frac{b}{2}, \frac{c}{2}\right)$ and $N = \left(\frac{a + b}{2}, \frac{c}{2}\right)$. The slope of $\overline{MN} = 0 =$ slope of $\overline{OP}$, so $\overline{MN} \parallel \overline{OP}$. By the Distance Formula, $MN = \frac{a}{2} = \frac{1}{2} \times OP$. This proves that the segment that joins the midpoints of two sides of a triangle is parallel to the third side and half as long as the third side.
17. $E(a, 0)$, $F(a + b, c)$, $G(b + d, c + e)$, $H(d, e)$ **19.** The segments joining the midpoints of the opposite sides of a quadrilateral bisect each other. **29.** $(x - 7)^2 + y^2 = 36$
31. $(x + 1)^2 + (y - 4)^2 = 49$ **33.** 720° **35.** $\sqrt{89} \approx 9.43$
37. 5

Page 204 Checking Key Concepts

1. (4, 0, 2) **3.** (0, 3, –1) **5.** (–1, 4, –4) **7.** $\left(2, -3, -\frac{1}{2}\right)$
9. $\sqrt{14} \approx 3.74$

Pages 205–207 Exercises and Applications

1. (4, 0, 3) **3.** (4, 6, 3) **5.** (0, 0, 0) **7.** the xy-plane
9. the yz-plane **11.** the xz-plane **13.** (5, –2, 4.5)
15. (4, –0.5, –1) **17.** $\sqrt{21} \approx 4.58$ **19.** isosceles
21. scalene **31.** If all the sides of a triangle are congruent, then the triangle is equilateral. **33.** Answers may vary. An example is given.

Page 207   Assess Your Progress

1. 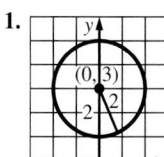   **2.**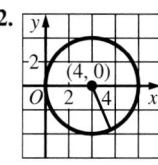

3. $D\left(-\dfrac{k}{2}, \dfrac{3k}{2}\right)$; $E\left(\dfrac{k}{2}, \dfrac{3k}{2}\right)$ **4.** slope of $\overline{AE} = 1$;

slope of $\overline{CD} = -1$ **5.** -1 **6.** If the product of the slopes of two lines is -1, then the lines are perpendicular.
7. equilateral   **8.** isosceles

Pages 212–213   Chapter 4 Assessment

1. B   **2.** E   **3.** C   **4.** D   **5.** A   **6. a.** $PR = 3\sqrt{4}$;
$QR = 3\sqrt{4}$; $PQ = 2\sqrt{2}$ **b.** isosceles triangle
7. a. $AB = 4\sqrt{2}$; $CD = 4\sqrt{2}$; $BC = 2\sqrt{2}$; $AD = 2\sqrt{2}$
b. rectangle   **8.** $-\dfrac{1}{2}$ **9.** -4 **10.** undefined   **11.** 3; -12

12. $\dfrac{3}{4}$; 1   **13.** -1; 4   **14.** $y = 2x - 11$ **15.** $y = -\dfrac{3}{2}x$

16. $x = 1$ **17.** $y = -1$ **18.** The slope of $\overline{JK}$ is -3 and the
slope of $\overline{KL}$ is $\dfrac{1}{3}$. Since $(-3)\left(\dfrac{1}{3}\right) = -1$, $\overline{JK} \perp \overline{KL}$, so $\triangle JKL$ is a
right triangle.   **20.** $x^2 + y^2 = 9$ **21.** $x^2 + y^2 = 5$
22. $(x - 4)^2 + (y - 2)^2 = 81$ **23.** $(x - 6)^2 + (y + 2)^2 = 1.25$
24. a. 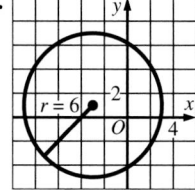   **b.** $(-3, 1)$ **c.** See part (a).

25. $(-2a, 0)$ **26.** $S(a, b)$; $T(-a, b)$; $O(0, 0)$ **27. a.** $TS = 2a$;
$TO = SO = \sqrt{a^2 + b^2}$ **b.** $TO = SO$; $\triangle STO$ is isosceles by
definition.   **29.** $(-1, 6, -3)$ **30.** $MN = 2\sqrt{14}$

Pages 214–215   Algebra Review/Preview

1. 2   **3.** 0   **5.** $\dfrac{12}{7}$ **7.** -8 **9.** 1   **11.** 3   **13.** $\dfrac{97}{11}$

15. $(10, 20)$ **17.** $\left(-1, \dfrac{2}{3}\right)$ **19.** no solution

21. 1550 adults and 2700 children

23. $z < 18$ (number line: 0, 6, 12, 18, 24; open circle at 18)

25. $p \geq 1$ (number line: $-4, -2, 0, 2, 4$; closed dot at 1)

27. $k \geq -7$ (number line: $-12, -8, -4, 0, 4$; closed dot at -7)

CHAPTER 5

Page 223   Checking Key Concepts

1. j **3.** $\angle 3$ and $\angle 6$, $\angle 4$ and $\angle 5$ **5. a.** $60°$ **b.** $60°$
c. 58

Pages 223–225   Exercises and Applications

1. $\angle 2$ and $\angle 5$, $\angle 3$ and $\angle 8$ **3.** $\angle 1$ and $\angle 5$, $\angle 2$ and $\angle 6$,
$\angle 3$ and $\angle 7$, $\angle 4$ and $\angle 8$ **5.** corresponding angles
7. corresponding angles   **9.** same-side interior angles
17. $x = 55$; $y = 70$; $z = 27$ **19.** same-side interior angles
21. none of these   **23.** none of these   **25.** same-side interior
angles   **29.** $\overline{AB}$ and $\overline{DE}$; $\overline{BC}$ and $\overline{EF}$; $\overline{CD}$ and $\overline{FA}$
33. $(2.5, -3.5, 5)$; $3\sqrt{6}$ **35.** False; answers may vary. An
example is given.

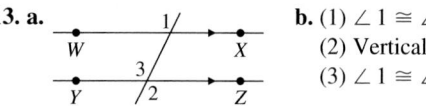

Page 229   Checking Key Concepts

1. $z = 50$; $m\angle 1 = 130°$ **3.** Answers may vary. An
example is given.

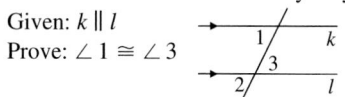

Pages 230–233   Exercises and Applications

1. $m\angle 1 = m\angle 2 = 125°$ **3.** $m\angle 5 = m\angle 7 = m\angle 9 = 140°$;
$m\angle 6 = m\angle 8 = 40°$ **5.** 26   **7.** always   **9.** sometimes
11. $115°$
13. a. (diagram with lines W–X, Y–Z, angles 1, 2, 3)   **b.** (1) $\angle 1 \cong \angle 3$
(2) Vertical $\angle\!\!\angle$ are $\cong$.
(3) $\angle 1 \cong \angle 2$

15. Answers may vary. An example is given.

17. not possible; Answers may vary. For example, Example
2 on page 229 showed that the angles of a trapezoid consist
of two pairs of supplementary angles. Two acute angles
cannot be supplementary, so a trapezoid cannot have three
acute angles.   **23.** $\angle 2$, $\angle 4$, and $\angle 6$ **25.** Answers may
vary. An example is given. The sum of the measures is the
same for both pairs of angles.   **31.** $m\angle 1 = 140°$;
$m\angle 2 = 40°$ **33.** Answers may vary. An example is given.
Given: $k \parallel l$
Prove: $\angle 1 \cong \angle 3$

Statements	Reasons
1. $k \parallel l$	1. Given
2. $\angle 1 \cong \angle 2$	2. If two $\parallel$ lines are intersected by a transversal, then corresponding $\angle\!\!\angle$ are $\cong$.
3. $\angle 2 \cong \angle 3$	3. Vertical $\angle\!\!\angle$ are $\cong$.
4. $\angle 1 \cong \angle 3$	4. Transitive Property (Steps 2 and 3)

37. $w° = 115°$; $z° = 65°$; $y° = 68°$

Page 233   Assess Your Progress

1. same-side interior angles: $\angle 3$ and $\angle 7$, $\angle 2$ and $\angle 6$;
alternate interior angles: $\angle 3$ and $\angle 6$, $\angle 2$ and $\angle 7$;
corresponding angles: $\angle 1$ and $\angle 7$, $\angle 3$ and $\angle 5$, $\angle 4$ and
$\angle 6$, $\angle 2$ and $\angle 8$ **2.** 29   **3.** $y = 135$; $z = 38$

4. Answers may vary. For the example shown, $\overline{JK}$ and $\overline{ML}$ are the bases and $\overline{KL}$ and $\overline{JM}$ are the legs.

Page 238 Checking Key Concepts

1. $m \angle 1 + m \angle 2 + m \angle 4 = 180°$

Pages 239–241 Exercises and Applications

1. (1) $\angle A$ and $\angle B$ are complements; (3) $m \angle A + m \angle B + m \angle C = 180°$; (4) 180°; Substitution Property (Steps 2 and 3), (5) $m \angle C = 90°$; Subtraction Property. **7.** If a triangle contained more than one angle with measure 90° or greater, then the sum of the measures of the three angles would be greater than 180°. Therefore, if one angle is right or obtuse, the other two angles must be acute in order for the angle sum to be exactly 180°.

13. 1. $m \angle 1 + m \angle 2 = 180°$; $m \angle 4 + m \angle 5 = 180°$
(The $\angle$s form a linear pair.)
2. $m \angle 1 + m \angle 2 = m \angle 4 + m \angle 5$ (Substitution)
3. $m \angle 1 = m \angle 5$ (Given)
4. $\angle 2 \cong \angle 4$ (Subtraction Property)

17. Methods of proof may vary. An example is given. Since $\overline{AB} \parallel \overline{CD}$, $\angle ABC \cong \angle BCD$. (If two $\parallel$ lines are intersected by a transversal, then alternate interior $\angle$s are $\cong$.) Similarly, since $\overline{BC} \parallel \overline{DE}$, $\angle BCD \cong \angle CDE$. By the Transitive Property, $\angle ABC \cong \angle CDE$. **21.** If a triangle is isosceles, then the triangle is equilateral; False.

Page 245 Checking Key Concepts

1. If two lines are intersected by a transversal and same-side interior angles are supplementary, then the lines are parallel.

Pages 245–248 Exercises and Applications

1. –32 **3.** 13 **5.** parallel; Answers may vary. An example is given. $m \angle YTV = 36°$, so $\angle XYW$ and $\angle YTV$ are congruent, and since corresponding angles are congruent, the lines are parallel. **7.** not parallel; The labeled corresponding angles cannot be congruent. **9.** $\overline{AD}$ and $\overline{BC}$

11. $\overline{AB}$ and $\overline{DC}$

21.

Statements	Reasons
1. $\angle 1 \cong \angle 3$	1. Given
2. $\angle 2 \cong \angle 1$	2. Vertical $\angle$s are $\cong$.
3. $\angle 2 \cong \angle 3$	3. Transitive Property
4. $j \parallel k$	4. If two lines are intersected by a transversal and corresponding $\angle$s are $\cong$, then the lines are $\parallel$.
5. $\angle 4$ and $\angle 5$ are supplements.	5. If two $\parallel$ lines are intersected by a transversal, then same-side interior $\angle$s are supplementary.

25. Proofs may vary. An example is given.

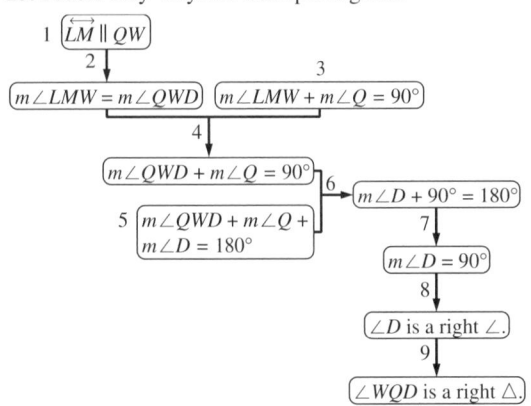

Reasons

1. Given
2. If two $\parallel$ lines are intersected by a transversal, then corresponding $\angle$s are $\cong$.
3. Given
4. Substitution Property (Steps 2 and 3)
5. The sum of the angle measures of a $\triangle$ is 180°.
6. Substitution Property (Steps 4 and 5)
7. Subtraction Property
8. Def. of right $\angle$
9. Def. of right $\triangle$

27. $y = \frac{1}{3}x - 6$

Page 251 Checking Key Concepts

1. $\overleftrightarrow{CE} \parallel \overleftrightarrow{FD}$ **3.** $\overleftrightarrow{CE} \parallel \overleftrightarrow{AB}$; If two lines are both parallel to a third line, then the lines are parallel.

Pages 252–255 Exercises and Applications

1. They are parallel; in a plane, if two lines are both perpendicular to a third line, then the two lines are parallel. **3.** $x = 15$; $m \angle LPQ = 65°$ **11.** $4\sqrt{6}$ **17. b.** 4 units **23.** (1) Given; (2) $\angle 3 \cong \angle 4$; (3) If two lines are intersected by a transversal and corresponding $\angle$s are $\cong$, then the lines are $\parallel$; (4) $p \parallel m$; (5) $n \parallel p$ (If two lines are both $\parallel$ to a third line, then the two lines are $\parallel$.) **25, 27.** Answers may vary. Examples are given.

25. **27.**

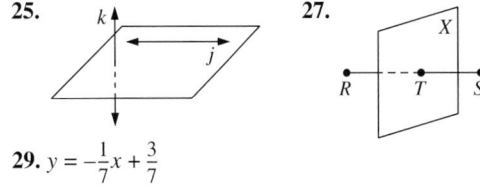

29. $y = -\frac{1}{7}x + \frac{3}{7}$

Page 255 Assess Your Progress

1. Since the sum of the angle measures of a triangle is 180°, $m \angle J + m \angle K + m \angle L = 180°$. It is given that $m \angle J + m \angle L = m \angle K$, so by the Substitution Property, $m \angle K + m \angle K = 180°$. $2 \cdot m \angle K = 180°$, and so $m \angle K = 90°$. $\angle K$ is a right angle by the definition of right angle and JKL is a right triangle by the definition of right triangle.

2. Statements | Reasons

Statements	Reasons
1. $\angle 1 \cong \angle 2$	1. Given
2. $\overleftrightarrow{FE} \parallel \overleftrightarrow{DC}$	2. If two lines are intersected by a transversal and corresponding $\angle$s are $\cong$, then the lines are $\parallel$.
3. $\overleftrightarrow{DC} \parallel \overleftrightarrow{BA}$	3. Given
4. $\overleftrightarrow{FE} \parallel \overleftrightarrow{BA}$	4. If two lines are both $\parallel$ to a third line, then the two lines are $\parallel$.

3. $\overline{MN} \cong \overline{PO}$; If two lines are intersected by a transversal and same-side interior angles are supplementary, then the lines are parallel.   **4.** $\overline{RQ} \parallel \overline{ST}$; Answers may vary. An example is given. $\angle RQT$ and $\angle T$ are both right angles. In a plane, if two lines are both perpendicular to a third line, then the lines are parallel.   **5.** 36   **6.** 42

Page 258   Checking Key Concepts

1. a. $\overline{AE}$, $\overline{AD}$, $\overline{BC}$, and $\overline{BF}$ **b.** Answers may vary. An example is given. Planes ABCD and EFGH are parallel and $\overleftrightarrow{AB}$ and $\overleftrightarrow{HG}$ are the lines of intersection when ABCD and EFGH are intersected by plane ABGH. By the Intersecting Planes Theorem, $\overline{AB} \parallel \overline{HG}$.

Pages 259–262   Exercises and Applications

1, 3. Answers may vary. Examples are given.

1. 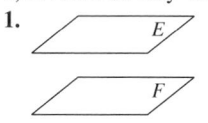   **3.**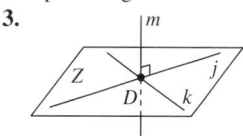

5. a. They are parallel; if two parallel planes are intersected by a third plane, the lines of intersection are parallel.
b. a trapezoid   **7.** $\angle FDE$ and $\angle BDC$
11. $m\angle QNR = 45°$; $m\angle MPN = 30°$
13. Answers may vary. An example is given.

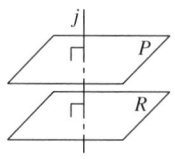
Given: plane $P \perp$ line j;
plane $R \perp$ line j
Prove: plane $P \parallel$ plane R

19. a. 60°   **b.** No; since $m\angle AED \neq 90°$, $\overleftrightarrow{AE}$ is not perpendicular to $\overleftrightarrow{ED}$, so $\overleftrightarrow{AE}$ cannot be perpendicular to plane CDE.   **25.** D   **27.** $\sqrt{97}$ **33.** Yes.

Page 265   Checking Key Concepts

1, 3. Use the indicated method.   **1.** the Parallel Lines Construction   **3.** the Congruent Angles Construction

Pages 266–269   Exercises and Applications

1, 3. Use the indicated method.   **1.** the Congruent Angles Construction   **3.** the Perpendicular Bisector Construction

5. Constructions may vary. In the example, $m\angle PQR = x°$, $m\angle SQR = y°$, and $m\angle PQS = (x - y)°$.

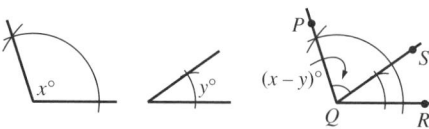

7. a. The same radius was used to draw the arcs used to locate points E, D, and H.   **13. a.** Using M and N as centers and a radius greater than $\frac{1}{2}MN$, draw arcs that intersect below line g at a point R. Draw $\overleftrightarrow{QR}$. $\overleftrightarrow{QR} \perp g$ **23. a.** Use the Congruent Angles Construction and the Perpendicular Bisector Construction.   **25.**

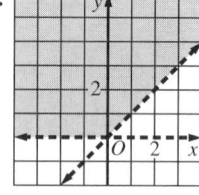

Page 269   Assess Your Progress

1. plane $P \parallel$ plane Q **2.** They are parallel; if two parallel planes are intersected by a third plane, the lines of intersection are parallel.   **3.** 122°; 58°   **4–6.** Use the indicated method.   **4.** the Congruent Angles Construction   **5.** the Perpendicular Bisector Construction   **6.** the Parallel Lines Construction

Pages 274–275   Chapter 5 Assessment

1. trapezoid; bases   **2.** transversal   **3.** flow proof   **4.** $\angle 1$ and $\angle 3$ **5.** $\angle 2$ and $\angle 4$ **6.** $\angle 2$ and $\angle 3$ **8.** $z = 59$; $y = 17$; $x = 112$ **9.** $p = r = 95°$ **10.** $s = 55$; $q = 35$
11. a. $\angle X$ and $\angle Y$ are supplementary since $\overline{WX} \parallel \overline{YZ}$; if two parallel lines are intersected by a transversal, then same-side interior angles are supplementary.   **b.** Yes; $\angle Y$ and $\angle Z$ are not supplementary. If they were, $\overline{WZ}$ and $\overline{XY}$ would have to be parallel and WXYZ would not be a trapezoid. (They may, however, be congruent.)
12. Given: $\overleftrightarrow{AY} \parallel \overleftrightarrow{CZ}$; $\overrightarrow{AB}$ bisects $\angle XAY$; $\overrightarrow{CD}$ bisects $\angle ACZ$.
Prove: $\overleftrightarrow{AB} \parallel \overleftrightarrow{CD}$
13.

```
1 [ AY ∥ CZ ]
     │2
     ▼
[ ∠XAY ≅ ∠ACZ ]      5 [ AB bisects ∠XAY.
     │3                   CD bisects ∠ACZ. ]
     ▼                         │6
[ m∠XAY = m∠ACZ ]              ▼
     │4            [ m∠XAB = ½ m∠XAY
     ▼               m∠ACD = ½ m∠ACZ ]
[ ½ m∠XAY = ½ m∠ACZ ]
                    7│
                     ▼
[ m∠XAB = m∠ACD ]─8→[ ∠XAB ≅ ∠ACD ]
                         │9
                         ▼
                    [ AB ∥ CD ]
```

Reasons

1. Given
2. If two ∥ lines are intersected by a transversal, then corresponding ⦞ are ≅.
3. Def. of ≅
4. Multiplication Property
5. Given
6. Def. of bisector
7. Substitution Property
8. Def. of ≅
9. If two lines are intersected by a transversal and corresponding ⦞ are ≅, then the lines are ∥.

14. If two lines are intersected by a transversal and alternate interior angles are congruent, then the lines are parallel.
15. In a plane, two lines perpendicular to the same line are parallel.   **16.** If two lines are intersected by a transversal and corresponding angles are congruent, then the lines are parallel.   **17.** If two lines are intersected by a transversal and same-side interior angles are supplementary, then the lines are parallel.   **18.** If two lines are both parallel to a third line, then they are parallel to each other.   **19.** j is parallel to l; thus, l must be parallel to k.

CHAPTER 6

Page 281   Checking Key Concepts

1. Yes.   **3.** Yes.   **5.** It is between 6 ft long and 32 ft long.
7. It is less than 16 m long, and like every length, greater than 0 m long.

Pages 282–284   Exercises and Applications

1. Yes.   **3.** Yes.   **5.** No.   **7.** It is between 1 ft long and 7 ft long.   **9.** It is between 3 cm long and 21 cm long.   **11.** It is between $\frac{1}{2}$ in. long and 14 in. long.   **13.** $5 < AB < 27$
15. $4 < AB < 6$ **17.** $AB > 4$ **19.** B   **21, 23.** Answers may vary. Examples are given.   **21.** Yes; the three points are not collinear.   **23.** No; all three points lie on the line $y = x$.
27. Answers may vary. Examples are given.

Student	Lengths	Triangle?
1	7, 5, 2	No
1	5, 4, 2	Yes
1	2, 2, 6	No
2	3, 11, 6	No
2	4, 10, 3	No
2	10, 8, 1	No
3	9, 2, 1	No
3	11, 2, 1	No
3	6, 1, 10	No
4	7, 2, 11	No
4	7, 8, 9	Yes
4	9, 11, 10	Yes

41. 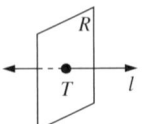   **43.** 0; 4

Page 288   Checking Key Concepts

1. △CBA   **3. a.** 18   **b.** 5

Pages 289–291   Exercises and Applications

1. Yes; △FME ≅ △ELF.   **3.** Yes; △WXY ≅ △WVU.
5. $\overline{BC}$ **7.** $WC = 12$ **9.** 21   **13.** △CEF   **15. a.** No.
b. Yes.   **25.** Reflexive Property

Page 291   Assess Your Progress

1. No.   **2.** No.   **3.** Yes.   **4.** It is greater than 3 in. and less than 5 in.   **5.** It is greater than 2 ft and less than 6 ft.
6. It is greater than 6 cm and less than 30 cm.   **7.** △AJC ≅ △BJG and △ACG ≅ △BGC   **8. a.** 15   **b.** 3.5

Page 295   Checking Key Concepts

1. Yes; SAS.   **3.** Yes; SSS.   **5.** $\overline{RQ} \cong \overline{RS}$

7.

Statements	Reasons
1. $\overline{XY} \cong \overline{ZW}$; $\overline{XY} \parallel \overline{ZW}$	1. Given
2. ∠WZX ≅ ∠YXZ	2. If two ∥ lines are intersected by a transversal, then alternate interior ⦞ are ≅.
3. $XZ \cong ZX$	3. Reflexive Property
4. △XYZ ≅ △ZWX	4. SAS Postulate

Pages 295–298   Exercises and Applications

1. ∠ACB   **5.** Yes; △ABD ≅ △CBD; SAS Postulate.
7. No.   **9.** No.

11.

Statements	Reasons
1. $\overline{AB} \cong \overline{CB}$; $\overline{BE} \cong \overline{BD}$	1. Given
2. ∠ABE ≅ ∠CBD	2. Vertical ⦞ are ≅.
3. △ABE ≅ △CBD	3. SAS Postulate

19. No.   **21.** Answers may vary. An example is given. It is given that $\overline{PQ} \cong \overline{RQ}$ and that $\overline{PS} \cong \overline{RS}$. By the Reflexive Property, $\overline{QS} \cong \overline{QS}$. Then, by the SSS Postulate, △PQS ≅ △RQS.

23. a. 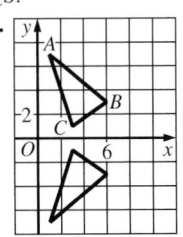   The coordinates of the vertices of the image of △ABC are $(1, -7)$, $(6, -3)$, and $(3, -1)$.

b. 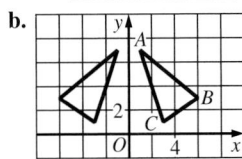   The coordinates of the vertices of the image of △ABC are $(-1, 7)$, $(-6, 3)$, and $(-3, 1)$.

c. Yes; the lengths of the sides of each image of △ABC are the same as the lengths of the sides of △ABC, so by the SSS Postulate, each image is congruent to △ABC.
25. △MON ≅ △QOP and △MQN ≅ △QMP **27.** 1 : 120
29. x° = 38°

Page 302 Checking Key Concepts

1. In any △ABC, $\overline{AB}$ is the side included between ∠A and ∠B. The non-included sides for ∠A and ∠B are $\overline{AC}$ and $\overline{BC}$. **3.** SAS Postulate **5.** ASA Postulate, AAS Theorem
7. none

Pages 303–305 Exercises and Applications

1. AAS Theorem, ASA Postulate **3.** SAS Postulate
5. SSS Postulate **7.** △PRU ≅ △PTQ; AAS Theorem, ASA Postulate **9.** △XWZ ≅ △YZW; AAS Theorem, ASA Postulate **13.** Answers may vary. An example is given. Since JKLM is a rectangle, ∠M and ∠L are right angles. Therefore, △JMP and △KLP are right triangles. A rectangle is a parallelogram, and the opposite sides of a parallelogram are congruent, so $\overline{JM} ≅ \overline{KL}$. $\overline{JP} ≅ \overline{KP}$ (given), so △JMP ≅ △KLP by the HL Theorem. **15.** Since $\overline{PQ} \parallel \overline{VS}$, ∠P ≅ ∠SVT. (If two ∥ lines are intersected by a transversal, then corresponding ∡ are ≅.) Since $\overline{QU} \parallel \overline{ST}$, ∠QUP ≅ ∠T. (If two ∥ lines are intersected by a transversal, then corresponding ∡ are ≅.) $\overline{PQ} ≅ \overline{VS}$ (Given), so △PQU ≅ △VST by the AAS Theorem. **17. a.** Answers may vary. There may be one or two ways to do Step 4. When there are two ways, only one of them results in a triangle that is congruent to △ABC. **b.** There is just one way to do Step 4, and that way gives a triangle that is congruent to △ABC. **21.** $\overline{JL}$
23. $\overline{CA}$ **25.** $128\frac{4}{7}°$; $51\frac{3}{7}°$

Page 309 Checking Key Concepts

1. △DGH and △EGF **3.** △DEF and △EDH **5.** △XVY and △ZVW or △WVX and △YVZ **7.** P, R **9.** isosceles

Pages 309–312 Exercises and Applications

1. △ABE and △DCE **3.** △ABC and △DCB or △ABD and △DCA

5.
Statements	Reasons
1. $\overline{PS} \parallel \overline{QR}$	1. Given
2. ∠PSQ ≅ ∠RQS	2. If two ∥ lines are intersected by a transversal, then alternate interior ∡ are ≅.
3. ∠P ≅ ∠R	3. Given
4. $\overline{QS} ≅ \overline{QS}$	4. Reflexive Property
5. △PSQ ≅ △RQS	5. AAS Theorem
6. $\overline{PQ} ≅ \overline{RS}$	6. Def. of ≅ △s

7. (1) $\overline{PB}$; exactly one; (2) perpendicular bisector; (3) SAS Postulate; (4) PA = PB; Definition of congruent triangles
9. 5

17.
Statements	Reasons
1. ∠WQX ≅ ∠YQZ	1. Vertical ∡ are ≅.
2. $\overline{WQ} ≅ \overline{YQ}$ and $\overline{XQ} ≅ \overline{ZQ}$	2. Given
3. △WQX ≅ △YQZ	3. SAS Postulate
4. ∠W ≅ ∠Y	4. Def. of ≅ △s
5. $\overline{WX} \parallel \overline{YZ}$	5. If two lines are intersected by a transversal and alternate interior ∡ are ≅, the lines are ∥.

19. D **21.** 1.8 **23.** 2.7 **25.** 43°

Page 312 Assess Your Progress

1. No. **2.** Yes; SSS Postulate or SAS Postulate. **3.** Yes; SSS Postulate. **4.** Yes; ASA Postulate or AAS Theorem.
5. Yes; HL Theorem or SSS Postulate. **6.** Yes; ASA Postulate or AAS Theorem. **7.** △YVZ and △XVW
8. △XWZ and △YZW **9.** Answers may vary. An example is given.

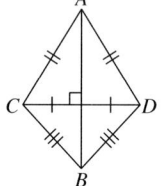

Page 315 Checking Key Concepts

1. 9 **3.** 42°

Pages 316–318 Exercises and Applications

3. ∠PKD and ∠PDK **5.** the roofs in Lithuania, Japan, and Germany **7.** (1) $\overline{AB} ≅ \overline{AC}$; (2) Reflexive Property; (3) SAS Postulate; (4) ∠B ≅ ∠C and Def. of ≅ △s **9.** 2
11. 35° **13.** 5.5 **15.** 12; 120° **19.** As shown in Ex. 7 on page 316, if you could pick up △PQR and flip it over so that $\overline{PQ}$ lands on $\overline{PR}$, the resulting triangle would be congruent to △PQR. "Flipping the triangle over" is equivalent to reflecting △PQR over the perpendicular bisector of $\overline{QR}$, which is the line of symmetry. (This line also bisects ∠P.)
21. Draw the bisector of ∠XZY. (Every angle has a bisector.) Let P be the intersection of the bisector and $\overline{XY}$. ∠XZP ≅ ∠YZP by the definition of angle bisector. ∠X ≅ ∠Y (given) and $\overline{ZP} ≅ \overline{ZP}$ (Reflexive Property). △XPZ ≅ △YPZ (AAS Theorem), and so $\overline{XZ} ≅ \overline{YZ}$ by the definition of congruent triangles. **25.** If a point is on the perpendicular bisector of a segment, then the point is equidistant from the endpoints of the segment. **27.** $\overline{DC}$ and $\overline{DE}$

Page 321 Checking Key Concepts

1. $\overline{EQ}$ **3.** Yes; if $\overline{ED} ≅ \overline{EF}$, then $\overline{EQ}$ is both an altitude from E and the bisector of ∠DEF. **5.** 11 cm

Pages 322–325 Exercises and Applications

1. a. 90° **b.** 14 **3. a.** In an isosceles triangle, the altitude to the base of the triangle is also the bisector of the vertex angle. **b.** Answers may vary. An example is given. Plan of Proof: Use the diagram on page 320. Use the HL Theorem to prove that $\triangle DAB \cong \triangle DAC$. Then $\angle ADB \cong \angle ADC$, so $\overrightarrow{DA}$ bisects $\angle BDC$. **7.** $y = -x + 6$ **11.** 8 **13. a.** No; each angle bisector lies inside the angle it bisects, that is, inside the triangle in which it is drawn. **b.** The circle touches each side of the triangle in just one point; the incenter of a triangle is equidistant from its sides. **15.** Yes; if a triangle is obtuse, then the lines containing the altitudes intersect outside the triangle. **19. a.** 130° **b.** 22 **21. a.** 12 **b.** 8

Page 325 Assess Your Progress

1. 6 **2.** 62° **3.** line n **4.** $\overline{JS}$ **5.** 47° **6.** 94

Pages 330–331 Chapter 6 Assessment

1. A **2.** C, E **3.** C, D, F **4.** E **5.** B **6.** No. **7.** It is between 14 cm long and 2 cm long. **8.** $\overline{AB}$; $\overline{AC}$ **9.** $\triangle GCD$, $\triangle GFD$; $\triangle DCG$, $\triangle DFG$; $\triangle BCD$, $\triangle BFD$; $\triangle BCG$, $\triangle BFG$ **10.** polygon $KVMR$ **11.** $x = 15$; $y = \frac{25}{7}$; $z = 22$ **12.** $\triangle XTY \cong \triangle ZTW$; SAS Postulate **13.** $\triangle TUV \cong \triangle TWV$; SSS Postulate **14.** none **15.** $\triangle SPR \cong \triangle QRP$; ASA Postulate or AAS Theorem **16.** none **17.** $\triangle EKF \cong \triangle GHF$; ASA Postulate or AAS Theorem

19.

Statements	Reasons
1. Line m is the perpendicular bisector of $\overline{CD}$.	1. Given
2. $\overline{FC} \cong \overline{FD}$	2. If a point is on the perpendicular bisector of a segement, then the point is equidistant from the endpoints of the segment.
3. $\angle C \cong \angle D$	3. If two sides of a triangle are $\cong$, then the $\angle$ opposite the sides are $\cong$.

20. 120° **21. a.** $\overline{KN}$ **b.** $\overline{JM}$ **22.** $\overline{KN}$ is the perpendicular bisector of $\overline{JL}$.

Pages 332–333 Algebra Review/Preview

1. 37.68 **3.** 1.84 **5.** 6 **7.** 400 ft **9.** $l = \frac{A}{w}$ **11.** $w = \frac{1}{2}P - l$ **13.** $r = \sqrt[3]{\frac{3V}{4\pi}}$ **15.** $\frac{1}{8}$ **17.** $\frac{25}{2}$ **19.** $\frac{10}{21}$ **21.** $3\frac{1}{3}$ **23.** 6 **25.** $(-3, 4)$ **27.** no solution **29.** $\left(-1\frac{1}{18}, \frac{5}{12}\right)$

CHAPTERS 4–6

Pages 334–335 Cumulative Assessment

1. $5\sqrt{5} \approx 11.2$; $\left(\frac{1}{2}, 0\right)$ **3.** $y = \frac{1}{2}x - 4$ **5.** $(x - 3)^2 + (y + 5)^2 = 125$

7.
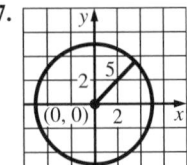

11. 0; $RA = \sqrt{(-a - 0)^2 + (0 - b)^2} = \sqrt{a^2 + b^2}$; $RB = \sqrt{(a - 0)^2 + (0 - b)^2} = \sqrt{a^2 + b^2}$; $RA = RB$ **13.** 25

15.
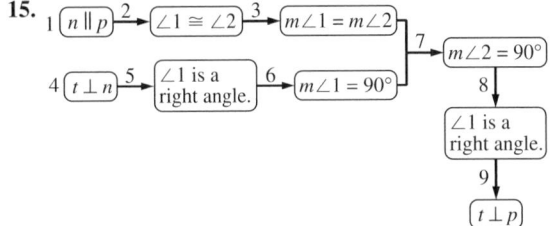

Reasons

1. Given
2. If two $\parallel$ lines are intersected by a transversal, then corresponding $\angle$ are $\cong$.
3. Def. of congruent angles
4. Given
5. Def. of perpendicular lines
6. Def. of right angle
7. Substitution Property
8. Def. of right angle
9. Def. of perpendicular lines

17. $\angle 1 \cong \angle 2$ (given), $\angle 2 \cong \angle 3$ (vertical angles), and $\angle 3 \cong \angle 4$ (given). Thus, $\angle 1 \cong \angle 4$ (Transitive Property), and so $\overleftrightarrow{AB} \parallel \overleftrightarrow{DE}$ (alternate interior angles are congruent). **19. a.** $\overline{MR}$ **b.** $\angle Z$ **c.** $\triangle MAR$ **21.** SSS Postulate **23.** Answers may vary. An example is given. Since $\angle 1 \cong \angle 2$, $\overline{PT} \cong \overline{ST}$ (converse of the Isosceles Triangle Theorem). $\angle 3 \cong \angle 4$ (given). $\triangle PTR \cong \triangle STQ$ (AAS Theorem), so $\overline{PR} \cong \overline{SQ}$ (def. of congruent triangles). Therefore, $\overline{PQ} \cong \overline{RS}$ (Segment Addition Postulate and Subtraction Property). **25.** never

CHAPTER 7

Page 342 Checking Key Concepts

1. Answers may vary. An example is given.

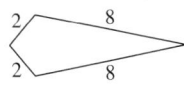

3. a. 7.5 **b.** 15

Pages 343–345 Exercises and Applications

1. $AD = 4$; $CD = 5$ **3. a.** $\overline{WX} \cong \overline{WZ}$ because they are radii of the same circle. Similarly, $\overline{YX} \cong \overline{YZ}$. **b.** Draw two circles with centers A and C and different radii that intersect in two points, B and D. Draw $\overline{AB}$, $\overline{BC}$, $\overline{CD}$, and $\overline{AD}$. $ABCD$ has two pairs of congruent sides, but opposite sides are not

congruent. Two adjacent sides are radii of circle A and two are radii of circle C and A and C have different radii. *ABCD* is a kite.   **5.** If two rectangles are congruent, then their diagonals are congruent. (Given rectangle *PQRS* ≅ rectangle *TUVW*, if you draw diagonals $\overline{PR}$ and $\overline{TV}$, $\triangle RQP \cong \triangle VUT$ by the SAS Postulate.) Then since the diagonals of any rectangle are congruent, $\overline{BD}$, $\overline{BF}$, $\overline{HD}$, and $\overline{HF}$ are congruent and *DBFH* is equilateral and, therefore, a rhombus.
7. a. $\sqrt{89} \approx 9.4$ **b.** 5 **9. a.** 3 **b.** 6
11. Answers may vary. An example is given.
Given: *ABCD* is a rhombus.
Prove: $\overline{AC} \perp \overline{BD}$

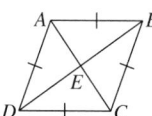

ABCD is a rhombus, so *ABCD* is a parallelogram. Then $\overline{EA} \cong \overline{EC}$ since the diagonals of a parallelogram bisect each other. $\triangle ABE \cong \triangle CBE$ by the SSS Postulate. $\triangle AEB \cong \triangle CEB$, so $\overline{AC}$ and $\overline{BD}$ form congruent adjacent angles; therefore, $\overline{AC} \perp \overline{BD}$.
13. Construct a line through A parallel to $\overline{BC}$ and a line through C parallel to $\overline{AB}$. (Through a point not on a line, there is exactly one line parallel to the given line.) Label E, the point where the constructed lines intersect. *ABCE* is a rectangle. Since *ABCE* is a parallelogram, its diagonals bisect each other. Since *ABCE* is a rectangle, its diagonals are congruent. Then $AD = BD = CD$.   **15.** 8 **17.** 5
19. $\sqrt{73} \approx 8.5$ **31. a.** 9 **b.** 75° **33. a.** 90° **b.** 28°

Page 349 Checking Key Concepts

1. *VWXY* is a parallelogram because $\overline{XY}$ and $\overline{WV}$ are both parallel and congruent. **a.** 14 **b.** 6 **c.** 10.3 **3.** Answers may vary. An example is given. Let *PQRS* be a parallelogram and point T be the intersection of the diagonals. Since the diagonals bisect each other and vertical angles are congruent, $\triangle PTQ \cong \triangle RTS$ and $\triangle PTS \cong \triangle RTQ$ (SAS Postulate). Then $\overline{PQ} \cong \overline{SR}$ and $\overline{PS} \cong \overline{QR}$, so *PQRS* is a parallelogram.

Pages 349–352 Exercises and Applications

1. *ABCD* is a parallelogram because both pairs of opposite angles are congruent. **a.** 130° **b.** 3 **c.** 5
3. 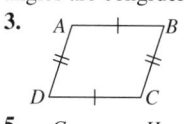 *ABCD* must be a parallelogram because both pairs of opposite sides are congruent.

5. 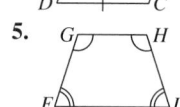 *FGHI* may or may not be a parallelogram. In the example, *FGHI* is not a parallelogram.

17, 19. Answers may vary. Examples are given.
17. 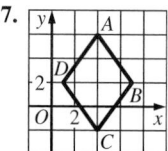 (4, 2) is the midpoint of both $\overline{AC}$ and $\overline{DB}$. *ABCD* is a parallelogram because its diagonals bisect each other.

19. 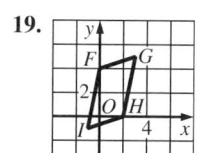 slope of $\overline{FG}$ = slope of $\overline{IH} = \frac{1}{3}$;
$FG = IH = \sqrt{10}$; *FGHI* is a parallelogram because one pair of sides, $\overline{FG}$ and $\overline{IH}$, are both congruent and parallel.

21.  Given: *JKLM*; $\angle J \cong \angle L$, $\angle K \cong \angle M$
Prove: *JKLM* is a parallelogram.

Statements	Reasons
1. $\angle J \cong \angle L$ and $\angle K \cong \angle M$	1. Given
2. $m \angle J = m \angle L$ and $m \angle K = m \angle M$	2. Def. of ≅
3. $m \angle J + m \angle L + m \angle K + m \angle M = 360°$	3. The sum of the angle measures of a quad. is 360°.
4. $2m \angle J + 2m \angle K = 360°$ $2m \angle L + 2m \angle M = 360°$	4. Addition Property
5. $m \angle J + m \angle K = 180°$ $m \angle L + m \angle M = 180°$	5. Algebra
6. $\overline{JK} \parallel \overline{ML}$ and $\overline{KL} \parallel \overline{JM}$	6. If two lines are intersected by a transversal and the same-side interior ∠s are supplementary, then the lines are ∥.
7. *JKLM* is a parallelogram.	7. Def. of parallelogram

23. *JKLM* is a kite with $\overline{JK} \cong \overline{JM}$. Suppose that $\overline{JK}$ is congruent to $\overline{KL}$ or to $\overline{ML}$ as well. Both possibilities present a contradiction. If $\overline{JK} \cong \overline{ML}$, then two opposite sides of *JKLM* are congruent and *JKLM* is not a kite. Similarly, if $\overline{JK} \cong \overline{KL}$, then $\overline{JM} \cong \overline{KL}$ and *JKLM* is not a kite. Therefore, $\overline{JK}$ is congruent to exactly one other side of the kite.
27. a. $(c, 0)$ **b.** b **c.** $a + c$ **d.** If one pair of sides of a quadrilateral is both parallel and congruent, the quadrilateral is a parallelogram. $\overline{MN}$ and $\overline{OP}$ are both parallel and congruent. **31.** $12\sqrt{2} \approx 17.0$ **33.** 6; $(-1, 4)$
35. $\sqrt{13} \approx 3.6$; $\left(3, 2\frac{1}{2}\right)$ **37.** $\sqrt{73} \approx 8.5$; $\left(0, -\frac{1}{2}\right)$

Page 355 Checking Key Concepts

1. parallelogram **3.** square

Pages 356–359 Exercises and Applications

1. The quadrilaterals are not necessarily parallelograms.
3. square **5.** rhombus **7.** rectangle **9.** rectangle
15, 17, 19. Answers may vary. Numerical examples should follow the given form, where each variable represents a positive number. **15.** $(0, a)$, $(0, -a)$, $(-b, 0)$, $(b, 0)$ $(a \neq b)$
17. If the diagonals of a rectangle are perpendicular, the rectangle must be a square. **19.** If the diagonals of a parallelogram are perpendicular, the parallelogram must be a rhombus. See Exs. 15 and 16. (The rhombus may or may not be a square.)

21. Given: quadrilateral $ABCD$; $\overline{AC}$ is the perpendicular bisector of $\overline{DB}$; $\overline{DB}$ is not the perpendicular bisector of $\overline{AC}$.

Prove: $ABCD$ is a kite.

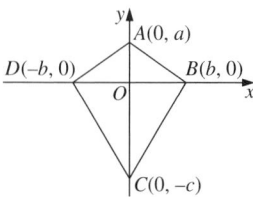

$AD = \sqrt{(-b-0)^2 + (0-a)^2} = \sqrt{b^2 + a^2}$;
$AB = \sqrt{(b-0)^2 + (0-a)^2} = \sqrt{b^2 + a^2}$;
$CD = \sqrt{(-b-0)^2 + (0-(-c))^2} = \sqrt{b^2 + c^2}$;
$CB = \sqrt{(-b-0)^2 + (0-(-c))^2} = \sqrt{b^2 + c^2}$;
Since $\overline{DB}$ is not the perpendicular bisector of $\overline{AC}$, $|a| \neq |c|$, so $a^2 \neq c^2$ and $\sqrt{b^2 + a^2} \neq \sqrt{b^2 + c^2}$. Then $ABCD$ has two pair of congruent sides but opposite sides are not congruent. $ABCD$ is a kite.   **27.** $\overline{AC} \cong \overline{DB}$ (They are both diameters of the circle.)

29. a.

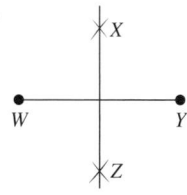

Draw a segment $\overline{WY}$. When you construct its perpendicular bisector, let X and Y be the points where the two pairs of arcs intersect. $\overline{WY}$ and $\overline{XZ}$ are perpendicular bisectors of each other. (You may choose different points X and Y by drawing any two arcs with the same radius and center at the midpoint of $\overline{WY}$.)
b. rhombus; The diagonals bisect each other, so $WXYZ$ is a parallelogram. The diagonals are perpendicular, so $WXYZ$ is a rhombus.   **35.** $50°$ **37.** $y = -3x + 20$

Page 359   Assess Your Progress

1. a. 5   **b.** 10   **c.** $2\sqrt{21} \approx 9.2$ **2. a.** 15   **b.** 9   **c.** 18
3. a. 8   **b.** 16   **c.** $8\sqrt{2} \approx 11.3$ **4.** Suppose that $\angle W \cong \angle Y$. Since it is given that $\overline{WZ} \cong \overline{WX}$ and $\overline{YZ} \cong \overline{YX}$, and $\overline{WY} \cong \overline{WY}$ by the Reflexive Property, $\triangle WZY \cong \triangle WXY$ by the SSS Postulate. Then $\angle Z \cong \angle X$. However, if $\angle W \cong \angle Y$, then $WXYZ$ is a parallelogram since both pairs of opposite angles are congruent. This is a contradiction, since opposite sides of a kite are not parallel. The assumption that $\angle W \cong \angle Y$ must not be true. So $\angle W \not\cong \angle Y$.   **5–7.** Answers may vary. Examples are given.   **5.** rhombus; Both pairs of opposite sides are parallel, so the quadrilateral is a parallelogram by definition; it is a rhombus because the diagonals are perpendicular.   **6.** rectangle; The diagonals bisect each other, so the quadrilateral is a parallelogram; it is a rectangle because the diagonals are congruent.   **7.** square; It is given that the quadrilateral is equiangular. Then both pairs of opposite angles are congruent and the quadrilateral is an equiangular parallelogram, that is, a rectangle. Since two consecutive sides are congruent, it is a rhombus. A rectangular rhombus is a square.

1. 12   **3.** 64   **5.** 234 ft^2 **7.** $56\sqrt{11} \approx 185.7$ in.^{2}

Pages 364–366   Exercises and Applications

1. 180.5   **3.** 16   **5.** 36   **11.** 18   **13.** 60 in.^{2}
15. $m = \frac{1}{2}(b_1 + b_2)$ **17.** rectangle; 21   **19.** triangle; 28
21. parallelogram; 15   **23.** If a rectangle and a parallelogram that is not a rectangle have the same base and height, they have the same area. To see why, picture cutting the parallelogram along the perpendicular segment and positioning the resulting triangle as shown.

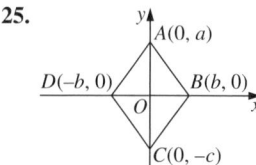

25.

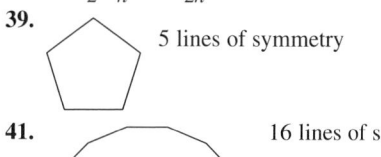

$ABCD$ is a rhombus; it is equilateral and $\overline{AC}$ and $\overline{BD}$ are perpendicular and bisect each other. Then $\overline{AC}$ and $\overline{BD}$ divide the rhombus into four congruent right triangles, each with area $\frac{1}{2}ab$. The area of the rhombus is $2ab$.

$AC = \sqrt{(0-0)^2 + (a-(-a))^2} = \sqrt{4a^2} = 2a$,
$BD = \sqrt{(b-(-b))^2 + (0-0)^2} = \sqrt{4b^2} = 2b$,
so $\frac{1}{2}AC \cdot BD = \frac{1}{2}(2a)(2b) = 2ab$. Then the area of the rhombus is half the product of the lengths of the diagonals.

27. a. 48; 48   **b.** Let n be any positive integer. To divide a triangle into n smaller triangles whose areas are equal, divide the base of the triangle into n congruent segments. The triangles all have the same height, h, as the original triangle. The base of each is $\frac{1}{n}$ times that of the original triangle, so each has area $\frac{1}{2} \cdot \frac{b}{n} \cdot h = \frac{bh}{2n}$.   **37.** rhombus

39.

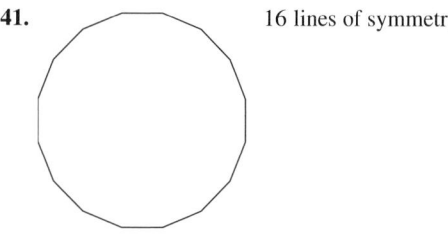

5 lines of symmetry

41.

16 lines of symmetry

Page 370   Checking Key Concepts

1. 36   **3.** $65\sqrt{90.21} \approx 617.54$ **5.** $\frac{169\pi}{4} \approx 133$

Pages 371–373   Exercises and Applications

1. 480   **3.** $98\sqrt{3} \approx 170$ **5.** $121\pi \approx 380$
7. $\frac{25\pi}{2} - 25 \approx 14$ **9.** $7\pi \approx 22$

15. Each side was constructed using the same compass radius. **17.** Connect only vertices A, C, and E or only vertices B, D, and F. To show that $\triangle ACE$ is equilateral, draw $\overline{AC}$, $\overline{AE}$, and $\overline{EC}$. Since $ABCDEF$ is a regular hexagon, $\triangle FEA$, $\triangle BAC$, and $\triangle DCE$ are congruent by the SAS Postulate, and $\overline{AC} \cong \overline{AE} \cong \overline{EC}$. **21.** about 44.7 **23.** 359.0 **29.** 49 **31, 33.** Answers may vary. Examples are given.

31. 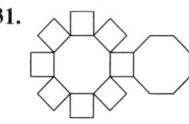 **33.**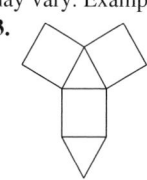

Page 378 **Checking Key Concepts**

1. 147; 182 **3.** $576\sqrt{3} \approx 997.7$; $192\sqrt{3} + 288 \approx 620.6$

5. 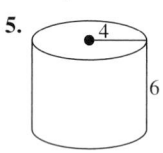 $96\pi \approx 301.6$; $80\pi \approx 251.3$

Pages 378–381 **Exercises and Applications**

1. $\frac{891\pi}{4} \approx 699.8$; $\frac{279\pi}{2} \approx 438.3$ **3.** 307.2; 281.6

5. 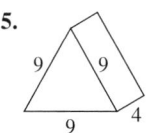 $81\sqrt{3} \approx 140.3$; $\frac{81\sqrt{3} + 216}{2} \approx 178.1$

9. $\frac{45\pi}{2} \approx 70.7$ m³ **11.** 60 in.³

21. $V = 21$ **23.** $V = 192$

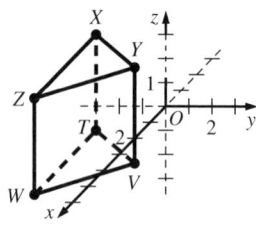 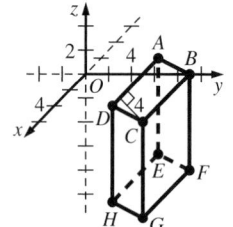

25. 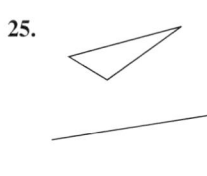 **27.**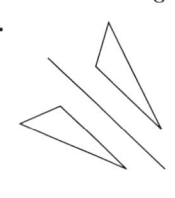

Page 381 **Assess Your Progress**

1. $2\sqrt{21} \approx 9.2$ **2.** 32.5 **3.** $18\sqrt{10} \approx 56.9$
4. $54\sqrt{3} \approx 93.5$ **5.** $\frac{65\sqrt{78.75}}{2} \approx 288.4$ **6.** $64\pi \approx 201.1$
7. $128\pi \approx 402.1$; $96\pi \approx 301.6$ **8.** 1750; 990
9. $240\sqrt{5} \approx 536.7$; $92\sqrt{5} + 300 \approx 505.7$

Pages 386–387 **Chapter 7 Assessment**

2. indirect **3.** base **5. a.** 2 **b.** 100° **6. a.** 10 **b.** 8
7. a. 6 **b.** $3\sqrt{2}$

8. 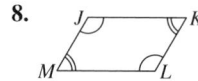 If both pairs of opposite angles of a quadrilateral are congruent, the quadrilateral is a parallelogram.

9. 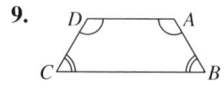 not necessarily

10. 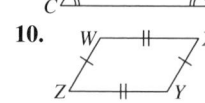 If both pairs of opposite sides of a quadrilateral are congruent, the quadrilateral is a parallelogram.

11. Answers may vary. An example is given. Suppose that a kite is a parallelogram. Then both pairs of opposite sides are congruent. But opposite sides of a kite are not congruent, so a kite is not a parallelogram. **12.** rectangle **13.** square
14. rhombus

15.

Statements	Reasons
1. $\overline{AB} \parallel \overline{DC}$; $\overline{AB} \cong \overline{DC}$	1. Given
2. $ABCD$ is a parallelogram.	2. If one pair of opposite sides of a quadrilateral is both parallel and congruent, the quadrilateral is a parallelogram.
3. $\overline{AC}$ and $\overline{BD}$ bisect each other.	3. The diagonals of a parallelogram bisect each other.
4. $\overline{EA} \cong \overline{EC}$; $\overline{EB} \cong \overline{ED}$	4. Def. of bisector
5. $\angle AED \cong \angle CEB$	5. Vertical $\angle$s are $\cong$.
6. $\triangle AED \cong \triangle CEB$	6. SAS Postulate

16. 28 **17.** 16 **18.** 36 **19–24.** Answers are given to the nearest tenth. **19.** 158.7 **20.** 8.8 **21.** 160.2
22. a. 51.8 ft² **b.** 1021.7 ft³ **23. a.** 21 cm³ **b.** 42.3 cm²
24. a. 332.6 in.³ **b.** 192 in.²

CHAPTER 8

Page 394 **Checking Key Concepts**
1. $\overline{P'Q'}$; $\overline{R'S'}$ **3.** $\triangle S'P'Q'$ **5.** quadrilateral $PQRS$

Pages 394–397 **Exercises and Applications**

1. 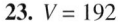 **3.**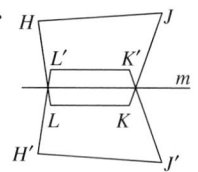

5.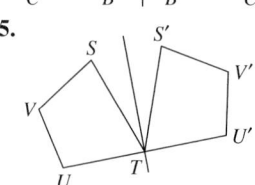

7. $a = 90$; $b = 3$; $c = 4$; $d = 2\frac{1}{2}$ **9.** $x = 30$; $y = 100$

15. H, O, T, U, V, W, X, Y; Answers may vary. Examples are given.

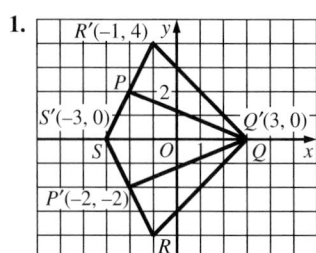

17. F, G, J, L, N, P, Q, R, S, Z   **27.** $18\pi \approx 56.5$
29. $\triangle DCE$ **31.** $\triangle GLF$ **33.** $y = 2x - 2$ **35.** $y = 7$

Page 401   Checking Key Concepts

1.

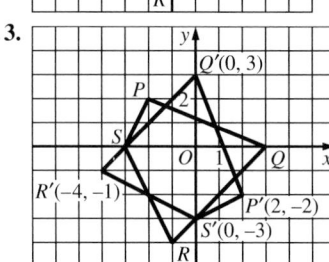

3.

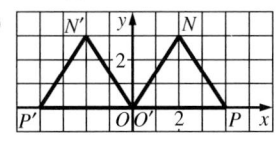

Pages 401–404   Exercises and Applications

1. $N'(-2, 3)$; $O'(0, 0)$; $P'(-4, 0)$

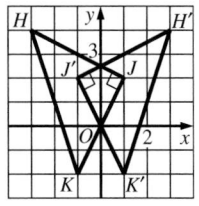

3. $H'(3, 4)$; $J'(-1, 2)$; $K'(1, -2)$

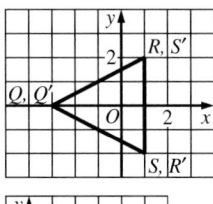

5. $Q'(-3, 0)$; $R'(1, -2)$; $S'(1, 2)$

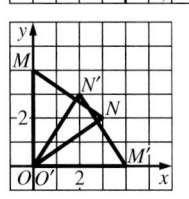

7. $M'(4, 0)$; $N'(2, 3)$; $O'(0, 0)$

9. $D'(2, 0)$; $E'(4, 2)$; $F'(4, 4)$; $G'(-1, 1)$

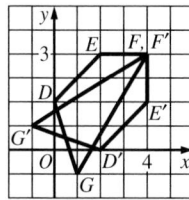

17. $(8, 0)$

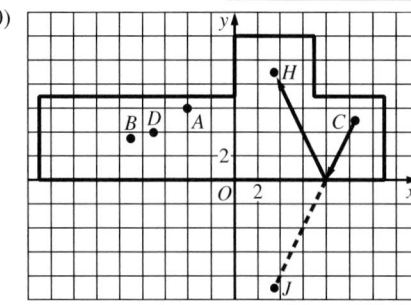

23.

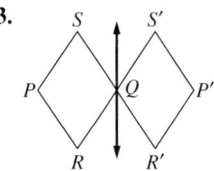

25.

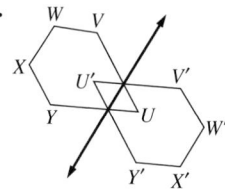

29, 31. Answers may vary. Examples are given.
29.

31.

Page 407   Checking Key Concepts

1.

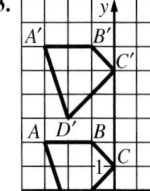

3.

5. $(a, b) \to (a - 3, b + 4)$

Pages 408–411   Exercises and Applications

1.

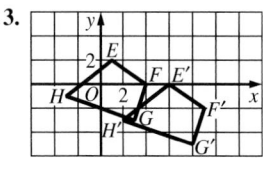

3.

5. $(a, b) \to (a - 7, b + 7)$ **7.** $(a, b) \to (a, b + 4)$ **9.** Every point moves to the right 3 units; $(a, b) \to (a + 3, b)$.
11. Every point moves up 5 units; $(a, b) \to (a, b + 5)$.
13. D   **17.** The top flight can be translated down by a distance equal to the distance between the floors.

19. Answers may vary. An example is given. lines *q* and *s*
25. a, b. Answers may vary. An example is given. △*ABC* has vertices *A*(0, 1), *B*(2, 4), and *C*(5, 5). **a.** *A*′(3, 6); *B*′(5, 9); *C*′(8, 10) **b.** *A*″(−1, 8); *B*″(1, 11); *C*″(4, 12) **c.** (*a, b*) → (*a* − 1, *b* + 7) **27.** acute **29.** Answers may vary. An example is given.

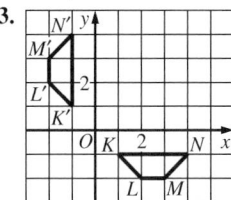

Page 411 Assess Your Progress

1. *L*′*M*′; *K*′*L*′ **2.** △*K*′*M*′*N*′
3.

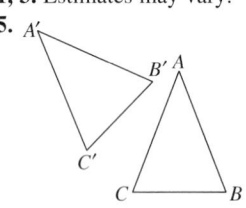

4.

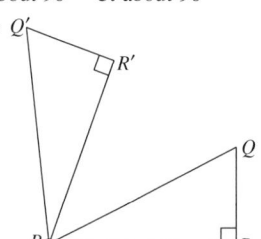

5. (*a, b*) → (*a* + 5, *b* − 7) **6.** (*a, b*) → (*a* − 3, *b* + 4)

Page 414 Checking Key Concepts

1. △*DEF* **3.** △*GHI* **5.** *D*(−2, −1); *E*(−4, 1); *F*(−1, 4)
7. *J*(2, 1); *K*(4, −1); *L*(1, −4)

Pages 415–418 Exercises and Applications

1, 3. Estimates may vary. **1.** about 90° **3.** about 90°
5.

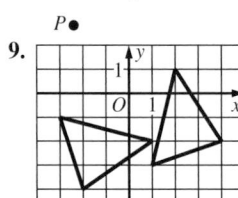

7.

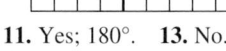

9.

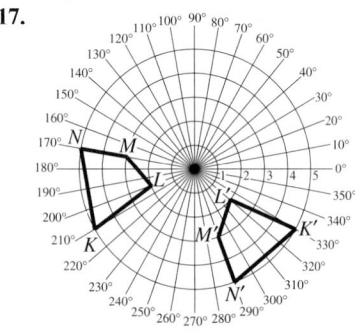

11. Yes; 180°. **13.** No.
17.

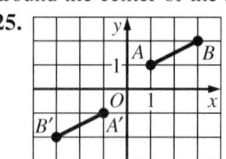 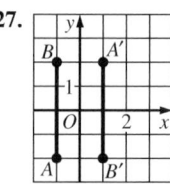

19, 21. The center of each rotation is the center of the square. **19.** 90° rotation **21.** no rotation
23. a–c. Answers may vary. Students may choose any two lines that intersect at an angle of 45°. Reflecting Triangle 1 over one of the lines and reflecting its image over the other has the same effect as rotating Triangle 1 either 90° or −90° around the center of the square.
25.

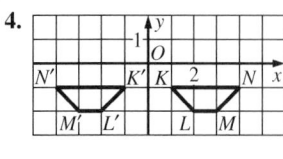

27.

29. a 180° rotation around the origin **31.** Yes; a 90° rotation of quadrant *D* has the same effect as a reflection of quadrant *D* over the *x*-axis. **35.** (*a, b*) → (*a* − 3, *b* + 6)
37. △*DFG* and △*EFH* **39.** △*ABH* and △*CBG*
41. (5, 0) **43.** $\left(0, 1\frac{1}{2}\right)$

Page 422 Checking Key Concepts

1. Yes. **3.**

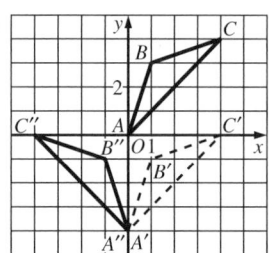

Pages 423–425 Exercises and Applications

1.

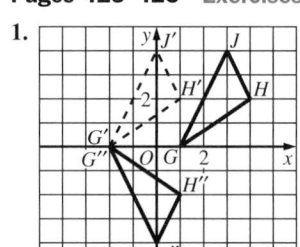

3.

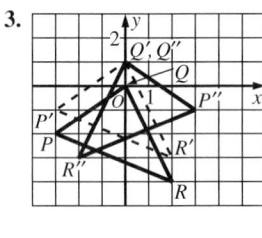

7, 9. Answers may vary. Examples are given.
7.

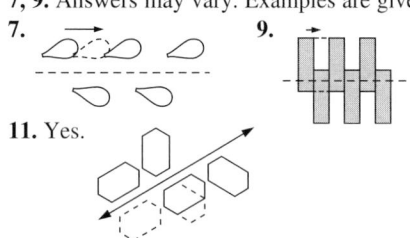

9.

11. Yes.

15. D **17.** generation 2; generation 3
21.

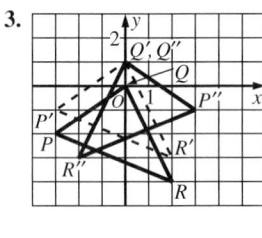

SA20 Selected Answers

A translation left 1 unit followed by a reflection over a horizontal line moves generation 0 to generation 2.

23. A glide reflection is a translation followed by a reflection over a line parallel to the translation. Then, if *j* and *k* are parallel lines and line *m* is perpendicular to both *j* and *k*, reflection over *j*, then *k*, then *m* has the same effect as a glide reflection.

25. translation length: 0.6 cm (6 mm)

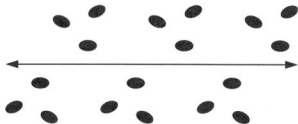

27. translation length: 11 in.

31.

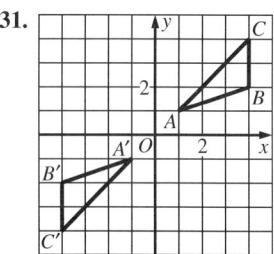

35. The third side is horizontal and is 4 units long. The other midpoint of the segment is (2, 2), so its length is 2, which is half the length of the third side. The segment is horizontal, so it is parallel to the third side.

Page 429   Checking Key Concepts

1.

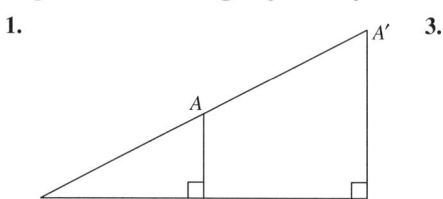

3. $B; \frac{1}{2}$

Pages 430–433   Exercises and Applications

1.

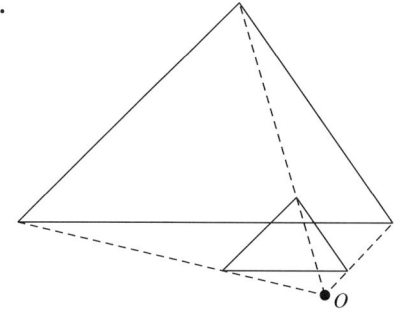

3.

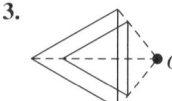

5.

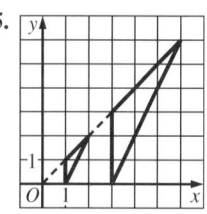

7. a.

Figure	Scale	Image
(2, 0)	3	(6, 0)
(3, 5)	3	(9, 15)
(−2, 4)	3	(−6, 12)

b. (ka, kb)

c.

Figure	Scale (Ex. 5)	Image	Figure	Scale (Ex. 6)	Image
(3, 0)	$\frac{1}{3}$	(1, 0)	(−1, −1)	4	(−4, −4))
(3, 3)	$\frac{1}{3}$	(1, 1)	(−2, 3)	4	(−8, 12)
(6, 6)	$\frac{1}{3}$	(2, 2)	(−4, −3)	4	(−16, −12)

11. $O; 4$ **13.**

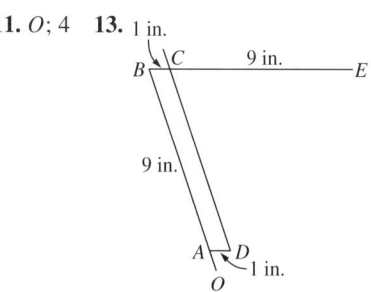

15. center: P; scale factor: $\frac{8}{5}$

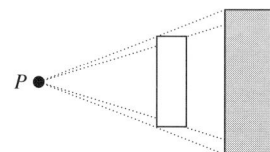

25. $XY = 2\sqrt{2} \approx 2.8; XZ = YZ = \sqrt{10} \approx 3.2$

Page 433   Assess Your Progress

1–4. The image of a point P after a 90° rotation around the origin is labeled P'. The image after a 180° rotation around the origin is labeled P''.

1.

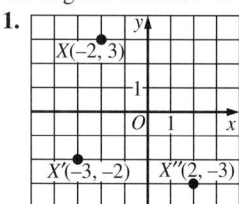

2.

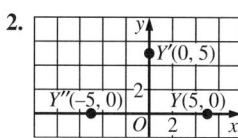

Selected Answers   **SA21**

SA21

3.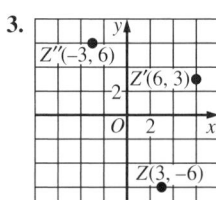

4.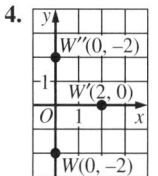

5.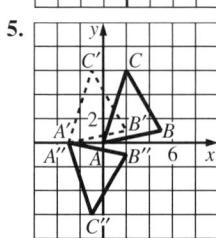

6.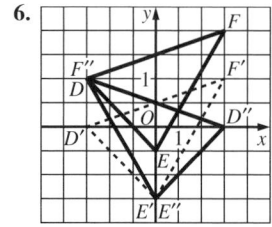

7.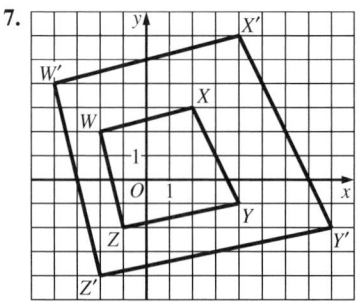

8.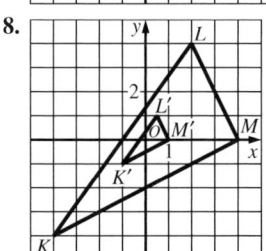

Pages 438–439 Chapter 8 Assessment

1–4.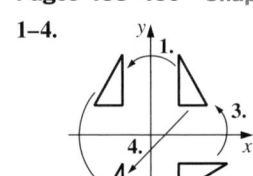

1. reflection
2. translation
3. rotation
4. glide reflection

5.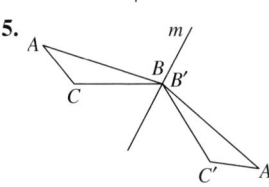

6.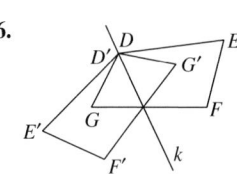

7.

8.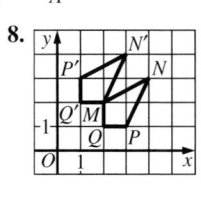

$M'(2, 2)$;
$N'(3, 4)$;
$P'(1, 3)$;
$Q'(1, 2)$

9. a. 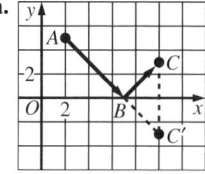 **b.** (7, 0)

10. $(a, b) \rightarrow (a + 3, b - 1)$ **11.** $(a, b) \rightarrow (a - 1, b - 1)$

12.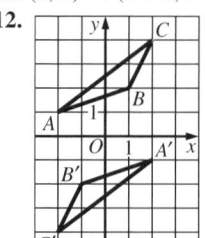

14. 6.2 cm

$D''(-1, 4)$; $E''(0, -1)$; $F''(3, -5)$

15.

16.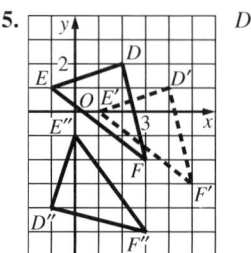

17.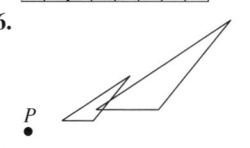

Pages 440–441 Algebra Review/Preview

1. cannot be simplified   **3.** cannot be simplified

5. $7n^2 - 3n$ **7.** $\frac{13}{12}x^2$ **9.** $-1.4x^2$ **11.** $7\sqrt{6}$ **13.** 37

15. $\frac{3\sqrt{5}}{4}$ **17.** 11   **19.** 105   **21.** $36\sqrt{2}$ **23.** ± 10

25. $\pm 3\sqrt{6}$ **27.** $\pm\sqrt{13}$ **29.** $\pm\sqrt{21}$ **31.** $\pm 3\sqrt{2}$ **33.** 4

35. 12   **37.** 20.5   **39.** 15   **41.** $\pm\sqrt{15}$ **43.** ± 21 **45.** ± 6

CHAPTER 9

Page 449 Checking Key Concepts

1. Yes; corresponding angles are congruent.
(The measure of the third angle of the larger triangle is
$180° - (112° + 30°) = 38°$.) Lengths of corresponding sides
are in proportion. $\frac{12}{8} = \frac{15}{10} = \frac{22.5}{15} = \frac{3}{2}$ **3.** $\triangle TUV \sim \triangle CAB$;
$\frac{TU}{CA} = \frac{UV}{AB} = \frac{TV}{CB}$ **5.** $\frac{1}{4}$

1. 8   **3.** 2   **5.** 4   **7.** No; since *PQRS* is a parallelogram, $m\angle P = 120°$ and $m\angle S = m\angle Q = 60°$. Similarly, since *TUVW* is a parallelogram, $m\angle U = 45°$ and $m\angle T = m\angle V = 135°$. Then no two angles of the figures are congruent.   **9.** Yes; $\angle AEB \cong \angle CED$ since vertical angles are congruent. Since $\overleftrightarrow{AB} \parallel \overleftrightarrow{DC}$, $\angle B \cong \angle D$ and $\angle A \cong \angle C$. So corresponding angles are congruent. Also, $\dfrac{AE}{CE} = \dfrac{EB}{ED} = \dfrac{AB}{CD} = \dfrac{3}{4}$.   **11.** $x = 10.4$; $y = 90$

13. always

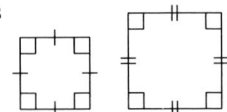

15. sometimes

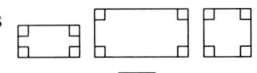

17. sometimes

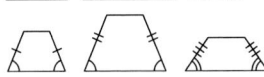

19. $11 : 1{,}400{,}000{,}000$ **23. a.** $1 : 1200$; CN Tower: 18.15 in.; TMG Offices: 7.93 in.; Washington Monument: 5.55 in.; Empire State Building: 12.5 in.   **27.** D

31. $1 : 35\frac{3}{16}$; Workers can multiply any length on the model by $35\frac{3}{16}$ to find the appropriate length on the monument.

33. none   **35.** none

Page 456   Checking Key Concepts

1. by the SAS Similarity Theorem; Vertical angles are congruent and $\dfrac{7}{10} = \dfrac{14}{20}$.   **3.** Yes; since the ratio of the lengths of any two corresponding sides is $\dfrac{5}{8}$, the triangles are similar by the SSS Similarity Theorem.   **5.** They are similar.   **7.** They are similar by the SSS Similarity Theorem.

Pages 456–457   Exercises and Applications

1. Yes; the triangles are similar by the SAS Similarity Theorem since the two vertical angles are congruent and the sides including the angles are in proportion.   **3.** Yes; it can be shown the triangles are similar by the AA Similarity Postulate. Because the two lines are parallel, it can be shown that two angles of one triangle are congruent to two angles of the other triangle.   **5.** by the SAS Similarity Theorem; The sides including the congruent angles are in proportion; $\dfrac{20}{32} = \dfrac{15}{24} = \dfrac{5}{8}$.   **7.** Since $\overline{BC} \parallel \overline{ED}$ and $\overline{AB} \parallel \overline{DC}$, $\angle BCA \cong \angle DEC$ and $\angle A \cong \angle ECD$. (If two $\parallel$ lines are intersected by a transversal, then alternate interior $\measuredangle$ are $\cong$.) Then $\triangle ABC \sim \triangle CDE$ by the AA Similarity Postulate.   **15. a.** Since the right angles are congruent and the sides including the right angles are in proportion $\left(\dfrac{3}{6} = \dfrac{4}{8} = \dfrac{1}{2}\right)$, the triangles are similar. The ratio of the lengths of corresponding sides is $\dfrac{1}{2}$, so $\dfrac{5}{c} = \dfrac{1}{2}$ and $c = 10$.   **b.** $d = 50$; $e = 20$; $f = 24$

17. Answers may vary. An example is given.

Statements	Reasons
1. $\overline{AB} \perp \overline{AE}$ and $\overline{ED} \perp \overline{AE}$.	1. Given
2. $\overline{AB} \parallel \overline{ED}$	2. In a plane, two lines $\perp$ to the same line are $\parallel$.
3. $\angle A \cong \angle E$ and $\angle B \cong \angle D$.	3. If two $\parallel$ lines are intersected by a transversal, alternate interior $\measuredangle$ are $\cong$.
4. $\triangle ABC \sim \triangle EDC$	4. AA Similarity Postulate

25. Sketches may vary. Examples are given.
a. not necessarily

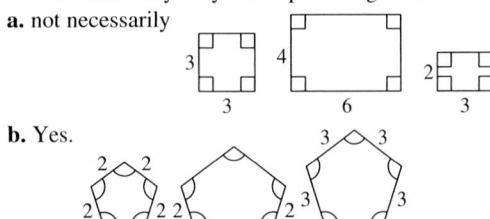

b. Yes.

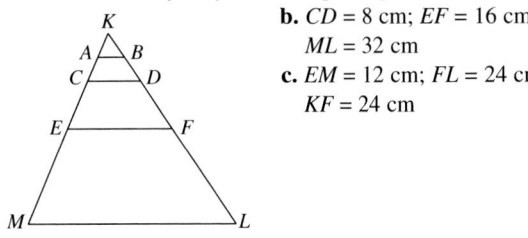

29. Yes; corresponding angles are congruent and the lengths of corresponding sides are in proportion.   **31.** No; $\dfrac{4}{6} \neq \dfrac{6}{10}$.

Page 464   Checking Key Concepts

1. True.   **3.** True.   **5. a.** 1   **b.** 3   **c.** 4

Pages 464–467   Exercises and Applications

1. $x = 3\dfrac{3}{4}$ **3.** $z = 2\dfrac{2}{9}$ **5.** $\dfrac{3}{x}$ **7.** 2 cm; 2.25 cm; 1.5 cm

9. a. Answers may vary. An example is given.

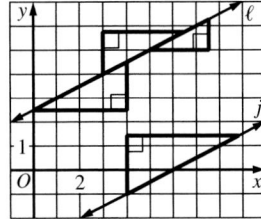

b. $CD = 8$ cm; $EF = 16$ cm; $ML = 32$ cm
c. $EM = 12$ cm; $FL = 24$ cm; $KF = 24$ cm

15. $\dfrac{AB}{FE}$; $\dfrac{AC}{FD}$

17. Answers may vary. An example is given.

a. In each pair of triangles, each leg of one triangle is parallel to a leg of the other. The parallel lines can be used to show that two angles of one triangle are congruent to two angles of the other. Then the triangles are similar by the AA Similarity Postulate. (Alternatively, the parallel lines can be used to show that one angle of one triangle is congruent to one angle of the other. Then, since one angle of each triangle is a right angle, the triangles are similar by the AA Similarity Postulate.) **b.** Yes; the reasoning is similar to that described in part (a). However, additional parallel lines (horizontals and verticals) may be needed to show the triangles are similar. **19.** Answers may vary. An example is given.

If $\frac{a}{b} = \frac{c}{d}$, then $1 \div \frac{a}{b} = 1 \div \frac{c}{d}$ or $\frac{b}{a} = \frac{d}{c}$. **21.** C **25.** 104 ft^2 **27.** 60 in.^{2} **29.** 0.001 **31.** 0.04

Page 467 Assess Your Progress

1. 10 **2.** 4.2 **3.** 8 **4.** No. **5.** Yes. **6.** $x = 29$; $y = 18\frac{1}{3}$
7. $x = 2\frac{2}{3}$; $y = 5\frac{7}{9}$ **8.** $\frac{5}{2}$ **9.** $\frac{5}{y}$ **10.** $\frac{2+x}{x}$

Page 471 Checking Key Concepts

1. $\frac{4}{3}$ **3.** $\frac{4}{3}$ **5.** 250

Pages 471–474 Exercises and Applications

1. small triangle: $A = 25.2$; large triangle: $P = 37.8$; $A \approx 56.7$ **3.** $EF = 6$; $A = 63$ **5.** 9 **7. a.** $3:5$; $9:25$ **b.** $3\sqrt{15}$; $5\sqrt{15}$ **9.** not similar **11.** $\frac{13}{6}$; $\frac{13}{6}$; $\frac{169}{368}$ **13.** $2:5$; $4:25$ **15.** doubled; doubled; quadrupled **21. a.** No; the ratios of the lengths of corresponding sides are not proportional. **b.** The front faces of the boxes are what you see when you look at the shelf. If they are similar, you might make assumptions about the volumes of the boxes that are not necessarily true. **25.** The length of a side of the image square is kx. Its area is $(kx)^2 = k^2x^2$.
29.

Statements	Reasons
1. $\overline{AB} \parallel \overline{CD}$	1. Given
2. $\angle ECD \cong \angle EAB$; $\angle ABE \cong \angle EDC$	2. If two $\parallel$ lines are intersected by a transversal, then alternate interior $\angle$s are $\cong$.
3. $\triangle ABE \sim \triangle CDE$	3. AA Similarity Postulate

31. $\frac{32}{43} \approx 74\%$ **33.** $114\frac{2}{7}\% \approx 114.3\%$ **35.** 375% **37.** $\frac{1}{200}$
39. $\frac{19}{20}$

Page 478 Checking Key Concepts

1. $\frac{16}{49}$ **3.** $\frac{w}{z}$ **5.** $\frac{1}{2}$

Pages 478–481 Exercises and Applications

1. $\frac{6}{25}$ **3.** $\frac{4-\pi}{4} \approx 21.5\%$ **5.** $\frac{1}{2}$ **9.** $\frac{4}{23}$ **11. a.** $\frac{25\pi}{1296} \approx 6\%$
b. $\frac{125\pi}{1296} \approx 30\%$ **19.** B **23.** $\frac{x+y}{y}$ **25.** right **27.** obtuse

Page 481 Assess Your Progress

1. large trapezoid: $A = 102$, $P = 49$; small trapezoid: $A = 18\frac{36}{49}$ **2.** $P = 324$; $A = 12{,}285$ **3.** $\frac{3}{5}$ **4.** $\frac{2}{3}$ **5.** $\frac{4}{9}$ **6.** $\frac{4}{9}$

Pages 486–487 Chapter 9 Assessment

3. Yes; AA Postulate or SAS Similarity Theorem. **4.** No. **5.** Yes; SSS Similarity Theorem. **6.** Yes; SAS Similarity Theorem. **7.** $r = 14$, $s = 18$ **8.** $a = 4.5$, $b = 10$
9. $n = \frac{40}{9}$; $m = \frac{98}{9}$ **10.** $x = 5.2$; $y = 4.8$ **11.** 6.48
12. $A = 51.2$ m^2, $f = 3.6$ m **13.** $V \approx 202.2$ cm^3, $h = 12$ cm
14. a. 24 cm^2 **b.** 243 cm^3; $\frac{2^3}{3^3} = \frac{72}{x}$ **15.** $\frac{1}{4}$ **16.** $\frac{4}{49}$

CHAPTERS 7–9

Pages 488–489 Cumulative Assessment

1. Since $\overline{WX} \parallel \overline{YZ}$ and $\overline{WX} \cong \overline{YZ}$, $WXYZ$ must be a parallelogram. **3.** 60°; 120° **5.** $2\pi \approx 6.28$ **7.** 128
9. Answer is given to the nearest whole number.
$2880\pi \approx 9048$; $768\pi \approx 2413$ **13. a.** $D(2, 4)$, $E(-1, -3)$, $F(-2, 0)$ **b.** $D(4, -2)$, $E(-3, 1)$, $F(0, 2)$ **15.** $D(-6, -2)$, $E(1, 1)$, $F(-2, 2)$ **21.** Yes; SAS Similarity Theorem.
23. $3:5$ **25.** 37.5 square units **27.** about 0.59

CHAPTER 10

Page 496 Checking Key Concepts

1. $\angle HGF$ **3.** FH **5.** 15

Pages 497–499 Exercises and Applications

1. $\triangle JKL$, $\triangle JMK$, $\triangle KML$ **3.** 15 **5.** $6\sqrt{5}$ **7.** 1 **11.** 18
13. $x = 5\sqrt{5}$; $y = 10\sqrt{5}$ **15.** $u = 5.4$; $v = 9.6$; $w = 7.2$
17. $\triangle ACD \sim \triangle CBD$. (If the altitude is drawn to the hypotenuse of a right triangle, then the two triangles formed are similar to the original triangle and to each other.) Then, since corresponding sides of similar triangles are in proportion, $\frac{BD}{CD} = \frac{CD}{AD}$, or $\frac{e}{d} = \frac{d}{f}$. **19.** (2) If an altitude is drawn to the hypotenuse of a right triangle, then the length of the altitude is the geometric mean of the lengths of the hypotenuse and the segment of the hypotenuse adjacent to that leg. (3) a^2; b^2; a property of proportions (4) $a^2 + b^2$; Addition Property (5) $a^2 + b^2$; Distributive Property (6) c; Segment Addition Postulate (7) Substitution Property
27. $(3, -1)$ **29.** $\left(\frac{7}{2}, -\frac{1}{2}\right)$ **31.** $b = 4\sqrt{6} \approx 9.80$

Page 502 Checking Key Concepts

1. 3; $3\sqrt{2}$ **3.** 5; $5\sqrt{3}$ **5.** $4\sqrt{2}$; $4\sqrt{2}$

Pages 503–506 Exercises and Applications

1. 10 **3.** 4 **5.** $n = \frac{5}{\sqrt{3}}$; $m = \frac{10}{\sqrt{3}}$
11. about 13,317.16 mm^2 **13.** 11,032 mm^2 to the nearest square millimeter **15.** $6\sqrt{3} \approx 10.4$ in.; The height is the length of an altitude of an equilateral triangle. The altitude determines two 30-60-90 triangles with shorter leg 6 in. long. The altitude is the longer leg of each triangle.
17. about 18 ft 2 in. **19.** $x = 45$; $y = 12$ **21.** $h = 2\sqrt{2}$; $f = \sqrt{6}$; $g = \sqrt{3}$ **23.** C

25. a. $AB = \frac{1}{2}x$ because $\overline{AC}$ was folded in half; $AD = x$ because D was chosen so that $AD = AC$. **b.** 30-60-90; $AD = x$ and $AB = \frac{1}{2}x$, so by the Pythagorean theorem, $DB = \frac{\sqrt{3}}{2}x$. Then $\triangle ADB$ is a 30-60-90 triangle because it is similar to any 30-60-90 triangle by the SSS Similarity Theorem. **c.** $\overrightarrow{AE}$ bisects $\angle DAC$ as $\triangle EAD$ and $\triangle EAC$ are congruent. **d.** Given: $\triangle ADB$ obtained by paperfolding as described.

Prove: $m\angle EAC = 30°$

$\overline{AD} \cong \overline{AC}$ and $\overline{DB}$ bisects $\overline{AC}$, so $AB = \frac{1}{2}AD$. By the Pythagorean theorem, $DB = \frac{AB\sqrt{3}}{2}$, so $\triangle ADB$ is a 30-60-90 triangle with $m\angle DAC = 60°$. From the way the paper was folded, $\triangle EAD \cong \triangle EAC$. Thus, by the definition of congruent triangles, $\angle EAD \cong \angle EAC$. But $m\angle EAD + m\angle EAC = m\angle DAC$. So $m\angle EAC = \frac{1}{2}m\angle DAC = \frac{1}{2}(60°) = 30°$.

27. Given: Equilateral $\triangle ABC$;
$\overrightarrow{BD}$ bisects $\angle ABC$.
Prove: $AD = \frac{x}{2}$ and $BD = \frac{\sqrt{3}}{2}x$.

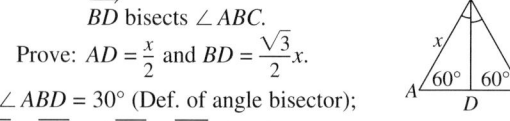

$m\angle ABD = 30°$ (Def. of angle bisector); $\overline{BD} \perp \overline{AC}$ and $AD = DC$ (the bisector of the vertex angle of an isosceles triangle is the perpendicular bisector of the base). Therefore, $\triangle BAD$ is a 30-60-90 triangle. Since $AB = AC$, $AD = \frac{1}{2}x$. By the Pythagorean theorem, $x^2 = \left(\frac{1}{2}x\right)^2 + (BD)^2$. So $(BD)^2 = x^2 - \frac{1}{4}x^2 = \frac{3}{4}x^2$. Thus, $BD = \frac{\sqrt{3}}{2}x$. **31.** AA Similarity Postulate **33.** $\frac{7}{4}$ **35.** $-\frac{4}{3}$

Page 506   Assess Your Progress

1. 9   **2.** $2\sqrt{10}$ **3.** $7\sqrt{15}$ **4.** $c = 7$; $d = 7\sqrt{5}$

5. $t = \frac{11}{\sqrt{2}}$; $s = \frac{22}{\sqrt{2}}$ **6.** 36   **7.** $n = 30$; $q = 5$; $p = 5\sqrt{3}$

8. $j = 9$; $k = \frac{9}{\sqrt{2}}$ **9.** $a = 2\sqrt{3}$; $b = 2\sqrt{3} - 2$; $c = 4$

Page 510   Checking Key Concepts

1. $\frac{1}{2\sqrt{2}}$; $2\sqrt{2}$ **3. a.** about 31,569 ft   **b.** about 7.4°

Pages 511–513   Exercises and Applications

1. $\frac{3}{4}$; $\frac{4}{3}$ **3.** $\frac{5}{4}$; $\frac{4}{5}$ **5.** 2.4751   **7.** 572.9572   **9.** 15.9°
11. 51.7°   **13.** 7.3   **15.** $s = 5.1$; $r = 40.4°$
19. $m\angle P = 18.9°$; $m\angle Q = 71.1°$ **21.** $m\angle Y = m\angle Z = 63.4°$; $m\angle X = 53.2°$ **25.** about 51.3 ft

27. In $\triangle ABC$, $0 < m\angle A < 45°$. Since $m\angle A + m\angle B = 90°$, $45° < m\angle B < 90°$. This means that $b > a$ because the side opposite the larger angle is longer than the side opposite the smaller angle. Since $\tan A = \frac{a}{b}$ and $\frac{a}{b} < 1$, $\tan A < 1$.
31. Yes.   **33.** No.

Page 517   Checking Key Concepts

1. $\frac{3}{\sqrt{13}} \approx 0.8321$; $\frac{2}{\sqrt{13}} \approx 0.5547$; 56°   **3.** about 234 ft

Pages 517–520   Exercises and Applications

1. $\sin A = \cos B = \frac{5}{13} \approx 0.3846$; $\cos A = \sin B = \frac{12}{13} \approx 0.9231$

3. $\sin A = \frac{5}{7} \approx 0.7143$; $\cos A = \frac{2\sqrt{6}}{7} \approx 0.6999$;

$\sin B = \frac{5}{\sqrt{34}} \approx 0.8575$; $\cos B = \frac{3}{5} = 0.6$

5. 0.0628   **7.** 0.3387   **9.** 51.9°   **11.** 16.3°   **13.** Answers are given to two decimal places. $r = 5.25$; $s = 7.85$

15. $m\angle A = 36.9°$; $m\angle B = 53.1°$ **17.** $m\angle X = m\angle Z = 50.0°$; $m\angle Y = 80°$ **21. a.** about 24.7 in.   **b.** $x \approx 35.2$ in.; $y \approx 30.8$ in.   **c.** about 39.5 in.   **23.** $\frac{1}{\sqrt{2}}$ **25.** $\frac{1}{2}$ **27.** $\frac{1}{2}$

29. a. $d = 300t$ **b.** $h = d\cos 15° = 300t\cos 15° \approx 289.78t$; $v = d\sin 15° = 300t\sin 15° \approx 77.65t$ **c.** about 2897.8 ft; about 776.5 ft   **31.** B   **33.** $(\sin A)^2 + (\cos A)^2 = \left(\frac{a}{c}\right)^2 + \left(\frac{b}{c}\right)^2 = \frac{a^2 + b^2}{c^2} = 1$ (By the Pythagorean theorem, $a^2 + b^2 = c^2$.)

37. 71.6°   **39.** 55.2°   **41.** $\left(\frac{3}{2}, 2\right)$ **43.** $(2, -2)$

Page 525   Checking Key Concepts

1–3. Answers may vary.

1. $|\overrightarrow{AB}| = 5$

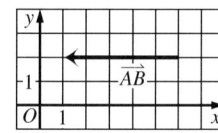

3. $|\overrightarrow{EF}| = 3\sqrt{5} \approx 6.7$

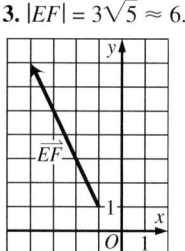

5. Answer is given to the nearest tenth. 56.3   **7.** $(-7, -14)$

Pages 525–528   Exercises and Applications

1, 3. Values of the variables are given to the nearest tenth.
1. $(-4, -2)$; 26.6   **3.** $(250, 100)$; 21.8
5. $|\overrightarrow{KL}| = \sqrt{73} \approx 8.5$

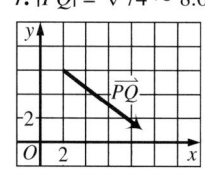

7. $|\overrightarrow{PQ}| = \sqrt{74} \approx 8.6$

9. $(2, 7)$ **11.** $(3, 10)$ **13.** $(7, 11)$ **15. a.** $(3600, -1100)$ **b.** $(3600, -1100)$ **c.** Both represent paths from checkpoint P to S; any path from checkpoint P to S can be represented by a vector with initial checkpoint P and terminal checkpoint S.
17. 91 ft

19. a. $\vec{AB} + \vec{CD} = (9, 0)$

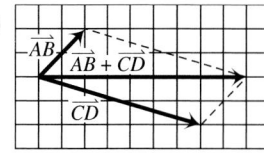

b. $\vec{AB} + \vec{CD} = (9, 0)$

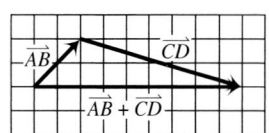

c. In the parallelogram method, the length of the side opposite $\vec{CD}$ equals $|\vec{CD}|$. Since $\vec{AB}$ is the same in both diagrams, the diagonal of the parallelogram has the same length as the third side of the triangle in the second diagram.

23. $a = 60\sqrt{3}$; $b = 60$; $(60\sqrt{3}, 60)$

25. $e = f = 2\sqrt{2}$; $(-2\sqrt{2}, -2\sqrt{2})$ **27. a.** about 219.2 mi/h
b. about 50.6 mi/h   **31.** 33.7°   **33.** 83.7°   **35.** 45.0°
37. $V = Bh$, where B is the area of a base and h is the height of the prism.

Page 528   Assess Your Progress

1. 85.8°   **2.** 11.5°   **3.** 41.4°   **4.** 49.5   **5.** 8.6   **6.** 9.1
7. $|\vec{PQ}| = \sqrt{58} \approx 7.6$ **8.** $|\vec{RS}| = 5\sqrt{89} \approx 47.2$

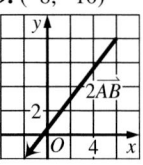
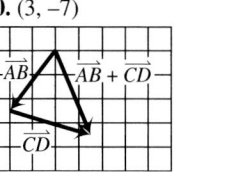

9. $(-8, -10)$ **10.** $(3, -7)$

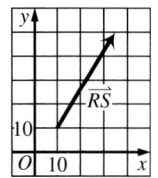

11. $(10, -1)$

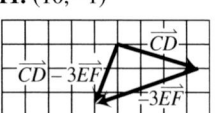

Page 531   Checking Key Concepts

1, 3, 5. Answers are given to the nearest tenth.   **1.** 43.2
3. 404.8   **5.** 232.0 in.^{2}

Pages 532–534   Exercises and Applications

Answers are given to the nearest tenth.   **1.** 33.5   **3.** 31.2
5. 110.0   **7.** 522 cm^2 **9.** 210 m^2 **11. a.** 6.4; 7.7   **b.** 13.7
c. 68.5

13. a. Answers may vary. An example is given.
b. about 0.025 in.^{2}

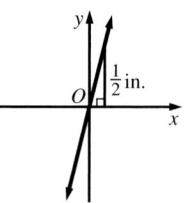

15. 138.9 m^3 **17.** 1211.2 cm^3

23. The area of $\triangle XYZ$ is $A = \frac{1}{2}(XZ)(YZ)$.

Since $\tan X = \dfrac{YZ}{XZ}$, $YZ = XZ \cdot \tan X$.

By substitution, $A = \frac{1}{2}(XZ)^2 \tan X$.

25. E   **29.** $\sqrt{2} \approx 1.4$ **31.** $432\pi \approx 1357.2$ cm^2

Page 538   Checking Key Concepts

1. 5.2; 14.0   **3.** 51.4 cm^3; 98.8 cm^2

Pages 539–541   Exercises and Applications

1. 37.7; 75.4   **3.** 314.2; 282.7   **5.** 593.8; 497.1
7, 9. Answers may vary. Examples are given.
7.   **9.**

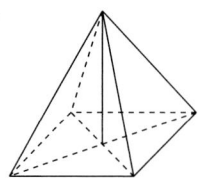
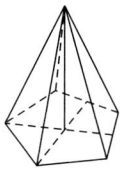

15. $r = 6$; $h = 12$; $s = 6\sqrt{5} \approx 13.4$ **21. a.** 8.0   **b.** 49.0

23. $|\vec{AB}| = 7$ **25.** $|\vec{EF}| = \sqrt{462,500} \approx 680.0$

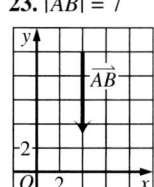
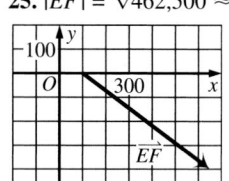

27. 100

Page 541   Assess Your Progress

1. 36.9   **2.** 186.2   **3.** 344.4   **4.** 615.8; 459.6   **5.** 326.7;
320.0   **6.** 5641.5; 2299.0

Pages 546–547   Chapter 10 Assessment

1. Answers may vary. Examples are given.

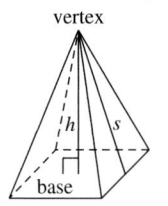
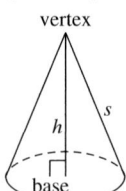

2. $\sin R = \dfrac{TS}{RS}$; $\cos R = \dfrac{RT}{RS}$; $\tan R = \dfrac{TS}{RT}$; $\sin S = \dfrac{RT}{RS}$; $\cos S = \dfrac{ST}{RS}$;

$\tan S = \dfrac{RT}{TS}$ **3.** Find the square root of their product.

4. No; a vector has both magnitude and direction.   **5.** An angle of elevation is formed with the horizontal and the line of sight to a point above the viewing point. An angle of depression is formed with the horizontal and the line of sight to a point below the viewing point.   **6.** $x = 2\sqrt{10}$; $y = 2\sqrt{14}$; $z = 2\sqrt{35}$ **7.** $m = \frac{121}{21}$; $r = \sqrt{\frac{68{,}002}{441}}$

8. $f = 4\sqrt{7}$; $g = 12$ **9.** $s = \frac{95}{7}$; $t = \sqrt{95}$ **10.** $a = 4$; $b = 4\sqrt{3}$; $c = 4\sqrt{3}$; $d = 4\sqrt{6}$ **11.** $x = 30$; $y = 6\sqrt{3}$

12. a. $\sqrt{\frac{5}{12}}$ **b.** $10\sqrt{10}$ **c.** 8   **13.** 128   **14.** $\frac{25}{4}\sqrt{3}$

15. $\frac{49}{4}\sqrt{3}$ **16.** 0.6101   **17.** 0.7923   **18.** 0.7701

19. about 1.4 ft   **20.** about 32.5 ft tall   **21.** $m\angle B \approx 97.2°$, $m\angle A = m\angle C \approx 41.4°$

22. 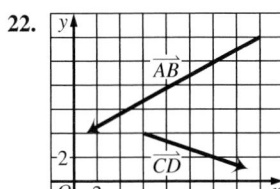 $|\vec{AB}| = 17$; $|\vec{CD}| = 3\sqrt{10}$

23. $(-6, -11)$ **25–29.** Answers may vary slightly due to rounding. Answers are given to the nearest tenth.
25. 296.1   **26.** 225.38 cm²   **27.** 1279.5; 2520   **28.** 3078.8; 8796.5   **29.** 562.9; 1299.0

Pages 548–549   Algebra Review/Preview

1. $2y^3$ **3.** $5\pi r^2$ **5.** $\frac{5}{2}s^2$ **7.** cannot be simplified

9. $-x^3 + x^2 - 2x$ **11.** 7×3 **13.** 23 losses   **15. a.** not possible   **b.** $\begin{bmatrix} 5 \\ -9 \\ 8 \end{bmatrix}$ **c.** not possible   **d.** $\begin{bmatrix} 6 & -2 \\ 3 & -4 \\ 0 & 4 \end{bmatrix}$

19. $\pm\sqrt{x^2 - r^2}$ **21.** 2   **23.** ± 6 **25.** 5   **27.** $\frac{3V}{\pi r^2}$

CHAPTER 11

Page 556   Checking Key Concepts

1. $\overline{AD}$ or $\overline{DC}$ **3.** $\angle ADC$ **5.** $\overset{\frown}{AC}$ and $\overset{\frown}{CDA}$, or $\overset{\frown}{AD}$ and $\overset{\frown}{ACD}$
7. 60°

Pages 557–559   Exercises and Applications

1. 50°   **3.** 25°   **5.** 310°   **7.** 136°; 104°   **9.** 62°; 127°
13. $x = 15$; $z = 16$; $(4x + 20)° = 80°$; $(10z + 120)° = 280°$
15. $x = 22$; $(x + 100)° = 122°$; $(5x + 10)° = 120°$; $(4x + 30)° = 118°$ **17.** 28.5   **19.** 50   **29.** Answers are given to the nearest tenth. 183.3; 213.7   **31.** $\frac{23}{36} \approx 63.9\%$

Page 562   Checking Key Concepts

1. $x = 40$; $y = 60$ **3.** $x = 55$; $y = 80$

Pages 563–566   Exercises and Applications

1, 3, 5. Answers may vary. Examples are given.   **1.** $\overline{DA}$
3. $\angle BFE$ **5.** 35   **7.** $y = 90$; $z = 20$ **9.** $x = 30$; $z = 160$

11.

Statements	Reasons
1. Secants $\overrightarrow{AP}$ and $\overrightarrow{CP}$ intersect as shown.	1. Given
2. Draw chord $\overline{BC}$.	2. For any two points, there is exactly one line through the two points.
3. $m\angle 3 = \frac{1}{2}m\overset{\frown}{AC}$; $m\angle 2 = \frac{1}{2}m\overset{\frown}{BD}$	3. The measure of an inscribed angle is half the measure of the intercepted arc.
4. $m\angle 3 = m\angle 2 + m\angle 1$	4. The measure of an exterior angle of a triangle is equal to the sum of the measures of the two interior angles that are not adjacent to it.
5. $m\angle 1 = m\angle 3 - m\angle 2$	5. Subtraction Property
6. $m\angle 1 = \frac{1}{2}m\overset{\frown}{AC} - \frac{1}{2}m\overset{\frown}{BD}$	6. Substitution Property (Steps 3 and 5)
7. $m\angle 1 = \frac{1}{2}(m\overset{\frown}{AC} - m\overset{\frown}{BD})$	7. Distributive Property
8. $m\angle 1 = \frac{m\overset{\frown}{AC} - m\overset{\frown}{BD}}{2}$	8. Algebra

13. 43°

15.

Statements	Reasons
1. Draw chord $\overline{AB}$.	1. For any two points, there is exactly one line through the two points.
2. $m\angle PAB = \frac{1}{2}m\overset{\frown}{AB}$	2. The measure of an angle formed by a tangent and a chord is half the measure of the intercepted arc.
3. $m\angle ABC = \frac{1}{2}m\overset{\frown}{AC}$	3. The measure of an inscribed angle is half the measure of the intercepted arc.
4. $m\angle ABC = m\angle PAB + m\angle P$	4. The measure of an exterior angle of a triangle is equal to the sum of the measures of the two interior angles that are not adjacent to it.
5. $m\angle P = m\angle ABC - m\angle PAB$	5. Subtraction Property
6. $m\angle P = \frac{1}{2}m\overset{\frown}{AC} - \frac{1}{2}m\overset{\frown}{AB}$	6. Substitution Property (Steps 2, 3, and 5)
7. $m\angle P = \frac{m\overset{\frown}{AC} - m\overset{\frown}{AB}}{2}$	7. Distributive Property

25. $(a, b) \rightarrow (a - 5, b + 2)$

27. Answers may vary. An example is given.

Page 566   Assess Your Progress

1. 228°   **2.** 43°   **3.** 73°   **4.** Each angle of the triangle is an inscribed angle with measure 60°. Then each of the arcs has measure 120°. (The measure of an inscribed angle is equal to half the measure of the intercepted arc.) **5.** $3x° = 75°$; $(2x + 5)° = 55°$ **6.** $t° = 40°$ **7.** $(6b - 10)° = 158°$; $(2b + 14)° = 70°$

Page 570 Checking Key Concepts

1. 49° **3.** 262° **5.** $x = 20$ **7.** $x = y = 15$

Pages 570–572 Exercises and Applications

1. $t = 18$ **3.** $z = 75$; $y = 15$ **5.** $t = 21$; $x = 42$
9. $AB \cong CD$, so $m\ \overset{\frown}{AB} = m\ \overset{\frown}{CD}$. The measure of a central angle is equal to the measure of its intercepted arc, so $m\angle APB = m\ \overset{\frown}{AB}$ and $m\angle DPC = m\ \overset{\frown}{CD}$. Then $m\angle APB = m\angle DPC$ or $\angle APB \cong \angle DPC$. $\overline{PA}$, $\overline{PB}$, $\overline{PD}$, and $\overline{PC}$ are all congruent since they are radii of the same circle. By the SAS Postulate, $\triangle PAB \cong \triangle PDC$, so corresponding sides $\overline{AB}$ and $\overline{CD}$ are congruent. **17.** (1) Given; (2) Def. of perpendicular lines; (3) Reflexive Property; (4) For any two points, there is exactly one line through the points; (5) Def. of radius; (6) HL Theorem; (7) Def. of congruent triangles
21. $\angle ABE$; $\angle BAE$ **23.** $\overrightarrow{CD}$ **25.** 0.3907

Page 575 Checking Key Concepts

1. $BC \cdot BE = BA^2$; $EG \cdot GC = FG \cdot GD$; $BA = BD$; $BC \cdot BE = BD^2$ **3.** $2z = 30$

Pages 576–578 Exercises and Applications

1. True. **3.** False. **5.** False. **7.** $CD \cdot CG = CE \cdot CF$, so $\dfrac{CD \cdot CG}{CE \cdot CG} = \dfrac{CE \cdot CF}{CE \cdot CG}$ and $\dfrac{CD}{CE} = \dfrac{CF}{CG}$. **9.** $CB^2 = CE \cdot CF$, so $\dfrac{CB^2}{CB} = \dfrac{CE \cdot CF}{CB}$, or $CB = \dfrac{CE \cdot CF}{CB}$. **11.** $x = 6\sqrt{2} \approx 8.5$
13. $y = 4$; $y + 3 = 7$ **15.** $x = 12$ **17.** (1) For any two points, there is exactly one line through the points; (2) AEB; (3) $\angle C$; (5) Def. of similar figures; (6) $CA \cdot CD$; a property of proportions **19.** about 26,100 mi

21.

Statements	Reasons
1. Draw $\overline{AB}$ and $\overline{BD}$.	1. For any two points, there is exactly one line through the points.
2. $\angle C \cong \angle C$	2. Reflexive Property
3. $m\angle CBA = \frac{1}{2}m\ \overset{\frown}{BA}$	3. The measure of an angle formed by a tangent and a chord is half the measure of the intercepted arc.
4. $m\angle D = \frac{1}{2}m\ \overset{\frown}{BA}$	4. The measure of an inscribed angle is equal to half the measure of the intercepted arc.
5. $m\angle CBA = m\angle D$ or $\angle CBA \cong \angle D$.	5. Substitution Property (Steps 3 and 4)
6. $\triangle ABC \sim \triangle BCD$	6. AA Similarity Theorem
7. $\dfrac{AC}{CB} = \dfrac{CB}{CD}$	7. Def. of similar figures
8. $AC \cdot CD = CB^2$	8. A property of proportions

29. $\dfrac{81}{\pi}$ cm$^2 \approx 25.8$ cm^2 **31.** 70° **33.** 220°

Page 581 Checking Key Concepts

1. 74 in.^{2}; 15 in. **3.** 236 m^2; 63 m **5.** 64 in.

Pages 582–585 Exercises and Applications

1. 24 in.^{2} **3.** 56 in.^{2} **5.** 19 ft^2 **9.** 8 cm **11.** 19 in.
13. 23 cm **15.** 5.6 in. **23.** about 20.9 ft
31. about 416 ft^2; 500 ft^3

Page 585 Assess Your Progress

1. 15 **2.** 4 **3.** 1.2 **4.** 101 m^2; 59 m

Page 588 Checking Key Concepts

1. $\overline{CT}$ **3.** $\overleftrightarrow{BD}$ **5.** $\overline{VT}$, $\overline{VC}$, or $\overline{VH}$ **7.** about 113 in.^{3}
9. about 1 in.

Pages 588–591 Exercises and Applications

1. B **3.** C **5.** A **7.** 314 in.^{2}; 524 in.^{3} **9.** 129 cm^2; 137 cm^3 **13, 15.** Answers are given to the nearest tenth.
13. 1.7 ft **15.** 3.0 ft **17. a.** about 201,062,000 mi^2
b. about 58,308,000 mi^2 **21.** about 4.25 in.
23. about 449 in.^{3} **29. a.** 8 : 1 **b.** 4 : 1 **c.** The thickness is divided by 4. **33.** about 5 in.^{2} **35.** about 94 ft^2 **37.** $\dfrac{7}{5}$

Page 594 Checking Key Concepts

1. 300,000 ft^2 **3.** 4 : 9 **5.** 16 : 81

Pages 594–596 Exercises and Applications

1. a. 3.7 : 1 **b.** 13.4 : 1 **c.** 49 : 1 **3.** about 8 oz
5. about 37 oz **7. a.** $a : b$; $a^3 : b^3$ **b.** smaller solid: $\frac{2}{3}\pi a^3$; larger solid: $\frac{2}{3}\pi b^3$ **c.** $\dfrac{\frac{2}{3}\pi a^3}{\frac{2}{3}\pi b^3} = \dfrac{a^3}{b^3}$; Yes. **9. a.** 2 : 1

b, c. Answers may vary. **b.** about 1.26 : 1 **c.** about 1.59 : 1
d. about 1870 mm^2 **11.** As the size increases, the ratio of surface area to volume decreases. **17.** the Parallel Postulate (Through a point not on a line, there is exactly one line parallel to the given line.)

Page 599 Checking Key Concepts

1.

3. perpendicular lines m and l

5. No; the figure for Ex. 4 provides a counterexample. In fact, through a point not on a line, there are infinitely many lines perpendicular to the given line.

Pages 600–603 Exercises and Applications

3. No; if you tried to do so, you would end up drawing a line. On a sphere, a ray could not extend in only one direction.

5. The figure shows a quadrilateral that has four congruent angles. Yes; a square or rectangle is a quadrilateral with four congruent angles.

7.

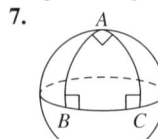

9. The triangle in Ex. 7 is a counterexample. Considering two sides of the triangle as two lines and the third side as the transversal, corresponding right angles are formed, but the lines are not parallel.   **11. a.** 120°; 108°   **b.** No; the sum of the measures of the angles at each vertex would be 348°, which would leave a gap at each vertex.   **13.** about 111.4°; Since the sum of the measures of the angles at each vertex on a soccer ball is 360°, the measure of each interior angle of a pentagon is about 360° − 2(124.3°) = 111.4°.   **15.** 100°20′
17. a. 4200′   **b.** about 4200 nautical miles
c. about 4830 mi   **25.** 100 : 49   **27.** $A'(-2, 1)$; $B'(0, -3)$; $C'(5, 3)$

Page 603   Assess Your Progress

1. Answers may vary. An example is given.

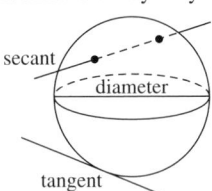

2. about 113 in.^{2}; about 113 in.^{3} **3.** about 105.7 cm^2; about 102.2 cm^3 **4. a.** 25 : 144   **b.** 125 : 1728   **5.** No; if you draw a diagonal, you divide the quadrilateral into two triangles. The sum of the measures of the interior angles of the each triangle is greater than 180°. Then the sum of the measures of the interior angles of the quadrilateral is greater than 360°.

Pages 608–609   Chapter 11 Assessment

1. arcs in congruent circles whose measures are equal

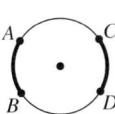

2. two points on a circle and all points of the circle not on the minor arc between the two given points

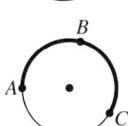

3. an angle whose vertex is on a circle and whose sides contain chords of the circle

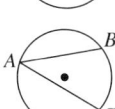

4. a line, ray, or segment that contains a chord

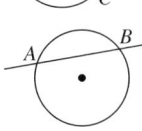

5. a line in the same plane as a circle that intersects the circle in only one point

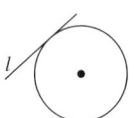

6. the intersection of a sphere with a plane containing the center of the sphere

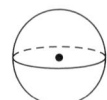

7. 59°   **8.** 52°   **9.** 116°

10.

Statements	Reasons
1. $\overarc{AB} \cong \overarc{CD}$	1. Given
2. $\angle ADB \cong \angle CBD$	2. In a circle or congruent circles, angles that intercept the same arc or congruent arcs are congruent.
3. $\overline{AD} \parallel \overline{BC}$	3. If two lines are intersected by a transversal and alternate interior angles are congruent, then the lines are parallel.

11. $\sqrt{119} \approx 10.9$ **12.** 24   **13.** $x + 1 = 8$; $x - 1 = 6$

14.

Statements	Reasons
1. $\angle MNP \cong \angle RSP$; $\angle NMP \cong \angle SRP$	1. In a circle or congruent circles, angles that intercept the same arc or congruent arcs are congruent.
2. $\overarc{MN} \cong \overarc{RS}$	2. Given
3. $\overline{MN} \cong \overline{RS}$	3. Congruent arcs have congruent corresponding chords.
4. $\triangle MNP \cong \triangle RSP$	4. ASA Postulate

15–19. Answers are given to the nearest tenth.   **15.** 40.8 ft^2; 13.6 ft   **16.** 351.9 in.^{2}; 58.6 in.   **17.** 47.0 m^2; 13.4 m
18. 113.1 m^2; 113.1 m^3 **19.** 132.7 cm^2; 143.8 cm^3
20. 28 cm   **21. a.** 3 : 5   **b.** 9 : 25   **c.** 27 : 125   **22.** In spherical geometry, the sum of the measures of the interior angles of a triangle is greater than 180°. In Euclidean geometry, the sum of the measures of the interior angles of a triangle is equal to 180°.

CHAPTER 12

Page 616   Checking Key Concepts

1. 12; a hexagon

3. $\begin{bmatrix} 4.5 & 3 & -6 & -4.5 \\ 6 & 1.5 & 1.5 & 6 \end{bmatrix}$

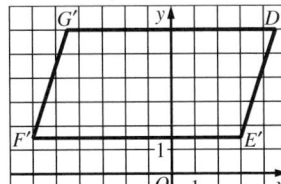

1. $\begin{bmatrix} -3 \\ 2 \end{bmatrix}$ **3.** $\begin{bmatrix} -2 & 0 \\ -2 & 4 \end{bmatrix}$ **5.** $\begin{bmatrix} 0 & 3 & -2 \\ 4 & 0 & -2 \end{bmatrix}$ **7.** $\begin{bmatrix} 15 \\ -5 \end{bmatrix}$

9. $\begin{bmatrix} 0 & 9 & -6 \\ 0 & -3 & -12 \end{bmatrix}$ **11.** $\begin{bmatrix} 10 & 5 & 15 \\ 5 & 5 & 10 \end{bmatrix}$

13. (7) dilation of a point with center (0, 0) and scale factor 5; (8) dilation of a line segment, with center (0, 0) and scale factor 4; (9) dilation of a triangle, with center (0, 0) and scale factor 3; (10) dilation of a quadrilateral, with center (0, 0) and scale factor 2.5; (11) no transformation (the same figure before and after); (12) dilation of a line segment, with center (0, 0) and scale factor $\frac{2}{3}$.

15. $4\begin{bmatrix} 2 & 3 & 0 \\ -1 & 4 & \frac{1}{4} \end{bmatrix} = \begin{bmatrix} 8 & 12 & 0 \\ -4 & 16 & 1 \end{bmatrix}$

17. $\frac{1}{2}\begin{bmatrix} -4 & 0 & 4 \\ 0 & 6 & 0 \end{bmatrix} = \begin{bmatrix} -2 & 0 & 2 \\ 0 & 3 & 0 \end{bmatrix}$

19. $2\begin{bmatrix} -2 & 0 & 3 & 0 \\ 0 & 2 & 0 & -2 \end{bmatrix}$

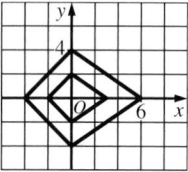

29. a. 2×8 **b.** 2×8 **33.** $\sqrt[3]{\frac{9}{4}}$ **35.** $x = 12$; $y = 15$; $z = 20$

37. 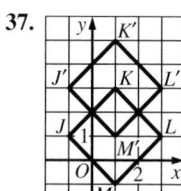 **39.**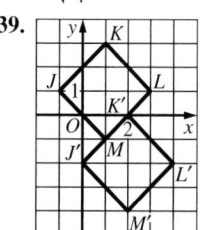

1. $\begin{bmatrix} 4 \\ 4 \end{bmatrix}$ **3.** $\begin{bmatrix} 4 & 7 \\ 3 & 0 \end{bmatrix}$

5. (1) The point (0, 3) is translated to the right 4 units and up 1 unit, to image point (4, 4). (2) The triangle with vertices (0, 3), (3, −4), and (7, −2) is translated to the left 1 unit and up 5 units. The image triangle has vertices (−1, 8), (2, 1), and (6, 3). (3) The line segment connecting (1, 5) and (4, 2) is translated to the right 3 units and down 2 units. The image segment has endpoints (4, 3) and (7, 0). (4) A quadrilateral with vertices (−2, 0), (3, −1), (2, 6), and (−5, 5) is translated up 4 units. The image quadrilateral has vertices (−2, 4), (3, 3), (2, 10), and (−5, 9).

1. $\begin{bmatrix} 2 \\ 4 \end{bmatrix}$ **3.** $\begin{bmatrix} 0 & 4 \\ 3 & -1 \end{bmatrix}$

5. (1) The point (5, −1) is translated to the left 3 units and up 5 units. (2) A triangle with vertices (1, −3), (2, 4), and (−1, 6) is translated to the right 3 units and up 2 units. (3) A line segment with endpoints (−1, 3) and (3, −1) is translated to the right 1 unit. (4) A quadrilateral with vertices (0, −2), (3, 2), (0, 3), and (−5, 1) is translated to the right 2 units and down 4 units.

7. $\begin{bmatrix} -2 & -2 \\ 0 & 0 \end{bmatrix}$ **9.** $\begin{bmatrix} 6 & 6 & 6 & 6 \\ -1 & -1 & -1 & -1 \end{bmatrix}$ **11. a.** $\begin{bmatrix} -4 & -4 & -4 \\ 4 & 4 & 4 \end{bmatrix}$

b. $\begin{bmatrix} -8 & -12 & -4 \\ 1 & 6 & 7 \end{bmatrix}$

c.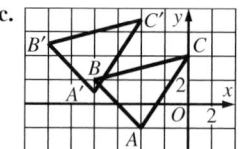

13. a. $\begin{bmatrix} 1 & 3 & 5 & 3 & 1 \\ 5 & 5 & 3 & 1 & 1 \end{bmatrix}$ **b.** $\begin{bmatrix} -4 & -4 & -4 & -4 & -4 \\ -8 & -8 & -8 & -8 & -8 \end{bmatrix}$

c. $\begin{bmatrix} -3 & -1 & 1 & -1 & -3 \\ -3 & -3 & -5 & -7 & -7 \end{bmatrix}$

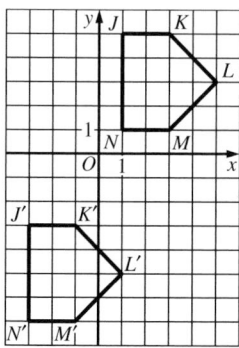

15. $\begin{bmatrix} 4 & 4 & 4 \\ -3 & -3 & -3 \end{bmatrix} + \begin{bmatrix} -1 & 1 & 0 \\ 4 & 4 & 1 \end{bmatrix} = \begin{bmatrix} 3 & 5 & 4 \\ 1 & 1 & -2 \end{bmatrix}$

17. $\begin{bmatrix} 0 & 0 & 0 & 0 \\ -4 & -4 & -4 & -4 \end{bmatrix} + \begin{bmatrix} -5 & -4 & -1 & -2 \\ -1 & 1 & 1 & -1 \end{bmatrix} =$

$\begin{bmatrix} -5 & -4 & -1 & -2 \\ -5 & -3 & -3 & -5 \end{bmatrix}$ **19.** $\begin{bmatrix} -2 & 0 & 6 & 4 \\ 6 & 3 & 6 & 9 \end{bmatrix}$ **31.** about 2.93

1. $\begin{bmatrix} 0 & 20 \\ 25 & 15 \end{bmatrix}$ **2.** $\begin{bmatrix} 1.5 & 4.5 & 6 \\ 0 & 3 & 4.5 \end{bmatrix}$ **3.** $\begin{bmatrix} 1 & 2 & 1.5 & 0.5 \\ 0 & 2 & 3 & -1 \end{bmatrix}$

4. $\begin{bmatrix} 0 & 50 & 10 \\ 0 & -20 & -30 \end{bmatrix}$ **5.** $\begin{bmatrix} -1 & -1 & -1 & -1 & -1 \\ 6 & 6 & 6 & 6 & 6 \end{bmatrix}$ **6.** $\begin{bmatrix} 5 & 5 & 5 & 5 \\ 0 & 0 & 0 & 0 \end{bmatrix}$

1. $\begin{bmatrix} 16 & -24 & 65 \\ 26 & -16 & 5 \end{bmatrix}$

3. The transformation matrix should be first, so the matrices cannot be multiplied in the order given.

Pages 630–633 Exercises and Applications

1. a. $\begin{bmatrix} 11 \\ 1 \end{bmatrix}$

3. a. $\begin{bmatrix} 1 & 3 & -1 & -5 & -4 \\ 7 & -1 & -3 & 3 & 10 \end{bmatrix}$

b.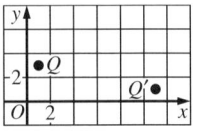

b.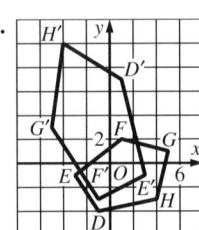

5.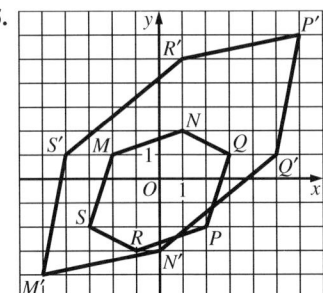

7.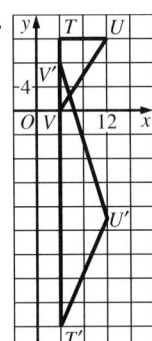

9. a. $\begin{bmatrix} 4 & 8 & -8 \\ 6 & -6 & 0 \end{bmatrix}; \begin{bmatrix} 6 & 12 & -12 \\ 9 & -9 & 0 \end{bmatrix}; \begin{bmatrix} 8 & 16 & -16 \\ 12 & -12 & 0 \end{bmatrix}$

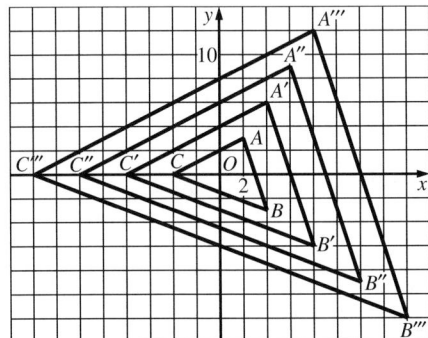

b. a dilation about center $(0, 0)$ with scale factor n;
$\begin{bmatrix} n & 0 \\ 0 & n \end{bmatrix} = n \begin{bmatrix} 1 & 0 \\ 0 & 1 \end{bmatrix}$, so $\begin{bmatrix} n & 0 \\ 0 & n \end{bmatrix} A = nA.$

13. a. $\begin{bmatrix} 181 & 103 & 146 \\ 485 & 443 & 362 \end{bmatrix}$ **b.** $\begin{bmatrix} 290 & 212 & 255 \\ 328 & 286 & 205 \end{bmatrix}$

c. a translation 109 units right and 157 units down **17. B**

21. $\begin{bmatrix} 2 & 6 & 8 \\ 0 & 4 & 6 \end{bmatrix}$ **23. a.** $\begin{bmatrix} -2 & 4 & 4 & -2 \\ -1 & -1 & -3 & -3 \end{bmatrix}$ **b.** $\begin{bmatrix} 2 & -4 & -4 & 2 \\ 1 & 1 & 3 & 3 \end{bmatrix}$

c. $\begin{bmatrix} 1 & 1 & 3 & 3 \\ -2 & 4 & 4 & -2 \end{bmatrix}$

Page 636 Checking Key Concepts

1. $\begin{bmatrix} 1 & 0 \\ 0 & -1 \end{bmatrix}$ **3.** $\begin{bmatrix} 3 & 1 & 1 & 3 \\ 0 & -1 & -3 & -3 \end{bmatrix}$ **5.** $\begin{bmatrix} 0 & 1 & 3 & 3 \\ 3 & 1 & 1 & 3 \end{bmatrix}$

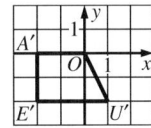

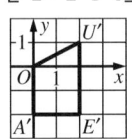

Pages 637–640 Exercises and Applications

1. $\begin{bmatrix} 7 \\ 2 \end{bmatrix}$; the x-axis; **3.** $\begin{bmatrix} -2 \\ 7 \end{bmatrix}$; line $y = x$ **5.** $\begin{bmatrix} -1 & 0 \\ 0 & 1 \end{bmatrix}$

7. $\begin{bmatrix} -2 & -2 & 1 & 0 \\ 0 & -2 & -2 & 0 \end{bmatrix}$ **9.** $\begin{bmatrix} 0 & 2 & 2 & 0 \\ -2 & -2 & 1 & 0 \end{bmatrix}$

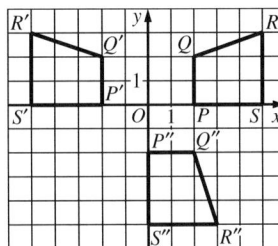

11. $A'(24, -10)$; $A'(-24, 10)$ **13.** $C'(-15, 27)$; $C'(15, -27)$

17. $\begin{bmatrix} -1 & 0 & 3 \\ -2 & 3 & 0 \end{bmatrix}; \begin{bmatrix} 1 & 0 & -3 \\ -2 & 3 & 0 \end{bmatrix}$ **19.** $\begin{bmatrix} -3 & -1 & -3 & -5 \\ 3 & 3 & -1 & -1 \end{bmatrix};$

$\begin{bmatrix} -3 & -3 & 1 & 1 \\ 3 & 1 & 3 & 5 \end{bmatrix}$ **21. a.** (a) $\begin{bmatrix} 2 & 2 & 5 & 5 \\ 0 & 2 & 3 & 0 \end{bmatrix}$ (b) $\begin{bmatrix} -2 & -2 & -5 & -5 \\ 0 & 2 & 3 & 0 \end{bmatrix}$

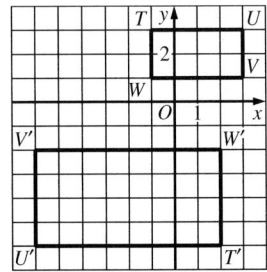

(c) $\begin{bmatrix} 0 & 2 & 3 & 0 \\ -2 & -2 & -5 & -5 \end{bmatrix}$

(d) $P''Q''R''S''$ is the rotation of $PQRS$ about $(0, 0)$ by $270°$.

25. $A'(-237, 572)$ **27.** $C'(365, 374)$

29. $\begin{bmatrix} -85 & -69 & -27 & 27 & 69 & 85 \\ 0 & 50 & 81 & 81 & 50 & 0 \end{bmatrix}$

33. $\begin{bmatrix} -1 & 0 \\ 0 & 1 \end{bmatrix} \begin{bmatrix} 3 & 3 & 0 \\ -3 & 4 & -1 \end{bmatrix} = \begin{bmatrix} -3 & -3 & 0 \\ -3 & 4 & -1 \end{bmatrix}$

37. $\begin{bmatrix} 2 & -6 & -6 & 2 \\ -6 & -6 & -2 & -2 \end{bmatrix}$

39. $t = 6$ **41.**

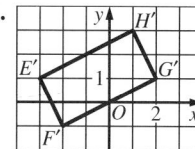

Page 644   Checking Key Concepts

1. a. $\begin{bmatrix} -1 & 2 & 2 & -3 \\ 5 & 5 & 3 & 3 \end{bmatrix}$

b.

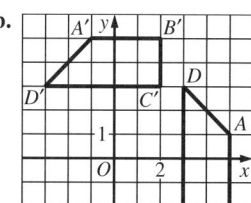

90° rotation

Pages 644–647   Exercises and Applications

1. $\begin{bmatrix} 1 & 0 & -3 \\ 4 & 2 & 5 \end{bmatrix}$

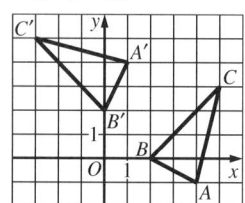

3. $\begin{bmatrix} -1 & -3 & -5 & -2 & 0 \\ 0 & 0 & 1 & 3 & 3 \end{bmatrix}$

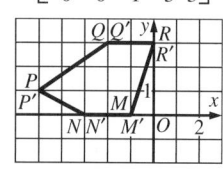

5. $\begin{bmatrix} 0 & 10 & -20 \\ -15 & 0 & 0 \end{bmatrix}$

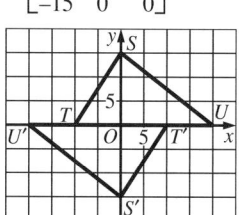

17. 0.9659   **19.** $x = 7\sqrt{2}$ **21.** $x = 20$

Page 647   Assess Your Progress

1. $x = 2;\ y = 12$ **2.** $g = -2;\ h = -4$

3. $\begin{bmatrix} 1 & 3 & 1 & -2 \\ 2 & 1 & -2 & 1 \end{bmatrix};\ \begin{bmatrix} 2 & 1 & -2 & 1 \\ 1 & 3 & 1 & -2 \end{bmatrix}$

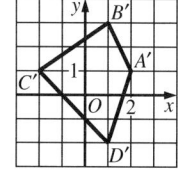

4. $\begin{bmatrix} -1 & 3 & -1 \\ -1 & -2 & -4 \end{bmatrix};\ \begin{bmatrix} 1 & 2 & 4 \\ 1 & -3 & 1 \end{bmatrix}$

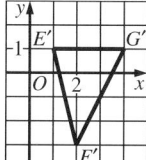

5. $\begin{bmatrix} 1 & 2 & 4 \\ -1 & 3 & -1 \end{bmatrix}$

6. $\begin{bmatrix} -1 & -2 & -4 \\ 1 & -3 & 1 \end{bmatrix}$

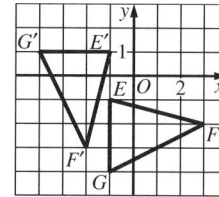

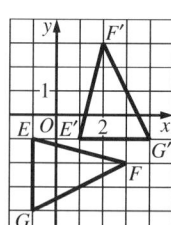

Pages 652–653   Chapter 12 Assessment

1. $\begin{bmatrix} 3 & 2 & 1 \\ 0 & 5 & 1 \end{bmatrix};\ 2 \times 3$ **2.** $3 \cdot \begin{bmatrix} 1 & 6 \\ 5 & 2 \end{bmatrix} = \begin{bmatrix} 3 & 18 \\ 15 & 6 \end{bmatrix}$

3. $\begin{bmatrix} 1 & 1 \\ 2 & 3 \end{bmatrix} + \begin{bmatrix} 2 & 1 \\ 0 & 1 \end{bmatrix} = \begin{bmatrix} 3 & 2 \\ 2 & 4 \end{bmatrix}$ **4.** $\begin{bmatrix} 2 \\ 1 \end{bmatrix}$ **5.** $\begin{bmatrix} -3 \\ 0 \end{bmatrix}$

6. $\begin{bmatrix} 2 & 3 \\ 1 & -2 \end{bmatrix}$ **7.** $\begin{bmatrix} -2 & -3 \\ -2 & 0 \end{bmatrix}$ **8.** $\begin{bmatrix} -3 & 2 & 3 \\ 2 & 1 & -2 \end{bmatrix}$

9. $\begin{bmatrix} -3 & 2 & 3 & -2 & -3 \\ 2 & 1 & -2 & -2 & 0 \end{bmatrix}$ **10.** $\begin{bmatrix} -10 & 5 & 25 \\ 0 & 15 & -5 \end{bmatrix}$ **11.** $\begin{bmatrix} 40 & -16 \\ 24 & 72 \end{bmatrix}$

12. $\begin{bmatrix} 0 & 0 & 0 \\ 0 & 0 & 0 \end{bmatrix}$ **13.** $\begin{bmatrix} 11 \\ -9 \end{bmatrix}$ **14.** $\begin{bmatrix} -3 & -2 & -1 \\ 1 & -2 & 2 \end{bmatrix}$

15. $\begin{bmatrix} -7 & -4 & -2 & -4 \\ 2 & 4 & 2 & 0 \end{bmatrix}$ **16.** $\begin{bmatrix} 2 & 2 & 2 & 2 \\ 0 & 0 & 0 & 0 \end{bmatrix} + \begin{bmatrix} -1 & 0 & -2 & -3 \\ 3 & 1 & -2 & 2 \end{bmatrix}$

17. $\begin{bmatrix} -3 & -3 & -3 \\ 2 & 2 & 2 \end{bmatrix} + \begin{bmatrix} 3 & 2 & 1 \\ 1 & -3 & 0 \end{bmatrix}$

18. $\begin{bmatrix} -1 & -1 & -1 & -1 & -1 \\ -3 & -3 & -3 & -3 & -3 \end{bmatrix} + \begin{bmatrix} 0 & -1 & -1 & -1 & 2 \\ 0 & 0 & 1 & 3 & 3 \end{bmatrix}$

19. $\begin{bmatrix} 3 & 18 \\ -9 & -38 \end{bmatrix}$ **20.** $\begin{bmatrix} 5 & 0 & -5 & -1 & 3 \\ -20 & 8 & 12 & -20 & -20 \end{bmatrix}$

21. $\begin{bmatrix} -3 & -2 & 2 & 1 \\ -8 & 0 & 7 & -2 \end{bmatrix}$

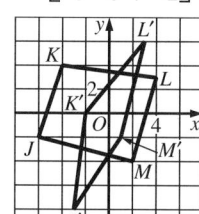

22. a. $\begin{bmatrix} -1 & 4 \\ -2 & 1 \end{bmatrix}$ **b.** $\begin{bmatrix} -2 & 1 \\ -1 & 4 \end{bmatrix}$ **c.** a reflection over $y = x$

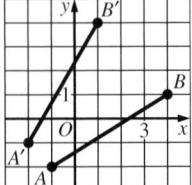

23. $\begin{bmatrix} 0 & -3 & 1 & 2 \\ 4 & -1 & -3 & 0 \end{bmatrix}$

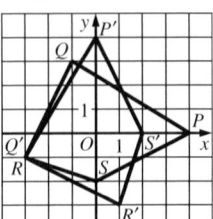

24. $\begin{bmatrix} 0 & 0 & 3 & 5 \\ -1 & -3 & -2 & 1 \end{bmatrix}$

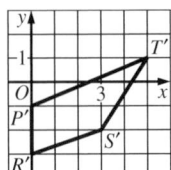

reflection over the
x-axis

25. $\begin{bmatrix} -1 & -3 & -2 & 1 \\ 0 & 0 & -3 & -5 \end{bmatrix}$

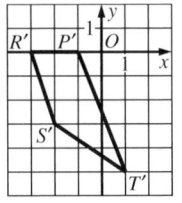

reflection over $y = -x$

26. $\begin{bmatrix} 0 & 0 & -3 & -5 \\ -1 & -3 & -2 & 1 \end{bmatrix}$

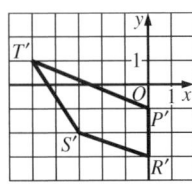

rotation of 180°
about the origin

27. $\begin{bmatrix} 0 & 0 & 3 & 5 \\ 1 & 3 & 2 & -1 \end{bmatrix}$

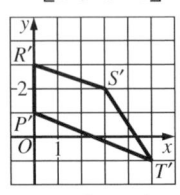

no transformation

28. $\begin{bmatrix} -1 & -3 & -2 & 1 \\ 0 & 0 & 3 & 5 \end{bmatrix}$

rotation of 90°
about the origin

29. $\begin{bmatrix} 1 & 3 & 2 & -1 \\ 0 & 0 & -3 & -5 \end{bmatrix}$

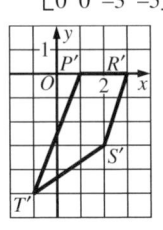

rotation of 270°
about the origin

CHAPTERS 10–12

Pages 656–657 Cumulative Assessment

1. *JMK, KML* **3.** 31°; 59° **5. a.** 41.0°, 139.0°, 41.0°,
139.0° **b.** about 262 cm² **9.** about 610; about 784
11. 13,824; 82,944 **13.** $x = 90$; $y = 90$ **15.** $t = 7\sqrt{2} \approx$
9.90 **17.** $16\pi \approx 50.3$; $20\pi \approx 62.8$; $144\pi \approx 452$

21. $\begin{bmatrix} 4 & -1 & -3 & 0 \\ 0 & 3 & -1 & -2 \end{bmatrix}$

EXTRA PRACTICE

Pages 659–661 Chapter 1

1. −14, −17 **3.** 16, 25 **5.** 112, −224 **7.** ⟵

9. $|n - 3|$ **11.**

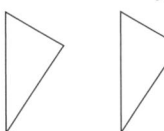

13. Answers may vary. An example is given.

15. reflection symmetry **17.** translation symmetry
19. If I forget to fill the tank, then my car runs out of gas.
21. hypothesis: the weather gets cold; conclusion: birds fly
south **23.** True. **25.** False; Carmine is in Huma. **27.** F
29. $\overleftrightarrow{DH}$ **31.** *AEH* and *AEF* **33.** 4 **35.** 2 **37.** 5 **39.** 10
41. 6 **43.** 70° **45.** 110° **47.** 90° **49.** True. **51.** True.
53. 6 **55.** 14 **57.** $\left(-\frac{1}{2}, 3\right)$ **59.** $\left(\frac{1}{2}, -1\right)$ **61.** (−3, −2)

Pages 661–663 Chapter 2

1. $\angle AFH$ **3.** $\angle JGD$ **5.** 70° **7.** 30° **9.** 60° **11.** $x = 54$
13. $w = 48$, $z = 44$, $2z = 88$ **15.** $b = 36$, $c = 50$ **17.** $x = 60$
19. $a = 40$, $b = 35$ **21.** equilateral quadrilateral
23. equilateral pentagon **25.** regular quadrilateral or square
27. 1260° **29.** $x° = 65°$ **31.** $z° = 135°$; $(z - 15)° = 120°$
33. 144°; 36° **35. a.** 8 cm **b.** 8 cm **c.** 125° **37. a.** 40°
b. 55° **c.** 125° **d.** 85° **39.** 9.8 m **41.** 90° **43.** 40°
45. pentagonal prism **47.** 3 **49.** 10

Pages 663–665 Chapter 3

1. inductive **3.** deductive **5.** 17; Each term is 3 more than
the previous term. **7.** $\frac{1}{27}$; Each term is $\frac{1}{3}$ of the previous
term. **9.** Reflexive Property **11.** Definition of congruent
segments **13.** Definition of congruent segments
15. Segment Addition Postulate; Addition Property
17. False. **19.** 145° **21.** 71° **23.** Two angles that form
a linear pair are supplementary. **25.** If a wildflower is
yellow, then it is goldenrod. False. **27.** If I take my
umbrella, it will rain. False. **29. a.** If two angles are
congruent, then their complements are congruent.

b. If the complements of two angles are congruent, then the angles are congruent. **31.** 4.47; $2\sqrt{5}$ **33.** ± 5.66; $\pm 4\sqrt{2}$ **35.** -12.25; $-5\sqrt{6}$ **37.** Yes. **39.** No. **41.** No. **43.** $b = 5$ **45.** $a = 16$ **47.** $c = 169$ **49.** obtuse **51.** right **53.** right **55.** Inverse: If Jo does not study hard, then she will not do well on the test. Contrapositive: If Jo does not do well on the test, then she does not study hard. **57.** Inverse: If Bill was not born in France, then he cannot speak French. Contrapositive: If Bill cannot speak French, then he was not born in France.

Pages 665–667   Chapter 4

1. 5 **3.** $\sqrt{85} \approx 9.22$ **5.** $3\sqrt{5} \approx 6.71$ **7.** (4, 4) **9.** $(-4, 5)$ **11.** $(5, -1)$ **13.** $AB = 5$, $BC = 5$, $AC = 8$; isosceles triangle **15.** $AB = 2\sqrt{5} \approx 4.47$, $BC = 5$, $CD = 2\sqrt{5} \approx 4.47$, $DA = 5$; parallelogram **17.** -2 **19.** -2 **21.** 3 **23.** 1 **25.** $y = 2x + 1$ **27.** $y = \frac{1}{4}x - \frac{11}{4}$ **29.** $x = 5$ **31.** $y = \frac{7}{2}x - 20$ **33.** perpendicular **35.** parallel **37.** 0 **39.** undefined **41.** 3 **43.** Yes. **45.** No. **47.** $(x - 2)^2 + (y - 5)^2 = 16$ **49.** $(x + 2)^2 + (y + 5)^2 = 20.25$

51. 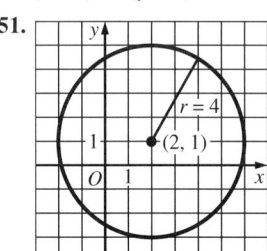 **53.**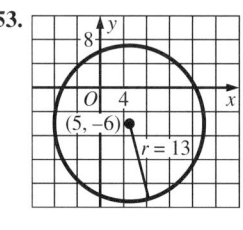

55. $H(c - a, 0)$ **57.** $J(a, 0)$ **59.** $M(\sqrt{3}, 1)$; $MA = 2$, $MB = 2$, $MO = 2$ **61.** $(-3, 0, 6)$ **63.** 5 **65.** $2\sqrt{42} \approx 12.96$

Pages 667–670   Chapter 5

1. alternate interior angles **3.** same-side interior angles **5.** alternate interior angles **7.** same-side interior angles **9.** $x = 60$, $y = 22$, $z = 15$ **11.** $m\angle 1 = 45°$, $m\angle 2 = 45°$ **13.** $m\angle 5 = 105°$, $m\angle 6 = 75°$ **15.** $\overline{LM}$ and $\overline{KN}$ **17.** $105°$ **19. a.** corresponding angles **b.** $\angle 2 \cong \angle 3$ **c.** $\angle 1 \cong \angle 3$ **21.** alternate interior; Substitution **23.** $t = 40$ **25.** parallel; $78° + 102° = 180°$ **27.** not parallel; $x + 10 \neq x + 6$ **29.** $\overline{PM}$ **31.** $66°$ **33.** 3 **35.** $104°$

37. 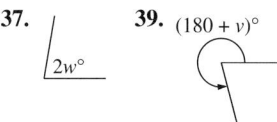 **39.** $(180 + v)°$

Pages 670–672   Chapter 6

1. Yes. **3.** Yes. **5.** No. **7.** Yes. **9.** No. **11.** It is between $1\frac{1}{4}$ in. long and $7\frac{3}{4}$ in. long. **13.** It is between 1 ft long and 11 ft long. **15.** It is between 6 cm long and 22 cm long. **17.** It is between 9 cm long and 31 cm long. **19.** $6 < AB < 14$ **21.** $AB = CB$ **23.** $AB < 12$ **25.** $\triangle FCE$

27. $\angle FEC$ **29.** $\overline{FE}$ **31.** $\triangle YXW$ **33.** $\angle Y$ **35.** $x = 57$ **37.** SSS Postulate **39.** No. **41.** SAS Postulate **43.** (1) Given; (2) midpoint; (3) Given; (4) $\overline{AE} \cong \overline{ED}$; Def. of midpoint; (5) Vertical angles are $\cong$; (6) SAS Postulate **45.** none **47.** ASA Postulate **49.** SSS Postulate **51.** $\triangle BAD \cong \triangle ABC$ **53.** $\triangle ACB \cong \triangle BDA$ or $\triangle ACE \cong \triangle BDE$ **55.** $\overline{PT} \cong \overline{QT}$ **57.** $\angle 2 \cong \angle 3$ **59.** $90°$ **61.** 8

Pages 672–675   Chapter 7

1. a. 5 **b.** 5 **c.** 5 **3. a.** 6 **b.** 12 **c.** 10 **5. a.** 12 **b.** 13 **c.** 20 **d.** 11 **e.** 20 **7.** Both pairs of opposite sides are parallel. (Def. of parallelogram) **a.** $67°$ **b.** 10 **c.** 20 **9.** Both pairs of opposite sides are congruent. **a.** $90°$ **b.** $45°$ **c.** 4 **11.** rectangle **13.** square **15.** rectangle **17.** 20 **19.** 216 **21.** 36 **23.** 242 cm^2 **25.** $16\sqrt{3} \approx 27.7$ in.^{2} **27.** 16 **29.** $9\pi \approx 28.27$ **31.** $16.81\pi \approx 52.81$ **33.** $36\pi - 72 \approx 41.10$ **35.** $V = 170$; $S.A. = 193$ **37.** $V = 250\sqrt{3} \approx 433.01$; $S.A. = 300 + 50\sqrt{3} \approx 386.60$ **39.** 96 m^3

Pages 675–677   Chapter 8

1.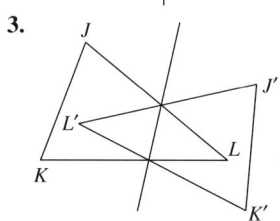

3.

5.

7. $A'(-2, 3)$, $B'(-3, -1)$, $C'(-1, 0)$ **9.** $G'(1, 1)$, $H'(-1, 3)$, $J'(-3, 1)$, $K'(-2, -1)$ **11.** $D'(-1, 0)$, $E'(-2, -2)$, $F'(1, -3)$, $G'(2, 0)$ **13.** $P'(2, 0)$, $Q'(3, 2)$, $R'(0, 1)$ **15.** $W'(-1, -1)$, $X'(2, -1)$, $Y'(3, 1)$, $Z'(2, 2)$

17. 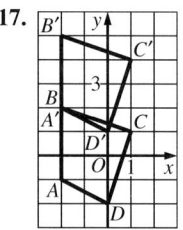 **19.**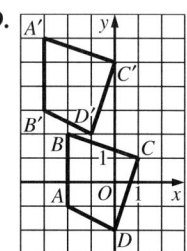

21. $(a, b) \rightarrow (a + 2, b + 1)$

SA34   Selected Answers

23.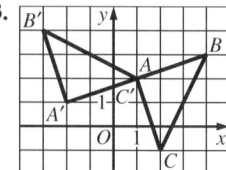

25.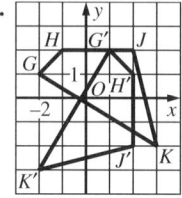

27.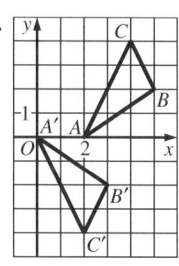

29.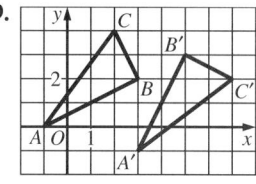

31.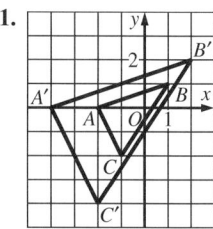

33.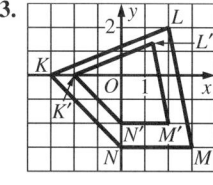

59.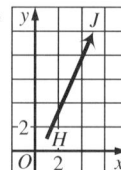

$|\overrightarrow{HJ}| = \sqrt{97} \approx 9.8$

61.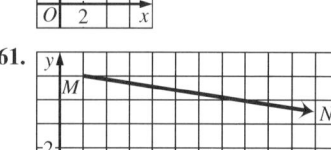

$|\overrightarrow{MN}| = \sqrt{409} \approx 20.2$

63. $(-1, 3)$ **65.** $(-4, 10)$ **67.** about 105   **69.** 640

71. 1728   **73.** $V = \dfrac{500\sqrt{2}}{3} \approx 235.7$;

$S.A. = 100\sqrt{3} + 100 \approx 273.2$

Pages 683–685   Chapter 11

1. $270°$ **3.** $120°$ **5.** $310°$ **7.** $305°$ **9.** $218°$ **11.** $n = 35$
13. $x = 25$ **15.** $z = 70$ **17.** $x = 20$ **19.** $x = 105$
21. $x = 3, y = 4$ **23.** $x = 3, y = 4$ **25.** $w = 15$
27. $y \approx 2.34$ **29.** $m = 4$ **31.** $w = 7$ **33.** $s = 3.6$
35. $\dfrac{2}{3}\pi$ ft^2 ≈ 2.1 ft^2 **37.** 18π cm^2 ≈ 56.5 cm^2
39. $54\dfrac{4}{9}$ cm^2 ≈ 54.4 cm^2 **41.** $3\dfrac{1}{9}\pi$ in. ≈ 9.8 in.
43. $S.A. = 576\pi$ in.^{2} ≈ 1809.6 in.^{2}; $V = 2304\pi$ in.^{3} $\approx$ 7238.2 in.^{3} **45.** $S.A. = 256\pi$ m^2 ≈ 804.2 m^2;
$V = 682\dfrac{2}{3}\pi$ m^3 ≈ 244.7 m^3 **47.** $\dfrac{128\pi}{3}$ ft^3 **49.** False.

Pages 685–686   Chapter 12

1. $\begin{bmatrix} -2 \\ 1 \end{bmatrix}$ **3.** $\begin{bmatrix} -2 & 1 \\ 1 & 0 \end{bmatrix}$ **5.** $\begin{bmatrix} 8 \\ -6 \end{bmatrix}$ **7.** $\begin{bmatrix} 4 & 3 & -3 \\ -2 & 1 & 5 \end{bmatrix}$

9. $\begin{bmatrix} 20 & -40 \\ 15 & 30 \end{bmatrix}$ **11.** $\begin{bmatrix} -3 \\ 9 \end{bmatrix}$ **13.** $\begin{bmatrix} 2 & 4 & 5 \\ -4 & 0 & -1 \end{bmatrix}$ **15.** $\begin{bmatrix} -3 & -3 \\ 0 & 0 \end{bmatrix}$

17. $\begin{bmatrix} 0 & 0 & 0 & 0 \\ -4 & -4 & -4 & -4 \end{bmatrix}$ **19.** $\begin{bmatrix} -4 & -1 \\ 10 & 3 \end{bmatrix}$

21. $\begin{bmatrix} 0 & -0.5 & 1.5 \\ 2.4 & 3.8 & 12.6 \end{bmatrix}$

23.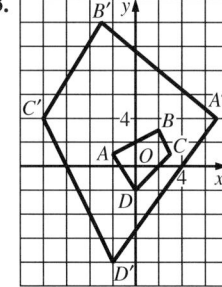

25. $(-1, 2)$ **27.** $(1, -2)$

Pages 677–679   Chapter 9

1. $x = 12$ **3.** $z = 3\sqrt{5}$ **5.** $x = 9$ **7.** Yes; corresponding sides are proportional.   **9.** No; corresponding sides are not proportional.   **11.** $x = 8.25, y = 2.75$ **13.** $x = 3\dfrac{2}{3}$ **15.** No; only one pair of corresponding angles are congruent.
17. No; only two pairs of corresponding sides are in proportion.   **19.** SAS Similarity Theorem   **21.** SAS Similarity Theorem or SSS Similarity Theorem   **23.** AA Similarity Postulate   **25.** $y = 6$ **27.** $w = 12.5$ **29.** $y = 9.6$
31. $\dfrac{y}{5}$ **33.** $2\dfrac{2}{9}$ **35.** $12\pi \approx 37.70$ **37.** 22.5   **39.** $\dfrac{1}{2} = 0.5$
41. $\dfrac{16}{49} \approx 0.33$ **43.** $\dfrac{5}{14} \approx 0.36$ **45.** $\dfrac{3}{15} \approx 0.2$

Pages 680–682   Chapter 10

1. SQR **3.** SQ **5.** 9   **7.** $1\dfrac{1}{2}$ **9.** 1   **11.** $x = 2\sqrt{2} \approx 2.8$
13. $u = 8\dfrac{1}{3}$ **15.** $x = 8$ **17.** $x = 10$ **19.** $z = 5\sqrt{2} \approx 7.1$
21. $x = 6\sqrt{2} \approx 8.5$ **23.** $p = 10.5$ **25.** $x = 45$,
$y = 6\sqrt{2} \approx 8.5$ **27.** $\tan A = \dfrac{4}{3}, \tan B = \dfrac{3}{4}$ **29.** 0.5317
31. 11.4301   **33.** $y \approx 5.5$
35. $\sin A = \dfrac{2\sqrt{5}}{5}, \cos A = \dfrac{\sqrt{5}}{5}, \sin B = \dfrac{\sqrt{5}}{5}, \cos B = \dfrac{2\sqrt{5}}{5}$
37. $\sin A = \dfrac{3}{5}, \cos A = \dfrac{4}{5}, \sin B = \dfrac{4}{5}, \cos B = \dfrac{3}{5}$
39. 0.1478   **41.** 0.9781   **43.** 75.2°   **45.** 11.6°   **47.** 50.3°
49. 56.3°   **51.** $m \approx 3.53, n \approx 7.52$ **53.** $x \approx 53.1$
55. $z \approx 41.4$ **57.** (2, 3); $y \approx 56.3$

TOOLBOX

Page 687   Using a Protractor

1. $45°$ **2.** $105°$ **3.** $90°$ **4.** $25°$ **5.** $70°$ **6.** $135°$ **7.** $60°$
8. $115°$

Page 688   Using a Compass

1–8. Answers may vary.

Page 689   Transformations

1. 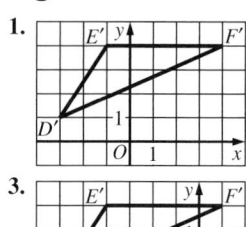   **2.**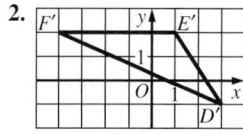

3. 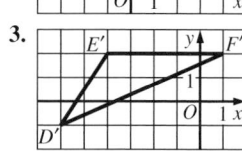   **4.**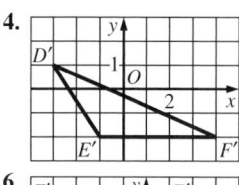

5. 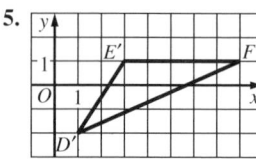   **6.**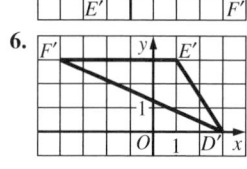

Page 690   Recognizing Symmetry

1. no symmetry   **2.** reflection and rotational symmetry
3. rotational symmetry   **4.** reflection symmetry
5–7. Answers may vary. Examples are given.

5.   **6.**   **7.**

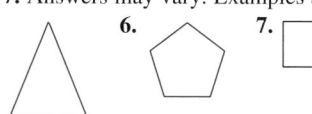

Page 692   Simplifying and Evaluating Expressions

1. -8 **2.** 49 **3.** -168 **4.** 14 **5.** -9 **6.** -18 **7.** -1
8. 53 **9.** 3 **10.** 2 **11.** $-\frac{5}{9}$ **12.** 13 **13.** $3x - 1$
14. $4x^2 - 4x^3$ **15.** $4x^3 - 10x$ **16.** $11x + 21$
17. $t^2 + 9t + 36$ **18.** $9x^2 + 15x$ **19.** x **20.** $-23m$
21. $5x^3 + 7x^2 + 7x + 7$ **22.** $x + 1$ **23.** $-7x + 12$
24. $7x^2 - 17x$ **25.** -23 **26.** 239 **27.** -56 **28.** 32 **29.** 10
30. 25 **31.** -45 **32.** 45 **33.** 2 **34.** 8 **35.** -13 **36.** 45
37. 23 **38.** 12 **39.** 5 **40.** 23

Page 693   Translating Phrases into Variable
Expressions

1. A child y years old must pick up $3y$ toys.   **2.** The cost of m main dishes, s side dishes, and d drinks is $4.25m + 0.75s + d$.   **3.** If there is d amount of orange juice, each glass has $\frac{d}{8}$ amount.   **4.** Monica will charge $\$20h$ for working h hours.   **5.** The hurricane will move $45h$ miles in h hours.

6. m mosaics will require $44m$ yellow tiles, $125m$ white tiles, $90m$ blue tiles, $78m$ red tiles, and $13m$ green tiles.

Page 693   Algebraic Properties

1. a. commutative **b.** distributive **2. a.** associative
b. commutative **c.** associative **3. a.** distributive
b. commutative

Page 694   Evaluating Equations for Given Values

1. $y = -4$ **2.** $y = 1$ **3.** $y = 0$ **4.** $y = 12$ **5.** $y = 11$
6. $y = 14$ **7.** $y = -\frac{1}{4}$ **8.** $y = 24$

Page 695   Translating Sentences into Equations

1. The cost c of soda in dollars equals 0.80 times the number of bottles b. $c = 0.80b$ **2.** The total number of pieces p is 17 times the number of weeks w. $p = 17w$ **3.** Elaine's speed s in feet per second is the distance run d in feet divided by 15 s. $s = \frac{d}{15}$ **4.** The cost of using the program c in dollars is 1 plus 0.05 times the number of minutes the program is used, m. $c = 1 + 0.05m$ **5.** The total weight w of a container of food from the deli is the weight of the food f plus 0.01 lb. $w = f + 0.01$ **6.** The number n of oxygen atoms required is half the number of hydrogen atoms, h. $n = \frac{h}{2}$

Page 696   Solving One-Step Equations

1. $y = 5$ **2.** $x = 13$ **3.** $y = 9$ **4.** $t = -4$ **5.** $x = \frac{8}{3}$
6. $z = 45$ **7.** $t = 32$ **8.** $x = -1$ **9.** $y = 5$ **10.** $w = \frac{9}{8}$
11. $y = 4$ **12.** $x = -3$

Page 696   Solving Two-Step Equations

1. $x = \frac{1}{2}$ **2.** $x = 1$ **3.** $t = 4$ **4.** $x = -\frac{7}{6}$ **5.** $y = 2$ **6.** $x = 9$
7. $x = 9$ **8.** $y = -4$ **9.** $x = -6.416$ **10.** $x = \frac{23}{3} = 7.6$
11. $t = 5.4583$ **12.** $y = 0.44$

Page 697   Graphing Points

1–8. 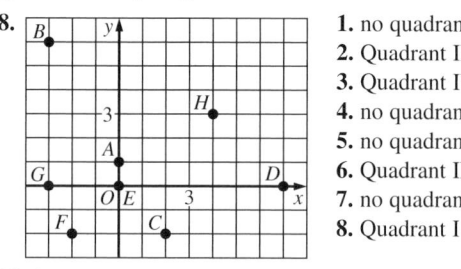   **1.** no quadrant
2. Quadrant II
3. Quadrant IV
4. no quadrant
5. no quadrant
6. Quadrant III
7. no quadrant
8. Quadrant I

9, 10. Answers may vary. Examples are given.   **9.** $(2, -1)$
10. $(-2, 0)$

Pages 699–700   Graphing Linear Equations

1. slope 3; y-intercept 0 **2.** slope -2; y-intercept 0
3. slope -1; y-intercept 2 **4.** slope 0; y-intercept -4
5. slope -5; y-intercept -1 **6.** slope 8; y-intercept -3
7. slope 1; y-intercept 2 **8.** slope 13.8; y-intercept 122

SA36   Selected Answers

9. slope 3; y-intercept -3 **10.** slope $-\dfrac{5}{4}$; y-intercept $-\dfrac{3}{2}$

11. slope $\dfrac{1}{2}$; y-intercept $\dfrac{1}{2}$ **12.** slope $-\dfrac{2}{3}$; y-intercept 3

13. 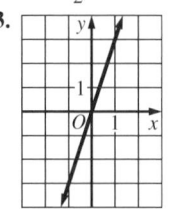   **14.**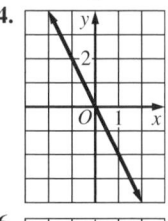

15. 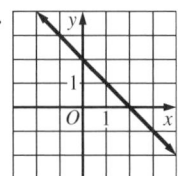   **16.**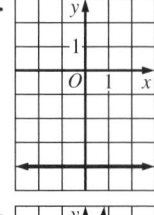

17. 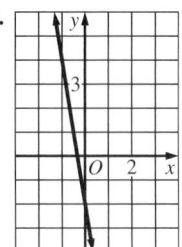   **18.**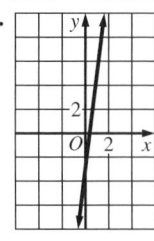

19. 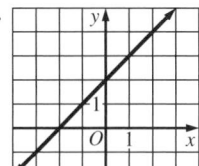   **20.**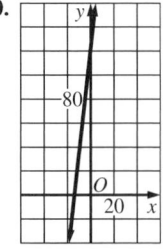

21. 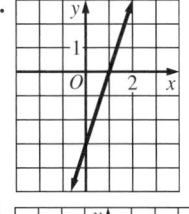   **22.**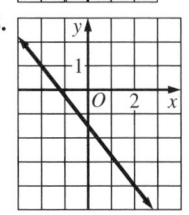

23. 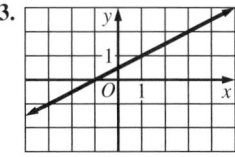   **24.**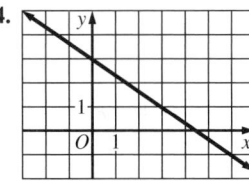

25. slope 0; y-intercept 2   **26.** slope 1; y-intercept 12

27. slope $-\dfrac{1}{2}$; y-intercept -2 **28.** slope 2; y-intercept -2

29. slope $\dfrac{1}{3}$; y-intercept -5 **30.** slope -5; y-intercept -12

31. slope $\dfrac{4}{3}$; y-intercept -8 **32.** slope $\dfrac{1}{3}$; y-intercept 3

33. slope $-\dfrac{1}{3}$; y-intercept 6

Page 701   Simplifying Radicals

1. 4   **2.** $5\sqrt{3}$ **3.** $\sqrt{2}$ **4.** 6   **5.** $\dfrac{2\sqrt{2}}{3}$ **6.** $\dfrac{1}{8}$ **7.** $\dfrac{\sqrt{15}}{2}$

8. $\dfrac{2\sqrt{3}}{5}$ **9.** $5\sqrt{2}$ **10.** $5\sqrt{5}$ **11.** $-5\sqrt{5}$ **12.** -6 **13.** 49

14. 9   **15.** $\dfrac{1}{2}$ **16.** $\dfrac{\sqrt{6}}{2}$

Page 702   Solving Simple Quadratic Equations

1. $x = 5$ or $x = -5$ **2.** $x = 5\sqrt{2}$ or $x = -5\sqrt{2}$ **3.** $x = \sqrt{6}$ or $x = -\sqrt{6}$ **4.** $x = 6$ or $x = -6$ **5.** $x = 1$ or $x = -1$ **6.** $x = 4$ or $x = -4$ **7.** $x = \sqrt{10}$ or $x = -\sqrt{10}$ **8.** $x = 0$ **9.** $x = 2$ or $x = -2$ **10.** $x = 3$ or $x = -3$ **11.** $x = \sqrt{29}$ or $x = -\sqrt{29}$ **12.** $x = 3$ or $x = -3$

Page 703   Working with Formulas

1. $t = \dfrac{1}{3}$ h   **2.** $d = 25$ mi   **3.** $d = 66\dfrac{2}{3}$ mi/h   **4.** $s = 25$ mi

5. $A = 0.16\pi$ **6.** $r = 3$ **7. a.** $c = \dfrac{ad}{4b}$ **b.** $a = 16$ in.

Page 703   Finding Perimeter

1. 19 ft   **2.** 26   **3.** 36   **4.** 16 in.   **5.** 14 ft   **6.** 21 units

Page 704   Finding Circumference

1. 16π in.   **2.** 4π cm   **3.** 2π in.   **4.** 24π in.   **5.** 4π

6. 28π m   **7.** 60π **8.** 9π **9.** 9 in.   **10.** 17   **11.** $\dfrac{1}{2}$

12. $\dfrac{1}{\pi}$ mi

Page 705   Finding Area and Volume

1. 12   **2.** 2   **3.** 168   **4.** 3.5   **5.** 2.25   **6.** 60   **7.** 4   **8.** 20
9. 10   **10.** 27   **11.** 1400   **12.** 6   **13.** 96   **14.** 0.125
15. $l = 2$ **16.** $w = 3.75$ **17.** $w = \dfrac{1}{3}$ **18.** $A = 50$

Page 705   Solving Inequalities

1. $y < 16$

2. $n > \dfrac{9}{7}$

3. $n \le 2$

4. $n > 0$

5. $x \le 1$

6. $x < \dfrac{15}{14}$

7. $y < -8$

8. $n \le -1$

Pages 706–707 Graphing Systems of Equations and Inequalities

1.
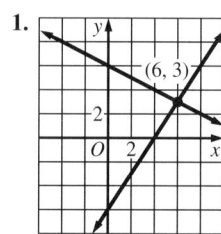
(6, 3)

2.
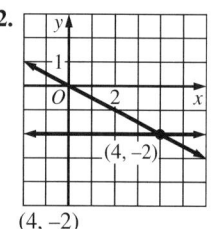
(4, −2)

3.
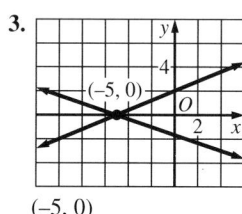
(−5, 0)

4.
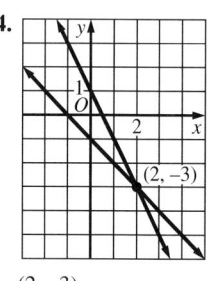
(2, −3)

5.
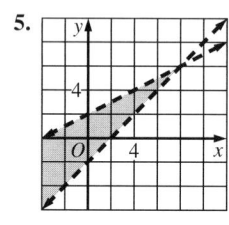

6.
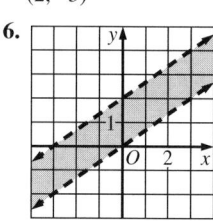

7.
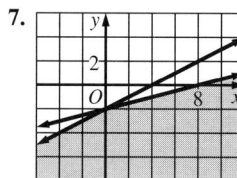

8.
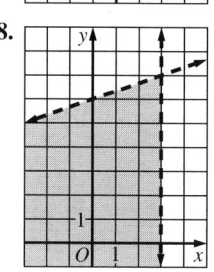

Page 707 Creating a Ratio

1. $\frac{4}{1}$ **2.** $\frac{9}{20}$ **3.** $\frac{1}{5}$ **4.** $\frac{3}{7}$ **5.** $\frac{1}{3}$ **6.** $\frac{1}{8}$ **7.** $\frac{12}{5}$ **8.** $\frac{1}{100}$ **9.** $\frac{1}{6}$
10. about 3.3 miles per hour **11.** $40 per hour **12.** $7500 per year **13.** about 6.4 hours per day **14.** 3.25 cats per hour **15.** 24 miles per day **16.** 120 cycles per second **17.** 0.25 inches per minute **18.** 30 apples per basket

Page 708 Solving Proportions

1. $x = 2$ **2.** $x = 7$ **3.** $n = \frac{1}{3}$ **4.** $y = 26$ **5.** $x = 27$
6. $x = 20$ **7.** $m = \frac{4}{5}$ **8.** $x = 35$

Page 709 Finding Averages

1. mean: 7.25; median: 7 **2.** mean: 19.4; median: 20
3. mean: 5.125; median: 6 **4.** mean: 11.75; median: 11.5
5. mean: 14.16; median: 16 **6.** mean: 1.857142; median: 2
7. mean: −1; median: −3 **8.** mean: 19; median: 18

Page 709 Scientific Notation

1. 0.000018 **2.** 0.0000000000455 **3.** 5,200,000,000
4. 381.2 **5.** 98.4 **6.** 0.000000057 **7.** 0.39 **8.** 42,000
9. 8.8×10^{-4} **10.** 1.2×10^{-1} **11.** 1.245×10^{6}
12. 1.31×10^{-2} **13.** 6×10^{-11} **14.** 4.5×10^{4}
15. 1.58×10^{8} **16.** 2.3×10^{5} **17.** 6.2×10^{3}
18. 5×10^{-3} **19.** 1.1×10^{-12} **20.** 3.48651×10^{6}

Page 710 Probability

1. $\frac{3}{4}$, or 0.75 **2.** $\frac{10}{29} \approx 0.34$ **3.** $\frac{2}{5}$, or 0.4 **4.** $\frac{4}{7} \approx 0.57$
5. $\frac{3}{5}$, or 0.6 **6.** $\frac{1}{2}$, or 0.5 **7.** $\frac{1}{13} \approx 0.08$ **8.** $\frac{1}{2}$, or 0.5
9. $\frac{3}{20}$, or 0.15 **10.** 0

Page 711 Matrices

1. $\begin{bmatrix} 6 & 5 \\ 2 & 13 \end{bmatrix}$ **2.** $\begin{bmatrix} 7 \\ 10 \end{bmatrix}$ **3.** $\begin{bmatrix} 2 & -4 \\ 2 & 1 \end{bmatrix}$ **4.** $\begin{bmatrix} 1 \\ -5 \\ -5 \end{bmatrix}$ **5.** $\begin{bmatrix} 10 & 7 \\ 8 & 16 \end{bmatrix}$

6. $\begin{bmatrix} 1 & 11 & -6 \\ 2 & -2 & -4 \end{bmatrix}$ **7.** $\begin{bmatrix} 2 & 0 \\ 0 & 2 \end{bmatrix}$ **8.** $\begin{bmatrix} -4 & \frac{3}{2} \\ 3 & -2 \end{bmatrix}$ **9.** $\begin{bmatrix} 72 \\ 40 \end{bmatrix}$

10. $\begin{bmatrix} 3 & 12 & -6 \\ -12 & -3 & 3 \end{bmatrix}$ **11.** $\begin{bmatrix} -3 & 0 \\ 0 & -3 \end{bmatrix}$ **12.** $\begin{bmatrix} -6 & 0 & 10 \\ 16 & -4 & -2 \\ 18 & 14 & 0 \end{bmatrix}$

13. $\begin{bmatrix} -\frac{25}{3} & -5 \\ -5 & 5 \\ \frac{5}{3} & -10 \end{bmatrix}$ **14.** $\begin{bmatrix} \frac{9}{2} \\ 3 \\ -12 \end{bmatrix}$

TECHNICAL HANDBOOK

Pages 722–725 Calculator Practice

1. −12 **2.** 31.85 **3.** 3.4 **4.** about 10.3 **5.** 4096
6. about 1436.8 **7.** 1.804 **8.** 51.5° **9.** The shape of the scatter plot is similar to half of a parabola.

10. $\begin{bmatrix} -3 & 0 & 5 \\ 4 & 8 & 3 \end{bmatrix}$ **11.** $\begin{bmatrix} 5 & 20 & 45 \\ 15 & 35 & 10 \end{bmatrix}$ **12.** $\begin{bmatrix} -3 & -7 & -2 \\ -1 & -4 & -9 \end{bmatrix}$

13. $\begin{bmatrix} -3 & -7 & -2 \\ 1 & 4 & 9 \end{bmatrix}$

SA38 Selected Answers

Page 725 **Spreadsheet Practice**

14, 15.

Type of ball	r(cm)	V(cm^3)		
Golf ball	2.1	38.79238609		
Tennis ball	3.3	150.5325536		
Softball	4.8	463.2466863		
Volleyball	10.5	4849.048261		
Basketball	11.9	7058.777513		
		Pythagorean	Triples	
n		n^2–1	2n	n^2+1
2		3	4	5
3		8	6	10
4		15	8	17
5		24	10	26
6		35	12	37
7		48	14	50
8		63	16	65
9		80	18	82
10		99	20	101

Pages 728–729 **Software Practice**

16.
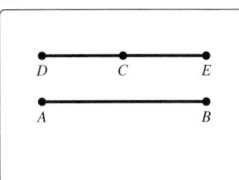

17.
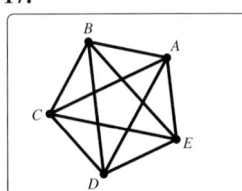

18.
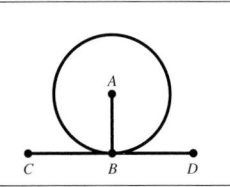

19.
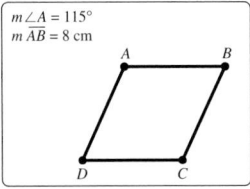
$m\angle A = 115°$
$m\,\overline{AB} = 8$ cm

c. The point at which the three angle bisectors of a triangle intersect is equidistant from the sides of the triangle.

20. a, b.
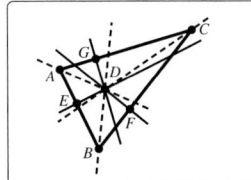

21.
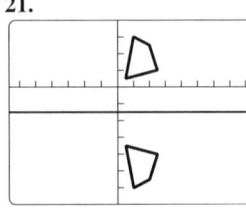

22.
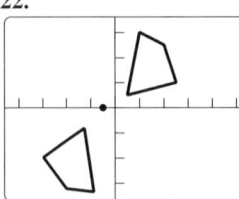

23.
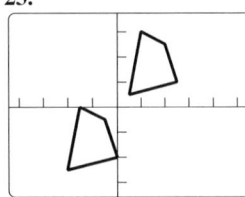

24.
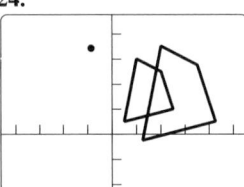

25.
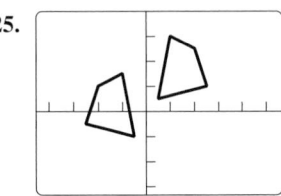

APPENDIX 2

Pages 740–741 **Exercises and Applications**

1. I win and you win.   **3.** I do not win and you do not win.
5. If you win or they win, then I do not win.   **7.** I win or both you and they win.   **9.** $q \rightarrow r$ **11.** $\sim p \rightarrow \sim(s \vee r)$

13.

p	$\sim p$	$p \wedge \sim p$
T	F	F
F	T	F

contradiction

15.

p	$\sim p$	$\sim(\sim p)$
T	F	T
F	T	F

17.

p	q	$\sim p$	$\sim p \vee q$
T	T	F	T
T	F	F	F
F	T	T	T
F	F	T	T

19.

p	q	$\sim p$	$\sim p \wedge q$
T	T	F	F
T	F	F	F
F	T	T	T
F	F	T	F

21. (1) "It is not the case that p or q is true" has the same meaning as "It is not the case that p is true, and it is not the case that q is true." (2) "It is not the case that both p and q are true" has the same meaning as "It is not the case that p is true or it is not the case that q is true."

23.

p	q	$p \rightarrow q$	$q \rightarrow p$	$(p \rightarrow q) \wedge (q \rightarrow p)$
T	T	T	T	T
T	F	F	T	F
F	T	T	F	F
F	F	T	T	T

25. a contradiction, since there are all F's in the final column of the truth table:

p	q	$p \wedge q$	$\sim p$	$\sim q$	$\sim p \vee \sim q$	$(p \wedge q) \wedge (\sim p \vee \sim q)$
T	T	T	F	F	F	F
T	F	F	F	T	T	F
F	T	F	T	F	T	F
F	F	F	T	T	T	F

27. a tautology, since there are all T's in the final column of the truth table:

p	q	$\sim p$	$p \vee q$	$(p \vee q) \wedge \sim p$	$[(p \vee q) \wedge \sim p] \to q$
T	T	F	T	F	T
T	F	F	T	F	T
F	T	T	T	T	T
F	F	T	F	F	T

29.

p	q	$p \to q$	$\sim q$	$(p \to q) \wedge \sim q$	$\sim p$	$[(p \to q) \wedge \sim q] \to \sim p$
T	T	T	F	F	F	T
T	F	F	T	F	F	T
F	T	T	F	F	T	T
F	F	T	T	T	T	T

31.

p	q	r	s	$p \to q$	$q \to r$	$r \to s$	$(p \to q) \wedge (q \to r) \wedge (r \to s)$
T	T	T	T	T	T	T	T
T	T	T	F	T	T	F	F
T	T	F	T	T	F	T	F
T	T	F	F	T	F	T	F
T	F	T	T	F	T	T	F
T	F	T	F	F	T	F	F
T	F	F	T	F	T	T	F
T	F	F	F	F	T	T	F
F	T	T	T	T	T	T	T
F	T	T	F	T	T	F	F
F	T	F	T	T	F	T	F
F	T	F	F	T	F	T	F
F	F	T	T	T	T	T	T
F	F	T	F	T	T	F	F
F	F	F	T	T	T	T	T
F	F	F	F	T	T	T	T

$p \to s$	$[(p \to q) \wedge (q \to r) \wedge (r \to s)] \to (p \to s)$
T	T
F	T
F	T
F	T
F	T
F	T
F	T
F	T
T	T
F	T
F	T
F	T
T	T
F	T
T	T
T	T

16 rows

33. Answers may vary. An example is given. Relate circuits placed in series to truth tables by labeling a closed switch T and an open switch F and using the truth table for $p \wedge q$. Relate circuits placed in parallel to truth tables by labeling a current that will flow in a circuit T and current that will not flow F and use the truth table for $p \vee q$.

Additional Answers

CHAPTER 1

Page 3 Exploration

1–3. Check students' work.

4, 5. The table values for the first seven steps are given.

Step	Number of new segments	Total number of segments
1	1	1
2	2	3
3	4	7
4	8	15
5	16	31
6	32	63
7	64	127

6. The figure shown is Step 2 is repeated over and over again, on a smaller and smaller scale.

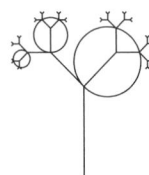

7. The number of new segments doubles at each step; the total number of segments increases by a power of two at each step. At Step n, the number of new segments is 2^{n-1}; the total number of segments is $2^n - 1$.

8. 1023 segments; $2^{10} - 1 = 1023$

Page 11 Exercises and Applications

7. **8.**

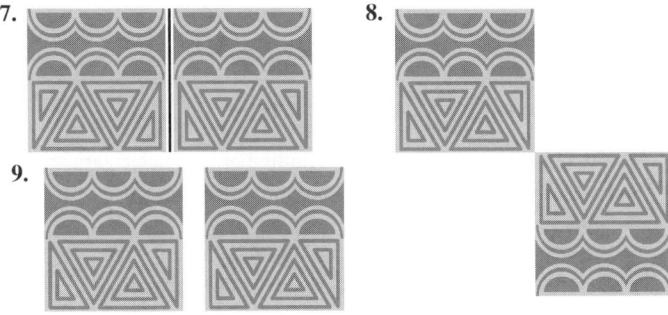

9.

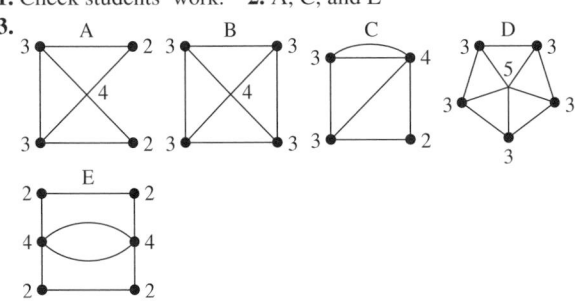

Page 14 Exploration

1. Check students' work. **2.** A, C, and E

3.

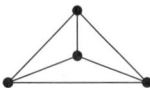

4.

Network	Number of odd vertices	Traceable?
House	2	Yes
A	2	Yes
B	4	No
C	2	Yes
D	6	No
E	0	Yes

Networks with 0 or 2 odd vertices are traceable. Networks with 4 or 6 odd vertices are not traceable.

5. Sketches may vary. An example is given.

The network is not traceable. Any network with more than 2 odd vertices is not traceable.

6. Answers may vary. An example is given. A network is traceable if it has 0 or 2 odd vertices.

Page 19 Exercises and Applications

23–27. Answers may vary. Examples are given. **23.** If today is January 1, then tomorrow is January 2. **24.** If $5x + 7 = -3x - 5$, then $x = -\frac{3}{2}$. **25.** If a point is located on the x-axis, then the y-coordinate of the point is 0. **26.** If a shape has reflection symmetry, then it has a line of symmetry. **27.** If you translate a point up one unit on a coordinate plane, then the y-coordinate of the point increases by 1. **28.** D **29.** Answers may vary. An example is given. If I am not awake by 7:00 A.M., I will be late for school. Counterexample: Today is Sunday.

Page 20 Assess Your Progress

5. a. **b.** **c.**

Page 53 Chapter 1 Assessment

24. Answers may vary. Examples are given. $\overline{ST} \cong \overline{VU}$, $\overline{SV} \parallel \overline{TU}$; Angles S, T, U, and V are all congruent to one another. **25. a.** 5 **b.** 10 **c.** 8

CHAPTER 2

Page 78 Exercises and Applications

25. Sketches may vary. Examples are given.

Polygon	Triangle	Quadrilateral	Pentagon
Concave	not possible		
Convex			
Regular			
Equilateral only	not possible		
Equiangular only	not possible		

Page 79 Exploration

1–2. Sketches may vary. Examples are given.

Type of polygon	Number of sides	Number of triangles formed	Sum of ∠ measures of triangles	Sum of ∠ measures of polygon
triangle	3	1	180°	180°
quadrilateral	4	2	360°	360°
pentagon	5	3	540°	540°
hexagon	6	4	720°	720°
heptagon	7	5	900°	900°

For each polygon, the number of triangles formed is 2 less than the number of sides. The sum of the angle measures in the triangles is 180° times the number of triangles. The sum of the angle measures in the polygon is the same as the sum of the angle measures of the triangles.

3.

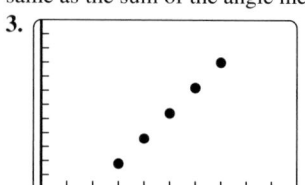

```
WINDOW    FORMAT
  Xmin = 0
  Xmax = 10
  Xscl = 1
  Ymin = 0
  Ymax = 1200
  Yscl = 100
```

The data points appear to lie along a line.

4. For an n-gon, the number of sides is n, the number of triangles formed is $n - 2$, the sum of the angle measures of the triangles is $(n - 2)180°$. The sum of the angle measures of a polygon is found by subtracting 2 from the number of sides, then multiplying by 180°.

Page 97 Exercises and Applications

23. a. 5; 4   **b.** Answers may vary. An example is given.

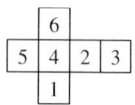

24. Answers may vary. An example is given. If the lateral faces of the prism are rectangles, then the prism is right. If the lateral faces are parallelograms that are not equiangular, then the prism is oblique.

CHAPTER 3

Page 138 Checking Key Concepts

1. If you are in Guyana, then the official language is English; True; True.
2. If an official language is Quechua, then you are in Peru; True; False.
3. If the official language is Portuguese, then you are in Brazil; True; True.
4. Answers may vary. An example is given. If you are in Suriname, then the official language is Dutch. If the official language is Dutch, then you are in Suriname; True.   **5. a.** If a quadrilateral is a square, then it is a rectangle.
b. If a quadrilateral is a rectangle, then it is a square.   **6. a.** If I earn enough money, then I will buy a car.   **b.** If I buy a car, then I earn enough money.
7. a. If a figure is a triangle, then the sum of the angle measures is 180°.
b. If the sum of the angle measures of a figure is 180°, then the figure is a triangle.

Page 138 Exercises and Applications

6. a. If an angle is obtuse, then a supplement of the angle is acute. **b.** If a supplement of an angle is acute, then the angle is obtuse.   **7. a.** If a triangle is isosceles, then it is not scalene.   **b.** If a triangle is not scalene, then it is isosceles.   **8. a.** If a quadrilateral is a parallelogram, then its diagonals bisect each other.   **b.** If the diagonals of a quadrilateral bisect each other, then the quadrilateral is a parallelogram.   **9. a.** If a quadrilateral is a square, then its diagonals are congruent and perpendicular.   **b.** If the diagonals of a quadrilateral are congruent and perpendicular, then the quadrilateral is a square.

Page 139 Exercises and Applications

12. The order of the statements may vary.
Given: In $\triangle PQR$, $\angle P$ and $\angle Q$ are complementary angles.
Prove: $\angle R$ is a right angle.

Statements	Reasons
1. $\angle P$ and $\angle Q$ are complementary angles.	1. Given
2. $m\angle P + m\angle Q = 90°$	2. Definition of complementary angles
3. $m\angle P + m\angle Q + m\angle R = 180°$	3. The sum of the angle measures of a triangle is 180°.
4. $90° + m\angle R = 180°$	4. Substitution Property (Steps 2 and 3)
5. $m\angle R = 90°$	5. Subtraction Property
6. $\angle R$ is a right angle.	6. Definition of a right angle

13. a. sometimes true   **b.** If an English translation uses the word "live," a Japanese sentence would use the word "sumu"; False.

3. Table values may vary. Examples are given.

Length of sides	$a^2 + b^2$	c^2	Type of triangle
3, 4, 5	25	25	right
4, 4, 5	32	25	acute
3, 6, 7	45	49	obtuse
5, 6, 7	61	49	acute
5, 12, 13	169	169	right
4, 12, 13	160	169	obtuse
6, 12, 13	180	169	acute
4, 6, 8	52	64	obtuse
6, 8, 10	100	100	right
8, 10, 12	164	144	acute
10, 12, 14	244	196	acute

39. Answers may vary. An example is given. True conditional statement: If I plan to go sailing today, I will put on sunscreen before I leave my house. Inverse: If I do not plan to go sailing today, I will not put on sunscreen before I leave my house. (False.) Counterexample: I may put on sunscreen before I leave the house because I plan to go swimming, not sailing. Converse: If I put on sunscreen before I leave my house, then I plan to go sailing today. (False.) Counterexample: I might put on sunscreen before I leave my house because I plan to go for a walk. Contrapositive: If I do not put on sunscreen before I leave my house, then I do not plan to go sailing today. (True.) False conditional statement: If $AM = MB$, then M is the midpoint of $\overline{AB}$. (This is false because A, M, and B could be the vertices of a triangle with $AM = MB$.) Inverse: If $AM \neq MB$, then M is not the midpoint of $\overline{AB}$. (True.) Converse: If M is the midpoint of $\overline{AB}$, then $AM = MB$. (True.) Contrapositive: If M is not the midpoint of $\overline{AB}$, then $AM \neq MB$. (False.) Counterexample: A, M, and B could be the vertices of a triangle with $AM = MB$.

CHAPTER 4

1. a.

lane	swimmer position (S)	timer position (T)	PS (m)	PT (m)	$\dfrac{PS}{346}$	$\dfrac{PT}{346}$	advantage (s)
8	$S(0, 19.00)$	$T(50.00, 19.00)$	24.21	39.82	0.070	0.115	0.045
7	$S(0, 16.60)$	$T(50.00, 16.60)$	22.37	38.74	0.065	0.112	0.047
6	$S(0, 14.20)$	$T(50.00, 14.20)$	20.66	37.77	0.060	0.109	0.049
5	$S(0, 11.80)$	$T(50.00, 11.80)$	19.09	36.94	0.055	0.107	0.052
4	$S(0, 9.40)$	$T(50.00, 9.40)$	17.70	36.24	0.051	0.105	0.054
3	$S(0, 7.00)$	$T(50.00, 7.00)$	16.55	35.69	0.048	0.103	0.055
2	$S(0, 4.60)$	$T(50.00, 4.60)$	15.69	35.30	0.045	0.102	0.057
1	$S(0, 2.20)$	$T(50.00, 2.20)$	15.16	35.07	0.044	0.101	0.057

b. Answers may vary. An example is given. The swimmers in the first two lanes have the greatest time advantage. The time advantages decrease as the lane number increases. The swimmer in lane 8 has the least advantage.

24, 25.

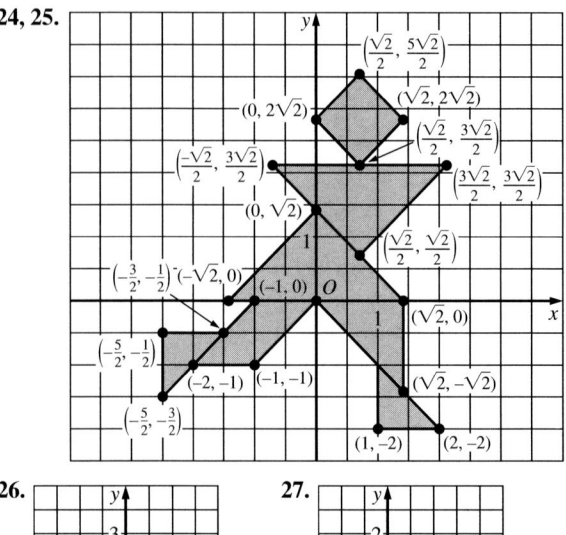

26. **27.**

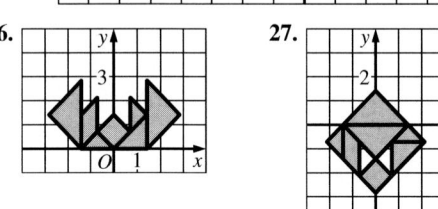

7. Answers may vary. An example is given. Find the midpoint of the segment, $M(a, b)$. If the segment is neither horizontal nor vertical and has slope k, then the slope of the perpendicular bisector is $-\frac{1}{k}$. Substitute the slope and the coordinates (a, b) in the slope-intercept equation $y = mx + b$ and simplify to find the value of b. Write the equation of the perpendicular bisector in slope-intercept form. (If the segment is vertical, then an equation for the perpendicular bisector is $y = b$. If the segment is horizontal, then an equation for the perpendicular bisector is $x = a$.) Sketches may vary.

14. **15.**

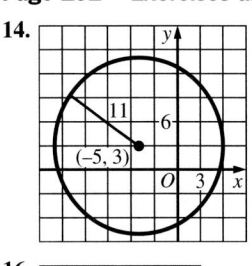
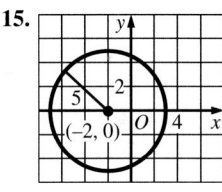

16.

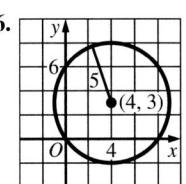

17–20. Answers may vary. Examples are given.

17. **18.**

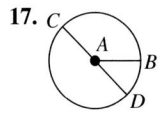
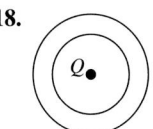

19. **20.**

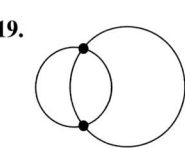
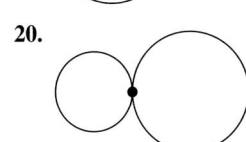

24. Answers may vary. An example is given. Use the diagram given in Exs. 17–20. The coordinates of points E, F, G, and H are $E(a, 0)$, $F(a + b, c)$, $G(b + d, c + e)$, and $H(d, e)$. The slope of $\overline{EF} = \frac{c}{b}$ = slope of $\overline{GH}$ and the slope of $\overline{FG} = \frac{e}{d - a}$ = slope of $\overline{EH}$. Since the slopes of the opposite sides of quadrilateral $EFGH$ are equal, the opposite sides are parallel. $EFGH$ is a parallelogram by definition.

25. a. slope of $\overline{OY} = \frac{r}{p + q}$; slope of $\overline{XZ} = \frac{r}{q - p}$ **b.** $\frac{r}{p + q} \cdot \frac{r}{q - p} = -1$

c. $p = \sqrt{q^2 + r^2}$ **d.** $OX = \sqrt{(p - 0)^2 + (0 - 0)^2} = p$; $XY = \sqrt{(p + q - p)^2 + (r - 0)^2} = \sqrt{q^2 + r^2} = p$; $YZ = \sqrt{(p + q - q)^2 + (r - r)^2} = p$; $OZ = \sqrt{(q - 0)^2 + (r - 0)^2} = \sqrt{q^2 + r^2} = p$. Therefore, $OXYZ$ is a rhombus by definition.

28. a. $PR = QS = \sqrt{(a + b)^2 + c^2}$, so the diagonals are congruent.

b. $M = \left(\frac{b - a}{2}, \frac{c}{2}\right)$ and $N = \left(\frac{a - b}{2}, \frac{c}{2}\right)$. **c.** Use the Distance Formula; $MN = a - b$ and $\frac{1}{2}(PS - QR) = \frac{1}{2}(2a - 2b) = a - b$. Therefore, $MN = \frac{1}{2}(PS - QR)$.

d. Answers may vary. An example is given. Given points $P(-a, 0)$, $Q(-b, c)$, $R(b, c)$, and $S(a, 0)$, use the Midpoint Formula to find the coordinates of M, the midpoint of $\overline{PR}$, and the coordinates of N, the midpoint of $\overline{QS}$. Then use the Distance Formula to find the lengths MN, PS, and QR. Show that $MN = \frac{1}{2}(PS - QR)$. This proves that the length of the segment whose endpoints are the midpoints of the diagonals of an isosceles trapezoid is equal to half the difference of the lengths of the bases.

9. Answers may vary. An example is given. I want to prove that the opposite sides of a parallelogram are congruent. I can use the diagram in Example 1 on page 195. Using the Distance Formula, $EF = GH = a$, and $FG = EH = \sqrt{b^2 + c^2}$. Since opposite sides have equal lengths, the opposite sides are congruent.

CHAPTER 5

5. Answers may vary. An example is given, based on the diagram provided for Exs. 22 and 23 on page 231. The angles that are supplementary to $\angle 2$ are $\angle 1$, $\angle 3$, $\angle 5$, and $\angle 7$. $\angle 1$ and $\angle 3$ are supplementary to $\angle 2$ because $\angle 2$ forms a linear pair with both of those angles. $\angle 2$ and $\angle 5$ are supplementary because they are same-side interior angles formed by two parallel lines and a transversal. $\angle 5 \cong \angle 7$ since they are vertical angles, so by the Substitution Property, $\angle 2$ and $\angle 7$ are supplementary.

12. Check students' work.

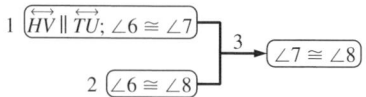

Reasons: (1) Given; (2) If two ∥ lines are intersected by a transversal, then corresponding ∡ are ≅; (3) Transitive Property (Steps 1 and 2)

14. Check students' work.

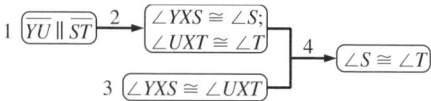

Reasons: (1) Given; (2) If two ∥ lines are intersected by a transversal, then alternate interior ∡ are ≅; (3) Given; (4) Transitive Property (Steps 2 and 3)

25. Proofs may vary. An example is given.

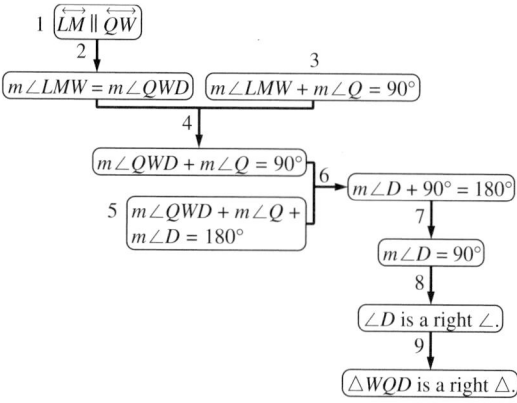

Reasons: (1) Given; (2) If two ∥ lines are intersected by a transversal, then corresponding ∡ are ≅; (3) Given; (4) Substitution Property (Steps 2 and 3); (5) The sum of the angle measures of a △ is 180°; (6) Substitution Property (Steps 4 and 5); (7) Subtraction Property; (8) Def. of right ∠; (9) Def. of right △

7. Answers may vary. An example is given. A paragraph proof focuses on the key ideas and omits minor details. Because of this, a paragraph proof is a concise form of proof, but it may be hard to follow. A two-column proof is easy to understand because it is well organized and contains all the necessary information, but it may be quite long and involved. A flow proof emphasizes the relationships between the steps, but may be hard to follow since the reasons are separated from the steps and the arrows sometimes make a proof look confusing. Writing the key steps of a proof may be desirable when you want to be informal and to focus on the most important steps, but it may be hard to understand since some of the details are omitted.

1. $\overleftrightarrow{HG} \parallel \overleftrightarrow{DC}$ since $\overleftrightarrow{HG}$ and $\overleftrightarrow{DC}$ are the lines of intersection formed when the parallel planes $ABCD$ and $EFGH$ are intersected by plane $CDHG$.

2. It does not state that faces $ABFE$ and $DCGH$ lie in parallel planes, so the Intersecting Planes Theorem cannot be used.

13.

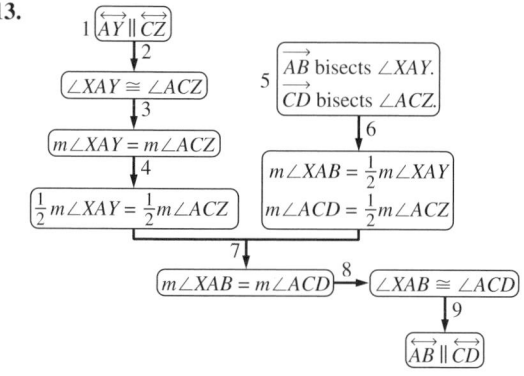

Reasons: (1) Given; (2) If two ∥ lines are intersected by a transversal, then corresponding ∡ are ≅; (3) Def. of ≅; (4) Multiplication Property; (5) Given; (6) Def. of bisector; (7) Substitution Property; (8) Def. of ≅; (9) If two lines are intersected by a transversal and corresponding ∡ are ≅, then the lines are ∥.

CHAPTER 6

Page 283 Exercises and Applications

27. Answers may vary. Examples are given.

Student	Lengths	Triangle?	Student	Lengths	Triangle?
1	7, 5, 2	No	2	3, 11, 6	No
1	5, 4, 2	Yes	2	4, 10, 3	No
1	2, 2, 6	No	2	10, 8, 1	No
3	9, 2, 1	No	4	7, 2, 11	No
3	11, 2, 1	No	4	7, 8, 9	Yes
3	6, 1, 10	No	4	9, 11, 10	Yes

Page 287 Exploration

1. a right triangle   2. The corresponding sides and angles of the triangles are congruent.   3. $\triangle BCA \cong \triangle XZY$

4. Diagrams may vary. Examples are given. In each diagram below, $\triangle ABC \cong \triangle YXZ$.

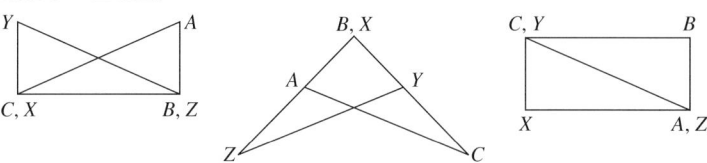

5. Answers may vary. Examples are given. In each diagram below, $\triangle ABC \cong \triangle YXZ$.

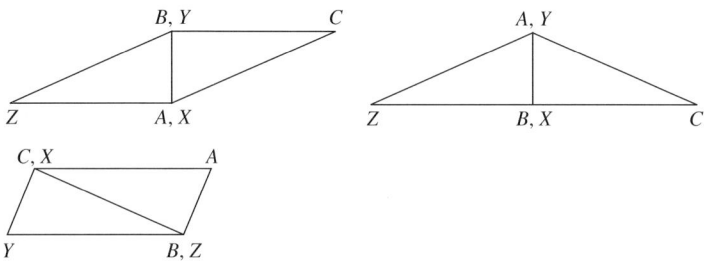

Page 298 Exercises and Applications

24. You can prove that two triangles are congruent by definition, that is, by showing that all six parts of one triangle are congruent to the corresponding six parts of another triangle, by showing that the three sides of one triangle are congruent to the three sides of another triangle (SSS Postulate), or by showing that two sides and the included angle of one triangle are congruent to two sides and the included angle of another triangle (SAS Postulate). Examples may vary.   25. $\triangle MON \cong \triangle QOP$ and $\triangle MQN \cong \triangle QMP$
26. $\triangle ABC \cong \triangle AED$ and $\triangle ABD \cong \triangle AEC$ 27. $1 : 120$ 28. $x° = 29°$
29. $x° = 38°$ 30. $x° = 37°$; $(2x - 6)° = 68°$

Page 304 Exercises and Applications

16. Since $\overline{NH} \perp$ plane EHG, $\overline{NH} \perp \overline{HG}$ and $\overline{NH} \perp \overline{HE}$. Then $\triangle GNH$ and $\triangle ENH$ are right triangles. $\overline{GN} \cong \overline{EN}$ (given) and $\overline{NH} \cong \overline{NH}$ (Reflexive Property), so $\triangle GNH \cong \triangle ENH$ by the HL Theorem.   17. Check students' work.   a. Answers may vary. There may be one or two ways to do Step 4. When there are two ways, only one of them results in a triangle that is congruent to $\triangle ABC$.   b. There is just one way to do Step 4, and that way gives a triangle that is congruent to $\triangle ABC$.   c. This investigation shows that if the hypotenuse and a leg of one right triangle are congruent to the corresponding parts of another right triangle, the triangles must be congruent. However, if two sides and a non-included angle of one triangle are congruent to two sides and a non-included angle of another triangle, the triangles may or may not be congruent.

Page 318 Exercises and Applications

22. Answers may vary. An example is given. Since $\overline{SY} \cong \overline{SW}$, $\angle SYW \cong \angle SWY$. Then, since $\overline{WR}$ and $\overline{YT}$ are angle bisectors, $\angle SYT$, $\angle TYW$, $\angle SWR$, and $\angle RWY$ are all congruent. $\triangle YWR \cong \triangle WYT$ by the ASA Postulate, and $\triangle SYT \cong \triangle SWR$ by the ASA Postulate. $\overline{YR} \cong \overline{WT}$ by the definition of con-

gruent triangles, so $\triangle YRX \cong \triangle WTX$ by the AAS Theorem.
23. Answers may vary. An example is given. An open pair of scissors forms two pairs of triangles. The handle end is usually like a scalene triangle; the cutting end is like an isosceles triangle. The cutting blades are the legs and the angle at the hinge of the scissors is the vertex angle. It would not make sense for the triangle to be scalene because the scissors only cut where the two blades meet. The longer blade part would serve no purpose.

CHAPTER 7

Page 345 Exercises and Applications

26, 27. Answers may vary. Examples are given.
26. 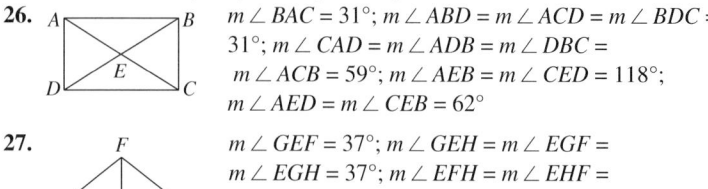 $m \angle BAC = 31°$; $m \angle ABD = m \angle ACD = m \angle BDC = 31°$; $m \angle CAD = m \angle ADB = m \angle DBC =$
$m \angle ACB = 59°$; $m \angle AEB = m \angle CED = 118°$;
$m \angle AED = m \angle CEB = 62°$

27. $m \angle GEF = 37°$; $m \angle GEH = m \angle EGF =$
$m \angle EGH = 37°$; $m \angle EFH = m \angle EHF =$
$m \angle GFH = m \angle GHF = 53°$

28. E
29. Given: rectangle $JKLM$; $\overline{JM} \cong \overline{NL}$
Prove: $m \angle LJK + m \angle LMK = m \angle KLN$
$JKLM$ is a rectangle (and, so, a parallelogram). The diagonals of a parallelogram bisect each other; the diagonals of a rectangle are congruent. Then $\triangle KNL$ and $\triangle JNM$ are equilateral triangles and $m \angle KLN = m \angle NJM = m \angle NMJ = 60°$. $\angle LJK$ and $\angle LMK$ are complements of $60°$ angles, so the measure of each angle is $30°$. Then $m \angle LJK + m \angle LMK = m \angle KLN$.
30. The diagonals of a parallelogram with no special properties form two distinct pairs of congruent scalene triangles that are not right triangles. Since the diagonals of a rhombus are perpendicular and bisect each other, if a rhombus is not a square, the triangles form two distinct pairs of congruent scalene right triangles. Since the diagonals of a rectangle are congruent and bisect each other, if the rectangle is not a square, the triangles form two distinct pairs of isosceles triangles that are not right triangles. Since a square is both a rhombus and a rectangle, the four triangles are congruent isosceles right triangles. For a kite, the triangles form two distinct pairs of congruent right triangles that are, in general, scalene. (It is possible that two of the triangles are isosceles.) For an isosceles trapezoid, two of the triangles are congruent and scalene and two are non-congruent and scalene. For a general quadrilateral, the triangles are generally non-congruent and scalene. If the diagonals are perpendicular but neither bisects the other, the four triangles will be right triangles.   31. a. 9
b. $75°$ 32. a. $40°$ b. $100°$ 33. a. $90°$ b. $28°$

Page 349 Exercises and Applications

6. 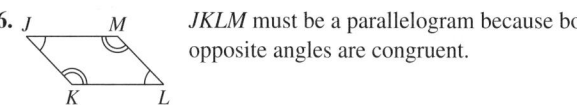 $JKLM$ must be a parallelogram because both pairs of opposite angles are congruent.

7. Answers may vary. An example is given. Draw a pair of congruent parallel segments $\overline{AB}$ and $\overline{DC}$. Draw $\overline{AD}$ and $\overline{BC}$. $ABCD$ is a parallelogram since one pair of opposite sides is both parallel and congruent.

Page 351 Exercises and Applications

23. $JKLM$ is a kite with $\overline{JK} \cong \overline{JM}$. Suppose that $\overline{JK}$ is congruent to $\overline{KL}$ or to $\overline{ML}$ as well. Both possibilities present a contradiction. If $\overline{JK} \cong \overline{ML}$, then two opposite sides of $JKLM$ are congruent and $JKLM$ is not a kite. Similarly, if $\overline{JK} \cong \overline{KL}$, then $\overline{JM} \cong \overline{KL}$ and $JKLM$ is not a kite. Therefore, $\overline{JK}$ is congruent to exactly one other side of the kite.   24. Let $WXYZ$ be a trapezoid with right $\angle X$. If it is not true that $WXYZ$ has exactly two right angles, then it has either one, three, or four right angles. If it has four right angles, both pairs of opposite angles are congruent and $WXYZ$ is a parallelogram. This is a contradiction, since a trapezoid has exactly one pair of parallel sides. If three angles of $WXYZ$ are right angles, so is the fourth, which we have already seen leads to a contradiction. (The measure of the third angle is $360° - 3(90°) = 90°$.) If

WXYZ has only one right angle, then neither angle adjacent to ∠ *X* is a right angle. This is a contradiction. One of the two must be a right angle. If $\overline{WX} \parallel \overline{ZY}$, then ∠ *Y* must be a right angle. (If two parallel lines are intersected by a transversal, then same-side interior angles are supplementary.) Similarly, if $\overline{WZ} \parallel \overline{XY}$, then ∠ *W* must be a right angle. The assumption that *WXYZ* does not have exactly two right angles must be incorrect. Therefore, if a trapezoid has a right angle, then it has exactly two right angles.

25. a. congruent rhombuses; Since the congruent metal pieces bisect each other, the twelve pieces that form the sides of the three quadrilaterals are all congruent. Therefore, each is an equilateral parallelogram, or a rhombus. (If both pairs of opposite sides of a quadrilateral are congruent, then the quadrilateral is a parallelogram.) **b.** Since the lengths of the sides do not change, the four quadrilaterals remain rhombuses. While the lift is in motion, vertical angles ∠*NMO* and ∠*KML* remain congruent. ∠*KML* and ∠*KJL* are opposite angles of a parallelogram, so they must remain congruent. By similar reasoning, all of the following angles remain congruent: ∠*NMO*; ∠*KML*, ∠*KJL*, ∠*GJH*, ∠*GFH*, ∠*DFE*, ∠*DCE*, and ∠*ACB*. Then ∠*K*, ∠*G*, and ∠*D* remain congruent as do ∠*L*, ∠*H*, and ∠*E*. (They are supplements of congruent angles.) **26. a.** The diagonals bisect each other. **b.** *GKLH* is a parallelogram, so $\overline{GH} \parallel \overline{KL}$. Similarly, *DGHE* and *ADEB* are parallelograms, so $\overline{DE} \parallel \overline{HG}$ and $\overline{AB} \parallel \overline{DE}$. Then if $\overline{KL}$ is horizontal, all of the indicated segments are horizontal and the lift stays horizontal.

Page 356 Exercises and Applications

12.

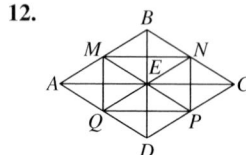

Answers may vary. An example is given. *ABCD* is a rhombus and *M, N, P,* and *Q* are the midpoints of consecutive sides. Then △ *AMQ* ≅ △ *CNP* and △ *BMN* ≅ △ *DPQ* by SAS, so $\overline{MN} \cong \overline{QP}$ and $\overline{MQ} \cong \overline{NP}$. Since both pairs of opposite sides of *MNQP* are congruent, *MNQP* is a parallelogram. Moreover, *AMPD* is a parallelogram. ($AM = \frac{1}{2}AB = \frac{1}{2}DC = DP$, so $\overline{AM}$ and $\overline{DP}$ are both parallel and congruent.) Similarly, *CNQD* is a parallelogram. Then *MP* = *AD* = *CD* = *NQ*. Since the diagonals of parallelogram *MNPQ* are congruent, *MNPQ* is a rectangle.

Page 357 Exercises and Applications

20. Answers may vary. An example is given. Since *WXYZ* is a parallelogram, $\overline{WZ} \parallel \overline{XY}$. Then ∠ *W* and ∠ *X* are supplementary. (If two parallel lines are intersected by a transversal, then same-side interior angles are supplementary.) Since ∠ *W* and ∠ *X* are both supplementary and congruent, they are right angles. Then *WXYZ* is a rectangle. (If one angle of a parallelogram is a right angle, then it is a rectangle.)

21. Given: quadrilateral *ABCD*; $\overline{AC}$ is the perpendicular bisector of $\overline{DB}$; $\overline{DB}$ is not the perpendicular bisector of $\overline{AC}$.
Prove: *ABCD* is a kite.

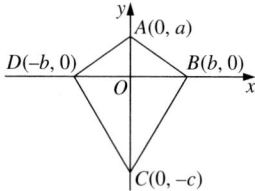

$AD = \sqrt{(-b-0)^2 + (0-a)^2} = \sqrt{b^2 + a^2}$; $AB = \sqrt{(b-0)^2 + (0-a)^2} = \sqrt{b^2 + a^2}$; $CD = \sqrt{(-b-0)^2 + (0-(-c))^2} = \sqrt{b^2 + c^2}$; $CB = \sqrt{(-b-0)^2 + (0-(-c))^2} = \sqrt{b^2 + c^2}$; Since $\overline{DB}$ is not the perpendicular bisector of $\overline{AC}$, $|a| \neq |c|$, so $a^2 \neq c^2$ and $\sqrt{b^2 + a^2} \neq \sqrt{b^2 + c^2}$. Then *ABCD* has two pair of congruent sides but opposite sides are not congruent. *ABCD* is a kite.

22. Given: □*JKLM*; $\overline{JL} \cong \overline{KM}$
Prove: *JKLM* is a rectangle.

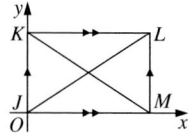

Proof: $\overline{JM}$ and $\overline{KL}$ are parallel, so ∠ *JML* and ∠ *KLM* are supplementary; that is, $m \angle JML + m \angle KLM = 180°$. $\overline{JM}$ and $\overline{KL}$ are congruent, as are $\overline{JL}$ and $\overline{KM}$ (given) and also $\overline{ML}$ and $\overline{ML}$ (Reflexive Property). Then △ *JML* ≅ △ *KLM* (SSS Postulate) and ∠ *JML* ≅ ∠ *KLM* or $m \angle JML = m \angle KLM$. Substituting, $2m \angle JML = 180°$ or $m \angle JML = 90°$. Then □*JKLM* is a rectangle. (If one ∠ of a parallelogram is a right ∠, then it is a rectangle.)

Page 358 Exercises and Applications

30. Construct the perpendicular bisector of $\overline{RS}$ to locate its midpoint, *P*. Follow Steps 1–3, using $\overline{PR}$ as the radius for circle *E*. **31.** By the Linear Pair Theorem, $2x + 2y = 180$ and $x + y = 90$. Consider △ *EDC*. Let $m \angle EDC = z°$; then $m \angle ECD = z°$ and $2z + 2y = 180$ by the Triangle Angle Sum Theorem. By substitution, $2x + 2y = 2z + 2y$ and $x = z$. Similarly, using △ *ECB*, $m \angle ECB = m \angle EBC = y$. Since ∠ *ABD* ≅ ∠ *BDC*, $\overline{AB} \parallel \overline{DC}$ and since ∠ *ADB* ≅ ∠ *DBC*, $\overline{AD} \parallel \overline{BC}$. By definition, *ABCD* is a parallelogram. Moreover, since $m \angle BCD = x° + y° = 90°$, ∠ *BCD* is a right angle and *ABCD* is a rectangle. **32.** D **33.** Answers may vary. Examples are given.

Method (1):

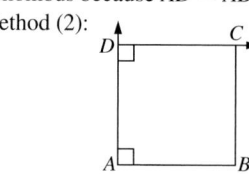

Draw $\overline{AB}$. Use the protractor to draw lines perpendicular to $\overline{AB}$ at *A* and *B*. Use the ruler to draw $\overline{AD}$ and $\overline{BC}$ on the perpendiculars, labeling points *D* and *C* so $\overline{AD} \cong \overline{AB} \cong \overline{BC}$. Draw $\overline{DC}$. *ABCD* is a parallelogram because $\overline{AD}$ and $\overline{BC}$ are both parallel and congruent. It is a rectangle because ∠ *DAB* is a right angle. It is a rhombus because $\overline{AD} \cong \overline{AB}$.

Method (2):

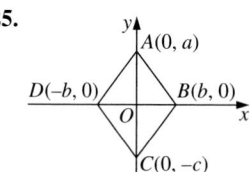

Draw $\overline{AB}$. Use the protractor to draw a line perpendicular to $\overline{AB}$ at *A*. Use the ruler to draw $\overline{AD}$ congruent to $\overline{AB}$ on the perpendicular. Use the protractor to draw a line perpendicular to $\overline{AD}$ at *D*. Use the ruler to draw $\overline{DC}$ on the perpendicular congruent to $\overline{AB}$. Draw $\overline{BC}$. *ABCD* is a parallelogram because $\overline{AB} \parallel \overline{DC}$. It is a rectangle because ∠ *DAB* is a right angle. It is a rhombus because $\overline{AD} \cong \overline{AB}$.

Page 359 Assess Your Progress

8. A quadrilateral is a parallelogram if both pairs of opposite sides are parallel, both pairs of opposite sides are congruent, one pair of opposite sides are both parallel and congruent, both pairs of opposite angles are congruent, or the diagonals bisect each other. A parallelogram is a rhombus if two consecutive sides are congruent or if the diagonals are perpendicular. A parallelogram is a rectangle if the diagonals are congruent or if one angle is a right angle. A parallelogram is a square if it is both a rectangle and a rhombus. A quadrilateral is a trapezoid if exactly one pair of opposite sides are parallel. A quadrilateral is a kite if it has two pairs of congruent sides, but opposite sides are not congruent or if exactly one diagonal is the perpendicular bisector of the other.

Page 365 Exercises and Applications

25.

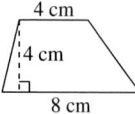

ABCD is a rhombus; it is equilateral and $\overline{AC}$ and $\overline{BD}$ are perpendicular and bisect each other. Then $\overline{AC}$ and $\overline{BD}$ divide the rhombus into four congruent right triangles, each with area $\frac{1}{2}ab$. The area of the rhombus is $2ab$.

$AC = \sqrt{(0-0)^2 + (a-(-a))^2} = \sqrt{4a^2} = 2a$, $BD = \sqrt{(b-(-b))^2 + (0-0)^2} = \sqrt{4b^2} = 2b$, so $\frac{1}{2}AC \cdot BD = \frac{1}{2}(2a)(2b) = 2ab$.

Then the area of the rhombus is half the product of the lengths of the diagonals. **26.** Answers may vary. An example is given.

4 cm
4 cm
8 cm

27. a. 48; 48 **b.** Let *n* be any positive integer. To divide a triangle into *n* smaller triangles whose areas are equal, divide the base of the triangle into *n* congruent segments. The triangles all have the same height, *h*, as the original triangle. The base of each is $\frac{1}{n}$ times that of the original triangle, so each has

area $\frac{1}{2} \cdot \frac{b}{h} \cdot h = \frac{bh}{2n}$.

Page 372 Exercises and Applications

14. a. The table gives the perimeter and area of an *n*-sided regular polygon ($3 \le n \le 8$) inscribed in a circle whose radius is 1. Perimeter and area are given to the nearest thousandth.

Number of sides	Perimeter	Area
3	5.196	1.299
4	5.657	2.000
5	5.878	2.378
6	6.000	2.598
7	6.074	2.737
8	6.123	2.828

b.

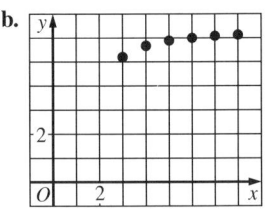 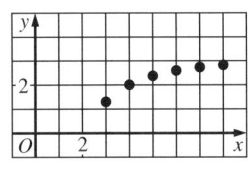

c. As the number of sides increases, both the perimeter and the area increase. The rate of increase for both quickly levels off, however. As *n* gets very large, the perimeter of the polygon gets very close to the circumference of the circle ($2\pi \approx 6.283$) and the area of the polygon gets very close to the area of the circle ($\pi \approx 3.142$).

Page 379 Exercises and Applications

15. a. The amount of plastic needed is determined by the surface area of the bottle. I calculated the surface area of each bottle assuming one is a cylinder and one a triangular prism, ignoring the neck of each bottle. A cylindrical bottle requires about 51 in.^{2} of plastic. A bottle with a triangular base requires about 63.5 in.^{2} of plastic. **b.** Answers may vary. Examples are given. I think the cylindrical bottle costs less; less plastic is required. The bottle shaped like a triangular prism might be easy to stack, but it would be difficult to hold in your hand. I think the cylindrical bottle is more practical.

Page 381 Assess Your Progress

10. Formulas are given in the tables. For figures, see pages 362, 368, 369, 375, and 376 of the text.

Figure	Formula
rectangle with base *b* and height *h*	$A = bh$
square with side lengths *s*	$A = s^2$
triangle with base *b* and height *h*	$A = \frac{1}{2}bh$
triangle with side lengths *a*, *b*, and *c* and $s = \frac{1}{2}(a + b + c)$ (Ex. 7, page 364)	$A = \sqrt{s(s-a)(s-b)(s-c)}$
parallelogram with base *b* and height *h*	$A = bh$
trapezoid with bases b_1 and b_2 and height *h*	$A = \frac{1}{2}(b_1 + b_2)h$
rhombus with diagonals d_1 and d_2 (Ex. 25, page 365)	$A = \frac{1}{2}d_1 \cdot d_2$
regular polygon with perimeter *p* and apothem *a*	$A = \frac{1}{2}ap$
circle with radius *r*	$A = \pi r^2$
prism with base perimeter *p*, base area *B*, and height *h*	$V = Bh$; S.A. $= ph + 2B$
cylinder with radius *r* and height *h*	$V = \pi r^2 h$; S.A. $= 2\pi rh + 2\pi r^2$

The formula for the area of a triangle can be derived from the formula for the area of a parallelogram. The formula for the area of a trapezoid can be derived from the formulas for the area of a rectangle and a triangle.

CHAPTER 8

Page 401 Exercises and Applications

3. $H'(3, 4)$; $J'(-1, 2)$; $K'(1, -2)$

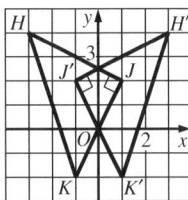

4. $L'(-1, 0)$; $M'(1, -3)$; $N'(4, -3)$; $P'(5, -1)$

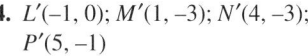 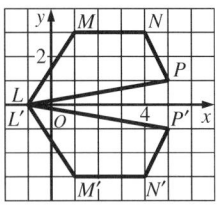

5. $Q'(-3, 0)$; $R'(1, -2)$; $S'(1, 2)$

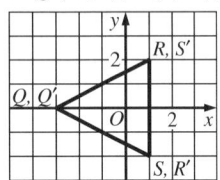

6. $T'(-2, -3)$; $U'(2, 0)$; $V'(-3, 2)$; $W'(-1, -1)$

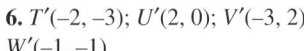 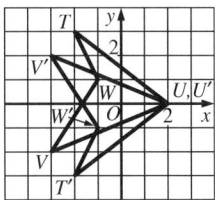

7. $M'(4, 0)$; $N'(2, 3)$; $O'(0, 0)$

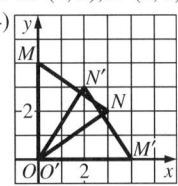

8. $O'(0, 0)$; $A'(2, 0)$; $B'(2, 4)$; $C'(0, 4)$

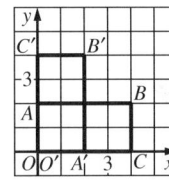

9. $D'(2, 0)$; $E'(4, 2)$; $F'(4, 4)$; $G'(-1, 1)$

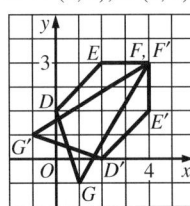

Page 403 Exercises and Applications

18. $(0, 8)$

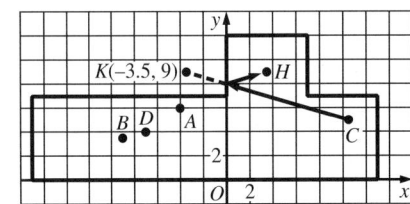

19. Reflect *H* over line *m*. The image of *H*, *H'* has coordinates $(10.5, 9)$. If Dianne aims at *H'*, the ball will bounce off the bumpboard that lies along line *m* and will reach *H*. The ball will hit the bumpboard at $(7, 8)$.

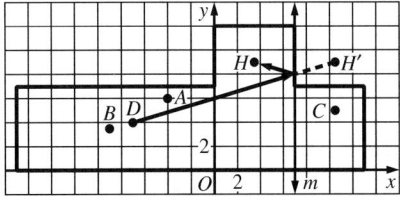

Page 404 Exercises and Applications

29–32. Answers may vary. Examples are given.

29. **30.**

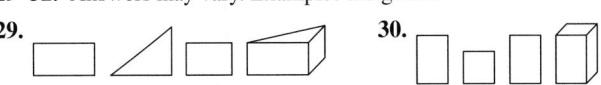

A-7

31. 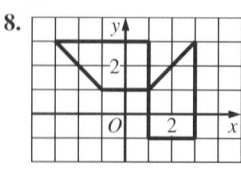   **32.**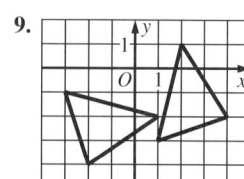

Page 415   Exercises and Applications

8. 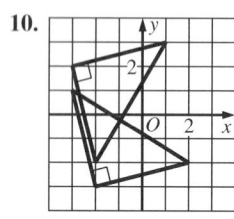   **9.**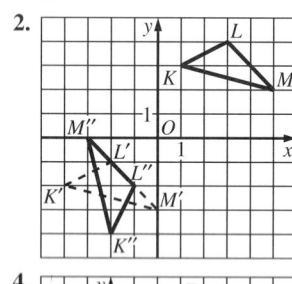

10.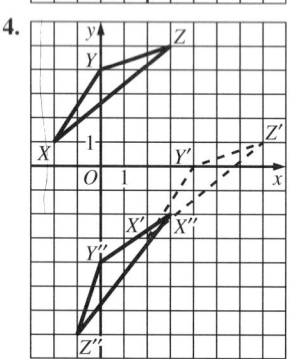

11. Yes; 180°.   **12.** Yes; the cooking surface of the waffle iron has rotational symmetry; 72°, 144°, 216°, 288°.   **13.** No.   **14.** Check students' work.

Page 416   Exercises and Applications

15. a. $\left(-9\frac{1}{4}, 6\frac{1}{2}\right); \left(-6\frac{1}{2}, -9\frac{1}{4}\right); \left(9\frac{1}{4}, -6\frac{1}{2}\right)$

b. $\left(9\frac{1}{4}, 6\frac{1}{2}\right); \left(-6\frac{1}{2}, 9\frac{1}{4}\right); \left(-9\frac{1}{4}, -6\frac{1}{2}\right); \left(6\frac{1}{2}, -9\frac{1}{4}\right)$ **c.** $x = 20; y = 70$

Page 423   Exercises and Applications

2.   **3.**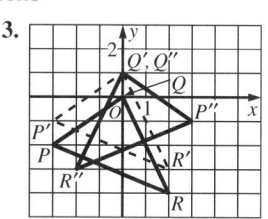

4.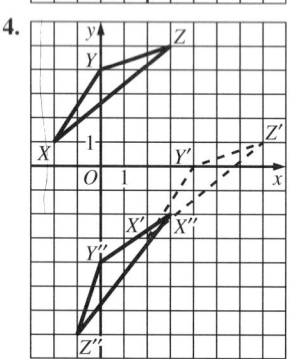

Page 424   Exercises and Applications

20.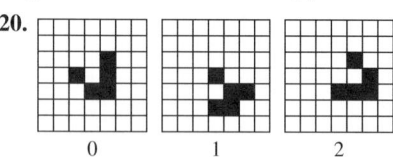

A translation down 1 unit followed by a reflection over the diagonal from the upper left corner of the grid moves generation 0 to generation 2.

21.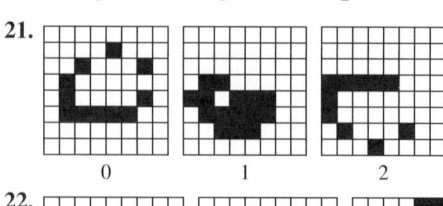

A translation left 1 unit followed by a reflection over a horizontal line moves generation 0 to generation 2.

22.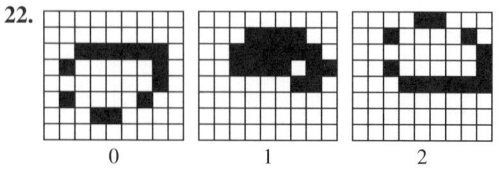

A translation right 1 unit followed by a reflection over a horizontal line moves generation 0 to generation 2.

Page 432   Exercises and Applications

18. Answers may vary. Examples are given.

a. 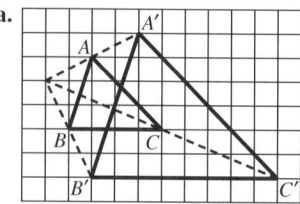   **b.**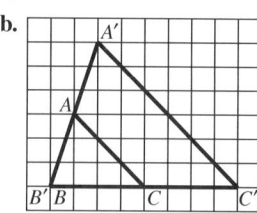

c. 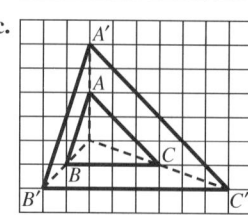   **d.** The images of △ABC under all three dilations are congruent. If the center of dilation is inside or outside the triangle, each side of the image is parallel to the corresponding side of the original triangle. If the center of dilation is a vertex, the lines containing two sides of the image are concurrent with the lines containing the corresponding sides of the original triangle. The third pair of corresponding sides are parallel.

Page 433   Assess Your Progress

7. 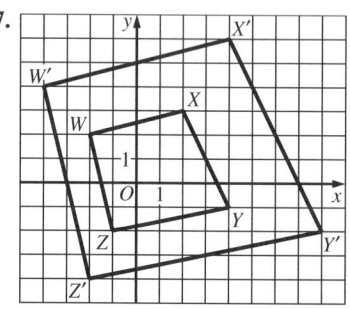   **8.**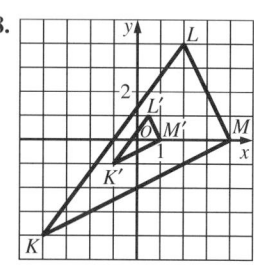

9. The following transformations can be described using coordinate notation: reflections over the y-axis, the x-axis, and the lines $y = x$ and $y = -x$, rotations of 90°, 180°, and 270° around the origin, dilations with center at the origin. The coordinates of the image of point (a, b) after the indicated transformation are given in the table.

Transformation	Coordinates of image
reflection over the x-axis	$(a, -b)$
reflection over the y-axis	$(-a, b)$
reflection over the line $y = x$	(b, a)
reflection over the line $y = -x$	$(-b, -a)$
90° rotation around the origin	$(-b, a)$
180° rotation around the origin	$(-a, -b)$
270° rotation around the origin	$(b, -a)$
dilation with center O and scale factor k	(ka, kb)

CHAPTER 9

Page 457   Exercises and Applications

8. Since $\overline{PT} \parallel \overline{SR}$, $\angle P \cong \angle S$ and $\angle T \cong \angle R$. Then $\triangle QTP \sim \triangle QRS$ by the AA Similarity Postulate. (The fact that vertical angles $\angle PQT$ and $\angle SQR$ are congruent could also be used.)   **9. a.** about 41 ft   **b.** If the sun is not at the same height when both shadows are measured, the triangles are not similar. The method only works if both shadows are clearly visible and measurable.

10. a. $\angle ACB$ and $\angle ADE$ are both right angles, so $\angle ACB \cong \angle ADE$. $\angle BAC \cong \angle EAD$ because vertical angles are congruent. Then $\triangle ABC \sim \triangle AED$ by the AA Similarity Postulate. **b.**

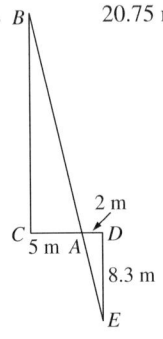

Page 460 Exercises and Applications

27. a. Answers may vary. An example is given. Consider $\triangle ABC$ with $m\angle A = 40°$ and $m\angle B = m\angle C = 70°$, and $\triangle DEF$ with $m\angle D = m\angle E = 40°$ and $m\angle F = 100°$. $\angle A$ and $\angle D$ are congruent, but the triangles are not similar. **b.** If the vertex angle of one isosceles triangle is congruent to the vertex angle of another isosceles triangle or if a base angle of one of the triangles is congruent to a base angle of the other, then the triangles are similar. The first part of the conjecture was proved in Ex. 14. Suppose a base angle of one triangle is congruent to a base angle of the other. Since base angles of an isosceles triangle are congruent, the triangles are similar by the AA Similarity Postulate.

29.

Statements	Reasons
1. $\overline{AD} \cong \overline{DC}$; $\angle ADB \cong \angle BDC$	1. Given
2. $\overline{BD} \cong \overline{BD}$	2. Reflexive Property
3. $\triangle ABD \cong \triangle CBD$	3. SAS Postulate

Page 472 Exercises and Applications

16. a. Answers may vary. The table below gives values for integer lengths from 2 cm to 10 cm.

	A	B	C
1	Edge length (cm)	Surface Area (cm²)	Volume (cm³)
2	2	24	8
3	3	54	27
4	4	96	64
5	5	150	125
6	6	216	216
7	7	294	343
8	8	384	512
9	9	486	729
10	10	600	1000

b.

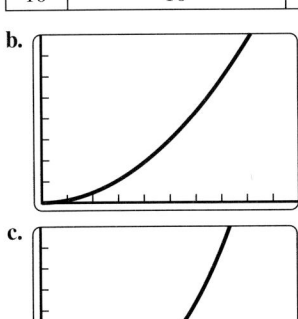

The graph is part of a parabola; it curves upward to the right. The surface area increases more rapidly than the edge length.

c. The graph curves upward to the right. The volume increases more rapidly than the edge length.

d. Answers may vary. An example is given. For $x \leq 6$, as edge length increases, surface area increases more rapidly than volume. For $x > 6$, the opposite is true.

Page 474 Exercises and Applications

28. a. Check students' work. **b.** Explanations may vary. An example is given. Each side of each smaller triangle is half as long as a side of the original triangle. (Two sides of each outer small triangle have the midpoint of a side as a vertex. The third sides of the outer small triangles and all three sides of the inner triangle are half as long as a side of the original triangle by the Midsegment Theorem.) Then the four triangles are congruent by the SSS Postulate; each is similar to the original triangle by the SSS Similarity Theorem. **c.** Answers may vary. Examples are given. (1) The four triangles are congruent and the sum of their areas is the area of the original triangle, so each has area $\frac{1}{4}$ that of the original triangle. (2) Since each triangle is similar to the original triangle and the ratio of the lengths of the sides is $1:2$, the ratio of the areas is $1:4$. So each new triangle has area $\frac{1}{4}$ that of the original triangle. (3) The area of the original triangle is $\frac{1}{2}bh$. Since each smaller triangle is similar to the original triangle, all linear measures are in proportion. Then the area of each smaller triangle is $\frac{1}{2} \cdot \frac{1}{2}b \cdot \frac{1}{2}h = \frac{1}{4}\left(\frac{1}{2}bh\right)$.

29.

Statements	Reasons
1. $\overline{AB} \parallel \overline{CD}$	1. Given
2. $\angle ECD \cong \angle EAB$; $\angle ABE \cong \angle EDC$	2. If two ∥ lines are intersected by a transversal, then alternate interior ∠ are ≅.
3. $\triangle ABE \sim \triangle CDE$	3. AA Similarity Postulate

30. $LK = 4\frac{1}{2}$

CHAPTER 10

Page 499 Exercises and Applications

25. a. In the table, n is the number of the pipe and l is the length of the pipe to the nearest tenth of a centimeter. Values of l may vary due to rounding. In the table, rounded values of l were used in subsequent calculations.

n	1	2	3	4	5	6	7	8	9	10	11	12	13
l	16.4	15.5	14.6	13.8	13.0	12.2	11.4	10.7	10.0	9.3	8.6	8.0	7.4

b. Check students' work.

Page 506 Exercises and Applications

29. If the length of a diagonal of a square is x, the length of a side is $\frac{x}{\sqrt{2}}$ and the area is $\left(\frac{x}{\sqrt{2}}\right)^2 = \frac{x^2}{2}$. If the length of one side of an equilateral triangle is x, a median of the triangle forms a 30-60-90 triangle with base length $\frac{x}{2}$ and height $\frac{x\sqrt{3}}{2}$. The area of the original triangle is $\frac{x^2\sqrt{3}}{4}$.

Page 506 Assess Your Progress

10. Answers may vary. Examples are given.
Similar Right Triangles Theorem

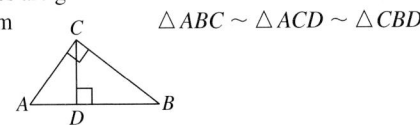

$\triangle ABC \sim \triangle ACD \sim \triangle CBD$

Geometric Mean Theorems:
In the figure above, $\frac{AD}{CD} = \frac{CD}{BD}$, $\frac{AB}{AC} = \frac{AC}{AD}$, and $\frac{AB}{CB} = \frac{CB}{DB}$.

45-45-90 Triangles

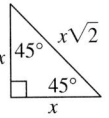

30-60-90 Triangles

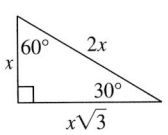

36. Answers may vary. An example is given.

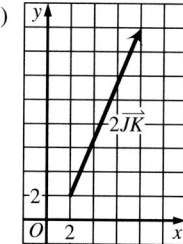

$\cos 50° = \frac{AC}{6}$; $AC \approx 3.9$; $\sin 50° = \frac{BC}{6}$; $BC \approx 4.6$

6. $2\overrightarrow{JK} = (6, 14)$

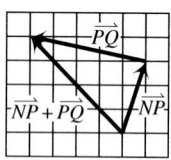

7. $(-7, -14)$

8. $\overrightarrow{NP} + \overrightarrow{PQ} = (-4, 4)$

6. $|\overrightarrow{MN}| = 4\sqrt{2} \approx 5.7$

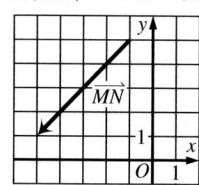

7. $|\overrightarrow{PQ}| = \sqrt{74} \approx 8.6$

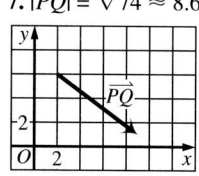

8.

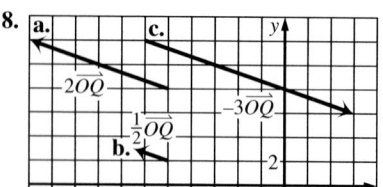

a. $4\sqrt{10} \approx 12.6$
b. $\sqrt{10} \approx 3.2$
c. $6\sqrt{10} \approx 19.0$

20. Check students' work. In the table, B represents the area of a base, h represents height, r represents the radius of a base, p represents perimeter, and s represents slant height.

Figure	Surface Area	Volume
prism	$ph + 2B$	Bh
cylinder	$2\pi rh + 2\pi r^2$	$\pi r^2 h$
pyramid	$\frac{1}{2}ps + B$	$\frac{1}{3}Bh$
cone	$\pi rs + \pi r^2$	$\frac{1}{3}\pi r^2 h$

CHAPTER 11

21. Case 1 (The center of the circle is outside the inscribed angle.)

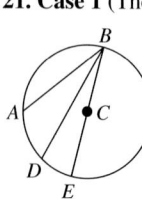

Draw diameter $\overline{BE}$. The center of the circle is on a side of inscribed angles $\angle ABE$ and $\angle DBE$. By the proof in Ex. 20, $m \angle ABE = \frac{1}{2}m\,\widehat{AE}$ and $m \angle DBE = \frac{1}{2}m\,\widehat{DE}$. By the Angle Addition Postulate, $m \angle ABD = m \angle ABE - m \angle DBE$. By the Substitution Property, $m \angle ABD = \frac{1}{2}m\,\widehat{AE} - \frac{1}{2}m\,\widehat{DE} = \frac{1}{2}m\,\widehat{AD}$ (Arc Addition Postulate).

Case 2 (The center of the circle is in the interior of the inscribed angle.)

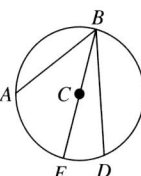

Draw diameter $\overline{BE}$. The center of the circle is on a side of inscribed angles $\angle ABE$ and $\angle DBE$. By the proof in Ex. 20, $m \angle ABE = \frac{1}{2}m\,\widehat{AE}$ and $m \angle DBE = \frac{1}{2}m\,\widehat{DE}$. By the Angle Addition Postulate, $m \angle ABD = m \angle ABE + m \angle DBE$. By the Substitution Property, $m \angle ABD = \frac{1}{2}m\,\widehat{AE} + \frac{1}{2}m\,\widehat{DE} = \frac{1}{2}m\,\widehat{AD}$ (Arc Addition Postulate).

25. a. $\odot C$ was constructed so that $m \angle C = 2m \angle MTB$ (Ex. 24). Then $m\,\widehat{MB} = 2m \angle MTB$, so $m \angle MTB = \frac{1}{2}m\,\widehat{MB}$, which is true for an inscribed angle of $\odot C$ that intercepts $\widehat{MB}$ (first Inscribed Angle Theorem). Moreover, any inscribed angle of $\odot C$ that intercepts $\widehat{MB}$ is congruent to $\angle MTB$ (second Inscribed Angle Theorem). Note that for $\angle MTB$ to intercept $\widehat{MB}$, T cannot be on $\widehat{MB}$.

b. No; since the boat is in the ocean, it is on $\odot C$. (There is another circle of position, constructed in the same manner as $\odot C$, with center C' on the opposite side of $\overleftrightarrow{MB}$ that describes possible positions of T on land or closer to land than the buoy.) This can be shown using an indirect proof. Suppose T is not on $\odot C$. Draw $\overleftrightarrow{TM}$ intersecting C at X.

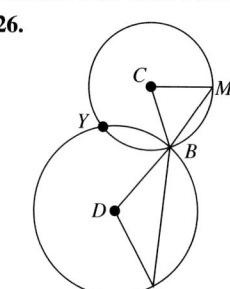

$\angle MXB$ intercepts arc $\widehat{MB}$, so $\angle MXB = \frac{1}{2}m\,\widehat{MB} = m \angle MTB$. Then, since corresponding angles formed by $\overleftrightarrow{TB}$, $\overleftrightarrow{XB}$, and transversal $\overleftrightarrow{MT}$ are congruent, $\overleftrightarrow{TB}$ and $\overleftrightarrow{XB}$ are parallel. Clearly, this is a contradiction, since $\overleftrightarrow{TB}$ and $\overleftrightarrow{XB}$ intersect at B. Then T is on $\odot C$.

26.

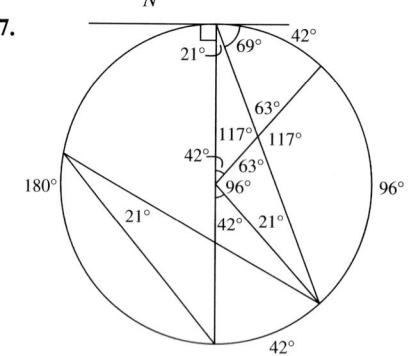

a. Since T is on both circles, C and D, T is one of the points of intersection. Since it cannot be at B, the buoy, it must be at Y.
b. Check students' work.

27.

(circle figure with arc measures: 21°, 69°, 42°, 63°, 117°, 117°, 42°, 63°, 180°, 96°, 21°, 42°, 21°, 96°, 42°)

24. Answers may vary. An example is given. Dear Andre: In this lesson, we learned about the measures of angles formed by tangents, secants, and chords. We learned that the measure of an angle formed by the intersection of two tangents, two secants, or a secant and a tangent, at a point outside a circle, is half the difference of the measures of the intercepted arcs. The measure of an angle formed by two chords is equal to half the sum of the measures of the intercepted arcs. In the figure, if you know $m \angle APB$ and $m\,\widehat{BC}$, you can find

$m\,\widehat{AB}$ because $m\angle APB = \frac{m\,\widehat{BC} - m\,\widehat{AB}}{2}$. $\angle C$ is an inscribed angle, so to find $m\angle C$, you just need to know $m\,\widehat{AE}$, since $m\angle C = \frac{1}{2}m\,\widehat{AE}$. If you know $m\,\widehat{CD}$ and $m\,\widehat{AE}$, you can find $m\angle CPD$, because $m\angle CPD = \frac{m\,\widehat{CD} - m\,\widehat{AE}}{2}$. I hope this will help you learn what you missed.

Page 570   **Exercises and Applications**

8. Statements (Reasons)

1. Draw radii $\overline{CP}$ and $\overline{CE}$. (For any two points, there is exactly one line between the points.)
2. $\overline{CP} \cong \overline{CE}$ (Def. of radius)
3. $\overline{CQ} \perp \overline{PR}$; $\overline{CF} \perp \overline{EG}$ (Given)
4. $\angle CQP$ and $\angle CFE$ are right angles; $\triangle CQP$ and $\triangle CFE$ are right triangles. (Def. of perpendicular lines and right triangles)
5. $\overline{CQ}$ bisects $\overline{PR}$; $\overline{CF}$ bisects $\overline{EG}$. (A diameter that is perpendicular to a chord bisects the chord and its corresponding arc.)
6. $QP = \frac{1}{2}PR$; $EF = \frac{1}{2}EG$ (Def. of segment bisector)
7. $\overline{PR} \cong \overline{EG}$ or $PR = EG$ (Given; def. of congruence)
8. $\frac{1}{2}PR = \frac{1}{2}EG$ (Multiplication Property)
9. $QP = EF$ or $\overline{QP} \cong \overline{EF}$ (Substitution, Steps 6 and 8; def. of congruence)
10. $\triangle CQP \cong \triangle CFE$ (HL Theorem)
11. $\overline{CQ} \cong \overline{CF}$ (Def. of congruent triangles)

9. $\widehat{AB} \cong \widehat{CD}$, so $m\,\widehat{AB} = m\,\widehat{CD}$. The measure of a central angle is equal to the measure of its intercepted arc, so $m\angle APB = m\,\widehat{AB}$ and $m\angle DPC = m\,\widehat{CD}$. Then $m\angle APB = m\angle DPC$ or $\angle APB \cong \angle DPC$. $\overline{PA}$, $\overline{PB}$, $\overline{PD}$, and $\overline{PC}$ are all congruent since they are radii of the same circle. By the SAS Postulate, $\triangle PAB \cong \triangle PDC$, so corresponding sides $\overline{AB}$ and $\overline{CD}$ are congruent.

Page 571   **Exercises and Applications**

22. Answers may vary. An example is given. If you think of $\overline{AC}$ in the second theorem as a secant for which A and D are the same point, the length of the chord inside the circle is 0, and $AC(AC + CD) = AC(AC + 0) = AC^2$.

Page 572   **Exercises and Applications**

19. Answers may vary. A example is given. Using patty paper, draw a circle and a chord. Construct a diameter perpendicular to the chord. Fold the paper along the diameter. The two segments of the arc will correspond exactly, demonstrating that they are congruent. Using geometry software, draw a circle, a chord, and a diameter perpendicular to the chord. Measure the segments of the arc.

CHAPTER 12

Page 616   **Exercises and Applications**

13. (7) dilation of a point with center $(0, 0)$ and scale factor 5;
(8) dilation of a line segment, with center $(0, 0)$ and scale factor 4;
(9) dilation of a triangle, with center $(0, 0)$ and scale factor 3;
(10) dilation of a quadrilateral, with center $(0, 0)$ and scale factor 2.5;
(11) no transformation (the same figure before and after);
(12) dilation of a line segment, with center $(0, 0)$ and scale factor $\frac{2}{3}$

14. $\frac{1}{3}\begin{bmatrix} 9 & -3 \\ 21 & 0 \end{bmatrix} = \begin{bmatrix} 3 & -1 \\ 7 & 0 \end{bmatrix}$ **15.** $4\begin{bmatrix} 2 & 3 & 0 \\ -1 & 4 & \frac{1}{4} \end{bmatrix} = \begin{bmatrix} 8 & 12 & 0 \\ -4 & 16 & 1 \end{bmatrix}$

16. $\frac{5}{2}\begin{bmatrix} 0 & 0 & 2 & 2 \\ 0 & 2 & 2 & 0 \end{bmatrix} = \begin{bmatrix} 0 & 0 & 5 & 5 \\ 0 & 5 & 5 & 0 \end{bmatrix}$ **17.** $\frac{1}{2}\begin{bmatrix} -4 & 0 & 4 \\ 0 & 6 & 0 \end{bmatrix} = \begin{bmatrix} -2 & 0 & 2 \\ 0 & 3 & 0 \end{bmatrix}$

Page 618   **Exercises and Applications**

31. a. $\begin{bmatrix} 4 & 0 & 0 \\ 0 & 6 & 0 \\ 0 & 0 & 8 \end{bmatrix}$ **b.** $\begin{bmatrix} 1 & 0 & 0 \\ 0 & 1.5 & 0 \\ 0 & 0 & 2 \end{bmatrix}$

a dilation of the triangle about center $(0, 0, 0)$ by the scale factor

c.
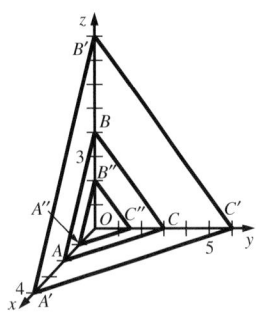

Page 622   **Exercises and Applications**

11. a. $\begin{bmatrix} -4 & -4 & -4 \\ 4 & 4 & 4 \end{bmatrix}$ **b.** $\begin{bmatrix} -8 & -12 & -4 \\ 1 & 6 & 7 \end{bmatrix}$

c.
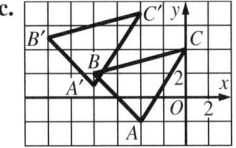

12. a. $\begin{bmatrix} -3 & -1 & 2 & 2 \\ 2 & 4 & 4 & 1 \end{bmatrix}$ **b.** $\begin{bmatrix} 5 & 5 & 5 & 5 \\ -3 & -3 & -3 & -3 \end{bmatrix}$ **c.** $\begin{bmatrix} 2 & 4 & 7 & 7 \\ -1 & 1 & 1 & -2 \end{bmatrix}$

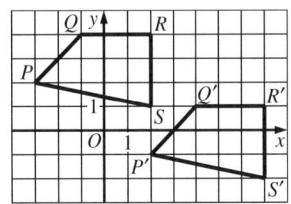

Page 623   **Exercises and Applications**

20. a. $\begin{bmatrix} 2.6 \times 10^5 & -1.0 \times 10^5 & 10.6 \times 10^5 \\ 0.4 \times 10^5 & -1.1 \times 10^5 & -0.7 \times 10^5 \\ 0.7 \times 10^5 & 0 & 3.5 \times 10^5 \end{bmatrix}$

b. 12.1×10^5 km; 12.1×10^5 km   **c.** Both distances represent the distance between Europa and Callisto. They are the same, as the distance between these two moons is constant regardless of the coordinate system used.

Page 630   **Exercises and Applications**

6.
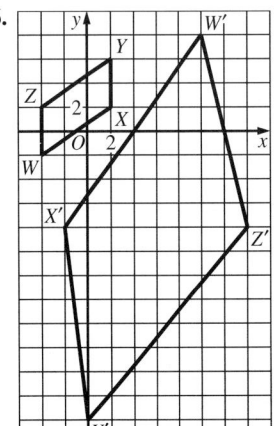

7.
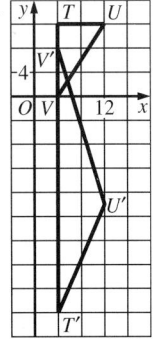

8. a. Answers may vary.   **b.** When you multiply a $2 \times n$ matrix by $\begin{bmatrix} 1 & 0 \\ 0 & 1 \end{bmatrix}$, you get the same matrix.

9. a. $\begin{bmatrix} 4 & 8 & -8 \\ 6 & -6 & 0 \end{bmatrix}$; $\begin{bmatrix} 6 & 12 & -12 \\ 9 & -9 & 0 \end{bmatrix}$; $\begin{bmatrix} 8 & 16 & -16 \\ 12 & -12 & 0 \end{bmatrix}$

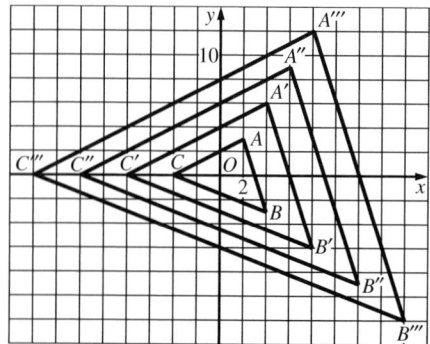

b. a dilation about center (0, 0) with scale factor n; $\begin{bmatrix} n & 0 \\ 0 & n \end{bmatrix} = n\begin{bmatrix} 1 & 0 \\ 0 & 1 \end{bmatrix}$, so $\begin{bmatrix} n & 0 \\ 0 & n \end{bmatrix}A = nA$.

Page 637   Exercises and Applications

8. $\begin{bmatrix} 2 & 2 & -1 & 0 \\ 0 & 2 & 2 & 0 \end{bmatrix}$

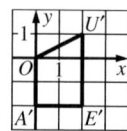

9. $\begin{bmatrix} 0 & 2 & 2 & 0 \\ -2 & -2 & 1 & 0 \end{bmatrix}$

Page 640   Exercises and Applications

36. $\begin{bmatrix} 6 & 0 & -3 & 3 \\ 6 & 3 & -3 & 0 \end{bmatrix}$

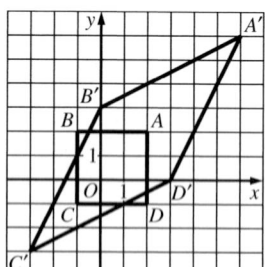

Page 641   Exploration

1. $\begin{bmatrix} 1 & 0 \\ 0 & 1 \end{bmatrix}$; $\begin{bmatrix} 1 & 0 \\ 0 & -1 \end{bmatrix}$; $\begin{bmatrix} -1 & 0 \\ 0 & 1 \end{bmatrix}$; $\begin{bmatrix} -1 & 0 \\ 0 & -1 \end{bmatrix}$; $\begin{bmatrix} 0 & 1 \\ 1 & 0 \end{bmatrix}$; $\begin{bmatrix} 0 & 1 \\ -1 & 0 \end{bmatrix}$; $\begin{bmatrix} 0 & -1 \\ 1 & 0 \end{bmatrix}$; $\begin{bmatrix} 0 & -1 \\ -1 & 0 \end{bmatrix}$

2. $\begin{bmatrix} 1 & 0 \\ 0 & 1 \end{bmatrix}\begin{bmatrix} 3 & 4 & 7 \\ 1 & 3 & 1 \end{bmatrix} = \begin{bmatrix} 3 & 4 & 7 \\ 1 & 3 & 1 \end{bmatrix}$; $\begin{bmatrix} 1 & 0 \\ 0 & -1 \end{bmatrix}\begin{bmatrix} 3 & 4 & 7 \\ 1 & 3 & 1 \end{bmatrix} = \begin{bmatrix} 3 & 4 & 7 \\ -1 & -3 & -1 \end{bmatrix}$;

$\begin{bmatrix} -1 & 0 \\ 0 & 1 \end{bmatrix}\begin{bmatrix} 3 & 4 & 7 \\ 1 & 3 & 1 \end{bmatrix} = \begin{bmatrix} -3 & -4 & -7 \\ 1 & 3 & 1 \end{bmatrix}$; $\begin{bmatrix} -1 & 0 \\ 0 & -1 \end{bmatrix}\begin{bmatrix} 3 & 4 & 7 \\ 1 & 3 & 1 \end{bmatrix} = \begin{bmatrix} -3 & -4 & -7 \\ -1 & -3 & -1 \end{bmatrix}$;

$\begin{bmatrix} 0 & 1 \\ 1 & 0 \end{bmatrix}\begin{bmatrix} 3 & 4 & 7 \\ 1 & 3 & 1 \end{bmatrix} = \begin{bmatrix} 1 & 3 & 1 \\ 3 & 4 & 7 \end{bmatrix}$; $\begin{bmatrix} 0 & 1 \\ -1 & 0 \end{bmatrix}\begin{bmatrix} 3 & 4 & 7 \\ 1 & 3 & 1 \end{bmatrix} = \begin{bmatrix} 1 & 3 & 1 \\ -3 & -4 & -7 \end{bmatrix}$;

$\begin{bmatrix} 0 & -1 \\ 1 & 0 \end{bmatrix}\begin{bmatrix} 3 & 4 & 7 \\ 1 & 3 & 1 \end{bmatrix} = \begin{bmatrix} -1 & -3 & -1 \\ 3 & 4 & 7 \end{bmatrix}$; $\begin{bmatrix} 0 & -1 \\ -1 & 0 \end{bmatrix}\begin{bmatrix} 3 & 4 & 7 \\ 1 & 3 & 1 \end{bmatrix} = \begin{bmatrix} -1 & -3 & -1 \\ -3 & -4 & -7 \end{bmatrix}$

3. $\begin{bmatrix} 1 & 0 \\ 0 & 1 \end{bmatrix}$; identity

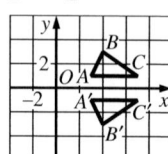

$\begin{bmatrix} 1 & 0 \\ 0 & -1 \end{bmatrix}$; reflection over x-axis

$\begin{bmatrix} -1 & 0 \\ 0 & 1 \end{bmatrix}$; reflection over y-axis

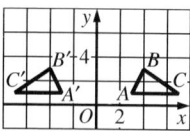

$\begin{bmatrix} 0 & 1 \\ 1 & 0 \end{bmatrix}$; reflection over $y = x$

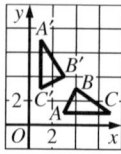

$\begin{bmatrix} 0 & -1 \\ 1 & 0 \end{bmatrix}$; 90° rotation

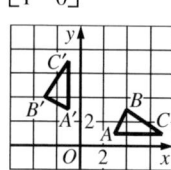

$\begin{bmatrix} -1 & 0 \\ 0 & -1 \end{bmatrix}$; 180° rotation

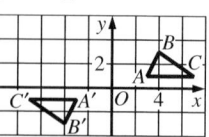

$\begin{bmatrix} 0 & 1 \\ -1 & 0 \end{bmatrix}$; 270° rotation

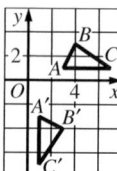

$\begin{bmatrix} 0 & -1 \\ -1 & 0 \end{bmatrix}$; reflection over $y = -x$

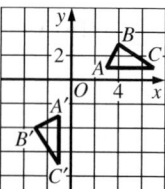

4.

Matrix	Transformation
$\begin{bmatrix} 1 & 0 \\ 0 & 1 \end{bmatrix}$	identity/360° rotation
$\begin{bmatrix} 0 & -1 \\ 1 & 0 \end{bmatrix}$	90° rotation
$\begin{bmatrix} -1 & 0 \\ 0 & -1 \end{bmatrix}$	180° rotation
$\begin{bmatrix} 0 & 1 \\ -1 & 0 \end{bmatrix}$	270° rotation
$\begin{bmatrix} 1 & 0 \\ 0 & -1 \end{bmatrix}$	reflection over x-axis
$\begin{bmatrix} -1 & 0 \\ 0 & 1 \end{bmatrix}$	reflection over y-axis
$\begin{bmatrix} 0 & 1 \\ 1 & 0 \end{bmatrix}$	reflection over $y = x$
$\begin{bmatrix} 0 & -1 \\ -1 & 0 \end{bmatrix}$	reflection over $y = -x$

Page 644   Exercises and Applications

2. $\begin{bmatrix} -3 & -1 & -2 & -3 \\ -2 & -2 & 3 & 1 \end{bmatrix}$

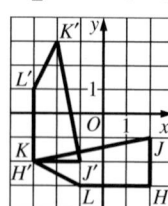

3. $\begin{bmatrix} -1 & -3 & -5 & -2 & 0 \\ 0 & 0 & 1 & 3 & 3 \end{bmatrix}$

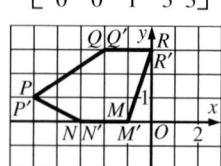

4. $\begin{bmatrix} 12 & 9 & 0 \\ 0 & -9 & -6 \end{bmatrix}$

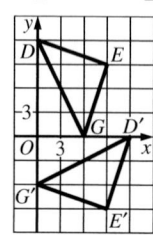

5. $\begin{bmatrix} 0 & 10 & -20 \\ -15 & 0 & 0 \end{bmatrix}$

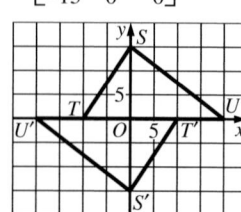

6. $\begin{bmatrix} -1 & 1 & 3 & 1 & -1 \\ 3 & 3 & 1 & -2 & -1 \end{bmatrix}$

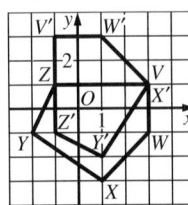

Page 645 **Exercises and Applications**

9. in order to aim cameras and other recording equipment correctly

10. a. $\begin{bmatrix} -5 & -5 & -3 \\ 0 & -4 & -4 \end{bmatrix}$ b. $\begin{bmatrix} 5 & 5 & 3 \\ 0 & 4 & 4 \end{bmatrix}$

c.

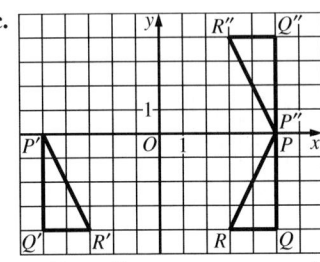

11. a. $\begin{bmatrix} 0 & 1 \\ -1 & 0 \end{bmatrix}$; 270° rotation b. $\begin{bmatrix} 0 & 1 \\ -1 & 0 \end{bmatrix}$

Page 654 **Chapter 12 Assessment**

28. $\begin{bmatrix} -1 & -3 & -2 & 1 \\ 0 & 0 & 3 & 5 \end{bmatrix}$

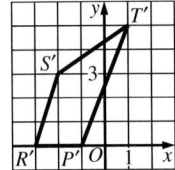 rotation of 90° about the origin

TECHNOLOGY HANDBOOK

Page 725 **Spreadsheet Practice**

14, 15.

Type of ball	r(cm)	V(cm^3)		
Golf ball	2.1	38.79238609		
Tennis ball	3.3	150.5325536		
Softball	4.8	463.2466863		
Volleyball	10.5	4849.048261		
Basketball	11.9	7058.777513		
		Pythagorean	Triples	
n		n^2–1	2n	n^2+1
2		3	4	5
3		8	6	10
4		15	8	17
5		24	10	26
6		35	12	37
7		48	14	50
8		63	16	65
9		80	18	82
10		99	20	101

APPENDIX 2

Page 741 **Exercises and Applications**

25. a contradiction, since there are all F's in the final column of the truth table:

p	q	$p \wedge q$	$\sim p$	$\sim q$	$\sim p \vee \sim q$	$(p \wedge q) \wedge (\sim p \vee \sim q)$
T	T	T	F	F	F	F
T	F	F	F	T	T	F
F	T	F	T	F	T	F
F	F	F	T	T	T	F

26. neither, since there are both T's and F's in the final column of the truth table:

p	q	$p \vee q$	$\sim q$	$(p \vee q) \rightarrow \sim q$
T	T	T	F	F
T	F	T	T	T
F	T	T	F	F
F	F	F	T	T

27. a tautology, since there are all T's in the final column of the truth table:

p	q	$\sim p$	$p \vee q$	$(p \vee q) \wedge \sim p$	$[(p \vee q) \wedge \sim p] \rightarrow q$
T	T	F	T	F	T
T	F	F	T	F	T
F	T	T	T	T	T
F	F	T	F	F	T

29.

p	q	$p \rightarrow q$	$\sim q$	$(p \rightarrow q) \wedge \sim q$	$\sim p$	$[(p \rightarrow q) \wedge \sim q] \rightarrow \sim p$
T	T	T	F	F	F	T
T	F	F	T	F	F	T
F	T	T	F	F	T	T
F	F	T	T	T	T	T

30.

p	q	r	$p \rightarrow q$	$q \rightarrow r$	$(p \rightarrow q) \wedge (q \rightarrow r)$	$p \rightarrow r$	$[(p \rightarrow q) \wedge (q \rightarrow r)] \rightarrow (p \rightarrow r)$
T	T	T	T	T	T	T	T
T	T	F	T	F	F	F	T
T	F	T	F	T	F	T	T
T	F	F	F	T	F	F	T
F	T	T	T	T	T	T	T
F	T	F	T	F	F	T	T
F	F	T	T	T	T	T	T
F	F	F	T	T	T	T	T

31.

p	q	r	s	$p \to q$	$q \to r$	$r \to s$	$(p \to q) \wedge (q \to r) \wedge (r \to s)$
T	T	T	T	T	T	T	T
T	T	T	F	T	T	F	F
T	T	F	T	T	F	T	F
T	T	F	F	T	F	T	F
T	F	T	T	F	T	T	F
T	F	T	F	F	T	F	F
T	F	F	T	F	T	T	F
T	F	F	F	F	T	T	F
F	T	T	T	T	T	T	T
F	T	T	F	T	T	F	F
F	T	F	T	T	F	T	F
F	T	F	F	T	F	T	F
F	F	T	T	T	T	T	T
F	F	T	F	T	T	F	F
F	F	F	T	T	T	T	T
F	F	F	F	T	T	T	T

$p \to s$	$[(p \to q) \wedge (q \to r) \wedge (r \to s)] \to (p \to s)$
T	T
F	T
F	T
F	T
F	T
F	T
F	T
F	T
T	T
F	T
F	T
F	T
T	T
F	T
T	T
T	T

16 rows

32.

r	p	t	$r \to p$	$t \to p$	$(r \to p) \wedge (t \to p)$	$r \to t$	$[(r \to p) \wedge (t \to p)] \to (r \to t)$
T	T	T	T	T	T	T	T
T	T	F	T	T	T	F	F
T	F	T	F	F	F	T	T
T	F	F	F	T	F	F	T
F	T	T	T	T	T	T	T
F	T	F	T	T	T	T	T
F	F	T	T	F	F	T	T
F	F	F	T	T	T	T	T

The argument is not logically valid because the last column of the truth table does not contain only T's.